Lecture Notes in Civil Engineering

Volume 209

Lecture Notes in Civil Engineering (LNCE) publishes the latest developments in Civil Engineering - quickly, informally and in top quality. Though original research reported in proceedings and post-proceedings represents the core of LNCE, edited volumes of exceptionally high quality and interest may also be considered for publication. Volumes published in LNCE embrace all aspects and subfields of, as well as new challenges in, Civil Engineering. Topics in the series include:

- Construction and Structural Mechanics
- Building Materials
- Concrete, Steel and Timber Structures
- Geotechnical Engineering
- Earthquake Engineering
- Coastal Engineering
- Ocean and Offshore Engineering; Ships and Floating Structures
- Hydraulics, Hydrology and Water Resources Engineering
- Environmental Engineering and Sustainability
- Structural Health and Monitoring
- Surveying and Geographical Information Systems
- Indoor Environments
- Transportation and Traffic
- Risk Analysis
- Safety and Security

To submit a proposal or request further information, please contact the appropriate Springer Editor:

- Pierpaolo Riva at pierpaolo.riva@springer.com (Europe and Americas);
- Swati Meherishi at swati.meherishi@springer.com (Asia - except China, and Australia, New Zealand);
- Wayne Hu at wayne.hu@springer.com (China).

All books in the series now indexed by Scopus and EI Compendex database!

More information about this series at http://www.springer.com/series/15087

Ioannis Vayas · Federico M. Mazzolani
Editors

Protection of Historical Constructions

Proceedings of PROHITECH 2021

Springer

Editors
Ioannis Vayas
National Technical University of Athens
Zografou, Athens, Greece

Federico M. Mazzolani
University of Naples Federico II
Napoli, Italy

ISSN 2366-2557 ISSN 2366-2565 (electronic)
Lecture Notes in Civil Engineering
ISBN 978-3-030-90787-7 ISBN 978-3-030-90788-4 (eBook)
https://doi.org/10.1007/978-3-030-90788-4

This Springer imprint is published by the registered company Springer Nature Switzerland AG
The registered company address is: Gewerbestrasse 11, 6330 Cham, Switzerland

Preface

It was long ago recognized that the construction environment is an important component of the cultural heritage that needs preservation and protection. In previous centuries, entire monuments or parts of them were often dismantled from their original places, transported and exhibited in museums of the metropoles, mainly in Europe, which became cultural centres. This changed during the formation of national states that considered the monuments on their soil as their own cultural identity and undertook considerable restoration efforts by themselves. In any case, independently of property issues, the existing construction expresses a set of values: artistic, architectural, technical, historical, sacred, symbolic, functional or educational. Protection of the monuments means protections of these values, which are indispensable for the human kind as the results of its own creation. This is the reason why nowadays the destruction of monuments, wherever it takes place, is considered an act of barbarism.

Restoration works at their beginnings were sometimes made, without the required knowledge or experience. They were consequently associated with structural, architectural or aesthetic errors that, sometimes, modified the original form or had detrimental consequences due to lack of durability. As an example, restorations with the use of steel reinforcements in antique temples resulted in destruction of the marbles due to corrosion of embedded steel bars. Having this recognized, international bodies agreed with the objective to draft general intervention principles for the preservation of the values connected to the historical buildings, by ensuing their protection over the time. According to these principles, intervention criteria have been established, based on fundamental properties, such as reversibility, compatibility, durability and/or robustness.

The objectives of interventions may differ and can be classified through different types, like restauration, reconstruction, retrofitting, repair or strengthening. Interventions refer to the building itself but also to separate parts of it, e.g. covering roof, for its protection. To comply with the established criteria, the interventions require interdisciplinary scientific approaches and team work in all stages of the project, including preparatory exploration works, design or execution. New technologies for digital documentation are increasingly introduced, such as image matching and 3D reconstruction, point cloud processing and analysis, virtual and augmented reality for complex architectures, and BIM for cultural heritage (HBIM). Another important area of research and practical applications deals with material science, characterization of old and traditional material and development of new structural materials and reconstruction techniques. Architecture and urban planning are involved in terms of architectural heritage, industrial heritage, religious heritage, urban transformation and cultural heritage. Intervention strategies for anamnesis, diagnosis and therapy have to be established, by taking into due account the adaptation to climate change, energy efficiency and sustainability. Geotechnical engineering is important to recognize and find solutions for buildings and sites on hazardous soils. Structural engineering is involved for recognizing the response of antique static systems to gravity, dynamic and seismic loading, taking into consideration that many past structures were not monolithic as the modern ones and using methods of experimental testing, numerical modelling and analysis of ancient systems, safety requirements, vulnerability and risk assessment of old structures, or monitoring the performance under the real loading conditions.

The International Conference on Protection of Historical Constructions, PROHITECH 2020, was rescheduled due to COVID-19 for 25–27 October 2021 in Athens, Greece. It is, after 2009 in Rome, Italy, 2014, in Antalya, Turkey, and 2017 in Lisbon, Portugal, the 4th edition of this series of relevant conferences, that followed the important EU-supported research project (2004–2008), with participation of 12 Mediterranean and South European Countries. The PROHITECH Conference series provides a valuable platform for presentation of the latest scientific and practical developments in the protection of historic and monumental structures. The pandemic situation made the organization of this edition extremely difficult. However, it was the strong belief of the chairmen that the physical presence is indispensable for this kind of conference, since, beside its scientific component, it gives a unique opportunity for meeting and exchanging ideas of experts of different disciplines working together in this field. This resulted in, unfortunately but expectedly, a leaner conference than before, as the pandemic does not seem to stop despite the worldwide vaccination policy.

This volume contains around 100 papers authored by participants from all over the world. This includes many countries from near to far East, South to North America and Europe, mostly from the South. The papers are devoted to a variety of subjects of scientific and technical nature, which includes a large number of case studies of protection measures illustrating the approaches followed in practical applications. The scientific subjects include HBIM, 3D modelling, point cloud processing and analysis, urban transformation and cultural heritage, architectural, industrial and religious heritage, anamnesis, diagnosis, therapy and controls, numerical modelling, re-use and valorization of built heritage, experimental tests, traditional materials and techniques, vulnerability and risk assessment and structural behaviour of various types of constructions (stone, masonry, timber, iron, etc.).

Each paper is assigned to one of the five main chapters of the volume. Chapter 1 with the title Properties, Investigation and Monitoring has 18 papers, most of which devoted to experimental tests performed on-site or in the laboratory with application of various techniques in order to define the properties of constitutive materials of entire structures. Chapter 2 with the title Modelling and Analysis is the most extended one with 28 papers. It starts with papers on image matching and 3D reconstruction, point cloud processing and analysis, after it proceeds with architectural heritage, then it goes to numerical modelling and it ends with contributions on protection and repair. Chapter 3 with the title Assessment includes 20 papers dealing with vulnerability and risk assessment, safety requirements and code provisions. Chapter 4 has the title Restoration Techniques. It includes 23 papers that report specific case studies of rehabilitation already carried out or in progress. The case studies refer to residential or religious buildings (temples, churches, mosques, etc.) and other types of cultural heritage. The last Chapter 5 with the title Protection and Preservation strategies has 21 papers. They are dealing with HBIM, rehabilitation interventions, re-use and valorization of built heritage, mostly in the context of case studies of various shape and use, ranging from the Temples of Abu Simple in Egypt to royal roads in Mexico.

We thank all participants for their contributions, Editor Mr. Pierpaolo Riva and SPRINGER for their commitment and effort that allowed the realization of this volume.

Federico Mazzolani
Ioannis Vayas
The chairmen of PROHITECH 2020

Organization

Conference Chairmen

Theodosios P. Tasios (Honorary Chairman)	National Technical University of Athens, Greece
Federico M. Mazzolani	University of Naples Federico II, Italy
Ioannis Vayas	National Technical University of Athens, Greece

Organizing Committee

Maria-Eleni Dasiou	National Technical University of Athens, Greece
Stavros Kourkoulis	National Technical University of Athens, Greece
Ioannis Psycharis	National Technical University of Athens, Greece
Andreas Spiliopoulos	National Technical University of Athens, Greece
Ioannis Vayas	National Technical University of Athens, Greece
Elizabeth Vintzileou	National Technical University of Athens, Greece

International Scientific Committee

Zeynep Ahunbay, Turkey
Amalia Androulidaki, Greece
Goran Arun, Turkey
Stathis Bousias, Greece
Maria Lucia Bressan, Brazil
Bruno Calderoni, Italy
Carlo Castiglioni, Italy
Zekai Celep, Turkey
Christis Chrysostomou, Cyprus
Gianfranco De Matteis, Italy
Florea Dinu, Romania
Matjaz Dolsek, Slovenia
Eirini Efesiou, Greece
Vasiliki Eleftheriou, Greece
Antonio Formisano, Italy
Antonio Sousa Gago, Portugal
Andreas Georgopoulos, Greece
Alper Ilki, Turkey
Charalambos Ioannidis, Greece
Arif Kamal, India
Yordanka Kandulkova, Bulgaria
Konstantinos Karadimas, Greece
Fyllitsa Karantoni, Greece
Petros Komodromos, Cyprus

International Advisory Committee

Contents

Properties Investigation and Monitoring

Strength Grading of Structural Timber in Existing Structures – Study on the Apparatus Supported Grading with the Ultrasonic Time-of-Flight Measurement

Gunter Linke[1](✉), Wolfgang Rug[2], and Hartmut Pasternak[3]

[1] Brandenburg University of Technology, Cottbus, Germany
linke@holzbau-statik.de
[2] University of Sustainable Development, Eberswalde, Germany
wolfgang.rug@hnee.de
[3] Brandenburg University of Technology, Cottbus, Germany
Hartmut.Pasternak@btu.de

Abstract. The material properties of timber are determined by growth characetristics of this natural material what results in a significant variability of its parameters. Their limitation is required for its application as construction material. This is achieved by strength grading. The grading methods developed for new timber can only be applied on timber members in existing historic buildings with major restrictions. Therefore, a strength grading of timber members in existing structures is rarely carried out nowadays. Though, exact evaluation of the structural state and the planning of a substance careful, professional redevelopment is only possible when the present material quality is determined. Here, non- and semi-destructive test methods – e.g. ultrasonic time-of-flight measurement – can be used. This contribution describes comparative material studies on the applicability of this test method for the on-site strength grading. Two alternatives are investigated to determine how the ultrasonic time-of-flight measurement can be integrated in the present grading standards. The results show a significant improvement of the grading yield and the accuracy of the grading process can be achieved by the combined use of visual grading and ultrasonic measurements.

Keywords: Historic timber structures · Strength grading · Ultrasonic time-of-flight measurement

1 Introduction

The material properties of structural timber show significant variation which result mainly from the wood structure itself. Additional variation is caused by local growth conditions. Their limitation is an unconditional requirement for its application as regulated construction material. This is achieved by a strength grading.

I. Vayas and F. M. Mazzolani (Eds.): PROHITECH 2021, LNCE 209, pp. 3–19, 2022.
https://doi.org/10.1007/978-3-030-90788-4_1

The requirements of strength grading of structural timber are regulated on European level by EN 14081–1, which are met by the German grading standards DIN 4074–1/-5. In general, strength grading is divided into visual and mechanical grading procedures.

The visual grading concentrates on superficial visible and visually determinable growth properties. The timber is classified into three grades (coniferous wood: “S”-classes, deciduous wood: “LS”-classes). Hereby, the timber is divided into structural timber with low (S7, LS7), average (S10, LS 10) and high load-bearing capacity (S1, LS13). If the visual strength grading is combined with non-/semi-destructive test methods, the timber can also be classified into grade S15/LS15 (see DIN 4074–1 & DIN 4074–5, chapter 7.3.1). The assignment of the visually determined classes to the strength classes according EN 338 is accomplished according to EN 1912 based on the provenance, the wood species and the applied grading standard.

This assignment process is not necessary if the timber is machine graded. The results of the applied nondestructive measurements are used to directly assign the timber to strength classes according to EN 338. Besides deflection measurements, optical methods and radiography/microwaves dynamic measurements are applied for mechanical grading (see Sandomeer and Steiger 2009).

The grading methods which have been developed for new structural timber can only be applied with great limitations on timber members in existing structures. This concerns basically the limited accessibility and visibility of the timber elements, the non-existing personnel qualification as well as the lack of in-situ flexible manageable and certified grading apparatuses (see Lißner and Rug 2018). Therefore, a strength grading of timber members in existing structures is rarely carried out. The present load-bearing capacity of the structural timber is at most intuitively estimated. Static calculations are then performed under the consideration that the structural timber belongs to the grades S10 and LS10 according to DIN 4074–1/-5 respectively C24 and D30 according EN 338. In doing so, load-bearing capacity reserves and deficits cannot be revealed. This can lead to a less substance-carefully and unprofessional redevelopment. Nevertheless, the determination of the present material properties by a reliable on-site grading enables a more reliable assessment of the structural stability and functional capability. This enables substance-careful and efficient redevelopment measures.

2 Strength Grading of Timber in Existing Structures

The strength grading of structural timber members in existing structures in combination with the application of non-/semi-destructive test methods allows the exact and reliable determination of material properties. This would not be possible with the visual strength grading alone.

The visually observable and measurable grading criteria show only a weak correlation to the strength properties of structural timber (see Glos 1995; Sandomeer and Steiger 2009; Denzler and Glos 2009). This leads to a low degree of distinctiveness, efficiency and significance. The combination of the visual grading with non-/semi-destructive measurements and test methods enables a significant enhancement of the efficiency, as shown in Table 1.

In the last decades many non-/semi destructive test methods for the in-situ evaluation of structural timber have been developed, investigated and tested (see Lißner and Rug

Table 1. Relation between non-destructive measurable indicating properties (IP) of the strength and the actual, destructive measurable strength properties (taken from Sandomeer und Steiger, 2009)

Indicating properties (IP)	Coefficient of determination (R^2)
Annual ring width	0,15 … 0,35
Knots	0,15 … 0,35
Density	0,20 … 0,40
Natural frequency, ultrasonic velocity	0,30 … 0,55
Static modulus of elasticity	0,40 … 0,65
Dynamic modulus of elasticity	0,30 … 0,55
Knots & density	0,40 … 0,60
Knots & modulus of elasticity	0,55 … 0,75
Knots, density & modulus of elasticity	0,55 … 0,80

2018). Although this is rarely possible, the laboratory testing of samples taken semi-destructively certainly enables the exact determination of material properties (see Rug und Seemann 1988) - especially in structures which are listed as national heritage. In such cases the non-destructive determination of material properties is only possible with sclerometers and dynamic test methods. The dynamic test methods include the measurement of the natural frequency (see Görlacher 1984) and the ultrasonic measurement (see Steiger 1996; Sandoz 1989; Augustin 2004). Both methods are nowadays state of the art and are used e.g. for the grading of timber in sawmills.

3 The Ultrasonic Test Method

The ultrasonic test method is based on the strong correlation between the velocity of an ultra-sonic pulse and the stiffness and density of the material. It is divided into the ultrasonic echo method and the time-of-flight measurement. The ultrasonic echo method uses the reflection of an ultrasonic pulse induced perpendicular to the grain on surfaces (i.e. surfaces or imperfections). This method is mainly used for the detection of imperfection and damage (see Linke et al. 2017). The time-of-flight measurement uses the time which is required to send an ultrasonic pulse from transmitter to receiver and is subdivided according to the application of the direction of measurement (see Fig. 1). This method suitable for the determination of material properties (see Linke et al. 2017).

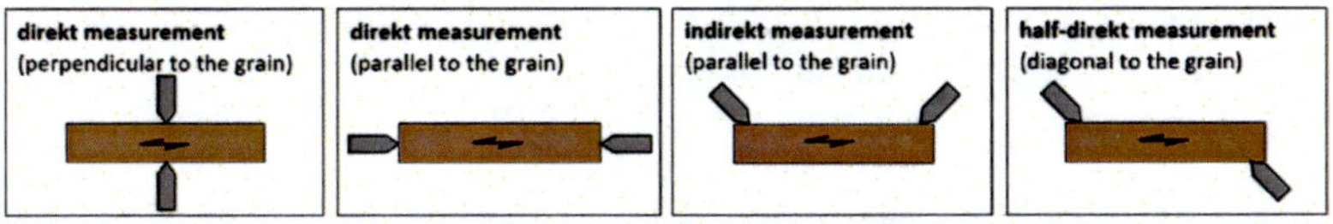

Fig. 1. Measurement methods for ultrasonic test taken from Linke et al. (2017)

Besides the investigation of the basic applicability and crucial influences – e.g. moisture content, temperature – the strength grading of timber with the ultrasonic time-of-flight measurement has been studied. This includes the relation between the ultrasonic velocity and the material properties which are relevant for strength grading – i.e. density, bending strength, modulus of elasticity. A detailed summary of the ultrasonic time-of-flight measurement's state of the art is given in Linke et al. (2017).

The predominant part of the previous studies focused on the application on new structural timber. However, single studies showed, that there is no significant difference between new and old timber (see Kranitz et al. 2010). Therefore, the application on old timber is possible. This has been the case in the last decade, although these studies focused mainly on single structures with a relatively small extent. Extensive systematic studies on old timber are missing hitherto.

4 Comparative Material Tests

4.1 Aim and Subject

The material tests described hereinafter are part of a systematical studies on new and old timber concerning the applicability of the ultrasonic time-of-flight measurement as a non-destructive method for the determination of the material properties of structural timber in existing structures. The aim of the study is the evaluation of the efficiency and reliability of the ultrasonic time-of-flight measurement as a grading tool. The subject of this sub study were specimen from spruce (Picea abies), pine (Pinus sylvestris) and oak (Quercus robur/petraea) with approximately 300 specimens of each species. The sample sizes and specimen dimensions are listed in Table 2.

Table 2. Sample sizes and specimen dimensions

Species	Sample size n	Specimen dimensions b/h/ℓ
Spruce (Picea abies)	303	50/80/1520 mm
Pine (Pinus sylvestris)	300	
Oak (Quercus robur/petraea)	301	

4.2 Methods

The comparative material tests are divided into three parts:

1. Visual grading according to EN 14081–1:2016 and DIN 4074–1:2012/-5:2008
2. Ultrasonic time-of-flight measurements
3. Destructive bending test according to EN 408:2012

Additionally, the density was determined according to EN 408:2012 and the moisture content was measured according to EN 13183–1/-2:2002.

4.3 Visual Strength Grading

The visual strength grading of the specimen was carried out according to DIN 4074–1:2012 and DIN 4074–5:2008. Herein, the criteria listed in chapter 5 of the respective standards were applied.

Several further features of the specimen, which do not account for one of the criteria listed above (e.g. finger joints, smaller damages due to production/transportation) were documented but not taken into consideration for the classification according to DIN 4074–1:2012/–5:2008.

4.4 Ultrasonic Time-of-Flight Measurements

After the specimen were graded visually the time-of-flight and the ultrasonic velocity were measured with the apparatus Sylvatest Trio (CBT CBS Lausanne, CH, see Fig. 2).

The measurements were carried out as direct (probes are placed end to end) and indirect (probes are placed sideways in an angle of approx. 30°) measurement parallel to the grain. On each specimen, the measurement was performed on the upper and lower third of the specimen's height (direct measurement) and on the top and bottom side of the specimen respectively (see Fig. 2). For each measurement the time-of-flight as well as the velocity of the ultrasonic impulse was documented.

Fig. 2. Direct (top) and indirect (bottom) ultrasonic time-of-flight measurements (1 ... transmitter/receiver probe, 2 ... test apparatus Sylvatest Trio, Fa. CBT CBS, Lausanne/CH)

Additionally, the climatie conditions (GANN Hydromette BlueLine Compact) and the moisture content (GANN Hydromette HT 85 with insulated electrodes, depth of measurement t = 15mm) was measured. The test results were adjusted to a moisture content of $\omega = 12\%$ and a temperature of $\upsilon = 20$ °C for better comparability. For this, the adjustment equations according to Sandoz (1993) were applied (see Eqs. (1) and (2)).

$$v_{12} = v_{\omega} + 29 \cdot (\omega - 12) \mathit{for}\, \omega \leq 32\% \tag{1}$$

where v_{12} is the velocity at $\omega = 12\%$, v_{ω} is the velocity at $\omega \neq 12\%$ and ω is the moisture content.

$$v_{20} = v_{\upsilon} - 3,9 \cdot (\upsilon - 20) \mathit{for}\, \omega = 12\% \tag{2}$$

where v_{20} is the velocity at $\upsilon = 20$ °C, v_υ is the velocity at $\upsilon \neq 20$ °C and υ is the temperature.

4.5 Destructive Bending Tests

The global modulus of elasticity and the modulus of rupture (i.e. bending strength) were determined in bending tests according to EN 408:2012, chapters 10 & 19. The used test setup is depicted in Fig. 3.

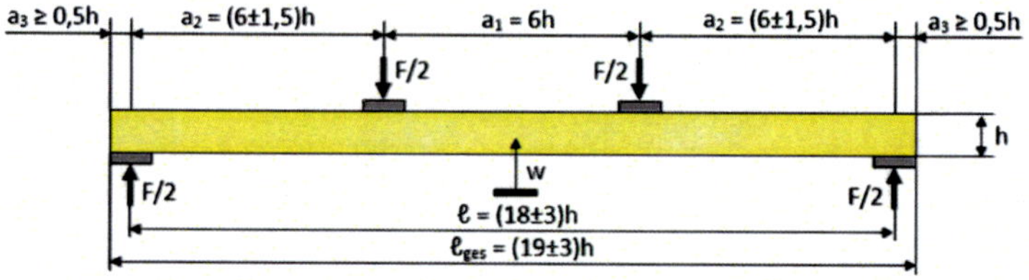

Fig. 3. Bending test setup ($\ell = 1440$ mm, $\ell ges = 1520$ mm, a1 = a2 = 480 mm, a3 = 40 mm)

The test load was applicated with a hydraulic press (max. load: 500 kN). The deflection was measured over the crosshead travel (with stiffness correction) with an external sensor (ASM position sensor WS11–2000).

The density was determined according to EN 408:2012 on samples which were cut out of the bending specimen (8 samples for each specimen). Additionally, the moisture content was determined on the same samples according to EN 13183–1:2002.

5 Results and Discussion

5.1 Visual Strength Grading

The results of the visual strength grading of the tested specimen is depicted in Fig. 4. The predominant part of the sample material was assigned to the grading classes S7/LS7 (low load-bearing capacity) and S10/LS10 (average load-bearing capacity). Only 11% (spruce and pine timber) and 19% (oak timber) met the requirements for the classification in the highest visual grading class S13/LS13 (high load-bearing capacity). What is more, 13–23% (spruce and pine timber) and 21% (oak timber) could not be assigned to any grading classes according to DIN 4074–1/-5. The main decisive grading criteria for all specimen (coniferous and deciduous wood) were knots, slope of grain and cracks.

The relatively large amount of timber with low and average load-bearing capacity (grading classes S7/LS7 and S10/LS10) as well as the quite high rate of rejected specimen is a common result for a visual strength grading of structural timber. The observed grading yield is mainly caused by the relatively week correlation between the visual determinable and measurable grading criteria and the actual strength and stiffness properties of timber (compare Table 1). These results prove the fact, that timber with a high load bearing capacity (i.e. S13/LS13 acc. DIN 4074–1/-5 respectively > C30/D30 acc. EN 338) cannot be graded visually in a reliable manner.

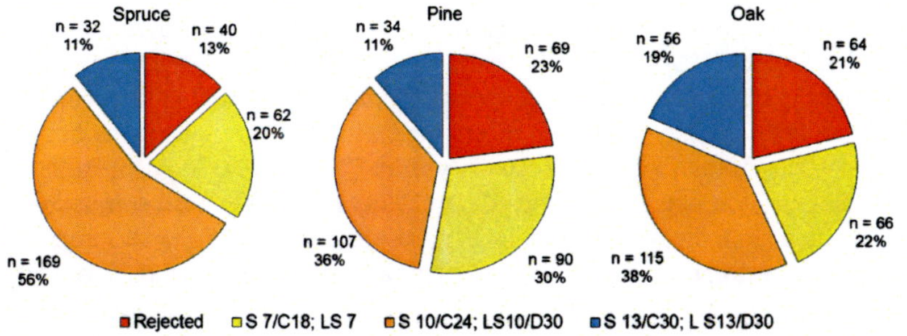

Fig. 4. Results of the visual strength grading according DIN 4074–1:2012/-5:2008 (n = sample size)

A significant improvement of the grading yield can be achieved by the additional application of non- and semi-destructive test methods. For example, the additional determination of the modulus of elasticity with the help of dynamic test methods leads to a significantly higher correlation (compare Table 1 & Fig. 5).

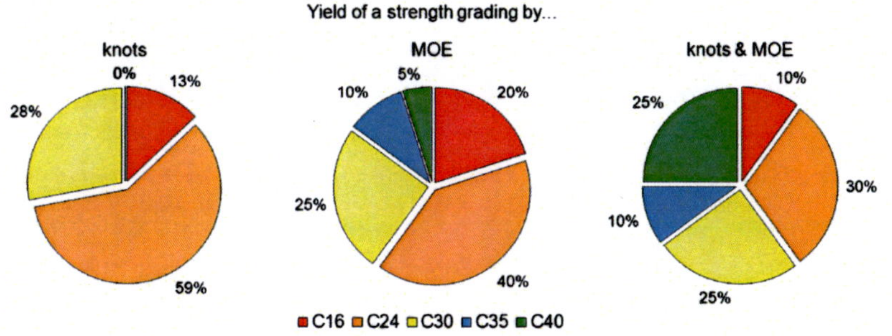

Fig. 5. Improvement of the grading yield by application of different grading criteria (taken from Blaß and Görlacher 1996)

The applicability of an apparatus-supported strength grading is proven amongst others by the studies of Blaß and Frese (2002). The proposed grading procedure includes the determination of the dynamic modulus of elasticity with the natural frequency measurement according to Görlacher (1984) and the density as well as a visual grading according to the requirements of grading class S10 according to DIN 4074–1. Interestingly, a grading solely based on non-destructive measurements - i.e. the waiver of the visual grading – led to a decrease of the characteristic values. The grading procedure proposed by Blaß/Frese was introduced into the German grading standard in 2008.

5.2 Grading with the Ultrasonic Time-of-Flight Method

The ultrasonic time-of-flight measurement gives the opportunity to determine the material values of timber with relatively low effort. The applicability of this test method for the non-destructive testing and strength grading of structural timber was partially investigated in the past. In particular, the studies on Swiss and Austrian spruce timber presented in Steiger (1996), Sandoz (1989) and Augustin (2004) must be mentioned. In order to use the ultrasonic velocity as a grading criterion, limiting values were derived from the test results. They have been determined with the regression equations and the normative values of the modulus of elasticity from EN 338 (e.g. Augustin 1998).

In analogy to the approach of Augustin (1998) several regression equations have been calculated from the results of the own ultrasonic time-of-flight measurements and the destructive bending tests. Herein, the relation between the ultrasonic velocity and the bending strength as well as the static modulus of elasticity was observed. The results are shown in Table 3.

Table 3. Relation between the ultrasonic velocity and the bending strength as well as the static modulus of elasticity of spruce, pine and oak (v_{dir} – direct measurement, v_{ind} – indirect measurement, r – coefficient of correlation, r^2 - coefficient of determination)

Relation	Spruce	Pine	Oak
$v_{dir} \leftrightarrow f_m$	$f_m = 0{,}02v_{dir} - 84$ ($r = 0{,}425$; $r^2 = 0{,}181$)	$f_m = 0{,}02v_{dir} - 26$ ($r = 0{,}227$; $r^2 = 0{,}052$)	$f_m = 0{,}03v_{dir} - 81$ ($r = 0{,}686$; $r^2 = 0{,}470$)
$v_{ind} \leftrightarrow f_m$	$f_m = 0{,}02v_{ind} - 76$ ($r = 0{,}445$; $r^2 = 0{,}198$)	$f_m = 0{,}02v_{ind} - 34$ ($r = 0{,}274$; $r^2 = 0{,}075$)	$f_m = 0{,}03v_{ind} - 86$ ($r = 0{,}717$; $r^2 = 0{,}514$)
$v_{dir} \leftrightarrow E_m$	$E_m = 4{,}6v_{dir} - 14267$ ($r = 0{,}543$; $r^2 = 0{,}295$)	$E_m = 2{,}8{*}v_{dir} - 1688$ ($r = 0{,}284$; $r^2 = 0{,}081$)	$E_m = 6{,}04v_{dir} - 15374$ ($r = 0{,}829$; $r^2 = 0{,}687$)
$v_{ind} \leftrightarrow E_m$	$E_m = 4{,}4v_{ind} - 12467$ ($r = 0{,}564$; $r^2 = 0{,}318$)	$E_m = 2{,}9v_{ind} - 2026$ ($r = 0{,}317$; $r^2 = 0{,}100$)	$E_m = 6{,}2v_{ind} - 15629$ ($r = 0{,}844$; $r^2 = 0{,}712$)

Table 3 shows, that there is no significant difference between the direct and indirect measuring method. Nevertheless, the relations to the modulus of elasticity are significantly stronger than the relations to the bending strength. In general, the observed relations differ significantly. While the observed coefficients of correlation for spruce and oak have an average to high value, the relations for pine are relatively weak.

The determined regression equations were used to calculate limiting values for the ultrasonic velocity. These were based on the characteristic bending strength and the mean value of the modulus of elasticity for coniferous and deciduous timber according to EN 338. The reliability and applicability of the derived limiting values was verified by comparing the classification according to the ultrasonic velocity with the classification according to the determined material properties density, bending strength and modulus of elasticity. In order to reach a high level of accordance the limiting values have been adjusted empirically. The results of this approach are shown in Table 4.

Table 4. Accordance between the grading results by grading according the ultrasonic velocity and the material properties density, bending strength and static modulus of elasticity (v_{dir} – direct measurement, v_{ind} – indirect measurement)

Relation	Spruce	Pine	Oak
$v_{dir} \leftrightarrow f_m$	57%	48%	54%
$v_{ind} \leftrightarrow f_m$	53%	50%	42%
$v_{dir} \leftrightarrow E_m$	84%	79%	94%
$v_{ind} \leftrightarrow E_m$	74%	55%	65%

The grading with the limiting values which were calculated on basis of the static modulus of elasticity shows a higher level of accordance than the approach based on the characteristic bending strength. Furthermore, the accordance based on the direct measurement is higher than on basis of the indirect measurement. This observation applies on all the three investigated species.

Secondly, the grading yield of the ultrasonic grading is improved when compared to the visual grading. The rate of timber with a high load-bearing capacity (>C30/D30 acc. EN 338) is significantly higher (see for example Fig. 6).

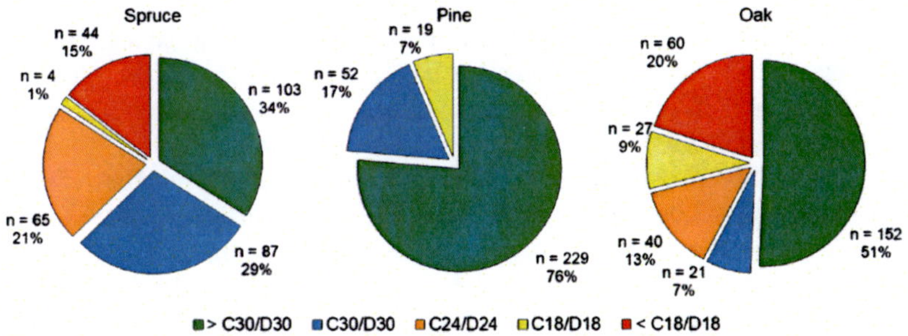

Fig. 6. Results of the grading based on the directly measured ultrasonic velocity and the static modulus of elasticity (n = sample size)

These results show that the ultrasonic velocity can principally be used as an additional grading criterion. To achieve a high level of accordance, the limiting values based on the relation between the ultrasonic velocity and the static modulus of elasticity should be applied. The limiting values, which were derived from the results of this study are shown in Table 5.

Table 5. Proposal for limiting values for the grading criterion "ultrasonic velocity" (v_{dir} – direct measurement, v_{ind} – indirect measurement)

Strength class	Spruce		Pine		Strength class	Oak	
	v_{dir}	v_{ind}	v_{dir}	v_{ind}		v_{dir}	v_{ind}
C18	5100	4900	3900	3900	D24	4300	4200
C24	5500	5350	4600	4600	D30	4500	4400
C30	5750	5550	5000	4900	D35	4600	4600
C35	5950	5800	5300	5200	D40	4800	4700
C40	6150	6050	5700	5600	D50	5000	4900

5.3 Grading Based on Indicating Properties

The results of the grading based on the ultrasonic velocity show, that thereliability must be enhanced since the observed accordance is partially low (see Table 4). Therefore, an alternative approach was investigated.

The machine grading uses a so-called grading parameter or indicating property. This parameter is derived from multivariate regression models and is used as an estimated value for the target values (i.e. strength and stiffness properties). Hereby, different test methods can be applied and considered in the grading decision.

According to this approach the results of the comparative material test were used to set up several multivariate linear regression models. In order to take the particularities of existing structures and the possible implementation of adjusted analysis methods into account the regression models were set up in a multi-level system.

The basis of this consideration are the so-called knowledge levels for the structural analysis proposed by Loebjinski et al. (2018). The proposal differentiates four levels which range from extremely limited (knowledge level KL 0) to high (knowledge level KL 3). Depending on the level of investigation and the gathered information on the respective structure the application of adjusted analysis methods is possible. Nevertheless, the determination of the present material quality with non- and semi-destructive test methods is a strict requirement for their application. Table 6 shows the knowledge level as well as the corresponding minimum strength grading levels.

The strength grading procedure required in the knowledge level KL 1 is comprehensively regulated in the present grading standards (EN 14081–1 & DIN 4074–1/-5). For the knowledge levels 2 & 3 the regulations of the DIN 4074–1/-5 concerning the apparatus supported visual grading can be considered. However, these normative regulations do not contain any concrete provisions on how the test methods are to be applied together with the visual grading regulations. Therefore, for further considerations the approach normatively regulated for machine grading is used.

The multivariate linear regression models were set up separately for the strength grading level SGL 2 & 3. Visual grading criteria as well as non- and semi-destructively measurable material properties were considered. The visual grading was taken into account by the criteria knots, slope of grain and cracks. The other visual grading criteria were not considered since they are often not measurable on-site (e.g. pith and annual ring

Table 6. Strength grading level (SGL) and the corresponding knowledge level (KL) according to Loebjinski et al. (2018)

Strength grading level	Required strength grading procedure	Corresponding Knowledge Level
SGL 0	No strength grading required	KL 0
SGL 1	Visual grading according EN 14081–1	KL 1
SGL 2	Apparatus supported visual grading *(visual grading and non-destructive test methods)*	KL 2
SGL 3	Apparatus supported visual grading *(visual grading and non-/semi-destructive test methods)*	KL 2/3

width), can be taken into account in the structural analysis (e.g. wane as reduced cross section, deformation as eccentricity) or can be eliminated in the course of redevelopment (e.g. decay, insects). The ultrasonic velocity as well as the transferred signal voltage were used as non-destructive criteria in the strength grading level SGL 2. Additionally, the dynamic modulus of elasticity and the density were considered in strength grading level SGL 3.

The relation between the grading criteria and the target property were investigated in different combinations. As a preliminary target property, the static modulus of elasticity was used. This was considered due to the generally high correlation between the static modulus of elasticity and the other material properties of timber. Furthermore, the investigated apparatus supported grading method is based on the dynamic properties of timber. For each species a total of 103 regression models were set up. The coefficients of correlation and determination of the strongest regressions of each species are listed in Table 7 for the strength grading levels SGL 1–3.

Table 7 shows, that the relations of strength grading level SGL 1 are relatively weak. The implementation of non-destructively measurable material properties (e.g. ultrasonic velocity) improves the relation significantly. The additional implementation of semi-destructively measurable properties (e.g. density determined on small core drill samples) raises the coefficients of correlation and determination further. Though, depending on the species only a minor improvement was found.

In order to use the investigated relations for the apparatus supported visual grading method appropriate limiting values must be determined for the indicating property. This was done following the regulations of the EN 14081–2:2018, Annex B. Therefore, one-dimensional linear regressions must be set up between the indication property (IP) and the target property (MOE) as shown in Eq. (3).

$$MOE = a_{MOE} \cdot IP + b_{MOE} \tag{3}$$

Table 7. Coefficients of correlation and determination of the investigated regressions

Strength grading level	Grading criteria	Coefficients of correlation r & determination r^2		
		Spruce	Pine	Oak
SGL 1	knots/slope of grain/cracks	$r = 0{,}294$ $r^2 = 0{,}086$	$r = 0{,}535$ $r^2 = 0{,}286$	$r = 0{,}310$ $r^2 = 0{,}096$
SGL 2	knots /slope of grain/cracks/ ultrasonic velocity/transferred voltage (direct measurement)	$r = 0{,}790$ $r^2 = 0{,}624$	$r = 0{,}817$ $r^2 = 0{,}667$	$r = 0{,}869$ $r^2 = 0{,}755$
	knots /slope of grain/cracks/ ultrasonic velocity/transferred voltage (indirect measurement)	$r = 0{,}799$ $r^2 = 0{,}638$	$r = 0{,}816$ $r^2 = 0{,}666$	$r = 0{,}878$ $r^2 = 0{,}771$
SGL 3	knots /slope of grain/cracks/ dynamic MOE/ density (direct measurement)	$r = 0{,}803$ $r^2 = 0{,}645$	$r = 0{,}931$ $r^2 = 0{,}867$	$r = 0{,}885$ $r^2 = 0{,}783$
	knots /slope of grain/cracks/ dynamic MOE/ density (indirect measurement)	$r = 0{,}815$ $r^2 = 0{,}664$	$r = 0{,}916$ $r^2 = 0{,}839$	$r = 0{,}899$ $r^2 = 0{,}808$

where MOE is the target value (here: static modulus of elasticity), IP is the indicating property, a_{MOE} is the regression coefficient of the target value and b_{MOE} is the intercept of the target value.

The limiting values of the indication property can then be calculated by converting the regression equation and using the characteristic values of the modulus of elasticity according to EN 338 as required target value as shown in Eq. (4).

$$S_{MOE} = \frac{MOE_{req} - b_{MOE}}{a_{MOE}} \tag{4}$$

where S_{MOE} is the limiting value of the indicating property and MOE_{req} is the required value of the target value (here: characteristic value according to EN 338).

The calculated limiting values must be verified. According to the regulations of EN 14081–3:2016 for output-controlled grading systems this must be done with the help of the characteristic values. They must be determined according to EN 384 based on material test according to EN 408. The limiting values are verified if the characteristic values of the sample are at least equal to the characteristic values given in EN 338. Otherwise the limiting values must be modified and verified again.

The determination of the limiting values for the indicating property for the investigated apparatus supported grading method was only necessary for the strength grading level 2 & 3. For the strength grading level SGL 1 the limiting values are not required since the classification must be done based on the limiting values given in DIN 4074–1/-5. The regression equations which were used for the determination of the limiting values are listed in Table 8.

Table 8. Regression equations for the determination of limiting values for the indicating property

Strength grading level	species	Regression equation	Coefficients of correlation r & determination r^2
SGL 2	Spruce Pine	$E_m = IP - 2,18 \cdot 10^{-11}$	$r = 0,790$; $r^2 = 0,624$
		$E_m = IP - 2,91 \cdot 10^{-11}$	$r = 0,799$; $r^2 = 0,638$
	Oak Spruce	$E_m = IP - 2,55 \cdot 10^{-11}$	$r = 0,817$; $r^2 = 0,667$
		$E_m = IP - 2,55 \cdot 10^{-11}$	$r = 0,816$; $r^2 = 0,666$
	Pine	$E_m = IP - 1,64 \cdot 10^{-11}$	$r = 0,869$; $r^2 = 0,755$
		$E_m = IP - 3,64 \cdot 10^{-12}$	$r = 0,878$; $r^2 = 0,771$
SGL 3	Spruce Pine	$E_m = IP + 7,26 \cdot 10^{-12}$	$r = 0,803$; $r^2 = 0,645$
		$E_m = IP - 1,27 \cdot 10^{-12}$	$r = 0,815$; $r^2 = 0,664$
	Oak Spruce	$E_m = IP + 3,64 \cdot 10^{-12}$	$r = 0,931$; $r^2 = 0,867$
		$E_m = IP - 3,64 \cdot 10^{-11}$	$r = 0,916$; $r^2 = 0,839$
	Pine	$E_m = IP + 5,46 \cdot 10^{-12}$	$r = 0,885$; $r^2 = 0,783$
		$E_m = IP - 3,64 \cdot 10^{-11}$	$r = 0,899$; $r^2 = 0,808$

The limiting values of the indicating property were only calculated for selected strength classes according EN 338:2016, as listed below:

- Coniferous timber (Spruce/Pine): C18, C24, C30, C35, C40
- Decidious timber (Oak): D18, D24, D30, D35, D40

The determination of limiting values for the strength classes higher than C40/D40 was waived, since experience has shown that these values do not produce any significant improvements in the verification procedure or any improvements that can be used in reality.

The verification as well as a potentially necessary empirical modification of the limiting values was carried out in two steps. Firstly, the characteristic values of the static modulus of elasticity $E_{0,mean}$ were calculated for each class. Secondly, the classification based on the indicating property as well as based on the static modulus of elasticity determined in destructive bending test were compared.

The characteristic values for the static modulus of elasticity were calculated according to EN 384:2019, chapter 5.5.2.1 and EN 14358:2016, chapter 3.3. The classification based on the static modulus of elasticity which was determined in destructive bending tests was carried out under consideration of a measurement correction of 3%. Furthermore, a limiting value undercut of 5% was considered acceptable for the classification based on the indicating property following the regulations of EN 384:2019, Eq. (10).

Table 9 shows the derived limiting values. It can be seen that - due to the relatively strong relation (see Table 8) - the characteristic values of the static modulus of elasticity which are given in EN 338:2016 can be applied as limiting values for the indicating properties without modification.

The comparison between the classification based on the indicating property and the classification based on the static modulus of elasticity determined in destructive bending test shows relatively high level of correspondence of approximately 85–95% depending on the species (see also Table 10). The empirical modification of the limiting values did not lead to a significant improvement of the level of correspondence.

Table 9. Limiting values for the indicating property for strength grading level SGL 2 & 3

Strength class acc. EN 338	Spruce [N/mm^2]	Pine [N/mm^2]	Strength class acc. EN 338	Oak [N/mm^2]
< C18	< 9000	< 9000	< D18	< 9500
C18	9000	9000	D18	9500
C24	11000	11000	D24	10000
C30	12000	12000	D30	11000
C35	13000	13000	D35	12000
C40	14000	14000	D40	13000

Table 10. Level of Correspondence between the classification based on Table 8 and the results of the destructive material tests

Strength grading level	Grading criteria	Level of correspondence		
		Spruce	Pine	Oak
SGL 2	knots /slope of grain/cracks/ ultrasonic velocity/transferred voltage (direct measurement)	85,1%	91,3%	91,3%
	knots /slope of grain/cracks/ ultrasonic velocity/transferred voltage (indirect measurement)	84,8%	92,3%	91,3%
SGL 3	knots /slope of grain/cracks/ dynamic MOE/ density (direct measurement)	85,1%	94,0%	91,6%
	knots /slope of grain/cracks/ dynamic MOE/ density (indirect measurement)	84,5%	94,0%	90,6%

The resulting grading yield is shown in Fig. 7. In General, there is a significant increase of timber with a very high load-bearing capacity (>C30/D30 according EN 338) compared to the visual grading as well as to the grading based exclusively on the ultrasonic time-of-flight measurement.

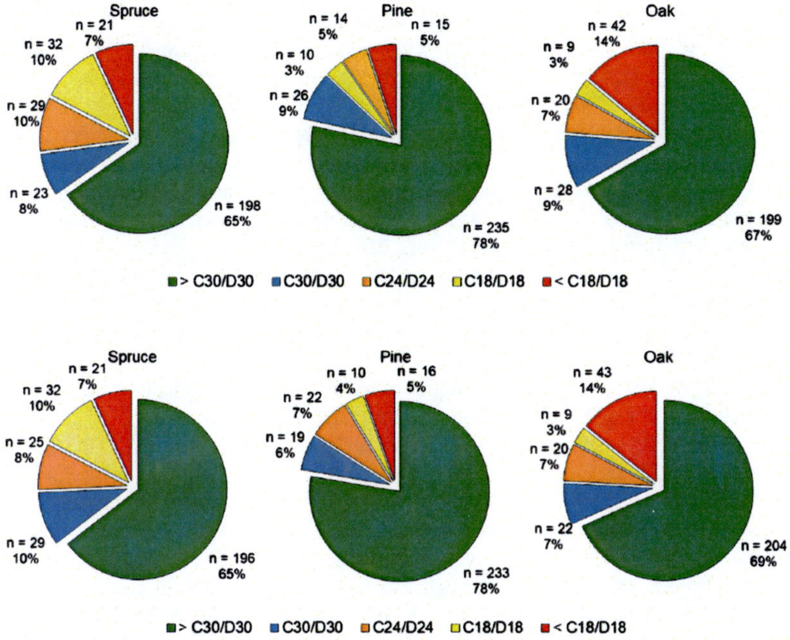

Fig. 7. Grading yield for the strength grading level – top: SGL 2; bottom: SGL 3 (indirect measurement; n = sample size)

6 Conclusion

The results of the study show that the application of the ultrasonic time-of-flight measurement for the on-site strength grading of timber members in existing structures in general is possible. The grading should not be carried out exclusively with non-destructive methods due to the partially weak correlation. The additional implementation of non- and semi-destructively measurable material properties – e.g. knots, slope of grain, cracks, density – leads to a significant improvement of the correlation. At the same time a higher level of correspondence with the destructive material tests was reached. The proposed grading method must be tested and verified in material tests on old timber on-site as well as in the laboratory. This will be the subject of future studies.

Acknowledgments. The studies have been financed exclusively by private donations. The authors want to thank all donors for the generous support.

References

Augustin, M.: Ermittlung der Auswirkungen der neu geregelten Sortiervorschriften für Kantholzquerschnitte für österreichische Sägebetriebe und Holzbauunternehmen. Grazer Holzbau-Fachtagung, Graz, Österreich (1998)

Augustin, M.: Eine zusammenfassende Darstellung der Festigkeitssortierung von Schnittholz. Technische Universität Graz, Lehrstuhl für Holzbau (2004)

Blaß, H.J., Frese, M.: Entwicklung eines Sortierverfahrens für die kombinierte maschi-nelle und visuelle Festigkeitssortierung. Holz und Steine, Abteilung Ingenieurholzbau, Universität Frideridiana Karlsruhe, Versuchsanstalt für Stahl (2002)

Blaß, H.J., Görlacher, R.: Visuelle und maschinelle Festigkeitssortierung von Vollholz. Mikado **5**, 64–71 (1996)

Denzler, J., Glos, P.: Maschinelle Festigkeitssortierung von Schnittholz. Zukunft Holz - Statusbericht zum aktuellen Stand der Verwendung von Holz und Holzprodukten im Bau-wesen und Evaluierung künftiger Entwicklungspotential, 355–359 (2009)

Glos, P.: Holz im technischen Einsatz, Perspektiven der Sortierung und Qualitätssicher-ung. Internationales Holzbauforum, Garmisch-Partenkirchen, Germany (1995)

Görlacher, R.: Ein neues Meßverfahren zur Bestimmung des Elastizitätsmoduls von Holz. Holz als Roh- und Werkstoff **42**, 219–222 (1984)

Linke, G., Rug, W., Pasternak, H.: Festigkeitssortierung von Bauholz in historischen Gebäuden – Bericht zum Stand der Technik. Bauingenieur **92**, 229–236 (2017)

Lißner, K., Rug, W.: Holzbausanierung beim Bauen im Bestand. Second edn. VDI Springer Verlag, Berlin (2018)

Loebjinski, M., Köhler, J., Rug, W., Pasternak, H.: Development of an optimisation-based and practice orientated assessment scheme for the evaluation of existing timber structures. In: 6th International Symposium on Life-Cycle Civil Engineering, pp. 28-31, Ghent, Belgium (2018)

Kranitz, K., Deublin, M., Niemz, P.: Strength estimation of aged wood by means of ultra-sonic devices. In: The Future of Quality Control for Wood & Wood Products, The Final Confer-ence of COST Action E53, 4-7 May, Edinburgh, United Kingdom (2010)

Rug, W., Seemann, A.: Festigkeit von Altholz. Holztechnologie **29**, 186–190 (1988)

Sandomeer, M.K., Steiger, R.: Potenzial der maschinellen Festigkeitssortierung von Schnittholz. Zukunft Holz - Statusbericht zum aktuellen Stand der Verwendung von Holz und Holzprodukten im Bauwesen und Evaluierung künftiger Entwicklungspotential, 361–370 (2009)

Sandoz, J.-L.: Grading of construction timber by ultrasound. Wood Sci. Technol. **23**, 95–108 (1989)

Sandoz, J.-L.: Moisture content and temperature effect on ultrasound timber grading. Wood Sci. Technol. **27**, 373–380 (1993)

Steiger, R.: Mechanische Eigenschaften von Schweizer Fichten-Bauholz bei Biege-, Zug-, Druck- und kombinierte M/N-Beanspruchung – Sortierung von Rund- und Schnittholz mittels Ultraschall. Institut für Baustatik und Konstruktion, Eidgenössische Technische Hochschule Zürich, Switzerland (1996)

DIN 4074–1:2012–06 - Sortierung von Holz nach der Tragfähigkeit – Teil 1: Nadelschnittholz

DIN 4074–5:2008–12 - Sortierung von Holz nach der Tragfähigkeit – Teil 5: Laubschnittholz

DIN EN 338:2016–07 – Bauholz für tragende Zwecke – Festigkeitsklassen

DIN EN 384:2019–02 – Bauholz für tragende Zwecke – Bestimmung charackteristischer Werte für mechanische Eignenschaften und Rohdichte

DIN EN 408:2012–10 - Holzbauwerke – Bauholz für tragende Zwecke und Brettschichtholz – Bes-timmung einiger physikalischer und mechanischer Eigenschaften

DIN EN 1912:2013–10 – Bauholz für tragende Zwecke – Festigkeitsklassen – Zuordnung von visuellen Sortierklassen und Holzarten

DIN EN 13183–1:2002–07 – Feuchtegehalt eines Stückes Schnittholzes – Teil 1: Bestimmung durch Darrverfahren
DIN EN 13183–2:2002–07 – Feuchtegehalt eines Stückes Schnittholzes – Teil 2: Schätzung durch elektrisches Widerstands-Messverfahren
DIN EN 14081–1:2016–06 – Holzbauwerke – Nach Festigkeit sortiertes Bauholz für tragende Zwecke mit rechteckigem Querschnitt – Teil 1: Allgemeine Anforderungen
DIN EN 14081–2:2018–12 – Holzbauwerke – Nach Festigkeit sortiertes Bauholz für tragende Zwecke mit rechteckigem Querschnitt – Teil 2: Maschinelle Sortierung; zusätzliche Anforderungen an die Erstprüfung
DIN EN 14081–3:2018–12 – Holzbauwerke – Nach Festigkeit sortiertes Bauholz für tragende Zwecke mit rechteckigem Querschnitt – Teil 3: Maschinelle Sortierung; zusätzliche Anforderungen an die werkseigene Produktionskontrolle
DIN EN 14358:2016–11 – Holzbauwerke – Berechnung und Kontrolle charakteristischer Werte

In Situ Tests Procedures for the Evaluation of Masonry Mechanical Parameters

Pietro Croce(✉), Filippo Landi, Paolo Formichi, Maria Luisa Beconcini, Beconcini Puccini, and Vincenzo Zotti

Department of Civil and Industrial Engineering (DICI), Università di Pisa, Largo Lucio Lazzarino 1, 56122 Pisa, Italy
{p.croce,filippo.landi}@ing.unipi.it

Abstract. The evaluation of seismic vulnerability of existing masonry building is a first fundamental step for the preservation of historical buildings in seismic prone area.

However, the definition of masonry's mechanical behavior, which is characterized by a complex, nonhomogeneous, anisotropic, non-linear behavior, is still a critical issue that significantly influences the structural analysis and the evaluation of seismic risk index. Different test procedures have been proposed in the past such as double flat jack, diagonal compression and shear compression tests to characterize the masonry behavior; but their outcomes are often contradictory and can differ significantly for the same class of masonry or even for the same wall.

In the paper, a new test procedure for the characterization of masonry mechanical parameters and especially the shear behavior of masonry walls is presented together with the experimental test results obtained for a stone masonry wall and a brick masonry wall. The procedure consists of a combination of three different in situ tests to be performed on the investigated wall. First, a single flat jack test is done to derive the normal compressive stress acting on the wall, then a double flat jack test is carried out to estimate the elastic modulus and finally a shear test is performed to directly assess the capacity curve and estimate the shear modulus and the shear strength.

The first results obtained in the experimental campaign carried out by the authors show the capability of the proposed methodology that allows a complete assessment of masonry mechanical parameters reducing the uncertainty in the estimation of shear behavior and consequently on the evaluation of seismic vulnerability of the investigated buildings.

Keywords: Existing masonry buildings · Masonry mechanical parameter · Flat jack test · Shear test · Shear behavior

1 Introduction

The majority of existing masonry buildings in seismic prone areas is characterized by a high vulnerability, mainly due to the lack of seismic design provisions at the time of construction and the typological and structural characteristics of the existing buildings. The

I. Vayas and F. M. Mazzolani (Eds.): PROHITECH 2021, LNCE 209, pp. 20–33, 2022.
https://doi.org/10.1007/978-3-030-90788-4_2

safety assessment and the design of strengthening interventions for masonry buildings is a standing challenge and the characterization of mechanical properties of masonry is undoubtedly one of the most delicate issues in the process, in particular in the field of heritage structures.

Particular attention should be devoted to the evaluation of the shear strength and the shear modulus, which are, among the mechanical parameters of masonry, those that most control the determination of the structural response to seismic actions (Beconcini et al. 2019) (Croce et al. 2019). Numerical experiences carried out by the authors showed that the seismic risk index (I_R), i.e. the value which summarizes the seismic performance of the structure, could significantly vary based on the assumption on its shear behavior (Croce et al. 2018b, 2019a and 2019b). Furthermore, the identification of shear strength and the shear modulus is still critical, since the available results from experimental campaigns show values that can significantly differ within the same masonry class or even within the same wall (Bosiljkov et al. 2005).

It is thus highlighted the necessity of new testing procedures that could allow a proper, faster but still accurate, estimation of masonry parameters for a suitable definition of the capacity curves of masonry walls (Beconcini et al. 2019).

In the framework of an important research project regarding the seismic assessment of more than 80 school masonry buildings in Florence, a robust campaign of in situ tests has been carried out by the authors on different masonry typologies. The experimental campaign allowed the evaluation of different test methods on a large number of masonry buildings, aiming to define a methodology for a quick and reliable estimation of masonry parameters, to be used in the structural analysis (Croce et al. 2018a), also identifying valid innovative non-destructive in-situ tests. In fact, in addition to the common testing procedures, an innovative in-situ shear test is experimented by the authors, which, together with the test with the single and double flat jacks, provides all the needed mechanical parameters to model the investigated masonry wall.

Once carried out the in-situ investigations and an adequate knowledge of the building obtained, it is possible to implement a mathematical model, aiming to represent the actual behavior of the structure.

The proposed testing methodology and the preliminary results obtained from the survey campaign in two case studies, a brick masonry panel in the primary school "Francesco Baracca" and a stone masonry panel in the kindergarten and primary school "Giotto", both located in Florence, are presented in the following sections.

2 Masonry Mechanical Characterization

2.1 Masonry Types

The most common types of masonry present in the Italian built environment are stone masonry, which can be irregular or partially dressed, and brick masonry with lime mortar (Croce et al. 2018a).

Numerous experimental campaigns were carried out on these and on other types of masonry for the evaluation of their mechanical properties (Corradi et al. 2003; Croce et al. 2021), leading to a classification of existing masonry typology, valid in Italy. A review of experimental results, especially concerning the shear behavior can be found in (Croce et al. 2018a).

More in detail, eight different types of masonry are identified according to the Italian Guidelines for the application of Italian Building Code 2018 (Italian Public Works Council 2019). In Table C8.5.I, for each masonry type, ranges of values are provided for mechanical parameters, such as the average compressive strength f_d, the average shear strength τ_0, the elastic modulus E, the shear modulus G and the specific weight w. These ranges are given as reference values for the mechanical definition of other masonry similar to those shown in the Table 1 and on which it is not possible to carry out more specific investigations. Moreover, they refer to low quality masonry characterized by mortar of poor characteristics, lacking of masonry courses, badly connected wall leaves and unconsolidated masonry. The presence of higher quality masonry is acknowledged by means of appropriate coefficients varying for each masonry type.

Table 1. Table C8. 5. I of the Italian guidelines 2019: reference values of the mechanical parameters of the masonry and average specific weight for different types of masonry.

Masonry Type	f_d (N/mm^2)	τ_0 (N/mm^2)	E (N/mm^2)	G (N/mm^2)	w (kN/m^3)
Rubble or Irregular stone	1.00–2.00	0.018–0.032	690–1050	230–350	19
Partially dressed stone	2.00	0.035–0.051	1020–1440	340–480	20
Fully dressed stone	2.6–3.8	0.056–0.074	1500–1980	500–660	21
Irregular soft stone	1.4–2.2	0.028–0.042	900–1260	300–420	13–16
Regular soft stone	2.0–3.2	0.04–0.08	1200–1620	400–500	13–16
Rectangular stone blocks	5.8–8.2	0.09–0.12	2400–3300	800–1100	22
Solid brick and lime mortar	2.6–4.3	0.05–0.13	1200–1800	400–600	18
Hollow bricks and cement mortar	5.0–8.0	0.08–0.17	3500–5600	875–1400	15

2.2 Tests Methods

In addition to the average values provided by the Italian Guidelines, the mechanical characteristics can be determined experimentally both in the laboratory and in situ to achieve an ever higher knowledge level for the materials (Beconcini 2018). Depending on the type of test, one or more mechanical parameters of the material under examination are estimated.

Material characterization tests, carried out in a laboratory, may concern, for example, direct compression tests on a masonry samples, compression tests on single resistant blocks, tests on mortar samples, such as Darmstadt test (Henzel and Karl 1987), PNT-G (Gucci and Barsotti 1995) or direct compression.

In situ tests are instead carried out, for example, with single and double flat jacks (ASTM 1991), pull-out tests (Croce et al. 2017), microseismic tests (Marchisio et al.

2003), sliding tests, diagonal compression tests (ASTM E 519–81, 1981), shear tests (Turnšek and Sheppard 1980), shear-compression tests (Corradi et al. 2003).

Destructive in-situ tests, such as diagonal shear tests, can result to be highly problematic, due to their significant invasiveness and they are commonly substituted by semi-destructive tests. Moreover, different types of tests can often lead to contradictory results even for the same type of masonry (Croce et al. 2018a) thus calling for a specific attention to the correct interpretation of the test results.

For this reasons, in the next chapter a reliable and effective test procedure is proposed, which allows a complete characterization of masonry walls, by means of the combination of semi-destructive tests. In this way, the four basic mechanical parameters, the elastic modulus E, the compressive strength f_d, the shear modulus G and the shear strength τ_k, can be obtained.

3 A Test Procedure for the Experimental Characterization of Masonry Walls

The proposed procedure is based on the combination of three experimental in situ tests to be carried out on an investigated wall:

- a single flat jack, which allows to evaluate the state of compression on the wall σ_0;
- a double flat jack, which allows to determine the elastic modulus, E, the Poisson ratio, v, and the compressive strength, f_d;
- a shear compression test (Sheppard 1980), to identify the shear strength τ_k and the shear modulus G

The following paragraphs briefly describe how the tests are carried out and the experimental results are elaborated.

3.1 In-situ Flat Jack Tests

The most common type of test used to estimate compressive behavior of masonry is the in-situ flat jack test (Gregorczyk and Lourenço 2000). The tests with in-situ flat jacks are semi-destructive tests, not too expensive and with very short execution times.

The flat jack can be applied either as a single or as a double jack (ASTM 1991). In the first case, the test is carried out to estimate the normal stress in the masonry, σ_0.

Three pairs of gauge points are fixed in the area where the cut is made. The distances between each pair is read both before and after cutting the wall, using millesimal deformometers (see Fig. 1). Once the cut is made and the flat jack is inserted, pressure is applied. When the initial distances between the sensors are found, the corresponding value of the pressure inside the jack indicates the stress which was present in the masonry before the cut.

The test with two flat jacks, instead, allows to obtain the behavior of an in-situ sample of masonry under uniaxial compression, from which to estimate compressive strength, the elastic modulus E and the Poisson ration v (calculated as the ratio between horizontal and longitudinal displacement). From the fundamental law linking material constants of isotropic materials an equivalent shear modulus G can be also derived.

$$G = \frac{E}{2(1+\nu)} \tag{1}$$

The test is performed by making a second cut at about 50 cm from the previous one and installing a total of four pairs of measurement bases, three longitudinal and one transversal, placed between the two jacks (see Fig. 1).

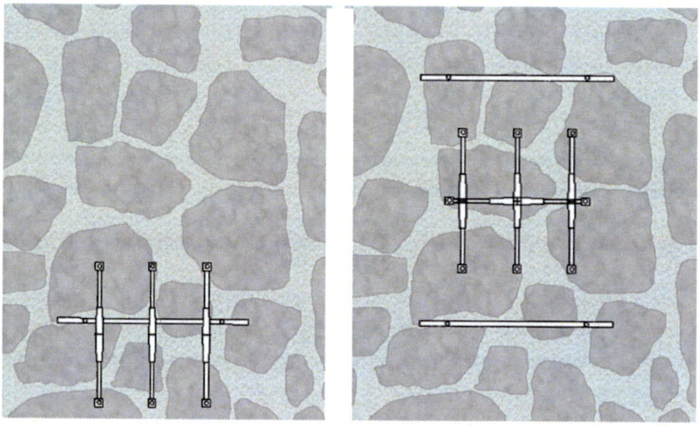

Fig. 1. Single and double flat jacks test on a generic masonry wall.

A gradual increase of pressure is applied, recording at each step the distances between the measurement bases, up to the pressure under which the first crack in the specimen appears. At the end of the test, particular attention should be paid to the shape and width of the cracks that occurred during the execution of the test.

Once both stress and deformation values have been determined, it is possible to draw the σ-ε diagram for the tested masonry and to determine the elastic modulus E.

Flat jack tests provide acceptable results, but require particular attention regarding the position of the cuts, the placement of the measurement bases, and the identification of opening cracks around the tested area.

In order to estimate the elastic modulus E, in different conditions, it was decided to evaluate it interpolating the points on the graph $\sigma-\varepsilon$ with three straight lines considering a percentage of load first ranging from 10% to 40% (which takes into account the elastic section of the masonry behavior), then an intermediate section from 40% to 70% of maximum load and finally from 70% to 100% (which takes into account the plastic section of the masonry behavior). In this way, the curve σ $-\varepsilon$ is approximated by a tri-linear one (see Fig. 2).

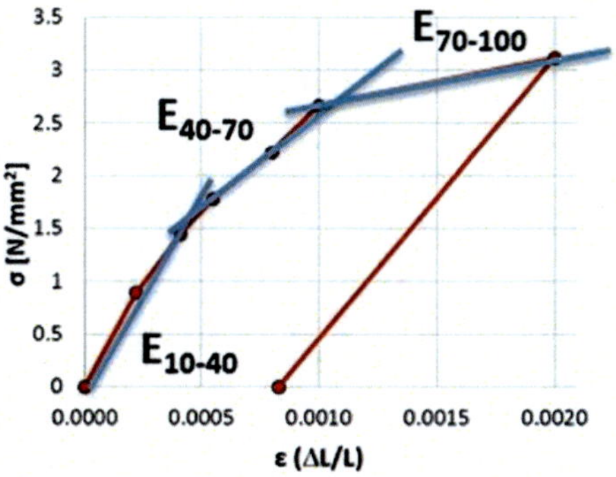

Fig. 2. Stress-deformation, $\sigma - \varepsilon$, diagram resulting from double flat jacks tests and approximated tri-linear curve.

3.2 An Innovative In-situ Shear Test

The shear strength τ_k and the shear modulus G are fundamental parameters for the definition of the seismic response of the construction. Indeed, they represent the major sources of uncertainty in the evaluation of seismic performance of masonry buildings (Croce et al. 2019).

For this reason the authors have developed an innovative in-situ shear test (Beconcini 2018), which have been tested on two masonry buildings in the Municipality of Florence.

To carry out the shear test it is first necessary to isolate laterally a masonry wall of adequate size, close to an opening, cutting at about 80 cm from the opening and as high as the opening (see Fig. 3). The wall which is under compression for the effects of the existing vertical loads acting on the structure, is subjected to an incremental horizontal load applied at the centerline of the panel, by means of two hydraulic jacks, stored in a hole in the masonry. Considering the presence of a fixed constraint for both ends of the panel, the jack identifies two parts of the panel, in which there is a state of compression and shear. During the test, the horizontal load is gradually increased until the maximum horizontal load is achieved, and diagonal cracks occur in one of the two parts.

The deformations are measured by using eight strain gauges placed along the diagonals of the two parts of the wall with a measuring base of about 70 cm, and six horizontal displacement transducers placed on the free lateral face, with a symmetrical arrangement with respect to the centerline (see Fig. 3).

The test thus provides the shear-displacement curve for the wall subjected to both compression and shear and allows to derive, combined with the previously determined parameters, the shear strength τ_k and the shear modulus G.

In particular, to determine the shear strength τ_k, the relationship developed in (Turnšek and Ĉaĉoviĉ 1971), which considers diagonal shear failure, has been used:

$$T_u = A\tau_k\sqrt{1 + \frac{\sigma_0}{b\tau_k}} \quad (2)$$

in which b, is the shear stress distribution factor that considers the variability of the shear action along the horizontal surface of the wall face and may be set equal to 1.5, for aspect ratio $h/l \geq 1.5$ (Tomazevic 2009).

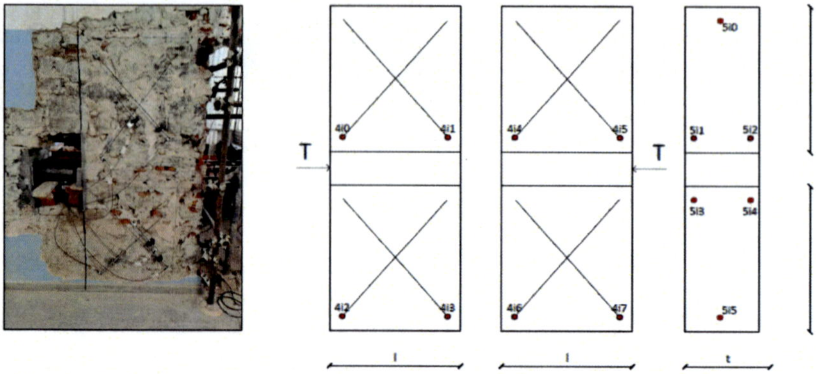

Fig. 3. In-situ shear test carried out on a stone masonry wall, load arrangement and position of measurement devices.

The ultimate horizontal load T is obtained from the test and assuming $T_u = T/2$, the c shear strength τ_k is determined by means of Eq. (2).

Starting from the shear displacement curve, the elastic stiffness k_{el} is determined taking as reference a horizontal load equal to 2/3 of T_u (Zimmermann 2011).

$$k_{el} = \frac{\frac{2}{3}T_u}{\delta_{\left(\frac{2}{3}T_u\right)}} \tag{3}$$

Modeling the masonry panel as a Timoshenko's beam, i.e. taking into account bending and shear deformations, considering the hypothesis of rigid spandrel, the stiffness of the wall can be defined as:

$$k_{el} = \frac{1}{\frac{h^3}{3EJ} + \frac{1.2h}{GA}} \tag{4}$$

Therefore, once calculated the elastic stiffness, it is possible to determine the shear modulus G of the wall, which correspond to the value of elastic modulus E obtained from the double flat jack.

4 Case Studies

4.1 Case Study A

The first masonry panel under study is located in the primary school "Francesco Baracca", a masonry building in the Municipality of Florence built in 1930. It is a three-story building, with a total height of about 16 m, an area of each floor of about 800 m^2 and a basement of 160 m^2.

The building is made up of different types of masonry walls, including masonry in disordered stones and bricks with lime mortar. During the experimental campaign, tests on masonry components and walls were carried out with different methodologies of different degrees of invasiveness, in order to evaluate the performance of different techniques and highlight the correlations between the results. The tests were carried out both in situ and in the Laboratory of the Department of Civil and Industrial Engineering of the University of Pisa.

With reference to solid bricks and lime mortar masonry: laboratory tests such as Darmstadt (Henzel and Karl 1987), PNT-G (Gucci and Barsotti 1995) and direct compression tests were performed on the mortar; two direct compression tests were also carried out, one on a masonry block and the other on a single brick. Further tests were done in situ: a single flat jack test, a double flat jack test (ASTM 1991), the proposed shear test (Beconcini 2018) and a microseismic test (Marchisio et al. 2003).

All the above tests were concentrated on the same wall, chosen in the basement, where there is a high state of compression.

Focusing on the in-situ tests, the results obtained by the single and double flat jack are: the normal stress σ_0 (value close to that determined with the structural analysis), compressive strength f_d and elastic modulus estimated on different sections of the curve with relative coefficients of Poisson and shear modulus G.

The results of the flat jack test for the case study A are reported in the following Table 2 and in Table 3.

The in-situ shear test is carried out on the wall with a section of 73 × 48 cm and a height of 175 cm, including cement plaster on both sides (see Fig. 4). The test was interrupted when cracks occurred in the upper half of the wall.

Below the graphs of the four strain gauges placed on the diagonals of the upper half of the panel (Fig. 5a) and the lateral displacements of the section at half height of the masonry wall analyzed are illustrated (Fig. 5b).

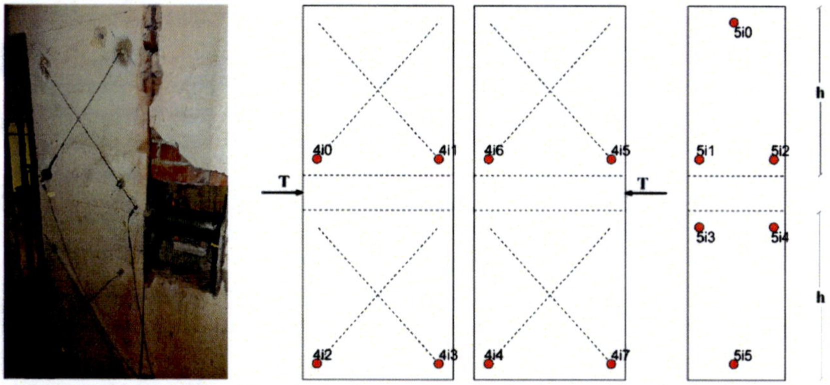

Fig. 4. Position of the instruments for the case study A: four strain gauges on the diagonals on each side and six horizontal displacement transducers (Fig. 4a, Fig. 4b)

In this first case study, the test determined an ultimate horizontal load T_u, from which it is derived the shear strength τ_k. The elastic stiffness k_{el} of the T-δ curve is estimated according to Eq. (3). So, by determining k_{el}, the shear modulus G of the wall is obtained

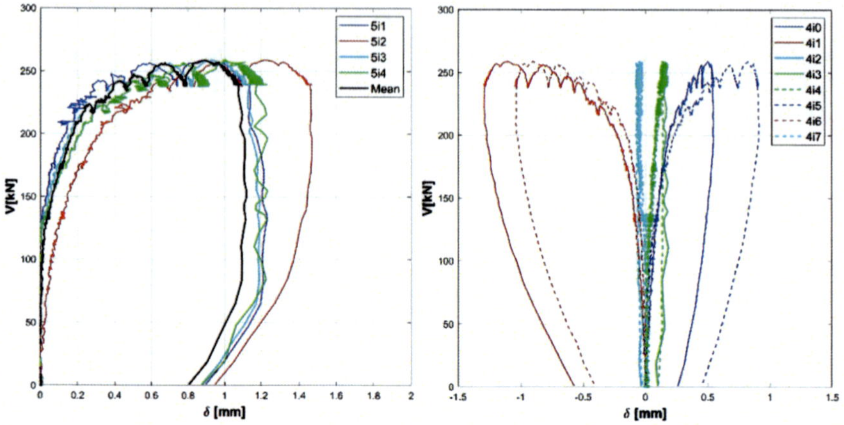

Fig. 5. Load displacement diagram of the four horizontal bases (Fig. 5a) and of the eight strain gauges on the diagonals (Fig. 5b), case study A.

using the value of the elastic modulus of the double flat jack ($E_{10\%–40\%}$). The values of the parameters of the shear test are reported in Table 4.

4.2 Case Study B

The second masonry panel belongs to the three-story masonry building housing the kindergarten and primary school "Giotto" in Florence, built in 1907–1908, with a U-shaped plan of approximately 38 × 25 m and an area of about 780 m^2 on the ground and first floor and 620 m^2 on the second floor. The load-bearing structure of the building is mainly made of irregular stone masonry.

Three tests have been carried out on this structure with single and double flat jacks together with the proposed in-situ shear test. The wall chosen for tests determined a normal stress σ_0 (value confirmed by the load analysis carried out), a maximum value of compressive strength σ and an elastic modulus calculated on three sections of the curve $\sigma - \varepsilon$ with relative coefficients of Poisson and shear modules G, all basic parameters for the following shear test. The results of the flat jacks for the case study B are reported in Tables 2 and 3. Subsequently, the wall measuring 90 × 48 × 195 cm^3 is isolated using a circular saw, on which the shear action, generated by two hydraulic jacks arranged in parallel inside the wall thickness in a hole in the wall, is applied in the middle (see Fig. 3). Two identical panels are determined, with dimensions of 90 × 48 × 90 cm^3, in which there is a state of shear-compression and with the same conditions of constraint at the ends, shaped like a fixed end.

Again, during the test, the shear action is implemented gradually, until one of the two parts of the wall reached breakage. The deformations are measured using 16 LVDT transducers, placed as illustrated in Fig. 3. The test was interrupted when cracks occurred in the top panel of the wall.

The results obtained from the diagonal and lateral transducers for the upper half of the panel, as the action increases, are shown in the T-δ diagrams in Fig. 6.

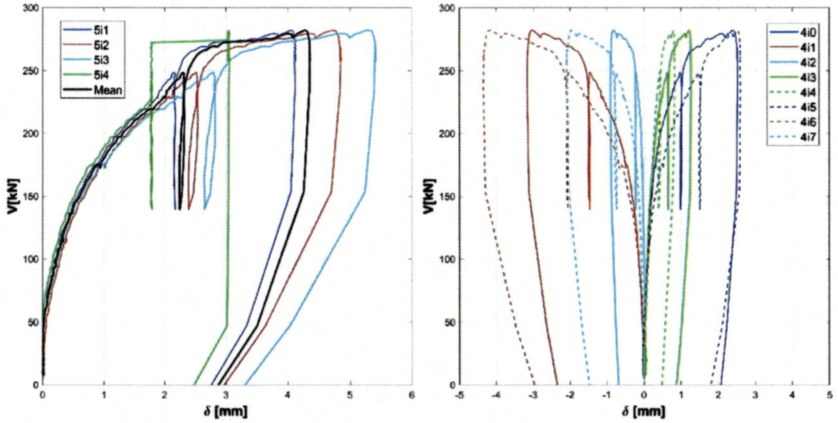

Fig. 6. Load displacement diagram of the four horizontal bases (Fig. 6a) and of the eight strain gauges on the diagonals (Fig. 6b), case study B.

From the shear action exerted by the jack is then possible to derive the capacity curve of the wall. The parameters obtained for the case study B are summarized in Table 4.

4.3 Analysis of Results and Comparisons

The results of the described in-situ tests, supplemented by previous studies reported in literature, define the values of mechanical parameters for the types of masonry walls under study that should be used for the assessment of the seismic performance of the investigated buildings.

The tests carried out with the double flat jacks provide information on the compressive strength of the load-bearing masonry wall of the building, comparing the values with those proposed in Table C8.5.I of the Italian Guidelines for the application of Italian Building Code 2018 (Table 1). The first cracking stresses obtained from the tests with the double flat jacks are used to determine the average compressive strength f_d of the masonry to be used in the analysis according to the methodology described in the Italian Guidelines. From these tests it is also possible to determine the elastic modulus *E* and the corresponding shear modulus *G*.

In the following Tables, the parameters obtained from the tests with single and double flat jacks for the two case studies are reported.

The experimental values of the ultimate shear forces, the ultimate strength, the shear resistances in the absence of compression and the elastic stiffness, from which the *G* shear modules are obtained using $E_{10\%–40\%}$ of the flat jacks, that are derived by means of the in-situ shear tests are reported in Table 3.

In accordance with the results presented in Croce et al. (2018a), it is possible to derive a procedure for determining the masonry mechanical parameters to be used for seismic analysis. The test campaign conducted by Tomažević (1999) leads to *G* values in the range $1000\tau_k \leq G \leq 700\tau_k$ and, as shown in Tomažević (1999) and Tomažević (2009), the experimental values of the *G/E* ratio are considered between 0.06 and 0.25,

Table 2. Values of normal stress σ_0 and compressive strength σ of flat jacks, average compressive strength $f_{d(TabC8.5.I)}$ for Italian Guidelines.

Case study	Flat jack test	σ_0 (N/mm^2)	σ (N/mm^2)	*fd (TabC8.5.I)* (N/mm^2)
A	**A.FJ1**	0.51	3.65	3.45
B	**B.FJ1**	0.168	2.23	2.0
	B.FJ2	0.31	1.34	2.0
	B.FJ3	0.465	3.56	2.0

Table 3. Values of elastic modulus *E* and shear modulus *G* calculated in three part of the curve σ−ε of double flat jacks and according to the average values proposed by Italian Guidelines.

Case study	Flat jack test	*E10–40%*	*E40–70%*	*E70–100%*	*G10–40%*	*G40–70%*	*G70–100%*	*E(TabC8.5.I)*	*G(TabC8.5.I)*
		(N/mm^2)			(N/mm^2)			(N/mm^2)	
A	**A.FJ1**	3740	2384	1824	1533	740	546	1500	500
B	**B.FJ1**	2180	983	718	858	354	226	1230	410
	B.FJ2	1654	1097	527	326	223	98		
	B.FJ3	3022	1455	807	414	127	71		

Table 4. Values of the ultimate shear forces T_u, the ultimate strength τ_u, the shear strength τ_k, the elastic stiffness k_{el} and the relative shear modules G, average shear strength $\tau_{k\,(TabC8.5.I)}$ for Italian Guidelines.

	Shear test	T_u (N)	τ_u (N/mm^2)	τ_k (N/mm^2)	k_{el} (N/mm)	G (N/mm^2)	$\tau_{k(TabC8.5.I)}$ (N/mm^2)
Case study A	**A.S1**	140000	0.40	0.28	729167	2918	0.09
Case study B	**B.S1**	141000	0.33	0.20	88479	292	0.043

with a value most commonly assumed in the seismic evaluation of a masonry building of about 1/6 ≈ 0.17.

The study of Croce et al. (2018a), comparing the results obtained from tests carried out with double flat jacks, diagonal compression jacks and shear tests and based on Tomažević's considerations, defines the shear modulus *G* and the elastic modulus *E* by means of the following formulas:

$$G = 2000 \cdot \tau_k \tag{5}$$

$$E = 6 \cdot G \tag{6}$$

Once identified the capacity curves of the two parts of the panels through the experimental in-situ shear test, it is possible to compare them with the trilinear curves obtained from the values of E and G of the flat jacks and with the bilinear capacity curves used for the non-linear static analysis, which are based on the relations between G and E defined above.

$$T_u = A\tau_k\sqrt{1 + \frac{\sigma_0}{1.5\tau_k}} \tag{7}$$

$$k = \frac{GA}{1.2h}\left(1 + \frac{G}{1.2E}\left(\frac{h}{l}\right)^2\right)^{-1} \tag{8}$$

$$\delta_e = \frac{T_u}{k} = 1.2\frac{\tau_k}{G}h\sqrt{1 + \frac{\sigma_0}{1.5\tau_k}} \tag{9}$$

The final displacement is calculated both controlling the ductility $\delta_u = \mu\ \delta_{el}$, as well as controlling the drift: $\delta_u = \gamma_u h$, where h is the wall height, More specifically, here it is assumed $\mu = 1.5$and, according the Italian Code, $\gamma_u = 0.04$.

In the following diagrams (Fig. 7), the experimental capacity curve (in blue) is compared with those obtained from the experimental values of the shear strength and assuming the relationships in Eq. (5) and (6) to evaluate E and G (bilinear curve in green). A further comparison is made with the trilinear (in red in the graphs) obtained using the values of E and G of the various sections of the curve $\sigma - \varepsilon$.

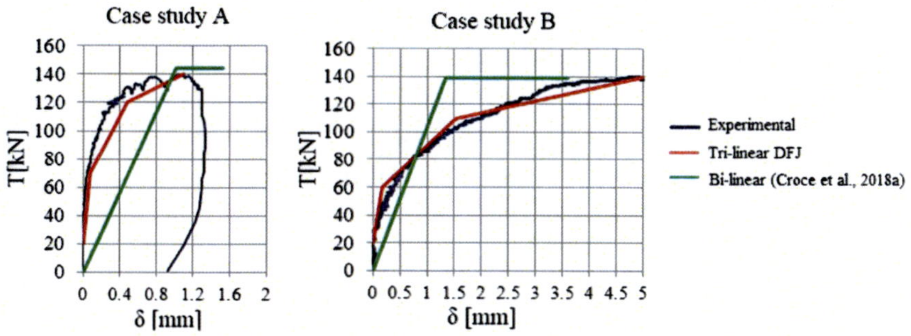

Fig. 7. Comparison of different capacity curves for the case study A and for the case study B.

In the case study A, the shear behavior obtained by the in-situ shear test (in blue) is close to that obtained with the results of the double flat jacks (in red) but more rigid than the bilinear one (in green) while in the "Giotto" school the tri-linear built according to the double flat jacks results(in red) is very close to the *T-δδ* curve of the shear test and the theoretical bilinear (in green) remains within the range of the investigated values. Considering that the tri-linear curves are built by calculating the stiffness of the various sections identified between 10% and 40%, 40% and 70%, and 70% and 100% of the maximum shear action achieved in the in-situ shear test, it is possible to affirm that they do not deviate from the experimental curves, well reproducing the experimental behavior in the various phases of the analysis.

Therefore, the results highlight the possibility to derive from double flat jack tests, a good description of the stiffness of the wall in different phases.

5 Conclusions

The definition of the masonry mechanical parameters is a critical issue in the assessment of the seismic performance of existing masonry buildings. In particular, the identification of the shear behavior of masonry wall is a fundamental step to provide a proper modeling of masonry walls in the structural analysis.

In order to achieve an adequate knowledge of the characteristics of materials and their degradation, in addition to the available documentation, extensive in-situ visual verifications and targeted experimental investigations are necessary.

A procedure to derive the fundamental mechanical parameters, elastic modulus, compressive strength, shear modulus and shear strength, to be used for the structural analysis of masonry buildings is proposed.

The procedure, which consists of the combination of three in-situ tests, single flat jack, double flat jack and an innovative shear test, has been applied in two case studies: a brick masonry panel and a stone masonry panel.

Preliminary results show the ability of the procedure to derive the required parameters in an efficient way, reducing the uncertainty in the estimation of shear behavior and, consequently, on the evaluation of seismic performance of the investigated buildings.

Moreover, the results highlight the possibility to obtain relevant information about the stiffness of the walls, in terms of both elastic and shear directly from in-situ double flat jacks tests.

References

ASTM, American Society for Testing and Materials, C 1197–91 .Standard test method for in-situ measurement of masonry deformability properties using the flat jack method (1991)

ASTM, American Society for Testing and Materials, E 519–81. Standard Test Method for Diagonal Tension (Shear) in Masonry Assemblages, American Society for Testing Materials (1981)

Beconcini, M.L.: Laboratorio Delta: indagini sui materiali sulle strutture e sul sottosuolo, Pacini Fazi, Lucca (2018). (in Italian)

Beconcini, M.L., Cioni, P., Croce, P., Formichi, P., Landi, F., Mochi, C.: Influence of shear modulus and drift capacity on non-linear static analysis of masonry buildings. In: IABSE Symposium, Guimaraes 2019: Towards a Resilient Built Environment Risk and Asset Management – Report, pp. 876–888 (2019)

Bosiljkov, V.Z., Totoev, Y.Z., Nichols, J.M.: Shear modulus and stiffness of brickwork masonry: an experimental perspective. Struct. Eng. Mech. **20**(1), 21–43 (2005)

Italian Public Works Council. Guidelines for Application of Italian Building Code; Istituto Poligrafico e Zecca dello Stato, Roma, Italy (2019). (In Italian)

Corradi, M., Borri, A., Vignoli, A.: Experimental study on the determination of strength of masonry walls. Constr. Build. Mater. **17**, 325–337 (2003)

Croce, P., Beconcini, M.L., Formichi, P., Cioni, P., Landi, F., Mochi, C., Castelluccio, R.: Expeditious assessment of seismic vulnerability of existing buildings from statistic sample of school buildings, property of the Comune di Firenze. In: XVII Conference ANIDIS, Seismic Engineering in Italy, September, Pistoia (2017)

Croce, P., et al.: Shear modulus of masonry walls: a critical review. Procedia Struct. Integrity **11**, 339–346 (2018)

Croce, P., et al.: Influence of mechanical parameters on non-linear static analysis of masonry buildings: a relevant case-study. Procedia Struct. Integrity **11**, 331–338 (2018)

Croce, P., Landi, F., Formichi, P.: Probabilistic seismic assessment of masonry buildings. Buildings **9**, 237 (2019)

Croce, P., Beconcini, M.L., Formichi, P., Landi, F., Puccini, B., Zotti, V.: Seismic risk evaluation of existing masonry buildings: methods and uncertainties. In: IABSE Congress New York: The Evolving Metropolis; IABSE: Zurich, Switzerland, pp. 2510–2515 (2019b)

Gregorczyk, P., Lourenco, P.B.: A review on flatjack testing. Engenharia Civil **9**, 39–50 (2000)

Gucci, N., Barsotti, R.: A non-destructive technique for the determination of mortar load capacity in situ. Mater. Struct. **28**, 276–283 (1995)

Henzel, J., Karl, S.: Determination of strength of mortar in the joints of masonry by compression tests on small specimens. Darmstadt Concr. **2**, 123–136 (1987)

Marchisio, M., D'Onofrio, L., De Falco, A., Frediani, L., Guidoni, F.: New tomographic techniques (micro-seismical and geoelectrical) for the non-destructive testing on masonry structures. In: International Symposium NDT-CE (Non-destructive Testing in Civil Engineering), Berlin (2003)

Sheppard, P.F.: In situ test of the shear strength and deformability of an 18th century stone and brick masonry wall. In: Proceedings of 7th International Brick Masonry Conference, vol. 1, pp. 149–160 (1985)

Tomaževic, M.: Laboratory and in-situ tests of the efficacy of grouting and tying of stone masonry walls. In: Proceedings International Workshop on the Effectiveness of Injection Techniques for Retrofitting of Stone and Brick Masonry Walls in Seismic Areas, Milano (1992)

Tomaževič, M.: Earthquake-Resistant Design of Masonry Buildings. Series on Innovation in Structure and Construction, vol.1, London (1999)

Turnšek, V., Ĉaĉoviĉ, F.: Some experimental results on the strength of brick masonry walls. In: Proceedings of the 2nd International Brick Masonry Conference, Stoke-on-Trent, pp. 149–156 (1971)

Turnšek, V., Sheppard, P.F.: The shear and the flexural resistance of masonry walls. In: Proceedings of the Research Conference on Earthquake Engineering, Skopje, pp. 517–573 (1980)

Zimmermann, T., Strauss, A.: Shear bearing behavior of old unreinforced masonry under seismic load (in German). Bautechnik **89**(8), 553–563 (2012)

Zimmermann, T., Strauss, A., Wendner, R.: Old masonry under seismic loads: stiffness identification and degradation. In: Structures Congress 2011 © ASCE 2011, pp. 1736–1747 (2011)

Croce, P., Beconcini, M.L., Formichi, P., Landi, F., Puccini, B., Zotti, V.: Bayesian methodology for probabilistic description of mechanical parameters of masonry walls. ASCE-ASME J. Risk Uncertain. Eng. Syst. Part A: Civ. Eng. **7**, 04021008 (2021). https://doi.org/10.1061/AJRUA6.0001110

Application of Non-destructive Techniques for the Investigation of Old Masonry Structures

Erato Ch. Oikonomopoulou(✉), Vasiliki A. Palieraki, Vasiliki S. Nikolopoulou, and Elizabeth Vintzileou

Laboratory of Reinforced Concrete, Department of Civil Engineering, National Technical University of Athens, Iroon Polytechniou 9, 15780 Athens, Greece
eratooikonomopoulou@central.ntua.gr

Abstract. This study aims at the acquisition of a reliable image of the non-visible parts of historical masonry buildings in Greece by in-situ application of non-destructive techniques (NDTs). To serve the purpose of obtaining overall conclusions on the documentation of the monuments' internal condition, the deriving results are examined and combined together with the in-situ observations. The structures under investigation are the Katholikon of Osios Loucas Monastery (Boeteia), the Monastery of Kaisariani (Attica), the Agia Paraskevi Church (Chalkida) and the Nativity of Christ's Church (Athens). The main investigation method applied on all monuments was that of the Ground Penetrating Radar. Additionally, the methods of Boroscopy and Schmidt Hammer Testing were also used when further investigation was required. The techniques were applied on targeted areas, representative for the entire examined structure. Thanks to the assessment of the results, it was possible to identify data relevant to the dimensions and the type of stones used in the piers and the three-leaf masonries, the thickness and the condition of the mortar and the infill, the coexistence of different construction types in a structure, as well as the approximate compressive strength of stones.

Keywords: Monuments · Historical buildings · Non-destructive testing · Structural documentation

1 Introduction

All four examined structures (Fig. 1) are Christian historical buildings, listed by the Ministry of Culture for their historical, architectural and artistic value. Although they are constructed in different areas of Greece, during different historical periods, they exhibit similarities in terms of construction types and materials. Furthermore, they either continue to serve the purpose of their construction (Fig. 1a, c, d) or are monuments with open access to the public (Fig. 1a, b). At the same time, they are prone to ageing, to earthquakes and environmental risks, and have suffered severe damages during their lifetime. Their preservation and maintenance, essential for their structural safety and functional performance, depend on the alleviation of potential risks. Towards this direction, the

I. Vayas and F. M. Mazzolani (Eds.): PROHITECH 2021, LNCE 209, pp. 34–45, 2022.
https://doi.org/10.1007/978-3-030-90788-4_3

identification of the structural system and the mechanical behavior of the monument is crucial.

To effectively interpret the behavior of existing structures and, hence, to design appropriate intervention techniques for their repair or strengthening, the most important step is the investigation of their current condition, which is the purpose of the present study. Thanks to the evaluation and the combination of the results deriving from in-situ non-destructive techniques (NDT), it was possible to identify data relevant to: (a) the thickness of stones used for the construction of piers, as well as the exterior leaves in perimeter masonry walls; thus, the thickness of the internal filling material for the case of three-leaf stone masonry could be estimated, (b) the coexistence of different construction types in a structure, (c) the mortar's condition and the existence of voids and discontinuities between adjoining stones could be identified, as well as (d) the strength of the stones, in one of the examined cases, with an approximate estimation.

Fig. 1. The investigated monuments: (a) the Katholikon of Osios Loucas Monastery (Boeteia), (b) the Monastery of Kaisariani (Attica), (c) the Agia Paraskevi Church (Chalkida) and (d) the Nativity of Christ's Church (Athens).

2 Non-destructive Techniques Used

The acquisition of a reliable image of the non-visible parts of the structures was performed by applying three NDTs, namely the Ground Penetrating Radar (GPR), Boroscopy and Schmidt hammer testing (briefly described in the respective section). The equipment of the NTUA Laboratory of Reinforced Concrete was used, Ground Penetrating Radar being the main method that was applied. Since no single technique can sufficiently provide the required reliability, and taking into account the limitations of each method, the two other techniques were used in those cases where the need for further investigation was apparent and safe access was feasible.

2.1 Ground Penetrating Radar

The non-destructive technique of GPR, which is based on the transmission of electromagnetic waves, was applied on various areas of all the examined monuments. The system consists of an antenna, serving as the transmitter and the receiver of the pulse, and a control unit (Daniels 1996; Utsi 2017). When the pulse reaches an interface of materials with a different dielectric constant, a part of it returns to the transmitter, furnishing the system with this information, whereas its remaining part continues to penetrate the material in greater depth while attenuated (GSSI 2009). The final radar image is a section of the examined area. Given the scope of these applications, where the thickness of the masonry is relatively small (up to 100 cm approximately), a rather high frequency antenna (1600 MHz) was selected, offering the twofold advantage of the high image quality and the small size, which subsequently facilitates scanning on areas with limited dimensions.

To use the GPR results on a quantitative basis, rather than a qualitative one only, the calibration of the device is a prerequisite. In the examined monuments, the calibration was carried out in locations where the dimensions of single stones or the entire width of a pier or a masonry wall was already known or could be easily measured. The dielectric constant that results in correct depth evaluations is adopted and, hence, the vertical axis of the radar files presented herein expresses the depth (in [m]), whereas the horizontal axis indicates the length of the path (also in [m]).

2.2 Boroscopy

In some of the cases, following the radar method, and to verify the reliability of the images taken, the slightly destructive method of boroscopy was applied. In this method, the borescope is inserted into a (pre-existing or drilled on purpose) hole of sufficient diameter (12 mm–25 mm, depending on the width of the mortar joint in which boroscopy is applied) and length (usually up to 500 mm), allowing for direct visual observation of the masonry's interior and for videos or pictures to be taken at various depths, offering the ability to detect stone surfaces, discontinuities, gaps and timber or metal elements, as well as their position inside the masonry (Palieraki et al. 2007; Palieraki et al. 2013). In the present study, boroscopy was applied with the aim to confirm the findings of the GPR technique and to acquire also qualitative information about the in-depth condition of each examined area.

2.3 Schmidt Hammer Testing

For the estimation of the mechanical properties of rock material, the Schmidt hammer non-destructive method provides a quick measure of surface hardness (Aydin and Basu 2005). A metal mass hits the tested surface and, based on its average rebound value, the compressive strength is estimated by means of calibration diagrams, depending on the material of the tested element. To enhance the reliability of the results, at least ten measurements per application site were conducted, the repeated application of the method at close points and near the edges of the stones were avoided, whereas the plaster was carefully removed and the stone surface was of acceptable roughness.

Since the acquisition of stones from the monuments is -in general- not preferred or even not permitted, the calibration of the hammer through in-laboratory testing of stone specimens under compression is usually not possible. Therefore, in this study, the strength assessment was based on calibration curves available at bibliography. It is emphasized that the deriving strengths are approximate, indicative estimations.

3 Application of the Techniques and Test Results

The suitable NDT or a combination of NDTs is selected on the basis of the specific features of each monument and the scope of its investigation. To obtain accurate and -as much as possible- conclusive results, both horizontal and vertical radar scans were performed in the examined areas, in all cases. The results of NDTs were combined together and decisively contributed towards drawing conclusions on the construction type of masonry. Hereafter, the concise presentation and the commentary of the most significant results offered by the applied techniques is given for every structure.

3.1 Katholikon of Osios Loucas Monastery in Boeteia

The Katholikon of Osios Loucas, dating back to the 11th–12th centuries AD, consists of two storeys and is one of the most efficiently preserved Middle-Byzantine Churches in Greece. Due to its dominant significance, many researchers have devoted efforts towards understanding the monument from the structural perspective, encountering, however, the lack of adequate relative written sources. This fact renders the need for in-situ investigation of monument's targeted areas even greater. In the recent years, several investigation campaigns of the structure have taken place, with the most recent one conducted during 2016–2017 within the context of an NTUA Research Programme, where the authors participated. In this programme, 93 radar scans and observation with the boroscope at 15 positions were performed in total.

The main goal of the recent investigation campaign was the correlation of the gathered results with those that derived from previous investigations (Vintzileou 2007; Vintzileou et al. 2008). The scope of the investigation was fulfilled, as the observations reported in the aforementioned papers were confirmed; more specifically, the fact that the piers in the Yperoon demonstrate a solid construction, with extensive use of large stones and joints of limited volume was verified. Also, the findings of both the present and the previous investigation campaigns on the examination of the perimeter walls in the ground floor revealed the existence of three-leaf masonry, with the external leaves being constructed with cut stones of significant dimensions.

The most important findings of one typical pier of the gallery (Yperoon-second storey of the monument) are reported in the present study. The fact that the entire surface of the sides was covered with plaster was the main problem faced. To allow for a reliable calibration of the technique, and for an assessment of the plaster's effect on the suitable value of the dielectric constant, the plaster of two sides was removed up to a specific height of the pier (80 cm approximately, measured from the floor) (Fig. 2a). Thanks to the plaster removal, the exact position of the structural units placed at the lowest part of the pier was locally revealed. The value of the dielectric constant both for the covered and

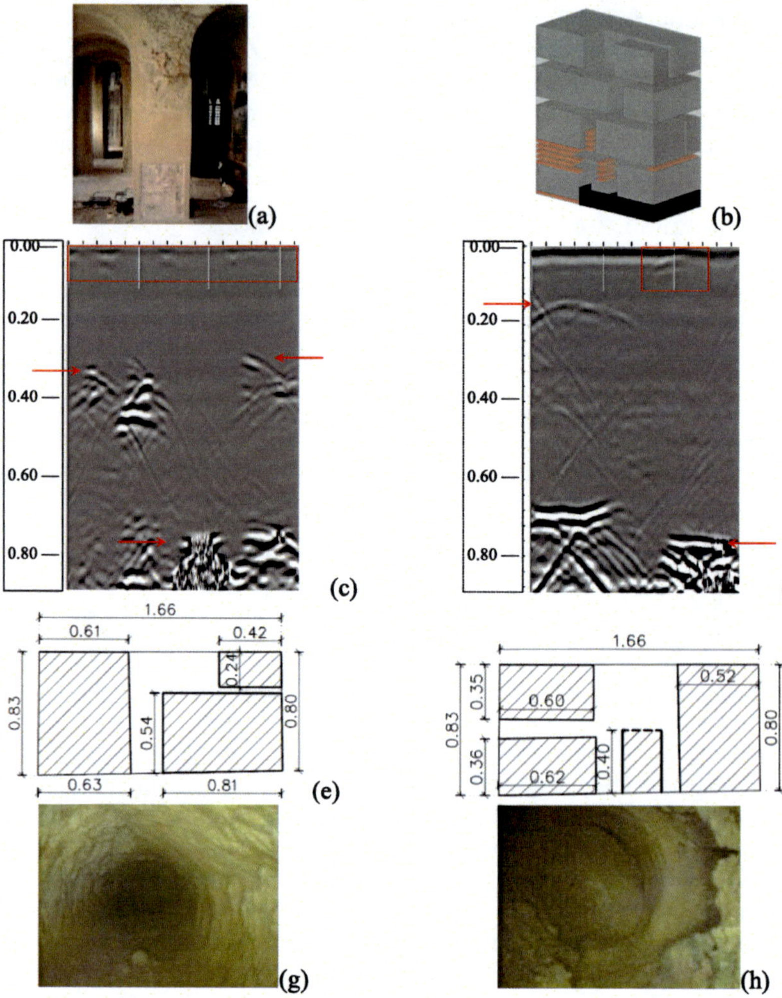

Fig. 2. (a) Pier Y25 of the Yperoon, (b) Three-dimensional drawing of the pier, depicting the overall findings of the visual inspection and the applied NDTs, (c) Radar profiles of vertical and (d) horizontal path along the pier: the arrows indicate the thickness of the identified distinct stones, whereas the frame denotes reflections created due to the presence of bricks, (e)– f) Horizontal sections at two levels (0.30 m and 0.80 m respectively), measured from the floor level, as derived from the GPR findings and the visual confirmations (where possible). Only the stones, and not the bricks, are depicted in the sections. (g),(h) Images taken at various depths (45 cm and 49 cm, respectively) of the borescope's path along the hole drilled in mortar area, between stones placed up and down.

the uncovered parts remained the same. This is attributed to the insignificant thickness of the plaster compared to that of the entire pier, and the nature of the plaster's material, as well. As all four sides of the pier were available for scanning without obstacles, a more detailed investigation was feasible.

As observed in Fig. 2(c) and (d), the reflections corresponding to the adjacent stones are strong and very clear. Thanks to the penetrating capacity of the antenna, the entire thickness of the pier could also be identified, in some cases. By combining the scans of the grid, the horizontal sections in two different levels are drawn (Fig. 2e, f), revealing a solid construction, with large-size cut stones, interconnected along the vertical axis of the pier. As visually noticed, also bricks were used at specific areas of the pier at its lowest part, depicted only in the three-dimensional drawing (Fig. 2b). The findings correlate well with those reported in Vintzileou et al. 2008, where similar observations yielded regarding pier Y32 of the gallery.

As for the nature and the quality of the mortar, its investigation was attempted via boroscopy. It seems to be rather stable and uniform along the hole, as illustrated in Fig. 2 (g) and (h). The perimeter of the hole provides the observer with information relative to a stone surface (up and right- Fig. 2g). The stone is visible until a depth that is lower than its thickness, as estimated with the radar. This could indicate that the stone is semi-cut, so its surface does not follow the axis of the horizontal hole, and therefore, its entire dimension cannot be identified with the borescope. Also, the existence of some voids of insignificant magnitude (right- Fig. 2 h) can be identified.

3.2 Monastery of Kaisariani in Attica

The in-situ measurements in the Kaisariani Monastery have been executed in the framework of the Research Programme "Thales-NTUA: Antiseismic Protection of Monuments and Historical Structures" (Vintzileou et al. 2015). The Kaisariani Monastery is situated in the area of Imittos, in Attica. The Katholikon (main Church) of the Monastery has been constructed in the 11th century and it has been subjected to various interventions and modifications during its lifetime. The masonry of the Katholikon follows a construction type typical for Byzantine Churches: Masonry is constructed with stones of rather small dimensions, while solid bricks are used both within the horizontal and several vertical mortar joints. The Narthex, as well as the small church of Saint Antonios in the south part of the Katholikon, have been added during the 17th century. The initial paintings of the interior of the Church have been lost. The older painting still existing in the Katholikon dates back in the 14th century.

The aim of the in-situ investigations was the documentation and the assessment of the current state of the Katholikon and the Saint Antonios Church. The radar technique, as well as the boroscopy, have been used in order to identify the masonry construction type, without disturbing the masonry materials. Almost 60 radar scans over more than 120 m of masonry and observation with the boroscope at three positions were performed. The paths for the application of the radar technique have been selected according to the information to be obtained and the ease of safe access. Scans have been performed along the perimeter masonry, along the piers and the marble columns of the interior, as well as along the pavement of the Church.

It is noted that in some areas the obtained radar images are not reliable: The stones are of small dimensions, leading to the acquisition of multiple reflections, the humidity in the lower part of the masonry may be high, decreasing the accuracy of the obtained image, while, on the other hand, the radar images differ between parts of the masonry constructed in different phases (Fig. 3a). Nevertheless, the systematic radar measurements together with the application of the boroscopy gave the possibility of measuring the width of the exterior masonry leaf in most of the areas of the monument (Fig. 3b), and drawing conclusions about the construction type of the piers (Fig. 3c).

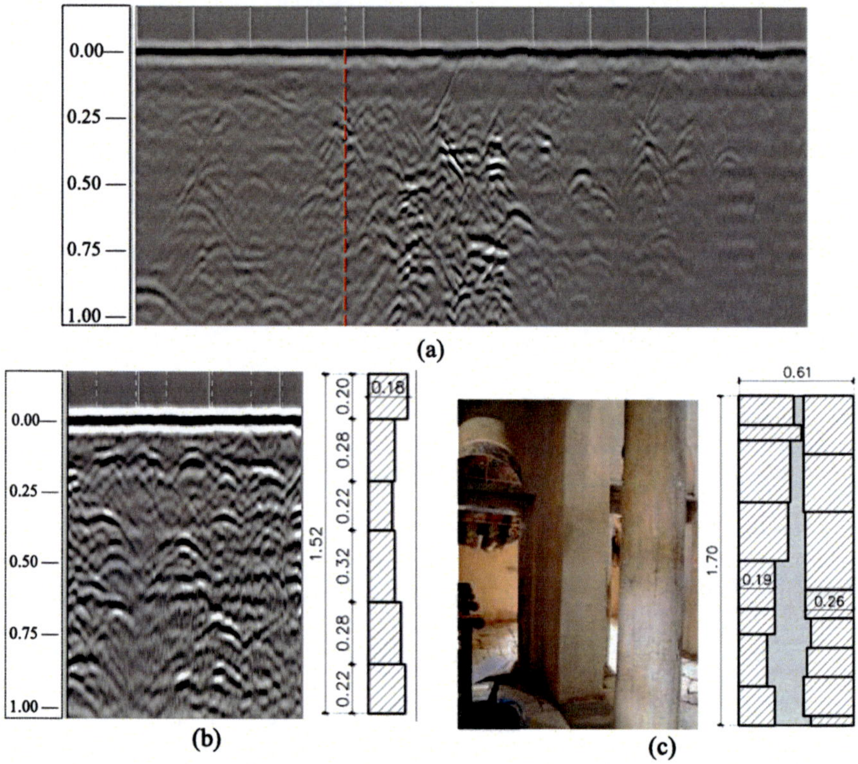

Fig. 3. (a) Radar profile of horizontal scan at the exterior (Northern façade). The red line stands for the limit between the two different construction phases, (b) Radar profile of vertical scan at the east façade of the Katholikon and the respective masonry section, (c) The scanned masonry pier and the vertical section deriving from the radar measurements. Only indicative dimensions are provided in the sections.

3.3 Agia Paraskevi Church in Chalkida

The Agia Paraskevi Church is a dominant monument of the Frankish occupation period (13th century). Parts of the church were constructed again during the 14th or the 15th century, whereas the west façade had to be reconstructed after the 1853 major earthquake. The historical building is considered to be one of the most important ones of its era (Bertolini Cestari et al. 2007). NDTs were applied on both the interior and the exterior of the monument, in selected areas. In total, 139 radar scans and observation with the boroscope at four positions were performed.

As different construction phases of the monument throughout the years have been reported, the present research intended to identify areas of different construction types. The goal was accomplished, since the historical data concerning the reconstructed parts of the monument guided the in-situ investigation strategy and, the combined findings deriving from the visual inspection and the GPR and boroscopy application led to the conclusion that the different construction types of distinct areas can be indeed recognized.

The most important findings from representative areas of the west and south perimeter walls are presented herein. Since the same number, level and direction of scans was kept for the interior and the exterior leaf of the three-leaf masonries, the direct combination of the findings became feasible. Regarding the west examined area of the monument, the façade is covered with plaster, whereas the interior is not, allowing thus for the positions of the stones and bricks to be known. Despite the presence of plaster at the façade, the obtained image has great clarity, with strong reflections indicating the thickness of the stones comprising the exterior leaf (Fig. 4a). The image of the interior is also clear, and the vertical short dotted lines of the file (Fig. 4 b) denote the joints between adjacent stones. Local horizontal sections depicting the geometry of the three-leaf masonry can be reproduced and, as observed in Fig. 4 (c), the average thickness of the infill is not negligible; a fact that demonstrates that further investigation is required.

This is valid for the south examined area (Fig. 4f) as well, where the thickness of the infill is estimated to be slightly smaller. The radar image exhibited less clarity (Fig. 4d), as the existence of numerous interfaces made the accurate interpretation of the files rather questionable, especially at areas constructed extensively with bricks. As structural units with various characteristics (cut-stones, rubble stone, oblong stones, bricks) have been used at the examined areas and different construction phases coexist in the structure, the acquired images significantly vary from region to region. For an overall survey of the south masonry, two interior positions were also examined with the borescope: the captures of Fig. 4(e). The first image corresponds to hole positioned right by gothic stone of the interior south wall, in proximity to the Sanctuary. The surface of the stone can be clearly identified at the left side of the hole's perimeter, as well as the drilled mortar of reddish shade. In a short distance within the same area, another hole is drilled (Fig. 4e-second image) right over a horizontally placed long timber element, at area where a buttress exists at the exterior. The images taken from this position are significantly different compared to the previous one: the light-colored mortar is unstable, powdered due to drilling, and several voids can be identified. In addition, near the end of the hole, a rectangular cross-section can be clearly observed together with a dark-colored fragment, which could be an indication of timber element position.

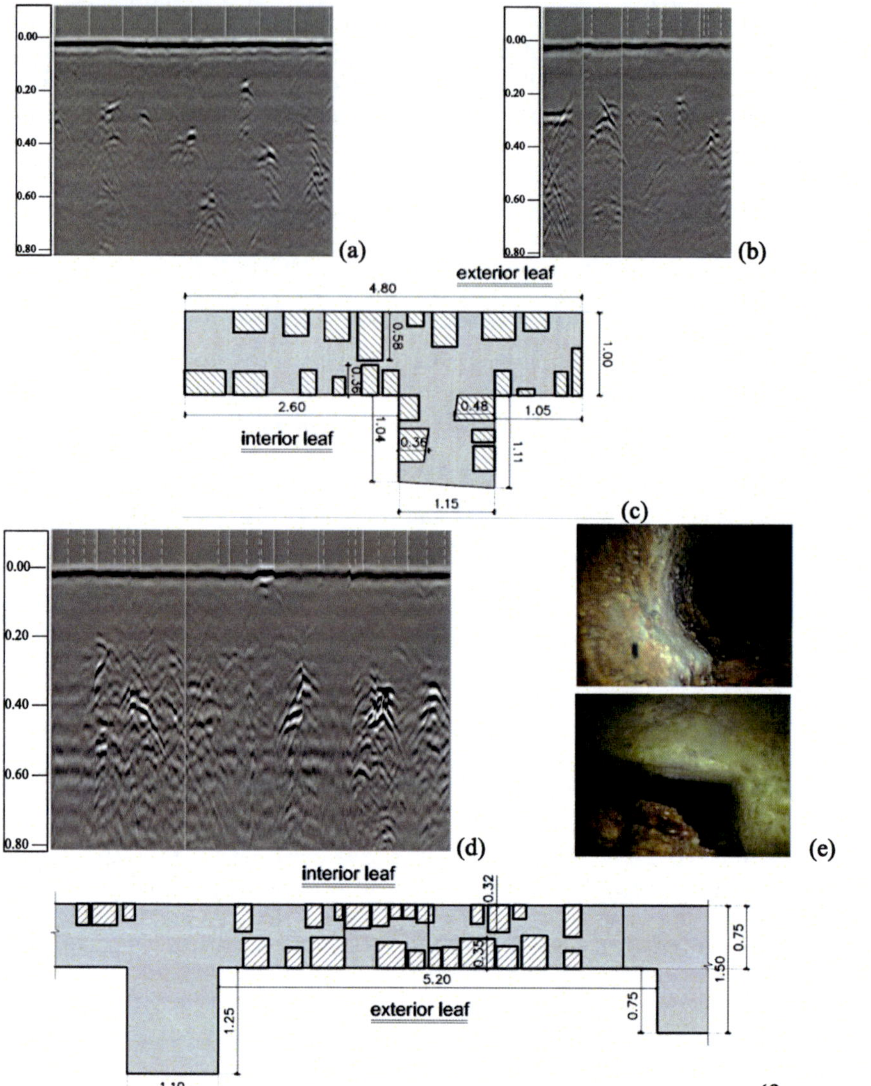

Fig. 4. (a) Radar profiles of horizontal scan at the exterior and (b) respective horizontal scan at the interior of the examined west area: the clear reflections correspond to the thickness of the identified distinct stones, (c) Horizontal section of the west wall, at the level of +2.0 m measured from the ground floor, as derived from the GPR and the visual findings, (d) Radar profile of horizontal scan at the exterior of the examined south area, (e) Images taken at various depths (24 cm and 95 cm respectively) of the borescope's path along two different holes in the examined south area, (f) Horizontal section of the south wall, at the level of + 1.70 m measured from the ground, as derived from the GPR and boroscopy findings. Only indicative dimensions are provided in the sections.

3.4 Nativity of Christ's Church in Athens

The position of the Nativity of Christ's Church, alternatively known as "Christocopidis Church," was occupied by a temple owed by the Copidis Athenian family, who gave the name to the monument. The monument at its current scheme was constructed with cut-stones in 1860 or 1880 and has experienced two main earthquake incidents in 1981 and 1999, which caused major structural damages that can easily be observed nowadays. The GPR technique was applied on both the interior and the exterior of the monument and, in total, 100 radar scans were performed. Schmidt hammer testing was applied on twelve stones of the south perimeter masonry wall, in limited area with removed plaster.

The investigation campaign for this monument was conducted as the first prerequisite step in the context of an extensive study on the assessment of the monument's existing condition, ultimately aiming at the design of suitable structural interventions. Towards this direction, the sound simulation of the structure is required and, hence, the collection of adequate and accurate data related to the construction type of distinct parts of the monument (perimeter walls and piers). In addition, for the complete modelling of the structure, sufficient estimation about the strength of the materials used (stones, mortar) was also required. Thanks to the application of the described NDTs, the goals were achieved, decisively assisting the structural design towards the definition of the needed interventions. The most important findings from representative areas are reported herein.

The great effect of the frescoes on the radar image is depicted in Fig. 5(b): when the antenna crosses the area where gold color has been used in the frescoes (Fig. 5a), the signal propagation is totally obstructed and no information can be acquired by the image. When the antenna is moved in other areas of the same side of the pier, a very clear profile is obtained, as observed in the vertical scan of Fig. 5(c), where the thickness of the adjoining stones is depicted with clarity, revealing the consistent interconnection between them. Thanks to the existence of adequate surfaces without frescoes, the geometry of the pier can be represented in indicative horizontal sections (Fig. 5d).

The results obtained from Schmidt hammer testing (Fig. 6) indicate the existence of stones that can be distinguished in two categories based on their strength: (a) stones of relatively low strength, varying between 8 MPa and 17 MPa (i.e. average rebound values between 18 and 30), and (b) stones of higher strength, ranging from 33 MPa to 41 MPa (i.e. average rebound values between 39 and 42). The estimated values reported correspond to the average of the strengths resulting from two calibration curves available at bibliography (Tassios and Mamillan 1985; Aliabdo and Elmoaty 2012). The findings from the macroscopic observation of the stones correlate well with the estimated strength values. In particular, stones of lower strength are friable and porous. On the contrary, stones of higher strength do not exhibit the aforementioned characteristics, while being slightly different in color and having reddish veins. As anticipated, stones of high strength are concentrated in the lower zone of the examined area.

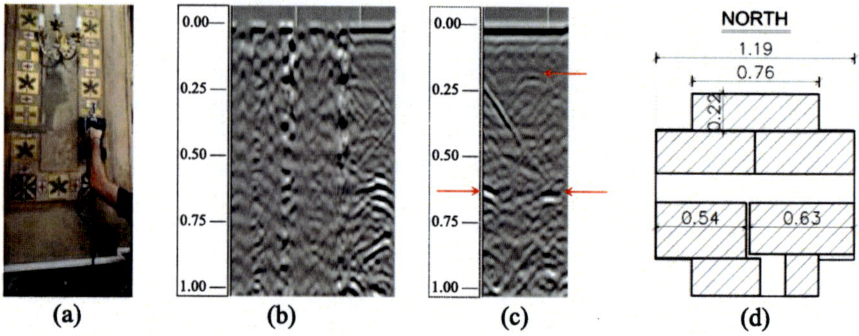

Fig. 5. (a) Scanning on the frescoes of the pier, (b) Radar profile of vertical path along the pier: the gold color of the frescoes does not allow for the propagation of the signal, (c) Radar profile of vertical scan along the pier, path without frescoes: the arrows indicate the thickness of the identified stones. (d) Horizontal section, as derived from the GPR findings: only indicative dimensions are provided.

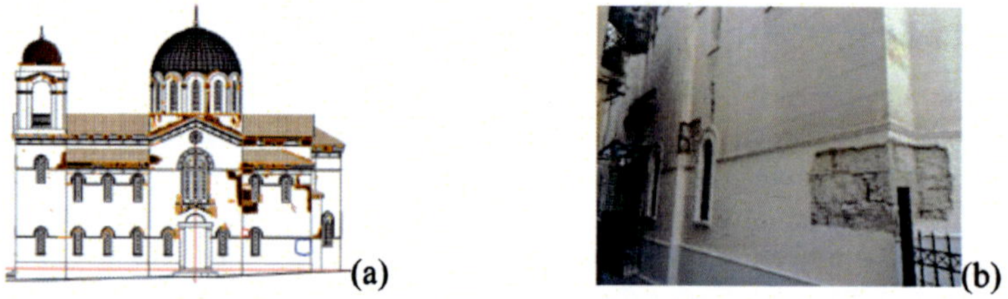

Fig. 6. (a) Side view of the church (drawing provided by E. Tavouktsi, Civil Engineer, Designer of the interventions): the blue frame indicates the presented examined area, (b) Photo of the examined area: the plaster was removed and the stone surfaces were revealed.

4 Conclusions

Based on the material presented, it is concluded that the application of non-destructive techniques in gingerly selected areas, organized in specific steps and preceded by careful inspection, can yield reliable and in-detail results regarding the structure of the monuments. Thanks to the combination of the information emerging from each method, the outcome is enhanced and the final results provide sufficient data for the visible and non-visible parts of the examined structures.

The evaluation of the mechanical properties of the materials and the comprehension of the geometry of the structure are the prerequisite for a sound analysis, which is strictly linked to the decision-making process on formulating proposals towards efficient interventions in monumental structures of high historical and cultural value. All the methods presented herein are applicable in the field of monument investigation, on condition that constrains arising by the characteristics of each unique structure are considered.

References

Aliabdo, A.A.E., Elmoaty, A.E.M.A.: Reliability of using nondestructive tests to estimate compressive strength of building stones and bricks. Alexandria Eng. J. **51**(3), 193–203 (2012)

Aydin, A., Basu, A.: The Schmidt hammer in rock material characterization. Eng. Geol. **81**(1), 1–14 (2005)

Bertolini Cestari, C., et al.: The timber roof of Hagia Paraskevi Basilica in Halkida, Greece: multi-disciplinary methodological approach for the understanding of the structural behaviour. analysis and diagnosis. In: Proceedings of XVI International Conference and Symposium of the IIWC From Material to Structure, Florence (2007)

Daniels, D.J.: Surface-penetrating radar, The Institution of Electrical Engineers, Radar, Sonar, Navigation and Avionics Series, Editors Prof. E.D.R.Shearman and P.Bradsell, London, p. 300. (1996).

Geophysical Survey Systems INC., SIR-3000 Manual, USA, p. 91 (2009)

Palieraki, V., Vintzilaiou E., Miltiadou-Fezans, A.: The use of radar technique and boroscopy in investigating historic masonry: Application of the techniques in Byzantine monuments in Greece. In: Proceedings of the International Symposium: Studies on Historical Heritage (SHH07) (2007)

Palieraki, V., Adami, Ch.-E., Vintzileou, E.: In-situ measurements using radar and boroscopy techniques: case study -Hagia Sophia museum of Trabzon, Turkey. In: Proceedings of the Conference BH2013 (Built Heritage 2013, Monitoring Conservation Management), 18-20 November, Milan, Italy (2013)

Tassios, T.P., Mamillan, M.: Valutazione strutturale dei monumenti antichi (1985)

Utsi, E.C.: Ground Penetrating Radar: Theory and Practice. Butterworth-Heinemann (2017)

Vintzileou, E.: Investigation of the construction type and mechanical behavior of Byzan-tine masonry before and after interventions, with application in the case of the Katholikon of Osios Loucas Monastery (in Greek), Report for the Hellenic Ministry of Culture, Athens, p. 59 (2007)

Vintzileou, E., Konteas, I., Manoledaki, A.-A., Palieraki, V., Miltiadou-Fezans, A.: The Katholikon of Osios Loucas Monastery: In: situ investigations. RILEM Symposium on Site Assessment of Concrete, Masonry and Timber Structures-SACoMaTiS 2008. RILEM Publications SARL (2008)

Vintzileou, E., Palieraki, V., Nikolopoulou, V.: Thales-NTUA: Antiseismic Protection of Monuments and Historical Structures, Work Package 2: Development and application of methodology, Deliverable 2.2.1: Report on the application of the methodology on the selected historical buildings or monument and specification text for the application of the methodology (in Greek), Report for the Thales Programme, Athens, p. 142 (2015)

In-situ Experimental Research to Reinforce, Restore and Enhance the Historical Monument EVANGELICAL CHURCH C.A. SIBIU

Claudiu Lucian Matei(✉), Cora-Maria Stamate, and Mariana Cioncu–Puenea

National Research and Development Institute, URBAN-INCERC, Sos. Pantelimon nr. 266, Sector2, Bucharest, Romania

Abstract. The entry into force of Romanian Code P 100–3 / 2009 (updated in 2019) regulating the expertise of existing, seismically vulnerable constructions requires the assessment of the seismic performance of structures having the structure of structural masonry. The code also applies to the assessment of historical heritage buildings. The Evangelical Cathedral in Sibiu (in German Evangelische Stadtpfarrkirche) is one of the oldest and most impressive Gothic buildings in Transylvania (Romania).The building is dominated by the 7-level tower, with four turrets on the corner of the roof, a sign that the city has a right to condemnation (in Latin and Gladiers, literally "the right of the sword").

According to the requirements of normative P 100–3 "Evaluation and design code of the consolidation works on existing buildings, seismically vulnerable" in chapter 4.4.3.- Materials, are presented in-situ investigation methods for such constructions. The tests will be in terms of the factors of confidence in the knowledge levels of type: KL2 - normal knowledge and in the category KL3 - complete knowledge. From the point of view of the level of inspection and testing, the tests fall into the category of limited type tests.

Keywords: Masonry · In-situ tests · Multi-cycle test · Flat presses

1 Introduction

The seismic assessment of buildings, classified as monuments, requires a complex of activities that highlight the structural vulnerability.

The seismic assessment of buildings, classified as monuments, requires a complex of activities that highlight the structural vulnerability. In the operations that make up the assessment process, an important role is to know the characteristics of the masonry.

The application of knowledge level involves procedures simplified or more complex to investigation materials. Normal level KL2 knowledge as well as full knowledge KL3 requires extensive, in-situ, tests on masonry building materials on a representative number of elements. Determination of component materials of masonry, for old buildings, is a complex process of investigation, evaluation, sampling, testing, interpretation and conclusion that is completed by a qualitative assessment of parameters necessary for

I. Vayas and F. M. Mazzolani (Eds.): PROHITECH 2021, LNCE 209, pp. 46–55, 2022.
https://doi.org/10.1007/978-3-030-90788-4_4

the structural calculation to verify satisfy requirements limit state associated to seismic action.

Considering the importance of construction and belonging to the cultural heritage (being on the list of historical monuments), a closer analysis of the resistance characteristics was decided, taking the decision to evaluate the characteristics by the method of destructive investigation, determining the data of the realities of the materials components of the masonry, in this case is being use: determination of the strength and deformability characteristics of the masonry by the use of *flat presses*.

2 Experimental Results

In order to have a more accurate picture of the resistance structure, the decision was made to conduct some surveys. Following the investigations it was found that the masonry construction was made of ceramic elements of full pressed brick type (cpp) built with mortar on lime base combined with masonry mixed with natural stone elements or even natural stone masonry.

The masonry is made with ceramic elements full of different dimensions depending on the stages of manufacture, being made with weaving performed quite negligently, on rather large masonry thicknesses, the resulting thickness being a massive masonry walls.

Considering the importance of the construction and belonging to the cultural heritage (being on the list of historical monuments), a more careful analysis on the resistance characteristics was decided, taking the decision to evaluate the characteristics through the destructive investigation method, determining the material characteristics as real as possible. Components of the masonry, in this case asking: determining the strength and deformability characteristics of the masonry by using flat presses.

Photo 1. The mixed construction of the masonry

2.1 Determinations of Masonry Elements

The determination of the compressive strength of the ceramic elements was performed for elements taken from several areas of the resistance structure, so as to make an overview of the materials used, during the construction.

The results of the compressive strength for each of the ceramic elements taken under test were presented according to the Table 1.

Table 1. Tests to determine the compressive strength for ceramic masonry elements

Index	Compression section dimensions (mm)	Force (N)	Compression strength (N/mm^2)	Specimen after the test
Ez-1	144x130	156800	8,38	
Average value of compressive strength				***8,38***

2.2 Determinations of the Mortar Characteristics

In order to determine the compressive strength of the mortar used for the masonry, mortar samples were taken from the investigated areas.

The mortar samples were processed in the laboratory by cutting with a thin diamond blade, in order to obtain specimens that must fit the standard dimensions for the test, respectively ½ of the standard 40 × 40 × 160 mm prism.

Photo 2. Mortar samples taken to determine the resistance to compression

Photo 3. Mortar samples processed to determine the resistance to compression

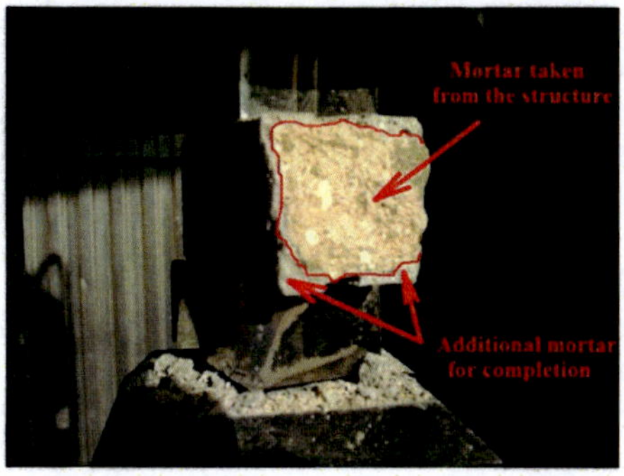

Photo 4. Mortar samples after the tests

The contact surfaces were obtained by correcting the samples with the addition mortar, ensuring the prismatic form. The mortar layer was very thin and did not influence essential the test itself.

The obtained prisms were tested for compression under standard conditions, as for new mortars.

The results of the compressive strength of the mortar, taken from the masonry of the Evangelical Church C.A. Sibiu, from the level of the surveys are determinate with the "Eq. (1) values mortar strengths".

$$fm,\ real_{=}K * fm \tag{1}$$

Where *fm, real* is the "real in-situ" compressive strength, *K* is the correction factor (area ratio between the real and the laboratory section) and *fm* is the compressive strength obtained in the laboratory. K depends on the ratio between the mortar taken from the site and the mortar added in the laboratory.

2.3 Determination of the Compressive Strength of the Masonry in Situ Using Flat Presses

There is a particular interest in the study of masonry as a composite material, especially since in a very large proportion of the world heritage and civil buildings have the structures of resistance predominantly constructed from this material.

There is a continuing need for information on how best to maintain a good state of preservation of the masonry and intervention techniques for repairing and restoring the damaged masonry.

In order to evaluate the materials used in both new and old structures, in laboratory accelerated durability tests are required. Similarly, an in-situ, non-destructive range or semi-invasive tests are required to assess the condition and state of the structures on the ground and to allow quality control of repair systems.

This recommendation specifies an in-situ method for determining local force-displacement behavior as a function of compressive condition, of built masonry, by using flat presses reacting against the main body of the masonry (as a reaction) to apply a force at a small sample volume. Guidance is provided on the main features involved in the characterization of the masonry, the preparation for the test, the apparatus, the test method, the calculation method and the content in the test report.

The choice of specimens depends on the type of masonry, the dimensions of the test, the capacity of the flat presses, the possibility of measuring the deformations as well as avoiding the effects of concentration of efforts.

Tests involves performing several cycles that take into account the behavior at different loading stages, the expected (predictable strength of the masonry), the masonry and its integrity. Test will be avoided the sections with local degradation, unfinished joints and cracked masonry elements.

The size and shape (and position) of the specimen (the area of the masonry) is determined by the shape of the masonry and the type of flat press used and it is recommended that the specimen tested should fit into a square (the sides being equal to the length of the presses used).

The test technique is based on the use of two presses / flat presses connected to a common oil source to apply a pressure field of a masonry volume between them. It is obvious that the tendency to deform the presses to the masonry creates a pressure that must react against the rest of the structure in order to exert a reaction force and the applied compression level which will be limited to the test area.

The type of flat presses will be chosen according to the structure of the masonry and the efforts expected to be obtained.

Because the masonry in the area of the test gap will continue to be physically bound (also by mortar adhesion) to the rest of the wall the edges are not detached from the rest of the structure. It was not chosen for the test of the four sections the detachment of the masonry by vertical decoupling.

The value of Young's module measured by this technique together with the measured or calculated data of the specimens are probably sufficiently accurate for estimating the design values of the masonry at complex demands.

In some cases, the test can be used as a real test of compressive strength, but it is not recommended as a safe and accurate technique and it is necessary to use the other results obtained from the material tests.

The schematic diagram of the flat presses is given in Fig. 1

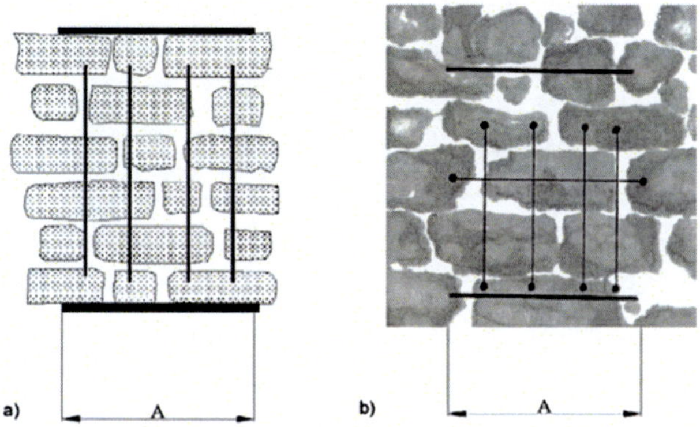

Fig. 1. Installation of flat presses

2.3.1 Determination of the Compressive Strength of the Masonry in the Chosen Sections

The test sections have been chosen so that it is possible to have an exact analysis of the masonry at representative points in the structure plan. The test sections were materialized at the base of the construction, the masonry being required for gravitational loads, as close as possible to the base using the masonry ballast as a reaction of the flat presses.

In each section, four zones for measuring the deformation on the verticals and a horizontal section were measured.

For each loading sections the pairs of values were measured so that the S-E curve can be drawn for each of the sections.

Before starting the test, zero measurements were made as a distance between the metal points. The deformations were measured using a 250 mm base deform meter.

For each loading step, the averages of the deformations were calculated for the four vertical sections and the characteristic curves were drawn.

To draw the curves, a test protocol was decided in two or three cycles, so that the best calibration of the results is possible.

After each cycle it was considered as a deformed starting point translated into origin.

The characteristic curves for the test no.1 and 2, on each loading cycle, are shown in Fig. 2

The characteristic curves for the test made on inhomogeneous sections, on each loading cycle, are shown in Fig. 3

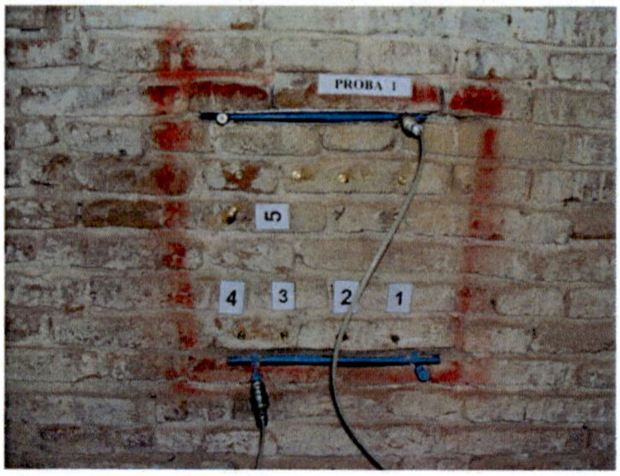

Photo 5. Section 1 homogeneous masonry made of ceramic elements

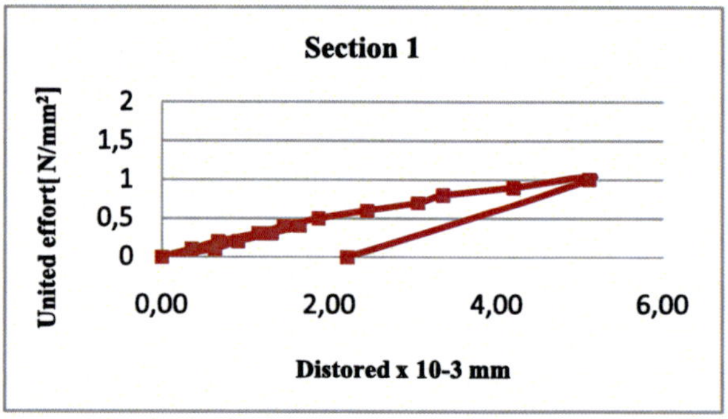

Fig. 2. The characteristic curves for section made only with ceramic elements

Photo 6. Section 2 homogeneous masonry made of ceramic elements

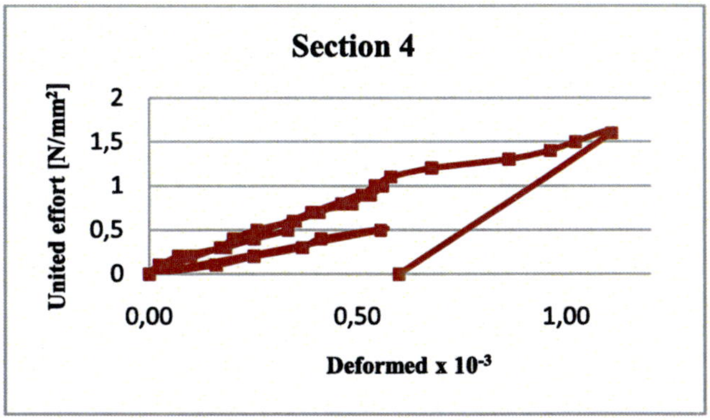

Fig. 3. The characteristic curves for compression section

3 Conclusion

The analysis of the results obtained on the component elements of the masonry, as well as of the results obtained on the compression tests, using flat presses, led us to the conclusion that during its operation the construction has often undergone modifications of functionality and consequently of structures.

In the course of history, some of the structural elements have been modified by restoring, filling some gaps or practicing a new one, using different types of ceramic elements or natural stone.

Photo 7. The characteristic curves for masonry, a) in situ curve b) test made in laboratory

The results achieved in situ were compared with the tests performed in the laboratory, the values obtained by using flat presses being confirmed by the tests performed in the laboratory for masonry with similar structures to the materials (mortar and masonry elements) taken from the structure.

The obtained values were used in the analysis and the intervention decision for the structural consolidation taking into account that intervention will be made on a heritage building.

The intervention measures consisted of the realization of vertical reinforcement introduced in drilled channels, vertically and horizontally inside the massive masonry structure. The joists are anchored at the level of the infrastructure in isolated foundations made under the masonry of the structure.

Before the drilled holes were made, the neo-homogeneous cracks and masonry were injected with injection mortars.

In Fig. 4 the results obtained are comparatively presented.

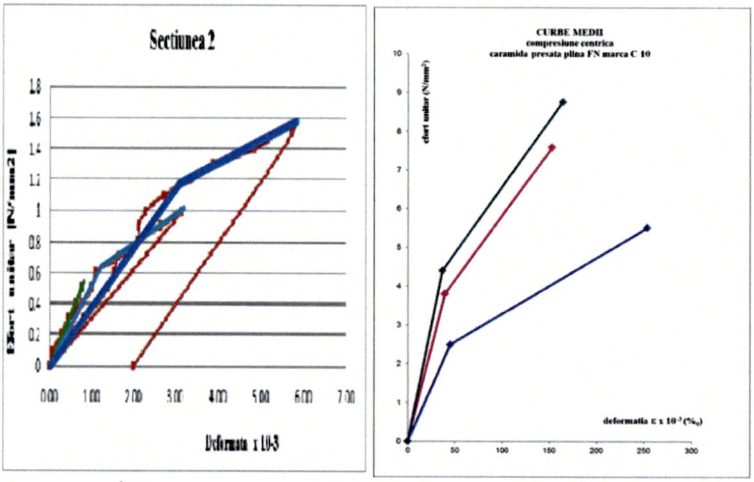

a) b)

Fig. 4. The characteristic curves for masonry, a) in situ curve b) test made in laboratory

Acknowledgements. This program is part of the activities for the rehabilitation and put safety of the resistance structure of the evangelical parish church "St. Mary" in Sibiu, which was included in the list of historical monuments in Sibiu County in 2004, having the classification code SB-II -I-12078. On this list the constructions is dated as the period 1350–1520.

References

Franchi, A., Crespi, P., Ronca, P., Pizzamiglio, F.: An in situ diagonal compression test for brick walls with displacement control on the two external layers Int. J. Hous. Sci. Appl. **38**(1), 25–35 (2014)

Steil, R.O., Calçada, L.M.L, Oliveira, A.L., Martins, V.C., PrudêncioJr, L.R.: Influên-cia do tipo de argamassa no fator de eficiência e nadeformabilidade de alvenarias de blocos de concreto. In: Anais do IV Simpósio Brasileiro de Tecnologia de Argamassas, Brasília, 423–434 (2001)

Cunha, E.H., Guimarães, G.N., Carasek, H.: Influência do tipo de argamassanaresistência à compressão da alvenariaestrutural. In: Anais do IV Simpósio Brasileiro de Tecnologia de Argamassas, Brasília, pp. 397–408 (2001)

Zejak, D., Vatin, N., Murgul, V.: Analysis of the Masonry Structure Calculation with vertical ring beams according to European standards Appl. Mech. Mater. 725–726, 111–117 (2015)

Matei, C.L., Matei, R.C.: Study to evaluate the characteristics of masonry for Stirbey family chapel to retrofit the structure. Structures and Architecture: Concepts, Applications and Challenges – 263–264ISBN 978–0–415–66195–9, CRC Press /Balkema Cruz (ed) 2013 Taylor & Francis Group, London (2012)

Abdunur, C.: Stress and deformability in concrete and masonry. In: IABSE Symposium on Strengthening of Building Structures - Diagnostic and Therapy, Venice, Italy (1983)

Noland, J.L., Atkinson, R.H., Schaller, M.P.: A review of the flat-jack method for Nonde structive evaluation, Proc. Nondestr. Eval. Civ. Struct. Mater. Boulder (1990)

Scrivener, J.C.: Old Masonry Buildings: Earthquake Performance and Materials Testing, Internal Report No. 624, Institute for Research in Construction, National Research Council, Canada (1992)

ASTM, In-situ compressive stress within solid unit masonry estimated using flat-jack Measurements, ASTM Standard C 1196–91 (1991)

ASTM, In-situ measurement of masonry deformability properties using flat-jack method, ASTM Standard C 1197–91 (1991)

RILEM, LUM.D.3, In-situ strength/elasticity tests on masonry based on the flat-jack (1990)

Model Tests on the Stability of Retaining Walls with Gabions Under Earthquake and Rainfall Conditions

Hiroshi Nakazawa[1(✉)], Tomohiro Ishizawa[2], Toru Danjo[2], Yasuhiro Onoue[2], Daisuke Suetsugu[3], and Tadashi Hara[4]

[1] Earthquake Disaster Mitigation Research Division, National Research Institute for Earth Science and Disaster Resilience Tennodai, 3-1, Tsukuba 3050006, Japan
nakazawa@bosai.go.jp

[2] Storm, Flood and Landslide Resarch Division, National Research Institute for Earth Science and Disaster Resilience Tennodai, 3-1, Tsukuba 3050006, Japan

[3] Department of Civil and Environmental Engineering, University of Miyazaki Kibanadai-Nishi, Gakuen 1-1, Miyazaki 8892192, Japan

[4] Department of Global Environment and Disaster Prevention, Kochi University, Akebono-cho 2-5-1, Kochi 7808520, Japan

Abstract. In this study, based on the damage survey in Nepal, shake table tests and rainfall experiments using a typical gabion retaining wall model were conducted to investigate the influence on earthquake and rainfall basically. In a series of experiments, two same models with retaining wall with slope ground same as typical gabion retaining wall in Nepal were constructed in rigid soil container. Then, the model experiments containing shake table test after preceding rainfall and rainfall after preceding shake table test were simultaneously performed in order to confirm difference of the damage aspects caused by pattern of earthquake and rainfall. As main results, it was confirmed that the deformation tendency of the gabion retaining wall was different by the order of rainfall and earthquake. Especially, the deformation of the retaining wall in Case of preceding shake table test was minor compared to preceding rainfall. Regardless of the order by rainfall and earthquake, the gabion retaining wall showed the same tendency on deformation mode.

Keywords: Gabion · Retaining wall · Earthquake · Rainfall · Model test · Residual deformation

1 Introduction

As a traditional construction method, there are seen many structures utilizing gabion in many countries around the world. Though gabion is a historical technology cultivated in river construction for a long time (Higaki 2015), it is thought that research on gabion construction method has not been progressed. However, in developing countries, because of its ease of understanding, easy construction and low cost, gabion is widely used in

I. Vayas and F. M. Mazzolani (Eds.): PROHITECH 2021, LNCE 209, pp. 56–67, 2022.
https://doi.org/10.1007/978-3-030-90788-4_5

construction works such as retaining wall and slope stability as well as river levee. According to damage survey results on gabion retaining walls for road after the 2015 Gorkha earthquake in Nepal, it was already reported that different types of damages such as tilting or deformation depended on gabion structure types and ground conditions behind the gabion retaining walls (Hara et al. 2018). Although these deformations of gabion retaining walls were mainly caused by the earthquake, it was suggested that many damaged gabion retaining walls showed structural tenacities without causing collapse because of their high flexibilities.

At present in Nepal, it is recognized that the restoration in the damaged area has not progressed after the earthquake (Ishizaki 1987). In addition, progress of deformations of the damaged gabion retaining walls and loss of stability of the ground behind gabion retaining walls are concerned. Moreover, it can be thought that these risks will be increased due to heavy rain during the rainy season every year. The problems on restoration there include lack of the knowledge for a design, construction and evaluation method on gabion structure because of their empirical rules (Association of Nationwide Disaster Prevention 1963).

In this study, based on the damage survey in Nepal, shake table tests and rainfall experiments using a typical gabion retaining wall model were conducted to investigate the influence on earthquake and rainfall basically.

2 Previous Study on Gabion Structures

The earthquake (Mw.7.3) that struck Nepal on April 25, 2015, caused damage to many civil engineering and architectural structures. While several road gabion retaining walls in mountainous regions incurred damage, there is very little information that can be used to draw up earthquake countermeasures in Nepal, since there have been few construction cases, case studies or experimental or analytical studies of gabion structures.

2.1 Damage Survey

Hara et al. (2018) conducted three field surveys, in July and November, 2015, and No-vember, 2016, of the damage to the gabion road retaining walls along the Araniko Highway in the mountainous regions, and have previously reported the findings in detail. The usage of gabions reported by Hara et al. (2018) and Nakazawa et al. (2018), shown in Fig. 1, indicates that, of the 115 locations, 49%, or approximately half, were used as road retaining walls, followed by 19% as crash barriers.

As shown in Fig. 1, the damage level of the retaining walls is divided into the three categories of A, B, and C. Categories A, B, and C indicate no damage, bulging, and collapse, respectively. In this previous survey, one third of the road retaining walls belonged to category A and a half was categorized in B. In addition, vertical types as gabion structure constituted roughly half of category A, while the stepwise type exceeded the vertical type in categories B and C. Although vertical type gabion structure construction is rarely used at Japanese construction sites due to its instability as compared to the stepwise type, it is characteristic of Nepal and is often used there.

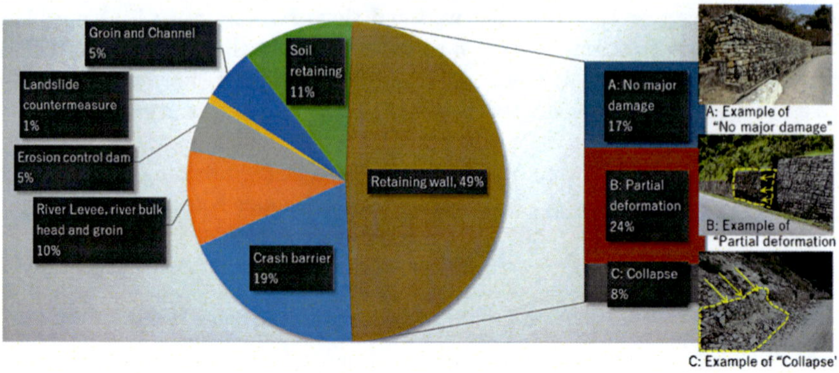

Fig. 1. Breakdown of gabion structures along Araniko highway in Nepal (adapted from Nakazawa et al. (2018)).

2.2 Full-Scale Experiments of Gabion Retaining Wall

Nakazawa et al. (2018; 2019) reported that the retaining wall model was a three-layered structure which height was 3 m and each layer was built in different stacking ways to imitate a retaining wall actually damaged in the 2015 earthquake. In addition, the ground was constructed behind the retaining wall. In the shake table tests, a sinusoidal wave of 8 s (3Hz) was used. Intensity of the input motion was gradually increased up to the maximum acceleration of about 300 Gal. In this series of experiments, we compared the three cases; a typical vertical retaining wall in Nepal (Case 1), a stepwise type (Case 2), and gravity type added a number of gabions (Case 3).

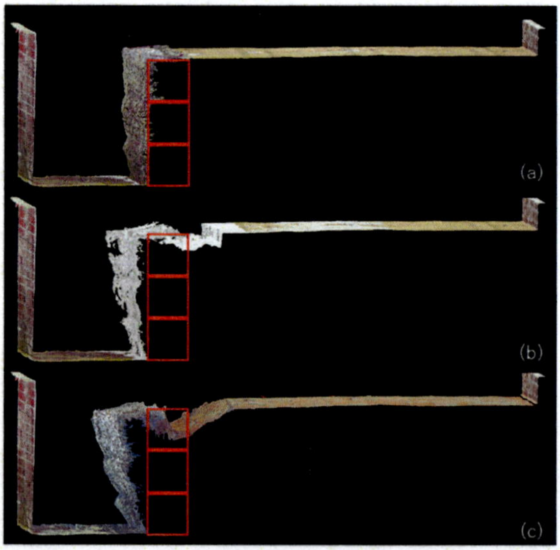

Fig. 2. Results of 3D measurements. (a) Before shake test. (b) After shaking at 203 Gals. (c) Residual deformation after final shake test (adapted from Nakazawa et al. (2018)).

According to the results of 3D territorial laser measurement of Case1 shown in Fig. 2, although the wall was not collapsed after shaking, there was confirmed significant inclination of 18°. This type of retaining wall deformation was also commonly observed in Nepal. This durability based on soft structure, was believed to be the reason why the wall did not eventually collapse. In the other two cases, the deformation was slight and stable. Based on these results, the recommended structure, and design and construction method suitable for Nepal were summarized in the guidelines (Hara et al. 2019).

3 Outline of Model Test

This chapter describes the outline of a siriease of experiments carried out in this study as an experimental bulletin. In previous studies, the seismic resistance of the gabion retaining wall was verified using a full-scale gabion retaining wall. On the other hand, the damage caused by heavy rain has not become negligible recently. However, there is no method to quantitatively evaluate the combined influence against heavy rainfalls and earthquakes. Therefore, it is difficult to improve the maintenance level while considering both heavy rainfall countermeasures and earthquake resistance at the same time.

In this study, a series of rainfall and shake table tests using a same model was carried out in order to confirm that the residual deformations and collapse patterns differ depending on the occurrence of preceding damage reproducing rainfall or earthquake.

3.1 Experimental Model, Materials and Experimental Device

In a series of experiments, two same models with retaining wall with slope ground were constructed in rigid soil container which size was 1.5 m high, 1.0 m wide and 4.0 m long shown as Fig. 3. The model on the left shown in Fig. 3 shows Case A for the effect of preceding rainfall, while the model on the right is illustrated as Case B to estimate the effects of preceding earthquake. The model was scaled down to one fifth compared with full scale experimental model with reference to typical gabion retaining wall in Nepal and which height was 0.6 m and the slope of model ground behind the retaining wall model was about 25°. Since it was indicated that gabion connection was important by Suetsugu et al. (2018), the upper and lower gabion unit were tightly connected by steel wire each other during the experiment in this study. The sensors like accelerometers and pore water pressure gauges and deformation transducers for shake tests and soil moisture meters and earth pressure meters for rainfall tests were installed in the model.

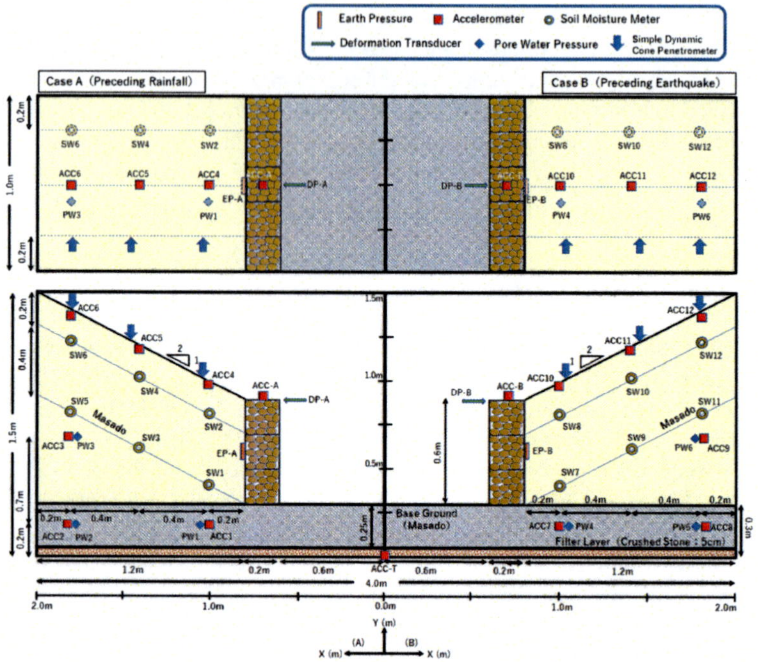

Fig. 3. Plan view and cross section.

A series of experiments were performed at Large-Scale Earthquake Simulator of National Research Institute for Earth Science and Disaster Resilience, Tsukuba, Japan. The features of this facility are shown in Table 1. Photo 1 shows the situation of shake table. As indicated in Table 1, the shake table has a size of 14.5 m × 5.0 m (the area to be loaded is 12.0 m × 2.0 m), the maximum load capacity is 4,900 kN, its performance provided an acceleration of approximately 0.5 G (490 Gal) with load of 4,900 kN and approximately 0.8 G (784 Gal) with load of 2,450 kN; the stroke is ± 22 cm.

Table 1. Specifications of shake table

Loading capacity	500 ton
Table size (area)	14.5 m × 15 m (217.5 m^2)
Power supply	Hydraulic pump system
Shaking direction	Horizontal, 1- dimensional
Excitation force	3,600 kN (four 900 kN actuators)
Maximum acceleration	940 cm/s^2 for 200 ton, 500 cm/s^2 for 500 ton
Maximum velocity	100 cm/s
Maximum displacement	±22 cm

Photo 2 shows the appearance of the sprinkler for rainfall test. This device has a structure that straddles the outer shape of the soil container used in the experiment and can generate uniform rainfall within the area of 1m by 1.8 m. The watering mechanism is a system by simultaneously vibrating 500 rain needles with a motor and it generates rain from a height of 2.0 m above the soil container. In addition, the raindrop diameter is variable in the range of Ø 1.7 to 3.0 mm and the rainfall intensity can be controlled from 10 to 80 mm/h.

Photo 1. Situation of shake table.

Photo 2. Soil container with rainfall device.

Decomposed granite soil called as Masado was used for the ground model. Figure 4 and Table 2 show the grain size distribution curve and the characteristics on physical properties and shear strength, respectively. Most grain sizes are those of gravel and sand. The maximum dry density $\rho_{d\max}$ used in preparing the ground model is 1.884 g/cm^3. The water content, w_c of the ground material during the construction of models was 8.9% and the average densities of the back slope behind the retaining wall were 1.568 t/m^3 for Case A and 1.572 t/m^3 for Case B respectively.

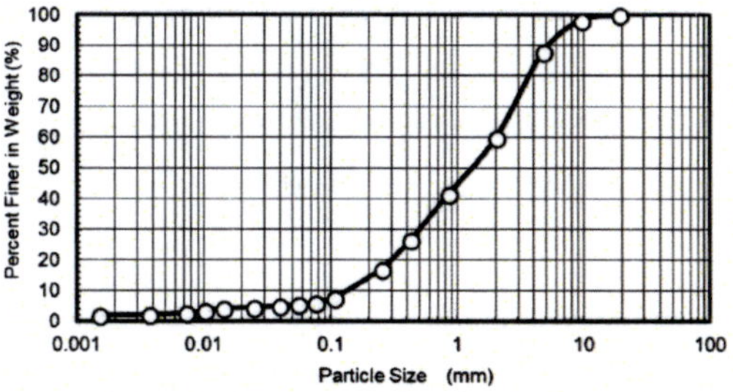

Fig. 4. Grain size distribution of decomposed granite soil.

In Nepal, the wire diameter ranges from 3 to 5 mm, and various mesh shapes are used such as hexagon, square, and rhombus. The minimum and maximum mesh sizes for the gabion retaining wall were 9 cm and 18 cm, respectively. As for the dimensions of gabion units, the dimensions 1 m × 1 m × 1 m are commonly prevalent in Nepal (Hara et al., 2018). On the other hand, the size of the model gabion used in this experiment is 20 cm × 20 cm × 20 cm shown in Photo. 3. The filling material used was crushed stone with a particle size of 37 mm to 19 mm. The maximum and minimum densities calculated with reference to the Japan Geotechnical Society Standard (2009) were ρ_{dmax} = 1.590 g/cm^3 and ρ_{dmin} = 1.251 g/cm^3 respectively. The average densities of Gabion retaining walls were 1.73 t/m^3 for Case A and 1.84 t/m^3 for Case B respectively.

Table 2. Physical and shear characteristics of material for ground model.

Density of soil particles	r_d	(g/cm^3)	2.657
Maximum grain size	D_{max}	(mm)	19.0
Fine content (<0.075 mm)	F_c	(%)	6.1
Mean particle size	D_{50}	(mm)	1.61
Uniform coefficient	U_c	–	19.5
Optimum water content	w_{opt}	(%)	11.6
Maximum density	r_{dmax}	(g/cm^3)	1.880
Plastisity index	I_p	–	*NP*
Cohesion	c_d	(kN/m^2)	20.5
Angle of shear resistance	$\emptyset_d$	(°)	38

Photo 3. Scaled gabion model.

3.2 Experimental Case

As shown in Table 3, the model experiments containing two cases of shake table test after preceding rainfall (Case A) and a rainfall test after preceding shake table tests (Case B) were simultaneously performed in order to confirm difference of the damage aspects caused by pattern of earthquake and rainfall. As a general outline of the experiment, a rainfall test for Case 1, shake table tests with three stages of acceleration for both cases, a rainfall test for Case 2, and finally shake tests with three stages of acceleration for both cases were performed.

The rainfall test conditions were 10 mm/h during 20 min, and shake table tests were targeted for 100 Gal, 200 Gal and 300 Gal. As for the waveform, a 3 Hz sinusoidal wave containing 2 s for each of the gradually increasing and decreasing portions was input to the shake table for 8 s. The actual accelerations acting on the shake table and at the bottom of the model were respectively 134 Gal, 259.4 Gal and 405.2 Gal for No. 2 to 4 shown in Table 3.

Table 3. Test cases and summary of observation results.

No	Case A	Case B	Event
1	Rainfall of 10 mm/h (20 min)	–	Erosion occurs on the back slope of Case 1
2	Shaking by Sinusoidal wave with 100 Gal acceleration		No deformation or damage in both cases
3	Shaking by Sinusoidal wave with 200 Gal acceleration		Inclination occurs in the retaining wall of Case 1
4	Shaking by Sinusoidal wave with 300 Gal acceleration		Retaining walls of both cases are inclined and deformation occurs on the back slopes in both cases. The retaining wall of Case 1 completely collapsed 32 s after the end of shaking
5	-	Rainfall of 10 mm/h (20 min)	Erosion occurs on the back slope of Case 2
6	Shaking by Sinusoidal wave with 100 Gal acceleration		No deformation or damage in both cases
7	Shaking by Sinusoidal wave with 200 Gal acceleration		During shaking, the retaining wall of Case 2 collapsed. No change on the slope of Case 1
8	Shaking by Sinusoidal wave with 300 Gal acceleration		The retaining walls of both cases had already collapsed, and no changes were seen in the models

4 Results of Experiments

4.1 Effects of Preceding Rainfall on Dynamic Behavior

In order to describe the effects of preceding rainfall, The data is explained focused on No. 1 to 4 experimental results in Table 3. In No. 1, some erosion was observed on the slope behind the retaining wall, but there was no deformation in both the retaining wall and the slope.

For shake table tests from No. 2 to 4, time histories of the acceleration response (ACC-A and B) and horizontal displacements (DP-A and B) measured at the top of the retaining wall are shown as representative data in Fig. 5. In No. 2, deformations of the retaining wall regardless of histories of the preceding rainfall in both case A and B are hardly observed and it can be seen that the acceleration response did not change. At the time of No.3 when deformation began to be occurred during shaking, the increase in a horizontal displacement with the progress of shaking in case A is remarkable. It is thought that its reason was due to the mass increase of the back slope due to rainfall and the softening of the ground material with high water content by the occurrence of excess pore water pressure. On the other hand, in case B, the reason is not clear, but it was confirmed that spiked behavior appeared in the acceleration.

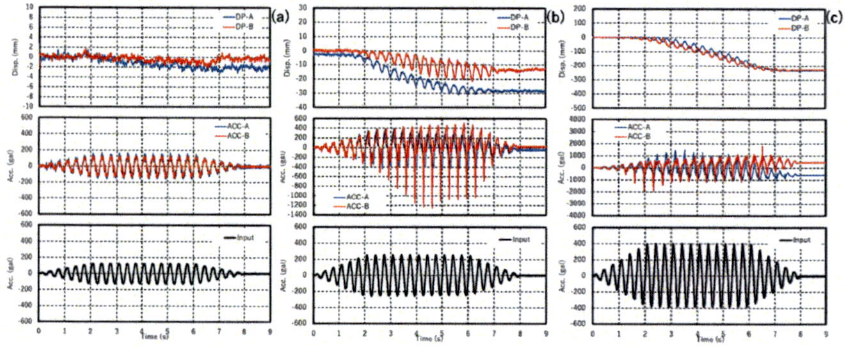

Fig. 5. Time history data obtained from (a) No. 2, (b) No. 3 and (c) No. 4 shake test shown in Table 3.

Focused on No.4, there was seen no significant difference in acceleration and horizontal displacement increments in both cases. As shown in Photo 4, a large deformation occurred in the retaining wall and back slope in Case A after a shake test in No. 4 was applied. Though it was barely escaped collapse of the retaining wall, it was observed that it slowly collapsed at about 32 s after the end of the shake test of No. 4. For the deformation tendency of the retaining wall during shaking, it was confirmed that the residual deformation before the collapse of Case A was roughly the same as case B. Finally, as shown in Photo 5, the retaining wall collapsed in Case A, while the deformation of retaining wall in Case B remained and it was self-supporting.

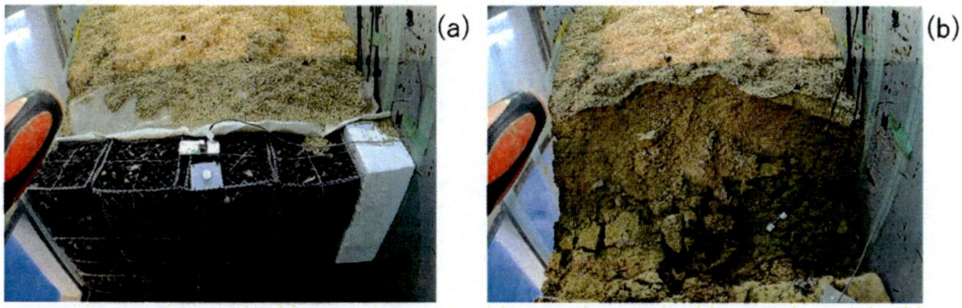

Photo 4. Deformation of retaining of Case A. (a) is immediately after shaking in No. 4 and (b) shows the collapse at 32 s after shaking.

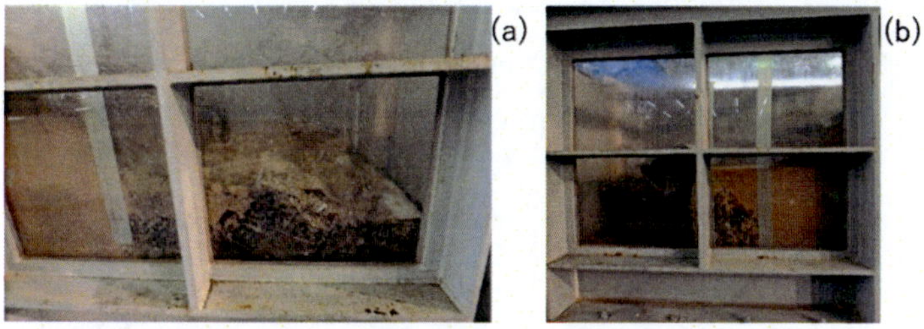

Photo 5. Comparison between (a) Case A and (b) Case B after No. 4.

4.2 Effects of Combined Effects of Earthquake and Rainfall

The residual deformations in each experimental phase of both cases are indicated in Fig. 6 and 7 respectively. A point of interest here is the difference of residual deformation tendency due to the different order of rainfall and earthquakes.

Compared with the residual deformations in both cases each other at the situation of No. 4, in Case A which was affected by preceding rainfall, the retaining wall was finally collapsed after the shake table testat No. 4. While in Case B affected by the preceding shaking, although a significant deformation occurred in the retaining wall by shake table tests from No. 2 to 4, it was confirmed that the collapse was escaped though the deformation progressed further by subsequent rainfall test at No. 5.

As described above, it could be confirmed from the experiment that the damage situation differs depending on the order of event.

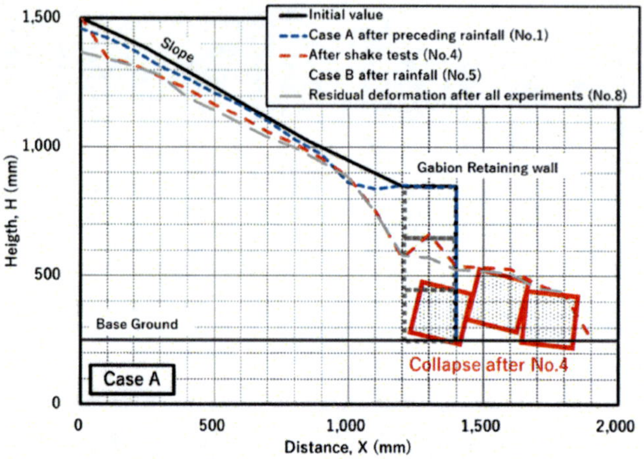

Fig. 6. Residual deformation in case A at No. 1, 4, 5 and 8.

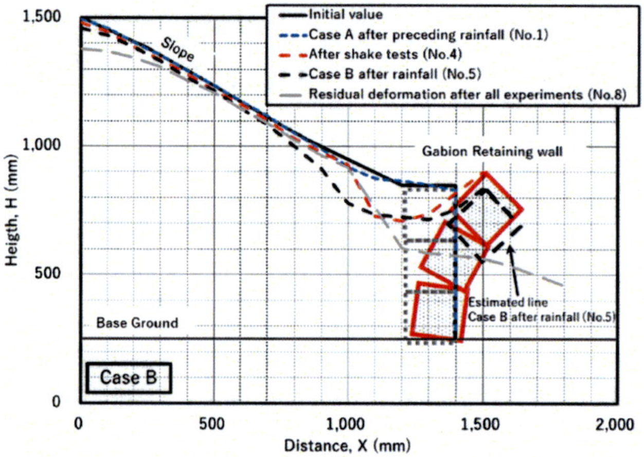

Fig. 7. Residual deformation in case B at No. 1, 4, 5 and 8.

5 Summary

As results from a series of experiments, the followings were confirmed.

1) The deformation tendency of the gabion retaining wall was different by the order of rainfall and earthquake. Especially, the deformation of the retaining wall in Case B was minor compared to Case A.
2) Regardless of the order by rainfall and shake test, the gabion retaining wall showed the same tendency on deformation mode.

3) A collapse of the retaining wall in Case A formed for a while after finishing shake table test. It is thought that this phenomenon was tenacity of gabion retaining wall and suggested a merit of flexible structures.

Acknowledgements. The authors would like to express their gratitude to the relevant parties such as Mr. Nakamura of Nexus Corporation who helped their experiment.

References

Higaki, D., Yagi, K. Wakai, A.: Sediment-related disasters caused by the 2015 Nepal Gorka earthquake. Jpn Landslide Soc. Emerg. Surv. Team Rep. 1–11 (2015)

Hara, T., et al.: Field survey on damages of Gabion structures caused by the 2015 Nepal Gorkha earthquake and examination of specific measures for earthquake resistaance improvement. J. Jpn. Soc. Civ. Eng. Ser. A1 (Struct. Eng. Earthq. Eng.) **74**(4), I_586–597 (in Japanese) (2018)

Ishizaki, M.: A historical study on Gabion. In: Proceedings of the Japan Society for Civil Engineering History, vol. 7, pp. 253-258 (1987). (in Japanese)

Association of Nationwide Disaster Prevention Knowledge of gabion, Japan River Association (1963). (in Japanese)

Nakazawa, H., et al.: Experimental evaluation on earthquake-resistance of road retaining wall using Gabion. J. Disaster Res. **13**(5), 897–916 (2018)

Nakazawa, H., et al.: Problems in earthquake resistance evaluation of gabion retaining wall based on shake table test with full-scale model. J. Disaster Res. **14**(9), 1154–1169 (2019)

Hara, T., et al.: Gabion Plan / Design / Construction Manual. First edn. Yusuhara Town and Kochi University (2019)

Suetsugu, D., Hara, T., Nakazawa, H., tadokoro, Y., Kuribayashi, K., Nishi, T.: Evaluation of earthquake resistance of retaining wall using gabions by laboratory tests -Part.2 Lateral loading test for model gabion retaining wall-. In: 53rd Japan National Conference on Geotechnical Engineering, pp. 1798–1799 (2018). (in Japanese)

The Japanese Geotechnical Society. Test method for minimum and maximum densities of gravels (JGS 0162–2009). Japanese Geotechnical Society Standards Laboratory Testing Standards of Geomaterials 3 (2009)

Technology, Diagnosis and Structural Assessment on Timber Composite Beams in Venice Arsenal at the Biennale

Clara Bertolini-Cestari[1](✉), Luciana Cestari[1], Stefano Invernizzi[2], Tanja Marzi[1], Olivia Pignatelli[3], and Mariangela Rossino[1]

[1] Department of Architecture and Design (DAD), Politecnico di Torino, Viale Mattioli 39, 10125 Turin, Italy
clara.bertolini@polito.it
[2] Department of Structural, Geotechnical and Building Engineering (DISEG), Politecnico di Torino, Corso Duca degli Abruzzi 24, 10124 Turin, Italy
[3] Laboratorio Dendrodata, Verona, Italy

Abstract. In the architectural and cultural heritage of the Arsenale of Venice, the authors have worked on several interventions for diagnosis and recovery of roofs, foundations and floors of the so-called "woods of Venice". The work refers to the complex of the "Sale d'Armi" located in the central area of the Arsenale, currently used as exhibition space for the Biennale of Venice. The original nucleus (1460) was used as weapons deposit of the Serenissima. The South "Sala d'Armi", three buildings on two levels realized in the 17th century, were rebuilt after the fire of 1728. This article describes the diagnostic assessment, including dendrochronological and mechanical assessment, carried out in one of these buildings, the largest one, a large room at full height, which currently preserves original timber elements left in place after the fire: composite beams belonging to a timber floor supported by timber columns. The paper focuses on a very interesting construction system such as the historical composite wooden beams. This technology, although in frequent use in different parts of Italy for the realization of tie-beams of the trusses that were covering large spans, can be found specifically in the area of Venice for the construction of timber floors during the 17th century.

Keywords: Constructive technology · Composite beams · Timber floor · Burned structures · Diagnosis NDT · Dendrochronology

1 Introduction

The present study focuses on a particular construction technology adopted in 17th-century carpentry which consists on the compositions of wooden beams, also called composite beams. This construction system was designed to cover large spans. The use of this technology in the tie-beams of large span wooden trusses is not uncommon. Of particular interest are those of the Arsenale of Venice. The construction technique analyzed in this study, on the other hand, concerns its application in timber floors in

I. Vayas and F. M. Mazzolani (Eds.): PROHITECH 2021, LNCE 209, pp. 68–79, 2022.
https://doi.org/10.1007/978-3-030-90788-4_6

buildings belonging to cultural heritage. The case of the Sale d'Armi beams is reflected in the peculiarity of the construction system present in the city of Venice and in other parts of Italy, as for example in the cities of Ferrara and Florence, in the Renaissance buildings. This article, after an analysis of these constructive systems, describes an unusual intervention on one of the buildings of the "Sale d'Armi" of the Venice Arsenal which is the largest building in the system, a single large room at full height, which currently has timber elements left in place after the fire. These are beams belonging to a timber floor supported by timber columns. The diagnostic assessment carried out were necessary both to understand the behavior of the beams and - mainly - for the extensive area of charred wood resulting after the fire. The research activities, aimed at evaluating the state of preservation and identifying the mechanical strength of the beams, were carried out according to the indications of the Italian Technical Standards (UNI 11119: 2004 and UNI 11138: 2004). This preliminary assessment before the design of the consolidation intervention was presented with a multi-disciplinary approach that made it possible to optimize the use of the available resources and to avoid further damage to the existing structures that remained as evidence of the cultural heritage and of the techniques and technologies present in Venice. This work has constituted the base for a complex restoration of the building, currently used as exhibition space for the Biennale of Venice and as multipurpose cultural space.

2 Historical Background

The Venice Arsenal, with its 48 acres of extension and a history that goes back more than 800 years, is an extraordinary subject of study of constructive techniques and of experimentation in the development of restoration methods. The wooden roof systems are among the most precious example of almost 700 years of building technology that was strictly linked and influenced by the development of naval shipbuilding industry. Since 1983 the Venice Superintendence (National Board of Antiquities) has been working and investing many resources on a program of study of the Arsenal, concentrating especially on roofs structures (Menichelli et al. 2009; Bertolini, Marzi 2018).

One of the aspects that contribute in determining the extraordinary cultural interest of the Venice Arsenal is the dimension and richness of the material, constituting a witness of architecture and constructive technologies for naval shipbuilding.

The assessment was mainly focused on the roofing structures of numerous buildings located in the Arsenale area: the tie-beam was a constant. This element, the longest within the truss system, looks like a composite beam, with different connection systems used to connect different wooden elements.

The paper focuses on the South Sale d'Armi, in particular the "E" building (Fig. 1). The complex of the South and North Halls consists of two imposing two-story buildings. These buildings, whose original nucleus dates back around 1460, constituted the repository of the most important military armaments of the Serenissima Republic and were also used for representation purposes in occasion of visits of particularly important guests. With the adjacent Artillery building, the Halls of Arms, Sale d'Armi, constituted the symbol of the war power of Venice "by land and sea". It was accessible only through the monumental Artillery Gate, built around 1560 - attributed to Sanmicheli - to separate

Fig. 1. Aerial view of the city of Venice (a), the Arsenale complex (b) and the Sale di'Armi (Room E highlighted in orange) (c)

the artillery sector from the other functions of the Arsenale. The Sale d'Armi South, built after the North Hall facing the dock, presumably date back to the mid-17th century. They were rebuilt in the mid-1700s, after a fire dating back to 1728. They are divided into three buildings (E, F, G). The overall dimensions of the South "E" sector, on two floors, are 2300 m^2 (Fig. 2 and 3).

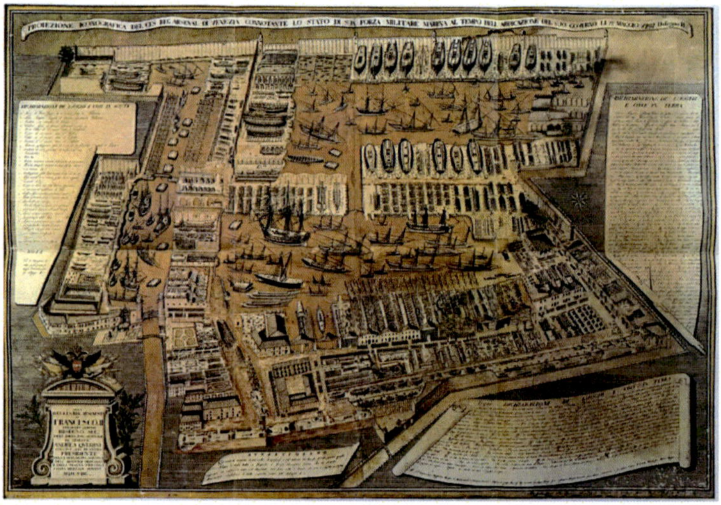

Fig. 2. Iconographic representation of the Arsenale of Venice, G. Maffioleti, 1797 (from U. Pizzarello, V. Fontana, *Pietre e legni dell'Arsenale di Venezia*, L'altra Riva, Venice 1983)

In 1964 the Arsenale ceased all activity and the Sale d'Armi were at first used as warehouses by the Navy and then abandoned.

Since 2007, having survived almost intact in the load-bearing structures, after the passage of management to the Venice Biennale, the block of buildings has undergone a complete restoration which foresees its new destination as exhibition, show and cultural spaces linked to the activities of the Biennale (Fig. 4).

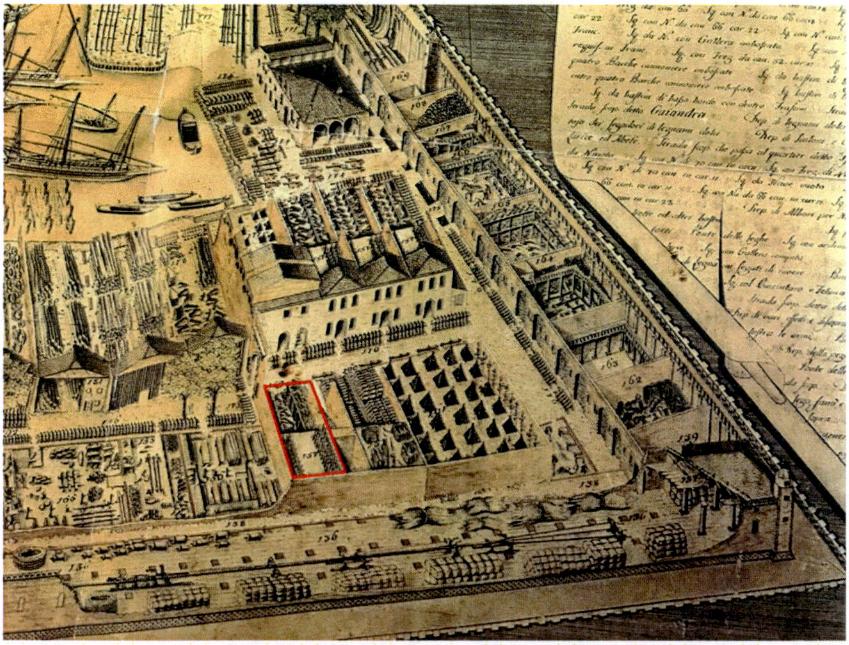

Fig. 3. Iconographic representation of the Arsenale of Venice, G. Maffioleti, 1797 (from Pizzarello, Fontana 1983). Highlighted in red the Sala d'Armi South, building E.

Fig. 4. View of the Sale d'Armi South, 1980's (left) (from Pizzarello, Fontana, 1983); Sale d'Armi North and South during the recent restoration.

The analyzed "E" building is the largest block in the sector. It currently looks like a single large full-height nave, due to the collapse of the intermediate floor due to the 18th-century fire.

As evidence of the architectural importance of the building, the two large wooden beams, object of the present study, were still preserved on site (Fig. 5).

Fig. 5. Interior view of the Sala d'Armi "E" (left), detail of the burned wooden beam with wooden column (right). (Photo D. Ferrante 2019)

3 Constructive Technology

The definition of "composite beam" can be found in historical documents or treatises with reference to a generic term of "*chiavon*", or big beam. With this term we mean a wooden element made through an appropriate assembly of different wooden elements that allow to cover large spans with correct structural behavior.

Composite beams are quite frequent in the typology of large wooden trusses with particular regard to the longest element of the construction system which is the tie-beam. At the Arsenale of Venice, in many "Tesoni" buildings, the tie-beams are made with large composite beams that simultaneously served to perform the function of a tie-beam. In some cases, these elements also functioned as beams to mast ships.

The singularity of the beams analyzed in building "E" is that they are composite beams used as support for the main floor.

These beams are made of five overlapping wooden elements: a tie-beam made of two elements side by side, two rafters elements positioned symmetrically to the extrados of the beam, an element that functions as a block or key to close the system.

Another characteristic of the analyzed composite beams is that the elements that compose them are made with different wood species.

The general characteristic of composite beams is the curvature of the tie-beam which corresponds to a counter-deflection imposed with respect to that resulting from the loads. Therefore, there is a curvature towards the upper part.

The connections between the different pieces that make up the construction type are made with cutting indentations. These shapes between the parts in contact prevent the overlapping elements from sliding. Through nails (metal nails inserted from the extrados and riveted) confirm the robustness of the system. These rivets are often joined by metal straps (*reggette*) that collaborate in their assembly (Fig. 6).

Each beam covered a span of about 11.80 m, had a section of 30 (base) × 68 (height) and originally had two wooden columns placed approximately at the distance of 1/3 from the support.

The two composite beams, which have remained in the "Sala d'Armi" building "E" building today, have extended traces of fire.

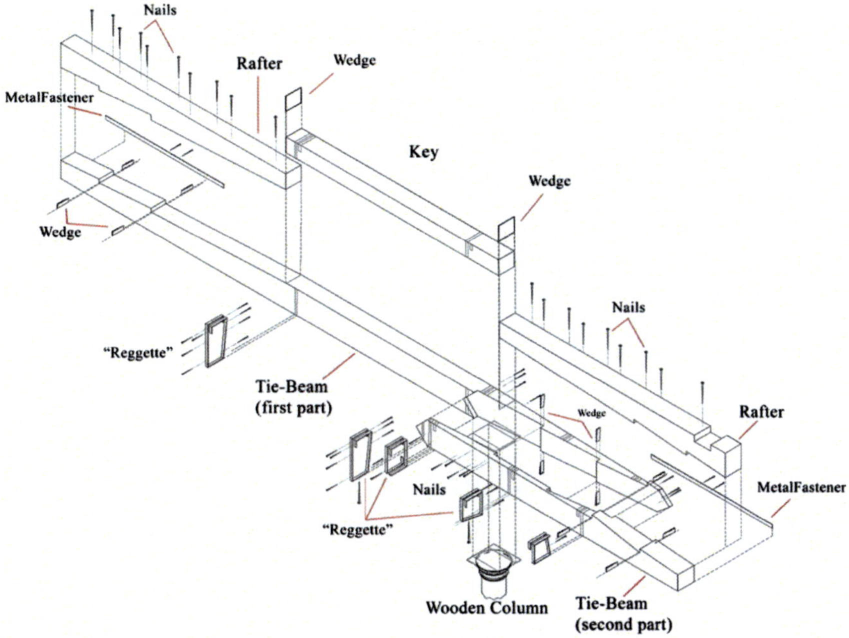

Fig. 6. Axonometric scheme of the different parts constituting the composite beam with the wooden column.

A single wooden column with Doric capital is currently present in the first beam, with a height of about 4.30 m (Fig. 7).

In the tradition of construction consolidated within the Venetian territory, horizontal wooden structures assume specific aspects, material, structural, formal and the whole construction organism connection, forming a complex technological device. The field-work of a large number of existing floor structures, dating back to the 14th and 18th century, has allowed to highlight the characters of both the individual elements and the "floor system" as a whole (Scappin 2015).

In Italy, this construction technology is very widely reflected in the wooden floors of the Renaissance city of Ferrara (Giunchi et al. 2004). The Ferrara composite beam, which we find widespread within important buildings of the 15th–16th centuries, is often made of four or six pieces. The singularity of these beams is their slimness. In particular, the six-piece beam consists of the tie-beam, the two rafters, the dowel or key inserted centrally, and to complete two "shoulders", they contribute to characterize the geometry.

Fig. 7. Interior view of the Sala d'Armi "E", composite beams and wooden column (Photo D. Ferrante 2019)

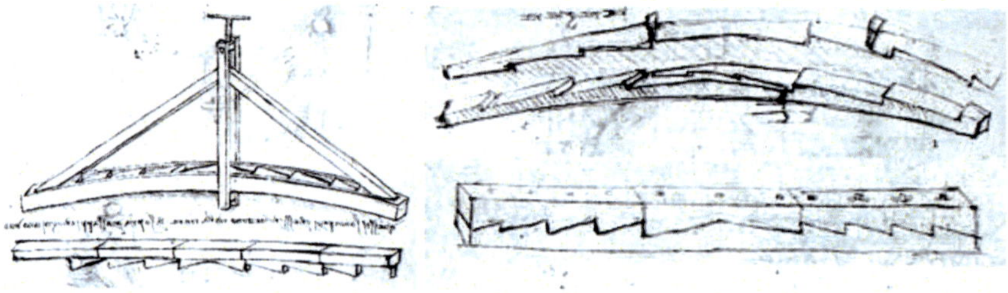

Fig. 8. Leonardo's "trestle" for bending composite beams interior (left); Leonardo's sketches (right) from the Atlantic Codex of Leonardo da Vinci, 15th–16th cent. (from Leonardo Da Vinci 2000)

A wide documentation of the composite beams can be found in Italian treatises. It is interesting to recall how in the 15th–16th centuries in the Atlantic Codex of Leonardo da Vinci the aspects of construction technique and technology are investigated, such as the possible solutions to impart the curvature to the tie-beam and to create cuts with adequate connections between the different components (Fig. 8).

4 Diagnosis

4.1 Technological Survey

The technological survey for assessing the state of conservation and geometry of connections has basically regarded:

- diagnosis, intended as the complex of on-site and/or laboratory techniques, which have as their object the survey of the original natural characteristics of wood (wood species, physical-mechanical properties, defects) and those acquired after the construction (biotic decay, breakages, deformations);
- timber grading according to resistance, using the results of the diagnostic phase to define the structural quality of the wooden elements according to "categories" from which the values of the main physical-mechanical characteristics used in the calculation of structural safety are obtained;
- assessment of the functionality of the joints, which is of fundamental importance for the proper functioning of traditional timber structures, since structural stability problems often stem from the inadequate capacity of the joints to transfer the forces from one element to the other;

These types of surveys went alongside with instrumental investigations, with the help of resistance drilling for wooden elements whose state of conservation needed to be verified. Furthermore, also the constructive survey was integrated with the use of resistance drilling to verify the thickness of the different parts constituting the joint, a very important indication for the assessment of the state of conservation and mainly of the stability of the structure.

For the Venice Arsenal, survey charts were used, which constitute a "reading grid" for the identification of all the features that allow to fully defining the technical solutions originally adopted for the construction of the structure and for the identification of all the decays (these obtained through the recalled diagnostic activity).

The diagnostic assessment mainly concerned the timber elements of the main supporting structure of the trusses, to which the investigations on secondary elements were added (Fig. 9).

The purpose of the inspection was to obtain the information listed below related to the wooden elements being investigated, in order to assess their integrity and mechanical performance. In particular, data was collected regarding:

a) wood species;
b) wood moisture content and any humidity gradients;
c) biological risk class for wood, according to UNI EN 335-1 and UNI EN 335-2;
d) geometry and morphology of the wooden element, with indication of the position and extent of the main defects, decay and any damage present;
e) position, shape and dimensions of critical areas and sections;
f) classification according to the resistance of the element as a whole and/or of the individual critical areas, according to the UNI 11119 standard.

The first three phases allowed to establish if the structure was subject to biotic degradation by specific xylophagous organisms. All the other evaluations allowed the assignment of a resistant profile for each of the elements constituting the examined structure. This has constituted the base of the restoration works.

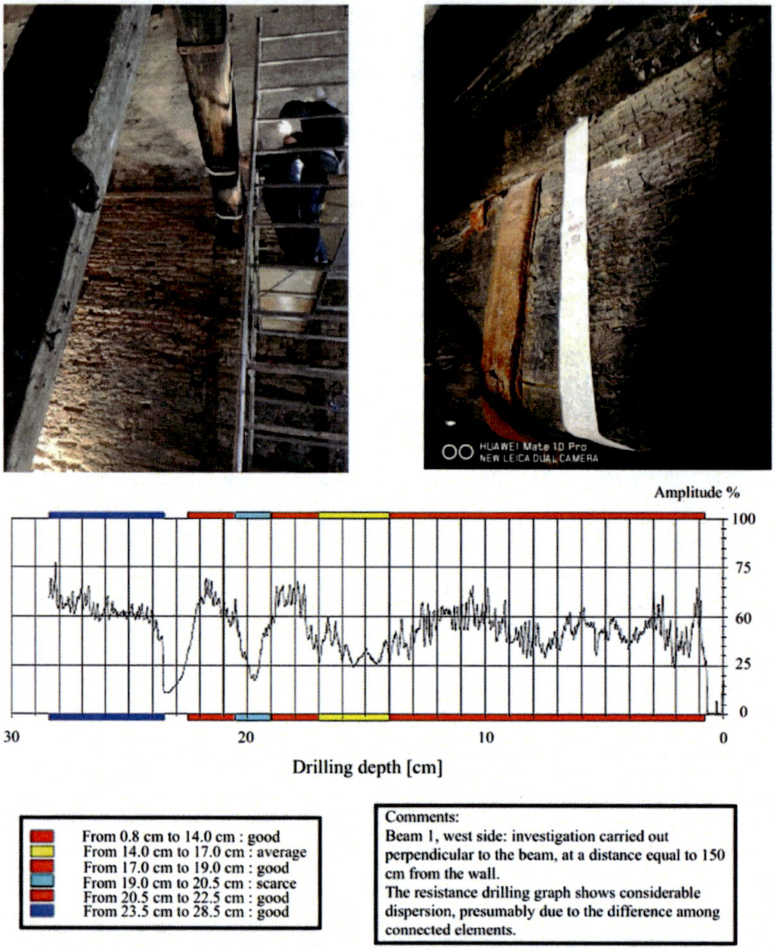

Fig. 9. Diagnostic survey on the composite beam. Overlapping of the different wooden elements composing the beam. Example of resistance drilling carried out on the beam.

4.2 Dendrochronological Investigations

Xylotomy (UNI 11118: 2004) and dendrochronological (UNI 111141: 2004) investigations aimed at identifying the wood species used and dating the two composite beams, involved almost all the elements, with the exception of the key of the beam T2 (Table 1).

Table 1. Characteristic of the examined elements

Composite beam	Element	Species	Lenght	sapwood	Last tree-ring
T1	Rafter side W	*Abies alba* Mill.	172 rings	undifferenciated wood	AD 1579
T2	Rafter side W	*Abies alba* Mill.	57 rings	undifferenciated wood	AD 1501
T2	Rafter side E	*Picea abies* Karst.	185 rings	undifferenciated wood	undated
T1	Rafter side E	*Abies alba* Mill.	87 rings	undifferenciated wood	AD 1550
T1	Key	*Larix decidua* Mill.	88 rings	absent	undated
T1	Tie-beam side E	*Larix decidua* Mill.	109 rings	28 rings	undated
T1	Tie-beam side W	*Larix decidua* Mill.	124 rings	25 rings	undated
T2	Tie-beam side E	*Larix decidua* Mill.	73 rings	22 rings	undated
T2	Tie-beam side W	*Larix decidua* Mill.	86 rings	18 rings	undated

The tie beams were all made of larch wood (*Larix decidua* Mill.), as well as the only key examined; for the rafters, instead, fir (*Abies alba* Mill.) and spruce (*Picea abies* Karst.) were used. The dendrochronological investigations carried out following the classic procedures (Baillie 1982, Fritts 1976, Schweingruber 1982), led to the date of the only three rafters in fir. These elements, all missing the outermost rings, partly removed with the processing of the trunks, and partly deteriorated as a result of the carbonization, provided only the date after which the matrix plants were felled (Table 1). The year AD 1579 constitutes the common *terminus ante quem non* for the felling. No dating has been identified for the other examined wooden elements. Unfortunately, the absence of dating for the tie-beams, all with sapwood, and, in two cases, also with traces of the bark, does not allow us to establish from a dendrochronological point of view whether the elements of fir represent reused wood.

However, it must be underlined that the larch series do not match with the other dated series in the Arsenale area. This seems to suggest that the larches used for the tie-beams of the composite beams do not refer to the felling phases identified so far for the Arsenale compound.

5 Structural Assessment

The two beams are composed by four main structural elements. The intrados portion is a two-pieces composed tie-beam. The connection between the parts is provided by means of horizontal steel pins and several notches, in addition to steel stirrups nailed to the beam. The tie-beam is subjected to tensile stresses. The tie-beam does not show, contrarily to other examples, a significant curvature. In addition, it can be noticed that each beam was supported by two vertical wooden pillars, although only one of them is survived. The extrados of the beam is composed of a central rectangular key piece surrounded by two symmetrical rafters. These three parts are subjected to compressive stresses. The extrados plan shows a series of regularly spaced big nails (rivets) which, together with notches, guaranties the transmission of shear stresses at the level of the neutral plane. A three-dimensional finite element model has been set up to assess the level of stress and amount of deformations of the two timber beams. The model account for the presence of the different structural elements of the composed beams, and for the presence of the survived pillar. The interaction between the parts has been accounted for by means of no-tension structural interface special elements. The structure is analyzed under the effect of dead weight only, in order to assess the feasibility of conservation of the beams, accounting also for possible degradation due to partially burned surfaces. The state of coaction in the structure mainly depends on the actual procedure followed during the assembling of the composed beams, in addition to the rheologic behavior of the parts and interfaces. Although it is very difficult to provide a precise picture of the present state, some parametric analyses have been carried out by considering different imposed deformation to the rafters and central key-part.

As a result, given the level of stresses and deformation of the beams, their conservation is strongly advised (Fig. 10).

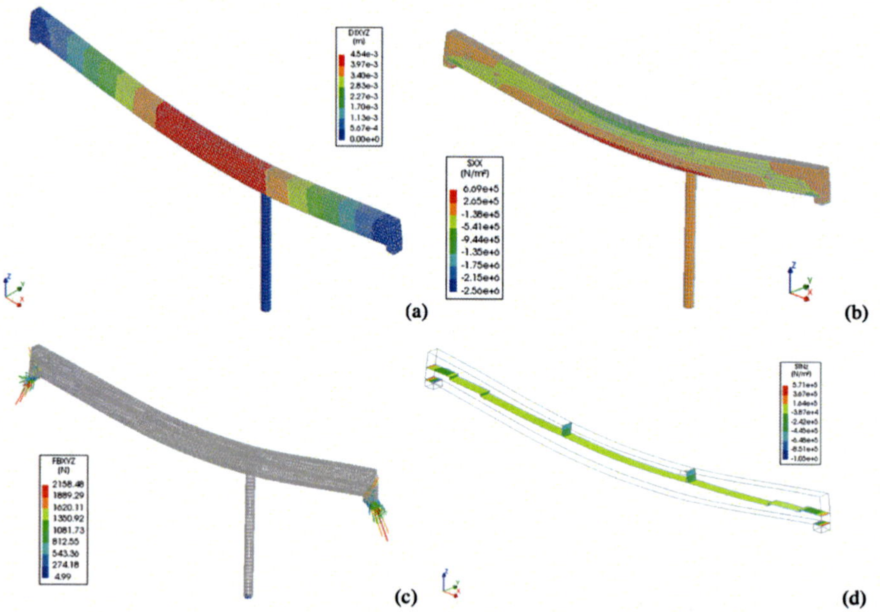

Fig. 10. Finite element analysis results: deformed mesh (a); normal stress component along the beam axis (b); reaction forces at the boundary (c); tractions in correspondence of the interfaces (d).

6 Conclusions

The main assessment conducted on the timber elements of the composite beams of the floor of the Sala d'Armi "E" at the Venice Arsenal concerned: the technological assessment of the components and their mechanical resistance, the dendrochronological investigation, the structural evaluation. The results of these multi-disciplinary activities have led to the validation of the various timber members in good condition and mechanical response. The diagnostic survey has constituted the base for the following restoration phase.

The proposed case-study represents an example of great interest among the timber floor structures realized with the composite beams' constructive technology.

Acknowledgements. The Biennale of Venice is gratefully acknowledged.

References

Baillie, M.G.L.: Tree-ring Dating and Archaeology, London – Canberra, pp. 1–274 (1982)

Bertolini-Cestari, C., Marzi, T.: Conservation of historic timber roof structures of Italian architectural heritage: diagnosis, assessment, and intervention. Int. J. Arch. Heritage **12**(4), 632–665 (2018)

Fritts, H.C.: Tree Ring and Climate, pp. 1–576. Academic press, London-New York-San Francisco (1976)

Giunchi, E., Malvezzi, R., Russo, M.: (2001–2002), Travi composte in legno nella Ferrara rinascimentale: indagini conoscitive e proposte di intervento. Thesis. Università di Ferrara

Giunchi, E., Malvezzi, R., Russo, M., Alessandri, C., Fabbri, R.: High technology and structural insight in the wooden composite beams of the renaissance architecture in Ferrara. In: Bertolini-Cestari, C., Marzi, T., Seip, E., Touliatos, P. (eds.) Interaction between Science, Technology and Architecture in Timber Construction, pp. 513–544. Elsevier, Paris (2004)

Da Vinci, L.: Il codice atlantico della Biblioteca Ambrosiana di Milano, trascrizione critica di Augusto Marinoni, Giunti, Firenze (2000)

Menichelli, C., et al.: Le strutture lignee dell'Arsenale di Venezia. Studi e restauri. XXV Convegno Internazionale Scienza e Beni Culturali - Conservare e restaurare il legno, Bressanone, Arcadia Ricerche, 1163–216 (2009)

Pizzarello, U., Fontana, V.: Pietre e legni dell'Arsenale di Venezia, L'altra Riva, Venice (1983)

Scappin, L.: Il solaio ligneo nel 'congegno' dell'edificio storico veneziano: elementi, connessioni e dimensionamenti. In: Marzo, M., Rakowitz, G., Nella ricerca. Città/Venezia, Università IUAV, Venezia, pp. 85–109 (2015)

Schweingruber, F.H.: Tree Rings, pp. 1–276. Reidel Publishing Company, Basics and applications of dendrochronology (1988)

UNI 11118:2004. Beni Culturali, Manufatti lignei. Criteri per l'identificazione delle specie legnose, UNI, Milano

UNI 11119:2004 Beni culturali - Manufatti lignei - Strutture portanti degli edifici – Ispezione in situ per la diagnosi degli elementi in opera, UNI, Milano

UNI 11141:2004. Beni Culturali. Manufatti lignei – Linee guida per la datazione dendrocronologica del legno, UNI, Milano

Characterization of Old Concrete from a Heritage Structure of Inousses Cluster of Islands

Konstantinos Sotiriadis[1,2], Konstantinos Aspiotis[3(✉)], Anton Mazur[4], Peter Tolstoy[4], Efstratios Badogiannis[5], and Sotirios Tsivilis[3]

[1] Institute of Theoretical and Applied Mechanics of the Czech Academy of Sciences, Prosecká 809/76, 19000 Prague 9, Czechia

[2] Department of Building Materials and Products, South Ural State University (National Research University), pr. Lenina 76, 454080 Chelyabinsk, Russia

[3] School of Chemical Engineering, National Technical University of Athens, Heroon Polytechniou 9, 15773 Athens, Greece

aspiotis@mail.ntua.gr

[4] Center for Magnetic Resonance, St. Petersburg State University, Universitetsky pr. 26, Petergof, 198504 St. Petersburg, Russia

[5] School of Civil Engineering, National Technical University of Athens, Heroon Polytechniou 5, 15773 Athens, Greece

Abstract. In this work, the amorphous part of samples extracted from damaged concrete structural elements of a heritage church of Inousses cluster of islands (Greece) was studied. The samples were obtained from both the interior (dome) and the exterior (two walls and one column) of the church. The techniques employed include simultaneous thermal analysis and solid state nuclear magnetic resonance spectroscopy. Concrete degradation was mainly due to carbonation and it was more intense for the structural elements exposed directly to the environment (external walls and column). The components identified in the sample coming from the external column indicate that a different concrete composition was likely used for its construction.

Keywords: Heritage concrete · Deterioration · Marine environment · Amorphous phase · Solid state nuclear magnetic resonance

1 Introduction

Cultural heritage is the link between the past and the present of people living in a specific area, and it concerns both tangible and intangible characteristics (ICOMOS 2002). In this aspect, considerable attention has been given at a European level since 1985, when the so-called Convention for the Protection of the Architectural Heritage of Europe was adopted in Granada (Spain) and came into force in 1987 (Council of Europe 1985).

In Greece, the Granada convention was approved in 1992 (Official Journal of the Hellenic Republic 1992), however the protection and conservation of traditional buildings and communities was anticipated already since 1973 within the Hellenic Building

I. Vayas and F. M. Mazzolani (Eds.): PROHITECH 2021, LNCE 209, pp. 80–89, 2022.
https://doi.org/10.1007/978-3-030-90788-4_7

Code. The earlier referred as "preserved settlements" were recognized as "traditional settlements" after the amendment of 1977. By today, a total number of 924 settlements all over the territory of Greece are characterized as such (Papageorgiou and Pozoukidou 2014).

The present study focuses on a church of traditional architecture built on Agios Panteleimon islet which is part of the Inousses cluster of islands at the Aegean Sea, situated between the Greek island of Chios and the western Turkish coast (Fig. 1). Since 1999, the settlement of the Municipality of Inousses (Chios Regional unit) is identified as traditional with a Presidential Decree (Official Journal of the Hellenic Republic 1999).

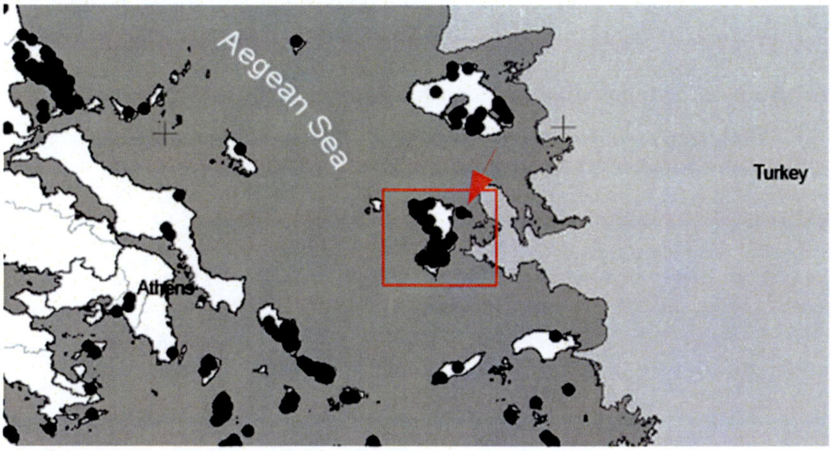

Fig. 1. Extract from the map showing the traditional settlements in Greece (black dots) (Source: Papageorgiou and Pozoukidou 2014). The red arrow points to the Municipality of Inousses within the Chios Regional unit (Greek territory within the red frame)

According to the provisions, prior to any intervention in a heritage object, the pathogenies of the structure should be recognized first and subsequently the damage observed has to be characterized. Steel reinforcement was revealed in several parts of the examined church; the diameter of the steel rebars was reduced due to corrosion. This type of attack is common in reinforced structures exposed to aggressive environments, and occurs mainly as a result of concrete carbonation and the effect of chlorides (Bertolini et al. 2008) (Cao et al. 2019). Several tests are usually applied to monitor the state of a damaged structure, including tests for concrete (ultrasonic pulse velocity, chloride and sulfate contents, porosity, petrography, X-ray diffraction) and steel reinforcement (strength, elongation, corrosion tests) (Marcos et al. 2016).

Here the aim is the characterization of the amorphous part of the hardened cement paste as altered due to the effect of the marine environment which is in close proximity to the structure. This will enable us to draw conclusions about the extent of concrete's damage and reveal the structural features of its aluminosilicate nature.

2 History and Topography

The diverse history of Inousses left its traces on the local architecture, developed mainly from late 18th to early 19th century, and it is reflected on two-storey masonry houses built with extensive employment of metal, wooden and tiled parts. Reinforced concrete became, at the second half of the 20th century, the dominant material in newer structures, which can be currently considered also as architectural heritage objects. Their construction involved natural marine aggregates and brackish water both bearing a multi-ion load potentially affecting the hardened cement paste and the steel reinforcement.

The traditional church under investigation is situated on an elevation of about 16 m at the southern edge of Agios Panteleimon islet (coordinates of central point: 692657.124, 4265136.683, according to the Hellenic Geodetic Reference System 1987), as shown in Fig. 2. It was originally constructed in 1938 and completely reconstructed in 1980s (Laimos 1965). Testimonies of local inhabitants refer to the use of crushed aggregates in concrete.

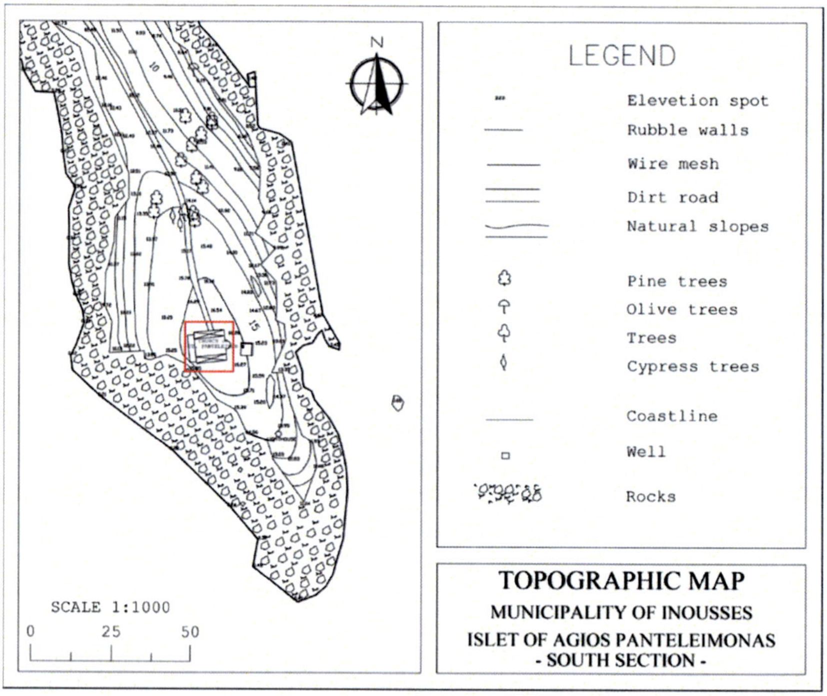

Fig. 2. Extract from the original topographic map of Inousses; here only the islet of Agios Panteleimon is presented. The church is indicated within the red frame. The map was produced during the Field Course in Geodesy II of the School of Rural and Surveying Engineering of the National Technical University of Athens, under the supervision of Professors E. Lambrou† and G. Pantazis, from 2001 to 2008.

3 Experimental

Damage was mainly observed on dome, columns and beams. The long-term exposure of the church to the local environment (periods of high humidity, rapid drying due to sunlight, deleterious species supplied by the sea breeze) is assumed as the main factor triggering the deterioration. Representative samples of damaged concrete were extracted from four structural elements of the church shown in Fig. 3, i.e. from dome (sample 1), from southern external wall (sample 2), from northern external wall (sample 3), and from southern external column (sample 4).

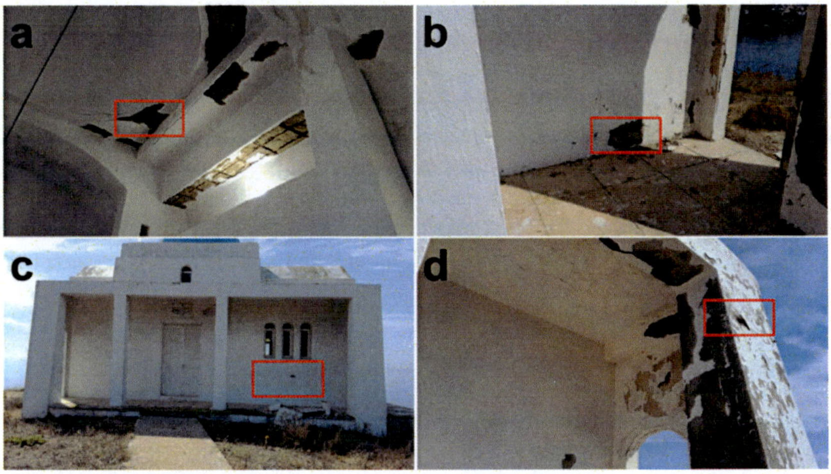

Fig. 3. Sampling spots indicated with red frame: (a) dome, (b) southern external wall, (c) northern external wall, (d) southern external column

Each sample was initially gently ground by hand and sieved for removing the aggregates. The resulting material was milled for 3 min in a Mini-Mill Pulverisette 23 (Fritch) at 35 oscillations/min until it became possible to pass through a sieve of 63 μm mesh size. The obtained powders were investigated with the following techniques.

Simultaneous thermal analysis was performed with a Discovery SDT 650 instrument (TA Instruments), coupled with a Discovery MS mass spectrometer (TA Instruments) used for the analysis of the generated gases. Samples (~20 mg each) were loaded on alumina crucible and heated in the temperature range 20–1000 °C with 20 °C/min heating rate and N_2 gas flow.

NMR experiments were performed with a Bruker Avance III 400 WB NMR spectrometer (magnetic field of 9.4 T), using a 4-mm CP/MAS probe. Larmor frequencies for the investigated ^{1}H, ^{13}C, ^{29}Si and ^{27}Al nuclei are 400.1, 100.6, 79.5 and 104.3 MHz, respectively. Samples were loaded into a zirconium oxide rotor of 4 mm in diameter spinning at 12.5 kHz. The chemical shifts obtained for each nucleus were calibrated using appropriate external reference compounds. One-pulse ^{1}H, ^{13}C, ^{29}Si and ^{27}Al MAS NMR spectra were acquired at 25 °C, recording 8, 64, 1024 and 2048 scans, with 3, 3.2, 4 and

1.6 μs pulse length, and 20, 30, 5 and 1 s relaxation delay, respectively. Q^n notation was used for the interpretation of ^{29}Si MAS NMR spectra (Engelhard 2007).

^{27}Al triple-quantum MAS (3Q MAS) NMR spectra were collected applying a three-pulse sequence with z-filter (5.4 μs excitation pulse, 2.2 μs conversion pulse, 20 μs weak selective 90° pulse) with rotor synchronization. 32 points in the f1 dimension (20 ms increments) were obtained with each point corresponding to 288 scans. Recycle time of 0.5 s was used for the acquisition of spectra. A shear transformation in the indirect dimension with line-broadening of 50 Hz was applied for data processing.

4 Results and Discussion

4.1 Thermal Analysis Results

In Fig. 4, the TGA, DTG and DSC curves obtained for sample 1 are presented. The main event observed is the decomposition of calcium carbonates, initiating at ~640 °C and showing a steep weight loss at ~800 °C. Two endothermic events at ~85 °C and ~130 °C, are related to removal of water from cement hydrates and gypsum, respectively (Scrivener et al. 2016) (Földvári 2011). The continuous weight loss up to ~400 °C is related to weight loss from cement hydrates, and especially from calcium silicate hydrate, which occurs in the range 50–600 °C (Scrivener et al. 2016) (Kunther et al. 2015). Similar curves were obtained for samples 2, 3 and 4.

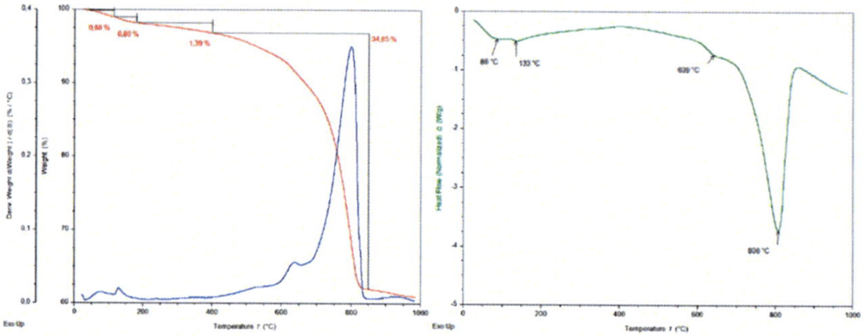

Fig. 4. Weight loss and derivative weight loss curves (left) and heat flow curve (right) obtained for sample 1

In Table 1, the weight losses determined at each temperature interval for each sample, are summarized. They indicate extensive carbonation of concrete and significant deterioration of C–S–H phase. Both phenomena were more intense for the external structural elements (samples 2, 3 and 4) and especially for those oriented to the South (samples 2 and 4).

Table 1. Weight loss (wt%) obtained from thermogravimetric analysis.

Temperature interval	Sample 1	Sample 2	Sample 3	Sample 4
20–120 °C	0.88	0.47	0.55	0.28
120–190 °C	0.80	0.62	0.40	0.37
190–410 °C	1.39	1.13	1.05	0.94
410–850 °C	34.85	36.81	37.28	31.78

4.2 NMR Results

Figure 5 illustrates the ^{1}H and ^{13}C MAS NMR spectra obtained for the four investigated samples; the relevant ^{29}Si and ^{27}Al MAS NMR spectra are presented in Fig. 6.

The single peak observed at around 4.7 ppm in ^{1}H MAS NMR spectra (Fig. 5a) is attributed to water present in capillary channels which is associated with structural water (Sotiriadis et al. 2018). This assessment is supported by the fact that increasing the rotation speed of the sample, the isotropic chemical shift decreases, as shown in the graph provided as inset, regarding sample 3. The line broadening observed for sample 4 is due to the higher relaxation speed of water. This indicates that free water is more rigidly bound to structural water in this sample, which is likely because of narrower capillary channels.

In ^{13}C MAS NMR spectra (Fig. 5b), a single peak at around 168.6 ppm was detected which is characteristic for calcium carbonates, including all possible polymorphs formed (Sevelsted and Skibsted 2015).

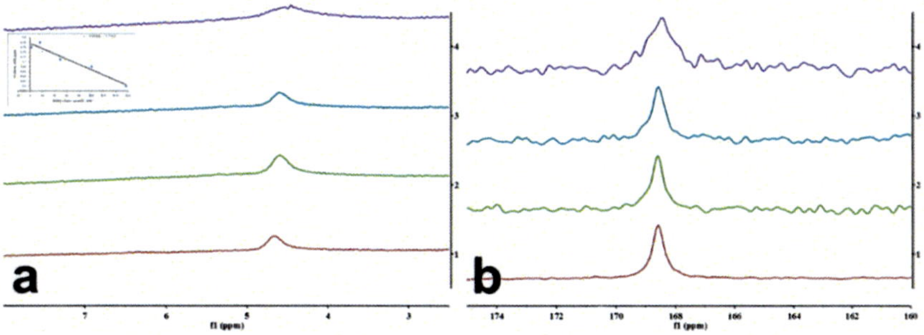

Fig. 5. ^{1}H (a) and ^{13}C (b) MAS NMR spectra obtained for all samples

Broad ^{29}Si MAS NMR spectra are illustrated in Fig. 6a, which are substantially same for all samples. According to the literature (Mendes et al. 2011) (Walkley and Provis 2019) (Kunther et al. 2016) (Bernard et al. 2017), the interpretation of the observed signals is as follows. Low signal intensities at isotropic chemical shift values around –79 ppm and –85 ppm, corresponding to Q^1 and Q^2 species that are characteristic for C–S–H, are observed. On the contrary, the signal intensity is significantly higher in the isotropic chemical shift range $-100 < \delta_{iso} < -90$ ppm, indicating the presence of condensed silicate and/or aluminosilicate structures. Both observations point to considerable degradation of C–S–H phase, confirming the results of thermal analysis. The signal for $\delta_{iso} < -100$ ppm denotes polymerized silicates (Q^4 species), which are most likely inherited from the aggregates used in concrete mix.

Two structural environments are observed in the range $-50 < \delta_{iso} < 100$ ppm of all ^{27}Al MAS NMR spectra (Fig. 6b), corresponding to six-coordinated ($-10 < \delta_{iso} <$ 20 ppm) and four-coordinated ($50 < \delta_{iso} < 80$ ppm) aluminum (Justnes et al. 1990). The intense distinct signal at ~57 ppm observed in all samples indicates the association of AlO_4 species with Q^3 silicate structures (Sun et al. 2006) (Houston et al. 2009). It should be noted that sample 4 shows an additional broad shoulder of the line in AlO_4 region (~70 ppm), attributed to Q^2 bridging sites charge-balanced by six coordinated and penta-coordinated Al. This finding suggests that likely a different concrete composition was used for the preparation of the column.

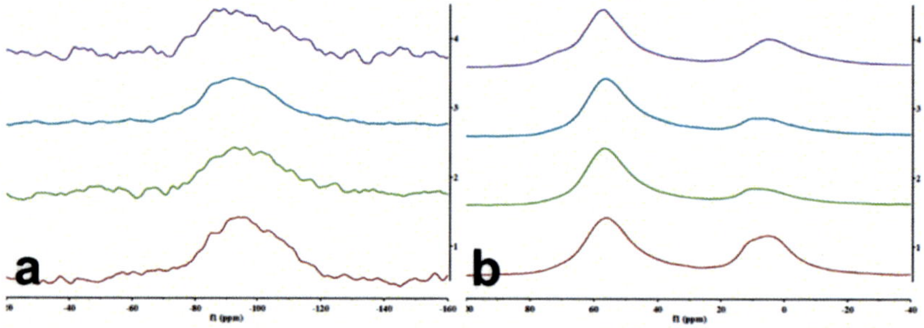

Fig. 6. ^{29}Si (a) and ^{27}Al (b) MAS NMR spectra obtained for all samples

The signal in AlO_6 region is quite broad (Fig. 6b) for all samples, thus the ^{27}Al 3Q MAS NMR technique was used to distinguish the coexisting species. In samples 1 and 4 (Figs. 7a and d, respectively), two components are identified at 4 ppm and 10 ppm, corresponding to disordered aluminate hydrate and AFm phases, respectively (Andersen et al. 2006) (Skibsted et al. 1993). The absence of the component at 4 ppm in samples 2 and 3 (Figs. 7b and c, respectively) is likely due to its low amount. Longer spectra acquisition times are required to reveal low content components with ^{27}Al 3Q MAS NMR technique (Webb, 2016). In general, the lines corresponding to samples 2 and 3 are quite similar and their intensities indicate lower content of the identified environments, compared to samples 1 and 4.

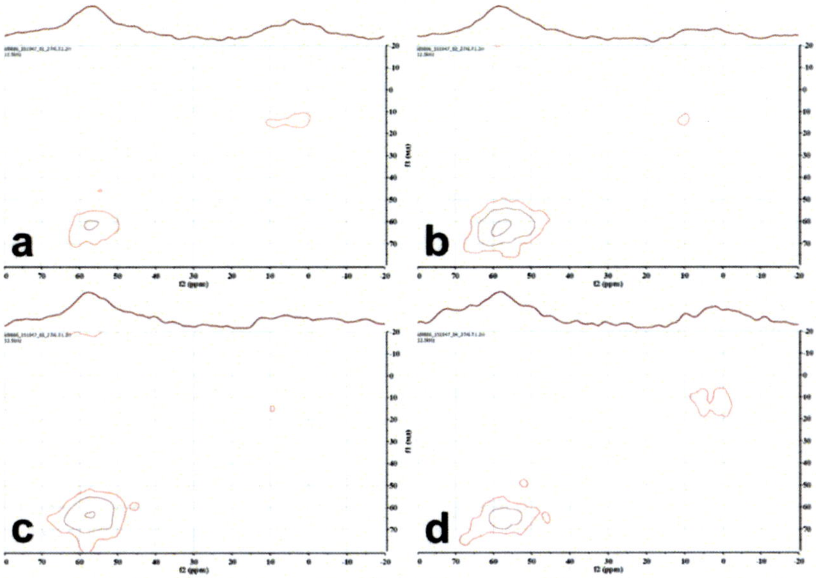

Fig. 7. ^{27}Al 3Q MAS NMR spectra obtained for all samples

5 Conclusions

In the present study, old concrete samples extracted from a heritage church of Inousses cluster of islands was characterized in terms of cement hydrate composition, in order to reveal the extent of degradation and possible differences in the original concrete mix used during construction. It was found that concrete suffered considerable degradation, owing mainly to carbonation. The structural elements exposed directly to the environment (external walls and column) suffered greater deterioration, compared to the internal part of the dome. The components identified in the sample coming from the external column indicate that a different concrete composition was likely used for its construction.

Acknowledgements. Solid state NMR spectroscopy experiments were implemented within the framework of the cooperation agreement between the St. Petersburg State University and the Institute of Theoretical and Applied Mechanics of the Czech Academy of Sciences.

References

Andersen, M.D., Jakobsen, H.J., Skibsted, J.: A new aluminium-hydrate species in hydrated portland cements characterized by ^{27}Al and ^{29}Si MAS NMR spectroscopy. Cem. Concr. Res. **36**, 3–17 (2006)

Bernard, E., Lothenbach, B., Rentsch, D., Pochard, I., Dauzères, A.: Formation of magnesium silicate hydrates (M-S-H). Phys. Chem. Earth **99**, 142–157 (2017)

Bertolini, L., Carsana, M., Radaelli, E.: Conservation of historical reinforced concrete structures damaged by carbonation induced corrosion by means of electrochemical realkalisation. J. Cult. Herit. **9**, 376–385 (2008)

Cao, Y., Gehlen, C., Angst, U., Wang, L., Wang, Z., Yao, Y.: Critical chloride content in reinforced concrete – an updated review considering Chinese experience. Cem. Concr. Res. **117**, 58–68 (2019)

Council of Europe: Convention for the Protection of the Architectural Heritage of Europe. European Treaty Series – No. 121 (1985)

Engelhardt, G.: Silicon-29 NMR of Solid Silicates, in Encyclopedia of Nuclear Magnetic Resonance. John Wiley & Sons, Ltd. (2007)

Földvári, M.: Handbook of Thermogravimetric System of Minerlas and its Use in Geological Practice. Geological Institute of Hungary (2011)

Houston, J.R., Maxwell, R.S., Carroll, S.A.: Transformation of meta-stable calcium silicate hydrates to tobermotite: reaction kinetics and molecular structure from XRD and NMR spectroscopy. Geochem. Trans. **10**, 1–14 (2009)

International Council on Monuments and Sites: Principles and Guidelines for Managing Tourism at Places of Cultural and Heritage Significance. ICOMOS International Cultural Tourism Committee (2002)

Justnes, H., Meland, I., Bjoergum, J.O., Krane, J., Skjetne, T.: Nuclear magnetic resonance (NMR) – a powerful tool in cement and concrete research. Adv. Cem. Res. **3**, 105–110 (1990)

Kunther, W., Dai, Z., Skibsted, J.: Thermodynamic modeling of hydrated white portland cement-metakaolin-limestone blends utilizing hydration kinetics from ^{29}Si MAS NMR spectroscopy. Cem. Concr. Res. **86**, 29–41 (2016)

Kunther, W., Lothenbach, B., Skibsted, J.: Influence of the Ca/Si ratio of the C-S-H phase on the interaction with sulfate ions and its impact on the ettringite crystallization pressure. Cem. Concr. Res. **69**, 37–49 (2015)

Laimos, A.G.: Aignousa Our Homeland. K. Tsikopoulos Publishing (1965)

Marcos, I., San-José, J.-T., Garmendia, L., Santamaría, A., Manso, J.M.: Central lessons from the historical analysis of 24 reinforced-concrete structures in Northern Spain. J. Cult. Herit. **20**, 649–659 (2016)

Mendes, A., Gates, W.P., Sanjayan, J.G., Collins, F.: NMR, XRD, IR and synchrotron NEXAFS spectroscopic studies of OPC and OPC/slag cement paste hydrates. Mater. Struct. **44**, 1773–1791 (2011)

Official Journal of the Hellenic Republic: Characterization of the Inousses Settlement of the Municipality of Inousses of the Prefecture of Chios and Determination of Special Terms and Limitations for Construction. Series D Issue No. 381, 27 May 1999. National Printing House (1999). (in Greek)

Official Journal of the Hellenic Republic: Ratification of the Convention for the Protection of the Architectural Heritage of Europe. Series A Issue No. 61, 13 April 1992. National Printing House (1992). (in Greek)

Papageorgiou, M., Pozoukidou, G.: The Traditional settlements of greece: planning and protection isuues. Geografies **24**, 107–125 (2014). (in Greek)

Scrivener, K.L., Snellings, R., Lothenbach, B.: A Practical Guide to Microstructural Analysis of Cementitious Materials. CRC Press, Taylor & Francis Group (2016)

Sevelsted, T.F., Skibsted, J.: Carbonation of C-S-H and C-A-S-H samples studied by ^{13}C, ^{27}Al and ^{29}Si MAS NMR spectroscopy. Cem. Concr. Res. **71**, 56–65 (2015)

Skibsted, J., Henderson, E., Jakobsen, H.J.: Characterization of calcium aluminate phases in cements by ^{27}Al MAS NMR spectroscopy. Inorg. Chem. **32**, 1013–1027 (1993)

Sotiriadis, K., Mácová, P., Mazur, A.S., Tolstoy, P.M., Viani, A.: A solid state NMR in-situ infrared spectroscopy study on the setting reaction of magnesium sodium phosphate cement. J. Non-Cryst. Solids **498**, 49–59 (2018)

Sun, G.K., Young, J.F., Kirkpatrick, R.J.: The role of Al in C-S-H: NMR, XRD, and compositional results for precipitated samples. Cem. Concr. Res. **36**, 18–29 (2006)

Walkley, B., Provis, J.L.: Solid-state nuclear magnetic resonance spectroscopy of cements. Mater. Today Adv. **1**, 100007 (2019)
Webb, G.A.: Annual Reports on NMR Spectroscopy 88. Academic Press (2016)

Check for updates

Technological Evolution of Historic Mortars: From Lime-Based Mortars to Roman Opus Caementicium

Dafni Kyropoulou(✉), Eustratios Heliades, Petros Karalis, George Diamantopoulos, Sophia Gougoura, and Elissavet Dotsika

Institute of Nanosciences and Nanotechnology, Stable Isotope Unit, National Centre for Scientific Research, Demokritos Agia Paraskevi, 15310 Attiki, Greece
{d.kyropoulou,e.heliades,p.karalis,g.diamantopoulos,s.gougoura, e.dotsika}@inn.demokritos.gr

Abstract. Bulk samples of mortars were collected from archaeological sites dated from Hellenistic to Roman period in order to investigate the technological evolution of historic structures from simple lime based systems to advanced hydraulic materials. Roman opus caementicium enabled the construction of vaults with a mechanical behaviour of an inert monolith; the rapidly hardened mortar with its concrete-like strength provided a monolithic effect on lighter vaulted structures. The samples were characterized in terms of their isotopic, chemical and mineralogical composition. Stable isotope analysis (^{13}C and ^{18}O) provided information relative to the origin of CO_2 and water during calcite formation making possible to distinguish different mortar technologies. The results of micro-morphological and petrographic examination elucidate the technological continuity of historic mortars. Hellenistic mortars are composed of lime enhanced with quartz aggregates. Roman mortars are composed of lime, pozzolan and various aggregates such as quartz, feldspar, ceramic and rock fragments. The isotopic values comprise a range of $\delta^{13}C$ and $\delta^{18}O$ values from −17.6‰ to 3.6‰ and −25.9‰ to 0.4‰ very different from that of local limestones used for mortar production. This study indicated that stable isotope analysis is an excellent tool to fingerprint the origin of carbonate and therefore indicate the variations in mortar's technology.

Keywords: Stable isotopes · ^{13}C · $\delta^{18}O$ · Mortars · Mineralogy · Transition · Concrete

1 Introduction

Although in Hellenistic times the capacities of mortars mixed with ground pottery not only were known but the technique had also been spread across the Mediterranean, architects continued - in general terms - the classical tradition of building in stone and were mostly interested in evolving the tectonic structural and aesthetic values of cut-stone masonry. The use of mortar in this period did not triggered a significant change to architectural design (Malacrino 2010, Robertson 1959). During Roman times, the use

I. Vayas and F. M. Mazzolani (Eds.): PROHITECH 2021, LNCE 209, pp. 90–96, 2022.
https://doi.org/10.1007/978-3-030-90788-4_8

of local materials such as the dark volcanic sand (pozzolana) allowed the development of a very high-quality mortar, known as opus caementicium, that during the phase of its crystallization was transformed into a kind of "artificial stone", immensely tough, thus providing walls with significant static capacities (Malacrino 2010, Lamprecht 1984, Lechtman and Hobbs 1986, MacDonald 1965). New wall construction methods were employed using this revolutionary mortar. The values of the new material allowed the construction of very thick strong walls that were able to support heavy structural burdens. As the concrete needed a long time to dry out, its successive layers and materials were incorporated and united into a single absolutely homogeneous mass that ultimately presented the behavior of a structural unit with elasticity and strength. The semi fluidity of the mortar and the use of centering during installation phase enabled the construction of a range of vaults with a mechanical behavior of an inert monolith (Taylor 2003, Wright 2000). The values of the new material contributed to a true revolution in architectural thought concerning vaulted constructions, circular lines and surfaces and their effect on the concept of the organization of interior space. Far more economical and without need of the working hours and skills required for the squared stone techniques, Roman cement served as the ideal material to support the political aim of "universality". The new technique and forms were easily exported thus spreading the architectural vocabulary of the new era throughout the empire (Taylor 2003, MacDonald 1965, Robertson 1959). The main objective of this study is to investigate the technological transition of historic mortars from Hellenistic, to Roman period using microscopy, mineralogical, chemical and isotopic analysis.

2 Materials and Methods

Thirty samples were obtained from Hellenistic and Roman historic constructions located at Makrygialos and Kavala, Greece. Mortar samples were obtained from 4thcentury BC Hellenistic tombs in the area of Pydna in North Greece. Ancient Pydna lies in the North of Pieria, two kilometers away from South Makrygialos. Roman mortars dated at 2nd century AD were obtained from Anaktoroupoli, Kavala in North Greece. All mortar samples were analyzed using Scanning Electron Microscopy (SEM) with Energy Dispersive X-ray (EDX) analysis. Major and minor elements for polished mortar samples were analyzed using a FEI/Quanta Inspect D8334 SEM fitted with a low vacuum chamber and an energy dispersive X-ray analyzer at NCSR Demokritos. Mineralogical analysis was achieved using optical microscopy. Optical microscopy was performed using a polarising microscope at the Department of Mineralogy, Petrology and Economic Geology in Aristotle University of Thessaloniki. The isotopic analyses took place in the Stable Isotope Unit of Institute of Materials Science (NCSR Demokritos) on a Thermo Delta V Plus IRMS equipped with GasBench II device. The $\delta^{13}C$ and $\delta^{18}O$ was measured with an accuracy of 0.1‰ estimated by the reproducibility of international standards. The international PDB (Pee Dee Belemmitella) was used as standard to calculate $\delta^{13}C$ and $\delta^{18}O$ respectively.

3 Results and Discussion

3.1 Mineralogy, Petrography, Microstructure and Mineralogy

Backscattered electron photomicrographs of sample A4 (Fig. 1a), illustrates the morphology of a typical mortar sample from a Hellenistic tomb at Makrygialos, Greece. The photomicrograph depicts large irregularly shaped grains acting as aggregates composed mainly of quartz and feldspar. The matrix is composed of fine calcitic particles with small portions of quartz grains. Hellenistic binding material is composed of 96% calcite according to oxides analysis (SEM).

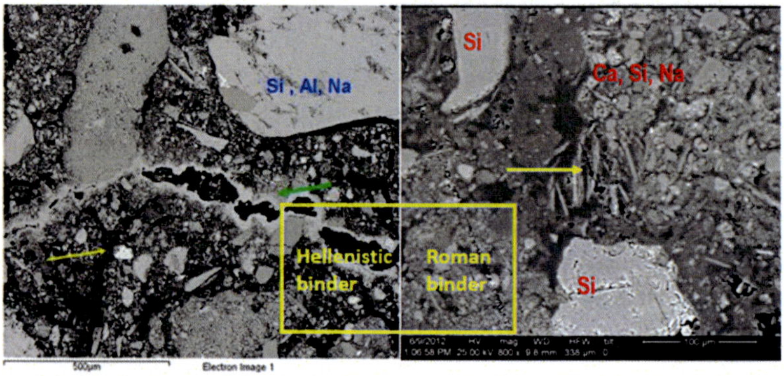

Fig. 1. 1a (left) BSE photomicrograph shows the technology of Hellenistic binders and 1b (right) shows the technology of Roman binders.

Backscattered electron photomicrographs of sample AN1b obtained from Anaktoroupoli, Nea Peramos, illustrates the morphology of a Roman mortar (Fig. 1b). The matrix is composed of lime and pozzolan. Compositional analysis of Roman mortars showed that the silica content in the matrix ranges from 30% to 90%. The high silica content is attributed to the addition of pozzolanic materials. The addition of pozzolans enhances the durability and strength of mortar structures. The use of dark volcanic sand (pozzolana) allowed the development of a very high-quality mortar, known as opus caementicium (Malacrino 2010, Lamprecht 1984). The innovative characteristic of this newly induced mortars, is that during the phase of its crystallization was transformed into a kind of "artificial stone", immensely tough, thus providing walls with significant static capacities (Lechtman and Hobbs 1986, MacDonald 1965).

Petrographic observation confirmed the results of microscopic observations under the sem. Figure (2a) elucidates a textured Hellenistic mortar with mineral components sized from 100 μm to 4 mm. The matrix is composed of relatively homogeneous powdery calcite that is interrupted by angular aggregates and pores. Hellenistic mortars are composed of microcystalline calcite with main addition of quartz, and plagioclase and feldspar minerals as aggregates (Fig. 3a). Calcite appears as the main mineral phase among other mineral components (size up to 6 mm). Quartz is the most abundant mineral after calcite and its size rises up to 2 mm. Petrographic investigations of Roman mortars

(Fig. 2b) elucidated the mineralogical composition and petrography of the examined samples. The matrix is composed of powdery calcite together with silicate minerals, such as quartz, plagioclase, orthoclase, biotite and muscovite. Microscopic analysis also detected grains of volcanic rocks. Quartz is the most abundant mineral after calcite and its size rises up to 2 mm. Quartz is textured and demonstrates a strong recrystallization matrix and is surrounded with reaction rims (Fig. 3b) and appears as the main aggregate material. The morphology of aggregates changes from Hellenistic to Roman time. During the Hellenistic period the aggregates are composed of quartz, plagioclase and potassium feldspar, whereas during Roman period new material was used as aggregates such as crushed ceramic and rock fragments adding coherence and strength in the newly formed mortar (Fig. 3 a, b).

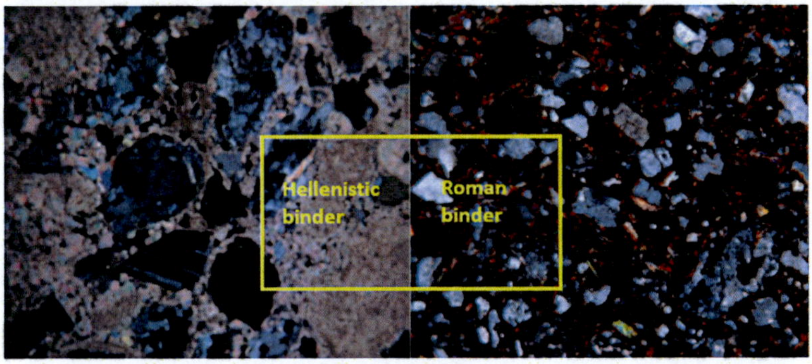

Fig. 2. 2a (left) optical microscopy photograph showing the minerology of Hellenistic binder and 2b (right) showing the minerology of a Roman binder.

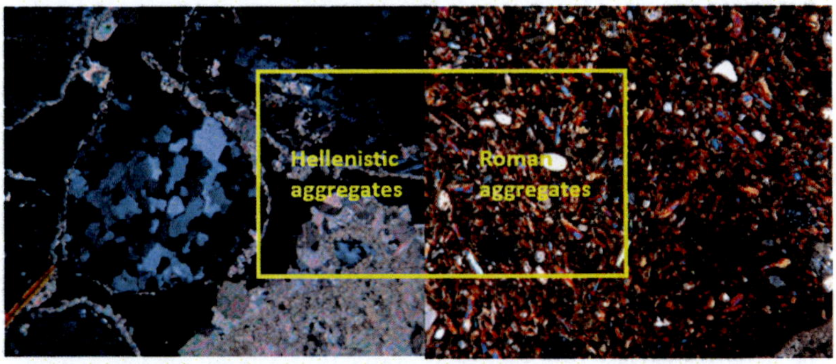

Fig. 3. 3a (left) optical microscopy photograph showing the minerology of Hellenistic aggregates and 3b (right) showing the minerology of Roman aggregates.

3.2 Stable Isotope Analysis

The stable isotope (^{13}C and ^{18}O) analysis is highly useful in providing extra parameters, associated with the conditions of formation and the origin of the carbonates and therefore suitable to differentiate different mortar technologies. Mineralogical and chemical analysis indicated that Hellenistic mortars differ from Roman mortars on the basis of silica content in the matrix. The technological differences and the impact of raw materials are expected to affect isotopic composition of historic mortars. The isotopic value of carbonates depends mainly on the isotopic composition of atmospheric CO_2 and water and on the degree to which the isotopic equilibrium is reached and therefore the factor of fractionation of the components ($\delta^{13}C_{CaCO3}$, $\delta^{13}C_{CO2}$, $\delta^{18}O_{CO2}$ and $\delta^{18}O_{H2O}$). Isotope data of the calcite from mortar represent non isotopic equilibrium.

Isotopic values for mortar samples comprise a wide range of δ13Cmatrix and δ18Omatrix values, from −4.2 to −9.6‰ and from −2.1 to −9.7‰, for Hellenistic mortars, from −6.4 to −15.4 ‰ and from −7.6 to −19.7 ‰ for Roman mortars and finally from −11.3 to −15.6‰ and from −16.1 to −18.0 ‰ respectively for Late Roman mortars. All the samples are plotted in Fig. 4.

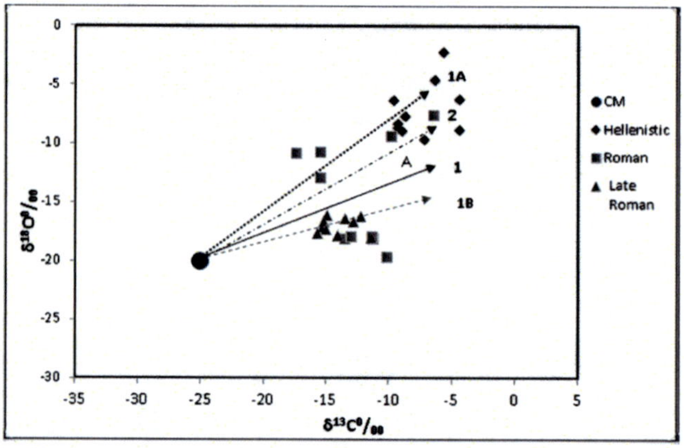

Fig. 4. Isotopic diagram showing the various sources of CO_2 and H_2O (lines) and mechanisms (areas).

Figure 4 shows also the general trend for potential changes in the primary environmental setting behavior of lime mortar. Line 1 is the Regression line of equation δ^{18}Ocalcite = 0.67 × δ^{13}Ccalcite–6.4 (Dotsika et al. 2018); Point CM represents the $\delta^{13}C$ and $\delta^{18}O$ composition of the calcite formed directly by the absorption of CO_2 from the atmosphere using the local isotopic composition of Macedonia water and atmospheric CO_2. The regression of Roman era lime mortar line δ^{18}Ocalcite = 0.61 × δ^{13}Ccalcite–3.3 (Dotsika et al. 2018), (line 2) is also depicted. Relicts of local limestone, from insufficient calcinations during the ancient burning or contamination from limestone aggregate due to insufficient separation during sampling preparation conduce a deviation (lines 1A and 1B) from line 2, depending on the δ of local limestone and on the isotopic behaviour (line 2) of the calcite along the setting path of a lime mortar layer (Undoski and Hoefs 1993, Dietzel 1992). Most of the data lie close or on this line, indicating different setting environment and secondary effects.

For Hellenistic mortars, the contamination of relicts from heavy limestone is confirmed by SEM analysis. The impact of silicate minerals in mortar matrix affects the isotopic composition of Roman and Byzantine mortars, since it results in more negative isotopic values (Hutcheon and Abercrombie 1990). The silicates matrix enhances mortar durability and the potential to maintain their properties overtime. Therefore the concentration of the majority of Roman mortars close to line 1B demonstrates that these samples have maintained their physicochemical properties and they appear greater resistance to environmental degradation. Despite this, the isotopic composition of Hellenistic mortar samples position along line 1A reflects the $\delta^{13}C_{CO2}$ enrichment relative to δ12CCO2 in the gas phase with continuous calcite formation along the lime mortar layer and the presence of impurities of relicts from isotopically heavy Greek limestone of (δ^{13}Climestone = 2‰, δ^{18}Olimestone = −5 + 2‰; Dotsika et al. 2009). The location of these mortar samples in the plot reveals a strong impact of technology in isotopic composition.

4 Conclusion

The observation of the use of mortars from the Hellenistic to Roman time shows that in many cases, along with any aesthetical reasons, symbolisms and the general concept of style prevailing, the architectural forms employed are also a result of constructional and technical conditions. Thus, various uses of materials, building practices, elements of construction and techniques employed constitute important factors that may play a significant role in the alterations of the architectural vocabulary of an era in terms of both typology and morphology. The results of compositional, isotopic and mineralogical analysis confirmed that the evolution in the technology of mortars. The simple lime based systems were replaced by powerful hydraulic mortars enabling the development of glorious historic constructions. This technological evolution is depicted in the mineralogical and chemical composition of the binding materials and aggregates. The content of silicate minerals is increased in Roman mortars in Greece resulting in a negative shift in their isotopic composition, providing coherence and strength to mortar samples. The technological developments in the consistency of mortar, the evolution of its composition and the differences in its qualities are related to the generation of new structural systems during each period of time.

References

Dietzel, M., Usdowski, E., Hoefs, J.: Chemical and 13C/12C and 18O/16O-isotope evolution of alkaline drainage waters and the precipitation of calcite. Appl. Geochem. **7**, 177–184 (1992)

Dotsika, E., Psomiadis, D., Poutoukis, D., Raco, B., Gamaletsos, P.: Isotopic analysis for degradation diagnosis of calcite matrix in mortar and plaster. Anal. Bioanal. Chem. **395**, 2227–2234 (2009)

Dotsika, E., Kyropoulou, D., Christaras, V., Diamantopoulos, G.: δ13 C and δ18O Stable isotope analysis applied to detect technological variations and weathering processes of ancient lime and hydraulic mortars. Geosciences **8**(9), 339 (2018)

Hutcheon, I., Abercrombie, H.: Carbon and dioxide in clastic rocks and silicate hydrolysis. Geology **18**, 541–544 (1990)

Lamprecht H.O.: Opus Caementicium, Beton – Verlag, pp. 21–40 (1984)

Lechtman H.N., Hobbs L.W.: Roman concrete and the Roman architectural revolution, ceramics and civilization VII: high technology ceramics, past, present and future, the nature of innovation and change in ceramic technology, Kingery (eds), American ceramic society Inc, USA, pp. 81–128 (1986)

MacDonald W.L.: The architecture of the Roman Empire, pp. 152–168. Yale University Press (1965)

Malacrino C.G.: Constructing the ancient world. Architectural techniques of the Greeks and Romans, pp. 72–137. Getty Publications (2010)

Robertson, D.S.: A Handbook of Greek and Roman Architecture, pp. 147–266. Cambridge University Press (1959)

Taylor R.M.: Roman Builders. A Study in Architectural Process, pp. 175–178. Cambridge University Press (2003)

Usdowski, E., Hoefs, J.: Oxygen isotope exchange between carbonic acid, bicarbonate, carbonate, and water: a re-examination of the data of McCrea (1950) and an expression for the overall partitioning of oxygen isotopes between the carbonate species and water. Geochimica and Cosmochimica Acta **57**, 3815–3818 (1993)

Wright, G.R.H.: Ancient building technology, pp. 115–116. Historical Background, Brill (2000)

Stable Isotopic Composition of Carbonate Materials for Determination of the Origin of Marble Artifacts

Petros Karalis[1(✉)], Maria Tassi[1], Eustathios Heliades[1], Georgios Diamantopoulos[1], Vassilios Christaras[2], Panagiotis Spathis[3], and Elissavet Dotsika[1]

[1] Institute of Nanoscience and Nanotechnology, Stable Isotopes Unit, National Center for Scientific Research DEMOKRITOS, Neapoleos 27, 153 10 Aghia Paraskevi, Attiki, Greece
{p.karalis,m.tassi,e.heliadis,g.diamantopoulos, e.dotsika}@inn.demokritos.gr

[2] School of Geology, Faculty of Sciences, Aristotle University of Thessaloniki, GR-541 Thessaloniki, Greece
christar@auth.gr

[3] Faculty of Chemistry, Aristotle University of Thessaloniki, Thessaloniki, Greece
spathis@chem.auth.gr

Abstract. In this study we present the isotopic signatures of quarries located in North-East (Tempi, Tisaion, Pangaio), Southern (Hymittos, Penteli) Greece in order to enrich the existing databases used in ancient marble provenance. This expansion of the databases will enhance our understanding in the archaeological characterization of ancient monuments and potentially determine economic and cultural associations and trade routes. Furthermore, it will reduce the overlap of the isotopic signals of marbles in the up to now databases, enhancing the dissemination and facilitate the provenance studies of the ancient marble monuments .

Keywords: Stable isotopes · Marbles · Provenance · Oxygen isotope · Carbon isotope

1 Introduction

Due to its white colour and speckled appearance as well as to its suitable physical and chemical properties for building and sculpturing, marble was particularly appreciated in antiquity, especially in ancient Greek and Roman times when the quantity of marble employed and the quality of workmanship involved often reached quite high levels. Carbonate rocks, except for their use in cultural works, have been widely used as structural material and as base of wall paintings from antiquity until modern time. The scientific determination of the origin of marble, from which archaeological objects are constructed, has contributed largely to the archaeological and art-historical studies and interpretations, to the identification of trade and technology evolution and provides guidelines for restorations and reconstructions studies.

I. Vayas and F. M. Mazzolani (Eds.): PROHITECH 2021, LNCE 209, pp. 97–101, 2022.
https://doi.org/10.1007/978-3-030-90788-4_9

A number of techniques and methods (only few of which are non-destructive) have been tried and developed over the last decades (Petrographic Examination, Electron Paramagnetic Resonance Spectroscopy, Maximum Grain Size Analysis, Cathodeluminescence, Stable Isotope Analysis). These methods are associated with extensive databases from ancient marble quarries. For the determination of the quarries' origin, the isotopic results of the samples were compared with the isotopic database diagram for the most known ancient marble quarries produced by L. Moens (Moens et al. 1992a, Moens et al. 1992b, Moens 2003). However the large overlapping of the isotopic data fields from different quarries renders $\delta^{13}C$-$\delta^{18}O$ plots often ineffective in discriminating the marble sources. For this reason it is very important to complete the database for the most important marbles of antiquity and to create data for new marble quarries.

The aim of this study is to expand the existing databases with isotopic measurements of other known ancient marble quarries of Greece. This expansion of the databases and the application of developed isotopic techniques will reduce the overlap of the isotopic signals and facilitate the determination of the provenance of ancient marbles of unknown origin, providing invaluable information concerning the archaeological characterization of the samples.

The present study focuses on five white marble quarries in Tempi, Tisaion Mountain in Magnesia, Pangaio Mountain in Kavala, Hymittos and Penteli from Attica. The results of this study provide additional data and complete the databases presented in previous work of (Craig and Craig 1972, Coleman and Walker 1979, Germann et al. 1980, Herz 1987, Tsirambides 1996, Gorgoni et al. 2002, Capedri et al. 2004, Melfos 2004, Melfos et al. 2010).

2 Method

The isotope analysis took place in the Stable Isotope Unit of Institute of Nanoscience and Nanotechnology (NCSR Demokritos) on a Thermo Delta V Plus IRMS equipped with GasBench II device. Small quantities of clean (leached with acetic acid) carbonate material were separated and after grinding to a fine powder were diluted in ortho-phosphoric acid and the CO_2 produced was measured in the isotope ratio mass spectrometer. The $\delta^{13}C$ and $\delta^{18}O$ were measured with an accuracy of 0.2‰ estimated by the reproducibility of international standards and are expressed in delta notation ($\delta^{13}C$ and $\delta^{18}O$) as ‰ deviation relative to the international PDB (Pee Dee Belemnite) standard:

$$\delta = \left[(\text{Rsample} - \text{Rstandard})/\text{Rstandard}\right] \times 1000 \tag{1}$$

where R is the ratio between the heavy and the light isotope, in this case $^{18}O/^{16}O$ or $^{13}C/^{12}C$. The reported values are the mean of two or more consistent measurements of each sample.

Measurements of Maximum Grain Size (MGS) were performed at the Department of Mineralogy, Petrology and Economic Geology at Aristotle University of Thessaloniki.

3 Samples and Results

Ten samples (white marble) from each of the ancient quarries (dating from 3 to 7 b.c.) from the five sites (Tempi, Tisaion Mountain in Magnesia, Pangaio Mountain in Kavala,

Hymittos and Penteli Mountain in Attica) were collected. Measurements of Maximum Grain Size for all the samples are presented in Fig. 1.

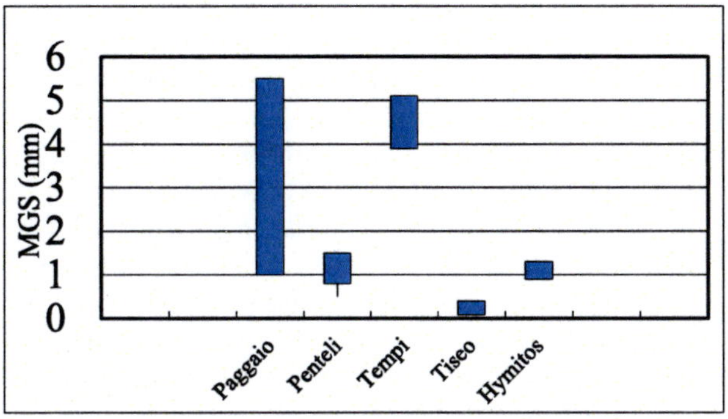

Fig. 1. MGS of all the studied samples.

All the samples were isotopically analysed and the results (mean values) are presented in the following $\delta^{13}C$-$\delta^{18}O$ plot (Fig. 2). In the same figure we present the available bibliographic data (rectangular areas) of the same quarries, for comparison.

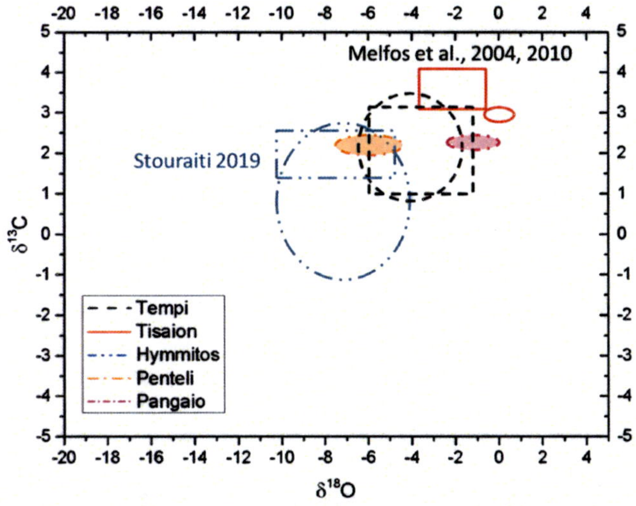

Fig. 2. $\delta^{13}C$-$\delta^{18}O$ plot of marble samples from six regions of Greece (circles). The rectangular are the available bibliographic data for the same quarries [Hymittos, (Stouraiti et al. 2019) and Tempi, Tisaion, (Melfos et al. 2010)].

Specifically, the marble of Tempi consists only of middle- to coarse-grained calcite crystals. Its colour is from greyish white to grey. The quarry was used from 5–7 B.C. The marble consists of calcite crystals of 4 to 5 mm of MGS. The isotopic results of 10 marble samples collected from different sites of the quarry range from 0.9 to 3.4‰ and from −1.5 to −6.5‰ for carbon and oxygen respectively.

The quarries of the Tisaion Mountain extent up to 100m above the sea level. The marble in the Tisaion Mountain has a white to greyish white colour. This type is called "Lafkos White". The marble is characterized as a fine-grained marble, with grain size varying from 10 μm to 100 μm. Only rarely there are calcite crystals with a diameter up to 1mm (MGS). The quarry was used from 5–7 B.C. Ten samples were analysed for carbon and oxygen isotopes. The results are homogeneous and the $\delta^{13}C$ values range from 2.8 to 3.1‰ and the $\delta^{18}O$ range from −0.4 to 0.2‰.

The rocks of northern Hymittos Mt. comprise impure dolomitic marbles, grey or white in colour, and mica-schist, in successive alternations. This unit is known as the "Upper Marble" (Lepsius 1893). The quarry was used from 6–7 B.C. The marble consists of crystals 1mm of MGS. Ten marble samples were isotopically analysed and their values range from −1.2 to +2.6‰, $\delta^{13}C$ and for $\delta^{18}O$ the values ranging from −10.0 to –5.0‰.

The marble of Penteli Mountain has white fine grain marble with schist veins. The quarry was used from 5 B.C. The marble consists of crystals 1 to 1.3mm of MGS. Ten samples of Penteli calcite were isotopically analysed. $\delta^{13}C$ values were from 1.8 to 2.6‰, and $\delta^{18}O$ values from −7.5 to −4.5‰.

The marble of Pangaio Mountain has white fine- to coarse-grained calcite crystals. The quarry was used from 3 to 7 B.C. The marble consists of crystals 1 to 5.5mm of MGS. Ten samples of Pangaio calcite were isotopically analysed. $\delta^{13}C$ values were from 2.0 to 2.5‰, and $\delta^{18}O$ values from −2.3 to 0.0‰

4 Discussion-Conclusions

The O-C isotopic results presented here complete the isotopic databases of Greek marbles. The isotopic values of the marble coming from five different ancient quarries show distinct differences and with the combination of the petrographic results may be used to identify the provenance of archaeological marbles.

According to their stable isotope compositions most of the samples plot within the same area, although some groupings are observed (Fig. 2).

Globally, $\delta^{18}O$ values in marine carbonates range from −10‰ to +5‰, but most marine carbonate sediments and limestones are in the range of −10‰ to +2‰ (Nelson and Smith 1996). The values of the marbles from this study vary from −19.3‰ to 0.0‰ and fall inside this range. More specific the oxygen isotopes can be compared with the isotopes of the Mesozoic marine limestones which range from −7 to 0 ‰ (Veizer, Ala et al. 1999), suggesting that they were relatively affected by very low grade metamorphism, which resulted in the recrystallization of the carbonates without any affect in the O-isotopes.

The carbon isotope values of the studied marbles (−1.2 to +3.9 ‰) demonstrate a relatively narrow range, which probably shows that the Carbon source was rather homogeneous for the carbonates which were incorporated within all the marble occurrences. These values represent seawater composition during carbonate deposition and are comparable to values of the non-metamorphic Mesozoic pelagic limestones worldwide which have $\delta^{13}C$ values mainly from 0 to 3 ‰ (Veizer et al. 1999).

References

Capedri, S., et al.: Accessory minerals and δ18O and δ13C of marbles from the Mediterranean area. J. Cult. Herit. **5**(1), 27–47 (2004)

Coleman, M., Walker, S.: Stable isotope identification of Greek and Turkish marbles. Archaeometry **21**(pt. 1), 107–112 (1979)

Craig, H., Craig, V.: Greek marbles: determination of provenance by isotopic analysis. Science **176**(4033), 401–403 (1972)

Germann, K., et al.: Determination of marble provenance: limits of isotopic analysis. Archaeometry **22**(pt. 1), 99–106 (1980)

Gorgoni, C., et al.: An updated and detailed mineropetrographic and CO stable isotopic reference database for the main Mediterranean marbles used in antiquity. Asmosia **5**, 115–131 (2002)

Herz, N.: Carbon and oxygen isotopic ratios: a data base for classical Greek and Roman marble. Archaeometry **29**(1), 35–43 (1987)

Lepsius, R.: "über die Geologie von Attika." Zeitschrift der Deutschen Geologischen Gesellschaft, pp. 530–533 (1893)

Melfos, V.: MIneralogical and stable isotopic study of ancient white marble quarries in Larisa, Thessaly, Greece. Bull. Geol. Soc. Greece **36**(3), 1164–1172 (2004)

Melfos, V., et al.: Mineralogical, petrographic and stable isotopic study of ancient white marble quarries in Thessaly, Greece-II. Chasanbali, Tempi, Atrax, Tisaion mountain. Bull. Geol. Soc. Greece **43**(2), 845–855 (2010)

Moens, L.: He Kindly Provided the Updated Isotope Field Diagram (earlier version published in Moens et al. 1992). Faculty of Sciences, Ghent University, Belgium (2003)

Moens, L., et al.: Multidisciplinary research and cooperation: keys to a successful provenance determination of white marbles. Ancient Stones: Quarrying, Trade Provenance - Acta Archaeologica Lovaniensia Monographiae. **4**, 247–252 (1992a). M. Waelkens, N. Herz and L. Moens. Leuven, Leuven University Press

Moens, L., et al.: Provenance determination of white marble sculptures from the Allard Pierson Museum in Amsterdam, based on chemical, microscopic and isotopic criteria. Ancient Stones: Quarrying, Trade Provenance - Acta Archaeologica Lovaniensia Monographiae. **4**, 269–276 (1992b). M. Waelkens, N. Herz and L. Moens. Leuven, Leuven University Press

Nelson, C.S., Smith, A.M.: Stable oxygen and carbon isotope compositional fields for skeletal and diagenetic components in New Zealand Cenozoic nontropical carbonate sediments and limestones: a synthesis and review. NZ J. Geol. Geophys. **39**(1), 93–107 (1996)

Stouraiti, C., et al.: Silver-rich sulfide mineralization in the northwestern termination of the Western Cycladic Detachment System, at Agios Ioannis Kynigos, Hymittos Mt. (Attica, Greece): a mineralogical, geochemical and stable isotope study. Ore Geol. Rev. **111**, 102992 (2019)

Tsirambides, A.: Provenance determination of Greek white marbles by isotopic analysis of oxygen and carbon. In: Proceedings of the 2nd Symposium of the Hellenic Archaeometrical Society (1996)

Veizer, J., et al.: 87Sr/86Sr, δ13C and δ18O evolution of Phanerozoic seawater. Chem. Geol. **161**(1–3), 59–88 (1999)

Stable Isotope for Tracing the Salt Involved in the Degradation of Stone Monuments

Petros Karalis(✉), Maria Tassi, Sofia Gougoura, George Diamantopoulos, Dafni Kyropoulou, Eustratios Heliades, Emmanouil Palaigeorgiou, and Elissavet Dotsika

Institute of Nanoscience and Nanotechnology, Stable Isotopes Unit, National Center for Scientific Research DEMOKRITOS, Neapoleos 27, 153 10 Aghia Paraskevi, Attiki, Greece
{p.karalis,m.tassi,g.diamantopoulos,e.heliadis, e.dotsika}@inn.demokritos.gr

Abstract. Salts crystallizing in the pore space of building stones cause stone monument degradation. The origin of these salts can be variable: marine, air pollution, together with marine aerosols, building or restoration material. Previous studies using sulfur and nitrogen isotopes demonstrated the important role of atmospheric sulfur and nitrogen to black crust formation and determining it as a principal component of salt weathering of stone monuments. In this study we present isotopic measurements of black crusts from monuments located in North Greece and Attica. The applied methodology is capable to distinguish the sources of the sulfate salts (atmospheric pollution, marine aerosols, bio-pollution).

Keywords: Stable isotopes · Sulfur · Oxygen · Pollution · Black crusts

1 Introduction

Calcium sulfate ($CaSO_4$) black crusts are the major cause of deterioration of cultural heritage monuments. Gypsum black crusts are formed on marble or other carbonate rocks as a result of reaction of atmospheric sulfur with the calcium of the stone causing erosion of its surface. These salts can be originated from different natural (soils, rain and marine water) and anthropogenic (pollution) sources. Understanding this deterioration processes is crucial for restoration/conservation of cultural heritage monuments. The natural sources include biological sources (Skrzypek et al. 2008), construction materials (Vallet et al. 2006, Dotsika et al. 2010, Schleicher and Hernández 2010, Kloppmann et al. 2011), rainwater, marine spray (Kloppmann et al. 2011) and pyrite oxidation within the rock substrate (Kramar et al. 2011). Anthropogenic sources include the pollution from fossil fuels (Vallet et al. 2006, Schweigstillová et al. 2009, Schleicher and Hernández 2010, Kloppmann et al. 2011, Kramar et al. 2011). The oxygen and sulfur isotope can help to discriminate the S-origin of black crusts.

I. Vayas and F. M. Mazzolani (Eds.): PROHITECH 2021, LNCE 209, pp. 102–107, 2022.
https://doi.org/10.1007/978-3-030-90788-4_10

2 Sampling

Samples of crusts were collected by scraping off about 1 g of the layer of decayed carbonate rocks. A total of 16 samples from four Hellenistic, Late Roman, Byzantine and Ottoman historic constructions, located at Anaktoroupoli (n = 4) and Marmarion tower in Kavala (n = 4), Fortification walls in Drama (n = 4) and the Funerary monuments in Makrygialos (n = 4) in North Greece. From the historical area of Athens, four more samples were collected.

Anaktoroupoli is in the region of Nea Peramos in Kavala, Greece and is the remaining of a Byzantine town located in the west coast of Kavala near the marine coast. It is dated at 14th century AD based on the characteristics of the masonry and the features observed in the coins excavated in the region. The Marmarion tower was founded in 1367 and belongs to the complex of the Ancient Monastery of Pantokratoras. The Fortification walls in Drama are one of the most important monuments of the region (Dadaki 2010). The walls were founded in Late Roman times (3th–6th century AD) and later building phases are ascribed to the Frankish period. The Funerary monuments in Makrygialos are of 4th century BC. They are located in the area of Pydna in North Greece which lies in the North of Pieria, two kilometres away from South Makrygialos. The monuments were discovered in rescue excavations under the supervision of KZ´. The historical area in Athens is heavily affected by atmospheric pollution. Four samples were taken from fallen stones in order to study the "influence" of atmospheric pollution to the buildings .

3 Methods

The isotope analysis was conducted in the Stable Isotope Unit of Institute of Nanoscience and Nanotechnology (NCSR Demokritos).

For S isotope analyses, sulfate minerals were dissolved in deionized water and subsequently precipitated as $BaSO_4$. Sulfur isotopic compositions were measured after conversion of $BaSO_4$ to SO_2 using an elemental analyzer (Flash EA device) coupled with an isotope ratio mass spectrometer.

Sulfur isotope measurements were performed with an analytical error of better than ±0.3‰ and results reported in delta notation ($\delta^{34}S$) as ‰ deviation relative to the Canyon Diablo Troilite (CDT) standard. Oxygen isotope measurement results were performed with an analytical error of better than ±0.5‰ and are expressed in delta notation ($\delta^{18}O$) as ‰ deviation relative to Vienna Standard Mean Ocean Water (V-SMOW), according to Eq. (1):

$$d = \left[\left(R_{sample} - R_{standard}\right)/R_{standard}\right] \times 1000 \quad (1)$$

where R is the ratio between the heavy and the light isotope, in this case $^{18}O/^{16}O$ or $^{34}S/^{32}S$. The reported values are the mean of two or more consistent measurements of each sample.

4 Results and Discussion

The sampled black crusts consist of gypsum with minor contents of calcite derived from the building stone. Chemical analysis shows that all analysed black crusts are dominated by Ca and SO_4 ions, (up to 80 wt.%). In Fig. 1 we present the isotopic values of oxygen ($\delta^{18}O$) and sulfur ($\delta^{34}S$) of black crusts from the five sampled areas. The shaded areas are the mean isotopic values of black crust samples from different sampled monument of France (Kloppmann et al. 2011), Italy (Longinelli and Bartelloni 1978) and Slovenia (Kramar et al. 2011) as gathered from the literature.

The black crusts show $\delta^{34}S$ values from −1.2 to +7.5‰ and $\delta^{18}O$ from +3.2 to + 11.9‰, and these values stay within the ranges of black crusts observed in literature (Table 1).

Table 1. Isotopic values of black crust samples.

Material	Location	Sites	$\delta^{34}S$	$\delta^{18}O$
Black crusts	Greece	Anaktoroupoli, Kavala	4.7	3.5
Black crusts	Greece	Marmarion tower, Kavala	7.5	5.4
Black crusts	Greece	Fortification walls, Drama	2.1	3.2
Black crusts	Greece	Funerary monuments, Makrygialos	2.2	4.7
Black crusts	Greece	Athens	−1.2	11.9
Black crusts	France	Bourges, Chenonceau, Versailles, Marseille, Chartes	−1.4 to 7.4	6.3 to 12.4
Black crusts	Slovenia	Ljubljana	−9.0 to 3.1	−1 to 5
Black crusts	Italy	San Marco, Venice	0 ± 2	6 ± 2

Both isotopes in the black crusts are strongly depleted in $\delta^{34}S$ and $\delta^{18}O$ in relation to those of seawater (20 and 9.5‰ respectively, (Longinelli 1989, Dotsika et al. 2009). These significant differences between the study sites are attributed to mixing values from several environmental influences (pollution, marine aerosol). In fact, there are several types of sulfur and oxygen isotope signatures within the black crusts:

1. $\delta^{34}S$ values for atmospheric SO_2 range from −40‰ to +30‰, with urban air lying between 0‰ and +20‰ (Newman et al. 1991).
2. $\delta^{34}S$ values of industrial pollution coming from coal (−30 to 30‰; (Faure 1986)) and petroleum (−8 to 32‰ (Faure 1986)) ranging from slightly negative to +10‰ (Pankina 1991, Torfs et al. 1997).
3. $\delta^{34}S$ values of rainwater range between −2.5 to 19.5‰ depending from the seawater distance (Rafter and Mizutani 1967, Cortecci and Longinelli 1970).
4. $\delta^{34}S$ value of marine aerosol with value around to 20‰. (Siedel and Klemm 2000).
5. $\delta^{34}S$ value of biological sources with value from 4 to10‰ (Skrzypek et al. 2008)
6. $\delta^{34}S$ value originating from the oxidation of sedimentary sulfides present in the black limestone, with value ranging between −12‰ to −40 ‰ (Holser and Kaplan 1966, Pye and Schiavon 1989, Dowuona et al. 1992).
7. $\delta^{34}S$ value originating from carbonate materials (ex. Carrara marbles: 26 to 28‰)(Cortecci and Orlandi 1975).
8. $\delta^{34}S$ value of atmospheric dust from agricultural activities (fertilizers and phytosanitary applications ($\delta^{34}S_{CDT}$ values range between −6.5‰ and 11.5‰ (Moncaster et al. 2000, Soler et al. 2002, Vitòria et al. 2004, Otero et al. 2008).

There are significant differences between the study sites. All $\delta^{34}S$ values except for Athens are positive from almost 2 to 7.5 ‰, showing that this progressive enrichment in sulfur-34 may reflect different percentage of marine aerosols. This is coherent with the location of the Marmaryon tower monuments that is found very close to coastal. The $\delta^{34}S$ value from Athens monuments $\delta^{34}S$ can be interpreted in terms of mixing with anthropogenic sources from SO_2 emitters as different hydrocarbons (coals, lignite, fuels, and gas). This $\delta^{34}S$ trend illustrates an isotope mixing model involving generally two S sources a low $\delta^{34}S$ values member from the carbonate construction materials and another with higher $\delta^{34}S$ values from atmospheric S.

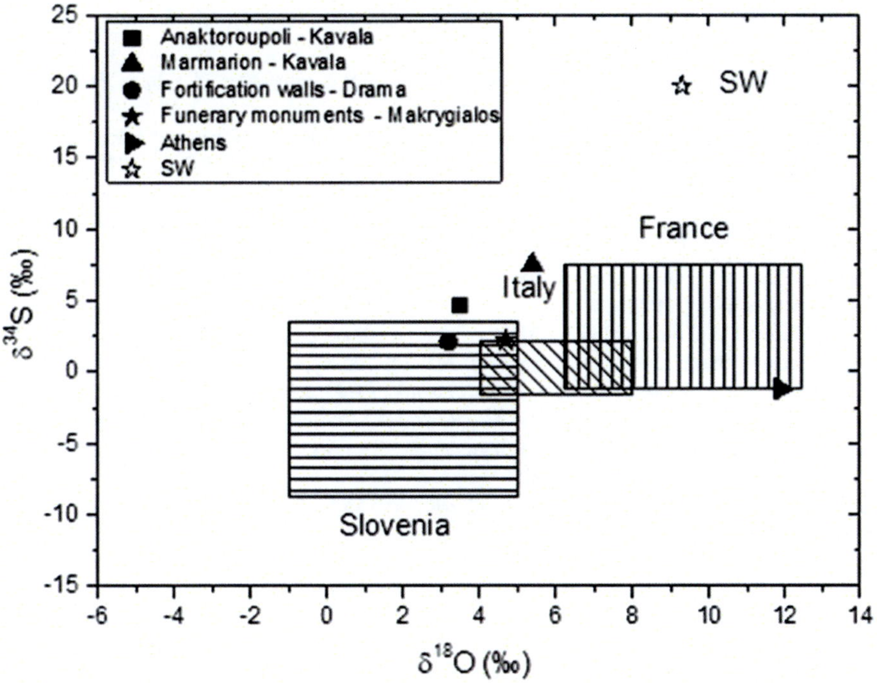

Fig. 1. Isotopic values of oxygen ($\delta^{18}O$) and sulfur ($\delta^{34}S$) of black crusts. The shaded areas are the isotopic values ($\delta^{18}O$ and $\delta^{34}S$) of black crusts from monuments in Slovenia, Italy and France as found in the literature.

The values of oxygen discriminate also Athens from the other sites. The oxygen of sulfates depends from the oxygen of air (+23‰ for gaseous O_2), water (from −11 to −5‰, (Dotsika et al. 2010, Dotsika et al. 2018)) and oxygen of diesel combustion residues (24.5‰; (Rivas et al. 2014)) and the relative contribution of each reservoir. Thus, the differences in the $\delta^{18}O$ of the black crusts could be the result of two Oxygen sources: low $\delta^{18}O$ values of rainwater and high $\delta^{18}O$ values of sulfates contained in dusts.

5 Conclusions

This is a preliminary study and it is evident that the stable isotope analysis of the $\delta^{34}S$ and $\delta^{18}O$ can discriminate the origin of the black crust on the deteriorated surface of the studied monuments. The competitive role of the sulfates of different origin was revealed (marine versus pollution). A systematic research is currently conducted that will also account for the orientation of the monuments. Furthermore, carrot samples will also be extracted and isotopically analyzed, in order to examine the "in depth" effect of the salts to the monuments.

References

Cortecci, G., Longinelli, A.: Isotopic composition of sulfate in rain water, Pisa, Italy. Earth Planet. Sci. Lett. **8**(1), 36–40 (1970)

Cortecci, G., Orlandi, P.: Oxygen-and carbon-isotopic composition of gypsum-calcite-dolomite crystals and metamorphic marble assemblages. Chem. Geol. **15**(4), 309–314 (1975)

Dadaki, S.: (Greek Ministry of Culture, Athens, Greece). Historical overview of the antiquities of 12th EBA, 12th Ephoreia of Byzantine antiquities, Archaeological report. Unpublished work (2010)

Dotsika, E., et al.: Isotopic composition of spring water in Greece: spring waters isoscapes. Geosciences **8**(7), 238 (2018)

Dotsika, E., et al.: Spatial distribution of the isotopic composition of precipitation and spring water in Greece. Global Planet. Change **71**(3–4), 141–149 (2010)

Dotsika, E., et al.: The use of O, H, B, Sr and S isotopes for tracing the origin of dissolved boron in groundwater in Central Macedonia, Greece. Appl. Geochem. **25**(11), 1783–1796 (2010)

Dotsika, E., et al.: Isotopic analysis for degradation diagnosis of calcite matrix in mortar. Anal. Bioanal. Chem. **395**(7), 2227–2234 (2009)

Dowuona, G., et al.: Isotopic composition of salt crusts in Saskatchewan, Canada. Chem. Geol. **94**(3), 205–213 (1992)

Faure, G.: Principles of Isotope Geology. John Wiley & Sons Inc, New York (1986)

Holser, W., Kaplan, I.: Isotope geochemistry of sedimentary sulfates. Chem. Geol. **1**, 93–135 (1966)

Kloppmann, W., et al.: Building materials as intrinsic sources of sulphate: a hidden face of salt weathering of historical monuments investigated through multi-isotope tracing (B, O, S). Sci. Total Environ. **409**(9), 1658–1669 (2011)

Kramar, S., et al.: Weathering of the black limestone of historical monuments (Ljubljana, Slovenia): oxygen and sulfur isotope composition of sulfate salts. Appl. Geochem. **26**(9–10), 1632–1638 (2011)

Longinelli, A.: Oxygen-18 and sulphur-34 in dissolved oceanic sulphate and phosphate. Marine Environ. **3**, 219–255 (1989)

Longinelli, A., Bartelloni, M.: Atmospheric pollution in Venice, Italy, as indicated by isotopic analyses. Water Air Soil Pollut. **10**(3), 335–341 (1978)

Moncaster, S., et al.: Migration and attenuation of agrochemical pollutants: insights from isotopic analysis of groundwater sulphate. J. Contam. Hydrol. **43**(2), 147–163 (2000)

Newman, L., et al.: Sulphur isotope variations in the atmosphere. In: Stable Isotopes: Natural and Anthropogenic Sulphur in the Environment (1991)

Otero, N., et al.: Controls of δ34S and δ18O in dissolved sulphate: learning from a detailed survey in the Llobregat river (Spain). Appl. Geochem. **23**(5), 1166–1185 (2008)

Pankina, R.: Sulfur isotopic composition of fossil fuels: oil. In: Stable Isotopes: Natural and Anthropogenic Sulphur in the Environment. Scope, pp. 100–108 (1991)

Pye, K., Schiavon, N.: Cause of sulphate attack on concrete, render and stone indicated by sulphur isotope ratios. Nature **342**(6250), 663–664 (1989)

Rafter, T.A., Mizutani, Y.: Preliminary study of variations of oxygen and sulphur isotope in natural sulphates. Nature **216**(5119), 1000 (1967)

Rivas, T., et al.: Sulphur and oxygen isotope analysis to identify sources of sulphur in gypsum-rich black crusts developed on granites. Sci. Total Environ. **482**, 137–147 (2014)

Schleicher, N., Hernández, C.R.: Source identification of sulphate forming salts on sandstones from monuments in Salamanca, Spain—a stable isotope approach. Environ. Sci. Pollut. Res. **17**(3), 770–778 (2010)

Schweigstillová, J., Přikryl, R., Novotná, M.: Isotopic composition of salt efflorescence from the sandstone castellated rocks of the Bohemian Cretaceous Basin (Czech Republic). Environ. Geol. **58**(1), 217–225 (2008). https://doi.org/10.1007/s00254-008-1510-y

Siedel, H., Klemm, W.: Evaluation of the environmental influence on sulphate salt formation at monuments in Dresden (Germany) by sulphur isotope measurements. In: Proceedings of the 9th International Congress on Deterioration and Conservation of Stone, Elsevier (2000)

Skrzypek, G., et al.: Stable isotope studies of moss sulfur and sulfate from bog surface waters. Geochem. J. **42**(6), 481–492 (2008)

Soler, A., et al.: Sulfur and strontium isotope composition of the Llobregat River (NE Spain): tracers of natural and anthropogenic chemicals in stream waters. Water Air Soil Pollut. **136**(1–4), 207–224 (2002)

Torfs, K.M., et al.: Use of stable isotope measurements to evaluate the origin of sulfur in gypsum layers on limestone buildings. Environ. Sci. Technol. **31**(9), 2650–2655 (1997)

Vallet, J.-M., et al.: Origin of salts in stone monument degradation using sulphur and oxygen isotopes: first results of the Bourges cathedral (France). J. Geochem. Explor. **88**(1–3), 358–362 (2006)

Vitòria, L., et al.: Fertilizer characterization: isotopic data (N, S, O, C, and Sr). Environ. Sci. Technol. **38**(12), 3254–3262 (2004)

Displacements with Remote Sensing and Standard Ground Survey for the Benedictine Basilica of S. Angelo in Formis (Southern Italy)

Gianpiero Russo[1](✉), Luca Di Girolamo[1], Diego Di Martire[2], Donato Infante[3], Gabriella Marone[1], and Massimo Ramondini[1]

[1] Department of Civil, Architectural and Environmental Engineering, Federico II University of Napoli, Naples, Italy
gianpiero.russo@unina.it

[2] Department of Earth, Environmental and Resources Sciences, Federico II University of Napoli, Naples, Italy

[3] SINTEMA Engineering, Spin-Off Federico II University of Napoli, Naples, Italy

Abstract. Cultural and historic buildings, such as ancient palaces and churches are rare and irreplaceable wealth of human civilization. Unfortunately, they are often impacted by natural calamities and anthropogenic activities. Consequently, the monitoring and preventive diagnosis of structure's deformation, as well as their surroundings, are essential for their safeguarding and conservation. High-precision and efficient monitoring facilitates the early recognition of potential risks and enables preventive diagnosis of heritage sites.

This work shows the potentiality of remote sensing technique as a complementary tool for the analysis of the movements of Benedictine Basilica of S. Angelo in Formis, an important monumental building in the area north of Napoli in Southern Italy.

Geological survey of the area, subsoil investigations and the integrated analysis of remote sensing and geodetic data, allowed to perform a very careful and detailed investigation about the structural performance of the Basilica. In detail, Differential Synthetic Aperture Radar (DInSAR) data have permitted a global interpretation of the Basilica's deformation and its surrounding areas: C-band products, available starting from 2003, have been integrated with topographic measurements, providing building displacement time-series which confirmed the occurrence of differential settlements of the monument.

Keywords: Forensic analysis · Differential synthetic aperture radar interferometry · Ground movements · Historic building monitoring

1 Introduction

Cultural heritage and archeological sites are inherited from past generations and serves in the understanding of cultural diversity as well as the evolutionary relationship between

I. Vayas and F. M. Mazzolani (Eds.): PROHITECH 2021, LNCE 209, pp. 108–119, 2022.
https://doi.org/10.1007/978-3-030-90788-4_11

earth and humans. Protection of historic sites (Russo et al. 2017) involves multi-temporal monitoring, aimed to detect geological, geotechnical or structural changes and identify potential threats (Russo and Alterio 2017).

Commonly, historic masonry structures are subjected to potential elastic deformation, damage susceptibility and brittle collapses, due to natural events such as ground instability (slow-moving landslides or subsidence phenomena) or related to structural performance mainly due to both intrinsic properties and to the concomitance with several modifications occurring over time (current state of maintenance, presence of irregularities and possible alterations occurred).

To this purpose, monitoring activities are significant for their preservation and safeguarding. Movements assessment of structures as well as their surroundings facilitates the early recognition of potential risks and enables effective conservation planning. Conventional investigations, including topographic leveling and GPS surveying are restricted by their limited spatial/temporal coverage as well as the need for considerable human and financial resources (Zhou et al. 2015).

In the last decade, a growing interest has been directed towards the exploitation of remote sensing approach as a tool for displacements monitoring of monuments (Tomás et al. 2012; Cigna et al. 2016; Di Martire et al. 2016; Tapete Cigna 2019): in particular, Differential Interferometry Synthetic Aperture Radar (DInSAR), based on processing of multi-temporal repeat-pass SAR images, has demonstrated to be a powerful technique to measure displacement annual rates and evolution of scatterers identified on Earth's surface. The availability of non-operating (ERS, ENVISAT-ASAR) and operating (COSMO-SkyMed, SENTINEL-1) satellite images, acquired with spatial resolution up to 3×3 m and revisiting time up to 6 days, allows to detect spatial and temporal historical movements since 1992, from large to local-site investigations, also at the scale of a single building (Bianchini et al. 2015; Cerchiello et al. 2016; Infante et al. 2019).

This study focuses on the assessment of stability conditions of the Benedictine Basilica of S. Angelo in Formis, in Campania Region (Italy), carried out by the cross-comparison of geological and geotechnical data, and results of displacements monitoring performed with geodetic surveys (Russo et al. 2013) and DInSAR processing.

The article is organized as follows: Sect. 2 describes available historical structural and architectonic characteristics of the Basilica; methodology and datasets, including geological and geotechnical settings, and displacement monitoring, are described in Sect. 3. Then, analysis and discussion of the results are provided in Sect. 4; finally, Sect. 5 summarizes the main conclusions of the paper.

2 The Benedictine Basilica of S. Angelo in Formis

A temple dedicated to Diana Tifatina had been erected in V century B.C. at the base of the South West side of the Tifata calcareous mountains in Southern Italy (De Franciscis 1956). In A.D. 595 a Longobard church dedicated to the Archangel Michael was constructed over the ruins of the Roman temple, as reported in the "Regesto" of S. Angelo in Formis", in the library of the great Montecassino Abbey.

In 1072 the Norman Richard the 1st, prince of Capua and count of Aversa, presented the church to the Abbott Desiderio, rector of Montecassino (Jacobitti and Abita 1992).

At the site of the ancient church, Desiderio erected the present Basilica whose floor is still lined with the white marble mosaic of the stylobates of the Diana temple, dating back to 150–170 B.C. The Basilica, in a simple and appealing Romanesque style, contains an outstanding cycle of frescos over the walls of its three naves. They represent stories of the Old Testament in the lateral naves, and of the New Testament in the central one. These frescoes are one of the reasons for the inclusion of the Basilica in the Italia recent tentative list for admission in the Unesco heritage.

The church has a plan divided in three naves, each one ending in semicircular apse (Fig. 1a). The columns that divide the naves are made of different materials (granite, white marble and green marble) and have Corinthian capitals from the temple of Diana. The façade is graced with a delicate portico or narthex of five arches, upheld by four Corinthian columns (Fig. 1b).

(a) (b)

Fig. 1. The Benedictine Basilica of S. Angelo in Formis: schematic geometrical plan (a) and the Western façade (b).

The foundations of the lateral walls, the façade and the apses are very shallow and of the same thickness of the overlying masonry structures. The columns are founded over stone blocks with a square shape in plan with a side of 0.7 to 1 m and a height of about 0.5 m. As it will be shown in the next sections, the apses and most of the lateral walls and columns rest on calcareous rock, while the façade, the portico and the front part of the sidewall are founded on debris or made soil.

3 Methodology and Datasets

Aimed to investigate structural performance of cultural heritage sites, as first step, a preliminary characterization of subsoil geological and geotechnical settings is required. In particular, the combined study of geological and geomorphological maps allows to evaluate if the area is prone to landsliding and the influence of geological properties on the distribution of potential mass movements.

Subsequently, the characterization of occurred damage by *in situ* survey represents an important step of building performance evaluation approach: cracks and ruptures occurring on structures or in the surrounding areas can be considered a useful indicator of processes involving existing constructions.

Finally, the integrated analysis of conventional and innovative monitoring data, performed with geodetic and DInSAR techniques, allows to achieve a validation of the results and to investigate preliminary cause-effect relationship between ground instability kinematics and building structural behavior.

3.1 Geological and Geomorphological Settings

The slope is geologically characterized by the presence of limestone and dolomite belonging to Trias–Cretacic, heavily fractured, somewhere brecciated and even interested by diffuse karst phenomena (Fig. 2a). The limestones are superimposed to Oligocene–Miocene deposits, composed by sandstones and variegated clays with a chaotic structure. Limestone slipped over the Miocene formation, and such a tectonic superposition, generated the fracture system affecting it. The rock formation is crossed by significant fracture lines in the North–South direction and become thinner towards the plain.

Later on (30,000 to 35,000 years B.P.) the plain and the slopes have been covered by the grey Campanian tuff erupted by the Phlegrean volcanoes (Fig. 2b). Since then, the morphological and structural pattern of the area remained unchanged, except for the anthropic actions. Among them the retaining walls and the fill to realize, in successive stages, the courtyard of the Basilica, and the opening of stone quarries. At least three of such quarries are located within a few hundred meters from the Basilica; they have been cultivated till the 1980s with extensive use of explosives.

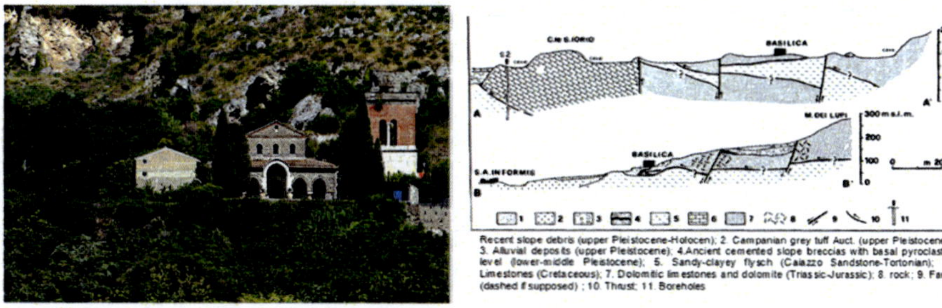

(a) (b)

Fig. 2. The Basilica located in the calcareous dolomitic slope (a) and geological section of the slope along North-South (above) and East-West (below) directions (b) (Corniello and Santo 1995).

The water table is generally found at the boundary between the limestone, very permeable by fractures, and the underlying complex of sandstones and clays.

Geophysical investigations allowed a detailed reconstruction of the subsoil within the zone surrounding the Basilica, which includes three horizons. The upper one is composed by made ground, for a thickness ranging from a few decimeters to some meters. Below the made ground a layer of fractured rock is found with a thickness variable between 15 and 30 m; it includes dolomite, slightly cemented dolomitic limestone and intensely fractured dolomite, besides cemented calcareous debris. Finally, the base formation of sandstones and variegated shaly clays is found.

The geological section (Fig. 3a) shows clearly that the basilica is located across a strati-graphic discontinuity, with the apses and the back-ward part of the naves founded on rock, and the front on the debris cover or even on the made ground.

As regards the slope instability phenomena, "IFFI" project, dedicated to landslide inventory and mapping in Italy, identified in the surrounding area two landslides classified as slide and complex, respectively (ISPRA 2008) (Fig. 3b).

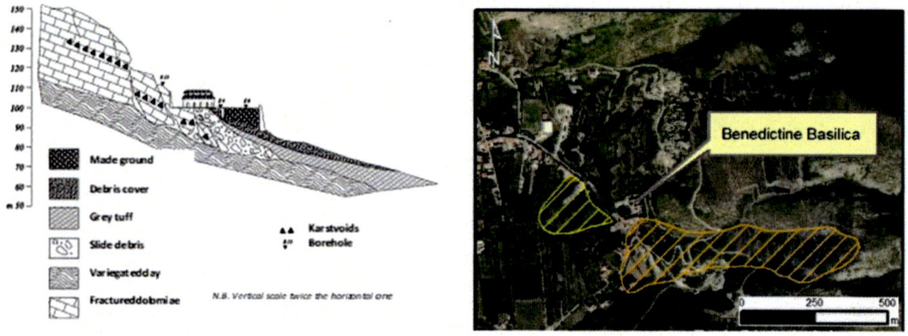

(a) (b)

Fig. 3. Geological section of the area (a) and landslide inventory map (b) within the area of Basilica (after ISPRA 2008).

3.2 Previous Damage Description

A stone in the right apse of the Basilica refers to repair works carried out in 1732 under the Cardinal Giuseppe Renato. Nothing is known about the damage requiring such interventions; perhaps they were consequence of an earthquake occurred in the same year. In 1930 an earthquake produced the collapse of the wooden roof. A number of fissures in the walls were repaired and the frescoes restored "with the help of German experts". Further earthquakes registered in 1962, 1970 and 1980 produced only minor effects,

In 1969 some fissures appeared in the walls above the columns of the central nave; in the following years the fissures gradually opened and extended to other parts of the Basilica, with a concentration in the right side of the façade and in the right nave near the façade. The occurrence of fissures was interpreted as the consequence of differential settlement of the foundations and a monitoring activity was organized for about one year, in 1972/73.

At the end of the 1970s, following repeated alarms on the safety of the Basilica, a Committee formed by a geotechnical engineer, a structural engineer and a geologist was installed, promoting a geological survey of the area, some further subsoil investigations, a geodetic monitoring of a number of points both inside the Basilica and outside and some further measurements of the fissures opening. Results of investigation and monitoring are discussed in Adriani et al. (1980) and in Pellegrino and Pescatore (1981) documents. The Committee conclusions related the damage mainly to the activity of the nearby quarries, where explosives were systematically used. Accordingly, the quarrying has been completely stopped since 1981. Unfortunately, fissures in the Basilica went on increasing, endangering the integrity of the frescoes; as consequence, a new cycle of investigation and monitoring has thus been started in 2012, with the aim of planning remedial and consolidation works.

3.3 Monitoring Data

To investigate the structural behavior of the Basilica, the opening of fissures, the offset of the columns and general movements of the structure occurred during different time spans, have been taken into account.

Two series of measurements of width of cracks in the walls are available: the first one in 1974/75 and the second one in 1980/81. The increase in the width of fissures was up to 0.5 mm in 1974/75 and to 0.8 mm in 1980/81, but unfortunately, location of observed cracks in 1980/81 is not known. Remarkable is the fact that the trend of the fissures measured in 1974/75 is compatible with a hogging deformation of the church.

Furthermore, in January 1980, the offset of the columns, measured with reference to the vertical direction, has been accurately determined. The offsets range from 10 to 89 mm and are small enough that they could be attributed to construction defects. The measurements have been repeated until January 1981, without showing any detectable variation.

As to regards the monitoring of displacements, geodetic and satellite data have been carried out.

In the periods February 1980-March 1981, May 2012 – January 2013 and 2018/2019, vertical displacements of some points inside the Basilica, on the base of each column (Fig. 4a) have been measured through 4 and 6 geodetic surveys, respectively. In the first period, the maximum settlement was about 3.5 mm, while thirty years later, a cumulated value of about 60 mm has been measured. In both cases, a hogging mode of deformation was observed (Fig. 4b).

In addition, C-band interferometric data have been used: firstly, ENVISAT ASAR descending products, just available on the Geoportal of the Italian Environmental Ministry in the period June 2003 – June 2010, have been analyzed. Subsequently, 91 SENTINEL-1 descending images, acquired in the period January 2017 – January 2020 have been processed by means of Coherent Pixels Technique (CPT, Mora et al. 2003).

As a result, for each dataset, displacement mean velocity maps of identified targets, measured along Line of Sight (LoS), have been obtained (Fig. 5).

Moreover, starting from 'LoS' displacement value, vertical component of movement has been evaluated according to Di Martire et al. (2013), as function of satellite incidence angle in descending geometry. Such analysis allows to obtain settlements time series of some targets identified on the Basilica, as indicator of structure deformations (Fig. 6). Maximum value of measured settlement is about 30 mm in the time span June 2003 – June 2010 and about 20 mm in the time span January 2017 – January 2020.

4 Discussion

All the available monitoring data have shown that the Basilica is affected by a continuing settlement. Geodetic surveys have recorded an average rate of displacement of about 1.4 mm/year in the period 1980–1981. The restart of the survey in the 2012/2013 have confirmed a long term movement rate of about 1.5 to 2 mm/year. The check of the recent surveys in the 2019 is still on the same more or less constant path even with some acceleration. An increase of movements rate has been detected also by interferometric data: a settlement means velocities of about 3 mm/yr and of little less than 10 mm/yr were measured in the periods 2003–2010 and 2017–2020, respectively.

Furthermore, these investigations highlighted the occurrence of differential settlements of the monument. As it is possible to note by displacement profiles along the bases of the columns (Fig. 4b) and DInSAR time series of scatterers located in different parts of the Basilica (Fig. 6), different cumulated displacements have occurred, generating deflection ratios and angular distortions, in other word classical damage parameters, as a consequence. Such analysis shows a good agreement between measured differential displacements and real distribution of damage surveyed. These effects could depend by complex geological pattern of the soil on which the Basilica was built: in fact, the apses and most of the lateral walls and columns is founded on calcareous rock, while the principal façade, the portico and the front part of the sidewall are founded on debris or made soil.

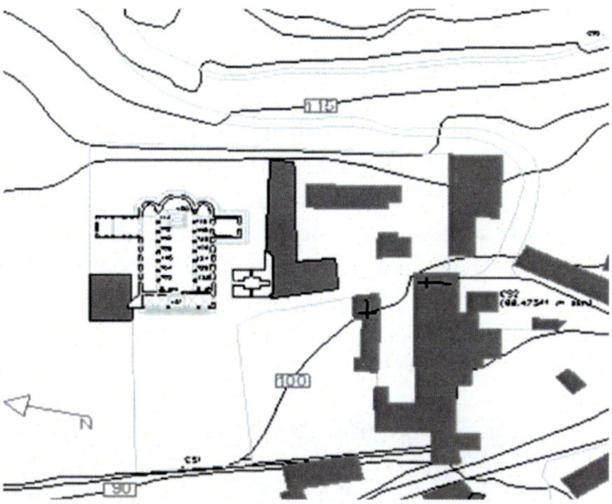

a) Plan of benchmarks used for topographic survey

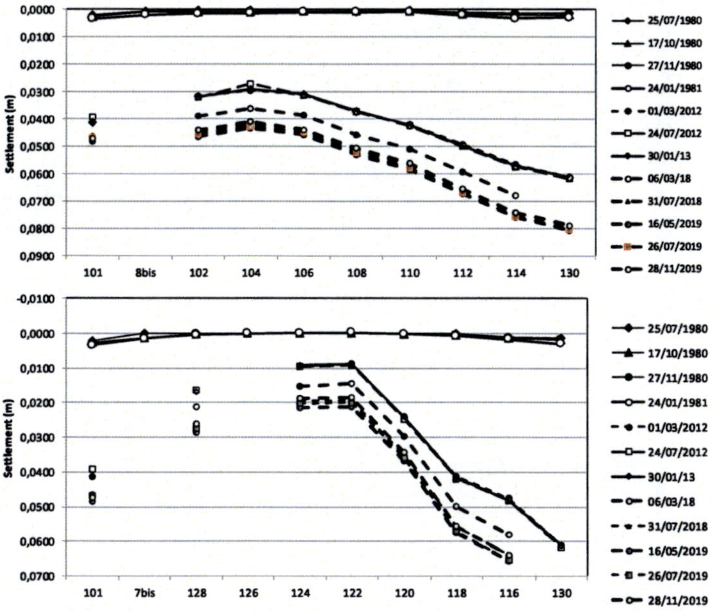

b) Settlement profile measured with optical survey at the base of the columns

Fig. 4. Plan of leveling points inside and outside the Basilica (a) and settlement profiles along the bases of the columns in the period 1980–2019 (b).

Different causes generating structure distortions and consequently cracks occurrence, have been considered in the past by various Authors (Russo et al. 2013): (i) the extensive use of explosives in the nearby stone quarries; (ii) the effect at the surface of karst phenomena in the limestone; (iii) residual slow tectonic movements of the slope. None of these suggestions seems totally convincing. As mentioned above, the quarrying activity has been stopped since 1980, while the movements are continuing. There are no evident reasons why the karst phenomena should begin at a time of some decades ago. Di Nocera (2013) believes that the movements are probably connected with a tectonic activity, triggered either by recent seismic activity or by the explosions in the quarries. The discussion is still open and the recent added data confirm the trends of the movements observed in the last 40 years.

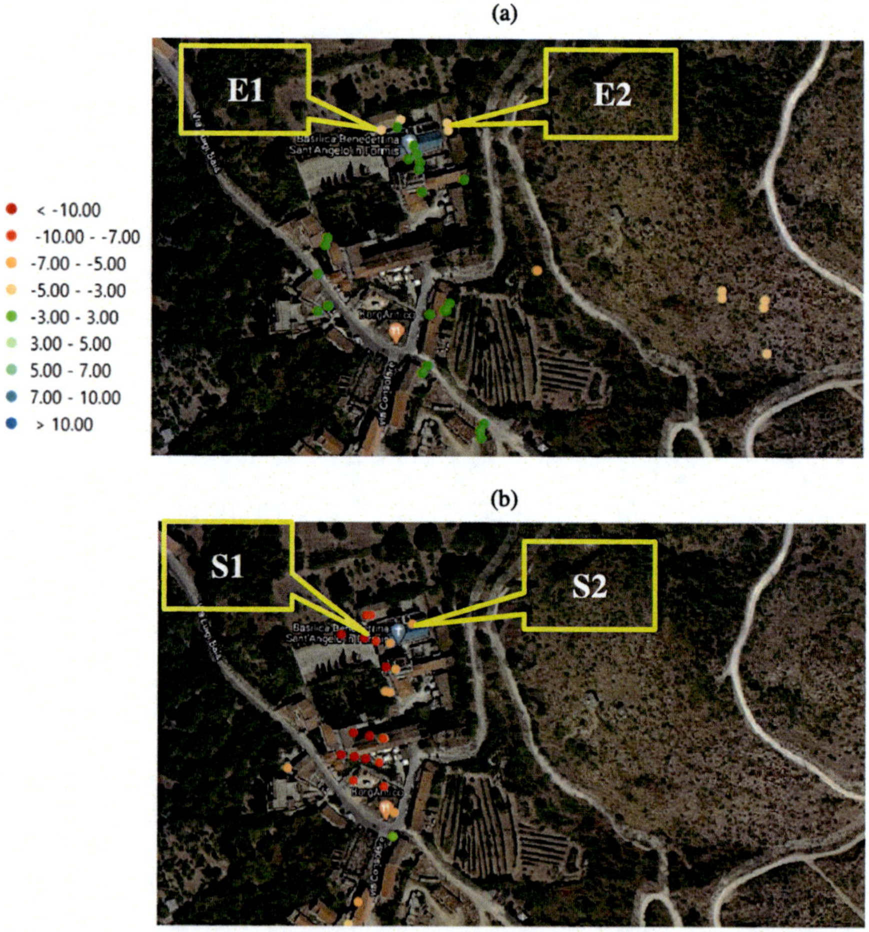

Fig. 5. "LoS" displacement velocity map (mm/yr) of targets identified by descending ENVISAT ASAR (a) and SENTINEL-1 (b) interferometric products.

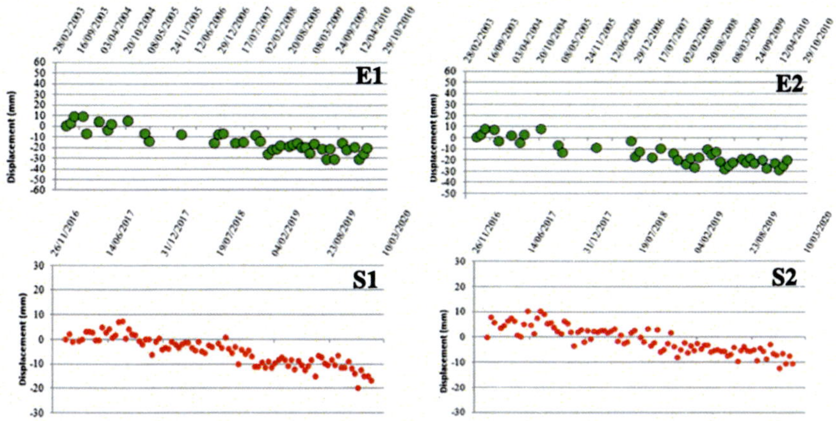

Fig. 6. Settlements time series of some targets identified on the Benedictine Basilica by ENVISAT ASAR (a) and SENTINEL-1 (b) interferometric products.

5 Conclusions

In the paper ground survey by optical precision levels is compared with satellite monitoring data obtained via DInSAR techniques with reference to a very ancient building, the Benedictine Basilica of Sant'Angelo in Formis which is largely founded over the ruins and ancient masonry of Roman temple built several centuries b.c. The comparison between the two techniques and the integration over different periods of time allow to describe a picture of the kinematics going on in the last 40 years approximately (Fig. 7). Movement are detected and confirmed by both the techniques even if with variable rates. Rates of 1–2 mm/year are confirmed with both techniques since the monitoring activities started in the late'70. A recent observed acceleration need some further and future check in order to prevent damage to the recently partially restored Basilica. The trend of the settlement pattern is also one of hogging type which is widely recognized as one of the worst possible for masonry buildings.

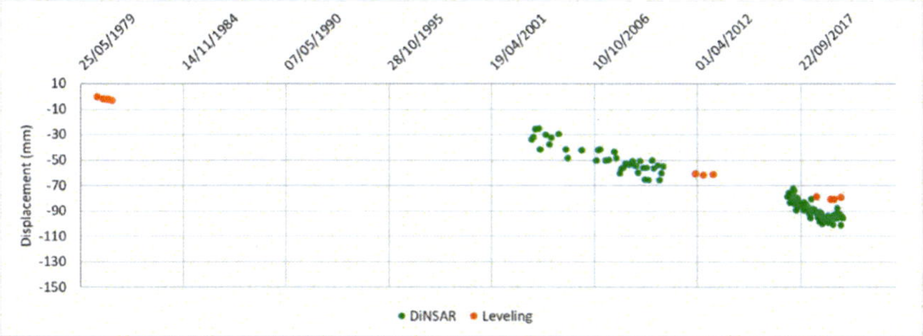

Fig. 7. Settlements time series recorded in the last 40 years.

Acknowledgements. The authors would like to thank the Segretariato Regionale Campania of the Mibact for their constant technical and financial support.

References

Adriani, L., Pellegrino, A., Bescatore, T.S.: La Basilica Benedettina di S. Angelo in Formis: prima relazione. Indagini sui terreni e sulle strutture del monumento per lo studio dei dissesti. Unpublished report, Soprintendenza ai Monumenti della Campania, Napoli (1980)

Bianchini, S., Pratesi, F., Nolesini, T., Casagli, N.: Building deformation assessment by means of persistent scatterer interferometry analysis on a landslide-affected area: the Volterra (Italy) case study. Remote Sens. **7**(4), 4678–4701 (2015)

Cerchiello, V., Tessari, G., Velterop, E., Riccard, P., Defilippi, M., Pasquali, P.: Risk of building damage by modeling interferometric time series. In: 2016 IEEE International Geoscience and Remote Sensing Symposium (IGARSS), pp. 7334–7337. IEEE, July 2016

Cigna, F., et al.: 25 years of satellite InSAR monitoring of ground instability and coastal geohazards in the archaeological site of Capo Colonna, Italy. In: Proceedings of SPIE 10003, SAR Image Analysis, Modeling, and Techniques XVI, 100030Q, 18 October 2016). https://doi.org/10.1117/12.2242095; http://dx.doi.org/https://doi.org/10.1117/12.2242095

Corniello, A., Santo, A.: I dissesti della Basilica Benedettina (XI secolo) di Sant'Angelo in Formis (Capua) ed il complesso assetto geologico dell'area. Geologia Applicata e Idrogeologia, XXX I, 125–137 (1995)

De Franciscis, A.: Templum Dianae Tifatinae. Soc. Di Storia Patria di Terra di Lavoro, Caserta, p. 60 (1956)

Martire, D.D., De Luca, G., Ramondini, M., Calcaterra, D.: Landslide-related PS data interpretation by means of different techniques. In: Margottini, C., Canuti, P., Sassa, K. (eds.) Landslide science and practice, pp. 347–355. Springer, Heidelberg (2013). https://doi.org/10.1007/978-3-642-31445-2_45

Di Martire, D., Novellino, A., Ramondini, M., Calcaterra, D.: A-differential synthetic aperture radar interferometry analysis of a deep seated gravitational slope deformation occurring at Bisaccia (Italy). Sci. The Total Environ. **550**, 556–573 (2016). https://doi.org/10.1016/j.scitotenv.2016.01.102

Di Nocera, S.: Personal communication (2013)

Jacobitti, G.M., Abita, S.: La Basilica Benedettina di Sant'Angelo in Formis, p. 97. ESI, Napoli (1992)

Infante, D., et al.: Assessment of building behavior in slow-moving landslide-affected areas through DInSAR data and structural analysis. Eng. Struct. **199**, 109638 (2019)

Italian National Institute for Environmental Protection and Research, ISPRA: Landslides in Italy. Special Report 2008. ISPRA, Rapporti, 83/2008 (2008)

Mora, O., Mallorquí, J.J., Broquetas, A.: Linear and nonlinear terrain deformation maps from a reduced set of interferometric SAR images. IEEE Trans. Geosci. Remote Sens. **41**, 2243–2253 (2003). https://doi.org/10.1109/TGRS.2003.814657

Pellegrino, A., Pescatore, T.S.: La Basilica di S. Angelo in Formis. Indagini sui terreni di fondazione ed analisi dei dissesti: relazione finale. Unpublished report, Soprintendenza per i Beni Ambientali e Architettonici della Campania, Napoli (1981)

Russo, G., Viggiani, C., Cammarota, A., Candela, M.: The Benedictine Basilica of S. Angelo in Formis (Southern Italy): a therapy without diagnosis? In: Geotechnical Engineering for the Preservation of Monuments and Historic Site, International Conference TC. 301, vol. 1, pp.225–232 (2013)

Russo, G., Alterio, L., Silvestri, F.: Seismic vulnerability reduction for historical buildings with non-invasive subsoil treatments: the case study of the mosaics palace at herculaneum. Int. J. Arch. Heritage **11**(3), 382–398 (2017). https://doi.org/10.1080/15583058.2016.1238969

Russo, G., Alterio, L.: Reduction of peak ground acceleration for monumental buildings via subsoil treatments the case study of Villa Favorita. In: Mazzolani, F., Calado, L. (eds.) 3rd International Conference on Protection of Historical Constructions – Lisbona 2017, pp. 1–12 (2017). ISBN:978–989–8481–58–0

Tapete, D., Cigna, F.: COSMO-SkyMed SAR for detection and monitoring of archaeological and cultural heritage sites. Remote Sens. **11**(11), 1326 (2019)

Tomás, R., et al.: Subsidence damage assessment of a gothic church using differential interferometry and field data. Struct. Health Monit. **11**(6), 751–762 (2012)

Zhou, W., Chen, F.L., Guo, H.D.: Differential radar interferometry for structural and ground deformation monitoring: a new tool for the conservation and sustainability of cultural heritage sites. Sustainability **7**, 1712–1729 (2015)

Achieving a Suitable Hygro-Mechanical Compatibility Between the Substrate and the Repairing Mortars in Historic Mansory Constructions

María Domizio[1,2](✉), Noemí Maldonado[1], and Luis Fernández Luco[3]

[1] CeReDeTeC, Universidad Tecnológica Nacional Facultad Regional Mendoza, Rodriguez 273, Ciudad, Mendoza, Argentina
mdomizio@frm.utn.edu.ar

[2] CONICET Consejo Nacional de Investigaciones Científicas y Tecnológicas, Buenos Aires, Argentina

[3] FI-UBA Facultad de Ingeniería, Universidad de Buenos Aires, Av. Paseo Colón 850, Buenos Aires, Argentina

Abstract. It is essential the study of restoration mortars on historical construction, because these must provide protection against harmful agents and must be sacrificable, not to cause damage on the primitive materials in case of removal by a new valorization. Repair mortars must have physical, chemical and mechanical compatibility with the substrate where they are applied and their mechanical strength must be similar to that of the wall where they are applied. The presence of water generates serious deterioration in the constructions, affecting its durability and causing greater maintenance and replacement costs for damaged materials. The characteristics of the repair mortars must be such that they do not allow water to enter the external faces and allow the evaporation of the water on internal faces. In this work the characteristics of compressive strength of different different composition mortars have been evaluated; absorption, speed of capillary suction and air permeability. The results obtained can facilitate the choice of mortars for the rehabilitation of heritage buildings.

Keywords: Mortar · Compatibility · Strength · Capillarity · Permeability · Masonry

1 Introduction

In the studies of value of heritage buildings, existing and repair materials must be analyzed, so that there are no incompatibilities between them, and they do not cause damages greater than those of normal behavior. In the restoration of historic buildings, preserve the original materials; and that replacement materials do not harm primitive construction is paramount argue Válek et al. (2012).

In the case of historical mortars, the originals should be kept as possible, because in them the technology, technique and construction history of the time in which they were

I. Vayas and F. M. Mazzolani (Eds.): PROHITECH 2021, LNCE 209, pp. 120–128, 2022.
https://doi.org/10.1007/978-3-030-90788-4_12

made are documented. If it is necessary to replace them in some places, they must have similar characteristics to the primitives, and have chemical, physical and mechanical compatibility with the support material. Likewise, restoration mortars must be designed so that they do not damage the historical construction, they must be "sacrificable", that is, they must be degraded before the support material (Gisbert et al. 2011, Domizio et al. 2016).

Masonry mortars have important and distinct properties depending on whether they are fresh or hardened. In the fresh state the properties that determine its aptitude in work correspond to the workability and water retention. In the hardened state the properties that influence the behavior of the masonry are: adhesion, durability, elasticity and compressive strength.

Methods have not always been developed to quantitatively evaluate all the behavior properties of the mortar, so that traditional specifications are applied as guidance in value enhancement.

A case study that has been monitored since its enhancement in 2015 is a heritage school of cooked ceramic masonry built between the end of the 19th century and the first decade of the 20th century and that has suffered the effect of earthquakes and lack of maintenance (Maldonado et al. 2011). Since its rehabilitation, a follow-up of the constructive pathologies is maintained and since 2017 an environmental monitoring of ambient temperature and humidity inside the premises and outside the school, which has allowed a mapping of the surface dampness in all the walls of the school as indicated in Domizio et al. (2018a), observing the importance of new uses and air conditioning technologies that impact historic masonry and its mortars.

2 Materials and Methods

2.1 Physical-Mechanical Characterization of Mortars

After the analysis of the pathologies found in the restoration mortars, the proportionsof mortars were studied. Due to the lack of national regulations for restoration mortars, is applied national regulations CIRSOC 501 (2007) and 501 – E (2007), INPRES CIRSOC 103 Part III (2016) and the local regulation CCSR'87 (1987). Mortar cubic specimens have been made, with normalized volume proportioning; then, they have been tested for compressive strength at 7, 28 and 56 days. The cement used is Pozzolanic Portland Cement (PPC40), with wide local availability. The water content was determined experimentally, until a workable mixture was obtained; with water/cement ratios ranging from 0.56 to 3.02. In addition to the normalizedproportioning, three more dosages were prepared for mortars without lime, increasing a part of sand, from the typical 1: 3 dosage, until reaching 1: 6, to contrast the dosage with lime, 1 cement: 1 lime: 6 sand.

Table 1 shows the proportions used, the w/c ratio, the designation according to the regulations analyzed and the test identification; and finally, the compressive strengths grouped by ranges, according to the values given in the INPRES CIRSOC 103-Part III (2016) regulation and in Fig. 1, the results obtained are plotted.

Table 1. Summary dosages mortar volume and test results.

Test	Proportions				Standard Quality		Compressive Strength (MPa)		
	Cement	Lime	Sand	w/c ratio	CIRSOC 501 IC 103-III	CCSR 1987	7 days	28 days	56 days
PPC40 1:3	1	–	3	0.56	Elevate	M3	18.2	29.9	33.1
PPC40 1:4	1	–	4	0.78	–	–	11.5	15.2	19.0
PPC40 1:5	1	–	5	0.89	–	–	8.9	13.3	17.4
PPC40 1:6	1	–	6	1.11	–	–	5.3	8.3	9.5
PPC40 1:0.5:4	1	0.5	4	0.89	Intermediate	–	6.5	11.3	16.7
PPC40 1:1:6	1	1	4	1.00	Normal	M2	7.3	12.2	15.4
PPC40 1:1.5:9	1	1.5	9	1.78	–	–	2.7	4.7	6.0
PPC40 1:3:9	1	3	9	2.22	–	–	1.7	3.2	4.6
PPC40 1:4:9	1	4	12	3.02	–	M3	1.3	2.1	3.1

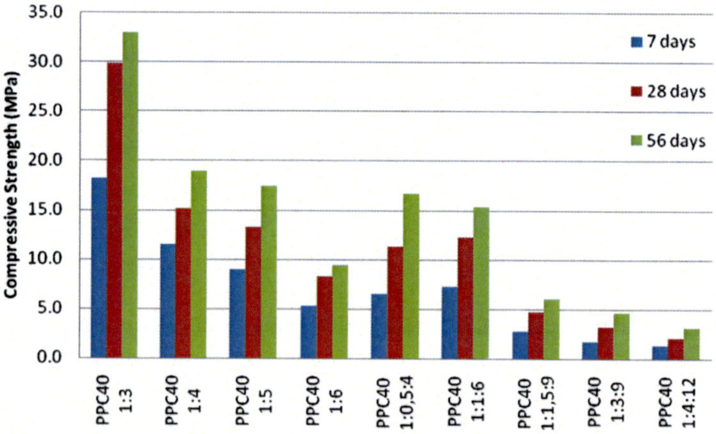

Fig. 1. Results of the compressive strength of different mortars studied

To characterize the behavior of repair mortar and substrate, the capillary suction test was conducted that guides the level of impermeability of the mortar (IRAM 2004). Figure 2a shows the laboratory test layout and the result of it in Fig. 3.

Two types of mortar samples were made, molded discs (M) and sawn discs (S) of 100, 110 and 150 mm in diameter and 50 mm high. The sorption slope is the same for all specimens, but the sorptivity increases more in the molded samples than the sawn ones and decreases with the diameter.

The air permeability test studied by Ebensperger and Torrent (2009) and Romer (2005), which allows guidance on the quality of the mortar used as indicated by Dobel et al. (2013), has been used to verify the behavior of the mortar-mortar interface (Fig. 2b and 4).

a) b)

Fig. 2. Provision of specimens for suction test (a) and air permeability test (b).

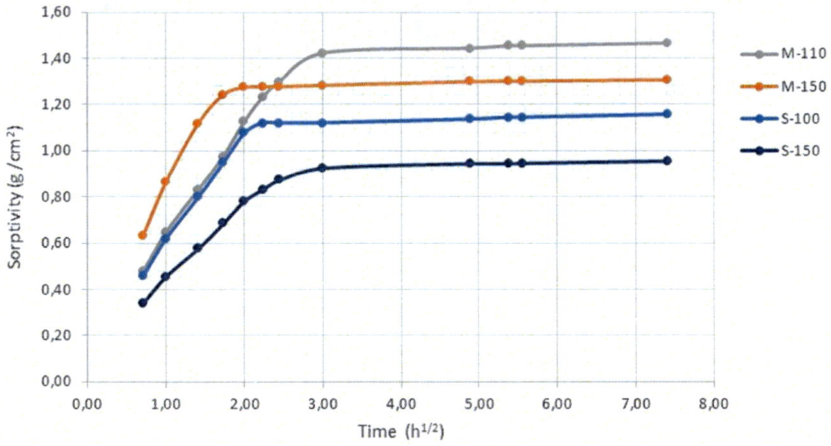

Fig. 3. Results of suction test

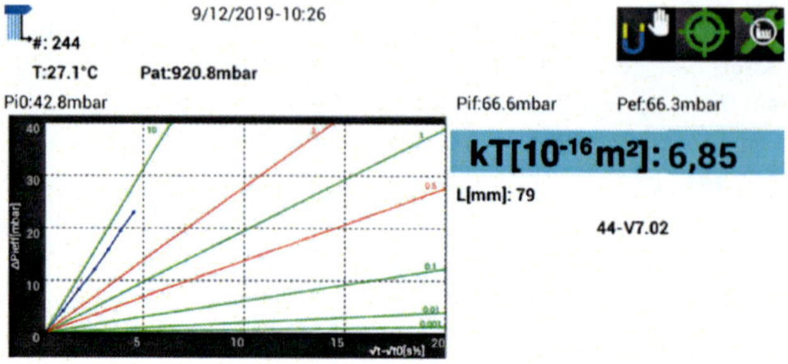

Fig. 4. One measuring test on the screen of equipment (b).

2.2 Environmental Monitoring and Surface Moisture Mapping

The use of cementitious plasters, generated new problems in the rehabilitated building, because there are chemical and physical incompatibilities with the heritage masonry. Over the years, large concentrations of sulfate salts were deposited inside the walls. The combination of these salts present in the masonry, the cement hydration compounds of the new plasters, high humidity conditions and low temperatures during their placement; made possible the formation of thaumasite and ettringite, two salts that cause the appearance of efflorescence; and then, expansion, cracking and disintegration in mortars (Maldonado et al. 2017) (Figs. 5 and 6).

a) b) c)

Fig. 5. South Wall #3 State in a) 2017 b) 2018 c) 2019

The monitoring of temperatures and humidity was carried out with HOBO brand dataloggers, inside the premises, inside the ventilation chambers, and also outside. It was verified that the relative humidity in the ventilation chambers of the premises was around 80–90% (Fig. 7a).The reason for the high humidity in the ventilation chamber of a room, after a visual inspection of it, was that the floor was very wet, close to the foundation beam of the south wall, which borders an old construction and in poor building conditions. The chemical analysis of the soil detected the presence of escherichiacolii that corresponds to contamination due to loss of sewage effluents.

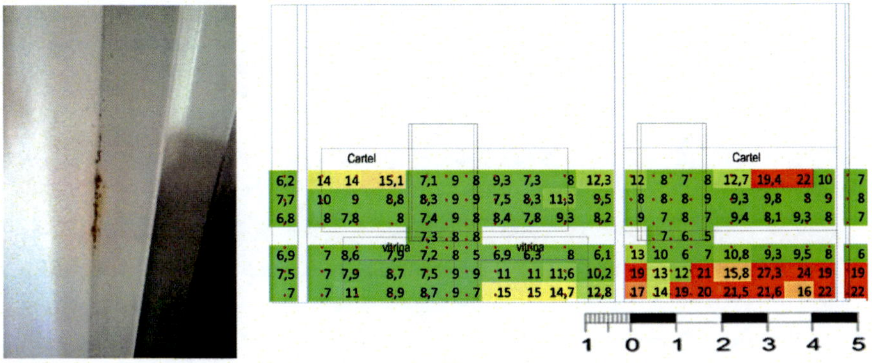

a) b)

Fig. 6. Corrosion in metal reinforcement profiles (a). View of the south wall Local 3, with surface moisture mapping (b)

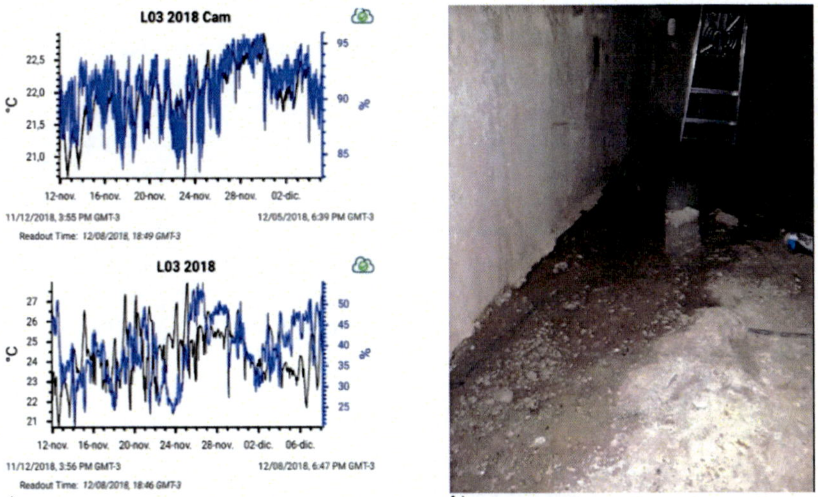

a) b)

Fig. 7. Humidity and temperature data in the ventilation chamber of room 3 and in the habitable zone of the same room (a). Photograph of the interior of the ventilation chamber of room 3 (b).

The value of the internal humidity of the party walls with Tramex humidimeter has also been verified, exceeding the average values of 99% relative humidity, which indicates the saturation of the masonry and the consequent problem that affects the mortars of coatings, since the presence of soluble salts is located above the level + 1.50 m (Fig. 8).

a) b)

Fig. 8. Provision of probes to measure internal wall moisture and measurement (a). Internal humidity measuring point in room 3 (b)

3 Results and Discussions

In the present study, the analysis of the environmental monitoring in the patrimonial construction and the mapping of surface moisture in the walls has been carried out to evaluate the appearance of the new pathologies, after the putting in value and second use of the building. The natural ventilation of the humidity of the chambers is not possible due to the location of the school: between buildings of many floors and with different orientations, and because of the low wind in the city of Mendoza. As a consequence, that moisture contained in the chambers, rises by capillarity in the walls and affects the plasters, in the places where there is concentration of salts, and to the metal profiles, in the areas of greatest humidity in the walls.

The opening of the mouths of the air chambers in summer has shown an effective reduction of the moisture content, a 5% decrease in five days, but they are not compatible with the current uses of the building (museum) such as the use of natural cross ventilation (Fig. 9).

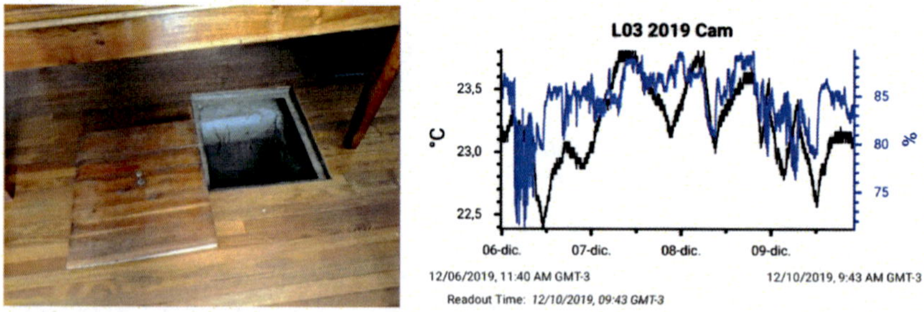

a) b)

Fig. 9. Basement air chamber cover opening (a). Measure of humidity and temperature data in the ventilation chamber of room 3, after opening of the mouths of the air chambers in summer (b).

The study of the compressive strengths of the standarized mortars, dosed in volume, allowed to determine: 29.9 MPa at 28 days, for the pure cementitious mortar, with a dosage of 1: 3 (the minimum resistance of 15 MPa, required in the regulation IC 103-III is doubled). For dosages 1: 4 and 1: 5, the resistance values are similar to each other, close to 15 MPa; and the 1: 6 dosage, has resistance below 10 MPa, at 28 and 56 days.

Lime mortars, Intermediate Quality of CIRSOC 501 and IC 103-III (1 cement: 0.5 lime: 4 sand), M2 of CCSR 1987 and Normal Quality mortar of CIRSOC 501 and IC 103-III (1 cement: 1 lime: 6 sand), have a resistance close to 12 MPa at 28 days, but exceed 15 MPa at 56 days. The M1 dosages of CCSR 1987 (1 cement: 4 lime: 12 sand) and own proposals with 1 cement: 1.5 to 3 lime: 9 sand, obtained resistance less than 5 MPa.

The study of capillary suction and air permeability in mortars made with Pozzolan Portland Cement according to the standard dosages, allowed to obtained the capillary suction speed, around 24 g/m^2 $s^{1/2}$,and the air permeability coefficient kT, that was greater than 2×10^{-16} m^2; these values indicate high permeability, in relation to results for normal concrete.

4 Conclusions

From the results obtained, it is concluded that:

- The environmental monitoring of a rehabilitated building allows its preservation and helps to understand the appearance of new pathologies.
- In cases of walls that have a capillary elevation of the water, it is not appropriate to use cementitious restoration mortars, especially with the 1: 3 dosage, due to their high resistance, which implies low permeability, with respect to the other dosages but with a great capacity of capillary absorption. It is better in these cases, the use of resistance mortars similar to the original masonry, and that are as permeable as possible, to allow evaporation of the water contained inside the walls to the outside.
- For the case study, if forced ventilation is implemented in the chambers, it is possible to evacuate the moisture contained therein. With this, the ascent of water to the walls would be cut and the appearance of efflorescence and new pitting of corrosion in the profiles of the structural reinforcement would be avoided.

Acknowledgements. This work has been part of Program PICT 2015-761 supported by the Technological National University of Argentina, National Agency for Promotion of Science and Technology of Argentina and Government of Mendoza. The authors want to thank the University staff, the professionals of preservation of Heritage Bureau of Province of Mendoza and CONICET, that they had made the development of the research program possible.

References

CIRSOC 501-2007. Reglamento Argentino de Estructuras de Mampostería (2007)
CIRSOC 501E-2007. Reglamento Empírico para Construcciones de Mampostería de Bajo Compromiso Estructural (2007)

Código de Construcciones Sismorresistentes de la Provincia de Mendoza, (CCSR) (1987)
Dobel, T., Torrent, R.J., Fernandez Luco, L.: Detección de sustratosheterogéneos mediante la medición de la permebilidad al aire. *1°* Congreso ALCONPAT Argentina 2013 Patología, control de calidad y rehabilitación en la construcción. Paper 44 (2013)
Domizio, M.C., Maldonado, N.G., Pablo, M., Fernandez Luco, L., Affranchino, G.: Incompatibilidad de morteros en la puesta en valor patrimonial: un caso de estudio. 21 Reunión Técnica AATH. 28–30 September 2016, Salta, Argentina (2016)
Domizio, M., Maldonado, N., Fernandez Luco, L.: Evaluación de morteros para la puesta en valor de mampostería patrimonial. 22 Reunión Técnica de la AATH - 5 al 9 de Noviembre de 2018. Olavarría, Bs. As., Argentina (2018a)
Domizio, M., Maldonado, N., Fernandez Luco, L., Arena, P.: Importance of environmental monitoring after rehabilitation of heritage masonry buildings in seismic zone: case study. In: 11th International Conference on Structural Analysis of Historical Constructions (SAHC 2018), 11–13 September. Cusco, Perú (2018b)
Ebensperger, L., Torrent, R.: Medición 'in situ' de la permeabilidad al aire del hormigón: Status quo. In: Proceedings of CONPAT 2009, Valparaíso, Chile, 29 de Septiembre al 2 de Octubre, 2009
Gisbert Aguilar, J., Royo, I.M., Somovilla de Miguel, I.A.: Morteros de restauración. Revista Zabaglia, del Colegio Oficial de Aparejadores, Arquitectos Técnicos e Ingenieros de Edificación de Huesca y UTA Huesca, Aragón, España. Parte I en Revista N°8, febrero de 2011. Parte II en Revista N°9, setiembre de 2011 (2011)
IRAM 1871-2004: Hormigón. Método de ensayo para determinar la capacidad y la velocidad de succióncapilar de agua del hormigónendurecido (2004)
Maldonado, N., Martín, P., Maldonado, I.: Seismic Mitigation of a Historic Masonry Building. Open Constr. Build. Technol. J. **5**(1), 61–70 (2011)
Maldonado, N., Martín, P., Maldonado, I., Domizio, M., González del Solar, G., Calderón, F.: Behaviour and durability of ceramic heritage masonry in near source fault zone. In: Proceedings 16th World Conference on Earthquake (16WCEE); 9–13 January 2017; Santiago de Chile: Paper 3371 (2017)

Shaking Table Tests Determining Frequency and Stiffness Before and After Strengthening of a Confined Masonry Wall Without an Opening

Emin Mahmud(✉) and Emad Abdulahad(✉)

Department Reinforced Concrete Structures, Faculty of Structural Engineering, University of Architecture, Civil Engineering and Geodesy (UACEG), 1 Hristo Smirnenski Blvd., 1046 Sofia, Bulgaria
mahmud_fce@uacg.bg

Abstract. Shaking table tests are planned in order to deepen the understanding of the behavior of confined masonry structures with or without openings. This paper presents the observations from a series of shaking-table tests done on a 1:1 scaled confined masonry wall model without openings (model CMF). Frequency and stiffness changes before and after GFRP wall strengthening are analyzed. The main targets of this research are assessment of the influence of location and area of the opening on the load-bearing capacity, rigidity and ductility of confined masonry, definition of appropriate systems for improvement of the stiffness, load capacity and ductility of the system and contribution in the assessment procedures proposed for masonry building in Eurocode 8. The initial frequency of the undamaged model CMFuS is 24,63 Hz, while at the end of the testing, the frequency decreased to 10,60 Hz. This emphasizes the reduction of the initial stiffness of the model due to damage, especially in the masonry and tie-beam to tie-column connection. After strengthening of the damaged wall, the natural frequency increases to 19,78 Hz. This highlights the beneficial effect of the strengthening. After completion of dynamic testing at CMFS, the natural frequency is reduced to 12,48 Hz.

Keywords: Behavior of masonry structures · Confined masonry · Eurocode · Frequency · Shaking table test · Strengthening

1 Introduction

Natural disasters are an effect of natural hazards such as tornados, volcanic eruptions, landslides, tsunamis, or earthquakes, which cause millions of deaths and severe socio-economic impacts, affecting the development of many countries (Mendes et al. 2014). From this perspective, earthquakes are one of the most devastating natural hazards on Earth. However, earthquakes hardly kill people, being the collapse of the buildings the main reason of the deaths. This means that efforts should be conducted to reduce the seismic vulnerability of buildings.

I. Vayas and F. M. Mazzolani (Eds.): PROHITECH 2021, LNCE 209, pp. 129–145, 2022.
https://doi.org/10.1007/978-3-030-90788-4_13

Masonry systems have a wide variety of forms and have been used as structural material for thousands of years. Existing masonry buildings were built for many centuries taking into account mostly vertical static loads according to the experience of the builder and without reference to any particular seismic code (Tomaževič et al. 1996). For their importance and value, many of those building have been classified into mankind's historical and cultural heritage of highest category. Thus, the study of the vulnerability of existing buildings is receiving much attention in the recent decades due to the increasing interest in the conservation of the built heritage and the awareness that life and property must be preserved. Considering that many of these buildings were built in the past, most of them do not meet the requirements of the Eurocode 8 (2005) recommendations and should be properly reinforced. Furthermore, from past earthquakes, it has been observed that the behavior of masonry systems under seismic impacts is substantially influenced by the presence and location of the openings in them.

2 Outline of the Research and Main Targets

Experimental research is part of the dissertation of PhD student Emin Mahmud. In his dissertation, original (unstrengthen) and strengthen models of confined masonry walls with different position and size of openings were tested. As defined, confined masonry is a construction system where reinforced concrete vertical tie-columns connected together with reinforced concrete horizontal tie-beam, confine plain brick or block masonry structural walls, thus forming a framed masonry panel. The main targets of this research are assessment of the influence of location and area of the opening on the load-bearing capacity, rigidity and ductility of confined masonry, definition of appropriate systems for improvement of the stiffness, load capacity and ductility of the system and contribution in the assessment procedures proposed for masonry building in Eurocode 8.3.

Shaking table tests are planned in order to deepen the understanding of the behavior of confined masonry structures with or without openings. In order to fulfill the purposes of the project three confined masonry walls with the same geometrical and material characteristics but with different configuration of the openings were tested. The first specimen was designed without an opening (specimen CMF), the configuration of the second specimen was with a window opening (specimen CMW) while the third specimen configuration was with a door opening (specimen CMD). Specimens at the beginning of the tests are shown in Fig. 1. The models were built on a scale of 1:1, in the laboratory of the Institute of Earthquake Engineering and Engineering Seismology. Moreover, the instrumentation scheme for the models was conceived in a way to obtain as many significant and valid experimental results, defining the types of measurement instruments, optimal location and number of measuring points, according to the available capacity of instruments and systems acquisition in the laboratory of the Institute of Earthquake Engineering and Engineering Seismology (IZIIS) - Skopje. The specimens were examined separately on the shaking table, with uniaxial, in-plane excitation following a defined testing methodology containing two phases - definition of the dynamic characteristics of the specimens and definition of the dynamic behavior of the specimens under earthquake/dynamic excitation. After testing, the entire surface of the damaged walls was coated with a three-component thixotropic epoxy mortar, whereas the larger cracks were further strengthened with the externally glued GFRP (Glass Fiber Reinforced Plastic) plate. The retrofitted specimens were tested following the defined testing procedure for the original models, respectively. The use of externally bonded glass fiber reinforced plastic (GFRP) materials exhibit considerable potential as an alternative solution to conventional strengthening techniques because of the low installation costs, high durability, negligible influence on the global mass, and excellent mechanical properties of these materials (Ceroni et al. 2014a, 2014b).

This paper presents the observations from a series of shaking-table tests done on a 1:1 scaled confined masonry wall model, without openings. Frequency and stiffness changes before and after GFRP wall strengthening are analyzed. Definition of dynamic properties of model CMF was the first step of the experimental testing, which enabled the acquiring of important information about the achieved stiffness (natural frequencies) of the model. The natural frequency was defined in the Y direction of the model by applying resonant frequency search tests. It is important to mention that both specimens CMFuS and CMFS are subject to the same effects.

Fig. 1. Specimens CMFuS, CMWuS and CMDuS at the beginning of the tests

Shaking table tests are able to simulate most accurately the excitation of structures under seismic loading. Basically, the shaking table is a rigid platform, where models are fixed, moved by hydraulic actuators. The most complex shaking tables have six degree of freedom (three translational and three rotational), which require a complex control system. The costs to build this type of facility and the cost of making the models are themselves high.

3 General Methodology for Testing of the Specimens – Phases, Ground Motion and Instrumentation

The testing procedure consisted of two main phases. Phase 1: Tests for definition of the dynamic characteristics of the models, in order to check the stiffness degradation of the model produced by the micro or macro cracks developed during the tests – resonant frequency search tests. Phase 2: Seismic testing by a selected earthquake record until heavy damage. The tests were performed in several steps, by increasing the input intensity of the earthquake, (Table 1 and Table 2), in order to obtain the response in the linear range, as well as to define the initial crack state, the development of the failure mechanism and the possible collapse of the model – seismic response tests. In such a way, the complete seismic performance of the structures starting from the linear range, the appearance of the first cracks in the walls up to the development of the failure mechanisms was captured. The testing has been performed on 5,00 × 5,00 m 5 DOF (Degrees-of-Freedom) MTS seismic shake table at IZIIS Laboratory (Fig. 2). Detailed information on the shaking table in IZIIS can be found in Corbi and Rakicevic (2013). Moreover, for simulating the foreseen real load in the exploitation period of a structure, additional load of 5,20 kN/m^2 (4 × 400 kg) was placed on the walls (Fig. 3). Also, in order in-plane excitation to be provided, special system for lateral support of the models was constructed (Fig. 3).

Fig. 2. The shaking table (a) and the relevant hydraulic system (b) at the Dynamic Testing Laboratory at IZIIS, Skopje (Corbi and Rakicevic 2013).

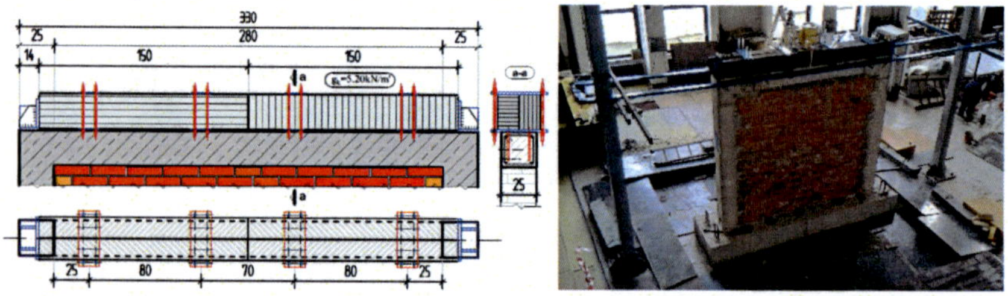

Fig. 3. Additional load for the walls (technical drawing) and system for lateral support

3.1 Definition of the Dynamic Characteristics of the Specimens

Determination of dynamic characteristic of the specimens, in both horizontal directions, was done using harmonic sine sweep excitation and random excitation, generated by the seismic shaking table. The harmonic (sine sweep) motions were performed with in-plane, uniaxial excitation, before testing of the model (initial state), after certain tests and after all performed tests (final state). The assumption that the test structure behaves linearly is essential to attaining accurate FRF measurement.

The experimental determination of the natural frequency of an element/structure is of particular importance in view of their trouble-free serviceability. The fundamental frequency is determined by applying the principles of structural dynamics. Modal testing is an experimental technique used to derive the modal model of a linear time-invariant vibratory system (He and Fu 2001). The theoretical basis of the technique is secured upon establishing the relationship between the vibration response at one location and excitation at the same or another location as a function of excitation frequency. This relationship, which is often a complex mathematical function, is known as frequency response function, or FRF in short. Combinations of excitation and response at different locations lead to a complete set of frequency response functions (FRFs) which can be collectively represented by an FRF matrix of the system. This matrix is usually symmetric, reflecting the structural reciprocity of the system. On the measurement of frequency response function data, (Ewins 2000; He and Fu 2001) prove a brief coverage of all aspects involved in conducting a successful frequency response function measurement. This includes preparation of the tested structure, general measurement set-up, selection of excitation forces and locations, selection of measurement locations, calibration of measurement set-up and acquisition of FRF data.

The performed resonant frequency search tests were with the following characteristics: random tests – horizontal, in plane direction with frequency range 1– 50 Hz and peak excitation level 0,02g–0,05g and sine sweep tests – horizontal, in plane direction with frequency range 1–50 Hz, peak excitation level 0.02g–0.05g and sweep rate 2,00 octave/min.

3.2 Definition of the Dynamic Behavior of the Specimens Under Earthquake/dynamic Excitation

To determine the dynamic response of the specimens under earthquake excitation, in-plane, uniaxial tests were performed. As representative real earthquake excitation, El-Centro time history record was used. Since the El-Centro time history record has different dominant frequencies compared with the natural frequency of the specimens, it had been frequently scaled with factor 8.0. The tests were conducted by gradually increasing the intensity of the excitation. Because the specimens could not be significantly damaged with the representative earthquake, both with the original and the scaled one, the models were further excited with random and sine sweep excitation in a previously specified frequency range in a correspondence with the natural frequencies for each model separately. Firstly, random vibrations in defined frequency range were performed, with increase in the intensity of the excitation. Afterwards, sine sweep excitation was carried out with frequency range changed almost in every test, following the decrease in the natural frequency. The intensity of the sine sweep excitations was also progressively raised. The results of these tests will be presented in other future publications.

3.3 Description of the Specimen

Specimen CMFuS is a full masonry wall, scale 1:1, confined with reinforced concrete tie-column, reinforced concrete tie-beam and reinforced concrete foundation. The masonry wall is 2.60 m height, with length of 2.80 m and thickness of 25 cm, whereas the dimensions of the confined wall are 2.85 m, 3.30 m and 25 cm, respectively (Fig. 4). The CMFuS specimen is one-open and one-story. The strengthened model (CMFS) presents the same geometry of the unstrengthened one (CMFuS). The size of reinforced concrete tie-column and tie-beam are 25 × 25 cm and the foundation is 400 × 95 × 40 cm. The foundation provides 8 holes with a diameter of 50 mm for fixing the specimens to the shake table. The wall was built with materials typical of the Republic of Bulgaria and the region. Detailed information on the specimens can be found in Abdulahad and Mahmud (2019).

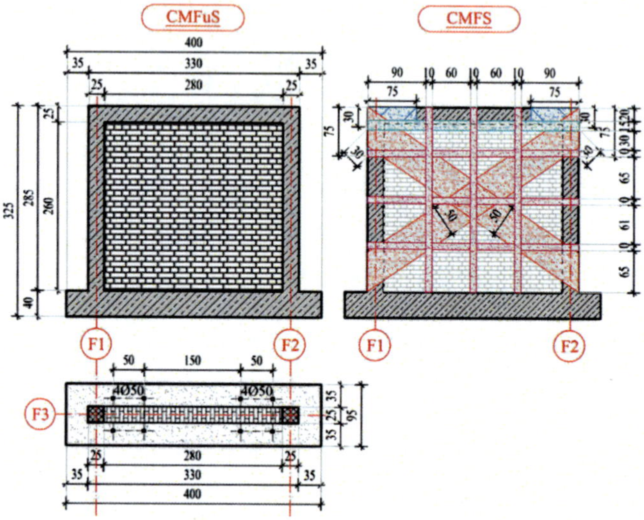

Fig. 4. Geometrical characteristics of specimens CMFuS and CMFS

The following sequence was followed in the built of the wall: the build of the foundation, then the masonry and finally the reinforced concrete tie-column and tie-beam. Eurocode 1998 (Eurocode 8 2005) requires that in order to obtain an effective connection between the tie-elements and the masonry, the concrete in the tie-elements must be poured after the masonry has been built. Good bonding between a masonry wall and adjacent RC tie-columns is important for satisfactory earthquake performance, and for delaying undesirable cracking and separation at the wall-to-tie-column interface. Bonding is an essential feature of confined masonry construction and it can be achieved by toothing at the wall-to-tie-column interface. Horizontal reinforcement anchored into tie-columns, also known as dowels, can be used instead of toothing (Brzev and Meli 2012; Jain et al. 2015; Singhal and Rai 2014a, 2014b). In the implementation of CMFuS, the connection between the RC tie- elements and the masonry is by toothing, and the concrete in the restraining elements is poured after the masonry has been built.

The implementation of the samples complies with the requirements of Eurocode 6 (2005) and Eurocode 8 (2005). The vertical joints are completely filled with mortar. The horizontal and perpendicular joints made of general-purpose masonry mortar are between 6 mm and 15 mm thick. Implementation of the specimen CMFuS and strengthening of the CMFS is presented in Fig. 5.

According to Eurocode 8 (2005) the longitudinal reinforcement of confining elements may not have a cross sectional area less than 300 mm^2, nor than 1% of the cross-sectional area of the confining element. The longitudinal reinforcement in the confining element is 4Ø14. Lap splices is 60 bar diameters in length. Stirrups 8 mm in diameter spaced in 100 and 150 mm are provided around the longitudinal reinforcement. Reinforcing steel is of Class B in accordance with (Eurocode 2 2005).

Fig. 5. Implementation of the specimen CMFuS and strengthening of the CMFS

3.4 Instrumentation Set-Up

The model response was monitored by a high-speed data acquisition system consisting of 3 accelerometers (ACC), 14 linear variable differential transformers (LVDT), 3 linear potentiometers (LP) and 8 strain gages (SG), providing information about accelerations at different levels and points, relative displacements and deformations and strains at selected points. Also, input parameters are obtained from the fixed instruments under the shaking table. The position and the general information for the instruments are given in Fig. 6. The strain gauges were placed on the reinforcement, before concreting. Detailed information on the instrumentation can be found in Abdulahad and Mahmud (2019).

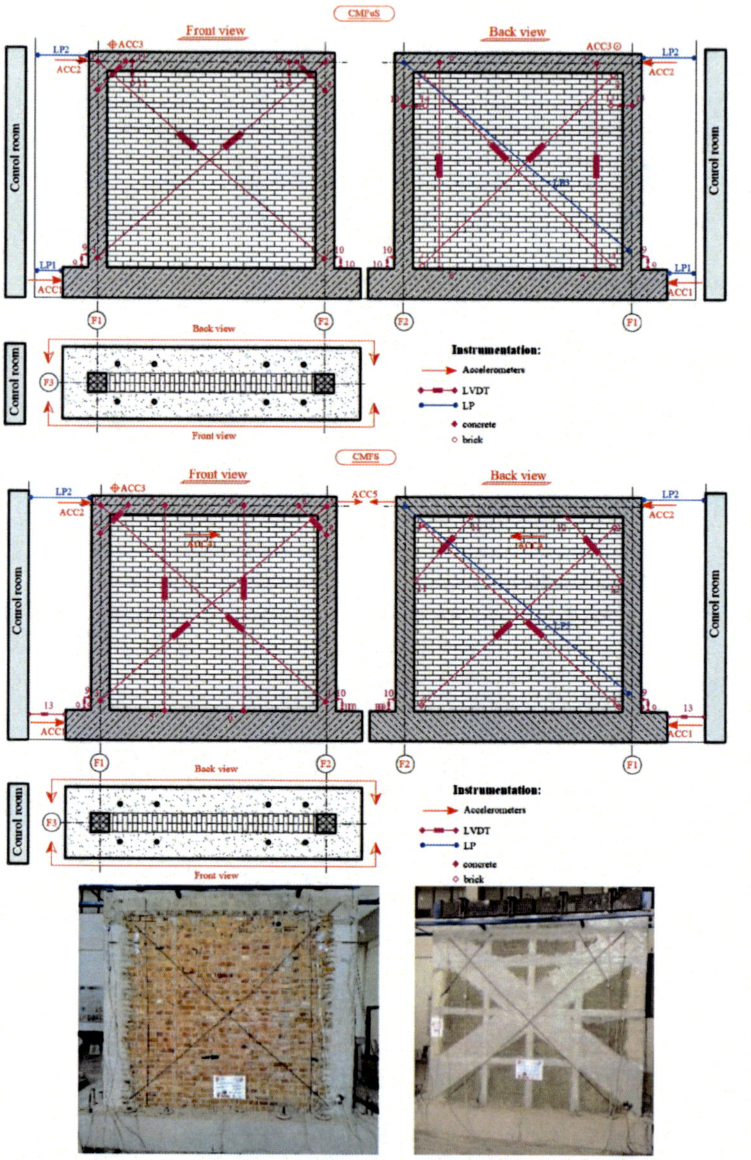

Fig. 6. Instrumentation scheme of specimens CMFuS and CMFS – technical drawing (front and back view) and photo documentation

4 Shaking Table Tests for Definition of the Dynamic Characteristics of the Specimens

In this section, the results of the tests carried out at IZIIS shaking table are presented. The methodology used for analyzing the results is first presented, followed by the discussion on the different results obtained in the dynamic identification tests. The dynamic identification tests aim at estimating the dynamic properties, namely the frequencies. The change of the dynamic properties, such as the decrease of the frequencies, is related to the stiffness of the structure and, consequently, to the damage concept usually used to define the degradation of the mechanical properties of materials. Thus, the main objective of these tests is to define a damage indicator based on the decrease of frequencies and to compare the seismic performance of the models (strengthened versus unstrengthened). Equations (1) to (4) are used, which are described in detail in Mendes et al. (2014).

$$K_{i,n} = \left(1 - d_{2,i,n}\right).K_{i,0}. \tag{1}$$

$$d_{2,i,n} = 1 - \left(\frac{f_{i,n}}{f_{i,o}}\right)^2. \tag{2}$$

$$d_{2/3,i,n} = 1 - \left(\frac{f_{i,n}}{f_{i,o}}\right)^{\frac{2}{3}}. \tag{3}$$

$$d_{i,n} = 1 - \left(\frac{f_{i,n}}{f_{i,o}}\right). \tag{4}$$

where K is the generalized stiffness; f is the frequency of mode n in the dynamic identification test i; d_2, $d_{2/3}$ and d damage indicator of the mode i in the dynamic identification.

Further on, results for frequencies and stiffness from specimens CMFuS and CMFS are presented, while the investigations of walls with openings are ongoing and will be presented in other future publications.

4.1 Unstrengthening Wall - Specimen CMFuS

On March 6, 2019, a specimen CMFuS was tested at IZIIS. A total of 20 tests were performed, of which 7 were to determine the fundamental frequency and the remaining 13 were to investigate seismic behavior on the wall. Table 1 presents a list of performed tests for specimen CMFuS. Testing of specimen CMFuS was performed by following the test general methodology described in Sect. 3.

Table 1. List of performed tests for specimen CMFuS

Test number	Name of the test	Excitation type	Input frequency range	Input acceleration	Measured Frequency [Hz]
TuS 01	Test01_Random_002	Random	1–50 Hz	0,066.g	/
TuS 02	Test02_Sweep_002	Sine sweep	1–50 Hz	0,030.g	24,63
TuS 03	Test03_Elcentro _50	Time history	/	0,147.g	/
TuS 04	Test04_Elcentr_100	Time history	/	0,340.g	/
TuS 05	Test05_Elcentro_50_X8	Time history	/	0,142.g	/
TuS 06	Test06_Elcentro_200_X8	Time history	/	0,568.g	/
TuS 07	Test07_Elcentro_400_X8	Time history	/	1,088.g	/
TuS 08	Test08_Elcentro_800_X8	Time history	/	1,739.g	/
TuS 09	Test09_Random_005	Random	1–50 Hz	0,169.g	21,18 Hz
TuS 10	Test10_Random_015	Random	8–28 Hz	0,537.g	/
TuS 11	Test11_Random_05	Random	8–28 Hz	1,464.g	/
TuS 12	Test12_Random_09	Random	8–28 Hz	2,274.g	/
TuS 13	Test13_Sweep_005	Sine sweep	1–50 Hz	0,062.g	18,89 Hz
TuS 14	Test14_Sweep_03	Sine sweep	15–22 Hz	0,331.g	/
TuS 15	Test15_Sweep_100	Sine sweep	14–20 Hz	1,089.g	/
TuS 16	Test16_Sweep_005	Sine sweep	1–50 Hz	0,062.g	12,99 Hz
TuS 17	Test17_Sweep_100_8-16	Sine sweep	8–16 Hz	1,247.g	/
TuS 18	Test18_Sweep_005	Sine sweep	1–50 Hz	0,061.g	11,65 Hz
TuS 19	Test19_Sweep_03_5-12	Sine sweep	5–12 Hz	0,388.g	/
TuS 20	Test20_Sweep_005	Sine sweep	1–50 Hz	0,062.g	10,60 Hz

Definition of the dynamic properties of specimen CMFuS was the first step of the experimental testing, which enabled the acquiring of important information about the achieved stiffness (natural frequencies) of the model. The natural frequency was defined in the Y direction of the model, by applying resonant frequency search tests (Fig. 7). The frequencies obtained before testing of the model (initial state, Test02), after certain tests and after all performed tests (final state Test20) of the specimen CMFuS are presented in Fig. 8.

As the damage/cracks in the sample increase with increasing number of tests, the natural frequency begins to gradually decrease. The change in the natural frequency is due to the degradation of the stiffness of the specimen and the damages obtained therein. The first fundamental initial frequency before the start of dynamic testing is 24.63 Hz as determined by Test02. Following the completion of the dynamic tests using the El-Centro earthquake acceleration (unscaled and scaled) whereby the PGA reaches 1,008.g, the fundamental frequency drops to 21,18 Hz as determined by Test09. The frequency decreased by 16.4% compared to the initial frequency. Current stiffness represents 74% of the initial stiffness Ko on the specimen CMFuS. After completing a series of three tests using random excitation when the PGA reaches 2.274.g, the frequency drops to 18.89 Hz as determined by Test13. The frequency decreased by 15.5% by the frequency determined

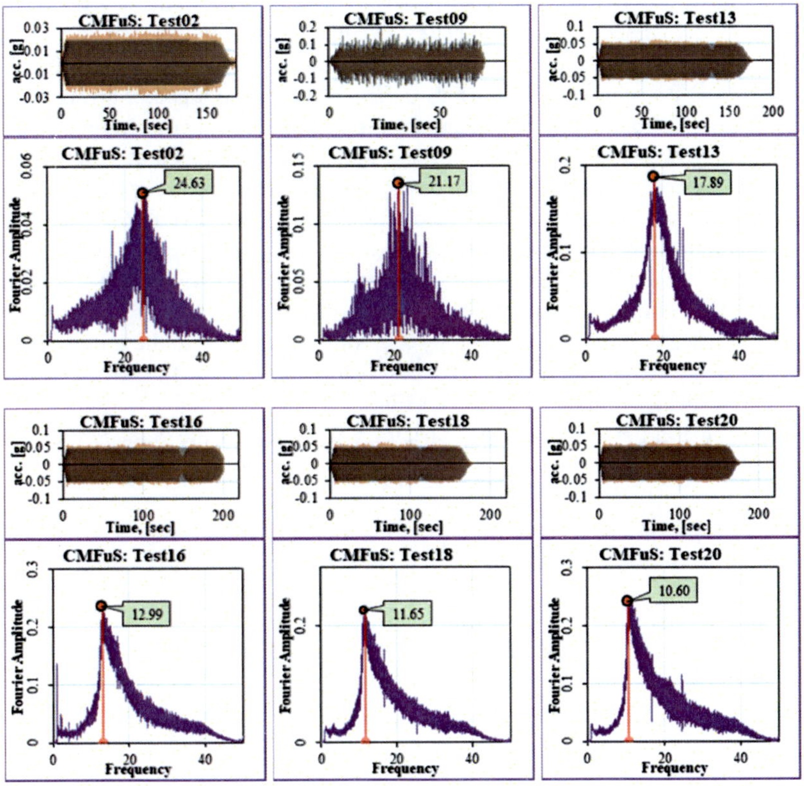

Fig. 7. Input signal und Fourier amplitude spectrum - specimen CMFuS

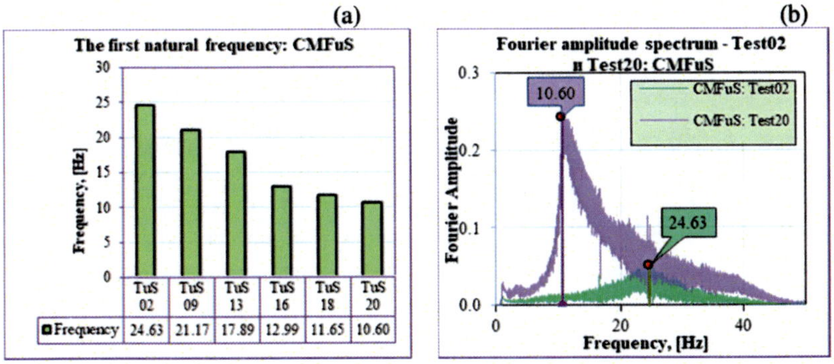

Fig. 8. The first natural frequency (a) and Fourier amplitude spectrum for Test02 and Test20 (b) of the specimen CMFuS

at Test09 and 1.38 times initial frequency. Stiffness is reduced by 29% from the stiffness determined in Test 09 and represents 53% of the initial stiffness Ko on specimen CMFuS. After completing a series of two tests using a sine sweep, with a PGA of 1.089 g, the frequency drops to 12.99 Hz as determined by Test16. The frequency decreased by 27,40% by the frequency determined at Test13 and 1.89 times initial frequency. Stiffness is reduced by 47% from the stiffness determined in Test 13 and represents 28% of the initial stiffness Ko on specimen CMFuS. After completion of another test using a sine sweep, with a PGA of 1.247 g, the frequency drops to 11,65 Hz as determined by Test18. The frequency decreased by 10,30% by the frequency determined at Test16 and 2,12 times initial frequency. Stiffness is reduced by 20% from the stiffness determined in Test 16 and represents 22% of the initial stiffness Ko on specimen CMFuS. After completion of another test using a sine sweep, with a PGA of 0,388g, the frequency drops to 10,60 Hz as determined by Test20. The frequency decreased by 9% by the frequency determined at Test18. After completing all 13-dynamic specimen CMFuS tests, the final frequency is 10.60 Hz. The frequency decreased by 2.33 times compared to the initial frequency. Stiffness is reduced by 18% from the stiffness determined in Test 18 and represents 19% of the initial stiffness Ko on specimen CMFuS.

The change in the stiffness of the specimen CMFuS and the damage indicator are presented in Fig. 9.

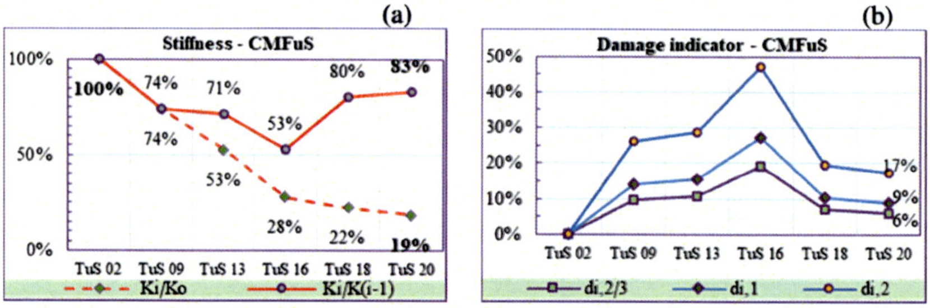

Fig. 9. Degradation of stiffness (a) and damage indicator for (b) of the specimen CMFuS

4.2 Strengthening Wall - Specimen CMFS

On March 26, 2019, specimen CMFS was tested at IZIIS. A total of 12 tests were performed, of which 7 were to determine the fundamental frequency and the remaining 5 were to investigate seismic behavior on the wall. Table 2 presents a list of performed tests for specimen CMFS. Testing of specimen CMFuS was performed by following the test general methodology described in Sect. 3.

Table 2. List of performed tests for specimen CMFS

Test number	Name of the test	Excitation type	Input frequency range	Input acceleration	Measured Frequency [Hz]
TS 01	Test01_Random_005	Random	1–50 Hz	0,191.g	19,78 Hz
TS 02	Test02_Sweep_005	Sine sweep	1–50 Hz	0,063.g	/
TS 03	Test03_Random_005	Random	1–50 Hz	0,175.g	/
TS 04	Test04_Random_005	Random	1–50 Hz	0,117.g	/
TS 05	Test05_Elcentro_100	Time history	/	0,333.g	/
TS 06	Test06_Elcentro_400_X8	Time history	/	1,007.g	/
TS 07	Test07_Elcentro_800_X8	Time history	/	1,733.g	/
TS 08	Test08_Random_005	Random	1–50 Hz	0,142.g	18,44 Hz
TS 09	Test09_Random_09	Random	12–22 Hz	2,058.g	/
TS 10	Test10_Sweep_005	Sine sweep	1–50 Hz	0,063.g	13,91 Hz
TS 11	Test11_Sweep_1g	Sine sweep	10–16 Hz	1,179.g	/
TS 12	Test12_Sweep_005	Sine sweep	1–50 Hz	0,063.g	12,48 Hz

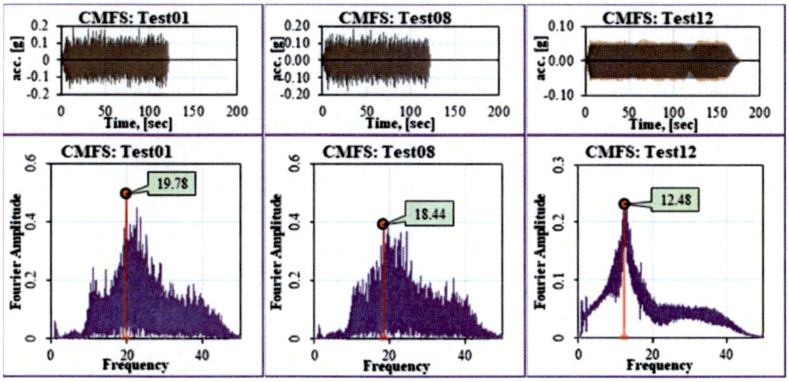

Fig. 10. Input signal und Fourier amplitude spectrum - specimen CMFS

Definition of the dynamic properties of specimen CMFS was the first step of the experimental testing, which enabled acquiring of important information about the achieved stiffness (natural frequencies) of the model. The natural frequency was defined in the Y direction of the model, by applying resonant frequency search tests (Fig. 10). The frequencies obtained before testing of the model (initial state, Test01), after certain tests and after all performed tests (final state Test12) of the specimen CMFS are presented in Fig. 11.

As the damage/cracks in the sample increase with increasing number of tests, the natural frequency begins to gradually decrease. The change in the natural frequency is due to the degradation of the stiffness of the specimen and the damages obtained therein. The first fundamental initial frequency before the start of dynamic testing is 19.78 Hz as determined by Test01. The determination of the initial frequency is to obtain information to what extent the stiffness of the unstrengthening specimen CMFuS tested after its testing has been restored. Following the completion of the dynamic tests using the El-Centro earthquake acceleration (unscaled and scaled) whereby the PGA reaches 1,733.g, the fundamental frequency drops to 18.44 Hz as determined by Test08. The frequency decreased by 6.78% compared to the initial frequency. Current stiffness represents 87% of the initial stiffness Ko on the specimen CMFS. After completing a series tests using random excitation when the PGA reaches 2.058.g, the frequency drops to 13.91 Hz as determined by Test10. The frequency decreased by 24.57% by the frequency determined at Test08 and 1.42 times initial frequency. Stiffness is reduced by 43% from the stiffness determined in Test08 and represents 79% of the initial stiffness Ko on specimen CMFuS. After completion of another test using a sine sweep, with a PGA of 1.189g, the frequency drops to 12.50 Hz as determined by Test12. The frequency decreased by 10.14% by the frequency determined at Test10. After completing all 5-dynamic specimen CMFS tests, the final frequency is 12.50 Hz. The frequency decreased by 1.58 times compared to the initial frequency. Stiffness is reduced by 20% from the stiffness determined in Test 10 and represents 40% of the initial stiffness Ko on specimen CMFS (Fig. 14).

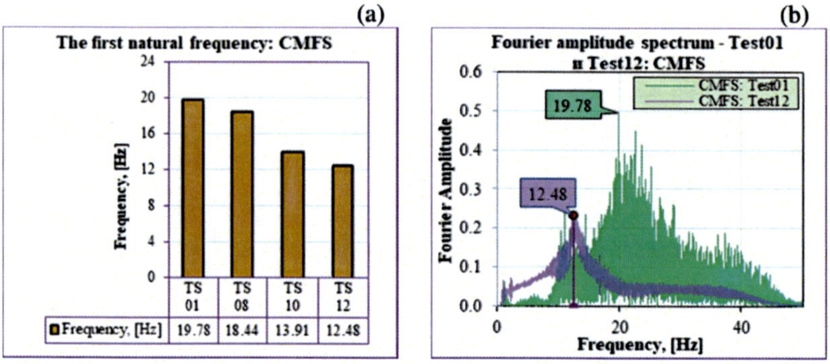

Fig. 11. The first natural frequency (a) and Fourier amplitude spectrum for Test01 and Test12 (b) of the specimen CMFS

The change in the stiffness of the specimen CMFS and the damage indicator are presented in Fig. 12.

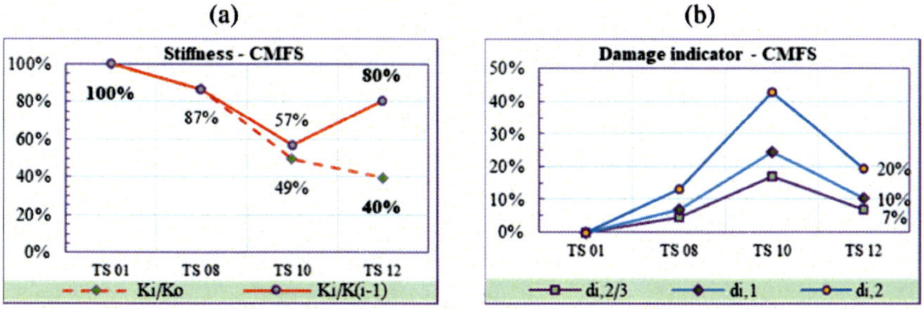

Fig. 12. Degradation of stiffness compared to the initial state (a) and damage indicator for (b) of the specimen CMFS

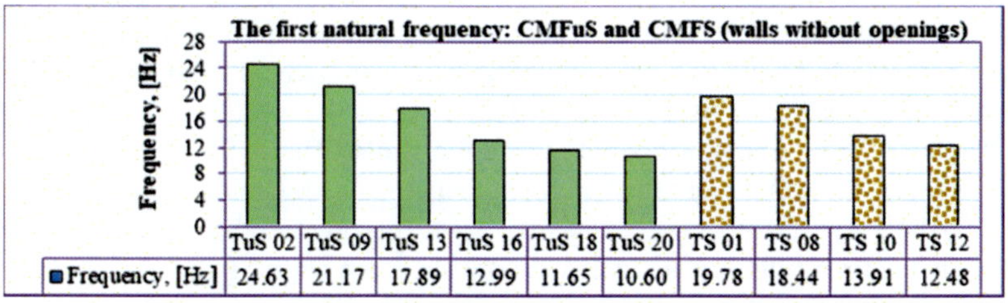

Fig. 13. The first natural frequency of the specimens CMFuS and CMFS

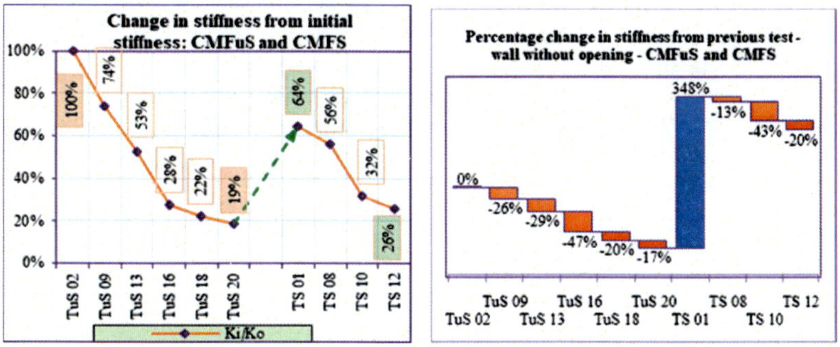

Fig. 14. Percentage change in stiffness from initial and previous test - CMFuS and CMFS

5 Conclusions

This paper discusses the results of the shaking table tests of a masonry wall without opening, before and after strengthening. The methodology used for analyzing the results is first presented, followed by the discussion on the different results obtained in the dynamic identification tests. The dynamic identification tests aim at estimating the dynamic properties, namely the frequencies.

The specimens were examined separately on the shaking table, with uniaxial, in-plane excitation following a defined testing methodology. After testing, the entire surface of the damaged walls was coated with a three-component thixotropic epoxy mortar, whereas the larger cracks were further strengthened with the externally glued GFRP (Glass Fiber Reinforced Plastic) plate. The retrofitted specimens were tested following the defined testing procedure for the original models, respectively.

The variation of the first wall frequency without a hole is shown graphically in Fig. 12, and in the graphs of Fig. 13 shows the change in stiffness. The following results can be summarized. The initial frequency of the unstrengthening wall without the opening is 24.63 Hz and after the realized tests it is reduced to 10.60 Hz, i.e. the degradation is 2.32 times. As a result of the strengthening, the first frequency of the wall increases by 1.87 times and reaches 19.78 Hz. The frequency after completion of the wall CMFS test is 12.48 Hz. Substantial degradation is observed after dynamic tests in the stiffness of the wall. After completion of the tests, the stiffness of the unstrengthening wall be 19% of the initial stiffness. As a result of the strengthening, the stiffness of the wall increases by 3.48 times, so that the stiffness after strengthening represents 64% of the initial stiffness of the wall CMFuS. Stiffness after completion of the strengthening wall tests represents 26% of the initial stiffness of the wall CMFuS.

The chosen strengthening system allows the wall stiffness to be restored to 64%, which indicates the beneficial effect of strengthening. The seismic behavior of the specimens CMFuS and CMFS under earthquake will be presented in other publications. Further on, the investigations of walls with openings are ongoing and will be presented in other future publications.

Acknowledgements. This research project under contract Д-111/2018 was supported financially by the Center for Research and Design at the University of Architecture, Civil Engineering and Geodesy (UACEG). The authors would like to express their deepest and appreciation to the staff of the Center for Research and Design.

The authors would like to express their deep gratitude and appreciation to the team of the following companies for their support for the successful implementation of the experiments: Baumit Bulgaria Ltd., Isomat International Ltd., Shpici stroi Ltd., Arcon Construction Ltd., A&K Engineering Ltd. and Emil Kroumov & Co. Ltd.

The authors would also like to thank all participants who contributed to the realization of the experimental tests and to the staff of the Dynamic Testing Laboratory at IZIIS, Skopje.

References

Abdulahad, E., Mahmud, E.: Implementation of samples and preparatory activities before dynamic tests. Annual of the University of Architecture, Civil Engineering and Geodesy, Sofia, Bulgaria, vol. 52, no. (3), pp. 891–900 (2019)

Brzev, S., Meli, R.: International guideline for seismic design of low-rise confined masonry buildings in regions of high seismic risk. In: The 15th World Conference on Earthquake Engineering (15WCEE), Lisbon, Portugal (2012)

Ceroni, F., Garofano, A., Pecce, M.: FE modelling of masonry panels externally bonded with FRP grid. In: PROHITECH 2014 Proceedings of the 2nd International Conference on Protection of Historical Constructions, Antalya, Turkey (2014a)

Ceroni, F., Garofano, A., Pecce, M.: Modelling of the bond behaviour of tuff elements externally bonded with FRP sheets. Compos. Part B **59**, 248–59 (2014b)

Corbi, I., Rakicevic, Z.: Shaking table testing for structural analysis. Int. J. Mech. **7**(4), 459–466 (2013)

Eurocode 2: Design of concrete structures. Bulgarian Institute for Standardization. BDS, Sofia (2005)

Eurocode 6: Design of masonry structures. Bulgarian Institute for Standardization. BDS, Sofia (2005)

Eurocode 8: Design of structures for earthquake resistance., Bulgarian Institute for Standardization. BDS, Sofia (2005)

Ewins, D.J.: modal testing: theory, practice and application. 2nd edn. Research Studies Press Ltd., Baldock (2000)

He, J., Fu, Z.-F.: Modal Analysis. Butterworth-Heinemann (2001)

Jain, S., et al.: Confined Masonry for Residential Buildings. Indian Institute of Technology Gandhinagar, India (2015)

Mendes, N., Lourenço, P., Campos-Costa, A.: Shaking table testing of an existing masonry building: assessment and improvement of the seismic performance. Earthq. Eng. Struct. Dynam. **43**, 247–266 (2014)

Rakicevic, Z, Garevski, M., Naumovski, N., Markovski, I., Golubovskic, R., Filipovski, D.: Upgrading of 5 DOF seismic simulation system with the newest real-time three variable digital control system. In: 15th World Conference on Earthquake Engineering, Lisboa (2011)

Singhal, V., Rai, D.: Seismic behavior of confined masonry walls when subjected to in-plane and out-of-plane loading. In: Tenth U.S. National Conference on Earthquake Engineering. Anchorage, Alaska (2014a)

Singhal, V., Rai, D.: Role of toothing on in-plane and out-of-plane behavior of confined masonry walls. J. Struct. Eng. (2014b)

Tomaževič, M., Lutman, M., Weiss, P.: Seismic upgrading of old brick-masonry urban houses: tying of walls with steel ties. Earthq. Spectra **12**(3), 599–622 (1996)

Parameterization of Stainless Steel Rebars to Improve Bonding Strength in Masonry Repairing

Beatriz Hortigon[1](✉), Fernando Ancio[1], and Esperanza Rodriguez-Mayorga[2]

[1] Departamento de Mecánica de Medios Continuos y Teoría de Estructuras, Universidad de Sevilla, C/ Virgen de Africa, 7, 41011 Sevilla, Spain
bhortigon@us.es

[2] Departamento de Estructuras de Edificación e Ingeniería del Terreno, Universidad de Sevilla, Avda. Reina Mercedes, 2, 41012 Sevilla, Spain

Abstract. Transversally tying and bed joint structural repointing are well-known techniques when speaking about repairing and reinforcing masonry walls and columns. Both techniques are widespread applied since they have given excellent and durable results. The introduction of stainless steel rebars or, more recently, polymeric fibre strips, along the joints or transversally tying the external masonry leaves of structural elements improves the behaviour of the conjunct and its mechanical properties.

The current shape of stainless steel rebars is the legacy of the rebars used to reinforce concrete. Many researches can be found about the influence of shape in bonding of carbon steel rebars and concrete. However, no research has been found about stainless steel rebars. The context of repairing masonry structures is significantly different of reinforced concrete mainly since so, consequently, a deeper study is compulsory to improve the design of these rebars when are used with this target.

In this paper, the preliminary studies about modelling and numerical simulations are exposed. To accomplish these goals, the rebars have been geometrically parametrised and analysed by means of the Finite Element Method. This way, the influence of several geometrical parameters have been researched and the results compared.

Keywords: Masonry repair · Transversal tying · Bonding strength · Stainless steel · Ribbed bars

1 Introduction

Historic masonry walls are usually composed by external leafs of brick or blockmasonry infilled with rubble masonry and joints of lime poor mortar. This characteristic morphology, containing a high percentage of voids, is quite sensitive to the effect of damp. The passing of time and weather phenomena usually drive in this types of structures to the weakening of structural elements and, consequently, to the appearance of cracks. In this

I. Vayas and F. M. Mazzolani (Eds.): PROHITECH 2021, LNCE 209, pp. 146–155, 2022.
https://doi.org/10.1007/978-3-030-90788-4_14

line, many researches have been conducted to develop and improve consolidation and reinforcing techniques for historic masonry walls (Binda et al. 1997; Valluzzi et al. 2005). Among the most used and tested are grout injection (Rodriguez-Mayorga et al. 2013; Rodriguez-Mayorga et al. 2019) and external or internal reinforcement (Mazzolani and Mandara 2002).

Transversal bars in walls improve substantially the load-bearing capacity of walls, mainly due to the effect of bracing. Stainless steel is usually used due to its excellent behaviour regarding oxidation and corrosion. The results are even better when the reinforcement is applied together with grout injection since the inner layer, with low resistance and usually degraded, is consolidated. On the other hand, if bed joints are structurally repointed, results are similar and the technique is less invasive, but no consolidation of the core of the walls is accomplished. Despite these reinforcement bars have been widely used for many years, with or without grouts, giving successful results, their shapes have not been deeply researched yet. Both designs used for bars of fiber reinforced polymers and stainless steel rebars (Molnàr and Larsson 2016) are not in fact optimised for an efficient bonding behaviour bar-grout.

Several designs of ribbed bars can be found depending on the country. Regarding Spanish specific legislation (UNE 36068 2011; UNE 36065 2011), a minimum value of bonding-strength values between concrete and steel are stablished, but no shape rib is described for Tempcore steel (EHE 2008; UNE-EN 10080 2006). On the other hand, the identification of ribbed bars in the country is done according to the early Tempcore ribbed steel standards, dated to years 1994 for weldable steel and 2000 for weldable steel with high ductility.

Regarding stainless ribbed bars, the disposition and shape of ribs is similar to those of Tempcore ribbed steel in case of their diameter are equal or higher than 12 mm. In case of lower diameters, the ribbed bars profile is determinate by the specific standard for stainless ribbed wires (UNE 36067 2017) with three areas and parallel and equidistant ribs along its entire length (Fig. 1). The reason of these two different geometries is only due to manufacturing criteria, with no structural justification.

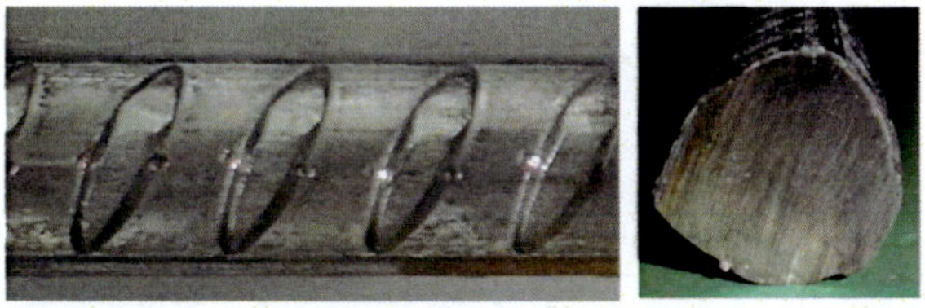

Fig. 1. Ribs geometry of stainless steel bars with diameter lower than 12 mm

Currently, several researches can be found about the influence of the shape of the ribs on the tensile and fatigue behaviour of the reinforcement Tempcore steel, in addition to the bonding strength concrete-steel (Lorrain et al. 2010; Tao et al. 2017; Barbosa et al. 2008; Majumdar et al. 2018; Hortigón et al. 2019). These authors conclude that the ribs cause a reduction of the mechanical properties of steel but it is worth to note that no references to the improvement of their designs is found. Hence, this research is focused on improving the design of stainless ribbed bars with minor diameters in order to enhance their performance when they are used as masonry-reinforcement, especially their bonding strength to grouts of lime or microcement. As both transversally tying and bed joint structural repointing, these ribbed bars are introduced in a different material, with lower cover than concrete as well, affecting all these factors to the bonding behaviour.

In line with this, the research project referenced as PGC2018-098185-A-I00, granted by Ministerio de Innovación, Ciencia y Universidades of Spanish government, which is being developed since 2019 to 2021, is focused on the optimization of rib shape of stainless ribbed bars used as reinforcement of masonry in order to improve their bonding behaviour. Despite the use of polymers is progressively more widespread, it is worth to note the high performance of steel as reinforcement of masonry and the significant weight within the Spanish economy. Both issues are sufficient to justify an in-depth research of the design of the bars, especially of the ribs shape, in order to enhance the bonding strength to lime, cement or microcement grout. The description of the steps of this project and the advances made is the main subject of this article.

2 Project Steps and Working Plain

The work plain to develop during the three years of the research project is structured in 4 main steps. These steps are divided in 6 sub-steps. The global structure of the project can be observed in Table 1.

Table 1. Global estructure of the project

Step 1	Finite-Element Simulation of prototypes	
	Sub-step 1	Obtaining of new designs of ribbed bars
	Sub-step 2	Modelling of prototypes
	Sub-step 3	Analysis of the simulation-results. Choice of prototypes
Step 2	Manufacturing of prototypes	
	Sub-step 4	Printing of prototypes in 3D
Step 3	Experimental validation of prototypes	
	Sub-step 5	Tests-carring out
Step 4	Validation of experimental results	
	Sub-step 6	Analysis of experimental results and correlation with those of FEM simulation

2.1 Finite-Element Simulation of Prototypes

In order to obtain initial data to compare subsequently with results obtained from the designs analysed by the research team, as a first phase, ribbed bars of stainless steel with reduced diameter will be considered. These bars have been manufactured by the Spanish company ACERINOX S.A., and their design, according to UNE 36067 (2017) is shown in Fig. 1. The further prototypes will be the result of modifications of this first design, taking into account several parameters: (a) angle between the rib and the longitudinal axis of the bar, (b) radius at the root of the rib, (c) height of the rib and (d) changes in rib shape. The 3D Finite-Element Simulations of the different designs lead to the knowledge about the theoretic values of bond strength between bar and grout for both static and dynamic pull-out test. Therefore, a first selection can be done by means of a comparative analysis of results.

2.2 Step 2: Manufacturing of Prototypes

The ribbed-bar prototypes selected will be printed with carbon fiber or Kevlar by means of a 3D printer. This procedure allows a more quickly and economic manufacturing of prototypes than traditional processes such as engraving or molding. Subsequently, printed designs will be tested according to the methodology described in Step 3. It should be noted that results obtained from the tests will be qualitative, not quantitative, since the mechanical properties of the prototypes material are very different to those of stainless steel. From these results, the most optimum ribbed bar design will be manufactured with stainless steel.

2.3 Step 3: Experimental Validation of Prototypes

The experimental validation of prototypes will be carry out from the "pull-out" test, according to ACI 440.3R-12 (2017). The system necessary to perform this test in a machine of static or dynamic test is equipped with 4 digital probes for the measure of displacements.

2.4 Step 4: Validation of Experimental Results

Results obtained by means of Finite-Elements Analysis will be compared with those obtained from pull-out tests described previously. This fact will allow the validation of the first hypothesis and, consequently, the application of them to further researches both in this field and similar ones, such as reinforced concrete structures.

3 Advances of the Project

Currently the project is at the first step. The works are focused on two main blocks: the generation of the models and their analysis by means of Finite-Elements simulation. It should be note that the diameter of the bars used in the project is 5 mm, a usual diameter used in masonry repair.

3.1 Generation of Models

Ribbed bars have been parameterized by means of the software Grasshopper, modulus integrated in design software (Rhinoceros v. 6 2019). In particular, three models of bars with several rib shape have been parameterized from the geometry of ribbed stainless steel bars established in UNE 36067 (2017). Geometrical properties of the models have in common the following parameters: (a) height of ribs equal to 0.45 mm, (b) distance between ribs equal to 4 mm, and (c) angle between rib and landscape axes equal to 55°. From this parameterization, several models will be generated subsequently by iterative modifications of these parameters.

3.2 Finite-Elements Analysis of the Models

From the solid models, the Finite-Elements analysis is being carried out by means of the software (ANSYS v. 19 2019). The simulation of these models implies the analysis of the three following issues: the modelling of the steel or the grout in addition to the modeling of the contact between them. Due to the steel is an isotropic material widely known, its modelling is relatively straightforward. However, the modelling of the grout is more complex as a consequence of: (a) different values of tensile and compressive strength, (b) elastoplastic behaviour, (c) possibility of tensile failure which implies its separation and, therefore, a discontinuity in the material and a reduction of the stiffness and (d) possibility of failure due to crushing, with changes of the stiffness and strength values of the interface.

Instead of the more usual modeling procedure from the application of the damage-model known as CONCR (based on the failure criterion of Willam and Warnke (1975) and the element SOLID 65 (based on the model of plasticity of Drucker and Prager (1952), in this research the MPLANE damage-model has been used. Both CORN model and SOLID 65 element present problems of convergence in case of complicated geometries like that of ribbed bars. Furthermore, different results can be found due to changes of the mesh.

The MPLANE damage-model is developed from the researches of Bazant and Gambarova (1984), Bazant and Oh (1985) and it is based on the Microplane Theory. The microplane formulation is based on a volumetric–deviatoric split (V–D) in a macroscopic viewpoint. The material behaviour is characterized by constitutive laws formulated on the individual microplanes and the anisotropic material behaviour can be taken into account simply and effectively. The microplane model formulates the constitutive equations between single stress and strain components on each of individual microplane. The microplane formulation is related to well-known macroscopic models but its application is more effective, especially in the case of non-isotropic materials. Nevertheless, some of isotropic materials have an anisotropic response under inelastic regimes. The need to propose continuum damage mechanics to these materials leads to complex formulations that can be simplified under the microplane theory. The MPLANE model is based on the modelling of the reduction of the stiffness in one specific direction from the uniaxial damage laws, applied at 42 different planes, which implies an anisotropic and macroscopic formulation of the damage. Among the several available options, an elastic model for each microplane associated with the SOLID 185 element has been chosen.

Regarding the interface of contact, a cohesive model has been implied using both CONTA 173 and TARGE 170 element in addition to the major tensile strength and slide values and the tensile and slide fracture energies. It is supposed that this model will simulate successfully the grout-steel join.

Grout Modeling from MPLANE Formulation
According to MPLANE model, the microscopic free energy (ψ^{mic}) on each of 42 planes can be integrated, obtaining the Helmholtz´s microscopic free energy (ψ^{mac}).

$$\psi^{mac} = \frac{3}{4\pi}\int_{\Omega} \psi^{mic} d\Omega \tag{1}$$

It states, that the energy of a material point can be understood as the integral of the microplane-based free energy over the entire solid angle Ω. Therefore, the free energy at the planes is integrated in order to obtain the initial free energy on the sphere. Due to the symmetry, the integration is done only at 21 planes. The failure criterion of microplane modeling assume an energetic criterion rather than the usual stress-strain relationship. The study of damage is done in every symmetry plane per each point of integration and not only in three perpendicular directions as usually, which implies a wider estimation of the damage and assure more successful simulation results. In order to take into account the degradation and the damage of the material, the function of the free energy includes a damage parameter.

$$\psi^{mic} = \left(1 - d^{mic}\right)\psi^{mic} \tag{2}$$

The damage parameter d^{mic} (Eq. (3)) is a standardized variable $\{0 \leq d^{mic} \leq 1\}$.

$$d^{mic} = 1 - \frac{\gamma_0^{mic}}{\eta^{mic}}\left[1 - \alpha^{mic} + \alpha^{mic} exp\left(\beta^{mic}\left(\gamma_0^{mic} - \eta^{mic}\right)\right)\right] \tag{3}$$

where α^{mic} defines the maximal degradation, β^{mic} determines the rate of damage evolution and, $\gamma_0{}^{mic}$ characterizes the equivalent strain energy on which the material damage starts (damage starting boundary).

The parameter η^{mic} is the equivalent strain energy, which characterizes the damage evolution law and is defined as

$$\eta^{mic} = k_0 I_1 + \sqrt{k_1^2 I_1^2 + k_2 J_2} \tag{4}$$

where I_1 is the first invariant of the strain tensor ε, J_2 is the second invariant of the deviatory part of the strain tensor ε, (Eq. (5)) and k_0, k_1, and k_2 are the material parameters which characterize the damage-function, depending on the compressive strength (*fc*) and the tensile strength (*ft*) of the grout (Eq. (6)).

$$I_1 = \varepsilon_1 + \varepsilon_2 + \varepsilon_3 \tag{5}$$

$$J_2 = \frac{1}{6}\left[(\varepsilon_1 + \varepsilon_2)^2 + (\varepsilon_2 + \varepsilon_3)^2 + (\varepsilon_3 + \varepsilon_1)^2\right]$$

$$k_0 = k_1 = k - 1/2k(1 - 2\nu) \quad (6)$$

$$k_2 = 3/k(1 + \nu)^2$$

$$k = f_c/f_t$$

where υ is the Poisson's coefficient.

Modelling and Boundary Conditions

In order to simulate correctly the behaviour of the bar, the grout has been modeled with rectangular shape. Bars, with a diameter of 5 mm, have been embedded in joints with a thickness of 15 mm, modelled as a parallelogram surrounding them. Three of the four faces of the parallelogram have been considered fixed as the fourth has free moving in order to reproduce the free area of the joint between rough stones or bricks (Fig. 2).

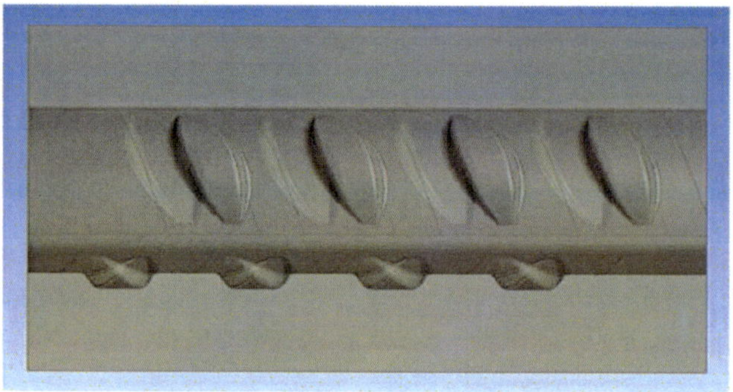

Fig. 2. Ribbed bar embedded in grout

The bar has been subjected to a longitudinal displacement equivalent to the tensile-force value supported. As it is common in this type of simulation, the displacement control has been chosen instead of force control in order to assure the convergence of the model. Figure 3 shows an inner section of the grout (without the bar) where the stress in the interface grout-bar can be seen clearly reflecting the transmission of the tensile force from the bar to the grout.

The parameter used to determinate the suitability of the rib design is the total force applied to the bar in order to reach this displacement value, calculated as the sum of all the force in the interface. Higher force value implies more optimal rib design. Figure 4 shows the three ribs models which are being analyzed currently.

For a similar displacement value, higher force values have been reached in models b) and c), over 3% higher than model a). Similar force values have been obtained for models b) and c). This difference of values is not relevant which is going to conclude that the existence of longitudinal ribs doesn't have influence on bonding behavior. Therefore, longitudinal ribs have been refused in further studies in order to reduce de computational time.

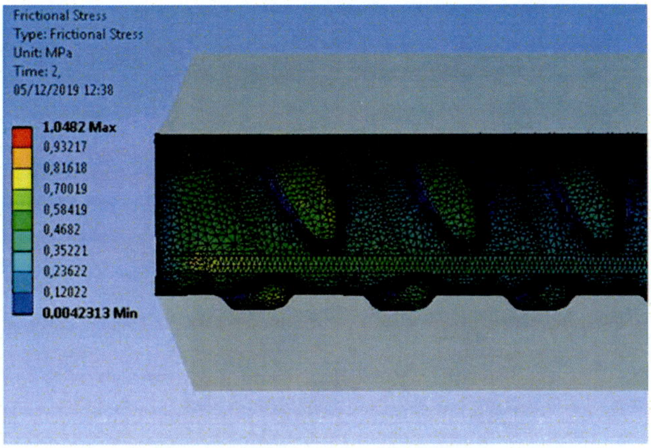

Fig. 3. Distribution of stresses in the interface bar-grout

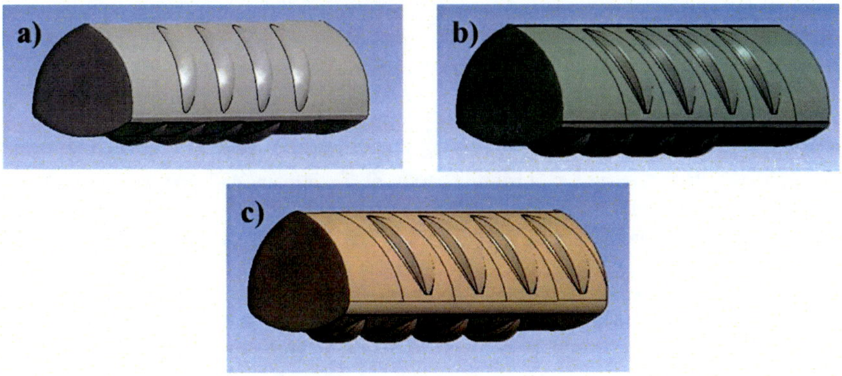

Fig. 4. Geometry of the three models analysed. a) Circular profile of transversal rib with longitudinal rib, b) trapezoidal profile of transversal rib with longitudinal rib, c) trapezoidal rib without longitudinal rib

4 Conclusions

A research about the influence of rib shape on the bonding strength between stainless steel and grout have been carried out in order to recover the use of the steel in masonry repair rather than the fiber reinforced polymers since, on the one hand, the steel industry is one of the most important within the Spanish economy and, on the other hand, excellent results have been obtained when it has been used in these cases.

In view of the first analyses, the optimal design seems a ribbed bar without longitudinal rib and a trapezoidal profile of transversal ribs. Further researches will be necessary in order to determinate the optimal shape of transversal ribs instead of the distance between them and their inclination angle.

Acknowledgements. Authors of this article thank to Ministerio de Ciencia, Innovación y Universidades of the Spanish government the economic support to this research by the project PGC2018-098185-A-I00.

References

ACI 440.3R-12: Guide test methods for fiber-reinforced polymer (FRP) composites for reinforcing or strengthening concrete and masonry structures, pp. 65–90. American Concrete Institute (2017)

Ansys: ANSYS, Inc. (2019)

Barbosa, M.T.G., de Souza Sánchez Filho, E., de Oliveira, T.M., dos Santos, W.J.: Analysis of the relative rib area of reinforcing bars pull out tests. Mater. Res. **11**(4), 453–457 (2008)

Bazant, Z.P., Gamborava, P.G.: Crack shear in concrete: crack band microplane model. J. Struct. Eng. **110**, 2015–2036 (1984)

Binda, L., Modena, C., Baronio, G., Abbaneo, S.: Repair and investigation techniques for stone masonry walls. Constr. Build. Mater. **11**(3), 133–142 (1997)

Drucker, D.C., Prager, W.: Soil mechanics and plastic analysis for limit design. Q. Appl. Math. **10**(2), 157–165 (1952)

EHE-08: Instrucción de hormigón Estructural. Ministerio de Fomento, Madrid (2008)

Hortigón, B., Gallardo, J.M., Nieto-García, E.J., López, J.A.: Strain hardening exponent and strain at maximum stress: steel rebar case. Constr. Build. Mater. **196**, 175–184 (2019)

Lorrain, M.S., Caetano, L.F., Silva, B.V., Gomes, L.E.S., Barbosa, M.P., Silva Filho, L.C.P.: Bond strength and rib geometry: a comparative study of the influence of deformation patterns on anchorage bond strength. In: 3rd International Fib Congress and Exhibition, Incorporating the PCI Annual Convention and Bridge Conference: Think Globally, Build Locally, Proceedings (2010)

Majumdar, S., Ali, R., Kumar, A., Deb, A.: Optimum rib design in TMT rebars to enhance fatigue life while retaining bond strength. J. Mater. Civ. Eng. **30**(3), 04017313 (2018)

Mazzolani, F.M., Mandara, A.: Modern trends in the use of special metals for the improvement of historical and monumental structures. Eng. Struct. **24**(7), 843 (2002)

Molnár, M., Larsson Ivanov, O.: Clay brick masonry facades with cracks caused by corroding bed joint reinforcement – findings from field survey and laboratory study. Constr. Build. Mater. **125**, 775–783 (2016)

Rhinoceros: Mc Neel & Associates (2019)

Rodríguez-Mayorga, E., Yanes, E., Compán, V., Sáez-Pérez, A.: La restauración del templo parroquial de San Dionisio (Jerez de la Frontera, España). La inyección como método de reparación de estructuras de fábrica. Informes de la Construcción **65**(529), 5–16 (2013)

Rodríguez-Mayorga, A., Cobo, E., Yanes, E., Sáez-Pérez, A.: The repair of the structure of Santiago's church (Jerez De La Frontera, Spain) using grout-injection. Int. J. Architect. Herit. **13**(8), 1234–1251 (2019)

Tao, W., Chen, C., Jun, H., Ting, R.: Effect of bolt rib spacing on load transfer mechanism. Int. J. Min. Sci. Technol. **27**(3), 431–434 (2017)

UNE 36068: Barras corrugadas de acero soldable para armaduras de hormigón armado. Aenor, Madrid (2011)

UNE 36065: Barras corrugadas de acero soldable con características especiales de ductilidad para armaduras de hormigón armado. Aenor, Madrid (2011)

UNE-EN 10080: Acero para el hormigón armado. Acero soldable para armaduras de hormigón armado. Generalidades. Aenor, Madrid (2006)

UNE 36067: Alambres de acero inoxidable para armaduras de hormigón armado. Aenor, Madrid (2017)
Valluzzi, M.R., Binda, L., Modena, C.: Mechanical behaviour of historic masonry structures strengthened by bed joints structural repointing. Constr. Build. Mater. **19**(1), 63–73 (2005)
Willam, K.J., Warnke, E.D.: Constitutive model for the triaxial behavior of concrete. In: Proceedings of IABSE, Structural Engineering Report 19, Section III, vol. 19, pp. 1–30 (1975)

Monitoring of Changes of External Climate Influence on Inner Microclimate of Sacral Building

Lukas Balik(✉), Lucie Kudrnacova, and Klara Nedvedova

Klokner Institute Czech Technical University, Solinova 7, 160 00 Prague, Czech Republic
lukas.balik@cvut.cz

Abstract. The paper deals with the internal environment of the side chapel of the important sacral building in Kutna Hora and its changes in relation to changes of selected parameters of the external environment. One of the important phenomena affecting the durability of construction materials (especially plasters and their surface treatments) is the existence and frequency of water vapor condensation on the surfaces of these structures. The formation and presence of condensation on interior plasters is, in a simplified view, dependent on the parameters of the internal environment (air temperature and relative humidity) and the surface temperature of the structure itself. The aim of the article is to map the state of the inner microclimate of the sacral building at a selected location of the side chapel in close proximity to the stained glass at the interface of the window and the window sill. Attention is paid to the existence and frequency of condensation and condensation zones and changes of specific air humidity from the position of long-term empirical monitoring of selected parameters of indoor microclimate.

Keywords: Inner microclimate · Condensation · Construction durability · Long term monitoring

1 Introduction

Maintenance of sacral buildings and approach to their repair and reconstruction cannot be done without an objective and comprehensive view of the technical condition of the building and of the conditions in which the building is located. Years of experience point to the fact that each historical sacral building has its specific characteristics and therefore it is necessary to approach each of them individually in terms of maintenance and operation. The technical condition of individual structural elements and their service life is undoubtedly influenced also by the environment in which they are located, both the external (conditions under the external terrain and the influence of outdoor climate), as well as the internal. This is an complex issue in which the internal microclimate is influenced by changes in the exterior microclimate but also by the technical condition of the structural elements (the presence of moisture, the existence of leaks, the presence of salts in the masonry) and structural arrangement of the building (the presence of building openings and the links between levels etc.). The internal microclimate in turn influences the structural elements and the interior furnishings of the space.

I. Vayas and F. M. Mazzolani (Eds.): PROHITECH 2021, LNCE 209, pp. 156–163, 2022.
https://doi.org/10.1007/978-3-030-90788-4_15

2 Aim of the Experiment

The aim of the performed long-term measurement was to evaluate the characteristics of the inner microclimate of the side chapel with attention paid in particular to the reference comparison of the microclimate in the space of the chapel itself and the microclimate in close proximity to the window panel. The stated objective was based on the assumption that the behavior of the indoor microclimate in the vicinity of the window panel will be significantly influenced by solar radiation or sunlight, its thermal component that passes through the glass panel. Expected severe temperature fluctuations should affect the formation and development of condensation zones as well as changes in specific humidity. To a certain extent, the specific air circulation along the window panels also contributes to the specific characteristics of the microclimate at this point.

3 Method of the Experiment

For a purpose of the experiment to obtain the values of the microclimate parameters the monitoring devices sensing selected parameters of indoor microclimate were installed in the side chapel of the Cathedral of St. Barbara in Kutna Hora. It is a chapel in the southwest corner of the cathedral. The sensors were placed so as to monitor selected parameters of the indoor microclimate both in the interior of the inner chapel and in the micro-space, which is located in close proximity to the contact of the glass window panel (stained glass) and the adjacent stone block (window sill). The overview of sensors and sensing points is shown in Table 1 below. These were measuring instruments (data loggers), measuring relative humidity and air temperature. Data were taken from 24. 7. 2018 to 19. 11. 2019 and was automatically stored in the device at regular 4-h intervals. Outdoor temperature sensors were connected to these data loggers (Fig. 1).

Table 1. Description of measuring instruments

Sensor number	Sensor type	Placement height above the church nave floor (m)	Further placement specification	Measured parameter
1	Data logger Comet 3631	3	Interior	Temperature °C, relative humidity % rel.
2	Probe	3	Interior by glass window	Temperature °C
3	Weather station Vantage Pro2	Tower of the Church of All Saints in Sedlec near Kutná Hora	Exterior	Temperature and humidity °C, % rel.

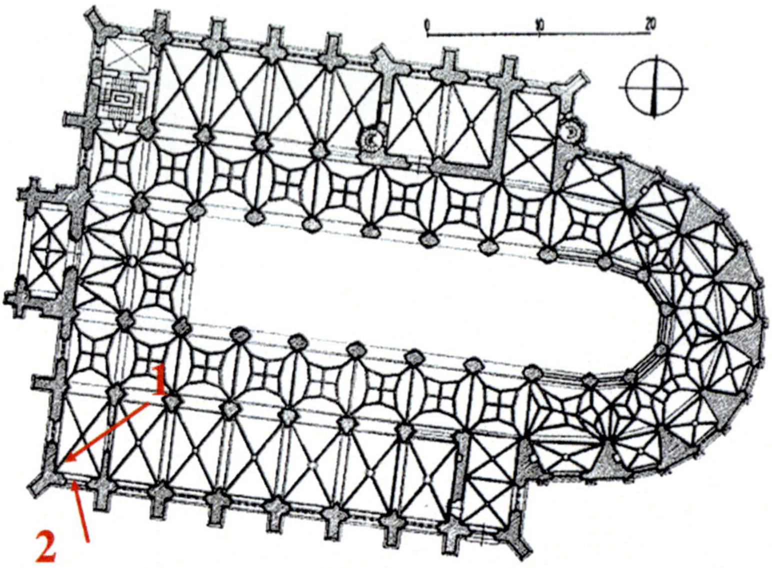

Fig. 1. Drawing of sensors placement in the temple plan

4 Results of Experiment

Direct monitored parameters were temperature, relative humidity and dew point temperature. Parameter derived was specific air humidity. Correlations of temperature, relative humidity and specific humidity in the comparison between exterior and interior were also calculated. The graphs of temperatures and specific humidity in the whole monitored period are shown in the following graphs Fig. 2 and 5. Detailed reference comparison of temperatures and dew points of the measured points of selected (critical) time periods is shown in graphs Fig. 3 and 4. Detailed reference comparisons of specific humidity of measured points of selected (critical) time periods are shown in Fig. 7. The summarizing information on the measured parameters is contained in Table 2. Table 3 clearly summarizes the frequency of condensation occurring at the point of contact between the stained glass and the stone sill.

On the basis of measurements and comparative analyzes carried out it can be stated:

1) the characteristics of the indoor microclimate in the immediate vicinity of the window panel and stone sill (hereinafter referred to as detail) are significantly different from the microclimate in the open space of the chapel (Fig. 2 and 5);
2) temperatures close to detail oscillated between –14.3 to 37 °C with average value of 12.5 °C in the two-year cycle and in the open space between −0.4 to 13.9 °C with average value of 13.9 °C;
3) no significant condensation process has been recorded in the side chapel area (on the inner wall surface) (Fig. 4);
4) a process of periodically occurring water vapor condensation has been recorded on the perimeter walls near the detail (Fig. 4);

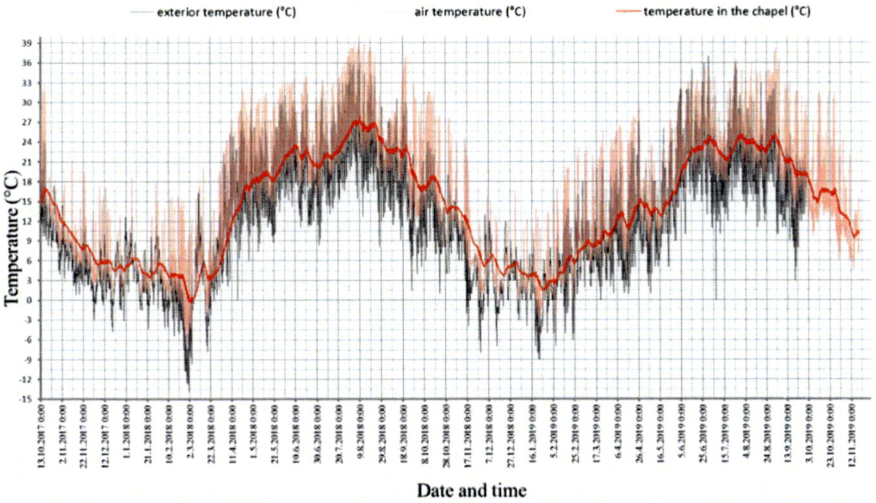

Fig. 2. Course of temperatures in the chapel (space and immediate proximity of the window panel) and exterior in the period October 2017 to November 2019

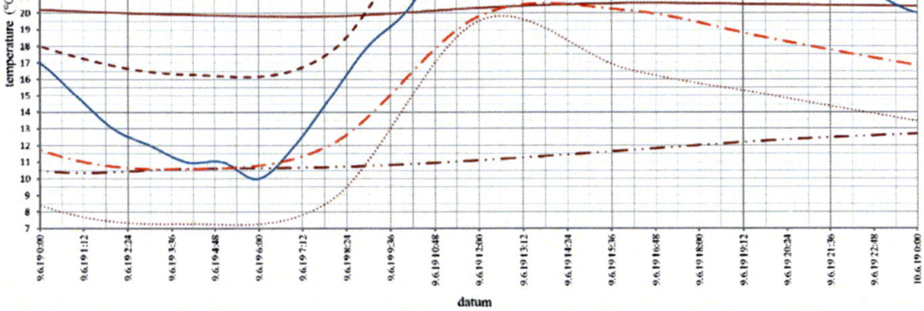

Fig. 3. Graphs of selected temperatures on the surface of the stone window sill of the stained glass window as of 9.6. 2018

5) the most risky period in terms of the frequency of water vapor condensation in the vicinity of the detail was recorded in both years of measurement during the March to June (Table 3);

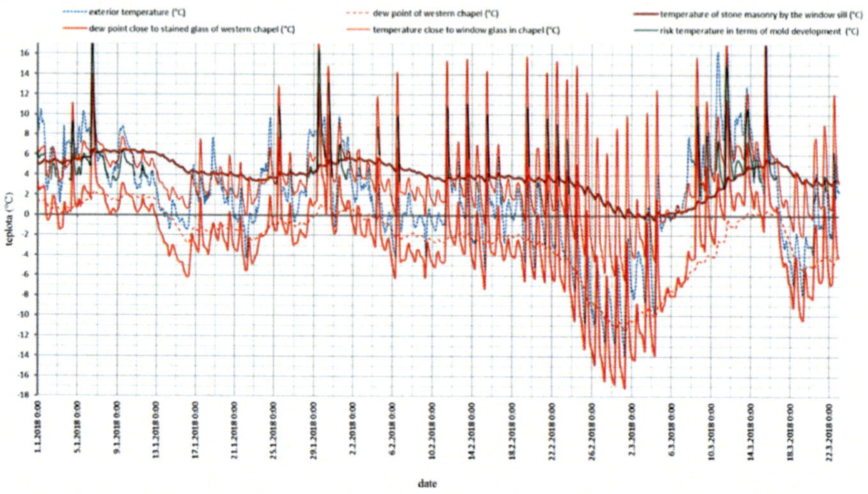

Fig. 4. Waveforms of selected temperatures and dew point near the contact between the glass window and the window sill

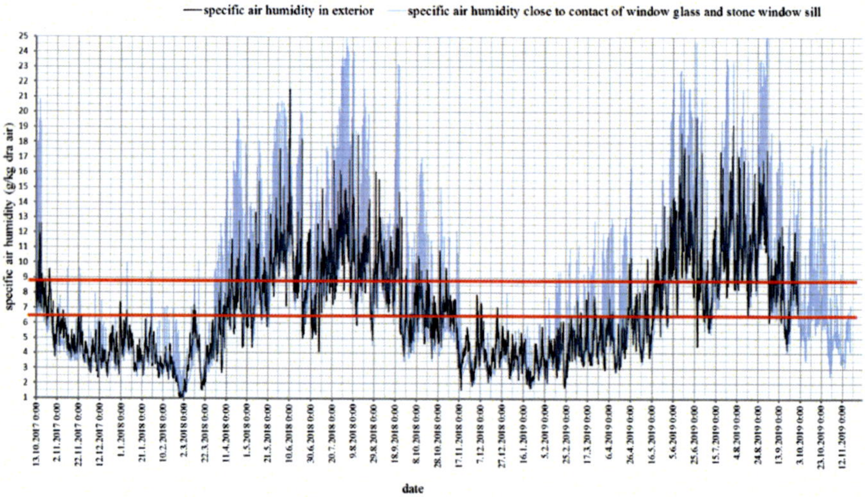

Fig. 5. Specific humidity of selected areas from October 2017 to November 2019

6) the most intense condensation was recorded in vicinity of the detail in October and November 2017, January, March, April and October 2018 and February to June and October 2019;
7) in the area of detail, there is a sharp periodic change of the specific air humidity (Fig. 5, 6 and 7) compared to the chapel space, where its changes are significantly attenuated;

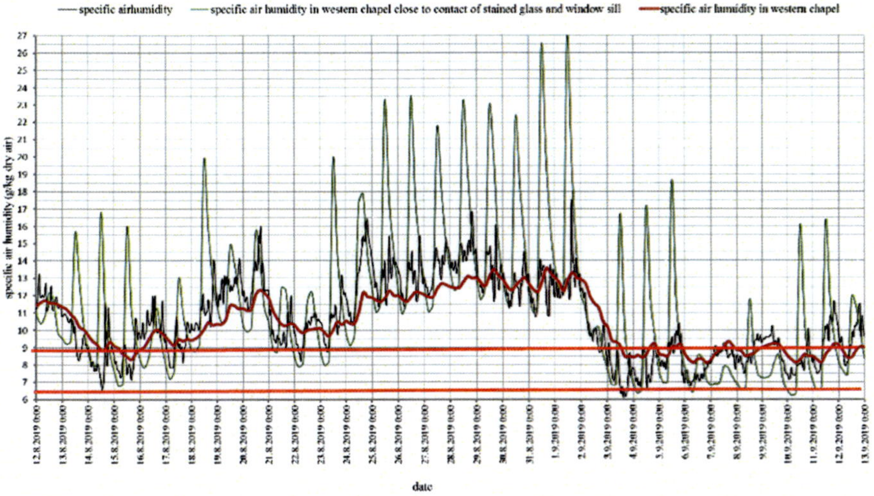

Fig. 6. Specific humidity of the western chapel in the immediate vicinity of the contact between the window panel (stained glass) and the stone sill in the period from 12 August to 13 September 2019

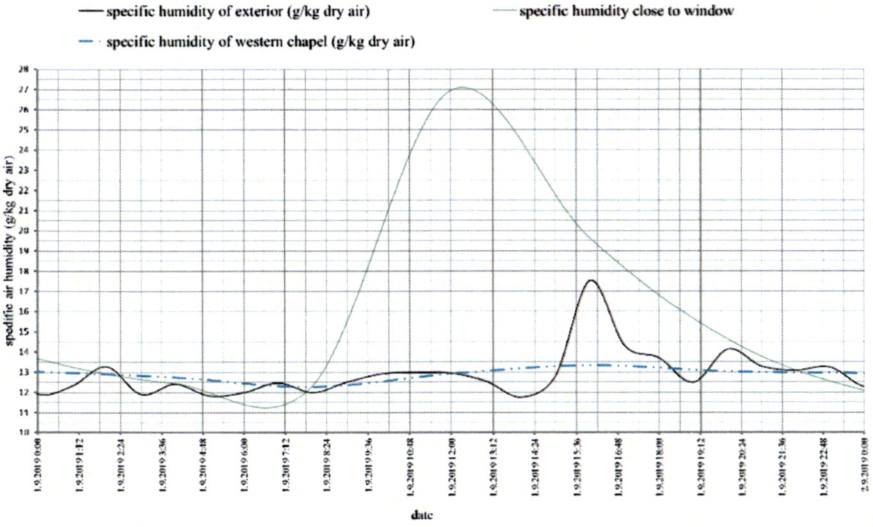

Fig. 7. Specific humidity of the western chapel in the immediate vicinity of the window panel (stained glass) and the stone sill in the period from 12 August to 13 September 2019

8) variations in specific air humidity in the detail area reached up to 16 g/kg dry air at a time interval of 5 h (Fig. 7);

9) correlation of functional curves of the relative humidity of the interior of the side chapel, even in the vicinity of detail has a lower value (below 0.5) than in the case of correlations of temperature functions of interior and exterior (above 0.8).

Table 2. Condensation of water vapor in place of window stained glass and stone sill

		Space of western chapel		Immediate proximity to the glass window	
Obtained values	Temperature	Dew point	Relative humidity	Specific humidity	Temperature
	[°C]	[°C]	[%]	[g/kg d. a.]	[°C]
Min.	−0,4	– 17,1	31,8	1,6	−14,3
Max.	27,9	29,7	80,2	13,6	37
Average	13,9	6,5	59,3	6,4	12,5

Table 3. Frequency of water vapor condensation at the point of contact of window stained glass and stone window sill

Year	January	February	March	April	May	June	July	August	September	October	November	December
2017	X	X	X	X	X	X	X	X	X			
2018												
2019												X

5 Conclusions

The measurements showed that the micro-space in the immediate vicinity of the window panels of churches and chapels has significantly different climatic properties than the open and ventilated space of church navels and chapels. This micro-space is characterized by a frequent and periodic increase of the condensation point above the temperature of the surfaces of related structures (glass, stained glass, plaster or stone elements of the sill). The likely major influence that causes external thermal-humidity changes in this micro-space is the sunlight acting on the window glass panels. Structural elements (plasters, stone, bricks) are absorbing moisture not only by the absorption of condensation water, but also by the absorption of moisture from the air according to the characteristics of the sorption isotherms for the material. The change in the moisture conditions in the sill also causes a change in the thermal insulation characteristics of the masonry and destruction due to the change of the water state in the ice in winter. We consider the most harmful phenomenon to wall paintings, valuable plasters and wooden window elements to be rapid changes in the specific humidity of the microclimate, which cause different dilatation movements of the materials present and stresses at their mutual contacts. The found points pointed out that it is necessary to pay special attention to the behavior and influencing of the microclimate in the micro-space of window panels and related sill in order to solve the question of the interior microclimate of sacral buildings.

Acknowledgements. The activity presented in the paper is part of the research grant NAKI II DG18P02OVV012 project of the Ministry of Culture of the Czech Republic.

References

ČSN EN ISO 12572: Tepelně-vlhkostní chování stavebních materiálů a výrobků – Stanovení prostupu vodní páry (2002)

ČSN EN ISO 7783-2: Paints and varnishes – Coating materials and coating systems for exterior masonry and concrete – Part 2: Determination and classification of water-vapor transmission rate.

Černý, M., Němeček, M.: Mikroklima v historických interiérech. NPÚ Praha (2011). ISBN 978-80-87104-82-8

Pavlík, Z., Balík, L., Kudrnáčová, L., Maděra, J., Černý, R.: Chapel of cemetry church of all saints in sedlec–long-term analysis of hygrothermal conditions. In: Thermophysics. AIP, Santa Fe (2017)

Kudrnáčová, L, Balík, L.: Sledování mikroklimatu v prostorách kostela. In: Sanace a rekonstrukce 2017, WTA Praha, pp. 71–77 (2017). ISBN 978-80-01-06347-7

Balík, L., Kolísko, J.: The Influence of Incoming Parameters on Accuracy of Determination Water Vapour Resistance Factor of Lime and Modified Mortars. Klokner Institute CTU in Prague (2009)

Balík, L., Nenadálová, Š., Kolísko, J.: Dependence of brickwork natural drying speed on surroundings conditions. In: Sanace a rekonstrukce staveb 2012. Brno, 06.11.2012–07.11.2012. Vědeckotechnická společnost pro sanace staveb a péči o památky WTA CZ, Praha, p. 219 (2012). ISBN 978-80-02-02414-9

Structural Behaviour of Clay Brick Lime Mortar Masonry Walls Under Lateral Cyclic Loading in Dry and Wet Conditions

A. Y. Elghazouli[1] (✉), D. V. Bompa[1,2], S. A. Mourad[3], and A. Elyamani[4]

[1] Department of Civil and Environmental Engineering, Imperial College London, London, UK
a.elghazouli@imperial.ac.uk

[2] Department of Civil and Environmental Engineering, University of Surrey, Guildford, UK

[3] Department of Structural Engineering, Cairo University, Giza, Egypt

[4] Department of Archeological Conservation, Cairo University, Giza, Egypt

Abstract. This paper examines the experimental structural response of clay brick lime mortar masonry walls in wet and ambient-dry conditions. The properties of fired-clay bricks and hydraulic lime-mortar materials are selected to resemble those of existing heritage masonry structures in Historic Cairo. The investigation includes tests on square panels under diagonal compression, and large-scale walls subjected to gravity loading and in-plane lateral cyclic displacements. In addition to the conditioning type, the effectiveness of strengthening with helical bars in horizontal bed joints is also investigated. Implications of embedding helical bars in lime mortar as well as the provision of end anchorages are assessed. The complete load-deformation response of the large-scale members is also evaluated, including the main behavioural characteristics and failure modes. The results show that moisture has a notable effect on the main mechanical properties and overall structural response of such masonry components. For the panels subjected to diagonal compression, the strength reduction under wet conditions is shown to be more than 40% compared to the dry counterparts. For the large-scale walls, subjected to combined lateral loading and precompression, this reduction is significantly lower but can exceed 10%. It is also shown that the provision of helical bars can, depending on their end anchorage and arrangement, double the diagonal tension strength of masonry and offset the adverse effects occurring due to moisture.

Keywords: Masonry walls · Lime mortar · Clay bricks · Wet conditions · Diagonal tension · Lateral deformation · Cyclic loading

1 Introduction

Heritage masonry is characterised by high material heterogeneity, irregular bearing elements and degradation of constituents due to environmental effects. It is recognised that environmental wet-dry cycling can affect both the mechanical characteristics of masonry as well as plaster and rendering (Cotic et al. 2013; Gentilini et al. 2015). Unreinforced masonry walls can develop shear- or flexure-governed failures depending on

I. Vayas and F. M. Mazzolani (Eds.): PROHITECH 2021, LNCE 209, pp. 164–174, 2022.
https://doi.org/10.1007/978-3-030-90788-4_16

their geometry, aspect ratio and boundary conditions. The diagonal shear and sliding strength of URM is greatly dependent on the mortar-brick interaction. In general, the interface mortar-brick shear response greatly depends on the moisture content, porosity, mortar strength, and conditioning type (Mazzotti et al. 2014; Bompa and Elghazouli 2020a), with saturation causing a deterioration in strength and stiffness (Franzoni et al. 2015; Bompa and Elghazouli 2020b).

The ductility and robustness of lateral-resisting masonry elements are of particular importance for structures subjected to seismic loading (Elghazouli 2016). To this end, previous studies have shown that the geometry, coupling level, and material properties, influence not only the strength but also the deformation capacity and cyclic degradation of masonry (Vasconcelos and Lourenço 2009). Comparative cyclic tests on dry low-strength mortar masonry also indicated a modification in failure mode from combined rocking-shear to rocking when the height-to-depth aspect ratio increased from 0.7 to 1.4, representative of squat and slender elements, respectively (ElGawady et al. 2005). Other tests on small-scale dry historic masonry exhibited diagonal-cracking failures, reaching ultimate drifts below 0.4% (Capozucca 2011).

Although experimental assessments on ambient-dry masonry walls representative of heritage structures have been carried out (Calvi et al. 1996), studies to investigate the lateral cyclic performance of wet large-scale masonry are lacking, and the influence of moisture on the diagonal cracking response has also not been examined. Such masonry structures are typically also not specifically designed for resisting lateral seismic loading.

This paper therefore presents an experimental investigation into the material and structural response of ambient-dry and wet masonry elements consisting of fired clay brick and hydraulic lime mortar, which are used in many historic structures. The properties of the fired-clay bricks and lime-mortar materials were selected to resemble those of masonry components investigated in a wider research programme on the management and conservation of heritage masonry structures in Historic Cairo.

2 Experimental Programme

2.1 Material Properties

Commercial fired clay facing solid bricks with measured sizes of 229 × 110 × 66 mm were used for the construction of all specimens. The measured average compressive strength of brick units parallel to the testing bed, obtained from a minimum of three tests, was about 15.5 MPa in both dry and wet conditions, with the latter corresponding to full submersion in water for 48 h. From readily available materials, this type of fired-clay bricks has the closest physical and mechanical properties to those from the Mausoleum of Fatima Khatun (Umm al-Salih) built in the 13th century in Cairo, which is assessed in the project. Site surveys indicated that: (i) 'red' bricks (used for the foundation) have a compressive strength (f_b) of about 5.2 MPa and water absorption $wa = 27.5\%$, (ii) 'light brown' bricks have an $f_b = 14.7$ MPa and $w_a = 18.13\%$ and (iii) 'dark brown' bricks have an $f_b = 22.7$ MPa and $w_a = 13.4\%$.

The characteristics of the selected fired-clay bricks are hence within the range of those from the site survey, typically found in such heritage masonry, and can be used for comparative assessments and structural repair studies. The moisture content of both conditioning cases for bricks was assessed. The ambient-dry samples and those submersed in tap water for a minimum of 48 h, were dried in an oven for 6 h at 60 °C and another for 18 h at 105 °C until the sample mass was relatively constant. The moisture content of ambient-dry bricks was 0.07% by weight (wt.%), and 10.46 wt.% for those submersed in water.

Mortars incorporating natural hydraulic lime (NHL5) with a binder-to-aggregate ratio of 1:3 were selected for the study as these are typically used for conservation works. These are singular binders which combine hydraulic with air setting, obtained by carbonation with atmospheric CO2. The free lime Ca(OH)2 is above 15%, whilst the sulphates content is below 2%.

Compressive strengths on a minimum of 5 samples for each masonry specimen were determined from compression tests on ambient-dry and wet cubes (50 × 50 × 50 mm). The resulting strengths of the mortars in ambient-dry conditions were between $f_{lm} = 2.08 - 5.11$ MPa, and in wet conditions were between $f_{lm} = 0.70 - 2.37$ MPa (Table 1 and 2). An average reduction in elastic modulus in the range of 50% was also typically observed due to moisture. Additionally, the moisture content of both conditioning cases for NHL mortar samples was assessed following the same procedure as for the bricks. The moisture content of the lime mortars was 2.54% by weight (wt.%) for ambient-dry samples and 10.80 (wt.%) for those submersed in water.

2.2 Diagonal Panels

Diagonal panel tests on single leaf square panels of $b \times h \times t = 710 \times 710 \times 110$ mm in wet and air-dry conditions were carried out in order to assess the diagonal tensile strength of masonry (Fig. 1). These were either non-strengthened or strengthened with helical bars. The specimen reference adopts the following formats: DB-Xx for non-strengthened specimens, and DHx-Yy for strengthened specimens. For non-strengthened members (DB-Xy), X refers to conditioning D-dry or W-wet, and y to testing sequence, whilst for the strengthened specimens (DHx-Yy), x stands for end anchorage (N – not present, A – end anchorage), Y refers to conditioning as above, and y is for the testing sequence.

The panels for diagonal testing, as depicted in Fig. 1a, b, had both horizontal and vertical lime mortar joints with an average thickness of 9 ± 1.5 mm. After the last course of bricks was laid, the specimens were kept in laboratory conditions. Plastic sheets were used to cover the specimens at early curing, and they were then tested at an age of 30–35 days. Helical stainless-steel reinforcing bars for masonry repair were used in four diagonal panels and two large-scale tests.

Four bars of 4.5 mm and proof stress around 1100 N/mm^2 were used per specimen. These were directly incorporated in the third and the sixth bed joints of the panels as shown in yellow in Fig. 1b. Two bars were embedded in the lime mortar laid in each of the two horizontal joints, at 30 mm from the face of the panel. The helical bars from Specimens DHA were provided with end anchorage systems to prevent slip, whilst Specimens DHN had only straight bars.

Prior to testing, the wet specimens were submersed 3/5 of depth in water and were sprinkled with a hose from the top. After testing, the walls were dismantled, and samples were extracted to assess the moisture content. After each brick and mortar joints were weighed, all components were dried in an oven for 6h at 60 °C for at least 18h at 105 °C

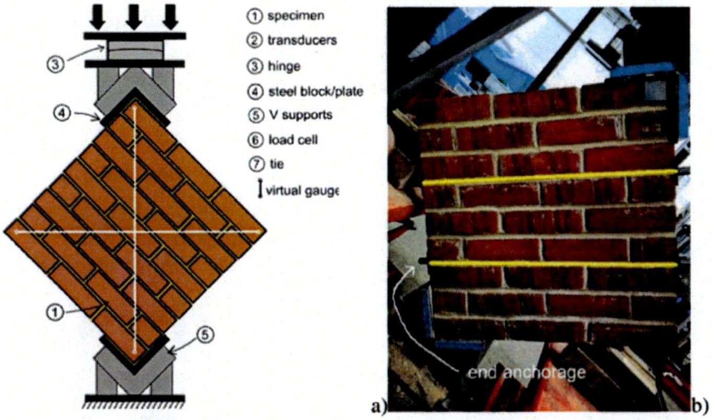

Fig. 1. (a) Arrangement of diagonal tests, (b) Strengthening details

until reaching a constant mass. The moisture distribution results indicated that the same moisture content of 10.7% ± 0.2 wt was consistently obtained in all nine brick courses. Cylinders of diameter d = 69 mm and height of around h = 145 mm were extracted, after testing, from undamaged areas of the diagonal panels to assess the compression properties of the masonry. The compressive strengths of masonry (f_m) from cylinder testing are given in Tables 1 and 2.

The tests on diagonal panels were carried out in a rig which included a main loading transfer frame with a 1000 kN Instron actuator and a connected load cell. As shown in Fig. 1a, the specimens were positioned and loaded through V-shaped supports. External transducers were used for secondary measurements alongside the displacement recordings provided by the machine, coupled with detailed data from a Digital Image Correlation (DIC) system (Elghazouli et al. 2021).

2.3 Large Walls Under Combined Loading

Single leaf masonry walls of 1910 mm length, 1300 mm height and 110 mm width were built using fired-clay bricks and lime mortars, as described above. A total of four selected tests on walls under combined loading are presented in this section. All specimens were tested under lateral cyclic loading and a representative gravity load of about 1.0 MPa.

The specimen reference adopts the format W*X*-*Y*, where W stands for wall, *X* represents the specimen type: (B for non-strengthened specimens or H for members with helical bars), and *Y* represents the conditioning (D for air-dry or W for wet). For example, WB-D is a non-strengthened ambient-dry wall. From the two non-strengthened walls, one was tested in wet conditions and the other in dry conditions. The same approach was applied to the two strengthened walls.

Stretcher/running bond was used to construct the 16 course high walls (Fig. 2). To achieve the desired specimen height, the mortar joint thickness was in the range of 8–10 mm. For the strengthened walls, a total of four bars were inserted in two bed joints located above the fifth and eleventh brick course from the bottom to the top of the wall. The helical bars were positioned directly in the lime mortar layer. All bars were provided with end anchorages as a secondary force transfer mechanism along with the bar-mortar bond. The properties of the helical bars used in the large walls were identical to those used for the diagonal panels described above.

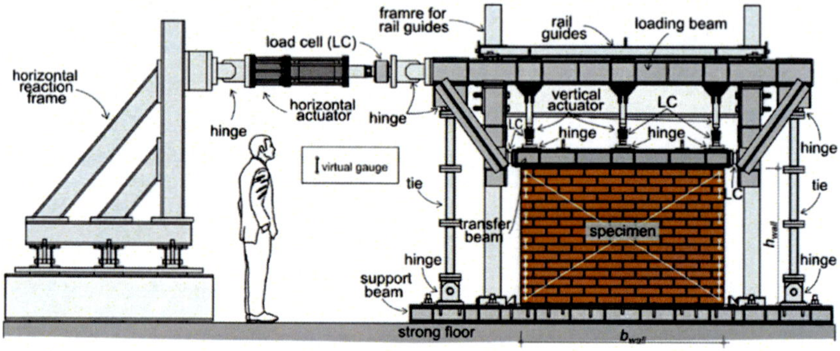

Fig. 2. Testing arrangement for large scale walls

To enable wetting through capillary absorption of wet walls, after the specimens were placed in the testing rig and the prestressing ties were removed, a water tank was built in-place which was connected to the first course of bricks. In addition to the water tank, a pipe/sprinkler system mounted at the top of the wall was manufactured and was used to accelerate the wetting process. The conditioning procedure was identical for both non-strengthened walls and those with helical bars. The moisture content of samples from the tested walls was assessed similarly to the diagonal panels. The average moisture content was 11.1% by weight with a standard deviation was 0.34%, indicating that the moisture was evenly spread.

The rig shown schematically in Fig. 2 was designed to enable realistic experimental assessment of the ultimate behaviour of large-scale members subjected to lateral cyclic displacements and co-existing axial loading. The specimens were directly supported by a 25 mm thick plate and connected by means of prestressed Ø20 mm bolts to a supporting steel beam. The latter was connected to the strong floor by means of 4 × Ø33 mm prestressing ties to avoid sliding and overturning under lateral loads. At the top of the wall, three 120 kN Enerpac actuators were connected to a steel loading beam and were used to apply the gravity load by means of unidirectional hinges to the specimen through a transfer steel beam.

For the application of the lateral loading, another 250 kN Instron actuator was placed horizontally and connected to the reaction frame. After the application of the constant vertical load, corresponding to an axial stress around 1.0 MPa, the lateral deformations were applied based on a pre-defined quasi-static cyclic history. A set of three cycles were applied for each deformation level, corresponding to a drift of 0.025, 0.050, 0.075, 0.10, 0.15, 0.20, 0.25, 0.30, 0.35, 0.40, 0.50, 0.80, 1.0, 1.25, 1.50 (%). The displacement rate and loading procedure were chosen based on recommendations available in the literature (Petry and Beyer 2015; Magenes et al. 2008).

3 Test Results and Observations

3.1 Diagonal Panels

Table 1 depicts the main results of the tested panels, whilst Fig. 3 illustrates a typical crack pattern and failure characteristics for selected members. In the diagonal panel tests, when the principal tensile stresses perpendicular to the compressed strut reached the maximum tensile strength, failure occurred due to bond failure between the brick unit and the mortar in the joints. In terms of the crack kinematics, similar behaviour was observed during tests on both air-dry and wet specimens. Close inspection of the results showed that the moisture reduced both the stiffness and the specimen strength, albeit with different extents as a function of mortar strengths (Elghazouli et al. 2021).

Table 1. Main results of diagonal panel tests

Specimen	P(kN)	α(°)	f_m(MPa)	f_{lm}(MPa)	Specimen	P(kN)	α(°)	f_m(MPa)	f_{lm}(MPa)
DB-D1	48.0	39.7	5.70	2.53	DHN-D1	66.2	31.3	5.70	2.13
DB-D2	53.1	38.4	5.70	2.25	DHN-D2	63.4	35.6	5.70	2.09
DB-D3	52.3	40.3	6.06	2.56	DHA-D1	128.3	42.7	6.06	2.09
DB-W1	26.5	39.3	4.28	0.70	DHA-W1	62.9	34.6	4.78	0.98
DB-W2	31.5	39.8	4.28	0.70					

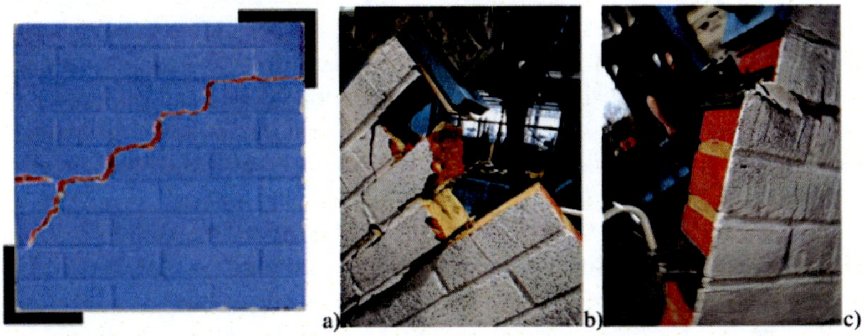

Fig. 3. (a) typical crack pattern, (b) failure characteristics of DHN Specimens, (c) failure characteristics of DHA Specimens

The response of dry panels was generally characterised by loss of bond at a horizontal bed joint at the centre of the panel at around 80–90% of the peak capacity Pu. The critical crack propagated following the top and bottom plates through vertical and horizontal bed joints in a stepped mode. As shown in Fig. 4a, failure largely occurred due to a single diagonal crack. For wet panels, the failure patterns resembled those of their corresponding dry counterparts. However, the first strain concentrations associated with loss of bond at the brick-to-mortar interface occurred at 73–75% of Pu.

Dry specimens provided with helical bars and no end anchorage (DHN-D) showed a modest strength increase of about 12–17% compared to their non-strengthened counterparts, as indicated in Table 1. The strength increase was modest because the critical crack passed only through one of the reinforced bed joints, and the bond between the lime mortar and helical bars was relatively low. In contrast, when the bars were provided with end anchorages the strength increase was by a factor of 2.28 (DHA-D) compared to the non-strengthened dry specimens. When helical bars were used with anchorage and in wet conditions, the contribution of the strengthening system was largely offset by the presence of moisture (Specimen DHA-W1).

3.2 Large Scale Walls

The complete load-drift (P-Δ) curves of the tested walls are shown in Fig. 4. The crack patterns at failure obtained from DIC recordings are illustrated in Fig. 5, and the main test results are given in Table 2. The P-Δ response of the air-dry Specimen WB-D and its envelope are illustrated in Fig. 4a, and the failure patterns are shown in Fig. 5a. The lateral deformation at peak corresponded to a drift of $\Delta_{peak+} = 0.53\%$ and $\Delta_{peak-} = 0.53\%$ in the push (positive) and pull (negative) cycle (Table 2). The corresponding maximum lateral load carrying capacity was $P_{u+} = 146.3$ kN and $P_{u-} = 142.2$ kN, respectively. The transition between elastic and inelastic stiffness occurred at a lateral load (P) around 80 kN in both loading directions. This corresponded to a drift ratio of about $\Delta = 0.1\%$ and was associated with initial signs of cracking. At this stage, the behaviour was largely governed by flexure.

The first visible diagonal crack occurred in the negative (pull) cycle at a drift level $\Delta = 0.53\%$ and closed during unloading. At the same drift level in the push (positive cycle), a diagonal crack nearly perpendicular to that occurring from the pull cycle was initiated. This corresponded to the maximum capacity of the specimen. At the third displacement cycle in the positive direction, the specimen failed and was unable to reach the load attained in the previous cycles. The ultimate drifts, assumed to correspond to 20% reduction in load carrying capacity (Bompa and Elghazouli 2019), were $\Delta_{u+} = 0.61\%$ and $\Delta_{u-} = 0.58\%$.

Specimen WB-W was tested in wet conditions. The full load-drift (P-Δ) and envelope curves are depicted in Fig. 4b, while the crack patterns at failure are shown in Fig. 5b. A noticeable reduction in stiffness started at about 35kN ($\Delta = 0.07\%$) then at 95 kN ($\Delta = 0.25\%$). The initial stiffness was $K_0 = 75.6$ kN/mm. The maximum lateral load was $P_{u+} = 127.0$ kN and $P_{u-} = 129.2$ kN in the two directions, corresponding to drifts of $\Delta_{peak+} = 0.35\%$ and $\Delta_{peak-} = 0.31\%$ in the push (positive) and pull (negative) cycles, respectively. The behaviour of WB-W was initially governed by flexure. The first signs of diagonal cracking were observed at around $\Delta = 0.20\%$, with the maximum capacity reached closely after. Failure was characterised by sliding along the diagonal cracks.

Specimen WH-D had helical bars and was tested under combined axial load and lateral cyclic displacement, in dry condition. The complete load-drift (P-Δ) and crack patterns at failure are illustrated in Fig. 4c and Fig. 5c. The maximum lateral load carrying capacity was $P_{max,pos} = 134.1$ kN and $P_{max,neg} = -142.2$ kN, respectively. This corresponded to a drift of $\Delta_+ = 0.55\%$ and $\Delta = -0.55\%$ in the push (positive) and pull (negative) cycle. The first inclined crack occurred in the second pull (negative) cycle

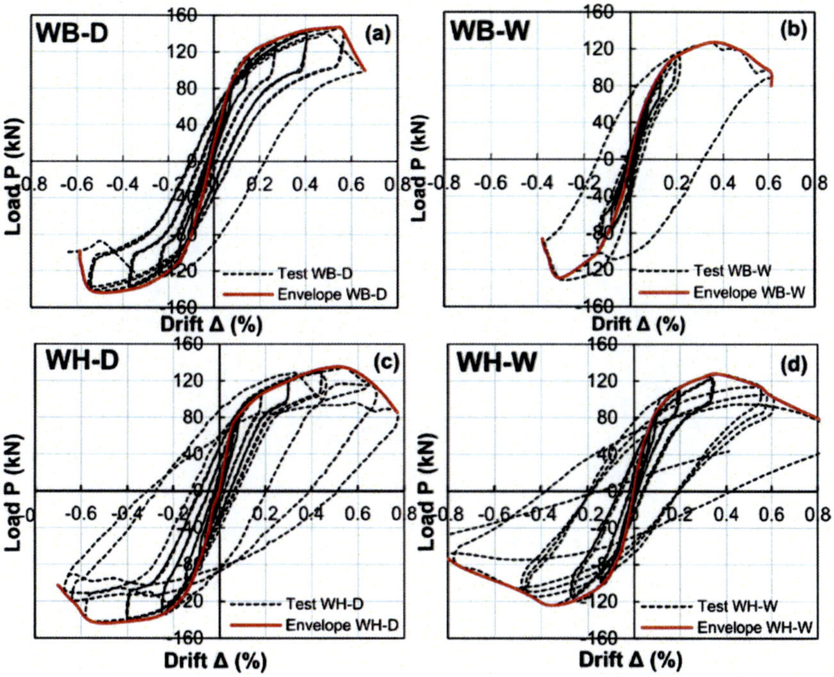

Fig. 4. Load-drift (P-Δ) curves of (a) WB-D, (b) WB-W, (c) WH-D, (d) WH-W

Table 2. Main test results from tests on large scale walls

	P_{u+}(kN)	P_{u-}(kN)	P_u(kN)	Δ_{peak+}(kN)	Δ_{peak-}(kN)	Δ_{u+}(%)	Δ_{u-}(%)	K_{test}(kN/ mm)	f_m(MPa)	f_{lm}(MPa)
WB-D	146.3	142.2	144.3	0.53	0.53	0.61	0.58	68.4	6.59	5.11
WB-W	127.0	129.2	128.1	0.35	0.31	0.57	0.36	54.6	6.04	2.36
WH-D	134.1	142.2	138.2	0.55	0.55	0.70	0.67	89.6	6.06	3.12
WH-W	127.6	124.5	126.1	0.37	0.37	0.61	0.58	96.0	4.87	1.62

at drift around a drift of 0.20%. As the helical bars kept the crack closed at displacement reversal, weaker regions developed outside of the shear-reinforced regions. Hence, a sliding crack occurred below the third brick course from top to bottom. This was followed by an inclined crack in the push (positive) cycle. Ultimately, failure was due to the sliding of the top non-reinforced region. The ultimate drifts were $\Delta_{u+} = 0.70$ and $\Delta_{u-} = 0.67$, in the positive and negative loading direction, respectively.

Specimen WH-W had helical bars and was tested in wet condition. The complete load-drift (P-Δ) and crack patterns at failure are illustrated in Fig. 4d and 5d, respectively. The maximum lateral load carrying capacity was $P_{max,pos} = 127.6$ kN and $P_{max,neg} = -$ 124.5 kN, respectively. This corresponded to drifts of $\Delta_+ = 0.37\%$ and $\Delta = -0.37\%$ in the positive and negative cycles. The first inclined crack occurred in the third negative cycle at a drift of 0.38%. For the wet wall, the presence of the helical bars seems to have

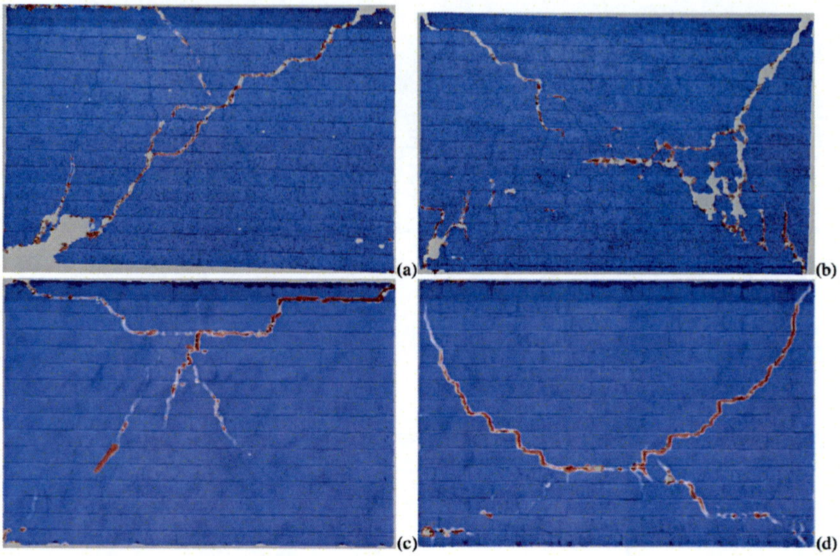

Fig. 5. Crack patterns at failure for Specimens: (a) WB-D, (b) WB-W, (c) WH-D, (d) WH-W

created a weak region at the bottom bed joint. The horizontal crack where the two inclined cracks intersect correspond to the position of the helical bars.

Comparing the response of the wet (WB-W and WH-W) and dry (WB-D and WH-D) specimens, the influence of moisture on the capacity of the members becomes evident. For the wet walls, the average reduction in lateral strength was about 11%, whilst the reduction in stiffness was around 20% compared to the dry walls. These ranges are significantly lower both in terms of stiffness and strength than the differences between wet and dry properties obtained from the diagonal panel tests. This is attributed to the different loading and failure conditions with respect to mortar and brick interaction.

Specimens WB-D and WB-W had a flexure-governed response characterised by opening of an interface gap between the supporting plate and the first brick course. The response of both WH-D and WH-W walls was largely symmetric and governed by shear, ultimately failing in diagonal tension with the two diagonal cracks intersecting outside of the region reinforced with helical bars. Unreinforced masonry can develop such response, as often explained by aspects of non-linear elastic behaviour (Petry and Beyer 2015). This is mainly attributed to the reduction of the compression zone with the increase in lateral displacement and the associated decrease in effective stiffness of the member. The contribution of the tension side of the wall also gradually diminishes as cracks open along the horizontal bed joints.

As discussed above, the ductility of the wet walls, in terms of ultimate drift, was broadly similar or lower than that of their dry counterparts, primarily influenced by the post-peak kinematics. Given the influence of moisture on the stiffness and effective yield, perhaps a more representative measure of ductility would be an ultimate-to-yield drift ratio ($\mu_\Delta = \Delta_u/\Delta_y$). The yield drift can be estimated from a simplified bilinear P-Δ curve by equating areas under the average test P-Δ envelope and the bilinear P-Δ

idealisation. Using this approach, the drift ductility ratios (μ_{Δ}) are estimated as 4.0 for WB-D, 3.5 for WB-W, 4.8 for WH-D and 4.2 for WH-W. Overall, direct experimental comparison between the cyclic response of the wet and dry specimens shows the negative effect of moisture on the stiffness, capacity and ductility of masonry walls subjected to initial gravity and increasing lateral loads.

4 Concluding Remarks

This paper presented an experimental investigation into the response of ambient-dry and wet clay-brick/lime-mortar masonry elements representative of those in some historic structures. Several tests on large-scale walls subjected to gravity load and lateral displacements as well as square panels under diagonal compression, were described. Although, as expected, there is significant inherent variability in the properties of masonry elements, depending on the constituent materials and construction conditions, this investigation offered an insight into the influence of moisture on the behaviour.

A direct comparison between strength properties of dry and wet specimens tested under diagonal compression showed that moisture reduced the strength on average by about 47% of the dry specimens, respectively. Full-field DIC measurements also showed that the cracking load was reduced by moisture with the brick-to-mortar interface bond loss occurring at around 75% and 85% of ultimate for the wet and dry specimens, respectively. Specimens provided with helical bars and no end anchorage showed a modest strength increase compared to their non-strengthened counterparts. In contrast, when the bars were provided with end anchorages, the strength increase was by a factor of 2.28. For the cylindrical cores tested in this study, the reduction in the elastic modulus and compressive strength was on average in the range of 13% and 15%, respectively, between the dry and wet cases.

Large-scale tests on masonry walls under cyclic loading and practical levels of gravity load indicated up to 11% reduction in lateral strength in the presence of moisture. Although these ranges are significantly lower than those obtained from the diagonal panel tests, due to the different loading and failure conditions, the combined results show that notable reductions in elastic and ultimate properties would occur when the masonry is wet. All non-strengthened specimens tested under cyclic loading had a brittle failure in diagonal tension. For the dry strengthened wall, the helical bars kept the crack closed at displacement reversal, and weaker regions developed outside of the reinforced regions. For the wet wall, the presence of the helical bars created a weak region at the bottom bed joint, similarly shifting the critical cracks outside the reinforced region. The results and failure mode point to the need for further examination of various effective arrangements and extent of the helical bars in such large elements. Overall, the ultimate deformations of the wet cyclic walls were either broadly similar or lower than that of the dry counterparts. The wet walls also had ductility ratios about 10% lower than their dry counterparts.

Acknowledgements. The study was supported by the Arts and Humanities Research Council of the UK Research and Innovation agency, within the project "Interdisciplinary approach for the management and conservation of UNESCO World Heritage Site of Historic Cairo - Application to Al-Ashraf Street", Grant No. AH/R00787X/1. The authors would like to acknowledge the support provided by the technical staff of the Structures Laboratories at Imperial College London, particularly Mr. P. Crudge, Mr. L. Clark, and Mr. T. Stickland.

References

Bompa, D.V., Elghazouli, A.Y.: Inelastic cyclic behaviour of RC members incorporating threaded reinforcement couplers. Eng. Struct. **180**, 468–483 (2019). https://doi.org/10.1016/j.engstruct.2018.11.053

Bompa, D.V., Elghazouli, A.Y.: Experimental and numerical assessment of the shear behaviour of lime mortar clay brick masonry triplets. Constr. Build. Mater. **262**, 120571 (2020a). https://doi.org/10.1016/j.conbuildmat.2020.120571

Bompa, D.V., Elghazouli, A.Y.: Compressive behaviour of fired-clay brick and lime mortar masonry components in dry and wet conditions. Mater. Struct. **53**, 60 (2020b). https://doi.org/10.1617/s11527-020-01493-w

Calvi, G.M., Kingsley, G.R., Magenes, G.: Testing of masonry structures for seismic assessment. Earthq. Spectra **12**(1), 145–162 (1996). https://doi.org/10.1193/1.1585872

Capozucca, R.: Experimental analysis of historic masonry walls reinforced by CFRP under in-plane cyclic loading. Compos. Struct. **94**(1), 277–289 (2011). https://doi.org/10.1016/j.compstruct.2011.06.007

Cotic, P., et al.: Effect of moisture on the reliability of void detection in brickwork masonry using radar, ultrasonic and complex resistivity tomography. Mater. Struct. **46**(10), 1723–1735 (2013). https://doi.org/10.1617/s11527-012-0011-3

ElGawady, M.A., Lestuzzi, P., Badoux, M.: In-plane seismic response of URM walls upgraded with FRP. J. Compos. Constr. **9**(6), 524–535 (2005). https://doi.org/10.1061/(ASCE)1090-0268(2005)9:6(524)

Elghazouli, A. (ed.): Seismic Design of Buildings to Eurocode 8. CRC Press, Boca Raton (2016)

Elghazouli, A.Y., Bompa, D.V., Mourad, S.A., Elyamani, A.: In-plane lateral cyclic behaviour of lime-mortar and clay-brick masonry walls in dry and wet conditions. Bull. Earthq. Eng. **19**, 5525–5563 (2021). https://doi.org/10.1007/s10518-021-01170-5

Franzoni, E., Gentilini, C., Graziani, G., Bandini, S.: Compressive behaviour of brick masonry triplets in wet and dry conditions. Constr. Build. Mater. **82**, 45–52 (2015). https://doi.org/10.1016/j.conbuildmat.2015.02.052

Gentilini, C., Franzoni, E., Graziani, G., Bandini, S.: Mechanical properties of fired-clay brick masonry models in moist and dry conditions. In: Key Engineering Materials, vol. 624, pp. 307–312. Trans Tech Publications Ltd. (2015). https://doi.org/10.4028/www.scientific.net/KEM.624.307

Magenes, G., Morandi, P., Penna, A.: D 7.1 c Test results on the behaviour of masonry under static cyclic in plane lateral loads. Test report, ESECMaSE Poject, University of Pavia, EURCENTRE, Italy (2008)

Mazzotti, C., Sassoni, E., Pagliai, G.: Determination of shear strength of historic masonries by moderately destructive testing of masonry cores. Constr. Build. Mater. **54**, 421–431 (2014). https://doi.org/10.1016/j.conbuildmat.2013.12.039

Petry, S., Beyer, K.: Force–displacement response of in-plane-loaded URM walls with a dominating flexural mode. Earthq. Eng. Struct. Dynam. **44**(14), 2551–2573 (2015). https://doi.org/10.1002/eqe.2597

Vasconcelos, G., Lourenço, P.B.: In-plane experimental behavior of stone masonry walls under cyclic loading. J. Struct. Eng. **135**(10), 1269–1277 (2009). https://doi.org/10.1061/(ASCE)ST.1943-541X.0000053

Shear-Compression Failure Envelopes for Clay Brick Lime Mortar Masonry Under Wet and Dry Conditions

D. V. Bompa[1,2] and A. Y. Elghazouli[2(✉)]

[1] Department of Civil and Environmental Engineering, University of Surrey, Guildford, UK
d.bompa@surrey.ac.uk

[2] Department of Civil and Environmental Engineering, Imperial College London, London, UK
a.elghazouli@imperial.ac.uk

Abstract. This study describes an experimental investigation into the fundamental mechanical properties of masonry incorporating fired-clay bricks and hydraulic lime mortars under ambient-dry and wet conditions, used in many historic structures. The investigation includes tests on cylindrical cores in compression, square panels in diagonal compression and triplets in shear. Apart from conventional measurements, the use of digital image correlation techniques enables a detailed assessment of the influence of moisture on the constitutive properties and on the response kinematics of the member, including behaviour characteristics of the brick-mortar interfaces, crack propagation in tension and confinement effects in compression. The tests results show that on average the uniaxial compressive strength of wet brick–mortar cylinders is about 15% lower than their dry counterparts, whilst the shear strength of wet triplets is about 20% lower than of those in dry conditions. The diagonal tension strength of wet panels depends largely on the mortar properties and is about 30% lower on average than that of the dry panels. Based on the test results, the material shear-compression strength envelopes for wet and dry masonry are evaluated and compared with existing assessment models.

Keywords: Clay bricks · Lime mortar · Compression · Shear response · Diagonal tension

1 Introduction

Clay bricks and stone blocks have been widely used as the main construction materials in historic structures. These are rarely provided with protection systems against moisture or rising damp, nor against significant temperature variations. Amongst the main causes of degradation of heritage masonry is the presence of high moisture content resulting from capillary absorption of groundwater. It is recognised that moisture can affect both the mechanical characteristics of masonry, as well as the plaster and rendering, and its effects need to be identified in combination with other structural problems (Cotic

I. Vayas and F. M. Mazzolani (Eds.): PROHITECH 2021, LNCE 209, pp. 175–185, 2022.
https://doi.org/10.1007/978-3-030-90788-4_17

et al. 2013). A fundamental understanding and quantification of moisture effects on the mechanical properties of porous building materials such as masonry is hence required.

The behaviour of masonry components under combined normal and shear actions can be largely characterised by three independent mechanisms: shear sliding, diagonal cracking, and compressive crushing. These mechanisms are typically assessed through material tests to define the limit strength domain of masonry in the form of shear-precompression diagrams (Malyszko 2004). The interface brick-mortar bed joint response is characterised by a proportionality between shear and precompression in a Mohr-Coulomb type representation and is often obtained from small-scale triplet tests (Pelà et al. 2017). It has been shown that the mortar-brick interface shear response is dependent on the moisture content, porosity, mortar strength, and conditioning type (Franzoni et al. 2014). A recent comparative study on wet and dry triplets also indicated that the shear strength of wet triplets is, on average, about 20% lower than in dry conditions for practical ranges of precompression (Bompa and Elghazouli 2020a).

The diagonal cracking mechanism, on the other hand, is characterised by splitting or stepped sliding along the mortar joints, with the ultimate condition depending on the properties and geometry of the brick and mortar as well as their interaction (Malyszko 2004). Although various representative diagonal panel tests are reported in the literature, particularly for comparing strengthened and non-strengthened cases (Koutas et al. 2014), such tests on wet conditions seem to be lacking. The third mechanism defining the shear-compression stress space is masonry crushing. Whilst there is general agreement that moisture reduces the compressive strength and elastic modulus of masonry, this varies with the mortar properties and brick unit porosity (Sathiparan and Rumeshkumar 2018; Bompa and Elghazouli 2020b).

In this paper, the performance of fired-clay brick-lime mortar components in ambient-dry and wet conditions, corresponding to 48 h submersion in water, are investigated through a series of compressive tests on square diagonal panels, cylindrical cores and shear tests on triplets. The bricks and mortars were selected to resemble the properties of masonry components investigated in a wider research programme on the management and conservation of heritage masonry structures in Historic Cairo (Elgahzouli et al. 2018). This paper focuses on the assessment and quantification of the influence of moisture on the constitutive response and mechanical properties of masonry components and enable the assessment of the material shear-compression strength envelopes for the masonry investigated in this paper.

2 Constituent Materials

Commercial fired clay facing solid bricks with measured sizes of 229 × 110 × 66 mm were used for the construction of all specimens. The measured average compressive strength of brick units parallel to the testing bed, obtained from a minimum of three tests, was about 15.5 MPa in both dry and wet conditions, with the latter corresponding to full submersion in water for 48 h. From readily available materials, this type of fired-clay bricks has the closest physical and mechanical properties to those from the Mausoleum of Fatima Khatun (Umm al-Salih) built in the 13th century in Cairo, which is assessed in the project.

The characteristics of the selected fired-clay bricks are hence within the range of those from the site survey, typically found in such heritage masonry, and can be used for comparative assessments and structural repair studies. The moisture content of both conditioning cases for bricks was assessed. The ambient-dry samples and those submersed in tap water for a minimum of 48 h, were dried in an oven for 6 h at 60 °C and another for 18 h at 105 °C until the sample mass was relatively constant. The moisture content of ambient-dry bricks was 0.07% by weight (wt.%), and 10.46 wt.% for those submersed in water.

Mortars incorporating natural hydraulic lime (NHL5) with a binder-to-aggregate ratio of 1:3 were selected for the study as these are typically used for conservation works. These are singular binders which combine hydraulic with air setting, obtained by carbonation with atmospheric CO_2. The free lime $Ca(OH)_2$ is above 15%, whilst the sulphates content is below 2%. Mortars with a consistency of 180–200 mm were prepared in 15 L buckets and mixed until becoming homogenous using a hand-held electric paddle mixer. Samples of fresh mortar were taken selectively, covered with a plastic sheet and removed from moulds after five days.

Compressive strengths were determined from compression tests on a total of 95 ambient-dry and wet cubes (50 × 50 × 50 mm). The resulting strengths of the mortars in ambient-dry conditions were in the range of f_{lm} = 2.08–5.11 MPa, reducing to f_{lm} = 0.70−2.37 MPa under wet conditions. An average reduction in elastic modulus in the range of 50% was typically observed due to moisture. Additionally, the moisture content of both conditioning cases for NHL mortar samples was assessed following the same procedure as for the bricks. The moisture content of the lime mortars was 2.54% by weight (wt.%) for ambient-dry samples and 10.80 (wt.%) for those in water.

3 Experimental Assessment

3.1 Shear Triplet Tests

A total of 30 specimens of b × h × t = 227 × 229 × 111 mm (±2.8 × 2.0 × 2.9 mm), were subjected to varying levels of pre-compression (σ_c = 0–2.5 MPa). The specimen reference adopts the format *Txyyz*, where x indicates the specimen conditioning (D for dry, and W for wet); yy represents the pre-compression stress (e.g. 00 for 0 MPa, and 15 for 1.5 MPa), and *z* represents the specimen sequence (a, b, c, etc.). From the total number of specimens, 15 were kept in laboratory conditions (T = 24–30 °C, RH = 30–50%), whilst the remaining 15 specimens were wet.

A stiff four-post hydraulic servo-controlled machine with a capacity of 750 kN was used for testing the triplets in shear. The two lateral bricks of each triplet were positioned on 65 mm wide and 115 mm deep steel blocks, which were connected to the machine bed through a series of hinges (Fig. 1). A 65 mm wide steel block and a hinge were used to transfer the load from the actuator to the central brick. Prior to applying the monotonic displacement procedure to the central brick, a preload equivalent to axial stresses of up to 2.5 MPa was applied through four threaded ties and recorded by a load cell (denoted as '5' in Fig. 1a). Besides the transducers used to record overall deformations, a digital image correlation (DIC) system was used to obtain strain fields from all triplet tests (Fig. 1b).

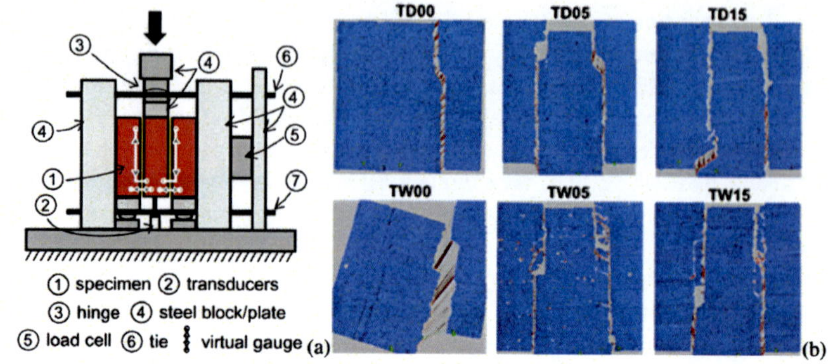

Fig. 1. (a) Specimen details and testing arrangement, (b) strain fields for selected triplets

Figure 2 illustrate the average shear stress versus displacement ($\tau - \delta_s$) curves for various values of pre-compression (σ_c) for the dry (TD) and wet (TW) specimens. As observed, the shape of the $\tau - \delta_s$ curves depends on the level of pre-compression $\sigma_{c,}$ with the shear strength τ_m increasing proportionally with σ_c. Additionally, the test results indicate a reduction in τ_m for wet (TW) specimens in comparison with the dry (TD) counterparts for the same level of σ_c. By comparing the $\tau - \delta_s$ curves in Fig. 2, it can be observed that depending on σ_c, 7–32% reductions in τ_m occur between dry and wet cases. A larger reduction of τ_m around 48% is obtained for cases with $\sigma_c = 0$ MPa.

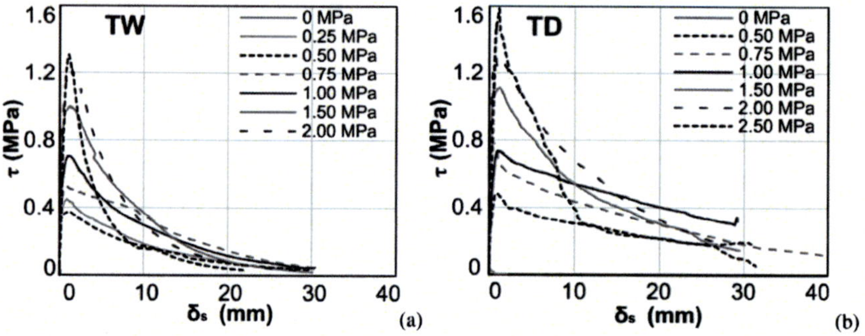

Fig. 2. Shear stress versus displacement curves for: (a) dry triplets TD, (b) wet triplets TW

For both dry and wet tests, when $\sigma_c = 0$ MPa, the failure initiated at one of the four brick-mortar interfaces propagating abruptly throughout the depth of the specimen. After the peak shear stress τ_m was reached, as shown in Fig. 2b, the specimen separated into two distinct bodies and τ dropped instantaneously to zero. For triplets with 0.25 $\geq \sigma_c \geq$ 1.5 MPa, failure occurred on the faces of both sides of the mortar layer with multiple cracks forming inside each layer. Test observations showed that the mortar from dry specimens had less cohesion, breaking in multiple lumps, whilst the mortar from wet specimens had less fracture surfaces due to paste-like consistency.

3.2 Diagonal Compression Tests

The diagonal tests on single leaf square panels had b × h × t = 710 × 710 × 110 mm (Fig. 1b, d). The specimen reference adopts the format D-*Y*z, where D stands for diagonal panel), *Y* represents the conditioning (D for air-dry or W for wet) and z represents the specimen sequence (1, 2, 3, etc.). Prior to testing, the wet specimens were submersed 3/5 of depth in water and were sprinkled with a hose from the top. The tests on diagonal panels were carried out in a rig which included a main loading transfer frame with a 1000 kN Instron actuator and a connected load cell and a hinge. As shown in Fig. 3a, b, the specimens were positioned and loaded through V-shaped supports. The load was applied from the actuator to the top support. To ensure uniform contact between the loading/support plates and the specimen, a thin timber panel was used.

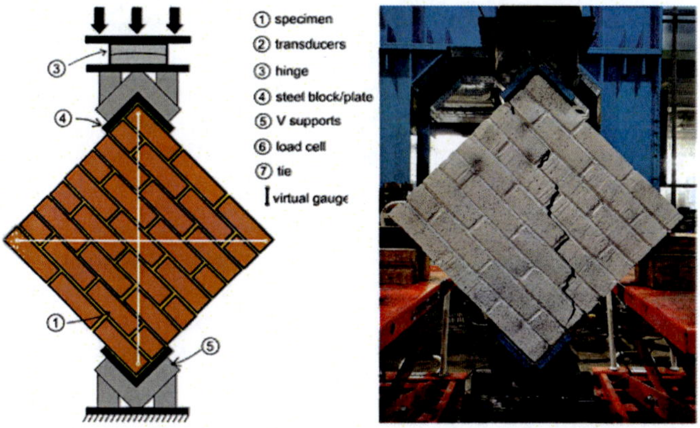

Fig. 3. (a) Specimen details and testing arrangement (b) specimen at failure

Figure 4a depicts the average stress-strain behaviour of the tested panels. In these tests, the ultimate condition occurs when the principal tensile stresses perpendicular to the compressed strut reaches the maximum tensile strength of the masonry. The shear stress was determined using Eq. (1), in which P is the applied load, α is the angle between the panel diagonal and horizontal axis, parallel to the bed joint.

$$\tau_u = \frac{P\cos\alpha}{0.5(b+h)t\cdot n} \tag{1}$$

$$\gamma = |\varepsilon_v| + |\varepsilon_h| \tag{2}$$

The parameters b, h and t are the panel width, height and thickness, respectively, whilst n is a factor which depends on the perforations (n = 1 for solid bricks). The shear strain γ, is the sum of the absolute vertical and horizontal strains, ε_v and ε_h, respectively (Eq. 2). The shear stiffness G of the wet specimens was reduced by about 39% and 33%, compared to the corresponding dry panels, respectively. As for the triplet tests, a DIC system was used to assess strain fields (Fig. 4b).

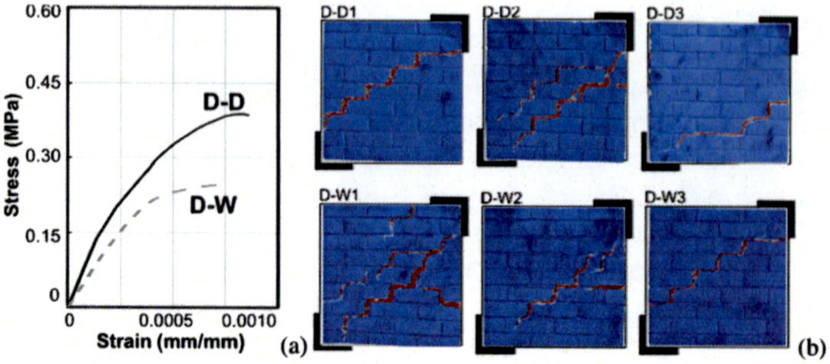

Fig. 4. (a) Average stress-strain curves of tested diagonal panels (b) crack patterns at failure of selected specimens

The typical crack patterns illustrated in Fig. 4b, obtained from full-field DIC measurements, indicate a typical brittle panel response with failure surfaces following the principal tensile stresses within the specimen. The response of Specimen D-D1 was characterised by loss of bond at a horizontal bed joint at the centre of the panel at around 90% of the peak capacity τ_u. For the wet Specimen D-W3, the failure patterns illustrated in Fig. 4b resembled those of their corresponding dry counterparts. However, the first strain concentrations associated with loss of bond at the brick-to-mortar interface occurred between 73–75% of their ultimate strength. In addition to stiffness and strength reduction due to moisture, the above results indicate that the bond properties are also influenced by the moisture content. concentration at a horizontal bed joint was observed around 85% of peak load.

3.3 Compression Tests

The cylinders were divided equally into two groups: wet and dry. Half of the specimens were kept in laboratory conditions (T = 24–30 °C, RH = 30–50%), whilst the remaining specimens were in wet conditions by full submersion in water for 48h. To assess the compressive strength of the masonry, cylinders made of two vertical brick cores, with a mortar joint of about 10 mm in between, were extracted from undamaged masonry walls (Elghazouli et al. 2021). The cylinders had a diameter d = 69 mm and height of around h = 145 mm, representing an aspect ratio h/d ≈ 2.10. Such geometries (h/d ≥ 2.0) are required to ensure a uniaxial stress state at mid-height of the specimen and to obtain the material strength rather than the response of the specimen (Oliveira et al. 2006). As illustrated in Fig. 5a, the specimens were tested in a four-post Instron 3500 kN machine and the testing arrangement included top and bottom high strength steel transfer plates, with the actuator at the top. A DIC system was used to assess surface deformations and strain fields (Fig. 5b).

As shown in Fig. 6a, b, the response of cylinders in compression exhibits a typical brittle response with significant softening immediately after the peak. Based on the stress-strain (σ-ε) curves from Fig. 6a, b, it can be observed that both the elastic modulus and the compression strength are notably influenced by the moisture. The dry specimens had, on average, about 18% and 16% higher stiffness and strength, respectively, compared

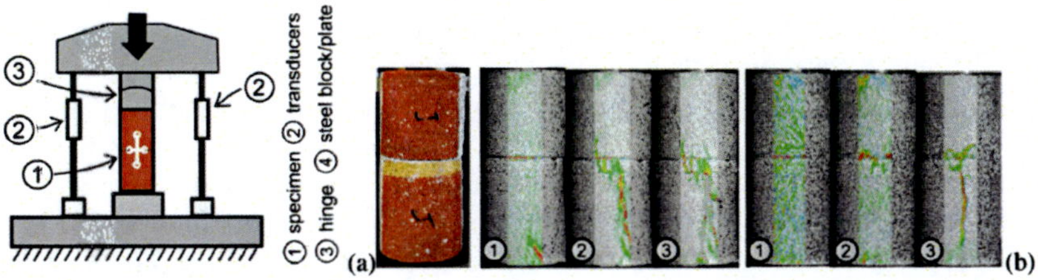

Fig. 5. (a) Specimen details and testing arrangement (b) strain fields for selected cylinders

to their wet counterparts. Additionally, the volumetric strain (ε_{vol}) is illustrated against the stress-to-strength ratio (σ/f_m) in Fig. 6c as an average of the test curves for each configuration (Bompa and Elghazouli 2020c). As shown in the figure, with the increase in σ/f_m, the mortar developed dilation earlier than the brick components, and this effect is stronger for the wet specimen in comparison to the ambient-dry counterpart.

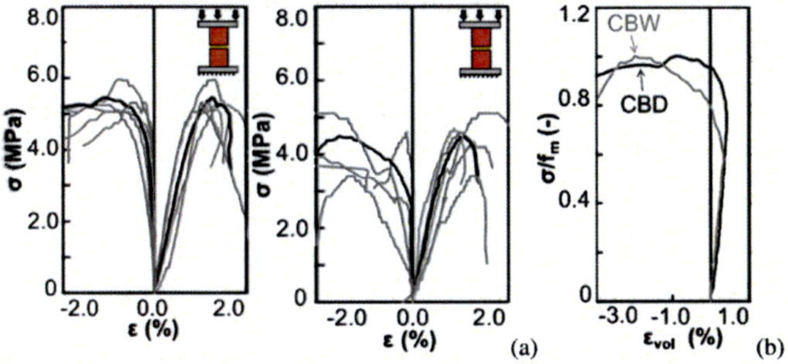

Fig. 6. Selected stress-strain curves for: (a) dry cylinders, (b) wet cylinders; (c) volumetric response

Close inspection of other results from the literature, on masonry elements made of solid clay bricks and lime mortar (Sathiparan and Rumeshkumar 2018), indicate that the compressive strength is relatively constant for values of moisture between 0% and 2–3%, which correspond tao oven-dry and air-dry conditions, respectively. Beyond this value, a gradual reduction in compression strength in the range of 15% occurs with the increase in moisture up to values around 10–13%. It is worth noting that the moisture content is dependent on the porosity of the material, a parameter which governs the total water absorption. Hence, the above comments are specific to the masonry components investigated in this paper.

4 Shear-Compression Envelopes

The tests described herein can be used to evaluate the shear-compression (τ-σ) failure envelopes. This enables a detailed insight into the moisture effects on the mechanical properties of masonry. The shear sliding properties of masonry joints can be determined from triplet tests and the diagonal tension strength can be assessed from diagonal panel tests. The failure envelope is then capped by compression crushing (Li et al. 2005).

The shear sliding is generally represented by a Mohr-Coulomb criterion (Eq. 3) with model coefficients c and μ. As given in Eq. 4, the cohesion c is referred to in codified provisions as the initial shear strength f_{v0} and is representative of cases without normal load ($\sigma = 0$) (BS EN 1996-1-1 2005). The slope of the Mohr-Coulomb criterion $\mu = \tan\phi$ is typically considered in codes as a value of 0.4. However, as shown by the tests from Sect. 3.1, this is about 50% higher than that suggested by codified guidance, with $\mu_{dry} = 0.62$ and $\mu_{wet} = 0.58$.

The linear representation of τ-σ for shear sliding, using Eq. 5, is marked with '(1)' and represented by an inclined dashed line in Fig. 7. To represent ranges of the sliding failure mode, the dashed line is offset to intersect the average of the two triplet test results. It is worth noting that Eq. 5 applies for singular brick-mortar interfaces.

$$\tau_s = c + \mu\sigma \tag{3}$$

$$f_v = \min(f_{v0} + 0.4\sigma, 0.065 f_b) \tag{4}$$

$$f_{v1} = f_{v0} + \mu\sigma \tag{5}$$

$$\mu' = \frac{\mu}{1 + 2\mu(h_b/l_b)} \text{ and } c' = \frac{c}{1 + 2\mu(h_b/l_b)} \tag{6}$$

$$f_{v2} = \frac{f_t}{2.3}\sqrt{1 + \frac{\sigma}{f_t}} \tag{7}$$

$$f_{v3} = n_c f_m \sqrt{1 - \frac{\sigma}{f_m}} \tag{8}$$

To obtain these parameters for masonry elements of larger sizes, the interface parameters c and μ would need to be replaced with c' and μ' using Eq. 6, in which h_b and l_b are the height and length of the brick unit, respectively (Mann and Muller 1982). By considering the size effect, the shear stress limit condition moves to lower stress levels. In other terms, the inclined dashed line, marked with '(1)' is shifted downwards in Fig. 7a to '(1*)'. The same approach was applied to obtain the τ-σ for shear sliding for wet conditions, as shown in Fig. 7b.

The response of the diagonal panel tests is characterised by principal stresses acting in the two main specimen directions: a compression field for the direction of load, and a perpendicular tension field. This failure criterion is referred to as 'tensile stress', and requires the diagonal tensile strength as an input parameter f_t. This parameter corresponds to the unit strength obtained from diagonal tests τ_u (Eq. 1). The average test results from Fig. 4 are used as input in Eq. (7) (Li et al. 2005) to plot the second curve '(2)' of the shear-compression τ-σ envelope shown in Fig. 7a, b. The grey areas represent the test ranges.

The third failure mode, denoted '(3)' in Fig. 7a, b, and estimated using Eq. 8, is used to cap the τ-σ failure envelope. This third mode is related to the principal compressive

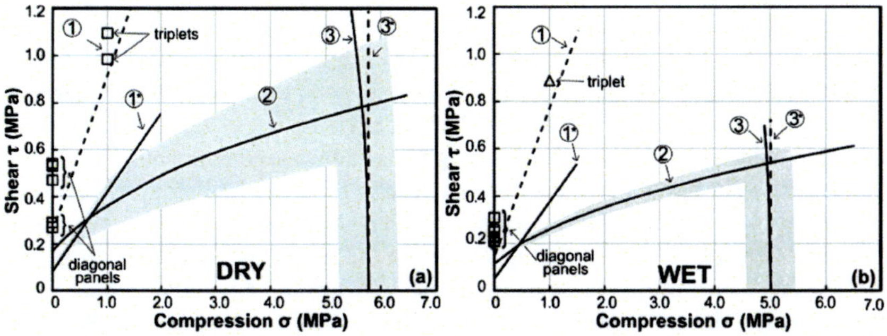

Fig. 7. Shear-compression envelopes for (a) dry masonry, (b) wet masonry

stress criterion for the case of biaxial compression (Malyszko 2004), in which f_m is the compressive strength of masonry obtained from cylindrical cores with the bed joint perpendicular to the loading direction. The parameter n_c accounts for the compressive strengths parallel and normal to the bed joints (n_c = 1.0 herein). As for the diagonal tension case, the grey areas adjacent to the average τ-σ for '(3)' correspond to the test strength domain obtained from cylinder testing. The dashed grey and black vertical lines from Fig. 7a, b, marked with '(3^*)', represent the uniaxial compression strength ranges. The intersection between the three failure criteria determine the τ-σ failure envelope for a given type of masonry. When a pair of stresses (τ, σ) falls on or outside the envelope, this would represent failure.

5 Concluding Remarks

This paper described an experimental investigation into the fundamental mechanical properties of masonry elements incorporating fired-clay bricks and hydraulic lime mortars under ambient-dry and wet conditions, corresponding to 48h of submersion in water. Tests on shear triplets, square panels under diagonal compression, and cylindrical cores in compression, were included. After describing the constituent materials, specimen details and instrumentation, the main test results and key observations resulting from detailed test measurements were reported.

Shear triplet tests showed that for specimens without pre-compression the failure was characterised by specimen separation into two distinct bodies, with the shear stress dropping instantaneously. On the other hand, for tests with pre-compression, failure occurred at the brick-mortar interface and propagated through the mortar layer. Moreover, the mortar from dry specimens had less cohesion, breaking in multiple lumps, whilst the mortar from wet specimens had fewer fracture surfaces due to paste-like consistency. The results show that the shear strength of wet triplets was about 20% lower on average than of those in dry conditions.

A direct comparison between the average load displacement curves of dry and wet specimens tested under diagonal compression showed that moisture reduced both the elastic stiffness and the strength on average by about 35% and 29% of the dry specimens, respectively. Full-field DIC measurements also showed that the cracking load was reduced by moisture with the brick-to-mortar interface bond loss occurring at around 75% and 85% of ultimate for the wet and dry specimens, respectively. Compression tests on cylindrical cores indicated that all uniaxial compression properties were lower when moisture was present. For the configurations tested in this study, the reduction in the elastic modulus and compressive strength was on average in the range of 13% and 15%, respectively, between the dry and wet cases.

The tests were used to assess the shear-compression failure envelope, which enables an in-depth understanding of moisture effects on the mechanical properties of masonry. The clay-brick/lime-mortar masonry envelope, which comprises three regimes, namely shear sliding, diagonal tension, and compression crushing, showed that the envelope for wet masonry lies consistently within the limit condition for dry masonry, indicating reductions in strength across all shear-precompression ranges as a function of the governing response.

Acknowledgements. The study was supported by the Arts and Humanities Research Council of the UK Research and Innovation agency, within the project "Interdisciplinary approach for the management and conservation of UNESCO World Heritage Site of Historic Cairo - Application to Al-Ashraf Street", Grant No. AH/R00787X/1. The authors would like to acknowledge the support provided by the technical staff of the Structures Laboratories at Imperial College London.

References

Bompa, D.V., Elghazouli, A.Y.: Experimental and numerical assessment of the shear behaviour of lime mortar clay brick masonry triplets. Constr. Build. Mater. **262**, 120571 (2020a). https://doi.org/10.1016/j.conbuildmat.2020.120571

Bompa, D.V., Elghazouli, A.Y.: Compressive behaviour of fired-clay brick and lime mortar masonry components in dry and wet conditions. Mater. Struct. **53**, 60 (2020b). https://doi.org/10.1617/s11527-020-01493-w

Bompa, D.V., Elghazouli, A.Y.: Stress–strain response and practical design expressions for FRP-confined recycled tyre rubber concrete. Constr. Build. Mater. **237**, 117633 (2020c). https://doi.org/10.1016/j.conbuildmat.2019.117633

BS EN 1996-1-1: Eurocode 6: Design of masonry structures—General rules for reinforced and unreinforced masonry structures. British Standards Institution, UK (2005)

Cotic, P., et al.: Effect of moisture on the reliability of void detection in brickwork masonry using radar, ultrasonic and complex resistivity tomography. Mater. Struct. **46**(10), 1723–1735 (2013). https://doi.org/10.1617/s11527-012-0011-3

Elghazouli, A., Butler, A., Mourad, S., Cheeseman, C., Elyamani, A.: Interdisciplinary approach for the management and conservation of UNESCO World Heritage Site of Historic Cairo. Application to Al-Ashraf Street. UK Research and Innovation, 7 July 2019 (2018)

Elghazouli, A.Y., Bompa, D.V., Mourad, S.A., Elyamani, A.: In-plane lateral cyclic behaviour of lime-mortar and clay-brick masonry walls in dry and wet conditions. Bull. Earthq. Eng. **19**, 5525–5563 (2021). https://doi.org/10.1007/s10518-021-01170-5

Franzoni, E., Gentilini, C., Graziani, G., Bandini, S.: Towards the assessment of the shear behaviour of masonry in on-site conditions: a study on dry and salt/water conditioned brick masonry triplets. Constr. Build. Mater. **65**, 405–416 (2014). https://doi.org/10.1016/j.conbuildmat.2014.05.002

Koutas, L., Bousias, S.N., Triantafillou, T.C.: Seismic strengthening of masonry-infilled RC frames with TRM: experimental study. J. Compos. Constr. **19**(2), 04014048 (2014). https://doi.org/10.1061/(ASCE)CC.1943-5614.0000507

Li, T., Galati, N., Tumialan, J.G., Nanni, A.: Analysis of unreinforced masonry concrete walls strengthened with glass fiber-reinforced polymer bars. ACI Struct. J. **102**(4), 569 (2005)

Malyszko, L.: In-plane shear and tensile strength tests of small brickwork specimens. In: Structural Analysis of Historical Constructions-2 Proceedings of the IVth Int. Seminar on Structural Analysis of Historical Constructions, 10–13 November 2004, Padova, Italy (2004)

Mann, W., Muller, H.: Failure of shear-stressed masonry. An enlarged theory, tests and application to shear walls. Proc. Br. Ceram. Soc. **30**, 223–235 (1982)

Oliveira, D.V., Lourenço, P.B., Roca, P.: Cyclic behaviour of stone and brick masonry under uniaxial compressive loading. Mater. Struct. **39**(2), 247–257 (2006). https://doi.org/10.1617/s11527-005-9050-3

Pelà, L., Kasioumi, K., Roca, P.: Experimental evaluation of the shear strength of aerial lime mortar brickwork by standard tests on triplets and non-standard tests on core samples. Eng. Struct. **136**, 441–453 (2017). https://doi.org/10.1016/j.engstruct.2017.01.028

Sathiparan, N., Rumeshkumar, U.: Effect of moisture condition on mechanical behavior of low strength brick masonry. J. Build. Eng. **17**, 23–31 (2018). https://doi.org/10.1016/j.jobe.2018.01.015

Rediscover Traditional Gypsum: Territory, Traditional Architecture and Rediscovery of Traditional Gypsum

Belén González-Sánchez[1], Antonia Navarro Ezquerra[1](✉), Manuel Julià Comas[2], Judit Ramirez-Casas[1], and Mariana Palumbo Fernandez[1]

[1] Department of Architectural Technology, Universitat Politècnica de Catalunya, Av. Doctor Marañon, 44-50, 08028 Barcelona, Spain
antonia.navarro@upc.edu
[2] Architect and Archaeologists, Center for the Study of Gypsum (CEG), Carrer Major, 23, 08750 Molins de Rei, Spain

Abstract. In the region of La Segarra, in the province of Lleida (Spain), there is a geological formation of the Oligocene age, called the "Talavera Gypsum Formation". The traditional buildings located in this area are built with this material, and its use ranges from ornamental elements to complex and structural construction systems. Within the traditional patrimony of the area, we can find in a generalized way the use of gypsum in defensive towers, churches, castles and private houses, among other patrimonial elements of the territory. In this article, some representative buildings of the area are analysed, indicating their geographical and geological location. Moreover, with the aim of obtaining a material suitable for future restoration of La Segarra region's built heritage, and by extension, of all the heritage built using this material, a traditional gypsum furnace was built, the firing temperatures within the furnace were monitored during the firing process, and the product obtained was characterized by mineralogical.

Keywords: Traditional architecture · Gypsum · Construction systems · Materials · Territory · Built heritage

1 Introduction

As architectural constructions have progressed, increasing in simultaneously and reducing the thickness of their walls, building materials have also changed. In fact, there is still a debate about if architecture has conditioned the search for new building materials, or if the finding of these has contributed to the evolution of architecture.

Up to the 20th century, the construction use of materials and the architectural techniques was linked to the resources available in the near territory, which explains why, in the built heritage, the same material is used in a wide diversity of construction systems (La Spina 2016).

Among the binding materials used in traditional construction, lime is the best known. However, another binder with visual characteristics similar to lime; gypsum, is frequently

I. Vayas and F. M. Mazzolani (Eds.): PROHITECH 2021, LNCE 209, pp. 186–195, 2022.
https://doi.org/10.1007/978-3-030-90788-4_18

found in such constructions too. This binding materials currently limited to indoor use, in linings and plasterboard, and can be acquired in the market within a variability of products limited by the existing regulations (AENOR 2010; AENOR 2012; AENOR 2009).

Nonetheless, traditionally, and especially in geological gypsum areas, gypsum has been used in a wider range of applications, either in floors, ceilings, beams, walls, partitions, and even in external coatings. The versatility of this material depends essentially on its purity and the firing temperature (Navarro Ezquerra and Julià 2019; La Spina and Grau 2020).

In the region of La Segarra, in the province of Lleida (Spain), there is a geological formation of the Oligocene age, called the "Talavera Gypsum Formation" (Pla et al. 2016). The traditional buildings located in this area are built with this material, and its use ranges from ornamental elements to complex and structural construction systems. Within the traditional patrimony of the area, we can find in a generalized way the use of gypsum in defensive towers, churches, castles and private houses, among other patrimonial elements of the territory (Garganté Llanes 2006; Llobet i Portella 2012; Martí Falip 2018).

In this article, some representative buildings of the area are analyzed. Moreover, with the aim of obtaining a material suitable for future restoration of La Segarra region's built heritage, and by extension, of all the heritage built using this material, a traditional gypsum furnace was built, when the firing temperatures within the furnace were monitored during the firing process, and the product obtained was characterized.

2 Arquitectura de la Segarra

According to the documents kept in the Arxiu Comarcal de la Segarra (Cervera), in 1486, plaster made by the farmers of Sant Pere dels Arquells i Vergós de la Ribera (two villages in La Segarra) was supplied for the construction of the church of Santa Maria de Cervera, and years later for various private constructions. It seems that in the 16th century there was a strong demand for gypsum in the area, which led to an increase in the price of plaster, until in the 17th century a selling price was set by the local authorities. Some historical studies explain that in the 19th century the use of lime and gypsum was very abundant in the towns of La Segarra (Garganté Llanes 2006).

This use of plaster can be clearly documented in the different constructions that are spread throughout the area today. We find examples in both private and public architecture, and in various municipalities such as those described in the study carried out by the architect Antoni Martí in Torà, Biosca, Cervera or Talavera, among others (Martí Falip 2018).

Although gypsum is conceived as a building material from the firing and grinding of gypsum stone, there are also examples of the use of gypsum stone in its natural state for the construction of cyclopean walls, such as those found at Timor Castle (in the municipality of Sant Pere dels Arquells). This construction is documented from the 12th century onwards (Pladevall i Font 1997), and a farmhouse was built on top of it in the 17th or 18th century, which was inhabited until the second half of the 20th century (Fig. 1). Nowadays it is in ruins, as well as the nearby church of Sant Jaume de Timor where there are also many constructive elements made of plaster in its natural state and as a building material.

Just walking through the aforementioned Segarra villages, one can see the extensive use of plaster in construction in general. Thus, plaster as an exterior rendering gives buildings that still remain a particular characteristic, both in terms of shape and appearance, and gives them a roughness that is difficult to achieve with other materials.

Fig. 1. Cyclopean plaster wall (left) and plaster stone pillar supporting two plaster stone arches taken with plaster mortar (right), Timor Castle

There are many examples in the construction of the walls, ranging from masonry with gypsum mortar as a bonding agent, to walls made of rammed earth with gypsum concrete, to thin walls made of mass plaster or those with stone slabs inside.

The masonry walls were usually covered with plaster, although few remains today, especially in the door jambs, edges of doors and windows. These were made of a thinner and harder plaster, which also served as a master to facilitate the execution of the plastering. The existence of these edges allows us to see the evolution of the façades, since in certain cases we can see, for example, how what was once a door later became a window, and the conservation of the profiles of the openings.

On the façades of the houses, there are also cornices and cantilevers made of plaster that finish off the upper part of the walls, and door decorations made of plaster are also documented.

We find interesting examples of covered passages in the streets of La Segarra made in the form of a vault or with wooden beam slabs and interlaced with plaster vaults. An example of a flat plaster ceiling with molded decoration can be found in a covered walkway in Torà. This is a fairly widespread technique in Catalonia and has a varied iconography that is repeated in different areas.

The use of plaster on the joists allows for the creation of the upper part of the floors, which are also mostly plaster-based, but with different characteristics to those of the vault itself (Fig. 2).

In Cervera and Talavera there are examples of basements with barrel vaults made of mass-produced gypsum concrete or with brickwork and also with groin vaults and fajon arches (Fig. 3).

Perhaps one of the most representative examples of the use of plaster in architecture from a structural point of view is the tower of Ivorra. With a circular floor plan and a height of about 20 m, it is dated to around the year one thousand (Martí Falip 2018). After carrying out some complete previous studies, it has been concluded that this tower is built with mass-produced plaster, and for this reason in the restoration carried out in 2013 and 2014 the original aspect was recovered with the application of plaster in the deteriorated and disappeared parts (Fig. 4).

Fig. 2. Flat roof in covered passage, Torà

Fig. 3. Basement with barrel vault, Talavera

Fig. 4. Outside Ivorra tower

3 The Geology of the Segarra

Geologically, the Segarra region is located in a land of geological formation from the Oligocene period, dated between 40 and 25 million years B.C., where the formation known as the "Talavera Gypsum" emerges, interspersed with large layers of marl (Pla et al. 2016).

According to the information available at the "Institut Cartogràfic i Geològic de Catalunya" (ICGC), the towns of Cervera, Vergós, Sant Pere dels Arquells, Timor, Sant Antolí i Vilanova, and Talavera, are located on land composed of alternating gypsum and marl, known as the "Talavera Gypsum" (POxm) formation. The towns of Biosca and Ivorra are located close to soils composed of gray marl and sandstone with intercalation of gypsum (PEox). Finally, the town of Torà is located in nearby land composed of gypsum and marl, known as the "Balbastre gypsum" formation (Fig. 5).

Currently, the company Cemento Natural Tigre, located in the town of Cervera, exploits large layers of marl to produce NHL hydraulic limes and natural cements. However, in the quarries where it exploits its raw material there are outcrops of gypsum, interspersed between the marls, which are not productive and are stockpiled to later fill in the finished quarry fronts. The composition of these gypsums varies in terms of their degree of purity and associated minerals (calcite, dolomite, clays, quartz).

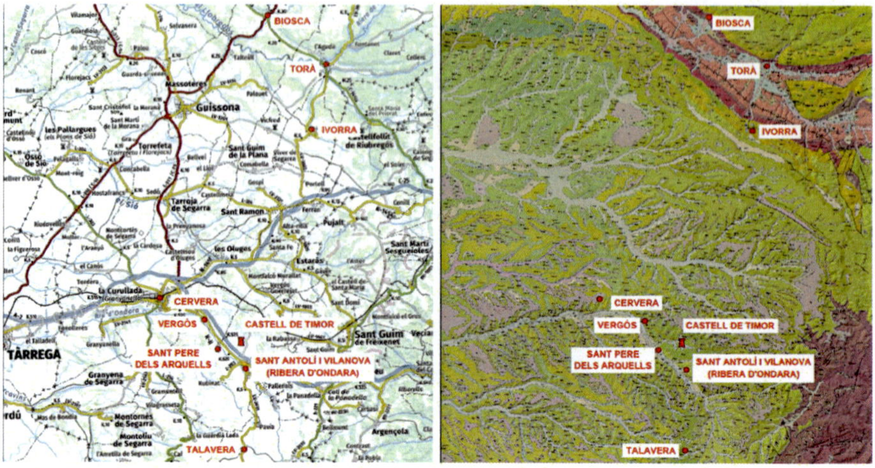

Fig. 5. Topographical map (left) and geological map (right) of the Segarra area with indications of the towns studied

4 Recovery of a Traditional Gypsum Oven

At the end of November 2018 the congress "El guix tradicional en el patrimoni construït" organized by the Centre d'Estudis del Guix (CEG), took place in Ribera d'Ondara, located to the east of the city of Cervera. This meeting brought together professionals, academics, students and amateurs, to share knowledge and recover popular wisdom regarding the construction of a traditional gypsum oven.

The morphology of the kiln built was characterized by a square floor plan, with a cylindrical section, open, uncovered, medium-sized, half-buried, isolated, built on one side of the quarry front of Cemento Natural Tigre, finished off on three of its four sides with a brickwork with exposed ceramic bricks, and with two entrances that formed a false vault (La Spina and Grau 2020).

The oven was lighted with pine wood due to the difficulties in obtaining other combustible material that was sufficient for the correct development of the cooking of the gypsum, although traditionally native vegetation was used. In the last phase of burning the oven, gypsum vegetation from the area was used. The oven was shut down by sealing the two entrances (CEG 2020a, 2020b, 2020c).

The particularity of this traditional plaster oven was the incorporation of temperature sensors in different areas in order to monitor its thermal evolution during the plaster firing process, and the comparative study of the samples collected.

5 Materials and Methodology

5.1 Mineralogical Composition

Before the experimental development of the traditional gypsum furnace, two samples of gypsum stone were collected from the Cemento Natural Tigre quarry. The selection criterion of these samples was their visual heterogeneity. Once the oven was fired, the resulting material was homogenized and samples were also collected.

These two groups of samples were analysed in the laboratory to determine their mineralogical composition by X-ray diffraction (XDR), comparing the results obtained with the data already analysed in previous laboratory work on gypsum mortars in the Ivorra Tower and the Timor Castle.

The diffractometer used is a PANalytical X'Pert PRO MPD Alphal power diffractometer in Bragg-Brentano $\theta/2\theta$ geometry of 240 mm of radius, nickel filtered Cu Kα radiation and sample spinning at 2 revolutions per second.

5.2 Evolution of the Temperature of the Traditional Plaster Oven

The traditional gypsum oven was monitored, at 21 different points distributed homogeneously over the entire surface of the interior structure of the oven made with the gypsum stone itself, with type K high-temperature sensors (1400 ºC). Readings were taken with the Applent AT4532 multi-channel temperature meter, which was connected to a battery, powered by a 150W Poly-crystalline silicon siries solar panel with an energy transformer (DC 12 V to AC 220 V) during the entire plaster firing process, from the time the oven was switched on to the time it was switched off and the resulting material was cooled (48 h in total).

6 Results

6.1 Characterization of the Gypsum

The study of the two gypsum stone samples collected before the experimental development of the traditional gypsum oven [Gypsum Stone "A" and Gypsum Sone "B"], reflect that in both cases the presence of hydrated gypsum was detected, although in a very variable proportion, ranging from 40% to 95%. The soil studied is rich in calcite and dolomite, which is found in both samples with variable concentrations, and the most predominant clay in the area is Muscovite.

Table 1. Semi-quantitative (%) X-ray diffraction (XRD) analyses of the samples analysed

Sample	G_D	GH	GA	Qz	C	D	M	A	C_2S	SCH
Gypsum stone "A"	95	–	–	2	3	–	–	–	–	–
Gypsum stone "B"	40	–	–	6	28	6	17	3	–	–
Traditional gypsum	2	9	77	–	2	–	–	–	10	–
Ivorra tower gypsum	51	–	–	8	32	–	–	–	–	9
Timor castle gypsum	85	–	9	–	4	–	–	–	–	2

G_D = calcium sulphate dihydrate; G_H = calcium sulphate hemihydrate; G_A = anhydrite; Qz = quartz; C = calcite; D = dolomite; A = Analcime; C_2S = Belite; SCH= Tobermorite.

Analysis of samples of traditional gypsum [Traditional gypsum] indicates that in addition to anhydrous gypsum (77%), hemihydrate (9%) and dihydrate (2%), phases of calcium silicates, which are characteristic of hydraulic limes, have also been detected in a proportion of 10%.

The mineralogical composition of some samples taken from the tower of Ivorra and the castle of Timor was also studied [Ivorra tower gypsum and Timor castle gypsum respectively]. These samples indicate that the gypsum is fully hydrated, although the presence of non-reactive anhydrite gypsum has also been detected in Timor Castle. The presence of carbonate phases (calcite) and hydrated hydraulic phases (hydrated bicalcium silicate) was also detected (Table 1).

6.2 Thermal Analysis of the Traditional Gypsum Oven

Based on the thermal data collected in situ, this study has only analysed 10 reading points, located according to Figs. 6 and 7, corresponding to the data of the false left vault. The graphic representation is highlighted in colours according to its spatial location, where the lilac colours represent the lower part of the furnace, the blue ones the lower-middle part, the green ones the central area and the red ones the upper area.

Fig. 6. Plant of the traditional gypsum oven

Fig. 7. Elevation of the traditional gypsum oven

In the following graph (Fig. 8) we can see that the temperatures with the greatest oscillation correspond to the lower zone of the kiln (lilac colour), due to its proximity to the flames produced by the burning of the fuel, in this case pine wood, by the incorporation of new material that rapidly decreases its temperature, and by the production of embers at the base of the kiln. The highest average temperatures reached in this area would be around 700 °C, with peaks of over 800 °C.

The highest temperatures are obtained in the lower-middle zone of the oven (blue colour) with punctual temperature peaks above 1000 °C, and with an average thermal accumulation of between 600 °C and 900 °C for more than 150 min.

The temperatures with less oscillation and with some curves, of ascent and descent, more moderate are in the superior zone of the furnace (red colour). The average temperature of the area during the furnace burning process is between 300 °C and 400 °C, with two specific maximum peaks of 650 °C and 750 °C respectively.

The central zone of the oven (green colour), despite what might initially be expected from the thermal behaviour of this zone, has a very similar evolution to the data collected in the upper zone of the oven. This phenomenon could be produced by the generation of internal current which generates a loss of heat from the central zone of the oven.

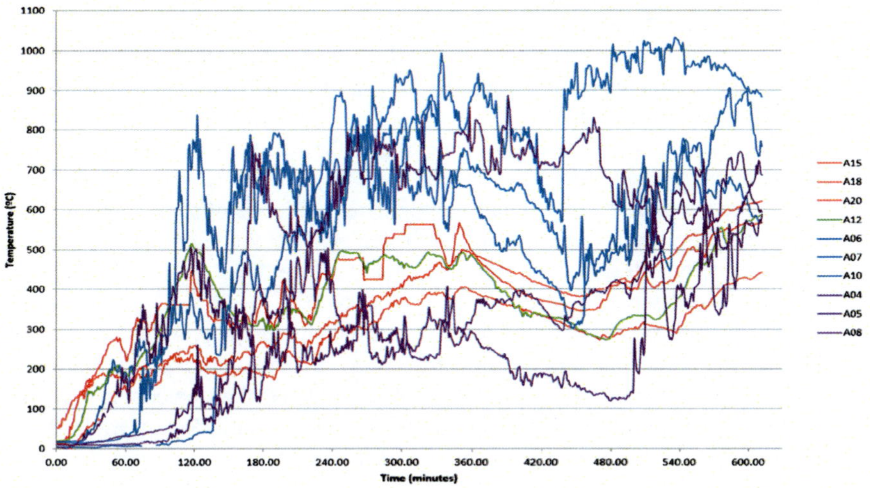

Fig. 8. Graph of the thermal evolution of traditional gypsum oven firing

The Table 2 shows the firing temperature values of the traditional plaster oven according to the literature consulted (Ruiz de Gauna 1973; Sanz Arauz 2009; Bel-Anzué 2017), with the different compounds found in the mineralogical analysis of the samples analysed by X-ray diffraction.

Table 2. Relationship between processing temperature and generated compounds

	Name	Mineral phase	Temp	XDR
G	Calcium sulphate dihydrate	$CaSO_4 + 2\ H_2O$	Ambient	2%
	Calcium sulphate hemihydrate	$CaSO_4 + 0.5\ H_2O + \uparrow 1.5\ H_2O$	100–170 °C	9%
	Anhydrite III	$CaSO_4$	170–300 °C	77%
	Anhydrite II	$CaSO_4$	300–700 °C	
	Anhydrite I	$CaSO_4$	+1200 °C	
	Calcium oxide	$CaO + \uparrow SO_3$	+800 °C	
C	Clay	$mSiO_2 \cdot nAl_2O_3 \cdot pFe_2O_3 \cdot qH_2O$	Ambient	
	Alumina	$Al_2O_3 + \uparrow 2\ H_2O$	+600 °C	
	Silica	$SiO_2 + \uparrow 2\ H_2O$	+600 °C	
	Iron oxide	$Fe_2O_3 + \uparrow 2\ H_2O$	+600 °C	
L	Calcite	$CaCO_3$	Ambient	2%
	Calcium oxide	$CaO + \uparrow CO_2$	+850 °C	
X	Calcium silicate	2CaO·SiO2	+850 °C	2%

G = gypsum; C = clay; L = limestone; X = combination mineral phase.

As can be seen, the highest percentage (77%) corresponds to anhydrite, which requires a temperature between 300 °C and 700 °C for its transformation. If we analyse the temperature graph, only the gypsum located in the lower-middle zone of the oven has acquired these temperatures.

Both the calcium sulphate dehydrate (2%) and the calcite (2%) resulting from the mineralogical analysis correspond to unfired material, in other words, the firing in the gypsum furnace is very heterogeneous and the innermost zone of some gypsum stones has not received any heat. This is also the explanation for the appearance of calcium sulphate hemihydrate (9%), which requires a temperature of between 100 °C and 170 °C for its transformation.

The calcium silicate (10%) detected in the mineralogical analysis needs temperatures above 850 °C, and is formed by the combination of calcium oxide and silicon oxide, coming from the decomposition of the limestone and clays present in the quarry face studied. The appearance of this compound generates what is known as "hydraulic gypsum", and gives better performance to the final product.

7 Conclusions

In this study it has been possible to demonstrate that in the manufacture of traditional gypsum a mono-mineral stone is not used, but that gypsum is associated with other minerals such as clays, limestone, quartz, etc. As we have seen in the temperature graph, when a heterogeneous firing of the material takes place in a traditional gypsum kiln, with maximum temperature peaks above 1000 °C, the associated minerals are also transformed into new compounds that combine with each other, such as belite (bicalcium silicate), obtaining the formation of hydraulic phases. When this happens, it is popularly said that we have "hydraulic gypsum".

The study of the existing gypsums in the constructions of the tower of Ivorra and the castle of Timor, reflects that these have similar compositions to those obtained in the baked stone of our traditional gypsum furnace, which does not happen with the industrial gypsums used in the construction sector. That is why we propose the use of a traditionally baked plaster to rehabilitate this type of building, to guarantee the compatibility of materials and the durability of the applied material.

Acknowledgements. The authors of this study would like to thank the Centre d'Estudis del Guix for organizing the congress "El guix tradicional en el patrimoni construït", as well as all the participants who took part in the course; the company Cemento Natural Tigre for providing its facilities, machinery and raw materials to build the traditional gypsum oven; Professor Felipe Buill for his active participation in the documentation process of the experimental part; the EPSEB-UPC for financing part of the experimental monitoring material. Finally, we would like to thank the UPC for awarding the FPI-UPC 2018 grant to hire research staff in training to carry out the doctoral thesis.

References

La Spina, V.: Estudio del yeso tradicional en España. Yacimientos, canteras, hornos y la arquitectura tradicional, su estado de conservación y propuestas de itinerarios visitables para su revalorización y difusión (Informe). Instituto del Patrimonio Cultural de España, Ministerio de Educación, Cultura y Deporte, Madrid (2016)

Asociación Española de Normalización y Certificación. UNE-EN 520:2005+A1: Placas de yeso laminado. Definiciones, especificaciones y métodos de ensayo. AENOR, Madrid, February 2010

Asociación Española de Normalización y Certificación. UNE-EN 12859: Paneles de yeso. Definiciones, especificaciones y métodos de ensayo. AENOR, Madrid, April 2012

Asociación Española de Normalización y Certificación. UNE-EN 13279-1: Yesos de construcción y conglomerantes a base de yeso para la construcción. Parte 1: Definiciones y especificaciones. AENOR, Madrid, July 2009

Navarro Ezquerra, A., Julià, M.: El guix tradicional. L'Informatiu del Col·legi d'Aparelladors i Arquitectes de Barcelona, vol. 360, p. 80–83 (2019)

La Spina, V.; Grau Giménez, C.J.: La diversidad tipológica de los hornos tradicionales de calcinación de yeso en España. Informes de la Construcción 72, 557, e334 (2020)

Pla, O., Ferrer García, J.O., Gratacós Torrà, Ò., Muñoz, J.A., Roca i Abella, E.: Influencia de los niveles evaporíticos sin-orogénicos en la geometría del frente surpirenaico central: Anticlinales de Sanaüja y de La Sentiu-Almenara. Geo-Temas **16**(1), 65–68 (2016)

Garganté Llanes, M.: Arquitectura religiosa del segle XVIII a la Segarra i l'Urgell. Condicionants, artífexs i pràctica constructiva. Pagès editors, S.L., Barcelona (2006)

Llobet i Portella, J.M.: Els materials de construcció a Cervera al llarg dels temps i la seva procedència. Els materials de l'arquitectura popular **10**, 116 (2012)

Martí Falip, A.: El guix en la construcció tradicional de la Segarra. El guix tradicional en el patrimoni construït, Ribera d'Ondara, 25–27 October 2018

Pladevall i Font, A.: Catalunya Romànica. Segrià, les Garrigues, el Pla la Segarra, l'Urgell, vol. 24. Enciclopèdia Catalana, Barcelona (1997)

Centre d'Estudis del Guix: Time-lapse del proceso de carga del horno de yeso tradicional construido durante las II Jornadas de Yeso Tradicional en Ribera d'Ondara. UPC, Barcelona (2020a). https://youtu.be/uEA4jZIPMY4. Accessed 10 Dec 2020

Centre d'Estudis del Guix: Time-lapse del encendido y quema del horno de yeso tradicional construido durante las II Jornadas de Yeso Tradicional en Ribera d'Ondara. UPC, Barcelona (2020b). https://youtu.be/Evevzz-_5r4. Accessed 10 Dec 2020

Centre d'Estudis del Guix: Time-lapse de la descarga del horno de yeso tradicional construido durante las II Jornadas de Yeso Tradicional en Ribera d'Ondara. UPC, Barcelona (2020c). https://youtu.be/XlvDmcbL4qY. Accessed 10 Dec 2020

Ruiz de Gauna, A.: La determinación del contenido en yeso hemihidrato y dihidrato en los Cementos por anàlisis térmico. Mater. Constr. **23**(152), 5–25 (1973)

Sanz Arauz, D.: Análisis del yeso empleado en revestimientos exteriores mediante técnicas geológicas (doctoral thesis). Universidad Politécnica de Madrid (2009)

Bel-Anzué, P., Almagro, A., Sáez Pérez, M.P., Rodríguez-Navarro, C.: Influence of the calcination process in traditional gypsum with structural behavior. Ge-conservación **11**, 79–85 (2017)

Modelling and Analysis

Response Simulation of Aseismic Retrofit for a Reinforced Concrete Historic Building Structure Using a Variable Friction Damper

Kazutaka Shirai[1](✉), Tomoaki Ito[2], and Takeshi Sano[3]

[1] Faculty of Engineering, Hokkaido University, Kita 13, Nishi 8, Kita-ku, Sapporo 060-8628, Hokkaido, Japan
shirai.kazutaka@eng.hokudai.ac.jp

[2] Graduate School of Engineering, Hokkaido University, Kita 13, Nishi 8, Kita-ku, Sapporo 060-8628, Hokkaido, Japan

[3] Technology Division, Obayashi Corporation, Shinagawa Intercity Tower B, 2-15-2 Konan, Minato-ku, Tokyo 108-8502, Japan

Abstract. For the preservation of historic architecture, historic building structures in areas prone to seismic excitations should have sufficient aseismic performance. A useful approach for enhancing the aseismic performance of historic buildings is aseismic retrofitting, where energy dissipation elements, such as frictional devices, are installed. However, for conventional friction dampers (FDs), an increase in damper force increases the story shear force and response acceleration under earthquakes. The authors previously developed a passive variable friction damper (VFD) that produces a decreased frictional force when the device displacement exceeds a predetermined value. In the present study, a time history response simulation was conducted using a nonlinear analytical model assuming a historic building made of a reinforced concrete structure retrofitted by the VFD subjected to earthquakes. The seismic responses were compared to those obtained with the FD. The results show that the VFD reduced the peak response story shear force and acceleration compared with those for the FD.

Keywords: Historic architecture · Aseismic performance · Passive variable friction damper · Frictional force · Reinforced concrete structure · Seismic response

1 Introduction

The preservation of historical architecture is important to society. Historic building structures in areas prone to seismic excitations thus need sufficient aseismic performance. The seismic performance of historic buildings can be enhanced by aseismic retrofitting by installing energy dissipation elements, such as frictional devices (Shirai et al. 2017). To date, many friction devices have been proposed (Pall and Marsh 1982; Mualla and Belev 2002; Sano et al. 2019). However, for conventional friction dampers, an increase in

I. Vayas and F. M. Mazzolani (Eds.): PROHITECH 2021, LNCE 209, pp. 199–205, 2022.
https://doi.org/10.1007/978-3-030-90788-4_19

damper force increases the story shear force and response acceleration under earthquakes (Shirai et al. 2018).

The authors previously developed a passive variable friction damper (VFD) that produces a decreased frictional force when the device displacement exceeds the predetermined value (Shirai et al. 2013.b). It is expected that the VFD will mitigate the peak response of historic buildings without significantly increasing the burden due to the damper force.

In the present study, a nonlinear time history response simulation is conducted using an analytical model assuming a historical building made of a reinforced concrete (RC) structure retrofitted by the VFD subjected to earthquakes. The earthquake responses are compared to those for a conventional friction damper (FD). This paper is a revision of a previous paper presented by the authors (Shirai et al. 2013a).

2 Variable Friction Damper

The basic mechanism of the VFD is shown in Fig. 1. A unit of the VFD mainly consists of disc springs, a high-tension bolt, a friction material, a stainless steel plate, and a variable height component. Conceptual diagrams of the hysteresis loops of the FD with rigid-plastic damping force characteristics and those of the VFD are depicted in Fig. 2.

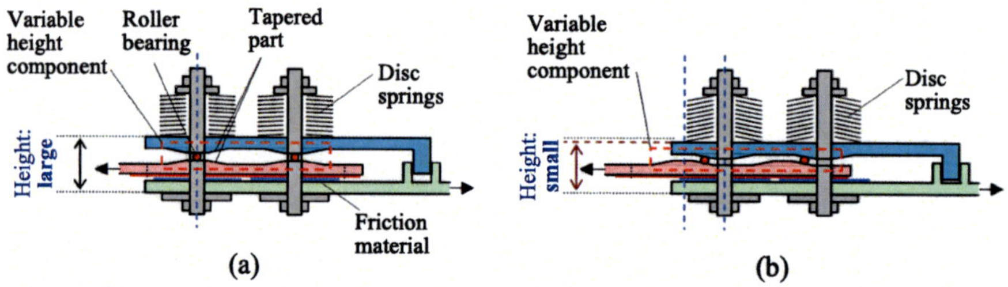

Fig. 1. Basic mechanism of VFD in (a) the neutral position and (b) when displaced

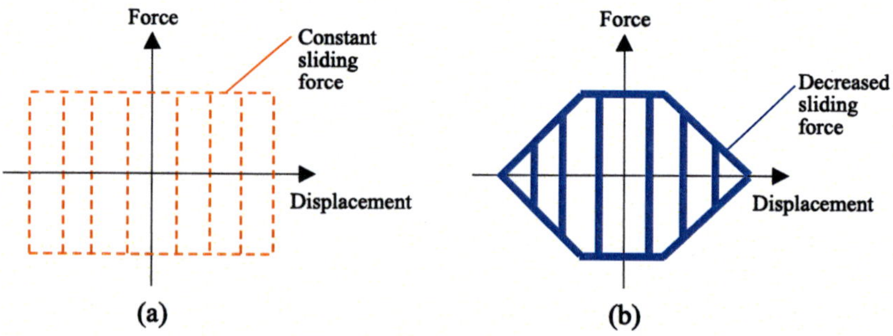

Fig. 2. Conceptual diagrams of hysteresis loops of (a) FD and (b) VFD

Owing to the function of the variable height component, the VFD produces a decreased sliding force when the damper displacement exceeds a predetermined value.

3 Analysis Methods

3.1 Analytical Model of Main Frame

In this study, a nonlinear seismic response simulation was carried out using a single-degree-of-freedom (SDOF) model, representing a main frame that incorporates the FD or VFD for retrofitting, or lacks a damper. Figure 3(a) shows the analytical SDOF model. The properties of the main frame SDOF model are shown in Table 1, where g is acceleration due to gravity. For the main frame, a fictional historic RC building was assumed. A yield shear coefficient C_{Fy} smaller than that of a typical newly built low-rise RC building in Japan was adopted. A degrading trilinear restoring force characteristic was given for the nonlinear spring element of the main frame, as shown in Fig. 3(b).

Table 1. Properties of RC main frame

Property	Symbol	Value
Mass	m	2000 t
Natural period of main frame	T_{F0}	0.3 s
Yield shear coefficient	C_{Fy}	0.2
Yield shear force	F_{Fy}	mgC_{Fy}
Cracking force	F_{Fc}	$(1/3)F_{Fy}$
Secant yield stiffness ratio	α_y	0.25
Stiffness ratio after yielding	α_3	0.03
Unloading stiffness coefficient	γ	0.4

3.2 Analytical Model of Damper

The damper properties for the models that incorporate a damper (FD or VFD) are shown in Table 2. The FD and VFD have the same values of the sliding (or initial sliding) force, initial damper stiffness, and stiffness of damper support member in series.

3.3 Input Motions and Other Conditions

Two simulated earthquake waves, Wave-M and Wave-S, were used as the input ground motions. These waves were prepared in a previous study (Shirai and Inoue 2014). The time history acceleration and response acceleration spectra (damping factor: 5%) for the two waves are shown in Fig. 4.

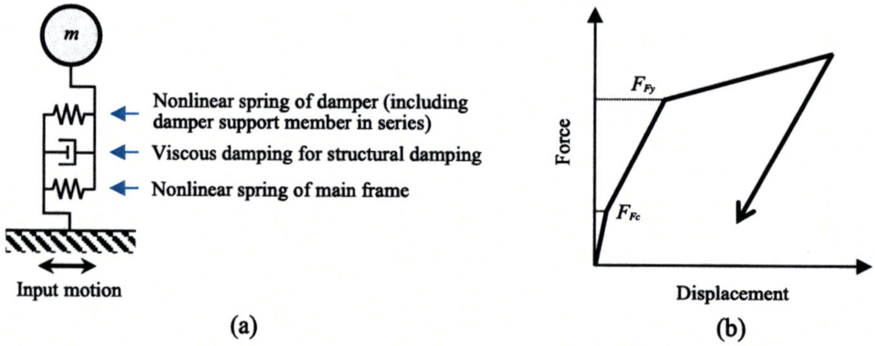

Fig. 3. Analytical model: (a) SDOF model and (b) restoring force characteristics of main frame

Table 2. Properties of FD and VFD

Property	Symbol	Value
Ratio of damper sliding force (FD)	F_{Ds}/F_{Fy}	0.3
Ratio of damper initial sliding force (VFD)	F_{Ds}/F_{Fy}	0.3
Initial stiffness ratio of damper (FD and VFD)	K_{D0}/K_{F0}	Almost infinity
Stiffness ratio of damper stiffness during sliding (FD)	K_{Ds}/K_{D0}	0
Stiffness ratio of damper support member in series (FD and VFD)	K_B/K_{F0}	0.2
Damper displacement at oriented point (VFD)	$D_{VFD,o}$	0.1 m
Damper force at oriented point (VFD)	$F_{VFD,o}$	Almost zero
Damper displacement when force decrease begins (VFD)	$D_{VFD,d}$	Almost zero

The Newmark-β method ($\beta = 0.25$) was adopted for numerical integration. A tangent stiffness proportional damping with a damping factor of 0.03 was given. The software SNAP version 7 (Kozo System 2015) was used for the simulation.

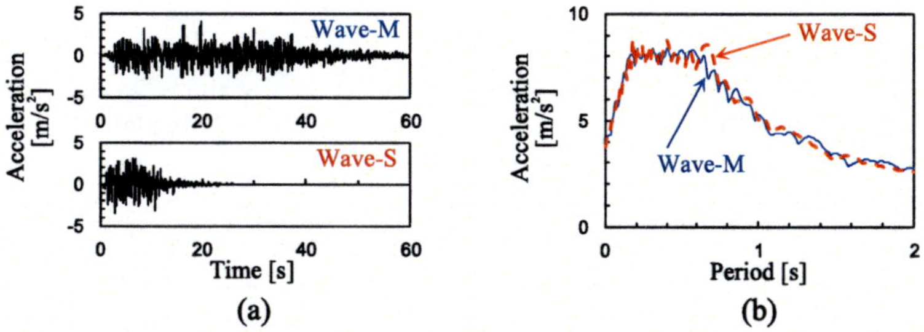

Fig. 4. Input seismic motions: (a) time history accelerations and (b) response acceleration spectra with damping factor of 5%

4 Analysis Results

4.1 Response Hysteresis Loops

The response hysteresis loops obtained without a damper, with the FD, and with the VFD from the analysis (input motion: Wave-M) are compared in Fig. 5. The damper force of the FD (Fig. 5(d)) was constant when the damper displacement varied, whereas that of the VFD (Fig. 5(e)) decreased as the damper displacement increased. The peak response displacements of the main frame with the FD (Fig. 5(b)) and VFD (Fig. 5(c)) were both obviously reduced compared to that without a damper (Fig. 5(a)). This is due to the energy dissipation effect and the initial stiffness enhancement achieved by the friction dampers.

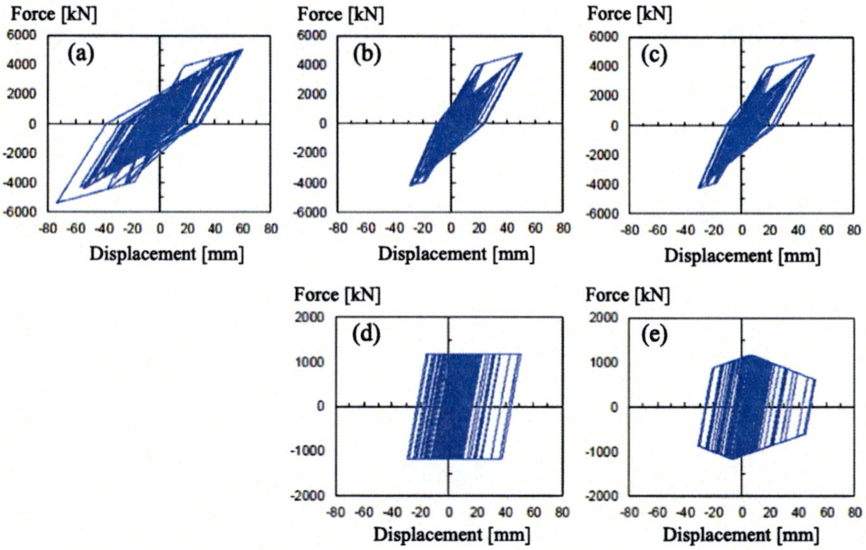

Fig. 5. Response hysteresis loops under Wave-M input at (a) main frame without damper, (b) main frame with FD, (c) main frame with VFD, (d) damper with FD, and (e) damper with VFD

4.2 Peak Responses

Figure 6 shows the obtained peak response displacement of the main frame for each analytical case (i.e., without a damper, with the FD, and with the VFD) under Wave-M and Wave-S inputs. The symbol N denotes the analytical case without a damper. The displacement responses of the two cases with a damper (FD and VFD) were significantly decreased compared to that for the case without a damper (N). The case with the VFD showed almost the same displacement as or slightly larger displacement than that with the FD.

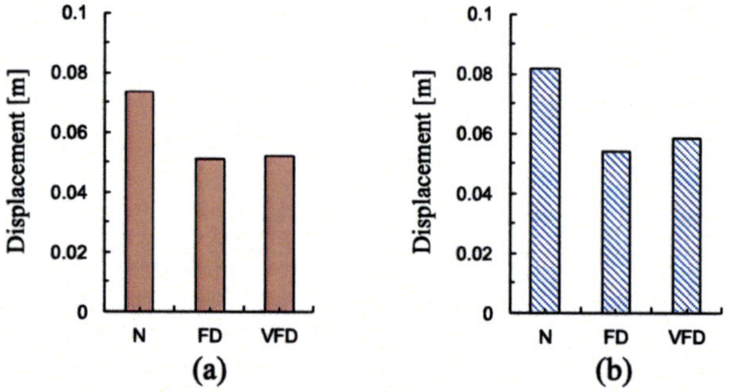

Fig. 6. Peak response displacement under (a) Wave-M and (b) Wave-S

Figure 7 plots the obtained peak response shear force of the structure, including both the main frame and damper, for each analytical case. The VFD shows a smaller force compared to that for the FD. Moreover, even though the damper element was installed, the VFD shows almost the same peak shear force as that for the case without a damper (N).

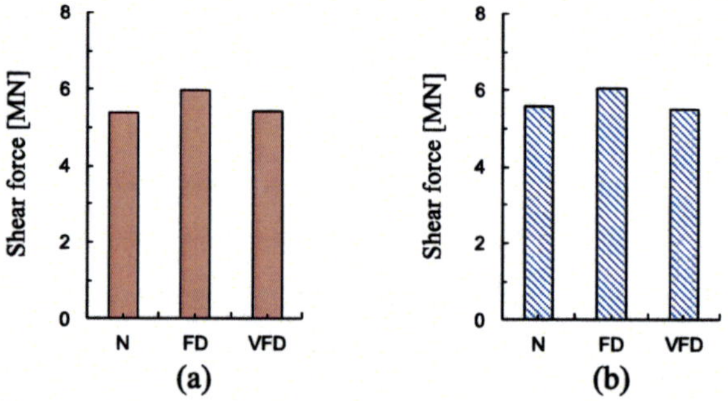

Fig. 7. Peak response shear force under (a) Wave-M and (b) Wave-S

Similar to the shear force response, the obtained peak acceleration response of the main frame for the VFD was mitigated compared to that for the FD, as depicted in Fig. 8.

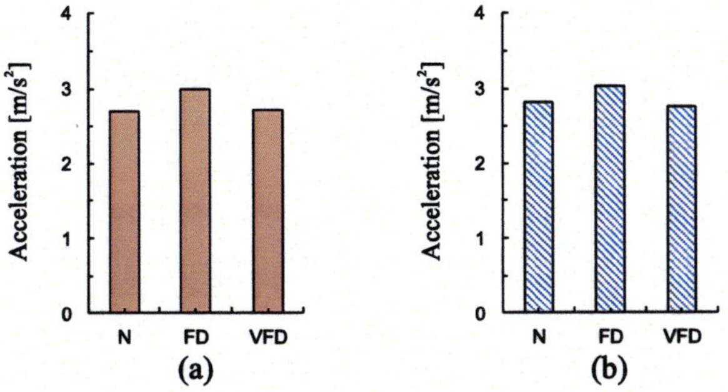

Fig. 8. Peak response acceleration under (a) Wave-M and (b) Wave-S

5 Conclusions

In this study, a nonlinear time history seismic response analysis was conducted using an analytical model assuming a fictional historic RC building that incorporates the VFD for aseismic retrofitting subjected to earthquakes. The responses were compared to those obtained with the FD or without a damper. The results show that the VFD can reduce the peak response displacement compared to that obtained without a damper, and mitigate the peak response story shear force and acceleration compared with those for the FD.

Acknowledgements. The research presented in this paper was in part supported by a research grant from the Japan Society for the Promotion of Science, KAKENHI grant number JP19H02279.

References

Kozo System: SNAP Ver. 7 Technical manual (2015). (in Japanese)

Mualla, I.H., Belev, B.: Performance of steel frames with a new friction damper device under earthquake excitation. Eng. Struct. **24**(3), 365–371 (2002)

Pall, A.S., Marsh, C.: Response of friction damped braced frames. J. Struct. Div. (ASCE) **108**(6), 1313–1323 (1982)

Sano, T., Shirai, K., Suzui, Y., Utsumi, Y.: Loading tests of a brace-type multi-unit friction damper using coned disc springs and numerical assessment of its seismic response control effects. Bull. Earthq. Eng. **17**, 5365–5391 (2019)

Shirai, K., Sano, T., Kikuchi, M.: Seismic response analysis of RC structure using SDOF model with variable friction damper. Proc. Japan Concrete Inst. **35**(2), 961–966 (2013). (in Japanese)

Shirai, K., Sano, T., Suzui, Y.: Development of a passive variable friction damper with displacement-dependent damping force characteristics. In: 13th East Asia-Pacific Conference on Structural Engineering and Construction, September 11–13, Sapporo, Japan (2013b)

Shirai, K., Inoue, N.: A seismic response estimation method for RC structures using random vibration theory. J. Adv. Concr. Technol. **12**(2), 62–72 (2014)

Shirai, K., Kikuchi, M., Ito, T., Ishii, K.: Analytical study on the seismic retrofitting of a historic reinforced concrete temple using friction dampers. In: 3rd International Conference on Protection of Historical Constructions, July 12–15, Lisbon, Portugal (2017)

Shirai, K., Kikuchi, M., Ito, T., Ishii, K.: Earthquake response analysis of the historic reinforced concrete temple Otaniha Hakodate Betsuin after seismic retrofitting with friction dampers. Int. J. Architect. Heritage **13**(1), 47–57 (2018)

Seismic Damage Evaluation and Decisions on Interventions Supported by UAV-Based Surveys

Valeria Croce[1,3](✉), Dimitris Diamantidis[2,3], and Miroslav Sýkora[1,3]

[1] ASTRO Laboratory, Department of Civil and Industrial Engineering (DICI), University of Pisa, Largo Lucio Lazzarino, 56122 Pisa, Italy
valeria.croce@unifi.it

[2] Faculty of Civil Engineering, Ostbayerische Technische Hochschule, Galgenbergstrasse 30, 93053 Regensburg, Germany
dimitris.diamantidis@oth-regensburg.de

[3] Klokner Institute, Czech Technical University in Prague, Solinova 7, 16608 Prague 6, Czech Republic
miroslav.sykora@cvut.cz

Abstract. The contribution demonstrates the benefits of Unmanned Aerial Vehicles (UAV)-based surveys for building damage evaluation and decisions on interventions after an earthquake. The case study focuses on the historic village of Castelluccio di Norcia (Italy), heavily damaged by the 2016 earthquake. By constructing a 3D model based on UAV flights, the damage levels are estimated for buildings in the village and an optimal set of intervention measures is proposed. For the decision analysis, a pre-posterior Bayesian technique is utilized. The efficiency of the UAV-based survey is highlighted through a cost-benefit analysis and by considering parameters such as the time constraints, the quality of obtained data and the risk of inspecting persons. The proposed method seems to be efficient in organizing rescuing activities, managing emergency measures and specifying interventions, following the occurrence of an accidental situation.

Keywords: Post-seismic areas · Natural hazards · UAV survey · Decision analysis

1 Introduction

The study illustrates the application of Unmanned Aerial Vehicles (UAVs, i.e. drones) for surveys of damaged areas, supporting decisions regarding interventions after a seismic event. The proposed methodology is applied to the village of Castelluccio di Norcia, Umbria, Italy that was heavily damaged by the 6.5 magnitude earthquake on October 30th, 2016. Shortly after this earthquake, the Fire Brigade utilized UAVs to get a first insight into the extent of the seismic damage. Through the acquisition of videos taken by the drones and by extracting frames from videos, the authors reconstruct the 3D model of the damaged village after the earthquake. The modern techniques of Structure from

I. Vayas and F. M. Mazzolani (Eds.): PROHITECH 2021, LNCE 209, pp. 206–221, 2022.
https://doi.org/10.1007/978-3-030-90788-4_20

Motion and Multi-View Stereo are applied for photogrammetric surveys. Starting from these 3D data, a map showing the level of damage of each building in the village is developed, using the European Macro-seismic scale EMS-98.

The integrated digital methodology allows consequently deriving a 3D model when terrestrial photogrammetry cannot be applied, and the locations of interest are inaccessible.

The procedure developed in this contribution makes it possible to evaluate degrees of damage and propose an optimal set of intervention measures. The efficiency of the UAV-based is discussed through a pre-posterior Bayesian cost benefit analysis.

The paper starts off from Sect. 2 where a brief overview of the digital technologies applied to seismic damaged sites is provided. Section 3 presents the seismicity of the Apennines region, with a particular reference to the area of Castelluccio. In Sect. 4, the steps of the workflow that led to the construction of the two 3D study models are described. Section 5 shows how these 3D models were used for the construction of damage maps and highlights the connection between geomatic survey and disaster management. In Sect. 6, the reliability level of existing structures is discussed and a relationship between the target reliability and damage level is proposed. In Sect. 7, the case study is presented, and its results are discussed. Section 8 summarises main findings of this contribution and provides an outlook of further research.

2 Digital Technologies in Post-earthquake Scenarios

Nowadays, digital technologies are more and more used to enhance knowledge and documentation on built heritage structures (Croce et al. 2019; Croce et al. 2020a). With the integration of photogrammetry and computer vision, it is possible to develop in a coherent way 3D reality-based models of a site, building or complex of buildings, significantly reducing the time needed for *in situ* data acquisition, and optimizing subsequent field surveys.

Recent studies have witnessed the application of these restitution techniques even in the case of damage scenarios, in which the presence of risk areas and the uncertainty over the inspectable areas require changes in the timing and methods of on-site surveys (Bevilacqua et al. 2018; Calvano and Guadagnoli 2016).

As an example, 3D restitution and modeling techniques have been largely applied for the detailed description of damaged buildings and aggregates in Italy, following earthquakes of considerable intensity (Croce et al. 2020.b). Italian villages and hamlets affected or threatened by earthquakes are mostly historical buildings, built from stone or brick masonry and even by earth masonry in a few cases, before the definition of rules and methods to guide design and execution of the structures in seismic zones (Sisti et al. 2019; Guidoboni 2017; Zampilli 2017; Indirli et al. 2013).

This is why it becomes more and more necessary to introduce numerical evaluation parameters, even expeditious ones, to assess the level of damage to buildings following an earthquake. These evaluation metrics should guide the choice of subsequent investigations to be carried out, even directly on site, also for the definition of intervention and recovery plans (Croce et al. 2020b). In the phases of the post-intervention emergency, the assessment of damage to buildings plays a key role, making it possible to identify the most critical situations in terms of safety accessibility for rescue teams and experts on surveys.

To evaluate the damage level, different scales have been proposed over time (Calderoni et al. 2020; Sisti et al. 2019; Goretti et al. 2002). In this work, the classification scale provided by the European Macroseismic Scale EMS98 (Grunthal 1998) is applied since it is considered an indispensable reference tool, at least at European level, to quantify the entity of earthquake-induced damages on masonry buildings. The damage level assessment is carried out considering primarily in-plane failure mechanisms, which are obviously associated to well manifested crack patterns. Note that the proposed method for damage evaluation can be efficiently applied in detecting out-of-plane failures. In the latter case, the mechanisms are identified and analysed when the failures are associated to crack openings; for a general discussion see Asteris et al. (2014).

Nonetheless, it may happen that the observed damages are not sufficient to directly detect the failure mechanism, or it might be unclear which elements are affected by such mechanism. This may be for instance the case of multiple façades connected on a same building aggregate. In these cases, the proposed method should require the integration with a more detailed analysis, for example with remote sensing techniques such as laser scanning surveys.

3 Seismicity in the Apennines Region

Seismicity refers to how often earthquakes of a certain magnitude occur in a particular region. The usual expression relating earthquake magnitudes with their rates of occurrence is the magnitude-frequency relationship by Gutenberg and Richter (1954). Considering the seismicity history and a domain of engineering interest, the magnitude M in a region can be modelled by a truncated exponential distribution (Campbell 1985). The cumulative probability distribution is given by:

$$F(M) = \frac{1 - e^{-\beta(M - M_L)}}{1 - e^{-\beta(M_U - M_L)}} \tag{1}$$

where F(M) is the probability of not exceeding magnitude M, M_{L} is the smallest magnitude of engineering interest (usually $M_{\mathrm{L}} = 4.0$), M_{U} is the largest magnitude, which can be expected in the region, and β is a site-specific constant. The upper bound M_{U} may be a random variable. The distribution in Eq. (1) is used in reliability analyses of structures subjected to seismic hazard.

In order to obtain the ground motion at the site it is necessary to adopt a relationship between the seismicity information (magnitudes) and peak ground acceleration (PGA), in terms of hypocentral (or epicentral) distance. If fault-lines are well defined, it may be assumed that the earthquake occurs anywhere along the fault-line with equal probability. In many geographical regions, fault locations and their activity are not known. In such cases the distribution of the horizontal distance R from the focus to the site can be calculated under the assumption that focus locations are uniformly distributed over regions of homogeneous seismicity (seismotectonic province).

Based on this, the first part of the workflow adopted here consists of an analysis of data obtained from seismicity catalogues (Luzi et al. 2016; Rovida et al. 2015), geographically referring to the Italian Regions of Umbria, Lazio, Marche and Abruzzi. These data are used as basic input data to estimate earthquake recurrence rates. The estimates are refined by incorporating regional seismicity models and site-specific fault data. Finally, locally adjusted attenuation laws for the horizontal PGA are considered and included in the hazard assessment (Bindi et al. 2011; Ambraseys et al. 2005; Sabetta and Pugliese 1997).

Figure 1 shows the location and year of seismic events that occurred in the Apennine area, which has an active tectonic regime running from west to east responsible for the presence of many fault systems in the area (INGV 2016). The squares and colors refer to the observed values of M while the blue stars identify the location of the 2016 earthquakes epicenters. Figure 2 displays the cumulative distribution function of the magnitude, taking into account the earthquakes that occurred in the area from 1972 to 2018 with a magnitude $M_L = 4.0 \leq M \leq M_U = 7.0$.

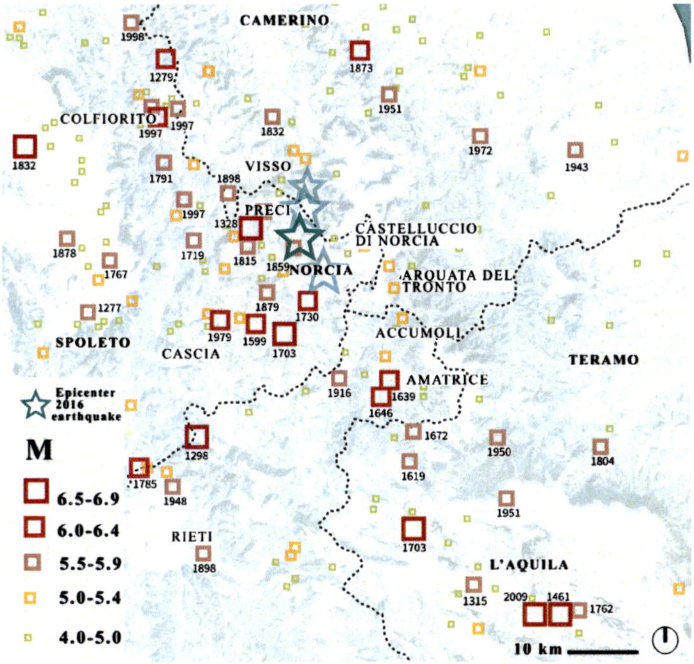

Fig. 1. The Apennine ridge earthquakes (period 1000–2018).

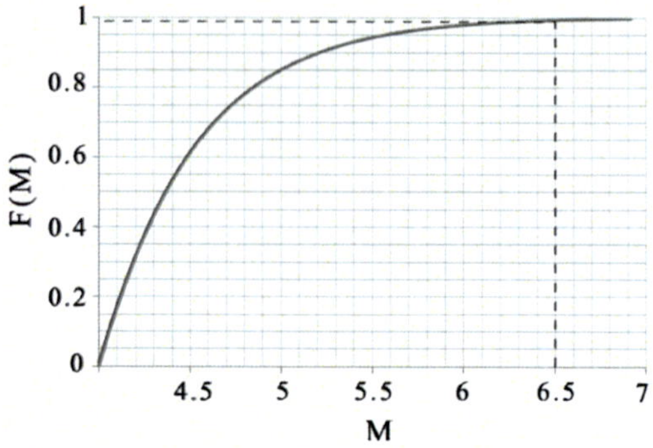

Fig. 2. Cumulative distribution function of the Apennine ridge earthquake magnitude (during the period 1972–2018).

4 Monitoring Information and Construction of 3D Study Models

In recent years, several studies have been focused on the need to apply geomatics studies in damaged and risky areas. Photogrammetry studies the use of photography in surveying and mapping for the 3D reconstruction of existing objects. This discipline has undergone considerable developments due to the integration of the techniques of Structure from Motion and Multiple View Stereo, derived from Computer Vision, and has been applied in post-catastrophe situations (Remondino et al. 2011; Caroti et al. 2015).

The use of remotely controlled systems allows to acquire images of damaged zones without the performance of direct inspections in situ, and consequently to reconstruct—through the photogrammetric process—models and maps of disrupted areas. Recent studies on this topic have highlighted the advantages of these techniques, and the importance of geomatics studies to gather aerial information on dangerous and inaccessible sites and to have an accurate and reliable reconstruction of the damage extent (Murtiyoso et al. 2018; Zaragoza et al. 2017; Nex and Remondino 2014).

The case study of Castelluccio di Norcia represents an emblematic example of application of photogrammetric techniques to damaged scenarios: few hours after the occurrence of the seismic event, the Civil Protection and the Fire Department carried out an aerial reconnaissance of the damage in the village, through the use of UAVs. The results of these flights are videos, which were published few days after the earthquake on the web-site of the Fire Department <www.vigilfuoco.tv>. These videos along with another recording dated to many days before the earthquake provided the authors of the paper with the basis for 3D reconstruction. Even though these videos were originally not recorded for specific survey purposes, the idea is to test their reliability in view of studies of damage identification and subsequent design of intervention measures, using photogrammetry.

The process led to the construction of the 3D study models (Croce et al. 2019.a; Croce and Zaragoza 2018); it consists of the following steps:

1. Video download and consequent extraction of frames: as a first step, the videos are downloaded and frames are extracted, with an average rate of 2 s/frame. In total,

146 digital images are obtained for the pre-earthquake model and 334 images for the post-earthquake model.

2. Insertion of Ground Control Points (GCPs): points of known coordinates—GCPs—are inserted on the basis of prior information on the area. In particular, reference is made to a 2004 technical map drafted by the Region of Umbria, in order to detect the X, Y, Z attributes of the GCPs.
3. Camera alignment: the photogrammetric procedures allows to process the digital images.
4. Accuracy assessment: the resulting models present a maximum accepted error of 0.40 m (compliant with the graphic error) in the determination of GCPs.
5. Dense point cloud generation, mesh generation and texturing: the dense point cloud is generated, then a polygonal mesh model and a texture atlas are constructed both for pre-earthquake (Fig. 3) and post-earthquake (Fig. 4) models.
6. Comparison between the resulting 3D models. The computation of the distances is performed through the *CloudCompare*, by a cloud-to-mesh comparison tool; the results are displayed in Fig. 5. For most damaged areas, the difference between the two models is higher than 0.9 m, corresponding to collapsed buildings.

Fig. 3. 3D model of the pre-earthquake phase.

Fig. 4. 3D model of the post-earthquake phase.

Fig. 5. Cloud-to-mesh comparison: computed distances (Croce et al. 2019.a).

5 Importance of Geomatics Studies for Disaster Management

Strategies for the management and planning of intervention after the occurrence of a disaster are based on the in-depth evaluation of building methods, architectural elements and vulnerability assessment on existing assets. The strategies should also take into account the urban scale and the general layout of cities, villages or minor historical centers; considering these aspects several studies (D'Amico and Currà 2018; Zampilli 2017) focused on the impact of catastrophic events at urban level and on the investigation of the characteristics of *resilience* of urban environments, intended as the capacity of a system to overcome the crisis.

The implementation and development of devices for the management of crisis and emergency phases has been extensively addressed by national and international programs: the United Nations International Strategy for Disaster Risk Reduction (UNISDR 2012) and the International Recovery Platform discuss the concept of Pre-Disaster Recovery Planning, i.e. the series of decision and actions taken before and after a disaster, in order to:

- Identify and establish shared recovery goals, objectives and strategies to guide post disaster decision-making, ensure that survey and recovery activities align with long-term development goals, address actual needs, and enhance resilience to future disasters
- Develop and have ready the capacity to plan, initiate and manage; to achieve an efficient, adaptive, and well-coordinated recovery effort that progresses towards the recovery goals.

The Pre-Disaster Recovery Planning is a tool directed towards the planning of recovery and reconstruction programs, in order to direct the communities to *build back better* after the occurrence of natural hazards. This concept is linked to the structuring of the disaster life cycle (Fig. 6), which includes mitigation and preparedness phases (for prevention) and recovery and response phases (for reconstruction).

In this regard, the latest developments in the disciplines of Information and Communication Technologies offer a great potential in supporting the different phases of the disaster life cycle. Aerial acquisition systems, image-based reconstruction, satellite

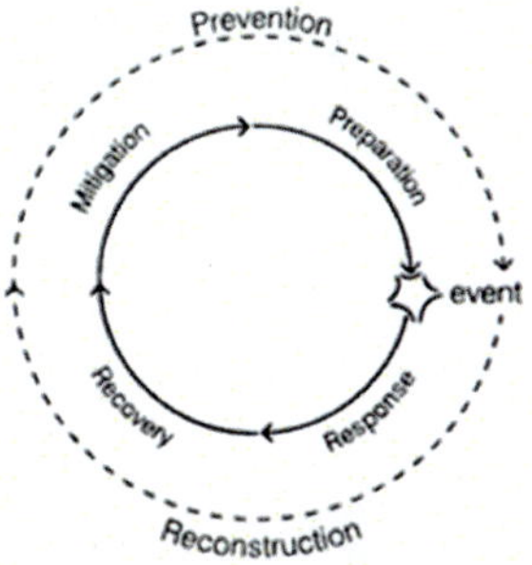

Fig. 6. Disaster life cycle (D'Amico and Currà 2018)

imagery and sensor networks represent nowadays means to support disaster management. They allow a constant and quick monitoring of the areas at risk and environmental hazards and make it possible to gather a huge amount of spatial data even in areas that present safety issues or that are difficult to access and to inspect.

6 Damage and Reliability Levels

Damage to structural elements may be classified into several levels. In order to forecast damage due to an earthquake, it is necessary to learn how various types of structures have behaved when exposed to ground motions of different intensities. The level of damage after a given earthquake depends on vulnerability of the building and on the intensity and duration of the ground motion.

The levels of damage to buildings adopted in this study follow EMS98 (Grünthal 1998); see Fig. 7. The damage level can be associated to volume and human loss ratios reaching values from 0% to 100%; see Sect. 6.

The damage level can be related to structural reliability and to post-earthquake intervention strategies. A common indicator of the reliability level is the reliability index β (indicator equivalent to failure probability).

The reliability of the structure or of its member is sufficient if the reliability index is equal or greater than a target value, normally adopted from relevant standards. EN 1990:2004 for basis of structural design provides the target reliability indexes for three consequence classes CC (small failure consequences CC1, medium CC2 and large CC3) and for two reference periods (1 year and 50 years) without any explicit link to design service life.

The present study deals with existing structures in historical centers. The reliability assessment of existing structures differs from new structures in a number of aspects, including possibly lower target reliability levels when safety measure costs are much higher for existing structures in comparison to new structures.

(Caspeele et al. 2017) indicated:

- Minimum target reliability levels β_{ex} below which the existing structure should be upgraded
- Target reliability levels β_{up} for an optimal upgrade strategy

The target reliability levels for different consequence classes, adopted in the case study in Sect. 6, are provided in Table 1.

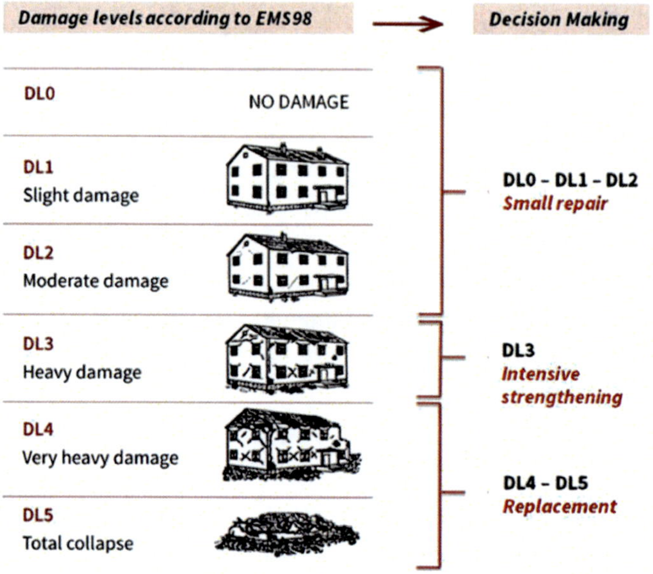

Fig. 7. Damage levels according to (Grünthal 1998) and associated intervention strategies.

Table 1. Target reliability indices for different consequence classes, (Caspeele et al. 2017; Steenbergen et al. 2015)

Consequence class	β_{code} (50 y.)	β_{up}a	β_{ex}a
CC1	3.3	2.8	1.8
CC2	3.8	3.3	2.3
CC3	4.3	3.8	2.8

[a]For a reference period equal to a remaining service life.

7 Case Study Implementation Proposal

In the case study of Castelluccio, the operational procedure to manage the recovery and response phases of the disaster life cycle is based on the definition of a damage level map that is obtained through the outcomes of the UAV-based survey. Several classes of interventions are proposed based on an estimated damage level (Fig. 7).

The data acquired by an UAV-based procedure only are inevitably affected by uncertainties since:

- Aerial acquisitions were originally not aimed at specific needs of building surveys.
- It was impossible to perform checks by physical inspections in situ.

It is thus reasonable to assume that more accurate results would be achieved by combining the UAV-based survey with a terrestrial survey. The distinction between three different approaches to the survey can then be made (Croce 2018):

*Approach 1.*UAV-based survey only.

*Approach 2.*Terrestrial survey (physical inspection on the site) only.

*Approach 3.*UAVs combined with terrestrial survey.

For the sake of brevity, this study discusses in detail *Approach 1* (the only available approach in Castelluccio) and *Approach 3* (deemed to provide the best information) only.

In the post-earthquake phase, it is normally necessary to estimate EMS 98 damage levels for affected buildings by performing many surveys in situ. Depending on an estimated damage level, operational decision on appropriate risk reduction measures has then to be made. The study treats both the afore-mentioned aspects.

The approach to survey (*Approaches 1 to 3*) may affect the results of the damage assessment and consequently decision on interventions. The aspects to be considered when selecting the approach include:

- Costs of survey (inspection), repair, strengthening or replacement
- Time constraints
- Quality of data and possibility to take a wrong decision (based on either underestimation or overestimation of the damage level)
- Risk to experts conducting the survey

The UAV-based survey (*Approach 1*) is associated with lower costs and times of inspection and does not cause risk to people, but its outcomes may be affected by uncertainties as experts do not physically access the area. The terrestrial survey provides higher quality information, but it is more time consuming and it is related to higher costs since the inspection is made by experts that have to be insured.

This might suggest that the combined methodology (*Approach 3*) represents the optimum strategy. To confirm this, the so-called pre-posterior cost-benefit analysis is conducted to quantify the value of monitoring and inspections in buildings subjected to seismic hazards. For background information see the guidelines for operators, engineers and scientists available at <cost-tu1402.eu/action/deliverables/guidelines>.

The reference period is equal to a remaining service life (in the case of repair of an existing building) or to 50 years (upgraded building); the two are considered to be similar and no distinction is further made between them.

Initially, *Approach 3* is considered. The related decision tree displayed in Fig. 8 shows that the analysis starts off with estimating occurrence probability of damage level DL_i, considering prior information about expected consequences of a seismic event; obtained for instance from a Damage Probability Matrix. It is assumed that DL_i is assigned without uncertainty using *Approach 3*. Buildings with *DL5*—total collapse—do not require inspections and they are thus disregarded in the following analysis.

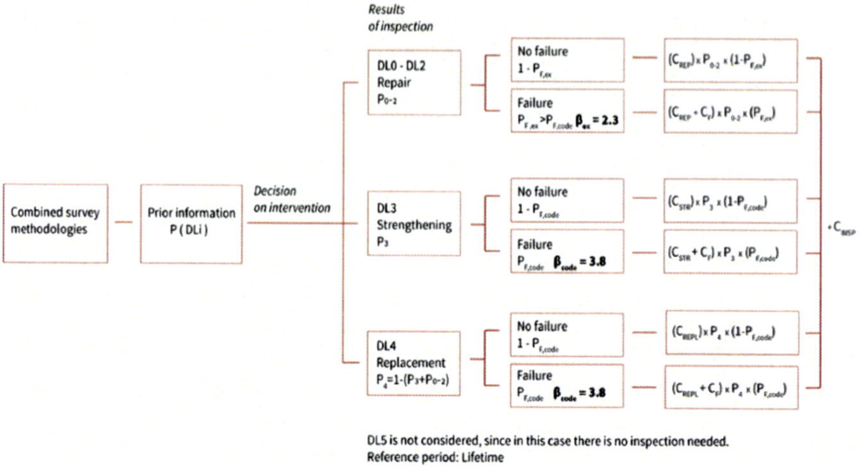

Fig. 8. Decision tree for *Approach 3.*

Depending on the identified damage level, a decision about intervention is made. For the sake of clarity, the following simplifications are adopted:

- For *DL0* to *DL2*—as identified by the survey—only small repairs are made and a minimum reliability index, $\beta_{ex} = 2.3$ (Table 1), is assumed to be achieved.
- Buildings with *DL3* must undergo intensive strengthening that is assumed to ensure $\beta_{code} = 3.8$ for new structures (Table 1).
- Buildings classified into *DL4* must be replaced and again $\beta_{code} = 3.8$ applies.

After interventions, failure may occur during the reference period with probability P_F corresponding to the aforementioned reliability indices β. Each of the six scenarios in Fig. 8 can then be associated with different costs; the following notation is used:

Pi: occurrence probability of damage level P(DL_i) based on prior information considering previous experience
C_{REP}: cost of repairs
C_{STR}: cost of intensive strengthening
C_{REPL}: cost of replacement (including demolition and reconstruction costs)
C_{INSP}: cost of inspection

For *Approach 1* imperfect results of a survey and related uncertainty should be considered. A survey may thus over- or underestimate the true *DL*. Probabilities of identifying a (correct or false) damage level DL_i given a true DL_j should be based on previous experience with applications of the technique in similar cases; expert judgement might be needed in the absence of data. The required input could be a table that provides the likelihoods of correct and false identifications; see the example in Table 2 based on the authors' judgement.

Table 2. Probability of an outcome of the survey given a true DL_j (false indications marked – underestimation of the true DL in **bold**, overestimation in grey).

True damage level, DL_j	Damage level obtained by survey, DL_i	Probability of identifying DL_i by a survey given true DL_j, $P_{i,UAV}\|DL_j$
DL0	*DL0*	100%
	DL0	**15%**
DL1	*DL1*	80%
	DL2	5%
	DL1	**15%**
DL2	*DL2*	80%
	DL3	5%
	DL2	**15%**
DL3	*DL3*	80%
	DL4	5%
DL4	*DL4*	100%

Table 2 indicates likelihoods for false or correct identification of the true damage level. For instance when a particular building suffered moderate damage *DL2*, there is:

- 80% probability that the UAV-survey would reveal the true level is *DL2*.
- 15% probability that *DL1* is identified by the UAV-survey; thus a true DL is underestimated.
- 5% probability that the true damage level is overestimated (*DL3*) by the survey and the decision is to strengthen (while repair should be sufficient).

A pre-posterior cost-benefit analysis can provide the recommendation on an appropriate survey method under particular conditions:

1. The prior occurrence probabilities of a true damage level, P(DL_j) = P_j
2. Information about the imprecision of a survey method, $P_{i,\mathrm{UAV}}|DL_j$
3. Costs related to decisions based on survey outcomes.

In comparison to *Approach 3* the decision tree for *Approach 1* becomes more complex as it is deemed to be imperfect and aspect 2 needs to be newly introduced. Following the notation in Table 2, the probabilities of identifying a damage level DL_i in Fig. 9 are obtained as:

$$P_{0-1}* = P_{0,\mathrm{UAV}}|DL0 \times P_0 + \left(P_{0,\mathrm{UAV}}\middle|DL1 + P_{1,\mathrm{UAV}}\middle|DL1\right) \times P_1 + P_{1,\mathrm{UAV}}|DL2 \times P_2 \tag{2}$$

$$P_2* = P_{2,\mathrm{UAV}}|DL1 \times P_1 + P_{2,\mathrm{UAV}}|DL2 \times P_2 + P_{2,\mathrm{UAV}}|DL3 \times P_3 \tag{3}$$

$$P_3* = P_{3,\mathrm{UAV}}|DL2 \times P_2 + P_{3,\mathrm{UAV}}|DL3 \times P_3 \tag{4}$$

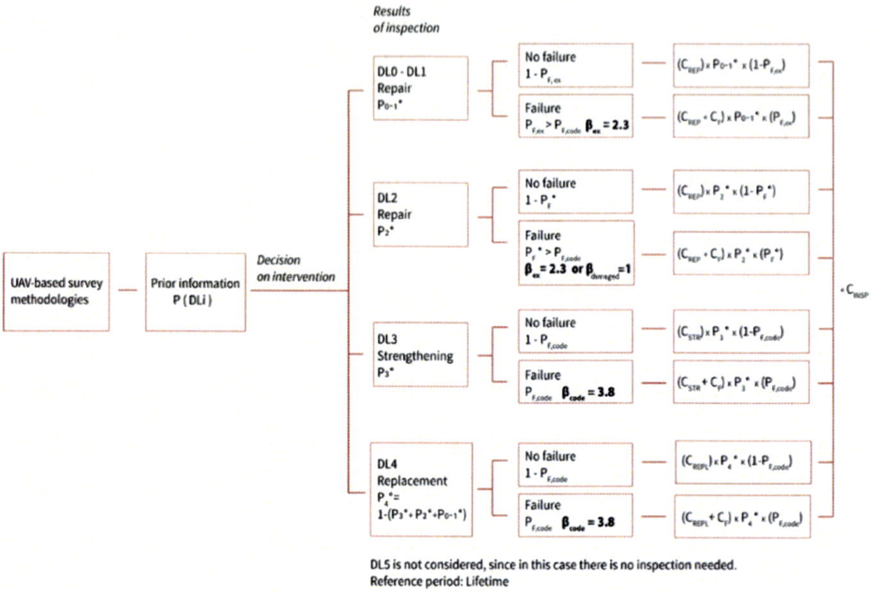

Fig. 9. Decision tree for *Approach 1*.

$$P_4* = P_{4,\mathrm{UAV}}|DL3 \times P_3 + P_{4,\mathrm{UAV}}|DL4 \times P_4 \tag{5}$$

The decision tree in Fig. 9 contains eight scenarios. The most interesting case is when.

For building(s) under consideration, the costs in Fig. 9 need to be finally estimated and the decision to take *Approach 1* or *Approach 3* should be based on the comparison of the related total costs.

8 Conclusions

This contribution discusses the benefits and drawbacks of Unmanned Aerial Vehicle (UAV)-based surveys in damage evaluation and intervention decisions for residential areas after an earthquake. The case study focuses on the village of Castelluccio di Norcia (Italy), heavily damaged by a 2016 earthquake. Videos taken during UAV flights provide essential information for estimating the damage levels for buildings in the village.

The study proposes the methodology linking occurrence probability of a damage level for a particular building, reliability index (or failure probability) for a damaged building and the decision-making on interventions for a particular type of survey. The selection of an optimum survey strategy is supported and the efficiency of the UAV-based survey is highlighted by the results of a pre-posterior Bayesian cost-benefit analysis. The analysis reflects all the key parameters such as the time constraints, the quality of obtained data and the risk of inspecting persons.

Detailed analysis of two different survey strategies reveals that photogrammetric methods along with UAV acquisition makes it possible to acquire spatial data on areas damaged by an earthquake. If necessary, the UAV-based photogrammetry can be supplemented by in-depth in-situ inspections that, however, imply risk to inspectors.

Depending on expected damage, the drawbacks of UAV surveys—limited information on real conditions in the damaged area with possibility to under- or overestimate a damage level—may or may not be traded off by low survey costs and low time demands.

The proposed method supports efficient organization of rescuing activities, managing emergency measures and design of interventions in the post-earthquake phases. The methodology can be utilized to manage other accidental situations of historical centers or areas hit by catastrophic events. For important and safety critical structures such as public theatres or hospitals, more detailed studies would be needed. The presented results will be utilized within further research of a disaster life cycle and resilience assessment.

Acknowledgements. The present contribution was supported by the Regensburg Centre of Energy and Resources (RCER) of OTH Regensburg, by the Technology and Science Network Oberpfalz (TWO), and by the Ministry of Education, Youth and Sports of the Czech Republic under Grant LTT18003. It is partly based on the thesis by V. Croce, supervised by M. G. Bevilacqua, E. Bascherini, L. Lanini and I. Martinez-Espejo Zaragoza from University of Pisa, and by D. Diamantidis from OTH Regensburg.

References

Ambraseys, N.N., Douglas, J., Sarma, S.K., Smit, P.M.: Equations for the estimation of strong ground motions from shallow crustal earthquakes using data from Europe and the Middle East: vertical PGA and spectral acceleration. Bull Earthquake Eng. **3**, 1–53 (2005)

Asteris, P.G., et al.: Seismic vulnerability assessment of historical masonry structural systems. Eng. Struct. **62–63**, 118–134 (2014)

Bevilacqua, M.G., Caroti, G., Piemonte, A., Terranova, A.A.: Digital Technology and Mechatronic Systems for the Architectural 3D Metric Survey. In: Ottaviano, E., Pelliccio, A., Gattulli, V. (eds.) Mechatronics for Cultural Heritage and Civil Engineering. ISCASE, vol. 92, pp. 161–180. Springer, Cham (2018). https://doi.org/10.1007/978-3-319-68646-2_7

Bindi, D., et al.: Ground motion prediction equations derived from the Italian strong motion database. Bull. Earthq. Eng. **9**, 1899–1920 (2011)

Calderoni, B., Cordasco, E.A., Del Zoppo, M., Prota, A.: Damage assessment of modern masonry buildings after the L'Aquila earthquake. Bull. Earthq. Eng. **18**(5), 2275–2301 (2020). https://doi.org/10.1007/s10518-020-00784-5

Campbell, K.W.: Strong motion attenuation relations: a ten-year perspective. Earthq. Spectra **1**(1985), 759–804 (1985)

Calvano, M., Guadagnoli, F.: 3D reconstruction of the city of Amatrice. An "instant modelling" operation, DisegnareCon, vol. 9, no. 17 (2016)

Caroti, G., Zaragoza, I.M.-E., Piemonte, A.: Accuracy assessment in structure from motion 3D reconstruction from UAV-born images: The influence of the data processing methods, ISPRS Int. Arch. Photogramm. Remote Sens. Spat. Inf. Sci. **XL-1-W4**, 103–109 (2015)

Caspeele R., et al.: Partial Factor Methods for Existing Concrete Structures, (fib bulletin 80). Fédération internationale du béton (fib), fib COM3 TG3.1 (2017)

Croce, V.: Post-earthquake reconstruction and renewal of urban centres. Case study: Castelluccio di Norcia. Master thesis, Ingegneria Edile-Architettura, Università di Pisa (2018)

Croce, V., Zaragoza, I.M.-E.: UAV-based 3D photogrammetry for post-earthquake studies on seismic damaged cities - a case study: Castelluccio di Norcia. Proc. IMSCI **2018**, 79–84 (2018)

Croce V., Caroti G., Piemonte, A. (2019). Assessment of earthquake-induced damage level on buildings: analysis of two different survey methods for a case study. Int. Arch. Photogramm. Remote Sens. Spatial Inf. Sci. **XLII-2/W15**, 351–358 (2019). https://doi.org/10.5194/isprs-archives-XLII-2-W15-351-2019

Croce, V., Caroti, G., Piemonte, A., Bevilacqua, M.G.B.: Geomatics for cultural heritage conservation: integrated survey and 3D modeling. In: Proceedings of the IMEKO TC4 International Conference on Metrology for Archaeology and Cultural Heritage, MetroArchaeo, 4–6 December 2019, Florence, Italy, pp. 271–276. MetroArcheo Florence (2019)

Croce, V., Caroti, G., De Luca, L., Piemonte, A., Véron, P.: Semantic annotations on heritage models: 2D/3D approaches and future research challenges. ISPRS - Int. Arch. Photogramm. Remote Sens. Spat. Inf. Sci, TCII/WG8, XLIII-B2–2020, 829–836 (2020a). https://doi.org/10.5194/isprs-archives-XLIII-B2-2020-829-2020

Croce, V., Diamantidis, D., Sýkora, M.: Planning post-earthquake surveys: assessment and reconstruction of small historical centers. In: Proceedings of the IABSE Symposium, Synergy of Culture and Civil Engineering, Wroclaw, Poland, 7–9 October 2020, pp. 332–339 (2020b)

D'amico, A., Currà, E.: Resilienza urbana dei centri storici italiani. Strategie di pianificazione preventiva, Techne. Ricerca e sperimentazione 15, (2018)

Goretti, A., Di Pasquale, G.: An overview of post-earthquake damage assessment in Italy. In: EERI Invitational workshop, An action plan to develop earthquake damage and loss data protocol, 19–20 September 2002, Pasadena, California (2002)

Grünthal, G.: European Macroseismic Scale 1998 (EMS-98), Center Europèen de Géodynamique et de Séismologie, Luxembourg (1998)

Gruppo di lavoro INGV sul terremoto in Centro Italia, Summary report on the October 30th, 2016 earthquake in central Italy Mw 6.5. https://doi.org/10.5281/zenodo.16623

Guidoboni, E.: Il valore della memoria. Terremoti e ricostruzioni in Italia nel lungo periodo, Quellen und Forschungen aus italienischen Archiven und Bibliotheken, vol. 96, no. 1, pp. 415–444 (2017). https://doi.org/10.1515/qfiab-2016-0020

Gutenberg, B., Richter, C.F.: Seismicity of the Earth and Associataed Phenomena, Princeton University Press, Princeton, NJ, USA (1954)

Indirli, M., Kouris, S., Formisano, A., Borg, R.P., Mazzolani, F.M.: Seismic damage assessment of unreinforced masonry structures after the abruzzo 2009 earthquake: the case study of the historical centers of L'Aquila and Castelvecchio Subequo. Int J Archit Herit **7**(5), 536–578 (2013)

Luzi, L., Puglia, R., Russo, E.: ORFEUS WG5, Engineering Strong Motion Database, INGV, Observatories & Research Facilities for European Seismology (2016)

Murtiyoso, A., Grussenmeyer, P., Suwardhi, D., Awalludin, R.: Multi-scale and multi-sensor 3D documentation of heritage complexes in urban areas. ISPRS Int. Geo-Inf. **7**(12), 483 (2018). https://doi.org/10.3390/ijgi7120483

Nex, F., Remondino, F.: UAV for 3D mapping applications: a review. Appl. Geom. **6**(1), 1–15 (2014). https://doi.org/10.1007/s12518-013-0120-x

Pugliano, A.: La ricostruzione del patrimonio immateriale nei luoghi danneggiati dai terremoti. In: Ricerche di storie dell'arte, 2, Maggio-Agosto 2017, pp. 23–35. Società editrice Il Mulino, Bologna (2017)

Remondino, F., Barazzetti, L., Nex, F., Scaioni, M., Sarazzi, D.: UAV photogrammetry for mapping and 3D modeling. ISPRS Int. Arch. Photogramm. Remote Sens. Spat. Inf. Sci. **XXXVIII-1/C22**, 25–31 (2011) https://doi.org/10.5194/isprsarchives-XXXVIII-1-C22-25-2011

Rovida, A., Locati, M., Camassi, R., Lolli, B., Gasperini, P.: CPTI15, Parametric Catalogue of Italian Earthquakes, Istituto Nazionale di Geofisica e Vulcanologia INGV (2015)

Sabetta, F., Pugliese, A.: Estimation of response spectra and simulation of non-stationary earthquake ground motion. Bull. Seismol. Soc. Amb. **86**(2), 337–352 (1997)

Sisti, R., Di Ludovico, M., Borri, A., Prota, A.: Damage assessment and the effectiveness of prevention: the response of ordinary unreinforced masonry buildings in Norcia during the Central Italy 2016–2017 seismic sequence. Bull. Earthq. Eng. **17**(10), 5609–5629 (2019). https://doi.org/10.1007/s10518-018-0448-z

Steenbergen, M., Sýkora, M., Diamantidis, D., Holický, M., Vrouwenvelder, T.: Economic and human safety reliability levels for existing structures. Struct. Concrete. J. Fib **16**(3), 323–332 (2015)

Sýkora, M., Diamantidis, D., Holický, M., Jung, K.: Target reliability for existing structures considering economic and societal aspects. Struct. Infrastruct. Eng. **13**(1), 181–194 (2017)

UNITED NATIONS OFFICE FOR DISASTER RISK REDUCTION (UNISDR, 2012). Annual report 2012, Geneva, Switzerland (2012)

Zampilli, M.: Come affrontare il processo di ricostruzione dei centri storici, *Ricerche di storia dell'arte*, 2, Maggio-Agosto 2017. Società editrice Il Mulino, Bologna (2017)

Zaragoza, I.M.-E., Caroti, G., Piemonte, A., Riedel, B., Tengen, D., Niemeier, W.: Structure from motion (SfM) processing of UAV images and combination with terrestrial laser scanning, applied for a 3D-documentation in a hazardous situation. Geomat. Nat. Haz. Risk **8**(2), 1492–1504 (2017). https://doi.org/10.1080/19475705.2017.1345796

System Identification and Finite Element Modelling of Damaged Bal Mandir Monument in Kathmandu After the 2015 Gorkha Earthquake

Rajesh Rupakhety[1](✉), Dipendra Gautam[2,3], Rabindra Adhikari[2,3], Pratyush Jha[3,4], Lalit Bhatt[3], and Rewati Baruwal[3]

[1] Earthquake Engineering Research Center, Faculty of Civil and Environmental Engineering, School of Engineering and Natural Sciences, University of Iceland, Austurvegur 2a, 800 Selfoss, Iceland
rajesh@hi.is
[2] Department of Civil Engineering, Cosmos College of Management and Technology, 44601 Lalitpur, Nepal
[3] Structure and Infrastructure Dynamics Laboratory (SID-Lab), Interdisciplinary Research Institute for Sustainability, 44600 Kathmandu, Nepal
[4] Digicon Engineering Consult, 44601 Lalitpur, Nepal

Abstract. Bal Mandir is a Palladian style three-storied monument located in Kathmandu, Nepal. The monument was damaged by the 2015 Gorkha earthquake and part of the aggregate building was declared inoperable after post-earthquake damage assessment. Bal Mandir serves an orphanage and educational center for the orphans and homeless children; seismic performance of the monument is a crucial concern. Aiming to quantify the damage level of the aggregate masonry building, field reconnaissance was performed in preliminary to detailed level. Furthermore, ambient vibration tests were carried out to perform system identification of the building. The results of system identification were compared with that of finite element analysis and the results show that damage has caused detectable change in the dynamic properties of the structure. This paper reports some representative damage scenarios and the results of the finite element analysis as well system identification.

Keywords: System identification · Finite element modeling · Greco-Roman monument · Ambient vibration · Gorkha earthquake

1 Introduction

Commonly, monumental buildings are centuries old and their dynamic characteristics are uncertain. It is primarily because the historic construction system was largely practice-oriented rather than guideline-based. As monumental buildings have unique construction system, it is not difficult to generalize the characteristics of buildings falling under the same class of construction. To this end, understanding the behavior of such buildings

I. Vayas and F. M. Mazzolani (Eds.): PROHITECH 2021, LNCE 209, pp. 222–231, 2022.
https://doi.org/10.1007/978-3-030-90788-4_21

remains a challenging job for structural engineers. The recent development and advances in non-destructive in-situ measurement techniques have facilitated in understanding the behavior of such complex structures greatly. Structural system identification based on recorded earthquake response or ambient vibration is another popular method to estimate dynamic properties of such structures. Parametric and non-parametric system identification (Ködeström and Stoica 1989; Jenkins and Watts, 1968) and the various algorithms of these methods have been used in many studies (e.g. Guleret al. 2008; Sawaki et al. 2019 among others). Masciotta et al. (2017) performed dynamic characterization of historic masonry chimney in Portugal that was damaged by a lightening. They monitored the chimney before and after the lightening event and concluded that the disappearance of several modes occurred due to repair activities. Similarly, Diaferio et al. (2016) performed operational modal analysis (OMA) to estimate the modal frequencies and mode shapes of a masonry bell tower in Greece. They concluded that the material properties assumed after the damage may be varied significantly so dynamic properties are required to be determined using dynamic identification. Cimellaro, Piantà, and De Stefano (2012) performed ambient vibration measurement of the Margherita Palace in L'Aquila, Italy that was damaged by the 2009 L'Aquila earthquake. Using frequency-domain decomposition, random decrement combined with eigensystem realization algorithm, and natural excitation technique, they determined the dynamic characteristics of the structure. Potenza et al. (2015) performed three-year long monitoring of the Basilica S. Maria di Collemaggio that was damaged by the 2009 L'Aquila earthquake. Using stochastic subspace identification technique, they presented discussions on the near- and far-field micro-earthquake effects on the behavior of historic masonry structure.

Measurement of response during real earthquake as reported by Sawaki et al. (2019) are rare, and system identification often needs to rely on ambient vibration measurements. This paper reports and compares the dynamic characteristics of Balmandir inferred from ambient vibration measurements and finite element modelling. Of interest is examination in changes in the dynamic properties due to damage caused by the Gorkha Earthquake.

2 Bal Mandir

Bal Mandir is a Greco-Roman monument located at central Kathmandu (27°42′48.86"N 85°19′51.92"E). It is a load-bearing masonry construction system. Bal Mandir has three plus attic stories and two courtyards with wing projections in the north. The second and third floor heights are less than that of the first floor by up to 1m. The total ground coverage is 3894 m^2. Total floor area is 12282 m^2 excluding attic. The wall thickness is 900 mm at the ground floor, 800 mm at the first floor, 750 mm at the second floor, and 650 mm at the top floor. Some of the partitions were found to be made of up single brick or plywood. The main walls are brick in mud mortar. The floor structure is comprised of timber girder and joists, steel girder and timber joist, steel joist and brick/clay tiles, and steel I-beam and jack arch. The bricks found in the building show that the construction of the building took place between 1932–1936. The underlying soil is silty fine soil at foundation level with backfill of coarse sandy soil. Bal Mandir was used as Children home that housed around 300 orphan students until the 2015 Gorkha earthquake. However, it was used by the reining Rana dynasty for many years as a palatial building. The plan of the building is shown in Fig. 1. In Fig. 1, the locations indicated by circles show the locations of accelerometers used for ambient vibration measurement. Three instruments were installed, one each on ground, first, and second floors.

During the Gorkha earthquake of 25 April 2015, Bal Mandir sustained DS-3 damage grade per the EMS-98 damage grading system. The southern extension of the Bal Mandir collapsed as shown in Fig. 2a. Similarly, the eastern extension was heavily damaged as shown in Fig. 2b. Figures 2c, 2d, and 2e depict some local damages. In particular, the corners sustained tension cracks and subsequent separations. As shown in Figs. 2d and 2e, the wall portion above the openings were extensively damaged throughout the building.

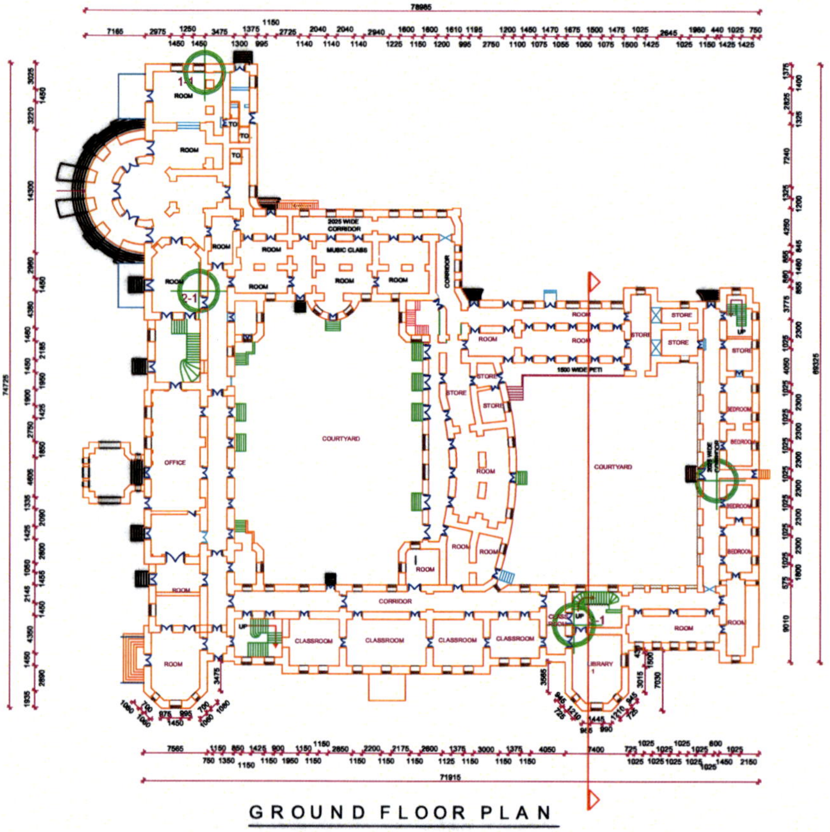

Fig. 1. Ground floor plan and instrument location (marked by circles). Horizontal and Vertical directions in this plan are denoted as X and Y axis, respectively.

Fig. 2. a) Collapsed southern extension of the building, b) damaged east face of the building, c) typical corner separation, d) damage above window, e) cracks above the arch and damaged floor-wall region

3 System Identification

Modal identification was performed by using ambient vibration measurement. Three digital accelerometers, one at each floor of the building were installed and ambient vibrations were measured at 100 Hz for 15 min. The accelerometers are ETNA-2 manufactured by Kinemetrics Inc. The location of the accelerometers is shown in Fig. 1. Each of the accelerometers recorded time through a GPS connection. The measured vibrations at the three floors were synchronized before being further processed.

Only the two horizontal components of vibration were used in system identification. The recorded signals were detrended before further processing. Each signal was divided into 10 segments. Each of these segments were checked for transients. The transients were removed by windowing the segments of the signal which looked stationary by initial visual inspection. Then stationarity was checked by looking at the cumulative power build-up of the signal. Roughly uniform slope of the power build-up was considered to imply stationarity. The windowed signals were low-pass filtered using a fourth order Butterworth filter with a corner frequency of 20 Hz. The filtered signals were then tapered with a Tukey Window. System Identification was then carried out using two methods described as follows.

3.1 Non Parametric Method

This method was based on the power spectral densities (PSD) of the recorded signals. PSD of the signals were computed by using the Welch's averaged periodogram considering 10 windows of equal length and 50% overlap between the windows. The signals at the first and the second floor are considered as response and those at the ground floor is considered as excitation. Denoting the PSD of response and excitation as $S_{yy}(f)$ and $S_{uu}(f)$, respectively, with f denoting frequency, an estimate of the squared amplitude of the frequency response function of the structure is given by:

$$|H(f)|^2 = \frac{S_{yy}(f)}{S_{uu}(f)} \tag{1}$$

The peaks of this function can be interpreted as the modal frequencies of the structure. The PSD of ambient measurements and the corresponding transfer functions are shown in Figs. 3 and 4 for X and Y directions respectively. All the 10 segments of the records show similar PSDF except at very low frequencies. This shows that the measurements are roughly stationary. The PSDF of ground acceleration is roughly constant over the range of frequencies of interest and may thus be regarded as a white noise process. The PSDF of accelerations at the first and the second floor show clear peaks at 2.1 Hz in the X direction and 2.7 Hz in the Y direction. These are the estimates of the fundamental frequencies of translational modes in X and Y directions. A second peak around 8Hz is visible in some of the transfer functions and might indicate the second mode. However, this mode is not resolved as confidently as the first mode. The peak at the fundamental mode is rather narrow, indicating small damping ratio, which will be further investigated in the following sections.

3.2 Parametric Method

The numerical algorithms for subspace space state system identification (N4SID, see for example, Ljung (1999)) was used for identification of the structure. This algorithm allows estimation of state matrices of the system from measured input and output. The state matrices can then be used to compute natural frequencies, damping ratios, and mode shapes of the structure, as well as its frequency response function. N4SID algorithm was used separately for the 10 segments of measurements. For each segment two state-space models were calibrated, one for X direction and the other for Y direction. The response was represented by accelerations at the first and the second floor, and the excitation was considered as unknown. Due to the almost white noise nature of the ground acceleration as discussed above, selection of this output only method is justified.

The magnitude of frequency response function expressed as dynamic flexibility in the X and Y directions are shown in Figs. 5 and 6 respectively. Each figure shows dynamic flexibility in the first and the second floor. Estimates from the 10 segments of the measurement and their average are shown with grey and black lines, respectively. These functions show clear peaks in both the floors at frequencies of approximately 2.5 Hz in the X direction and 2.1 Hz in the Y direction. The higher mode frequencies are not identifiable as clearly as the first mode frequencies, but some peaks in the range 5–10 Hz seem to suggest existence of higher modes of vibration in that range.

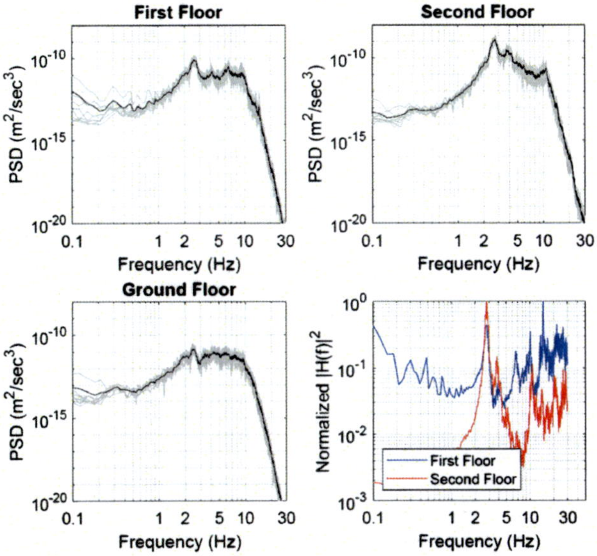

Fig. 3. Power spectral density functions of ambient vibration measured at the ground, first, and second floor of Bal Mandir and the corresponding amplitudes of frequency response function. X component of the measurements is used in this figure. The grey lines indicate the PSD of 10 windows of the signals and the black line is the mean PSD. The transfer function amplitudes are estimated according to Eq. 1 where the mean PSD at different floors are used. The squared amplitudes of transfer functions are normalized by their respective maximum values for easy visualization.

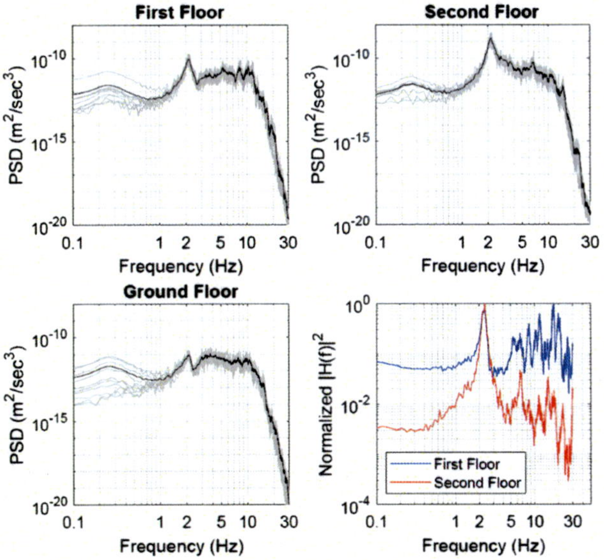

Fig. 4. Same as in Fig. 3 but of Y component.

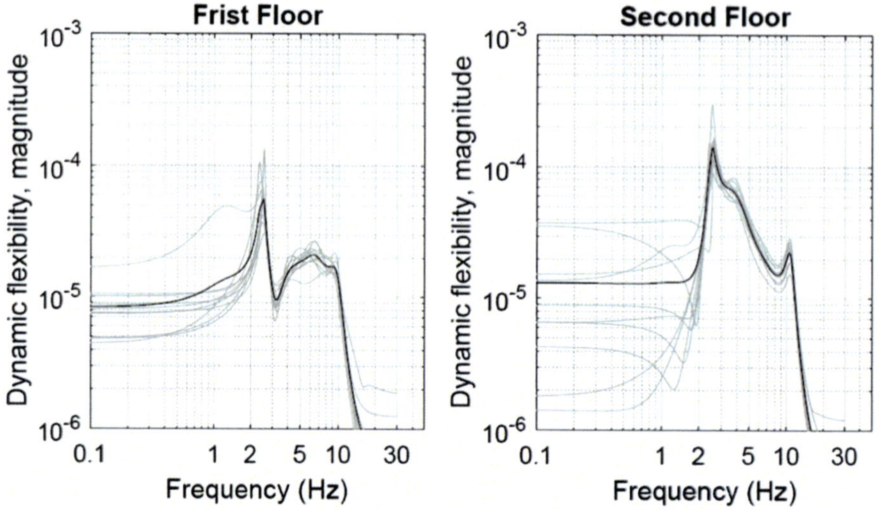

Fig. 5. Magnitude of frequency response function of the 10 segments of ambient measurements at the first and the second floor are drawn in grey and their corresponding averages are drawn in black lines (for X axis)

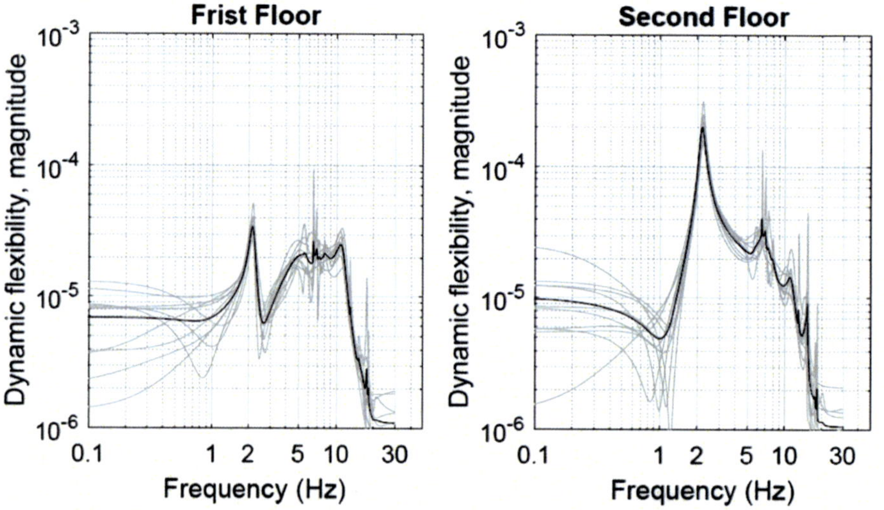

Fig. 6. Magnitude of frequency response function of the 10 segments of ambient measurements at the first and the second floor are drawn in grey and their corresponding averages are drawn in black lines (for Y axis)

The modal frequencies and damping ratios of the identified systems was computed by using the lease squares complex exponential method (see Brandt (2011)). The estimated periods of vibration and damping ratio of the 10 segments are presented in Table 1 along with their average and standard deviations. The average fundamental vibration period in the Y and X directions are 0.46 s and 0.40 s, respectively, and their variation across different segments is negligible. The corresponding frequencies are 2.2 Hz and 2.5 Hz, which are consistent with those estimated from the non-parametric method described above. The corresponding damping ratios are 5.2% and 5.7%. Variation in estimated damping ratio is larger than that in the natural periods. The results show that the first translational mode of the structure is in the Y direction.

Table 1. Natural periods of vibration and damping ratios estimated from the identified state space models

Segment	Period (s)		Damping ratio (%)	
	X	Y	X	Y
1	0.393	0.452	5.973	4.748
2	0.393	0.456	5.120	4.707
3	0.392	0.456	7.668	3.336
4	0.377	0.462	7.750	5.803
5	0.393	0.456	7.328	4.965
6	0.414	0.459	7.411	5.922
7	0.389	0.466	1.407	4.091
8	0.423	0.453	3.791	5.940
9	0.391	0.463	7.445	6.926
10	0.398	0.464	2.766	5.062
Mean	**0.396**	**0.459**	**5.666**	**5.150**
Standard deviation	**0.013**	**0.005**	**2.304**	**1.033**

4 Finite Element Modeling

Three-dimensional finite element model of the monument was prepared in ETABS program (Computers and Structures Inc. 2018). The model is shown in Fig. 7. The walls are modelled by shell elements and pinned support is used at the base.

The redistribution of stress once the tension cracks are formed near the plinth undergirds a more realistic assumption of the fixed base model. Iso-parametric 4-nodded and 3-nodded thin elements were used to model the masonry walls in the building. Timber joists and beams were modeled using line elements, whereas, the timber flooring was modeled using thin shells equivalent to timber planks which provide partial in-plane rigidity. The mechanical properties of materials were determined by laboratory testing. The average compressive strength was adopted as 1 MPa as recommended by the Indian Standard for the brick strength of 0.25 MPa. The Young's modulus was taken as 1000 MPa. The Poison's ratio of masonry was taken as 0.2. The height to width ratio of

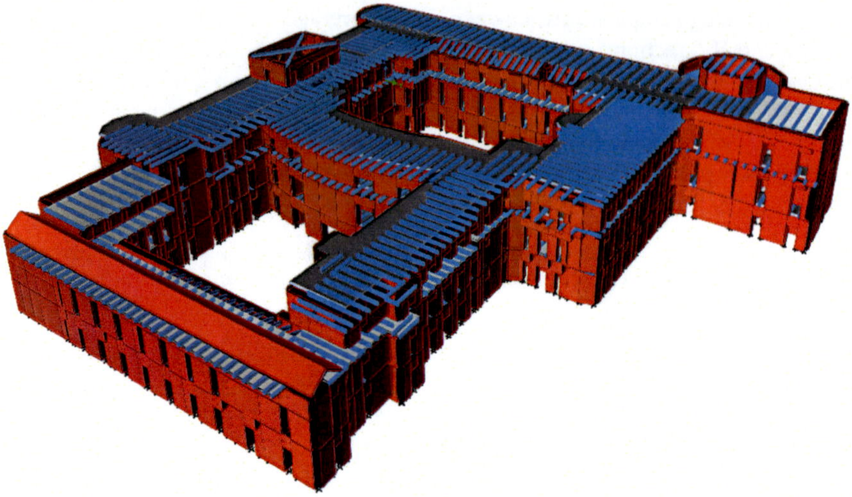

Fig. 7. Three-dimensional finite element model of Bal Mandir

brick units is less than 0.75, slenderness ratio of walls is below 6; and no portion of the wall was less than 0.2 m^2; thus, per the IS 1905:1987 recommendations, the permissible stress in masonry equals basic compressive strength of masonry, which is one fourth of the average compressive strength of masonry. The results eigen value analysis of the model shows that the first mode was dominant in Y direction with the period of 0.30 s. Similarly, the second, third, and fourth modes have periods of 0.29 s, 0.27 s, and 0.25 s respectively. It should be noted that a lot of local modes were present in the building. The modal mass participation was nearly 50% in the first 12 modes.

5 Concluding Remarks

The Bal Mandir, an aggregate masonry complex, was considered for system identification using ambient vibration records taken after the Gorkha earthquake that damaged the structure significantly. Ambient vibration records were processed using non-parametric and parametric system identification algorithms. Welch's Algorithm and N4SID methods. Each of the records was divided into 10 segments to estimate uncertainties in the estimation of natural frequencies and damping ratios. The results obtained from the non-parametric method using Welch's averaged periodogram estimate of frequency response function and the N4SID were similar. The period of vibration of the first mode estimated from different segments of the measurements showed very little variation. Variation in damping ratios was larger but acceptable.

A finite element model of the building was also prepared to compare the dynamic characteristics of the building before and after the earthquake. The finite element model is meant to represent the undamaged structure. The average fundamental periods in the Y and X directions from system identification are 0.46 s and 0.40 s respectively. The corresponding periods of the finite element model are 0.30 s and 0.26 s. The results show significant elongation of the vibration frequencies of the structure due to the damage caused by the earthquake. Both system identification and finite element analysis consistently showed that the first translational mode of the structure was in the Y direction. It is interesting to note that the building depicted 5.67% and 5.15% damping ratios in X and Y directions respectively, which give close approximation to the commonly used 5% damping in finite element models. Despite several uncertainties in mechanical modelling such as material properties, geometry, interfaces, connections, among others, the system identification results, the finite element model seems to be consistent with the measured vibration periods of the structure. The results can be used to update the finite element model of the structure. Such updated models can be used to access seismic vulnerability of the building and design appropriate intervention measures. Similar dynamic measurements after proper retrofitting can be useful to quantify the effects of retrofitting in the vibration periods of the structure. This study uses one sensor per floor; however, dense sensor network in each floor and a greater number of sensors would improve the outputs with better accuracy. So, future researchers are recommended to increase the number of accelerometers in each floor.

Acknowledgements. Authors acknowledge the support by Ram Udar Yadav, Gangesh Joshi, Binay Sah, Gokarna Motra, and others for supports. Rajesh Rupakhety acknowledges financial support from the University of Iceland Research Fund.

References

Brandt, A.: Noise and Vibration Analysis: Signal Analysis and Experimental Procedures. Wiley, Chichester (2011)

Cimellaro, G.P., Piantà, S., De Stefano, A.: Output-only modal identification of ancient L'Aquila city hall and civic tower. J. Struct. Eng. (2012). https://doi.org/10.1061/(ASCE)ST.1943-541X.0000494

Computers and Structures Inc.: "ETABS v.17." California (2018)

Diaferio, M., Foti, D., Giannoccaro, N.I.: Modal parameters identification on environmental tests of an ancient tower and validation of its FE model. Int. J. Mech. (2016)

Guler, K., Yuksel, E., Kocak, A.: Estimation of the fundamental vibration period of Existing RC buildings in Turkey utilizing ambient vibration records. J. Earthquake Eng. (2008). https://doi.org/10.1080/13632460802013909

Jenkins, G.M., Watts, D.G.: Spectral Analysis and Its Application. Holden-Day, San Francisco (1968)

Masciotta, M.G., Ramos, L.F., Lourenço, P.B., Vasta, M.: Damage identification and seismic vulnerability assessment of a historic masonry chimney. Ann. Geophys. (2017). https://doi.org/10.4401/ag-7126

Potenza, F., Federici, F., Lepidi, M., Gattulli, V., Graziosi, F., Colarieti, A.: Long-term structural monitoring of the damaged Basilica S. Maria di Collemaggio through a low-cost wireless sensor network. J. Civ. Struct. Heal. Monit. **5**(5), 655–676 (2015). https://doi.org/10.1007/s13349-015-0146-3

Sawaki, Y., Rupakhety, R., Ólafsson, S., Gautam, D.: System identification of a residential building in Kathmandu using aftershocks of 2015 Gorkha earthquake and triggered noise data. In: Rupakhety, R., Olafsson, S., Bessason, B. (eds.) ICESD 2017. GGEE, vol. 47, pp. 233–247. Springer, Cham (2019). https://doi.org/10.1007/978-3-319-78187-7_18

Söderström, T., Stoica, P.: System Identification. Prentice Hall International, Cambridge (1989)

The Urban Refugee Settlements of Xanthi (1922–1930): Recording and Typological-Morphological Analysis of the Existing Refugee Dwellings

Eirini Chara Tsetine(✉) and Nikolaos Lianos

Department of Architectural Engineering, Democritus University of Thrace (DUTH), Kimmeria Campus, 67100 Xanthi, Greece
{etsetine,nlianos}@arch.duth.gr

Abstract. This paper aims to present the first results of a research program(In the context of the doctoral thesis, conducted by the architect Mrs. Eirini Tsetine, with a scholarship from the Hellenic Foundation for Research and Innovation (HFRI), supervised by Prof. Mr. N. Lianos and is part of the research programs of the Morphology-Rhythmology Laboratory of DUTH.) which examines the refugee settlements that were established in the prefecture of Xanthi, in northern Greece, after the end of the Asia Minor Campaign (1919–1922), to permanently accommodate the refugees coming from the regions of Asia Minor, Eastern Thrace and Pontus. The refugee flow towards Greece, which was originally initiated before the armed conflict, continued during the events of the Asia Minor Campaign (1919–1922) and was finally concluded with the exchange of population between Greece and Turkey, as dictated by the Treaty of Lausanne (Lianos 2013, p. 1). The permanent settlement of the refugees was carried out under the auspices of the League of Nations, via the establishment of the Refugee Settlement Commission (RSC). The refugee settlements were divided into two main categories, rural and urban. Urban refugee settlements were created in cities such as Athens and Piraeus and regions such as Thessaly, central Greece, the Aegean islands, Macedonia and Western Thrace. Especially in the last area, where Xanthi is located, the need for housing was higher, as the Muslim population had been excluded from the compulsory exchange and for this reason, there were not enough empty properties to be reused for the settlement of the refugees. The urban refugee settlements of Xanthi belong in this wide network, that was established throughout Greece, and they consist of four areas known as "Stratonon", "Kyplselis" and "Gymnastiriou" settlement and part of the "Assa" district. The main features of this work are the presentation of the process of identifying and recording the existing urban refugee dwellings in Xanthi, as well as their classification in building types and their morphological analysis.

Keywords: Urban refugee dwellings · Urban refugee settlements · Xanthi

I. Vayas and F. M. Mazzolani (Eds.): PROHITECH 2021, LNCE 209, pp. 232–257, 2022.
https://doi.org/10.1007/978-3-030-90788-4_22

1 Introduction

1.1 General Approach and Brief Historical Events

The study and preservation of historic urban areas have become an important aspect of both the professional and academic sector that occupies with the conservation of architectural heritage.

This academic work focuses on the case of Xanthi; a city with a rich historical background that is located in northern Greece (Fig. 1). One of the elements that give Xanthi a special architectural character is the Old Town[1]. This act of statutory protection is very important for the preservation of the historic fabric and the formation of the local cultural identity. However, there is a large number of historic buildings, located outside this conservation area, which are not protected by any similar framework. A type of historic buildings that belong to this wide category is the refugee dwellings, which constituted the urban refugee settlements, built in the context of the permanent settlement of refugees in the Prefecture of Xanthi (1925–1930)[2].

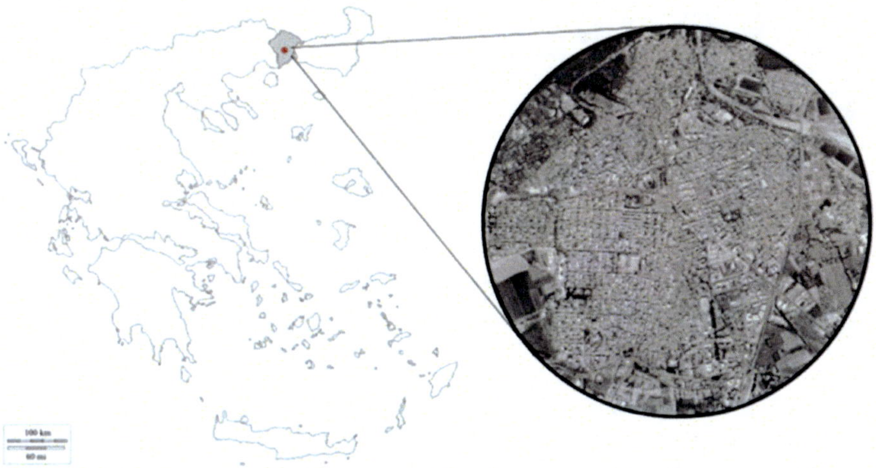

Fig. 1. Location of Xanthi within Greece (author's own).

As far as the exact number of refugees that arrived in Greece during the period 1919–1924 is concerned, one should mention that its identification is one of the most

[1] The historic center that was designated as a protected traditional settlement in 1976 with FEK 661/B/17-5-1976.

[2] The displaced population arrived in Greece firstly during the Asia Minor Campaign (1919–1922) and later with the mandatory population exchange between Greece and Turkey, as dictated by the Treaty of Lausanne in 1923 This agreement involved the Turkish citizens of the Greek Orthodox religion who lived in Turkey and the Greek citizens of the Islamic religion that lived in Greece. Moreover, two groups were excluded from the exchange; the Greek residents of Constantinople, who had settled before the 30th of October 1918, and the Muslim citizens of Western Thrace (Treaty of Lausanne, 1923).

perplexing issues, that historians have dealt with throughout the years. Most of them fled to Greece before the population exchange, under conditions that did not allow the systematic recording of the incoming population. For this reason, several researchers have presented different statistics and numbers. For example, Pallis (1977, p.78–79) estimated the number of displaced individuals to be 1.293.600 people, based on the 1928 national census (Table 1). However, one should consider that by 1928 the real number of refugees had been altered due to childbirths and deaths from various diseases (Lampidis 1992, p. 91). Finally, according to the League of Nations (LN), the Greek refugees were approximately 1.5 million (League of Nations 1924a, p. 1). As stated by Yildirim (2006, pp. 129), only 150.000 Greeks migrated according to the terms of the population exchange agreement. This brief discussion reveals the proportions of the problem Greece had to tackle, as the country received a vast number of refugees, almost equal to one-quarter of its local population (Maurogordatos 2008, p. 101).

Table 1. Table showing the number of refugees that arrived from each area according to the national census of 1928 (Lampsidis 1992, p. 92).

Area	Total	Before 1922	After 1922
Asia Minor	626.954	37.728	589.226
Eastern Thrace	265.635	27.057	229.578
Pontus	182.169	17.528	164.641
Constantinople	38.458	4.109	35.349

The issue of housing and integrating the refugees into the Greek host society was added to a long list of financial, social, and political problems (Apostolopoulos 2008, p. 26). During the early stages of refugee influx foreign organizations, such as the *American Red Cross*, the *Near East Relief* and the *All-British Appeal* provided the necessary humanitarian aid (League of Nations 1924a, p. 5). One of the measures taken by the Greek government was the requisition of existing public buildings and the mandatory housing of refugees within local households. Moreover, the Greek government established the Refugee Fund (RF), which undertook the creation of temporary refugee settlements in the areas around Athens and several provinces, such as Xanthi (Paralikis 2006). This institution functioned from November 1923 to May 1925. Within this period, it constructed 4.000 dwellings, while the construction of another 2.500 buildings had started before its dissolution (Gizeli 1984, p. 136).

The permanent settlement of refugees, although it was a priority, proved to be a rather difficult task for the Greek government, which in February 1923 asked the LN for its advisory and financial assistance (League of Nations 1924, p. 16–22). Thus, the Refugee Settlement Commission (RSC) was established on the 29th of September 1923, as an autonomous organization to undertake the permanent settlement of the Greek refugees and handle the funding provided by the loan (Ibid). In Thrace four RSC Headquarters were being operated, one of which was located in Xanthi (Koutsoukos 2012).

The RSC continued the work started by the RF with the difference that it created permanent settlements and did not provide temporary humanitarian relief. According to the RSC's 1st Quarterly Report, the purpose of the organisation was the *"...establishment of refugees in productive work..."* (League of Nations 1924a, p. 5), which means that commission aimed to provide not only a permanent shelter to the refugees but also the base for their new life, making them self-sufficient.

The required land was put at the disposal of the RSC by the Greek state and consisted of no less than 500.000 hectares (Ibid, p. 4). The Greek government expropriated the large agricultural estates, known as *chifliks*, as well as 500.000 rural dwellings, which were abandoned by the exchanged Muslim population (Exarhou 2011, p. 158–159).

The settlements were divided into two major categories: rural and urban, and the organisation attempted to allocate the refugees according to their previous financial and social status, whenever this was possible (Gizeli 1984, p. 130). By January 1929 the RSC had established 59 urban refugee settlements in the areas of Athens, Piraeus, Thessaly, Volos, Central Greece, Peloponnese, the islands, Macedonia and Western Thrace (Paralikis 2006). Especially in the latter, the need for housing was even more urgent as the Muslim population of the area was excluded from the Population Exchange and there were not enough empty properties to be reused (Exarhou 2010, p. 21). These refugee suburbs were constructed deliberately on the outskirts of existing cities so that the settlement of the new social group would not trigger any kind of conflict (Lampsidis 1992), following in several cases the provisions of the Legislative Decree *"On plans of cities, towns and settlements of the State and their construction"* of 1923 (FEK 228/A/1923).

To this wide network belong the areas of Xanthi, which were created based on a new city plan and consist of the Stratonon, Kypselis, Gymnastiriou settlements and part of the Assa district (Archive of Eleftherios Venizelos, File 213, 03–09).

2 City of Xanthi: Brief History and Urban Development

Xanthi existed as a walled settlement since the Byzantine period and developed into an important town during the 16th century (Patrikios et al. 2008, p. 209). In 1829 two major earthquakes demolished a great number of buildings, which were reconstructed during the period 1829–1860 (Pentazos 1979, p. 14). After this stage, the town was still enclosed within its previous limits that correspond almost entirely to those of Xanthi's Old Town nowadays.

During its second phase of development (1860–1912) (Fig. 2), the city expanded out of its former boundaries and towards the southern plains (Ibid). More specifically, in 1872 Xanthi became the headquarters of the local Ottoman administration and its urban centre was organised according to the typical Ottoman plans, with a Kulliye at the location of the modern central square (Melkidi 1991, p. 16). A Kulliye usually combines a variety of administrative, religious, and social functions such as a mosque, a fountain, both outdoor and indoor market, an inn (han), a school (madrasa), a Turkish bath (Hamam), a hospital, a poorhouse (imaret), a cemetery and other auxiliary uses (Goodwin 1987). At the same time, the construction of the railway line Thessaloniki-Constantinople benefited the financial transactions and commercial activities (Melkidi 1991, p. 16), especially the export of tobacco (Bakirtzis 2008, p.31–32). Xanthi was then led into the most prosperous phase of its history, a period during which the industrial zone of tobacco warehouses was constructed following the Ottoman city planning regulations of 1882 and the Hatzistavrou neighbourhood by tobacco workers (Patrikios et al. 2008, p. 212).

The third period of urban development was characterised by a series of turbulent political and social events. On November 8th 1912, during the First Balkan War (October 1912 - May 1913) Xanthi was occupied by Bulgarian forces. Subsequently, in 1913, with the end of the First Balkan War and the signing of the Treaty of London (17/30 May 1913), Western Thrace, and consequently Xanthi, was found under Bulgarian rule (Kerkidis 2006, p. 30). Within a few months, the Second Balkan War broke out (June 29 - July 31, 1913) and on July 13th 1913, Xanthi was liberated by the Greek army for only 15 days. The Second Balkan War ended with the signing of the Treaty of Bucharest (July 28/August 10, 1913), which defined the borders of the Balkan countries with Xanthi being once again under Bulgarian administration (Ioannidis 1980).

The situation of the city remained the same until the end of World War I (July 28, 1914 - November 11, 1918), in which Greece participated on the side of the Entente. In September 1919, the Supreme Council of the Allies in Paris imposed the evacuation of Western Thrace by the Bulgarians and on October 4, 1919, the 9th Greek Army Division entered Xanthi, under Lt. Georgios Leonardopoulos, by order of the Great Powers. Then forces of the French army occupied the rest of Thrace under the command of General Sarpi, who from October 10, 1919, settled in Komotini. On October 20th of the same year, the provisional Inter-Allied Administration was established with Komotini as its capital and General Sarpi took over the military and political administration (Kerkidis 2006, p. 30).

The allied forces divided the area into 3 districts: 1) Xanthi, 2) Komotini and 3) Karagats (Orestiada). In contrast to the other two, where French military commanders were appointed, in Xanthi George Leonardopoulos took over as commander. On November 27th, 1919, the Treaty of Neighue was signed, according to which Bulgaria relinquished all its sovereign rights in Western Thrace (Bakirtzis 2008, p. 32). The inter-allied administration of the region lasted for seven months, until May 1920, when Greece was given the right to establish its sovereignty. Finally, on July 28th/August 10th, 1920, the Treaty of Sevres was signed, by which Western Thrace and of course Xanthi were ceded to Greece.

Amid these unrests, Xanthi was called to face the influx of refugees, both those who fled to Greece before the Asia Minor campaign (1919–1922) and those who moved with the Exchange of Populations and the evacuation of Eastern Thrace, as it was suggested by the Treaty of Lausanne. The creation of rural and urban refugee settlements in Xanthi and its surrounding regions took place in this historical context.

Fig. 2. View of Xanthi around 1900, before the settlement of refugees (personal archive of Tefronidis A.).

3 The Urban Refugee Settlements of Xanthi and the Recording of Existing Buildings

As has been mentioned above, four urban settlements were built in Xanthi, from 1925 to 1930 (Fig. 3, 4). Archival material has provided information on the number of urban refugee houses, as shown in Table 2. Moreover, the same document, entitled *"List of urban houses that have been built and will be built in Thrace"* (1930) sheds light on the cost of these buildings which reached the amount of 188.496.894 drachmas in total for the area of Western Thrace.

Table 2. Table showing the number of refugee houses in the urban settlements of Xanthi (Eleutherios Venizelos Institution Archives, Folder 213, 03–09).

Settlement name	Number of houses
Stratonon	1228
Gimnastiriou	200
Kipselis	350
Assa	90

The construction of the urban refugee areas resulted in the expansion of the city's limits and contributed to the increase of Xanthi's population. A document from 7th July 1923, found in the Archive of the Ministry of Foreign Affairs, mentions that in 1923 the number of refugees settled in Xanthi was 8.832 individuals while the local population counted 16.584 people. According to the same document the total number of refugees in the Prefecture of Xanthi was 18.613, while a few months later this number increased to 23.000 (Newspaper "Fos", 29th September 1923).

Fig. 3. View of Xanthi in 1952. In the background the Stratonon and Kypselis settlements are visible (personal archive of Tefronidis A.).

The documentation of the existing historic buildings was initially carried out employing a photographic recording, along with the identification of their position on the modern city plans. The next stage of the process included the matching of the old plans (Fig. 5 a, b), found in archives, with the current situation. This was necessary to understand to which of the original refugee dwellings correspond the surviving structures. Subsequently, the numbering system of the old documents was used to name the existing buildings via the following code: initial letter of refugee settlement (S for Stratonon, K for Kipselis, G for Gymnastyriou) - block number - plot number-letter of property (A or B). For example, the name S_12_102_A means that this specific building is located in the Stratonon settlement, in Block No 12, plot No 102 and is property A. As there was no numbering for the Assa district plots, a new one was created using a similar system: A_ building number 1-n.

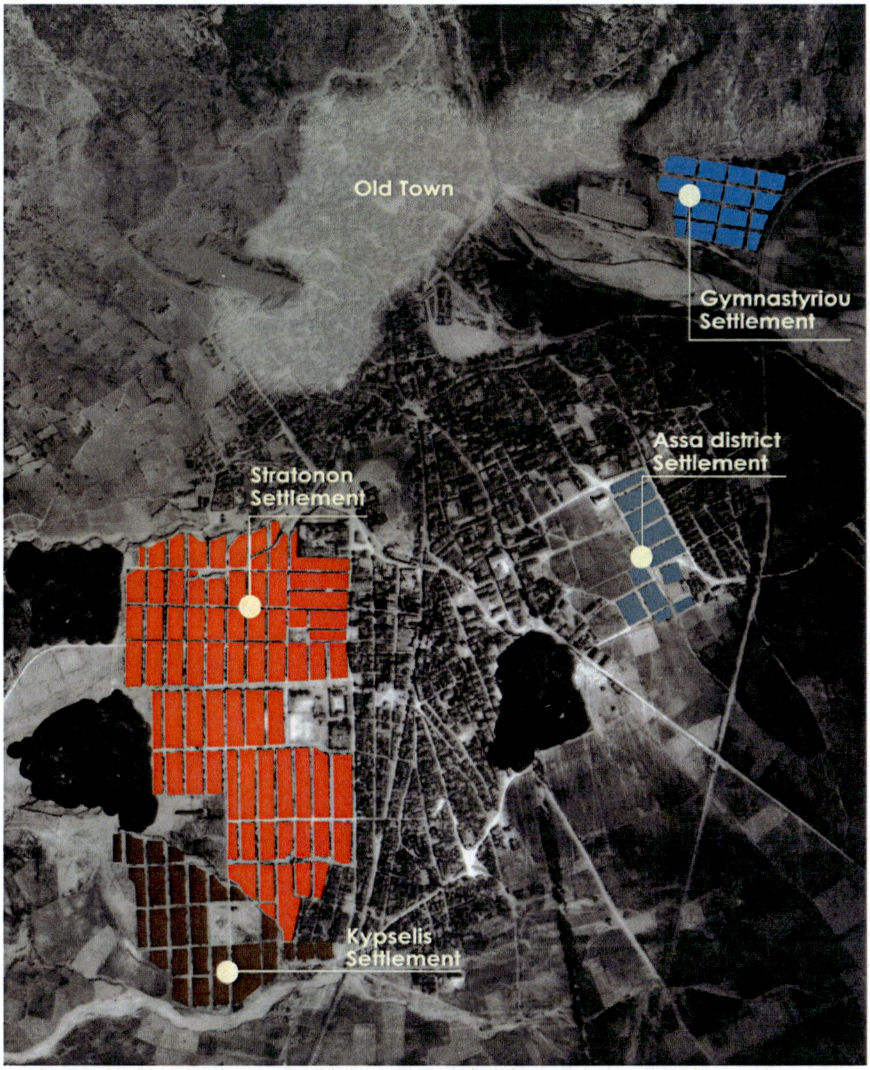

Fig. 4. Aerial photograph of Xanthi in 1938 (Hellenic Military Geographical Service). In color are the locations of the four urban refugee settlements (author's own).

During this process, one was able to obtain quantitative results regarding the number of existing buildings in comparison to the original plans, which are presented in Table 3.

a

b

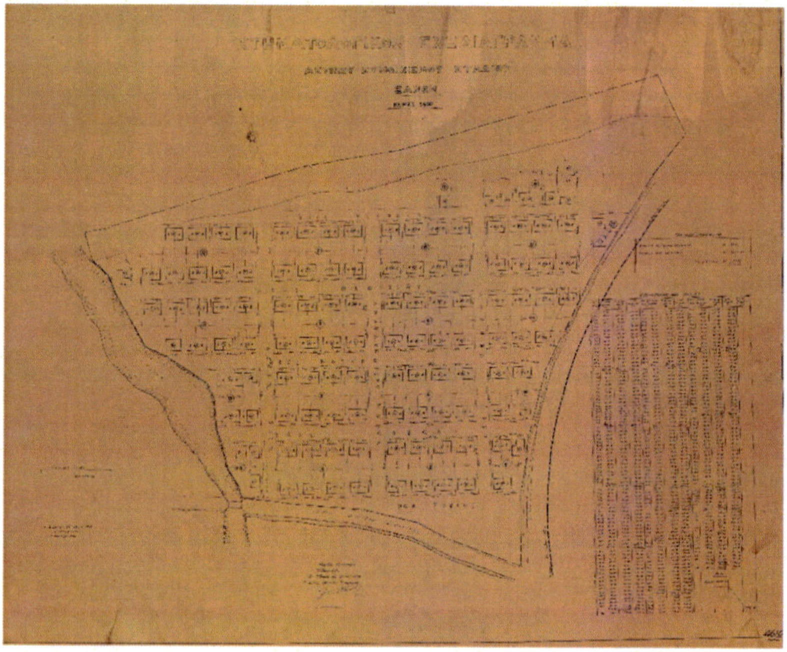

Fig. 5. a. Original old plan of Stratonon and Kispelis settlemetns (1928), b. Original old plan of Gymnastyriou settlement (1928) (General State Archives, Rodopi division).

Moreover, the current condition of the area was documented using as background the city plan of 1985, which was provided by the Municipality of Xanthi. The modern alterations were identified using a series of orthophotographs, that constitutes part of a survey conducted by the Greek Topographical Service, and extensive in situ observations. The outcome of this process is several maps that show the evolution of the built environment and its gradual domination over the open space and original structure of the refugee settlements. The geometrical characteristics such as the road grid and the shape and size of the blocks remain the same nowadays, except for slight differences, which regard mainly the creation of new blocks and parks. One notes that apartment blocks and other modern structures have replaced the original dwellings, disturbing the equilibrium of the original urban plan, whose basic principle "street-building-garden" nowadays has almost entirely vanished. Another impressive fact is that the urban fabric has raised vertically instead of spreading. In this way, the city maintained its strict original limits, while new urban areas could have been created without altering the historic refugee settlements (Fig. 6).

The massive construction of apartment blocks and internal migration of the rural population to the city characterised Xanthi during the 70s (Giannopoulou- Roukouni et al 1984, p. 188). The first regions where the new building activity took place were the urban refugee settlements, as there existed well-defined blocks, a clear status of ownership and a high building coefficient (Patrikios et al. 2008, p. 217). It is important to mention that at that time the areas under examination were not considered historic and from this point of view it was anticipated that the residents would seek the modernisation of their daily life. Progress and evolution are not negative on their own, but in this case, the urban fabric evolved without proper control and planning.

Table 3. Table showing the number of existing refugee houses in the four urban settlements of Xanthi (author's own).

Settlement name	Original number of properties	Number of existing properties
Stratonon	1228	284
Gimnastiriou	225	112
Kipselis	350	48
Assa	90	54
Total	1893	498

Fig. 6. Views of refugee houses in 2021 and 1930s: the high-rise apartment blocks have replaced the old refugee dwellings (author's own and personal archive of Tefronidis A.).

3.1 The Stratonon and Kypselis Settlements

The Stratonon and Kypselis settlements are examined and presented together because they constitute a single unit within the modern urban fabric, located in the southwest part of Xanthi.

The main source of information regarding this area was a topographic plan, made in 1947 by the engineer Konstantinos Benakis, which is an exact copy of the original RSC plan dating from 1928 (General State Archives). It provides the exact limits of the area given to the RSC and the RF, while at the same time depicts some buildings of particular importance. These include the old Ottoman barracks, demolished nowadays, which give the settlement Stratonon its name (Stratonon means "of the barracks"), the 4th Elementary School and a gunpowder warehouse, which occupied, probably for security reasons, an area of eight plots (Fig. 7). According to the testimonies of locals, this warehouse existed until the 1960s when it was finally demolished, and the refugee residences were built with almost forty years of delay.

One may also notice that the limits of the two settlements were surrounded by streams, until 1968 when the local authorities started the process of desiccation (Patrikios et al. 2008, p. 216).

An interesting fact regarding the names of the settlements is that some of Stratonon houses are known, even nowadays, as *Tsimentenia*, taking their name from their basic material, cement blocks. Moreover, the RF houses were also called *Chirika*, named after the Greek word *chira*, meaning widow, because that land was intended to house the widow refugees and their children (Kanellopoulos 1950).

From the original 1578 properties, today one may identify 332 existing refugee houses (Fig. 8).

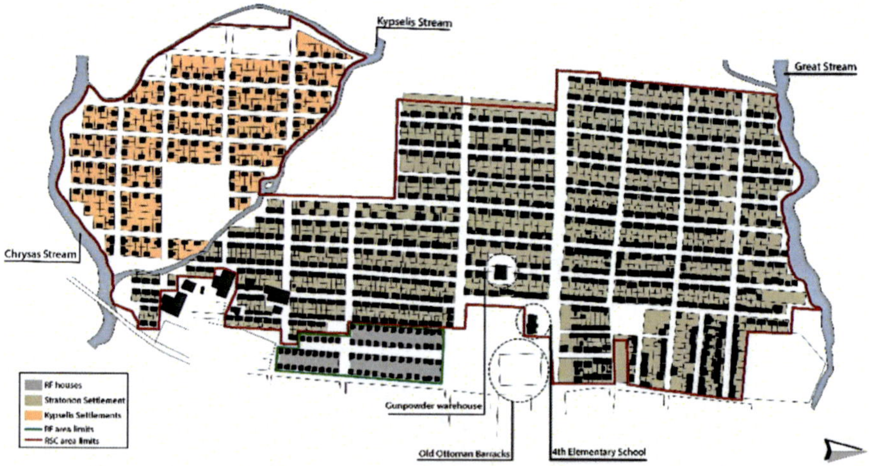

Fig. 7. Map showing the division of the refugee settlements according to the 1947 plan, the limits of the RSC and RF land and the extra buildings (author's own).

Fig. 8. Map showing the existing refugee houses in Stratonon and Kypselis settlements highlighted in yellow (author's own).

3.2 The Gymnastyriou Settlement

The Gymnastiriou settlement is located in the northeast part of Xanthi, in proximity to the Old Town, Kosynthos river and the railway. This settlement takes its name from an old stadium that existed in the area (Fig. 9), which has now been replaced by modern sports facilities. In later years, the settlement has been known to the locals also as "Konstantinou and Elenis", this time taking the name of a nearby small church. For this research, the name Gymnastyriou has been selected as it is the one mentioned in archival sources. From the originally 225 properties, today 112 still exist (Fig. 10).

Fig. 9. Panoramic view of Xanthi circa 1925. The stadium "Gymnastyrion" is visible at the bottom left (personal archive of Tefronidis A.).

Fig. 10. Map showing the existing refugee houses in Gymnastyriou settlement highlighted in yellow (author's own).

3.3 The Assa District Settlement

This settlement constitutes the expansion of the pre-existing Assa neighbourhood and is located in the eastern part of the city. It was the smallest urban settlement of Xanthi consisting only of 90 properties, that correspond to ten blocks. The most interesting fact about this area is that the refugees were settled exactly next to the local population and not in the outskirts of the city as it usually happened. Nowadays 54 of the original structures remain close to a great number of old houses that belong to Xanthi's second stage of urban development (Fig. 11).

4 Typological Analyses of the Urban Refugee Houses

A new urban model was introduced in Greek cities with the establishment of refugee settlements. In the case of Xanthi, the urban model was simple, and it consisted of a Hippodamian grid that formed rectangular blocks. More specifically, Stratonon settlement is organised in 68 blocks, Kipselis settlement in 28, Gymnastiriou in 16 and Assa in 10 blocks. In many cases, the blocks have a slightly irregular shape due to elements such as roads or the railway.

Fig. 11. Map showing the existing refugee houses in Assa settlement highlighted in yellow along with other old houses in light grey (author's own).

A typical block was divided into rectangular plots of the same size, with the dwelling being usually in contact with the building line, while a small garden existed behind each house (Fig. 12). As stated by Zardalidis, in 1925 the urban settlements of Xanthi were an impressive sight (Exarhou 2013, p.140). Plans that were found in the National Archives of Greece reveal that this type of city planning was also applied in other provincial refugee settlements.

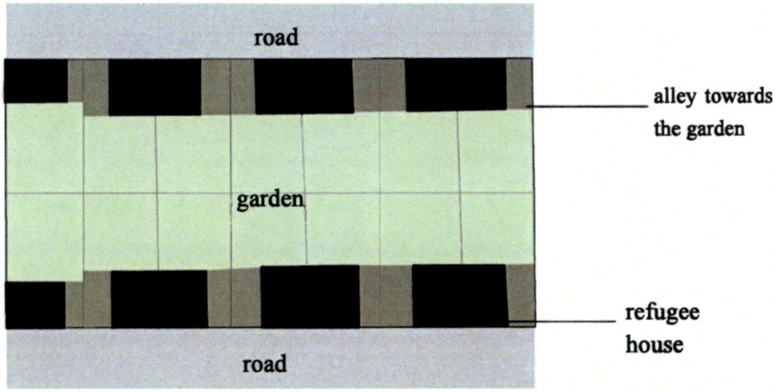

Fig. 12. Diagram showing the division of a typical block (author's own).

As mentioned earlier, the RF was the first organisation that constructed houses to relieve the refugees. Multiple bibliographical sources agree that those dwellings had temporary character and for this reason, cheap building materials were applied (Morgenthau and Strother 1994, passim.). The load-bearing structure was made of timber; the roofs were covered with corrugated metal sheets or bituminous paper and common lavatories existed in the centre of each block. Those early buildings did not have a particular typology because they were constructed very quickly to tackle the issue of homeless refugees. Additionally, during that period one could notice numerous cases of self-housing as the refugees had built houses, using for example mudbricks that had dried out in the sun[3], in their attempt to protect themselves and their families (Ibid).

Subsequently, the RSC took over the permanent settlement of refugees. In the early stages, the committee also constructed low-quality shelters, but after signing contracts with foreign industries, it started using materials of higher standards (League of Nations 1924b, p.4). The construction of the settlements was realised via direct assignments to construction companies, both Greek and foreign, and by the refugees themselves (Arabatzi 2005). In the second case, this decision was taken on the 10th of March 1925, during the Refugee Conference of Thessaloniki, as the refugees were disappointed by the delayed delivery of houses (Exarhou 2011b, p. 88). The responsible authorities provided all the necessary materials while local surveyors supervised the building process (Fig. 13).

Fig. 13. Refugees build a house on Perikleous street in Xanthi, 1926, Stratonon settlement (personal archive of Ramadan Oglou S.).

[3] Well known technique from their homeland, applied also in other refuge settlements (p.e. Kaisariani) cnf. N. Lianos, The housing settlement of refugees (in Greek: Η στεγαστική αποκατάσταση των προσφύγων)), LIFO 16.12.2016.

The engineer in chief was named Sgoutas and had already cooperated with the RF (Morgenthau and Strother 1994). Together with his team, they ended up with a combination of local stone or bricks, with an external layer of lime plaster and a roof covered with local ceramic roof tiles, while towards the end of the committee's operation they started using concrete blocks (Paloukis 2011). It is noted that in in many buildings of Xanthi they also used French ceramic roof tiles. Moreover, the ground floor was most often made of wood beam, covered with beaten earth that seems to have been the most affordable solution, but the refugees used to replace it with ceramic tiles or mosaic (Ibid). This fact reveals that the refugees used to have a different lifestyle and even if they were found in a very difficult situation, they tried to increase the quality of their accommodation.

The refugee houses were in most cases based on typical plans and were the result of mass production (Fig. 14) (Vasiliou 1944, p. 74). Generally, there were two basic types of urban dwellings: a) double dwelling with one floor for two families and b) double dwelling with two floors for four families. Each family inhabited three rooms, the larger of which served a dual function; it was both living room and bedroom, while both building types had a garden (League of Nations 1924b, p. 4–5). Several researchers mention the following types as well: a) small single dwellings with one floor that had one living room and an auxiliary space divided into kitchen and lavatory, b) single dwellings with two floors, usually for wealthier families, c) houses in a row with one or two floors that were inhabited by 6–12 families per floor, d) single dwellings with one floor that had one room, hall and kitchen, e) corner houses that were intended for larger families, f) *germanika;* timber frame houses with nine rooms whose construction was funded by the German war compensations, and g) refugee apartment blocks that were built after

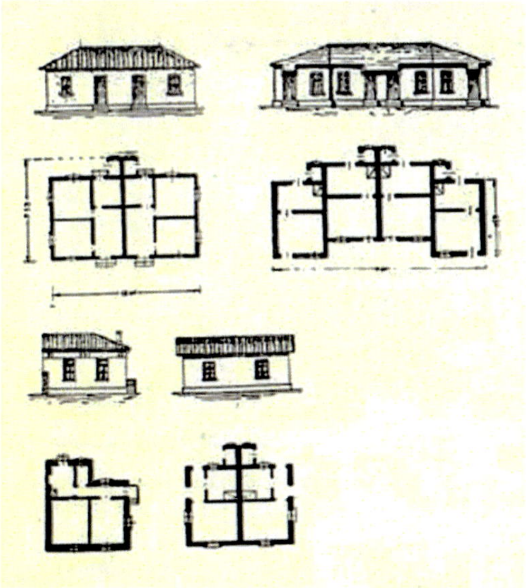

Fig. 14. Typical plans of the RSC urban dwellings (Vasiliou 1944, pp. 72).

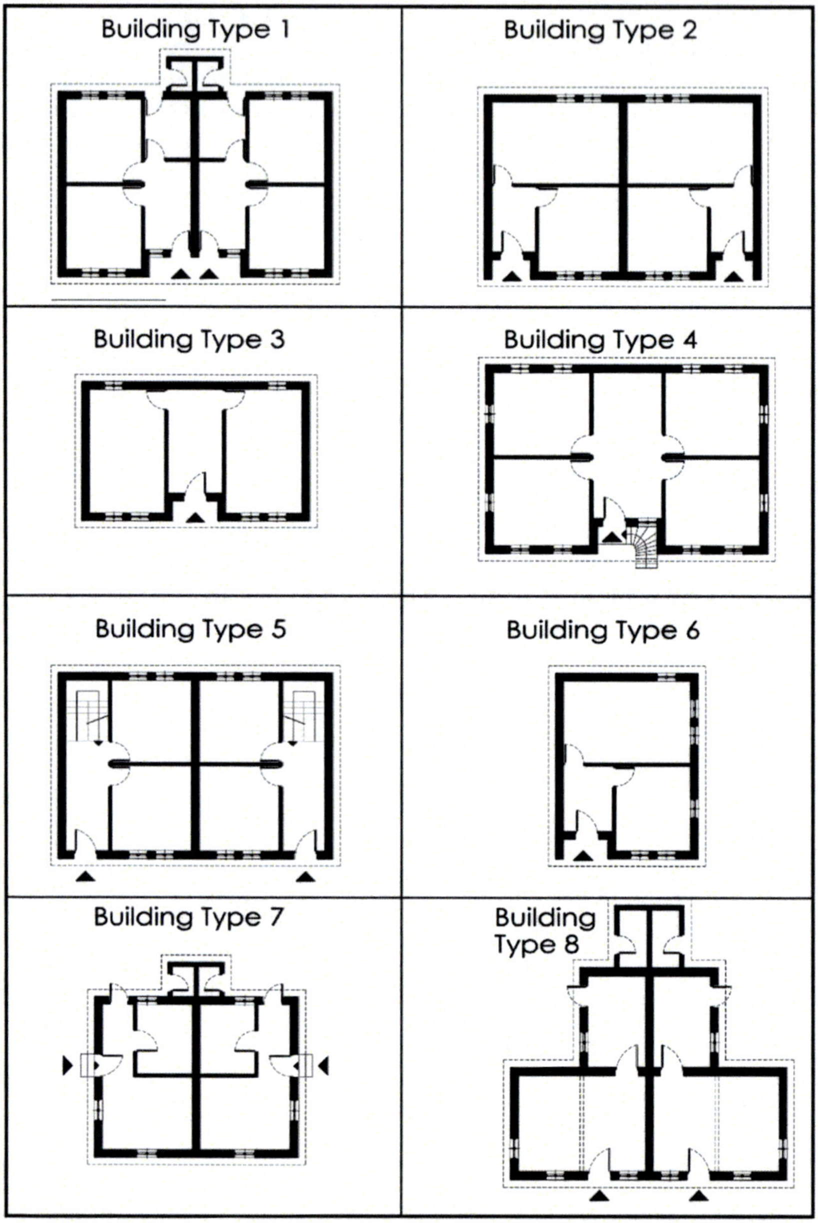

Fig. 15. Types of refugee houses found in Xanthi (author's own 2021).

the 1930s (Papaioannou 1975; Gouvousi 2011). According to Filippidis (1984), despite the obvious typological repetition, by 1930 over a hundred different variations had been used.

In the case of Xanthi's settlements, one may notice several building types and their variations. The houses were built mainly by the refugees themselves with materials provided by the RSC. Only in the Kypselis settlement were the refugee houses built almost exclusively by contractors (Exarhou 2011, pp. 157).

The basic building types in Xanthi are the following (Fig. 15):

a) Type 1: double dwelling with one floor/one floor and semi-basement, for two families, with hallway, three small rooms and auxiliary space in the rear, two separate central entrances in recess.
b) Type 2: double dwelling with one floor/one floor and semi-basement, for two families, with hallway and two rooms, entrances in a recess at the ends of the facade.
c) Type 3: double dwelling with one floor/one floor and semi-basement for two families, with hallway and at least one room for each family, one central entrance in a recess.
d) Type 4: double dwelling with two floors/ one elevated ground floor and sei-basement, for four families, with hallway and two rooms for each family, one central entrance in a recess usually with external staircase.
e) Type 5: double dwelling with two floors, for four families, with hallway and two rooms for each family, two entrances at the ends of the façade, with internal staircase.
f) Type 6: single dwelling with two floors/ one elevated ground floor and semi-basement, with hallway and at least two rooms, grand entrance in a corner recess.
g) Type 7: double dwelling with one floor, for two families, with hallway, two small rooms and auxiliary space in the rear, two separate entrances at the side facades.
h) Type 8: double dwelling with one floor, for two families, with hallway, at least two small rooms and auxiliary space in the rear, T shaped plan, two separate central entrances.

Types 1–7 have rectangular floor plans, either square or orthogonal, in contrast to type 8 that has a T- shaped plan. An important feature is the position of the entrance to the street, its position within the façade and the number of doorways. They are very important as they affect not only the interior arrangement of the building but also the relationship between the residents. For example, in buildings of type 3, with a single entrance, the division of the two properties is internal, which means that the users interact daily. On the contrary, in dwellings of type 1, there are two separate doorways, and the properties are distinguishable from the exterior.

Also, the doorway has a significant role, as it is formed either in a recess or directly on the facade. When it is located in a recess, it creates a transitional space, a type of threshold between the street and the dwelling, which was most important for daily social life, as shown in old photographs (Fig. 16 a, b). This fact is important for another reason as well. The formation of the entrance in a recess was a common practice in Eastern Thrace, a location where a great number of refugees derived from. Thus one may assume that the refugees adapted the typical plans according to the building practices they were familiar with (Fig. 17 a,b).

Types 1–6 can be found throughout the Stratonon settlement. In the Gimnastiriou settlement, one may notice types 1, 2, 3 and 6, while in the Assa neighbourhood the houses belong to types 1, 2 and 3. Types 7 and 8 are found only in the Kysplelis settlement.

Fig. 16. Women t aking dressmaking lessons in a dwelling's t hreshold, circa 1952, Themistokleous st reet, Xanthi, St ratonon set t lement (personal archive of Baltazi S.).

a

b

Fig. 17. A) House in Eastern Thrace with middle entrance in recess (Mavridis 2006, pp. 57), b) Refugee dwelling in Xanthi, with middle entrance in recess (author's own).

5 Architectural Forms Analyses of the Urban Refugee Houses

Xanthi is an outstanding example of architectural expression, as far as the morphology of the buildings is concerned. The city, due to its rich historical background, constitutes nowadays a hybrid of architectural rhythms, ranging from traditional architecture to neoclassical and eclectic styles. During the city's great financial development between the late 19th and early 20th centuries, entrepreneurs, and other wealthy citizens, who travelled abroad for business or leisure purposes, brought to Xanthi all the European architectural trends that impressed them and incorporated them into their manor houses (Lianos 2008, p. 82).

One should consider that this variety of architectural styles existed when the refugees arrived in the 1920s. Despite the consistency indicated by the old plans and the uniformity, the refugee settlements seem to have in old photographs taken from a long distance, the reality is much different (Fig. 18 a, b). Only a limited number of dwellings look alike.

Fig. 18. A) Old photograph of Gymnastyriou settlement, the houses used to be quite similar (personal archive of Tefronidis A.), b) Two separate properties with different facades, refugee dwelling S_49_211_AB (author's own).

The structures might belong to the same building type, but their external appearance is entirely different. The first explanation for this is the division of double dwellings into separate properties. The owners treated their properties as two different buildings and thus used diverse decorative elements. One may notice many cases where one property is highly decorated or painted in a bright colour and the other is plain.

Fig. 19. Examples of refugee houses in Xanthi with interesting morphological features (author's own).

In the 19th RSC Quarterly Report, it is mentioned that "*While on the subject of uniformity, we should mention that most of the urban quarters established in the centres of Western Thrace, particularly a t Comotini and Xanthi, form an exception to this rule. We are not referring to several quarters built by private enterprise but to hundreds of six- to eight-roomed houses which our Thrace Direction has caused to be constructed by providing loans or building material up to the value of 30,000 drachma. The refugee carries out the constructional work himself, or at any rate supervises it, and makes any additions he thinks fit a t his own expense. Thus quarters have sprung up in which each house has a style and individuality of its own. Some of the houses indeed are in excellent taste; and as every owner gives free play to his imagination or desire for decorative effect by a bright touch here or there to the exterior or interior of his house, the general aspect of these quarters finally becomes most agreeable"* (League of Nations 1928, pp. 22).

The existing old historic centre and their homelands constituted sources of inspiration for the refugees. They started imitating the style of Xanthi's manor houses and at the same time incorporating traditional elements and craft techniques from their own culture. Consequently, one might suggest that the refugee dwellings demonstrate features of vernacular architecture even though they were products of typical plans.

One may point out a list of morphological elements that are found in many dwellings, in a variety of combinations (Fig. 19): a) cornices around the openings, b) parapets of various forms on the roof level, c) roofs with gables or pediments, d) quoins of various forms, e) dripstones above windows, f) canopies above doorways or balconies, g) decorative engravings made of plaster, h) projecting bands on the facades, either plain or with ornamental motifs, i) traditional decorative techniques.

The buildings in the Kypselis settlement seem to have a simple morphology without any decorations, combined with simple typology. On the contrary, the dwellings of Stratonon settlement have more elaborate facades, incorporating decorative elements, regardless of their dimensions. This is an indication of the background the families had when they were settled in Xanthi. Wealthier families seem to have inhabited Stratonon settlement as many houses resemble small manors and not typical refugee dwellings.

6 Conclusion

Taking into consideration all the information presented in the previous sections, it becomes apparent that the districts of the refugee settlements are of great importance for the history and evolution of the city of Xanthi. They are part of a broad network of both rural and urban settlements throughout Greece. Consequently, they were not the result of a local necessity and have an additional group value, which reinforces their historical significance. The study of the historical events and the complex international relations of that era has shown that the refugee settlements including the ones in Xanthi are inextricably associated with events of national and global importance.

Moreover, the refugee settlements were the result of the work conducted mainly by the RSC, under the aegis of the LN. This is of paramount importance as the project of housing the Greek refugees of 1922 was one of the first humanitarian acts of this scale and kind.

Also, it is important to keep in mind that the arrival of the refugees and their interaction with the local population constituted the base on which modern Greek society was formed. Their extroverted way of daily life could give an alternative to the modern lifestyle. People in that era had a strong sense of community and were in constant interaction and interdependence with each other. They cared for their neighbors as if they were their families and helped one another to overcome their difficult situations in life. The refugee settlements did not have all the materialistic comforts of modern housing but their spatial organization as model villages would be interesting if applied again in modern cities. The single or two-story dwelling with a garden seems to be an effective alternative to the apartment blocks, where one often notices minor interaction and social alienation among the residents.

Subsequently, in the city of Xanthi, the refugee settlements are highly significant from a sociological point of view. The city had always been a multi-cultural society and the arrival and permanent settlement of a new population added a vibrant group to the existing social structure. For these reasons, it is very important to preserve and protect the remaining historic urban fabric as it is tangible evidence of the refugee presence in the city.

Most importantly the research has shown that these buildings do not have solely historical, social and emotional value but also demonstrate interesting typological and morphological characteristics, which prove that the area has also architectural significance.

Acknowledgements. The research work presented in this paper was supported by the Hellenic Foundation for Research and Innovation (HFRI) under the HFRI PhD Fellowship grant (Fellowship Number: 1037).

H.F.R.I.
Hellenic Foundation for
Research & Innovation

References

Apostolopoulos, F.D.: 1920–1922 The Greco-Turkish Collision, TA NEA, The Asia Minor Catastrophe, in collaboration with the Centre for Asia Minor Studies, pp. 16–32 (2008)

Arabatzi, C.: Refugee Settlements in Chalkidiki in the First Half of the 20th Century, PhD. Aristotle University of Thessaloniki, Department of History and Archaeology, Thessaloniki (2005)

Archive of Eleftherios Venizelos, File 213, 03–09

Bakirtzis, G.: Modern History of Xanthi, in Mavridis, D., ed.,2008, Xanthi: The City of the Thousand Colors. Published by the Municipality of Xanthi and PAKETHRA, Xanthi (2008)

Exarhou, Th.P.: Xanthi 1925: Everyday Life and Historical data. Self-publication, Xanthi (2013)

Exarhou, T.: Xanthi 1924: Everyday Life and Historical Data. Published by the Municipality of Xanthi, Xanthi (2011)

Exarhou, T.: Xanthi 1925: Everyday Life and Historical Data. Published by the Municipality of Xanthi, Xanthi (2011)

Exarhou, T.P.: Xanthi 1923: Everyday Life and Historical Data. Published by the Municipality of Xanthi, Xanthi (2010)

FEK 661/B/17–5–1976

FEK 228 / A / 1923

General State Archives, Rodopi division: folder GRGSA_ROD_ADM021_F000001 (1928)

Roukouni, G., M., et al.: The open cities of the county of Xanthi, Thracian Chronicles Annual Edition 1984, 39, 186. Published by Stefanos Ioannidis, Xanthi (1984)

Gizeli, V.D.: Social alterations and the origins of social housing in Greece 1920–1930. Epikairotita Publications, Athens (1984)

Goodwin, G.: A History of Ottoman Architecture. Thames & Hudson, London (1987)

Gouvousi, S.: Residential Complexes in Greece: Potential of Upgrade of the OEK Settlements, PhD, Thessaloniki: Aristotle University of Thessaloniki, School of Architectural Engineering, Department of Architectural Design and Architectural Technology (2011)

Ioannidis, S.: Xanthi 1870–1920: Elements from its history, Thracian Chronicles Annual Edition 1980–1981, No36, p. 11. Published by Stefanos Ioannidis, Xanthi (1980)

Kanellopoulos, K.: Letter to the Ministry of Social Welfare- General Administration of Refugee Settlement, Komotini: National Archives of Greece, Department of Rodopi, Folder GRGSA_ROD_ADM021.01. F000104 (1950)

Kerkidis, S.: Thrace from the Ottoman conquest to its integration in Greece: a summary of its historical course, in Varvounis, M., (ed.) (2006), Thrace: historical and folklore approach of its culture. Alitheia Publications, Athens (2006)

Koutsoukos, V.: From an Empire to a National State: Spatial Aspects of the Integration of Thrace, PhD, Greece, Rethimno: University of Crete - School of Philosophy, Department of History and Archaeology (2012)

Lampsidis, G.N.: The Refugees of '22: their Contribution to the Country's Development, 3rd edn. Kiriakidi Brothers Publications, Thessaloniki (1992)

League of Nations: Nineteenth quarterly report of the refugee settlement commission (1928)

League of Nations: Report on the operations of the refugees settlement commission for the first three months (1924a)

League of Nations: Third quarterly report on the operations of the refugee settlement commission for the third three months (1924b)

Lianos, N.: The housing settlement of refugees (1922), in Lifo (2016)

Lianos, N.: Yunanistan'da Mülteciler İçin Oluşturulan Yeni Toplukonut Alanları ve Bunların Kentsel Gelişmeye Etkisi: Kavala Kentinden Örnek Alan İncelemesi (The new residential complexes for accommodating refuges in Greece and their consequences for urban development: A case study from the city of Kavala), in "Turkish – Greek Compulsory Population Exchange in its 90th year: New Approaches, New Findings", Koç University, İstanbul, November, pp. 16–17 (2013)

Lianos, N.: Morphological characteristics of the buildings in the Old Town of Xanthi. In: Mavridis, D. (ed.) (2008), Xanthi: The City of the Thousand Colors, pp. 79–90. Published by the Municipality of Xanthi and PAKETHRA, Xanthi (2008)

Maurogordatos, G.: The unrepeatable achievement, TA NEA, The Asia Minor Catastrophe, in collaboration with the Centre for Asia Minor Studies, pp. 101–106 (2008)

Melkidi, C.: Social rearrangements in the urban development of Xanthi, Thracian Chronicles Annual Edition 1991, No45, p. 13. Published by Stefanos Ioannidis, Xanthi (1991)

Morgenthau, H., Strother, F.: An International Drama, Athens: Trochalia Publishing, 1st published 1930. Jarrolds Ltd, London (1994)

Newspaper "Fos", 29th September 1923.

Pallis, A.: Racial immigration in the Balkans and persecutions of Hellenism (1912–1924) [in Greek: Φυλετικές μεταναστεύσεις στα Βαλκάνια και διωγμοί του ελληνισμού (1912–1924], Bulletin of the Centre for Asia Minor Studies. A', Athens (1977)

Paloukis, K.: The Urban Settlement, in the History of Asia Minor, Volume 7: The Refugees in Greece; Settlement and Integration. Tegopoulos Publishing, Athens (2011)

Papaioannou, I.: Part I: Housing in Greece: Public Initiatives. Technical Chamber of Greece, Athens (1975)

Paralikis, S.N., (2006), History of the Municipality of New Smyrna 1922–1940: Establishment and Urban Development, PhD, Athens: NTUA, School of Architectural Engineering, Department of City Planning.

Patrikios, G., et al.: City planning development of Xanthi. in Mavridis, D.: Xanthi: The City of the Thousand Colors. Published by the Municipality of Xanthi and PAKETHRA, Xanthi (2008)

Pentazos, E.: Xanthi: the Chronicle of the Listing, Tackling the Problems of the Traditional Settlement, Thracian Chronicles Annual Edition 1979, No35, p. 14. Published by Stefanos Ioannidis, Xanthi (1979)

Treaty of Lausanne (1923). http://treaties.fco.gov.uk/docs/pdf/1923/ts0016-1.pdf. Accessed 30 Aug 2021

Vasiliou, I.: Traditional House: Social, Technical and Economic Aspects, the Traditional House in various Foreign Countries and in Greece. P.A. Dialisma, Athens (1944)

Yildirim, O.: Diplomacy and Displacement: Reconsidering the Turco-Greek Exchange of Populations 1922–1934. Routledge (2006)

Cultural Heritage Exposed to Natural Hazards: The Case Study of the Convent of San Domenico in Maiori

Raffaele Landolfo(✉), Cristiana Tarantino, Lucrezia Cascini, and Francesco Portioli

Department of Structures for Engineering and Architecture, University of Naples "Federico II", via Forno Vecchio 36, 80134 Naples, Italy
{raffaele.landolfo,lucrezia.cascini,fportiol}@unina.it
https://www.docenti.unina.it/raffaele.landolfo,
https://www.docenti.unina.it/lucrezia.cascini,
https://www.docenti.unina.it/francescopaoloantonio.portioli

Abstract. It is nowadays acknowledged that the minimum intervention principle shall guide any structural interventions on Cultural Heritage buildings. The enhancement of the structural capacity shall not result in a weakening of the values and the authenticity of the monument. Hence, the design of an efficient structural intervention has to be based on a correct interpretation of the current damage. Within this framework, this paper presents the preliminary analysis carried out to assess and evaluate the structural behaviour of a complex case study affected by several deficiencies. The case study is the convent of Saint Domenico, a monumental seventeenth century's masonry structure, belonging to the traditional architectural typology of the court building and characterized by the presence of a large variety of vaulted systems.

The building is located in Maiori, a small town in the Amalfi Coast (Italy), a site of great cultural and naturalistic interest, included in the UNESCO World Heritage List since 1997.

The structure, heavily damaged by two seismic events occurred in the 80s, is abandoned and currently shows an extensive and diffuse crack pattern that is the result of the interaction of several causes. In this paper, the multidisciplinary approach adopted to acquire a deep knowledge of the Convent is presented first. The constructive evolution of the building, traced through a critical historical analysis, is presented, providing a special focus on the transformations that increased the structure vulnerability. An accurate survey of the building geometry, with the aid of photogrammetric technology, and of the current crack pattern are presented. Lastly a first interpretation of the crack pattern is provided based on the kinematic analysis of the corresponding potential collapse mechanisms.

Keywords: Cultural heritage · Masonry building · Differential settlement · Photogrammetric survey · Damage analysis · Kinematic analysis

I. Vayas and F. M. Mazzolani (Eds.): PROHITECH 2021, LNCE 209, pp. 258–265, 2022.
https://doi.org/10.1007/978-3-030-90788-4_23

1 Introduction

In the field of structural retrofit, it is widely recognized that the selection of a suitable intervention stems from a proper interpretation of the damage and the identification of its actual causes. This is even more true when the construction is a cultural heritage building, which requires a special effort to define the minimal necessary actions to solve the problems, while respecting the authenticity and the value of the monument.

As a matter of fact, the damage analysis of a historical building it's a tricky issue and requires a deep investigation of the multiple factors affecting the structural capacity (MiBAC 2011).

In this perspective, this paper presents a multidisciplinary analysis carried out to perform a damage analysis and vulnerability assessment of a seventeenth century's masonry building.

The building is the Ex Convent of Saint Domenico (Fig. 1), a typical courtyard construction, usually employed in monastic architecture, abandoned since the 80s and currently in a deep state of decay.

The Convent is located in the town of Maiori, a small touristic city in the Campania region. It is one of the 15 towns sited along the Amalfi Coast, a site of great cultural and natural interest, included, since the 1997, in the UNESCO World Heritage List.

Fig. 1. The Ex Convent of Saint Domenico (aerial photo)

The importance of the building is related to its uniqueness features. Indeed, it is one of the few monuments documenting the construction activity during the seventeenth century on the Amalfi Coast and contains interesting details of baroque architecture and remarkable artistic works, such as a fresco by the local painter Gaetano Capone. Very few buildings along the coast can compete with its dimensions, whereby it represents an important spatial resource for the city and the entire coastal area. Furthermore, it is one of the few surviving elements of the historical urban fabric of Maiori, which has been almost completely destroyed in the past by several floods.

The Convent is threated by several natural hazards. It lies on an area of high landslide risk, in a site exposed to floods, and it has been already strongly hit by strong earthquakes in the recent past.

An additional risk factor is the poor quality of the soil, worsened by the nearness of the Reghinna river, that flows on the west side of the convent.

Currently, the building shows a severe state of deterioration due to the long period of neglect and the missing of any intervention needed to repair the damages produced by the 80's earthquakes.

This status, together with the deficiencies of the structure resulting from the transformations in time, defines a high vulnerability condition of the building, which increases the risk related to all the aforementioned phenomena (Lagomarsino 2008).

2 History of the Building

The Convent of S. Domenico in Maiori (SA) is a porticoed courtyard building, characterized by a complex plan configuration, which is the consequence of several manipulations and adaptation, carried out during the centuries, to solve unexpected problems and adapt the construction to new functions.

The construction process was long and complicated, also because of several structural deficiencies, occurred even in the construction stage, due to poor quality of foundation soils. As an example, the position of the Church, originally located on the South-West Wing of the monastery, was moved on the South-East side and different foundation typologies were built in different parts of the building to minimize the effect of settlements (D'Aprile 2003). In the 1809, after the suppression of the religious Orders, the Convent was abandoned and left without any use for about 50 years. In 1865, the local Authorities assigned three different functions to the building: the ground floor, on the East side, became a courthouse, on the West side, a jail and the whole first floor became a school. Consequently, very invasive restoration works were carried out to adapt the old building to the new functions: lofts were created at the ground floors and new windows were opened in the walls. At the first floor, large openings were created in the transverse walls to define the classrooms (Fig. 2), enlarging the old cell monks.

In 1980, a strong earthquake heavy damaged the structures and subsequently the building was definitively closed and abandoned. In 1985, a new project of restoration was launched but only partially completed, due to the lack of funds. Nowadays, the building is still closed, abandoned and in critical condition, used by the administration as a warehouse.

Fig. 2. The large openings in the transverse walls realized to define the classrooms from two cells.

3 Geometric Survey and Description of the Building

The geometrical survey was carried out adopting the range imaging technique structure-from-motion (De Luca 2011). The capturing of the pictures was made by two different modes and instruments: by air, with a Mavic Pro drone with integrated camera; from the ground, using a Nikon camera, model D7100. A set of 598 pictures of the building was modeled in the Agisoft software Photoscan and the obtained point cloud model was then processed in cad environment to extract all the geometrical data and the necessary graphs.

The Convent of San Domenico is composed by four bodies of different width and height, placed around a square courtyard of 32 m side length. Two of them (the East and the North blocks) develop for two levels, while the last two (the South and the West blocks) feature a single above-ground level and only consists of a portico covered by a panoramic terrace facing the sea (see Fig. 3).

The Church is located on the South East corner, connected to the cloister along the edge, and sharing with it a bearing wall. There is a single entrance of the building, located near the church's facade, which directly leads to the portico, in the long sequence of regular squared spans of 3, 6 m side length. Along the North and the East sides big barrel vaulted halls open off, while the South and West porticos have blind back walls. On the first level a long corridor runs the whole building and serves all the rooms once being the monks cells. Two stairways are located on the extremes of the North block. The one on the East side is a three ramps stairway, covered with flying barrel vaults. That on the West side was cut after the first ramp by the nineteenth century's transformations, thus it is no longer serviceable.

The thickness of the masonry walls ranges between 1, 2 m (at the ground floor) and 0,6 m (on the last level).

At the ground floor there are three-leaf load-bearing walls, made of two exterior layers in calcareous stone, tuff and clay elements, of limited thickness, and a core in between, made of poor quality mortar and large size aggregates. The upper floors structures are ashlar masonry walls, made of regular squared blocks in grey tuff from Nocera.

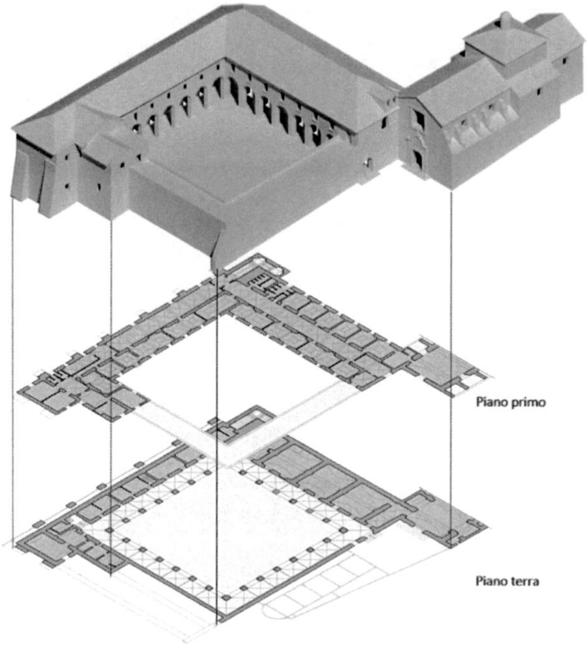

Fig. 3. The architectural layout of the building.

The horizontal structures consist of a great variety of vaulted structures made of yellow tuff: big barrel vaults cover the rooms at the ground floor, while cloister vaults roof the sequence of monastery cells; groin vaults span along the portico and the corridor at the first floor. The roof is made by double slopes timber structures, covered by clay roman-style tiles.

The foundation structures are inhomogeneous and placed on different laying surfaces. The geognostic study reports three different typologies of footings: superficial continuous masonry foundation, in calcareous stone; deep masonry foundation of pillars connected by arches, in calcareous stone; deep masonry foundation of pillars connected by walls, in calcareous stone.

At the present, the building shows a complex crack-pattern, emphasized by the decay condition, due to the last 40 years of neglect.

4 The Damage Survey and Preliminary Interpretation

The 3D point cloud model of the building, with its high level of detail and geometrical precision, made it possible to move the study environment of the building's damage from the real site into a virtual space. Here, by the aid of software's tools, the deformations invisible to the naked eye can be investigated.

Generally speaking, the analysis of the model revealed a slight out of plumb of all the facades on the courtyard, and a non-perfect line-up of the columns edges of the East wing portico, with a maximum shift of about 8 cm.

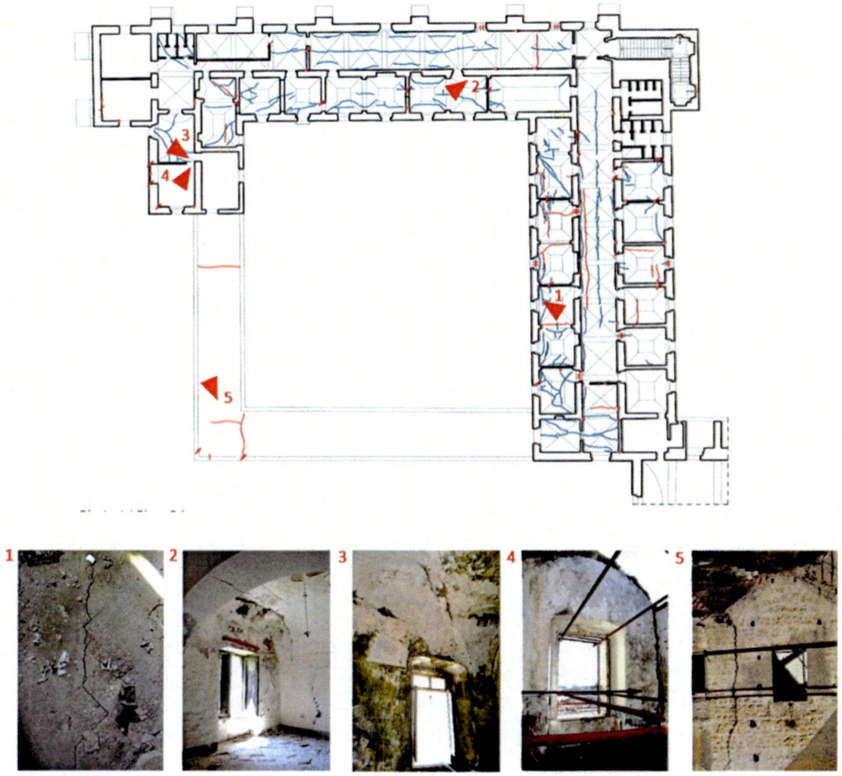

Fig. 4. The first floor crack pattern survey.

The building currently shows an extensive and diffuse crack pattern (Fig. 4), which is the result of the interaction of several causes such as: the natural aging of material, the lack of maintenance, the modifications occurred during the centuries, the seismic events occurred in the past but also the poor quality of the foundation soil, susceptible of producing differential settlements.

The groin vaults of the portico and of the corridors at the first floor show fissures in the centerline and at the attachment to the walls, corresponding to the typical three-hinged mechanism. The big barrel vaults of the ground floor halls are completely detached from the head wall separating them from the courtyard. The trough vaults of the monks cells exhibit diagonal fissures and multiple dense cracks, parallel to the courtside wall.

The most concerning situation is that of the head of the west building, where a vertical "passing through" crack takes the whole height of the floor, reaching the width of 7 cm at the top, and where the ceiling vault and the roof in the recent years have collapsed.

Generally speaking, most of the building exhibits a damage state that is typically related to the effect of seismic actions (Mastrodicasa 1993; D'Ayala and Speranza 2003). A set of collapse local mechanisms were analyzed, by means of linear kinematic analysis, and found congruent with the damage scenario: the cracks detected in the vaulted ceilings, increasing from the ground towards the top, the rotations of the walls and the diffused collapses of parts of the roof, mainly sited in the middle spans of the blocks, where the horizontal displacement is larger.

In the case of the East wing of the building, some inconsistencies raised in the crack pattern have suggested to study additional collapse mechanisms, considering not only the the overturning of the façade, referable to the earthquake but also a ground displacement. Figure 5 reports the two kinematic chains corresponding to the two mechanisms that revealed a potential compatibility of existing crack pattern with a foundation settlement (Santiago Huerta 2008; Roca et al. 2010). Based on the current information, both of the two mechanisms are feasible and more detailed and advanced analysis are underway to establish the correct diagnosis.

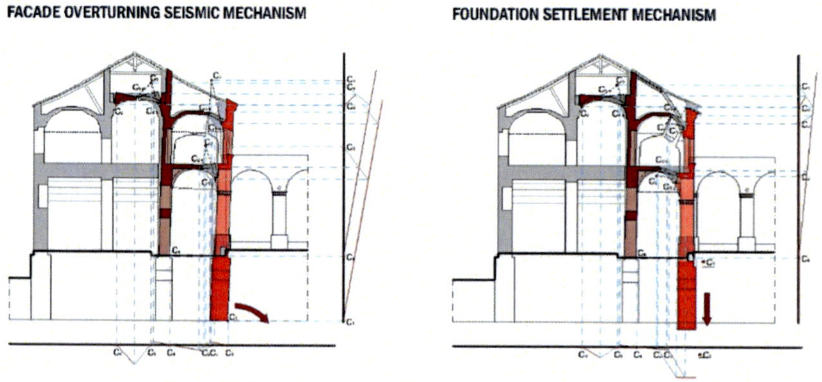

Fig. 5. The two collapse mechanisms compatible with the crack pattern.

5 Conclusions

In this paper, the first results of the damage analysis of the Ex Convent of Saint Domenico were presented. An integrated and multidisciplinary approach was adopted to acquire a deep knowledge of the building and its context. A kinematic analysis supported by the evidences of the crack pattern was conducted and revealed that, in addition to seismic damage, the current crack pattern can be referred to ground instabilities as well. Further development of the present study will consider the development of numerical models to proper address the structural capacity and identify the actual source of damages.

Acknowledgements. The activity presented in the paper were carried out in the framework of the PRIN 2015 PERICLES (Protecting Cultural Heritage from water-soil interaction related threats), coordinated by Prof.Raffaele Landolfo.

References

D'Aprile, M., Giannattasio, C.: L'architettura religiosa di Maiori e Minori tra XVII e XVIII secolo ed il Complesso domenicano del SS.Rosario, in G.Fiengo (a cura di), (1998). La Costa d'Amalfi nel secolo XVII, atti del Convegno di Studi, Centro di Storia e Cultura Amalfitana, Amalfi (2003)

D'Ayala, D., Speranza, E.: Definition of collapse mechanisms and seismic vulnerability of historic masonry buildings. Earthquake Spectra **19**(3) (2003)

De Luca, L.: La fotomodellazione architettonica, Dario Flaccovio Editore, Palermo (2011)

Lagomarsino, S.: Vulnerability assessment of historical buildings. In: Oliveira, C.S., Roca, A., Goula, X. (eds.) Assessing and Managing Earthquake Risk. Geotechnical, Geological And Earthquake Engineering, vol 2. Springer, Dordrecht (2008). https://doi.org/10.1007/s10518-006-9025-y

Mastrodicasa, S.: Dissesti statici delle strutture edilizie: diagnosi e consolidamento, Hoepli, Milano (1993)

MiBAC (Italian Ministry for Cultural Heritage and Activities): Guidelines for the seismic risk evaluation and reduction on Cultural Heritage – alignment to the technical standards for constructions, Italy (2011)

Huerta, S.: The analysis of masonry architecture: a historical approach. Archit. Sci. Rev. **51**(4), 297–328 (2008)

Roca, P., Cervera, M., Gariup, G., Pelà, L.: Structural analysis of masonry historical constructions. classical and advanced approaches. Arch. Comput. Methods Eng. **17**, 299–325 (2010)

Check for updates

The Dynamic Response of a Vertical Dry-Stone Masonry Assembly as a Mock-Up of Masonry Wall Fortifications

George C. Manos(✉), Lazaros Melidis, Olympia Felekidou, and Konstantinos Katakalos

Laboratory of Strength of Materials and Structures, Aristotle University, Thessaloniki, Greece
gcmanos@civil.auth.gr

Abstract. The dynamic response of an ancient defense wall built with large monoliths as a dry-stone-masonry is examined here. For this purpose a mock-up was formed employing five rectangular stones having a cross-section of 150 mm × 300 mm and a height of 200 mm. The horizontal surfaces of each stone was made plane in order to fit each other without the use of mortar. This stone assembly was subjected to one-direction sinusoidal horizontal base motions at the shaking table of Aristotle University. Acceleration sensors were provided at mid-height of each stone and at the moving base aimed at recording the base horizontal excitation and the acceleration response of each stone. A numerical simulation of this mock-up was also formed in order to reproduce numerically the geometric non-linear sliding and rocking response of the stones at their contact surfaces. A summary of the obtained experimental behaviour is presented and compared with the corresponding numerical results. It is concluded that the numerical predictions of the stable-unstable dynamic response of this 5-stone assembly resemble the observed response. It is also concluded that the presence of wire links has a stabilizing effect on the dynamic response of this 5-stone mock-up of wall fortifications.

Keywords: Ancient walls · Monoliths · Dynamic response · Macedonian Palace · Soil-structure Interaction

1 Introduction

Ancient Greek and Roman walls are usually composed of large heavy members (monoliths) that simply lie on top of each other in an almost perfect-fit construction without the use of connecting mortar (Figs. 1 and 2). In this way they are distinctly different from other types of fortifications that use connecting mortar and different type of masonry construction. This type of dry stone masonry is also used as part of the typical structural form of ancient Greek or Roman temples of the peripteral form. The columns and the dry stone masonry are connected at the top with the epistyle (entablature), also composed of monolithic orthogonal blocks, spanning the distance the columns and the stone masonry wall. The earthquake response of free standing columns has been researched by the first author (Manos et al. 1990, 1996, 1997, 1999, 2000, 2001, 2002, 2011, 2013, 2015) in

I. Vayas and F. M. Mazzolani (Eds.): PROHITECH 2021, LNCE 209, pp. 266–283, 2022.
https://doi.org/10.1007/978-3-030-90788-4_24

the past as well as by other researchers (Abrazeys et al. 2011 Aslam et al. 1978, Demosthenous 1994, Koh et al. Koh and Mustafa 1990, Mouzakis et al. 2002, Papantonopoulos et al. Papantonopulos et al. 2002, Psycharis et al. 1993, 1990, 2000, 2003, and Spanos et al. 1984). This paper focuses on the response of square monoliths that lie on top of each other without connecting mortar at their contact horizontal surfaces, as it is typical in the mentioned before dry stone masonry construction.

Fig. 1. Greek ancient walls at the ancient Palace of Tyrins, Greece.

(a) (b)

Fig. 2. a. Mycenae, Greece, Outer wall b. Part of the Hadrian wall, North of England

The seismic response mechanisms that develop on this solid block structural system during strong ground motions can include sliding and rocking of the various monoliths, thus dissipating the seismic energy in a different way from that of other forms of masonry. This paper presents results and conclusions from an experimental study that examines the dynamic response of such prismatic stone formations. Towards this objective the dynamic response of a simple dry stone masonry assembly that is composed by five (5) stones is studied. This stone dry-masonry assembly is simply supported on the bottom of the first block (Figs. 7 and 8). It is then subjected to various types of horizontal base motions (including sinusoidal as well as earth-quake base motions), reproduced by the Earthquake Simulator Facility of Aristotle University, as will be described in Sect. 3.

This study is partially linked with the North ancient walls of the Macedonian Palace of Vergina Greece. At present, these walls are in a state of ruin; however, the restoration of these walls in their original formation is planned together with similar repair and restoration works for the Macedonian Palace itself. The restoration of the walls will utilize only those original monoliths that are in sound condition (Figs. 3a and 3b). However, because such ancient monoliths are in small numbers the construction of new monoliths, having the same dimensions as the ancient ones is planned.

(a) (b)

Fig. 3. a. The remains of the stone monoliths of the ancient North wall of the Macedonian Palace at Vergina Greece. b. The reconstruction of this ancient wall by using recovered ancient monoliths as well as new monoliths formed for this purpose

2 Study of the New Monoliths

Specimens of new travertine material, taken from a quarry located at the prefecture of Serres-Greece, were part of an experimental investigation that was conducted at the laboratory of Strength of Materials and Structures of Aristotle University, exploring the possibility of forming monoliths of this new travertine material. Initially, cubic specimens were taken from the ancient as well as from the new travertine material with dimensions 150 mm × 150 mm × 150 mm and were subjected to axial compression. The average compressive strength of the specimens taken from the ancient monoliths was found to be equal to $\sigma comp = 10.92$ MPa whereas the corresponding value for the specimens taken from the new monoliths was equal to $\sigma comp = 31.02$ MPa. Figure 4 depicts the measured compressive behaviour of the specimens taken from the ancient and the new travertine monoliths. As can be seen, the specimens taken from the ancient travertine monoliths have lower values of compressive strength (average value $\sigma comp = 10.92$ MPa) as well as lower values of Young's modulus ($E = 13.3$ GPa) than the corresponding values for the specimens taken from the new travertine monoliths ($\sigma comp = 31.02$ MPa and $E = 50.0$ GPa). These low compressive strength and the Young's modulus values measured for the specimens taken from the ancient travertine monoliths must be attributed to the weathering of these monoliths for such a long time interval (more than 2000 years). The possibility of constructing artificial stone monoliths of the same shape and the same dimensions as the ancient monoliths, by pouring appropriate white cement - sand mixtures in special forms in-situ, was also investigated. Two such artificial stone monoliths, constructed in-situ at the archaeological site of Vergina-Greece, were transported at the premises of the laboratory at Aristotle University (Figs. 5a and 5b). They were placed on a test rig having as an objective to determine the coefficient of friction between these monoliths when sliding response developed between them at their horizontal contact area.

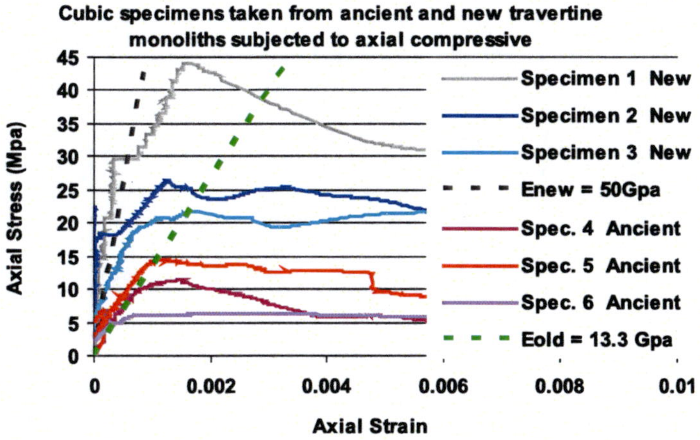

Fig. 4. Axial compressive behaviour of the specimens taken from ancient and new travertine monoliths.

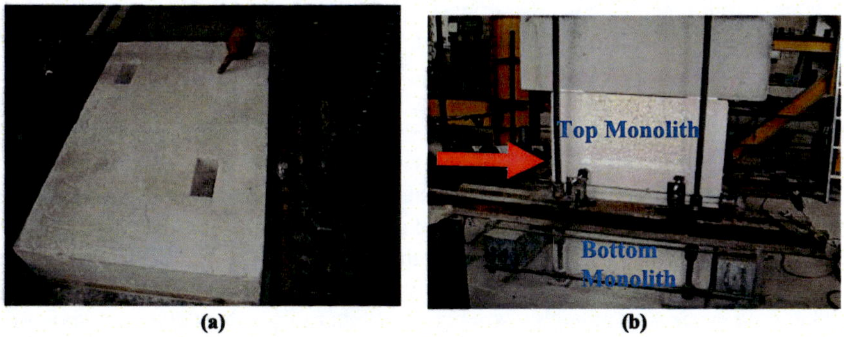

Fig. 5. a. Detail of the horizontal contact surface of the new monoliths b. Arranging two monoliths on top of each other at the laboratory of Strength of Materials and Structure at Aristotle University

The loading arrangement for these tests is shown in Fig. 5b. More information on the experimental set-up can be found in Manos et al. 2015. The obtained sliding response in terms of imposed load and resulting sliding displacement is depicted in Fig. 6. From the stable cyclic sliding response of the two monoliths, depicted in Fig. 6, the value of the dynamic coefficient of friction was found to be equal to $\mu = 0.72$. This value of the friction coefficient remained almost constant for all the range of compressive stress amplitudes ($\sigma n = 0.0126$ MPa, 0.0265 MPa, 0.047 MPa, 0.0686 MPa and 0.0831 MPa) that were applied normal to the sliding surface during this experimental sequence.

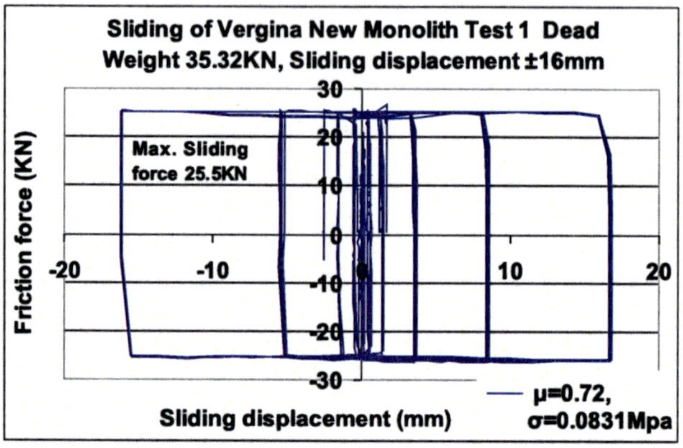

Fig. 6. Cyclic friction load/sliding displacement cyclic response obtained for the Vergina new monoliths.

3 A 6-Stone Dry-Masonry Assembly Wall Mock-Up Subjected to Horizontal Base Motions

In this section summary details of a series of tests dealing with the dynamic response of a dry-masonry 5-stone assembly are presented. These five stones were of almost identical geometry; their horizontal cross section was 280 mm × 160 mm and their height equal to 200 mm. Their top and bottom cross sections were plane and parallel to each other; they were placed one on top of each other forming in this way the 5-stone assembly. The friction coefficient for the contact surfaces between these five stones was found equal to 0.57 from a series of tests with prismatic stone samples taken from these blocks. The supporting base was the top plane-surface of a reinforced concrete (R/C) footing. Each pair of stones was in contact to each other at the intermediate supporting contact surface without the inclusion of any type of mortar (dry masonry). The R/C foundation footing together with the 5-stone assembly was placed within a shear-stack aluminum box filled with dry sand, having part of the R/C footing embedded within the volume of the sand. In turn the base of this sand-box was rigidly secured on top of an earthquake simulator steel platform (shaking table), as is shown in Fig. 7; it was then subjected to uni-directional horizontal base motions, as indicated in the same figure. The dynamic response of this 5-stone assembly was studied in two different configurations. First, each one of the five stones, when subjected to the horizontal base motion, was free to rock and slide without any restrictions, simply resting on each other. A number of tests were performed with this type of 5-stone configuration increasing the intensity of the dynamic horizontal base motion from test to test till this 5-stone assembly became geometrically unstable and partially collapsed. Next, the same sequence of horizontal base motions was repeated having the top two stones of this assembly connected at their back with a stiff steel T-beam, using metallic wire connectors (Figs. 7 and 8). One end of each metallic wire connector was attached at mid-height to the corresponding stone on the vertical plane of symmetry of the 5-stone assembly; then this wire was extended horizontally for 30 mm having its other end secured on a stiff steel T-beam as is shown in Figs. 7 and 8. This steel T-beam was rigidly fixed on the R/C foundation footing, having its vertical plane of symmetry coinciding with the corresponding vertical plane of symmetry of the 5-stone assembly. Due to the large flexural stiffness of the steel T-beam and its small mass the most significant T-beam eigen-frequencies were of much larger values than those of the horizontal base motion. Therefore, the horizontal displacement response of all the points of this steel T-Beam in the direction of motion, relative to the moving base, was approximately equal to zero.

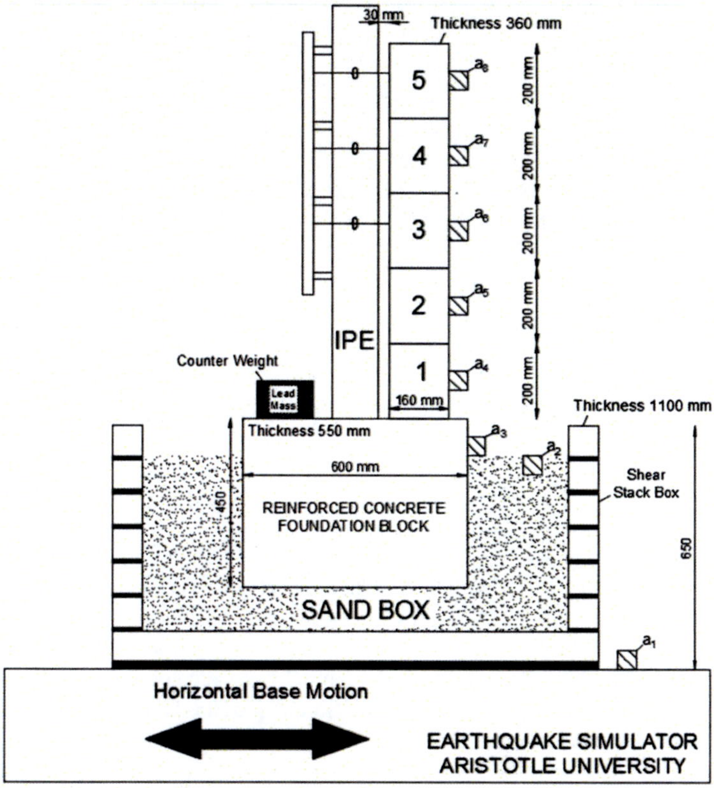

Fig. 7. Testing arrangement of the five stone dry-masonry assembly

Consequently, each metallic wire exercised a form of restrain in the horizontal sliding displacement of the stone, which was connected through this wire to the steel T-beam. Through this mock-up an effort was made to resemble, up to a point, the conditions of a wall formed with prototype monoliths resting on soil. Acceleration sensors were rigidly attached at mid-height on each one of the five stones as well as on the foundation block (a4, a5, a6, a7, a8, from bottom to top, Figs. 7 and 8). Moreover, additional acceleration sensors were also placed near the surface of the sand box (a3, a2) as well as on the steel platform of the shaking table (a1) as shown in these figures.

Fig. 8. Testing arrangement of the five stone dry-masonry assembly

4 Measured Dynamic Response of the Tested 5-Stone Dry-Masonry Assembly

4.1 Dynamic Response of the Five-Stone Assembly Without Wires When Subjected to Horizontal Sinusoidal Base Motions

A number of tests were performed with horizontal sinusoidal base motions keeping the frequency of the excitation constant (1.0 Hz) and gradually increasing the amplitude of the base motion from test to test. Results from two tests are included here. Small rocking could be measured when the maximum horizontal base acceleration reached during a test the value of 0.15 g, where g the acceleration of gravity. During the next test (Test 6a), despite the relative large amplitude of the base motion, the five-stone assembly did not develop any form of large rocking or sliding displacements leading to instability. However, for the next test (Test 7a, 0.173 g) with larger base excitation than Test 6a, the five-stone assembly developed large rocking displacement response (Fig. 9a).

The acceleration response along the height of the five-stone assembly for Test 7a is depicted in Fig. 9b. As can be seen in this figure, the top 5th-stone develops a peak acceleration of approximately 0.68 g. It was observed in the past during the experimental sequences with columns having drums that for a given frequency of excitation there is a

(a)

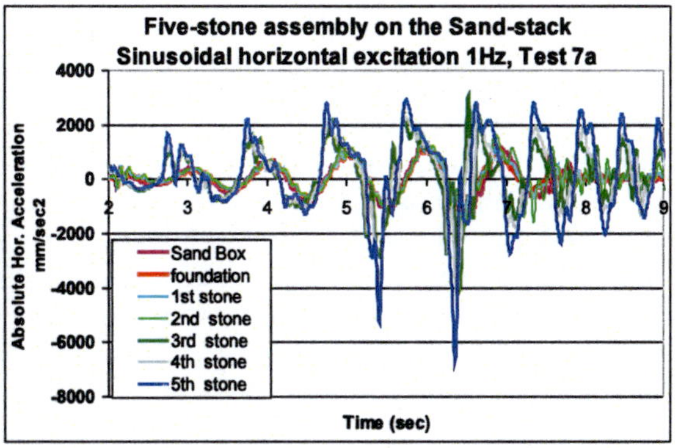

(b)

Fig. 9. a. Rocking displacement response of the unattached five stone dry-masonry assembly, being unattached from the steel beam b. Acceleration response along the height of the unattached five-stone assembly (peak base horizontal acceleration 0.173 g, Test 7a)

boundary between the stable rocking response and the overturning (Manos et al. 1990, 1996, 1997, 1999, 2000, 2001, 2002, 2011, 2013 2015). This boundary has relatively high values of non-dimensional excitation amplitude for high frequency values and relatively low values of non-dimensional excitation amplitude for low frequency values (Manos et al. 2015). In addition, sliding between the various stones at their contact interface became also noticeable (Fig. 10).

Fig. 10. Sliding between stone-blocks at the contact interface of the unattached five-stone assembly

4.2 Dynamic Response of the Five-Stone Assembly Having Wire Connectors When Subjected to Horizontal Sinusoidal Base Motions

This time the top two stones of the assembly were provided with metallic wire attachments. Again, a number of dynamic tests were carried out keeping as before the frequency of the excitation constant (equal to 2.0 Hz) and gradually increasing the amplitude of the base motion from test to test. Results from two tests are included here. Small rocking could be measured when the maximum horizontal base acceleration reached during a test the value of 0.2 g, where g the acceleration of gravity. During the next tests, despite the relative large amplitude of the base motion, the five-stone assembly did not develop any form of unstable rocking or excessive sliding displacements leading to instability (Fig. 11, Test 7b). Unstable rocking developed during test 8b as can be seen in Fig. 12.

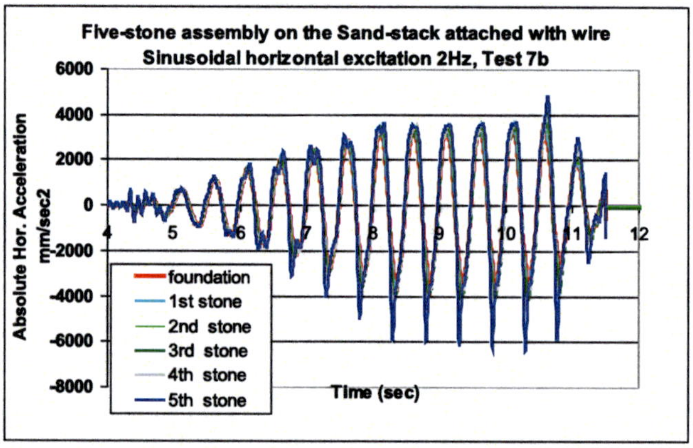

Fig. 11. Acceleration response along the height of the five-stone assembly attached with wires (peak base horizontal acceleration approximately equal to 0.40 g, Test 7b)

The acceleration response along the height of the five-stone assembly being attached with metallic wires to the steel beam for Test 7b and Test 8b is depicted in Figs. 11 and 12, respectively.

As already described (Fig. 7), the acceleration sensor a3 recorded the horizontal acceleration response of the reinforced concrete (R/C) footing which was embedded within the sand and was used to provide the base support of the 5-stone assembly. Figure 13 depicts the horizontal acceleration recorded with this sensor at the R/C footing during test 7a (5-stone assembly without metallic wires) as well as during tests 7b and 8b whereby metallic wires were utilized to attach the 5-stone assembly to the steel T-beam (Sect. 3, Figs. 7 and 8). As can be seen in Fig. 13, the peak recorded R/C footing horizontal acceleration during tests 7b and 8b (with wires) was twice as large as the peak horizontal acceleration recorded during test 7a (no wires). Figure 14 depicts the horizontal displacement of the mid-height of the top 5th-stone of the assembly relatively to the R/C footing for test 7a (5-stone assembly without metallic wires) as well as for tests 7b and 8b whereby wires were utilized to attach the 5-stone assembly to the steel

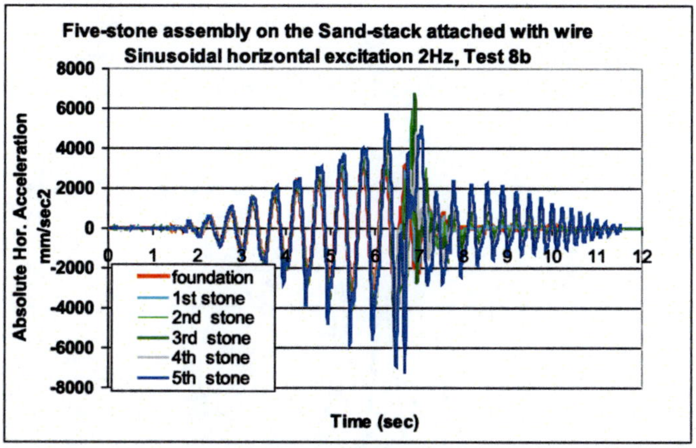

Fig. 12. Acceleration response along the height of the five-stone assembly attached with wires (peak base horizontal acceleration >0.40 g, Test 8b)

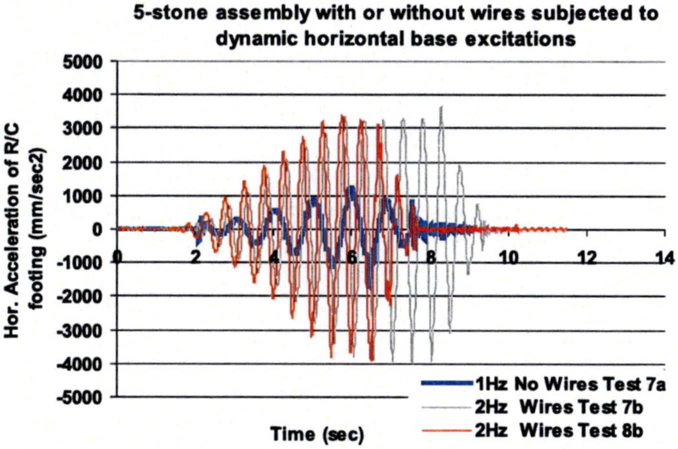

Fig. 13. The horizontal excitation in terms of acceleration of the R/C footing supporting the 5-stone assembly.

T-beam (Sect. 3, Figs. 7 and 8). Figure 15 depicts the horizontal acceleration recorded at mid-height of the top 5th-stone of the assembly by acceleration sensor a7 (Figs. 7 and 8), during test 7a (5-stone assembly without metallic wires) as well as during tests 7b and 8b whereby wires were utilized to attach the 5-stone assembly to the steel T-beam (Sect. 3, Figs. 7 and 8). The relative horizontal displacement response curves in Fig. 14 were derived from the numerical integration of the relevant absolute acceleration recordings (Fig. 15). As can be seen in Fig. 14, the relative displacement response of the top 5th-stone during test 7a (no wires) becomes excessive, with large rocking and sliding displacements leading to instability. On the contrary, the relative displacement response of the top 5th-stone during test 7b (with wires) corresponds to a stable rocking response; this occurs despite the fact that the excitation for this tests, in terms of R/C footing horizontal acceleration, is much larger than the excitation for test 7a (no wires).

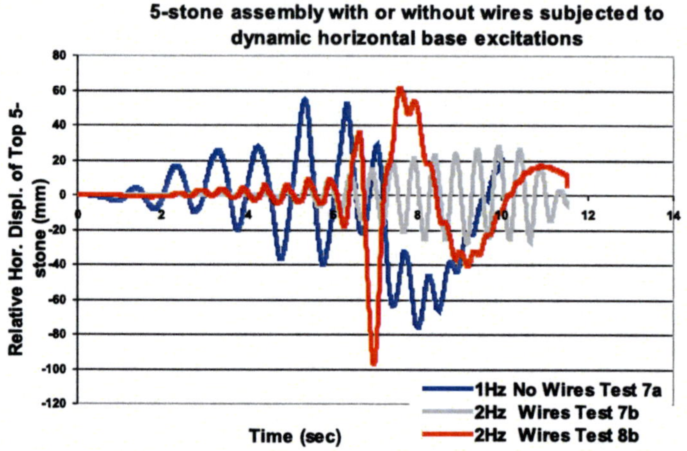

Fig. 14. Horizontal displacement response of the 5th-stone relative to the R/C footing. Test 7a without wires, Tests 7b and 8b with wires.

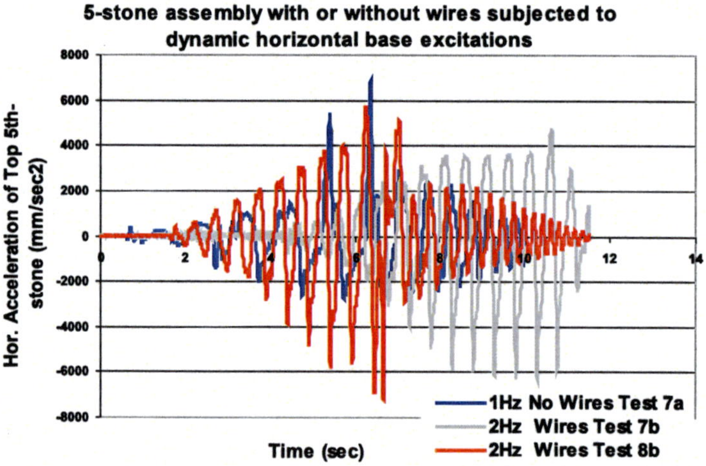

Fig. 15. Horizontal acceleration response of the 5th-stone. Test 7a without wires, Tests 7b and 8b with wires.

During test 8b, with R/C footing horizontal acceleration of the similar amplitude as test 7b (Fig. 13), the wires were less effective because they were elongated during the previous test (test 7b). This explains the fact that the relative displacement response

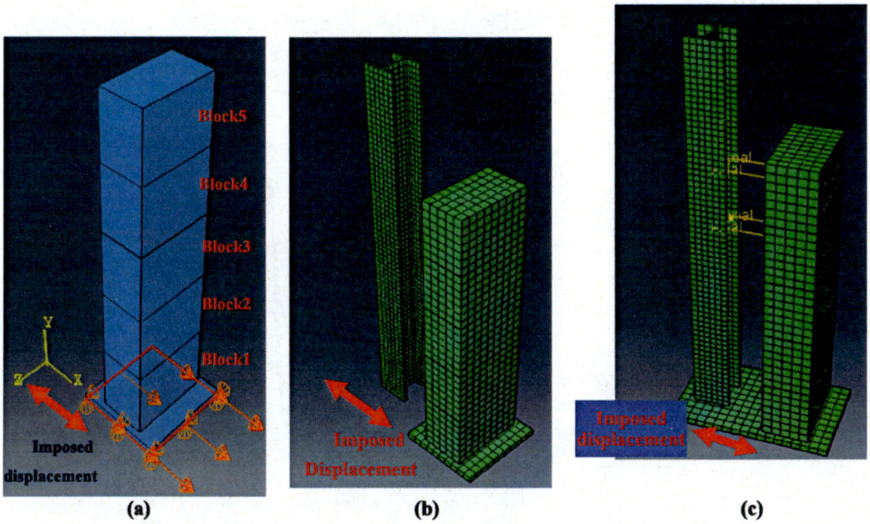

Fig. 16. a. Numerical simulation of the 5-stone assembly without the steel T-beam b. Numerical simulation of the 5-stone assembly with the steel T-beam (without wires). c. Numerical simulation of the 5-stone assembly with the steel T-beam (with wires).

during this test becomes excessive (Fig. 14). It must be underlined that the horizontal excitation for test 8b was again much larger than for test 7a (no wires, Fig. 13). From these observations it can be concluded that the presence of the metallic wires leads to the desired control of the sliding and rocking acceleration and displacement response of the 5-stone assembly prohibiting up to a point unstable response accompanied with partial or total collapse. This is also evident in Fig. 15 whereby the horizontal acceleration response of the 5th-stone during test 7b indicates stable rocking.

5 Numerically Predicted Dynamic Response of the Tested 5-Stone Dry-Masonry Assembly

The numerical process with finite element (F.E.) discretization within the ABAQUS soft-ware package used before by Manos et al. (2013) is also applied here. It was demonstrated, by comparing the obtained numerical results with the corresponding observed response as well as with analytical predictions, that these numerical predictions of the rocking response have an acceptable degree of approximation. More details are given by Manos et al. (2013) and only basic information is included here. As was done during the experimental investigation, the influence of the presence or the absence of the metallic wires on the dynamic response of the assembly was again studied here. As already described in Sect. 3, these metallic wires were used to attach the top stone blocks of the assembly to the steel T-beam. The value of the friction coefficient used in the numerical simulation was the same with the one measured during the experimental sequence ($\mu = 0.57$).

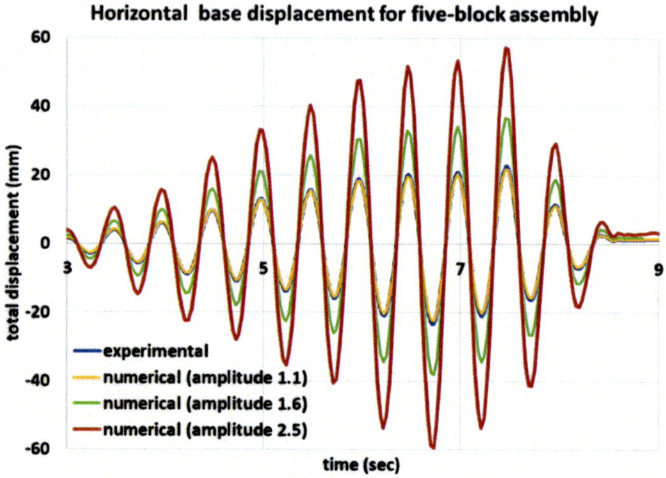

Fig. 17. Comparison between the displacement base motion recorded during the experiments and the base motion that occurred during the numerical simulations of the 5-stone assembly dynamic response.

Figures 16a, 16b and 16c depict the formed numerical simulations, one without the steel T-beam (without wires) and the second with the steel T-beam (with or without wires).

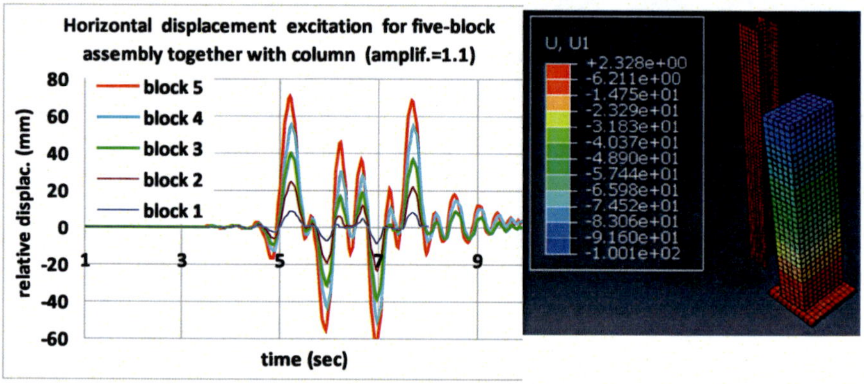

Fig. 18. Hor. displacement response along the height of the five-stone assembly (without wires). Numerical simulation with the steel T-beam. Amplification value = 1.1

All these numerical models (Figs. 16a, 16b and 16c) were subjected to a uni-directional horizontal excitation, as is indicated, which was obtained by numerically integrating the acceleration recorded at the R/C footing during test 7a of the experimental sequence (Sect. 4, Fig. 13). This basic horizontal excitation (sinusoidal variation with a frequency of 1Hz, with the code name Base-Sin1Hz)) was used for all subsequent numerical simulations in the following way. A number of horizontal base excitations were produced by multiplying this basic excitation with an amplitude factor multiplier (amplification) ranging from 1.1 to 2.5 as is shown in Fig. 17. The excitation curves plotted in this figure as numerical with a constant amplitude value (e.g. 1.6) represent the horizontal displacement of the moving base as a result of the numerical solution for a corresponding desired input motion equal with the displacement amplitude of the moving based measured during the experimental sequence (Base-Sin1Hz) multiplied by the amplification factor 1.1. As can be seen in Fig. 17, despite the use of the amplification factor with the value 1.1 the base motion during this numerical simulation is somewhat smaller that the experimental base motion.

5.1 Numerical Simulation of the Dynamic Response of the Five-Stone Assembly Without Wires When Subjected to Horizontal Sinusoidal Base Motions

Figure 18 depicts the numerical predictions of the horizontal response of the 5-stone assembly relatively to the moving base. An amplification factor equal to 1.1 was employed in this numerical simulation. As can be seen in this figure, the predicted response is that of stable rocking which does not agree with the corresponding experimental observation whereby this base motion resulted in unstable response (Figs. 9a, and 14). In order to reach unstable response by this numerical simulation the base excitation was gradually increased till the amplification factor reached the value of 1.6. (Fig. 19).

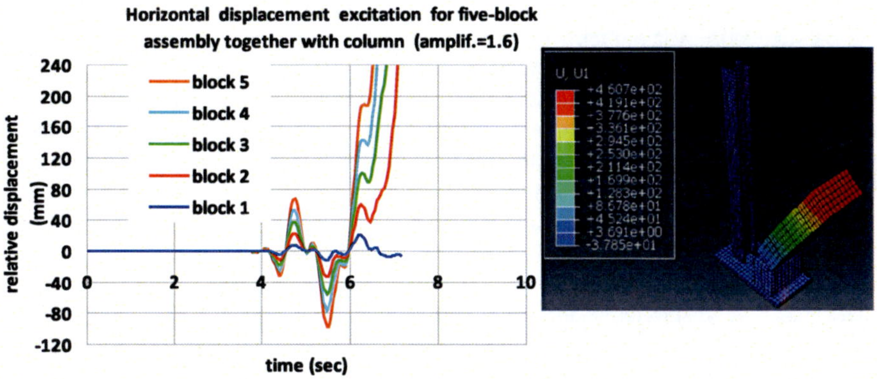

Fig. 19. Hor. displacement response along the height of the five-stone assembly (without wires). Numerical simulation with the steel T-beam. Amplification value = 1.6

5.2 Numerical Simulation of the Dynamic Response of the Five-Stone Assembly Having Wire Connectors When Subjected to Horizontal Sinusoidal Base Motions

In this section the numerical predictions of the dynamic response of the 5-stone assembly, having its top two stones attached with wires to the steel T-beam, as was done during the experimental sequence. The axial behaviour of each wire was governed by an elasto-plastic constitutive law with a yield force equal to 60 N, which represents a numerical approximation of the axial behaviour of the metallic wires used during the experiments. Figures 20 and 21 depict the numerical displacement response for a horizontal excitation with an amplification factor equal to 1.6 and 2.5, respectively.

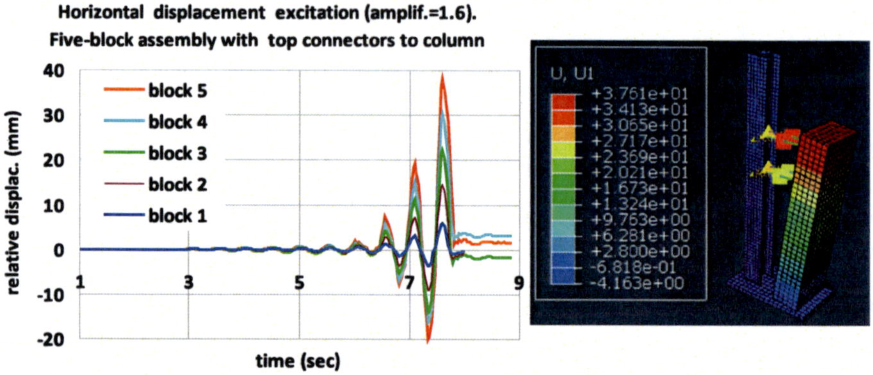

Fig. 20. Hor. displacement response along the height of the five-stone assembly (with wires). Numerical simulation with the steel T-beam. Amplification value = 1.6

By comparing the numerical horizontal displacement response of Figs. 19 and 20 of the 5-stone assembly with the wires in place with the corresponding response of Fig. 18 for the assembly without the wires, the stabilizing effect of the presence of the used wires can again be ascertained. In particular, for the same base excitation (basic sinusoidal motion multiplied by the amplification factor equal to 1.6), the numerical simulation of the assembly with the wires rocks with a stable response (Fig. 20) whereas the corresponding numerical simulation of the assembly without the wires overturns (Fig. 19). For the assembly with the wires to reach an unstable response the amplification factor must attain the value of 2.5, which represents a 55% increase to the base excitation that caused the unstable response of the assembly without the wires. This observation agrees with a similar observation made during the experimental sequence (Sect. 4). The numerically predicted displacement response of the 5-stone assembly with the wires in place appears to combine rocking as well as sliding response, mainly for the top three stones (Fig. 21). This response resembles the displacement response of the individual stoned of the assembly without the wires rather than the one with the wires (Figs. 9a, 9b and 10).

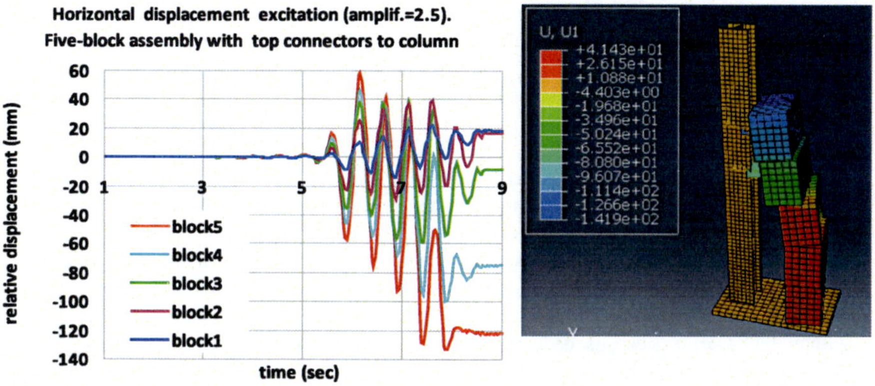

Fig. 21. Hor. displacement response along the height of the five-stone assembly (with wires). Numerical simulation with the steel T-beam. Amplification value = 2.5

6 Conclusions

- The mock-up of a five stone dry masonry assembly without any wire attachments developed stable rocking and unstable rocking dynamic response that was combined with sliding at the contact interfaces. This stable-unstable rocking response of the mock-up of a five stone dry masonry assembly resembles similar behaviour that was observed for the response of mock-ups of ancient free standing columns and colonnades.
- Metallic wire links, which exercise horizontal displacement restraints on such monoliths, can have, up to a point, beneficial influence on this stable-unstable rocking response for this type of stone dry masonry. Stable rocking response was observed for much larger base acceleration amplitudes when the metallic wire links were present than when the used mock-up of the five-stone assembly was left unattached without any such links.
- The numerical simulation of the dynamic response of the 5-stone assembly confirmed this stabilizing influence resulting from the presence of such metallic wire attachments.

Acknowledgements. The authors would like to thank Ms. A. Kottaridi, the Director of the archaeological studies and restoration works at the Vergina Macedonian Palace and the Head of the 17th Prehistoric and Classical Archaeological Sites Agency of Greece.

References

Ambraseys, N., Psycharis, I.N.: Earthquake stability of columns and statues. J. Earthquake Eng. **15**, 685–710 (2011)

Aslam, M.M., et al.: Rocking and overturning response of rigid bodies to earthquake motions, Report No. LBL-7539, L. B. Lab. Univ. Of California, Berkeley (1978)

Demosthenous, M.: Experimental and numerical study of the dynamic response of solid or sliced rigid bodies, Ph. D. Thesis, Dep. Civil Eng., Aristotle University (1994)
Koh, A.S., Mustafa, G.: Free rocking of cylindrical structures. J. Eng. Mech. **116**(1)35–54 (1990)
Manos, G.C., Demosthenous, M.: The behavior of solid or sliced rigid bodies when subjected to horizontal base motion, Proc. 4th U.S. Nat. Confer. Earth. Eng **3**, 41–50 (1990)
Manos, G.C., Demosthenous, M.: Study of the dynamic response of models of ancient columns or colonnades subjected to horizontal base motions, 11WCEE, Session 6.3.3. of Abstract Volume Analysis of Masonry Structures, Acapulco, Mexico (1996)
Manos G.C., Demosthenous, M.: Models of ancient columns or colonnades sub-jected to horizontal base motions - study of their dynamic and earthquake behavior. In: STREMA97, pp. 289–298, San Sebastian Spain (1997)
Manos, G.C.: The dynamic performance of models of ancient columns and colonnades with and without the insertion of wires made of shape memory alloy. In: STREMAH 1999, pp. 281–290, Dresden, Germany (1999)
Manos, G.C., Demosthenous, M., Hatzigeorgiou, A., Manou-Evison J.: Study of the dynamic and earthquake behavior of models of ancient columns and colonnades with and without the inclusion of wires with energy dissipation characteristics. In: 12th WCEE, Auckland, New Zealand (2000)
Manos, G.C., Demosthenous, M., Kourtides, V., Hatzigeorgiou A.: The dynamic and earthquake behavior of ancient columns and colonnades with and without shape memory alloy devices. In: STREMAH 2001, Florence, Italy (2001a)
Manos, G.C. et al.: Study of the dynamic and earthquake behavior of rigid bodies with or without the inclusion of shape memory alloy devices. In: 16SMIRT, Washington D.C (2001b)
Manos, G.C., Demosthenous, M., Kourtides, V., Hatzigeorgiou, A.: Study of the dynamic and earthquake behavior of models of ancient columns and colonnades with and without the inclusion of wires with energy dissipation characteristics. In: 9th Canadian Conference on Masonry Structures, Univ. New Braunswick, Canada (2001c)
Manos, G.C., Demosthenous, M., Kourtides, V.: The Response of models of ancient columns and colonnades under horizontal forces with or without SMAD's. In: 2nd International Conference Studies in Ancient Structures, Constantinoupol, Turkey (2001d)
Manos, G.C., Demosthenous, M., Kourtides, V.: The Dynamic and Earthquake Response of Models of Ancient Columns and Colonnades with and without Energy Dissipation Devices, 12th European Earthquake Engineering Conference. England, London (2002)
Manos, G.C., Koidis, G., Demosthenous, M.: Investigation of the sliding response of a rigid body system to uni-direction horizontal dynamic and earth-quake excitations. In: COMPDYN 2011, Corfu, Greece, 26–28 May 2011a
Manos, G.C, Koidis, G., Demosthenous, M.: The sliding response of a rigid body system subjected to uni-directional horizontal dynamic and earthquake excitations. In: EURODYN 2011, Leuven, Belgium, 4–6 July 2011b
Manos, G.C, Koidis, G., Demosthenous, M.: The response of a two-degree-of-freedom dynamic sliding system subjected to uni-direction horizontal dynamic and seismic excitations. Int. J. Comput. Methods Exp. Measur. **1**(3), 221–237 (2013a)
Manos, G.C., Petalas, A., Demosthenous, M.: Numerical and experimental study of the rocking response of unachored body to horizontal base excitation. In: 4th ECOMASS Thematic Conference on Computational Methods in Structural Dynamics and Earth-quake Engineering – COMPDYN 2013, 12–14 June 2013, Kos Island, Greece (2013a)
Manos, G.C., Katakalos, K., Kotoulas, L., Felekidou O.: The Dynamic response of a mock-up of the restored ancient north wall scheme for the macedonian palace at Vergina-Greece. In: COMPDYN 2015, 5th ECCOMAS Thematic Conference on Compu-tational Methods in Structural Dynamics and Earthquake Engineering, Crete Island, Greece, 25–27 May 2015 (2015)

Mouzakis, H., Psycharis, I.N., Papastamatiou, D.Y., Carydis, P.G., Papantonopoulos, C., Zambas, C.: Experimental investigation of the earthquake response of a model of a marble classical column. Earthquake Eng. Struct. Dynam. **31**, 1681–1698 (2002)

Papantonopulos, C., Psycharis, I.N., Papastamatiou, D.Y., Lemos, J.V., Mouzakis, H.: Numerical prediction of the earthquake response of classical columns us-ing the distinct element method. Earthquake Eng. Struct. Dynam. **31**, 1699–1717 (2002)

Psycharis, I.N., Jennings, P.C.: Rocking of slender rigid bodies allowed to uplift. Earthquake Eng. Struct. Dynam. **11**, 57–76 (1983)

Psycharis, I.N.: Dynamic behaviour of rocking two-block assemblies. Earthquake Eng. Struct. Dynam. **19**, 555–575 (1990)

Psycharis, I.N., Papastamatiou, D.Y., Alexandris, A.: Parametric investiga-tion of the stability of classical columns under harmonic and earthquake excitations. Earthquake Eng. Struct. Dynam. **29**, 1093–1109 (2000)

Psycharis, I.N., Lemos, J.V., Papastamatiou, D.Y., Zambas, C., Papantonopoulos, C.: Numerical study of the seismic behaviour of a part of the parthenon Pro-naos. Earthquake Eng. Struct. Dynam. **32**, 2063–2084 (2003)

Spanos, P.D., Koh, A.S.: Rocking of rigid blocks due to harmonic shaking. J. Engin. Mech. ASCE **110**(11), 1627–1642 (1984)

Preserving a Modern Neighborhood: Koşuyolu

Yıldız Salman(✉)

Department of Architecture, İstanbul Technical University, Taşkışla Campus, Taşkışla Caddesi, 34367 İstanbul, Turkey
salman@itu.edu.tr

Abstract. The concept of a "designed neighborhood became widespread following the foundation of the young Turkish Republic, particularly in regard to the works of invited urban planners and architects from Germany focused on creating modern environments. Concurrent economic problems and the lack of building materials limited construction activity and raised the prices. Koşuyolu Neighborhood was planned as one of the pioneer examples of municipality's initiatives in 1948 for constructing low-cost houses in order to meet the demand. Leading planners and architects of the period like Kemal Ahmet Aru, Rebii Gorbon, Seyfi Arkan, and Leyla Turgut were involved in the planning process between the years 1948–1956. Regarding contemporary preservation concepts, Koşuyolu Neighborhood is a heritage place as one of the first examples of its type and should have a legal conservation status Planning decisions and implementations within the site should consider its modern architectural heritage values. The paper will discuss the architectural and urban features of a "modern neighborhood" as well as its heritage values and preservation possibilities, in regard to challenging conditions of the site.

Keywords: İstanbul · Koşuyolu · Low-cost housing · Modern architectural heritage

1 Introduction

The decade around and after the WWII was a period of large population shifts from country to big cities, growing housing demands for the newly established republic. As in the 50s, affected by the agricultural and industrial policies, social and political changes, İstanbul has become one of the most preferred cities, Faced with the difficulties of providing low-cost housing for the new citizens and increasing worker population in the big cities, urban planning and housing studies has become significant. Construction activities in the urban scale were carried out in two ways, planning new cities and modernizing the existing ones with partial plans and new settlements. The idea of "designed neighborhood" (Akcan 2005) became widespread mostly with the contribution of architects and urban planners invited from Germany. The urgent need for modern, healthy environment and low-cost housing for the public was the dominant factor for all types of construction and development activities.

I. Vayas and F. M. Mazzolani (Eds.): PROHITECH 2021, LNCE 209, pp. 284–296, 2022.
https://doi.org/10.1007/978-3-030-90788-4_25

The novel idea of "low-cost housing" appears to be the most important development of this rationalist and functionalist period in the architecture of Turkey (Yöney and Salman 2010). On the other hand, economic conditions of the period, land values, difficulties in providing qualified construction materials and prices, prompted for different financial models. Thus, public initiated solutions included land and housing co-operatives, in many cities that was more effective solving the urban housing problem at long last.

2 Housing Initiatives in the Young Turkish Republic

Right after the foundation of the Turkish Republic three state agencies were founded to meet the financial and technical needs of public constructions including dwellings. Land and Orphan Bank (Emlak Eytam Bankası) in 1926, Bank for Reconstruction (İmar Bankası) in 1929 and Bank of Municipalities (Belediyeler Bankası) in 1933 have become key actors for public constructions. In 1946, Land and Orphan Bank has changed its name as Emlak Credit Bank (Emlak Kredi Bankası) and was restructured with the aim of providing long-term low-interest loans to middle class families especially for housing constructions and invested many projects in various cities including Ankara and İstanbul.

Like many other planners and architects who studied in Germany or worked with Germans Kemal Ahmet Aru and Seyfi Arkan, have become prominent actors in various housing projects for different social and economic groups The only woman architect in the designers group, Leyla Turgut would later design İETT Housing Blocks in1962.

Professor Kemal Ahmet Aru who has directed urban planning studies in many cities, has stated that in the50s İstanbul was on a verge of a dramatic transformation in regard to social and urban structure therefore was in an urgent need of 40.000 houses as per estimated by the commission of which he was a member as well (Aru 1953). Koşuyolu Low-Cost Housing Neighborhood was one of the earliest examples of social housing enterprises with the objective to solve housing shortage in the developing city with a high increase in population. In 1950s architectural circles in Turkey have had various debates on housing policy, social housing and low-cost housing. Implementations financed by Emlak Credit Bank which, has been founded to give financial support for housing constructions planned by the government and the municipalities were criticized as to be quite wide of the mark and did not satisfy the demand for low-cost housing. Zeki Sayar, architect and the owner of the architectural journal, named Arkitekt, in many of his articles (Sayar 1946, 1952 and 1953) underlines that low-cost and social housing policies of the government were not sufficient to solve the problems. Sayar (1958) stated that, housing blocks on Atatürk Boulevard, planned and constructed by İstanbul Municipality Dwelling Office in partnership with Emlak Bank Technical Committee between the years of 1954–57 likewise IVth Levent Neighborhood planned by Kemal Ahmet Aru and financed by Emlak Credit Bank were far from being included in low-cost housing concept due to the internationally developed minimum standards and norms as well as. According to Sayar (1958), both housing projects could be accepted as luxurious even for middle income families thus, they went beyond the scope of "habitation of low-income citizens". The only exception for widely criticized housing implementations is the Koşuyolu Neighborhood.

3 Koşuyolu Neighborhood: A Low-Cost Housing Project

Free lands at the outskirts of the city were relatively cheaper to establish new housing settlements for the low and middle income groups. Mecidiyeköy on the European site, like Kadıköy Koşuyolu (Fig. 1a, b) and Üsküdar Selamsız on the Anatolian site were selected for the pioneer cases of low-cost housing in İstanbul.

(a) **(b)**

Fig. 1. a. Street Plan of Üsküdar and Kadıköy by Carl Lörcher, 1924–26. b. Aerial view of the site in 1946

These "designed neighborhoods" were the new suburbs of the period. Koşuyolu Neighborhood which, completed mainly in four phases between the years of 1949–59 include more than 400 house units for low and middle income families, shops, school, day nursery, cinema, police station, green areas and public spaces. The first phase designed and constructed between 1949 and 1952 includes two parts located on both sides of the main road named Koşuyolu Street crossing the site from north to south. Although the whose site is planned between 1948–49 the construction and competition of all houses were realized in a long period. The part located on the east side was named Municipal Houses and the other group of houses located on the west side is called Bank Houses (Banka Evleri). The second phase established between 1955 and 1957 is called Worker's Houses while Credit Edifice Fund Houses (Kredi Yapı Sandığı Evleri) built between 1957 and 1958 was added to the site as the third phase. The last phase namely Dock Workers' Houses was the latest and a rather late addition to the neighborhood in 1962 (Fig. 2).

İstanbul Municipality issued and published a bylaw including financial and legal terms as well as basic characteristics of the houses like base area limitations, room numbers, floor heights, stair width, etc. According to the bylaw text (Anon 1950), a house may have maximum three rooms, the maximum area for single storey houses is 50 m^2 while two storey types might be 60 m^2. Considering the topographical conditions, rows of six unit blocks have been provisioned. Article 12 of the bylaw, stated that the

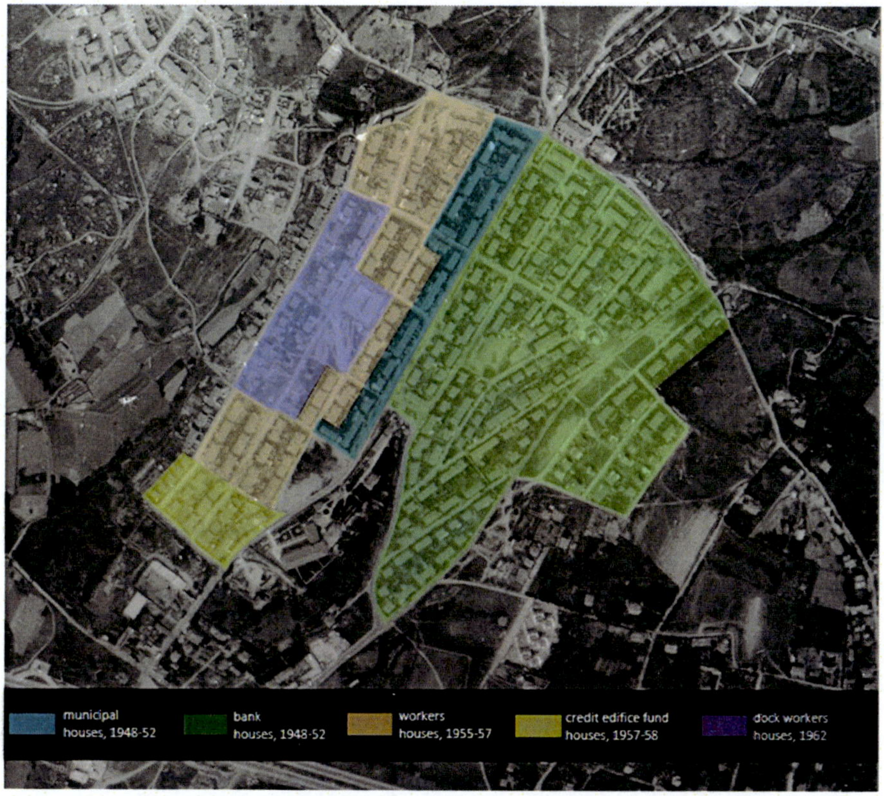

Fig. 2. Koşuyolu Low-Cost Housing Neighborhood, phases

municipality either might develop typologies in reference to international experiences or organize a competition. However, in 1950, the municipality opened an architectural competition for different typologies of public low-cost houses including one-storey with two rooms, one-storey with three rooms and two-storey with three rooms (Özdoğan et al. 1950) but unfortunately there is no clear information whether the winning projects has been implemented or not.

The first implemented stage of Koşuyolu Low-Cost Housing Neighborhood was included in the *International Housing Architecture Plans and Photographs Exhibition* along with other selected low-cost housing examples from 17 countries, opened at Emlak Bank Pavilion, in 1952 İzmir International Fair.

The first part of the first stage, named roughly Municipal Houses has been developed for the municipal workers and dock workers by the Urbanism Office of Municipality Development Directorate. The site plan including housing units, green spaces and an elementary school prepared in 1948 by Kemal Uralman, an urban planner in the Urbanism Department and approved in 1949 (Fig. 3).

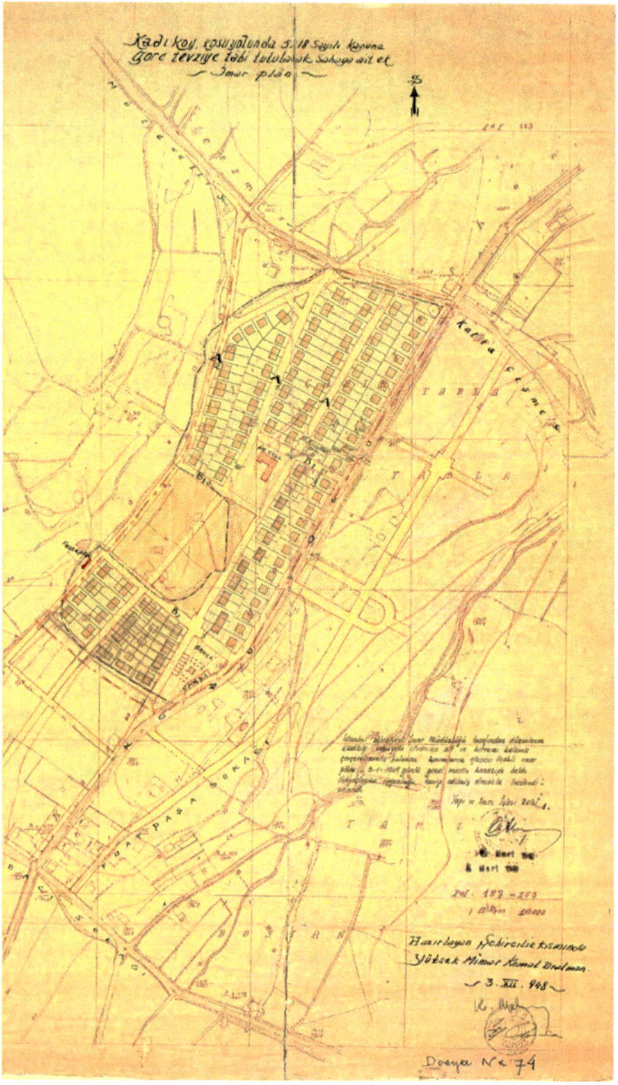

Fig. 3. Development plan for low-cost housing at Koşuyolu, 1948

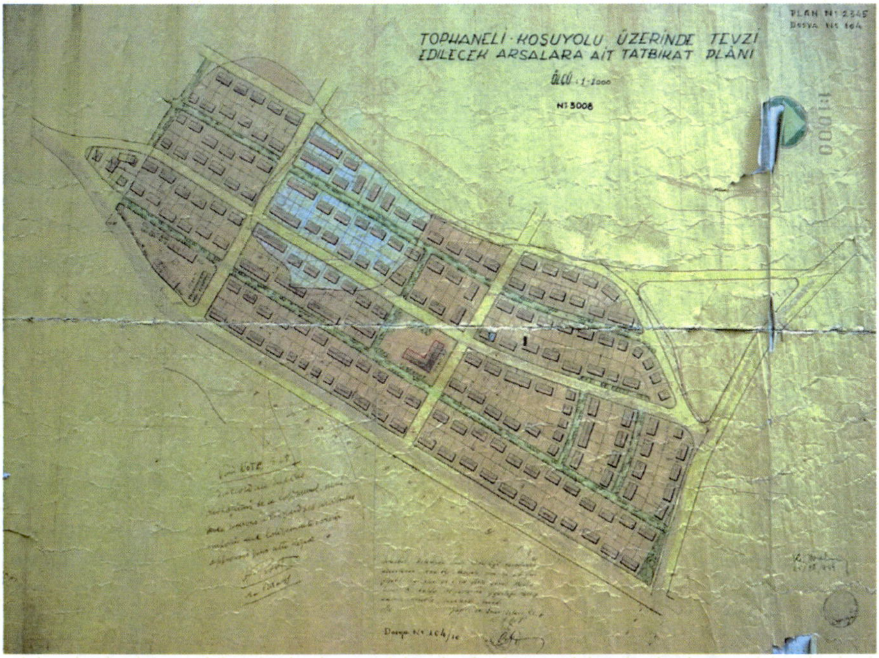

Fig. 4. Plan revision of the Municipal Houses, 1949

Although there is very limited document about the architectural details, house types designed and developed by Sait Özden and Leyla Turgut (Broek 1964) and it is known that this typology was also implemented in another concurrent housing project at Üsküdar-Selamsız. Soon after the plan had been revised so as to include more housing units and alteration of street organization and approved by Henri Prost, the chief of the Municipal Planning Office (Fig. 4). The construction of the planned 100 houses was financed by Development Limited Partnership (İmar Limited Ortaklığı) which, established with the accession of İstanbul Municipality and Emlak Kredi Bank has been completed in 1952 (Milliyet 1952). After the competition of the first stage, other housing projects have been developed for different worker cooperatives with different economic models on the rest of the planned site.

Group of houses in smaller size located in the second phase of the neighborhood named as Workers Houses were designed in 1955. The set of drawings archived in Kadıköy Municipality include two storey houses with cantilevers having grid glass facades in reference to traditional houses are dramatically different from other houses in the neighborhood (Fig. 5a). However, these facades were quite likely have not been implemented in regard to economic inefficiency of the design and minimum standards for low-cost housing. Instead for the façade design, below and above the window openings, horizontal lines formed with cement mortar visually referring to timber weather boarding have been used (Fig. 5b).

Credit Edifice Fund Houses located at the south end of the Municipal Housing were designed constructed as the third phase of the site. In a booklet published by Credit Edifice Fund in 1957, it has been emphasized that Koşuyolu Neighborhood has already become an elite area of the city with sufficient infrastructure,

a

b

Fig. 5. a. Workers Houses, drawing. b. Workers' Houses facades, 2017.

It is also stated that the construction of the houses will be finalized by June 1958 and will be available to move in (Suoğlu 2009). Seyfi Arkan one of the prominent names of modernist architecture in Turkey is mentioned as the designer of the urban plan prepared for this stage (Cengizkan et al. 2012). Although no original drawings with his signature is included in the archives, the architectural language of the movie theater building included in the booklet reminds of his style. The booklet included eight different types of houses and a movie theater building facing the main street (Fig. 6). Unlike other residential units, in this phase house types with shops on the ground floor are added to the choices. It is also mentioned that, maximum comfort with minimum cost is the main objective and thus, in the houses small but functional spaces with possible maximum storage has been designed.

Fig. 6. Credit Edifice Fund booklet, cover and drawings, 1957

Second part of the first phase of the neighborhood, Bank Houses were designed by two prominent names of the period, urban planner Kemal Ahmet Aru and architect Rebii Gorbon in 1950. Settlement plans prepared both in 1/1000 and 1/500 scales include rows of shops, a school, a coffee house and an elementary school in addition to 413 housing units located on 14 plots situated in a field of 140.000 m^2 (Fig. 7).

The settlement including five different types of one or two-storey semi-detached houses with gardens designed in rows aligned along the streets. On the slopes basic grid plan modified in accordance with the topography thus all houses might have a panorama of the valley. The advertisement flyers published in 1952 and 1953 by Emlak Bank, hass detailed information about different types of houses, credit amounts and implementation stages (Fig. 8). However, rich variety of house types on site probably refer to later revisions of the project.

The fourth and the last phase called Dock Workers Houses located on Cenap Şahabettin Street, has been implemented rather late in 1962 by the İstanbul Dockers Labor Union Cooperative (İstanbul Liman ve Dokları Gemi Sanayii İşçileri Sendikası Kooperatifi) (Suoğlu 2009). A tender notice by Turkish Trade Union about the construction of 25 houses on according to the existing project agreement has been published in the newspapers (Milliyet 1961) The housing units in this section designed and constructed in accordance to low-cost housing principles, have plain prismatic masses with hipped roofs.

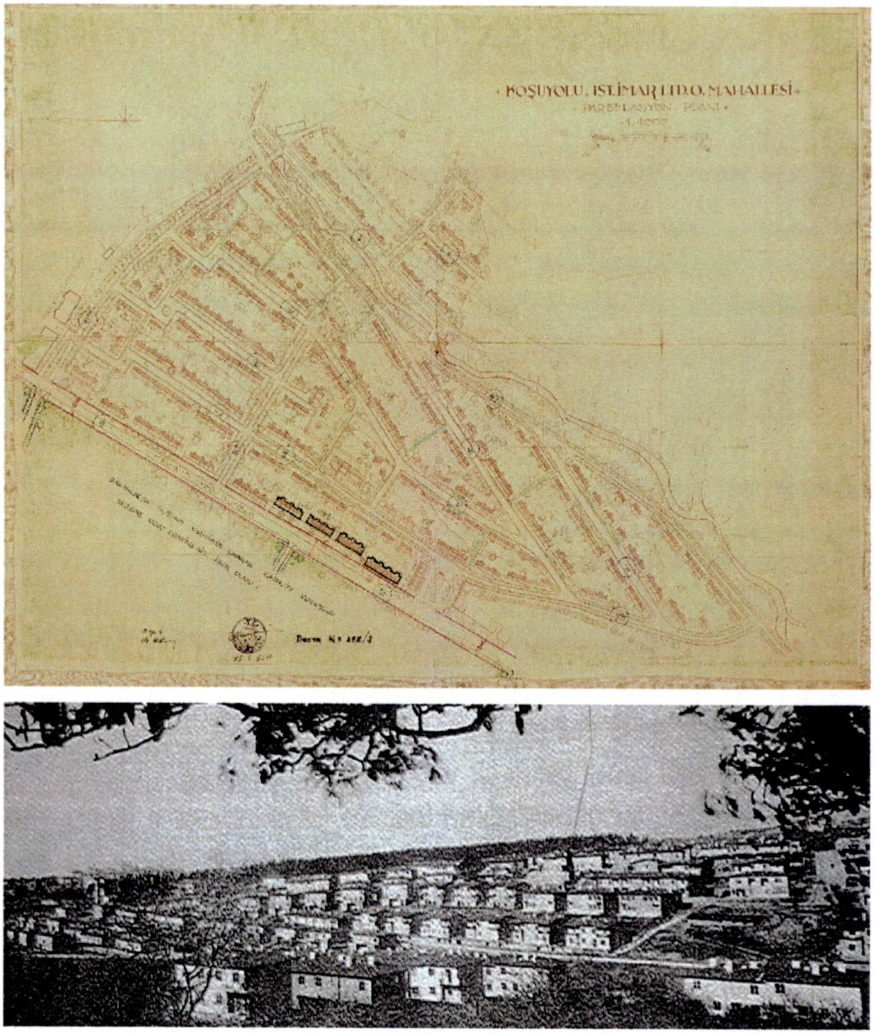

Fig. 7. Plan and view of Bank Houses

Fig. 8. Booklet of Bank Houses, covers and different house types, 1952–1953

4 Neighborhood Transformed

Construction of the Boğaziçi Bridge in 1973 has been the triggering factor for the transformation of Koşuyolu Neighborhood. Its close location to the main arterial road increased the land values and functional transformation from residential use to offices. This continuous functional transformation had also accompanied with physical alterations. The Master Plan approved in 1997 is the second turn point for the site. As a solution for insufficient parking lots based on the increase of car ownership, the plan permitted underground parking spaces in the gardens and gave additional storey permissions for the lots facing Koşuyolu Street. Thus, in a very short time period construction of underground parking spaces has led to landlessness throughout the whole site while nearly most of the houses facing Koşuyolu Street completely lost their architectural characteristics. As another factor, earthquake in 1999 accelerated the demand for low-rise single family houses located in the city center. Last but not least, after the designation of Levent Housing Neighborhood as an urban conservation site in 2008, many offices left Levent and moved to Koşuyolu. Thus a very pleasing progressive step for the conservation of a contemporaneous ex-suburb became a threat for Koşuyolu Neighborhood. Today, nearly 45% of the total building stock maintains original residential use, yet the rest 55% is transformed for various commercial activities including offices, cafes, restaurants and other services. This functional change obviously affected the physical character of the neighborhood negatively and many houses went through a rapid transformation. On the other hand the functional discontinuity is not the only problem since most of the houses within the neighborhood constructed additional spaces, floors due to daily needs and changed original materials with new ones (Fig. 9a, b and c).

a

b

c

Fig. 9. a. Bank Houses, well preserved units, 2018. b. Municipal Houses, improper alterations 2000–2008. c. Dockers' Houses, additional car parking spaces, 2018.

5 Heritage Values of the Place

Theoretical approach for the preservation of modern architectural heritage covered an remarkable distance in the last 30 years. From Eindhoven Statement in 1990 to Madrid-New Delhi Document in 2017, the importance of preserving the heritage places and sites of the 20th century came to gain acceptance in the world. However many places are still at the risk from a lack of appreciation and care. Preservation of modernist housing projects commonly threatened by rapid urbanization and the changing life style. Designed and constructed in a different urban context, for a completely different society, social and cultural groups, it could be extremely challenging to set a conservation approach and convince the related stakeholders for conservation. Managing the limits of acceptable change becomes a crucial issue.

Designed as one of the few surviving low-cost housing settlements in İstanbul, Koşuyolu Neighborhood should definitely be accepted and evaluated as a heritage place. Prominent architects and urban planners of the period who have been involved in the planning and designing processes constitutes the canonic value of the place. Although the near vicinity and urban context has been radically changed over time, the settlement is still a reminiscent of the functionalist approach to city planning in the 50s and reflect the changing social patterns of the mid-20th century life. Henceforth, it should be accepted as one of the significant elements of the historical layers of İstanbul, when all those fields were empty lawns. However, the authenticity and integrity of the place is highly at risk in terms of tangible and intangible attributes. Tangible attributes suffered a lot from adverse effects of the urban dynamics in the last two decades and urgent seek for a solution is needed. On the other hand, intangible attributes like the collective memory of a small group of people living in the neighborhood from their childhood is both precious but fragile at the same time.

6 Conclusion: Threats, Potentialities and Suggestions

Most of the housing units have been altered in time and lost many of their design features amongst are open space organization, shared roofs of attached units, façade finishing, and architectural elements like doors or windows. This negatively dramatic change in the outer appearance of the buildings has constructed a barrier against the existing values to be represented. Due to the analyses carried out between the years 2016 and 2017 an evaluation method was developed in order to unveil the invisible values of the site. Plot -street, plot-building, plot -garden wall, façade character, and mass configuration analyses resulted that on the side of Municipal Houses, well preserved houses are only 15%, while houses preserved in the medium level are of 18%. Thus at the Bank Houses well preserved units had 29% yet the ones have medium level preservation remained at 9%. These percentage ratios are very frustrating at the first glance. However, with an optimistic look a total potential of 71% could be provisioned considering that a certain amount of the negative alterations can be reversed and hidden features can be revealed by a proper approach.

Madrid-New Delhi Document, the latest international text on the preservation of 20th century cultural heritage should be accepted as the basic guideline both for evaluation of the heritage values, establishing limits of change and for any type of interventions. Undoubtedly the very first step should be the acceptance and designation of the neighborhood as an urban conservation site and this should be followed by a new conservation plan including a comprehensive inventory of the buildings and guidelines for radically altered ones. Although it is not easy to reverse the functional change, appropriate short and medium term planning and preserving decisions may hopefully act as a tool to stop the demand for functional change.

Koşuyolu Neighborhood like few other similar places in the city, merits an integrated planning and preservation approach to present its existing but invisible character as a representative of the 50s İstanbul.

Acknowledgements. The statistical data included in the paper is part of an ongoing research based on the collective term study within the scope of the course *Conservation of Modern Architectural Heritage*, given by the author at İTÜ Graduate School of Science Engineering and Technology, Restoration Ph.D. program.

References

Akbulut, M.R.: Koşuyolu, Dünden Bugüne İstanbul Ansiklopedisi, Kültür Bakanlığı ve Tarih Vakfı Yayını, pp. 78–79 (1994)

Akcan, E.: Modern Komşuluğun İç İçe Geçmiş Tarihleri (2005). www.eurozine.com/modern-komsulugun/. Accessed 20 Feb 2019

Akcan, E.: Architecture in Translation Germany Turkey and the Modern House. Duke University Press, Durham (2012)

Anon: İstanbul Belediyesi Tarafından Taksitle Satılacak Ucuz Evler Yönetmeliği, İstanbul Belediye Matbaası (1950)

Anon: Levend ve Koşuyolu Evleri, Emlak Bankası Broşürü (1953)

Anon: Kredi Yapı Sandığı Koşuyolu Evleri Broşürü (1957)

Aru, K.A.: Avrupa'da Mesken Problemi. Arkitekt, sayı **259–260**, 100–102 (1953)

van den Broek, J.H.: Habitation 3 U.I.A.: Programme, Design and Production. Elsevier Publishing Company, Amsterdam (1964)

Özdoğan, F., Balkan, A., Arpat, A.: İstanbul Belediyesi'nin Ucuz Evler Müsabakası Münasebetiyle. Arkitekt **221–222**, 135–136 (1950)

Sayar, Z.: Mesken Davası. Arkitekt **171–172**, 49–50 (1946)

Sayar, Z.: Mesken Davasında Teşkilat. Arkitekt **253–254**, 213–214 (1952)

Sayar, Z.: Mesken Davasında Devletin Yardım ve Rolü. Arkitekt **259–260**, 69–70 (1953)

Sayar, Z.: Şu Mesken Davamız. Arkitekt **283**, 3–4 (1956)

Sayar, Z.: İmar Vekaletinden Beklediklerimiz. Arkitekt **290**, 3–4 (1958)

Suoğlu, R.: Koşuyolu Yerleşmesinin Oluşumu ve Gelişimi, İTÜ FBE (2009). Unpublished master thesis

Şener, H., Yıldız, D.: Dünden Bugüne Koşuyolu. Mimarlık **291**, 26–37 (2000). TMMOB Mimarlar Odası

Yıldız, D.: Modern Dönemin Konut Alanlarının Değişen Yüzü: Koşuyolu Konutları Örneği. TMMOB Şehir Plancıları Odası Haber Bülteni **227**, 25–27 (2014)

Yöney, N., Salman, Y.: Mass housing development by a government agency and the politics of urbanization. In: IPHS 2010 Istanbul Conference Proceedings, 12–15 July, pp. 441–456 (2010)

The Female Presence in the Brazilian Modern Architecture: A View Through Acrópole Magazine's Pages

Tabosa Mayara(✉)

Faculty of Architecture and Urbanism (FAU), University of Brasília, Campus Universitário Darcy Ribeiro, ICC Norte, Brasília 70904-970, Brazil

Abstract. This article is a result of the concern about the absence of female references during academic education, which resulted in a research still in progress. Nonetheless, the paper is structured in a bibliographical research on the presence of women in architecture, contextualizing the Brazilian reality and in the screening in the 391 editions published along 33 years, which result in more than 23 thousand pages. The main result of this paper consists in the listing of the authors' names and their respective works, as well as the analysis of the data collected and a closer presentation about three of these women: Lygia Fernandes, Odiléa Toscano and Rosa Kliass. Aware that this is only one of thousands of ways to begin to raise the names of women active in modern architecture, I intend to provoke debate on these contributions and provide material on various architects whose built heritage remain in the shadows of the modern movement history, often told through the name of the "big stars" of this period.

Keywords: Women in modern architecture · Brazilian modern architecture · Acrópole magazine · Lygia Fernandes · Odiléa Toscano · Rosa Kliass

1 Introduction

As Cohen (2013) points out, history of modern architecture told in bibliography published until the 1970s propagated "a mythology of modern architecture that privileged the radical character of its innovations" (CASTEX cited in Cohen 2013, p. 15) and, consequently, the great masters of this period. This apparent linearity in history has left many important contributions to the Brazilian modern architecture in the shadows.

In Brazil woman are already majority in architecture. Dara from de Architecture and Urbanism Consil of Brazil (CAU/BR) concludes that this percentage which has been growing over the years, tends to consolidate more and more, given that women are 67% of the country's architecture students. The same survey gives us inquiring statistics about federal positions held by women: they don't even reach 20%. Why do women still minority in prominent positions? How did woman become majority and where are our built heritage designed by them? Do it still exist to tell its (and our) history?

I. Vayas and F. M. Mazzolani (Eds.): PROHITECH 2021, LNCE 209, pp. 297–316, 2022.
https://doi.org/10.1007/978-3-030-90788-4_26

As a consequence of the lack of knowledge and – even less - of the recognition of the production of the female architects, many of the examples found during this research are completely depersonalized or have already been demolished. So, in addition to losing such an important part of a still current and constant struggle, works rich in cultural significance are being quietly wiped out. The risk of not preserving it for our generation and for the future, is to erase an important part of our history that was built due to the excellent work of our precursors in a scenario not conducive to their professional performance. And those registers are being erased as if it they had never existed.

Lygia Fernandes, an architect of whom we will talk about in this article, had the house Dr. José Lyra published in different national journals. It was also published in the French magazine Architecture d'Aujourd'hui and was a postcard at the time. The work had elements typical of a residential Brazilian modern architecture architecture: a large wooden lattice panel, muxarabis, shutters, concrete brises-soleil used for thermal and solar control inside the house, and the butterfly roof - widely used by architects such as Acácio Gil Borsoi and others from that region of the country. In its spatiality, the prevailing winds define the position of the rooms on the upper floor. Which it's layout form on the floor below, a covered area under circular *pilotis* - integrated with the internal patio, allowing the ventilation of other rooms on the ground floor. In addition to having other very particular elements.

Images of the house in the 50s can be seen in Fernandes (1955) and the current images of the property are available at "416 Av. Fernandes Lima" on Google Street View, in accordance with the links available in the bibliographic references. The reality observed in the comparison is an alert to the sad constancy of loss that our historical heritage continues to suffer.

Considering this brief contextualization, this article aims to contribute to a debate on female legacy in Brazilian modern architecture, and to honor these professionals who were some of the precursors in our professional field. The spatiotemporal outline is the Acrópole magazine, an important specialized journal which covered most of the architecture produced in national modern movement. It was published between 1932 and 1971, and compose a group of magazines that "reflect the period of transition to the affirmation of modern architecture in Brazil" (Miranda 2010, p. 9).

This paper is product of a graduation final project entitled "Arquitetas da modernidade na perspectiva da revista Acrópole, 1932–1971 (Female architects on the perspective of Acrópole magazine, 1932–1971) and consists in the collection and analysis of data, followed by the elaboration of an overview of the trajectory of three of these women. The project seeks to broaden the look on the feminine performance in architecture and raise some questions about a production that was an effective part of Brazilian modern architecture in order to provoke curiosity about the "whys" and "hows", which can result in overwhelming results for the academic community. The consolidation of the current professional scenario, in which – in Brazil – women are majority, runs through a recent past trod by women is the 20th century, whose contribution in still threatened by the lack of recognition.

To better understand the reader, this article was structured as follows: in the session we are in, I present the topic and approach the relation of women with the architecture field, followed by problem delimitation and justification for the research; Then, the contextualization session brings a brief historic about women's access to education in Brazil, when they began to show up in architecture programs, and the presentation of the Acrópole magazine, main source for this research. After familiarizing ourselves with the context and topics relevant to its comprehension, the last session, called "Women in the Acrópole", summarizes the results found in this research, together with a brief presentation of three of these women: Lygia Fernandes, Odiléa Toscano and Rosa Kliass.

Therefore, I invite you to walk through these pages and to let you be sharpened by questions and curiosities that have arisen along the way, which can - and should - result in new discoveries about our architectural heritage.

1.1 Women and the Architecture

> *Academic studies of women's history have been expanding since the mid-1970s, notably in Europe and the United States. This was largely due to feminist movements and the increased entry of women into universities [...] in the developed countries of the West, this work is well developed. In Latin America, we still have a long way to go (Lima* 2013*, p. 10).*

In Brazil, the efforts to bring these names up have resulted in interesting works and names hitherto little or unknown to the majority of student and professional class. Noteworthy is the research group "Arquitetas Invisíveis" (Invisible [female] architects, in free translation), created in 2014 by students of the Architecture and Urbanism University of Brasilia, in order to "promote gender equality within the scope of architecture and urbanism, through the recognition and spreading the life and work of architects who are disgraced by history"[1]. Based on the efforts of these researchers, a website was created containing the results of these studies, exhibitions and lectures were given, as well as the publication of magazines with more than 20 names - national and international - found over the years.

Other professionals have also been studying over the years about female architects who had until then no bibliography about their work. Like Goldchmit (2008), whose thesis on Odiléa Setti Toscano's work was the primary source for this research, and Andrea Gáti (2014), who has dedicated her studies to Janete Costa, an architect and designer from Pernambuco, state of northeastern Brazil. Besides her thesis, Gáti also has other papers on the topic, including "Wives: Acts in Architecture, Interiors, and Design" (free translation) in which she addresses architects who were wives and partners of renowned architects, but who did not have the same professional recognition. This scenario highlighted by Gáti happens with some recurrence in 20th century Brazil and still has a lot do revel us about our recent past (or also about our present). Lima (2013), recurrently referenced in this paper, outlines a perspective of the Latin American female architects of the 20th century, which thesis brings a pioneer book about the topic in Brazil. Therefore, although the efforts initiated by these and other authors are noteworthy and relevant steps, there is still a long way to go regarding the recognition of female personalities of Brazilian architecture and the insertion of these women as part of history of our national architecture. That said, this paper seeks to contribute to the theme,

[1] Arquitetas Invisíveis. *Trajetória.* Available in: arquitatasinvisiveis.com/. Free translation.

presenting itself as an initial effort from this author, who is committed to continue research on the topic, aiming to delve into the subject and help make this heritage known and recognized by professionals and social factors involved.

For a fuller understanding, I will briefly address the trajectory of women and the access to education and their insertion in the architecture market in Brazil. According to Stamatto (2002) girls were excluded in the schooling process until 1758, when slowly they began to have permission to attend school, and a few decades later a woman had access to higher education in Brazil. The rise of women in professional training of architects in the country and the subsequent development of a predominantly female scenario are also discussed, in order to support the theoretical framework and the context in question.

1.2 Problem Delimitation and Justification

On the international stage, the Pritzker Prize is considered the most important in the field and reached 78 awards in 2019. Of these 78, only 3 were earmarked to women, and only Zaha Hadid was individually awarded, in 2004. Kazuyo Sejima was nominated along with her partner Ryūe Nishizawa in 2010, and Carme Pigem with Rafael Aranda and Ramon Vilalta, with whom she compose the 2017 RCR Arquetects team. In contrast, 75 architects have already been individually awarded by the Pritzker Prize (Fig. 1).

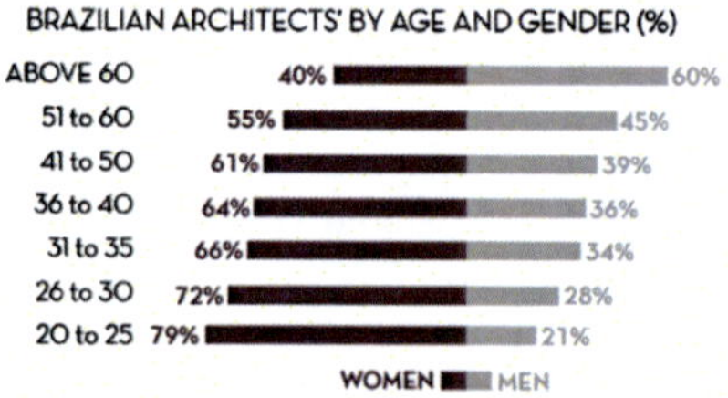

Fig. 1. Brazilian architects' by age and gender. Source: CAU/BR (2019), edited by the author.

Besides this award, the numbers are even more disproportionate if we take a look at American Institute of Architects' (AIA)[2] Gold Medal laureates, in which the first woman was awarded only in 2014, more than 105 years after the annual award began. Following the forerunner Julia Morgan[3], Denise Scott Brown received the prize in 2016, along with Robert Venturi, composing the only two AIA Gold Medal laureates, both professionals who were belatedly acknowledged for their contributions. Although the AIA Gold Medal is an award focused primarily on the United States, in addition to

[2] According to the AIA, the Gold Medal is the institute's largest annual award, considered one of the most prestigious in the architect's world, and is aimed at individuals whose work has had a significant influence on architectural practice or theory. (AIA - American Institute of Architects. Gold Medal. Available at: <https://www.aia.org/awards/7046-gold-medal>).

[3] Julian Morgan (1952–1957) born in the state of California, was the first woman to enter the Paris School of Fine Arts and the first licensed woman to work in architecture, as stated by the Invisible Architects (2015).

inconsistency in the number of female and male laureates, the award is held by a country where research on women in architecture is older and more developed, what is supposed to be a breakthrough in the professional recognition of these women, however, as we can see, this recognition is not reflected in the country main award.

In Brazil, data from the (CAU/BR) indicates that women are the majority of active professionals in the country. But even they represents more than 63% of architects with active registration, what means that they are the only ones allowed to have technical liability record, women's work does not receive the same visibility and recognition of male colleagues, as stated by Lima (2013).

Approaching the Brazilian modern architecture scenario, according to Miranda (2010), the Acrópole magazine (1938–1971), main source of research of this work, along with Arquitetura e Urbanismo (RJ, 1936–1941) and the Federal District's city hall magazine Revista do Clube de Engenharia (RJ, 1932–1932), portrays the affirmation and recognition of the movement in Brazil. Moreover, the author emphasizes the privileged immersion of the architecture journals in the diffusion of the national cultural quality, which was accentuated in the 1950s with the investment of cultural producers, such as critics and artists, mainly in the cities of Rio de Janeiro and São Paulo.

The author states that until 1970, cultural magazines were small or medium sized and faced market difficulties. In the case of architecture magazines, there was an aggravation of the lack of public, once it is estimated that the number of active architects active in Brazil in the 1950s was 1077. With the opening of new architectural schools and consequent increase in the number of architects in the country added to the interest of Brazilian architecture abroad, as stated by Miranda (2010), the periodicals gain more visibility and currently "Acrópole magazine is considered by historiography and criticism as one of the most important periodicals of Brazilian architecture, characterized by a high frequency of publications, of a monthly nature" (Almeida 2008, p. 98).

In more than 30 years of publication, names such as Paulo Mendes da Rocha, Niemeyer, Vilanova Artigas, Warchavchik and several others that fit in what Lima (2013) identifies as "superlative figures that have influenced, changed and forever marked directions of the profession" (Lima 2013, p. 10), appear frequently in its pages, as well as appear in the academy and in the most diverse bibliographies about modern Brazilian architecture. However, as Roth (2017, p. 8) states, "architecture is also the physical record of human activities and aspirations," which means that architecture is directly linked to its history. This connection between architecture and history does not tell us only about the evolution of architecture, but also the changes and advances of human society. So, to understand the present, we need to know the process and, as Lima (2013) states, not only from the point of view of big names, but with the social and broader vision of those who contributed to the construction of professional and social scenario that we find ourselves today.

Thus, to understand our history we must understand and recognize the precursors of a predominantly female scenario that is also the result of competent "anonymous" professionals of the 20th century, who contributed to important changes in the context in which they were inserted.

2 Contextualization

In the second half of the 18th century, when girls started attend school, boys and girls were taught separately: male teachers taught boys and female teachers – who could only lecture with their father or husband's permission – taught in schools for girls. This scenario has changed only in the late 19th century with the emergence of mixed schools in Brazil, where girls and boys started sharing the same classroom. At this point, female teachers began to be allowed to lecture to children of the opposite sex up to a certain age – usually from 12 to 14 years old, while male would still exclusively teach boys, as pointed out by Stamatto (2002). By the end of the19th century the proportion was of one girl out of every three boys in school.

In regard to higher education, it was only in 1881 that we had the first female diplomaed in a Brazilian college: Abrosina de Magalhães graduated as a doctor from the Faculty of Medicine, in Rio de Janeiro. Two years after her, this same medicine school had four female students of which, according to Freire (1989, p. 104), "one attended the classes accompanied by her father, another by an old lady, the others exempt protection and all were banned certain classes like anatomy and physiology"[4].

What seems like a "natural" process when told in a nutshell, was the result of hard struggles for women's rights. Here it is worth mentioning the importance of the União Universitária Feminina (Women's University Union, in free translation), chaired and cofounded by Carmem Velasco Portinho, which had an important role for the insertion and permanence of women in colleges, as Motta (2014) pointed out. In addition, Carmem Portinho was a prominent personality in modern Brazilian architecture, being the third woman to graduate in Civil Engineering at the Polytechnic School (of the currently called Universidade do Rio de Janeiro) and the first female urban planner in the country.

Although these changes have occurred since the late 19th century, only in 1962, with the Married Woman Statute[5], women were legally granted with several rights, including the right to practice profession without the need of her husband's permission.

2.1 The Labor Market and Its Feminization

Regarding architecture academic education in Brazil, although the female insertion occurred belatedly, Oliveira (2011) states that the formation of architects took place since the colonial period. It was made through the "Military Classes" responsible for training "military engineers" or "military architects"; and in the "mechanical craft corporations" where knowledge was transmitted from master to apprentice.

However, Sá (2010) argues that whilst women's entry into academia has been delayed as a result of social and cultural difficulties of the time, there is no evidence that women's access to architecture schools has been clearly restricted at no time in our country.

[4] Free translation. Original quote: "uma frequentava as aulas acompanhada pelo pai, outra por uma velha dama, as restantes dispensavam proteção e todas eram interditadas a certas aulas como anatomia e fisiologia".

[5] Law No. 4,121, AUGUST 27, 1962 defined by the House of Representatives as an Ordinary Law that "regulates the civil rights of married women and other measures" (free translation). Given that the Statute has not been revoked, it is valid in that which does not conflict with the Civil Code and the Code of Civil Procedure, amended in 2002 and 2015, respectively.

Taking a look at Sá's (2010) data between the early 1950s and early 2000s on the percentage of male and female graduates of the Faculdade de Arquitetura e Urbanismo of Universidade de São Paulo (FAU-USP) (Fig. 2a) and of Universidade Federal do Rio Grande do Sul (FA-UFRGS) (Fig. 2b), two of the country's oldest architectural colleges, it is observed that both more than tripled the number of women graduates, going from approximately 15% to over 50% over a period of about 30 years.

Salvatori (2003) points out that although in the first 50 years of FAU-URGS (1952–2003) it formed 53.4% men, women became major graduates from 1973 onwards, except in 1981, 1987 and 1993. The author further states that:

> *"The same tendency is detected by Durand (1989) in the Brazilian context: between 1970 and 1980, while the total number of architects increased by 1.5 times (1.5), female participation within the professional category grew seven (7.0) times. There are many hypotheses that intend to explain the phenomenon; probably, the fact originates from a combination of these factors, and may, depending on the context, tend to one or the other" (Salvatori 2003, p. 1, free translation).*

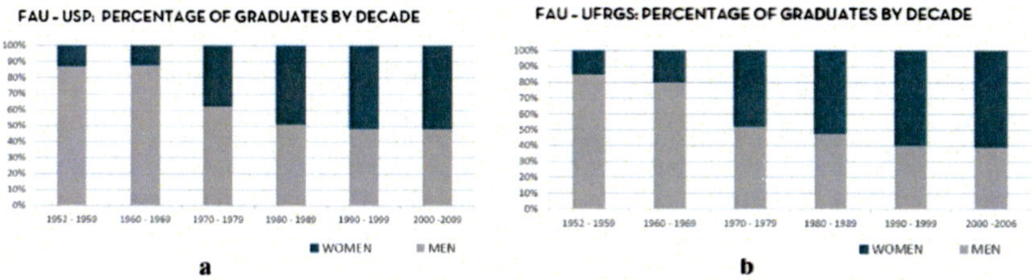

a b

Fig. 2. a. Percentage of graduates by decade on FAU-USP. Source: data from Sá (2010), edited by the author. b. Percentage of graduates by decade on FAU-UFRGS. Source: data from Sá (2010), edited by the author.

According to the author, the first of these reasons would be the new role of women after World War II, when the demand for services related to education, health and culture increased, areas where female insertion was more prominent. The author also adds that there was an absorption of female labor also by trade and bureaucratic activities, as well as women also began to exercise liberal professions, mainly because of the ease in reconciling family obligations and motherhood. "The educated and professionalized woman, with improved aesthetic dispositions, becomes part of an audience that consumes various types of cultural goods, which eventually generated a circular increase in educational demand." (Salvatori 2003, p. 1, free translation).

Concerning the choice of women by architecture, Durand (1989) raises some hypotheses. Among them a change in social perception of architecture as linked more to arts and decoration, which may have been a result of the 1962 Teaching Reform, pioneered in the FAU-UFRGS. Its objective was to distinguish architecture by the specificity of architectural design, that is, to detach a little from the idea of a construction site practitioner – where they competed with several other professionals – to refer to the project's design and conception, more suited to their intellectual status.

"The problems of this meaning are discussed at FAU-FRGS in the late 70's and end up resulting in a new curriculum reform in 1982, which seeks to reincorporate the construction as indispensable to the living space, as well as the project itself - besides trying to minimize the effects of the University Reform of 1969" (Salvatori 2003).

The University Reform of 1969 refers to changes in the entrance exam that reduced or ever extinguishing the credit of freehand drawing, which would have allowed the admission of culturally disadvantaged classes. According to Durand (1989), this fact would have come into convergence with the devaluation of the profession, a consequence of the accelerated development of civil construction and the decrease in the quality of buildings during the period of "economic miracle" after the Military Dictatorship. In other words, at the same time as the devaluation of the profession due to the University Reform of 1969 and the reduction of opportunities in the labor market occurred, the social vision becomes to associates architecture with the arts. These facts happen at a time when women are been accepted in other segments of labor market, and would probably have favored the insertion of women in the architecture field; given that, the female figure was already well admitted and recognized in the fine arts:

"Well, by the time the first reform occurs, women of the middle strata of the population are in the process of integration into the labor market. But this insertion seems to come from its own logic, as an expansion of its own female inclinations and attributions and domestic skills. Durand's studies show a rapid numerical and percentage increase in female participation not only in architecture, but in the field of visual arts in general in Brazil between 1950 and 1980 - including photography, decoration and cinematography, sculpture and painting" (Salvatori 2003).

However, while women have shown a significant amount in both academia and professionals in labor market, research shows that prominent positions and projects are mostly held by men[6] - not only in architecture but in several areas. A report from the O Globo newspaper, dated March 2017, among other data, confirmed that women occupy is only "37% of the leadership and management positions. At the top, in the executive committees of large companies, they are only 10% in Brazil". Besides the fact that this subject requires an in-depth debate, which is not the focus of this paper, it is important to note that:

"The visible and invisible barriers that keep women out of the most qualified and highest paid positions are numerous: the feminization of certain professions and their subsequent devaluation, social resilience, motherhood and inequality in the division of household chores, lack of critical mass of women in organizations, and others. Only through the acquisition of power this fact can be transformed, in other words, this fight involves power and it will certainly be fought in the

[6] Arcilla, P. (2015). AJ's 2015 Women in Architecture Survey Says "Pay Gap" is Slowly Closing. *ArchDaily*. Available in: https://www.archdaily.com/591058/aj-s-2015-women-in-architecture-survey-says-pay-gap-is-slowly-closing/.

political spheres, at home and in the companies" (ARROIO e RÉGNIER, 2002 cited in Maldaner 2005, p. 34, free translation).

Lima (2013) emphasizes in her thesis disproportion of the visibility and recognition of women active in professional field of architecture, considering the fact that there is currently a gender balance in the exercise of profession and academia, where, in fact, women are often noticeably more numerous in the latter. These finding made in various sources of research is difficult to explain and there is, until then, no definitive conclusion on the subject.

2.2 Acrópole Magazine: Introducing the Journal

"In the 1950s, and especially in the 1960s, there was no architect or architecture student who was effectively engaged in the profession who did not consult or collect the magazine. (...) or an architecture student in the second half of the 1960s who signed the Acropolis and consulted it as it was a treatise on architecture. There will be many episodes of this nature that I have heard and will still hear. Forming a mythology that grows as an imaginary for many generations: from those who eagerly awaited the new edition each month, to those that today we leafed with the reverence of a valuable historical document." (Segawa 2014).

In the presentation text for the website of the Acrópole magazine, developed by FAU-USP, Segawa, while emphasizing the importance of the journal, states that although the journal survived from publicity and that it was not a vanguard magazine, trends[7], ideologies or principles defined, it may have been all at the same time.

Tue magazine has been under the direction of Roberto A. Corrêa de Brito from its launch until 1953, when it passed to Max Gruenwald, who directed the magazine until its last edition in 1971. However, according to Segawa, it was not only its editors that were responsible for the journal's editorial quality, but the various "consultant-architects and collaborators who throughout the long-lived existence (...) have outlined the Acrópole's various personalities over more than three decades, making an impressive and astonishing testimony of the time" (2014, free translation).

According to Corona (1971), Acrópole magazine emerged from the enthusiasm of the architect Eduardo Kneese de Mello who, along with Roberto Corrêa de Brito, sought the author to publish a small printed portfolio of his works. The architects Henrique Mindlin and Alfredo Ernesto Becker were invited by Corrêa de Brito to participate in magazine formulation, which inaugural edition receives Ramos de Azevedo's monument on the cover and an article by Kneese de Mello on the Acropolis of Athenas – what names to the journal.

Manfredo Gruenwald bought the Acrópole magazine along with his father Max in late 1952. He points out the main changes the magazine has undergone "related to the character of its publications, its graphic communication, the participation of its collaborators, its publicity and the way it relates with the architects and the Architects

[7] "Understood as one that is editorially limited to a set of related values, shared by a closed group" (Segawa 2014. Free translation).

Institute of Brazil" (Instituto dos Arquitetos do Brasil – IAB)" (GRUENWALD *cited in* Almeida 2008, free translation). So in the new administration, the journal began to explore reflections on the architecture being produced at the time, which was not explored in previous phase of its production.

In another interview, given to the magazine Projeto Design, Gruenwald was asked about how the selection of texts and works was made for publication, and he states that the architects were usually sought after by the magazine and not the other way around.

> *"It wasn't restricting, but we selected more. Those young architects were different from the old firms, which made that everyday architecture... In fact, after we took over, the selection was automatic, because the old ones were no longer in demand" (Gruenwald 2006, free translation).*

In 1971 the magazine ceased the circulation for financial reasons, according to former editor. It was also during this period that the Brazilian modern architecture began to decline due to the post-dictatorship economic boom that generated a large production in the construction industry, whose focus turned to the quantity of its works, leaving behind the quality of what was being produced. Therefore, it can be said that the Acrópole magazine followed almost completely the process of the introduction of modern architecture in Brazil, from the debate and establishment of a national modern expression, to its peak production – marked by Brasilia –, and stopped circulating approximately in the period when Brazilian modern architecture was no longer as strong as in the previous decades.

3 Women in the Acrópole

3.1 The Search

It is in the second period of the magazine, from 1954 to 1971, that we come across the publications of female authored projects. The first three will be in 1954 (ed. 190, 192 and 195) on landscape designs by the American landscape architects Roberto Coelho Cardozo and **Susana Osborn Coelho**. Following we have the first architectural design of female authorship, published in 1955, with some peculiarities: signed individually by a northeastern architect from São Luís of Maranhão, **Lygia Fernandes**, with a building located outside the state of São Paulo, where most of the works published in the magazine were located. In addition to Lygia Fernandes, the only architect whose published project is under individual authorship is **Miranda Martinelli**, whose work named by the Acrópole as "Residence in Boaçava" was published in 1957 in the edition 229, bringing the fifth female name printed in the magazine.

Maria Laura Osser appears as co-author of the "Ducol Stores in Rio de Janeiro" in edition 229 of 1957 and, in the same year, in the 229 edition, **Rosa Grena Kliass** first appears in the journal: she is mentioned by a project developed with some architects, including Jorge Wilheim and Roberto Coelho Cardozo. In the following year she is cited twice in edition number 234, for preliminary designs of residences made in partnership with her husband, the architect Wlademir Kliass.

In 1959 (ed. 247), **Paola Pestalozzi** and her husband Walter Pestalozzi sign the authorship of the "Campos de Jordão Municipal Market" and in the following year

Deisy Igel is mentioned as a collaborator in the project of a "House in Jardim Paulistano", published in edition number 260. In the same year, **Rosa Kliass** appears as the architect responsible for the landscape design of the "Itaim Residence" published in issue 272, and appears again in editions 294 (1963), 315 (1966), 334 (1966) and 337 (1966) with her signature in the landscaping of residential design signed by other colleagues. In edition 294, Rosa Kliass and **Miranda Martinelli** appear with the landscape design of "Residence in Alto de Pinheiros". Rosa's last project published in 1969, in edition 361, was the "Green Recreation Areas" also developed with **Miranda Martinelli** and is characterized as an unusual publication for being a large urban project that took five pages of the journal.

In 1961 **Ruth Kok de Sá Moreira** with Fábio Kok de Sá Moreira have their project for the Acoustic Shell and auditorium of the Universidade Federal do Ceará (UFC), published in the 277 edition. In the same year, the project for Panair in Brasilia, by **Maria Elisa Costa** and Elias Kaufman, prints the pages of edition 268. In the volume 281 a residential design by **Nina Albu** and Bernardo Vaisman is published, which also have their name in a building garage published in release 333 of 1968.

Marlene Picarelli is cited for the first time in the journal in 1963 as coauthor of the "Residence in Santa Bárbara do Oeste", published in the 298th edition, and has two other projects published in 1963 (294th edition) and 1964 (302th edition). These projects were developed in partnership with a group of architects of what Marlene was part of, and they tribute in the 314th edition, in 1965, with the publication of several school projects designed by the office.

In 1963 **Heleny A. Marques Lins** had the project "SESI Gymnasium in Recife", design along with Forismundo Marques Lins Sobrinho published in issue 297. In sequence, **Maria Lucia Motta de Athayde** and José Fernandes Nunes de Carvalho's proposer for the "Contest for hospital in Recife", which won in second place was published in 308th edition released in 1964. **Ana Maria Gottardi** appears in the magazine in 1966 (331th edition) and **Maria Cláudia Repetto de Boccara** in 1967, in the 340 and 344 editions, both for residential projects designed alongside their spouses and partners.

Odiléa Setti Toscano, as well as some of her previously mentioned colleagues, appeared as co-author of projects developed with her husband and also architect, João Walter Toscano. Their first work was published in 1965 with the project "Public Square and Garden" and just like Lygia Fernandesis a bit out of what was usually published in the magazine; the issuing also have some peculiarities: although located in São Paulo, public spaces did not receive the same prominence as residential and other private projects in Acrópole magazine, thus being one of only two urbanistic projects found in this research signed by an female architect. **Odiléa** and João Walter Toscano also have projects in pages on the editions 345 and 349, on 1967 and 1968, and the publication "Tiles and Tiles" in 1967 (ed. 338). Although the latter is not about an architectural design project, but such publication arouses curiosity since the journal dedicates a few pages to the fine arts and graphic design works developed by Odiléa, which characterizes her as the only architect with non-project publication, together with **Maria Cláudia Repetto de Boccara**, who writes an article with her husband for the 1960 issue 261.

Alice KonderComparato has her name published as researcher for a design for "School Expansions: Emergency Plan" published in issue 364, and **Ana Maria de Biase**

along with **Maria Helena Fynn** appear in issue 361, 1969, as co-authors of the project for the contest of the "Osaka Pavilion", in which it was fourth placed. In 1969 **Maria Estela Mélega** and **Cristina Jucá** appear as co-authors in architectural designs published respectively in editions 378 (1970) and 381 (1971). Cristina Jucá is the second co-author of a residential project in Recife and the last woman mentioned in the journal.

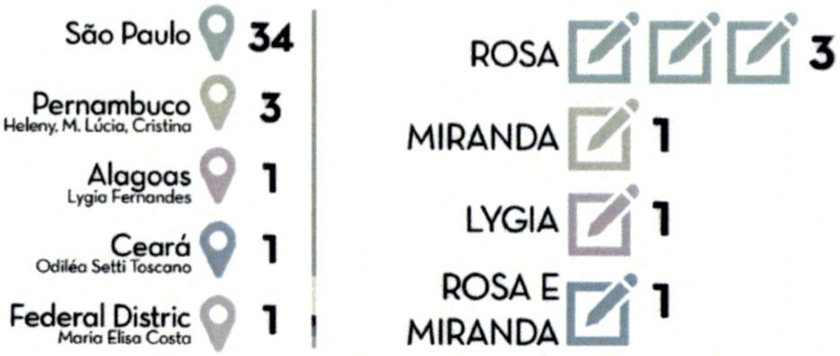

Fig. 3. a. Number of projects buplished under female authorship by State. b. Projects under exclusively female authorship.

These 22 women were identified in a total of 40 published projects (Fig. 3a). This group had to be reduced to a smaller number of personalities to be presented in more detail. The three architects, **Lygia Fernandes**, **Odiléa Setti Toscano** and **Rosa Grena Kliass**, were chosen according to some criteria evolving their peculiarities: Rosa Kliass for notoriously having the most publications, Odiléa Setti Toscano for second architect with most published projects, including a public space project, that it was less common to appear on the pages of the Acropole, and with an additional publication about its artwork; and Lygia Fernandes for being the first architect who had an architectural project published in the magazine, which she signs individually (Fig. 3b) and whose location is outside the state of São Paulo.

3.2 Lygia Fernandes

Lygia Fernandes was born in the northeastern State of Maranhão and in addition to being the first Brazilian to print the pages of the journal, in 23 years of publication, she is one of only three women who have projects exclusively female authored in the magazine, one of the few who is not from the Rio-São Paulo axis.

According to Silva (1991), Lygia was born in São Luís, capital of her State, and graduated from the first class of the Faculty of Architecture of the School of Fine Arts[8], in 1945, when the name of the professional class changed from "engineer-architect" to just "architect". Thus, her academic formation took place in a period of intense debate and about modern architecture and she was influenced by both teachers with eclectic, neoclassical guidelines and by teachers with modern guidelines, being the latter the one that directed the studies of Lygia Fernandes.

[8] Currently named as Universidade Federal do Rio de Janeiro (UFRJ).

She interned with Jorge Moreira, an important Brazilian architect who was also present in the team of architect who designed the Ministry of Education and Health building in Rio de Janeiro, with whom she later worked with. Lygia also accompanied Affonso Eduardo Reidy and Carmem Portinho in the works carried out in the early 1950s - including the construction of the Pedregulho building - besides having professional relations with architect Henrique Mindlin.

In the 40's and 50's, for example, the Brazilian Lygia Fernandes worked in the architectural modernist means with resourcefulness. She participated, for example, in the contest for the architectural project of the headquarters of the Brazilian Jockey Club, with Francisco Bologna, Giuseppina Pirro and Israel Correia, ranking in 2nd place. Although not built, the project merited publication in the Architecture d'Aujourd'hui magazine (n° 21) and in the Arquitetura e Urbanismo magazine in July/August 1947 edition (Lima 2013, p. 91).

Although she professionally recognized, Lygia Fernandes opted for the public service: she worked in the Department of Popular Housing, from where she was transferred in 1960 to the Department of Parks and Gardens and then to the Department of Roads, where she remained until her retirement, in 1989, as stated by her in an interview granted in 2002.

The works of her authorship published in the bibliography are dated from the 50's and located in the city of Maceió, capital of Alagoas (AL), where the architect had family ties, according to Silva (1991). The works demonstrate a mastery of the formal language of Brazilian modern architecture, with responsive solutions and adaptations related to the local culture - the ceramic tile butterfly roof, widely used by Borsoi, muxarabi, stone cladding, among other attributes. Lygia Fernandes also shows in these projects a strong technician mastery, considering that the electric and hydraulic projects, the construction detailing, landscaping, tile design, besides and all the furniture are signed by her.

Information about Lygia, her professional career or her works is still quite scarce. The architect is cited by Cortês (2011) and Lima (2013), and mentioned on some online pages that speak little about her. But the most accurate information about this professional is found in the two pages of Mindlin's book (1999), Silva's book (1991) and in an interview with Costa (2002).

Lygia Fernandes's "invisible" production shows unquestionable relevance in the context of Brazilian modern architecture, especially considering modernization of the context in which it was inserted – the northeastern city of Maceió – and her professional recognition at the time, cited by the main sources of search. The fact that Lygia had this recognition is already an important fact considering the professional scenario in Brazil in the early second half of 20th century. However the lack of knowledge and recognition nowadays about her trajectory and work is seriously compromising her contributions – that are also a thankworthy heritage –that are almost completely uncharacterized, and many of it are still unknown.

3.3 Odiléa Toscano

Odiléa Toscano was born in 1934 in a family greatly influenced by culture and the arts, in the city of São Bernardo do Campo, inland of the state of São Paulo. She draws attention for being one of the most published female architects in the magazine: appears for the first time in the periodical in 1965 with a project of "Square and public garden" and again in 1967 and 1969 with two private projects, the three being carried out in partnership with João Walter Toscano, her partner and husband.

As previously mentioned, the magazine in 1967, dedicated a few pages in edition 338 for Odiléa's design Works on the title "Tiles and Ladders; oil panels; graphic arts", which indicates her versatility but also the relevance of her work. This publication is the only of a few of women designers who had their work on the magazine.

According to Goldchmit (2008), Odiléa Toscano started the architecture school at the FAU-USP in 1953, in a class with 4 female colleagues and 25 male colleagues immersed in a "pluralistic formation" characteristic of their college. Odiléa took classes with of prominent names in Brazilian modern architecture, such as Vilanova Artigas, Rino Levi, Abelardo de Souza and Carlos Lemos, as well as with professionals like Renina Katz, a Brazilian artist who stood out as an illustrator, designer and label, which illustrates the incentive for interdisciplinarity with the fine arts.

She began to be known for the quality of her work while still in college, through FAU's "Sunday Artists" exhibitions and others outside of academia, which earned her works such to draw Jorge Wilheim's proposal for Brasilia's Pilot Plan contest. She graduated in 1958, same year she married João Walter Toscano, who was also her professional partner.

Her work was first published in 1957, in the fine arts section of O Estado de São Paulo's Literary Supplement, and became an illustrator and effective designer of Editora Brasiliense from 1959 to 1966, where she developed over 40 covers for the "Young People Worldwide" collection, worthy of international diffusion, as stated by Goldchmit (2008). In parallel, she began teaching at the Drawing Teacher Training Course of the Armando Álvares Penteado Foundation (FAAP) in 1960, and in the following years she joined other institutions. She joined the FAUUSP faculty in at the Visual Programming Disciplines Group of the Architectural Design Department, where she taught until her retirement in the 2000s at the age of 66.

Throughout her career, she developed several architectural design works, both in private properties and in public spaces, where she was most prominent. In the thesis "Odiléa Setti Toscano: from drawing to design", Sara Miram Goldchmit studies in detail many of these works, their principles and correlations. Regarding the murals for the São Paulo subway stations, which were part of the initiative Art in the Subway and of great importance in Odiléa's career, the author points out some of the principles of the architect's work:

> *In some ways, the production of visual design projects for public use was already a daily challenge for Odiléa, who, since the first illustrations for the O Estado de S. Paulo newspaper Literary Supplement, always had her drawings reproduced graphically in large runs, that is, directed to the whole of society. It can then be said that the city - in addition to often appearing as a theme in printed design work, showing its concerns as an architect - is once again the recipient of its visual message, this time in the form of chromatic interventions in architecture (Goldchmit 2008, p. 118, free translation).*

About her production in architecture, were not found by this research works signed exclusively by the architect. However, analyzing the chronology of João Walter Toscano's works in Artigas (2002), it is observed that in many of his architectural design projects, the landscaping is under the name of Odiléa Toscano, which shows once again the relationship of the architect with issues related to city. But little is known about Odiléa's performance in this field of architectural design. In addition, some of the projects that this professional coauthored were winners of contests and awards, such as the Londrina Yacht Club and Largo 13 de Maio Station.

Regarding the projects carried out with João Walter, in which Odiléa Toscano developed the color study and panels or murals for some, the architect comments on the joint way in which the conception of the projects took place:

> *[...] Most of the work that I developed in this regard related to our projects, which I followed from the beginning until the construction, where the chromatic proposals were born together with the first sketches, at least in terms of intention. Under these conditions, from the preliminary study, it was defined, with a good approximation, the supports that should receive painting and those elements that would remain uncoated, revealing their own material, such as concrete, steel, brick, glass (TOSCANO cited in Goldchmit 2008, p. 119, free translation).*

The graphic design works developed by her are undoubtedly a great differential in her work not only for its graphic quality, but for the concern as an architect in the quality of space and use of graphic arts as an extension of the architectural work, which is applied strategically, with studies that direct their colors and shapes according to the needs of the space and the user. However, her decades-long acting as landscape architect, in urban design as well as in restoration and architectural projects, shows a still silent versatility about Odiléa Toscano's performance.

In the award-winning Largo 13 Station project, the differential is largely credited to the station's color study and panels, which were also defining in design decisions. But Odiléa is rarely mentioned in the publications and, when it is, usually almost querulously. This is a strong example of several other cases that remain hidden in our modern heritage.

3.4 Rosa Kliass

Born in São Roque, also in the interior of São Paulo, in 1932, Rosa Kliass is the female architect who accumulates the largest number of publications in the journal. She signs ten projectual publications and two publications in the "Living Drawing" session, which featured drawings of constructive details. In addition, Rosa Kliass covers some pages of the 1969 edition, 361, with an exclusively female project with Miranda Martinelli, for squares in the city of São Paulo, entitled "Green Recreation Areas".

She joined FAU-USP in 1951 and in her last year of college it was added to the curriculum the course of landscape architecture, taught by the American landscape architect Roberto Coelho Cardozo. It was then that Rosa Kliass immediately understood this as her mission (CAU/RJ 2016), and became the first landscape female architect in Brazil.

She graduated in 1956, same year she married FAU fellow architect and colleague Wlademir Kliass (1929–1985), with whom she developed several projects, including the first published by the magazine. At the time of entering the job market, the only landscape architect in Brazil was Roberto Burle Marx, based in Rio de Janeiro and active in a field that is still little explored in Brazil, in which Rosa Kliass soon entered, doing projects in several states of the country.

She internship with Abelardo de Souza, Rino Levi and also Jorge Wilheim, with whom developed several large projects during his career: among them the Angelica Master Plan (1954), considered the first master plan of Brazil, the proposal for the Brasilia Plan Pilot (1957), the Curitiba Master Plan (1964), Joinville Urban Plan (1965), the winning project for the revitalization of the Anhangabaú Valley (1970), among others.

At the Acropole, Rosa Kliass' first published projects are privately owned and designed together with her husband. However, the following publications already have their landscape architecture in the design of other architects, which indicates a recognition in the professional environment and an expansion of their performance beyond the office that had with Wlademir Kliass.

In the 1960s, she was invited by the then mayor of São Paulo, Faria Lima, to develop the Parque of Morumbi project, "the first city park project designed by a professional and done by the city with someone outside the staff" (Kliass 2012); she was then hired again with Miranda Martinelli to develop the 44 city squares project in São Paulo - the first landscape planning for the city, published on the Acrópole under the title "green recreation areas" and winner of the Carlos Milan award, also mentioned in the journal.

In 1970 she worked at the Vale do Paraíba Municipalities Integrated Development Consortium (Codivap), which resulted in the compilation of the book "Characterization of the Paraíba Valley Knowledge, Codivap 1971". About it, Rosa Kliass states in an interview with AU magazine: "from it derived the policy of regional planning. It was very beautiful, it was the greatest moment of knowledge production I ever had".

From 1974 to 1977 she was professor at the Mackenzie and in the 1980s she joined the Pontifícia Universidade Católica do Paraná (PUC-PR) faculty, where she started the postgraduate course in landscape architecture with Fernando Magalhães Chacel. She Founded the Brazilian Association of Landscape Architects (ABAP) in 1970 in order to "assist, promote, support, encourage and develop scientific, technological, educational, cultural, social and environmental actions aimed at the development, dissemination and professional enhancement of Landscape Architecture in the country" (ABAP 2017), principles that have always been linked to Rosa Kliass' struggle for recognition of the profession in the country.

In 1982 she was invited by Wilheim to take over the board of the Urban Planning Secretariat of São Paulo, and in the 1990s, the parks that he developed mainly in Salvador, Belém and São Paulo stand out in her production, besides the landscaping of the Brasilia airport and of the Youth Park, in São Paulo.

Rosa Kliass was not only a pioneer in landscape architecture, but achieved a professional projection at a time when many female architects worked with their husbands and ended up not receiving the same recognition as their partners, such as Odiléa Toscano herself, or others like Denise Scott Brown or Myriam de Melo Cordeiro, wife of renowned Pernambucan architect Vital Pessoa de Melo. Even so, although Rosa worked in partnership with Wlademir Kliass, her name overcame her husband's at the time when they shared the same office.

To this day she is recognized for her work developed by the office that bears her name: Rosa Grena Kliass - Landscape Architecture, Planning and Projects, whose trajectory of more than 50 years "blends with the very origin of landscaping in Brazil" (PEREIRA e PEDRONI 2016).

4 Final Considerations

Given what was presented throughout the paper, it is understand that female performance in architecture has been comprehensive – given that in a brief analysis of three of these architects we see their performance in various fields of architecture – but are still silenced and in the "shadows" of the history of the Brazilian modern architecture.

This fact is easily verified when the disparity between the projection of men and women in the journal is put in numbers: based on the data from Almeida (2008) and the research that gave rise to this present work, 520 residential projects are signed by men, with 13 of them coauthored by women while only 2 projects are exclusively female authored. That is, women represent only 2.8% of the residential projects published (Fig. 4), data that make us more attentive to the analyzes of Lima (2013) which states that by analyzing specialized publications in architecture until the late 1990s, it is notable devalued female authorship and veiled in architectural projects, especially large ones.

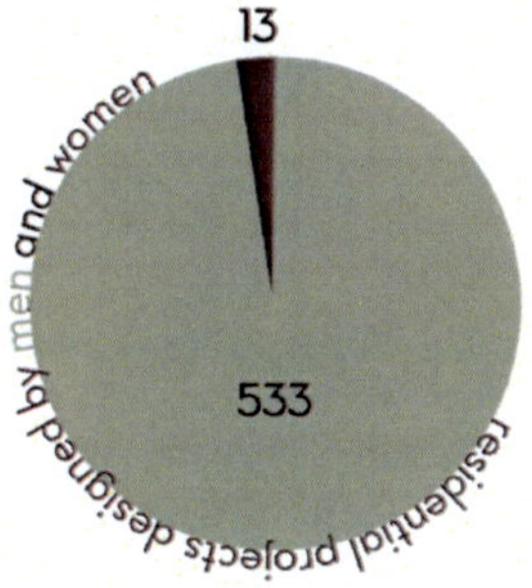

Fig. 4. Residential projects published by the Acrópole from 1954 to 1971: authorship quantitative from a gender perspective

Another question that arouses curiosity is Lygia Fernandes is one of the only two female architects, and the other is Lina Bo Bardi, who appears in Mindlin's book "Modern Architecture in Brazil". As previously stated, she had professional relations with Mindlin, and her work is published in the same edition of the magazine where some works by Henrique also are published. Just like Rosa Kliass appears for the first time in the edited journal where Jorge Wilheim also print pages of the magazine. Although there is no data that includes the relationship between business contacts and the opportunity to be in such a restricted environment for women, I believe it is a matter for further consideration and may perhaps give us some answers regarding the market opportunities for women in twentieth century architecture.

Fig. 5. List of female architects published in the Acrópole.

As a result, the professional relationships between spouses, as in the case of Rosa Kliass, Odiléa Toscano and other professionals who appear in, is a fact already noted by other researchers such as Gáti (2016), whose complexity and importance are likely to be further investigated.

Finally, we can say that the three professionals found (Fig. 5) bring us some concerns: Lygia Fernandes, the oldest of them, is the architect with the greatest bibliographical scarcity about her, but the little knowledge about her career indicates that she has conquered a space in labor market, which was even more restricted to female performance than at the time of its successors, besides indicating great technical and formal mastery of Brazilian modern architecture. Odiléa Toscano, who, although is best known for her wide performance in design, shows a versatility in her often "silent" architectural production in the face of João Walter Toscano's signature and a singularity in her constant concern with the chromatic study of architectural works, as well as the communication between architecture and design. And Rosa Kliass, a pioneer in landscape architecture, also shows a wealth of knowledge about urbanism and architectural projects and the achievement of an acclaimed space in the professional market, whose name is still little known to many. Nevertheless, their work shows us that, like other professionals, they actively participated in the production of national modern architecture.

Although there is still a long way to go in terms of women's contribution to modern architecture in Latin America, as Lima (2013) states, the recognition of the role and production of these women in national architecture is not only of importance for the historical context of Brazilian modern architecture, but also in the recognition of the relevance of this production, whose conservation is increasingly threatened by the 'silence' about its value, as well as for the contemporaneity considering that the feminine productions and contributions, saving the due proportions, continue to be minimized in the present times.

References

416 Avenida Fernandes Lima, Maceió. Google Street View. https://www.google.com.br/maps/@-9.6469341,-35.7340062,3a,75y,82.88h,92.97t/data=!3m6!1e1!3m4!1sWZGDBoZy8LwaZl5Ozle4NQ!2e0!7i16384!8i8192/. Accessed Apr 2021

Acrópole. Acrópole Magazine Online Collection. http://www.acropole.fau.usp.br/. Accessed 10 Dec 2019

Acrópole. Acrópole Magazine Online Collection. Apresentação. Hugo Segawa. http://www.acropole.fau.usp.br/. Accessed 10 Dec 2019

Andersen, A., Persavento, U., Wang, Z.J.: Unsteady aerodynamics of fluttering and tumbling plates. J. Fluid Mech. **541**, 65–90 (2005)

Almeira, C.: Mulheres estão em apenas 37% dos cargos de chefia nas empresas: no setor público, elas são apenas 21,7%. Jornal O Globo, São Paulo (2017). https://oglobo.globo.com/economia/mulheres-estao-em-apenas-37-dos-cargos-de-chefia-nas-empresas-21013908

de Almeida, M.F.: Revista Acrópole publica residências modernas – análise da revista Acrópole e sua publicação de residências unifamiliares modernas entre os anos de 1952 a 1971. 456f. Masters thesis – Escola de Engenharia de São Carlos, Universidade de São Paulo, São Carlos (2008)

Conselho de Arquitetura e Urbanismo do Brasil - CAU/BR: Inédito: visão completa sobre a presença da mulher na Arquitetura e Urbanismo (2019). https://www.caubr.gov.br/inedito-visao-completa-sobre-a-presenca-da-mulher-na-arquitetura-e-urbanismo/

Cohen, J.-L.: O futuro da arquitetura desde 1889: uma história mundial. São Paulo: Cosac Naify, p. 594 (2013)

de O. Cortês, C.M.: O Moderno brasileiro em Moçambique, 1950–1975: a importação de uma imagem. Masters thesis, Universidade Federal de Pernambuco, CAC. Arquitetura. Recife. 252° (2011)

Durand, J.C.: Arte, privilégio e distinção: artes plásticas, arquitetura e classe dirigente no Brasil, 1855/1985. São Paulo: Perspectiva (1989)

Fernandes, L.: Residência em Alagoas. Acrópole, São Paulo, no. 204, set. 1955c, pp. 542–545. http://www.acropole.fau.usp.br/edicao/204/50. Accessed 05 Apr 2021

Gáti, A.H.: Arte e artesanato na arquitetura de interiores moderna de Janele Costa. Recife, vol. 2014, p. 188. Masters thesis. Universidade Federal de Pernambuco, CAC (2014)

Gáti, A.H.: ESPOSAS: atuação em Arquitetura, Interiores e Design. In: IV enanparq – Encontro da Associação Nacional de Pesquisa e Pós-Graduação em Arquitetura e Urbanismo, 9 ed., Porto Alegre. Sessão temática: arquitetura, gênero e sexualidade, p. 18 (2016)

Goldchmit, S.M.: Odiléa Setti Toscano: do desenho ao design. São Paulo, p. 156. Masters thesis. Faculdade de Arquitetura e Urbanismo – Universidade de São Paulo (2008)

Goldchmit, S.M., Loschiavo, M.C., Santos, L.R.: Odiléa toscano: design visual, espaços públicos e ensino. Pós. São Paulo **22**(38), 38–57 (2015)

Lima, A.G.G.: Arquitetas e arquiteturas na América Latina do século XX. Altamira Editorial. Online book (2013)

Maldaner, I.S.: Discriminação por gênero no mercado de trabalho paranaense. IPEA. Honorable mention. Awarded monographs (2007). http://www.en.ipea.gov.br/agencia/images/stories/PDFs/livros/mencao_iandrasouza_23.pdf

Miranda, C.L.: A circulação das teorias artísticas nas revistas brasileiras de arquitetura nos anos 1950. Thematic topics. Páginas de arquitetura moderna brasileira nas revistas especializadas, Rio de Janeiro (2010)

Mindlin, H.E.: Arquitetura Moderna no Brasil, p. 188. Aeroplano Editora, Rio de Janeiro (1999)

Motta, D.: Pesquisa analisa a trajetória de inserção das mulheres no ensino superior. Rio de Janeiro: Fundação Carlos Chagas Filho de Amparo à Pesquisa do Estado do Rio de Janeiro (FAPERJ). Government of Rio de Janeiro (2014). http://www.faperj.br/?id=2748.2.6

de Oliveira, A.F.: A regulamentação do exercício profissional da arquitetura no brasil, p. 421. Doctorate thesis. Faculdade de Arquitetura da Universidade Federal da Bahia. Salvador (2011)

Pereira, M., Pedrotti, G.: Rosa Kliass: Poeta da paisagem. Arch Daily Basil (2017). https://www.archdaily.com.br/br/880958/rosa-kliass-poeta-da-paisagem

de Sá, F.C.: Profissão: arquiteta. formação profissional, mercado de trabalho e projeto arquitetônico na perspectiva das relações de gênero. Masters thesis. FAU-USP. São Paulo, p. 196 (2010)

Salvatori, E.: Análise crítica das relações entre o campo profissional da arquitetura e o âmbito acadêmico de formação de arquiteto no Brasil: questões de gênero no campo profissional da arquitetura. InfoIAB-RS. Porto Alegre (2003). www.iabrs.org.brcolunasartigo.php?art=14

Stamatto, M.I.S.: Um olhar na história: a mulher na escola (Brasil: 1549–1910). II Congresso Brasileiro de História da Educação, Natal (2002)

In-plane Behavior of Iron-Framed Masonry Panels: Numerical Analyses

Thomas Celano[1], Luca Argiento[2], Francesca Ceroni[1(✉)], and Claudia Casapulla[2]

[1] Department of Engineering, University of Naples "Parthenope", Centro Direzionale is C.4, 80143 Napoli, Italy
thomas.celano001@studenti.uniparthenope.it, francesca.ceroni@uniparthenope.it

[2] Department of Engineering, University of Naples "Federico II", Via Forno Vecchio, 80134 Napoli, Italy
{lucaumberto.argiento,casacla}@unina.it

Abstract. The 'baraccato' system is a typical building technique for masonry structures with clever earthquake resilient features, used for the reconstruction of historical city centers in Europe and especially in the South of Italy after some seismic events occurred in the 18th–19th centuries. An atypical and very innovative 'baraccato' system was realized in the Ischia Island at the end of the 19th century, when it was characterized by masonry walls encaged in iron frames other than timber elements.

The paper presents the results of extensive numerical analyses on the behavior of such iron-framed masonry walls, with reference to both slender and squat geometrical configurations. The analyses are mainly focused on the assessment of: a) the more appropriate modeling strategies for simulating the connection between the iron frames and the masonry walls, b) the contribution of the iron frames on the overall nonlinear behavior of the masonry walls by means of push-over analyses. To achieve these goals, two models and corresponding software are used and compared for the numerical analyses, i.e. DIANA for the finite element model (FEM) and 3DMacro for the discrete macro-element model (DMEM).

Keywords: 'Baraccato' system · Masonry · Pushover analysis · Finite element analysis · Discrete element analysis · Structural modeling

1 Introduction

For millennia, masonry and timber were two of the most used construction materials in the Mediterranean, thanks to their high availability and easiness to use. Several historical masonry buildings are still exploited in all parts of the world and need adequate prevention against seismic events. The main differences among these masonry buildings are related to their location, i.e. type of material, masonry system texture, morphology, number of layers in the thickness of the wall, etc.

Nowadays, several intervention techniques can be realized in order to improve the seismic performance of existing masonry buildings, which are intrinsically vulnerable

I. Vayas and F. M. Mazzolani (Eds.): PROHITECH 2021, LNCE 209, pp. 317–333, 2022.
https://doi.org/10.1007/978-3-030-90788-4_27

to horizontal actions; one of the most common in the last decades is the use of fiber reinforced plastic (FRP) or fiber reinforced cement based mortar (FRCM) materials, externally bonded on masonry panels, arches, vaults (Bilotta et al. 2017). However, the issue of increasing the seismic performance of masonry buildings was also addressed in the past with different systems. The mixed timber-masonry constructions represent a resilient building system consisting of masonry walls encaged in one or more layers of timber frames, suitable to guarantee an adequate resistance to earthquakes. In Europe two of the most important examples of this system are the Portuguese 'Pombalino system' (Fig. 1a) and the Italian 'baraccato system' (Fig. 1b).

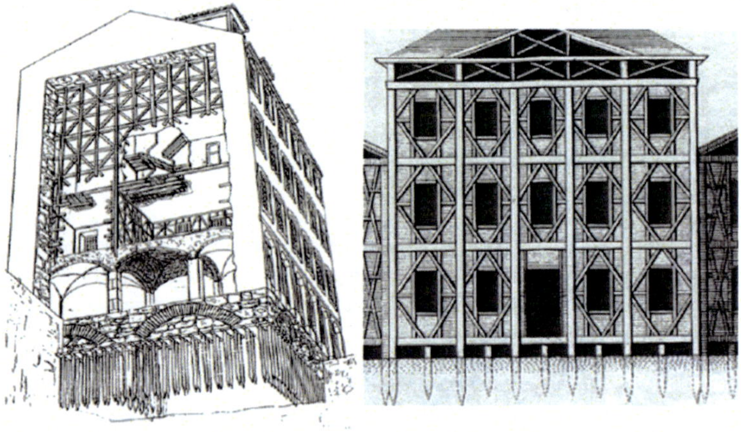

(a) (b)

Fig. 1. (a) Portuguese 'Pombalino' system (Meireles et al. 2012), (b) Italian 'baraccato' system (Vivenzio 1788).

In Portugal, after the devastating 1755 Lisbon earthquake, several masonry buildings were constructed with a timber framework (also called 'gaiola-pombalina'), very similar to the 'baraccato' one. The latter (also called "Borbone system") was introduced in Italy by Ferdinando IV of Borbone, in the 18th century, after the 1783 catastrophic seismic events. The main difference in the seismic behavior of these two composite structures is related to the different function of the timber elements. In the 'Borbone' system, the timber frames are present from the foundation to the upper floor (Fig. 1b), strength and stiffness are significantly provided by the masonry walls since the aim was to realize a masonry system strengthened with timber elements (Stellacci et al. 2016). Conversely, in the 'Pombalino' system, the ground floor was often made of masonry walls and vaults, since the timber elements start being present from the first floor only. In this system the masonry had a primary role of insulation (Poletti and Vasconcelos 2015) and the composite structure mainly behaves as a timber frame filled with masonry.

More in general, under dynamic loads, in a composite timber-masonry structure, timber frames allow increasing the ductility significantly, improving the tensile strength of the system, and guaranteeing a box-like behavior of the building. The latter is achieved through good connections among structural elements, and, in particular, thanks to the connection among timber frames in the three dimensions of the building.

Other similar types of mixed timber-masonry building systems are distributed and traceable throughout the world, such as the Dhajji Dewari in Central Asia or the Himis system in Turkey (Poletti and Vasconcelos 2015). The substantial differences among several types of mixed timber-masonry buildings around the world are the masonry typology (depending on the availability on each territory), the configuration of the reinforcement system (i.e. one or more layers, quadratic or rectangular mesh frame, presence of a single or double diagonal), and their dimensions.

However, although there are numerous studies in the literature on the mixed timber-masonry constructions, especially related to the Portuguese 'Pombalino' system (Mendes and Lourenço 2010; Vasconcelos et al. 2013; Poletti and Vasconcelos 2015), there is still a great lack of knowledge about their structural behaviour.

Besides, in the Ischia Island there are two particular buildings where the masonry walls are engaged in frames made of slender iron elements. After the catastrophic 1883 earthquake that hit the Ischia Island and, in particular, the area of Casamicciola Terme, the Church of Santa Maria Maddalena (Fig. 2a) and the large building of "Pio Monte della Misericordia" were both built with an iron 'baraccato' system, very innovative for that period. This atypical system is constituted of two external iron frames, connected each other through transversal flat plates (Fig. 2b) and embedded in the masonry walls (Fig. 2c). The masonry walls, made of squared yellow tuff blocks with a regular texture, are continuous between the two iron frames, and, thus, unlike the case of the more traditional 'baraccato' system realized with timber elements, the masonry has not only a filling function within the frame.

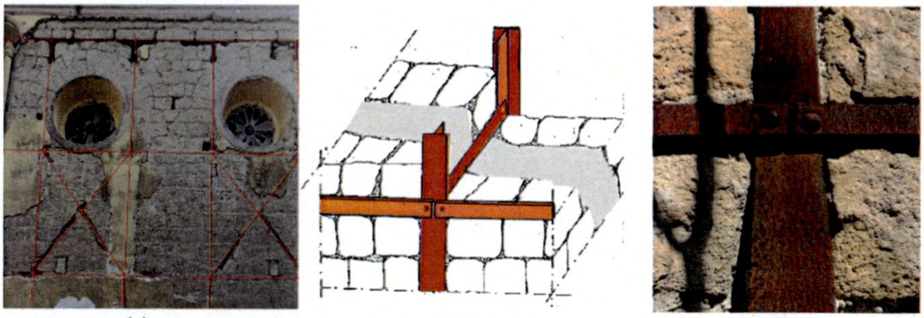

(a) (b) (c)

Fig. 2. (a) Detail of iron 'baraccato' system in the Santa Maria Maddalena Church (Casapulla et al. 2019), (b) particular of the connection between the two iron frames, (c) detail of the joint between iron elements embedded in the masonry.

If there is still a lack of knowledge on the behavior of the more traditional 'baraccato' system, there is a dramatic gap in the literature for the system made of iron elements, which represents a 'unicum'. A first step in this line is represented by the recent work of Casapulla et al. (2019), mainly describing the damage status and the in situ dynamic test performed on the Santa Maria Maddalena Church in the Ischia Island after the seismic event of August 2017, and the Msc thesis of Di Napoli (2019) focused on its 3D modeling and safety assessment. The analysis of the damage in the church after the 2017 earthquake confirmed the effectiveness of such a construction technique since the church did not exhibit significant damage and out-of-plane mechanisms did not occur. However, an improvement of knowledge on such an earthquake-resistant system, aimed to understand the effect of the iron frames on the global behavior of masonry buildings, still is an interesting and important issue to be addressed.

This paper starts to fill this gap by presenting the results of several numerical analyses carried out on masonry panels strengthened according to the iron 'baraccato' system. The analyses were mainly focused on the understanding of the mechanical behavior of the panels under horizontal actions and on the assessment of the best modeling strategies for simulating the interaction between the iron frame and the masonry panel (perfectly adherent to or completely independent of the masonry). The numerical analyses were performed both through a finite element model in the software Diana FEA (DIANA FEA bv © 2019) and a discrete element model in the software 3DMacro (Gruppo Sismica © 2019) in order to simulate a shear-compression test.

Since no experimental data on iron-framed masonry systems are present in the literature, the paper first took advantage of the preliminary study of unreinforced masonry panels experimentally tested by Anthoine et al. (1995) in order to calibrate both the numerical models. Successively, the numerical analyses were carried out on the same panels reinforced by the external iron frames according to the same lay-out observed in the Church of Santa Maria Maddalena in the Ischia Island (Casapulla et al. 2019). The numerical analyses are also aimed to furnish indications about the range of variability of the maximum loads and displacements expected in the steel strengthened panels, in order to design a reliable set-up for experimental tests.

2 Numerical Analyses of Unreinforced Masonry Walls

Two different modeling approaches are used in this paper for both the unreinforced and reinforced masonry panels under study. The first one is based on a finite element model (FEM) that simulates the masonry as a homogenous material (Lourenço 1996), with nonlinear response in a detailed mesh refinement. This model was implemented in the DIANA FEA software.

The second approach, implemented in 3DMacro software, is based on the discrete macro-element model (DMEM) proposed by Caliò et al. (2012). Such an approach considers the masonry building as an assemblage of macro-elements and is capable of assessing the nonlinear response of the structure with a limited computational effort, thus making it suitable also for professional aims. In particular, the in-plane behavior of a masonry panel can be studied by means of a two-dimensional "macro-element" made of four hinges connecting four rigid edges and two diagonal nonlinear springs. Such a 2D macro-element is capable of catching the main in-plane failure mechanisms of a masonry panel: flexural, shear-diagonal and shear-sliding failure. The activation of the flexural failure is controlled by the orthogonal springs at the interface elements, while the shear-diagonal and the shear-sliding failure modes are governed by the diagonal nonlinear springs and the additional longitudinal nonlinear spring at the interface elements, respectively.

Numerical analyses using such models were firstly performed on the two unreinforced brick masonry panels with different slenderness ratio $\lambda = H/B$ (Fig. 3a) tested by Anthoine et al. (1995) according to quasi-static in-plane cyclic loads (Fig. 3b). The first panel, named in the following "squat", has $\lambda = 1.35$, with the base of 1 m and height of 1.35 m, while the second, named "slender", has $\lambda = 2.00$, with the same base of 1 m, but height of 2 m, both panels having the thickness of 0.25 m. The geometric parameters of the panels are gathered in Table 1. The two slenderness ratios were originally chosen by Anthoine et al. (1995) in order to catch different failure mechanisms in the panels.

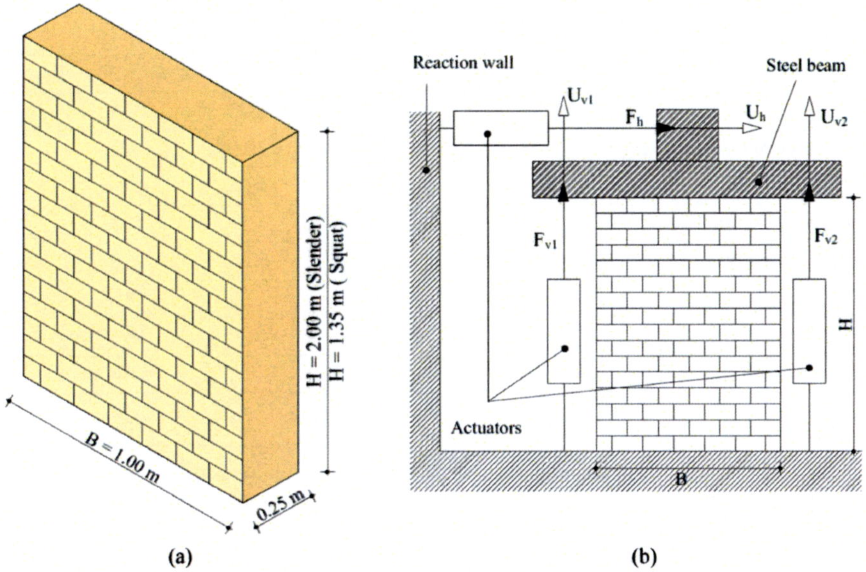

Fig. 3. (a) Geometric parameters of the unreinforced masonry panels, (b) schematic view of the set-up for the quasi-static in-plane cyclic test (Anthoine et al. 1995).

Table 1. Geometric parameters of the unreinforced masonry panels.

Panel	Thickness	Base	Height	Slenderness ratio
	t [m]	B [m]	H [m]	λ [-]
Squat	0.25	1.00	1.35	1.35
Slender			2.00	2.00

The tested walls were composed of brick units and hydraulic lime mortar, arranged in an 'English' bond pattern. Experimental tests were carried out to determine the mechanical properties of the single materials, providing average tensile strengths of 2.44 MPa and 0.58 MPa and compressive strengths of 26.9 MPa and 3.31 MPa for the bricks and the mortar, respectively (Anthoine et al. 1995; Magenes and Calvi 1997). The values of the Young's modulus, E, compressive strength, f_c, tensile strength, f_t, and Poisson's ratio, ν, listed in the FEM column of Table 2 refer to the brick masonry and were derived by Liberatore and Addessi (2015) via a homogenization procedure based on the brick and mortar properties.

Table 2. Mechanical masonry properties

			FEM	DMEM
Young's modulus	E	[MPa]	1700	1943
Poisson's ratio	ν	–	0.19	0.19
Compressive strength	f_c	[MPa]	6.20	6.20
Tensile strength	f_t	[MPa]	0.25	0.25
Compressive fracture energy	G_c	[N/mm]	10.0	–
Tensile fracture energy	G_t	[N/mm]	0.012	–
Compressive ductility	μ_c	–	–	4.6
Tensile ductility	μ_t	–	–	3.3
Shear modulus	G	[MPa]	–	816
Shear strength	τ_0	[MPa]	–	0.17
Shear strain capacity	γ_u	[%]	–	0.4

The walls tested by Anthoine et al. (1995) were loaded under double clamped boundary conditions in order to better reproduce the seismic behavior of the piers connected with the rigid nodes in a regular masonry structure. The top rotations were blocked using a steel beam constrained with two electronically controlled actuators. A horizontal actuator was used to apply an alternated lateral top displacement of increasing amplitude quasi-statically imposed under a constant vertical pre-compression stress, i.e. $\sigma_c = 0.6$ MPa ($F_v = 150$ kN). Since no crack openings were taken into account, only a monotonic horizontal displacement was considered in the numerical analyses.

In DIANA FEA, the masonry was modeled through the "total strain crack model" using an exponential law for the constitutive behavior in tension (Fig. 4a) and a parabolic law in compression (Fig. 4b). Such a modeling strategy also requires the definition of fracture energy in tension, G_t, and compression, G_c that were fixed based on the commonly used values given by Lourenço (2009) and listed in Table 2.

An optimization analysis of the mesh was carried out in order to evaluate the better value of the mesh size, which resulted to be 0.05 m. The type of the element mesh used for masonry is CHX60 (Fig. 4c), which is a twenty-node isoparametric solid brick element based on the quadratic interpolation and Gauss integration. The total number of solid elements used in the model is 420 for the squat panel and 600 for the slender panel.

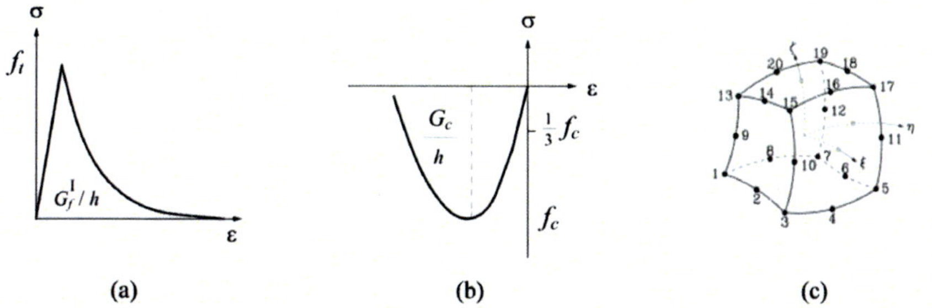

Fig. 4. FE Model implemented in DIANA: (a) Exponential tensile law of masonry, (b) parabolic compressive law of masonry; (c) CHX60 solid element mesh type.

The secant (quasi-Newton) method with the maximum number of iterations equal to 100 and the normalized energy with the tolerance of 0.001 were respectively used as iteration method and convergence criterion for carrying out nonlinear static analyses (push-over). The horizontal displacement was applied at top of the masonry and increased of 0.1 mm each step.

On the other hand, the computational discrete model (DMEM) implemented in the software 3DMacro was obtained by properly dividing the structural objects in order to produce the mesh of elements (wall panels) used in the calculation routine. In this work, the minimum mesh was set to 1 m. Nonlinear springs were used to take into account the mechanical properties of the masonry and the spacing among the springs at the interface elements was calibrated equal to 20 mm and 100 mm for unreinforced and reinforced panels, respectively. The nonlinear static analysis of this model was performed under force control in the first phase and under displacement control in the second phase.

Since different constitutive models of the masonry were used for the FE and the DME models, it was necessary to calibrate some parameters of the DME model in order to compare the results correctly. The elastic-plastic law for masonry used in the DME model (Fig. 5a) required, indeed, the definition of μ_c and μ_t, which respectively represents the ductility in compression and tension of the masonry, expressed as the ratio of the ultimate strain (ε_{cr} and ε_{tr}) to the limit elastic one (ε_{ce} and ε_{te}).

The Young's modulus E and the ductility in tension and compression, μ_t and μ_c, were calibrated in the DME model, while for the tensile and compressive strengths the same values used in the FE model and listed in Table 2 were used. The shear-diagonal stiffness was defined through the shear modulus G, based on the values of E and ν, and the Turnsěk and Čačovič criterion for the shear-diagonal mechanism was adopted. Thus, the pure shear strength, τ_0, was calculated by dividing the tensile strength, f_t, to 1.5 and the shear strain capacity γ_u (Fig. 5b) was fixed equal to 0.4% of the height of the panel, in compliance with the ultimate shear drift provided by the current Italian Technical Standards for Constructions (NTC 2018). The mechanical parameters of masonry used in the DME model are listed in Table 2.

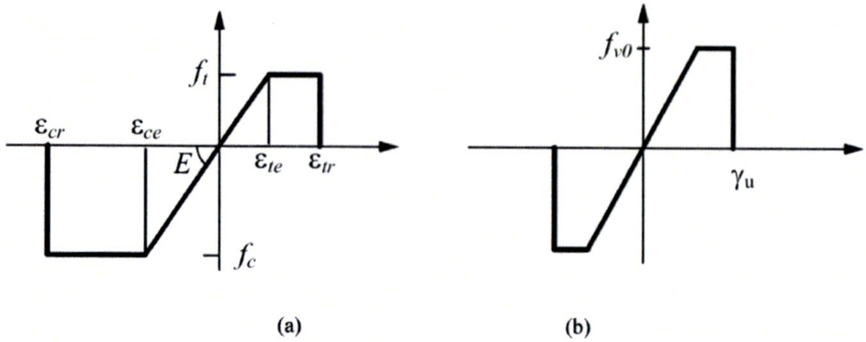

Fig. 5. Constitutive laws adopted in the macro-element model in: (a) tensile/compression and (b) diagonal shear.

2.1 Comparisons Between the Results of FE and DME Models for the Unreinforced Panels

The experimental curves in terms of shear-displacement relation for the unreinforced panels tested by Anthoine et al. (1995) were compared with the numerical curves obtained with the FEM and the DMEM, with reference to the slender (Fig. 6a) and squat (Fig. 6b) configurations.

For the slender panel (Fig. 6a), the numerical curves of both models are in very good agreement with the experimental values, with some differences in the initial stiffness, which in the experimental curve is slightly higher than the numerical ones. The failure was related to a flexural mechanism of the panel in both models, in agreement with the experimental results. It is important to underline that, since no apparent degradation of the resistance was observed for the slender panel, the analyses were interrupted at the displacement of 12 mm, which corresponds to a drift of 0.6%, as in the experimental test.

The behavior of the squat panel (Fig. 6b) is governed by a shear-diagonal failure, as highlighted by the brittle decay observed both in the experimental and in the numerical push-over curves. The agreement between the numerical and experimental curves is very good in terms of initial stiffness, while the plastic phase is quite different due to the different constitutive laws used for modeling the masonry behavior and the different strategies for modeling the shear behavior. Modeling of shear behavior is, indeed, not necessary in DIANA, if the rotating total strain crack model is used (DIANA manual v. 10.3 2019), since the failure is related to the achievement of the tensile strength along the principal direction. Conversely, in the DMEM the failure occurs when the shear springs collapse. A slight greater capacity can be observed in the curve given by the FEM with respect to both the experimental and DMEM curves.

Figure 7 and Fig. 8 show the comparison between the crack patterns provided by both models for the slender and squat panel, respectively. It can be observed that similar failure mechanisms are predicted by the two models.

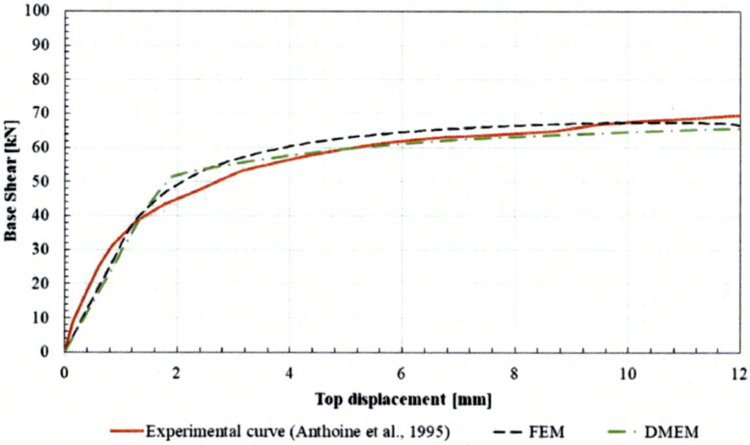

(a)

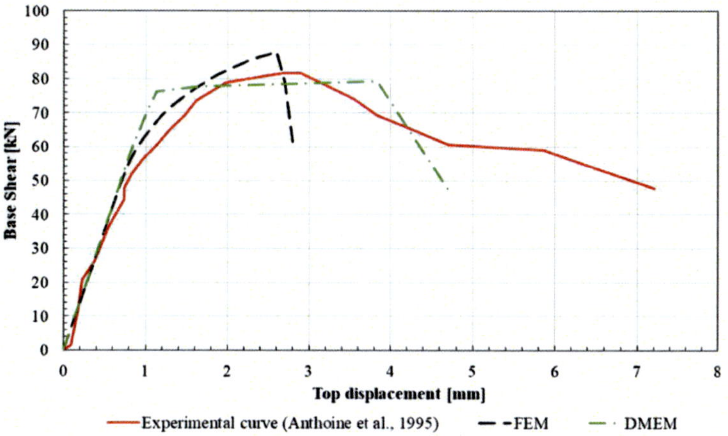

(b)

Fig. 6. Experimental vs. numerical curves given by FEM and DMEM for: (a) slender and (b) squat panel.

3 Numerical Analyses of Reinforced Masonry Walls

One of the most relevant issues for the analysis of the iron 'baraccato' system is the modeling strategy for the iron elements and for their interaction with the masonry panel. The iron elements can be modeled with beam, truss or embedded elements both in the FEM and DMEM. The main differences among the three modeling strategies are related to the type of connection with masonry and the blocked degrees of freedom.

Beam elements can guarantee axial and flexional stiffness, but the nodes of the beams representing the iron elements have to be coincident with the nodes of the solid elements simulating the masonry, leading, thus, to a perfect adherence between the iron and masonry in all the common nodes.

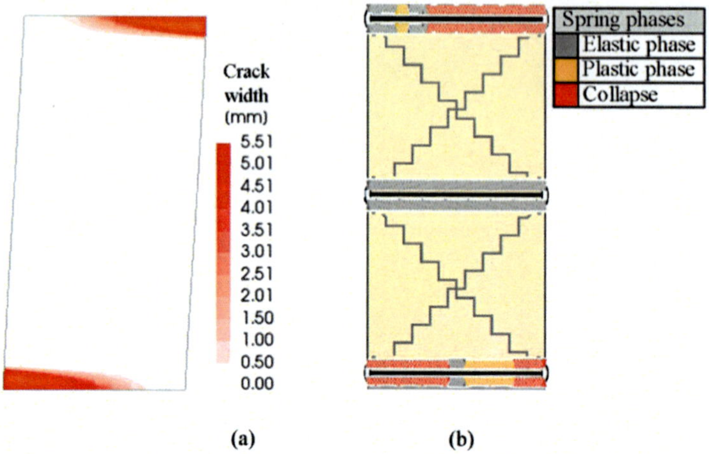

Fig. 7. Crack patterns at the ultimate displacement for the slender panel: (a) FEM and (b) DMEM.

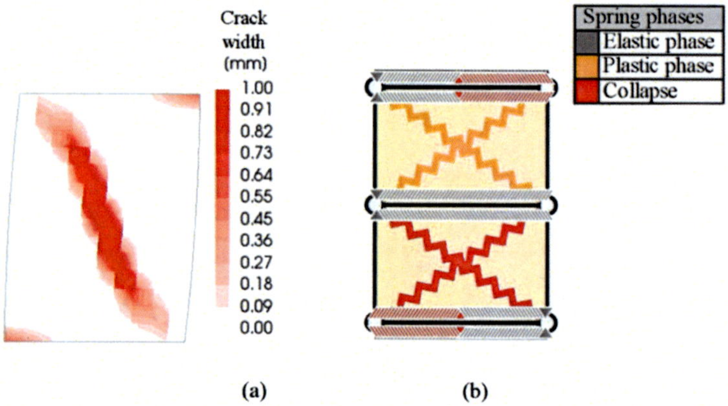

Fig. 8. Crack patterns at the ultimate displacement for the squat panel: (a) FEM and (b) DMEM.

Conversely, truss elements can guarantee only axial stiffness, i.e. they are not able to transfer bending moment, and can be connected to the masonry substrate only in the end nodes. This means that there is no adhesion between iron elements and masonry. Lastly, embedded elements are mainly used for simulating bars in reinforced concrete structures, since they warrant perfect adherence, even if the nodes of the embedded element do not necessarily coincide with those of the solid element representing the masonry.

These modeling strategies can be differently combined in order to simulate different elements of the iron frame.

In beam elements, it is possible to define the shape and size of the cross section, while for truss or embedded elements only the area of the cross section necessary for the axial stiffness can be assigned. This means that the geometric nonlinearity due to the buckling of the elements can be taken into account only in the case of beam elements.

Based on the iron 'baraccato' system observed in the Church of 'Santa Maria Maddalena' (Fig. 2a and Fig. 9), two modeling solutions were considered for the single masonry panel:

1) all the elements of the iron frame were modeled as beam elements (Beam model);
2) the horizontal and vertical elements of the iron frame were modeled as beam elements, while the diagonals as truss elements (Beam+Truss model).

The transversal connecting elements between the two steel frames are always modeled as embedded elements, since they are actually embedded in the masonry.

Truss elements were considered suitable for the diagonals since they can be considered hinged, thanks to the simple bolted connections observed in the church.

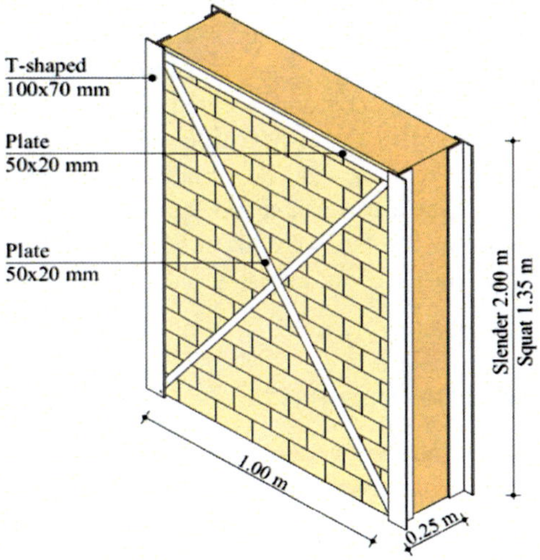

Fig. 9. Geometric parameters of the reinforced masonry panels.

Lacking experimental information about the properties of materials used in the Church of Santa Maria Maddalena, a commercial S235 steel was used for the iron elements, which were, thus, characterized by a nominal yielding strength of 235 MPa and Young's modulus of 205000 MPa (NTC 2018). The S235 steel will probably be used in future experimental tests planned by the authors on masonry panels strengthened similarly to the 'baraccato' technique observed in the Santa Maria Maddalena Church, since its strength and Young's modulus are not far from the range of values found in the literature for iron elements (Brussel1997; Di Napoli 2019).

The frame is made by vertical T-shaped profiles, with the flange of 100 mm, web of 70 mm and their thickness of 10 mm, and horizontal, diagonal and transversal rectangular plates with the base of 20 mm and height of 50 mm.

In the FEM implemented in DIANA, the three-node beam element CL18B and the two-node truss element L2TRU were used and the steel was modeled as an elastic-perfectly plastic material. These elements were applied to both faces of the panel since the steel frame consists in a 3D structure.

In 3DMacro, an equivalent 2D model was adopted for the frame with the same elastic-plastic behavior. In fact, the steel frame was modeled with equivalent inertial elements located in the vertical mid-plane of the panel.

In order to take into account possible buckling phenomena for the compressed steel elements, which could occur when damage in masonry elements or deterioration of bond behavior along the interfaces become significant, the numerical analyses were carried out according to two limiting schemes:
1) all the steel elements active until the global failure of the reinforced panel, which corresponds to the assumption of no buckling phenomena in the compressed elements; 2) compressed elements not active from beginning of loading because of buckling phenomena.

With the aim of giving upper and lower bounds of the global shear strength of the panels, the second case was modeled without considering the compressed elements from the beginning of the analysis, since it is not known if and when buckling phenomena may occur. With reference to the reinforced panel reported in Fig. 9, for both squat and slender configurations, it was verified that the critical value of the axial load for the diagonal plates is very low, meaning that the capacity curves are immediately affected by buckling phenomena. Conversely, it was also verified that the vertical elements remained in the elastic field and the maximum stresses were lower than the critical values for buckling. Thus, in the model taking into account buckling phenomena, only the diagonal plates were neglected.

3.1 Comparison Between the Results of the Two Models of the Reinforced Panels

Figure 10 and Fig. 11 show the comparisons among several numerical curves obtained for the iron 'baraccato' panels according to the slender (Fig. 10a and Fig. 11a) and the squat (Fig. 10b and Fig. 11b) configurations. The following cases are considered for each geometry:

- "Beam+Truss" approach in FEM and DMEM, with and without buckling phenomena;
- "Beam" approach in FEM and DMEM, with and without buckling phenomena.

In addition, the results related to the unreinforced masonry panels and to the only steel frame, with and without buckling, given by the FEM are also reported in Fig. 10 and Fig. 11 in order to better understand the contribution of the steel elements in terms of both stiffness and strength.

Focusing the attention on the results in Fig. 10, it is worth noting that the steel frame plays a very significant role on increasing both the strength and the stiffness of the masonry panel. In particular, the strength increases of about 4 and 8 times, if the buckling phenomena are considered or not in the slender configuration (Fig. 10a), and of about 5 and 8 times in the squat one (Fig. 10b). Moreover, it is interesting to note that the diagonal plates give a relevant strength contribution, since, when they are not included in the model, both the strength of the only frame and of the reinforced panels are sensibly lower (−42% for the slender and −38% for the squat panel). For both FE and DME models, the slender reinforced panel still shows a flexural failure as the unstrengthened one. Conversely, for the squat panel without buckling effect in both models the failure switches from the shear-diagonal mechanism of the unreinforced panel to a flexural failure.

The steel frame also allows increasing the ductility of the masonry panel for both geometrical configurations. It is worth noting that no limitations were assumed for the deformability of the steel. For this reason, the analyses were interrupted up to the displacements of 20 mm and 15 mm, which correspond to the drifts of 1% and 1.2% for the slender and the squat panels, respectively.

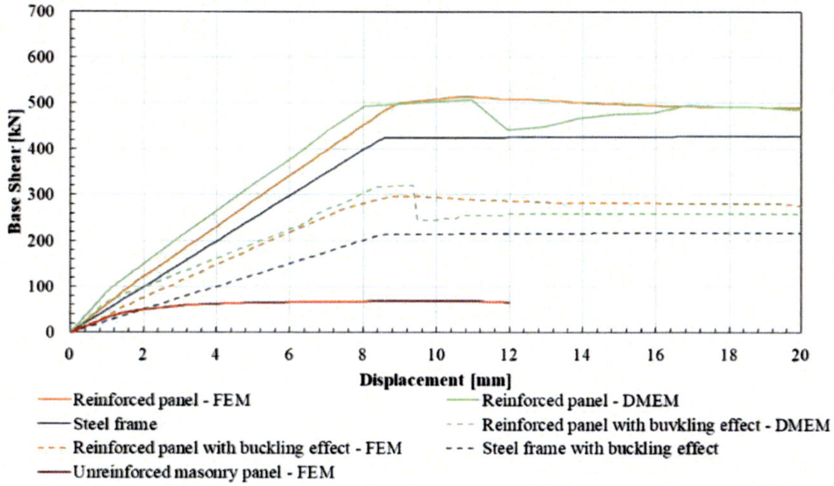

(a)

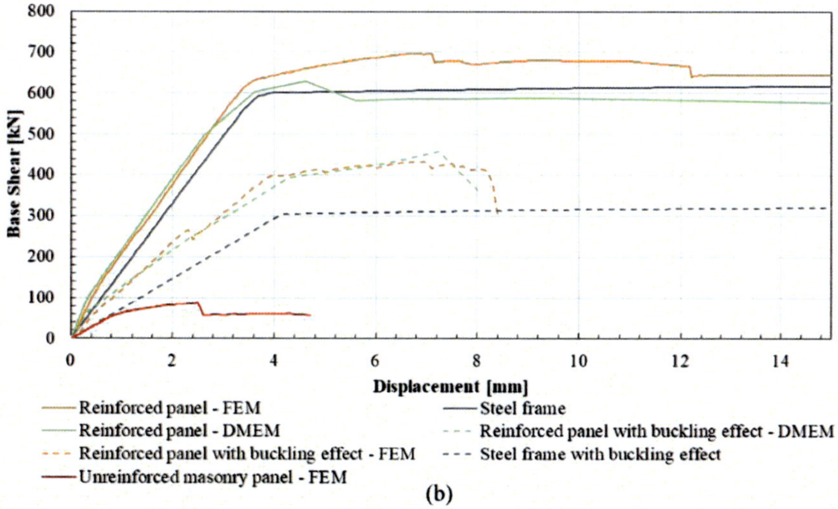

(b)

Fig. 10. Numerical curves of strengthened panels for 'Beam+Truss' model: (a) slender and (b) squat panel.

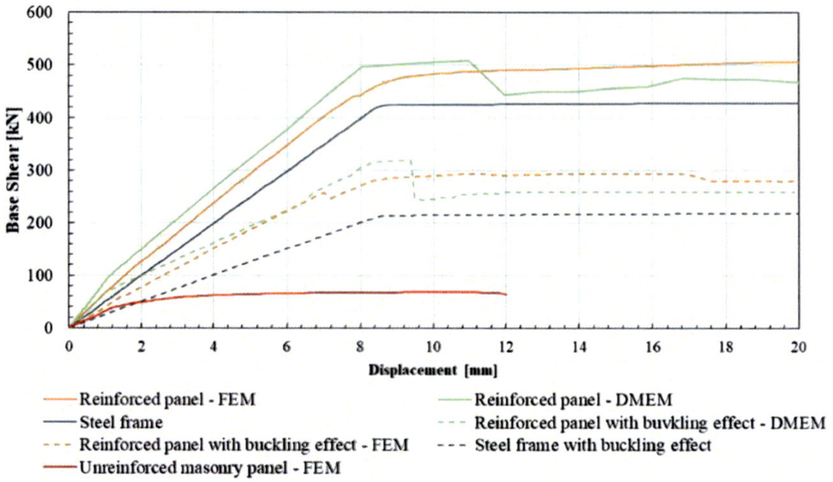

(a)

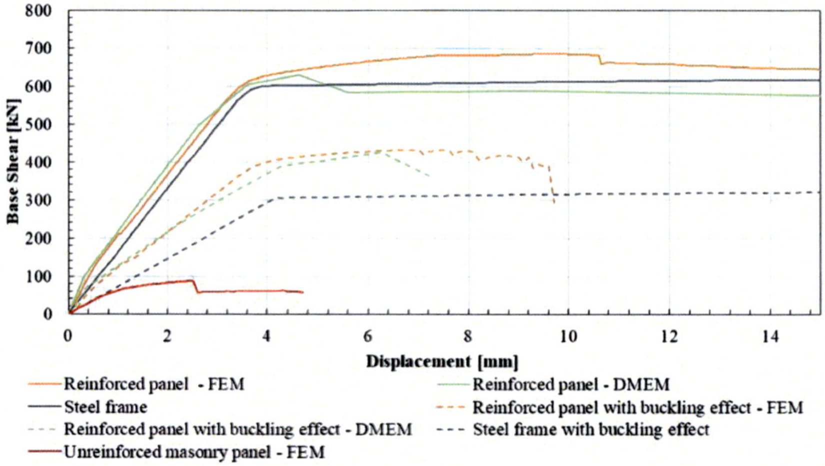

(b)

Fig. 11. Numerical curves of strengthened panels for 'Beam' model: (a) slender and (b) squat panel.

Figure 10b shows a good agreement for the squat panel in terms of stiffness and ultimate load between the FE and DME models when the compressed diagonal element is neglected, while a slight difference is observed in terms of strength when it is included. Conversely, Fig. 10a shows a good agreement for the slender panel between the two models in terms of strength and a slight difference in terms of stiffness in all cases.

Figure 11 shows that, when beams elements are adopted for all the irons profiles (Beam model), the results are very similar to those obtained in the 'Beam+Truss' model for both the FEM and DMEM with a negligible difference in terms of capacity (2% higher for the 'Beam+Truss' case compared to the 'Beam' one). Thus, the connection between diagonal elements and masonry solid does not play a crucial role in the in-plane behavior of these masonry panels.

Figures 12 and 13 show the crack patterns given by the FEM and DMEM approaches for the strengthened slender and squat panels, respectively, in the case of 'Beam' model. The cracks are mainly distributed along the steel elements showing, thus, the effectiveness of the connection between the masonry and steel reinforcement. Clearly, in order to better understand the behavior of the reinforced panels, the numerical results should be further validated against other numerical outcomes and experimental tests.

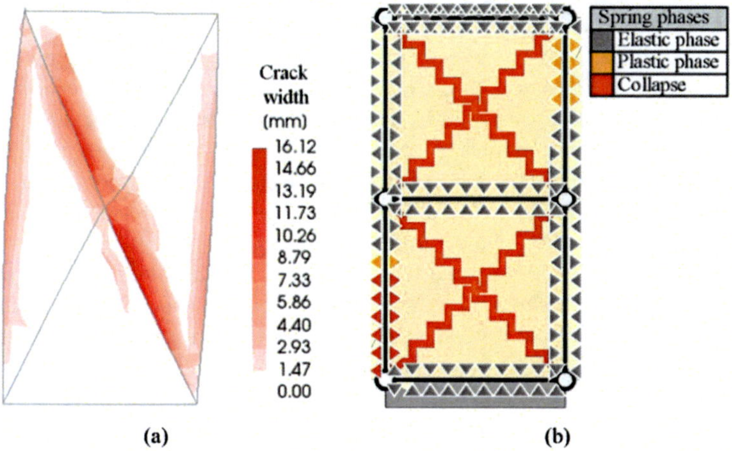

Fig. 12. Numerical crack pattern of the iron 'baraccato' slender panel: (a) FE 'Beam' model; (b) DME 'Beam' model.

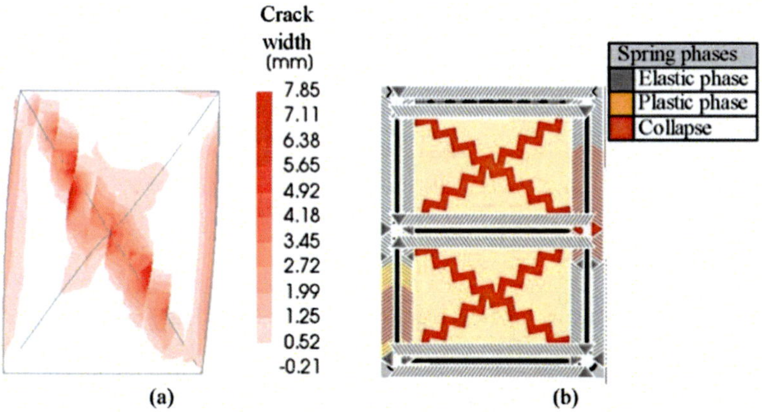

Fig. 13. Numerical crack pattern of the iron 'baraccato' squat panel: (a) FE 'Beam' model; (b) DME 'Beam' model.

4 Conclusions and Future Developments

Although still poorly investigated, the 'baraccato' system is an interesting building technique for masonry structures which started to be diffused in Italy and in Europe at the end of 18th century in order to improve the seismic performance of such structures.

This paper presents several numerical analyses carried out on masonry panels arranged with a particular 'baraccato' system found in the Ischia Island and characterized by tuff masonry walls engaged in iron frames. Two types of numerical models were considered: 1) a finite element model based on the simulation of masonry as homogenous material, and 2) a discrete element approach where macro-elements are used for simulating the shear and flexural behavior of masonry walls.

Firstly, nonlinear analyses were carried out by means of the two approaches and were successfully compared with the experimental results in terms of capacity curves of two unreinforced masonry panels having different slenderness ratios, i.e. 1.35 and 2.

Successively, the two modeling strategies were used to simulate the behavior of the masonry panels with the same geometries, but reinforced according to the iron 'baraccato' technique found in the Ischia Island. The numerical analyses evidenced a very significant contribution of the profiles in terms of both strength and initial stiffness in comparison with the unreinforced masonry panel. The behavior of the compressed diagonal elements was influenced by buckling phenomena and, thus, upper and lower bounds of the shear resistances of the reinforced walls were assessed considering or not the contribution of the compressed elements.

However, despite the good agreement between the results given by the two approaches for all the investigated panels, the reliability of the two models for the strengthened panels will be definitely checked by means of the comparison with the outcomes of experimental tests that the authors are planning. The numerical analyses presented in this paper for the strengthened panels are, indeed, also aimed to assess the range of variation of both maximum loads and displacements and, thus, correctly design the set-up for carrying out the experimental tests.

Acknowledgements. The authors would like to thank Gruppo Sismica s.r.l. and Prof. Paulo B. Lourenço for providing the licenses of the software 3D-Macro and DIANA, respectively. They also wish to thank Dr. Bartolomeo Pantò and UMINHO research group for the technical/scientific support provided in the use of these software.

References

Anthoine, A., Magonette, G., Magenes, G.: Shear-compression testing and analysis of brick masonry walls. In: 10th European Conference on Earthquake Engineering, August 28–September 2, 1994, Vienna, Austria (1995)

Bilotta, A., Ceroni, F., Nigro, E., Pecce, M.: Experimental tests on FRCM strengthening systems for tuff masonry elements. Constr. Build. Mater. **138**, 114–133 (2017)

Bussell, M.N.: Appraisal of Existing Iron and Steel Structures. The Steel Construction Institute, Ascot -Berkshire (1997)

Caliò, I., Marletta, M., Pantò, B.: A new discrete element model for the evaluation of the seismic behaviour of unreinforced masonry buildings. Eng. Struct. **40**, 327–338 (2012)

Casapulla, C., Ceroni, F., Rainieri, C., Argiento, L.U., Arcamone, P., Fabbrocino, G.: Structural assessment of Santa Maria Maddalena church in Ischia (Italy) by experimental modal analysis under operational conditions. In: 7th International Conference on Computational Methods in Structural Dynamics and Earthquake Engineering, COMPDYN 2019, June 24–26, 2019, Crete, Greece (2019)

DIANA FEA BV: Diana User's Manual. Release 10.3 (2019)

Di Napoli, B.: Modelling and Safety Assessment of the Santa Maria Maddalena Church, Ischia, Italy. Msc Thesis, Universidade do Minho, Portugal (2019)

Liberatore, D., Addessi, D.: strength domains and return algorithm for the lumped plasticity equivalent frame model of masonry structures. Eng. Struct. **91**, 167–181 (2015)

Lourenço, P.B.: Recent advances in masonry modelling: micro-modelling and homogenization. Comput. Exp. Methods Struct. **3**, 251–294 (2009)

Lourenço, P.B.: Computational Strategies for Masonry Structures. Ph.D. thesis, Technische Universiteit Delft, The Netherlands (1996)

Magenes, G., Calvi, G.M.: In-plane seismic response of brick masonry walls. Earthq. Eng. Struct. Dyn. **26**, 1091–1112 (1997)

Meireles, H., Bento, R., Cattari, S., Lagomarsino, S.: A hysteretic model for 'frontal' walls in Pombalino buildings. Bull. Earthq. Eng. **10**(5), 1481–1502 (2012)

Mendes, N., Lourenco, P.B.: Seismic assessment of Masonry Gaioleiro buildings in Lisbon, Portugal. J. Earthq. Eng. **14**(1), 80–101 (2010)

NTC 2018: Ministero delle Infrastrutture e dei Trasporti. Norme Tecniche per le Costruzioni di cui al D.M. 17/01/2018. Gazzetta Ufficiale N. 42 del 20/02/2018 (2018). (in Italian)

Poletti, E., Vasconcelos, G.: Seismic behaviour of traditional timber frame walls: experimental results on unreinforced walls. Bull. Earthq. Eng. **13**(3), 885–916 (2014). https://doi.org/10.1007/s10518-014-9650-9

Gruppo sismica S.r.l.: 3DMacro. Il Software per Le Murature (3D Computer Program for the Seismic Assessment of Masonry Buildings). Technical report (2019)

Stellacci, S., Ruggieri, N., Rato, V.: Gaiola vs. Borbone system: a comparison between 18th century anti-seismic case studies. Int. J. Archit. Herit. **10**(6), 817–828 (2016)

Vasconcelos, G., Poletti, E., Salavessa, E., Jesus, A.M.P., Lourenço, P.B., Pilaon, P.: In-plane shear behaviour of traditional timber walls. Eng. Struct. **56**, 1028–1048 (2013)

Vivenzio, G.: Istoria e Teoria de' Tremuoti in Generale ed in Particolare di Quelli della Calabria, e di Messina del MDCCLXXXIII. Nella Stamperia Regale, Napoli, Italy (1788)

Assessment of Unmanned Aerial System (UAS) Post-processing Techniques Applied to the Auckland War Memorial Cenotaph

Silvia Porcarelli[1], Dmytro Dizhur[2], Davide Forcellini[2](✉), and Jason M. Ingham[2]

[1] Department of Civil and Environmental Engineering, Università della Repubblica di San Marino, San Marino, San Marino

[2] Department of Civil and Environmental Engineering, University of Auckland, Auckland, New Zealand

dfor295@aucklanduni.ac.nz

Abstract. The use of Unmanned Aerial Systems (UAS) in aerial photography allows photogrammetry to be an affordable and fast technology for 3D map generation. In order to process the number of images acquired by UAS and to generate automatic 3D model reproduction, different platforms for post-processing have been recently developed. Several studies have been previously conducted to provide partial assistance in the selection of software, through comparisons of 3D models of physical objects. Following such contributions, the historical cenotaph located in front of the Auckland War Memorial Museum in Auckland domain park (New Zealand) was chosen as a case study. Four commercial software (Agisoft PhotoScan, Recap Pro, VisualSFM and Pix4D) were compared.

Keywords: Unmanned Aerial Systems (UAS) · 3D models · Post-processing · Auckland War Memorial Museum Cenotaph

1 Introduction

In recent decades the conservation of historical heritage has been improved by adopting various technologies to achieve the objectives of protection and valorization of artefacts (Pieraccini et al. 2017). Among them, aerial photography has received much interest as an accessible and cost-efficient technique due to the use of Unmanned Aerial Systems (UAS) (Remondino et al. 2011; Uysal et al. 2015; Ruiz Sabina et al. 2015; Boccardo et al. 2015; Chiabrando et al. 2016; Georgopoulos et al. 2016; Guerra et al. 2017; Forcellini et al. 2019). This technique is based on the acquisition of significant quantities of images and the generation of automatic 3D models (Colomina et al. 2008; Eisenbeiss 2009 and Haala et al. 2011). For example, Structure-from-Motion (SfM) is one of the most common techniques used to automatically reproduce three-dimensional scenes from a set of two-dimensional digital images (Lowe 1999 and Triggs et al. 2000; Westoby et al. 2012 and Bevilacqua et al. 2016). Different software for post-processing have recently been developed and commercialized and 3D models were compared in order to select

I. Vayas and F. M. Mazzolani (Eds.): PROHITECH 2021, LNCE 209, pp. 334–344, 2022.
https://doi.org/10.1007/978-3-030-90788-4_28

the best-performing software (Remondino et al. 2014; Gross 2015; Niederheiser et al. 2016; Alidoost and Arefi 2017). To further contribute to this pursuit, a 3D-case study is herein presented in order to apply four commercial software (Agisoft PhotoScan, Recap Pro, VisualSFM and Pix4D) and assess the different capacities and potentialities of each software. The chosen case study consists of the historical cenotaph (Fig. 1), constructed in 1929 in remembrance of the war dead of the First and Second World Wars and located in front of the Auckland War Memorial Museum in Auckland Domain. The cenotaph presents a regular geometry with specific details, such as three steel ropes, stone crowns, and several engravings.

(a) Cenotaph 3D view **b) Cenotaph details**

Fig. 1. Auckland war memorial cenotaph

2 The Survey

A survey was performed using a Phantom 3 Professional (manufactured by DJI Company), which is a drone that is equipped with a 4K camera and a 3-axis gimbal that allowed stabilized video and images to be captured. The drone camera had a 1/2.3" CMOS sensor, 12 MP effective, 20 mm fixed lens and 94° field of view and supported single or burst shooting, offering the ability to collect images or video from nadir and vertical positions. An application called DJI Go (https://www.dji.com/goapp) was used to acquire the images and to control the flight path, the camera inclination and the calibration of image exposure. Three circular drone flight paths with a tilted camera were performed. Specifically, a nadiral acquisition was used for the upper part of the object with one path, while a closer flight was performed along both sides of the memorial to capture the cenotaph details from different angles. The cenotaph was geo-referenced to provide its correct position, scale and orientation and a local coordinate system was defined by placing four visible targets on the four corners at the base of the memorial (Fig. 2). Being a comparative study concerning a physical object, the introduction of a global reference system was not necessary for evaluating the Ground Control Point (GCP) coordinates and therefore three distances (horizontal, vertical and a diagonal) were acquired from two reciprocal targets. Light conditions considerably varied during the acquisition phase, affecting the images processing and the final modelling.

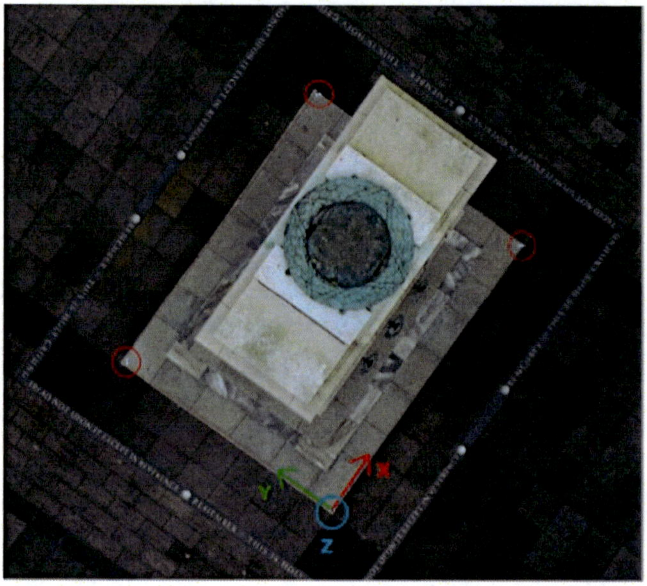

Fig. 2. Local coordinate system with four targets at the base of the Cenotaph.

3 Post Processing Comparison

A comparison of the 4 considered open-source point cloud software is presented in this section, which was undertaken using an open-free software called CloudCompare that can be used to generate 3D point clouds, edit 3D meshes, and compute relative errors in terms of distance between two-point clouds.

3.1 Point Cloud

A comparison of the four 3D point cloud models generated using each of the four different open-source point cloud software is shown in Fig. 3. VisualSFM was the only software that failed to adequately reproduce the Cenotaph, resulting in voids on both lateral facades and in the upper part of the Cenotaph where reproduction was difficult because of the variability of light conditions. The model generated using Pix4D had the most dense point cloud (22 million number of points), followed by PhotoScan, VisualSFM and Recap Pro.

No. Points	5 million	170,000	22 million	850,000
Computation time	3 h 24 min	55 min (Whole model)	3 h 35 min	3 h 20 min
Geometry Reconstruction	🙂	🙂	🙂	🙁

Fig. 3. Quality comparison of the four dense point clouds.

Recap Pro was the fastest software, with the model generated in approximately 55 min when compared to approximately 3 h and 20 min for the other three software.

The models generated using Pix4D and VisualSFM adequately reproduced the three steel ropes located on the two lateral facades, while voids occurred up the height of the other two models (Fig. 4). In addition, the details of the crown located on the upper part of the memorial were only adequately reproduced when applying PhotoScan and Pix4D.

In order to evaluate the quality of the models generated by the four software, each model was oriented with respect to an external relative coordinate system by using the four markers placed on the base of the Cenotaph. The comparisons of the point clouds and the correlations between the number of points and the relative distance are shown in Figs. 5, 6 and 7. In Fig. 5 the peak points of relative distance were used to compare PhotoScan and VisualSFM, where it is shown that the first peak consisted of more than 9000 blue points having 3 mm distance, while the second peak consisted of 8000 blue points with a distance of 40 mm. As noted in the table shown in Fig. 5, the average value and the standard deviation were equal to 30 mm and 0.022, respectively.

Figure 6 reports results from a comparison of the models generated using Pix4D and Visual SFM, with two peaks again shown similar to Fig. 5 and with a similar average value (28 mm). A comparison of the models generated using Recap Pro and VisualSFM is shown in Fig. 7, where the first peak consists of more than 7000 points having a distance of 20 mm distance and the second peak had a distance of 50 mm.

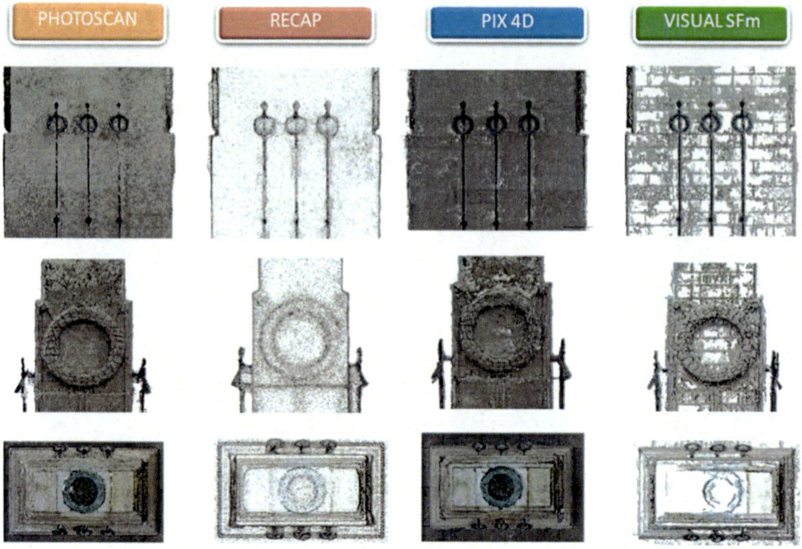

Fig. 4. Quality comparison of the cenotaph details.

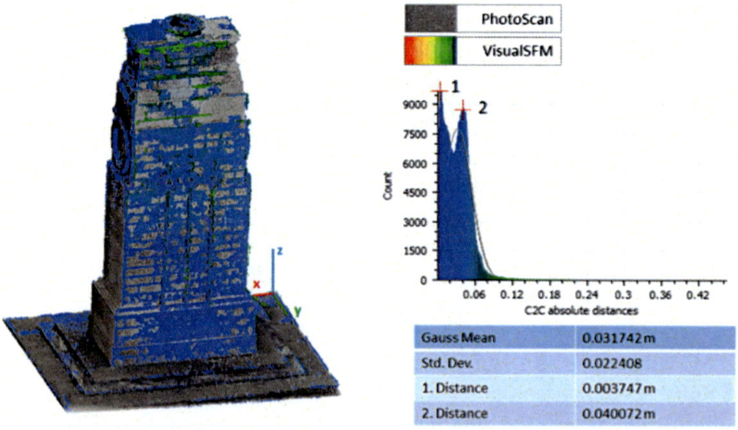

Fig. 5. Point cloud comparison between PhotoScan and VisualSFM.

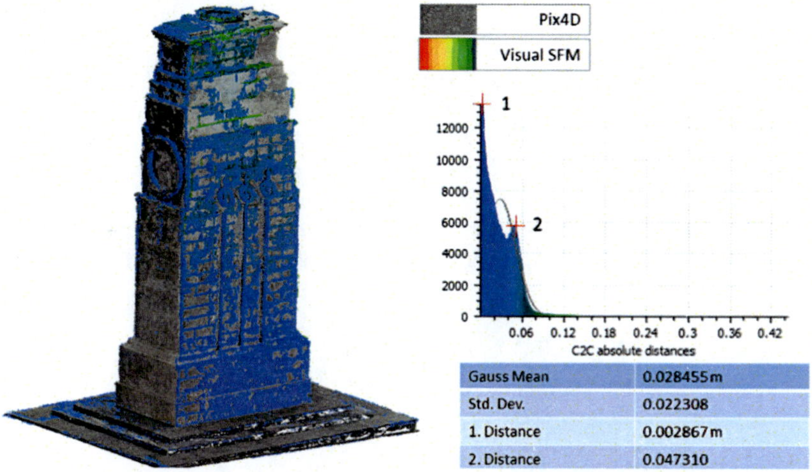

Gauss Mean	0.028455 m
Std. Dev.	0.022308
1. Distance	0.002867 m
2. Distance	0.047310

Fig. 6. Point cloud comparison between Pix4D and VisualSFM.

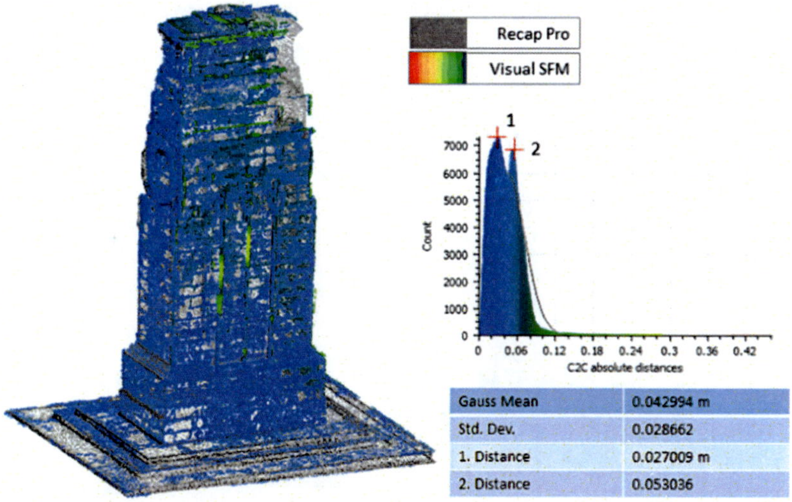

Gauss Mean	0.042994 m
Std. Dev.	0.028662
1. Distance	0.027009 m
2. Distance	0.053036

Fig. 7. Point cloud comparison between Recap Pro and VisualSFM.

3.2 Mesh

In Fig. 8 a comparison is shown for the results from the Pix4D model (represented with the textured colour) and the Recap Pro model (represented with the coloured scalar field) using the relative coordinate system defined by using the four markers placed at the base of the cenotaph (Fig. 8). The final models had similar geometries, with 12000 faces and a relative distance equal to 0.2 mm with an average value of 3 mm. The differences between the PhotoScan and Recap models (Fig. 9) were primarily due to difficulties encountered in reproducing some features of the cenotaph, such as the three steel ropes.

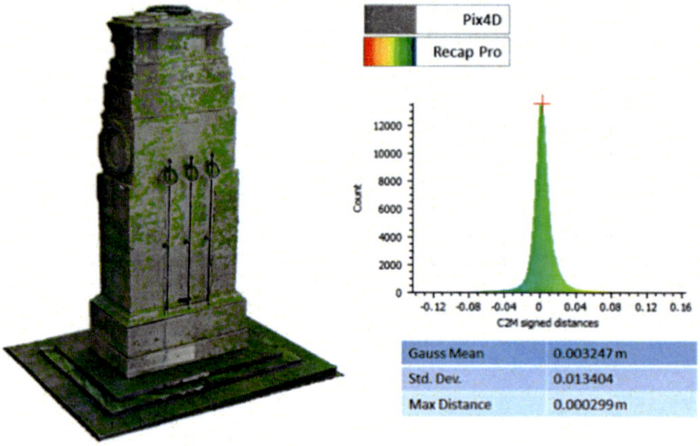

Fig. 8. Mesh comparison between Pix4D and Recap Pro.

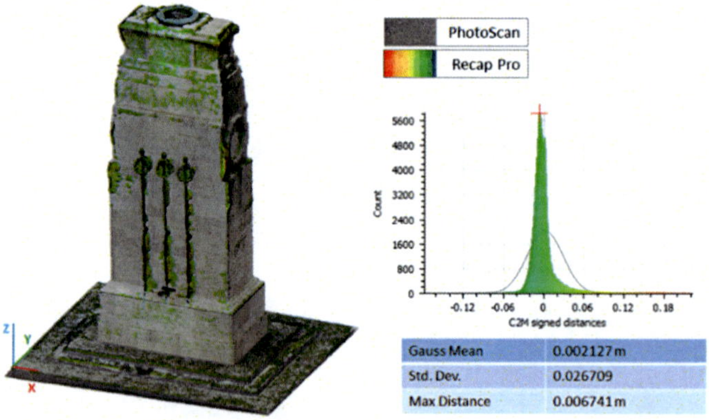

Fig. 9. Mesh comparison between PhotoScan and Recap Pro.

3.3 Geometry

In order to test the accuracy of the geometry of the four point cloud models, each model was oriented along the local coordinate system and scaled by referring to manually recorded measurements (Fig. 10). The geometry obtained by the dense-point cloud process is represented with different colours (black represents Recap, green Visual SFM, blue PhotoScan and red Pix4DRecap). The errors between the measured width dimension (4.255 m) and the dimensions obtained with the 3D models are reported in Table 1, with Recap presenting the highest error (7.4%).

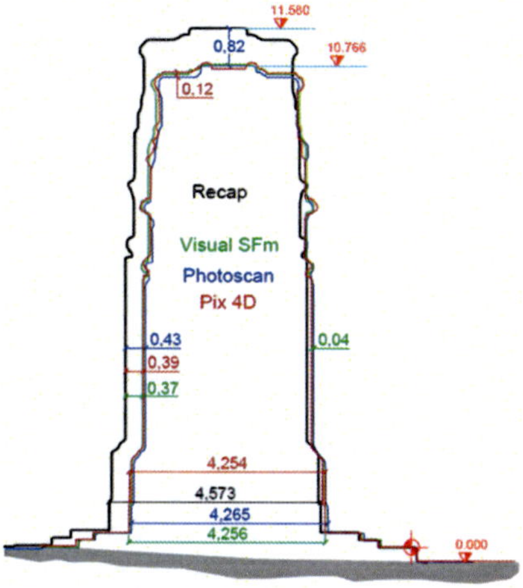

Fig. 10. Model geometry after the dense cloud reproduction

Table 1. Error evaluation between the four final models

Sofware	PhotoScan	RecapPro	Pix4D	VisualSFM
Measurement (m)	4.265	4.573	4.254	4.256
Error (%)	0.23	7.4	0.02	0.02

3.4 Overall Comparison

In Table 2 a comparison of the models is presented based on coloured smiles (green for the best results, yellow for a good performance and red for the worst results), by considering several parameters related to the models. When considering mesh reproduction, VisualSFM was shown to be the weakest performing while the model generated using Pix4D reproduced the cenotaph with the highest quality of details. RecapPro was the

Table 2. Model parameters

Software	PhotoScan	RecapPro	Pix4D	VisualSFM
Point clouds	🙂	😐	🙂	🙁
Mesh	🙂	😐	🙂	Not permitted
General Reproduction	🙂	🙂	🙂	🙁
Detail Reproduction	🙁	🙁	🙂	😐
Computing time	😐	🙂	😐	😐

best software in terms of computation time of the entire process including generation of the dense point cloud and mesh.

General parameters associated with the use of each software considered in this study are shown in Table 3. RecapPro was the most user-friendly software because the entire final model can be generated in one step and automatically processed picture is uploaded on the cloud server. However, RecapPro can be classified as the least reliable because the user does not control each step of the process. Pix4D and PhotoScan provide the possibility to choose the accuracy of each processing step, while VisualSFM requires a binary code to run the dense point cloud.

Table 3. General parameters

Software	PhotoScan	RecapPro	Pix4D	VisualSFM
Workflow complexity	😐	🙂	😐	😐
User accuracy	🙂	🙁	🙂	🙁
All in one	🙁	🙂	🙁	🙁
Possibility to scale	🙂	🙁	🙂	🙁
User friendliness	😐	🙂	😐	🙁
Output control	🙂	🙁	🙂	🙂

4 Conclusion

The performance of four post-processing software (Agisoft PhotoScan, Recap Pro, VisualSFM and Pix4D) was assessed by analyzing the historical cenotaph located in front of the Auckland War Memorial Museum in Auckland Domain park (New Zealand). Potentialities of the various software were evaluated by considering several parameters such as mesh, detail reproduction, computing time, overflow complexity, accuracy, user-friendliness and output control. It is anticipated that future studies will be conducted on other monuments in New Zealand.

References

Agisoft PhotoScan User Manual Professional Edition version 1.4. n.d. Agisoft PhotoScan. http://www.agisoft.com/pdf/photoscan-pro_1_4_en.pdf. Accessed 15 Nov 2017

Alidoost, F., Arefi, H.: Comparison of UAS-based photogrammetry software for 3D point cloud generation: a survey over a historical site. In: 4th International GeoAdvances Workshop, Turkey (2017)

Bevilacqua, V., et al.: Photogrammetric meshes and 3D points cloud reconstruction: a genetic algorithm optimization procedure. In: Rossi, F., Piotto, S., Concilio, S. (eds.) WIVACE 2016. CCIS, vol. 708, pp. 65–76. Springer, Cham (2017). https://doi.org/10.1007/978-3-319-57711-1_6

Boccardo, P., Chiabrando, F., Dutto, F., Tonolo, F.G., Lingua, A.: UAV deployment exercise for mapping purposes: evaluation of emergency response applications. Sensors **15**(7), 15717–15737 (2015)

Chiabrando, F., Di Lolli, A., Patrucco, G., Spanò, A., Sammartano, G., Teppati Losè, L.: Multitemporal 3D modelling for cultural heritage emergency during seismic events: damage assessment of S. Agostino church in Amatrice (RI). In: The International Archives of the Photogrammetry, Remote Sensing and Spatial Information Sciences, Volume XLII-5/W1, 2017, GEOMATICS & RESTORATION – Conservation of Cultural Heritage in the Digital Era, 22–24 May 2017, Florence, Italy (2017)

Chiabrando, F., D'Andria, F., Sammartano, G., Spanò, A.: 3D modelling from UAV data in Hierapolis of Phrigia (TK). In: 8th International Congress on Archaeology, Computer Graphics, Cultural Heritage and Innovation 'ARQUEOLÓGICA 2.0', Valencia (Spain), 5–7 September 2016, pp. 347–349 (2016)

CloudCompare. n.d. User Manual version 2.6.1. http://www.dan-ielgm.net/cc/doc/qCC/CloudCompare%20v2.6.1%20-%20User%20manual.pdf. Accessed 15 Dec 2017

Colomina, I., Blázquez, M., Molina, P., Parés, M.E., Wis, M.: Towards a new paradigm for high-resolution low-cost photogrammetry and remote sensing. In: IAPRS&SIS 2008, pp. 1201–1206 (2008)

Eisenbeiss, H.: UAV Photogrammetry, Diss. ETH No. 18515, Institute of Geodesy and Photogrammetry", ETH Zurich, 2009, p. 235 (2009)

Forcellini, D., Giardi, F., Tanganelli, M.: Seismic assessment of the historical third tower in San Marino based on a 3D laser scanner survey (3D-LSS). Innov. Infrastruct. Solut. **4**(1), 1–20 (2019). https://doi.org/10.1007/s41062-019-0207-2

Georgopoulos, A., Oikonomou, C., Adamopoulos, E., Stathopoulou, E.K.: Evaluating unmanned aerial platforms for cultural heritage large scale mapping. The International Archives of the Photogrammetry, Remote Sensing and Spatial Information Science, Volume XLI-B5, pp. 355–362 (2016)

Gross, J.W.: A Comparison of Orthomosaic Software for Use with Ultra High Resolution Imagery of a Wetland Environment (2017)

Guerra, G., et al.: 3D numerical simulations of the San Marino third tower (Montale) based on 3D Mapping. In: Computational Methods in Structural Dynamics and Earthquake Engineering (COMPDYN), Rhodes, Greece (2017)

Haala, N., Cramer, M., Weimer, F., Trittler, M.: Performance test on UAV-based photogrammetric data collection, ISPRS ICWG I/V UAV-g. In: Conference, Zurich, Switzerland, 2011 (2011)

Lowe, D.G.: Object recognition from local scale-invariant features. In: Proceedings of the seventh IEEE international conf. on Computer vision, vol. 2, pp. 1150–1157. IEEE (1999)

Niederheiser, R., Mokroš, M., Lange, J., Petschko H., Prasicek, G., Elberink, S.O.: Deriving 3D point clouds from terrestrial photographs comparison of different sensors and software. In: The International Archives of the Photogrammetry, Remote Sensing and Spatial Information Sciences, Prague, Czech Republic, Vol. XLI-B5, pp. 685–692 (2016)

Pieraccini, M., et al.: Radar detection of pedestrian-induced vibrations on Michelangelo's David. PlosONE (2017). https://doi.org/10.1371/journal.pone.0174480.g004

Pix4D, April 2019. https://cloud.pix4d.com

ReCap PRO, November 2019. https://www.autodesk.com/products/recap/overview

Remondino, F., Barazzetti, L., Nex, F., Scaioni, M., Sarazzi, D.: UAV Photogrammetry for mapping and 3D modeling current status and future perspectives. Int. Archives Photogram. Remote Sens. Spatial Inf. Sci. **38**(1), C22 (2011)

Remondino, F., Spera, M.C., Nocerino, E., Menna, F., Nex, F.: State of the art in high density image matching. Photogram. Rec. **29**(146), 144–166 (2014)

Sabina, J.R., Valle, D.G., Ruiz, P.C., García, J.M., Gómez Laguna, A.: Aerial Photogrammetry by drone in archaeological sites with large structures. Virtual Archaeol. Rev. **6**(13), 5–19 (2015)

Triggs, B., McLauchlan, P.F., Hartley, R.I., Fitzgibbon, A.W.: Bundle adjustment — a modern synthesis. In: Triggs, B., Zisserman, A., Szeliski, R. (eds.) IWVA 1999. LNCS, vol. 1883, pp. 298–372. Springer, Heidelberg (2000). https://doi.org/10.1007/3-540-44480-7_21

Uysal, M., Toprak, A.S., Polat, N.: DEM generation with UAV Photogrammetry and accuracy analysis in Sahitler hill. Measurement **73**(2015), 539–543 (2015)

VisualSFM, November 2019. http://ccwu.me/vsfm/

Westoby, M.J., Brasington, J., Glasser, N.F., Hambrey, M.J., Reynolds, J.M.: Structure-from-Motion photogrammetry: a low-cost, effective tool for geoscience applications. Geomorphology **179**, 300–314 (2012)

Numerical Evaluation of Seismic Retrofitting of Heritage Building by Friction Dampers Based on Geometry Obtained by Laser Scans

Shakhzod M. Takhirov[1](✉), Amir Gilani[2], Brian Quigley[3], Liliya Myagkova[4], and Sultan Kudratov[5]

[1] Structures Laboratory, Civil and Environmental Engineering Department, University of California - Berkeley, 337 Davis Hall, Berkeley, CA, USA
takhirov@berkeley.edu
[2] Miyamoto International, Inc., Sacramento, CA, USA
[3] BNZ, Tashkent, Uzbekistan
[4] Smart Scanning Solutions LLC, Tashkent, Uzbekistan
[5] Department of Informatics, Tashkent University of Information Technologies, Tashkent, Uzbekistan

Abstract. The main objective of this paper is to investigate the feasibility of using friction dampers for a seismic retrofit of a historic building in Uzbekistan. This study is part of the extensive ongoing project dealing with the structural assessment of heritage buildings along the ancient Great Silk Road by means of laser scanning and other advanced methods of structural health monitoring. In an earlier study, it was shown that the existing structural reinforcement of the portal in the Kok-Gumbaz monument in Shakhrisabz was deficient and had a strong potential of poor performance in case of an earthquake. As one of the conclusions of the earlier study, it was shown that the building's existing reinforcement needs to be modified. The structural assessment and numerical analysis was limited to the portal of the movement while the main building with a globe was not considered. This was related to the fact that there is a continuous crack separating the portal from the main building. A complex numerical model of the portal was generated, which was examined for its seismic performance and stability. The numerical model of the historic monument was developed from the point clouds collected with a terrestrial laser scanner to ensure that it closely replicates all imperfections of the actual monument. As one of the best retrofitting alternatives, friction dampers were introduced into the model to study their efficiency. The numerical model was subjected to a ground shaking recommended by the structural code of Uzbekistan. A nonlinear time history analysis of the numerical model protected with the friction dampers clearly demonstrated the benefits of the friction dampers in the seismic protection of the historic monument. Based on the results of this study, the friction dampers are recommended for practical application in the future retrofitting projects of the monument.

Keywords: Structural health monitoring · Seismic assessment · Laser scanning · Friction dampers · Seismic protection · Seismic retrofit strategies

I. Vayas and F. M. Mazzolani (Eds.): PROHITECH 2021, LNCE 209, pp. 345–353, 2022.
https://doi.org/10.1007/978-3-030-90788-4_29

1 Introduction

Seismic retrofitting of historic monuments has several challenging requirements. One of the main requirements is that the modification of the structure or its components should not be invasive and should be not visible. This is difficult to achieve since very little modifications are permitted. This paper is focused on a study of friction dampers those are proposed to be installed in line with the existing reinforcement. Since they combine both stiffness and damping properties in one device and do not require large spaces, the friction dampers can be one of the best candidates for structural reinforcement of historic monuments.

2 Laser Scans and Numerical Modelling of Monument

The Kok-Gumbaz monument in Shakhrisabz was studied earlier (Takhirov et al. 2018). Laser scans of the monument were collected from multiple stations and a registration of those is presented in the left image of Fig. 1. The point cloud manipulation was conducted in Cyclone (Leica Geosystems 2019). During field studies and laser scanning it was discovered that there is a continuous crack between the main building and the portal. Based on this discovery it was assumed that the seismic performance of the portal can be separated from that of the main structure. An accurate numerical model in SAP2000 (CSI 2014) replicating all imperfections (including uneven inclination of the portal) was developed.

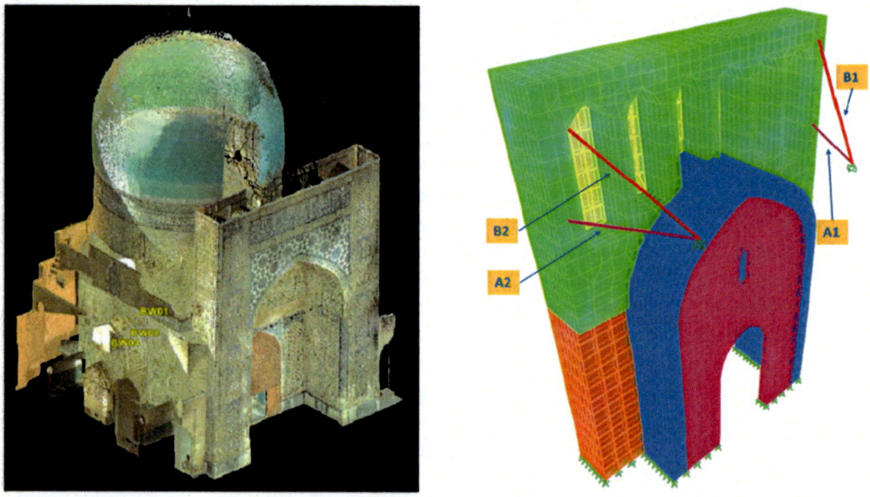

Fig. 1. Point cloud of the monument (left) and a numerical model of the portal (right) with additional braces B1 and B2 proposed earlier (Takhirov et al. 2018)

It was shown that the existing reinforcement is deficient in providing an adequate seismic performance of the main portal. To improve its performance, two more reinforcement braces (B1 and B2) similar to the existing ones (A1 and A2) were proposed as shown in the right image of Fig. 1.

In addition to the braces (A1 and A2) securing the portal to the main structure, the portal was also reinforced with a tie-rod (A3) as shown in Fig. 2. Although the seismic performance of the portal improved significantly, the new braces (B1 and B2) would be visible from the ground level, which is not preferable. This study is focused on an alternative that would not require new braces and only two friction dampers would need to be installed in-line with the existing braces (A1 and A2). Since they provide both stiffness and damping in a single device, this main advantage was utilized. These dampers are popular in the seismic protection of nonstructural components (Takhirov et al. 2019a; Takhirov et al. 2019b). The friction dampers are also commonly used for structural applications with results discussed in many publications, see (Bishay-Girges and Carr 2014) as an example.

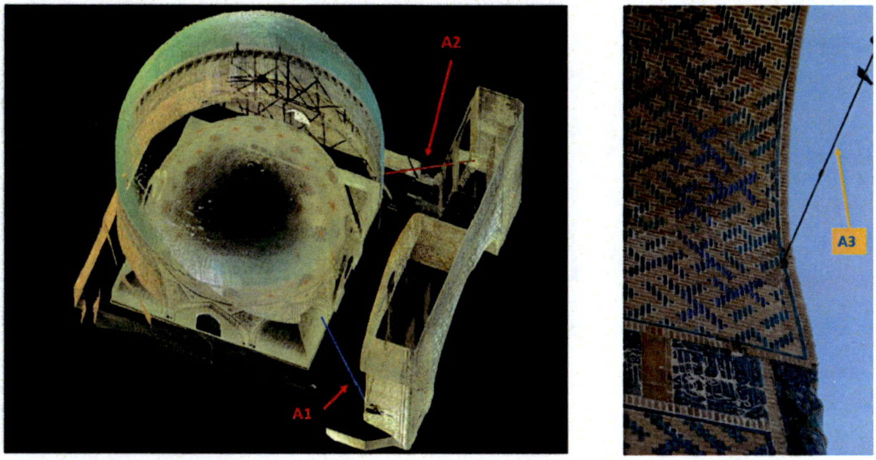

Fig. 2. Existing reinforcement: two braces between the portal and the main structure (left) and a tie-rod between the arches of portal (right)

3 Experimental Evaluation of Friction Dampers

A special test setup was designed and fabricated for component testing of the friction dampers. It is presented in Fig. 3 and it simulates loading of the friction damper in a typical installation where the damper is installed as a stiffness and damping component of a retrofit system. An actual implementation of the design idea shown in Fig. 3 is presented in Fig. 4.

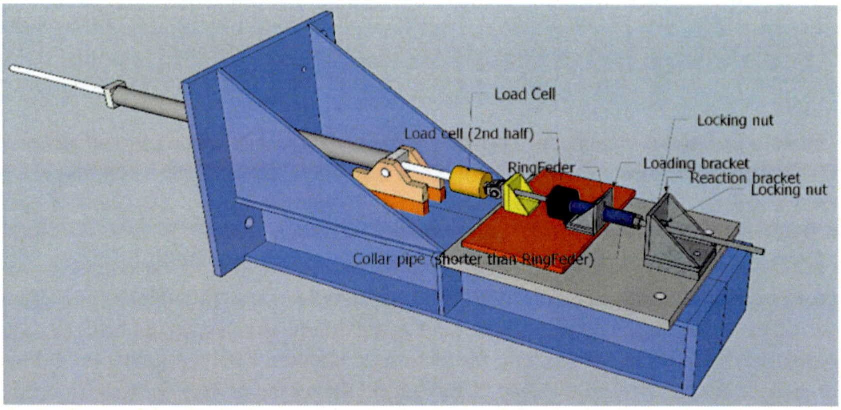

Fig. 3. Schematic 3D drawing of the self-reacting setup designed for friction damper tests

Fig. 4. Overall view of the experimental setup

As one of possible friction damping devices to be used, friction dampers from Ringfeder (Ringfeder 2016) were used in this study. Since the friction dampers work in compression only, there are several ways of using friction dampers as seismic protection devices. For example, if only one friction damper is used in the so-called single-acting configuration, the force will be developed in compression only as presented in the left side image of Fig. 5. In this case, another spring or friction damper needs to be used on the tension side that corresponds to a positive force and displacement to resist the tension forces. When there is one damper on each side of the loading point, they create the so-called double-acting configuration. The force versus displacement curve is more symmetric in this case as presented in the right image of Fig. 5.

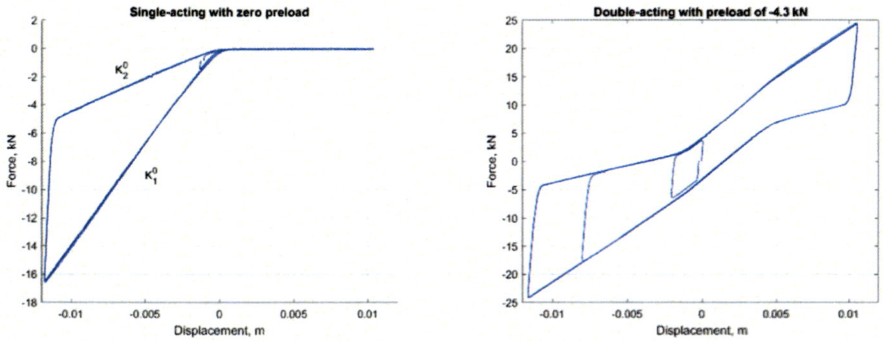

Fig. 5. Force vs. displacement from component testing in single-acting with no preload (left) and double-acting with preload (right) configurations

The dampers used in this study were installed in a double-acting configuration, where there is a damper on each side of the loading point. The latter results in symmetric behavior in tension and compression. A testing protocol comprised of several groups of cycles with different amplitudes was followed, which is presented in Fig. 6 on the left. A typical result for the cyclic tests with increasing amplitude is presented in the right image of Fig. 6.

It is worth noting that the maximum forces measured during the component testing of the friction dampers were much lower than those to be developed in the braces of the monument. For the purpose of this evaluation study, it was assumed that a scaled-up version of the friction damper with properties proportional to those measured during the component tests can be developed. This assumption reflects reality because the friction dampers can be engineered to meet certain force, displacement, and damping requirements. Since all the pieces of the friction damper are machined with high accuracy, the anticipated and as-found properties of the friction dampers are close to each other. The predictability of their performance was demonstrated during the component tests.

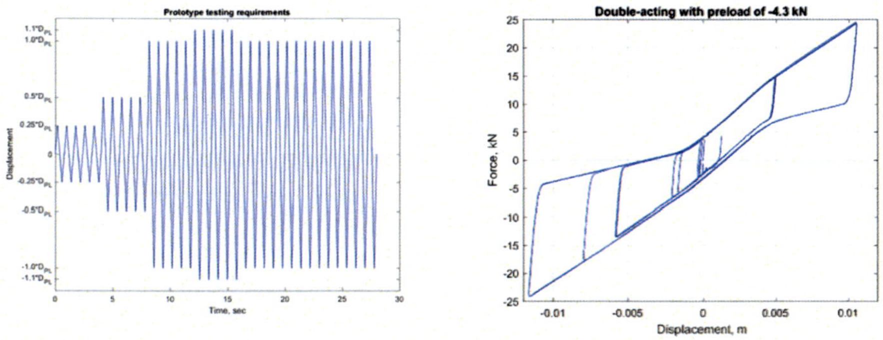

Fig. 6. Component testing: protocol (left) and sample test results (right)

4 Numerical Modelling of Monument with Friction Dampers

The numerical model of the monument (Takhirov et al. 2018) was updated by introducing friction damper models available in SAP2000 (CSI 2014). In addition to that, the nonstationary analysis was changed from linear to nonlinear analysis to capture the performance of the friction dampers. Since the structural code of Uzbekistan (KMK 2.01.03-96) requires to use Gazli strong motion time histories for a seismic excitation, they were downloaded from PEER's NGA-West2 database (PEER 2019) and utilized in this study. The friction damper was in line with the existing braces as presented in Fig. 7 (left). A typical result showing the force versus displacement diagram in one of the braces (A1 is shown as a typical example) is presented in the right image in Fig. 7.

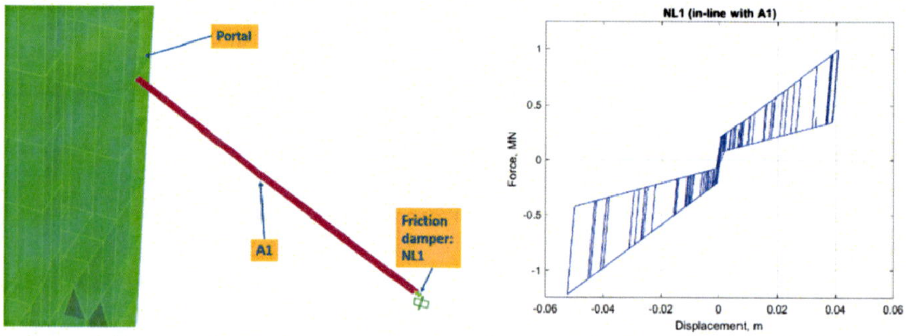

Fig. 7. Modelling of friction damper (left) and force vs. displacement diagram in damper NL1 installed in-line with A1 brace

It is worth noting that the force vs. displacement diagram depicted in Fig. 7 (right) closely matches the actual performance of a friction damper presented in Fig. 6 (right). As a typical example of the reaction forces in the braces, the reaction force along the axis of the A1 brace is presented in Fig. 8. The left plot shows a comparison of the time history of the reaction force with friction damper (blue line) to that without it (red line). On the right plot, the Fast Fourier Transforms (FFTs) of the forces are presented. In the latter plot FFTs of the reaction forces were normalized by the maximum FFT of the case without the friction damper. Both plots in Fig. 8 show a significant reduction of the reaction forces.

The results of the nonlinear time history analysis are presented in Fig. 9. The plots compare the case with the friction damper to that without it. For comparison purposes the results are presented in a normalized form. The normalization was conducted in the following way. All peak values obtained from the model with the friction dampers were divided by the corresponding peak values obtained from the model without the dampers that corresponds to Model A in the terminology of the earlier publication (Takhirov et al. 2018). The benefits of using the friction dampers can be observed from these plots and they are as follows. The plots on the left side of Fig. 9 show that tension stresses at the bottom of the portal are reduced by about 30%, which was the most critical reduction that helps to avoid masonry cracking at the bottom of the portal.

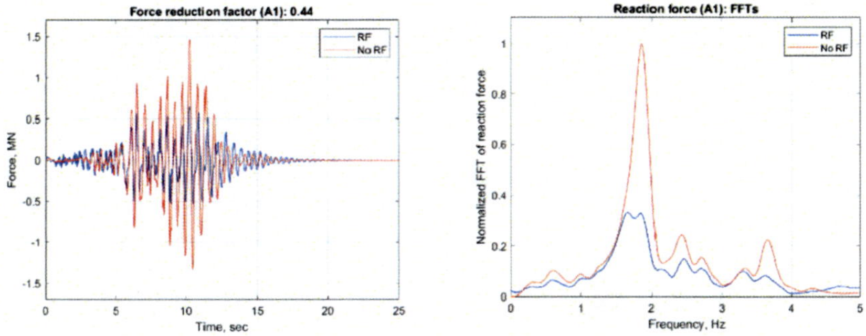

Fig. 8. Reaction forces in A1 brace: time histories (left) and FFTs (right)

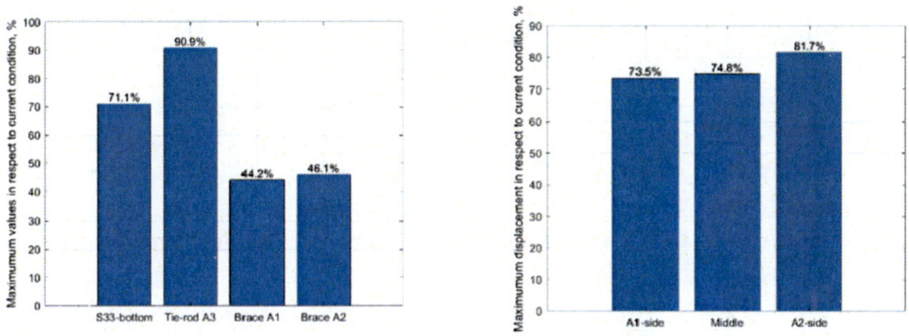

Fig. 9. Normalized changes of peak force related values (left) and displacements on the portal's top (right)

The forces in the tie-rod A3 are reduced by 10%, which was not too critical because they were well within the elastic range of the tie-rod. More importantly, the forces in braces A1 and A2 are reduced by a large percentage (about 56% and 54%, respectively). The latter means that the concentrated forces imposed by the braces at their attachment points to the portal were reduced dramatically. This reduction can help in the future reductions of the present and ongoing localized failures of masonry around the locations where the brace meets the portal. These failures zones are presented in Fig. 10.

In addition to reduction of all these force related values, the normalized peak displacements on top of the portal are also reduced due to the introduction of the friction dampers as shown in right image of Fig. 8. As it can be seen from the plot, the reduction is not symmetric, which is related to the geometrical imperfections of the model generated from the point clouds. The relatively small reduction in the displacements is related to the fact that some travel in the friction damper needs to be allowed to maximize its damping performance.

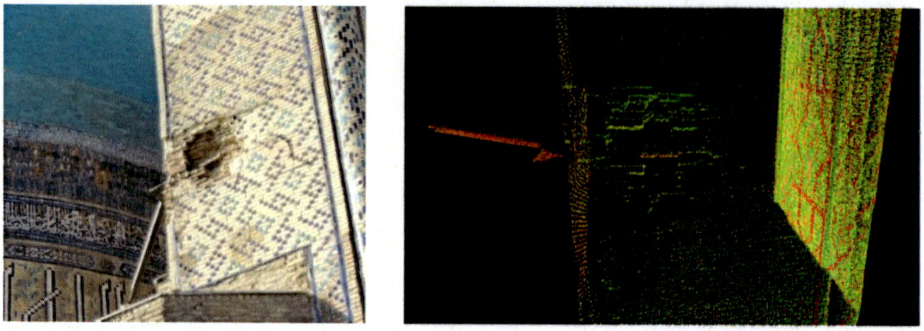

Fig. 10. Localized failures at the locations where concentrated forces from the reinforcement transferred to masonry: photo (left) and laser scan (right)

5 Conclusions

In the present study, a component testing of the friction dampers and their evaluation in seismic protection of a historic monument was conducted. The results of the study discussed in this paper clearly show the benefits of friction dampers for seismic protection of the monument. Since the friction damper combines both stiffness and damping characteristics, its utilization does not require any additional reinforcement (in opposite to a viscous damper that lucks a stiffness component). Hence the installation of the friction damper is more compact and less visible, which is very important for historic monuments. The results of the numerical nonlinear time history analysis clearly show several benefits of the friction dampers. First, the dampers reduced tension stresses at the bottom of the portal's masonry by about 30%. Second, their introduction resulted in significant reduction of forces in the existing braces, the concentrated forces from which are causing localized masonry failures at the attachment points. This reduction was about 55% on average for both braces. Third, some reduction in the displacements at the top of the portal was observed. Based on the results of this study, the friction dampers are recommended for practical application in future retrofitting projects of the monument.

Acknowledgements. Special thanks are due to BNZ and Smart Scanning Solutions (SSS) (both from Uzbekistan) for providing a laser scanner and partial financial support of this research program. Support from staff of Ministry of Culture and Physical Education of Uzbekistan in providing access to the historic monument is greatly appreciated. The component testing of friction dampers was sponsored by the Electric Power Research Institute (EPRI) whose support is greatly appreciated. The research team thanks Mr. Robert Cochran of Seattle City Light for donating Ringfeder friction dampers. Special thanks are due to the staff of the Structures Laboratory at the University of California, Berkeley. Special thanks are due to Mr. Llyr Griffith, Mr. Phillip Wong, and Mr. Matthew Cataleta. The successful completion of the testing phase of the project would not be possible without their technical support. Technical help by Ms. Sage Shamsai (CEE, UC Berkeley) on preparing 3D drawings and testing is greatly appreciated.

References

Bishay-Girges, N.W., Carr, A.: Ring spring dampers: passive control system for seismic protection of structures. Bull. New Zealand Soc. Earthq. Eng. **47**(3), 173–180 (2014)

Computers and Structures, Inc. (CSI). SAP2000 Ultimate Version 16.1.1. Structural Analysis Program (2014)

Leica Geosystems, AG. Cyclone Version 9.4.2 (2019)

Pacific Earthquake Engineering Research Center (PEER) (2019). http://ngawest2.berkeley.edu/site

Ringfeder Power Transmission. Ringfeder Damping Technology (2016). http://www.Ringfeder.com/Documents/Products/Ringfeder/Damping_US.pdf

Seismic code of Uzbekistan, KMK 2.01.03-96 "Norms and Regulations for Construction in Seismic Zones", Tashkent, Uzbekistan, 125 p. (1996). (in Uzbek and Russian)

Takhirov, S., Gilani, A., Quigley, B., Myagkova, L.: Assessment of seismic vulnerability and retrofit strategies of a historic building on the great silk road based on its current condition captured by laser scans. Int. J. Archit. Heritage 1–13 (2018). https://doi.org/10.1080/15583058.2018.1497231

Takhirov, S., Kempner, L., Riley, M., Fujisaki, E., Low, B.: Seismic protection of high-voltage equipment by friction dampers: numerical modelling correlated with full-scale component tests. In: Papadrakakis, M., Fragiadakis, M. (eds.) COMPDYN-2019, 7th ECCOMAS Thematic Conference on Computational Methods in Structural Dynamics and Earthquake Engineering, Crete, Greece, 24–26 June 2019 (2019a)

Takhirov, S., Kuse, H., Yoshida, K., Murase, S., Fujisaki, E.: Numerical optimization of seismic performance of tall and slender system protected by seismic protection device. In: Papadrakakis, M., Fragiadakis, M. (eds.) COMPDYN-2019, 7th ECCOMAS Thematic Conference on Computational Methods in Structural Dynamics and Earthquake Engineering. Crete, Greece, 24–26 June 2019 (2019b)

Check for updates

Progressive Collapse Retrofit of a RC Hospital Building Using Steel Braces

Massimiliano Ferraioli, Angelo Lavino, and Alberto Mandara(✉)

Department of Engineering (DI), Università Della Campania "Luigi Vanvitelli", Via Roma 9, 81031 Aversa, Italy
{massimiliano.ferraioli,angelo.lavino, alberto.mandara}@unicampania.it

Abstract. The progressive collapse retrofit of a RC hospital building making use of steel braces is described in the paper. The vulnerability of the construction when subjected to specific threat scenarios is preliminary analysed by removing critical column and applying the alternate load path method to evaluate the capability of the damaged structure to bridge across the removed column. A detailed three-dimensional model of the building has been created to carry out both nonlinear static and nonlinear dynamic analyses. Both geometrical and material nonlinearities have been considered in the analysis in order to account for the inelastic behaviour following the sudden column removal and the possible subsequent catenary effects. In order to retrofit the building against progressive collapse the use of steel braces in some bays has been proposed, designed on the basis on simple equilibrium conditions under vertical loads. The retrofitting bracing system proved to be effective in improving the global building behaviour under several accidental scenarios.

Keywords: Progressive collapse · Retrofit · RC buildings · Steel bracing

1 Introduction

Recent historic events have demonstrated that constructions designed according to standardized building codes might not necessarily be able to resist to abnormal loads such as fire, explosion or vehicle impact. When these quite unusual loading conditions lead to the destruction of one or more bearing columns of the frame structure, then the collapse of the whole building may result. This kind of vulnerability is particularly important for public buildings such as schools and hospitals, whose collapse consequences could be very severe in terms of human casualties. Recently erected buildings designed according the current seismic codes usually have high ductility and redundancy properties, so as to show generally a good capacity to contrast the global collapse in case of column loss when compared to buildings designed for gravity loads only. In addition, reliability-based resistance factors implicitly increase the level of structural safety. Nevertheless, due to the inherent design philosophy based on ductility and capacity design, the design

I. Vayas and F. M. Mazzolani (Eds.): PROHITECH 2021, LNCE 209, pp. 354–368, 2022.
https://doi.org/10.1007/978-3-030-90788-4_30

base shear can be relatively small even in high seismicity sites. In addition, this approach does not take into consideration extreme loading events that may cause a critical column to suddenly lose its load bearing capacity and lead to progressive collapse. As a consequence, the current seismic design, even though carried out according to modern codes, not always guarantees an adequate level of robustness, which is why a specific assessment of progressive collapse potential is necessary in most cases.

Most of papers about progressive collapse assessment published in the latest years were concerned with steel structures (Chen et al. 2016; Kim and Park 2008; Kim et al. 2009, 2011; Ferraioli et al. 2014, 2018a), whereas less contributions devoted to reinforced concrete structures are present in literature (Sasani and Sagiroglu 2008; Ferraioli et al. 2012; Ferraioli 2019a). Even though accurate analyses based on finite-element models have shown in general good simulation capabilities (Keyvani et al. 2014), the validity of these complex models has not been checked for large structural systems, yet. This results from some commonly adopted simplifying assumptions, which neglect specific local effects such as axial-flexure-shear interaction, joint failure, and bond failure in the assessment of global structural response. In addition to this, simplified linear analyses are often carried out with their inability to include geometric and material nonlinearities, second order and large displacement effects, stiffness and strength degradation, beam-catenary action and so on. This may lead to great inaccuracy in the prediction of behaviour under progressive collapse as, for example, the catenary action is an important resisting mechanism allowing the beams to carry further vertical loads beyond the flexural limit at large displacements. Such mechanism, which occurs only if the beam-and-slab substructures are effectively restrained by adjacent lateral boundary elements and the beam reinforcement bars are adequately arranged and satisfactorily anchored, can be pointed out only by a fully non-linear large displacement analysis. Eventually, just a few if any numerical modelling guidelines dealing with simple and reliable rules for progressive collapse analysis of buildings are available. This paper is focused the progressive collapse retrofit of a RC hospital building making use of steel braces. The construction under consideration is the hospital of Avezzano, located in the district of L'Aquila, Central Italy, in a high seismicity area. The study is focused on the "alternate load path" approach, which aims at analysing possible stress redistributions throughout the structure after the sudden loss of a vertical bearing element. This column loss scenario is widely considered in the literature as it undoubtedly leads to an upper bound of the deformations related to an event dependent simulation analysis. In order to simulate the complex dynamic effects following the column removal, a suitable dynamic increase factor (DIF) is used when a static approach is applied. A detailed three-dimensional model of the hospital has been created to perform both nonlinear static and nonlinear dynamic analyses, accounting for both geometrical and material nonlinearities. The retrofitting bracing system proved to be effective in improving the global building behaviour against both earthquake and progressive collapse under several accidental scenarios.

2 The Case Study

The hospital of Avezzano consists of several adjacent structures separated by seismic joints (Fig. 1). The study described herein refers to a Reinforced Concrete wall-frame body) erected in the 70's and designed according to the Italian standards of the time (Italian Ministerial Decree N.1684 November 25th, 1962) (Fig. 2). A general plan of the basement floor of the building is plotted in Fig. 3. The building, which has planar irregularity due to its T-shaped plan geometry, has seven storeys with a global height of about 20 m. Floors consist of 21 cm or 25 cm thick slabs, with mixed reinforced concrete and hollow tiles typology. The foundation of the building relies upon a beam gridwork. Typical cross-section properties for columns and beams are illustrated in Fig. 4.

Before the earthquake occurred in 2009 the building underwent extensive measurements and tests (both in-situ and in the laboratory) for the diagnosis and assessment of seismic vulnerability. In particular, the following investigations were carried out: 1) geometrical survey; 2) soil analysis including sampling and testing; 3) assessment of mechanical properties of materials by testing of samples taken from the structure. The campaign comprised a large number of tests, including 3 monotonic compressive tests on cylindrical specimens, 24 tensile tests on steel rebars, ultrasonic tests, 40 carbonation depth measurement test, 163 Schmidt rebound hammer tests, 180 radiographic tests. The concrete compressive strength was finally estimated by the "SonReb" methodology (SONic + REBound). According to Italian Standard (NTC 2018) and Eurocode 8 (CEN 2004) the concrete compressive strength f_c and the steel yield strength f_{yw} were calculated as mean values from in-situ tests ($f_{cm} = 23.4$ MPa; $f_{ym} = 402.6$ MPa), appropriately divided by the confidence factors CF = 1.20, accounting for the level of knowledge attained (KL2 Normal knowledge).

Additional in-situ experimental tests were performed by means of Environmental Testing Methods, based on the measurement of environmental vibrations. Both floor noise and low frequency vibrations from wind and traffic loading were measured without direct excitation of the building while in use. The environmental vibration measurements under wind and traffic vibration were carried in July 2008, (before the 2009 L'Aquila earthquake), leading to the structural identification by means of techniques of modal extraction in the frequency domain (Ferraioli 2010a, 2010b, 2015).

The mechanical properties of soil and the ground type according to soil classification of Eurocode 8 (CEN 2004) were obtained from geological and geotechnical tests such as N.1 soil profile test, N.3 Standard Penetration Tests (SPT) and N.1 Down-Hole Test. The last one estimates the soil stiffness properties by analysing direct compression and shear waves along a borehole down to about 30 m. The average value obtained for the propagation velocity of S waves ($V_{S,30}$) in the upper 30 m of the foundation soil is 306.4 m/s for S-waves in X-direction and 329.7 m/s for S-waves in Y-direction. These results led to classify the ground as type C ($180 < V_{S,30} < 360$; $15 < N_{SPT} < 50$). As the building is located on a flat ground, topographic conditions have been assumed not to affect the seismic behaviour.

The experimental data have been exploited to set up a linear-elastic model of the structure (Fig. 5) implemented in SAP2000 (2019). This model has been then extended to include the strength of structural elements and their post-elastic behaviour, according to Eurocode 8 (CEN 2004) provisions.

In particular, a Coupled PMM hinge model for members of the framed structures has been used for the analyses, by assuming the plastic hinge generalised load-deformation diagrams as bilinear. The elastic stiffness of the bilinear force-deformation relation

Fig. 1. Aerial view of the Hospital of Avezzano (Italy)

Fig. 2. External view of the Hospital building structure considered in the analysis

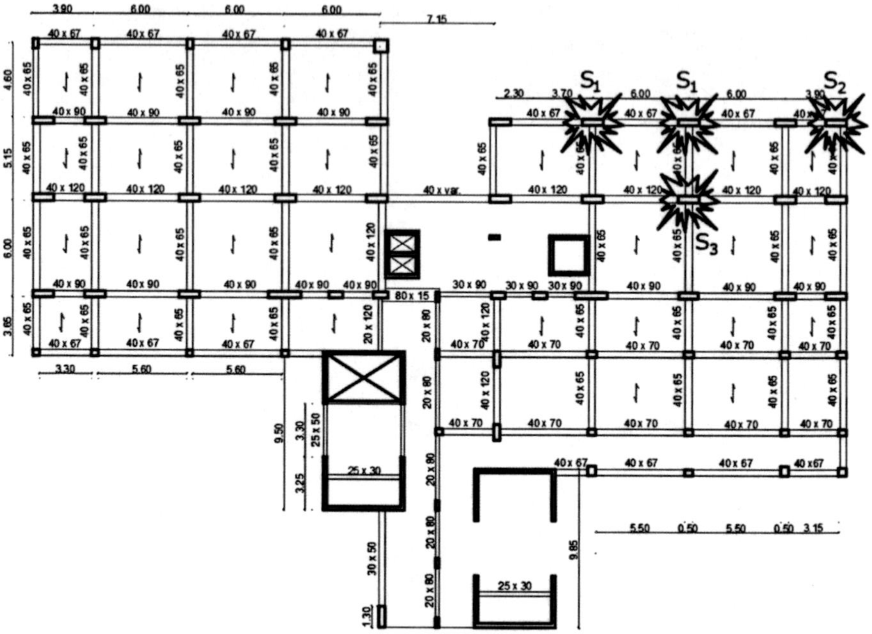

Fig. 3. Plan view of the Hospital building structure with column removal scenarios

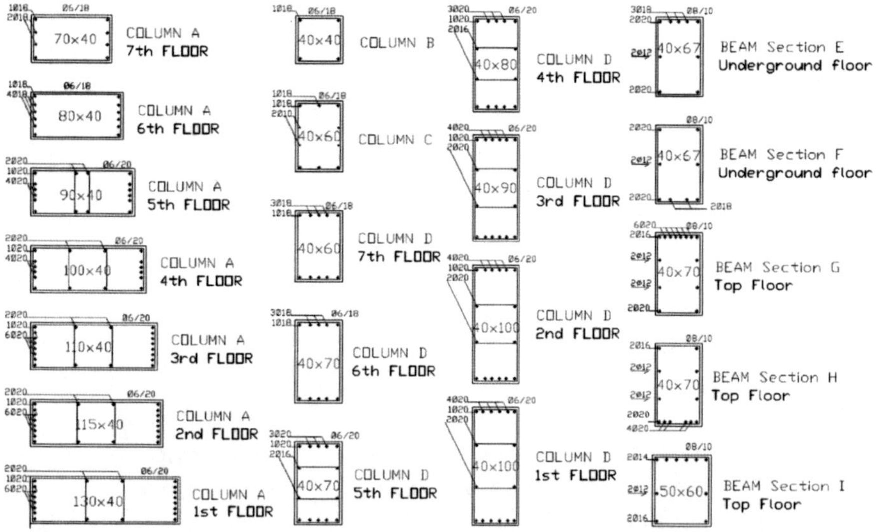

Fig. 4. Typical cross-sections properties of columns and beams

has been evaluated taking into account the effect of cracking. In particular, the elastic flexural stiffness properties of concrete elements have been taken equal to the secant stiffness corresponding to the initiation of reinforcement yielding. More details about the nonlinear modelling and the inelastic seismic behaviour of the building structure may be found in Ferraioli (2015).

Fig. 5. Tridimensional FEM model of the Hospital building structure

3 Progressive Collapse Analysis of the Existing Structure

3.1 Nonlinear Static (Pushdown) Analysis

The sudden removal of a structural member is a strongly nonlinear and dynamic phenomenon, which as a rule requires sophisticated, time consuming finite element modelling and computationally intensive analyses. For this reason, even though losing a part of accuracy, simpler equivalent nonlinear static analyses are commonly used in practice (GSA 2013; UFC 2013; Ruth et al. 2006; Stevens et al. 2008; Izzuddin et al. 2008; Ellingwood and Dusenberry 2009; Mirtaheri and Zoghi 2016; Xu and Ellingwood 2011). In this context, the equivalent static load method, also known as "pushdown analysis", aims at reproducing these dynamic effects applying increasing gravity loads to the structure in which one or more elements have been suppressed. In this procedure, a gravity load multiplier named dynamic increase factor (DIF) is applied on the bays that are directly involved in the column removal in order to account for the dynamic effects. The DIF value initially recommended in the first guidelines on progressive collapse was in the order of 2. However, some subsequent studies (McKay et al. 2012; Liu 2013; Mashhadi

2016, 2017; Mashhadi and Saffari 2017; Ferraioli 2016, 2019b) demonstrated that this value leads to an overconservative estimation of progressive collapse resistance. In fact, the amplification factor of 2 would be rigorously correct only if the response after the column removal were purely elastic. This is not the case during this extreme loading event, when the structure in most cases well enters into plastic range. Generally speaking, the DIF decreases as long as the flexural ductility of beams is exploited. An infinite plastic deformation capacity would ideally yield the theoretical lower bound value of 1.0. For this reason, many researchers suggested increase factors ranging between 1.3 and 1.5 (Marchand and Alfawakhiri 2004; Ferraioli 2018b) or proposed alternative formulations for DIF (Liu 2013; Ferraioli 2019c).

In this paper, the standard load factor approach was applied for pushdown analysis. This procedure is based on the application of the load combination DIF×LF×(DL + 0.25LL) to the spans where the column has been removed, whereas the load combination LF×(DL + 0.25LL) is applied to the remaining spans (Fig. 6a). LF is the load factor that is the ratio of the applied load to the GSA-specified load of DL + 0.25LL (DL and LL being dead loads and live loads). The DIF has been defined using the formula originally proposed by Marchand et al. (2009) for RC structures and then adopted in the latest version of both UFC (2013) and GSA (2013):

$$\mathrm{DIF} = 1.04 + \frac{0.45}{\theta_{pra}/\theta_y + 0.48} \tag{1}$$

θ_{pra} being the prescribed maximum acceptable plastic hinge rotation angle and θ_y the yield rotation angle. θ_{pra} is calculated using the acceptance criteria of Eurocode 8 (CEN 2004) for the appropriate structural response level (Collapse Prevention or Life Safety).

The chord rotation for the Limit State (LS) of Collapse Prevention (CP) is estimated using the formula A.1 from EN 1998-3 (CEN 2004). The chord rotation for the Limit State (LS) of Life Safety (LS) is assumed as 3/4 of the chord rotation for the Limit State (LS) of Collapse Prevention (CP). The chord rotation at yielding θ_y, is evaluated by the formula (A.10b) and (A.11b) from EN 1998-3 (CEN 2004). Its value defines the achievement of the Immediate Occupancy Limit State. In Eq. (1), θ_{pra}/θ_y is the smallest ratio between the structural components (columns excluded) contributing to progressive collapse resistance and being within the immediately affected bays.

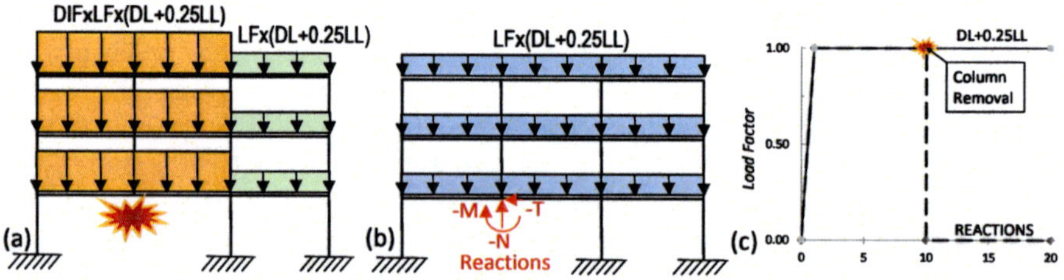

Fig. 6. (a) Loads for nonlinear static analysis; (b) Loads for nonlinear dynamic analysis; (c) Time history function for nonlinear dynamic analysis.

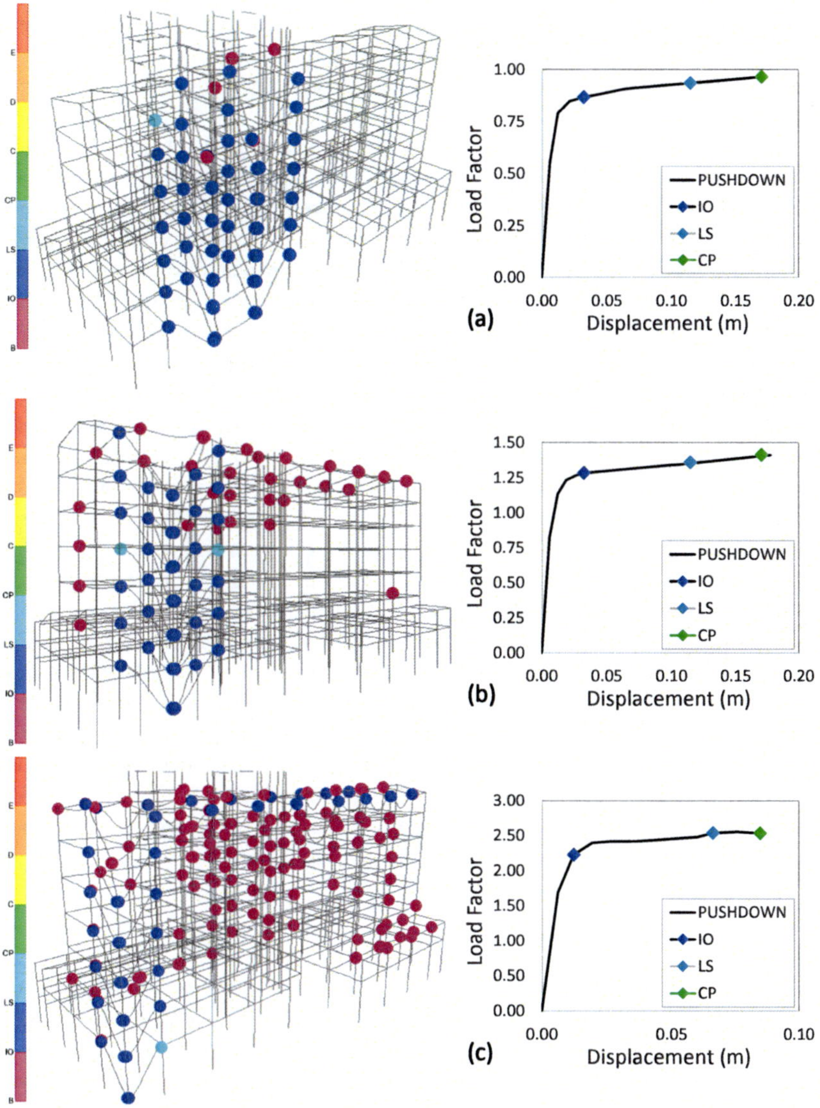

Fig. 7. Plastic hinges at Life Safety Limit State and pushdown curves under different column removal scenarios: (a) Scenario S_1; (b) Scenario S_2; (c) Scenario S_3.

Equation (1) is governed by the mechanical properties of the structural members and acceptance criteria, only. This leads to the same DIF value regardless of the specific gravity loading. In addition, further parameters playing an important role, such as the structural configuration and axial forces in beams (i.e. catenary effect), are not taken into account. Moreover, even though this formulation for DIF is an improvement over the standard load factor approach that assumes a constant DIF, the monotonic decreasing of DIF with ductility is now a matter of debate among researchers.

This paper uses the displacement controlled pushdown analysis to evaluate the response of the structure following the removal of a column. In this scenario, structural members (designed to resist seismic loads in flexure) are subjected to large displacements which may cause them to develop axial forces. In this procedure, gravity loads, amplified within the affected bays, are increased incrementally after the column removal. This increases displacement to an arbitrary level in the location of the removed column, whose maximum value is governed by the concentrate hinge deformation limits and acceptance criteria determined according to the Eurocode 8 (CEN 2004). In Fig. 7 the "pushdown curve" (load-factor versus vertical displacement of the column-removed joint) is plotted for each column removal scenario. In the same figure, the points corresponding to the limit states of Immediate Occupancy (IO), Life Safety (LS) and Collapse Prevention (CP) are plotted according to the Eurocode 8 (CEN 2004) acceptance criteria. As shown in the pushdown figures, in many cases the collapse is reached under a load factor lower than 1 for the S_1 column removal scenario and higher than 1 for the S_2 and S_3 column removal scenario. This means that the building structure has adequate progressive collapse resistance under the single column removal scenarios S_2 and S_3. However, it should be observed that these results are obtained under the hypothesis of perfect adherence of the steel bars within the reinforced concrete of beams. This hypothesis might not be verified since smooth steel bars are used as internal reinforcement. Figure 7 also shows the plastic hinges developed at the Life Safety limit state of the building and their stages represented by different colours. This allows checking if enough residual capacity or alternate load paths take place to prevent the propagation of failure. It is interesting to note that the S_1 scenario essentially involves the beams in the two bays above the removed column. On the contrary, under the S_2 and S_3 scenarios the plastic hinges develop in many of the beams of the higher floors that are characterized by smaller cross sections.

3.2 Nonlinear Dynamic Analysis

The pushdown analysis is an approximated nonlinear static procedure and tries to simulate the progressive collapse of the structure that is a dynamic phenomenon. In order to calibrate and validate the pushdown procedure, a nonlinear response-history analysis has been carried out by analysing the process that dynamically simulates the sudden loss of a column. This procedure encompasses the following steps:

1) Vertical loads are statically applied on the undamaged model, from which axial force N, shear force V and bending moment M in the column to be removed are calculated.
2) The column is replaced by the corresponding reaction forces and then the gravity loads are statically applied together with the calculated end forces acting in the opposite directions (i.e. $-N$, $-V$, $-M$) (Fig. 6b). This application takes 1s (during which loads are increased linearly until their full value) and then kept unchanged for 9s (Fig. 6c), so that the structure can reach a stable condition that reproduces the conditions before the column loss. Under these circumstances, the dynamic effects

caused by the loads applied to the structure so far virtually vanish, so as the state of stress immediately prior to column removal can be considered static.

3) The reaction forces are simultaneously and abruptly nulled. Practically, at 10s, after the frame has reached a static equilibrium, the recorded end forces in original directions (i.e. *N*, *V*, *M*) are applied over a very short time period (in the order of 10 ms, but not higher than 1/10 of the period of the fundamental response mode) (Fig. 6c).

The tangent stiffness-proportional Rayleigh damping is assumed to rule the nonlinear dynamic analysis. The mass-proportional and the stiffness-proportional damping coefficients are calculated to achieve the real critical damping ratio of 5% for both the first and second mode shapes. The direct-integration dynamic procedure including both P-delta and large displacements effects is used in the analyses. Figure 8 shows the vertical displacement time-histories of the column-removed joint. The time-history plot starts at the instant of the ideal column removal. The structural response after the column removal is essentially oscillatory and is primarily controlled by the first vertical mode of vibration. The residual damaged structure oscillates vertically around a position where very large deformations occur. No plastic hinge reach the Life Safety limit state even though as aforementioned this result is obtained under the hypothesis of perfect adherence of the steel bars within the beams.

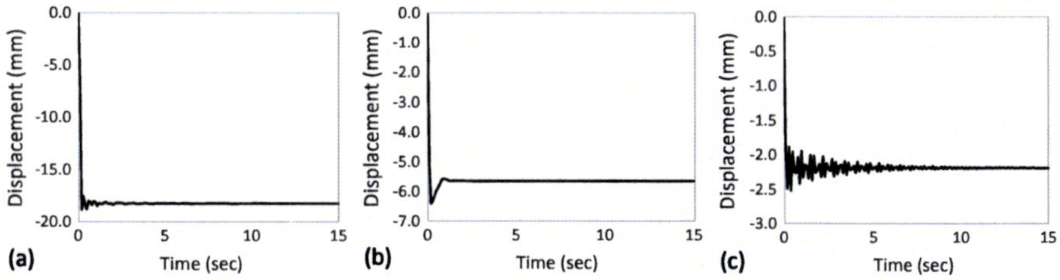

Fig. 8. Displacement time-history from nonlinear dynamic analysis under different column removal scenarios: (a) Scenario S_1; (b) Scenario S_2; (c) Scenario S_3.

4 Progressive Collapse Retrofit with Steel Braces

It should be observed that the progressive collapse resistance was carried out under the hypothesis of perfect adherence between steel bars and concrete. In the case study, smooth steel bars are installed across the transverse joint thus providing a not satisfying mechanical connection. Therefore, the calculated collapse load factor should be considered as an upper limit. The progressive collapse retrofit has been carried out using steel braces provided in some bays (Fig. 9). The braces have been designed following simple equilibrium conditions. The axial load in the column to be removed is distributed between the bare frame structure and the steel braces. In particular, only 20% of the vertical loads is assigned to the frame. Figure 10 shows the plastic hinges developed at the Life Safety state. The results show that the Life Safety limit state is reached in the plastic hinges at the upper storey, that is very far from the column removed spans. This is due to the steel braces that activate alternate load paths. In Fig. 10, also the pushdown curves after retrofit are plotted. The results show that the steel bracing gives a significant increase of progressive collapse resistance for all the column removal scenarios.

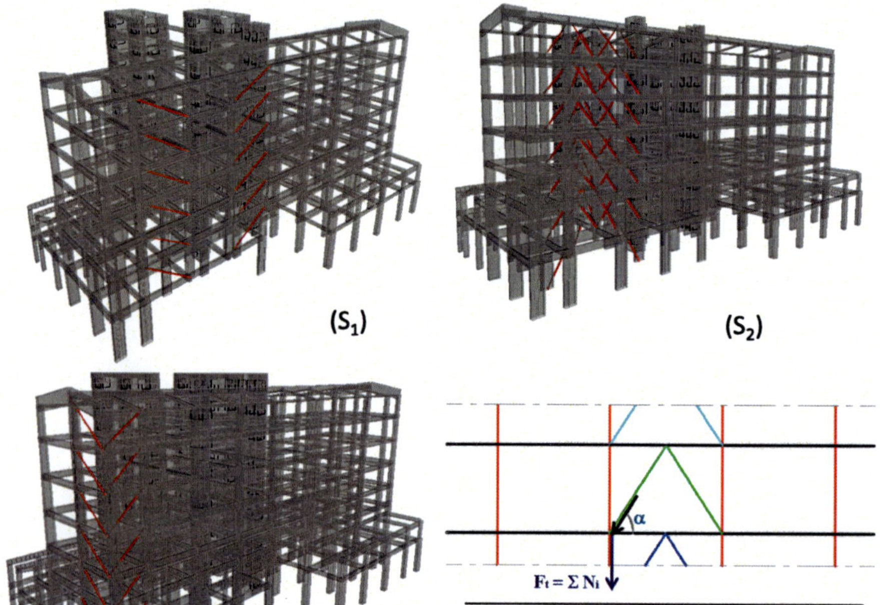

Scen.	S_1	S_2	S_3
Brace	CHS 139,7x10	CHS 114,3x8	CHS 193,7x10

Fig. 9. Building structure retrofitted with steel braces for different column removal scenarios.

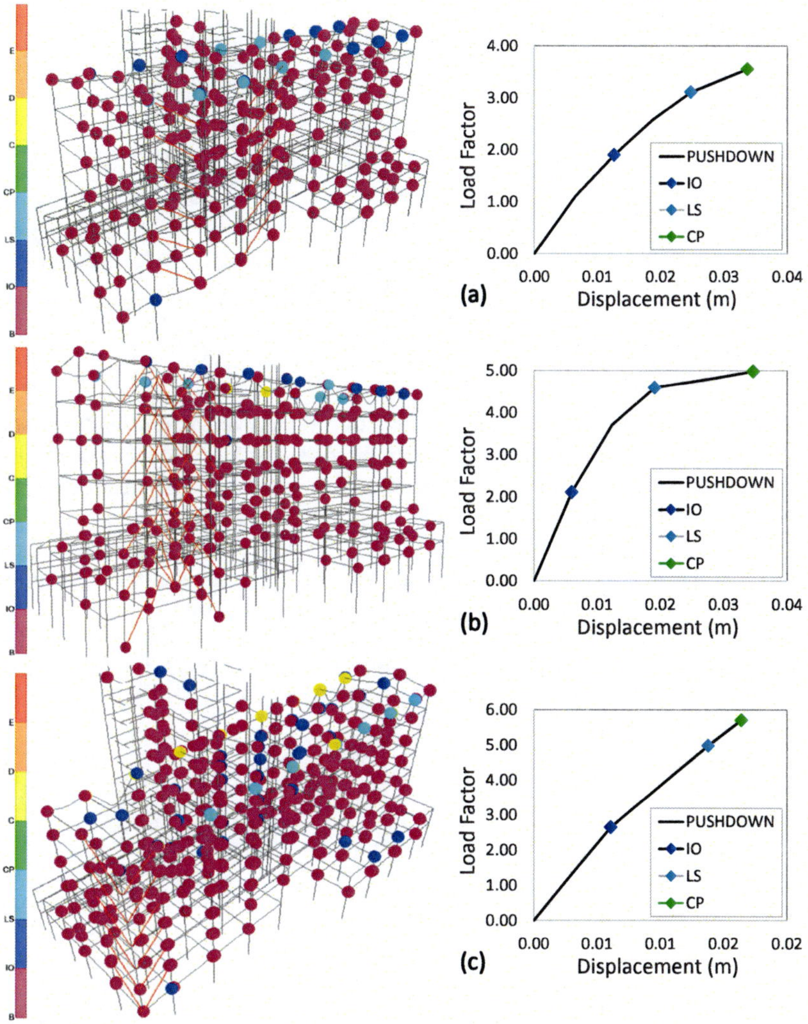

Fig. 10. Plastic hinges at Life Safety Limit State and pushdown curves under different column removal scenarios: (a) Scenario S_1; (b) Scenario S_2; (c) Scenario S_3.

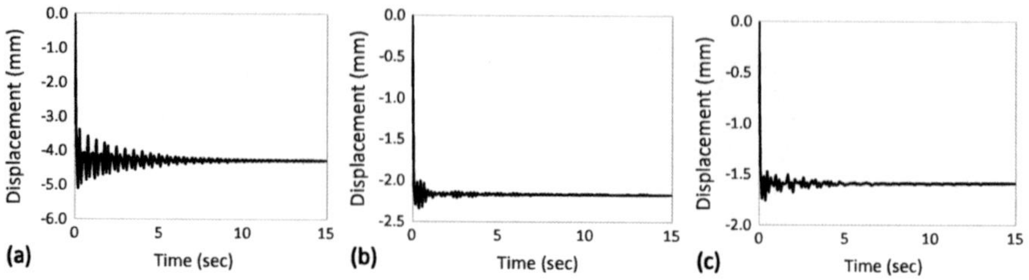

Fig. 11. Displacement time-history from nonlinear dynamic analysis under different column removal scenarios: (a) Scenario S_1; (b) Scenario S_2; (c) Scenario S_3.

5 Conclusions

A retrofit design strategy to mitigate the risk of progressive collapse under abnormal actions was investigated. To this aim, within the framework of the recurrent structural typologies, a particularly representative building was selected and its inelastic behaviour following the sudden column removal was investigated. A design solution based on vertical steel bracing was considered for progressive collapse retrofit. The results obtained showed that the steel bracing might provide an effective and economical solution for strengthening existing RC buildings against progressive collapse.

Acknowledgements. The activity presented in the paper is partly funded by the Italian "Dipartimento della Protezione Civile", through Project RELUIS 2019–2021 "Steel and composite steel-concrete structures".

References

CEN European standard EN1998-1. Eurocode 8: design of structures for earthquake resistance. Part 1: general rules, seismic action and rules for buildings. European Committee for Standardisation, Brussels (2004)

Chen, C.H., Zhu, Y.F., Yao, Y., Huang, Y.: Progressive collapse analysis of steel frames structure based on energy principle. Steel Compos. Struct. **21**(3), 553–571 (2016)

Ellingwood, B.R., Dusenberry, D.O.: Building design for abnormal loads and progressive collapse. Comput-Aided Civ. Inf. **20**(3), 194–205 (2009)

Ferraioli, M.: Inelastic torsional response of an asymmetric-plan hospital building in Italy. COST In: ACTION C26: Urban Habitat Constructions under Catastrophic Events - Proceedings of the Final Conference, pp. 365–370 (2010a)

Ferraioli, M., Abruzzese, D., Miccoli, L., Vari, A., Di Lauro, G.: Structural identification from environmental vibration testing of an asymmetric-plan hospital building in Italy. In: COST ACTION C26: Urban Habitat Constructions under Catastrophic Events - Proceedings of the Final Conference, pp. 981–986 (2010b)

Ferraioli, M., Avossa, A.M.: Progressive collapse of seismic resistant multistory frame buildings. In: 3rd International Symposium on Life-Cycle Civil Engineering. IALCCE 2012, Vienna, Austria (2012)

Ferraioli, M., Avossa, A.M., Mandara, A.: Assessment of progressive collapse capacity of earthquake-resistant steel moment frames using pushdown analysis. Open Constr. Build. Technol. J. **8**, 324–336 (2014)

Ferraioli, M.: Case study of seismic performance assessment of irregular RC buildings: hospital structure of Avezzano (L'Aquila, Italy). Earthq. Eng. Eng. Vib. **14**(1), 141–156 (2015). https://doi.org/10.1007/s11803-015-0012-7

Ferraioli, M.: Dynamic increase factor for pushdown analysis of seismically designed steel moment-resisting frames. Int. J. Steel Struct. **16**(3), 857–875 (2016). https://doi.org/10.1007/s13296-015-0056-6

Ferraioli, M., Lavino, A., Mandara, A., Donciglio, M., Formisano, A.: Seismic and robustness design of steel frame buildings. Key Eng. Mater. **763**, 116–123 (2018a)

Ferraioli, M., Lavino, A., Mandara, A.: Assessment of dynamic increase factors for progressive collapse analysis of steel frames subjected to column failure. Ingegneria Sismica **35**(2), 67–77 (2018b)

Ferraioli, M.: Dynamic increase factor for nonlinear static analysis of RC frame buildings against progressive collapse. Int. J. Civil Eng. **17**(3), 281–303 (2019a)

Ferraioli, M.: Evaluation of dynamic increase factor in progressive collapse analysis of steel frame structures considering catenary action. Steel Compos. Struct. **30**(3), 253–269 (2019b)

Ferraioli, M.: A modal pushdown procedure for progressive collapse analysis of steel frame structures. J. Constr. Steel Res. **156**, 227–241 (2019c)

GSA. Alternate Path Analysis and Design Guidelines for Progressive Collapse Resistance. Washington, DC: U.S. General Service Administration (2013)

Izzuddin, B.A., Vlassis, A.G., Elghazouli, A.Y., Nethercot, D.A.: Progressive collapse of multi-storey buildings due to sudden column loss — part I: simplified assessment framework. Eng. Struct. **30**(5), 1308–1318 (2008)

Keyvani, L., Sasani, M., Mirzaei, Y.: Compressive membrane action in progressive collapse resistance of RC flat plates. Eng. Struct. **59**, 554–564 (2014)

Kim, J., Park, J.: Design of steel moment frames considering progressive collapse. Steel Compos. Struct. **8**(1), 85–98 (2008)

Kim, T., Kim, J., Park, J.: Investigation of progressive collapse-resisting capability of steel moment frames using push-down analysis. J. Perform. Constr. Facilities ASCE **23**(5), 327–335 (2009)

Kim, J., Park, J.H., Lee, T.H.: Sensitivity analysis of steel buildings subjected to column loss. Eng. Struct. **33**, 421–432 (2011)

Italian Ministerial Decree N.1684 November 25th, (1962). (in Italian)

Liu, M.: A new dynamic increase factor for nonlinear static alternate path analysis of building frames against progressive collapse. Eng. Struct. **48**, 666–673 (2013)

Marchand, K.A., Alfawakhiri, F.: Blast and progressive collapse. In: Facts for Steel Buildings 2, USA: AISC (2004)

Marchand, K., McKay, A., Stevens, D.J.: Development and application of linear and nonlinear static approaches in UFC 4-023-03. In: Proceedings of Structures Congress 2009, Austin, Texas, 30 April–2 May (2009)

Mashhadi, J.J.: Effects of damping ratio on dynamic increase factor in progressive collapse. Steel Compos. Struct. **22**(3), 677–690 (2016)

Mashhadi, J.J.: Effects of postelastic stiffness ratio on dynamic increase factor in progressive collapse. J. Perform. Constr. Facilities ASCE **31**(6), 04017107 (2017)

Mashhadi, J.J., Saffari, H.: Dynamic increase factor based on residual strength to assess progressive collapse. Steel Compos. Struct. **25**(5), 617–624 (2017)

McKay, A., Marchand, K., Diaz, M.: Alternate path method in progressive collapse analysis: variation of dynamic and nonlinear load increase factors. Pract. Period. Struct. Des. Constr. ASCE **17**(4), 152–160 (2012)

Mirtaheri, M., Zoghi, M.: Design guides to resist progressive collapse for steel structures. Steel Compos. Struct. **20**(2), 357–378 (2016)

NTC-2018 Guidelines. Technical Standards for Constructions. Official Journal of the Italian Republic, 20.02.2018, Rome, Italy (2018). (in Italian)

Ruth, P., Marchand, K.A., Williamson, E.B.: Static equivalency in progressive collapse alternative path analysis reducing conservatism while retaining structural integrity. J. Perform. Constr. Facilities ASCE **20**(4), 349–364 (2006)

SAP2000: Linear and nonlinear static and dynamic analysis of three-dimensional structures. Research Ultimate Version 21.0, Analysis Ref. Manual, Computer and Structures, Berkeley, CA (2019)

Sasani, M., Sagiroglu, S.: Progressive collapse of reinforced concrete structures: a multihazard perspective. ACI Struct. J. **105**(1), 96–103 (2008)

Stevens, D.J., Crowder, B., Hall, B., Marchand, K.: Unified progressive collapse design requirements for DoD and GSA. ASCE Struct. Congress 1–10 (2008)

UFC: Design of Buildings to Resist Progressive Collapse. Department of Defense. Unified Facilities Criteria. UFC 4-023-03 Including Change 2 – 1 June 2013, Washington, DC, USA (2013)

Xu, G., Ellingwood, B.R.: An energy-based partial pushdown analysis procedure for assessment of disproportionate collapse potential. J. Constr. Steel Res. **67**, 547–555 (2011)

Investigation of the Pathology Causes of Xana, Greece by the Use of Nonlinear Finite Element Analyses

Vasiliki G. Terzi(✉)

Department of Civil Engineering, Aristotle University of Thessaloniki, 54124 Thessaloniki, Greece
terziv@civil.auth.gr

Abstract. The monument of Xana, a typical caravanserai, is located in the archeological site of the municipality of Traianoupolis, Evros, Greece and is one of the earliest khans in the Balkan area. Nowadays, the roof that once covered the monument and almost the entire first part that was used for travelers do not exist. The monument suffers by a vertical fracture on the northern wall, tensile cracks which run along the interior surface of the dome and vertical tensile cracks located at the openings of the western wall. Furthermore, hair cracks, loss of stone material, exfoliation and efflorescence worsen the pathology image of Xana.

The target of the present study is the investigation of the monument's pathology by the use of nonlinear finite element analyses. A three-dimensional finite element model of the Xana's initial state is constructed. Soil-structure interaction effects are taken into account by the use of Winkler foundation. Various possible loading scenarios are investigated: (i) absence of wooden tendons; (ii) destruction of heavy roof; (iii) earthquake action; (iv) foundation settlement. The results of the nonlinear finite element analyses in terms of displacements and crack patterns can explain some of the main causes of the monument's pathology.

Keywords: Finite element method · Nonlinear masonry · Pathology · Earthquake · Settlement

1 Introduction

Monuments are an important part of the cultural heritage and the history of a country. Not only the past is alive by their preservation but also the future development of a country is enhanced by cultural heritage tourism (Mortezaei and Mortezaei 2012). Therefore, the study of their static and dynamic behavior is vital. However, the aforementioned task is very demanding and difficult to achieve for the following reasons: geometry and morphology complexities; variation of materials and their mechanical properties; complex nature of masonry materials; existing pathology; missing information of usage, loading, restoration history; limited financial and computational resources (Betti and Vingoli 2011; Sanchez-Aparicio et al. 2014; Lourenco 2002). The creation of a sophisticated

I. Vayas and F. M. Mazzolani (Eds.): PROHITECH 2021, LNCE 209, pp. 369–388, 2022.
https://doi.org/10.1007/978-3-030-90788-4_31

numerical model is necessary in order to investigate the pathology causes and to propose the most appropriate restoration solution (Lourenco 2005).

The target of the present study is the investigation of the monument's pathology by the use of nonlinear finite element analyses. A three-dimensional finite element model of the Xana's initial state is constructed. The masonry is simulated by 3D solid elements while the wooden beams and tendons are simulated by beam elements. Two different cases are considered for the foundation's boundary conditions: (i) fixed base; (ii) flexible base by taking into account soil-structure interaction effects. The structural response under dead and live loads is studied by staged construction. Therefore, by implementing the birth option for finite elements the finite element model of the initial state is concluded. Cracks on the key stone area of the arches and at the openings of the northern wall are entirely attributed to the large self-weight. Since the response of the monument under normal operating conditions is established, various possible scenarios may be further investigated: (i) deterioration of the wooden beams and tendons (implementation of death option for beam elements); (ii) destruction of the wooden span roof (cancelation of the corresponding loading function); (iii) soil subsidence at the northern wall due to the combination of clay soil foundation and high water level (selection of a proper displacement distribution pattern); (iv) earthquake action (implementation of pseudo-dynamic analysis by lateral loading). The results of the nonlinear finite element analyses in terms of displacements and crack patterns can explain some of the causes of the monument's pathology.

2 Case Study, Xana: Structure and Pathology

The monument (Fig. 1a) is located 13 km far from the city of Alexandroupolis, prefecture of Evros in Greece. Xana was constructed by Ghazi Evrenos Bey, circa at 1375–1385 in order to meet the needs of an important geographical point situated close to via Egnatia, the thermal springs of Loutra village and the Roman city of Traianoupolis (Kiel 1990). The specific area was characterized in 1964 by the Hellenic Ministry of Culture and Tourism as an archeological site.

The plan of monument is rectangular with dimensions 39.62 m × 13.92 m (Kalafatzis 2008). The circumferential walls rise to 4.97 m; consist of a three-leaf masonry and have a thickness of 1.4 m. The exterior leaves correspond to cloissone masonry while the interior ones to rumble stones and mortar. A hemicylindrical dome rises to 7.26 m; is constructed by bricks and mortar; was initially protected by a wooden roof with red tiles and covers the monument. The interior of the structure is divided by a thick inner wall (1.40 m) into two parts (Fig. 1b). As is typical at caravanserais, the first part was used for travelers lodging and the second part was used as a stable for the animals. Six fire places, F1 to F6 were constructed at the first part for cooking and heating. Two stones arches (A1, A2) divide almost equally the second part. A third stone arch (A3) is in contact with the northern wall while a fourth one (A4) is in contact with the inner wall. At the height of 4.30 m, four wooden tendons (W1-W4) were constructed in order to tie down the span roof while two others (W5, W6) were constructed at the height of 2.60 m at the two stone arches (A1, A2). At the height of 2.60 m, wooden beams were placed all along the length of the circumferential walls, taking into account the two square holes (0.15 m × 0.15 m) that are visible. Physical lighting was achieved by three windows on the western side and another three at the northern side.

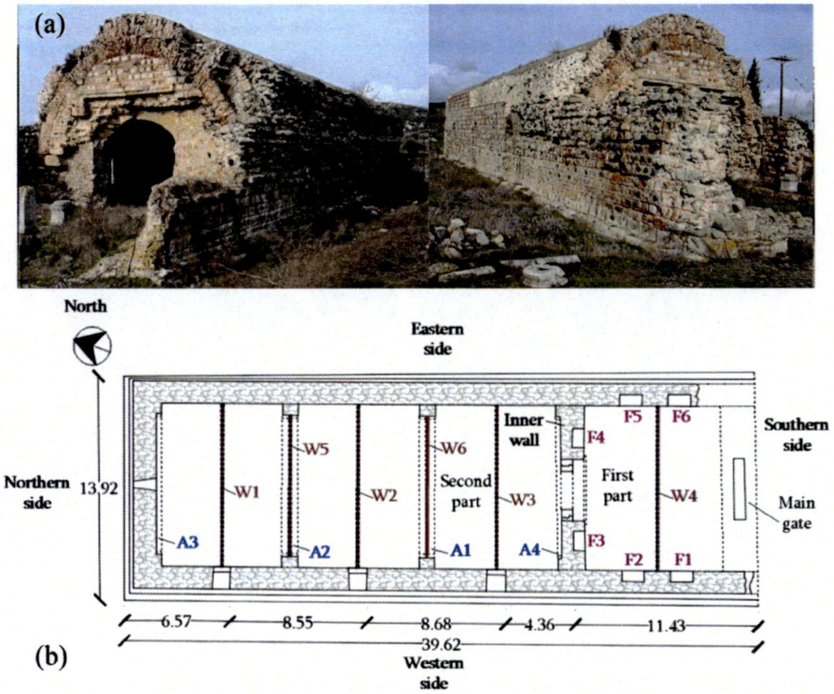

Fig. 1. Xana: (a) present state: east and west depiction (b) plan depiction

The first part of the monument is completely destroyed. The dome collapsed in the early 1900s due to its heavy weight (Kiel 1990). In 1939, during the construction of a military road nearby, the main gate and the highest parts of the walls of the monument collapsed (Minopoulos 2001). All the wooden elements, which used to link the lateral walls or were place in the masonry body are absent. The monument suffers by extreme tensile cracks in two positions: (i) on the northern wall there is a vertical fracture, which can possibly be attributed to soil subsidence taking into account the clayey soil foundation in combination with high water level (nearby thermal baths); (ii) along the key line of the barrel vault and in the vertical direction, at locations where the windows are constructed. Furthermore, the pathology picture fill out: hair cracks; loss of stone material; mortar pulverization; efflorescence and blistering; exfoliation that can be attributed to biological colonization; drainage problems; unsuitable repair materials; soil moisture and the action of marine aerosols taking into account that the monument is situated 6 km far from the Aegean sea (Fig. 2).

Fig. 2. Pathology images: (a) loss of material, location of fireplaces at the first part; (b) absence of wooden tendons; (c) vegetation; (d) biological colonization.

Restoration works that aimed mainly at the healing of cracks and protection from rain took place in 1968 and 1969 (Varvitsas 1968; Pentazos, 1969). Furthermore, in 1998 one part of the western masonry was reconstructed (Terzi 2010). Nowadays, the entrance to the monument is prohibited.

3 Finite Element Modeling

The process of creating a finite element model of an existing structure consists of the following stages: geometrical imprint; creation of numerical model's geometry in 3D; selection of appropriate finite element types; numerical model's meshing; laboratory or in situ testing for the identification of material properties; selection and implementation of appropriate material models; application of loading functions; application of boundary conditions; solution; interpretation of the results (Ramirez et al. 2012). The finite element model must represent the overall spatial configuration of mass; self-weight; stiffness; architectural elements; openings or wall thickness (Betti and Vignoli 2011; Fratini et al. 2011). Therefore, the initial geometry of the monument, architectural image and structural configuration were the outcome of historical survey; bibliographical references; old time photographs; existing drawing; and direct in situ surveys (Terzi 2010) (Fig. 3).

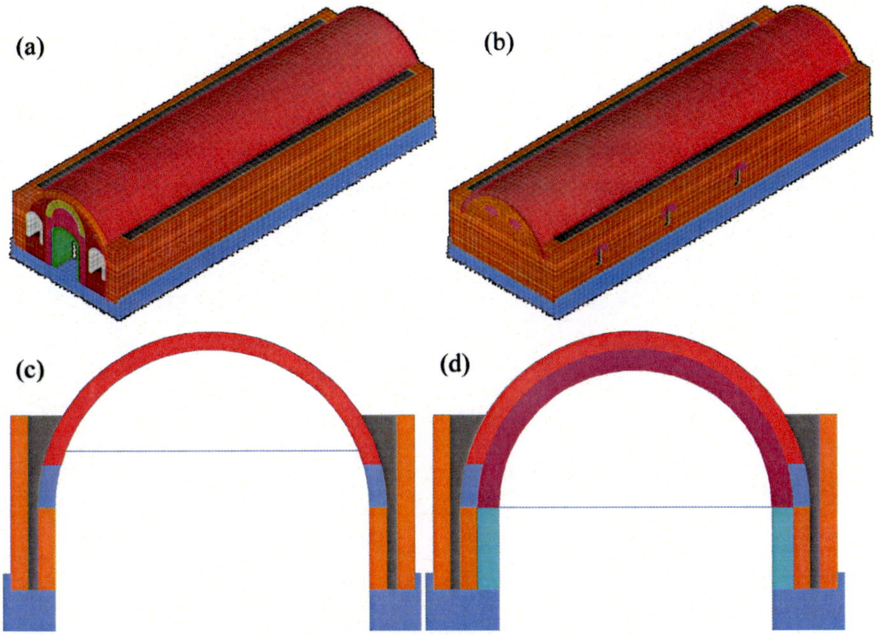

Fig. 3. Finite element model: (a) eastern elevation; (b) western elevation; (c) wooden tendon at + 4.30 m; (d) wooden tendon at +2.60 m

Nonlinear behavior is the most important characteristic of masonry structures. Therefore, the gathering of experimental data for the estimation of masonry mechanical characteristics is considered to be mandatory (Sanchez-Aparicio et al. 2014). However, due lack of financial and technical resources, only six brick and four stone samples were taken from debris and tested in the laboratory. The results refer to density and compressive strength (Table 1). The significant diversity of the compressive strength of bricks can be attributed to the lack of production standards in the old days (Eslami et al. 2012).

The nonlinear mechanical characteristics of all masonry types were determined by the application of method of homogenization in combination (Psycharis et al. 2010) with the EC6 norm (EC6, 1996). Homogenized properties are used by various researchers in the study of masonry structures (Luis et al. 2018; Di Nino and Luongo 2019; Peng et al. 2018; Malena et al. 2019; Kujawa et al. 2020).

In order to describe the nonlinear behavior of masonry in finite element code, the "concrete model" was used (Acito et al. 2014). Therefore, the characteristic compressive strength is larger than the characteristic compressive strength ($f_{tk} = f_{ck}/20$).

Wooden tendons were considered to be elastic and their Young's modulus was estimated at 10 GPa. Their cross-section was estimated according to the dimensions of the holes detected in the masonry body.

Table 1. Mechanical characteristics of brick and stone samples

Samples	Bricks		Stones	
	Density [kgr/m^3]	Compressive strength [MPa]	Density [kgr/m^3]	Compressive strength [MPa]
1	1244.963	8.706	1419.316	17.248
2	1406.211	13.011	2679.477	66.095
3	1607.322	8.690	2609.037	50.726
4	1632.837	24.140	1461.326	27.812
5	1352.545	15.836		
6	1516.894	19.815		

Table 2. Characteristic strength values for the description of "concrete model"

Masonry type	f_{ck} [MPa]	f_{tk} [MPa]	$E_{tangent}$ [MPa]
Circumferential masonry	6.15	0.31	5531
Arches stone masonry	13.06	0.65	11756
Foundation's stone masonry	6.22	0.31	5598
Brick masonry	5.02	0.25	4521

In order to estimate soil-structure interaction effects. Two different cases of boundary conditions were taken into account concerning the foundation of the monument: fixed; and flexible base. According to the first assumption all the finite element nodes that belong to the lowest level of the foundation have zero displacements, in terms of translations and rotations. Furthermore, according to the second assumption, all the aforementioned nodes have the ability to translate only at the vertical direction. Therefore, the Winkler model was applied (Winkler 1867). The modulus of subgrade reaction was estimated to 70 MN/m^3, which corresponds to neither dry nor saturated clayey soils. Finally, the vertical stiffness corresponding to each node was based on its tributary areas. In order to ensure the proper function between beam and 3D solid elements, the rotational degrees of freedom of beam elements are prohibited at nodes common with solid elements.

Three basic construction stages are taken into account: foundation level; level of 2.60 m; and barrel vault. The construction sequence is achieved by the birth element option included in the finite element code. During each construction stage, the self-weight of the structure is activated gradually. The self-weight is applied by mass-proportional loading at the gravity direction, therefore each construction stage is considered as a separate loading function (Ayensa et al. 2015). The heavy roof is not simulated by the use of finite elements, but its load is taken into account as a separate loading function which is activated after the third construction stage. Foundation settlement is taken into account by imposing a displacement distribution pattern at the vertical direction. Finally, earthquake action is applied as a mass-proportional loading at the horizontal direction under the prism of a push over analysis. All finite elements are submitted to linearly increasing horizontal acceleration. Two different loading combinations are determined: 1.35G + 1.5Q; and G + 0.3Q, where G is the dead and Q is the live load. The first combination represents the static loading due to self-weight of the structure and the live loading of the gabled roof. The second combination is used during the pseudo-analysis and corresponds to reduced live loads taking into account the probabilistic assumption that during a seismic event live loads are not fully present (Mallardo et al. 2008). Each loading case is applied incrementally to the structure due to the non-linearity of the masonry material (Bernardeschi et al. 2004). The gradual application of loadings is essential especially in cases of near collapse behavior (Lucchesi et al. 1995).

The finite element model includes 17,263 3D solid elements; 384 Hermitian beam elements; and 1,072 linear spring elements. The solid elements are applied for the simulation of the masonry; the beam elements for the wooden tendons and the spring elements for the soil flexibility.

4 Results

4.1 1st Loading Scenario: Gravity Loads, Function of Tendons

The first loading scenario refers to the structural behavior of the monument under the action of gravity loads. Figure 4 depicts the three construction stages of the masonry body plus the fourth which corresponds to the roof loading. The loading combination is determined by 1.35G + 1.5Q. Two different cases are studied concerning the foundation base: fixed; and flexible.

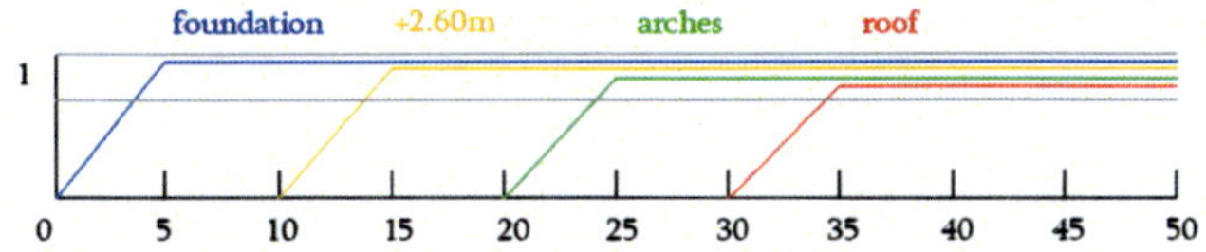

Fig. 4. 1st loading scenario: gravity loads

Figure 5 depicts the deformed shape and the maximum displacements of the monument for the two different boundary condition assumptions. The points, where the maximum values are located, are the same for the two models, but their values are different. In particular, the fixed base model presents smaller displacements than the flexible base model. The most characteristic displacement is at the vertical direction and it is attributed to the very heavy roof loading. Along the longitudinal direction, X, the dome follows the vertical displacement while along the direction Y, the out-of-plane deformation of the longitudinal walls is characteristic.

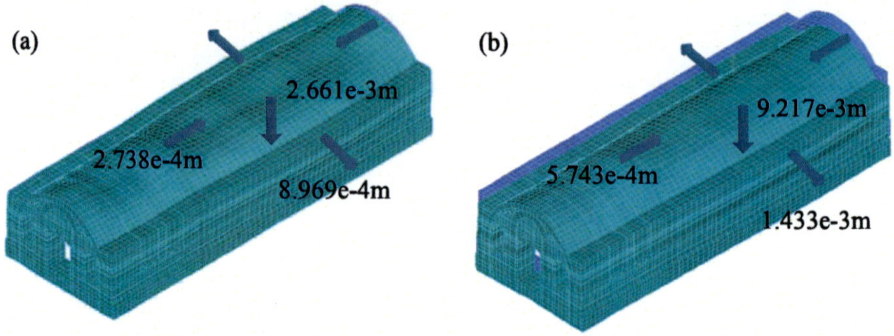

Fig. 5. Deformed shape due to 1st loading scenario: (a) fixed base; (b) flexible base

Figure 6a and 6b depicts the crack pattern for each boundary condition assumption. No cracks appear at the upper surface of the barrel vault. For both cases cracks at the key stones of arches A1 and A2 are characteristic and attributed solely to the large roof loading. For the case of the flexible base, additional tensile cracks are observed at the highest level of the foundation and at the windows of the northern and southern wall. Both crack patterns are expected in the case of a masonry structure due its large self-weight. However, taking soil-structure interaction effects, even under the prism of Winkler foundation, leads to results closer to the real structural behavior.

For the fixed base condition only open cracks appear.

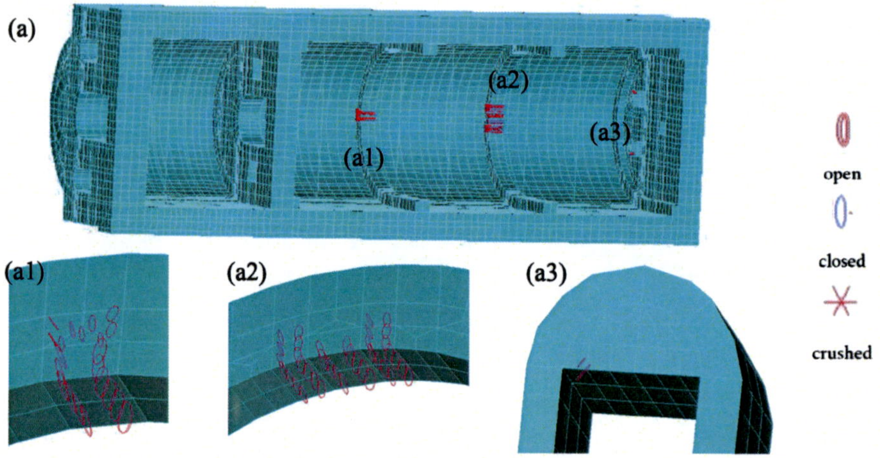

Fig. 6a. Crack pattern due to 1st loading scenario for the fixed base condition.

For the flexible base condition all the cracks are open except for the case of A2, where closed cracks are also visible (light blue circles).

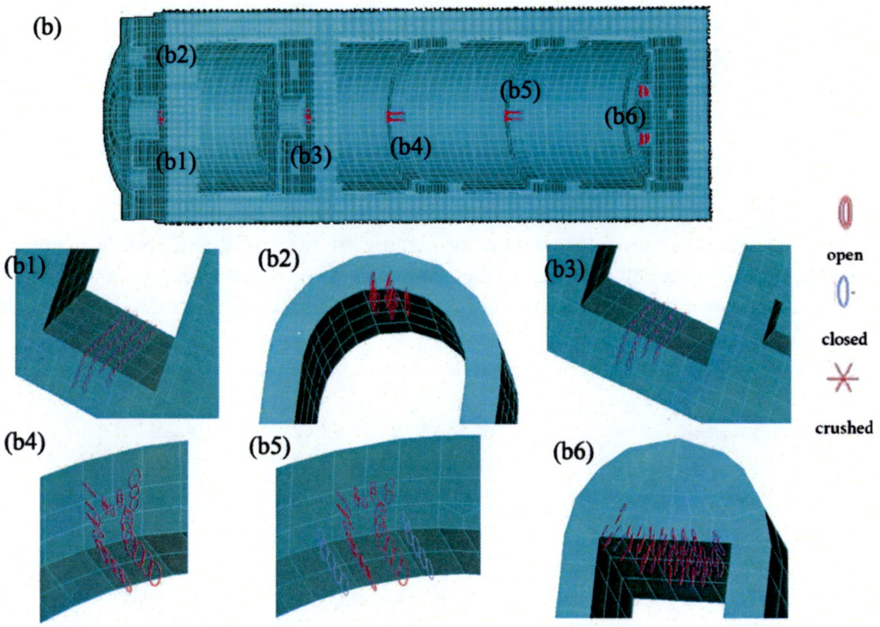

Fig. 6b. Crack pattern due to 1st loading scenario for flexible base condition

Table 3 presents the internal forces of the six wooden tendons. R denotes the axial force while S,T denote the lateral forces.

Table 3. Wooden tendons forces: R = axial, S,T = lateral.

Tendon	R [kN]		S [kN]		T [kN]	
	Fixed	Flexible	Fixed	Flexible	Fixed	Flexible
W1	24.531	39.783	1.981	1.981	1.316	2.058
W2	38.644	66.296	2.368	2.368	0.228	0.228
W3	23.467	36.569	2.366	2.366	0.248	0.284
W4	11.398	14.437	2.373	2.373	0.156	0.156
W5	27.362	55.445	3.270	3.370	0.020	0.033
W6	27.222	53.629	3.217	3.302	0.019	0.024

The axial forces are much larger than the lateral forces. Under the action of gravity loading, the wooden tendons prevent the out-of-plane deformation of the longitudinal walls. Furthermore, the flexible base assumption intensifies the aforementioned deformation and the basic outcome is the increase of the tendons axial forces, especially in the second part of the monument. The soil-structure interaction effect does not alter significantly the lateral internal forces of the wooden tendons.

The comparison of the numerical analyses results between the fixed and flexible base reveals that soil-structure interaction effects cannot be ignored. Therefore, all the preceding loading scenarios are under the assumption of flexible soil.

4.2 2nd Loading Scenario: Gravity Loads, Absence of Tendons

The second loading scenario refers to gravity loading, but under the absence of the wooden tendons. Taking into account the ageing of materials, the present scenario assumes that the wooden tendons do not function. The aforementioned is simulated by the "death" option of the Hermitian beam elements that correspond to the wooden tendons (Fig. 7).

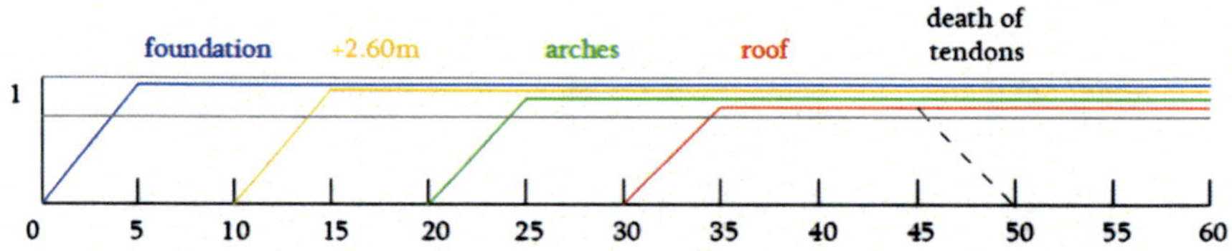

Fig. 7. 2nd loading scenario: gravity loads, absence of wooden tendons

Figure 8 depicts the deformed shape of the monument under the absence of the wooden tendons. The displacements in all directions increase in comparison to the operational state of the wooden tendons. In particular, the increase at the vertical direction is equal to 10%, in the longitudinal direction is 40% while in the transverse direction is 55%. Therefore, the out-of-plane deformation of the two longitudinal walls is more pronounced than the other two directions.

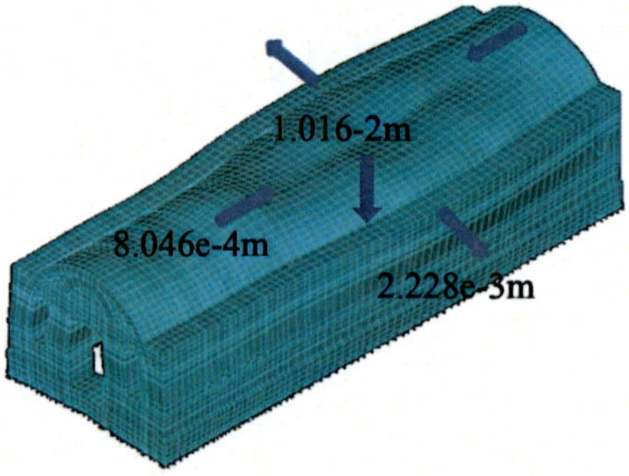

Fig. 8. Deformed shape due to 2nd loading scenario

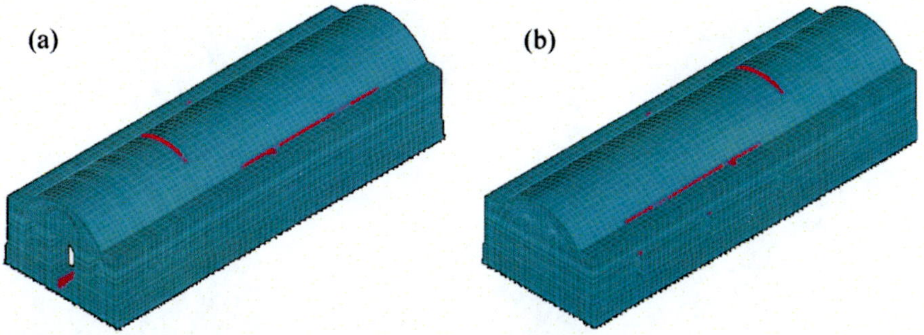

Fig. 9. Crack pattern due to 2nd loading scenario, external view: (a) eastern elevation; (b) western elevation

Figure 9 depicts the crack pattern due to the absence of wooden tendons. Tensile cracks appear in a linear distribution pattern along the interfaces between the barrel vault and the longitudinal walls and are attributed to the out-of-plane bending of the walls. This failure figure must be examined in comparison with Fig. 1, which depicts the present state of the monument. Therefore, the present pathology, and in particular the loss of material of the external masonry leaf is attributed to the out-of-plane bending of the eastern and western walls. Furthermore, tensile cracks also develop on the exterior of the dome in the transition zone between the second part and the thick masonry between the first and second part of the monument (Terzi and Ignatakis, 2018).

Figure 10 depicts the crack pattern at the key stone of the dome. The absence of the wooden tendons is the most important reason for the linear distribution crack pattern that develops at the key stone area. Since the two longitudinal walls move away from each other, the vertical displacements of the dome increase due to the heavy roof loading. The finite element results under the assumption that the wooden tendons do not function, capture the described phenomenon adequately.

Part of the aforementioned linear crack pattern is depicted in Fig. 11 and refers to the key stone of arches A1 and A2 that divide the second part of the monument almost equally. In particular, in Fig. 11(a) cracks between the stones can be detected while in Fig. 11(b) a previous restoration attempt by the use of new stone masonry is evident.

Furthermore, tensile cracks appear at the highest level of the windows of the northern wall. In real state, not only the absence of wooden beams can be visually confirmed (Fig. 12(a)) but also the reconstruction of the window's arch is also obvious as a restoration attempt (Fig. 12(b)).

Therefore, the combination of soil-structure interaction effects; important self-weight loading and the absence of wooden tendons can explain some basic and significant elements of the monument's pathology.

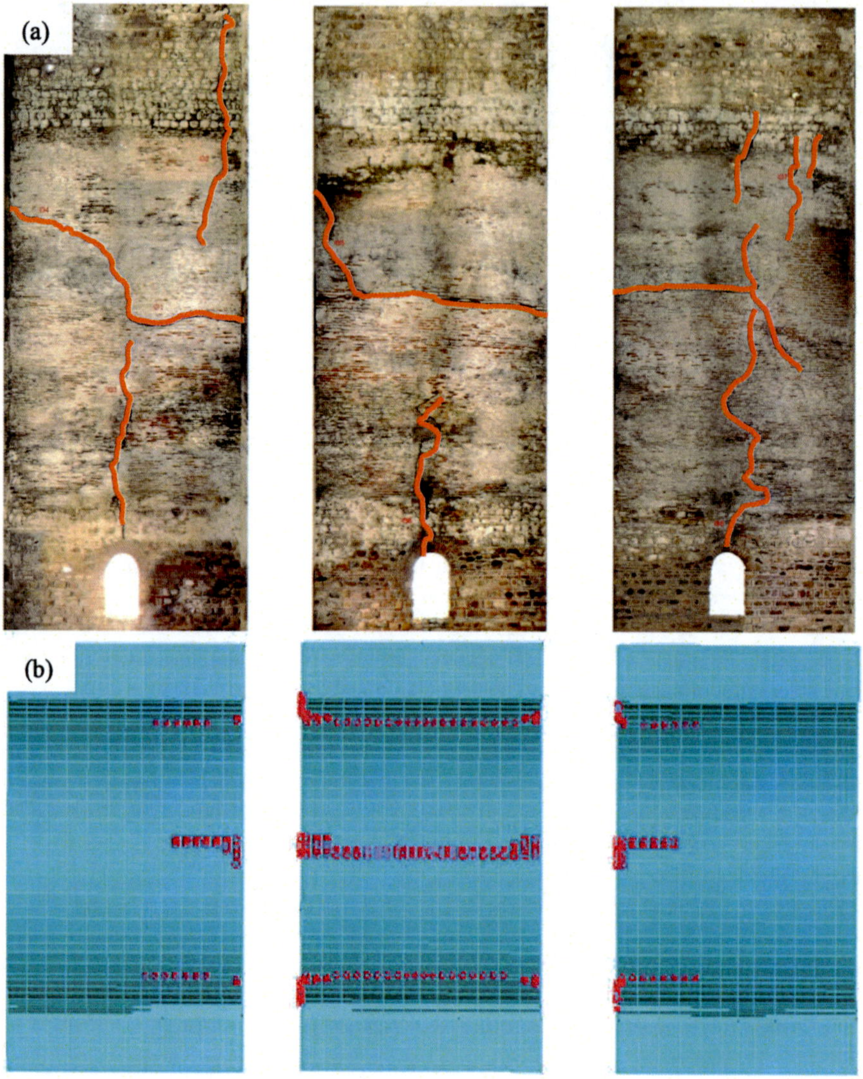

Fig. 10. Crack pattern due to 2nd loading scenario, view of the key stone of arches A1, A2: (a) nowadays condition (b) finite element analyses results.

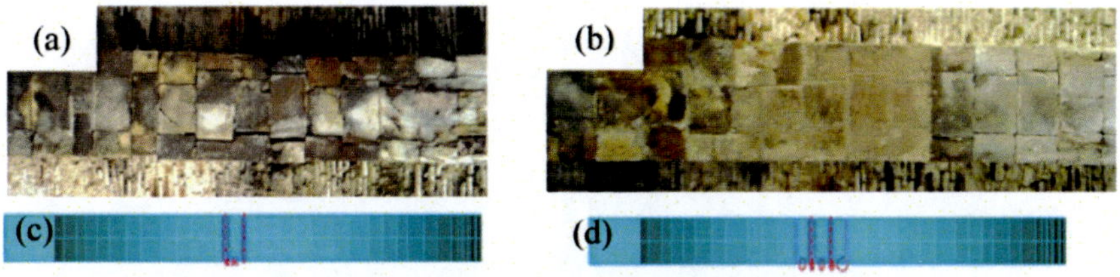

Fig. 11. Crack pattern due to 2nd loading scenario, view of the key stone arch at the second part: (a, b) nowadays condition (c, d) finite element analyses results.

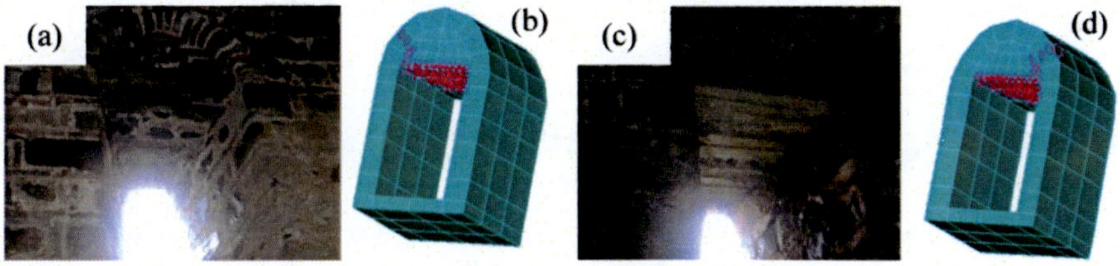

Fig. 12. Crack pattern due to 2nd loading scenario, windows at the northern wall: (a, c) nowadays condition (b, d) finite element analyses results.

4.3 3rd Loading Scenario: Gravity Loads, Absence of Heavy Roof

The third loading scenario refers to the action of gravity loads, the function of wooden tendons and the cancellation of the roof loading. In particular, the time function that describes the roof loading, at a certain time step starts to decrease and finally, receives zero values (Fig. 13). The above assumption is based on the hypothesis that is partially or totally destroyed.

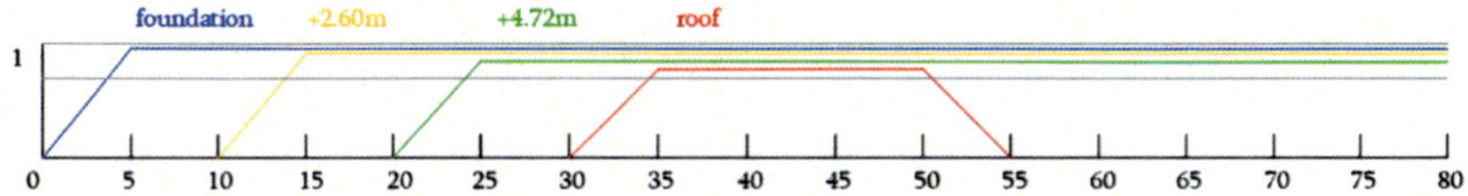

Fig. 13. 3rd loading scenario: gravity loads, death of heavy roof

Figure 14 depicts the deformed shape of the monument after the "death" of the heavy roof. The displacements along vertical, longitudinal and transverse axes are reduced by 11%, 17% and 21%, respectively in comparison to the 1st loading scenario. The crack pattern does not change at all. Therefore, the cancellation of the heavy roof loading acts as a relief measure of the structural behavior.

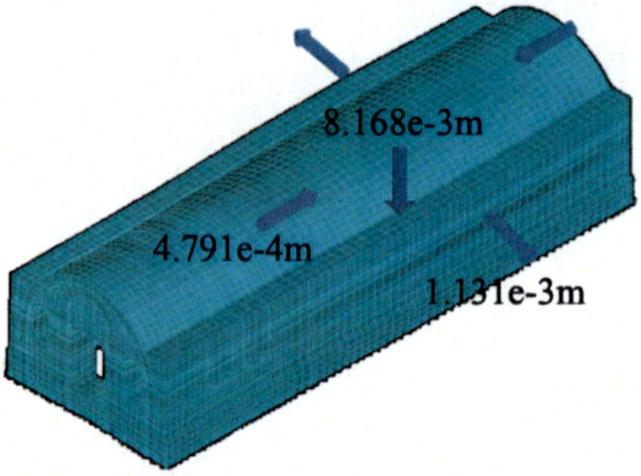

Fig. 14. Deformed shape due to 3rd loading scenario

4.4 4th loading Scenario: Gravity Loads, Absence of Tendons, Absence of Heavy Roof

The fourth loading scenario corresponds to the action of the self-weight under the absence of the wooden tendons and the heavy roof (Fig. 15).

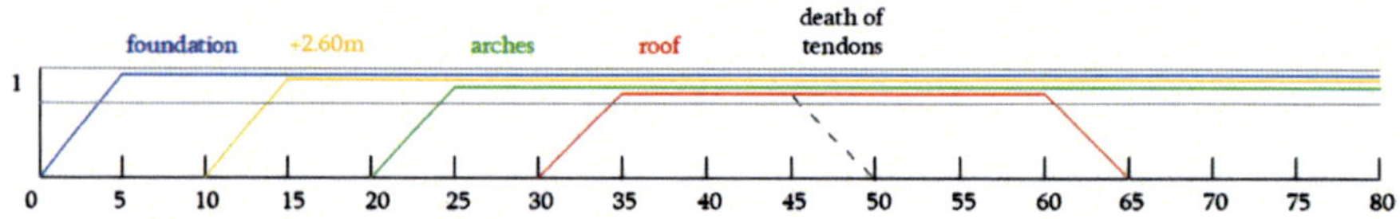

Fig. 15. 4th loading scenario: gravity loads, absence of wooden tendons, death of heavy roof

The deformation of the monument follows the same shape as in the 2nd loading scenario. However, the displacement at the vertical, longitudinal and transverse axes are reduced by 13%, 19% and 22%, respectively. Furthermore, the crack pattern remains the same as in 2nd loading scenario. Therefore, the cancellation of the heavy roof mitigates the out-of-plane bending of the eastern and western masonry walls (Fig. 16).

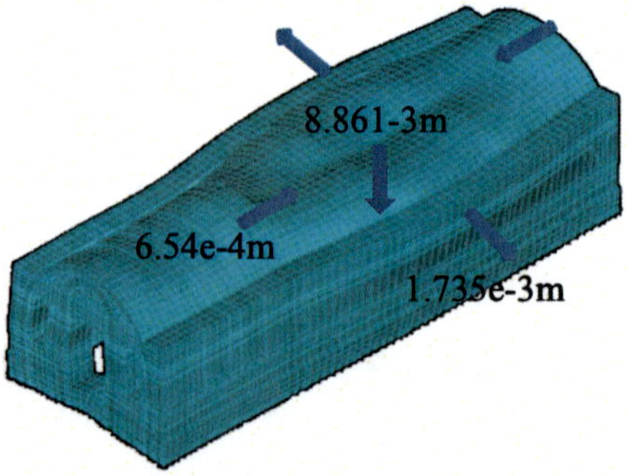

Fig. 16. Deformed shape due to 4th loading scenario

4.5 5th Loading Scenario: Earthquake Loading

Figure 17 depicts the fifth loading scenario, which consists of the three construction stages, the application of roof loading and finally the application of earthquake action. In particular, the whole finite element model is imposed in a mass-proportional loading at the transverse axis, which increases linearly. Once the model becomes unstable the analysis stops. Therefore, the particular earthquake loading follows the push over analysis pattern.

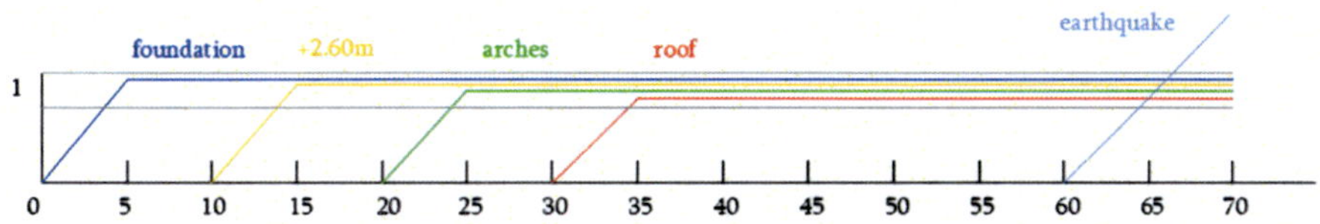

Fig. 17. 5th loading scenario: earthquake loads, function of wooden tendons

The pseudo-dynamic loading is directed from the western to the eastern wall. Therefore, tensile cracks accumulate on the eastern side of the monument (Fig. 18). Once the horizontal acceleration becomes equal to 0.235g, the analysis stops. Although, no real acceleration records are used, even this type of loading achieves to reveal the parts of the monument that are susceptible to a possible earthquake action or even the possible damage patterns. In particular, the earthquake action amplifies the out-of-plane bending of masonry walls which the main cause of pathology. Furthermore, the stone arches (A1-A4) receive an important percentage of the horizontal load and therefore, the cracks concentrate in areas close these elements.

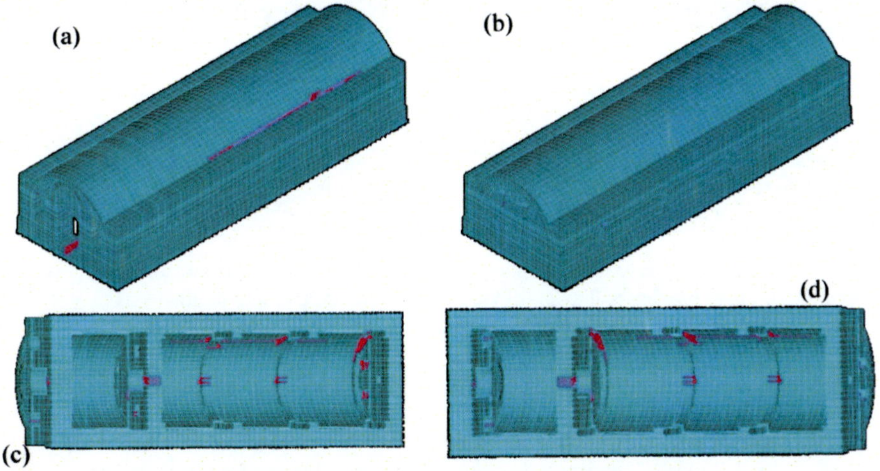

Fig. 18. Crack pattern due to the 5th loading scenario: (a) eastern wall; (b) western wall; (c) view of dome at negative longitudinal axis; (d) view of dome at positive longitudinal axis

Therefore, some of the present cracks at the aforementioned areas can be attributed to earthquake action, especially if the alternating direction movement is taken into account.

4.6 6th Loading Scenario: Earthquake Loading, Absence of Wooden Tendons

The sixth loading scenario refers to the pseudo-dynamic analysis but under the absence of wooden tendons (Fig. 19).

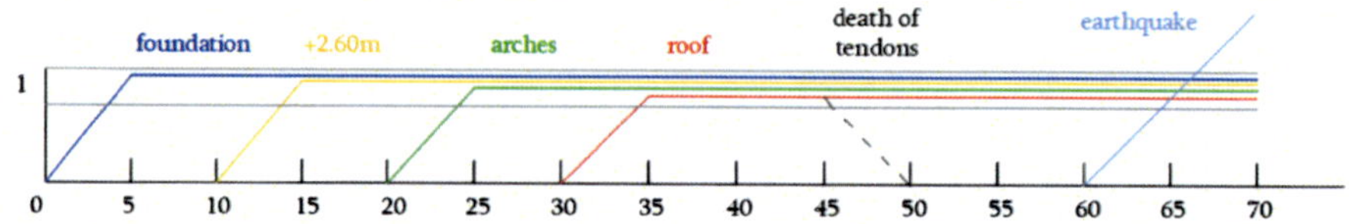

Fig. 19. 6th loading scenario: earthquake loads, absence of wooden tendons.

The crack pattern retains the pathology due to gravity loading and additional cracks appear on the eastern walls due to the earthquake action (Fig. 20).

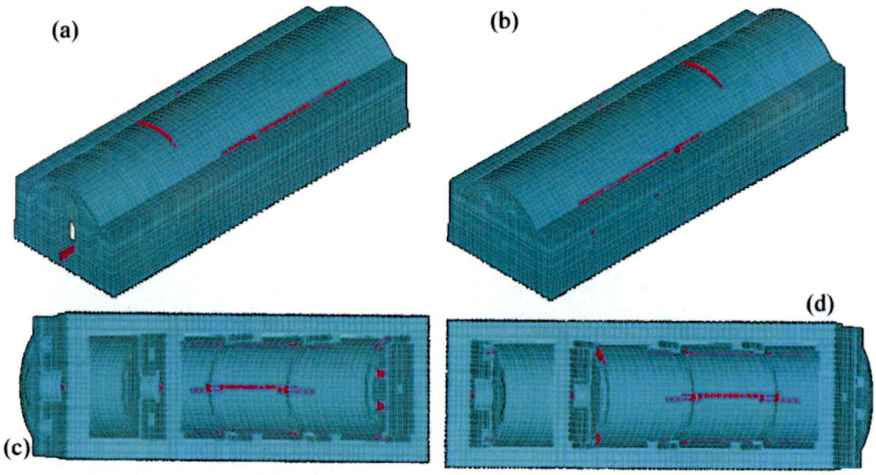

Fig. 20. Crack pattern due to the 6th loading scenario: (a) eastern wall; (b) western wall; (c) view of dome at negative longitudinal axis; (d) view of dome at positive longitudinal axis

The analysis stops when the horizontal acceleration becomes equal to 0.17g, which is less than in the case of the presence of wooden tendons in the system. Therefore, the restraining action of the tendons on the out-of-plane bending of masonry walls is important to all types of loadings. In conclusion, once the tendons are destroyed or do not function properly, the structure is susceptible to dynamic loadings acting on the transverse axis. Furthermore, the stone arches (A1–4) and the areas near them are more vulnerable to seismic actions.

4.7 7th Loading Scenario: Gravity Loading, Absence of Wooden Tendons, Foundation Settlement

The seventh loading scenario attempts to explain the pathology image of an inclined crack that appears on the northern wall of the monument. The image is typical of foundation settlement. Therefore, vertical displacements which increase linearly are applied on the foundation of the northern wall. In order to enhance the failure pattern, the loading is imposed under the assumption of the absence of the tendons (Fig. 21).

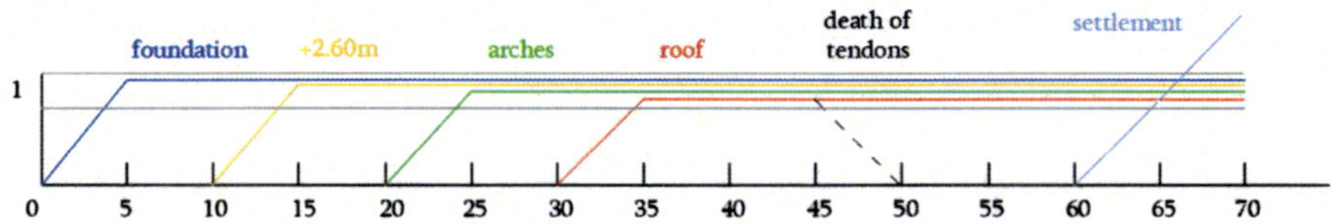

Fig. 21. 7th loading scenario: foundation settlement, absence of wooden tendons

Figure 22 depicts the crack pattern after the end of the analysis.

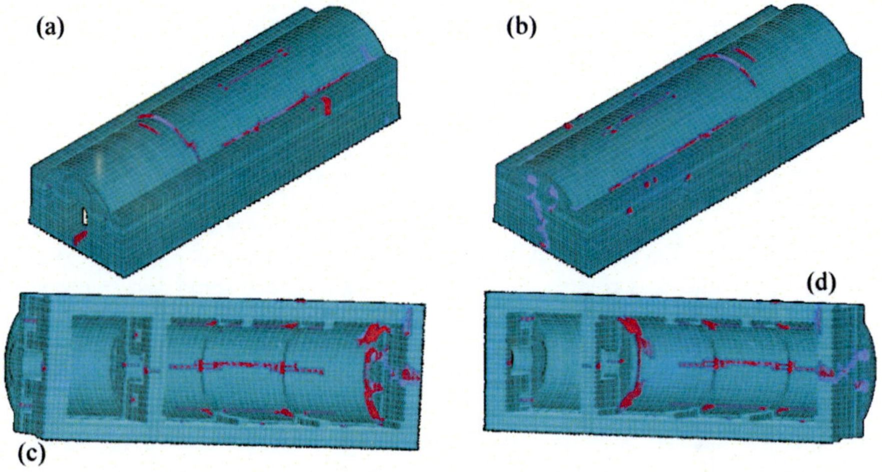

Fig. 22. Crack pattern due to the 7th loading scenario: (a) eastern wall; (b) western wall; (c) view of dome at negative longitudinal axis;(d) view of dome at positive longitudinal axis

One interesting observation is the line of tensile cracks that appear at the outer surface of the dome. In other words, the cracks at the key stone of the barrel vault extend to the exterior face. Furthermore, additional cracks develop at the intermediate thick wall that divides the monument into two parts. Finally, the most important failure figure refers to the northern wall, where the foundation settlement has been applied.

Figure 23 compares the crack pattern of the northern wall to the real state. The failure image resembles the real condition.

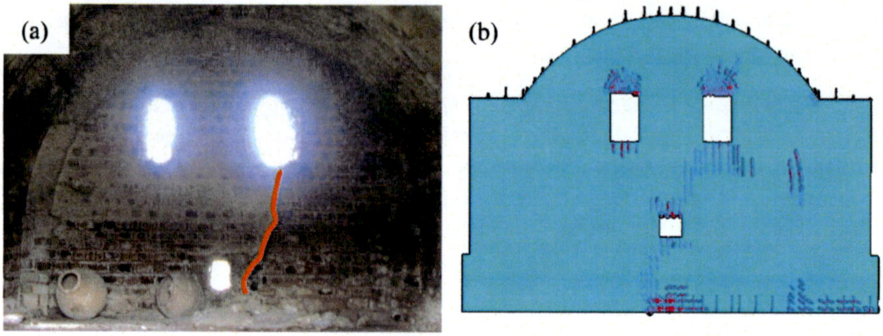

Fig. 23. Crack pattern due to 7th loading scenario: (a) real state; (b) finite element analysis results

The crack line based on the finite element analysis results crosses the small window which is located near the foundation and ends to the right upper window whereas in the real state the crack line does not crosses the small window. This difference can be attributed to the real damage history of the monument which cannot be captured by the finite element analyses, mainly due to lack of information or simulation details, such as masonry parts that are already missing. However, the most important conclusion is that the crack observed on the northern masonry wall has been caused by foundation settlement, which is in accordance with finite element results.

5 Conclusions

The purpose of the present study was the investigation of the causes of pathology of Xana's monument. All the possible information sources concerning the monument's history were taken into account. A three-dimensional finite element model of the monument's initial state was created. Laboratory testing provided some basic mechanical characteristics of the masonry material which were included in the finite element model as nonlinear material behavior. Soil-structure interaction was included in the model in the form of Winkler foundation. Various loading scenarios were examined under different hypotheses: function/absence of wooden tendons; existence/absence of heavy roof loading; earthquake action; foundation settlement. The results of each of the aforementioned scenarios reveal vital information regarding the monument's structural behavior and provide solid explanation for the monument's nowadays condition.

References

Mortezaei, A., Mortezaei, K.: Seismic evaluation and rehabilitation of the historical masonry structure of arg gate in semnan. In: 15th World Conference of Earthquake Engineering, Lisboa (2012)

Betti, M., Vignoli, A.: Numerical assessment of the static and seismic behavior of the ba-silica of Santa Maria all' Impruneta (Italy). Constr. Build. Mater. **25**, 4308–4324 (2011)

Sanchez-Aparicio, L.J., Riveiro, B., Gonzalez-Aguilera, D., Ramos, L.F.: The combination of geomatic approaches and operational modal analysis to improve calibration of finite element models: a case of study in Saint Torcato church (Guimaraes, Portugal). Constr. Build. Mater. **70**, 118–129 (2014)

Lourenco, P.B.: Computations of historical masonry constructions. Prog. Struct. Mat. Eng. **4**, 301–319 (2002)

Lourenco, P.B.: Assessment, diagnosis and strengthening of Outeiro Church, Portugal. Constr. Build. Mater. **19**, 634–645 (2005)

Kiel, M.: Studies in the Ottoman Architecture of the Balkans. Gower Publishing Company, Varorium (1990)

Kalafatzis, I.: The thermal baths complex and Xana in Traianoupolis, Evros. Diploma Thesis, Architectural Department, Polytechnic School, Aristotle University of Thessaloniki (2008)

Minopoulos, M.D.: Loutros. In the steps of time and space. Cultural and Educational Association of Loutra, Traianoupolis (2001)

Vavritsas, A.K.: Ancientries and monuments of Thrace. Archeological Bull. **23**, 364–366 (1968)

Pentazos, E.: Ancientries and monuments of Thrace. Archeological Bull. **24**, 362–364 (1969)

Terzi, V.: Restoration study of Xana, Traianoupolis municipality, Evros prefecture. Estimation of bearing capacity. Evaluation of restoration measures. MSc Thesis, Civil Engineering Department, Polytechnic School, Polytechnic School, Aristotle University of Thessaloniki (2010)

Ramirez, J.C., Lozano, J., Ferrer, H.O., Rojas, J., Vazques, R., Mijares, H.: Dynamic behavior of puebla City Cathedral. In: 15th World Conference of Earthquake Engineering, Lisboa (2012)

Fratini, M., Pieraccini, M., Atzeni, C., Betti, M., Bartoli, G.: Assessment of vibration reduc-tion on the Baptistery of San Giovanni in Florence (Italy) after vehicular traffic block. J. Cult. Heritage **12**, 323–328 (2011)

Eslami, A., Ronagh, H.R., Mahini, S.S., Morshed, R.: Experimental investigation and non-linear FE analysis of historical masonry buildings – a case study. Constr. Build. Mater. **35**, 251–260 (2012)

Psycharis, I.N., Mouzakis, H.P., Miltiadou, A., Pavlopoulos, E.P., Taflampas, I.M.: Investigation of the seismic response of a byzantine church and evaluation of the effectiveness of proposed interventions. Earthq. Spectra **26**, 473–497 (2010)

EC6. BS EN 1996 -1–1. Rules for reinforced and unreinforced masonry (1996)

Silva, L.C., Lourenco, P.B., Milani, G.: Derivation of the out-of-plane behavior of masonry through homogenization strategies: micro-scale level. Comput. Struct. **209**, 30–43 (2018)

Di Nino, S., Luongo, A.: A simple homogenized orthotropic model for regular masonry walls. Int. J. Solids Struct. **167**, 156–169 (2019)

Peng, B., Wang, D., Zong, G., Zhang, Y.: Homogenization strategy for brick masonry walls under in-plane loading. Constr. Build. Mater. **163**, 656–667 (2018)

Malena, M., Portioli, F., Gagliardo, R., Tomaselli, G., de Felice, G.: Collapse mechanism analysis of historic masonry structures subjected to lateral loads: a comparison between con-tinuous and discrete models. Comput. Struct. **220**, 14–31 (2019)

Kujawa, M., Lubowiecka, I., Szymczak, C.: Finite element modelling a historic church structure in the context of a masonry damage analysis. Eng. Fail. Anal. **107**, 104233 (2020)

Acito, M., Bocciarelli, M., Chesi, C., Milani, G.: Collapse of the clock tower in finale Emilia after the May 2012 Emilia Romagna earthquake sequence: numerical insight. Eng. Struct. **72**, 70–91 (2014)

Winkler E.: Vorträge über Eisenhanbau. Heft 1, 2. H. Dominicus, Prag (1867)

Ayensa, A., Beltran, B., Ibarz, E., Gracia, L.: Application of a new methodology based on Eurocodes and finite element simulation to the assessment of a Romanesque church. Constr. Build. Mater. **101**, 287–297 (2015)

Mallardo, V., Malvezzi, R., Milani, E., Milani, G.: Seismic vulnerability of historical masonry buildings: a case study in Ferrara. Eng. Struct. **30**, 2223–2241 (2008)

Bernardeschi, K., Padovani, C., Pasquinelli, G.: Numerical modelling of the structural behavior of Buti's bell tower. J. Cult. Herit. **5**, 371–378 (2004)

Lucchesi, M., Padovani, C., Pasquinelli, G.: On the numerical solution of equilibrium problems of elastic solids with bounded tensile strength. Comput. Meth. Appl. Mech. Eng. **127**, 37–56 (1995)

Terzi, V.G., Ignatakis, C.E.: Nonlinear finite element analyses for the restoration study of Xana, Greece. Eng. Struct. **167**, 96–107 (2018)

Presentation and Evaluation of the Loading and Unloading Mechanisms of the Industrial French Wharf of Lavrion

Nikolaos Rogkas(✉), Georgios Vasileiou, Vasilios Spitas, and Ioannis A. Antoniadis

Department of Mechanical Constructions and Automatic Control, National Technical University of Athens, NTUA Zografou Campus 9 Heroon Polytechniou Str., 15780 Zografou, Athens, Greece
nrogkas@mail.ntua.gr

Abstract. Industrial French Wharf of Lavrion was built in 1888 by French Mining Company of Lavrion and it was used by the ships in order to load and unload their carriage which mainly were mining materials, coal or commercial everyday products. Today it stands as a monument of significant industrial heritage. Referring to its operation, a different loading and unloading spot on the top of the wharf was used, depending on the type of the transported product. The loading of the ships with minerals was accomplished by means of carriages filled with the product, manually guided on railway lines in the top of the wharf by the workers. When the transportation of commercial products was desired, two steam cranes located to the left and right of the wharf were used, which around late 50s were substituted by a conveyor belt system to transport the minerals from the loading area to the end of the wharf in a continuous way using electrical power. It can be assumed that through the years of the wharf's operation, various mechanical devices and mechanisms were used and therefore understanding and describing their operation became of great interest form a historical and industrial point of view. This work aims to present, explain and evaluate the mechanical loading and unloading mechanisms of the wharf, which were found to be: a) carriages and rail way equipment b) a belt conveyor and c) one steam load lifting crane. Regarding the train-related equipment, this study focuses into understanding the operation of the carriages transportation as the design of the corresponding railways showed some remarkable characteristics, while the steam cranes were truly inovative at that time. Finally, the preliminary design of a particular lifting platform device for people with special needs is proposed in this study.

Keywords: French Wharf of Lavrion · Momentum restoration · Lifting mechanisms

I. Vayas and F. M. Mazzolani (Eds.): PROHITECH 2021, LNCE 209, pp. 389–401, 2022.
https://doi.org/10.1007/978-3-030-90788-4_32

1 Introduction and Historical Perspective

The present work was conducted by the Mechanical Engineering team of the National Technical University of Athens (NTUA) as part of the study for the restoration of the French Wharf of Lavrion. The industrial French Wharf of Lavrion was built in 1888 and named after the French Mining Company of Lavrion (FMCL), which had a key role at that time in the wider Lavreotiki area as referred by Cheirchanteri (2019), by exploiting the Lavrion mines (Katerinopoulos 2010). Its construction was completed just three years after the establishment of Athens-Lavrion Railway (1885) whose terminal station lied very close to the Wharf, while the launch station was in Lavrion Square in downtown Athens as reported in Markouli-Mponioti's (2008) and Papanikolaou's (2008) work. The construction of the Wharf was aiming to facilitate ships loading and unloading with minerals and commercial every-day goods respectively. The minerals included silver and lead ores (Skarpelis 2007), while according to historical reports, coal was often transported towards and from the Wharf. Loading of the ships was accomplished by means of wagons filled with ore, which were pushed by workers on rails on top of the Wharf and the ore was deposited on the ship using a hopper located at the end. When the transportation of unloaded commercial products was desired, two steam cranes located at the left and right extremities of the Wharf were used to lift them from the ship to the Wharf deck. In that case a rotating disc was employed so that the wagon could change its direction by 90° and thus put within the cranes' area of operation. Around late 50 s the manual transportation of the wagons ceased and instead a conveyor belt system was built in order to convey the ore from the loading area to the end of the Wharf in a continuous fashion using electrical power. Also, the two steam cranes were replaced by an electric one at the end of the Wharf.

Figure 1, obtained from Markouli-Mponioti's dissertation (NTUA 2008), shows an overview of the Wharf during crane operation as an anchored ship was unloaded. Figure 2 shows a schematic drawn by FMCL where the rails, the rotating discs, the hopper and the two rotating cranes with their operating range visible in dash-dotted lines.

Fig. 1. View of the wharf, taken from an old photograph [1].

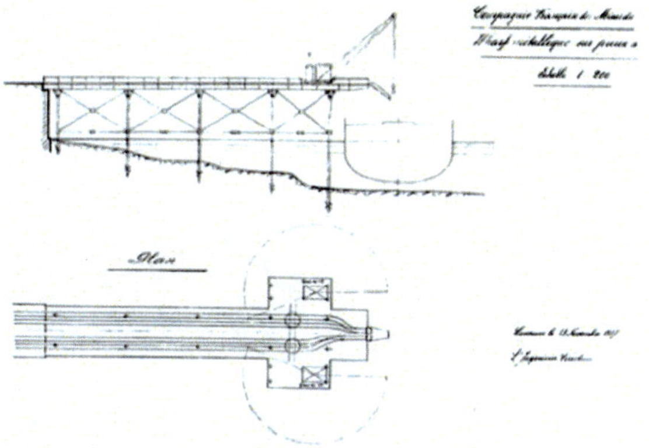

Fig. 2. Schematic by FMCL [1].

2 Wharf Mapping of Mechanical Equipment and Carriages Movement Infrastructure

Within the scope of this study, the NTUA Mechanical Engineering team visited the remains of the Lavrion Warf in order to map the mechanical loading and unloading mechanisms of the Wharf and assess their condition. The remaining related equipment comprised of the following items: a) railways on the deck of the wharf b) a belt conveyor and ancillary equipment and c) remains of the electric crane. The three discrete items along with their associated infrastructures are presented in the photographs taken below (Fig. 3 a to h).

3 Explanation of Wharf Infrastructures

Out of the three individual ore carrying systems, the most controversial was found to be the rail-way equipment due to the corrosion sustained to the system over the years. Parts of them have either disappeared completely or have been widely displaced, therefore causing much ambiguity compared to the original schematics by FMCL (Fig. 2).

After carrying out the mapping of the wharf deck the following observations were made: two sets (one right and one left) of four rails each are placed symmetrically along the Wharf, which, contrary to FMCL's plan (Fig. 2), do not reach up to the end of the wharf but are interrupted at their intersection with the two rotating discs. (Fig. 1-(c) to (f)). The pair of the transverse railways shown in the drawing, as well as the mountings of the two cranes (Fig. 3 (e) to (f)) are missing and finally the distances between the tracks were measured and found to be similar per pair but not exactly equal, as expected due to the displacements they have undergone over the years.

(a) (b) (c) (d) (e) (f) (g) (h)

Fig. 3. Autopsy of the wharf. General view from the bottom of the wharf (a). Top place of the wharf (starting point of railways) (b). View of the railways (conveyor system to the left, railways, and rotating disc in center/right) (c). Details of rotating disc and hopper (d). Rotating disc. In the background the place where the steam crane was based can be seen (e). The later-installed electric motor system (f), (g). Ending point of conveyor-belt system (h).

The rotating discs were investigated with regard to their rotational mechanism and it was found to consist of the following parts: a) a main manually rotating shaft transmitting motion and torque to the disc and supporting it, b) a set of tapered rollers placed radially beneath each disc, connected externally with a rim and internally with radially placed spokes mounted on the shaft using a tapered bearing, and c) a circular raceway fixed at the bottom on which the mechanism was mounted and the tapered rollers were able to roll. The mechanism is presented in Fig. 4.

(a) (b)

Fig. 4. Rotating mechanism. Rollers and retaining hoop (a), Vertical rotating shaft, rollers' spokes and circular raceway.

Regarding the belt-conveyor system the following remaining components were found: a) a hopper, attached on an iron frame and located at the east side of the Wharf, used for the supply of minerals onto the conveyor belt, b) a conveyor belt which was supported by successive sets of three upper rollers and one bottom one (for the feed and return respectively) and c) a lower drum for the wrapping and the reverse of rotation of the drive-belt (Fig. 5).

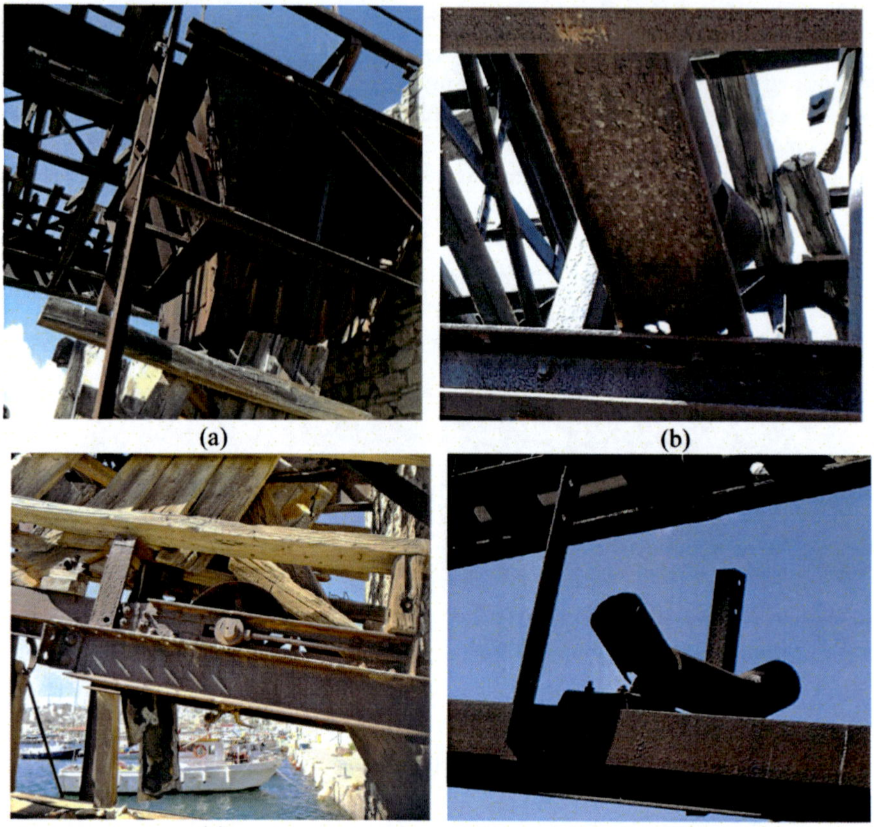

(a) (b) (c) (d)

Fig. 5. Belt conveyor system. Hoper (a) Supporting frame (b)–(c). Roller (d).

All the aforementioned structures and their components are assumed to be constructed of cast iron due to their morphology and their surface condition after the corrosion they have undergone over time.

3.1 Investigation of the Railways

The design of the railways prevails some remarkable characteristics, bearing in mind the fact that they were built before 1900. The main modes of operation were the loading of minerals from the wharf onto the ships using the hopper and unloading of every-day goods using the steam cranes. Those modes are illustrated schematically in Fig. 6

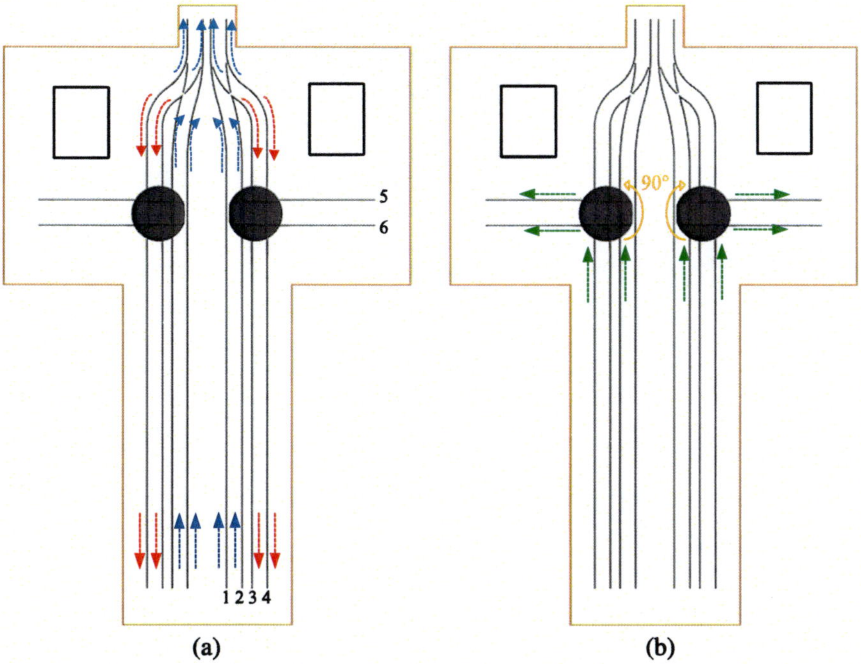

Fig. 6. Schematic representation of wagons tracks. Loading of minerals-The blue arrows indicate the direction of filled wagons towards the hoper (paths 1, 2) and the reds indicate the direction of return of the empty wagons (paths 3, 4) (a). Unloading of commercial goods.-Tracks 2–4 and 5–6 are used (b).

In case (a) of Fig. 6, the wagons are driven to the end of the Wharf to unload their contents into the hopper and afterwards return to their starting point after the ore has been deposited.

In case (b) the carriages are driven to the unloading area on the side of the Wharf which falls into the operating zone of the steam-powered cranes as shown in Fig. 2. To achieve that, the rotating discs are activated enabling the access of the wagons within the cranes' operating range.

After analyzing and simulating the kinematic operation of the mechanism in a CAD environment, it has been concluded that the width of the wagons used in case (b) is different from that of case (a) to allow the same pairs of tracks to be used in different combinations. An illustration of the mineral wagon is presented in Fig. 7.

3.2 Phase I – Loading of Minerals

In the first mode of operation the carriages filled with minerals drive on the rails 1, 2 towards the hopper (Fig. 8-(a), (b)) while lines 3.4 are used by the empty wagons returning to the starting point in order to be refilled (Fig. 8-(c)). Therefore, the distances between the railways 1–2 and 3–4 are identical and they were found approximately equal to 60 cm. Just before the hoper the four rails are merged into two.

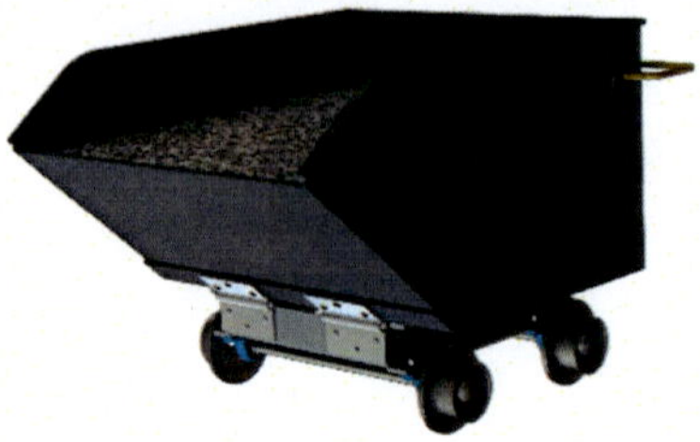

Fig. 7. Illustration of wagon. Pivot mechanism in the front part.

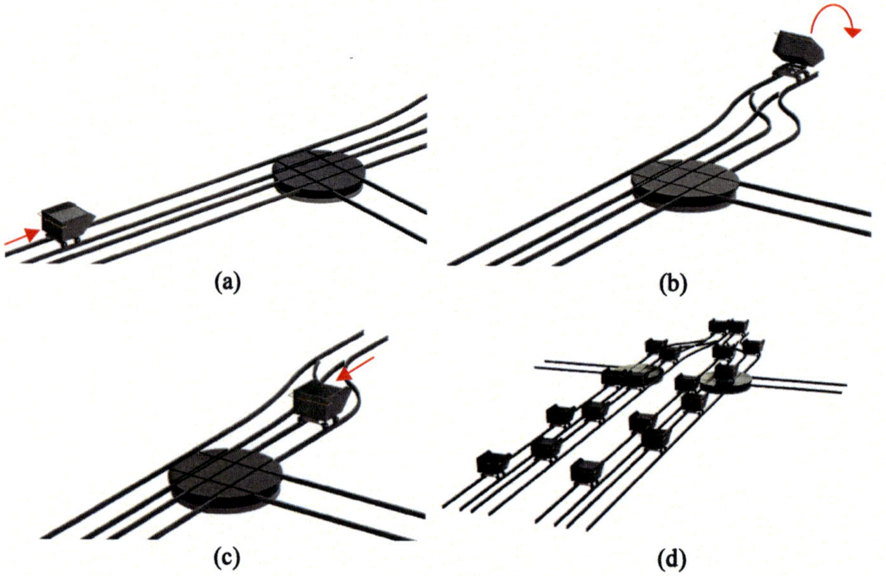

Fig. 8. Carriages used for the transportation of mining material. Phase sequence of a single wagon (a)–(c). Work cycle of wagons (d).

In addition, the use of lines 1, 2 for loading of minerals is likely to occur as they are less curved than 3, 4 and therefore the danger of derailment is eliminated. The sequence of movements for a single wagon is shown in Fig. 8 (a) to (c). The complete work cycle is illustrated in Fig. 8 (d).

3.3 Phase II – Unloading of Commercial Products

The wagons carrying commercial products travel onto the rails 2, 4 and 5, 6. The accordant distances were measured and found equal to 100 cm. The corresponding sequence of operating phases is shown in Fig. 9 (a) to (d).

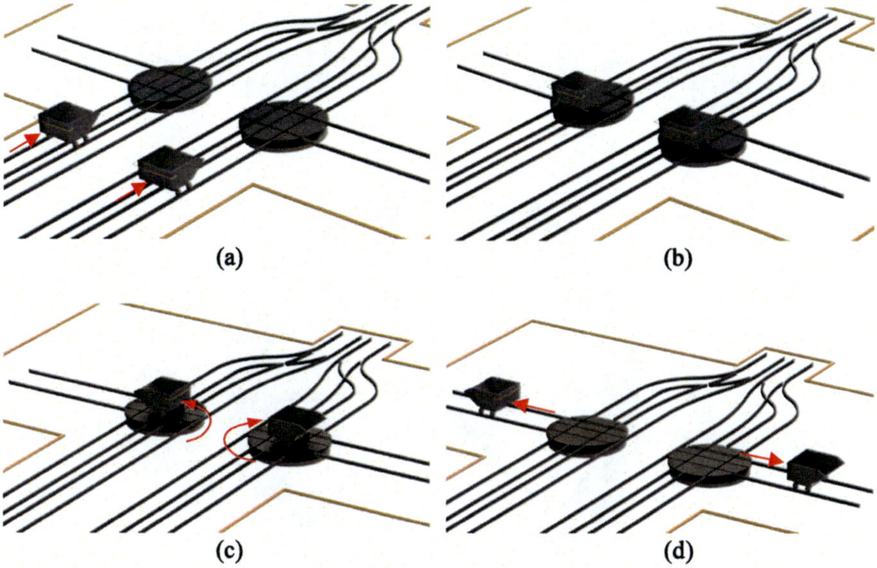

Fig. 9. Sequence of operating phases of wagons carrying every-day goods. Wagons traveling towards the discs (a). They reach the discs (b). Rotation of discs is applied and the wagons change direction by 90° (c). Arrival to the cranes' operating place (d).

3.4 Analysis of Rotating Disc

The rotating discs permit the access of carriage to the cranes' operating range. Each disc carries a set of recesses on its top which prevent the derailment of the carriage as the flanges of its wheels pass through them while the carriage travels above the disc. The recesses and their dimensions are shown in Fig. 10.

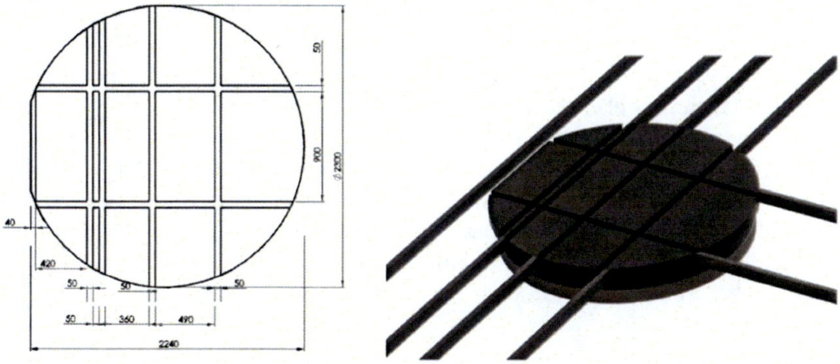

Fig. 10. Rotating disc recesses (all dimensions in mm).

The rotation of the shaft is accomplished by means of a bearing-type mechanism by intervening a brass ring inside the hub allowing the shaft to slice with the minimum friction loses. To achieve non-sliding motion the rollers should be tapered with a half-cone angle of 4° (Fig. 11).

Fig. 11. Rotating disk, top and section view.

4 Lifting Platform

The second objective conducted by the Mechanical Engineering team regarding the restoration of Lavrion Wharf and its use as an outer space museum, was to investigate how people with special needs could access the Wharf deck. Since there is no other way of access to the deck (e.g. with perimeter ramps) it is necessary to have a lift-type installation which may also serves other purposes (e.g. lifting loads).

The proposed platform is shown in Figs. 12, 13. The platform is scissor-type, similar to the ones proposed by Hongyu and Ziyi (2011) and Sauli et al. (2019). The scissor mechanism is preferred over other lifting-type mechanisms, as when not used, it will remain hidden, buried inside a cavity at a depth of 2.5 m. Also, the lift guarding rails are mechanically retractable, and when the mechanism reaches the wharf level, they open towards the Wharf forming a ramp, thereby securing the users access to onto the deck.

The energy required to ascend and descend the platform is provided by a typical hydraulic system (motor, pump, accumulator and one single-way hydraulic cylinder). Through renewable energy sources (RES) proposed to be implemented with vertical axis wind turbines and solar tree panels, a UPS (24 V) battery is charged which can power the motor of the hydraulic pump in cases of power outages.

The hydraulic elements are concentrated inside the cavity at the bottom and are accessible for inspection and maintenance operations through an open hatch and ladder. At the ground level there is a control panel that implements the overall automation of the lifting mechanism and the rest electro-mechanical systems (wind turbines, solar trees, lighting, etc.).

Fig. 12. Photorealistic view of the Wharf and the lifting platform at its extended position.

Regarding the lifting mechanism it is consisted of six scissors and its total height at the retracted and extended position is 2m and 7 m, respectively. The platform's dimensions are 1.5 m × 15 m in order to allow two individuals using it simultaneously (the person with special needs and his personal escort). Finally, the platform can ascend and descend a total mass up to 300 kg.

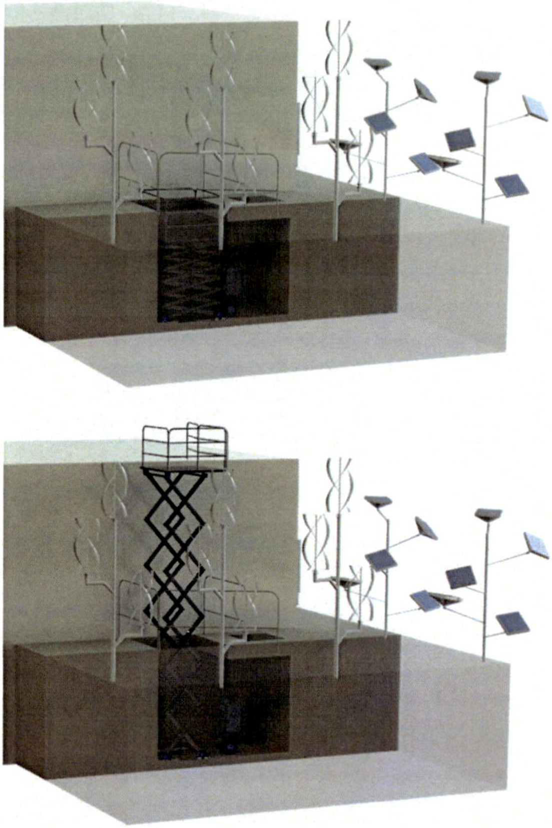

Fig. 13. Details of the cavity where the hydraulic components can be spotted. Top, the platform remains hidden at the retracted position and bottom, the platform at the extended position reaches Wharf deck

5 Conclusions

In this study, the mechanical mechanisms used in the Industrial French Wharf of Lavrion were presented and explained. The study focused in depicting the various devices and mechanisms used for the transportation of wagons carrying minerals or every-day goods and within this scope the design of the railways on the Wharf deck was investigated. Finally, a scissor-type lifting platform was proposed to facilitate people with special needs for their access onto the deck.

References

Cheirchanteri, G.: Industrial cultural heritage as an alternative tourism model: case study of the Wider Lavreotiki area in Southeastern Attica. In: IOP Conference Series: Materials Science and Engineering, vol. 603, no. 2, p. 022001. IOP Publishing (2019)

Katerinopoulos, A.: The Lavrion mines. In: Evelpidou, N., Figueiredo, T., Mauro, F., Tecim, V., Vassilopoulos, A. (eds.) Natural Heritage from East to West, pp. 27–33. Springer, Heidelberg (2010). https://doi.org/10.1007/978-3-642-01577-9_3

Skarpelis, N.: The Lavrion deposit (SE Attica, Greece): geology, mineralogy and minor elements chemistry. Neues Jahrbuch für Mineralogie-Abhandlungen J. Mineral. Geochem. **183**(3), 227–249 (2007)

Markouli-Mponioti, A.: The creation and the development of French Mining Company of Layrio (F.M.C.L.) shows new evolutions to the economical and industrial section of Greek society of 19th and 20th century, Doctoral dissertation (2008)

Papanikolaou, M.: The construction of the Athens-Lavrion railway line and its socio-economic impact on the development of the South East Attica region (1885–1938). In: Proceedings of the IB Scientific Meeting of Southwest Attica (2008)

Hongyu, T., Ziyi, Z.: Design and simulation based on Pro/E for a hydraulic lift platform in scissors type. Procedia Eng. **16**, 772–781 (2011)

Sauli, S.A., Ishak, M.R., Mustapha, F., Yidris, N., Hamat, S.: Hybridization of TRIZ and CAD-analysis at the conceptual design stage. Int. J. Comput. Integr. Manuf. **32**(9), 890–899 (2019)

Sanusi, H., Ishak, M.R., Hussin, M.S., Yuzairi, A.R., Imran, M.S., Sauli, S.A.: Optimisation of scissor lifting machine structures using finite element analysis (FEA). In: AIP Conference Proceedings, vol. 2129, no. 1, p. 020003. AIP Publishing LLC (2019)

Hemp-FRP for Seismic Retrofitting of Existing Masonry Buildings

Antonio Formisano(✉), Generoso Vaiano, and Nicola Junior Petrucci

Department of Structures for Engineering and Architecture, University of Naples "Federico II", Piazzale Tecchio 80, 80125 Naples, Italy
antoform@unina.it, generoso.vaiano@inwind.it

Abstract. Nowadays, renewable technologies and reduced environmental impacts are important concepts of the modern construction systems, especially in relation to the production of building materials. In this framework, the seismic performance of H-FRPs (Hemp - Fibre Reinforced Polymers) for retrofitting masonry structures are studied in the current paper, where both analytical and experimental researches on laminated composite green materials are reported and discussed. With respect to analytical studies, proper equations ruling the percentages of fibres and matrixes to be used in green composites are examined. Moreover, full scale experimental tests in order to evaluate the hemp fibres mechanical behaviour and the seismic behaviour of masonry panels are taken into account. The numerical calibration of the above experimental tests are conducted with the ABAQUS non linear numerical software in order to know the behaviour of hemp fibres and masonry walls under the combined effect of vertical and horizontal actions. This activity is used to apply hemp fibres under form of a grid of strips (with variable pitch) as seismic retrofitting system applied on one or both sides of an existing masonry panel. The behaviour of hemp strips in terms of resistance and ductility is compared to that of C-FRP (Carbon - Fibre Reinforced Polymer) ones aiming at both evaluating the effectiveness of using green fibres and finding the best retrofitting system of the analyzed masonry panel.

The analysis results have shown the convenience of using hemp fibres, which allow to perform low cost and minimally invasive seismic interventions without the building disruption of use, offering at the same time strength and ductility comparable to those of the most famous C-FRP materials.

Keywords: Hemp · Natural fibres · Green composites · Seismic retrofitting · FE analysis

1 Introduction

Italy is a highly seismic country where most of masonry buildings can suffer very serious damage following seismic events. Masonry is characterized by anisotropy, inhomogeneity, non-linearity, different behaviour in elastic field with a good compression strength and a poor tensile one. The main problem is represented by the poor mechanical properties, which increase building seismic vulnerability even for low seismic excitations

I. Vayas and F. M. Mazzolani (Eds.): PROHITECH 2021, LNCE 209, pp. 402–417, 2022.
https://doi.org/10.1007/978-3-030-90788-4_33

(Como 2013; Anthoine 1992; Parisi and Augenti 2011; Fonti et al. 2017; Betti and Vignoli 2011; Clementi et al. 2017; Valente and Milani 2017; Valente and Milani 2016; Milani and Valente 2015a, 2015b). Furthermore, on the basis of the past earthquake experiences, it has been observed that masonry buildings show the repetition of different collapse mechanisms related to the loss of equilibrium. In-plane and/or out-of-plane failure mechanisms interest masonry buildings characterised by some seismic deficiencies, such as poor connections between walls or walls and floors, pushing roofs, wall excessive slenderness, poor mechanical properties, etc. (Leite et al. 2017; Gattulli et al. 2014; Milani and Valente 2015a, 2015b; Valente et al. 2016; Stockdale 2016; Lagomarsino and Podestà 2004; Milani et al. 2018). For these reasons, it is necessary to perform seismic upgrading or retrofitting interventions on these buildings. To this purpose, many traditional techniques, i.e. reinforced plaster, steel pre-stressed tie rods, mortar injections, steel connectors, etc., can be used (Maragna et al. 2016; Modena et al. 2002; Valluzzi et al. 2005; Preciado et al. 2016; Formisano and Marzo 2017; Faggiano et al. 2009). Recently, great importance is given to the application of C-FRP (Carbon-Fibre Reinforced Polymers), which is an innovative, reversible e few invasive material for consolidation interventions (CNR DT 200 2013; Milani et al. 2016; Milani et al. 2017; Mosoarca et al. 2017; La Manna et al. 2017; Bertolesi et al. 2017, Formisano et al. 2015).

In recent years, scientific community has turned its interest towards the use of natural composite materials in the construction field with the aim to use recyclable, renewable and eco-sustainable materials in place of synthetic ones consisting of carbon or glass fibres. For this reason, the aim of this paper is to analyze the Hemp-Fibre3 Reinforced Polymer (H-FRP) as reinforcement system of masonry constructions. In fact, hemp fibres are biodegradable, recyclable and eco-sustainable and have, therefore, behaviour contrary to synthetic fibres, which require for dismantlement and disposal operations a significant quantity of energy, so leading towards significant CO_2 emissions.

Hemp fibres are found in the plant stems, that gives them good mechanical characteristics. Through experimental studies it has been found that mechanical properties of hemp fibres are comparable to those of glass ones (Cristescu et al. 2003; Teodorescu-Draghicescu et al. 2011; Vlase et al. 2009, 2011). Hemp fibres composites made of a thermoplastic matrix, which are characterised by a strong anisotropy, show fair mechanical properties, which can be improved by surface treatments.

Starting from these hypotheses, this paper shows how H-FRP can increase masonry buildings mechanical features. A fundamental part of this research is the analytical study on laminated green composite materials, which have led to define the optimal percentage between fibres and matrix and the behaviour of the composite material, starting from the numerical calibration by the ABAQUS software of experimental test results on both hemp laminates (Scutaru et al. 2013) and masonry walls (Ganz and Thürlimann 1984; Lourenço 1996). After these experimental tests have been numerically reproduced by FEM analyses, a new FEM model of the before studied existing masonry wall reinforced with the investigated H-FRP system has been setup. Non-linear static analyses have been conducted with the ultimate goal to evaluate the masonry wall performance increase in term of resistance and ductility, as well as the difference of behaviour between H-FRP and C-FRP.

2 Experimental Reference Tests

The first experimental reference test (Scutaru et al. 2013) was carried out in order to evaluate mechanical properties of hemp fibres reinforced by the "Derakane 411-350" thermoplastic resin. The composite laminate was obtained by a manual lay-up process using an impregnating roller until reaching a thickness of 5 mm. This process was characterized by the following phases: formwork construction, mold realization, application of the release on the mold, cutting, stratification, polymerization, extraction and finishing. Eight specimens having "dog bone" shape (Fig. 1) were cut from the laminate using a diamond point in order to avoid additional internal stress formation.

Tensile tests were carried out using the "LS100 Plus Lloyd Instruments materials" testing machine. Figure 2 shows the stress-strain (σ–ε) curves obtained from tensile tests performed on all of the tested specimens.

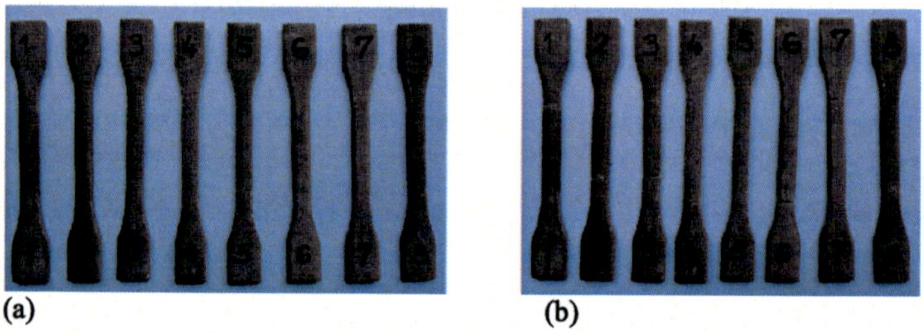

Fig. 1. Hemp fabric specimens before (a) and after (b) tensile testing (Scutaru et al. 2013).

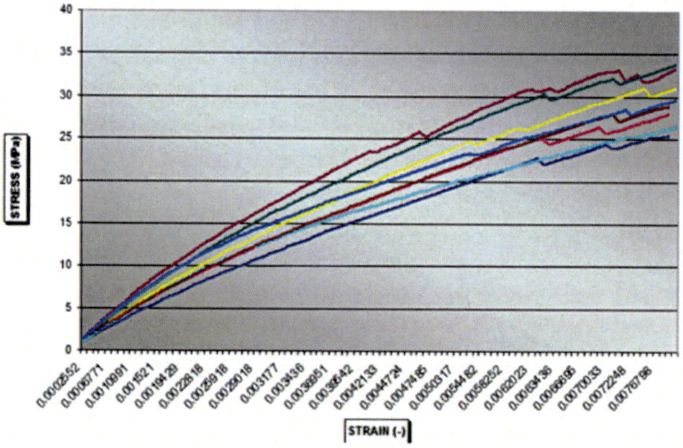

Fig. 2. Stress-strain curves of eight specimens.

The second experimental reference test (Ganz and Thürlimann 1984) was carried out in order to evaluate mechanical characteristics of a masonry wall experienced a shear failure mode.

The test was performed on a full-scale specimen, as shown in Fig. 3, where geometric dimensions of the masonry wall are provided. Four different loading conditions have been applied to the wall. However, in the current paper, only the first loading condition has been considered. In particular, an initial vertical load of 415 kN (equivalent to 0.61 MPa) has been applied to the top surface of the wall, which has been then horizontally loaded. After experimental test, a very ductile response of the wall, showing shear failure with diagonal cracks (Fig. 4), has been noticed. Figure 5 shows the experimental test result in term of force-displacement curve.

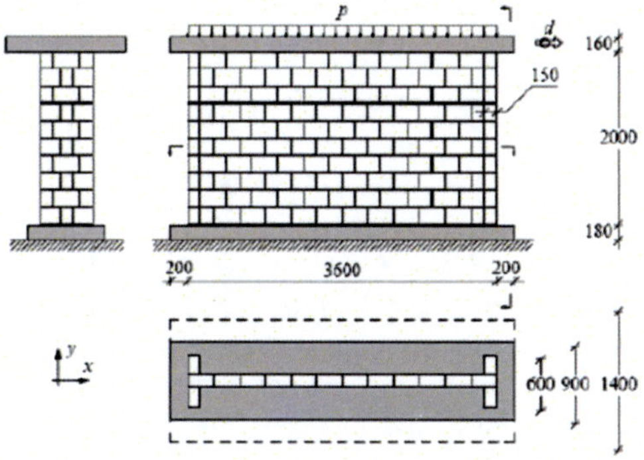

Fig. 3. Geometric dimensions of the tested wall specimen (Lourenco 1996).

Fig. 4. Wall experimental failure pattern (Ganz and Thurlimann 1984).

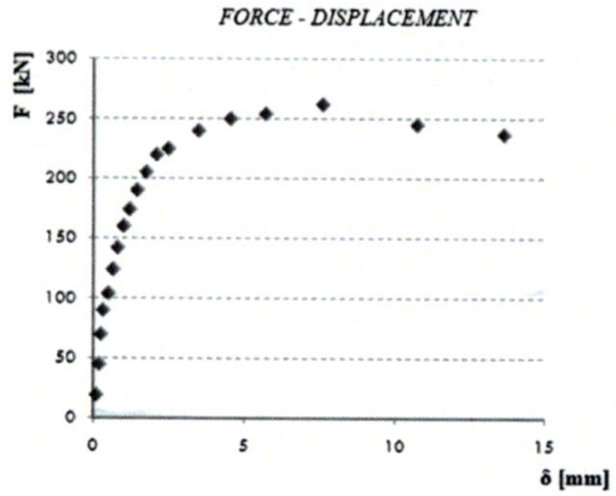

Fig. 5. Experimental force-displacement curve of the wall (Lourenco 1996).

3 Numerical Analysis

The experimental activities presented in Sect. 2 has been numerically simulated through the finite element software ABAQUS 6.14.

According to the experimental tests of Scutaru et al. (2013), the composite laminate specimens have been modelled based on the real mechanical and geometric features of H-FRP samples (Fig. 6a). About boundary condition, the specimen has been fully restrained at the base. After, a tensile force has been applied at the top of the sample (Fig. 6b). The choice of the mesh size has been done taking the best accuracy results with the less computational time. In particular, the following three different analysis cases have been considered:

- Case 1: mesh size of 10 mm (model with 838 seeds) and time analysis of 30″;
- Case 2: mesh size of 5 mm (model with 2918 seeds) and time analysis of 5′;
- Case 3: mesh size of 3 mm (model with 7569 seeds) and time analysis of 30′.

As depicted in Fig. 7, reducing the mesh size from 5 to 3 mm, the obtained curve is almost the same, but the analysis time is drastically increased. For this reason, the mesh with seed length of 5 mm can be considered as the best analysis case. Figure 8 illustrates the stress configuration of the specimen at the end of the numerical test. The stress-strain numerical curve has been compared with experimental curves in Fig. 9, where it is noticed how the former is very close to the average experimental curve represented by that of specimen no. 5.

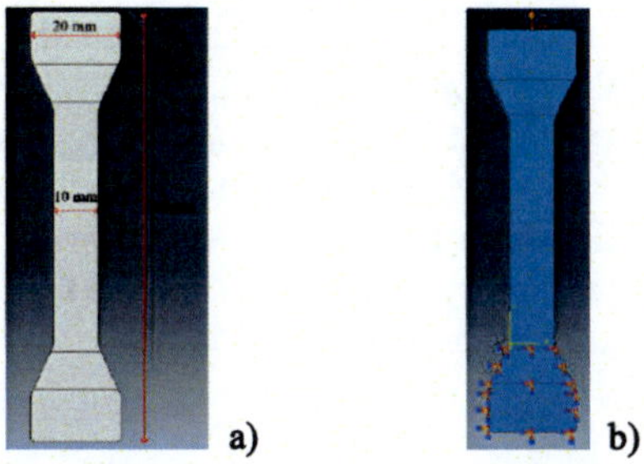

Fig. 6. The ABAQUS FEM model: geometric features (a) and boundary conditions (b).

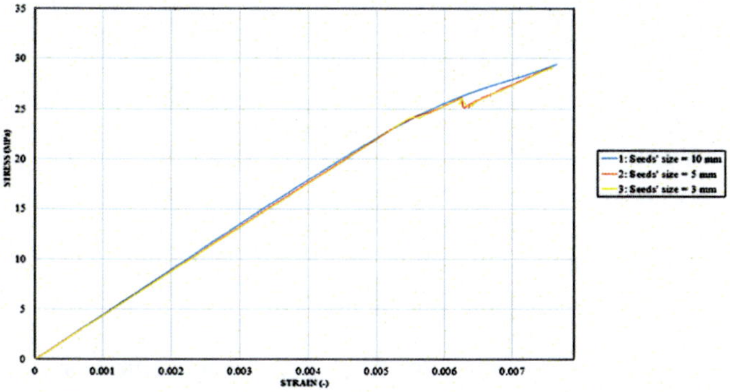

Fig. 7. Sensitivity analysis for assessment of the best mesh size.

Fig. 8. Stress state in the specimen at the numerical test end.

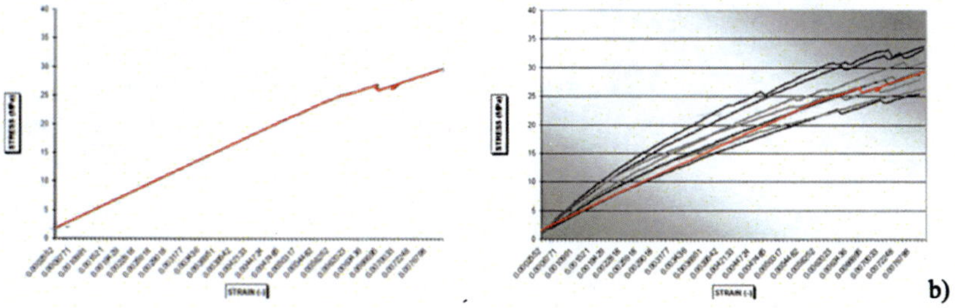

Fig. 9. Numerical stress-strain curve (a) and comparison with results of experimental tests (b).

Subsequently, the numerical calibration of Ganz and Thurlimann's experimental tests (1984) has been performed. Isotropic Damage Plasticity Model (Lubliner et al. 1989; Lee and Fenves 1998) has been used for masonry modelling. Originally, this model was formulated to describe the concrete behaviour, but it is easily extendable to other brittle materials under tensile actions like masonry, provides excellent results in the case of structures subject to cyclic and/or dynamic loads. The used modelling approach is used for materials having the following properties:

- isotropic behaviour;
- diverse tensile and compression stresses;
- different behaviour, with softening in the tensile field and hardening followed by softening until failure in the compression field;
- dissimilar degradation of tensile and compression elastic stiffness;
- stiffness recovery in the case of cyclic loading.

Figure 10a shows the numerical model obtained by deformable 3D solid elements, which are assembled with different constraints. In particular, the "Tie constraint" is employed to assemble orthogonal masonry panels, whereas the "surface interaction" is used to connect masonry walls with the lower and upper reinforced concrete (rc) tie-beams. About boundary condition, the lower rc tie-beam surface is fixed. The load application develops in two phases (Fig. 10b):

- phase 1: vertical load (415 kN or 0.61 MPa) on the upper rc tie-beam surface;
- phase 2: lateral displacement of 15 mm.

Figure 11 shows the wall deformation and the related tensile stress state at the end of the numerical test. Numerical curve in terms of force-displacement is compared with the experimental one in Fig. 12. From the comparison, it can be observed that, since the two curves are very similar to each other, the numerical calibration has been carried out accurately.

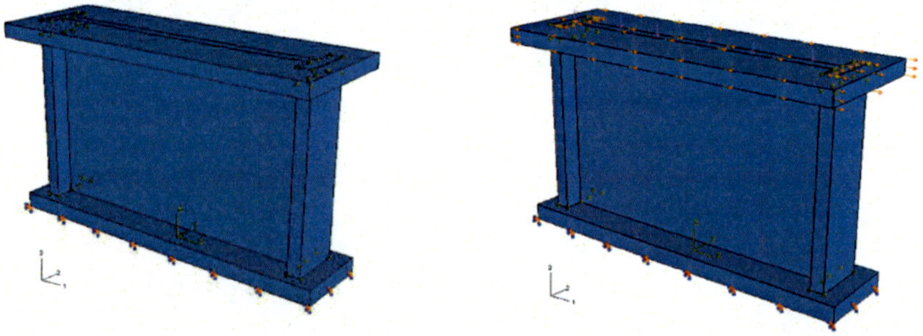

a) b)

Fig. 10. Numerical model (a) and boundary conditions (b) of the tested masonry wall.

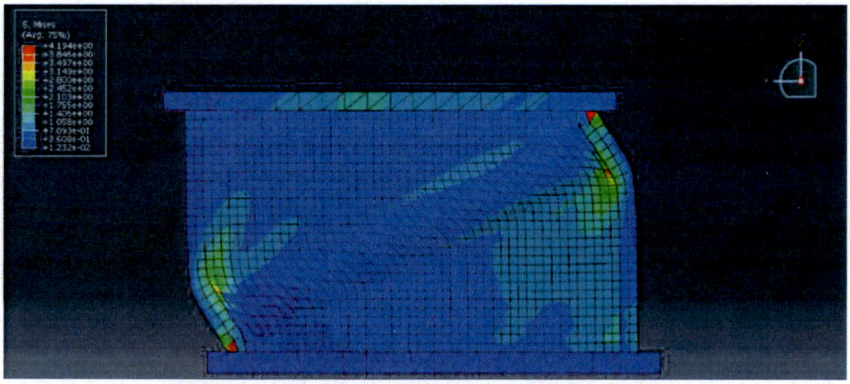

Fig. 11. Wall deformation at the end of numerical test.

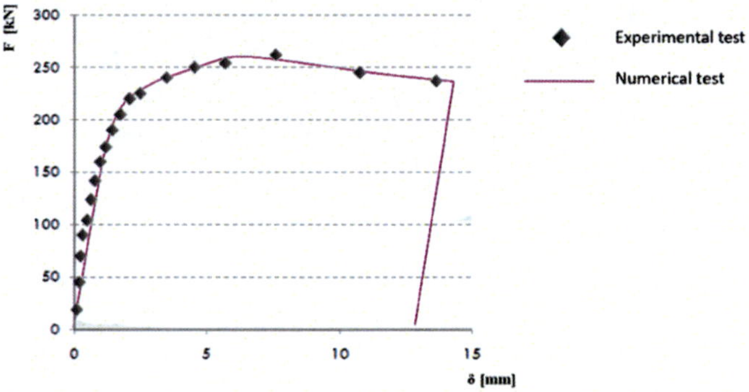

Fig. 12. Experimental-numerical comparison in terms of the force-displacement curve.

4 Seismic Upgrading of Masonry Walls by H-FRP

The seismic behaviour of the masonry wall studied in Sect. 3 has been herein improved by applying H-FRP strips having width of 10 cm.

These strips, which have been glued on the masonry surface as occurs with the usually adopted C-FRP analogous elements, have been placed in horizontal position at a pitch of 50 cm, while vertically three different pitches, namely 110 (4 strips), 50 (7 strips) and 40 cm (10 strips), have been considered (Fig. 13).

The analysis has been carried out considering the application of strips either on one side or on both sides of the wall. In the FEM model, the strip application on one side of the wall has been carried out through a "TIE constraint", since it is considered that bonding by thermoplastic resin avoids the occurrence of frictional sliding.

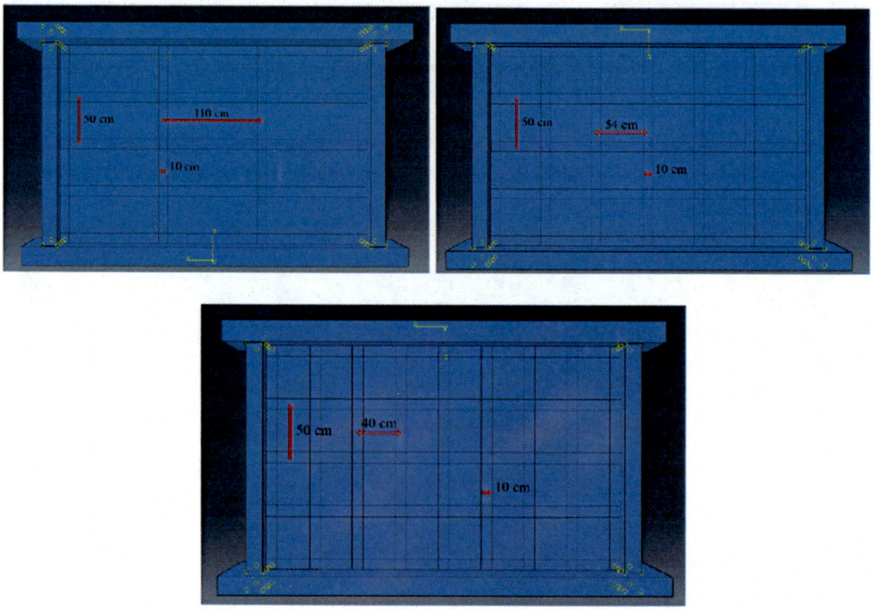

Fig. 13. Different arrangements of H-FRP strips on the masonry wall

On the other hand, the application of strips on panel both sides has been made through a connector constraint. In particular, the corresponding points on the two surfaces of the masonry panel have been connected by a proper connector placed at the intersection between longitudinal strips and transversal ones (Fig. 14).

Restraints and loads are the same used by Ganz and Thurlimann in their experimental test shown in the previous section. Figure 15 shows the final configuration of examined walls at the end of the tests. It is observed that the highest compression values are arranged almost in diagonal direction. The obtained force-displacement curves (Fig. 16) have allowed for an immediate understanding about benefits provided to the masonry wall by H-FRP in terms of resistance and ductility. In Table 1 the six numerical results have been compared in terms of maximum force (F_{max}), ultimate force (F_u), maximum displacement (δ_1 or δ_{max}) and ultimate displacement (δ_2). Furthermore, Table 2 shows how for the reinforced wall the curve resistance and displacements increase with respect to the same values achieved on the original wall. In particular, it has been seen that the proposed interventions entail small benefits in terms of resistance (percentage increase from 2.42% to 12.33%), but large improvements in terms of ductility occur (percentage increase from 14.20% to 36.68%).

Fig. 14. Numerical modelling of connectors among H-FRP strips.

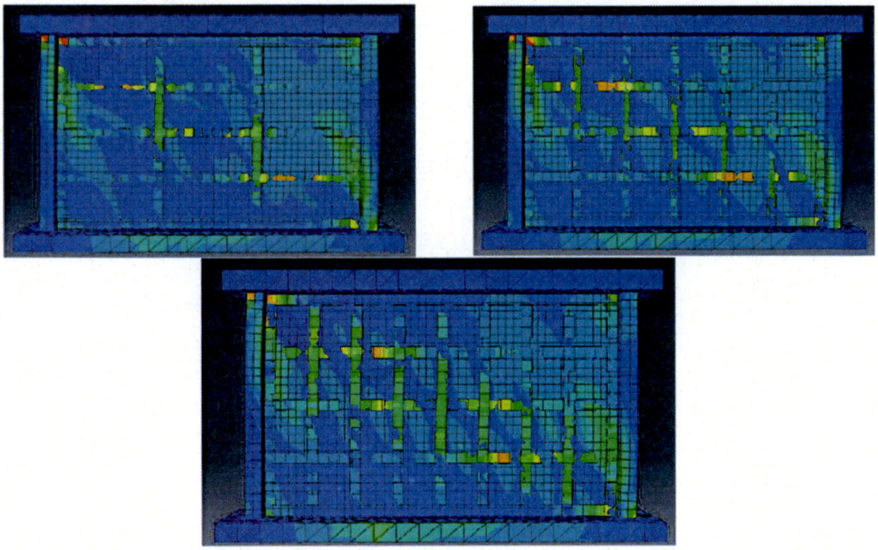

Fig. 15. Final configuration of the walls at the end of the numerical simulation tests.

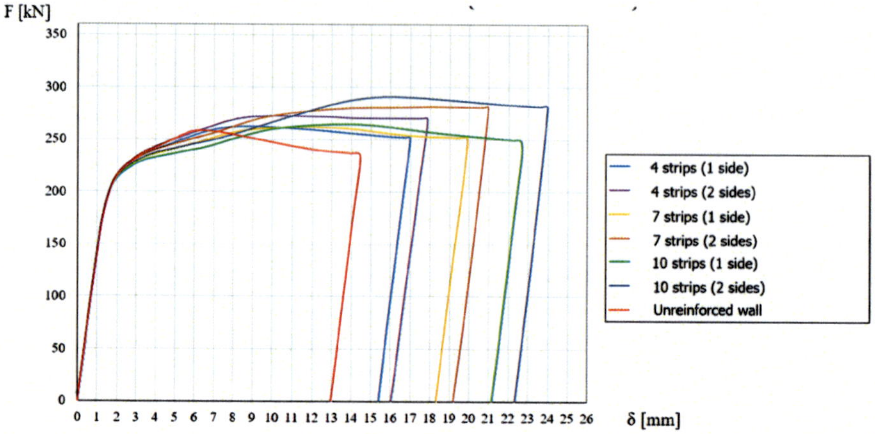

Fig. 16. Force-displacement curves of the wall reinforced by different configuration of H-FRP.

Table 1. Punctual interpretation of numerical test results.

Curve	F_{max}[kN]	F_u[kN]	δ_1[mm]	δ_2[mm]
Unreinforced wall	258.2	234.2	14.5	13.0
4 strips_1 side	264.6	253.9	16.9	15.3
4 strips_2 side	273.3	252.2	19.5	18.1
7 strips_1 side	267.7	251.1	22.5	20.8
7 strips_2 side	275.2	273.8	17.0	15.4
10 strips_1 side	283.1	281.9	19.9	18.3
10 strips_2 side	294.5	290.6	22.9	21.1

Table 2. Percentage increments due to use of different H-FRP systems.

Curve	F_{max}[%]	δ_{max}[%]
Unreinforced wall	/	/
4 strips_1 side	2.42	14.20
4 strips_2 side	5.52	14.71
7 strips_1 side	3.51	25.64
7 strips_2 side	6.18	27.14
10 strips_1 side	8.80	35.56
10 strips_2 side	12.33	36.68

5 Comparison Between H-FRP and C-FRP

The last group of analyses have been carried out using C-FRP (Carbon-Fibre Reinforced Polymers) strips as reinforcement systems of masonry panels in order to have a comparison with the H-FRP ones. Numerical test layout and C-FRP strips arrangement are the same used for H-FRP systems shown in Sect. 4. C-FRP strips have density of 1.82 g/cm^3 and mechanical properties implemented in the constant elastic matrix according to the parameters shown in Table 3.

Table 3. Mechanical C-FRP features: E = elastic modulus, G tangential modulus, N_u ultimate compression strength, 1, 2, 3 are x, y and z directions, respectively.

E_1 [MPa]	E_2 [MPa]	E_3 [MPa]	Nu_{12} [MPa]	Nu_{13} [MPa]	Nu_{23} [MPa]	G_{12} [MPa]	G_{12} [MPa]	G_{12} [MPa]
4700	4700	92316	0.15	0.12	0.12	2400	2100	2100

Figure 17 and Table 4 show the numerical results in terms of force-displacement curves and maximum force and displacement, respectively, achieved on masonry panel reinforced with C-FRP strips on either one or two sides. Table 5 shows the performance percentage increases of the two used technologies with respect to the behaviour of the unreinforced panel. From obtained results, it is clear that C-FRP systems offer the greatest improvements in terms of strength and ductility. Nevertheless, even if a better behaviour improvement is given by C-FRP strips, interventions with H-FRP systems are preferable, since the latter however satisfies the seismic retrofitting requirements with natural materials having lower environmental impact and costs.

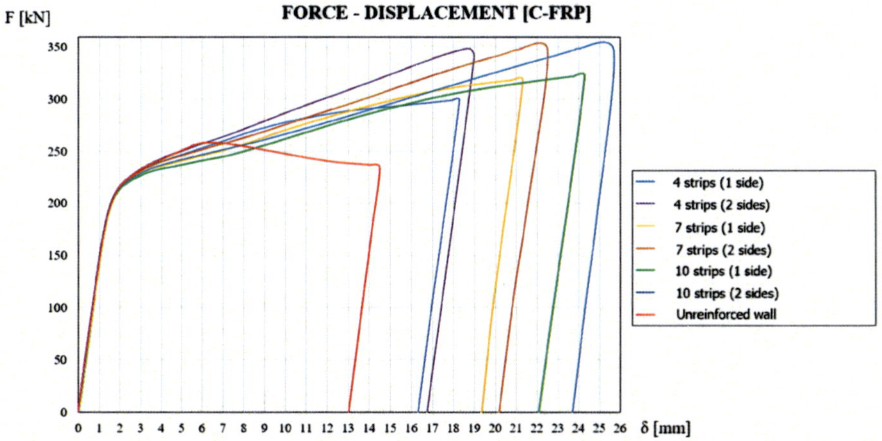

Fig. 17. Force-displacement curves of the masonry wall reinforced with C-FRP systems.

Table 4. Comparison of C-FRP solutions with the unreinforced wall behaviour.

Curve	F_{max}[MPa]	δ_{max}[mm]
Unreinforced wall	258.2	13.00
4 strips_1 side	298.7	18.30
4 strips_2 side	348.6	19.34
7 strips_1 side	318.7	21.30
7 strips_2 side	348.9	22.49
10 strips_1 side	322.0	24.30
10 strips_2 side	349.5	25.68

Table 5. Percentage increments of performances deriving from all studied seismic upgrading techniques.

Curve	H-FRP		C-FRP	
	F_{max} [%]	δ_{max} [%]	F_{max} [%]	δ_{max} [%]
Unreinforced wall	/	/	/	/
4 strips_1 side	2.42	14.20	13.55	28.96
4 strips_2 side	5.52	14.71	25.93	32.78
7 strips_1 side	3.51	25.64	18.98	38.97
7 strips_2 side	6.18	27.14	25.99	42.19
10 strips_1 side	8.80	35.56	19.81	46.50
10 strips_2 side	12.33	36.68	26.12	51.27

6 Conclusions

The goal of this paper was the use of Hemp-Fibre Reinforced Polymers (H-FRP) for seismic upgrading and retrofitting of masonry buildings. The starting point of the research was the numerical calibration carried out using the finite element software ABAQUS of literature experimental tests on both hemp specimens and unreinforced masonry walls. The very satisfactory simulation of experimental results allowed to proceed with the parametric numerical analysis on a masonry wall reinforced with different systems made of hemp fibres and thermoplastic resin aiming at finding the best strengthening configuration. By comparing the obtained numerical results with those obtained on the unreinforced wall it was detected a modest percentage increase in terms of resistance (from 2.42 to 12.33%), but an important improvement in terms of ductility (from 14.20 to 36.68%). The last step of the research program was devoted to the numerical investigation of Carbon-Fibre Reinforced Polymers (C-FRP) strips as an alternative technique of H-FRP for reinforcing masonry structures. C-FRP strips were arranged analogously to the H-FRP ones on the masonry panel examined as a benchmark structure. The gotten numerical results revealed that in the configuration with the major number of elements (10 strips on one or two sides) C-FRP are able to offer the best improvement of the unreinforced wall behaviour in terms of both ductility (more than two times) and resistance (almost two times). Nevertheless, even if C-FRP offered the highest behaviour improvement, H-FRP system can be considered as the best interventions since it satisfies the seismic retrofitting checks with the use of natural materials having environmental impact and costs lower than those of the other investigated synthetic materials. However, considering that in Italy this kind of green reinforcing technique of existing buildings is not allowed by current standards, further numerical studies and experimental campaigns should be performed in order to highlight much more the benefits of the proposed upgrading and retrofitting system.

References

Anthoine, A.: In-plane behaviour of masonry: a literature review. Report EUR 13840 EN, Comission of the European Communities, JRC - Institute for Safety Technology, Ispra, Italy (1992)

Bertolesi, E., Fabbrocino, F., Formisano, A., Grande, E., Milani, G.: FRP-strengthening of curved masonry structures: local bond behavior and global response. Key Eng. Mater. **747**, 134–141 (2017). https://doi.org/10.4028/www.scientific.net/KEM.747.134

Betti, M., Vignoli, A.: Numerical assessment of the static and seismic behaviour of the basilica of Santa Maria all'Impruneta (Italy). Constr. Build. Mater. **25**(12), 4308–4324 (2011)

Clementi, F., Gazzani, V., Poiani, M., Mezzapelle, P.A., Lenci, S.: Seismic assessment of a monumental building through nonlinear analyses of a 3D solid model. J. Earthq. Eng. (2017)

CNR DT 200: Instructions for the Design, Execution and Control of Static Consolidation Interventions through the use of Fiber-Reinforced Composites (in Italian). Italian National Research Council (2013)

Como, M.: Statics of Historic Masonry Constructions. Springer, Heidelberg (2013). https://doi.org/10.1007/978-3-642-30132-2

Cristescu, N.D., Craciun, E.M., Soos, E.: Mechanics of Elastic Composites. Chapman & Hall/CRC (2003)

Faggiano, B., Marzo, A., Formisano, A., Mazzolani, F.M.: Innovative steel connections for the retrofit of timber floors in ancient buildings: a numerical investigation. Comput. Struct. **87**(1–2), 1–13 (2009). https://doi.org/10.1016/j.compstruc.2008.07.005

Fonti, R., Borri, A., Barthel, R., Candela, M., Formisano, A.: Rubble masonry response under cyclic actions: experimental tests and theoretical models. Int. J. Mason. Res. Innov. **2**(1), 30–60 (2017)

Formisano, A., Marzo, A.: Simplified and refined methods for seismic vulnerability assessment and retrofitting of an Italian cultural heritage masonry building. Comput. Struct. **180**, 13–26 (2017)

Formisano, A., Iaquinandi, A., Mazzolani, F.M.: Seismic retrofitting by FRP of a school building damaged by Emilia-Romagna earthquake. Key Eng. Mater. **624**, 106–113 (2015). https://doi.org/10.4028/www.scientific.net/KEM.624.106

Ganz, H.R., Thurlimann, B.: Plastic strength of masonry shear walls. Institute of structural engineering, Swiss Federal Institute of Thecnology, Zurich, Switzerland (1984)

Gattulli, V., Antonacci, E., Vestroni, F.: Field observations and failure analysis of the basilica S. Maria di Collemaggio after the 2009 L'Aquila earthquake. Eng. Fail. Anal. **34**, 715–734 (2014)

La Manna Ambrosino, G., Brigante, D., Mauro, A., Formisano, A.: Concept, prototyping and application of a tensioning system for FRP ties into masonry structures. Key Eng. Mater. **747**, 298–304 (2017). https://doi.org/10.4028/www.scientific.net/KEM.747.298

Lagomarsino, S., Podestà, S.: Seismic vulnerability of ancient churches: II. Statistical analysis of surveyed data and methods for risk analysis. Earthq. Spectra **20**, 395–412 (2004)

Lee, J., Fenves, L.: Plastic-damage model for cycling loading of concrete structure. J. Eng. Mech. **124**(8) (1998)

Leite, J., Lourenco, P.B., Ingham, J.M.: Statistical assessment of damage to churches affected by the 2010–2011 Canterbury (New Zealand) earthquake sequence. J. Earthq. Eng. **17**, 73–97 (2017)

Lourenço, P.B.: Computational strategies for masonry structures. Doctoral thesis, Delft University Press, Netherlands (1996)

Lubliner, J., Oliver, J., Oller, S., Onate, E.: A plastic-damage model for concrete. Int. J. Solids Struct. **25**(3), 299–326 (1989). https://doi.org/10.1016/0020-7683(89)90050-4

Maragna, M., Casacci, S., Gentilini, C.: In-plane shear behaviour of masonry wall panels strengthened by structural repointing. Int. J. Mason. Res. Innov. **1**(3), 253–276 (2016)

Milani, G., Valente, M.: Failure analysis of seven masonry churches severely damaged during the 2012 Emilia-Romagna (Italy) earthquake: non-linear dynamic analyses vs conventional static approaches. Eng. Fail. Anal. **54**, 13–56 (2015a)

Milani, G., Valente, M.: Comparative pushover and limit analyses on seven masonry churches damaged by the 2012 Emilia-Romagna (Italy) seismic events: possibilities of non-linear finite elements compared with pre-assigned failure mechanisms. Eng. Fail. Anal. **47**, 129–161 (2015b)

Milani, G., Valente, M., Alessandri, C.: The narthex of the Church of the Nativity in Bethlehem: a non-linear finite element approach to predict the structural damage. Comput. Struct. **207**, 3–18 (2018)

Milani, G., Shehu, R., Valente, M.: Seismic upgrading of a masonry church with FRP composites. Mater. Sci. Forum **866**, 119–123 (2016)

Milani, G., Shehu, R., Valente, M.: Possibilities and limitations of innovative retro-fitting for masonry churches: advanced computations on three case studies. Constr. Build. Mater. **147**, 239–263 (2017)

Modena, C., Valluzzi, M.R., Tongini Folli, R., Binda, L.: Design choices and interven-tion techniques for repairing and strengthening of the Monza cathedral bell-tower. Constr. Build. Mater. **16**(7), 385–395 (2002)

Mosoarca, M., Apostol, I., Keller, A., Formisano, A.: Consolidation methods of Romanian historical building with composite materials. Key Eng. Mater. **747**, 406–413 (2017). https://doi.org/10.4028/www.scientific.net/KEM.747.406

Parisi, F., Augenti, N.: Evolutionary strength domains of unreinforced masonry spandrel panels including strain softening. In: Proceedings of the 9th Pacific Conference on Earthquake Engineering, Auckland, New Zealand, 14–16 April 2011

Preciado, A., Sperbeck, S.T., Ramirez-Gaytan, A.: Seismic vulnerability enhancement of medieval and masonry bell towers externally prestressed with unbonded smart tendons. Eng. Struct. **122**, 50–61 (2016)

Scutaru, M.L., Cofaru, C., Teodorescu-Draghicescu, H., Timar, J.: Properties of advanced new hemp fiber materials used in automotive engineering. In: Advances in Production, Automation and Transportation Systems (2013). ISBN: 978-1-61804-193-7

Stockdale, G.: Reinforced stability-based design: a theoretical introduction through a mechanically reinforced masonry arch. Int. J. Mason. Res. Innov. **1**, 101–141 (2016)

Teodorescu-Draghicescu, H., Vlase, S.: Homogenization and averaging methods to predict elastic properties of pre-impregnated composite materials. Comput. Mater. Sci. **50**, 1310–1314 (2011)

Valente, M., Milani, G.: Non-linear dynamic and static analyses on eight historical masonry towers in the North-East of Italy. Eng. Struct. **114**, 241–270 (2016)

Valente, M., Milani, G.: Effects of geometrical features on the seismic response of historical masonry towers. J. Earthq. Eng. **22**, 2–34 (2017)

Valente, M., Barbieri, G., Biolzi, L.: Damage assessment of three medieval churches after the 2012 Emilia earthquake. Bull. Earthq. Eng. **15**(7), 2939–2980 (2016). https://doi.org/10.1007/s10518-016-0073-7

Valluzzi, M.R., Binda, L., Modena, C.: Mechanical behaviour of historic masonry structures strengthened by bed joints structural repointing. Constr. Build. Mater. **19**(1), 63–67 (2005)

Vlase, S., Teodorescu-Draghicescu, H.: Finite element method used in the multibody systems with elastic elements. In: 4th IASME/WSEAS International Conference on Continuum Mechanic (CM 09), Univ Cambridge, Cambridge, England, 24–26 February 2009

Vlase, S., Teodorescu-Draghicescu, H., Calin, M.R., Serbina, L.: Simulation of the elastic properties of some fibre-reinforced composite laminates under off-axis loading system. Optoelectron. Adv. Mater. Rapid Commun. **5**(3–4), 424–429 (2011)

Restoration and Structural Analysis of an Irregular Byzantine Monument – A Discussion on a Simplifying Analysis Procedure

Styliani Papatzani[1](✉), Nikolaos Pnevmatikos[2], Eleni Zarogianni[1], Konstantinos Dimitroulias[1], Georgios Tzamalis[3], Georgios Sapiridis[4], and Vasilakis Flentzouris[4]

[1] Hellenic Ministry of Culture and Sports, Directorate for the Restoration of Byzantine and Post-byzantine Monuments, 8-10 Tzireon Street, 117 42 Athens, Greece

[2] School of Engineering, Department of Civil Engineering, University of West Attica, 250 P. Ralli & Thivon, 122 44 Egaleo, Greece

[3] G. Tzamalis & Associates Engineering and Consultancy, 4, Kiprion Ethnomartiron Sq., 113 63 Athens, Greece

[4] Proentasi - George D. Sapiridis & Associates - Design and Construction Company, 4 Skopou Street, 546 36 Thessaloniki, Greece

Abstract. The present papers explores the response of a rare monument situated on Mount Athos under several loading combinations. The structure comprises of two connected blocks; a two-storey masonry building which served as a workmen's house and a single-storey rectangular masonry block, which served as an ox stable. The masonry rectangular building is founded on a rock with masonry pillar footings of different height. Therefore, the elevation irregularity and the different elevations of the footings of the structure presents an additional challenge for the structural analysis. A brief architectural analysis of the building is presented, as well as a structural analysis with a finite element (FE) model of the structure. Performing seismic analysis, lateral force and response spectrum analysis, stresses and deformations at critical points of the structure were calculated. Maximum normal and shear stresses were compared with the stress capacity of the masonry. The latter was calculated according to the Eurocode 6 and empirical formulas. It was concluded that simplifying structural checks can be adequate for preliminary analysis since in that state time does not allow for a full FE analysis of the entire shell, although local failure mechanisms will be the subject of future research of the team.

Keywords: Irregular masonry monument · Finite element analysis

I. Vayas and F. M. Mazzolani (Eds.): PROHITECH 2021, LNCE 209, pp. 418–430, 2022.
https://doi.org/10.1007/978-3-030-90788-4_34

1 Introduction

In the current paper the ox stable, "Vourdounario" of the Pantokratoros Monastery, a monastery located at the east coast of the Athos peninsula is presented. It is a rare, representative specimen of an extinct type of building, as most of the monasteries' stall buildings have been reused, changing function or demolished due to the non-use of animals nowadays in various works.

The date of construction is not known, but it presents a typical form of such type of buildings in Mount Athos during the 18th and 19th century. In an engraved inscription, which is preserved on a stone on the staircase, the year 1863 is quoted. Moreover the structure is also depicted in an 1871 photograph (Fig. 1).

Fig. 1. Photograph taken in 1871 (Paschalides 2005)

The monument combines two distinct uses and hence, consists of two parts, a stable and a workhouse, that form a single unit (Fig. 2). The building is elongated with a rectangular floor plan and it is constructed by stone masonry, stone pillars, wooden vertical posts and wooden internal columns and it is covered by a wooden roof with sheet metal. The entire structure is founded on natural rock with steep inclination. Hence, a significant foundation height irregularity is prominent and although the main volume of the structure in the south shows a two-floor structure, in the north it is a semi-basement.

The first part of the building has a floor plan of 9,30 m × 22,60 m and a height of 5.70 m. It hosts the stall and occupies the western part of the complex. It is a ground floor building with a basement, which due to the steep slope of the ground, appears as a ground floor on the south side (Fig. 3). Coverage is provided with a three-sided roof. Despite its deterioration, the monument retains its originality and the typical elements of its function (wood flooring, wooden roof, feeders, benches, etc.) (Fig. 4).

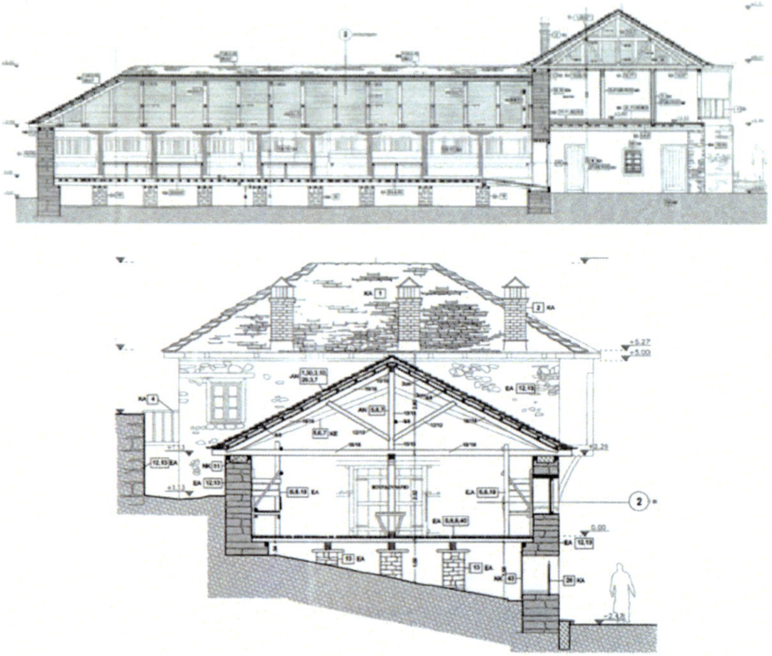

Fig. 2. Sections of the structure

Fig. 3. Recent photographs of different views of the structure; SE view shows the two-floor workman's house and the south side of the "vourdounario" stable. West view showing the west side of the "vourdounario" stable. South view showing the south side of the single-floor "vourdounario" and the two-floor workman's house towards the East.

The geometry of the building openings varies according to the use of each space. Typical are the openings of stable's basement, which are arched, formed with bricks. The openings of the stable are closed with wooden covers on the ground floor and simple safety rails in the basement (Fig. 4).

Fig. 4. "Vourdounario" interiors; (a) foundation level supporting the wooden floor of the ground level; (b) masonry pillar foundation on natural rock; (c) ground floor interiors – feeders; (d) wooden columns, floors and roof beams and frames and (e) wooden central column and roof frames

The second part has a floor plan of 10.80 m × 7.80 m and a height of 7.40 m. It houses the workmen's house and forms the eastern part of the complex (Fig. 5). It is a two-floor construction with a four-sided roof. In the northeast part of the ground floor a room is placed, while south of it a cobblestone semi-open area provides access to the stable. The access to the first floor was through an arched staircase on the east side of the building, as well as by a small wooden bridge to the north, leading to a covered balcony that occupies the entire east side of the level. From this balcony, one enters to the three rooms of the floor. The two extreme rooms housed dormitories, while the middle one hosted the kitchen. The room on the south side extends out of the ground floor, forming a "sahnisi" with timber struts.

Fig. 5. "Workmen's house; (a) east view; (b) south view with the sahnisi; (c) north view and (d) interior of the timber roof

2 Pathology of the Structure

The stable is maintained in a mediocre condition, with deterioration attributed to aging of the building materials, environmental and biological degradation and poor maintenance. The proximity to the sea has induced the additional deterioration caused by salts. In brief the following was observed:

- Moisture has caused severe deterioration of all wooden elements; the timber reinforcement (wooden connections embedded in masonry) have been severely deteriorated (Fig. 6a). Furthermore, the timber floors and roofs (Fig. 6b) have been additionally attacked by wood-eating insects (Fig. 6c). This observation is valid for both the "vourdounario" and workmen's house.
- Masonry walls (Fig. 6d) and pillar footings present loss of stone material and connecting mortar, while extensive vegetation is present in a number of locations (Fig. 6e).
- The brick elements which form the arcs of the openings present extensive peeling (Fig. 6f).

Fig. 6. Characteristic pathology of the structure

3 Description of the Structural Model

The structure was simulated using finite elements and was studied as a continuous structure formed by two parts the "vourdounario" and the workmen's house, sharing a wall with an opening which allows the communication of the two distinct areas (Fig. 7). The software used was SAP2000.

The walls were modelled using quadrilateral shell elements. Two different section properties were defined; (i) one with a thickness of 0.75 m (mostly belonging to the "vourdounario" and the ground floor of the workmen's house), (ii) walls had a thickness of 0.25 m belonging to the first floor of the workmen's house – front wall and all partition walls.

The actual mechanical properties of the masonry were derived for both types of walls by the empirical formulae. It should be noted that the thickness of the walls varied with height. In fact, at the base the walls had a thickness of 0.90 m, whereas at the top a thickness of 0.65 m. The average thickness equal to 0.75 m was used for the simulations.

The masonry pillar footings were of 0.60 $\times$ 0.60 m cross sections but of different heights, due to the ground inclination. The exact location of the pillars assigned to the model, was determined by a through surveying plot. The mechanical properties were assigned as per the walls.

The timber roofs were modelled using linear elements, which formed plain frames at specific distances forming the 3-dimensional wooden structure. Two different 3-D roof structures were modelled; one for the "vourdounario" and the other for the workmen's house. Primary beams were of 0.15 $\times$ 0.15 m and the secondary beams were of 0.05 $\times$ 0.05 m. The wooden triangular frames were simply supported on the side walls, while at midspan were connected with a wooden column. All timber columns transfer the loads to the underlying mid-line masonry pillar footings which were perfectly aligned.

The floors comprised of beams at specific spaces for the “vourdounario”, whereas for the workmen’s house the floors were modelled as a diaphragm with shell elements with the equivalent thickness, due to different construction conditions.

The boundary condition for the building was considered as fixed to the ground.

Lateral force (LFX in the X direction and LFY in the Y direction) and response spectrum seismic (RFX in the X direction and RFY in the Y direction) analysis were performed in both directions and compared.

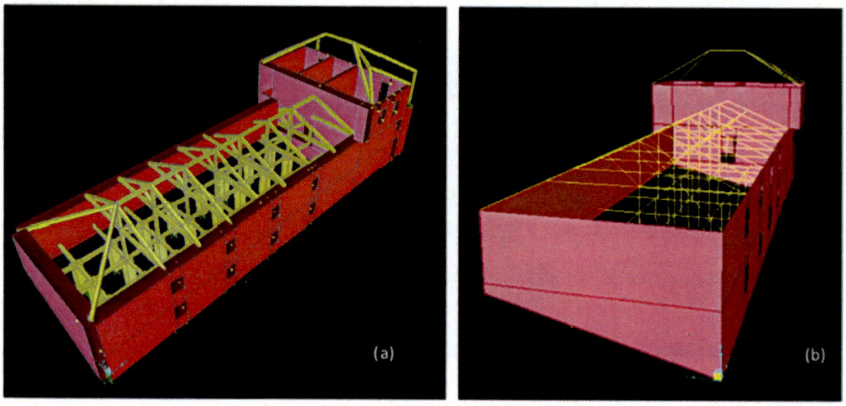

Fig. 7. Geometric layout of the finite element model (a) general layout, (b) different height of foundations

It should be noted that, typically, for lateral force analysis, software applies calculations on the first mode shape. However, in the case of this structure, the first mode shape marginally activated the wooden beams of the roof. Therefore instead of allowing the software to run the lateral force analysis using the first mode shape, the total mass was calculated using the seismic design response spectrum ($\Phi_{d(T)}$) and equivalent forces were applied manually along the walls of the structure.

In fact, using the Greek seismic code (Earthquake Planning and Protection Organization 2001), for the horizontal section of the spectrum:

$$\Phi_d = \gamma_I \cdot A \cdot \frac{\eta \cdot \theta \cdot \beta_0}{q} = 1{,}30 \cdot 0{,}24g \cdot \frac{1{,}0 \cdot 1{,}0 \cdot 2{,}5}{1{,}5} = 0{,}52g$$

where: $\gamma_I = 1{,}30$ (Table 2.3 - for category 4).

A = 0,24g (Table 2.1 & 2.2 for the Athos Peninsula)

$$\eta = \sqrt{\frac{7}{2+\zeta}} = \sqrt{\frac{7}{2+5}} = 1 \quad \text{(Eq 2.2 for } \zeta=5\% \text{ Table. 2.8)}$$

$\theta = 1$ (§2.3.7)

$\beta_o = 2.5$

$q = 1.5$ (Table 2.6)

$$V_o = \mathrm{M} \cdot \Phi_{d(\mathrm{T})}$$

The total mass of the structure is:

$$\mathrm{M} = \frac{N_{sd} \cdot l_{tot}}{g} = \frac{130{,}33 \cdot (22{,}3 + 8{,}3) \cdot 2}{g} = \frac{7.976{,}2}{g} kN$$

Therefore:

$$V_o = \frac{7.976{,}2}{g} \cdot 0{,}52g = 4.147{,}62kN$$

4 Simplifying Calculations

Given the time required to create a finite element model, the following simplifying calculations were made in order to check the results with the ones given by the FEA.

The mechanical properties of masonry were initially calculated using the Eurocode formulae (CEN 2005):

Compressive strength According to EC.6.1 § 3.6.1:

$$f_k = K \cdot f_b^{0,65} \cdot f_m^{0,25} \tag{3.1}$$

where:

f_k: is the characteristic compressive strength of the masonry, in N/mm^2.

K: is a constant.

f_b: is the normalized mean compressive strength of the units, in the direction of the applied action effect, in N/mm^2.

fm is the compressive strength of the mortar, in N/mm^2.

Therefore,

$$f_k = K \cdot f_b^{0,65} \cdot f_m^{0,25} = 0{,}45 \cdot 5{,}0^{0,65} \cdot 5{,}0^{0,25} = 1{,}915\,\mathrm{MN/m^2} = 1.915\,\mathrm{kN/m^2}$$

Shear strength (EC.6.1 § 3.6.2)

$$f_{vk} = f_{vk0} + 0,40 \cdot \sigma_d \leq 0,065 \cdot f_b \tag{3.5}$$

where:

f_{vk0}: is the characteristic initial shear strength, under zero compressive stress;

f_{vlt}: is a limit to the value of f_{vk};

σ_d: is the design compressive stress perpendicular to the shear in the member at the level under consideration, using the appropriate load combination based on the average vertical stress over the compressed part of the wall that is providing shear resistance;

f_b: is the normalized compressive strength of the masonry units, as described in 3.1.2.1, for the direction of application of the load on the test specimens being perpendicular to the bed face.

From Table 3.4, $f_{vk0} = 0{,}10$ N/mm^2 (MPa).

Moreover: $\sigma_d = N_{sd}/A_{TOIX,HOR} = 130{,}33/(1{,}00 \cdot 1{,}00) = 130{,}33$ kPa $= 130{,}33$ * 10–3 MPa.

Therefore,

$$f_{vk} = 0,10 + 0,40 \cdot 130,33 \cdot 10^{-3} = 0,152\,\text{MPa} \leq 0,065 \cdot 5 = 0,325\,\text{MPa}$$

However, since this type of structures does not conform with the Eurocodes because in old structures stone and mortar are not specified, it is preferred to use Tasios and Vitzilaiou giving compressible and shear strength according to the mortar and masonry characteristics.

Simplifying calculations for the structure subjected to vertical loading, shear loading and lateral loading have also been carried out but lie beyond the scope of the current paper and are therefore not shown.

5 FEA Model Results and Discussion

With respect to the response spectrum analysis it is worth noting that 300 mode shapes had to be considered in order to satisfy the EC8 rule which states that the number of mode shapes should be enough in order to excite at least 90% of the total mass. This

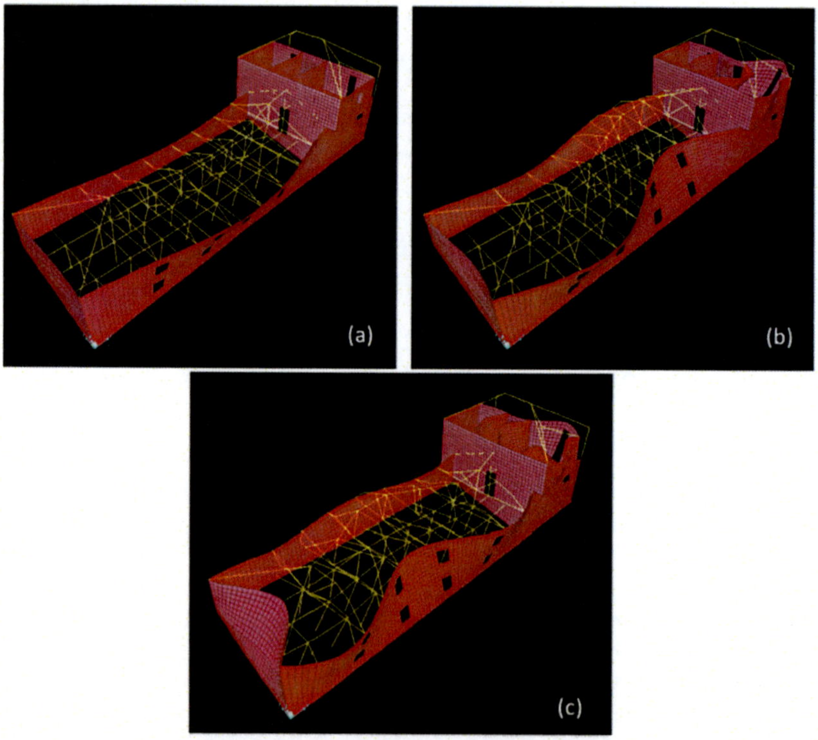

Fig. 8. Eigen modes of the finite element model (a) Mode 15th T = 0.069 s, (b) Mode 17th T = 0.059 s and (c) Mode 29th T = 0.048 s

significant number of mode shapes shows how robust this type of structure is. Figure 8 shows the main translational eigen modes at different periods (T).

5.1 Displacements

Figure 9 shows maximum displacements at specific joints. The maximum displacement for excitation at both directions for seismic action (X and Y) have been calculated for both lateral force analysis (LFX and LFY) and response spectrum analysis (RSX and RSY).

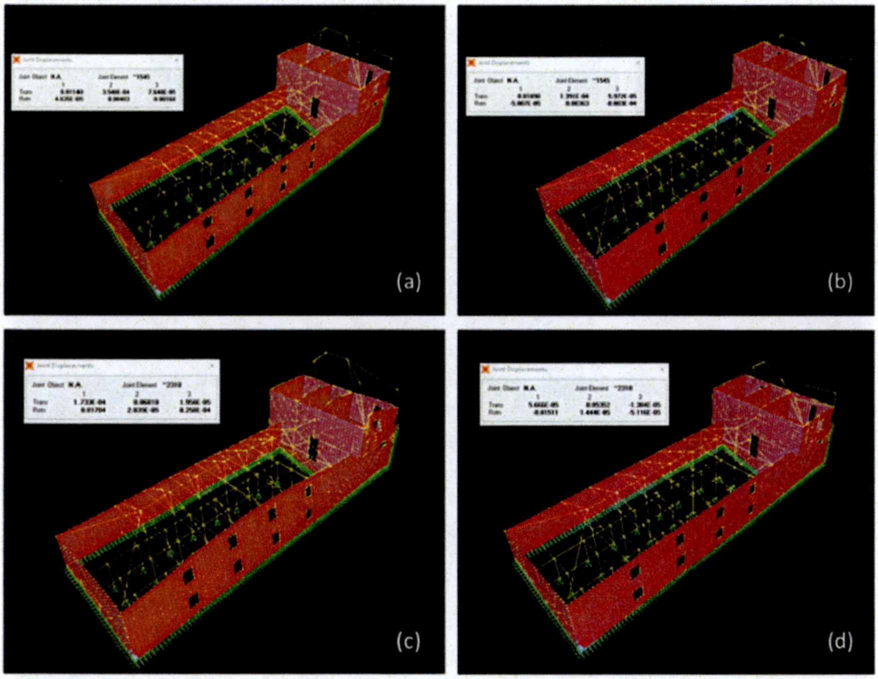

Fig. 9. Displacement results of the finite element model (a) RSX, (b) LFX, (c) RSY and (d) LFY

For the response spectrum analysis, the maximum displacement recorded for the RSY excitation was 6.01 cm in the middle of the south wall of "vourdounario". Due to the irregularity in section the workmen's house undergoes the highest displacements and the greatest stress as will be shown in the next paragraph. The maximum displacement recorded for the RSX excitation was 1.14 cm in the middle of the west wall of "vourdounario".

For the lateral force analysis, the maximum displacement recorded for the LFY excitation was 5.35 cm in the middle of the south wall of "vourdounario". Due to the irregularity in section the workmen's house undergoes the highest displacements and the greatest stress as will be shown in the next paragraph. The maximum displacement recorded for the LFX excitation was 1.09 cm in the middle of the west wall of "vourdounario".

5.2 Stresses

Figure 10 shows the normal (S11 and S22) and shear stress (S12) distribution in the structure. Maximum normal compressive stress calculated by the analysis does not exceed the masonry stress capacity which is 1.9 MPa, calculated using the Eurocode formulas. The stresses were calculated for seismic action in the x direction (RSX) and for seismic action in the y direction (RSY). Combinations with earthquake, permanent and variable actions have been considered, but are not shown for brevity.

Maximum stresses are identified by blue colour. Again, due to the irregularity in section the workmen's house undergoes the greatest stress as discussed previous with respect to displacements. The same goes for all corners of "vourdounario" which are the most vulnerable (Fig. 11).

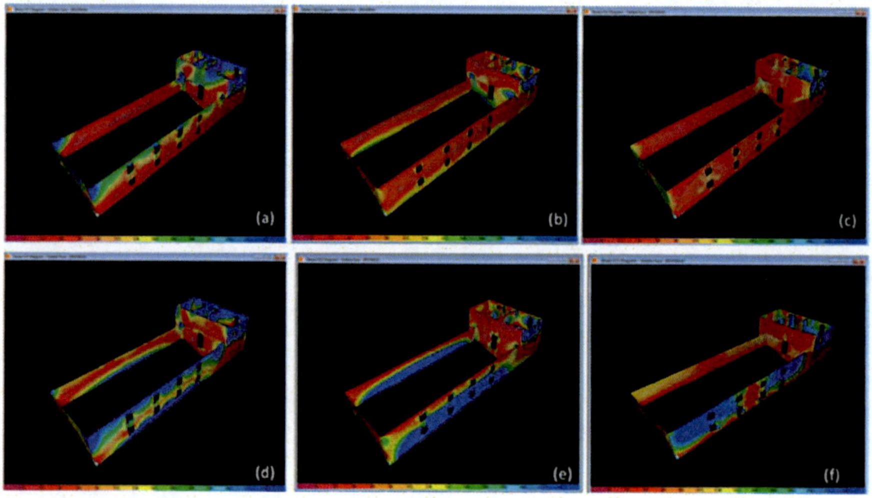

Fig. 10. Stress results of the finite element model (a) Stress 11 RSX, (b) Stress 22 RSX, (c) Stress 12 RSX, (d) Stress 11 RSY, (e) Stress 22 RSY and (f) Stress 12 RSY

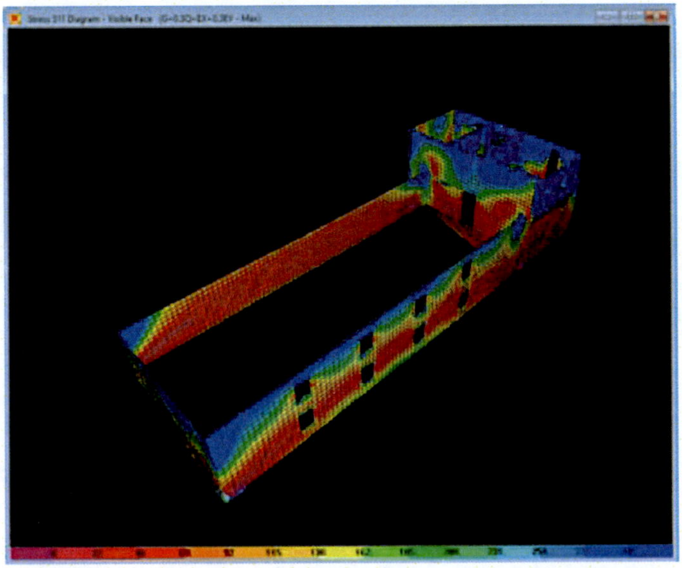

Fig. 11. Stress results (S11) of the finite element model Seismic combination (X direction)

6 Conclusions

In the present study, an extinct type of structure has been presented in terms of historical and typological characteristics. The current pathology of the monument has been briefly described and a set of simplifying calculations for the load bearing capacity of the structure have been recorded. Then a numerical finite element analysis model was designed to evaluate the displacements and consequently the stresses which are developed in the structure under various loading conditions. It was found that the elaborate analytical model yielded stresses which did not exceed the bearing capacity calculated using the simplified procedure. Therefore, this procedure even in irregular monuments can yield acceptable results if the time and resources are not sufficient for the development of a finite element model.

It is well known that, with finite element modelling, a better picture of the stress and deformation environment due to loading is produced. However, in many cases of masonry structures, the local failure mechanisms are proven more critical (Κωνσταντίνος Σπυράκος 2019). These local mechanisms cannot be detected by finite element modelling software and will be the subject of future investigation so that both analyses will be compared.

References

CEN (2005): BS EN 1996–1–1:2005 - Eurocode 6. Design of masonry structures. General rules for reinforced and unreinforced masonry structures (2005)

Earthquake Planning and Protection Organization (EPPO): National Greek seismic code (2001)

Paschalides, S.: Ιερά Μονή Παντοκράτορος. Προσκυνηματικός Οδηγός (Άγιο Όρος 2005) (In Greek: The Holy Monastery of Pantokrator: A Pilgrim's Guide). The Holy Monastery of Pantokrator (2005)

Spyrakos, K.: Κατασκευές από Τοιχοποιία. Αποτίμηση και επεμβάσεις για σεισμικά φορτία. Εργονόμος ΙΚΕ. (In Greek: Masonry structures. Evaluation and interventions for seismic loading). Ergonomos IKE (2019)

Three Dimensional Modeling of a Multiblock Roman Column Under Harmonic Excitations

Mazen Tabbara, Gebran Karam(✉), Charbel Beaino, and Josiane Jello

Department of Civil Engineering, Lebanese American University, Byblos, Lebanon
gkaram@lau.edu.lb

Abstract. A three dimensional DEM model was developed for a single multiblock Corinthian column from the Roman temple of Jupiter in Baalbek and used to investigate the response of the full size column as well as 0.5 m and 1 m reduced models that are used in shaking table experiments. 25 cycles of harmonic pulses of varying periods and accelerations were applied until collapse was observed. The critical combinations causing collapse were identified. The displacement history of the top column block was plotted alongside the harmonic excitation signal applied to the base of the column in order to visualize the rocking and/or wobbling response. The in-plane rocking and the out-of-plane wobbling were identified. Numerical stability charts were plotted showing safe-unsafe boundaries. Column responses included stable and unstable rocking of the whole column, stable and unstable multi-block rocking with toppling of upper blocks and combinations of rocking and wobbling. The type of column response, resulting structural disorder and collapse modes are shown to be dependent on the excitation period. The reliability of the results and the predictive capabilities of the model are discussed.

Keywords: Rocking · Overturning · Wobbling · Harmonic excitation · Distinct element method · Multiblock column

1 Introduction

The stability and seismic response of classical columns has been researched for the last twenty years in order to predict their response, develop strategies for preserving them and to understand the events that lead to their damage and overturning from an archeoseismological perspective. Seismic events represent one of the major hazards that endanger the conservation of large masonry structures such as encountered in archeological and historical sites. These monuments represent an important part of the world cultural and historical heritage requiring protection. Lebanon is particularly rich in these heritage sites such as Baalbek, Tyr, Saida, Tripoli, and Byblos. In addition to the immediate objective of strengthening and conservation, studying the damage and structural collapse caused to these archeological structures by historically recorded seismic events can help estimate the magnitude and nature of these events making contributions to archaeoseismology (Galadini et al. 2006; Hinzen 2009a, Sintubin 2011).

I. Vayas and F. M. Mazzolani (Eds.): PROHITECH 2021, LNCE 209, pp. 431–442, 2022.
https://doi.org/10.1007/978-3-030-90788-4_35

A UNESCO World Heritage site since 1984, the temple of Jupiter at Baalbek in the Bekaa valley of Lebanon is the largest temple complex built by the Romans relative to its size, its complexity and the height of its colonnade. The temple was originally surrounded by a colonnade of 54 external Corinthian columns. The site has been devastated by earthquakes over the last ten centuries notably the major earthquakes that occurred in the 12th century, in 1138–39 and 1157–59, 1170 and 1202 (Ambraseys and Jackson 1998; Guidoboni et al. 2004). The last destructive earthquake that hit Baalbek was in 1759. Only 6 damaged columns survived the November 25th, 1759 Bekaa valley earthquake (Ms ~7) (Ambraseys and Barazangi 1989; Ambraseys and Jackson 1998; Lewis, 1999). These large multiblock columns are 2.2 m in diameter and 20 m in height. They are constituted by five distinct blocks: a base block 1m high, three cylindrical drums of heights 7.5 m, 5.5 m and 4 m respectively, on top of each other constituting the shaft, and one capital block 2 m high. The 6 columns are connected at the top by an entablature, 4 m high, that consists of two rows of blocks only, the lower row combining architrave and frieze, the upper row forming the cornice (Raguette 1980). The colonnade sits on an elevated foundation made of large masonry blocks. Figure 1 shows a photograph of the colonnade.

The first major numerical effort in investigating the stability of classical columns was carried out by Psycharis et al. (2000) who used a 2D Distinct Elements Method, DEM, code (UDEC) to simulate the response of multi-drum columns to harmonic and earthquake excitations identifying numerically the existence of a safe-unsafe boundary for free standing columns in the acceleration-period domain for harmonic excitations. Psycharis et al. (2000) also investigated qualitatively the modes of overturning of columns noting that for earthquake motions the acceleration level required for collapse was similar to the one imposed by around three cycles of harmonic excitations close to the predominant period. Mouzakis et al. (2002) then reported on an experimental program subjecting a scaled down column model to 19 tests corresponding to earthquake records. The high sensitivity of the experiments was noted as well as the frequency effect of the base motion. Residual displacements were measured but no major rocking or overturning was observed. UDEC was also used by Papantonopoulos et al. (2002) to model numerically in 2D 4 of these tests showing qualitative agreement without predicting any overturning. Komodromos et al. (2008) simulated the response of ancient columns to harmonic and earthquake excitations in a DEM parametric study with a very small experimental program. They observed that the required accelerations to initiate rocking or sliding were inversely related to the excitation frequencies, with rocking at low frequency harmonic excitations and sliding at very high frequencies. Hinzen (2009b) used numeric rigid block models of columns representing those fallen at the Cathedral of Sussita to study the columns' response. Dimitri et al. (2011) carried out a systematic parametric study based on a DEM model to evaluate the dynamic behavior and resistance of multi-drum columns and arches on buttresses subjected to two different base motions such as step and harmonic impulses. They also presented a detailed investigation on failure domains and modes of collapse.

Toumbakari and Psycharis (2010) used 3DEC, a three dimensional DEM, to model parametrically the three dimensional seismic response of a free standing column of the Aphrodite Temple at Amathus by subjecting the column to four different earthquakes, reporting stability in all cases. Ambraseys and Psycharis (2011) studied the earthquake stability of columns and statues using 3DEC to model the statue of Apollo at the Academy of Athens. Psycharis et al. (2013) and Stefanou et al. (2015) presented a methodology for the seismic risk assessment of classical columns using 3DEC to model a multi-drum column subjected to synthetic ground motions.

Frontal view of the colonnade showing the bases of the three fallen columns in the 1759 earthquake (two on the right side and one on the left side above).

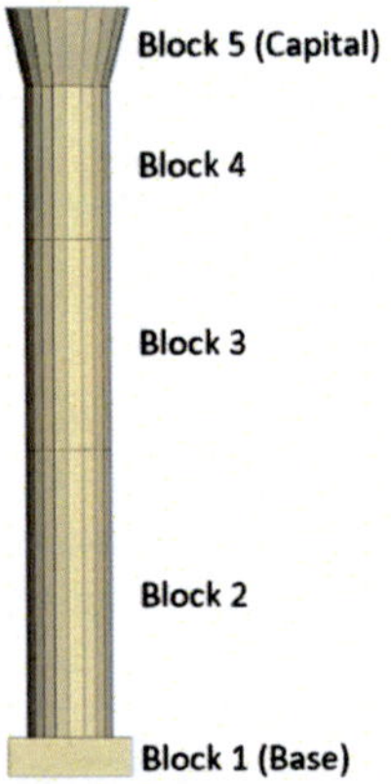

The 3DEC single column five blocks computational model.

Fig. 1. Photo of Baalbek colonnade and 3DEC single column model.

Drosos and Anastosopoulos (2014; 2015) presented the results of an interesting experimental program on scaled down columns subjected to Ricker wavelets and real seismic records. In total 25 tests were run on a free standing columns and another 24 on a portal of two columns. Minor rocking and residual displacements were observed but no overturning or large scale rocking. The authors also reported the high sensitivity of the results to minor changes in experimental parameters as well the fact that spectral displacement and the length scale were more effective intensity indicators that peak ground accelerations.

More recently Pappas et al. (2017) modeled numerically the dynamic response of a free-standing ancient column in the Roman Agora of Thessaloniki using 3DEC and subjecting the column to multiple seismic records selected based on the seismic hazard for the region of interest to identify the various impacts in the relative importance of earthquake intensity measures.

Following previous numerical and experimental work by the authors on the seismic stability of large columns with variable size drums such as those of Baalbek (Tabbara and Karam 2010, 2018 and 2019), it was found necessary to develop a 3D numerical model to capture the out of plane movements as well as the full response of the various blocks that constitute the columns with respect to each other.

The ultimate goal is to develop robust tools to analyze the response of the Baalbek columns and similar rocking blocks structures to complex earthquake excitations, keeping in mind that earthquake time histories and acceleration records can be treated as a series of pulses or pulse trains in the analysis of rocking blocks; hence the appeal of developing a modeling approach based on simple periodic pulses at first then combining them into a complex earthquake record analysis. The work presented in this paper is part of a larger research effort to that effect by the authors.

2 Numerical Modeling

2.1 Method

A commercial DEM software, 3DEC (Itasca 2019), was used to model the response of the reduced models and to establish the validity of the numerical approach in modeling the full size colonnade. A numerical model was developed for 0.5 m and 1.0 m reduced column models that were used in shaking table experiments (Tabbara and Karam 2018 and 2019) and for the full sized 20 m Baalbek column. All DEM blocks were modeled as rigid with non-penetrating contact with density = 2720 kg/m^3 and the joint material was assumed to follow a Mohr Coulomb model with a friction angle = 26°, a dilation angle = 0°, cohesion = 0 and tension cutoff = 0. Damping was not considered in the model given that the strong motion part of an earthquake signal response, which controls the stability of the structure, is not sensitive to the actual value of damping. Since the proposed numerical analysis is concerned with the stability threshold during strong shaking leading to overturning, the damping coefficient was set, for simplicity, equal to zero (Psycharis et al. 2000).

Two types of loads were applied to the model: (i) the self-weight of the blocks and (ii) a repeated harmonic cosine acceleration pulse with a prescribed amplitude and period. All pulses were applied for up to 25 cycles, and for a given period the acceleration was incremented until overturning of the whole column or part of it was observed within the 25 cycles of signal application. This approach allows the determination of a conservative value for the critical acceleration required for a given period. The period, velocity amplitude and corresponding acceleration amplitude were noted for all the critical combinations causing overturning as well as the mode of overturning. The displacements and rotations in three dimensions of each of the five blocks were tracked to identify the rocking movements as well as the out of plane movements and wobble.

2.2 Results

The results of the numerical modeling are presented in Figs. 2, 3, 4, 5, 6, 7 and 8. Figures 2, 3 and 4 present the in-plane and out-of-plane displacements' of the top of the columns versus the applied cosine pulses at selected critical values that caused collapse. The displacement of the bottom of the column is provided for reference in black, while the in-plane displacement and the out-of-plane displacements of the top of the column are shown in red and green respectively. A flat green line indicates that no out-of-plane displacements are taking place. Overlapping between the red and black signals indicates that the column is moving in one piece with no rocking. Phased out red and black signals indicates that rocking is taking place and the displacement of the top with respect to the bottom is in-plane and can be calculated from the difference between the two displacements' values at the same time. A non-zero green line indicates that out-of-plane displacement is taking place simultaneously with an in-plane displacement indicating wobbling.

Figures 5, 6 and 7 present the collapse/overturning modes of the columns modeled under the critical cosine pulse accelerations. A snapshot side view and a top view are provided of the columns at the instant of collapse. The side view shows the mode of collapse, the number of overturning blocks and the status of disorder of the different blocks. The top view allows the determination of whether the overturning is taking place in the plane or out of the plane of the movement. Out-of-plane overturning is indicative of wobble.

Figure 8 presents the stability charts showing the boundary between the safe and unsafe domains for the three column sizes. Harmonic pulses that have accelerations and periods that plot above these boundaries will cause collapse and overturning of the columns for each of the three sizes respectively.

3 Discussion of the Results

The three columns sizes studied, 0.5 m and 1 m shaking table models, and the 20 m full sized column exhibited consistently the same response patterns to the applied harmonic pulses but with a shift in the time domain as expected according to the applicable similitude laws. With reference to Figs. 2, 3, 4, 5, 6 and 7 three response modes are observed for columns:

(a) At large periods, roughly greater than 0.6 s, 0.8 s and 3 s for the 0.5 m, 1 m and 20 m columns respectively, the movement of the top of the column follows closely that of the base with a minor lag, indicative of the inception rocking with a very small amplitude that increases suddenly resulting in overturning of the whole column after one or two impacts in the plane of the applied pulse. No wobbling is observed.

(b) At intermediate periods, roughly in the intervals of [0.3 s; 0.6 s], [0.4 s; 0.8] and [1.5 s; 3 s] for the 0.5 m, 1 m and 20 m columns respectively, multiple rocking is observed prior to overturning; the rocking response has a time interval between impacts, or an apparent period, that is distinct from the period of the harmonic excitation with a resonance like response culminating within a few cycles in overturning of the

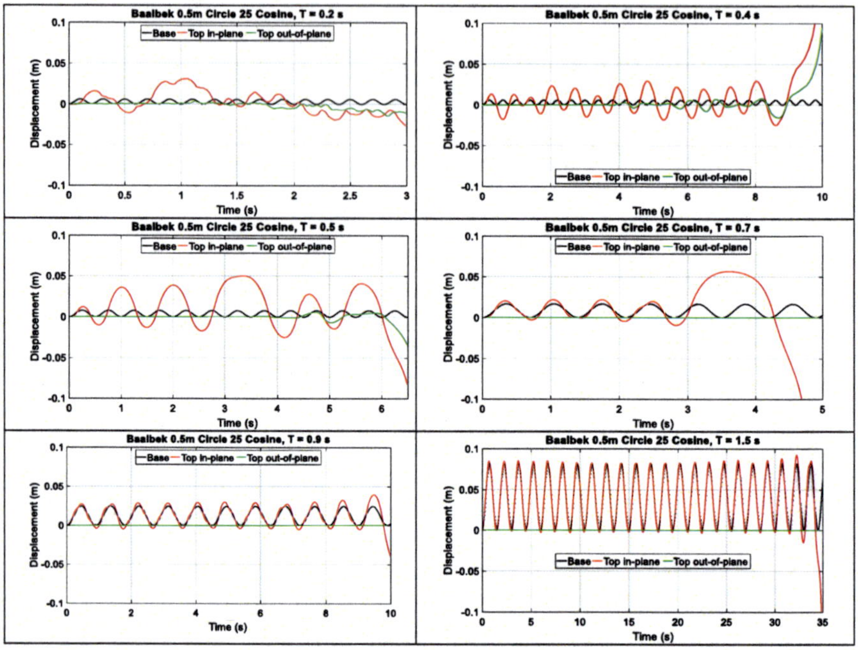

Fig. 2. Displacements of the top of the 0.5 m column for selected critical cosine pulses.

whole column out of the plane of the applied pulse. An out-of-plane displacement is observed with an apparent period similar to that of the in-plane displacement indicating wobbling. Wobbling intervenes and takes over after a few cycles of rocking; the in-plane displacement observed becomes controlled by the rolling movement. The cylindrical column can roll on its edge under small inclination angles and, contrary to rocking, the pole of rotation does not change abruptly from one to the other corner of its base as in rocking, but rather it follows a smooth translation along that curved edge (Stefanou et al. 2011).

(c) At small periods, roughly smaller than 0.3 s, 0.4 s and 1.5 s for the 0.5 m, 1 m and 20 m columns respectively, large accelerations are required to cause collapse. The movement of the top of the columns follows a complex response, totally different from the forcing harmonic excitation. The observed displacements are indicative of multiple inter block rocking and probably wobbling movements resulting in toppling of one or more of the top blocks alone. The large accelerations reached are sufficient to activate the rocking modes of shorter columns constituted of 3, 2, and 1 blocks.

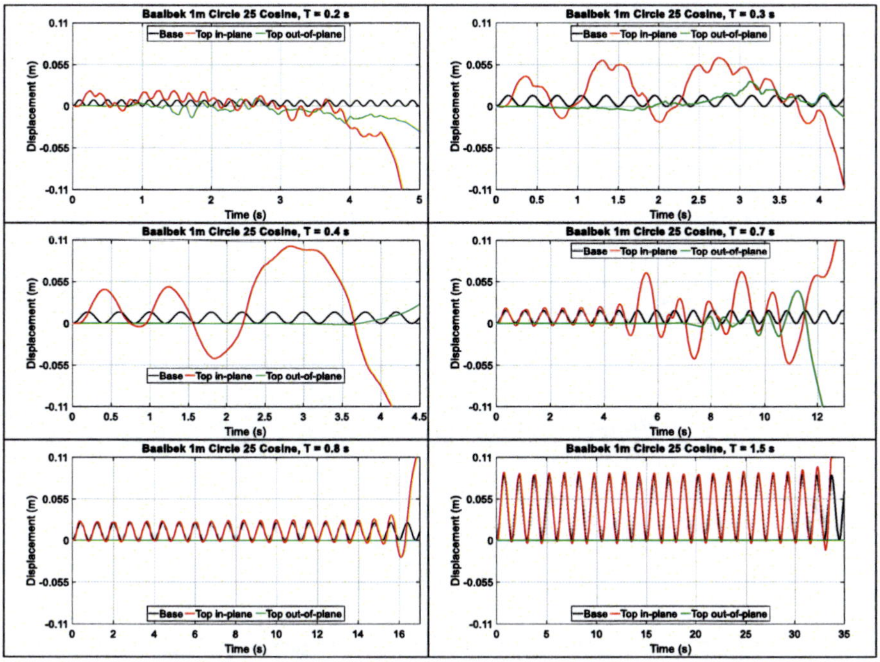

Fig. 3. Displacements of the top of the 1 m column for selected critical cosine pulses.

With reference to Fig. 8, the numerically compiled safe-unsafe boundaries follow at large periods a more or less horizontal line that rises sharply in a hyperbolic fashion as periods drop below a threshold value. This characteristic threshold period is unique for each column and related to its size. For periods above 2.5 s the 20 m column stability chart falls on a flat horizontal line that follows the theoretical minimal required acceleration for overturning: $a = \alpha g$, where α is the critical stability angle given by $\alpha = \cot^{-1}(H/B)$ with H = height of the column or block and B = width of the column or block (Housner 1963). For Baalbek $\alpha = 0.10956$. Whereas the horizontal part of the 20 m column stability chart fall exactly on $\alpha = 0.10956$, the horizontal parts of the 0.5 m and 1 m columns' stability charts fall below this value indicating a size related lower stability or a higher sensitivity of numerical parameters.

At very short periods of the order of 0.1 s to 0.3 s the required accelerations for overturning grow hyperbolically and may exceed those required to overturn less slender columns with shape factors $\alpha 2$, $\alpha 3$, $\alpha 4$ greater than $\alpha 1 = \alpha = 0.10956$. These shape factors ($\alpha 2$, $\alpha 3$, $\alpha 4$) correspond to those of columns made up of the top single block, two top blocks, and then the three top blocks respectively. At these accelerations the rocking of the top parts of the multiblock columns can be activated by the harmonic excitation signal applied and multiple rocking modes with overturning or toppling of one or more of the upper blocks are observed such as in Figs. 5, 6 and 7.

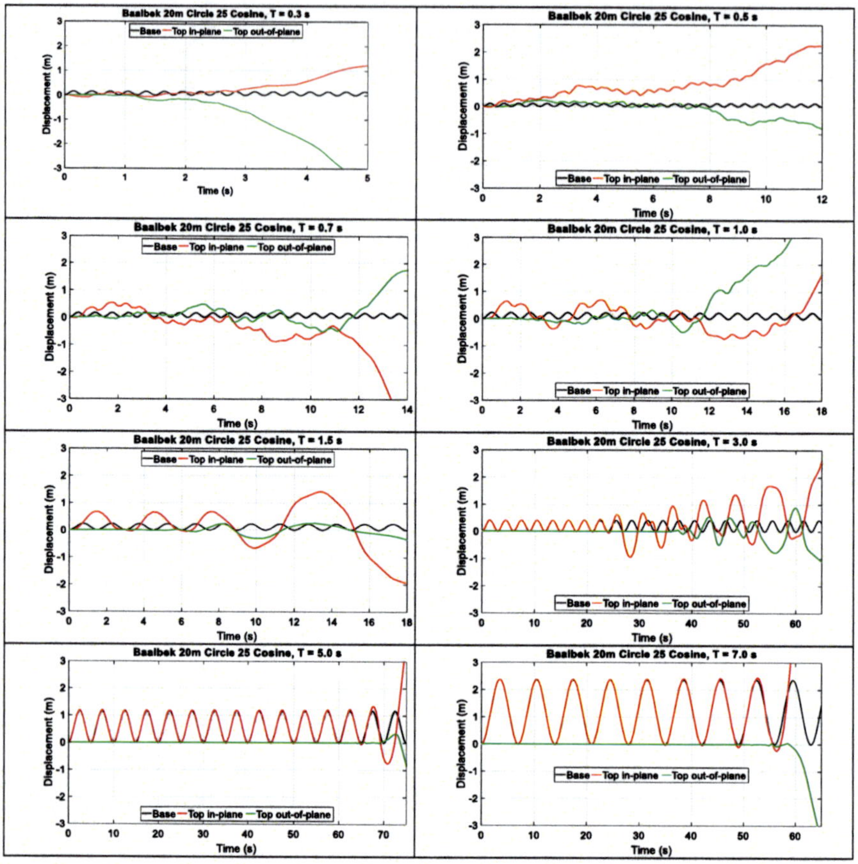

Fig. 4. Displacements of the top of the 20 m column for selected critical cosine pulses.

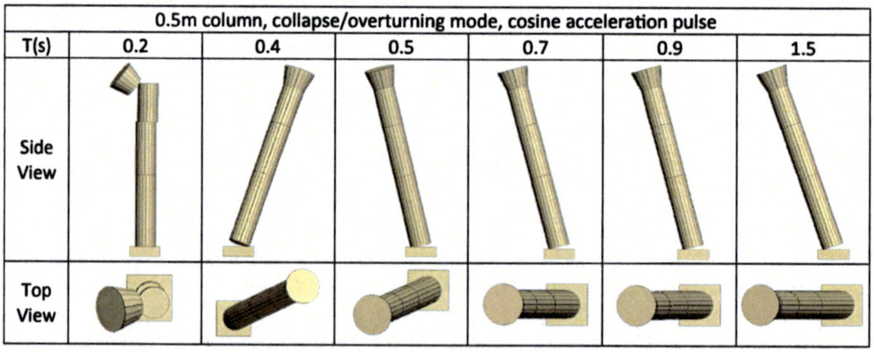

Fig. 5. Collapse/overturning modes of the 0.5 m column under critical cosine pulse accelerations.

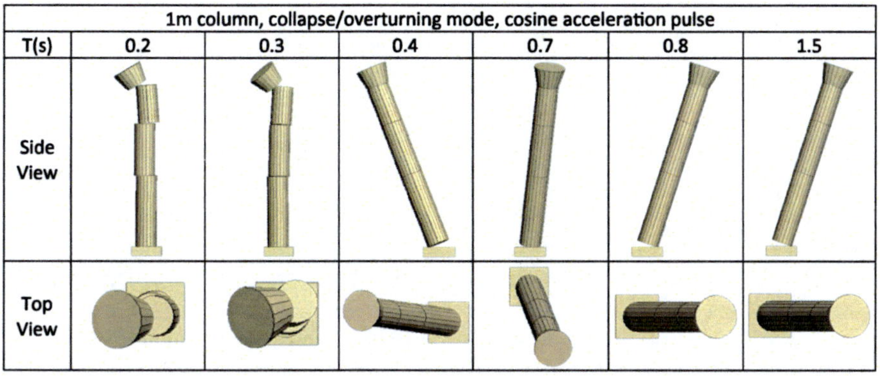

Fig. 6. Collapse/overturning modes of the 1 m column under critical cosine pulse accelerations.

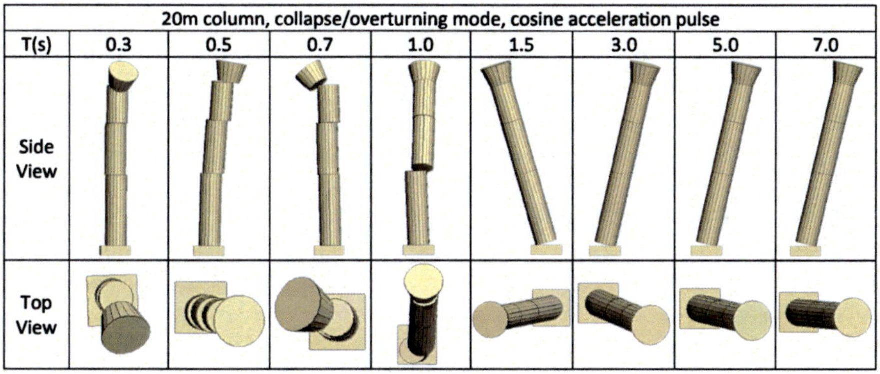

Fig. 7. Collapse/overturning modes of the 20 m column under critical cosine pulse accelerations.

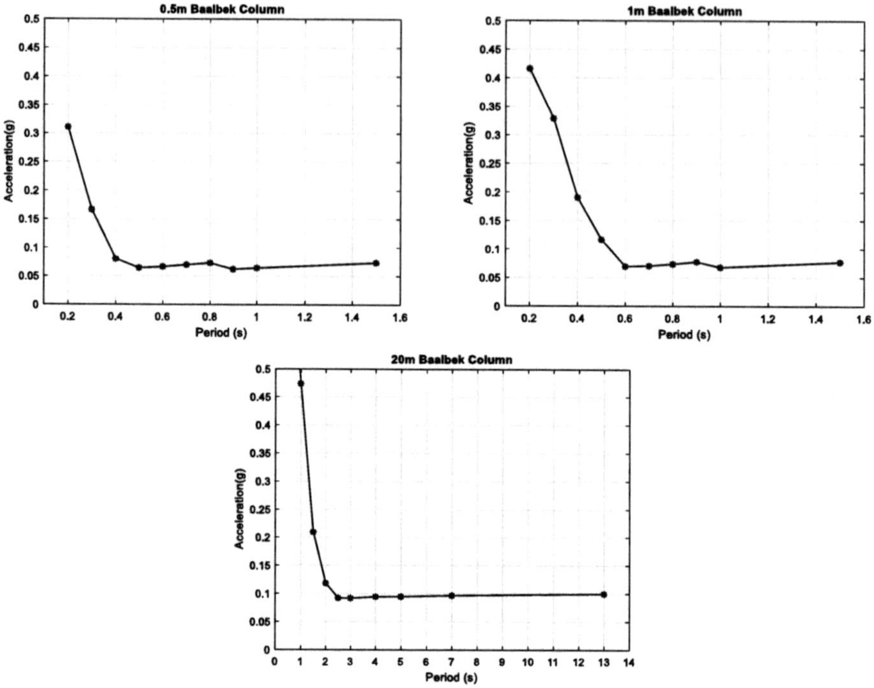

Fig. 8. Stability charts showing the boundary between the safe and unsafe domains for the three column sizes.

4 Conclusions

The numerical 3DEC simulations identified the same type of response and collapse modes for all multiblock column sizes investigated but with a shift in the time domain which agrees with similitude rules. The 1/40 and the 1/20 scaled models are shown to be qualitatively equivalent to the full sized column while being less stable.

This work confirmed the existence of three regimes for the response of multiblock columns to harmonic pulses/excitations in the acceleration period domain:

(i) at large excitation periods the multiblock column behaves as a single block with very little rocking if at all; failure is by overturning of the whole column; (ii) at intermediate excitation periods rocking then wobbling are observed prior to overturning; failure is by overturning of the whole column out of the plane of the applied excitation because of wobble; (iii) at small excitation periods it is observed that the top blocks may enter into rocking movements of their own, with or without accompanying wobbling; the movement of the top of the column follows a complex response, totally different from the forcing harmonic excitation, failure is mostly by toppling of the top blocks alone.

Wobbling, a particular type of instability of inverted pendulum, was shown to be confined to a definite region in the acceleration-period response domain. The conditions for the bifurcation from rocking into wobbling have yet to be confirmed.

The work presented here has confirmed the capabilities of the proposed 3D numerical approaches in simulating test results and real life events. The results presented are part of a larger ongoing research effort to investigate the response of colonnades to complex seismic events.

References

Ambraseys, N.N., Barazangi, M.: The 1759 earthquake in the Bekaa Valley: implications for earthquake hazard assessment in the eastern mediterranean region. J. Geophys. Res. **94**(B4), 4007–4013 (1989)

Ambraseys, N.N., Jackson, J.A.: Faulting associated with historical and recent earthquakes in the eastern mediterranean region. Geophys. J. Int. **133**, 390–406 (1998)

Ambraseys, N., Psycharis, I.N.: Earthquake stability of columns and statues. J. Earthq. Eng. **15**(5), 685–710 (2011)

Dimitri, R., Lorenzis, L., Zavarise, G.: Numerical study on the dynamic behavior of masonry columns and arches on buttresses with the discrete element method. Eng. Struct. **33**, 3172–3188 (2011)

Drosos, V., Anastasopoulos, I.: Shaking table testing of multidrum columns and portals. Earthq. Eng. Struct. Dyn. **43**, 1703–1723 (2014)

Drosos, V.A., Anastasopoulos, I.: Experimental investigation of the seismic response of classical temple columns. Bull. Earthq. Eng. **13**(1), 299–310 (2015)

Galadini, F., Hinzen, K.G., Stiros, S.: Archaeoseismology: methodological issues and procedure. J. Seismolog. **10**(4), 395–414 (2006)

Guidoboni, E., Bernardini, F., Comastri, A., Boschi, E.: The large earthquake on 29 June 1170 (Syria, Lebanon, and central southern Turkey). J. Geophys. Res. Solid Earth **109**(B7) (2004)

Hinzen, K.G.: Simulation of toppling columns in archaeoseismology. Bull. Seismol. Soc. Am. **99**(5), 2855–75 (2009a)

Hinzen, K.G.: Sensitivity of earthquake toppled columns to small changes in ground motion and geometry. Israel J. Earth Sci. **58** (2009b)

Housner, G.W.: The behavior of inverted pendulum structures during earthquakes. Bull. Seismol. Soc. Am. **53**(2), 403–417 (1963)

Itasca Consulting Group Inc.: 3DEC-3 Dimensional Distinct Element Code, version 5.2 –User's Guide, Minneapolis, USA (2019)

Komodromos, P., Papaloizou, L., Polycarpou, P.: Simulation of the response of ancient columns under harmonic and earthquake excitations. Eng. Struct. **30**(8), 2154–2164 (2008)

Lewis, N.N.: Baalbek before and after the earthquake of 1759: the drawings of James Bruce. Levant **31**(1), 241–253 (1999)

Mouzakis, H.P., Psycharis, I.N., Papastamatiou, D.Y., Carydis, P.G., Papantonopoulos, C., Zambas, C.: Experimental investigation of the earthquake response of a model of a marble classical column. Earthq. Eng. Struct. Dyn. **31**(9), 1681–1698 (2002)

Papantonopoulos, C., Psycharis, I.N., Papastamatiou, D.Y., Lemos, J.V., Mouzakis, H.P.: Numerical prediction of the earthquake response of classical columns using the distinct element method. Earthq. Eng. Struct. Dyn. **31**(9), 1699–1717 (2002)

Pappas, A., Sextos, A., Da Porto, F., Modena, C.: Efficiency of alternative intensity measures for the seismic assessment of monolithic free-standing columns. Bull. Earthq. Eng. **15**(4), 1635–59 (2017)

Psycharis, I., Papastamatiou, D., Alexandris, A.: Parametric investigation of the stability of classical columns under harmonic and earthquake excitations. Earthq. Eng. Struct. Dyn. **29**, 1093–1109 (2000)

Psycharis, I.N., Fragiadakis, M., Stefanou, I.: Seismic reliability assessment of classical columns subjected to near fault ground motions. Earthq. Eng. Struct. Dyn. **42**(14), 2061–79 (2013)

Ragette, F.: Baalbek. Chatto & Windus, London (1980)

Sintubin, M.: Archaeoseismology: past, present and future. Quat. Int. **242**(1), 4–10 (2011)

Stefanou, I., Psycharis, I., Georgopoulos, I.: Dynamic response of reinforced masonry columns in classical monuments. Constr. Build. Mater. **25**, 4325–4337 (2011)

Stefanou, I., Fragiadakis, M., Psycharis, I.N.: Seismic reliability assessment of classical columns subjected to near source ground motions. In: Psycharis, I., Pantazopoulou, S., Papadrakakis, M. (eds.) Seismic Assessment, Behavior and Retrofit of Heritage Buildings and Monuments. Computational Methods in Applied Sciences, vol. 37, pp. 61–82. Springer, Cham (2015). https://doi.org/10.1007/978-3-319-16130-3_3

Tabbara, M.R., Karam, G.K.: The use of numerical methods to investigate the seismic response of a large block masonry colonnade. In: Proceedings of the 9th U.S. National & 10th Canadian Conference on Earthquake Engineering (9USN/10CCEE): Reaching Beyond Borders, Toronto, Canada, 25–29 July, pp. 5148–5157 (2010)

Tabbara, M., Karam, G.: Experimental investigation of the stability of colonnades under harmonic excitation. In: 16th European Conference on Earthquake Engineering (ECEE) 2018 (2018)

Tabbara, M., Karam, G.: Experimental, numerical and theoretical investigation of the rocking response of Baalbek columns under harmonic excitations. J. Earthq. Eng. (2019). https://doi.org/10.1080/13632469.2019.1693448

Toumbakari, E., Psycharis, I.N.: Parametric investigation of the seismic response of a column of the Aphrodite Temple in Amathus, Cyprus. In: 14th European Conference on Earthquake Engineering (ECEE) 2010 (2010)

Assessment of Archaeological Theories for Monuments on the Basis of Sophisticated Structural Analysis Numerical Simulations

Konstantinos A. Papadopoulos(✉)

Directorate of Prehistoric and Classical Antiquities, Hellenic Ministry of Culture and Sports, Trivonianou 62, 11636 Athens, Greece
k.papadopoulos@culture.gr

Abstract. This paper highlights the benefits of using sophisticated numerical simulations to assess theories related to archaeological and historical aspects of monuments, if these theories are based on evidence that can be evaluated through structural analysis. The main advantage of this method of assessment is that it permits the necessary scrutiny through the conduction of numerous parametric numerical analyses (which can be characterized as series of "virtual experiments") without significant cost in time or money. An important side-benefit of the method is that when it reaches satisfactory results, it includes verification and calibration of numerical models which can be used also for the diagnosis of damages of the monument under analysis, and potentially for the design of a structural intervention on the monument. To demonstrate the capacities of the method, a case study is presented, regarding the cause of the unusual damages of a column in the classical monument of the Propylaea on the Acropolis of Athens.

Keywords: Monuments · Diagnosis of damages · FE simulations · Numerical reproduction of damages

1 Introduction

Structural analysis and particularly diagnosis of damages when they concern monuments are usually quite complicated and difficult tasks. This is because nearly each monument constitutes a singular case which requires additional efforts for the acquisition of data on material properties, internal morphology and damage, and for the adequate interpretation of structural arrangements, overall organization and historical facts (Icomos/Iscarsah Committee 2005). In their efforts to address these complex issues, the scientists who work for the protection of monuments relay, increasingly during the recent years, on sophisticated numerical simulations.

Advanced computer methods started to be applied for the analysis of historical structures in the mid-1990's, when the potentials of this approach was demonstrated by the pioneered work of Macchi et al. (1993), Papastamatiou and Psycharis (1993), Croci (1995), and Mola and Vitaliani (1995). Since then, numerous numerical studies have been carried out for the better understanding of the structural behaviour of monuments

I. Vayas and F. M. Mazzolani (Eds.): PROHITECH 2021, LNCE 209, pp. 443–457, 2022.
https://doi.org/10.1007/978-3-030-90788-4_36

(e.g. Roca et al. 2004; Psycharis et al. 2000, 2003; Lourenço and Mourão 2001; Konstantinidis and Makris 2005; Chiorino et al. 2008; Peña and Chávez 2014; Bartoli et al. 2015; Papadopoulos and Vintzileou 2016; Papadopoulos et al. 2019). Obviously, most (if not all) of these studies were ultimately aiming to be found useful during the decision-making process upon intervention in the monuments concerned.

Moreover, there are cases where numerical analyses have been used directly for the verification of the diagnosis of damages for a monument or for the investigation of a specific theory concerning the cause(s) of collapse (or of the severe damages) of a monument. One of the earliest relative examples is the work of Zambas (1994) who investigate numerically the hypothesis that the general deformations of the western part of the Parthenon walls were caused by the fire that occurred in the interior of the temple in the 3rd century A.D. Zambas used the Finite Element software Sofistik, and the derived results strengthen his hypothesis. Quite later the author of the present paper used the FE software Abaqus to verify that most of the severe deformations of the classical temple of Apollo Epikourios at Bassai, Peloponnese, are due to the settlements of its foundation (Papadopoulos 2010). Soon after, with the use of the same software, the possibility that the missing epistyle beams from the north façade of the Bassai temple were collapse due to an earthquake, was investigated (Papadopoulos and Vintzileou 2014), however this investigation did not yield definitive results. Similarly, Alexandris et al. (2014) used the Distinct Element software 3-DEC to examine in a quantitative way the widely accepted theory that the columns of the classical Temple of Zeus at Olympia were overturned by the seismic shaking of a damaging earthquake, revealing that this is rather unlikely. In all these cases the drawn conclusions are equally interesting not only to engineers and architects but also to archaeologists and historians.

In this paper the latter-mentioned use of engineering software is discussed. More specifically, it is proposed herein that sophisticated numerical simulations can be found useful in a broader context: for the assessment of theories related to archaeological and historical aspects of monuments, if these theories are based on evidence that can be evaluated through structural analysis. To demonstrate the capacities of this method of assessment, a case study is also presented herein in a concise manner, which regards the cause of the unusual deformations of a column in the classical monument of the Propylaia on the Acropolis of Athens.

2 The Method

An assessment of a theory usually follows predefined steps, the first of which always being the collection of the data which led to the formulation of the theory and the arguments supporting it. The study of the data comes next, for the necessary full understanding of the theory. That is why relative discussions of the analyst(s) with the scientist(s) which proposed the theory, if possible, are always helpful (obviously, not only in the initial step but throughout the process of assessment).

In case of a theory related to the structural behaviour of a monument, the gathered data must involve information not only (i) structural (e.g., arrangement of bearing elements, material properties, deformations and damages) and (ii) architectural (general form and dimensions, shape and dimensions of distinct architectural members, etc.), but also (iii) archaeological (the conclusions from excavations in and/or around the monument, the knowledge about the building techniques at the time of monument's construction and of its later repairs, etc.), and (iv) historical (e.g. date or period of construction, original use and change in uses throughout the lifetime of the monument, and catastrophic events in the area such as battles, earthquakes, landslides, etc.), as well as (v) any other type of information that, according to the researchers, can be found useful for the analyses (e.g., the available seismological and climatic data).

The gathering and study of the information must be followed (if possible) by an in-situ survey and, in some cases (according to the judgment of the analyst) by the formulation of alternative hypotheses which seem as possible as the theory to be assessed. The likelihood of the alternative scenarios must be investigated along with the theory under assessment.

At some point the numerical tool which will be used for the assessment must be selected. It is useful for this selection to be after the understanding of the theory and the formulation of the various hypotheses, as it must be derived on the basis of the software's credibility to simulate phenomena of the same or similar kind to the ones needed to be simulated for the assessment of the theory (and the hypotheses) in question. Obviously, there are other factors which must be taken into account (e.g. which software are available to the analyst, which one is expected to present the lesser computation time etc.), but they must have secondary influence on the selection (e.g. for the final choice between two codes which are expected to perform equally well during the assessment).

In any case the verification and suitable calibration of the selected software and/or the various numerical models should precede any simulation directly related to the assessment. Afterwards, the actual assessment should be through the design and the execution of (most likely numerous) parametric numerical simulations, carried out in phases. The conclusions from the evaluation of the results of each phase should determine the input for the simulations of the next phase. Sometimes the results could lead to the formulation of other scenarios and the process must start again from the initial phase.

Last but not least, any conclusion derived from the simulations not only has to be evaluated according to the analyst engineering judgment, but also it must be cross-checked with all the data available to the analyst. That is why, before the final formulation of the assessment, another survey in the monument must be carried out, for checking if there is any evidence or traces which were initially overlooked and could be contradictive to the final conclusion(s).

3 Case Study

3.1 The Monument: The Propylaea of the Athenian Acropolis

3.1.1 Brief Description and Historical Data

Across the top of the western slope of the Acropolis of Athens stands a building constructed in the classical period as a monumental gateway, the famous Propylaea (Fig. 1). The monument is entirely of marble and its architect was Mnesikles. Its construction started in 437 BC and continued till 432 BC after which the remainder of the project had to be abandoned because of the beginning the Peloponnesian War, the conflict between

Athens and Sparta. So far as it was completed, it consists of a very wide central building/passage with porches outward and inward, and an outer wing on either side (Fig. 2); the northern was quite large and had a large room, whereas the southern one was a mere stoa leading to the nearby sanctuary of Athena Nike (Fig. 1). The central building had six Doric columns at both its facades and was divided into two unequal parts (the western hall and the eastern porch) by a transverse wall, pierced with five doors. A ramp along the central axis was flanked, in the western hall, by three pairs of fine Ionic columns, supporting a magnificent ceiling (Tanoulas and Ioannidou 2002).

Fig. 1. The Acropolis of Athens from W. and above (2009, from the archive of the Acropolis Restoration Service). The Propylaea are shown in the center of the photograph. In front and to the right the temple of Athena Nike is shown, whereas the Parthenon and the Erechtheum are shown in the back.

The monument was remodeled in the 15th century to house the palace of the Florentine dukes of Athens (though this modification consisted mainly of additions). Lateron during the period of Turkish domination of Athens, the monument underwent major changes. In 1640, a gunpowder explosion blew up a large part of the ceiling and other features of the upper parts of the building, mainly in the east part of the western hall and probably also in the eastern porch. In the early 18th century, the Turks dismantled the upper parts in the west part of the western hall in order to use the material for the repair and strengthening of the Acropolis fortification, whereas towards the end of the 18th century, they removed the capitals and upper drums from the four central columns

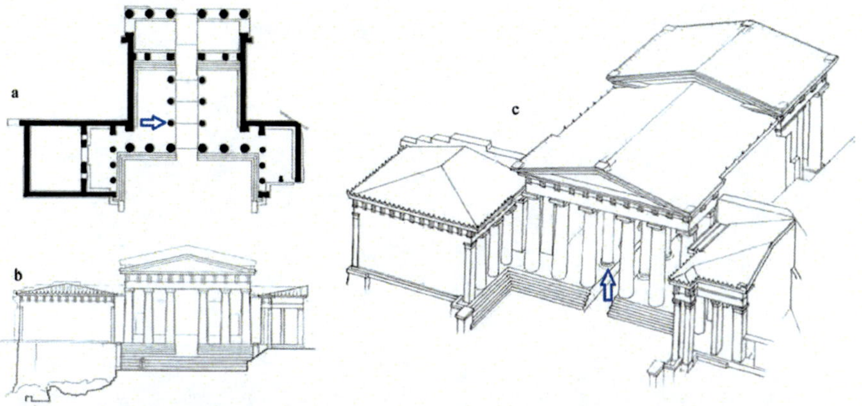

Fig. 2. The Propylaea in antiquity: (a) plan; (b) western façade; and (c) aximetric view from SW (Tanoulas 2012). The blue arrows indicate the Ionic column under discussion in the present paper

of the west façade (Tanoulas 1987). The first restoration of the Propylaea was carried out between 1909–1917, under the direction of Nikolaos Balanos. Since 1981, a modern restoration of the monument is being carried out, under the supervision of the Committee for the Conservation of the Acropolis Monuments.

3.1.2 The Theory Under Assessment

In 1994, the 3rd International Meeting for the Restorations of the Athenian Acropolis Monuments was held. In its Proceedings an additional paper was included (as part of the answer of the author to the questionnaire posed to the Meeting participants), in which the major structural damages of the Propylaea were discussed in a comprehensive way (Korres 1995). In that paper, for one very particular case: the damages of the first, to the left from the entrance, Ionic column in the western hall (Fig. 3), the author wrote that are due to special reasons (and not due to an earthquake nor an explosion) and apposed only one relative, yet quite "talkative", drawing (Fig. 4).

The column in question, in its current (non-restored) condition, has only seven drums in place, from its second drum up is very dislocated, its second drum has lost a quite large part, and its base has a local, limited, damage at the circumference. The theory of Manolis Korres about how these damages occurred can be summarized, on the basis of the interpretation of his drawing, as follows:

At some point when the column was still in full height (10.29 m) and supporting the ceiling, the overlying cross beam (6.4 m long and weighing ~10 t) was somehow fractured close to its end that was supported by the column; consequently, the large fragment of the beam started to fall and followed a course that led to a collision with the second drum of the column. The results of the impact were the fracture of a large part of the drum and the severe dislocation of the column. Afterwards, the beam fragment continued to fall and damaged the column base, before rested on the floor.

Fig. 3. The Propylaea (left) and the Athena Nike temple from NW (2011, from the archive of the Acropolis Restoration Service). The blue arrow indicates the Ionic column under discussion in the present paper

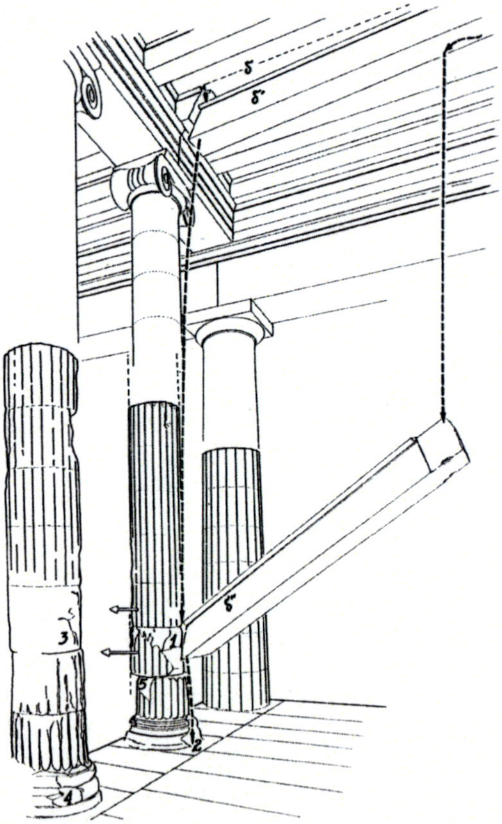

Fig. 4. Drawing of Manolis Korres (1995) presenting his theory about how an Ionic column of the Athens Acropolis Propylaea was severely damaged

3.2 The Assessment

3.2.1 Preliminary Approach

For the assessment of the above-mentioned theory, the working steps presented in Sect. 2 were followed. The initial estimations derived by studying the theory, visiting the monument, and collecting and studying various relative data, are the following. (a) Probably the damages in the 2nd drum came from a severe strike of some kind. (b) The strike was not exactly perpendicular to the colonnade, i.e. in the direction from N. to S., but slightly deviated towards the southwest (Fig. 5).

Fig. 5. The severely damaged lower part of the Propylaea ionic column; view from N. (left) and SE.

(c) If this strike came indeed from a falling fragment of the overlying cross beam, the ceiling probably was not in its place as it would have resisted (perhaps even prevented) the fall through various interlocking of fragments and blocks.

Hence, the scenarios formulated from the preliminary steps, to be later investigated numerically, were four: (i, ii) the beam (without overlying members) after being fractured close to its support over the column (or close to its other support) fell and hit the 2nd drum of the column, and (iii, iv) the beam intact (without overlying members) after being pushed (or displaced), either from the side of the column or the other side of the wall, towards the interior of the Hall, in a deliberate effort to be overthrown, fell and hit the column's 2nd drum.

3.2.2 Verification of Numerical Tool

The numerical tool selected for the numerical investigations was the FE software Abaqus/Explicit (Abaqus 2015). This choice was based not only on the fact that Abaqus is one of the most advanced engineering computer codes, commercially available, but also because it is generally considered one of the most suitable for highly nonlinear problems involving changing contact conditions. In fact, its ability to predict the (nonlinear) rocking bevaviour of stone-blocks under gravitational load was verified in a recent, among others, research (Papadopoulos et al. 2019), as with this software, the experimentally recorded free rocking response of two granite blocks with different slenderness was numerically reproduced very accurately (Fig. 6).

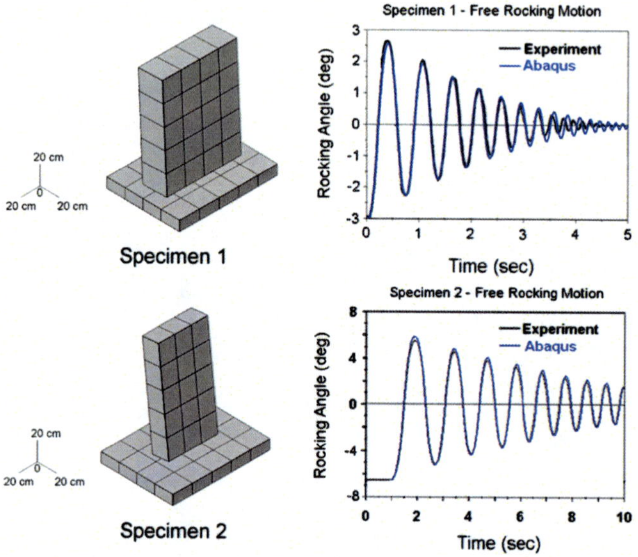

Fig. 6. Simulation of two free rocking motion experiments with granite specimens: numerical models (left) and comparison of main experimental data with numerical results (Papadopoulos et al. 2019)

The programmed numerical analyses, however, had to include simulations with also nonlinearity in material behaviour. From the various plasticity models for materials, available in Abaqus/Explicit, the most suitable for marble are: (a) the brittle cracking model which is most accurate in applications where the brittle behavior dominates such that the assumption that the material is linear elastic in compression is adequate; and (b) the damaged plasticity model, developed for the simulation of concrete and other quasi-brittle materials, which combines isotropic damaged elasticity with isotropic tensile and compressive plasticity to represent the material inelastic behaviour.

To examine whether these two models simulate satisfactory the plastic behaviour of marble, a preliminary numerical investigation was carried out. More specifically, using Abaqus/Explicit, a 3-point (3-P) bending test (Fig. 7) on a marble specimen was simulated two times, each time with a different model for the material behaviour. In both cases the mechanical properties of Dionysos-Pentelicon (D-P) marble (the building material of the Athens Acropolis monuments), known from the work of Vardoulakis and Kourkoulis (1997), were introduced in the model: modulus of elasticity equal to 84.5 GPa, Poisson's ratio equal to 0.26, tensile and compressive strength, equal to 10.8 MPa, and

80 MPa, respectively (all these were measured in marble's strong direction), and density equal to 2730 kgr/m^3. The nominal fracture stress of a D-P marble specimen in a 3-P bending test with parameters as presented in Fig. 7 is very close to 18 MPa (Kourkoulis et al. 1999). This value corresponds to critical load (P) ~30 kN.

In the simulation of the test with the brittle cracking model, the marble "beam" was fractured when the load was equal to ~35 kN (Fig. 8 left), whereas in the simulation with the damaged plasticity model, plastic deformation appeared when the load reached 29 kN and total failure occurred for load equal to kN (Fig. 8 right). Hence, the preliminary numerical investigation yielded satisfactory results.

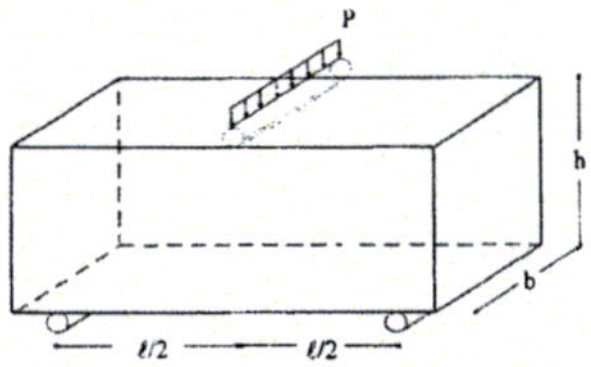

Fig. 7. Sketch presenting the three point bending test. In the case of the simulated test, its parameters were $l = 0.40$ m, $h = b = 0.10$ m, total length of specimen equal to 0.42 m and estimated contact width between the rollers and the specimen equal to 0.8 mm

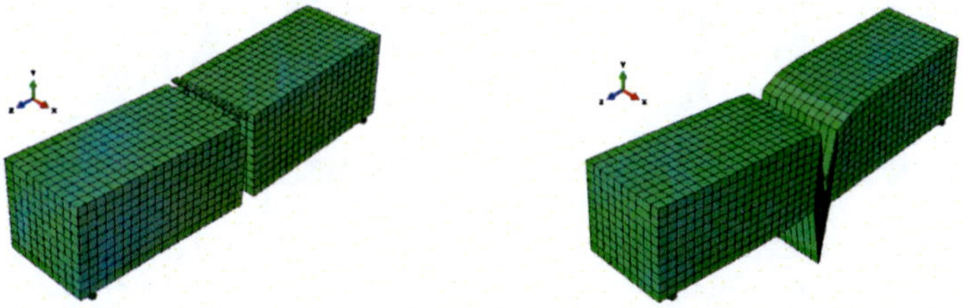

Fig. 8. The deformed model of the specimen at the instance of total failure in the simulations in which the brittle cracking model and the damaged plasticity model were used (left and right, respectively)

3.2.3 Numerical Simulation with Optimum Results

For the investigation of the aforementioned in Sect. 3.2.1 four scenarios, numerous numerical parametric analyses were carried out. According to the numerical results no one leads to a direct strike of the beam to the 2nd drum of the column (they lead firstly to an impact with the floor, Fig. 9), except for a variation of one scenario, derived from further parametric investigations. In this alternative scenario the beam intact, in an effort to be removed (descended), was displaced by 76 cm towards the interior of the hall (the distance between its northern end and the edge of its support was 31 cm) and then somehow was slipped and end-up impacting the 2nd drum of the column (Fig. 10). The simulation of this additional scenario led to results for the column dislocation and drum damages very similar to the actual ones (Figs. 11 and 12).

The meshing and the main dimensions of the numerical model in this "optimum" simulation are shown in Fig. 10a and b. The model was formatted in such a way so that all marble architectural members were in the state and position that they were in the initial, undamaged condition of the monument (i.e., as distinct members, in full contact, with no relative displacement, and in, practically, vertical or horizontal positions) apart from the beam which was displaced by 76 cm towards the interior of the hall. For the formulation of each marble member, the actual shape and dimensions of the respective member in the monument were followed to a great extend; the only note-worthy difference was in the circumferences of the column drums which in the model were polygonal and not as the real circular with flutes.

For the interactive behaviour of the model members normal to their interfaces a 'hard' contact model was determined, which allows when two surfaces are in contact, any pressure to be transmitted between them and, when the surfaces separate, it reduces the contact pressure to zero. In the tangential direction of the interfaces a classical friction model was used, with friction coefficient of 0.70.

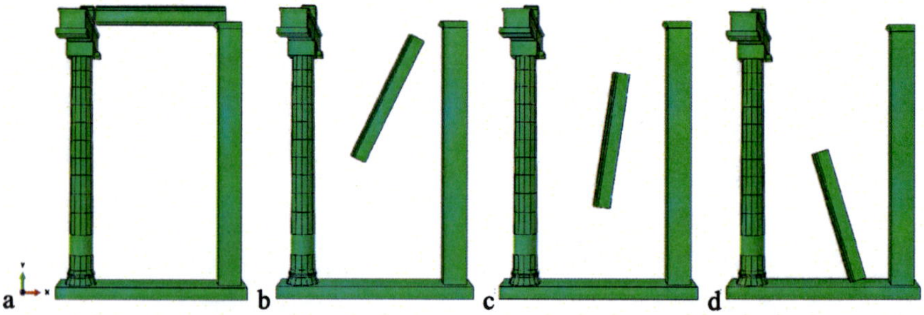

Fig. 9. Numerical results for various moments in the simulation very close to the theory under assessment: (a) the model before the beginning of the analysis (t = 0); (b, c) 1 s and 1.2 s after analysis started; and (d) instant just before the impact of the beam with the floor (t = 1.5 s)

To simulate the material behaviour of the beam the brittle cracking model was used, whereas for the material behaviour of the 2nd drum the damaged plasticity model was used. This choice was derived from a series of parametric preliminary analyses which showed that this leads to the best numerical results from the physical point of view. For the material behaviour of the rest marble members, a linear elastic model was adopted. The various parameters in the brittle cracking model and the linear elastic model were defined according to the results of the work of Vardoulakis and Kourkoulis (1997) for the strong direction of D-P marble (see Sect. 3.2.2). However, the values defined for the damaged plasticity model (for the simulation of material behaviour of the 2nd drum) were the following: modulus of elasticity equal to 79.5 GPa, Poisson's ratio equal to 0.26, tensile strength equal to 9.5 MPa, and compressive strength equal 65 MPa, which were taken from the results for the intermediate direction of D-P marble (Vardoulakis and Kourkoulis 1997). The latter choice was based on the fact that the orientation of material layers in the column drums is different from the one in the beams, in a way for the latter-mentioned strength values to be considered closer to reality.

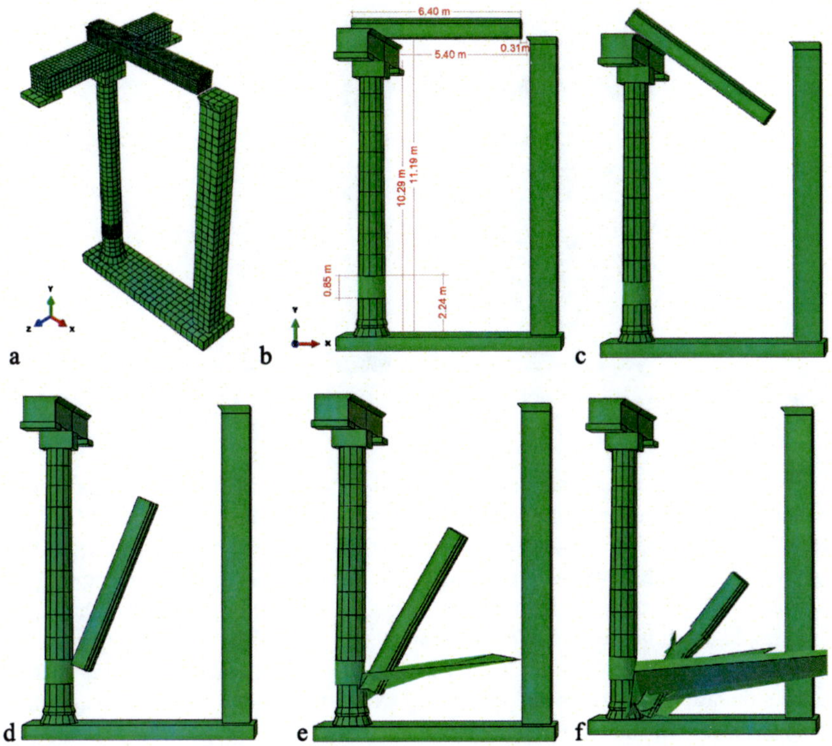

Fig. 10. (a) The meshing of numerical model in the "optimum" simulation and (b) its main dimensions. (c to f) Numerical results for crucial moments, specifically: (c) t = 0.7 s after the start of beam's fall; (d) beginning of impact between the beam and the column's 2nd drum (t = 1.45 s); (e) end of impact (t = 1.55 s); and (f) the beam hits the floor (t = 1.68 s)

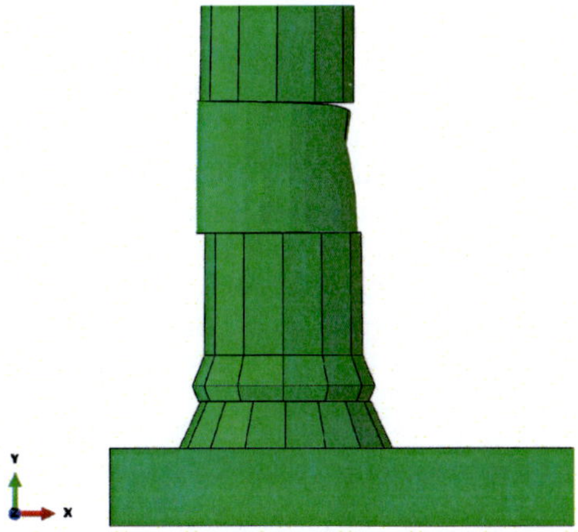

Fig. 11. The profile of the damaged 2nd drum of the column, after the completion of the simulation

Fig. 12. The damages of the 2nd drum of the Ionic column from SE, and detail of Fig. 11

Obviously, there is a certain degree of uncertainty in the above-described simulation. The uncertainty comes from the fact that some basic parameters are unknown and therefore had to be reasonably estimated (position of the beam at the beginning of its fall, state of preservation of the overlying part of the monument, and material strengths for the beam and the 2nd drum) and from the assumption (with small possibility) that at some point people attempted to remove the beam in one piece. That is why any conclusion drawn from the numerical results can be considered only as estimation. However, the fact that the simulation was conducted apparently without any major simplifying assumption for the definition of the model, increases the likelihood of the drawn estimations to be correct. These estimations are presented in the following paragraph.

3.2.4 Conclusions

The numerical analyses, carried out for the assessment of the theory of Manolis Korres that the unusual damages of the 2nd drum of an Ionic column in the Propylaea on the Acropolis of Athens were caused by an impact with a falling beam (Korres 1995), did not yield definitive results; however, some results were favorable to the theory's basic concept. The simulation which results were corroborative for the theory indicated that the beam (without any overlying stone member) fell in one piece, after being slipped during an effort to be descended (while it was quite high, very close to its original level), ending-up impacting the 2nd drum of the column which was supporting its northern end. If this is what actually took place, then the event probably happened in the early 18th century, during the dismantling of the west-upper parts of the western hall of the Propylaea by the Turks, in their efforts to use building material of the monument for the repair and strengthening of the near-by parts of the Acropolis fortification.

4 Concluding Remarks

In this paper the use of sophisticated numerical simulations to assess theories related to archaeological and historical aspects of monuments, if these theories are based on evidence that can be evaluated through structural analysis, is discussed. It is obvious that the main advantage of this method of assessment is that it permits the necessary scrutiny through numerous parametric numerical analyses. In each analysis the alteration in the parameter(s) can be from very small to quite large, depending on the previous results. Hence, the detailed examination of various influential (physical or numerical) factors is possible, as well as the evaluation of various hypotheses related to the theory under assessment. In this way the optimum results are being gradually derived, and consequently the theory is assessed. This process (which can be characterized as the conduction of a series of "virtual experiments") has no significant cost in time or money.

Another benefit of the method is that when it reaches successful results, it includes the verification and calibration of numerical models that can be used also for the diagnosis of damages of the monument in case. This becomes highly important in cases of monuments that need restoration, as the calibrated models can be used with confidence (at a later step) for the design of the structural intervention on the monuments.

To demonstrate the capacities of the method, a case study was presented, regarding the cause of the unusual damages of a column in the classical monument of the Propylaea on the Acropolis of Athens. According to the theory proposed by Manolis Korres, the column was damaged when one of the overlying cross-beams was fractured and fell, ending-up striking the 2nd drum of the column (Korres 1995). For the assessment of this theory, numerous numerical analyses were carried out, included calibration and verification of various numerical models, some results of which were favorable to the theory's basic concept.

It is believed that in the near future, the continuous progress of computer science and technology will further facilitate research approaches similar to the one described in the present paper, and eventually will lead to a significant increase of the cases where theories about archaeological and historical aspects of monuments are being assessed and with structural analysis numerical simulations.

Acknowledgements. Communications of the author with Dr. Konstantinos Karanasos, architect, about various metric data of the Athenian Acropolis Propylaea are much appreciated. The permission from both the Director of the Acropolis Restoration Service, Dr. Vasiliki Eleftheriou, and the Head of the Propylaea Restoration Project, Dr. Tasos Tanoulas, to the author to have free access to the Propylaea in order to study the western hall of the monument, is gratefully acknowledged. Comprehensive discussions with Prof. Manolis Korres about his theory and relative aspects of the history of the Propylaea are also gratefully acknowledged.

References

Abaqus 2016: Theory guide. Simulia, Dassault Systèmes (2015)

Alexandris, A., Psyharis, I., Protopapa, E.: The collapse of the ancient temple of Zeus at Olympia revisited. In: 2nd European Conference on Earthquake and Seismology, August 25–29, Istanbul, Turkey (2014)

Bartoli, G., Betti, M., Borri, C.: Numerical modeling of the structural behavior of Brunelleschi's dome of Santa Maria del Fiore. Int. J. Archit. Herit. **9**(4), 408–429 (2015)

Croci, G.: The colosseum: safety evaluation and preliminary criteria of intervention. In: Structural Analysis of Historical Constructions I, Barcelona, Spain (1995)

Chiorino, M.A., Spadafora, A., Calderini, C., Lagomarsino, S.: Modeling strategies for the world's largest elliptical dome at Vicoforte. Int. J. Archit. Herit. **2**(3), 274–303 (2008)
ICOMOS: International scientific committee for analysis and restoration of structures and architectural heritage (ISCARSAH). In: Recommendations for the Analysis, Conservation and Structural Restoration of Architectural Heritage. Document approved on 15/6/2005 in Barcelona, Spain (2005)
Konstantinidis, D., Makris, N.: Seismic response analysis of multidrum classical columns. Earthquake Eng. Struct. Dynam. **34**(10), 1243–1270 (2005)
Korres, M.: Major structural damages in the Propylaea. In: Proceedings of the 4th International Meeting for the Restoration of the Acropolis Monuments, May 27–29, 1994, Athens, Greece (In Greek) (1995)
Kourkoulis, S.K., Exadaktylos, G.E., Vardoulakis, I.: U-notched dionysos-pentelicon marble in three point bending: the effect of nonlinearity, anisotropy and microstructure. Int. J. Fract. **98**, 369–392 (1999)
Lourenço, P.B., Mourão, S.: Safety assessment of Monastery of Jerónimos. In: Lourenço, P.B., Roca, P. (eds.) Historical Constructions, pp. 697–706, Guimarães, Portugal (2001)
Macchi, G., Ruggeri, M., Eusebio, M., Moncecchi, M.: Structural assessment of the leaning tower of Pisa. In: Structural Preservation of the Architectural Heritage, Rome, Italy (1993)
Mola, F., Vitaliani, R.: Analysis, diagnosis and preservation of ancient monuments: the St. Mark's Basilica in Venice. In: Structural Analysis of Historical Constructions I, Barcelona, Spain (1995)
Papadopoulos, K.: The restoration of the north-side foundation of the temple of Apollo Epikourios. Int. J. Archit. Herit. **4**(3), 246–269 (2010)
Papadopoulos, K., Vintzileou, E.: The collapse of epistyle beams from the north side of the classical temple of Apollo Epikourios: analysis and numerical approach. In: 2nd International Conference on Protection of Historical Buildings, May 7–9, Antalya, Turkey (2014)
Papadopoulos, K, Vintzileou, E.: New titanium connectors for the columns capitals of the classical temple of Apollo Epikourios. Int. J. Archit. Herit. **10**(6), 749–765 (2016)
Papadopoulos, K, Vintzileou, E., Psycharis, I.N.: Finite element analysis of the seismic response of ancient columns. Earthquake Eng. Struct. Dynam. **48**(13), 1432–1450 (2019)
Papastamatiou, D., Psycharis, I.: Seismic response of classical monuments - a numerical perspective developed at the temple of apollo in Bassae Greece. Terra Nova **5**, 591–601 (1993)
Peña, F., Chávez, M.M.: Assessment of the seismic vulnerability of typical Mexican colonial churches. In: SAHC2014 - 9th International Conference of Structural Analysis of Historical Constructions, October 14–17, Mexico City, Mexico (2014)
Psycharis, I.N., Papastamatiou, D.Y., Alexandris, A.P.: Parametric investigation of the stability of classical columns under harmonic and earthquake excitations. Earthquake Eng. Struct. Dynam. **29**, 1093–1109 (2000)
Psycharis, I.N., Lemos, J.V., Papastamatiou, D.Y., Zambas, C., Papantonopoulos, C.: Numerical study of the seismic behavior of a part of the Parthenon Pronaos. Earthquake Eng. Struct. Dynam. **32**(13), 2063–2084 (2003)
Roca, P., Massanas, M., Cervera, M., Arun, G.: Structural analysis of Küçük Ayasofya Mosque in İstanbul. In: Structural Analysis of Historical Constructions IV, November 10–13, Padova, Italy (2004)
Tanoulas, T.: The propylaea of the Acropolis at Athens since the seventeenth century. Jahrbuch des deutschen archaologishen Instituts **102**, 413–483 (1987)
Tanoulas, T., Ioannidou M.: Study for the restoration of the superstructure of the central building of the Propylaia. In: Committee for the Conservation of the Acropolis Monuments (in Greek) (2002)
Tanoulas, T.: The propylaea. In: Bouras, C., Ioannidou, M., Jenkins, I. (eds.) Acropolis Restored. The British Museum, London, UK (2012)

Vardoulakis, I., Kourkoulis, S.K.: Mechanical properties of dionysos marble. In: Final Report of the Environment Project EV5V-CT93–0300. Monuments under Seismic Action. National Technical University of Athens (1997)
Zambas, C.: Study for the restoration of the parthenon, vol. 3b. In: Committee for the Conservation of the Acropolis Monuments (in Greek) (1994)

Parametric Investigation of the Seismic Response of Ancient Isodomic Masonry Walls

Antigone Vrouva[1,2](✉), Ioannis Psycharis[1], and Vasiliki Eleftheriou[2]

[1] Department of Structural Engineering, Laboratory of Earthquake Engineering, School of Civil Engineering, National Technical University of Athens, Polytechnic Campus, Heroon Polytechniou 9, 15780 Zografou, Greece
ipsych@central.ntua.gr

[2] Hellenic Ministry of Culture and Sports, Acropolis Restoration Service, Polygnotou 10, 10555 Athens, Greece
veleftheriou@culture.gr

Abstract. In this paper the seismic behavior of ancient isodomic masonry walls is investigated through the case study of the walls of the Parthenon's cella. The isodomic masonry wall is formed of an orthostate, consecutive alternating courses of headers and strertchers and an external entablature. The material of the examined wall is Pentelikon marble. The blocks rest one upon another and side by side in total absence of mortars but with a meticulous plot and are elaborately connected with metal connectors both in the horizontal and the vertical direction. The structure is modelled using the code 3DEC by Itasca and the analyses are carried out using the discrete element method with the blocks modelling the individual stones assumed to be rigid. The models are subjected to specific seismic actions that correspond to the maximum possible earthquake for the site under consideration within a return period of more than 2000 years. The effect of several features of the wall on the seismic behavior is discussed, as the inclination of the wall towards the inside of the cella, the presence and function of the connecting elements and the damages of the ancient blocks.

Keywords: Seismic response · Ancient walls · Isodomic masonry · Restoration · Connectors · Damages

1 Introduction

The walls of the cella of Greek and Roman temples are usually built of meticulously formed isodomic masonry blocks, closely fitted together and put one upon another in total absence of mortars. The arrangement of the blocks in the isodomic system is such to provide increased interlocking. Monumental structures of that type have proven to comport very well in seismic actions, provided that they maintain their structural integrity. However, strength and stability problems arise if the structure starts lacking parts and gradually turns into a ruin. The outline of the ruin usually self-degrades into

I. Vayas and F. M. Mazzolani (Eds.): PROHITECH 2021, LNCE 209, pp. 458–476, 2022.
https://doi.org/10.1007/978-3-030-90788-4_37

stable parts following nature's mechanisms depending on the original dimensions, the materials and the actions sustained, mainly earthquakes in high seismicity regions. The restoration of such ruins is a complex and difficult task, as it has to achieve maximum exploitation of the remnant material and an outline that can withstand in safety the seismic actions.

The seismic response of such a restoration is examined in this paper, namely the north wall of the cella of the Parthenon (Fig. 1) on the Acropolis of Athens, Greece. The Parthenon was constructed in 438 BC as a temple devoted to Goddess Athena. In 267AD, the building was severely damaged after put in fire by the Erouli. However, the thermal fracture of the blocks during the fire must have left their inner surface almost intact, since the interlocking shape of the independent fractions kept them together [1]. A second large destruction took place in 1687, when an explosion occurred during the siege of the Acropolis by the Venetians. During the explosion, three of the four walls of the cella collapsed. The collapsed blocks lost their integrity as the thermal fractures were released and more raptures were provoked from the fall (Fig. 2). In the meantime, the temple had undergone through repairs and remodelling to suit the various functions it served for more than 1400 years. A number of manly imposed damages followed before the attempts made in modern years to bring the scattered material of the longitudinal walls back into place, first in 1841–1844 and later in 1928 during a more radical intervention that revealed the ancient material. In these restorations, numerous misplacements of stone blocks took place and, for this reason, the Acropolis Restoration Service has decided to dismantle the restored parts of the walls (Fig. 3) and restore them again, putting the blocks at their initial position in the walls' structure [2–4] and also including additional blocks found scattered around the hill. Considering the seismicity of the Athens area and the damaged condition of the ancient blocks, the question of the seismic capacity of several proposed restoration scenarios raised, in conjunction with the different proposals made for the support of the inner damaged façade [2–4].

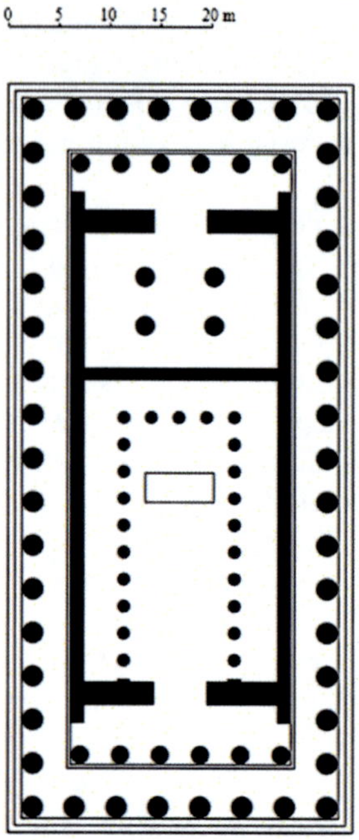

Fig. 1. Plan view of the Parthenon [1].

Fig. 2. View of part of the stone blocks of the Parthenon's cella walls to be restored [2]. Most stones present significant damage.

Fig. 3. View of the Parthenon's cella in its present condition (YSMA Photo Archive).

In this paper, based on the case study of the approved restoration proposals for the Parthenon's cella [6], the response of ancient isodomic masonry walls under seismic loading is investigated and the impact of several simplifications in the numerical model of the structure are evaluated. Several issues are examined, as the effect of the inclination of the wall, the importance of the connecting elements to the seismic stability and the need for the exact representation of the geometry of the blocks and their respective seat interfaces.

2 Structural Models Considered

2.1 Construction of the Original Walls

In the original structure, the isodomic masonry walls of the Parthenon's cella were formed of an orthostate, 17 consecutive alternating courses of headers and strertchers and an external entablature consisting of four courses, all made of Pentelikon marble. The blocks rested one on top of the other and side by side, in total absence of mortars but with a meticulous arrangement, as depicted at the right plot of Fig. 4 [1]. The placement of the blocks at the crossing of the longitudinal walls (north and south) with the transversal walls (east, west and partition walls) are shown in Fig. 5. Today, only part of the walls are still standing (Fig. 3).

The blocks were elaborately connected with iron elements, both in the horizontal and the vertical direction, aiming at preventing shear and longitudinal relative displacements. The horizontal connectors had a typical double T shape while vertical dowels were used for the shear connection of the blocks (Fig. 6).

Indicatively, it is mentioned that the width of the north wall at the base of the orthostate is 1.165 m, while the dimensions of the typical blocks of the orthostate are approximately 2.44 × 0.50 × 1.16, the ones of the typical headers are 1.22 × 1.15 × 0.52 and the dimensions of the typical stretchers are 1.22 × 0.55 × 0.52 (dimensions in m) [6]. Interesting to note that, unlike the transeversal walls (east, west and partition wall) which are vertical, the north and south walls were constructed with an inclination towards the inside of the cella, equal to 1.455% [6, 7].

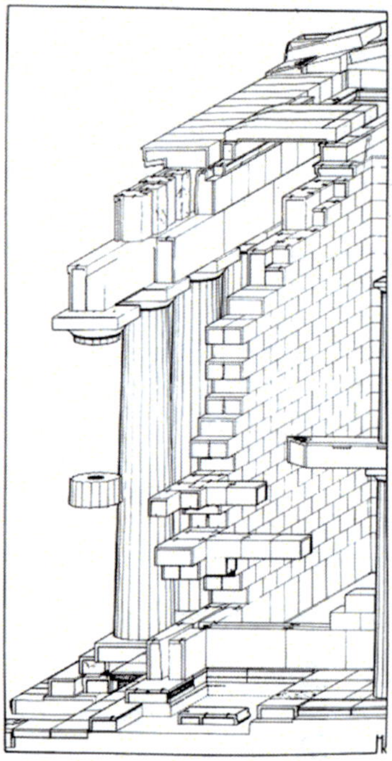

Fig. 4. Vertical section showing the structure of the temple and a detailed representation of the placement of the wall blocks [1].

Fig. 5. Left: Connection of the partition with the south wall [5]; Right: Connection of the south wall with the west wall of the cella [5].

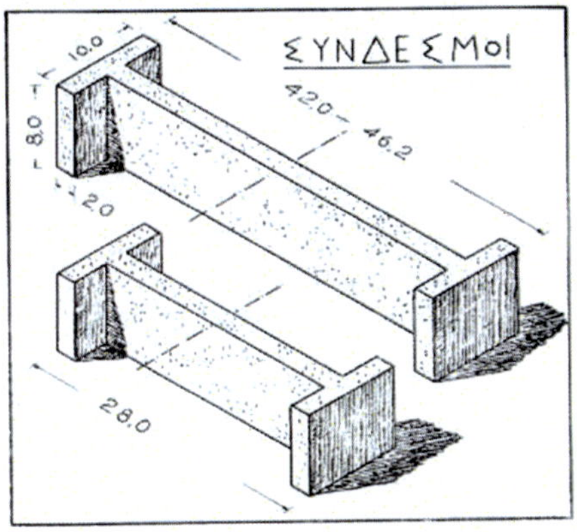

Fig. 6. Left: Typical double T-shaped clamps used as horizontal connectors of the blocks of the walls (dimensions in cm) [5].

2.2 Structure Analyzed

The geometry of structure analyzed was based on the restoration proposal for the north cella wall of the Parhtenon [6] and the in situ blocks of the west and south walls (Fig. 7). For simplicity reasons, the foundation and the connection with the Opisthonaos colonnade were excluded from the structural model and the restoration proposal for the south wall was not considered. Four alternative geometries of the structure were analysed in order to study the effect of various geometric and structural features of the walls, as the inclination of the long walls to the inside of the cella, the presence and function of the connecting elements and the imperfections of the ancient blocks. Four models considered are described in the following.

Model w1 (Reference Model)

Model w1 (Fig. 7) had the most simplified geometry, as the long walls were assumed vertical and the blocks were assumed to be intact with their initial dimensions. Also no connection were considered between the blocks, neither in the horizontal (clamps) nor in the vertical (dowels) direction. This model was used as reference.

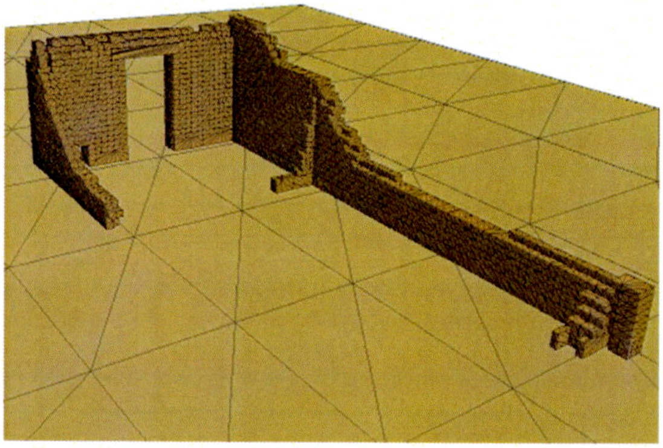

Fig. 7. Numerical representation of model w1.

Model w2

The model w2 had the very same geometry and materials with the reference model w1, but both the long walls were considered with their actual inclination towards the inside of the temple.

Model w3

The third model w3 had exactly the same geometry with model w2, but the horizontal and the vertical connection elements (clamps and dowels, respectively) were considered in accordance with the ancient construction. The importance of the connection elements of the wall has been already pointed out in other studies [8].

Model w4

The fourth model w4 was similar to model w3 but the existing damage of the blocks was also incorporated (Fig. 8). The imperfections of the blocks consist mainly of missing parts and edge cutoffs, which were caused mainly by the fire of the 3rd century AD and the action of the connection elements. Due to the considered imperfections of the stones, model w4 had a more sparse distribution of the connection elements compared with model w3.

It is noted that, apart from missing parts, many stones are also severely cracked due to thermal fractures that occurred during the fire [1] and also caused by the action of the connection elements as pointed out in [8]. However, these cracks were not considered in model w4 and only the missing parts of the blocks were incorporated in the numerical model.

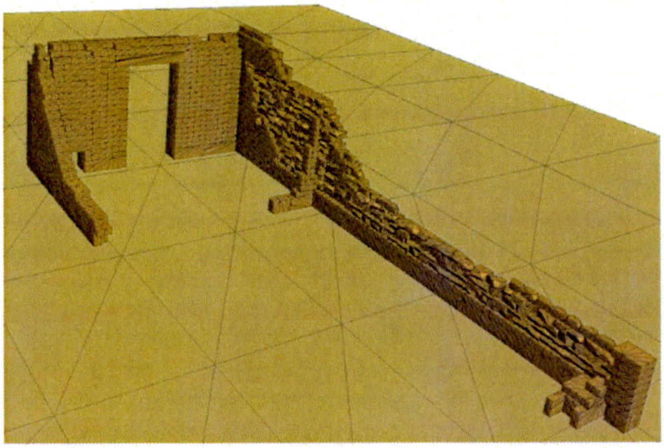

Fig. 8. Numerical representation of model w4 showing the considered imperfections of the blocks.

3 Numerical Analysis

3.1 Numerical Model of the Structure

The structural system of ancient temples differentiates significantly from the one of modern structures, since it consists of individual structural elements (blocks) with the friction and the actions in the iron connections (if they exist) being the sole forces preventing dismantling during an earthquake. Energy is dissipated through sliding and rocking of the independent blocks that take place at the joints and the response is strongly nonlinear and sensitive to the properties of the structure and the characteristics of the excitation.

Numerical analyses of such structures can be performed with the use of the Distinct (or Discrete) Element Method (DEM). Previous studies [9–13] have shown that DEM can reliably predict the seismic response of such spinal structures. The method deals with the problem of discontinuities allowing large displacements and rotations between adjacent blocks, even full detachment, while it searches for new contacts during the numerical process. The analyses presented herein were performed with the code 3DEC by Itasca [14] which is based on DEM.

The analyses were performed assuming that the marble blocks are rigid and that all deformations take place at the joints, where elastic contacts are assumed in the normal (only in compression) and the tangential direction. Appropriate values for the joint parameters were considered, based on previous studies for similar structures [9–13], in which 3DEC was calibrated and its accuracy was verified through comparisons with test results on the shake table response of multi-drum marble columns and colonnades performed at the Laboratory of Earthquake Engineering of NTUA [9, 10, 12]. Based on these studies, the parameters shown in Table 1 were adopted in the numerical model.

Table 1. Model properties considered

Property	Value
Blocks	
Density	2780 kg/m^3
Damping (Rayleigh)	10% mass proportional at frequency 0.3 cps
Joints	
Normal stiffness, *kn*	5 GPa/m
Tangential stiffness, k_s	1 GPa/m
Friction angle	37°
Clamp connectors	
Axial stiffness	1.19e8 N/m
Shear stiffness	1.36e6 N/m
Yield strength	45.2 kN
Shear strength	6.8 kN
Dowel connectors	
Axial stiffness	Null (1 N/m)
Shear stiffness	1.0e6 N/m
Yield strength	60 kN
Shear strength	10 kN

Each marble stone was modelled as an individual block [15], either with its original dimensions (models w1, w2, w3) or with its current imperfections (model w4). The connectors (clamps and dowels) incorporated in models w3 and w4 were modeled with nonlinear springs placed at the positions of the connectors in the original structure [16]. Their properties are given in Table 1 and were derived from experimental studies conducted for the axial and the shear behavior of metal connections in marble and concrete [17–21].

3.2 Seismic Excitations

In the investigation of the seismic response of ancient monuments it is of outmost importance to use base excitations compatible with the seismotectonic environment of the area and covering a wide range of the expected frequency content (especially the predominant period) of past and anticipated earthquakes. In the performed analyses, suitable accellerograms were chosen from international strong motion databases satisfying these criteria, which were properly amplified to correspond to the worst-case earthquake that could have occurred in Athens in more than 2000 years according to [22]. The corresponding amplification factor for each ground motion was determined so that, marginally, the base excitation does not cause the collapse of the two free-standing columns on the south flank of the Acropolis of Athens, which were built more than 2000 years ago and are still standing. It is noted, though, that the corresponding worst-case earthquakes are upper limits derived on the basis that the columns are still standing and it is possible that such strong ground motions have not occurred yet in Athens.

Five ground motions with varying predominant periods were selected from [22], the characteristics of which are given in Table 2. In the last column of Table 2, the amplification factor applied to each record to scale it to the worst-case earthquake is shown.

Table 2. Ground motions considered

No	Earthquake	Date	Magnit. M_w	Record	Dist. from rapt. (km)	Soil type	*pga* (g)	*pgv* (cm/s)	T_p(*) (s)	Ampl. Factor
1	Campano Lugano, Italy	23/11/80	6.9	Bisaccia (bsc)	19	Rock	L: 0.079 T: 0.092 V: 0.055	L: 14.7 T 16.3 V: 9.9	1.02 2.02 3.51	1.1
2	Campano Lugano, Italy	23/11/80	6.9	Bagnoli–Irpino (bgi)	6	Rock	L: 0.139 T: 0.181 V: 0.104	L: 20.5 T: 30.6 V: 14.6	1.26 1.58 1.50	1.4
3	Northridge, California USA	17/1/94	6.7	Wonderland Ave., L.A (won)	23	A	L: 0.112 T: 0.172 V: 0.106	L: 8.7 T: 11.8 V: 3.6	1.17 0.84 0.70	3.2
4	Kozani, Greece	13/5/97	6.5	Kozani, Prefecture Blg (koz)	14	A	L: 0.208 T: 0.143 V: 0.084	L: 8.6 T: 6.6 V: 4.5	0.23 0.39 1.30	3.2
5	Athens, Greece	7/9/99	5.9	Syntagma, Metro B (sgmb)	10	Stiff Soil	L: 0.109 T: 0.086 V: 0.087	L: 9.8 T: 10.6 V: 3.5	0.89 0.52 1.71	4.4

(*) Predominant period derived from the peak of the *PSV* spectrum for 5% damping.

4 Results and Discussion

In order to investigate the effect of the considered parameters (inclination of the wall, connectors, imperfections), the response of models w1 to w4 was derived and compared under the above-mentioned ground motions. Similarly to [23], maximum displacements during the ground shaking and residual ones corresponding to the deformed shape at the end of each earthquake motion were calculated along the height of the long wall at selected places, which are shown in Fig. 9. The displacements were measured at the centroid of the blocks shown in red. It is noted that the plots depict the maximum displacements along the height of the wall and thus they are not a record of simultaneous displacements.

Although results were obtained at all positions shown in Fig. 9, the discussion is focused mainly on the deformation at positions P3 (connection with the reconstructed partition wall) and P4 (the mid-length between the two transversal walls of the structure – partition and west wall), where the reconstructed wall attains its maximum height (over 10 m high) and the displacements are significantly larger than the ones at places P1-P2.

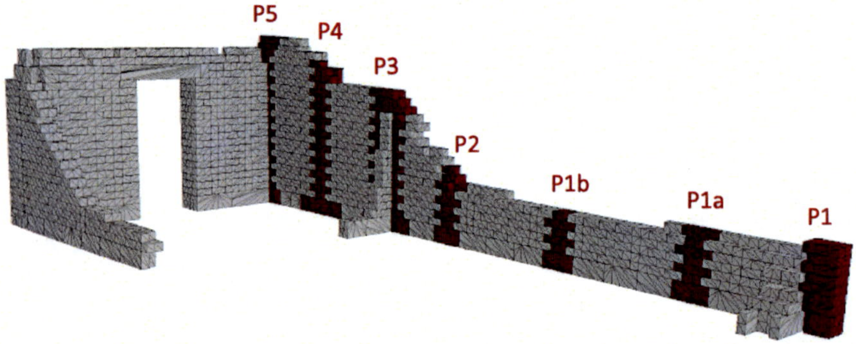

Fig. 9. Positions at which maximum and residual displacements are presented. The blocks considered at each position for the calculation of the deformation of the wall are shown in red.

4.1 Effect of the Wall Inclination

The effect of the wall inclination was examined by comparing the results for models w1 (without the inclination of the wall) and w2 (with the inclined wall) at position P3. The results are shown in Figs. 10 and 11, which depict the out-of-plane and the in-plane residual (left) and maximum (right) displacements, respectively, along the height of the wall for the five ground motions considered. The relative displacements with respect to the centroid of the reference vertical wall (model w1) are shown, while the inclined axis of wall w2 is depicted with a black line. Similar results were obtained for the other positions considered.

The maximum out-of-plane displacements measure reached 0.25 m (22% of the wall's thickness) at the upper course of the blocks and the residual displacements were up to 0.08 m (7% of the wall's thickness). Note that the inclination of wall w2 is 0.15 m at the top (13% of the wall's thickness). Both models w1 and w2 exhibited a rather similar deformation, but with the inclined wall (model w2) showing larger displacements: For model w1, the deviation of the upper block from the base of the section was 0.265 m, while for model w2, it was 0.39 m. It is obvious, therefore, that the inclination of the model might be crucial, as the inclined model inherently has a deficit of stability to the leaning side, showing, thus, larger vulnerability to collapse. It is notable that the inclination is almost twice the value of the maximum permanent displacement of all seismic events considered in this study, and only in one seismic excitation the maximum displacement on the upper block of the structure exceeded the deviation from verticality caused by the inclination.

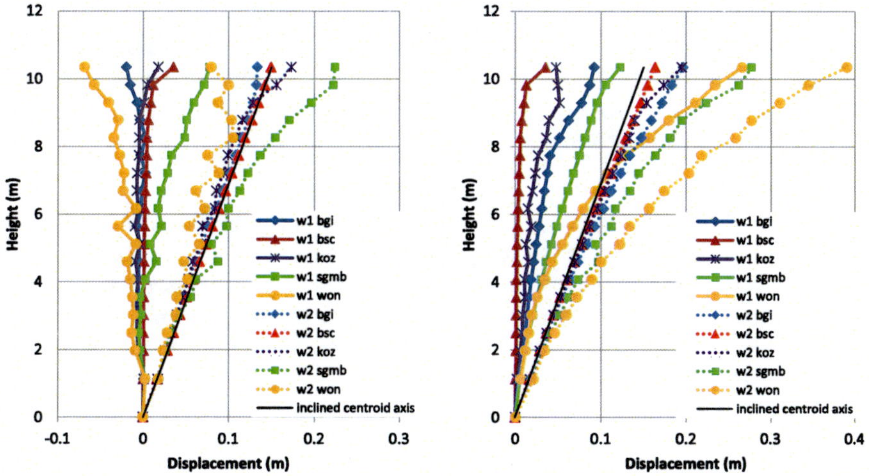

Fig. 10. Displacements at position P3 in the out-of-plane direction of the wall. Left: residual displacement; right: maximum displacement during the excitations.

In the in-plane direction, the inclined model (w2) developed smaller displacements, in general, than the vertical wall (w1) (Fig. 11). It is noted that significant slippage of the blocks occurred during some ground motions, especially at the upper part of the wall.

4.2 The Effect of the Connectors

The results for models w2 (without connectors) and w3 (with connectors) at position P4 are shown in Figs. 12 and 13, which depict the out-of-plane and the in-plane displacements (residual and maximum) of the centroid of each block, respectively. In this case, the out-of-plane displacements are measured from the inclined axis.

It is evident that the connectors resulted in the detainment of the displacements, especially above the mid-height of the wall, in both the out-of-plane (Fig. 12) and the in-plane (Fig. 13) directions. In the out-of-plane direction, the slips that occurred at the upper blocks of model w2 (without the connectors), which caused large permanent

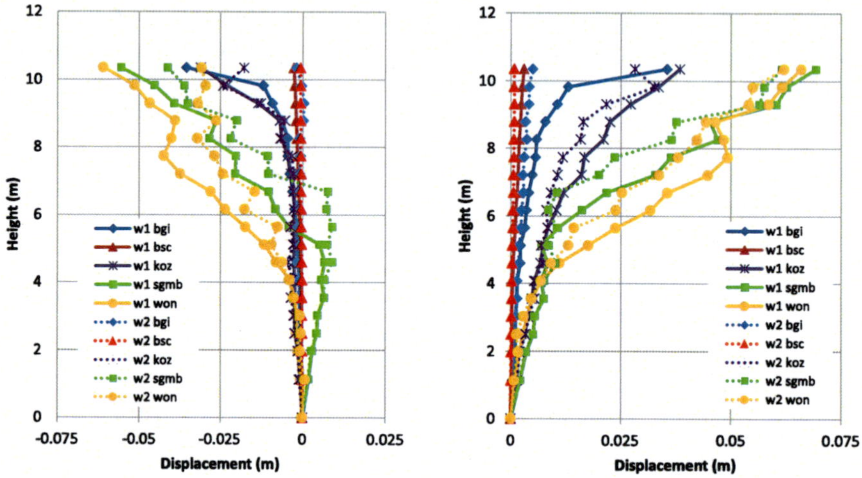

Fig. 11. Displacements at position P3 in the in-plane direction of the wall. Left: residual displacement; right: maximum displacement during the excitations.

displacements, were constrained in the w3 model. Maximum displacements were also constrained for the model with the connectors, with the effect of the connectors being more obvious above the mid-height of the wall. Same observations can be made for the in-plane direction. An exception to this behavior can be seen in the maximum displacements for the Athens 1999 (sgmb) ground motion, where model w2 (without the connectors) exhibited smaller maximum displacements above the mid-height of the wall than model w3 (with the connectors).

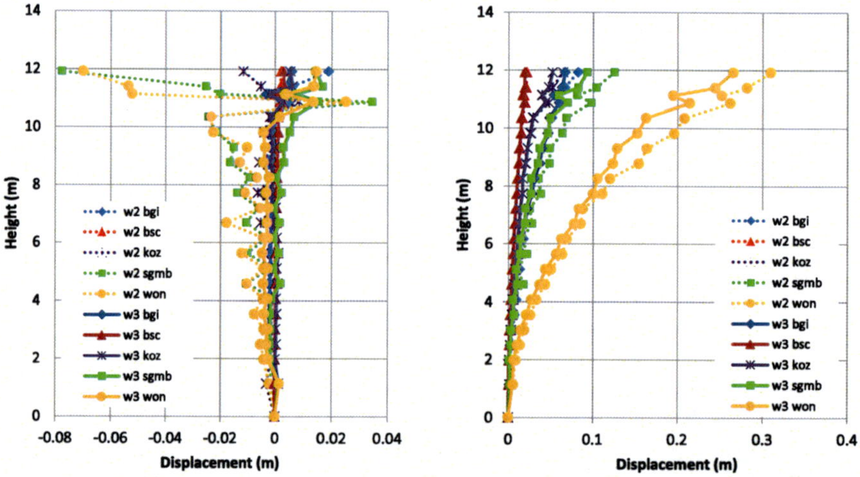

Fig. 12. Envelop of maximum displacements at position P4 in the out-of-plane direction of the wall. Left: residual displacement; right: maximum displacement during the excitations.

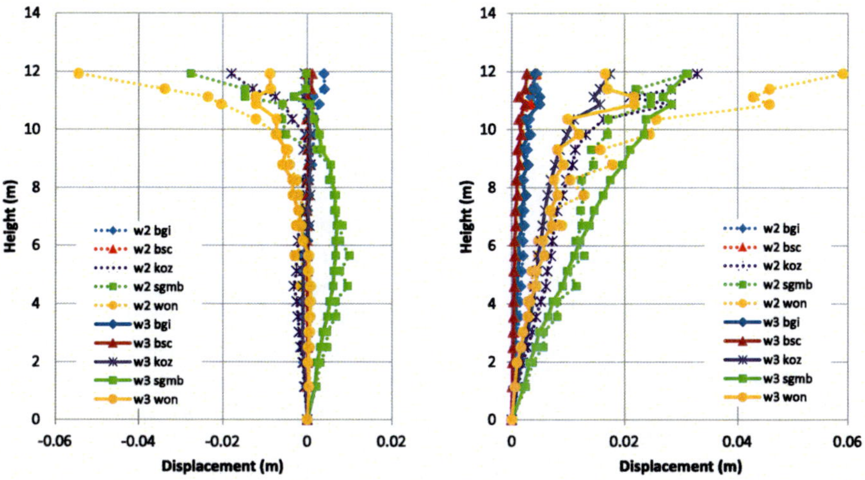

Fig. 13. Envelop of maximum displacements at position P4 in the in-plane direction of the wall. Left: residual displacement; right: maximum displacement during the excitations.

At position P3, where the response of the wall is affected from the connection with the partition wall (Fig. 9), the response in the out-of-plane direction was somehow different, as the large stiffness of the partition wall affected the overall behavior. Thus, as shown in Fig. 14, the residual displacements were slightly decreased compared to position P4, especially above the mid-height of the wall. Note that, in these plots, the relative displacements with respect to the centroid of the reference vertical wall (model w1) are shown, while the inclined axis of walls w2 and w3 is depicted with a black line.

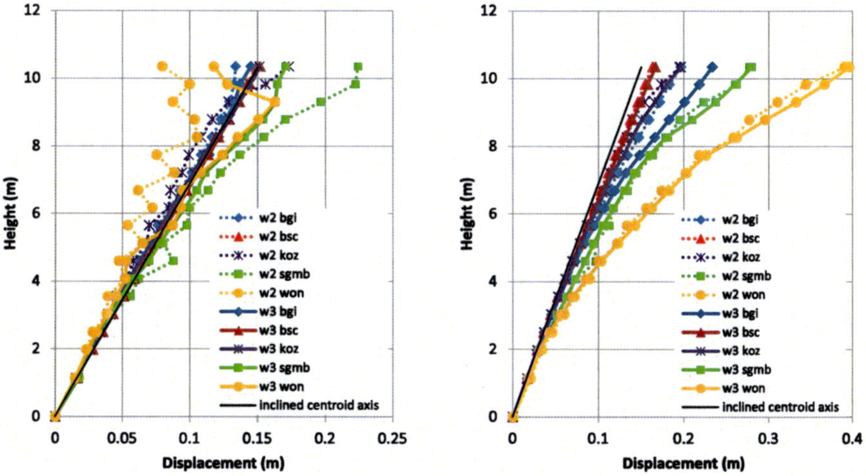

Fig. 14. Envelop of maximum displacements at position P3 in the out-of-plane direction of the wall. Left: residual displacement; right: maximum displacement during the excitations.

4.3 The Effect of Damage

The effect of the blocks' damages was examined from the comparison of models w3 (without damages) and w4 (with damages). The damages were introduced with alterations in the geometry of the blocks and in the allocation of the connectors. The results show that blocks' damages (model w4) are of outmost importance and might undermine the stability of the structure. The worst case proved to be the Wonderland Ave record of the Northridge 1994 earthquake (won), for which the numerical model predicted collapse of the central part of the wall above the 8th layer (Fig. 15). Localised collapses (3 to 4 blocks) also occurred when the model was subjected to the Athens 1999 (sgmb), the Kozani 1997 (koz) and the Bagnoli – Irpino 1980 (bgi) ground motions. The reconstructed part of the wall with height up to the 6th course of blocks (positions P1, P1a P1b, P2) did not present any collapse failure under the seismic actions considered. The reconstructed partition wall seems to provide some support at position P3, but cannot restrain the failure at P4 (Figs. 16 and 17).

In Fig. 18, the out-of-plane displacements (residual and maximum during the events) at position P1a, i.e. at the lowest reconstructed part of the wall, are plotted for models w3 and w4. For the model without the damages (w3), the maximum residual displacement of the upper block did not exceed 1 cm from the inclined axis, whereas the corresponding displacement for the model with the damages (w4) was approximately 3.4 cm. The maximum displacements during the seismic events were also, in general, detained for model w3.

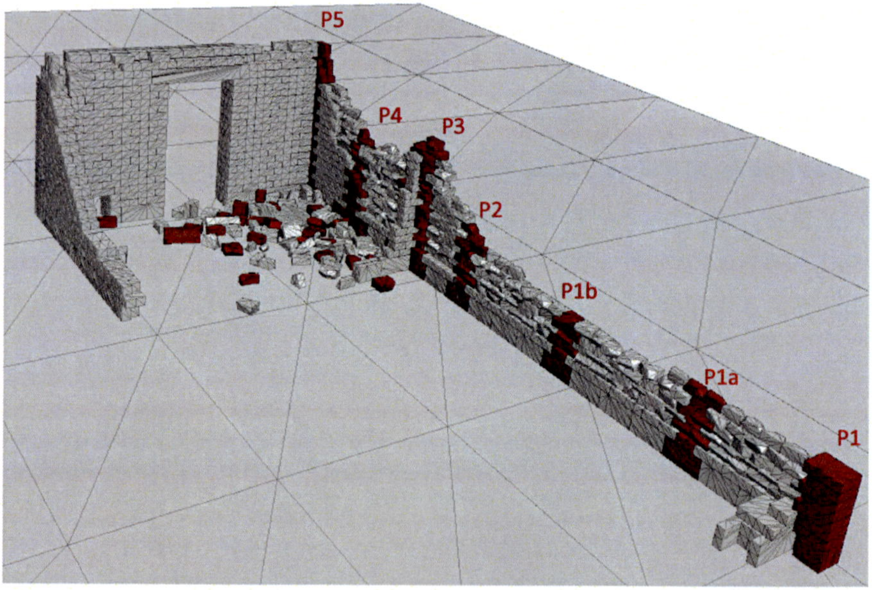

Fig. 15. Partial collapse of the wall when model w4 is subjected to the Northridge, California, 1994 earthquake.

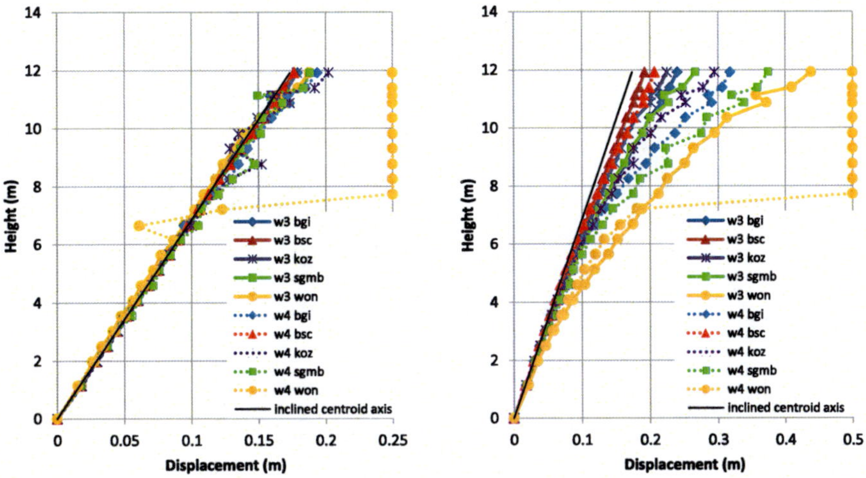

Fig. 16. Displacements at position P4 in the out-of-plane direction of the wall. Left: residual displacement; right: maximum displacement during the excitations. Residual displacements equal to 0.25 m and maximum displacements equal to 0.5 m indicate collapse of the respective blocks.

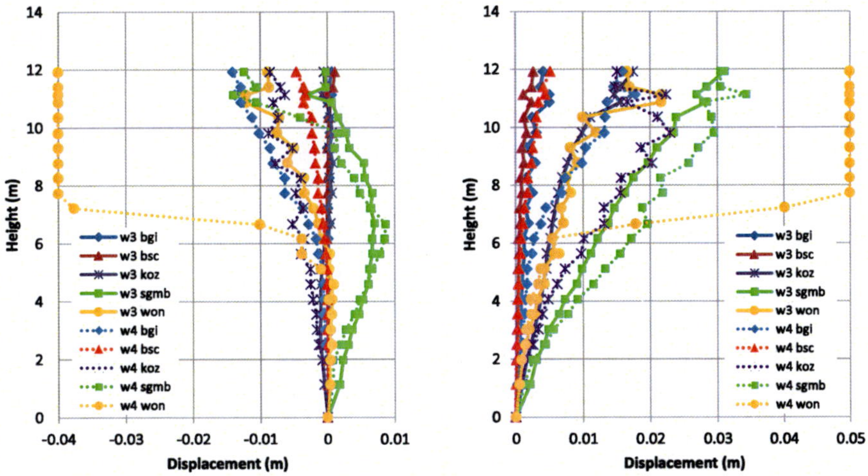

Fig. 17. Displacements at position P4 in the in-plane direction of the wall. Left: residual displacement; right: maximum displacement during the excitations. Residual displacements equal to − 0.04 m and maximum displacements equal to 0.05 m indicate collapse of the respective blocks.

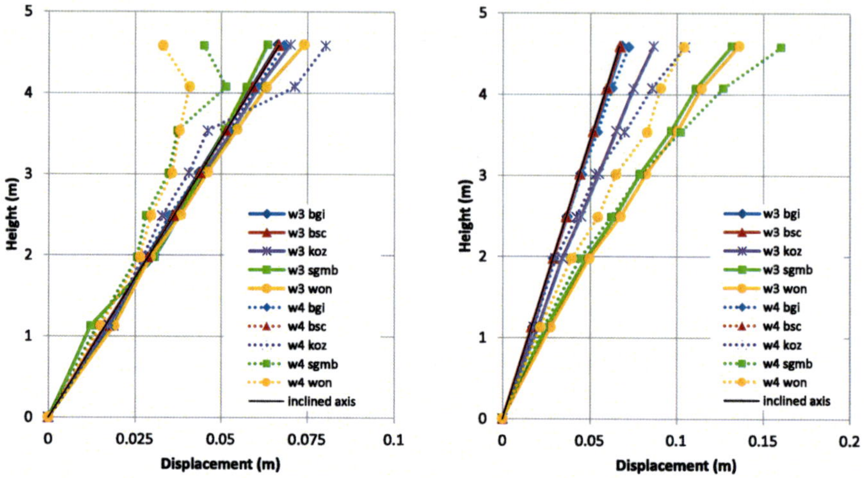

Fig. 18. Displacements at position P1a in the out-of-plane direction of the wall. Left: residual displacement; right: maximum displacement during the excitations.

5 Conclusions

In this paper, different numerical models of the Parthenon's north wall were analysed in an effort to estimate the effect of specific features of the structure, passing, step by step, from the simplest geometrical model (w1) to a finer detailed one (w4), in which damage was also included. Five earthquake motions were considered and the following results were drawn:

- The largest displacements were calculated for the upper blocks at position P4 in the out-of-plane direction. For models w1 and w2, the maximum displacements during the considered seismic events at the upper course of blocks slightly exceeded 0.3 m (26% of the thickness of the wall), while the residual displacements were less than 0.1 m (8% of the thickness). For model w3, the same blocks suffered about 20% decreased displacements. For model w4, collapse occurred above the mid-height of the wall for one of the imposed excitations.
- The inclination of the wall leads to larger displacements and might be crucial, as the inclined wall inherently has a deficit of stability to the leaning side, showing, thus, larger vulnerability to collapse.
- Concerning the effect of the connectors, the analysis showed a restriction of the displacements recorded for the model with connectors (w3). This restriction was larger above the mid-height of the wall and in the in-plane direction of the wall. An exception was observed at position P3 where the wall is connected with the reconstructed part of the partition wall at this place, the connectors slightly increase the maximum displacements observed in the out-of-plane direction, as the behavior of the partition wall prevails, but they manage to decrease almost in half the maximum residual displacements at the upper blocks of the wall.

- The inclusion of the blocks' damages in the numerical model (model w4) proved to have the most significant impact on the results. For the worst case scenario investigated, the numerical model predicted collapse of a significant part of the wall.

Acknowledgments. The authors are indebted to architect Mr Kosmas Skaris of the Acropolis Restoration Service. This study was based on his drawings for the proposed restoration of the north wall of the Parthenon cella, and the 3d models he created after the surveys of the ancient damaged blocks.

References

1. Korres, M., Bouras, Ch.: Study for the restoration of the Parthenon Vol. 1. Committee for the Preservation of the Acropolis Monuments, Ministry of Culture and Science, Athens, in Greek with a Summary in English (1983)
2. Toganides, N.: Study for the restoration of the Parthenon Vol. 5. Committee for the Preservation of the Acropolis Monuments, Ministry of Culture, Athens, in Greek with a Summary in English (1994)
3. Paraschi, C., Toganides, N.: Study for the restoration of the Parthenon Vol. 6. Committee for the Preservation of the Acropolis Monuments, Ministry of Culture, Athens, in Greek with a Summary in English (2002)
4. Toganides, N., Matala, C.: Study for the restoration of the Parthenon's north cella wall. Ministry of Culture, Committee for the Preservation of the Acropolis Monuments, Athens, in Greek with a Summary in English (2002)
5. Orlandos, A.: Architecture of the Parthenon Vol. b, Athens, in Greek (1977)
6. Skaris, K.: Study for the restoration of the north cella wall. Unpublished approved study by the Committee for the preservation of the Acropolis Monuments (2016)
7. Korres, M.: Study for the restoration of the Parthenon Vol. 4. Committee for the preservation of the Acropolis Monuments, Ministry of Culture, Athens, in Greek with a Summary in English (1994)
8. Toumbakari, E.E.: The Athens Parthenon: analysis and interpretation of the structural failures in the orthostate of the northern wall. In: Structural Analysis of Historic Construction, vol. 1. Taylor & Francis, London (2008)
9. Dasiou, M.E., Mouzakis, H.P., Psycharis, I.N., Papantonopoulos C., Vayas I.: Experimental investigation of the seismic response of parts of ancient temples. In: Proceedings of the International Conference "Protection of Historical Buildings" (PROHITECH 09), Rome, 21–24 June 2009, vol. 2, pp. 1671–1622. CRC Press/Balkema, Leiden, The Netherlands. ISBN: 978-0-415-55805-1 (2009a)
10. Dasiou, M.E., Psycharis, I.N., Vayas, I.: Verification of numerical models used for the analysis of ancient temples. In: Proceedings of the International Conference "Protection of Historical Buildings" (PROHITECH 09), Rome, 21–24 June 2009, vol. 2, pp. 1269–1274. CRC Press/Balkema, Leiden, The Netherlands. ISBN: 978-0-415-55805-1 (2009b)
11. Psycharis, I.N., Papastamatiou, D.Y., Alexandris, A.: Parametric investigation of the stability of classical columns under harmonic and Earthquake excitations. Earthquake Eng. Struct. Dynam. **29**, 1093–1109 (2000)
12. Papantonopoulos, C., Psycharis, I.N., Papastamatiou, D.Y., Lemos, J., Mouzakis, H.: Numerical prediction of the earthquake response of classical columns using the distinct element method. Earthquake Eng. Struct. Dynam. **31**, 1699–1717 (2002)

13. Psycharis, I.N., Lemos, J.V., Papastamatiou, D.Y., Zambas, C., Papantonopoulos, C.: Numerical study of the seismic behavior of a part of the Pronaos. Earthquake Eng. Struct. Dynam. **32**, 2063–2084 (2003)
14. 3DEC: 3-Dimensional Distinct Element Code: Theory and Background, Itasca Consulting group.
15. Kourkoulis, S.K., Exadaktlyos, G.E., Vardoulakis, I.: Unotched dionysos-pentelicon marble beams in three point bending: the effect of non-linearity, anisotropy and microstucture. Int. J. Fract. **98**, 369–392 (1999)
16. Zambas, K.: Study for the restoration of the Parthenon (Vol. 3b). Committee for the Preservation of the Acropolis Monuments, Ministry of Culture and Science, Athens, in Greek with a Summary in English (1994)
17. Psycharis, I.N., Mouzakis, H.P.: Shear resistance of pinned connections of precast members to monotonic and cyclic loading. Eng. Struct. **41**(2012), 413–427 (2012)
18. Vintzileou, E., Papadopoulos, K.: Dowel action of titanium bars connecting marble elements In: Rilem International Symposium on Connections Between Steel and Concrete, 10–12/09/2001 (2001)
19. Vintzileou, E., Toumbakari, E.E.: Dowel action of titanium bars connecting marble fragments at different angles. In: 13th International Brick and Block Masonry Conference, Amsterdam, 4–7/07/2004 (2004)
20. Vintzileou, E, Toumbakari, E.E., Papadopoulos, K.: Dowel action of titanium bars connecting marble fragments. Masonry Soc. J. **24**, 9–22 (2006)
21. Pasiou, E.D., Stavrakas, I., Triantis, D., Kourkoulis, S.K.: Marble epistyles under shear: an experimental study of the role of "Relieving Space." Front. Struct. Civ. Eng. **13**(4), 767–786 (2019). https://doi.org/10.1007/s11709-019-0515-9
22. Psycharis, I.N.: Technical report of the research programme "Selection of appropriate accellerograms for the Acropolis monuments dynamic analysis". Acropolis Restoration Service and Laboratory of Earthquake Engineering, NTUA, Athens, in Greek (2015)
23. Psycharis, I.N., Dasiou, M.E., Pavlopoulou, E., Vrouva, A., Ioannidou, M.: Final Report of the research project "Investigation of restoration scenarios for the north wall of the Parthenon cella". Acropolis Restoration Service and Laboratory of Earthquake Engineering, in Greek (2008)

The Application of 3D Image-Based Photogrammetry in Thai Heritage Sites: A Case Study of Wat Chai Wattanaram

Krisada Chaiyasarn(✉), Shekhar Mahat, and Supawit Prukpratin

Department of Civil Engineering (DI), Thammasat University, Pathumthani 12120, Khlong Luang, Thailand
ckrisada@engr.tu.ac.th

Abstract. Many historical structures from Ayutthaya Historical Park have been deteriorating due to aging and human-made activities. Therefore, it is necessary to find a systematic approach in order to maintain and preserve valuable archaeological ruins. This paper presents a novel approach in collecting information about historic structures, which is applied in Wat Chai Wattanaram in Ayutthaya as a case example. The method involves the 3D model creation of a historical building using the technique called Structure from Motion, which can create a detailed 3D model from images collected via a drone. The model can be used for many purposes, such as for archiving and structural analysis. This paper shows an example used of the model by applying the 3D model in Finite Element analysis. As shown in this paper, the 3D model is useful for structural analysis, and the presented approach can be applied in order to archive and manage these archeological sites sustainably.

Keywords: Historical buildings · Image-based 3D modelling · Structure from motion

1 Background and Importance

Thailand has several important historical sites and is considered as a cultural heritage by the Science Education Organization and United Nations Culture (UNESCO). These archaeological sites create a reputation and play an integral part of income for the country. At present, the ancient structure has been affected by damage caused by various environmental phenomena such as floods, human activities, and longevity. Even though restoring ancient structures is maintained regularly, but such damage still appears periodically, such as cracks. Wat Chai Watthanaram in Ayutthaya province is used for the current study, as shown in Fig. 1, which is a significant site for Thailand. The site has many stupas, including the main pagoda located in the middle of the site, and children pagodas (called Meruthit-Meru pagodas), located around the main pagoda. All children's units are connected by terraces to form the entire complex.

I. Vayas and F. M. Mazzolani (Eds.): PROHITECH 2021, LNCE 209, pp. 477–488, 2022.
https://doi.org/10.1007/978-3-030-90788-4_38

Fig. 1. Wat Chai Watthanaram

Fig. 2. Crack on Wat Chai Watthanar

3D reconstruction technologies can be applied to various engineering works. There has been a rapid development in 3D technologies, namely LiDAR and image-based Photogrammetry, that can significantly help to facilitate surveying works. The data obtained from the 3D technologies can be used to create a 3D model that provides information about the geometry of historic structures. This information can be used for structural restoration, progress monitoring, predicting structural behavior by the Finite Element Method (FEM). LiDAR technology has many benefits over the photogrammetry method, but its output lack texture due to the limited number of color that can be displayed. Also, Lidar cannot be used to collect data at a high altitude easily, and to obtain information from such inaccessible areas can become expensive. Therefore, an alternative method such as unmanned aerial vehicles, together with photogrammetry technology, can be an excellent combination for obtaining 3D models of historic structures. The photogrammetry technology can provide an output that is rich in color texture, and the drone can be used to obtain images at high altitude. With these benefits and capabilities, the photogrammetry technique can then be applied to examine and analysis of damage to ancient structures. Figure 2 shows an example of cracks on an ancient structure that requires constant inspection and monitoring. In this study, an unmanned aerial vehicle or drone technology is used to collect the images of Wat Chai Wattanaram. Then the images are processed by the photogrammetry technique to create a three-dimensional model of the temple, which is used to convert into a solid model for structural analysis by the FEM method. The contribution of this paper is to a new process for structural analysis for Thai structures that utilizes photogrammetry technology as part of its core process.

2 Literature Review and Related Theory

3D modeling technology, for example, a laser scanner technology, is applied for structural inspection and preserving historic sites. Armesto-González et al. (2010) investigated structure through the application of three-dimensional laser scan technology in the masonry bridge structure to predict the deformation by assuming a symmetrical parabolic curve of the bridge, which can then be analyzed to see the deviation from the curve. Fregonese et al. (2013) applied a 3D laser scanner to control the deformation of old buildings by monitoring its change from a 3D point cloud from the sets of laser scans combined with the general geographic data. Pieraccini et al. (2014) measured the tilt angle of the "Torre del Mangia" (Mangia's tower) in Siena, Italy, using the data from a laser scan in three dimensions. Costanzo et al. (2015) proposed a methodology by combining data obtained from a laser scanner and infrared thermal imaging for inspection at the St. Augustine Monumental location in Calabria, Italy. Pesci et al. (2012) applied three-dimensional laser scanning technology with digital photos to check the regeneration of the Palazzo d'Accursio in Bologna, Italy. Kouimtzoglou et al. (2017) inspected the structure of the Plaka Bridge in Epila, Northwestern Greece, by photogrammetry method, in which the 3D model was created using the Agisoft Photoscan program to analyze the causes of the bridge collapse. Photogrammetric survey via unmanned aerial vehicle (UAV) technology can be used to collect data that give the details of elements in the structure at current conditions. The collected photos can be used to create a three-dimensional model for reporting and analyzing the structure. Achille et al. (2015) demonstrated the application of UAV for exploring in the historic structure "Santa Barbara" in Italy. Bhadrakom and Chaiyasarn (2016) implemented collected photos to create a 3D model via structure from motion (SFM) in order to assess the tilt angle of Wat Yai Chai Mongkhon in Phra Nakhon Si Ayutthaya province. Chaiyasarn et al. (2017) also implemented the photogrammetry technique by using unmanned aerial vehicles to study an ancient structure.

Structural analysis using the finite element method (FEM) is a very commonly used method for analyzing and examining various effects that occurred in the structures. Generally, the analysis results depends greatly on geometry information, which in the past, this information was is normally from drawings that can be inaccurate. With the photogrammetry technique, as-is geometry information can be obtained to enable more accurate prediction of a structural model. Macchi et al (1993) studied the Tower of Pisa to compare models with force-free conditions using the finite element technique using the photogrammetry technique. Meli and Sánchez-Ramírez and (2007) studied the response of the Corpus Christi archaeological site in Mexico City using the finite element method without considering the effects of damage on the structure but they applied the 3D technology to obtain the true geometry information of the structure. Roca et al (2010) conducted the analysis of masonry structures with complex geometries using the 3D photogrammetry techniques for finite element modelling. It can be seen that the photogrammetry technique is a useful tool to obtain geometry information of a structure for a more accurate prediction of structural stability for ancient structures.

3 Methodology

Figure 3 shows the process summarized in the study. The outline of the section is as follows. Section 3.1 explains how images are collected by a drone using various collection strategies. Section 3.2 illustrates the formation of a three-dimensional model, which is followed by a generation of 3D geometry solid model in Sect. 3.3. Finally, the details about structural analysis using Finite Element Method is explained in Sect. 3.4.

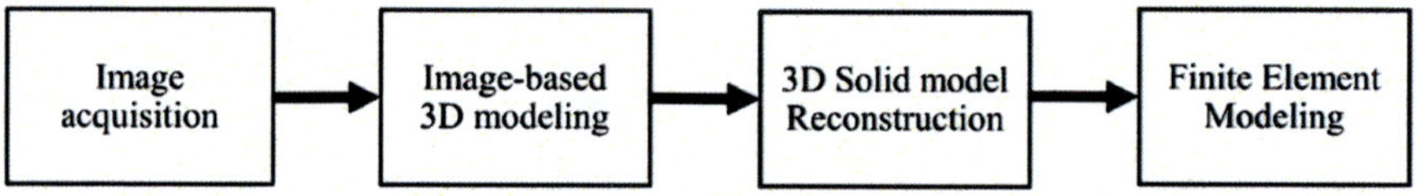

Fig. 3. The study process

3.1 Image Acquisition

In this study, image dataset was used to create a three-dimensional model for analyzing the structure. To obtain images, a drone is used for data collection. Table 1 shows the specification of the drone used in this study. There are three collection strategies used to collect the data in this study. The first strategy is called a Point-Of-Interest (POI) strategy, in which a drone is flown to collect data of interest in a circular motion as shown in Fig. 4. This strategy can be done automatically in many drone flight applications by specifying various parameters, such as radius, high, and speed. The second method is called Scanning, in which a drone flies in a horizontal plane at a specified elevation as shown in Fig. 5. This collection strategy is useful to obtain the overview of study area. Lastly, sometimes the UAV is unable to collect information at the base area of a structure, therefore images are collected manually from the ground level.

Table 1. Properties of unmanned aircraft

Specs	DJI Phantom 4
Camera sensor	1/2.3" CMOS, 12.4 Megapixel
Max Speed	20 m/s
Controller range	5 km
Satellite Positioning	GPS/GLONASS
Fight time	Approximately 28 min
Weight	1380 g

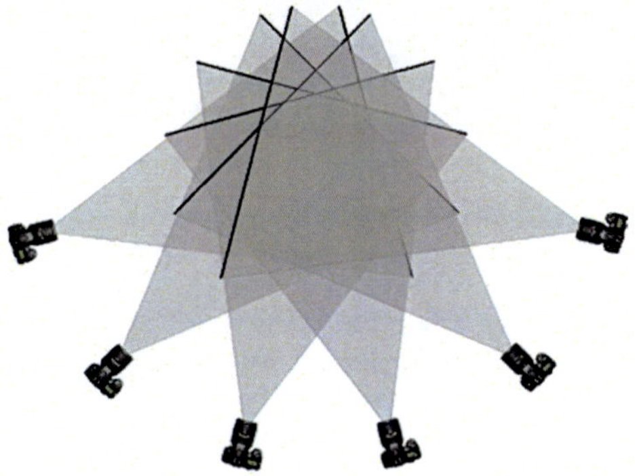

Fig. 4. Point of interest (POI)

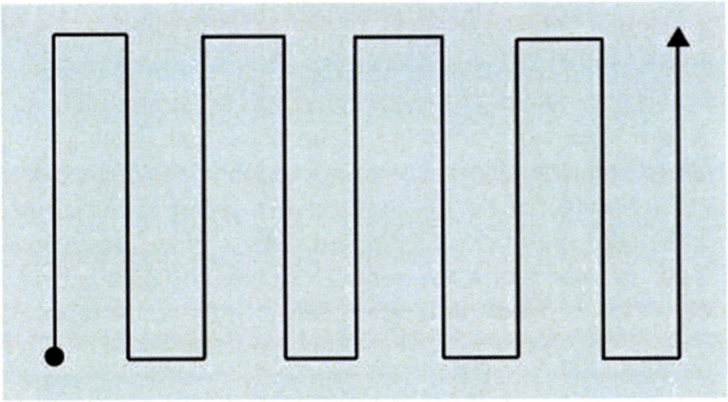

Fig. 5. Scanning strategy

3.2 Image-Based 3D Photogrammetry

After the images are collected using the method from Sect. 3.1, the next process is to create a three-dimensional model based on the structure from motion (SFM) and multiview stereo (MVS) techniques. The process finds the relationship between the photos by automatic key point matching. The key points in each image are used to orient the images through the ray intersections. These intersections are used to initialize the camera calibration (if required) and orientation by the process called Triangulation. The whole network is adjusted using the bundle adjustment algorithm at once. The sparse point cloud is produced, which later is densified by the MVS technique. The example model of a sparse point cloud is shown in Fig. 6.

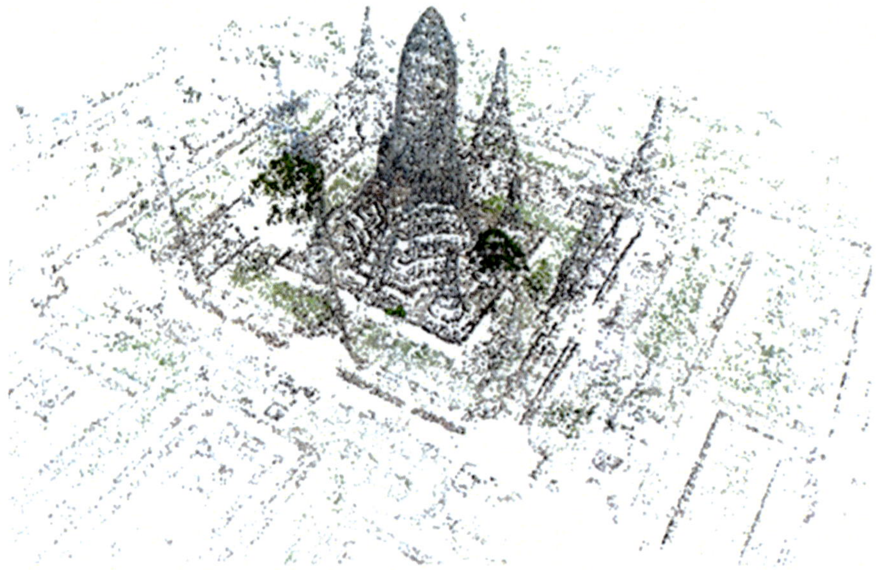

Fig. 6. A sparse point cloud of a structure

A dense cloud is then obtained after the sparse point cloud from Fig. 6. This process creates more details of the cloud point to a higher resolution, as shown in Fig. 7a. This step can be quite computationally expensive when greater detail is required. The dense point cloud is then used to create a 3D solid model for FEM, which is explained in Sect. 3.2. Once the dense point cloud is created, a mesh model and texture are created to obtain a 3D texture model of a structure. This model offers a very realistic model that can be used for inspection of the surface damage of a structure, as exemplified in Figs. 7(b) and 8.

3.3 Solid Model Creation

The solid model is required for the FEM model. This model is created from a series of geometry to create a watertight model. Currently, no software package can automatically convert the mesh or point cloud model into a solid model for FEM. Therefore, in this study, we used Autodesk Recap software to create the solid model. The process starts by importing a dense cloud model into the program. Then, we use the point cloud as a reference to create the geometry of the temple. We start by sectioning the point cloud

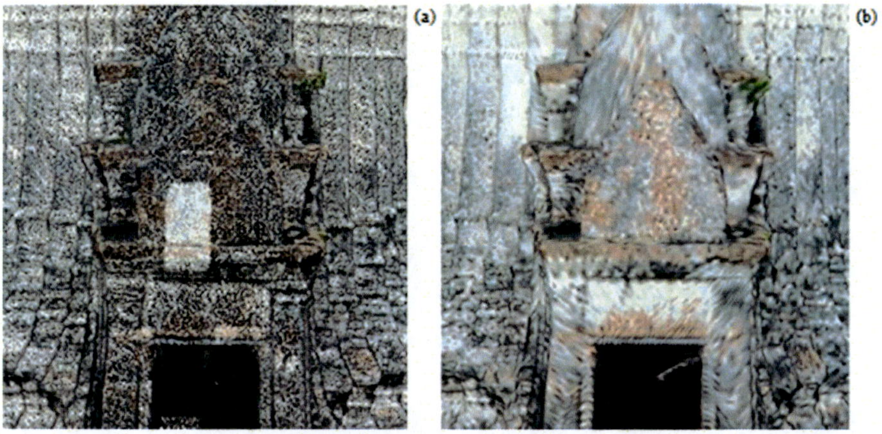

Fig. 7. (a) Dense cloud model (b) Texture model

into different layers at various height. We then used the sectioned point cloud to draw an outline of the model and then extrude the model at the same height as the sectioned point cloud. By this method, the solid model is a collection of extruded layers that are then combined to form a solid model. In this study, we create a hollow of the structure using the laser scanner data. We carved the interior of the model to get the most realistic geometry of the structure. The process of creating a solid is exemplified in Figs. 9, 10, and 11.

Figure 10 shows that the dense point cloud has to be aligned with the physical world coordinates before sectioning. Once the data is aligned, we can then section the point cloud in the Autodesk Recap and then create a solid model. The solid model is then converted into the format that can be accepted by Autodesk Inventor for Finite Element Modelling.

Fig. 8. A texture model of Wat Chai Wattanaram

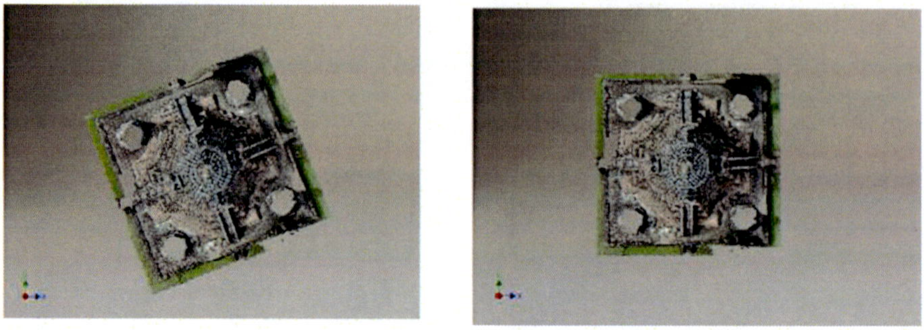

Fig. 9. Aligning positions of the point cloud model

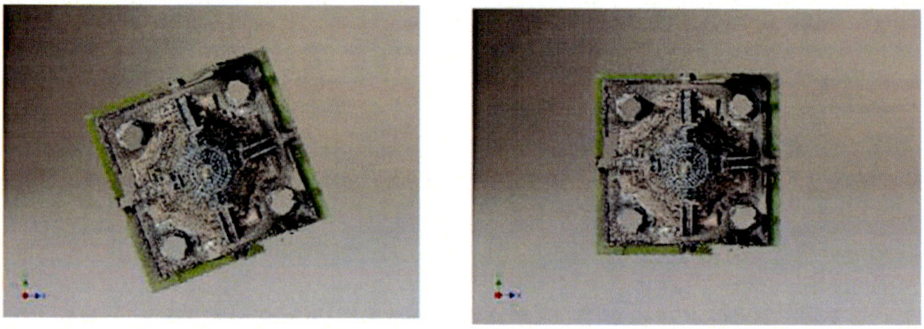

Fig. 10. Sectioned point cloud is used to guide the solid model

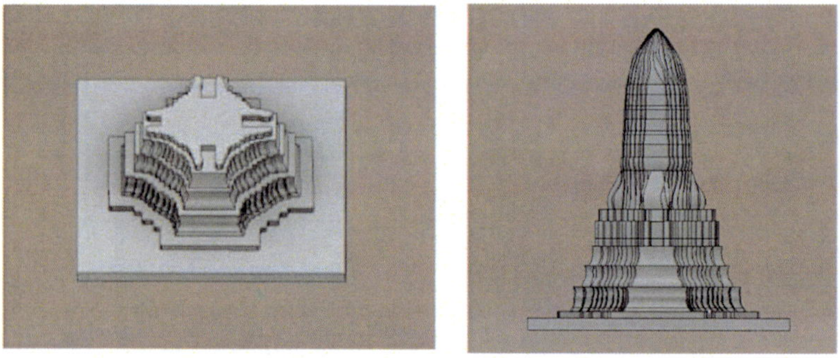

Fig. 11. The solid 3D model by Autodesk Inventor

3.4 Analysis of the Three-Dimensional Model by Finite Element Method

The finite element method is a method for solving numerical differential equations based on estimating solutions from algebraic equations systems. This method assumes a solid material model. In this model, we model the effect of the self-weight of the structure using the Autodesk Inventor software program using the solid three-dimensional model. We obtained the variables of the temple using experimental results of the bricks from the site, and the variables include the modulus of the body and the density. The parameters used for modeling is explained in Sect. 4.

4 Experimental Results and Discussion

The DJI Phantom 4 unmanned aircraft is used to collect the images from Wat Chai Watthanaram Temple. The images must be at least 50% overlapping. In this study, we collect images from 3 main locations, which are the central pagoda with 240 pictures, Merutid-Merurai or the children pagoda of 615 pictures, and the base of the main pagoda is 163 pictures.

The three-dimensional model of the temple is divided into sub-model as follows: the main pagoda has a total of 103,195 sparse point cloud and 168,926 points for the children pagodas and 107,637 points for the pagoda base. The total sparse point cloud is 379,758 points. The sparse points cloud is then used to obtain the dense point cloud, which becomes 35,425,647 dense point cloud models. The three-dimensional model of the point cloud model is a complete model that is suitable for creating a geometric model. The accuracy of the 3D point cloud can significantly affect the accuracy of the solid model since the incorrect 3D model can lead to a wrong solid model.

The solid model is analyzed in the Autodesk Inventor program; the geometry model of Wat Chai Watthanaram is shown in Fig. 12. The parameters used for the FEM model is summarized in Table 2.

Table 2. The parameters used for modelling the temple obtained from a laboratory experiment.

Parameters	Values
Young modulus	980 MPA
Poisson ratio	0.216
Density	1800 kg/m^3

The solid model is divided into 51,893 elements, as shown in Fig. 13. The structure is subjected to a self-weight alone, and in this experiment, we assume fixed support at the base. As shown in Fig. 14, the 3rd principal stress is uniform and almost 0 MPa. In the central hall of the pagoda, it is found that there is higher compressive stress as expected, as shown in Figs. 14 and 15, respectively.

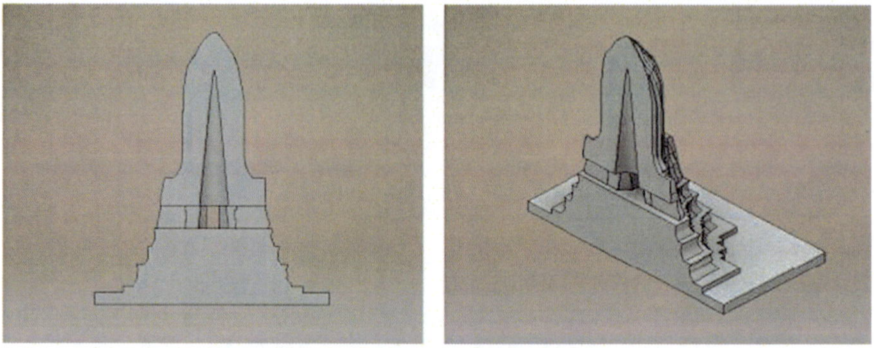

Fig. 12. The details of interior of the 3D model

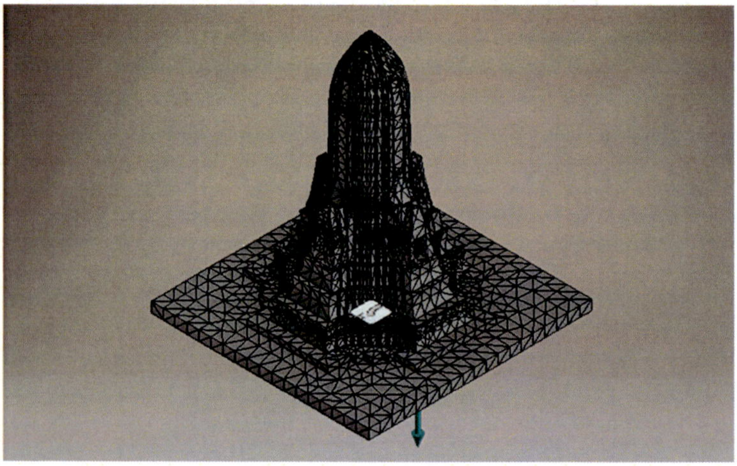

Fig. 13. Elements and nodes of the temple FEM model

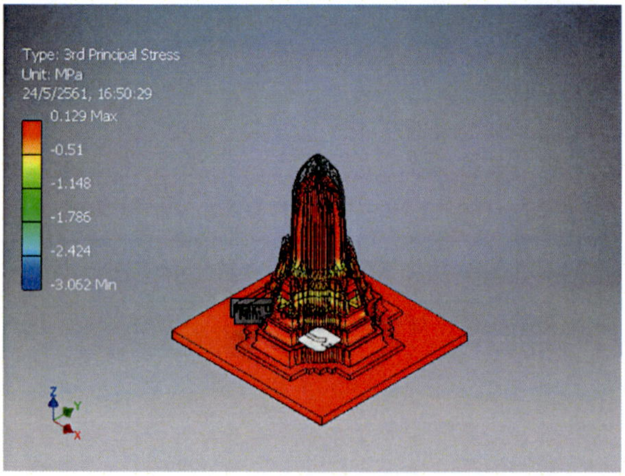

Fig. 14. The 3rd principal stress of the FEM model

Fig. 15. The vertical stress of the FEM model

4.1 Discussion

This work presents the process of structural analysis of a historical temple using the combination of the image-based photogrammetry technique and a drone to create a three-dimensional model. With this technology, it is also possible to obtain the as-is condition of the temple. The texture model of the temple can also be used for damage inspection on the surface. The stability of the temple is analyzed by the FEM method. It is found that the main pagoda of Wat Chai Watthanaram has high compressive stress in the central hall. However, stress is not critical to cause further damage to the stability of the structure. However, since the structure is still symmetrical, and the stress is distributed throughout the structure, therefore, no critical stress is of concern. This study exemplifies that an unmanned aerial vehicle can be used to obtain information from inaccessible areas, which makes the 3D geometry of the temple much easier to create. To improve the accuracy of the 3D model, GPS may be used to obtain more accurate ground control points.

The accuracy of the FEM model depends much on the solid model. Currently, this process is done manually, which can be challenging to model any imperfection in geometry, although this method is a much-improved process than the traditional FEM model that creates a 3D model from drawing. Furthermore, the interior of the model may not be accurate in this study, as there may be other unknown voids in the structure, which is hard to obtain. To obtain the real as-in condition is an on-going research which will be published in the future.

5 Conclusion

This work presents the framework that can be used to obtain a structural analysis of a historical temple in Thailand. The process utilized modern technology to obtain the as-is geometry of the model using the drone and image-based geometry. The results from this study suggest that the 3D model can be created with high accuracy, and the temple does not have any stress that is of concern for failure.

Acknowledgements. This research is funded by the Young Researcher Grant, Faculty of Engineering, Thammasat University.

References

Armesto-González, J., Riveiro-Rodríguez, B., González-Aguilera, D., Rivas-Brea, M.T.: Terrestrial laser scanning intensity data applied to damage detection for historical buildings. J. Archaeol. Sci. **37**(12), 3037–3047 (2010)

Fregonese, L., Barbieri, G., Biolzi, L., Bocciarelli, M., Frigeri, A., Taffurelli, L.: Surveying and monitoring for vulnerability assessment of an ancient building. Sensors **13**(8), 9747–9773 (2013)

Pieraccini, M., Dei, D., Betti, M., Bartoli, G., Tucci, G., Guardini, N.: Dynamic identification of historic masonry towers through an expeditious and no-contact approach: application to the "Torre del Mangia" in Siena (Italy). J. Cult. Herit. **15**(3), 275–282 (2014)

Costanzo, A., Minasi, M., Casula, G., Musacchio, M., Buongiorno, M.: Combined use of terrestrial laser scanning and IR thermography applied to a historical building. Sensors **15**(1), 194–213 (2015)

Pesci, A., Bonali, E., Galli, C., Boschi, E.: Laser scanning and digital imaging for the investigation of an ancient building: Palazzo d'Accursio study case (Bologna, Italy). J. Cult. Herit. **13**(2), 215–220 (2012)

Kouimtzoglou, T., Stathopoulou, E.K., Georgopoulos, A.: Image-based 3d reconstruction data as an analysis and documentation tool for architects: the case of Plaka bridge in Greece (2017)

Achille, C., et al.: UAV-based photogrammetry and integrated technologies for architectural applications—methodological strategies for the after-quake survey of vertical structures in Mantua (Italy). Sensors **15**(7), 15520–15539 (2015)

Bhadrakom, B., Chaiyasarn, K.: As-built 3D modeling based on structure from motion for deformation assessment of historical buildings. Int. J. Geomate **11**(24), 2378–2384 (2016)

Chaiyasarn, K., Bhadrakom, B., Buatik, A., Poovarodom, N.: The structural analysis of historical structures based on image-based as-built 3d modelling using unmanned aerial vehicles. In: 3rd International Conference on Protection of Historical Constructions, Lisbon, Portugal (2017)

Macchi, G., Ruggeri, G., Eusebio, M., Moncecchi, M.: Structural assessment of the leaning tower of Pisa. In: IABSE reports, pp. 401–401 (1993)

Meli, R., Sánchez-Ramírez, R.: Criteria and experiences on structural rehabilitation of stone masonry buildings in Mexico City. Int. J. Archit. Herit. **1**(1), 3–28 (2007)

Roca, P., Cervera, M., Gariup, G.: Structural analysis of masonry historical constructions. Classical and advanced approaches. Arch. Comput. Methods Eng. **17**(3), 299–325 (2010). https://doi.org/10.1007/s11831-010-9046-1

Generalized Algorithm for Segmental Topology of Rampant Arch

Vaishnavi Yogesh Totla[1](✉), Pallavi Varma[2](✉), and Prathamesh Varma[3](✉)

[1] Department of Civil Engineering, Government College of Engineering, Aurangabad, India
[2] Nandadeep Designer's and Valuers Pvt. Ltd., 21, Ramdurg, Raghuveer Nagar, Jalna Road, Aurangabad, Maharashtra 431005, India
arpallavi@nandadeep.org
[3] Saraswati Bhuvan College of Science, Aurangabad, Maharashtra 431005, India

Abstract. Gothic architecture inspired by the Romanesque architecture flourished in the Middle Ages between the 12th to 15th century. The principal structural characteristics including barrel vaults, domes and flying buttresses (external rampant arches) of Gothic architecture arose out of medieval masons' efforts to support heavy masonry ceiling over wider spans. Vault topology, keeping all interfaces of blocks in compression allowed these medieval masons to construct larger spans using small blocks, without any reinforcement. This concept is further extended in construction of flying buttresses to support the thrust of masonry dome in tall Gothic structures, and in the staircase to commute the height difference between two floors. This arch whose support on one side is having greater elevation than the support on the other side is called a rampant arch.

The rampant arch built using two arch segments were used in gothic architecture to shape the flying buttresses and to build the stairs. Various examples of construction of stairs over rampant arch are observed in historical structures, for example, Monastery of Ucles in Cuenca (Spain), Beauvais Cathedral in France and La Lonja Valencia (Spain).

Rampant arch is widely used as a composition of two segments of circular arch in such a way that these curves have a common tangent at the meeting point. Cases in which common tangent is parallel to slope line connecting two supports of rampant arch and referred to as optimum geometry for rampant in some literatures. In general, infinite possibilities exist for different topology of rampant arch made from two segments and having common tangent at the meeting point. Possibilities of having different slope of tangents at either springing point further entangle the problem.

Historical structures exhibit many topologies for rampant arch construction. To understand the effect of these different topology on load path and stability need development of generalized algorithm for plotting rampant arch geometry. This paper demonstrates the generalized algorithm for plotting varying rampant arch geometry.

Keywords: Rampant arch · Flying buttresses · Gothic architecture · Geometry

I. Vayas and F. M. Mazzolani (Eds.): PROHITECH 2021, LNCE 209, pp. 489–499, 2022.
https://doi.org/10.1007/978-3-030-90788-4_39

1 Introduction

Rampant arch is an arch whose support is higher on one side as compared to the other and with considerable difference in height. An arch carry loads around an opening, transferring them around the profile of the arch to abutments, jambs or piers on either side. They exhibit structural strength and span large areas by resolving forces into compressive stresses and eliminating tensile stresses this is referred to as arch action [1]. This makes them very stable and efficient, capable of supporting greater loads and larger spans compared to horizontal beams.

Arches have been being built from as long ago as the second millennium but were not used for a variety of structures until the roman took advantage of their structural capabilities. Romanesque architecture dating back to the 10th century utilized internal buttresses in the form of arch as supporting elements for the inside of the church walls [2]. Towards the end of the Romanesque period, architects were aspiring to build churches and cathedrals on an ambitious scale. With the arrival of the Gothic period in the 12th century from the idea of internal buttresses used in Romanesque architecture flying buttresses were introduced as means of making such buildings taller [3].

Rampant arches were mostly witnessed in the form of flying buttresses. Flying buttresses are built projecting from the walls of a structure down to the foundation. As the forces in an arch are carried toward the ground, the arch will push outward at the base, called thrust [4]. The purpose of such projections is to support the weight and horizontal thrust of the high arches and domes spanning the interior space. The flying buttress serves as a bridge, carrying the lateral thrust produced at the base of the arches and domes due to their weight, across to the outer buttress, which is massive enough to absorb the pressure [5]. The stability of the entire building depends upon the balance of pressures and with the existence of flying buttresses, cathedrals were able to be built taller and more glorious than ever before.

One of the earliest, cathedrals to incorporate the use of flying buttresses was the Notre Dame Cathedral in France shown in Fig. 1. Its construction began in 1163 and the cathedral was finally completed around the year 1345. Many different architects and

Fig. 1. Notre Dame Cathedral depicting Rampant Arches

ideals went into the construction of Notre Dame. Flying buttresses were incorporated into the architecture as shown in Fig. 2 for the primary use of balancing the pressure produced by such vaulted spaces. Another very important reason flying buttresses were used in the Notre Dame Cathedral was to allow adequate sunlight into the building [6].

With such high walls and lack of windows, the cathedral proved to be quite dark. Architects then realized that with the addition of more flying buttresses, they could place large stained glass windows along the walls of the cathedral which provided great luminosity to the interior of the cathedral and give symbolic value to Gothic architecture, which contrasts with the heaviness of Romanesque architecture [7]. The incorporation of these large windows would further weaken the stability of the walls; however, the strength of the flying buttresses solved such issues. It showed that the flying buttress was important from a structural point of view.

Fig. 2. Zoomed version of Notre Dame Cathedral exhibiting Rampant Arch

The rampant arch of a single circumferential arch played an essential role in structural and formal conception of Gothic architecture; the number of rows of buttresses in the external volume of cathedrals increased the spatial richness of building. Rampant arch is also widely used as a composition of two segments of circular arch in such a way that these curves have a common tangent at the meeting point [8]. In Gothic architecture to shape the flying buttresses and also to build stairs this type of arches were used. Examples include Flying Buttresses of Monastery of Batalha in Fig. 3 and the Cathedral of Palma de Mallorca, Spain in Fig. 4 is cited. Rampant arch in the staircase of the Monastery of Ucles, Cuenca as shown in Fig. 5 In France, Amiens Cathedral, etc

A buttress is a structure built against another structure in order to strengthen or support it. Historically, they have been used to strengthen large walls or buildings such as churches. The rampant arches were widely used again at the end of the 19th century with the appearance of neo-Gothic architecture. Many churches were built throughout Europe. One of the prominent structure Washington National Cathedral in Fig. 6 which uses several rows of rampant arches which increases spatial richness of building [9]. Also, they continue to be used in large modern structures such as retaining walls and dams also in residential buildings in present. Figure 7 shows a rampant arch over the staircase in a residential building located in Maharashtra, India.

Fig. 3. Monastery of Batalha

Fig. 4. Flying Buttresses of the Cathedral of Palma de Mallorca, Spain

Design of rampant arch of single arch is not difficult, but it is difficult to design rampant arches consisting of two arches, were throughout history calculated using graphical methods, in some literatures rampant arch in which common tangent is parallel to slope line connecting two supports of rampant arch, and referred to as optimum geometry for rampant in some literatures [8]. In general, infinite possibilities exist for different topology of rampant arch made from two segments and having common tangent at the meeting point. Possibilities of having different slope of tangents at either springing point further entangle the problem. The objective of this work is to develop a generalized algorithm for plotting varying rampant arch geometry. Work is further extended to optimize the quantity of filling under the stairway built over the rampant arch.

Fig. 5. Rampant arch in the staircase of the Monastery of Ucles, Cuenca

Fig. 6. Washington National Cathedral

Fig. 7. Rampant arch before steps construction, Maharashtra, India.

2 Generalized Algorithm for Plotting Rampant Arch Geometry

Where,

θ - Angle subtended at the centre by common tangent for two segmental arches.
α - Angle subtended by tangent at the start of arch 1 with horizontal in clockwise direction.
β - Angle subtended by tangent at start of arch 2 with horizontal in anti-clockwise direction.
S - Horizontal distance between two supports.
h - Vertical distance between two supports.
R_1 - Radius of arch 1
R_2 - Radius of arch 2

In Fig. 8, the rampant arch as composition of two circular arches is shown. With angles θ, α *and* β known along with span(S) and h. For modelling of rampant arch with given data radius R_1 and R_2need to be evaluated which further gives us location of point O_1 (centre of arch 2), point A (start of arch 2),point O_2 (centre of arch 1), point B (start of arch 1) and also C (meeting point of two arches where common tangent exists).

R_1 and R_2 are two unknowns so we need two equations to solve.

In the Δ AO_1P_1,

By sine rule,

$$\frac{P_1A}{\sin(90+\beta-\theta)}=\frac{R_2}{\sin\theta}$$

$$P_1A=\frac{R_2\cos(\theta-\beta)}{\sin\theta} \tag{1}$$

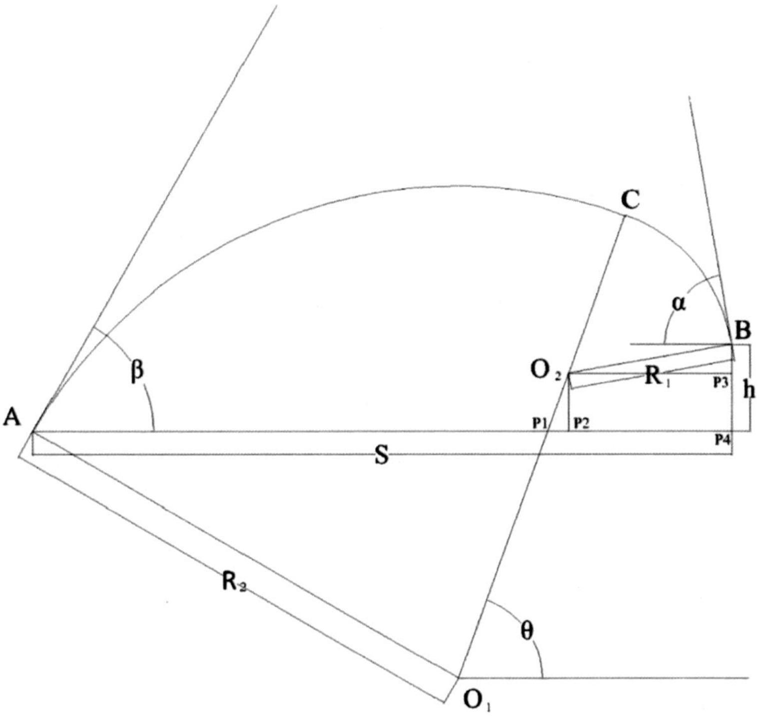

Fig. 8. Rampant arch as two segments of circular arch

$$\frac{O_1P_1}{\sin(90-\beta)} = \frac{R_2}{\sin\theta}$$

$$O_1P_1 = \frac{R_2\cos\beta}{\sin\theta} \tag{2}$$

$$P_2P_4 = R_1\sin(\alpha)$$

$$o_1o_2 = R_2 - R_1$$

In the Δ AO_1P_1,

$$P_1P_2 = (R_2 - R_1 - \frac{R_2 cos\beta}{sin\theta})cos\theta$$

$$\text{Span } = AP_1 + P_1P_2 + P_2P_4$$

$$\text{S} = \frac{R_2 cos(\theta-\beta)}{sin\theta} + (R_2 - R_1 - \frac{R_2 cos\beta}{sin\theta})cos\theta + R_1 sin(\alpha) \tag{3}$$

Similarly,

$$\text{Height} = P_4P_3 + P_3B$$

$$P_3B = R_1cos(\alpha)$$

$$P_3P_4 = (O_1O_2 - P_1O_1)\sin\theta$$

Therefore,

$$h = R_1cos\alpha + \left(R_2 - R_1 - \frac{Rcos\beta}{sin\theta}\right)\sin\theta \tag{4}$$

Even when there is change in location of centre of arch 2(O_2) with respect to point A or point O_1 as shown in Fig. 8. The equation calculated for span(S) and h will not alter that is they will remain same it can be crosschecked and concluded that Eqs. (3) and (4) are valid for all the possible geometries with given valid inputs such that both R_1 or R_2 values are positive (>0). This happens either due to less span available or greater height difference between two supports for given data.

2.1 Drawing of Rampant Arch

From Eqs. (3) and (4) for given inputs we can calculate R_1 and R_2 for two arches.

Consider O_1 as reference point, for example set O_1 as (0, 0) in Cartesian plane.

Co-ordinates for all other points in Cartesian plane will be as follows:

O_1 (0,0)

O_2 (($R_2 - R_1$)cos θ, ($R_2 - R_1$)sin θ)

A ($-R_2$sin β, R_2cos β)

B (($-R_2$sin β + s), (R_2cos β + h))

C (R_2cos θ, R_2sin θ)

Draw arch 1 with starting point (B), ending point (C) and centre as (O_2).

And draw arch 2 with starting point (C), ending point (A) and centre as (O_1).

We observe point O_1, O_2 and C lie in straight line this shows both arches have common tangent existing at point C.

Rampant arch geometry is generated.

3 Computer Software

Rampant arches are modelled in AutoCAD using AUTOLISP and Microsoft Excel (spreadsheet program) in this work. AUTOLISP is full version of AutoCAD's inbuilt programming language. AUTOLISP Is used to create programs that can generate customized command; AUTOLISP have rich set of functions that can be used to perform any activity in AutoCAD automatically [10]. Programs are developed in AUTOLISP and Microsoft Excel for generating rampant arch in AutoCAD by giving required inputs. This program calculates R_1 and R_2 for two arches and further provides commands for drawing rampant arch. Using program file named "RAMPANT" of both AUTOLISP and Microsoft Excel which is attached at the end of work using which rampant arch geometries for infinite inputs can be generated. Some of the program with varying inputs is shown in Fig. 9 with inputs mentioned respectively. In Fig. 9a all three angles θ, α *and* β

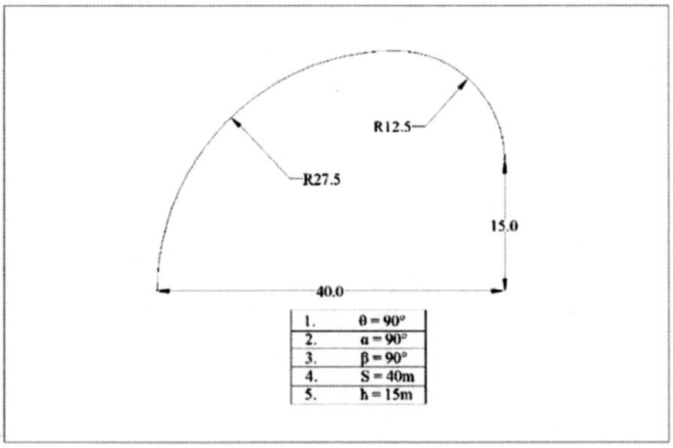

(a)

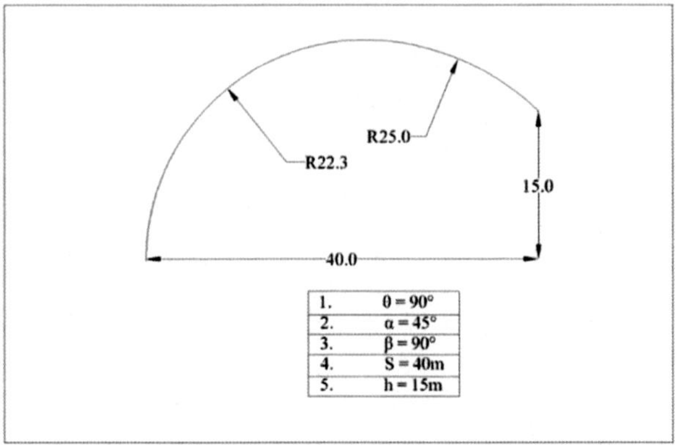

(b)

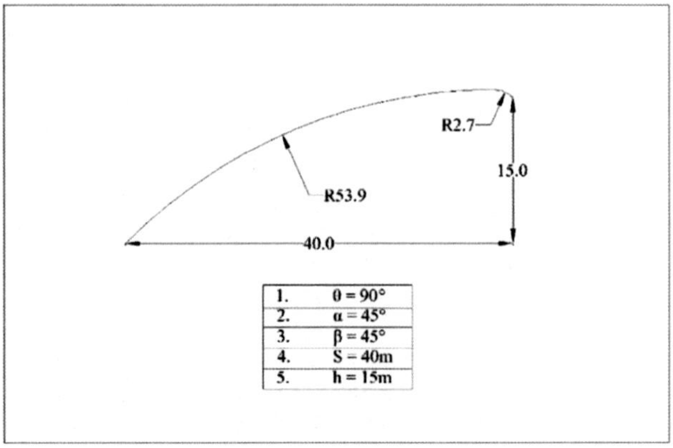

(c)

Fig. 9. a to e depicts Rampant arches generated for given inputs with AUTOLISP and Microsoft Excel

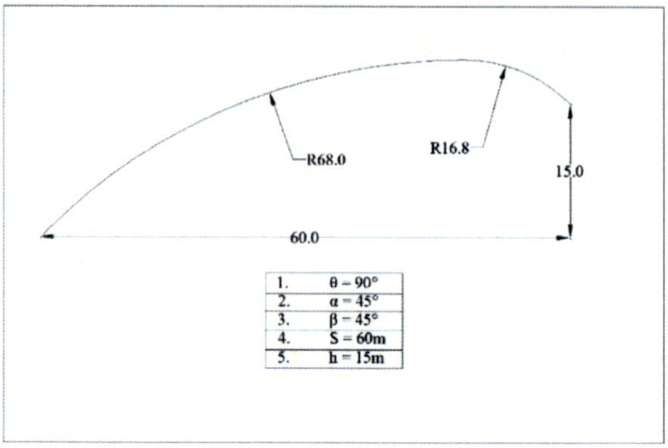

(d)

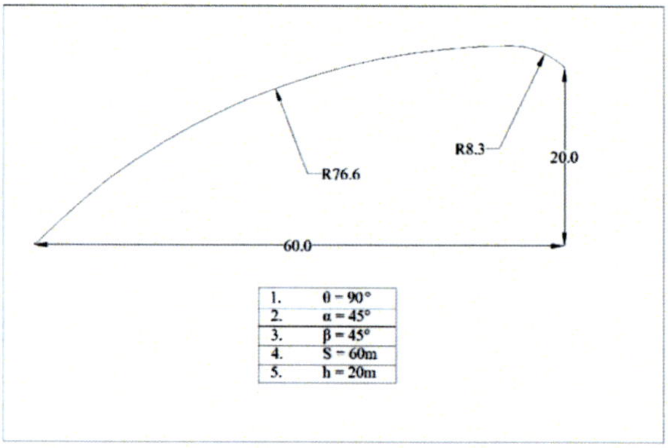

(e)

Fig. 9. continued

are 90^0 each with span of 40 m and elevation difference between two supports as 15 m, the rampant arch is generated with $R_1 = 12.5$ m and $R_2 = 27.5$ m. In Fig. 9b angle α is changed to 45^0 and all other quantities are same as in Fig. 9a rampant arch is generated with $R_1 = 25$ m and $R_2 = 22.3$ m. Further in Fig. 9c angle β is changed to 45^0 all other quantities are same as in Fig. 9b rampant arch is generated with $R_1 = 2.7$ m and $R_2 = 53.9$ m. Further span of rampant is changed to 60 m in Fig. 9d rampant arch with $R_1 = 16.8$ m and $R_2 = 68$ m is generated and further h is changed to 20 m from 15 m in Fig. 9e rampant arch with $R_1 = 8.3$ m and $R_2 = 76.6$ m is generated.

4 Conclusion

In Gothic architecture's initial phase, rampant arches used were generally circular. With passing of time and thoroughly understanding structural requirements of improving thrust resistance rampant arches involving two circular arches were introduced. Whose centers are at different levels and there exist a common tangent point of their meeting.

There are many variations possible in inputs of rampant arch in order to optimize rampant arch from different aspects such as structural, architectural, economical, etc. we need a generalized algorithm for geometry of rampant arch. Under this work, generalized geometry of rampant arch is modeled by developing an algorithm using trigonometry. Further programs for rampant arch geometry generation in AutoCAD are developed using AUTOLISP and Microsoft Excel. So this work opens up a new perspective for understanding the optimum rampant arch construction and also can further serve as an initial point of verification of existing structure, in spite of their stability they may require intervention which can be understood.

References

1. Viola, E., Panzacchi, L.: Structural analysis of historical masonary arches. In: International Conference of Restoration, Recycling and Rejuvenation Technology for Engineering and Architectural Application (2004, June)
2. DannyRand: Romanesque and gothic architecture. https://ivypanda.com/essays/romanesque-and-gothic-architecture/
3. West, G.H.: Gothic Architecture in England and France, vol. I. G. Bell and Sons, London (1927
4. Ricci, E., Sacco, E.: A method of analysis of masonry arches. In: 10th International Conference of Structural Analysis of Historical Construction (2016, September)
5. Varma, M., Jangid, R., Ghosh, S.: Thrust line using linear elastic finite element analysis for Masonry structures. Adv. Mater. Res. **133–134**, 503–508 (2010). https://doi.org/10.4028/www.scientific.net/amr.133-134.503
6. Bruzelius, C.: The construction of Notre-Dame in Paris. Art Bull. **69**, 540–569 (1987)
7. Chilton, J., Isler, H.: The Engineer's Contribution to Contemporary Architecture, p. 32. Thomas Telford, London (2000). ISBN 9780727728784
8. Velilla, C., Alcayde, A., San-Antonio-Gómez, C., Montoya, F.G., Zavala, I., Manzano-Agugliaro, F.: Rampant arch and its optimum geometrical generation. Symmetry **11**(5), 627 (2019). https://doi.org/10.3390/sym11050627
9. Wilsmaft, H.: Washington National Cathedral. Retrieved April 23, 2018 (2016)
10. Nikolinakou, M.-K., Tallon, A.J., Ochsendorf, J.A.: Structure and form of early Gothic flying buttresses. Revue Européenne De Génie Civil **9**(9–10), 1191–1217 (2005). https://doi.org/10.3166/regc.9.1191-1217
11. Bony, J.: French Gothic Architecture of the Twelfth and Thirteenth Centuries. University of California press, Berkeley (1983)

Protection of Historic Stone Masonry Walls Against Blast Loads

Laura Ciornei and Murat Saatcioglu(✉)

Department of Civil Engineering, University of Ottawa, 161 Louis Pasteur, Ottawa, Canada
{lcior058,Murat.Saatcioglu}@uottawa.ca

Abstract. Historic masonry buildings often have cultural and heritage significance becoming iconic buildings of a nation. As such, they may become targets for maliciously intended acts of terror in the form of bomb blasts. Vehicle bombs at close proximity of masonry buildings can cause significant damage to structural and non-structural building components. Blast performance of stone masonry buildings has not been researched sufficiently to develop blast retrofit technologies. An experimental research program is currently underway at the University of Ottawa to fill this research gap. Scaled models of stone masonry infill walls are tested under simulated blast loads. A new retrofit methodology is developed, consisting of tension braces that provide reactive forces to externally applied blast overpressures. The test walls consist of 2.0 m^2 geometry with a wall thickness of 100 mm. Blast shock waves are applied through a shock tube. The results indicate that the blast capacity of walls can be increased multiple times by the implementation of the retrofit strategy, protecting historic stone masonry walls.

Keywords: Blast loads · Bomb blast · Blast retrofit · Masonry wall · Stone masonry · Tension brace

1 Introduction

A large number of historic buildings were built using stone masonry walls, worldwide. These buildings have significant heritage, architectural, and social meaning; often reflecting history and character of a nation. In Canada, most of the historic buildings were built in the late 19th and early 20th century. The Federal Canadian Parliament Building, Provincial Legislature Buildings and a large number of museums and art centres, in addition to many government office buildings are examples of historic stone masonry construction that are preserved in Canada. These buildings either have structural load bearing walls or non-structural infill wall panels, consisting of single or double wythe panels, with or without rubble infills. They also differ in terms of coursing of the stones, as well as the properties of stone and mortar used. Because of their historical, heritage and political significance they become iconic buildings and hence potential targets for maliciously intended acts of terror, which often manifests itself in the form of a bomb attack. The performance of historic stone masonry buildings in Canada, subjected to blast loads

I. Vayas and F. M. Mazzolani (Eds.): PROHITECH 2021, LNCE 209, pp. 500–511, 2022.
https://doi.org/10.1007/978-3-030-90788-4_40

is not well understood. Therefore, their protection becomes a challenge for structural engineers.

A combined experimental and analytical research project has been underway in the Blast Research Laboratory of the University of Ottawa in Canada. Large scale stone masonry walls were tested using a large-capacity shock tube that could simulate blast over-pressures. The wall specimens represented scaled models of a typical single-wythe stone masonry wall, with or without gravity loads. The specimen geometry consisted of 2.0 m by 2.0 m surface area with a wall thickness of 100 mm. Two walls were tested as representatives of the actual prototype building and served as control specimens. Companion walls were also tested after hardening against blast loads. Two hardening techniques were developed and tested as a blast-retrofit strategy. The first technique had tension braces that consisted of threaded steel bars and turnbuckles resting on spreader beams while supported at the top and bottom slab systems. This system showed superior load resistance, increasing the capacity of the wall significantly. However, at later stages of loading, fragmentation of stones between the braces limited further improvements in load capacity. The same system was then used with a spray-on polyurea as the second hardening technique, which maintains the integrity of the wall as the braces further increase lateral load resistance. Single-degree of freedom dynamic analysis was conducted utilizing the resistance function developed earlier by a set of walls tested under incrementally increasing static pressure by means of an air-bag. The paper presents the early results of the research project, offering an innovative solution for the protection of historic stone masonry walls.

2 Experimental Research

2.1 Test Specimens

Two large-scale stone masonry walls were built and tested at the Blast Research Laboratory of the University of Ottawa in Canada under simulated blast shock waves. The walls represented exterior infill stone masonry walls, enclosed in structural frames. As such, the wall specimens were built on a bottom supporting element with free sides, representing an interior segment of a continuous wall with a gap at the top that simulated the actual construction practice where an allowance is often provided for the deflection of top beam of the enclosing frame. The University of Ottawa Shock Tube was used to test the walls with a test area of 2.0 m by 2.0 m. They walls were built on a 152 mm by 152 mm (6 in $\times$ 6 in) hallow steel section (HSS) having a 6.4 mm thickness. The top gap was subsequently filled with grout to activate arching mechanism under out-of-plane blast pressure. The walls were further supported by shear end plates against out-of-plane sliding, generating simple supports at the ends. The geometry of the test walls is illustrated in Fig. 1.

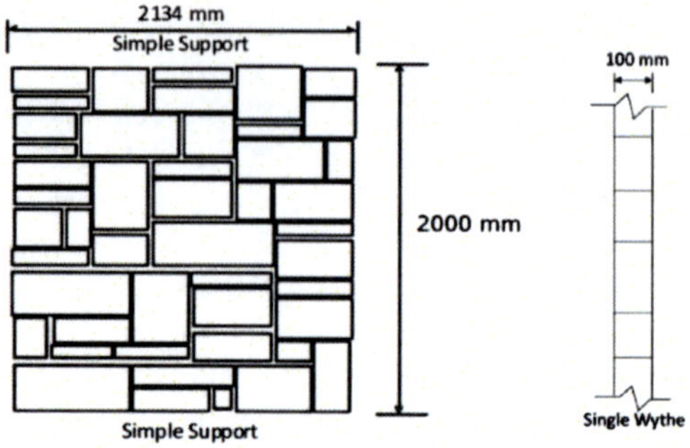

(a) Elevation View (b) Side View

Fig. 1. Stone masonry wall specimens tested in the experimental program

Stone masonry prisms were prepared from the same batch of stone and mortar used for constructing the wall specimens. Two-stone, three-stone and four-stone prisms were tested with respective compressive strengths of 17.0 MPa, 11.4 MPa and 13.4 MPa, resulting in an average compressive strength of 14.0 MPa. Figure 2 shows stone masonry walls test specimens after construction, during curing.

Fig. 2. Sandstone single wythe walls prior to testing, during curing

2.2 Test Equipment (Shock Tube) and Test Setup

The tests were conducted using the University of Ottawa Shock Tube as the blast simulator. The shock tube, located in the Blast Research Laboratory of the University of Ottawa, is one of the very few facilities where blast simulation tests can be performed on structural components. Experimental tests could be carried out on individual structural members such as beams, columns, slabs and walls as well as non-structural building components such as windows, masonry walls and curtain walls. The shock can simulate the effects of blast shock wave mechanism by using a compression chamber, also called a driver section that rapidly releases compressed air into an expansion chamber where it travels along its length until it interacts with the test specimen. The shock tube, shown in Fig. 3a and 3b, consists of a driver section, a double diaphragm spool section with an aluminum diaphragm at each end, and an expansion chamber section that terminates in a rigid test frame. Operation of the shock tube involves simultaneously charging the driver and the spool sections to the desired pressures, then suddenly releasing the air in the spool section causing the aluminum diaphragm to rupture. The intensity and duration of the load are controlled by varying the driver section pressure and the driver section length, respectively. The benefits of using a shock tube over live explosives testing include low cost of testing, predictability of shock wave parameters, ease of monitoring and instrumentation, as well as ease of operation and safety of operation (Lloyd et al. 2011).

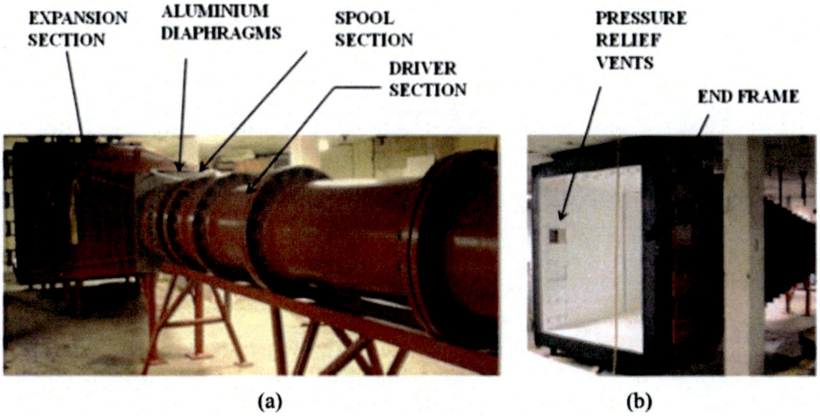

Fig. 3. University of Ottawa Shock Tube (a) driver section, spool and expansion chamber (b) rigid end frame

The overall experimental research program consists of tests of multiple masonry walls, including those made with stone and concrete block masonry units, with or without simulated gravity loads for load bearing masonry and infill masonry walls, respectively. In addition, walls with spray on polyurea for fragmentation control and walls braced against out-of-plane blast forces by means tension braces are considered as blast retrofit techniques. The current paper only reports on the results and comparisons of two infill stone masonry walls, one without any blast retrofit (control wall) and the other with tension braces used as blast retrofit against out-of-plane blast pressures.

All walls were tested using the same test set-up. The walls were positioned in front of the shock tube for testing. The bottom HSS section, on which each wall had been built, was secured to the rigid frame of the shock tube. For the infill walls, another set of HSS section was placed and secured to the shock tube frame with a gap above the wall to simulate the actual construction practice. This gap was filled with grout to promote arching action of the wall between the framing elements (top and bottom beams in the prototype building) during response to blast loads. This filler grout is present in most existing buildings. The wall was restrained against out-of-plane shear sliding by means of horizontal plates, one at each end, while it was free to move along its side edges. These two retrofit measures, i.e., filling the top gap with grout and the prevention of sliding shear may be viewed as simple but limited blast retrofit measures, without which the wall was expected to have limited, or negligible blast resistance. They also facilitate comparison with the second wall, which also had the same features in addition to lateral tension braces as the main retrofit strategy. The first wall was labelled as the "control wall," and the second wall was labelled as the "retrofitted wall." Figure 4 illustrates the control wall.

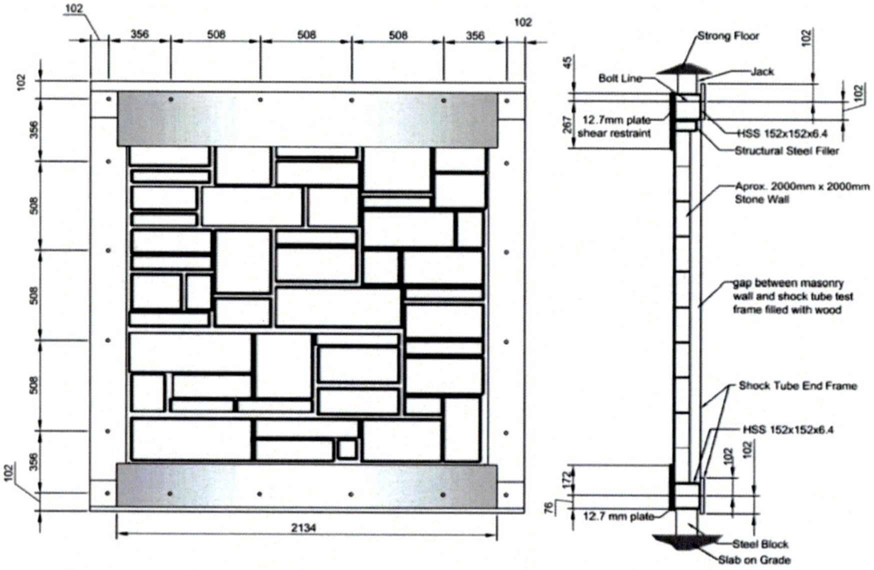

Fig. 4. Geometric details of the control wall and the test setup

The retrofitted wall had two vertical tension braces, placed approximately 1.0 m (1016 mm) apart on the inside face (protected side) reacting to external blast pressures. The braces consisted of 4 threaded rods connected together with 4 hinges and a turnbuckle, forming three straight segment. The top and bottom segments are connected to the top and bottom frame elements (welded to the top and bottom HSS sections, which represent top and bottom beams/slabs. These segments had linearly increasing eccentricity relative to the wall centerline, reaching to a maximum end eccentricity of 121 mm in the mid-height region of the wall, where the third straight segment of the brace was attached. The resulting tension brace with a maximum eccentricity of about 120 mm in the mid- one-third segment of the wall developed horizontal forces reacting to the applied blast pressure. Figure 5 illustrates the geometric details of the retrofitted wall with the tension braces. The braces distributed the reactive forces through two spreader beams.

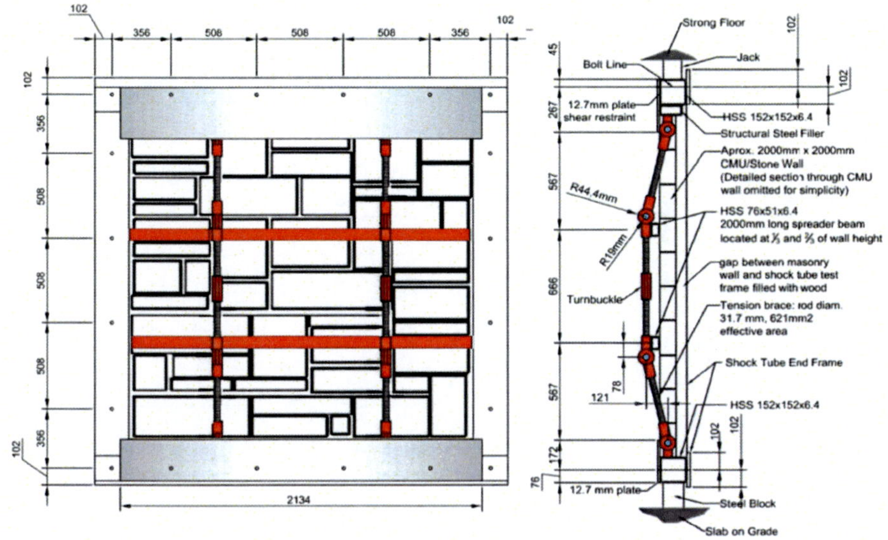

Fig. 5. Geometric details of the retrofitted wall and the test setup

2.3 Blast Loading and Test Results

The walls were subjected to simulated blast overpressures using the shock tube. Applied pressure – impulse combinations are indicated in Table 1. The applied pressure was uniform over the exterior surface of the walls and they were measured by means of pressure transducers near the wall surface, giving reflected blast pressures. The same transducers also provided time history of applied pressure. The impulse was computed as the area under the positive duration of blast pressures. The walls behaved in one-way mode, with top and bottom supports providing restraints against translations (simple supports) and the wall sides being unrestrained and free to deflect. Two high-speed cameras and an image processing technique was used to record the wall deflected shape. Maximum deflection was recorded at wall mid-height. The walls were also instrumented with linear variable transducers (LVDs) for backup deflection measurements. Table 1 also includes the recorded values for wall mid-height horizontal (out-of-plane) deflections. The walls were subjected to progressively increasing blast loads, as indicated in Table 1. Figures 6 and 7 illustrate recorded pressure and impulse time-histories for both walls at selected stages of loading. Also plotted in the same figures are the mid-height horizontal deflections.

The results indicate that the control wall was able to sustain a reflected pressure-impulse combination of $P_r = 14.5$ kPa and $I_r = 157$ kPa-ms with little damage in the wall in the form of cracking taking place along mortar joints. The wall developed maximum mid-height deflection of 55 mm and permanent mid-height deflection of 12 mm. The subsequent load stage involved $P_r = 27.4$ kPa and $I_r = 247$ kPa-ms, at which stage there was a global collapse. Figures 8 and 9 illustrates the performance of the Control Wall.

Table 1. Reflected blast pressure (P_r) and impulse (I_r) combinations applied during tests

Wall	P_r (kPa)	I_r (kPa-ms)	Mid-height Deflection (mm)
Control	14.5	157	55
	27.4	247	n/a
Retrofitted	18.3	221	8
	27.3	276	12
	46.2	488	17
	59.3	691	24
	66.1	727	31
	70.2	740	34
	97	761	40

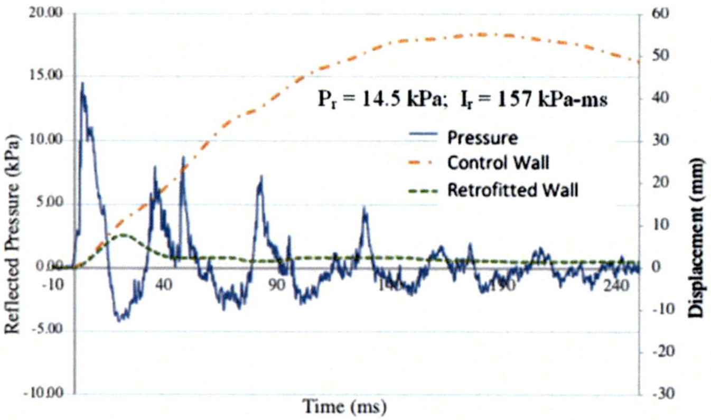

Fig. 6. Pressure and displacement time histories under $P_r = 14.5$ kPa and $I_r = 157$ kPa-ms

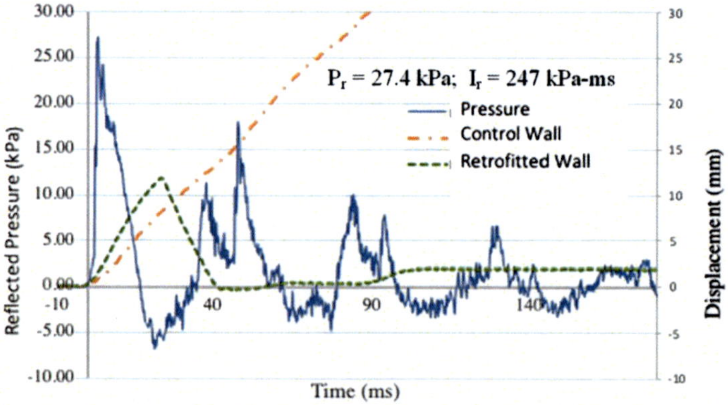

Fig. 7. Pressure and displacement time histories under $P_r = 27.4$ kPa and $I_r = 247$ kPa-ms

Fig. 8. Pressure and displacement time histories under $P_r = 14.5$ kPa and $I_r = 157$ kPa-ms, control wall

Fig. 9. Pressure and displacement time histories under $P_r = 27.4$ kPa and $I_r = 247$ kPa-ms, control wall

The retrofitted wall showed significantly more rigid response, as expected. The lateral bracing controlled out-of-plane deflections. The first shot was at 18.3 kPa reflected pressure and 221 kPa-ms of impulse. The resulting mid-height deflection was only 8 mm, in comparison to the first shot of the control wall, which developed 55 mm deflection at 20% lower reflected pressure and 29% lower impulse. The second shot was at approximately the same pressure and impulse level as those for the control wall, but only generated 12 mm mid-height deflection, though the companion control wall suffered global collapse at this level of blast loading. The gradual increase in reflected pressure and impulse showed almost linear increase in deflections with load. At early stages of loading only a few hairline cracks were observed in some of the mortar joints. The cracks have become more visible under increasing blast loads and eventually the stones started moving. The retrofitted wall survived 7 shots up to a reflected pressure-impulse combination of 97 kPa and 761 kPa-ms without failure. The braces remained elastic and the wall experienced only 40 mm deflection at mid-height, less than the 55 mm deflection experienced during the first shot of the control wall. However, at the last shot, top and bottom left stones became loose, and some of the masonry stones fell off the wall. At this stage the test was discontinued. Figure 10 shows the comparisons of the performance of the two walls tested and the performance of the retrofitted wall under increasing reflected pressures and impulses. Figures 11 shows the retrofitted wall at the end of the test.

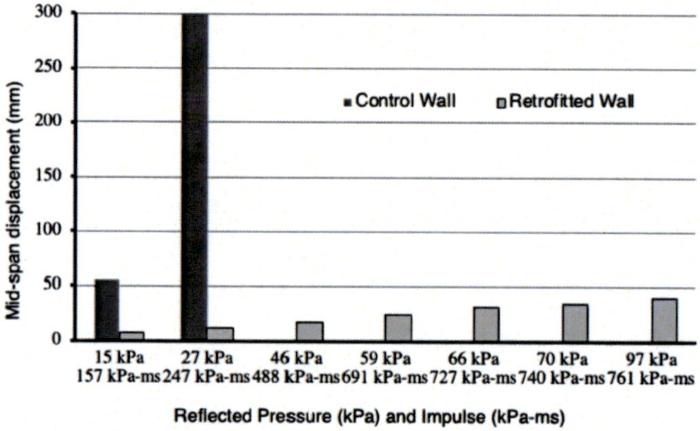

Fig. 10. Performance and comparisons of masonry walls tested

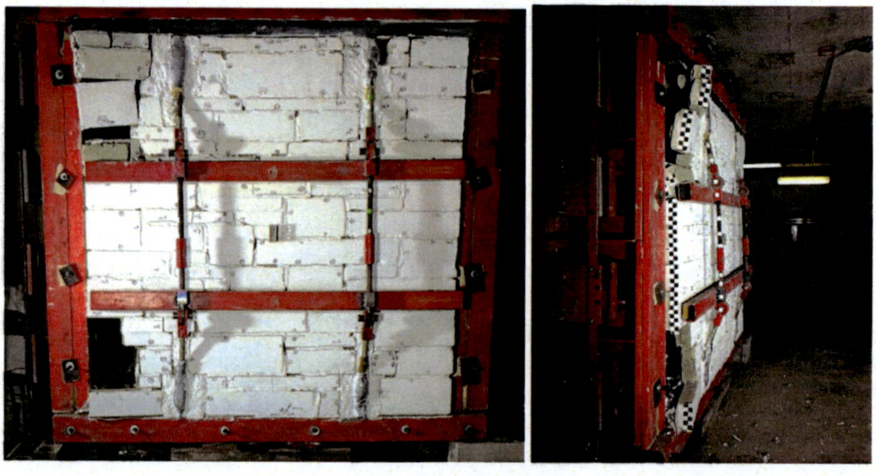

Fig. 11. Damage to retrofitted wall at the end of the test (after shot 7)

It was clear that while the tension braces performed well in terms of resisting high levels of blast pressure, individual masonry units were not able to maintain the integrity of the wall. Although the stones did not fragment and fly in as projectiles, and fell on the floor near the wall, it is not difficult to anticipate fragmentation as a potentially lethal failure mode under stronger blast loads. A potential improvement to the tension-brace retrofit technique would be a protective coating on the wall that could keep the masonry units intact. In an earlier phase of the same project, spray-on polyurea coating was applied on concrete block masonry walls of the same size as the stone masonry walls tested in the current phase. The details of the performance of polyurea retrofitted walls are presented elsewhere (Ciornei et al. 2011; Ciornei 2012). Figure 12 shows the performance of polyurea sprayed masonry wall at a maximum deflection of 160 mm, which was 4 times the 40 mm deflection at which the stones became unstable in the tension-braced masonry wall, indicating that polyurea spray can improve wall integrity up to very large deflections, over and above those that are resisted by tension braces under high level of blast loading. Hence, it may be concluded that polyurea sprayed stone masonry walls, strengthened by tension braces can be used as a blast retrofit technique.

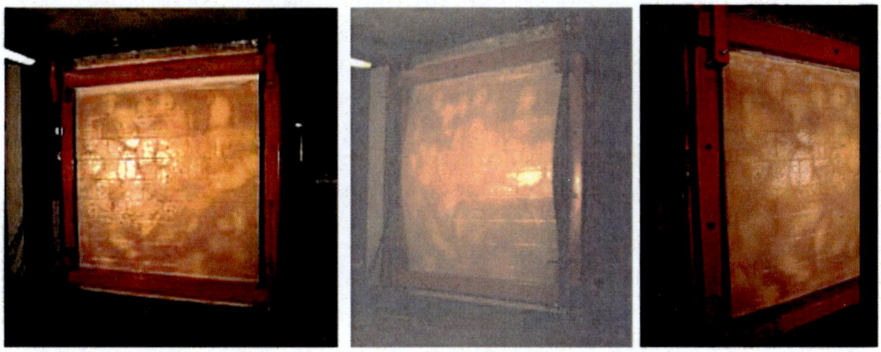

(a) (b) (c)

Fig. 12. Performance of polyurea sprayed concrete block infill masonry wall at $P_r = 84.4$ kPa and $I_r = 597$ kPa-ms (Ciornei 2012), (a) prior to testing (b) at maximum deflection of 160 mm (c) after testing

3 Retrofit Design Recommendations

The retrofit technique developed in the current investigation is based on a number of improvements in blast resistance of infill stone masonry walls. The first step involves improving the arching action that the wall develops between the end restraints at the top and bottom, provided by the framing element that enclose the wall. Infill walls typically have gaps at the top between the wall and the framing element (beam or slab). This is provided to allow for long term vertical deflections of the top structural elements. Filling this gap in existing buildings, which may have already experienced a significant portion of their long term deflections, ensures full arching action, increasing out-off plane bending capacity of unreinforced masonry. Gapped arching, which occurs when the gap is not completely filled, results in limited improvement in the compression membrane action (arching). The other retrofit component is the end plates against sliding shear, which ensures stability of the end support regions while preventing out-off plane shear sliding. The third and the major component of the retrofit technique involves tension braces that are secured to the top and bottom structural elements at the floor levels with built-in eccentricity near the mid-height region of the wall relative to the wall centerline. This eccentricity results in horizontal reactive forces on the walls to resist externally applied blast pressures. As the eccentricity increases, the horizontal force components of tension braces acting on the wall surface also increases. These reactive forces are governed by two factors; the eccentricity of brace and the axial tension force that develops in the brace. The bracing increases out-off plane wall capacity and wall stiffness. If the elastic force capacity of the steel brace is exceeded, it may also provide ductility to the retrofitted wall. Finally, the integrity of individual masonry units during blast response can be attained through polyurea spray, which has been proven to increase the out-off plane deformation capacity of unreinforced masonry without loosing the integrity of the masonry units. It was experimentally observed that a 3.0 mm thick single layer of polyurea was sufficient to prevent wall fragmentation for the 2.0 m by 2.0 m wall panels that were tested (Ciornei et al. 2011; Ciornei 2012). This may have to be increased to two layers of polyurea with a total thickness of 6 mm, which is considered by the technical experts of the material to be the practical upper limit for uniform application and curing of the material (Line-X of Ottawa 2012).

The retrofitted wall may be modelled as a single-degree-of-freedom system for design purposes. The wall resistance function can be established with due considerations given to the wall arching action and the presence of lateral bracing forces. It is recommended to keep the braces remain elastic during blast response to minimize lateral deflections. The elastic stiffness of the wall-brace assembly and the total mass of the system can be used to compute the period of vibration. Load-mass factors are used to account for the distribution of the applied pressure. The fundamental period of the system, relative to the duration of idealized triangular impulsive forcing function helps compute the maximum dynamic load factor (DLF), which can be used to compute the capacity of the retrofitted wall under dynamic loading conditions as affected by the brace capacity and eccentricity. Depending on the level of threat, the parameters of design can be established in terms of the diameter, strength and eccentricity of tension braces that are needed for the system.

4 Conclusions

The following conclusions can be made from the experimental research reported in this paper:

- Lateral bracing of stone masonry infill walls provides substantial improvements to the blast resistance of unreinforced masonry walls provided that the braces have sufficient axial tension capacity and eccentricity to develop horizontal force components acting perpendicular to the surface of the wall.

- The lateral bracing system, supported at top and bottom floor levels, and the attached spreader beams is sufficient to improve blast resistance multiple times (the 2.0 m^2 walls tested in the current investigation experienced a factor of 3 improvement in reflected pressure and impulse resistances). However, further increase in blast resistance and the elimination of premature fragmentation requires keeping the integrity of masonry units. This may be achieved by the application of spray-on polyurea.
- Filling of any potential gaps between the wall and the top structural frame element improves arching action and the resulting wall capacity. Similarly, providing top and bottom steel plate restraints against potential shear sliding eliminates premature support failures in regions of small brace eccentricity near the ends.
- A single-degree-of freedom analysis can be used for design purposes to establish brace capacity and eccentricity required for a given threat level.

References

Ciornei, L., Saatcioglu, M., Tikka, T.K.: Blast retrofit of unreinforced masonry infill walls with polyurea. In: Proceedings of the 9th International Conference on Shock and Impact Loads on Structures, Fukuoka, Japan, pp. 241–250 (2011)

Ciornei, L.: Performance of polyurea retrofitted unreinforced Concrete masonry walls under blast loading. MASc thesis, Department of Civil Engineering, University of Ottawa, ON, Canada (2012)

Lloyd, A., Jacques, E., Saatcioglu, M., Palermo, D., Nistor, I., Tikka, T.: Capabilities and effectiveness of using a shock tube to simulate blast loading on structures and structural components. ACI-SP 281 Blast Resistance Concrete Structures **281**(3), 1–20 (2011)

Line-X of Ottawa: Personal conversations in the Structures Laboratory of the University of Ottawa, Canada (2012)

Numerical Modelling for Assessing the Structural Efficiency of CAM® Reinforcement System for Masonry Walls

Mattia Zizi[1], Alessandro Vari[2], Nino Spinella[3], Piero Colajanni[4], and Gianfranco De Matteis[1](✉)

[1] Department of Architecture and Industrial Design, Università of Campania "Luigi Vanvitelli" Via San Lorenzo 31, 81031 Aversa, Italy
gianfranco.dematteis@unicampania.it
[2] EDILCAM® Sistemi Srl, Via Dei Genieri, 39, 00143 Rome, Italy
[3] Department of Engineering, University of Messina, C.da di Dio 1, 98166 Sant'Agata Messina, Italy
[4] Department of Engineering, University of Palermo Viale Delle Scienze, 90128 Palermo, Italy

Abstract. A large portion of the Italian building heritage is made of masonry constructions, which were erected in the first decades of the last century and were conceived to support gravitational loads only. Many research activities have been carried out in order to propose retrofitting techniques aimed at improving seismic behaviour of existing masonry buildings. The present paper focuses on the CAM® reinforcing system. Such a retrofitting technique, which is widely used in Italy, consists in the application of pre-tensioned stainless steel ribbons on existing masonry walls, conferring to them additional strength, ductility and beneficial confinement effects. The preliminary outcomes of a common research activity developed in cooperation between the University of Campania "Luigi Vanvitelli", the University of Palermo, the University of Messina and the EDILCAM® Sistemi Srl, aimed at setting up appropriate design criteria, are provided in this paper. In particular, the calibration procedure of a specific FEM numerical model developed by Abaqus is presented, which then has been applied to interpret the cyclic response of unreinforced and reinforced masonry walls previously tested within the research project In.CAM.M.I.N.O..

Keywords: Masonry walls · Stainless steel · Ribbons · Shear · Numerical analysis

1 Introduction

The protection of historical masonry buildings is a complex challenge which is promoting the development of many research activities. The need to preserve such valuable constructions retrieves many difficulties due to their inefficiency to face earthquake loads. In fact, many masonry buildings, due to their structural complexity and the poor mechanical features of the base material (i.e. lack of tensile strength, non-linear behaviour and reduced ductility), are particularly vulnerable against seismic actions.

I. Vayas and F. M. Mazzolani (Eds.): PROHITECH 2021, LNCE 209, pp. 512–523, 2022.
https://doi.org/10.1007/978-3-030-90788-4_41

To define the extent of the problem, it would be enough to mention the notable damage occurred to masonry buildings (both monumental and ordinary ones) after the 2016 Central-Italy earthquake, despite it was not particularly intense in terms of magnitude.

Many research activities have been carried out by various authors, by means of experimental tests and numerical models, with the aim to propose innovative reinforcement techniques suitable to improve the seismic behaviour of masonry structures (e.g. ElGawady et al. 2006, Campitiello et al. 2007, Bischof et al. 2014, Gattesco et al. 2015, Zizi et al. 2017a). In this context, steel-based retrofitting techniques may result very effective. In fact, the introduction of metallic materials, which are characterized by high mechanical characteristics in terms of ductility, strength and stiffness, could strongly enhance the seismic behavior of masonry structures.

In particular, the CAM® system, which consists in the application of stainless steel ribbons on existing structural elements, is analysed in this paper. Such a reinforcement system is easy to be applied and since it could be embedded in a small layer of external plaster, its application may be suitable for a wide range of masonry buildings, including monumental ones. Although such a system has been already studied in previous works (e.g. Dolce et al. 2002, Spinella et al. 2014), nowadays specific design criteria are not still available.

Based on these premises, in the present paper a specific numerical model, which has been calibrated on available results obtained from an experimental campaign carried out on the CAM® system, is presented. The aim is to lay the groundwork for the definition of more appropriate design criteria.

2 The CAM® System: Description of the System and Reference Experimental Tests

2.1 Description of the CAM® System

The CAM® system is recognized as an efficient retrofitting technique for existing masonry and reinforced concrete structural elements. With regard to masonry buildings, the principles at the base of the CAM® system are essentially: i. to provide additional tensile strength, ductility and stiffness by introducing steel ribbons; ii. to offer a beneficial confinement effect on the existing material due to pre-tensioning of the metallic elements. In practice, it consists of a three-dimensional mesh of pre-tensioned steel ribbons that pierce the masonry throughout transversal holes, which are realized before the installation of the system (Fig. 1).

The pre-tension to the steel elements is conferred by connecting their extremities with specific seals during the closing operations, which are carried out by using a special device. At transversal holes, on the masonry surface, small metallic plates (125 × 125 × 4 mm) are applied, in order to distribute and diffuse the stresses transmitted by the steel strips, avoiding local stress peaks in the base material. In addition, angular steel plates are used in the corners where the steel elements and the masonry wall are overlapped. By applying this system, the seismic behaviour of masonry structural elements (i.e. walls or columns) may be strongly improved, as additional stiffness, strength and ductility are provided.

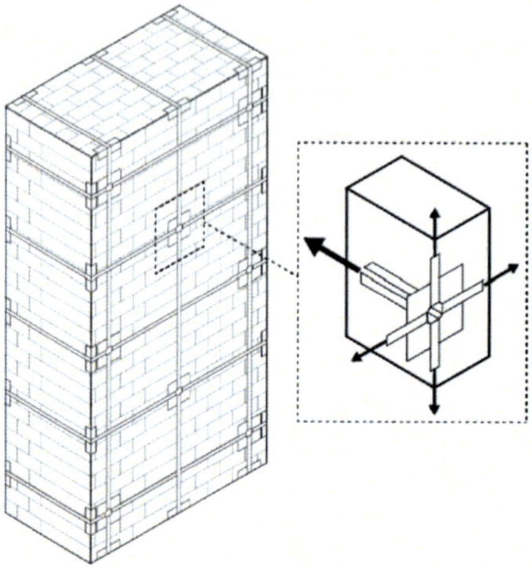

Fig. 1. Example of application of the CAM® system on an existing masonry wall

2.2 The Reference Experimental Test

The experimental campaign adopted as reference test for this study has been carried out in Messina and was part of the framework of the In.CAM.M.I.N.O. research project, which involved various institutes and private companies (namely CHIMETEC s.a.s., ABI s.r.l. of Ragusa and University of Messina). The activities were aimed at investigating the efficiency of the CAM® system on existing masonry panels. The tested masonry specimens had a thickness of 0.55 m and were obtained from a retaining wall, which was shaped in order to define seven different panels with a height of 2.15 m and a width of 1 m for the first panel and 1.1 m for the others. The quality of the masonry was very poor and its arrangement was characterized by stones of different size and crumbly lime mortar.

All tested masonry panels were divided in two parts by inserting two steel beams in the middle height (i.e. UPN 200 with two transversal bolts ϕ24). In particular, on the first panel, double flat-jacks (downer half-panel) and a compression test (upper half-panel) were carried out. The remaining six panels were tested in unreinforced (one) and reinforced (five) configurations by means of Sheppard tests, i.e. in condition of constant vertical compression and increasing horizontal force applied in the middle height of the specimen until the collapse (Turnsek and Sheppard 1980). The experimental campaign included also a preliminary verification of the mechanical characteristics of the masonry material.

With reference to the reinforced configurations, the CAM® system has been applied in combination with simple external plaster or with polymeric net embedded in it. For the aims of this study, the masonry panel with the reinforcement and simple external plaster are considered. More details on the experimental campaign may be found in Spinella et al. (2014).

2.3 Experimental Results

The preliminary mechanical characterization of the masonry revealed the poor quality of the base material, as it can be also deduced by observing the irregular texture of the wall (Fig. 2a). In particular, by means of a double flat-jack test, an initial Elastic Modulus of E = 633 MPa and a compressive strength of $f_k = 1.07$ MPa were obtained, while the compressive test, whose stress-strain results are showed in Fig. 2b, provided a compressive strength $f_k = 0.89$ MPa with an ultimate strain of $\varepsilon_u = 0.37\%$.

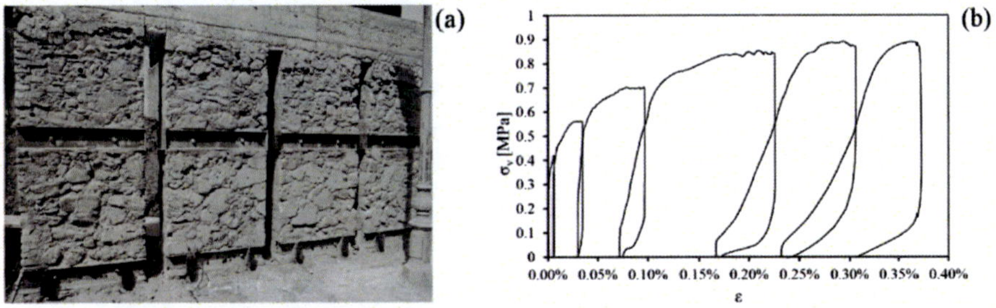

Fig. 2. The tested panel (a) and result of preliminary compressive test (b).

For the Sheppard test set-up of the unreinforced panel (Fig. 3a), the masonry panel was connected to a rigid reinforced concrete beam (50 × 55 cm) at the top, while two UPN 200 steel beams with bolted connections were applied at the basement, similarly to the middle height section. In such a way, perfect restraining condition could be supposed, without allowing rotational and translational movements. A compressive force of 200 kN has been firstly applied on the masonry panel, by imposing tensile stress to dywidag bars (ϕ32, two for each side of the panel), which were connected to the upper and downer rigid beams.

The unreinforced reference panel (URM), whose surface was preliminary treated with a simple external plaster, attained the collapse condition due to a shear failure (diagonal cracking). At the end of the test a more serious damage on the upper half-panel respect to the downer ones was observed, evidently due to the different boundary conditions (Fig. 3c). In fact, since the translational stiffness of the upper beams was higher than the downer system, small displacements in the bolt-masonry interface, as well as rotations of the base section, probably occurred. The panel reached the collapse for maximum horizontal load of 180 kN and ultimate lateral displacement of 18 mm (Fig. 3b).

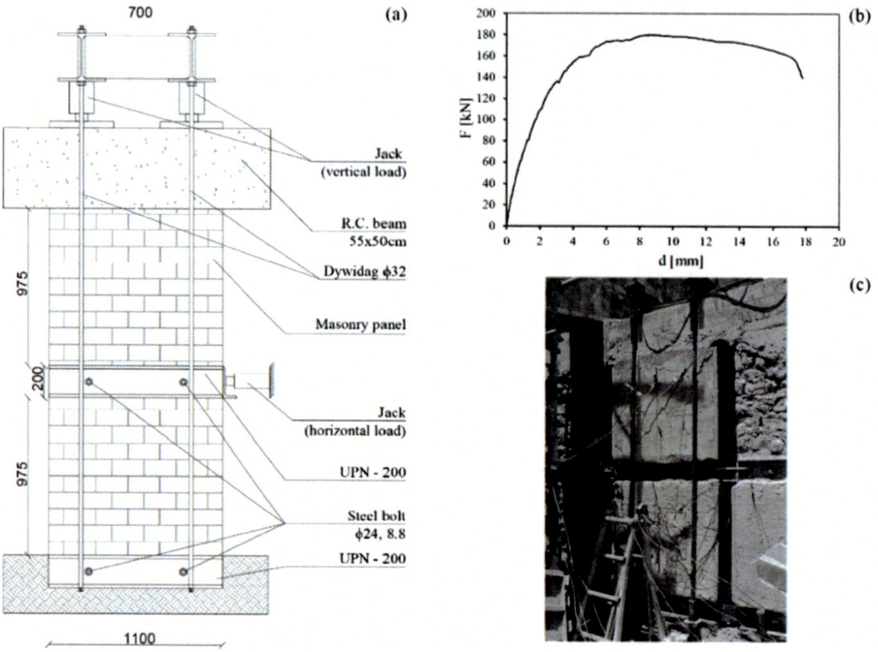

Fig. 3. Test set-up (a), force-displacement curve (b) and cracking pattern (c) of the unreinforced reference panel.

Among the reinforced configurations, the panel reinforced with the CAM® system and with simple external plaster has been selected by the authors for the aims of this study. The test set-up (Fig. 4a) was similar to the one adopted for the unreinforced panel, with rigid elements placed at the top, at the base and in the middle height sections. Nevertheless, in order to avoid displacements and rotations of the downer UPN-masonry system, steel plates welded to the webs of the base beams were introduced during the experimental campaign on the reinforced panels. The panel under consideration was assembled with two overlapped stainless steel ribbons AISI 304 (19×1 mm^2) disposed with a vertical and horizontal distance of 40 cm. At the end of the test a significant increase of the maximum lateral load respect to the unreinforced configuration (about 45%), as well as a lateral stiffness increment were noticed (Fig. 4b). Nevertheless, due to some problems, the test was stopped before reaching the global panel collapse. In particular, the test provided that the horizontal jack was directly supported by the adjacent panel, at whose interface a steel element as contrast was interposed. Therefore, since during the test the adjacent panel showed small cracks and movements of the contrast steel element occurred, the test was interrupted at a lateral displacement of about 17 mm. According to the crack pattern (Fig. 4c), it could be asserted that a maximum horizontal load of 262 kN was reached. Also in this case, the upper part of the panel absorbed a larger force, due to the free rotation that probably occurred at the base.

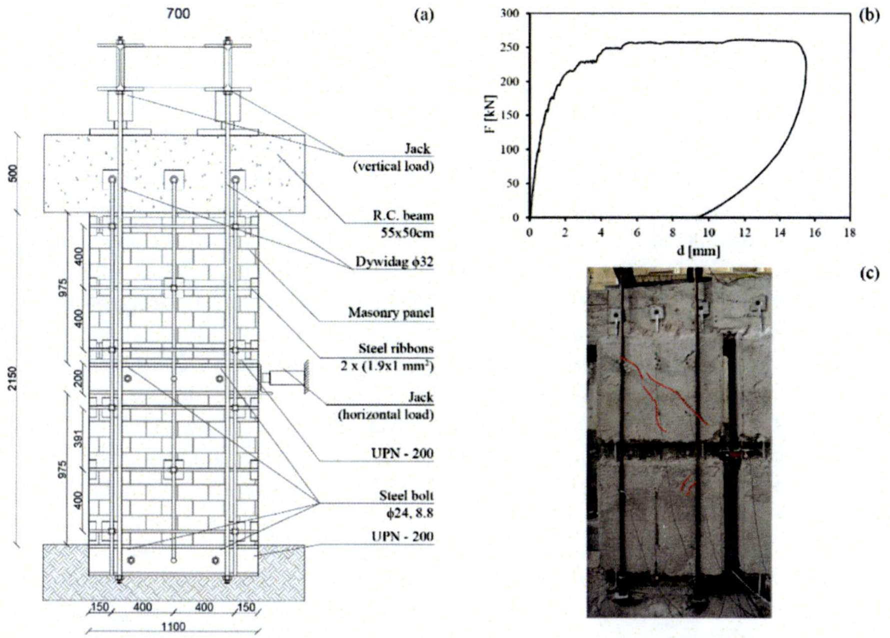

Fig. 4. Test set-up (a), force-displacement curve (b) and cracking pattern (c) of the reinforced reference panel.

3 Numerical Modelling

3.1 Introduction

A numerical model have been developed in order to define a tool suitable to interpret and predict the effect of the CAM® system when applied on existing masonry walls. The Abaqus software with standard solution has been used for the FE Models, which have been calibrated on the basis of the aforementioned experimental tests. The adopted modelling technique and the results related to the calibration procedure with a real test referred to both unreinforced and reinforced configurations are shown in the following.

3.2 The FE Model for the Unreinforced Panel URM

The unreinforced panel has been modelled by using 8-node linear brick with reduced integration (C3D8R). Since the contribution of the external plaster has been neglected in terms of both strength and stiffness, it has not been included in the numerical model. The masonry in the unreinforced FE Model (URM) has been modelled by adopting a macro-modelling approach, which is suitable for the description of the global response of masonry walls (Lourenço 1996). In order to faithfully reproduce the set-up test and simulate the different boundary conditions, the upper and the downer nodes of the panel have been rigidly linked to control nodes (reference points RPs) by means of internal constraints (i.e. rigid body). In such a way external constraints and loading conditions applied to the RPs are directly transmitted to the nodes of the panel. Moreover, the same technique has been adopted for taking into account the presence of the intermediate steel beams. The vertical dywidag bars have been also considered in the numerical model by introducing between the upper and downer nodes four axial springs, whose stiffness $K = 78.5$ kN/mm has been evaluated according to real axial stiffness of the bars.

The numerical analysis has been developed in two different steps, which correspond to the application of vertical loads (first step) and horizontal displacement (second step). In particular, during the first step, to the upper RP only vertical degree of freedom (DOF) was left free, while the others movements were constrained; then vertical concentrated forces ($F = 50$ kN) have been applied to the upper nodes in correspondence of the four axial springs, in order to simulate the total compressive load of 200 kN. In the second step, the boundary condition of the top RP was changed: an entire DOF restriction was introduced in order to fix the vertical displacements obtained in the first step. Hence, an increasing horizontal displacement was applied to the central RP. The boundary condition of the bottom RP has been assumed equal for both steps. In particular, only rotations in the plane of the panel and horizontal translations have been permitted, while the other DOFs have been blocked. In order to take into account that small horizontal movements could occur due to the not infinite stiffness offered by the bolts-masonry system, a translational external spring has been introduced. Since in the bolt-masonry a shear-out failure did not occur during the test, the stiffness of the introduced spring has been evaluated considering an elastic behaviour. Therefore, according to Gelfi and Giuriani (1987), who proposed a theoretical model for the concrete-steel bolt connections, the translational stiffness K_t of the connection has been defined as $K_t = 7.6$ kN/mm.

A picture of the URM FE Model is provided in Fig. 5.

The masonry has been modelled by adopting a plastic-damage model material, namely concrete damage plasticity (CDP), similarly to other research experiences carried out by the authors of this study (Campitiello et al. 2007, De Matteis and Mazzolani 2010, Zizi et al. 2017a). In particular, the CDP, which is available in the software material library, refers to the amendments proposed by Lee and Fenves (1998) to the plastic-damage model originally implemented by Lubiner et al. (1989).

Since they do not influence significantly the global response at the end of the analysis (Zizi et al. 2017b), the base parameters of the adopted material model have been defined according to the default values proposed in software guide (Abaqus Inc. 2014).

Therefore, the following data have been assumed: i. dilation angle $\psi = 35°$; ii. ratio between biaxial and uniaxial compressive strength $f_{bo}/f_{co} = 1.16$; iii. eccentricity $\varepsilon = 0.1$; iv. parameter that influence the yield surfaces in the deviatoric plane $K_c = 2/3$.

The uniaxial behaviour of the masonry has been defined according to the results obtained by the preliminary material tests and following an iterative calibration procedure. In particular, an initial Elastic Modulus E = 633 MPa has been assumed, as obtained in the double flat-jack test. Hence, a parabola-rectangular compressive stress-strain law has been adopted. The compressive strength (σ_k) has been evaluated according to the results obtained in the preliminary tests ($f_k = 0.89$ MPa and $f_k = 1.07$ MPa for double flat-jack and compressive test, respectively), while an elastic limit stress (σ_y) corresponding to the 30% of the ultimate one has been assumed. Therefore $\sigma_k = 1$ MPa and $\sigma_y = 0.3$ MPa, with an ultimate strain $\varepsilon_u = 0.37\%$, have been fixed (Fig. 6a). The

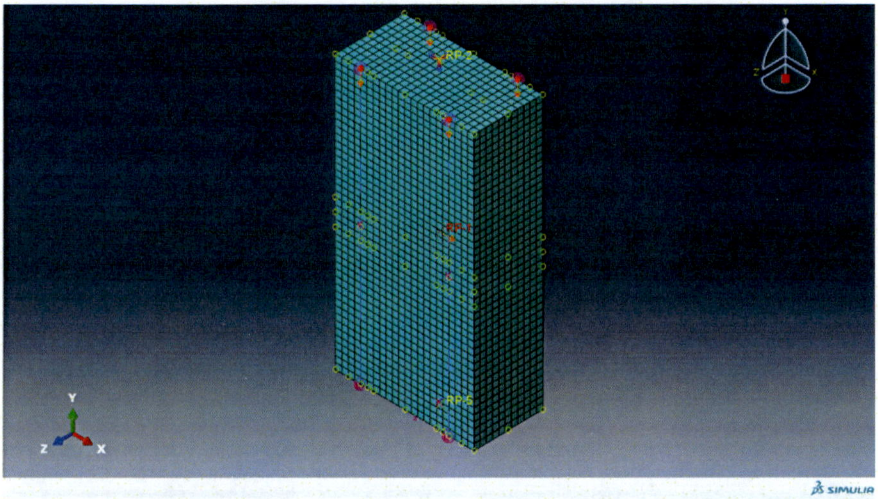

Fig. 5. The FE Model for the unreinforced system (URM)

uniaxial behaviour in tension has been defined according to an analytical and then an iterative procedure, which for the sake of brevity are not reported in the present paper. In particular, a fracture energy approach has been used, with a strength of $\sigma_t = 0.23$MPa and a stress-displacement linear constitutive law with an ultimate displacement $u_{t0} =$ 2.5mm (Fig. 6b).

At the end of the analysis, the adopted URM allowed a correct interpretation of the experimental results both in terms of quantitative behaviour and failure mode.

In fact, the obtained force-displacement curve (Fig. 7a) is able to reproduce the experimental test result with a good accuracy. Furthermore, as shown in Fig. 7b, 7c and 7d, the crack pattern at the failure point, which has been conventionally assumed at the maximum displacement of 18mm, revealed also a good agreement with the reference experimental test. Indeed, the numerical analysis revealed a more serious damage level of the upper half panel.

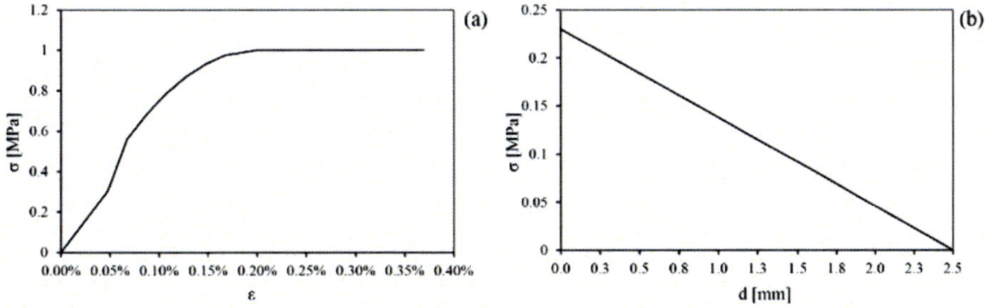

Fig. 6. Uniaxial stress-strain law in compression (a) and uniaxial stress-displacement post-elastic law in tension (b).

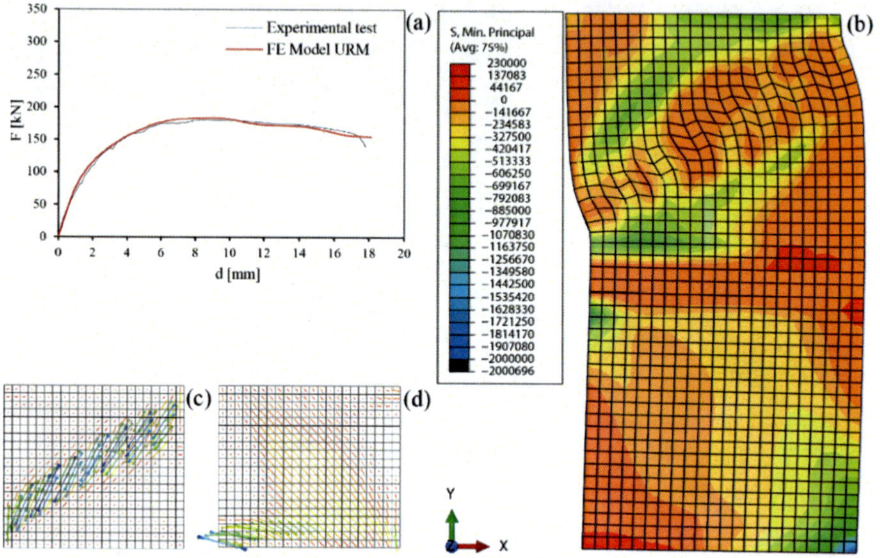

Fig. 7. Results of the FE Model analysis for the unreinforced panel (URM): force-displacement curve (a), compressive stress at the collapse point (deformed shape amplified 5 times) (b) and plastic compressive strain of the upper (c) and downer (d) half panels

3.3 The FE Model for the Reinforced Panel RM

The reinforced RM FE Model has been defined on the same basis of the URM FE model. Also in this case the presence of the external plaster has been neglected, while the masonry solid elements have been modified in order to simulate the presence of holes for the application of steel ribbons. The overlapped stainless ribbons have been considered as a unique element, considering bi-dimensional reduced integration 4 node elements (CPSR4) with a thickness of 2 mm.

The material model adopted for the reinforcement elements has been defined according to the experimental tensile tests preliminary carried out (Spinella et al. 2014). In particular a typical bi-linear material law has been adopted, with an Elastic Modulus $E = 212000$ MPa, a yielding stress of $f_y = 317$ MPa, an ultimate strength $f_k = 567.3$ MPa and an ultimate strain $\varepsilon_u = 5.4\%$.

Moreover, the pre-tension to the stainless steel ribbons, which could be approximatively estimated equal to $\sigma_p = 200$ MPa, has been conferred by means of thermal loads. Therefore, an *expansion* behaviour has been settled for the material assuming the typical thermal coefficient of the steel $\alpha_c = 1.2 \cdot 10^{-5}\,°\mathrm{C}^{-1}$. The difference of temperature ΔT to assume in order to obtain the aforementioned stress state has been evaluated according to Eq. (1):

$$\sigma_p = \alpha_c \cdot E \cdot \Delta T. \tag{1}$$

Hence, an additional step has been introduced in the RM before the application of the compressive load, in which the temperature of the steel elements has been varied from a value of 0 °T of the initial state, to −80 °C.

The *interaction* between the CAM® system and the existing masonry panel has been defined with a *hard contact* for the normal behaviour, and a *penalty* with a friction coefficient of 0.3 for the tangential one. Moreover, rigid bodies have been introduced in correspondence of the small metallic plates at masonry holes. Aimed at simulating the perfect connection of the ribbons with the top, middle and bottom rigid beams, the terminal nodes of the steel elements have been included in the existing rigid bodies.

In order to take into account the additional contribute offered by the steel plates welded to the UPN 200, the horizontal stiffness of the translational spring adopted for the downer reference node has been modified in the RM. In particular a value $K_t = 10$ kN/mm has been adopted on the basis of a preliminary calibration procedure carried out with the aim to obtain the same lateral stiffness of the real test.

In Fig. 8 a picture of the implemented RM FE Model is provided.

At the end of the analysis, the RM revealed a good correspondence with the experimental test in terms of global response, initial stiffness and maximum horizontal force (Fig. 9a).

Since the real test was interrupted before the complete collapse of the wall, the failure point in the numerical analysis has been conventionally assumed when the horizontal force decreased more than 20%. This condition corresponded to a maximum displacement of 30 mm.

It is important to note that after the elastic branch of the obtained force-displacement curve, there is a disagreement between the two curves. This is probably due to some local failures that occurred in the experimental test. In fact, in the experimental curve, it is possible to observe some parts where there is increment of displacement with a constant force. The stress and strain states at the failure point plotted in Fig. 9b, 9c and 9d showed a very similar damage state of the both half masonry panels with small rotation at the base.

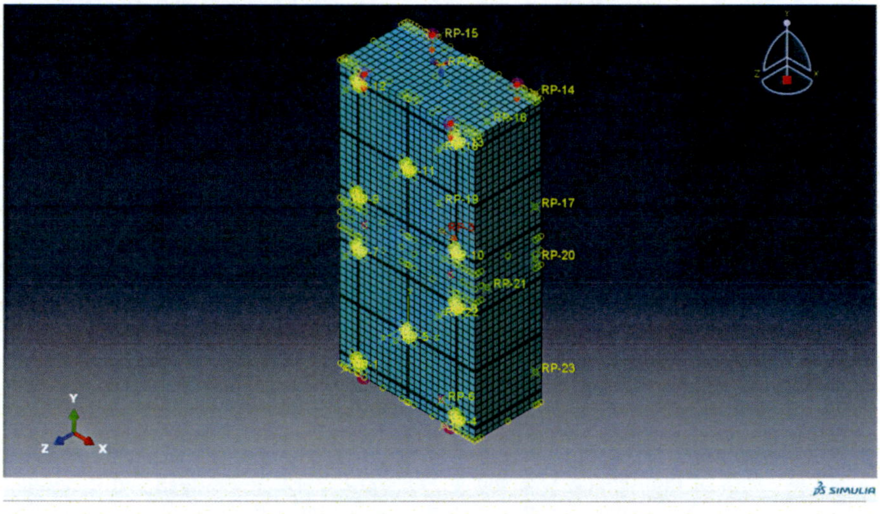

Fig. 8. The FE Model for the reinforced system (RM)

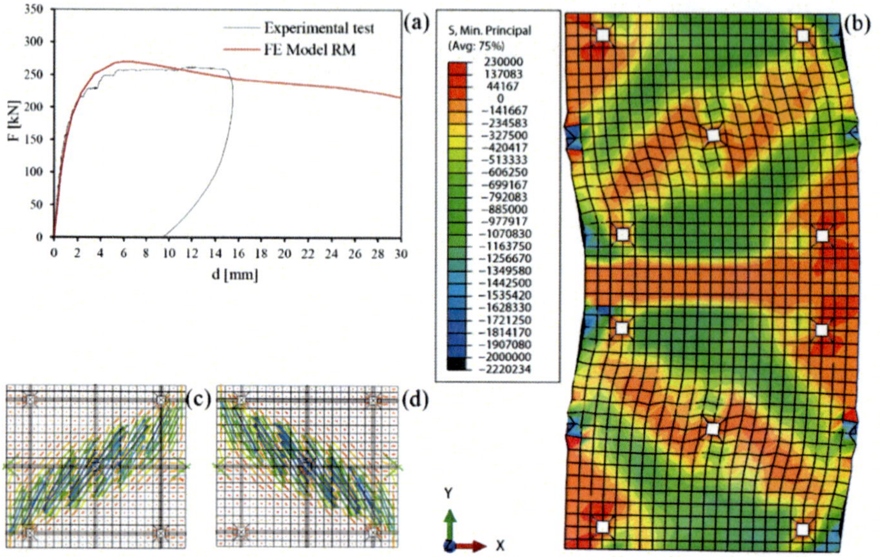

Fig. 9. Results of the FE Model analysis for the reinforced panel (RM): force-displacement curve (a), compressive stress at the collapse point (deformed shape amplified 2.5 times) (b) and plastic compressive strain of the upper (c) and downer (d) half panels

4 Conclusions and Future Developments

In this paper, the problem of structural preservation of the existing masonry buildings has been faced by analysing innovative steel based retrofitting techniques suitable for improving the seismic behaviour of masonry walls under compressive and shear forces. In particular, the potentialities of the CAM® system, patented by the Italian company EDILCAM® Sistemi Srl, has been examined based on a previous experimental and research activity.

To this purpose, a FEM model able to reproduce the shear behaviour of a masonry walls retrofitted with the CAM® system has been set up, considering as reference tests the experimental results on full scale masonry panels obtained within the research project In.CAM.M.I.N.O..

The obtained outcomes show that there is a very good agreement between numerical and experimental results, in terms of strength, ductility and stiffness, evidencing that the implemented numerical tool could be profitably adopted in the future in order to define suitable design criteria for this retrofitting technique when applied to simple and complex structures.

Acknowledgements. Part of this study has been developed within the scholarship funded in the framework of the National Operational Program ESF-ESFR Research and In-novation (PON RI 2014–2020), Action I.1 related to Innovative Industrial Ph.D (Project Code – CUP - B25D18000010006) of which EDILCAM® Sistemi Srl is the industrial partner.

References

ABAQUS Inc.: Abaqus analysis User's guide, v. 6.14. Dassault Systemes Simulia Corp., Providence, RI, USA (2014)

Betti, M., Galano, L., Petracchi, M., Vignoli, A.: Diagonal cracking shear strength of unreinforced masonry panels: a correction proposal of the b shape factor. Bull. Earthq. Eng. **13**(10), 3151–3186 (2015)

Bischof, P., Suter, R., Chatzi, E., Lestuzzi, P.: On the use of CFRP sheets for the seismic retrofitting of masonry walls and the influence of mechanical anchorage. Polymers **6**, 1972–1998 (2014)

Campitiello, F., De Matteis, G. and Mazzolani, F. M.: Risposta a taglio di pannelli murari tozzi rinforzati mediante elementi metallici: analisi numeriche mediante un approccio "smeared cracking". In: Wondermasonry 2007 workshop, October 11–12. Lacco Ameno, Ischia, Italy (2007)

De Matteis, G., Mazzolani, F.M.: The fossanova church: seismic vulnerability assessment by numeric and physical testing. Int. J. Arch. Herit. **4**, 222–2245 (2010)

Dolce, M, Cacosso, A., Ponzo, F.C., Marnetto, R.: New Technologies for the Structural Rehabilitation of Masonry Constructions: Concept, Experimental Validation and Application of the CAM System. In: Seminar: "the intervention on built heritage: conservation and rehabilitation practices" (Invited lecture), Oct. 2–4, Porto, Portugal (2002)

ElGawady, M., Lestuzzi, P., Badoux, M.: Shear strength of URM walls retrofitted using FRP. Eng. Struct. **28**, 1658–1670 (2006)

Gattesco, N., Amadio, C., Bedon, C.: Experimental and numerical study on the shear behavior of stone masonry walls strengthened with GFRP reinforced mortar coating and steel-cord reinforced repointing. Eng. Struct. **90**, 143–157 (2015)

Gelfi, P., Giuriani, E.: Modello teorico del legame costitutivo per le connessioni a piolo. Studi e ricerche **9**, 323–341 (1987)

Lee, J., Fenves, G.L.: Plastic-damage model for cyclic loading of concrete structures. J. Eng. Mech. **124**(8), 892–900 (1998)

Lourenço, P.B.: Computational Strategies for Masonry Structures. PhD Thesis, University of Porto, Porto, Portugal (1996)

Lubliner, J., Oliver, J., Oller, S., Onate, E.: A plastic-damage model for concrete. Int. J. Solids Struct. **25**, 299–326 (1989)

Spinella, N., Colajanni, P., Recupero, A.: Experimental in situ behaviour of unreinforced masonry elements retrofitted by pre-tensioned stainless steel ribbons. Constr. Build. Mater. **73**, 740–753 (2014)

Turnsek, V., Cacovic, F.: Some experimental results on the strength of brick masonry walls. In: 2nd International Brick Masonry Conference. Apr. 12–15. Stoke-on-Trent, England (1970)

Turnsek, V., Sheppard, P.: The shear and flexural resistance of masonry walls. In: International Research Conference on Earthquake Engineering, Jun. 30–Jul.3, Skopje, Yugoslavia (1980)

Zizi, M., Campitiello, F., Dogariu, A. De Matteis, G.: Cyclic response of brick-cement mortar masonry shear-walls retrofitted with steel grids. In: 3rd International Conference on Protection of Historical Constructions PROHITECH 2017, Jul. 12–15, Lisbon, Portugal (2017a)

Zizi, M., Campitiello, F., De Matteis, G.: Non-Linear FE Model for cyclic response of brick-cement mortar masonry shear-walls, In: XV International Forum Le Vie dei Mercanti. Jun. 15–17, Naples-Capri, Italy (2017b)

Comprehensive Management Systems of the Quito Heritage. Church the Compañía de Jesús

Juan Enrique Nieto[1](✉), M. Lenin Lara[2], and Juan Moyano[1]

[1] Universidad de Sevilla – Departamento de Expresión Gráfica e Ingeniería en la Edificación, Sevilla, Spain
jenieto@us.es

[2] Universidad Internacional del Ecuador / Universidad Politécnica de Madrid, CipArq – Facultad para la Ciudad, Arquitectura y Paisaje, DTCA Departamento de Construcciones y Tecnología Arquitectónica, Quito, Ecuador
mlara@uide.edu.ec, lenin.lara.calderon@alumnos.upm.es
https://investigacion.us.es/sisius/sis_showpub.php?idpers=5737, https://www.uide.edu.ec/, https://www.dcta.upm.es

Abstract. Currently, the plans for the protection of historical groups tend to be sustainable in the management of Heritage, covering the conservation, intervention and dissemination phases. The interoperability of operators in rehabilitation - archaeologists, architects, historians, engineers, restorers - is essential and for this it is necessary that there is a workflow. In addition, new data acquisition technologies, scanning and photogrammetry techniques, are helping to capture the real state of historical constructions, allowing an analysis of processes, materials and pathologies, essential for prevention and maintenance. of the historical building. These standardized operations provide us with data that must be continuously recorded. A para-loyal commitment is the management of the information collected by the different experts, its sharing and the deposit in an alphanumeric data container model. It is here that the H-BIM project takes a fundamental role, having a virtual 3D model of multidisciplinary information.

In this investigation a workflow is developed in which the numerical models obtained through laser and photogrammetry scans and the parametric model based on the orthophoto sequences generated in a previous restoration are superimposed. For this, the Church of the Society of Jesus is taken as a study, emblem of that Latin American baroque that the Jesuit order built between 1605 - 1765 in Quito (Ecuador). The preparation phases are structured to adequately model a historical building of these characteristics, breaking down the different stages and presenting its Level of Development -LOD-, detail and information.

Keywords: Religious architecture of Quito · Scanning · Photogrammetry · HBIM · Intervention in architectural heritage

1 Introduction

The AEC sector (Architecture, Engineering, Construction) is moving in a context that demands increasing coordination, integration and productivity operations. Aspects that

I. Vayas and F. M. Mazzolani (Eds.): PROHITECH 2021, LNCE 209, pp. 524–538, 2022.
https://doi.org/10.1007/978-3-030-90788-4_42

have to do with the production process and the different information, documentation and management systems. Therefore, the Cultural Heritage environment must take advantage of the advantages and opportunities offered by the industry to activate these processes throughout the entire life cycle of the heritage itself. Building Information Modeling (BIM) technology makes it possible to contribute to technological development and the continuous advances in interdisciplinary operations that occur in this area. In another sense, the new data acquisition technologies, such as scanning techniques and photogrammetry, are helping to capture geometries of historical constructions, allowing an analysis of processes, materials and pathologies, essential for prevention and maintenance of the historic building. A new area of knowledge arises here from these new technologies, the Heritage-BIM or H-BIM, which aim to work with a tool available to operators of the heritage set that transmits past historical information and the current modifications to future generations.

2 Related Works

The scientific contributions to the emergence of these new technologies applied to heritage - an H-BIM project - can be detected in works dedicated to the review of the state of the art (Zalama and Lerones 2018), (Pérez-Lombard et al. 2008), (Anagnostopoulos et al. 2015), and under the perspective of diagnosis and performance evaluation (Bruno et al. 2018) and implementation (Bruno and Roncella 2019). The particularity of this new paradigm in digital design is the possibility of creating parametric objects that manage and manage information of any kind and work area.

3D digital reconstruction systems have benefited from the development of massive data acquisition technologies and the advancement of BIM platforms. But this has not always been the case, the last century ended the year with technologies that aspired to lift 3D scans and manipulate the data to obtain floor plans and building elevations. Previous work was proposed to conduct a survey and compare the results with the data obtained from the massive point cloud (Serrano et al. 2014). Other works have been carried out with the photogrammetry technique especially focused on restitution (Astorqui 2003) and photogrammetry dedicated to real estate (Almagro 2000); and more recently extract from the cloud of orthophoto-points of various planes, in JPG or TIFF format with color inversion, (Puche Fontanilles et al. 2017). Advancing in technology, the plans were obtained by means of transfer of vector support, on which the measures were taken. Nowadays, the operability between BIM and the techniques of massive data acquisition opens up new scenarios (Wang et al. 2014) aimed at enabling workflows to be shorter and cheaper. In this area, it should be known that the point cloud registers geometric, chromatic and material characteristics information (Attenni 2019) and also information related to visible pathologies (Santagati and Lo Turco 2016). It is, therefore, that as technological development advances the data is more complete. Looking at this panorama, a research paper is proposed that superimposes numerical models obtained through laser and photogrammetry scans and the parametric model based on the orthophoto sequences generated in a restoration of the late twentieth century, when the planimetry was obtained through orthophotography.

3 Case Study

To develop the study, the Church of the Society of Jesus is taken, a baroque temple that is the most representative and important of Colonial America. A construction that lasted 150 years -1605–1765-, having become one of the most important religious monuments in the historic center of the city of Quito. Declared Cultural Heritage of Humanity by UNESCO in 1978 is one of the works of art of Colonial America. This document presents the results of the analysis carried out on several of the exquisite elements of the patrimonial joint of the Company, highlighting the pulpit of the Church, worked in carved, openwork and gilded cedar wood in gold leaf (Fig. 1).

Fig. 1. a. Current image of the Facade of the Church of the Society of Jesus in Quito. b. Post-processed facade of the Church of the Company, explored from the Leica 360° JetStream Viewer software.

4 Methodology

The methodology of this work aims to properly structure the phases of modeling and incorporation of data into a historical information model, in order to effectively manage an H-BIM project. In a first stage it is essential to carry out a work of gathering historical and bi-bliographic information of the building. The Church of the Society of Jesus is the most representative and important Bavarian temple in Colonial America, whose construction lasted approximately a century and a half. As such a building, very few plans and details are preserved from this era. As relevant and updated information, a 2D CAD planimetry is currently available, derived from a photogrammetric restitution on the occasion of the last restoration of the beginning of the 21st century. Archeology documentation is also important for the knowledge about the constructive stages carried out throughout history, for the purposes of stratigraphic analysis, although the data is very scarce.

At a later stage, the data set based on dimensional metrology is obtained. This means obtaining, through the new technologies, the acquisition of geometric data and textures, the cloud of points that allows a precise and faithful survey of reality (Fig. 2).

Fig. 2. Sketch of the patrimonial complex of the Society of Jesus in Quito. Author: Darío Domosos, 1983, page 57.

4.1 Data Acquisition Procedure

The data acquisition process generally begins with a survey of three-dimensional terrestrial laser scanner (TLS) or photogrammetry structure from motion (SFM) in order to acquire the coordinates of the points in a massive way, obtaining at the same time a three-dimensional representation of the captured geometry and the precise measurements. In any topographic campaign two processes are required, the initial field work and, secondly, the postprocessing. With the information of the complete point cloud, a relationship of points of the building with values (xi, yi, zi) that will support the vector segments that will be introduced and coupled in the BIM software to be used as a reference to modeling will make up the 3D model.

Scan The Church of the Company was going to be a new challenge, due to the richness of details in architectural, sculptural and pictorial decoration. We did not want to lose any important details, and take advantage of all the existing means available for the most complete capture of the building's geometry. Thus, a previous rethinking of the three ships was made and to obtain an adequate position of the scanner equipment, BLK360, that no corners were left without scanning, with a complete 360° scan and choosing for this purpose some points that will define a grid with a compensated distance, varying between 4 and 4.50 m. To complete the staking of the first level or wooden floor of the Church, complementary scans were carried out of the annexed rooms: Chapel of Santa Mariana de Jesús, Sacristy, Ante-sacristy Room, Multipurpose Room and Anteroom. The sum of scans for level 1 gave a total of 70 valid and 4 nulls. The time of each scan for a 360° scan, HDR image and maximum quality, did not exceed 5′30″(Fig. 3).

Fig. 3. Staking Scanners plant with the position of the BLK scanner parking lots at level 1 (+1.00). HBIM project. ArchiCAD Software 23. October 2019.

The facade was swept from two levels, the first with respect to the exterior pavement, with the height of the equipment, distributed in 7 positions; and a second level at 8.10 m, with another 6 parking lots (Table 1).

Table 1. Summary of the scanning campaign in the Church of the Company. October 2019.

Scans	Level1 (+1.00)	Facade
Valid	74	13
Null	4	0
Total processed	70	13

4.2 First Photogrammetric Studies of the Church of the Company

In 2002, studies were carried out on the rich heritage of the Churches of the city of Quito. Two of its main elements "the pulpit and the facade", representatives of the maximum example of the Jesuit Latin American baroque, were analyzed by means of the stereoscopic photogrammetric system. The easiest way to create the perception of depth in the brain is to provide the viewer's eyes with two different images, representing two perspectives of the same object, with a minor deviation from the perspectives that both eyes naturally receive in the binocular vision Traditional stereoscopic photography consists of creating a 3-D illusion that begins with a pair of 2-D images (Fig. 4).

Fig. 4. Assembly of scaffolding on the facade of the Church of the Company in Quito and taking reference measures prior to stereoscopic photographic shots. Images contributed by Lenin Lara, 2002.

Its use was indispensable in the case of study since the method allows the visualization of reliefs, such as roughnesses and faults, facilitating the taking of data on the surfaces and reducing the time of the graphic production. In the case described, a similar 60 mm Rollei camera was used.

It is coincidental that in the first decade of 2000 the analytical period of photogrammetry became fashionable (Azkarate Garai-Olaun 2002) until the new algorithms developed a more technically developed plot. Currently, photogrammetric pairs are being used for vector restitution (Rodríguez Miranda et al. 2015) in order to create 3D models with photographic texture and orthophotographs (Fig. 5).

This first study relied on pairs of 2 Mp photographs for the creation of photomosaics. The resulting product was planimetric representations in 2-D as plants, sections and elevations of the elements, which served as graphic input for the maintenance of expost and book editing (Figs. 6 and 7).

Fig. 5. Plan of the Church of the Company, planimetry 2D at + 1.70 m. The areas explored with the stereoscopic photogrammetric system are marked. Document of the Church of the Company Foundation.

4.3 Study of the Pulpit. Photogrammetry Structure from Motion (SFM)

In a later and recent phase, the SFM was initially used, which has been used as a basis for the new studies initiated, a project that emerges from the call for Research Residences of the International University of Ecuador, where they participate the research group of the University of Seville "Technological Innovation, 3D Modeling Systems and Energy Diagnosis in Heritage and Building" in collaboration with the UIDE CipArq Architecture Faculty.

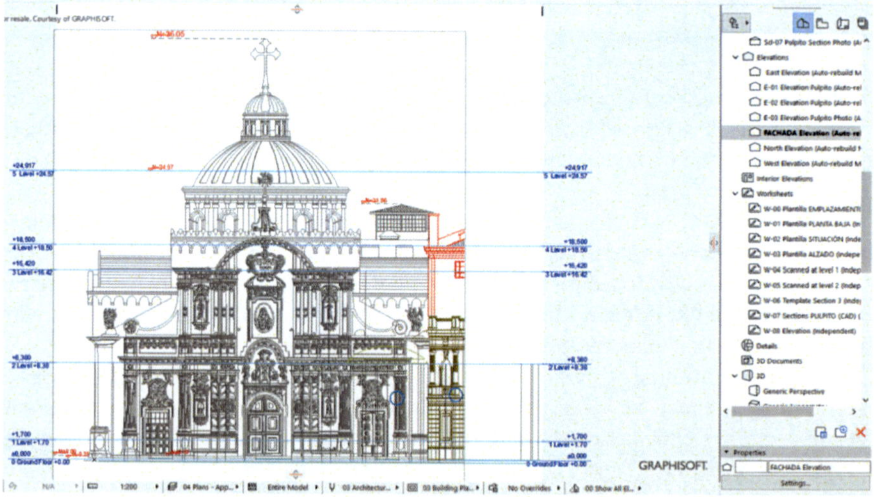

Fig. 6. 2D elevation of the facade inserted in the HBIM Church of the Company project, ArchiCAD software. Document provided by the Church of the Company Foundation

Fig. 7. Side elevation and cross section of the pulpit of the Church of the Company. Author: Lenin Lara, arch-tecto. Executive Director: Carlos Palleres Sevilla. December 2002.

The use of SFM photogrammetry has been accompanied by a data collection to obtain control point measurements. A Canon EOS650D camera has been used for the pulpit, with 4.38 × 4.38 μm of pixel size and 18–55 mm lens supported under a tripod. A set of 244 images were obtained in raw and jpg formats, revealed before processing in the Agisoft Photoscan software. The photogrammetry technique has served to obtain specific details such as the pulpit of the church.

In a third phase, the digital model was built from the set of data obtained. The orthophotos generated by point cloud algorithms that flow from photogrammetry can be integrated into HBIM models, as well as the TLS mass data set, which serves as an auxiliary window for lifting walls, pillars and vaults (Fig. 8).

Fig. 8. Perspective of the SFM model of the pulpit, point cloud from the photogrammetric survey, and its fit in the HBIM project. ArchiCAD Software 23. October 2019.

The last phase of the methodological process will try to configure the model to be a true reflection of the real model.

In the new pulpit data collection, using the SFM technique, a series of metric information has been obtained that has allowed the contrast of the results obtained in stereoscopic photogrammetry with the current algorithms (Table 2).

Table 2. Camera calibration

Camera model	Resolution	Focal length	Pixel Size	Nº images
Canon EOS650D	5184 × 3456	18 mm	4.38 × 4.38 μm	244 images
Type:	Frame	F:	4321.51	
Cx:	34.6119	B1:	−2.25942	
Cy:	54.0962	B2:	1.64978	
K1:	−0.171156	P1:	0.00044325	
K2:	0.164935	P2:	0.00250693	
K3:	−0.0225315	P3:	0	
K4:	0	P4:	0	

For the photogrammetric procedure, numerous control points with natural targets and marker targets were available, showing the results of Table 3 (Fig. 9, Fig. 11).

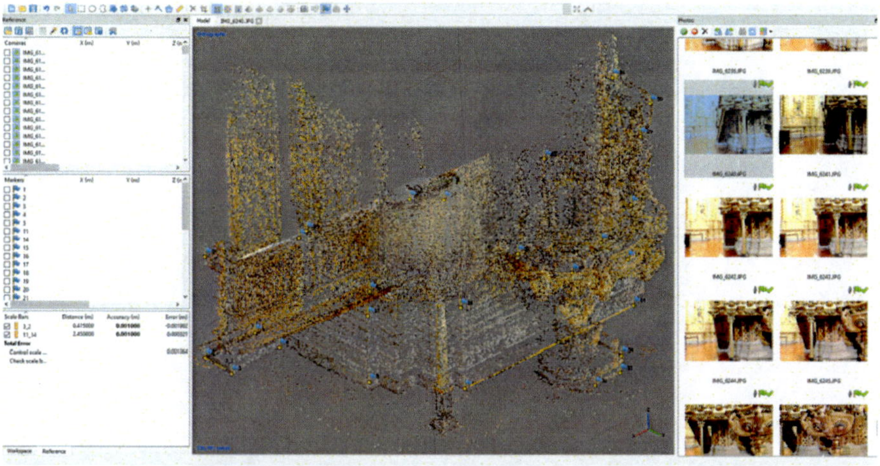

Fig. 9. Perspective of the SFM point cloud of the pulpit where you can see the markers introduced. Agisoft Photoscan. October 2019.

Table 3. Control scale bars

Label	Distance (m)	Error (m)
11–14	2.45032	0.000320686
3–2	0.413098	−0.001902
Total		**0.0013639**

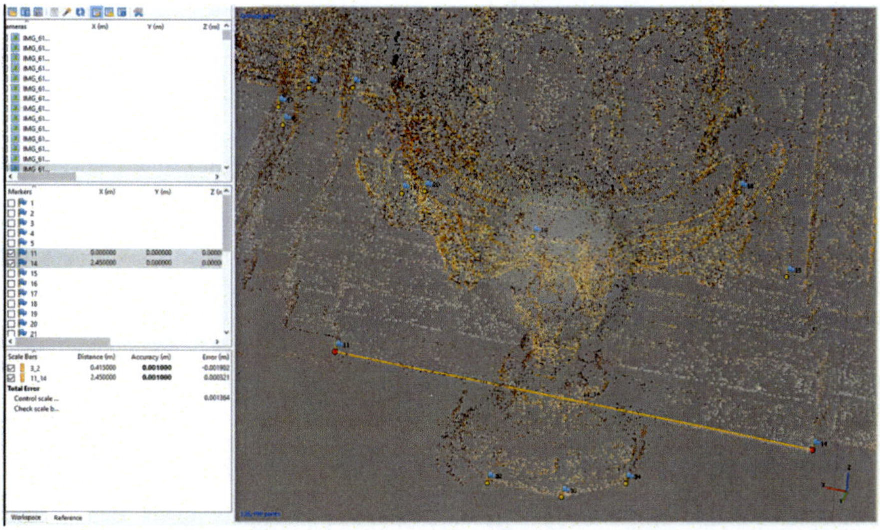

Fig. 10. Introduction of Control scale bars of the Pulpit. Agisoft Photoscan. October 2019.

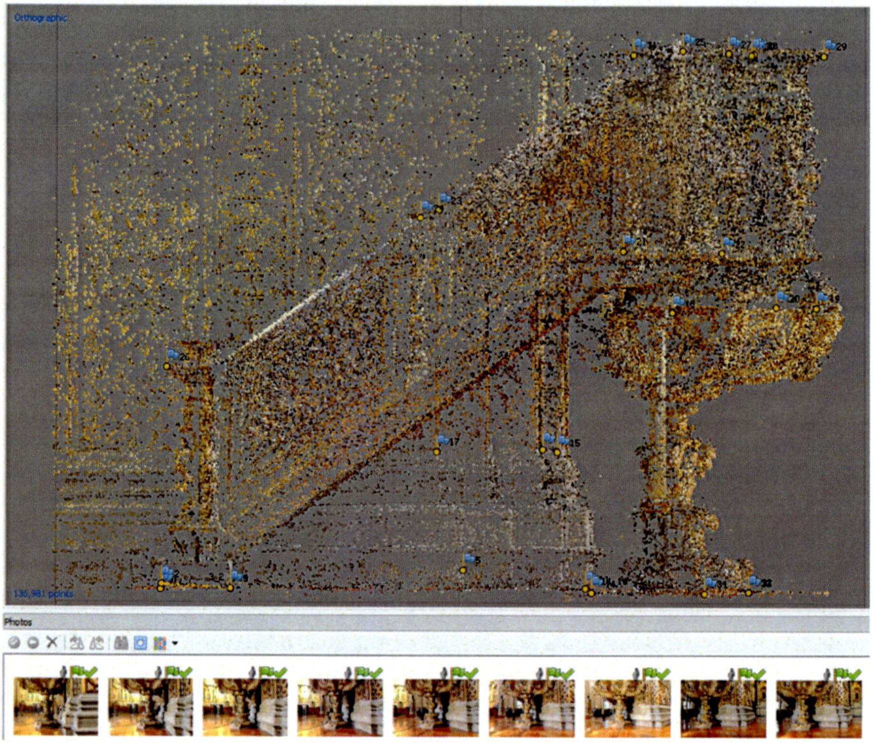

Fig. 11. Side elevation of the Cloud of points of the Pulpit with the markers. Agisoft Photoscan. 2019.

5 Results

As a more relevant result of the study carried out, the great evolution of the photogrammetric technique and the precision of the algorithms implemented in the SFM system should be highlighted when contrasted with the three-dimensional terrestrial laser scanner (TLS) survey.

The final textured point cloud of the pulpit, obtained by the SFM system, has been contrasted with the CAD planimetry (Fig. 8), some vector drawings that had as an reference template an orthophoto derived from a restoration, coming from stereoscopic photogrammetry.

The precision of the process of creating a dense point cloud has been endorsed based on the use of "scale bar control" (Fig. 10), thus obtaining a scaled object according to the real pulpit. Next, we have also proceeded to mesh from the Agisoft Photoscan itself, although due to the large number of polygons generated, it has been proven that it entails many difficulties for efficient management. The overall mesh of the entire pulpit is very heavy once the parametric object has been generated and inserted into the ArchiCAD project library (PULPITO COMPAÑIA R1–181019.lcf, 1008328 kb); Therefore, its analysis in the HBIM project is not fruitful due to the great difficulties that the graph of the PC-Collapse MSI i7-7700 HQ CPU, 2.80 GHz, 16 Gb RAM- computer has for its management. Finally it has been decided to work only with the point cloud of the pulpit. As it has been shown in the images presented, the quality of the colored point cloud has been quite high once inserted in the H-BIM project, with an excellent texture and shortage of holes, which has allowed an exploration Efficient and fruitful without generating the mesh.

It has been possible to certify, in the comparison between the SFM model and the digital projections in vector format (type *.dwg), that there is a slight deviation not very important both in its width and in the heights of the different sections of the pulpit, showing in the images with red and blue lines -front elevation (Fig. 12) and sections (Fig. 13) -. But if you look at the side elevation that faces the stairs (Fig. 14), this image shows that the deviation is very considerable, exceeding 140 mm at several points of its edges. This could be due to an error in the survey or because the drafting operator, not having an orthophoto as a reference, has drawn the vector drawing by analogies with the other elevations.

The H-BIM Church of the Company project has served as a container for all the information collected: Plant, Elevations and Sections of the CAD planimetry from the 2002 survey. But also of the SFM point cloud of the pulpit, which has functioned as an independent parametric object that can be introduced, in property files, all the data coming from a multidisciplinary investigation, in its structural, sculptural, material facets,…, to serve as a basis for a future restoration intervention.

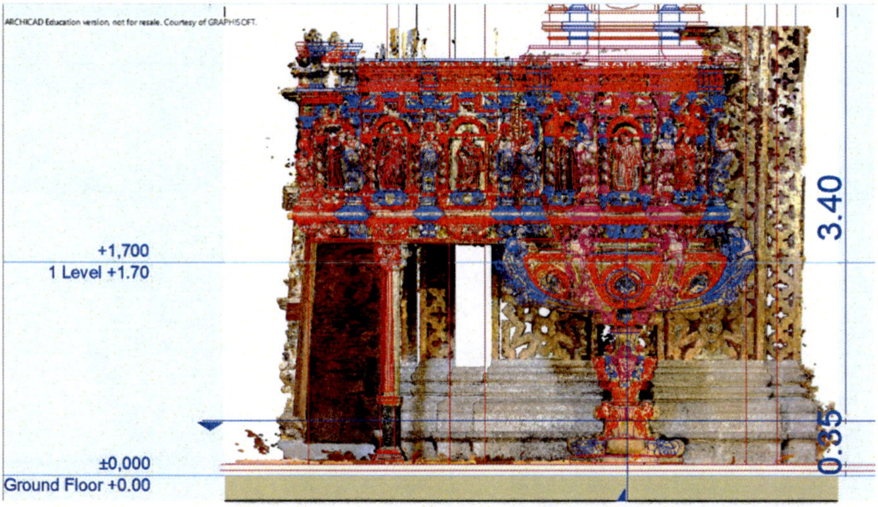

Fig. 12. Front elevation of the pulpit. Comparison between the SFM model and the vector elevation (dwg) based on a photogrammetric ortho-situation.

Finally, although this article has not been studied in depth, underline that the H-BIM project is also a container for the point clouds of the Church as a whole coming from laser scans, which have been introduced by sectors for facilitate a more fluid work in the use

Fig. 13. Central sections by the pulpit. Comparison between the SFM model and the vector section based on a photogrammetric ortho-situation.

Fig. 14. Left lateral elevation of the pulpit. Comparison between the SFM model and the vector elevation, in red and blue lines (dwg), showing a significant deviation in the upper front of the pulpit.

of graphics and the RAM memory of the PC. For the post-processing of this information, the Leica Register 360 software has been used, generating two global clouds by levels. First, the.e57 format was used for export in complete point clouds, obtaining a file of 37.037.014 kb (level 1).

Subsequently, to reduce the abusive size (greater than 8 million points), the point clouds have been cleaned and exported by sectors, eliminating noise and unwanted objects such as banks and moving people. This procedure has been fundamental for the subsequent insertion in the HBIM project. The point cloud imported in .xyz format is transformed into an ArchiCAD parametric object once inserted into the HBIM project. At the end of the process, 6 sectors of the global scan of the Church of the Company have been linked, being managed by different layers to facilitate the work and performance of the PC graphics card.

6 Conclusions

New information technologies allow the dissemination of cultural heritage with increasing rigor and precision. This is, in large part, thanks to the scientific work of the institutions that put a large number of electronic media and equipment at the service. What we know today as an architectural survey is a way to recreate a reduced model of the object, with sufficient information that allows the transmission of knowledge for its proper restoration (Gorbea 2011). Photogrammetric techniques represent the combination of the cheapest and easiest instruments to handle. It allows us to transfer information from the projected space 2D to 3D (Pérez García et al. 2011). In this way, complex objects such as those in this study (pulpit) can be combined with the new 3D platforms by importing the point cloud into BIM software.

This article presents the main ideas to develop a workflow in which numerical models are superimposed, obtained through laser scans (TLS) and photogrammetry (SFM), and the parametric sequence-based model of the orthophotos generated in older times, where photogrammetry based on current algorithms allow millimeter accuracies in architectural ensembles (Sapirstein 2016). These measurements are possible by the comparison between the different massive point clouds that define the spatial coordinates (x, y, z). In the first decade of the twentieth century, as a more advanced method, the extraction of the geometry of the buildings studied from the same point cloud processed was used using traditional 2D representation systems, that is, the generation of system-based projections dihedral. The comparison now made between the elevations and the sections and the 3D point cloud generated by the STM system, -Figs. 12, 13 and 14-, shows the metric deviation that exists between 2D projections based on a photogrammetric ortho-restitution and the precise surveys using the current photogrammetry.

Acknowledgments. The activity presented in this document is the result of the work included in the call for Research Residences of the International University of Ecuador for the year 2019. To thank the entire team of technicians, assistants and maintenance personnel of the Foundation of The Society of Jesus supported the researchers, especially for the great help in logistic media that has greatly facilitated the exploration of the heritage as a whole.

References

Almagro, A.: Fotogrametría para arquitectos. El estado de la cuestión. In: Las nuevas tecnologías de la representación gráfica arquitectónica en el siglo XXI: Actas del VIII Congreso EGA, pp. 277–280 (2000). https://dialnet.unirioja.es/servlet/libro?codigo=529751

Anagnostopoulos, I., Brilakis, I., Vela, P. a. A Review on methods for generating as-built building information models. In: Proceedings of the 32nd CIB W78 Conference 2015, (Trimble 2014) (2015)

Astorqui, J.S.C.: La fotogrametría digital en el levantamiento de planos de edificios | Santa Cruz Astorqui | Informes de la Construcción (2003). Retrieved January 10, 2020, from http://informesdelaconstruccion.revistas.csic.es/index.php/informesdelaconstruccion/article/view/539/614

Attenni, M.: Informative models for architectural heritage. Heritage **2**(3), 2067–2089 (2019). https://doi.org/10.3390/heritage2030125

Azkarate Garai-Olaun, A.: Intereses cognoscitivos y praxis social en Arqueología de la Arquitectura. Arqueología de La Arquitectura **1**, 55 (2002). https://doi.org/10.3989/arq.arqt.2002.6

Bruno, N., Roncella, R.: HBIM for conservation: a new proposal for information modeling. Remote Sens. **11**(15), 1751 (2019). https://doi.org/10.3390/rs11151751

Bruno, S., De Fino, M., Fatiguso, F.: Historic building information modelling: performance assessment for diagnosis-aided information modelling and management. Autom. Constr. **86**, 256–276 (2018). https://doi.org/10.1016/j.autcon.2017.11.009

Gorbea, A.A.: Veintidós años de experiencia de fotogrametría arquitectónica en la Escuela de Estudios Árabes, CSIC (2011). Retrieved January 13, 2020, from Ministerio de Cultura website: https://digital.csic.es/handle/10261/49861

Pérez-Lombard, L., Ortiz, J., Pout, C.: A review on buildings energy consumption information. Energy Build. **40**(3), 394–398 (2008). https://doi.org/10.1016/j.enbuild.2007.03.007

Pérez García, J.L., Mozas Calvache, A.T., Cardenal Escarcena, F.J., López Arenas, A.: Fotogrametría de bajo coste para la modelización de edificios históricos TT - Low-cost photogrammetry for the modeling of historic buildings. Virtual Archaeol. Rev. **2**(3), 121–125 (2011). https://doi.org/10.4995/var.2011.4633

Puche, J.M., Fontanilles, J.M., Solé, M., Toldrà, J.M., Domingo, P.D., Solà-Morales,: Más allá de la métrica. Las nubes de puntos como expresión gráfica semántica. EGA Revista de expresión gráfica arquitectónica **22**(31), 228 (2017). https://doi.org/10.4995/ega.2017.6781

Rodríguez Miranda, Á., Pérez Vidiella, P., Martínez Lázaro, R., Valle Melón, J.M., Melón, J.M.V.: Reutilización de pares fotogramétricos de elementos arquitectónicos para la obtención de modelos 3D y ortofotografías a partir de técnicas SFM. Arqueología de La Arquitectura **0**(12), e024 (2015). https://doi.org/10.3989/arq.arqt.2015.004

Santagati, C., Lo Turco, M.: From structure from motion to historical building information modeling: populating a semantic-aware library of architectural elements. J. Electron. Imaging **26**(1), 011007 (2016). https://doi.org/10.1117/1.jei.26.1.011007

Sapirstein, P.: Accurate measurement with photogrammetry at large sites. J. Archaeol. Sci. **66**, 137–145 (2016). https://doi.org/10.1016/j.jas.2016.01.002

Serrano, F., Saumell, J., Berenguer, F.: Análisis de resultados métricos de una nube de puntos y una medición directa en el patrimonio edificado. El Santuario de La Montaña en Cáceres. Inf. Constr. **66**(534), e016 (2014). https://doi.org/10.3989/ic.12.059

Wang, J., et al.: Integrating BIM and LiDAR for real-time construction quality control. J. Intell. Rob. Syst. **79**(3–4), 417–432 (2014). https://doi.org/10.1007/s10846-014-0116-8

Zalama, E., Lerones, P.M.: A review of heritage building information modeling (H-BIM). Multimodal Technol. Interac. **2**(21), 1–29 (2018). https://doi.org/10.3390/mti2020021

Seismic Behavior of Historical Royal Cenotaphs (Chhatris) in Gujarat, India During the 2001 Bhuj Earthquake

Satwant Rihal[1](✉), William Koh[2], and Anjali Mehrotra[3]

[1] Architectural Engineering Department, College of Architecture and Environmental Design, Cal Poly State University, San Luis Obispo, CA, USA
[2] Consulting Structural Engineer, Los Angeles, CA, USA
[3] ISISE, Department of Civil Engineering, University of Minho, Guimaraes, Portugal

Abstract. The state of Gujarat in western India has a rich architectural heritage, as evidenced by the abundance of historical monuments found to extend from Bhuj in the west to Ahmedabad in the east. A unique feature of the cultural heritage of Bhuj in particular are the cenotaphs (Chhatris) built during the 1700s under the Jadeja dynasty that ruled the region from 1540–1948. These cenotaphs are canopy-like structures, and are traditional symbols of Hindu architecture. Typically constructed out of local stone, they consist of an open simple structure on a raised platform, with a central dome or multiple domes supported by stone pillars. This paper presents the results of a study comprising the documentation, assessment, and restoration of historical royal cenotaphs following the 2001 Bhuj earthquake. Using SAP2000, a 3D finite element model of a representative cenotaph structure was first developed, based on approximate observed geometry and assumed material properties for the structure. A structural analysis was then carried out under the action of gravity and lateral (seismic) loads. The results are presented in terms of seismic base shears and expected displacement response behavior, while an assessment of the overall seismic safety and stability of the structure is also carried out.

Keywords: Cultural heritage · Cenotaphs · Earthquakes · Structural assessment · Seismic analysis · Restoration

1 Background

The rich cultural heritage landscape of India includes beautiful cenotaphs that are found scattered all over north, central, western and eastern parts of the country. These cenotaphs are canopy-like structures, and are traditional symbols of Hindu architecture. Also known as "Chhatris", their name is derived from the Hindi/Urdu word for umbrella. Typically constructed out of local stone, they consist of an open simple structure on a raised platform, with a central dome or multiple domes supported by an appropriate number of stone pillars, while some cenotaphs also have partial-height stone infill walls to provide a sense of enclosure.

I. Vayas and F. M. Mazzolani (Eds.): PROHITECH 2021, LNCE 209, pp. 539–555, 2022.
https://doi.org/10.1007/978-3-030-90788-4_43

These cenotaphs were built as memorials to members of royalty, as well as to those belonging to wealthy and the powerful families such as prominent landlords and traders. They are commonly found in Rajasthan, some parts of Madhya Pradesh, and Gujarat which are connected to the Rajput lineage. In Rajasthan, many Rajput dynasties constructed cenotaphs as memorials for their royal family members at traditional cremation sites. Representative examples include the Ahar cenotaph complex in Udaipur (Figs. 1a and b), Bada Bagh near Jaisalmer (Fig. 1c), and the Poddar cenotaph complex in the city of Ramgarh (Fig. 1d).

Fig. 1. (a-b) Ahar Cenotaph Complex in Udaipur, Rajasthan, (c) Cenotaphs at the Bada Bagh in Jaisalmer, Rajasthan and (d) Poddar Cenotaph Complex in Ramgarh, Rajasthan

Further southwest, these cenotaphs also make up part of the cultural heritage in the region of Bhuj, Gujarat. A number of these cenotaphs (Chhatris) were built during 18th century, under the Jadeja dynasty that ruled the princely state of Kutch from 1540 to 1948. Typical heritage cenotaphs (chhatris) in Bhuj are presented in Fig. 2.

Fig. 2. Typical cenotaph complex in Bhuj, Gujarat

2 Seismicity of the Region

The Gujarat region of India is one of the most active seismic regions of the world, as evidenced by historical earthquakes dating back to the Great Rann earthquake of 1819 as presented in Table 1. A composite map showing the historical earthquakes that struck the Bhuj area, and other seismic prone regions of India is presented in Fig. 3.

Table 1. Bhuj – Kutch Region, Gujarat – Historical Earthquakes [10]

Year	Location	Magnitude	MMI
1819	Great Rann of Kutch	7.8	XI
1845	Southeast of Lakhpat	6.3	VII
1956	North of Anjar	6.3	IX
2001	Near Bhachau	7.6–7.7	IX-XII

2.1 2001 Bhuj Earthquake

The 7.7 magnitude 2001 Bhuj earthquake caused a catastrophic amount of destruction, resulting in a significant loss of both life and property. More than 20,000 people were reported to have died in the earthquake, over 167,000 were injured, while economic losses were estimated to be approximately $5 billion [3]. The highest amount of damage was sustained in the Kutch region of Gujarat, with approximately 60% of the casualties due to the earthquake concentrated there. The observed damage included collapse of building systems including recently-built multistory reinforced concrete frame buildings, failure of masonry arch and reinforced concrete bridges, as well as extensive geotechnical damage such as liquefaction and failure of earth dams [3].

Fig. 3. Major historical earthquakes in the Bhuj region and other seismically-active regions in India [10]

The epicenter of the earthquake was about 20 km ENE of Bhuj, however shaking effects were felt as far away as Rajasthan and Pakistan – as illustrated by Fig. 4. For more detailed information on the tectonics of the earthquake, fault mechanisms, resulting ground motions and the effects of this catastrophic earthquake, please refer to [1, 3, 5, 7, 10], and [11].

2.2 Performance of Heritage Structures - Cenotaphs

The 2001 Bhuj earthquake caused widespread devastation and damage to the rich cultural and architectural heritage of Bhuj and the Kutch region of Gujarat. Extensive damage reconnaissance surveys of heritage and historical structures were carried out and reported by EERI [1], INTACH [2, 4], and UNESCO [3] amongst others. A detailed study of the performance of heritage structures during the 2001 Bhuj earthquake was reported by EERI [1], based on a detailed condition assessment of heritage and historical structures carried out by INTACH [2] and EERI reconnaissance teams. According to the detailed EERI report [1], it was reported that there were close to 15,000 heritage structures spread over 250 towns in the Kutch region, out which 10,000 were very badly damaged or destroyed. The EERI/INTACH damage survey of heritage structures covered the districts of Kutch, Jamnagar, Junagarh and Rajkot; and Mehsana, Surendranagar and Ahmedabad.

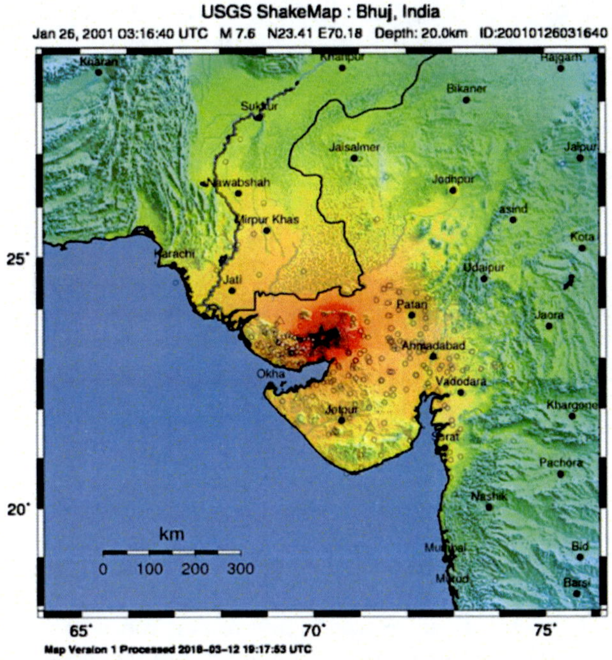

PERCEIVED SHAKING	Not felt	Weak	Light	Moderate	Strong	Very strong	Severe	Violent	Extreme
POTENTIAL DAMAGE	none	none	none	Very light	Light	Moderate	Mod./Heavy	Heavy	Very Heavy
PEAK ACC.(%g)	<0.007	0.08	1.0	5.0	8.8	15	27	47	>83
PEAK VEL.(cm/s)	<0.003	0.04	0.5	3.0	6.5	14	30	63	>136
INSTRUMENTAL INTENSITY	I	II–III	IV	V	VI	VII	VIII	IX	X+

Scale based upon Atkinson & Kaka, 2007

Fig. 4. Shake map of Bhuj and the affected region. Source: USGS.

The heritage cenotaphs (chhatris) in Bhuj suffered severe damage and collapse during the earthquake as shown in Fig. 5. The damage consisted of failure of the stone columns mainly in shear, accompanied by cracking and collapse of the domes supported on the stone columns due to the large displacements they experienced during the earthquake. For example, the main dome and supporting structure of the historic cenotaph (chhatri) of Rao Lakhpat Ji in Bhuj (Fig. 5a) totally collapsed during this devastating earthquake as shown in Fig. 5b, and reported by Langenbach [6].

Additional damage modes included failure of stone slabs overhanging the structural framing, failure of partial-height infill walls, as shown in Figs. 5c and d.

Fig. 5. Rao Lakhpat Ji Chhatri in Bhuj, Gujarat (a) before and (b) after the 2001 earthquake, (c and d) failure of partial height infill walls and (d) failure of structural framing, both shored with temporary brick pillars

3 Cenotaphs (Chhatri) Geometry and Material Properties

Based on available records and augmented by information from a personal visit to Bhuj, an attempt was made to estimate the approximate dimensions of a representative cenotaph structure in Bhuj. From photographic records, based on a fixed known dimension for the columns supporting the cenotaphs, all other remaining dimensions were estimated relative to this one fixed dimension. A reconstructed schematic plan and section of a representative cenotaph structure is presented in Fig. 6.

As the cenotaphs are typically built of red sandstone, material properties of the same are used in this analysis. These properties are listed in Table 2, and are based on available published data [9].

Table 2. Material properties for red sandstone used in this analysis [9]

Parameter	Value
Unit weight	21.78 kN/m^3
Modulus of elasticity	5.51 GPa
Poisson's ratio	0.3
Tensile strength	6.3 MPa
Compressive strength	58.69 MPa

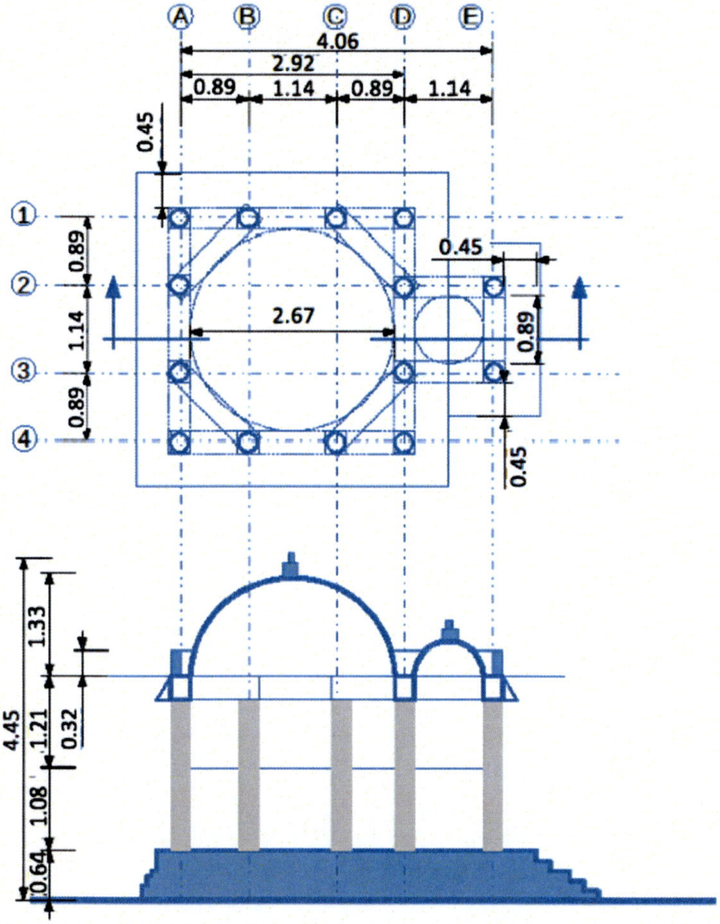

Fig. 6. Schematic plan and section of the representative cenotaph structure (all dimensions in meters)

4 Structural Analysis

Using the geometric dimensions and material properties as defined in the previous section, a 3D finite element (FE) model of the representative cenotaph was developed in SAP2000 [8] as shown in Fig. 7. As Fig. 7 illustrates, shell elements were used to model the domes, while frame elements were used for the supporting stone structure. As the columns are typically connected to the base using stone or metal dowels, the columns in the model were assumed to be pinned at the base in order to reproduce these connections.

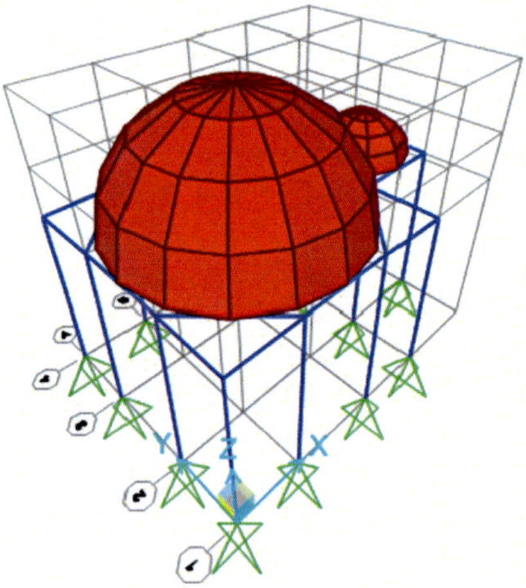

Fig. 7. FE model used for analysis

Structural analyses were carried out under the action of gravity as well as three different seismic load cases (applied in the primary direction of deformation of the structure, in this case the global Y direction) which included a static lateral load corresponding to a PGA of 0.2 g, response spectrum analysis as well as a full time-history analysis. A modal analysis was also conducted in order to determine the natural frequencies and mode shapes of the structure.

In the case of the response spectrum analysis, the response spectrum curve was obtained directly from the Indian Code IS 1893–2016 based on the site and structure-specific parameters listed in Table 3. Note that in the absence of a full geotechnical investigation of the site, soil type II (i.e. medium soil) was assumed. The resulting response spectrum function can be found in Fig. 8.

Table 3. Site and structure specific parameters used for generation of response spectrum curve

Parameter	Value
Seismic zone	V
Seismic zone factor	0.36
Importance factor *I*	1.5 (for monumental structures)
Soil type	II (medium soil assumed)
Response reduction factor *R*	1.5 (for unreinforced masonry without seismic strengthening)
Damping	5%

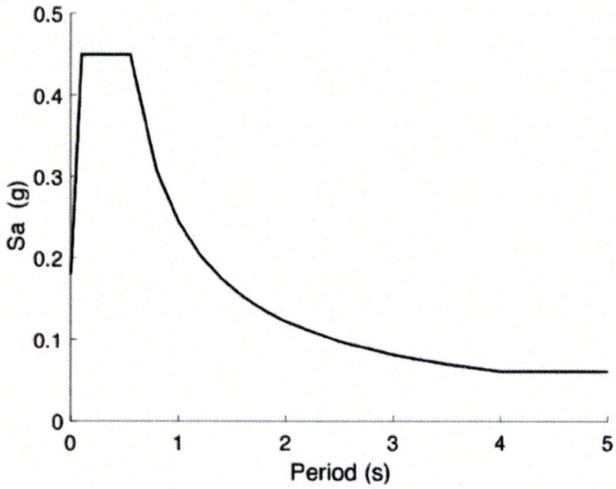

Fig. 8. Response spectrum used in the analysis, as defined in IS 1893–2016

Similarly, the earthquake excitation used for full time-history analysis is the Bhuj earthquake time history data recorded at the IIT Roorkee station in Ahmedabad, as presented in Fig. 9. The actual recorded PGA for this time history record is 1.04 m/s^2 (0.106 g). However, as this is less than the site-specific PGA of 0.36 g as defined in the Indian Building Code (IS-1893 2016), a scaling factor of 3.4 was applied to the time-history record. Damping was assumed to be 10% (as recommended for masonry), specified at the first and third modal periods, as determined from modal analysis which was also carried out.

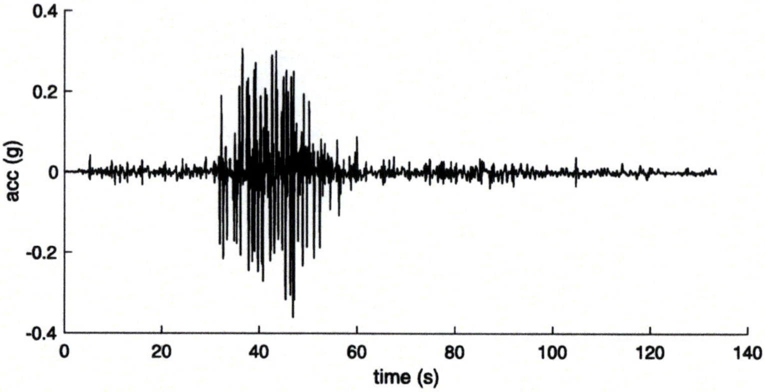

Fig. 9. Bhuj earthquake ground motion as recorded at IITR-Ahmedabad station, scaled by a factor of 3.4

5 Results

5.1 Gravity Load

Under the influence of gravity alone, the structure was found to undergo a maximum displacement of 0.2 mm in the vertical direction, with most deformation taking place in the domed structure as illustrated by Fig. 10. The meridional force and stress distributions in the dome can be found in Fig. 11, from where it can be seen that the highest force/stress concentration in the dome is at the end supports. The total vertical reaction under this load case was found to be 43.1 kN at the base of the large dome, and 6.3 kN at the base of the small dome. Similarly, the total vertical reaction at the base of the frame was found to be 150.6 kN (which also corresponds to the total weight of the structure), distributed among the 14 stone supporting columns, with the two columns between the large and small domes sustaining the largest vertical reactions of 16.13 kN each, as well as shear forces of 0.19 kN, as illustrated by Fig. 12. This corresponds to an axial stress in the

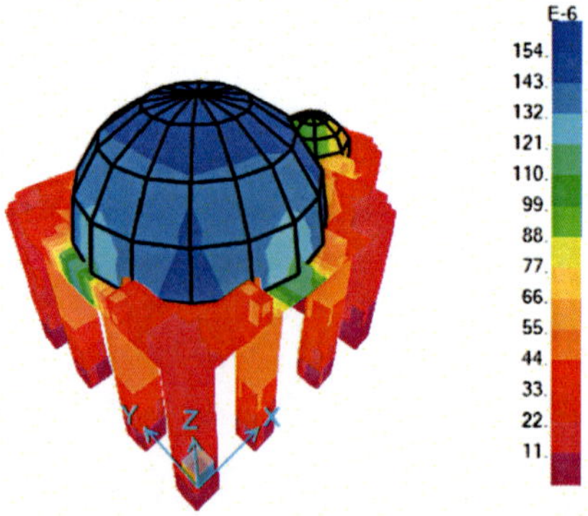

Fig. 10. Deformed shape of the structure showing resultant displacements (m) under gravity load

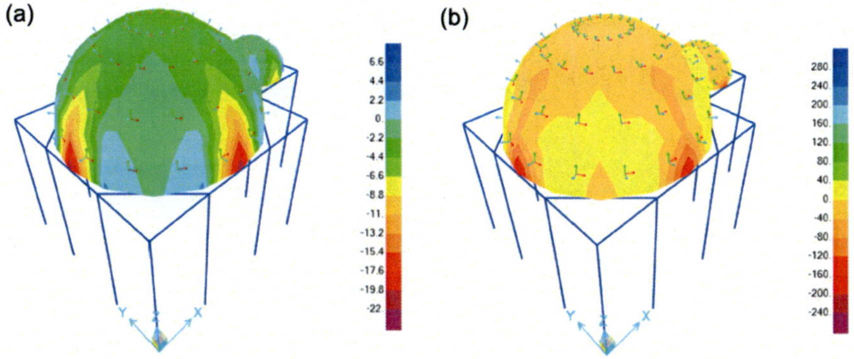

Fig. 11. Meridional (a) force (kN) and (b) stress (kPa) distributions in the domes - DL

columns of 173.62 kPa and a shear stress at the base of 2.05 kPa (Table 5). It should be noted that the weight of the two domes make up nearly a third of the weight of the entire structure.

5.2 Modal Analysis

The first ten mode periods and frequencies as extracted from the modal analysis are listed in Table 4, while the first three mode shapes can be found in Fig. 13. From this analysis it was found that the first mode has a period of 0.201 s with a mass UY participation factor of 99.47% indicating it is a primary mode of deformation in the global Y direction. Similarly, the second mode, with a period of 0.196 s and UX mass participation factor of 99.99% is the primary mode of deformation in the global X direction. Finally, the third mode has a period of 0.164 s and RZ mass participation factor of 99.4%, indicating this is the fundamental torsional mode (as also illustrated by Fig. 13).

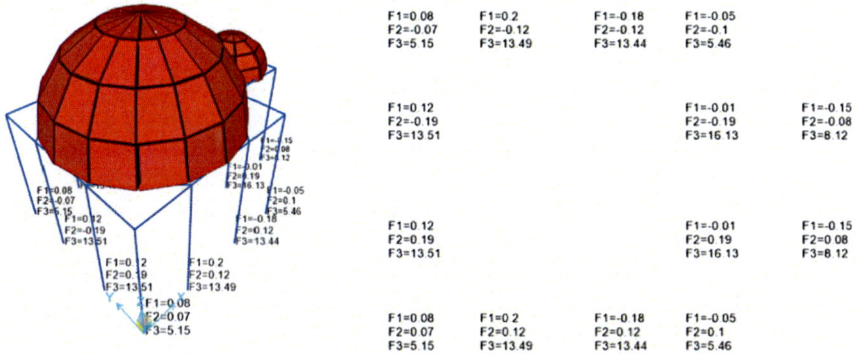

Fig. 12. Joint reactions at base of the structure DL – 3D view (left) and plan view (right) (in kN). Note that F1, F2 and F3 in this figure correspond to the forces in the global X, Y and Z directions respectively.

Table 4. Modal periods and frequencies

Mode	Period (sec)	Frequency (Hz)
1	0.201	4.96
2	0.196	5.11
3	0.164	6.12
4	0.045	22.13
5	0.034	29.51
6	0.027	37.48
7	0.026	38.91
8	0.025	39.99
9	0.023	44.43
10	0.020	50.30

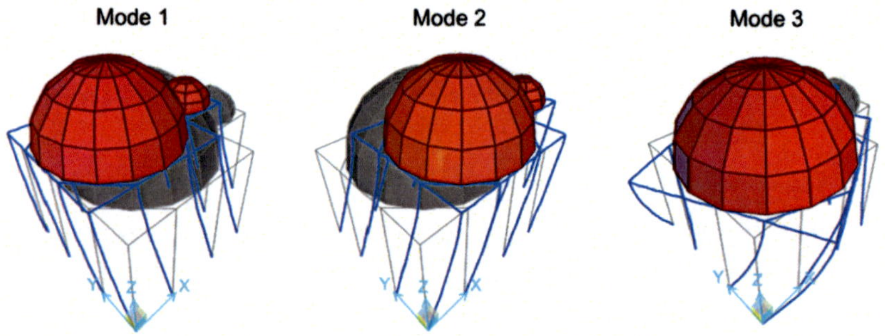

Fig. 13. First three modes shapes of the cenotaph structure

5.3 Earthquake (Static) Loading (EQY)

Under the influence of the static lateral loading based on a PGA = 0.2 g, the structure was found to undergo a maximum displacement of 2.1 mm as illustrated by Fig. 14. The total static base shear under this load case was found to be 23.64 kN, while total story shear at the base of the domes was found to be 9.77 kN. The critical column under this load case corresponds to the column A4 (ref: Fig. 6) which is found on the southwest corner of the structure, under the large dome. This column was found to experience a maximum axial force of 6.99 kN under the lateral loading alone, and a maximum shear (in the direction of the load) of 1.66 kN, corresponding to axial and shear stresses of 75.24 kPa and 17.87 kPa respectively (Table 5).

Fig. 14. Deformed shape of the structure showing resultant displacements (m) under lateral loading

5.4 Response Spectrum Analysis (RSY)

Similarly, in the case of the response spectrum analysis, the cenotaph structure was found to undergo a maximum probable displacement of 4.6 mm as illustrated by Fig. 15. The maximum probable base shear under this load case was found to be 53.84 kN, while total story shear at the base of the domes was found to be 61.97 kN. The critical column once again corresponds to the column A4 (Fig. 6) which is found to experience a maximum probable axial force of 16.14 kN and a maximum probable shear of 3.92 kN, corresponding to axial and shear stresses of 173.73 kPa and 42.19 kPa respectively (Table 5).

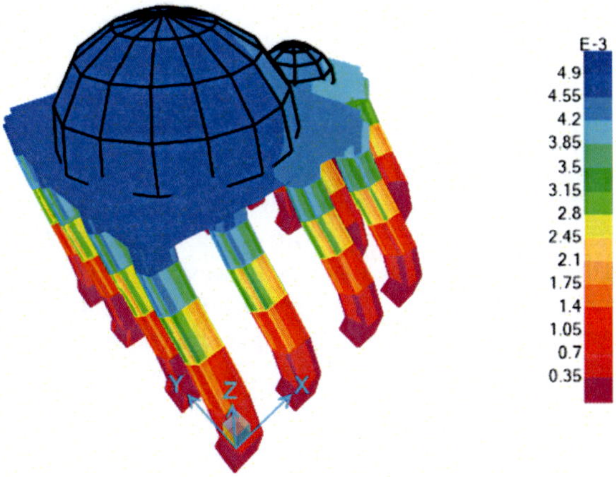

Fig. 15. Deformed shape of the structure showing maximum probable displacements (m) during the response spectrum analysis

5.5 Full Time-History Analysis (THY)

In the case of the full time-history analysis, the largest dynamic displacement of the structure was recorded at the top of the large dome, with a maximum magnitude of 6.71 mm as illustrated by Fig. 16. The total maximum base shear under this load case was found to be 78.72 kN (Fig. 17), while the maximum probable story shear at the base of the domes was found to be 82.27 kN.

As in the case of the static loading and response spectrum analysis, the critical column once again corresponds to column A4 (Fig. 6) - which is found to experience a maximum axial force of 23.40 kN and a maximum probable shear of 5.54 kN, which correspond to axial and shear stresses of 251.88 kPa and 59.63 kPa respectively. These results have been tabulated in Table 5 in order to facilitate comparison with the results of the different load cases.

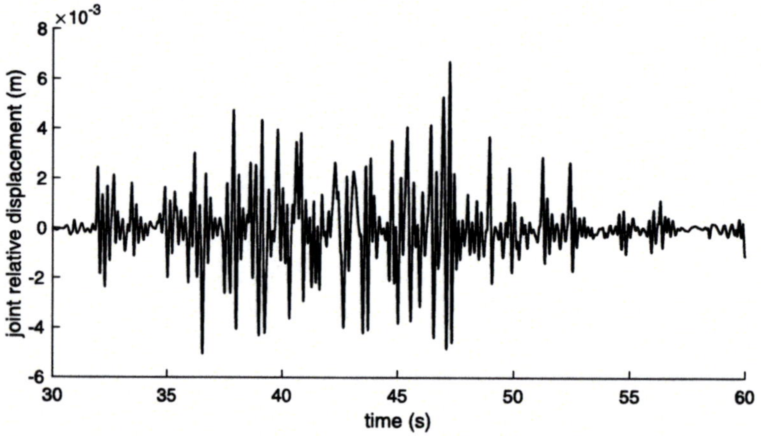

Fig. 16. Displacement time history as recorded at the top of the large dome

Finally, in order to compare the overall response of the structure under the different loading conditions, Fig. 18 presents the. maximum base shears (normalized by the total weight of the structure) for the different seismic load cases. As Fig. 18 illustrates, the structure experiences the smallest base shears under the static lateral loading and the largest shears for the full time-history analyses. However, this is due in part to the smaller PGA associated with this load case (0.2 g compared to the 0.45 g and 0.36 g used for the response spectrum and time-history loadings respectively). Similar trends are also observed upon comparison of the maximum displacements, with the static, response spectrum and time-history loading resulting in displacements of 2.1 mm, 4.6 mm and 6.7 mm respectively.

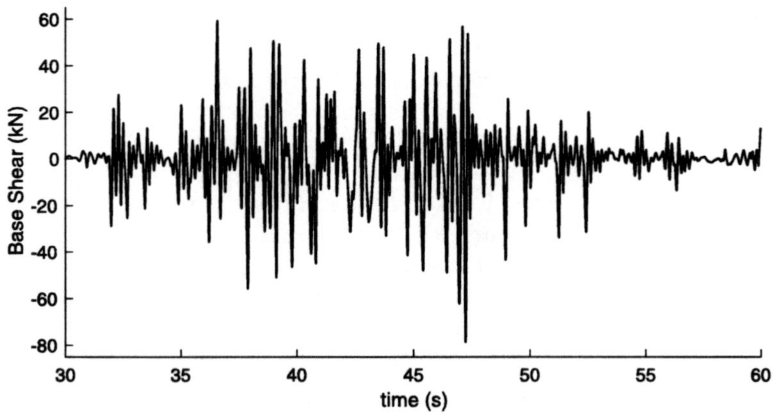

Fig. 17. Total base shear time history

Table 5. Comparison of the axial and shear forces/stresses under the different load cases (Column A4). Note that EQY: Lateral earthquake load, RSY: response spectrum loading and THY: time-history load case

	Dead load	EQY (static)	RSY	THY
Axial force (kN)	5.15	6.99	16.14	23.40
Axial stress (kPa)	55.4	75.24	173.73	251.88
Shear force (kN)	0.07	1.66	3.92	5.54
Shear stress (kPa)	0.75	17.87	42.19	59.63

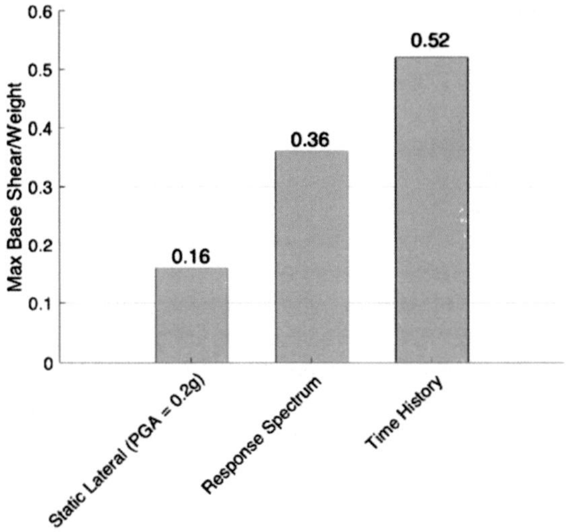

Fig. 18. Comparison of the maximum total base shear normalized by the total weight of the structure, for all seismic load cases

6 Restoration

The heritage cenotaph (chhatri) complex in Bhuj, Gujarat is under the protection of the Archaeological Survey of India (ASI). While the cenotaph (chhatri) of Rao Lakhpat Ji was totally destroyed including the collapse of its dome during the 2001 Bhuj earthquake, restoration works on other damaged cenotaphs in the complex have been in progress at a very slow pace. The site was visited in 2014, and it appears that not much progress has been made. The temporary brick pillars that were erected to shore up the damaged cenotaphs (chhatris) against further damage and collapse (Figs. 5c and d) are still in the same state as before. Systematic studies are needed to develop cost-effective restoration solutions for the different components of the sandstone cenotaph structures, and their connections, based on local building traditions and practices, to ensure their overall structural integrity. A deeper study is needed to investigate the challenges and bottlenecks in the process of restoration of this cultural heritage complex, so that it may continue to be enjoyed by all future generations.

7 Conclusions

The results of a study on the seismic behavior of heritage cenotaphs (chhatris) in Gujarat during the 2001 Bhuj earthquake have been presented. These cenotaphs have a structural form dominated by a heavy domed roof supported on an open layout of red sandstone columns, which are connected to the base with stone or metal dowels. This type of structural system is itself problematic, as it relies on its own mass, and friction between components to provide a minimal resistance to lateral loads and torsional movements.

The first torsional mode for this heritage cenotaph structure is mode 3 with a period of 0.164 s. The importance of torsional response of the selected cenotaph structure is consistent with the observed damage of cenotaphs (chhatris) in Bhuj which showed clear evidence of significant torsional response displacements during the 2001 Bhuj earthquake.

Results of max base shears under different earthquake load cases, presented in Fig. 18 show that selected heritage cenotaph (chhatri) structure appears to have a safety factor of $1/0.52 = 1.9$, and may be considered safe under earthquake loading. However, this factor is based on underlying assumptions about material properties, connections between different components, assumed linear elastic analysis, and a ground excitation recorded relatively far away from the actual location of the structure which thus does not account for local site effects.

Based on the results for maximum axial forces and shears in the critical column as presented in Table 5 and Fig. 12, it is found that in the critical column under the combined dead and time-history loading, the maximum axial stress is 307 kPa, and maximum shear stress is 60.4 kPa which are very low and well below the compressive and tensile strength of sandstone (Table 2). Similarly, the maximum meridional tension stress in the dome under earthquake loading was found to be about 1.36 MPa due to the bulging behavior experienced by such domes during earthquakes. However, this is still much less than the 6.3 MPa tensile strength of sandstone. In reality, it appears the cenotaphs in Bhuj failed due to excessive lateral and torsional displacements imposed on them by the earthquake ground motions, which could not be accommodated by these structures.

References

1. Jain, S., et al., (ed.): Bhuj, India Earthquake of January 26, 2001 Reconnaissance Report. In: Earthquake Spectra Vol. 18, Earthquake Engineering Research Institute (EERI), Oakland, California (2002)
2. INTACH: "Field Report – 26th. January 2001 Gujarat (India) Earthquake", Indian National Trust for Art and Cultural Heritage, New Delhi, India (2001)
3. Jain, S., et al.: The republic day earthquake in the land of M.K. Gandhi – the Father of the Nation, In: NICEE, IIT – Kanpur, India (2001)
4. Booth, E., Vasavada, R.: Earthquake of 26 January 2001 and its effects on architectural heritage of kutch and saurashtra, Prepared for INTACH, New Delhi, May (2001)
5. Mishra, P.: The Kutchh Earthquake of 2001. National Institute of Disaster Management, NIDM, New Delhi (www.nidm.gov.in)
6. Langenbach, R.: UNESCO mission to the earthquake damage district in Gujarat, India – for the conservation of cultural properties (2001)
7. D'Ayala, D., Kansal, A.: An analysis of the seismic vulnerability of the architectural heritage of Bhuj, Gujarat, India. In: SAHC (2005)
8. SAP2000: Computers and Structures Inc., Walnut Creek, California (2020)
9. Shams Alam, M., et al.: Characterization of kota sandstone under different strain rates in uniaxial loading. Geotech. Geol. Eng. **33**(1), 143–152 (2014). https://doi.org/10.1007/s10706-014-9810-3
10. Edinger J., (ed.): Gujarat (Kutch) India earthquake of January 26, 2001 – lifeline performance. In: Technical Council on Lifeline Engineering, Monograph no. 19 (2001)
11. Krinitzsky, E.L., Hynes, M.E.: The Bhuj, India, earthquake: lessons learned for earthquake safety of dams on alluvium. Eng. Geol. **66**(3–4), 163–196 (2002). https://doi.org/10.1016/S0013-7952(02)00049-2

Estimation of Seismic Behaviour of Historic Masonry Structures by Shake Table Testing

Lidija S. Krstevska(✉)

Dynamic Testing Laboratory (DYNLAB), Institute of Earthquake Engineering and Engineering Seismology-IZIIS, University Ss. Cyril and Methodius, Todor Aleksandrov 165, 1000 Skopje, Republic of North Macedonia
lidija@iziis.ukim.edu.mk

Abstract. Evaluation of seismic stability and behavior of structures involves several important aspects and activities that should be considered. Among these is testing of scaled models on a shaking table. Within several complex international projects related to the problem of seismic stability of historic masonry structures, a number of tests were performed on the 5DOF shake table in the Dynamic Testing Laboratory of the Institute of Earthquake Engineering and Engineering Seismology in Skopje. Most of these tests were aimed at experimental verification of new proposed methodologies for repair and seismic strengthening by testing models of representative existing prototype structures for specific types of monuments. Some models were tested mainly for the purpose of assessing the vulnerability of a particular monument or to estimate the intensity of a past earthquake as a reason for existing damage. The applied testing procedure, the obtained experimental results and the main conclusions on selected and characteristic historic masonry structure models tested in the past, as well as recently tested models of the stone masonry arch bridge – the Old Bridge in Mostar and the ancient dry masonry Protiron structure in Split will be presented.

Keywords: Historic structure · Monument · Shake table · Seismic behavior · Failure mechanism

1 Introduction

Historic structures are of permanent interest in research projects worldwide, as they are witnesses of history and culture at the time of their construction. Most of them are under protection and particular care toward increase of their resistance to damage caused by various sources. Historic structures located in seismic active regions are usually significantly damaged during strong earthquakes due to their stiff and brittle structural components and lack of ductility. They are not designed in compliance with seismic rules and codes. The way of their protection has continuously been a subject of many discussions, but what is clear is that strengthening of these structures must be in a way that requires the least intervention and maximum preservation of authenticity, (Feilden 1982; Tassios 2010; Oliveira and Costa 2010). For many reasons, each historic monument

I. Vayas and F. M. Mazzolani (Eds.): PROHITECH 2021, LNCE 209, pp. 556–575, 2022.
https://doi.org/10.1007/978-3-030-90788-4_44

represents a case for itself. Even though the only way to look into the problem of finding the best way for protection is to analyze these structures one at a time, considering the original structure and all interventions for its repair done in the course of time, we are trying to generalize the problem by representative examples of specific types of monuments. This requires development of a scientifically based methodology for repair and seismic strengthening of such structures in seismically active regions, a significant part of which is the need for experimental tests of the main dynamic characteristics of structures (ambient and forced vibration technique), as well as laboratory testing of physical models on shaking table, (Modena et al. 1992; Carydis et al. 1996; Krstevska and Tashkov 1998; De Canio et al. 2008;). Based on experimental data obtained from in-situ testing of structures, the laboratory testing of models provides very reliable data on the seismic behaviour and stability, pointing out the weak points of the structures. These data are of great importance for further analysis, i.e., development of an appropriate methodology for seismic strengthening and development of sophisticated numerical models of monuments (Nikolic et al. 2019a,2019b).

Within the framework of extensive research activities for estimation and evaluation of seismic stability and behavior of different types of historic structures, the Institute of Earthquake Engineering and Engineering Seismology has realized important scientific research projects involving experimental and analytical research. The most representative seismic shaking table experiments on scaled models of historical buildings are presented herein.

2 Testing Procedure

For design, construction and shake table testing of an appropriate scaled physical model of a structure and estimation of its seismic behavior and stability by shake table testing, a study of the expected seismic hazard, the local soil conditions,the dynamic characteristics of the structure and the strength and deformability characteristics of the built-in materials should be performed (Krstevska et al. 2007). The selection of appropriate model type is of crucial importance that will enable drawing of correct and valuable conclusion on the seismic behavior of the prototype structure.

The shake table testing procedure applied to scaled models consists of two main phases:

1. Tests for definition of the dynamic characteristics of the model, before and after performance of seismic tests, in order to check the stiffness degradation of the model produced by micro or macro cracks developed during the tests.
2. Seismic testing by selected earthquake record until collapse. The tests are usually performed in several steps, increasing the input intensity of excitation in order to obtain the response in the linear range, as well as to define the initial crack state, development of failure mechanism and collapse of the model.
3. The recorded response parameters are checked after each test and visual inspection of the tested model is performed in order to mark the cracks.

Fig. 1. IZIIS Shaking Table – top view

Table 1. Kinematic quantities for zero pay load

Direction	Displacement	Velocity	Acceleration
Horizontal	± 125 mm	± 1.0 m/s	± 3.0 g
Vertical	± 60 mm	± 0.5 m/s	± 1.5 g

3 IZIIS Shaking Table

The shaking table in the Dynamic Testing Laboratory of IZIIS is proportioned 5 m × 5 m, while its total mass is 330 kN, Fig. 1. The newest state-of-the-art digital controller 469D from MTS was installed later in March 2011. The shaking system controls five degrees of freedom of the table, two translations and three rotations, and is capable of generating translational vibrations in horizontal and vertical direction. The maximum weight of the models can be up to 400 kN with the height of 6 m. The maximum displacements which can be achieved are 125 mm in horizontal direction, and 60 mm in vertical direction. The maximum acceleration in horizontal direction is 3 g, and 1.5 g in vertical direction. The operating frequency ranges from 0 to 50 Hz. Table 1 presents the main kinematic quantities for zero pay load.

4 Shaking Table Tests of Historic Masonry Structures - Examples

4.1 Models of Historic Adobe Structures in California

Within the cooperation project between IZIIS and the Getty Conservation Institute (GCI) during the 90-ties, a shaking table testing of two scaled models representing existing historic adobe buildings in California was carried out (Gavrilovic et al. 1996; Tolles et al. 2000). The objective of the experimental investigation was to contribute to preparation of engineering guidelines for design of seismic retrofit measures for adobe buildings. The models were designed as typical south-western American adobe structures with floor and roof systems as well as highly vulnerable gable end walls (Fig. 2).

The first model was an original (control) model while the second one was retrofitted by combination of horizontal and vertical straps, center cores, and partial plywood diaphragms.Unretrofitted and retrofitted models were built to a scale of 1:2 and tested under seismic excitation with increasing intensities applying the record of Taft earthquake, California. The objective of the testing was to investigate the models' behaviour both in linear and nonlinear range, until reaching the heavy damage state.

Fig. 2. Characteristic adobe structure

During the seismic tests, the gable end wall of the original unretrofitted model building collapsed. In the case of the retrofitted model, even though the vibration was very intensive, the applied retrofitting proved very effective in improving stability and preventing collapse, Fig. 3.

Fig. 3. Gable end wall – failure of the unretrofited (original) model (left) and damage to the retrofitted model (right)

It is worthy to mention that the use of the survey of 19 historic adobes in the Los Angeles area conducted after the 1994 Northridge earthquake evidently showed that the damage to the real structures and the damage to the original model were very similar. Based on all the results and the GCI' 104 Engineering Guidelines, a number of historic adobe structures in South California were seismically upgraded applying this experimentally verified methodology (Toles et al. 2000).

4.2 Model of Representative Byzantine Church

For the experimental investigations planned within the research project "Study for Seismic Strengthening, Conservation and Restoration of Churches Dating from the Byzantine period (ninth–fourteenth century) in Macedonia", realized by IZIIS, the Republic

Institute for Protection of Cultural Monuments (RZZSK) in Skopje and the Getty Conservation Institute (GCI) in the USA, the church of St. Nikita near Skopje was selected as a representative prototype structure, Fig. 4.

Fig. 4. Prototype (left) and model (right) of St. Nikita church

For experimental verification of the proposed methodology for repair and seismic strengthening, a 1:2.75 scale model of this church was constructed, model M-SN-EXIST (Shendova et al. 2012, 2014). To satisfy the criteria for realistic reproduction of the global nonlinear behavior and failure mechanisms, the "gravity forces neglected" modelling principle was adopted. The two main goals of the shake table tests were: assessment of the vulnerability of such monument structures and selection of the most appropriate procedures for repair and strengthening of damaged structures in post-earthquake protection. The testing by gradually increasing the intensity of input earthquake excitations enabled monitoring of the progressive development of cracks and failure mechanism, modification of the dynamic characteristics as well as definition of the elasticity limit (occurrence of the first cracks).

The investigation was performed by simulating two main types of earthquake: the 1976 Friuli (IT) earthquake (Breginj record) as a local one, and the 1979 Montenegro (Petrovac record) and 1940 El Centro earthquakes as distant ones. From the general behavior of the model, it was concluded that it behaved as a rigid body in the elastic range, but when the first larger cracks occurred, separation of the bearing walls and damage up to a state close to complete failure took place. This was proved by the decrease in the models' natural frequency from 11.0 to 6.6 Hz.

The model was then repaired by injection of lime-based mixture and structurally strengthened following the proposed methodology: incorporation of strengthening elements - horizontal and vertical belt courses formed by incorporating steel reinforcement into the wall mass and filling the area around them with an appropriate injection mixture that enabled contact with the surrounding existing masonry, model M-SN-STR.

The response of the strengthened model was considerably different from that of the original model, Table 2. The increased elasticity limit and reduction of displacements at the top were characteristic. Although there was a considerable deterioration in the load bearing capacity under maximum seismic effect, the complete stability of the model structure was not disturbed due to the presence of ductile elements, while the damage was such that it was repairable. The structure of the strengthened model was in a state of deep nonlinearity, but still far from failure. A qualitatively different type of failure mechanism was also observed. The strengthened model did not suffer separation of bearing walls and vertical cracks, but failure was transferred to the lower zone and resulted in occurrence of diagonal cracks (Fig. 5).

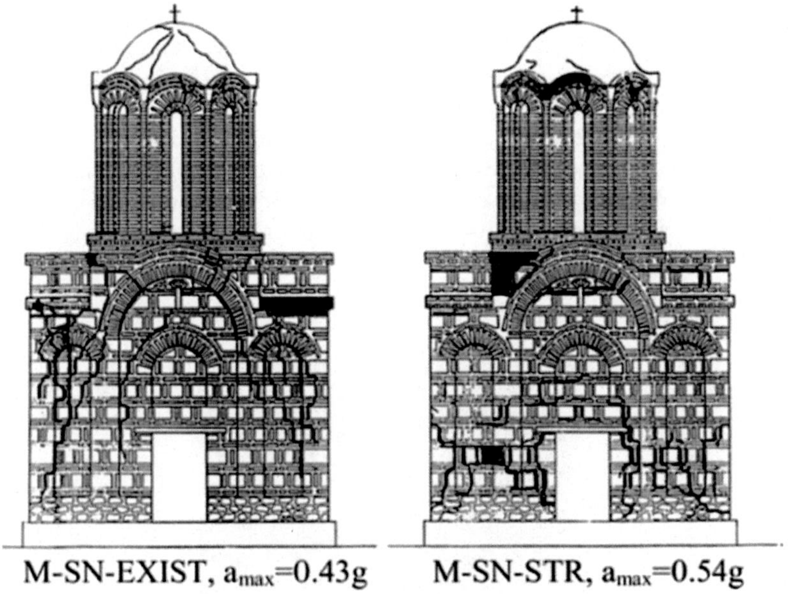

Fig. 5. Failure mechanisms - west façade, original model (M-SN-EXIST) and strengthened model (M-SN-STR)

Table 2. Comparison between the experimental results obtained for the models

Earthquake	Input acc(g)	Output acceleration for the church models			
		M-SN-EXIST		M-SN-STR	
		Level1	Level2	Level1	Level2
El Centro	0.17	0.29	0.55	0.20	0.47
El Centro	0.30	–	–	0.65	1.10
El Centro	0.49	–	–	0.91	1.59
El Centro	0.54	–	–	0.77	1.41
El Centro	0.63	–	–	–	–
Petrovac	0.19	0.39	0.76	0.27	0.48
Petrovac	0.40	–	–	0.77	1.36
Breginj	0.17	0.22	0.52	0.30	0.55
Breginj	0.28	–	–	0.20	0.40
Breginj	0.38	–	–	0.34	0.79

Level1 Base of the tambour, *Level2* Top of the dome

This proposed and experimentally verified methodology using "ties and injection" was successfully implemented in the process of reconstruction and seismic strengthening of real historic monuments.

As a continuation of previous activities, experimental and analytical investigations as well as shaking table testing of a base isolated church model were performed to develop a methodology for application of seismic isolation as a way of seismic protection of a large number of similar cultural monuments. These investigations were realized at IZIIS within the framework of the joint US-Macedonian research project and PHARE Cultural Development Program. For the needs of testing the isolated church model, Fig. 6, eight seismic isolators of the type of rubber bearings were specially designed and produced.

The performed seismic tests showed that the base isolated model behaved as a rigid body without any visible cracks. It did not suffer damage under low and moderate earthquake intensities and the damages under maximum expected accelerations with a return period of 1,000 years were minimal and repairable. The tests proved that the technology of seismic base isolation of historic monuments offers absolute safety and protection and that its application should become an imperative in earthquake protection of monuments in future (Tashkov et al. 2010, 2014).

Fig. 6. Base isolated model on the shaking table

5 Model of Marcus Aurelius Column in Rome

The main objective of the cooperation project between Istituto Nazionale di Geofisica, Rome and IZIIS realized in the period March 1994 – July 1995 was to find out the reasons for the existing damage to the Marcus Aurelius column in Rome (also known as Antonina column, Fig. 7 left), as well as to estimate the seismic resistance and provide an insight into its seismic behavior. This column is one of the oldest monuments of Ancient Rome, located in the historical center of Rome. The column is constructed of cylindrical marble blocks, placed over massive foundation. Significant local damages to the column were evident, manifested by cracks and dislocations particularly in the upper part of the structure.

The seismic behaviour of Marcus Aurelius column was investigated by testing of a model designed as a true replica model to the scale of 1:6 (Fig. 7, middle). A special water-cement mixture was applied as a material for the construction of the model. The model generally properly reproduced the dynamic behavior of the actual structure, which was proved by comparison of the model and the prototype dynamic characteristics (Krstevska et al. 1996).

The seismic input for the shake table testing was the accelerogram recorded at Mercato San Severino Station, Irpinia earthquake, 23 Nov. 1980, E-W component and the peak ground acceleration was taken to be up to 0.15 g. The behavior of the model was linear under low level of excitation, while nonlinearity started under excitation of 0.10 g. Under earthquake intensity of 0.15 g the model was near collapse, vibrating very intensively. A lot of cracks appeared due to compression stresses and dislocation between the cylindrical blocks above the foundation as well as sliding of the blocks in the upper third (blocks 17, 18, 19 – Fig. 7 right). The instability of the model was mainly caused by the lack of connecting elements between the blocks. It was obvious that, in the case of low frequency content of the input, i.e., in the case of close values of input and column model frequencies, the displacements and uplifting between the elements became large and critical for the stability of the model.

Fig. 7. The column of Marcus Aurelius in Rome (left), the model (in the middle) and dislocation of the model elements after the test (right)

After the seismic testing with different intensities until failure of the model, the crack pattern as well as the dislocation of the blocks along the upper third of the column were very similar to those existing on the column prototype in Rome. So, the explanation was that the damage was caused by earthquakes that occurred in the past centuries and the soft soil conditions at the site of the column. The general conclusion based on shaking table test was that the column should be strengthened in some way in order to achieve continuity between the blocks and prevent sliding between them (Tashkov et al. 1995).

5.1 Model of Masonry Walls at Beauharnois Powerhouse

In 2006, a bi-lateral project for experimental and analytical investigation of the seismic stability of the brick masonry walls of the Beauharnois powerhouse located in Quebec, Canada, was initiated. The project was realized in cooperation between "Hydroquebec" (Montreal, Canada) and the Institute of Earthquake Engineering and Engineering Seismology-IZIIS (Skopje, Republic of North Macedonia). The main objective of the investigation was to evaluate the seismic stability of the masonry walls of the machinery building at the Beauharnois powerhouse according to the new design codes in Canada (Krstevska et al. 2010a, 2010b, 2010c).

The powerhouse of the Beauharnois power plant is a typical example of a composite steel frame-masonry curtain wall structure built at the beginning of the twentieth century.The structure is considered a historical heritage and is protected as such. More than 50 powerhouses of dams and sub-stations in Quebec are constructed by application of the same mixed system.

Fig. 8. Beauharnois powerhouse (left) and the model of the wall on the IZIIS' shake table (right)

The superstructure of the Beauharnois powerhouse consists of single bay steel frames spaced at 5.1 m along the length of 826 m (Fig. 8 left). The masonry brick walls, incorporating the columns of the steel frames, serve as enclosures at the downstream side. The height of the walls is 20 m. Their thickness varies form 0.3 to 0.6 m and they accommodate a series of large window frames. The walls were built after completion of the steel structure and they are integrated with the majority of the steel frames. The masonry consists of bricks and cement mortar with a very specific way of construction.

The physical model of the selected portion of the Beauharnois powerhouse measured by in-situ ambient vibrations was designed according to the similitude requirements defined for model testing on a shake-table. Considering the prototype structure dimensions, as well as the dimensions and the load capacity of the shake-table, the model was designed to the geometrical scale of 1:3 (Fig. 8 right). An adequate model (combined true replica-artificial mass simulation model) was adopted.

The out-of-plane testing was chosen as the most critical testing direction for the wall, considering its impressive height. The seismic excitation selected for the shake-table testing of the model was the representative (modified) accelerogram recorded during the 1985 Nahanni, NWT earthquake of M6.8, horizontal component H1, with peak acceleration of 0.2 g.The accelerogram represents a typical earthquake time history for the hard rock sites in Montreal, Quebec. It is compatible with the uniform hazard spectra (UHS) provided in the Canada's national seismic hazard maps. Time histories of ground motion that match the UHS were developed.

The main conclusion after the performed out-of-plane seismic tests on the model was that, under the expected intensity level for the site where the Beauharnois powerhouse is located, i.e., 0.2 g, the behaviour of the model was stable, without any cracking. The dynamic amplification factor (top/base) was < 2. The first cracks occurred under an intensity of 0.7 g, i.e., under very intensive excitations, mainly in the part of the wall next to the openings. Some of the bricks below the level of the first opening were broken. The part of the structure without openings was practically undamaged, with only a few micro-cracks developed during the most intensive excitation. The relative displacements between the wall and the steel columns were also monitored during the seismic tests. They were relatively small and within the range between 0.2 and 1.8 mm (in reality: 0.6–5.4 mm), for the excitation level of 0.2–0.7 g. The reason for this was the very good connection between the steel structure and the wall, achieved by the steel bracings incorporated into the wall and connected to the steel columns.

The numerical model calibrated by these experimental tests represented the basis for analytical evaluation of the seismic stability of the real structure and prediction of the safety of the powerhouse, which was done later.

5.2 Models of Historical Monuments Tested Within PROHITECH Project

In the period 2006–2009, within the framework of FP6 PROHITECH project "Earthquake Protection of Historical Buildings by Reversible Mixed Technologies" experimental shaking table tests on scaled models of three representative historical monuments – mosque, cathedral and church, were carried out. The main objective of the project was:

- Experimental investigation of both traditional and innovative materials and techniques for seismic protection of existing structures, with special emphasis on the Cultural Heritage.
- Experimental contribution to the assessment and setting up reversible mixed technologies for repair and strengthening of reinforced concrete, masonry, steel and wooden buildings, monuments and other structures typical for the European and Mediterranean building heritage.

The results from the shake table tests have shown the efficiency of all applied retrofitting systems.

Model of Mustafa Pasha Mosque in Skopje

Mustafa Pasha mosque is one of the biggest monuments of the Ottoman sacral architecture in Skopje, built at the end of the 15th century (Fig. 9 left). It is a one-dome structure consisting of massive double-leave walls with a thickness of 2 m and has one minaret. (Mazzolani et al. 2007; Krstevska et al. 2010a, 2010b, 2010c). The model of the mosque was designed according to the "gravity forces neglected" modeling principle and it was constructed in the IZIIS' Laboratory (Fig. 9 right).

The seismic testing was performed by use of the Montenegro Petrovac 1979 earthquake record that was scaled according to the similitude requirements by a factor of 6 and applied in horizontal direction. The testing was performed in three main phases: (1) Testing of the original model under low intensity level, with the aim of provoking damage to the minaret only; (2) Testing of the model with a strengthened minaret, with the aim of provoking its collapse and damage to the mosque and (3) Testing of the strengthened mosque model until reaching heavy damage.

The strengthening solution for the minaret consisted of application of CFRP wrap upon a layer of epoxy glue on four sides along the length of the minaret, in vertical direction, at four levels along the height of the minaret in horizontal direction and at the minaret base. For the main mosque structure, the strengthening consisted of formation of a horizontal belt course around the bearing walls by CFRP rods as well as around the tambour and at the base of the dome by a CFRP wrap.

Fig. 9. Mustafa Pasha mosque (left) and its model (right)

After the performed seismic tests, the strengthened model was heavily damaged, Fig. 10 left). The push-over curves were constructed based on the different intensity level of the repeated seismic test, within an input acceleration of 0.05 g - 1.50 g (Fig. 10 right).

The shaking table testing of both the original and strengthened model showed that the strengthening of the minaret by application of a CFRP wrap enabled stiffening and increasing of its bending resistance. The mosque model's behaviour, after strengthening, was evidently different in respect to that of the original model. During the intensive tests, the failure mechanism was transferred to the lower zone of the bearing walls where typical diagonal cracks occurred due to shear stress. Comparing the obtained results, it was concluded that the applied strengthening technique significantly improved the seismic resistance of the monument.

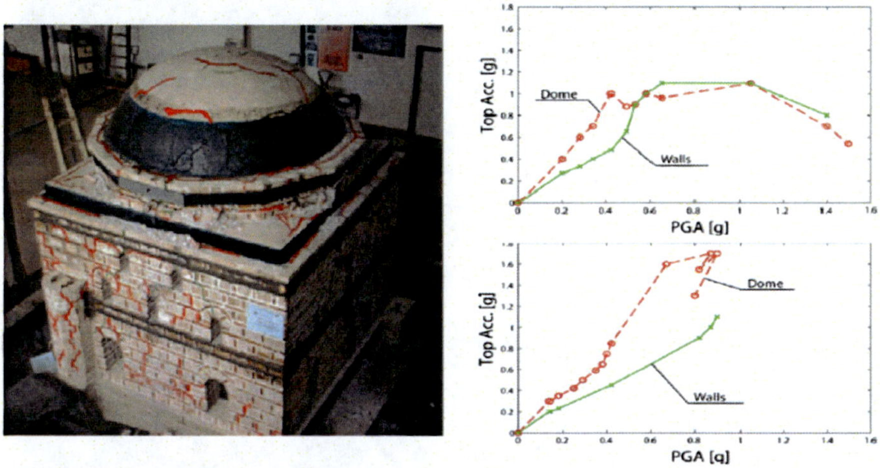

Fig. 10. Damaged model of the mosque after testing in Phase 3 (left) and response of the model: acceleration measured at the top of the shear walls and of the dome as a function of input acceleration, phase 2 and phase 3 (right)

Model of Gothic Cathedral – Fossanova Abbey, Italy

The Fossanova Gothic cathedral is located in Priverno, Lazio, Italy. It was constructed in the fifteenth century. (Fig. 11 left). The model for seismic shake table testing was designed as a true replica, to a scale of 1:5.5, Fig. 11 right. A very detailed model was conceived in order to take into account the complexity of the prototype and keep correct the scale factor in terms of its dynamic response. Equivalent materials were considered, having the same weight, but reduced strength according to the adopted scale factor (Tashkov et al. 2009).

Seismic investigation of the model was performed by simulating the Calitri earthquake selected as a representative one and applied in horizontal direction only. After testing of the original model and development of failure mechanism, the cracks were repaired by expansive aluminum-cement mortar.

The proposed methodology for strengthening consisted of incorporation of prestressed vertical and horizontal carbon fiber ties at several levels of the model (Fig. 12 left). During the final test, with intensity of 0.4 g, the strengthened model was severely damaged, with completely developed failure mechanism (Fig. 12 right).

Fig. 11. The Fossanova Cathedral, Italy – prototype (left) and model (right)

The main conclusion was that the integrity of the model was significantly improved by the applied strengthening. For the original model, the critical input intensity was 0.14 g, while for the strengthened model, it was 0.40 g. These parameters clearly showed the effectiveness of the applied strengthening, enabling confining of the structure.

Fig. 12. The strengthened model of the cathedral (left) and damage after the seismic test (right)

Model of St. Nicholas church, Psacha, Republic of North Macedonia

St. Nicholas church was built in the mid-fourteenth century. The model of the church was designed to the length scale of 1:3.5 according to the "gravity force neglected" modelling principles, Fig. 13 (Tashkov et al. 2009).

Testing of the model was performed by simulating the 1979 Montenegro Petrovac earthquake. According to the "gravity force neglected" modelling principle, the original earthquake record was scaled by 3.5 in time domain (compressed).The shake-table testing was performed in two main phases: testing of the base-isolated model with the innovative ALSC floating-sliding system and testing of the original fixed-base model. The model with the sliding base was tested first because it was expected that no damage would occur.

Fig. 13. St. Nicholas church, prototype (left) and its model (right)

The comparative test of the base-isolated model by ALSC system and classical fixed base model, clearly showed the superior behaviour of the ALSC floating-sliding base-isolation system. The tests showed that system could protect the structure in any frequency and/or amplitude range and under the strongest earthquake because of the specific property: limited transmissibility (0.2–0.3 g) for the scaled model i.e., 0.07–0.08 g for the real structure, which cannot produce damage when sliding as a rigid body with amplification factor of 1.0, along the total height of the structure, Fig. 14. On the contrary, the fixed base case is very vulnerable. The amplification factor ranged from 1.5 for the walls to 3.0 for the dome. The consequence of that was early damage to the dome (0.45 g) and serious damage to the walls.

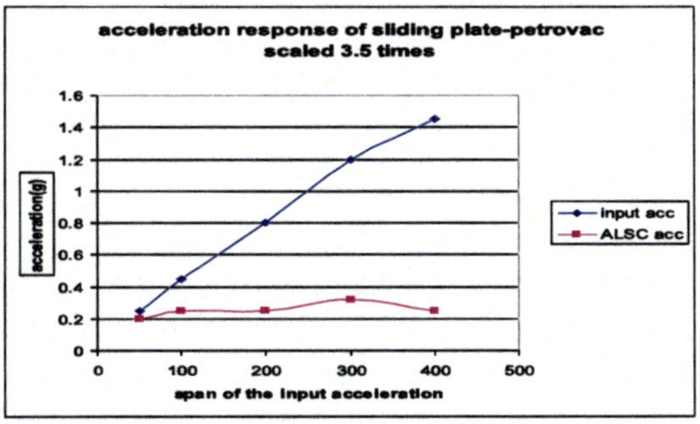

Fig. 14. The efficiency of the ALSC system for base isolation

5.3 Model of the Old Bridge in Mostar

The Old Bridge in Mostar, Bosnia and Herzegovina, was originally built in 1566. During the war in 1993, the bridge was demolished and then reconstructed in 2004. This bridge attracts the attention of the wider public with its appearance, especially because of its extremely narrow arch, Fig. 15, left. The structure is made mostly of a local Tenelija stone with a span of approximately 28.80 m, arch thickness of approximately 80 cm and arch width of approximately 4 m. The principles of construction during the bridge reconstruction in 2004 completely complied with the original construction method. The method of sealing clamps and dowels with liquid lead was used for assembling and fixing the connections with the stone blocks.

Investigation of the seismic stability of the bridge model was performed for the purpose of experimental research within the doctoral thesis (Kustura, 2018). The model was designed and constructed to the geometrical scale of 1:9 as a gravity force neglected model, where stone, mortar and steel connecting element materials were used with the same or sufficiently similar characteristics as in the original structure. The same method of assembling the vault used for the reconstruction of the prototype was applied also in the construction of the model - clamps and dowels, as well as liquid lead was used in connecting the stone blocks (Kustura et al. 2018).

Fig. 15. The Old Bridge in Mostar (left) and the scaled model placed on the shake table (right)

For the seismic testing of the Old Bridge model, the accelerogram of the Montenegro Petrovac earthquake of 15 April 1979 was applied as a characteristic earthquake. The excitation was applied simultaneously in longitudinal (Y) and vertical (Z) directions. The accelerogram was scaled according to the similitude requirements for the first series of seismic tests by a scaled factor (SF) of 1:9 in the time domain. Considering the model frequencies and the predominant frequencies of the applied earthquake, in order to excite the resonant conditions during the model testing and to provoke intensive vibrations until its collapse, the accelerogram of the Petrovac earthquake scaled to 1:6 and 1:3 was applied in the final series of tests.

The experimental testing was conducted for the purpose of determining the global behaviour and failure mechanism of the masonry arch under intense vibration conditions until collapse, by monitoring all response values – displacement, acceleration and strains at characteristic points. The instrumentation set-up included 31 measuring points.

During the first series of tests, it was observed that the connections between the vault elements of the model remained strong and unchanged and occurrence of cracks was not detected (scaling factor 1:9). Such unchanged connections enabled direct transfer of vibrations from the foundation to the top of the model. In the case of the most intensive tests, cracks appeared, and their development led to collapse of the model, as a result of loss of stability due to large strains and failure of connections between the elements (clamps and dowels). The failure of the model was due to loosening of the connections between the elements in the bridge model vault manifested by pulling out of clamps and dowels from their original position (Fig. 16). Breaking of steel connecting elements was not detected. The most vulnerable parts were the third points of the arch as well as its top.

The global stability of the bridge was not affected by the expected earthquake intensity for the area where the prototype is located but it will cause some damage to the bridge, manifested by cracks in the mortar. Clamps and dowels as reinforcing elements improve the seismic stability by significantly increasing the strain capacity of the model.

The experimental research of the model showed that the assembling technique was quite effective in increasing the seismic resistance of historical masonry structures. The presented experimental investigation was focused on analysis of the global behavior of such structures and their collapse mechanisms. Unlimited prospects for further research can be found in the analysis of local effects of connectors considering their types, geometry, dimensions, assembling methods, material of the blocks, connectors and infill, and the interaction between the stone, infill and connectors.

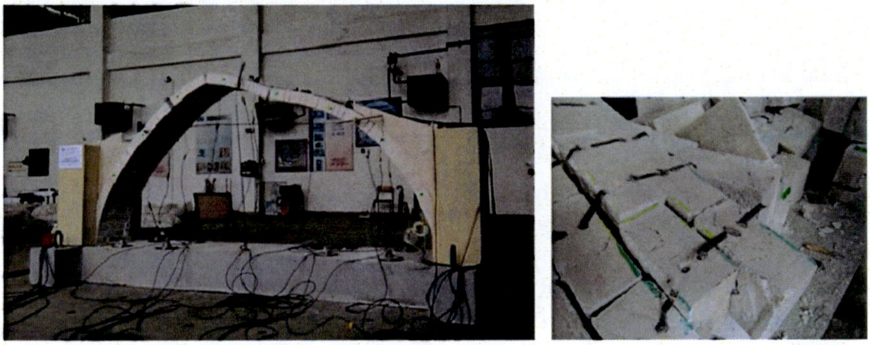

Fig.16. Failure mechanism of the model (left) and pulling-out of connection elements (right)

5.4 Model of Protiron Structure in Split

The experimental study focusing on the response of the Protiron model, an ancient dry-stone masonry structure located in Split, Croatia, to seismic excitation was aimed at providing experimental data for verification and calibration of the 3D numerical model and investigating the seismic resistance of this dry-stone ancient monument strengthened with clamps and dowels, which is representative for that period (Nikolic et al. 2019a,2019b). The research work was fully supported by the Croatian Science Foundation under the project "Development of Numerical Models for Reinforced-Concrete and Stone Masonry Structures under Seismic Loading Based on Discrete Cracks" (IP-2014-09-2319).

Protiron is a dominant structure in the Peristyle square of the Diocletian's Palace in Split (Fig. 17, left), built from ca. 295 to 305 AD by the Roman Emperor Gaius Aurelius

Valerius Diocletianus. The structure is composed of four massive granite columns with Dorian capitals supporting a broad gable. Originally, the structure was built of stone blocks with dry joints and dowels embedded between the base, the columns, the capitals and the beams of the broad gable. The original Protiron was a separate structure, while the surrounding buildings were built later. Due to the movements of the stone blocks and the columns, extensive restoration works were carried out at the beginning of the 20th century, when the stone blocks in the broad gable were connected by clamps. The dowels and clamps have kept the structural integrity throughout history so Protiron is one of the ancient structures that is very well preserved to date.

Fig. 17. Protiron structure in Split (left) and the model ready for testing (right)

The physical model was designed as a true replica model, Fig. 17, right, at a scale 1:4. According to the similitude requirements for testing of models on a shake table, the scaling factors were defined considering all relevant characteristics: shape and geometry, mass density, modulus of elasticity and seismic excitation characteristics. An equivalent material for stone was chosen to fulfill the required scaling factors, i.e., micro-concrete.

In order to overcome all unknown parameters regarding the characteristics of the connections applied in the Protiron structure and having in mind the main objective of the project, it was decided to use copper for all model connections – clamps and dowels. The real 1979 Petrovac (Montenegro) earthquake N-S component recorded on April 15th, 1979 was used as input acceleration.

Seismic tests were performed by gradual increase of intensity of input acceleration, ranging from 0.017 g to 1.9 g, until collapse of the model. Considering the similarity rules for the true replica model, these intensities are valid both for the model and for the real structure.

Due to the horizontal in-plane seismic loading of the model, stone blocks of the arch of the broad gable tended to collapse and space two central columns. The damage to the real structure was very similar to that of the tested model. Namely, the same damage mechanism, which included separation of the blocks and deformation of the clamps placed in the lower part of the arch, was visible on the real structure as well as on the tested model.

The collapse of the model occurred due to reaching the critical opening of the contacts at the ends of the columns accompanied by their rotation and loss of geometrical stability when the large mass pediment moved to one side causing complete loss of connections with the columns, Fig. 18.

The experiment provided valuable data for verification and calibration of numerical models for strengthened stone masonry structures. Those models are very important for understanding the mechanisms that allowed stone masonry structures avoid structural collapse and destruction during strong earthquakes, since laboratory tests are very difficult and costly to perform. The experiment showed that strengthening of a stone masonry structure with clamps and dowels could increase the collapse acceleration and, generally, seismic resistance of the structure.

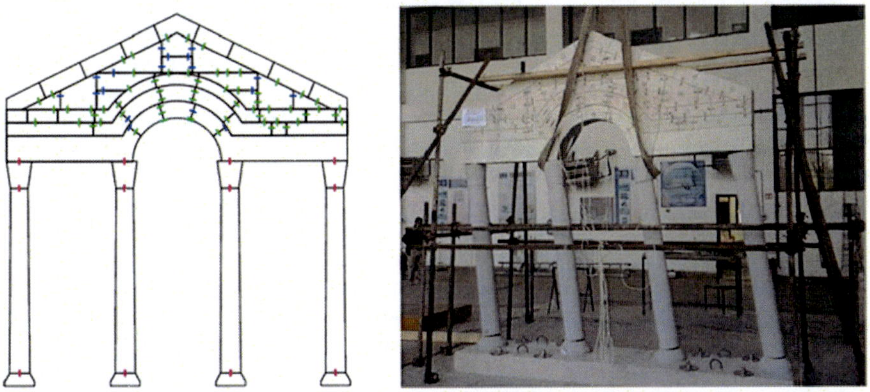

Fig. 18. Disposition of the clamps and dowels (left) and loss of the models' stability – pre-collapse state (right)

6 Conclusions

Shake table testing of scaled models is an important experimental research activity particularly when seismic behavior and stability of specific structures like historical buildings and monuments are considered. Verification of new methodologies for improving their stability as well as calibration of developed numerical models for analysis should be based on results obtained by such testing. For this purpose, it is important to adopt, design and construct an appropriate physical model of the prototype structure according to similitude requirements and to apply a proper testing procedure. Based on the extensive experience of the IZIIS' Dynamic Testing Laboratory, this paper provides a review of selected examples of shake table testing of models realized within several international projects. Some of the experimentally verified methodologies presented herein were successfully implemented for real historic monuments. Newly developed, advanced numerical models, have successfully been calibrated by these experimental tests, representing the basis for analytical evaluation of the seismic stability of existing historical masonry structures.

Acknowledgements. The author wishes to express her gratitude to all institutions that financially supported the realized projects and all principal investigators of the projects whose vision and creativity contributed to their extraordinary success. Gratitude is also accorded to the colleagues from the IZIIS' Dynamic Testing Laboratory for the successful realization of the presented experimental investigations.

References

Carydis, P.G., et al.: Experimental and numerical investigations of earthquake response of classical monuments. In: 11WCEE, Acapulco, Paper 1388 (1996)

De Canio, G., Muscolino, G., Palmeri, A., Poggi, M., Clemente, P.: Shaking table tests validating two strengthening interventions on masonry buildings. In: Seismic engineering conference, MERCEA08, Reggio Calabria (2008)

Feilden, B.M.: Conservation of Historic Buildings. Butterworth Scientific, London (1982)

Gavrilovic, P., Sendova, V., Tashkov, Lj., Krstevska, L., Ginell, W., Tolles, L.: Shaking Table Tests of Adobe Structures. Report IZIIS 96–36 (1996)

Krstevska, L., Mihailov, V., Bosci, E., Rovelli, A.: Experimental dynamic testing of prototype and model of the Marcus Aurelius column in Rome. In: 11th World Conference of Earthquake Engineering (11 WCEE). June 1996, Acapulco, Mexico, Paper No. 546 (1996)

Krstevska, L., Tashkov, Lj.: Testing methodology and practical applications for evaluating the seismic behaviour of historical monuments. In: Monument-98, Workshop on Seismic Perform of Monuments, November 12–14, 1998, Lisbon, Portugal, pp. 187–195 (1998)

Krstevska, L., Tashkov, Lj., Arun, G., Akoz, F.: Evaluation of seismic behaviour of historical monuments. In: SHH07 Studies on Historical Heritage - International Simposium, Antalya, Turkey, September 17–21, 2007, pp. 411–418 (2007)

Krstevska, L., Taskov, Lj.: Verification of effectiveness of seismic protection and retrofit techniques by experimental testing. In: Malta Symposium - COST Action C26 Urban Habitat Construction under Catastrophic Events, Malta, 2008, pp. 223–228 (2008)

Krstevska, L., Tashkov, Lj., Gramatikov, K., Landolfo, R., Mammana, O., Portioli, F., Mazzolani, F.: Large-scale experimental investigation on mustafa pasha mosque. J. Earthquake Eng. **14**(6), 842–873 (2010). https://doi.org/10.1080/13632460903338528

Krstevska, l., Tashkov, Lj., Gocevski, V., Garevski, M.: Experimental and analytical investigation of seismic stability of masonry walls at Beauharnois powerhouse. In: Bulletin of Earthquake Engineering, Springer, Netherlands, Vol. 8, Issue 2, p 421 (2010). https://doi.org/10.1007/s10518-009-9142-5

Krstevska, L., Taskov, Lj., Garevski, M., Gocevski, V.: Experimental investigation of seismic stability of power plants in Canada, In: 9th US National and 10Th Canadian Conference on Earthquake Engineering: Reaching Beyond Borders, Toronto, July 25–29, 2010. Paper ID 1285 (2010)

Kustura, M., Krstevska, L., Ćubela, D.: Experimental testing of scaled model and arch segment of the old bridge in mostar. In: E-Proceedings No. 16, December 2018, Faculty of Civil Engineering, University of Mostar, Mostar (2018)

Kustura, M.: Investigation of Seismic Behaviour of Stone Arch Bridges Including the Effect of Connecting Elements. *Doctoral Disertation.* IZIIS, Skopje (2018)

Mazzolani, FM., Krstevska, L., Tashkov, Lj., Gramatikov, K., Landolfo, R.: Shaking table testing of Mustafa Pasha Mosque model. FP6-PROHITECH Final Report, IZIIS, Skopje (2007)

Modena, C., La Mendola, Terrusi, A.: Shaking table study of a reinforced Masonry building Model. In: 10WCEE, vol. 6, Madrid, pp. 3523–3526 (1992)

Nikolić, Ž., Krstevska, L., Marović, P., Smoljanović, H.: Experimental investigation of seismic behaviour of the ancient Protiron monument model. Earthq. Eng. Struct. Dyn. **2019**, 1–21 (2019). https://doi.org/10.1002/eqe.3149

Nikolic, Ž., Krstevska, L., Smoljanovic, H., Zivaljic, N.: Modelling of the influence of metal connectors on the resistance of historical dry-stone masonry structures. Int. J. Arch. Herit. **14**(10), 1468–1483 (2019). Taylor & Francis. https://doi.org/10.1080/15583058.2019.1613455.

Oliveira, C.S., Costa, A.: Reflections on the rehabilitation and the retrofit of historical constructions. In: Garevski, M., Ansal, A. (eds.) Earthquake engineering in Europe, pp. 199–221. Springer Netherlands, Dordrecht (2010). https://doi.org/10.1007/978-90-481-9544-2_9

Shendova, V., Rakicevic, Z.T., Krstevska, L., Tashkov, L., Gavrilovic, P.: Shaking table testing of models of historic buildings and monuments – IZIIS' experience. In: Fardis, M.N., Rakicevic, Z.T. (eds.) Role of Seismic Testing Facilities in Performance-Based Earthquake Engineering, pp. 221–245. Springer Netherlands, Dordrecht (2012). https://doi.org/10.1007/978-94-007-1977-4_12

Shendova, V., Rakicevic, Z., Garevski, M., Apostolska, R., Bozinovski, Z.: Implementation of experimentally developed methodology for seismic strengthening and repair of historic monuments seismic evaluation and rehabilitation of structures. Geotech. Geol. Earthquake Eng. **26**, 115–154 (2014). https://doi.org/10.1007/978-3-319-00458-7_8

Tashkov, Lj., et al.: Shaking table test of 1/6 scale model of the column of Marcus Aurelius in Rome. In: SECED Conference - European Seismic Design and Practice, October 1995, Chester, UK (1995)

Tashkov, Lj., Krstevska, l., De Matteis, G., Gramatikov, K., Mazzolani, F.: Shake table test of a model of Fossanova Church in reduced scale. In: Protection of Historical Buildings, Conference Prohitech 09, Rome, Italy, 2009, Taylor & Francis Group, London, ISBN978-0-415-55803-7, Vol 2, 1683–1690 (2009)

Tashkov, Lj., Krstevska, L., Gramatikov, K., Mazzolani, F.: Shake-table test of model of st. nicholas church in reduced scale 1/3.5. In: Protection of Historical Buildings, Conference Prohitech 09, Rome, Italy, 2009, Taylor & Francis Group, London, ISBN978-0-415–55803-7, Vol 2, pp 1691–1697 (2009)

Tashkov, Lj., Shendova, V., Krstevska, L., Gavrilovic, P.: Application of two base isolated techniques for seismic protection of byzantine churches - shake table test studies. In: Proceedings of the 14th European Conference on Earthquake Engineering, 28 August-3 September 2010, Ohrid, Macedonia (2010)

Tashkov, Lj., Krstevska, L.: Seismic shake-table testing of base-isolated structures-IZIIS experience. In: 2ECEES & 15ECEE, 24–28 August 2014, Istanbul Turkey, Paper ID 2297 (2014)

Tolles, L., Kimbro, E., Webster, F., Ginell, W.: Seismic stabilization of historic adobe structures. In: Final Report of the Getty Seismic Adobe Project, Los Angeles (2000)

Tashkov, Lj., Krstevska, L.: Application of two base isolation systems for seismic protection of historical structures – shake table test studies. In: Workshop on Seismic Risk of Historical Structures, November 3–5 2014, Istanbul, Turkey, 107–112 (2014)

Tassios, T.P.: Seismic engineering of monuments, earthquake engineering in Europe. In: 14ECEE, Skopje (2010)

Tolles, L., Kimbro, E., Webster, F., Ginell, W.: Seismic Stabilization of Historic Adobe Structures. Final Report of the Getty Seismic Adobe Project, Los Angeles (2000)

Structural Model Updating of a Historical Stone Masonry Tower in Tønsberg, Norway

Amirhosein Shabani(✉), Agon Ademi, and Mahdi Kioumarsi

Department of Civil Engineering and Energy Technology, Oslo Metropolitan University, Pilestreet 35, 0167 Oslo, Norway
amirhose@oslomet.no

Abstract. Conservation of historical structures plays a pivotal role in every nation, and numerical modeling of historic buildings is an essential part of a conservation methodology. Since deriving the mechanical properties of construction materials using destructive tests is usually not allowed, and most of the cultural heritage assets are so complicated in architecture, numerical modeling of historic buildings is challenging for the analysts. Therefore, in this study, a non-invasive approach for numerical simulation of historic structures has been proposed and applied on a historical stone masonry tower (Slottsfjell tower) in the city of Tønsberg, Norway. Two types of sensors were utilized for finite element modeling of the stone masonry tower. Firstly, three-dimensional (3D) laser scanners were used as a fast and precise tool to provide a finite element model of the case study with and without considering soil-structure interaction. Secondly, operational modal analysis (OMA) was performed using accelerometers to derive the tower's dynamic characteristics. Finally, as a preliminary result of the case study in the context of a European project, the produced finite element models' mechanical properties are updated based on the OMA results, and the effect of considering soil-structure interaction has been investigated.

Keywords: Finite element model · Numerical modeling · Historical towers · Masonry tower · 3D laser scanner · Accelerometer · Operational modal analysis

1 Introduction

Comprehensive knowledge about the geometry, material properties, and boundary conditions is required to develop a robust numerical model of historical structures in a non-invasive way. In addition, to assess the vulnerability of a structure, different types of risks, and their effects on the structure's performance should be investigated and applied to the numerical model (Shabani et al. 2020).

I. Vayas and F. M. Mazzolani (Eds.): PROHITECH 2021, LNCE 209, pp. 576–585, 2022.
https://doi.org/10.1007/978-3-030-90788-4_45

In recent years, contemporary digital methods help structural engineers with structures' geometric documentation (Stampouloglou et al. 2020). A Three-dimensional (3D) laser scanner is a tool for converting objects to reliable and accurate 3D digital models. 3D laser scanners can be easily exploited even by non-experts in new technologies, and 3D geometries result from them can be converted to 3D finite element models (FEM) for structural analysis (Barazzetti et al. 2015, Castellazzi et al. 2015).

Soil-structure interaction (SSI) is an effective parameter on the results from the structural analyses of the numerical models of structures, including historical masonry towers (Anand and Kumar 2018). In order to simplify the modeling procedure, SSI was neglected in some studies, but the basement level structural elements were modeled (Shabani and Erfani 2020, Zaherdannak et al. 2020). Various methods have been presented for considering SSI effects, i.e., modeling springs, dashpots, or the soil layers with solid elements (Çarhoğlu and Usta 2018, Fathi et al. 2020, Hökelekli and Al-Helwani 2020).

Operational modal analysis (OMA) of a case study using accelerometer sensors is a tool to constrain and calibrate the FEMs (Shimpi et al. 2019). FEM tuning of historical structures has been done by dynamic identification of historical structures, including masonry towers, as a non-destructive procedure (Lacanna et al. 2020, Standoli et al. 2020). Optimization algorithms were employed for accurate calibration of the masonry towers so that a tower was divided into different parts, and the material properties of each part were updated based on the results from ambient vibration testing (Milani and Clementi 2019).

Although the effect of SSI on the dynamic characteristics of a structure is not negligible, for the structural model updating of historical towers, the SSI was usually neglected, and only a few studies have been done that the FEMs were updated by considering SSI (Ramos et al. 2013, Ceroni et al. 2014).

In the present study, a preliminary result of a European project (Hyperion project) focusing on improving historic areas and monuments' resiliency is presented. Two FEMs of a historical stone masonry tower (Slottsfjell tower) in Tønsberg, Norway, were developed using 3D laser scanners. SSI is neglected in one of the FEMs, but the soil is modeled in the second model considering SSI effects. Finally, all two models are updated based on OMA results, and the impact of considering SSI is investigated.

2 An Overview of the Case Study

The Slottsfjell tower is located on a rocky hill, which dates from 1888 with architectural characteristics of the middle age era (see Fig. 1). The Slottsfjell tower has a significant historical value to both city of Tønsberg and the county of Vestfold and Telemark because of the region's identification with the Viking Age. The tower's total height is 21 m with one basement and three stories on top of the soil level. Each storey is like a box, and the width of each box is decreased from the basement level to the top floor, as is shown in Fig. 1c.

Fig. 1. (a) The location of Tønsberg in Norway, (b) the place of the Slottsfjell and (c) Slottsfjell tower, a historical stone masonry tower in Tønsberg

3 Numerical Modeling

A Topcon 2000 3D laser scanner was utilized for deriving the dense point clouds of the structure. Totally 20 scans were performed inside and outside the tower, and the data was imported into Autodesk Recap Pro software to combine all the raw point clouds of the shots into one complete model. The combined dense point clouds were then imported to Autodesk Revit for 3D modeling of the tower, as illustrated in Fig. 2a.

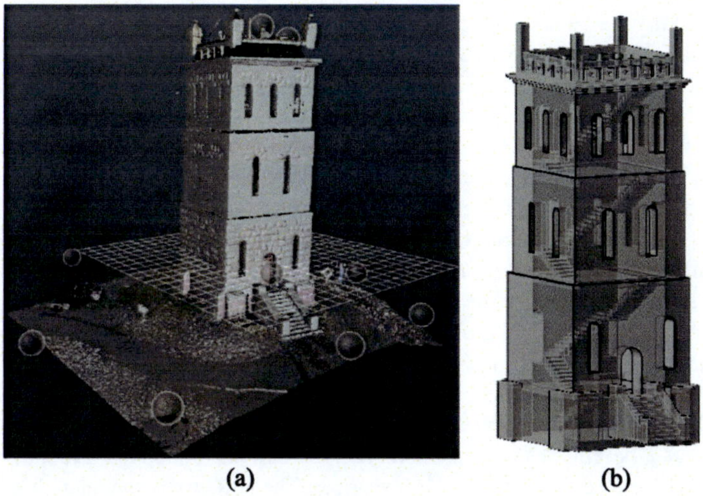

Fig. 2. (a) Point clouds derived from the 3D laser scanner and position of the scanners during the data acquisition and (b) 3D drawing of the tower

As it is shown in Fig. 2b, after developing an accurate 3D model of the tower, the drawing file has been imported to the DIANA FEA software package (DIANA 2020). In this software, the tower's 3D geometry has been converted to the FEM by considering a homogenized behavior for the masonry units, which is a robust and prevalent method for numerical modeling of the full-scale masonry structures (D'Altri et al. 2019, Sotoudeh et al. 2020).

For importing the 3D drawing files from the Revit software to the FEM, the industry foundation classes (IFC) format of the model has been provided, and by the CAD exchanger software, the IFC format file has been converted to the standard for the exchange of product model data (STEP) format which is suitable for importing the model in DIANA with solid elements. Figure 3 shows the methodology of converting the point clouds from the 3D laser scanners to the FEM.

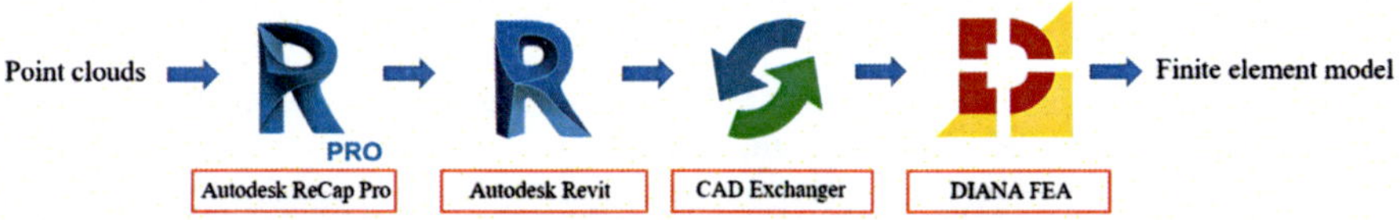

Fig. 3. The workflow of converting point clouds from the 3D laser scanners to FEM and the involved software packages

Based on the previous studies and due to the masonry's nature, an orthotropic material has been considered for the stone masonry with a shear modulus, 0.33 times of modulus of elasticity (E) (Gentile et al. 2015).

Two FEMs have been developed in such a way that in model 1, the SSI is neglected, but in model 2, the soil is modeled, and SSI effects are considered. For model 1, restrained boundary conditions have been considered for the bottom of the tower in all directions. The tower has a basement floor, and, in the model, the soil loads are neglected, as shown in Fig. 4a (Gentile et al. 2015). For model 2, the soil and SSI are considered for the FEM of the tower. Since no exact geotechnical data has been provided for the selected area, and due to the tower's location on a rocky hill based on the onsite inspection of the site, the hard soil's material properties are considered for the box soil element. Restrained boundary conditions have been considered for the soil box endings in different directions, as is illustrated in Fig. 4b. An interface element with a linear material type is modeled between the soil and the tower for considering the soil's no-tension connection behavior and the structure in the normal direction. Besides, high constant stiffness has been considered for the interface element to model the hard shear connection behavior of the soil and the tower (Hökelekli and Al-Helwani 2020). Considering the point that the tower has a basement floor, in the model, the soil surrounds the structure until the elevation was derived from the 3D laser scanner. The soil box is modeled as a box with a square plan with the dimensions of $35 \times 35 \times 25$ m^3 considering isotropic material characteristics.

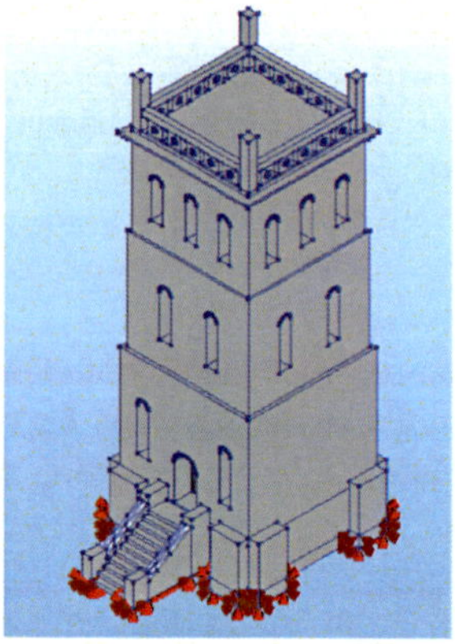

(a)

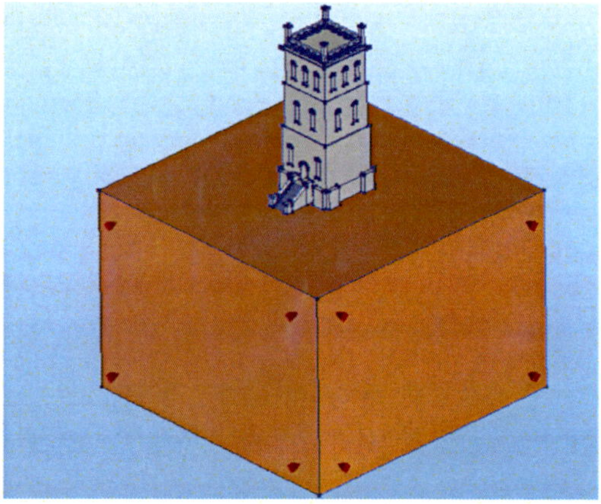

(b)

Fig. 4. (a) FEM of the case study converted from 3D drawing without considering SSI (model 1) and (b) FEM of the tower considering SSI (model 2)

Eight-nodes 3D quad mesh with adaptive element size has been considered for meshing the FEM. The number of elements for the tower model with a mesh size of 0.15 m is more than two times the number of the elements for the model with a mesh size of 0.2 m. Furthermore, the modal analysis results showed around a 1% deviation for the two types of meshing. Therefore, a maximum mesh size of 0.2 m has been employed for the tower as it is shown in Fig. 5a, and for the soil maximum mesh size of 4 m is considered as illustrated in Fig. 5b.

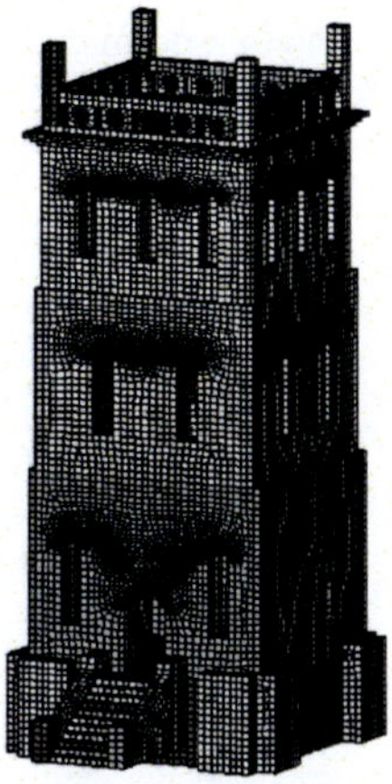

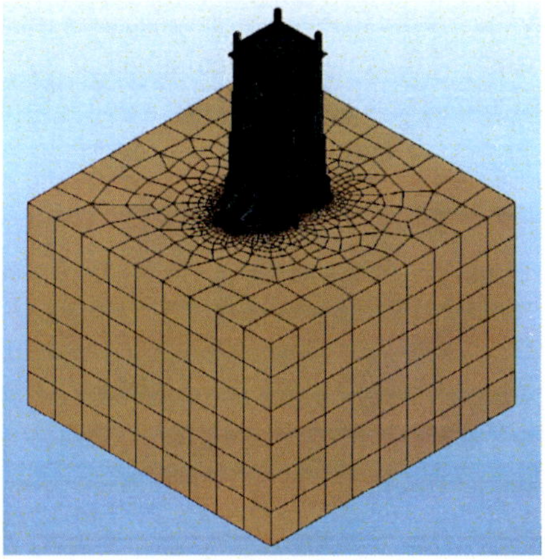

(a) (b)

Fig. 5. Meshing of FEM of the tower for (a) model 1 without considering SSI and (b) model 2 considering SSI

4 Dynamic Identification

OMA has been done to derive the dynamic characteristics of the tower, including the natural frequencies. For OMA, 3-Axis MEMS digital Unquake accelerometers with the maximum sensitivity of 8 g were installed on the second and the third floors with heights of 7.9 m and 13.3 m, respectively. Figure 6 shows the test setup of the accelerometer.

Fig. 6. Test setup for AVT, the accelerometer and the data logger are in the blue and yellow circles, respectively

Ambient vibration testing was performed for three days, and Fast Fourier Transform (FFT) has been utilized for each sensor to investigate the natural frequency of the structure (Brincker and Ventura 2015). Different types of more sophisticated methods exist to derive the exact frequencies and the mode shapes by employing a number of sensors (Brincker 2014). However, in this study, FFT as a fast method is utilized to get just the first dominant frequency of the case study from a sensor. Another sensor also was installed to verify the results from the first sensor. As illustrated in Fig. 7, the first dominant natural frequency of the tower in each direction for two sensors is 5.7 Hz showing the regular box like the tower's behavior.

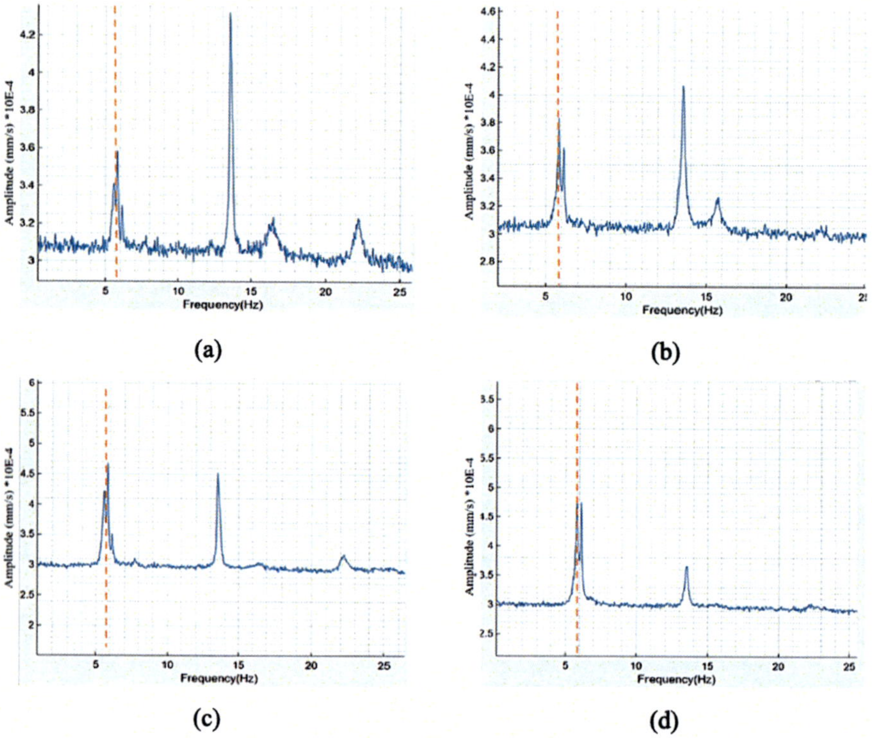

Fig. 7. FFT of the raw data from (a) X direction of the second floor, (b) Y direction of the second floor, (c) X direction of the third floor, and (d) Y direction of the third floor

5 FEM Updating

After providing natural frequencies of the tower from OMA, the FEM of model 1 has been updated by performing the sensitivity analysis to investigate the tower's exact material properties, including modulus of elasticity. Lower and upper boundary limits of 6 and 15 GPa have been considered respectively for the stone masonry prisms (Ip 1999). The final updated material properties of the FEM have been illustrated in Table 1.

Table 1. Updated material properties of FEM of model 1

	E-modulus	Poisson's ratio	Shear modulus	Density
Masonry	7.25 GPa	0.2	2.4 MPa	2000 kg/m^3

For updating model 2, the modulus of elasticity of the soil is updated as well as the masonry material. Lower and upper boundary limits of 5 and 20 GPa have been considered for the hard soil (Kramer 1996). The updated properties after the sensitivity analysis are illustrated in Table 2.

Table 2. Updated material properties of FEM of model 2

	E-modulus	Poisson's ratio	Shear modulus	Density
Masonry	11.2 GPa	0.2	3.7 MPa	2000 kg/m^3
Soil	6 GPa	0.25	–	2000 kg/m^3

As illustrated in Tables 1 and 2, the modulus of elasticity of the material of the case study in model 2, including the SSI, is 1.54 times more than the modulus of elasticity of the masonry material in model 1, which shows a significant difference in the initial elastic stiffness and structural behavior of the model. As it is investigated in this study by updating a case study without considering the SSI effect, the model's initial stiffness is underestimated. Therefore, conservative results will be derived from the updated models for further structural vulnerability assessment studies compared to the updated model in which the SSI effects are considered.

6 Conclusion

A 3D laser scanner is considered a reliable and fast tool for providing the 3D models of the structures that can be easily converted to FEM and accelerometer sensors as tools for calibrating the FEM are two sides of a coin for the approach presented in this paper. In this paper, a reliable FEM has been developed for numerical modeling of the case study utilizing 3D laser scanners and accelerometers, which can be used for further numerical analysis to assess the structure's vulnerability to different types of risks. Converting 3D geometries derived from 3D laser scanners to FEM is possible in some FEM software packages, i.e., DIANA FEA, which can accelerate the modeling procedure. Natural frequencies of structures can be derived using FFT, a fast method that can be done with one accelerometer sensor. When SSI is considered in the FEM, it is observed that for the masonry material of the tower, the modulus of elasticity and the shear modulus, which is a multiple of the modulus of elasticity in orthotropic materials, was 1.54 times more than the model without SSI effects. Therefore, conservative results are derived from future structural analysis of an updated FEM without considering SSI; however, for decreasing the uncertainties related to SSI consideration for structural health monitoring purposes, the numerical model needs to be modeled with SSI effects. This study should be done by performing other OMA methods, and the FEM should be updated for different natural frequencies of the case study by considering the mode shapes for each frequency.

Acknowledgements. This work is a part of the HYPERION project. HYPERION has received funding from the European Union's Framework Programme for Research and Innovation (Horizon 2020) under grant agreement no. 821054. The contents of this publication are the sole responsibility of Oslo Metropolitan University (Work Package 5, Task 2) and do not necessarily reflect the opinion of the European Union. Special thanks to Haidar Hosamo for his help during the measurements.

References

Anand, V., Kumar, S.S.: Seismic soil-structure interaction: a state-of-the-art review. Struct. Elsevier (2018). https://doi.org/10.1016/j.istruc.2018.10.009

Barazzetti, L., et al.: BIM from laser clouds and finite element analysis: combining structural analysis and geometric complexity. In: ISPRS - International Archives of the Photogrammetry, Remote Sensing and Spatial Information Sciences XL-5/W4, pp. 345–350 (2015)

Brincker, R.: Some elements of operational modal analysis. Shock. Vib. **2014**, 1–11 (2014). https://doi.org/10.1155/2014/325839

Brincker, R., Ventura, C.: Introduction to Operational Modal Analysis. John Wiley & Sons (2015)

Çarhoğlu, A., Usta, P.: Investigation of soil structure interaction problem for a masonry building. Acad. Persp. Proc. **1**, 1010–1018 (2018)

Castellazzi, G., D'Altri, A., Bitelli, G., Selvaggi, I., Lambertini, A.: From laser scanning to finite element analysis of complex buildings by using a semi-automatic procedure. Sensors **15**, 18360 (2015)

Ceroni, F., Sica, S., Pecce, M.R., Garofano, A.: Evaluation of the natural vibration frequencies of a historical masonry building accounting for SSI. Soil Dyn. Earthq. Eng. **64**, 95–101 (2014)

D'Altri, A.M., et al.: Modeling strategies for the computational analysis of unreinforced masonry structures: review and classification. Arch. Comput. Methods Eng. **27**(4), 1153–1185 (2019). https://doi.org/10.1007/s11831-019-09351-x

DIANA- Finite Element Analysis User Manual Version 10.4. Delft, Netherlands (2020)

Fathi, A., Sadeghi, A., Emami Azadi, M., Hoveidaie, N.: Assessing seismic behavior of a masonry historic building considering soil-foundation-structure interaction (case study of Arge-Tabriz). Int. J. Arch. Herit. **14**(6), 795–810 (2020). https://doi.org/10.1080/15583058.2019.1568615

Gentile, C., Saisi, A., Cabboi, A.: Structural identification of a masonry tower based on operational modal analysis. Int. J. Arch. Herit. **9**(2), 98–110 (2015)

Hökelekli, E., Al-Helwani, A.: Effect of soil properties on the seismic damage assessment of historical masonry minaret–soil interaction systems. Struct. Design Tall Spec. Build. **29**(2), e1694 (2020). https://doi.org/10.1002/tal.1694

Ip, F.: Compressive Strength and Modulus of Elasticity of Masonry Prisms, Carleton University (1999)

Kramer, S.: Geotechnical Earthquake Engineering. University of Washington. SL Prentice-Hall, Inc., United States of America (1996)

Lacanna, G., Betti, M., Ripepe, M., Bartoli, G.: Dynamic identification as a tool to constrain numerical models for structural analysis of historical buildings. Front. Built Environ. **6**(40) (2020). https://doi.org/10.3389/fbuil.2020.00040

Milani, G., Clementi, F.: Advanced seismic assessment of four masonry bell towers in Italy after operation modal analysis (OMA) identification. Int. J. Arch. Herit. **15**, 1–30 (2019)

Ramos, L.F., Aguilar, R., Lourenço, P.B., Moreira, S.: Dynamic structural health monitoring of Saint Torcato church. Mech. Syst. Sig. Process. **35**(1–2), 1–15 (2013)

Shabani, A., Erfani, S.: Seismic performance evaluation of SSMF with simple beam–column connections under the base Level. Int. J. Steel Struct. **20**(1), 89–100 (2020). https://doi.org/10.1007/s13296-019-00273-9

Shabani, A., Kioumarsi, M., Plevris, V., Stamatopoulos, H.: Structural vulnerability assessment of heritage timber buildings: a methodological proposal. Forests **11**(8), 881 (2020). https://doi.org/10.3390/f11080881

Shimpi, V., Sivasubramanian, M., Singh, S.: System identification of heritage structures through AVT and OMA: a review. Struct. Durab. Health Monit. **13**, 1–40 (2019)

Sotoudeh, S., Jahangiri, M., Ranjbarnia, M., Zakeri, J.-A.: Three-dimensional modeling of an old masonry bridge and assessing its current capacity. Periodica Polytechn. Civil Eng. (2020). https://doi.org/10.3311/PPci.14942

Stampouloglou, M., Toska, O., Tapinaki, S., Kontogianni, G., Skamantzari, M., Georgopoulos, A.: Archaeological anastylosis of two Macedonian tombs in a 3D virtual environment. Virtual Archaeol. Rev. **11**(22), 26 (2020). https://doi.org/10.4995/var.2020.11877

Standoli, G., Giordano, E., Milani, G., Clementi, F.: Model updating of historical belfries based on OMA identification techniques. Int. J. Arch. Herit. **15**, 1–25 (2020)

Zaherdannak, A., Shabani, A., Erfani, S.: Seismic performance evaluation of special RC frames with gravity steel columns under the base level. Shock. Vib. **2020**, 8825258 (2020). https://doi.org/10.1155/2020/8825258

Digital Documentation and 3D Modelling in the Restoration of the First Ancient Theatre of Larissa, Greece

Dimitrios L. Karagkounis[1] (✉) and Sofia D. Tsanaktsidou[2]

[1] Department of Archaeological Projects and Studies, Ephorate of Antiquities of Larissa, Ministry of Culture and Sports Diachronic Museum of Larissa, Mezourlo, 41500 Larissa, Greece
[2] Ancient Theatre of Larissa, Ministry of Culture and Sports Ancient Theatre of Larissa, Ephorate of Antiquities of Larissa, 10-12 Mitropolitou Arseniou, 41223 Larissa, Greece

Abstract. Digital documentation and 3D modelling find a wide application in the protection, preservation and restoration of Cultural Heritage Monuments. In the current paper, the use of these technologies is demonstrated in the case of the First Ancient Theatre of Larissa, one of the largest and most important theatres in Greece, the restoration of which has been ongoing for more than 20 years. In the NSRF 2014–2020 funded project "Restoration of the first Ancient Theatre of Larissa – Phase E", part of the koilon (cavea) is restored by repositioning 300 ancient marble seats, "edolia", to their original or corresponding positions, and by the addition of 240 new "edolia", according to the approved restoration study. The koilon is documented through 3d laser scanning and photogrammetry methods to produce a three dimensional reconstructed model. Then, an as-designed model of the koilon is created, depicting the curvature, elevation and the new position of both ancient and new "edolia". Separate 3d models are created for each new "edolio" that incorporate all the information needed to produce them on a CNC machine, ensuring the best fit. The paper also presents the implementation of digital documentation in the project's execution where modern technology meets ancient techniques.

Keywords: Ancient theater · 3d modelling · Restoration · Monument documentation · Project management

1 Introduction

The First Ancient Theatre of Larissa, Greece, (Fig. 1) is one of the largest and most important theatres of Greece and it is situated on the southern foot of the "Frourio" (Fortress) hill, in the centre of the modern city of Larissa, in the Region of Thessaly. According to long time Ephor, archaeologist Athanasios Tziafalias, (Tziafalias 1985 and Tziafalias 2000) the theatre had a lifespan of six centuries, from the early 3rd century BC until the late 3rd or early 4th century AD. During the first centuries, apart from theatrical performances, the assemblies of the senior regional authority were hosted, the so called "Koinon" of the Thessalians. In the 2nd c. BC the Romans converted it into an arena,

I. Vayas and F. M. Mazzolani (Eds.): PROHITECH 2021, LNCE 209, pp. 586–593, 2022.
https://doi.org/10.1007/978-3-030-90788-4_46

reserved for official celebrations. The following centuries it was gradually buried. In the 1980's, the first compulsory purchases took place, and a gradual unearthing began that lasted until the mid-2000s.

The theatre's unearthing and restoration have been the objective of several European funded projects, as early as the 2nd Community support framework. The current project "Restoration of the first Ancient Theater of Larissa – Phase E" project, is co-financed by the European Union and Greek national funds through the Regional Operational Program Thessaly 2014–2020 of the National Strategic Reference Framework (NSRF) 2014 – 2020.

In this paper, project manager and main supervisor Karagkounis D., architect, and assistant Tsanaktsidou S., civil engineer, present the project's outline and its progress since its beginning in 2019, emphasizing on the implementation of digital means in project management and documentation, as well as the design of subproject 3, which involves the 3d modelling of the main theatre.

Fig. 1. Aerial view of the First Ancient Theatre of Larissa (Karagkounis et al. 2017)

2 Restoration of the First Ancient Theater of Larissa – Phase E

The project "Restoration of the first Ancient Theater of Larissa – Phase E" has a total budget of 3.400.000 € and is subdivided into 6 subprojects:

- Subproject 1 "Restoration works of Ancient Theatre of Larissa" is procured via the direct labour system. It entails the construction of a stone masonry retaining wall with

total length of 61 m, on the foot of the main theatre for its support and the installation of ancient and new marble seats ("edolia"). 300 ancient edolia will undergo conservation before being reinstalled to their original or corresponding ("homologous") positions, and 240 new edolia will be installed, according to the approved restoration study (Karagkounis et al 2017).
- Subproject 2 involves the Procurement of marble blocks
- Subproject 3 involves the scanning and 3d modelling of the main theatre and production of 3d models for each new edolio.
- Subproject 4 involves the transport of marble blocks from quarry to CNC and ancient theatre, and the rental of crane trucks to aid in Subproject 1 heavy lifting works.
- Subproject 5 involves the creation of new edolia by processing marble blocks with CNC machine according to subproject 3 designs
- Subproject 6 involves the construction of entrance staircase for pedestrians and people with disabilities

As of November 2020, the first two stages of conservation works have been completed and preliminary works for the construction of the retaining wall have begun. Subproject 3, which is discussed in Sect. 5, has just been contracted and will be complete in the following two months. Also, a proposal for the restoration of the Skene (Karagkounis et al. 2020) was drafted, which has been approved by the Central Archaeological Council and the Minister of Culture and Sports. The restoration of the Skene will be incorporated in the current project, as the Region of Thessaly authorities have announced their willingness to double the project's total budget for this purpose.

3 Digital Documentation: Implementation in Project Management

Archaeological projects, including restoration ones, are notorious for being unpredictable and therefore difficult to plan and track. For this reason, and especially for NSRF projects, the Ephorates of Antiquities utilize pre-made forms included in the Manual of Procedures of Archaeological Works that are published by the Ministry of Culture and Sports. In the current project, apart from the use of said pre-existing documents, such as site diaries, complementary project documentation is implemented.

An example is the documentation of interventions on architectural members, as in the case of the conservation of ancient edolia. All works are documented by photographing each edolio before and after intervention (Fig. 2) and reports on the completion of objectives are drafted, including quantities and statistics.

Apart from the reports and photographic records, an MS Access database is used. This database contains entries of the architectural members that are involved in the restoration process, and is updated at a regular basis. Figure 3 is a screenshot of the Access database showing a typical tab. Information on the general attributes of the member are documented, as well as its pathology, drawings and photographs at different conservation stages. A similar methodology is followed for other works in the project's scope.

Another example is the use of Gantt charts in excel sheets (Fig. 4) for monitoring the progress of subprojects, spanning from the preparation of the tender documentation for

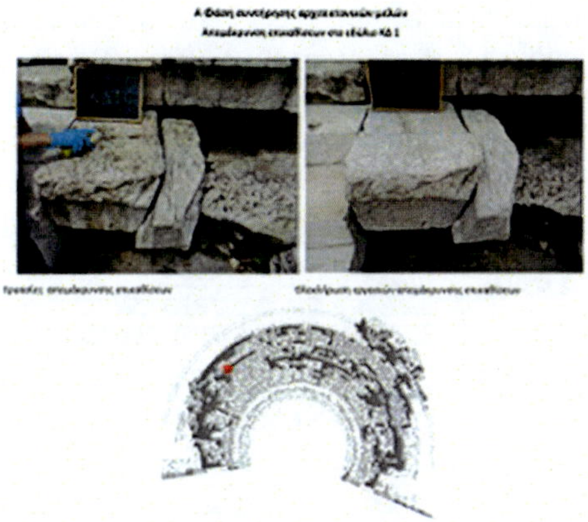

Fig. 2. Conservation documentation. Before and after pictures of an edolio during the 1st phase of conservation and its location on the main theatre.

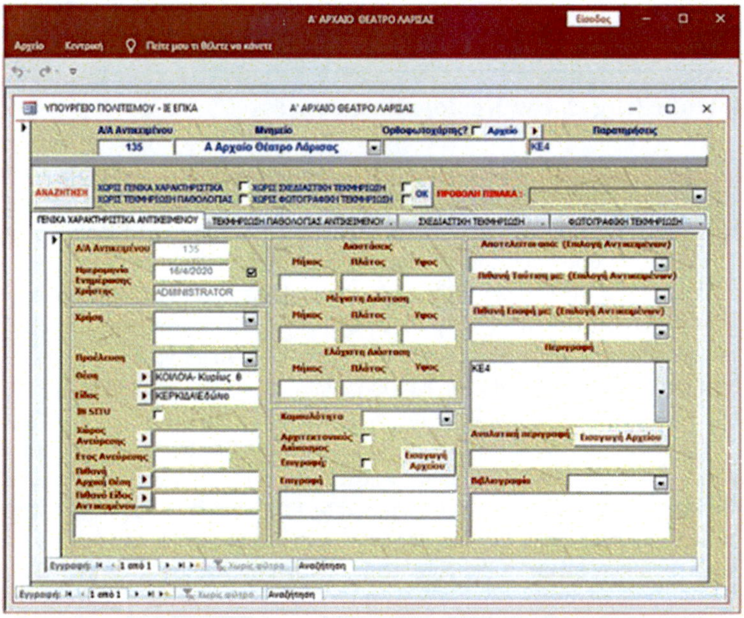

Fig. 3. Screenshot of the MS Access database

the open public procurements to the subproject's completion and delivery. In the future, the authors aspire to utilize project management software such as MS Project in order to test its application on archaeological projects of this scale.

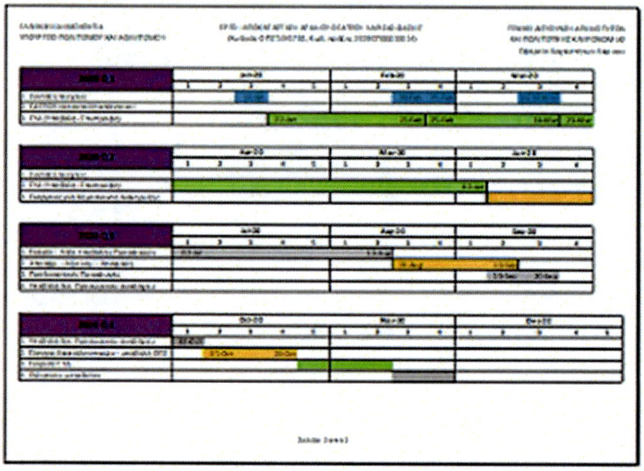

Fig. 4. Typical Gantt chart for monitoring subprojects

4 3D Modelling in the Restoration Process

The use of photogrammetry methods and 3d modelling in the documentation and restoration of monuments is becoming increasingly popular in recent years. These methods allow for detailed documentation of our cultural heritage and have opened new possibilities in their study, restoration and preservation.

In the case of the west frieze of the Parthenon (Bouzakis et al. 2016) it was possible to visualise and consider it in its complete form, by digitally assembling dispersed parts.

In other instances, it can be a useful tool in decision making as to which approach would be appropriate when restoring a monument.

In the process of drafting the Restoration of the Skene of the First Ancient Theatre of Larissa – Phase A proposal (Karagkounis et al., 2020), a 3d reconstruction of the scene building was performed (Fig. 5). This allowed the closer inspection of every part of it and from every angle, without the limitations posed by accessibility issues. It was especially useful in decision-making, as it was possible to draft several restoration scenarios. A similar approach was taken in the case of the roman theatre of Nikopolis (Bilis et al. 2017).

In other instances, 3d modelling is used for the reconstruction and manufacturing of accurate copies of archaeological findings (Efstathiou et al. 2013).

Fig. 5. 3D reconstruction of the Scene building (bottom) and stereometric presentation of the restoration proposal (top). Northwest view. (Karagkounis et al. 2020).

5 3D Modelling and Digital Restoration of the Main Theatre

As it has been described in Sect. 2, Subproject 3 entails the digital restoration of the main theatre of the Ancient Theater of Larissa by creating three-dimensional models of the seats (edolia) in the locations where those are missing using CAD/CAM software.

In order to create 3d models for all 240 new marble seats, laser scanning technologies will be used along with digital photogrammetry methods. The main theatre will be scanned and existing ancient edolia will be digitized.

The subproject is executed in three stages:

1. Surveying and 3D representation of the wider area based on the suggested positions of new seats according to the approved study (Fig. 6).
2. 3D Editing the existing state by

    a. Editing point clouds - creating 3D models using the appropriate techniques (target-to-target, cloud-to-cloud or surface-to-surface registration)
    b. Definition and 3D update of the restoration study with presentation of areas where the installment of new seats is required
    c. Evaluation of the surviving neighboring members and selection study of the horizontal and altitude starting points.
    d. Dimensioning of new edolia.

3. 3D design - representation of each new member in the form of 3D models, in its optimal position.

After the approval of the deliverables, the models will be used for the production of new members, by processing marble blocks with a CNC machine under subproject 5.

An example of the process in shown in Figs. 7a) to 7f) with the 3D representation of part of the monument and the addition of a new fragment.

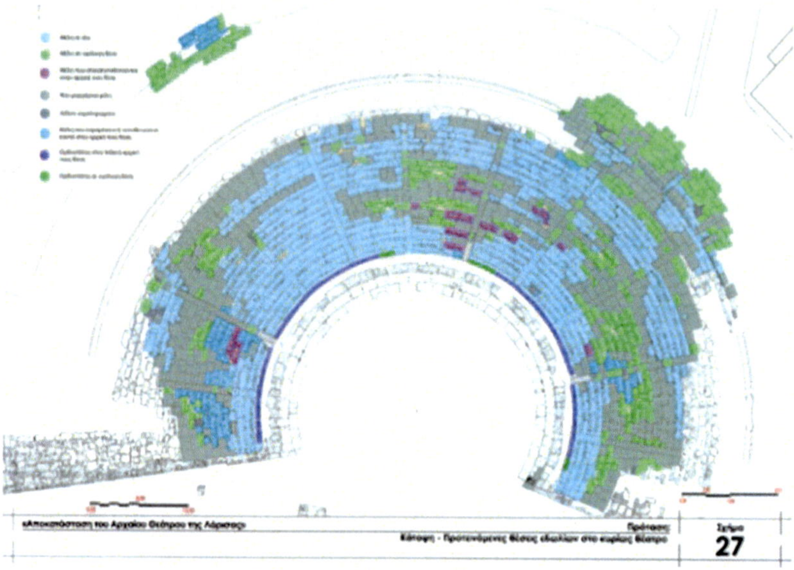

Fig. 6. Suggested positions of ancient and new marble seats (Karagkounis et al 2017)

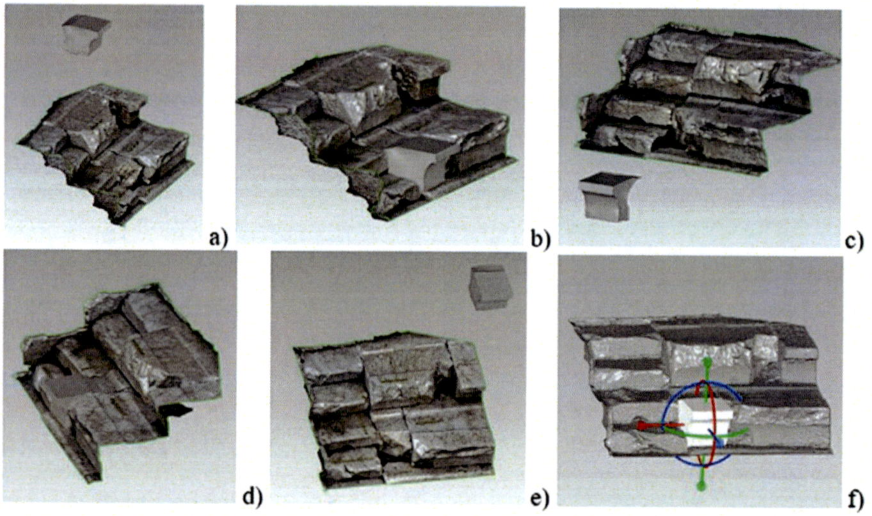

Fig. 7. 3d reconstruction of part of the main theatre and 3d modelling of a new marble fragment

6 Conclusions

Modern technology has provided several tools for the documentation, preservation and restoration of the monuments of our cultural heritage. This paper presents the implementation of digital documentation and 3d modelling in the current project "Restoration of the first Ancient Theater of Larissa – Phase E". Simple digital tools are used to document, monitor and plan the course of restoration works. State of the art techniques will be used to digitize the main theatre and create three dimensional models 240 new architectural members. This allows the accurate and detailed reproduction of these members on CNC machines, saving time in the project's execution. On the other hand, the Ephorate's database is enriched with useful information that will allow the better understanding of the monument and provide new insights to researchers of various specialties.

Acknowledgements. The activity presented has been conducted under the Project "Restoration of the first Ancient Theater of Larissa – Phase E" that is co-financed by the European Union (European Regional Development Fund – ERDF) and Greek national funds through the Regional Operational Program Thessaly 2014–2020 of the National Strategic Reference Framework (NSRF) 2014 – 2020.

References

Bouzakis, K.-D., et al.: 3D-laser scanning of the Parthenon west frieze blocks and their digital assembly based on extracted characteristic geometrical details. J. Archaeol. Sci. Rep. **6**, 94–108 (2016). https://doi.org/10.1016/j.jasrep.2016.01.030

Bilis, T., Kouimtzoglou, T., Magnisali, M., Tokmakidis, P.: The use of 3d scanning and photogrammetry techniques in the case study of the roman theatre of nikopolis. surveying, virtual reconstruction and restoration study. Int. Arch. Photogrammetry Remote Sens. Spat. Inf. Sci. **XLII-2/W3**, 97–103 (2017). https://doi.org/10.5194/isprs-archives-XLII-2-W3-97-2017

Efstathiou, K., Bouzakis, K., Efstathiou, C., Basiakoulis, A.: Investigation, reconstruction and manufacturing of accurate copies of archaeological findings, based on innovated technologies as: 3D scanners, photogrammetrie, X-ray tomography, neutron tomography, CAD systems. In: Symposium: New Technologies In Service To Our Knowledge Of Antiquity, Academy of Institutions and Cultures, Thessaloniki (2013)

Karagkounis, D., Chatzidakis, N., Angeli, E.A., Plakotaris, I., Karnavezos, Ch., Tziafalias, A.: Αποκατάσταση του Αρχαίου Θεάτρου της Λάρισας (Restoration of the First Ancient Theatre of Larissa), unpublished study (2017)

Karagkounis, D., Matsarokos, E., Tsanaktsidou, S.: Αποκατάσταση Σκηνικού Οικοδομήματος του Α' Αρχαίου Θεάτρου Λάρισας – Α' Φάση. (Restoration of the Skene of the First Ancient Theatre of Larissa – Phase A), unpublished study Ephorate of Antiquities of Larissa, Ministry of Culture and Sports (2020)

Tziafalias, A.: Το αρχαίο θέατρο της Λάρισας, *Πρακτικά του Α΄ Ιστορικού-Αρχαιολογικού Συμποσίου*. Larissa **1985**, 162–185 (1985)

Tziafalias A.: Το έργο της ΙΕ΄ ΕΠΚΑ Λάρισας, *Το έργο των Εφορείων Αρχαιοτήτων και Νεωτέρων Μνημείων του ΥΠΠΟ στη Θεσσαλία και την ευρύτερη περιοχή της (1990–1998), 1η Επιστημονική Συνάντηση*, Volos, 91–96 (2000)

Assessment

The Restoration of the Eupalinos Tunnel: The Many Faces of a Composite Project

Costas Papantonopoulos(✉)

25 Martiou 10, 16233 Vyron, Greece

Abstract. The Eupalinos tunnel built in Samos island, Greece, in the 6th century BC, is considered as one of the most important engineering achievements of antiquity. The tunnel, which is the central part of an aqueduct, was excavated from two portals, set about 1 km apart: the two bores met with high degree of accuracy at a depth of 175 m under the ancient fortifications of the Kastro Hill. It should be noted that the tunnel is not only an archaeological monument, but among others a geological one as well. The present paper deals with some of the various aspects of the restoration interventions undertaken between 2013 and 2016, namely: structural restoration, geological consolidation, monument's enhancement, and illumination. With the completion of the project the visitor can now admire the design and the construction of the tunnel in its full length, and understand the ingenious engineering of Eupalinos.

Keywords: Aqueduct · Geological consolidation · Eupalinos · Restoration · Tunnel

1 Introduction

Samos island lies in the eastern part of the Aegean Sea, and of the coast of Asia Minor, from which it is separated by the 1.6 km-wide Mycale Strait. In ancient times Samos was an especially rich and powerful city-state, well known among others for exceptional engineering achievements. The "father of history" Herodotus (481–425 BC) considered the tunnel of Eupalinos, built in the mid-sixth century BC, as the foremost of the three largest technical works in ancient Greece [2]:

> "*I have written at such length of the Samians, because the three greatest works of all the Greeks were engineered by them. The first of these is the tunnel with a mouth at either end driven through the base of a hill nine hundred feet high; the whole tunnel is forty-two hundred feet long, eight feet high and eight feet wide; and throughout the whole of its length there runs a channel thirty feet deep and three feet wide, through which the water coming from an abundant spring is carried by pipes to the city of Samos. The designer of this work was Eupalinus son of Naustrophus, a Megarian. This is one of the three works; the second is a breakwater in the sea enclosing the harbor, sunk one hundred and twenty feet, and more than twelve hundred feet in length. The third Samian work is the temple,*

I. Vayas and F. M. Mazzolani (Eds.): PROHITECH 2021, LNCE 209, pp. 597–614, 2022.
https://doi.org/10.1007/978-3-030-90788-4_47

which is the greatest of all the temples of which we know; its first builder was Rhoecus son of Philes, a Samian."

Although the ruins of the breakwater and the great temple of Hera remained visible until our time, the traces of the tunnel had been lost for eleven centuries. It was as late as 1853 that Victor Guérin discovered the underground collector cistern under the old church of Aghios Ioannis and a part of the trench near the spring. The finding of the tunnel was made by the monks Kyrillos and Theofanis in 1882 under the guidance of the Hegemon (ruler) of Samos Konstantinos Adosidis. The monument was cleaned in a systematic and archaeological way and was partially consolidated by the German Archaeological Institute (DAI) between 1972 and 1973. The explanatory documentation was published in 1995 by Hermann Kienast [3] and the archaeological findings were presented in 2004 by Ulf Jantzen.

The aqueduct consisted of three parts as follows (Fig. 1):

a. North channel

From the cistern, where fifteen large stone pillars support a roof of massive stone slabs, a buried channel winds for 890 m along the hillside to the northern portal of the tunnel. The channel is 60–70 cm wide and about 5.0 m deep. After it had been cut out of the bedrock, it was covered over with stone slabs and burried. The last 150 m of this channel pass as an underground excavation under a small hill using the well-known at that time, shafts and gallery (or qanat) method. According to this method vertical shafts were dug from the surface at intervals of 30–50 m and then linked by a gallery.

b. Central part

The tunnel, which is the central part of an aqueduct, was excavated from two portals, set about 1 km apart: the two bores met with high degree of accuracy at a depth of 175 m under the ancient fortifications of the Kastro Hill. The tunnel consists of a dry corridor having internal dimensions 1.80 × 1.80 m and of a side ditch 0.60 m wide. The depth of the ditch ranged from 4.0 m in its northern section to 8.9 m in its southern one. The gradual differentiation of the ditch's depth provided the necessary longitudinal inclination to convey the water by means of gravitational flow. The ditch itself wasn't excavated as an open trench at its full length, but portions of it are galleries running underneath the main tunnel (Fig. 2). The majority of tunnel's length is unlined, although, parts of it of a cumulative length −230 m were lined along tunnel's sections of weak or potential unstable rock mass.

c. Southern channel

Shortly before the southern portal of the tunnel the water channel diverges eastwards, for more than 1000 m, bringing fresh water to the ancient capital of Samos (today called Pythagoreion). The digging technique for the southern part is similar to that of the northern part, i.e. the above-mentioned shafts and gallery (or qanat) method.

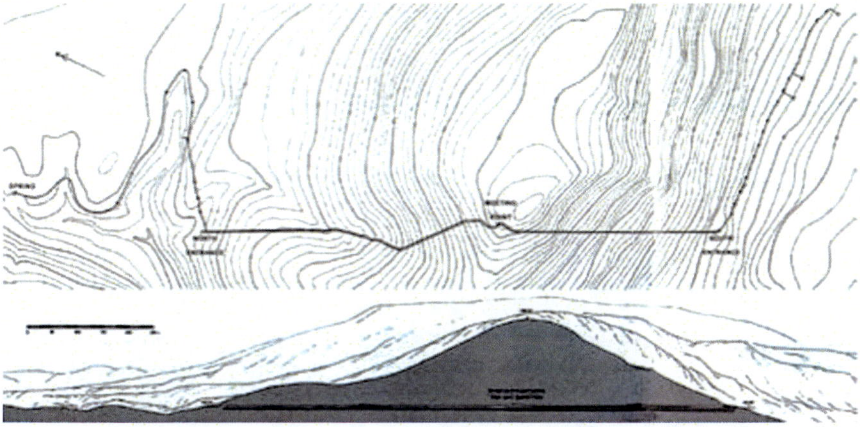

Fig. 1. Plan and longitudinal section of the aqueduct from the spring (left) to the city (right) [4].

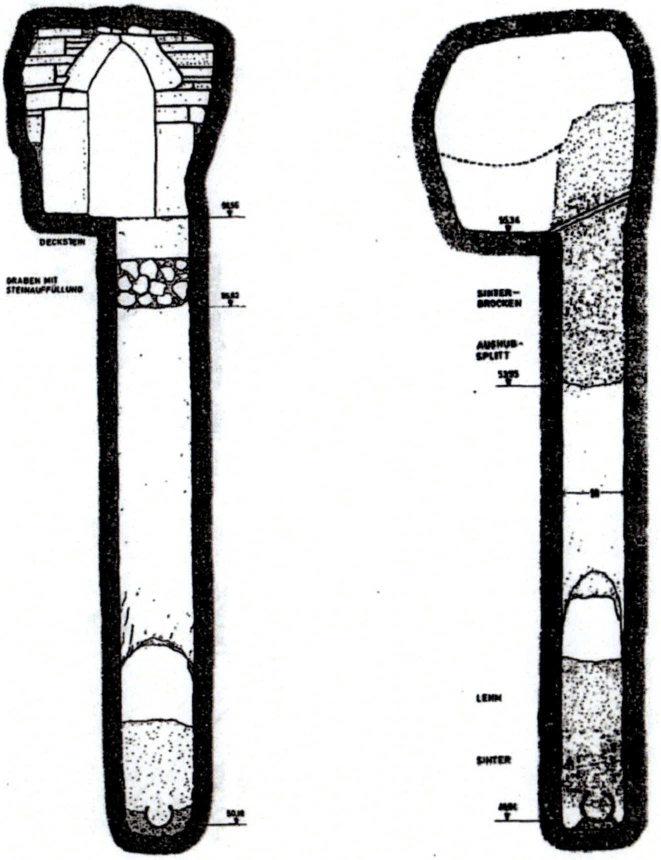

Fig. 2. Cross sections of the tunnel, showing the corridor, the trench and the galleries below [3].

Setting the zero point of the tunnel's chainages (Ch.) on the lowest step of the modern stair of the north portal, the paper deals with some of the various aspects of the restoration interventions undertaken between 2013 and 2016. The presentation is structured in consecutive sections, namely:

a. Structural restoration (Sect. 2).
b. Geological consolidation (Sect. 3).
c. Monument's enhancement (Sect. 4).
d. Illumination (Sect. 5).

2 Structural Restoration

In less than 64.0 m (Ch. 63.9) after entering into the north portal starts the Archaic lining running for a sum length of 153.6 m. There are two parts: one with a length of 133.6 m up to Ch. 197.5, another with a length of 20 m from Ch. 243.5 to Ch. 263.5. It is a structure of great aesthetic value with an admirable durability and has survived in an excellent state of preservation. The lining consists of two vertical walls with a height of 1.3–1.4 m, at a distance 0.60 to 0.68 m. apart; a pointed barrel vault was set upon them, with its highest point at a distance between 1.8 to 1.9 m from the ground surface. The vault is formed by single slabs set opposite to each other, seated on the walls. Their visible surfaces have a curvilinear section with a chord between 56 and 59 cm and a sagitta of 3.5 cm. For such an underground project this subtle curvilinear appearance is astonishing. It should be noted that only four parts of the Archaic lining suffered serious deformations, namely in [6]:

- Ch. 96–100, where partial reconstruction of the east wall with bricks took place during the interventions by the DAI,
- Ch. 109–117, where the two walls have become deformed inwards and the minimum width of the tunnel diminished to about 43 cm,
- Ch. 130–137, Ch. 178.6–183, where the hazardous roof slabs (Fig. 3) were dismantled and reassembled during the restoration.

(a) (b)

Fig. 3. The deformed roof slabs in: (a) Ch. 130–137, and b) Ch. 178.6–183 [5].

For the first time the dismantling of the above mentioned areas gave the opportunity to examine the inner sides of the lining in quite extended parts. The quality of the rock mass in both regions is poor: platy limestone, marls and clay shales. Due to the consecutive failures above the roof slabs, cavities of variable dimensions were formed (Fig. 4). It is obvious that the broken plates of the rock fell on the slabs of the roof and produced the deformations of the lining. Just between Ch. 130 and 136 the width of the tunnel varies significantly with remarkable curvature on the west side (Fig. 4a). The walls were built of stone blocks with all their sites carefully dressed, except the back one, which was left as it was in the quarry. The dressing is of high quality, especially on their front faces, where the variety of the traces of the tools produces an impressive aesthetic effect [6]. The thickness of the walls varies from 0.5 to 1.1 m and that of the slabs from 10 to 10 cm. The space between the stone blocks facing the walls and the surface of the rock behind was filled with stone-plates in a systematic manner and the joints between them were filled with soil. As a result, the wall was built by erecting the stone blocks in front and at the same time filling the gap between and the rock with these stone-plates. Therefore, the procedure of construction was as follows [6]:

- Completion of the excavation and removal of all the material so created.
- A remarkable quantity of building material, that is stone blocks and stone-plates for the walls, and curved slabs for the roof were transported from a nearby quarry through the north portal to the site for the completion of the lining.

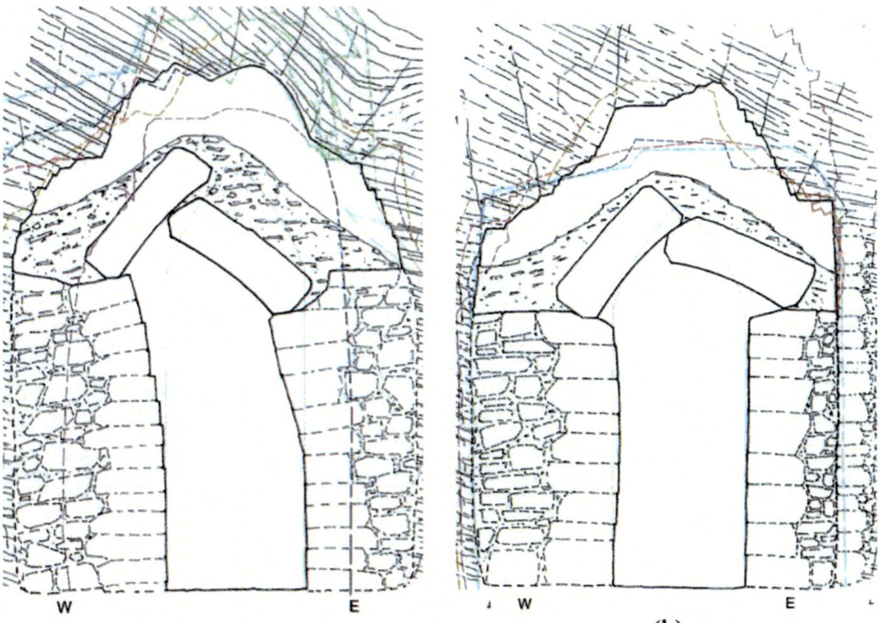

(a) (b)

Fig. 4. Cross sections in: (a) Ch. 130–137, and (b) Ch. 178.6–183 [6].

In the first segment between Ch. 130 to Ch. 137, the following interventions were performed:

- Safe dismantle of 20 slabs, 10 from each side. The slabs were repaired with the aid of titanium pins. One only slab whose a few fragments of the face had survived was replaced with a new stone, in which the survived fragments were incorporated.
- Correction of the deformation of the west wall (Fig. 5a) with partial dismantling, but mainly with the application of lateral forces.
- Installation of 2.0 m long rock anchors Φ16 in a shotcrete mantle reinforced with stainless steel bars.
- Erection of the supporting steel frames (Fig. 5b), and final restitution of the roof (Fig. 7a).

(a)

(b)

Fig. 5. Ch. 130–137: (a) The curvature of the west wall, (b) Steel frames and reinforcement.

(a)

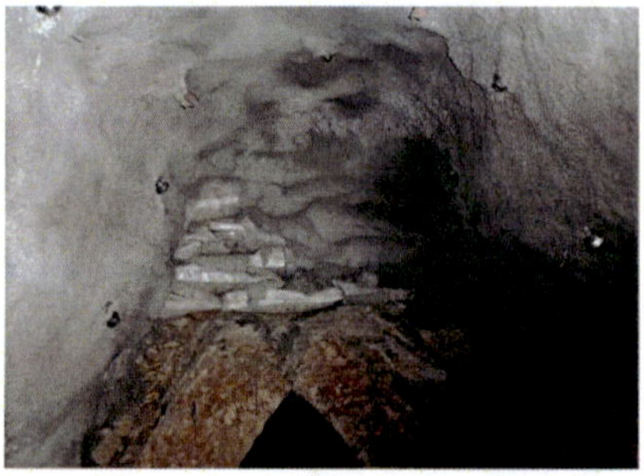

(b)

Fig. 6. Ch. 178.6–183: (a) The ancient wall and the stainless steel mesh, (b) Shotcrete mantle and rock anchors.

In the second segment between Ch. 178.6 to Ch. 183, the following interventions were performed:

- Safe dismantle of 15 slabs, 7 from the east side and 8 from the west.
- Because of the lack of space, the steel frames were omitted in this case. Instead a dense layout of rock anchors in a double layer shotcrete mantle was installed (Fig. 6).
- Final restitution of the dismantled slabs (Fig. 7b).

(a) (b)

Fig. 7. Final restitution: (a) Ch. 130–137, (b) Ch. 178.6–183.

Further downwards to the south, another Archaic lining exists between Ch. 1023 and 1036. This segment, with a length of 13 m., is for crucial importance since it is the principal access path to the tunnel. However, it has remarkable morphological differences from the north lining. The walls are built with voluminous polygonal blocks and the roof is formed by thick slabs with level surfaces. Furthermore, the walls had suffered serious deformations inwards and the slabs had severe structural damages. Many repairs had also been conducted since the time of the discovery of the tunnel and some of the slabs had been reconstructed. The first attempt to proceed with the procedure adopted in north segments, that is dismantling from below, resulted in failures, showing that the overlying material was very loose. For this reason the decision was made to continue with a systematic archaeological excavation from above (Fig. 8). The soil material removed by hand and showed Byzantine ceramic fragments, which means that it had already been moved in the past. Near the bottom of the excavation burnt material was found, which is shown on the left side of Fig. 9a. Since the thickness of the slabs of the roof reaching 48cm is remarkably greater than that of the north lining, it is obvious that the lining of the south portal was a heavy structure, constructed with the 'cut and cover' method.

A total of 27 slabs were disassembled and repaired using titanium rods and new small stone supplements (Fig. 9b). The reassembled slabs were protected against the earth pressure with a new invisible concrete vault supported on a group of micropiles, and reinforced with stainless steel bars (Fig. 10). The distortion of the eastern wall was corrected with lateral pressure and partial dismantling of the upper parts of the wall, thus restoring the original width of the walkway.

(a) (b)

Fig. 8. The excavation in Ch. 1023–1036: (a) Plan [6], and (b) Dismantling the lining.

(a) (b)

Fig. 9. Ch. 1023–1036: (a) The roof slabs during disassembling, (b) Restoration of a slab.

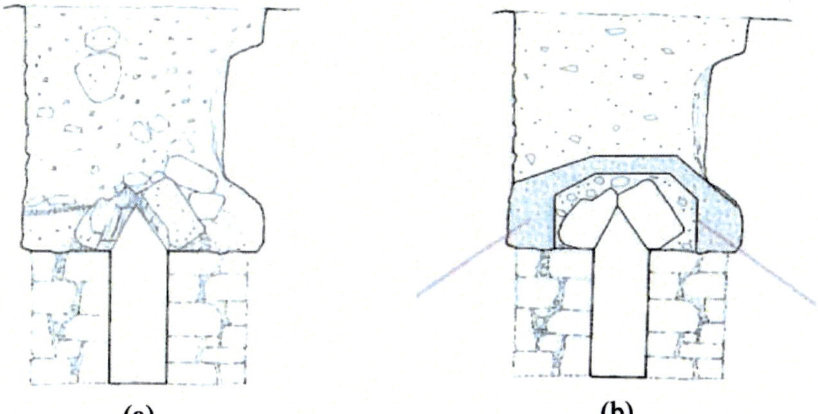

Fig. 10. Ch. 1023–1036: Cross sections (a) before, and (b) after the restoration [6].

3 Geological Consolidation

The consolidation and support of the vulnerable parts of the rock surfaces was made with rock anchors with especially designed spherical ends and a fine wire-rope mesh made of stainless steel AISI316. Depending on the rigorousness of the geological conditions, three sizes of rock anchors were used, with diameters of Φ12 to 16 mm respectively, with a length from 0.5 to 2.5 m. Furthermore three types of flexible wire-rope mesh were also used according to the severity of the instabilities (diameter of rope) × (mesh width) in mm: Type I: 1 × 30, Type II: 2 × 40 Type III: 3 × 60. The wire-rope meshes are connected with cables of Φ3 τo 5 mm, to the anchor plates with special end connectors and clamps (Fig. 11a). It should be underlined that the main concern in fixing the wire-rope mesh was to be under light tension and installed as close as possible to the rock surface, without protruding elements (Fig. 11b). For this reason some special techniques were locally invented and implemented. However, for areas of less vulnerability, the thin and platy rock slabs of the roof were consolidated with pins, injections and the sealing over of open cracks. The great advantage in supporting the progressively collapsed areas with a combination of anchors and thin wire meshes is that all the surfaces remain visible (Figs. 12–13). Hence, a spacious, high-altitude space between Ch. 1009 and 1017, which is a product of constant falls that had become necessary formerly to construct a scaffolding protected corridor, is now free of any obstacles (Fig. 14). Keeping in mind that the tunnel is not only an archaeological monument, but a geological as well, even unworked rock surfaces exposed through damage, which are important only from the geological point of view remain visible [6]. As a result of all the above mentioned procedures at present time the visitor can safely transverse the tunnel and admire both the initial surfaces with the traces of the tools and at the same time the broken natural surfaces with their geological strata. Moreover in invisible parts, e.g. above or behind the linings, the support was made with rock anchors, stainless steel mesh and a shotcrete mantle applied with controlled low pressure.

(a) (b)

Fig. 11. Fixing the wire-rope mesh: (a) end connectors and tensioners, and (b) auxiliary rod anchor for the closest adjustment of the mesh to the rock.

Fig. 12. The wire-rope mesh in Ch. 1009.

Fig. 13. The wire-rope mesh in Ch. 200.

(a) (b)

Fig. 14. Ch. 1009–1013: The site (a) before [5], and (b) after the restoration [1].

It is worth noting the intervention between Ch. 14 and 49 above the cylindrical vault, the so-called Roman lining. This lining was considered by Kienast [3] as a structure of Roman times, taking into account its structural characteristics: plate-like stones built with wide joints, filled with mortar. The space above the vault, which was inaccessible during the preparation of the restoration study, was cleared and it was found that the height of the rock roof was much more than had been estimated. A large cave and a rock chimney above it had formed because of the collapse, reaching a maximum height of about 8 m above the vault (Figs. 15a-b). The thickness of the layers of the fallen materials was up to 6 m., whereas the remaining thickness of the rock mass up to the surface level of the hill had decreased to a low overburden of only 8 m. It should be noted that this type of progressive failure leading to formation of a chimney is typical for this type of rock mass. A large part of the fallen material was removed in an archaeological way: it was found that the lower layer included finds of the Byzantine period, and even a construction with stone steps, revealing the use of the tunnel as a refuge in the seventh century AD. Finally after the removal of the fallen material, the roof and the exposed sides of the rock was supported with rock anchors and reinforced shotcrete mantle (Fig. 15c).

(a)

(b)

(c)

Fig. 15. Roman lining Ch. 14–49: (a) cross section [6] (b) Cavity, and (c) shotcrete mantle.

4 Monument's Enhancement

For the enhancement of the monument the side trench was covered with heavy duty removable gratings made of stainless steel. The bearing system of the gratings consists of transverse beams securely fixed into the east and west sides and of pairs of longitudinal UNP-beams resting on them. The west end of the trench, which was broken and had been temporarily restored with concrete, was cleaned and carefully reconstituted with fresh hydraulic lime mortar, thus preserving and improving the ancient cuttings.

The implementation of the gratings has some distinctive advantages:

- In horizontal level enlarges the floor space available for the movement of the visitors.
- Because the level of the gratings is the same as the ground level of the so-called "bridges" over the galleries, the free height for the visitor is 10-12 cm more than that above rock floor.

- Because all the necessary electrical equipment, i.e. wires, drivers, junction boxes etc. were hided under the gratings, the tunnel is now free for any sighting obstacles (Fig. 16).
- Permits the contemplation of the ancient ceramic water pipes beneath.

(a) (b)

Fig. 16. (a) Electrical equipment supported on the bearing system of the gratings, and (b) Final position of the gratings.

In conclusion, the whole appearance of the tunnel has drastically changed; one can have now an unimpeded view of it, after the replacement of the old projecting and rusted steel bars by the new inox gratings (Fig. 17).

(a) (b)

Fig. 17. Gratings (a) before, and (b) after the restoration [1].

For places of extreme vulnerability an additional structure for the protection of the visitors was provided: the protective canopies. They are light and 'transparent' constructions formed of steel beams and tubes and covered with steel gratings (Fig. 18b). The longest protective canopy was installed, between Ch. 49 and 63.9, which is the intermediate segment between the Roman and Archaic lining. The rock mass in this part of the tunnel is of highly unstable condition; it was characterized as "shaly-thinly bedded alternations", comprising a weak and heterogeneous unit consisting of alternating platy limestones, marls and clay shales. Formerly, the poor condition of the rock forced the DAI excavators to construct a provisional protected corridor from steel frames and sheets (Fig. 18a).

(a) (b)

Fig. 18. Ch. 49–63.9: (a) the old protective corridor [5], and (b) the new canopy [1].

For the appropriate reception of the visitors the whole area was transformed with new lined paths, retaining walls, drainage system, and renovation of the ticket office and the WC, and more with new rest areas, information booths and screens. The historic (1883) building of the south entrance was repaired with a variety of interventions: stone wall grouting, replacement of the old wooden roof, restoration of the broken marbles, new lining on the floor and on the access staircase to the tunnel etc. (Fig. 19). Moreover, in contact with the western wall of the entrance building, a new small house was built housing the new electrical generator.

(a) (b)

Fig. 19. The entrance building: (a) before [5], and (b) after the restoration.

Finally, the origin of the aqueduct has been restored with the rehabilitation of the facade of the cistern, by replacing the old shaft by a new window that allows viewing the archaic stone pillars and the massive stone slabs of the roof (Fig. 20).

(a) (b)

Fig. 20. The spring of the aqueduct: (a) the old shaft leading to the archaic cistern, and (b) the new facade with the window.

5 Illumination

Electrical and mechanical facilities such as communication systems and lighting were installed into the tunnel. At the present time the whole tunnel is illuminated with soft line

LED lights, which respect the underground character of the monument without losing their functionality (Figs. 21 and 22a). Locally at the places where Archaic red marks and inscriptions are found, LED spots have been used (Fig. 23). The trench and the water pipes are also illuminated with strong LED spots, installed under the gratings (Fig. 22b). As a result the visitor can now understand the main point of the construction of the tunnel, which is the installation of the pipes of the aqueduct at the bottom of the trench. In the ancient linings all the wires leading to the spots were installed into inox tubes, very close to the eastern wall (Fig. 24).

(a) (b)

Fig. 21. The Archaic lining of the entrance (Ch. 1023–1036). Illumination (a) before [5], and (b) after the restoration [1].

(a) (b)

Fig. 22. Illumination of: (a) the tunnel, and (b) the trench.

(a) (b)

Fig. 23. The new illumination of (a) an archaic red mark, and (b) the famous inscription ΠΑΡΑΔΕΓΜΑ (Ch. 174) [1].

(a) (b)

Fig. 24. The lighting in: (a) the Roman, and (b) the Archaic lining.

6 Conclusions

The Eupalinos tunnel built in Samos island, Greece, in the 6th century BC, is considered as one of the most important engineering achievements of antiquity. The tunnel, which is the central part of an aqueduct, was excavated from two portals, set about 1km apart: the two bores met with high degree of accuracy at a depth of 175 m under the ancient fortifications of the Kastro Hill. It should be noted that the tunnel is not only an archaeological monument, but among others a geological one as well. The present paper deals with some of the various aspects of the restoration interventions undertaken by an interdisciplinary group between 2013 and 2016, namely: structural restoration, geological consolidation, monument's enhancement, and illumination. With the completion of the project the visitor can now admire the design and the construction of the tunnel in its full length, and understand the ingenious engineering of Eupalinos [6].

Acknowledgments. I would like to express my thanks to Dr. Eng. C. Zambas and Dr. Eng. G. Dounias, who both gave me the opportunity to work in situ for the project of restoration of Eupalinos tunnel. It should be mentioned that the presentation is mainly based on the final technical report of the project [1], and the studies of C. Zambas [5, 6].

References

1. Dounias G., Zambas C.: Παροχή Υπηρεσιών Τεχνικού Συμβούλου για την επίβλεψη του έργου «Αποκατάσταση και ανάδειξη του Ευπαλίνειου Υδραγωγείου στο Πυθαγόρειο Σάμου». Τελική αναφορά παροχής υπηρεσιών (unpublished final report of the project, Athens 2016) (2016)
2. Herodotus: Histories 3.60, with an English translation by A. D. Godley. Cambridge. Harvard University Press (1920)
3. Kienast H.: Die Wasserleitung des Eupalinos auf Samos, Samos 19 (Bonn 1995) (1995)
4. Tokmakides, K.: Surveying the Eupalinian Aqueduct in Samos Island. In: 22nd CIPA Symposium, October 11–15, 2009, Kyoto, Japan (2009)
5. Zambas C., Thomas G., Doudoumi E.: Οριστική μελέτη αποκατάστασης, οριστική Αρχιτεκτονική μελέτη διαμόρφωσης χώρων και μελέτη εφαρμογής του Ευπαλινείου Υδρα-γωγίου στη νήσο Σάμο. Μελέτη εφαρμογής αποκατάστασης και ανάδειξης της σήραγγας (unpublished report, Thessaloniki 2010) (2010)
6. Zambas, C.: More light in the tunnel of Eupalinos, Athenische Mitteilungen DAI, Sonderdruck, Band 131/132, 2016–2017, 100–145 (2017)

Preliminary Seismic Risk Assessment of Monolithic Columns of the Aphaia Temple in Aegina

Vasileios E. Melissianos(✉), Christos G. Lachanas, and Dimitrios Vamvatsikos

Institute of Steel Structures, School of Civil Engineering,
National Technical University of Athens, 9, Iroon Polytecheiou str., Zografou Campus,
GR-15780 Athens, Greece
{melissia,divamva}@mail.ntua.gr, lahanasch@central.ntua.gr

Abstract. Studies on the seismic assessment of monumental structures of antiquity are mostly focused on the estimation of the structural behavior and do not involve the pertinent uncertainties. Towards filling this research gap, a preliminary seismic risk assessment of monolithic columns of the Aphaia Temple in Aegina island, Greece, is presented. A comprehensive application of the framework of Performance-Based Earthquake Engineering is carried out. Site-specific seismic hazard estimation is performed at first by employing Probabilistic Seismic Hazard Analysis. The rocking column is analyzed under seismic excitation by numerically solving the equation of motion in order to extract the fragility curves. Finally, the convolving of seismic hazard and structural response yields the seismic risk of the column that is used to estimate the mean annual rate of exceeding predefined limit states that are associated with damage and collapse of the column.

Keywords: Seismic hazard · Record selection · Rocking block · Limit states · Fragility curves · Seismic risk assessment

1 Introduction

The protection of cultural heritage has drawn the attention of engineers, scientists, and the general public during the last decades. In Greece, numerous monuments of classical antiquity are scattered throughout its territory. Most of these monuments have been exposed to various environmental actions during their lifetime and consequently, they are already in some stage of deterioration. The evaluation of their seismic performance is a very complex mathematical and computational problem as the marble structural elements of classical antiquity monuments are typically not rigidly connected and thus susceptible to rocking.

Numerous studies have been conducted on the seismic assessment of monuments, indicatively, Dasiou et al. (2009), Psycharis et al. (2013), Makris and Vassiliou (2013), Psycharis (2018). However, most of these studies are limited to the estimation of structural behavior and do not involve the incorporation of pertinent uncertainties. The aim of the present study is to contribute towards the estimation of the seismic risk of monuments

I. Vayas and F. M. Mazzolani (Eds.): PROHITECH 2021, LNCE 209, pp. 615–624, 2022.
https://doi.org/10.1007/978-3-030-90788-4_48

of classical antiquity. To that effect, a comprehensive application of Performance-Based Earthquake Engineering, initially developed by Cornell and Krawinkler (2000), is carried out, comprising four inter-related steps: (1) estimation of the seismic hazard in terms of a scalar intensity measure via a comprehensive seismic source model, (2) use simplified structural models to enable a multi-fidelity estimation of the response distribution, as characterized by appropriate engineering demand parameters that can be connected to damage, (3) definition of appropriate limit-state thresholds and the corresponding aleatory and epistemic uncertainties to allow determining discrete damage states and associated fragility curves, and (4) assessment of the seismic risk in terms of the mean annual frequency of exceeding a limit state via the convolution of fragility and seismic hazard. The methodology is applied to monolithic columns of the Temple of Aphaia, Greece.

The Temple of Aphaia is located within a sanctuary complex in Aegina island, Greece, which is located 27 km southwest of Athens in the Saronic Gulf. The Temple of Aphaia (Fig. 1) was dedicated to the mother-goddess Aphaia and was founded circa 500 BC in the Late Archaic phase (510 to 470 BC) of the sanctuary. The Temple was a hexastyle peripteral Doric order structure of a 6 × 12 column plan on a 15.5 m × 30.5 m platform (www.wikipedia.org). All columns, except three, were monolithic. The remains of the Temple today include free-standing columns, colonnades of two or more columns with epistyles, and an internal segment with a two-story colonnade. In general, the remnants of the Temple have suffered non-negligible damage.

Fig. 1. North view of the Temple of Aphaia, Aegina Island, Greece (courtesy of the authors)

The selected monolithic free-standing column is located in the southeast corner of the Aphaia Temple and has a total height of 5.29 m including the capital. The column is presented in Fig. 2(a) and its drawing is shown in Fig. 2(b). It is noted that within the present preliminary seismic risk assessment, fractures and damages of the column are not considered.

(a) (b)

Fig. 2. Selected free-standing monolithic column of the Aphaia Temple: (a) column photo (courtesy of the authors) and (b) drawing

2 Seismic Hazard

Ancient monuments are structures of significant social and cultural importance and when it comes to seismic assessment, the application of probabilistic methods is practically a one-way street. The appropriate tool to estimate the seismic hazard is the Probabilistic Seismic Hazard Analysis (PSHA) (Baker 2008). In PSHA, two approaches are available, the so-called classic (Cornell 1968) and the event-based. Herein, the classical approach is employed, as only a single site is considered. Cotemporaneous consequences over multiple sites can be evaluated via an event-by-event view of seismic ruptures, which is beyond the scope of our study.

PSHA is performed for the IM of *AvgSA*, which is the geometric mean of the log spectral acceleration $Sa(T_1)$ at a set of periods of interest (Bianchini et al. 2009). The range of periods is set from 0.1 s to 1.5 s. It is noted that ancient columns do not possess natural modes in the classical sense since the period of free vibrations is amplitude-dependent (Psycharis 2018) and consequently T_1 is not defined. Still, the lower periods tend to govern early damage, as the peak ground acceleration determines uplift, while longer periods (e.g., moderate periods indicative of peak ground velocity) tend to correlate well with overturning (Giouvanidis and Dimitrakopoulos 2018). Thus, the range of periods selected stands as a reasonable engineering assumption.

The seismic hazard analysis for the site of interest (Temple of Aphaia) is performed using the open-source platform OpenQuake (GEM 2016) and using the latest European seismic source model (SHARE) (Woessner et al. 2015). A comprehensive logic tree is available in SHARE to account for epistemic uncertainties. In this study, only the area source model is employed together with the Ground Motion Prediction Equation (GMPE) of Boore and Atkinson (2008). Calculations were performed for soil type "rock" ($V_{s30} = 800$ m/s) and the resulting hazard curve is shown in Fig. 3.

Then, 16 hazard levels are considered, corresponding to probabilities of exceedance ranging from 0.0001 to 0.90 in 50 years, or equivalently to mean annual frequencies ranging from 0.046052 to 0.000020. The hazard levels and the corresponding values of *AvgSA* obtained from the hazard curve (Fig. 3) are tabulated in Table 1.

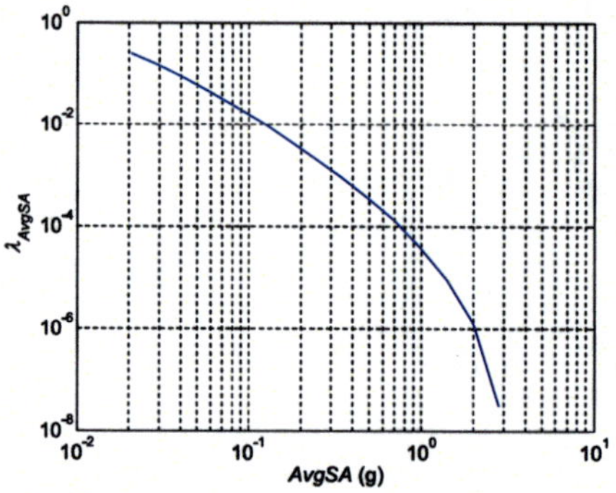

Fig. 3. Seismic hazard curve for the Temple of Aphaia site (island of Aegina), displaying the mean annual frequency (λ_{AvgSA}) of exceeding *AvgSA* (intensity measure)

Table 1. Hazard levels obtained from seismic hazard curve

No	Mean annual frequency	*AvgSA* (g)	No	Mean annual frequency	*AvgSA* (g)
1	0.046052	0.057	9	0.003250	0.212
2	0.032189	0.069	10	0.002107	0.247
3	0.024079	0.081	11	0.001026	0.333
4	0.018326	0.094	12	0.000404	0.477
5	0.013863	0.108	13	0.000201	0.632
6	0.008616	0.136	14	0.000100	0.797
7	0.005754	0.162	15	0.000040	0.991
8	0.004463	0.182	16	0.000020	1.253

Subsequently, hazard disaggregation (Bazzuro and Cornell 1999) is performed using OpenQuake for the predefined hazard levels in order to obtain the magnitude – distance – epsilon distribution of the events that contribute to each hazard level. On the basis of this data, the Conditional Spectrum methodology (Lin et al. 2013) was employed to perform hazard-consistent record selection based on AvgSA (Kohrangi et al. 2017) from the PEER database (https://ngawest2.berkeley.edu/site). In particular, 30 records have been selected for each hazard level.

3 Structural Analysis

The response of classical columns to earthquake excitation follows the dynamics of rocking rigid blocks. This rocking response is highly nonlinear and extremely sensitive to initial conditions and system parameters. Housner (1963) was the first to carry out a systematic investigation of the problem and developed the equation for the rocking response of a rigid block:

$$I_o\ddot{\theta} + mgR\sin[\alpha - \theta\text{sgn}(\theta)]\text{sgn}(\theta) = +m\ddot{u}_g R\cos[\alpha - \theta\text{sgn}(\theta)] \tag{1}$$

where I_o is the moment of inertia of the block with respect to its point of rotation, $\ddot{u}_g$ is the earthquake excitation, $\text{sgn}(\theta)$ denotes the signum function, R is the radial distance of the center of mass to the center of rotation, and α is the block (or column) slenderness. Equation (1) is applied for the structural analysis of the column by assuming that neither sliding nor bouncing takes place. Housner (1963) has developed Eq. (1) for rocking rigid blocks.

The shape of the monolithic column including the capital is geometrically transformed into an equivalent rectangular shape, considering that the height difference between the center of mass of the original column and the equivalent shape is lower than 5%, as schematically illustrated in Fig. 4(a), (b). It is noted that this is a typical transformation carried out by researchers in case of examining the rocking response of columns using analytical expressions. Under the assumption that the capital is rigidly connected to the column, Makris and Kampas (2016) and Kavvadias et al. (2017) have, also, examined a monolithic column of the Aphaia Temple using the same approach. Furthermore and regarding the selection of the shape, it has been found that the difference between the rectangular and the cylindrical shape on the mass moment of inertia is below 0.21%. If a truncated cone shape was adopted, as indicatively by Manos and Demosthenous (1996), then the capital would have to be ignored.

The main geometric characteristics of the equivalent block are $R = 2.687$ and $\alpha = 0.178$. For the rocking response, the coefficient of restitution is equal to $1\text{–}3/2\ \sin^2\alpha \approx 0.96 < 1$ (Housner 1963) and is defined as the ratio between the post-impact and the pre-impact angular velocity. Housner's coefficient of restitution is a geometric property and can be calculated a priori for the rocking block if there is no material dissipation (Ĉeh et al. 2018). The response of the column is evaluated via the Engineering Demand Parameter (EDP), which is the quantity used to predict the damage. The selected EDP is the peak displacement at the capital (i.e. at the top) normalized by the base diameter (Psycharis et al. 2013):

$$u_{top} = \frac{\max_t(u_{top})}{D_{\text{base}}}. \tag{2}$$

The EDP definition is shown in Fig. 4(c), where D_{base} is the diameter at the column base, θ is the column rotation, R is the radial distance of the center of mass to the center of rotation, h is the column height and α is the column slenderness. The adopted EDP after Psycharis et al. (2013) has a physical meaning in terms that collapse is probable to occur for $u_{\text{top}} > 1$ because the column's weight becomes a pure overturning force, while

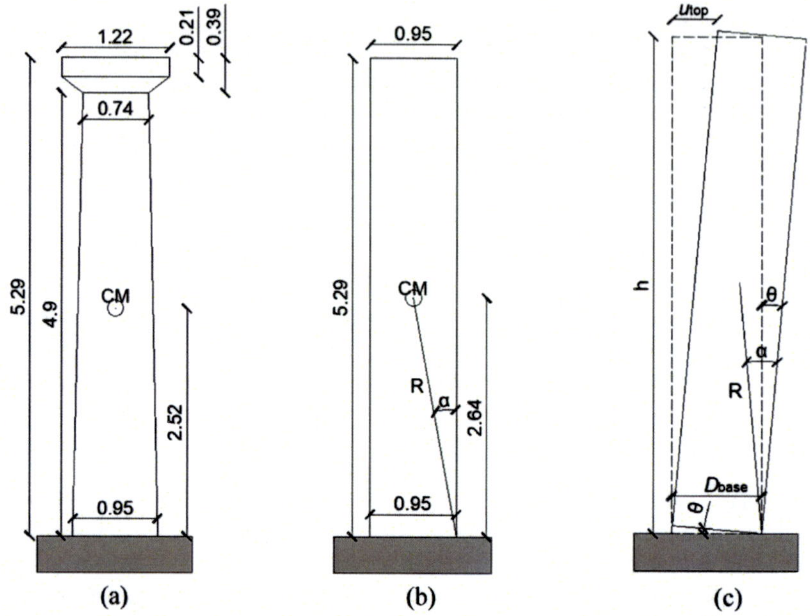

Fig. 4. (a) Original column shape and (b) equivalent rectangular shape, (c) EDP definition for the monolithic column

for $u_{top} = 0.30$ the maximum displacement at the top is one-third of the base diameter, indicating no danger of collapse.

In order to characterize the damage of a structure, it is necessary to introduce discrete damage states or limit states (LS), each with distinct consequences to the structure (Bakalis and Vamvatsikos 2018). The performance criteria (limit states) proposed by Psycharis et al. (2013) for the collapse risk of ancient columns are adopted (Table 2). It is noted that these performance criteria have been proposed for multidrum columns, rather than the monolithic one employed here, yet they are still indicative of similar levels of damage.

Table 2. Performance criteria and associated limit states for single classical antiquity columns after Psycharis et al. (2013)

LS	u_{top}	Performance level
LS1	0.15	Damage limitation
LS2	0.35	Significant damage
LS3	1.00	Near collapse

Structural analysis of the column is performed by solving Eq. (1) for every record selected (30 in total) at each hazard level (16 in total). The numerical solution of Eq. (1) has been carried out using the software of Prof. Michalis Vassiliou at ETH Zurich (https://n.ethz.ch/~mvassili/scripts.html). In fact, a stripe analysis is performed, where each of the 30 records selected per IM level is scaled to match said IM level and used to run response history analysis. The results are shown in Fig. 5, where the intensity measure (*AvgSA*) is plotted on the vertical axis and the engineering demand parameter (u_{top}) on the horizontal axis. Each one of the 16 stripes (horizontal discrete set of results) corresponds to a single hazard level and it comprises 30 EDP values from the 30 ground motion records employed.

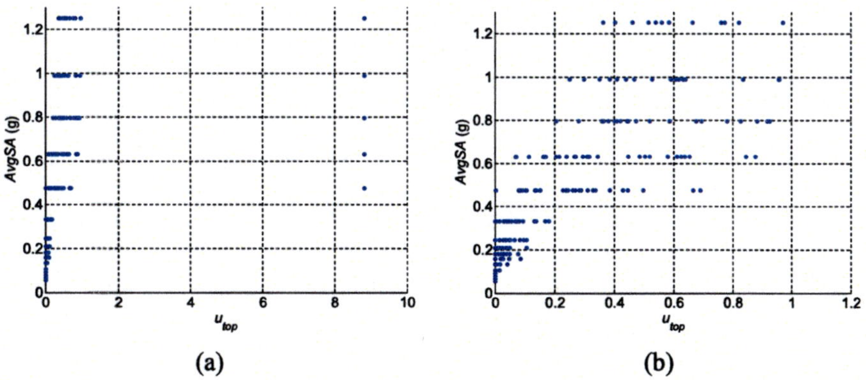

Fig. 5. Stripe analysis results for (a) the full range of EDP values, where the values at $u_{top} = 9$ correspond to collapse, and (b) curtailed range of EDP values, excluding collapse points

After performing the structural analysis of the column, the next step is the fragility definition. A fragility curve is defined as the probability function of violating a predefined limit state given the IM (Bakalis and Vamvatsikos 2018) and is mathematically expressed as follows taking into account that LS violation is defined only by u_{top} that is a single EDP:

$$FS_{LS}(IM) = P[LS\text{violated}|IM] = P[D > C|IM] \quad (3)$$

where *IM* is the intensity measure, while *D* and *C* are the structure's demand and capacity, respectively. The fragility curves are depicted in Fig. 6 for the limit states of Table 2. At each hazard level, a set of 30 analyses has been carried out, introducing record-to-record variability in the structural response. This source of uncertainty is incorporated via the lognormal fitting of the discrete probability data points back into a single continuous cumulative distribution function (CDF). The *AvgSA* (intensity measure) is plotted on the horizontal axis and the probability of limit state violation via Eq. (3) is plotted on the vertical axis as *P(EDP > edp | AvgSA)*, where *EDP* is the demand and *edp* is the threshold (capacity).

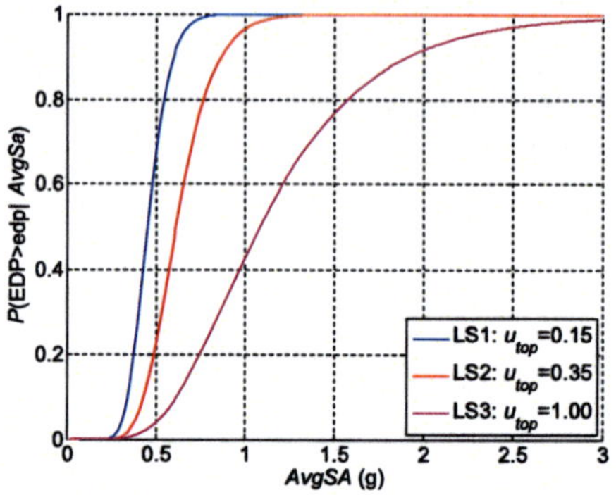

Fig. 6. Column fragility curves

4 Seismic Risk

The seismic risk assessment is based on the simplified expression of the Cornell-Krawinkler framing equation adopted by the Pacific Earthquake Engineering Research (PEER) Center (Cornell and Krawinkler 2000), which can be reduced to a single integral equation (Vamvatsikos and Cornell 2006) for convolving seismic hazard with the fragility in order to estimate the mean annual frequency (MAF):

$$\lambda_{LS} = \int P(EDP > edp|IM)|d\lambda(IM)| \tag{4}$$

where EDP is u_{top}, IM is *AvgSA* and $P(EDP > edp \mid IM)$ is the fragility corresponding to limit state LS. The resulting MAF values are listed in Table 3. It is noted that demand uncertainty is indirectly incorporated via the lognormal fitting at the fragility curves, while capacity uncertainty is not considered due to the lack of data. When more information is available, this source of uncertainty can also be incorporated by appropriately inflating the dispersion of the fragility curve (Bakalis and Vamvatsikos 2018).

In order to make results more comprehensible, the MAF of exceeding LS is converted into the probability of exceedance (POE) in 50 years (P_{50}) using the exponential distribution CDF and assuming a Poisson process:

$$P_{50} = 1 - e^{-50MAF} \tag{5}$$

Table 3. Mean annual frequency and probability of exceeding limit states

LS	Performance level	P_{50}	MAF of exceeding LS
LS1	Damage limitation	2.59%	0.000524
LS2	Significant damage	1.20%	0.000251
LS3	Near collapse	0.37%	0.000074

5 Conclusions

Seismic risk assessment within the framework of Performance-based Earthquake engineering is the appropriate tool to estimate the seismic vulnerability of ancient monuments. This approach can provide the necessary information in quantified terms regarding the actual seismic risk of monumental structures, helping to support decision making for any given monument. Herein, this was employed to study a free-standing monolithic column from the Temple of Aphaia in Aegina island, Greece. Its application can be based on a simple model of response, but it requires comprehensive incorporation of all sources of uncertainty, including both epistemic and aleatory sources influencing demand and capacity. Most importantly, though, to control bias, one needs to carefully choose ground motion records, as well as employ an efficient intensity measure to connect hazard and fragility.

The same framework can be extended to cover all adjacent monuments, e.g., the entire Temple and Sanctuary and the nearby museum, where identical ground motions and fully correlated hazard is expected. Extension to multiple sites over an extended region can be based on the same fragility analyses yet also requires further considerations of spatial variability of hazard, to be treated by the more comprehensive event-based approach of PSHA.

Acknowledgments. This research has been co-financed by the European Regional Development Fund of the European Union and Greek national funds through the Operational Program Competitiveness, Entrepreneurship and Innovation, under the call RESEARCH – CREATE – INNOVATE (project code: T1EDK-00956), project: "ARCHYTAS: Archetypal telemetry and decision support system for the protection of monumental structures". The authors thank, also, Dr. Michalis Vassiliou, Assistant Professor at ETH Zurich, for his support on modeling the rocking response of columns.

References

Bakalis, K., Vamvatsikos, D.: Seismic fragility functions via nonlinear response history analysis. J. Struct. Eng. **144**(10), 04018181 (2018)

Baker, J.W.: An introduction to probabilistic seismic hazard analysis. Version 1.3 (2008). https://web.stanford.edu/~bakerjw/Publications/Baker_(2008)_Intro_to_PSHA_v1_3.pdf

Bazzuro, P., Cornell, C.A.: Disaggregation of seismic hazard. Bull. Seismol. Soc. Am. **89**(2), 501–520 (1999)

Bianchini, M., Diotallevi, P., Baker, J.W.: Prediction of inelastic structural response using an average of spectral accelerations. In: 10th International Conference on Structural Safety and Reliability (ICOSSAR09), Osaka, Japan, 13–17 Sept (2009)

Boore, D.M., Atkinson, G.M.: Ground-motion prediction equations for the average horizontal component of PGA, PGV, and 5% damped PSA at spectral periods between 0.01s and 10.0s. Earthquake Spectra **24**(1), 99–138 (2008)

Ĉeh, N., Jelenić, G., Bićanić, N.: Analysis of restitution in rocking of single blocks. Acta Mech. **229**(11), 4623–4642 (2018)

Cornell, C.A.: Engineering seismic risk analysis. Bull. Seismol. Soc. Am. **58**(5), 1583–1606 (1968)

Cornell, C.A., Krawinkler, H.: Progress and challenges in seismic performance assessment. PEER Center News **3**(2), 1–4 (2000)

Dasiou, M.-E., Psycharis, I., Vayas, I.: Numerical investigation of the seismic response of Parthenon, Greece. In: Protection of Historical Buildings: PROHITECH 09, Rome, Italy, 21–24 Jun (2009)

Global Earthquake Model (GEM). OpenQuake Engine User Instruction Manual, Version 1.9.0, Pavia (2016)

Giouvanidis, A.I., Dimitrakopoulos, E.G.: Rocking amplification and strong-motion duration. Earthquake Eng. Struct. Dynam. **47**, 2094–2116 (2018)

Housner, H.H.: The behavior of inverted pendulum structures during earthquakes. Bull. Seismol. Soc. Am. **53**, 404–417 (1963)

Kavvadias, I.-E., Vasiliadis, L.K., Elenas, A.: Seismic response parametric study of ancient rocking columns. Int. J. Architect. Heritag. **11**(6), 791–804 (2017)

Kohrangi, M., Bazzurro, P., Vamvatsikos, D., Spillatura, A.: Conditional spectrum based ground motion record selection using average spectral acceleration. Earthquake Eng. Struct. Dynam. **46**(10), 1667–1685 (2017)

Lin, T., Harmsen, S.C., Baker, J.W., Luco, N.: Conditional spectrum computation incorporating multiple causal earthquakes and ground motion prediction models. Bull. Seismol. Soc. Am. **103**(2A), 1103–1116 (2013)

Makris, N., Kampas, G.: Size versus slenderness: two competing parameters in the seismic stability of free-standing rocking columns. Bull. Seismol. Soc. Am. **106**(1), 104–122 (2016)

Makris, N., Vassiliou, M.F.: Planar rocking response and stability analysis of an array of free-standing columns capped with a freely supported rigid beam. Earthquake Eng. Struct. Dynam. **42**(3), 431–449 (2013)

Manos, G.C., Demosthenous, M.: Study of the dynamic response of models of ancient columns or colonnades subjected to horizontal base motions. Eleventh World Conference on Earthquake Engineering, Acapulco, Mexico, 23–28 Jun (1996)

Psycharis, I.N.: Seismic vulnerability of classical monuments. In: Pitilakis, K. (ed.) Recent Advances in Earthquake Engineering in Europe, pp. 563–582. Springer International Publishing, Cham (2018). https://doi.org/10.1007/978-3-319-75741-4_24

Psycharis, I.N., Fragiadakis, M., Stefanou, I.: Seismic reliability assessment of classical columns subjected to near-fault ground motions. Earthquake Eng. Struct. Dynam. **42**(14), 2061–2079 (2013)

Vamvatsikos, D., Cornell, C.A.: Direct estimation of the seismic demand and capacity of oscillators with multi-linear static pushovers through incremental dynamic analysis. Earthquake Eng. Struct. Dynam. **35**(9), 1097–1117 (2006)

Woessner, J., et al.: The 2013 European Seismic Hazard Model: key components and results. Bull. Earthq. Eng. **13**(12), 3553–3596 (2015)

Spatial Variability of Ground Motion Hazard and Preliminary Regional Damage Assessment of Ancient Monuments

Christos G. Lachanas(✉), Vasileios E. Melissianos, and Dimitrios Vamvatsikos

Institute of Steel Structures, School of Civil Enginnering, National Technical University of Athens, 9, Iroon Polytecheiou Street, Zografou Campus, 15780 Athens, Greece
lahanasch@central.ntua.gr, {melissia,divamva}@mail.ntua.gr

Abstract. The seismic hazard is a spatial variable and its distribution depends on the site and soil conditions, the distance from the influencing faults, their geometric characteristics etc. To characterize a future seismic event, multiple correlated Intensity Measure (IM) fields are employed. In Greece and especially in the Attica region, there are numerous monuments of classical antiquity that are spread throughout the territory, e.g., the Temple of Poseidon in Sounio, the Ancient Walls of Piraeus etc., and not only in the center of the city of Athens, where the well-known Parthenon, the Temple of Olympian Zeus, etc. lie. To capture the ensemble risk of such widely distributed antiquities a regional risk assessment approach is adopted, based on the principles of the Performance-Based Earthquake Engineering. Event-Based Probabilistic Seismic Hazard Analysis (PSHA) is used as basis, where the distribution of the IM fields is evaluated for a stochastic event set, spanning thousands of realizations, allowing the estimation of damages over an entire region on an event-by-event basis. A basic application is offered by considering a typical single multi-drum column at each site of interest to assess its performance. Different fragility representation schemes are employed, and their consequences are compared for both the event-based and the classical approach of PSHA as a first decisive step towards the formulation of a comprehensive methodology for the assessment of seismic damages of ancient monuments within a region.

Keywords: Seismic hazard · Risk assessment · Classic PSHA · Event-based PSHA · Spatial variability · Performance-based earthquake engineering · Ancient column

1 Introduction

The seismic risk assessment of monumental structures is a complex task that involves considerations of seismic hazard, structural fragility and resulting consequences, compounded by multiple sources of uncertainty. As a first step towards simplification, the usual practice is to view each site and structure separately. However, in regions where multiple cultural heritage sites are within a few kilometers of each other, they cannot be viewed separately. Attica is such a region, roughly 50 by 70 km with numerous

I. Vayas and F. M. Mazzolani (Eds.): PROHITECH 2021, LNCE 209, pp. 625–637, 2022.
https://doi.org/10.1007/978-3-030-90788-4_49

monuments of classical antiquity, as well as the accompanying museums. Most lie at the center of Athens with the Parthenon being the most emblematic one, while a fair number is spread throughout the surrounding territory of Attica. There are well-known monuments in the west and southwest side (Elefsina, the Temple of Aphaia etc.), in the east and southeast side (the Temple of Artemis, the Temple of Poseidon etc.) and in the north side (the Amphiareio at Oropos). They are all influenced by the same seismic events, each site to a different degree depending on parameters such as the soil conditions, the distance from the rupture, etc.

Thus, while long-term averages of losses per site can still be considered at each site separately, losses per event will be compounded with the simultaneous occurrence of damages severely stretching available resources. So, there is need for a simple, but accurate at the same time, computational tool for risk estimation and hierarchization of the needs for potential remedial actions. Before fully implementing such a tool, though, we want to verify our approach for assessing risk on an event-to-event basis by comparing it against the classical "all-event" approach based on the principals of Performance Based Earthquake Engineering (PBEE) (Cornell and Krawinkler 2000) and the Probabilistic Seismic Hazard Analysis (PSHA) (Cornell 1968; Bommer 2002; Baker 2008).

On the front of monument fragility and vulnerability, one may develop his/her own functions or employ data available in the literature. In both cases, one needs to be aware of the consequences of the finer details in fragility modeling to the overall resulting risk. This fact becomes more pronounced when structures of classical antiquity are investigated, thus some ground rules need to be set, especially regarding rocking columns that comprise most of the cases of interest. Thus, we want to test an approach for processing literature fragility functions (Psycharis et al. 2013) to provide homogeneous and rational results.

So, this entire paper is just an exploration of how a stakeholder (ephorate of antiquities) can plan, expect and understand the actual risk posed by temples or monuments in general under seismic threat and how one is going to quantify this for an entire region, rather than a single site.

2 Hazard Definition

Both classic and event-based PSHA with spatial correlation (Jayaram and Baker 2009) approaches were employed to calculate the seismic hazard. The classic PSHA method is based on the direct estimation of the mean annual frequency (MAF) of exceeding each value x of a selected Intensity Measure (IM), $\lambda(IM > x)$. This is performed by considering cumulatively all the potential ruptures and events for a specific site using the causative faults and associated distributions of the IM given the rupture and site characteristics, as provided by Ground Motion Prediction Equations (GMPE). On the other hand, the event-based approach considers a stochastic set of possible ruptures during an extended time period on a one-by-one basis. Each rupture represents a possible seismic event and generates spatially distributed values of the IM, leading to a set of so-called ground motion IM fields (Fig. 3). Of the two approaches, the event-based needs more computational effort, as it requires the investigation of events spanning tens of thousands of years in order to include the rarer ground motions. At the same time, it offers more comprehensive data on a per event basis. While long term averages (i.e., estimates of the MAF) are more or less the same per each site regardless of the computational approach, the event-basis option can quantify the correlation of ground motion at adjacent sites as would be witnessed in any real individual event. Thus, individual events may produce either consistently higher or consistently lower damages over an extended area, which would otherwise be "smeared" to a common average per classic PSHA. Thus, higher variability in the ensemble results may be expected (Weatherill et al. 2015).

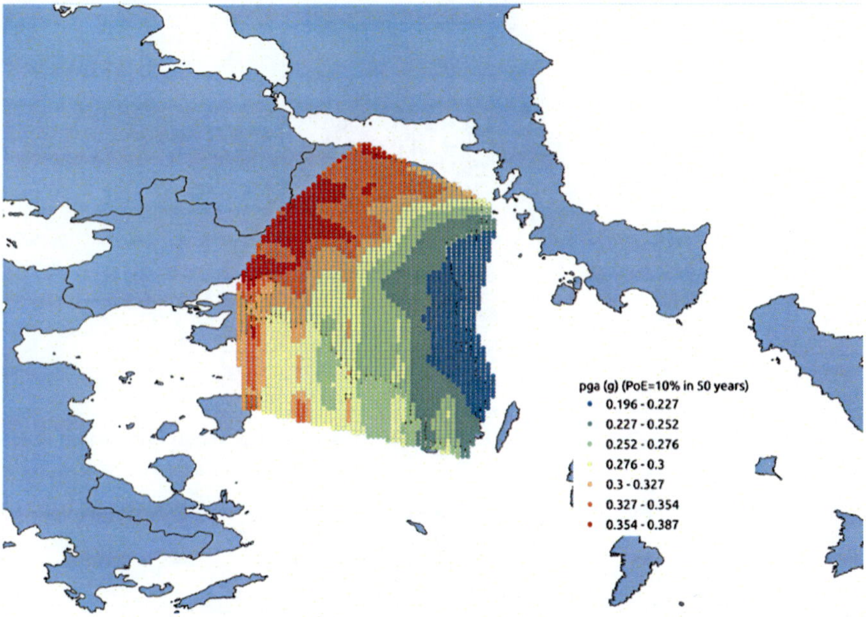

Fig. 1. Classic PSHA results for Attica's region for *pga* with PoE equal to 10% in 50 years.

The open source platform OpenQuake (GEM 2016), where the computational tools for PSHA are provided, was used to calculate the seismic hazard of Attica. All the calculations were made using the latest European seismic source model (SHARE) (Woessner et al. 2015). Of the logic tree branches available, only the area source model and one GMPE (Boore and Atkinson 2008) were employed, whereas all the calculations were made for soil type "rock" ($V_{s30} = 800$ m/s). Peak ground acceleration (*pga*) was selected as the IM for compatibility with literature fragility curves (Psycharis et al. 2013). A 1 km × 1 km grid covering the whole Attica region was set for the seismic hazard definition. Figure 1 shows the results of classic PSHA by using the QGIS (2009) platform for plotting the mean *pga* values with a MAF of $\lambda = 0.0021$ or a probability of exceedance (PoE) equal to 10% in 50 years that is the typical MAF proposed by modern code provisions for the design of new buildings. As illustrated, west Attica shows *pga* values ranging from 0.28–0.39 g higher of those of east Attica (0.19–0.24 g), whereas both estimates exceed the code (EAK 2000) proposed *pga* values. As a further example, Fig. 2 captures the mean hazard curve for the site of the Acropolis as estimated from classic versus event-based PSHA, the latter for two different investigation time periods. In all cases, the frequent events (low *pga*-high MAF) are well captured by all approaches. For rarer events, the event-based PSHA with an effective investigation period of 200,000 years obviously matches better with the classic PSHA than the 10,000 years approach.

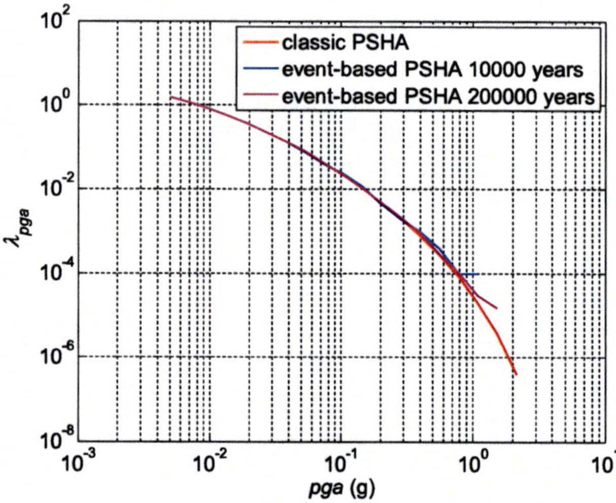

Fig. 2. Mean hazard curve for *pga* at the Acropolis site from classical versus event-based PSHA

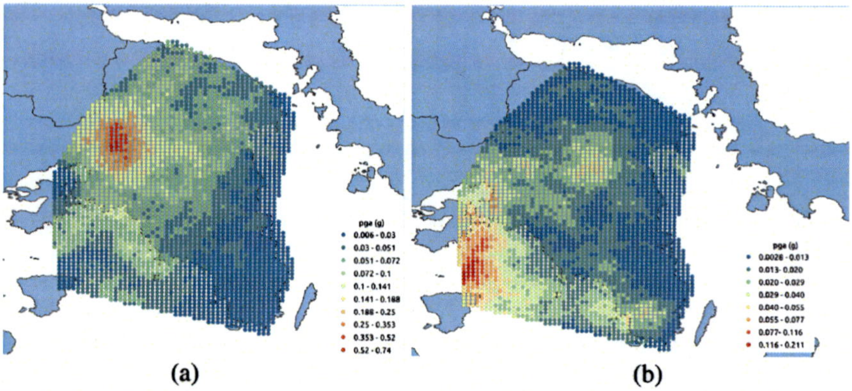

Fig. 3. IM fields for event-based PSHA, a) a possible rupture in the northwest and b) a possible rupture in the southwest

3 Fragility Assessment

The next step of the risk assessment procedure is the fragility definition or, in other words, *the susceptibility of a structure to seismic damage* (Bakalis and Vamvatsikos

2018). A single fragility curve (Fig. 4b) was adopted from Psycharis et al. (2013) to represent a typical multidrum column of classical antiquity, originating from the Pronaos of Parthenon (Fig. 4a). As an Engineering Demand Parameter (EDP) the maximum displacement at the height of the capital normalized by the bottom drum diameter (u_{top}) was employed. Three different Limit States (LS1 to LS3), as defined at associated threshold u_{top}, were employed to define four Damage States (DS0–DS3). Specifically, LS1 to LS3 correspond to the definitions of Damage Limitation, Significant Damage, and Near Collapse by Psycharis et al. (2013), with threshold values of $u_{top} = 0.15, 0.35, 1.00$, respectively.

In general, *pga* values between 0.1 and 1.5 g were employed for the derivation of the fragilities, as evident in Fig. 4b. Linear interpolation was originally employed to approximate the lower tail between 0.0 and 0.1 g. At the same time, the upper tail could not be fully captured for the higher limit-states as reaching 100% probability of exceedance requires *pga* > 1.5 g for LS2 and LS3. Still, the upper and lower tails of the resulting fragility curves are not of equal significance. This fact affects the risk estimations mainly for low *pga* values (lower tail) where most of the events stack. On the other hand, the upper tail refers to the rare events and high *pga* values so it missing does not adversely affect our estimations. However, they may prove useful to determine the effects of a large magnitude event that can effect a wide catastrophe in a scenario assessment.

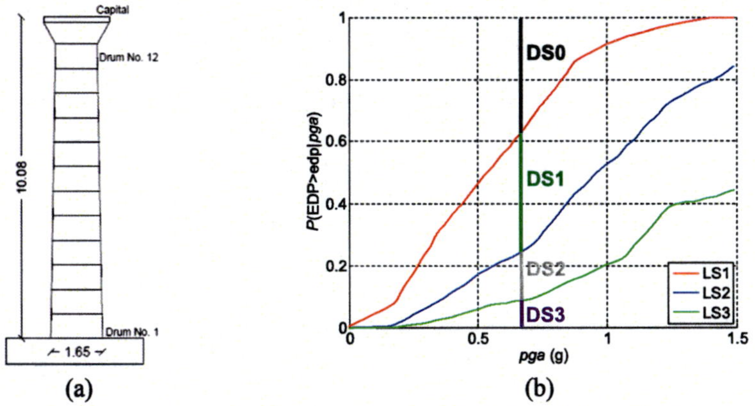

Fig. 4. a) Typical ancient multidrum column and b) fragility curves, both adopted from Psycharis et al. (2013).

4 Risk Assessment

The PBEE methodology (Cornell and Krawinkler 2000) was employed to utilize the classical PSHA results for the risk estimation. It is an application of the total variability theorem for computing the MAF of a predefined *Decision Variable* (*DV*). When only damages are of interest, it can be reduced to a single integral equation (Vamvatsikos and Cornell 2004) for estimating the MAF of violating a LS by convolving the hazard (Fig. 5b) with the fragility (Fig. 5a) as:

$$\lambda(LS) = \int_{IM} G(EDP|IM)\ |d\lambda(IM)| \tag{1}$$

where $G(EDP|IM)$ is the conditional complementary cumulative distribution function (CCDF), which can be defined for each i-th LS and j-th value of the IM as the probability

$$P\big(EDP > EDP_{LSi}|IM = IM_j\big) \tag{2}$$

The event-based approach can also be employed in the same way, via Eq. (1), for estimating the MAF of violating an LS or being in a DS (Fig. 5a), as long as the hazard curve is estimated on an event-to-event basis as in Fig. 2. Essentially, this is equivalent to a simple treatment of the different possible IM fields (Fig. 3) as produced from event-based PSHA whereby the MAF of the column being in each i-th DS at a given site, can be calculated as:

$$\lambda_{being}(DSi) = \frac{\sum_{j=1}^{n} P_{being}(DSi|IMj)}{t_{eff}} \tag{3}$$

where n is the number of the ground motion fields produced from event-based PSHA, P_{being} the probability of being in each DS given the j-th IM field value at the monument site, and t_{eff} the effective investigation time. Equally the MAF of exceeding each of the j limit-states, $\lambda(LSj)$, is calculated by adding the MAFs of higher damage states occurring:

$$\lambda(LSj) = \sum_{i=j}^{3} \lambda_{being}(DSi) \tag{4}$$

The MAF of exceedance is converted into PoE in 50 years (P_{50}) using the exponential distribution CDF characterizing the probability of exceeding given event-to-event interarrival times of the underlying Poisson process as:

$$P_{50} = 1 - e^{-50\lambda} \tag{5}$$

As a first step to a more comprehensive evaluation, we considered the same generic column at the classical antiquity heritage sites of Attica listed in Table 1. To compare the fidelity of the two approaches, the MAF for each LS is computed both by numerical integration (Eq. 1) from Classic PSHA results and from event-based PSHA results (Eq. 3). Taking the full fragility curves from Psycharis et al. (2013) (Fig. 5a), high MAF values, especially for LS1 ($P_{50} \approx 0.70$), were obtained. This comes from the low *pga* levels where there are no analysis results. The seemingly innocuous adoption of a linear relationship between EDP and IM values of 0 to 0.1 g, are poorly combined with the high hazard MAF values characterizing low values of *pga* (Fig. 5).

Instead, to avoid this error source and to offer a more reliable estimation, a lower tail cut-off needs to be employed, otherwise even for values of acceleration caused by vehicle traffic, one obtains non-zero probabilities of damage. Herein, some knowledge of the structure is indispensable. Specifically, the geometric characteristics of the column (Fig. 4a) suggest that rocking response (i.e., uplift) initiates at $pga > \mathrm{g}\tan a \approx 0.1637$ g (Housner 1963) where the block slenderness angle a was approximated as $\arctan(b/h) = \arctan[(1.65/2)/(10.08/2))] = 0.1637$ rad. Additionally, using the friction angle φ from Psycharis et al. (2013), the slipping of drums initiates at $pga = \mathrm{g}\tan\varphi = \tan 37^{\circ} = 0.753$ g. Thus, the minimum pga value for damage of any kind is approximately 0.16 g. The resulting modified fragility curves appear in Fig. 6a. To also add an upper tail, where missing, a lognormal fitting procedure was employed (Fig. 6b) using a moment fitting method (Baker 2015). The lognormal fit is an obviously imperfect approximation that can still acceptably accommodate both the (cut-off) lower and the (often missing) upper tail.

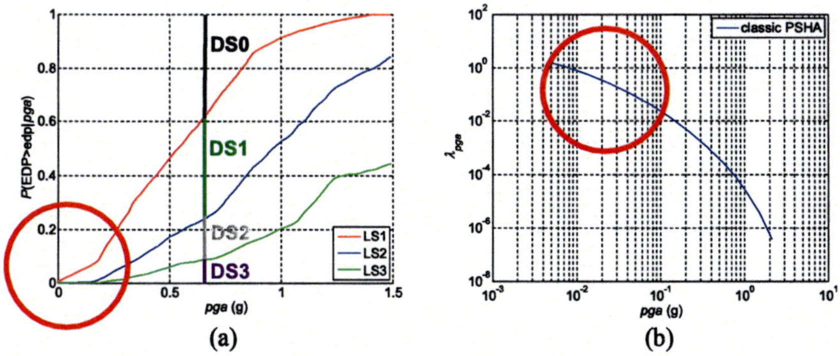

Fig. 5. a) Fragility curves (Psycharis et al. 2013) and b) mean hazard curve for pga at the Parthenon.

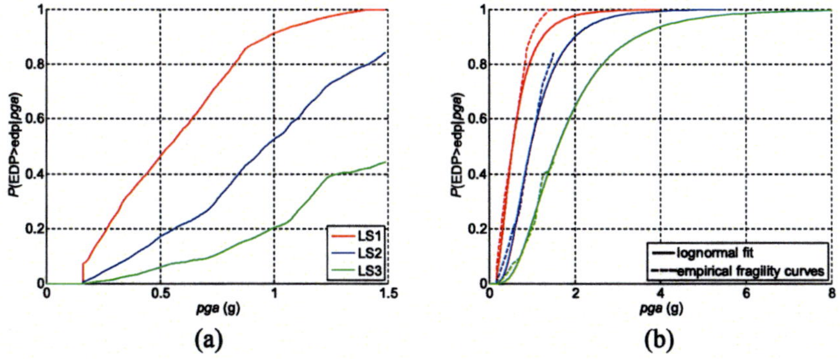

Fig. 6. a) Lower tail cut-off empirical CDF fragility curves and b) their lognormal fit.

Table 1 presents the P_{50} values for the three LS (Fig. 5) as defined by the numerical integration (Eq. 1). As illustrated, the monuments sited in the north, west or southwest Attica (Elefsina, the Temple of Aphaia, the Amphiareio at Oropos, etc.) show higher

P_{50} values than those in the center of Athens (the Parthenon, the Temple of Olympian Zeus, etc.) and those on east or southeast Attica (the Temple of Artemis, the Temple of Poseidon etc.). For instance, $P_{50} \approx 0.106, 0.030, 0.009$ for violating the three LS for the Temple of Aphaia in the southwest, $P_{50} \approx 0.073, 0.020, 0.005$ for the Parthenon in the center, and $P_{50} \approx 0.035, 0.009, 0.002$ at the Temple of Artemis in the east. Lognormal fitted curves capture fairly well the P_{50} showing lower probabilities than the (cut-off) empirical CDF fragilities because of the slight differences due to fitting at low pga values (Fig. 6b).

Table 1. P_{50} values of the three LS from Eqs. 1, 5 for the cut-off fragility curves and their lognormal fit for 12 monuments spread throughout Attica's region. Results from the fitted lognormal distribution are shown in *italics*.

No.	Place	Lon.	Lat.	LS1	LS2	LS3
1	Temple of Aphaia (Aegina)	23.533	37.754	0.1061	0.0305	0.0090
				0.0827	*0.0170*	*0.0048*
2	Archaeological Site of Elefsina	23.539	38.042	0.1205	0.0364	0.0111
				0.0960	*0.0221*	*0.0066*
3	Ancient Walls of Piraeus	23.636	37.946	0.0848	0.0235	0.0067
				0.0649	*0.0125*	*0.0035*
4	Parthenon	23.727	37.972	0.0726	0.0196	0.0055
				0.0549	*0.0100*	*0.0027*
5	Horologion of Andronikos Kyrrhestes	23.727	37.974	0.0730	0.0198	0.0056
				0.0553	*0.0101*	*0.0027*
6	National Archaeological Museum of Athens	23.733	37.989	0.0746	0.0203	0.0058
				0.0567	*0.0105*	*0.0029*
7	Temple of Olympian Zeus	23.733	37.969	0.0726	0.0197	0.0056
				0.0550	*0.0101*	*0.0027*
8	Amphiareio, Oropos	23.845	38.291	0.0995	0.0300	0.0092
				0.0793	*0.0184*	*0.0056*
9	Temple of Artemis (Brauron)	23.993	37.926	0.0348	0.0089	0.0024
				0.0256	*0.0043*	*0.0011*
10	Temple of Poseidon (Sounion)	24.025	37.650	0.0518	0.0137	0.0038
				0.0388	*0.0068*	*0.0018*
11	Archaeological Site of Ramnous	24.027	38.223	0.0829	0.0248	0.0076
				0.0660	*0.0151*	*0.0045*
12	Ancient Theatre of Thoricus	24.054	37.738	0.0404	0.0105	0.0029
				0.0300	*0.0051*	*0.0014*

Table 2. P_{50} values of the three LS from Eqs. 3, 4, 5 for the cut-off fragility curves and their lognormal fit for 12 monuments spread throughout Attica's region. Event-Based PSHA approach for an effective investigation time of 1,000 years. Results from the fitted lognormal distribution are shown in *italics.*

No.	Place	Lon.	Lat.	LS1	LS2	LS3
1	Temple of Aphaia (Aegina)	23.533	37.754	0.0687	0.0181	0.0049
				0.0530	*0.0085*	*0.0018*
2	Archaeological Site of Elefsina	23.539	38.042	0.0689	0.0163	0.0041
				0.0488	*0.0057*	*0.0012*
3	Ancient Walls of Piraeus	23.636	37.946	0.0914	0.0285	0.0091
				0.0774	*0.0196*	*0.0049*
4	Parthenon	23.727	37.972	0.0864	0.0240	0.0070
				0.0680	*0.0136*	*0.0033*
5	Horologion of Andronikos Kyrrhestes	23.727	37.974	0.0842	0.0233	0.0066
				0.0662	*0.0121*	*0.0028*
6	National Archaeological Museum of Athens	23.733	37.989	0.0693	0.0182	0.0051
				0.0535	*0.0086*	*0.0018*
7	Temple of Olympian Zeus	23.733	37.969	0.0826	0.0228	0.0064
				0.0650	*0.0121*	*0.0028*
8	Amphiareio, Oropos	23.845	38.291	0.1251	0.0372	0.0113
				0.1038	*0.0207*	*0.0046*
9	Temple of Artemis (Brauron)	23.993	37.926	0.0492	0.0112	0.0027
				0.0333	*0.0028*	*0.0005*
10	Temple of Poseidon (Sounion)	24.025	37.650	0.0278	0.0077	0.0024
				0.0233	*0.0045*	*0.0010*
11	Archaeological Site of Ramnous	24.027	38.223	0.0570	0.0139	0.0034
				0.0426	*0.0053*	*0.0009*
12	Ancient Theatre of Thoricus	24.054	37.738	0.0230	0.0051	0.0013
				0.0164	*0.0018*	*0.0003*

Tables 2, 3, and 4 shows the resulting P_{50} for the three LS as defined via the event-based PSHA approach (Eqs. 3–5). Three different realizations of event-based PSHA are illustrated in terms of effective investigation time from 1,000 years (Table 2) to 10,000 years (Table 3) and finally 200,000 years (Table 4). As it is shown in Table 2, the simulation of 1,000 years gives different results than the numerical integration method (Table 1), yet not unacceptably so. For instance, for the Temple of Aphaia in Aegina $P_{50} \approx 0.069, 0.018, 0.005$ where for the Parthenon $P_{50} \approx 0.086, 0.024, 0.007$, per each of the three LS. This is happening despite the 1,000 years estimate of the hazard curve being less than ideal in Fig. 2, as this mismatch is confined to the rare events (high *pga* - low MAF), that do not dominate the risk. On a site-by-site basis, the results of a short investigation time may be either conservative or unconservative, depending on the randomness of the individual IM fields. When simulating a short time period,

Table 3. P_{50} values of the three LS from Eqs. 3, 4, 5 for the cut-off fragility curves and their lognormal fit for 12 monuments spread throughout Attica's region. Event-Based PSHA approach for an effective investigation time of 10,000 years. Results from the fitted lognormal distribution are shown in *italics*.

No.	Place	Lon.	Lat.	LS1	LS2	LS3
1	Temple of Aphaia (Aegina)	23.533	37.754	0.1166	0.0349	0.0111
				0.0890	*0.0194*	*0.0063*
2	Archaeological Site of Elefsina	23.539	38.042	–	–	–
				0.0912	*0.0245*	*0.0085*
3	Ancient Walls of Piraeus	23.636	37.946	0.0947	0.0298	0.0091
				0.0754	*0.0191*	*0.0058*
4	Parthenon	23.727	37.972	0.0751	0.0212	0.0063
				0.0571	*0.0124*	*0.0037*
5	Horologion of Andronikos Kyrrhestes	23.727	37.974	0.0749	0.0212	0.0062
				0.0573	*0.0124*	*0.0036*
6	National Archaeological Museum of Athens	23.733	37.989	0.0854	0.0228	0.0063
				0.0644	*0.0115*	*0.0029*
7	Temple of Olympian Zeus	23.733	37.969	0.0750	0.0204	0.0057
				0.0561	*0.0108*	*0.0031*
8	Amphiareio, Oropos	23.845	38.291	0.1124	0.0368	0.0125
				0.0919	*0.0240*	*0.0076*
9	Temple of Artemis (Brauron)	23.993	37.926	0.0346	0.0092	0.0024
				0.0253	*0.0043*	*0.0012*
10	Temple of Poseidon (Sounion)	24.025	37.650	0.0579	0.0144	0.0038
				0.0422	*0.0061*	*0.0014*
11	Archaeological Site of Ramnous	24.027	38.223	0.0716	0.0194	0.0055
				0.0553	*0.0103*	*0.0025*
12	Ancient Theatre of Thoricus	24.054	37.738	0.0424	0.0105	0.0027
				0.0311	*0.0044*	*0.0010*

any individual strong event may generate too high or too low damages for a site and consequently influence (or bias) the risk estimate vis-à-vis the classic approach.

Contrarily, the simulation of a long time period (Table 3) results into a risk estimation on par with the classical approach (Table 1). As it shown, for the Temple of Aphaia $P_{50} \approx 0.117, 0.035, 0.011$, for the Parthenon $P_{50} \approx 0.075, 0.021, 0.006$ and for the Temple of Artemis $P_{50} \approx 0.035, 0.009, 0.002$ at the 3 LS, values that correlate well with those of the classic approach. But, as the effective investigation time is increased, the inclusion of rarer events (high *pga* values) limits the use of the empirical CDF fragility curves

(Fig. 5a). Thus, no results for the Elefsina's Archaeological site are computed from the 10,000 years approach, where potential IM fields produce *pga* values beyond the limits of the empirical CDF fragility curves (≈1.5 g). The empirical CDF fragilities become insufficient for the majority of the monuments at the 200,000 years simulation. This fact necessities the use of a fitting approach to treat the rare events and to get unbiased estimations. Here, similarly to the aforementioned classical method, the lognormal fit curves give a relatively adequate approximation of the empirical CDF ones with slightly lower P_{50} values.

Table 4. P_{50} values of the three LS from Eqs. 3, 4, 5 for the cut-off fragility curves and their lognormal fit for 12 monuments spread throughout Attica's region. Event-Based PSHA approach for an effective investigation time of 200,000 years. Results from the fitted lognormal distribution are shown in *italics*.

No.	Place	Lon.	Lat.	LS1	LS2	LS3
1	Temple of Aphaia (Aegina)	23.533	37.754	–	–	–
				0.0879	*0.0190*	*0.0055*
2	Archaeological Site of Elefsina	23.539	38.042	–	–	–
				0.0976	*0.0242*	*0.0078*
3	Ancient Walls of Piraeus	23.636	37.946	–	–	–
				0.0662	*0.0134*	*0.0041*
4	Parthenon	23.727	37.972	–	–	–
				0.0568	*0.0110*	*0.0031*
5	Horologion of Andronikos Kyrrhestes	23.727	37.974	–	–	–
				0.0573	*0.0113*	*0.0032*
6	National Archaeological Museum of Athens	23.733	37.989	–	–	–
				0.0568	*0.0110*	*0.0032*
7	Temple of Olympian Zeus	23.733	37.969	–	–	–
				0.0567	*0.0107*	*0.0030*
8	Amphiareio, Oropos	23.845	38.291	–	–	–
				0.0788	*0.0178*	*0.0054*
9	Temple of Artemis (Brauron)	23.993	37.926	–	–	–
				0.0271	*0.0048*	*0.0013*
10	Temple of Poseidon (Sounion)	24.025	37.650	–	–	–
				0.0407	*0.0068*	*0.0017*
11	Archaeological Site of Ramnous	24.027	38.223	0.0831	0.0241	0.0072
				0.0650	*0.0135*	*0.0037*
12	Ancient Theatre of Thoricus	24.054	37.738	–	–	–
				0.0317	*0.0053*	*0.0014*

5 Conclusions

The seismic risk assessment of cultural heritage sits hinges upon the accurate representation of the hazard and the monuments' fragility. On the first front, comprehensive consideration of an ensemble of monuments requires an event-by-event view of seismic hazard, whereby individual events are represented by spatial fields of intensity that incorporate within-event correlation at adjacent sites. On the fragility front, careful consideration of literature fragility functions is an excellent way to incorporate the results of true experts, yet may require some additional post-processing to remove lower-tail artefacts that can introduce unwanted bias.

After such treatment, the probability of exceeding a low damage state for the column was estimated for a period of 50 years to be 0.09–0.12 at the north, northwest, west and southwest Attica, while it was reduced to 0.07–0.08 for the center of Athens and to 0.03–0.05 for the east side. This is in line with the estimation of hazard and the prevalence of large faults in the north and west of Athens (Parnitha, Atalanti and the Corinth Gulf).

The next step forward is the enhancement of the vulnerability (fragility) model by using structure-specific structural models and therefore individual fragility functions for each single monument, as well as the consideration of damages on an event-by-event basis over the entire region. Thus, more accurate estimates can be provided to the authorities for deciding on potential remedial actions and efficiently allocating funds, as the moral and the historical cost of potential damages at a monument of classical antiquity are irreparable.

Acknowledgements. This research has been co-financed by the European Regional Development Fund of the European Union and Greek national funds through the Operational Program Competitiveness, Entrepreneurship and Innovation, under the call RESEARCH – CREATE - INNOVATE (project code: T1EDK-00956), project: "ARCHYTAS: Archetypal telemetry and decision support system for the protection of monumental structures".

References

Abrahamson, N.A., Bommer, J.J.: Probability and uncertainty in seismic hazard analysis. Earthq. Spectra **21**(2), 603–607 (2005)

Bakalis, K., Vamvatsikos, D.: Seismic fragility functions via nonlinear response history analysis. J. Struct. Eng. **144**(10), 04018181 (2018)

Baker, J.W.: An introduction to probabilistic seismic hazard analysis. Version 1.3 (2008). https://web.stanford.edu/~bakerjw/Publications/Baker_(2008)_Intro_to_PSHA_v1_3.pdf

Baker, J.W.: Efficient analytical fragility function fitting using dynamic structural analysis. Earthq. Spectra **31**(1), 579–599 (2015)

Bazzuro, P., Cornell, C.A.: Disaggregation of seismic hazard. Bull. Seismol. Soc. Am. **89**(2), 501–520 (1999)

Bazzuro, P., Cornell, C.A.: Vector-valued probabilistic seismic hazard analysis (VPSHA). In: Proceedings of the 7th U.S. National Conference on Earthquake Engineering. EERI, Boston (2002)

Bommer, J.J.: Deterministic vs. probabilistic seismic hazard assessment: an exaggerated and obstructive dichotomy. J. Earthquake Eng. **6**, 43–73 (2002)

Bommer, J.J., Scherbaum, F.: The use and misuse of logic trees in probabilistic seismic hazard analysis. Earthq. Spectra **24**(4), 997–1009 (2008)

Boore, D.M., Atkinson, G.M.: Ground-motion prediction equations for the average horizontal component of PGA, PGV, and 5% damped PSA at spectral periods between 0.01s and 10.0s. Earthquake Spectra **24**(1), 99–138 (2008)

Cornell, C.A.: Engineering seismic risk analysis. Bull. Seismol. Soc. Am. **58**(5), 1583–1606 (1968)

Cornell, C.A., Krawinkler, H.: Progress and challenges in seismic performance assessment. PEER Center News **3**(2), 1–4 (2000)

Global Earthquake Model (GEM). OpenQuake Engine User Instruction Manual, Version 1.9.0, Pavia (2016)

Gutenberg, B., Richter, C.F.: Frequency of earthquakes in California. Bull. Seismol. Soc. Am. **34**(4), 185–188 (1944)

Housner, H.H.: The behavior of inverted pendulum structures during earthquakes. Bull. Seismol. Soc. Am. **53**(2), 404–417 (1963)

Jayaram, N., Baker, J.W.: Correlation model for spatially distributed ground-motion intensities Earthquake. Earthquake Eng. Struct. Dynam. **38**(15), 1687–1708 (2009)

McGuire, R.K.: Computations of seismic hazard. Ann. Geofis. **36**(3–4), 181–200 (1993)

McGuire, R.K.: Probabilistic seismic hazard analysis and design earthquakes: closing the loop. Bull. Seismol. Soc. Am. **85**(5), 1275–1284 (1995)

McGuire, R.K.: Probabilistic seismic hazard analysis: early history. Earthquake Eng. Struct. Dynam. **37**, 329–339 (2008)

Organization for Earthquake Resistant Planning and Protection, Ministry of Environment Planning and Public Works, Greece (OASP). Greek code for Seismic Resistant Structures - EAK2000 (enhanced in 2003) (2000)

Psycharis, I.N., Fragiadakis, M., Stefanou, I.: Seismic reliability assessment of classical columns subjected to near-fault ground motions. Earthquake Eng. Struct. Dynam. **42**(14), 2061–2079 (2013)

QGIS Development Team. QGIS Geographic Information System. Open Source Geospatial Foundation Project. http://qgis.osgeo.org (2009)

Vamvatsikos, D., Cornell, C.A.: Applied incremental dynamic analysis. Earthq. Spectra **20**(2), 523–553 (2004)

Weatherill, G.A., Silva, V., Crowley, H., Bazzurro, P.: Exploring the impact of spatial correlations and uncertainties for portfolio analysis in probabilistic seismic loss estimation. Bull. Earthq. Eng. **13**(4), 957–981 (2015). https://doi.org/10.1007/s10518-015-9730-5

Woessner, J., et al.: The 2013 European Seismic Hazard Model: key components and results. Bull. Earthq. Eng. **13**(12), 3553–3596 (2015)

Evaluation of the Load-Bearing Capacity of Timber Members in Existing Structures Based on Information from a Qualified Investigation In Situ Within a Stepwise Procedure

Maria Loebjinski[1](✉), Wolfgang Rug[2], and Hartmut Pasternak[1]

[1] Chair of Steel and Timber Structures, Brandenburg University of Technology, Konrad-Wachsmann-Allee 2, 03046 Cottbus, Germany
Maria.Loebjinski@b-tu.de

[2] Department of Timber Engineering, Eberswalde University for Sustainable Development (University of Applied Sciences), Alfred-Möller-Street 1, 16225 Eberswalde, Germany

Abstract. Enhanced knowledge from a qualified survey in situ can help to model the load-bearing behaviour of an existing structures more realistically by considering specific knowledge on material parameters and conditions. A semi-probabilistic evaluation format with partial safety factors is used in common practice. Standardised options to include specific information into the design concept have to be worked out. These are improved strength grading using non-/semi-destructive means and adjustments within the safety concept itself. However, for everyday practice two problems arise. It is often not possible to determine strength parameters but reference properties like Young's modulus or density. Besides, an exhaustive determination of material parameters is cost-intensive and therefore often not possible for smaller projects. This contribution presents a procedure to include data gained in a qualified survey in situ into the design concept for the verification of load-bearing capacities of existing timber structures. The level of detail of the design format is increased stepwise considering the level of detail of the survey. These levels are named Knowledge Levels in accordance with the Science and Policy Report of the Joint Research Centre (JRC) from 2015. Knowledge Level 1 includes the evaluation based on current Eurocodes without adjustments. Knowledge Level 2 is divided into three sub-levels including adjustments of partial safety factors for target values derived for existing structures, an improved strength grading based on non-/semi-destructive technical means and the updated of the partial safety factor for the material strength based on measured reference properties. Knowledge Level 3 includes a probabilistic evaluation comparing the actual reliability of the current design situation with and without parameter update.

Keywords: Timber · Existing structures · Evaluation procedure · Semi-probabilistic safety concept · Partial safety factors

I. Vayas and F. M. Mazzolani (Eds.): PROHITECH 2021, LNCE 209, pp. 638–650, 2022.
https://doi.org/10.1007/978-3-030-90788-4_50

1 Introduction

The challenging task of investigating and evaluating existing structures has been addressed several times in current research work. These studies range from non-/semi-destructive investigation techniques in situ to probabilistic aspects to analyse and update the reliability of an existing structure. To evaluate the load-bearing capacity of an existing structure, its structural elements have to be investigated in detail. This embraces the aspects illustrated in Fig. 1 from (Lißner and Rug 2015).

The figure shows that for the investigation and evaluation of timber structures, numerous issues have to be studied in detail. To investigate these aspects named, different techniques and technical means are available and under continuous development and improvement. An overview concerning investigation techniques to be applied in situ can for example be studied in (Dietsch and Köhler 2010), (Kasal and Tannert 2011) and (Lißner and Rug 2015).

The attempt to standardise the investigation and evaluation of existing structures has been dealt with mainly on a national level. To be named here are for example the Italien standards UNI 11119 (UNI 2004a) and UNI 11138 (UNI 2004b), the Swiss standard SIA 269 (SIA 2011) and the German Leaflet (DBV 2013). In (Perria and Sieder 2019) and (Loebjinski et al. 2019c) the current state of the art concerning the investigation and evaluation of existing timber structures as given in codes, standards and selected research work is summarised.

On an international level work has been carried out to standardise national approaches, e.g. in (European Commission, Joint Research Centre 2015) independent from the building material. A European standard for the investigation of existing timber structures has been developed lately, EN 17121 (CEN 2017). However, whereas this standard provides a common approach for the investigation, no methods for an adjusted evaluation of the load-bearing capacity and serviceability including data collected in a qualified survey in situ are provided. For concrete, advices are given on an international level by the Fédération Internationale du Béton (fib) (FIB 2016). For timber, no standardised methods considering the special material properties of this natural grown and thus inhomogeneous material are available at state.

This contribution illustrates a proposal that has been worked out so far and summarizes the results of the research that has been carried out for the steps of the evaluation. This contribution focusses on the modification of the semi-probabilistic evaluation concept. For more detailed information on non- and semi-destructive investigation techniques available for timber, the interested reader may refer to the literature.

Existing Structures		
Cross sections	Material parameters	
• Size • Adherence of minimum cross sections • Influence on the load-bearing capacity • Influence on the serviceability	• Timber species • Density • Strength • Young's modulus • Moisture content • Number/ location of knots • Core and sapwood • Cracks	Grading & strength class obtained from decisive criteria of DIN 4074-1 DIN 4074-5
Structure/ Structural system	Bracing	
• Date of erection • Load history/ load assumption • Static system/ changes • Geometry, cross sections, connections • Deformations • Alterations of positions of elements	• Construction principle • Sufficiently existent • Effective and including connections • Influence on the load-bearing capacity • Sufficient load-bearing capacity and serviceability	
Connections		
• Type, material strength • Load distributions • Serviceability • Type of damage • Load-bearing capacity • Structural facts influencing the load-bearing capacity and serviceability		

Evaluation of the load-bearing capacity and serviceability	
Evaluation of the static and structural functionality	
Load-bearing capacity	Serviceability

Damages			
Cracks		Deformations	
• Depth • Length • Static relevant/ irrelevant		• Size • Influence on the load-bearing capacity • Influence on the serviceability	
Insects and fungi		Corrosion	
• Type • Amount of damage • Influence on load-bearing capacity • Reconstruction measures • Treatment measures	Timber protection investigation	• Type • Amount of damage to timber & connections • Influence on load-bearing capacity and serviceability	Investigation of corrosion (e.g. corrosion of tension cracks for steel, maceration for timber)

Fig. 1. Aspects for the evaluating of load-bearing capacity and serviceability of timber structures in existing buildings from (Lißner and Rug 2015)

2 Proposal for an Evaluation Procedure

Whenever the evaluation of the load-bearing capacity of a structural element in an existing structure becomes necessary (e.g. due to load changes or structural damage), current design codes have to be applied to evaluate its sufficiently load-bearing behaviour. At state, it is not possible within the concept of the Eurocode, to include updated information on a critical member that can be gained through a qualified and detailed investigation in situ. However, as the decision on the amount of information gathered in situ based on e.g. material tests always aims for an optimal allocation of time and resources considering sufficient structural safety, it always forms a very individual optimisation problem of the task at hand. Thus, this research does not aim to replace current practice but to provide suggestions for its enrichment. Therefore the presented procedure is built on current practice providing methods to be applied if necessary or desired. The proposal is illustrated in Fig. 2 as presented in (Loebjinski et al. 2019a) and developed further in (Loebjinski et al. 2019b) and described in more detail hereinafter. This contributions summarizes results of different updating steps that have been presented bevor and informs about some recently developed achievements and open challenges as well as open research questions. The abbreviation *KL* stands for *Knowledge Level*, which is based on JRC Science and Policy Report 2015 (European Commission, Joint Research Centre 2015).

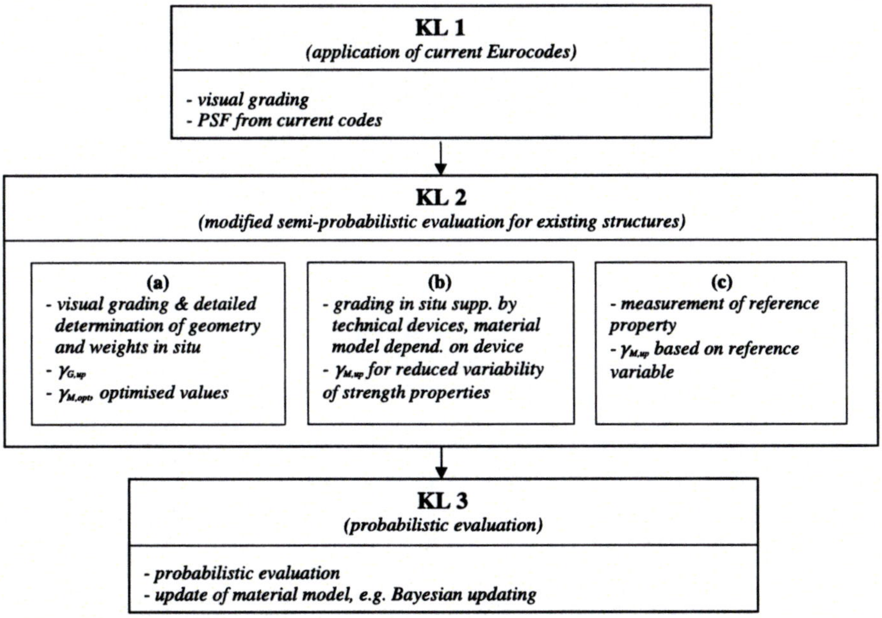

Fig. 2. Framework for the evaluation of the load-bearing capacity of existing structures (Loebjinski et al. 2019b).

3 Evaluation in Knowledge Level 1

This level represents the current state of the art applying EN 1990, EN 1991 and EN 1995 to analyse the structural behaviour and check sufficient structural safety. Partial safety factors (PSF) to be applied are summarised in Table 1.

Table 1. Partial Safety Factors from EN 1990 (CEN 2010a), EN 1995 (CEN 2010b) & EN 1995/NA (CEN 2010c)

γ_G	γ_Q	Code	γ_M		Code
Permanent act	Variable act.		Solid timber	Glulam	
1.35	1.50	EN 1990	1.30	1.25	EN 1995–1–1
			1.30	1.30	NA for GER

This concept is well developed and applicable for a variety of practical problems. However, the safety elements for a semi-probabilistic evaluation do not allow for a consideration of updated information concerning load and material parameters and is thus rather less flexible. Especially for existing timber structures a skilled selection, treatment and arrangement of structural elements are important to ensure that structures are reliable und durable. What is more, the grading of the material is often performed only by visual inspection, whereas grading supported by technical devices can significantly enhance knowledge about the material properties of the elements at hand, see e.g. (Sandomeer and Steiger 2009), (Kessel and Sandoz 1989), (Sandoz 1989), (Linke et al. 2019). This information shall be used by introducing a Knowledge Level 2 in the suggested approach.

4 Evaluation in Knowledge Level 2

4.1 General Remarks

This level embraces different options for an adjusted semi-probabilistic evaluation of existing structures. Current Eurocodes form the basis of structural design and verification of load-bearing capacity. EN 1990 (CEN 2010a) allows a probabilistic evaluation. However, the semi-probabilistic path is chosen in practice in most common cases, see Fig. 3 from (CEN 2010a). Optional, but not part of current versions of Eurocodes, are deterministic methods (method a). They are often part of older code formats and may be applicable to evaluate existing structures as e.g. permitted in Germany by (Fachkommission Bautechnik der Bauministerkonferenz (ARGEBAU) 2008).

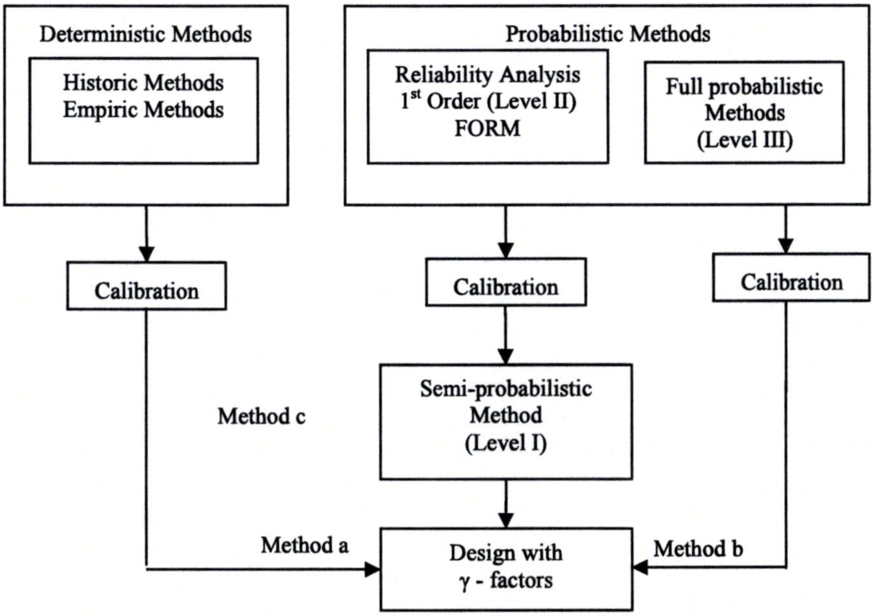

Fig. 3. Overview of methods of reliability analysis from EN 1990 Annex C (CEN 2010a)

The following sections summarize research work performed to update the semi-probabilistic design concept for the evaluation of existing structures. Case studies have been used to test the applicability of the developed concepts and presented in (Loebjinski et al. 2019b) and (Loebjinski et al. 2019d).

4.2 Evaluation in Knowledge Level 2a

In this level, the evaluation of the load-bearing capacity can be done with modified PSF for the material strength without directly updating the material parameters by technical means, but by a more qualitative evaluation. These factors are based on an adjusted target reliability level for maintenance measures in existing structures.

In ISO 2304:2015 target reliability indices are given considering consequences of failure and relative cost of life safety measures, see Table 2 from (ISO 2015).

In ISO 2394 (ISO 2015) it is emphasized, that these values have been determined considering economic optimisation and may be not acceptable concerning life safety risks. For existing structures, (Vrouwenvelder 2002) suggests to move from medium to large relative costs of safety measures.

The association of consequence classes given in Table 2 to the consequence classes of EN 1990:2010-12 Annex B is not straight forward even though both documents contain descriptions and examples for types of structures within different classes. Within EN 1990:2010-12 Annex C three classes are described whereas ISO 2394:2015 contains four classes of failure consequences. However, SIA 269:2011, the Swiss code for existing structures, contains the target reliabilities indicated in Table 2 indicated with *low consequences of failure* (equals values in *class 2*), *moderate consequences of failure* (equals values in *class 3*) and *high consequences of failure* (equals values in *class 4*).

Table 2. Tentative target reliabilities related to one year reference period and ultimate limit states, based on monetary optimization, from (ISO 2015)

Relative cost of safety measure	Consequences of failure					
	Class 2		Class 3		Class 4	
Ref. period	$T_{ref} = 1a$	$T_{ref} = 50a^*$	$T_{ref} = 1a$	$T_{ref} = 50a^*$	$T_{ref} = 1a$	$T_{ref} = 50a^*$
Large	3.1	1.7	3.3	1.8	3.7	2.7
Medium	3.7	2.7	4.2	3.2	4.4	3.5
Small	4.2	3.2	4.4	3.5	4.7	4.2

*The value has been calculated statistical independency of following years

(Steenbergen et al. 2015) suggest to reduce the target reliability for existing structures based on the target value for new structures by $\Delta\beta = 0.5$ and for a minimum reliability index by $\Delta\beta = 1.5$. The JRC Science and Policy Report 2015 (European Commission, Joint Research Centre 2015) indicates reliability indices for four consequence classes $\beta = 1.8$ (CC 0), $\beta = 1.8$ (CC 1), $\beta = 2.5$ (CC 2) and $\beta = 3.3$ (CC 3) for a reference period of 50 years. What is more, target reliability indices for CC 2 and CC 3 are illustrated depending on the affected area in case of collapse.

ISO 13822:2010 (ISO 2010) which contains guidelines for the assessment of existing structures refers to ISO 2394 (ISO 2015) for underlying principles of reliability analysis. Here, guidance to calculate partial safety factors with fixed sensivity factors is given. What is more, fib bulletin no. 80 (FIB 2016) describes two options to adjust partial safety factors for existing structures with fixed sensivity factors. These are the *Design Value Method* (DVM), which is also described in ISO 2394:2015 and the *Adjusted Partial Safety Factor Method* (APFM). The former allows a direct calculation of PSF with fixed sensivity factors and adjusted target reliability whereas the latter allows the calculation of factors to adjust given PSF for different requirements.

The determination of PSF with fixed sensitivity factors is a good, straight forward method. However, this method needs to be applied for load and resistance variables in a limit state consistently. Input parameters need to be chosen with special care considering especially the information that is currently available to model variable actions, see e.g. (Grünberg 2004) (Loebjinski et al. 2017). Thus, a sound option for the modification of one or more PSF is their calibration within a probabilistic optimisation. Results for such an optimisation for γ_M to be applied on structural timber material strength properties are under preparation within this research project. These values are optimised values that are only valid for different categories of measures (assurance of minimum reliability, evaluation of a member in service, new structural element) and different limit states, e.g. uniaxial strain induced by permanent action and live load. Similar studies have been published e.g. by (Fischer 2010) for reinforced concrete and by (Stauder 2015) for structures of hydraulic engineering. Parameter studies indicated a great scatter of the reliability among different design situations, see e.g. (Baravalle 2017), own studies support this. A high sensitivity of the coefficient of variation of the material strength could be observed.

4.3 Evaluation in Knowledge Level 2b

In this level information gained by strength grading supported by technical devices is available. Thus, the strength class can be updated. To generated a trustful result, different grading methods should be combined, i.e. visual grading ultrasonic measurements, extraction of core samples etc. By these means, and updated strength class can be determined, which is often higher than solely by visual grading, see e.g. (Linke et al. 2019).

The accuracy of grading significantly influences the material model in a strength class, see e.g. (Faber et al. 2004). In former studies it has been investigated, if the material model can be updated based on the grading technique applied. In (Loebjinski et al. 2019b) the stochastic grading model by (Pöhlmann and Rackwitz 1981) has been applied on test results from (Linke et al. 2019). It has been figured out, that the results depends on the timber species. For oak samples an improvement of the material model and reduction of the coefficient of variation (cov) of the assumed prior model could be observed, the results for softwood also depend on the timber species. The prior model was assumed based on (JCSS 2006). Baysian update has been applied to investigate the influence of the results of the calibration tests to update the prior model. Thus, an updated material model for different grading techniques could not be developed with certainty yet.

It has been figured out, that the assumption of a reduction of the variability of strength properties in a test based on different grading techniques is critical applying this model, as the uncertainty of the tests themselves has to be considered. Besides, it was identified, that multiple correlation coefficients to combine different parts of information are needed, see e.g. (Linke et al. 2019). Current studies including multiple regression analysis of different grading parameters gave good to very good results, see (Linke et al. 2020).

However, as the reliability and the calibrated PSF γ_M is highly sensitive against changes of the coefficient of variation cov_x of the material strength, this value has been studied. It is defined in Eq. (1)

$$cov_x = \frac{\sigma_x}{\mu_x} \tag{1}$$

where σ_x is the standard deviation and μ_x is the mean value of the variable x. It was assumed that with higher strength classes cov_x is reduced as the mean value increases and the standard deviation can be assumed equal in different strength classes. This was approved by the calibration test. Thus, high strength classes support a reduction of the PSF γ_M, although this is still difficult to quantify as this depends on the timber species.

4.4 Evaluation in Knowledge Level 2c

If it is decided to carry out material tests in situ, the PSF to be applied on the target variable (e.g. bending strength) can be updated based on the correlation relation of target variable and measured reference variable (e.g. Young's modulus from ultrasonic measurement). In (Loebjinski et al. 2019a) a formula has been developed to update the PSF γ_M based on a measured reference property. The updated PSF $\gamma_{m,up}$ (see Eq. (2))

without and $\gamma_{M,up}$ (see Eq. (3)) including model uncertainty is

$$\gamma_{m,up} = \exp\left(\frac{cov_{y,target} \cdot \sqrt{1-\rho_{x,y}^2}}{1+\rho_{x,y} \cdot cov_{y,target} \frac{x_{meas}-\mu_{x,ref}}{\mu_{x,ref} \cdot cov_{x,ref}}} \cdot \left(\alpha_R \cdot \beta + \Phi^{-1}(q)\right)\right) \quad (2)$$

$$\gamma_{M,up} = \gamma_{Rd} \cdot \gamma_{m,up} \quad (3)$$

with $cov_{y,target}$ the cov of the target variable y, $cov_{x,ref}$ the cov of the reference variable x, $\mu_{x,ref}$ the mean value of the reference variable x, $\rho_{x,y}$ the correlation coefficient, α_R the sensivity factor of the material resistance ($\alpha_R = 0.8$ from EN 1990:2010-12), β the target reliability, q the quantile for the definition of the characteristic value of the target variable used for design ($q = 0.05$) and γ_{Rd} the model factor that should be calculated from a normal distribution and the 50%-quantile considering the adjustment for a non-dominant variable (see EN 1990:2010-12 or ISO 2394:2015).

The results applying this approach have to be evaluated carefully, as they only refer to information based on one measured reference variable, which can differ locally. Thus, it does not release the engineer from a detailed investigation and careful evaluation of results.

5 Evaluation in Knowledge Level 3

As indicated above, EN 1990 which is the basis for current design concept of the Eurocodes, allows for probabilistic methods to verify sufficient structural safety. However, target values of the reliability index are still under discussion. In (Baravalle and Köhler 2017) it was figured out, that the reliability level differs significantly between different design situations depending on load shares and types of (variable) loads. What is more, the target value of $\beta = 3.8$ for a reference period $T_{ref} = 50a$ seems not entirely match the reliability level of current codes. What is more, target values for existing structures considering an enhanced knowledge and sufficient structural performance are not given ISO 2394:2015 (ISO 2015) gives target reliability indices depending on the relative costs of safety measures and consequences of failure.

This principle is also included in SIA 269:2011, the Swiss code giving guidelines for the evaluation of existing structures. This allows for the decision on an individually chosen target reliability. In this relation, (Vrouwenvelder 2002) suggest to move for existing structures into the range of high costs of safety measures.

Another approach is the distinction into minimum (disapproval) and target reliability (reconstruction) level as e.g. given in NEN 8700. Values are also discussed for example in (Diamantidis et al. 2017).

Knowledge Level 3 includes an update of the material model, e.g. by Bayesian Updating. It seems to be more appropriate to evaluate the reliability of a design situation and study the effects of changes of the random variable for the case at hand to indicate possible structural measures rather than to compare the result to a fixed target value. As mentioned above, the reliability level of different design situation differs depending on the type of loading, load shares and assumptions for material parameters.

6 Conclusion

The evaluation of the load-bearing capacity of an existing structure is fundamentally different compared to the design of new structures. The availability of information to update load and material parameters due to the existence of the former in tangible form is a big advantage. However, the quantification and utilisation of different forms of information in structural design is also a challenge. Thus, the major part of work consists of the development of standardised approaches (statistical methods, adjusted semi-probabilistic methods) within a flexible and widely applicable format, that is being approved by the building authorities. An attempt to structure different options has been presented in (Loebjinski et al. 2019b), see Fig. 4.

Figure 4 illustrates the parts of work and processes carried out within the research project and tries to structure ideas on how to combine information. However, these are some attempts without claiming to be complete. A lot of knowledge concerning statistical methods, good technical devices, a well based code format and expert knowledge experienced engineers and architects are already existent. All these fields need to be combined to satisfy the special requirements and challenges when evaluating existing structures. Thus, a lot of further work is necessary to develop good and practicable tools to include different types of information about an existing structure into its evaluation.

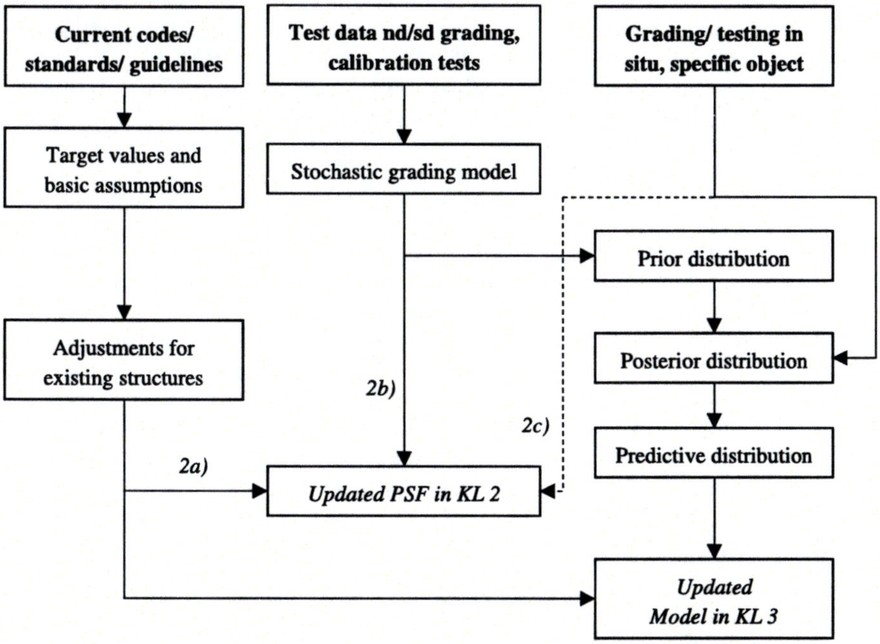

Fig. 4. Schematic updating procedure, extended from (Loebjinski et al. 2019b)

Acknowledgements. TThe research presented in the paper has been funded by The German Federal Environmental Foundation (Deutsche Bundesstiftung Umwelt - DBU).

References

Baravalle, M.: Risk and Reliability Based Calibration of Structural Design Codes. Dissertation. Trondheim (2017)

Baravalle, M., Köhler, J.: A framework for estimating the implicit safety level of existing design codes. In: 12th International Conference on Structural Safety and Reliability, pp. 1037–1046. TU-Verlag, Vienna (2017)

CEN. Basis of Structural Design. Beuth Verlag, Berlin (2010a). DIN EN 1990:2010-12

CEN. Design of Timber Structures – Part 1-1: General – Common Rules and Rules for Buildings. Beuth Verlag, Berlin (2010b). DIN EN 1995-1-1:2010-12

CEN. Nationaler Anhang – National Festgelegte Parameter – Eurocode 5: Bemessung und Konstruktion von Holzbauten – Teil 1-1: Allgemeines – Allgemeine Regeln und Regeln für den Hochbau. Beuth Verlag, Berlin (2010c). DIN EN 1995-1-1:2010-12/NA

CEN. Erhaltung des kulturellen Erbes – Historische Holzkonstruktionen – Leitlinien für die Bewertung vor Ort (2017). E DIN EN 17121:2017-06

DBV. Modifizierte Teilsicherheitsbeiwerte für Stahlbetonbauteile. Modified Partial Safety Factors for Reinforced Concrete Members. Berlin (2013)

Diamantidis, D., Holický, M., Sýkora, M.: Target reliability levels based on societal, economic and environmental consequences of structural failure. In: 12th International Conference on Structural Safety and Reliability, pp. 644–653. TU-Verlag, Vienna (2017)

Dietsch, P., Köhler, J.: Assessment of Structural Timber COST Action E55 "Modelling of the Performance of Timber Structures." Shaker Verlag, Aachen (2010)

European Commission, Joint Research Centre. New European Technical Rules for the Assessment and Retrofitting of Existing Structures. Luxembourg (2015)

Faber, M.H., Köhler, J., Sørensen, J.D.: Probabilistic modelling of graded timber material properties. Struct. Saf. **26**, 295–309 (2004)

Fachkommission Bautechnik der Bauministerkonferenz (ARGEBAU). Hinweise und Beispiele zum Vorgehen beim Nachweis der Standsicherheit beim Bauen im Bestand (2008). http://www.bvpi.de/bvpi-content/aktuelles/argebau.htm. Accessed 15 Jan 2016

FIB. Partial Factor Methods for Existing Concrete Structures. Recommendation. Fédération Internationale Du betón, Lausanne, Switzerland (2016)

Fischer, A.M.: Bestimmung modifizierter Teilsicherheitsbeiwerte zur semiprobabilistischen Bemessung von Stahlbetonkonstruktionen im Bestand. Dissertation. Kaiserslautern (2010). https://kluedo.ub.uni-kl.de/frontdoor/index/index/docId/2259. Accessed 29 Mar 2016

Grünberg, J.: Grundlagen der Tragwerksplanung - Sicherheitskonzept und Bemessungsregeln für den Konstruktiven Ingenieurbau. Erläuterungen zu DIN, pp. 1055–100. Beuth, Berlin (2004)

ISO. Bases for Design of Structures - Assessment of Existing Structures. ISO 13822:2010(E) (2010)

ISO. General Principles on Reliability for Structures. Schweiz, ISO 2394:2015(E) (2015)

JCSS. Probabilistic Model Code. Part 3 - Resistance Models (2006). http://www.jcss.byg.dtu.dk/. Accessed 08 Mar 2016

Kasal, B., Tannert, T.: In Situ Assessment of Structural Timber: State of the Art Report of the RILEM Technical Committee 215-AST. Springer, Netherlands (2011)

Kessel, M.H., Sandoz, J.L.: Zur Effizienz der Festigkeitssortierung von Fichtenkantholz. Holz als Roh- und Werkstoff **47**, 279–284 (1989)

Linke, G., Rug, W., Pasternak, H.: Strength grading of timber in historic structures - material testing concerning the application of the ultrasonic-time-of-flight measurement. In: 5th International Conference on Structural Health Assessment of Timber Structures, pp. 589–598. Guimarães, Portugal (2019)

Linke, G., Rug, W., Pasternak, H.: Strength grading of structural timber in existing structures – Study on the apparatus supported grading with the ultrasonic time-of-flight measurement (in preparation). In: Proceedings of the 4th International Conference on Protection of Historical Constructions, Athens, Greece (2020)

Lißner, K., Rug, W.: Ergänzung bzw. Präzisierung der für die Nachweisführung zur Stand- und Tragsicherheit sowie Gebrauchstauglichkeit von Holzkonstruktionen in der Altbausubstanz maßgebenden Abschnitt der DIN 1052: August 2004 (aktualisierte Fassung aus dem Jahre 2004 unter Berücksichtigung der Einführung der Eurocodes) (2015). http://www.holzbau-statik.de/ibr/downloads/05_bestand/Erg%C3%A4nzungen-Forschungsbericht_Holzkonstruktion%20-%20Altbausubstanz_03.12.pdf

Loebjinski, M., Köhler, J., Rug, W., Pasternak, H.: Development of an optimisation-based and practice orientated assessment scheme for the evaluation of existing timber structures. In: 6th International Symposium on Life-cycle Analysis and Assessment in Civil Engineering, pp. 353–360. CRC Press, London (2019a)

Loebjinski, M., Linke, G., Rug, W.: Instandsetzung einer denkmalgeschützten Dachkonstruktion in Holzbauweise. Bauingenieur **94**(10), 378–385 (2019b)

Loebjinski, M., Linke, G., Rug, W. Pasternak, H.: Evaluation of existing timber structures – Current standards for the assessment and evaluation in Germany and Europe. In: 5th International Conference on Structural Health Assessment of Timber Structures, pp. 884–893. Guimarães, Portugal (2019c)

Loebjinski, M., Linke, G., Rug, W., Pasternak, H.: Redevelopment of a wooden roof construction under preservation order. In: 5th International Conference on Structural Health Assessment of Timber Structures, pp. 912–921. Guimarães, Portugal (2019d)

Loebjinski, M., Rug, W., Pasternak, H.: Zuverlässigkeitsbewertung von Holzbauteilen im Bestand. Bauingenieur **92**, 65–73 (2017)

Loebjinski, M., Rug, W., Pasternak, H.: The influence of an improved strength grading in situ on modelling timber strength properties. In: 5th International Conference on Structural Health Assessment of Timber Structures, pp. 619–628. Guimarães, Portugal (2019e)

Perria, E., Sieder, M.: Process of structural assessment of heritage timber structures: definition based on the state of the art. In: 5th International Conference on Structural Health Assessment of Timber Structures, pp. 922–934. Guimarães, Portugal (2019)

Pöhlmann, S., Rackwitz, R.: Zur Verteilungsfunktion von Werkstoffeigenschaften bei kontinuierlich durchgeführten Sortierungen. Materialprüfung **23**(8), 277–278 (1981)

Sandomeer, M.K., Steiger, R.: Potential der maschinellen Festigkeitssortierung von Schnittholz. In: Holz, Z. (ed.) Statusbericht zum aktuellen Stand der Verwendung von Holz und Holzprodukten im Bauwesen und Evaluierung künftiger Entwicklungspotentiale. Kapitel 05 – Forst und Sägeindustrie, pp. 361–370 (2009). http://www.fva-bw.de/publikationen/zukunft_holz/zh_k05.pdf. Accessed 06 Mar 2017

Sandoz, J.L.: Grading of construction timber by ultrasound. Wood Sci. Technol. **23**, 95–108 (1989)

SIA. Grundlagen der Erhaltung von Tragwerken, vol. 269. SIA, Zürich (2011)

Stauder, F.: Zuverlässigkeitskonzept für bestehende Tragwerke im Wasserbau. Dissertation. Kaiserslautern (2015)

Steenbergen, R., Sýkora, M., Diamantidis, D., Holický, M., Vrouwenvelder, T.: Economic and human safety reliability levels for existing structures. Struct. Concr. **16**(3), 323–332 (2015)

UNI. Cultural Heritage. Wooden Artefacts. Load-bearing structures - On site inspections for the diagnosis of timber members, vol. 91.080.20, UNI 11119:2004 (2004a)

UNI. Cultural Heritage. Wooden Artefacts. Load-Bearing Structures of Buildings - Criteria for the Preliminary Evaluation, Design and Execution of Works, vol. 91.080.20, UNI 11138:2004 (2004b)

Vrouwenvelder, T.: Developments towards full probabilistic design codes. Struct. Saf. **24**, 417–432 (2002)

Non-linear Methods for the Assessment of Seismic Vulnerability of Masonry Historical Buildings

Pietro Croce(✉), Filippo Landi, Paolo Formichi, Maria Luisa Beconcini, Benedetta Puccini, and Vincenzo Zotti

Department of Civil and Industrial Engineering, (DICI), University of Pisa, Largo Lucio Lazzarino, 56 122 Pisa, Italy
p.croce@ing.unipi.it

Abstract. The evaluation of seismic performance of existing masonry structures is a fundamental issue for the preservation of cultural heritage in general and of historical buildings in particular, even in view of the adoption of suitable policies for risk mitigation. To assess the seismic performance of existing masonry buildings, several refined push-over procedures for non-linear static analysis can be found in literature, mainly based on the equivalent frame approach. However, these procedures are not only very time consuming, but also very sensitive to modelling hypotheses, so that results could vary in a large interval even when users are very skilled. To solve these problems, the authors, aiming to enhance the classical pushover programs for masonry building, especially in terms of easiness of use, simplicity of modelling and computational demand, developed a reliable and sound push-over procedure allowing a quick evaluation of seismic risk index. The so called E-PUSH program, which allows the verification of multi-story masonry buildings according to the most recent structural codes and relies on very simple structural models, has been validated considering several relevant benchmark studies and comparing the results with classical procedures implemented in commercial programs. In the paper, a review of macro-element methods for the seismic assessment of unreinforced masonry buildings is then discussed considering the main hypothesis on masonry modelling, choice of masonry mechanical parameters and failure criteria and evaluating their impact on the estimation of seismic risk index. The results of the comparison will be finally presented and discussed for a relevant case study of a four story masonry building investigated by the authors in the framework of a research program agreement between the Municipality of Florence and the Department of Civil and Industrial Engineering of the University of Pisa.

Keywords: Pushover · Unreinforced masonry · Equivalent frame model · Seismic risk

1 Introduction

The preservation of historical constructions in seismic prone areas is a critical issue, due to the high seismic vulnerability of the built environment. Recent earthquake events

I. Vayas and F. M. Mazzolani (Eds.): PROHITECH 2021, LNCE 209, pp. 651–670, 2022.
https://doi.org/10.1007/978-3-030-90788-4_51

confirmed the seismic vulnerability of historical towns, where most of masonry constructions were erected before any specific seismic provision. In order to set-up intervention strategies and to identify priorities for seismic retrofit and strengthening, it is fundamental to define reliable procedures for the assessment of the seismic performance of masonry structures.

A proper seismic evaluation requires, as a first input, an adequate knowledge of masonry mechanical properties that should be combined with consistent modeling strategies. Indeed, understanding and modelling the mechanical behavior of masonry elements is a longstanding challenge in civil engineering (Lourenco 2002).

Several modeling strategies can be adopted for masonry structures, depending on the field of application and the level of complexity that is intended to be reached, such as macro-modelling, simplified micro-modelling and detailed micro-modelling. Among them, macro-modeling is still the most common technique in engineering practice, due to the possibility to idealize masonry walls as one-dimensional elements and simulate the structural response by means of a frame analysis.

Different macro-modelling approaches and software were developed in the recent years, especially in Italy (Marques and Lourenco 2011), to be used for the assessment of the seismic performance of unreinforced masonry buildings. However, the different modelling hypotheses and the analysis methods adopted by each software may lead to significant differences in the results with important consequences for the final assessment.

In this work, a critical review of the most common commercial software adopted in Italy for the structural analysis of masonry constructions is performed. Moreover, the E-PUSH program, developed by the authors (Beconcini et al. 2018; Croce et al. 2019a) for the analysis of masonry structures, is presented and compared with the others.

The comparison is carried out on different case studies, and the results of the non-linear static analysis, in terms of capacity curves, are discussed highlighting the consequences of the different hypothesis behind each method.

2 Masonry Modelling and Structural Analysis

Modelling strategies and analysis methods are the key points in assessment of existing masonry buildings. Historical masonry buildings have been often built according to empirical rules, without following any codified provisions and it is thus difficult to properly identify the underlying structural model. Therefore, the analysis of existing buildings firstly requires a standard procedure for modelling, which should be, at the same time, reliable and user-friendly. Indeed, the definition of finite element models able to describe the non-linear mechanical behaviour of masonry panels requires complex constitutive laws, characterized by several uncertain parameters and significant computational costs. Moreover, since masonry is a composite material, detailed modeling strategy must then include units, mortar and interface (Lourenco 2002). For this reason, detailed modeling approaches are rarely applied to large structures and used for practical applications.

The choice of the most appropriate model to represent the structural behavior of masonry buildings is not univocal. Finite element method can be used to define the structural model and simulate the structural behavior, but accurate and robust constitutive models for masonry are needed.

An example of possible different modeling strategies for the same masonry wall is shown in Fig. 1, considering only mono-dimensional (Fig. 1b and c) and two-dimensional (Fig. 1d) elements, masonry being treated as a homogenous material in both cases. 2D continuum elements (Fig. 1a) are adopted by advanced FE software and require the definition of specific constitutive laws for the yield surface of the homogenized masonry material (Lourenco et al. 1998) and are generally not used in current practice.

Using mono-dimensional beam elements, the structure is conceived as an equivalent frame in which walls are divided into macro-elements: piers and spandrels connected by rigid nodes. This kind of modeling, known as Equivalent Frame Model, allows to create a three-dimensional outline of the structure and enable a global analysis. Each vertical beam has the characteristics of the corresponding pier in terms of mass, stiffness and strength, while the horizontal beams have the corresponding spandrels' ones. Most commercial software are based on the Equivalent Frame Model approach but, as it will be shown in the following section, they may significantly differ in the schematization of piers and spandrels (see also, Fig. 1b and c).

Once defined the structural model, the other key element for the analysis is the constitutive law, which is adopted to describe the non-linear behaviour of masonry.

In the following section, a detailed description of the modelling strategies and material assumptions adopted by the investigated softwares will be discussed.

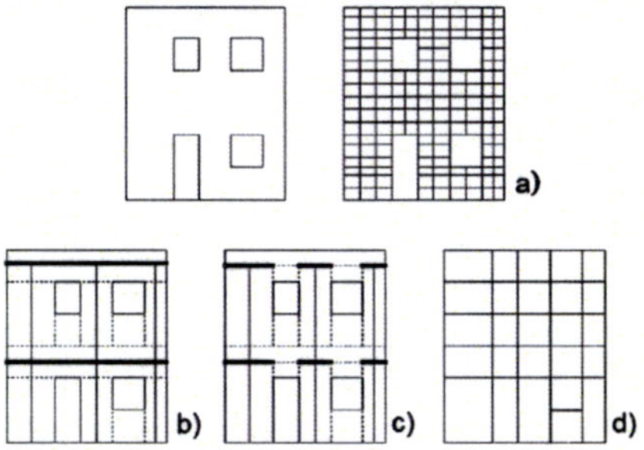

Fig. 1. Examples of different modelling strategies, a) 2D continuum elements, b) equivalent frame model with rigid spandrels, c) equivalent frame model and d) two dimensional macro-elements.

3 Comparison of Software Programs for the Analysis of Masonry Buildings

To discuss the features and the capabilities of the software programs for the analysis of masonry buildings commonly used in Italy, they have been applied to three relevant case studies:

- a single-story building which was the benchmark in the development of the POR method (Regione Autonoma Friuli Venezia Giulia 1980);
- a two-story building tested at the Georgia Tech laboratory (Yi et al. 2006);
- a real case study of a four-story masonry building in the Municipality of Florence.

3.1 Investigated Software Codes

Starting from the POR method (Tomazevic et al. 1978), the first computer program for the analysis of masonry structure, the most common codes for the analysis of masonry buildings in Italy have been analyzed and compared. In particular, the commercial software Aedes PCM (Aedes 2016), 3 MURI (Lagomarsino et al. 2013), Sismicad (Concrete 2018), Modest (Tecnisoft 2018; XFinest 2017) and 3D Macro (Caliò et al. 2011), have been considered together with the non-linear program developed by the authors in (Beconcini et al. 2018) and (Croce et al. 2019a). In the following, a brief description of each code is provided, focusing on the modeling assumptions on both the structural elements and the material behavior implemented.

In the discussion particular attention is devoted to the different material constitutive laws adopted in each software program (Croce et al. 2019b).

POR Method

The POR method was introduced by Tomaževič after the Friuli earthquake in 1976 (Tomazevic et al. 1978), and adopted by the Italian regulations published in 1978 (Regione Autonoma Friuli Venezia Giulia 1978) and 1981 (Italian Ministry of Public Works 1981) for the reparation and strengthening of existing masonry buildings.

The method, which was one of the first computer program regarding the seismic assessment of masonry building, can be seen as a simplified nonlinear, pushover-type method, in which a story failure mechanism for the whole building is assumed.

The global response of each story is evaluated in terms of base shear and story displacement. Each resisting wall is modelled via an elastic-plastic constitutive law with a limited plastic plateau (meaning limited ductility).

The shear resistance curves for each story of the investigated building are calculated assuming that the maximum shear capacity of the story itself corresponds to the first attainment of the ultimate displacement in a wall and the seismic assessment is fulfilled when the shear resistance is bigger than the design seismic action.

E-PUSH

The E-PUSH program is an innovative method for nonlinear static analysis of masonry building, simple and efficient from a computational point of view such as the POR Method, but without its limitations and inaccuracies.

The method, in fact, allows to take into account the ductility of the whole structure and can be applied to analyze mono or multi-story buildings, where stiffness of the floors can be assumed to be, in turn, negligible or rigid, depending on the characteristics of the floors themselves (Beconcini et al. 2018). It is based on the following assumptions:

- the stiffness of the shear walls in the relevant directions, is approximated considering only the lateral stiffness, disregarding the transverse stiffness;
- the capacity curve of each wall is approximated by a bi-linear curve, defined by the ultimate resistance given by the diagonal shear failure, with the plastic plateau limited by the elastic drift δ_e and the ultimate drift δ_u (which may be set both in terms of a ductility factor and inter-story drift);
- the story displacements are considered in order to perform the verification in terms of seismic capacity and demand, using the Acceleration Displacement Response Spectra (ADRS).

During the analysis, at each step of the procedure, the inter-story drift of each shear resistant wall is compared with the elastic drift, δ_e, and the ultimate drift, δ_u, considering 3 possible situations:

- the drift of the wall is lower than δ_e, then the wall is still in elastic phase;
- the drift of the wall is higher than δ_e and lower than δ_u, then the wall is in the plastic range, the shear force is equal to the resistance and the stiffness is reduced;
- the drift of the wall is higher than δ_u, then the wall is collapsed, its shear resistance and its stiffness are set to zero and the wall is assumed to sustain only vertical loads.

The lateral forces are increased until the base shear resistance reduces to approximately 80% of the relative maximum base shear resistance, or as soon as all the walls pertaining to the same floor collapse. In this way, the capacity curve of the whole structure is derived.

The E-PUSH algorithm has been validated focusing on the ability to reproduce the seismic response of masonry building prototypes tested in laboratory (Croce et al. 2019a) and has been also tested on real case studies, comparing the outcomes with those obtained by means of other commercial software (Croce et al. 2018b).

SISMICAD 12.12

SISMICAD is a complete software tool for different structural systems, such as reinforced concrete, steel, wood and masonry. Regarding masonry the program offers the possibility to use 2D elements or equivalent frame modelling. In this study, the attention has been devoted on the second option. The equivalent frame consists of elastic beams with elasto-plastic hinges at both ends, connected by rigid links. The spandrel is characterized by two horizontal elements: one for the masonry with rigid links at the nodes and elastic beams on top of openings, and the second one for the floor, which varies depending on the presence of r.c. curb (see Fig. 2). As an alternative, an equivalent frame model based on inelastic beams with diffuse plasticity can be adopted.

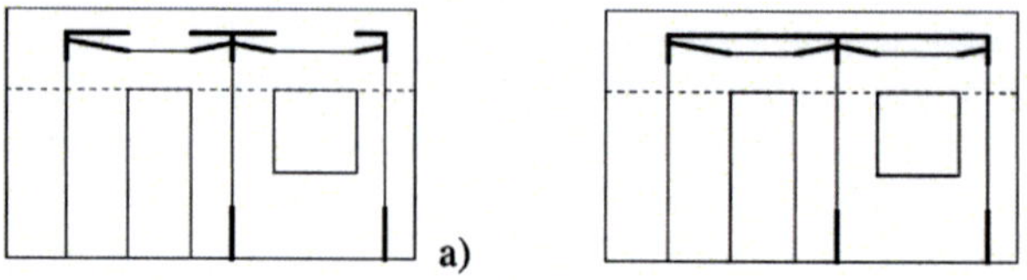

Fig. 2. Sismicad equivalent frame model, a) without r.c. curb and b) with r.c. curb.

Aedes PCM

Aedes PCM is a commercial software specific for the analysis of masonry buildings. It is based on the equivalent frame approach, the default modelling scheme is the one proposed by (Dolce 1991). The length of rigid links is defined assuming for the stiffness contribution from upper or bottom openings spandrels a 30° distribution (see Fig. 3). A peculiarity of this software is that it allows modifications of the generated structural elements, especially concerning the lengths of elastic beams and rigid links. The FEM model is generated replacing each wall with an equivalent frame with one dimensional beam elements, piers and spandrels, connected by rigid links.

As far as the constitutive law of panels is concerned, a bi-linear curve is adopted, with ultimate resistance defined according to the most vulnerable failure modes and ultimate displacement, δ_u, which can be set both in terms of a ductility factor, $\mu = 1.5$, and inter-story drift.

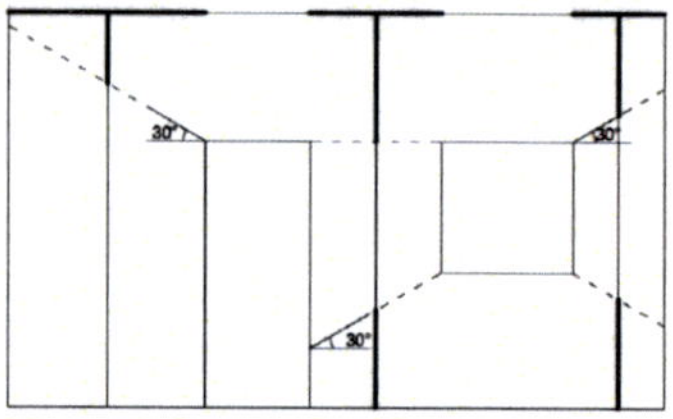

Fig. 3. Equivalent frame model according to (Dolce 1991).

MODEST 8.18 - Tecnisoft s.a.s.

Modest is a complete post-processor which can be adopted for all the main construction materials, it is combined with different solvers and for this study the Xfinest solver has been used.

The program, starting from the 3D building, defines an equivalent frame model according to (Dolce 1991), assuming a 30 degrees distribution for the stiffness contribution from upper or bottom spandrels (Fig. 3). Moreover, it offers two options for the definition of structural walls: continuous piers from the foundation to the floors; piers separated for each story, disregarding the continuity.

Bi-linear constitutive laws are adopted for each element considering the inter-story drift as ultimate displacement.

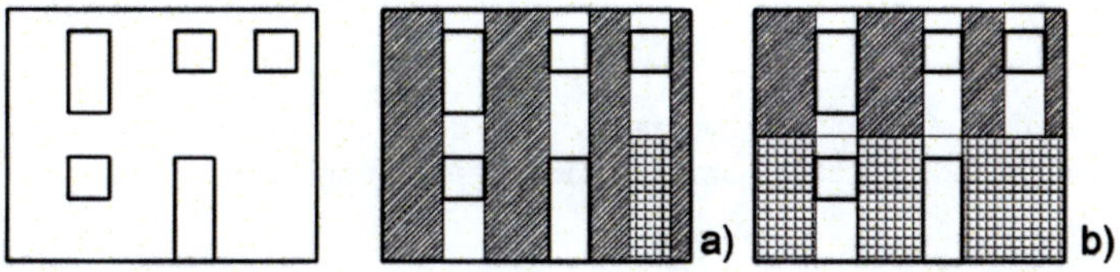

Fig. 4. Modest different modelling options of piers, a) continuous piers and b) single story piers.

3MURI S.T.A. Data

3MURI is a commercial sofware for the non-linear seismic analysis of masonry building, based on the equivalent frame approach, developed at the University of Genoa (Lagomarsino et al. 2013). In particular, two-dimensional non-linear macro-elements (see Fig. 5) are used to model structural components, i.e. piers and spandrels, characterized by the stiffness components reported in the following Table 1. The non-linear behaviour of 2D elements is modelled by a bi-linear constitutive law with cut-off in strength (without hardening) and stiffness decay in the nonlinear phase (Lagomarsino et al. 2013). A constant stiffness reduction coefficient is taken into account during pushover analysis.

Walls are modelled as plane frames which are connected to assemble the 3D model by means of 3-dimensional nodes at the intersection. Finally, the presence of flexible floor can be taken into account by modelling them as 3- or 4-nodes orthotropic membrane finite (plane stress) elements, with two displacement degrees of freedom at each node.

3D Macro

3D Macro is a commercial software for the evaluation of the seismic behaviour of unreinforced masonry buildings, developed by the University of Catania (Caliò et al. 2012). It is based on the macro-elements discretization where the elements consist of articulated quadrilaterals with rigid edges connected by two diagonal springs governing the shear behaviour. The flexural and sliding shear behaviour is instead modelled by discrete distributions of springs on the sides of the quadrilateral, governing the interaction between the adjacent macro-elements.

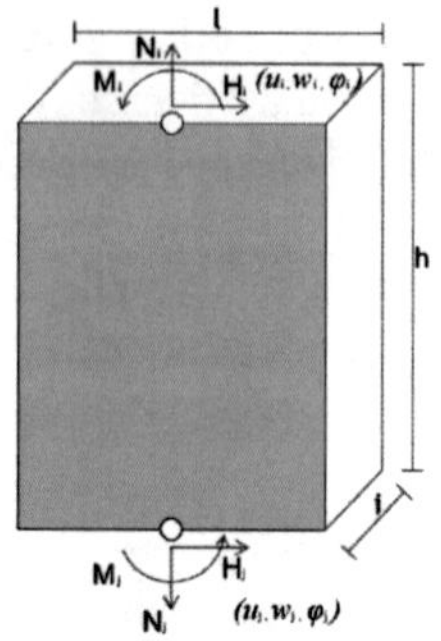

Fig. 5. Idealization of a single panel in 3MURI, macro-element with kinematic variables, generalized forces and geometrical properties according to (Lagomarsino et al. 2013).

Therefore, the shear deformability of the masonry panel is represented by the diagonal spring, while the axial and bending deformability is concentrated in the springs located at the interfaces; in addition, at the interfaces there is also a longitudinal spring to model the sliding, if relevant for the analysis.

In Fig. 6, an example of masonry wall discretization in macro-elements is illustrated. The mesh can be also refined by the user in order to obtain a more detailed description of the kinematics of the structure and the collapse mechanism.

The elasto-plactic behaviour of panels is modelled assigning to each spring (in all directions) a bi-linear law. The limiting displacement for diagonal springs is related to the inter-story height.

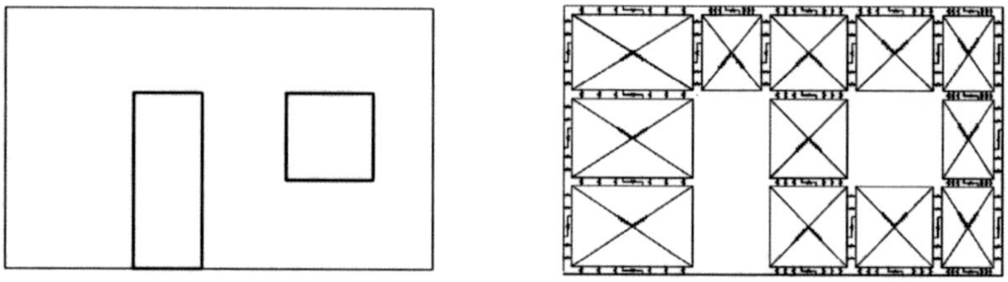

Fig. 6. Example of masonry wall discretization in 3D Macro.

In Table 1, a summary of the different modeling hypotheses in terms of both structural elements and constitutive laws for the masonry material is reported. In the formulae, h is the height of the piers, l is the length, t is the thickness, A is the area of the cross section, J is the in-plane moment of inertia of the wall, E and G are the elastic and shear modulus of masonry, σ_0 is the compressive stress in the wall, τ_k is the shear strength of masonry, f_d is the compressive strength and b is the shear resistance factor, which depends on the aspect ratio of the wall h/l.

Table 1. Main modelling hypotheses of the examined software.

<table>
<tr><th colspan="2">POR Method</th></tr>
<tr><td>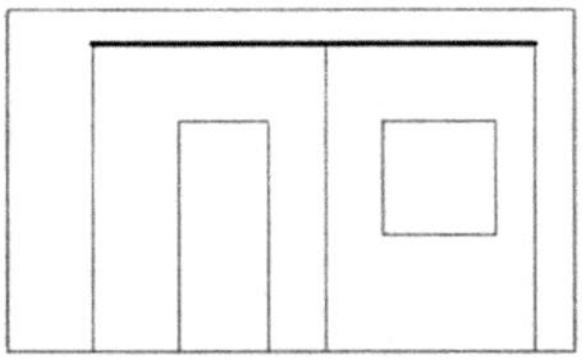
Equivalent frame, defined for each story</td><td>Lateral stiffness
$$k = \frac{GA}{1.2h}\left(1 + \frac{G}{1.2E}\left(\frac{h}{l}\right)^2\right)^{-1}$$
Diagonal shear resistance
$$H_{Rd} = A\frac{1.5\tau_k}{b}\sqrt{1 + \frac{\sigma_0}{1.5\tau_k}}$$
Ultimate displacement
$\delta_u = \mu\delta_e$ with $\mu = 1.5 \div 2$</td></tr>
<tr><th colspan="2">E-PUSH</th></tr>
<tr><td>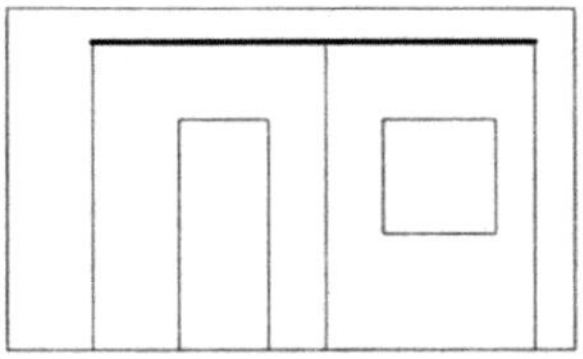
Equivalent frame</td><td>Lateral stiffness
$$k = \frac{GA}{1.2h}\left(1 + \frac{G}{1.2E}\left(\frac{h}{l}\right)^2\right)^{-1}$$
Diagonal shear resistance
$$H_{Rd} = A\frac{1.5\tau_k}{b}\sqrt{1 + \frac{\sigma_0}{1.5\tau_k}}$$
Ultimate displacement
• Ductility check: $\delta_u = \mu\delta_e$ with $\mu = 1.5 \div 2$
Drift check: $\delta_u = \gamma_u h$ with $\gamma_u = 0.004 \div 0.006$</td></tr>
<tr><th colspan="2">SISMICAD</th></tr>
<tr><td>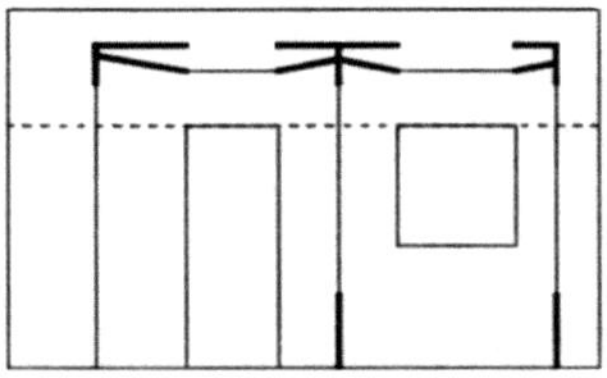
Equivalent frame</td><td>Lateral stiffness
To lateral force $k = \frac{12EJ}{h^3\left(1+\frac{12EJ}{GAh^2}\right)}$
To moment $k = \frac{6EJ}{h^2}$
Rotational stiffness
To moment $k = \frac{4EJ}{h}$
To lateral force $k = \frac{6EJ}{h^2}$
Diagonal shear resistance
$$H_{Rd} = A\frac{1.5\tau_k}{b}\sqrt{1 + \frac{\sigma_0}{1.5\tau_k}}$$
Rocking/crushing resistance
$$M_{Rd} = l^2 t\frac{\sigma_0}{2}\left(1 - \frac{\sigma_0}{0,85f_d}\right)$$
Ultimate displacement
Drift check: $\delta_u = \gamma_u h$ with $\gamma_u = 0.004 \div 0.006$</td></tr>
</table>

(*continued*)

Table 1. (*continued*)

<table>
<tr><th colspan="2">MODEST</th></tr>
<tr><td>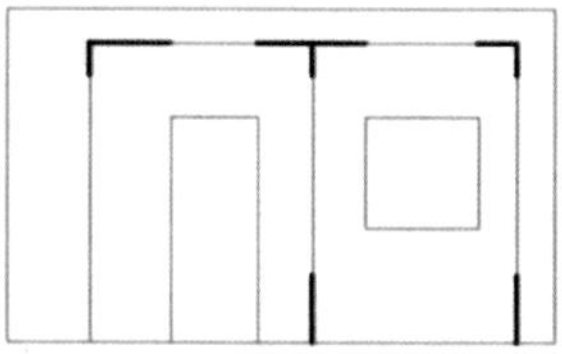
Equivalent frame</td><td>Lateral stiffness
To lateral force $k = \frac{1}{\frac{1,2h}{GA}+\frac{h^3}{3EJ}}$
Rotational stiffness
To moment $k = \frac{4EJ}{h}$
Diagonal shear resistance
$H_{Rd} = A\frac{1.5\tau_k}{b}\sqrt{1+\frac{\sigma_0}{1.5\tau_k}}$
Rocking/crushing resistance
$M_{Rd} = l^2 t\frac{\sigma_0}{2}\left(1-\frac{\sigma_0}{0,85f_d}\right)$
Ultimate displacement
Drift check: $\delta_u = \gamma_u h$ with $\gamma_u = 0.004 \div 0.006$</td></tr>
<tr><th colspan="2">Aedes PCM</th></tr>
<tr><td>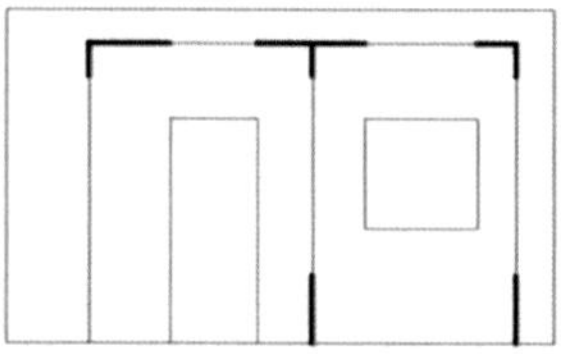
Equivalent frame</td><td>Lateral stiffness
To lateral force $k = \frac{12EJ}{h^3\left(1+\frac{12EJ}{GAh^2}\right)}$
To moment $k = \frac{6EJ}{h^2\left(1+\frac{12EJ}{GAh^2}\right)}$
Rotational stiffness
To moment (top end) $k = \frac{EJ\left(4+\frac{12EJ}{GAh^2}\right)}{h}$
To moment (bottom end) $k = \frac{EJ\left(2+\frac{12EJ}{GAh^2}\right)}{h}$
To lateral force $k = \frac{6EJ}{h^2\left(1+\frac{12EJ}{GAh^2}\right)}$
Diagonal shear resistance
$H_{Rd} = A\frac{1.5\tau_k}{b}\sqrt{1+\frac{\sigma_0}{1.5\tau_k}}$
Rocking/crushing resistance
$M_{Rd} = l^2 t\frac{\sigma_0}{2}\left(1-\frac{\sigma_0}{0,85f_d}\right)$
Ultimate displacement
• Ductility check: $\delta_u = \mu\delta_e$ with $\mu = 1.5 \div 2$
Drift check: $\delta_u = \gamma_u h$ with $\gamma_u = 0.004 \div 0.006$</td></tr>
</table>

(*continued*)

Table 1. (*continued*)

3 MURI	
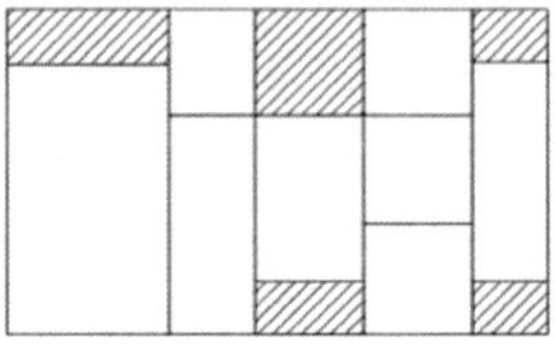 *2D macro-elements*	**Lateral stiffness** To lateral force $k=\frac{12EJ}{h^3\left(1+\frac{1.2EI^2}{Gh^2}\right)}$ To moment $k=\frac{6EJ}{h^2\left(1+\frac{1.2EI^2}{Gh^2}\right)}$ **Rotational stiffness** To moment $k=\frac{EJ\left(4+\frac{1.2EI^2}{Gh^2}\right)}{h\left(1+\frac{1.2EI^2}{Gh^2}\right)}$ To lateral force $k=\frac{6EJ}{h^2\left(1+\frac{1.2EI^2}{Gh^2}\right)}$ **Diagonal shear resistance** $H_{Rd}=A\frac{1.5\tau_k}{b}\sqrt{1+\frac{\sigma_0}{1.5\tau_k}}$ **Rocking/crushing resistance** $M_{Rd}=l^2t\frac{\sigma_0}{2}\left(1-\frac{\sigma_0}{0{,}85f_d}\right)$ **Ultimate displacement** Drift check: $\delta_u=\gamma_u h$
3D Macro	
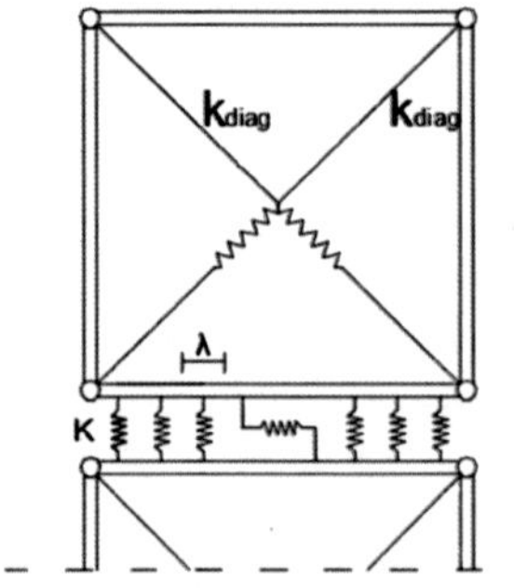 *Articulated quadrilaterals connected by non-linear springs*	**Axial behavior** Elasto-plastic spring which replaces two spring in series k_1 and k_2 for the two adjacent panels $k=\frac{k_1 k_2}{k_1+K_2}$ with $k_1=\frac{E\lambda t_1}{h_1}$ and $k_2=\frac{E\lambda t_2}{h_2}$ Ultimate displacement, assuming spring 1 yielded and 2 in elastic range $\delta_u=\frac{h_1}{2}\varepsilon_u+\frac{F_{y1}}{k_2}$ **Lateral stiffness** Two diagonal springs $k_{diag}=\frac{GA}{2h\cos^2\alpha}$ $\delta=\frac{H}{2\cos^2\alpha\ k_{diag}}$ **Diagonal shear resistance** $H_{Rd}=A\frac{1.5\tau_k}{b}\sqrt{1+\frac{\sigma_0}{1.5\tau_k}}$ **Rocking/crushing resistance** $M_{Rd}=l^2t\frac{\sigma_0}{2}\left(1-\frac{\sigma_0}{0{,}85f_d}\right)$ **Ultimate displacement** Drift check: $\delta_u=\gamma_u h$

3.2 Case Study – "Casolare DT2"

As a first case study, the simple masonry buildings, known as "Casolare DT2", illustrated in the Italian Guidelines (Regione Friuli Venezia Giulia 1978) for the application of the POR method has been considered.

The layout of the building is illustrated in Fig. 7, while the geometrical and mechanical properties of the masonry walls are reported in Table 2.

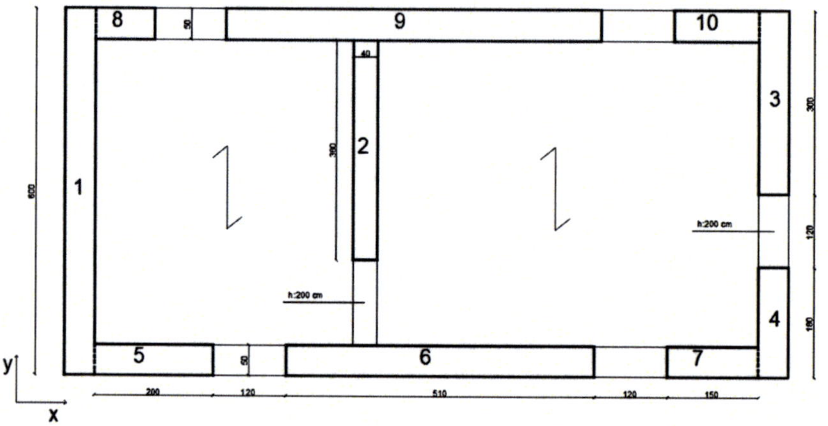

Fig. 7. Layout of the "DT2 Casolare" masonry building.

Table 2. Geometrical and mechanical properties of the masonry walls in the "DT2 Casolare".

Wall	*Lx* [m]	*Ly* [m]	*H* [m]	τk [N/mm^2]	*G* [N/mm^2]	*E* [N/mm^2]	*Fd* [N/mm^2]	γ [kN/m^3]	μ
1	0.50	5.50	3.00	0.112	123	740	3.2	18.5	1.5
2	0.40	3.60	3.00	0.245	269	1615	3.2	10	1.5
3	0.50	2.75	3.00	0.112	123	740	3.2	18.5	1.5
4	0.50	1.55	3.00	0.112	123	740	3.2	18.5	1.5
5	2.25	0.50	3.00	0.112	123	740	3.2	18.5	1.5
6	5.10	0.50	3.00	0.112	123	740	3.2	18.5	1.5
7	1.75	0.50	3.00	0.112	123	740	3.2	18.5	1.5
8	1.25	0.50	3.00	0.112	123	740	3.2	18.5	1.5
9	6.10	0.50	3.00	0.112	123	740	3.2	18.5	1.5
10	1.75	0.50	3.00	0.112	123	740	3.2	18.5	1.5

A non-linear static analysis has been carried out by means of the investigated software codes, the results are reported in terms of capacity curves in Fig. 8, for the seismic force acting along the horizontal direction, *x*. In order to perform the comparison, it has been assumed the presence of rigid and resistant spandrels together with a rigid floor. In case of PCM and E-PUSH program, the two different options for the definition of the ultimate displacement have been considered ("*drift*" solid lines and "*ductility*" dashed lines).

The intrinsic differences among different software can be clearly appreciated in terms of the spread of the capacity curves, even in this simple, but relevant, benchmark study.

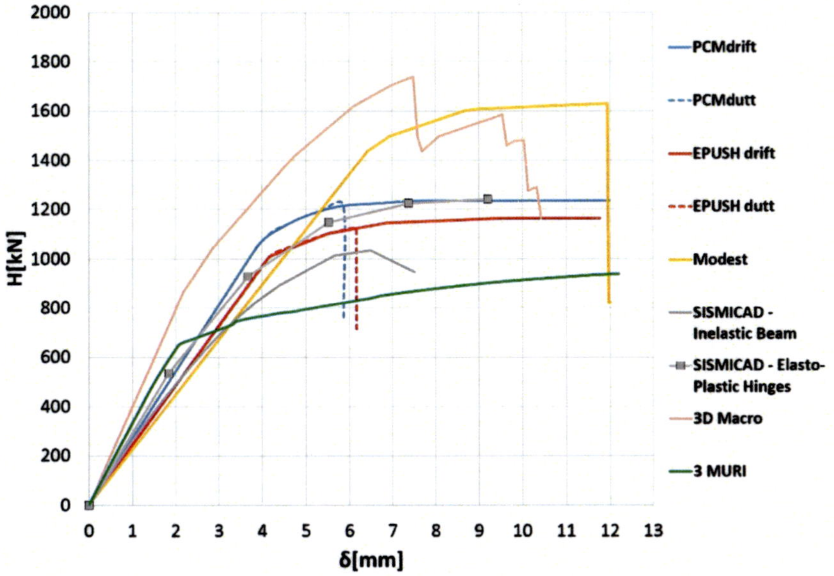

Fig. 8. Comparison of capacity curves obtained from the non-linear static analysis performed with different software, case study of "DT2 Casolare", x direction.

The capacity curves of the whole structure are influenced by the geometric proprieties, mechanical characteristics and displacement capacity associated to the each modelled element. Therefore, depending on the adopted modelling scheme, the first difference is detected in the elastic slope of the capacity curves; the length of flexible beams, the constraints, the presence and localization of rigid links, clearly influence the elements' stiffness.

Observing Fig. 8, we can divide the elastic slopes of different curves in two subgroups:

- software with equivalent frame models including Modest, Sismicad, PCM and E-PUSH, showing comparable initial stiffness. The slightly stiffer behaviour calculated with PCM and Sismicad is related to the presence of rigid offsets at the beam's ends. The Modest model exhibits the lower stiffness in the group because of the adoption of a cantilever beam for pier modelling instead of fixed ended beam. E-PUSH leads to intermediate results;
- a second group, including the remaining software, adopting a modelling scheme based on plane elements, 3MURI and 3DMacro.

Regarding the ultimate resistance, it can be noticed that the average estimate is approximately 1200 kN. 3DMacro and Modest lead to higher values because of the adoption of a shear resistance factor *b* equal to 1, instead of 1.5, for diagonal shear resistance (see Table 1). 3MURI shows the lowest resistance and this is a consequence of the flexural strength behaviour, that governs the failure of walls nr. 7, 8 and 10.

As expected, the ductility behaviour is significantly influenced by the criteria adopted for the definition of ultimate displacement, ductility factor μ or inter-story drift (Beconcini et al. 2019).

3.3 Two-Story Unreinforced Masonry Building Tested at the Georgia Tech Laboratory

The second step of the comparative study has been the evaluation of capacity curves for a two-story masonry building, which was tested under lateral loads at the Georgia Tech Laboratory (Yi et al. 2006a, b). It is a two-story unreinforced brick masonry structure, with timber floor and roof diaphragms, with in-plan dimensions of 7,32 m by 7,32 m and story height of 3,6 m for the ground story and 3,54 m for the first story (Fig. 9).

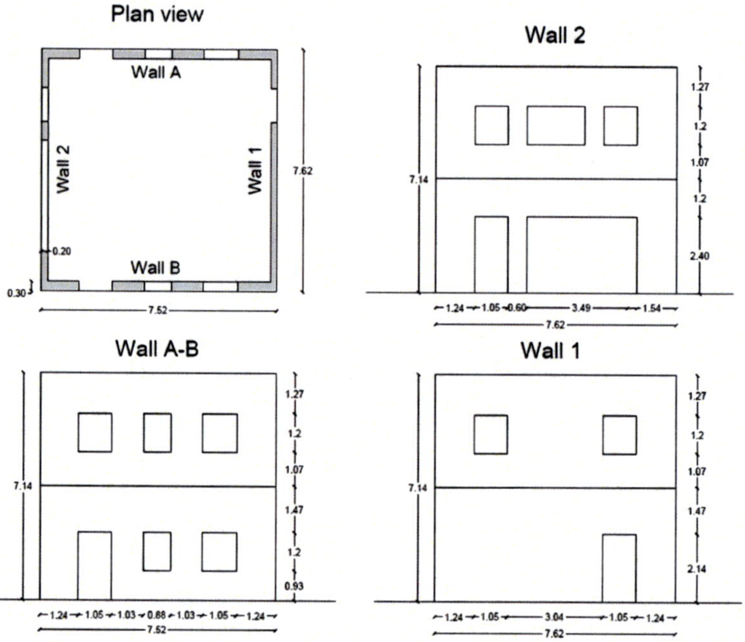

Fig. 9. Sketch of the two story masonry structure tested at Georgia Tech laboratory (Yi et al. 2006a), plan view and elevation.

The building is made-up of four unreinforced masonry walls connected at the corners, labelled as Walls A, B, 1, and 2, designed with different thickness and opening ratios to represent typical masonry walls. The different walls are composed as follows:

- Wall A and B: these walls are identical, 30 cm thick with moderate and well distributed openings;
- Wall 1, 20 cm thick, presents small openings and robust piers
- Wall 2, 20 cm thick, is characterized by a large opening at the ground floor, producing a discontinuous, "false", wall at the first story.

The purpose is to represent some common structural characteristics of existing buildings, the walls A-B have well distributed openings, Wall 1 represents a typical robust wall, and Wall 2, with his large and diffuse openings, is a kind of strong spandrel-weak pier type wall.

The experimental load displacement curves reported in (Yi et al. 2006a) for each wall have been compared with those obtained performing a non-linear static analysis by means of the E-PUSH, Aedes PCM and 3MURI and software, which were selected among those specifically designed for masonry buildings As far as mechanical properties of the brick masonry and the initial vertical stress levels acting on the walls are concerned, the same assumptions were used in the different software.

The results are summarized in Fig. 10 for the walls A and B, and for wall nr. 2, where they are shown the experimental curves (in black) and the capacity curve obtained by means of E-PUSH (in red), Aedes PCM (in blue) and 3MURI (in green).

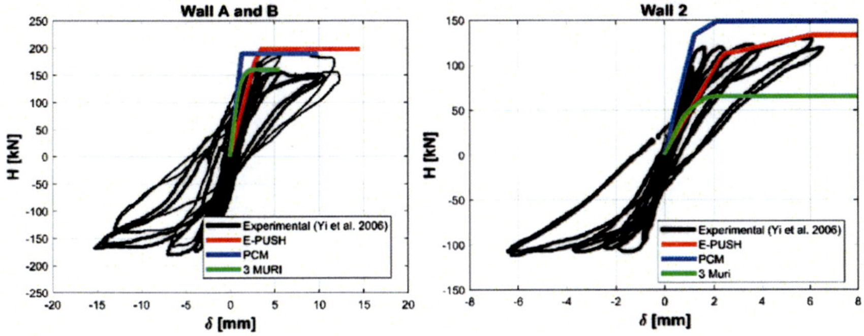

Fig. 10. Comparison of capacity curves of walls A and B and wall 2 with experimental test results.

The curves that are extrapolated from the global analysis can be evaluated considering the different hypotheses, underpinning each software.

The structural behaviour is strongly influenced by the geometric asymmetry of the walls along the *y*-direction. Along the *x*-axis, where the walls A-B provide a simpler and more regular model, the results are easier to evaluate. As expected, the slopes of the elastic branch of the curves are influenced by the length of flexible beams, while the difference in ultimate resistance is motivated by the resistance domain and the consequent failure mode, which is considered by the software. In E-PUSH and PCM, diagonal shear failure has been set, while in 3MURI the lower resistance arising from diagonal shear failure and the rocking/crushing failure is considered. Indeed, the lower values of resistance shown by the 3MURI curves are consequence of the flexural strength which dominates the failure of panels associated to high aspect ratios *h*/*l* (see Fig. 11a, for masonry panel in wall A and B having $l = 1.24$ m, $h = 3.2$ m and $t = 0.3$ m) and low axial load (see Fig. 11b, for masonry panel in wall 1 having $l = 3.04$ m, $h = 2.8$ m and $t = 0.3$ m).

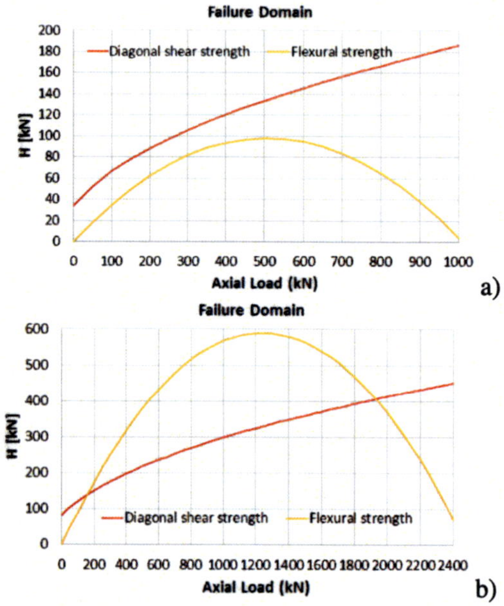

Diagonal Shear Failure

$$H_{Rd} = A\frac{1.5\tau_k}{b}\sqrt{1+\frac{\sigma_0}{1.5\tau_k}}$$

Rocking/crushing failure

$$H_{Rd} = \frac{2M_{Rd}}{h} = l^2 t\frac{\sigma_0}{2}\left(1-\frac{\sigma_0}{0{,}85f_d}\right)$$

Fig. 11. Influence of axial load and aspect ratio (h/l) on the failure domain for masonry walls.

The analysis of the behavior of curves along the *y*-direction is more complex. First of all, the lower stiffness of wall 2, which presents a large opening, leads to the global torsional rotation of the structure. Therefore, along this direction the differences between the software become more relevant.

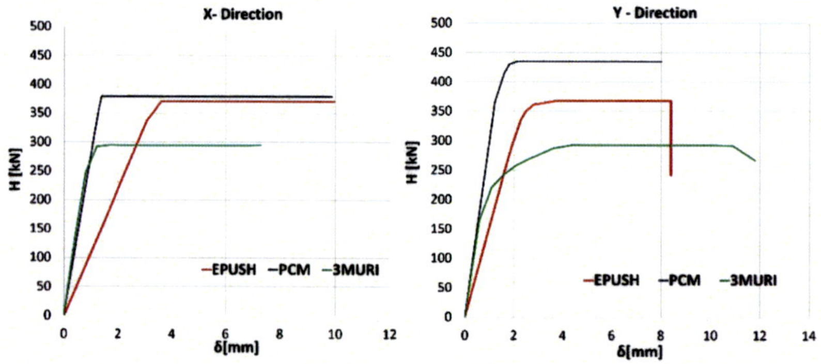

Fig. 12. Comparison of capacity curves of the whole building in the two relevant directions, *x* and *y*.

The results of the global pushover analysis are illustrated in Fig. 12, for the two main directions of the seismic action (x-direction along wall A and B, and y-direction along wall 1 and 2).

3.4 Case Study, The "Niccolini Primary School" in Florence

As the last step of this comparative study, a four-story masonry building has been investigated. The building is the primary school "G.B. Niccolini", which is a Neoclassic building located in the Isolotto, urban district of Florence, was built in 1910. It presents an in-plan L layout and covers a surface of about 1150 m^2 at each story, the inter-story height is 5 m, with the only exception of the ground floor where it is reduced to 3,5 m. The total height of the structure is approximately 18 m, which is significant for an unreinforced masonry building. The building shows a regular distribution of openings in the facades and the wall thickness varies from 70/75 cm for the ground floor to 45 cm in the upper story. Walls are mainly made of dressed stone masonry and only at the upper story there are some walls made of brick and lime mortar. Masonry walls are well connected to each other. Floors are made of one-way rigid slabs, justifying the global 3D modelling of the building.

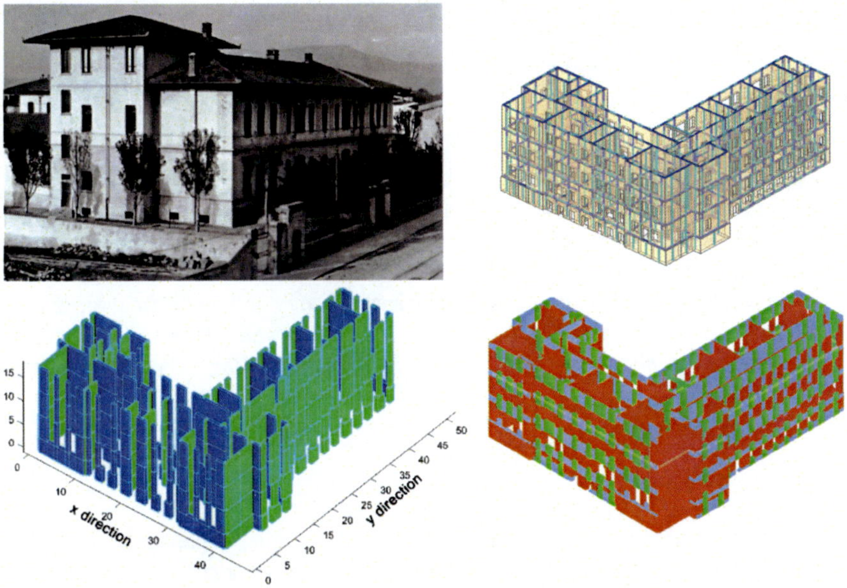

Fig. 13. The "Niccolini school" building and the different structural models: PCM (top right), E-PUSH (bottom left) and 3 Muri (bottom right).

The structure has been modelled by means of the same software programs (see Fig. 13) mentioned in the previous paragraph, assuming the material mechanical parameters reported in Table 3.

Table 3. Masonry mechanical properties adopted in the analysis.

Masonry Type	f_d [N/mm^2]	Tk [N/mm^2]	G [N/mm^2]	E [N/mm^2]	Γ [kN/m^3]
Stone	2,5	0,043	86	516	20
Bricks and lime mortar	3,2	0,076	152	912	18

The results of the global pushover analysis are illustrated in Fig. 14, for the two main directions of the seismic action.

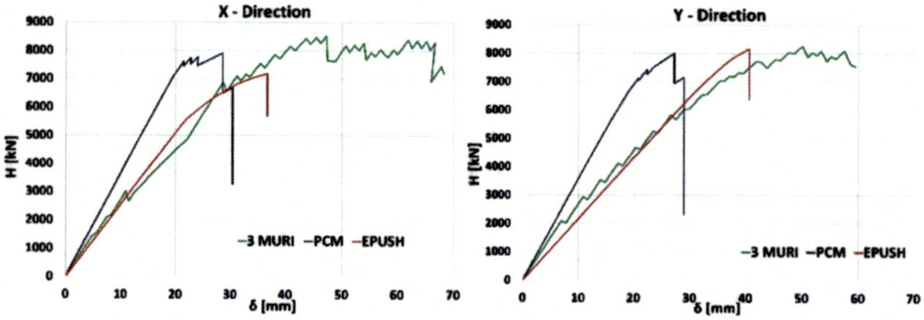

Fig. 14. Comparison of capacity curves for the whole building in the two relevant directions, x and y.

A detailed comparison of the curves confirms some of the results already observed before. In particular, the different slope of the curves obtained by E-PUSH and PCM is consistent with the different modelling assumption on the length of flexible beams, while 3MURI shows a model stiffness comparable to that in E-PUSH due to the absence of r.c. curb. Then, in this case, similar ultimate resistance are obtained in the three program since the failure is dominated by shear for most of the walls. However, the ultimate displacement results significantly higher in 3MURI, thus leading to better performance seismic indexes.

4 Conclusions

In the present work, a comparative study of different methods, implemented in different commercial softwares for the assessment of seismic performance of existing masonry buildings, has been carried out to evaluate the influence of the underlying modelling hypotheses.

Several refined push-over procedures for non-linear static analysis, which are based on structural component modeling, have been investigated together with the E-PUSH software, developed by the authors.

The results of the comparison are presented and discussed for three different case studies: a simple reference one-story building, a two-story prototype tested under lateral loads at the Georgia Tech Laboratory (Yi et al. 2006a), and a four story masonry school building in Florence. The differences between the investigated programs appear generally in the range of the uncertainties of the adopted mechanical models and there are no evidences to define a reference software.

The illustration and critical review of the obtained results clarify that the detected differences are mainly related to the different sources of uncertainties affecting the analyses and to the modelling criteria, implemented by different software.

Unfortunately, there are no enough experimental data available to identify one of the modelling techniques as the most appropriate for general purposes.

In general terms, more complex analyses require a higher number of input parameters, which are themselves unavoidably affected by significant uncertainties, making the increased computational effort not generally justified.

On the other hand, simplified models, which are commonly used in engineering practice, may suffer from unappropriated modelling when non-experienced users are facing complex design cases.

Evidently, it should be avoided to consider the outcome of a software analysis as an absolute measure of the actual ability of a structure to sustain seismic actions, especially when the mechanical model is so sophisticated and complex to require assumptions often based on subjective engineering judgements, Consequently, since the assumptions are user dependent, even the same software could lead to different outcomes. For that reason, the authors' opinion is that the results of the analysis should be considered in relative terms, to compare performances obtained with the same software under the same hypotheses. A typical application is thus the evaluation of the efficiency of different strengthening techniques, when comparing the performance of the same model with and without interventions.

This is a further motivation to mitigate some tendency in the engineering practice to use the outcomes of different programs to validate each other or to extrapolate the best estimate of seismic performance indicators from a sort of "mean" model, also considering that available experimental results are obtained on rather small and simple buildings, not fully representative of the real set of masonry buildings.

References

Aedes Software. Aedes PCM User's Manual; Aedes Software per Ingegneria Civile, San Miniato, Italy, (In Italian) (2016)

Beconcini, M.L., Cioni, P., Croce, P., Formichi, P., Landi, F., Mochi, C.: Non-linear static analysis of masonry buildings under seismic actions. In: IMSCI 2018 - 12th International Multi-Conference on Society, Cybernetics and Informatics, Proceedings, Orlando, United States, 8–11 July 2018, vol. 1, pp. 126–131 (2018)

Beconcini, M.L., Cioni, P., Croce, P., Formichi, P., Landi, F., Mochi, C.: Influence of shear modulus and drift capacity on non-linear static analysis of masonry buildings. In: IABSE Symposium, Guimaraes 2019: Towards a Resilient Built Environment Risk and Asset Management – Report, pp. 876–888 (2019)

Caliò, I., Marletta, M., Pantò, B.: A new discrete element model for the evaluation of the seismic behaviour of unreinforced masonry buildings. Eng. Struct. **40**, 327–338 (2012)

Concrete s.r.l. Manuale d'uso Sismicad 12.12. (in Italian) (2018)

Croce, P., et al.: Shear modulus of masonry walls: a critical review. Procedia Struct. Integr. **11**, 339–346 (2018)
Croce, P., et al.: Influence of mechanical parameters on non-linear static analysis of masonry buildings: a relevant case-study. Procedia Struct. Integr. **11**, 331–338 (2018)
Croce, P., Landi, F., Formichi, P.: Probabilistic seismic assessment of masonry buildings. Buildings **9**, 237 (2019)
Croce, P., Beconcini, M.L., Formichi, P., Landi, F., Puccini, B., Zotti, V.: Seismic risk evaluation of existing masonry buildings: methods and uncertainties. In: IABSE Congress New York: The Evolving Metropolis; IABSE: Zurich, Switzerland, pp. 2510–2515 (2019b)
Dolce, M.: Schematizzazione e modellazione degli edifici in muratura soggetti ad azioni sismiche. L'Industria delle Costruzioni **25**, 44–57 (1991). (in Italian)
Lagomarsino, S., Penna, A., Galasco, A., Cattari, S.: TREMURI program: an equivalent frame model for the non-linear seismic analysis of masonry buildings. Eng. Struct. **56**, 1787–1799 (2013)
Lourenco, P.B., Rots, J.G., Blaauwendraad, J.: Continuum model for masonry parameter: estimation and validation. J. Struct. Eng. **124**(6), 642–652 (1998)
Lourenço, P.B.: Computations on historic masonry structures. Prog. Struct. Eng. Mater. **4**, 301–319 (2002)
Marques, R., Lourenco, P.: Possibilities and comparison of structural component models for the seismic assessment of modern unreinforced masonry buildings. Comput. Struct. **89**, 2079–2091 (2011)
Galasco, A., Lagomarsino, S., Penna, A.: Analisi sismica non lineare a macroelementi di edifici in muratura, X Congresso ANIDIS, Potenza-Matera 9–13 Sep 2001 (2001)
Regione Autonoma Friuli Venezia Giulia. TD n.2—Guidelines for Strengthening and Repair of Masonry Buildings; Gruppo Interdisciplinare Centrale, Udine, Italy, 1980 (in Italian) (1980)
Ministry of Public Works. Guidelines for Application of Structural Code for Strengthening and Repair of Masonry Buildings; Istituto Poligrafico e Zecca dello Stato, Roma, Italy, 1981 (In Italian) (1981)
Tecnisoft. Manuale d'uso ModeSt 8.18 (in Italian) (2018)
Tomazevic, M., Turnsek, V., Tertelj, S.: Computation of the Shear Resistance of Masonry Buildings; Report ZRMK-ZK; Institute for Testing and Research in Materials and Structures: Ljubljana, Slovene. (In Slovene) (1978)
Yi, T., Moon, F.L., Leon, R.T., Kahn, L.F.: Lateral load tests on a two-story unreinforced masonry building. J. Struct. Eng. **132**, 643–652 (2006)
Yi, T., Moon, F.L., Leon, R.T., Kahn, L.F.: Analyses of a two-story unreinforced masonry building. J. Struct. Eng. **132**, 653–662 (2006)
XFinest. Manuale Utente: solutore ad elementi finiti analisi statica, dinamica, non-lineare, Instabilità e pushover, CeAS Centro di Analisi Strutturale, Milano (in Italian) (2017)

Investigating Stone Masonry Arch Bridges in Greece Employing In-situ Measurements and Numerical Predictions

George C. Manos[1](✉), Nick Simos[2], and G. Psyrras[1]

[1] Laboratory of Strength of Materials and Structures, Department of Civil Engineering, Aristotle University, Thessaloniki, Greece
gcmanos@civil.auth.gr
[2] Brookhaven National Laboratory, Upton, USA
simos@bnl.gov

Abstract. The present study presents a series of in-situ measurements conducted at selected old stone masonry bridges, using up-to-date system identification techniques, in an effort to identify their dynamic characteristics in terms of eigen-frequencies, eigen-modes and damping properties. All these information is part of a data base that can be used in the future as a reference for identifying noticeable changes in these dynamic characteristics as part of a structural health monitoring effort for these bridges. Moreover, this information provides a basis for build-ing realistic numerical simulations towards studying the structural behaviour of such stone masonry bridges and assessing their expected structural behaviour in extreme future seismic events. Selected in-situ measurements are presented together with their use in building numerical models of various levels of complexity. These numerical models are finally utilized in assessing the expected performance of specific case studies of stone masonry bridge structures in Greecetowards meeting the demands of extreme events that include design earth-quake loads. The described system identification technique can also be linked to specific actions, such as earthquake activity, and thus serve as warning for specific maintenance counter-measure.

Keywords: Stone masonry bridges · In-situ vibration measurements · System identification · Numerical simulation · Foundation deformability · Plaka bridge collapse

1 Introduction

In what follows selective results are presented from an extensive study, which focused on old stone masonry bridges that are located mainly in the prefectures of Western Macedonia and Epirus and Thessaly in Greece (Gialeridis 1995; Grassos 2007; Milas 2016; Psimarni et al. 2000). These bridges are examples of out-standing stone-masonry construction that was dominant for a long period in these parts of Greece (Fig. 1a and b).

The use of the stone masonry arch that is utilized in forming stone masonry bridges was extensively used in the times of the Roman Empire as part of the transportation

I. Vayas and F. M. Mazzolani (Eds.): PROHITECH 2021, LNCE 209, pp. 671–690, 2022.
https://doi.org/10.1007/978-3-030-90788-4_52

a. Konitsa Bridge, Ipiros, Greece.

b. Plaka Bridge, Ipiros, Greece. (Before its collapse in 1st February 2015)

Fig. 1. (**a**) Konitsa Bridge, Ipiros, Greece. (**b**) Plaka Bridge, Ipiros, Greece. (Before its collapse in 1st February 2015)

system that was established and linked the various provinces of the Roman Empire. Evidence of stone masonry arch bridges prior to Roman times is not known although stone masonry structures in the East Mediterranean area for other uses date to prehistoric times. A well known use of arch/vault stone masonry structural form is the one that can be seen at the royal tombs which have been excavated during the last 200 years in many places in Greece.

a. The royal tomb of Atraeus at Mycenae in Peloponnese Southern Greece.

b. Interior of a Macedonian, royal tomb, Greece.

Fig. 2. (**a**) The royal tomb of Atraeus at Mycenae in Peloponnese Southern Greece. (**b**) Interior of a Macedonian, royal tomb, Greece.

In Fig. 2a the royal tomb of Atraeus at Mycenae-Greece is depicted where stone masonry is employed to form an underground vaulted structure with a diameter at its base of 14.60 m and a height of 13.30 m constructed with 33 subsequent series of stone masonry along the height. The use of such vaulted stone masonry structures demonstrates the efficient utilization of this structural form in order to bear efficiently the dead loads as well as the weight of the overlying soil volume in a state of stress dominated by compression. The use of the stone masonry arch/vault type formation is also evident in the structural system of the royal tombs of the Macedonian kings at Vergina in Northern Greece, dated from 350 B.C. and excavated during the last 30 years (Fig. 2b). Despite the use of arch/vaulted stone masonry structural formations for these underground Macedonian royal tombs at Vergina in Northern Greece there is no evidence of such structural formations being used for bridges at that time. Figure 3a shows the location of the Macedonian royal palaces at Vergina and Pella in Northern Greece (red arrows). In the same figure the location of the remains of an ancient Roman bridge (blue arrow) is also indicated. These remains correspond today to only one main arch with a span of 15 m and height of 7.5 m (Fig. 3b). This surviving part of a Roman stone masonry bridge is dated between 50 A.D. and 150 A.D. and, as can be seen in the map of Fig. 3a, is located at a close distance (25 km) from the Macedonian palaces of Vergina and Pella and 30 km from the city of Thessaloniki.

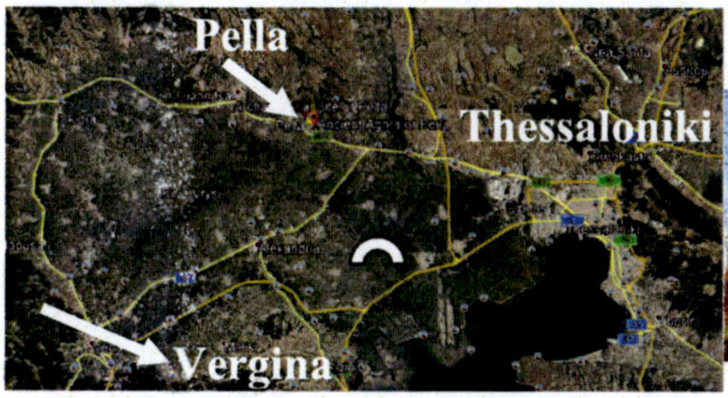

a. Map of Macedonia-Greecde near the gulf of Thessaloniki with the location of the royal palaces at Vergina and Pella, Greece. The location of the remains of the ancient Roman stone bridge is also indicated.

b. The remains of an ancient Roman bridge at a distance of 25km from the Macedonian palaces of Vergina and Pella.

Fig. 3. **(a)** Map of Macedonia-Greecde near the gulf of Thessaloniki with the location of the royal palaces at Vergina and Pella, Greece. The location of the remains of the ancient Roman stone bridge is also indicated. **(b)** The remains of an ancient Roman bridge at a distance of 25 km from the Macedonian palaces of Vergina and Pella.

The old stone masonry bridges that survive today have been built during the 18th and 19th century. They were used to connect villages located in rough mountainous terrain bridging currents that could be quite turbulent during part of the year. This type of transporting people, animals and goods was accomplished using a relatively narrow deck with width varying from 2.0 m to 3.0 m. On the contrary, their size spans from 8 m to 40 m when a single arch is employed (Konitsa, Fig. 1a) or over 70 m for multi-arch structures (Plaka, Fig. 1b). More information on the geometry, construction characteristics and mechanical properties of the employed materials are given by Manos et al. (2016). Today, these structures have retain but only a small part of this primary function, as new roads and bridges have been built to facilitate the contemporary transportation needs. Despite this fact they have recently attracted considerable attention as cultural heritage structures together with an effort to become parts of a network of mountain trails. Because a number of people use these structures as visitors a conservation effort was also initiated for their maintenance. Our study has also partly such an objective. That is to obtain information through in-situ instrumentation and measurements on the vibration characteristics of these structures at a given time in a health monitoring framework, as described in Sect. 2. In addition, these in-situ measurements are complemented with a numerical investigation that has as first target to form realistic numerical simulations of such old stone masonry structures. Moreover, to employ next such realistic numerical simulations in order to assess the behaviour of these old stone masonry bridges to a combination of loads that include accidental actions, such as forces generated from earthquakes or flooding.

a. The studied old stone masonry bridge of Saint Vissariona 23rd August, 2017. Span of main arch 28.9m and height 16.3m

b. The studied old stone masonry bridge of Aziz Aga 31st August, 2016. Span of main arch 28m and height 13.5m.

Fig. 4. (**a**) The studied old stone masonry bridge of Saint Vissariona 23rd August, 2017. Span of main arch 28.9 m and height 16.3 m. (**b**) The studied old stone masonry bridge of Aziz Aga 31st August, 2016. Span of main arch 28 m and height 13.5 m.

2 In-situ Measurements of the Dynamic Characteristics of the Studied Stone Bridges

In measuring the dynamic response of all the studied stone bridges two types of excitation were mobilized. The first, namely ambient excitation, mobilized the wind, despite the variation of the wind velocity in amplitude and orientation during the various tests. Due to the topography of the areas where these stone bridges are located, usually a relatively narrow gorge, the orientation of the wind resulted in a considerable component perpendicular to the longitudinal bridge axis. This fact combined with the resistance offered to this wind component by the façade of each bridge produced sufficient excitation source resulting in small amplitude vibrations that could be recorded by the employed instrumentation. The second type of excitation that was employed, namely vertical in-plane excitation, was produced from a sudden drop of a weight on the deck of each stone masonry bridge (Aoki et al. 2007; Manos et al. 2015a,2015b, 2016, 2017; Ozden et al 2012; Ruocci et al. 2013).

The level of this second type of excitation was capable of producing mainly vertical vibrations; however, depending on the location of the stone-bridge that such an excitation was applied, horizontal vibratory response components could also be recorded. From the recorded three-component response the in-plane and out-of-plane eigen-frequencies and eigen-modes of each studied stone masonry bridge structure could be identified. The employed tri-axial velocity sensors had a sensitivity of 0.001 mm/s and a data acquisition system with a sampling frequency of 800 Hz. The wind orientation relative to the geometry of each bridge structure coupled with the bridge stiffness properties could excite mainly the 1st symmetric out-of-plane eigen-modes. The variability of the wind orientation could also excite, although to a lesser extend, some of the other in-plane and out-of-plane eigen-modes. The drop weight excitation in various locations combined with the placement of the sensors in selected locations was capable of capturing the in-plane as well as out-of-plane modes of response. A careful study of the numerous in-situ response measurements together with the assistance of numerical simulation tools and a back analysis process resulted for such stone bridges in their dynamic system identification. This process is presented in what follows for two old masonry bridges, namely the Saint Vissariona Bridge at Pile, in Trikala prefecture (Fig. 4a) and the bridge of Aziz-Aga, at Grevena prefecture (Fig. 4b).

Due to space limitations only selective measurements of the out-of-plane response, which was recorded utilizing either the wind or the drop weight excitation are included here. More information is reported by Manos et al. (2016).

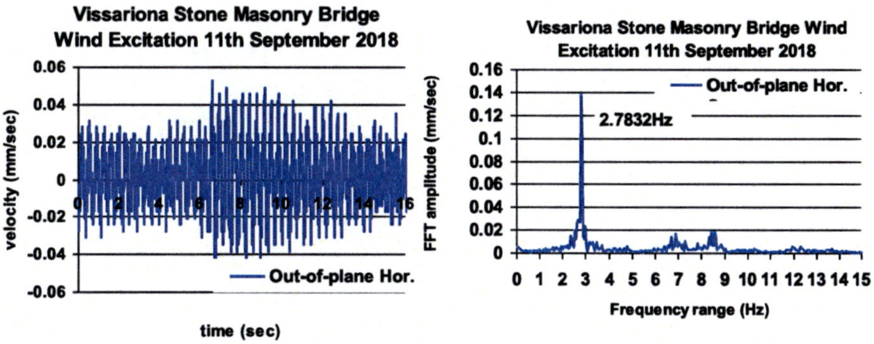

Fig. 5. Wind excitation Saint Vissariona Bridge 9th September, 2018. Out-of-plane horizontal component.

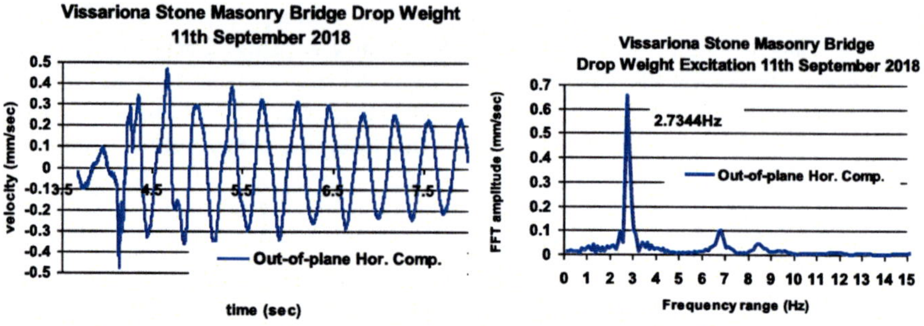

Fig. 6. Drop weight excitation Saint Vissariona Bridge 9th September, 2018. Out-of-plane horizontal component.

In Fig. 5, the out-of-plane horizontal component response of the Saint Vissariona bridge is presented due to wind excitation. The time history plot is shown at the left hand side of this figure with the corresponding FFT plot shown at the right hand side. As dominant frequency the value of 2.7832 Hz can be identified. Figure 6 depicts the corresponding out-of-plane horizontal component response, in terms of time history and FFT plots, for the same bridge due this time to the drop weight excitation. The dominant frequency value is this time equal to 2.7344 Hz. For the Aziz Aga stone bridge the out-of-plane horizontal component response, in terms of time history and FFT plots, due to the drop weight excitation is depicted in Fig. 7. As can be seen in this figure the dominant eigen-frequency is equal to 3.4668 Hz. By studying further the FFT plots of Figs. 4 and 5 one can also identify through the other two distinct peaks the eigen-frequencies of two more high-order out-of-plane response eigen-modes.

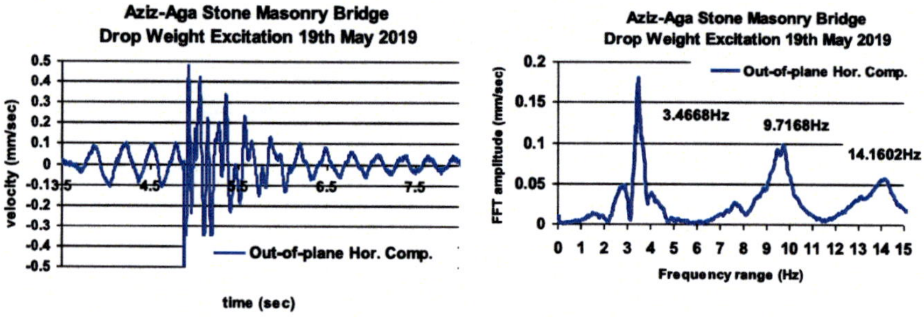

Fig. 7. Drop weight excitation Aziz-Aga Bridge19th May, 2019. Out-of-plane horizontal component

3 Numerical Simulation of the Dynamic Response Recorded In-Situ Utilizing Linear-Elastic Response Assumptions

In what follows an attempt is made to numerically simulate the eigen-frequencies and eigen-modes of Saint Vissariona and Aziz Aga stone masonry bridges, as they resulted from the dynamic system identification process, based on the measured in-situ numerous dynamic response records. At this initial stage the basic assumption of the numerical simulation is the linear elastic response approximation. The overall geometry of the various parts as well as their overall thickness was reproduced as was measured in-situ. The numerical simulation tried to reproduce the primary arch and the secondary arch separately. Moreover, separately were reproduced the abutments and the foundation footings. Thin layers of more flexible material (mortar layers) were introduced in regions where such main parts were connected. Reported by Manos et al. (2016) is a limited study of the mechanical characteristics of stone and mortar samples taken from a prototype bridge which collapsed in 2015 and is currently rebuilt shown in Fig. 6. More information of this bridge's performance is presented by Manos et al. (2019). Initially, for simplicity purposes, these numerical simulations are made in the 3-D domain representing the bridge structure with its mid-surface employing thick shell finite elements (SAP 2000).

The predicted eigen-frequency values obtained from this type of numerical simulation are listed in column 2 of Tables 1 and 2 for the stone masonry bridges of Saint Vissariona and Aziz Aga, respectively. The corresponding modal mass participation ratio values are also listed in this table. The out-of-plane and the in-plane horizontal directions are denoted with the axes y-y and x-x, respectively, whereby z-z is the vertical axis. At a second stage an alternative 3-D numerical was carried out whereby all the various parts of each bridge were represented employing solid prismatic finite elements. The predicted eigen-frequency values obtained from this type of numerical simulation are listed in column 3 of Tables 1 and 2 for the stone masonry bridges of Saint Vissariona and Aziz Aga, respectively.

The corresponding eigen-mode shapes are also included in these tables. The various main parts of this stone masonry bridge, that is the primary and the secondary arches, the abutments, the deck, the mandrel walls and the parapets were simulated in such a way that narrow contact surfaces could be introduced between them, representing in this way a different "softer" medium. All available information, measured during the in-situ campaign, on the geometry of each one of these parts was used in building up these numerical simulations. The mechanical property values obtained from the stone and mortar sample tests were utilized (see Manos et al. (2016). Moreover, there is important information that is needed in order to form with some real-ism the boundary conditions at the river bed and banks. The lack of specific studies towards clarifying in a systematic way all these uncertainties represents a serious limitation in the numerical simulation process. The approximation adopted in this study is a process of back simulation (Manos et al. 2015a,2015b; Simos et al. 2017, 2018). This is done by adopting values for these unknown mechanical stone masonry properties, respecting at the same time all the measured geometric details, which result in reasonably good agreement between the measured and predicted in this way eigen-frequency values. Following this approximate process the boundary conditions introduced in both numerical simulations at all the boundaries, either at the river bed or the river banks, were constraining to zero the translational displacement in the two horizontal and the vertical direction. It was observed from a sensitivity analysis that was performed by Manos et al. (2016) for a number of stone masonry bridges that introducing fixity boundary conditions for the first type of numerical simulation influences, as expected, the out-of-plane and not the in-plane stiffness of the studied stone masonry bridges. Moreover, for the Saint Vissariona and Aziz Aga Bridge whereby the main central arch is supported at the North end in adjacent arches rather than in a rocky river bank the influence of the variation of the boundary conditions, as expected, is again less pronounced.

Table 1. Measured and predicted eigen-frequencies for Saint Vissariona stone masonry bridge

Eigen-modes and measured eigen-frequencies (1)	3-D numerical simulation with **shell elements** representing the longitudinal mid-section (2)	3-D numerical simulation with **solid elements** representing the full geometry of the stone bridge (2)
1st Out-of-plane Measured eigen-frequency 2.832Hz	 Pred. eigen-frequency 2.784Hz, m_y=40.3%	 Computed eigen-frequency 2.719Hz
2nd Out-of-plane Measured eigen-frequency 5.029Hz	 Pred. eigen-frequency 5.270Hz, m_y=0.0	 Computed eigen-frequency 5.240Hz
1st In-plane Measured eigen-frequency 7.129Hz	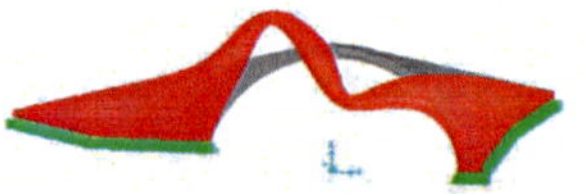 Pred. eigen-frequency 8.38Hz, m_x=9.27% m_z=0.4%	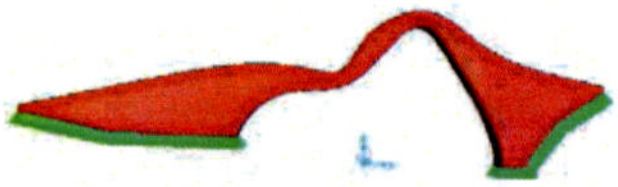 Computed eigen-frequency 8.392Hz
3rd Out-of-plane Measured eigen-frequency 8.839Hz	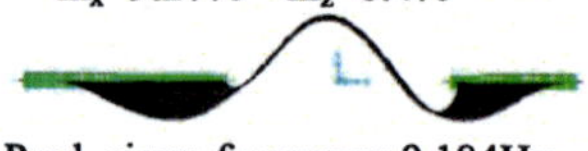 Pred. eigen-frequency 9.184Hz, m_y=20.5%	 Computed eigen-frequency 8.202Hz
2nd In-plane Measured eigen-frequency 9.570Hz	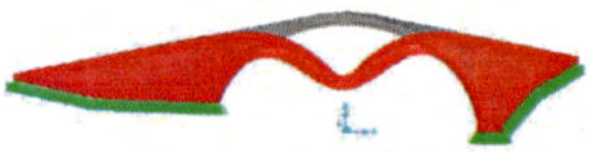 Pred. eigen-frequency 9.45Hz, m_x=0.002% m_z=9.7%	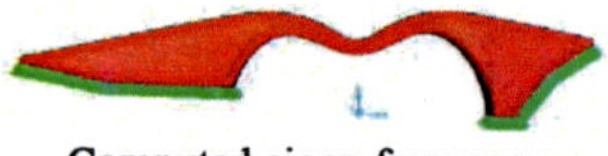 Computed eigen-frequency 9.475Hz
3rd In-plane Measured eigen-frequency 13.037Hz	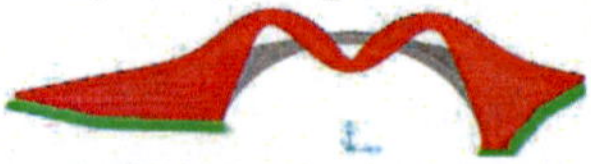 Pred. eigen-frequency 12.69Hz, m_x=0.01% m_z=14.1%	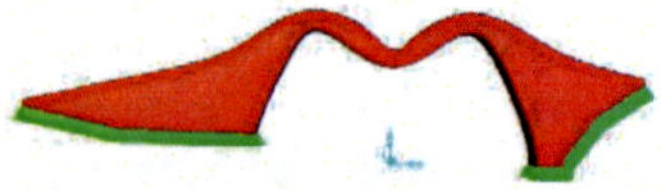 Computed eigen-frequency 12.712Hz

Table 2. Measured and predicted eigen-frequencies for Aziz Aga stone masonry bridge

Eigen-modes and measured eigen-frequencies (1)	3-D numerical simulation with **shell elements** representing the longitudinal mid-section (2)	3-D numerical simulation with **solid elements** representing the full geometry of the stone bridge (3)
1st Out-of-plane Measured eigen-frequency 3.4670Hz	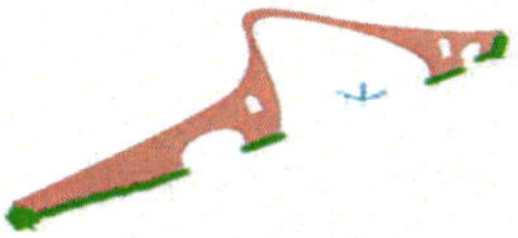 Computed eigen-frequency 3.360Hz	 Computed eigen-frequency 2.618Hz
2nd Out-of-plane Measured eigen-frequency 6.250Hz	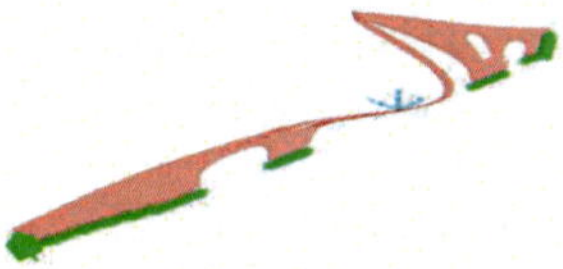 Computed eigen-frequency 5.857Hz	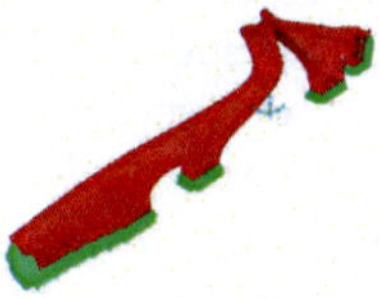 Computed eigen-frequency 4.399Hz
1st In-plane Measured eigen-frequency 8.398Hz	 Computed eigen-frequency 7.920Hz	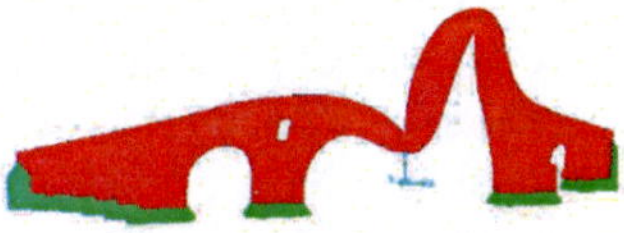 Computed eigen-frequency 7.776Hz
3rd Out-of-plane Measured eigen-frequency 7.764Hz	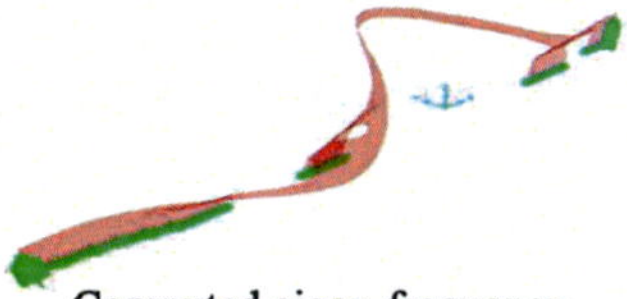 Computed eigen-frequency 8.024Hz	 Computed eigen-frequency 6.150Hz
2nd In-plane Measured eigen-frequency 9.375Hz	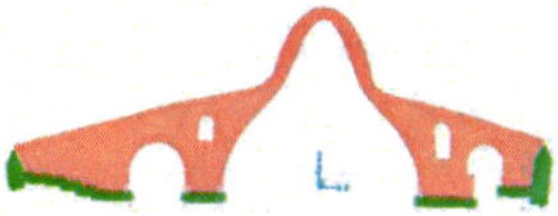 Computed eigen-frequency 9.440Hz	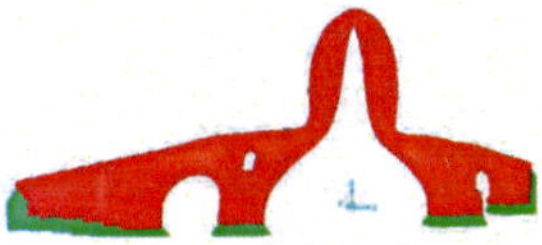 Computed eigen-frequency 9.332Hz

The value of the corresponding measured eigen-frequency values is included in the first column of Tables 1 and 2. These values, as already mentioned, were obtained from the system identification study for each bridge based on the numerous measurements of the vibratory response recorded in-situ utilizing either wind or drop weight excitations (see Figs. 5, 6 and 7). When the predicted eigen-frequencies obtained from the shell elements numerical simulation of Saint Vissariona bridge (Table 1 column 2) are compared with those obtained from the solid elements numerical simulation (Table 1 column 3) a good agreement is observed for all the listed numerically predicted eigen-modes apart for the third out-of-plane eigen-mode. When this comparison is made between the numerically predicted eigen-frequency values with the ones deduced from the in-situ measurements (Table 1 column 1), a good agreement can be observed for the 1st and 2nd out-of-plane and for the 2nd and 3rd in-plane eigen-modes, for both types of numerical simulations. For the 3rd out-of-plane eigen-mode the numerical simulation employing shell element is in better agreement with the measured value than the numerical simulation employing solid elements. Both types of numerical simulation predict an eigen-frequency value for the 1st in-plane eigen-mode that differs substantially for the measured value. This discrepancy is of need of further research. When the predicted eigen-frequencies obtained from the shell elements numerical simulation of Aziz Aga bridge (Table 2 column 2) are compared with those obtained from the solid elements numerical simulation (Table 2 column 3) a good agreement is observed for all three in-plane numerically predicted eigen-modes but not for the corresponding three out-of-plane eigen-modes. When this comparison is made between the numerically predicted eigen-frequency values with the ones deduced from the in-situ measurements (Table 2 column 1), a good agreement can be observed between the measured and predicted values for the 1st type of numerical simulation employing shell finite elements. The predicted values employing solid finite elements are in reasonable agreement with the measured values only for the in-plane eigen-modes but not for the out-of-plane eigen-modes. This must be attributed to the effect of the boundary conditions for the numerical simulation employing solid elements.

4 Seismic Behaviour of the Saint Vissariona Bridge

This section includes results from a series of simplified numerical simulations of the Saint Vissariona stone masonry bridge when it is subjected to a combination of actions that include the dead weight (D) combined with extreme actions generated from seismic ground motions (Sevim Barıs et al. 2011, Kiyono et al. 2012). The seismic forces were defined by making use of the current definition of the seismic forces by EURO-Code 8 (2004). Towards this horizontal design spectral curves were derived (Fig. 8) based on the horizontal design ground acceleration. This value, as it is defined by the zoning map of the current Seismic Code of Greece, is equal to 0.16 g (g the acceleration of gravity) for the location of the Saint Vissariona bridge located at Pyle Trikala Prefecture of Greece (EAK 2000; ITSAK 2019). Furthermore, it is assumed that the soil conditions belong to category B because of the bridge is founded near the banks of the river bed and not directly at a rocky sub-terrain. The importance factor was assigned to a value equal to 1.2 whereas the foundation coefficients was as-signed a value equal to one (1.0); the damping ratio is considered equal to 5% and the behaviour factor is equal to 1.5 (unreinforced masonry).

It must be pointed out that the flexibility of the foundation that was taken into account in the numerical simulation of stone masonry bridges in the past was not considered here. Two types of numerical analyses were performed. In the first type of numerical analysis,

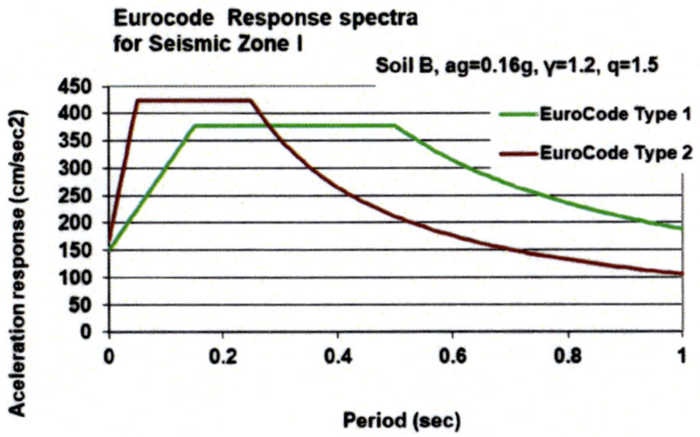

Fig. 8. Design acceleration spectral curves.

the dynamic spectral method was employed making use of the contribution of twelve modes, which include the dominant in-plane and out-of-plane eigen-modes. Moreover, an amplification factor was introduce in order to compensate for the fact that only a portion of the total mass is mobilized by these eigen-modes in the in-plane and out-of-plane directions. In the second type of numerical analysis equivalent horizontal seismic forces are applied either in the in-plane (Ex) or in the out-of-plane (Ey) directions. These forces are of an almost triangular distribution along the height of this bridge with the maximum value applied at the key of the central arch. Moreover, the base shear that results by applying these equivalent horizontal seismic forces is approximately equal to the base shear that results from the dynamic spectral analyses (Manos et al. 2015c). This is done by adjusting the amplitude of the equivalent horizontal seismic forces. In what follows results from this second type of analyses are presented and discussed. Load combinations which included the simultaneous application of the deal load (D) and the equivalent horizontal seismic forces either in the in-plane (D + Ex) or in the out-of-plane (D + Ey) directions have been considered. The obtained deformation and stress results for the Saint Vissariona stone masonry bridge as obtained from these load combinations are shown in the following Figs. 9, 10, and 11.

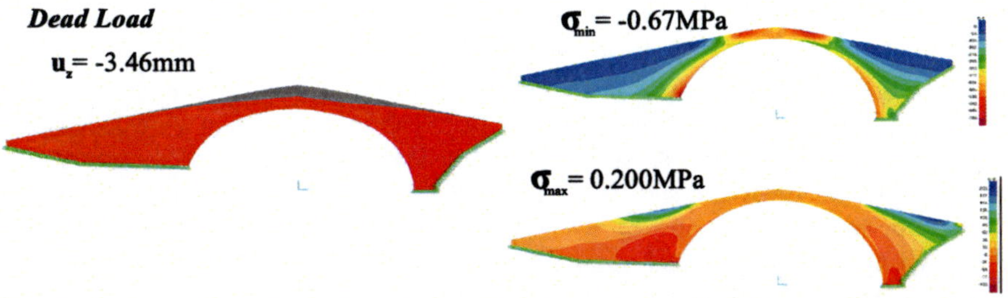

Fig. 9. Deformation and in-plane stress distribution patterns of Saint Vissariona bridge due to the dead load

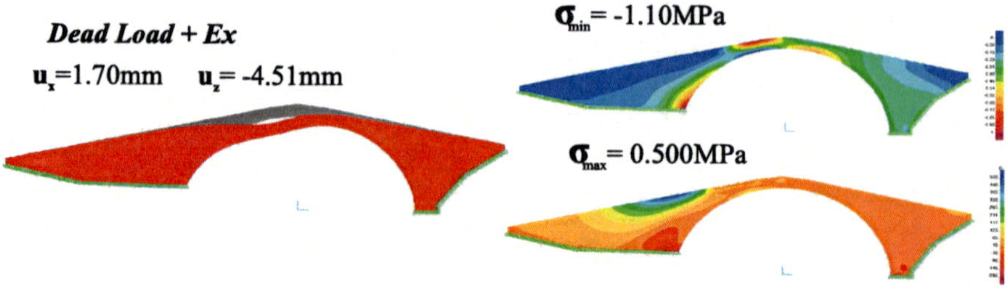

Fig. 10. Deformation and in-plane stress distribution patterns of Saint Vissariona bridge due to dead load and equivalent in-plane seismic horizontal forces

Figure 9 shows the deformation and in-plane stress distribution patterns of Saint Vissariona bridge due to its dead load. As can be seen in this figure, the deformation of the key of the central arch due to dead load is relatively small ($\mathbf{u_z} = -3.46$ mm). In addition the maximum tensile stress value is also relatively small ($\sigma_{max} = 0.200$ MPa) and develops at a confined region at the deck of the bridge above the left and right abutments. The minimum compressive stress value is of the order of $\sigma_{min} = -0.70$ MPa that can be easily sustained by such stone masonry construction. Figure 10 shows the deformation and in-plane stress distribution patterns of Saint Vissariona bridge due to the load combination including the dead load and the in-plane equivalent horizontal seismic forces. Again, the in-plane deformation of the key of the central arch due to this load combination is relatively small ($\mathbf{u_x} = 1.70$ mm $\mathbf{u_z} = -4.51$ mm). However, it can be seen that the region at the deck above the left abutment develops a considerable increase in the maximum tensile stress value ($\sigma_{max} = 0.500$ MPa). Such tensile stress values represent a relatively high tensile stress demand for stone masonry. The region that this high tensile stress value develops correlates with the deformation pattern of the bridge for this load combination. Again, the minimum compressive stress value is of the order of $\sigma_{min} = -0.70$ MPa that can be easily sustained by such stone masonry construction.

Figure 11 shows the deformation and out-of-plane stress distribution patterns at the two extreme faces (top and bottom) of Saint Vissariona stone masonry bridge due to the load combination of the dead load and the out-of-plane horizontal equivalent seismic forces. As can be seen in this figure, the out-of-plane horizontal deformation of the key of the central arch due to this load combination is relatively large ($\mathbf{u_y} = 35.0$ mm $\mathbf{u_z} = -3.88$ mm). As a result, large maximum tensile stress values develop at the central region of the main arch (Bottom face $\sigma_{max} = 2.07$ MPa) as well as at the region where the left abutment joins the foundation. Such tensile stress values represent a relatively very high tensile stress demand for stone masonry. Again, the minimum compressive stress value is of the order of $\sigma_{min} = -5.00$ that, despite the relatively large amplitude, can be easily sustained by such stone masonry construction. As can be seen from the results of such a simplified numerical analysis the Saint Vissariona bridge can be quite vulnerable for earthquake ground motions that are dictated from the cur-rent seismic code. According to historical information this stone masonry bridge was built on 1514 A.D. In 1967 this stone masonry bridge was declared cultural monument by the Greek

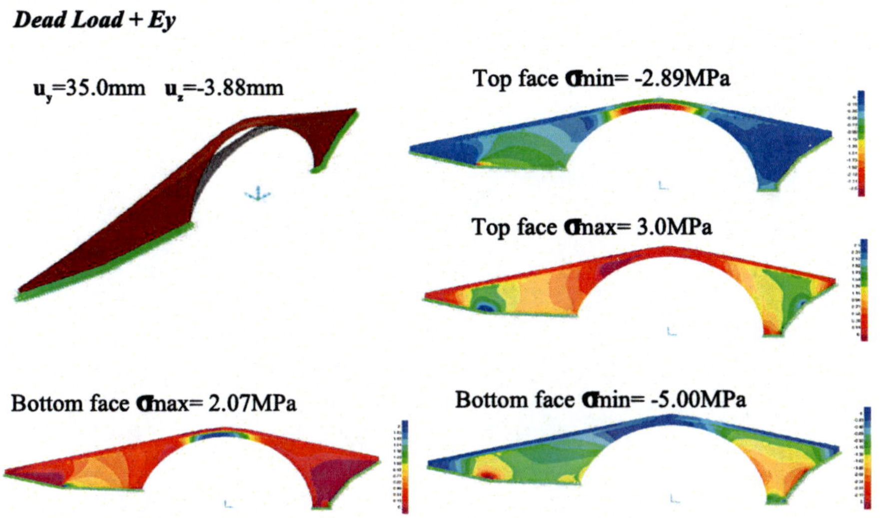

Fig. 11. Deformation and out-of-plane stress distribution patterns at the facades (top and bottom) of Saint Vissariona bridge due to dead load and equivalent out-of-plane seismic horizontal forces

Ministry of Culture and in 2000 the surrounding area was also declared as a region in need of special protection by the same Ministry (Greek Government Gazette 1967). During the last century the most intense earthquake activity related to this bridge must be considered the Sofades in Karditsa earth-quake sequence on 30th of April 1954 which is assigned by Papazachos and Papazachos (1989) a magnitude of 7.0 at the Richter scale with an epicenter approximately 50 km from the site of Saint Vissariona bridge. It was not possible up to now to find any record of structural damage relating this earthquake to Saint Vissariona bridge despite the fact that are reports of substantial structure damage which was widespread in the Thessalia prefecture.

5 Influence of the Foundation Deformability

In the preceding Sects. 3 and 4 the piers and the abutments of the examined stone masonry bridges were assumed to be supported in non-deformable soil-foundation interface. In what follows a preliminary study of the influence of deformable soil where a stone masonry is found is carried out. The Plaka stone masonry bridge collapsed on the 1st of February 2015 due to the flooding of the Arachtos river. Figure 12a depicts this bridge as it was standing before this collapse whereas Fig. 12b shows the remains of the abutments after the collapse of the bridge due to the flooded river which is also shown in the same figure. Figure 12c depicts the collapsed bridge ten months afterwards (12th December 2015) whereas Fig. 12d depicts the picture (19th October 2019) of the reconstruction of the old bridge in the same overall geometry and using materials and constructions techniques as close as possible to the original construction. Recently (7th December 2019), it was announced that the primary arch of the reconstructed new Plaka stone masonry bridge was completed by placing the top key stones at the top- middle of

Fig. 12. Collapse from flooding and reconstruction of the Plaka stone masonry bridge

this primary arch. Manos et al. (2017) investigated, through a simplified linear analysis approach, the effect of the flooding forces acting on the Plaka bridge structure adopting the assumption that the foundation of its mid-pier footing was on a relatively flexible soil.

In addition, the damage suffered by the main arch during an explosion that occurred in 1944, was also included in this study. More information can be found in Manos et al. (2017).

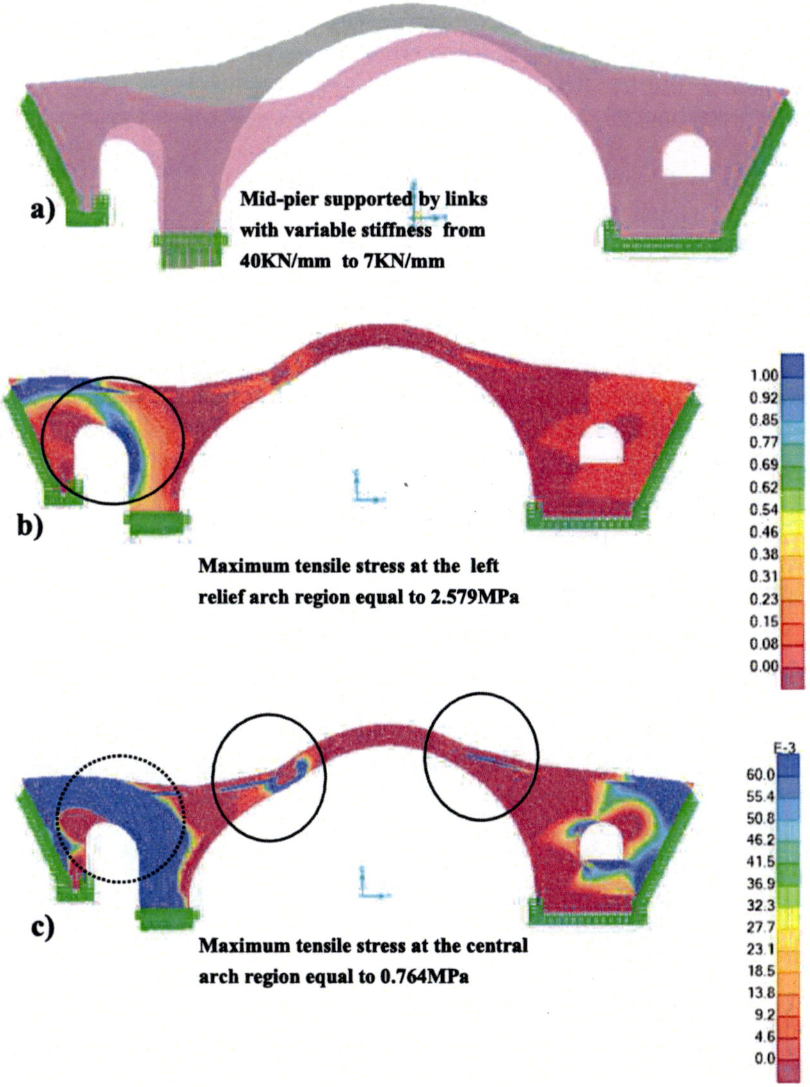

Fig. 13. Displacement and stress response of the Plaka stone masonry bridge

Figure 13a, b, c depict the displacement and stress response of the Plaka bridge under the combined action of the gravity and flooding forces, taking into account the arch damage as well as the deformability of the soil under the footing of the mid-pier. The maximum vertical deflection response, shown in Fig. 13a, is equal to 21.95 mm at the top middle of the central arch. At the same location, the horizontal in-plane displacement is equal to 4.5 mm (y-y direction), whereas the out-of-plane horizontal displacement (x-x direction, flow of river) is equal to 0.87 mm. The maximum (tensile) stress response is depicted in Fig. 13b and c. As can be seen, at the central arch region (Fig. 13c) the maximum stress value is equal to 0.764 MPa whereas at the left relief arch region the maximum stress value us equal to 2.579 MPa (Fig. 13b).

Simos et al. (2017). Investigated the collapse of the Plaka Stone Bridgedue to extreme flooding using high fidelity, three-dimensional non-linear numerical analysis and sensitivity studies. The potential of a number of postulated scenarios in initiating failure and leading to the eventual collapse of the bridge were explored aided by sensitivity analyses relying on non-linear finite element modeling, augmented by actual material properties of the structural elements deduced from laboratory tests following the collapse. More information can be found in Simos et al. (2017). The numerical analysis and the failure scenario postulated and presented in detail in this study tend to support that the triggering of the collapse and the most-likely cause was the undermining by the flood (scouring of the river bed and hydrodynamic loads) of the central pier. In what follows the Plaka Bridgeis numerically simulated this time together with a volume of soil under its foundation of the piers and the abutments within the river bed, as is shown in Fig. 14. Elastic properties are assigned to this volume in an effort to approximate the flexibility of the foundation under the footings of the piers located within the river bed and to further research on the collapse mechanisms.

Fig. 14. Numerical simulation of the Plaka bridge together with a volume of soil.

Apart from numerically simulating the flexibility of the foundation, seven critical cross sections, five at the central arch, one at the left relief arch and one at the right relief arch, were numerically simulated employing contact surfaces which could allow a non-linear palstification response to develop, according to an assumed cohesive constitutive law. In order to check how this numerical model reproduces the effect of the foundation flexibility, simulated as described, the gravity forces were gradually applied till they reached a final level equal to the gravity forces amplified by a factor of 2. The amplified displacement response is shown in. Figure 15a and b, for applied forces reaching the level of 80% and of 200% of gravity, respectively.

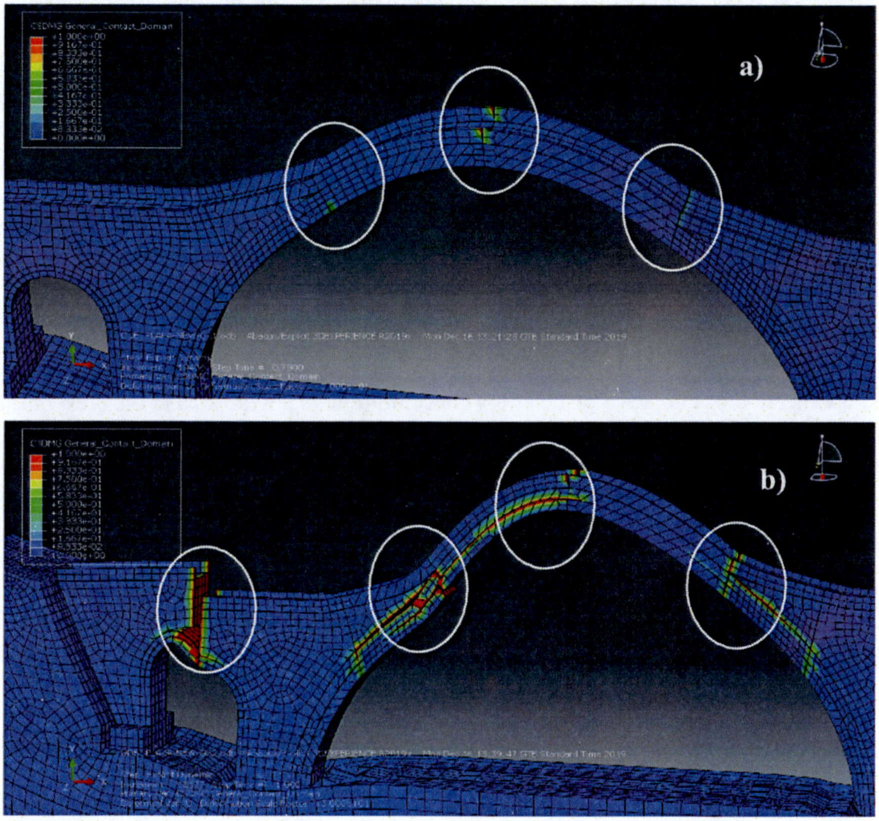

Fig. 15. Plastification of critical cross-sections of the Plaka stone bridge (a) for vertical forces 80% of gravity. (b) For vertical forces 200% of gravity

These plotted deformation pattern of the bridge bears resemblance to the collapse mechanism of the Plaka bridge depicted in Fig. 12b, c. Thus, the numerical response obtained at the end of the described gradual loading sequence shows signs of non-linear structural response which can be considered as realistic. This investigation is still under way by investigating in a parametric way the foundation deformability. Stone bridge behaviour employing advanced non-linear analyses tools were also utilized by Drosopoulos et al (2006), Korompilias (2015), Simos et al. (2018).

6 Conclusions

- Stone masonry construction has a long tradition in many places worldwide. Stone masonry bridges built many centuries ago are one such example. Despite the rigidity and resilience of stone masonry bridges they are in need of maintenance in order to preserve them as part of the built cultural heritage. Towards this end in-situ measurement campaigns were conducted on a number of stone masonry bridges in order to identify their dynamic characteristics in terms of eigen-frequencies, eigen-modes and damping properties. This information is believed to rep-resent a valuable basis for building realistic numerical simulations of the structural behaviour of such bridges as well as for their structural health monitoring.
- The simplified numerical analyses yielded numerical predictions of bridge deformations and stresses that are useful in understanding the structural behaviour and the structural dam-age potential for such masonry structures.
- The peak tensile values predicted when design seismic forces were applied for the Saint Vissariona bridge are many times above the tensile strength of the stone masonry thus indicating the severe potential structural damage for this old stone masonry bridge from the design earthquake. The most vulnerable parts of the bridge, as obtained from the location of the peak tensile stress concentration, is a wide area at the crest of the main central arch of the bridge as well as at the abutment near the foundation.
- The integrity of the stone masonry in various parts of the bridge is an additional maintenance issue of considerable importance. Intervention recommendations for such stone masonry bridges should include clauses for applying preparatory actions of measurements and analyses like the ones included here together with established principles that govern a major retrofitting/maintenance effort for cultural heritage together with effective retrofitting/maintenance techniques that are proven to be durable.
- The described system identification technique can also be utilized in identifying changes of stiffness that may be linked to specific actions, such as earthquake activity, and thus serve as warning for certain maintenance counter-measures.
- As shown from the numerical predictions, the flexibility of the foundation, which may be partly attributed to long or short term erosion of the bridge footings, results in detrimental response for the Plaka bridge. These predicted peak tensile stress values are well beyond the stone masonry strength. Consequently, they could well have contributed towards the February 2015 collapse.
- The peak tensile values predicted when design seismic forces were applied for the Plaka bridge are many times above the tensile strength of the stone masonry thus indicating the severe potential structural damage for this old stone masonry bridge from the design earthquake. The most vulnerable parts of the bridge, as obtained from the location of the peak tensile stress concentration, is a wide area at the crest of the main central arch of the bridge as well as the mid-pier internal side footing. Thus, these earthquake vulnerability predictions for the Plaka bridge indicate that had the bridge not collapsed from flooding the design earthquake would have led to its structural damage and its partial collapse.
- Results form advanced non-linear numerical analyses show that such numerical simulations can reproduce in a realistic way the collapse failure scenario and support the hypothesis that the most-likely cause of the collapse of the Plaka bridge.

References

Aoki T., et al.: Theoretical and experimental dynamic analysis of Rakanji Stone Arch Bridge. In: 7th International Conference on Motion and Vibration Control, MOvIC 04, Honyabakei, Oita, Japan (2007)

Drosopoulos, G.A., Stavroulakis, G.E., Massalas, C.V.: Limit analysis of a single span masonry bridge with unilateral frictional contact interfaces. Eng. Struct. **28**(2006), 1864–1873 (2006)

EAK. Provisions of Greek Seismic Code 2000, OASP, Athens, December 1999. Revisions of seismic zonation introduced in 2003 (2000)

Eurocode 8. Design of structures for earthquake resistance – Part 1 and Part 2: Bridges, DRAFT No 3. European Committee for Standardization, 2004 (2004)

Galeridis, A.: The Stone Masonry Bridges in Thessalia. Technical Chamber of Greece, Eptalofos, Athens, Greece (In Greek) (1995)

Gazette of the Greek Government. (ΦΕΚ 352/Β/31-5-1967) and (ΦΕΚ 194/Β/21-2-2000) (in Greek) (1967)

Grassos, G.: The stone masonry arch bridges of Greece. In: Center of Environmental Education Makrinitsas. Grassos, G. (ed.). ISBN: 978-960-98043-9-4 (in Greek) (2007). http://kpe-makrin.mag.sch.gr

ITSAK: Institute of Engineering Seismology and Earthquake Engineering (ITSAK). Data Base of Greek Earthquake Strong Motions (2019)

Kiyono, J., et al.: Seismic Assessment of Stone Arched Bridges. In: 15 WCEE, Lisboa, Portugal (2012)

Korompilias, D.: Study of the inelastic behaviour of the Konitsa bridge using an inelastic model for masonry and applying strengthening methods. Ph.D. Thesis, Univ. of Patras, Greece (in Greek) (2015)

Manos, G.C.: Consequences on the urban environment in Greece related to the recent intense earthquake activity. Int. J. Civil Eng. Architect. **5**(12) (Serial No. 49), 1065–1090 (2011)

Manos, G.C., et al.: Field experiments for monitoring the dynamic soil-structure-foundation response of model structures at a Test Site. J. Struct. Eng. Am. Soc. Civil Eng. Special Issue "Field Testing of Bridges and Buildings 141(1), D4014012 (2015a)

Manos, G.C., Kozikopoulos, E.: In-situ measured dynamic response of the bell tower of Agios Gerasimos in Lixouri-Kefalonia, Greece and its Utilization of the Numerical Predections of its Earthquake Response. In: COMPDYN 2015, Greece (2015b)

Manos, G.C., Kotoulas, L., Soulis, V., Felekidou, O.: Numerical simulation of the limit non-linear behaviour of unreinforced masonry under in-plane state of stress from gravitational and seismic actions. In: COMPDYN 2015, Greece (2015c)

Manos, G.C., Nick Simos, N., Kozikopoulos, E.: Chapter 4: The structural performance of stone-masonry bridges. In: Structural Bridge Engineering (2016). ISBN 978-953-51-2689-8, Print ISBN 978-953-51-2688-1. https://doi.org/10.5772/64752

Manos, G.C., Kozikopoulos, E., Kotoulas, L., Simos, N.: In-situ measurements related to the performance of stone masonry bridges in Greece. In: COMPDYN 2017, 6th ECCOMAS Thematic Conference on Computational Methods in Structural Dynamics and Earthquake Engineering. Papadrakakis, M., Fragiadakis, M. (eds.). Rhodes Island, Greece (2017)

Manos, G., Simos, N., Lambri-Gaitana, N.: Dynamic and seismic behaviour of stone masonry brisges in Greece utilizing in-situ measurements and numerical predictions. In: COMPDYN 2019, 7th ECCOMAS Thematic Conference on Computational Methods in Structural Dynamics and Earthquake Engineering. Papadrakakis, M., Fragiadakis, M. (eds.). Crete, Greece (2019)

Milas, I.: Stone Bridges. Ethnos Travel Book. Ethnos Publications. ISBN: 978-960-577-122-5 (in Greek) (2016)

Ozden, C.B., Ozakgul, K., Tezer, O.: Assessment of a concrete arch bridge using static and dynamic load tests. Struct. Eng. Mech. **41**(1), 83–94 (2012)

Papazachos, B.K., Papazachou, K.: Earthquakes in Greece. Ziti Publishers, Thessaloniki, Greece (1989).(In Greek)

Psimarni, K., Georgopoulos, A., Balodimos, D.D.: Development of a Geographic In-formation System for the Traditional Bridges of Central Zagori. Report to the Municipality of Zagori, in Greek (2000)

Ruocci, G., Quattrone, A., Zanotti Fragonara, L., Ceravolo, R., De Stefano, A.: Chapter 6: Experimental testing of a masonry arch bridge model subject to increasing level of damage IRIS. In: Industrial Safety and Life Cycle Engineering, VCE Vienna Consulting Engineers (2013). www.vce.at, ISBN 978-3-200-03179-1

SAP. Integrated Software for Structural Analysis and Design. Computers and Structures Inc. (2000)

Barıs, S., et al.: Finite element model calibration effects on the earthquake response of masonry arch bridges. Finite Elem. Anal. Des. **47**(2011), 621–634 (2011)

Simos, N., Manos, G.C.: Numerical analysis of seismic response of natural stone arch bridges-field observations and a case study. In: COMPDYN 2013 (2013). http://www.eccomasproceedings.org/cs2013/

Simos, N., Manos, G.C., Kozikopoulos, E., Kotoulas, L.: On the assessment of the Plaka Bridge Collapse using non-linear analysis preventable or doomed? In: COMPDYN 2017, 6th ECCOMAS Thematic Conference on Computational Methods in Structural Dynamics and Earthquake Engineering. Papadrakakis, M., Fragiadakis, M. (eds.). Rhodes Island, Greece, 15–17 June 2017 (2017)

Simos, N., Manos, G.C., Kozikopoulos, E.: Near- and far-field earthquake damage study of the Konitsa stone arch bridge. Eng. Struct. **177**(2018), 256–267 (2018). https://doi.org/10.1016/j.engstruct.2018.09.072

The Performance of Two Unreinforced Masonry Churches in Greece Under Gravitation and Earthquake Actions

George C. Manos[1(✉)], Lambros Kotoulas[2], V. Matsou[2], Olga Felekidou[2], J. Arnaoutis[2], and Konstantinos Katakalos[2]

[1] Laboratory of Strength of Materials and Structures, Aristotle University, Thessaloniki, Greece
gcmanos@civil.auth.gr

[2] Laboratory of Strength of Materials and Structures, Aristotle University, Thessaloniki, Greece

Abstract. The seismic performance of "Greek" churches made by low-strength unreinforced masonry is examined. These structures were damaged during recent strong seismic activity in Greece combined with long term effects from foundation settlement. Measurements from shear-sliding laboratory tests of stone masonry triplets are presented and discussed together with corresponding numerical simulation results in an effort to quantify the in-plane sliding shear failure criterion. The cohesive surface interaction constitutive law was employed in forming realistic limit-state criteria for such weak mortar stone masonry. Next, the performance of specific "Greek" churches is numerically simulated employing simplified dynamic linear elastic analysis and assumed limit-state criteria. This simplified approach yielded realistic predictions of the observed performance. However, the necessity to obtain a comprehensive set of measured strength properties for such type of masonry construction is underlined. From this simplified dynamic linear elastic analysis it can be concluded that the soil-foundation deformability results in a significant increase in the tensile stress demands at critical regions. This conclusion is also in agreement with the observed damage.

Keywords: Greek-churches · Unreinforced-masonry · Earthquake-performance · Foundation-settlement · Shear-sliding · Numerical-predictions · Laboratory-measurements

1 Introduction

During the last fifty years various parts of Greecehave been subjected to a number of damaging earthquakes ranging from Ms = 5.2 to Ms = 7.2 on the Richter scale. Some of these events, not necessarily the most intense, occurred near urban areas (Manos 2011). One of the most demanding tasks for counteracting the consequences of all these seismic events was the effort to ensure the structural integrity of old masonry structures which sustained considerable damage. In this framework, it was essential to study for a considerable number of old masonry Greek churches their structural system and to investigate the most significant causes of structural damage. From such

I. Vayas and F. M. Mazzolani (Eds.): PROHITECH 2021, LNCE 209, pp. 691–710, 2022.
https://doi.org/10.1007/978-3-030-90788-4_53

an investigation two main contributing factors come to light. The first factor is the pre-existing state of stress and strain either from previous earthquake events and/or from long term permanent foundation settlement. The effect of foundation deformability is significant for structures that are currently designed and constructed with contemporary materials and construction techniques. The effects of interaction between old masonry structures and deformable layers of supporting soil are far more significant. One of the most celebrated cases are masonry towers that are inclined, like the tower of Pisa in Italy, due to soil deformability; in some cases such inclination led to total collapse. An in-depth presentation of the causes of soil settlement and its effect on old masonry structures as well as remedial measures is given by Croci (1998) in the chapter with the relevant subtitle. Stone masonry bridges are another structural type that suffers from foundation settlement (Jones 1990). In this case foundation deformability, which results from long term river flow or short term turbulent river flow from flooding, also leads to collapse of such stone masonry structures (Manos et al. 2016; Simos et al. 2018). The worst case scenario for the various masonry structural elements is the accumulation of stress and strain from such long term effects and the absence of any appropriate counter-measures. The second factor is a strong seismic ground motion and the earthquake forces generated by it (GEER 2014; Manos 2011; Manos and Kramitsios 2013; Manos et al. 2015, 2016, 2019; Lagomarsino 2011; Modena et al. 2010). Such seismic events result in stress demands that many times exceed the capacities of unreinforced masonry structural elements and their connections; this is more likely to occur for masonry structures which have already accumulated considerable pre-existing state of stress and strain (Manos et al. 2015, 2019). The combination of these two contributing factors can lead unreinforced masonry structural members and their connection to a critical state that is accompanied by significant structural damage and partial collapse. The combination of these two actions must be seen in a relatively wide time window when one studies their effect on Cultural Heritage Structures. M. Cerone et al. (2001), by investigating the influence of the soil together with earthquake forces on the Colosseum in Rome (Italy), they concluded that the registered collapses are due to the combination of soil movement together with the earthquake activity over the centuries and the continued lack of maintenance. The most detrimental state of stress is the one resulting from uneven foundation settlement (Croci 1998). Heavy structural damage develops as a result of such actions because the resistance of unreinforced weak mortar masonry to tensile or shear stress demands is rather low and is accompanied with a brittle type behaviour. In many cases this resistance is in effect even lower than assumed due to poor maintenance conditions. The longitudinal and transverse walls of such structures are usually long and thick with very large in-plane stiffness. These walls are interconnected to form the main part of the total 3-D structural system, which mainly resists the horizontal earthquake forces. Moreover, any change of shape imposed by soil-foundation settlement, because of the large in-plane wall stiffness, leads to considerable stressing of these planar structural elements and their interconnections (Manos et al. 2015, 2019). In some cases, these planar masonry walls as well as the internal colonnades support a simple or complex relatively stiff masonry vaulting that is in turn protected by a wooden roof. The soil-foundation settlement and the resulting state of stress-strain that develops on these planar masonry walls spreads also to such masonry vaulting. Therefore, the resulting structural damage develops in either the planar walls or/and the vaulting. Such relatively heavy masonry vaulting also generates large inertia forces in case of a strong earthquake ground motion, which in turn result in large tensile and shear stress demands that can be detrimental for the structural performance of the masonry planar walls and/or the masonry vaulting (Manos and Karamitsios 2013).

2 Simplified Numerical Evaluation Process

In evaluating the static, dynamic and earthquake response of such masonry structures it is initially assumed that the various planar and vaulting masonry structural elements are well interconnected at their intersections as well as with the wooden roof and the foundation. At this "first stage" evaluation, the performance of each particular structural element is assessed individually, assuming that these interconnections are withstanding the imposed demands without any form of damage. Assumed elastic properties are adopted for each individual masonry structural element in order to approximate its in-plane and out-of-plane stiffness characteristics. The actual main architectural features are used to form a three dimensional (3-D) numerical model of the whole structural system. This linear elastic numerical model is further simplified by utilizing shell elements for numerically simulating each masonry structural element thus approximating the in-plane and flexural (out-of-plane) stiffness without having to actually portray the masonry structural thickness in this numerical approximation. Towards this objective, appropriate software packages are utilized in order to form this 3-D numerical model (Abaqus, SAP 2000). It is important at this stage to carefully check all the intersections in order to ensure that there is compatibility in the finite element representation of the 3-D actual structural system, despite the simplification introduced by the use of shell finite elements. The deformability of the foundation is also approximated in two different ways. First, deformable supports are placed under the foundation; these supports have elastic properties equivalent to the deformability properties of the underlying soil layers. Alternatively, layers of deformable soil are used, placed under the foundation of this numerical simulation. Relatively non-deformable soil is characterized when the shear wave velocity is higher than 400 m/s whereas shear wave velocity values smaller than 400 m/s and larger than 200 m/s characterize a soil of medium deformability. For shear wave velocity values lower than 100 m/s the soil is of considerable deformability. By studying the deformability behaviour of the upper layers of a soil volume the elastic stiffness properties of equivalent linear elastic link elements are defined. This numerical process to account for the soil deformability is presented in Manos and Kozikiopoulos (2015b).

Both basic assumptions, e.g. that "the various planar and vaulting masonry structural elements are well interconnected at their intersections as well as with the wooden roof and the foundation", and that "elastic properties are adopted for each individual masonry structural element in order to approximate its in-plane and out-of-plane stiffness characteristics", have a very debatable degree of realism at the same time such an elastic model also possess certain advantages:

(a) Elastic numerical solutions are relatively fast and the corresponding results easily obtained.

(b) The role of the various structural elements in transferring either the gravitational forces or forces generated by extreme actions (e.g. earthquake, snow, wind) can be easily identified.
(c) For low level dynamic/seismic excitations, which do not mobilize the various non-linear mechanisms inherent within such masonry structures, the most dominant dynamic eigen-modes can be measured. In this way, the validity the relevant dynamic system, formed in this way, can be checked and properly adjusted.
(d) Critical structural elements with stress concentration regions can be identified.
(e) Using parametric studies the influence of various parameters, either regarding the actions or the structural formation (e.g. adding or removing parts of the structure, examining the flexibility of the foundation etc.) can be studied with relative ease.
(f) The predicted demands on structural elements resulting from such a simplified elastic numerical analysis can be utilized in predicting plausible limit-state scenarios for the performance of the examined structural system, by adopting realistic strength limit state criteria. When in-situ or laboratory data are not yet available limit-state criteria included in EuroCode-6 may be employed with caution in this simplified evaluation process, assuming low strength mortar. More detailed description of this simplified evaluation process is presented by Manos and Karamitsios (2013), Manos et al. (2019).

In order to estimate the performance of each masonry structural element the following inequalities are employed:

$$Ri = SR_{di}/SE_{di} > 1 \quad (1)$$

or

$$Ri = SR_{di}/SE_{di} < 1 \quad (2)$$

SE_{di} represents the demand posed for each masonry structural element as it results from the simplified numerical simulation; SR_{di} is the corresponding capacity value which is obtained on the basis of assumed strength values for the stone masonry. Note that no safety coefficients are used in estimating the capacity values at this stage of the evaluation process. Inequality 1 signifies safe structural performance. Inequality 2 denotes that the predicted structural performance exceeds a certain limit state thus signifying the development of structural damage corresponding to the specific limit state that is exceeded. These corresponding capacity over demand ratio (Ri) values are used in this simplified numerical evaluation process; a ratio (Ri) value smaller than 1 indicates that a distinct limit state has been reached leading to the corresponding failure mode. The following five common structural damage scenarios are stated corresponding to five distinct relevant limit-states through the relevant ratio values ($Ri = SR_{di}/SE_{di}$). Scenario a1) addresses the in-plane shear limit state which corresponds to a sliding failure mode through the value of the ratio ($R\tau_{sli}$); scenario a2) addresses the in-plane shear limit state corresponding to a diagonal tension failure mode ($R\tau_{dia}$). Scenario b) corresponds to a compressive mode of failure ($R\varsigma$) whereas scenario c) corresponds to the in-plane tensile limit state ($R\sigma$). Finally, scenario d) corresponds to the out-of-plane tensile limit state (R_M). Both scenario c) and scenario d) use an assumed tensile strength value. The above linear-elastic approach is a reasonable approximation for a "first stage" evaluation of the actual behaviour and of predicting regions of structural damage, being both less complex and time consuming than the corresponding non-linear approach (Manos et al. 2007). A relevant expert system was also developed for assessing the various resisting capacities of vertical masonry structural elements (Manos et al. 2015; Manos and Kotoulas 2017). Despite the above stated advantages, the results from such an elastic numerical representation must be considered with sufficient skepticism bearing in mind that the validity of assuming elastic properties as well as assuming full interconnection of the various structural elements is an approximation with serious limitations. Moreover, in order to predict plausible limit state scenarios for the performance of the structural system, apart for the realism of the predicted demands the necessity to form realistic strength limit state criteria is also of obvious importance. Measurements form properly arranged and executed tests in-situ or at the laboratory level are very essential in validating the realism of such predictions of limit-state structural performance. A brief presentation of a series of laboratory tests is presented in Sect. 3. Manos and Kozikiopoulos (2015b) utilized in-situ measurements from a bell tower in Kefalonia island in order to approximate the stiffness characteristics of the underlying soil layers. Moreover, a 3-D finite element approximation of flexible soil layers was utilized to obtain the stiffness properties of equivalent link elements used to replace the soil layer in a 3-D finite element representation of the structure together with its flexible foundation. This 3-D numerical approximation is subjected to a variety of load combinations that include the gravitational forces as well as snow or earthquake loads, described by relevant design provisions. This "first stage" approximate evaluation process was successfully applied in many cases of earthquake damaged Greek churches like the two churches presented in Sect. 4.

3 Sliding-Shear Laboratory Tests

A process is presented that can be utilized to quantify a shear-sliding stone masonry failure criterion in the absence of verified masonry shear strength values. Masonry triplets can be built with materials that somehow resemble the stone masonry of a particular structure of interest. To demonstrate this process specimens are built employing stones with a cubic compressive strength of 60 MPa and medium strength mortar with a mean cubic compressive strength equal to 4.50 MPa; they do not resemble the stone masonry of the churches studied in Sect. 4. Each triplet was built with three stones and two mortar joints, each 20 mm thick. The dimensions of each stone were approximately 160 mm × 380 mm × 200 mm. Each specimen was placed in a testing rig hosting a vertical jack with a load cell and a flat sliding bearing resting at the top surface of each specimen (Fig. 1a). In addition, a horizontal actuator was attached at the mid of the central stone in order to apply a horizontal load in a gradually increasing manner, keeping the vertical load constant at a predetermined level. Thus each specimen was forced to fail in an almost horizontal sliding mode at either one of the two mortar joints. Two displacement transducers were attached at each one of the mortar joints in order to measure the sliding displacement of the central stone relatively to the top and bottom stones. The final objective of this test set-up is to be able to quantify the shear resistance offered by stone-mortar joint which is within a stress field with compressive stresses acting normal to the stone-mortar joint interface. The stone-mortar joint shear resistance involves both the cohesion mechanism between stone and mortar as well as the sliding friction mechanism. The resulting sliding response is depicted in Fig. 1b. The variation of the equivalent shear and normal stress values plotted in this figure were derived from variation of the recorded horizontal and vertical load, respectively, based on the assumption of uniform distribution along the two mortar joint surfaces.

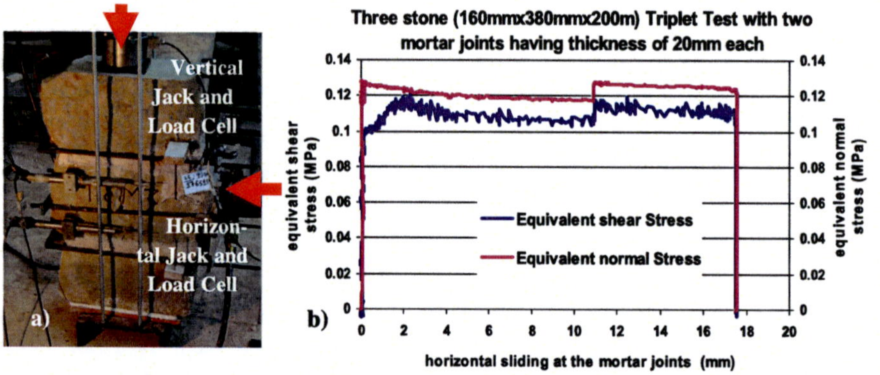

Fig. 1. Shear-sliding test. a) Experimental set-up (left) b) Obtained shear-sliding response under normal stress at the mortar joints (right).

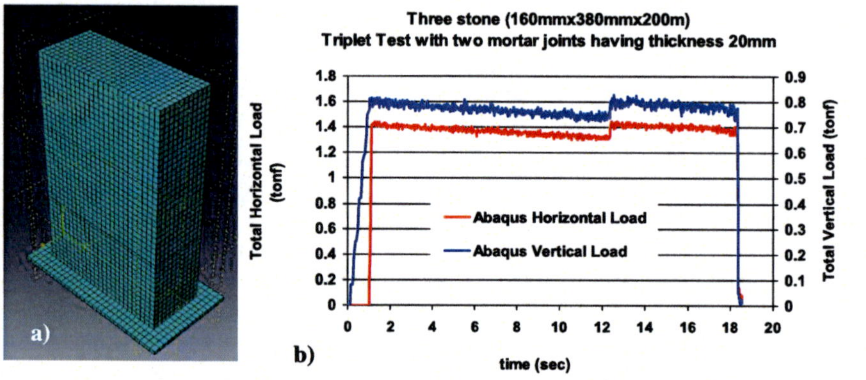

Fig. 2. Numerical simulation of the shear-sliding test. a) F.E. discretization of stones and mortar joints (left) b) Obtained shear-sliding response under normal stress at the mortar joints (right).

3.1 Numerical Simulation of the Triplet Test

A numerical model of the performed triplet test was formed (Abaqus) as shown in Fig. 2a. The three stones were assumed to behave elastically whereas the two mortar joint were provided with non-linear constitutive behaviour. Each mortar joint was numerically simulated with a contact surface joining the two adjacent stones governed by a cohesive surface interaction constitutive law. The most important parameters governing this numerical behaviour is the value of the cohesion as well as the value of the coefficient of friction.

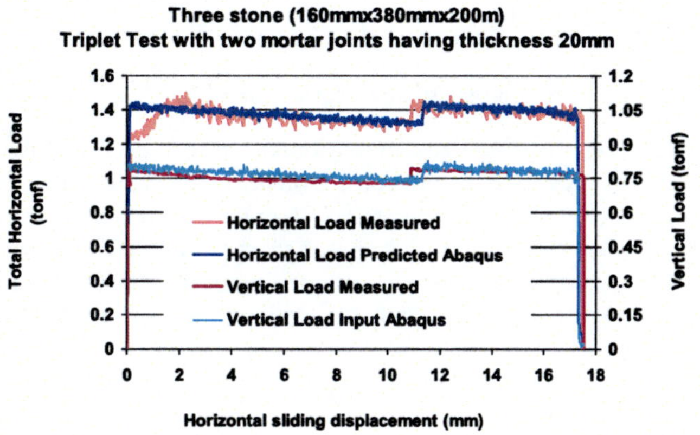

Fig. 3. Comparison between numerically predicted and measured shear-sliding response

In this attempt to simulate numerically the measured triplet response depicted in Fig. 2a the values of these parameters were varied in a parametric way using the measured response as a level of reference through a back analysis process. This non-linear numerical simulation was performed with a time step-by-step analysis with a very small time step that is adjusted by the software according to convergence requirements.

The loading sequence was also varied in time and it included at each time step the measured vertical load, which was applied on the numerical model in the same way as was applied during testing. Together with this vertical load a horizontal displacement was also imposed on the numerical central stone resembling the horizontal load which was applied on the triplet during testing. The resulting horizontal reaction is plotted in Fig. 2b against the corresponding horizontal displacement of the central stone relatively to the adjacent top and bottom stones. Figure 2b depicts the variation of the numerically predicted horizontal and vertical load (numerical), as it resulted from the numerical simulation of the tested stone triplet (Fig. 1a). In this numerical simulation the vertical load is initially applied till it reaches the desired level; then the imposed horizontal displacement is gradually applied in small steps (amplitude and time). This results in a corresponding sudden increase in the resulting horizontal load that soon reaches a peak (1.42 tonf). In Fig. 3 a comparison is made between the measured horizontal force horizontal displacement response of the three stone triplet of Fig. 1a with the corresponding numerical predictions obtained from the numerical simulation described in this section (Fig. 2). In the same figure the vertical load measured during the experiment and applied as input in the numerical simulation is also plotted and compared with the vertical reaction that results from the numerical simulation solution. As can be seen in this figure, good agreement is reached between the experimental measurement and the numerical predictions in this case, utilizing the back analysis process for choosing the values of the various parameters for the cohesive surface interaction non-linear constitutive law.

4 Assessing the Structural Performance of Two Greek Churches Applying the Simplified Numerical Evaluation Process

4.1 The Observed Performance of the Transfiguration of Saviour

This is a Byzantine church in Hortiatis, Thessaloniki, Greecedating since the 12th century A.D. This church is been listed as protected monument since 1967. Its dimensions are of 6.80 m × 8.56 m in plan and a height of 8.16 m with a foundation wall assumed to extend to approximately1m in depth. The South view of this church is shown in Fig. 4a whereas Fig. 4b represents the elastic numerical simulation of the structural system of this church above ground together with its foundation extending at a depth of 1 m below the ground surface.

a)

b)

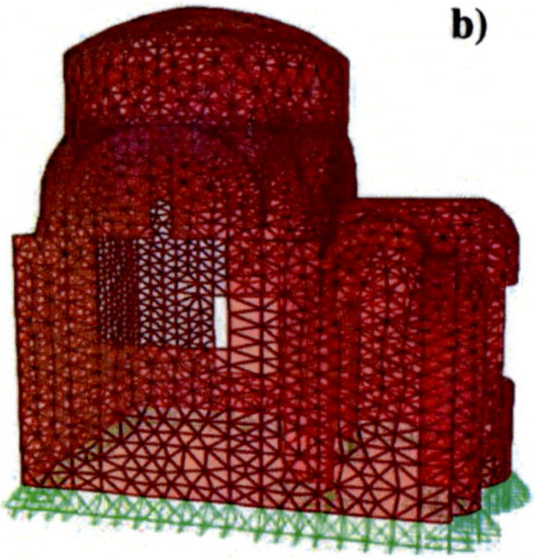

Fig. 4. (a) The Byzantine church of the Transfiguration of Saviour. (b) Numerical simulation

4.1.1 Foundation Settlement

This church has visible signs of accumulated strain (Fig. 5a) that can be attributed to past earthquake activity combined with settlement of its foundation, as described in the introduction. The simplified process described in Sect. 2 is utilized. The flexibility of the soil-foundation interface was approximated by utilizing equivalent linear link elements which were assigned axial low axial stiffness value to approximate in this way the potential for soil settlement in the vertical direction (Manos and Kozikopoulos 2015b; Manos et al. 2019). Two distinct cases of soil-foundation flexibility are considered. In the first case the soil foundation interface was considered as being practically non-deformable for the vertical supports of the peripheral external masonry walls and the internal masonry walls and columns. In the second case the supports of the masonry pilasters in the interior bearing part of the vaulting were considered to be deformable.

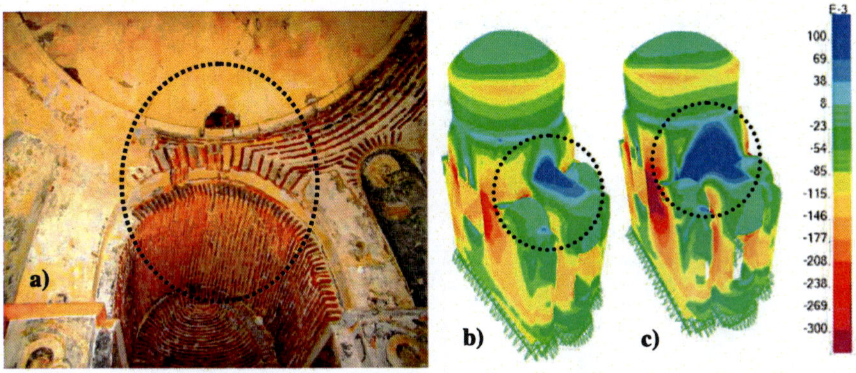

Fig. 5. (a) Observed damage at the East vault. (b) Numerical simulation assuming hard soil (c) Numerical simulation assuming soft soil. Axial σ_{22} stress distribution (MPa). For the vertical walls the axial σ_{22} stresses act normal to horizontal cross-section.

This was approximated in the numerical simulation by assuming flexible vertical supports for these masonry pilasters being assigned a stiffness of 10 KN/mm. The obtained internal fiber axial σ_{22} stress response for the external masonry walls and the dome is depicted in Fig. 5b and c. For the case of non-deformable soil (Fig. 5b) the extreme axial stress values are: Max value 0.18 MPa, Min value −0.25 MPa. For the case of deformable soil (Fig. 5c) the extreme axial stress values are: Max value 0.33 MPa, Min value −0.38 MPa. As can be seen, the considered soil-foundation deformability results in a significant increase in the tensile stresses at the region that the tympanum of the spherical dome is supported on the vaulting part that springs from the two interior masonry pilasters and the apse. This observation is in agreement with the damage that developed in the region of the structural system for this church (Fig. 5a).

4.1.2 Earthquake Action

In this section the earthquake activity is approximate through a linear elastic dynamic spectral analysis (SAP 2000) employing the Type-1 design spectrum of EuroCode-8 (ELOT 2005) for soil category B, with design base acceleration equal to 0.16 g (g is the acceleration of gravity, assuming that this church belongs to seismic zone I, of the new seismic zone map of Greece (EAK 2000). The value of the behaviour factor was assumed to be equal to 1.5.

Only the first two translational modes were considered in the numerical analysis, which mobilized approximately 73% and 68% of the total mass, respectively, compensating this fact with a corresponding amplification factor. The value of the 2nd eigen-period is equal to 0.121 s, corresponding to the East-West translational mode (longitudinal direction x-x from the main entrance located in the West wall towards the apse which is located at the East) with a 68% modal mass ratio (ignoring the mass of the foundation). The value of the 1st eigen-period is equal to 0.175 s, corresponding to the

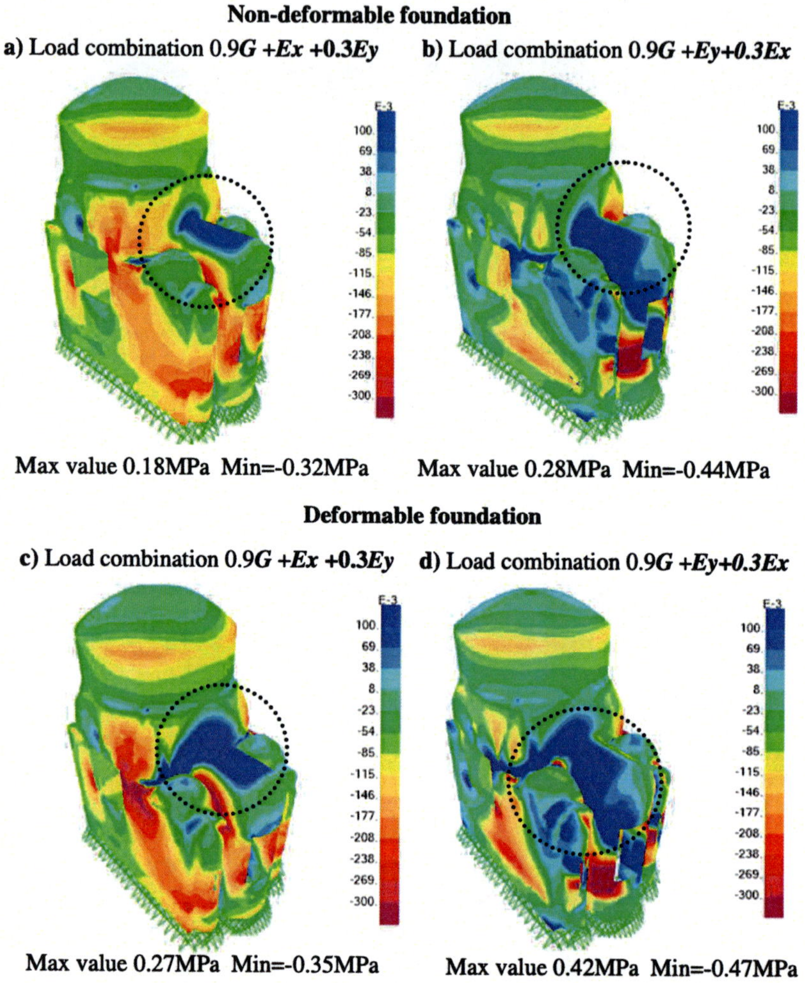

Fig. 6. Numerically predicted axial stress response. Axial σ_{22} stress distribution (MPa). For the vertical walls the axial σ_{22} stresses act normal to horizontal cross-section.

North-South translational mode (transverse direction y-y from South peripheral transverse wall towards the South peripheral transverse wall) with a 73% modal mass ratio. The resulting base shear, along the x-x and y-y directions, attains values approximately equal to 699 KN and 706 KN, respectively. Equivalent lateral static horizontal forcing (Ex and Ey), equally spread in all the masonry elements, resulting in the same base shear values as the dynamic spectral analyses were also considered. The results presented here correspond to the following loading combinations: 0.9G + Ex + 0.3Ey, 0.9G + Ey + 0.3Ex, where G is the dead load. An extended number of load combinations can also be employed. The obtained internal fiber axial σ_{22} stress response for the external masonry walls and the dome is depicted in Fig. 6a and b for the non-deformable foundation and in Fig. 6c and d for the deformable foundation. All these stress results (MPa) are plotted with the same scale in order to be easily comparable. As can be seen the stress response for combination 0.9G + 1.4Ey with deformable foundation results in a maximum tensile value equal to 0.42 MPa, which is by far the largest value obtained from this type of analysis and should be seen as exceeding the corresponding weak mortar masonry strength. It is also interesting to note that the region that such large axial tensile stress response develops coincides with the region of the observed damage (Fig. 5a). It must be underlined that complex numerical simulations in predicting the performance of old masonry structures are also currently available. However, they should not be considered a' priori with a high degree of confidence. They are also faced with the immense variability of old stone masonry in terms of materials and construction techniques and the subsequent lack of relevant in-situ or laboratory measurements. Therefore, it is essential not to rely on the complexity of the numerical modeling; instead to follow a number of limit state scenario, ensuring that these are based on a certain degree of realism for every particular case being studied (Manos et al. 2007).

4.2 The Observed Performance of Profitis Elias

This is a Post-Byzantine three nave Basilica built in 1701 A.D. on the top of a hill in the town of Siatistaof the prefectureof Kozani, Greece. It has a wooden roof without a vaulting system made of stone masonry. The horizontal dimensions of this church are 23.25 m in length and 16.60 m in width. The top of the roof lies at 7.1 m from the floor level of the interior of this church. The naves are formed by 4-column colonnades (Fig. 7a) built with stone masonry. The lower part of the key of each arch of these colonnades lies at 5.25 m from the floor level of the interior of this church. All the exterior walls are made of stone masonry. Apart from the main church a narthex was built at the North side at a later stage; this is of a relatively lower height than the main church. The West part of this church is allocated to the women' quarters that is separated from the rest of the interior by a mid-transverse wall The South longitudinal wall is supported by a system of wooden beams as is shown in Fig. 7b. These were installed after the Kozani 1995 earthquake sequence. Additional wooden supports are also placed at the mid-transverse wall as well as at the mid-span of a longitudinal beam that spans the women quarters from East to West. According to past records, the structural system of this church showed signs of distress from soil settlement sometime before the occurrence of this particular earthquake sequence. The records do not give information of any countermeasures being taken in the past up to the point of the earthquake occurrence. The main structural damage, as recorded after this earthquake sequence, is described below. Inclination of the South longitudinal wall outwards that is accompanied with extensive cracking at its joints with the East and West exterior masonry walls as well as with the mid-transverse wall. Cracking is evident at the arches of the colonnades. The structural rehabilitation of this church commenced almost twenty years after it was damaged by the Kozani earthquake sequence of 1995.

a. Longitudinal cross section

b. Wooden shoring of South longitudinal wall

Fig. 7. The church of Profitis Elias at Siatista – Kozani

4.2.1 Foundation Settlement

The current numerical investigation of this church included the following:

- Simulation of the behaviour assuming non-deformable supports at the foundation level. - Simulation of the behaviour assuming deformable supports at the foundation level, introducing at this level elastic springs with properties reflecting the actual soil deformability that was found from in-situ sampling. The soil consist of clay in its upper layers. For quantifying the stiffness of these soil layers use was made of the data from three boreholes drilled in the vicinity of the church relevant to the constitution of the soil deposits at a depth up to 15 m. All the numerical simulations assumed elastic behaviour with relatively low values of the modulus of elasticity for the stone masonry equal to E = 1660 MPa. The system of the wooden roof was modeled as well as all the wooden elements that connect the longitudinal and transverse walls with the interior colonnades and the mid-transverse wall. A small number of mortar samples were taken from the church to check the mortar composition. The natural stone used in the masonry elements is a type of slate with a compressive strength equal to 24 MPa, as given in the literature survey. The simplified numerical evaluation process, described in Sect. 2, is applied here. The following masonry strength values were assumed towards approximating the capacities SR_{di} (see Sect. 2) of the various structural elements (Euro-Code 6). Compressive strength fk = 1.0 MPa, shear strength fvko = 0.1 MPa, tensile strength normal to bed joints fxk1 = 0.1 MPa, tensile strength parallel to bed joints fxk2 = 0.25 MPa. The ratio of the in-plane tensile or shear strength value over the corresponding demand (see Sect. 2, inequalities 1 and 2) is signified by $\mathbf{R_\sigma}$ or $\mathbf{R_\tau}$ whereas $\mathbf{R_M}$ denotes the ratio of the out-of-plane tensile strength value over the corresponding demand. Ratio values smaller than one ($\mathbf{R\sigma}$, $\mathbf{R\tau}$,$\mathbf{R_M}$ < 1) predict a corresponding limit state conditions. The load combination used to investigate the flexibility of the foundation comprises of all the dead loads of the walls and the roof (G) plus part of the live loads from the ladies' quarters (0.3Q). Initially, the soil foundation interface was considered as rigid for all the vertical peripheral walls and the internal colonnades.

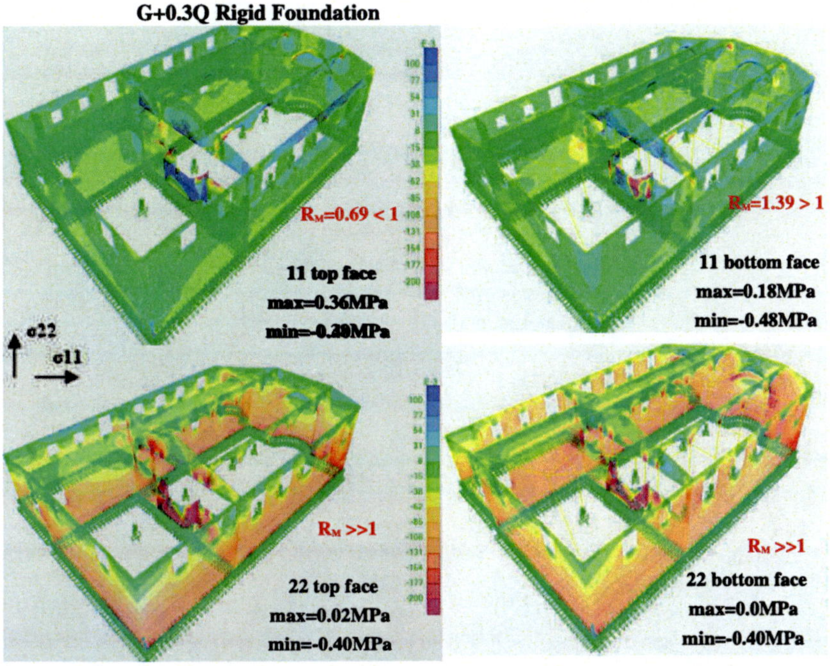

Fig. 8. Stress distribution response of the church of Profitis Elias at Siatista – Kozani for the load combination G + 0.3Q assuming rigid foundation. For the vertical walls the axial σ22 stresses act normal to horizontal cross-section.

The obtained stress distribution response is depicted in Fig. 8. As can be seen in this figure the stress response is dominated by compression. Tensile stresses of limited amplitude develop at the internal top transverse top girder. Therefore the values of the capacity/demand ratio R_M is well above 1 (see inequalities 1 and 2, Sect. 2). Next the same load combination is applied to a numerical simulation of the same overall characteristics as before; however, this time the flexibility of the soil-foundation is introduced at all the supports of the vertical walls (see Manos and Kozikopoulos 2015a, 2015b; Manos et al. 2019). This was done by using linear link elements with a finite stiffness property in the vertical direction. Due to the properties of the soil layers under the church, as they were obtained by a series of boreholes, the stiffness of these link elements of the South wall were assigned a value of 1/10 of the corresponding stiffness properties of the links for the rest of the peripheral walls and the internal colonnades. The obtained stress distribution response is depicted in Fig. 9.

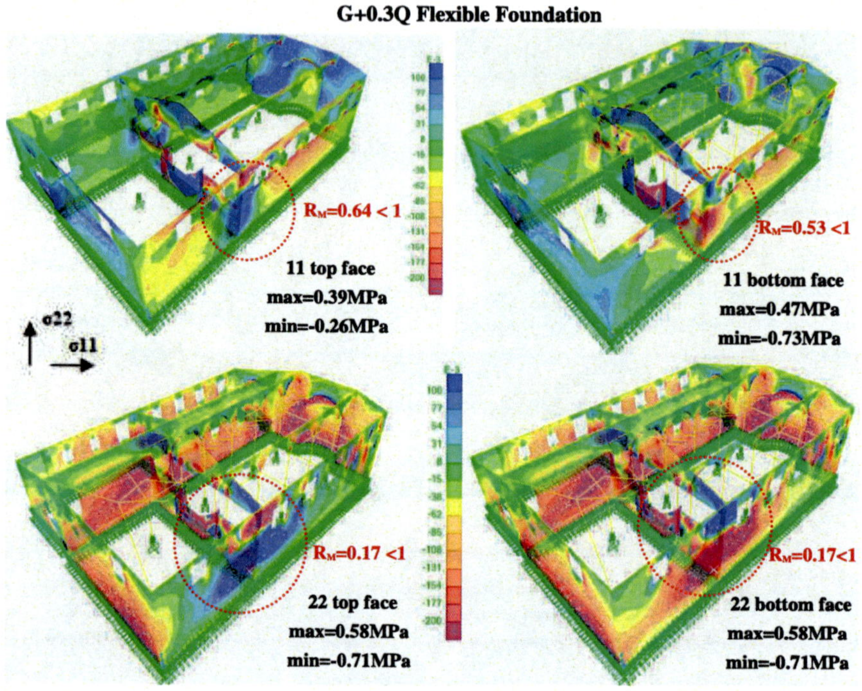

Fig. 9. Stress distribution response of the church of Profitis Elias at Siatista – Kozani for the load combination G + 0.3Q assuming flexible foundation. For the vertical walls the axial σ22 stresses act normal to horizontal cross-section.

As can be seen in Fig. 9, considerable tensile stress response develops at the junction of the South wall with the transverse wall and top beam (this region is encircled in the stress plots of Fig. 9). This transverse wall separates the main part of this 3-nave Basilica church from the women' quarters at its Western part. Based on these maximum tensile stress demands and the corresponding assumed tensile strength values (fxk1 = 0.1 MPa, tensile strength parallel to bed joints fxk2 = 0.25 MPa) this time for the flexible foundation the corresponding ratio values of the capacity/demand (R_M) are well below 1 (see inequalities 1 and 2, Sect. 2). These ratio (R_M) values are shown in Fig. 9, indicating structural damage (Manos et al. (2019).

Figure 10a and b depict the structural damage which was observed at the junction of the South wall with the transverse wall (a) and top beam (b) after the Kozani earthquake of 1995. There is no available data to document the pre-existing state of deformation and strain of this church prior to this seismic event. The way the numerical results are presented is to underline that the appearance of damage after a strong earthquake motion may be also due to a long-term pre-existing state of deformation and strain due to soil-deformability. A good correlation can be seen when the observed structural damage is compared with the numerical predictions of Fig. 9, obtained by applying the simplified numerical evaluation process. In this way the validity of this simplified evaluation process is demonstrated. Another part of this structure with considerable

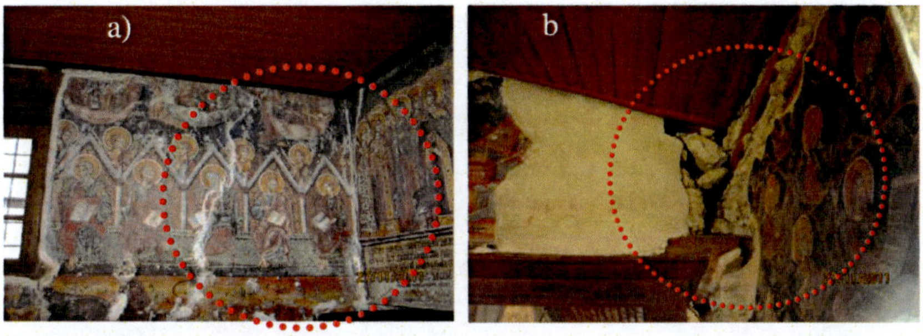

Fig. 10. Damage after the 1995 Kozani earthquake at the junction of the South wall with the transverse wall (a) and top beam (b) (Manos 2011).

tensile stress response is the region with the apses at the East wall. This also correlates with observed damage (Figs. 13 and 14). The corresponding stress response numerical predictions due to the seismic actions are presented in the next Sect. 4.2.2. (Figs. 11 and 12). It is shown again that by applying the simplified numerical evaluation process a reasonably good correlation can be obtained with observed damage for this type of action. A more complex non-linear type of numerical simulations are currently tried in an effort to combine long term effects with seismic actions.

4.2.2 Earthquake Action

In this section the earthquake activity is approximate through a linear elastic dynamic spectral analysis (SAP 2000) employing the design spectrum of the Greek Seismic Code (EAK 2000) for soil category B and behaviour factor equal to 1.5 (unreinforced masonry), with design base acceleration equal to 0.16 g (g is the acceleration of gravity, assuming that this church belongs to seismic zone I, of the new seismic zone map of Greece (EAK 2000). Only the first two translational modes were considered in the numerical simulation mobilizing approximately 40% and 60% of the total mass in the x-x (East-West, longitudinal direction) and y-y (North-South, transverse direction) axes, respectively. Thus, relevant amplification factors were used to account for the total mass. The resulting base shear values were approximately 5000 KN along either the x-x or y-y axes. Apart from dynamic spectral analyses along the x-x and y-y axes equivalent horizontal distributed force static analyses were also performed, Ex_{stat} and Ey_{sta}, having a base shear values equal to the ones resulting from the dynamic spectral analyses (5000 KN).This was done for a foundation of medium deformability. Figure 11 depicts the stress distribution response of the church of Profitis Elias for the load combination $0.9G + Ey_{stat} + 0.3\ Ex_{stat}$, whereas Fig. 12 for the load combination $0.9G + Ey_{stat} + 0.3\ Ex_{stat}$.

0.9G+ Ey$_{stat}$ + 0.3Ex$_{stat}$ Medium-Flexible Foundation

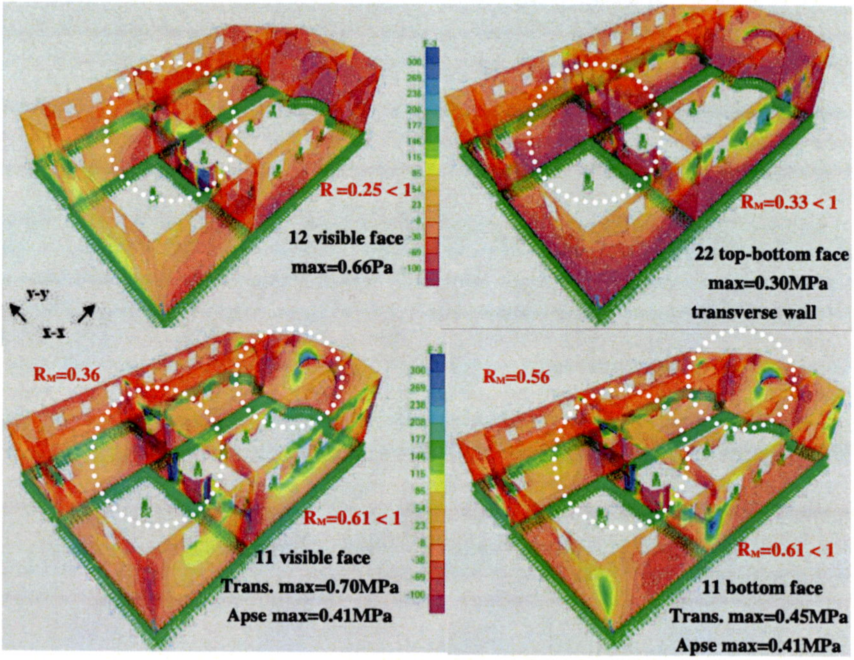

Fig. 11. Stress distribution response of the church of Profitis Elias at Siatista – Kozani for the load combination 0.9G + Ey$_{stat}$ + 0.3 Ex$_{stat}$, assuming foundation of medium flexibility. For the vertical walls the axial σ22 stresses act normal to horizontal cross-section.

Based on these numerically predicted stress demands and the corresponding tensile strength values mentioned before (fxk1 = 0.1 MPa, tensile strength parallel to bed joints fxk2 = 0.25 MPa) this time for the flexible foundation the corresponding ratio values of the capacity/demand (R_M) are well below 1 (see inequalities 1 and 2, Sect. 2). These ratio (R_M) values are shown in Fig. 9, indicating structural damage (Manos et al. (2019). Similarly, the corresponding ratio of the capacity/demand (Rτ) for the shear limit-state scenario takes values smaller than 1.

As can be seen in these figures the most vulnerable part of this structure, based on the results of the applied simplified numerical evaluation process, is the transverse wall separating the main part of this three nave Basilica church from the women' quarters located at its Western part. Moreover, the apses located at the East wall appear to develop considerable tensile stress response, as was also pointed studying the case of the foundation settlement. The predicted in this way structural damage correlates well with the in-situ observations depicted in Fig. 10a and b for the transverse wall and 13 and 14 for the apses. As can be seen in Fig. 14, repair works were carried out during 2014 involving deep repointing, mortar injections and consolidation of the foundation.

0.9G+ Ex_{stat} + 0.3Ey_{stat} Medium-Flexible Foundation

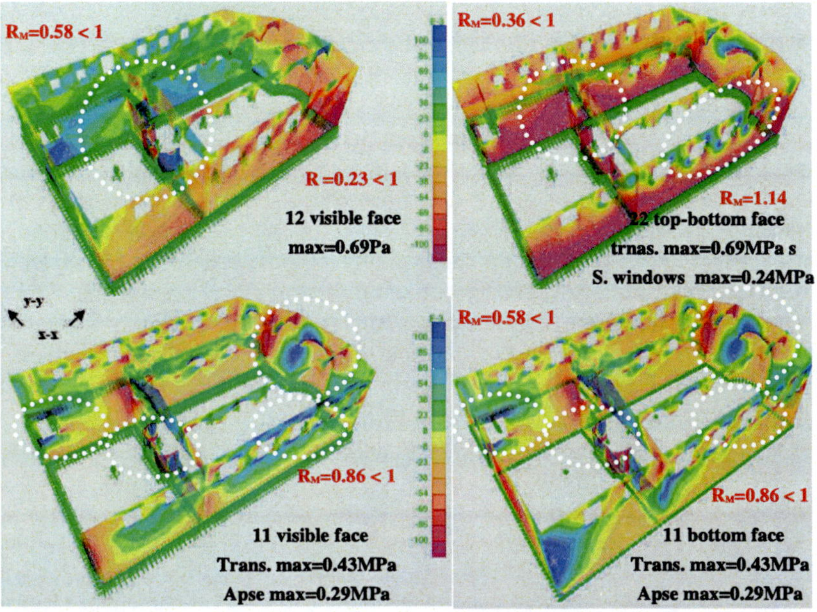

Fig. 12. Stress distribution response of the church of Profitis Elias at Siatista – Kozani for the load combination 0.9G + Ex_{stat} + 0.3 Ey_{stat}, assuming foundation of medium flexibility. For the vertical walls the axial σ22 stresses act normal to horizontal cross-section.

Fig. 13. Damage of the main apse at the East wall of the Profitis Elias of Siatista, Kozani, Greece

Fig. 14. Rapair works (2014) at the damaged parts Northern apse at the East wall of the Profitis Elias of Siatista, Kozani, Greece.

5 Conclusions

- A systematic study of the performance of damaged stone masonry structures representing "Greek" Christian churches substantiates two fundamental causes; the long term permanent uneven foundation settlement combined with seismic forces generated from relatively strong earthquake ground motions.
- A simplified evaluation process is summarized based on a dynamic linear elastic numerical simulation for obtaining the imposed demands on the various structural elements.
- Limited experimental results of the shear-sliding response of a three stone triplet with two mortar joints are also presented in an effort to provide measurements for the validation of relevant limit-state criteria.
- Non-linear numerical simulation of the shear-sliding response of the tested three stone triplet, based on the cohesive surface interaction constitutive law, is also presented and discussed. It can be argued that such a cohesive surface interaction constitutive law, with values of its parameters validated through the process of the specific triplet test, can be also employed in forming realistic limit-state criteria for stone masonry elements.
- It is believed that such a cohesive surface interaction constitutive law, with values of its parameters validated through the process of the specific triplet test, can also be used in forming more complex non-linear numerical model of structural elements, in an effort to simulating the behaviour of such stone masonry structural elements whereby the shear-sliding response is expected to prevail as the dominant response mechanism.
- For the numerical predictions resulting from this simplified dynamic linear elastic analysis of the church of the "Transfiguration of Saviour" it can be concluded that the soil-foundation deformability results in a significant increase in the tensile stresses at the region that the tympanum of the spherical dome which is supported on the vaulting part that springs from the two interior masonry pilasters and the apse. This predicted tensile stress response is well above the corresponding masonry strength values thus indicating structural damage. The same observation can be made based on the predicted structural response for earthquake forces. This conclusion is in agreement with the observed damage that developed in this region of the structural system for this church.
- For the three nave Basilica church of "Profitis Elias" the numerically predicted vulnerable parts of the structure, by applying the simplified evaluation process for either the foundation settlement or for the earthquake forces, correlate well with the observed in-situ damage. In this case the most vulnerable parts are the transverse wall separating in the main part of this three nave Basilica church from the women' quarters located at its Western part and the apses located at the East wall. Again, this conclusion is in agreement with the observed damage that developed in this region of the structural system for this church.
- It can be thus concluded that despite the fact that this simplified evaluation process is partly based on linear numerical simulations of the structural response, it can yield realistic predictions of expected structural performance of such masonry structures. This was presented here for two such "Greek" churches. However, the necessity to

obtain a more comprehensive set of measured strength properties for such type of masonry construction must be also underlined. This is necessary in order to increase the confidence on the validity of simple or complex numerical approximations.

Acknowledgements. The work which is presented in the paper is based on a long-term research effort performed at the Laboratory of Strength of Materials and Structures of Aristotle University. No financial support is directly linked with this work.

References

Abaqus Unified FEA - SIMULIA™ by *Dassault Systèmes*

Cerone, M., Viscovic, A., Carriero, A., Sabbadini, F., Capparela, L.: The soil stiffness in-fluence and the earthquake effects on the colosseum in Roma. In: 2nd International Conference on Studies in Ancient Structures, Istanbul, Turkey, pp. 421–426 (2001)

Croci, G.: The conservation and structural restoration of architectural heritage. Computational Mechanics Publication (1998). ISBN 1 85312 4826

EAK-2000 Greek Seismic Code

ELOT EN 1998-1/2005-05-12. Eurocode 8: Design of Structures for Earthquake Resistance - Part1: General Rules, Seismic Actions and Rules for Buildings (2005)

European Committee for Standardization. *Euro-code 6*; "Design of Masonry Structures, Part 1-1: General Rules for Building. Rules for Reinforced and Un-reinforced Masonry", EN 1996-1-1 (2005)

GEER - EERI - ATC. Cephalonia GREECE Earthquake Reconnaissance January 26th/February 2nd 2014. Report Version 1: 6 June 2014 (2014)

Jones, C.J.F.P.: Defects originating in the ground. Chapter 9 of the "The Maintenance of Brick and Stone Masonry Structures", Sowden, A.M. (ed.). (1990). ISBN 0419-14930-9

Sergio, L.: Damage assessment of churches after L'Aquila earthquake (2009). Bull. Earthq. Eng. (2011). https://doi.org/10.1007/s10518-011-9307-x

Manos, G.C., Soulis, V.J., Diagouma, A.: Numerical investigation of the behaviour of the church of Agia Triada, Drakotrypa, Greece. Adv. Eng. Softw. **39**(4), 284–300 (2007)

Manos, G.C.: Consequences on the urban environment in Greece related to the recent intense earthquake activity. Int. J. Civil Eng. Architect. **5**(12) (Serial No. 49), 1065–1090 (2011)

Manos, G.C., Karamitsios, N.: Numerical simulation of the dynamic and earthquake behavior of Greek post-Byzantine churches with and without base isolation. In: Syngellakis, S. (ed.). Earthquake Engineering Retrofitting of Heritage Structures, Design and Evaluation of Strengthening Techniques, pp. 171–186. Wessex Institute of Technology, UK (2013). ISBN: 978-1-84564-754-4, eISBN: 978-1-84564-755-1

Manos, G.C., Kozikopoulos, E.: The dynamic and earthquake response of Basilica Churches in Kefalonia-Greece including soil-foundation deformability and wall detachment. In: CompDyn 2015, Crete, Greece (2015a)

Manos, G.C., Kozikopoulos, E.: In-situ measured dynamic response of the bell tower of Agios Gerasimos in Lixouri-Kefalonia, Greece and its utilization in the numerical predictions of its earthquake response. In: CompDyn 2015, Crete, Greece (2015b)

Manos, G.C., Kotoulas, L., Felekidou, O, Vaccaro, S., Kozikopoulos E.: Earthquake damage to Christian Basilica Churches – application of an expert system for the preliminary in-plane design of stone masonry piers. In: International Conference of STREMAH 2015 (2015c)

Manos, G.C.: The seismic behaviour of stone masonry Greek orthodox churches. J. Architect. Eng. **1**(1), 44–53 (2016)

Manos, G.C., Nick Simos, N., Kozikopoulos, E.: Chapter 4: The structural performance of stone-masonry bridges. In: Structural Bridge Engineering (2016). ISBN 978-953-51-2689-8, Print ISBN 978-953-51-2688-1. https://doi.org/10.5772/64752

Manos, G.C., Kotoulas, L.: Unreinforced stone masonry under in-plane state of stress from gravitational and seismic actions. Measured and predicted behavior. In: CompDyn2017, Rhodes Island, Greece, 15–17 June 2017 (2017)

Manos, G.C., Kotoulas, L., Kozikopoulos, E.: Evaluation of the performance of unreinforced stone masonry Greek "Basilica" Churches when subjected to seismic forces and foundation settlement. Buildings **9**, 106 (2019). https://doi.org/10.3390/buildings9050106

Modena, C., Casarin, F., da Porto, F., Munari, M.: L'Aquila 6th April 2009 Earthquake: emergency and post-emergency activities on cultural heritage buildings. In: Garevski, M., Ansal, A. (eds.) Earthquake Engineering in Europe, pp. 495–521. Springer, Dordrecht (2010). https://doi.org/10.1007/978-90-481-9544-2_20

SAP. Integrated Software for Structural Analysis and Design. Computers and Structures Inc. (2000)

Simos, N., Manos, G.C., Kozikopoulos, E.: Near- and far-field earthquake damage study of the Konitsa stone arch bridge. Eng. Struct. **177**, 256–267 (2018). https://doi.org/10.1016/j.engstruct.2018.09.072

The Rehabilitation of the Historical Bridge of Konitsa: A Documentation-Based Investigation of Structural Behaviour

Christos Giannelos and Elizabeth Vintzileou(✉)

Laboratory of Reinforced Concrete, School of Civil Engineering, National Technical University of Athens, 5 Iroon Polytechniou, 15773 Zografou, Greece
elvintz@central.ntua.gr

Abstract. The Konitsa bridge was constructed in 1869 over the Aoos river. It is one of the most important arch bridges in the Balkan region and the larger in Greece, with a clear span of its semi-circular arch approximately equal to 37,0m. After an unsuccessful attempt of the Ottoman Army in retreat to destroy the bridge by explosives, it was repaired by the French Army the same year 1913 using unreinforced and reinforced concrete. Although the general state of the bridge is good, there is a need for interventions in various areas, including the part that was repaired in 1913.

A detailed architectural and structural survey of the monument, as well as the collection of historical data allowed for significant data to be gathered and interpreted related to the in-time behaviour of the bridge.

In this paper, the methodology applied with the purpose of investigating the structural behaviour of the bridge is presented. To this end, gradual application of loads and actions to the numerical model was adopted, in order to reach a numerical confirmation of the observed pathology. Furthermore, such a procedure has allowed for the assessment of the current state of the bridge, as well as for the identification of the remedial measures that are needed for its protection.

On the other hand, this investigation put in evidence the major significance of exhaustive documentation, as well as the inherent uncertainties related to all steps of this numerical investigation.

Keywords: Monuments · Historical stone bridges · Structural documentation · Interventions

1 Introduction

In this paper, the main steps followed during the study for the rehabilitation of the historical stone bridge of Konitsa (Fig. 1) are presented. This study was performed at the National Technical University of Athens (NTUA), in the framework of a Contract among the Prefecture of Epirus, the Municipality of Konitsa (owner of the bridge), the Ministry of Culture and NTUA. The work presented herein emphasizes on the significance of the documentation of the bridge and shows how its findings led to the final proposal for the rehabilitation of the bridge.

I. Vayas and F. M. Mazzolani (Eds.): PROHITECH 2021, LNCE 209, pp. 711–726, 2022.
https://doi.org/10.1007/978-3-030-90788-4_54

Fig. 1. The historical bridge of Konitsa – view from upstream (Vintzileou et al. 2017)

The need for rehabilitation of the historical bridge was felt by the Authorities due to the advanced corrosion and disintegration of the RC jacket, constituting an older intervention for the repair of the bridge. The jacket covers part of the intrados of the central arch, in the region of its key. For suitable interventions to be selected and adequately designed, it was necessary to identify the time of application of this intervention, to interpret its structural role and, hence, to judge on whether it would be preserved or removed from the monument.

The basic steps of the research are described herein, with the purpose of highlighting their effect on the adopted methodology for numerical calculations and, finally, on the measures that were proposed for the rehabilitation of the bridge.

2 Bridge Description

The bridge of Konitsa is a typical stone arch bridge for Epirus in the middle of the 19th century and it is listed by the Ministry of Culture since 1982. It consists of a central arch (36.90m in diameter) and two abutments (Fig. 1).

The total length of the bridge at the level of the parapets is equal to 60.5 m. Its width at the level of the pavement varies between 2.90 m at the key of the arch and 3.30m at the end of the abutments. The arch consists in two superimposed arches, the main arch (at the intrados) and the secondary one (at the extrados), as shown in Fig. 2a. The main arch is in recess (by 5.5 cm to 6.5 cm) with respect to the secondary one, whereas the construction of the abutments follows the width of the secondary arch.

The height of the main arch is equal to 75–80 cm. The arch is provided with eighteen (18) transverse metal connectors (Fig. 2b). The height of the secondary arch is equal to 60 cm approx. in the region between its support and an angle of approx. 45°, while it is reduced to approx. 50 cm in the central portion of the arch. Both the main and the secondary arch are made of schist. The thickness of the stones varies between 6 and 10 cm. The mortar joints are thin, with an average thickness of 10 mm.

In the main arch, both at its intrados and on its (upstream and downstream) faces, there are visible holes (Fig. 2c). The investigation through radar has revealed (Fig. 5a) the locations of timber grids (the timber elements being completely disintegrated) connecting the two arches. Those elements are quite similar to those identified in the Plaka bridge, after its partial collapse (Fig. 2d, NTUA 2016).

Fig. 2. Construction details (a) Portion of the arch (between 45° and 60° from its theoretical support): View of the N-abutment from downstream, (b) Detail of the anchorage of transverse metal connector, (c) View of the intrados of the arch at the location of a timber grid; the holes corresponding to the location of disintegrated timber elements are visible, and (d) timber grids, as documented in Plaka bridge (NTUA 2016).

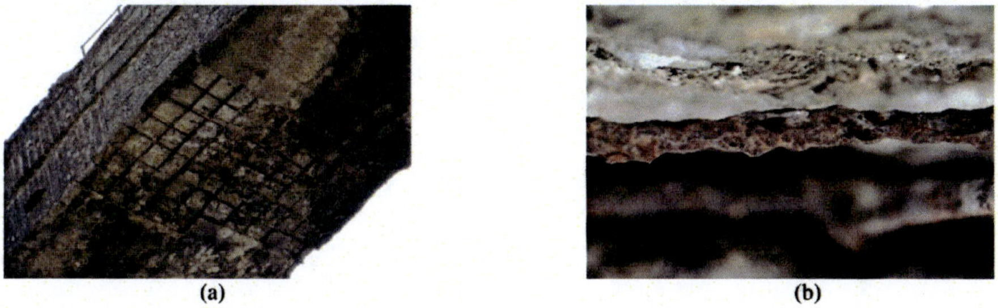

Fig. 3. Reinforced concrete jacket at the intrados of the arch and its condition: (a) Loss of the concrete cover and (b) significant reduction of the reinforcing bars section, due to corrosion.

3 The Need for Rehabilitation

The main damage of the bridge, which led to this study, is the poor condition of the RC jacket close to the key of the arch (Fig. 3). On the contrary, the unreinforced concrete identified at the extrados of the same location is in remarkably good condition.

The environmental conditions of the site have contributed to the corrosion of the steel, to the loss of the concrete cover and, with the time, to the in-depth disintegration of the concrete. The corrosion of most of the reinforcing bars is pronounced, as witnessed by the significant loss of their cross section (Fig. 3b). Under these conditions, it is clear that the bearing capacity of the RC jacket is reduced to a degree that does not allow it to behave as a structural component of the bridge. Thus, the question arose about the purpose of the construction of that RC jacket and, by way of consequence, about the need to preserve this intervention (after suitable interventions).

To provide an adequate reply to this question, one should acquire documentation on (a) when and why the RC jacket was constructed and (b) whether the unreinforced concrete at the extrados and the RC jacket at the intrados were cast simultaneously. Furthermore, (c) information was needed on the quality of materials. It has to be noted that during the first inspection, vaguely distinguished ribs were identified on the longitudinal reinforcement of the RC jacket, suggesting that the unreinforced concrete at the extrados (known to be cast by the French Army, see Sect. 4) and the RC jacket were cast at different times.

4 Documentation of the Bridge

Following the established procedure of assessing a monument and intervening to it, a systematic program for the documentation of the bridge was planned and executed. This investigation plan includes (a) Historical research, in order to identify the structural history of the monument as well, (b) geometrical survey, using conventional and photogrammetric techniques (Fig. 4, Michaelidis 2017), (c) survey of morphological characteristics and identification of the bearing system of the bridge, using georadar measurements and boroscopy (Fig. 5), (d) sampling of materials and laboratory tests to identify their chemical and mechanical properties, (e) geotechnical investigation, and (f) investigation of the hydraulic and seismic conditions of the region affecting the bridge.

Fig. 4. Photogrammetric model

(a) (b)

Fig. 5. Data obtained through georadar and boroscope applications are reported on the photogrammetric model (a) general view - grids of timber elements are depicted in brown and metal objects in blue (b) detail – 3D representation of cut stones and timber element, at the intrados, lower region of the bridge (Vintzileou et al. 2019).

(a) (b) (c)

Fig. 6. Historical photographs of the bridge: (a) Rhomaides-Zeitz 1913; the damage to the central part of the arch is visible, (b) Fr. Boissonnas 1913; picture taken during the repair of the bridge by the French Army and (c) M.Hrousaki; picture taken between the two World Wars: the 1913 repair is visible, along with the metal railing (at the top of the parapet).

According to the historical data, the construction of the bridge started in 1869 and it was mainly funded by the entrepreneur I. Loulis. After the liberation of Ioannina (the capital city of Epirus) on February 23rd, 1913, the Ottoman Army in retreat attempted to destroy the bridge using explosives. The bridge was severely damaged at the central region of the arch (Fig. 6a) and it was repaired in the summer of the same year by the Engineers of the French Army (Fig. 6b). The repair technique applied by the French Army, after the construction of a scaffolding (Fig. 6b) was to fill the holes caused by the explosion using a "stony" material, i.e. unreinforced concrete. However, there is neither historical data nor photographs documenting the construction of the RC jacket at the intrados, in the same region of damage. Thus, further investigation was needed.

The detailed survey of the region of damage and repair shows (Fig. 7) that the Ottoman Army has positioned explosives in two locations (one upstream, the other downstream) close to the key of the bridge (known to be of key significance for the stability of arches), with the obvious purpose to destroy the arch. Although the loss of mass at the central region of the arch was significant, its collapse was avoided.

The Engineers of the French Army have filled the holes using plain concrete, as documented through georadar investigations that did not detect any metallic object in the concrete. As for the RC jacket at the intrados, its thickness varies between 10 cm and 12 cm and it is provided with longitudinal reinforcement (diameter 24 mm@25 cm approximately) and transverse reinforcement (diameter 14 mm@22–28 cm). The jacket is anchored to the arch using smooth steel dowels. The inspection revealed on the longitudinal (pronouncedly corroded) bars small height bulges suggesting that deformed bars were used for the construction of the RC jacket. Therefore, one could assume that the RC jacket belongs to an intervention subsequent to that of the French Army and, more specifically, in the '60s, when the use of deformed steel bars became a rule for the country. However, as there were no archival data confirming this assumption, samples of both the longitudinal and the transverse bars were taken and analyzed with the purpose of identifying their chemical composition. The chemical analysis of all samples has shown that the same steel quality was used both for longitudinal and transverse bars. It is a steel of low carbon content (~0,10%-wrought iron), typical for the end of 19th-beginning of 20th century. On the other hand, the chemical analysis of concrete samples taken both from the extrados and the intrados has revealed their similarity (Vintzileou et al., 2018b).

As for the "bulges" in the longitudinal reinforcing bars, an effort was made to find a-as sound as possible-portion of longitudinal steel bar, by means of careful optical inspection (Fig. 8a). Indeed, a sample was taken and examined. It became obvious that the bar was smooth, with manually produced bulges (Fig. 8b) with the purpose of improving the bond properties of the bars. Afterwards, it was also observed that those manually roughened bars were used in the railing of the bridge (Fig. 8c). Furthermore, the photograph in Fig. 6a (taken immediately after the bridge was damaged) does not show any railing on top of the parapet. On the contrary, the railing is visible in the photograph of Fig. 6c, taken after the completion of the intervention.

It is, therefore, legitimate to assume that both the unreinforced concrete at the extrados and the RC jacket at the intrados belong to the same intervention, applied in 1913.

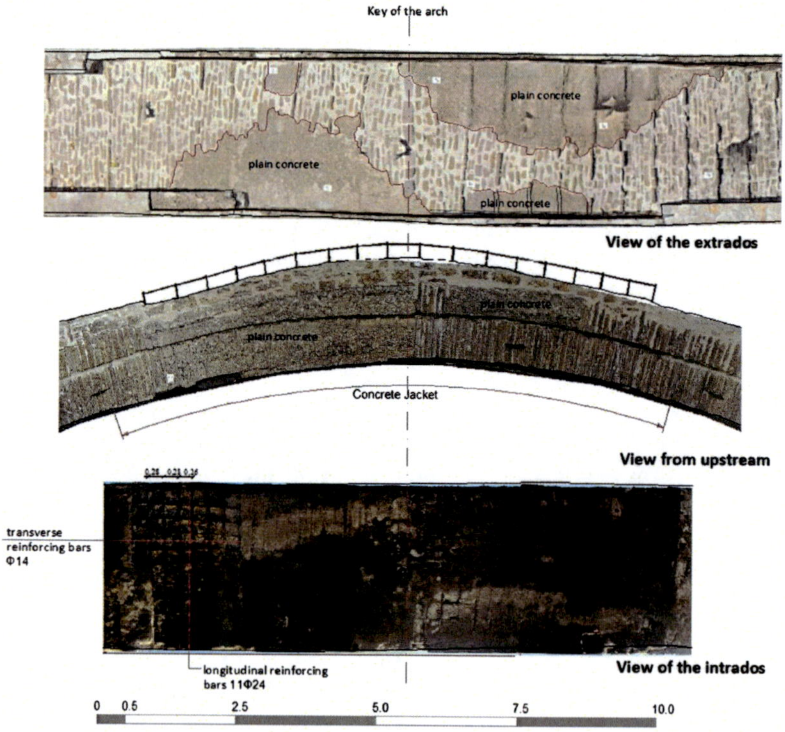

Fig. 7. The region of interventions to the bridge

(a) (b) (c)

Fig. 8. Visual inspection of the reinforcement of the RC jacket at the intrados of the arch: (a) Sampling, (b) Sample of longitudinal bar, (c) The railing fixed in the parapet.

5 Numerical Modeling - Calculation Methodology

Significant effort was devoted to numerical calculations with the purpose of acquiring a thorough understanding of the structural system of the bridge. Thus, the structural function of the timber elements found in the abutments of the bridge (see Fig. 5a) could be interpreted, the need for the RC jacket added at the intrados of the arch in the region of damage could be identified, the margins of safety at the current state of the bridge could be assessed, the efficiency of various intervention techniques could be estimated, etc. Static and time-history analyses and, in some cases, also parameter analyses were performed. Only part of this analytical work is presented herein.

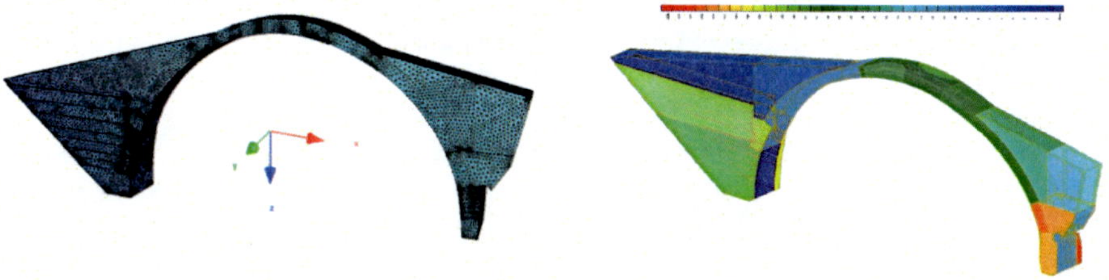

(a) (b)

Fig. 9. 3D numerical model of the bridge: (a) general view of the mesh, (b) groups of elements per region.

For the numerical calculations presented herein, the software SOFiSTiK (2014) was used. Volume finite elements were used for modelling the various parts of the bridge, with the exception of the RC jacket at the intrados, where shell elements were used (Fig. 9). The numerical model was based on the 3D photogrammetric model, to reach the maximum possible geometrical accuracy (Fig. 10). Grouping of finite elements per region of the bridge was also performed (Fig. 9b), to allow for a calculation procedure taking into account the construction phases of the bridge, as described in detail herein.

On the basis of the information collected during the investigation phase, regions of various types of masonry were identified (Fig. 11) and their mechanical properties were estimated accordingly (Vintzileou et al., 2018a and Table 1).

Fig. 10. View of the S-abutment and its numerical model

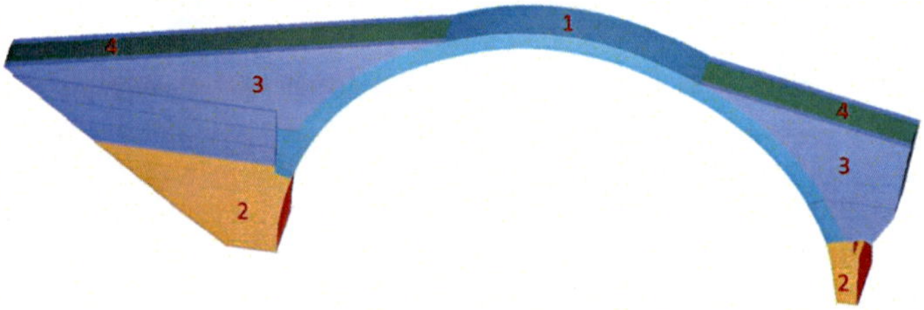

Fig. 11. The identified regions of masonry characterized by different geometrical and mechanical properties

Table 1. Mechanical properties of various regions of masonry used in numerical calculations

No	Region (Fig. 11)	Mean compressive strength of masonry f_{wc} (MPa)	Modulus of elasticity of Masonry E (MPa) = (600–1000)f_{wc}
1	Lower part of the abutments	14.50	8700–14500
2	Arch	8.90	5350–8900
3	Spandrels	2.50	1500–2500
4	Filling of the abutments	0.80	500–800

For the simulation of the foundation, elastic springs were used, their constants being estimated on the basis of the results of the geotechnical investigation (Vintzileou et al. 2018a). Thus, the values presented in Table 2 were adopted for the N- and the S-abutment respectively. For numerical analyses including the seismic action, those elastic constants were multiplied by three, as proposed in (Kavvadas 2016). However, based on the fact that high uncertainties are related both to the elastic properties of masonry and to the properties of foundation soil, the decision was made to perform parameter analyses, using adequate combinations of high and low values of those properties (Table 2).

Table 2. Values of modulus of elasticity and of the elastic constants of the soil adopted for parameter analyses of the bridge

a/a	Modulus of elasticity of masonry per region (MPa)	Elastic constants of foundation soil (MPa)
I	min{$E_1 = 8700$, $E_2 = 5350$, $E_3 = 1500$, E4 = 500}	soft{$K_{v,south} = 40.000$, $K_{v,north} = 100.000$}
II	min{$E_1 = 8700$, $E_2 = 5350$, $E_3 = 1500$, E4 = 500}	med{$K_{v,south} = 40.000 \times 3$, $K_{v,north} = 100.000 \times 3$}
III	max{$E_1 = 14500$, $E_2 = 8900$, $E_3 = 2500$, $E_4 = 800$}	soft{$K_{v,south} = 40.000$, $K_{v,north} = 100.000$}
IV	max{$E_1 = 14500$, $E_2 = 8900$, $E_3 = 2500$, $E_4 = 800$}	med{$K_{v,south} = 40.000 \times 3$, $K_{v,north} = 100.000 \times 3$}

As the documentation of the monument has revealed two construction phases, as well as phases related to its damage by explosives and to its repair (Table 3), the decision was made to follow the history of the bridge in the numerical calculations as well, in an effort to better assess its current state.

It should be noted that (a) the parapets (made of masonry) are not included in the numerical model; however, their self-weight is taken into account as a load on the pavement of the bridge, and (b) In this part of the numerical calculations, the timber elements identified in various regions of the bridge (Fig. 5) are not taken into account both because their state of preservation is very poor (in most of cases, they are completely disintegrated) and because their replacement in the body of the bridge is not feasible. Nonetheless, their structural role during the construction phase of the abutments of the bridge was numerically investigated and confirmed.

Table 3. Phases considered in the investigation of the structural behaviour of the bridge

Construction phase –Numerical model	Region of damage
LC9-construction of abutments 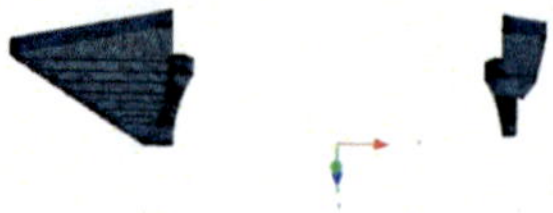	
LC10-construction of the arch/construction bridge completed, 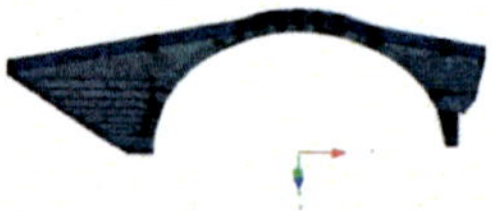	
LC11-Damage due to explosion (1913) 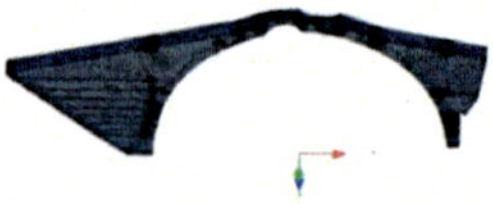	 Rhomaides-Zeitz 1913 
LC12: Repair of the bridge-Current condition 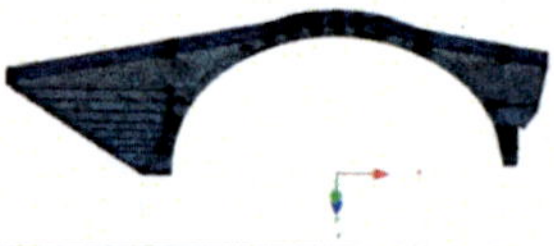	 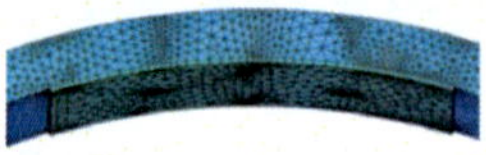

6 Investigation of the Structural Behaviour of the Bridge

In order to provide a qualitative understanding of the structural behaviour of the bridge at the phases listed in Table 3 (LC9 to LC12), selected results are presented in the form of deformations and principal (compressive and tensile) stresses.

During the construction of the abutments, (LC9), the maximum vertical deformation occurs at the south abutment (Fig. 12a) both because of the larger mass of this abutment and because of the softer soil in the south bank of the river. For the most adverse case (Case I, Table 3), a difference of 8mm in the vertical deformations at the base of the two abutments was calculated. Although any differential settlement occurred during that phase did not affect the behaviour of the completed bridge, as the abutments were independent from one another, it has to be noted that calculating the bridge after the completion of its construction, without accounting for the intermediate phase of construction of the abutments, leads to significant differences in the vertical deformations underwent by the bridge (Fig. 12c).

According to the adopted methodology, the self-weight of the arch and its connection with the abutments are imposed to the already deformed shape of the abutments. It is interesting to observe that at the state where the construction of the bridge is completed (LC10), the maximum vertical deformation (for all parameter analyses) is not located close to the key of the arch, but at the region of connection between the arch and the south abutment (Fig. 12b). This feature can be attributed to the asymmetry of the abutments of the bridge.

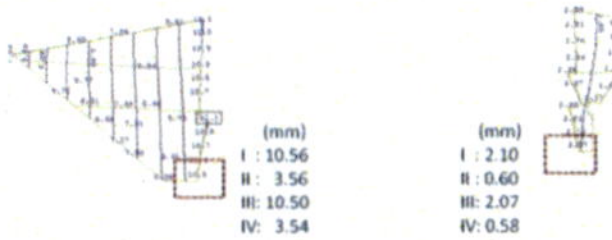

(a) Vertical deformations during the construction of the abutments (LC9).

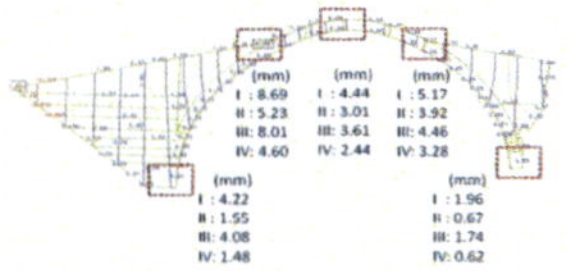

(b) Additional vertical deformations due to the construction of the arch (LC10).

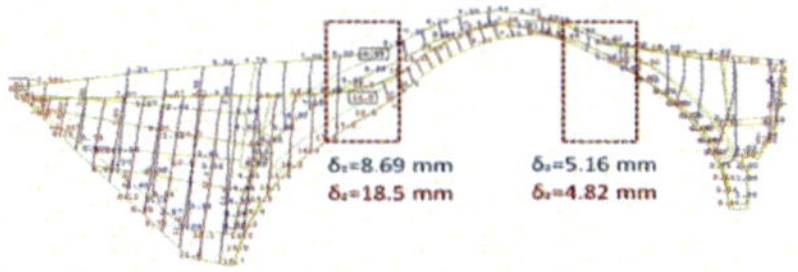

(c) Vertical deformations: in blue-considering phases, in red-without accounting for construction phases

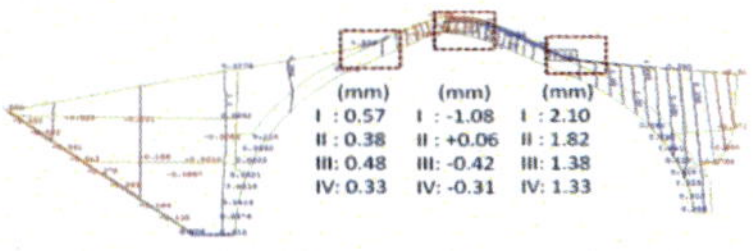

(d) Additional vertical deformations due to the loss of mass caused by the explosion (LC11).

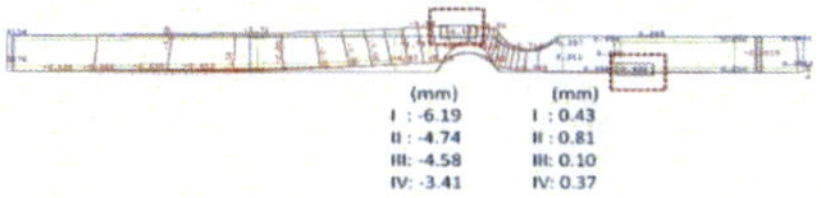

(e) Horizontal deformations (perpendicular to the longitudinal axis of the bridge) due to the explosion (LC11)

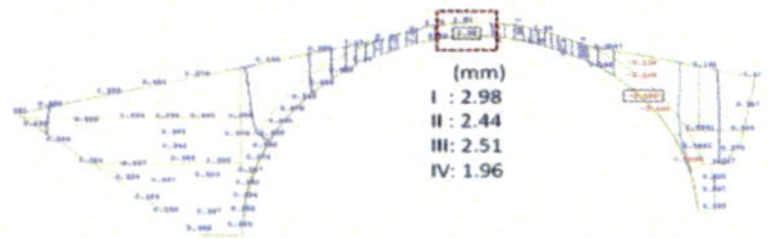

(f) Additional vertical deformations due to the self-weight of the repair concrete (LC12).

Fig. 12. Parameter analyses: Deformations at characteristic locations for various loading-cases

After the explosion (LC11), the significant loss of material in the region of the key and the resulting reduction of stiffness, leads to increased vertical deformations in the critical regions of the bridge (Fig. 12d). As expected, on the basis of the kinematics of the damaged region, a rotation occurred with respect to the key of the arch and, hence, horizontal deformations (perpendicular to the longitudinal axis of the bridge, Fig. 12e). An increase is observed also in the vertical deformations after repair of the damaged region using concrete (Fig. 12f). The out-of-plane deformation of the arch due to the explosion leads to additional compressive stresses in the downstream corner of the S-abutment and the upstream corner of the N-abutment (as evidenced in Fig. 13). On the contrary, additional tensile stresses occur in the upstream corner of the S-abutment and in the downstream corner of the N-abutment. As shown in Fig. 13, the results of the numerical calculations seem to be in agreement with the damages observed in the abutments, which-therefore-can be qualitatively interpreted and attributed to the explosion.

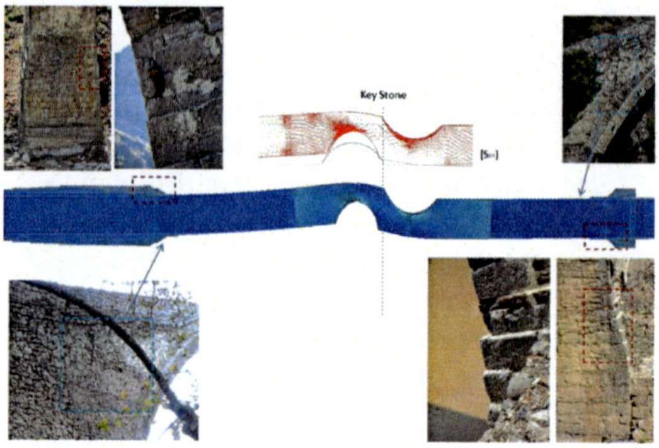

Fig. 13. Damage outside the damaged region of the arch, attributed to the explosion

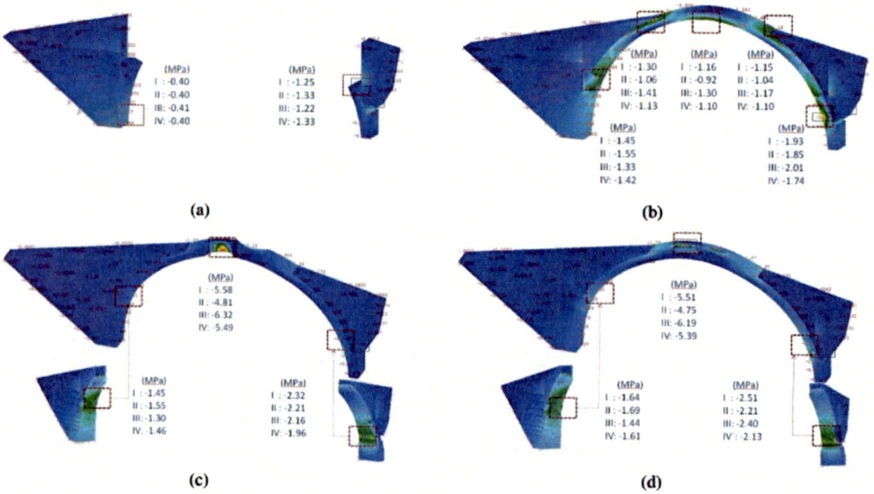

Fig. 14. Principal compressive stress contours and stress-values at characteristic locations for the examined phases: (a) LC9, (b) LC10, (c) LC11and (d) LC12 (see Table 3)

For the first construction phase (LC9-construction of the abutments), the values of principal compressive stresses (Fig. 14a) are very small compared to the estimated compressive strengths of masonry for various regions of the bridge. Similarly, very small principal tensile stresses seem to occur (Fig. 15a).

When the bridge is complete (LC10), expectedly, the principal compressive stresses increase (Fig. 14b). The maximum value (equal to 1.30 MPa) occurs at the most critical region (of connection between the arch and the south abutment). However, they remain in every location significantly smaller than the respective compressive strength of masonry. It should be noted that the parameter analyses did not show any significant effect of the moduli of elasticity of masonry or of the elastic constants of foundation soil on the values of the principal compressive stresses. This is not the case for principal tensile stresses

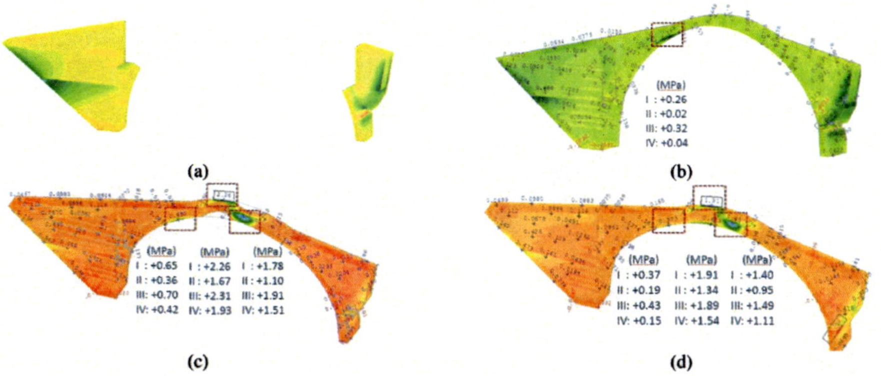

Fig. 15. Principal tensile stress contours and stress-values at characteristic locations for the examined phases: (a) LC9, (b) LC10, (c) LC11and (d) LC12 (see Table 3)

that seem to be affected by the values of the investigated parameters (Fig. 15b). In any case, the largest values of both compressive and tensile principal stresses (along with the largest values of vertical deformations) occur at the connection of the arch with the south abutment. As shown in Fig. 16, this location is nearly at the same vertical line with the centre of mass of the bridge.

As expected, the loss of material due to the 1913 explosion (LC11) caused a significant increase of the principal compressive stresses in the survived bridge (Fig. 14c). In the damaged region, the maximum compressive stress reaches the value of 6.32 MPa, still smaller than the estimated compressive strength of the arch (equal to 8.90 MPa). Although it has to be admitted that the performed elastic analyses are of qualitative value only, as they cannot simulate the complex dynamic event of the explosion and its effects, it is believed that the calculations provide a plausible interpretation of the survival of the bridge.

The high-value principal tensile stresses in the damaged region (Fig. 15c) are not commented upon, as they are considered only as an indication of the extensive cracking occurred in the damaged region leading to the urgent need for immediate repair. Indeed, as shown in Fig. 17, repair concrete is visible in the regions where the maximum tensile stresses are calculated.

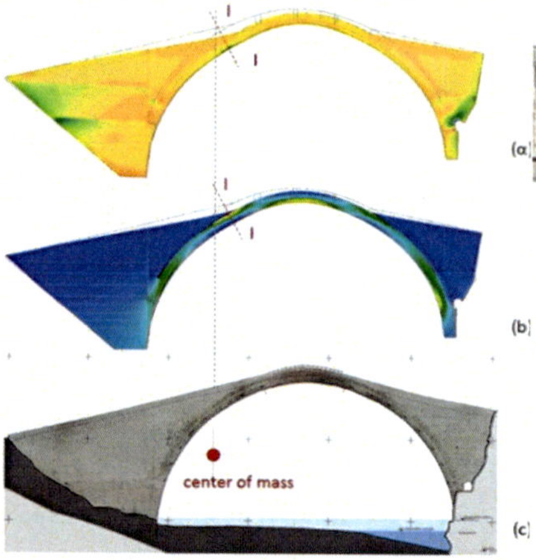

Fig. 16. Most stressed region of the bridge: (a) principal tensile stresses, (b) principal compressive stresses and (c) location of the centre of masses of the bridge.

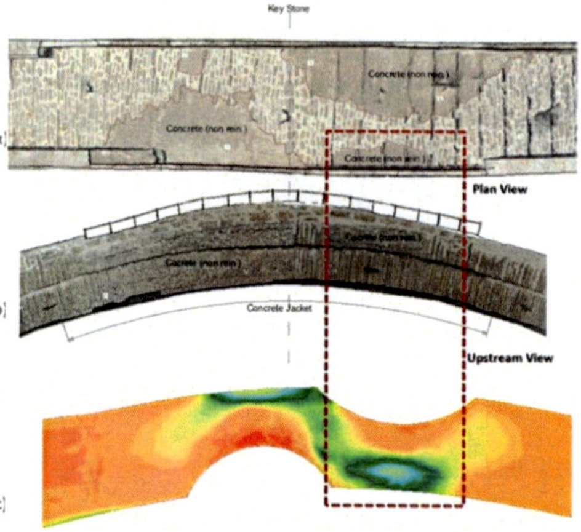

Fig. 17. Correlation between the most stressed in tension damaged region and the regions where interventions were applied.

7 The Structural Role of the RC Jacket at the Intrados

It has to be noted that the concrete added to the bridge for its repair does not contribute to its behaviour under self-weight alone. This is because the repair was done without shoring of the bridge and, hence, the added concrete remains practically free of stresses (Fig. 14d and Fig. 15d), as it is stressed only by the insignificant live loads of the bridge.

The rehabilitation of the damaged region, by filling the holes caused by the explosion with unreinforced concrete at the extrados, was of obvious necessity to reinstate the geometrical integrity of the arch. However, this repair does not contribute to the structural safety of the bridge against actions perpendicular to the longitudinal axis of the bridge, such as flood and earthquakes. Indeed, as shown in Fig. 18, the portion of the unreinforced concrete that belongs to the compressed zone contributes to the out-of-plane resistance of the bridge, whereas the portion within the tensioned zone tends to separate from the surrounding stone masonry of the arch. It seems therefore that the RC jacket at the extrados was needed to take up the (parallel to the axis of the bridge) tensile forces that cannot be sustained by the interface between the plain concrete at the extrados and the arch.

Thus, the decision was made to preserve the RC jacket at the intrados, as it contributes to the improvement of the behaviour of the bridge against accidental actions. Having assessed the current state of the concrete (both unreinforced and reinforced), the NTUA team had to propose suitable interventions for the rehabilitation of the 1913 repair. It is reminded here that (a) the unreinforced concrete at the extrados is in very good condition one century after its casting. Its compressive strength (estimated using the rebound technique) is higher than 28 MPa, whereas the depth of carbonated concrete does not exceed 3 cm. This is attributed to the fact that there is no pollution in the area and, as this concrete is at the extrados, it does not keep in its mass the rainwater that

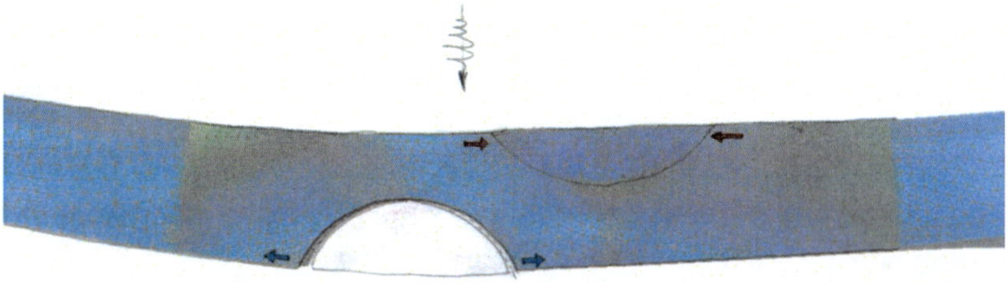

Fig. 18. The state of stresses of the unreinforced concrete at the extrados, due to actions perpendicular to the exis of the bridge (sketch)

might affect the integrity of the concrete during the freeze-thaw cycles. On the contrary, the RC concrete at the intrados is in so poor condition that it cannot play any structural role. Thus, the decision was made to remove the existing RC jacket and to replace it by a new one, using concrete and stainless steel reinforcing bars.

This proposal was approved by the other parties in the Contract, namely, the Prefecture of Epirus, the Municipality of Konitsa and, indeed, the Service of Modern Monuments and Technical Works of Epirus, North Ionian and West Macedonia (Ministry of Culture). It is to be noted that the NTUA proposal includes more interventions for the protection of the bridge. Those remedial measures are not mentioned here, as this paper focuses on some parts of the entire work of documentation, assessment and formulation of rehabilitation scenario, performed in the framework of the project.

8 Conclusions

This paper presents part of the work performed within a project for the rehabilitation of the historical arch stone bridge of Konitsa (Epirus) built in 1869 and listed by the Ministry of Culture. The following aspects of this work are highlighted herein, as concluding remarks:

1. The systematic documentation of the monument, in terms of historical data, exhaustive visual inspection assisted by in situ and in laboratory investigations, as well as by modelling and parameter analyses allowed for a thorough comprehension of the structural behaviour of the bridge, as well as for a reliable assessment of its current state. More specifically,
2. Construction and intervention phases were identified, the soil conditions (different for the two abutments) were investigated, the quality of materials and their properties were assessed by means of in situ and in laboratory tests. Furthermore,
3. Significant structural characteristics were detected, such as timber grids in the abutments, as well as connection between the primary and the secondary arch, by means of timber elements, along with various construction types of masonry in various areas of the bridge.
4. The major event that affected the monument (i.e. the explosion occurred in 1913, aiming at destruction of the bridge) was identified and its effect was studied. Finally,

5. The extent and the structural role of the intervention, applied after the explosion that destroyed part of the central arch in 1913, were identified and led to the formulation of remedial measures to preserve both the integrity of the bridge and its structural history.

References

Boisssonas, F.: Image of Greece: Epirus – Cradle of the Greeks, p. 1915. Editions d'Art Boissonnas, Geneva (1913)

NTUA: The Bridge of Plaka: The work of NTUA for its rehabilitation. Athens (in Greek) (2016)

Hroussaki, M. (2 *Photographs 1917–1958*, National Gallery and Alexandros Soutzos Museum, Athens (in Greek) (2002)

Kavvadas, M.: On the E-modulus of ground materials for seismic applications and its relation to the corresponding "static" values. Personal communication (2016)

Michaelidis, G.: Technical report of geodetic, topographic and photogrammetric survey on the arched bridge of Konitsa. Service of Modern Monuments and Technical Works of Epirus, North Ionian and West Macedonia (in Greek) (2017)

Rhomaides, A., Zeitz, F.: Balkan wars 1912–1913. Photographic album (1913)

SOFiSTiK: Version: 14.15–30, Servicepack, 2014–15, SOFiSTik AG (2014)

Vintzileou, E., Palieraki, V., Giannelos, C., Oikonomopoulou, E.: Rehabilitation of the Konitsa Historical Stone Bridge. Research report, work package 1, NTUA (in Greek) (2017)

Vintzileou, E., Palieraki, V., Giannelos, C., Oikonomopoulou, E.: Rehabilitation of the Konitsa Historical Stone Bridge. Research report, work package 2, NTUA (in Greek) (2018a)

Vintzileou, E., Palieraki, V., Giannelos, C., Oikonomopoulou, E.: Rehabilitation of the Konitsa Historical Stone Bridge. Research report, work package 3, NTUA (in Greek) (2018b)

Vintzileou, E., Palieraki, V., Nikolopoulou, V., Oikonomopoulou, E., Giannelos, C.: Documenting the bearing system of the Konitsa historical stone bridge. In: 2nd International Conference on Natural Hazards & Infrastructure, 23–26 June, Chania, Greece (2019)

Constructive Stages in the Architecture of Quito. XVII-XIX Centuries

Inés Del Pino[1](✉) and M. Lenin Lara[2,3](✉)

[1] Facultad de Arquitectura, Diseño y Artes, Pontificia Universidad Católica del Ecuador, Quito, Ecuador
idelpinom@puce.edu.ec
[2] DTCA, Departamento de Construcción y Tecnología Arquitectónica, Universidad Politécnica de Madrid, Madrid, Spain
lenin.lara.calderon@alumnos.upm.es
[3] CipArq – Facultad para la Ciudadla Arquitectura y Paisaje, Universidad Internacional del Ecuador, Quito, Ecuador
mlara@uide.edu.ec
https://www.puce.edu.ec, https://www.dcta.upm.es/, https://www.uide.edu.ec/

Abstract. The article incorporates the occurrence of seismic events in the study of the construction stages of Quito architecture, since these represent inflection points and mark a before and after in the history of the city, particularly earthquakes. During the 17th and 19th centuries Quito was shaken by 18 earthquakes, of these, five compromised the structures of the buildings: 1587, 1662, 1755, 1859 and 1868. The study of the architectural morphology of six buildings: La Catedral, La Compañía de Jesus, San Francisco, La Merced, Santo Domingo and San Agustín, describe the circumstances of their transformation and the most frequent risks, physical, economic and social reaction, which are interwoven in the construction, which until then used materials and similar techniques in the period studied. However, the review of the transformations happened in the buildings show that the monumental architecture of Quito's heritage is made up of parts, each representing a chapter of the city's history, necessary reforms, political projects, symbols of power, the local and the global. The earthquake contributes in this process to configure the proportions of the architecture and the characteristic urban profile of Quito.

Keywords: Earthquakes · Seismic vulnerability · Risk · Constructive stages · Historic buildings · Quito

1 Introduction

The religious buildings of the historic center of Quito, built between the seventeenth and nineteenth centuries, are urban landmarks that implanted in properties of one to three hectares, which generated a built area that varies between 1,500 sqm and 6,000 sqm.

At the end of the 19th century, the 1888 plan built by Gualberto Pérez identified 40 religious buildings in an area of 315 ha that are now part of the historic center of Quito

I. Vayas and F. M. Mazzolani (Eds.): PROHITECH 2021, LNCE 209, pp. 727–733, 2022.
https://doi.org/10.1007/978-3-030-90788-4_55

(IGM-ORSTOM 1992). These correspond to 6 convents and *recoletas*[1]; 6 monasteries; 20 churches, which include the cathedral and parish churches; 1 urban chapel and 5 *recoletas* at the edges of the city; and 2 seminars (Pérez 1888) (Fig. 1).

The common aspects that identify this type of buildings are the number, the concentration in a small area, its monumentality, architectural complexity and urban presence. Inside, each convent is an autonomous unit, and each church tower, a milestone for the population, they shape the urban profile of the historic city; composes the image of a temporality typical of Quito.

These buildings have up to now a constructive homogeneity despite their origin and the transformations produced during the three centuries of Hispanic presence in America. Until the nineteenth century the construction techniques of the sixteenth century were in use with few transformations, related architectural orders, proportion and scale of the buildings related to the configuration of the geographical environment; nevertheless, 1888 was a year in which the city had passed the onslaught of the war of independence, and a seismic period that was reflected in the presence of ruined buildings, city poverty and the diminished mood of its inhabitants.

This background suggests that Quito presents a series of obvious risks in its urban history. Taking some of the concepts of Gustavo Wilches-Chaux and Omar Cardona for the Colombian case (Cardona 2019), ten components or levels of global vulnerability in disasters are identified, among them, three could be cited, applicable to the subject of this article: the first one is of a physical nature (Wilches-Chaux 1989, p. 3.21), alternates the programmed growth of the buildings with the emergence of earthquakes. These destabilize continuity to integrate the logic of rupture, creation, adaptation and innovation: introduction of constructive forms such as the barrel vault reinforced with slab arches, both in architecture and in public works. That is, without fear of misunderstanding, the experience of the earthquake should be part of the tale of the history of cities and planning strategies, since the recurrence of seismic events and human action shape the image of the city.

In this scenario of indeterminacy, the second element of risk is of an economic nature (Wilches-Chaux 1989, p. 3.24), since the earthquake represents uncertainty, it is not known when the magnitude of the economic loss arrives: the projects are postponed without specific dates or deadlines; it reveals errors in construction, inadequate repairs with civil responsibilities, among others, however, when economic resources arrive, it is the opportunity to rebuild, demolish or transform. As an example, from the report on the state of ruin of the buildings and the successive earthquakes in Quito, the Italian reformers attracted funds for the restructuring of the interior design and the significant change of the image of the churches of Santo Domingo and San Agustín in the 19th century.

In this and other cases, and in the face of adverse political circumstances, economic resources for reconstruction were not sufficient, and for better or worse, religious communities were forced to sell part of their material assets. That is, the economy is a risk factor that when studied in the light of other factors such as political and social, acquires nuances that go beyond quantitative estimation and enriches the understanding of the vulnerability of the built heritage.

[1] Religious building that keeps a closer observance of the rule than what is commonly kept.

The third risk is the social reaction (Wilches-Chaux 1989, p. 3.26), given the loss of human lives and material goods, the period studied was strongly related to the action of the church in the emergency. In the earthquake of 1755 the closing of the convents was raised so that the friars and nuns can assist the wounded (Recio 1947, p. 382); the church gave its explanation about the meaning of the earthquake and acted to control collective fear; it accompanied the acceptance of the loss, in short, social control in circumstances of panic at a time of destabilization of order. On the other hand, the social reaction exposes interests of different social groups allowing us to understand the circumstances of the time; the appearance of expressions of faith and instruments of psychological support that exercised their power in the acceptance of the crisis: *novenarios*[2], images, processions. Once the earthquake is over, the diffusion of works of art exhibited in the churches and financed by "donor" families, which evoke miracles happened during the earthquake, feed the memory of the event, the urban imaginary and the history of the territory and the city. In this sense, the reaction of the population to the earthquake, and its link with the temple and public squares of Quito contribute to configure the symbols of the city, the collective recognition of intangible and material heritage.

2 Construction Stages and Earthquakes

The integrated analysis of the morphology of buildings and earthquakes explains the configuration of the building that we know today, is traversed by the risks mentioned earlier in the construction stages, in a constant process of expansion and reconstruction. The earthquakes that caused major damage to the architecture of Quito and that marked moments of temporary rupture took place in five years: 1587, 1662, 1755, 1859, 1868. That is to say that in the interval between one event and another there were several recorded aftershocks like tremors that sometimes destroyed the repairs made. The earthquakes of 1859 and 1868 activated previous fissures that were not repaired properly, these were usually masonry and not structural Fig. 1.

Thus, the risk is latent, the possibility of recurrence of an earthquake equal to or greater than 7MSK in Quito has 30-year cycles (Del Pino and Yepes 1990, p. 95). The above means that in a century, three generations experienced at least one earthquake, followed by a series of replicas or minor earthquakes, this means that this type of event is in the collective unconscious and is expressed unexpectedly, for example, with gestures of reciprocity that arise spontaneously from the population.

In the nineteenth century, on the one hand, the social dissatisfaction that ended with the independence of Spain and the formation of Republics; and on the other, the force of nature expressed in two earthquakes in a row that exposed the not visible damage in the masonry. With independence, the expropriation of church property around 1820, and occupation of a part of the convents for accommodation of the libertarian armies by Simón Bolívar and General Antonio José de Sucre. The Italian reform that began in 1863 until 1890, added to the social and economic crisis in the architecture of the Franciscan, Jesuit, Augustinian and Dominican groups. A difficult time in which the risk was represented in the architecture.

[2] Prayers dedicated to devotions and protection against diseases, unknown or uncertain situations.

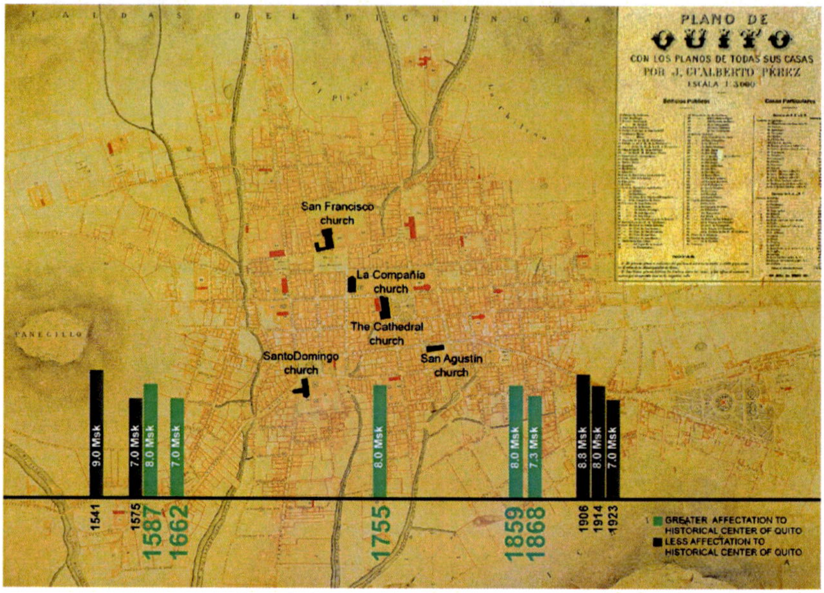

Fig. 1. The earthquakes that most affected Quito in the timeline. Map of Quito with the plans of all its houses. Author: Gualberto Pérez. Year 1888. Scale 1: 3000. Reserve of the Municipality of the Metropolitan District of Quito. Source: (Del Pino and Yepes 1990). Preparation: study team.

Economic and social instability is expressed in architecture in a differentiated way in buildings as illustrated in two examples: The Cathedral had its first church in 1543. In 1545, Quito was the seat of the Bishopric and required a more comfortable building. The church was reconstructed until 1572. In 1587 the city suffers an earthquake that destroys the tower in the part of the bell tower and the breakage of six cedar beams of the roof (National Institute of Cultural Heritage of Ecuador 1977). The repairs take until 1630 when it is decided to extend the church behind the main altar for which several houses were expropriated, the works had not advanced when the earthquake and eruption of 1660 opened cracks in the walls. This fact led to expanding the church again and orienting the main side towards the square with a side door and "took a street" to build the parapet. In 1698 and 1704 new earthquakes affected the roof, the repairs that were made until 1733 took advantage of this repair to include functional changes such as the relocation of the baptistery, the control of the entrance to the Tabernacle, the access to the snail-shaped tower and the brickwork from the floor of the church. The earthquake of 1755 destroyed the church again and the priests proposed to abandon it and build another one in the main square, a project that did not prosper (Instituto Nacional de Patrimonio Cultural del Ecuador 1977, pág. 56). That is, the Cathedral church we know today is the fourth in time, with significant transformations in its architecture. The earthquakes and the change of their hierarchy were the occasion for the expansion and renovation of the temple without taking into account the important risks for its conservation: the sinking due to the proximity to one of the largest ravines in the city (Peltre 1989, págs. 45–91) and the poor maintenance of the building that finally weakened it. Hence, the mention of leaks and cracks in the walls is recurrent, complaints of lack of budget for repairs follow the same order. However, being the most important church in the city, it was a space of ritual, meeting and citizen recognition during the earthquakes, and at the same time, a collection center for economic donations and assets of the population of Quito.

Another case, the building of the Compañía de Jesús. The congregation arrived in Quito when the main lots had already been assigned. Through swaps and land purchases, they managed to unify the largest property in the city, located in the center. The construction of its cloisters and church were carried out under a systematic program, they resolved on the way the possible damages that occurred in the building that was completed in 1765. In 1767 the congregation was expelled from America, the buildings passed to the government or others religious orders. Hereinafter, the earthquakes of 1797 and 1859, and the severity of the abandonment, were the cause of the material damage that deepened in 1868. Information on the state of deterioration due to abandonment is the situation described in 1862 by the archbishop of Quito, who delivered the goods of the Jesuits upon their return: the tower had no stairs nor bells (Ortiz 2008, p. 203). What has been described suggests that the tower had fissures, rain flooded the interior, architectural elements such as stairs, which support the stability of the tower were destroyed putting its stability at risk.

The repair of damages and the high cost of the same demanded of extraordinary budgets that were not foreseen; the tower was rebuilt in 1868 and the same year it was cracked again. The decision of the Architect of the Republic, Thomas Reed was the demolition of the tower at half of its height (Saldarriaga 2005, p. 118) as it is known until today. This physical loss was accompanied by the loss of memory of the tower and the clock where the people of Quito saw the time until the nineteenth century, a detail that at the time of proposing its restitution in the twentieth century, was a reason for discussion and a strong argument from those who opposed the project.

3 Conclusions

The religious buildings of Quito, built between the sixteenth and nineteenth centuries were transformed by three causes: the first motivated by operational needs, the second due to the incidence of earthquakes, and the third due to risk factors not adequately addressed: abandonment, lack of maintenance or economic resources.

The type of buildings, symmetrical, bearing structure, have been efficient to resist the series of earthquakes and tremors that have affected the city, however, the towers were severely affected, among other reasons, for their slenderness; the successive destruction of these elements, walls and constructive forms underwent a selective process of use and adaptation, after which it could be affirmed that the earthquakes contributed to mold the proportions of the architecture to prevent future disasters and determine the urban profile of Quito.

The study conducted in six buildings of the period of Hispanic presence in the city of Quito: San Francisco, La Merced, San Agustín, Santo Domingo, La Catedral and the Compañía de Jesús, reveal a complex constructive history: an architecture composed of parts. The destruction by the earthquake was the opportunity to demolish and rebuild or renovate spaces that probably did not work, without losing the urban prominence. In this way it could be ensured that the architecture of Quito kept an architectural unit until the 19th century: the constructive and functional structure around a courtyard was not changed, construction techniques and similar materials were used, behind which there is a workforce with high degree of specialization and social prestige.

The construction stages of religious architecture from the seventeenth to the nineteenth centuries can be typified in three: the first, the overlapping of buildings on the same site destroying the former to reuse materials such as La Catedral and the second church of San Francisco. The second, the construction of cloisters in a systematic expansion program: The Compañía de Jesús, Santo Domingo and San Agustín. In these buildings the transformation is the result of other factors: in the first building, abandonment; in the second and third, the Catholic reform applied in the nineteenth century whose result was the drastic transformation of the interior of the church of San Agustín and Santo Domingo. In the first with the incorporation of European decorative painting, false vaults made with local traditional techniques; in the second, the transformation of the baroque church into neoclassical, the apse and the neo-Gothic altar.

On the other hand, the physical growth of the buildings is not linear or constant, it depends on the circumstances of adaptation to each place, and especially on the earthquakes and tremors that break with the continuity of the growth of the buildings. The history of the architecture seen from the changes in the institutions and the earthquakes provides a chronological vision with negative moments by the loss, but also positive moments and of bonanza. These peaks realize the vulnerability of the entire city and reveal its fragility in the face of future events.

Looking ahead, the quality of traditional materials and constructive forms that were efficient in the long term is proven. The study of mortar components, proportions of the masonry, cooking temperatures and application in the architectural restoration will validate non-invasive procedures and results that guarantee the permanence of the architectural asset. On the other hand, it is necessary to evaluate the results obtained from the application of restoration and structural reinforcement techniques carried out in the 1980s and 1990s, their compatibility and behavior in the last thirty years could develop codes for architectural rehabilitation or eliminate practices that represent new risks in the future.

In a retrospective view since the 21st century, it can be said that the mechanisms of social control of the emergency and economic risk have changed in some way, nowadays, the institutions specialized in risk and prevention issues have approached social management since the field of psychology, however, the church has not lost its space since its role in the rituality of death, the acceptance of loss is mitigated by faith, which comes to life in circumstances of population crisis. On the other hand, spontaneous social expressions of reciprocity emerge, organized from social networks and in parallel with the action of the institutions, although at one point they entered into a dispute, they were positive in the early attention during the 2016 earthquake in the province of Manabí.

Until the nineteenth century there were no banks, the convents became custodians of private property, the audience and the town hall, which weaves a chain of circumstances and conflicts of legal responsibility. In that sense, the economic risk for a part of the population has been reduced, which contributes to a guarantee for physical reconstruction.

References

Cardona, O.: (20 de diciembre de 2019). Obtenido de http://scholar.google.com.ec/scholar?q=vulnerabilidad+y+riesgo&hl=es&as_sdt=0&as_vis=1&oi=scholart: https://repositorio.gestiondelriesgo.gov.co/bitstream/handle/20.500.11762/19852/VulnerabilidadRiesgoHolistico%28Cardona_2002%29.pdf?sequence=1&isAllowed=y

Del Pino, I., Yepes, H.: Apuntes para una historia sísmica de Quito. In: Quito, M.D. (ed.) Centro Histórico de Quito. Problemática y perspectivas, pp. 67–100. Municipio de Quito-Junta de Andalucía, Quito (1990)

IGM-ORSTOM: Atlas infografico de Quito. Quito: IGM (1992)

Instituto Nacional de Patrimonio Cultural del Ecuador: Salvaguarda de la Catedral Primada de Quito. Corporación Andina de Fomento, Quito (1977)

Ortiz, A.: Guía de arquitectura de Quito. Junta de Andalucía, Sevilla (2004)

Ortiz, A.: Desde la primera piedra hasta la expulsión: 160 años de historia constructiva. In: Moreno, J. (ed.) Radiografía de la piedra, pp. 173–211. FONSAL, Quito (2008)

Peltre, P.: Quedradas y riesgos naturales en Quito, periodo 1900–1988. In: Peltre, P.C. (ed.) Riesgos naturales en Quito: lahares, aluviones y derrumbes del Pichincha y del Cotopaxi, pp. 32–53. Corporación Editora Nacional, Quito (1989)

Pérez, G.: Plano de Quito con todas sus casas. Municipio del Distrito Metropolitano de Quito. Quito: Municipio del Distrito Metropolitano de Quito (1888)

Recio, S.J., B.: Compendiosa relación de la cristiandad de Quito. Instituto Santo Toribio de Mogrovejo, Madrid (1947)

Saldarriaga, A.O.: En busca de Thomas Reed. Panamericana Formas e Impresos, Bogotá (2005)

Wilches-Chaux, G.: Desastres, ecologismo y formación profesional. Servicio Nacional de Aprendizaje (SENA), Bogota (1989)

Hospital Building Performance Assessment Through Walkthrough Methodology Focusing on Technical Assessment

Alyria Donegá(✉) and João Pantoja(✉)

Faculty of Architecture and Urbanism (FAU), University of Brasilia, UnB, Brasilia 70910-900, Brazil

Abstract. This article aims to present a qualitative assessment on the Sarah Brasília Hospital with the walkthrough tool, based on the post-occupation evaluation method developed by the Fluminense Federal University - UFF in partnership with the Oswaldo Cruz Foundation - FIOCRUZ, adapted to the need of case study in question. It includes the evaluation of the elements that make the building, through their characteristics through existing projects, photographic surveys, on-site surveys, images, schemes, and comparative tables. The study aims to focus on the verification of critical points and those in need of maintenance, as well as compliance with the Ministry of Health standards, accessibility, emergency exits, maintenance and other related. The results identify the adequacy needs of the building and the adaptations are mainly related to safety and accessibility.

Keywords: Performance evaluation · Building inspection · Hospital architecture · Maintenance · Health care facilities

1 Introduction

As Souza and Ripper (1998) remind us, the life cycle of a building comprises three stages: the project, the execution, and the maintenance - which must be coordinated by the project manager. In the case of maintenance, it should be considered that the buildings have a period of "useful life" and, therefore, should be monitored to verify their durability, as well as identify the presence of pathologies as soon as possible. As in health, it is well known that the pathology, the earlier it is investigated, the more likely it is to have a more viable and less invasive and therefore more economical solution. It is necessary to emphasize that the maintenance is foreseen in the Brazilian legislation (ABNT 2012) as a means of conservation of the buildings, as well as of relevant impact for the performance evaluation of the buildings and their elements (ABNT 2013).

It is noteworthy that in the case of hospital maintenance, according to Karman (1994), this is a "necessary evil" to prevent an "undesirable evil", a means of prevention against a "sick hospital", which includes predictive maintenance, preventive, corrective and operational continuity. These types of maintenance, according to Karman and Fiorentini (2006), are directly related to competitiveness in the hospital architecture, because the

I. Vayas and F. M. Mazzolani (Eds.): PROHITECH 2021, LNCE 209, pp. 734–743, 2022.
https://doi.org/10.1007/978-3-030-90788-4_56

"healthier" building is - in all aspects related to it - the more functional and optimized it tends to become. Seppanen et al. (2018) clarify that building facilities directly influence patient care, so there are studies in this regard in the United States and Europe.

The authors Souza and Ripper (1998), in turn, present a simplified classification concerning Karman (1994), which divides into strategic maintenance and sporadic maintenance. The strategic maintenance includes the inspections conditioned and the periodic and the sporadic maintenance is generally related to emergency maintenance.

In case of conditioned inspections are related the cases of damage with the highest risk potential and should be analyzed by specialized technical personnel, checking the conditions of the buildings disease progression, whether active or stationary. In the case of periodic inspection is indispensable for preventive maintenance, and should be applied specifically to the structure to be analyzed, since its purpose is the identification of damage and pathologies of the building, and can be applied by a non-specialized but a trained team for this procedure.

In this context, it is valid to clarify that the study of pathology, according to Milititsky et al. (2005), seeks to identify the forms of manifestation, degradation, the mechanism of failures and, consequently, their origins, and can be divided into simple problems and the complexes ones. The last one usually needs more detailed analysis and to latter to have a clearer diagnosis and prophylaxis. Thus, simple pathologies are subject to standardization and definition of a script for corresponding maintenance, while complex ones require analysis by a highly specialized professional, not applying conventional inspection mechanisms.

The authors Souza and Ripper (1998) also clarify that the periodic inspection should evaluate the following situations: absence of damage, minor damage, major damage, emergency damage, and alarm. These situations characterized, respectively, by the need for: no attitude, correction by unskilled personnel, correction by small or medium-sized companies with technical manager, call for specialists and partial or total support - subject to ruin.

Thus, it is concluded that the maintenance and inspections, should verify mainly the places subject to potential deterioration and wear to evaluate the performance of the components that compose the building and monitor its state of conservation to prevent the problems and pathologies of the building. The building will diminish the building's ability to perform.

There are several works focused on health buildings that highlight the importance of establishing, or seeking to establish, methodologies for the development of architecture and its maintenance and those applied in other countries of which it is worth mentioning Araújo (2008), Martin et al. (2015), Jones (2018), Horden (2010), Barros and Martinez-Giralt (2009) and Peters (2018). It is worth noting that Kirkeby et al. (2015) seeks improve the design techniques mainly by comparing the nowadays methodologies.

In this context, it was decided to apply the performance evaluation method developed by FIOCRUZ in field research. The method applied consisted of "identifying the main problems and pathologies of the buildings under analysis and their origins (design, execution or use), with a view to the elaboration of a building and maintenance planning policy for the buildings".

That is, the analysis performed is based on the evaluation method developed by FIOCRUZ based on the information and data exposed by Penna et al. (2002), and Castro, Lacerda, and Penna (2004) divided into three evaluations: technical, qualitative and opinion of the user. The technical evaluation seeks to verify, mainly, the usability aspects of the facilities applied to the building under study. In the qualitative evaluation research, we try to evaluate the performance of the built environment, through the walkthrough procedure - preliminary analysis with visual analysis, verifying the functionality, the flows, the layouts, and the ergonomic aspects, aiming to identify the problem and provide the recommendations.

2 The Performance Appraisal

The evaluation was carried out through technical visits to the site to analyze the present pathologies, through the walkthrough tool, as well as to verify the conformity of the building elements. Analyzing their characteristics through plans, photographic surveys, on-site surveys, images, schemes and comparative tables. In addition to information provided by hospital maintenance staff and reports from staff and patients. Focusing on the verification of critical points and those that need maintenance and the elaboration of a technical report containing the recommendations for damage control and recovery, considering the current norms.

3 Case Study: Sarah Centro Hospital, Brasilia - Brazil

The Sarah Brasília Hospital, Fig. 1a, consists of a hospital complex with approximately 68.000 m^2 and capacity for 300 beds, focused on specialized locomotor care, has all the basic structure of a specialized hospital housing the support and care sectors, except immediate care - urgency and emergency, according to the classification of RDC 050 (ANVISA 2002). This building was chosen as the object of this study due to its relevance in the field of architecture, construction method, and efficiency in care. The Hospital was designed to optimize space, high productivity and efficiency, high technology, humanized environments and the application of natural methods to provide environmental comfort. Implemented through an industrialized building system with precast parts and the project was designed by the architect João Filgueiras Lima and was inaugurated in 1974.

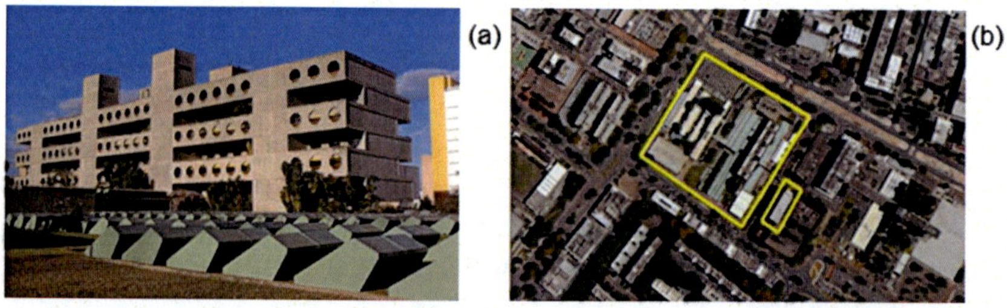

Fig. 1. Main Facade Sarah Brasília Building (a) and Location of the lot in relation to the surroundings (b).

Sarah Brasília is located in the Southern Hospital Medical Sector (SMHS), central region of Brasília. In a land of approximately 58.000,00 m^2, designed with compact volumes in order to offer a safe free outdoor areas and space for future expansions, such as the auditorium and Sarinha (LIMA 2012). The building is located between three collecting ways - in a predominantly commercial zone with more than six floors, with the presence of little vegetation as shown in Fig. 1b.

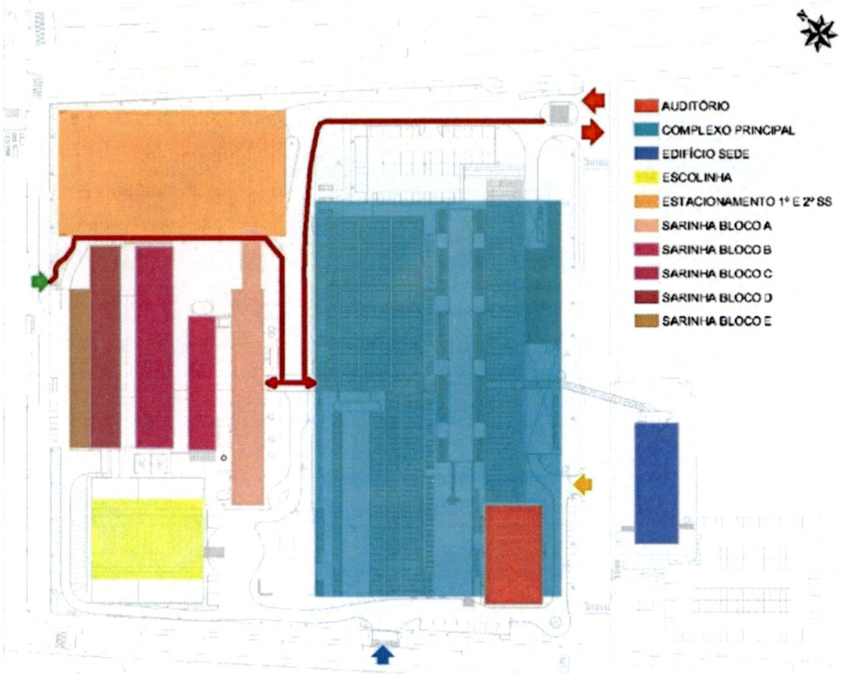

Fig. 2. Identification of the units that make up the Sarah Centro hospital complex.

This hospital complex consists of ten volumes (Fig. 3) comprising the buried parking lot, Sarinha, block A to D, Escolinha - which has been deactivated, the Main Complex, the Auditorium and the Headquarters Building. Each block has specific property functions being used independently of each other.

As for the internal plastic composition, the predominant features are internal walls of exposed concrete, vinyl floor tiles, ceramic tiles on exterior floors and workshops, metal frames with colorless glass, with no ceiling lining - most of all complex - and facilities embedded in the gutter niches. At some points in the Sarah Brasília complex, it is also possible to identify decorative elements of the artist Atos Bulcão, such as Sarinha block A and the ground floor of the main complex. As for the coverings, the walls are mostly removable in white color, apparent structure and facilities, wooden stretcher and vinyl flooring - green for general circulation and beige for the technique mostly.

Fig. 3. Main complex sectorization, 1st underground.

4 Results

In almost the entire complex, the vinyl flooring has seams with projections of up to 1 cm, which from accessibility does not cause inconvenience, but can disturb the transfer of patients on stretchers, also causing small accumulations of dirt. In the courtyard at the entrance of Sarinha and in the workshops located in blocks D and E, however, it was found damages in the flooring and types of floor in disagreement with the RDC 050 (ANVISA 2002). The standard in question specifies that floors in healthcare facilities should not contain joints. Regarding the structures, fixed walls, semi-movable walls, visitable and movable accesses, no apparent pathologies were found.

In the headquarters building were identified inadequate emergency signaling, undersized distribution of extinguishers and/or inadequate type, inadequacy of the 11th-floor library railing, need for resizing of the Technical Fire Reserve, deficiency in emergency lighting, no presence of automatic showers (required for buildings above 12 m in height for the area) and inadequate escape routes.

Other problem identified, was the walkway that connects the headquarters building to the main complex showed inadequate stairway, located between the buildings, regarding safety standards, and the "wind tunnel" factor. The situation was due to the installation of a mechanized ventilation system too strong for the structure that aimed to cool the necessary walkway because its constituent materials, polycarbonate and steel, were overheating the space. In the auditorium, it was verified the presence of infiltration and consequent damage to the floor and furniture, besides the inadequacy regarding the signaling norms and fire extinguishers (ABNT 2001, 2015).

In the main complex, underground, the inadequacy of escape route stairs, emergency signaling, emergency lighting, fire extinguishers and problems with some sprinkler showers has been verified. In addition to failures in the protective structure to the x-ray technicians - it was not fully isolated - and obstruction of access to safety equipment, extinguishers, and fire hydrants. The ramp that connects the basements to the ground floor also presented problems with inadequate floor and handrail for accessibility and safety. In the hospitalization tower, on the first floor, the medical dormitories were converted into dormitories for caregivers by their staff, since it is the pediatric nursing floor, the caregivers have the right to stay with the patient and the hospital does not have this type of structure.

Regarding the setorization, it was identified that the hospital has seven of the eight sectors recommended by RDC 050 (ANVISA 2002), which are administrative support, diagnostic support and therapy, logistic support, technical support, outpatient care, inpatient care, and teaching and research, with the addition of internal gardens. About the main complex, the second underground consists of areas of logistical support, technical support, and exclusively technical circulation. On this pavement, the logistic support consists of the maintenance rooms, dietary and storage sectors and technical support corresponds to sterilization plants.

The first underground consists of areas of logistical support, technical support, and support for diagnosis and therapy, corresponding to the changing rooms, operating room, and examination sector, respectively. Circulation is predominantly general with the presence of critical circulations in pathological examination areas and operating room - due to the risk of contamination. The ground floor comprises administrative, outpatient and technical support sectors, which represent the medical offices, consulting rooms and doctors' rooms respectively. Circulation is for general use and is restricted to technicians only on the metal walkway to the headquarters building.

About the upper floors, the 1st to 5th floors are divided into the service, hospitalization and logistic support sectors, which represent the hospitalization areas on the sides and medical dormitories in the center, in this order. The volumes shifted abroad house the restrooms and support environments for hospitalization. Each floor has a specific specialty and from the 1st to the 3rd house the pediatric ward, oncology and neurosurgical orthopedics. The 4th and 5th floors have the specialties of neurological rehabilitation and muscle injury.

On the 6th floor, the sectors were redistributed for teaching and research and logistical support. The teaching and research sector represents the meeting rooms, while the logistical support area to the dormitories, used by staff and/or guests visiting Sarah Brasília for training, conferences or seminars. Hospitalizations are located in the upper floors while in the lower are the larger sectors. Besides, it can be seen that the building houses the main sectors of Sarah Brasilia, such as the maintenance areas, changing rooms, sterilization centers, exams, and hospitalization.

Finally, there is the so-called Sarinha, which consists of five blocks - A to D. Sarinha's blocks A and B consist of technical support areas (block A) and diagnostic support and therapy (block B), which include the daycare center to children's employees and rehabilitation environments. In block A, there are also leased lines connecting the block A to the other blocks and the vertical circulations that give access to the main complex (external volume to the building) and the parking lot of employees on the right. In the courtyard near the elevators and stairs, vertical circulations - at the right end of the building, outdoor activities are performed for hospitalization and recovery patients. On the second floor of the block A of Sarinha is located another sector of diagnostic and therapy support, which houses the patient recovery academy and some logistic support environments for the activities developed. Blocks B, C, and D consist of ground blocks where logistic support and diagnostic support and therapy activities are located, the latter consisting of pulmonary and exercise testing environments. The logistics areas of these blocks house Sarah Brasilia's hospital support workshops, where they manufacture or maintain in-patient clothes, canes, prosthetics, among others, as well as the presence of a warehouse dedicated to these activities.

The headquarters building is mainly composed of the administrative support and teaching and research sectors, with two vertical circulation towers at its extremities, being the tallest building of the hospital complex that comprises the Sarah Brasília with its 11 upper floors. It is a typical building of offices and administrative rooms in vain, adapted with furniture and removable partitions according to the needs. On its first floor, it is connected to the main complex through the catwalk and in this are located the administrative support and teaching and research environments with a small auditorium for events. The ground floor also houses administrative support environments consisting of the depot and screen-printing sectors. In the basement, the generator equipment and emergency pumps are located, making up the logistics sector.

From the second to the tenth floor of the headquarters building, there are the typical administrative support environments for hospitals, with rooms and offices and, in this case, they also develop support activities for the other Sarah Network member units in Brazil. The floors are identified as type floors, with few changes between floors through removable modules and furniture. On the 11th floor and mezzanine, additionally, is located the library of Sarah Brasilia, thus constituting sectors of education and research of the building, as well as logistical support areas that include the pump house and reservoir of drinking water of the building.

As for the flows of the hospital complex as a whole it can be classified as critical circulation, technical circulation and general circulation, where critical consists of areas with a higher risk of contamination, technically refer to areas of exclusive access to employees, and general as those with common access. In the hospital complex, the vertical circulation is mainly carried out in three towers, one to the north and two to the south, giving access to Sarinha and the hospitalization tower, respectively, as well as some additional circulation between the basements.

In the 2nd underground, the circulation is predominantly technical and in the 1st underground and ground floor, the circulation is predominantly general with the presence of examination rooms. Access to the area is controlled by the balcony located on the ground floor near the elevators of the south tower. On the ground floor, the predominant

Table 1. Comparative table of results.

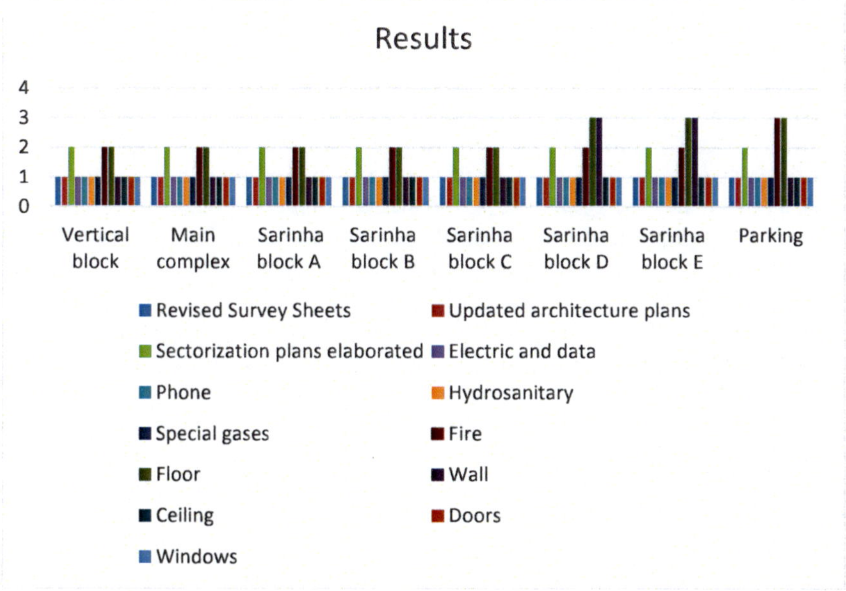

circulation is general, with access controlled by the balcony located near the main access of the complex, near the entrance of block A of Sarinha and the north vertical circulation tower. In Sarinha Blocks A to E, the predominant circulation in general, due to the predominance of examination and rehabilitation rooms, and the headquarters building has a circulation classified as technical due to the functional characterization of the departments located in this building.

In the Table 1, it is the data summary of all anamnesis and diagnosis identified in the case study in question, where the value 1 indicates items by the current normative, the value 2 indicates items in disagreement with the normative current and the items with value 3 indicate areas where urgent repair is needed. As well as it is visually possible to identify, through this table, that Blocks D, E, and parking are the locals, where it is necessary a larger intervention. The obtained results indicate, therefore, that the interventions should be initiated by Block D, E and parking as they consist of areas of a greater problem. Considering that in other locations the frequency of problems identified indicates less urgency and the no pathologies identification that would substantially compromise the safety and functionality of the spaces.

5 Conclusion

Regarding the technical evaluation, it is noted that, in practically the entire hospital complex, there are inadequacies regarding emergency signs, emergency lighting, extinguisher sizing, obstruction of access to safety equipment, inadequate escape routes -

including railing and handrail inadequacy - and leaks on roofs and balconies structures - except sheds. Therefore, there is a need to adjust the security, accessibility and maintenance issues of the building as the main point, as well as a more in-depth study on the issue of infiltration in the roof slabs of the hospitalization block and in the headquarters building, in addition to the study of correcting the floors of the hospital complex - as for the amendments - to comply with current legislation.

As for the users' opinion, there were no complaints regarding the cleanliness of the entire complex or bathrooms, nor the state of conservation as a whole. About the environments, they said they had good lighting and thermal comfort - except the hottest months in Brasilia, thus requiring a more in-depth study on the issue. Only the dissatisfaction regarding public parking outside the hospital complex and the absence of a snack bar in the complex - for companions and visitors - were evidenced.

Regarding the qualitative evaluation, based on the data collected, there were no problem points presented beyond the opinion of users and the technical evaluation, which makes it clear that the maintenance recommendations regarding safety issues of Sarah Brasília refer to the a more technical aspect, since its design has not been reviewed for approximately 40 years and the applicable standards have been updated since then.

In conclusion as to the walkthrough methodology, it was identified that it consists of a method that seeks the global assessment of the building. Adopting the items of technical assessment and user opinion that make up the post-occupation assessment established by the authors, which was a tendency to prioritize the user's opinion in relation to the evaluation of the building as a whole.

Regarding the technical evaluation, it was noted that the evaluation items suggested by the method mainly include the evaluation of architectural elements of the building and some types of facilities. Thus, with the present study and its application, a gap was identified regarding more technical and specific items to be evaluated in a hospital, with emphasis on complex facilities, structures and accessibility. These items are more specific and of considerable weight for the proper planning and management of a complex building as a hospital. It should be noted that the evaluation of these studies was based on publications on the method applied to hospitals in the period from 2002 to 2004 and that it was already adapted to the studies in which it would be applied. Therefore, it must be concluded that the method itself tends to be a targeted tool and as such needs some adaptation to each case. In view of the proposal to be a simplified assessment method, but which could be improved to include more technical items of great relevance in the case of hospitals, especially regarding the requirements of today regarding accessibility and fire prevention systems and escape routes.

References

ABNT: Associação Brasileira de Normas Técnicas. NBR 15575: Norma de Desempenho (2013)

ABNT: Associação Brasileira de Normas Técnicas. NBR 5674: Manutenção de edificações — Requisitos para o sistema de gestão de manutenção (2012)

ABNT: Associação Brasileira de Normas Técnicas. NBR 9050: Acessibilidade a edificações, mobiliário, espaços e equipamentos urbanos (2015)

ABNT: Associação Brasileira de Normas Técnicas. NBR 9077: Saídas de emergência em edifícios. . Rio de Janeiro: [s.n.] (2001)

ANVISA: Agência Nacional de Vigilância Sanitária. Resolução (RDC) 50, de 21 de fevereiro de 2002. Dispõe sobre o Regulamento Técnico para planejamento, programação, elaboração e avaliação de projetos físicos de estabelecimentos assistenciais de saúde (2002)
Araujo, E.P.: Avaliação crítica de ambientes em estabelecimentos assistenciais de saúde (2008)
Barros, P., Martinez-Giralt, X.: Contractual design and PPPs for hospitals: lessons for the Portuguese model. Eur. J. Health Econ. **10**(4), 437–453 (2009)
Castro, J., Lacerda, L., Penna, A.C. (org): Avaliação pós-ocupação APO: saúde nas edificações da FIOCRUZ. FIOCRUZ (2004)
Horden, P.: Hôpitaux au Moyen Âge et aux Temps modernes: France, Allemagne et Italie: une histoire comparée. The English Historical Review (2010)
Jones, P.: Private finance initiative hospital architecture: towards a political economy of the Royal Liverpool University Hospital. Sociol. Health Illn. **40**(2), 327–339 (2018)
Karman, J.: Manutenção hospitalar preditiva. 1. ed. PINI (1994)
Karman, J., Fiorentini, D.: Conceitos de arquitetura manutente e de arquitetura voltária. Exacta, Engenharias I B4 QualisInterdisciplinar B2 Qualis, pp. 159–168 (2006)
Kirkeby, I.M., et al.: Designing for health in school buildings: between research and practice. Scandinavian J. Public Health **43**(3), 260–268 (2015)
Lima, J.F.: Arquitetura: uma experiência na área da saúde, 1. ed. Romano Guerra Editora (2012)
MARTIN, Daryl, et al. 2015 Martin, D., et al.: Architecture and health care: a place for sociology. Sociol. Health Illn. **37**(7), 1007–1022 (2015)
Milititsky, J., Schnaid, F., Consoli, N.C.: Patologia das fundações. Oficina de Textos (2005)
Penna, A.C.M., et al.: Avaliação pós-ocupação (APO) em edificações da Fundação Oswaldo Cruz (FIOCRUZ): o caso do Instituto Fernandes Figueira (IFF). USP, p. 10 (2002)
Peters, A., et al.: Keeping hospitals clean and safe without breaking the bank. Antimicrob. Resist. Infect. Control **7**(1) (2018)
Seppanen, A., et al.: Modern forensic psychiatric hospital design: clinical, legal and structural aspects. Int. J. Ment. Syst. **12**(1) (2018)
Souza, V.C.M., Ripper, T.: Patologia, recuperação e reforço de estruturas de concreto. PINI (1998)

Conservation of Cultural Heritage. Evaluation of Archaeological Digs Effect on the Degradation of the Structures Found

Veronica Vitiello(✉), Roberto Castelluccio, and Massimo Ramondini

Department of Civil, Construction and Environment Engineering (DICEA), Università degli Studi di Napoli Federico II, Via Claudio 21 - Piazzale V, Tecchio 80, 80125 Napoli, Italy
{veronica.vitiello,roberto.castelluccio, massimo.ramondini}@unina.it

Abstract. The restoration of Archaeological Heritage gives a remarkable importance to the features of formal and material preservation of elements in their context pursuing the enhancement of its value as historical document. The diagnosis phases must be structured in a complete and complex way to find correct methods of digging, considering that the archaeological digs provide a sudden alteration of the equilibrium condition that the underground structures have reached with the filling layers over the centuries, causing degradation phenomena or accelerated deterioration of the unearthed parts. Some studies have shown a pushed aging for stones subjected to saturation and rapid drying cycles, both in terms of lowering mechanical resistance and activation of pathogenic dynamics. Starting from these studies, a laboratory survey has been carried out on samples of Neapolitan yellow tuff aimed at reproducing, at the small scale, the alteration due by digging phases on natural stones. The paper shows the first results of this survey and proposes intervention strategies to ensure preservation criteria during the archeological digging phases, starting from indirect geometric relief of underground structures, including a thermo-igrometric characterization of the soil-artifact system and developing an intervention method for digging and protecting the Archeological Heritage.

Keywords: Archaeological heritage · Archeological digs · Preservation · Diagnosis · Intervention strategies · Thermo-igrometric characterization

1 Introduction

The restoration of the Architectural Heritage pursues the goals of functional recovery, formal and material conservation, as well as the valorization of the monument as a historical document. In the more restricted and sensitive area of the restoration of the Archaeological Heritage, the component of the functional recovery is rarely taken into account while it is necessary to develop, in an even more complete and complex way, the strategies of conservation and enhancement of the element in its own context.

Planning the conservation needs of archaeological structures must begin on site, from the first moment of planning the dig phase which, according to Stanley Price (1986),

I. Vayas and F. M. Mazzolani (Eds.): PROHITECH 2021, LNCE 209, pp. 744–754, 2022.
https://doi.org/10.1007/978-3-030-90788-4_57

is crucial: *"first because it is the moment when the archaeologist has the maximum opportunity to collect information related to the context of the finding and its materials; second, for the potentially disastrous consequences of a lack of environmental control on chemically and mechanically unstable findings"*.

Price introduces the theme of archaeological conservation on site, distinguishing it from archaeological conservation in the laboratory, understood as the set of techniques to be applied on the remains during and immediately after their exposition to the air.

The dig operations thus represent the first strategic phase in which measures must be taken to safeguard the monument and they must be conducted by ensuring that the underground structures enter the new exposure condition in a time compatible with the materials they are composed with.

The archaeological ruins, in fact, remain for centuries in underground conditions, reaching a thermo-hygrometric equilibrium with the covering layers. The excavation phases lead to a sudden alteration of this equilibrium as they expose the structures into contact with the external environment in a short time. As a result of the difference in pressure, temperature and ventilation with the atmosphere, thermal and water flows are triggered out of the excavated elements causing the deterioration of the building materials and of the decorations that may be present.

The effects of this sudden dynamics can be studied by laboratory investigations, considering that the alteration produced by the dig phases is similar to that induced by subjecting natural building materials to cycles of aging, with wetting and drying. The parameter that is tested acts on the durability of the material, which represents its ability to resist external agents while maintaining its original shape, resistance and appearance for a long time and depends on its petro-physical properties, such as porosity, hydraulic behaviour and mechanical strength.

Research on the change of the petro-physical properties of some granites subjected to heating and cooling cycles, showed a performance decay of the material in terms of reduction of apparent density, Young's modulus, surface hardness, effective porosity and, on the other hand, an increase in internal microfractures (Freire-Lista et al. 2016).

Other studies have shown that degradation phenomena, closely linked to the dynamics of water absorption by porous materials and their evaporation, increase with the progressive ageing of the material, also leading to a reduction in their mechanical performance (Colback et al. 1965).

Experimental surveys developed on sandstone rocks by the researchers of the Department of Geology of the University of Bristol (Hawkins et al. 1992) have shown a substantial reduction in the uniaxial compressive strength of samples tested under completely soaked conditions compared to the values recorded for similar samples tested under dry conditions (after drying in the oven at 50 °C). The maximum performance decay, equal to 78%, was recorded in rock samples containing large amounts of clay or fragments of sedimentary rocks. For the first one, the reduction of mechanical resistance in saturated conditions was attributed to the volumetric expansion of clay minerals when of a moist content is present.

This phenomenon is also found in bricks and can be described by the mathematical relation known as *Power-Law of Moisture Expansion Kinetics* (Hall and Hoff 2012).

Similar tests, made by at the University of Sannio and the University of Naples Federico II (Ceroni Ceroni et al. 2004) on Neapolitan yellow tuff, have recorded a reduction of the mechanical resistances from dry to saturated condition of 40%.

The relationship between the degree of saturation of samples of Neapolitan yellow tuff and its moisture content was studied by Evangelista (1980) who has shown that the mechanical behaviour of this material is influenced by its degree of saturation and also, that the variation of the water content causes not negligible deformations of the tufaceous mass. The results of the mono-axial compressive tests, in analogy with the results obtained in this study, had shown a reduction in the compressive strength of tests from dry to saturated conditions.

Similar results were highlighted in the research developed by the University of Bologna (Franzoni et al. 2015) on bricks and walls manufactured with cement and lime mortars.

In addition to the influence on the mechanical resistances, the degradation related to the presence of water also determines a variation on their thermal behaviour, both for their volumetric expansion (Gordejuela 1994) and for their thermal conductivity (Baker 2011, Doran 2001, Evangelisti 2015; Foucquier et al. 2013; Gaspar 2016; Kosmina 2016; Marshall 2017; Pehlivanli 2010; Troppová et al. 2015; Vitiello et al. 2020).

The evaluation of the variation of the thermal parameters, however, can be careless in the context of those interventions on the Archaeological Heritage that do not promote a refunctioning of the ruin and, therefore, prefer conservation to efficiency actions.

Very important, finally, is the evaluation of the influence of salts in the activation of the degradation phenomena for the stones and their finishes. In its movement within the porous system of the building materials, the water carries salts in solution which, in the evaporation phase of the fluid, move on the surface layers. The crystallization and activation of the relative degradation phenomena are closely linked to the interaction that the underground masonry has with the external environment when the covering layers are removed; at this stage, in fact, the supersaturation solution, which occupies the reduced volume of the capillary pores, develops a hydrostatic pressure which determines the precipitation and crystallization of the salts, with increase in volume and formation of micro-tensions which cause the degradation and spraying of the superficial layers (Charola 2000; Correns 1949).

Possible degradation phenomena depend on the chemical composition of the building materials and the nature of the salts dissolved in the water which is absorbed and conveyed by the soil (Maguregui 2007). The scientific research examines with great interest the techniques to be adopted for the rehabilitation of the walls; among these, the most widespread involves the realization of clay or cellulose wraps; however, they can lead to the formation of second order decay phenomena related to the growth of mould or activation of inverse osmosis-dynamic (Delgado et al. 2016).

In the final analysis, the phenomena of degradation that occurred as a result of incongruous technological solutions adopted for the realization of museum itineraries or for the covering of archaeological ruins cannot be neglected. Sometimes, the choice of materials and closures that are not very breathable, feeds the dynamics of the flow of water into the wall elements altering the finishes and the decorations. As an example, the phenomena of moisture from rising damp, with the formation of moulds and saline efflorescence, on the mosaics of the Pavilions of the Archaeological site of Piazza Armerina after the realization of non-breathable covers, can be mentioned (Castelluccio and Vitiello 2016).

Considering the above, it is interesting that scientific research identifies a correct methodology for carrying out archaeological campaigns which will make it possible to limit the formation of degradation of the elements during the digs.

Aimed at studying alterations in the performance of building materials recurrent in archaeological structures, a research is being carried out at the Department of Civil, Construction and Environmental Engineering of the University of Naples Federico II (DICEA), to quantify the variation of the mechanical resistance of the stone materials in the various phases of the ruin: from the underground condition, in which it is in thermo-hygrometric equilibrium with the covering ground, to the archaeological excavation, which activates a high evaporation of the moisture content.

The contribution presents the first results of the experimental survey which reproduces, at the small scale, the alterations caused on the material in the phase in which the walls are buried and then completely saturated.

2 Materials and Methods

The experiment was conducted on cubes of dimensions 10cm × 10cm × 10cm of Neapolitan yellow tuff, a soft and workable rock historically used in the Neapolitan construction tradition and strongly recurrent in the archaeological sites of the ancient city for its characteristics of good resistance in relation to the specific weight and for the low cost. This material is the product of the third cycle of volcanic activity of the Campi Flegrei, the second for importance in the area of Campania. The rock originated from a rather rapid eruptive dynamic that caused the formation of a very warm and mobile suspension that flooded vast areas giving the solidified material its typical chaotic nature, the partial devitrification of its pumices and its firmness (Rittmann 1950; Savino 2000).

Over the years, this rock has been widely used as a building material, because it offers excellent strength characteristics that have allowed it, over the centuries, to be easily quarried and reduced in small blocks (Ceroni et al. 2004).

The widespread use of chaotic yellow tuff in the Neapolitan construction tradition gave it the name of Neapolitan yellow tuff, indicating that this material is representative of the local constructive culture (Nicotera et al. 1967).

To the volcanic nature of the Neapolitan yellow tuff is connected the high porosity of this material which makes it strongly vulnerable to degradation phenomena caused by physical, chemical and biological attacks.

Physical phenomena linked to frost and thaw of the water contained in the porous structure can produce lesions and crumbling of the material; chemical alterations linked to the crystallization of dissolved salts in the water, on the other hand, can lead to the formation of subflorescences and, as a result, to the exfoliation and detachment of surface crusts; the presence of moisture, finally, favours the occurrence of pathologies of biological origin linked to the formation of moulds and lichens which can cause damage to the health of those who live the environments in which these phenomena occur.

2.1 Physical Characterization of the Material

The experimental survey, conducted at the DICEA, was developed using yellow tuff extracted from S. Rocco cave near Capodimonte hill in 1970–1980 and used for previous geotechnical test surveys. The material was kept for years in one of the courtyards of the Department. This condition of long-term exposure to atmospheric agents has proved particularly interesting to assimilate the degree of natural ageing of the material, subjected to drying and bleaching cycles by pouring rain, to that of the underground walls of an archaeological ruin.

To achieve a simulation closer to the condition of the on-site ruin, the cube samples obtained from each block have been further wetted until they've reached the complete saturation. A total of 15 samples of material were tested (Fig. 1).

Fig. 1. Material before cutting, dimensional check of samples, cataloguing of specimens

The first phase of characterization of the material provided the measurement of its specific weight using a pycnometer, according to the methods prescribed by ASTM D854. The test is based on the measurement of the volume of water contained in a pycnometer which capacity is known, which is moved from the volume of material inserted into it. Knowing the density of the liquid and the initial weight of the material, which is baked in the oven before the test to remove the physiological moisture content, it is possible to determine the density of the investigated sample (Fig. 2).

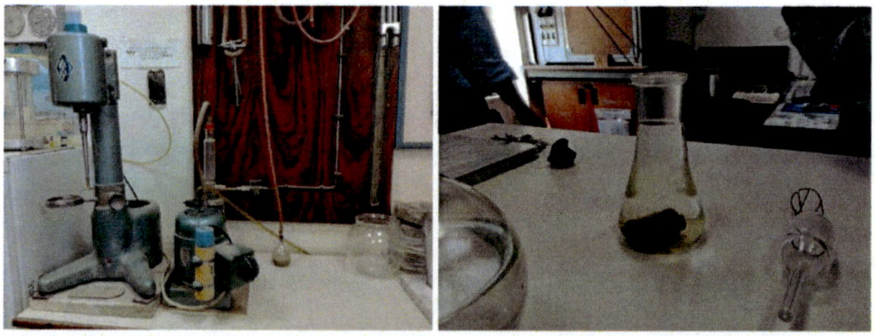

Fig. 2. Test steps on a powder sample (left) and solid sample (right)

Whereas a piece of stone material completely immersed in water retains a very small percentage of uninhibited voids, relating to the pores not communicating with each other which are not allowed to pass through by the fluid, it was decided to carry out the measurement using both pulverized and leaked material and solid material. In the two tests, therefore, it was necessary to use pycnometers with different capacities of 100cc for the powder test and 250cc for the solid test.

As was easily guessed, the measurement results showed a density 12% higher than the solid sample, because of the effect of pulverisation in the elimination of residual pores (Table 1).

Table 1. Specific gravity measurement by water pycnometer method

DISINTEGRATED AND SIEVED			SOLID		
$\mathbf{P_{pic}}$ [gr] pycnometer weight 100cc	$\mathbf{P_{mat}}$ [gr] material weight	$\mathbf{P_{mat}/V_{mat}}$ [gr/cm^3]	$\mathbf{P_{pic}}$ [gr] pycnometer weight 250cc	$\mathbf{P_{mat}}$ [gr] material weight	$\mathbf{P_{mat}/V_{mat}}$ [gr/cm^3]
59,77	16,93	**2,55**	189,94	20,27	**2,24**

2.2 Conditioning of Samples

The samples were tested in three conditioning classes corresponding to dry (S = 0%), semi-saturated (S = 50%) and saturated (S = 100%).

The conditioning of the test samples was carried out following the weight method prescribed by UNI 11085, which is based on the principle of loss of mass of a sample after drying by heating, and was applied using the following equipment:

- Balance with accuracy of 0,01 gr
- Electric heating laboratory heater capable of maintaining drying temperature with a stability and uniformity of 2 °C
- Dryer containing silica gel with cobalt chloride indicator

It should be noted that the evaluation of saturated weight of the stone elements in the laboratory is not immediate and sometimes requires complex procedures, with application of vacuum, pressure or boiling to achieve the total imbibition of the blocks, (Verduch 1975), which reduces the time taken to carry out experiments but requires more laboratory control. In this study, the total immersion method was adopted, as described by UNI EN 772-11; however, for the sake of completeness, reference should be made to the simplified method proposed by Z. Zaretzky (1972), which provides comparable results obtained by household instruments.

After recording the initial weight of the test pieces in environmental conditions, they were all subjected to 4-h drying cycles in the oven at 105 °C and cooled in the dryer until the ambient temperature was reached. At each stage the samples were weighed. The test was completed when the samples have reached a stable mass, determining the dry weight of the samples (P_d). A first group of samples was tested in this saturation conditions.

Subsequently, the samples were completely immersed in water until the constant mass was reached, thus determining the wet weight (P_w) of each sample and, by difference with the dry weight, the percentage of the absorbed water content C_w, applying the relation (Eq. 1):

$$C_w = \frac{(P_w - P_d)}{P_d} \times 100 \quad (1)$$

A second group of samples was tested in this condition of complete saturation (S = 100%); a third one was putted to air-dry in a ventilated environment shielded by direct solar radiation until the intermediate values of moisture were reached, determined according to the quantity of known water to be disposed of. In order to ensure a uniform distribution of the moisture content within the elements investigated, the samples were placed in thermo-hygrometric equilibrium prior to each test, to allow the expansion of the moisture content within the capillary system of the material, thus hindering evaporation to the outside.

The samples have been identified with the acronym TGN (Tufo Giallo Napoletano), then the classes of samples corresponding to each degree of saturation where found, each containing 5 specimens on which the tests were carried out (Table 2):

Table 2. Grouping of samples

Saturation degree	Samples	
	Group	Elements
$S = 0\%$	TGN $_{1-5}$	$TGN_1 - TGN_2 - TGN_3 - TGN_4 - TGN_5$
$S \approx 50\%$	TGN $_{6-10}$	$TGN_6 - TGN_7 - TGN_8 - TGN_9 - TGN_{10}$
$S = 100\%$	TGN $_{11-15}$	$TGN_{11} - TGN_{12} - TGN_{13} - TGN_{14} - TGN_{15}$

The five samples of each saturation class were subjected to mechanical uniaxial compression tests carried out in accordance with UNI 9724-3. The test was made using a monotone load at a speed of 0,5 MPa/s. At the same time the values of the axial de-formations were read from 4 measures (Fig. 3).

Fig. 3. Test steps before beaking (left) and after (right)

3 Results and Discussion

The analysis of the data collected shows a reduction in the mechanical compressive resistances when the degree of saturation increases.

In fact, for samples tested under dry conditions (S = 0) a mean of the five maximum diverting strain of 2.3 MPa was recorded; for samples tested in semi-saturated conditions (S = 50%) an average of the five maximum diverting strain of 1,6 MPa was recorded; finally, for the samples tested under full saturation conditions (S = 100%), an average of the five maximum diverting strain of 1,3 MPa (Table 3) was recorded.

Table 3. Mechanical measurements at different saturation degrees

Sample	Saturation [%]	Maximum compressive strength [MPa]	Medium compressive strength [MPa]	Standard deviation [%]
TGN 1	S = 0	2,53	2,30	11,2%
TGN 2		2,02		
TGN 3		2,59		
TGN 4		2,07		
TGN 5		2,28		
TGN 6	S ≈ 50%	1,37	1,64	10,2%
TGN 7		1,61		
TGN 8		1,69		
TGN 9		1,72		
TGN 10		1,81		
TGN 11	S = 100%	1,35	1,30	20,4%
TGN 12		1,56		
TGN 13		1,42		
TGN 14		1,33		
TGN 15		0,86		

There was a 29% reduction in compressive strength from dry to semi-saturated conditions and 43% from dry to saturated conditions.

The diagram between the stresses and deformations is shown in the graph (Fig. 4).

The test results showed a 37% reduction in mechanical resistance from the dry condition (S = 0%) to the semi-saturated condition (S = 50%) and 45% from the dry condition to the condition (S = 100%). These reductions are greater than those recorded in Evangelista's results (1980), this can be attributed to the ageing effect of the material due to its long exposure to air and weather conditions.

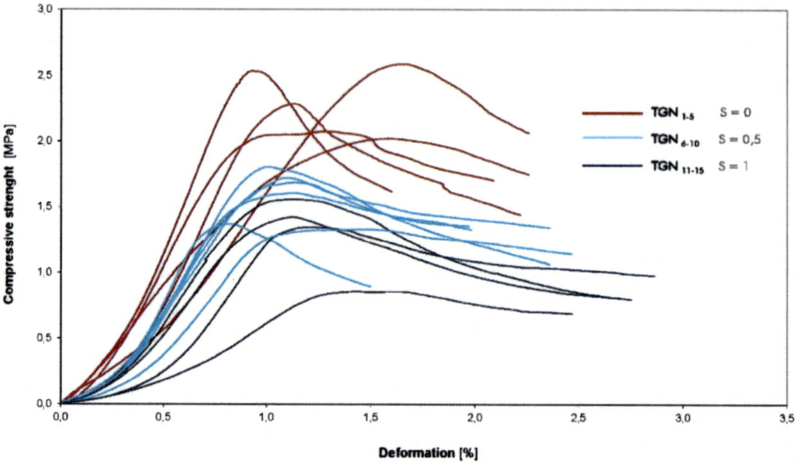

Fig. 4. Compressive strength values at different saturation degrees

However, the independence of the reduction of the mechanical resistance of the specimens with respect to the number of cycles of wetting and drying of sample, showed by the Author in his study, must be interpreted in the light of the method of conditioning used in the experiment. The research, in fact, had been carried out using the air sample drying method, which represents a condition close to the behaviour of the wet masonry of an above-ground building.

In the case of underground archaeological structures which are brought to light during excavation digs, the thermal stress produced by the sudden exposure to the air of masonry must be considered, which is closely related to the high value of the temperature gradient to which the structure is subjected-to. Therefore, it would be extremely limiting to interpret the dynamics of the evaporation of the wet content only in its positive meaning, considering as the only consequence the recovery of the mechanical resistance of the material from the saturated to the dry condition.

A more truthful schematization can be obtained by assimilating the temperature gradient to which the walls are subjected during the excavation to a drying cycle of the damp material in the oven.

Therefore, in a second experimental phase, already partially underway, the variation of the mechanical resistances of the wetted materials is evaluated, considering the comparison between drying cycles to air and drying cycles in the oven.

The first results of this experiment clearly show that the application of a fast and suitable drying process on samples of saturated stone material produces the formation of internal micro-cracking which further exacerbate the reduction in mechanical performance of the samples tested. Some test pieces, dried in the oven and then completely saturated, burst before being subjected to the test.

4 Conclusions Remarks

The issue of the conservation of archaeological ruins must take into account the alterations that can be induced on the building materials by the combination of various decay factors. Many of the dynamics of alteration and deterioration of the material are already activated in the dig phases, due to the abrupt change of the thermo-hygrometric parameters to which the walls are subjected: from a condition of equilibrium with the cover layers of ground to a sudden exposure to air and atmospheric environments.

The experimental survey conducted at the DICEA has highlighted the alterations produced on the material in the first phase of the life of the ruin, the one underground. The results have confirmed studies and surveys made also by other researchers on the reduction of mechanical resistance of a porous-so material with the increase of its degree of saturation.

In addition, further development of the test procedure has led to the hypothesis of simulating the effects on building materials during excavation, when the structures are suddenly exposed to the air. The results of the tests will provide the parameters needed to determine the precautions to be taken during the execution of archaeological digs.

Acknowledgements. The activity presented in the paper is part of the research founded by the M.I.U.R. (Italian Ministry of Education, University and Research) through a PhD scholarship grant for "Studies with Industrial characterization" (PON RI) to the Department of Civil, Building and Environment Engineering (DICEA) of the University of Naples Federico II.

References

ASTM D854 – 14: Standard test methods for specific gravity of soil solids by water pycnometer

Baker, P.: U-Values and Traditional Buildings. In: Situ Measurements and their comparison to calculated values. Historic Scotland Technical Paper 10, Dinburgh, UK (2011)

Castelluccio, R., Vitiello, V.: Deleting of rising damp in the archaeological site of Piazza Armerina through the application of the technology by neutralizing electrical charge T.N.C. World Heritage and Degradation, Le Vie dei Mercanti - XIV Forum Internazionale di Studi, pp. 1457–1466 (2016). ISBN: 978-88-6542-257-1

Charola, A.E.: Salt in the deterioration of porousmaterials: an overview. J. Am. Inst. Conserv. **39**, 327–343 (2000)

Ceroni, R., et al.: Analisi e caratterizzazione meccanica di murature di tufo. In: Proceedings of the 15° Congresso C.T.E. Bari (2004)

Colback, P., et al.: The influence of moisture content on the compressive strength of rocks. In: Proceedings of the 3rd Canadian Symposium on Rock Mechanics, pp. 65–83 (1965)

Correns, C.W.: Growth and dissolution of crystals under linear pressure. Discuss. Faraday Soc. **5**, 267–271 (1949)

Delgado, J.M.P.Q., et al.: Salt damage and rising damp treatment in building structures. Adv. Mater. Sci. Eng. (2016)

Doran, S.: Field Investigations of the thermal performance of construction elements as built. Building Research Establishment (BRE), East Kilbride, UK (2001)

Evangelista, A.: Influenza del contenuto d'acqua sul comportamento del tufo giallo napoletano. Atti del XIV Convegno Nazionale di Geotecnica, Firenze **2**, 63–68 (1980)

Evangelisti, L., et al.: In situ thermal transmittance measurements for investigating differences between wall models and actual building performance. Sustainability, 10388–10398 (2015)

Foucquier, A.J., et al.: State of the art in building modelling and energy performances prediction: a review. Renew. Sustain. Energy Rev. **23**, 272–288 (2013)

Franzoni, E., et al.: Compressive behaviour of brick masonry triplets in wet and dry conditions. Constr. Build. Mater. **82**, 45–52 (2015)

Freire-Lista, D.M., Fort, R., Varas-Muriel, M.J.: Thermal stress-induced microcracking in building granite. Eng. Geol. **206**, 83–93 (2016)

Gaspar, K., et al.: A comparison of standardized calculation methods for in situ measurements of facades U-value. Energy Build., 592–599 (2016)

Gordejuela, F.: Expansíon por la humedad de los productos cerámicos españoles: revision de la normativa, Congreso Nacional de Tecnología en la Arquitectura (1994)

Hall, C., Hoff, W.D.: Moisture expansivity of fired-clay ceramics. J. Am. Ceram. Soc. **95**, 1204–1207 (2012)

Hawkins, A.B., et al.: Sensitivity of sandstone strength and deformability to changes in moisture content. Q. J. Eng. Geol. **25**, 115–130 (1992)

Kosmina, L.: In-situ measurement of U-value. Building Research Establishment (BRE), East Kilbride, UK (2016)

Maguregui, M., et al.: Identification of historical bricks degradation products by means of combined Raman spectroscopy and scanning electron microscopy. Conserv.ation Sci., 212–219 (2007)

Marshall, A.: Domestic building fabric performance: closing the gap between thein situ measured and modelled performance. Energy Build. (2017)

Nicotera, P., et al.: La costituzione geologica, in AA. VV. "Il sottosuolo di Napoli", Relazione della commissione di studio, Comune di Napoli (1967)

Pehlivanli, Z.: Effect of moisture and temperature on thermal conductivity of G2/04 class autoclaved aerated concrete. Asian J. Chem. **22**, 4104–4110 (2010)

Rittmann, A.: Sintesi geologica dei Campi Flegrei, in Bollettino della Società Geologica Italiana, LXIX-II, 117–128 (1950)

Savino, E.: Pompei il Vesuvio e la Penisola Sorrentina, Atti del secondo ciclo di conferenze di geologia, storia e archeologia. Pompei, Istituto B. Longo, ottobre 1997-febbraio 1998, vol. 11, pp. 297–299 (2000)

Stanley Price, N.: Scavo e conservazione. In: La conservazione sullo scavo archeologico, ICCROM e CCA, pp. 1–12 (1986)

Troppová, E., Švehlík, M., Tippner, J., Wimmer, R.: Influence of temperature and moisture content on the thermal conductivity of wood-based fibreboards. Mater. Struct. **48**(12), 4077–4083 (2015). https://doi.org/10.1617/s11527-014-0467-4

UNI 9724-3:1990: Materiali lapidei. Determinazione della resistenza a compressione (1990)

UNI 11085:2003: Materiali lapidei naturali ed artificiali. Determinazione del contenuto d'acqua: metodo ponderale (2003)

UNI EN 772-11:2011: Metodi di prova per elementi per muratura - Parte 11: Determinazione dell'assorbimento d'acqua per muratura di calcestruzzo, di calcestruzzo areato autoclavato, si materiale lapideo agglomerato e naturale dovuto alla capillarità ed al tasso iniziale di assorbimenti d'acqua degli elementi per muratura di laterizio (2011)

Verduch, G.: Método para la determinación de agua de los ladrillos. Materiales de construcción, vol. 159, pp. 97–100 (1975). ISSN 0465-2746

Vitiello, V., Castelluccio, R., Del Rio Merino, M.: Experimental research to evaluate the percentage change of thermal and mechanical performances of bricks in historical buildings due to moisture. Constr. Build. Mater. (2020). https://doi.org/10.1016/j.conbuildmat.2020.118107

Zaretzky, Z.: Porosidad aparente y otros datos - Control rápido en refractarios (1972)

Traction Resistance in Chemical Anchors Resistencia a Tracción En Anclajes Químicos

Diego Fernando Páez Moreno[1,2,3](✉) and Jeniffer Hamón Caicedo[3,4]

[1] Universidad Nacional de Colombia, Bogotá, Colombia
diego.paez@uptc.edu.co
[2] Ingeniería Civil, Universidad de los Andes, Bogotá, Colombia
[3] Universidad Pedagógica y Tecnológica de Colombia, Tunja, Colombia
[4] Ingeniería Civil, Pontificia Universidad Javeriana, Bogotá, Colombia

Abstract. In researches done on chemical anchors, also known as post installed structural anchors, they have shown a link between embedded depth and diameter of drilling with maximum resistance to traction. The present article shows the results of an investigation that analyzed the link from the traction resistance variation and type of failure in prototypes of structural anchors with changes in the embedded depth and drilling diameter. The prototypes were made in concrete with different compressive strengths, epoxy adhesive, and various types of reinforcement bars, subjected to pure tensile in universal machine. The results showed that tensile strength and type of failure are on function of a greater embedded depth and the diameter of the reinforcement bar anchored.

Keywords: Chemical anchors · Traction

1 Introduction

In Colombia, the existence of historical buildings built before the effective date of technical documents such as the NSR-10 (COLOMBIAN SEISMIC-RESISTANT REGIME OF CONSTRUCTION NSR-10) (Ministerio de ambiente, vivienda y desarrollo territorial 2010) is well known, therefore, it is evident the need to reinforce and restore this type of buildings in order to update them regarding new functionality and service requirements. Among the constructive methods for reinforcement and restoration of structures, there are chemical anchors with epoxy adhesive (Rendón 2014). As specified by Committee ACI 503 (American Concrete Institute 1998) such anchors consist of threaded or shaped bars that are cemented using chemical components from two compounds formed. These are mainly loaded through devices for fixing the embedded anchor and the solicitation can be flexion, traction, cutting or a combination of the mentioned above.

Analytical and experimental investigations have been developed upon this subject in order to make clarity on the interaction of the variables that take action on the anchors with epoxy adhesive, and the influence they have on their mechanical behavior. Among the analytics, the one elaborated by Gesoğlu et al. (Gesoğlu et al. 2014) in which through the use of gene expression programming (GEP) a model of prediction of shear capacity

I. Vayas and F. M. Mazzolani (Eds.): PROHITECH 2021, LNCE 209, pp. 755–763, 2022.
https://doi.org/10.1007/978-3-030-90788-4_58

was developed in post-installed anchors in hardened concrete; (Obata and Goto 1998) which experimentally and analytically studied the influence of the free edge on the tensile strength of the anchors exhibiting a relationship between the length of embedment and the failure; and (Upadhyaya and Kumar 2015) which emphasized in predicting the stress that would cause failure by shear or traction in chemical anchors, determining that due to the increase in the length of embedment, the maximum shear stress is reduced.

Regarding the experimental ones, the research carried out by (Çalışkan et al. 2013) evaluated structural anchors with concrete of low compressive strength applied to shear, concluding that by increasing the diameter of the anchor, its shear strength was reduced; (Yang et al. 2008) analyzed the tearing of the anchor in a system composed of anchoring-mortar-concrete establishing that the maximum load increases linearly with the length of embedment. Regarding the structural anchors with high strength concrete, (Çolak 2001) focused mainly on the behavior of the epoxy adhesive, establishing the importance in the dosage of this one and moreover, in this type of anchors, better results are not obtained with a depth of embedment greater than 10 times the diameter of the bar. Regarding the types of failure, (McVay et al. 1996) analyzed the characteristics of the failure process determining that in the concrete the "cone" type of failure is due to traction, while the failure in the interface of concrete - adhesive is due to the cutting. From the researches referenced, the influence of the length of embedment and perforation diameter on the behavior of the post-installed structural anchors is clear. Other investigations such as the one developed by (Contrafatto and Cosenza 2014) demonstrated the possibility of using these structural systems in basalt, sandstone, and limestone, as well as new applications according to (Puigvert et al. 2014) such as the replacement of corrugated rebar by Carbon fiber reinforced polymer (CFRP). In regard to the constructive process of chemical anchors, the document prepared by (Gerber and Ekenel 2013) exposes the contrast between the requirements established by different standards for the inspection of the aforementioned process.

Furthermore, technical documents such as (Hilti North America 2016) compile guidelines to be considered regarding some other procedures for the design and installation of structural anchors. In particular, this document includes a chapter with the description of the design equations gathered in (American Concrete Institute 2014) to establish the length of embedment (underrun) in anchors. However, in the technical sheets of manufacturers of Epoxy adhesive, such as (Sika Colombia 2014), it is established a specific value for the length of the embedment and the diameter of the perforation according to the diameter of the reinforcement bar.

Based on the previously described context, and due to the need to continue evaluating the influence of the variables of perforation diameter and length of embedment on the chemical anchors, the present investigation is structured with chemical anchors applied to direct traction, constructed from concrete specimens with five types of compressive strength and three types of rebar. In this way, the parameters to be evaluated were the last load resisted by each structural anchoring system and its respective type of failure. It should be noted that this investigation does not consider the conditions of confinement of reinforced concrete.

2 Materials and Methods

The materials, experimental design and procedure used in this investigation were established as it follows.

2.1 Materials

2.1.1 Concrete

125 cylindrical concrete specimens were used to develop the resistance tests of the chemical anchors, in addition, to evaluate the resistance from each type of resistance, were failed 8 specimens for the compressive strength of 14 MPa, 8 for the compressive strength of 17.5 MPa, 6 for the compressive strength of 28 MPa, 3 for the compressive strength of 34 MPa, and finally 3 specimens for the compressive strength of 42 MPa. All the specimens were failed at the age of 28 days, and for the characterization of the type of concrete was used the average result depending on the compressive strength evaluated. The compressive strength test was made according to the (ICONTEC 2012) and the characterization of the concrete is shown in Table 1.

Table 1. Characteristics chemical anchors elements.

Physical Characteristics				Mechanical Characteristics		
Material		Dimensions (m)		Parameter	Source	
		Ø	Length		Fabricator	Laboratory
Concrete		0.15	0.30	Specified compressive strength of concrete f'c		13.45 MPa
		0.15	0.30			17.51 MPa
		0.15	0.30			28.71 MPa
		0.15	0.30			34.38 MPa
		0.20	0.40			42.11 MPa
Reinforcing bar	Steel rebar	0.009	0.15	Last breaking load	–	35.20 kN
				Yield load	–	26.39 kN
				Specified yield strength	420 MPa	-
		0.012	0.20	Last breaking load	–	64.55 kN
				Yield load	–	55.73 kN
				Specified yield strength	420 MPa	-
		0.015	0.25	Last breaking load	–	85.36 kN
				Yield load	–	81.73 kN
				Specified yield strength	420 MPa	-

Source: Authors own creation.

2.1.2 Reinforcement Bar

Steel rebars of diameter 3/8″ (9.5 mm), 4/8″ (12.7 mm), and 5/8″ (15.9 mm) type Grade 60 were used. Of the latter, from each diameter two units were tested, based on the (ICONTEC 1995) their last breaking load, yield, and the graphic load vs. elongation were obtained. The characterization of the reinforcing bars is shown in Table 1.

2.1.3 Epoxy Adhesive

The adhesive used is a two-component-100% solid epoxy system. Once the components are mixed, a soft paste with high adhesion and mechanical strength is obtained to be used in post-installed structural anchors. It meets the requirements of ASTM STANDARD C-881-99, type IV, grade 3 (Sika Colombia 2014).

2.2 Method

2.2.1 Experimental Design

The experimental design is based on both the variation of the length of embedment (Le) defined as n times the diameter of the reinforcement bar to be anchored, and the variation of the diameter of perforation (ØP) in which 3 measurements are established (under ØB, medium ØM, and high ØA).

2.2.2 Preparation of the Specimens for the Test

For the development of the chemical anchors, the instructions of the manufacturer of the epoxy (Sika Colombia 2014) were taken into account, following the order stated:

- Perforation of concrete specimens (after 28 days of age)
- Cleaning the perforation with air pressure
- Application of the epoxy adhesive mixed in the nozzle to approximately half of the perforation
- Insertion of the reinforcement bar by rotary movements
- Cure of two days

2.2.3 Direct Tensile Test

The chemical anchors were subjected to a direct tensile test in universal machine as shown in Fig. 1.

Based on the guidelines of (ASTM 2014) to guarantee the stability of the structural anchor to be tested, an area free for the development of the failure, and that the forces that were applied through the reinforcement bar were perpendicular to the surface of the section of the test element, a complement of a laminar frame and a plate was used in the universal machine with a circular hole arranged in the geometric center of 0.10 m or 0.20 m according to the diameter of the concrete cylinders. This complement is shown in Fig. 2.

The loading speed used in the test was 5 MPa / s on an ongoing basis until reaching the final loading point of the element. Immediately after the trial, a film and photographic

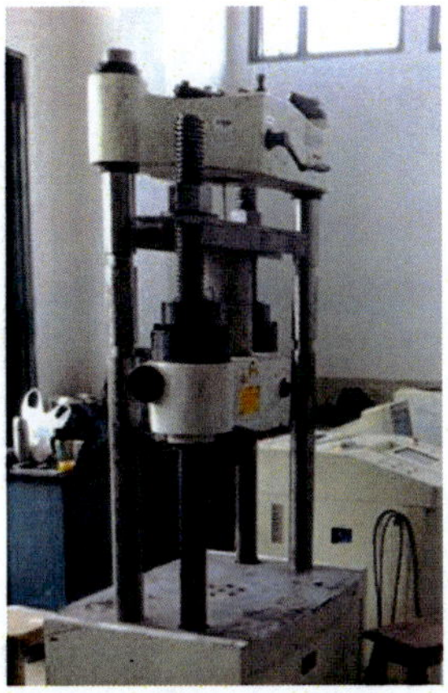

Fig. 1. Direct tensile test in universal machine Source: Authors own creation.

Fig. 2. Complement of a laminar frame and plate in universal machine. Source: Authors own creation.

record of the type of failure presented for further characterization was taken according to (American Concrete Institute 1997).

3 Results

The analysis is based on a quantitative and qualitative evaluation of the results. The qualitative one is given by the characterization of the type of failure classifying it in failure by pull-out, failure of steel or failure by concrete breakage. In contrast, the quantitative one is based on the premise that in the development of the chemical anchors the yield of the bar precedes the system failure, for this analysis of the results the values of yield and last load were taken as reference.

For the group of chemical anchors with 3/8″ (9.0 mm) diameter steel rebar, the qualitative analysis allowed to establish that in most anchors the failure was due to breakage of the concrete as shown in Fig. 3. In contrast, it is highlighted that the type of failure presented in the anchors with greater length of embedment (Lex15) was by steel which is shown in Fig. 4. This last failure is characterized by the fact that the embedded length of the anchor was sufficient for the resistance of the steel to control the break thus excluding concrete failures.

Regarding the group of chemical anchors with 4/8″ (12.0 mm) diameter steel rebar, it can be noticed that there were two main failures on those structural anchors, the pull-out failure and the breakage of the concrete. Taken into account that these anchors also had variations on their embedment length, the failures presented can be related with the compressive resistance of the concrete, which is considered insufficient to permit that the steel assume the imposed traction, and as a result, the concrete of the structural anchor remains subjected to the traction and quickly begins to develop cracks to the point that causes the failure. That can be seen in the Fig. 5.

Fig. 3. Breakage of the concrete failure. Source: Authors own creation

Fig. 4. Steel failure. Source: Authors own creation.

Fig. 5. Pull-out failure. Source: Authors own creation.

Continuing with the group of chemical anchors with 5/8″ (15.0 mm) diameter steel rebar, from the qualitative analysis it can be seen that the principal failure was because of the breakage of the concrete. For this case in particular, the type of failure can be related with the high compressive strength that does not allow the performance of the yield on the steel.

As for the quantitative analysis, from the results obtained can be notice two important facts. The first one, is that the capacity of the structural anchor to develop the steel yield is directly related with the increasing of the embedment length, and the second one, is that the concrete resistance does not make a significant improve on the perform of the structural anchor. To simplified the analysis of the data collected from the tests, four different types of graphics were created, in each one of them one specific variable of the anchors was evaluated. Those graphics are shown as it follows.

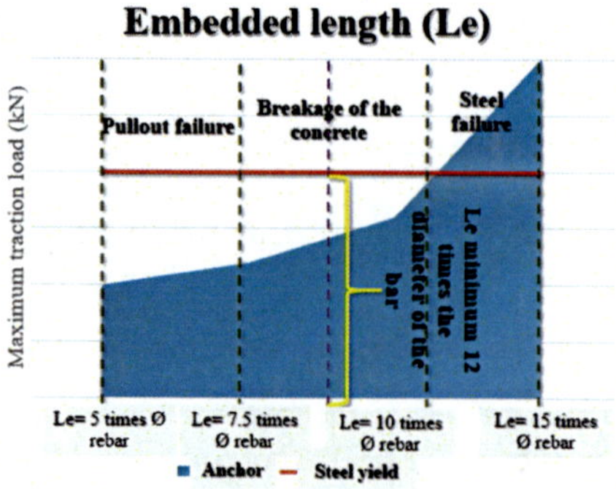

Fig. 6. Embedded length. Source: Authors own creation.

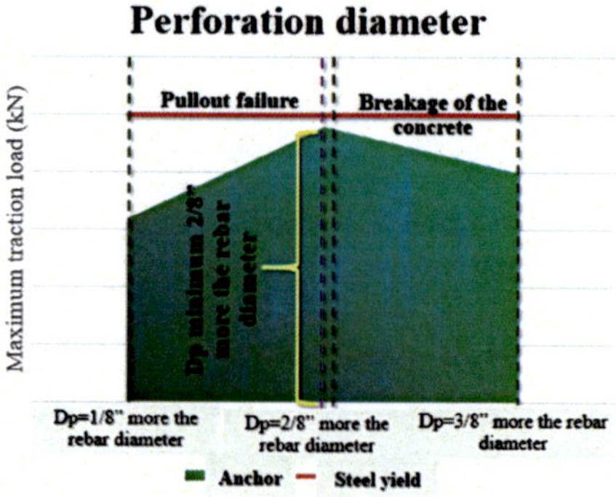

Fig. 7. Perforation diameter. Source: Authors own creation.

Based on the Fig. 6, it can be noticed that a greater embedment length improves significantly the performance of the anchors, because the anchors with embedment length further than 12 times the diameter or the rebar, were able to be subjected to loads beyond the equivalent to the steel yield. The mean type of failure related to this variable was the breakage of the concrete, it can be assumed that it happens because most of the anchors were made with embedment length below 10 times the diameter of the rebar, and that generates that the steel doesn't have the sufficient embedment to develop its yield. On the other hand, for the Fig. 7 it's clear that regard the perforation diameter this variable doesn't must be used below the value recommended from the fabricant of the adhesive, been this value 1/8″ times the diameter the rebar to be anchored.

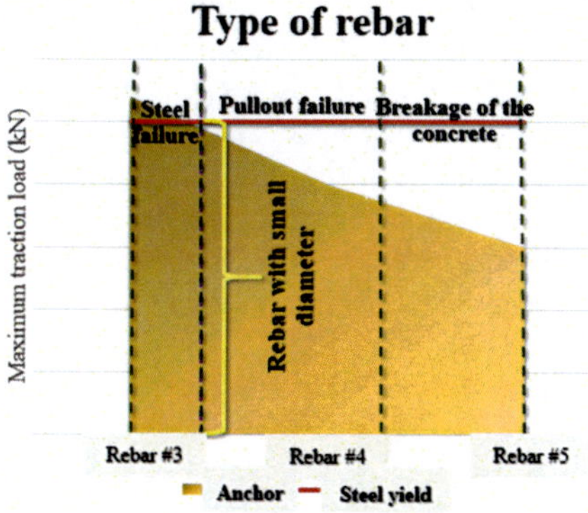

Fig. 8. Type of rebar. Source: Authors own creation.

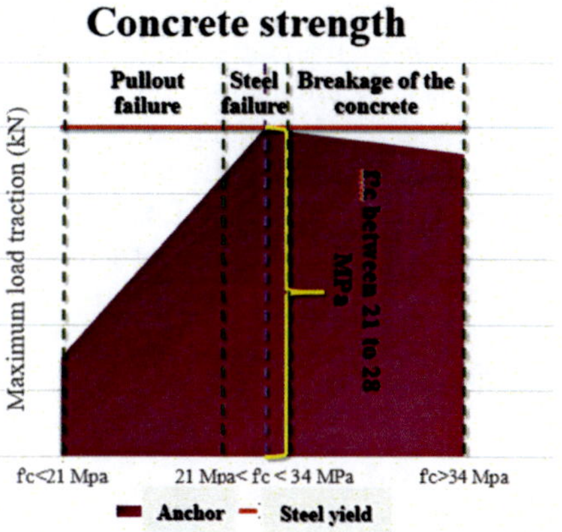

Fig. 9. Concrete strength. Source: Authors own creation.

As to the Fig. 8 it can be seen that for this type of structures it's much better to use smaller diameters rebars which implies more rebars to afford the required steel area. Also, it can be noticed that the use of rebars with diameters beyond 4/8″ doesn't improves the performance of the anchors. By the way, with the smallest rebar the yield of the steel during the performance of the anchor can be guaranteed. Finally, for the Fig. 9 it can be estimated that a higher compressive strength of the concrete does not gives better conditions for the development of the strength in the anchors.

4 Conclusions

Considering the materials, experimental design and analysis used in this study, the following conclusions can be obtained.

- In chemical anchors, the use of an embedment length (Le) equivalent to 12 times the diameter of the rebar to be anchored will cause that the traction to be assumed by the reinforcing steel, until its creep develops.
- In view of the results of the tensile tests carried out on the anchors, it is recommended to use the drilling diameter of 2/8" more than the diameter of the bar used to obtain a good maximum load.
- By analyzing the anchor length with the diameter of the rod, it was observed that with the smaller diameter of the rebar, the post-installed anchoring system develops greater tensile strength. The above confirms that for a defined reinforcement area will be more convenient to use a larger amount of steel of smaller diameter.

- The use of concrete with an effective compressive strength (f'c) greater than 34 MPa does not significantly influence the increase in traction strength of the structural anchoring system. Also, it is not recommended to do chemical anchors in concrete with a compressive strength (f'c) less than 21 MPa.
- In concrete of high and moderate resistance, systems with anchor length of 12 times the diameter of the rod, with reinforcement of equal diameter 3/8 "and drilling diameter of 2/8" more than the diameter of the bar to be used, they will be able to satisfy the resistance requirements, given that it was found that with this depth it is possible to reach to develop a minimum of 85% of the strength of the reinforcing steel.
- When the strength of the structural anchor is significantly lower than the maximum yield strength of the steel, there is pull-out failure, when the resistance is close to the maximum yield value the failure that occurs is the breakage of the concrete, and when the strength of the anchor reaches the ultimate breakage value of the reinforcing bar, a breakage failure of the reinforcing bar occurs without breaking of the adjacent concrete.

References

American Concrete Institute: Report on Anchorage to Concrete, ACI 355 1R-91 (Reapproved 1997). American Concrete Institute, Michigan (1997)

American Concrete Institute: Use of epoxy compounds with concrete, ACI 503 R-93. ACI, Michigan (1998)

American Concrete Institute: Building Code Requirements for Structural Concrete and Commentary, ACI 318-14. ACI, Michigan (2014)

ASTM: ASTM E488/488M – 10, Standard Test Methods for Strength of Anchors in Concrete Elements. ASTM International, West Conshohocken (2014)

Çalışkan, Ö., Yılmaz, S., Kaplan, H., Kıraç, N.: Shear strength of epoxy anchors embedded into low strength concrete. Constr. Build. Mater. **38**, 723–730 (2013)

Çolak, A.: Parametric study of factors affecting the pull-out strength of steel rods bonded into precast concrete panels. Int. J. Adhes. Adhes. **21**(6), 487–493 (2001)

Contrafatto, L., Cosenza, R.: Behaviour of post-installed adhesive anchors in natural stone. Constr. Build. Mater. **68**, 355–369 (2014)

Gerber, B., Ekenel, M.: Building code requirements for inspection of adhesive anchors in concrete. J. Constr. Eng. Manag. **139**(10), 06013001 (2013)

Gesoğlu, M., Güneyisi, E., Güneyisi, E., Mermerdaş, M.Y.: Modeling and analysis of the shear capacity of adhesive anchors post-installed into uncracked concrete. Compos. B Eng. **60**, 716–724 (2014)

Hilti North America: Post-Installed Reinforcing Bar Guide. Hilti Group, Schaan (2016)

ICONTEC: NTC 2, SIDERURGIA. ENSAYO DE TRACCIÓN PARA MATERIALES METÁLICOS. MÉTODO DE ENSAYO A TEMPERATURA AMBIENTE. Instituto Colombiano de Normas Técnicas y Certificación, Bogotá D.C. (1995)

ICONTEC: NTC 673, CONCRETOS. ENSAYO RESISTENCIA A LA COMPRESIÓN DE ESPECIMENES CILÍNDRICOS DE CONCRETO. Instituto Colombiano de Normas Técnicas y Certificación, Bogotá (2012)

McVay, M., Cook, R., Krishnamurthy, K.: Pullout simulation of postinstalled chemically bonded anchors. J. Struct. Eng. **122**(9), 1016–102 (1996)

Ministerio de ambiente, vivienda y desarrollo territorial. (2010). REGLAMENTO COLOMBIANO DE CONSTRUCCIÓN SISMO RESISTENTE DEL 2010, NSR-10. Colombia: Asociación Colombiana de Ingeniería Sísmica.
Obata, M., Goto, M.I.: The failure mechanism and the pull-out strength of a bond-type anchor near a free edge. Mech. Mater. **28**(1–4), 113–122 (1998)
Puigvert, F., Crocombe, A., Gil, L.: Static analysis of adhesively bonded anchorages for CFRP tendons. Constr. Build. Mater. **61**, 206–215 (2014)
Rendón, J.: Los anclajes adhesivos en la construcción. Noticreto **126**, 43–47 (2014)
Sika Colombia, S.A.: Hoja técnica Sikadur AnchorFix-4. Sika Colombia, Bogotá D.C. (2014)
Upadhyaya, P., Kumar, S.: Pull-out capacity of adhesive anchors: an analytical solutionl. Int. J. Adhes. Adhes. **60**, 54–62 (2015)
Yang, S., Wu, Z., Hu, X., Zheng, J.: Theoretical analysis on pullout of anchor from anchor–mortar–concrete anchorage system. Eng. Fract. Mech. **75**(5), 961–985 (2008)

Influence of Joints on the Seismic Response of Traditional Timber Frames in Turkey

Güzide Aslankaya[1](✉), Cenk Ustundag[2], and Albert Albareda-Valls[1]

[1] Department of Architectural Technology, Polytechnic University of Catalonia, Avenida Diagonal, 649, 08028 Barcelona, Spain
guzide.aslankaya@upc.edu

[2] Department of Architecture, Istanbul Technical University, Taskisla, 2, 34367 Istanbul, Turkey

Abstract. Timber frame structures which constitute an important cultural heritage of many countries, are well known as efficient seismic resistant structures worldwide and are worth to be preserved. Hımış is one common traditional Turkish timber system, which consists of a simple timber frame filled with masonry (such as bricks, adobes or stones with mortar), and a masonry ground floor, built on continuous stone foundations. These buildings are usually located in seismic areas. This study aims to make a review of the structural performance of Hımış timber system under seismic loading, with specific emphasis on joints and following strengthening of joints with CFRP (Carbon fiber reinforced polymer). Due to the seismic demands these timber structures mostly depend on connections, so that the joints have to be evaluated accurately in terms of translational and rotational stiffness and moment resistance. Subsequently, a series of experimental tests on two different types of timber joints (lap joint and mortise-tenon) which are common in Turkish timber structures have been carried out under monotonic bending loading. The numerical analysis, FEM (the finite element method) has been performed in order to calibrate the results from experiments. Finally, a numerical analysis considering semi-rigid joints in traditional timber connections has been performed globally.

Keywords: Hımış timber frame · Traditional joints · Stiffness · Carbon fiber reinforced polymer

1 Introduction

Timber is one of the most frequently used building materials in both traditional and modern engineering constructions. Thanks to the low weight and high load-bearing capacity, timber frame buildings can stand horizontal forces imposed during earthquakes and thus are well suited for seismic zones. Traditional timber frame structures are characterized by a timber frame filled with an infill which is mainly masonry acting as shear walls. The masonry behaves very well under compressive stresses while wood elements act as ties resisting to tensile stresses.

Timber framed constructions spread not only throughout European countries, such as Portugal (edificios pombalinos), Italy (casa baraccata), Germany (fachwerk), Greece

I. Vayas and F. M. Mazzolani (Eds.): PROHITECH 2021, LNCE 209, pp. 764–783, 2022.
https://doi.org/10.1007/978-3-030-90788-4_59

(ksilopikti tichopiia), France (colombage), Scandinavia (bindingverk), United Kingdom (half-timber), Spain (entramados) etc., but also in India (dhaji-dewari), Turkey (hımış and bagdadi), Peru (quincha), USA (balloon frame in Chicago), Haiti (gingerbread houses) (Poletti and Vasconcelos 2014). Even the evident difference in traditional construction worldwide, there is always a common idea in all them: timber resists tension, so that timber frames show better performance under seismic loading rather than masonry and provide ductility.

Hımış timber structure became a well-established construction technique under seismic loading in part of the Balkans and Turkey, especially from the sixteenth century. Hımış constructions show essential qualities that become beneficial under seismic loading. This research starts with a brief review of the current state of the art on structural performance of hımış buildings under seismic loading, with specific emphasis on joint details and discuss how these affect the overall structural behaviour under earthquakes. In Turkey since 1960s after major earthquakes, timber buildings remained intact, but suffered partial damages, such as: partial or complete collapse of infill material (stone or adobe), damages of ground floor and roof material and failure of joints. Joints bring together timber structure members. In other words, connections between timber frames are important to avoid a loss of physical integrity and to keep the box behaviour in place.

It is needed that these structures continue resisting future earthquakes without suffering serious damages.

Main objective is to preserve these structures by reinforcing with fiber for recovering and increasing the mechanical properties of traditional timber elements subjected to seismic loads.

2 State of Art

The traditional Turkish timber structural system, and therefore, timber joints against seismic loading, become the main focal point of this study.

There are many different typologies of traditional timber structures in Turkey, resulting from cultural preferences determined by material availability and the climate. Traditional timber houses in Turkey can be mainly classified in: (a) log houses, (b) timber frames (*hımış, bagdadi, dizeme*) and (c) combined construction depending on the structural system. This study focuses on the most common timber typology, known as *hımış*. *Hımış* timber structure simply consists of a simple timber frame filled with masonry, bricks, adobes or stones. The basement is always made of masonry at the ground floor (Fig. 1). Foundations are generally made of stone, with a system which is composed of posts, studs, wall plates, joists, ledgers, braces, knee braces, windowsills and lintels (Fig. 2). Typical typology includes diagonal bracing members to reinforce global frames.

The timber frame is usually connected to the masonry basement through base wall plates by overlapping one to each other at the corners, and nailed together. The cross-section of the wall plate is usually 10 × 10 cm (Şahin 2017). The studs and window studs divide the space between the main posts at constant intervals, being the cross section of the studs smaller than the posts (approximately 5 × 10 cm). The order of posts changes depending on the typologies of walls. For walls without windows, the posts are placed every 100–140 cm. Studs are usually varied in intervals depending on infill materials.

Fig. 1. Turkish timber house (Turkish Timber Association 2018)

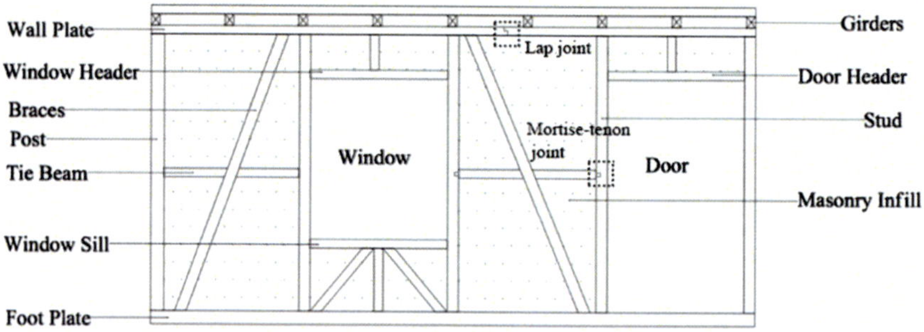

Fig. 2. Timber frame members

Horizontal members including the tie beam, windowsill, and knee braces are inserted between the studs in order to support the timber frame and to maintain the infill material in place. Cantilever is one of the peculiar features of the Turkish traditional timber house. Structural elements forming cantilevers are built together with the upper floor, as an extension of the joists. They are generally supported by bracings. The shape of the roof system is selected after completing the whole timber frame system. Hipped roof system with 4-side slope is usually preferred. Rafters are extended 50 to 60 cm outwards, in order to form the eaves. Tiles are generally used for roofing, by overlapping one tile on another.

Joints constitute usually a significant part of these structures. Many variables, such as the way of loading of timber elements, the type of fasteners, the existence of knots or even moisture content have a direct influence on timber joint design. From a structural point of view, joints are classified according to the type of acting forces (shear, compression, tension and bending). However, joints can also be classified depending on the type of fastener: bolted, doweled, nailed, plate components or glued connections.

Among all these types of joints, the mortise-tenon, the preferred joint for beam-column connections (a) in Turkish traditional timber house. Another common typology is, the lap or half-lap joint (b), which is suitable for beam-beam connections (Fig. 3–4).

The connection between braces (diagonal) to the bottom and (angle brace of) cantilevers to the beams is achieved using tongued-grooved (c) and notch joints (d) (Fig. 5), (Oztank 2008). Nailed connections have been widely used in Turkish traditional timber frame structures. Nailed connections provide the ability to absorb and dissipate energy during severe earthquakes (Dogangun et al. 2015).

Fig. 3. Timber joints in frame (Turkish Timber Association 2018)

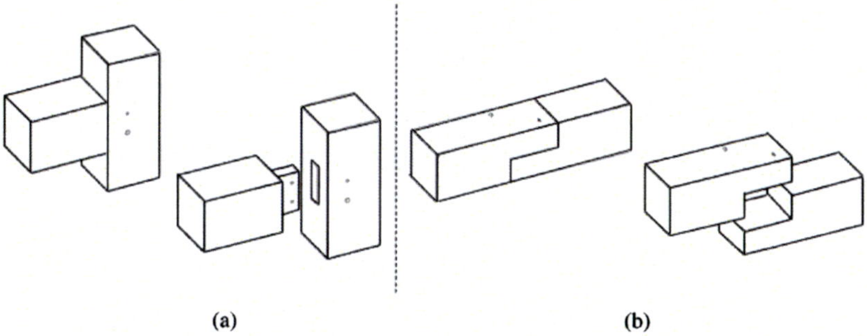

Fig. 4. Timber joints, mortise-tenon (a) and half-lap (b) joints

3 Experimental Analysis of Timber Joints

Some characterization tests, usually used for timber joints, have been carried out. These tests have been done under compression in both directions and bending. Compressive strength parallel to the fiber and perpendicular to the fiber, bending strength, global modulus of elasticity have been calculated with the expressions provided by standard. Two different types of timber joints (lap joint and mortise-tenon joints) have been analyzed under monotonic loading. Subsequently, joints which locally have been strengthened with carbon fiber textile were tested under monotonic and cyclic loadings. The purpose of these tests were to increase the flexural strength and load-bearing capacity of joint.

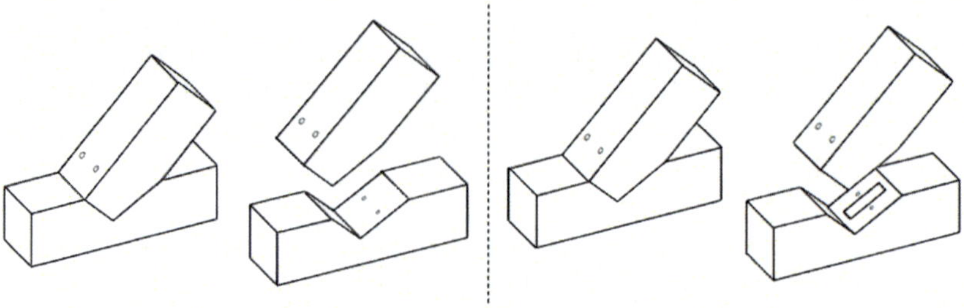

Fig. 5. Timber joints, tongued-grooved (c) and notch (d) joints

Comparative results between reinforced and un-reinforced specimens are given below graphics. Besides, the failure pattern of specimens have been examined in detail.

3.1 Timber Characterization Tests

The type of wood which has been used for specimens belongs to softwood species, namely pine. The size of timber specimens were manufactured according to the standard EN 408:2010. A total of 3 specimens with 45 × 70 mm of cross section (b × l) and height 90 mm (h), were subjected under compression parallel to the grain. A total of 3 specimens with a cross section of 45 × 70 mm (bxl) and 90 mm height (h), have been subjected under compression test perpendicular to the grain.

Experimental tests have been performed with an automatic press with maximum capacity of 300 kN and strain rate sensitivity about 0,01 mm. According to EN 408:2010, load was applied at a constant loading-head movement so that the maximum load is reached within (300 ± 120) s. The load was applied continuously with constant velocity 3 kN/min until failure (Fig. 6).

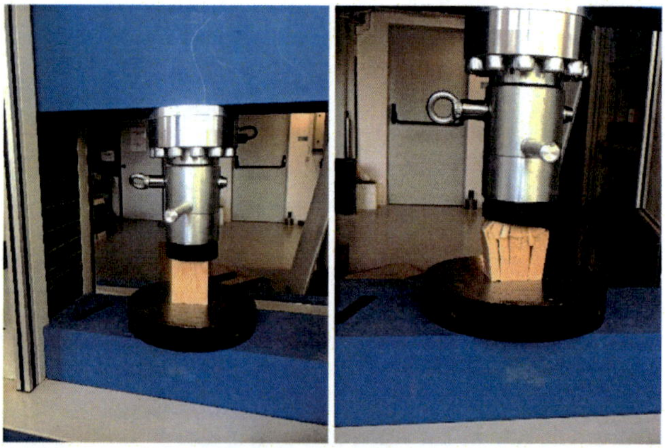

Fig. 6. Compression test

Load-deformation curve for compression parallel to the grain for three specimens are given in Fig. 7. The stress-strain behavior under compression parallel to the grain is characterized by a decrease after reaching the ultimate load. At the ultimate load, the weakest cell starts to collapse followed by the adjacent cells to guarantee that crushing band occurs. The collapse, which is a stability failure of the cell walls, leads to the loss of the load capacity of the cell. The load drops down to a level between 40 and 50% of the ultimate load. On average, the load slightly decreases in a ductile manner followed by a significant softening. Specimens 1 and 3 show less load capacity than specimen 2 due to the existence of knot in timber.

In order to determine the compressive strength of timber, Eq. (1) from EN 408:2010 was used. The average compressive strength of three specimens was calculated as 39 N/mm^2.

$$fc, o = \frac{F\max}{A} \tag{1}$$

$fc, 0$ compressive strength parallel to the grain, in newtons per square millimetre;
A cross-sectional area, in square millimetres;
$Fmax$ maximum load, in newtons.

Load-deformation curve for compression perpendicular to the grain for three specimens are given in Fig. 8. Timber is markedly ductile with a continual increment of load after yielding and an additional hardening after 45 to 55% deformation. Specimen 3 shows less load capacity than specimens 1 and 2, due to the existence of a knot in timber. The average compressive strength perpendicular to the grain of three specimens has been calculated as 4,12 N/mm^2 which is 9 times less than compressive strength parallel to the grain.

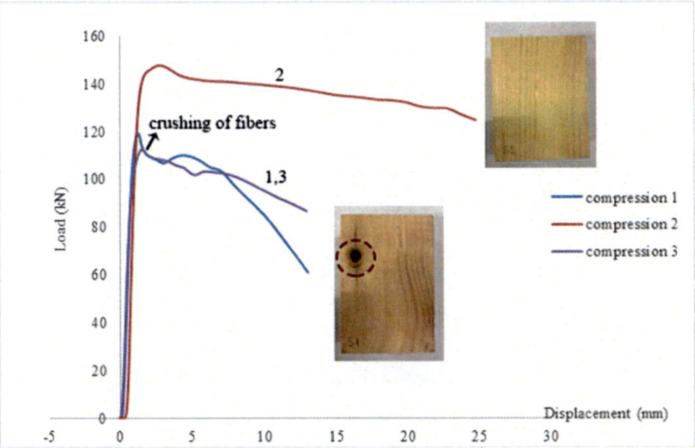

Fig. 7. Load-deformation curve for compression test parallel to the grain

Moreover, in order to determine the global modulus of elasticity and static bending (flexural) strength of wood, four point bending tests have been carried out according to

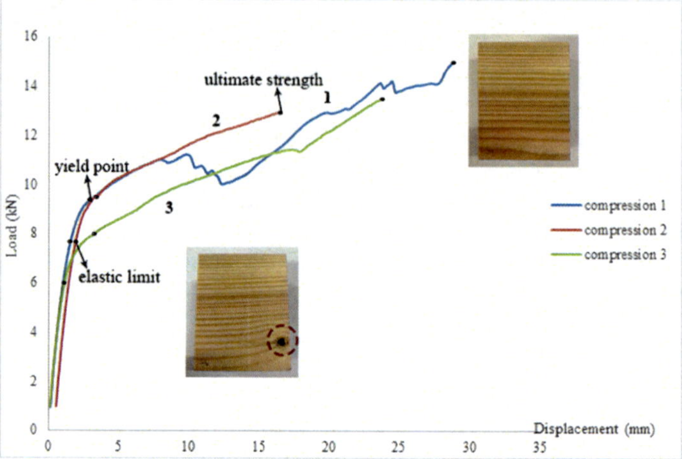

Fig. 8. Load-deformation curve for compression test perpendicular to the grain

EN 408:2010. According to the standard, the specimen shall have a minimum length of 19 times the depth of the section. A total of 3 specimens were tested, with a cross section of 90 mm × 90 mm (b × h) and 1800 mm length (Fig. 9). The deformation was measured at the center of the span and from the center of the tension edge. Constant velocity of load application has been imposed to 12 mm/min.

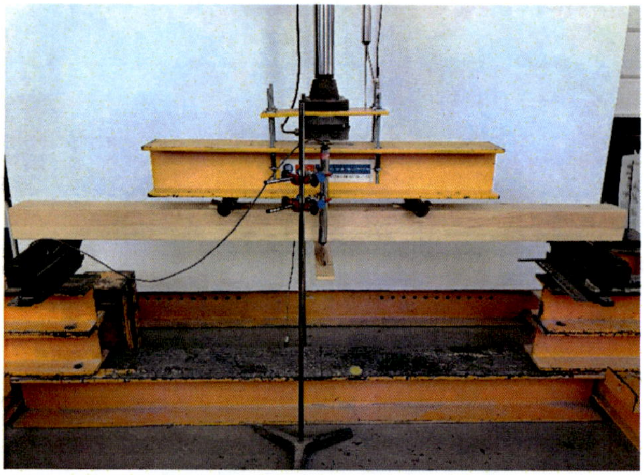

Fig. 9. The set up of bending test

The load-deformation graph was obtained from test results (Fig. 10). The obtained diagram mainly consists in three principal phases: A first phase with an initial linear elastic behavior followed by a second phase of nonlinear behavior in which the maximum loading is reached. It can be seen the upper surface starts to crush while the lower part

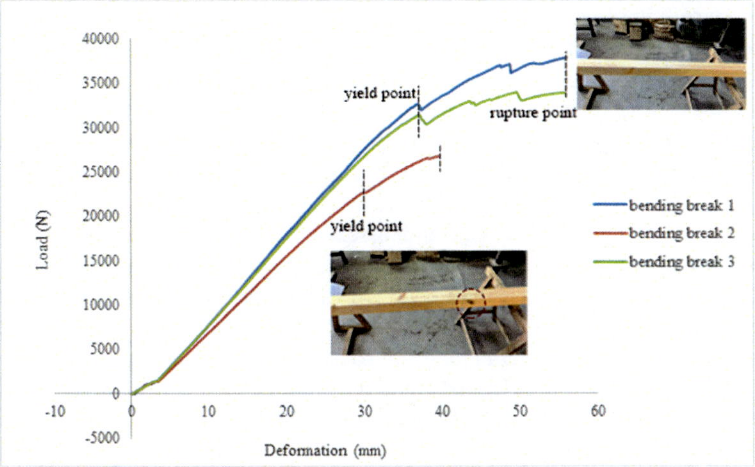

Fig. 10. The set up of bending test

starts to crack. In the last phase, a reduction of the load which is applied, is observed until the total rupture of the specimens. The strain is at an important rate when the failure occurs.

The slope of the linear part of the curves is defining the bending stiffness; the higher slope is the higher bending stiffness. Ultimate load-bearing capacity is defined by the load at which the curve turns into horizontal. Specimen 2 shows the smallest bending stiffness and load-bearing capacity than other specimens due to the existence of knot. Knot led to reduce the properties of timber.

According to the characterization test results, mechanical properties of selected timber, which is pine, are given from the test results (Table 1).

Table 1. The mechanical properties of timber from experiment results

Properties	Values
Density (p)	500 kg/m3
Compressive strength (fc, 0)	39 N/mm^2
Compressive strength (fc, 0)	4.2 N/mm2
Bending strength (fm)	72 N/mm2
Global modulus of elasticity (Em, g)	13647 N/mm^2

3.2 Lap Joint Tests

A set of monotonic tests of unreinforced specimens have been performed in order to describe the behavior of lap joints which is used for beam-beam connections in Turkish timber house. Subsequently, joints locally strengthened with carbon fiber textile have been tested under monotonic loadings.

To determine the load-deformation behavior of lap joint under monotonic loading, four point bending tests have been carried out according to BS EN 26891 (Timber structures- Joints made with mechanical fasteners- General principles for the determination of strength and deformation characteristics). A total of 2 lap joint specimens have been tested, by using the dimensions of 90 mm × 90 mm in cross-section (b × h) and 1800 mm length. Two screws with dimensions of Ø4.5 mm × h:80 mm have been used for connection introduced with an angle of 450 angles (Fig. 11a–b). The loading procedure of test has been obtained from BS EN 26891:1991 (Fig. 12). According to the standard, the load is applied up to 0,4 Fest and maintained for 30 s. Then the load is reduced to 0,1 Fest and maintained for 30 s. Thereafter, the load is increased until the ultimate load. The test is stopped when the ultimate load is reached. Following, the estimated maximum load, *F*est, 1000 N has been taken on the basis of previous bending experiments. The load has been applied up to 0.4 *F*est, which corresponds to 400 N, maintained for 30 s. Then, the load has been reduced to 0.1 *F*est, which corresponds to 100 N, maintained for 30 s at this value. Thereafter, the load has been increased until the ultimate load. Constant velocity of load application has been imposed to 10 mm/min.

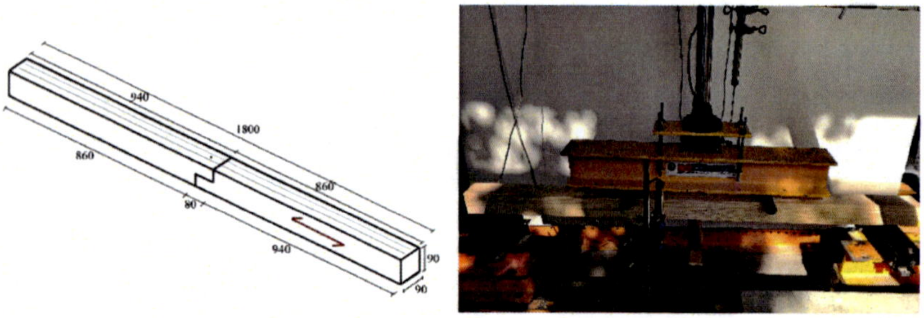

Fig. 11. The dimension of specimens (a), set up of test on lap joint (b)

When the timber beam deforms under bending, the deformation has been caused by moment and changing geometry. In addition to tensile and compressive stresses, shear also take place. The deformation is a result of normal and shear stresses in the beam cross section. The load-deformation diagram has been obtained directly from test results (Fig. 13). The curves show a quite elastic behavior until the maximum strength; this point when a slip occurs, there is loss of friction by inducing a rapid decrease of resistance. Thus, the brittle behavior is followed by an inelastic phase. Finally, a total loss of friction occurs with the failure of the connection. The screws have bent under stresses and at the same time have been pulled out from the wood.

A total of 2 lap joint specimens, with dimensions of 90 mm × 90 mm of cross-section (b × h) and 1800 mm length, have been reinforced with undirectional carbon fiber textile (Fig. 14a). High strength carbon fiber textile has been wrapped all around the timber specimen, as 400 mm wideness. It has been bonded parallel to the longitudinal direction of the beam in one layer with two component epoxy. The epoxy, named MasterBrace P

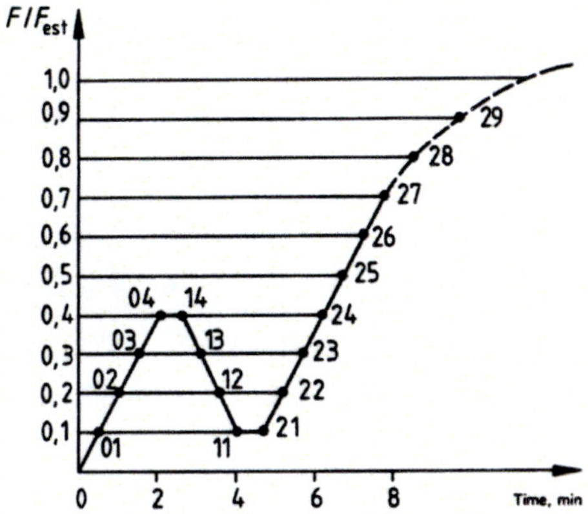

Fig. 12. The loading procedure of monotonic test (BS EN 26891:1991)

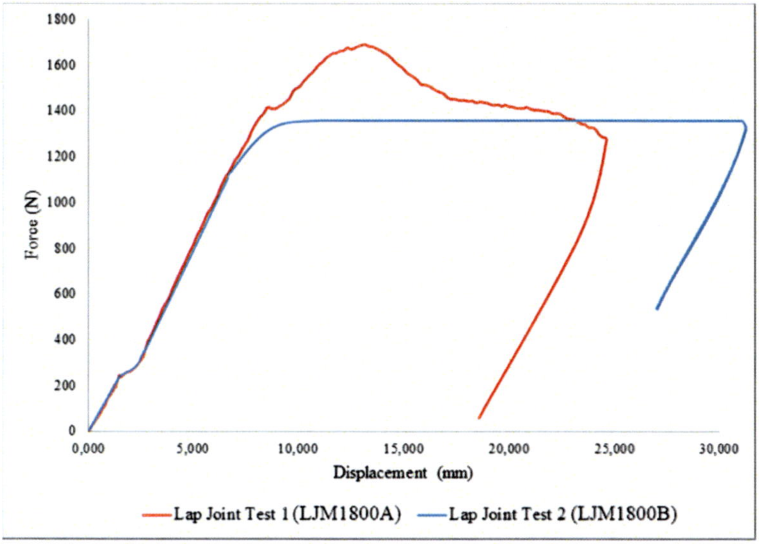

Fig. 13. Load-deformation curve of monotonic tests on unreinforced specimens

3500, is composed by two parts, A (resin) and B (hardener). The mix ratio is 3:1 (Part A to Part B) by volume. Each component were carefully measured and then added part B (hardener) to part A (resin).

Timber reinforced specimens have been subjected to four point flexural loading, using a 1000 kN displacement control hydraulic jack (Fig. 14b). The loading steps are similar to monotonic tests on unreinforced specimens.

The load-deformation diagram has been obtained from test results (Fig. 15). The curves show that beam exhibited more essentially linear elastic behavior up to the failure. The averaged maximum force is approximately 9 times more than in the unreinforced

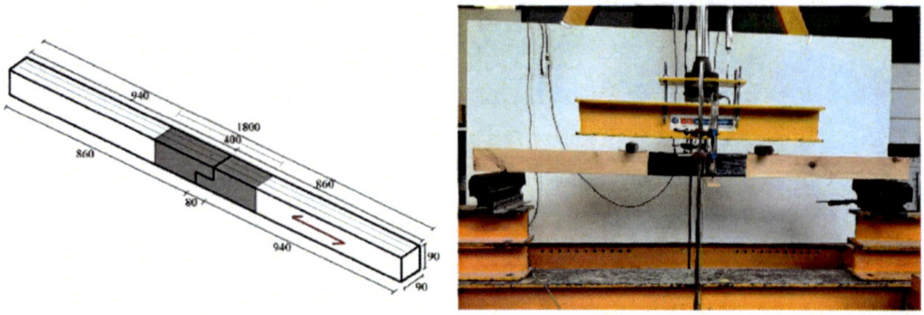

Fig. 14. The dimension of specimens (a), set up of test on reinforced lap joint (b)

specimen. After a loss of friction, a rapid decrease of the resistance takes place. Thus, the brittle behavior is replaced by an inelastic phase. Finally, a total loss of friction occurs with the general failure of the connection. Carbon fiber textile reinforced beams revealed less ductile behavior compared to the un-reinforced beams. The CFRP reinforcement derived into a global increment of the maximum load at failure, from 1400 N to 13000 N, which represents an increase of 830%.

It should be noticed that the rupture of the strengthened timber beams occurred due to the first crack of the solid timber in the tensile region. Failure has been initiated at the joint in tension zone due to screw buckling.

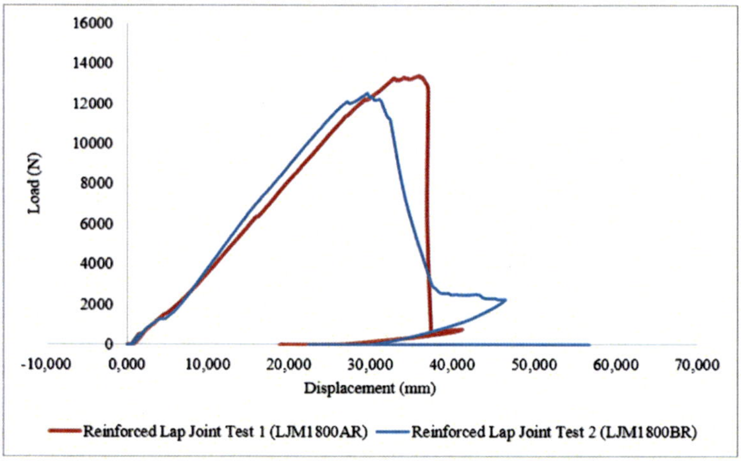

Fig. 15. Load-deformation curve of monotonic tests on reinforced specimens

3.3 Mortise-Tenon Joint Tests

A set of monotonic tests of unreinforced specimens have been performed in order to describe the behaviour of mortise tenon joint which is commonly used between beam and column members in Turkish traditional timber house.

The joint comprises two components: the mortise hole and the tenon tongue. These joints were implemented with metal fasteners such as; nails, screws or bolts and their ability to carry the loads was achieved through friction. Moreover, various reinforcement techniques such as; metal plates (strips, stirrup), glued composites (glass or carbon fibres textiles) and glued-in rods are used. In this study, joints which are locally strengthened with carbon fiber textile have been tested under monotonic loading.

A total of 3 pine specimens with 10% moisture content, have been used in T-type mortise-tenon joint, with the dimension of 90 × 90 × 500 mm beam and 90 × 90 × x1000 mm post member (bxhxl) (Fig. 16a). The dimension of 30 × 40 × 50 mm tenon (bxhxl) is connected to mortise hole with 2 lateral screws (Ø4.5 mm × h: 80 mm). The post of each specimens has been horizontally placed and bolted to steel reaction wall. The load has been concentrated in one point through a rectangle metal plate (Fig. 16b). The loading procedure of test is similar with lap joint test.

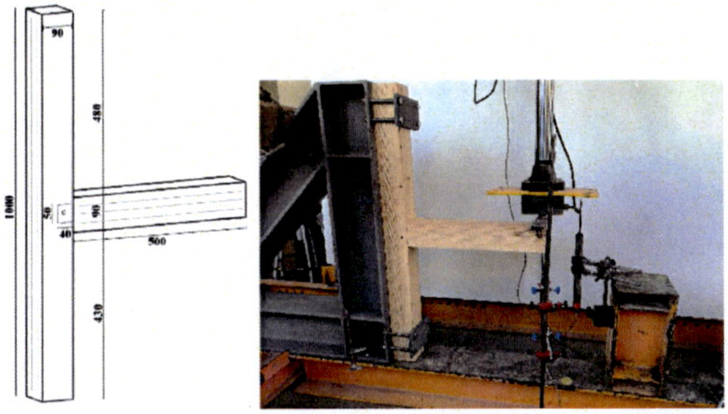

Fig. 16. The dimension of specimens (a), set up of test on mortise-tenon joint (b)

Typical load-displacement curves in the three tested specimens measured is shown in Fig. 17. Initially, the response is linear and elastic, where a linear increment of displacement corresponds to a linear increment of load. When the load reached 800 N, yielding took place in three specimens. Later, a nonlinear load-displacement curve has occurred and smooth plateau associated with tenon end crushing of mortise. The seperation between tenon and mortise has gradually seen and screws have bent under combined stresses. All specimens showed ductile behaviour under bending loading. Finally, a total loss of friction occurred with the global failure of the connection.

The experiments revealed that at early stages of loading, tenon and mortise have squeezed each other on contact surfaces, thus the specimens squeaked. Then, the tenon member has started to rotate. The upper tenon surface has slipped outside the mortise and has moved downwards, while the bottom of tenon surface has slipped inside of the mortise. With the increase of loading, the nonlinear compressive deformation has been observed on the interfaces.

Furthermore, a total of 3 mortise tenon joint specimens, with the same dimension of beam and post member have been reinforced with undirectional carbon fiber textile and they have been subjected to monotonic vertical loads. High strength carbon fiber textiles have been bonded to the upper surface of joint, as 90 × 200 mm with L shaped. It has been bonded parallel to the longitudinal direction of the beam in single layer with two components epoxy. Moreover, two CFRP textiles have been bonded with 45° angle to two lateral surfaces of the specimens with a dimension of 100 × 200 mm (Fig. 18a). After the preparation of specimens, the post of each specimens was vertically placed and bolted to steel reaction wall. The loads concentrated on one point with rectangle metal plate. Timber reinforced specimens have been subjected to one point flexural loading, at the end of beam the distance of 480 mm from the post. The load has been applied by one loading cell (Fig. 18b). Loading steps have been chosen similar to monotonic tests on unreinforced specimens.

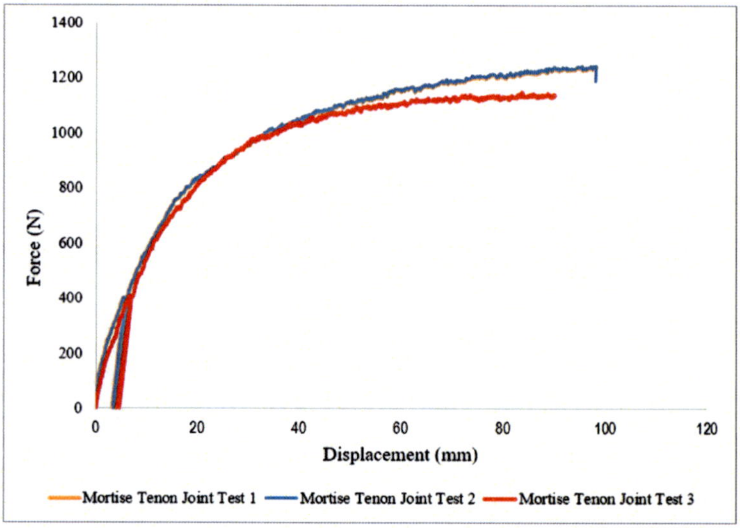

Fig. 17. Load-deformation curve of monotonic tests on unreinforced specimens

The load-deformation curve has been obtained from test results (Fig. 19). The behaviour of the three specimens has been quite similar, even though the ultimate values of strength varied. The curves show how beams exhibited more essentially linear elastic behaviour up the failure. The average maximum force is approximately 3 times more than unreinforced specimen. After a loss of friction, there is rapid decrease of capacity. Thus, the brittle behaviour is replaced by an inelastic phase. Finally, a total loss of friction takes place with the failure of the connection. Carbon fiber textile reinforced beams revealed more brittle behavior compared to the un-reinforced beams. The CFRP reinforcement caused an increment of the average maximum load at failure from 1200 N to 3600 N, which represents an increase of 300%.

It is worth to highlight that the failure of the strengthened timber joints occurred due to the separation of joint components in the tensile region. Failure has been initiated at the joint in tension zone due to rotation of tenon member. Even though joint components have been separated in tension zone, it still resisted to the increment of load by help of

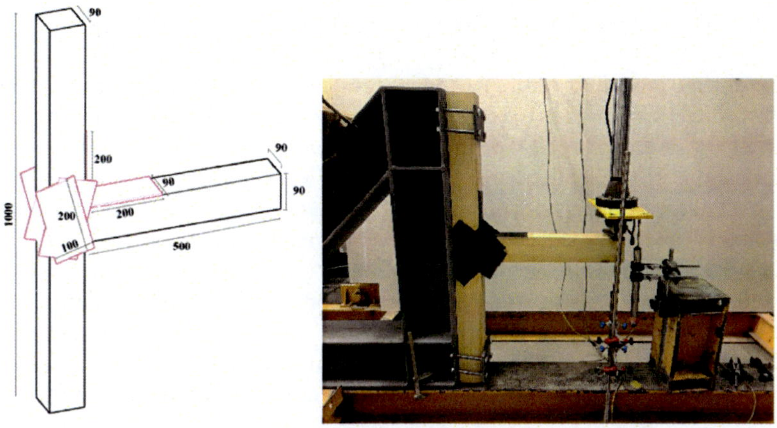

Fig. 18. The dimension of specimens (a), set up of test on reinforced mortise-tenon joint (b)

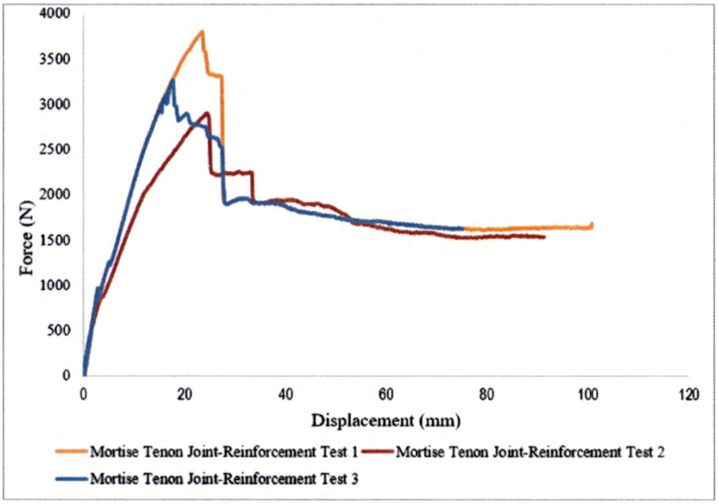

Fig. 19. Load-deformation curve of monotonic tests on reinforced specimens

high tensile strength of carbon fiber textile. When the specimen reaches the maximum peak of load, the carbon fiber textile yielded and separated from the timber surface. CFRP worked as a binder holding two timber members and provided continuity together until the ultimate deformation that was average of three specimens, 90 mm. In other words, CFRP reinforcements have led to progressive/gradual failure of joint rather than sharp failure.

4 Global Semi-rigid Analysis

The semi-rigid modelling of timber connections, using nonlinear moment-rotation laws and hysteretic rules, intends to represent the behaviour of timber structures with a comparable level of detail for all the structural components. The original and strengthened traditional timber connections are modelled using a nonlinear spring element available in a structural frame analysis software, RSTAB in order to analyze the internal forces, deformations and support reactions of frame in terms of global scale.

In order to evaluate the internal forces of members and deformation in existing structure and reinforced structure, a selected frame configuration has been analyzed under lateral loads (Fig. 20). The timber posts and beams have been modelled as isotropic linear elastic bars. The mechanical characterization of the materials is obtained through experimental testing on two types of joint. The reference material properties were obtained from experimental tests.

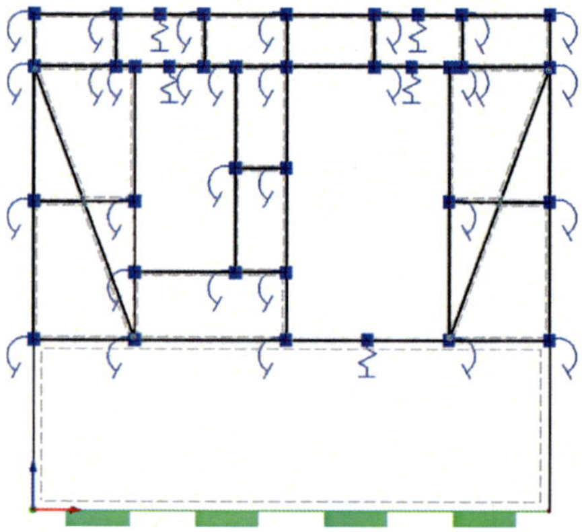

Fig. 20. The frame configuration for analysis

In order to represent the behaviour of semi-rigid joints, spring models have been derived from the average of force-displacement curves of three lap joint specimens and the average of moment-rotation curves of three mortise-tenon joint specimens under monotonic bending load. Non-linearity effects concentrated hinges were applied to the joints. Stiffnesses of connections which have been obtained from experiments, were progressively added at the joints. In total, three spring stiffness values were adopted to the nodals:

1. Hinge1 where lap joints between the beam and beam have been considered as semi-rigid.
2. Hinge2 where mortise-tenon connections between the beam and column have been considered as semi-rigid.
3. Hinge3 where connections between the diagonal (brace) and the main frame have been considered pinned.

The translational spring is introduced at Hinge1 and rotational spring for the nonlinear analysis of Hinge2. Properties of nonlinear hinges have been defined according to the force displacement and the moment-curvature, following diagrams presented at Fig. 21 and Fig. 22.

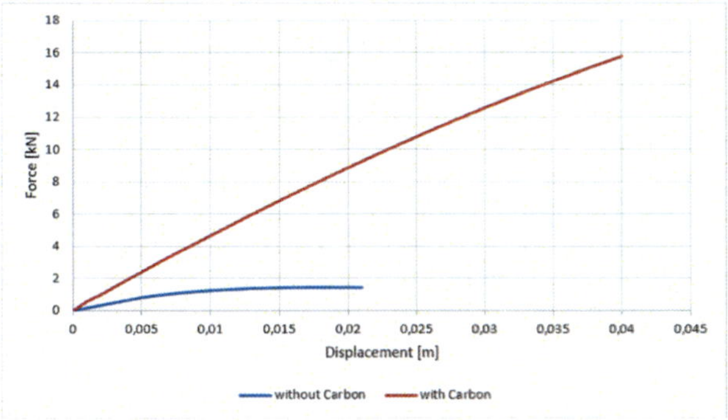

Fig. 21. The translational spring of lap joint

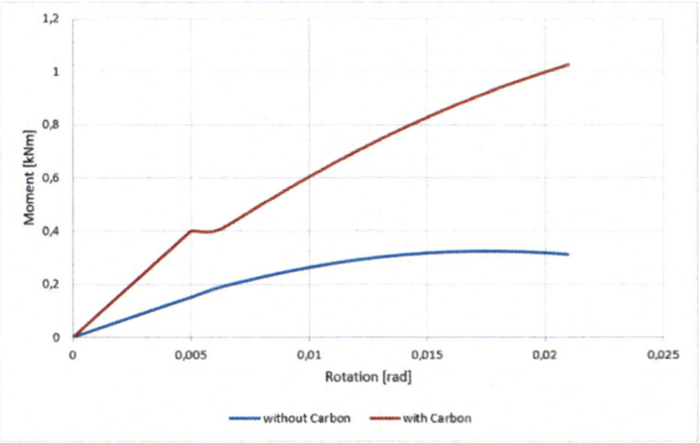

Fig. 22. The rotational spring of mortise-tenon joint

Different values of horizontal loads have applied at the top of frame until maximum 144 kN. As a boundary conditions, the masonry base has been chosen as fixed support, which restrained in vertical and horizontal directions.

Maximum deformation and normal forces took place at the upper storey, to direction of the applied lateral loads in both unreinforced and reinforced conditions. High bending moment has been seen at the left of base timber plate, where lap joints connect the beams. Also, the corners of door and window which are exposed to high bending, particularly in the place of mortise-tenon joints, that connect column and beam members. The reinforcement of joints with CFRP, has reduced the bending moment of joints, approximately %25 (Fig. 23–24).

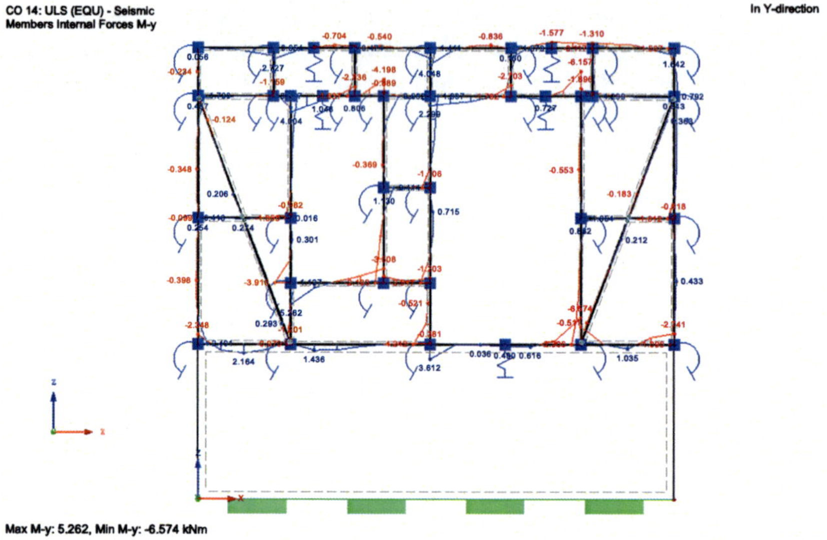

Fig. 23. Bending moment of unreinforced frame

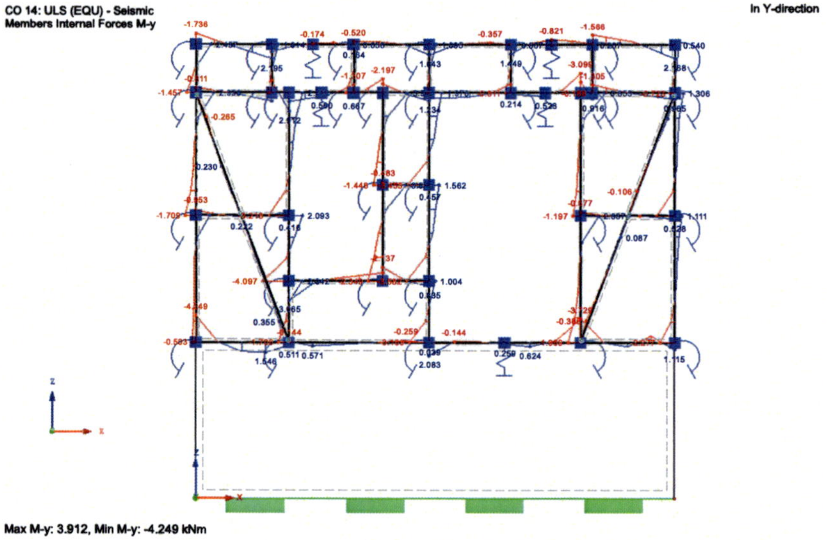

Fig. 24. Bending moment of reinforced frame

The results show that reinforcement of the joints with CFRP, significantly enhances the stiffness of joints and flexural strength of structure. Also, analysis indicates that reinforced frame with CFRP lead to improve nonlinear behavior of timber joints under high stresses and reduces deformation in an approximately %35 at same level of load.

Besides, the unreinforced frame endures until the load of 108 kN, while the reinforced frame bears until 144 kN (Fig. 25).

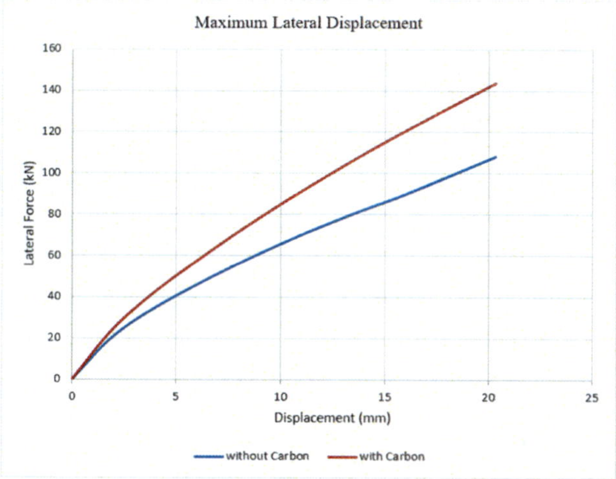

Fig. 25. The comprasion of global analysis results between unreinforced and reinforced frames

Minimum displacement and stresses are obtained in the cases where the frame is made up of reinforced with CFRP. After reinforcement, the deformation of system decreased as %35 in frame with infill while %43 decreasement was seen in frame without infill. In either case, CFRP increases the strength properties of timber elements subjected to lateral loads. Besides, strain was detected along the CFRP reinforced joints, it was seen that carbon fibers were subjected to high stress. This showed the high performance of CFRP in terms of bonding during stress-transfer process.

5 Conclusions

This study aims to assess the influence of the joint stiffness in the monotonic behaviour of Turkish traditional timber system (Hımış) identifying and evaluating new strengthening technique with carbon fiber reinforced polymers (CFRP).

Some characterization tests on the selected type of timber, which used in experiments related to timber joints, are carried out. In Turkish traditional timber house, as a type of timber mostly pine is used. These tests were compression in both directions and bending test of timber. The compressive strength in parallel to fiber and perpendicular to fibers, bending strength, global modulus of elasticity are calculated with the formula according the standard. Then, two different types of timber joints (lap joint and mortise-tenon joints) are analyzed under monotonic and cyclic bending loadings. Subsequently, the joints locally strengthened with carbon fiber textile have been tested in order to increase the flexural strength and load-bearing capacity of joint have been tested. The comparative result between reinforced and un-reinforced specimens show that the rotational stiffness of reinforced mortise tenon joint specimens and translational stiffness of reinforced lap joint specimens have been higher than the unreinforced specimens which led to more brittle behaviour. The rotational behavior of the degraded mortise tenon joints and the translational behavior of lap joints are semi-rigid. After the reinforcement, specimens reached to higher load. It can be observed that CFRP has provided the reduction of deformation as 23% and the increasement of load bearing capacity 325% under cyclic bending loading. Experiments gave an insight of the behaviour of two joints under bending load and they have proved that the reinforced specimens showed more ductile behaviour.

Consequently, according to the analyses and literature researches of earthquake damages of himis structure, it has been seen that these types of structures are less deformed after reinforced the joints of frame. Briefly, it is possible to conclude that:

- Timber connections in traditional Turkish structures are always complemented with nails to provide resistance against tensile or shear forces. Furthermore, nails contribute the ductility and the ability of dissipating energy in the structure.
- Most vulnerable points of the structure are the lap joints (beam-beam connections) and mortise tenon joints (beam-column connections), where nonlinear hinges are noted at early stage of the analysis and also progressively reach the status of failure. Besides, the diagonals concentrate high tension and compression stresses and as a result, hinges near the failure level are obvious.
- The failure mode of the specimens indicated that the joints are weak, but the timber members are strong. In any case, the timber beam bending capacity is dependent on the bearing capacity of connection, this controls the stability of lateral force-deformation response of system under lateral loads.
- When the specimens are subjected to lateral loading, the semi-rigid joint connections are experienced bending moments, which create tension perpendicular to the wood grain and a longitudinal shear stress, it causes the premature splitting at the joints, thus connections are easily damaged under lateral loading. Reinforcement with CFRP prevents the seperation of joint and specimen is able to resist the high lateral loads.
- The strengthening joints concerns the behaviour of the friction-based connection in its own plane, and is intended to avoid the detachment of the connected members. Particularly, reinforcement can prevent loss of capacity and possible separation of friction surfaces due to the reduction of compression forces under lateral loading, the application of strengthening solution can maintain a stable structural behaviour.

The global strength and ductility that are the two most important parameters for structure in seismic area and it should be aware of the seismic behaviour of traditional structure. In case of the structure was not constructed properly to resist earthquakes, it requires the rehabilitation for better performance. Innovative techniques can play an important role in the rehabilitation of the traditional structures. Whilst respecting the original structure concept and, therefore, their authenticity, CFRP strengthening technique is capable of improving the global strength, ductility and energy dissipation capacity of the structures. In particular, ease of handling and application, their light weight are some factors that are advantageous in the strengthening of timber joints.

References

Dogangun, A., Tuluk, I., Livaoglu, R.: Traditional wooden buildings and their damages during earthquakes in Turkey. Karadeniz Technical University, December 2015

Oztank, N.: Traditional timber Turkish houses and structural details. Dokuz Eylul University, Izmir (2008)

Poletti, E., Vasconcelos, G.: Seismic behavior of traditional timber frame walls: experimental results on unreinforced walls. Bull Earthquake Eng. **2015**(13), 885–916 (2014)

Şahin, G.N.: Observations on earthquake resistance of traditional timber framed houses in Turkey. Build Environ. **42**, 840–851 (2017)

Turkish Timber Association, March 2018. http://www.ahsap.org/bilgi/yapi-sistemleri

EN 408:2010: Timber structures-Structural timber and glued laminated timber-Determination of some physical and mechanical properties. Brussels, 16 July 2012

BS EN 26891:1991: Timber structures –Joints made with mechanical fasteners- General principles for the determination of strength and deformation characteristics. Brussels, 31 July 1991

Check for updates

$\delta^{13}C$ and $\delta^{18}O$ Stable Isotope Analysis Applied to Indicate Sources of Water-Induced Degradation in Historic Monuments

Dafni Kyropoulou(✉), Eustratios Heliades, Petros Karalis, George Diamantopoulos, Sophia Gougoura, and Elissavet Dotsika

Stable Isotope Unit, National Centre for Scientific Research, Institute of Nanosciences and Nanotechnology, Demokritos Agia Paraskevi, 15310 Attiki, Greece
{d.kyropoulou,e.heliades,p.karalis,g.diamantopoulos,s.gougoura,e.dotsika}@inn.demokritos.g

Abstract. Samples of mortars were collected from lime and hydraulic mortars affected by environmental, mainly water induced degradation. Moisture and humidity possesses a main threat for the preservation of historic monuments. Agents of decay related with water such as acid rain, sulfate attack, acid attack, leaching action, salts attack, damage due to frost and many more can cause extensive cracks and total disintegration of historic constructions. 50 samples were obtained from historic constructions located in North Greece. Isotopic data will make possible to provide weathering gradients. The samples were characterized in terms of their isotopic, chemical and mineralogical composition. Compositional and morphological analyses were achieved using energy dispersive X-ray analysis in the scanning electron microscope. The results of micro-morphological and petrographic examination elucidate the different sources of degradation of historic mortars. Stable isotope analysis (^{13}C and ^{18}O) provided information relative to the origin of CO_2 and water making possible to distinguish different sources of water-induced degradation. This study indicated that stable isotope analysis is an excellent tool to fingerprint the origin of water –induced degradation and to determine the weathering depth and the potential secondary degradation mechanisms.

Keywords: Stable isotopes · ^{13}C · $\delta^{18}O$ · Mortars · Mineralogy · Degradation

1 Introduction

Hydraulic and lime mortars constitute main structural components in architectural constructions. Water induced and environmental degradation of historic mortars possesses a main threat for the preservation of historic monuments (Sabbioni et al. 2001; Yates 2003; Zappia et al. 1994). Agents of decay related with water such as sulfate attack, acid attack, leaching action, salts attack, damage due to frost and many more can cause extensive cracks and total disintegration of historic constructions (El-Turki et al. 2007; Larbi 2004). In order to determine the above degradation mechanisms and propose suitable conservation strategies, it is essential to examine the calcium carbonate origin and to

I. Vayas and F. M. Mazzolani (Eds.): PROHITECH 2021, LNCE 209, pp. 784–789, 2022.
https://doi.org/10.1007/978-3-030-90788-4_60

identify the possible causes that change the initial composition. Isotope method provides information of the primary sources of raw material and offer considerable potential in the investigation of lime mortar and plaster.

2 Experimental and Sampling

Major and minor elements for polished mortar samples were analyzed using a FEI/Quanta Inspect D8334 SEM fitted with a low vacuum chamber and an energy dispersive X-ray analyzer at the NCSR Demokritos Greece. The operating conditions were a 25 kV accelerating potential and a 107 μA emission current, giving a count rate of 400cps on metallic cobalt, the working distance was normally 10 mm. The detector was calibrated with a range of pure elements, synthetic oxides and well characterized minerals. The elements present in mortars samples were determined using X-ray measurements of each particle. A backscattered electron image was obtained in every sample and qualitative analysis of 10–20 individual particles was performed. The imaging was performed using a four-quadrant backscattered electron (BEI) and a secondary electron detector (SEI). The isotopic analyses took place in the Stable Isotope Unit of Institute of Materials Science (NCSR Demokritos) on a Thermo Delta V Plus IRMS equipped with GasBench II device. Small quantities of clean carbonate material were separated and after grounding to a fine-ground powder were diluted in ortho-phosphoric acid and the CO_2 produced was measured in the isotope ratio mass spectrometer. The isotopic ratios of $^{13}C/^{12}C$ and $^{18}O/^{16}O$ were determined and expressed as $\delta^{13}C$ and $\delta^{18}O$ relative to the international PDB standard.

3 Results

3.1 Micromorphological and Elemental Analysis

Backscattered electron photomicrographs of sample AN1b obtained from Anaktoroupoli, Nea Peramos, illustrates the morphology of a Roman mortar. The photomicrograph (Fig. 1) depicts large irregularly shaped grains acting as aggregates composed mainly of quartz and feldspar minerals. The matrix is composed of hydraulic lime. The black holes in the pictures are pores in the mortar structure. The sample represents the main building phase of Anaktoroupoli. Fibrous crystals were attributed to degradation products of hydraulic mortars; the mortar's surface is entirely covered with salts.

Backscattered electron photomicrographs of sample M1 (Fig. 2), illustrating the morphology of typical mortar sample from Makrygialos, Greece. The photomicrograph (Fig. 3) depicts large irregularly shaped grains acting as aggregates composed mainly of quartz and feldspar minerals. The matrix is composed of fine calcitic particles with small portions of quartz grains. The black holes in the pictures (Fig. 2) are pores in the mortar structure.

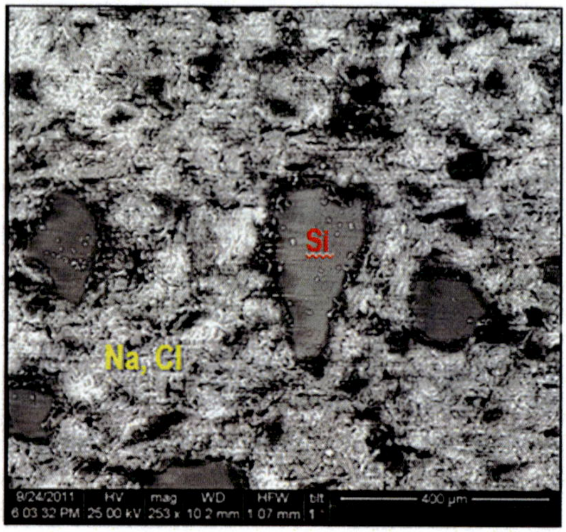

Fig. 1. SEM/BSE image showing the deposition of salts in mortar surface.

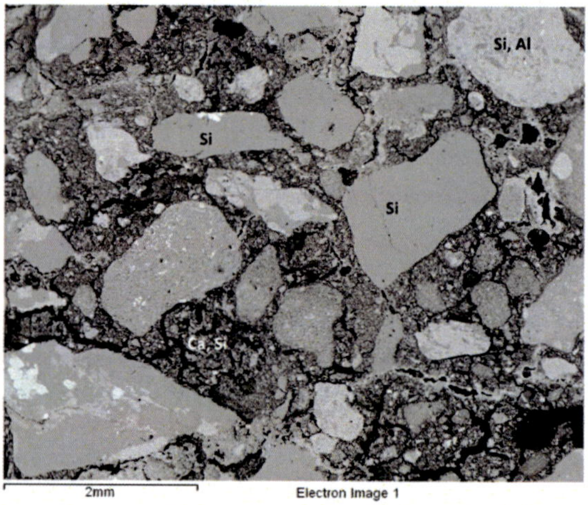

Fig. 2. SEM/BSE image showing degradation features in Hellenistic mortar.

In this sample, it is possible to have two types of pores. The elongate cracks are attributed to shrinkage and the rounded pores are derived from bubbles during setting. In places where heating was more intense, the half-burnt limestone fragments demonstrate an edge along the area of a crack (pore). It represents deposition of calcite on the edge of the crack due to the movement of solutions during burial. The black hole in Fig. 2 is lined with a very light grey line composed of calcite. This pore lining requires a source of dissolved carbonate. The dissolved carbonate could be attributed to the dissolution of calcite matrix or dissolution of the calcitic particles in the aggregate. In sample A4 the matrix is mostly composed of calcite, therefore the pore lining is attributed to the dissolution of calcite matrix.

3.2 Stable Isotope Analysis

The stable isotope (^{13}C and ^{18}O) analysis is highly useful in providing extra parameters, associated with the conditions of formation and the origin of the carbonates and therefore diagnostic of the mechanisms and processes that cause material degradation. The carbon and oxygen isotopes in mortar were analyzed in order to evidence the carbonate origin and to identify the potential sources of mortar decay related with water such as salts attack, signs of sulfation and dissolution/recrystallization processes. Mineralogical and chemical analysis indicated that the hydraulic mortars are attacked by salts while the lime-based mortars are decomposed because of the leaching action that causes material dissolution. Apart from material degradation the technological differences and the impact of raw materials are expected to affect isotopic composition of historic mortars. The isotopic value of carbonates depends mainly on the isotopic composition of atmospheric CO_2 and water and on the degree to which the isotopic equilibrium is reached and therefore the factor of fractionation of the components ($\delta^{13}C_{CaCO3}$, of $\delta^{13}C$ versus $\delta^{12}C$ of the calcite precipitated in the CO_2 absorption experiments is explained by a continuous enrichment of $\delta^{13}C$ relative to $\delta^{12}C$ of the gaseous CO_2 due to kinetic isotope fractionation during calcite formation. The isotopic values for C and O within the precipitation of calcite by atmospheric CO_2 absorption using the local isotopic composition of water and atmospheric CO_2 has been detailed in previous work of Dotsika et al. (2009). The isotopic values of C and O are calculated using the same mathematical equations modifying the isotopic composition for water of West Macedonia ($\delta^{18}O_{H2O} = -7‰$), $\delta^{18}O_{calcite} = -20‰$. Consequently according to the study of Dotsika et al. (2010)) and modifying the O value for West Macedonia, isotopic values for $\delta^{13}C$ and $\delta^{18}O$ are ($\delta^{13}C_{calcite} = -27‰$; $\delta^{18}O_{calcite} = -20‰$). Such values express the calcite precipitation that formed with the absorption of atmospheric CO_2.. Isotopic values for mortar samples comprise a wide range of δ^{13}Cmatrix and δ^{18}Omatrix values, from −17.3 to −6.4‰ and from −18.1 to −4.6‰, respectively. All the samples are plotted in Fig. 3.

Figure 3 shows the various sources of CO_2 and H_2O (lines) and mechanisms (areas). Point CM indicates that precipitated calcite is formed directly by the absorption of atmospheric CO_2 in strong alkaline aqueous environment (Dotsika et al. 2018); its values change to more enriched isotopic composition due to continuous enrichment of $\delta^{13}C$ of CO_2 during calcite precipitation; area A indicates that precipitated calcite is formed from atmospheric CO_2 and contaminated by residual natural limestone; line 1A reflects that the primary source of water used for setting of the lime mortar is coming from heavy source, evaporation effect; area B indicates that precipitated calcite is formed from atmospheric CO_2 and heavy, evaporated water. Enrichment of C and O isotopes indicates recrystallisation of calcite with water of meteoric origin and atmospheric CO_2; area C shows the enrichment of $\delta^{18}O$ of calcite due to the re- equilibrium with the silicate minerals; line 1B reflects that the primary source of water used for setting of the lime mortar is coming from light source, condensation effect; area D indicates that precipitated calcite is formed from atmospheric CO_2 and isotopically light local meteoric water or

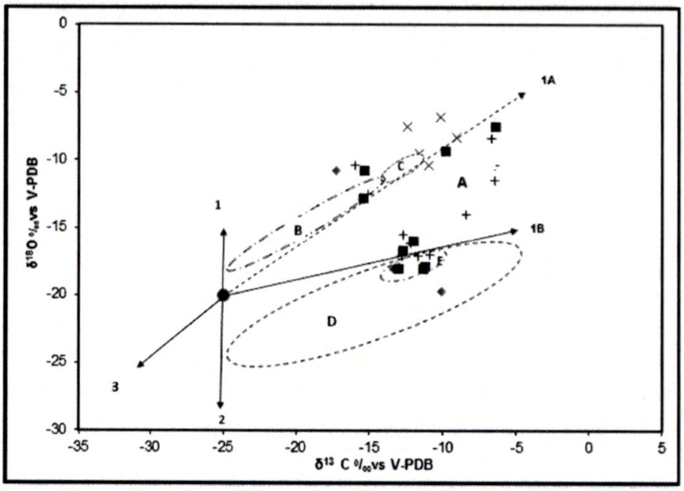

Fig. 3. Schematic diagram of exogenic processes causing changes in $\delta^{13}C$ and $\delta^{18}O$ of calcite of mortar samples.

isotopically light re-condensed primary water; line 3 depicts the depletion of C and O isotopes indicating recrystallisation of calcite with light water and CO_2 of mixed origin (atmospheric and soil origin); and area E, depletion of C and O isotopes, indicates also human influence (surface treatment) and biological growth. Samples position along line 1A reflects the $\delta^{13}C_{CO2}$ enrichment relative to $\delta^{12}C_{CO2}$ in the gas phase with continuous calcite formation along the lime mortar layer and the presence of impurities of relict from isotopically heavy limestone (Craig 1961; Dotsika et al. 2010). Samples from lime mortars and Roman mortars are plotted along line 1A. For lime mortars contamination of relicts from heavy limestone is possible, as indicated by SEM analysis. For Roman mortars this enrichment is attributable to silicates in the matrix that increase water porosity.

The severely decayed mortars (AN1a-T5b) indicate different values in comparison to the less weathered material (samples M1-M7). In the case of O isotopic composition, when the secondary solution during the alteration of calcite is meteoric water, like in our case, the δ^{18}Ocalcite should shift to heavier values (El-Turki et al. 2007). So, a secondary recrystallisation of calcite with porewater of meteoric origin and atmospheric CO_2 gives heavier values. The isotopic composition of samples from Makrygialos (M1-M7) is increased according to line 1 demonstrating a shift towards heavier values. This is attributed to secondary recrystallisation of calcite with porewater of meteoric origin. Mortar samples obtained from Anakroroupoli are affected by salts according to the results of mineralogical and elemental analysis. The location of these mortar samples in the plot reveals that the ^{18}O for these samples shows a shift towards more negative values along line 2. Also the samples of Anaktoroupoli are concentrated in area E a location that demonstrates extensive biodegradation. The late Roamn mortars are plotted between lines 1A and 1B in area A. The conditions for the calcite formation for samples plotted in area A indicate that precipitated calcite is formed from atmospheric

CO_2 and contaminated by residual natural limestone. The hydraulic mortars (samples from Anaktoroupoli) show a shift towards more negative values as compared to lime based mortars. This shift is attributed to re-equilibrium with the silicate (Hutcheon and Abercrombie 1990) at high CO_2 pressures and at high $\delta^{18}O$ (Dotsika et al. 2010) of local water and very low porewater quantity.

4 Conclusion

The aim of this work was to determine the sources of degradation of a historic mortars dating from Hellenistic to Roman period. The mineralogical, microscopic and elemental analysis provided information on the different typologies of mortar degradation. Mortars degradation was attributable to water related factors such as salts attack, sulfation and leaching action. Stable isotopes of $\delta^{18}O$ and $\delta^{13}C$ showed that in general there are extended dissolution/reprecipitation processes that take place on the surface layers of the material. The recrystallisation of calcite took place with meteoric water and CO_2 from various sources: atmospheric and biogenic CO_2. Also, human influence (surface treatment) and biological growth are candidates between the exogenic processes giving change in $\delta^{18}O$ and $\delta^{13}C$ in lime mortar. The diagnostic contribution of stable isotopic study indicated that environmental conditions were severe on all of historic mortars, resulting in total disintegration of masonry walls. In general, the isotopic values appear to be a useful tool to diagnose mortar degradation and determine the different sources of decay.

References

Craig, H.: Isotopic variations in meteoric waters. Science **133**, 1702–1703 (1961)

Dotsika, E., Lykoudis, S., Poutoukis, D.: Spatial distribution of the isotopic composition of precipitation and spring water in Greece. Global Planet. Change **71**, 141–149 (2010)

Dotsika, E., et al.: Isotopic analysis for degradation diagnosis of calcite matrix in mortar and plaster. Anal. Bioanal. Chem. **395**, 2227–2234 (2009)

Dotsika, E., Kyropoulou, D., Christaras, V., Diamantopoulos, G.: $\delta^{13}C$ and $\delta^{18}O$ Stable isotope analysis applied to detect technological variations and weathering processes of ancient lime and hydraulic mortars. Geosciences **8**(9), 339 (2018)

El-Turki, A., Ball, R.J., Allen, G.C.: The influence of relative humidity on structural and chemical changes during carbonation of hydraulic lime. Cem. Concr. Res. **37**, 1233–1240 (2007)

Hutcheon, I., Abercrombie, H.: Carbon dioxide in clastic rocks and silicate hydrolsis. Geology **18**, 541–544 (1990)

Larbi, J.A.: Microscopy applied to the diagnosis of the deterioration of brick masonry. Constr. Build. Mater. **18**, 299 (2004)

Sabbioni, C., Zappia, G., Riontino, C., Blanco-Varela, M.T., Aguilera, J., Puertas, F., et al.: Atmospheric deterioration of ancient and modern hydraulic mortars. Atmos. Environ. **35**, 539–548 (2001)

Yates, T.: Mechanisms of air pollution damage to brick concrete and mortar. In: Brimblecombe, P. (ed.) Air Pollution Reviews: The Effects of Air Pollution on the Built Environment, pp. 107–132. Imperial College Press, London (2003)

Zappia, G., Sabbioni, C., Pauri, M.G., Gobbi, G.: Mortar damage to airborne sulfur compounds in a simulation chamber. Mater. Struct. **27**, 469–473 (1994)

The Restoration Proposal for the 4-storey Load Bearing Timber Structure of Matsopoulos Wheat Mill in Trikala (Greece)

Eleftheria Tsakanika(✉) and Sonia Charalambides

School of Architecture, National Technical University of Athens, Patission 42, 10682 Athens, Greece
eletsaka@central.ntua.gr, scharalambides@arch.ntua.gr

Abstract. The paper focuses on the restoration proposal for the 4-storey timber structure of the central building that belongs to Matsopoulos Wheat Mill complex, built in 1884. The study was conducted in the frame of a research program of the School of Architecture of the National Technical University of Athens and the Municipality of Trikala, for the restoration of the industrial complex and the creation of a museum that will host again the original function of this 'machine building', since the timber load bearing structure is an integral part of the main production equipment of the Mill. An excessive survey and assessment of the structure and the original machinery was conducted. The 3D structural model was a valuable tool for both the assessment procedure and the interventions for a huge and complicated timber structure. A new multi-task steel structure was designed, forming corridors for the control of the visitors, and the suspension of information equipment. This structure, reduced also the loads more than 50% and consequently the relevant reinforcements. In parallel, steel elements were placed next to the timber ones in specific places in order to carry the additional loads when needed. The main principles of the restoration proposal were: local, distinguishable and reversible interventions, with minimum impact on the original timber fabric, since this fabric along with the machinery were the main exhibits of the museum. The reinforcing interventions in this case served quite successfully both, museological, museographical, architectural, structural and mechanical requirements.

Keywords: Timber wheat mill · Timber industrial building · Structural analysis as assessment tool · Steel reinforcements of timber structures · Industrial museum

1 Introduction

The Wheat Mill was built in 1884, in the periphery of the town of Trikala, in Thessaly, by the Agiamoniotis river, an affluent of the Lithaios river that crosses the town. It's a great complex in an area about 90 acres (Fig. 1), that was donated to the Municipality by the owner I. Matsopoulos. It was declared a cultural heritage monument in 1995.

I. Vayas and F. M. Mazzolani (Eds.): PROHITECH 2021, LNCE 209, pp. 790–803, 2022.
https://doi.org/10.1007/978-3-030-90788-4_61

Intervention works had been carried out several years ago (1998), by the Municipality of Trikala (structural engineer: V. Papaioannou). Thankfully, the buildings kept most of their original architectural and structural features and the production machinery 'in situ', with interventions mainly at the external masonry walls and only minor ones at the timber fabric. The Mill complex now is called the Mill of the Elves, accepting millions of visitors, constituting a very important cultural center of the town, a thematic museum and the greatest Christmas Park of Greece.

Fig. 1. The whole complex of the Wheat Mill (left). The main building of the Wheat Mill (right).

This paper focuses mainly on the restoration proposal for the internal 4-storey timber framed structure of the central building of the Mill complex (Fig. 2). The study was conducted in the frame of a research project of the Architecture School of the National Technical University of Athens and the Municipality of Trikala, for the restoration of one of the most important historical industrial complexes in Greece and the creation of a museum that will host again the original function of a '*machine building*'.

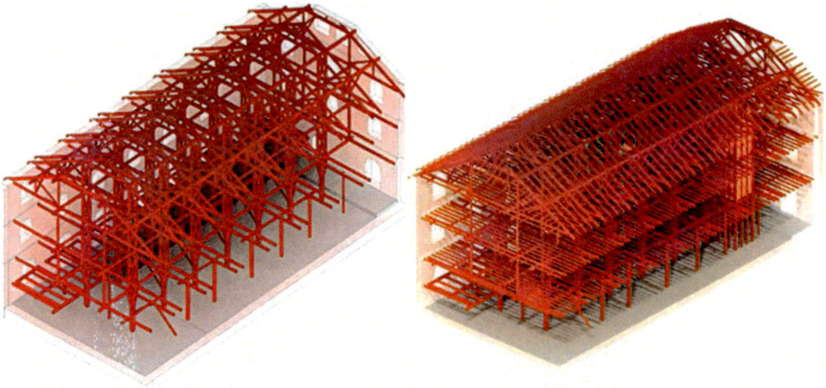

Fig. 2. Axonometric drawings of the timber load bearing structure.

The structural system, the pathology, the numerical analysis and the restoration proposal of the timber part of the building, are presented briefly, always in the frame that the above are considered an integral part of the whole restoration project (museological, museographical, architectural, mechanical).

Fig. 3. Old photograph during the production process of wheat. Representation of the architectural proposal (ground floor).

2 Description of the Timber Load Bearing Structure

The dimension in plan of the main building is ~33.80 × 16.80 m, and its height is ~18.0 m. The external walls are built of stone and brick masonry (Fig. 1, 4, 5). Their width at the ground floor is 85 cm, at the 1st floor 75 cm, and at the 2nd floor 57 cm. The 4-storey timber frame of the main building is a rare example of a multi-level industrial complex of the 19th century, bearing the 3 floors and the roof. It is also an integral part of the main production equipment and process of the Mill, that develops also along the whole height of the building, and is considered the most prominent feature of the museum as an exhibit.

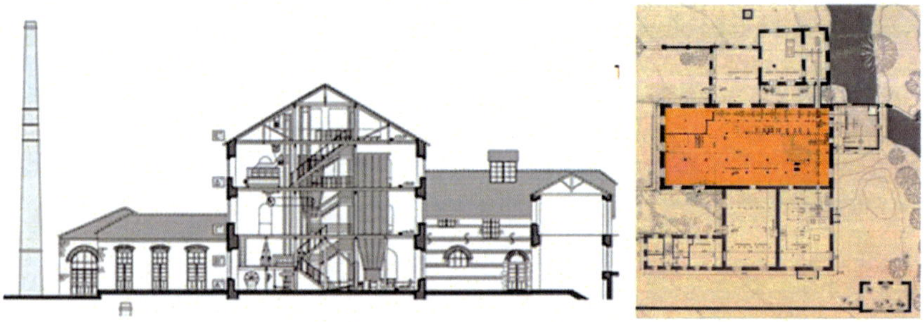

Fig. 4. Section and plan of the main building of the Wheat Mill.

The columns of the timber structure of the 1st, 2nd and 3rd floor (roof level) are double (~12–14 × 19–22 cm), placed in a distance of 1.12 m, reinforced by cross diagonal timber elements as bracings (14 × 12 cm), oriented along the longitudinal axis of the building (Fig. 5). The columns are constructed differently at the ground floor. One piece of timber (17–21 × 22 cm) is reinforced at the upper part by double struts (9 × 23 cm) and at the lower part by the addition of double timber pieces of 6cm width, that increase its section to ~ 22 × 33 cm (Fig. 3, 5).

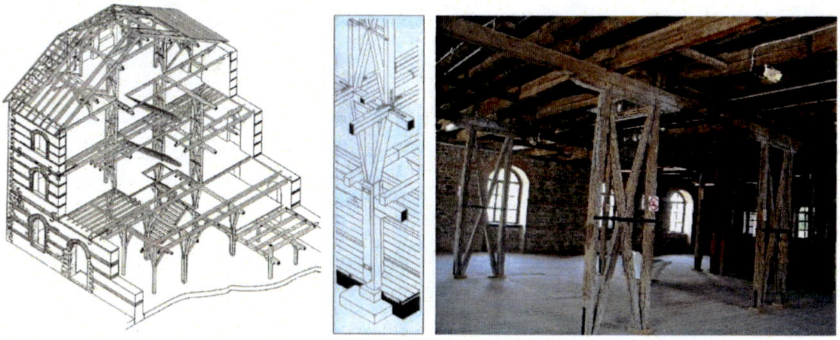

Fig. 5. Axonometric drawings[1] and photo of the overall structural system, the columns along the height of the buildings and their joints.

The loads from the timber planks of the floors are transferred on the joists (12 × 12 cm, Fig. 6, brown color) and the double beams (green color) made of 2 timbers 12 × 12 cm, placed one over the other and connected with vertical timber dowels. The joists are resting on the masonry walls, the transversal beams (12 × 12 cm, blue color) and the main longitudinal beam (17 × 21 cm, yellow color), placed over the timber columns, with the addition of a long timber capital (Fig. 5). The transversal timbers (blue color) are transferring the biggest part of the loads to the double beams (green color), with a very weak connection, poorly constructed, not possible to transfer the loads (Fig. 7).

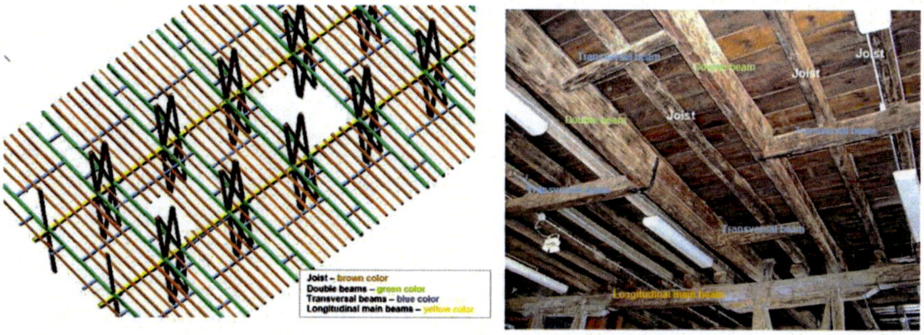

Fig. 6. The load bearing system of the floors. The structural model (left) as key drawing for the recording of the structural system

[1] These axonometric drawings were conducted during an older research program of NTUA (1990) and the Municipality of Trikala, for the constructional analysis of the Wheat Mill, with Head of the research Prof. P. Touliatos.

Fig. 7. The connection of the transversal beams (blue color in Fig. 6), to the double beams (green color) through a wedge cutting and a steel angle, is extremely weak and poorly constructed.

The load bearing structure of the roof is constructed following similar principles with the floor and is presenting the same problems (see next chapter, pathology). The rafters rest on longitudinal beams (at the ridge and 2 other points), that through small double posts with diagonal bracings and struts are transferring the loads to double beams which finally transfer the loads to the central double columns (Fig. 8).

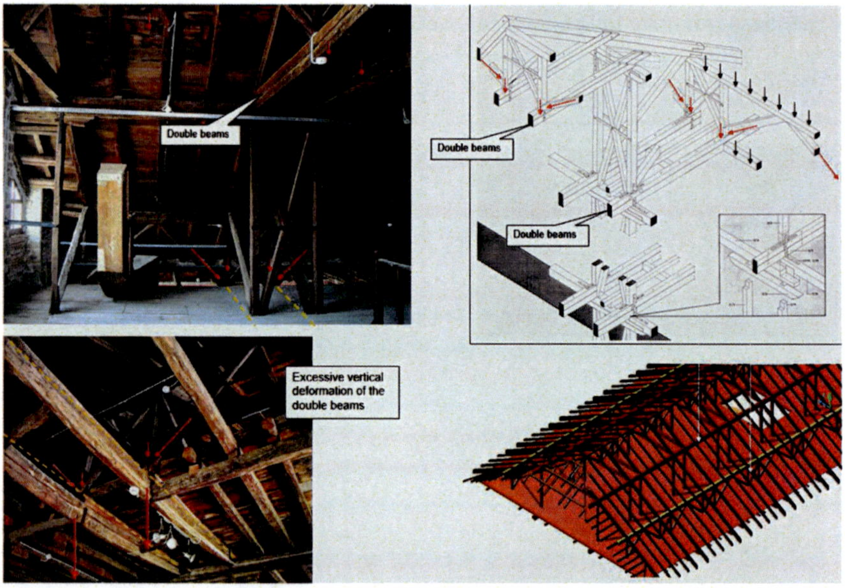

Fig. 8. The load bearing structure of the roof. The large vertical deformation of the double beams, is the biggest problem not only of the roof, but of the whole structural system.

3 Recording - Pathology - Assessment

3.1 General

A visual inspection for the recording of the structure, its pathology and vulnerability was carried out on site. An excessive documentation and survey of the overall timber structure and its connections was conducted with 2d and 3d drawings (Fig. 2, 8).

The information concerning the recognition of the structural system, its behavior and pathology (damages and interpretation of their causes), was organized in Tables with photos and drawings, a methodology quite useful especially when hundreds of elements and joints need to be assessed (Fig. 8, 9, 11).

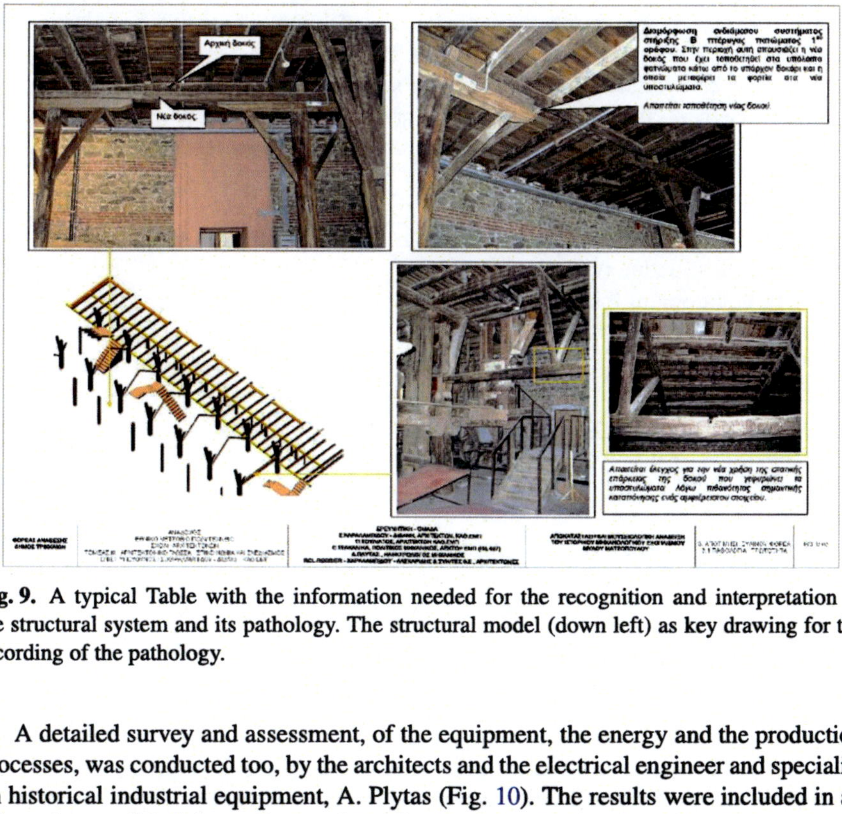

Fig. 9. A typical Table with the information needed for the recognition and interpretation of the structural system and its pathology. The structural model (down left) as key drawing for the recording of the pathology.

A detailed survey and assessment, of the equipment, the energy and the production processes, was conducted too, by the architects and the electrical engineer and specialist on historical industrial equipment, A. Plytas (Fig. 10). The results were included in an exhaustive, codified for each member, diagnostic report. The repair proposals by A. Plytas for the functioning in a museum mode of the machines, was based on the above recording and assessment procedure of the machinery and its interaction with the timber load bearing structure of the building.

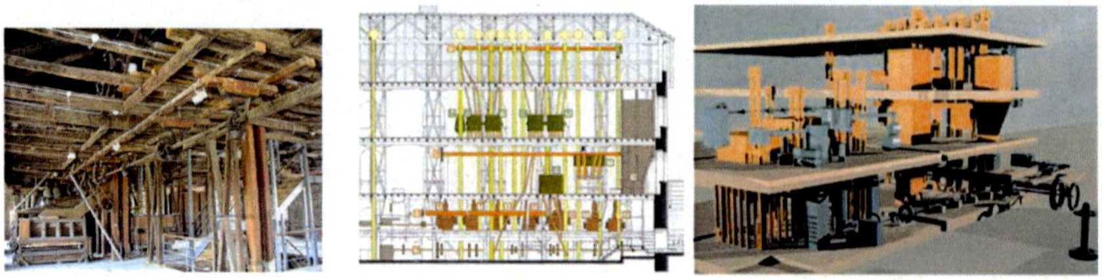

Fig. 10. Photos, 2d and 3d drawings of the recording and survey of the mechanical equipment.

3.2 The Structural Model as an Assessment Tool

It must be highlighted that the recording of the pathology and the assessment of all the timbers and joints that consist the timber structure of the Mill, with hundreds of members and joints (Fig. 2) is actually impossible. For this, Mill's assessment was guided by the mathematical analysis of the existing timber structural system. A 3D structural model, using finite elements that included the components of the timber structure (vertical timber frames, floors and roof), became the most valuable tool (Fig. 6, 9) in order to determine:

- The critical areas and the areas of high stresses, where more attention is required during survey, limiting and guiding the assessment procedure of a huge and very complicated structure (Cruz et al. 2015).

  The definition of the critical areas gave also the possibility the decided interventions to be local and less heavy.
- The type of stresses for each element. The identification of the different structural member types is critical for the strength assessment (grading) of the timbers[2].
- The overall level of forces, quite important for controlling and calculating the load bearing capacity of the existing structure.
- The verification of the pathology (failures and deformations). The detailed analysis of the contribution (good or bad) of the existing timbers revealed the causes of the problems and damages and the problematic points and areas (Fig. 8, 11).
- Moreover, the structural model became an important methodological tool for the recording of the existing condition of a very complex structure, a forest of timbers. It was used as a key drawing and reference, in order to document the timber members and joints, to comment and interpret the pathology (Fig. 6, 9).

[2] Current grading rules used for new timber structures are not well suited to historic timbers for a number of reasons. The commercial grading of timbers into strength classes takes no account of the conditions of use. In contrast, the location and stress conditions of historic timbers are known as are the positions of specific growth characteristics, thus allowing the importance of defects to be evaluated in relation to the applied stresses (Cruz et al. 2015). As simple examples, a crack that divides the whole section of a member is not important if the member is in tension. On the contrary, if the member is subjected to bending, this crack may diminish significally its load bearing capacity. Where compression members are large, as the main columns of timber frames, stresses will be low and the effect of knots will usually be small. Where compression members are small, as in stud walls, knots may well have a serious effect on the stiffness of a member.

It must be mentioned, that non-linear analysis was conducted, in order to take into account the connections that cannot transfer tension (Tsakanika 2007). The strength class of the timber that was used for the calculations according to EC5, was C22.

3.3 The Pathology

The most important problem of the timber structure was the excessive vertical deflection of the double beams and the joists, at the level of the floor and the roof (Fig. 8, 11), caused mainly by errors of the original conception and construction of the structural system. The inefficiency of the connections of the joists to the double beams (Fig. 7), and the inefficiency of the connections in several other cases too, due to poor construction and old interventions, caused several shear failures and problems (Fig. 11, right).

Decay problems were few and negligible since restoration works that had been conducted in 1998, faced some of this problems.

Fig. 11. Excessive deformation of the double beams at the floor level (left). Shear failure of timber beams due to careless cuttings (original or due to later interventions) at the connection areas.

4 The Restoration Proposal

4.1 Principles of the Proposal

The architectural, museological and museographical aim was to:

- show the cultural dimension of the industrial heritage, the history of the local industrial evolution, the technical and the technological aspects of the production processes. The architecture of industrial buildings presents interesting solutions to particular problems resulting from the production process requirements.
- revive the original use of the building and restart the function of the machinery creating a place of discovery, knowledge, education and entertainment,
- relate the technical equipment with the human factor and the social and local conditions of the creation and evolution of the industrial activities.

The important axes of the design process were defined by the following assumptions:

- The organization of the production process is inscribed in the layout of the industrial plans and the architecture of the buildings.

- New technologies that may be included in the museological concept and introduced in the museum project, as actual energy production, or information technologies, are in the trend of the continuous innovation process that defined the industrial production.

The space planning and the architecture of the buildings are considered as the main 'exhibit', along with the mechanical equipment. This important industrial heritage construction is actually 'a machine building'.

The main principles concerning the restoration of the building were:

- Preservation as much as possible of the authentic materials (e.g. minimization of replacements of timbers or other materials either of the load bearing or non-load bearing elements, or the machinery).
- Preservation of the authenticity and integrity of the original timber structural system, since in this case this exceptional fabric, along with the machinery, as mentioned before, are the main 'exhibits' of the museum. Some parts of the building, are a machine themselves, that interacts with the real machinery.
- Minimum interventions with local repairs is preferred rather than total structural replacements.
- The interventions (repairs or reinforcements) for the improvement of the load bearing capacity of the joints, must not change their stiffness and consequently the overall behavior of the original structural system.
- All necessary new infrastructures must be as invisible as possible, or clearly distinguished from the historical machinery.
- New architectural and structural elements, necessary also for the rehabilitation and restoration/reinforcement process of the building must be clearly distinguished too. For this all new structures was decided to be from steel elements.
- The reinforcements of the original timber structural system due to the fact that its use will change (an industrial building will become a museum, which means higher level of safety for the building and the visitors), must have minimum impact on authentic structural system.
- The interventions must be reversible and compatible with the original parts of the buildings.

4.2 Structural Analysis and Evaluation of the Intervention

New 3D structural models were used that included except the existing timber elements, every component of the new steel structure of the 'corridor' (Fig. 12, 15). The structural analysis verified the positive role of the 'corridor-bridge' (Fig. 12) and became again an important methodological tool, guiding, evaluating and optimizing the intervention proposal, step by step. The enlarged areas that were created along the corridor (Fig. 13–15), were determined through the mathematical analysis, that was used to find the position and the scale of these areas in order to have the minimum impact on the timber structural system. In each model, the new steel elements were added gradually, in order to control their contribution, and to ensure the minimum of interventions.

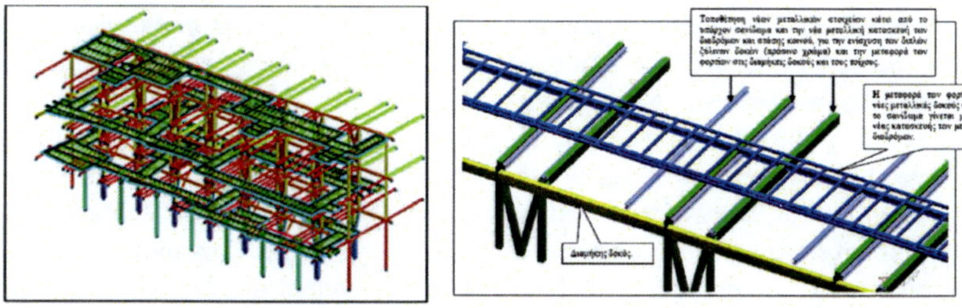

Fig. 12. Structural models (left. M. Tsitsinaki). The 'corridor' working as a 'bridge' element.

4.3 Description of the Reinforcement Proposal

The main design idea that served the above principles and the museological, museographical, architectural, structural and mechanical requirements, was the creation of a multi-task solution, a "corridor' with few areas of enlarged space along its length (Fig. 13–15 plan), product of a multi-disciplinary collaboration.

In order to conform with museum requirements and specifications, especially to the number and specific flow of visitors, a path was created through the machinery, that made possible to trace the production process, limiting the movements allowed, thus protecting the visitors from the machinery, especially when functioning, and the exhibits from manipulation. The rail that delimits the corridor, is supporting at the same time, an infrastructure for the suspension of information equipment. The "corridor", a slightly elevated steel structure, was designed to be placed over the existing timber planks of the floor, which will be maintained and preserved (Fig. 13).

Fig. 13. Plan and 3d representations of the architectural proposal of the "corridor".

This new structure, makes also possible to fulfill the goal of minimal visual interference of the new systems in the museum, allowing, the 4-storey timber load bearing structure and the machinery to be the main exhibit.

The necessary infrastructure for the electromechanical systems is designed in the space created under the new wooden flooring of the "corridor", between the steel structure, which for this purpose was formed by steel elements (HEA160) with holes (Fig. 14). The few enlarged spaces that were created along the length of the "corridor" can be used for rest and for the accommodation of additional permanent or not, information and exhibition areas (Fig. 14).

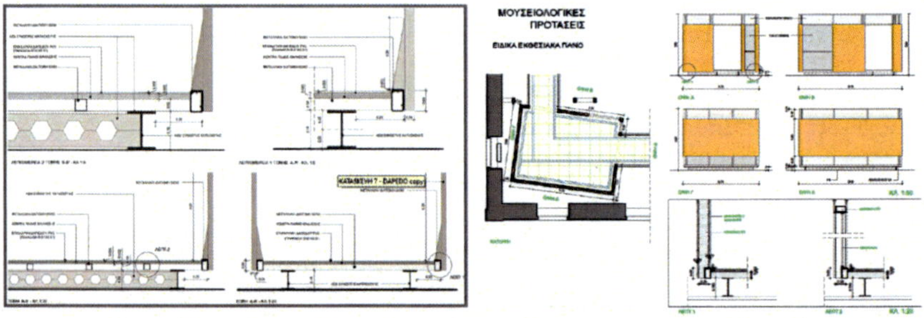

Fig. 14. Construction details of the "corridor" and the enlarged areas with exhibition panels.

The corridor consists a very effective structural solution for many reasons too.

- The decision of limiting, and organizing the areas of the visitors, reduced the area that the loads are applied more than 50%.
- The decision of controlling the flow of visitors reduced even more the variable loads to 3.5 KN/m^2, and consequently, the relevant reinforcements which otherwise would be extremely heavy on the original timber structural system as it was verified by the structural analysis.
- The 'corridor', was also designed to work as a 'bridge' (Fig. 12), transferring the loads to the main load bearing double timber beams next to the columns, avoiding the secondary joists, that as mentioned before in pathology, had already exhibited excessive deformation, their joints were poorly constructed, and their section was not adequate as it is verified by calculation according to EC5 (Fig. 15).
- The introduction of supplementary steel elements under the 'corridor' will relieve the existing timber structure of loads (Fig. 15, red color).

The steel structure of the 'corridor' was formed by transversal tubular steel joists (5.0 × 5.0 × 0.4 cm, Fig. 14, 15, 16 yellow color) that rest on longitudinal beams (HEA160) with hollows (Fig. 14, 15, 16 blue color)[3]. The longitudinal beams act as

[3] The final dimensions, the position and the number of the steel elements were provided by the structural study of M. Tsitsinaki, that was conducted with the collaboration and supervision, of the research team of NTUA.

bridging elements, transferring the loads to the double timber main beams next to the timber columns and to the new steel ones that were added (red color).

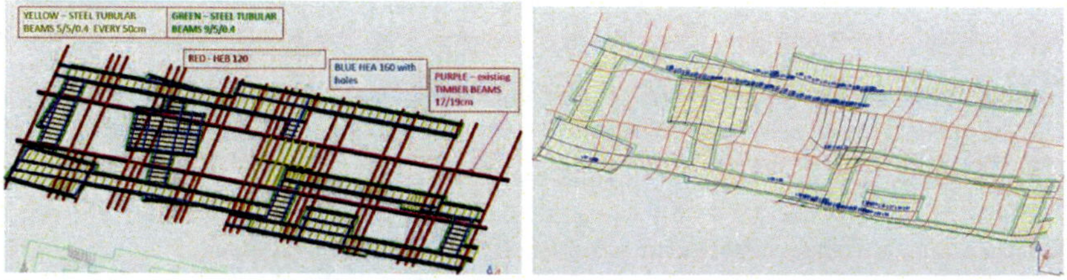

Fig. 15. Structural model at the level of the floor. The new steel structure of the 'corridor', working as a 'bridge' (yellow, green and blue members), is resting on the original double timber beams and the new steel ones (red color) that have been added next to the timber ones. The results of the structural analysis is showing the most loaded areas, evaluating the position and the efficacy of the interventions.

As mentioned before in pathology, the double timber beams and the joists were severely deformed. In order to increase the flexural strength and stiffness of these beams, the improvement of the vertical connection of these 2 beams was proposed, using steel dowels and screws. Moreover, additional steel beams (HEB120) was proposed to be placed next to them, without connection, in order to act supplementary, diminishing even more their loads (Fig. 15, 16 red color). The use of these steel beams is as the 'corridor', a reversible and distinguishable intervention.

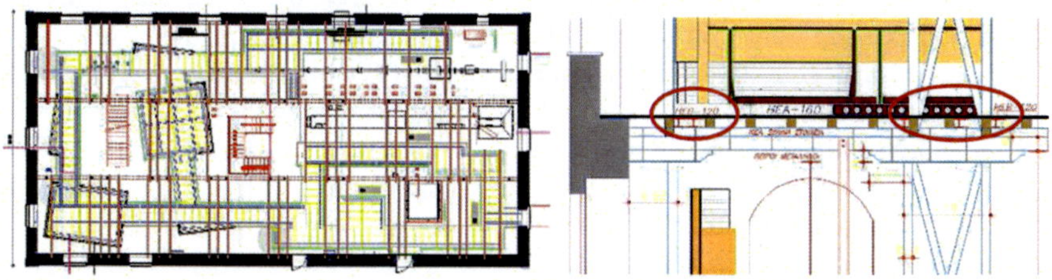

Fig. 16. Plan and section with the proposed position of the 'corridor' and the new steel beams, near the timber double ones.

The decision of using new steel beams close to the double timber beams, was dictated by the need to transfer the loads directly to the columns, relieving the longitudinal beams (Fig. 15, purple color, 16). In areas that this was not possible, or their shear load bearing capacity needed to be improved the reinforcement presented in Fig. 17 was proposed.

Finally, measurements of the natural frequencies of the building using the microtremor method on-site were proposed, due to the vibrations during the functioning of the machinery and its interaction with the whole load bearing system of the building (timber and masonry). The measurements should be conducted in 3 phases, before the interventions, after the interventions and during the functioning of the Wheat Mill machinery.

Fig. 17. Addition of timber elements in areas that the shear load bearing capacity of the beams is not adequate and connection with the original ones with bolts. Reinforcement of the original joints can be made using self-threaded screws (left). Addition of new metal columns in limited areas that excessive stresses are occuring in longitudinal beams or columns (center). Interaction of the timber load bearing system with the machinery. Because of this, the machinery in several areas will transfere vibrations on the load bearing system of the building, when it is functioning (right).

5 Conclusions

In the frame of the research program conducted by NTUA, a multi-disciplinary team (architects, specialist in museology, civil engineer, electrical engineer), proposed a multi-functional intervention. The team worked closely from the beginning, trying to understand and evaluate an exceptional 4-storey historic timber load-bearing system of an industrial building of the 19th century, ~570 m^2 per level, interconnected and inter- acting with the mechanical equipment of the Mill, also developed along the plan and the height (18 m) of the building.

It must be highlighted that the mathematical analysis and the 3D structural model that included the existing timber structure and the new steel one, became the most valuable tool for the recording and the assessment of a huge very complicated timber structure, with hundreds of members and joints and moreover, evaluated, guided and optimized the intervention proposal, step by step.

The main principles of the restoration proposal were: local, distinguishable and reversible interventions, with minimum impact on the original timber fabric, saving its authenticity and integrity, since this complex timber structure-machine, along with the whole mechanical equipment of the Wheat Mill, were the main ‘exhibits’ of the museum. The reinforcing interventions were based on an idea that in this case served quite successfully architectural, museological, museographical, structural and mechanical requirements.

Acknowledgements. The research team was: S. Charalambides-Divani Architect Prof. Emeritus NTUA Head of the project, P. Touliatos, Architect Prof. Emeritus, NTUA, E. Tsakanika Civil Engineer Ass. Professor NTUA, A. Plytas Electrical Engineer Historical Industrial Equipment specialist, RCL-M.Rossier, I. Charalambides, C. Laskaridis, Architects Urban Planners.

References

Branco, J.M., Descamps, T., Tsakanika, E.: Repair and strengthening of traditional timber roof and floor structures. In: Costa, A., Arêde, A., Varum, H. (eds) Strengthening and Retrofitting of Existing Structures. Building Pathology and Rehabilitation, vol 9, pp. 113–138. Springer, Singapore (2018)

Cruz, H., et al.: Guidelines for the on-site assessment of historic timber structures. In: International Journal of Architectural Heritage: Conservation, Analysis, and Restoration, vol. 9, no. 3 (2015)

IIWC ICOMOS Wood Committee: Principles for the preservation of Historic Timber Strucures (2017)

Tsakanika, E.: Byzantine and Post-Byzantine Historical Timber Roofs in Greece. Typical failures, misunderstanding of their structural behaviour, restoration proposals. "From Material to Structure", XVI International Conference and Symposium of the IIWC, Florence. Online (2007). http://www.icomos.org/iiwc/2007.htm

Adhesive Materials for the Conservation Interventions in the Acropolis Monuments

Eleni Aggelakopoulou[1], Chrissy-Elpida Adami[2(✉)], Anthi Tsimereki[1], Elizabeth Vintzileou[2], and Vasiliki Eleftheriou[1]

[1] The Acropolis Restoration Service, Hellenic Ministry of Culture and Sports, Athens, Greece
{eaggelakopoulou,atsimereki,veleftheriou}@culture.gr
[2] National Technical University of Athens, Athens, Greece
{adamis,elvintz}@central.ntua.gr

Abstract. During the conservation interventions that occur in the Acropolis monuments, there is a need to develop compatible adhesive materials that will bond small fragments to the stone substrate of architectural members. For this purpose, different hydraulic pastes and mortars were designed with the aim to assess their adhesive efficacy. For comparative reasons, syntheses of white cement and an epoxy resin were evaluated as well. "Sandwich" specimens were prepared by using stone (marble and limestone) and the adhesive materials and subjected in shear tests, in 1,6 and 12 months of curing time. The higher shear bond values are ensured for cement and epoxy resin, accompanied by a substrate failure. On the other hand, sufficient shear strength was acquired by hydraulic materials. Thus, they may consider an attractive alternative for binding small stone fragments on the ancient blocks of the Acropolis monuments.

Keywords: Acropolis · Marble · Limestone · Adhesive materials · Hydraulic mortars/pastes · Shear bond strength

1 Introduction

The Acropolis of Athens, located on a rocky hill above the city, contains several ancient buildings of great architectural and historic significance. Pentelic marble, a fine-grained calcitic marble from the quarries at Penteli, north of Athens, has been used for the construction of classical period's monuments (Parthenon, Propylaia, Erechtheion and Athena Nike temple). A sedimentary limestone derived from Piraeus peninsula was used for the construction of the monuments of the archaic period, the foundations of the classical monuments and the circuit walls.

Pentelic marble is a metamorphic rock composed mainly by calcite (approx. 98%) and other minerals such as muscovite, chlorite, dolomite, quartz, hematite, pyrite, albite (Exadaktylos et al. 2001), (Skoulikidis 2000). It presents a low porosity value (approx. equal to 0.4%), a low pore radius (approx. equal to 0.12 μm) and specific surface area (approx. equal to 0.03 m^2/g) in sound condition (Moropoulou et al. 2003). The sound pentelic marble presents high mechanical strength values: compressive strength (f_c)

I. Vayas and F. M. Mazzolani (Eds.): PROHITECH 2021, LNCE 209, pp. 804–815, 2022.
https://doi.org/10.1007/978-3-030-90788-4_62

varying from 59 to 96 MPa, tensile strength (f_{fl}) varying from 11 to 22 MPa, while the static modulus of elasticity (E_s) and Poisson (ν) ratio are equal to 43 ÷ 89 GPa and 0.33 ÷ 0.35 respectively (Zambas 1994).

Regarding the limestone derived from the Piraeus peninsula (Eleftheriou 2015), it mainly consists of dolomite, calcite and aluminosilicates. Depending on the area that the limestone was mined, the Piraeus stone may be either Aktitis stone, coming from the west coast or Mounychia stone, coming from the hill of Mounychia (Theoulakis et al. 2000). Aktitis stone is a hard dolomitic stone having a compressive strength of the range of 20 to 107 MPa. Mounychia stone is a soft marly stone ~~(clayey)~~ with a compressive strength of the order of 10–20 MPa. The scatter in the values of the stones' characteristics may be attributed to the materials' inhomogeneity caused by their sedimentary nature and their different mineralogy (Ksinopoulou and Aggelakopoulou 2018).

During the conservation interventions that occur in the Acropolis monuments, there is a need to develop compatible adhesive materials that will bond small stone fragments to the stone substrate of the architectural members. These small stone fragments (in centimeters scale) were detached ~~and~~ or peeled off the ancient blocks due to various factors such as earthquakes, fires, bombardments, environmental causes, etc. Furthermore, these actions led to a great reduction of the structural stones' original mechanical properties ~~of the monument~~. Until recently, these small parts were re-attached in their original position on the stone substrate with white cement paste. It is well known that, the use of cement as an adhesive material in monuments has several disadvantages: a) ~~in case it is~~ when used to bind soft stones, such as limestone, the damage occurs in the substrate rather than in the adhesive material [for example (Adami and Vintzileou 2008), (Adami 2011)] and b) it is difficult or practically impossible for cement pastes to be removed without harming the original material.

Wishing to take a new approach to the conservation interventions of the Acropolis monuments and taking into account the evolution of conservation materials science, a need to investigate the bonding capability of other hydraulic mortars has emerged (Tsimereki 2016). According to this approach, the following design criteria were adopted for the proposed adhesive materials:

- sufficient bond strength
- physico-mechanical compatibility with the authentic materials and
- in case of failure, the damage will occur on the adhesive material, whereas the authentic material will remain intact.

Eight syntheses and one epoxy resin were considered. In order to evaluate the bonding capability of the materials, composite specimens (sandwich) were prepared by joining two stone prisms (marble or limestone) with the proposed adhesive materials. The sandwich specimens were tested in shear. Cement materials and an epoxy resin were tested in one month while the hydraulic materials were tested in 12 months of curing.

2 Experimental Program

2.1 Material and Properties

The syntheses of adhesive pastes and mortars were proposed by the conservation department of the Acropolis' Restoration Service (YSMA) and are presented in Table 1. Five syntheses of pastes were considered: one of pure white cement, one of white cement and hydrated lime, three syntheses of hydrated lime with metakaolin in different ratios and one of natural hydraulic lime (NHL3.5-Z). In addition two syntheses of mortars were designed by using natural hydrated lime or a mixture of hydrated lime and metakaolin with a fine calcitic sand. The grain size distribution of this sand is concentrated in two grain size ranges (0.063 μm and 0.125 μm) while the contained fraction < 0.063 μm is less than 7% p.w. Lastly, an epoxy resin composed of two components (EPOXOL 2874, Neotex S.A.), addressed to marble, was used for comparison reasons. More specifically this epoxy resin was used, after the application of a layer of an acrylic resin (PARALOID B-72) on the stone substrate. This technique is referred in the literature (Riccardelli et al. 2014), in order to assure the reversibility of the intervention since the acrylic resin (and therefore the resin) can be removed by using an organic solvent.

Table 1. Mix proportions of the studied mixes

Mix	Composition [%-wt]						
	White cement	Hydrated lime	Natural Hydraulic Lime (NHL3.5-Z)	Metakaolin	Calcitic sand (0.063–0.250 μm)	Epoxy-acrylic resin	w/c
C	100						0.50
LC	50	50					0.62
LMK40		60		40			0.72
LMK50		50		50			0.68
LMK60		40		60			0.66
NHL			100				0.45
NHLSc			60		40		0.30
LMK30Sc		30		30	40		0.46
Epoxy Resin						100	

w/c: water to binder ratio

In order to measure the main mechanical properties of the mortars/pastes, specimens with dimensions 40 × 40 × 160 mm^3 were prepared and tested in the Laboratory of Concrete/NTUA. Table 2, presents the available results of the compressive strength of the hydrated lime-metakaolin and hydraulic lime based mortars/pastes. These properties were measured at the age of 28 days and according to EN196–1. In general, one may observe that the studied syntheses develop compressive strengths that vary between 7.00 and 14.60 MPa. Higher compressive strength values are recorded for the hydrated lime-metakaolin mixes, as compared to those of the natural hydraulic lime ones. Moreover, the addition of sand seems to have a positive effect on their compressive strength. This effect is more pronounced for the hydrated lime-metakaolin mixes.

Table 2. Physico-mechanical properties of the mixes containing metakaolin and natural hydraulic lime (28 days)

Mix	Compressive strength (MPa)	Density (gr/cm^3)
LMK 40	10.36	1.23
LMK50	12.88	1.25
LMK60	16.65	1.32
NHL	6.34	1.45
NHLSc	7.07	1.62
LMK30Sc	14.60	1.56

For the substrates, stones corresponding to those of the Acropolis walls were used. Dionysos marble was used instead of the original pentelic marble (substrate type M). As Piraeus limestone (Aktitis) from the ancient quarry of Hadjikyriakio had been used for the restoration project of the Theater of Dionysus, three stones were kindly offered to the Acropolis' Restoration Service by the Ephorate of Antiquities of the City of Athens. Those stones (substrate types A, B and C) were selected due to their different hardness. However, it was impossible to find stone from the area of Mounychia, which was originally used in the inner elevations of the circuit walls. For this reason, with the permission of the Committee for the Conservation of the Acropolis Monuments (ESMA), an amorphous fragment from the inner elevation of the south wall that was collected after a small-scale failure was used (substrate type D). Figure 1 presents the four stone types, namely type A, B, C and D, which were considered here.

Fig. 1. Stone types A, B, C and D prisms

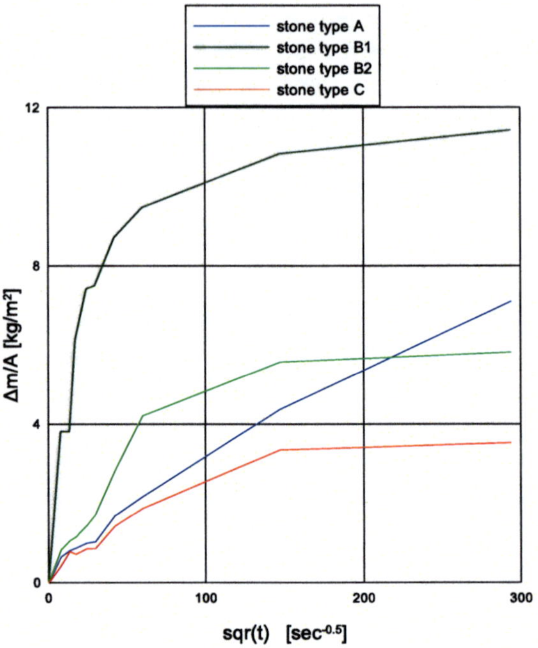

Fig. 2. Absorbed water over time by capillary suction in the three stone types A, B and C.

The compressive strength of the Dionysos marble (type M) and of the four types of Piraeus limestone (type A, type B, type C and type D) was measured on 50 × 50 × 50 mm^3 cubes in the Laboratory of Concrete/NTUA. Table 3 presents the compressive strength of the stones, as well their density values. In general, Dionysos marble, stone type A and type B present higher compressive strength values as compared to those of Stone type C and D. As mentioned above, the deviation of the compressive strength for each stone type is high due to the inhomogeneity of the parent rock. These characteristics are, therefore, expected to affect the failure mode and the value of the interfacial shear bond strength of the respective composite specimens (Fig. 2).

Additionally to the mechanical characterization of the substrates, the absorbed water over time was measured for stone types A, B and C. For stone B, two specimens with different macroscopic properties were used. Stone type B1 presented the highest percentage of absorbed water (8.07%) and stone type C the lowest value (2.35%) while the relevant values of the other two stone types lie in between (Type A:4.97%, Type B2:5.45%). Analogous behaviour is observed for the rate of absorbing water. Stone type B1 presented the higher value (0.24 kg/(m^2/$s^{0.5}$)) and stone type C the lowest one (0.035 kg/(m^2/$s^{0.5}$)). The relevant values of absorbing water rate are 0.044 kg/(m^2/$s^{0.5}$) for stone type A and 0.062 kg/(m^2/$s^{0.5}$) for stone type B2. The difference in the hygric behavior between the two specimens of stone type B is ascribed to the great inhomogeneity that this stone type presents.

Table 3. Physico-mechanical properties of the stones

Type of stone	Compressive strength (MPa)	Density (gr/cm^3)
Dionysos marble	59 ÷ 96*	2.70**
Stone Type A	81.2 (39.9%)	2.49
Stone Type B	83.9 (17.2%)	2.46
Stone Type C	31.7 (63.4%)	2.11
Stone Type D	10 ÷ 20***	n/a

(*) (Exadaktylos et al. 2001)
(**) (Moropoulou et al 2003).
(***) (Ksinopoulou and Aggelakopoulou 2018).
(****) In the parenthesis, the coefficients of variation ratios are given.

The composite specimens (sandwich) were prepared in the conservation department of the Acropolis' Restoration Service (YSMA). Initially, tiles (40 × 40 × 20 mm) were cut out of each of the five stone categories (Dionysos marble and Piraeus limestone type A, type B, type C and type D). Then, composite (sandwich) specimens were produced by adhering two stone prisms. In order to bring the in-field experience into the laboratory, decision was made over the optimal values of water content and bond line thickness over a range of 0.39–0.42 mm for each of the six pastes and two mortars. These two parameters (water content and bond line thickness) were constant for the total of the sandwich specimens of each category of stone.

During the jointing process a digital calliper was used in order to measure the specimens before and after placing the adhesive material. The jointing surfaces were initially roughened with a chisel and the stone tiles were hydrated until saturation before the application of the adhesives. The specimens were allowed to cure covered with wet compresses in order to maintain the moisture level.

The parameters investigated within this part of the research are the following: (a) the type of the adhesive material, (b) the type of the substrate and (c) the age of the interface. The selected combinations of parameters are shown in Table 4. In total, sixty six specimens were tested in shear. The shear tests were performed using a shear box device (Adami and Vintzileou 2008), (Vintzileou and Adami 2009) at a speed of 0.5 mm/min (Fig. 3). During testing, the value of the normal compressive stress, σ, on the interface was kept constant for all sandwich specimens (equal to 0.1 MPa). The horizontal displacement (s), the lateral dilatancy (w) and the corresponding resisting shear stress (τ) were continuously recorded during testing. The cement and epoxy resin composite specimens were tested in 1 month of curing, two syntheses of hydrated lime/metkaolin were tested in 6 months while the other specimens were tested in 12 months of curing, taking into account the chemical reactions completion in the adhesive materials (Aggelakopoulou et al 2011), (Bakolas et al 2008).

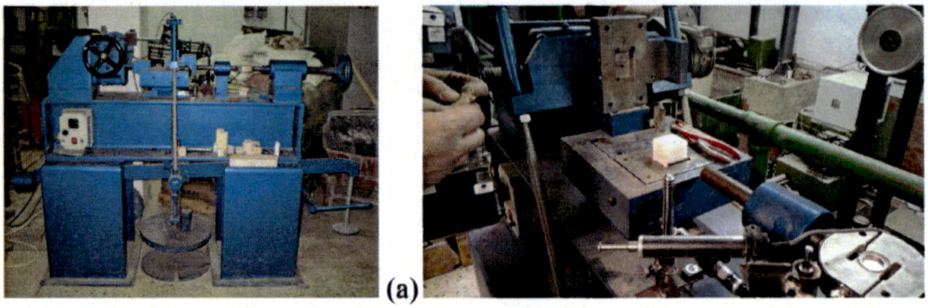

(a) (b)

Fig. 3. Pictures of the shear box device.

3 Experimental Results

This section summarizes the experimental results obtained to date for the hydraulic mixes (Table 4). Although testing is still in progress, the results allow for several key features to be discussed. They also allow for comparison with the results obtained from resin and cement paste adhesive materials.

Table 4. Adhesive materials to substrate interfaces: Summary of shear test results (mean values of two or three specimens)

Days	Ref. No of the mix	Substrate type	τ_u (MPa)	s_u (mm)	w_u (mm)
28	C	Marble (M)	7.157 (2)	3.474	2.390
	Resin	Marble (M)	5.933 (2), (4)	1.942	2.140
		Stone Type A (A)	6.917 (1), (4)	3.200	2.471
		Stone Type B (B)	3.745 (2), (4)	2.154	2.243
		Stone Type C (C)	8.253 (2), (4)	2.624	2.303
90	LMK50	Marble (M)	2.116(1), (3)	1.498	0.899
		Stone Type A (A)	0.899(1), (3)	0.608	0.294
		Stone Type B (B)	2.202(1), (3)	1.598	0.632
		Stone Type C (C)	2.598(1), (3)	1.396	0.631
	LMK60	Marble (M)	1.514(1), (3)	1.033	0.897
		Stone Type A (A)	0.797(1), (3)	0.690	1.312
365	LC	Marble (M)	7.730 (2), (3)	2.690	1.890
		Stone Type A (A)	3.810 (2)	1.930	1.850
		Stone Type B (B)	4.350 (1), (2)	2.620	1.700
	LMK40	Stone Type A (A)	0.200 (1)	1.090	1.890
		Stone Type B (B)	0.320 (1)	0.800	1.290

(continued)

Table 4. (*continued*)

Days	Ref. No of the mix	Substrate type	τ_u (MPa)	s_u (mm)	w_u (mm)
		Stone Type C (C)	0.170 (1)	0.490	0.200
	NHL	Marble (M)	2.180 (1), (3)	1.610	1.430
		Stone Type A (A)	0.240 (1), (3)	0.470	0.070
		Stone Type B (B)	1.710 (1)	1.710	1.180
		Stone Type C (C)	0.770 (1)	0.890	1.220
	NHLSc	Marble (M)	0.530 (1)	0.850	1.370

(1) Specimens failed due to failure mode ITZ
(2) Specimens failed due to a combination of failure modes ITZ and S
(3) Specimens failed due to failure mode Z
(4) Specimens failed due to failure mode S

3.1 Failure Modes

As in case of grouts to substrate interfaces presented in (Adami and Vintzileou 2008) and (Vintzileou and Adami 2009) adhesive material to stone interfaces subjected to shear presented the following failure modes (Fig. 4):

a) Mode ITZ: Shear failure along one adhesive material to substrate interface.
b) Mode Z: Similar to (a). Practically half of the joint remains attached to the one substrate prism, whereas the other half remains attached to the second one.
c) Mode S: Shear failure of the substrate.

(a) (b) (c)

Fig. 4. Failure modes a) ITZ, b) Z and c) S.

3.2 Marble to Mixes Interfaces

Marble to stronger adhesive materials (cement or resin) at the age of one month failed due to shear failure of the substrate. Similarly, marble to cement-hydrated lime composite specimens presented mainly substrate failure at the age of 1 year. On the contrary, medium strength metakaolin and natural hydraulic lime mixes failed along the substrate to adhesive material interface (failure modes ITZ and Z) at the age of testing.

Regarding the values of the maximum recorded shear resistance (τ_u), one may observe that the stronger adhesive materials ensure significantly higher values of τ_u, as compared to the values reached by the hydraulic lime based mixes. Actually, cement and resin interfaces exhibit a mean value of τ_u equal to 7.157 MPa and 5.933 MPa at the age of one month, whereas slightly increased values are recorded for the lime-cement interfaces (7.730 MPa). Since the composite specimens of these materials (cement, lime-cement and resin) failed at the substrate, one may say that the observed differences on the values of τ_u may attributed to the scatter of the substrate properties. Moreover, it should be noted that the slip (τ_u) and lateral dilatancy (w_u) values corresponding to τ_u of the abovementioned mixes are of the same order. Thus, one may say that the cement, lime-cement and resin mixes present practically an identical response.

As far as the natural hydraulic lime compositions are concerned, one may observe that NHL paste specimens reached high values of τ_u (2.180 MPa), whereas the addition of marble powder in the mix (NHLSc composition) led to a significant reduction of the maximum attained shear resistance ($\tau_u \sim 0.530$ MPa). Given that the compressive strength is of the same order for both mixes (Table 2), this reduction shows that the addition of sand reduces the bonding capability and this behaviour needs to be investigated. As expected, the slip and lateral dilatancy values corresponding to τ_u of the abovementioned mixes are accordingly lower than those obtained for the stronger mixes.

Regarding the hydrated lime materials, the best adhesive behavior is reported for the paste produced by mixing hydrated lime and metakaolin in equal percentages (50% -wt). Actually, the maximum attained shear resistance (2.116 MPa) is practically identical to that recorded for the NHL paste.

3.3 Stone Type A-To-Mixes Interfaces

The behaviour of resin to stone type A interfaces is similar to that recorded for resin to marble interfaces (in terms of failure mode, shear strength and corresponding slip and dilatancy values). However, hydraulic adhesive materials to stone A present a drastic reduction (varying from 50–900%) in their shear strength (Table 4) as compared to those of marble interfaces. It should be noted that the slip (τ_u) and lateral dilatancy (w_u) values corresponding to τ_u of the abovementioned specimens are reduced accordingly. Actually, the maximum shear resistance, τ_u, for LC interfaces is equal to 3.810 MPa, whereas LC composite specimens fail mainly due to a combination of failure modes ITZ and S. Similarly, for LMK50 and LMK60 specimens the maximum shear resistance equals to approx. 0.8–0.9 MPa. The reduction of τ_u is more noticeable for NHL interfaces; the maximum shear resistance, τ_u, for NHL specimens is equal to 0.240 MPa. Similarly, low values of τ_u are recorded for the metakaolin based composition LMK40 composite specimens (~0.200 MPa). It should be noted that both NHL and LMK40 sandwich specimens failed along the interface. The abovementioned behaviour may be attributed to the substrate properties, which led to an extensive microcracking in NHL and LMK40 joints (Fig. 5a, 5b).

(a)

(b)

Fig. 5. Microcracking of the joint of stone A to a) NHL and b) LMK40 composite specimens.

3.4 Stone Type B-to-Mixes Interfaces

Except NHL specimens, all stone type B-to-mixes specimens present similar behaviour as compared to those of stone A interfaces at the age of 1 year. Composite specimens with joints filled resin and LC paste develop shear resistances that are approx. equal to 3.750 MPa and 4.350 MPa, respectively. These specimens failed mainly due to substrate failure. Sufficient shear resistance values are presented by the LMK50 (~2.202 MPa) and NHL mix (~1.700 MPa), which are accompanied by a ITZ failure. On the contrary, metakaolin paste LMK40 was unable to ensure sufficient τ_u (approxim. equal to 0.170 MPa). It should be noted that all LMK40 to stone B specimens failed along the interface.

3.5 Stone Type C-to-Mixes Interfaces

Besides resin composite specimens, LMK50 synthesis presents the best adhesive behavior for limestone type C. Moreover, the latter synthesis fulfils all the design criteria that were adopted for this investigation (i.e. sufficient bond strength, physico-mechanical compatibility with the authentic materials and failure of the repairing material). NHL and LMK 40 specimens at the age of one year show reduced shear strength capacity. Actually, for NHL mixes τu values equal to 0.770 MPa are recorded, whereas for LMK40 the relevant values are of the order of 0.170 MPa. As in the case of stone A specimens, this behaviour may be attributed to the extensive microcracking that was observed in NHL and LMK40 joints (Fig. 4a, 4b) (Fig. 6).

(a) (b)

Fig. 6. Microcracking of the joint of stone A to a) NHL and b) LMK40 composite specimens.

4 Conclusions

With this paper, the adhesive capacity of nine materials was investigated. For this purpose, shear tests were performed on composite substrate/adhesive materials specimens in 1,6 and 12 months of curing time. The main conclusions of this research are:

- Cement paste, hydrated lime/cement and epoxy resin provided an excellent bonding between the stone fragments (~~independently~~ regardless of the stone type), expressed by high values of shear strength (varying between 3.700–7.700 MPa). However, in all cases of substrate, the stone substrate has failed during the experiment.
- Natural hydraulic lime paste presented a sufficient shear strength and therefore a bonding strength for marble substrates (shear strength of the order of 2.2 MPa), without failure of the stone substrate. Analogous behavior was reported in the case of hydrated lime/metakaolin in mixing ratio 1/1 per weight (shear strength equal to 2.1 MPa).
- Regarding the Piraeus stone types B and C, these were sufficiently bonded together by using the paste of hydrated lime/metakaolin in mixing ratio 1/1 per weight. The shear strength measured was up to 2.20 MPa and 2.60 MPa for type B and type C respectively.
- Natural hydraulic lime and hydrated lime/metakaolin pastes presented low shear strength (<1MPa) in the case of the stone type A. For this type of stone cement, lime/cement and epoxy resin presented sufficient shear values but the stone substrate was failed during testing.
- Finally, the pastes presented a better adhesive capacity than the relevant mortars.

Acknowledgements. The present study was funded by the Acropolis Restoration Service (Hellenic Ministry of Culture and Sports). The stones (types A, B and C) were offered by the Ephorate of Antiquities of the City of Athens. The composite specimens were prepared by marble sculptor L. Michalakos and conservator A. Tsimereki. The contribution of N. Seferi and M. Skaza to the execution of the shear tests is gratefully acknowledged.

References

Adami, C.E.: Experimental and analytical investigation of high injectability hydraulic grout to stone or brick interfaces, PhD Thesis, National Technical University of Athens, p. 552 (2011). (in Greek)

Adami, C.E., Vintzileou, E.: Interventions to historic masonries: Investigation of the bond mechanism between stones or bricks and grouts. Mater. Struct. **41**(2), 255–267 (2008)

Adami, C.E., Toumbakari Vintzileou, E.: Lime-pozzolan-cement compositions for the repair and strengthening of historic structure. HMC08 Historical Mortars Conference, Lisbon, Portugal (2008)

Adami, C.E., Vintzileou, E., Moutselis, M.: Shear behaviour of interfaces between masonry units and hydraulic-lime grouts. HMC08 Historical Mortars Conference, Lisbon, Portugal (2008)

Aggelakopoulou, E.: Design and evaluation of restoration mortars for the Acropolis monuments. The Acropolis Restoration News. https://ysma.gr/en/open-access/publications/newsletter, pp. 13–17 (2013)

Aggelakopoulou, E., Bakolas, A., Moropoulou, A.: Properties of lime-metakaolin mortars for the restoration of historic masonries. Appl. Clay Sci. **53**, 15–19 (2011)

Bakolas, A., Aggelakopoulou, E., Moropoulou, A.: Evaluation of pozzolanic activity and physico-mechanical characteristics in metakaolin-lime paste. J. Therm. Anal. Calorim. **92**(1), 345–351 (2008)

EN 196–1:2016. Methods of testing cement. Determination of strength (2016)

Eleftheriou, V.: Protection of the rock and the circuit walls of the Acropolis: Works accomplished and future projects, The Acropolis restoration news, The Acropolis Restoration News, pp. 11–14 (2015). https://ysma.gr/en/open-access/publications/newsletter

Exadaktylos, G.E., Vardoulakis, I., Kourkoulis, S.K.: Influence of nonlinearity and double elasticity on flexure of rock beams, characterization of dionysos marble. Intern. J. Solids Struct. **38**, 4119–4145 (2001)

Ksinopoulou, E., Aggelakopoulou, E.: Investigation of the physico-chemical and the mechanical characteristics of porous stones - Compatible stone selection for the Acropolis wall - Revised version, Technical report, The Acropolis Restoration Service (YSMA) files (2018). (in Greek)

Moropoulou, A., Bakolas, A., Aggelakopoulou, E., Pineli, T., Prasianakis, I.: Estimation of elastic constants of stones, used in historic monuments, using ultrasonic technique and correlation to their microstructure parameters. In: 3rd International Conference on Non-Destructive Testing, pp. 242–245. Hellenic Society for NDT, Chania (2003)

Riccardelli, C., Soultanian, J., Morris, M., Becker, L., Wheeler, G., Street, R.: The treatment of tullio lombardo's adam: a new approach to the conservation of monumental marble sculpture, metropolitan museum journal. The Metropolitan Museum of Art, New York, pp. 49–116 (2014)

Skoulikidis, T.N.: Corrosion and Conservation of monuments' building materials. Crete University Press (2000). (in Greek)

Theoulakis, P., Bardanis, M.: The stone of piraeus at the monuments of the acropolis of Athens. In: Proceedings of the 9th International Congress on Deterioration and Conservation of stone, Venice, pp. 255–264 (2000)

Tsimereki, A.: Interventions in the Acropolis Walls – Conservation issues, Specialised research and implementation issues in the Acropolis restoration works carried out during the period of 2010–2015, Nov. 18–19, Athens, pp. 343–353 (2016). (in Greek)

Vintzileou, E., Adami, C.E.: The bond mechanism in stone or brick to grout interfaces. Strain **45**(5), 400–409 (2009)

Zambas, C.: Mechanical Properties of Pentelic Marble. Committee for the Restoration of Parthenon Publications, Athens (1994)

http://en.wikipedia.org/wiki/Sandstone

Condition Assessment of Cultural Heritage Buildings in the Historic Strand District of Galveston

Saadet Toker Beeson[1(✉)], Angela Lombardi[1], and Kasim A. Korkmaz[2]

[1] College of Architecture, Construction and Planning, Department of Architecture, University of Texas at San Antonio, San Antonio, TX, USA
Saadet.Beeson@utsa.edu

[2] College of Engineering and Technology, School of Visual and Built Envrionemnts, Eastern Michigan University, Ypsilanti, MI, USA

Abstract. The city of Galveston is located by the coast on the Galveston Island in Texas. Despite its flat and narrow land, the island became an important port in the history for thousands of people to enter the country and a significant market for trade and supplied labor to cotton planters. The city gained its architectural characteristics in the late 19th century as it grew to become a city of expanding commercial and shipping center. However, the hurricane of September 8, 1900, which is known as the deadliest natural disaster in North America, devastated the island by killing over six thousand people and destroying about one-third of the buildings. Hurricanes and tropical storms striking the coast along the Gulf of Mexico are known to have caused severe damage. Among the all, Harvey was the last destructive hurricane that affected the Galveston area in 2017. The community is still vulnerable, and in need of resilient design based on sound structures against hurricanes and storms. After the disasters, the economy has been adversely affected as well. The community has been in serious need for the revival to support the economy. The tourism industry of Galveston is based on leisure travel, conventions, and tourism. The city has an invaluable cultural asset. The majority of the cultural heritage is located at the East End Historic District, Silk Stocking National Historic District, Lost Bayou Historic District and Strand/Mechanic Historic District. Among these districts, the Strand hosts a great many number of historic buildings that are still in use today. Most of these buildings were damaged in the 1885 fire and numerous hurricanes and tropical storms. This research presents an investigation of the historical building profile of the Strand District. Historical structures have been investigated via the Rapid Visual Inspection approach. Based on the results, a condition assessment is obtained and a behavioral pattern is defined for the historic structures of the Stand District of Galveston.

Keywords: Historical building profile in Strand-Galveston · Iron facade buildings · Assessment of historic structures · Rapid visual inspection

I. Vayas and F. M. Mazzolani (Eds.): PROHITECH 2021, LNCE 209, pp. 816–827, 2022.
https://doi.org/10.1007/978-3-030-90788-4_63

1 Introduction

Galveston has one of the finest collections of historical buildings from the 19th century. Taking advantage of the deep waters of the gulf, the island became an important port of entry for thousands of people and a significant market for trade and supplied labor to cotton planters.

The last quarter of the 19th century, the growth of Galveston was exponential. With a population of 37,000, the city was the center of trade, banks, churches and other public buildings (Alexander 2010). It became the second richest city only after Providence, Rhode Island (Shanks 1998). However, several hurricanes and big fires thwarted the victory of the city several times handing the port trade over to Houston. The fire in 1885 destroyed more than 40 blocks of buildings. The big hurricane in 1900 killed more than 6,000 people and destroyed many buildings. An important portion of the population abandoned the city after this big disaster. It also led to construction of 5 m tall and 6 km long sea wall, later extended for another six and a half miles. The initial construction was completed in July 1904. The process required altering the entire topography of the island and raising all the buildings from the shore to downtown. The Seawall performed well for the next hurricanes to protect the water from penetrating into the island; but the water, running around the Seawall, flooded the downtown area for an extended time due to the bowl shape that the city resumed after the radical change in topography.

The 1900 hurricane was not the only disaster that hit the region. The storms and hurricanes that happened in 1915, 1943, 1961, 1983, 2001, 2002, 2003, 2005, 2008 and 2017 damaged or destroyed an important portion of the buildings as well, some of which are irreplaceable due to their historic value. Table 1 summarizes the recent hurricanes affecting Galveston. In the meantime, the city also went through the Civil War Battle of Galveston and the Great Depression.

Table 1. List of recent hurricanes in the gulf area affecting galveston

Name	Date
Harvey	August-September 2017
Cindy	June 2017
Ike	September 2008
Edouard	August 2008
Rita	September 2005
Grace	August-September 2003
Claudette	July 2003
Fay	September 2002
Bertha	August 2002
Allison	June 2001

Galveston holds an important place in the historical and cultural context. The harbor provided a great advantage for the growth and development of the city in a short time. Though the city plan developed in the 1838 defined arrow lots for residential units, mansions were built for wealthier residents. There was also an increased interest in beautifying the business district by adding iron facades to the buildings. They were mostly used for decorative purposes in an attempt to characterize the business district, which is mainly known as the Strand.

This study provides an investigation of the historical building profile located Strand, Galveston. Since the city is constantly under hurricane and tropical storm risk, historical building preservation is very important. All of the buildings in the area sustained various loading cases. Cracking in walls can be detrimental to the overall expected performance of these types of buildings. Diagonal "stair-step" or through-brick cracking with widths greater than 1/8″ would be captured for additional investigation by checking the "Significant Damage/Deterioration to the Structural System" per FEMA (FEMA P-154 2015). Cracks over at least 25% of the length of a member are sufficient to warrant a Detailed Structural Evaluation. In investigated buildings, there is no cracks were defined with their current conditions. However, these buildings have been repaired in various cases over the years.

With rapid and detailed investigations for existing buildings, protecting the historical buildings, and helping them maintaining their original characteristics, and preparing them for future disasters is more important than ever. In the present research, historical buildings located in the Strand have been assessed and structurally investigated using the Rapid Visual Inspection Method. Based on such investigation, a condition assessment for the selected historic buildings has been defined.

2 Characteristics of Galveston and the Historical Strand District

Following the independence from Spain, Congress of Mexico established the Port of Galveston in 1825. Having been named after the Count of Galvez, Bernardo de Galvez, and the city served as the main port for Texas Navy during the Texas Revolution (World Heritage Encyclopedia 2019).

The first European settlements on the island were constructed in the early decades of the 19th century. In 1838, a city plan with very precise grid pattern was developed as given in Fig. 1 (Drie 1871). The narrow lots, which were probably the result of scarce land, shaped the architectural characteristics of the city. The city also hosts large mansions built for the wealthier population that owned the businesses. However, the big fire in 1885 destroyed hundreds of buildings in about forty blocks. Majority of the houses were built again within a year, bringing a more uniform look to the entire town (Beasley and Fox 1996).

The second half of the 19th century, Galveston as an expanding commercial and shipping center flourished with its economy. The built environment is the first place to see the effects of the rapid growth with a wide collection of buildings representing the Victorian, early Renaissance, Classical and Gothic periods. Iron became very popular in the construction of commercial buildings very soon after its use in the northeast part of the country. Though the applications were decorative rather than structural, it was still an indication of the wealth and refinement that the city reached in such a short period of time.

Contributing to the architectural style of the city in their own way, these commercial buildings has been cohesive and adaptable in all these years despite the major hurricanes and fires. Some of these occasions caused buildings getting in serious need of help, most of these structures have been keeping their conditions representing the glorious times of the city even though they are now in use for completely different functions than what they were built for (Shanks 1998).

Fig. 1. Bird's eye view of the city of galveston, 1871 (Drie 1871)

The Strand is the area between the Water Street on the north, alley between the Avenues C and D on the south, 20th on the east and 25th on the west. The main street in the Strand District holds many buildings of shops, boutiques, retail outlets and art galleries. Majority of the buildings hosts bars, restaurants, stores, and old-fashioned ice cream and candy shops at the entrance levels have and provide studios, lofts or offices on the upper floors (Buff 2003).

These buildings evolved in a very short time after iron was started to be used in the northeastern part of the country. Using iron in construction was advantageous in so many ways. It could be added to the facade while construction was going on at the other parts of the structure. When used structurally, it let more light in and provided more room due to slender member sizes. Structural use of iron was limited to the first floor only while decorative use focused on the first floor shop windows, decorative members such as lentils, medallions and bandings in combination with iron columns and cornices (Shanks 1998). It is still possible to see original iron members, whether decorative or structural in most of these structures even after all the disasters and after the buildings have been repurposed as the street still demonstrates the glorious days of the city of Galveston (National Park Service 1976).

3 Historical Building Profile in the Strand

30 buildings in total have been investigated to determine the overall condition of historic structures in the Strand. For the presented research, 9 buildings were selected as a case study. Selection of the buildings were based on their size, current overall conditions based on the damage level from previous storms and floods, and base-to-height ratios to determine their possible resistance to wind and flood. The names, locations and the construction dates of the selected buildings are given in Table 2.

Table 2. Selected buildings on the strand

No	Name	Date	Location
1	Hutchings, Sealy and Company Bld.	1895–97	2326–2328 Strand
2	Hendley Row	1859	2002–2016 Strand
3	Rosenberg Bld.	1871	2309–2311 Strand
4	George Schneider and Co. Bld.	1878	2101–2107 Strand
5	Clara Lang Bld.	1878	2109 Strand
6	J.S. Brown and Co. Bld.	1878	2111–2113 Strand
7	Thomas Jefferson League Bld.	1871	2301–2307 Strand
8	Greenleve Block and Co. Bld.	1882	2310–2314 Strand
9	First National Bank Bld.	1878	2127 Strand

3.1 Hutchings, Sealy and Company Building

Built in 1895, The Hutchings, Sealy and Company Building is among the finest designs of Nicholas Clayton, who is one of the first local professional architects in Texas. The three-story structure comprises of two adjoining buildings. It has an atrium and a skylight inside. The interior of the building is heavily remodeled.

Fortunately the vaulted ceilings, one of the features that make the structure unique, are still visible despite the modern partitions and additions. The pink colored exterior bricks are flaking due to the damage from the previous flood damage (Fig. 2).

Fig. 2. The hutchings, sealy and company building

3.2 Hendley Row

One of the first businesses on the Strand, Hendley Row comprises of four buildings separated by fire walls. This building is an early example of iron use in the interior spaces. The exterior columns at the façade are granite while the interior columns are cast iron. The upper floors on the façade are covered by brick veneered walls creating an uninterrupted horizontal look on the entire block (Shanks 1998). The granite columns on the façade sustain longitudinal cracks through the entire height of the first floor (Fig. 3).

Fig. 3. The Hendley Row

3.3 Rosenberg Building

The Rosenberg Building has square pillars supporting the round arches on the façade. The pillars are brick covered with iron plates on three sides. The cornice separates the iron from the masonry at the top. Iron and brick are the interior structural members. There is also an atrium to let the sunlight in. The upper floors host residential lofts (Shanks 1998) (Fig. 4).

Fig. 4. The rosenberg building

3.4 Schneider Building

Schneider Building is another example that uses cast iron columns at the interior for structural support (Alexander 2010). The building is covered in plaster. It is possible to see cracks on the plaster, which are classified as cosmetic damage (Fig. 5).

Fig. 5. The schneider building

3.5 Clara Lang Building

The Clara Lang Building was constructed as a three-story building. However, the third floor was destroyed by the 1900 Storm. It features a little iron decoration at the front façade and it is designed to be in harmony with the building next door (Shanks 1998; Alexander 2010). The Clara Lang Building sustains material deterioration and discoloration due to water penetration (Fig. 6).

Fig. 6. The clara lang building

3.6 J. S. Brown and Co. Bld.

This building shows heavy use of iron to decorate its façade. Heavy vegetation has been observed at the cornice level. There are also several vertical cracks on the iron pillars and architrave (Fig. 7).

Fig. 7. J.S. Brown and Co. Building

3.7 Thomas Jefferson League Building

Built in 1872, the Thomas Jefferson League Building is one of the four buildings that made up the Insurance Square (Shanks 1998). It features iron arcades on the first floor level, which has sidewalk-level storefronts (Fig. 8).

Fig. 8. Thomas jefferson league building

3.8 Greenleve, Block and Company

Greenleve, Block and Company building is another structure designed by Nicholas Clayton in 1882. It is built of stone and brick, and it features iron columns (Fig. 9). The façade has oval and quatrefoil cutouts. The elaborate cornice that the building used to have over the 4th floor was destroyed in the 1900 hurricane. Also the 4th floor was removed after Hurricane Carla of 1961.

Fig. 9. Greenleve, block and company building

3.9 First National Bank Building

The First National Bank Building was Texas's oldest chartered bank. It was built in 1878 as a corner building. It has Corinthian cast iron columns on the front façade. Today it serves as the building for Galveston Art Center. It is heavily remodeled at the interior but still displaying the original design (Fig. 10).

Fig. 10. First national bank building

4 Historical Building Profile Determination in the Strand

To investigate the historical building profile of the buildings in Strand, Galveston, Rapid Visual Inspection (RVI) method was used. There are several of inspection techniques are available from various sources as given in ATC 21 (2001) and 156 (2001) other references. In the present study, ATC 21 has been considered as the base for the RVI methodology for evaluation of historic structures of the Strand.

The objective of this methodology is to assess conditions of the specified historic structures in a given area quickly and with minimum labor and effort. Rapid evaluation is designed to utilize the talents and experience of building inspectors and other people with similar experiences. The rapid evaluation method is designed for the use of individuals with some experience in general building design and construction techniques. Material knowledge is also important for historical buildings. Inspectors are generally selected from architects, civil engineers, construction engineers.

Rapid evaluation procedure begins with the inspection of the subjected buildings area. Then the envelope of the structure in question should be assessed. The inspectors need to walk around the entire structure. Ordinarily, only the exterior of the building is inspected at this phase, unless there is a suspected or reported problem. This is done primarily to maximize the number of the inspections in the immediate post-event period.

In Table 2, Matrix of RVI Applications is given (ATC 21 2001). Table 3 gives RVI Probability Scale in the historic structures on the Strand. In Table 3, probability damage is given through three parameters. These parameters are condition of structural members, structural risk and visual strength (Karanth and Mishra 2014). For the selected buildings Application levels are chosen as 4 for all buildings that indicating importance of Developing Building Vulnerability Information. Probability Scale for scaling the buildings are given in Table 4. In Table 5, Condition of Structural Members, Structural Risk and Visual Strength are given in chart for strand (Karanth and Mishra 2014).

Table 3. Matrix of RVI applications

Application	RVI
Ranking seismic rehabilitation needs	1
Designing seismic hazard mitigation programs	2
Developing inventories for Planning post-disaster safety	3
Developing building vulnerability information	4

Table 4. RVI probability scale in the historic structures on strand

RVI Probability Scale			
RVI	C	B	A
Condition of structural members	0–59	60–79	80–100
Structural risk	0–59	60–79	80–100
Visual strength	0–59	60–79	80–100

Table 5. Evaluation of the buildings on the strand

No	Building name	Condition of structural members	Structural risk	Visual strength
1	Hutchings, Sealy and Company Bld.	B	B	B
2	Hendley Row	C	B	B
3	Rosenberg Bld.	A	A	A
4	George Schneider and Co. Bld.	B	B	B
5	Clara Lang Bld.	B	B	B
6	J. S. Brown and Co. Bld.	C	B	B
7	Thomas Jefferson League Bld.	B	B	B
8	Greenleve Block and Co. Bld.	A	A	A
9	First National Bank Bld.	A	A	A

After defining structures in Strand in Table 5, Scale for Critical Scores are defined in Table 6. In Table 6, (a) human exposure, (b) building importance and (c) environmental effects on the buildings. Degree is defined throughout Low-critical to Moderate-critical for Critical-Benchmark Score Modifiers in Moderate seismicity. Other parameters such as (d) Design Value, (e) Over strength, (f) ductility, (g) Demand, (h) probability factor have similarly degraded in segmental trends with varying ratios of the associated Critical value. Table 7 summarizes assigned scores for buildings in Strand.

Table 6. Scale for critical scores

Critical scores	Scale				
		I	II	III	IV
Building importance	PA	0–25	26–50	51–75	76–100
Design Value	PB	0–25	26–50	51–75	76–100
Strength	PC	0–25	26–50	51–75	76–100
Probability factor	PD	0–25	26–50	51–75	76–100

Table 7. Critical score evaluation for buildings on strand

No	Building name	PA	PB	PC	PD
1	Hutchings, Sealy and Company Bld.	II	II	II	III
2	Hendley Row	III	III	II	III
3	Rosenberg Bld.	III	III	II	III
4	George Schneider and Co. Bld.	III	II	III	III
5	Clara Lang Bld	III	III	III	III
6	J. S. Brown and Co. Bld.	III	II	II	III
7	Thomas Jefferson League Bld.	II	III	II	III
8	Greenleve Block and Co. Bld.	II	III	III	II
9	First National Bank Bld.	II	III	III	II

5 Conclusions

In this study, significance of historical structures in Galveston-Strand area regarding the use of iron as a modern material of the 19th century has been discussed. Strand Street is the home to more than 30 historic structures built in the 19th century. These buildings have sustained damages due to fires and various hurricanes and storms in the region. Since the area is prone to storms and hurricanes the historic structures in question are under high risk and their safety is a concern.

In this study, the historic building profile of the Strand has been investigated. Throughout the investigation, RVI was used to evaluate the selected buildings. Table 5 summarizes Condition of Structural Members, Structural Risk and Visual Strength are given in chart and Table 7 summarizes assigned scores for buildings in Strand (Karanth and Mishra 2014). As a result of RVI evaluation, most of the existing buildings are providing enough resistance for external extreme loading. However, building should be evaluated in detail with Finite Element Analysis to define the risk of structural damage. Since these buildings are important, it is important to carry these buildings to next generations.

References

Alexander, D.: Galveston's Historic Downtown and Strand District. Arcadia Publishing, Charleston, South Carolina (2010)

ATC-21 and ATC-156: Applied Technology Council, California, USA (2001)

Buff, B.: The strand national historic landmark district. Galveston historical foundation. Publ. Historian **25**(4), 118–121 (2003)

Beasley, E., Fox, S.: Galveston Architecture Guide Book, Rice University Press, Houston, and Galveston Historic Foundation (1996)

City of Galveston, Historic Districts. http://www.cityofgalveston.org/DocumentCenter/Home/View/338. Accessed December 2019

City of Galveston, Galveston, Texas History with Tropical Systems. http://www.hurricanecity.com/city/galveston.htm. Accessed December 2019

Drie, C.N.: Bird's Eye View of the City of Galveston (1871). http://www.birdseyeviews.org/zoom.php?city=Galveston&year=1871&extra_info=. Accessed January 2020

FEMA P-154: Rapid Visual Screening of Buildings for Potential Seismic Hazards (2015). https://www.fema.gov/media-library-data/1426210695633-d9a280e72b32872161efab26a602283b/FEMAP-154_508.pdf

Karanth, A., Mishra, S.: TARU, A Guide Book for Integrated Rapid Visual Screenings of Buildings, Integrated Rapid Visual Screening of Buildings for Seismic Hazard, Gurgaon, India (2014)

United States, Department of the Interior, National Park Services, National Register of Historic Places Inventory Nomination Form. https://npgallery.nps.gov/NRHP/GetAsset/NHLS/70000748_text. Accessed January 2020

United States, Department of Commerce, NOAA. https://coast.noaa.gov/hurricanes/. Accessed January 2020

World Heritage Encyclopedia, Galveston, Texas. http://www.worldheritage.org/article/WHEBN0000053840/GalvestonTexas. Accessed 05 Dec 2019

Structural Evaluation of Historical Churches in Gumushane

Kasim A. Korkmaz[1] and Saadet Toker Beeson[2](✉)

[1] College of Engineering and Technology, School of Visual and Built Environments, Eastern Michigan University, Ypsilanti, MI, USA
[2] College of Architecture, Construction and Planning, Department of Architecture, University of Texas at San Antonio, San Antonio, TX, USA
saadet.beeson@utsa.edu

Abstract. Anatolia is a rich and a vibrant land with many historical buildings including churches. The historical churches in Northeast Anatolia are under risk due to frequent, high-magnitude earthquakes. This part of Anatolia is a secluded mountainous region with several plateaus and rugged terrain. Historical churches are important cultural heritage for all of the world. They often lack sufficient structural performance in seismic resistance. In recent earthquakes, some of these churches sustained damages in various parts. In majority of the cases, poor seismic performance was attributed to the degrading effects of aging along with poor maintenance. Frequency of earthquakes and currently poor seismic performance of these historic buildings necessitate the seismic risk assessment. Such seismic risk assessment is also essential to proper restoration and renovation projects. The present research provides a detailed investigation of the historical churches in North East Anatolia. The study includes the application of a rapid visual inspection methodology on selected churches located in Gumushane. ATC 21 methodology was used for seismic risk assessment and evaluation of selected churches in Gumushane. Eight historical churches were investigated in the research using ATC approach. As a result of the investigation, structural assessment has been carried out for the historical buildings.

Keywords: Seismic risk assessment · Historical churches · North east anatolia · Rapid visual inspection

1 Introduction

Anatolia is a land rich with a diverse historical background. Many civilizations have left various traces over the centuries. Churches, mosques and all other historical buildings comprise the main part of the historical heritage. Christian civilizations have built churches in various locations of the country in different eras from the history. A few of these buildings are located at the northeastern part of Anatolia from historical era dating back to the 10th century BC.

I. Vayas and F. M. Mazzolani (Eds.): PROHITECH 2021, LNCE 209, pp. 828–838, 2022.
https://doi.org/10.1007/978-3-030-90788-4_64

Gumushane, which is seen in the records as Gyumushkhana Arghyropolis, is both the name of the city and the capital district of Gumushane Province in the Black Sea region. The city lies along the Harshit River, at an elevation of 1,500 m, about 60 km south-west of Trabzon. In the recent history, the Arghyropolis was recorded as to belong to the Diocese of Chaldie. Historical traces from the past civilizations revealed the social, archaeological, aesthetic, economic, political, architectural and technical features of their times. The region hosted Hurri, Urartu, and Scythian civilizations before it was conquered multiple times by Arabs and Byzantines. It is documented that Seljouk Turks entered the region in the 11th century though there is no information about the date of the Ottoman defeat. Russians and consequently Georgians dominated the region for short periods until 1921. Due to the variety in the cultures and civilizations hosted by the region, the historical heritage in the region is invaluable.

The historical background of Gumushane dates back to 3000 BC due to rich silver mines. Despite the very old history, the amount of architectural works in and around the city is quite limited. The reason for the lack of architectural examples of early ages could be seen as the fact that the province remained as a buffer zone because of its geographical conditions. On the other hand, it might not be pretentious to claim that Gumushane has the highest number of churches among the other cities in Anatolia. The city first hosted the Hittites. Enjoying considerable power in the Middle Anatolia, Urartu lasted in the region from 860 BC to 550 BC until the dominancy of Alexander the Great. The Persians brought the end of Alexander the Great, but they could not reach until Gumushane. From 301 BC to 20 AC, Gumushane was the territory of the Pontus Empire. From 20 to 395 it was the time of the Roman Empire. The city seems to have hosted either the Arabians or the Byzantines from 7th to 11th centuries. In 1016 the province was taken by the Turks. The Mongol Empire took the domain from the Seljouks in the 13th century. The Ottoman Empire lasted in the region from 1514 to 1916.

2 Seismic Characteristics of Gumushane

Gumushane is a typical Black Sea city covered with hazelnuts that extend for kilometers within the greenery. It is very historical town famous for its natural beauties and historical heritage (Black Sea Region 2020). On the other hand, the region is seismically very active. North Anatolian fault line is passing through the region. There are many historical earthquake records affecting the historical buildings. In 1999, two major earthquakes hit that region and damaged and destroyed many existing buildings along with more than 50,000 life lost. North Anatolian Fault Line, which is one of the most important seismic sources in Anatolia, constitutes the northern boundary of the Anatolian plate, which constantly moves. Most of the devastating earthquakes in this region occur close by this fault line. Accordingly, seismic safety of the historical buildings in this region also is of concern for earthquake engineers (Modena et al. 2002). Figure 1a gives aerial view of the region. Figure 1b gives North Anatolian Fault Line lying on the region (Google Map 2020; DEMP 2019).

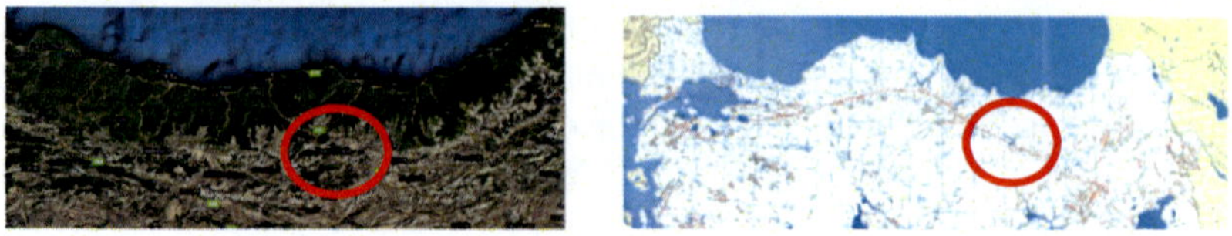

(a) (b)

Fig. 1. Aerial View of Gumushane (a) and North Anatolian Fault Line (b)

3 Risk Assessment for the Churches in Gumushane

This study provides an investigation of the historical building profile mainly formed with churches and located in Gumushane and its neighborhood. During the study, we investigated various churches in the region and structurally investigated these buildings. All of these historical buildings in the area are rigid masonry buildings built with stone and have sustained various earthquakes over the centuries. We detected many cracks and failures in structural walls of the historical buildings. Cracking in walls can be detrimental to the overall expected performance of these types of buildings. Diagonal "stair-step" or through-brick cracking with widths greater than 1/8″ would be captured for additional investigation by checking the "Significant Damage/Deterioration to the Structural System" per FEMA (FEMA P-154 2015). Cracks over at least 25% of the length of a member are sufficient to warrant a Detailed Structural Evaluation.

With rapid and detailed investigations for existing historical buildings in Gumushane, a heritage protection is targeted for the region. It is very critical to protect these historical buildings as a world heritage. Once they are investigated, they will be defined with their original characteristics. With further studies, they will be prepared for future possible earthquakes, which is more important to carry them for future generations. In summary, in the present research, historical buildings located in Gumushane have been assessed and structurally investigated using the Rapid Visual Inspection Method. Based on such investigation, a risk assessment for these historic buildings has been defined.

4 Rapid Visual Inspection for Selected Historical Churches

In this study, Seismic Risk Assessment was carried out for historical churches in North East part of Anatolia. Rapid Visual Inspection (RVI) methodology from ATC 21 was carried out for historical buildings. The method was applied to the selected churches. Eight different churches are considered and RVI methodology is applied on these buildings to understand the structural condition. This study evaluates an earthquake risk regarding with structural conditions of the existing historical churches. In the existing literature, there are various Rapid Visual Inspection (RVI) methods available from various sources as given in ATC 21 and 156 some other references (ATC-21,156 2001).

The objective of the methodology is to inspect and evaluate historical churches in the subjected area quickly and with minimum time and effort. Rapid evaluation is designed to utilize the talents and experience of building inspectors and other people with similar experience. This does not preclude the possibility of using experienced structural engineers. Once all buildings in a given area have been inspected and those are apparently safe or obviously unsafe have been posted, the remaining structures are left for detailed evaluation by a structural engineer. The rapid evaluation method is designed for use by individuals with some experience in general building design. This includes building inspectors in particular as well as volunteer civil engineers and other who have been involved in building design and construction. In this study, ATC 21 was used as RVI methodology for seismic risk assessment of historical churches in North East part of Anatolia. Churches were selected from the region based on their existing conditions. Each church was evaluated using six basic RVI criteria given below. These are used to rate the building condition with respect to nature of building especially in case the building has the historical value (ATC-21 2001). Rapid Visual Inspection Methodology Procedure could be given as below and also in Fig. 2 (ATC-21 2001);

1. Examine the outside of the structure entirely,
2. Examine the ground in general area of the structure for fissures, bulged ground, or signs of slope movement,
3. Ordinarily enter a building only when the structure cannot be viewed sufficiently from the outside,
4. Evaluate the structure according to condition of the structure and complete the RVI. Doubtful buildings should be left to detailed evaluation,
5. Post the structure according to the result of the evaluation. Post every entrance to a building classified limited entry or unsafe,
6. Explain the significance of limited entry or unsafe postings to building occupants and advise them to leave immediately.

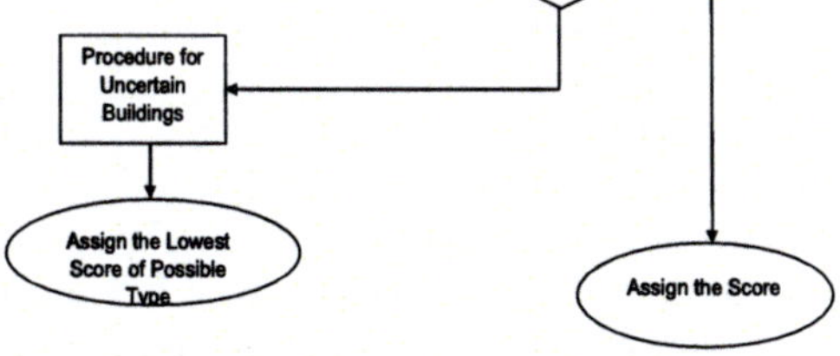

Fig. 2. Flow Chart of RVI methodology

Gumushane Mother Maria Church: The church (Fig. 3), which is located in the Kurtun Province-Yaylali Village is a work of 19th century. It is rectangular in plan and has one nave and one abscissa. Constructed of cut-stones, the structure is covered by a steep triangular roof at the top. There are no any ornaments in the church.

Gumushane Sariun Church: The Sariun Church (Fig. 4), which was constructed in the 19th century, is located in Yaglidere Village, Krom Valley, Sariun District. The building is rectangular in plan with three naves and three abscissas. Constructed on a scenery hill, the top cover of the church was demolished years ago. The only means of ornament of the church is seen to be the motif of a cross on the entrance door.

Gumushane Harmancik Church: The building (Fig. 5) is located in the Torul Province, Harmancık Village, Mesehor District. It is an Armenian church, which is believed to have been built in the second half of 19th century. The building is constructed of regular cut stones on the western wall, and rough stones. It has a rectangular layout.

Gumushane Ugurtas Church: The building (Fig. 6) is in Torul Province, Ugurtas Village. The construction date is snot known. The building is rectangular in plan and has three naves and three abscissas. There are 14 windows around the building. The top cover consists of a barrel vault and a hexagonal dome, which is partially demolished. There take place 6 columns and arches in between, to support the roof.

Gumushane Mandiri Church: Mandiri Church (Fig. 7) is located in Mescitli village at 3 km drive from city center. It was built in 19th century as a Greek church. Plan of the church is rectangular. It has a triangular roof over the rectangular shape which is along with environmental conditions of the city. Entrance is a simple rectangular. This church is a good representative of minimalism in that era. There is a narthex at the entrance.

Gumushane Zemberek Church: Zemberek Church (Fig. 8) of Gumushane is located in Yaglidere village. This one was constructed in the 19th century as a Greek church. It has a rectangular shape with its single Nave and apsidal. The roof is triangular shape in its original design however it is not existing in its current form.

Gumushane Ayvalos Church: Ayvalos Church (Fig. 9) is in the city center. It is located on the way towards to Arnavutlu from Mecitli. It is a small one with rectangular shape. The entrance is rectangular and small. It is structurally stable but does not have any sectional properties inside of the building.

Gumushane Hagios Stephanos Church: The church (Fig. 10) is located in Suleymaniye village and was constructed in the 18th century. It was built with three Naves Basilica plan. This church has architectural similarities with other churches in old downtown of Gumushane. The building has apsidal outing to the main body. The roof has two sided roof to support its main body which is considered as a big lay over for its era.

Fig. 3. Gumushane Mother Maria Church

Fig. 4. Gumushane Sariun Church

Fig. 5. Gumushane Harmacik Church

Fig. 6. Gumushane Ugurtas Church

Fig. 7. Gumushane Mandiri Church

Fig. 8. Gumushane Zemberek Church

Fig. 9. Gumushane Ayvalos Church

Fig. 10. Gumushane Hagios Stephanos Church

These selected eight historical buildings were investigated with Rapid Visual Inspection and results gathered from the inspection is summarized in Table 1. Table 1 summarizes risk of collapse for the historical buildings. This risk is defined based on the structural assessment and structural members of the buildings. Majority of the buildings got damages over the years and their main body is exiting with severe damages. This damage occurrence should be taken care of for future earthquakes for all buildings.

Table 1. Rapid visual inspection for selected historical buildings

No	Churches	Damage levels	Under protection	Restoration	Entrance control	Structural members	Risk of collapse
1	Gumushane Mother Maria	Mid	No	No	No	Existing	Mid
2	Gumushane Sariun	Mid	No	No	No	Existing	Mid
3	Gumushane Harmancik	High	No	No	No	Damaged	High
4	Gumushane Ugurtas	Mid	No	No	No	Damaged	High
5	Gumushane Mandiri	High	No	No	No	Damaged	High
6	Gumushane Zembelek	Mid	No	No	No	Existing	Mid
7	Gumushane Ayvalos	Mid	No	No	No	Existing	Mid
8	Gumushane Hagios Stephanos	High	No	No	No	Damaged	High

5 Structural Behavior Estimation for the Churches

To define structural behavior of historical buildings in Gumushane, Rapid Visual Inspection (RVI) methodology was used. ATC 21 (2001) and 156 (2001) are mainly used for this definition. The objective of this methodology is to assess conditions of the specified historic buildings in a given area quickly and with minimum labor and effort. Rapid evaluation is designed to utilize the talents and experience of building inspectors and other people with similar experiences. The rapid evaluation method is designed for the use of individuals with some experience in general building design and construction techniques. Material knowledge is also important for historical buildings (Sorace and

Terenzi 2002). Inspectors are generally selected from architects, civil engineers, construction engineers. Table 2 gives RVI Probability Scale. In Table 3, probability damage is given through three parameters. These parameters are condition of structural members, structural risk and structural strength (Karanth and Mishra 2014).

Table 2. RVI probability scale in the historical buildings

RVI	C	B	A
Condition of structural members	0–59	60–79	80–100
Structural risk	0–59	60–79	80–100
Visual strength	0–59	60–79	80–100

Table 3. Evaluation of the historical buildings

No	Building name	Condition of structural members	Structural risk	Structural strength
1	Gumushane Mother Maria	B	C	B
2	Gumushane Sariun	B	C	B
3	Gumushane Harmancik	C	C	C
4	Gumushane Ugurtas	C	C	C
5	Gumushane Mandiri	C	C	C
6	Gumushane Zembelek	B	C	B
7	Gumushane Ayvalos	B	C	B
8	Gumushane Hagios Stephanos	C	C	C

After Evaluation of Historical building, structural behavior estimation is determined by using Table 4. Table 4 gives RVI Probability Scale in Masonry Buildings. In Table 4, probability damage is given through 4 parameters. These parameters are Damageability Grading, Structural Risks, Visual Strength, and Spectral Hazard Risk for masonry buildings. These parameters are explained in Table 5 and Table 6. Table 5 gives Damage Grading and Structural Risks for masonry buildings in RVI Methodology. Spectral Hazard Risk for masonry buildings in RVI Methodology is also given in Table 5. Table 6 gives Visual Strength for masonry buildings in RVI Methodology. Table 7 presents final RVI Probability Matrix for Selected Churches.

Table 4. RVI structural behavior estimation scale in masonry buildings

Masonry buildings	RVI probability scale				
Damageability grading	G5	G4	G3	G2	G1
Structural risk	B5	B4	B3	B2	B1
Spectral hazard risk	R5	R4	R3	R2	R1
Visual strength	–	–	–/+	+	+

Table 5. Damage grading and structural risk faktors for masonry buildings

Damage grading	Levels	Structural risk	Levels	Spectral hazard risk	Levels
G5	Severe damage	B5	Severe risk	R5	Severe spectral hazard risk
G4	Heavy damage	B4	High risk	R4	High spectral hazard risk
G3	Moderate damage	B3	Moderate risk	R3	Moderate spectral hazard risk
G2	Slight damage	B2	Slight risk	R2	Slight spectral hazard risk
G1	No damage	B1	No risk	R1	No spectral hazard risk

Table 6. Visual strength for masonry buildings in RVI methodology

Visual strength	Levels
–	No strength
–/+	Low strength
+	Acceptable level of strength

Table 7. RVI probability matrix for selected churches

No	Churches	Damage grading	Structural risk	Spectral hazard risk	Visual strength
1	Gumushane Mother Maria	G2	B2	R2	+
2	Gumushane Sariun	G2	B2	R2	+
3	Gumushane Harmancik	G3	B3	R2	–
4	Gumushane Uguras	G3	B3	R2	–
5	Gumushane Mandiri	G3	B3	R2	–
6	Gumushane Zembelek	G2	B2	R2	–/+
7	Gumushane Ayvalos	G2	B2	R2	–/+
8	Gumushane Hagios Stephanos	G4	B3	R2	–

6 Conclusions

In this study, the importance of the historical buildings in North East part of Anatolia was emphasized. This region is very important in terms of historical and natural values. Since the region is seismically very active at the same time, the existing historical buildings in North East part of Anatolia are under high risk and safety of these historical buildings is a concern. It is important to investigate these buildings' safety against earthquake and disaster risks. There are significant structural damages on the walls and heavy diagonal cracks on the wall structural systems. These churches' structural members are pretty old and got severe damages over the years. Throughout the investigation, RVI was used to evaluate the selected buildings. Tables 1 and 3 summarizes Condition of Structural Members, Structural Risk and Visual Strength. Table 7 presents final RVI Probability Matrix for Selected Churches.

As a conclusion, with RVI evaluation, most of the existing buildings are providing insufficient resistance for seismic loading. Historical churches in North East part of Anatolia under structural damage risk should be consider as a cultural heritage to protect for future generations. Some of them are in need of structural strengthening and must be checked with structural analytical damage assessment methodologies. Moreover, these buildings should be evaluated in detail with Finite Element Analysis to define the risk of structural damage (Mele et al. 2003). Since these buildings are important for the world, it is essential to keep these buildings standing for future.

References

ATC-21 and ATC-156: Applied Technology Council, California, USA (2001)

Black Sea Region (2020). https://en.wikipedia.org/wiki/Black_Sea_Region

DEMP, Disaster and Emergency Management Presidency, Prime Ministry (2019). http://www.deprem.gov.tr/sarbis/depbolge/artvin.gif

FEMA P-154: Rapid Visual Screening of Buildings for Potential Seismic Hazards (2015). https://www.fema.gov/media-library-data/1426210695633-d9a280e72b32872161efab26a602283b/FEMAP-154_508.pdf

Google Map (2020). https://www.google.com/maps/place/Turkey/@40.599042,38.2984601,583965m/data=!3m1!1e3!4m5!3m4!1s0x14b0155c964f2671:0x40d9dbd42a625f2a!8m2!3d38.963745!4d35.243322

Karanth, A., Mishra, S.: TARU, A Guide Book for Integrated Rapid Visual Screenings of Buildings, Integrated Rapid Visual Screening of Buildings for Seismic Hazard, Gurgaon, India (2014)

Modena, C., Valluzzi, M.R., Folli, R.T., ve Binda, L.: Design choices and intervention techniques for repairing and strengthening of the Monza cathedral bell-tower. Constr. Build. Mater. **16**, 385–395 (2002)

Mele, E., Luca, A.D., ve Giordano, A.: Modelling and analysis of a basilica under earthquake loading. J. Cult. Heritage **V4**, 355–367 (2003)

Sorace, S., Terenzi, G.: Seismic evaluation and retrofit of historical churches. Struct. Eng. Int. Sci. Technol. **12**, 283–288 (2002)

Seismic Retrofit of RC Buildings Using Metal Shear Panels

Gianfranco De Matteis[1(✉)], Massimiliano Ferraioli[2], and Angelo Lavino[2]

[1] Department of Architecture and Industrial Design, Università Della Campania "Luigi Vanvitelli", Via San Lorenzo, 81031 Aversa, Italy
gianfranco.dematteis@unicampania.it

[2] Department of Engineering (DI), Università Della Campania "Luigi Vanvitelli", Via Roma 9, 81031 Aversa, Italy
{massimiliano.ferraioli,angelo.lavino}@unicampania.it

Abstract. The application of damped bracing systems is seriously limited by the lack of a standardized design procedure. The traditional methods appropriate for conventional structures (such as the use of a behaviour factor) become worthless. On the other side, the nonlinear response history analysis is complex and computationally demanding and its implementation is generally avoided in current practice. In this paper, an appropriate design methodology is developed to design the metal shear panels in relation to the structural capacity of the RC building. The metal shear panels are designed such that the distributions of drift angle and ductility demand of the retrofitted building become uniform, although those of the existing RC building may be non-uniform. In order to prove the effectiveness of the design procedure, a three-dimensional RC framed building is considered in the analysis. The reliability of the proposed procedure is finally checked by evaluating the nonlinear response of the building under earthquake strong ground motions.

Keywords: Seismic retrofit · Metal shear panels · RC buildings

1 Introduction

The past earthquake experiences demonstrated that many existing buildings designed without earthquake-resistance requirements or following outdated seismic codes are dramatically vulnerable to earthquakes. This is particularly important for public buildings such as hospitals or schools in view of the possible consequences of failure in terms of risk to life, injury, potential economic losses (Ferraioli 2010a, 2010b, 2015). Such buildings, designed without consideration on seismic loading, often have a seismically weak direction or are irregular in plan and/or elevation. The consequent lateral-torsional coupling leads inevitably to increase their seismic vulnerability. Many techniques have been developed for seismic retrofitting of buildings and have proved to be effective in increasing the capacity of the structure (Kasai et at. 1998, 2004, 2010, JSSI Manual 2007, Ito and Kasai 2006, Ferraioli et al. 2018, 2019) and/or reducing the seismic demand (Ferraioli and Mandara 2016, 2017). Among them, a suitable seismic retrofit solution is the

I. Vayas and F. M. Mazzolani (Eds.): PROHITECH 2021, LNCE 209, pp. 839–856, 2022.
https://doi.org/10.1007/978-3-030-90788-4_65

installation of Metal Shear Panels (MSPs) that have been effectively used as seismic resisting system for new steel buildings (De Matteis et al. 2003, 2005). Many studies available in the literature focused on experiment tests and analytical results and proved the excellent hysteretic behaviour of MSPs (De Matteis et al. 2011, Brando et al. 2014). Well-designed MSPs exhibited large ductility and energy dissipation capacity as well as high strength under earthquake ground motions. Recently, innovative methodologies for seismic retrofitting of existing RC buildings using aluminium shear panel were proposed in the literature (Mistakidis et al. 2007, De Matteis et al. 2009, 2018, Formisano et al. 2010).

2 Retrofit Design Using Aluminium Shear Panels

2.1 Pushover Curves and Trilinear Idealization

The sizing of the metal shear type was carried out generalizing a design procedure proposed in literature for seismic retrofit of RC buildings with hysteretic dampers (Pu and Kasai 2012, Sutcu et al. 2014). At first, a preliminary pushover analysis of the existing building was carried out using the first-mode distribution of lateral forces. This conventional analysis gives the pushover curves of the bare frame structure where, for each storey, the i-th storey shear force is plotted as a function of the i-th interstorey drift. The pushover curves are then idealized using a degrading trilinear behaviour following the Takeda model (Takeda and Sozen 1970) (Fig. 1). With reference to symbols in Fig. 1:

- $Q^f{}_{y,i}$, $\delta^f{}_{y,i}$ and $K^f{}_{y,i}$ are the shear force, the interstorey displacement and the secant stiffness of the i-th storey at yielding;
- $K^f{}_{0,i}$, $\alpha_1 K^f{}_{0,i}$ and $\alpha_2 K^f{}_{0,i}$ are the elastic, post-crack and post-yield stiffnesses;
- $Q^f{}_{c,i}$ and $\delta^f{}_{c,i}$ are the shear force and the interstorey displacement corresponding to cracking of concrete;
- $\delta^f{}_i = \mu^f{}_i\, \delta^f{}_i$ is the maximum interstorey displacement;
- $K^f{}_{\mu,i}$ is the secant stiffness corresponding to the maximum interstorey displacement.

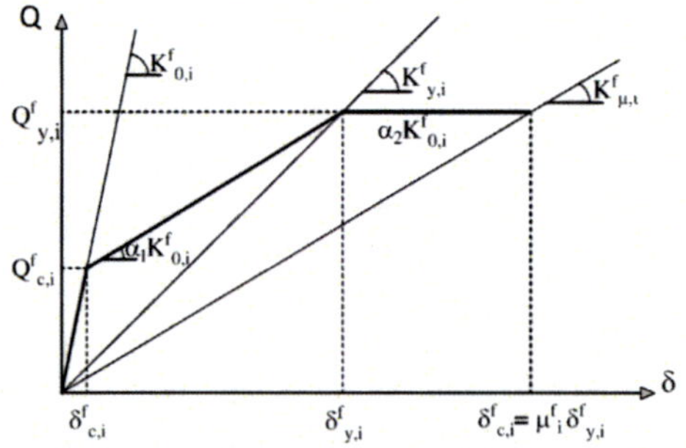

Fig. 1. Trilinear idealization of the pushover curve (i-th storey shear force vs i-th interstorey drift)

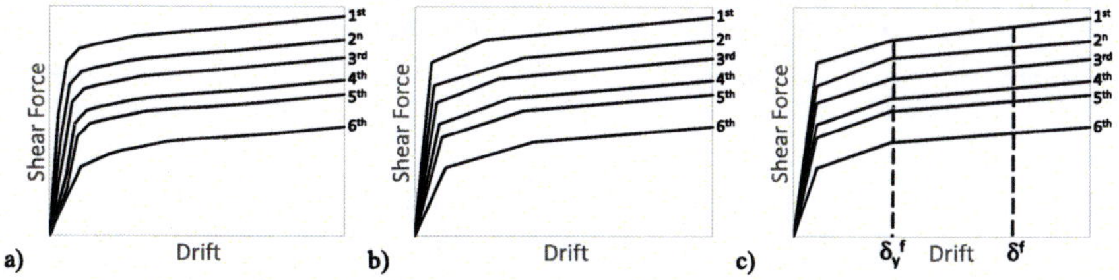

Fig. 2. a) Pushover curves; b) Trilinear idealization; c) Simplified trilinear behaviour

Some hypothesis are made for the sake of simplicity. First, the elastic stiffness $K^f{}_{0,i}$ is calculated as the tangent stiffness of the pushover curve. Then, the post-yield stiffness is assumed to be zero (i.e. $\alpha_2 = 0$), and this gives $\delta^f{}_{c,i} = Q^f{}_{c,i}/K^f{}_{0,i}$. The yielding displacement $\delta^f{}_{y,i}$ remains the only unknown parameter of the problem. Sutcu et al. (2014) calculated $\delta^f{}_{y,i}$ using the following condition:

$$\mu^f_{c,i} = \frac{\delta^f_{c,i}}{\delta^f_{y,i}} = \frac{1}{10} \tag{1}$$

which gives:

$$\delta^f_{y,i} = 10\delta^f_{c,i} \quad \alpha_1 = 0.22 \quad \alpha_{y,i} = \frac{Q^f_{y,i}}{\delta^f_{y,i}} = 0.30 \tag{2}$$

As an alternative, in this paper $\delta^f{}_{y,i}$ was calculated using the equal energy criterion between the original pushover curve (Fig. 2a) and the tri-linearized model (Fig. 2b). This gives a trilinear idealization of pushover curve for each storey of the building (Fig. 2b).

2.2 Equivalent Single-Degree System of Bare Frame

The equivalent SDOF system is defined under the hypothesis that all the storeys have the same yield displacement, calculated using the mean value as follows:

$$\delta^f_y = \frac{\sum_{i=1}^{N} \delta^f_{y,i}}{N} \tag{3}$$

where N is number of the storeys of the building. Moreover, the same maximum inter-storey displacement and corresponding ductility ratio is considered for all the storeys, as follows (Fig. 2c):

$$\delta^f_i = \delta^f \tag{4}$$

$$\mu^f_i = \frac{\delta^f_i}{\delta^f_{y,i}} = \frac{\delta^f}{\delta^f_y} = \mu^f \tag{5}$$

where μ^f is the maximum ductility ratio of the bare frame.

The equivalent height (H_{eq}), mass (M_{eq}) and period ($T_{\mu}{}^{f}$) of the single-degree system (SDOF) are defined as follows:

$$H_{eq} = \frac{\sum_{i=1}^{N} (m_i \cdot u_i \cdot H_i)}{\sum_{i=1}^{N} (m_i \cdot u_i)} \tag{6}$$

$$M_{eq} = \frac{\left(\sum_{i=1}^{N} (m_i \cdot u_i)\right)^2}{\sum_{i=1}^{N} (m_i \cdot u_i)} \tag{7}$$

$$T_{\mu}^{f} = \sqrt{\frac{\mu^f}{\alpha_y}} T^f \tag{8}$$

where m_i is the i-th storey mass, u_i is the first mode displacement of the building without dampers at height H_i while T^f is the fundamental vibration period of the bare frame building. Once the fundamental vibration period of the equivalent SDOF system is known, the drift angle demand is calculated as follows:

$$\theta^f = \frac{S_D\left(T_{\mu}^{f}, \xi_0^f\right)}{H_{eq}} \cong S_A\left(T_{\mu}^{f}, \xi_0^f\right)\left(\frac{T_f}{2\pi}\right)^2 \tag{9}$$

where S_D and S_A are the values of spectral displacement and acceleration of the equivalent SDOF system of the bare frame structure without damped braces as a function of the corresponding fundamental vibration period $T_{\mu}{}^{f}$ and the equivalent viscous damping $\xi_0{}^f$ of the bare frame.

2.3 Design of Metal Shear Panels

The equivalent SDOF system of the metal shear panels to be used to retrofit the existing building is designed to meet the performance criteria based on the target drift angle θ_t. This gives the following closed form expression (JSSI 2007) of the panels to bare frame stiffness ratio $r_p = K^p/K^f$:

$$r_p = \frac{p\left(\left(\frac{\theta^f}{\theta_t}\right)^2 - 1\right)\left(1 + a\left(\xi_0^f + \frac{1}{\pi} \cdot \frac{\alpha_y + \mu^c - \alpha_y^{\lambda}(1+\mu^c)}{\alpha_y + \mu^c} \cdot R\right)\right)}{\left(1 + a\xi_0^f\right) \cdot \frac{1}{\mu^p} + \frac{2aR}{\pi \mu^p}\left(1 - \frac{1}{\mu^p}\right)} \tag{10}$$

where $p = \alpha_y/\mu^f$, μ^c is the cracking ductility ratio of the bare frame, $a = 25$, $R = 0.6$ and $\lambda = 0.5$.

The optimal shear panel distribution is calculated under the following assumptions:

1) the equivalent viscous damping of the MDOF system is the same as the SDOF system;
2) the distributions of drift angle and ductility demand of panel-frame system become uniform, although those of the bare frame is non-uniform;

3) yield drift angle of shear panels for each story is uniform.

This gives the following distribution rule of the lateral stiffness $K_{p,i}$ of the metal shear panels (Takeuchi and Wada 2017):

$$K_{p,i} = \frac{Q_i}{H_i}\frac{\sum_{i=1}^{N}\left(K_{\mu,i}^{f}H_i^2\right)}{\sum_{i=1}^{N}(Q_iH_i)}\left(\mu^p + \frac{\mu^f}{\alpha_y}\cdot r_p\right) - \mu^p K_{\mu,i}^{f} \tag{11}$$

where Q_i is the design shear force at the i-th storey.

The overall design strength of the metal shear panels at the i-th storey is given by:

$$V_{p,i} = K_{p,i}\cdot\frac{\theta_t\cdot H_i}{\mu^p} \tag{12}$$

The multi-stiffened bracing-type pure aluminium shear panels tested by Brando and De Matteis (2014) were considered in this paper. These panels were designed by fixing the spacing of the ribs according to the slenderness limit value provided by Eurocode 9 (2007) for girders with longitudinal and/or transversal web stiffeners, ensuring yielding strength lower than the elastic buckling strength. In particular, with reference to rectangular plates, for web with both rigid and non-rigid transverse stiffeners at support only, Eurocode 9 (2007) provides the following design resistance:

$$V_{Rd} = \rho_v\cdot t_w\cdot h_w\frac{f_0}{\sqrt{3}\cdot\gamma_{M1}} \tag{13}$$

According to Fig. 3, t_w is the thickness of the web panel of the girder, h_w is the panel height, f_0 is the conventional yield strength of the base material, γ_{M1} is the safety factor and ρ_v is a factor for shear buckling which is obtained according to the following equations (Brando and De Matteis 2011):

$$\begin{aligned}
&\rho_v = \eta && \lambda_w < \frac{0,48}{\eta}\\
&\rho_v = \frac{2,774}{\lambda_w^{0,2175}} && \frac{0,48}{\eta}\le\lambda_w < 0,95\\
&\rho_v = \frac{2,8163}{\lambda_w^{0,1047}} && 0,95\le\lambda_w < 2,29\\
&\rho_v = \frac{2,6176}{\lambda_w^{0,0122}} && \lambda_w \ge 2,29
\end{aligned} \tag{14}$$

where η is factor that accounts for the influence of hardening on the buckling behaviour in the inelastic field and depends on the hardening ratio $\alpha = (f_{aw}/f_{ow})$, as follows (Eurocode 9 2007)

$$\eta = 0,7 + 0,35\cdot\frac{f_{aw}}{f_{ow}} \tag{15}$$

and λ_w is the slenderness parameter given by:

$$\lambda_w = \sqrt{\frac{f_v}{\tau_{cr}}} = \frac{0,81}{\sqrt{k_\tau}} \cdot \frac{\mathrm{a}}{t_w} \cdot \sqrt{\frac{f_{0,2}}{E}} \tag{16}$$

In such equation, f_v represents the yielding shear strength of the base material, τ_{cr} is the Eulerian shear buckling stress, E is the elastic modulus, a is the distance between the intermediate stiffeners. Since squared subpanels and rigid stiffeners are applied, the shear buckling coefficient k_τ may be calculated as follows:

$$k_\tau = 4,00 + 5,34 \cdot (b_w/\mathrm{a})^2 + k_{\tau st} \tag{17}$$

The sum of the first two addenda of Eq. (17) represents the shear buckling coefficient according to the Timoshenko theory for simple supported web panels (1934), determined by a critical load analysis as a function of the web geometry and boundary conditions. The term $k_{\tau st}$ accounts for the presence of longitudinal stiffeners, according to the following equation:

$$k_{\tau st} = 9,00 \cdot \left(\frac{b_w}{\mathrm{a}}\right)^2 \cdot \left(\frac{I_{st}}{t_w^3 \cdot b_w}\right)^{\frac{3}{4}} \tag{18}$$

where b_w is the plate depth, while I_{st} is the sum of the values of the second moment of area of individual longitudinal stiffeners, considering, for each of them, a plate collaborating area with a width of $15t_w$.

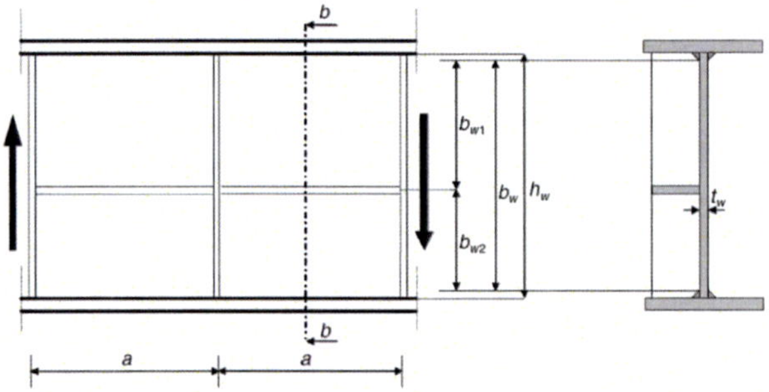

Fig. 3. Web with transverse and longitudinal stiffeners (Eurocode 9 2007)

Finally, it should be underlined that the design of the metal shear panels requires an iterative procedure. In fact, once the overall resistance is calculated from Eq. (12) and the number of shear panels for each storey is fixed, the required design resistance V_{Rd} is known. However, the factor for shear buckling ρ_v in Eq. (13) is a function of the numbers and geometry of applied stiffeners, that depend on the panel height h_w that is unknown. The iterative procedure is based on the following steps:

1) The thickness t_w of the web panel and numbers of applied stiffeners are fixed;
2) The attempt value of the panel height h_w is calculated using Eq. (13) with $\rho_v = \eta$;
3) $k_{\tau st}$, k_τ, λ_w and η and, finally, ρ_v are calculated from Eqs. (14)–(18)
4) The panel height h_w is estimated using Eq. (13).
5) The procedure is iterated until the initially assumed value and calculated value coincide with the prescribed tolerance.

3 Case Study

3.1 Seismic Assessment of the Existing Building

The retrofit design procedure is applied to a regular in plan and elevation RC framed building (Fig. 4). It is assumed that: a) the interstorey height is $h = 3.0$ m; b) at each storey all the columns have the same cross-section $b_C \times h_C$; c) all the beams have the same cross-section $b_B \times h_B$; d) $b_C = b_B = b$; e) the columns are tapered every two plans with taper ratio $r = \Delta h_C/h_C = 0.10$; f) the beam span is $L_B = 5.0$ m.

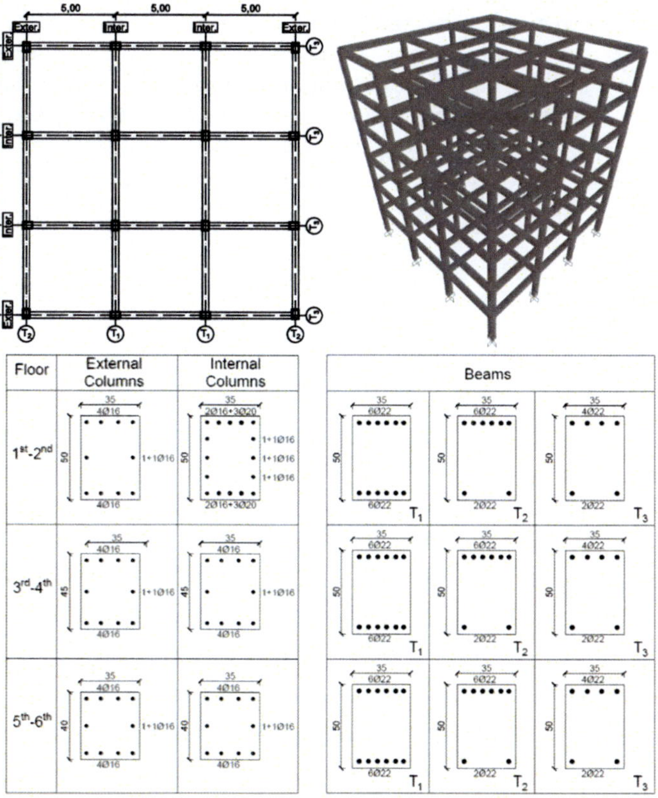

Fig. 4. Plan view of the 6-storey building, 3D model and cross-sections of beams and columns

As far as the material properties, Italian Feb38k steel and C20/25 concrete were used. The building was designed in compliance with an old Italian seismic code (DM96 1996) for a medium-risk zone. As far as the material properties were concerned, C20/25 concrete and Italian FeB38k steel were considered in the analyses. The nonlinear analyses were carried using a multi-axial spring model implemented in SAP2000 (2019). This model allows reproducing the behaviour of flexural and axial deformation of column element and representing the interaction among bi-directional bending moments and axial load. Each fibre was modelled with the nonlinear stress-strain relationship of the material it was made of. In particular, the cross section is composed of fibres of three different material types: confined concrete for the core, unconfined concrete for the cover, and steel for the longitudinal bars. The steel was modelled with an elastic-plastic relationship while the confined concrete was modelled with the stress-strain model originally proposed by Mander et al. (1988). The plastic hinge length is defined according to NTC-Instructions (2019). The seismic assessment of the building was carried out using the nonlinear static analysis under two different vertical distributions of lateral loads: a) Uniform distribution; b) First mode distribution. In the case study, the pushover analysis under the first mode distribution gives the lowest seismic performance. Figure 5 shows the pushover curves and the performance points corresponding to the limit states of Immediate Occupancy (IO), Damage Limitation (DL), Life Safety (LS) and Collapse Prevention (CP) defined according to the Italian Code (NTC 2018) acceptance criteria. Table 1 shows the coordinates of the performance points (i.e. Base shear /Weight vs top displacement), the capacity peak ground acceleration (PGA^c), the demand peak ground acceleration (PGA^d) and the safety index ($\zeta_{PGA} = PGA^c / PGA^d$). The results illustrate that the limit states of Life Safey and Collapse Prevention are not satisfied and, thus, the building needs seismic retrofit.

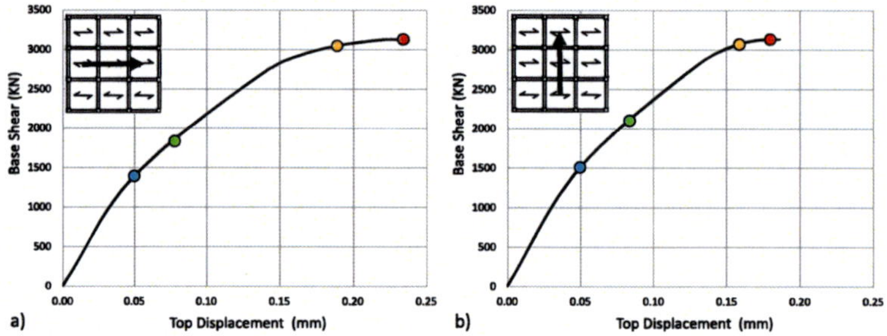

Fig. 5. Pushover curves and performance points a) X-Direction. b) Y-Direction.

Table 1. Performance points (T/W vs d). Capacity and demand in terms of PGA. Safety index.

X-Direction						Y-Direction				
Limit state	d (cm)	$\frac{T}{W}$	$\frac{PGA^c}{g}$	$\frac{PGA^d}{g}$	ζ_{PGA}	d (cm)	$\frac{T}{W}$	$\frac{PGA^c}{g}$	$\frac{PGA^d}{g}$	ζ_{PGA}
IO	0.050	0.090	0.056	0.048	1.18	0.050	0.098	0.060	0.048	1.25
DL	0.078	0.117	0.088	0.066	1.33	0.084	0.133	0.105	0.066	1.59
LS	0.189	0.198	0.218	0.256	0.85	0.159	0.199	0.198	0.256	0.77
CP	0.234	0.203	0.244	0.366	0.67	0.180	0.205	0.215	0.366	0.58

3.2 Design of Aluminiun Shear Panels

In general, the rehabilitation strategy aims to improve the seismic behaviour of the building by: 1) increasing the structural capacity of columns, beams and joints; 2) reducing the seismic demand through isolation and/or energy dissipation; 3) changing the system by inserting steel bracing or infilling the reinforced concrete frames with RC shear walls. In this paper, the seismic retrofit was carried out using the aluminium shear panels. In particular, the aluminium alloy AW1050AH24 was selected as base material for the ongoing study. The following parameters were considered in the analysis: Elastic conventional yielding strength $f_{0.2} = 21$ N/mm^2, ultimate deformation $\varepsilon_u = 13\%$, ultimate strength $f_u = 97.44$ N/mm^2, elastic modulus $E = 70000$ N/mm^2.

The aluminium shear panels were designed according to the procedure previously described using the target response spectrum at the Collapse Prevention Limit State plotted in Fig. 6. At first, a pushover analysis was carried out under the first mode distribution of the lateral loads and the storey pushover curves (storey shear force vs interstorey drift) were plotted (Fig. 7). Then, the storey pushover curves were modelled with an idealized trilinear behaviour using procedure described in 2.2 (Fig. 8, Tables 2–3). An equivalent SDOF system with height (H_{eq}) and mass (M_{eq}) was defined using Eqs. (6)–(7) (Table 4). The periods T^f and $T_\mu{}^f$, the spectral displacement S_D and the drift angle demand $\theta_\mu{}^f$ were calculated using Eqs. (8)–(10) as follows: $T^f = 0.95$ s, $T_\mu{}^f = 2.33$ s, $S_D = 39$ mm and $\theta_\mu{}^f = 0.031$ in X-Direction; $T^f = 0.92$ s, $T_\mu{}^f = 2.51$ s, $S_D = 42$ mm and $\theta_\mu{}^f = 0.033$ in Y-Direction. Once fixed the target drift angle $\theta_t = 0.02$, Eq. 10 gives the panels to bare frame stiffness ratio that is $r_p = 0.56$ in X-Direction and $r_p = 0.62$ in Y-Direction. Finally, the aluminium shear panels were designed with the procedure described in 2.3. The spacing $a/b_w = 0.25$ of the stiffeners was selected to avoid the detrimental effects caused by shear buckling. The details of the design parameters of the aluminium shear panels is shown in Tables 5–6. The surrounding steel frame columns were supposed to be connected directly to the RC beams. The RC beams were jacketed by two coupled UPN200 profiles to allow the transfer of the forces absorbed by the panels (Fig. 9).

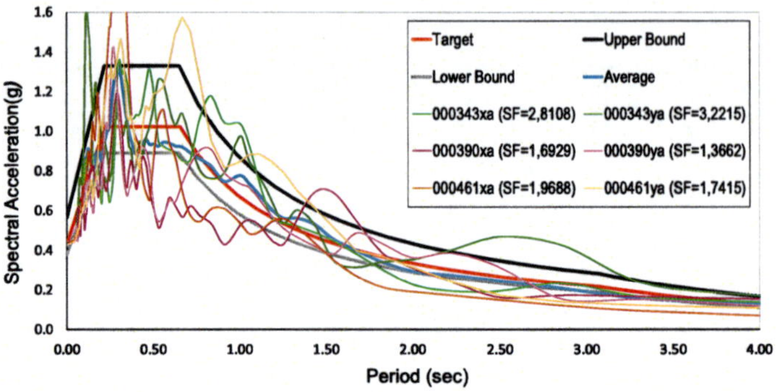

Fig. 6. Target response spectrum for the Collapse Prevention Limit State. Spectrum compatibility for the selected records (Upper bound and Lower bound according to NTC-2018)

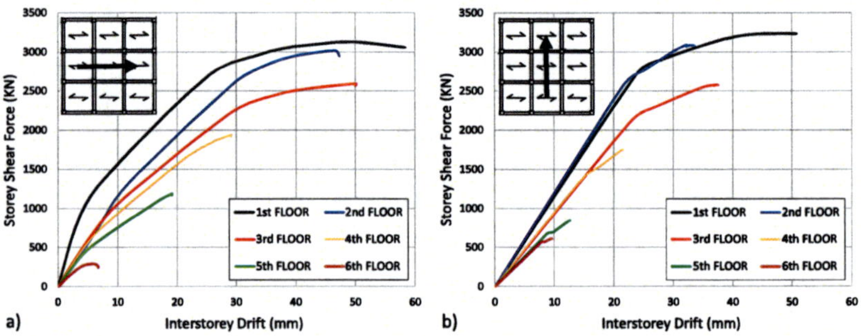

Fig. 7. Pushover curves for model building

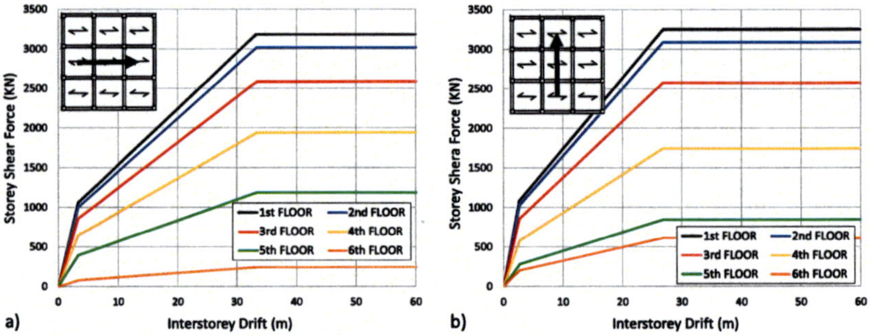

Fig. 8. Simplified trilinear idealization of pushover curves

Table 2. Parameters of trilinear idealization of pushover curves (X-Direction)

Floor	δ_y [mm]	Q_y [kN]	δc [mm]	Q_c [kN]	K_f [kN/mm]	$\alpha_1 K_f$ [kN/mm]	$\mu_f \delta_y$ [mm]	μ_f	$K_\mu{}^f$ [kN/mm]	p	$\alpha_y K_f$ [kN/mm]
1	33.3	3181.89	3.33	1060.63	318.52	70.78	60	1.80	53.032	0.166	95.56
2	33.3	3017.11	3.33	1005.70	302.03	67.12	60	1.80	50.285	0.166	90.61
3	33.3	2590.47	3.33	863.49	259.32	57.63	60	1.80	43.174	0.166	77.80
4	33.3	1942.67	3.33	647.56	194.47	43.22	60	1.80	32.378	0.166	58.34
5	33.3	1186.60	3.33	395.53	118.78	26.40	60	1.80	19.777	0.166	35.64
6	33.3	245.65	3.33	81.88	24.59	5.46	60	1.80	4.094	0.166	7.38

Table 3. Parameters of trilinear idealization of pushover curves (Y-Direction)

Floor	δ_y [mm]	Q_y [kN]	δc [mm]	Q_c [kN]	K_f [kN/mm]	$\alpha_1 K_f$ [kN/mm]	$\mu_f \delta_y$ [mm]	μ_f	$K_\mu{}^f$ [kN/mm]	p	$\alpha_y K_f$ [kN/mm]
1	26.9	3248.93	2.69	1082.98	402.98	89.55	60	2.23	54.149	0.166	120.89
2	26.9	3088.52	2.69	1029.51	383.08	85.13	60	2.23	51.475	0.166	114.92
3	26.9	2575.42	2.69	858.47	319.44	70.99	60	2.23	42.924	0.166	95.83
4	26.9	1744.11	2.69	581.37	216.33	48.07	60	2.23	29.069	0.166	64.90
5	26.9	845.40	2.69	281.80	104.86	23.30	60	2.23	14.090	0.166	31.46
6	26.9	613.07	2.69	204.36	76.04	16.90	60	2.23	10.218	0.166	22.81

Table 4. Storey mass and height

Floor	Mass [kNs^2/m]	Hi [m]	$m_i \times H_i^2$ [kNs^2m]	$m_i \times H_i$ [kNs^2]		
1	269.14	3	2422.26	807.42		
2	268.09	6	9651.24	1608.54		
3	267.03	9	21629.43	2403.27		
4	265.98	12	38301.12	3191.76		
5	264.93	15	59609.25	3973.95		
6	200.68	18	65020.32	3612.24		
			196633.62	15597.18	Heq = 12.6m	M_{eq} = 1237.18 kNs^2/m

Table 5. Design parameters of the aluminium shear panels (X-Direction)

N.	K_{pi} [kN/mm]	N_i	$K_{pi}^{(1)}$ [kN/mm]	V_{Rdi} [kN]	t_w [mm]	b_w [mm]	I_{st} [mm^4]	a [mm]	$k_{\tau st}$	k_τ	λ_w	ρ_v
1	128.77	4	32.19	275.90	5	995	814687.5	249	589.59	679.03	0.11	4.577
2	122.11	4	30.53	261.63	5	930	814687.5	233	620.18	709.62	0.10	4.640
3	117.34	4	29.34	251.41	5	894	814687.5	223	638.99	728.43	0.09	4.640
4	111.11	4	27.78	238.07	5	846	814687.5	212	665.67	755.11	0.09	4.640
5	93.83	4	23.46	201.04	5	715	814687.5	179	755.66	845.10	0.07	4.640
6	93.51	4	23.38	200.34	5	712	814687.5	178	757.62	847.06	0.07	4.640

Table 6. Design parameters of the aluminium shear panels (Y-Direction)

N.	K_{pi} [kN/mm]	N_i	$K_{pi}^{(1)}$ [kN/mm]	V_{Rdi} [kN]	t_w [mm]	b_w [mm]	I_{st} [mm^4]	a [mm]	$k_{\tau st}$	k_τ	λ_w	ρ_v
1	167.53	4	41.88	358.94	5	1441	814687.5	360	446.5	535.9	0.17	4.110
2	158.18	4	39.55	338.92	5	1330	814687.5	332	474.4	563.8	0.16	4.207
3	156.29	4	39.07	334.87	5	1307	814687.5	327	480.4	569.9	0.15	4.228
4	159.06	4	39.76	340.80	5	1340	814687.5	335	471.6	561.1	0.16	4.198
5	144.99	4	36.25	310.64	5	1176	814687.5	294	520.1	609.5	0.13	4.359
6	103.82	4	25.96	222.45	5	791	814687.5	198	700.4	789.86	0.08	4.640

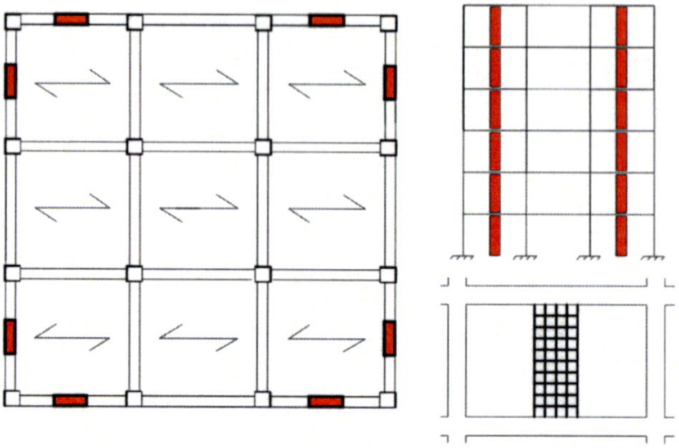

Fig. 9. Seismic retrofitting of the existing RC building by metal shear panels

3.3 Seismic Assessment of Retrofitted Building

The design procedure gives the strength and stiffness of the aluminium shear panels that can be used for the nonlinear static analysis of the building. However, it should be observed that the pushover procedure is not the most suitable method to evaluate the seismic response of such a kind of buildings. In fact, the metal shear panels have very high dissipation capacity and complex hysteretic behaviour with stiffness and strength degradation and pinching effects. Thus, a nonlinear response history analysis should be performed. To this aim, the behaviour of the metal shear panels should be characterized with an effective hysteretic model that is representative of the relationship between the load and displacement of panel under progressive alternate loading. When the accumulated plastic deformation increases, the local buckling of aluminium web plates due to the residual strain causes the slip type hysteresis. This slip phenomenon varies depending on the local buckling stiffeners provided. In order to account for this effect, three different hysteretic models were considered in the analyses: a) bilinear cinematic model. b) semi slip model. c) slip model. It should be observed that the local buckling stiffeners should be designed to avoid the slip type hysteresis, so that the aluminium shear panels are used in a region that shows the spindle-shaped relationship between the load and displacement. Thus, the real load-displacement relationship of the panels should be between cases a) and b). The nonlinear response history analysis was carried out using two simultaneously acting accelerograms along both horizontal directions. A group of three pairs of time-histories was applied. The envelope of the response quantities from all the analyses was used as the design value of the seismic effect. The seismic motion was made by using recorded accelerograms selected with the rules recommended in seismic codes (NTC-2018). The European Strong-motion Database (ESD), the SIMBAD database and the Italian Accelerometric archive (ITACA) were used for selecting the recorded accelerograms (Iervolino et al. 2010, Smerzini and Paolucci 2011). The spectrum compatibility for the selected acceleration records is represented in Fig. 6. In the range of periods of interest no value of the mean elastic spectrum, calculated from all time histories, is less than 90% (lower bound) and more that 130% (Upper bound) of the corresponding value of the target elastic spectrum. In Fig. 11, the comparison between capacity and demand in terms of maximum interstorey drift of the corner columns is shown. In the same figure also the limit domain corresponding to the target drift (θ_t = 0.02) is plotted. The demand is given by the interstorey drift time-histories for the selected earthquake records. The capacity is represented by the limit domain for the limit state of Collapse Prevention (CP) calculated from the corresponding chord rotation. In columns under seismic ground motion the nodal rotation is low with respect to the drift Δ of the equivalent cantilever. Thus, the chord rotation θ may be reasonably defined as $\theta = \Delta/L_s$, where L_s is the shear span length considered placed at the middle of the column. Considering the bi-axial bending, the drift capacity depends on the direction of bending axle. However, studies available in literature (Fardis 2009) show that, if the components of chord rotation along the sides of the section in bi-axial bending are normalized to the corresponding chord rotations in uniaxial loading, then a circular interaction diagram may be considered. The results shows that the ultimate chord rotation is exceeded only in the first-storey corner column if the bilinear cinematic model or the semi slip model is used for the aluminium shear panels.

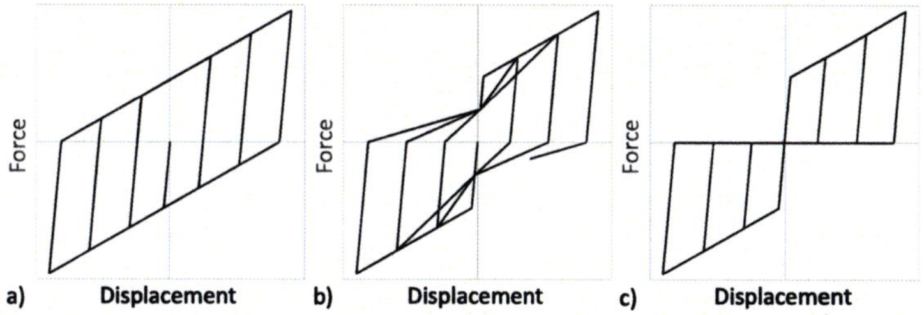

Fig. 10. a) Bilinear cinematic model. b) Semi slip model. c) Slip model

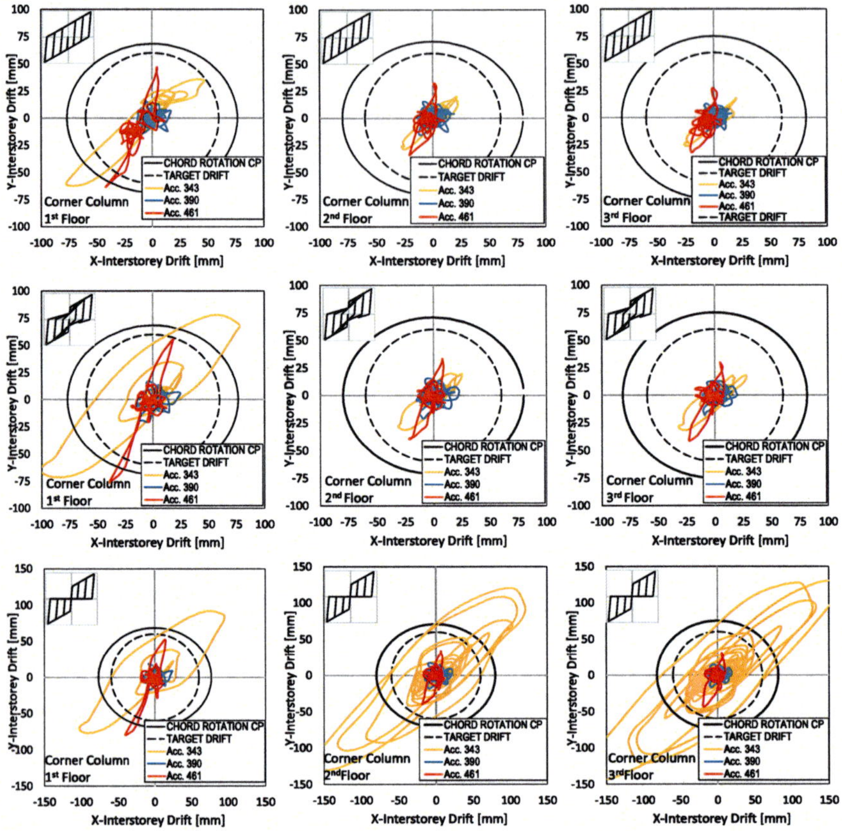

Fig. 11. Drift time-histories and limit domains for the Collapse Prevention Limit State

On the contrary, the ultimate chord rotation is exceeded even in the columns of the other floors if the slip model is considered. These results confirm the importance of the local buckling stiffeners that should be designed to avoid the slip type hysteresis.

4 Conclusions

In this paper, an effective design procedure for seismic retrofit of RC buildings using metal shear panels was developed to meet a certain performance requirement. The metal shear panels were designed based on the target drift angle such that the distributions of drift angle and ductility demand of the retrofitted building become uniform, although those of the existing RC building is non-uniform. The proposed procedure was applied successfully on a three-dimensional RC framed building. Its effectiveness was demonstrated by a series of nonlinear response history analysis under different earthquake strong ground motions. The influence of the type of hysteresis loop due to the local buckling of aluminium web plates was investigated. The results have shown the importance of local buckling stiffeners that should be designed to avoid the slip type hysteresis.

In fact, this type of hysteresis loop significantly decreases the energy dissipation capacity of the aluminium shear panels and this can significantly reduce the effectiveness of the seismic retrofit of the building.

References

DM96. Norme tecniche per le costruzioni in zona sismica. D.M. 16 Gennaio (January) (1996). (in Italian)

Brando, G., De Matteis, G.: Design of low strength-high hardening metal multi-stiffened shear plates. Eng. Struct. **60**, 2–10 (2014)

De Matteis, G., Landolfo, R., Mazzolani, F.M.: Seismic response of MR steel frames with low-yield steel shear panels. Eng. Struct. **25**(2), 155–168 (2003)

De Matteis, G.: Effect of lightweight cladding panels on the seismic performance of moment resisting steel frames. Eng. Struct. **27**(11), 1662–1676 (2005)

De Matteis, G., Brando, G., Mazzolani, F.M.: Hysteretic behaviour of bracing-type pure aluminium shear panels by experimental tests. Earthquake Eng. Struct. Dynam. **40**, 1143–1162 (2011)

De Matteis, G., Formisano, A., Mazzolani, F.M.: An innovative methodology for seismic retrofitting of existing RC buildings by metal shear panels. Earthquake Eng. Struct. Dynam. **38**, 61–78 (2009)

Formisano, A., De Matteis, G., Mazzolani, F.M.: Numerical and experimental behaviour of a full-scale RC structure upgraded with steel and aluminium shear panels. Comput. Struct. **88**, 1348–1360 (2010)

De Matteis, G., Ferraioli, M.: Metal shear panels for seismic upgrading of RC buildings: a case study. Key Eng. Mater. **763**, 1058–1066 (2018)

Eurocode 9. Design of Aluminium Structures. EN1999–1–1 (2007)

Fardis, M.N.: Seismic Design, Assessment and Retrofitting of Concrete Buildings, Springer, Cham (2009). https://doi.org/10.1007/978-1-4020-9842-0

Ferraioli, M.: Inelastic torsional response of an asymmetric-plan hospital building in Italy. In: COST ACTION C26: Urban Habitat Constructions under Catastrophic Events - Proceedings of the Final Conference, pp. 365–370 (2010a)

Ferraioli, M., Abruzzese, D., Miccoli, L., Vari, A., Di Lauro, G.: Structural identification from environmental vibration testing of an asymmetric-plan hospital building in Italy. In: COST ACTION C26: Urban Habitat Constructions under Catastrophic Events - Proceedings of the Final Conference, pp. 981–986 (2010b)

Ferraioli, M.: Case study of seismic performance assessment of irregular RC buildings: hospital structure of Avezzano (L'Aquila, Italy). Earthq. Eng. Eng. Vib. **14**(1), 141–156 (2015). https://doi.org/10.1007/s11803-015-0012-7

Ferraioli, M., Mandara, A.: Base isolation for seismic retrofitting of a multiple building structure: evaluation of equivalent linearization method. Math. Probl. Eng. **2016**, 8934196 (2016)

Ferraioli, M., Mandara, A.: Base isolation for seismic retrofitting of a multiple building structure: design, construction, and assessment. Math. Probl. Eng. **2017**, 4645834 (2017)

Ferraioli, M., Lavino, A.: A displacement-based design method for seismic retrofit of RC buildings using dissipative braces. Mathematical Problems in Engineering, p. 5364564 (2018)

Ferraioli, M., Nuzzo, D., Concilio, A.: Shape memory alloys for earthquake building protection. In: Proceedings of SPIE - The International Society for Optical Engineering, p. 10970, 109701M (2019)

Iervolino, I., Galasso, C., Cosenza, E.: REXEL: computer aided record selection for code-based seismic structural analysis. Bull. Earthq. Eng. **8**, 339–362 (2010)

Ito, H., Kasai, K.: Passive control design for elasto-plastically damped building structure. In: Proceedings of STESSA 2006 – Mazzolani & Wada (eds). Taylor & Francis Group, London (2006)

JSSI Manual. Design and Construction Manual for Passively Controlled Buildings. Japan Society of Seismic Isolation (JSSI), Tokyo, Japan (2007)

Kasai, K., Fu, Y., Watanabe, A.: Passive control systems for seismic damage mitigation. J. Struct. Eng. **124**(5), 501–512 (1998)

Kasai, K., Ito, H.: JSSI Manual for building passive control technology Part-8 peak response evaluation and design for elasto-plastically damped system. In: Proceedings of 13th World Conference on Earthquake Engineering, Vancouver, B.C., Canada (2004)

Kasai, K., Pu, W.: Passive control design method for MDOF slip-hysteretic structure added with visco-elastic damper. J. Struct. Construct. Eng. **650**, 781–790 (2010)

Mander, J.B., Priestley, M.J.N., Park, R.: Theoretical stress-strain model for confined concrete. J. Struct. Eng. **114**(8), 1804–1826 (1988)

Mistakidis, E.S., De Matteis, G., Formisano, A.: Low-yield metal shear panels as an alternative for seismic upgrading of concrete structures. Adv. Eng. Softw. **38**(8–9), 626–636 (2007)

NTC-2018 Guidelines. Technical Standards for Constructions, Official Journal of the Italian Republic. 20.02.2018, Rome, Italy (2018). (in Italian)

NTC-Instructions. Instructions for the application of the new technical norms on constructions (in Italian), Official Journal of the Italian Republic. Circ. 21.01.2019, no.7 (2019). (in Italian)

Pu, W., Kasai, K.: Design method for RC building structure controlled by elasto-plastic dampers using performance curve. In: Proceedings of the 15th World Conference on Earthquake Engineering, Lisbon, Portugal (2012)

SAP2000. Linear and nonlinear static and dynamic analysis of three-dimensional structures. Research Ultimate Version 21.0, Analysis Ref. Manual, Computer and Structures, Berkeley, CA (2019)

Smerzini, C., Paolucci, R.: Research Project DPC – RELUIS 2010–2013, SIMBAD: a database with Selected Input Motions for displacement-Based Assessment and Design – 2nd release, Department of Structural Engineering, Politecnico di Milano, Italy (2011)

Sutcu, F., Takeuchi, T., Matsui, R.: Seismic retrofit design method for RC buildings using buckling-restrained braces and steel frames. J. Construct. Steel Res. **101**, 304–313 (2014)

Takeuchi, T., Wada, A.: Buckling-Restrained Braces and Applications, Japan Society of Seismic Isolation (JSSI), Tokyo, Japan (2017)

Takeda, T., Sozen, M.A., Nielsen, N.N.: Reinforced concrete response to simulated earthquakes. J. Struct. Div. (ASCE) **96**(12), 2557–2573 (1970)

Timoshenko, S.: Stability of the webs of plate girders. Engineering **138**, 207–209 (1934)

Emergency Interventions in the Castle Church of Monreal De Ariza (Spain)

Rosa Bustamante[1](✉), Juan Monjo[1], Pilar Rodríguez-Monteverde[2], and José Mª Valero[3]

[1] Department of Construction and Architectural Technology, Universidad Politécnica de Madrid, Av. Juan de Herrera, 4, 28040 Madrid, Spain
{rosa.bustamante,juan.monjo}@upm.es

[2] Department of Structures and Building Physics, Universidad Politécnica de Madrid, Av. Juan de Herrera, 4, 28040 Madrid, Spain
pilar.r.monteverde@upm.es

[3] Diputación Provincia de Zaragoza, Plaza de España, 2, 50071 Zaragoza, Spain
valerojm@urbantran.com

Abstract. The church of Santa María de la Asunción, in Monreal de Ariza (Spain), is located in the medieval castle of Mont Regal, in the defense line of the Christian territories during the Muslim occupation of the Iberian Peninsula. The first church, around 1200, is Romanesque, with stone masonry walls and main entrance with archivolts of this style; the addition of the sacristy, representative of the late Gothic, occurs in the sixteenth century. The current church comes from the Baroque reform of 1619, with basket-handle transverse arches and half-barrel vaults of double layers of ceramic bricks with lunettes. Its abandonment after 1970 generated a phase of ruin that required several interventions to stop its collapse. Despite this, there was a partial collapse of the 4th transverse arch and the collapse of some vaults. In July 2016, a pre-diagnosis was carried out, due to its dangerous situation. In September 2018, an emergency intervention was initiated to consolidate the building. The most important aspects of this emergency action are described below, analyzing the different types of usual interventions in historic buildings.

Keywords: Emergency · Damage · Intervention · Consolidation · Castle church

1 Introduction

The response to an emergency situation constitutes a series of interventions planned or not, taken in the context of such emergency (EN 15898 2019). The main causes of the ruin of the old religious and defensive structures are not different from those that occur in other types of immovable cultural heritage, which are: a) the incidence of the time involved in alterations or aging in the building and its surroundings, b) the unpredictable natural disasters such as earthquakes and c) violent aggressions caused by man (Bonazza et al. 2018). In Spain, the damages caused by the Civil War (1936–39) to the churches demanded several decades of restoration and repairing (Bustamante 1999).

I. Vayas and F. M. Mazzolani (Eds.): PROHITECH 2021, LNCE 209, pp. 857–869, 2022.
https://doi.org/10.1007/978-3-030-90788-4_66

Any of these causes can produce some of the 28 damage mechanisms in the macro elements of churches: facades, arches, vaults, towers, etc. (Lagomarsino 2012). Anagnostopoulos and Moretti (2008) classified the severity of masonry damage according to the width of the cracks (*d*), i.e.1) no damage with fine cracks visible from one side only; 2) slight; 3) moderate; 4) severe-total damage when $d > 5.0$ mm and visible from both sides of the wall, partial or total failure of bearing walls, floors and/or roof.

The evaluation on the site after an earthquake is a research opportunity (Binda et al. 2008). Modena et al. (2011) expose the main emergency interventions: bracing, shoring, reinforcement of masonry by substitution or addition of material, injections and even reconstruction with provisional materials recovering the traditional image of churches. It should be added the loss of the original function of the building causes abandonment.

This paper establishes that emergency work incorporates interventions that do not differ much from those that are usual in the conservation of existing structures and those that give priority to structural consolidation. This may involve the addition of reinforcement materials and elements and, sometimes, the demolition of harmful elements, to maintain the stability of the building.

In this sense we consider that the emergency works in the castle church of Santa María de la Asunción, in Monreal de Ariza (Spain) is an example of the classification of the emergency interventions carried out. This castle church is located inside the third enclosure of the medieval castle of Mont Regal, located in the defense line of the Christian territories during the Muslim occupation of the Iberian Peninsula at the southern end of the province of Zaragoza (Fig. 1a). This walled castle church (Fig. 1b) has undergone several alterations throughout its history.

(a)

(b)

Fig. 1. Location map (a); rampart in front of the church from the north (b)

The current church from the Baroque reform of 1619-1631 is built with stone masonry walls inserting spoliation ashlars; rectangular single nave (28.25 m × 9.10 m) and 11.50 m high, divided into five sections plus a semicircular apse with small vaults on the side chapels; sandstone transverse basket-handle arches and vaults between them of double layer of ceramic brick. Windows between semi buttresses rise from the Romanesque church walls of the year 1200. This reform caused the blinding of the windows of the Romanesque apse not only because of the growth of the walls but also of the original level of the floor of the nave. The addition of the sacristy, representative of the late Gothic, occurs in the sixteenth century.

After its abandonment towards 1970, to stop its collapse some interventions were performed with tension bars in three of the arches at an approximate height of 80 cm above the springing, as well as their shoring in the 1990s. Despite this, there was a partial collapse of the fourth arch and vaults of the second and fourth section, and the total collapse of the vault of the fifth section next to the apse. Therefore, interior scaffolding was placed as provisional propping. On the other hand, the defensive tower-gate in the NE corner, belonging to the fortified enclosure, has undergone several transformations parallel to the evolution of the church to become a bell tower with brick bulrush in the seventeenth century, increasing its height.

A pre-diagnosis was carried out (AIPA 2016), due to its dangerous situation, with deformations of the wall of the fourth enclosure, important collapses of the north facade of the church, with numerous vertical cracks in this facade and in the apse, as well as important cracks also in the arches and vaults, in the bell tower, in the sacristy and the aforementioned collapse of the fourth arch. In September 2018, an emergency intervention was initiated to consolidate the building. The most important aspects of the action are described below, analyzing the different types of usual interventions in historic buildings.

2 Emergency Interventions

Emergency works can be classified as: a. *Temporary* (structure shoring) and *definitive* (integral consolidation); b. *Preventive* (actions to prevent future damage or invasive interventions) and *repairing* (interventions to recover functionality and stability of the building); c. *Innovative* (incorporation of new and different materials to existing ones) and *traditional* (artisanal masonry techniques); d. *Reversible* (intervention that can be removed without causing damage) and *irreversible* (permanent work).

In this castle church, actions of all kinds have been carried out, which we can fit into these binomials taking into account that some of them have a place in several.

2.1 Temporary or Definitive Interventions

The planned action sought to carry out definitive interventions, in order to consolidate the ensemble built to avoid its collapse and prepare a complete future restoration. However, the rampart corresponding to the fourth enclosure (located to the north of the castle) presented a very unstable situation and needed a more complete diagnostic study. Consequently, the need to carry out a temporary intervention in this part of the ensemble was estimated.

2.1.1 Temporary Interventions

a) *Rampart propping*

The defensive rampart, limiting one of the streets of the village, showed a clear collapse towards the slope, in its entire length and in the extreme buttresses, with vertical cracks, both in the center of the intermediate section as in the union of the buttresses with the rampart, which threatened to collapse. So that the action had to be immediate and did not allow us to carry out an exhaustive study of the causes to take definitive measures. Consequently, it was decided to make a temporary metallic propping (Fig. 2a) to support the rampart by an indefinite period of time and a couple of cracks were monitored to check its stabilization. Other measurements (Fig. 2b) and materials sampling were done for research purposes. Moreover, preventive conservation and maintenance of the interventions of defensive structures is recommended by national conservation charter of *Baños de la Encina* (2006).

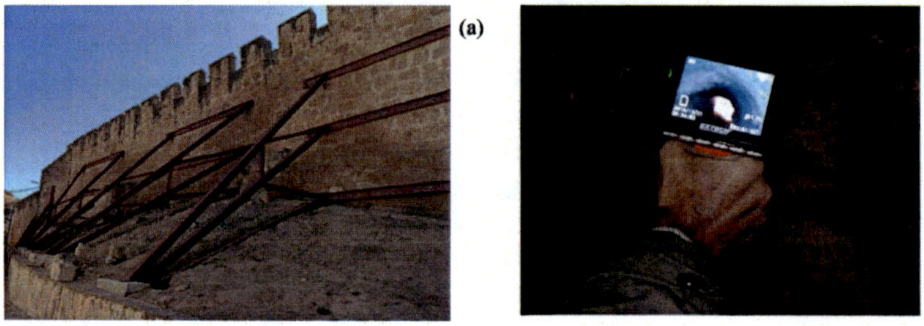
(a) (b)

Fig. 2. Provisional rampart propping (a); endoscope to check the core of the wall (b)

2.1.2 Definitive Interventions

b) *General consolidation of the church*
The consolidation works of the church can be considered definitive, although each one also corresponds to one of the other binomials.

The instability of the church was significant; scaffolding and provisional struts were raised inside. The most affected construction and structural elements are the following:

- The north wall of the church had a considerable collapse towards the north, about 37 cm out of the plumb-line, with important and ancient cracks, especially in its encounter with the apse wall. All this due, probably, to the limited support of the wall on the ground and the horizontal thrust of the roof structure. In addition, vertical cracks were visible on the inside, which indicated the separation of the pilasters with respect to the wall, probably due to the collapse of the wall indicated above and the lack of firm union between the outer wall and the interior pilasters.
- The transverse basket-handle arches had suffered the sinking of their keystones, probably as a result of the opening of their supports, which, in turn, may be due to the collapse of the north wall seen above, and the horizontal thrust of the roof structure. As a result, arches had cracks and separation of voussoirs in the middle of the intrados and in the transverse diaphragms by the extrados. One of them, the fourth arch, had collapsed. As a result, the vaults supported by these arches had suffered sinking and breaking, reaching the collapse in three of them.
- In the bell tower, many walls out of plumb-line, both north and west, causing numerous vertical cracks in the highest area, as well as in the union with the church walls, plus intermediate deformations with new vertical cracks, probably caused by the tower's own weight and by deficient joining of its ashlars.

- In the sacristy there was a slight collapse of the walls to the west and south and, especially, of the buttress located in the SO corner, with vertical cracks in both walls, in the middle and in its union with the south wall of the nave. Also the roof structure suffered collapse.

As a result of this situation, the following definitive consolidation measures were taken, which are described in the following sections, depending on their correspondence to the indicated binomials:

- Underpinning and containment of the north foundations;
- Bracing of the walls on its upper line;
- Beam attached to the inside of the upper part of the north wall of the church;
- Tied up from the walls and semi buttresses to the internal pilasters;
- Stitching of wall cracks once the deformations have been stabilized;
- Reconstruction of the 4th arch and the vaults of 4th and 5th sections;
- Reinforcement of the rest of arches and vaults with stitching of cracks;
- Tension bars in the 4th and 5th arches;
- Static cross ties in the bell tower; and
- Repairing of wood structure.

2.2 Preventive or Repairing Works

Among the listed actions there are both types.

2.2.1 Preventive Works

c) *Steel tension bars in the arches*
As has been said, the 1st, 2nd and 3rd transverse arches already had steel rods to control the opening of the supports. They were attached at their ends to the semi buttresses, by nut and locknut, to large steel plates, clearly visible from the outside. These rods have probably prevented these arches from collapsing (Fig. 3a).

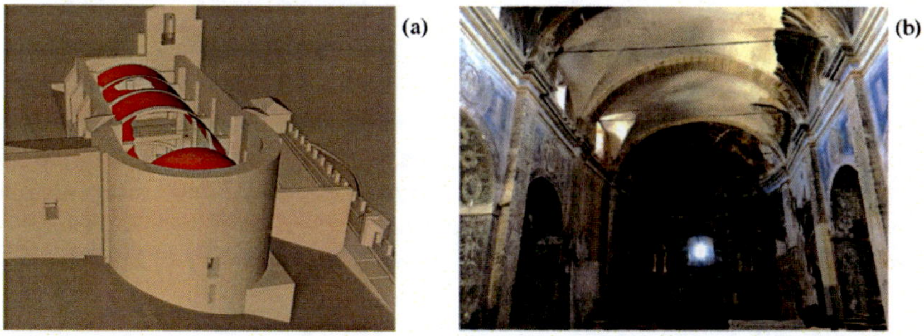

Fig. 3. Collapse of vaults before intervention (a); new steel tension bars of 4th and 5th arches (b).

Once the transverse arches and the vaults supported by them have been rebuilt and reinforced, tension bars have also been placed in the 4th and 5th arches (Fig. 3b) which, in addition to helping their structural consolidation, visually unifies the whole of the arches from the inside. These new tension bars have been made with elements of similar characteristics to those of the existing braces, that is, with smooth steel bars of 25 mm diameter, fastened by their ends with nut and locknut against steel plates of dimensions 500 mm/500 mm/20 mm.

2.2.2 Repairing Works

This section includes two different interventions, one in the arches and vaults and another in the roofing.

d) *Stone transverse arch and brick vaults*
   As indicated, the 4th arch was collapsed along with the brick vaults on both sides of it. It has been reassembled. Some original voussoirs were on the floor of the nave, many of them in good condition. For its reassembly a formwork was executed, copying the geometry from the existing arches and the voussoirs in good condition were placed, completing with replicas to restore the entire arch (Fig. 4a).

For the stone replicas beige sandstone with medium grain size was selected of density 2.1 g/cm^3 and Munsell color 10YR 6,5/2,4; physical and mechanical properties were tested. When placing the voussoirs, their extrados was reinforced with two bars 12 mm diameter of fiberglass polyester connected with perpendicular short bars inserted between the voussoirs, all received with hydraulic lime mortar.

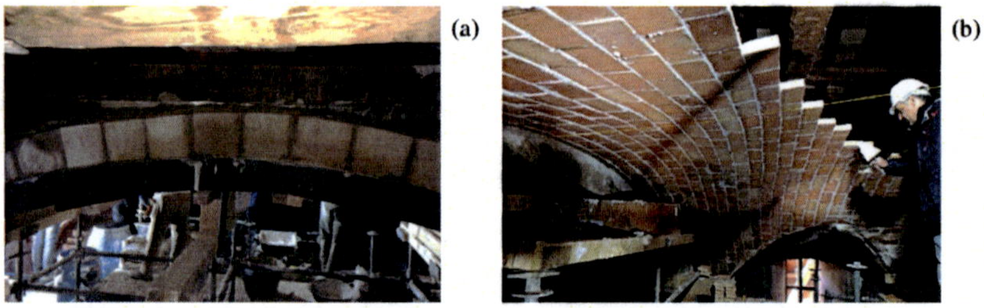

Fig. 4. Reassembled arch (a) and new brick vault (b)

As for the vaults, complete 5th section, 80% of the 4th section and the central part of the 2nd section, have been reconstructed with a double layer of solid ceramic brick 230 mm × 110 mm × 37 mm of density 1.6g/cm^3, received with plaster including polyester mesh between the two sheets and another reinforcement mesh with plaster on the extrados (Fig. 4b).

e) *Roof structure*
   The roofing of the main nave was in apparent good condition, since the elements of the structure were intact and without striking deformations. However, it suffered

a local subsidence on the north side as a result of the failure of a series of rafters and their support beam on the wall, due to the collapse of the north wall already commented. Consequently, only the broken elements and some of those adjacent to them have been changed, replacing them with similar ones of pine wood (Fig. 5a). Likewise, a double anti-xylophage treatment (in depth and surface) has been applied to all elements (existing and new).

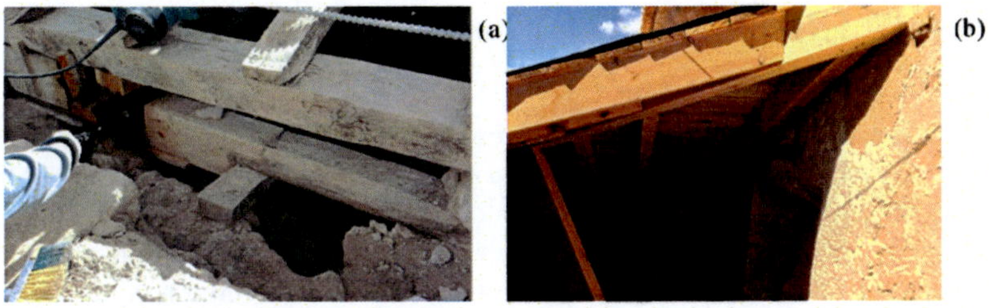

Fig. 5. Punctual repair of roof structure of the main nave (a) and new sacristy roof structure (b)

The roof of the sacristy was completely collapsed and resting on the vault, as a result of the failure of its support structure that was very poor. For its repair, a new structure has been executed which, in addition, has allowed it to improve its configuration in order to avoid the maximum risk of rainwater filtration (Fig. 5b).

The new structure has been made with pieces of sawn wood and treated against xylophages, with the following elements: a central beam arranged on the NE-SO diagonal, which divides it into two parts; rafters every 60 cm supported by the central diagonal beam and the wood beams on the south and west walls.

2.3 Innovative or Traditional Works

In any intervention there are some materials and techniques that are accepted even if there is no great experience in its use, while others can be considered consolidated over time.

2.3.1 Innovative Interventions

Those reinforcements based on more or less innovative materials are included below.

f) *Stitching of walls and vaults cracks with fiberglass reinforced polyester bars*
On the walls, once the vertical structural elements (walls and semi buttresses) have been stabilized and their movements controlled, the outside cracks have been stitched. To do this, corrugated bars of fiberglass reinforced polyester 12 mm diameter have been introduced, in a substantially horizontal plane and rotated about 45° respect to the facade in both directions, so that they intersect at the midpoint of the crack in the middle of the wall (Fig. 6a). The bars have been placed approximately every 40 cm vertically, and received with hydraulic lime mortar, and whenever possible, two bars have been placed on each side.

Fig. 6. Preparation of apse vault stitching (a); flat ridge for roof maintenance

Vaults cracks are the result of the deformations suffered, some longitudinal and other transversal. They have been stitched in some cases by the extrados (Fig. 6b) before executing the surface reinforcement and in others by the intrados. The stitch has been carried out by corrugated bars of fiberglass reinforced polyester 8mm diameter, placed perpendicular to the crack, overlapping on both sides of it a minimum of 50 cm, and received with plaster. On the other hand, the vaults that remained were also reinforced by their extrados by plaster reinforced with polyester meshes, with an average thickness of 40 mm.

g) *Tying up of walls and semibuttresses to the interior pilasters*
The presence of vertical cracks coinciding with the semi buttresses on the outside, thus on both sides of the inner of side chapels, implied that the connection between the outer wall, the semi buttresses and the internal pilasters, was very poor. Consequently, these elements have been tied using corrugated bars of fiberglass reinforced polyester 16 mm diameter, perpendicular to the facade crossing from side to side, placed horizontally every meter in height, bounded with hydraulic lime mortar.
h) *Flat ridge*
Execution of a flat ridge to facilitate the maintenance of the roof was an innovative change so to help the passage of people (Fig. 6c).

2.3.2 Traditional Interventions

Traditional interventions are referring to roof system and repointing masonry.

i) *Tile roof replacement*
The roof was damaged by breakage and detachment of tiles, especially in the small roofs of the side chapels, sunk and with numerous broken tiles and plant growth, and in the sacristy, where its configuration facilitated the entry of rainwater in its NE corner; all this increased by the actions of birds, wind and frost.

To solve these problems, a total renovation of the coverage of the main nave (Fig. 6a) and the sacristy has been carried out, with the following actions:

- Disassembly of all existing tile coverage until reaching the shingle, including grouting mortar;

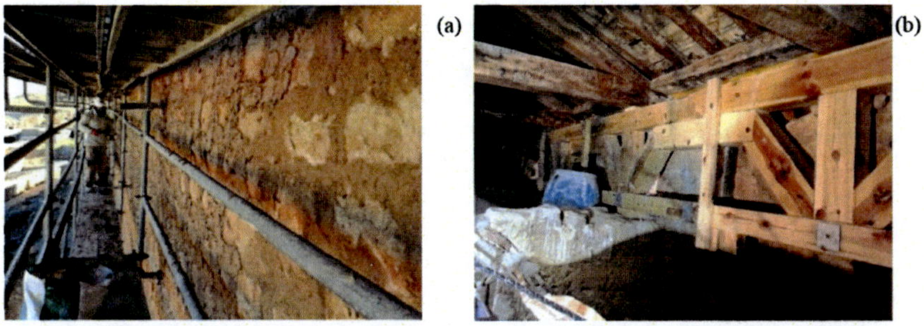

Fig. 7. Masonry repointing (a); wooden latticework for interior reinforcement of north wall

- Cleaning and renovation of the church's shingle board, replacing the broken pieces with new ones of the same characteristics;
- Placing corrugated fibro-bitumen plate in all the gables, nailed to the shingle board, to ensure the water tightness of the gables;
- Reposition of the curved tile recovering those in good condition and placement of new pieces like existing ones;
- New valleys of zinc in all the encounters of the gables.

j) *Masonry repointing*
General erosion was observed in all exterior surfaces, especially in the protruding elements, due to leaching and frost. The aim of the action was to leave the masonry firmly and repointing the most deteriorated joints, which can be summarized as:

- Elimination of disintegrated mortars and inadequate coatings, which had been detached due to lack of cohesion, faulty repairs or altering the masonry, especially Portland cement mortars.
- Repointing of open joints once the cleaning of deposits or vegetation is done. Masonry mortars of natural hydraulic lime (NHL5) and siliceous aggregates were used (Fig. 7a).

2.4 Reversible or Irreversible Interventions

It is an important binomial in restoration but partially debatable, especially when we talk about structural consolidation. In this sense, some of the actions carried out are reversible and others are not.

2.4.1 Reversible Interventions

Clearly reversible actions are indicated. If deemed necessary, all can be replaced by other consolidation techniques, given that the elements used can be clearly identified.

k) *North wall interior reinforcement*
The north wall of the church out of plumb-line moved outward had caused the partial loss of support of the roof structure. This had caused the breakage of the wood beam

on the wall and some of the rafters. A wooden lattice beam (Fig. 7b) has been placed inside the wall, which helps to recover the support of the timber elements of the roof structure. Said beam has been executed with a latticework of sawn wood treated against xylophages, 1m high, which has been fastened to the wall itself through steel bolts of 12 mm diameter screwed on both ends with double nut on 100 mm/100 mm/8 mm plates, all stainless steel.

l) *Tying of the bell tower walls*
The bell tower suffered various deformations, with the opening of joints and the appearance of vertical cracks in its walls, especially those facing west and north. To ensure its integrity, a series of static transverse ties have been introduced, placed at different heights of the tower, crossing it from side to side in both directions. They have been executed with steel bars Gewi type, 20 mm diameter and have been fastened at their ends by stainless steel plates of dimensions 150 mm/150 mm/10 mm and nut and locknut of the same material (Fig. 8a). A total of 9 braces of various dimensions have been placed depending on their location.

Fig. 8. Steel plate of tied walls of the bell tower (a); pre-consolidation of carved wood of altarpiece (b)

m) *Pre-consolidation of altarpiece*
The main altarpiece of good execution corresponds to the Baroque phase in the middle or late 17th century. The structural consolidation has been carried out by incorporating metal angles, oak strips and galvanized screws, metal squares and plates and corresponding screws in various parts of the altarpiece. The surface consolidation, by impregnating the stratum to fix the raised areas with 12g Japan paper and 50% Plextol B500 synthetic adhesive in distilled water (Fig. 8b).

2.4.2 Irreversible Interventions

As it has been said, there are consolidation actions that are necessary for the deteriorated building and that cannot be reversible. In our case we can mention two.

n) *Reinforced concrete for the bracing of the walls of nave and sacristy on its upper line*

The masonry wall of the nave and sacristy are basic structural elements to receive vertical loads, as well as possible horizontal thrusts, provided by the roof. Therefore, they need a horizontal bracing in their upper line, which, unfortunately, they do not have, which facilitates their opening movements. Therefore, it is frequent to incorporate these type of bracing in consolidation interventions and be done in an irreversible manner as in this case.

Therefore, an upper line bracing of reinforced concrete has been executed on the walls of the nave and sacristy. Its shape is irregular, adapting to the resulting gap between the carved sandstone blocks of cornice and the inner wall where rest the wood beam that receives the roof structure. It has a part of the section that is substantially rectangular, with the reinforcement of six bars 12 mm diameter, with its corresponding stirrups made with 8 mm diameter, which is wider on the north wall, and another triangular part on the interior area of the blocks cornice (Fig. 9a). It is anchored to all ashlars of the cornice on the inside, by means of corrugated bars of 8 mm diameter of fiberglass reinforced polyester, as well as to the wall itself by steel bars of 12mm diameter every meter, with a minimum depth of 60cm (Fig. 9b).

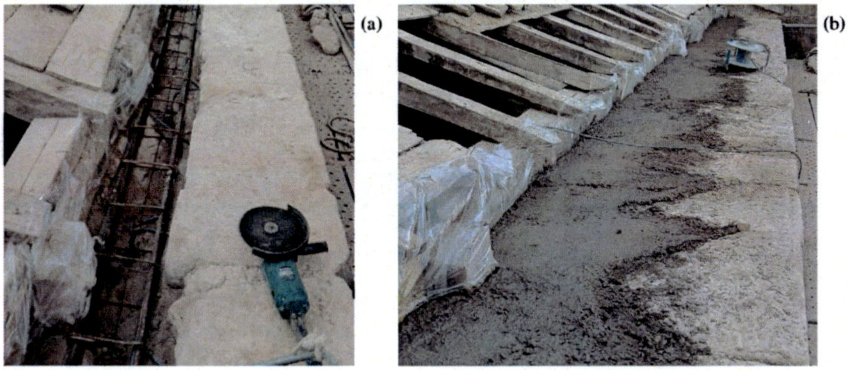

Fig. 9. Reinforcing bars of upper bracing of walls (a) and concrete cast (b)

o) *Containment-underpinning of the north facade wall and the apse*
One of the main problems of the church was the collapse of the north wall of the church as a result, among other reasons, of a poor foundation. In fact, there is no foundation, but the masonry wall rests directly on the natural rock, on the edge of it, which entails the risk of barefoot and turn. In addition, the rock was partially in the open. Therefore, it was decided to place a containment-underpinning beam, which holds and protects the supporting rock and contains the wall base (Fig. 10a).

It is carried out in three sections, corresponding to two different areas of the north wall and the apse area:

- In the first section of the north façade, a beam of variable height between 1.20 m and 1.50 m, adapting to the natural rock of the base, about 35 cm to 40 cm of edge, and overlapping the base of the wall at an average height of 30 cm. It is reinforced with a double mesh of 12 mm diameter, with 20 cm × 20 cm grid, and is anchored to the ground with Gewi bars of 25 mm diameter, approximately 3 m long, placed three times per meter, in two horizontal courses, and with an inclination of approximately 15° (Fig. 10b).
- In the second section of the same facade, a retaining wall of about 2.5 m of average height, about 40 cm thick and with characteristics similar to that of the first section, both in reinforcements and in ground anchors (Fig. 10c).
- In the third section, a longitudinal containment wall, following the trace of the previous one, to exceed the end of the apse, in order to contain the ground on which the base of the wall apse rests. This wall has an average height of 2 m, with similar characteristics to the previous one, and has been covered by the earth that was moved for its execution.

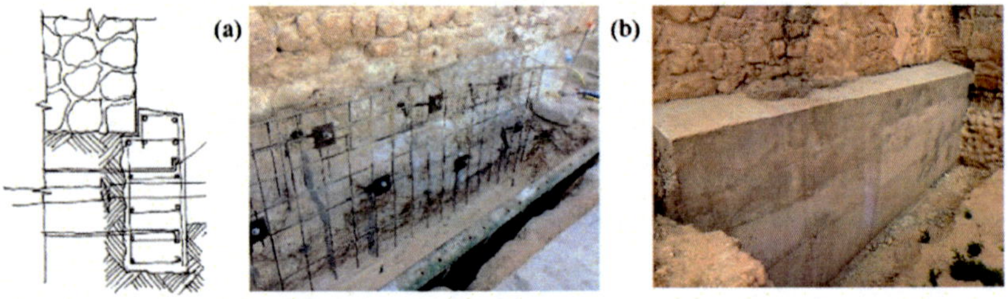

Fig. 10. Schema (a), reinforcing bars (b) and concrete containment-foundation (c)

3 Conclusions

Emergency works of a building without use and therefore lack of maintenance over several decades are similar to those necessary because of the deterioration due to aging, natural disasters (earthquakes) and violent events caused by man.

In view of the different actions that have been carried out in the emergency intervention for the structural consolidation of the church of Santa María de la Asunción in Monreal de Ariza, different types of actions are necessary that make up the binomials that have been stated, namely:

- Temporary and definitive – Preventive and repair
- Innovative and traditional, and – Reversible and irreversible.

And each of them must be adequately justified from a diagnosis as complete as possible, covering the three basic aspects of any cultural heritage building, that is:

- recovery of its constructive integrity ("firmitas"),
- functional possibilities *("utilitas"),*
- historical-artistic value *("venustas").*

In addition, it is important to preserve the site memory as a contextual value of the monument *(intangible values).*

This will allow us to select the appropriate intervention technique in each case to ensure all the values of the building.

Acknowledgements. To Elena Simon, Deputy Director of Coordination and Administrative Management of the General Directorate of Architecture, Housing and Land of Ministry of Development of Spain and EMR-Estudio Métodos de la Restauración S.L. that has carried out the emergency works, in particular to its director Mercè Ortiz.

References

AIPA, Research Group Análisis e Intervención en el Patrimonio Arquitectónico, Juan Monjo Carrió, Pilar Rodríguez Monteverde, Rosa Bustamante Montoro, Luis Cardona Lorente, Prediagnóstico y propuesta de actuación de emergencia en la Iglesia de Nuestra Señora de la Asunción de Monreal de Ariza, Madrid, october 2016

Anagnostopoulos, S., Moretti, M.: Post-earthquake emergency assessment of building damage, safety and usability—Part 1: Technical issues. Soil Dyn. Earthq. Eng. **28**(2008), 223–232 (2008)

Bonazza, A., Maxwell, I., Drdácký, M., Vintzileou, E., Hanus, C.: European Commission. Safeguarding Cultural Heritage from Natural and Man-Made Disasters. A comparative analysis of risk management in the EU. Publications Office of the EuropeanUnion, Luxembourg (2018)

Bustamante, R.: Salvaguardia y Trabajos de Emergencia durante la Guerra Civil (1936–1939). Tratado de Rehabilitación, vol. 1, pp. 79–129, ed. Munilla-Lería, Madrid (1999)

Binda, L., Modena, C., Casarin, F., Lorenzoni, F., Cantini, L., Munda, S.: Emergency actions and investigations on cultural heritage after the L'Aquila earthquake: the case of the Spanish Fortress. Bull. Earthquake Eng. **9**, 105–138 (2011). https://doi.org/10.1007/s10518-010-9217-3

CEN. EN 15898 Conservation of cultural heritage – Main general terms and definitions (2029)

Lagomarsino, S.: Damage assessment of churches after L'Aquila earthquake. Bull. Earthquake Eng. **10**(1), 73–92 (2012)

Ministry of Culture. Charter of Baños de la Encina for Conservation of Defensive Architecture in Spain. Reference documents of National Plan of Defensive Architecture (2006)

Modena, C., Valluzzi, M.R., Da Porto, F., Casarin, F.: Structural aspects of the conservation of historic masonry constructions in seismic areas: remedial measures and emergency actions. Int. J. Architect. Heritage **5**(4–5), 539–558 (2011). https://doi.org/10.1080/15583058.2011.569632

Restoration Techniques

Experimental Verification of New Seismic Reinforcing Method for Temples and Shrines Built with Traditional Japanese Timber Construction

Atsuo Takino[1](✉), Hiroyuki Masuda[2], Sara Yagami[3], Yusuke Ehara[2], Shunji Yamamoto[2], and Kazuo Aoki[2]

[1] Department of Architecture, Osaka Institute of Technology, 5-16-1 Omiya, Asahi-ku, Osaka 535-8585, Japan
atsuo.takino@oit.ac.jp

[2] Design Department, Takenaka Corporation Co., Ltd., 1-13, 4-chome, Hommachi, Chuo-ku, Osaka 541-0053, Japan

[3] KAP Co., Ltd., 2-4-9, Chiyoda-fujimi, Chiyoda-ku, Tokyo 102-0071, Japan

Abstract. Because seismic walls or braces cannot be added to temple and shrine buildings designated as cultural properties, seismic reinforcements are often installed under the floor. However, because many temple buildings have traditional Japanese construction, many column bases rest directly on foundation stones and are often decayed in old buildings. When only the decayed column base is cut off and replaced with new wood, a traditional method, called *Netsugi* ("grafting") in Japanese, is used to repair columns. However, seismic reinforcement is required when *Netsugi* is employed because *Netsugi* joints transmit very little bending moment. Therefore, in this study, we proposed a new seismic reinforcement construction method suitable for traditional Japanese wooden buildings and verified its reinforcement effect by experiments. This method, which uses carbon fiber, provided sufficient reinforcement, and it was possible to adopt it for the seismic repair work in an actual building.

Keywords: Seismic strengthening · Traditional Japanese wooden structure · Temple · Shrine · Carbon fiber · Experimental verification

1 Introduction

Japan contains many traditional wooden buildings, such as shrines and temples, many of which have been designated as cultural properties. Because earthquakes occur frequently in Japan, the seismic risk diagnosis has been increasing for buildings that are designated as cultural properties. Traditional wooden buildings tend to be very old, and most of them do not satisfy current laws and standards, but due to the seismic risk, it is necessary to implement seismic reinforcement. In general, adding a brace or a seismic wall is the simplest seismic retrofitting method [1], but there are often few walls in the types of buildings addressed by this paper, and it is difficult to add a brace or a seismic wall

I. Vayas and F. M. Mazzolani (Eds.): PROHITECH 2021, LNCE 209, pp. 873–891, 2022.
https://doi.org/10.1007/978-3-030-90788-4_67

to buildings designated as cultural properties because the existing structure cannot be changed significantly. For this reason, it is desirable to install seismic reinforcement under the floor, where there are fewer design restrictions.

However, many temple buildings use traditional construction methods, in which the support columns rest on foundation stones, and the column bases are often rotted, especially in old buildings. In Japan, such column bases are repaired using a method called *Netsugi* in Japanese (which roughly translates to "grafting" in English), in which only the rotted column base is cut out and replaced with a new member (Fig. 1). However, the traditional joint can transmit very little bending moment. If seismic reinforcement is applied to the column base with a *Netsugi* joint, some kind of reinforcement for the joint is required.

Therefore, in this study, we proposed a new under-floor seismic reinforcement construction method and conducted a verification experiment. Then, we also proposed a reinforcement construction method for the *Netsugi* joint and performed additional verification experiments.

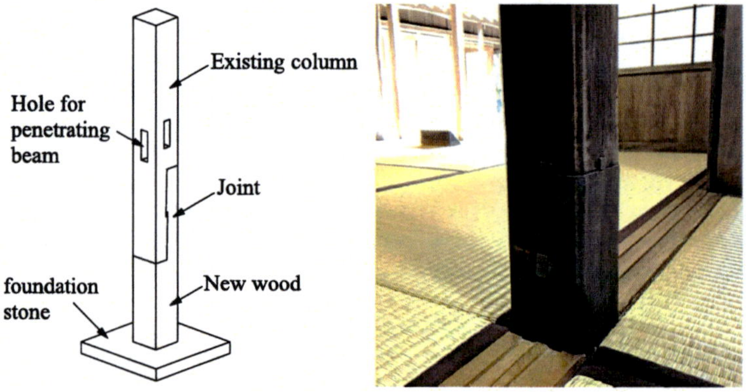

Fig. 1. *Netsugi* repair method for a column base

2 Proposal and Experimental Verification of the New Reinforcement Construction Method Using *Nageshi* Beam

2.1 Outline of the New Reinforcement Construction Method

Several methods have been reported for reinforcing structures under a floor, such as reinforcing a column-beam joint with a steel connection [2] and using an improved *Nageshi* beam [3, 4], a tie beam generally used as a decorative member (Fig. 2). In particular, a timber member is used for the reinforcement method using a *Nageshi* beam, and therefore this method has the characteristic that it does not damage existing columns because wood screws are not used for joining the reinforcement with existing members. In this *Nageshi* beam reinforcement method, because the reinforcement effect usually cannot be obtained unless two columns are tied, when the building is reinforced in both

X and Y directions in plan view, it is necessary to install two *Nageshi* beams on one column as shown in Fig. 3(a), and then it is difficult to fit the reinforcement under the floor. Therefore, as shown in Fig. 3(b), by joining and integrating two *Nageshi* beams orthogonal to each other with steel connections at the joint, the *Nageshi* beam in the Y direction can provide reinforcement even when force is applied in the X direction.

Fig. 2. Standard *Nageshi* member

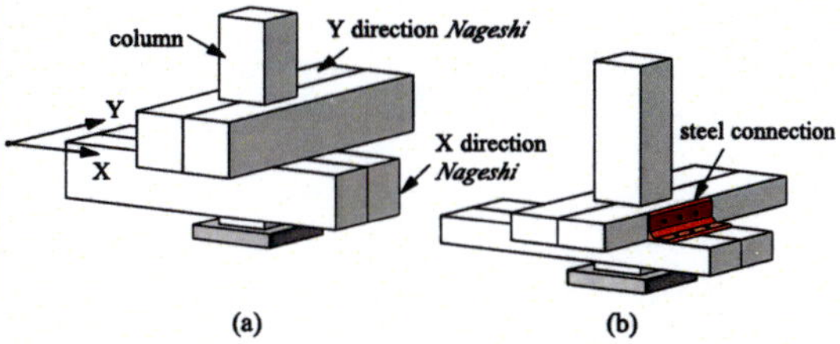

Fig. 3. Image of strengthening method using *Nageshi* beam

2.2 Nageshi Beam Specimens

An experiment was conducted to verify the reinforcement effect of the *Nageshi* beam construction method. Specimens and details of the joints are shown in Figs. 4 and 5, respectively. The basic configurations of the specimens were determined with reference to an existing temple. In general, the cross-sectional dimensions of columns used in temple buildings are very large, but for the convenience of the test equipment, the cross-sectional dimensions of the columns of the test specimen were determined to be 150 mm square. Specimen 1 consisted of only a frame with columns, base, beam, and penetrating beams for comparing the reinforcing effect of the *Nageshi* beam. Specimen 2 was a standard reinforced specimen effective in one direction. In specimens 3 and 4, an orthogonal *Nageshi* beam (75 × 150 mm) was joined by steel connections with a second *Nageshi* beam (75 × 150 mm) in the direction of the force. The same reinforcement effect as in specimen 2 was expected because the apparent height at the joint was 150 mm. In specimen 5, two *Nageshi* beams (75 × 150 mm) were stacked in the direction of force, the upper and lower beams were joined with a steel plate using bolts, and wooden dowels were embedded between the *Nageshi* beams. In specimens 6 and 7, two layers of orthogonal *Nageshi* beams (75 × 150 mm) were joined by steel connections with a third *Nageshi* beam (75 × 150 mm) in the direction of force, to form a joint with an apparent height of 225 mm.

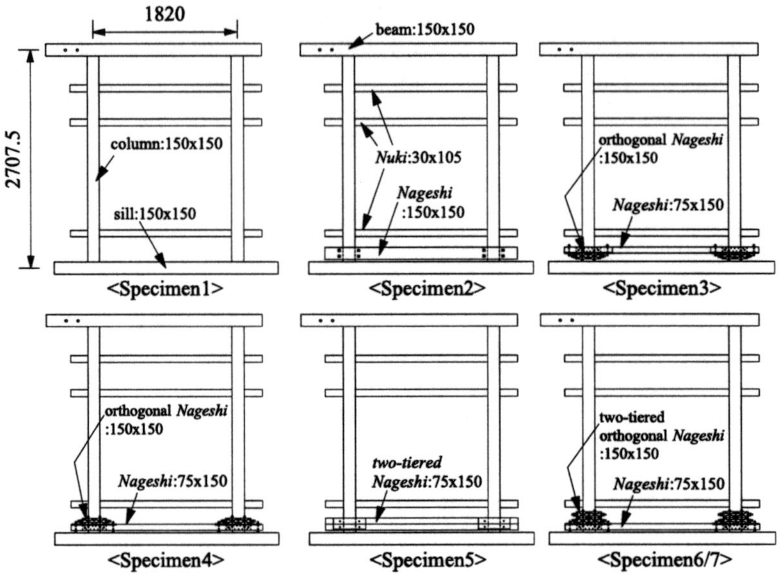

Fig. 4. Configurations of specimens

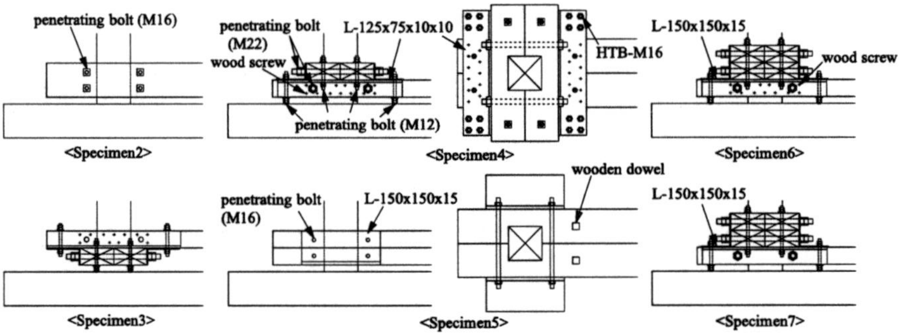

Fig. 5. Details of each joint type

2.3 Loading Method

The configuration of the loading setup is shown in Fig. 6. A specimen was placed on a steel frame, and a horizontal force was applied by a hydraulic jack connected to the beam at the top of the specimen. The applied force was controlled by displacement control, and a positive and negative alternating force was applied repeatedly to produce deformation angles of 1/300, 1/200, 1/150, 1/100, 1/75, 1/50, and 1/30 rad. The displacement cycle was repeated twice at each deformation angle, and finally, the force was applied to one side to produce a deformation of 1/10 rad.

Fig. 6. Loading setup

2.4 Results

The relationship between load and shear deformation angle in the *Nageshi* specimens is shown in Fig. 7. None of the specimens suffered significant damage or load reduction until 1/10 rad. Moreover, even in the specimen using the orthogonal *Nageshi* beams, no

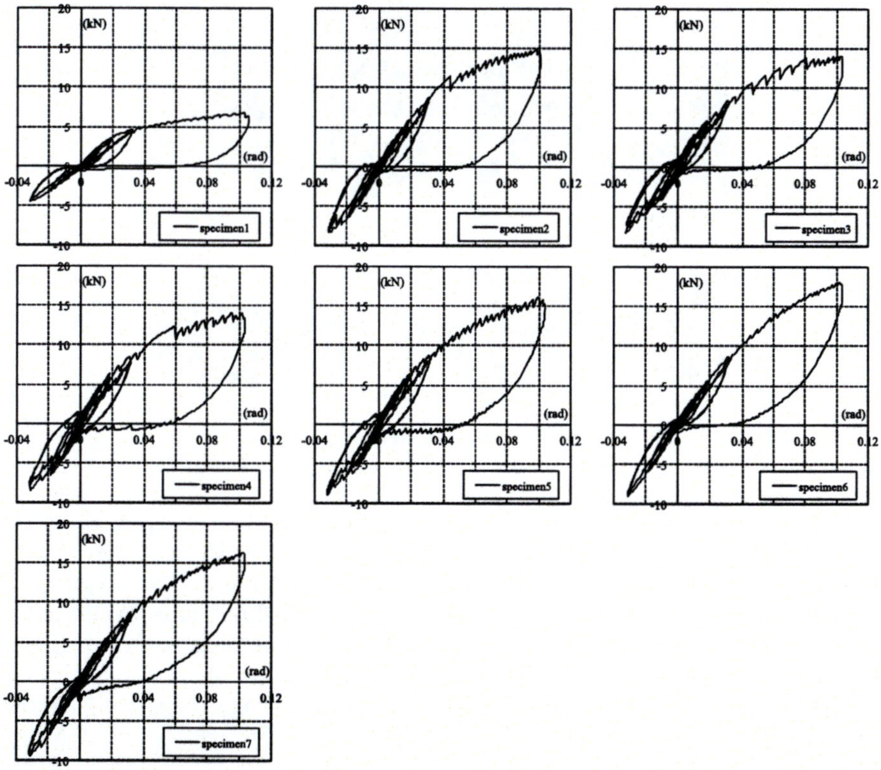

Fig. 7. Relationship between load and shear deformation angle in *Nageshi* specimens

significant damage was seen in the joint, and it seems that the effect of the reinforcement steel connection was maintained until the end.

An example of the deformation of *Nageshi* beams at 1/10 rad is shown in Fig. 8. In specimen 2 using a 150-mm *Nageshi* beam, the beam is almost unbent, but in specimens 3 and 6, where the height of *Nageshi* beam was 75 mm, it can be confirmed that the *Nageshi* beam is bent, especially in specimen 6, which had an apparent joint height of 225 mm. After the specimen was dismantled, it was possible to confirm the impression of the embedment of wood on the column side due to the *Nageshi* beam (Fig. 9). In addition, in the specimen using the orthogonal *Nageshi* beam, it was possible to confirm a slight trace of embedment of wood on the side of the orthogonal *Nageshi* beam. Although the joint was hardly damaged even at the end, in specimen 7, in which the wood screws were omitted, a gap was generated between the steel connection and the wood as the deformation progressed (Fig. 10). This is because only bolts were used to join the steel connection and wood.

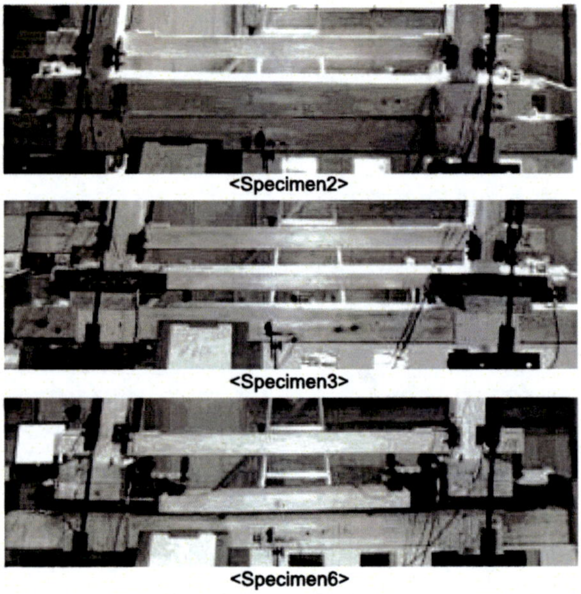

Fig. 8. Bending deformation of *Nageshi* beam at 1/10 rad

Figure 11 shows the column bending moments measured from strain gauges attached at the positions shown in Fig. 12. In each plot in Fig. 11, the vertical axis represents the bending moment of the column and the horizontal axis represents the shear deformation angle of the specimen. In all the specimens, the values of σ2 and σ4 on the lower side near the joint are large. Moreover, in specimens 6 and 7, the value of σ4 measured at the column on the compressive side is extremely large, and it can be seen that a large bending moment occurs at the joint. However, we concluded that the bending moment

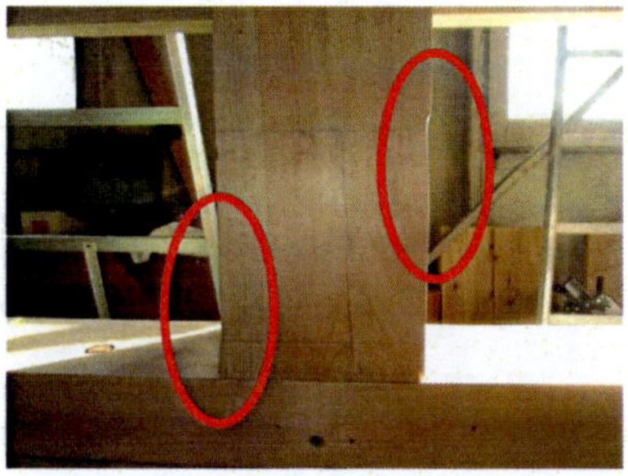

Fig. 9. Impression of the embedment of wood on the column in specimen 6

Fig. 10. Gap between the steel connection and the wood in specimen 7

of the column on the tensile side was less than that on the compressive side, because the column on the tensile side was slightly lifted.

The relative vertical deformation of the *Nageshi* beam from the base was measured at the displacement meter positions shown in Fig. 12, and the result of the bending deformation of the *Nageshi* beam in specimens 2, 3, and 6 at 1/10 rad is shown in Fig. 13. In Fig. 13, the horizontal axis represents the positions where displacement meters were installed, and the vertical axis represents the value obtained from the meters. The overall graph shows a downward slope because the column on the tensile side was lifted. In addition, the curve for specimen 2 with a 150-mm *Nageshi* beam has a linear distribution, whereas the curves for specimens 3 and 6 using a 75-mm *Nageshi* beam are not linear, which means that the *Nageshi* beam was bent. In particular, the bending deformation of the *Nageshi* beam is remarkable in specimen 6. This is the same as the result confirmed visually by Fig. 8.

A comparison of the positive envelope of the relationship between load and shear deformation angle is shown in Fig. 14. Comparison of all the reinforced specimens with specimen 1 shows that reinforcement effect of the *Nageshi* beam increased both the initial rigidity and maximum strength. In addition, the reinforcement effect of the orthogonal *Nageshi* beam can be confirmed, because the initial rigidity and maximum strength of specimens 3 and 4 with 75-mm orthogonal *Nageshi* beams were almost the same compared with those of specimen 2. However, in specimens 6 and 7, the initial rigidity and strength were not considerably increased, but the ultimate strength at 1/10 rad was

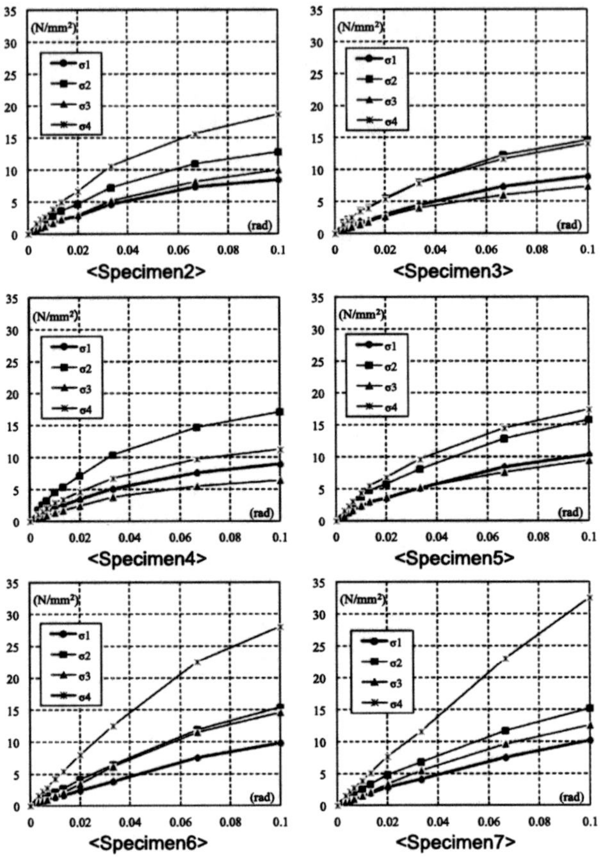

Fig. 11. Bending moments of columns reinforced with *Nageshi* beams

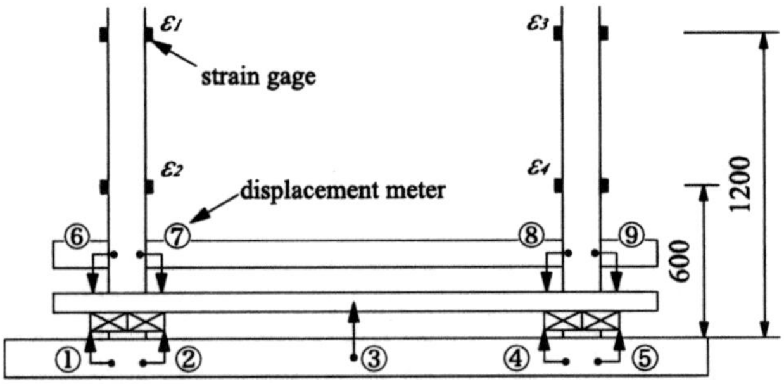

Fig. 12. Setup of column strain gages and displacement meters for *Nageshi* beam

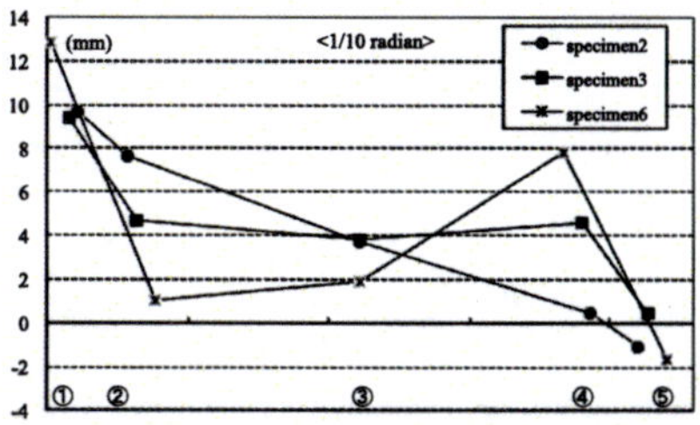

Fig. 13. Bending deformation of the *Nageshi* beam in specimens 2, 3, and 6 at 1/10 rad

slightly increased. As described above, this is because the bending moment generated in the column is increased due to the increased rigidity of the joint, and also the bending deformation of the column and *Nageshi* beam is increased.

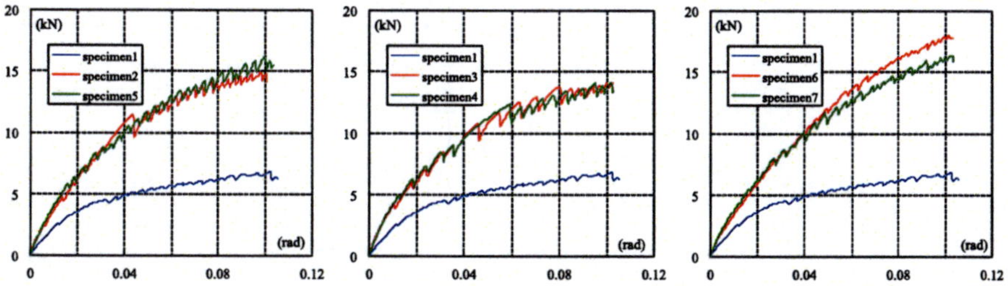

Fig. 14. Comparison of the positive envelope of the relationship between load and shear deformation angle in specimens reinforced with *Nageshi* beams

3 Bending Experiments of Timber *Netsugi* Joints Reinforced with Car-Bon Fiber

3.1 Kanawa-Tsugi Joint Specimens

Timbers 180 mm square were used as test specimens and the *Kanawa-tsugi* joint (also known as a "scarf" joint, Fig. 15) was used as the joining method. In addition, in some specimens, a through-hole for a penetrating beam was installed near the joint in consideration of the actual construction state (Fig. 15). The reinforcing material used was a unidirectional carbon fiber (CF) cloth [5] with a CF weight of 200 g/m^2. There are two types of *Kanawa-tsugi* joint, the strong axis direction and the weak axis direction. However, because there is a possibility that a force acts in both directions on the column, in most of the reinforcement specimens the load was applied in the weak axis direction.

A list of specimens is given in Table 1, and the specimen configurations are shown in Fig. 16. In addition, because some specimens were loaded by a different device, experimental conditions, such as the specimen lengths, were slightly different.

The CF cloth was cut to a width of 100 mm along the fiber direction. Then, it was attached to the joint area in the member axis direction with epoxy adhesive, and further

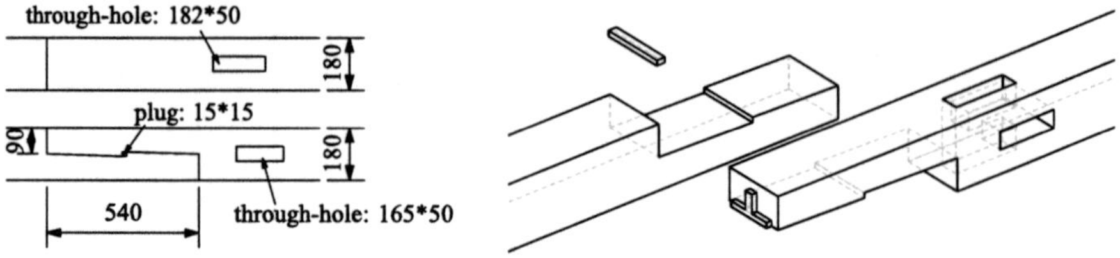

Fig. 15. Detail of *Kanawa-tsugi* joint and through-hole for a penetrating beam

Table 1. List of *Kanawa-tsugi* joint specimens

Type	Name	Length	*Kanawa-tsugi*	Through-hole	CF cloth	Number of layers	Attaching method of CF cloth		L_1 (Length of C cloth in member axis direction)	Loading method	Direction of applied force
							Direction				
		m					Member axis	Orthogonal	mm		
Kanawa-tsugi joint / unreinforced	K4-1	4.2	○	-	-	-	-	-	-	A	Weak axis
	K6-2	6.0	○	-	-	-	-	-	-	B	Strong axis
Through-hole / unreinforced	N6	6.0	-	○	-		-	-	-	C	-
Kanawa-tsugi joint / reinforced	K6-1-C-F-O-3	6.0	○	-	○	3	Center	Orthgonal	1040	B	Weak axis
	K6-2-C-F-O-3	6.0	○	-	○	3	Center	Orthgonal	1040	B	Strong axis
	K6-1-C-F-O-5	6.0	○	-	○	5	Center	Orthgonal	1240	B	Weak axis
	K6-2-C-F-O-5	6.0	○	-	○	5	Center	Orthgonal	1240	B	Strong axis
	K6-1-C-S-O-5	6.0	○	-	○	5	Corner	Orthgonal	1290	C	Weak axis
Kanawa-tsugi joint and through-hole / reinforced	KN6-1-C-F-O-5	6.0	○	○	○	5	Center	Orthgonal	1290	C	Weak axis
	KN6-1-C-S-O-5	6.0	○	○	○	5	Corner	Orthgonal	1290	C	Weak axis
	KN6-1-C-S-ON-5	6.0	○	○	○	5	Corner	Orthgonal	1290	C	Weak axis
	KN6-1-C-S-R-5	6.0	○	○	○	5	Corner	Spiral	1290	C	Weak axis

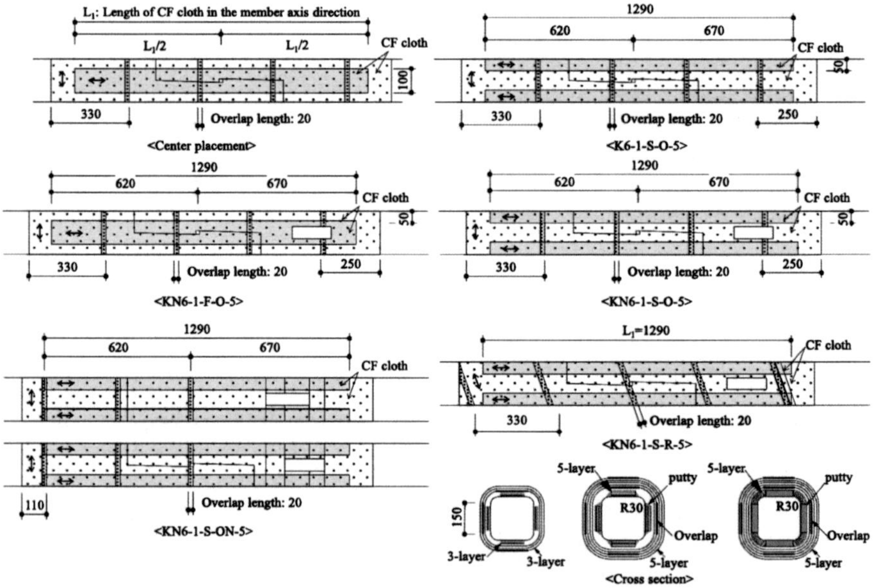

Fig. 16. Configuration of reinforced *Kanawa-tsugi* joint specimens

attached in the orthogonal direction of the member axis. Three or five layers were used, as indicated in Table 1. In all cases, the layers were attached in both the axial and orthogonal directions.

For specimens with a through-hole for the penetrating beam, an opening in the CF cloth must be cut at the through-hole when it is attached around the joint. For this reason, we prepared specimens in which the CF cloth at the through-hole was cut after the CF cloth was attached and specimens in which the CF cloth was applied to the corners of the columns to avoid the through-hole. In addition, several types of attaching methods in the orthogonal direction were provided, and a specimen in which a CF cloth was attached inside the through-hole and a specimen in which the CF cloth was attached in a spiral manner was prepared.

For comparison with reinforced specimens, three unreinforced specimens were prepared. The two specimens were with a *Kanawa-tsugi* joint only, and force was applied in the strong axis direction and the weak axis direction. The other was a specimen with only a through-hole for the penetrating beam.

3.2 Loading Method

Figure 17 shows the method used to apply force. Two force application methods were used depending on the availability of experimental equipment, but in both cases, the same bending moment and shear force were applied to the joint. The load at the applied point and the vertical displacement at the position shown in Fig. 17 were measured.

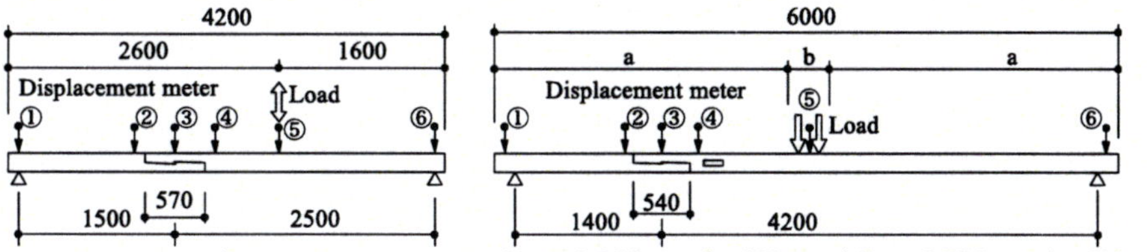

Fig. 17. Loading methods for *Kanawa-tsugi* joint specimens

3.3 Results

An example of the *Kanawa-tsugi* joint fracture behavior is shown in Fig. 18. Detailed fracture and deformation characteristics of each specimen are described below.

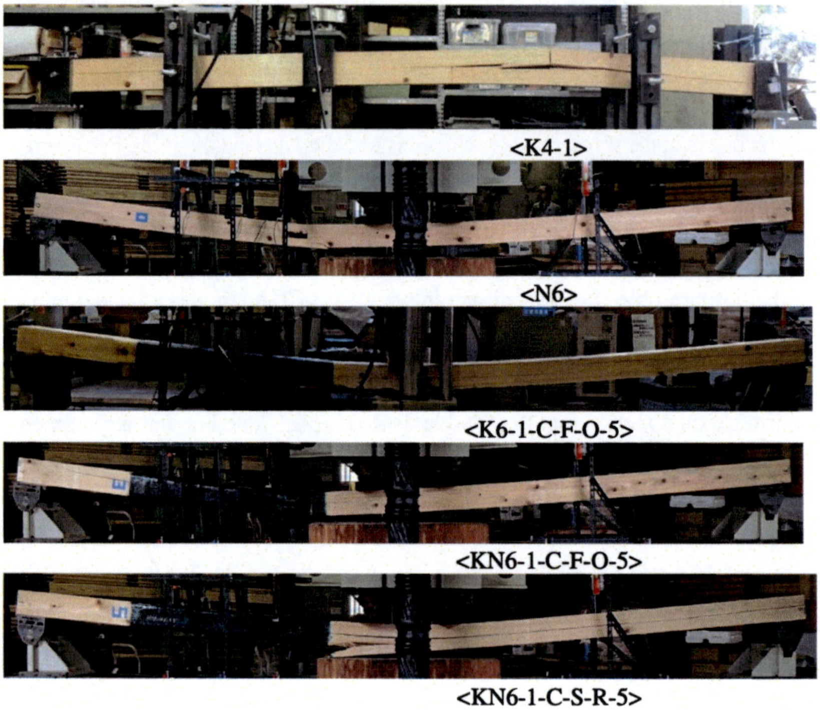

Fig. 18. Examples of fracture behavior

3.3.1 Unreinforced Specimens

Figure 19 shows the relationship between the load and vertical deflection at the center and joint position of the beam. The "E11" lines in the figure represent the elastic deflection of the beam calculated with a Young's modulus of 11.0 kN/mm^2 and the assumption that the beam is a single solid material without joints. Calculations were made using two lengths, L = 4.0 m and 6.2 m.

In all specimens with a *Kanawa-tsugi* joint, the rigidity decreased immediately after force was applied, and the joint was broken (Fig. 20). In addition, in the specimens with a through-hole, the rigidity remained almost linear until just before the maximum load was reached, but after reaching the maximum load, the load gradually decreased and compressive fracture around the through-hole occurred, as shown in Fig. 21, and finally, the specimen broke at the knot on the tensile side.

The distribution of vertical deflection at a specific load (P = 10 kN) is shown in Fig. 22. The vertical axis in the figure represents the amount of displacement, and the horizontal axis represents the positions of the displacement meters, which correspond to the numbers in Fig. 17. In addition, the results of other specimens described below are also shown in the figure. For the specimen with only the *Kanawa-tsugi* joint, it can be seen that the deflection at the joint is very large and the beam is bent at the joint. In contrast, it can be seen that in the specimens with only a through-hole, the influence of the through-hole is almost negligible in the initial stage of force application, and the bending deformation is generated smoothly.

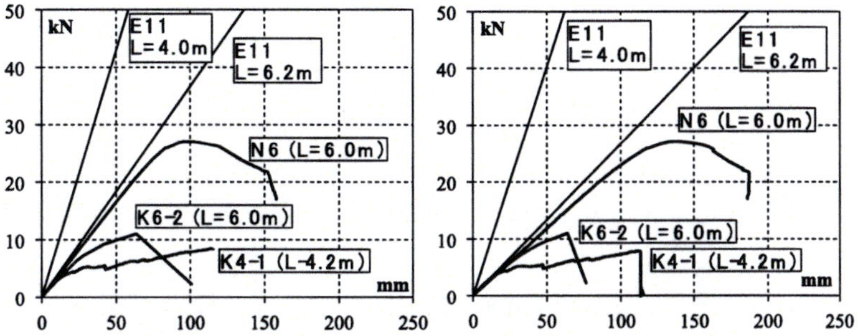

<at the joint position of the beam> <at the center of the beam>

Fig. 19. Relationship between load and vertical deflection in *Kanawa-tsugi* joints

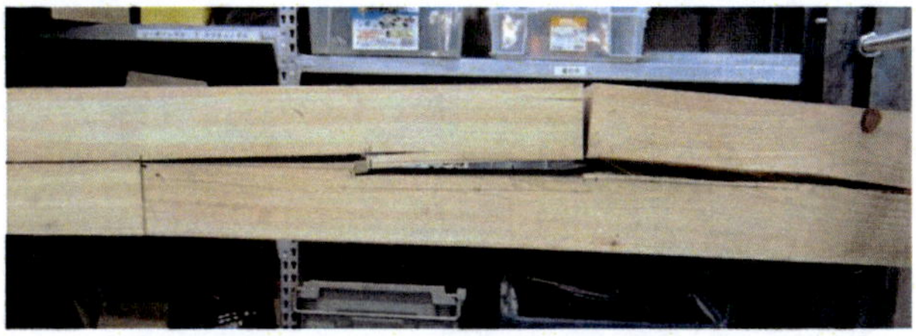

Fig. 20. Fracture of a *Kanawa-tsugi* joint

Fig. 21. Compressive fracture around a through-hole

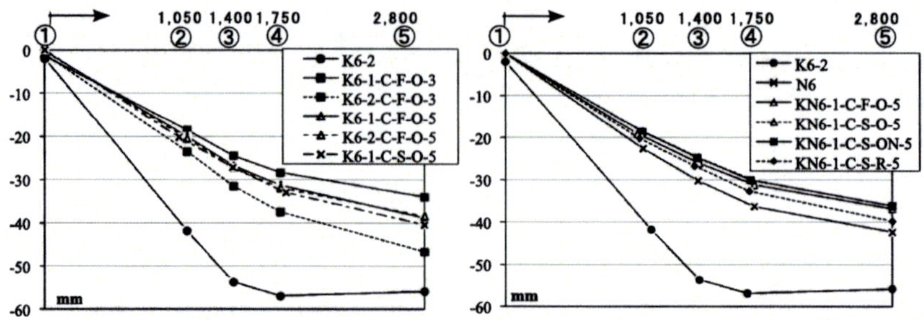

Fig. 22. Distribution of vertical deflection of *Kanawa-tsugi* joints at the load P = 10 kN

3.3.2 Effect of CF Cloth on Kanawa-Tsugi Joint Specimens Without Through-Hole

The relationship between the load and vertical deflection at the beam center and the joint position in beams without through-holes is shown in Fig. 23, and the distribution of vertical deflection at a specific load is shown in Fig. 22. For specimens K6-2-C-F-O-3, K6-1-C-F-O-5, and K6-2-C-F-O-5, results after displacement of 147 mm are not shown because the displacement at the center of the beam became impossible to measure during the experiment. In addition, because force method C was applied to specimen K6-1-C-S-O-5, the distance from the support to the center of the beam is different from the condition

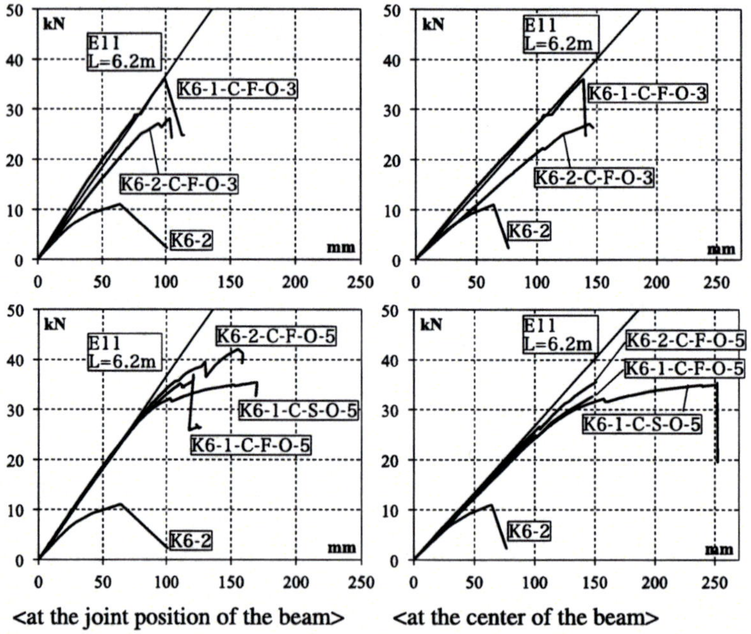

Fig. 23. Relationship between load and vertical deflection in *Kanawa-tsugi* joint specimens without through-hole

of force method B, but because the difference in the span is slight, we decided to compare the results within the same graph. Until the ultimate fracture, significant deformation of the joint was not observed, and the deformation characteristics are almost the same as those of the solid material without a joint. Ultimately, bending failure occurred in the wood part at the center of the beam, and the load decreased rapidly. In addition, in the load–vertical deflection relationship, almost no decrease in rigidity was observed until the maximum load was reached, and the initial stiffness was similar to that of the elastic deflection line (E11) assumed to represent a single solid material. For specimen K6-1-CFO-5 only, it was confirmed that the CF cloth was broken by the impact when the specimen was bent and fractured (Fig. 24). However, for the others, there was no noticeable damage to the CF cloth. In addition, specimen K6-1-C-S-O-5, in which the CF cloth was attached at the corners of the cross section, had results not significantly different from those of the other specimens. In this experiment, there was no difference in the results due to differences in the length of the CF cloth, the number of stacked layers, and the direction of the applied force.

Fig. 24. CF cloth was broken by the impact when the specimen was bent and fractured

3.3.3 Effect of CF Cloth on Kanawa-Tsugi Joint Specimens with Through-Hole

The relationship between the load and vertical deflection at the beam center and the joint position of beams with through-holes is shown in Fig. 25, and the distribution of vertical deflection at a specific load is shown in Fig. 22. In all the specimens, significant deformation around the joint and through-hole was not observed until immediately before the fracture, and the deformation properties were almost the same as those of a solid material without the joint. However, the fracture properties at the ultimate state were different for each specimen; in KN6-1-C-S-ON-5 with reinforced through-holes and KN6-C-S-R-5 with CF cloth spirally attached, the specimen was broken in the center of the beam due to the bending moment; however, the other specimens (KN6-1-C-F-O-5 and KN6-1-C-S-O-5) were bent and broken at the joint, and a shear failure of the joint was confirmed inside the through-hole, as shown in Fig. 26. This is thought to be due to through-hole formation making the wood discontinuous, and it could not withstand the shearing force transmitted from the *Kanawa-tsugi* joint. In contrast, the specimen with reinforcement inside the through-hole (KN6-1-C-S-ON-5) is bent and broken at a large knot under the load application point, as shown in Fig. 27, and because the maximum load was low, it is highly probable that there was no fracture at the joint.

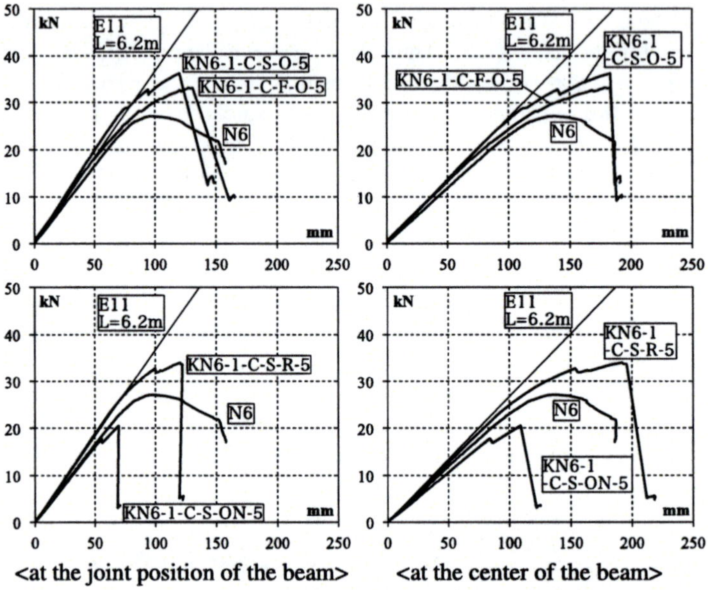

Fig. 25. Relationship between load and vertical deflection in *Kanawa-tsugi* joint specimens with through-hole

Fig. 26. Shear failure of the joint in specimen KN6-1-C-S-O-5

Fig. 27. Fracture from a large knot in specimen KN6-1-C-S-ON-5

3.3.4 Bending Stress

Table 2 summarizes the maximum load, fracture location, and bending stress at maximum load of all specimens. The bending stress shown in the table was obtained from the bending moment, which was calculated from the maximum load at the joint and center of the beam, divided by the section modulus ($Z = 972\ cm^3$) calculated with the assumption of a solid material 180 mm square.

Table 2. Summary of maximum load, fracture location, and bending stress at maximum load

Type	Name	Maximum load	Location of fracture	Bending stress at maximum load	
		kN		Joint	Center
Kanawa-tsugi joint / unreinforced	K4-1	8.47	Joint	4.9	6.1
	K6-2	11.03	Joint	7.9	14.8
Through-hole / unreinforced	N6	27.11	Throuth-hole	19.5	34.9
Kanawa-tsugi joint / reinforced	K6-1-C-F-O-3	36.12	Center	26.0	48.3
	K6-2-C-F-O-3	28.15	Center	20.3	37.7
	K6-1-C-F-O-5	36.87	Center	26.6	49.3
	K6-2-C-F-O-5	42.08	Center	30.3	56.3
	K6-1-C-S-O-5	35.42	Center	25.5	45.6
Kanawa-tsugi joint and through-hole / reinforced	KN6-1-C-F-O-5	36.32	Joint	26.2	46.7
	KN6-1-C-S-O-5	33.25	Joint	23.9	42.8
	KN6-1-C-S-ON-5	20.56	Center	14.8	26.4
	KN6-1-C-S-R-5	34.02	Center	24.5	43.8

The bending stress at the joint of the unreinforced specimens with only the *Kanawa-tsugi* joint (K4-1 and K6-2) was significantly lower than that of the bending standard material strength (26.7 N/mm^2) assumed for a single solid material without joints. In addition, in the unreinforced specimen with only a through-hole (N6), the bending stress at the joint (19.5 N/mm^2) was lower than the bending standard material strength.

All the specimens with only the *Kanawa-tsugi* joint reinforced with CF cloth were broken at the center of the beam, and the bending stress at the center was clearly higher than the bending standard material strength. The bending stress at the joint is not necessarily the maximum value because the fracture position is not at the joint, but all of the results exceeded 20.0 N/mm^2, and it can be said that a sufficient reinforcing effect was obtained. In addition, because differences depending on the direction of force (weak axis vs. strong axis) cannot be confirmed, as in the case of the unreinforced specimens, it can be seen that the reinforcing method has sufficiently contributed to the proof stress of the column with the *Kanawa-tsugi* joint.

In the specimens with the *Kanawa-tsugi* joint and through-hole reinforced with CF cloth, except for KN6-1-C-S-ON-5 (which had reinforcement inside the through-hole), because the bending stress of the joint exceeded 20 N/mm^2, it is considered that a sufficient reinforcing effect could be obtained. However, the reinforcing effect in specimen KN6-1-C-S-ON-5 could not be confirmed in this experiment because a bending failure occurred at a large knot just below the force application point, as shown in Fig. 27.

4 Conclusion

In this study, we proposed a new seismic reinforcement construction method suitable for traditional Japanese wooden buildings, such as shrines and temples, and verified the reinforcement effect by experiments. In addition, the method of reinforcing the *Netsugi* joint, which is a traditional Japanese method to repair column bases, using CF was verified by experiments. For all reinforcement methods tested, sufficient reinforcement effect was obtained, and it was possible to adopt it for the seismic repair work in an actual building of a cultural property (Fig. 28).

Fig. 28. Seismic repair work in an actual building

Acknowledgements. This work was supported by the Japan Society for the Promotion of Science, KAKENHI Grant Number JP15H05537. We would also like to thank the religious corporation of Shounen-ji Temple and the Nara Prefectural Board of Education, Office of Cultural Properties Preservation.

References

1. Architectural Institute of Japan: Standard for Structural Design of Timber Structures (2006). (in Japanese)
2. Editorial board of seismic design manual for wooden post-and-beam construction building: Seismic design manual for wooden post-and-beam construction building based on traditional construction method. Gakugei Shuppan-Sha (2004). (in Japanese)
3. Sato, H., Fujiwara, I., Fujita, K.: Research on the Chronology of the Form and the Structural Performance of NAGESHI. Summaries of Technical Papers of Annual Meeting, Structures-III. Architectural Institute of Japan, pp. 413–414 (2003) (in Japanese)
4. Takeshi, K., Osamu, S., Makoto, K., Kazuo, K.: Experimental tests on the improved Nageshi-column joint in traditional wood structures. AIJ J. Technol. Des. **19**(42), 531–536 (2013). (in Japanese)
5. Hyunjung, C.: The use of carbon fiber in historic wooden architecture. AIJ J. Technol. Des. **16**(34), 1175–1180 (2010). (in Japanese)

Strengthening Measures for Upgrading the Seismic Response of Gazi Hasan Pasha Mosque, Kos Island

Fillitsa V. Karantoni(✉) and Dionysia Dimakopoulou

Department of Civil Engineering, University of Patras, Patras, Greece
karmar@upatras.gr

Abstract. On 21/7/2017 a strong earthquake with epicentre between the Greek island Kos and the Turkey city Bodrum shook the area. The magnitude of the earthquake was Mw = 6,6 and it caused damage both to build environment and to humans, two people were killed. Older masonry buildings and newer ones of reinforced concrete, as well as roads and the port, developed damage. The Gazi Hasan Pasha mosque, a structure built during the 17th century developed very limited damage during this seismic event. In 1933 an earthquake of the same size had also shaken Kos Island and nevertheless the damage was extensive, the load-bearing walls responded well without cracks or extensive deformations; the mosque developed extensive damage to the minaret and to timber members as is the roof and the floor. In this paper both the capacity of the structure under the seismic loads of EN1998-1 as well as the effectiveness of recommended strengthening measures are studied. Two strengthening schemes provided the insertion of steel tendons are proposed. In order to assess the effectiveness of these measures, the assessment was done according to the Greek Code for Interventions to Masonry Structures (KADET) that suggests performance base criteria.

Keywords: Masonry · Mosque · Earthquake · Strengthening · Performance base criteria

1 Introduction

In June 2017 a strong earthquake of magnitude Mw = 6,6 and epicentre area the sea between the Greek island Kos and the Turkish city Bodrum, struck the region. In Fig. 1 from Ganas et al. 2017 the epicentre area as well as the nearby faults are presented. There were not recorded accelerations in Kos (15 km apart from the epicentre) and as it is mentioned in Kiratzi and Koskosidi 2018, to estimate the level of ground shaking at Kos town from the records of the three Turkish stations is not straightforward. Psyharis and Taflampas 2017 estimated the response acceleration spectrum shown in Fig. 2 considering the recorded spectrum accelerations at Bodrum and Norcia (Italy) as the upper and lower limits of the ground movements at the town of Kos relating the soil conditions. The earthquake caused significant damage to both historic structures and modern ones of reinforced concrete of the town of Kos and the death of two persons as well. Gazi

I. Vayas and F. M. Mazzolani (Eds.): PROHITECH 2021, LNCE 209, pp. 892–902, 2022.
https://doi.org/10.1007/978-3-030-90788-4_68

Pasha Mosque, having small piers and many openings as it is seen in Fig. 3 seems to be vulnerable to seismic actions but surprisingly developed light damage. As seen in Fig. 4(b) both the low percentage and the large horizontal span of the load-bearing walls contribute to the vulnerability of the structure. Instead of the expected severe damage, it did not develop but partially separation of the cladding wall of the south load-bearing wall and of the minaret as well. According to remembrances, after the large displacements of the minaret in 1933, a new one reinforced with concrete jacketing and cladding with a wall of limestones was erected; the cladding wall partially collapsed during the 2017 seismic event.

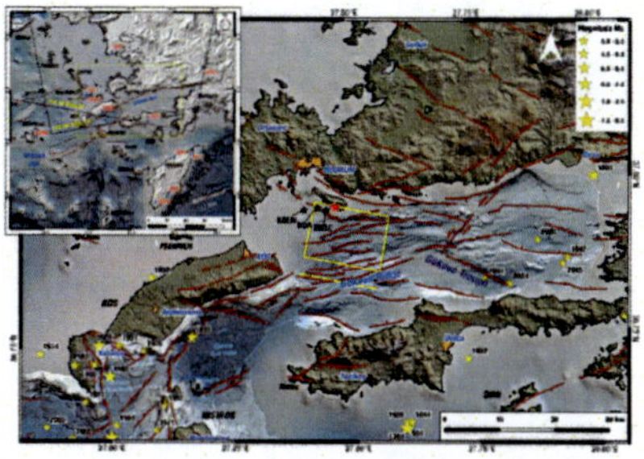

Fig. 1. Location map and relief model of the region affected by the Kos earthquake, *(from Ganas et al.* 2017*)*

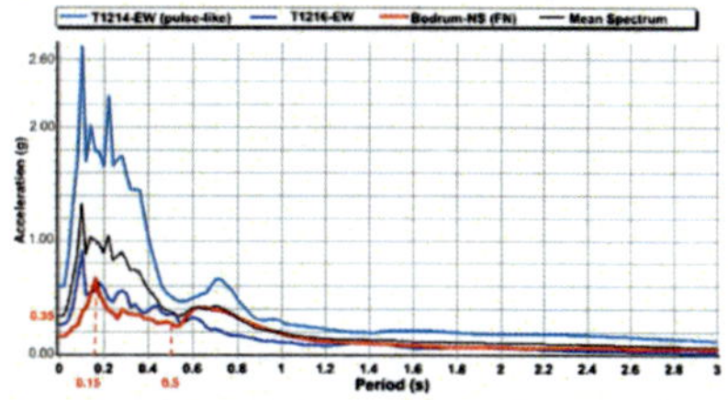

Fig. 2. Estimation of response acceleration spectrum, (*after* Psycharis and Taflampas 2017*)*

Fig. 3. The Gazi Hasan Pasha Mosque from East-North

For understanding this particular behavior, assessment of the seismic performance of the mosque using the estimated response spectrum of Fig. 2 performed. (Karantoni and Dimakopoulou 2020). The assessment implemented with terms of deformation capacity, namely with performance-based criteria. The seismic assessment of masonry structures used to be carried out in terms of forces derived from linear elastic analyses because the non-linear characteristics of each type of masonry are not known to the level necessary to get reliable output. The calculation of stiffness degradation after cracking is practically impossible and even after experiments, as Mojsilovic 2019 shown, there is not a trustworthy function to describe it. Nevertheless, the non-linearity of the bearing system, and not of the material can be considered if it is assumed that the structure behaves linearly until the yielding deformation and it cracks after that to a grade depending on the ultimate deformation capacity of the system (or the deformation just before it collapses). This method well known as a performance base method is not of wide use in masonry structures but is largely used to the seismic assessment of frame systems of reinforced concrete or steel members.

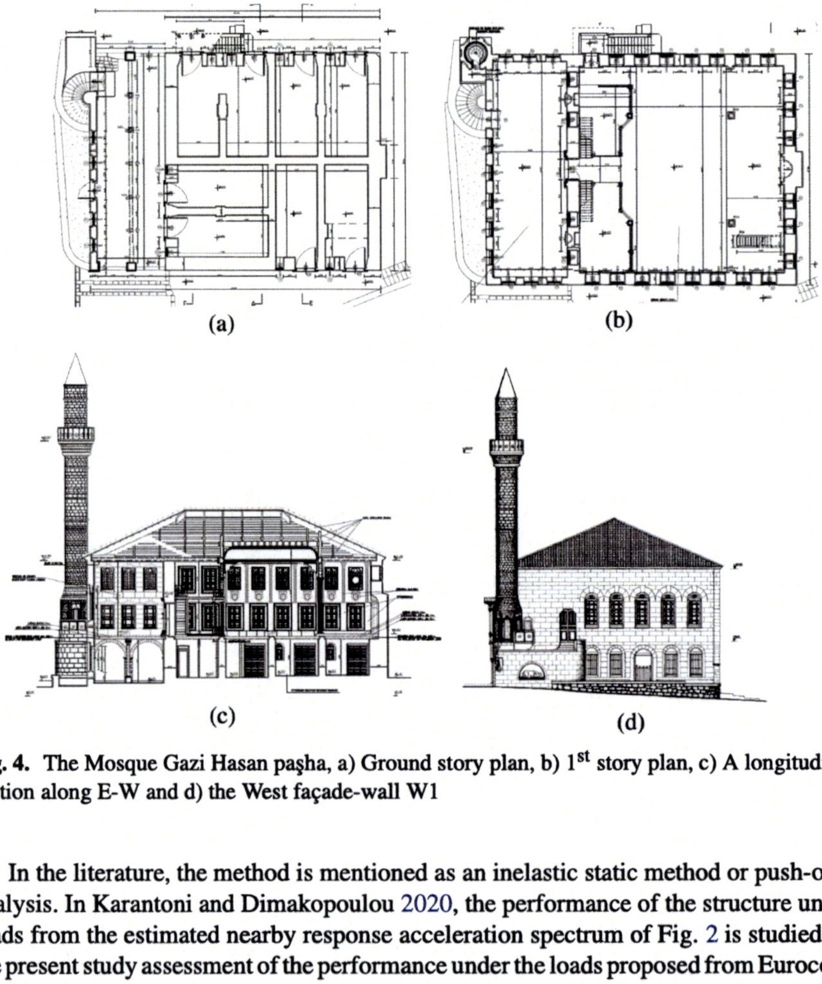

Fig. 4. The Mosque Gazi Hasan paşha, a) Ground story plan, b) 1st story plan, c) A longitudinal section along E-W and d) the West façade-wall W1

In the literature, the method is mentioned as an inelastic static method or push-over analysis. In Karantoni and Dimakopoulou 2020, the performance of the structure under loads from the estimated nearby response acceleration spectrum of Fig. 2 is studied. In the present study assessment of the performance under the loads proposed from Eurocode 1998-1-1 Type I elastic acceleration spectrum is implemented and two low- cost strengthening schemes to resist the demanded deformations are examined. The method is the insertion of steel tendons at two levels: a) at the crest of the longitudinal walls of narthex which are the most deformed ones, and b) at both the crest and the top of ground story of the same walls connecting also the colonnade of the ground story to the parallel walls. The first intervention assures a performance level of “immediate occupancy” to most of the masonry elements but is the application of the second intervention that ensures to all of them this performance level because then the demanded relative drift ratios remain under the capacity deformation of the masonry.

2 The Assessment Method of KADET

The structure analyzed for the seismic loads from the plateau of the elastic acceleration spectrum of Type I of EN 1998-1 (2004), because it was impossible the fundamental period to be estimated due to the low participated mass in each vibrated mode; the mode superposition is not recommended to the type of structures under consideration because the produced deflected shape of structures with distributed over the height mass is far away from its real deformation profile. For the assessment, the provisions of the draft of the Greek Code for Structural Interventions on Masonry Structures (KADET) are applied for each wall and pier too. Details on the method are in Pardalopoulos et al. 2019. From the displacements of the structure the relative drifts θ_{in}, $\theta_{out,h}$, $\theta_{out,v}$ of Fig. 5 are calculated and compared with the limits of the drift ratios of Table 1. In case they are smaller than the values corresponding to Performance Level (PL) the structure exhibits Performance Level of Immediate Occupancy (IO), meaning light or absence of damage. In case the relative drift ratios are higher, then the type of failure is considered and depending on bending or shear failure the PL from the above- mentioned Table 1 is determined. The procedure applies to walls and piers as well. The type of failure depends on the minimum shear derived from Eqs. (1).

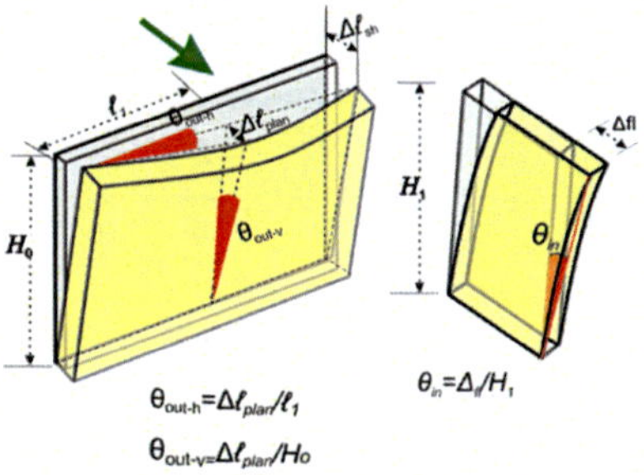

Fig. 5. Determination of the relative drift ratios

In structures without diaphragm function of the floors the out of plane bending of lintels of the top storey, namely the $\theta_{out,h}$ is critical for the walls whereas in case there are no openings in the walls and for the piers, critical is the bending deflection $\theta_{out,v}$. The shear failure due to θ_{in}, is frequent to piers of structures with diaphragm function of the floors.

The relative drift ratio at failure $\theta_{u,in}$, equals $0{,}008H_o/\ell$ for elements controlled by flexure and $0{,}004$ for elements controlled by shear. For in-plane member response, the strength $V_y = \min\{Vf, Vs\}$ depends on the dominated failure mechanism namely the flexural and sliding shear strengths of the member; V_f and V_s that are determined respectively from Eqs. (1).

Table 1. The Performance level criteria of KADET

Seismic response	Relative drift ratio limits	Performance level
In-plane	$\theta_{in} < 0.15\%$	Operational/Immediate occupancy
	$0.15\% \leq \theta_{in} < 0.75 \cdot \theta_{u,in}$	Significant but repairable damage
	$0.75 \cdot \theta_{u,in} \leq \theta_{in} < 1.33 \cdot \theta_{u,in}$	Life-Safe/No collapse
Out-of-plane	$\{\theta_{out\text{-}v}; \theta_{out\text{-}h}\} < 0.20\%$	Operational/Immediate occupancy
	$0.20\% \leq \{\theta_{out\text{-}v}; \theta_{out\text{-}h}\} < 0.75 \cdot \theta_{u,out}$	Significant but repairable damage
	$0.75 \cdot \theta_{u,out} \leq \{\theta_{out\text{-}v}; \theta_{out\text{-}h}\} < 1.33 \cdot \theta_{u,out}$	Life-Safe/No collapse

$$V_f = \frac{\ell \cdot N}{2 \cdot H_o} \cdot \left(1 - 1.15 \cdot \frac{N}{\ell \cdot t \cdot f_d}\right); \; V_s = 0.4 \cdot N < 0.065 \cdot f_b \cdot \ell_c \cdot t \quad (1)$$

Where: ℓ is the length of the element's cross-section orthogonal to the axis of bending, t is its thickness, H_o is the distance from the point about which rotation of the element occurs to the point where the bending moment attains peak value. Also, N is the element's axial load, measured at its base, f_d is the mean compressive strength of masonry and ℓ_c is the length of the element under compression.

The relative drift ratio $\theta_{u,out}$ is the minimum of $\{\theta_{u,1}\theta_{u,2}\}$derived from the Eqs. 2 and 3.

$$\theta_{u,1} = 0.003 \cdot \frac{H_o}{t} \quad (2)$$

$$\theta_{u,2} = \theta_{Ru} \cdot \left(1 - \frac{M_y}{M_{Rd}}\right) \text{ where, } \theta_{Ru} = \frac{t}{H_o} \quad (3)$$

With M_y being the minimum of the "yielding" moments from Eqs. 4 and M_{Rd} from Eq. 5.

$$M_y = f_t \frac{\ell \cdot t^2}{6}; \; M_y = \left(f_t + \frac{N}{2H_o \cdot t}\right)\frac{\ell \cdot t^2}{6} \quad (4)$$

Where f_t is the tensile strength of the masonry, usually taken equals 10% of the compressive strength, and ℓ, and H_o as in Fig. 5, t being the width of the wall and N the axial force for the load combination considered.

$$M_{Rd} = F_{Rd} \frac{H_o}{2}; \; F_{Rd} = \lambda \cdot S_a(T_1) \cdot t \cdot \gamma \cdot A_{L,w} \quad (5)$$

where, γ is the weight per unit volume of the masonry and $A_{L,w}$ being the vertical area of the wall excluding the openings if any. λ accounts for the degree of fixity of the wall ($\lambda = 2$ for cantilever walls and $\lambda = 1$ for walls pined or fixed at the upper end).

3 The Gazi Hasan Pasha Mosque

The Gazi Hasan Pasha mosque shown in Figs. 3 and 4 was built in 1786 on the Greek island Kos while it was under Ottomans' occupation. In 1933, under Italian occupation, a strong earthquake caused severe damage to the building stoke of the Island as well as of the Mosque and the authorities repaired the mosque and the minaret, a reinforced concrete jacket was erected to the last one cladding with a limestone masonry wall. As a strengthening measure of the mosque, the Italians constructed a reinforced concrete tie-beam at the crest of the perimeter load-bearing masonry walls. In 1962 with Kos as a territory of Greece from 1948, the timber roof replaced by a new one. In 1994, the Greek authorities decided to protect the monument from further damage owing to the bad condition of the roof and the reinforced concrete parts of the load-bearing walls according to Touliatos et al. 2009. Nowadays internal timber parts have decayed nevertheless the monument looks externally in good condition.

The mosque is at the first story of a rectangular structure of dimensions about 23.0 m $\times$17.0 m in plan (Fig. 4) and of total height 10,48 m excluding the roof. It is consisting of the prayer hall at east and of the narthex at the west. The north and south walls at ground story under narthex have two large arches as shown in Fig. 4c, to permit the free pass parallelly the short west side of the building. The rest of the ground story, the space under the prayer hall, by construction, was used as rooms for the pilgrims and later as storage space, as it is mentioned in Sideris (2009). Two rows of a total of 61 windows let the light lighten the two parts of the mosque, the narthex, and the prayer hall. In the storage space, many load-bearing walls were constructed to support the timber floor of the prayer hall. Inside the passage, a colonnade of circular columns is running along the NS direction and helps the support of the floor of the narthex. The perimeter load-bearing walls of the prayer hall are of width about 1,06 m in the ground floor and 0,90 m in the 1st story while the walls of the narthex are of a width 0,55 m in their whole height.

The load-bearing walls are of local rubble masonry, some of them used previously in a church located at the same site. A façade wall of 0,10 m thickness of marble stones covers the external face of the walls while the internal face is covered by plaster.

4 Seismic Assessment

For the purpose of seismic assessment, the mosque analyzed first for the accelerations estimated developed from the 2017 Bondrum earthquake to ensure that the assumptions made are correct. In the lack of experimental works to give data for mechanical properties and the structure of masonry, this step was fundamental. The results of the analysis are in Karantoni and Dimakopoulou 2020 and coincide impressively with the real behavior of the monument. After that, the assessment using loads from EN 1998-1 and the procedure of KADET was followed.

For the analysis, a finite element model of the whole structure excluding the minaret that it does not share structural elements with mosque was created using the code *Acord bat*. Three models analyzed using the seismic loads from the response spectrum of type I of Eurocode 8; one model for the "as is" condition and one model for each strengthening scheme. The walls are discretized with Mindlin type finite elements so that both the in-plane and out-of-plane bending to be accounted for. The thickness of each element has the real thickness of masonry they model. The compressive strength of masonry was taken $f_c = 2{,}65$ MPa. The Modulus of Elasticity $E = 2650$ MPa was taken equal to $1000 f_c$. The gravity loads G, as well as the live loads Q, are considered in the analysis with the factors presented in Table 2. The live load was taken equal to 5.0 KN/m^2 because people are expected to be gathered in normal use of the mosque. Seismic loads Ex, Ey for each one of the two horizontal directions x and y respectively were applied as acceleration. The spectral acceleration was calculated from the plateau of the spectrum of Type I for soil of category B; for seismicity Zone I and importance factor III, the ground acceleration a(g) = 0,16 g gives response acceleration $R(t) = 7{,}06$ m/sec^2. This acceleration was applied distributed uniformly along the height of the structure and combined with the dead and live loads. The analyses performed for the eight seismic load combinations (LC) of Table 2; the previously mentioned performance base criteria used for the seismic assessment. In Fig. 6 (a) the finite element mesh is shown; the green-colored finite elements represent the foundation and the blue ones the reinforced concrete tie-beam.

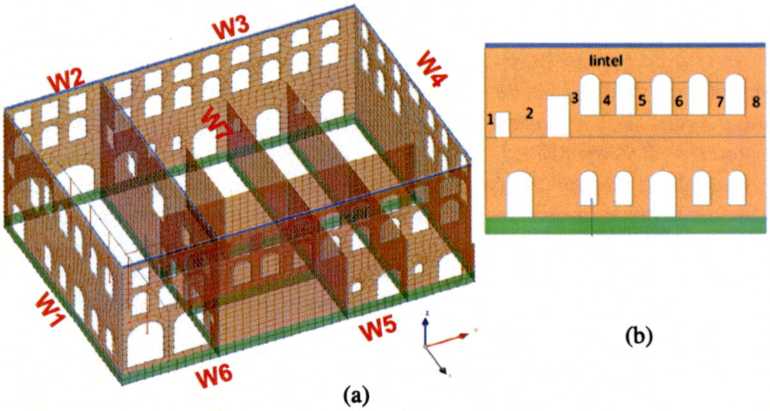

Fig. 6. (a) The finite element discretization of the mosque and notation of walls and (b) the wall W1

Table 2. Coefficients of the equation G + Q + Ex + Ey for each loading combination (LC)

LC	1	2	3	4	5	6	7	8
G	1,0	1,0	1,0	1,0	1,0	1,0	1,0	1,0
Q	0,30	0,30	0,30	0,30	0,30	0,30	0,30	0,30
Ex	1,0	1,0	−1,0	−1,0	0,30	−0,30	0,30	−0,30
Ey	0,30	−0,30	0,30	−0,30	1,0	1,0	−1,0	−1,0

The analysis showed that the walls will develop significant out of plane displacements mainly the west wall of narthex, W1, due to the seismic combinations 5–8 where the coefficient of the transverse to the wall seismic load Ey equals to 1.0 (see Table 2). In Fig. 7 the displacements for four of the eight load combinations are presented in plan as well as the displacement of the wall W1 along the height. The out of plane bending of all the walls is clearly observed and the shape of the displaced walls along the height, too; this shape is more like a parabola and certainly far away from a straight line.

The Table 3 presents the maxima relative drifts ratios of the eight load combinations from the most displaced elements of the monument, which is wall W1, and of the wall W1 the lintel A and the piers 7 and 8 as well, that are shown in Fig. 6 (b). These values correspond to the performance levels of Table 4. Note that the tables present also the relative results after the strengthening of the mosque which will be described in the next section of this paper. From this table is shown that the most deflected elements of the most displaced wall will develop Significand Damage (SD) and repairs will be needed. In order to prevent the structure from damage, strengthening is suggested.

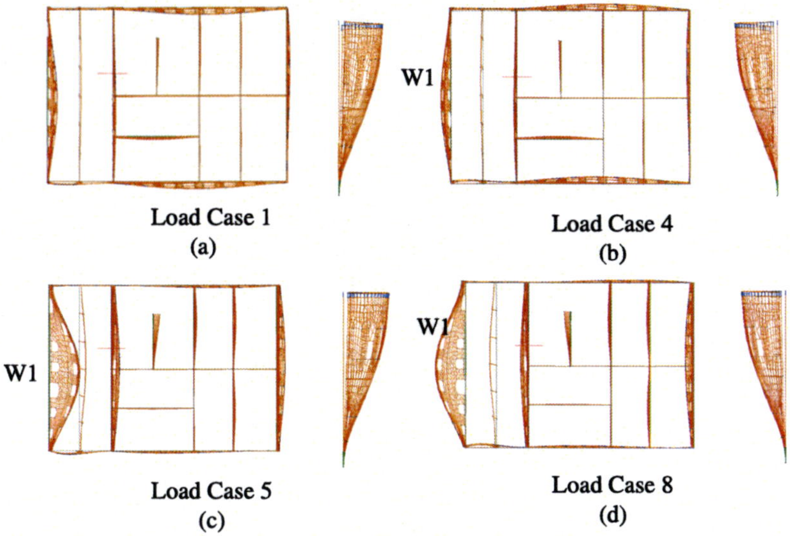

Fig. 7. Displacement of the mosque for selective load cases; in plan on the left and along height on the right.

Table 3. Maxima Relative drift ratios from the unstrengthened and strengthened structure

Element	Unstrengthed			Strengthening scheme 1			Strengthening scheme 2		
	θ_{in}%	$\theta_{out, v}$%	$\theta_{out, h}$%	θ_{in}%	$\theta_{out, v}$%	$\theta_{out, h}$%	θ_{in}%	$\theta_{out, v}$%	$\theta_{out, h}$%
Wall 1	0,042	0,535	0,46	0,040	0,165	0,11	0,040	0,158	0,11
Lintel	0,016	0,013	0,50	0,011	0,019	0,24	0,019	0,043	0,14
Pier 7	0,056	0,161	0,69	0,055	0,148	0,13	0,055	0,17	0,12
Pier 8	0,043	0,051	0,51	0,042	0,001	0,15	0,045	0,053	0,10

5 Strengthening Schemes

As it was proved in the previous section, significant damage will be exhibited to some elements of the structure owing to the out of plane bending of the walls. To eliminate these effects strengthening measures that will be simultaneously effective and of low cost are recommended and their effectiveness is examined. The idea is the establishment of steel tendons to connect the wall W1 with the parallel wall W7 to minimize the out of plane horizontal displacements. Two schemes of strengthening are examined: a) the scheme 1 consists the insertion of five steel tendons of diameter 80mm at the top of the wall W1 to join it with the wall W7 as seen in Fig. 8 (a) and, b) the scheme 2 provides five more tendons at the floor of first story level as seen in Fig. 8 (b).

Table 4. Performance levels for the relative drifts of Table 3

	Performance Level								
	Unstrengthen			Strengthening Scheme 1			Strengthening Scheme 2		
element	In	Out,v	Out,h	In	Out,v	Out,h	In	Out,v	Out,h
Wall 1	IO	SD	SD	IO	IO	IO	IO	IO	IO
Lintel	IO	IO	SD	IO	IO	SD	IO	IO	IO
Pier 7	IO	IO	SD	IO	IO	IO	IO	IO	IO
Pier 8	IO	IO	SD	IO	IO	IO	IO	IO	IO

In=in-plane
Out,v = Out-of plane, Vertical
Out,h= Out-of-plane, Horizontal

The insertion of the tendons at the crest of W1 and W7 reduces both the out of plane relative drift ratios $\theta_{out\text{-}v}$ and $\theta_{out\text{-}h}$ at the most displaced part of the wall W1 as seen in Table 3, keeping almost the same the ratio θ_{in}. The reduce of the displacement of W1 is shown in Fig. 9 where one can also see the change of the displaced profile of the wall. After the intervention the expected most displaced elements are the middle of the piers 4 and 5 of W1 but their relative drift ratios are smaller than that of the middle of the piers 7 and 8, which exhibits a performance level of Immediate Occupancy. Nevertheless, at the lintel, although the out of plane $\theta_{out,h}$ is reduced 50% and equals 0,24% (Table 3), the performance level of Significant Damage is sustained, as seen in Table 4.

By applying the strengthening scheme 2 we achieve the entire structure under the considered seismic loads to develop almost no damage or to exhibit Immediate Occupancy performance level. The relative drifts of all the walls and the piers of the structure, even the most displaced presented in Table 3 are lower than the cracking limits of Table 2. In Fig. 9 the isovalues of displacements, as well as the out of plane deformation of wall W1, are presented.

By applying the strengthening scheme 2, namely the insertion of steel tendons at two height levels at 10,40m and at 4,82m that is the top of each floor, the structure is behaving elastic and each relative drift ratio remains at the performance level of Immediate Occupancy. The cost of installation of the tendons is low making more attractive the considered intervention.

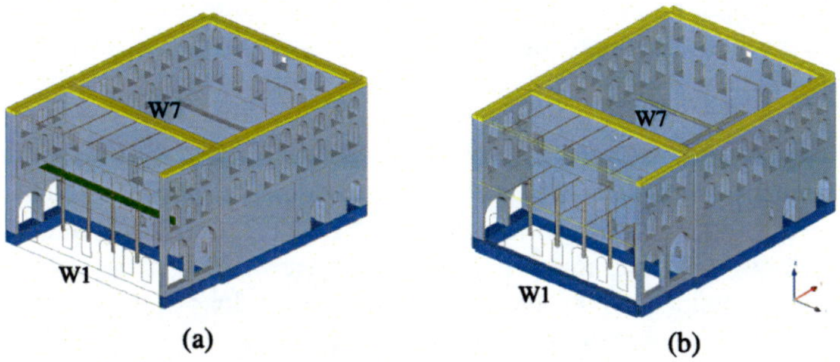

Fig. 8. Insertion of tendons connecting the walls W1 and W7 a) at the top and b) at the top and at the floor of the first story level.

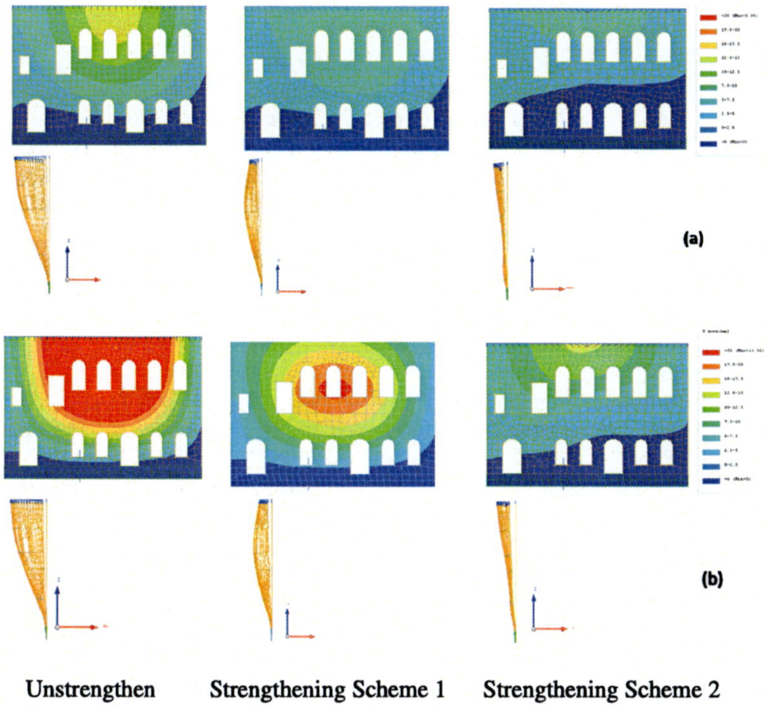

Fig. 9. Isovalues of displacement and deflected shape of wall W1 for LC 4, G + 0,30Q-Ex-0,30Ey (top) and LC 8, G + 0,30Q-0,30Ex-Ey (bottom). The color palette ranges from 0,0 to 20,0 mm for all the three cases for the comparison.

6 Concluding Remarks

Sometimes masonry structures that seem to be vulnerable, behave very well during an earthquake, as Gazi Hasan Pasha Mosque in Kos did in two earthquakes, of 1933 and 2017. For its seismic assessment according to codes in force, a model of the structure subjected to the applied seismic loads from the 2017 earthquake was studied as the first step. The predictions of this first step absolutely confirmed the assumptions made for the model, the method of analysis as well as the masonry properties in lack of laboratory or in site tests. In the next step, the assessment of Gazi Hasan Pasha Mosque performed using the seismic loads from Eurocode 8 and the performance-based method of the Greek Code for Interventions to Masonry Structures (KADET). The analysis predicted that the structure will not withstand the design loads and needs strengthening. Two low-cost proposals are recommended; a) the insertion of steel tendons at the crest of the most displaced wall that reduces but does not eliminate the damage and b) the addition of insertion of tendons and at the level of the top of the ground story wall. The latter one is proved very effective as after that all the walls and its elements behave at a performance level of Immediate Occupancy.

References

CEN. EN 1998-1 : Design of Structures for Earthquake Resistance-Part 1: General Rules, Seismic Actions and Rules for Buildings. Europ. Comm. for Standardization, Brussels (2004)

Ganas, A., et al.: The July 20, M6.6 Kos earthquake: "seismic and geodetic evidence for an active northdipping normal fault at the western end of the gulf of go kova (SE Aegean Sea) (2017). http://194.177.194.200/GPS/0_BGSG/Ganas2019_Article_TheJuly202017M66KosEarthquakeS.pdf

Karantoni, F., Dimakopoulou, D.: The response of Gazi Hasan Pasha Mosque (Kos Island, Greece) to 2017 Mw 6,6 Earthquake. In: Proceedings of REHABEND 2020, ID400, Granada (2020)

Kiratzi, A., Koskosidi, A.: Constraints on the near-source motions of the Kos-Bodrum 20 July 2017 Mw6.6 earthquake. In: Proceedings of the 16th European Conference on Earthquake Engineering, 18–21 June 2018, Thessaloniki, pp.13, ID11450 (2018)

Mojsilovic, N.: Stiffness of unreinforced masonry walls subjected to static-cyclic shear. In: Advances in Engineering Materials, Structures and Systems: Innovations, Mechanics and Applications, Cape Town(2019). https://doi.org/10.1201/9780429426506-290

Pardalopoulos, S.I, Karantoni, F.V., Pantazopoulou, S.J.: Practical assessment of the seismic behavior of a confined masonry system. Soil Dyn. Earthq. Eng. **127**, 105831 (2019). https://doi.org/10.1016/j.soildyn.2019.105831

Psycharis, I., Taflampas, I.: Preliminary estimate of the ground motion at the city of Kos at the 20 July 2017 earthquake [in Greek] p 20, Athens (2017)

Sideris, G.: Progetto di Restauro della Moschea Gazi Hassan Pascia, a Kos, Grecia. Universita degli Studi di Palermo, Agathon (2009). (in Italian)

Touliatos, P., Tsakanika, E., Tsouras, V.: Gazi Hasan paşa mosque on the island of Kos in Greece. Study and reconstruction of the timber roof. In: Conservation and repairing of Ottoman Monuments in Greece, University Studio Press, Thessaloniki (2009)

Seismic Protection of Historical Artefacts Using an Improved Spring Hanged Three-Dimensional Pendulum Isolation Method

Wen Bai(✉), Junwu Dai, and Yongqiang Yang

Key Laboratory of Earthquake Engineering and Engineering Vibration, Institute of Engineering Mechanics, China Earthquake Administration, Harbin 150080, China

Abstract. Along with historical structures, historical artefacts are also treasures left by our ancestors. Some artefacts are slender and easy to overturn. Besides, many artefacts are very fragile due to aging and erosion problems. To protect these invaluable treasures, different methods have been proposed. Base isolation is one of these methods and has been proved effective. Non-extensible cable hanged pendulum system is a simple and stable way to fulfil isolation. This method is mainly useful for lateral isolation. As more and more researches emphasize the importance of vertical isolation, especially at the epicentral areas. This paper tries to replace the non-extensible hanging cables with extensible springs to fulfil three-dimensional isolation. To validate the effectiveness of this method, dynamic history analyses of one typical artefact under a set of artificial ground motion without control, with traditional cable hanged pendulum isolation and with spring hanged pendulum isolation methods are compared. Since most artefacts are freestanding items and are not anchored to the base. Sliding and rocking are the two main indexes used to assess isolation effectiveness. Preliminary results show that both the extensible and non-extensible cable hanged pendulum methods are effective at lateral isolation and have similar trends. Meanwhile, the spring hanged pendulum method, which is the extensible cable method, can also fulfil vertical isolation and can help further reduce the artefact sliding distances than the non-extensible cable method.

Keywords: Artefacts · Overturn · Pendulum · Seismic protection · Three-dimensional isolation

1 Introduction

Historical artefacts, as treasures to all human beings, have suffered damage during the past strong earthquakes (Gong et al. 2016; Ning et al. 2018). During the 2008 Wenchuan earthquake, 4379 pieces of artefacts damaged, including 208 pieces of high value. Most damaged artefacts are porcelain made relics. Many researchers studied the dynamic response characteristics of artefacts and related floating items under earthquakes (Di Sarno et al. 2015; Konstantinidis et al. 2009 and Neurohr et al. 2008). Some artefact damages are caused by structure collapse and some are caused by sliding and dropping from shelves. Others are caused by rocking and overturning. To protect historical

I. Vayas and F. M. Mazzolani (Eds.): PROHITECH 2021, LNCE 209, pp. 903–909, 2022.
https://doi.org/10.1007/978-3-030-90788-4_69

artefacts, many methods have been proposed. For example, extra supports are added to increase the artefact stability as shown in Fig. 1a. Besides, nylon lines, as shown in Fig. 1b, which are fine and transparent, are used to fix the artefacts and help avoiding artefact overturning. Besides these traditional protection methods, base isolation is now getting more and more popular (Koumousis et al. 2019; Calio et al. 2003 and Berto et al. 2013).

There are many isolation device types and pendulum isolation is an effective one, its isolation period is controlled by the cable length and is not affected by the weight of the artefacts. Generally, this method is expected to fulfil lateral isolation. Nowadays, three-dimensional isolation is getting increasingly more emphases. This paper replaces the pendulum hanging cables with springs, trying to fulfil three-dimensional isolation instead of lateral isolation only.

a. Side support b. Nylon lines

Fig. 1. Traditional methods to protect historical artefacts

Finite element method is used to validate the effectiveness of this method. A vase, as typical fragile artefact, is chosen as an example and studied in this paper. Dynamic responses of the artefact without-control, with traditional pendulum lateral isolation method and with this improved spring hanged three-dimensional pendulum isolation method under a set of artificial ground motion are compared. The acceleration reduction ratios, the sliding and rocking responses are compared.

2 Finite Element Models

The finite element models are shown in Fig. 2. The vase is placed at a squared exhibition stand. Figure 2a is the case without control. Figure 2b is the case with traditional pendulum isolation control. The exhibition stand is hanged by four non-extensible cables.

Figure 2c is the case with spring hanged pendulum isolation schedule. The exhibition stand is hanged by four extensible springs. In order to make case 2 and case 3 comparable, the pendulum hanging lengths of these two methods are set at 1 m. The hanging length shown in Fig. 2c is shorter than that of Fig. 2b is because Fig. 2c is the original state and gravity has now been applied. The spring will be extended to the same length as case 2 after gravity applied. Meanwhile, translational dampers are installed beneath the exhibition stand to absorb the energy during earthquake.

Shell element is used to build the vase and stand. Spring element is used to simulate the cable. In case 2, the spring stiffness is given a very big value to avoid axial deformation. In case 3, the spring stiffness is set as calculated to fulfil specific period.

Since the period of the pendulum system is determined by the hanging length, the lateral isolation period can be determined directly. The hanging length is 2 m and the period is 2 s at lateral direction. As for vertical period, the period is set as 1.5 s through adjusting the spring stiffness. The damping coefficient of the dampers is set as 145 N·s/m. The equivalent damping ratio is about 20% at vertical direction.

As for the input ground motion, the acceleration histories and related spectra are shown in Fig. 3 and Fig. 4, respectively. The amplitude ratios among the three dimensions are set as 1: 0.85: 0.65. The damping ratio of the acceleration spectra is 5%.

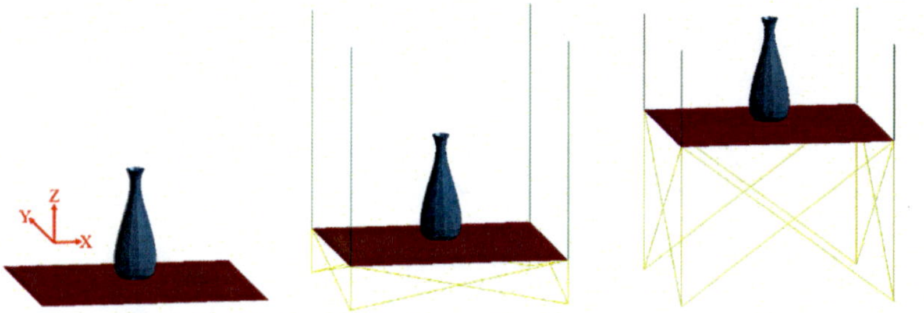

a. Non-control b. lateral isolation c. 3 Dimensional isolation

Fig. 2. Finite element models

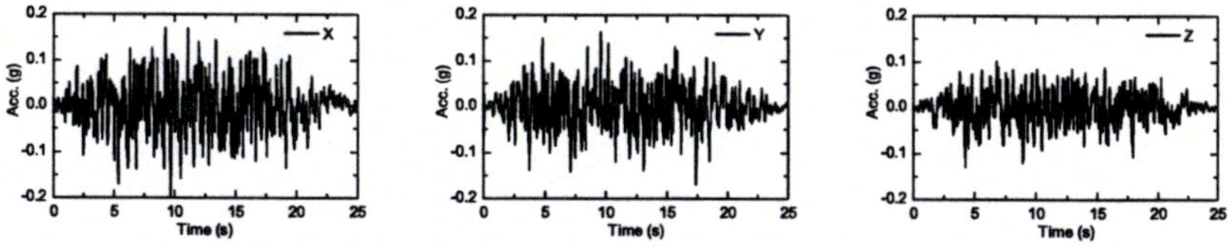

a. X direction b. Y direction c. Z direction

Fig. 3. Acceleration histories of the input artificial ground motion

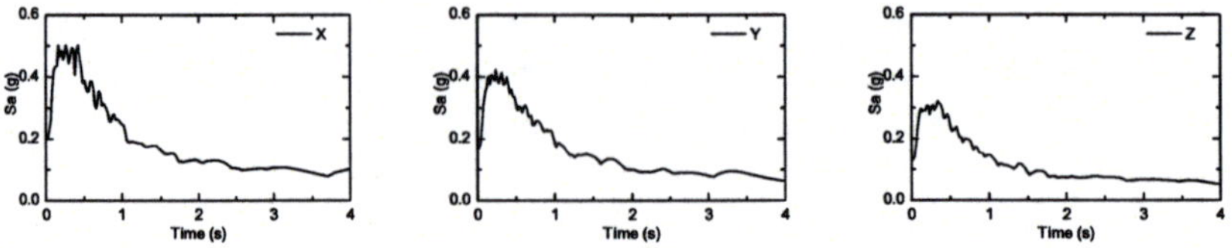

a. X direction b. Y direction c. Z direction

Fig. 4. Acceleration response spectra of the input artificial ground motion (damping ratio at 5%)

3 Exhibition Stand Dynamic Responses

The acceleration responses of the exhibition stand at X direction under the three schedules are plotted in Fig. 5 and corresponding responses at vertical direction are plotted in Fig. 6. The acceleration response of the exhibition stand under the non-control method is the same as the input ground motion.

The relative displacement responses at lateral and vertical direction are plotted in Fig. 7.

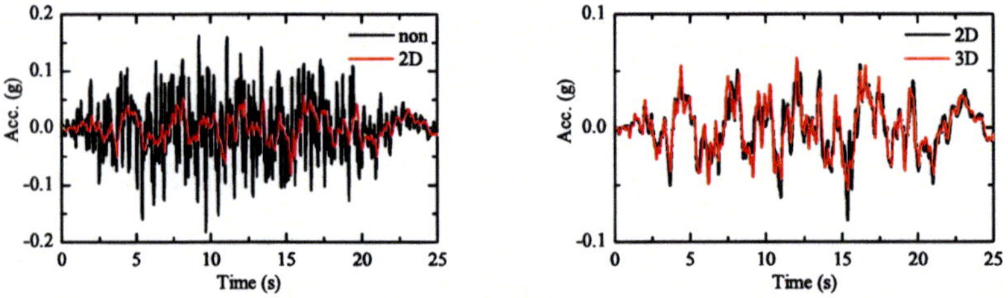

a. Without control and 2D control b. 2D control and 2D control

Fig. 5. Acceleration response comparisons at X direction among Non-control, 2-dimensional control and 3-dimensional control

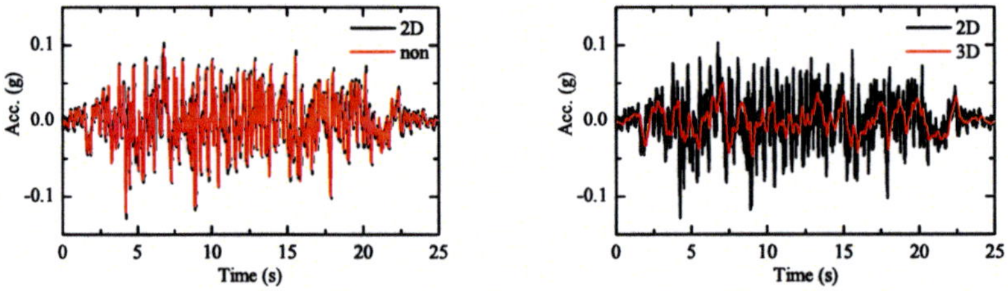

a. Without control and 2D control b. 2D control and 2D control

Fig. 6. Acceleration response comparisons at vertical direction among Non-control, 2-dimensional control and 3-dimensional control

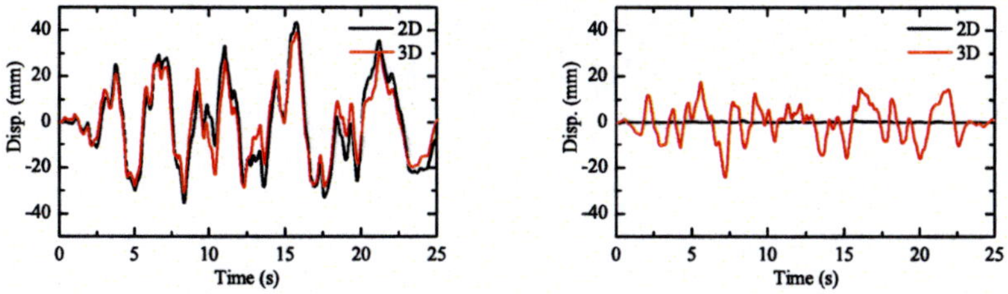

a. Lateral direction b. Vertical direction

Fig. 7. Relative displacement responses at lateral and vertical direction

It can be seen from the above figures that both the 2D and 3D methods are effective and show very similar trends at lateral isolation. The 2D method is not effective at vertical isolation, the vertical response of the stand is even a little bit bigger than that of the non-control method. The 3D method is effective at vertical isolation. The maximum relative displacement response of the 3D method at vertical direction is less than 20 mm.

4 Vase Rocking and Sliding

The sliding responses of the vase related to the stand at lateral X direction are plotted in Fig. 8. The rocking responses of the vase between top to bottom at X direction are plotted in Fig. 9.

It can been seen that both the 2D and 3D method are effective at reducing sliding. The 3D method works better than that of the 2D method. Both the 2D and 3D method are effective at reducing sliding. The 3D method works a little bit better than that of the 2D method.

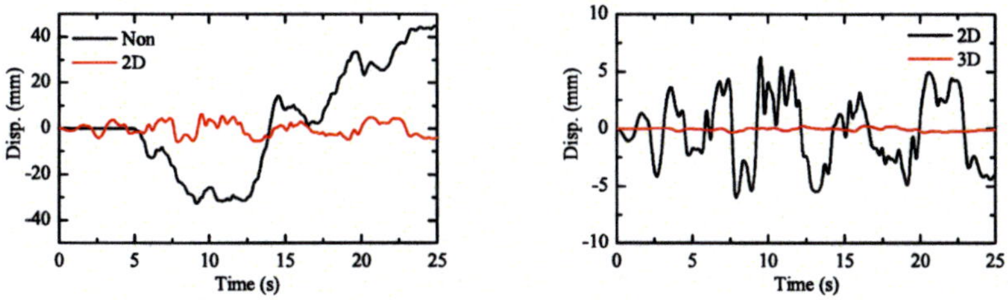

a. Without control and 2D b. 2D and 3D

Fig. 8. Vase sliding at X direction

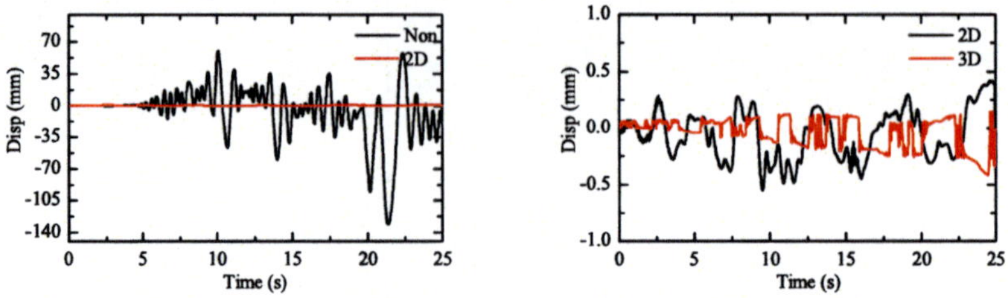

a. Without control and 2D

b. 2D and 3D

Fig. 9. Vase rocking at vertical direction

5 Conclusions

In order to protect floating artefacts, a three-dimensional isolation method is proposed. The cables of the traditional hanging pendulum isolation system are replaced by axial extensible springs, expecting to further fulfil vertical isolation. The effectiveness of this method is preliminarily validated using finite element method.

Results shows that this method can fulfil three-dimensional isolation. It can reduce the artefact rocking and sliding responses and performs better that the non-extensible cable hanged pendulum method. Meanwhile, the maximum vertical relative displacement, which is also the maximum spring axial deformation, is acceptable.

More simulations and tests are to be conducted in the future to further study the dynamic behavior of this improved spring hanged three-dimensional pendulum isolation method. Horizontal holding devices should also be used to avoid undesirable incline of the exhibition stand.

Acknowledgements. This work is supported by Earthquake Research Fund of CEA (2019C09, 2019B01), Heilongjiang Provincial Natural Science Foundation of China (LH2019E094), Program for Innovative Research Team in China Earthquake Administration.

References

Berto, L., Favaretto, T., Saetta, A.: Seismic risk mitigation technique for art objects: experimental evaluation and numerical modelling of double concave curved surface sliders. Bull. Earthq. Eng. **11**(5), 1817–1840 (2013). https://doi.org/10.1007/s10518-013-9441-8

Calio, I., Marletta, M.: Passive control of the seismic rocking response of art objects. Eng. Struct. **25**(8), 1009–1018 (2003)

Di Sarno, L., Petrone, C., Magliulo, G., Manfredi, G.: Dynamic properties of typical consultation room medical components. Eng. Struct. **100**, 442–454 (2015)

Gong, M., Wei, Q., Feng, P.: Survey on the damaged levels of the movable cultural relics and buildings during the 5.12 Wenchuan Earthquake in Sichuan Province. Sci. Conserv. Archaeol. **541**, 65–90 (2016)

Konstantinidis, D., Makris, N.: Experimental and analytical studies on the response of freestanding laboratory equipment to earthquake shaking. Earthq. Eng. Struct. Dyn. **38**(6), 827–848 (2009)

Koumousis, V., Moysidis, A.: On the dynamic behavior of nonlinear lightweight isolator for museum artifacts. Soil Dyn. Earthq. Eng. **117**, 251–262 (2019)

Neurohr, T., McClure, G.: Shake table testing of museum display cases. Can. J. Civ. Eng. **35**(12), 1353–1364 (2008)

Ning, X., Dai, J., Bai, W., Yang, Y., Zhang, L.: Seismic protection of cabinet stored cultural relics with silicone dampers. Shock Vib. **2018**, 1–15 (2018)

Check for updates

Conservation, Structural Consolidation and Seismic Strengthening of Ali Pasha Mosque in Ohrid, North Macedonia

Bekim Ceko[1](✉), Risto Petkovski[2], Onur Attar[3], Goran Jekic[4], and Predrag Gavrilovic[4]

[1] Faculty of Architecture and Space Planning, University of Business and Technology (UBT), Prishtina, Republic of Kosovo
bekim.ceko@ubt-uni.net
[2] Institute for Protection of Monuments of Culture and Museum - Stip, North Macedonia, Stip, North Macedonia
[3] Dogus 2000, Turkey, Ivedikkosb Mah. 1332 CAD. no. 78, Merkez Bucagi, Ankara, Turkey
[4] Ss. Cyril and Methodius University in Skopje, Institute of Earthquake Engineering and Engineering Seismology (IZIIS), 1000 Skopje, North Macedonia

Abstract. Ali Pasha Mosque is built during the 16-th centuries in the center of old part of Ohrid town with central dome, stone masonry walls, entrance hall covered with tree domes and minaret. As a monument beside used material represents the 16th century-built techniques in Ottoman period in the region, as one of the most distinguished building from the Ottoman era. During its existence monument has been partially damaged. In the First World War upper part of minaret has been destroyed. In 2015 Turkish foundation Vakuflar undertakes measures and funding for complete conservations, structural consolidations, repair, partial reconstruction of minaret and entrance hall and seismic strengthening for the whole monument structure. Investigations of present conditions, design of whole process with detailed analysis for seismic strengthening using modern techniques and composite materials, has been done by Institute for protection of monuments of culture and Museum – Ohrid. Complete field work of conservation, structural consolidation, repair, seismic strengthening and reconstruction has been done by Turkish company "Dogus 2000" Ankara, financed by Turkish Directorate General of Foundations (WAQUF). The mosque, as a monument of first category in the center of old town, represents a useful, singularly religion object, attractive touristic destination in Ohrid under the UNESCO protection. The conservation and rehabilitation of the mosque is good example of international cooperation in the field of cultural heritage protection.

Keywords: Conservation · Structural consolidation · Seismic strengthening · Monument

1 Introduction

A certain period of time (over five centuries) starting from 13th century till 19th century that ruled the Ottoman Empire in this area has left remarkable effects in many fields,

I. Vayas and F. M. Mazzolani (Eds.): PROHITECH 2021, LNCE 209, pp. 910–920, 2022.
https://doi.org/10.1007/978-3-030-90788-4_70

such as administration, economy, philology, architecture, culture etc. Architecture is one of the distinguished remarks that shaped throughout the Balkans and so the territory now known as North Macedonia was affected too. Within this, the typologies of the mosques were emblematic buildings in recognition of the Ottoman architecture that makes rich and divers the cultural heritage in general and built heritage of the North Macedonia in particular. The monuments in North Macedonia dating back to the Ottoman period are of a special interest having in mind their significance from historic and artistic point of view. The North Macedonia authorities in cooperation with Republic of Turkey Directorate General of Foundations (WAQUF) supported the conservation and restoration projects for Ali Pasha mosques, placed in city of Ohrid, North Macedonia. Having in mind the importance and the significance of the monuments, the present conditions and the damage state, the location of the monuments in zones of high seismicity, the modern principles of conservation and requirements, the criteria for protection of the historical monuments such as preservation of authenticity, reversibility and minimum interventions, a special study and a multidisciplinary approach have been implemented. Based on a detailed study of the structural stability and the expected seismic intensity at the considered locations, design of structural consolidation, repair and seismic strengthening has been performed and field work has been applied. For the monument, the general concept of seismic strengthening has been adopted based on experimental and analytical study performed for the Mustafa Pasha Mosque at the IZIIS' Dynamic Testing Laboratory in Skopje within the framework of the PROCHITECH Research Project. The use of the new selected composite materials such as carbon fiber with appropriate adhesives for historic monuments and old masonry structures represents a specific and new method of strengthening that has big advantages but, at the sometime, it requires special attention and caution during construction and design. (Shendova and Gavrilovic 2011).

2 History and Present Conditions of Monument

The Ali Pasha Mosque, (Fig. 1) is located in the center of the ancient city of Ohrid, close to the Bazaar. Extensive information about Ohrid is given by Evliya Celebi, who visited the city in 1670 (Kahraman 2011). The mosque, which was built by Süleyman Pasha in 1573, was named after him, in 1823 when it was repaired by Ali Pasha from Marash, Belgrade vizier, was known as Ali Pasha mosque (Fig. 1). That is mantioned in title deed of the ottoman empire of the 1190 year (Hicri year) (Ayverdi 1981).

Ali Pasha mosque is a typical representative of one-room sub-dome construction. Its base is in the shape of a square of 15 × 15 m, above which rises an octagonal drum and above it is an unusual sixteen-sided drum arched with a shallow dome. On each side of the tambour there is a bow arched window. The dome was originally covered with tile, but today it is covered with copper sheet. The transitional solution inside the mosque is solved with the help of tromps without the usual stalactite ornaments.

The interior of the mosque is characterized by unity of space, simplicity of form and volume dominance. Framed in a double profiled frame, it is housed on the southeast wall, and has simple stalactite ornamentation in the mihrab niche. To the south of the mihrab is a stone marble.

The porch occupies the entire length on the Northwest side of the building. The original porch is not preserved. But on the basis of analogies with similar buildings of

Fig. 1. Ali Paha Mosque, recorded in 1912

this period and the traces on the Northwest side wall, can be assumed that it was of the type of open porch with three small domes.

The remains of the minaret are to the west of the mosque. It was demolished in 1912 and recently rebuilt. According to the dimensions of its base it is a minaret which with its height and elegance dominated this part of the Ohrid Bazaar and was probably one of the higher minarets in Rumelia. It was built of brick and contained both the porch and the mahvil in the mosque. Written sources mention the existence of another minaret in the western part of the mosque. Evliya Çelebi states in this "Travelogue" that he reports that there was a mosque with two minarets in Ohrid, that is, the mosque originally had one minaret, and later Sultan Bayazit II (1481 - 1512) ordered another one to be erected. When performing the work on the east wall, a door was identified, at the level of the mahvil (most likely to enter the other minaret), - a counter on the west side entrance.

Ali Pasha Mosque is built of stone walls and stone dome, and it is among the first mosques to be built in Macedonia Fig. 2.

The base of the Minaret is made of quartz stone squares in lime mortar as well as the staircases which are one by one embedded in the minaret table. The ascertained thickness of the last row of stone quadrants certainly asserts that the minaret was built of brick just like the sheriff. The minaret was covered with a wooden structure, protected by a lead sheet. According to preserved old photography from the beginning of the twentieth century, the height of the minaret is set at 31.5 from the ground level.

The conservation research was carried out based on the Conservation Research Program, and within the framework of the Project "Conservation Restoration Works and Reconstruction of the Ali Pasha Mosque in Ohrid, Conservation and other previous interventions".

The first rebuilding of the mosque was made in 1823, when the shed was covered with tiles and a madrasa was built on the north ridge.

Minor repairs have been carried out several times, and in 1987/88 the roof construction and the dome were refurbished. In 1997 a project for remediation was made by the

Fig. 2. View of the Ali Pasha Mosque before the restoration

firm "Ziton" from Struga on the basis of which in 2000 construction works were carried out with the construction of a new closed porch on the north side and inside the building when a new mahvil was built.

During this period, the remains of the minaret were conserved and repaired in order to prevent the process of its further demolition, as well as the construction of a new timber made from reinforced concrete construction. In 2009, illegal interventions were carried out and stairs were constructed on the north-east side with inadequate building materials.

During these investigations it was concluded that the mosque was authentically preserved (the prayer space and part of the destroyed minaret). The roof cover, made of copper, sheet along with gutters originate from a more recent date, i.e. they were made during the interventions performed in 2000.

The floor in the mosque made of wooden boards is not authentic and was laid with the latest interventions, when a concrete floor was laid on the old stone slab flooring on which the floor was laid.

The ornament, built in the late 18th and early 20th century, is preserved in its original form. Interventions were carried out by plastering and painting the stairs.

The miharbeh is preserved in its original form, just as the mimbar is glazed and painted with oil paint.

The old mahvil at the main entrance on the north side has not been preserved. On the same site a new reinforced concrete mahvil is built.

3 Analysis of the Structure in Its Existing Conditions

Analysis of the structure has been carried out in accordance with the valid existing regulations and the Macedonian standards and codes for existing and strengthened conditions. A three-dimensional analysis of the structural system has been carried out using SAP2000 computer software.

For analysis of the existing structure under gravity and seismic load, the finite element method has been used, modeling the structure as 3D model. The walls were modeled applying solid elements and the dome applying shell elements. In the case of the strengthened structures, the strengthening elements have been modeled as frame tied elements with own characteristics. The numerical model is presented in Fig. 3 and the results of the analysis of the existing state are presented in Fig. 4. The both essential natural periods of vibrations in both orthogonal directions were calculated 0.26 s. with appropriate mode of vibrations. As to the results from the stress analysis for gravity load, it is clear that all the stresses are in normal conditions and value, the vertical stress in bearing walls is in the range 0.3–0.6 MPa and some tensions with low stresses of (0.2–0.5) MPa. The analysis results actually demonstrate the real conditions during serviceability life, leading to the following conclusion:

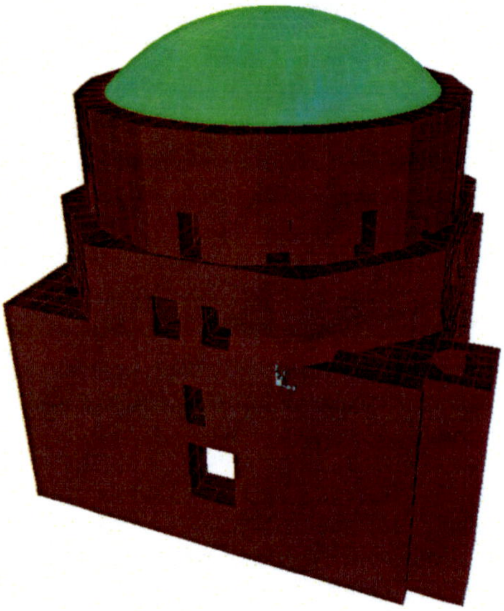

Fig. 3. Structural model of the non-strengthened structure of Ali Pasha Mosque

According to the observations and investigations of the structural conditions and stability, it is evident that the structure and the structural system is safe and satisfies the criteria for gravity and serviceability load. The mosque is built as masonry brittle structure located in zone of significant seismic risk. Having in mind that it represents a monument of the first category of importance, the seismic stability should be increased.

In Figs. 5 and 6 vertical tensile and shear stress distribution are shown. As to seismic conditions significant, analysis for expected seismic loads presented on Fig. 5 tensile stresses occur at the end of the walls and on the domes in the range of 1.2–1.5 MPa, meaning overstress and potential damage.

The analysis shows large tensile zones, below the dome and in the upper part of the structure. It has been concluded that the structure's seismic safety is insufficient, meaning that in the case of an earthquake damages are expected, and also a partial collapse. Strengthening of the structure is therefore a prerequisite.

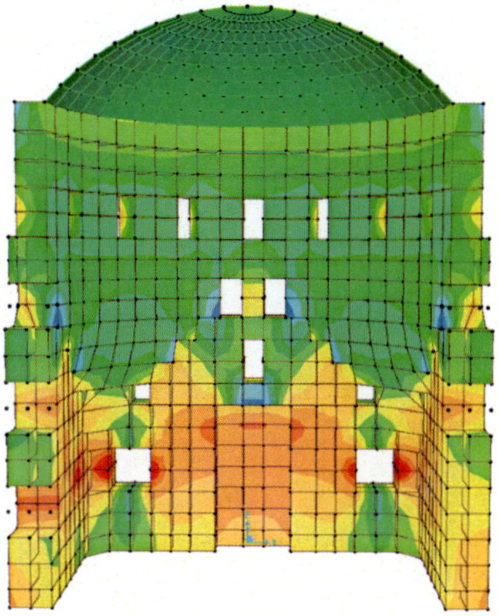

Fig. 4. Vertical stress distribution from dead load (non-strengthened structure)

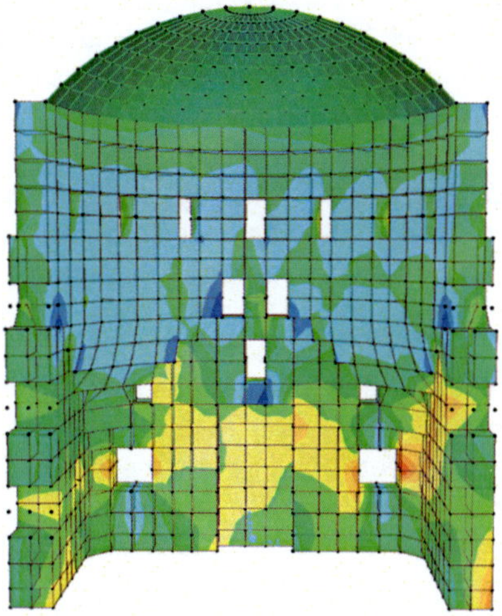

Fig. 5. Vertical stress distribution from dead and seismic loads (non-strengthened structure)

4 General Concept of Structural Strengthening and Analysis of the Strengthened Structure

During investigations it was concluded that the mosque was authentically preserved (the prayer space and part of the destroyed minaret). The roof cover, made of copper, sheet along with gutters originate from a more recent date, i.e. they were made during the interventions performed in 2000. The ornament, built in the late 18th and early 20th

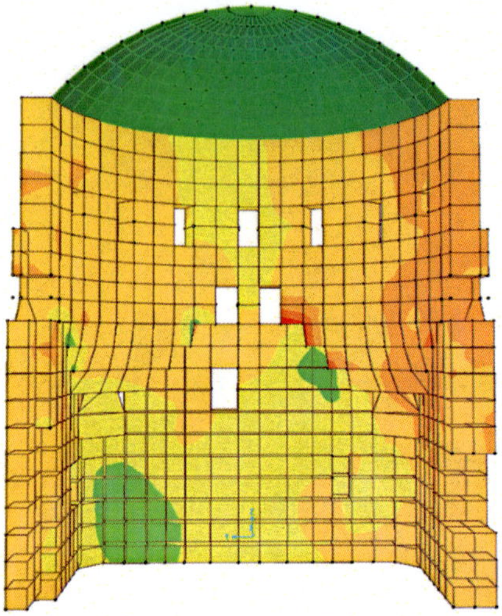

Fig. 6. Shear stress distribution from dead and seismic loads (non-strengthened structure)

century, is preserved in its original form. Interventions were carried out by plastering and painting the stairs. The miharbeh is preserved in its original form, just as the mimbar is glazed and painted with oil paint. The old mahvil at the main entrance on the north side has not been preserved. On the same site a new reinforced concrete mahvil is built. On the base of investigations and Project for architectural conservations fooling principal "minimum interventions - maximum protection" concept for structural interventions have been selected to used modern techniques and composite materials.

Considering the type of the structural system, the category and type of the monument and the required seismic stability criteria, structural consolidation and seismic strengthening has been proposed on the base of a study applying composite carbon materials. The carbon fiber and the other composite bonding materials used in the Project represent a relatively new technology and investigation (Mazzolani et al. 2006, 2007). The materials are characterized by a high tension-bearing capacity - 10 times higher than the tension capacity of the steel, relatively simple implementation, with minimal intervention on the existing structure. The general system of strengthening is as follows (Fig. 7, 8 and 9):

- strengthening of the dome in the lower part to improve their tensile strength by use of carbon WRAP;
- construction of a ring at the dome using a carbon fiber plate and carbon reinforcement;
- strengthening of the walls by using carbon reinforcements to sustain tensile stresses in the walls under seismic loads;
- rebuilding of the entrance part in its original shape and its strengthening by stainless steel ties;
- consolidation of the bearing walls by injecting cracks and strengthening of the foundations.

Before applications detail analysis has been performed according expected seismicity, seismic design parameters, category of monuments etc. Figure 10 represents the

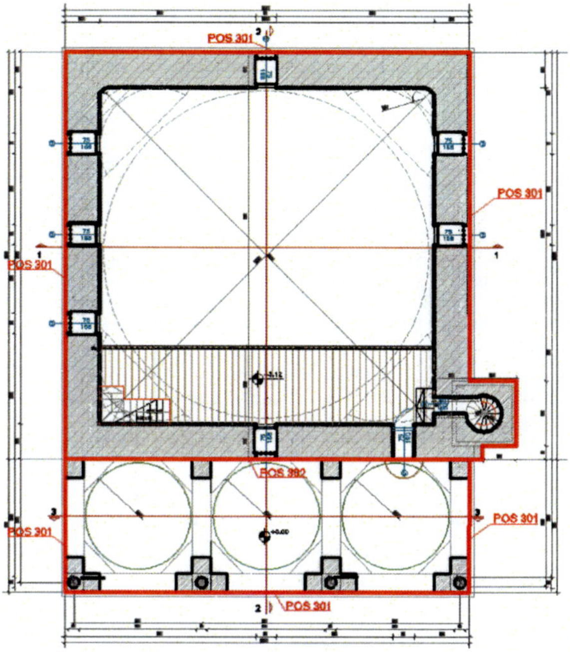

Fig. 7. Plane – seismic strengthening of the mosque

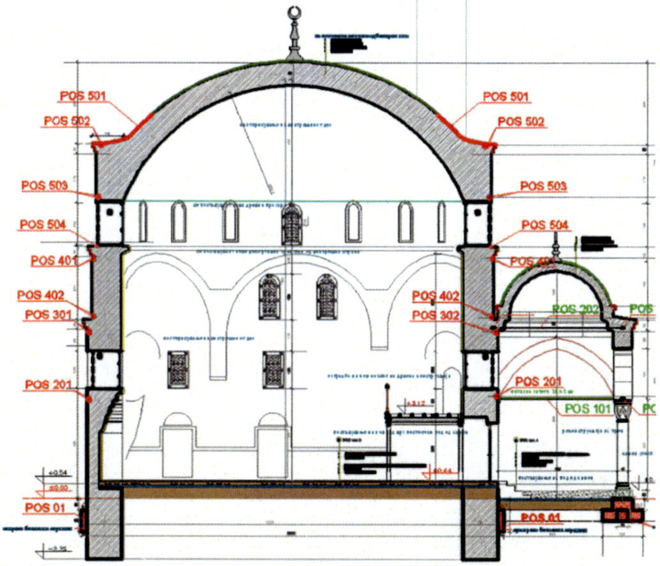

Fig. 8. Cross section – seismic strengthening of the mosque

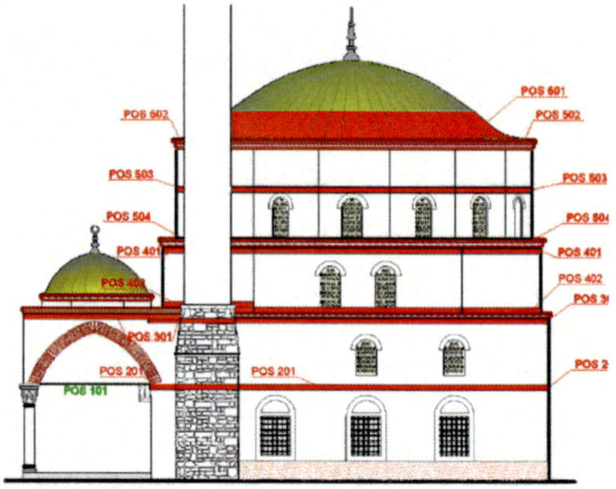

Fig. 9. Facade – seismic strengthening of the mosque

numerical model of the strengthening components. The results, obtained from the numerical analysis, show that the applied strengthening components provide more favorable shear stress distribution in the bearing walls (Fig. 11).

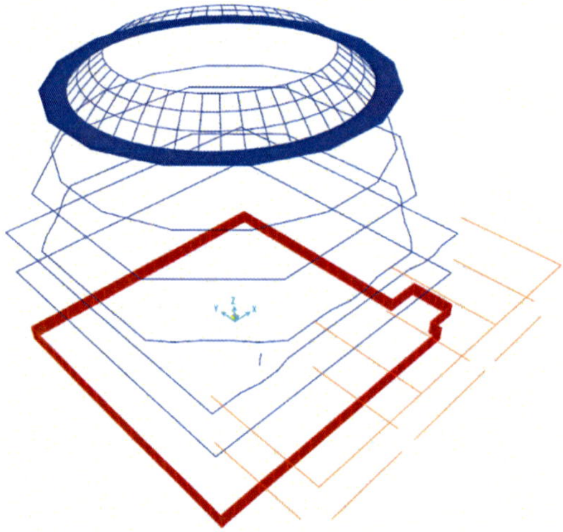

Fig. 10. 3D numerical model of the seismic strengthening

The results show that the level of tensile stresses in the strengthened structure is acceptable, deformations and, behavior in the case of the expected earthquakes satisfied safety criterions, the motions will cause some cracks and controlled level of damage, but not collapse of the structural elements, nor the whole structure, the general criterion of seismic safety and stability is satisfied. Figure 12 presents the Ohrid Ali Pasha Mosque after the restoration.

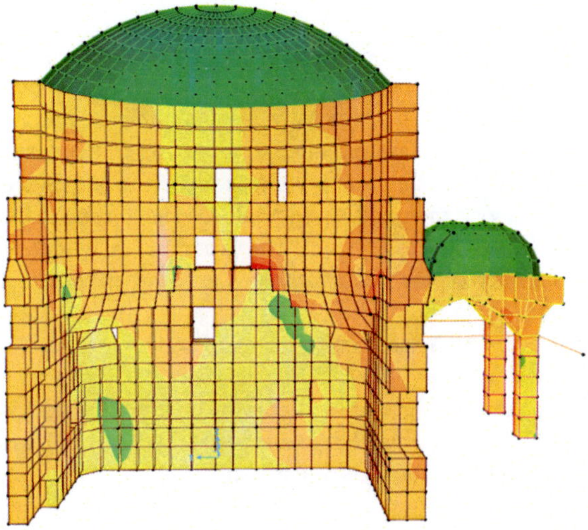

Fig. 11. Shear stress distribution from dead and seismic loads (strengthened structure)

Structural consolidations work, seismic strengthening and other field activities have been in cooperation with architectural conservations work and satisfied all criterions for protections of monuments, having in mind that Ohrid is on the UNESCO list of cultural heritage.

Fig. 12. The Ohrid Ali Pasha Mosque after the restoration

5 Conclusions

Using modern composite materials like Carbon fibers and others gives opportunity for minimum interventions on original structures as a monuments structure. The relevance of the design of such an important monument wisely suggested testing the validity of these techniques on a large scale model, which was first seriously damaged by a local simulated earthquake and then repaired and reinforced according to the design purposes by means of an innovative selected system. The experimental evidence was very helpful in proving the suitability of the design choices and the effectiveness of the applied consolidation system.

The design process, which has been followed as shown in this paper, should emphasize the suitable way to follow when facing the complex problem of protection and conservation of a cultural heritage construction. Construction field work is of special interest having in mind the specific characteristics and behavior of composite materials and it can be controlled all the time during construction.

References

Mazzolani, F.M.: Earthquake protection of historical buildings by reversible mixed technologies: the PROHITECH project. In: Proceedings of the International Conference STESSA 2006, Yokohama, Japan (2006)

Mazzolani, F.M., Krstevska, L., Tashkov, L., Gramatikov, K.: Shaking Table Testing of Mustafa Pasha Mosque Model. Final report, EU-FP6 PROHITECH project (2007)

Shendova, V., Gavrilovic, P.: Seismic upgrading of Mustafa Pasha mosque in Skopje. In: Sismic Protection of Cultural Heritage, Antalya, Turkey: WCCE-ECCE-TCCE Joint Conference 2, p. 47 (2011)

Kahraman, A.S., Yucel, D., Robert, D.: Evliya Celebi Seyhatnamesi. Yapi kredi Yayinlari. 8/2, Istanbul (2011)

Ayverdi, E.H.: Avrupa'da Osmanli Mimari Eserleri, Yugoslavya, II,/3, Istanbul (1981)

Check for updates

Application of the Methodology Developed Within the PROHITECH Project in Seismic Retrofitting of Mosques

Veronika Shendova(✉), Goran Jekic, Aleksandar Zlateski, and Predrag Gavrilovic

Institute of Earthquake Engineering and Engineering Seismology, IZIIS, Ss. Cyril and Methodius University, Skopje, P.O. Box 101, 1000 Skopje, Republic of North Macedonia
veronika@iziis.ukm.edu.mk

Abstract. Within the frame of the EU FP6 PROHITECH project a methodology for seismic retrofitting of monuments using CFRP elements has been developed and experimentally verified by shaking table tests on a model of the Mustafa Pasha Mosque in Skopje. The resulted efficiency of the applied methodology was crucial and starting point in making the decision about the concept of seismic upgrading of the real monument. Namely, the retrofitting methodology using CFRP was applied for the first time on the structure of Mustafa Pasha Mosque in Skopje in the period 2007–2010. The authors of this paper had the opportunity and the challenge to design and supervised the application of the seismic retrofitting methodology, followed by all the necessary solutions for in-situ technical detailing. Following this successful story, the same or slightly modified retrofitting methodology has been implemented in several other mosques on the territory of North Macedonia and Kosovo. The paper presents the main points of the specific process of pre-investigation, analysis, design of retrofitting and its application in the abovementioned case studies.

Keywords: Historic monuments · Mosques · Seismic retrofitting · CFRP · Design by testing

1 Introduction

Repair and/or retrofitting, as a fundamental part of preservation, conservation and restoration of monuments, has to be planned based on detailed analyses on the influence of seismic hazard, local soil conditions, dynamic properties of the structural systems, their strength and deformability as well as dynamic response of structures under the expected ground motions. Considering the importance of the above aspects in the evaluation of the earthquake response of historical monuments, as well as the fact that the seismic analysis cannot be performed using seismic design codes for modern buildings, the determination of criteria, methods and techniques for retrofitting and the process of structural restoration and preservation should be based on ad-hoc detailed studies, including design assisted by testing and taking into account the cost effectiveness of the alternative solutions by using advanced and innovative technologies.

I. Vayas and F. M. Mazzolani (Eds.): PROHITECH 2021, LNCE 209, pp. 921–937, 2022.
https://doi.org/10.1007/978-3-030-90788-4_71

The experimental investigation of models on a seismic shaking table is the most corresponding way of investigation from the aspect of dynamic structural behaviour during real earthquakes. Applying an appropriate modelling technique and according to the similarity laws, the models can be designed to different scales and tested under various seismic inputs. Based on experimental data obtained from site testing of the monuments, as well as the results of the seismic hazard analysis, the laboratory testing of models gives very reliable data on the seismic behaviour and stability, pointing out the weak points of the structures. These data are of a great importance for further analysis and development of an appropriate methodology for seismic retrofitting of monuments.

2 IZIIS' Experience in Earthquake Protection of Cultural Heritage

Within the frames of the IZIIS' research activities, in addition to seismic design and protection of modern structures, particularly noteworthy is the experience gathered in the field of protection of structures pertaining to the cultural historic heritage. During a period of more than 40 years of activities in this field, the Institute has realized important scientific research projects involving experimental and analytical research, field surveys of historic structures and application of knowledge during earthquake protection of important cultural historic structures and monuments, (Shendova et al. 2011).

2.1 The First Shaking Table Test of a Monument in the World

For the purpose of development of appropriate methods for repair and retrofitting of the Byzantine monuments in general, and particularly the Byzantine churches located within Macedonia, shaking table testing of a church model in a realistic geometrical scale was performed for the first time in the world within the frameworks of the scientific research projects realized at IZIIS in the period 1990–2000, (Gavrilovic et al. 1999). 1:2.75 scaled model of St. Nikita church was constructed and tested on the seismic shaking table in the IZIIS laboratory in its original and strengthened state by use of "ties and injection", (Fig. 1).

Preceding the shaking table testing, ample field and analytical studies, in situ and laboratory tests were performed for the existing structure of the St. Nikita church for the purpose of defining the physical-mechanic and chemical characteristics of the built-in materials, the dynamic characteristics of the structure and the seismicity of the terrain. A preliminary analysis of the seismic stability of the existing structure points to non-sufficient bearing and deformation capacity according to the seismic safety criteria. The proposed concept of retrofitting consists of incorporation of horizontal and vertical steel ties at the places formerly occupied with timer beams and filling the area around them with a lime-based mortar. So, after the testing of the church model in its original state it was strengthened in the proposed way and tested again.

Comparison of the results shows that the applied retrofitting is very effective since it increases the bearing capacity and deformability of the structure up to the level of the designed protection. The knowledge gained in this way was unique and it represents excellent background for further development of retrofitting methodologies.

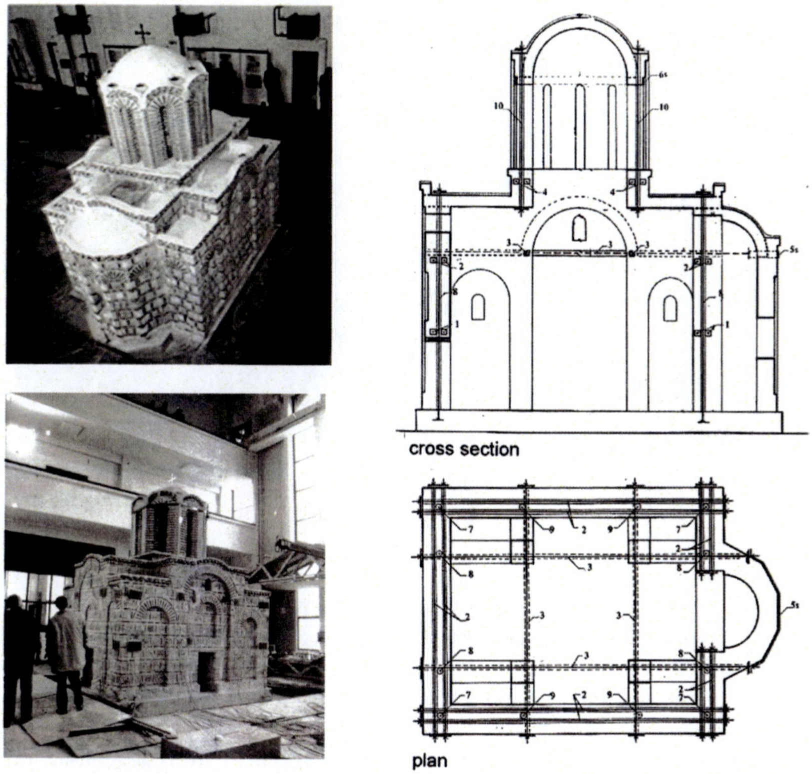

a) Original and strengthened church model b) Applied retrofitting using stainless steel ties

Fig. 1. The 1:2.75 scaled church model in its original and strengthened state

2.2 Shaking Table Testing of the Model of Mustapha Pasha Mosque in Skopje

Experimental shaking table tests on a model of the Mustafa Pasha Mosque were carried out at IZIIS within the frame of the EU project "Earthquake Protection of Historical Buildings by Reversible Mixed Technologies" - PROHITECH, (Mazzolani 2006). The 1:6 scaled model has been subjected to a series of earthquakes with increasing intensity, causing progressive damage (Fig. 2 left). Then the model was strengthened by application of CFRP elements (wraps and rods) and subjected to further tests (Fig. 4).

The experimental programme was performed in three main phases: (1) Testing of the original model under low intensity level, with the aim of provoking damage to the minaret only; (2) Testing of the model with strengthened minaret, with the aim of provoking its collapse and damage to the mosque, (Fig. 2 right), and (3) Testing of the strengthened mosque model until reaching heavy damage. The retrofitting solution is similar to that applied to the church model but this time using innovative composite materials. It consisted of formation of a horizontal belt course around the bearing walls

Fig. 2. The Mustapha Pasha model in its original state (left) and with retrofitted minaret (right)

by CFRP rods as well as around the tambour and at the base of the dome by CFRP wrap, (Fig. 4).

The seismic investigation of the model was performed by simulating the 1979 Montenegro Petrovac earthquake, selected as a representative one and applied in horizontal direction. Push-over curves were constructed based on the different intensity level of the repeated seismic test, within an input acceleration of 0.05 g–1.50 g, (Fig. 3).The results of the investigation have shown that the retrofitting system is efficient especially in providing the integrity of the structure under severe earthquake. The strengthened structure was able to resist a seismic acceleration quite three times the one corresponding to the original capacity (Mazzolani et al. 2007).

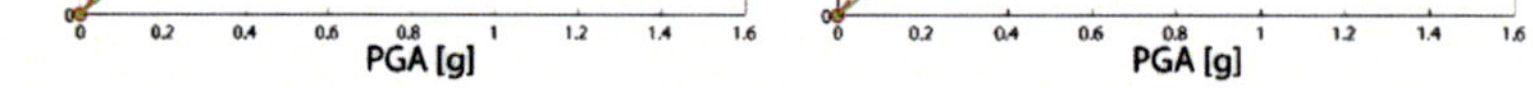

Fig. 3. Response of the model: acceleration measured at the top of the shear walls and of the dome as a function of input acceleration, phase 2 and phase 3.

Shaking table testing for both the original and strengthened model showed that the retrofitting of the minaret by application of a CFRP wrap enabled stiffening and increasing of its bending resistance. The mosque model's behaviour, after retrofitting, was evidently different in respect to that of the original model. During the intensive tests, the failure mechanism was transferred to the lower zone of the bearing walls where typical diagonal cracks occurred due to shear stress. Comparing the obtained results, it can be concluded that the applied retrofitting technique has significantly improved the seismic resistance of the monument.

Fig. 4. Retrofitted model before (left) and after the final test (right)

The knowledge gained in this way was unique and incomparable and hence necessary for seismic retrofitting of individual important cultural-historic structures where it is important to have an insight into the effect of the interventions upon the authenticity of the monument. Therefore, after the realization of these projects, IZIIS became partner of the National Conservation Center which enabled direct application of the gained knowledge in actual conditions and for specific historic monuments, (Shendova et al. 2014).

2.3 IZIIS' Integrated Approach to Repair and Retrofitting of Historic Monuments

Based on extensive research activities performed by IZIIS for the purpose of evaluation of a procedure for repair and retrofitting of valuable historic monuments an integrated approach to seismic protection of extraordinarily important cultural historic structures has been adopted and used in the process of reconstruction of the real monuments. This complies with the restoration and conservation requirements as well as procedures and regulations for high category structures, and encompass the following:

- Definition of expected seismic hazard;
- Definition of soil conditions and dynamic behaviour of soil media;
- Identification of structural characteristics, determination of main dynamic characteristics and bearing and deformation capacity of existing structures;
- Definition of safety criteria and selection of the concept for repair and/or retrofitting;
- Adoption of structural methods, techniques, materials and types of excitation;
- Analysis of dynamic response of repaired and/or strengthened structures and verification of their seismic stability;

- Definition of field works, execution and inspection.

Although the above stated seems to be the *normal procedure*, it is the only way of providing high quality in protection of cultural heritage. This task is certainly much more than simply listing of what is to be done since it requires a lot of knowledge and efforts.

3 Repair and Seismic Retrofitting of Mustafa Pasha Mosque

Mustafa Pasha's Mosque is one of the biggest and the best-preserved monuments of the Ottoman sacral architecture in Skopje and the Balkan. The building style belongs to the early Constantinople period at the beginning of the second half of the 15th century. The structural system of the mosque consists of massive peripheral walls in both orthogonal directions with a thickness of about 170 cm constructed partially of hewn stone and brick. The walls of the building are original in principle. They are constructed with two faces between which there is an infill of lime mortar and pieces of bricks and stone.

The catastrophic Skopje earthquake of 1963 inflicted damage to the mosque structure that dominantly affected the central dome and the domes of the porch, the east facade and the minaret. In 1968, these damages were repaired by injection of cement mortar based mixtures as well as incorporation of RC belt courses, (Fig. 5 and Fig. 6).

Fig. 5. Damage after the 1963 Skopje earthquake

Fig. 6. Retrofitting of the dome (1968)

Today, Mustafa Pasha's Mosque represents a cultural historic monument of an extraordinary importance for the city of Skopje and Republic of North Macedonia. As such, it is under the protection of the Law on Protection of Cultural Heritage and it is categorized as a structure belonging to the first category.

The authors of this paper had the opportunity and the challenge to design a system for seismic retrofitting of the structure, (financed by the Turkish foundation TIKA) for the needs of the conservation project on repair of Mustafa Pasha Mosque prepared by the Foundation of the University of Gazi in cooperation with the Ministry of Culture and Tourism of Turkey, the Ministry of Culture of Republic of North Macedonia and the National Conservation Centre and realized in 2006.

The starting point in making the decision about the concept and especially the selection of materials for retrofitting of the mosque was the shaking table investigation on a scaled mosque model. Respecting the modern requirements in the field of protection of historical monuments, as is the application of new technologies and materials, reversibility and invisibility of the applied technique, the authors have decided to choose a concept of repair and retrofitting involving the use of composite CFRP materials, which was for the first time in the region, (Shendova et al. 2007; Mazzolani and Shendova 2009).

3.1 Previous Investigation on the Site and the Monument

In realization of this project, the established IZIIS' integrated approach has thoroughly been respected, (Shendova et al. 2007). Taking into account that working at real site and real monument that have been existed several centuries is completely different in comparison with laboratory conditions, the concept of structural retrofitting aimed at reaching the designed level of earthquake protection has been selected based on following previous investigations:

(i) investigations of the characteristics of the built-in materials;
(ii) investigation of the main dynamic characteristics;
(iii) investigations of the soil conditions;
(iv) detailed geophysical surveys for definition of geotechnical and geodynamic models of the site;
(v) consequent relevant structural analysis

Investigation of the Characteristics of the Built-In Materials and the Main Dynamic Characteristics, (Tashkov and Krstevska 2006): In the course of 2006, within the bilateral scientific project with the Yildiz University, Turkey, the dynamic characteristics of the Mustafa Pasha Mosque structure were investigated by use of the ambient vibration technique. The obtained natural periods of the structure were $T^{N-S} = 0.33$ s and $T^{E-W} = 0.31$ s, whereas the fundamental period of the minaret structure is $T = 0.96$ s, (Fig. 7a).The mechanical characteristics of the built-in materials were also investigated through testing of samples of material taken from the structure, (Fig. 7b). These data are particularly important and necessary for relevant analysis of seismic stability of the monument.

Investigations of the Soil Conditions, (Sesov et al. 2007): To define the bearing capacity of the local soil and identify the foundation level of the structure, geomechanical drilling (3 sondages down to the depth of 8.00 m) was carried out in June 2007. These geomechanical investigations enabled sufficient definition of the lithological composition of the soil, definition of the geotechnical soil profiles below the structure as well as definition of the physical-mechanical characteristics of the present materials. It has been also concluded that the structure is founded at level −4.00 m on a layer of semi-bound to well compacted sandstone of high bearing capacity, (Fig. 8 left).

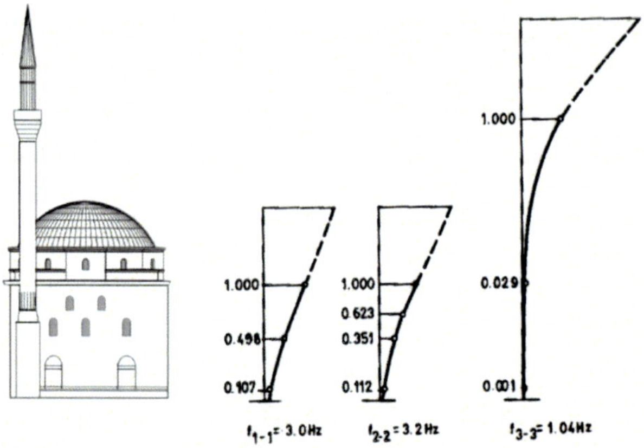

(a) Dynamic characteristics of the structure

b) testing of built-in materials

Fig. 7. Experimental testing of the mosque

Detailed Geophysical Surveys for Definition of Geotechnical and Geodynamic Models of the Site, (Sesov et al. 2007): The main purpose of these investigations has been to define the seismic parameters for evaluation of the seismic stability of the structure. Namely, for structures of extraordinary importance according to the valid technical regulations, it is necessary to define the seismic input concretely for the site of the structure in order to perform correct dynamic analysis. The investigations have been carried out in compliance with the latest achievements in the field of earthquake engineering. The main concept of the applied procedure is to consider the expected earthquake effect through a probabilistic approach, including also the local soil effects through nonlinear dynamic analysis of a representative geotechnical model, (Fig. 8 right). Table 1 shows the expected values of maximum acceleration for different return periods, whereas Table 2 shows the defined seismic design parameters through input acceleration and characteristic spectra for the design and maximal earthquake. For such defined input parameters, the characteristic records whose frequency content covers the frequency range of interest have been selected in the dynamic analysis.

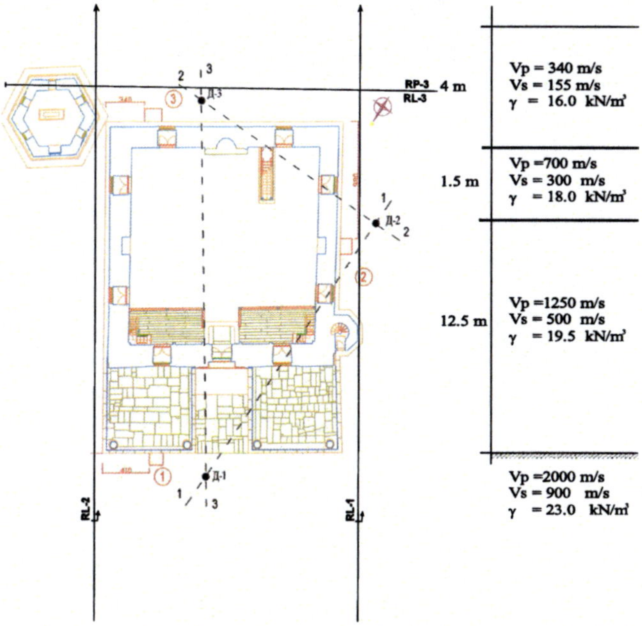

Fig. 8. Soil investigation with resulting geodynamic model

Table 1. Maximum acceleration (in g) for different return periods

	DAF	Return period (years)				
		50	100	200	500	1000
Bedrock	1.00	0.13	0.19	0.25	0.27	0.36
Foundation level	1.35	0.176	0.257	0.338	0.365	0.48

Table 2. Seismic design parameters

Serviceability life (years)	Level of seismic risk (%)	Earthquake	Max. acceleration a_{max} (g)
100 and more	30–40	Design	0.34
	10–20	Maximum	0.39

Concept for Seismic Retrofitting of Mustafa Pasha Mosque, (Shendova et al. 2007): Based on these investigations and the defined seismic parameters, as well as detailed analysis of the seismic stability of the structure, the solution of structural retrofitting has been accepted, (Fig. 9), that complies thoroughly with the conservation principles for repair and retrofitting of cultural historic monuments. It consists of incorporation of strengthening elements in the process of conservation, representing by CFRP bars in the walls and CFRP wrap on the dome, with the main purpose of providing integrity to the structure as well as simultaneous behavior of all the bearing walls.

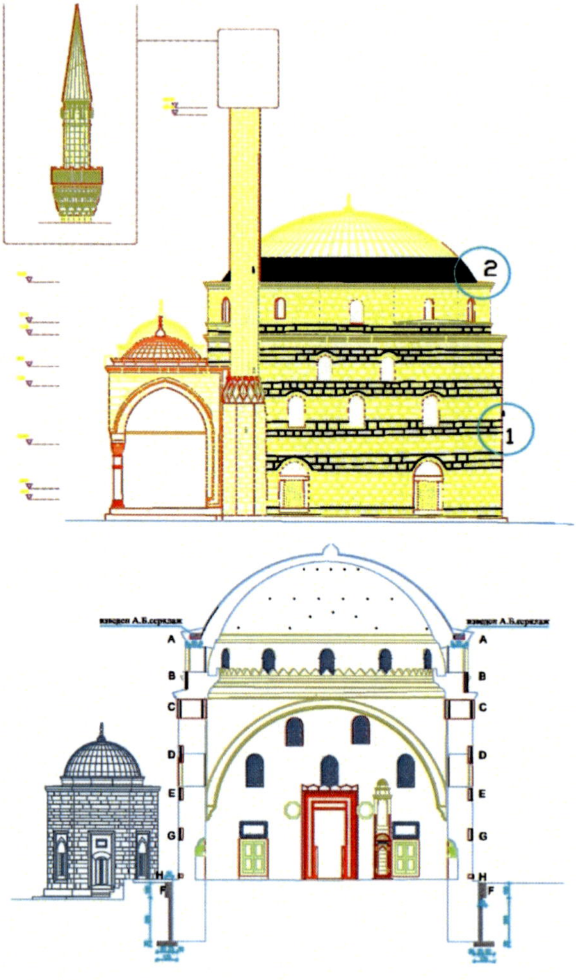

Fig. 9. Seismic retrofitting of Mustafa Pasha Mosque, (facade and cross section)

Structural Analysis: An analysis has been carried out in accordance with the valid existing regulations, (Shendova et al. 2007). Two methods were used: (1) Analysis of the bearing and deformability capacity of the structure in the form of ultimate storey shear force Qu, that compared to the equivalent seismic force Si yields the factor of safety against failure, F, (Table 3) and nonlinear dynamic analysis for the defined seismic parameters and (2) Static and equivalent seismic three-dimensional analysis of the structure using SAP 2000. The results show that the retrofitted structure possesses a sufficient bearing and deformation capacity up to the design level of seismic protection. It satisfies the prescribed requirements and the safety criteria for such type of historic structures, (F > 1).

Table 3. Results from bearing and deformation capacity analysis

Level	Stiffness Ki (kN/cm)	Ultimate transversal force Qu (kN)	Safety factor F = Qu/Si
4	38802	3798	1.184
3	140242	17534	1.382
2	100672	19343	1.868
1	49253	21276	1.748

3.2 Structural Retrofitting Process

Retrofitting of the mosque structure was realized in the period 2007–2009 in the following order:

Minaret: Anastylosis was anticipated for the structure of minaret. It was realized by removing and labeling in the same time of all the stone elements, checking the iron connections and replacing the deteriorated one with the new one and then rebuilding the minaret in opposite order, (Fig. 10).

Fig. 10. Anastylosis of the minaret structure

Foundation Structure: The solution for strengthening of foundation consists of construction of a reinforced concrete wall with a thickness of d = 25 cm along the perimeter of the foundation walls, on the external side, below the terrain level and down to the foundation level. From conservation reasons, this reinforced concrete wall was physically separated from the existing foundation walls by a polyurethane coating, (Fig. 11). However, to ensure interaction between the newly designed RC wall and the existing foundation structure, the solution anticipates placement of anchors made of chrome steel to an alternating length according to the reinforcement details, in previously formed openings filled with epoxy mortar (mixture of quartzite sand and epoxy) at an alternating length according to the reinforcement details.

(a) injection (b) coating

(b) anchors (d) reinforcing

Fig. 11. Strengthening of foundation structure

Tambour, Bearing Walls: After cleaning of all the joints on the outside with a depth of 8–10 cm, CFRP bars of defined mechanical characteristics were placed in an epoxy mortar layer and connected in the vertical joints. Then, the joints were filled by pointing lime mortar in accordance with the architectural-conservation project, (Fig. 12). Special detailing was designed for connecting the bars on the wall corners.

Dome: After removal of the cement mortar layer over the dome placed in 1968, the following was carried out: (i) coating of the formerly constructed reinforced concrete ring in the base of the main dome with an injection mixture based on lime mortar and (ii) placement of a 2.9 m wide CFRP wrap in a layer of epoxy glue along the perimeter of the dome. Then the entire dome was externally coated with a protective layer of lime mortar and covered in accordance with the architectural-conservation project, (Fig. 13).

In the entire part of the structure extending above the terrain level the cement mortar, injected in the cracks after the 1963 earthquake, was removed. During the retrofitting these cracks as well as all the cracks detected after the opening of the external joints were injected with lime mortar with defined mechanical characteristics.

Fig. 12. Placing CFRP bars in the wall's horizontal joints

Fig. 13. Retrofitting of the central dome structure, (injection, placing of CFRP wrap and roofing)

4 Retrofitting Using CFRP of Other Mosques in the Region

Following the successful seismic upgrading of Mustafa Pasha Mosque, the same or slightly modified retrofitting methodology by using CFRP has been so far implemented in several other mosques on the territory of North Macedonia and Kosovo, all of them from the Ottoman period: Isaki and Gazi Hajdar Kadi Mosques in Bitola, Ali Pasha Mosque in Ohrid, Yashar Pasha and Fatih Mosques in Pristina as well Sinan Pasha Mosque in Prizren, while for Orta Mosque in Strumica the process of pre-design investigation is currently under way. In each of these case studies, the retrofitting solution was designed in accordance with the specific on-site condition of the monument. Following are the main information for three of them.

Fatih Mosque: The Mosque of Sultan Mehmet Fatih, (Fig. 14, Gavrilovic et al. 2009) is located in the centre of the ancient city of Prishtina. It has been constructed upon the order of Sultan Mohammed II Fatih, the Conqueror in 1461. At the end of the 17th century it was converted into a Catholic church, while during Second World War, the German Army refurbished the interior of the mosque. For its monumentality, method of construction and decorative elements, this monument is listed among the most important buildings of the Islamic architecture in South-Eastern Europe. The wall thickness reaches 180 cm. The cupola is one of the biggest cupolas of the XV century and is one of the best architectural examples in South-East Europe. The damage to the Sultan Mohammed Fatih's Mosque is caused by climatic changes, ravages of time and permanent vibrations caused by traffic. The damage is manifested by separation of sandstones, vertical cracks in the walls indicating separation of rigid blocks as the first mode of failure, cracks in the arcade columns, vertical cracks in the corners and the cupola. The adopted retrofitting methodology, almost identical as in Mustafa Pasha Mosque except for the solution for strengthening of foundation, was applied in the period 2007–2009 by financial support of Turkish Foundations TIKA and Ministry of Culture of Kosovo.

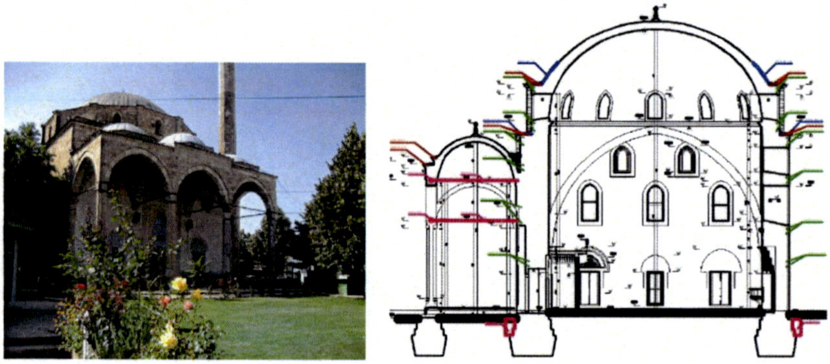

Fig. 14. Sultan Mehmet Fatih Mosque, (N-E view – left, retrofitting by CFRP – right)

Gazi Hajdar Kadi Mosque: This mosque was commissioned by Hajdar Bey, then beylerbey of Rumelia Eyalet, the largest Ottoman administrative territory, with Manastir, (today's Bitola) as a regional centre. It was built in 1561 as a domed building with massive walls of about 140 cm thick. In the 17th century, this mosque was considered as one of the most beautifully decorated. And the only one in the region with two minarets. In the past the mosque has been transformed into a warehouse by the local authorities. The damage, manifested by initial vertical cracks in the walls and deterioration of the wood and roofing elements, has been caused dominantly by ravages of time and by the number of local and short-distance earthquakes. A thorough rehabilitation including retrofitting by CFRP bars and wrap was accomplished in the period of 2013–2016.

(Fig. 15, Hadzi-Pecova et al. 2013) and financially supported by Turkish foundation Vakuflar and Ministry of Culture of North Macedonia. Because of the evident deterioration of the wooden built-in material, specific for this particulate monument is anticipation of rehabilitation of original wooden belt courses on the top of the bearing walls bellow the dome and their replacement with steel one.

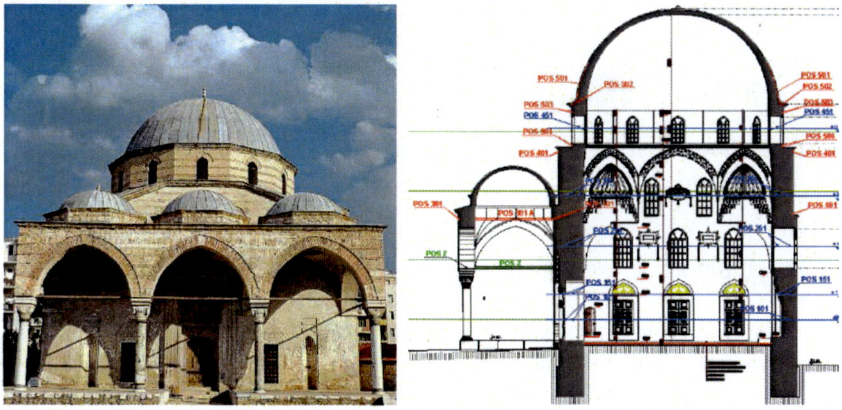

Fig. 15. Gazi Hajdar Kadi Mosque in Bitola, (N-E view – left, retrofitting by CFRP – right)

Ali Pasha Mosque: The mosque located in the center of Ohrid, was originally built by Süleyman Pasha in 1573, but since 1823, when it was repaired by Ali Pasha from Marash, Belgrade vizier, was known as Ali Pasha mosque. It is a typical representative of one-room sub-dome construction, with base in the shape of a square of 15 × 15 m, above which rises an octagonal drum and an unusual sixteen-sided drum arched with a shallow dome. The walls and the dome are in massive stone masonry. During its existence monument has been partially damaged. In the First World War upper part of minaret has been destroyed. In 2015 Turkish foundation Vakuflar undertakes measures and funding for complete conservations, structural consolidations, repair, partial reconstruction of minaret and entrance hall and seismic retrofitting for the whole monument structure. Investigations of present conditions, designing of whole process with detailed analysis has been done by the Institute for protection of monuments of culture and Museum – Ohrid and financially supported by Turkish Directorate General of Foundations. The structural retrofitting, which anticipated CFRP bars, laminates and wrap, ending with incorporation of horizontal RC belt course around the foundation wall bellow the terrain, was accomplished in 2019, (Fig. 16, Petkovski et al. 2016).

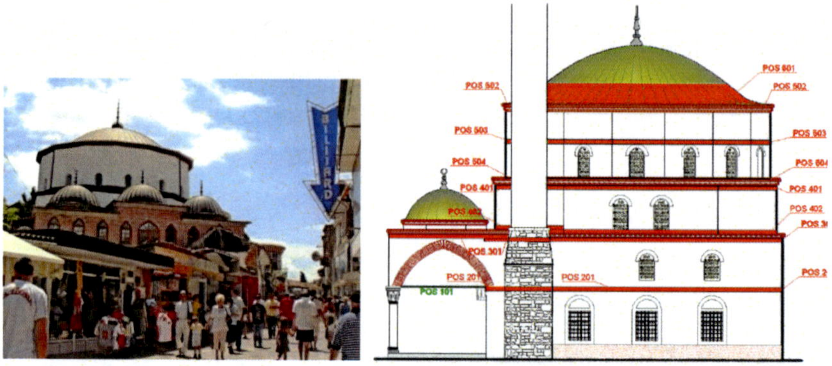

Fig. 16. Ali Pasha Mosque in Ohrid, (N-E view – left, applied retrofitting – right)

5 Conclusions

Shaking table tests on the models in a realistic geometric scale are excellent method for research investigation of dynamic properties and bearing characteristics, verification of technology and method for repair and retrofitting as well as for investigation of new retrofitting technologies in historical monuments. The experimental evidence is helpful in proving the suitability of the design choices and the effectiveness of the applied consolidation system. In this way the original methodology using CFRP for seismic retrofitting of monumental buildings was developed and verified through ample experimental and analytical investigation.

The developed and experimentally verified methodology was successfully implemented in the process of reconstruction and seismic retrofitting of structures of the real historic monuments. The seismic retrofitting of the Mustafa Pasha Mosque in Skopje, as well of several other mosques in the region, was accurately developed as far as possible according to the modern principles for the seismic protection of historical buildings, which require the use of reversible, low intrusive and possibly invisible technologies. The delicate problem of proving the effectiveness of the selected consolidation system has been successfully overcome by using the methodology of design assisted by testing, which represents a very powerful tool especially in a case of such complex structures, which is difficult, and therefore unsafe, to analyze by using the traditional methods.

However, it should be pointed out that, due to the enormous differences between laboratory and onsite conditions, special attention should be paid to the (i) inevitable pre-design detail investigation of monument and site, (ii) relevant structural analysis as well as (iii) specific technical detailing during implementation of retrofitting in order to provide complete realization of the project solution.

Acknowledgements. The authors of this paper and the participants involved in the realization of the projects, wish to extend their gratitude to all the mentioned institutions and companies who made the retrofitting of the mosques possible, for the entrusted task, their participation, collaboration and financial support.

References

Gavrilovic, P., Shendova, V., Ginell, W.S.: Seismic strengthening and repair of byzantine churches. J. Earthq. Eng. **3**(2), 199–235 (1999)

Tashkov, L., Krstevska, L.: Experimental testing of the historical monuments. Final Report. Bilateral scientific research project, Yildiz Technical University–IZIIS Skopje (2006)

Mazzolani, F.M.: Earthquake protection of historical buildings by reversible mixed technologies: the PROHITECH project. In: Proceedings of the International Conference STESSA 2006, 14–17 August, Yokohama, Japan (2006)

Mazzolani, F.M., Krstevska, L., Tashkov, L., Gramatikov, K.: Shaking table testing of Mustafa Pasha Mosque model, Final report, EU-FP6 PROHITECH project (2007)

Sesov, V., et al.: Definition of the seismic parameters for the evaluation of seismic stability of Mustafa Pasha Mosque in Skopje. IZIIS Report 2007-47, Skopje (2007)

Sendova, V., Gavrilovic, P., Stojanoski, B.: Main Project on Repair and Strengthening of the Mustafa Pasha Mosque in Skopje. IZIIS Report 2007-41, Vol. 1, 2, 3, Skopje (2007)

Mazzolani, F.M., Shendova, V., Gavrilovic, P.: Design by testing for the seismic restoration of Mustafa Pasha Mosque in Skopje. In: Proceeding of the International Conference PROHITECH 09, Rome, Italy (2009)

Gavrilovic, P., Jekic, G., Ceko, B., Ackay, M.: Structural consolidation and seismic strengthening of Fatih (Prishtina) and Pasha (Prizren) Mosques - Kosovo. In: Proceeding of the International Conference PROHITECH 09, Rome, Italy (2009)

Shendova, V., Rakicevic, Z., Krstevska, L., Tashkov, L., Gavrilovic, P.: Shaking table testing of models of historic buildings and monuments - IZIIS' experience. In: FP7 SERIES Workshop on the Role of Seismic Testing Facilities in Performance-Based Earthquake Engineering, pp. 221–245 (2011). Springer, Switzerland. https://doi.org/10.1007/978-94-007-1977-4_12. ISBN 978-94-00701976-7

Hadzi-Pecova, C., Gavrilovic, P.: Consolidation Repair and Strengthening of Hajdar Kadi Mosque in Bitola, Main Structural Design Project. National Conservation Center, Skopje (2013)

Shendova, V., Rakicevic, Z., Garevski, M., Apostolska, R., Bozinovski, Z.: Implementation of experimental developed methodology for seismic strengthening and repair in historic monuments. In: Seismic Evaluation and Rehabilitation of Structures, Geotechnical, Geological and Earthquake Engineering, vol. 26. Springer, Switzerland (2014). https://doi.org/10.1007/978-3-319-00458-7_8

Petkovski, R., Gavrilovic, P., Jekic, G., Kocanovska, V.: Main structural design project for conservation, structural consolidation an strengthening of Ali Pahsa Mosque in Ohrid. National Conservation Center, Skopje (2016)

Check for updates

Safety Analysis and Retrofitting of a Masonry Bell Tower

Massimiliano Ferraioli[1], Angelo Lavino[1], Donato Abruzzese[2], and Alberto Mandara[1](✉)

[1] Department of Engineering (DI), Università della Campania "Luigi Vanvitelli", Via Roma 9, 81031 Aversa, Italy
{massimiliano.ferraioli,angelo.lavino, alberto.mandara}@unicampania.it

[2] Department of Civil Engineering and Computer Science (DICII), Università of Roma Tor Vergata, Via Via del Politecnico 1, Roma, Italy
abruzzese@uniroma2.it

Abstract. The paper illustrates the performance assessment and strengthening of an existing masonry tower in Southern Italy. On-site diagnostic investigations including geometrical survey, laser scanning survey, flat jack tests, endoscopic tests, sonic pulse velocity tests, geognostic surveys and tie-rods tests performed on the tower are described in detail. The local collapse mechanisms are described through a kinematic limit analysis based on rigid block rotation, carried out according to the multilevel assessment path proposed by the Italian Guidelines for the assessment and mitigation of the seismic risk of the cultural heritage. Repair and strengthening interventions have been applied so as to repair the existing cracks, prevent the damage in the belfry and improve the seismic performance of the tower, keeping at the same time compatibility, durability, and reversibility of the intervention.

Keywords: Masonry towers · Seismic safety · Retrofit

1 Introduction

Masonry buildings such as bell towers and minarets are often characterized by very high state of stress due to gravity, which can easily produce severe damage up to collapse of the whole structure. In addition, these constructions are inherently very sensitive to horizontal actions, which make them particularly vulnerable to earthquakes. This is mostly the result of the poor tensile strength and ductility combined with the masonry heterogeneity and lack of wall connections, but also the consequence of the unfavorable constructional features of such buildings (namely high slenderness, openings, adjacent lower buildings, heavy and/or slender belfries on the top of the tower). Recent earthquakes occurred in Italy confirmed the great importance of above outlined problems and, at the same time, pointed out the necessity to define suitable strategies for the protection of the historical heritage. This process has led to a refinement of existing codification

I. Vayas and F. M. Mazzolani (Eds.): PROHITECH 2021, LNCE 209, pp. 938–950, 2022.
https://doi.org/10.1007/978-3-030-90788-4_72

at both national and international level in which a number of rules and recommendation were issued referring to general assessment, in situ and laboratory tests, structural analysis and, eventually, repairing and strengthening techniques (ASCE 2003, 2007, 2009; Eurocode 8 2004; FEMA 356 2000; FEMA 547 2006; Italian Code-NTC 2018, 2019). In this context, the importance of non-destructive assessment methodologies has been emphasized when historical constructions are dealt with (Bayraktar et al. 2009; Abruzzese et al. 2009a, 2009b, 2009c; Bartoli et al. 2013; Osmancikli 2012; Ferraioli et al. 2017, 2018) and many studies have been dedicated to full-scale ambient vibration tests and modal identification of civil structures (Ferraioli et al. 2010a, 2010b, 2015). This paper faces the case of the bell tower in Torre Orsaia (Salerno – Italy) in which, due to the widespread damage in the belfry, local or total failure was possible even under gravity loads. On-site diagnostic investigations based on nondestructive or slightly destructive test are described together with the structural analysis, which was carried out to evaluate the seismic capacity according to the multilevel approach of the Italian Guidelines on the assessment and reduction of seismic risk of cultural heritage (DPCM 2011). In the end, the strengthening intervention is presented, relying on a number of stainless steel reinforcing rings. The intervention was conceived and calibrated according to the compatibility, durability, ease of monitoring and reversibility properties required today for the restoration of cultural heritage buildings.

2 On-site Assessment of the Tower

The first construction of tower, formerly the civic tower of the old town of Torre Orsaia (Salerno), is today the bell tower of the church of San Lorenzo Martire (Fig. 1). It has square shape with four floors, was constructed with defensive purposes in the Lombard and Norman age and underwent several modifications over the centuries.

Fig. 1. Bell tower in Torre Orsaia. a) General view; b) North side view; c) North side cross-section

Originally built on three floors, two more floors were added in 1576 giving place to the belfry and resulting in the tower to reach the current height of about 30 m. The basic material used for the construction of the tower is limestone for the first floor and local sandstone for the two upper levels, whose windows and openings are surrounded by grey tuff elements. The upper part with octagonal section is made of plastered local stones and bricks, with the dome made of solid bricks and local stone inserts. The structure suffered extensive damage in the past due to both earthquake and lightning. Some traces of the latest repairing operations, made with injected mortar grouting and dating back after the Campano-Lucano earthquake of 1980, are still visible today. During this intervention, three of the ancient tie-rods connecting the belfry walls and destroyed by a lighting in 1911 were replaced. The investigation campaign on the bell tower was conceived in such a way to be less invasive as possible and included the geometrical survey of the building including cracking pattern and deformation phenomena, endoscopic tests, sonic pulse velocity tests and flat jack tests. Geognostic surveys for the assessment of both physical and mechanical properties of subsoil were also made. Restoration of the tower was mostly due to vegetation growth, mould and insect colonization. Structural damage in the form of widespread cracks was detected on the belfry, where several bricks of grey tuff had clearly detached from their support (Fig. 2).

Fig. 2. Details of the damage in the belfry

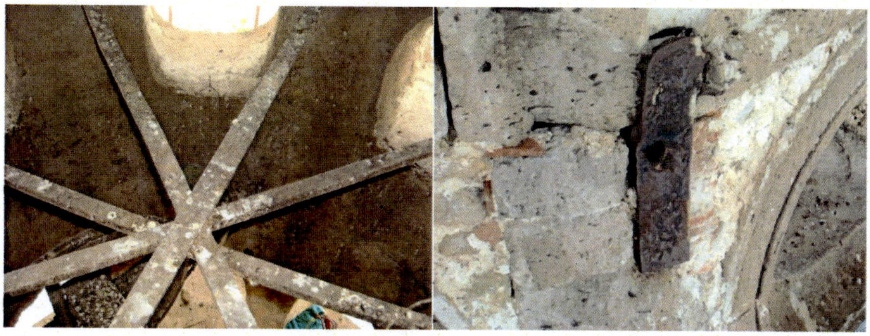

Fig. 3. Ancient tie-rods and external anchor plate

a) b)

Fig. 4. 3D views by laser scanning. a) East façade. b) Octagonal plan drum

Visual inspection suggested cracks to be consequence of the tension loss in the existing tie-rods, due to their degradation as well as to the lack of adequate anchor (Fig. 3). No damage was found at the lower level of the tower, thus excluding any possible foundation settlement. Due to both structural complexity and extensive degradation, laser scanning equipment "Faro Focus 3D" based on the phase-shift technology was conveniently exploited for the geometric survey of the belfry, greatly helping the material survey as well as the definition of the 3D model of the structure. This technique offers advantages such as ease of positioning even in restricted areas, rapid data acquisition and high precision. Due to the presence of scaffoldings around the tower, a higher number of scans was necessary to cover the whole structure adequately. In the end a very satisfactory 3D reconstruction of the entire tower was achieved, as shown in Fig. 4. A core drilling of foundation was carried out to determine the depth of the playing surface (Fig. 5a). The 40% leaning drilling on the masonry wall near the soil indicated the existence of a direct foundation with a basis at 1.4 m from the ground level. Morphology of the masonry was investigated by endoscopic examination into two existing voids in the upper part of the tower as well as into two 40 mm drilled holes at heights between 9 m and 13 m. Video-endoscopic images showed the great heterogeneity of the outer wall of the main façade at the upper part of the tower, whereas a good continuity of the masonry wall was found in the drilled cavities at lower level. This evidence was confirmed by the extensive visual inspections, which suggested that the stone masonry is relatively compact and well executed, in spite of a quite irregular texture and several local vertical joints. Sonic pulse velocity tests, performed on six selected areas of the outer wall of the façades to spot possible structural damage, showed solid brick sections, with sonic velocity values between 1200 m/s and 1400 m/s (Fig. 5b). Nevertheless, low velocity values were detected at some testing points, probably due to the degradation of the mortar joints and other local defects (insufficient stone interlocking, vertical cracks, etc.).

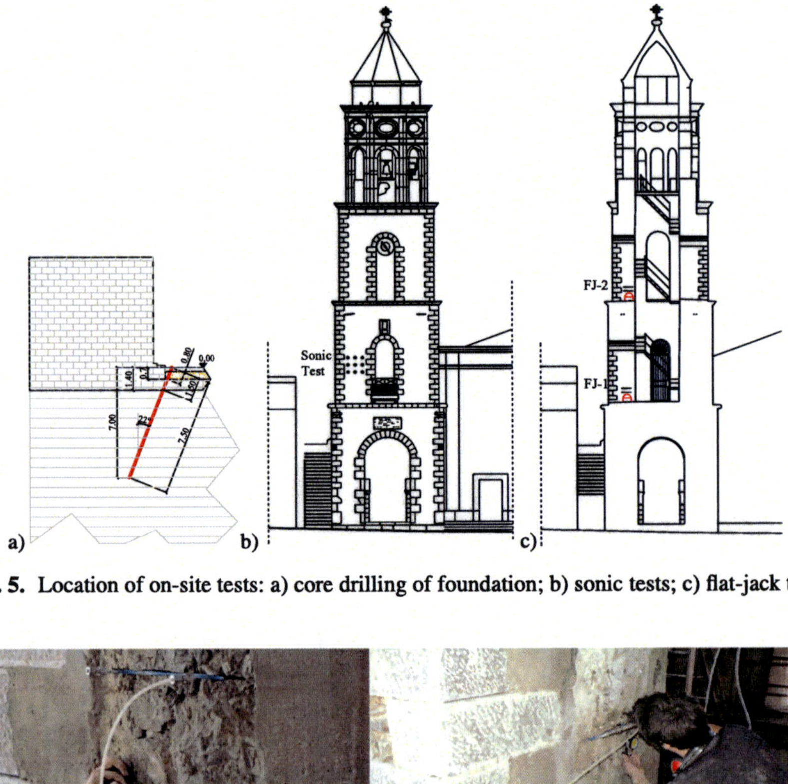

Fig. 5. Location of on-site tests: a) core drilling of foundation; b) sonic tests; c) flat-jack tests

Fig. 6. Double flat-jack tests

In order to assess the state of stress as well as the mechanical properties of the masonry, both single and double flat-jack tests (Fig. 6) were performed in the positions FJ-1 and FJ-2 shown in Fig. 5c. The state of stress from the two single tests resulted $\sigma_{Fj1} = 0.12$ MPa and $\sigma_{Fj2} = 0.11$ MPa, against a resistance at the first cracking of $\sigma_{Fj1} = 0.67$ MPa and $\sigma_{Fj2} = 0.71$ MPa obtained from double tests, respectively. These results show the conservative state of stress compared with the corresponding strength values which can be obtained from the current Italian codification (NTC 2018 and Instructions 2019). As noted above (Fig. 2), the cracking of the belfry comes from the tension loss in the existing tie-rods. A specific procedure was followed to estimate the tensile stress acting in the four existing tie-rods at the top of the belfry. Each tie-rod is encased into the walls of the belfry for about 1 m (Fig. 3). All of them are fastened to external clamps, except one (the N.1), which has no outside anchors. At first, the first experimental vibration frequency and the frequency response spectrum of the ties was determined through the Fourier transform based on a dynamic test carried out using four accelerometers for each rod. Both impact hammer and environmental vibration tests were performed. The axial force in the tie-rod was calculated from by theoretical formulations available in the literature (Shaker 1975), Graff 1991) on the basis of the geometry of each tie-rod and modelling the tie-rod as a single-span beam. Results show that only the rod N.2 can be considered actively working (Table 1). In the end, a geognostic investigation was made on the foundation. Two continuous core drilling and one Down-Hole survey (with the extraction of two undisturbed samples) were carried out. Some further data were collected in the surrounding area, obtained from N.22 continuous core drilling, N.12 continuous penetration tests (DL030), N.7 MASW tests, N.19 Down-Hole type tests, N.28 Dynamic penetration tests (SPT type) and laboratory tests on N.23 soil undisturbed samples. These tests revealed two layers immediately below the tower foundation. A first layer, up to a depth of about 19 m from the ground level, consisting of a limestone layer, and a second one, up to about 30 m from the ground level, consisting of a clay formation. In the surrounding areas, instead, MASW and Down-Hole tests detected the basic formation between 10 and 20 m from the ground (shear-wave velocity averaged over the top 30 m of soil $V_{s,30} > 800$ m/s and number of blows from Standard

Table 1. Geometrical and dynamic properties of tie-rods

Tie-rod number	1	2	3	4
Cross section width [mm]	46	38	36	43
Cross section height [mm]	14	23	22	14
Internal length [m]	2.82	2.81	2.80	2.85
External length [m]	4.82	4.81	4.80	4.85
First experimental frequency [Hz]	–	16	9.1	6.0
First theoretical freq. [Hz] (N = 0; pinned-pinned beam)	–	2.28	2.19	1.36
First theoretical freq. [Hz] (N = 0; fixed-fixed beam)		5.16	4.96	3.09
Axial force [kN] - pinned-pinned beam	–	19.8	8.52	2.81

Penetration Test $N_{SPT} > 50$), whereas at upper level soil consists of detrital materials of limestone in scarce clay matrix or altered clay materials ($V_{s,30} < 360$ and $N_{SPT} < 50$). In conclusion, subsoil can be classified of category E, according to Eurocode 8 (2004).

3 Structural Analysis of the Tower

The structural analysis of the tower has been carried out according to the Italian Building Code (2008), where the problem of the structural response under seismic action is faced via a multilevel approach is considered with three levels of analysis of increasing accuracy: LV1, LV2 and LV3. The level LV1 makes use of a simplified approach in which the structure is modelled as a free standing column. The level LV2 is based on the use of the linear kinematic analysis to study the collapse mechanism and determine the ultimate load of each single macro-element. The LV3 uses as a rule global analysis methods, such as the analysis of frame models or finite element analysis to investigate the seismic response of the structure. In the same code, three different levels are defined for the general knowledge of the construction: KL1: Limited knowledge, KL2: Normal knowledge, KL3: Full knowledge. Depending on the knowledge level achieved in the assessment of the building, a specific value is assumed of the confidence Factor *Fc*, defined as follows:

$$F_C = 1 + \sum_{k=1}^{4} F_{Ck} \tag{1}$$

where F_{Ck} are partial confidence factors accounting for: 1) geometric survey; 2) material types and structural details; 3) materials properties; 4) soil and foundation. Due to the exhaustive investigation carried out on the tower, in the case under consideration, the following values of the partial confidence factors have been assumed: $F_{C1} = 0.05$, because of the complete analysis of the geometrical characteristics; $F_{C2} = 0.12$, according to the lack of detailed information on structural details and materials; $F_{C3} = 0.12$, as the reference values indicated in the current Italian code (2018) have been used for materials properties; $F_{C4} = 0.03$, due to the relatively limited knowledge of the subsoil and foundation characteristics. This yields to a global confidence factor $F_C = 1.32$. Considering that destructive tests were limited to the minimum in order to preserve the historical value of the construction, and also that the assessment campaign was made on a limited budget, a limited knowledge (KL1) was prudentially assumed, in spite of the great amount of data collected. This led to assume values for masonry properties those given in the code for uneven stone masonry with poor mechanical resistance of the joints. Minimum resistance values recommended by the code were used, whereas the average value was taken for elasticity moduli. An additional strength reduction of 10% was assumed for masonry to account for the presence of an inner core with poor mechanical properties. Ultimately, the design compressive strength assumed in the analysis has been calculated as $f_d = 1 \times 0.90/1.32 = 0.682$ N/mm^2. The analysis has been carried out assuming the seismic parameters given in the Italian code for the construction site with values of PGA equal to 0.047 g and 0.116 g for damage limit state (DLS) and life safety limit state (LSLS), respectively. LV1 and LV2 analysis have been performed, only. The method outlined in Section 5.4.4 of Italian Guidelines (DPCM 2011) has been

Table 2. Parameters of elastic design response spectra

Parameter	SLS	DLS	LSLS	LSCP
P_{VR} (-)	0.81	0.63	0.10	0.05
T_R (yrs)	30	50	475	975
PGA (g)	0.037	0.047	0.116	0.149
F_o (-)	2.446	2.443	2.521	2.583

used for estimating the seismic vulnerability of the construction in the LV1 analysis. For this kind of structure, a collapse due to bending is assumed, considering the tower as a free standing column under seismic horizontal forces. The tower has been divided into three sections, according to the constructional features (Fig. 7). Under the assumption of plane sections up to failure, no tensile resistance of material and rectangular stress-block diagram for compressed areas, the resisting bending moment was calculated as (DPCM 2011):

$$M_u = \frac{\sigma_o A}{2}\left(b - \frac{\sigma_o A}{0.85\, a f_d}\right) \tag{2}$$

where σ_0 is the mean compressive stress, A is the cross-section area, f_d is the design compressive strength, a and b are the transversal and the longitudinal dimensions of the tower, respectively. The acting bending moment was calculated from linear static analysis considering a linearly distributed horizontal load and assuming the parameters of the elastic design response spectra given in Table 2. The Limit States of Immediate Occupancy (IO), Damage Limitation (DL), Life Safety (LS) and Collapse Prevention (CP) are defined according to Eurocode 8 (2004), whereas the behavior factor q was assumed equal to 2.8 as recommended in Italian Code (NTC 2018) for structures with stiffness irregularities along the height. In order to account for the effect of adjacent buildings on the vibration period, two limit conditions were considered: 1) isolated tower; 2) fixed tower. The first one assumes that no interaction with other buildings takes place (i.e. free height of the tower $h = 28.6$ m), whereas the second one assumes the neighboring church to be effective in constraining the tower fixed up to the height of 11.29 m (i.e. free height of the tower $H = 17.31$ m). Considering three different methods for calculating the first vibration period of the structure T_1, namely (1) the simplified formula $T_1 = 0.0187$ H, (2) the modal analysis of a shelf-type equivalent model and (3) ambient vibration measurements (Table 3), it has been found that the actual behavior of the tower is closer to the case of fully constrained structure, in which case the fundamental vibration period of the construction falls in the constant acceleration region of the spectrum.

In order to refine the results obtained from LV1 analysis, the LV2 analysis has been performed, in which the linear kinematic analysis has been used to determine the PGA corresponding to the ultimate limit state of the structure. The main assumptions of the linear kinematic analysis are that the structure consists of macro-elements (rigid bodies) hinged to each other and the materials have no tensile but infinite compressive strength.

Table 3. First period of vibration of the structure

	Simplified formula		Shelf-type equivalent model		Ambient vibration test	
	Period [sec]	Cracked period [sec]	Period [sec]	Cracked period [sec]	Period [sec]	Cracked period [sec]
Isolated tower	0.535	0.749	0.809	1.132	0.286	0.400
Non-isolated tower	0.323	0.453	0.360	0.504		

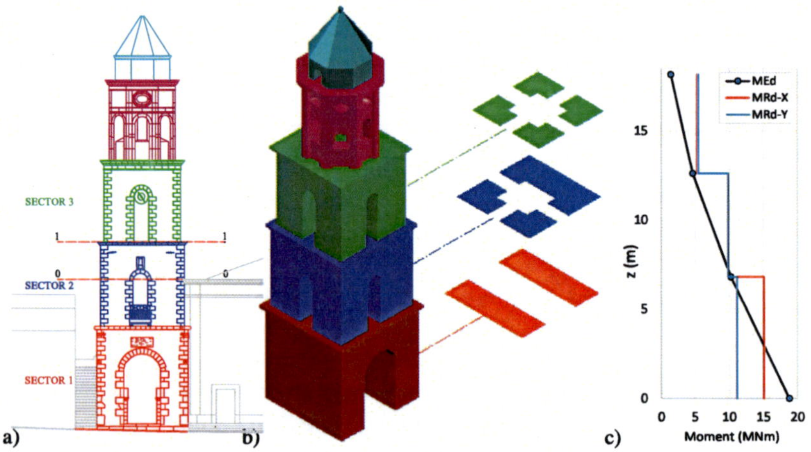

Fig. 7. a) Sectors of the tower. b) 3D view of sectors. c) Acting and resisting bending moments

Seismic loads are considered as static horizontal forces amplified by a multiplier α_0 and the corresponding kinematic multiplier at collapse for each possible mechanism is obtained by calculating the seismic spectral acceleration a^*_0 according to the procedure reported in DPCM (2011). Eight different collapse mechanisms were studied, chosen on the basis of the experience of past seismic events. The first three of them (M1, M2, M3) (Fig. 8) are local mechanisms related to the possible overturning of the belfry walls due to lack of internal connection among them, uneliminated thrusts, uneffectiveness of iron ties, etc. Mechanism M4, instead, assumes a soft story-mechanism of the masonry piers of the belfry. Mechanism M5 refers to the collapse of the whole belfry, whereas mechanisms M6, M7 and M8 consider the typical failure mode of the masonry tower by partial or global overturning. Eventually, the following seismic safety indexes were calculated (DPCM 2011):

$$f_{a,SLV} = \frac{a_{SLV}}{a_{g,SLV}} \quad I_{S,SLV} = \frac{T_{SLV}}{T_{R,SLV}} \tag{3}$$

The first one ($f_{a,SLV}$) is the acceleration safety index defined as the ratio between the peak ground acceleration (PGA) for soil of type A that leads the tower to the ultimate limit state (a_{SLV}) and the reference PGA of the site referred to a rigid soil of type A ($a_{g,SLV}$). The second one is the return period safety index which is the ratio between the return period T_{SLV} of the earthquake that leads the tower to the ultimate limit state and the expected return period $T_{R,SLV}$ of the design earthquake of the site. The values of the seismic safety indexes for the different constraints conditions are summarized in Table 4 and Table 5 for LV1 and LV2 analyses, respectively. Provided the local collapse mechanisms are avoided, the minimum value of the seismic index is equal to 0.716 if evaluated in terms of acceleration and 0.434 in terms of return period. Such value is associated to M7 mechanism.

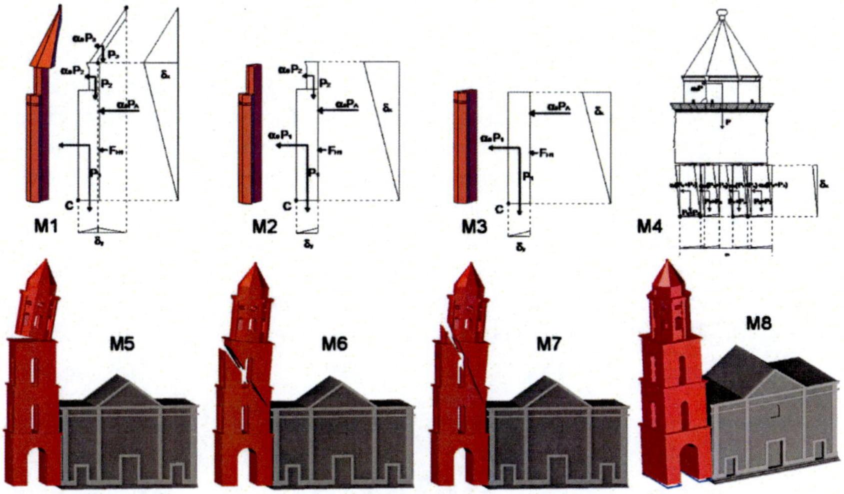

Fig. 8. Collapse mechanisms

Table 4. Acceleration safety index and return period safety index. Level LV1

Section	$f_{a,SLV}$			$I_{S,SLV}$		
	Isolated tower	Tower restrained by staircase	Tower restrained by church	Isolated tower	Tower restrained by staircase	Tower restrained by church
0	0.589	1.442	1.827	0.267	1.251	2.256
1	0.957	1.593	1.683	0.895	1.602	1.839
2	1.152	–	–	1.422	–	–

Table 5. Acceleration safety index and return period safety index. Level LV2

Collapse mechanism	M1	M2	M3	M4	M5	M6	M7	M8
$f_{a,SLV}$	0.411	0.371	0.305	2.738	1.673	0.900	0.716	1.210
$I_{S,SLV}$	0.109	0.084	0.052	12.302	3.606	0.769	0.434	1.608

4 Interventions on the Belfry

The intervention project has been intended as "repair or local intervention", after the Italian Guidelines (DPCM 2011). Operations on the tower were concentrated on its most vulnerable part that is the belfry. Damage in this element was due to the unrestrained roof thrust, which proved highly dangerous due to several factors, such as the slenderness of the masonry walls, the poor mechanical properties of masonry, the lack of effective tying or hooping and other constructional issues. In order to repair the belfry and improve its structural behavior, some stainless steel reinforcing rings consisting of a couple of Ø 30 mm bars were applied in four different horizontal sections of the tower (Fig. 9) so as to confine masonry and improve connection between the contiguous walls. A slight pre-tensioning, symmetrically applied from both ends of the rings, provides an active confinement which counteracts the creep damage occurring during local repair interventions. The effectiveness of pre-tensioning was monitored by means of an ad hoc

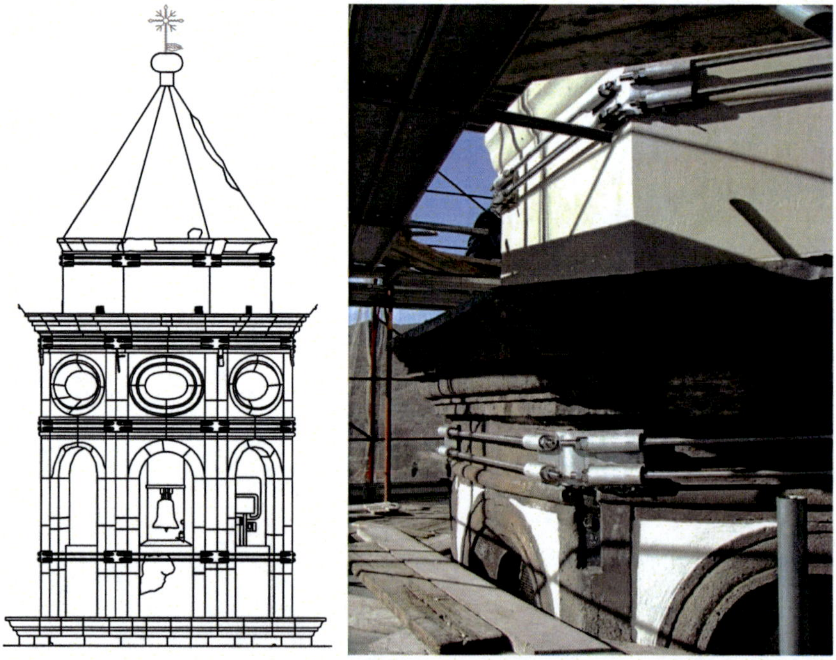

Fig. 9. The belfry after strengthening with stainless tie-rods

computerized acquisition system in such a way to limit the initial pre-tightening force to about 500 daN and, at the same time, to control the possible tension loss over time.

5 Conclusive Remarks

This paper illustrated the diagnostic investigation and performance assessment of an existing masonry tower, together with the subsequent repair and strengthening intervention. All steps of the general survey and on-site investigations including laser scanning, flat jack tests, endoscopic tests, sonic pulse velocity tests, geognostic assessment and tie-rods tests have been described in detail. Results obtained revealed an extensive damage of the belfry, mostly due to material degradation over time and loss of effectiveness of existing tie-rods. A comprehensive structural analysis, carried out at both LV1 and LV2 levels of investigation, demonstrated the great structural vulnerability of the tower, in particular from the seismic point of view. The repair and strengthening intervention were selected in order to satisfy the compatibility, durability, and reversibility issues currently required in the restoration of historical constructions. Moreover, it not only enhances resistance against vertical loads, but also improve the behavior under seismic actions, avoiding local collapse mechanisms that can involve the individual piers of the belfry or the walls supporting the roof. For this reason, even though referred to a specific study case, the approach used to assess the structural safety and define the retrofitting intervention can be intended as a generalized intervention policy for this type of buildings.

Acknowledgements. The author wishes to thank Prof. Lucio Olivares for geotechnical surveys and investigations; Alberto Maria Avossa, Phd, for supporting the numerical analysis; Luca Ferri, Phd, for on-site geometric and material surveys; Paola Paterna, PhD, for elaborating the results of the tie-bar tests; Konstantina Fleri for supporting the graphic representation; Maurizio Cice and SIA for geognostic surveys; Andrea Basile and the professional team belonging to Tecnolab for the structural monitoring activities.

References

Abruzzese, D., Miccoli, L., Yuan, J.L.: Mechanical behavior of leaning masonry Huzhu Pagoda. J. Cult. Herit. **10**, 480–486 (2009a)

Abruzzese, D., Miccoli, L., Vari, A., Ferraioli, M., Mandara, A., Froncillo, S.: Dynamic investigations on medieval masonry towers: vibration measurement and structural identification. In: 1st International Conference on Protection of Historical Buildings (PROHITECH 09), Rome, Italy (2009a)

Abruzzese, D., Miccoli, L., Vari, A., Ferraioli, M., Mandara, A., Froncillo, S.: Dynamic investigations on medieval masonry towers: seismic resistance and strengthening techniques. In: 1st International Conference on Protection of Historical Buildings (PROHITECH 09), Rome, Italy (2009c)

ASCE: Seismic Evaluation of Existing Buildings. Standard ASCE/SEI 31-03. American Society of Civil Engineers, Reston, Virginia, USA (2003)

ASCE: Seismic rehabilitation of existing buildings. Standard ASCE/SEI 41-06. American Society of Civil Engineers, Reston, Virginia, USA (2007)

ASCE: Supplement to ASCE 41-06, Reston, VA (2009)

Bartoli, G., Betti, M., Giordano, S.: In situ static and dynamic investigations on the "Torre Grossa" masonry tower. Eng. Struct. **52**, 718–733 (2013)

Bayraktar, A., Türker, T., Sevım, B., Altunisik, A.C., Yildirim, F.: Modal parameter identification of Hagia Sophia bell-tower via ambient vibration test. J. Nondestruct. Eval. **28**, 37–47 (2009)

DPCM: Valutazione e riduzione del rischio sismico del patrimonio culturale con riferimento alle norme tecniche per le costruzioni di cui al DM 14 gennaio 2008. Presidenza del Consiglio dei Ministri (Assessment and reduction of seismic risk of cultural heritage in relation to the Building Code included in the Ministerial decree of January 14th 2008. Presidency of the Council of Ministers), Rome (2011). (in Italian)

Eurocode 8, CEN EN 1998-1: 2004: Design of structures for earthquake resistance. Part 3: Assessment and retrofitting of buildings. European Committee for Normalization, Brussels (2004)

FEMA: Prestandard and Commentary for the Seismic Rehabilitation of Buildings. Report FEMA 356. Federal Emergency Management Agency, Washington, DC, USA (2000)

FEMA: Techniques for the Seismic Rehabilitation of Existing Buildings. Report FEMA 547. Federal Emergency Management Agency, Washington, DC, USA (2006)

Ferraioli, M., Abruzzese, D., Miccoli, L., Vari, A., Di Lauro, G.: Structural identification from environmental vibration testing of an asymmetric-plan hospital building in Italy. In: COST ACTION C26: Urban Habitat Constructions under Catastrophic Events - Proceedings of the Final Conference, pp. 981–986 (2010a)

Ferraioli, M.: Inelastic torsional response of an asymmetric plan hospital building in Italy. In: COST ACTION C26: Urban Habitat Constructions under Catastrophic Events - Proceedings of the Final Conference, pp. 365–370 (2010b)

Ferraioli, M.: Case study of seismic performance assessment of irregular RC buildings: hospital structure of Avezzano (L'Aquila, Italy). Earthq. Eng. Eng. Vib. **14**(1), 141–156 (2015). https://doi.org/10.1007/s11803-015-0012-7

Ferraioli, M., Miccoli, L., Abruzzese, D., Mandara, A.: Dynamic characterisation and seismic assessment of medieval masonry towers. Nat. Hazards **86**(2), 489–515 (2016). https://doi.org/10.1007/s11069-016-2519-2

Ferraioli, M., Miccoli, L., Abruzzese, D.: Dynamic characterisation of a historic bell-tower using a sensitivity-based technique for model tuning. J. Civ. Struct. Heal. Monit. **8**(2), 253–269 (2018). https://doi.org/10.1007/s13349-018-0272-9

Graff, K.F.: Wave Motion in Elastic Solids. Dover, New York (1991). 1975 re-issued 1991

Italian Code-NTC: Norme Tecniche per le Costruzioni, D.M. 17.01.18. G.U. No.42, February 4 (2018). (in Italian)

Italian Code NTC – Instruction: Circolare N.7 Istruzioni per l'applicazione dell'aggiornamento delle norme tecniche per le costruzioni di cui al decreto ministeriale 17 gennaio 2018 (2019). (in Italian)

Osmancikli, G., Uçak, S., Turan, F.N., Türker, T., Bayraktar, A.: Investigation of restoration effects on the dynamic characteristics of the Hagia Sophia bell-tower by ambient vibration test. Constr. Build. Mater. **29**, 564–572 (2012)

Shaker, F.J.: Effect of axial load on mode shapes and frequencies of beams, NASA Technical Note (NASA TN D-8109) (1975)

Ukraine's Cultural Heritage Objects Within Landslide Hazardous Sites

Oleksandr Trofymchuk[1], Oleksij Lebid[1], Viktoriia Berchun[1], Yaroslav Berchun[1], and Iurii Kaliukh[2](✉)

[1] Institute of Telecommunications and Global Information Space, NASU 13, Chokolivsky Blvd., Kyiv 03186, Ukraine
itelua@kv.ukrtel.net

[2] State Enterprise "Research Institute of Building Constructions", 5/2, Preobrajenskaya Street, Kyiv 03680, Ukraine

Abstract. During the 2012–2018 period LACUNHEN's (Margottini, Vilimek 2014) Ukrainian Department studied two Ukrainian Heritage Sites located in complicated geological and geotechnical conditions. The report presents examination results of two objects of cultural heritage of Ukraine: Livadia Palace located within active landslide systems and the Swallow's Nest castle situated on the top of the 40-m high Aurora Cliff of Cape Ai-Todor in the Black Sea. Both objects are located in Autonomous Republic of Crimea, Ukraine, and are a subject of intensive destruction during the recent years. The report demonstrates the analysis of application of monitoring and early warning system of Livadia Palace building constructions located within active Central Livadia Landslide system as an example of system approach for monitoring of World Heritage Sites placed on active landslides. The report also presents the results of visual and instrumental examination of the Swallow's Nest castle. In 1927, the Swallow's Nest survived a serious earthquake. At the end of 1950s the cracks indicating the threat of the castle collapse were detected. The castle was recognized as dangerous and was not used for a long time. There was the idea to disassemble the castle into pieces, to number the stones and slabs and to assemble on a new safe place.

Keywords: Livadia palace · Landslide · Monitoring · Early warning system

1 Introduction

The global climate changes and continued expansion of land use result in a tangible raise of the landslides frequency and intensity. Landslides are an important component of a number of major natural disasters and are charged with far greater losses than it is generally recognized. Usually, they are referred to in connection with floods, earthquakes or volcanoes eruptions with the losses from landslides often exceeding all other damages from a general disaster. During the last decade (from 2000 to 2009), the natural disasters damaged and destroyed about one million objects directly affecting nearly 2.5 billion people around the world. Every year in Europe about 20 major landslides occur, and this considerably exceeds the number of floods, earthquakes and hurricanes (Lacasse 2013).

I. Vayas and F. M. Mazzolani (Eds.): PROHITECH 2021, LNCE 209, pp. 951–961, 2022.
https://doi.org/10.1007/978-3-030-90788-4_73

Within this view, landslides and more generally slope instabilities are an important factor endangering cultural heritage sites and its degradation and requires additional protection measures, creation of the monitoring and early warning systems, etc.

The term "monitoring" originates from the English verb "to monitor" (to check, supervise, watch, or keep track of) derived from Latin monit - 'warned' or monere - to warn, admonish, remind. In a variety of scientific and practical activities, the method of observation has been used for a long time as a method of knowledge acquisition based on relatively long-term, purposeful and systematic perception of objects and phenomena of the surrounding environment. The brilliant patterns of the nature observation management were described in the first century AD in the Natural History (Naturalis Historia) by Pliny the Elder (Gaius Plinius Secundus). The work was divided into 37 books, organized into ten volumes. They covered topics including astronomy, geography, geology, zoology, botany, mineralogy etc. and became the most complete encyclopedia until the Middle Ages (Dolina 2002).

The organization of observations of the hydrocarbon oxide content in coal mines air in England and Belgium more than 100 years ago can be considered as the historical beginning of environmental monitoring. For those observations canaries, guinea pigs and cockroaches were used as some kinds of sensors (Dolina 2002). The systems for recording and accumulating information are the simplest forms of monitoring or of monitoring systems (MS). The historical sciences, which gather and analyze the data on historical facts and events, seem to be the oldest among such systems. Theoretical and methodological issues of these problems are analyzed in the works of modern scientists (Vayas et al. 2019; Tassios et al. 2003; Mazzolani and Calado 2019; Highland and Bobrowsky 2008; Kaliukh et al. 2018; Farenyuk et al. 2018, 2019; Slyusarenko et al. 2009, 2010, 2013, 2015; Shokarev et al. 2005, 2009; Mihalić et al. 2013; Trofymchuk et al. 2018, 2019) and others.

2 Concept of the Early Warning Systems New Integrated Methodology (EWS IM)

As defined by the UN International Strategy for Environmental Disaster Reduction (United Nations International Strategy for Disaster Reduction (UNISDR) 2009), the Early Warning System (EWS) is "the set of capacities needed to generate and disseminate timely and meaningful warning information to enable individuals, communities and organizations threatened by a hazard to prepare and to act appropriately and in sufficient time to reduce the possibility of harm or loss" (United Nations International Strategy for Disaster Reduction (UNISDR) 2009).This general definition can be applicable to any danger and does not contain a direct reference to landslides. Regardless of the definition and the hazard considered, EWS is used to reduce the risk by affecting the impact on exposed elements. The basic concept of EWS installed on landslides is that the elements at risk, especially people being away from the dangerous area, must have sufficient time to evacuate, if an imminent collapse is expected.

Therefore, an effective EWS shall include such four main sets of actions (Dixon and Spriggs 2007):

- Monitoring of the activity of the observed object, i.e. the data collection and transmission, as well as the equipment maintenance;
- The analysis and modeling of the observed and studied object;
- Warning, i.e. the dissemination of simple and clear information about the observed object;
- The effective response of risk exposed elements; full understanding of risks.

The key to the successful application of landslides early warning system (EWS) is the system ability to identify and measure in real time a limited number of important indicators called precursors that precede landslide catastrophic movements including disturbances and collapses. The recent advances in the development of control and measuring equipment in conjunction with GPS and photogrammetric techniques have increased the potential for obtaining the highly reliable measurements of various parameters, which then can be used to detect landslide activity preceding the entire slope breakage (Barla et al. 2010; Casagli et al. 2010; Intrieri et al. 2012).

It is quite obvious that whenever the mechanics and instability mechanism of a particular slope are ignored, it may be difficult or simply impossible to rely solely on the analysis based on the measurements of surface displacements and velocities. Therefore, it is necessary to describe the landslide forerunners for the purposes of early warning about soil movements (Di Biagio et al. 2007; O'Connor and Dowding 2000).

As EWS in its components is time-sensitive or stochastic errors susceptible, it is necessary to develop the EWS designing methodology, which will determine the methods for integrating the sources of monitoring information, identifying the potential hazard thresholds and assessing the associated risks within the frameworks of an explicit cause-and-effect analysis. Since both the corresponding forerunners and the landslides characteristic elements can vary depending on the type of a landslide and its location (urban, rural or mountainous areas), each EWS can be tailored for each particular landslide area to be investigated.

The EWS methodology includes four main elements as follows:

1. Special monitoring module;
2. Integrated monitoring module;
3. Module of characteristics description and mathematical modeling;
4. Module of analysis (verification).

Modules 1 through 3 are the sources of input data for the decision-making algorithm (orange dotted frame at the bottom of the flowchart). The algorithm allows the continuous assessment of hazard levels and identifies the appropriate actions to be taken to ensure an adequate level of safety for elements exposed to risk.

3 Example of EWS

The illustration of the EWS practical embodiment the project «System of monitoring the Central Livadia Landslide system (CLLS) and Livadia palace» (Trofymchuk et al. 2018) implemented during 2002–2014 can be taken. In this project unit 2, unit 3 and partially unit 4 of the four units of the EWS were implemented. To study the state of the geological environment of the CLLS the monitoring system was developed and technically implemented on the computer (ZSUV software). Heliogenic parameters included solar activity, changes in temperature and humidity regimes, the nature and intensity of precipitation, wind activity etc. The data were manually loaded into the computer. Lithogenic parameters were presented by a set of conditions and factors characterizing the mechanism and dynamics of changes in the equilibrium state of the CLLS slopes.

The system performed the following actions:

1. Control of the CLLS reference points displacements by means of landslide surface visual observations and subsequent manual loading of information into the PC).
2. The continuous real-time monitoring of the evolution of deviation angle changes for selected areas and zones within the landslide massif with the use of high precision electric inclinometers, filtering of electrical signals, converting of analog signals into a digital code by means of the analog-to-digital converter and data real-time downloading into the PC.

In order to improve the reliability and performance of the hardware component, the concept of the hardware-software complex open architecture was implemented in a form of separate external modules including the high-precision sensors of the angular displacements of the analog-to-digital converter, analog low-pass filters, a power supply unit and a computer. The software featured the possibility of long-term operation in real time without any operator assistance and the complete graphical representation of the statistical processing results on a display.

The processing of the measurement results showed the following:

1. The southeastern wing of the Livadia palace performs continuous waves relative to a certain intermediate position. These vibrations are of a noticeable periodic nature with a period being defined as twenty-four hours. The amplitude of daily vibrations varies within the range of approximately 1.5 angular minutes, that is, about 45 angular seconds to every side away from the intermediate position. The vibrations are directed relative to the transverse building axis.
2. Sometimes, the Livadia Palace tilt angle increase was recorded. In those cases, the amplitude increased to 6 angular minutes. The calculation results showed that the usual daily vibrations were 1.9 mm to each side from the intermediate position, but on the mentioned days the vibrations were about 4.2 mm to each side from the intermediate position.
3. The Livadia Palace daily vibrations can be caused by the difference of day and night temperatures when irregular heating and cooling of the building and the surrounding area occurs. If the third party access to sensors is excluded, the reasons for the increase of the amplitude of the Livadia Palace building facade deviation from the intermediate

position may include the following circumstances and actions: abundant precipitation with accompanying temporary rise of ground water levels; further CLLS landslide activation, seismic and anthropogenic dynamic impacts and other factors.

4 The Swallow's Nest Castle

The Swallow's Nest castle is a landmark of the whole Crimean coast (Ukraine) (Fig. 1).

Fig. 1. Swallow's Nest castle. South-west view. The crack in the cliff is obvious (the crack width is up to 0.5–0.7 м)

In 1927, the Swallow's Nest survived a serious earthquake. There were two shocks at night. The first one was weak but made people to leave their houses. The second one rated at 9 on the Richter scale. The castle was not damaged. However, the cliff itself developed a huge crack from its top to the middle so that the castle could break at any time (Fig. 2). Part of support cliff was thrown into the sea and observation platform was hanging over the precipice.

The building itself was not damaged apart from some small decorative items along with a small portion of the cliff under the bottom balcony. However, the cracks appeared on the walls and the castle was closed for uncertain term. At the end of 1950 s the cracks indicating the threat of the castle collapse were detected. The castle was recognized as dangerous and was not used for a long time. It was an idea to take the castle to pieces, to number the stones and slabs and assemble on a new safe place. In 1960–1970 s the castle was renovated and successfully used up to the present moment. However, the Aurora

Fig. 2. The northern view. The huge crack separa-ting some part of the rock from the main cliff part and a range of smaller cracks covering the whole Swallow's Nest castle base could be seen on the Fig. 1

Cliff – the Swallow's Nest foundation has been intensively destroyed during recent years (Figs. 1 and 2).

Therefore, State Research Institute of Building Constructions (RIBC) undertook scientific and research works on the above issue and developed recommendations as for the Aurora Cliff suspension as well as reconstruction and preservation of the castle structures. Visual and instrumental survey of the cliff Swallow's Nest castle foundation and castle itself performed by the group of RIBC specialists under the supervision of prof. Iurii Kaliukh has revealed that whole chalkstone rock mass has been affected by different types of destruction with perspective of cracks and cavernous porosity consolidation and extension. The above creates an immediate threat to the preservation of the unique monument of architecture, history and culture - the Swallow's Nest castle. On the photographs (Fig. 3), there are underwater damages that penetrate the rock massif-base of the castle from all sides are seen.

Visual inspection of building structures has allowed to ascertain the following. From the outside of the building, between the facing stones, there are vertical and horizontal cracks. Leads are installed at some cracks to control their condition. Inspecting the state of leads, crack opening doesn't happen. A crack from side of north-west facade is through. Number of defects is observed in the slab of the balcony of ground floor. In particular, there are cracks in the upper zone of this slab, as well as a horizontal crack on the butt of the slab (Fig. 4), the presence of which may indicate, that bonding between concrete of old and new slabs is destroyed. Thus, in accordance with results of a preliminary visual inspection of the building "Swallow's Nest" it is possible to conclude, that it necessary to per-form its detailed instrumental examination.

a) b)

Fig. 3. (a, b) Surface and underwater parts of the Swallow's Nest castle rock-base perforated with deep cracks, surface and underwater caves from all the sides

The purpose of vibratory-dynamic investigations - was to determine the actual dynamic parameters (period and frequency of oscillations, the logarithmic decrement of oscillations) of the structure and the foundation slab during oscillations.

Fig. 4. Horizontal crack along end of balcony slab

Before carrying of experimental studies, the possible dominant sources of oscillations were determined: microseismic action, the effects of sea surf, wind effects. In addition, the impulse actions were carried out on the rocky base. Vibro-dynamic observations were performed, which yielded the following results: frequencies of the natural dominant horizontal and vertical oscillations of the building are in the range 0.3 Hz–25 Hz; amplitudes of microseismic accelerations of foundation slab (rock base) along the X-axis range from 0.3 to 2.4 mm/sec^2; maximum horizontal acceleration amplitudes of floor of the ground floor in the direction of X- axis is 0.8–1.8 mm/sec^2; the maximum amplitudes of accelerations of wall of 1-nd floor are in the range 22.1–65.4 mm/sec^2; logarithmic decrement of oscillations of structures of the castle is recorded in the range $\Delta = 0.31 - 0.35$.

5 Conclusions

1. In recent decades, the concept of cultural heritage (Migon 2013) has evolved into one that encompasses an understanding of the history of humanity, together with scientific knowledge and intellectual attitudes. This changing concept has prompted a subsequent reevaluation of what constitutes the outstanding universal values of World Heritage sites and the operational methods for implementing the UNESCO World Heritage Convention (The World Heritage Convention 1972). The scope has broadened from studying a single monument in isolation to one that values a multidimensional, multiregional, and inter-disciplinary approach and encapsulates vast spans of human history, as demonstrated by the above: Livadia Palace and St. Andrew's Church, located within active landslide systems; the Swallow's Nest castle situated on the top of the 40-m high Aurora Cliff of Cape Ai- Todor in the Black Sea, Autonomous Republic of Crimea, Ukraine.
2. Up-to-date methods of geotechnical protection for historical monuments of architecture can provide them with reliable protection against adverse geological processes and ensure long-term reliable operation, as shown by the example of Livadia Palace and Swallow's Nest castle situated on the top of the 40-m high Aurora Cliff of Cape Ai-Todor in the Black Sea, Autonomous Republic of Crimea, Ukraine (temporary occupied in 2014 by Russian Federation).
3. Visual observations of the rock base of the Swallow's Nest show that in general the rocky massif of limestone is exposed to varying degree of destruction with prospect to consolidate and expand the cracking and karst, endangering a unique monument of architecture and history the "Swallow's Nest". To develop design solutions on strengthening of the rock base of the building "Swallow's Nest" it is necessary to perform complex of researches aimed at mapping of the surface cracks, clarification of their abyssal structure, the karst of studied massif of limestone. Taking into account the particular location Avrorin cliff of Cape Ai-Todor (shore of the Black Sea), these observations should be carried out both on the above-water part of rock and in underwater part.

Acknowledgements. Results presented herein have been obtained with the financial support from private consulting firm "Center of Science-Engineering Service", Yalta, Crimea, Ukraine (owner Michail Rijii). "Center of Science-Engineering Service" have been financing preparatory and installation works in 1998–2008 years on the first stages of project. National Academy of Science of Ukraine, the Institute of Telecommunications and Global Information Space have been financing next stages of the project in 2009–2014 years. City Office for Construction of the City, Utility

Services and Transport, City of Yalta enabled and financed supply of the system with electricity from public network. These supports are gratefully acknowledged. We would like to express our very great appreciation to the Ukrainian RIBC scientist, who have been directly involved in visual and instrumental survey The Swallow's Nest castle.

References

Barla, G., Antolini, F., Barla, M., Mensi, E., Piovano, G.: Monitoring of the Beauregard landslide (Aosta Valley, Italy) using advanced and conventional techniques. Eng. Geol. **116**, 218–235 (2010)

Di Biagio, E., Kjekstad, O.: Early warning, instrumentation and monitoring landslides. In: 2nd Regional Training Course, RECLAIM II, 29th January–3rd February 2007

Dixon, N., Spriggs, M.: Quantification of slope displacement rates using acoustic emission monitoring. Can. Geotech. J. **44**(8), 966–976 (2007)

Dolina, L.F.: Environmental monitoring and biosphere protection engineering. Part 1. Basics of monitoring. Continent L, Dnepropetrovsk (2002)

Casagli, N., Catani, F., Del Ventisette, C., Luzi, G.: Monitoring, prediction, and early warning using ground-based radar interferometry. Landslides **7**(3), 291–301 (2010)

Farenyuk, G., Kaliukh, I., Farenyuk, E., Kaliukh, T., Berchun, Y., Berchun, V.: Experimental and theoretical diagnostics of defects in ferroconcrete piles based on reflection of longitudinal and transverse waves. In: Hordijk, D.A., Luković, M. (eds.) High Tech Concrete: Where Technology and Engineering Meet, pp. 1307–1317. Springer, Cham (2018). https://doi.org/10.1007/978-3-319-59471-2_151

Highland, L., Bobrowsky, P.: The landslide handbook —a guide to understanding landslides: reston. U.S. Geological Survey Circular, Virginia (2008)

Intrieri, E., Gigli, G., Mugnai, F., Fanti, R., Casagli, N.: Design and implementation of a landslide early warning system. Eng. Geol. **147–148**, 124–136 (2012)

Kaliukh, I., Fareniuk, G., Trofymchuk, O., Fareniuk, I., Berchun, Y.: Identification of defects in reinforced concrete piles based on multi-wave reflection. In: Derkowski W., Gwoździewicz P., Hojdys Ł., Krajewski P. (eds.) Proceedings fib Symposium 2019: Concrete – Innovations in Materials, Design and Structures, Fédération Internationale du Béton (fib) – International Federation for Structural Concrete, pp. 991–998 (2019)

Kaliukh, I., Trofymchuk, O., Farenyuk, G., Ivanik, O., Shekhunova, S.: Practical measures fo landslide risk mitigation in the Ukrainian Carpathians. In First EAGE Workshop on Assessment of Landslide and Debris Flows Hazards in the Carpathians, vol. 2019, pp. 1–5. European Association of Geoscientists & Engineers (2019). https://doi.org/10.3997/2214-4609.201902165

Kaliukh, I., Dunin, V., Berchun, Y.: Decreasing service life of buildings under regular explosion loads. Cybern. Syst. Anal. **54**(6), 948–956 (2018). https://doi.org/10.1007/s10559-018-0098-9

Kaliukh, I., Fareniuk, G., Fareniuk, I.: Geotechnical issues of landslides in Ukraine: simulation, monitoring and protection. In: Wu, W., Yu, H.-S. (eds.) Proceedings of China-Europe Conference on Geotechnical Engineering. SSGG, pp. 1466–1469. Springer, Cham (2018). https://doi.org/10.1007/978-3-319-97115-5_124

Krivosheev, P., Slyusarenko, Y., Chervinsky, J.: Development of calculation methods of foundations on the pliable basis in Ukraine. In: Proceedings of the 17th International Conference on Soil Mechanics and Geotechnical Engineering: The Academia and Practice of Geotechnical Engineering, vol. 2, pp. 1818–1821 (2009)

Margottini, C., Vilimek, V.: The ICL network on "landslides and cultural & natural heritage (LACUNHEN)." Landslides **11**(3), 934–938 (2014)

Mazzolani, F.M., Calado, L.: Selected papers from the 3rd international conference on protection of historical constructions (Prohitech 2017). Int. J. Architectural Heritage **13**(1), 1 (2019). https://doi.org/10.1080/15583058.2018.1564591

Migon, P.: Cultural heritage and natural hazards. In: Bobrovsky (ed.) Encyclopedia of natural hazards. Springer Science + Business media, Dordrecht (2013)

Mihalić, S., Mihalić, A., Željko, A., Martin, K.: TXT-tool 2.385–1.2 a comprehensive landslide monitoring system: the Kostanjek landslide, Croatia (2013). https://scholar.google.fr/citations?view_op=view_Monitoring and Early Warning System507citation&hl=ru&user=7eoSoYYAAAAJ&citation_for_view=7eoSoYYAAAAJ:D03iK_w7-QYC

O'Connor, K., Dowding, C.: Comparison of TDR and inclinometers for slope monitoring. In: Proceedings of Geo-Denver 2000, Denver, Colorado, 5th–7th, August (2000)

Papadopoulos, K., Vayas, I.: Restoration and strengthening techniques for ancient-Greek monuments. Int. J. Architectural Heritage **13**(1), 33–46 (2019). https://doi.org/10.1080/15583058.2018.1497233

Slyusarenko, Y., Matveyev, I., Kisil, A., Ischenko, Y., Romanov, O., Kosheleva, N.: Solution of the geotechnical problems of the poshtova square reconstruction in Kiev.In: Proceedings of the XVI European Conference on Soil Mechanics and Geotechnical Engineering (ECSMGE 2015): Geotechnical Engineering for Infrastructure and Development, pp. 693–698, January 2015

Slyusarenko, Y., et al.: Modern methods of geotechnical defense of buildings in the difficult geological conditions of Ukraine. In: Proceedings 18th International Conference on Soil Mechanics and Geotechnical Engineering (Paris) (2013)

Slyusarenko, Y., Girshtel, G., Liseniy, O.: Optimization of multilayer repair systems of polymeric compositions for concrete structures at reconstruction of NSC Olimpiyskiy in Kiev. In: Proceedings of the fib Symposium Tel-Aviv 2013: Engineering a Concrete Future: Technology, Modeling and Construction, pp. 353–356 (2013)

Slyusarenko, Y., Kryvosheyev, P., Serjogin, J., Kozjavkin, I.: Concrete structures of the olympiyskiy national sports complex in Kiev. In: Proceedings of the 3rd International fib Congress and Exhibition, Incorporating the PCI Annual Convention and Bridge Conference: Think Globally, Build Locally (2010)

Shokarev, V., Shapoval, V., Chaplygin, V., Samchenko, R., Volkov, D.: Stress-strain state of the system base-strip foundation at elimination of excessive tilts of buildings. In: Proceedings of the 17th International Conference on Soil Mechanics and Geotechnical Engineering: The Academia and Practice of Geotechnical Engineering, vol. 1, pp. 680–683 (2009)

Trofymchuk, O., Kaliukh, I., Klymenkov, O.: TXT-tool 2.380-1.1: monitoring and early warning system of the building constructions of the Livadia palace, Ukraine. In: Sassa, K., et al. (eds.) Landslide Dynamics: ISDR-ICL Landslide Interactive Teaching Tools, pp. 491–508. Springer, Cham (2018). https://doi.org/10.1007/978-3-319-57774-6_37

Trofymchuk, O., et al.: Dynamic certification of landslide protection structures in a seismically hazardous region of Ukraine: experimental and analytical research. Earthquake geotechnical engineering for protection and development of environment and constructions. In Silvestri, F., Moraci, N. (eds.) Proceedings of the VII ICEGE 7th International Conference on Earthquake Geotechnical Engineering, Rome, Italy, 5337–5344, 17–20, June 2019

United Nations International Strategy for Disaster Reduction (UNISDR): Terminology on Disaster Risk Reduction (2009). http://www.unisdr.org

The World Heritage Convention (1972). http://whc.unesco.org/en/convention/

Zanakis, S., Theofanides, S., Kontaratos, A., Tassios, T.: Ancient Greeks' practices and contributions in public and entrepreneurship decision making. Interfaces, **33**(6), 72–88 (2003). Accessed 7 Oct 2020, http://www.jstor.org/stable/20141304

Zocenko, N., Matveyev, I., Rosenfeld, L., Shokarev, V., Kornienko, N.: Geotechnical aspects of construction on loessial soils. In: Proceedings of the 16th International Conference on Soil Mechanics and Geotechnical Engineering: Geotechnology in Harmony with the Global Environment, vol. 3, pp. 1307–1311 (2005)

Reinforcement of Dowel-Type Timber Joints with Self-tapping Screws

Jorge M. Branco[1,2](✉) and Douaa Elbashir[1,2]

[1] ISISE, Institute of Science and Innovation for Bio-Sustainability (IB-S), Department of Civil Engineering, University of Minho, Guimarães, Portugal
jbranco@civil.uminho.pt

[2] ISISE, Department of Civil Engineering, University of Minho, 4800-058 Guimarães, Azurém, Portugal

Abstract. An experimental campaign was carried out for dowel type joints made of two wood species, Maritime Pine (*Pinus pinaster*) and Scot's Pine (*Pinus sylvestris*). Monotonic (EN 26891) and cyclic (EN 12512) tests were performed on the joints, with and without reinforcement, and the effect of the reinforcement in the overall capacity of the joint is presented and analyzed. The effectiveness of the reinforcement was found to be linked directly to the slenderness of the dowel, which has proven to have an essential role in determining the failure mode of the joints. Reinforcement in general have projected different effects by means of the joints' load-carrying capacity, ductility and strength degradation, depending on many factors, such as the different densities and mechanical properties of each of the wood species, as well as the different loading procedures, as the effect was more evident with cyclic tests.

Keywords: Self-tapping screws · Dowel-type joints · Monotonic and cyclic tests · Reinforcement · Load-carrying capacity · Embedding strength

1 Introduction

A wide variety of timber joints are available, and their behavior and contribution to the structure's overall capacity and ductility is dependent on many factors such as, the type of the joint itself, its usage and position on the structure. Dias (2005) has portrayed the importance of stiffness and ductility in timber joints by studying the behavior of different types of joints, including dowel-type joints, pointing out that the load-slip behavior of these joints reflects on how stiffness and ductility in a connection are complementary parameters. Giving an example of glued and notched joints that showed almost linear behavior up to failure under monotonic loading, which projects high stiffness yet a lack of ductility, resulting in brittle failures. Opposed to joints with dowel-type fasteners that are ductile with non-linear behavior and a relatively lower stiffness. According to Dietsch and Brandner (2015), Lathuillière et al. (2015) and Chang et al. (2019), dowel-type joints can be subjected to brittle failures as well that can be a result of the concentration of high tensile forces perpendicular to the grain, as well as shear stresses. Wood as a material has

I. Vayas and F. M. Mazzolani (Eds.): PROHITECH 2021, LNCE 209, pp. 962–973, 2022.
https://doi.org/10.1007/978-3-030-90788-4_74

low resistance to these types of stresses, which increases the possibility of brittle failures, especially by splitting. Since it is not possible to avoid these forces from emerging, the design should be carried out in order to compensate for the wood's low resistance to these types of stresses. Ultimately, this compensation can be provided by reinforcing the connection area. As mentioned in Harte and Dietsch (2015), the capacity of the timber joint in this case is governed by the high tensile and shear capacity of the reinforcement instead of the low resistance of wood. Preventing such stresses to be taken entirely by the timber elements will result in increasing the overall load-carrying capacity of the joint. According to Blass et al. (2001), the load-carrying capacity in dowel-type joints is influenced by three main parameters, the joint's geometry, the dowel's yield moment, and the embedding strength of the timber elements. Therefore, several factors intervene in the design. Reynolds et al. (2014) carried out a study that addressed the stiffness of dowel-type joints under cyclic loads, and provided that the embedding strength of the timber elements contributes to the response of the timber joints under cyclic loading, and since it is a nonlinear process, the stiffness of the joints then depends on the nature of the applied load. This aspect leads to the role and importance of ductility, and the effect of embedding strength in achieving higher ductility. Jorissen et al. (2011) and Piazza et al. (2011) have defined ductility as the ability of joints to undergo plastic deformations, and that it's directly linked to the joint's ability to dissipate energy. It is a crucial parameter if earthquakes were a potential threat, as the seismic response of the structure depends mainly on the ductility of its joints.

The choice of reinforcement is governed by many factors including the type of the connection itself, wood species, service class, load duration, and the different types and geometry of fasteners. Each reinforcing technique has its own advantages as well as limitations. However, choosing self-tapping screws (STS) for this research project was decided to further investigate its potential as opposed to other reinforcing techniques with regards to the overall joint's capacity and ductility. Being less intrusive, economic and easy to install, added to its value as an effective reliable reinforcing technique. In a study that was carried out by Lam et al. (2010) on timber joints reinforced with STS perpendicular to the grain, STS can increase the moment resisting capacity of dowel-type joints under reverse cyclic loading, and effectively reduce the connection's tendency to split, due to their plasticity and high withdrawal capacity.

1.1 Aim and Subject

This experimental campaign is a part of a research project that addressed self-tapping screws (STS) as a reinforcement technique compared to other techniques. Aiming at studying the potential and effectiveness of STS as a reinforcement to dowel-type timber joints, and its contribution in increasing the overall load-carrying capacity of the joints. Moreover, studying the reinforcement's effect on the joints in terms of ductility, stiffness, energy dissipation and prevention against premature splitting.

1.2 Methods

The methodology followed in order to study the parameters mentioned above, comprises four main steps: 1) The estimation of the load-carrying capacity of the original joints prior to introducing any reinforcement, following the guidelines provided in Eurocode 5 (EN 2004); 2) Performing monotonic tests to characterize the mechanical properties of the joints, before and after reinforcement. The resulting elastic displacement and stiffness, assist in reaching a better judgment on the contribution of the dowels and reinforcement in increasing the capacity and ductility of the joints; 3) Performing cyclic tests as the behavior of the joints under compression and tension fluctuations gives an indication of the joint's ductility and ability to dissipate energy; and, 4) Evaluating the joints' behavior in correspondence to both tests in order to conclude on the effectiveness of the reinforcement by means of ductility and energy dissipation.

2 Material Characterization

Two wood species were chosen for testing within this project, Maritime Pine (*Pinus pinaster*) and Scot's Pine (*Pinus sylvestris*), in order to compare the effect of the reinforcement when adopted for different densities. The geometry of the joints was decided based on a previous study, (Matos 2011), for comparison purposes. The samples were stabilized prior to testing following the specification provided in (EN 1991). The density of the specimen was determined, according to the specifications provided within ISO 3130. The final density values for the wood specimen was 472 kg/m^3 for Scots Pine and 556 kg/m^3 for Maritime Pine. Regarding the yield moment value, a four-bending test for the dowels was carried out, and a value of the yield moment, M_y, was obtained accordingly. For the purpose of studying the influence of the dowel's slenderness on the overall load-carrying capacity of the joints, steel dowels with two different diameters, 12 and 8 mm, were used for monotonic testing in this project. The steel dowels are of grade S235 steel, supplied by Rothoblaas. The tests were performed following EN 409:1993. The characteristic value for the yield moment, $M_{y,RK,}$ was calculated prior conducting the tests according to the equation provided by Eurocode 5 (EN 2004), that depends directly on the value of the characteristic tensile strength $f_{\mathrm{u,k.}}$ This value for a galvanized structural steel dowel of grade S235 with a diameter between 3 to 16 mm, according to EN 10025, lies in the range of 350 to 510 MPa. Assuming $f_{u,k} = 360$ MPa, the corresponding characteristic yield moments were calculated.

3 Load-carrying Capacity of Original (Un-reinforced) Joints

In order to determine the load-carrying capacity for dowel-type joints in double shear (fastener is subjected to two shear planes), the possible failure mechanisms are driven and the load corresponding to each failure mode is determined directly from stress equilibrium conditions. Eurocode 5 (EN 2004) has provided four equations for each of the four failure modes that are developed based on Johansen's theory (Johansen 1949), which cater to the assumption of an ideal rigid-plastic behavior of the dowels up to failure under bending moment, and the wood under embedding stresses. Thus, the failure associated with the modes developed in this theory is dependent directly on the slenderness of the dowel. Hence, in order to fully benefit from the contribution of the dowel in increasing the load-carrying capacity of the joint, the maximum yield capacity of the dowel has to be reached and the entire dowel's cross-section has to be plasticized, forming two plastic hinges per shear plane as a result. The estimated values of the load-carrying capacity for both species, Sco's Pine and Maritime Pine, are shown in Table 1. It can be noticed that the corresponding failure mode for both species, is the 3rd mode (EN 2004), mode (j). Nonetheless, as mentioned earlier, that for the purpose of comparing the effect of dowels' slenderness, joints were tested with 8 mm dowels as well. Therefore, the joints capacity was similarly estimated prior testing, and the corresponding failure mode in this case is the 4th, mode (k). However, only Scot's Pine joints were tested with 8 mm dowels.

Table 1. Estimated failure modes and corresponding load-carrying capacity values for Scot's Pine and Maritime Pine joints

	Failure Mode	$F_{v,RK}$ (kN) Scot's Pine	$F_{v,RK}$ (kN) Maritime Pine		Failure Mode	$F_{v,RK}$ (kN) Scot's Pine
12 mm dowels	g 	49.04	57.78	8 mm dowels	g 	31.38
	h 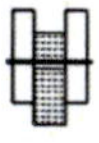	24.52	28.89		h 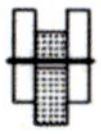	15.69
	j 	21.39	24.49		j 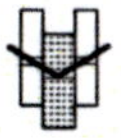	11.81
	k 	23.56	25.58		k 	8.16

4 Experimental Campaign

4.1 Monotonic Tests

The loading was applied in compression, parallel to the grain direction, following the protocol provided in EN 26891:1991. In order to control the procedure as described in EN 26891:1991, the test was performed under both force and displacement control. The first phase was carried out under force control with a rate of 0.058 kN/s, based on an estimated maximum force $F_{,est}$, equal to 20 kN. Thereafter, the test was performed under displacement control with a rate of 0.025 mm/s. A total of 12 specimens of each wood species were tested, 6 original joints and 6 with reinforcement. The reinforced joints were tested with 6 different scenarios in order to produce a configuration that allows for a ductile failure. Regarding original joints, Scot's Pine presented ductile failures, mode (j), while Maritime Pine joints have failed in embedment, which corresponds to mode (h). The average value of the maximum capacity reached, 18.22 kN for Scot's Pine, with a coefficient of variation (CoV) of 7.86%, and 19.17 kN, with a CoV of 9.61% for Maritime Pine. As for Scot's Pine specimen with 8 mm dowels, 6 specimen were tested, having a density, ρ_K, of 433.31 kg/m^3, and an embedment strength, $f_{h,k}$, of 32.69 N/mm^2. The failure was more ductile compared to the joints with 12 mm dowels. Figure 1 shows the specimen after testing, each with respect to the corresponding failure mode, and the force-slip monotonic curves.

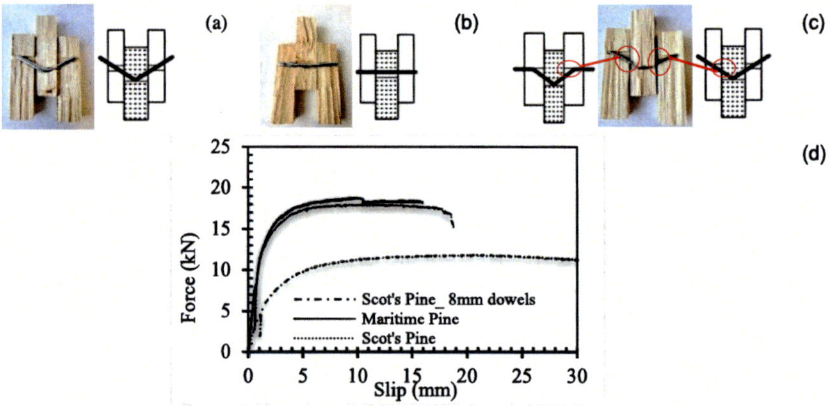

Fig. 1. Failure of Original Scot's Pine (12 mm dowel) (a) Failure of Original Maritime Pine (12 mm dowel) (b) Failure of Scot's Pine (8 mm dowel) (d) and Force-slip monotonic curves

As for the reinforced joints, two types of self-tapping screws (STS) were used for reinforcement, Rothoblass HBS 8 mm × 80 mm shear screws, and Rothoblass VGZ 5 mm × 80 mm fully threaded screws, manufactured in accordance with the provisions of ETA (European Technical Assessment, 2019). Six reinforcement schemes were developed, aiming at contributing to the formation of plastic hinges, where the plastic hinge position is then derived from the force and moment equilibrium in the shear planes (Bejtka et al. 2005). These schemes were tested on 6 specimens of Scot's Pine joints. The capacity of the joints has slightly increased yet the failure mode has remained the 3rd mode, mode (j). Therefore, the effect of the dowels' slenderness was essential. In light of the results, additional two schemes, (7) and (8), were developed and tested on joints with 8 mm dowels. Figure 2 shows the total 8 reinforcement schemes adopted. Scheme (5) originally provided higher results between the joints with 12 mm dowels, yet scheme (8) has provided better results.

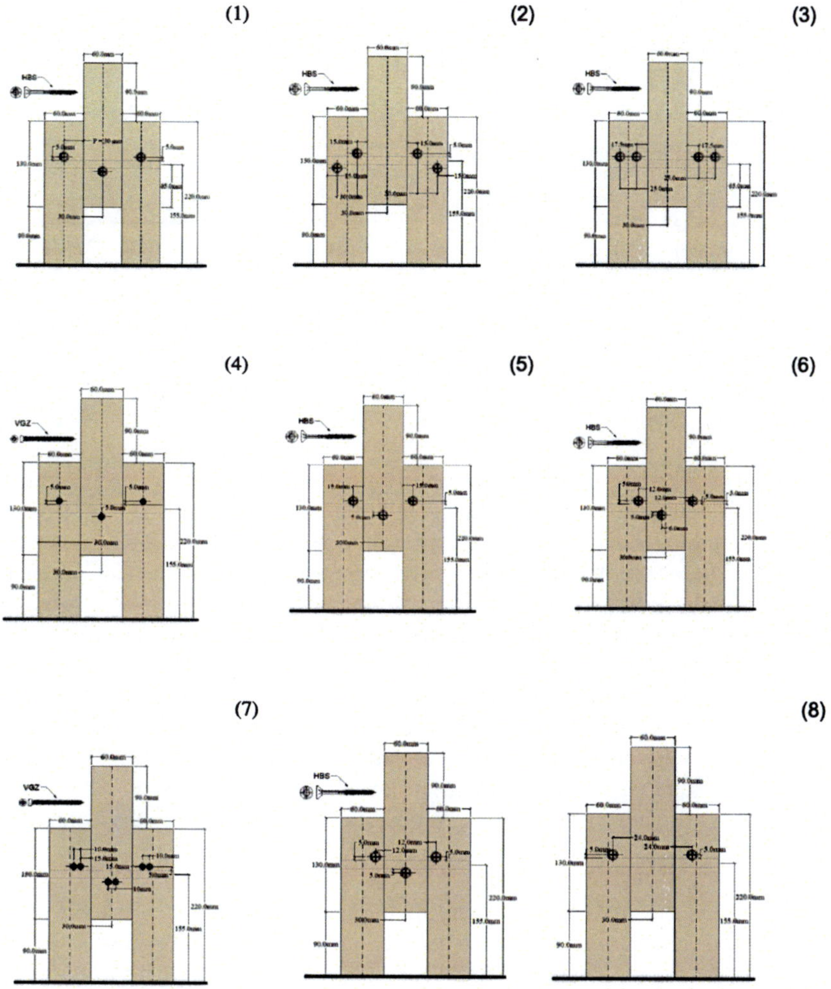

Fig. 2. Reinforcement schemes (1) to (8), with HBS and VGZ screws

In order to conclude on a scheme that is the most effective for both types of dowels, 3 specimens from each wood species were tested with scheme (8). The results have shown that scheme (8) provides higher capacity compared to the previous 7 schemes, for joints with 12 mm and 8 mm dowels. Taking Scot's Pine joints with 12 mm dowels, which are less slender, the mean value for the maximum force obtained with scheme (8) is 21.40 kN, with a CoV of 3.18%, and the mean value for the other 7 schemes is 18.36 kN, with a CoV of 9.29%. Figure 3 shows the force-slip curves for the 2 schemes with the highest results, (5) and (8), in comparison with the original.

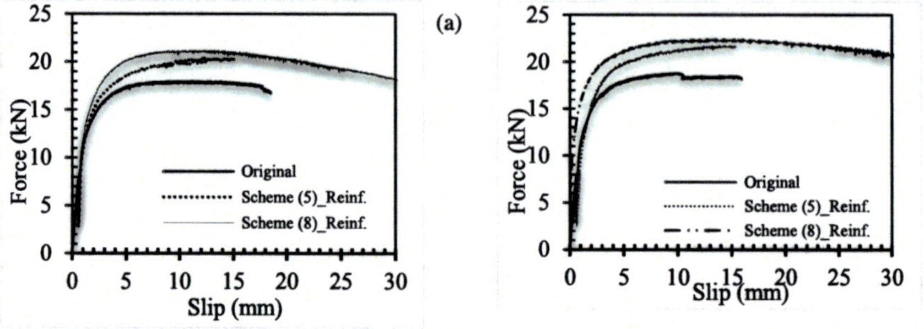

Fig. 3. Scot's Pine force-slip monotonic curves for scheme (5) and (8) opposed to the original (a) and Maritime Pine

4.2 Cyclic Tests

Cyclic tests were carried out following the new proposal for the European Standard EN 12512, which is currently under discussion within CEN TC 250 SC8/WG3. To start cyclic testing, the amplitudes of the cycles and the displacement rates are ought to be determined beforehand. These amplitudes are a percentage of the monotonic yield displacement of the joints, $V_{y,EEEP,m}$, that is obtained from the monotonic capacity curves. The value of the yield slip defined accordingly, $V_{y,EEEP,m}$, used for cyclic testing in this campaign is 3 mm for both wood species. Hence the displacement rates on the new draft are yet to be specified, the rates within this campaign were set with reference to Casagrande et al. (2020), who have studied the interaction between the strength degradation and the ductility capacity for dowel-type type joints through testing multiple cyclic protocols, employing different displacement rates. The study has provided that by increasing the number of steps, higher accuracy in the results can be achieved. Thus, the displacement rates adopted were specified taking into account 15 steps for each test, 0.5 mm/s and 1 mm/s for the first and second patterns, respectively. Providing reasonable testing time, bearing in mind the maximum capacity of the joints. Similar to monotonic tests, 6 non-reinforced joints from each wood species were tested as a primary step before the introduction of reinforcement. Both Scot's Pine and Maritime joints have failed by splitting, mainly in the middle element, yet Maritime joints have sustained all 15 cycles before failure, opposed to Scot's Pine joints. However, Scot's Pine joints failure was ductile, mode (j), while Maritime joints have failed in embedment, mode (g). Pinching effect was encountered for both wood species, which is expected to occur with cyclic loading, and it increases with the number of cycles. Regarding reinforcement, similar to monotonic tests, a total of 6 specimens, 3 of each species, were subjected to cyclic loading, reinforced with scheme (8), as it was the technique with the most potential. The reinforcement has provided better results for Scot's Pine joints, as it increased the overall capacity by 11.5%, opposed to only 2.66% for Maritime Pine. However, for both wood species, the splitting encountered after testing with reinforcement was limited and far less compared to original joints. Consequently, the ductility of the joints has significantly increased and brittle failures were prevented. Figure 4 shows the cyclic force-slip curves of reinforced joints compared to original joints.

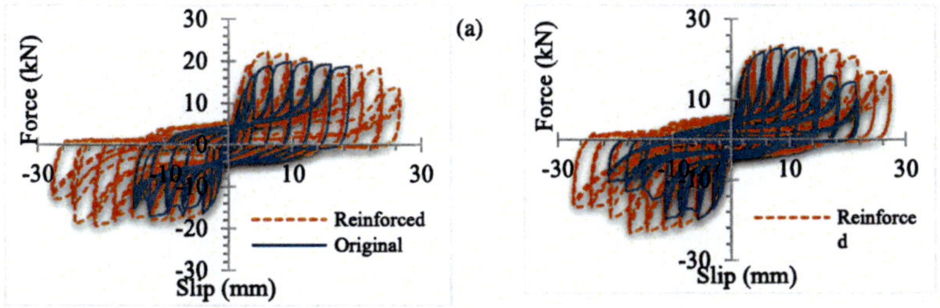

Fig. 4. Scot's Pine force-slip cyclic curves (a) and Maritime Pine force-slip cyclic curves

5 Results and Discussion

The new proposal for European standard EN 12512 differs from the previous version EN 12512:2005, in terms of testing procedure, protocol, and by means of data processing, which proposed a different approach to determine ductility, energy dissipation and strength impairment for the joints. The monotonic and cyclic data was analyzed by determining the Equivalent Energy Elastic-Plastic curves (EEEP). Following this method, the 1st and 3rd Load Envelope Curves (LEC) and the corresponding cyclic EEEP curves were obtained for both tension and compression, and the ductility was calculated in both cases. The design strength degradation factor, K_{deg}, was calculated in each case following Eq. (1):

$$k_{deg} = min\left[k_{deg}(v)\right] \geq k_{deg,min} \quad (1)$$

Where $K_{deg.min}$ is the minimum value of degradation factor, and $K_{deg(v)}$ is the strength degradation that is calculated using the Eq. (2) below:

$$k_{deg}(v) = \frac{F_3(v)}{F_1(v)} \leq 1 \quad (2)$$

Where $F1(v)$ and $F3(v)$ are the corresponding forces in the 1st and 3rd LEC, respectively.

The value of ductility was then determined by Eq. (3) below:

$$D_{EEEP,c} = \frac{V_{u,EEEP,c}}{V_{y,EEEP,c}} \quad (3)$$

Where $D_{EEEP,c}$ is the equivalent energy elastic-plastic ductility, $V_{u,EEEP,c}$ and $V_{y,EEEP,c}$ are the equivalent energy elastic-plastic ultimate and yield displacements for cyclic loading, respectively. Figure 5 shows the resulting EEEP curves and LEC for original and reinforced joints in comparison.

Table 2. Capacity, ductility and strength degradation calculations and results

				$F_{y,EEEP,c}$ (kN)	$V_{y,EEEP,c}$ (mm)	$F_1(v)$ (kN)	$F_3(v)$ (kN)	K_{deg}	$V_{u,EEEP,c}$ (mm)	$D_{,EEEP,c}$
PINE	Original	Comp	Avg	18.49	2.29	19.54	16.37	0.84	10.96	4.82
			CoV (%)	7.48	13.02	6.64	12.41	9.52	39.01	41.62
		Tens	Avg	6.28	2.13	17.2	15.57	0.91	10.66	5.48
			CoV (%)	11.28	22.51	7.84	8.03	1.15	39.87	58.50
	Reinforced	Comp	Avg	19.91	2.37	16.96	11.38	0.67	19.13	8.08
			CoV %	14.90%	4.60%	9.07%	4.75%	9.34%	16.74%	18.52%
		Tens	Avg	17.4	2.52	18.73	13.69	0.73	19.8	8.12
			CoV %	4.44%	10.57%	2.07%	17.12%	15.02%	41.11%	47.62%
PINE	Original	Comp	Avg	21.1	1.92	20.59	16.27	0.79	13.44	6.92
			CoV %	9.07%	27.03%	12.68%	21.59%	15.04%	56.21%	65.23%
		Tens	Avg	18.39	2.1	16.53	14.26	0.86	10.98	5.2
			CoV %	4.76%	20.64%	35.51%	38.19%	7.93%	62.06%	67.88%
	Reinforced	Comp	Avg	21.94	1.93	20.64	16.04	0.78	21.58	11.26
			CoV %	1.87%	10.69%	10.88%	9.45%	12.95%	2.74%	8.99%
		Tens	Avg	20.78	2.39	21.48	17.28	0.81	19.8	8.47
			CoV %	3.97%	16.96%	13.81%	18.21%	13.17%	13.23%	23.86%

Regarding un-reinforced joints, Maritime joints have shown 30.35% higher ductility in compression than Scot's Pine joints, yet 5.11% less in tension. Opposed to Scots Pine, Maritime Pine joints were 24.86% more ductile in compression, while Scot's Pine joints were found to be 12% more ductile in tension.

Reinforcement, on the other hand, has projected significant contribution to the ductility of the joints, as the value has increased by 40.35% with a coefficient of variation (CoV) of 18.52% in compression, and 32.52% with a CoV of 47.62% in tension for Scot's Pine joints. While for Maritime joints, the value has increased by 38.64% with a CoV of 23.86% in tension, and 38.51% with a CoV of 8.99% in compression. Detailed calculations and results are shown in Table 2.

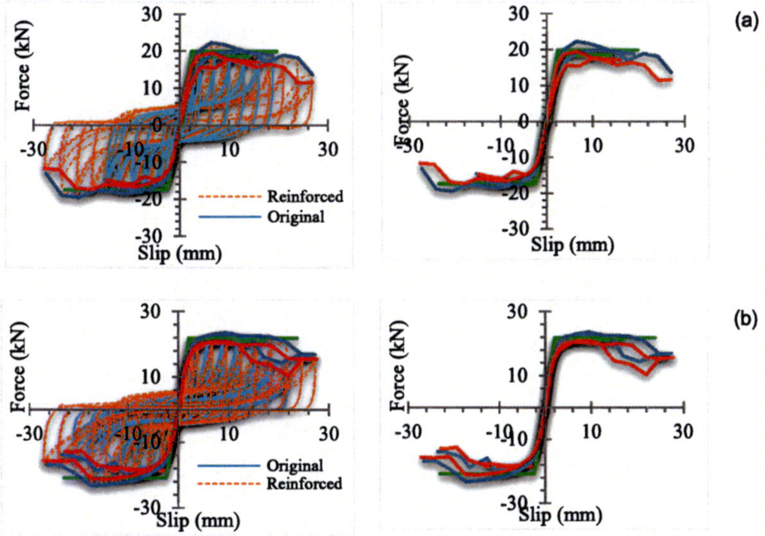

Fig. 5. Scot's Pine EEEP cyclic curves and LEC for original and reinforced joints (a) and Maritime Pine EEEP cyclic curves and LEC for original and reinforced joints

6 Conclusions

The dowel's slenderness was found to be an essential parameter in determining failure modes derived in EC5 (EN 2004), based on EYM and Johansen's theory. The equations described in the standard provide more accurate results for ductile failures. Regarding reinforcement schemes, the theory proposed in Bejtka et al. (2005), stating that placing the self-tapping screws (STS) in contact with the dowel, in between the shear plane and the plastic hinge position provides higher load-carrying capacity, was verified. Essentially, the contribution of reinforcement is primarily influenced by the mechanical properties of the wood. Therefore, different species cater for different response. Eventually, in both cases, the effect of reinforcement was more evident with cyclic loading.

Acknowledgements. This work was partly financed by FEDER funds through the Competitivity and Internationalization Operational Programme COMPETE, Portugal 2020, and by national funds through FCT – Foundation for Science and Technology within the scope of the project POCI-01-0145-FEDER-032031. Gratitude is extended towards Rothoblass, Portilame and AOF for material supply.

References

Bejtka, I., Blaß, H.J.: Self-tapping screws as reinforcements in connections with dowel-type fasteners. In: CIB-W18 Meeting, vol. 38 (2005)

Blass H.J., Bejtka, I.: Screws with continuous threads in timber connections. In: Aicher, S. (ed.) 'Joints in timber structures. The International RILEM Symposium, Stuttgart, Germany', vol. 22, pp. 193–201. 'RILEM Publications' (2001)

Blaß, H.J., Bienhaus, A., Kramer, V.: Effective bending capacity of dowel-type fasteners. In: The International RILEM Symposium: Joints in Timber Structures: Stuttgart, Germany, p. 71, 12–14 September 2001

Blaß, H.J., Schädle, P.: Ductility aspects of reinforced and non-reinforced timber joints. Eng. Struct. **33**(11), 3018–3026 (2011)

Casagrande, D., et al.: A methodology to determine the seismic low-cycle fatigue strength of timber connections. Constr. Build. Mater. **231**, 117026 (2020)

Harte, A.M., Dietsch, P.: Reinforcement of Timber Structures: A State-of-the-Art Report. COST: European Cooperation in Science and Technology, Shaker Verlag, Germany (2015)

Dias, A.M.P.G.: Mechanical behavior of timber-concrete joints. Ph.D. thesis, Delft University Press, Delft, Holland (2005)

Dietsch, P., Brandner, R.: Self-tapping screws and threaded rods as reinforcement for structural timber elements – A state-of-the-art report. Constr. Build. Mater. **97**, 78–89 (2015)

EN 1995–1–1: Eurocode 5: design of timber structures – part 1–1: General- common rules and rules for buildings. European Committee for Standardization (2004)

EN 12512: Timber structures - test methods - cyclic testing of joints made with mechanical fasteners. European Committee for Standardization, (2005)

EN 26891: Timber structures - joints made with mechanical fasteners - general principles for the determination of strength and deformation characteristics. European Committee for Standardization (1991)

EN 384: Structural timber - determination of characteristic values of mechanical properties and density. European Committee for Standardization (2010)

EN 338: Structural timber – strength classes. European Committee for Standardization (2009)

EOTA: European Technical Approval ETA-11/0030: Rotho Blaas Self-tapping screws. European Organization for Technical Approvals (2012)

ETA-11/0030, Rotho Blaas Self-tapping screws and threaded rods. European Technical Assessment. 8 Oct 2019

Gehloff, M., Closen, M., Lam, F.: Reduced edge distances in bolted timber moment connections with perpendicular to grain reinforcements. In: World Conference on Timber Engineering, vol. 230 (2010)

Guan, Z., Rodd, P.: DVW—local reinforcement for timber joints. J. Struct. Eng. **127**(8), 894–900 (2001)

ISO 13061-1: Physical and mechanical properties of wood - test methods for small clear wood specimens - part 1: determination of moisture content for physical and mechanical tests, International Organization for Standardization (2014)

ISO 13061-2: Physical and mechanical properties of wood - test methods for small clear wood specimens - part 2: determination of density for physical and mechanical tests, International Organization for Standardization (2014)

Johansen, K.W.: Theory of timber connections. Int. Assoc. Bridge Struct. Eng. **9**, 249–262 (1949)

Jorissen, A., Fragiacomo, M.: General notes on ductility in timber structures. Eng. Struct. **33**(11), 2987–2997 (2011)

Jorissen, A.J.M.: Double shear timber connections with dowel type fasteners. Delft, Neth. Delft Univ. Press **44**(3), 163–186 (1998)

Lam, F., Gehloff, M., Closen, M.: Moment-resisting bolted timber connections. Struct. Build. **163**(4), 267–274 (2010)

Lathuillière, D., Bléron, L., Descamps, T., Bocquet, J.F.: Reinforcement of dowel type connections. Constr. Build. Mater. **97**, 48–54 (2015)

Leijten, A.J.M., Brandon, D.: Advances in moment transferring dvw reinforced timber connections–analysis and experimental verification, part 1. Constr. Build. Mater. **43**, 614–622 (2013)

Marchi, L.: Innovative connection systems for timber structures. Ph.D. Thesis, University of Padua (2018)

Matos, S.: Avaliação experimental da ductilidade de ligações tipo cavilha em madeira. Universidade do Minho, Mestrado em Engenharia Civil (2011).(In Portuguese)

Mirdad, M.A., Hei Chui, Y.: Behavior of mass timber pane-concrete connections with inclined self-tapping screws and insulation layer. In: World Conference on Timber Engineering. Seoul (2018)

Pedersen, M.B.U.: Dowel Type Timber Connections Strength modelling. Ph.D. Thesis. Technical University of Denmark. Byg Rapport, no. R-039 (2001)

Piazza, M., Polastri, A., Tomasi, R.: Ductility of timber joints under static and cyclic loads. Struct. Build. **164**(2), 79–90 (2011)

Reynolds, T., Harris, R., Chang, W.S.: Stiffness and energy dissipation in dowel-type timber connections under in-service vibration. Eng. Struct. **65**, 21–29 (2014)

Rothoblaas: Declaration of performance STA_12_CPR_20130701, Cortaccia, Italy (2013)

Smith, I., Foliente, G., Nguyen, M., Syme, M.: Capacities of dowel-type fastener joints in Australian pine. J. Mater. Civ. Eng. **17**(6), 664–675 (2005)

Zhang, C., Guo, H., Jung, K., Harris, R., Chang, W.S.: Screw reinforcement on dowel-type moment-resisting connections with cracks. Constr. Build. Mater. **215**, 59–72 (2019)

Check for updates

Seismic Analysis and Rehabilitation by Steel Jacketing of a Historical Precast RC Industrial Building

Antonio Formisano(✉) and Antonio Davino

Department of Structure for Engineering and Architecture, School of Polytechnic and Basic Sciences, University of Naples "Federico II", P.le V. Tecchio, 80125 Naples, Italy
antoform@unina.it

Abstract. The field of seismic upgrading and retrofitting of historical buildings is an ever-developing discipline. The need of updated and ameliorative intervention techniques is fundamental, especially in highly seismic regions as Italy. The advantages of steel for the application of these techniques have been clearly stated. Nevertheless, its use as structural material is still limited both for new constructions and retrofitting of existing ones. The current study takes the inspiration from these premises, focusing the interest on the application of local steel interventions for rehabilitation of a historical precast RC industrial warehouse, that was struck by the Emilia-Romagna seismic sequence occurred in 2012. Two different local jacketing interventions of columns have been foreseen. Moreover, the efficiency of different beam-to-column joint types, namely hinge, semi-rigid and rigid joints, as well as the presence of a roof rigid diaphragm, have been examined and compared to each other by static non-linear analyses. Finally, from the achieved numerical analysis results it has been shown that the proposed local interventions also provide the global retrofitting of the building.

Keywords: Seismic assessment · Local interventions · Steel jacketing · Precast RC buildings · Industrial constructions · Joint types · Pushover analysis

1 Introduction

In 2012, Northern Italy was hit by two subsequent independent earthquakes on May 20th and 29th, spaced out by minor tremors, that struck the territory of Modena, Ferrara and Bologna in the region of Emilia-Romagna. The victims were 28, 17 of which lost their lives at work, 390 injured and 19 thousand evacuated people. These events had a strong repercussion on the industrial economics of the region (Di Giacomo, 2014). The Italian Civil Protection Department estimated that the seismic sequence provoked total losses for 13.2 billion euro. Many farmhouses, agricultural buildings and industrial warehouses collapsed. This peculiarity can be traced back not only to the diffusion of these construction typologies, but also to the late assessment of seismic hazard of the area and to the nature of the earthquake, which induced strong accelerations in high-rise warehouses (Giovanni and Vezzali, (2016).

I. Vayas and F. M. Mazzolani (Eds.): PROHITECH 2021, LNCE 209, pp. 974–985, 2022.
https://doi.org/10.1007/978-3-030-90788-4_75

In this scenario, the seismic assessment of existing buildings assumes fundamental importance. A study aimed at understanding the diffusion of the various technologies for seismic upgrading and retrofitting of existing buildings after the 2009 L'Aquila earthquake was carried out by Bertani et al. (2009). In this study, developed by the Italian Civil Protection Department in collaboration with the ReLUIS Consortium and Cineas, about 20,000 projects were examined. The recognition of these projects has highlighted as steel-based technologies are not commonly used by designers. In fact, even if global collapses of buildings and the consequent high number of victims are related to the use of a wide range of materials, several literature papers have shown that steel-based techniques are very effective in reducing or eliminating the negative consequences of earthquakes (Formisano et al., 2017).

In the Emilian territory, even if the buildings were designed considering the wind-induced accelerations only, the steel industrial structures, contrary to precast RC ones, have responded adequately to the induced seismic stresses. This excellent behaviour is prompted by the type of beam-to-column connections of this structural technology, as well as by the high strength and ductility of the material (Dall'Asta et al., 2009). In this study a well-designed local steel intervention is applied to an industrial precast RC warehouse in the Emilia-Romagna region, which is representative of similar buildings hit by the 2012 Northern Italy earthquake. Seismic assessment of this construction has been made by FEM pushover analyses developed in the SAP2000 environment. The efficiency of different beam-to-column joint types, namely hinge, semi-rigid and rigid joints, as well as the presence of a roof rigid diaphragm, have been examined and compared to each other by numerical analysis. Also two different local intervention techniques, represented by steel jacketing of existing RC columns, have been designed and implemented in SAP 2000, with the final goal to evaluate the benefits of the proposed rehabilitation technique, which has been also able to improve the global response of the entire building.

2 Case Study

The analysed case study is an industrial warehouse located in a district of Ferrara, a very productive area of the Emilia-Romagna Italian region. In order to have general results, it has been chosen a building representative of many other similar structures located in the area struck by the 2012 earthquake.

The structure has a rectangular plan of 120 × 72 m, for a total area of 8640 m2. The building has an internal height of 6.70 m under main beams and is covered by a pitched coverage supporting roof beams. In contrast with the actual seismic prescriptions, the beams are simply supported on the columns and the roof beams are simply supported on the main beams. Precast reinforced concrete (RC) was used for foundations and vertical structures, while prestressed RC was the material used for main and roof beams. All precast columns are oriented in the same direction and have a cross section of 54 x 44 cm. The amount of rebars in the columns depends on their location. Two sections are identified: type-A for the perimeter columns and type-B for the central ones. The less reinforcement detected in the perimeter columns is clearly the result of a design for vertical loads only. The roof was made of precast RC double-tapered beams with span of 10 m and variable cross-section ranging from 0.75 m at the supports to 1.65 m at the midpoint. The main beams are simply supported on 5 rows of columns, that create 4 bays with length of about 18 m. In the longitudinal direction, the perimeter beams have a L-shaped cross-section, whereas the central ones have an inverted-T profile. The structure is covered with double-T prestressed RC roof beams without any mechanical connection to the main beams. All prestressed RC elements are stretched using harmonic steel wires. The frames along the Y-direction have been named from Y1 to Y13, while the ones in the X-direction from X1 to X5 (Fig. 1).

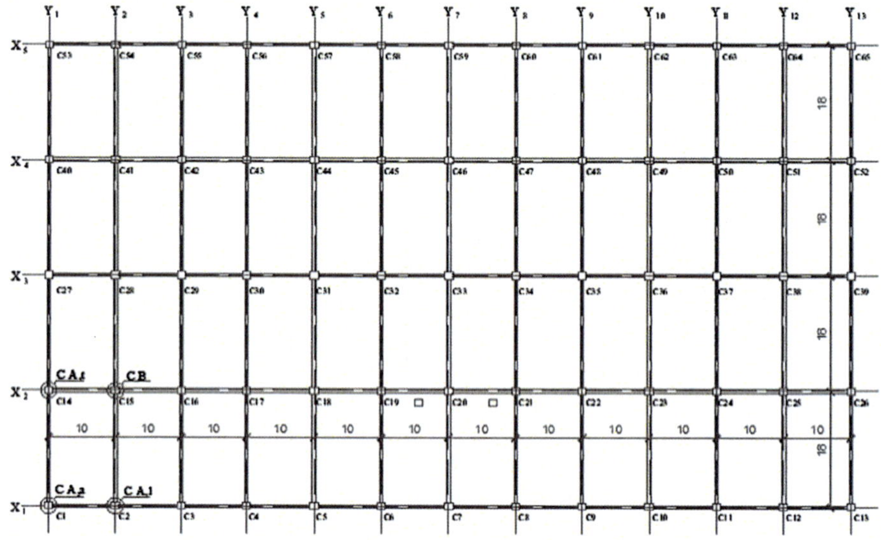

Fig. 1. Plan layout of the examined precast RC building

3 Numerical Structural Modelling

The structure was modelled using the software SAP2000 v.20 (Fig. 2). In a perspective of seismic assessment of precast RC structures, the type of connections assumes primary relevance (Lippi et al., 2018). Therefore, different connection types have been considered in order to understand how to reduce stresses in the structure. The achieved results, which will anticipate the global structural response evaluation, will address the best intervention to be carried out on the connections among elements aiming at eliminating the deficiencies found in the pure friction joints. To this purpose, hinge, semi-rigid and rigid connections among elements have been considered. Moreover, rigid joints together with the presence of the rigid diaphragm at the roof level has been taken into account. Finally, as a consequence of the four different schematizations of joints, the numerical FEM models related to these different constraint conditions have been developed.

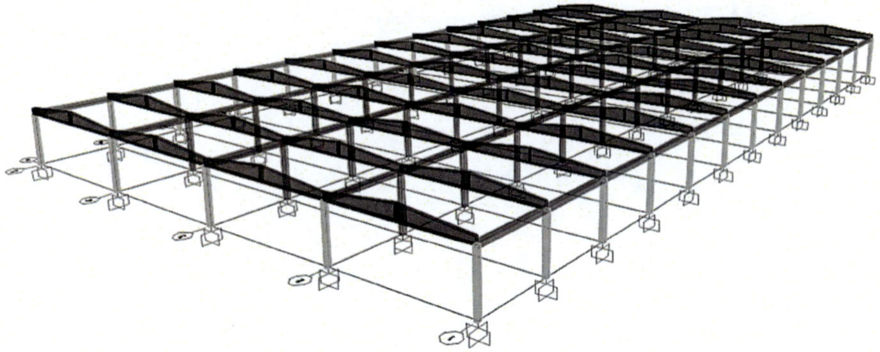

Fig. 2. SAP2000 3D model of the study building

4 Seismic Investigation

4.1 Linear Dynamic Analysis

The evaluation of the first three structural vibration modes for the four different constraint types considered among elements is shown in Table 1, where it is noticed that the vibration periods decrease as the rigidity of the structural system increases.

Table 1. Modal analysis results

	Joint type			
Vibration mode	Hinge	Semi-rigid	Rigid	Rigid+roof
First Percentage difference with respect to hinge scheme (%)	T = 2.29 s /	T = 1.37 s (–40)	T = 1.16 s (–50)	T = 1.13 s (–51)
Second Percentage difference with respect to hinge scheme (%)	T = 2.18 s /	T = 1.33 s (–39)	T = 1.11 s (–49)	T = 1.05 s (–52)
Third Percentage difference with respect to hinge scheme (%)	T = 2.04 s /	T = 1.27 s (–38)	T = 1.04 s (–49)	T = 1.03 s (–50)

The safety checks of beams have been all satisfied. On the other hand, with the help of the Software VcaSlu v.7.7., the resistance domains $M_{Rd,x}$ and $M_{Rd,y}$ have been plotted for a specific value of the acting normal stress (Ghersi, 2012), as shown in Fig. 3. Both type-A (perimetral) and type-B (central) columns have been examined. In order to consider the variation of the normal stress depending on the position of type-A columns, the verifications have been done for corner, longitudinal-perimeter and transverse-perimeter columns (Fig. 3). For type-B, as all columns have been subjected to the same normal stress, a unique safety check has been performed.

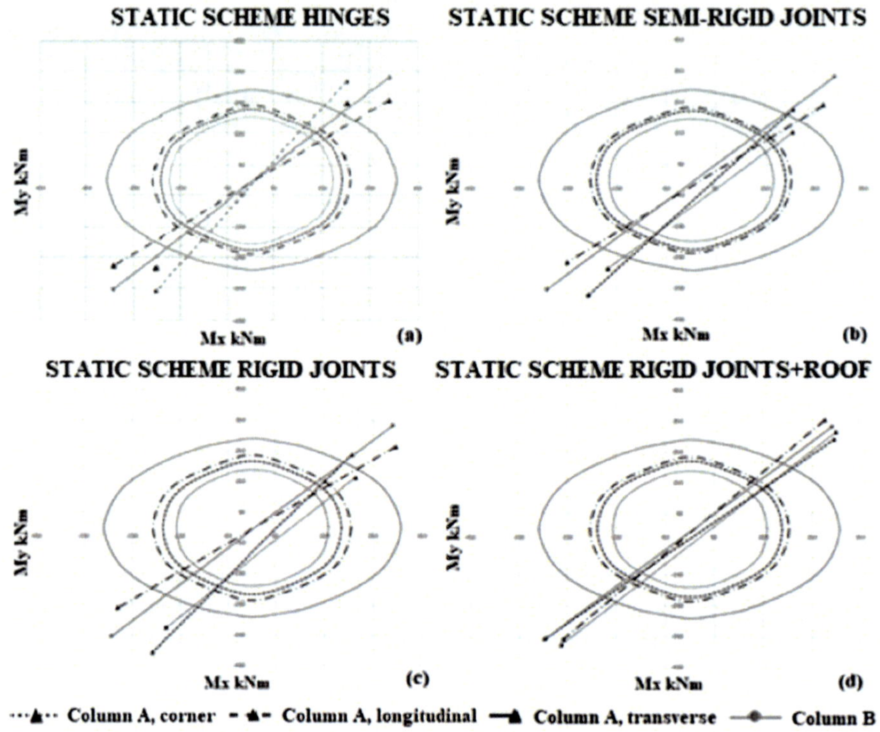

Fig. 3. Resistance domains for all investigated static schemes: hinges (a), semi-rigid joints (b), rigid joints (c) and rigid joints and roof (d)

4.2 Non-linear Static Analysis

The seismic global performance of the structure has been investigated in detail by pushover analyses (Verderame et al., 2012). In absence of infinitely or sufficiently rigid diaphragms, the distribution of forces has been chosen as proportional to structural masses. Therefore, the force applied to a single frame is obtained by dividing the mass of the considered frame by the total mass of the building. To better appreciate the variations of behaviour, the capacity curves of a reference frame for the different joint types have been plotted in the same diagram. The displacements are representative of the column that shifts the least, whereas the force is the sum of the base shears of the columns belonging to the considered alignment. For the sake of representation, Fig. 4 displays the pushover curves of only two frames, namely central (X3 and Y7) and perimeter (X1 and Y1), for each static scheme and direction considered.

The global safety checks of the structure have been done by considering risk indicators in terms of displacements and accelerations, the latter according to the provisions of the actual Italian technical code NTC18 (2018). The force distribution adopted in this work does not allow the application of the N2 method in a strict manner. In fact, considering the absence of a rigid roof diaphragm the structure is considered as the sum of different frames, each of them identifiable as a system with one degree of freedom.

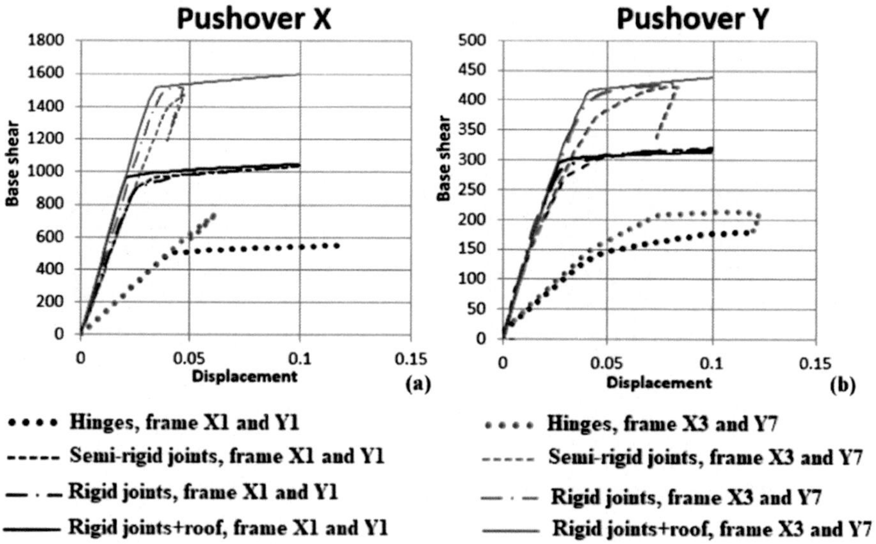

Fig. 4. Pushover curves in X (a) and Y (b) directions

Therefore, the calculated modal participation coefficient can be approximated to 1 with the assumption that the participating masses of each frame are identical.

The risk indicator in terms of displacement S_{LS} is defined as:

$$S_{LS} = d_{cap}/d_{dem} \quad (1)$$

where d_{cap} is the capacity displacement and d_{dem} is the demand one.

Instead, the risk indicator expressed in terms of acceleration ζ_{LS} is calculated as:

$$\zeta_{LS} = a_{g,max}/a_{g,NTC} \quad (2)$$

where $a_{g,max}$ is the capacity acceleration and $a_{,NTC}$ is the demand one.

All risk factors calculated according to the above relationships have been reported in the histograms of Fig. 5.

The above results have shown that in terms of displacement, the hinged structural scheme does not never fulfil the seismic check. The same occurs also with semi-rigid and rigid joints for the frame X3. Contrary, the structural schemes of frames X1, Y1 and Y7 characterised by semi-rigid joints, rigid joints and rigid joints plus roof satisfy the seismic checks in terms of displacement. On the other hand, it appears that in terms of acceleration, all frames, independently on the structural scheme considered, do not satisfy the seismic verifications imposed by the current standard. Therefore, the structure should be reinforced in order to improve its seismic behaviour.

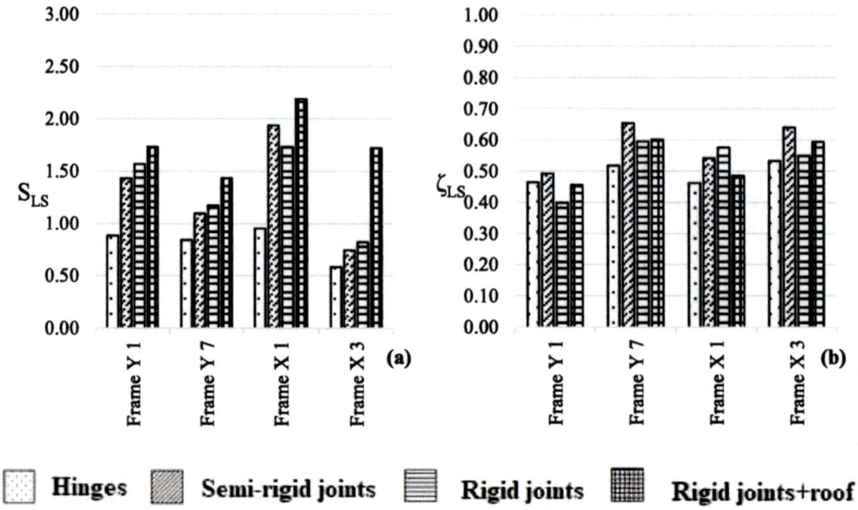

Fig. 5. Seismic indicators in terms of displacement (a) and acceleration (b)

5 Fragility Curves

The fragility curves of the study building with hinged structural scheme have been determined varying the PGA values based on different locations of the Italian territory. For the sake of example, the fragility curves of the frame Y7 with hinged joints is plotted in Fig. 6a. Afterwards, in order to assess the seismic performances of all investigated structures in their own site with respect to three different limit states, namely Immediate Occupancy (IO), Life Safety (LS) and Near Collapse (NC), the parameter μ, representing the ratio between demand and capacity in terms of acceleration, has been calculated for the above considered limit states, as depicted in Figures 6b and 6d.

From analysis results it is noticed that the building with hinged joints is not able to satisfy the standard checks with reference to all the three examined limit states. Therefore, the attention is herein focused on this structural scheme in order to design proper retrofitting intervention based on the use of steel profiles, as shown in the next section.

6 Seismic Retrofitting

Considering the results of linear dynamic analyses, it has been shown that the stress regimen applied on the columns is not compatible with their resistance domains, as shown in Fig. 3. This means that columns should be retrofitted against seismic actions.

Afterwards, based on the results of static non-linear analyses, the interventions on columns have been designed only for the hinged structural scheme. Among the various possible intervention technologies, steel has been herein chosen (Mazzolani et al., 2018). One of the potential interventions involving the use of steel is that based on steel plate shear walls (Formisano et al., 2006; 2008), which are used for global retrofitting

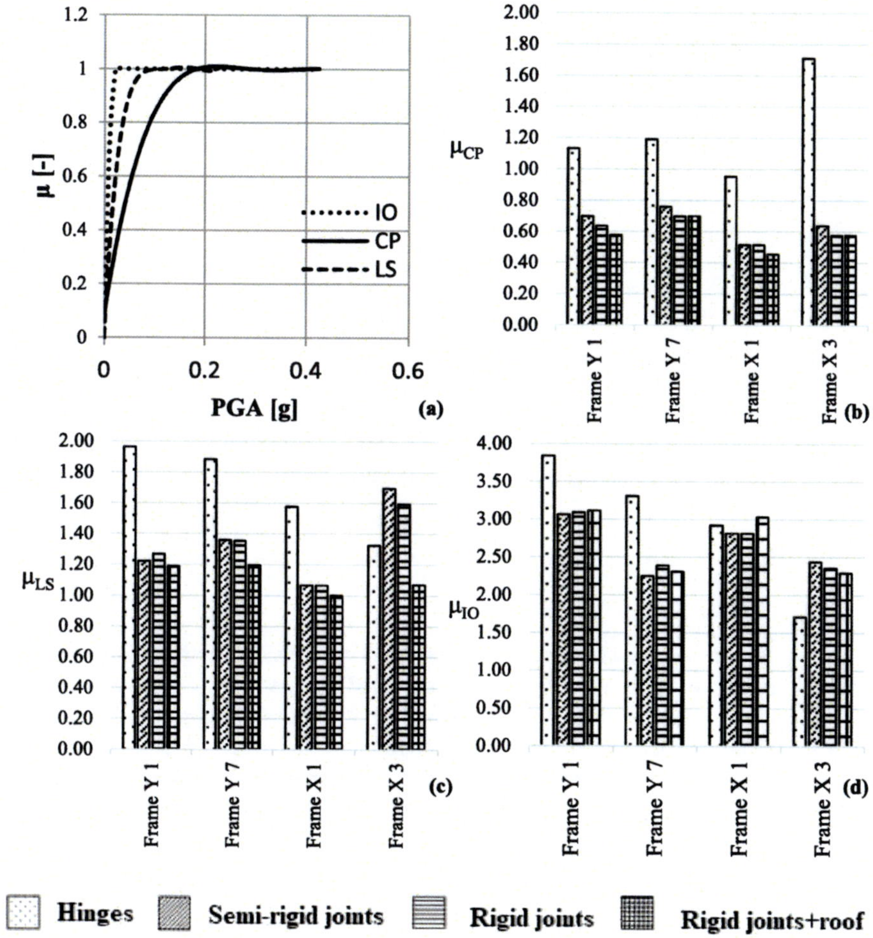

Fig. 6. Fragility curves of the frame Y7 with hinged joints (a) and demand-to-capacity ratios (m) in terms of acceleration at NC (b), LS (c) and IO (d) for the different structural frames located in their own site (Ferrara).

purpose, a type of intervention not required in the current case since only columns have demonstrated to have a deficient behaviour under seismic actions.

Therefore, local intervention by steelwork has been selected as restoration strategy. In particular, steel jacketing with angles connected to each other by welded batten plates has been considered as strengthening intervention of columns in order to increase the lateral confinement and the compressive strength of the concrete, thus minimizing their lateral deformations. The addition of the longitudinal steel reinforcement induces an increase of the flexural strength of columns and, at the same time, augments their ductility (Nunziata, 2011).

In a first reinforcement hypothesis, the steel angles used to reinforce existing columns have been designed in order to satisfy the minimum requirements in terms of safety, which is obtained by increasing the resistance domain of members up to have the point representative of the applied stress regime on the boundary of the new enlarged domain. The geometrical properties of employed steel angles are shown in Table 2, where b_s and h_s are base and height of the angles, respectively, while t_s is the corresponding thickness. Batten plates, which have been designed with the same thickness and height of angles, have been spaced to each other of 30 cm.

Table 2. Geometric parameters of steel angles in the first jacketing hypothesis

Column	b_s [mm]	h_s [mm]	t_s [mm]
A corner	60	60	5
A longitudinal	60	60	10
A transverse	60	60	10
B central	60	60	10

The pushover curves of single frames in directions X and Y retrofitted by the steel jacketing intervention above illustrated are depicted in Figs. 7 and 8, respectively.

After the first intervention, the structural response does not appear to be uniform, since the pushover curves of some frames in direction X have still a brittle behaviour, as shown in Fig. 7. Therefore, a second jacketing intervention has been hypothesized using unequal leg steel angles, whose geometrical characteristics are depicted in Table 3.

Table 3. Geometric parameters of unequal leg steel angles in the second jacketing hypothesis

Column	b_s [mm]	h_s [mm]	t_s [mm]
A corner	120	60	5
A longitudinal	60	60	10
A transverse	60	110	10
B central	80	110	10

As shown in Figs. 7 and 8, the second hypothesis of interventions with unequal leg steel angles provides a better seismic behaviour, with pushover curves more ductile than those achieved after the first jacketing intervention.

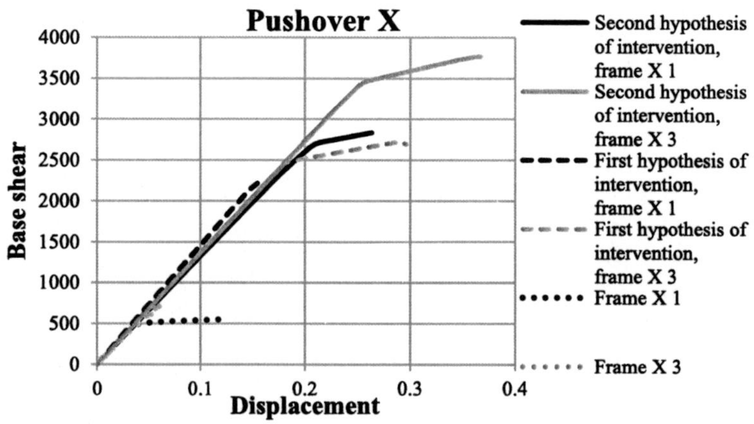

Fig. 7. Pushover curves in X direction before and after jacketing interventions

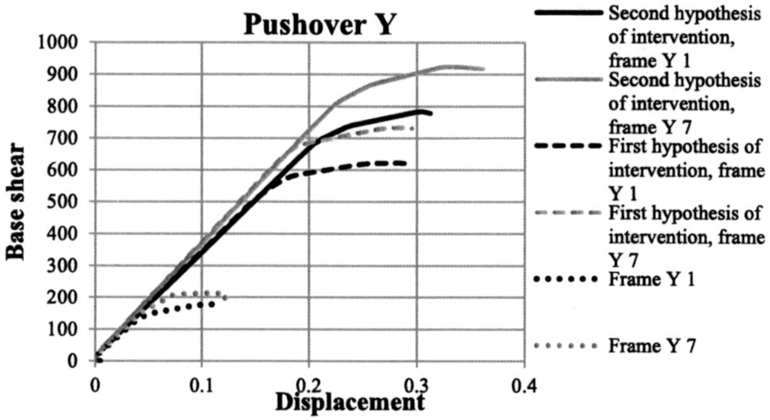

Fig. 8. Pushover curves in Y direction before and after jacketing interventions

Finally, the risk indicators in in terms of accelerations and displacements have been calculated after the two types of interventions performed (Fig. 9). The histograms of this figure clearly demonstrate how the second intervention type increases the safety coefficient of investigated frames, especially those in direction X.

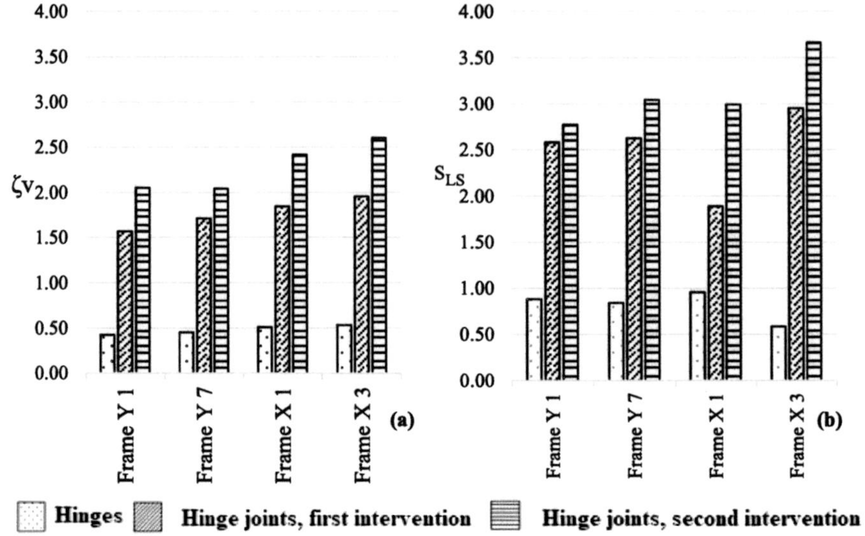

Fig. 9. Seismic safety indicators in terms of acceleration (a) and displacement (b) before and after jacketing interventions

7 Conclusions

The damage analysis and investigation of RC and steel buildings struck by the 2012 Northern Italy earthquake revealed the potentialities of steel as optimal material both for new constructions and retrofitting purpose. A broader use of this material could lead to less collapses, victims and economic losses. For what concerns the Emilia-Romagna region of Italy, it is to emphasize that this area entered the seismic classification only in 2003. Hence, almost all the buildings were designed in compliance with an outdated regulation, according to which it was enough to consider the wind-induced acceleration only, neglecting the seismic action effects. Starting from these premises, in this paper a historical precast RC industrial building with different structural schemes based on dissimilar joint types (hinges, semi-rigid and rigid joints), as well as considering rigid joints together with a rigid roof diaphragm, has been numerically investigated by the software SAP2000 in the linear dynamic and non-linear static field. The linear dynamic analysis has shown that as the structural rigidity increase, the vibration periods of the first three modes reduce of a quantity between 40 and 50% as well. In addition, while beams have been able to satisfy the design checks foreseen by the actual Italian standard, the existing columns have been subjected under seismic actions to a stress regimen not compatible with their resistance domain, so to require local strengthening interventions. In particular, the worst situation is achieved when the hinged joints scheme has been considered, it being characterised by risk indicators in terms of accelerations and displacements, intended as the ratio between capacity and demand, always less than the unit value. Two different local intervention hypotheses based on the use of steel jacketing with angles and batten plates have been foreseen. The first hypothesis has involved the use of the minimal transversal section of equal leg steel angles to be able the column to sustain the applied stresses. Despite the ameliorative operation, the new capacity curves of structural frames in the two analysis directions have displayed a different behaviour, with brittle performance experienced in direction X. Therefore, in order to improve the global structural behaviour in terms of pushover curve also in direction X, a second steel jacketing intervention of columns based on unequal leg angles has been designed. The pushover curves, as well as the risk indicators, have revealed the success of this second structural intervention, which has been able to transform the behaviour of frames in direction X from brittle to ductile. This outcome has demonstrated how well-designed local structural interventions can have a global impact on the structural performance, providing also the seismic retrofitting of the case study historical precast RC building.

References

Bertani, A., et al: White paper on private reconstruction outside historic centres in the municipalities affected by the 2009 April 6th Abruzzo earthquake (in Italian). Doppiavoce Editor, Naples, Italy (2009)

Dall'Asta, A., Landolfo, R., Salvatore, W., Barsacchi, G.: Single-storey steel buildings for industrial use (in Italian). Dario Flaccovio Editor, Palermo, Italy (2009)

Di Giacomo, M.: The costs of earthquakes in Italy (in Italian). Centro Studi Consiglio Nazionale Ingegneri, Ministero della Giustizia, Rome, Italy (2014)

Formisano, A., Di Lorenzo, G., Iannuzzi, I., Landolfo, R.: Seismic vulnerability and fragility of existing italian industrial steel buildings. Open Civil Eng. J. **11**, 1122–1137 (2017)

Formisano, A., De Matteis, G., Panico, S., Calderoni, B., Mazzolani, F.M.: Full-scale test on existing RC frame reinforced with slender shear steel plates. In: Proceedings of the 5th International Conference on Behaviour of Steel Structures in Seismic Areas – Stessa, pp. 827–834 (2006)

Formisano, A., De Matteis, G., Panico, S., Mazzolani, F.M.: Seismic upgrading of existing RC buildings by slender steel shear panels: a full-scale experimental investigation. Adv. Steel Construct. **4**(1), 26–45 (2008)

Ghersi, A.: The reinforced concrete (in Italian). Dario Flaccovio Editor, Palermo, Italy (2012)

Ghersi, A., Lenza, P.: Reinforced concrete antiseismic buildings (in Italian). Dario Flaccovio Editor, Palermo, Italy (2015)

Giovannini, D., Vezzali, L.: (eds). UNIMORE and the 2012 Emilia-Romagna earthquake. Interventions and Research to face emergency and sustain recovery (in Italian). University of Modena and Reggio Emilia, APM Editor, Carpi, Italy (2016)

Lippi, F.V., Mangini, F., Salvatore, W., Caprilli, S.: Steel in structural interventions on existing masonry and reinforced concrete buildings: examples and applications (in Italian). Dario Flacco Editor, Palermo, Italy (2018)

Mazzolani, F.M., Formisano, A., Vaiano, G.: Seismic upgrading of reiforced concrete buildings: BRB and FRP. Costruzioni Metalliche **1**, (2018)

Minister of infrastructures and transportations (NTC18). Upgrading of technical standards for constructions. Official Gazette n.20, Rome, Italy (2018)

Nunziata, V.: Theory and practice of steel structures (in Italian). Dario Flaccovio Editore (2011)

Verderame, G.M., Mariniello, C., Cosenza, E., Manfredi, G.: The Seismic Evaluation of Existing Reinforced Concrete Buildings: Comparison Between Linear And Nonlinear Analyses (in Italian). University of Naples Federico II, Naples, Italy (2012)

Check for updates

Restoration and Repair of the St. Nicola Church in Pisa

Pietro Croce[1(✉)], Nicola Croce[2], and Vladimiro Croce[2]

[1] Department of Civil and Industrial Engineering (DICI), University of Pisa, Largo L. Lazzarino, 1, 56122 Pisa, Italy
p.croce@ing.unipi.it

[2] Studio Croce Engineering, via G. Carducci, 47, Ghezzano, 56010 Pisa, Italy

Abstract. The paper deals with the restoration and repair of the façades of the St. Nicola Church in Pisa. The St. Nicola Church is one of the most important churches in Pisa. The construction of the Pisan Romanesque style church and of the adjacent monastery dates back to the 11th century; in the period 1297–1313 Augustinian fathers promoted significant enlargements sometime attributed to Giovanni Pisano and finally at the beginning of the 17th century it assumed the actual internal configuration. The main façade is characterized by a marble cladding in the inferior part and by exposed brick masonry in the upper part. During the years a significant crack pattern appeared on the upper part of the façade: despite in the past repair attempts were made with different techniques, as evident studying the texture of the façade itself, the propagation and reopening of the cracks were not effectively prevented.

The pathology of the façades is critically discussed and the repair solutions implemented in the restoration phase are illustrated, with particular emphasis on an innovative use of FRP (Carbon fiber) fabrics, which has been inserted in the horizontal mortar layers between adjacent brick rows of the exposed masonry to close and to bridge the cracks. Although risky from aesthetic point of view, this innovative technique was positively implemented. Moreover, since after some years from the intervention cracks are still closed, it can be concluded that the proposed technique is also very effective.

Keywords: Restoration · Repair · FRP · Carbon fiber fabric · Exposed masonry strengthening · Indent technique

1 Introduction

The paper deals with the restoration and repair of the facades of the St. Nicola Church in Pisa.

In 2004, during relevant restoration works, significant crack patterns were detected in the Church, mainly affecting the vaults, the façades and the external wall. For that reason specific studies were planned to diagnose the causes of the failures and to define the most appropriate intervention strategies.

I. Vayas and F. M. Mazzolani (Eds.): PROHITECH 2021, LNCE 209, pp. 986–998, 2022.
https://doi.org/10.1007/978-3-030-90788-4_76

The study of the texture of the masonry of the main façade of the Church demonstrated that in the past several repair attempts were performed with different techniques, but these interventions were not successful; in fact, the propagation and the reopening of the cracks were not effectively prevented.

As known, intervention on existing structures requires extensive investigations. Basic reference about this aspect is the ISO 13822 (2001), where a cyclic process is outlined allowing an increasingly refined appraisal of the structure itself. But further refinements have been introduced by Beconcini et al.(2016) who proposed an ad-hoc flow chart to plan extensive investigation on historic structures, previously adopted by Croce et al.(2018). A detailed discussion of the preliminary investigation carried out according the above mentioned procedures is out of the scope of the present paper, therefore only the main results are shortly summarized in the following. Anyhow, it must be stressed that the photographs reported in the paper were taken before or during the works, unless explicitly indicated.

The study presented here focuses on the repair and strengthening techniques adopted for the restoration of the façades of the Church, with particular emphasis on an innovative use of FRP fabrics, which has been inserted in the horizontal mortar layers between adjacent brick rows of the exposed masonry to close and to bridge the cracks.

Obviously, to mitigate the risks from the aesthetic point of view, the implementation of this technique on exposed historical masonry required highly skilled workmanship, but the results were extremely satisfactory.

2 The St. Nicola Church in Pisa

The St. Nicola Church in Pisa is one of the most relevant example of the Pisan Romanesque architectural style, which is the subject of huge literature (see, for example, Argan 1978a, Argan 1978b, Coroneo 1993, Ragghianti 1949, Salmi 1926, Sanpaolesi 1975, Tigler 2006, Willis 1835).

The Church, located in the historical center of the town, not far from the Arno River, is inserted in a building complex bounded at the West by St. Maria Street, at South by St. Nicola Street and at East by Carrara Place. The main façades is facing St. Maria Street, while the rear front is on Carrara Place.

On the right side of the main façade there is a succession of historical buildings facing St. Maria Street, in which the leaning bell tower is partly embedded, while the posterior part is connected with the Augustinian monastery.

The bell-tower (Figs. 1 and 2), which is considered a masterpiece of the Pisan Romanesque style, is 35.5 m high and it is inclined of about 1°13', corresponding to a maximum overhanging of about 1 m. The architectural organization of the bell tower is particularly complex: in fact, the cross section of the tower, which is circular at the base, with eight external ribs, becomes octagonal at higher orders and finally hexagonal at level of the belfry.

The most ancient part of the Church was erected in the 11th century, as it was firstly mentioned in the year 1097, but in in the period 1297–1313 Augustinian fathers promoted significant enlargements sometime attributed to Giovanni Pisano. At the beginning of the 17th century they were built the longitudinal barrel vault of the central nave and eight side chapels. Finally, the construction was completed in 1828, with the execution of the last two side chapels, adjacent to the main entrance.

The lower part of the main façade is characterized by an external cladding in black and white marble blocks, while the upper part is made by exposed brick masonry (Fig. 3).

Fig. 1. The bell tower of the St. Nicola Church seen from St. Maria Street

Fig. 2. The belfry of the bell tower of the St. Nicola Church

Fig. 3. The main façade of the Church

Finally, the South wall (Fig. 4a) and the rear façade (Fig. 4b) are made by brick masonry with a partial marble cladding.

(a) (b)

Fig. 4. Shout façade (a) and rear façade (b) of the Church

The actual plan of the Church (Fig. 5) can be roughly inscribed in a rectangle whose dimensions are about 23 × 47 m. The nave is divided into four bays with the eight chapels arranged along the longitudinal sides; other two chapels are located at the extremes of the transept, whose central part is occupied by the presbytery; while the choir occupies the central part of the nave, behind the presbytery. The nave is covered with a solid brick barrel vault, 0.12 m thick, stiffened by masonry arches, spaced about 2.4 m.

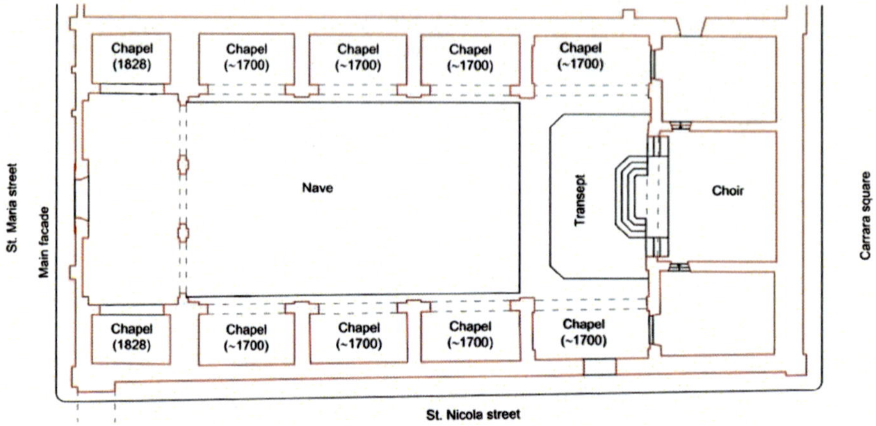

Fig. 5. Plan of the Church

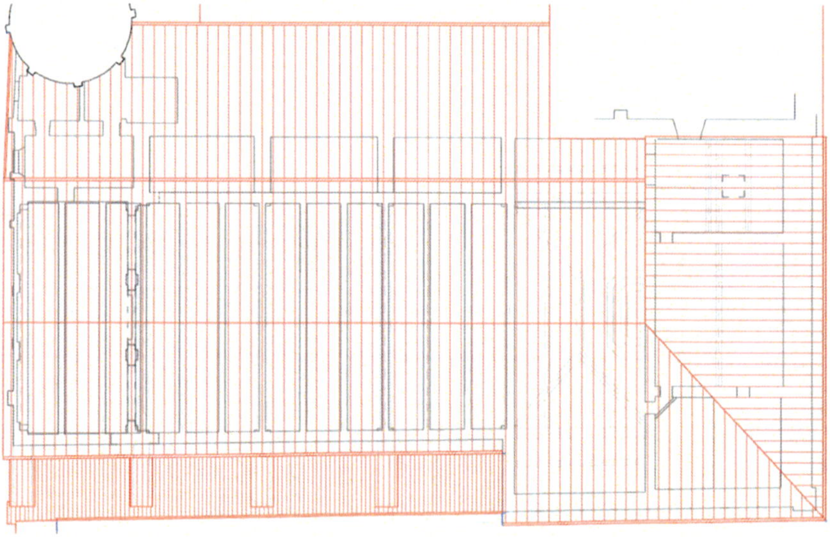

Fig. 6. Plan of the roof of the Church

The roof of the Church is mainly ridged, becoming hipped between the rear façade and part of the South façade (Fig. 6). The transverse walls of the South chapels come out from a lower mono-pitched roof in form of buttresses, sustaining the thrust of the vault (Fig. 7).

(a) (b)

Fig. 7. The roof of the South chapels (a) and a detail of the buttresses (b)

3 Crack Patterns and Diagnosis of Causes

A preliminary survey allowed to identify three main crack patterns, caused by two different failure mechanisms.

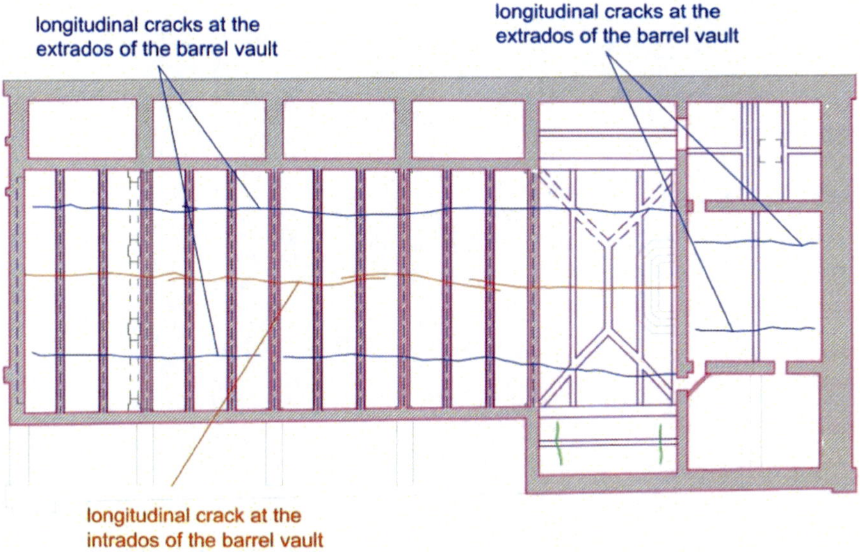

Fig. 8. The crack pattern in the vault and in the stiffening arches of the nave

The first crack pattern, summarized in Fig. 8, affected the longitudinal barrel vault and the stiffening arches of the nave. It was characterized by an exemplary and classical vault mechanism characterized by three longitudinal Pol Abraham's hinged cracks (Abraham 1934, Heyman 1995): a central one at the intrados of the crown (Fig. 9) and two lateral ones, nearly symmetric, at the extrados of the haunches of the vault (Fig. 10).

As clearly manifest in Figs. 3 and 4b, a second crack pattern was detected on the main façade and on the rear façade, diagrammatically represented in Fig. 11.

Fig. 9. Longitudinal crack at the intrados of the crown in the barrel vault of the nave

Fig. 10. Longitudinal crack at the extrados of the haunch in the barrel vault of the nave

The width of these vertical cracks, mainly concentrated near the rose windows on the main façade and around the second window on the rear façade, was increasing with the height.

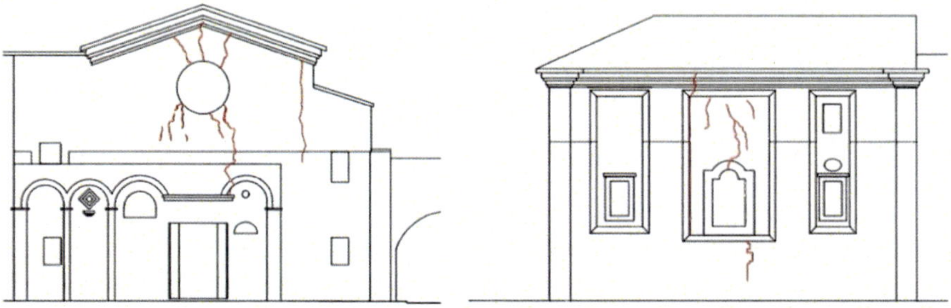

Fig. 11. Crack patterns on (a) the main façade and on the rear façade

These two crack patterns can be both ascribed to a unique failure mechanism: the relative horizontal displacement of the vault abutments caused by horizontal thrust and to the resulting rotation of the longitudinal walls around the base, so highlighting the modest restraining effects offered by the façades.

The above mentioned mechanism probably has quite remote origins. In fact, several attempts were carried out in the past to counteract it, as proven by the stone strips straddling the cracks on the main façade, presumably aiming to stitch the cracks themselves (Fig. 3). Unfortunately, all these interventions were unsuccessful; in fact, also stone strips were cracked.

Finally, a third significant crack pattern consisted in a relevant vertical crack on the longitudinal (South) façade (Fig. 12), at the attachment of the nave to the transept, characterized by opening increasing with the height.

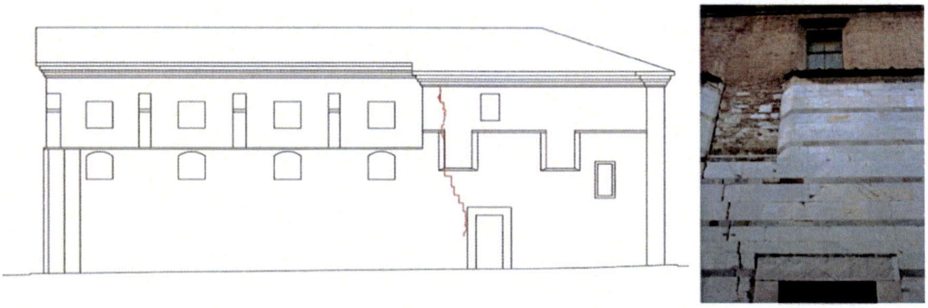

(a) (b)

Fig. 12. The vertical crack of the South façade (a) and a photo detail (b)

This second mechanism, initiated in fairly remote times, is due to the foundation settlement at the corner between the main façade and the South façade, which is also caused by the weight of the arch, partly visible in Fig. 7a, crossing St. Nicola Street and connecting the Church with the Royal Palace. But a suitable analysis of the evolution of this crack over time allowed to conclude that the settlements velocity was practically zero so that strengthening was judged not necessary.

For these reasons the interventions were concentrated on the crack patterns of the façades, which were reinforced with a rather innovative technique.

4 The Repair and Strengthening Interventions on the Façades

The rationale of the interventions on the façades was not only to close the cracks, but also to increase the efficiency and the tensile strength of the transversal connections of the masonry, in order to prevent further openings of the cracks. Moreover, interventions characterized by negligible bending stiffness were pursued, in order to minimize their sensitivity to relative vertical displacements.

With these objectives, two different types of interventions were setup.

The first type of intervention consisted in the strengthening of the edge joist at the top of the gable of the two façades by means of a composite steel-FRP strip, duly connected by stainless steel bars with the masonry; on the main façade the intervention was further supplemented by the insertion of a composite steel-FRP horizontal chain, as summarized in Fig. 13. The S355 steel plate was 600 × 3 mm. The fiber reinforcement was a bidirectional Carbon fiber fabric weighing around 1.6 N/m^2, connected to the steel plate by means of a low viscosity epoxy resin. An intermediate thin glass fiber net guaranteed the electrical insulation between the Carbon fiber and the steel.

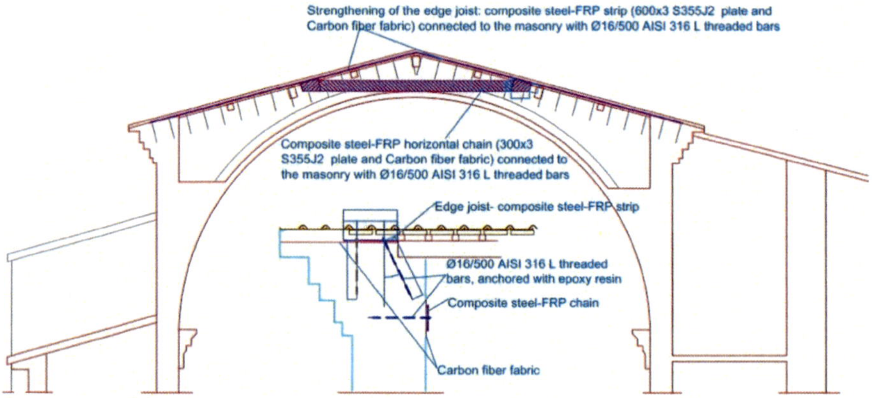

Fig. 13. Strengthening of the edge joist of the main façade

From the operational point of view, the intervention was carried out according to the following phases:

- cleaning of the masonry surface, in order to remove loose parts and any material potentially jeopardizing adhesion of the reinforcement;
- insertion of the anchoring bars;
- dust removal with a vacuum cleaner;
- application with a spatula of low viscosity epoxy resin on the surface;

- addition of the bidirectional Carbon fiber fabric and its impregnation with the low viscosity epoxy resin (Fig. 14a);
- addition of insulating glass fiber net;
- application of steel plate (Fig. 14b).

The second type of the intervention aimed to close the cracks on the exposed masonry, efficiently bridging the cracked surfaces.

This original technique basically relies on the insertion of continuous FRP strips in horizontal mortar layers, spaced 0.5–0.6 m around, by means of the so-called indent repair technique, consisting in a suitable and ordered removal and reinsertion of bricks of a given layer.

(a) (b)

Fig. 14. Strengthening of the edge joist: Carbon fiber fabric (a) and steel plate (b)

The extension of the brick course part affected by a single step of the indent (removal and reinsertion of the brick elements) was fixed around 1 m.

The operational sequence of the intervention was the following:

- preparation of the FRP strip characterized by a proper length, at least equal to total extension of the brick course interested by the intervention: a suitably wide piece of Carbon fiber fabric is folded one or several times in order to obtain the desired width and thickness, and subsequently rolled up;
- careful removal of the first stretch of the brick course (Fig. 15a) and dust removal with a vacuum cleaner;
- application with a paintbrush of low viscosity epoxy resin
- insertion of the Carbon fiber roll and its impregnation with low viscosity epoxy resin (Fig. 15b)
- reinsertion of the bricks of the stretch and reconstruction of the mortar layers (Fig. 16a)

- careful removal of the subsequent stretch of the bridge course and so on until the total extension of the course is completed (Fig. 16b).

Clearly, since applied to exposed masonry the intervention was extremely delicate and risky from the aesthetic point of view, but, thanks to highly skilled workmanship it was very successful.

Even if not covered by specific Codes, in designing the intervention reference was made to sound bibliographic references, subsequently summarized and updated in Ascione et al., 2016.

Moreover, it must be underlined that, after more than 15 years from the intervention, repaired cracks are still closed and that no new cracks have been detected. This is clearly demonstrated by the photo of the main façade, taken in February 2020 and reported in Fig. 17, which confirms the effectiveness of proposed technique.

(a) (b)

Fig. 15. Removal of the stretch of masonry course (a) and insertion and impregnation with low viscosity epoxy resin of the Carbon fiber roll (b)

(a) (b)

Fig. 16. Reinsertion and reconstruction of mortar layers (a) and completion of the intervention (b)

Fig. 17. The main façade in February 2020

5 Conclusions

In 2004, in the framework of the restoration works of the St. Nicola Church in Pisa, the crack patterns affecting the façades and the vaults of the church itself were analyzed in order to setup the most appropriate strengthening and repair interventions.

To close and to bridge the cracks affecting the exposed masonry preventing at the same time their reopening and their further propagation, an innovative solution was proposed, which is the main subject of the paper. The solution is based on the insertion of Carbon fiber fabrics in the horizontal mortar layers between adjacent brick rows. The insertion is carried out with the help of the indent repair technique, which is a rather classical method for intervention on masonry structures.

The intervention, even rather complex and risky from aesthetic point of view, thanks to skilled workmanship resulted very satisfactory and effective: in fact, at present, after more than 15 years, the cracks are still close and no new cracks appeared.

References

ISO 13822 (2001), Basis for design of structures. assessment of existing structures, ISO, Geneva, Switzerland (2001)

Beconcini, M.L., Croce, P., Marsili, F., Muzzi, M., Rosso, E.: Probabilistic reliability assessment of a heritage structure under horizontal loads. Probab. Eng. Mech. **45**, 198–211 (2016)

Sanpaolesi, P.: Il Duomo di Pisa e l'architettura romanica toscana delle origini, Pacini, Pisa, Italy (1975) (in Italian)

Salmi, M.: L'architettura romanica in Toscana, Bestetti and Tumminelli, Milano, Italy (1926) (in Italian)

Tigler, G.: Toscana Romanica, Jaca Book, Milano, Italy (2006)
Ragghianti, C.L.: Architettura lucchese e architettura pisana, Critica d'arte, **28** (1949) (in Italian)
Coroneo, R.: Architettura romanica dalla metà del Mille al primo '300, Ilisso, Nuoro, Italy (1993) (in Italian)
Willis, R.: Remarks on the Architecture of the Middle Ages (especially of Italy), Deighton, J and J.J., Cambridge, U.K. (1835)
Argan, G.C.: L'architettura protoromanica, preromanica e Cristiana. Dedalo, Bari, Italy (1978) (in Italian)
Argan, G.C.: L'architettura italiana del '200 e '300. Dedalo, Bari, Italy (1978) (in Italian)
Croce, N., Croce, P., Pelusi, M., Taccola, R.: Rehabilitation and seismic upgrading of the masonry arch bridge over the Magra river in Villafranca. Proc. Struct. Integr. **11**, 371–378 (2018)
Abraham, P.: Viollet-le-Duc et le rationalisme medieval, Vincent et Fréal, Paris, France (1934) (in French)
Heyman, J.: The Stone Skeleton. Cambridge University Press, Cambridge, United Kingdom (1995)
Ascione, L., et al.: Prospect for new guidance in the design of FRP (Eds.) Ascione, L., Gutierrez, E., Dimova, S., Pinto, A., Denton, S. JRC Report EUR 27666EN, JRC, Ispra, Italy (2016)

Deformation Resistance of the Gabion Retaining Wall Under Horizontal Loading

Daisuke Suetsugu[1](✉), Hiroshi Nakazawa[2], Tadashi Hara[3], Yashinori Fukubayashi[1], and Atshushi Koyama[1]

[1] Department of Civil and Environmental Engineering, University of Miyazaki, Gakuen-Kinabaidai-Nisi 1-1, Miyazaki 8892192, Japan
suetsugu@cc.miyazaki-u.ac.jp

[2] Earthquake Disaster Mitigation Research Division, National Research Institute for Earth Science and Disaster Resilience, Tsukuba 3-3, 3050006, Tennodai, Japan

[3] Department of Global Environment and Disaster Prevention, Kochi University, Akebono-Cho 2-5-1, Kochi 7808520, Japan

Abstract. This research investigated the effects of the connecting gabions, the packing density and the particle shape of the filling gravel, and the vertical pressure on the horizontal deformation resistance of the gabion retaining wall. The dimensions of the model gabion used in this experiment were 20 cm × 20 cm × 20 cm. The model gabion retaining wall was prepared by stacking three of them vertically. The horizontal loading tests were performed with a plane strain, in which a gabion in the middle of the retaining wall was pushed horizontally while applying constant vertical load. The obtained main findings are as follows. 1) The connecting gabions have the effect of increasing the deformation resistance due to the generation of tension in the gabion net due to the deformation of itself. 2) The particle shape of the filling gravel affects the deformation resistance of the gabion retaining wall. That is, the rigidity of the gabion retaining wall using the round gravel at a small deformation level was stiffer than in the case using the angular gravel. 3) The vertical stress acting on the gabion retaining wall has little influence on its deformation resistance.

Keywords: Gabion · Retaining wall · Fill material · Connecting · Model test

1 Introduction

Gabions are structural members that can be easily made using easily procured nets and stones and have been used as civil engineering structures such as waterworks since ancient times. In recent years, gabions using metal nets have been used, and are widely used to prevent erosion of river revetments and temporary retaining walls during disaster recovery.

The authors conducted a damage survey of the gabion structures along the main arterial road (Alaniko Highway) connecting the capital Kathmandu and the Chinese border after the Nepal Gorka Earthquake on April 25, 2015 (Hara et al. 2018). The gabion was

I. Vayas and F. M. Mazzolani (Eds.): PROHITECH 2021, LNCE 209, pp. 999–1007, 2022.
https://doi.org/10.1007/978-3-030-90788-4_77

used widely in Nepal, including retaining walls, crash barriers, and sabo dams. However, taking the example of a road retaining wall, its structural form was inconsistent, and most of the construction was done by engineer's experiences. In developing countries, gabions are valuable technology because gabion structures can be easily manufactured economically and manually. However, since there is no unified design and construction method for civil engineering structures using gabions including road retaining walls, it is necessary to establish and spread them. The authors are investigating the effects of gabion filling and gabion stacking on the behavior of gabion retaining walls during earthquakes to develop a design method for gabion retaining walls. (Nakazawa et al. 2018; Nakazawa et al. 2019).

On the other hand, gabions are often used to restore the remains and ensure the stability of the remains (Zhao et al. 2020). Similar to the above, the repair and reinforcement of archaeological sites using gabions largely depend on engineers' experience. The current situation is that designs based on the mechanical performance of gabions are rarely performed. Furthermore, preserving and utilizing archaeological sites requires passing on the value as cultural property to the next generation. Therefore, when repairing or preserving the remains, it is necessary to consider the short-term and long-term stability of such usage patterns of gabions.

As described above, the application of the gabion structure is widespread, but the design and construction in its use largely depend on engineers' experience. Even in Japan, there is no established method for designing gabions as retaining walls, so its use as an anti-earth pressure structure is limited. In order to use the gabion structure safely and rationally, it is necessary to clarify the stability of the gabion structure and the mechanical properties of the gabion itself.

This paper describes the results of a series of laboratory model tests that investigated the effects of the connecting gabions, the particle shape, and the packing density of the filling stone material on the deformation resistance of the gabion retaining wall.

2 Materials and Method of Horizontal Loading Test

2.1 Materials

In this study, the effects of connecting gabions with the steel wire, the bulk density of gabions, and the particle shape of the filling material on the horizontal deformation resistance of the gabion retaining wall were investigated by model tests. Therefore, two types of loading tests were performed. Run 1 was to examine the effects of the connecting gabions, and Run 2 was to investigate the influence of the particle shape of the filling stone.

The dimension of the model gabion used in this experiment was a cubic with 20 cm × 20 cm × 20 cm (Fig. 1). The filling stones applied in the series of model tests were angular gravel obtained by crushing rocks and around gravel collected from the river bed (Fig. 2). The range of the particle size was 37 mm to 19 mm. The maximum and minimum densities of the angular gravel was $\rho_{dmax} = 1.682$ g/cm^3 and $\rho_{dmin} = 1.376$ g/cm^3, and those of and the round gravel was ρ_{dmax} 1.773 g/cm^3 and $\rho_{dmin} = 1.470$ g/cm^3.

Fig. 1. A model gabion used in a series of laboratory tests

Fig. 2. Filling materials (left: crushed stone, right: riverbed gravel)

2.2 Method of Horizontal Loading Test

Figure 3 shows a schematic diagram of the horizontal loading test apparatus for the model gabion retaining wall. This apparatus pushes a model gabion retaining wall with one-row three-stage stacking vertically from two directions, horizontal and vertical with plane strain condition. The horizontal load was applied at a constant speed (1.5 mm/min) through the rigid plate (18 cm × 18 cm) on the back of the middle stage gabion, with a constant pressure applied vertically to the upper surface of the gabion via a rigid plate. The lower gabion and the upper gabion were fixed with stoppers as shown in this figure to prevent slip displacement. The behavior of the wall during loading was monitored by using horizontal displacement transducers attached in front of each gabion and taking digital imaging through the glass sidewall. Silicon grease was applied thinly on the inner surface of the sidewalls to cut the friction between the retaining wall and the sidewalls.

Table 1 shows a test condition (Run 1) for examining the effect of connecting gabions with using the steel wire. In this case, a test was performed to compare the horizontal resistance of gabion retaining walls with and without the connecting. A test was also conducted to compare the horizontal deformation resistance of the wall between the loosely and densely conditions of the filling gravels. Gabions were connected with two steel wires at the front and the back. The relative density in the dense condition exceeded 100%. Unlike the rigid frame, the gabion net has no rigidity and deformed at the time of filling gravels, and some of the angular gravel particles protrude from the mesh. Here, it was treated as the densest state.

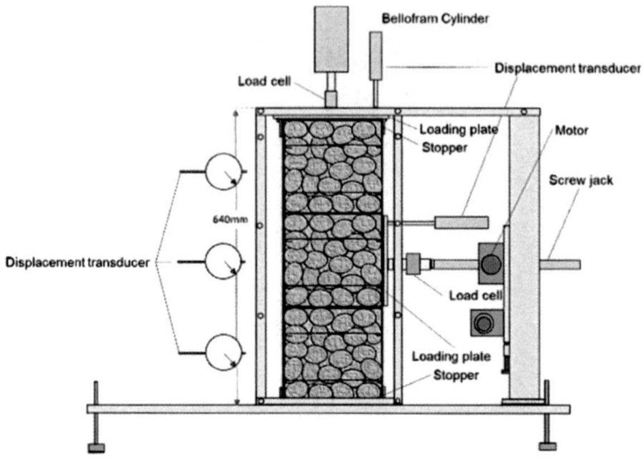

Fig. 3. Schematic diagram of the horizontal loading test apparatus

Table 1. Test condition of Run1

Test cases of Run 1	Material	Filling condition	Relative density (%)	Connecting condition	Vertical pressure (kPa)
Case 1-1	Crushed stone (Angular gravel)	Dense	111.4	connecting	6.35
Case 1-2		Dense	111.1	no connecting	
Case 1-3		Loose	57.3	connecting	

Table 2. Test condition of Run2

Test cases of Run 2	Material shape	Filling condition	Relative density (%)	Connecting	Vertical pressure (kPa)
Case 2-1r	Round	Dense	90.0	connecting	6.0
Case 2-1a	Angular		111.4		6.0
Case 2-2r	Round	Loose	88.5	no connecting	6.0
Case 2-2a	Angular		111.1		6.0
Case 2-3r	Round	Dense	21.3	connecting	6.0
Case 2-3a	Angular		57.3		6.0
Case 2-4r	Round	Dense	95.2	connecting	12.0
Case 2-4a	Angular		96.9		12.0

Table 2 shows a test condition (Run 2) for examining the influence of the particle shape of the filling material on the horizontal resistance of the gabion retaining wall.

In this case, a test was performed to compare the horizontal resistance using the angular gravel or the round gravel as the filling material of the gabions. A test comparing the bulk density of the gabions (dense or loose), and a test where the vertical load was different (6 kPa and 12 kPa) were also performed. The method of connecting gabions with using steel wires was the same as Run1.

3 Results and Discussions

3.1 Effects of Connecting Gabions and Filling Density

Figure 4 shows the relationship between the horizontal loading pressure and the horizontal displacement in each case. In Case 1–1 (with connecting and dense condition), the horizontal pressure increased immediately after the start of loading with small changes and showed a peak value of 30.6 kPa. After this, horizontal pressure decreased with increasing horizontal displacement. When the horizontal pressure showed the peak value, connecting wires at the front of the gabion broke.

In Case 1–2 (no connecting, dense condition), the rate of increase in the horizontal loading pressure immediately after the start of loading was smaller than that of Case 1–1 and tended to increase monotonically as the horizontal displacement increased. When the horizontal displacement reached 75 mm, the wire connecting the upper and middle gabions broke, and the horizontal loading pressure suddenly decreased and then increased again.

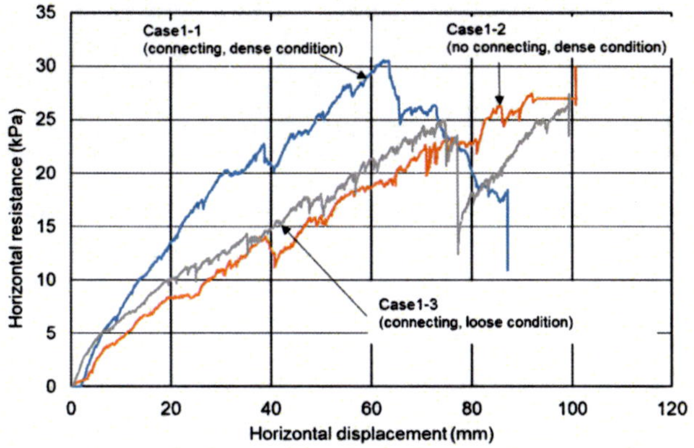

Fig. 4. Relationships between horizontal loading pressure and displacement

Figure 5 shows the deformation state of the model gabion retaining wall in each case at horizontal displacement was 100 mm. In Case1–1, the middle gabion was loaded almost horizontally, and the deformation of the gabion itself is small. Besides, the gabion in the upper and lower stages showed the behavior of rotating and falling in the loading direction.

On the other hand, in Case1–2, the middle gabion was displaced while rotating slightly counterclockwise while keeping its initial shape. The lower gabion followed the displacement of the middle gabion and caused greater shear deformation than that of Case 1–1. Furthermore, in Case1–3, the displacement and deformation of the middle and lower gabions were larger than in Case1–2.

In a state where the filling material was densely packed in the gabion and the gabions were connected to each other, the movement of the filling material is confined in the gabion net, and the gabion itself behaved like a rigid body. When some gabions are displaced or deformed, the tensile force also acts on the adjacent gabion net, so it was assumed that the rigidity of the entire gabion retaining wall increases. On the other hand, although the deformation resistance was smaller in the loosely filled condition than in the densely filled condition, it was considered that the rigidity increased because the entire gabion net of the retaining wall had a tensile force due to the connecting. Comparing Cases 1–2 and 1–3, the horizontal resistance is almost the same as in Case 1–2 until the connecting wire in Case 1–3 breaks. Therefore, the effect of connecting on the horizontal resistance of the gabion retaining wall is extremely large.

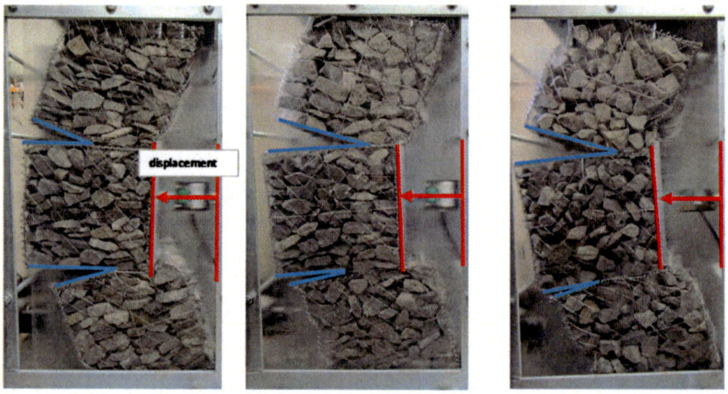

Fig. 5. Conditions of deformation at the end of loading

3.2 Effects of the Shape of Fill Materials

Figure 6(a) shows the relationship between horizontal resistance and displacement in Case 2–1 (connecting and dense condition). The horizontal resistance of the angular gravel showed a peak value and decreased sharply when the horizontal displacement exceeded 60 mm. The horizontal resistance of the round gravel increased monotonously without showing a peak. The sharp decrease in horizontal resistance in the case of the angular gravel was caused by the breakage of the wire at the front of the gabion. On the other hand, in the case of the round gravel, the wire at the connecting part did not break, but the gabion net was greatly extended. Comparing the stiffness of the angular and the round gravels immediately after the start of loading, the round gravel was stiffer than the round gravel. When the horizontal displacement exceeds 10 mm, the stiffness of the angular gravel was greater than that of the round one.

Figure 6(b) shows the results for Case2-2r and 2-2a (without connecting, dense condition). For both materials, the horizontal resistance monotonously increased with the increase of horizontal displacement. The rigidity of the round gravel immediately after the start of loading was larger than that of the round. Also, it can be confirmed that the horizontal resistance becomes small under no connecting condition from comparing to Fig. 6(a).

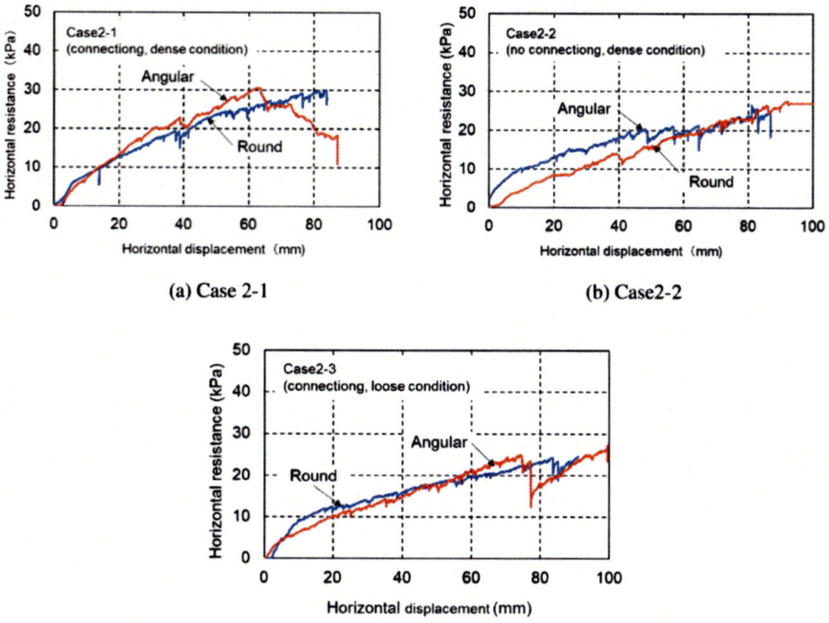

(a) Case 2-1

(b) Case2-2

(c) Case2-3

Fig. 6. Test results of Case 2–1, Case 2–2 and Case 2–3

Figure 6(c) shows the results of Case 2–3 (with connecting, loose condition). The horizontal resistance of the angular gravel decreased sharply after showing a peak value of 25.0 kPa. It was confirmed that the connecting wire at the front part of the gabion broke, similar to the result of the rubble shown in Fig. 6(a). However, the amount of horizontal displacement when the horizontal resistance showed a peak was larger than that of the dense condition of Case 2–1. The horizontal resistance of the round gravel increased monotonously without showing a peak. At the end of the test, the connecting at the front of the gabion was the same as in Case 2–1, the connecting wire did not break and the gabion net was greatly extended. In the case of loose filling, the rigidity of the round gravel immediately after loading is greater than that of the angular.

Figure 7 shows the results for Case 2–4 (vertical loading pressure 12 kPa with connecting). These figures also show results at 6 kPa. In the case of the round gravel shown in Fig. 7(a), the horizontal resistance and the initial stiffness hardly changed between 12.0 kPa and 6.0 kPa until the horizontal displacement reached 37 mm. As the displacement further progressed, the horizontal resistance became larger at 12 kPa than at 6 kPa. As shown in Fig. 7(b), the behavior of the angular gravel was similar to that of the round one.

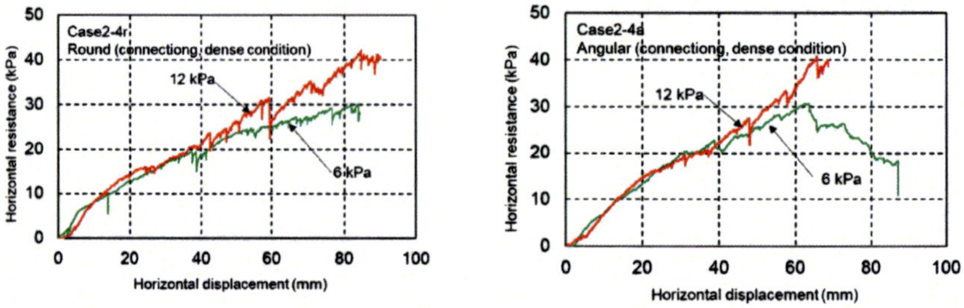

(a) Case 2-4r

(b) Case 2-4a

Fig. 7. Test results of Case 2–4

4 Conclusion

In this study, a model test was performed in a laboratory to examine the effects of the connecting gabions, the filling density of the filling material, the particle shape of the filling material, and the vertical pressure on the horizontal deformation resistance of the gabion retaining wall. The obtained findings are summarized as follows.

1) The connecting gabions have the effect of increasing the deformation resistance due to the generation of tension in the gabion net due to the deformation of itself.
2) The particle shape of the filling material affects the deformation resistance of the gabion retaining wall. That is, the rigidity of the gabion retaining wall at a small de-formation level is larger in the case of the gabion retaining wall using a round material than in the case using the angular material.
3) The vertical stress acting on the gabion retaining wall had little effect on the deformation resistance at the load level set in this study.

References

Hara, T., et al.: Field survey on damages of gabion structures caused by the 2015 Nepal Gorkha earthquake and examination of specific measures for earthquake resistance improvement. J. Jpn Soc. Civil Eng. Ser. A1 (Structural Engineering & Earthquake Engineering) **74**(4), I_586–597 (2018) (in Japanese)

Nakazawa, H., et al.: Experimental evaluation on earthquake-resistance of road retaining wall using Gabion. J. Disaster Res. **13**(5), 897–916 (2018)

Nakazawa, H., et al.: Problems in earthquake resistance evaluation of gabion retaining wall based on shake table test with full-scale model. J. Disaster Res. **14**(9), 1154–1169 (2019)

Nara National Research Institute for Cultural Properties. CAO NEWS Centre for Archaeological Operations No. 119 (2005). (in Japanese)

Zhao, W., You, Y., Chen, X., Liu, J., Chen, J.: Case study on debris-flow hazard mitigation at a world natural heritage site, Jiuzhaigou Valley Western China. Geomat. Nat. Hazards Risk **11**(1), 1782–1804 (2020)

Check for updates

Testing of Irregular Stone Masonry Strengthened with Cross-Laminated Timber

Ermes Rizzi, Ivan Giongo(✉), Daniele Riccadonna, and Maurizio Piazza

Department of Civil, Environmental and Mechanical Engineering, University of Trento, Via Mesiano 77, 38010 Trento, Italy
ivan.giongo@unitn.it

Abstract. The paper presents the outcomes of an experimental campaign investigating the in-plane behavior of stone masonry walls reinforced with cross-laminated timber (*CLT*) panels fixed to the masonry by using steel rods and adhesives. Timber panels are intended to be positioned only on the internal side of the building walls, to preserve the integrity of the external façades and for better durability. The campaign is part of a broader project that already saw testing (from connections to full-scale walls) of clay brick masonry strengthened with timber panels fixed to the masonry by using screw fasteners inserted dry. The research phase presented herein comprised 1) shear testing of timber-masonry connections fixed with epoxy-based resins and 2) in-plane diagonal compression testing of as-built and retrofitted masonry wallets to evaluate the shear capacity of the walls with and without the strengthening.

Keywords: Stone masonry · Earthquake · Retrofit · Cross-laminated timber · Adhesive connections

1 Introduction

Traditional unreinforced masonry (*URM*) buildings have frequently been reported as being particularly vulnerable when subject to seismic shaking (Augenti and Parisi 2010; Ingham and Griffith 2016; Penna et al. 2013; Dizhur et al. 2016). Following the catastrophic consequences of recent earthquakes, research on the topic of seismic retrofit of traditional URM buildings is attracting the attention of many academics worldwide (Bhattacharya et al. 2014; Borri et al. 2010; Lin et al. 2016). Within this context, an increasing number of studies has focused on the retrofit techniques based on the use of timber elements (Sustersic and Dujic 2014; Dizhur et al. 2017; Giongo et al. 2017; Riccadonna et al. 2019). The research presented herein was aimed at mechanically characterize the retrofit system proposed by Giongo et al. (2017) consisting in the use of cross laminated timber (*CLT*) panels that are connected to the masonry on the internal building façade by means of spread timber-masonry connections. An in-situ experimental campaign investigating the behavior of such retrofit system applied on brick masonry walls has already been presented by the authors in Riccadonna et al. (2019) and Giongo et al. (2021). The research work presented herein comprised shear testing of adhesive timber-stone masonry connection systems and diagonal compression testing of stone masonry wallets in the as-built and retrofitted configurations.

I. Vayas and F. M. Mazzolani (Eds.): PROHITECH 2021, LNCE 209, pp. 1008–1017, 2022.
https://doi.org/10.1007/978-3-030-90788-4_78

2 Materials

The timber-masonry connections were tested on the rubble stone (dolomite) and lime mortar masonry walls (referred to as *M1* masonry from here on) of a century-old building located in Northern Italy. Differently, eight double-leaf 40 cm thick masonry wallets were constructed from roughly-squared dolomite stone elements and hydraulic lime mortar for the purpose of being tested in diagonal compression according to ASTM E519-15 (ASTM 2015) (referred to as *M2* masonry from here on). The mechanical properties of the masonry components were characterized in the laboratory of the University of Trento (see Table 1). *M1* mortar and stone samples were extracted from the existing masonry walls and tested in compression, while specimens of the hydraulic lime mortar adopted for the *M2* masonry were prepared in the lab and then tested in bending and in compression according to EN 1015-11 (CEN 2019).

Table 1. Mechanical properties of masonry components

Material characteristic			n° samples	Mean	CoV
M1 Stone compression strength	$f_{sc,M1}$	[MPa]	5	64.30	0.33
M1 Stone modulus of elasticity	$E_{sc,M1}$	[MPa]	5	7660	0.40
M1 Mortar compression strength	$f_{mc,M1}$	[MPa]	46	2.95	0.95
M2 Mortar compression strength	$f_{mc,M2}$	[MPa]	12	5.40	0.07
M2 Mortar bending strength	$f_{mb,M2}$	[MPa]	6	2.0	0.26

For both timber-masonry connections and retrofitted wall specimens tested in diagonal compression, 60 mm-thick three-layered *CLT* panels [CIT n.03/2013] were used. Timber-masonry connections were installed using epoxy resin (*EX*) [ETA-10/0012] and hybrid *epoxy-vinyl ester* resin (*HY*) [ETA-02/0024] combined with various types of steel anchors (e.g. Class *8.8* threaded bars and *B450C* rebars) of different diameters (14 mm and 20 mm). Tested configurations are listed in Table 2. To assemble the timber-masonry connection specimens, 30 cm-deep holes were first drilled into the selected *M1* masonry locations. The hole diameter was equal to 22 mm for Ø 14 mm rods and equal to 30 mm for Ø 20 mm rods. The same hole diameters used for the masonry were adopted for the pre-drilling of the CLT specimens. Holes were cleared from drilling waste prior to the positioning of the timber specimens and the insertion of the metallic *injection anchor sleeve* that was subsequently filled with the resin. The anchor rods were then inserted into the holes and the specimens were tested after the prescribed curing times were elapsed.

The wallets tested in diagonal compression were divided into two groups of four specimens each (Table 3). Group 1 wallets were tested in the original configuration (*AsB*) while wallets of Group 2 were tested in the retrofitted configuration (*R*). The retrofit consisted in the application of a 110 × 110 cm^2 *CLT* panel to one side of the specimen and connected to the masonry by means of 5 *EX 14 TH* anchors, one per corner plus one in the middle of the panel (anchor distance from the wallet edges ≥20 cm; anchor distance from the *CLT* panel edges ≥15 cm).

Table 2. Tested grouted connection types

Connection ID	Repetitions	Resin	Anchor rod type	Anchor diameter
HY 14 TH	6	Hybrid	8.8 grade threaded	14 mm
HY 14 RB	3	Hybrid	B450C rebar	14 mm
HY 20 TH	7	Hybrid	8.8 grade threaded	20 mm
HY 20 RB	3	Hybrid	B450C rebar	20 mm
EX 14 TH	3	Epoxy	8.8 grade threaded	14 mm
EX 14 RB	3	Epoxy	B450C rebar	14 mm
EX 20 TH	4	Epoxy	8.8 grade threaded	20 mm
EX 20 RB	5	Epoxy	B450C rebar	20 mm

Table 3. Timber-masonry connection testing: test IDs and configurations

Group	Test ID	Wall	Configuration
1	AsB-1	1	As built
	AsB-2	2	As built
	AsB-3	3	As built
	AsB-4	4	As built
2	R-1	5	Retrofitted
	R-2	6	Retrofitted
	R-3	7	Retrofitted
	R-4	8	Retrofitted

3 Experimental Procedures

3.1 Connection Testing

The shear force was applied to the CLT elements parallel to the grain direction (with reference to the outer board layers) by means of the setup illustrated in Fig. 1. A 250 kN double-effect hydraulic actuator was connected to a timber reaction frame made of beech laminated veneer lumber which was anchored to the masonry by means of screw-type fasteners. A 75 kN load cell was interposed between the actuator and the steel bracket enabling the force transfer to the CLT specimen. Connection slip was measured by an *LVDT* supported on a rigid steel stand anchored to the floor.

Test protocol adopted for timber-masonry connection assemblages was defined according to EN 12512 (CEN 2001). To prevent out-of-plane buckling, only hemi cycles identified by positive displacements (i.e. with the actuator acting in tension) were performed.

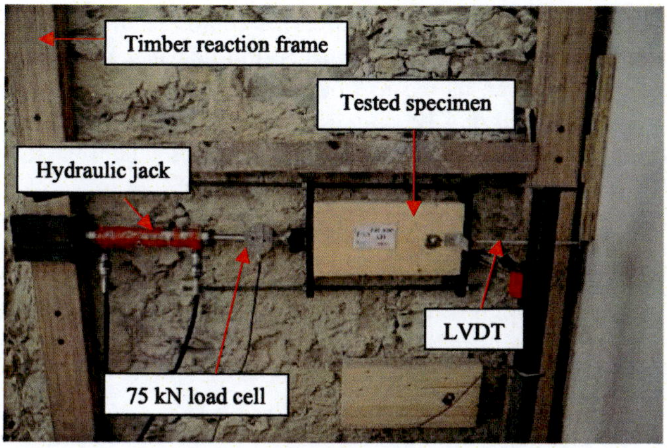

Fig. 1. Connection testing – test setup

3.2 Diagonal Compression Testing

The self-equilibrated test apparatus illustrated in Fig. 2 was designed according to instructions provided by ASTM E519-15 (ASTM 2015). Such system allowed for diagonal compression loads to be applied without the need of rotating the specimens. The test setup comprised: a) two steel beams, each one obtained by welding together two UPN 180 profiles, b) two welded steel shoes, c) a 600 kN single-effect hydraulic jack, d) and two Ø 36 mm 10.9 steel grade threaded rods. All carpentry elements were of S275 steel grade. To accommodate the steel shoes and to provide a uniform force distribution, an approximately 20 mm-thick M15 strength class mortar layer was applied

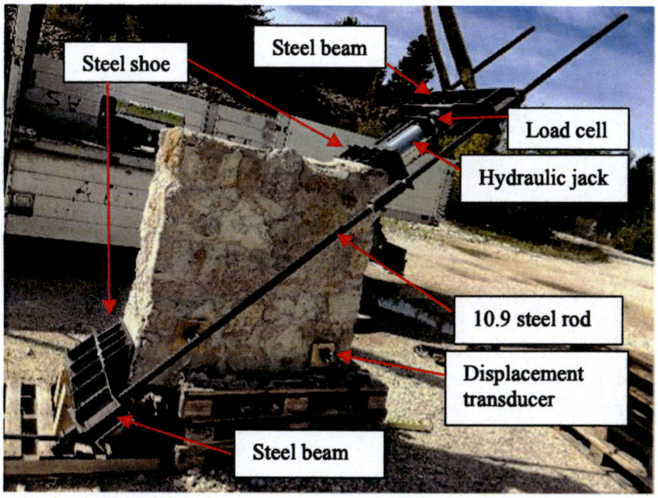

Fig. 2. Diagonal compression testing: test apparatus and instrumentation

to loaded corners. The steel shoes transferred the compression loads to the masonry only, without engaging the reinforcing elements (timber panel and reinforced plaster) directly. The jack was driven by means of a manual hydraulic pump. The specimens were instrumented by means of 4 100 mm wire displacement transducers located at the two diagonals on each side and a 600 kN load cell interposed between the jack and the upper steel beam.

A first 25 kN adjustment cycle was performed to get the steel shoes in contact with the specimen corners. Diagonal loads were then applied according to a hemi-cyclic force-controlled protocol (compression only). Hemi cycle amplitudes were gradually increased by 50 kN load steps up to the specimen capacity F_{max}. For retrofitted specimens test procedure continued with additional displacement-controlled hemi-cycles until the applied force fell below 0.8 F_{max}.

4 Experimental Results

4.1 Timber-Masonry Connections

After the test sequences were completed, the connection specimens were partially disassembled and the components were inspected (Fig. 3). Limited signs of local bearing failure were observed in the *CLT* elements while the anchors exhibited visible residual bending deformation highlighting the formation of a single plastic hinge located at timber-masonry interface. Tension and splitting cracks in the stone blocks were consistently observed, combined with the crushing of the surrounding mortar.

Mean values of connection strength and stiffness are listed in Table 4. Initial stiffness values k_1 are secant values measured at the peak of the 1st cycle and F_{max} refers to the maximum load recorded in the test. The irregular masonry texture (variability in stone element dimensions and shape, anisotropy of stone blocks, possible micro-cracking induced by drilling procedure) led to scattered results as highlighted by the relatively high values of the *CoV*s.

Fig. 3. Observed failure mechanisms: a) Tension cracking of stone elements and mortar crushing; b) Residual deformation of threaded steel rod and local timber bearing failure.

Despite the result variability, it was possible to estimate the influence of the investigated parameters (dowel diameter and type, grout formulation) on the connection behavior. As expected, the influence of the anchor diameter was observed to be relevant for both connection strength and initial stiffness, especially in the case of hybrid (*HY*) resin type (see Fig. 4). Specimens assembled with epoxy (*EX*) resin exhibited greater stiffness and strength compared to those assembled with *HY* resin (Fig. 5). As illustrated in Fig. 6 both connection stiffness and strength were observed to be minimally affected by anchor type and no clear trends can be reported.

Table 4. Observed mean strength and stiffness values for the tested connections

Connection ID	F_{max}		k_1	
	[kN]	CoV	[kN/mm]	CoV
HY 14 TH	15.84	0.23	3.01	0.71
HY 14 RB	15.77	0.28	2.43	0.14
HY 20 TH	27.62	0.08	4.75	0.44
HY 20 RB	28.83	0.15	5.90	0.26
EX 14 TH	25.97	0.23	8.97	0.29
EX 14 RB	22.88	0.16	8.56	0.47
EX 20 TH	35.90	0.20	8.39	0.41
EX 20 RB	34.51	0.16	10.26	0.64

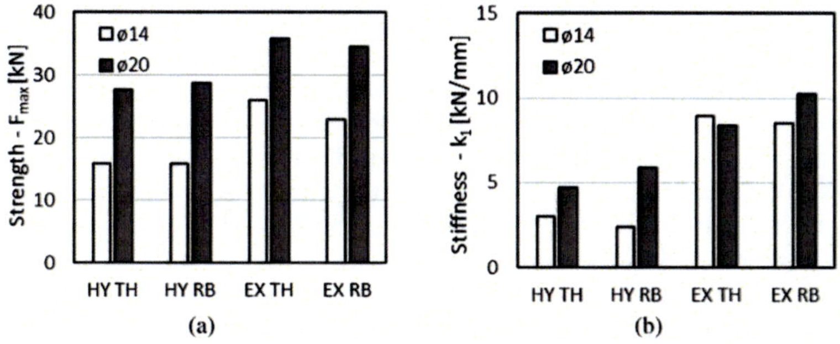

Fig. 4. Influence of dowel diameter: a) Connection strength *Fmax*; b) 1st cycle secant stiffness *kl*

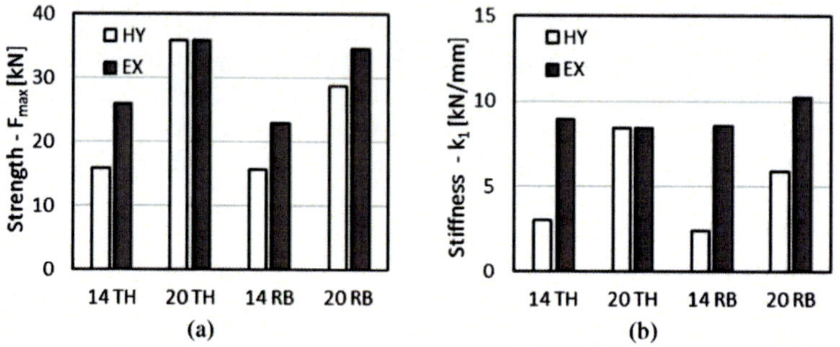

Fig. 5. Influence of grout formulation: a) Connection strength *Fmax*; b) 1st cycle secant stiffness *k1*

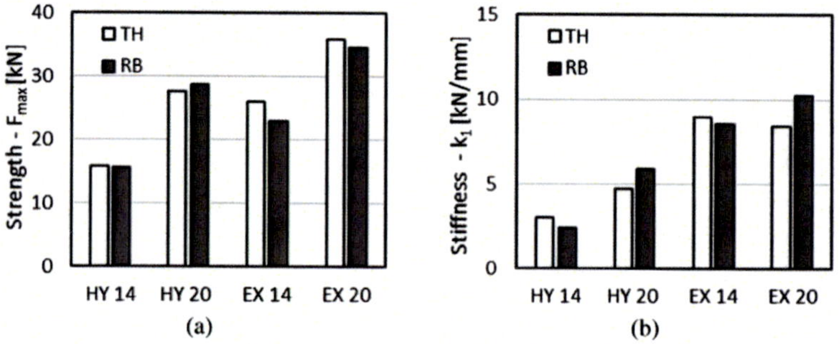

Fig. 6. Influence of dowel type: a) Connection strength *Fmax*; b) 1st cycle secant stiffness *k1*

4.2 Diagonal Compression Testing

Every specimen exhibited diagonal cracking failure. Diagonal cracks formed parallel to the loading direction as a result of the diagonal tensile stress perpendicular to the loading direction. Crack opening almost always interested the mortar joints except for few cases where also the stone elements were involved in the cracking mechanism. Masonry crack width reached remarkable amplitudes (>20 mm) while the *CLT* panels were observed to slightly deflect out-of-plane without undergoing damage (Fig. 7).

Test results are listed in Table 5 in terms of peak force F_{max} and equivalent shear strength s_s evaluated according to ASTM E519-15 (ASTM 2015).

The retrofitted specimens exhibited an average 14% strength increment with respect to the as-built configuration. Observed shear strength increments might have been mitigated by the modest number of timber-masonry connections due to the relatively small size of the wallets. Being the *CLT* reinforcement panel engaged solely by the timber-masonry connection shear transfer, testing of larger specimens featuring a higher number of connectors collaborating with each other (despite having the same number of connectors per squared meter of wall surface area as the specimens tested herein) is likely expected to provide greater strength increments when compared to the as-built wall configuration, consistently with the outcomes reported in Giongo et al. (2021).

Fig. 7. Retrofitted specimen during testing: a) un-reinforced side; b) reinforced side

Table 5. Diagonal compression testing: peak force F_{max} and equivalent shear strength s_s

Test ID	F_{max}	s_s
	[kN]	[MPa]
AsB-1	194.3	0.29
AsB-2	243.3	0.36
AsB-3	199.6	0.29
AsB-4	201.0	0.30
R-1	281.5	0.42
R-2	221.4	0.33
R-3	237.5	0.35
R-4	212.2	0.31

The experimental campaign also saw the diagonal compression testing of benchmark masonry wallets that were retrofitted on one side with a composite reinforced mortar (*CRM*) system made of ≈3 cm thick lime plaster (M10 hydraulic lime mortar) reinforced with a glass fiber mesh. In Fig. 8 the *CLT* retrofit technique is compared to the CRM one in terms of post-peak behavior in the dimensionless shear s/s_s vs. strain ductility $\varepsilon/\varepsilon(s_s)$ plane. From such comparison, the *CLT* retrofit solution seems to be the most effective in increasing the strain ductility of the retrofitted walls. Both retrofit techniques were applied to nominally identical masonry specimens that were tested according to the procedure reported in paragraph 4.2.

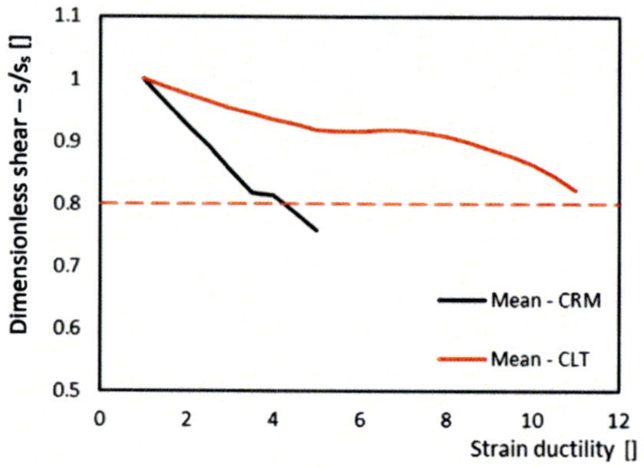

Fig. 8. Post peak behavior comparison, *CRM* vs. *CLT* retrofit solutions

5 Conclusions

The results of an experimental campaign investigating the mechanical behavior of a timber-based retrofit technique for URM walls were presented. The experimental program comprised shear testing of adhesive timber-masonry connection systems and diagonal compression testing of masonry wallets retrofitted by means of *CLT* panels.

Predictably, the shear behavior of the investigated connections was observed to be significantly influenced by the type of resin and anchor rod diameter, while the effects of modifying the type of anchor rod were found to be negligible. The shear capacity of the tested connection ranged from ≈16 kN for 14 mm anchors fixed using hybrid resin, to ≈36 kN for 20 mm anchors installed with epoxy resin.

The diagonal compression testing showed some promising results. The timber-based retrofit in fact, despite the non-symmetrical reinforcement configuration, resulted in remarkable increases in the post-peak deformation capacity and in non-negligible increases in the wall equivalent shear strength. Further testing is however recommended to fully evaluate the performance range of the retrofit technique.

Acknowledgements. The research work presented herein was carried out within the framework of the 2019–2021 ReLUIS-DPC network (Italian University Network of Seismic Engineering Laboratories and Italian Civil Protection Agency). The authors gratefully thank Terme di Comano (IT) thermal consortium for kindly providing access to the building and for allowing the experimental campaign on timber-masonry connection systems to be undertaken.

Miniera San Romedio S.r.l., Fischer Italia S.r.l., Rubner Holzbau S.r.l. are gratefully acknowledged for supporting the research and for supplying the materials required to assemble the specimens.

References

ASTM (American Society for Testing and Materials International). Standart test method for diagonal tension (shear) in masonry assemblages. In: ASTM E519-15/E519M-15, West Conshohocken, PA, USA (2015)

Augenti, N., Parisi, F.: Learning from construction failures due to the 2009 L'Aquila, Italy, Earthquake. J. Perform. Constr. Facil. **24**(6), 536–555 (2010). https://doi.org/10.1061/(ASCE)CF.1943-5509.0000122

Bhattacharya, S., Nayak, S., Dutta, S.C.: A critical review of retrofitting methods for unreinforced masonry structures. Int. J. Disast. Risk Reduct. **7**, 51–67 (2014). https://doi.org/10.1016/J.IJDRR.2013.12.004

Borri, A., Corradi, M., Speranzini, E., Giannantoni, A.: Reinforcement of historic masonry with high strength steel cords. Mason. Int. **23**(3), 79–90 (2010)

CEN (European Committee for Standardization). Methods of test for mortar for masonry – Part 11: Determination of flexural and compressive strength of hardened mortar. In: EN1015-11. Brussels, Belgium (2019)

Dizhur, D., Ingham, J.M., Biggs, D., Schultz, A.: Performance of unreinforced masonry and infilled RC buildings during the 2015 Gorkha, Nepal earthquake sequence. In: Proceedings, 16th International Brick and Block Masonry Conference, IBMAC 2016, pp. 2399–2408 (2016)

Dizhur, D., Giaretton, M., Giongo, I., Ingham, J.: Seismic retrofit of masonry walls using timber strong-backs. SESOC J. **30**(2) (2017)

European Technical Approval ETA-10/0012. Fischer injection system FIS EM. European Organisation for Technical Approvals (2013)

European Technical Approval ETA-02/0024. Fischer injection system FIS V, European Organisation for Technical Approvals (2016)

Giongo, I., Schiro, G., Piazza, M.: On the use of timber-based panels for the seismic retrofit of masonry structures. In: Proceedings of the 3rd International Conference on Protection of Historical Constructions (PROHITECH), Lisbon, Portugal (2017)

Ingham, J.M., Griffith, C.: Performance of unreinforced ma sonry buildings during the 2010 Darfield (Christchurch, NZ) earthquake. Aust. J. Struct. Eng. **11**(3), 207–224 (2016). https://doi.org/10.1080/13287982.2010.11465067

Italian Technical Approval CIT n.03/2013. Pannello in compensato di tavole RUBNER HOLZBAU (2013)

Lin, Y., Lawley, D., Wotherspoon, L., Ingham, J.M.: Out-of-plane testing of unreinforced masonry walls strengthened using ECC shotcrete. Structures **7**(2016), 33–42 (2016). https://doi.org/10.1016/j.istruc.2016.04.005

Penna, A., Morandi, P., Rota, M., Manzini, C.F., da Porto, F., Magenes, G.: Performance of masonry buildings during the Emilia 2012 earthquake. Bull. Earthq. Eng. **12**(5), 2255–2273 (2013). https://doi.org/10.1007/s10518-013-9496-6

Riccadonna, D., Giongo, I., Schiro, G., Rizzi, E., Parisi, M.A.: Experimental shear testing of timber-masonry dry connections for the seismic retrofit of unreinforced masonry shear walls. Constr. Build. Mater. **211**, 52–72 (2019)

Giongo, I., Rizzi, E., Riccadonna, D., Piazza, M.: On-site testing of masonry shear walls strengthened with timber panels. In: Proceedings of the Institution of Civil Engineers – Structures and Buildings (2021). https://doi.org/10.1680/jstbu.19.00179

Sustersic, I., Dujic, B.: Seismic shaking table testing of a reinforced concrete frame with masonry infill strengthened with cross laminated timber panels. In: Proceedings of the World Conference on Timber Engineering (WCTE), Quebec City, Canada (2014)

The Technical State Evaluation and Protection Measures Development for the Refectory Chamber Building with the Church of the National Kyiv-Pechersk Reserve

Yurii Melashenko(✉), Asia Kisil, Yurii Ishchenko, Yevheniia Pavliuk, and Nicolay Marienkov

State Enterprise "Research Institute of Building Constructions", 5/2, Preobrajenskaya Street, Kyiv 03680, Ukraine

Abstract. The paper presents the results of the technical condition survey and assessment by visual and design criteria carried out for one object of the National Kyiv-Pechersk Historical and Cultural Reserve, namely, the Refectory Chamber complex buildings (Upper Lavra, Fig. 1). The Refectory buildings include the Refectory Church, the Refectory Chamber and the Refectory Kitchen. To assess the Refectory buildings structures technical condition, the constructive solution features of the Refectory buildings individual elements and their structural scheme were analyzed. The Refectory buildings operating conditions were studied with highlighting the main factors affecting the materials and structures durability. The degree and types of damages were identified for building structural elements taking into account their functional purpose and location in a common spatial complex. The mathematical simulation of the Refectory Chamber building structures stress-strain state was carried out and based on its results the deformations forecast was made taking into account the base uneven settlement influence during an emergency soaking.

Keywords: National Kyiv-Pechersk reserve · Survey · Nondestructive examination · Design criteria · Stress-strain state

1 Introduction

More than 90% of the territory of Ukraine have complicated ground conditions, and about 120 000 sq. km of the Ukrainian territory are located in the zone of natural origin seismicity with a magnitude varying from 6 to 9. The ground failure processes are at the first place in Ukraine and the second in the world by caused damage volumes; the first place belongs only to earthquakes (Living with Risk 2002). Therefore, the unpredictable changes of natural geological and man-made factors governing ground conditions may lead to dangerous deformation processes in the heritage sites of Ukraine. These processes often end with accidents and disasters.

I. Vayas and F. M. Mazzolani (Eds.): PROHITECH 2021, LNCE 209, pp. 1018–1029, 2022.
https://doi.org/10.1007/978-3-030-90788-4_79

The landslides frequency of about 20 major events per year in Europe is the highest compared to floods, earthquakes and cyclones occurrences (Lacasse 2013). In this view, landslides and, more generally, slopes instabilities are the important factors, which endanger cultural heritage sites, cause their degradation and require the additional protection measures, monitoring, early warning systems creation etc. In recent decades, the cultural heritage concept (Margottini and Vilimek 2014; Migon 2013) has evolved into one that encompasses the humanity history understanding together with scientific knowledge and intellectual attitudes. This changing concept prompts a subsequent reevaluation of what constitutes the outstanding universal values of World Heritage sites and the operational methods for implementing the UNESCO World Heritage Convention (The World Heritage Convention 1972).

The fundamental principles of the World Heritage Convention are set forth in its Fourth and Sixth Articles. Article 4 recognizes that each state party has a duty "of ensuring the identification, protection, conservation, presentation and transmission to future generations of the cultural and natural heritage" situated in its territory. This duty "belongs primarily" to each state party. However, according to Article 6, "whilst fully respecting the sovereignty of the States on whose territory the cultural and natural heritage (The World Heritage Convention 1972) is situated, and without prejudice to property rights provided by national legislation, the States Parties to this Convention recognize that such heritage constitutes a world heritage for whose protection it is the duty of the international community as a whole to cooperate." To this end, "the States Parties undertake (The World Heritage Convention 1972) to give their help in the identification, protection, conservation and presentation of the cultural and natural heritage, if the States on whose territory it is situated so request." The conceptual scheme of the World Heritage Convention is clear. It entrusts each state party with a global function (the identification, protection, conservation, presentation, and transmission to future generations of the cultural and natural heritage), which must be performed in order to achieve an objective of the international community as a whole. Each state must manage a "world heritage," as Article 6 expressly defines it.

The theoretical and methodological issues of these problems are analyzed in the works of modern foreign scientists, including (Vayas et al. 2019; Mazzolani and Calado 2019; Tassios et al. 2003; Casagli et al. 2010; Highland and Bobrowsky 2008; Farenyuk et al. 2018, 2019; Kaliukh et al. 2018, 2019a, 2019b; Slyusarenko et al. 2009, 2010, 2013a, 2013b, 2015; Shokarev et al. 2005, 2009; Trofymchuk et al. 2019) and others.

In our paper "The technical state evaluation and protection measures development for the Refectory Chamber building with the church of the National Kyiv-Pechersk Reserve" the results are presented on the survey and assessment of the technical condition of one of the National Kyiv-Pechersk Historical and Cultural Reserve objects (Upper Lavra, Fig. 1), namely the Refectory buildings, based on the visual and design criteria. The Refectory buildings include the Refectory Church, the Refectory Chamber and the Refectory Chamber Kitchen (Figs. 2 and 3).

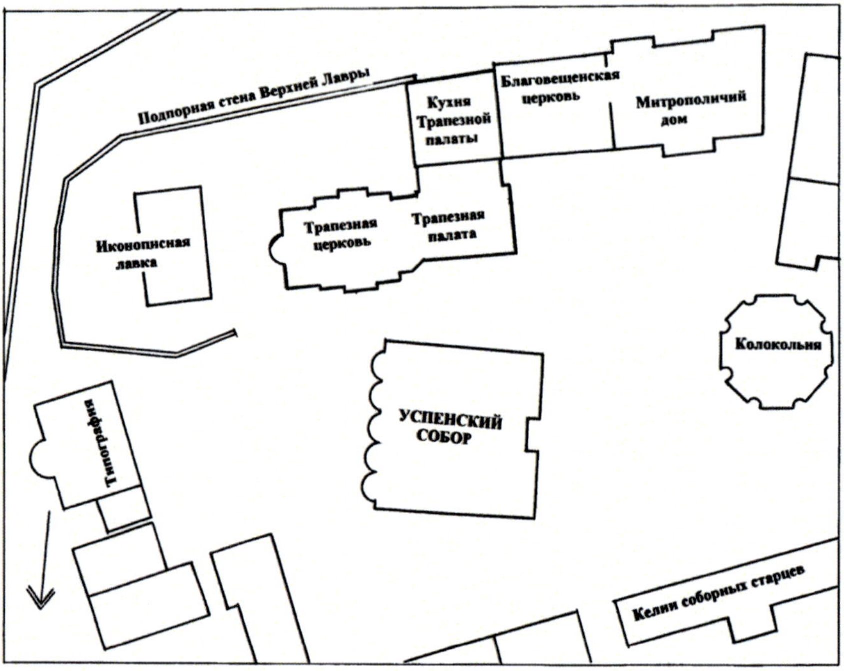

Fig. 1. The Upper Lavra site layout plan extract: 1 – the Holy Assumption Cathedral; 2 – the Great Bell tower; 3 – the Refectory Church; 4 – the Refectory Chamber; 5 – the Refectory Chamber Kitchen; 6 – the Annunciation Church; 7 – the Metropolitan's house; 8 – the Cells of the Cathedral Elders; 9 – the Printing house, 10 – the Upper Lavra retaining wall

Fig. 2. Refectory Chamber building

Fig. 3. Refectory Chamber buildingKindly note that the figure captions 2 and 3 seems to be same. Please check and and correct if necessary.

To assess the technical state of the Refectory Chamber buildings (RCB) structures the design features analysis of the building's individual elements and their structural scheme was envisaged. The Refectory buildings operating conditions were studied and the main factors affecting the materials and structures durability, as well as the extent and types of damage to structural elements were highlighted taking into account their functional purpose and the location in the RCB entire spatial complex. Mathematical simulation of the RCB structures stress-strain state was carried out and, based on its results, a deformations forecast was made taking into account the influence of the RCB base uneven settlement during emergency soaking.

2 Engineering-Geological and Structural Characteristics of the Study Object

At the depth investigated by geological excavations, the geological structure of the Refectory buildings development site is composed of a Quaternary and Neogene-Paleogene deposits system represented by sandy loams, loams, clays and sands, covered on the surface with bulk soil. Loess-like sandy loams with a thickness of 8–11 m, lying, as a rule, under the Refectory foundations bottom, have subsidence properties. According to archive materials, the first construction of the Refectory buildings was completed in 1108. Later, the buildings were rebuilt several times. In 1895, the Refectory complex was completely rebuilt. Structurally, the Refectory church is a one-story brick structure with a large spherical dome and five gold-plated domes (Fig. 2 and 3). The building internal dimensions in plan are 26.0–18.92 m. From the west, the Refectory Chamber adjoins the Refectory church. It is a rectangular two-storey brick building with internal plan dimensions of 27.69 m × 15.6 m. The Refectory kitchen building adjoins the Refectory Chamber from the south. These two buildings are connected by a passage. The Kitchen building is a two-storey, rectangular in plan. The building external dimensions are 20.3 m × 23.3 m. The building has no basement. The room of the Refectory building ground floor, which is a large hall, has two rows of supporting cast-iron columns with five columns in each row (Fig. 4). The ground floor middle span with a width of 6.9 m is covered with a brick cylindrical vault. The height of the cylindrical vault rise (internal size) is about 0.88 m.

Lateral spans are covered by coved vaults with a span of 4.6–4.65 m. Supporting beams are laid under the vaults along the building columns as follows: two rolled steel joists of I-section No. 24 are arranged in the longitudinal direction and two transverse bond beams of I-section No. 24 are perpendicular to the joists. The other ends of these beams rest on the external load-bearing walls. The second floor layout is solved in such a way that a transverse wall with a thickness of 1½ bricks divides the floor space into two spaces located between the building axes 1–2 and 2–7 (Fig. 4). Within the building length in axes 2–7 the second floor is divided into separate rooms by brick walls with a thickness of ½ brick. Rafters with intermediate support constitute the load-bearing roof structures on the part of the building length between axes 1 and 2. On the rest of the length, the building is covered by five metal gable trusses with a 16.6 m span supported by external longitudinal walls. Ceiling structures are suspended from the lower-chord panel points.

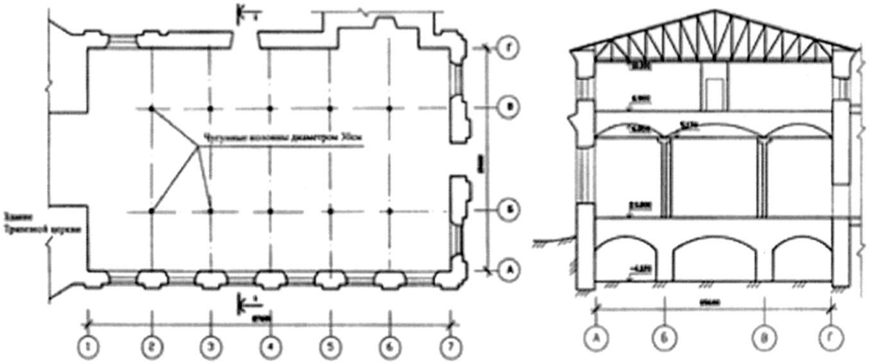

Fig. 4. Refectory chamber ground floor plan. Cross Section 1–1

3 Visual and Instrumental Examination

Under the Church and Chamber buildings there is a basement. The building structures surveys showed that there is a two-level ceiling over the basement. The lower level is composed of brick vaults reinforced with strengthened arches. These structures are visible from the basement room. The ceiling vaults and strengthened arches gaps are backfilled up to the vaults upper elevations level. On the smoothed with the screed horizontal surface the brick columns are arranged, through which a load is transferred to the arches of the lower level ceiling from the upper level ceiling. The latter is a system of vaults having a thickness of ½ brick and supported by steel longitudinal beams. The massive brick pillars and basement walls form the load-bearing structures, to which the load is transmitted from the ceiling and above situated structures.

The building walls foundations examination from the pits showed that the thicknesses of the basement outer walls of the northern and western facades (axes A and 7 in Fig. 4) are 145 cm and 73 cm, respectively, and the southern facade walls are 161 cm. The basement walls rest directly on the ground without any widening; their depth is 17–37 cm. Pilasters and poles below the basement floor level have widenings from 6 cm to 38 cm. The pilasters foundations bottoms depths are as follows: 54 cm along 7/Б axis, 73 cm along 3/A axis, 84 cm along the 4/G axis; the column depth along 6/V axis is 65 cm.The Refectory buildings, which have been in operation for a long time, were partially destroyed during the Second World War and were badly affected by changes in the foundations bases hydrogeological conditions.

In 1941, the explosion of the nearby Holy Assumption Cathedral caused a significant damage to the RCB structures (there is a version about the dynamic impact on the RCB structures from the base due to the Holy Assumption Cathedral structures collapse during the explosion). The Refectory Chamber roof was completely demolished, and the building itself received deformations in the form of the northern facade wall curving from its plane. It is assumed that this resulted in a span increase in the Refectory Chamber middle part. The vaults structures, which covered the Refectory Chamber ground floor premises, stood open until 1947–1948, when the building was restored. The RCB repair and restoration works after the World War II were completed only in 1956.

The building inspection in 1967 revealed that all cast-iron columns of the Refectory building ground floor deviated from the vertical towards the outer walls. The columns had vertical cracks for the depth of the column wall thickness and through-length cracks for the column entire height. The nature of the columns deviation from the vertical and deformations in the cylindrical vault indicated an arch thrust largeness and the reliable structure absence for this thrust perception. The building second floor inspection showed that there was a significant temporary load on the inter-flooring over the ground floor. All loads above the second floor were transferred through the partitions supporting structure to the vault covering the central span of the ground floor hall.

According to the specialists who examined the building, one of the reasons for the significant horizontal forces and the vault toes displacement from its center towards the outer walls, as well as the columns deviation from the vertical position, was a wrong constructive decision made when restoring the Refectory Chamber building. Based on these surveys results the following measures were taken: tie-bars were arranged between the central vault supporting structures; coupling metal clamps were mounted on the columns; the scheme of load transfer from the attic flooring and rafter system structures to the central vault was replaced by the load transfer to the building longitudinal walls. A number of cases of water leakage from communications and soaking of the base, the soils of which had the subsidence properties, have been recorded. This was accompanied by the cracks formation in the ceiling structures and buildings walls.

As a visual and instrumental examination result, it was found that the Refectory Chamber building walls had many cracks in the brickwork. The cracks of vertical orientation crossed the wall sections between window openings and continued on the building basement. It was possible to classify the cracks by their development nature as sedimentary, i.e. caused by uneven building settlements. At the time of surveying, most of the cracks visible on the facades walls were already repaired during the building current repair. The surveys of the walls and pillars technical condition in the basement and ceilings supporting structures over the basement revealed cracks in the brickwork of supporting arches and ceilings vaults. The crack opening width was 2–8 mm in the vaults and 2–4 mm in the arches. A probable cause of structural deformation was the walls and pillars uneven settlement.

In the Refectory Chamber building basement there were the biological damages to the masonry mortar of pillars and some vaulted structures. Here, the mortar destruction in the masonry joints to a depth up to 1.5 cm was observed. The roof covering was made of galvanized sheet iron. At the moment of examination, it was in unsatisfactory condition. In the flat arches of almost all window openings in walls, the vertical cracks with 1–2 mm opening widths were present on the render. There were soaking traces on the inner wall surface along the Γ axis; the paint layer peeled off in some places due to the biological corrosion present here. The measurements of the Refectory Chamber cast-iron columns deviations from the vertical plane were made using the geodetic vertical projection method. By measurements it was found that the columns had deviations from the vertical plane towards the longitudinal bearing walls. The deviation values were as follows: from 71 to 125 mm along the Б axis and from 21 to 41 mm along the В axis. The bricks samples were taken from the Refectory Chamber building walls. They were tested by means of a press in laboratory conditions.

The ultrasound method was used to study the bricks strength in brickwork separate sections. It was found that in the Refectory Chamber building the average values of the bricks ultimate compressive strength and bending limit were 12.34 and 2.5 MPa, respectively. Based on the brick and mortar tests results, the following theoretical assessment of the building walls brickwork strength was given: the brickwork ultimate compressive strength was 2.11 MPa. The non-destructive method allowed determining the stress levels in the ceiling vaults tie-bars above the Refectory Chamber ground floor.

The obtained results showed that the vaults tie-bars were included in the work. The stress levels in the tie-bars were in the range from 32.9 MPa to 88.3 MPa, which was less than the bars design tension resistance $R_s = 280$ MPa (reinforcement class AII). As mentioned above, during the building structures inspection the records were made of the cracks in the ceilings vaults and building walls, building northern wall deflection from the plane and ground floor columns deviation from the vertical. Some of the above types of deformations were due to the building structural design that existed before the repair and restoration works in 1973–1977, as well as to the adverse effects from the base. That referred to the effect caused by soils settlement during the base soaking and settlement during the compaction of bulk soils lying under the foundation bottom at the building south side. But the nature and value of the building ground floor walls and columns deviations were mainly influenced by the impact of an unknown origin (supposedly arising from the Holy Assumption Cathedral explosion in 1941).

4 Methodology of Mathematical Simulation

The results of the Refectory Chamber building visual and instrumental examination became the basis for the *Tasks No. 1* and *No. 2* formulation for the building stress-strain state mathematical simulation.

Task 1 determined the stress-strain state of the Refectory Chamber building structures using the structural scheme that existed before the repair and restoration works in 1973–1977. In this scheme, the loads from the roof, attic and partitions structures and live loads imposed on those structures were transferred to the central vault. The efforts of the central and side vaults thrust were perceived by the metal beams system and building longitudinal walls.

In *Task 2*, when determining the building structures stress-strain state, the same structural scheme as in the first task was considered. In that case, the influence on the Refectory Chamber building structures, including the ground floor columns, was taken into account the factor associated with the explosion of the nearby Assumption Cathedral (there is a version about the dynamic impact on the construction of the refectory chamber building from the base, which was caused by the fall of the construction of the Assumption Cathedral explosion). In that case, the influence of the factor associated with the explosion of the nearby Holy Assumption Cathedral (there is a version about the base dynamic impact on the Refectory Chamber building structures caused by the Holy Assumption Cathedral structures fall during the explosion) on the Refectory Chamber building structures, including the ground floor columns, was taken into account. The Refectory Chamber building structures stress-strain state and building deformations forecast were determined numerically, taking into account the influence of the base uneven settlement during its emergency soaking.

5 Results Analysis

The ground floor columns. The *Task 1* calculations results analysis showed the greatest horizontal displacements and greatest efforts in the ground floor columns. For one of the building cross sections, the total horizontal displacements of the wall and column top calculation model nodes were close to actual displacements. Structural stress analysis was performed for the most loaded columns. The stress determined by the calculation was less than the column material design resistance under tension and compression for $\sigma_B = 150$ MPa and was practically equal to the resistance for $\sigma_B = 120$ MPa. Columns accepted the loads that were imposed on them, but in the first case, the columns strength was at their strength limit; in the second case (assuming that the columns material design resistance was 107.1 MPa) there was a 25% margin of strength. The columns stability control showed that their stability margin exceeded the stability margin factor for cast iron, which was a negative factor. In such a situation, cast iron in the columns materials would begin to break down before the columns themselves reach the design stability limit.

Ceilings. The design scheme nodes displacements, which characterize the deformations of ceiling vaults over the ground floor and basement in *Task 1*, are shown in Fig. 5. The maximum value of the central vault middle vertical displacement is 25–31 mm relative to the nearby columns displacements and 3–10 mm relative to the extreme spans vaults displacements.

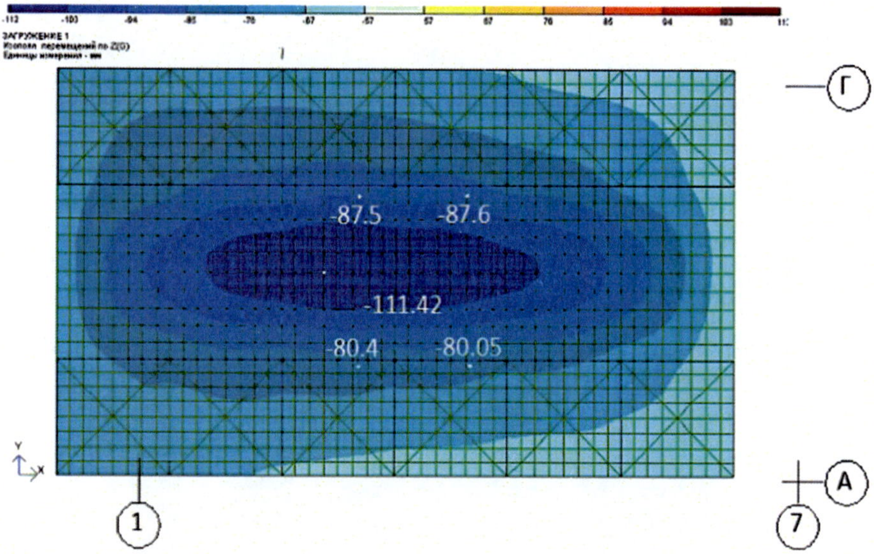

Fig. 5. Isofields of the ceiling vaults design scheme nodes vertical displacements over the ground floor

In Fig. 6 he main stresses distribution obtained in *Task 1* in the ground floor vaults brickwork middle layer is shown. The principal stresses in the lower, middle and upper layers of the Refectory Chamber building structure cross-section are calculated. As can be seen from Fig. 6, the central vault middle layer is in the compressed zone. The lower layer is located in the stretched area of the vault cross-section.

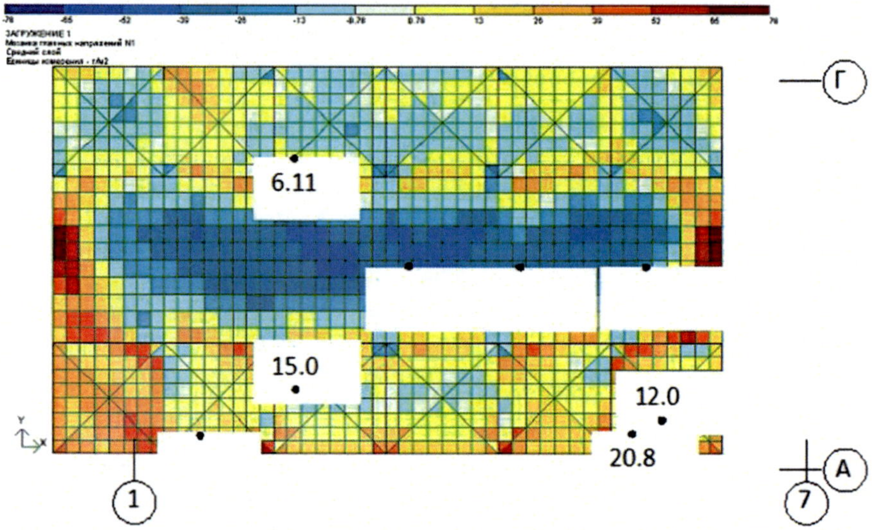

Fig. 6. Ceiling over the ground floor. Principal stresses N_1 (tf/m^2) distribution in vaults structures

The maximum tensile stresses in the building axes 3–4 are 1.1–1.28 MPa. In the building extreme spans vaults there are sections where the maximum tensile stress is 0.39 MPa. In the ceiling above the basement the supporting arches structures are in the most stressed state with the principal tensile stresses being 0.3–0.45 MPa. In the ceiling vaults the stresses are 0.04–0.165 MPa (Fig. 7).

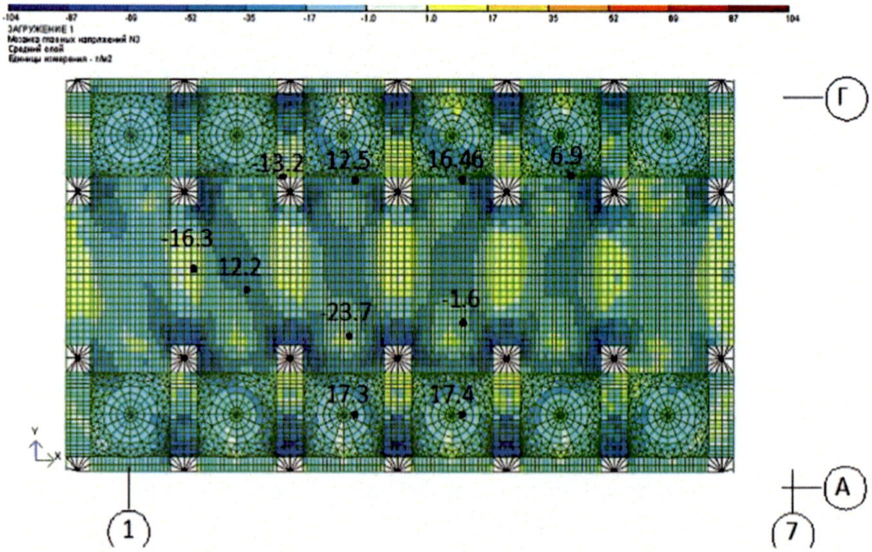

Fig. 7. Ceiling over the basement. Principal stresses N_3 (tf/m^2) distribution in the vaults structures middle layer

As can be seen from the stresses determination results, in the vaults design scheme the sections appeared where the tensile stresses in the vaults brickwork exceed the design resistance of 0.2–0.25 MPa to the principal tensile stresses. This indicates that the crack formation is possible in these areas. The results of the ceiling vaults actual state survey show that the pattern of crack formation in the examined structures in most cases coincides with the possible cracks formation zones obtained by calculations.

6 Conclusions

1. It was found that the building has damages caused by uneven foundation deformation, bricks and mortar weathering, biological damage to wooden structures and corrosion of metal. The structure of the base and physical and mechanical properties of layers, which compose it, have been established according to the engineering-geological survey's results. The samples of soils with special properties (loess-like sandy loams) that settle in the case of soaking were investigated in the laboratory. In addition, there is a filled soil layer of variable thickness, which form's the foundations base in the Refectory Chamber southern part. The both layers saturation under load results in their consolidation, causing the additional deformation of the aboveground structures.
2. If the results of the building calculation using the modern structural scheme are compared with the results of the building calculation based on the previously existing structural scheme, it should be noted that the forces in the walls, ceilings and columns structures have been redistributed due to the changed scheme of load transfer from the roof and attic, as well as the tie-bars arranged in the ceiling vaults. For example, the longitudinal force in the column along the 4/Б axis decreased by a factor of 1.4; the principal tensile stresses in the central vault decreased by a factor of 3–3.5 etc. The principal tensile stress in the most dangerous column section is 1.13 MPa (with a previously existing scheme, the stress was 6.95 MPa).
3. The total tensile stress in the cross-section of the most loaded column along the 4/Б axis obtained taking into account the influence factor associated with the Holy Assumption Cathedral explosion, is 79.5 MPa, which is practically the same as the tension strength of cast iron, equal to 85.7 MPa. In order to determine, if the columns have a more substantial strength reserve, it is necessary to experimentally find out the chemical composition of the column cast iron, and then qualify it in relation to modern cast iron brands.
4. The building calculation with an allowance for the soil base emergency soaking influence on the building deformation in the case of the breakthrough of heating network, which passes through the building basement, showed that the walls and columns base additional settlements by 29–55 mm are possible in such case.
5. The necessity of the individual structures strengthening and measures implementation for the foundation base protection against soaking as the main cause of damages to the structures is stated. The project for strengthening the existing foundations and structures of the National Kyiv-Pechersk Historical and Cultural Reserve Refectory Chamber and Church has been developed. The project implementation is currently in progress (according to the funding amount and timescale).

Acknowledgements. We would like to express our very great appreciation to the Ukrainian RIBC scientists, who have been directly involved in visual and instrumental survey and mathematical simulation of the building of the National Kyiv-Pechersk Reserve Refectory Chamber and Church.

References

Casagli, N., Catani, F., Del Ventisette, C., Luzi, G.: Monitoring, prediction, and early warning using ground-based radar interferometry. Landslides **7**(3), 291–301 (2010)

Farenyuk, G., Kaliukh, I., Farenyuk, E., Kaliukh, T., Berchun, Y., Berchun, V.: Experimental and theoretical diagnostics of defects in ferroconcrete piles based on reflection of longitudinal and transverse waves. In: Hordijk, D.A., Luković, M. (eds.) High Tech Concrete: Where Technology and Engineering Meet, pp. 1307–1317. Springer, Cham (2018). https://doi.org/10.1007/978-3-319-59471-2_151

Highland, L., Bobrowsky, P.: The Landslide Handbook —A Guide to Understanding Landslides: Reston. U.S. Geological Survey Circular, Virginia (2008)

Kaliukh, I., Dunin, V., Berchun, Y.: Decreasing service life of buildings under regular explosion loads. Cybern. Syst. Anal. **54**(6), 948–956 (2018). https://doi.org/10.1007/s10559-018-0098-9

Kaliukh, I., Fareniuk, G., Trofymchuk, O., Fareniuk, I., Berchun, Y.: Identification of defects in reinforced concrete piles based on multi-wave reflection. In: Derkowski, W., Gwoździewicz, P., Hojdys, Ł., Krajewski, P. (eds). In: Proceedings of fib Symp. 2019: Concrete – Innovations in Materials, Design and Structures, Fédération Internationale du Béton (fib) – International Federation for Structural Concrete, pp. 991–998 (2019)

Kaliukh, I., Trofymchuk, O., Farenyuk, G., Ivanik, O., Shekhunova, S.: Practical measures fo landslide risk mitigation in the Ukrainian Carpathians. In: First EAGE Workshop on Assessment of Landslide and Debris Flows Hazards in the Carpathians, vol. 2019, pp. 1–5. European Association of Geoscientists & Engineers (2019). https://doi.org/10.3997/2214-4609.201902165

Lacasse, S.: 8th Terzaghi Oration – Protecting society from landslides – the role of the geotechnical engineer. In: The 18th International Conference on Soil Mechanics and Geotechnical Engineering, Paris, pp. 15–34 (2013)

Living with Risk. A global review of disaster reduction initiatives. Preliminary version. ADRC, ISDR, UN, WMO, Geneva, vol. 384 (2002)

Margottini, C., Vilimek, V.: The ICL Network on "Landslides and Cultural & Natural Heritage (LACUNHEN)." Landslides **11**(3), 934–938 (2014)

Mazzolani, F.M., Calado, L.: Selected papers from the 3rd international conference on protection of historical constructions (Prohitech´17). Int. J. Architect. Heritage **13**(1), 1 (2019). https://doi.org/10.1080/15583058.2018.1564591

Migon, P.: Cultural heritage and natural hazards. In: Bobrovsky (ed.) Encyclopedia of natural hazards. Springer, Dordrecht (2013)

Slyusarenko, Y., Matveyev, I., Kisil, A., Ischenko, Y., Romanov, O., Kosheleva, N.: Solution of the geotechnical problems of the Poshtova Square reconstruction in Kiev. In: Proceedings of the XVI European Conference on Soil Mechanics and Geotechnical Engineering (ECSMGE 2015): Geotechnical Engineering for Infrastructure and Development, pp. 693–698 (2015)

Slyusarenko, Y., et al.: Modern methods of geotechnical defense of buildings in the difficult geological conditions of Ukraine. In: Proceedings of the 18th International Conference on Soil Mechanics and Geotechnical Engineering (Paris) (2013a)

Slyusarenko, Y., Girshtel, G., Liseniy, O.: Optimization of multilayer repair systems of polymeric compositions for concrete structures at reconstruction of NSC Olimpiyskiy in Kiev. In: Proceedings of the fib Symposium Tel-Aviv 2013: Engineering a Concrete Future: Technology, Modeling and Construction, pp. 353–356 (2013)

Slyusarenko, Y., Kryvosheyev, P., Serjogin, J., Kozjavkin, I.: Concrete structures of the Olympiyskiy National Sports Complex in Kiev. In: Proceedings of the 3rd International fib Congress and Exhibition, Incorporating the PCI Annual Convention and Bridge Conference: Think Globally, Build Locally (2010)

Shokarev, V., Shapoval, V., Chaplygin, V., Samchenko, R., Volkov, D.: Stress-strain state of the system base-strip foundation at elimination of excessive tilts of buildings. In: Proceedings of the 17th International Conference on Soil Mechanics and Geotechnical Engineering: The Academia and Practice of Geotechnical Engineering, vol. 1, pp. 680–683 (2009)

The World Heritage Convention (1972). [Online] http://whc.unesco.org/en/convention/

Trofymchuk, O., et al.: Dynamic certification of landslide protection structures in a seismically hazardous region of Ukraine: experimental and analytical research. Earthquake geotechnical engineering for protection and development of environment and constructions. In: Silvestri, F., Moraci, N. (eds.) Proceedings of the VII ICEGE 7th International Conference on Earthquake Geotechnical Engineering, Rome, Italy, 17–20 June 2019, pp. 5337–5344 (2019)

Check for updates

Research About the Techniques Sustainable to be Adopted in Conservation the Santa Teresinha Chapel

Camila R. Innocencio(✉) and Maria Teresa G. Barbosa

Department of Engineering, Federal University of Juiz de Fora, University Campus of Juiz de Fora, Juiz de Fora, MG, Brazil
camila.innocencio@arquitetura.ufjf.br,
teresa.barbosa@engenharia.ufjf.br

Abstract. The Santa Teresinha Chapel, built in 1927, is a city heritage, which guarantees its irreplaceable character. It is the first building built to worship this saint in the city, and the second building to worship the same saint in Brazil. Therefore, its conservation will preserve the collective memory of the Brazilian people. In this context, the present paper aims to show/deals with the relationship between religious vernacular heritage and sustainable conservation techniques. The highlights of this work are the guidelines for conservation and restoration of interventions on religious buildings from the 20th century, which guarantee the theoretical criteria, such as: legibility maintenance, authenticity maintenance and materiality maintenance of the heritage, as well as its relationship with sustainable development (social, economic and environmental dimensions). The literature review on heritage and sustainability, as well as their correlation was done after the diagnosis of the state of conservation of the building by visual inspection and non-destructive testing (thermography). The results and discussion show mitigating measures of pathological manifestations using sustainable techniques and materials. Finally, an intervention methodology will be proposed that enables in-depth analysis of the state of conservation of the built for the monitoring and implementation of future interventions its ensuring your maintenance of meaning for news generations.

Keywords: Santa Teresinha Chapel · Conservation · Historical heritage · Sustainability

1 Introduction

According to the authors Barranha (2016) and Araújo (2012), it is conceptually understood as cultural patrimony, the entirety of material, artistic and symbolic signs transmitted through the past to each culture and to humanity as a whole, being a constituent part of the affirmation and enrichment of cultural identities, as a legacy that belongs to all. About the importance of society in protecting its cultural heritage, the authors Aurélio and Scalabrini (2004) indicate that preserving is, above others things, a political

I. Vayas and F. M. Mazzolani (Eds.): PROHITECH 2021, LNCE 209, pp. 1030–1053, 2022.
https://doi.org/10.1007/978-3-030-90788-4_80

attitude and an essential factor for sustainable development, as it is a form of resistance to the transformation of space into merchandises and resistance to the homogenization of cities, which makes them more and more similar to each other. Therefore, the community has the role of being truly responsible for its cultural values, while preservation actions ought to be conceived having as principle its own interest and enjoyment.

With the recognition of the architectural, historical and cultural importance of any property, it is necessary to promote measures to ensure the maintenance of these values for future generations. Therefore, interventions of conservation and restoration, which have an intrinsic relationship with sustainability, are inserted in this context. The current panorama of preservation of the cultural heritage in the city of Juiz de Fora and in most of brazilian cities indicates a series of challenges that compromise the maintenance of historic buildings. There are several topics that contribute to the heritage in "stone and lime" being compromised, especially if we consider the temporal de-characterization and/or demolition (if not for the whole, at last for part of the building) and/or the abandonment that many times are associated with several factors. Among those factors we can mention: the inefficient protection of the public policies currently adopted; the limitation in the community's involvement in relation to its material and immaterial possessions resulting from the inexistence or insufficiency of education actions about heritage and diffusion; the lack of resources for maintenance, and in a very incisive way there is the real estate speculation that forcibly seeks for the demolishing of the remnants (which usually have a privileged location and dimensions) in order to obtain profits from new developments in the area.

According to the authors Castriota (2009) and Penteado et al. (2013), preservationist policies traditionally applied end up "sacralizing" the patrimony, incurring in the analysis of it as an isolated aesthetic instance, disconnecting it from more comprehensive urban policies, thus contributing to the creation of a supposed and untrue dichotomy between safeguarding cultural assets and sustainable development. The "tipping" itself, an instrument of legal protection of material assets, widely applied in the country during the 20th and 21st centuries, acts favoring, but also, in many situations, becoming a "bureaucratic factor" in the preservation process, because it imposes restrictions that may hinder maintenance and adaptation to new demands. According to Choay (2001), the dissemination of the use of this tool was driven by the internationalization of the theme, the strengthening of institutions in the field of preservation, the creation of specific laws, in addition to the expansion of the concept of heritage in this period through the typological, chronological and geographic extension.

However, the changes in society and in the dynamics of the cities have brought to light the necessity of an update of these practices looking to their effectiveness nowadays, with an extensive view of what is patrimony. Therefore, it became necessary to overcome the historical-stylistic approach and the attempt to replicate the safeguard practices adopted in European countries, being indispensable the integration with environmental, economic and social aspects of the place where the assets are inserted (Castriota 2009).

Considering this, we present this work's object of study: the Santa Teresinha Chapel, built in 1927, legally protected by the city of Juiz de Fora and important for the neighborhood that has the same name and where it is located, and for Brazil as the second place built in devotion to the saint in the country.

Note that Juiz de Fora is located in the Brazilian state of Minas Gerais. Its development took place on the sidelines of "Caminho Novo" ("New Path"), planned in 1698 and completed in 1725, at the request of the Portuguese Crown, to drain the gold extracted in the locality and to take it more safely and quickly to the then capital of the country, Rio de Janeiro. In order to encourage the occupation of the region around Caminho Novo, the king of Portugal began the policy of donating sesmarias for agricultural and settlement purposes (Esteves 1915).

The city developed intensely over the eighteenth and nineteenth centuries. Having an economy based primarily on coffee, and through investments from farmers, the city prospered. The installation of railways, factories and, in 1889, the first hydroelectric plant in Latin America, the Marmelos Plant, are highlights of this period (Fazolatto 2007). The intense development and industrial pioneering of Juiz de Fora in the late nineteenth century earned the city the nickname "Manchester Mineira", associating it with the English city, regarding the industrialization.

In this context, the scope of the present article is the elaboration of a diagnosis, which allows an in-depth analysis of the conservation status of the Santa Teresinha Chapel, serving as a subsidy for interventions that prioritize the application of sustainable techniques, considering the authentic and basic principles of the property; they may be reversible and distinguishable, but must also ensure the necessary durability and safety. Furthermore, this work is justified upon the presented scenario, by the need to deepen the discussions on: the preservation and the methodologies of intervention in the heritage in the scope of conservation and restoration, aiming at the maintenance of the authenticity in face of its constituent elements; the understanding of the values inherent to the asset that should be the basis for decision making aiming at the recovery of its integrity and the indispensability of analyzing and combining these heritage issues with the notion of sustainability.

2 The Santa Teresinha Chapel

The Santa Teresinha Chapel (Fig. 1) was built at the initiative of the military, with the initial objective of meeting their religious and family demands, but over time it was appropriated by the community that participated in the Sunday celebrations chaired by the military police chaplain. Built near Praça do Cruzeiro dos Militares, its implementation followed the tradition inherited from the Portuguese regarding the location of Catholic temples, since the building was visible at great distances and in topography that exalted its proximity to the sky (Juiz De Fora 1997).

Its inauguration took place on October 2, 1927. To have the "Saint of Modern Times" as a protector distinguished the chapel from the other military religious buildings that traditionally had São Jorge or Santo Expedito as protectors (Juiz De Fora 1997).

Fig. 1. Santa Teresinha Chapel and its location.

The building consists of two volumes: one referring to the tower and the other corresponding to the nave and the chancel. With simple features, its façades have pointed blind arcades under the eaves as ornamentation, resembling corbels and glass windows with full-arch markings and fixed rosette flags. Access to the interior of the chapel is through the tower's volume, with colorful circular stained-glass windows and metal roofing, which has a porch formed by porticos with columns carved in masonry, and asbestos roof, supported by stanchions, cross arm braces and bollards, interventions made after its construction and that interfere with its reading.

The interior of the chapel is of great relevance, with the choir located at the top near the nave entrance, following the colonial tradition; walls featuring parietal paintings by leading artists in the city, as well as curved stucco lining that also features paintings with religious motifs. The high altar with neo-Gothic throne has pictorial treatment in the crown and vault, representing angels and the Eternal Father.

The building was listed as a historical-cultural heritage at the municipal level on August 6, 1999, by Decree No. 6,501, but only its constructive volume and façades were indicated for legal protection and inscription in the Book of Tombo. However, its interior is in the process to be listed as a historical-cultural heritage since 2014.

3 Historical Heritage and Sustainability

According to Brandi (2005) restoration is to reestablish the potential unity of the work of art. Preserving the authenticity of the property is necessary, ensuring the certifying patina of its passage in time. The principles that guide the restoration services are: distiguibility (integrating and enabling recognition between the restored and the pre-existing), the constitution of the image aspect (to consider the material irreplaceable when it helps in the perception of the work image) and reversibility (any restoration intervention should be reversible to facilitate future changes).

Restoration, on the other hand, should be based on the use of materials and techniques compatible with those used in the property. Moreover, the unity of style should not be the goal of intervention, since the different eras regarding the "life" of the building must be respected. Finally, according to the Venice Heritage Charter (ICOMOS 1964), the insertion of elements intended to recompose missing parts should be harmoniously integrated, but distinguishable from the original parts, thus avoiding false artistic or false historical identification.

Considering the need to include buildings in the concept of sustainable development, including built heritage, Barbosa et al. (2018) points out that their behavior with the fundamental sustainability tripod stands out as:

- Environmental aspect: preservation as a tool to reduce the impacts of new construction. Therefore, maintenance and conservation processes need to be improved as a means of effectively extending the useful life and ensuring the integrity of the property.
- Economical aspect: It is important to apply a preventive maintenance program, which is economically more advantageous when compared to corrective maintenance operations. And in cases where rehabilitation and restoration interventions need to be performed, economically viable techniques and materials should be employed, especially regarding maintenance, durability and cost.
- Social aspect: It is understood that the preservation of the historical-cultural heritage of a given society contributes directly to its development, well-being, quality of life, and maintenance of the values of identity and collective memory.

In the regard of civil construction, it is well known that this is a sector with a strong impact in the increasing of the ecological footprint, mainly due to the amount of carbon incorporated in the production and transport of materials and its high rate of waste generation. Thus, the impact of construction materials differs based on their function at each stage of their production process and their life cycle. In this way, the so-called "green" materials with a low impact on the environment can decrease the amount of energy incorporated, CO2 emission levels and waste generation, reducing environmental burdens and also positively impacting the social and economic dimensions of sustainability (Khoshnava et al. 2018).

As a way to mitigate such impacts, the concept of "green buildings" is one of the most important elements in the requirement for sustainable construction. Therefore, it is important to obtain a complete understanding of what a green building really is, especially to promote and strengthen environmental and energy policies to meet the current demands. This brings the need for more effective policies to be implemented to reduce resource inefficiencies in the construction industry as a direct reflexion (Liu and Lin 2016).

In this context, there are positive aspects of the interventions for the conservation and restoration of historic heritage in reducing the ecological footprint and the potential for adapting old buildings to the concept of green buildings. This sustainable role of historical heritage is elucidated by Elefante (2005) when he says that the "greenest" building is the one that has already been built. And that because its protection and adaptation to sustainability criteria could contribute to the mitigation of impacts on the environment, something that can be achieved with interventions (in line with the theoretical principles of conservation and restoration) to make it more efficient in terms of energy demand, and considering the avoided emissions it represents (referring to demolition and construction), while it has already been built. In addition, originally in historic buildings are used local materials, as such: earth, lime-based mortar, stones and wood, reeds and straw that have less energy incorporated than modern materials such as glass, steel or aluminum. (Berg and Fuglseth 2018; Baracho 2013).

The authors Hammond and Jones (2008) conceptualize energy incorporated as being the energy consumed during the life of the building. However, according to Jalali and Torgal (2010), different approaches to this concept can be considered according to the period chosen for analysis. Thus, it can be based on: the start of the extraction of raw materials up to the factory doors (called a cradle to gate); the beginning of the inputs until the work (cradle to site), or from the beginning of the extraction until the phase of demolition and deposition (cradle to grave). Considering the "cradle to gate" approach, in Tables 1, 2, and 3 we have, respectively, the values of energy incorporated in different construction materials, as well as in the different types of mortar and masonry. Moreover, the restoration will allow the minimization of CO_2 emissions, as presented in Table 4.

Table 1. The values of energy incorporated in different construction materials (Source: Wellington 2005).

Material	MJ/Kg	MJ/m^3
Aggregates in general	0,1	150
River aggregates	0,02	36
Extruded aluminum	201	542.700
Anodized extruded aluminum	227	612.900
Recycled aluminum	8,1	21.870
Extruded recycled aluminum	17,3	46.710
Anodized recycled aluminum	42,9	115.830
Asphalt	3,4	7.140
Bitumen	44,1	45.420
Cement	7,8	15.210
Cement mortar	2,0	3.200

(*continued*)

Table 1. (*continued*)

Material	MJ/Kg	MJ/m^3
Ready mixed concrete (fc = 17,5 MPa)	1,0	2.350
Ready mixed concrete (fc = 30 MPa)	1,3	3.180
Ready mixed concrete (fc = 40 MPa)	1,6	3.890
Concrete block	0,94	–
Ceramic brick	2,5	5.170
Ceramic roof tiles	0,81	–
Adobe blocks stabilized with cement	0,42	–
Compressed earth block	0,42	–
Cement-stabilized Wattle and daub	0,8	–
Glass	15,9	40.060
Laminated glass	16,3	41.080
Stucco	4,5	6.460
Plasterboard	6,1	5890
Steel	32	251.200
Recycled steel	10,1	37.210
Local stone	0,79	1.890
Imported stone	6,8	1.890
Zinc	51	364.140
MDF	11,9	8330
Air-dried rough wood	0,3	165
Kiln dried raw wood	1,6	880
Air dried polished wood	1,16	638
Kiln dried polished wood	2,5	1380
Plywood	10,4	–
Polyester	53,7	7710
Polyurethane	74	44.400
PVC	70	93.620

Table 2. Energy incorporated in the different types of mortar (Source: Reddy and Jagadish 2003).

Types of mortar	Cement	Clay	Sand	Energy per m^3 (MJ)
Cement	1	0	6	1268
Cement + pozzolana	80%	0	6	918
Soil-cement	1	2	6	849
Lime-pozzolana (1:2)	–	0	3	732

Table 3. Energy incorporated in the different types of masonry (Source: Reddy and Jagadish 2003).

Types of masonry	Energy per m^3 (MJ)	Percentage of energy relative to brick masonry
Ceramic bricks	2141	100
Concrete blocks	819	38,3
Soil-cement blocks	646	30,2
Aircrete	1396	65,2

Fonte: Reddy and Jagadish (2003).

Table 4. Building materials and CO2 emissions per unit in the manufacturing process (Source: Lima et al. 2018).

Material	Unity of measurement	CO_2 amount per unit
Sand	m^3	17
Lime	20 kg (package)	95
Cement	50 kg (package)	120
Iron and Steel	kg	260
Wood	m^3	13
Gravel	m^3	12
Brick	Unity	8500
Roof Tile	Unity	2500

However, it is necessary to understand that an old building was designed to attend fundamentally different conditions than the current ones, where the climate and the greatest limitation of resources and construction materials have a relevant impact on the final result, differently from what happens today. On the other hand, these buildings can be considerably vulnerable to high wind loads, overload of the rainwater system, as well as fissures and cracks in the foundations due to the variation of soil moisture and its consequent expansion (Berg and Fuglseth 2018; Baracho 2013). In addition, it is well known that the materials used in construction are subject to numerous deterioration processes arising from natural and anthropogenic factors that cause a progressive loss of useful property during its useful lifetime, impairing its performance and for this reason, conservation and restoration interventions must be judicious and applied in conjunction with sustainable parameters in order to guarantee their physical and mechanical compatibility, effectiveness, durability and energy savings (Morillas et al. 2018).

Regarding the life cycle and the need for short and medium term results in minimizing the ecological footprint, Berg and Fulglseth (2018) report researchers indicating that careful interventions in a historic building contribute to the mitigation of climate change during a period of 60-year analysis. In contrast, a new building that meets the "green parameters" needs more than 50 years so that the initial emissions generated in its construction are overcome by the effects of lower energy consumption in use.

But, one of the main challenges of integrating the concept of sustainability into heritage preservation techniques is precisely the "sacralization" of the property as a monument. To advance these discussions, it is affirmed:

> "This is the great challenge of cultural heritage in the context of globalization: making traditional and modern compatible, while ensuring that cultural differences are respected and authentic. This will only be possible if citizenship is strengthened, guaranteeing to the people not only the right to cultural goods, but also the duty to participate in the construction of heritage, as well as preserving it so that present and future generations can enjoy it." (Penteado et al. 2013)

Despite the dichotomy and constant discussions about maintaining the materiality and authenticity of a property, applying similar techniques and materials to the ones originally used, and making interventions with safer, more efficient and sustainable materials and techniques for its preservation, it is possible to integrate these aspects. Regarding that:

> "In restoration, we have to take into account aspects such as: historical analysis, existing documentation and accumulated experience, since the available records are a process of transmission of technique and technology that was often unsuccessful. Conservation will be ineffective if the original technique used is ineffective, contradicting the logic of preservation." (Paschoalin and Barbosa 2013)

Thus, it is understood that in conservation and restoration interventions materiality should not be the only beacon element of authenticity and limitation of the work. It is necessary to consider issues such as culture, context, materials and techniques compatible with the demands verified in the building and, whenever possible, link them to sustainable solutions. Therefore, the ideal restoration or conservation work is one that seeks a respectful recovery of cultural heritage, but that also includes actions to reduce the ecological footprint. In addition, the adequacy of the policies for the renovation of historic buildings to the process of transition to a low-emission society must be carefully considered and adopted so that the values that constitute the significance of the property are not lost.

This is due to the fact that in order to meet strict environmental and energetic performance requirements, old buildings need to undergo adaptations, and in a context in which the importance of cultural heritage is undermined in favor of the environmental benefits related to new buildings, it can be caused a conflict in relation to the maintenance of historic buildings (Almas et al. 2011; Kohler and Hassler 2012; Norrström 2013). Therefore, it is necessary to expand the discussions on this subject and research on diagnostic methodologies and alternative materials in these types of intervention. Therefore, the present work seeks to advance in a practical and theoretical way in studies about the preservation of religious cultural property, considering the concepts cited here.

4 Materials and Methods

Initially, a historical survey of the building was made through documents and photographs from the municipal heritage protection agency and from the owner's collection; after that, a thorough literature review on the themes historical heritage and sustainability; thereafter the diagnosis of the state of conservation of the building by visual inspection along with the placement of plaster proof to determine the configuration and behavior of the fissures and cracks; thermography, as a non-destructive test, using "Flir" brand equipment, "One Pro" model; elaboration of damage mapping for allocation and measurement of the verified pathological manifestations; "Phantom 4" drone aerial inspection to analyze the conservation status of the roof and topographic conditions of the terrain, formulation of a GUT matrix (severity, urgency, and tendency) and, finally, the elaboration of intervention guidelines that take into consideration their relation with sustainability.

5 Results and Reviews

The complete diagnosis of the building's state of conservation was carried out in five stages, namely: visual inspection and monitoring of fissures; use of unmanned aerial vehicle (UAV); non-destructive testing (thermography); damage mapping; and elaboration of a GUT matrix (severity, urgency, and tendency). The use of this methodology aims to accurately diagnose the pathological manifestations (visible and still without evident signs) existing in the property, subsidizing the monitoring of the evolution of damages, as well as the proposition assertive mitigation measures.

5.1 Visual Inspection

Through visual inspection, a diagnosis was made to determine the main pathological manifestations and the real state of conservation of the building. Chart 1 lists the damage, the locations and their components.

Chart 1 Damage diagnosis

Place	Element	Photograph of the damage	Pathological manifestations
Front porch	Columns and capitals		Loss of section in the mortar coating.
	Cover slab		Loss of concrete overlay, carbonation, exposure and reinforcement corrosion.
	Bows	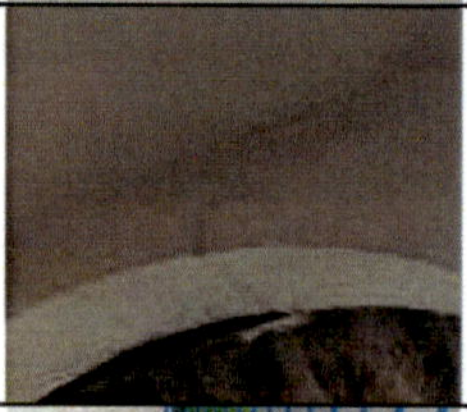	Cracks in the key point.
	Floors		Wear, adhered dirt, missing parts and small vegetation.
Façades	Front		Wear and loss of pictorial layer, humidity due to filtration, microorganisms and cracking of differential settlement.
	Left side		Wear and loss of the pictorial layer, detachment of the plaster, widespread dirt, rising moisture, mapped fissures and small vegetation.

	Right side		Wear and loss of the pictorial layer, detachment of the plaster, widespread dirt, rising moisture, mapped fissures and small vegetation
	Rear		Wear and loss of the pictorial layer, detachment of the plaster, widespread dirt, rising moisture, mapped fissures and small vegetation.
Nave	Inner walls		Loss of pictorial layer and parietal paintings, detachment of plaster, cracks from hygroscopic movement and adhered dirt
	Wall		Cracks from differential settlementand roof overload.
	Staircase, baluster, railing and choir		Presence of xylophagous insects and large amount of adherent bird droppings.
	Window frames		Cracked frames, loss of pictorial layer, presence of xylophagous insects and missing glass.

	Lining		Cracks from repression and overload, moisture stains.
Sacristy	Walls	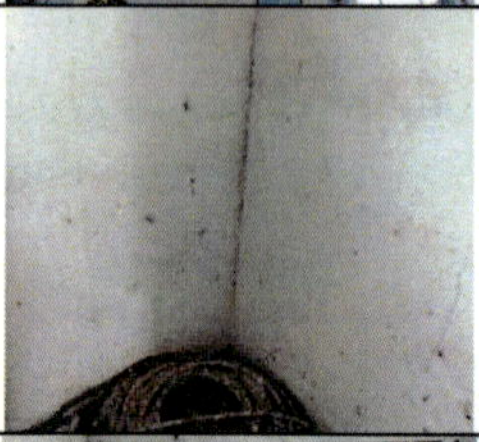	Cracks in the meeting of walls, cracks by hygroscopic movement and adhered dirt.
	Lining		Rotting, xylophagous insects, water infiltration and presence of numerous birds.
Roof	Tile roof		Bulging of the ridge, un-docked tiles and rust points on the tower.

5.2 Monitoring of Fissures

To determine the behavior regarding the activity and possible progression of fissures, plaster "tell-tales" were placed on those that are more representative and that indicate structural impairment of the self-supporting masonry arising from repression and overload, or by hygroscopic action, totaling five points, with monthly inspection. In all inspections carried out until July of 2020, all "tell-tales" remained intact, allowing the classification of fissures as passive. However, in the survey that took place on 08/17/2020, it was found that there was movement causing the tell tale to rupture. This behavior is due to the thermal variation suffered by the building during the transition of seasons (from atypical autumn with high temperatures to winter) causing the retraction of its components. In this way, this fissure is now considered to be active, while the other fissures remain stabilized and will therefore be treated as being passive.

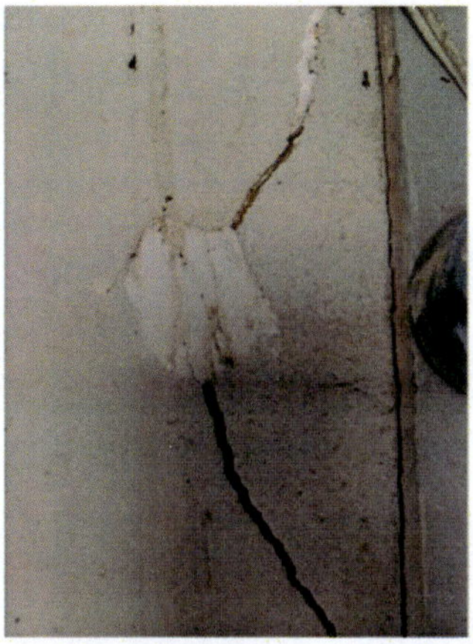

Fig. 2. Movimentation in the tell-tale.

Fig. 3. Crack where the tell-tail was fixed.

5.3 Thermography

Thermography is a type of non-destructive and non-invasive test with great potential for use in monitoring the evolution of pathological manifestations in buildings, which uses thermal imagers and radiometers to detect infrared radiation naturally emitted by materials with intensity directly proportional to their temperature. In this way, it is possible to determine the thermal variation of different surfaces in relation to a pre-established temperature pattern. Since no physical contact is necessary, this technique adds greater precision to the diagnosis of damage to a historic heritage, enabling the analysis of areas of difficult access through a sustainable approach, in an agile, safe and reliable way (Bauer 2013; Lerma et al. 2014).

In this research, after visual inspection to determine the state of conservation of the asset, thermography was used as a method of confirming the characteristics of the anomalies observed, as well as the location of other damages that, perhaps, did not yet have visible signs to the naked eye. For this purpose, a FLIR ThermaCAM b4os model camera was used, operating in the domain of the electromagnetic spectrum 7.5–13 lm, being classified in the high sensitivity category (0.10 °C), producing clear infrared images. The measurements were carried out in two moments: on dry days and, therefore, with all the construction elements with extremely low moisture content; and right after a rainy season with the saturated constituent materials. It was observed that the thermographic images obtained after saturation showed clearer results.

Through thermographic images (Fig. 2), it was possible to identify the points with presence of water infiltration in the stucco ceiling, as well as the presence of birds inside the lining (Figs. 3 and 4).

(a) (b) (c)

Fig. 4. Moisture in the ceiling (a) and (b), presence of birds inside the building (c).

5.4 Damage Mapping

In the damage mapping (Figs. 5, 6, 7, 8, 9, 10 and 11) it is possible to verify by the collected data:

i) The front façade (Fig. 5) shows wear and loss of the pictorial layer, humidity due to infiltration tand colonization of microorganisms near the stained glass window, as well as a crack from the differential settlement suffered by the building, which can also be observed from the inside of the wall. In addition, the corbels, sculptural elements around the roof, sculpted in masonry and coated with mass, present detachment of mortar and section loss in some points.
ii) The two lateral façades (Figs. 6 and 8) present similar pathological manifestations. It was observed: wear and loss of the pictorial layer, points with detachment of the plaster, generalized dirt, rising humidity (helped by the presence of the roughcast that makes it difficult for the walls to breathe), presence of microorganisms, small vegetation near the base and mapped fissures arising from the association of differentiated hygrothermal movements between the coating and the structure.
iii) The rear façade (Fig. 7), in addition to the damage mentioned above, faces the presence of microflora and oxidation of the grilles around the window spans, gaps in the glasses and section losses in the concrete shoulder in the building basement, leaving the masonry exposed.
iv) The inner walls of the building (Figs. 9 and 10) show loss of the pictorial layer and parts of the parietal paintings, detachment of the plaster, cracks arising from

hygroscopic movement and adhered dirt from the accumulation of bird droppings. In the wall of the main altar is possible to notice the existence of cracks near the spans and the arch that cross the self-supporting masonry structure. They can also be seen from the interior of the sacristy due to the differential settlement that was aggravated due to heavy vehicles moving around the chapel, causing vibrations, increased settlement and cracks to expand.

v) The nave's stucco ceiling (Fig. 11) has cracks in the four corners that come from the foundation's repression and the roof overload, whose structure is compromised. In addition, there is a bird's nest inside and over the cracks. Moisture patches with microflora from infiltration through the cover were also seen. The wooden lining of the sacristy is completely degraded due to the presence of xylophagous insects, dampness and the presence of numerous birds, whose droppings accelerate the rotting of the pieces.

Fig. 5. Front façade

Fig. 6. Left side façade

Fig. 7. Rear façade

Fig. 8. Right side façade

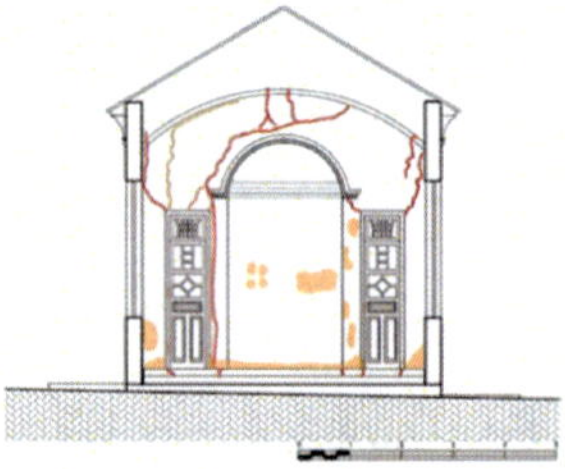

Fig. 9. Altar wall (Section AA)

Fig. 10. Inner-side wall (Section BB)

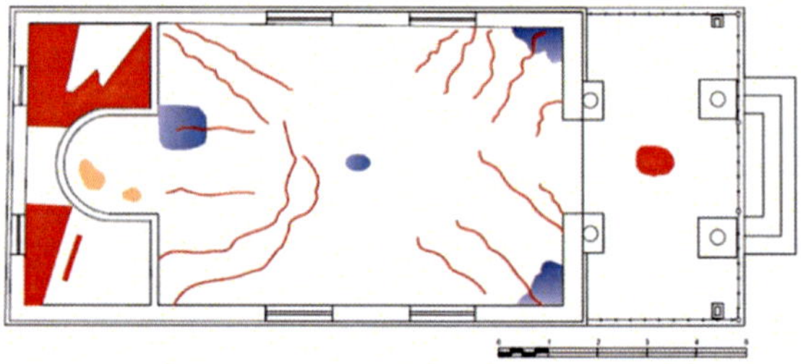

Fig. 11. Lining of nave and sacristy.

5.5 Aerial Inspection by Drone

The use of drones in civil construction is a promising technique, with several applications and which has become an important tool to assist professionals in the area, complementing inspection and diagnosis of the state of conservation of buildings. According to Falorca and Lanzinha (2018), among the advantages inherent to the use of unmanned aerial vehicles (UAVs) are: less time required to perform the tasks; possibility of inspection in places of difficult access; greater precision for analysis and diagnostics; operator safety; diversity and higher quality of the data generated. As camera that capture photos and shoots in high resolution, geolocation systems and other tools, can be coupled to their drones, their use enables the applicability in the diagnosis of pathological manifestations of great works of art, as well as the generation of 3D models that are useful at this stage, but also in the areas of design, maintenance and management.

Through drone inspection (Figs. 12, 13, and 14) it was possible to elaborate a 3D model. In it, it is possible to verify the relationship between the building and its surroundings, as well as its topography, a factor that directly contributes to the origination of the main structural pathological manifestations that the property presents. The existence of slope, together with inadequate fall and lack of surface drainage, associated with the high velocity of rainwater runoff, and loss of the concrete structure that protects the basement of the building, allow the infiltration of water into its base, carrying the soil and generating differential settlement at the four ends of the shallow foundation. The traffic of heavy vehicles using the building as a return generated vibrations that accentuated this movement and increased the size of existing cracks.

Fig. 12. Building and its surroundings.

Fig. 13. Coverage.

Fig. 14. NDVI map.

The inspection of the roof (Fig. 13) was also performed with the aid of the equipment, showing that it has missing tiles only in the ornamentation of the pediment, which are mostly in good condition, with only adhered dirt. However, they are undocked due to the settlement of the building, the excessive vibration generated by heavy vehicle traffic, rotten roof structure, and improper cleaning, all of which make it possible to infiltrate water in the building. Note also the bulging of the ridge and one of the slopes, indicative elements of the structural compromise of its timber. As for the roof of the tower, made of brass, there are oxidation points in its four slopes.

In addition, the equipment produced an NDVI (Normalized Difference Vegetation Index) map for monitoring the quality of the vegetation present in the talus (Fig. 14). It is a relevant analysis since its presence and integrity can interfere with element stability. Based on this, it was found that there is a predominance of undergrowth, with areas of exposed soil and some points with thicker vegetation, in general, with regular to good quality. Thus, it can be seen that the vegetation cover is of great importance for the control of possible erosive processes in the place.

5.6 GUT Method (Gravity, Urgency and Tendency)

The GUT method (Gravity, Urgency and Tendency) is a tool that helps in determining priority in a rational way. Through this matrix, one can analyze the severity of the problem for the people and operations involved; the urgency demanded for its resolution; and the trend to its positive or negative evolution. For this purpose, three decision columns are used to determine an order of priorities, and through the combination of them, it is possible to guide the making of decisions and consequently the resolution of problems. The scoring scales used can be seen in chart 5 (Table 5):

Table 5. GUT scoring scales (Source: Oliveira 1995).

Score	Gravity (G)	Urgency (U)	Tendency (T)
	Consequence if nothing is done	Deadline for decision making	Proportion of the problem in the future
5	Extremely serious	Immediate action	Immediate aggravation if nothing is done
4	Very serious	With urgency	Short-term worsening
3	Serious	As soon as possible	Worsening in the medium-term
2	A little bit serious	It can wait a bit	Long-term worsening
1	Weightless	There is no rush	There is no worsening

As the last stage of the diagnosis of the state of conservation, the GUT method was used to compile and analyze all the pathological manifestations verified with regard to its relationship with the impairment of the building. In addition, the method aims to subsidize the intervention guidelines to be carried out on the property according to the order of priorities obtained (Table 6).

Table 6. GUT matrix for analysis of pathological manifestations in the chapel

Pathological manifestation	G	U	T	GUT
Cracks in the key point of the balcony	2	3	3	18
Loss of section in the mortar coating	1	1	2	2
Absence of tiles on the balcony floor	2	2	2	8
Small vegetation on the balcony floor	2	3	4	24
Exposure and reinforcement corrosion	3	3	3	27
Cracked frames in the windows	2	3	2	12
Gap in the main door tramson	1	1	1	1
Attack of xylophagous insects on wooden frames	2	2	4	16
Oxidation in metal frames	1	2	2	4
Humidity due to capillarity on the facades	3	3	3	27
Microorganisms on the facades	1	1	2	2
Biological patina on the facades	1	1	3	3
Small vegetation on the building base	2	3	4	24
Loss of pictorial layer on the facades	2	2	3	12
Detachment of plaster at points on the facades	2	3	3	18
Mapped fissures on the facade cladding	2	2	3	12
Cracks from hygroscopic movement on the walls	3	3	3	27
Wear on the hydraulic floor tiles	1	1	2	2
Xylophages in the masonry (wall of the main altar)	4	3	4	48
Cracks from differential settlement and roof overload	5	4	3	60
Loss of pictorial layer and parietal paintings	1	1	2	2
Detachment of plaster on the internal walls	2	2	3	12
Damaged electrical installations	2	1	2	4
Damaged hydraulic installations	2	1	2	4
Rotting of the choir slats	2	2	4	16
Rotting floors on the wooden stairs	2	2	4	16
Infiltration and moisture stains on the stucco lining	4	5	5	100
Cracks on the stucco lining	4	4	5	80
Cracks in the meeting of walls	4	4	3	48
Wear of the burnt cement floor in the sacristy	1	1	4	4
Rotting of the joists and wooden slats in the sacristy lining	3	2	4	24
Deterioration of the wooden roof truss	4	5	5	100

(continued)

Table 6. (*continued*)

Pathological manifestation	G	U	T	GUT
Rust points on the tower	2	3	3	18
Widespread dirt on the walls and floors	1	1	4	4
Presence of numerous birds	4	4	4	48
Gap in the protective element at the base of the building	3	4	3	36
Loss of section in the misules	1	1	2	2
Widespread dirt on the roof tiles	1	1	2	2
Undocked tiles	3	4	3	36
Missing tiles in the ornamentation of the pediment	1	1	1	1

The determination of priorities in the resolution of pathological manifestations indicated that the most urgent interventions are those related to structural elements, following a vertical and descending logic (from the roof to the floor) such as the compromise of the roof lumber and the infiltration of rainwater in the lining of stucco. Factors that work as causes of other serious damage to the property.

Once the roofing problems have been solved, there must be a certain urgency to intervene in the factors that impact the bearing capacity and the performance of the structural masonry, such as: the slits, cracks and fissures in the walls arising from repression and overload, reestablishing their load capacity and enabling the receiving requesting efforts from coverage after the intervention. In addition, treatment for the removal of soil termites is also a priority whereas it is a pathological agent that quickly degrades the material, compromising the structure and degrading other parts of the building. The infestation of birds is also one of the priorities of intervention whereas these animals are co-responsible, due to their waste, for the degradation of various elements of the building, being necessary its removal so that all the resolutions of the other problems presented are in fact effective and lasting. In addition, it is a matter of public health to combat pigeon infestation that poses risks, due to dry excrement, to the neighboring population due to the possibility of inhaling Cryptococcus spores, culminating in diseases such as cryptococcosis and histoplasmosis.

6 Intervention Guidelines

Considering the above, the building has serious structural problems caused by the foundation's settlement and overload due to the uneven distribution of the roofing efforts. However, according to monitoring through the plaster proof placed at the critical points, the cracks appear to be consolidated. Thus, it is necessary to recover the fissures and cracks promoting their sealing and reinforcement and eliminating water infiltration points by injecting materials with adhesion and resistance compatible with existing ones, such as special Polyurethane resins. The indication of this technique, in detriment of the use of strong lime and sand mortar (materials usually applied in heritage interventions) is justified due to the size of the crack openings, the efficiency of the product, the guarantee of the durability of the intervention, and restoration of the masonry's carrying capacity.

In addition, the application of polyurethane injection is a less incisive alternative to structural repair interventions compared to, for example, stapling - which would be the other technique indicated in this case. However, based on the criterion of minimum intervention and seeking to preserve the materiality of the property as much as possible, the restoration of the clay coating mortar and the parietal paintings are essential.

As a corrective measure for roof overload problems, the wooden crate can be replaced by a metal crate, providing structure relief, longer part durability (with proper predictive maintenance), and lower cost. However, in this case, there are interferences with the materiality and authenticity of the property and, as an alternative to these criteria, one may maintain the wood structure making the necessary grafts and repairs, provided that the supporting masonry has its static capacity reestablished.

In order to avoid new movements in the foundation, it is extremely necessary to recompose the concrete elements that protect the building's foundation, promote surface drainage to rainwater and maintain restrictions on heavy vehicle traffic in the vicinity of the chapel. Finally, it is extremely important for the preservation of the building elements, as it is a matter of public health, to combat the infestation of pigeons present in the building that, besides degrading the property, offers risks to the neighboring population due to dry excrement that make possible the inhalation of Cryptococcus spores, culminating in diseases such as cryptococcosis and histoplasmosis.

7 Conclusions

Based on the historical and pathological surveys of the building, the present work points out that despite the duality between being marked by aspects of materiality and authenticity of the property and promoting more sustainable, efficient and safe interventions, the use of contemporary techniques enables the convergence of the two principles. Through the intervention guidelines indicated here for the restoration of the Santa Teresinha Chapel, we seek to ensure that the security of the building regarding its structural behavior, as well as the re-establishment of the reading of its elements are possible through the use of techniques and pre-existing materials that are economically viable in terms of execution and maintenance, and that enable the effectiveness of damage mitigation, without the reappearance of the same pathological manifestations in the short and medium terms, even though these materials are distinct from the original ones. In addition, considering sustainability in its social dimension, the restoration of the building represents the resumption of community values of identity, belonging and collective memory in relation to the property, that had been compromised due to the state of abandonment it was left. Thus, it is possible to resume its role as a symbol of the history of the neighborhood and as an important element in the celebrations that mark the collective experience of religiosity not only for the community, but for the city of Juiz de Fora as a whole.

References

Almås, A.-J., Lisø, K., Hygen, H., Flyen, C., et al.: An approach to impact assessments of buildings in a changing climate. Build. Res. Inf. **39**(3), 227–238 (2011)

Araújo, G.M.: Os valores do patrimônio cultural. Uma contribuição teórica para as políticas de conservação.*Arquitextos*, São Paulo, ano 13, n. 145.00, Vitruvius, Jun 2012. https://www.vitruvius.com.br/revistas/read/arquitextos/13.145/4387. Acessed 20 Mar 2020

Aurelio, C.R., Scalabrini, M.V.: Patrimônio e cidade. "Sobrevivências" do passado em Ribeirão Pires. *Arquitextos*, São Paulo, ano 04, n. 048.07, Vitruvius, maio 2004. http://www.vitruvius.com.br/revistas/read/arquitextos/04.048/587. Acessed 23 Mar 2020

Baracho, A.S.B.: ***Patrimônio sustentável:****reflexões sobre as melhores práticas anglo-saxônicas*. 2013. Dissertação (Dissertação em Ambiente Construído e Patrimônio Sustentável) – UFMG, Belo Horizonte, p. 27 (2013)

Barbosa, M.T.G., Silva, B.C.P., Oliveira, H.M., Fidelis, P.A.: Importância da sustentabilidade na restauração do patrimônio histórico. Estudo de caso: Pontes. *Construindo*, Belo Horizonte, vol. 10, pp. 1–9 (2018)

Barranha, H.: Património Cultural: Conceitos e Critérios Fundamentais. Lisboa (2016)

Bauer, E.: Condicionantes das medições termográficas para avaliação de temperatura em fachadas. In: X Simpósio Brasileiro de Tecnologia das Argamassas. Fortaleza, 2013. Anais... SBTA, Fortaleza (2013)

Brandi, C.: Theory of Restoration. Roma: Istituto centrale per il restauro; Firenze: Nardini (2005)

Berg, F., Fuglseth, M.: Life cycle assessment and historic buildings: energy-efficiency refurbishment versus new construction in Norway. J. Architect. Conserv. **24**(2), 152–167 (2018)

Castriota, L.B.: Patrimônio Cultural: Conceitos, Políticas, Instrumentos. Annablume, São Paulo (2009)

Elefante, C.: Historic Preservation and Sustainable Development: Lots To Learn, Lots to Teach. APT Bulletin, [S.l.: s.n] (2005)

Esteves, A.: Álbum do Município de Juiz de Fora. Imprensa Oficial, Belo Horizonte (1915)

Falorca, J.G.F., Lanzinha, J.C.G.: A utilização de drones como ferramenta tecnológica emergente para a inspeção técnica da envolvente de edifícios: revisão e ensaio de campo. In: Congresso Construção 2018 – Reabilitar e construir de forma sustentável. Anais...FEUP, Porto, 21–23 Nov (2018)

Fazolatto, D.: Juiz de Fora: imagens do passado. 4 edn. Panorama, Juiz de Fora (2007)

Hammond, G., Jones, C.: Inventory of carbon and Energy (ICE) Version 1,6a. [S. l.: s. n.] (2008)

ICOMOS. Conselho Internacional de Monumentos e sítios. Carta de Veneza, Veneza (1964)

Jalali, S., Torgal, F.P.A.: Sustentabilidade dos Materiais de Construção. [s. l.], Publindustria (2010)

Juiz De Fora. Prefeitura de Juiz de Fora. Processo Nº 4556-97. Tombamento de imóvel à rua Coronel Miranda, s/nº - Capela de Santa Teresinha. Requerente: Comissão Permanente Técnico-cultural. Interessado: Capela de Santa Teresinha. Juiz de Fora, v. 01 (1997)

Khoshnava, S.M., Rostami, R., Valipour, A., Ismail, M., et al.: Rank of green building material criteria based on the three pillars of sustainability using the hybrid multi criteria decision making method. J. Clean. Prod. **173**, 82–99 (2018)

Kohler, N., Hassler, U.: Alternative scenarios for energy conservation in the building stock. Build. Res. Inf. **40**, 401–416 (2012)

Lerma, C., Mas, A., Gil, E., Vercher, J., Peñalver, M.: Pathology of building materials in historic buildings. Relationship between laboratory testing and infrared thermography. Material. Construct **64**(313), e009 (2014)

Lima, E.M., Fernandes, R.T.V., Dantas, S.L.: Quantificação de CO2 emitido decorrente dos materiais empregados na construção de uma residência unifamiliar. In: Congresso Técnico Científico da Engenharia e da Agronomia CONTECC"2018. Maceió, 2018. Anais...CONTECC'2018, Maceió (2018)

Liu, H., Lin, B.: Ecological indicators for green building construction. Ecol. Indicator. **67**, 68–77 (2016)

Morillas, H., Vazquez, P., Maguregui, M., Marcaida, I., et al.: Composition and porosity study of original and restoration materials included in a coastal historical construction. Construct. Build. Mater. **178**, 384–392 (2018)
Norrström, H.: Sustainable and balanced energy efficiency and preservation in our built heritage. Sustainability **5**, 2623–2643 (2013)
Oliveira, S.T.: Ferramentas Para o Aprimoramento da Qualidade, 1st edn. Thomson Pioneira, São Paulo (1995)
Paschoalin, R.F., Barbosa, M.T.G.: Restauração da catedral de Brasília: desafios e conflitos da restauração da arquitetura moderna. Conserv. Patrimón. **18**, 45–53 (2013)
Penteado, F.C., Fabriani, C.B., Franco, L.F.D.R.: Patrimônio cultural, desenvolvimento sustentável E cidadania: O desafio das práticas preservacionistas. Rev. Direit. Cultur. **8**, 14 (2013)
Reddy, B., Jagadish, K.: Embodied energy of common and alternative building materials and technologies. Energy Build. **35**, 129–137 (2003)
Wellington, U. Of. Table of embodied energy coefficients. Centre for Building Performance, [s. n.] (2005)

Ambient Vibration Measurements: A Tool for the Preservation of the Heritage of Byblos City

Nisrine Makhoul[1] (✉) and Philippe Gueguen[2]

[1] Department of Architecture, Built Environment and Construction Engineering, Politecnico di Milano, Piazza Leonardo da Vinci, 32, 20133 Milan, Italy
nisrine.makhoul@polimi.it

[2] The Institute of Earth Sciences, Grenoble – Alpes University, Rue de la Piscine 1381, Grenoble, France

Abstract. Existing masonry buildings are a challenge for engineers. Each construction is endowed with different material, structural, and dynamic characteristics. This variability depends on the workforce expertise, which constructed the building, as well as on the history and aging of the construction, among others. The ambient noise vibration method can be used as a reliable tool providing information about the current state of a masonry building, helping thus, to assess the building's performance and vulnerability at the structural level as well as loss assessment at a city level, among others.

This paper deals more specifically with the masonry buildings in Byblos city. The city is at least 7000 years old, a UNESCO heritage as it is one of the oldest continuously inhabited cities and one of the 100 Resilient Cities Project. To this purpose, a measurement campaign already took place, and ambient noise recordings were acquired for many Byblos buildings typologies. This preliminary study presents the results of the ambient noise measurement campaign and identifies the benefits of the method.

Keywords: Ambient noise vibration · Cultural heritage · Dynamic characteristics · Seismic risk

1 Introduction

Byblos, known by its local name by Jbeil, is a UNESCO heritage site since it is one of the oldest continuously inhabited cities in the world. The coastal city is located around 40 km to the north of Beirut the capital of Lebanon. Byblos is divided into two main zones. The old town, where houses were built in masonry similarly to most of the Mediterranean Sea cities, their age span from 3000 BC to the early 20th century. And the relatively modern city, where reinforced concrete buildings are dominant since the late 20th century. Byblos is located about 40 km far from the Yammouneh fault and around 20 km from the Mount Lebanon Thrust. The two major faults can generate an earthquake of magnitude higher than 7, and thus the city is subjected to moderate-to-high seismic risk. Several

I. Vayas and F. M. Mazzolani (Eds.): PROHITECH 2021, LNCE 209, pp. 1054–1065, 2022.
https://doi.org/10.1007/978-3-030-90788-4_81

earthquakes already damaged the town significantly, we note the ones of 303 A.D., 551 A.D., and 1157 A.D. The characteristics of Byblos seismicity and buildings, as well as resilience characteristics, were detailed in Makhoul et al. (2016). To enhance the city's resilience and mitigate the earthquake risk it confronts, a series of studies were achieved (Makhoul et al. 2016, 2018; Makhoul 2018, 2019; Makhoul and Barakat 2017; Makhoul and Harb 2017; Makhoul and Mikhael 2016). Recently for the same purpose of enhancing the city's resilience, an ambient noise vibration recording campaign was conducted in spring 2018. Measurements were taken for 15 selected masonry buildings as a sample from a total of 205 it embeds (Makhoul et al. 2016). The ambient noise method is used to obtain dynamic characteristics of the building sample, such as the frequency, damping, and spectrum content. Moreover, the horizontal to vertical ratio method, HVNR or H/V, is used to determine and compare natural frequencies of the structure and the soil and thus to identify possible resonance occurrence during earthquakes.

This paper is a preliminary study, and it is limited to the study Byblos masonry buildings sample, which will help develop later the study of the fragility functions of Byblos masonry buildings. Therefore, in Sect. 1, we present Byblos buildings and the selected studied sample. In Sect. 2, we offer dynamic characteristics based on the ambient vibration method and show the identified patterns in the studied sample. In part 3, we discuss the possible resonance of those buildings using the H/V method in the event of a likely future earthquake. In Sect. 4, we conclude and present further work.

2 Byblos Buildings and Ambient Noise Vibration Measurements Campaign

In Makhoul et al. (2016), the total number of Byblos buildings with part of its suburb Blat was investigated. We have identified 829 buildings in total, including the 68 buildings located in Blat. The number of concrete structures is 623, and the number of masonry structures is 205. The number of stories ranged from 1 to 11. Nevertheless, few concrete constructions have eight stories and above and few masonry constructions have four stories and more. More specifically, one masonry building has five stories, two have four stories, and seven have three stories.

Table 1. Total available and studied masonry building sample

Stories number	1	2	3
Byblos	97	84	19
Sample	10	4	1

The ambient noise measurements were carried for 15 selected masonry buildings, as a sample from the total 205. Precisely, we have conducted measures for 10 one-story masonry buildings from the total 97, 4 two-stories masonry buildings from the total 84, and one three-stories masonry building from the total 19 buildings. The studied masonry buildings sample out of the entire Byblos available buildings is shown in Table 1. The ambient noise vibration measurement campaign for concrete buildings will be detailed in another paper.

The ambient vibration method (ANV) is used to determine the dynamic characteristics of the building structure and the horizontal to vertical ratio method, H/V, or HVNR of soil beneath. The measurements are carried out using a Lennartz seismometer, LE-3D, 5 s (Fig. 1b), and the Minishark recorder (Fig. 1a). The building ambient vibration recording of 10 min duration was conducted on the roof of the building for the sampling frequency of 200 Hz, while the soil recording was of 20 min in the vicinity of the buildings. The results of the method are detailed in Sect. 2.

(a) (b)

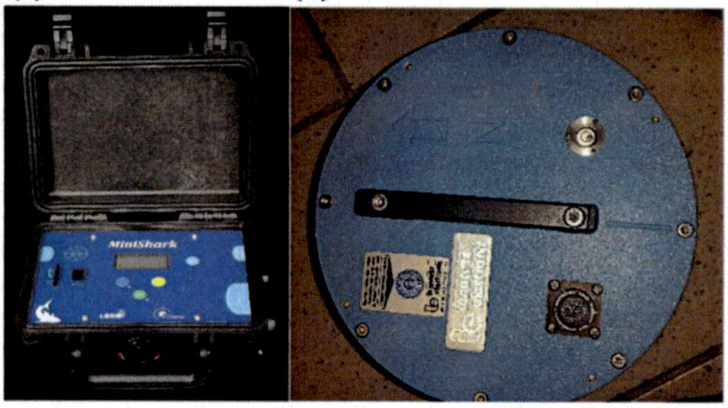

Fig. 1. Tools used for ANV (a) the recorder left figure and (b) the seismometer right figure

3 Dynamic Characteristics of the Municipality Building

The aim of the ambient noise vibration method (ANV) was to obtain, in a first step, for each building the frequencies, damping, spectrum content, and particle motion. In a second step, we aim to identify a pattern for the available masonry typologies in Byblos, in order to select and identify a representative prototype. In a third step, fragility functions will be built and represented in further publications. In this study, the Geopsy software is used for signal processing in the 0.1–50 Hz range.

3.1 Horizontal Spectrum and Frequencies

The frequencies for the longitudinal and transversal directions are identified by automatic picking of the highest frequency peak using the method of Fast Fourier Transform, FFT curves.

For the series of 10 one-story masonry buildings, we have identified the following frequencies: the fundamental frequency is whether 13 Hz for few times, otherwise it varies more frequently between 8–10 Hz. Moreover, whenever the 13 Hz is not a fundamental frequency, then it was often identified among other modes. Among different frequencies, we have repeatedly detected the frequency of 4 Hz and always noticed the 1.5 Hz.

For the four two-stories masonry buildings, we have identified the following frequencies: the fundamental is mainly 14 Hz. Otherwise, it varies between 8–9 Hz. Among other frequencies, we have always found the frequency ranging between 3–4 Hz and the one of 1.5 Hz regularly available. Moreover, whenever the fundamental 14 Hz is recognized, the 8–9 Hz is available among other modes and vice versa.

For the one 3-stories masonry building, the frequency is 8.5 Hz in the longitudinal direction and 7.2 Hz in the transversal direction. The frequencies of 4 Hz and 1.5 Hz are identified in both directions as well as the 13 Hz in the transversal direction.

Therefore, we may conclude that mainly all masonry buildings from one to 3 stories have expressed the same frequency ranges for the fundamental mode (13–14 Hz or 8–9 Hz) as well as for other higher modes (4 Hz and 1.5 Hz).

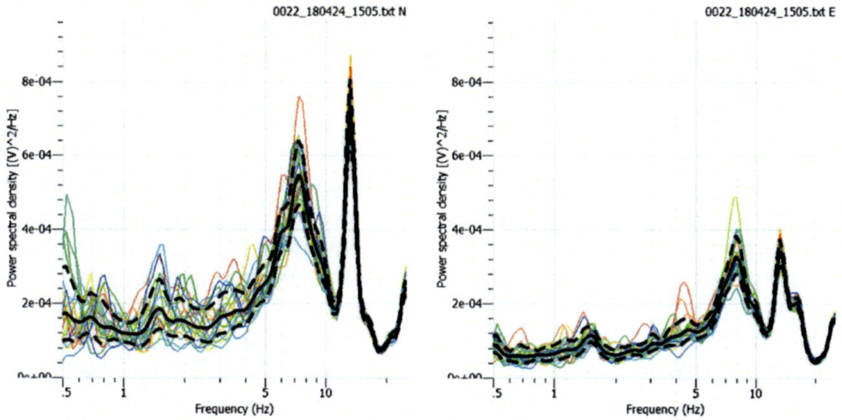

Fig. 2. Horizontal spectrum in the North and East direction for the one-story building

The chosen one-story building has a fundamental frequency of 13.285 Hz in both North and East directions. Then the frequencies of 1.5 Hz and 7.5 Hz are identified for other modes in both North and East directions. Finally, for the East direction, a small peak is detected for 16 Hz. All peaks 13.285 Hz, 7.5 Hz, and 1.5 Hz are identified in both directions, combined peaks; thus, the movement is rotation. The results of the horizontal spectrum for the one-story building are shown in Fig. 2.

The chosen two-stories building has a fundamental frequency of 14.8 Hz in both North and East directions. Then, the frequency of 1.5 Hz is identified for other modes in both North and East directions. For the East direction, the peaks of 3.383 Hz and 23.04 Hz are identified, and for the North direction, the peaks of 4.66 Hz, 7.31 Hz, and 19.34 Hz are found. All peaks 14.8 Hz and 1.5 Hz are identified in both directions, combined peaks; thus, the movement is rotation. The results of the horizontal spectrum for the two-story building are shown in Fig. 3.

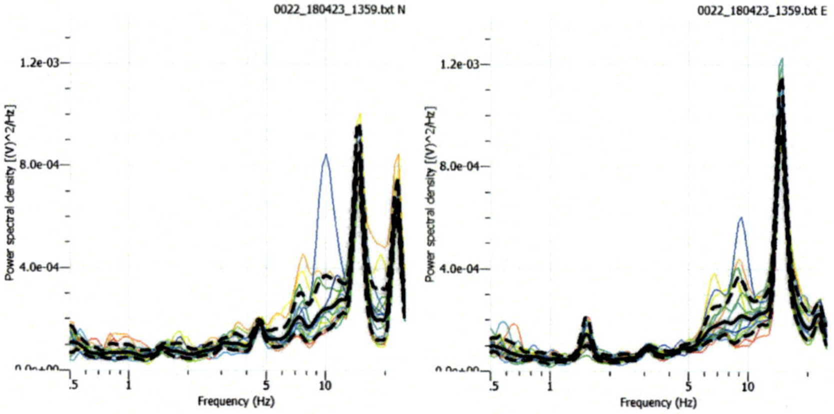

Fig. 3. Horizontal spectrum in the North and East direction for the two-story building

3.2 Damping

The random decrement method is followed to extract the damping utilizing the ambient noise measurements. It consists of decomposing the response of the oscillator into natural and forced responses. The obtained graphs include two curves (Figs. 4 and 5), where the mean is represented in red and mean of the random decrement ± one standard deviation is in black.

For the one-story building, for the fundamental frequency of 13 Hz in both North and East directions, we obtain that there is slight damping (0.2%); thus, the structure does not dissipate great energy while shaking in its fundamental period. On the other hand, for the different frequency modes, the energy dissipation is more significant. For the frequencies of 7.5 Hz in the North the damping is about 3.8% and in East directions, it is about 6.2%. For the 1.5 Hz, the damping is quite significant in the North direction (9.5%), and in the East direction (7.2%). The results of the horizontal damping for the one-story building are shown in Fig. 4.

For the two-story building, for the fundamental frequency of 14.75 Hz in both North and East directions, we obtain that there is quite no damping (0.09%); thus, the structure does not dissipate energy while shaking in its fundamental period. On the other hand, for the 1.5 Hz, the damping is quite essential in the North direction (14.3%), and thus the structure dissipates high energy while shaking in this frequency mode and this direction, while it dissipates less in the East direction (0.33%). For the frequencies of 4.65 Hz in North direction, the damping is 0.49% (very slight), while for the East direction, the damping is about 4.28%, thus within the normal range. The results of the horizontal damping for the two-story building are shown in Fig. 5.

Moreover, in the North direction, we note that for the frequency of 19 Hz, we obtain a damping value of 19.65% and for the 7.3 Hz a damping value of 14.99% thus in this direction the structure is considerably dissipating energy. For the East direction, for the frequency of 23 Hz, we obtain a damping value of 23.19%; thus, the structure is significantly dissipating energy as well.

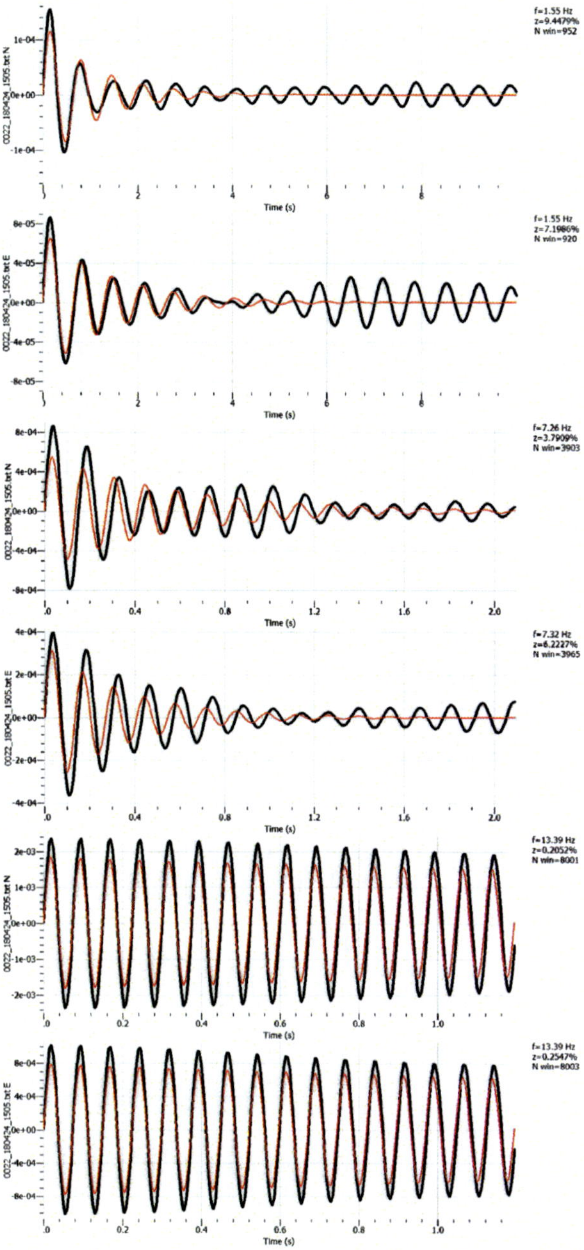

Fig. 4. Damping for the frequencies in the North and East directions for the one-story building

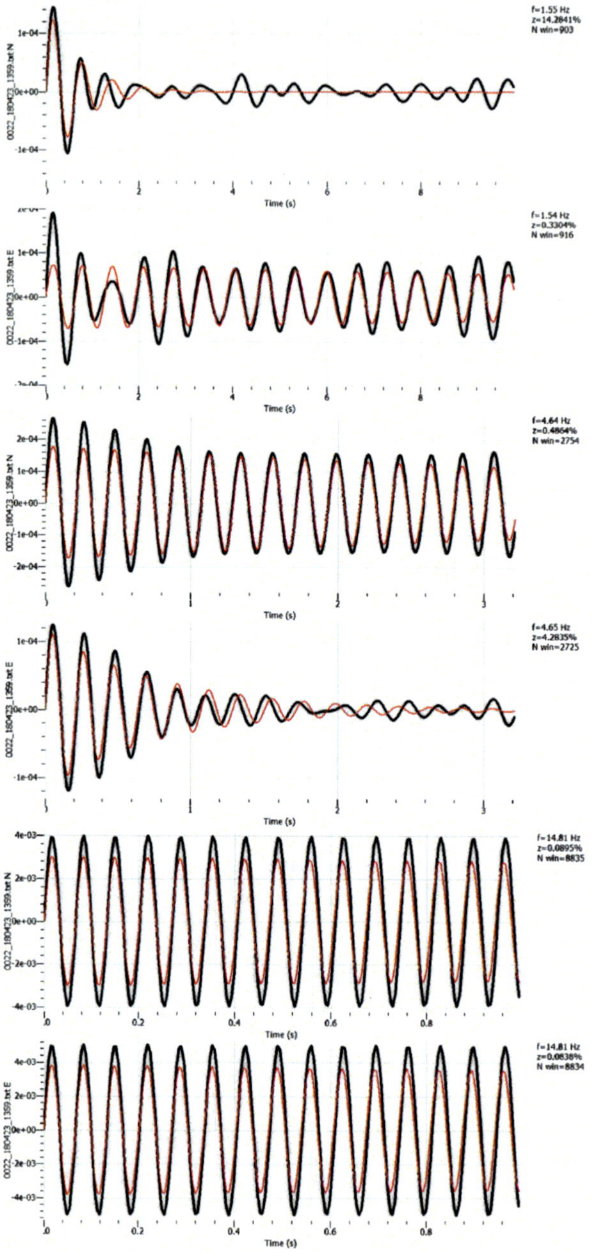

Fig. 5. Damping for the frequencies in the North and East direction for the two-stories building

Finally, we can note that for the two types of building one- and two-stories, we can identify similarities: for the fundamental frequency in the range of 13–14 Hz, there is no energy dissipation. While for the similarly identified frequency of 1.5 Hz, there is a significant energy dissipation in the North direction, while it is quite reasonable in the East direction.

3.3 Horizontal Spectrum Rotation and Energy Repartition

The energy repartition of the structure in the horizontal plane is obtained by the spectrum content. It allows representing the Fourier amplitude spectra in the horizontal plane, with respect to the azimuth or direction shown from 0° to 180°. For the one-story building, it is noted that for the frequency of 13.39 Hz, the movement is longitudinal. Moreover, we indicate that mainly energy is accumulated for this fundamental frequency. For the two-story building, we point that for the frequency of 14.75 Hz, the movement is transversal but also longitudinal. Moreover, we indicate that mainly energy is accumulated for this fundamental frequency (Figs. 6 and 7).

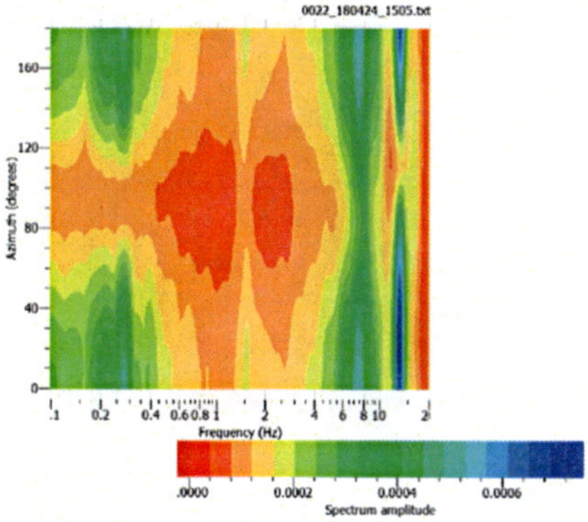

Fig. 6. Horizontal energy repartition with respect to the azimuth for the one-story building

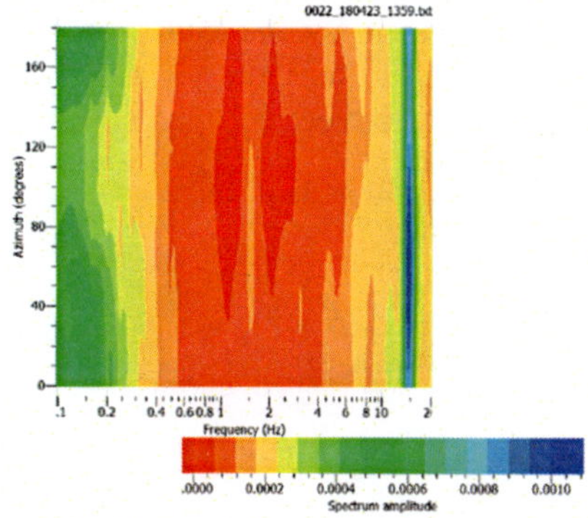

Fig. 7. Horizontal energy repartition with respect to the azimuth for the two-story building

3.4 Horizontal Particle Motion

The analysis of the horizontal particle motion confirms the results of the spectrum content. For the one-story building, it is deduced that for the fundamental frequency of 13.39 Hz, the movement is combined. Nevertheless, it is mainly along the longitudinal and leans towards the transversal direction, as shown in Fig. 8c.

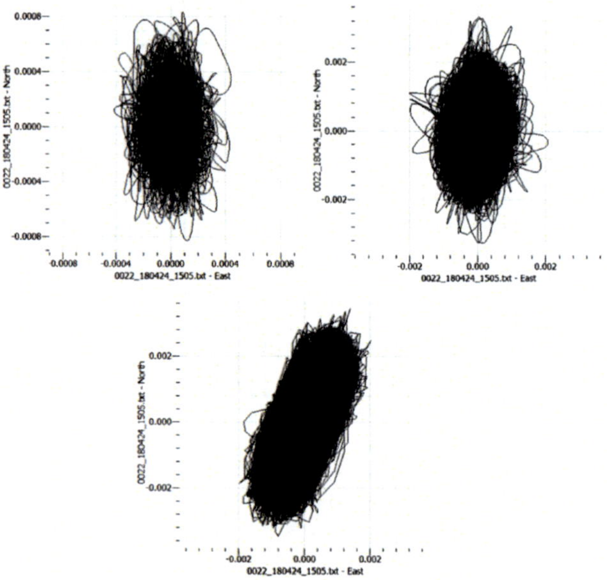

Fig. 8. Horizontal Particle Motion for the one-story building for a) 1.5 Hz; b) 7.5 Hz; c) 13.39 Hz

This result agrees with the horizontal Spectrum content in Sect. 3.1 and the horizontal spectrum rotate in Sect. 3.3. For the frequencies of 1.5 Hz and 7.5 Hz, the movement is combined (Fig. 8a and b).

For the two-story building, it is deduced that a fundamental frequency of 14.75 Hz, as well as for the 1.5 Hz, the movement is combined, as shown in Fig. 8a and d. For the frequencies of 3.12 Hz and 4.65 Hz, the movement is slightly combined (Fig. 8b and c). This result agrees with the horizontal Spectrum content in Sect. 3.1 and the horizontal spectrum rotate in Sect. 3.3 (Fig. 9).

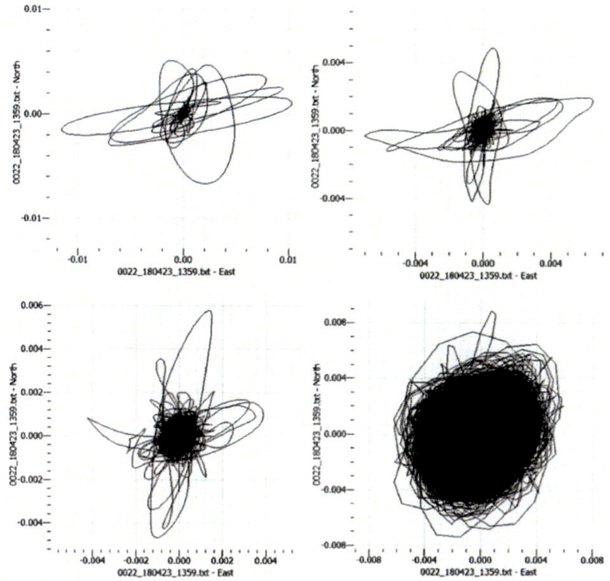

Fig. 9. Horizontal Particle Motion for the two-stories building for a) 1.5 Hz; b) 3.12 Hz; d) 4.65 Hz; d) 14.5 Hz

4 Geophysical Assessment and Resonance

The horizontal to vertical ratio method, HVNR, or H/V method consists of evaluating the ratio between the Fourier amplitude spectra of the horizontal component to the vertical component of the ambient noise vibrations recorded at one single station. It was initiated in Nogoshi and Igarashi (1971), expanded, propagated, and detailed in Konno and Ohmachi (1998).

In Makhoul and Harb (2017), the H/V for the soil in the nearby site of the Municipality building is obtained, and two peaks are observed: at 6.00 Hz and 10.22 Hz. Those same frequencies were obtained for many ambient noise vibrations measurements in different locations of the Byblos city. Therefore, we can note that in case the masonry buildings will be shaking in the same frequencies, it can go into resonance with the soil. Nevertheless, in all the series of 15 studied buildings, the frequency of 6 Hz was rarely noticed. However, the fundamental modes tend to have higher frequencies of 8–10 Hz and 13–14 Hz. Therefore, for the buildings with fundamental mode frequencies specifically of 8–10 Hz, there is a high probability that it will go in resonance with the soil (Fig. 10).

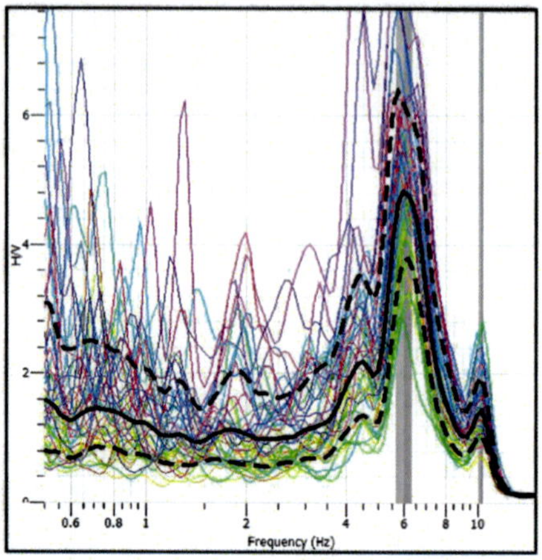

Fig. 10. H/V spectrum with respect to the frequency in Hz for the soil in the vicinity of the municipality building (Courtesy to Makhoul and Harb 2017)

5 Conclusions, Remarks, and Perspectives

In the present study, to enhance the Byblos resilience, an ambient noise vibration recording campaign was conducted specifically for the city's masonry buildings, which are considered as cultural heritage. Measurements were taken for a sample of 15 selected masonry buildings. The ambient noise method is used to obtain dynamic characteristics of the masonry buildings sample, the frequencies, damping, and spectrum content, among others. Moreover, the H/V method is used to determine and compare natural frequencies of the structure and the soil and thus to identify possible resonance occurrence during earthquakes.

For the chosen sample, we were able to identify a pattern. Essentially, most of the buildings have the same fundamental frequencies as well as many other similar modes frequencies are identified for this sample. Moreover, the structures show a somewhat identical dissipation pattern for the identified frequencies, as well as quite similar movements for the spotted modes.

Those results allow us in further steps: First, to broaden the measurement campaign to include a more significant number of masonry buildings to confirm further the theory; second, to identify several representative structures and develop detailed finite element models to study the response of the masonry buildings in terms of dynamic characteristics and then to develop accurate fragility functions for Byblos masonry buildings; and third to embed those fragilities in a robust loss assessment and resilience tool to study possible future losses and recoveries, once a likely earthquake occurs.

Acknowledgments. The work presented in the paper is part of the funded Agence Universitaire de la Francophonie Project (AUF-Patrimoine), to which we show gratitude. We show gratitude to Politecnico di Milano. Moreover, this work is willingly considered as part of the initiative of the Observatoire Libano-Français de l'Environnement (O-LIFE), thus this paper is allocated the number SA 2021 - 8.

References

Konno, K., Ohmachi, T.: Ground-motion characteristics estimated from spectral ratio between horizontal and vertical components of microtremor. Bull. Seismol. Soc. Am. **88**(1), 228–241 (1998)

Nogoshi, M., Igarashi, T.: On the amplitude characteristics of microtremor. J. Seismolog. Soc. Jpn **24**, 26–40 (1971)

Makhoul, N., Navarro, C., Lee, J., Abi-Younes, A.: Assessment of seismic damage to buildings in resilient Byblos city. Int. J. Disast. Risk Reduct. **18**, 12–22 (2016)

Makhoul, N., Harb, J.: Dynamic properties of Byblos municipality building with soil-structure interaction using geophysical methods. Procedia Eng. **199**, 206–211 (2017)

Makhoul, N., Navarro, C., Lee, J.: Earthquake damage estimations of Byblos potable water network. Nat. Hazards **93**(2), 627–659 (2018)

Makhoul, N., Mikhael, E.: Earthquake damage estimations of Byblos potable water network. In: Proceedings of the 19th IABSE Congress Stockholm: Challenges in Design and Construction of an Innovative and Sustainable Built Environment, Stockholm, Sweden, 21–23 Sep 2016 (2016)

Makhoul, N.: Performance-based assessment of Nahr Al-Fidar bridge. In: IABSE Symposium 2019 Guimarães, Towards a Resilient Built Environment - Risk and Asset Management, Guimarães, Portugal, 27–29 Mar 2019 (2019)

Makhoul, N.: Seismic loss estimation of Byblos city: a contribution to the "100 Resilient Cities" strategy. In: 16ECEE - Thessaloniki 16th European Conference on Earthquake Engineering, Thessaloniki, Greece, 18–21 Jun 2018 (2018)

Makhoul, N., Barakat, M.: Earthquake damage estimations of buildings in a typical north Lebanon village. In: 16th World Conference on Earthquake, 16WCEE 2017, Santiago, Chile, 9–13 Jan 2017 (2017)

The Role of Metallic Connections in the Global Behavior of Timber Frame Structural System of Tabique Walls

José Padrão[1,3](✉), Jorge Pinto[2], António Arede[3], and João Guedes[3]

[1] Department of Civil Engineering, School of Technology and Management, Polytechnic Institute of Viseu, University of Porto Campus Politécnico de Repeses, 3504-510 Viseu, Portugal
jpadrao@estgv.ipv.pt
[2] ECT - School of Science and Technology, University of Trás-os-Montes e Alto Douro, UTAD Quinta de Prados, 5000-801 Vila Real, Portugal
[3] CONSTRUCT-LESE, Department of Civil Engineering, Faculty of Engineering (FEUP), University of Porto, Rua Dr. Roberto Frias s/n, 4200-465 Porto, Portugal

Abstract. In order to perform a correct and efficient rehabilitation of old building, it is required to know the materials used, their main properties as well as the construction process. Tabique Walls (TW) are primary building components of many constructions in the historic center of Viseu and their function, in addition to being used as partitioning elements (the most common in Portugal), is to be part of the main structure of buildings. In this sense it is essential to understand how the TW interconnect with each other and with the other main structural elements of buildings, namely the floors and the stone masonry walls. The important role of metallic connectors in the realization of these bonds is emphasized. This paper presents the research carried out in several buildings of the historic center of Viseu. Several constructive solutions found are detailed, associating the shapes and dimensions of the connectors to their function, and results are presented out of a primary campaign of laboratory tests carried out to define their mechanical properties. The knowledge obtained allows a better understanding of the overall structural behavior of these buildings and the role of the TW in that behavior.

Keywords: Tabique walls · Metallic connections · Wood · Light-frame · Constructive details

1 Introduction

The vertical wooden boards are the main resisting member of Tabique Walls (TW). However, the way this type of construction is made gives metallic connections an important role. The overall structural behavior of the building depends significantly on the way the loads are conducted along all structural components. The connection between these

I. Vayas and F. M. Mazzolani (Eds.): PROHITECH 2021, LNCE 209, pp. 1066–1080, 2022.
https://doi.org/10.1007/978-3-030-90788-4_82

elements is generally ensured by metal connectors with different shapes and geometries, depending on their function.

Some of the major pathologies in constructions built with TW technique are related to the poor connection between their constituent elements and to the way these connections were built. So, it's essential to understand how the TW interconnect with each other and with the other main structural elements of buildings, namely the floors and the stone masonry walls.

In all these cases, metallic connectors (or the lack of them) play an important role. Therefore, the selection of correct and effective rehabilitation methods significantly depends on the increased knowledge about these important elements of the TW.

This research follows on from work already carried out to know the mechanical properties of constituent materials of TW, as described in Pinto R. and Padrão (2014), Pinto J. (2014) and Pinto J., et al. (2014). The specific objectives of the present study were to provide information about the construction process associated with the presence of metallic connectors and their elementary mechanical properties. In both cases, the *in situ* inspection carried out in several buildings in the historic center of Viseu was essential, as well as the collection of samples for subsequent laboratory tests.

2 The Tabique Walls – A Brief Summary

This study is based on the existing buildings in the historic center of Viseu (Fig. 1a and 1b). They are usually residential buildings with 2 to 4 floors. They are built using existing materials in the region: the masonry walls are made of granite, the timber (of wood floors, roof and walls) is *Pinus pinaster* (softwood) while the filling and coating mortar is made with earth-base material. Most were built between the early nineteenth century and the mid-twentieth century.

(a) (b)

Fig. 1. (a) and (b) Examples of dwellings with TW, Viseu

The use of TW is very common in Portugal. In most cases they appear inside buildings and are intended only for partitioning. In Viseu, in addition to this function, many assume a role of resistance, i.e., they are part of the main structure of the building, together with the stone masonry walls (exterior) and the floors and roofs (solid wood), which is opposing to what is often established, that the TW are considered secondary elements easily replaceable and without structural value. Therefore, following the proposal by the International Council on Monuments and Sites (ICOMOS 2003), an accurate assessment of the main elements that constitute this kind of buildings which are part of the Portuguese vernacular architecture should be performed. Knowing their location, the materials employed, their mechanical properties, etc., as well as the constructive details, is the right way to mitigate the errors of inappropriate and disproportionate rehabilitation interventions.

Existing research on TW is still relatively limited. However, it should be mentioned the work carried out at UTAD (Vila Real, Portugal) mostly based on its geometric and material characterization of buildings located in the northeast of Portugal (Carvalho et al. (2008), Cepeda et al. (2010), Martinho et al. (2010) and Pinto et al. (2010)) and at Polytechnic Institute of Viseu (Padrão and Pinto (2014), Pinto and Padrão (2014) and Padrão et al. (2016)).

In a TW construction technique (Fig. 2a and 2b), the timber boards are the main structural components.

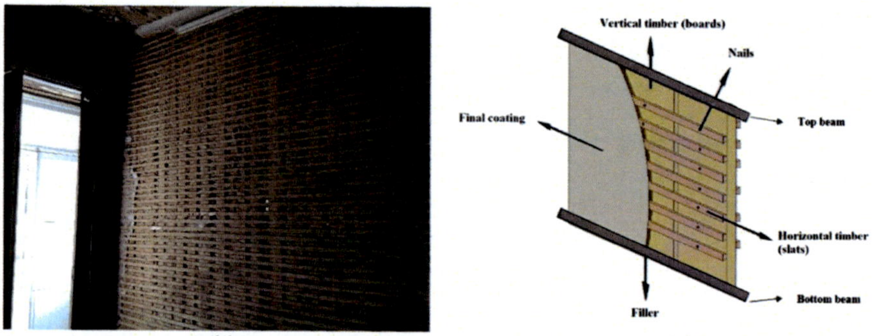

(a) (b)

Fig. 2. (a) Example of TW from case studies, (b) Scheme of the typical constitution of a TW

A traditional timber frame is formed by single layer or double layer with vertical boards and horizontal elements (slats), which are fastened to each other. The slats system joins the vertical boards and improves the adhesion of the finishing layer of lime mortar or earth-based render. Sometimes, the vertical boards are intersected by diagonal wood elements. All gaps are filled in by earth-based mortar. This material is considered a secondary element, having principally a filling and coating function. In fact, the earth mortar plays an important role because, besides providing good thermal and acoustic properties, it also protects the wood structure from the attack of biological agents and fire as Pinto (2014) refers. The vertical boards are attached to solid wood beams (members of the floor system) in the two points (top and bottom) through metal connectors. The tabique building technique is characterized as affordable, sustainable and low technology demanding.

The several research campaigns carried out in old buildings located in the historic center of Viseu allowed a better understanding of the construction process used as well as to draw some conclusions about its structural behavior, as described in the next paragraphs.

(1) There is a clear evolution in the construction system. Indeed, as shown in Fig. 3, in the older TW the vertical boards are more spaced and typically have a square section (background TW in Fig. 3a). The slats have larger width elements and rectangular section and are also more spaced. In more modern buildings the vertical boards are rectangular and generally wider than thick (foreground TW in Fig. 3a). The space between them is smaller. The slats have a trapezoidal section also with smaller spacing. Older buildings also include organic materials of vegetable origin in the mortar as cohesive elements (Fig. 3b). This material was later replaced by lime. Metal connectors also had an evolution. Initially, forged connectors with variable section were used, but newer ones were manufactured with constant section.

(a) (b)

Fig. 3. Evolution in TW construction process **(a)** in wood materials and **(b)** Filling mortar

(2) TW has good durability, as evidenced in many of the buildings still in use (many over one hundred years old) where TW were built. The protection of the less durable elements (wood and metal elements), provided by the filling and coating mortar, has not changed their properties.
(3) There is a great diversity in relation to the materials applied, their dimensions and even constructive details. Pinto and Padrão (2016) explains this variety by the local availability of building materials, the financial capacity of the developer, the builder's technical and empirical knowledge, weather conditions, etc.

3 Metallic Connections

The study of metallic connections as fundamental elements of TW is the main objective of this work. A research is presented concerning the constructive details associated with this construction technique and the results of some laboratory tests performed to characterize its mechanical properties.

On one hand, Viseu belongs to a region of Portugal where the seismic risk is moderately small. On the other hand, the studied buildings are in the city historic center (with narrow streets and large concentration of medium/large buildings) the reason why the wind action on the buildings is low. Therefore, the main actions to which the structures of these buildings are subject are essentially of gravitational origin such as dead and imposed loads. Considering these aspects, it can be stated that the metallic connections present in the TW plays a secondary role in their structural behavior. However, it is essential to ensure a connection between the constituent elements of TW and between these and the remaining structure.

Perhaps, that's why there is very little published literature on the subject, and most refer to the geometric study and some properties of the connectors.

As already mentioned, the success of this type of research involves carrying out exhaustive and complete inspection and diagnostic campaigns. Taking advantage of the rehabilitation works in some buildings with TW, it was possible to observe, measure and detail the presence and function of metallic connectors in these walls. Samples were also taken and used to determine their dimensions and the relationship between them and the function for which they were applied.

3.1 Constructive Details

The metal connectors are used in the TW with the purpose of acting as mechanical connectors between the different wooden elements, that is, in the connection between the vertical boards and the slats, between boards and the beams (floor and roof structural systems), between the TW elements and secondary elements such the lintels and door jambs and baseboards, between different TW, etc. In all these cases the metallic connectors come in the form of nails. In the connection between the TW and the stone masonry walls the metallic connectors are in the form of plates.

This research allowed to observe a wide variety of techniques. Such differences are associated with no criteria concerning the nailing angle or the distance between the nails and the edges of wood elements, for instance.

As already mentioned in the previous section, it is also possible to observe an evolution in the type of nails used, both in their geometric shape as in the type of manufacture

(a) (b)

Fig. 4. Evolution in the nails applied **(a)** in oldest TW and **(b)** in most recent TW as in the applied material. In older buildings it is observed that the nails have a variable rectangular section and made by local forgers (Fig. 4a). Newer ones have a rectangular or constant circular section and are manufactured (Fig. 4b).

Constructive solutions have been observed in which metallic connectors are applied to solve secondary but also important situations, namely for the durability of TW. One way the builders used to improve adhesion between the wood elements and the coating mortar was by nailing small metallic fasteners spaced along the wall without making apparent rule, as shown in Fig. 5a). In Fig. 5b) it can be observed the use of metallic connectors in external TW as fixing coating ceramic elements.

(a) (b)

Fig. 5. Nails application to **(a)** improve adhesion in TW and **(b)** fixing ceramic elements

Concerning mechanical connectors linking several wooden elements, the connection between the vertical boards and the slats (the most common link) is made by nailing one or two connectors, depending on the width of the boards, in an apparent vertical alignment (Fig. 6a). Following this rule, the main connections between wooden elements observed in the inspection campaign aim to provide a mechanical joint behavior of the connected elements (Fig. 6b). The main difference observed is the geometric dimension of the nails.

(a) (b)

Fig. 6. Nails connecting vertical boards to **(a)** Slats and **(b)** Door jambs

Table 1 aims to associate the dimensions of the connectors with their function. In general, all nails have square cross section (just like is head).

About 96 samples were measured. It's in the recognition of the connections between the main elements of the TW and the remaining main structure of the buildings that the role of the metallic connectors becomes more relevant. In the following, several figures show the main connections observed.

Table 1. Connection type vs. Connectors overall dimensions.

Rua Formosa, Praça D. Duarte, Travessa Escadinhas Buildings	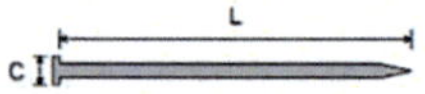 Manufactured with constant section (MSC)		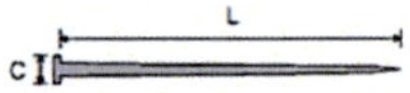 Forged with variable section (FVS)	
Connection Type:	Typology	Overall Dimensions - Average Values (mm)		
		Lenght (L)	Head Dimension (C)	Cross Section (D)
In wood elements to improve adhesion with the coating mortar	MSC	22,0	2,0	1,0
	MSC	24,4	2,2	1,0
In wood elements to fix coating ceramic elements	MSC	88,3	12,2	4,8
	MSC	64,1	11,1	3,6
Vertical Boards / Slats	MSC	47,2	5,8	2,4
Vertical Boards / Door lintels, baseboards,etc. (thicker secondary elements)	MSC	83,0	6,7	3,4
Vertical Boards / Door lintels, baseboards,etc. (thinner secondary elements)	MSC	67,7	6,5	3,0
Vertical Boards / Door jambs	MSC	83,0	6,2	3,7
Vertical Boards / beams (floor structural systems)	MSC	112,0	7,6	4,1
	FVS	94,7	6,0	5,4
	FVS	63,2	6,9	3,5
Between perpendicular TW	MSC	152,2	10,4	5,4
	FVS	177,5	17,6	7,3

(a) (b)

Fig. 7. Connection between TW and lower beams **(a)** in stone masonry walls and **(b)** between TW vertical boards

When a TW starts on stone masonry walls a solid wood beam is normally placed, slightly wider than the vertical boards of the TW which are fixed through metal connectors (Fig. 7a). Where TW develops over several floors in the same alignment, a solid wood beam is placed at the floor transition (Fig. 7b) and the vertical boards are nailed to this beam. In both cases, the nailing is performed diagonally, in a direction almost parallel to the vertical boards.

In the connection between the TW main boards and the upper beams (belonging to the floor construction system), the most common are those that have a continuous (centered) groove along the entire length of the wall or an individual groove (one for each vertical board) as shown in Fig. 8a) and 8b). These grooves have a width slightly larger than the thickness of the boards (to prevent any movement) and are fastened by nailing (one or two nails per board, depending on their width) and placed perpendicular to the boards. Examples were also found where the beams had no groove, that is, the beam had a rectangular section (Fig. 8c) or where the beams had a lateral groove to allow a better connection between the 2 elements (Fig. 8d). In the first case the nailing is performed diagonally and in the second it is perpendicular to the board direction.

a) b) c) d)

Fig. 8. Connection between TW and upper beams **(a)** with continuous groove **(b)** individual groove **(c)** with rectangular section and **(d)** with lateral groove

The constructive details used in the connections between TW are also very varied. In some cases, the walls had no connecting element, while large connectors (see Table 1) were used in other cases. The connection was made between the vertical boards, horizontally placing three to four nails divided by the total height of the wall (spacings between

0.5 and 1 m). The use of metal plates (rectangular or L-shaped) has been also observed, as shown in Fig. 9a and 9b). These plates were drilled to allow nailing of metallic connectors to the vertical boards (usually with three to four holes). The thickness and length of the plates were variable, with the most common dimensions being: thickness between 8 to 10 mm and length of 150 mm.

In the connection between TW and stone masonry walls there are two types of most common situations. First, with no effective metal connection where the slats extend beyond the vertical boards against stones or joint spaces between stones. The walls are then lined with mortar coating. The metallic plates were placed on each side of the TW, usually with rectangular shape. The plates were fixed to both the masonry wall and the TW. On the masonry wall, it was fixed with lead and, in the connection zone, the plate had grooves to increase the adhesion. On the TW side they had a zone with two to three holes to allow the connection with nails (Fig. 10a and 10b). There were usually one or two plates per link. In the examined cases, all the metallic elements had been forged.

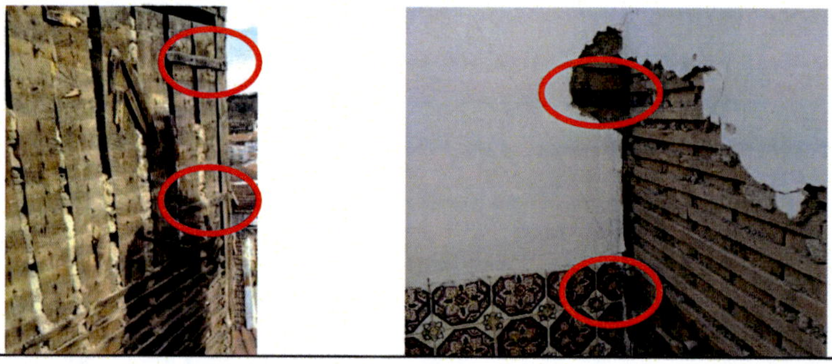

Fig. 9. Connection between TW with metal plates with **(a)** Rectangular shape and **(b)** L-shape

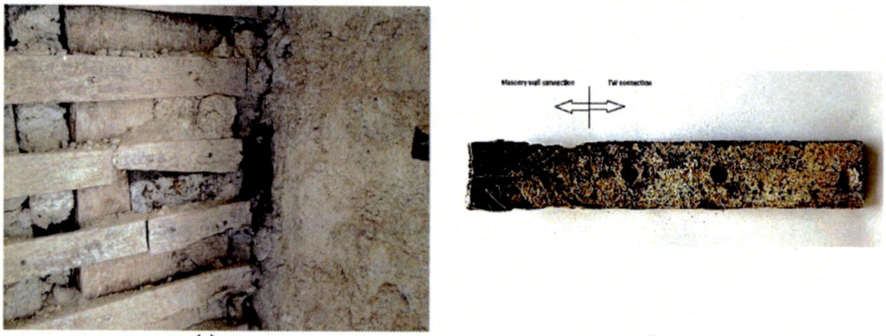

(a) (b)

Fig. 10. Connection between TW with stone masonry walls **(a)** *in situ* sample and **(b)** plate scheme

As an example, Table 2 presents the results obtained during the assessment of a building in the historic centre of Viseu. Twelve TW bonds were analysed. Three had two connectors per wall and the remaining had only one. Data was collected from 15 metallic plates. As in other inspections, the geometric dimensions of metallic plates are constant (with coefficients of variation (COV) of less than 10%). On the contrary, there is a great variability in the position of the connectors since the COV of the relative distance between the connectors and the lower floor is greater than 35%.

Once again, there is great variability and uncertainty concerning the adopted constructive solutions. Even though it is not the objective of this study, the inspection campaign showed that the use of diagonal boards inserted in the same layer of the vertical boards (see Sect. 2) was an alternative to the presence of metal connectors linking the walls.

Table 2. Connection type vs. Connectors overall dimensions.

Rua D. Duarte / Rua Dr. Luis Ferreira Building

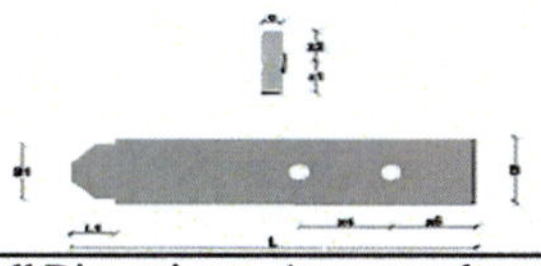

Plate geometry	Number of samples	Overall Dimensions – Average values (mm) Length (L)	Width (B)	Thickness (e)	Drilling distance (x4)	Edge distance (x5)
Speciments	15	163,10	37,10	6,40	52,10	21,20

3.2 Testing Details

The assessment described above allowed realizing that metallic connections are part of the construction process associated with the construction of TW and that they have an important influence on the way the various elements interconnect and on the overall structural behavior. Since the knowledge on the mechanical properties of metallic connectors is scarce, some laboratory tests were performed with the *in situ* collected samples. In this study the results obtained from the tensile test, the Vickers microhardness test and the microstructural analysis of some samples are presented. All tests were performed in the laboratories of the Department of Mechanical Engineering and Industrial Management of the Polytechnic Institute of Viseu.

The metallic nails and plates were carefully removed from the boards of TW structures without applying bending stress. Slight surface corrosion was present on all nails and plates tested. There was no significant loss of section due to corrosion.

The tensile test was performed according to NP EN 10002–1 (1990). Larger nails were selected to ensure the required dimensions for the good performance of the test (those used in connections between vertical boards and beams and between perpendicular TW, Table 1). Metallic plates were also prepared for the same purpose. Samples were extracted and prepared for the purpose (Table 2). It should be noted that fixing the samples (especially the nails) to the test straps was difficult to achieve. Specimens behavior results from tensile tests with nails and metallic plates are shown in Figs. 11 and 12, respectively.

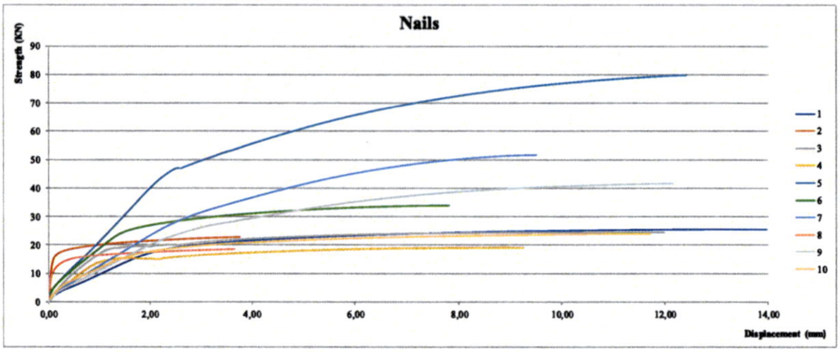

Fig. 11. Results of tensile tests with nails

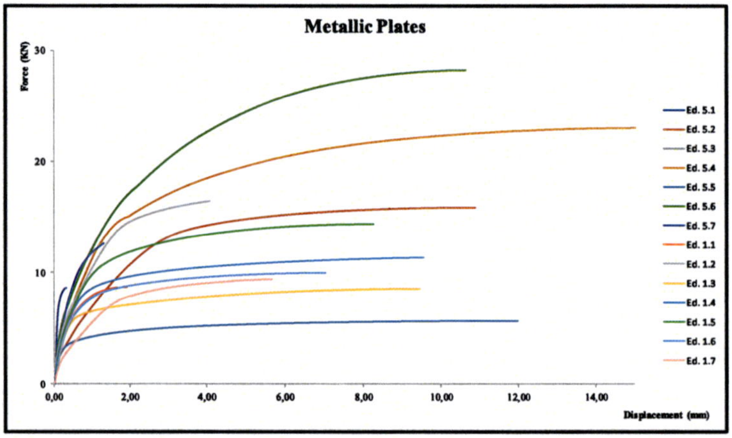

Fig. 12. Results of tensile tests with metallic plates

In Table 3, for each tensile test performed, the following characteristics are given:

- Fmax, the value of the maximum force and its corresponding displace-ment, Umax;
- Fmax, the corresponding maximum tensile stress.

By analysing the results, it can be concluded that the mean values of the maximum tensile stress is 11% higher in nail samples. However, the variation found cannot be considered high because these are materials with a manual manufacturing process without quality control.

Table 3. Results from tensile tests

Speciment	F_{max} [kN]	U_{max} [mm]	f_{max} [MPa]
Nails			
1	25,73	13,96	382,90
2	23,05	3,77	374,86
3	24,46	11,40	331,75
4	19,14	8,96	322,23
5	79,92	12,40	467,35
6	33,96	8,00	393,05
7	51,54	9,56	526,18
8	18,71	3,66	359,72
9	41,72	12,76	447,81
10	24,07	11,71	383,53
Mean			398,94
COV (%)			15,92%
Plates			
Ed. 5.1	8,63	0,31	313,85
Ed. 5.2	15,87	10,86	387,57
Ed. 5.3	8,61	6,56	337,57
Ed. 5.4	23,07	15,80	339,02
Ed. 5.5	5,68	11,96	354,86
Ed. 5.6	28,24	10,63	464,50
Ed. 5.7	12,67	1,29	335,42
Ed. 1.1	8,69	1,90	365,85
Ed. 1.2	16,46	4,06	351,18
Ed. 1.3	8,53	9,43	333,47
Ed. 1.4	11,37	9,50	389,78
Ed. 1.5	14,37	8,26	362,23
Ed.1.6	9,98	7,03	348,80
Ed. 1.7	9,43	5,66	348,67
Mean			359,48
COV (%)			10,16%

COV: Coefficient of variation

The Vickers micro-hardness test was performed according to NP 711–1 (1968). The samples were obtained from nails taken from the inspected buildings. The test surface was prepared to make it smooth, flat and free of oxides. Twelve samples were tested by measuring 5 distinct points per sample. Table 4 presents the main results obtained.

A large dispersion of results is observed, with values ranging from 87 to 228. This variability may be associated with the manufacturing process of the metal connectors mentioned above.

Table 4. Vickers Hardness Number (HV10).

Specimen	A	B	C	D	E	F	G	H	I	J	K	L
HV10	134	164	138	113	133	150	87	133	136	138	131	228

The test was performed in sections with and without inclusions. Perhaps this fact may explain the difference between the obtained values.

The microstructural analysis allowed to characterize the metallic alloy that constitutes the studied connectors. For this purpose, the nails were cut transversely and longitudinally, obtaining cubic samples with about 10 mm of edge (Fig. 13a). Once prepared according to NP 1467 (1977), the samples were observed and analysed under a reflection metallographic optical microscope at 5 to 50 times magnification. The observed surfaces were similar regarding grain size. It was also observed the presence of two phases, ferrite and perlite, which allows concluding to be in presence of a steel. High material heterogeneity was also observed with the observation of darker areas (inclusions) that must have been created in the forging process (Fig. 13b).

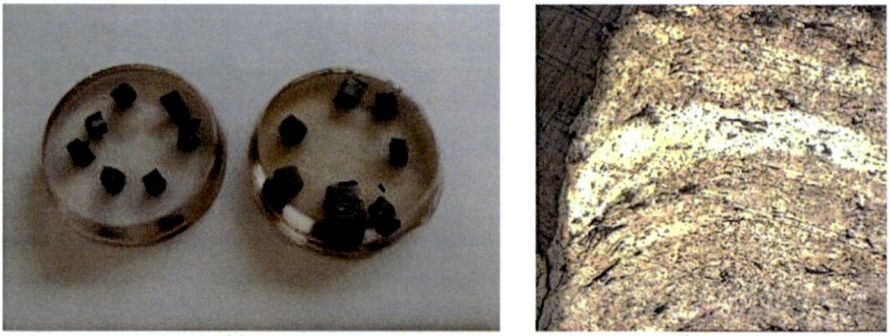

(a) (b)

Fig. 13. Microstructural analysis **(a)** preparation of test samples and **(b)** surface of a nail sample

4 Conclusions

Through an exhaustive assessment campaign on old buildings (more than one century old) located in the historic centre of Viseu, several situations were evaluated in which metal connectors associated with the TW construction technique are present. These buildings are considered to have representative construction solutions of TW buildings in the Viseu region. In each case observed, there is a great diversity of constructive solutions which seems to show that the construction in tabique depends largely on the builder's level of knowledge and his access to the applied materials. There is no doubt that in the Viseu region the TW are an integral part of the building's main structure which sets them apart from other regions of Portugal. Wood and metal elements, when properly protected by filling and coating mortar, have been found to be in a good state of conservation.

The role of metallic connectors is crucial, first to join all TW wood elements and second to link them to the rest of the building structure. For each type of connection there is a wide variety of connectors. Depending on the time of construction, the manufacturing process has advanced from forged to manufactured connectors. Due to the manual manufacturing process and without quality control they can have very different properties from each other, as can be seen from the results presented in Sect. 3.2. However, this first level of research into metallic connectors has led to the realization that these elements may and should deserve special care and be an integral part of any conservation and rehabilitation work in this type of building. The results obtained in the characterization of the materials studied here agree with the results obtained in other research carried out concerning this technique but in other regions of the country.

The constructive details and the results of the tests performed here are expected to contribute to the construction comprehensive databases. Research will continue with further material characterization testing and, more importantly, with further in situ and laboratory testing to better understand the overall behavior of TW (Padrão et al. 2019). Only in this way can it be rehabilitated more effectively, perhaps less costly and, thereby, mitigated the risk of losing (for lack of maintenance) unique buildings that are part of the vernacular heritage of Portugal and Europe.

Acknowledgements. This work was financially supported by: Base Funding - UIDB/04708/2020 and Programmatic Funding - UIDP/04708/2020 of the CONSTRUCT - Instituto de I&D em Estruturas e Construções - funded by national funds through the FCT/MCTES (PIDDAC).

References

Carvalho J., Pinto T., Varum H., Jesus A., Lousada J., Morais, J.: Construções em Tabique na região de Trás-os-Montes e Alto Douro. In: Varum, H., Carvalho, F., Costa, A., Bertini, A., Stepánek, P. (eds.) CINPAR 2008 – 4th International Conference on Structural Defects and Repair. Civil Engineering Department – University of Aveiro, Portugal, 25–28, June 2008. 978–989–95695–3–9

Cepeda A., et al.: Tabique construction in Alto Tâmega. In: Cruz, P. (ed.) Book of abstracts of the 1st International Conference on Structures and Architecture (ICSA2010), pp. 397–398. CRC Press, Taylor & Francis Group. London (2010). 978–0–415–49249–2

ICOMOS - International Scientific Committee for Analysis and Restoration of Structures of Architectural Heritage: Recommendations for the analysis, conservation and structural restoration of architectural heritage. Edited by International Scientific Committee for Analysis and Restoration of Structures of Architectural Heritage of ICOMOS. ICOMOS, Lisboa (2003)

Martinho J., Gonçalves C., Magalhães F., Lousada J., Silva P., Vieira J.: Study of the traditional tabique construction in the Alto Tâmega region. In: Brebbia, C.A. (ed.) The Sustainable World, pp.299–307. WIT Press 2011 (2010). 978–1–84564–504–5, 1746–448X (print), 1743–3541

NP EN 10002–1.: Materiais metálicos: ensaio de tração: parte 1 - método de ensaio (à temperatura ambiente). Instituto Português da Qualidade, Lisboa (1990)

NP 1467.: Aços e Ferros Fundidos – Preparação de provetes para metalografia. Portaria n.° 321/77, junho (1977)

NP 711.: Metais. Ensaio de dureza Vickers. Portaria n.° 23 589 de 6 de setembro de 1968 (1968)

Padrão, J., Pinto, R.: Caracterização Experimental das paredes de tabique do centro histórico de Viseu. In: CINPAR2014 - X Congresso Internacional sobre Patologia y Recuperatión de Estructuras, Santiago do Chile, Chile, junho 4–6 (2014)

Padrão, J., Fernandes, C., Miranda Guedes, J., Pinto, J., Arêde, A.: A contribution for tabique walls characterization. In: Proceedings of the 7th International Conference on Safety and Durability of Structures, ICOSADOS 2016. UTAD, Portugal, 10–12, May 2016

Padrão, J., Arêde, A., Guedes, J.M., Pinto, J.: Experimental characterization of mechanical behaviour of existing tabique walls under compressive and shear loading. In: Aguilar, R., Torrealva, D., Moreira, S., Pando, M.A., Ramos, L.F. (eds.) Structural Analysis of Historical Constructions. RB, vol. 18, pp. 568–576. Springer, Cham (2019). https://doi.org/10.1007/978-3-319-99441-3_61

Pinto J., et al.: Characterization of traditional tabique constructions in Douro north valley region. J. WSEAS Trans. Environ. Dev. **2**, 105–114 (2010). 1790-5079

Pinto, J. et al.: Save the *tabique* construction. In: Costa, A., Guedes, J., Varum, H. (eds) Structural Rehabilitation of Old Buildings. Building Pathology and Rehabilitation, vol. 2. Springer Heidelberg (2014). https://doi.org/10.1007/978-3-642-39686-1_6, 978–3–642–39685–4 (Print) 978–3–642–39686–1 (Online)

Pinto, J., et al.: An input on the characterization of timber frame structural system of tabique walls. In: Proceedings of the 6th International Conference on Safety and Durability of Structures, ICOSADOS 2014, Wroclaw, Poland, 13–15, May 2014

Pinto, R., Padrão, J.: Structural characterization of "tabique" walls from the historic centre of Viseu. In: 9th IMC – International Masonry Conference, Guimarães, Portugal, 4–9 July 2014

Critical Issues of Typical Existing Masonry Road Bridges Handled by the Province of Caserta

Gianfranco De Matteis[1](✉), Pasquale Bencivenga[1], Mattia Zizi[1], and Antonino Del Prete[2]

[1] Department of Architecture and Industrial Design, University of Campania "Luigi Vanvitelli", Via San Lorenzo 31, 81031 Aversa, Italy
gianfranco.dematteis@unicampania.it
[2] Province of Caserta, Settore Viabilità, Via Lubich 6, 81100 Caserta, Italy

Abstract. Due to bridge collapses recently occurred, it is now very clear that the conservation and the preservation of existing infrastructures is a real priority. On the other hand, many of the existing bridges were built before the first half of the last century on the basis of outdated design criteria. Therefore, they are affected by a significant ancientness, a bad state of conservation and structural deficiencies, also because of lack of maintenance. The present paper focuses on recurring structural issues affecting existing masonry bridges, which represent a large portion of the existing Italian infrastructural heritage. The main goal is to depict the current Italian scenario, in order to identify common structural deficiencies affecting such bridge typology. To this purpose, some typical bridges belonging to the roadway network of the Province of Caserta are examined. The study, which is part of a wider activity aimed at implement methodologies for the assessment and preservation of the existing bridges, shows that the considered bridges are affected by common structural deficiencies and defeats and therefore could set the stage for a wider investigation aimed at defining suitable intervention criteria that might be applied to existing European infrastructures.

Keywords: Masonry bridges · Infrastructures · Bridge safety · Structural degrade · Structural vulnerability

1 Introduction

Bridge collapses recently occurred in Italy but also in other European countries revealed that a large portion of the bridge heritage is affected by an inadequate structural safety, mainly due to lack of maintenance and material deterioration. The preservation of such strategic infrastructures is nowadays a significant challenge that must be seriously tackled, considering the important socio-economic defeats and the human life losses that might occur in case of bridge collapse.

I. Vayas and F. M. Mazzolani (Eds.): PROHITECH 2021, LNCE 209, pp. 1081–1092, 2022.
https://doi.org/10.1007/978-3-030-90788-4_83

From a qualitative analysis of the recent bridge failures, it has been pointed out that more than ten significant collapses occurred in Italy in the last five years. But, it is worth noting that the earthquake is not the prevailing cause of these failures. In fact, they were mainly due to environmental (e.g. material deterioration, flooding, etc.) and anthropological factors (e.g. excessive vertical loads, design or execution errors, lack of maintenance, etc.).

In Italy, but also in many other countries, an accurate and detailed census is still missing. Moreover, the complex management issues produce additional difficulties. For instance, it is interesting to note that in November 2019 a report of ANAS (*National Body for the Roads*) stressed that about one thousand of the existing Italian bridges did not belong to a well defined owner (ANAS 2019).

The bridge structures currently present in the Italian territory are managed by different administrative public and private bodies, namely Aiscat, ANAS, Rete Ferroviaria Italiana (*Italian Railway*), Regions, Provinces and Municipalities. According to some recent reports (ANAS 2019, Rapporto U.P.I. 2018, Aiscat 2018), it is known that:

- Associazione Italiana Società Concessionarie Autostrade e Trafori (Aiscat), which is the consortium for the management of the Italian highways, possesses 1626 viaducts and bridges (in the data overpasses are neglected);
- Autostrade per l'Italia, which is the most important society of the Aiscat consortium, deals with about 1.800 bridges, viaducts and overpasses;
- ANAS Society administers 14.644 bridges and viaducts;
- R.F.I. group manages about 19.000 viaducts, bridges and underpasses;
- Italian provinces, excluding Trentino Alto-Adige, handle more than 30.000 structures.

Thus, in this context, it is evident that an accurate priority strategy of bridge management aimed at preserving the existing bridge heritage ensuring an adequate structural safety is necessary.

Based on these premises, in the present paper, within a larger cooperative activity developed by the Department of Architecture and Industrial Design of the University of Campania "Luigi Vanvitelli" and the Province of Caserta, which handles about 1000 bridges, the scenario related to the road bridges of the Caserta district is described. In particular, the outcomes obtained from a record-keeping activity are presented, with a focus on masonry bridges, which, despite their ancientness, are still crucial and widespread in the infrastructural Italian network.

Then, moving from a large-scale approach to a more detailed one, the results obtained from visual inspections carried out with the aim to identify the recurrent critical issues are shown. Finally, 9 case studies are examined in order to emphasize critical structural deficiencies commonly present for such a bridge typology.

2 A Picture of the Bridge Scenario in the Caserta District

2.1 General

The current study deals with the bridges handled by the Caserta province, which is an Italian district located in the Campania Region. It has an extension of about 2.650 km^2 and boasts 104 municipalities. The administrative body, namely the Province of Caserta, manages 879 infrastructures spread all over its territory (Fig. 1).

Since about few years, a census of all these bridges has been carried out in collaboration with the Department of Architecture and Industrial Design of the University of Campania "Luigi Vanvitelli".

The total stock of the above manufacts includes typical bridges, viaducts, overpasses and underpasses. The record-keeping activity has been carried out considering exclusively cartographical, photographical, geometrical and typological parameters provided by the institution.

Hence, a priority scale for future specific inspections has been defined. Such surveys are aimed at assessing the real state of conservation of the considered bridges, in order to define possible provisions. Further details about the adopted methodology can be found in De Matteis et al. (2019a).

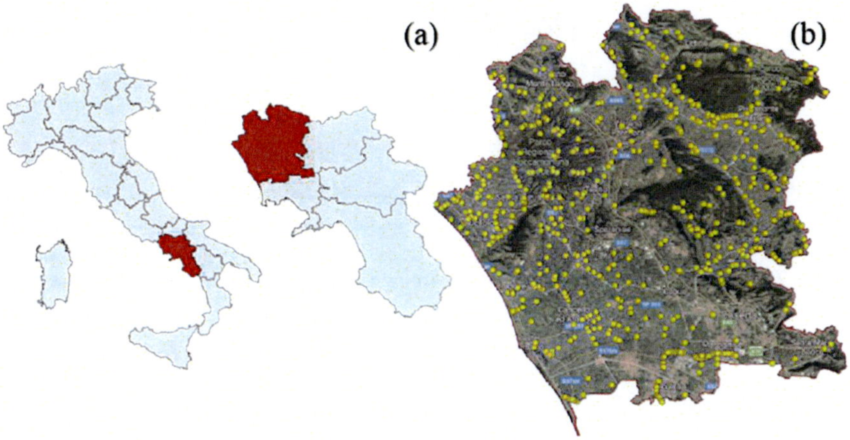

Fig. 1. The Caserta territory (a) and location of bridges handled by the Caserta province (b)

2.2 Characteristics of Masonry Bridges

Among the complete bridge stock managed by the Province of Caserta, 200 structures are masonry arch bridges (Fig. 2a). In few cases (about 5%), the structures present also reinforced concrete structural elements in combination with masonry structures. Based on the census activities carried out, masonry bridges are mainly characterized by short span length, i.e. either less than 5 m (55%) or in the range of 5 ÷ 10 m (33%). Therefore, only the 7% and the 4% of the masonry sample are characterized by total lengths in the range of 10 ÷ 20 m and more than 20 m, respectively. A complete picture of the situation is provided in Fig. 2b.

With reference to the number of spans (Fig. 2c), a wide portion of the masonry bridges present in this area is characterized by a single-span (85%). In addition to the few cases where the data were not available (5%), the remaining bridges are characterized by multi-spans, namely 2 ÷ 5 spans (18 structures) or more than 5 spans (3 structures).

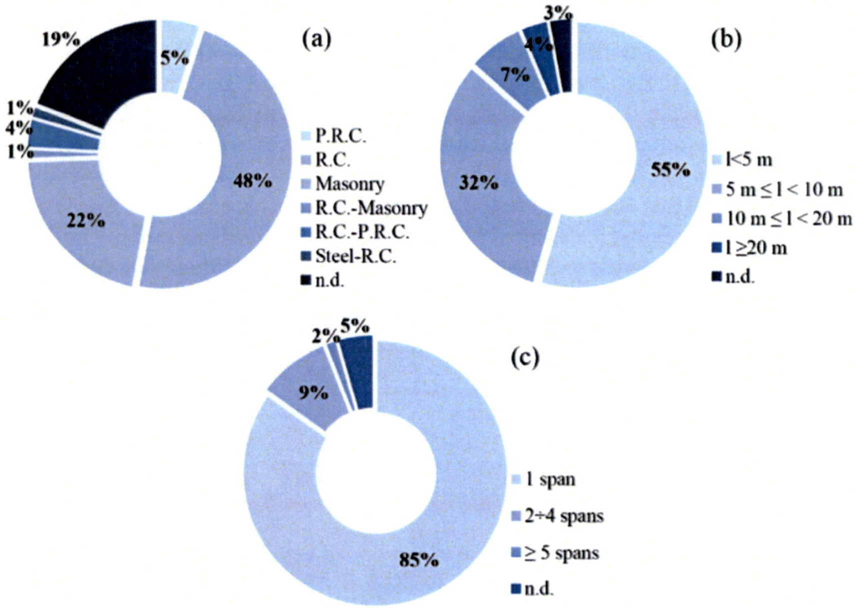

Fig. 2. Masonry bridge types: structural typologies, (a) Length (b) Number of spans (c)

2.3 Critical Issues of Inspected Masonry Bridges

The second level of the conducted activities consisted in specific visual inspections on bridges selected according to the ranking arisen from the previous phase. The main goals of these surveys were to provide a geometrical and typological description of the structures and define the critical structural issues and defects. Moreover, in some critical cases, immediate measures and provisions have been prescribed.

Among the 28 inspected bridges, 10 are characterized by a masonry structure. A summary of the bridges considered in this study is reported in Table 1, where the following geometrical features are provided: number of spans (*ns*), total length (*l*), arch span (*s*), arch height (*h*) and road width (*w*). Moreover in Fig. 3, a picture for each analyzed bridges is provided.

Table 1. The inspected bridges in the Caserta district

Code	Location	*ns*	*l* [m]	*s* [m]	*h* [m]	*w* [m]
330058	Pratella	1	13,00	9,90	2,50	6,80
336008	Caserta	3	28,00	7,00	4,30	8,50
085018	Roccamonfina	1	33,00	4,80	1,70	6,80
112001	Teano	3	38,00	7,00	3,10	7,50
330075	Ailano	1	12,00	10,70	4,50	8,60
014056	Mignano M.Lungo	5	83,00	11,40	7,30	8,40
111006	Teano	3	18,00	4,70	2,50	7,00
336064	Ruviano	8	241,00	22,00	4,60	7,80
329022	Teano	5	49,00	8,50	3,70	7,30
330066	Ailano	1	8,00	4,20	1,20	8,30

As far as the construction period is concerned, it can be supposed that the above bridges were realized in the first half of the last century. In fact, it has been ascertained that the considered territory resulted hugely damaged during the World War II and some of these bridges were built on the ruins of the previously existing ones.

Despite they are spread over a relatively large area, the use of recurrent dimensions of structural elements, as well as base materials, have been recognized. At this regard, it is worthy to note that half of the considered masonry arch bridges are characterized by a single-span structure, with a total length less than or equal to 10 m.

With reference to material properties, it has been generally verified a good workmanship of both constituents and assemblage techniques: a well-organized texture with joints of adequate thickness have been detected in almost all structures. Moreover, the inspected bridges commonly showed the presence of brick and stone masonry realized with the typical materials of Caserta territory. Generally, they are assembled with blocks in tuff or brick elements obtained from the material extracted by the surrounding mountainous areas.

Some of these manufacts present unoriginal interventions, such as the presence of ordinary or pre-stressed reinforced concrete elements (i.e. slabs at the intrados or additional beams) (e.g. bridges id. 330058, id. 336064 and id. 111006) and additional retaining walls (e.g. bridge id. 336008).

Moreover, in some cases, the actual bridge structure is constituted by two different substructures. For instance, bridges id. 085018, 330075, 111006 and 330066 have been subjected to enlargement interventions to adequate the road width to the modern traffic requirements.

The inspected structures revealed several recurring critical defects, which in some cases led the surveyors to prescribe precautionary provisions, such as traffic limitation or prohibition. The most critical case was the single-span bridge 085018 in Roccamonfina, where a partial collapse of the lateral abutments was evident. This bridge represents a very particular case of a stratified structure: in the actual state appears to be the result of

Fig. 3. The inspected structures: a) bridge id.: 330058, b) bridge id.: 336008, c) bridge id.: 085018, d) bridge id.: 112001, e) bridge id.: 330075, f) bridge id.: 014056, g) bridge id.: 11006, h) bridge id.: 336064, i) bridge id.: 329022, j) bridge id.: 330066

various ongoing adjustments carried out in the past. In particular, it could be considered as the union of two arch structures separated longitudinally and structurally independent from each other (Fig. 4a). In this structure, a significant failure of the both abutments of the oldest part occurred, probably due to excessive vertical loads, while, luckily, the new portion was not involved in the occurred deformation mechanism (Fig. 4b, 4c). A precautionary provision corresponding to complete traffic prohibition was assumed for this structure.

Another unsuitable situation has been detected for the bridge in Pratella (id.: 330058), where the presence of sub-vertical cracks on abutments, probably due to geological issues, have been observed (Fig. 5a, 5b). Similarly, for the three-span bridge id. 111006 in Teano, a significant longitudinal crack has been observed in one of the arches (Fig. 5c); in such a case, in order to check the possible damage evolution, a specific structural monitoring has been prescribed, in addition to an immediate reduction of traffic load acting

Fig. 4. Bridge in Roccamonfina (id.: 085018): stratification of the structure (a) and collapse of the abutments (b) and (c).

on the carriageway. Such phenomena can be ascribed to either the presence of differential settlements of foundation structures or unbalanced internal stresses, in addition to a progressive deterioration of the structural materials.

Fig. 5. Vertical cracks on abutment structures of bridge in Pratella (id.: 330058) (a) and (b) and longitudinal crack at the intrados of bridge in Teano (id.: 111006) (c)

A different critical situation commonly observed for inspected bridges is represented by the detaching of brick elements. In particular, the bridge id. 336008 in Caserta exhibited such a defect on piers (Fig. 6a, 6b). Although this situation does not compromise the global stability of the structures, it represents an alert for the structural safety, imposing the necessity of carrying out constant visual inspection of the bridge in order to verify the possible evolution of the degrade state.

In some other inspected structures, e.g. the bridge in Ailano (id. 330066), the consumption of mortar joints has been noticed (Fig. 6c). In these cases, the restoration of masonry surfaces has been prescribed.

Common defects surveyed during the inspections were due to the action of rainwater. In fact, many moisture stains and dark spots on various structural elements have been

Fig. 6. Detaching of brick elements on masonry piers of the bridge in Caserta (id.: 336008) (a) and (b) and consumption of mortar joints of the bridge in Ailano (id.: 330066) (c)

observed in almost all the examined bridges (Fig. 7a, 7b). In such case, in order to preserve the structures, the rehabilitation of rainwater drainage systems, in addition to an accurate cleaning of the deteriorated elements, was prescribed.

Finally, the presence of weeds, which in some cases penetrated into the masonry elements causing significant crack patterns, was observed (Fig. 7c, 7d).

Fig. 7. Moisture stains and dark spots of the bridge in Ruviano (id.: 336064) (a) and (b) and presence of infesting plants on the bridges in Ailano (id.: 330075) (c) and in Teano (id.: 112001) (d)

3 The Structural Safety Checks

3.1 General

In addition to the above surveying activities, structural vulnerability analyses have been carried out by external professionals, with the aim to check the actual structural performance of typical masonry bridges subjected to traffic and seismic loads according to the current Italian technical code (Italian Ministry of infrastructures and transport 2018). The results of such analyses, which are related to nine masonry arch bridges, have been examined in order to identify common structural deficiencies of this bridge typology.

3.2 The Analyzed Bridges

The analyzed structures are all located in the mountainous area of northern Caserta district and correspond to bridges having different geometrical characteristics and construction techniques. Such arch bridges belong to strategic roads since they represent routes which are the only connecting lines of many small villages. Therefore, all the examined bridges have a width ranging between 4.50 and 6.60 m. On the contrary, they present significant differences in terms of longitudinal length; in particular, the total length ranges between 9 and 140 m while the span of the arches varies in the range of 3.9 ÷ 18.50 m.

From a typological point of view, two main bridge types may be identified: single span (6 samples) and multi-span structures (3 samples).

Table 2. The analyzed bridges

Code	Location	ns	l [m]	s [m]	h [m]	w [m]
010007	Raviscanina	7	140	18,50	4,20	6,30
011012	Ciorlano	5	71	8,50 ÷ 12,40	4,20	6,30
083030	Valle Agricola	1	10,30	5,90	2,60	4,55
083031	Valle Agricola	1	9,00	6,00	3,00	4,80
089017	Gallo Matese	1	10,20	6,00	1,50	6,35
232001	Ciorlano	3	17,00	3,90	1,30	5,10
330047	Ciorlano	1	25,10	6,90	1,00	6,60
331017	S.Gregorio Matese	1	12,30	10,00	5,00	5,00
331020	S.Gregorio Matese	1	9,20	5,00	2,60	6,64

In particular, the single span bridges are characterized by arches that are directly supported by the foundation ground (bridge id. 083030, 083031, 089017, 330047 and 331017), while in one case only, high and massive lateral abutments supporting the arch are present (bridge id. 331020). On the contrary, multi-span bridges are characterized by very massive and stiff piers, which support the overlying masonry arches.

A summary of the aforementioned bridges is reported in Table 2, while in Fig. 8 some pictures of such bridges are provided.

3.3 Main Outcomes and Common Structural Deficiencies

The vulnerability studies showed that all the analyzed structures could be considered adequate to sustain vertical forces arisen from load schemes typical of 1st category bridges, despite their ancientness and obsolescence. This is mainly due to the fact that such structures were realized on the basis of empirical rules, which led to oversize the structural elements. Therefore, the evolution of traffic loads, which led to consider higher vertical loads in the current technical codes, could be considered negligible for these very massive structures, as the effect of self-weight is predominant.

On the contrary, when seismic actions are considered, only two of the analyzed bridges provided a sufficient performance according to the current Italian Technical Code (Italian Ministry of infrastructures and transport 2018).

Fig. 8. The analyzed bridges: a) bridge id.: 010007, b) bridge id.: 011012, c) bridge id.: 083030, d) bridge id.: 083031, e) bridge id.: 089017, f) bridge id.: 232001, g) bridge id.: 330047, h) bridge id.: 331017, i) bridge id.: 331020

A summary of the main outcomes is provided in Table 3, where seismic safety factors (SSF) evaluated for each structure as the ratio between the structural capacity (R_d) and the structural demand (E_d) together with the most influencing collapse mechanism are reported.

Table 3. Main outcomes of structural safety checks

Code	Location	SSF	Collapse mechanism
010007	Raviscanina	0,68	Limit load for foundation system
011012	Ciorlano	> 1,00	–
083030	Valle Agricola	0,22	Flexural crisis of the arch
083031	Valle Agricola	0,63	Overturning of the gable
089017	Gallo Matese	0,17	Flexural crisis of the arch
232001	Ciorlano	> 1,00	–
330047	Ciorlano	0,33	Flexural crisis of the arch
331017	S.Gregorio Matese	0,91	Flexural crisis of the arch
331020	S.Gregorio Matese	0,50	Flexural crisis of the abutments

It can be observed that the most recurring collapse mechanisms under seismic loads are arch or abutment crisis due to combined flexural and axial internal forces. Moreover, a particular case is represented by the bridge in Valle Agricola (id. 083031), which exhibited structural deficiencies due to possible overturning of the gable.

On the contrary, multi-span bridges, which are characterized by stiff and stocky piers, generally resulted adequate for seismic loads, with exception of bridge id. 010007, which showed geotechnical issues. This result confirms the outcomes of a previous study of the authors, which stated that the structural deficiencies of existing R.C. bridges are mainly due to vertical loads (De Matteis et al. 2019b). On the contrary, in case of single-span bridges, it is clear that the effects of horizontal actions are exclusively entrusted to the bearing capacity of the arch systems, which, as is well known, suffer this type of loads.

4 Concluding Remarks

The current study focused on masonry bridges located in the Caserta district, which have been surveyed within a cooperation between the Province of Caserta and the University of Campania "Luigi Vanvitelli". Typical structural issues and defects arisen from inspective activities have been analyzed. Then, the outcomes obtained by analytical structural analyses of 9 samples have been reported and commented.

The study revealed that despite their ancientness, such structures generally are adequate to sustain traffic loads, while on the contrary they are unsuitable to sustain design seismic forces. Although the considered bridges belong to a small territorial area, the considered typologies can be considered representative of a large portion of the Italian bridge heritage. Therefore, the conclusions of this study clearly point out that a general surveying and maintenance activity should be carried out in order to guarantee the functionality and also an adequate safety level of existing bridges.

Acknowledgments. The Authors would like to thank the President of the Caserta Province, Avv. Giorgio Magliocca, for the support in implementing the activity and geom. Carmine Sparano for the active cooperation provided.

Part of the present study is also framed within the national project ReLuis-DPC 2019–2021 financed by the Italian Department of Civil Protection.

References

ANAS: Focus sistema Anas di gestione e sorveglianza di ponti e viadotti (2018). https://www.stradeanas.it/sites/default/files/Focus%20Ponti%20e%20Viadotti_ottobre%202018.pdf

ANAS: Anas, censimento e ispezioni dei cavalcavia sulla rete stradale (2019). https://www.stradeanas.it/it/cavalcavia-senza-documentazione-di-proprieta

Associazione Italiana Società Concessionarie Autostrade e Trafori:. Aiscat in cifre 2018 (2018). http://www.aiscat.it/pubblicazioni/downloads/aiscat-in-cifre-2018.pdf

Chen, W., Duan, L.: Handbook of International Bridge Engineering (2014). 13: 978-1-4398-1030-9

De Matteis, G., Bencivenga, P., Zizi, M., Del Prete, A.: Rischio strutturale di ponti esistenti: considerazioni preliminari sullo scenario in provincia di Caserta. Atti del XVIII convegno ANIDIS L'ingegneria sismica in Italia (2019a). 978-88-3339-256-1

De Matteis, G., Zizi, M., Del Prete, A.: Structural features of typical Italian bridges built in the '50s: four case studies in the province of Caserta. In: Proceedings of the 7th Structural Engineers World Congress 2019, Istanbul, Turkey (2019b)

Dias, W.P.S.: Factors influencing the service life of buildings. J. Inst. Eng. **46**, 1–7 (2014). https://doi.org/10.4038/engineer.v46i4.680

Italian Ministry of infrastructures and transport: Circolare 2 febbraio 2009, n.617- Istruzioni per l'applicazione delle "Nuove norme tecniche per le costruzioni" di cui al decreto ministeriale 14 gennaio 2008 (2009)

Italian Ministry of infrastructures and transport: Aggiornamento delle Norme tecniche per le costruzioni (2018)

Jurina, L., Mazzoleni, M.: Verifiche statiche ed interventi di consolidamento nei ponti in muratura. Seminario CIAS: Ponti: sorveglianza, manutenzione ed interventi, Bolzano, Italy, 29, November 2002

Jurina, L., Mazzoleni, M.: Interventi di consolidamento su ponti in muratura. Seminario CIAS: Ponti e viadotti: ispezioni visive e tecniche di risanamento, Bolzano, Italy, 1–3, December 2005

Mandić Ivanković, A., Savor, Z.: Limit state of existing bridges. In Proceedings of Joint IABSE-fib Conference, Dubrovnik 2010, Zagreb, Croatia (2010)

ReLUIS: Linee guida e manuale applicativo per la valutazione della sicurezza sismica e il consolidamento dei ponti esistenti in c.a., Rome, Italy (2009)

Tonias. D.E., Zhao, J.J.: Bridge Engineering. Design, Rehabilitation and Maintenance of Modern Highway Bridges (2017). 13: 978-0-07-145003-7

Triglia, M., Maffei, G., Mari, D.: Innovazione nella manutenzione dell'infrastruttura in RFI, Quadrimestrale di Rete Ferroviaria Italiana (2005). http://www.rfi.it/cms-file/allegati/rfi/argomentinumero6.pdf

UPI - Unione Province d'Italia: Conferenza stampa Ponti: i risultati del monitoraggio delle Province (2018). https://www.provinceditalia.it/monitoraggio-ponti-upi-censite-5-931-opere-servono-3-miliardi-per-intervenire-su-sicurezza

Sustainability Potential of Intervention Strategies for Mortar Coating Repair: A Case Study in a Centenary Building

Camila Salvi Malacarne(✉), Anne Ketlelyn Domingos Biehl, Deividi Maurente G. Silva, Iago Lopes dos Santos, Luis Gabriel de Godoy, Angela Borges Masuero, and Mônica Regina Garcez

Department of Civil Engineering, Federal University of Rio Grande do Sul, Av. Osvaldo Aranha, 99, Porto Alegre, RS, Brazil

Abstract. Coatings play a fundamental role in the exterior wall system of a building and are directly related to its performance through life. Increased durability and lifespan can be achieved when coating's integrity can prevent the entrance of aggressive agents on the substrate. Therefore, the condition of this system should be periodically evaluated, considering not only technical and economic aspects in the selection of the materials for repair, but also sustainability aspects, in order to reduce the environmental impacts incorporated into the building. In this study, a Brazilian building dated from 1908 was evaluated with the purpose of detecting pathological manifestations in its outer coating. In a primarily inspection stage, cracks and coating detachments were identified as the pathological manifestations with higher incidence in the evaluated facades. Three maintenance strategies were assessed for crack repair and two for the purpose of resolving the loss of adhesion. Subsequently, a life cycle analysis (LCA) was performed for each intervention alternative. Results indicate that regarding the cracks correction, the solution that presented the lowest environmental impact potential was the model that uses lime paste for filling, while for detachments the indicated solution from an environmental point of view is the polymer-free model.

Keywords: Life cycle assessment · LCA · Pathological manifestations · Green maintenance · Historical building

1 Introduction

Historic buildings are an important resource to help understand humanity's temporal location, i.e., where we are and how we got here. Therefore, the choice of which buildings to turn into a historical heritage and its preservation can have a significant impact on the transmission of a building to future generations, as noted by Cherry (2007), and the understanding of it by the generation that receives it. In other words, the historical and artistic heritage of a nation are closely related to the historical evolution of a society and the cultural characteristics that compose it. Internationally, the importance of maintaining historical buildings is recognized as essential for the preservation of cultural heritage

I. Vayas and F. M. Mazzolani (Eds.): PROHITECH 2021, LNCE 209, pp. 1093–1110, 2022.
https://doi.org/10.1007/978-3-030-90788-4_84

and the capital (financial, social and environmental) associated with the infrastructure of a nation (Forster et al. 2011).

From an economic point of view, a country's expenditure on the maintenance of existing buildings can represent a significant percentage of the resources allocated to infrastructure. In the United Kingdom, for example, it is estimated that the amount spent on maintenance is up to half of the total allocated to the country's infrastructure (Balaras et al. 2005). In Brazil, unfortunately, specific statistics on maintenance activities of existing infrastructure are not recorded, but only general production data for the construction industry. The amount spent annually on maintenance may seem small, but accumulated over the useful life of buildings may be equivalent or even higher than the initial construction cost. According to Perret and Jouvent (1995), over a 50-year service life, usage and maintenance costs can represent up to 75 to 80% of total costs.

In recent years, a change in mindset has led to thinking about maintaining existing buildings not only from the perspective of preserving economic resources, but also considering the importance of the lower environmental impact associated with this practice, as opposed to the construction of new buildings. That's because the amount of resources needed for maintenance is far less than that needed for building a new one. This practice is stimulated by the signing of international agreements and cooperation on behalf of the reduction of emissions, following the example of Agenda 2030 for sustainable development, a document prepared in 2015 during the United Nations Conference in New York City that elucidates the importance of combining development with environmental conservation.

Information gathering about building materials by economic criteria, energy consumption or global warming potential can be used as a tool to help designers and builders make conscious decisions thus enabling more sustainable choices (Takano et al. 2014). This line of research has gained increasing attention in recent years, with Life Cycle Assessment (LCA) being the predominant methodology for quantifying impacts, as raised by Chen et al. (2019). This method consists in the evaluation of the environmental impact potential along all stages and processes of the life cycle of a product and its constituents, passing through the extraction of raw materials, manufacturing, distribution, use, repair and maintenance until its final disposal or recycling.

Within this context, the concept of green maintenance can be highlighted, as introduced by Forster et al. (2011), which prioritizes materials and repair techniques with lower CO2 emissions. The author establishes a relationship between maintenance and repair, giving preference to techniques and/or materials with higher longevity, which consequently incurs in fewer interventions and less impact incorporated during the useful life of the building, as illustrated in Fig. 1. After the end of construction, the building is in an optimal service condition, which decreases over the years of use and exposure to the environment. The maintenance works are aimed at returning the building to its initial service condition, and with this a new incorporation of carbon occurs. The deterioration rate after the intervention will be a consequence of the type of repair performed.

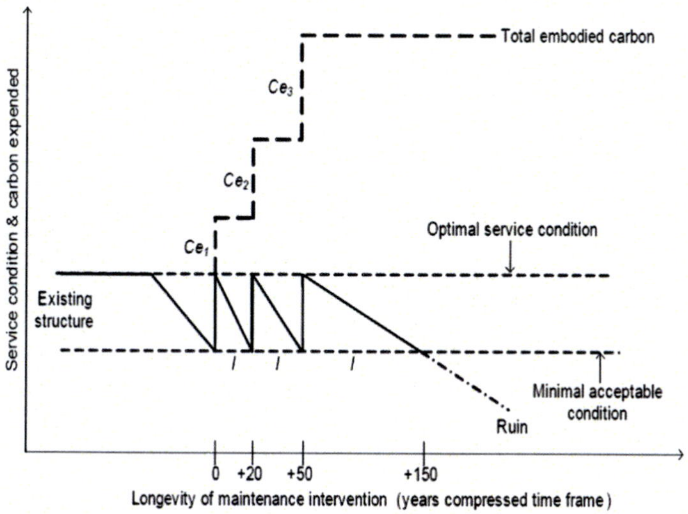

Fig. 1. Relation between longevity of the repair and carbon expenditure (Forster et al. 2011).

Maintenance has been considered a fundamental part of a building life cycle, since the structure deterioration of existing buildings has increased considerably across the globe (Kwon et al. 2020). That's because a building is made up of several systems that have the function of meeting users requirements, being these permanent or non-permanent occupants, human or non-human, living or inanimate beings, i.e., all the parts that generate requirements to be met by the building (CIB 1982). Among the different systems, the walls systems consist of delimiting internal and external environments, avoiding the action of undesirable agents, such as: rain, wind, noise, animals; among others. The external coating is an element of the system, being the one that, in most of the buildings, presents greater interface with the surroundings.

Sterflinger et al. (2018) define that coating's degradation may be associated with their interaction with the environment to which they are submitted or with the materials that compose them, either by chemical or physical phenomena or both (Bochen and Labus 2013). The evaluation of their constituents can be performed by destructive or non-destructive techniques, the latter being preferable due to the conservation of their integrity, but not always the use of this methodology is possible. Moreover, when it comes to historical buildings, extra care must be taken to preserve their structure and aesthetics, due to their cultural relevance. In this case, sensibility with the original structure is required: maintenance should involve repair techniques and compatible materials to the original ones (Mahmoud et al. 2015).

The purpose of this study arose from the interest to preserve and restore the integrity of the facade of a historic building made of mortar coating, taking into account the environmental impacts related to each alternative intervention. In the case study, a quantification of the most incident pathological manifestations was performed. This initial mapping supported the suggestion of different intervention alternatives, under the premise of maintaining the architectural characteristics of the building, using compatible materials with those originally used and other modern materials for comparison. The evaluation of the intervention strategies was made through the LCA methodology, considering the environmental impacts related to each alternative.

2 Methods

Figure 2 presents the main activity flow of this work, which will be detailed in the next topics. Briefly, in a preliminary inspection of the building, the most recurrent pathological manifestations were identified as cracks and detachment, which were than mapped and quantified. Three maintenance strategies were suggested and assessed for crack repair and two for the loss of adhesion. Finally, a Life Cycle Assessment (LCA) was conducted, with the objective of evaluating the alternative with the lowest environmental impact potential.

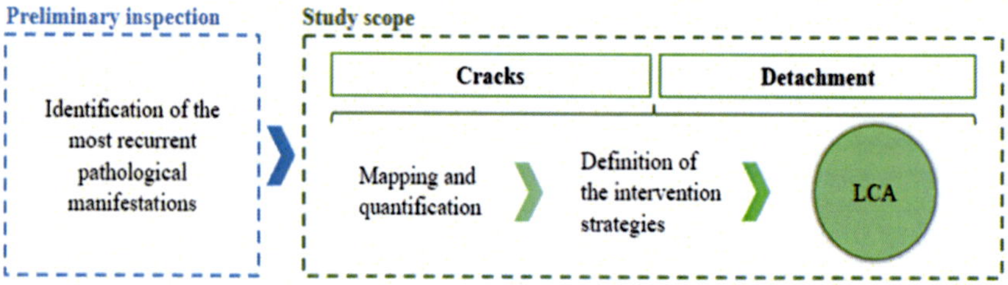

Fig. 2. Case study activity flow. LCA: Life Cýcle Assessment.

2.1 Case Study Description

The study was carried out in the building of the Electrotechnical Institute of the Federal University of Rio Grande do Sul, illustrated in Fig. 3(a), whose construction dates from 1908 and is located in the interior of the Federal University of Rio Grande do Sul (UFRGS) Central Campus, in the Historic Center district of the city of Porto Alegre/RS, Brazil. The building is positioned at the intersection of Osvaldo Aranha Avenue and Sarmento Leite Street. The main facades have east (1), northeast (2) and north (3) orientation, while the other ones, oriented to the interior of the UFRGS Central Campus, have west (4 and 6) and south (5 and 7) orientation, as shown in Fig. 3(b).

A power plant capable of supplying power to the Engineering School, the motor assembly and the development of studies related to these technologies were the purposes of the building when originally designed. The changes in the activities carried out in the building over the years have led to some changes in its structure and architectural design, as can be seen in Fig. 4.

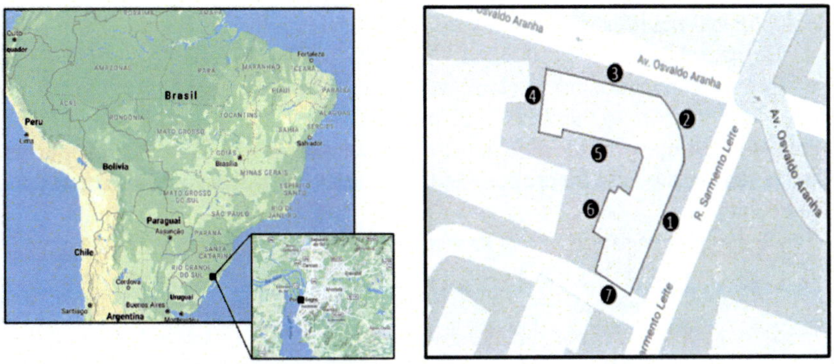

(a) (b)

Fig. 3. Electrotechnical Institute: (a) Building location and (b) Orientation and identification of the building's facades (Google Earth).

(a) (b) (c)

Fig. 4. Evolution of the building in the years (a) 1912, (b) 1929 and (c) 1990 (SPH 2007).

In 1951, all the upper ornaments and most side openings were removed, a third floor was built by replacing the French tile roof with a reinforced concrete slab. Later, in the 1970s, the chimney was removed. Currently, part of the ground floor headroom has been reduced through the insertion of mezzanines, extending the useful area, as well as metal railings were inserted in the internal areas to avoid access to the institute's premises by third parties (De Moraes, 2003, SPH 2007).

Considering the numerous modifications made and the changes in the initial architectural conception, an adopted delimitation of the study is the sole evaluation of the facades 1, 2 and 3, since they are the ones that most preserve the original building characteristics, besides having a greater visual impact on the community because they are oriented to the traffic routes.

2.2 Mapping and Quantification of Pathological Manifestation

Facades 1, 2 and 3 of the building were examined visually and with the aid of a photographic camera in order to map and quantify the cracks and detachments, identifying them in a sketch for recording and analysis of information obtained on site. Additionally, percussion tests were performed to increase the reliability and accuracy in identifying the pathological manifestations assessed. The elevations were obtained from the Historical Heritage Sector of the Federal University of Rio Grande do Sul (UFRGS) in digital format. The maximum working height in the studied facades was limited to two meters, with reference to the floor level around the building. The adoption of this limit is related to the impossibility of using specific equipment for the vertical displacement of people during the field inspection, and the checks are therefore limited to the reach of the operator.

The linear quantification of cracks was performed through image treatment with the aid of ImageJ software. In each facade, three regions were identified as a sample of the totality of cracks in this element. Each region was treated as illustrated in Fig. 5, and the linear length of cracks present in a given study area was quantified. Thus, the linear average of cracks per unit of area was multiplied by the whole cracked area of that facade (through the raised hatched area in the inspection stage) and a total length of cracks can be estimated.

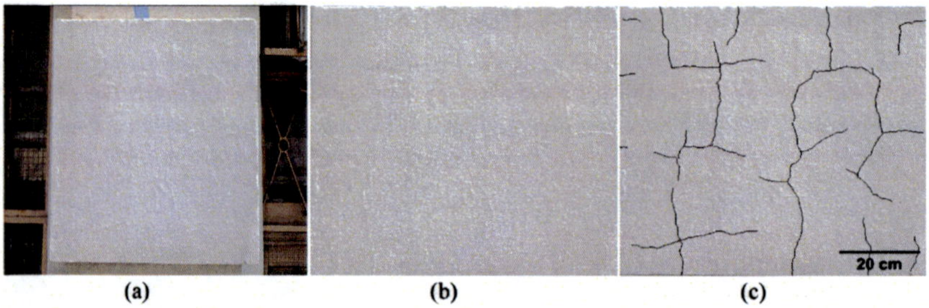

Fig. 5. Quantification of cracks by image analysis: (a) photograph of one of the analyzed areas, (b) Image treatment in ImageJ software for crack identification, (c) Sketch of cracks found, image area of $66 \times 76\ cm^2$.

A NIKON D5000 model digital camera was used to capture the image of the region under study, as shown in Fig. 5(a). The original image was treated in the imageJ software in 8 bits, i.e., a grayscale pattern was stipulated to identify the pixels representing the cracks, Fig. 5(b). In the last step, Fig. 5(c), a scale was assigned to the image after the study section dimension was known, and each crack was signaled by a line, which expresses its extension, enabling the determination of the total crack length.

As for the coating detached areas, its quantification was made with the percussion method, an expeditious technique for checking adhesion. Light impacts are produced with a rubber hammer or other instrument with inferior hardness to the coating, and the sound produced enables to identify the solidity between substrate and coating, according to NBR 13749 (ABNT 2013). The coating parts that presented hollow sounds were defined as points with loss of adhesion. An average of 0.4 mm thickness and 2.5 cm depth was considered to determine required materials for crack repair, based on average measurements performed on site. The specific mass used for the polymeric mortar, polyurethane and epoxy were based on technical specifications from local manufacturers.

2.3 Definition of Intervention Alternatives

The most common pathological manifestations in the building are cracks and coating detachment. Besides the aesthetic decline, however, these manifestations put at risk the functionality and durability of the building's coating, thus requiring correction. Therefore, the following repair interventions were proposed for:

- Cracks: polymeric mortar injection, lime paste injection and polyurethane injection;
- Detachment: bonding the detached coating with epoxy resin and total replacement of the detached coating.

Considering that the building is of historical interest, the repair technique must ensure compatibility between the substrate and the material used for repair. Therefore, the reconstitution of the coating mortar binder: aggregate proportion was carried out in a laboratory in order to estimate the proportion of its constituents. The method employed was proposed by Teutônico (1998), which allows the establishment of a relationship between aggregates and binder present in a cement-based sample in the hardened state through the chemical attack of a finely disaggregated mortar. The procedure is illustrated in Fig. 6.

(a) (b) (c)

Fig. 6. Procedure for determining the binder: aggregate proportion of a mortar sample: (a) Sample extraction, (b) Oven-dried sample and (d) Sample immersed in hydrochloric acid solution (Masuero et al. 2017).

Samples were taken from facades 1 and 3, in locations where the coating was not damaged. The painting layer was removed and the samples carefully comminuted with mortar and pestle, dried in an oven at a temperature of 75 °C until constant mass, and exposed to a solution of hydrochloric acid and distilled water (1:4) to consume the soluble fraction through a chemical neutralization reaction. The final solution was washed in distilled water and filtered several times until the total removal of the acid fraction and the materials solubilized by it. Subsequently, the solid fraction was dried until a constant mass. The difference between the initial and final masses corresponds to the amount of binder neutralized by the hydrochloric acid solution, the final mass being the fraction of aggregate present in the sample. The mix proportioning reconstituted are presented in Table 1.

Table 1. Reconstitution of the mix proportion in samples facades 1 and 3 (Masuero et al. 2017).

Sample	Binder	Aggregate
Facade 1	1.00	2.33
Facade 3	1.00	2.51

Coating cracks usually attract the attention of users and its recovery is the most frequently intervention in facade rehabilitation works. Three interventions were proposed as an alternative for crack repair:

- Model MI-F: polymeric mortar injection - selected due to the improvements provided by the polymeric additions in the mechanical and rheological mortar properties;
- Model MII-F: lime paste injection - material widely used for mortar coating crack repair, in order to restore monolithic properties and durability of building facades;
- Model MIII-F: polyurethane injection - elastic sealant that has the elasticity as main characteristic and it is a traditional recommended methodology for cracking closure in mortar coating.

Models MI-F, MII-F e MIII-F proposed above follow basically the same operational procedures. The first step is to prepare the substrate by cleaning with a compressed air flow. Subsequently, the materials are applied using the injection syringes technique.

The detachment of a building's coating is a pathological manifestation that greatly affects the appearance of the facade, favors the penetration of rainwater and, consequently, reduces the durability of the building. Its occurrence is a result of the loss of adhesion between mortar and substrate. In the studied building this damage is very widespread in all the facades analyzed, so two alternatives were proposed to recover them:

- Model MI-D: bonding the detached coating with epoxy resin - due to the historical character of the building, bonding the original mortar back to its place by applying a thin layer of low viscosity epoxy resin can be a simple and effective solution, meeting the recommendations of the material manufacturer, in order to preserve the originality of the facade (Torraca 2009);
- Model MII-D: total replacement of the detached coating – usual repair alternative for building facades. A new coating was made from the reconstituted mortar mix proportions of each facade, ensuring compatibility with the original mortar. The mortar binder: aggregate ratio used to the facade 2 was the same from facade 3 due to restrictions on sample extraction.

2.4 Life Cycle Assessment (LCA)

A Life Cycle Assessment (LCA) was carried out based on NBR ISO 14040 (ABNT 2009) and NBR ISO 14044 (ABNT 2009) in order to identify the environmental impacts of the suggested repair alternatives. Analysis was performed with OpenLCA v.1.9.0 software, Greendelta manufacturer, considering the process from cradle to grave. The

Functional Unit (FU) is 1 m^2 of repaired coating. Figure 7 represents the boundaries considered for the system analysis. The CML baseline method was used to assess environmental impacts. This method enables nine impact categories evaluation: acidification, eutrophication, freshwater ecotoxicity, climate change, human toxicity, marine aquatic ecotoxicity, ozone depletion, photo-oxidant formation and terrestrial ecotoxicity.

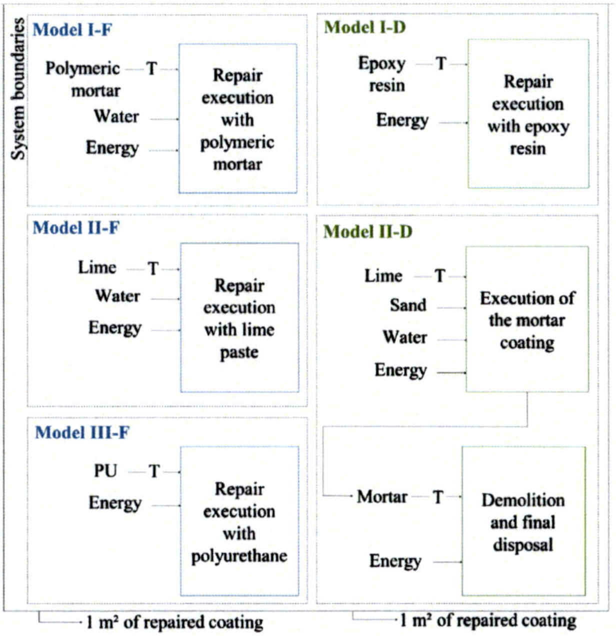

Fig. 7. System boundaries for the proposed intervention strategies. T: transport. MI-F: Polymeric mortar injection for crack closure, MII-F: Lime paste injection for crack closure, MIII-F: Polyurethane Injection for crack closure, MI-D: Bonding the detached coating with epoxy resin, MII-D: Total replacement of detached coating.

The database used in the study was obtained from Ecoinvent v.3.5 and adaptations were made to represent the local reality of the study. Due to the absence of data from the studied region, data from other regions were adapted, considering the best local scenario representation. Table 2 presents the amount of materials and energy necessary to repair all three facades during each process of the study, considering the FU. The energy consumption was adapted from CEMEX Sustainable Development Report (CEMEX 2015), with an 3.12 kwh/m^3 average consumption.

Table 2. Amounts of materials and energy to repair all three studied facades.

Process	Output flow	Input flow	Database	Model
Materials production	Mortar	Lime (kg)	Lime production, milled, packed I lime, packed I Cutoff, U - RoW	0.77 (B) 373.60 (E)
		Sand (kg)	Gravel and sand quarry operation I sand I cut-off, U - RoW	911.24 (E)
		Polymeric (kg)	Adhesive mortar production I adhesive mortar I Cutoff, U - RoW	1.47 (A)
	Epoxy resin	Epoxy resin (kg)	Epoxy resin production, liquid I epoxy resin, liquid I Cutoff, U - RoW	32.26 (D)
	Polyurethane (PU)	Polyurethane (PU) (kg)	Polyurethane production, flexible foam I polyurethane, flexible foam I Cutoff, U - RoW	1.15 (C)
Coating repair	Transport	Lime transportation to the building site (kg.km)	Transport, freight, lorry 1632 metric ton, EURO5 I transport, freight, lorry 1632 metric ton, EURO5 I cut-off, U - RoW	92.40 (B) 44,832.00 (E)
		Sand transportation to the building site (kg.km)	Transport, freight, lorry 1632 metric ton, EURO5 I transport, freight, lorry 1632 metric ton, EURO5 I cut-off, U - RoW	55,585.64 (E)
		Polymeric mortar transportation to the building site (kg.km)	Transport, freight, lorry 1632 metric ton, EURO5 I transport, freight, lorry 1632 metric ton, EURO5 I cut-off, U - RoW	154.35 (A)

(*continued*)

Table 2. (*continued*)

Process	Output flow	Input flow	Database	Model
		Epoxy transportation to the building site (kg.km)	Transport, freight, lorry 1632 metric ton, EURO5 I transport, freight, lorry 1632 metric ton, EURO5 I cut-off, U - RoW	3,645.40 (D)
		PU transportation to the building site (kg.km)	Transport, freight, lorry 1632 metric ton, EURO5 I transport, freight, lorry 1632 metric ton, EURO5 I cut-off, U - RoW	48.30 (C)
	Cleaning water	Cleaning water (kg)	Tap water production, conventional treatment I tap water I cut-off, U - RoW	1,827.00 (A) 1,827.00 (B) 1,827.00 (C)
	Mortar water	Mortar water (kg)	Tap water production, conventional treatment I tap water I cut-off, U - RoW	0.46 (B) 224.16 (E)
	Cleaning	Cleaning Electricity (kWh)	Electricity voltage transformation from high to medium voltage I electricity, medium voltage I cut-off, U - BR	7.57 (A) 7.57 (B) 7.57 (C) 5,220.00 (D)
	Mixture	Electricity for mortar mixing (kWh)	Electricity voltage transformation from high to medium voltage I electricity, medium voltage I cut-off, U - BR	2.35 (E)
End of life	Demolition and final disposal	Electricity for Demolition (kWh)	Electricity voltage transformation from high to medium voltage I electricity, medium voltage I cut-off, U - BR	352.00 (E)

(*continued*)

Table 2. (*continued*)

Process	Output flow	Input flow	Database	Model
		Mortar waste transportation to landfill (kg)	Transport, freight, lorry 1632 metric ton, EURO5 I transport, freight, lorry 1632 metric ton, EURO5 I cut-off, U - RoW	9,709.00 (E)

A: MI-F - Polymeric mortar injection for crack closure; **B: MII-F** - Lime paste injection for crack closure; **C: MIII-F** - Polyurethane Injection for crack closure; **D: MI-D** - Bonding the detached coating with epoxy resin; **E: MII-D** - Total replacement of detached coating.

Table 3 shows valid transport distances for this study, searched by means of a geographical possible suppliers mapping located close to the building and chosen the minimum displacement.

Table 3. Transport distances.

Activity	Transport	Distance (km)
Lime transportation to the building site	Truck	120
Sand transportation to the building site	Truck	61
Polymeric mortar transportation to the building site	Truck	105
Epoxy transportation to the building site	Truck	113
PU transportation to the building site	Truck	42
Mortar waste transportation to landfill	Truck	7

3 Results

3.1 Mapping and Quantification of the Pathological Manifestations

Based on visual analysis, on the photographic records and percussion test results, the pathological manifestations mapping of each building facade were obtained, to a height of up to 2 m having as reference the external level of the floor. Detachments are considerably evident in facades 1 and 3, while the cracks occurrence cracks quite pronounced in facade 3, with north orientation. These pathological manifestations may be the result of several facade degradation factors, such as: solar incidence, wind driven rain, rising damp, among others (Bauer et al. 2015). Table 4 shows linear length of cracks and detachment area per facade.

Table 4. Linear length of cracks and detached areas quantification.

Quantification	Facade 1	Facade 2
Total cracked area (cm^2)	28101.00	38955.00
Cracks density (cm/cm^2)	0.05	0.07
Total linear length of cracks (m)	1.41	2.84
Detached Area (m^2)	11.31	3.50

3.2 Life Cycle Assessment

In the environmental impact evaluation of different alternatives for the crack repair, model MIII-F was taken as reference, to which the value of 1.0 was attributed for all impact categories. The values referring to models MI-F and MII-F are expressed as a proportion of model MIII-F in Fig. 8.

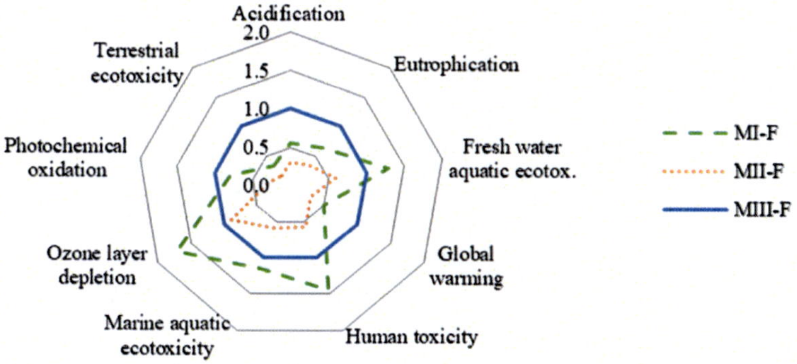

Fig. 8. Relative environmental impacts related to repair of 1 m^2 of repaired coating. MI-F: Polymeric mortar injection for crack closure, MII-F: Lime paste injection for crack closure, MIII-F: Polyurethane Injection for crack closure.

Model MIII-F, with polyurethane injection, presented the highest impact for the photochemical oxidation, terrestrial ecotoxicity, acidification, eutrophication and climate change categories, which also had the highest absolute impacts among all models. Regarding the model MI-F, with polymeric mortar injection, the highest impact were found in the fresh and marine water ecotoxicity, human toxicity and ozone depletion categories. The Model MII-F, with lime paste injection, presented lower impact in all categories, showing less significant impacts for the terrestrial ecotoxicity and climate change categories. There was also a trend to the same categories of the Model MI-F, however, the magnitude is significantly lower.

Figure 9 presents product manufacture and construction processes contribution in the environmental impacts generated by the different intervention alternatives for cracks closure. It should be noted that the construction process was equal for all models and, therefore, the impact of this process in absolute values is equal for models MI-F, MII-F and MIII-F.

Model MI-F, the environmental impacts attributed to the polymeric mortar production made a similar contribution to the construction process for all impact categories. Model MII-F, impacts contribution of the lime paste production process - which takes only water and decarbonated limestone - presents a impacts percentage lower than 3% in all impact categories. Thus, the Model II-F impacts are mostly attributed to the construction process. Finally, Model MIII-F, showed that polyurethane production process had a higher impacts percentage than construction process (close to 75%) for most impact categories, with the exception of human toxicity, marine ecotoxicity and depletion of the ozone layer.

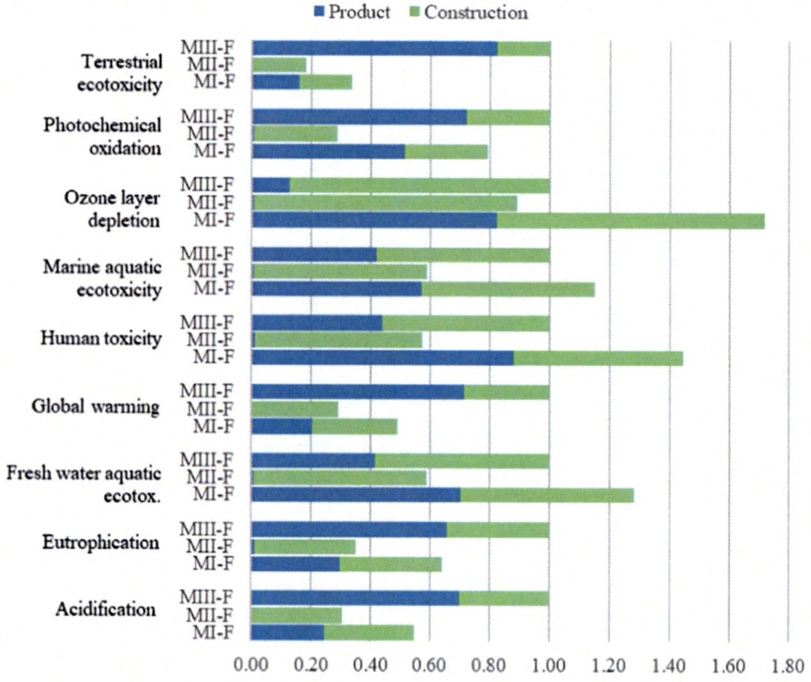

Fig. 9. Models MI-F, MII-F and MIII-F, divided into product manufacture and construction processes. MI-F: Polymeric mortar injection for crack closure, MII-F: Lime paste injection for crack closure, MIII-F: Polyurethane Injection for crack closure.

Figure 10 shows the relative impacts for the two repair alternatives related to coating detachment. Model MI-D was taken as reference, to which the value of 1.0 was attributed for all impact categories. The values referring to models MII-D are expressed as a proportion of model MI-D in Fig. 10.

It was observed that in all impact categories the environmental impacts generated by the model MII-D are 20% lower than those generated by the model MI-D. On the other hand, the quantities of materials used in the epoxy resin repair alternative are much lower than the model MII-D (Table 2). Considering the sum of all facades and models, it results in 32.26 kg of epoxy resin compared to 373.60 kg of lime, excluding water and sand, whose total sum exceeds 1500 kg materials. Thus, it is possible to predict, if the total impact was calculated per kilogram used for repair, the epoxy resin impact solution would be many times greater than the lime paste solution.

Figure 11 presents product manufacture and construction processes contribution in the environmental impacts generated by the different intervention alternatives for coating detachment.

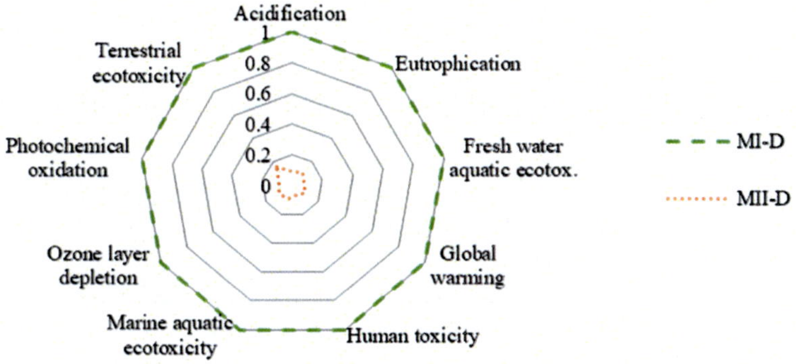

Fig. 10. Relative environmental impacts related to repair of 1 m^2 of repaired coating. MI-D: Bonding the detached coating with epoxy resin, MIID: Total replacement of detached coating.

Model MI-D product manufacture process impacts can be justified due to the epoxy material as the solution, the highest impact were human toxicity, with about 45% (389.53 kg1.4 DB) of the total impacts to this model, and freshwater ecotoxicity with 25% (261.87 kg1.4 DB). A similar result was obtained by La Rosa et al. (2014), where it was observed the contribution of 490.44 kg1.4 DB eq to the human toxicity and 246.50 kg1.4 DB eq for freshwater ecotoxicity for 1 tonne of epoxy. Moreover, it can be noted that the construction stage of model MI-D, which includes the epoxy resin transport to the building site and the energy used in the cleaning process, represents the majority impacts in all categories studied.

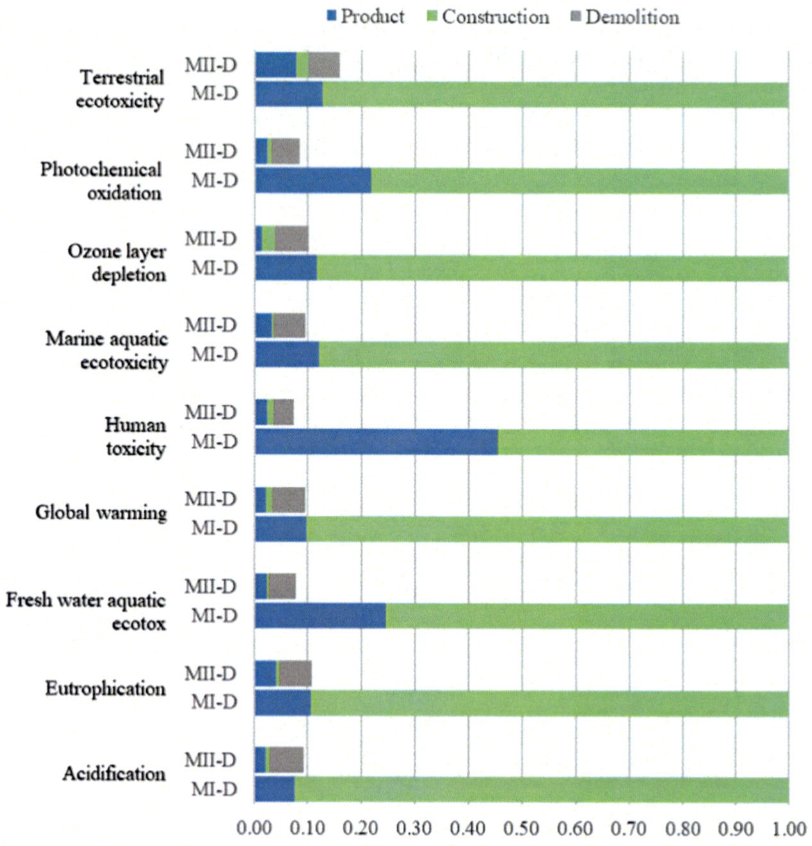

Fig. 11. Potential impacts related to models MI-D and MII-D, divided into the processes of product manufacture, construction and demolition. MI-D: Bonding the detached coating with epoxy resin and MII-D: Total replacement of detached coating.

4 Conclusions

Life cycle assessment is a methodology that can be used to estimate the potential environmental impact of building systems and components. Especially in civil construction, the concern with the preservation of existing buildings through maintenance and repair interventions has been increasing, especially in those of historical importance, requiring reliable data to properly evaluate the different repair solutions. Therefore, the present work evaluated, from the environmental perspective, different types of repair intervention for two recurrent pathological manifestations on the facades of a century-old building: cracks and lining detachments. The main findings of this research can be summarized as:

- The material choice and technique used to repair can significantly affect the environmental impacts of a building;

- The LCA tool proved to be appropriate and helpful for assessing the potential impacts of different intervention alternatives;
- For cracks, the solution with the lowest impact potential was model MII-F, which uses lime paste to repair cracks;
- From an environmental point the indicated alternative intervention for detachment coating is Model MII-D, without polymer-based products.

Acknowledgements. The authors are grateful for the financial support of CAPES and for the guidance and laboratory infrastructure of the research group NORIE/UFRGS.

References

ABNT - Associação Brasileira de Normas Técnicas: NBR ISO 14040: Gestão ambiental - avaliação do ciclo de vida - princípios e estrutura, Rio de Janeiro (2009)

ABNT - Associação Brasileira de Normas Técnicas: NBR ISO 14044: Gestão ambiental - Avaliação do ciclo de vida - requisitos e orientações, Rio de Janeiro (2009)

ABNT - Associação Brasileira de Normas Técnicas: NBR 13749: Revestimento de paredes e tetos de argamassas inorgânicas – especificação, Rio de Janeiro (2013)

Balaras, C.A., Droutsa, K., Dascalaki, E., Kontoyiannidis, S.: Deterioration of European apartment buildings. Energ. Build. **37**(5), 515–527 (2005)

Bauer, E., Castro, E.K., Silva, M.N.B.: Estimativa da degradação de fachadas com revestimento cerâmico: estudo de caso de edifícios de Brasília. Cerâmica **61**(358), 151–159 (2015)

Bochen, J., Labus, M.: Study on physical and chemical properties of external lime–sand plasters of some historical buildings. Constr. Build. Mater. **45**, 11–19 (2013)

CEMEX - Cementos Mexicanos: Concrete solutions for a sustainable future – sustainable development report, Mexico (2015)

Chen, Z., Martínez, L., Chang, J., Wang, X., Xionge, S., Chin, K.: Sustainable building material selection: a QFD-and ELECTRE III-embedded hybrid MCGDM approach with consensus building. Eng. Appl. Artif. Intell. **85**, 783–807 (2019)

Cherry, M.: Architectural history and conservation. In: Forsyth, M (ed.). Understanding Historic Building Conservation, pp. 9–25. John Wiley & Sons (2007)

CIB - International Council for Research and Innovation: CIB report (64): working with the performance approach in building (1982)

De Moraes, G.A.M.: A contribuição de Manoel Itaqui para a Arquitetura Gaúcha. Dissertação (Mestrado). Programa de Pesquisa e Pós-Graduação, Área de Concentração em Teoria, História e Crítica da Arquitetura (PROPAR) da UFRGS, Porto Alegre (2003)

Forster, A.M., Carter, K., Banfill, P.F.G., Kayan, B.: Green maintenance for historic masonry buildings: an emerging concept. Build. Res. Inf. **39**(6), 654–664 (2011)

Kwon, N., Song, K., Ahn, Y., Park, M., Jang, Y.: Maintenance cost prediction for aging residential buildings based on case-based reasoning and genetic algorithm. J. Build. Eng. **28**, 101006 (2020)

La Rosa, A.D., et al.: Environmental impacts and thermal insulation performance of innovative composite solutions for building applications. Constr. Build. Mater. **55**, 406–414 (2014)

Mahmoud, S., Khamidi, M.F., Idrus, A., Ashola, O.A.: Development of maintenance cost prediction model for heritage buildings. J. Teknol. **74**, 2 (2015)

Masuero, A.B., et al.: Instituto eletrotécnico – Diagnóstico das manifestações patológicas. Relatório técnico. Programa de Pós-Graduação em Engenharia Civil: Construção e Infraestrutura, Porto Alegre (2017)

Perret, J., Jouvent, M.: Guide de la maintenance des bâtiments: diagnostic d'un patrimoine bâti existant, prévention des désordres et actions pour y remédier 308 fiches tecniques de suivi des ouvrages, Le Moniteur (1995)

SPH - Setor de Patrimônio Histórico: Instituto Eletrotécnico. Universidade Federal do Rio Grande do Sul, Porto Alegre (2007). https://www.ufrgs.br/patrimoniohistorico/. Access 02 Nov 2019

Sterflinger, K., Little, B., Pinar, G., Pinzari, F., Rios, A., Gu, J.: Future directions and challenges in biodeterioration research on historic materials and cultural properties. Int. Biodeterior. Biodegradation **129**, 10–12 (2018)

Takano, A., Hughes, M., Winter, S.: A multidisciplinary approach to sustainable building material selection: a case study in a Finnish context. Build. Environ. **82**, 526–535 (2014)

Teutônico, J.M.: A laboratory manual for architectural conservators. International center for the study of preservation and restoration of cultural property (ICCROM), Rome (1998)

Torraca, G.: Lectures on material science for architectural conservation. The Getty Conservation Institute, Los Angeles (2009)

Structural Assessment of Historical Timber Roof Trusses: The Case of the Croce di Lucca Church in Naples

Gianfranco De Matteis[1(✉)], Luigi Ricci[2], Corrado Chisari[1], Alberto Mandara[2], and Simeone Panico[3]

[1] Department of Architecture and Industrial Design, University of Campania "Luigi Vanvitelli", Via San Lorenzo – Abazia San Lorenzo, 81031 Aversa, Italy
{gianfranco.dematteis,corrado.chisari}@unicampania.it

[2] Department of Engineering, University of Campania "Luigi Vanvitelli", Via Roma 9, 81031 Aversa, Italy
luigi.ricci@studenti.unicampania.it, alberto.mandara@unicampania.it

[3] Technical Office, University of Campania "Luigi Vanvitelli", Viale Lincoln, 81100 Caserta, Italy
simeone.panico@unicampania.it

Abstract. Large part of the historical heritage is characterised by the presence of timber elements for floors and roof, which need to sustain vertical load and comply with safety prescriptions of current building codes. One of the main stages in the assessment of historical timber structures is the evaluation of material properties. This is not a trivial task, as (i) timber is a complex material, (ii) the presence of defects strongly influences the overall bearing capacity of the element, and (iii) most times exhaustive destructive testing activities may not be feasible. In this respect, the use of resistographic non-destructive tests was proposed in past research as a means to estimate the strength of defect-free wood in longitudinal direction as well as a measure of the material local defectiveness. This approach is formalized in the present paper to derive a design procedure for the structural assessment of ancient timber structures. The procedure is applied to the case study of Croce di Lucca church in Naples. The evaluation of the bearing capacity of the roof structure by means of the proposed experimental methodology reveals significant structural issues in the system, in particular regarding the bottom tie beam.

Keywords: Structural assessment · Timber structures · Churches · Material characterisation · Resistographic test · Non-destructive tests

1 Introduction

The estimation of the mechanical properties of ancient timber elements, such as beams, trusses, floors, is critical for the structural assessment of historical buildings and monuments. Standards as UNI EN 14081 (2018) provide protocols for machine classification

I. Vayas and F. M. Mazzolani (Eds.): PROHITECH 2021, LNCE 209, pp. 1111–1125, 2022.
https://doi.org/10.1007/978-3-030-90788-4_85

of timber, based on which the current Italian and European standards framework define classes of resistance to be assigned to the structural elements (UNI EN 338, 2016). However, especially in the case of historical heritage, it is not always possible to extract samples on site. Alternatively, visual classification of in-situ structural timber, which in the past was carried out according to the UNI 11119 standard (2004), should be considered with due caution due to the large uncertainties involved and the subjectivity of the output.

For these reasons, non-destructive (NDT) or semi-destructive (SDT) testing methodologies for estimating material parameters of existing timber structures represent currently an active branch of research. Reviews of the state-of-the-art NDT-SDT methods are presented in Kloiber et al. (2015) and Feio and Machado (2015), where several techniques are critically assessed in their capability of estimating timber mechanical properties. In general, NDT or SDT techniques are effective insofar their output can be correlated with a minimum degree of uncertainty to the stiffness, strength and density parameters of the structural material, which very seldom can be measured directly. In this respect, De Matteis and co-workers (Calderoni et al., 2010; De Matteis et al., 2018; Guarino et al., 2018) investigated the use of resistographic tests to estimate timber strength parameters, performing extensive experimental activities including resistographic and compressive tests on real and defect-free specimens.

Building on these works, this paper proposes a framework including experimental testing by means of resistographic tests and appropriate data post-processing aimed at deriving the design strengths of ancient in-situ structural timber. This procedure is then applied to a real-world case study, namely the structural verification of the chestnut timber elements of the roof in Croce di Lucca church in Naples.

2 Evaluation of In-situ Design Strength

2.1 Design Strength Values According to the European and Italian Standards

Structural assessment of existing timber structures needs accurate estimation of strength properties of the elements. Following the approach proposed by UNI EN 1995-1-1 (2014) and NTC18 (2018), the design strength can be calculated through Eq. (1):

$$X_d = \frac{k_{mod} \cdot X_k}{\gamma_M} \tag{1}$$

where X_k is the characteristic value of the material property, γ_M is the partial safety factor and k_{mod} is the corrective factor that takes into account the effects of the duration load and humidity. Generally, wood does not degrade or significantly modify its strength characteristics over time unless there are alteration factors, such as xylophagous organisms, variations of humidity, mechanical actions, fire, etc., so the use of k_{mod} must be meant as a way to account for the viscous effects of timber on strength.

This relationship assumes knowledge of the characteristic strength X_k, which for new constructions can generally be obtained from the tables of the UNI EN 338 (2016), where the resistance classes are defined according to machine classification following UNI EN 14081–2 (2018).

As stated above, in the case of historical structures it is not always possible to perform destructive tests on in-situ structural timber elements in order to assign a resistance class and derive the characteristic values. The alternative is to estimate the design strength of in-situ structural timber from the output of non-destructive tests. The issue, in this case, is represented by the need of proved and reliable correlations between the NDT tests output and the strength parameters. The experimental test investigated in this paper and the correlation of its results with the unknown material properties is the topic of the following sections.

2.2 Experimental Test and Data Post-processing

The resistographic test uses a device, called resistograph, which measures the resistance given by the material to the advancement of a small tip. The resistance opposed to the perforation mainly depends on wood density and its variation can be related to the distribution of degraded and non-degraded areas. This test therefore provides a picture of internal degradation and thus an estimation of the quality of the examined material.

The resistographic test output consists of a graph in which the resistance to penetration on the ordinate axis is plotted against the drilling depth (Fig. 1). Generally, drilling is performed perpendicularly to the surface, although it is also possible to pierce the wooden elements with an angle other than 90° to evaluate covered areas of the structural elements; in particular cases, resistographic tests can also be carried out in the longitudinal direction, but this assumes that the element head is available for inspection (usually not in existing structures).

The number of tests to be performed is variable in that it is determined by the size of the element, its accessibility, the presumed extent of degradation and other issues. Generally, the operator starts with a minimum of three tests. Sometimes up to 10–15 tests are performed.

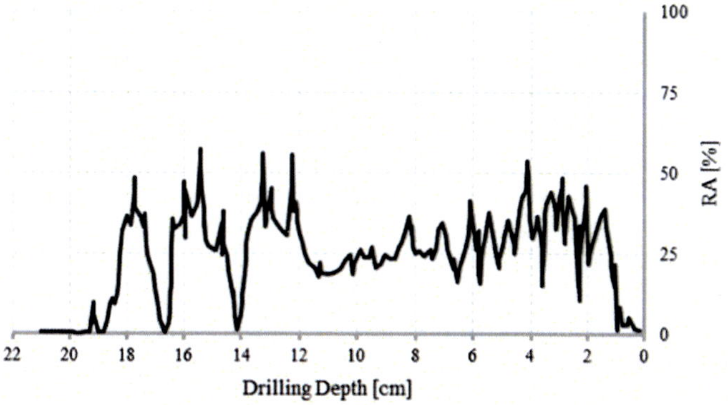

Fig. 1. Example of typical results obtained from a resistographic test.

As shown in Calderoni et al. (2010) and De Matteis et al. (2018), it is possible to correlate the results of the resistographic test to the compression strength in the grain direction $f_{c,0}$ obtained from destructive tests. To this purpose, it is necessary to calculate the mean value of relative resistance $AM_{(L,T)}$, representing the ratio between the integral of the diagram area and the drilling depth, evaluated in the transversal or longitudinal direction:

$$AM_{(L,T)} = \frac{\int_0^H RA \cdot dh}{H}[\%] \tag{2}$$

where RA is the relative resistance value, provided by resistographic test and H is the drilling depth.

The resistographic tests carried out in transversal direction, differently from the longitudinal ones, almost always have non-negligible variations along the drilling depth caused by the defects of materials. They can in turn be used to quantify the defectiveness of the material, by evaluating two different values of AM_T (Calderoni et al., 2010):

- $AM_{T,0}$ where only resistances RA greater than 5% are considered in Eq. (2), assuming thus that lower values are governed by defects (knots, cracks, etc.);
- $AM_{T,TOT}$ in which the RA values along the whole drilling depth are taken into account.

Following the approach proposed in Calderoni et al. (2010), the defectivity factor C_{DEF} is defined as:

$$C_{DEF} = \frac{AM_{T,0} - AM_{T,TOT}}{AM_{T,0}} \tag{3}$$

This coefficient allows to quantify the defectiveness of material, and will be used in the estimation of the resistance.

2.3 Derivation of the Compressive Strength in the Grain Direction

In previous research (De Matteis et al., 2018; Guarino et al., 2018), experimental tests were performed on ancient chestnut timber specimens, representing the wide majority of timber used in historic buildings, with the aim of elaborating useful abacuses relating compression strength $f_{c,0}$ to the results of in-situ resistographic tests (Fig. 2). While all the experimental results were considered in the evaluation of the regression law, it can be noticed from Fig. 2 that, depending on the value of defectiveness C_{DEF}, the individual points are differently located with respect to the bounds of the confidence interval.

The correlation between $f_{c,0}$ and AM must take into account the defectiveness of the material, represented by the C_{DEF} coefficient. Taking into account the experimental evidences, the following rules are herein proposed:

- the points characterized by defectivity factor less than 3%, are closer to the upper line. In this case it is possible to assume the maximum value, $f_{c,0,max}$, corresponding to the upper bound of the confidence interval.
- the points characterized by defectivity factor greater than 10%, are closer to the lower line. In this case it is possible to assume the minimum value, $f_{c,0,min}$, corresponding to the lower bound of the confidence interval.
- if the points are characterized by defectivity factor between 4% and 9%, a mean value between $f_{c,0,min}$ and $f_{c,0,max}$ can be assumed.

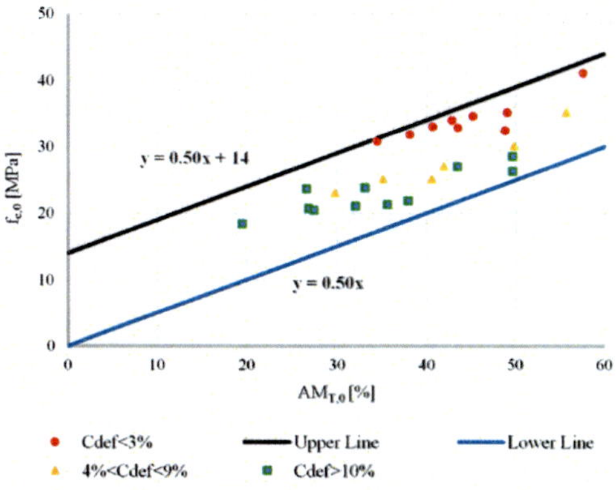

Fig. 2. Chestnut wood for real specimens: Correlation between $f_{c,0}$ and $AM_{T,0}$ and effect of defectiveness coefficient.

2.4 Derivation of the Characteristic and Design Compressive Strength

The population of n values of $f_{c,0}$ obtained for each structural element can be seen as sample of random variable X_i representing the material strength. According to the outcome of the extensive analysis carried out by Ravenshorst et al. (2004) statistical distribution of strengths in hardwood as chestnut can be well-represented by a normal distribution, provided that the defects of the elements such as knots and grain angle are limited. For this reason, assuming a parametric Gaussian distribution, it is possible to obtain the 5% fractile and thus determine the characteristic value of the compressive strength in the grain direction:

$$f_{c,0,k} = \mu - 1.645\sigma \tag{4}$$

where:

$$\mu = \frac{\sum_{i=1}^{n} X_i}{n} \tag{5}$$

$$\sigma = \sqrt{\frac{\sum_{i=1}^{n} (X_i - \mu)^2}{n}} \tag{6}$$

With this value of characteristic strength, it is possible to evaluate the design strength to be used in the structural checks by means of:

$$f_{c,0,d} = \frac{f_{c,0,k}}{\gamma_M} \tag{7}$$

where, according to the Italian standard, $\gamma_M = 1.5$ for massive wood. This equation differs from Eq. (1) for the k_{mod} coefficient, which is not present here. This has been deemed appropriate since the procedure considers in-situ material, in which viscous effects have been already accounted for. It is also important to note that the methodology has determined the relevant regression law as a function of the defectiveness C_{DEF}, and thus the material defects are included in the estimation.

In the next section, the proposed procedure will be applied to a case study of Croce di Lucca church in Naples. In particular, the design strength of ancient chestnut timber elements of the roof trusses will be obtained and used to perform a structural verification of the roof according to the limit states proposed in NTC18.

3 Case Study: Croce di Lucca Church in Naples

3.1 Historical References

Croce di Lucca church was built in 1537 in Naples, together with Carmelitane's monastery, and was dedicated to the crucifix venerated in Lucca Cathedral. During the XVII century, restoration and expansion works were carried out by architect Francesco Antonio Picchiatti but, in 1903, caused by University Polyclinic building, the monastery, cloister and some parts of church were demolished. The church is currently property of the University of Campania "Luigi Vanvitelli".

The front of the church consists of a piperno portal with marble inlaid decorations in which a painting depicting the Crucifix of Lucca was inserted (Fig. 3a); the internal structure consists of a single nave with side chapels and marble decorations executed by Pietro Barberiis.

The ceiling is made of ancient timber and its structure is realized with panels arranged above each other (Fig. 3b). This technique was often used in ancient structures to lighten the floor. The lower part is coated in gold and, in the central part, a paint of the Madonna del Carmine and some Saints, a 17th century work attributed to the school of Battistello Caracciolo, is present.

The roof structure has a rectangular plan with a double-pitched roof, which is supported by palladian chestnut-wood trusses (Fig. 3c) longitudinally arranged at approximately 2.0 m spacing. These, through the rafter-tie beam node, transfer the load on the lateral tuff walls. Moreover, in addition to absorbing the vertical load and consequent thrust of the roof, the lower tie-beams have the task of supporting the "timber coffered ceiling" by means of steel hangers arranged almost homogeneously over the entire roof (Fig. 3d), leading to a mixed tension-bending stress state. The resulting stress state induced significant deformations in the "timber coffered ceiling". In particular, through a laser scanner survey (Fig. 4), it was possible to observe that in the central part of the ceiling (points 3 to 10 in Fig. 4), deflections spanned from 11 to 20 cm. For this reason, the church is currently closed to the public.

Fig. 3. Croce di Lucca church: (a) Main entrance in piperno material; (b) Interior of the church with a view of the golden "coffered ceiling"; (c) Axonometric view of the roof trusses; (d) Palladian timber trusses supporting both the roof and the "coffered ceiling".

Fig. 4. Laser scanner survey of the ceiling and observation points for deflections.

3.2 Visual and Instrumental Inspections

To carry out structural checks on the timber elements composing the trusses of the church roof, visual and instrumental inspections were carried out according to UNI11119 (2004). In order to assign a category to the structural ancient timber elements, visual inspections were carried out by directly observing the external surface of elements under investigation. The defects taken into consideration were those affecting the mechanical properties of the wood such as knots, cracks, fibre deflection. The localization and extension of these were evaluated. As far as biological degradation is concerned, reference was mainly made to attacks by xylophagous insects and fungus, responsible for serious decreases strength of wood. According to UNI11119 (2004) , the structural elements were assigned category II as presenting a discrete state of conservation and a slight superficial degradation due to attack by insects.

The instrumental inspection (Fig. 5) consisted of the resistographic test described in the previous Section. An IML RESI F400-S resistograph with the following properties was used:

- resolution: 0.1 point per mm of drilling;
- drilling depth: 400 mm;
- feed speed: between 0 and 100 cm / min.

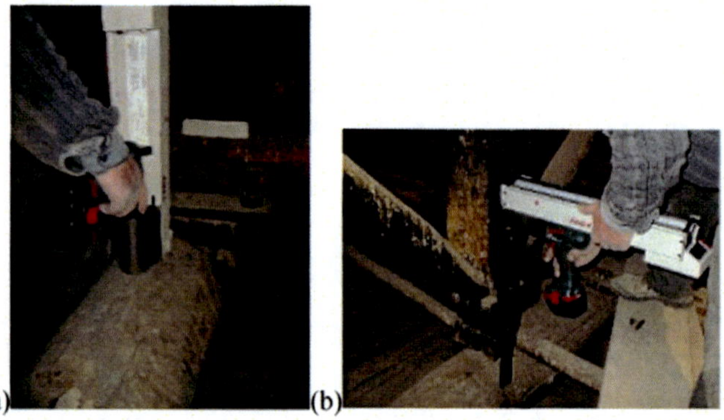

(a) (b)

Fig. 5. In-situ resistographic tests on the truss elements of the Croce di Lucca church.

Two trusses were analysed: truss n°5, identified with the code CA-5 and truss n°10 identified with the code CA-10 (Fig. 6). One resistographic test was performed on each element composing the Palladian truss, leading to a total of 9 tests.

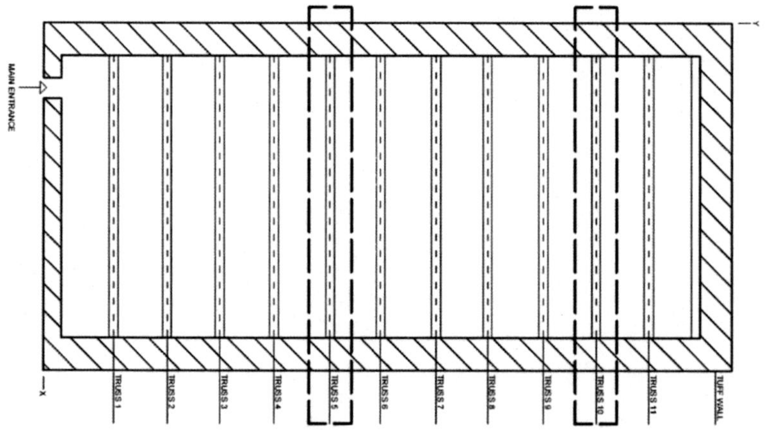

Fig. 6. Geometrical survey of Croce di Lucca church and investigated trusses.

3.3 Evaluation of Strengths

The 9 resistographic tests were performed in the transversal direction only, due to the impossibility of execution in the longitudinal direction. Starting from the data obtained from the graphs of the resistographic tests, it was possible to obtain the relative mean resistances $AM_{T,TOT}$ and $AM_{T,0}$, in the transversal direction, and the defectivity coefficient C_{DEF} (Table 1).

Table 1. Parameters derived from in-situ resistographic tests

Specimen	Elements	$AM_{T,TOT}$ [%]	$AM_{T,0}$ [%]	C_{DEF} [-]
CA-5	Left rafter	23.57	24.58	0.041
CA-5	King post	30.56	32.05	0.047
CA-5	Tie beam	31.82	33.02	0.036
CA-5	Left strut	34.33	36.91	0.070
CA-10	Right rafter	30.86	32.14	0.040
CA-10	Left rafter	28.62	29.70	0.036
CA-10	Tie beam	33.82	34.76	0.027
CA-10	King post	42.70	44.38	0.038
CA-10	Right strut	15.61	17.93	0.129

These were then used to define the compressive strength $f_{c,0}$ parallel to the grain direction after selecting the appropriate reference abacus. In this case, the reference abacus is that of Fig. 2, as NDT tests were carried out on real specimens in situ. The values of $f_{c,0}$ (Table 2, Fig. 7) were then estimated by considering the rules:

- $f_{c,0,min}$, if $C_{DEF} < 0.03$;
- $f_{c,0,max}$, if $C_{DEF} > 0.10$;
- $f_{c,0,mean}$, if $0.04 < C_{DEF} < 0.09$

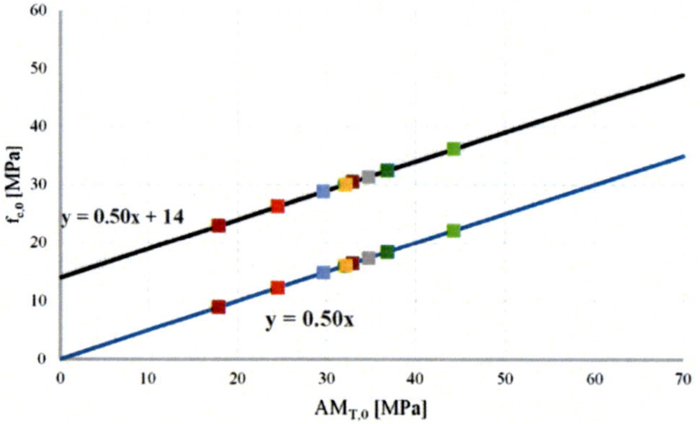

Fig. 7. Upper and lower values of $f_{c,0}$, for the 9 tests performed, starting from the mean relative resistance $AM_{T,0}$ in the transversal direction.

Table 2. Selection of the appropriate $f_{c,0}$ value, as a function of the defectiveness factor C_{DEF}.

Specimen	Elements	$AM_{T,0}$ [%]	C_{DEF} [-]	$f_{c,0}$ [MPa]			
				$f_{c,0,min}$	$f_{c,0,max}$	$f_{c,0,mean}$	$f_{c,0}$ assumed
CA-5	Left rafter	24.58	0.041	12.29	26.29	19.29	19.29
CA-5	King post	32.05	0.047	16.03	30.03	23.03	23.03
CA-5	Tie beam	33.02	0.036	16.51	30.51	23.51	23.51
CA-5	Left strut	36.91	0.070	18.45	32.45	25.45	25.45
CA-10	Right rafter	32.14	0.040	16.07	30.07	23.07	23.07
CA-10	Left rafter	29.70	0.036	14.85	28.85	21.85	21.85
CA-10	Tie beam	34.76	0.027	17.38	31.38	24.38	17.38
CA-10	King post	44.38	0.038	22.19	36.19	29.19	29.19
CA-10	Right strut	17.93	0.129	8.97	22.97	15.97	22.97

To derive the characteristic compressive strength, all the values evaluated in Table 2 were considered as samples of the same distribution. Then, it was possible to determine the mean μ, standard deviation σ and finally the 5% fractile $X_{0,05} = f_{c,0,k} = \mu - 1.645\sigma$, as reported in the following Table 3.

Once the characteristic strength in the grain direction $f_{c,0,k}$ was estimated, the other characteristic parameters of resistance were selected using UNI EN 384 (2019), which

Table 3. Characteristic value $f_{c,0,k}$ of compressive strength in the grain direction

Specimen	Elements	Xi = fc,0 [MPa]	Mean μ [MPa]	Dev. St σ [MPa]	$X_{0.05} = f_{c,0,k}$ [MPa]
CA-5	Rafter Leftt	19.29	22.86	3.37	17.31
CA-5	King post	23.03			
CA-5	Tie beam	23.51			
CA-5	Strut Left	25.45			
CA-10	Rafter Right	23.07			
CA-10	Rafter Left	21.85			
CA-10	Tie beam	17.38			
CA-10	King post	29.19			
CA-10	Strut Right	22.97			

also assumes knowledge of the average elastic modulus in the grain direction $E_{0,mean}$ and the characteristic value of the volume mass ρ_k. For this case study the following values were assumed:

- $E_{0,mean} = 9000\ MPa$. This value was obtained for chestnut wood with category II, according to the UNI 11119; (2004)
- $\rho_k = 395\ \text{kg/m}^3$. This value was obtained according to the UNI EN 11035–2 (2010) for chestnut wood.

In Table 4 the characteristic and corresponding design strength values of the truss elements are reported. The latter are obtained by applying Eq. (7).

Table 4. Characteristics and design strength values according to the proposed method and UNI EN 384 (2019).

Parameters	Characteristic values [MPa]	Design values [MPa]	Parameters	Characteristic values [MPa]	Design values [MPa]
$f_{m,k}$	16.20	10.80	$E_{0,mean}$	9000.00	9000.00
$f_{t,0,k}$	9.72	6.48	$E_{0,0.5}$	7560.00	7560.00
$f_{t,90,k}$	0.60	0.40	$E_{90,mean}$	600.00	600.00
$f_{c,0,k}$	17.31	11.54	G_{mean}	562.50	562.50
$f_{c,90,k}$	3.95	2.63	$\rho_k[\text{kg/m}^3]$	395.00	395.00
$f_{v,k}$	3.49	2.32			

With the design values obtained in Table 4, it is possible to carry out the checks on structural timber elements required by NTC18 and EC5. In particular, the following were performed:

- strength checks at the ultimate limit state (ULS);
- stability checks at the ultimate limit state (ULS);
- displacement checks at the serviceability limit state (SLS).

The analyses were carried out by means of SAP2000 Finite Element program (Figs. 8, 9).

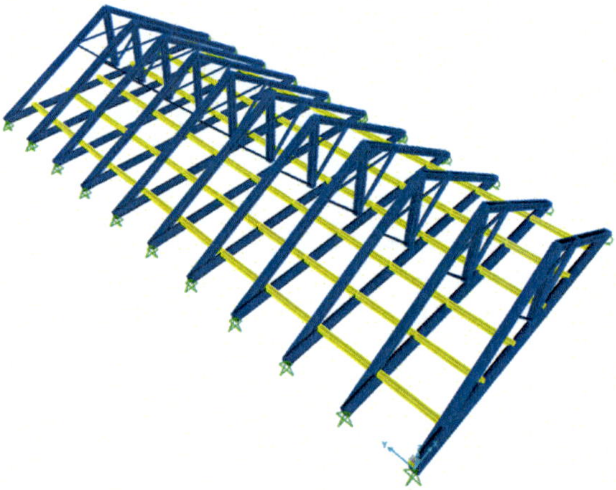

Fig. 8. Calculation model of the timber roof truss: 3D view.

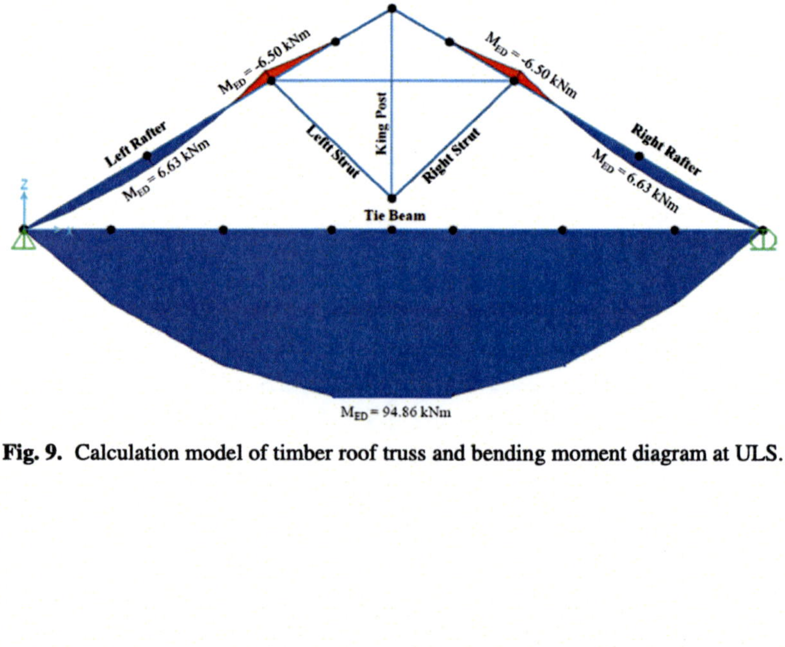

Fig. 9. Calculation model of timber roof truss and bending moment diagram at ULS.

In Table 5 the results of the verifications are reported. These highlight problems of strength, stability and deformation of the truss tie beam. Such problems will need to be solved by means of appropriate retrofitting actions aimed at reducing the stress state on the element. The design and application of such retrofitting action is underway.

Table 5. Check results according to NTC18 and EC5, with design strength values calculated according to the proposed method.

Strengths and displacement		Rafter	Tie beam	King post	Strut
Tensile	Resistance ULS	-	TRUE	TRUE	-
	Stability ULS	-	-	-	-
Compression	Resistance ULS	TRUE	-	-	TRUE
	Stability ULS	TRUE	-	-	TRUE
Bending	Resistance ULS	TRUE	**FALSE**	-	-
	Stability ULS	TRUE	**FALSE**	-	-
Shear	Resistance ULS	TRUE	TRUE	-	-
	Stability ULS	-	-	-	-
Buckling	Resistance ULS	TRUE	-	-	-
	Stability ULS	TRUE	-	-	-
Tensile-bending	Resistance ULS	-	**FALSE**	-	-
	Stability ULS	-	**FALSE**	-	-
Deformation	SLS	TRUE	**FALSE**	-	-

4 Conclusions

The determination of the mechanical characteristics of in-situ structural timber is a delicate process but critical for the structural assessment of ancient structures. For in-situ structural timber elements it is not always possible to perform destructive tests in order to determine the mechanical characteristics and the resistance class according to

UNI EN 338 (2016). It is evident that, in these cases, an alternative procedure is necessary for the designers to obtain the resistance values to be used for the structural checks.

In this paper, an experimental procedure aimed at the estimation of material strength of ancient timber is proposed. This is based on the outcomes of resistographic tests, which are non-invasive, simple to perform and for which reliable correlation with destructive tests is available in the literature. A formalised procedure aimed at establishing the design strength to be used in the structural assessment accounting for defectiveness of the material and dispersion of experimental results is described here. Since regression laws from past research are only provided for compressive strength in the grain direction, the remaining material parameters, such as specific weight, elasticity modules, bending strength, tensile strength and shear strength, have been selected starting from the former by using the relationships proposed in UNI EN 384 (2019).

The procedure has been applied to the case study of the timber roof of Croce di Lucca church in Naples. Through the use of the proposed procedure it was possible to check strength, stability and deformation of the components of the roof truss. In particular, it has been noted that the truss tie-beams require retrofitting interventions as they are extremely stressed by the load transferred by the coffered ceiling underneath. Future research will be focussed on further validation of the procedure by means of extensive experimental programmes, which will also aim to establish reliable regression relationships for different types of wood and all parameters involved in the structural checks.

Acknowledgements. The authors wish to thank the company GeoConsult srl for the technical support provided on site in the execution of visual and instrumental inspections.

References

Calderoni, C., De Matteis, G., Giubileo, C., Mazzolani, F.M.: Experimental correlations between destructive and non-destructive tests on ancient timber elements. Eng. Struct. **32**, 442–448 (2010)

De Matteis, G., Guarino, F., Mandara, A.: Correlation between NDT results and mechanical strength of structural timber, Le Vie dei Mercanti, XVI International Forum (2018)

Feio, A., Machado, J.S.: In-situ assessment of timber structural members: combining information from visual strength grading and NDT/SDT methods – a review. Constr. Build. Mater. **101**, 1157–1165 (2015)

Guarino, F., Mandara, A., De Matteis, G.:. Destructive tests for mechanical characterization of in situ structural timber, Le Vie dei Mercanti, XVI International Forum (2018)

Kloiber, M., Drdácky´, M., Machado, J.S., Piazza, M., Yamaguchi, N.: Prediction of mechanical properties by means of semi-destructive methods: a review, Construct. Build. Mater. **101**, 1215–1234 (2015)

NTC. Technical standards for construction, Superior Council of Public Works, Official Journal of the Italian Republic (2018)

Ravenshorst, G., Van der Linden, M., Van de Kuilen, T.J-W.: An economic method to determine the strength class of wood species, Heron **49** (2004)

UNI. Structural wood–Visible classification of timber according to mechanical strength–Part 2: Rules for visual classification according to mechanical strength and characteristic values for types of structural timber, European Standard, 11035–2 (2010)

UNI. Cultural heritage–Building load bearing structures–In situ inspection for the diagnosis of the elements in place, Italian Standard, 11119 (2004)
UNI EN. Timber structures–Strength graded structural timber with rectangular cross section Part 2: machine grading; additional requirements for type testing, European Standard, 14081–2 (2018)
UNI EN. Design of timber structures Part 1–1: general - Common rules and rules for buildings, European Standard (2014)
UNI EN. Structural wood–strength classes, European Standard, 338 (2016)
UNI EN. Structural timber–determination of characteristic values of mechanical properties and density, European Standard, 384 (2019)

Unmanned Aerial Vehicles for Mapping and Inspecting Historical Constructions

Alexandros Feretzakis(✉) and Efstratios Badogiannis

NTUA School of Civil Engineering, Zografou, Athens, Greece
info@inspedrone.com, badstrat@central.ntua.gr

Abstract. Structural inspection techniques of historical constructions are technologically developing in order to assist scientists with their protection and preservation. Nowadays UAV systems are widely used for the inspection of existing structures especially for all the difficult to reach areas, as they can provide an accurate and economic method for structural visual inspections. This paper describes a way to obtain various photogrammetric products from aerial photographs of the Manari Bridge located in Peloponnese, Greece, using a UAV system with photogrammetry techniques from the pix4D software. The results demonstrated the applicability of UAV's in engineering and herein their capabilities for the inspection of historic structures, as well as their potential to recognize defects and identify their causes.

1 Introduction

Historical structures are often defined as a single or a group of buildings with a universal value that is noticeable in historical, art or science perspective [1]. They may be buildings, landscapes, or any type of structure that has national significance, and they are usually protected by laws since they have been recognized by the official national historic agencies. Situated in a historic or heritage location where fragments of political, military, cultural, or social history have been preserved due to their cultural heritage value. Without proper and continuous conservation there is a concern that historical structures will disappear, overwhelmed by time, therefore we must explore new methods for their conservation. The effort to preserve and restore historic structures should be conducted so that the younger generation can have a chance to witness and enjoy the ancient history in its restored or in its current state. We have a duty for the next generations to preserve our history and present them in a state that is acceptable, before it is too late.

Recently we witness the growth of numerous research studies [2–6] aiming to develop new systems for the inspection of historic buildings and infrastructure in general. UAV's are used extensively in these studies as they can provide new cost-effective techniques. With regards to infrastructure inspections the research of Chan et al. [7] presented a detail comparison between drone-based and conventional inspection practices. Several considerations such as the various inspection requirements, the cost-benefit assessment, and the challenges of aerial platforms were presented with regards to the drone capabilities for bridge inspections. The conclusions presented was that UAVs have some important advantages over traditional inspection practices including reduced cost and

I. Vayas and F. M. Mazzolani (Eds.): PROHITECH 2021, LNCE 209, pp. 1126–1137, 2022.
https://doi.org/10.1007/978-3-030-90788-4_86

inspection time, minimal risk for the inspectors and improved quality of the inspection results.

Typical conservation methods consist from four different stages in the preservation and restoration of historical buildings: the preliminary technical studies, dilapidation studies, approved plan of action and finally the execution of the complete conservation program. This paper will focus on the first stage of the dilapidation studies which are more concerned with the process of identifying the overall condition of the buildings. It is carried out by conducting a detail visual inspection to collect digital images and document the condition of the structure and use the images to carry out further analysis in order to plan the action to be carried out. Visual inspections are probably the most important of all the non-destructive tests as they can provide valuable information to the well-trained eye. Normally, they are limited to the surfaces of the structure that are visually accessible, therefore the findings are mostly related to workmanship, structural serviceability, defects and material deterioration. The reliability of visual inspections as a method of NDT is a widely discussed topic in the preservation of historic structures. Despite of several limitations for visual inspections, it is one of the most widely accepted methods used in practice for asset management of buildings and bridges. Currently, due to advancement in computer technology and the use of UAVs, researchers are working towards automating the inspection process using digital image analysis.

Unnamed Aerial Vehicles have evolved beyond their military origin to become powerful business tools and while they already made the leap to the consumer market, and used for both commercial and civil government applications, creating therefore an opportunity to develop and carry out detail maintenance and structural inspections to all type of structures. Cultural heritage documentation is one of the major sectors where UAV systems can find a wide application, mainly due to their low cost, flexibility, availability and readiness to perform [8]. Access limitations can surely affect a visual inspection and in many cases, they result in poor image quality therefore the use of Unmanned Aerial Vehicles (UAVs) has recently become a very popular tool for acquiring aerial and close-range images.[9] This adds the advantage of aerial views over other classical terrestrial techniques, where traditionally special mechanisms, helicopters and airplanes, were necessary to rectify the poor access issues.. For our research we used state-of-the-art commercial UAVs to acquire the images of the Manari Arch Rail Bridge, build in the 19th century, located in the Peloponnese region of Greece.

The engineer conducting a visual inspection must be able to differentiate between the various signs of distress which may be encountered. These can include cracks, pop-outs, spalling, disintegration, color change, weathering, staining, surface blemishes and lack of uniformity. For example, it is very important whether the defects are random or appear in a specific pattern and whether the defect is confined to certain locations of members or is present all over the structure. Visual comparison of similar members is particularly valuable as a preliminary to testing to determine the extent of the problems in such cases. Extensive information about the structures condition must be gathered during the visual inspection in order to provide a preliminary indication of the condition of the structure and allow formulation of a subsequent testing program. It is important that the inspector properly document all observations related to the environmental exposure, durability, and the structure's performance in order to create a database of the structure and be able to review it when necessary. For large structures, such as bridges, which are normally difficult to inspect, the use of Photogrammetry (PG) offers much more than an alternative to a visual inspection because it captures more details than the human eye can detect, visualizes structural details from multiple angles and a 3D model is feasible to be created for a more thorough inspection [10]. This model can also be digitally stored for future inspections, setting a structure in a monitoring status that could show how the structure has changed over time and how this could be linked to changes in durability.

Proper documentation of heritage buildings can also assist engineers with the restoration process as the data collected will provide valuable information, especially when in most cases drawings and as build records are no longer available. A photogrammetry model can be analyzed, and accurate measurements can be extracted in terms of the structure's dimensions and morphology. These models are extremely useful for the evaluation of defects and the overall structural assessment because we can study any geometric anomalies, such as out movements, deformations and collapses. It is very important to highlight that if the same models are available prior to a natural disaster or an earthquake with the application of change detection techniques we can immediately evaluate in detail the before and after condition of the structure. Creating a digital database of all historic bridges is a very important step towards the preservation of the historic structures. As technology advances the development of optical sensors and dense matching algorithms assist with automatic defect detection techniques that are valuable for the documentation of heritage structures. Results based on available research show that although the dense matching of these UAV images may generate centimeter-level precision, a further increase in precision is often hampered by the quality of the onboard sensor [11].

2 Case Study

The Manaris Railway Bridge shown in 'Fig. 1' is the longest stone curved bridge in Greece and was constructed in the 19th century in order to support the railway system. The bridge was designed by the French engineer A. Cotteland for the "Société Anonyme Internationale de Construction et d'Entreprises de travaux publiques" and it was constructed during the period of 1891 and 1896. The bridge is no longer in service but remains in good condition and has been characterized as a historic structure part of the Greek railway legacy. The bridge's total curve span is equal to 116,00 m, its height is 20,00 m and its width is 4,25 m. It consists of 8 semicircular arches, the spans and the radius of which are 12,50 m and 5,75 m long, respectively. Under the train circulation deck, there is a 2-m infill of stones and the bridge itself is made of ashlar masonry with optimum construction quality and average block dimensions of 60 x 40 x 40 cm. Masonry arch bridges can be viewed as among the most sustainable structures ever to have been built and many have already been in service for hundreds of years without significant repair or strengthening works, exceeding the design life requirements of modern structures [12].

Manaris bridge is one of the most well-preserved stone bridges in Greece and while it appears to be in very good condition some defects have been observed. The structure can surely benefit from a fair amount of maintenance and some repairs are required in

Fig. 1. Manaris Railway Bridge in Peloponnese, Greece.

order to avoid the further expansion of the defects and preserve the bridge condition. It is a historic structure and one of the reasons that was selected for our study is to document the condition of the bridge because in the case of an earthquake engineers will be called to repair and restore the structure to its pre earthquake condition. In a country with many earthquakes like Greece documentation of historic structures is paramount.

3 Method and Technique

The DJI Phantom 4 Advance, equipped with a 12Megapixel camera, was used in order to collect images that will produce the photogrammetric model for our study. The flight was planned by using both the automatic 3D map option from the Pix4D platform and by flying manual in order to collect images from a close distance and from areas under the bridge arches. A total of 1233 images were captured with an overlap of 90% between them, however, obstacles such as trees and plantation in general reduced the quality of the 3D model and the final orthoimages. A photogrammetric model consists of thousands or millions of points, each with a 3D position and colored from the pixel in the digital image. This collection of points is referred to as the point cloud. Using sophisticated point cloud manipulation software, feature extraction routines can be run on the model to detect trouble areas based on the color of the pixel.

Data collected with the UAV was processed using the Pix4D photogrammetry software and the masking procedure was applied on the images aiming to exclude the background all through the photo alignment process to reduce the noise of the point cloud. During data processing parts of the 3D model, such as trees in our case, were masked out from the digital images and the 3D point cloud to reduce processing time and avoid errors. In order to increase the accuracy of our model we introduced four ground control points, (GCP's), which are points of known location used both to reference the photogrammetric model with an established local or national coordinate system as well as improve the overall accuracy of the model. Furthermore, we introduced 36 manual tie points (MTP's) in order to improve the reconstruction accuracy and the merging quality of the model. Due to collecting photos from both manual and automatic flights we had to first create two different models and then merge them together into the final model, therefore by introducing MTP's we improve the quality and the accuracy of the final product by identifying the same points in many different photos.

The imagery from the UAV was transmitted in real time to the operator, ensuring significant image overlap, while maneuvering the UAV to capture the skewed imagery from the object of interest. Automatic flight was performed at an elevation ranging from 30 m to 40 m and a manual flight was conducted for the bridge façades and under the arches. With this method the oblique images required for the 3D model were generated and all different angles of the structure in an efficient manner were managed to be covered. Unfortunately, the Pix4D software doesn't recognize or allows you to choose the type of structure to be inspected therefore the operator must conduct both automatic and manual flights to collect accurate images in order to minimize issues related to texture homogeneity.

The topographic data available by the Pix4D data acquisition process were used for the construction of a 3D photogrammetric model and further processed for the quantification of the damage. For example, area and volume of extensive cracking or spalling was calculated with high accuracy and speed compared to traditional methods. Furthermore, defects were indicatively identified and marked on the 3D model, making them available for future evaluation and comparison. One of the most important inputs when creating bridge inspection deliverables are the manual tie points (MTPs) and especially the ground control points (GCPs). The use of the ground control points can add absolute accuracy and place the inspection model in the exact global position but the addition of MTPs can improve the reconstruction accuracy which is more important for structural inspections. The absolute accuracy of a correctly reconstructed project is one to two GSD horizontally (X and Y) and one to three GSD vertically (Z) as reported by the Pix4D program. Accuracy depends not only on the precision of the actual GCP's, but also on the height at which the drone is flown and from the overlap of the captured photos. The lower the UAV flies, the higher image resolution is which improves the accuracy, but in that case the number of photos required to cover the entire bridge increases and so does the processing time and cost of inspection. Capturing images from low height will result to a low ground sampling distance (GSD) value which will provide high spatial resolution and clear visible details of the structure inspected.

4 Results and Discussion

An effective bridge inspection is comprised from the ability of the engineer to detect, characterize and document all deficiencies. To be successful, the bridge inspector's experience and knowledge play a key role, but also the access to the structure plays a very important role because it can affect the quality of the results. The results presented below demonstrated that a qualified engineer utilizing a UAV can improve the ability to detect deficiencies by alternative access and utilizing high-quality images and photogrammetry. The high quality, large-coverage images provided by the UAV inspection are a very important advantage for the quality results of the photogrammetry model and the overall inspection in general. We understood that the generation of orthoimages, 3D visualizations and the photogrammetry model can facilitate detection, classification, identification and measurement of any type of defects. Furthermore, they can provide a valuable archive file, allowing arbitrary revisiting, control and updating of the derived data, extraction of additional information, multitemporal analysis, etc. [13].

A digital orthophoto of the bridge's façade is illustrated in 'Fig. 2'. This kind of format is basically a large image, of very high quality, that is created by combining many individual images; it can be any planar surface such as the bridge's facade or any pier face. Orthofacades are very useful, especially for bridge and building inspections, because the entire façade is being presented in one photo and all defects can be marked in one file. As can be seen in 'Fig. 3' the quality of the orthofacades are very high and every single detail can be reviewed and investigated. Overall, the results were very positive and illustrated that the concept of using a UAV for a historic bridge inspection is a method that can provide accurate results in a fast manner and without any risks for the structure.

Fig. 2. Orthophoto of bridge's facade, derived from UAV imagery.

Based on the camera parameters and the flight altitude, approximately 40 m above the ground, we managed to achieve an average GSD value of 1.11 cm which is a very good indication of the accuracy accomplished. The GSD is the distance between the center of two pixels and this is how the Pix4D software measures distances in our bridge model, by drawing a polyline in our software the program will convert the pixels distance into a real-world distance. Our accuracy error is associated to the GSD value but if we try to measure a distance smaller than 1.11 cm we will not get the exact measurements because anything smaller than the GSD value will be contained into a pixel therefore it will lack definition. This is a very important factor that should be taken into consideration especially when the measurement of crack length and width is required for the visual structural inspection.

The photogrammetry 3D model presented below in 'Fig. 4' was created entirely from the UAV images and was used to measure deficiencies, cracks and to mark directly on the model all types of defects identified during the inspection. Once the model was generated, all the photos were located and referenced to the model which created an ideal environment for accurate measurements. Any point of interest can be selected directly on the bridge model and all the corresponding images of that area are displayed on the screen on right. With this feature we don't have to create a long photo log of the inspection, and this saved us a great amount of time during the report generation process. Also, with a photogrammetry model, we create a digital record of the bridge condition at a point in time so that as the bridge ages, deterioration rates can be observed and compared with future models. This can be particularly beneficial for post-earthquake and natural disaster inspections.

(a) (b)

Fig. 3. Orthophoto of the central arch, (a) facade (b) cross section of the pillar

Fig. 4. Measurements taken from Pix4D photogrammetry model

Main deficiencies presented in 'Figs. 5,8',were identified during UAV inspection are due to deterioration of materials, such as degradation and loss of stones, loss of mortar joints and salt efflorescence, all often due to inadequate rainwater drainage, freeze–thaw cycles and penetrating vegetation. Most of the defects identified are caused by moisture intrusion due to lack of maintenance and lack of inadequate water drainage provisions, both at the design and construction stage. Unfortunately, with unwanted moisture intrusion, the masonry stones and the mortar will crack; when water enters the natural stone, thermal expansion can cause the surface to peel, pop out, or flake off. This phenomenon, commonly known as spalling, is caused by excess moisture that exerts outward pressure. Eventually, spalling will cause large sections of the masonry to break and fall off, potentially leading to structural damage.

Fig. 5. UAV picture (a) Cracked stones due to corrosion of steel members, (b) Loss of mortar and loss of stones was a common defect

Fig. 6. UAV picture (a) Cracking and loss of stone, (b) Cracking, spalling and movement of stones

Degradation and loss of mortar in the joints of the masonry structure, 'Fig. 7', is an important defect that can constitute a conditioning factor of the mechanical behavior of the masonry. It is usually caused by the bad quality of mortar, but the material degradation is usually due to the environment of the bridge, the wind and the water erosion. It causes the acceleration of its aging and degradation and can provoke some very harmful modifications to its characteristics and loss of mechanical strength. Without the presence of mortar all the force stresses are directly supports by the stones and this can cause structural deficiencies.

The presence of salts was visible throughout the inspection, in many different areas, but in majority it was recorded under all the arches 'Fig. 8'. This is mainly because the water drainage pipes, that are situated under the arches, are blocked and the lack of maintenance is worsening the condition. Efflorescence is the fine, white, powdery deposit of water-soluble salts left on the surface of masonry as the water evaporates. Since efflorescence has been identified we can come to the following conclusions; water-soluble salts are present on the stones, sufficient amount of moisture exist in the masonry stones but most important it means that there is a path through the stones for the soluble salts to migrate to the surface where the moisture can evaporate, thus depositing the salts which then crystallize.

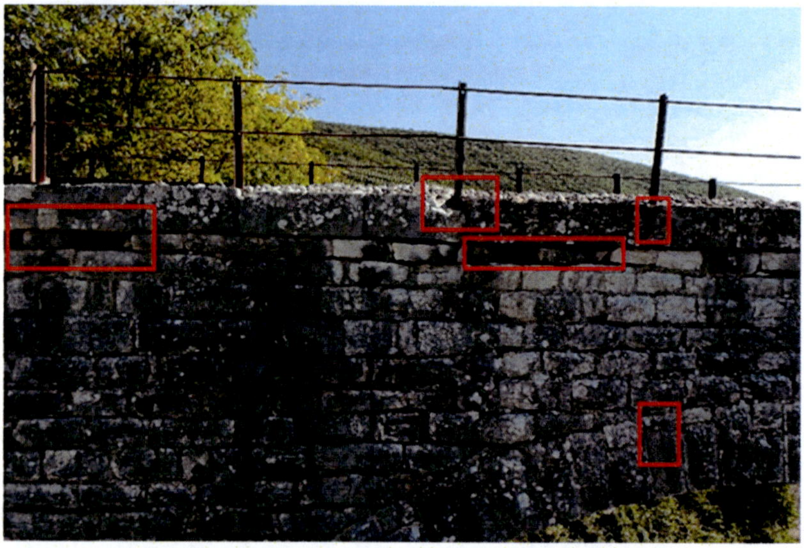

Fig. 7. Loss of mortar and stones above one of the arches

Fig. 8. Efflorescence and loss of mortar under an arch

We must declare that due to the presence of water and moisture the growth of vegetation will flourish on the structure and it can be responsible for important mechanical actions ‘Fig. 9’. As vegetation penetrates the existing cracks and structural joints, it is therefore contributing to the degradation of the structure. As the vegetation grows and expands the damage caused increases and sometimes the vegetation can hide existing damage hence making a visual inspection more difficult. Thus, the most efficient way of preventing this type of damage consists of adequately and regularly maintaining the structure.

Fig. 9. Vegetation growth between the joints

The table below presents a summary of the results derived from our inspection; a summary of the defects and their main causes as studied by the results of the UAV inspection (Table 1).

Table 1. Defects summary

Defects	Causes
Weathering	Natural disintegration and erosion of stone caused by wind and rain. Resulting in granular and rounded surfaces
Degradation and erosion of stones	Wearing of stones due to natural caused and lack of maintenance, salt crystallization, aqueous dissolution, frost damage and microbiological growth
Loss of mortar stones	Moisture and water intrusion, Growth Vegetation. It may be caused by an inherent weakness of the masonry and gradual dissolution of the binder, or it may be the result of external factors affecting the strength or durability of the masonry, such as salts or moisture entering the masonry
Salt efflorescence	Moisture and water intrusion and presence of salts. Efflorescence may also indicate salt accumulations under the surface of the masonry which are potentially damaging to the masonry
Chipping, Cracking and Spalling of stones	Freeze Thaw, Earthquakes, Corrosion of steel elements. Cracking may result from a variety of conditions, such as structural settlement of a building, too hard a repointing mortar, or it may be an inherent characteristic of the masonry itself
Steel corrosion	Lack of maintenance, expansion of steel elements, cracking of stones
Rising damp	Improper and blocked drainage causing wet stain marks and efflorescence

We should also mention that recent research [11] has shown that from the 3-D scaled model, we can import the data to a Finite Element Analysis (FEM) software and conduct a full structural analysis. This is not directly possible because the exported mesh that was created with the Photogrammetry software is affected by errors related to discrepancy between the various points constituting the mesh; for this reason, we cannot import directly into a FEM software. We must first make an intermediate step to eliminate the defects of the mesh. Once the model is imported to a FEM software, we can perform linear static and dynamic analyses to investigate the behavior of the structure under gravitational and earthquake loads.

5 Conclusions

- The technique used in this paper present the possibility of affordable photogrammetry for UAV inspections and digital documentation of historic structures. The procedure offered is capable to produce an accurate 3D model from digital images, with a high degree of accuracy.
- UAVs can offer great potential for the visual inspection of historic bridges eliminating the need for utilizing heavy bridge inspection units that can increase the imposed loads on the already defected structure.
- The technology has the capability to significantly reduce the overall inspection costs. Compared to a normal bridge inspection the UAV inspection will require less time to be spend on site, but photogrammetry will require more time to analyze the digital images.
- Before the inspection is carried out, time must be allocated for planning the inspection and determining the best positions for capturing the images.
- A defects visual inspection carried with the use of a UAV can produce accurate NDT results that can be utilized to evaluate the overall condition of a historic structure or any other structure in general.
- Photogrammetry documentation of a historic bridge can be efficiently developed using the Pix4D software and can provide measurement information associated with the historic structure inspection. In addition to this, the use of automated aerial image capture enables engineers to better understand a situation through the 3D spatial context offered by UAV systems.
- With the use of photogrammetry, we can produce an accurate 3D model, reducing the field time and cost of a conventional inspection. At the same time an increase in the processing model time can occur due to the large volume of data.

References

1. Savuti, N.S., Yusoff, N.M., Abu Bakar, N.A., Kaamin, M., Akbar, Z.A.: Visual inspection in dilapidation study of heritage structure using unmanned aerial vehicle (uav): case study as-solihin mosque, melaka. Politeknik Kolej Komuniti J. Lifelong Learn. **2** (2018)
2. Seoa, J., Duquea, L., Wacker, J.: Drone enabled bridge inspection methodology and application. Autom. Constr. **94**, 112–126 (2018)

3. Borin, P., Cavazzini, F.: Condition assessment of rc bridges. integrating machine learning, photogrammetry and BIM, Int. Arch. Photogramm. Remote Sens. Spatial Inf. Sci. Avilla, Spain. 201–208 (2019)
4. Karachaliou, E., Georgiou, E., Psaltis, D., Stylianidis, E.: UAV for mapping historic buildings: from 3D modelling to BIM. ISPRS Annal. Photogramm. Remote Sens. Spat. Inf. Sci. Bergamo, Italy, 397–402 (2019)
5. Colomina, I., Molina, P.: Unmanned aerial systems for photogrammetry and remote sensing: a review. ISPRS J. Photogramm. Remote. Sens. **92**, 79–97 (2014)
6. Barsanti, S.G., Remondino, F., Fenández-Palacios, B.J., Visintini, D.: Critical factors and guidelines for 3D surveying and modelling in cultural heritage. Int. J. Heritage Digital Era **3**, 141–158 (2014)
7. Chan, B., Guan, H., Jo, J., Blumenstein, M.: Towards UAV-based bridge inspection systems: a review and an application perspective. Struct. Monitor. Mainten. **2**, 283–300 (2015)
8. Caprioli, M., Mancini, F., Mazzone, F. Scarano, M., Trizzino, R.: UAV surveys for representing and document the cultural heritage. Herit. Technol. Mind Knowl. Experience-XIII Int. Forum, Aversa and Capri, Italy (2015)
9. Murtiyoso, A., Remondino, F., Rupnik, E., Nex, F., Grussenmeyer, P.: Oblique aerial photography tool for building inspection and damage assessment. Int. Archives Photogramm. Remote Sens. Spat. Inf. Sci. Denver, Colorado (2014)
10. Fernandez Galarreta, J., Kerle, N., Gerke, M.: UAV-based urban structural damage assessment using object-based image analysis and semantic reasoning. Nat. Hazard. Earth Syst. Sci. 1087–1101 (2015)
11. Barrile, V., Bilotta, G., Amore, E.D., Meduri, G.M., Trovato, S.: Structural modeling of a historic castle using close range photogrammetry. Int. J. Math. Comput. Simul. **10**, 370–380 (2016)
12. Baltsavias, E.P.: A comparison between photogrammetry and laser scanning. ISPRS J. Photogramm. Remote. Sens. **54**, 83–94 (1999)
13. Syrmakezis, C.A., Asteris, P.G., Antonopoulos, A.K., Mavrouli, O.A.: Stress-failure analysis of masonry structures under earthquake loading. Kourkoulis, S.K. (edn.) Fracture Failure Nat. Build. Stones, 157–166 (2006)

The Functional-Structural Rehabilitation of the Prince Caracciolo of Avellino Building Located in the Old Town of Naples "*UNESCO World Heritage Site*"

Beatrice Faggiano[1](✉), Luigi Fiorino[1], Vincenzo Macillo[1], Giulia Bonelli[2], Claudio Rossi[3], and Federico M. Mazzolani[1]

[1] Department Structures for Engineering and Architecture District, University of Naples Federico II, Via Claudio 21, 80125 Naples, Italy
beatrice.faggiano@unina.it
[2] Bonelli Studio Architettura, Via D.Cimarosa 37, 80129 Naples, Italy
[3] PSE srl - Progetti di Sviluppo Engineering, Via Martucci 48, 80121 Naples, Italy

Abstract. The paper presents the functional and structural rehabilitation of an historical masonry building located in the old town of Naples, which is an UNESCO World Heritage Site. The building was initially built in the 16th century as monastery and in the 17th century it was achieved by the Prince Caracciolo of Avellino family, who made major modifications. The building has a rectangular plan with a length of about 50 m and width of about 10 m. It has 6 stories, one of which is below the street level; the pitched roof reaches a maximum height of about 26 m. Before the renovation intervention the building was very much degraded, presenting widespread cracks in the masonry walls and vaults and deteriorated wooden floors. The rehabilitation was dedicated on one hand to the retrofit of existing vertical and horizontal structures, on the other hand to adapt the building to the new use, as a Foundation of Contemporary Art. The paper briefly describes the state of the construction before the intervention and its recovery and transformation, with particular reference to the structural aspects.

Keywords: Rehabilitation of historical masonry building · State of conservation of historical masonry building · Structural retrofitting interventions

1 Introduction

The Prince Caracciolo of Avellino building was built in the XIV century, then modified in the XVII century up to the current configuration, consisting in two buildings that overcome the Anticaglia decuman of the historical center of Naples, giving rise to the urban square of Largo Avellino Dominici (2003). In the following XVIII to XX centuries some extensions (Catalani et al. 1845) and valuable decorations (Sigismondo 1989) were introduced in the building.

I. Vayas and F. M. Mazzolani (Eds.): PROHITECH 2021, LNCE 209, pp. 1138–1143, 2022.
https://doi.org/10.1007/978-3-030-90788-4_87

The building is included in the heart of the Greek-Roman town, near to the Donnaregina Church, the Cathedral of Naples, as well as the ancient Greek Neapolis theatre then Roman forum. In the area worldwide renown museums are located, like the National Museum Mann, the M.a.d.r.e. Museum, as well as the Academy of Beaux Arts.

The building is protected by the Italian Ministry of Culture and listed by UNESCO as World Heritage Site.

The Morra Greco Foundation for Contemporary Art, established since 2003 by the building owner dr. Maurizio Morra Greco, is located in the building. Due to the extended state of degradation of the building, the Foundation decided to proceed to the restauration of the building for the valorization and enhancement of the museum and exhibition activities, as well as for hosting the permanent exposition of the collection of more than 200 artworks of young artists, patrimony of the Foundation itself.

The works were realized in four years from 2015 to 2019 by the Brancaccio Costruzioni S.p.A company (Naples, Italy), the overall design was coordinated by PSE srl Engineering (Naples, Italy), taking advantage of a financial contribution for culture and tourism of the Regional Office of Campania (FESR, 2007–2013; www.fondazionemorragreco.com). Prof. Mazzolani was the responsible of the structural restoration design.

The restoration and reuse of the building gives also a new impulse to the requalification of the whole Neapolitan area.

2 The Original Building and the State of Conservation Before the Rehabilitation

Prince Caracciolo of Avellino building historically belongs to a larger building complex, with an open courtyard. It consists of a large wing building facing the East open courtyard. The building plan sizes are about 50 m along the North-South direction and about 10 m along the East-West direction (Fig. 1a). The building has 6 stories and it is about 24 m high. The main load bearing structure is a tuff masonry (Fig. 1b). The ground floor and first floor consist of barrel vaults, while the second floor is made up of pavilion vaults. The second mezzanine floor and third floors, the latter having with a relatively large span of about 8.5 m, are made of a wooden structure. In particular, sail vault ceilings are suspended at the lower surfaces of the third floor. The double pitched roof is made of wooden trusses covered with tiles or corrugated steel sheets. The building has a large staircase located on the South side, which reaches all floors except the basement floor. In addition, a small and narrow internal staircase connects the building from the basement to the first floor. Finally, a spiral staircase, located on the West elevation connects the building from the first to the third floor.

Over the years, the building has undergone several transformations and different uses. The condition of degradation was widespread and it affected both the envelope and the interior side of the building. In particular, both vertical and horizontal (vaults) masonry structures were affected by cracks with variable width and all wooden structures were greatly damaged by the ravage of time (Fig. 2).

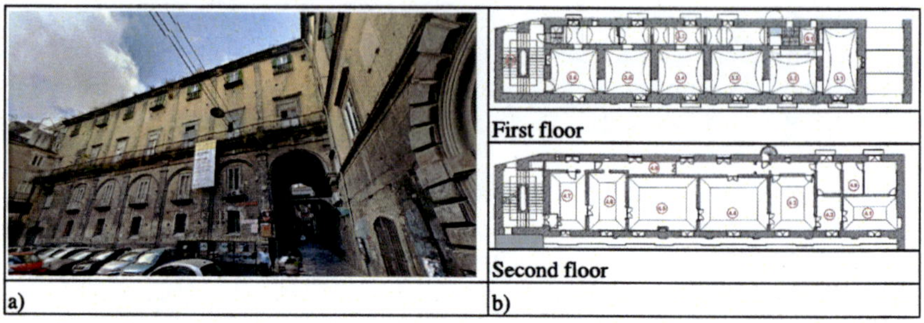

Fig. 1. The building before the rehabilitation: a) View from the East side; b) Some floor plans

Fig. 2. Typical damage in masonry and wooden structures.

3 The Functional-Structural Rehabilitation

The overall design was carried out according to the following phases: knowledge of the current state of the building through in situ geometrical and photographical surveys, including the site geotechnical investigations; evaluation of the structural capabilities of the building through numerical analysis on Finite Element models (FEM, before and after the interventions); identifications of the seismic vulnerability; selection of the type of intervention; design and execution.

In particular, the refurbishment of Prince Caracciolo of Avellino building had a twofold aim: the retrofitting with seismic upgrading of the existing structures and the integration of new structural parts related to the new functional purposes.

With regards to the retrofitting of existing structures the following interventions were designed and realized: repair of cracks in masonry based on their width; reinforcement of vaults; replacement of opening architraves with steel beams; installation of tie-rods in transversal structural walls; replacement or reinforcement of wood and concrete floors with new composite steel-concrete floors anchored to walls in order to obtain a rigid diaphragm effect; replacement of wood trusses of the roof with new steel trusses; reinforcement of spiral staircase by means of a bounding of steel plates; reinforcement with steel beams of the staircase between ground floor and basement; reinforcement of wood framed partition walls using steel frames.

Besides, the following new structural parts were designed and realized: two linear flying stairs, connecting the first to the third floor; steel frames to support new door openings in tuff masonry walls; a steel frame for the new lift shaft; a water reservoir and a transformer room both made of in reinforced concrete, located at the underground floor.

Some details of the new floor structure at the third level, in substitution of the wooden floor, are shown in Fig. 3. The new composite floors is made of steel beams collaborating with a lightweight concrete slab, cast on trapezoidal steel sheeting, by means of shear connectors. The diaphragm effect was obtained by a rigid ring along the perimetral structural walls, made of steel profile and anchored into the concrete slab.

Some details of the new linear flying stairs are shown in Fig. 4. They are made by a ribbed steel plate, which lightweight concrete steps are cast on; the stairs are also supported by steel profiles anchored in lateral existing masonry walls.

The new roof structure is shown in Fig. 5. The original wooden trusses were replaced by new ones made of CHS steel profiles, whereas the wooden beam floor was restored with particular attention to the supports to the masonries. It is now visible as ceiling of the third floor (see Fig. 7 b).

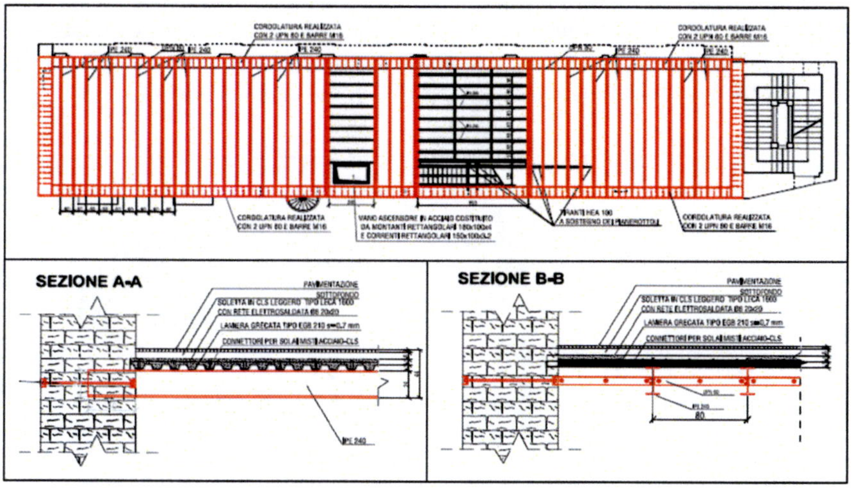

Fig. 3. Floor structures at the third level.

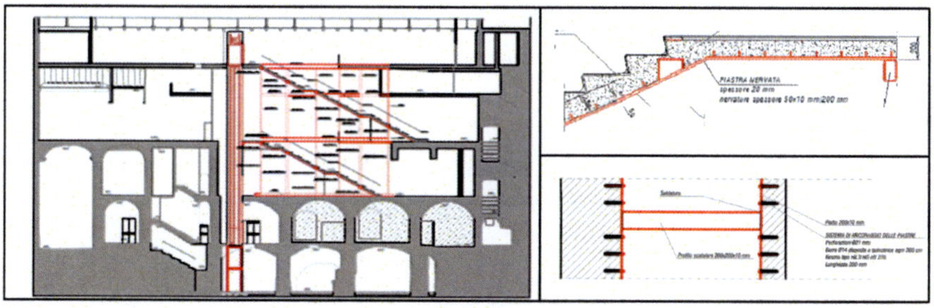

Fig. 4. New linear flying stairs.

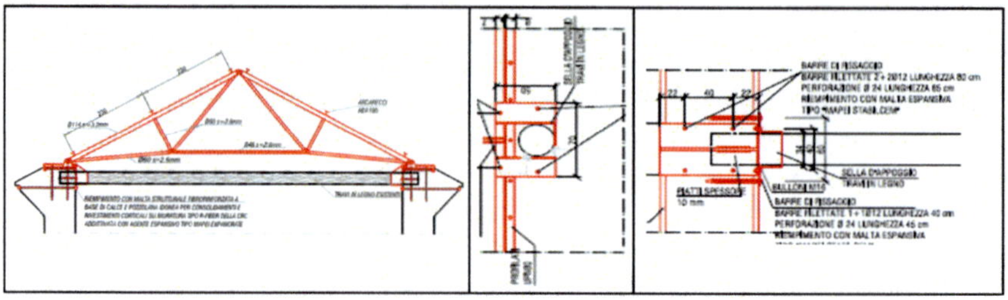

Fig. 5. New roof structure and detail of the support to masonry.

A linear dynamic analysis by a FEM model (Fig. 7), interpreting the structural behaviour before and after the interventions, was performed in order to evaluate the structural capabilities of the building in the current state and the enhancement of its seismic behavior after the interventions, with particular reference to the local safety checks in tuff masonry 'walls, to the evaluation of the global effects of rigid diaphragms given by new floors and to the effectiveness of the wall tie-rods (Fig. 6).

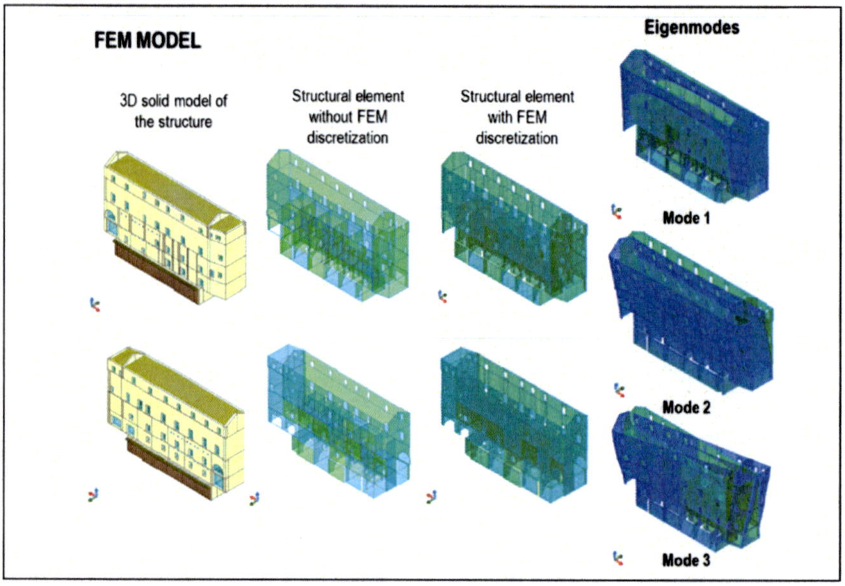

Fig. 6. FEM model and modes of vibration of the building.

4 The Building After the Rehabilitation Works

Hereafter some pictures of the renewal of the buildings are provided (Fig. 7).

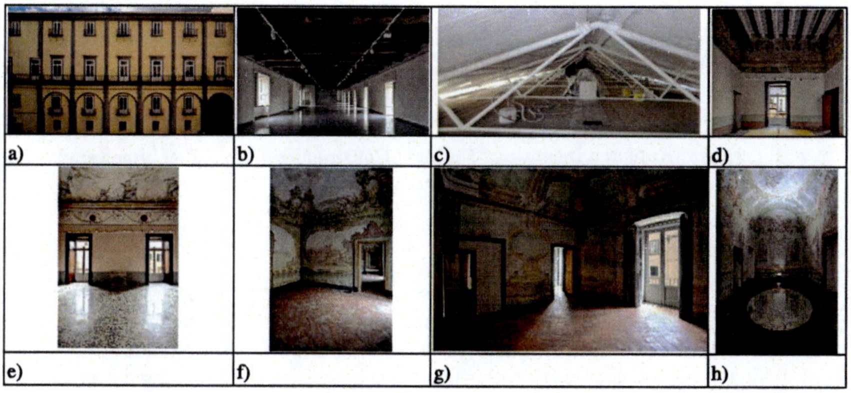

Fig. 7. a) The building East façade after restoration (fondazionemorragreco.com); b) the old timber beams in the upper hall (artribune.com); c) the new roof trusses; d) the ceiling of one of the halls (fondazionemorragreco.com); e, f, g, h) Some halls after restoration (fondazionemorragreco.com).

5 Conclusive Remarks

The functional-structural rehabilitation of the Prince Caracciolo of Avellino building is a very topical example of valorization of monumental built heritage, which combines principles of conservation and innovative technologies. This integration between ancient and modern environments, perfectly in line with the basic principles of the Restoration Charts, is even more apparent considering the destination of use to contemporary art exhibitions.

The comprehensive design allowed definitely to open two additional floors to the public for a total of 2000 square meters of exhibition space distributed over five levels, each revealing a remarkable variation of art-historical richness (www.fondazionemorragreco.com).

Acknowledgements. Fondazione Morra Greco is acknowledged for publishing.

References

Catalani, L.: I palazzi di Napoli, Napoli 1845, p. 64 (1845)

De Dominici, B.: Vite de pittori, scultori ed architetti napoletani, Napoli 2003, ed. a cura di F. Sricchia Santoro e A. Zezza, tomo I, p. 207 (2003)

Sigismondo, G.: Descrizione della città di Napoli e i suoi borghi, Napoli 1788-89, ristampa anastatica Forni (ed.) Bologna 1989, p. 137-138 (1989)

Protection of St. Andrew's Church Located Within Landslide Hazardous Sites and Ground Constraints

Yuriy Slyusarenko(✉), Volodymyr Tytarenko, Valeriy Shuminskiy, Anton Dvornyk, and Yaroslav Dombrovskyi

State Enterprise "Research Institute of Building Constructions", 5/2, Preobrajenskaya St., Kyiv 03680, Ukraine

Abstract. In recent decades, the concept of cultural heritage has evolved into one that encompasses an understanding of the history of humanity, together with scientific knowledge and intellectual attitudes. This changing concept has prompted a subsequent reevaluation of what constitutes the outstanding universal values of World Heritage sites and the operational methods for implementing the UNESCO World Heritage Convention. The scope has broadened from studying a single monument in isolation to one that values a multidimensional, multiregional, and inter-disciplinary approach and encapsulates vast spans of human history, as demonstrated by the St. Andrew's Church, located within active landslide systems, Ukraine. Andrew's Church is a unique historical and architectural monument of the eighteenth century built in 1747–1762 by I. Michurin according to B. Rastrelli's design. Since 1968 St. Andrew's Church is a museum, branch of the "Sofiya Kyivska" National reserve, nationally significant site, number 14 St. Andrew's Church was built in Baroque style. The church is located on a hill situated on the top of the Andriivskyi Descent (one of the most ancient streets in the Ukrainian capital) in the central historical district of Kyiv. The ground level varies from 181.7 m (area around the church) to 118.5 m (foot of the hill). The slopes of the hill are dissected by a thick girder-netting network where landslide processes were actively developing and are developing with active movements and erosion processes. Around the development site there is a complex of unfavorable physical and geological phenomena such as landslides, considerable thickness of fill-up ground, significant ground heterogeneity, mechanical suffixation of clay particles into an existing inactive gallery, and external erosion of the hill massif. Based on visual inspection of the superstructure constructions, general technical condition of the above structures can be classified as follows: superstructure building - as unsuitable for normal operation; runway construction - as unsuitable for normal operation. The technical condition of St. Andrew's Church building was assessed as unsuitable for normal operation as a whole that caused an urgent need for geotechnical research and restoration work undertaking.

Keywords: Andrew's church · Landslide · Visual inspection · Reliability · Restoration work

I. Vayas and F. M. Mazzolani (Eds.): PROHITECH 2021, LNCE 209, pp. 1144–1157, 2022.
https://doi.org/10.1007/978-3-030-90788-4_88

1 Pre-introduction

The first century AD. According to one of the legends, the first preacher of Christianity in the ancient Rus was the holy Apostle Andrew, nicknamed the First-Called. "Filled with the Holy Spirit, the Apostles, according to the covenant of our Lord Jesus Christ, went out to preach the Good News in all languages." According to a church tradition, Scythia was drawn to Saint Andrew the First-Called. Saint Andrew the Apostle sailed from Tauric Chersonese up the Dnipro-Slavutych River. Climbing the Kyiv mountains, he and his disciples set up a Wooden Cross and blessed the mountains over the Dnipro. He predicted that the faith of Christ will shine forth in the country, a great city will be there, and God will raise up many churches in it. Now on the place of the first Cross there is the best masterpiece of the famous architect Bartolomeo Rastrelli – St. Andrew's Church, one of the most marvelous Orthodox temples in the Ukrainian capital designed in the form of a cross. A beautiful painting on one of the church walls presents the pensive blond people listening to the holy Apostle's teaching against the background of a gloomy and austere landscape. The reality of this event raises doubts among many researchers of Kyivan Rus. However, there are arguments in favor of this concept. Before the Common Era, the trading routes existed between the Black and Baltic seas. Let us remember that amber was among the ten most important goods exported from Taurida by the ancient Greeks. Well, a preacher could also be a messenger from the Apostle. And since in the chronicle of the acts of St. Andrew the First-Called only Taurida and Colchis were mentioned, this version looks realistic. The amber merchants – Korels – could be the guides of the holy Apostle Andrew. The settlements of the Korels were recorded in the list of tribes that lived on the territory of Thrace (Bulgaria). Under this name they appeared in the Greek list of the third century BC among 30 tribes inhabiting Greece.

In recent decades, the concept of cultural heritage (Migon 2013) has evolved into one that encompasses an understanding of the history of humanity, together with scientific knowledge and intellectual attitudes. This changing concept has prompted a subsequent reevaluation of what constitutes the outstanding universal values of World Heritage sites and the operational methods for implementing the UNESCO World Heritage Convention (The World Heritage Convention 1972). The scope has broadened from studying a single monument in isolation to one that values a multi-dimensional, multiregional, and interdisciplinary approach and encapsulates vast spans of human history, as demonstrated by the St. Andrew's Church, located within active landslide systems, Ukraine.

The theoretical and methodological issues of these problems are analyzed in the works of modern foreign scientists, including (Vayas et al. 2019; Mazzolani and Calado 2019; Tassios et al. 2003; Barla et al. 2013; Casagli et al. 2010; Highland and Bobrowsky 2008; Dunin et al. 2018a; Farenyuk et al., 2018, 2019; Kaliukh and Berchun 2020; Lacasse 2013; Margottini et al. 2019; Migon 2013; Slyusarenko et al. 2009, 2010, 2013, 2015; Shokarev et al. 2005, 2009, 2020; Trofymchuk et al. 2018), and others.

In our report "Protection of St. Andrew's church located within landslide hazardous sites and ground constraints" the results are presented on the survey and assessment of the technical condition of the St. Andrew's church (Fig. 1).

2 Introduction

St. Andrew's Church is a unique historical and architectural monument of the eighteenth century built in 1747–1762 by I. Michurin according to B. Rastrelli's design. Since 1968 St. Andrew's Church is a museum, branch of the "Sofiya Kyivska" National reserve, Nationally significant site, number 14 (Fig. 1). St. Andrew's Church was built in Baroque style.

Fig. 1. St. Andrew's Church general view

This style is characterized by pageantry, dynamic architectural forms, riches of design, play of light and shadow. The church was built on the remains of the earth's fortress of the seventeenth century. The church building is completed with a central dome and four corner decorative towers. Building fronts are decorated with columns, pilasters, cornices of complex profile, cast iron and copper gilded details. The size of the church: length - 32 m, width - 20 m, height from the terrace to the top of the cross of the central dome - 50 m. The size of the superstructure is 33,5 × 9,5 m. The church is located on a hill situated on the top of the Andriivskyi Descent (one of the most ancient streets in the Ukrainian capital) in the central historical district of Kyiv. The ground level varies from 181.7 m (area around the church) to 118.5 m (foot of the hill (Fig. 2)). The slopes of the hill are dissected by a thick girder-netting network where landslide processes were actively developing and are developing with active movements and erosion processes (Fig. 3). Around the development site there is a complex of unfavorable physical and geological phenomena such as landslides, considerable thickness of fill-up ground, significant ground heterogeneity, mechanical suffixation of clay particles into an existing inactive gallery, and external erosion of the hill massif. The base of the foundations of the southern, western and northern Church facades is eolian-deluvial loess-like loamy sands, which have sagging properties. The base of the foundations of the eastern part is morainic loam. Hydrogeological conditions are characterized by the presence of two groundwater levels. The visible superstructure of the church is based on a slightly wider in terms of underground two-story foundation part of the church.

Fig. 2. Foot of the hill where the Church is situated

Fig. 3. An example of the active development of landslide processes with active developments and erosion processes on the slope of the Church

The base of the foundations of the eastern part is morainic loam. Hydrogeological conditions are characterized by the presence of two groundwater levels. The visible superstructure of the church is based on a slightly wider in terms of underground two-story foundation part of the church. The superstructure is joined to the underground foundation and is connected by a common entrance - a two-storey building, the covering of which is part of the church porch (see Fig. 4). The foundation of the superstructure is brick masonry in the section from 3 to 5 m. The church foundation base lays on different marks: in the western part (on the side of the Andriivsky Descent) - on the marks from 166.6 m to 165.8 m; in the eastern part that hangs over the hill - from 165.7 m to 167.8 m.

Fig. 4. General view of the church superstructure and the runway construction

3 Surveys of the Church Building and Adjacent Territory

During complex surveys by RIBC specialists it was discovered that the superstructure had significant damages and the main are: vertical and inclined cracks with a width of.

up to 2.0 cm in the walls of the superstructure; cracks width up to 2.0 cm (Figs. 5 and 6) in brick vaulted ceilings; walls and ceilings soaking and fungosity; deviation from the superstructure vertical wall.

Fig. 5. Vertical crack with a width of up to 2.0 cm

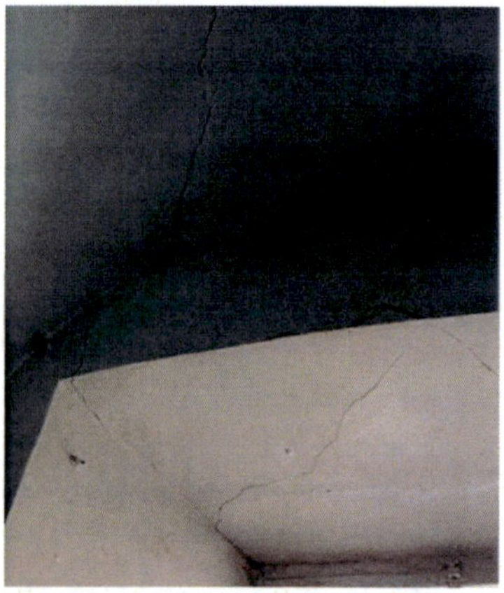

Fig. 6. Vertical and horizontal cracks in window lintels and vaulted ceilings with a width of up to 0.5 cm

The following damages were detected in the runway walls: diagonal cracks with a width of up to 3.0 cm; frost destruction of the brickwork and deviation from the vertical of reinforced concrete retaining wall, through cracks, local destruction of the outer layer and brickwork mortar (Fig. 7). The main source of detected damages was the uneven foundation subsidence, caused by the presence of subsiding ground, and sloping ground movements.

Fig. 7. Soaking and frost destruction of the top finishing

The above defects affect both the durability and operational reliability of St. Andrew's Church building structures. Based on visual inspection of the superstructure constructions, the construction of its runway in accordance with the current Ukrainian regulations in terms of inspection, certification, safe and reliable operation of industrial buildings and structures, general technical condition of the above structures can be classified as follows: superstructure building - as unsuitable for normal operation; runway construction - as unsuitable for normal operation. The technical condition of St. Andrew's Church building was assessed as unsuitable for normal operation as a whole that caused an urgent need for geotechnical research and restoration work undertaking.

4 Geotechnical Study of St. Andrew's Church

The following activities were carried out within geotechnical study: research and analysis of the church geological conditions; research, calculation and forecast of changes in geological and hydrogeological conditions of the adjacent territory along with hydrogeological model development. As a result, recommendations for improving of hydrogeological conditions were proposed. In addition to the existing drainage system a new drainage pipeline was built behind the retaining wall on the slope in order to hold the groundwater level down. The main purpose of the additional drainage system is groundwater interception and prevention of water saturation of the ground behind the retaining wall. Water saturation of the ground increases its pressure on the wall and reduces local slope stability near retaining wall. The new drainage system helped to avoid additional infiltration of atmospheric water and a rise of water level in the church foundation area. The new drainage system consists of two separate pipelines located along the retaining wall, which discharge water from opposite sides of the church building into the drain pipe. From the sewage pipe, water is discharged through the underground pipe-collector into the existing well of the drainage system gallery.

5 Analysis of the Slope's Stability Around St. Andrew's Church Site and Adjacent Territory

Study and analysis of slopes stability around St. Andrew's Church site and on the adjacent territory, as well as the study of current erosion processes on the surrounding slopes was carried out. According to Ukrainian State Construction Standards requirements, for normal operation of a structure built on a landslide slope, the value of the normative slope stability factor should not be less than 1.25. For the calculation of St. Andrew's Church slope stability, a software complex that has a wide range of opportunities for calculating and interpreting the results by 9 methods (Bishop, Yanbu, Spencer, Fellenius, etc.) was used. To improve results reliability, the calculations of slope stability were also carried out using software, developed based on the Terzaghi-Chugaev method. Comparison of the obtained results showed that they are mostly similar. Thus, the coefficients of soil stability in the lower and upper sections of the church landslide hazard slope due to the first variant were 1.015 and 1.229, and to the second one - 1.083 and 1.219. Based on historical materials study and visual survey of the slopes adjacent to the church building and the calculations performed, it was determined that a considerable part of the slopes of St. Andrew's Church hill is in a state close to the limit equilibrium. To increase slopes stability a new drainage system was designed in the form of concrete trays, which intercepts the atmospheric water from the church porch and conveys it into the drain pipe near the retaining wall. The main purpose of the new drainage system is to prevent ground saturation on the slopes of St. Andrew's Church hill that reduces the local stability of its slopes and leads to water and wind erosion of the hill ground.

6 Research and SDS Analysis of St. Andrew's Church Building Ground Base

The SDS (stress deformed state) analysis of the building was carried out taking into account the deformations of the ground base. As a result, the stresses and deformations of the bearing constructions of the building are obtained and compared with the strength values of the materials. The following groups of calculations of the "building-foundation" system were carried out:

Group 1 - calculation to determine the causes of cracking in the walls of the building and its substructure: with the values of soil properties in the natural state; with the water saturation of the foundation under the whole church building and its substructure; with water saturation of the base under the north-eastern part of the building; with the water saturation of the base under the superstructure and the southwestern part of the building; with water saturation of the base under the central part of the building (under the dome).

Group 2 - calculation to determine the actual structures SDS, taking into account the damages recorded during visual and instrumental survey.

Group 3 - analysis of prospecting SDS of building structure with possible changes in its foundation state. Various versions of water saturation of the base under the whole church building and its substructure were considered.

Modeling of water saturation of the ground in calculations was carried out by giving ground, lying at the base, values of properties corresponding to water saturation. It also took into account the appearance of zones of weakening and emptiness in the ground as a result of suffusion in the drainage system, which is located near the foundations. In order to avoid subsequent uneven deformations of the foundations and ground fixing at the baseline of the bearing walls, a jet grouting (the "jet"-columns device) was performed. Proportioning of the building with reinforcement elements was carried out on the basis of a mathematical and geometric model, in which existing cracks in the walls and overstressed sections of the foundations were taken into account. The results of the calculations gave a picture of actual SDS constructions after reinforcement. In the calculation model, the device "jet"-columns under the part of the foundations was taken into account by replacing the soil deformation module in the natural state by the average module of deformation of the natural ground "jet" columns. The calculation for determining the SDS of the ground mass and loads on the piles was carried out in an iterative way due to hardening ground model. This model is of elastic-plastic type, which is formed in the framework of plasticity with hardening during ground sliding, as well as compression during hardening for modeling of soil consolidation during the first compression loading. The model includes the following parameters: rigidity parameters (1), Poisson's coefficient (ν_{ur}) (2), clutch (c) (3), internal friction angle (ϕ) (4), and dilatation angle (ψ) (5). The program also takes into account the volumetric weight of the soil in the dry (γ_{unsat}) (6) and saturated water ((γ_{sat}) (7) states, the filtration coefficients Kx (8) and Ku (9). Thus, the model is nine-parameter. The calculation scheme is presented in Fig. 8.

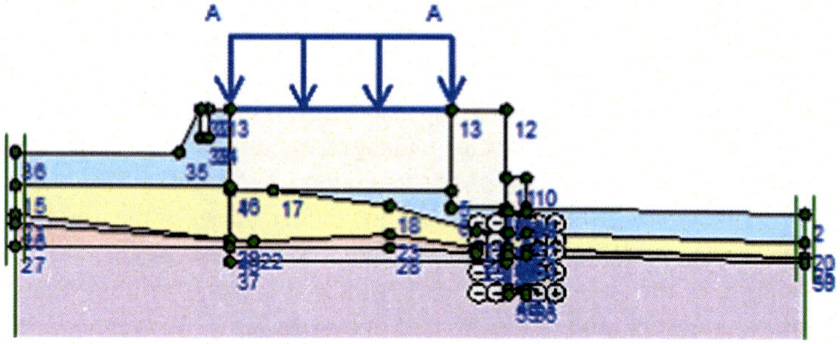

Fig. 8. Calculation scheme

Calculation in PLAXIS PC was divided into three phases: Phase 1 - assessment of stresses and displacements in the previous period of operation of the building; Phase 2 - assessment of stresses and displacements during the period of operation of the building during stabilization of the ground of the base; phase 3 - estimation of stresses and displacements during the period of operation of the building during stabilization of the foundation's grounds and arrangement of the superstructure runway. Some results of calculations of the stresses of the building are shown in Fig. 9. According to the results of geotechnical studies, design solutions and general concept of renovation works were developed.

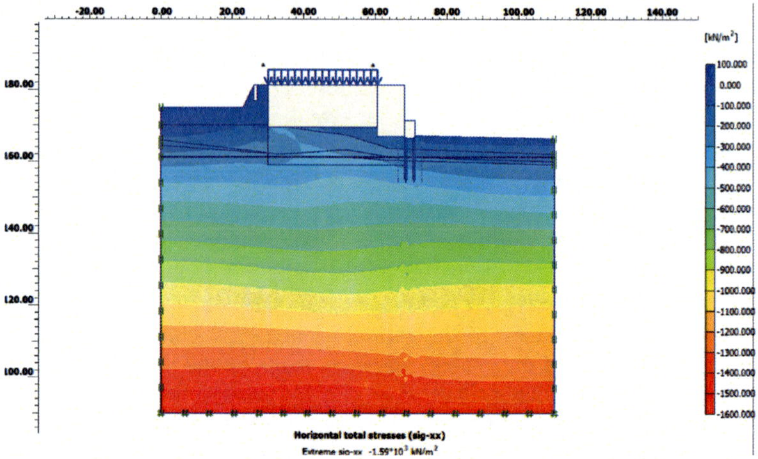

Fig. 9. Isofields of general horizontal stresses in soil massif during the period of operation and base soil stabilization and arrangement of the superstructure platform, PC PLAXIS

7 General Concept of Renovation Work

Renovation of St. Andrew's Church was carried out in the following areas: geotechnical activities; renovation and reinforcement of damaged building structures; restoration of facades and interiors; improvement of the technical condition of the surrounding area. In addition to geotechnical work the works on restoration of the integrity of building structures was carried out by reinforcing damaged sections of the masonry walls by injection methods and reinforcement of cracks (Figs. 5 and 6). The SDS monitoring of the building after renovation work showed that the deformations of the ground base have practically ceased and new damages in the building do not appear.

The simulation of the stabilization of St. Andrew's Church building foundation by the method of grounds jet grouting was carried out on the basis of the calculation scheme (Fig. 10) taking into account detected cross-sectional cracks in the masonry and the foundations base in wet state. In simulation model, the arrangement of "jet"-columns under the part of the church foundations was taken into account by changing of ground deformation module from the natural state to the weighted average module of deformation of natural ground and jet grouting "jet"-columns, which was determined by the formula:

$$\overline{E} = \frac{(A_{bn} \cdot E_{bn}) + (A_{jt} \cdot E_{jt})}{(A_{bn} \cdot A_{jt})}$$

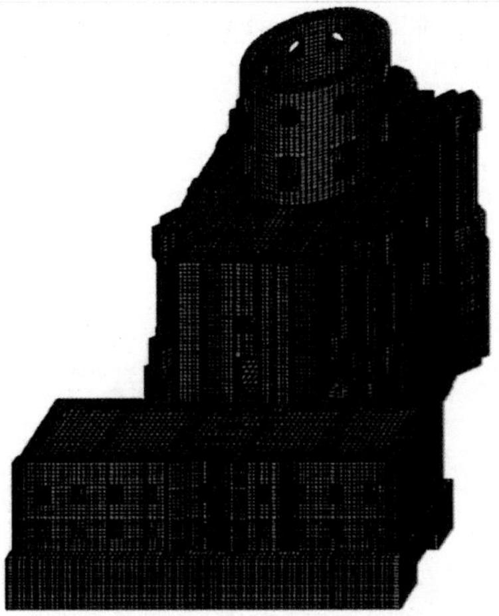

Fig. 10. General view of the computer model of the church building on the south-eastern side

where $\overline{E}$ – weighted average module of stabilized ground deformation under foundations site; A_{bn} – area of reinforced site under foundations without regard to "jet"-columns area; E_{bn} – module of ground deformation in natural state; A_{jt} – area of «jet»-columns on the site under foundations; E_{jt} – module of "jet"-columns material deformation. A brief characteristic of the calculation options is given in the table (Table 1).

Table 1. "Building foundation" system calculations options.

Option code	Brief description
1	Modeling of planned church foundations and its substructure reinforcement according to technical decisions developed using grounds jet grouting method by "jet"-columns
2	Modeling of actual reinforcement of the church building foundations and its substructure using basis jet grouting method by «jet»–columns
3	Modeling of actual reinforcement of the church substructure using basis jet grouting method and arrangement of over superstructure reinforced concrete slab by «jet»–columns

To assess effectiveness of actions on the church building reinforcement, several options of calculations were carried out to determine predicted constructions SDS under possible conditions of building's base operation in a soaked state. The calculation schemes take into account cross-cutting cracks that were recorded during visual inspection. Modeling of base watering was carried out only on areas of foundations that were not stabilized by the method of jet grouting of the base.

8 St. Andrew's Church Foundations Base Grounds Reinforcement by "Jet" Columns

On the basis of RIBK scientific and technical report the joint venture "OSNOVA-S SOLSIF" has reinforced the grounds of the basis by "jet" -columns according to the working design (Fig. 11).

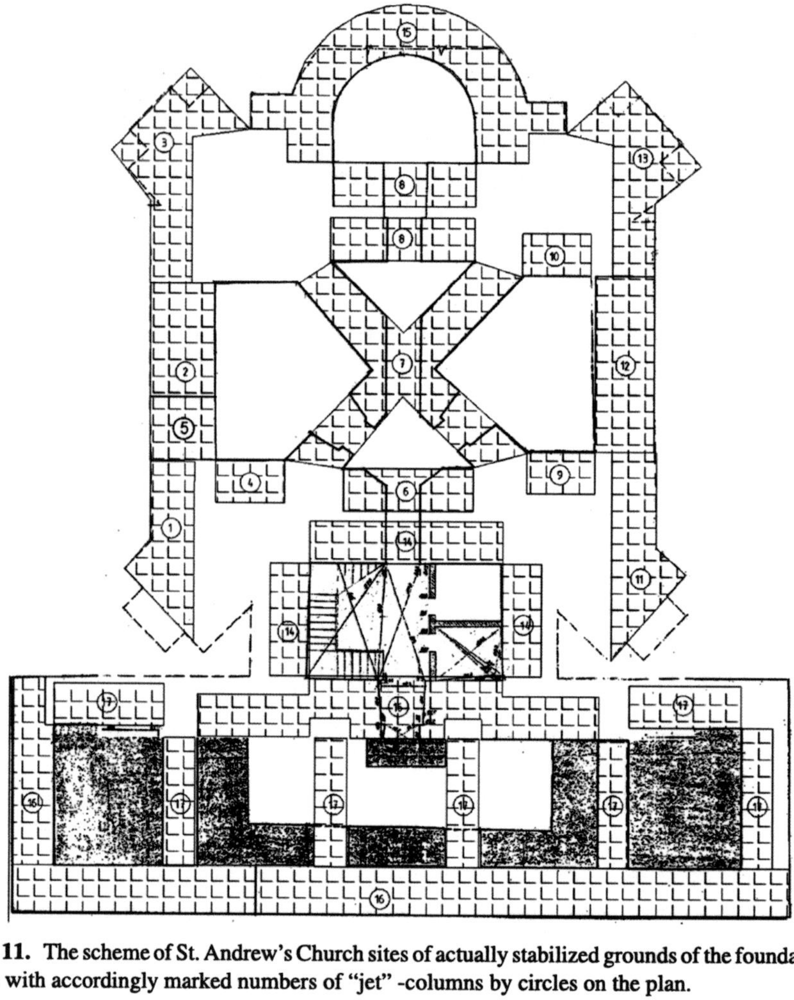

Fig. 11. The scheme of St. Andrew's Church sites of actually stabilized grounds of the foundations base with accordingly marked numbers of "jet" -columns by circles on the plan.

Jet grouting of the grounds has been made by the method of installation of "jet"-columns of 10,81–14,25 m long with an angle of 100–130 vertically, with a diameter of 0,6 m (with an 0,8 m extension under the foundation), the spacing is 1,0–2.7 m. The installation of "jet"-columns is performed outside and inside the church. The scheme of

St. Andrew's Church sites with actually stabilized grounds of the foundations base by "jet"-columns is shown in Fig. 11, where "jet"-columns are marked by circles on the plan. In addition to reinforcement of the grounds of foundations base by "jet"-columns the reinforced concrete slab with anchoring in its wall has been made to provide for operational reliability on the hazardous section of the substructure. The thickness of the superstructural plate is 200–550 mm (concrete grade B40, reinforcement net Ø16 A400C according to the standard DSTU 3760: 2006, the reinforcement bars spacing is 150 mm).

9 Conclusions

1. In recent decades, the concept of cultural heritage (Margottini and Vilimek 2014; Migon 2013) has evolved into one that encompasses an understanding of the history of humanity, together with scientific knowledge and intellectual attitudes. This changing concept has prompted a subsequent reevaluation of what constitutes the outstanding universal values of World Heritage sites and the operational methods for implementing the UNESCO World Heritage Convention (The World Heritage Convention 1972). The scope has broadened from studying a single monument in isolation to one that values a multidimensional, multiregional, and inter-disciplinary approach and encapsulates vast spans of human history, as demonstrated by the St. Andrew's Church, located within active landslide systems, Ukraine.
2. Up-to-date methods of geotechnical protection for historical monuments of architecture can provide them with reliable protection against adverse geological processes and ensure long-term reliable operation, as shown by the example of St. Andrew's Church.
3. Renovation of St. Andrew's Church was carried out in the following areas: geotechnical actions; renovation and reinforcement of damaged building structures; restoration of facades and interiors; improvement of technical condition of the surrounding landslide area. In addition to geotechnical work, the restoration of church building structures unity was carried out by means of reinforcement of the damaged sections of the stone masonry walls by cracks injection and reinforcement methods.
4. Based on the working design developed by "OSNOVA-SOLSIF" joint venture, the reinforcement of the church foundation by "jet" -columns was performed. Jet grouting of the grounds has been made by the method of installation of "jet" -columns of 10,81–14,25 m long with an angle of 100–130 vertically, with a diameter of 0,6 m (with an 0,8 m extension under the foundation), the spacing is 1,0–2.7 m. The installation of "jet"-columns is performed outside and inside the church.
5. The introduction of a new drainage system and a new water discharge system facilitated dewatering of the slopes grounds of St. Andrew's Church hill and stabilization of landslide processes on it.
6. Monitoring of SDS status of St. Andrew's Church building surrounding landslide hazardous area after the restoration work has approved that the deformation of the ground basis almost has stopped and new damages in the building do not arise.

Acknowledgements. Results presented herein have been obtained with the financial and scientific support from the RIBC, the private consulting firm "OSNOVA-SOLSIF", Kyiv, Ukraine.

We would like to express our very great appreciation to the Ukrainian RIBC scientist, who have been directly involved in visual and instrumental survey and mathematic simulation of SDS building structures of St. Andrew's Church and adjacent area. We would like to express our very great appreciation to the Ukrainian experts from "OSNOVA-SOLSIF" private Consultancy, who have performed not just working design for St. Andrew's Church foundations reinforcement, but also all the work on foundations reinforcement by "jet"- columns.

Our special thanks are extended to our colleagues Ph.D, now retiree, Iakov Chervinskii, Ph.D and Sergej Dvornik from the private Consultancy "OSNOVA-SOLSIF" for their support in the site and office work.

References

Barla, G., Antolini, F., Barla, M., Perino, A.: Key aspects in 2D and 3D modeling for stability assessment of a high rock slope. In: Workshops 'Failure Prediction', Austrian Society for Geomechanics, Salzburg, 9 October 2013

Casagli, N., Catani, F., Del Ventisette, C., Luzi, G.: Monitoring, prediction, and early warning using ground-based radar interferometry. Landslides **7**(3), 291–301 (2010)

Farenyuk, G., Kaliukh, I., Farenyuk, E., Kaliukh, T., Berchun, Y., Berchun, V.: Experimental and theoretical diagnostics of defects in ferroconcrete piles based on reflection of longitudinal and transverse waves. In: Hordijk, D.A., Luković, M. (eds.) High Tech Concrete: Where Technology and Engineering Meet, pp. 1307–1317. Springer, Cham (2018). https://doi.org/10.1007/978-3-319-59471-2_151

Highland, L., Bobrowsky, P.: The landslide handbook—a guide to understanding landslides: reston. U.S. Geological Survey Circular, Virginia (2008)

Kaliukh, I., Syvko, I., Chorna, K., Shokarev, V., Shokarev, I., Shokarev, A.: Assessment of residual service life for the residential buildings under systematical industrial explosions. In: Proceedings of IEEE 2nd International Conference on System Analysis and Intelligent Computing, SAIC 2020, 9239172 (2020). https://doi.org/10.1109/SAIC51296.2020.9239172

Kaliukh, I., Dunin, V., Berchun, Y.: Decreasing service life of buildings under regular explosion loads. Cybern. Syst. Anal. **54**(6), 948–956 (2018a). https://doi.org/10.1007/s10559-018-0098-9

Kaliukh, I., Fareniuk, G., Fareniuk, I.: Geotechnical issues of landslides in ukraine: simulation, monitoring and protection. In: Wu, W., Yu, H.-S. (eds.) Proceedings of China-Europe Conference on Geotechnical Engineering. SSGG, pp. 1466–1469. Springer, Cham (2018). https://doi.org/10.1007/978-3-319-97115-5_124

Kaliukh, I., Berchun, Y.: Four-mode model of dynamics of distributed systems. J. Autom. Inf. Sci. **52**(2), 1–12 (2020). https://doi.org/10.1615/JAutomatInfScien.v52.i2.10

Kaliukh, I., Fareniuk, G., Trofymchuk, O., Fareniuk, I., Berchun, Y.: Identification of defects in reinforced concrete piles based on multi-wave reflection. In: Derkowski, W., Gwoździewicz, P., Hojdys, Ł., Krajewski, P. (eds.) Proceedings of fib Symposium 2019: Concrete – Innovations in Materials, Design and Structures, Fédération Internationale du Béton (fib) – International Federation for Structural Concrete, pp. 991–998 (2019)

Kaliukh, I., Trofymchuk, O., Farenyuk, G., Ivanik, O., Shekhunova, S.: Practical measures fo landslide risk mitigation in the Ukrainian Carpathians. In: First EAGE Workshop on Assessment of Landslide and Debris Flows Hazards in the Carpathians, vol. 2019, pp. 1–5. European Association of Geoscientists and Engineers (2019). https://doi.org/10.3997/2214-4609.201902165

Krivosheev, P., Slyusarenko, Y., Chervinsky, J.: Development of calculation methods of foundations on the pliable basis in Ukraine. In: Proceedings of the 17th International Conference on Soil Mechanics and Geotechnical Engineering: The Academia and Practice of Geotechnical Engineering, vol. 2, pp. 1818–1821 (2009)

Lacasse, S.: 8th Terzaghi Oration – protecting society from landslides – the role of the geotechnical engineer. In: The 18th International Conference on Soil Mechanics and Geotechnical Engineering, Paris, pp. 15–34 (2013)

Margottini, C., Vilimek, V.: The ICL network on "Landslides and Cultural & Natural Heritage (LACUNHEN)." Landslides **11**(3), 934–938 (2014)

Mazzolani, F.M., Calado, L.: Selected papers from the 3rd international conference on protection of historical constructions (Prohitech'17). Int. J. Archit. Heritage **13**(1), 1 (2019). https://doi.org/10.1080/15583058.2018.1564591

Migon, P.: Cultural heritage and natural hazards. In: Bobrovsky (ed.) Encyclopedia of Natural Hazards. Springer, Dordrecht (2013)

Papadopoulos, K., Vayas, I.: Restoration and strengthening techniques for Ancient-Greek monuments. Int. J. Archit. Heritage **13**(1), 33–46 (2019). https://doi.org/10.1080/15583058.2018.1497233

Slyusarenko, Y., Matveyev, I., Kisil, A., Ischenko, Y., Romanov, O., Kosheleva, N.: Solution of the geotechnical problems of the Poshtova Square reconstruction in Kiev. In: Proceedings of the XVI European Conference on Soil Mechanics and Geotechnical Engineering (ECSMGE 2015): Geotechnical Engineering for Infrastructure and Development, January 2015, pp. 693–698 (2015)

Slyusarenko, Y., et al.: Modern methods of geotechnical defense of buildings in the difficult geological conditions of Ukraine. In: 18th International Conference on Soil Mechanics and Geotechnical Engineering (Paris) (2013)

Slyusarenko, Y., Girshtel, G., Liseniy, O.: Optimization of multilayer repair systems of polymeric compositions for concrete structures at reconstruction of NSC Olimpiyskiy in Kiev. In: Proceedings of the fib Symposium Tel-Aviv 2013: Engineering a Concrete Future: Technology, Modeling and Construction, pp. 353–356 (2013)

Slyusarenko, Y., Kryvosheyev, P., Serjogin, J., Kozjavkin, I.: Concrete structures of the Olympiyskiy National Sports Complex in Kiev. In: Proceedings of the 3rd International fib Congress and Exhibition, Incorporating the PCI Annual Convention and Bridge Conference: Think Globally, Build Locally (2010)

Shokarev, V., Shapoval, V., Chaplygin, V., Samchenko, R., Volkov, D.: Stress-strain state of the system base-strip foundation at elimination of excessive tilts of buildings. In: Proceedings of the 17th International Conference on Soil Mechanics and Geotechnical Engineering: The Academia and Practice of Geotechnical Engineering, vol. 1, pp. 680–683 (2009)

Trofymchuk, O.M., Kaliukh, Y.I., Dunin, V.A., Berchun, Y.A.: On the possibility of multi-wavelength identification of defects in piles. Cybern. Syst. Anal. **54**(4), 600–609 (2018b). https://doi.org/10.1007/s10559-018-0061-9

The World Heritage Convention (1972). http://whc.unesco.org/en/convention/

Zanakis, S., Theofanides, S., Kontaratos, A., Tassios, T.: Ancient Greeks' practices and contributions in public and entrepreneurship decision making. Interfaces **33**(6), 72–88 (2003). http://www.jstor.org/stable/20141304. Accessed 7 Oct 2020

Zocenko, N., Matveyev, I., Rosenfeld, L., Shokarev, V., Kornienko, N.: Geotechnical aspects of construction on loessial soils. In: Proceedings of the 16th International Conference on Soil Mechanics and Geotechnical Engineering: Geotechnology in Harmony with the Global Environment, vol. 3, pp. 1307–1311 (2005)

Proposal for the Restoration of the Skene of the First Ancient Theatre of Larissa, Greece

Dimitrios L. Karagkounis[1] (✉) and Sofia D. Tsanaktsidou[2]

[1] Department of Archaeological Projects and Studies, Ephorate of Antiquities of Larissa, Ministry of Culture and Sports, Diachronic Museum of Larissa, Mezourlo, 41500 Larissa, Greece

[2] Ephorate of Antiquities of Larissa, Ministry of Culture and Sports, Ancient Theatre of Larissa, 10–12 Mitropolitou Arseniou, 41223 Larissa, Greece

Abstract. The unearthing and restoration of the First Ancient Theatre of Larissa, Greece, built in the early 3rd c. BC, has been a painful and demanding process spanning almost half a century. Being one of the largest and most important theatres in Greece, its Skene is one of the best preserved, although it had been buried under blocks of flats for decades. Conservation and utility reasons prompted the currently presented proposal for the restoration of the Skene, which involves the addition of new limestone blocks to the four rooms consisting the scene-building, as well as restoring part of the proskenion, the colonnade in the front of the skene. The first was accomplished through a methodology of documenting the monument in situ, identifying ancient blocks to their original position and creating a parametric approach to decide on dimensions and position of each specific new block, validated by the monument's construction techniques (clamps, holes etc.). The proskenion will be restored by reassembling identified ancient marble column fragments using the necessary additions of new marble and titanium reinforcements.

Keywords: Ancient monuments · Ancient theatre · Anastylosis · Restoration of monuments

1 Introduction

The First Ancient Theatre of Larissa, Greece, is one of the largest and most important theatres of Greece and it is situated on the southern foot of the "Frourio" (Fortress) hill, in the centre of the modern city of Larissa, in the Region of Thessaly. It is oriented towards the south, where, according to archaeological evidence it faced the ancient free agora.

Its unearthing and restoration have been the objective of several European funded projects, as early as the 2nd Community support framework. The current restoration proposal was conducted in the context of the "Restoration of the first Ancient Theater of Larissa – Phase E" project, aiming to tackle structural and functional problems, as well as utilize a great number of ancient members that have been identified.

I. Vayas and F. M. Mazzolani (Eds.): PROHITECH 2021, LNCE 209, pp. 1158–1169, 2022.
https://doi.org/10.1007/978-3-030-90788-4_89

The drafting of the restoration proposal (Karagkounis et al. 2020), as well as the monument's documentation, were a particularly time-consuming and laborious intellectual task, because of the volume of the documentation elements and the many factors that had to be taken into account, but also because it was conducted simultaneously with the authors' other responsibilities in the aforementioned project. For this reason, the final proposal is a product of the collective work of all related disciplines. The study group consisted of D. Karagkounis, the architect who has been involved in the theatre's revelation and restoration for over 30 years, E. Matsarokos, architect and S. Tsanaktsidou, civil engineer, with the assistance of conservator K. Mpasiakou and stonemason O. Laiou, both of whom provided valuable insights on the monument's ancient construction techniques and materials. It also utilized various previous studies on the monument conducted by the Ephorate and other parties.

The study (Karagkounis et al. 2020) consists of the history and a description of the theatre focusing on the scene building, followed by the restoration proposal. A preliminary structural assessment report described the static and dynamic behaviour of the building, and specified the future steps for the analysis and design of titanium reinforcements. A detailed preliminary quantity report is included, calculating the percentages of new to ancient material, and four volumes where numerous ancient architectural members were catalogued and documented.

In the current paper, a brief description of the theatre and the scene building is presented, followed by a discussion of the restoration proposal. The methodology for dimensioning the new blocks is presented in the case of the north wall of Room 1.

2 The First Ancient Theatre of Larissa

According to long time Ephor and the monument's excavator, archaeologist Athanasios Tziafalias, (Tziafalias 1985 and Tziafalias 2000) the theatre had a lifespan of six centuries, from the early 3rd century BC until the late 3rd or early 4th century AD. During the first centuries, apart from theatrical performances, the assemblies of the senior regional authority were hosted, the so called "Koinon" of the Thessalians. In the 2nd c. BC the Romans converted it into an arena, reserved for the official celebrations. The following centuries it was gradually buried, notably after the 1868 earthquake.

In the early 20th century, the then Ephor of Antiquities Apostolos Arvanitopoulos began excavations that revealed part of the scene ("skene"). After the Second World War blocks of flats appeared, which, in many cases had their foundations constructed directly on the theatre's surfaces. In the 1980's, the first compulsory purchases took place, and a gradual unearthing began that lasted until the mid-2000s.

Several monuments from different eras are found in close proximity, such as the smaller second ancient theater, the basilica of St. Achilles, etc., witnesses to the city's continuous habitation throughout the centuries.

The First Ancient Theatre of Larissa (Fig. 1) has been unearthed almost in its entirety. Its cavea ("koilon") is built on a natural hill and is divided by the diazoma, a 2 m wide corridor, into the main theatre and the "epitheatron", the cavea's lower and upper section respectively. The main theatre was divided by 10 staircases ("klimakes") into 11 cunei, where each cuneus ("kerkida") consisted of 25 rows of seats ("edolia"). When the theatre was converted to an arena, the three first rows where removed and their marble seats where repurposed to retain the upper rows. To this day, only a small part of the epitheatron is preserved. The orchestra has an estimated diameter of 25 m and is surrounded by a closed conduit, the "evripos" that runs under the foundations of the eastern and western rooms of the scene building. In Fig. 2, a plan view of the marble slabs that cover the "evripos" can be seen. The retaining walls "anallimata" are maintained in excellent condition, although not fully unearthed.

Fig. 1. Aerial view of the First Ancient Theatre of Larissa (Karagkounis et al. 2017)

The Scene building ("Skene") is one of the best preserved and perhaps the most luxurious of the few examples of this category of Hellenistic buildings that survive. It is preserved in situ at a great height, retaining numerous architectural members of the colonnade that forms the Proskenion, and the dry stone masonry that forms the side rooms. The Proskenion is made of marble and the rooms are made of limestone and marble blocks without the use of mortar. The stones' origins are traced in the ancient marble quarry of Kastri (Melfos 2010) and limestone quarry of Tirnavos (Melfos 2011). Throughout the centuries several architectural members, especially limestone blocks were removed and used as building materials, and can now be found in nearby historic buildings of the early byzantine and ottoman eras.

3 Description of the Scene Building

3.1 Skene

The Skene is an elongated building of total dimensions of 7.70 × 35.41 m., with an east-west direction. It consists of four rooms in a row, separated transversely by corridors and a proskenion is placed in the north side (Fig. 2). Rooms are numbered from west to east from 1–4. The eastern room has internal dimensions of 5.60 × 5.45 m, and the western 5.55 × 5.30 m respectively. Each one is accessed by an opening to the south (fig). The middle rooms have dimensions of approximately 5.28 × 4.50 m. They are accessed by the main corridor. All openings preserve their marble threshold.

All corridors have a limestone floor that is level to the walls' foundations. The middle corridor is 2.70 m wide and the eastern and western corridors are 1.50 m wide. Three of the four original beams of the north entrance of the western corridor are preserved (not in their original position) (Fig. 2).

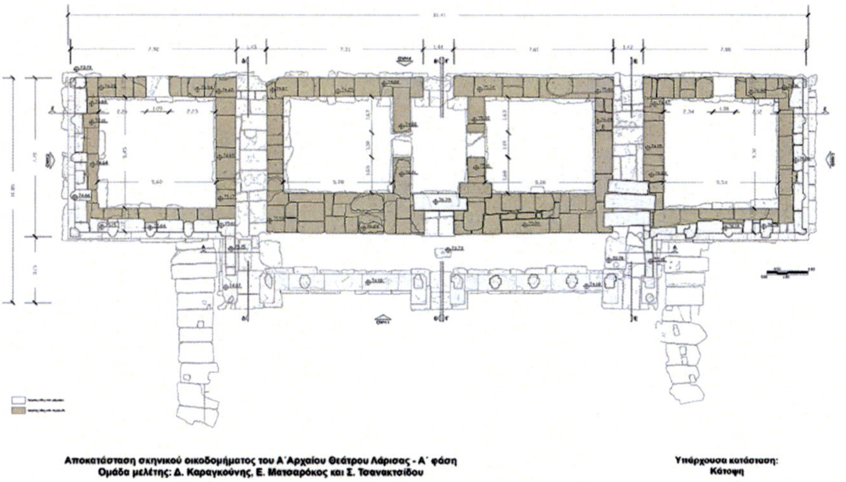

Fig. 2. Plan view of the scene building.

The southern exterior and all interior walls are made of large limestone blocks that are usually the width of the wall. In certain instances, the blocks are placed in pairs. The north wall of the middle rooms is wider than that of the outer rooms, with a width of 1.90 m, and is exclusively made of limestone blocks. The greater surviving height is found in the western side of the north wall, which is approximately 3.00 m. No connections are found between the limestone blocks, neither in the horizontal direction nor vertically, and no mortar was used. On their top faces, the holes that aided in the construction process are clearly visible. These holes (lever holes, "Mochlovothria") allowed the use of levers for the precise placement of the stone blocks of the upper row (Fig. 3a and 3b).

Fig. 3. a) Placement of limestone block 3NΓ6 using a lever b) lever holes on block 2NΓ1/2ΔΓ7.

The exterior walls of the outer rooms, except for the south wall, are made of limestone on the inner layer and marble on the outer. The marble facades display a doric decoration of inherent half-columns. The marble blocks are connected with the use of iron clamps, encased in molten lead. Joints between the marble blocks are dressed to ensure a tight fit using the method of "anathyrosis", which is dressing a part of the contact edges in a manner that resembles a doorframe ("thyra" is greek for door).

3.2 Proskenion

The Proskenion is a colonnade that forms a portico in the north side of the building, in front of the middle rooms of the Skene (Fig. 4). It is 20.72 m long and 3.07 m wide. On top of the Proskenion, in the second floor, was the "logeion" where the actors performed. The colonnade consists of 18 columns. There are 6 columns with a cross section of a rectangular shape with inherent half-column, 3 in front of each room. In the entrance of each corridor, there are 2 columns with an irregular cross-section and 2 of circular. The columns are made of single blocks with a separate doric capital, and they are placed on a marble platform, the "stylobate" which is 0.80 m wide and 0.30 m high.

In the current state of the Proskenion (Fig. 4), parts of 11 fragmented columns are preserved, mainly as a result of the 2002–2003 restoration works. The colonnade was connected to the north wall of the middle rooms of the Skene through beams with approximate dimensions 2.44 long, 0.60 m high and 0.38 m wide. One and one half of them are preserved. The architraves had a varying cross section of about 0.62 m × 0.27 m. On top of them would stand the frieze of triglyphs and metopes with an average height of 0.56 m. Many of them are found ex-situ. They have an inherent mutule ("geison") that protrudes by approximately 0.20 m and some of them bear inscriptions.

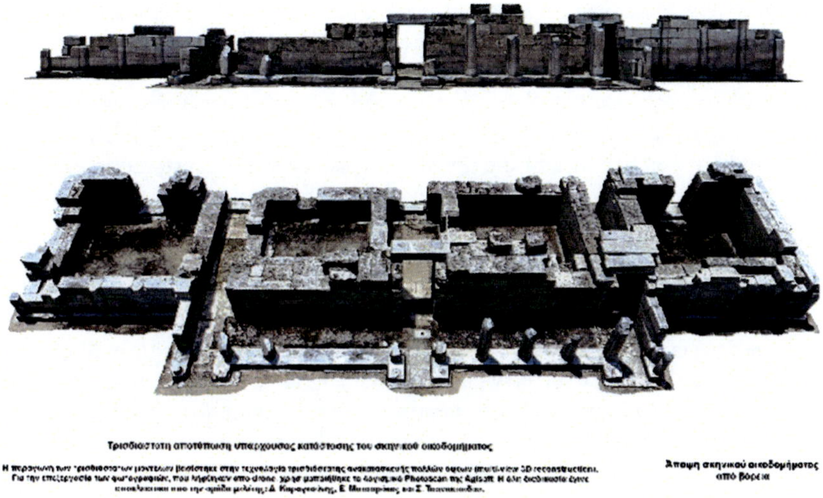

Fig. 4. 3D reconstruction of the Scene building in its present condition

3.3 Building Pathology

The building is in a relatively good structural condition, considering its centuries old life. During the 2002–2003 restoration works fragmented and cracked members where restored. In order to assess the building's pathology the following steps were taken:

- Study of historical records and bibliography
- Comparison of photogrammetric documentations
- In situ macroscopic inspection
- Damage mapping
- Photographic documentation

Some of the main causes of damages include: the materials' natural properties, anthropogenic interventions such as the removal of blocks to be used as building material and the construction of modern structures on top of the buried theater as well as in contact to the ancient members. Exposure to environmental conditions. Also, natural disasters, mainly two major earthquakes, the first during the pre-Christian period and the second circa 7th c. caused structural damages (Caputo et al 2011).

Damages include displacement and fracturing of blocks, laminations and local loss of material, superficial cracks, discoloring due to contact with oxidized iron reinforcements.

4 Aims and Principles

The aim of this restoration proposal is the Skene's structural and functional upgrade, its morphological restoration and the enhancement of the perception and readability of the Skene itself and the Ancient Theater in its entirety. The interventions concern the repositioning of ancient members of the Proskenion to their original or equivalent positions, the restoration of fragmented ancient members and also the placement of new members where necessary, in order to maintain the authenticity of the monument. In this way it will be possible to highlight the form of the Proskenion, showcasing the great number of original triglyphs and metopes but also give it the sense of height.

Additionally, certain rows of the dry stone masonry will be restored in each of the four rooms so that the form of the building is better understood and functionality is restored. This decision was made because the low height of the southeastern side of the scene building and the general condition of the dry stone masonry of its four rooms make it difficult for the visitor and passer-by to perceive its shape. This is a significant disadvantage, as this side of the Skene is the most prominent part of the Ancient Theater from the pedestrian street of Venizelos but also the first that the visitor will encounter when entering the site from the new entrance that will be located on the same street.

According to the International Declarations and Conventions on the protection and conservation of monuments and sites, architectural heritage is a living and invaluable testimony of the historical memory. The proposed interventions were based on the principles of the International Charters. The adherence to said principles ensures that the restoration complies with the internationally accepted ethics and codes of practice.

A key priority is to preserve the authenticity of the monument. In general, the criteria and principles of the restoration are: Respect for the original, reversibility of proposed interventions, financial costs of restoration works, completion time.

Specifically, the materials and techniques used will be compatible with the ones found to have been used on the monument. The authenticity of the ancient material will be preserved and its historic value revealed. The proposal respects all the historical phases and new work will be distinct from the original. All restoration works can be undone and the materials used can be removed. The restoration will allow the study of the monument by future researchers. The works will be accompanied by complete as-built drawings and photographic documentation. The intervention will be the result of choices and decisions, after examining the limitations, framework and priorities of the restoration process.

5 Restoration Proposal

5.1 Skene

As far as the dry stone masonry walls of the Skene are concerned, in room 1, three new marble blocks will be added to the outer layer of the north wall. In room 2, eight new limestone blocks will be added. In room 3, twenty-four new blocks will be placed and two ancient ones will be restored. In room 4, thirteen new marble blocks will be added and an ancient one will be restored. In addition, twenty-one new limestone blocks will be placed and ancient ones will be restored. In the opening of the central door, two marble lintels and three limestone blocks will be placed. In Figs. 5 and 6 the plan view and the front elevation and section of the Skene after the restoration are presented.

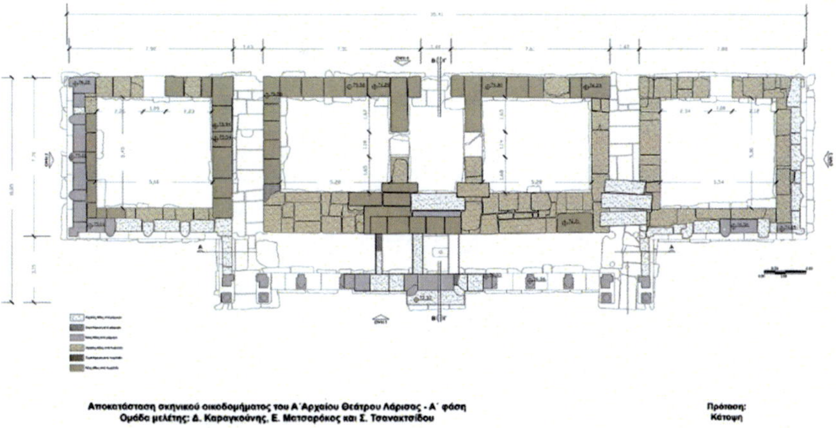

Fig. 5. Plan view of the Skene after restoration. White is ancient marble, grey is new marble. Beige is ancient limestone, dark beige is new limestone.

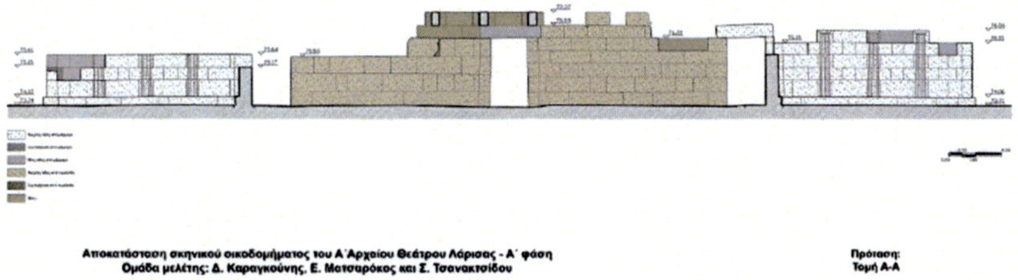

Fig. 6. Front elevation and section of the north side of the Skene. White is ancient marble, grey is new marble. Beige is ancient limestone, dark beige is new limestone.

For the dimensioning of the new blocks, the following were taken into account:

- The existence of holes for the use of levers
- The dimensions (length-width-height ratio) of the existing blocks. Each and every existing stone block was measured and catalogued by wall. The blocks where classified by their length, and the percentage in which they appeared was calculated in order to conclude as to the most common block sizes. This information helped decide on the potential dimensions of the new blocks
- The vertical joints of the surviving masonry

In the context of this paper, the dimensioning process is demonstrated in the case of the western room (Room 1).

In room 1, three new marble blocks will be added to the outer layer of the north wall, one in the fifth and two in the sixth row (Fig. 7).

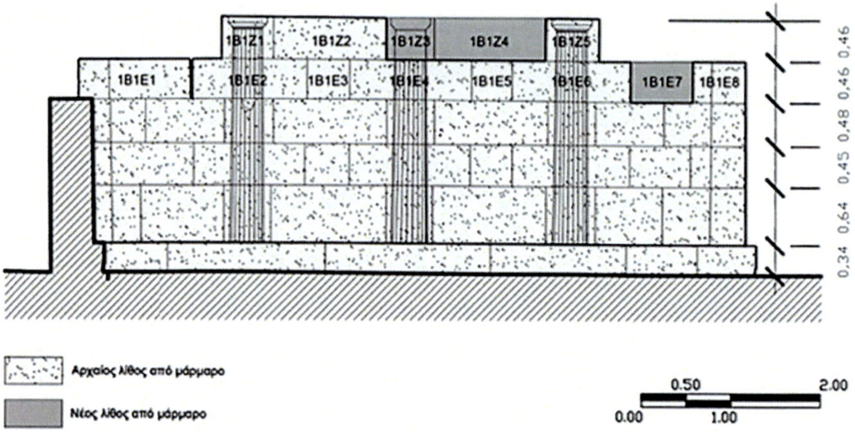

Fig. 7. The façade of the north wall of Room 1 after the restoration. White is ancient marble, grey is new.

Fifth row: New marble block 1B1E7, will be 0.70 m long and 0.51 m wide. The length is defined by the gap between the marble blocks 1B1E6 and 1B1E8. The width is defined by the presence of limestone blocks in the inner layer of the north wall. Obviously, its height is that of the adjacent blocks.

Sixth Row: New marble block 1B1Z3, will have an inherent half-column, and will be 0.56 m long. Its length is defined by the construction pattern of the masonry by repeating the corresponding block of the fourth row as well as by the existence of a hole on the underlying block 1B1E4 (Figs. 8a and 8b). Its width is defined by the presence of a limestone block in the inner layer of the wall.

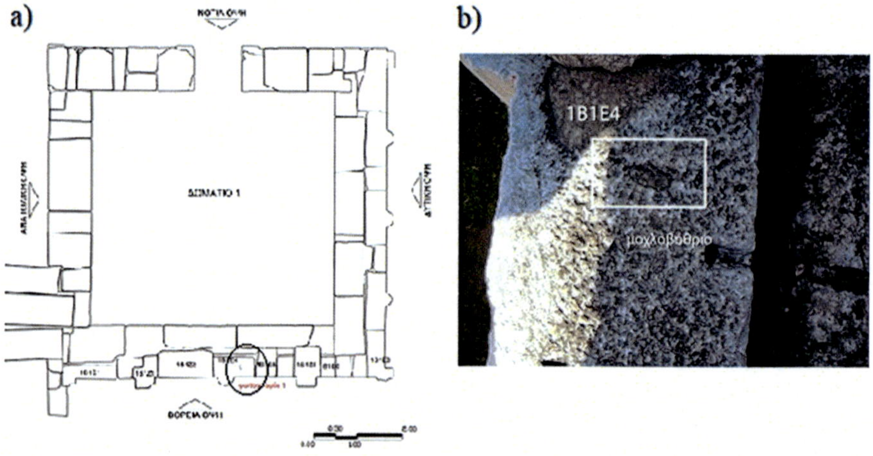

Fig. 8. a) Plan view of Room 1. b) Detail of the lever hole on block 1B1E4.

New marble block 1B1Z4, will be 1.26 m long, defined by the remaining gap between blocks 1B1Z3 and 1B1Z5, and by the corresponding blocks of the underlying rows. Its width (0.66 m) is defined by the presence of a limestone block in the inner layer of the wall. The height of the sixth row is 0.46 m, as defined by existing ancient blocks.

5.2 Proskenion

The current proposal suggested the restoration of the Proskenion in two phases, the first of which involves the columns and part of the frieze (Fig. 9). Three new circular columns would be installed and three ancient would be restored. Two rectangular columns with inherent half-column would be installed and four would be restored. Only the middle part of the frieze would be restored, therefore three new architraves would be installed and one would be restored. Five ancient and three new members of the frieze would be installed. One ancient beam would be placed and one would be restored.

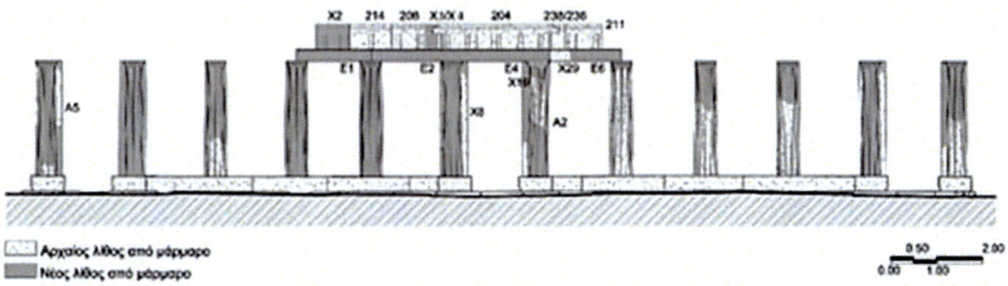

Fig. 9. Restoration of the Proskenion of the Skene of the First Ancient theater of Larissa. First phase

On its 4th August 2020 assembly, the Central Archaeological Council approved the restoration of the Proskenion in its entirety. In November 2020, the Region of Thessaly local authorities announced their willingness to fund the restoration of the Skene by including it in the current project, almost doubling its total budget. Consequently, at the time of this paper's submission, the study group is working on the detailed design for the restoration of the Proskenion as a whole (Fig. 10) according to the terms of the Greek Minister of Culture and Sports decision of approval (reference number ΥΠΠΟΑ/ΓΔΑΜΤΕ/ΔΑΑΜ/ΤΜΑΜ/551095/62244/2053/958).

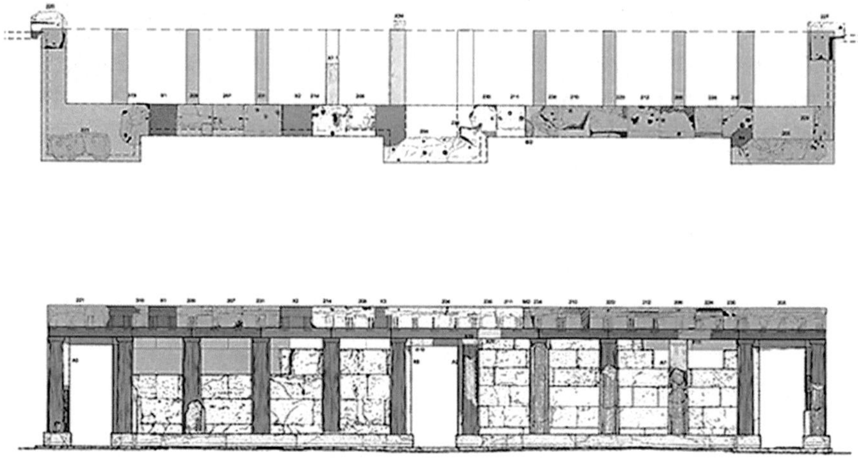

Fig. 10. Restoration of the Proskenion. Front and plan view. White is ancient members. Deep grey is new material. Medium grey is the outline of the second phase.

6 Conclusions

The restoration of an ancient monument is a demanding procedure that requires a multitude of factors to be taken into account, due to the uniqueness of every monument and its historical, architectural and aesthetic significance. Several sources were studied and processed in the course of this work, using a multidisciplinary approach.

In the present study, a proposal for the restoration of the Skene of the First Ancient Theatre of Larissa, Greece, a very important Hellenistic theatre, is discussed. A brief description of the monument is given for the understanding of its history and construction techniques. The restoration aims to solve structural and morphological issues, and restore a degree of functionality to the Skene. The proposal was conducted according to the international principles for the protection and conservation of monuments, respecting the original and ensuring the reversibility of interventions.

The proposal involves the addition of stone blocks to certain rows of the dry stone masonry of each of the four rooms of the Skene, as well as the restoration of the Proskenion, by reassembling identified ancient architectural fragments using the necessary additions of new marble and titanium reinforcements, installing new members and returning ancient members to their original positions.

The major outcomes will be the Skene's morphological restoration and the improvement of its readability, the promotion of its historical and architectural value. It is a major step in the restoration of the First Ancient Theatre, in the direction of rendering to the city of Larissa an accessible and functional theatre that is one of a kind in Greece.

Acknowledgements. The activity presented has been conducted under the Project "Restoration of the first Ancient Theater of Larissa – Phase E" that is co-financed by the European Union (European Regional Development Fund – ERDF) and Greek national funds through the Regional

Operational Program Thessaly 2014–2020 of the National Strategic Reference Framework (NSRF) 2014–2020.

References

Caputo, R., Hinzen, K.-G., Liberatore, D., Schreiber, S., Helly, B., Tziafalias, A.: Quantitative archaeoseismological investigation of the Great Theatre of Larissa, Greece. Bull. Earthquake Eng. **9**(2), 347–366 (2011)

ICOMOS: International charter for the conservation and restoration of monuments and sites (The Venice Charter 1964). In: 2nd International Congress of Architects and Technicians of Historic Monuments, Venice (1964)

Karagkounis, D., Chatzidakis, N., Angeli, E.A., Plakotaris, I., Karnavezos, Ch., Tziafalias, A.: Αποκατάσταση του Αρχαίου Θεάτρου της Λάρισας (Restoration of the First Ancient Theatre of Larissa), unpublished study (2017)

Karagkounis, D., Matsarokos, E., Tsanaktsidou, S.: Αποκατάσταση Σκηνικού Οικοδομήματος του Α' Αρχαίου Θεάτρου Λάρισας – Α' Φάση. (Restoration of the Skene of the First Ancient Theatre of Larissa – Phase A), unpublished study Ephorate of Antiquities of Larissa, Ministry of Culture and Sports (2020)

Melfos, V.: Προσδιορισμός της πηγής προέλευσης των πρώτων υλών από επιλεγμένα δείγματα πωρολίθου του Αρχαίου Θεάτρου Λάρισας (Determination of the source of origin of the raw materials from selected limestone samples of the Ancient Theater of Larissa), unpublished study (2011)

Melfos V. (2010). Προσδιορισμός της πηγής προέλευσης των πρώτων υλών από επιλεγμένα δείγματα μαρμάρου του Αρχαίου Θεάτρου Λάρισας (Determination of the source of origin of the raw materials from selected marble samples of the Ancient Theater of Larissa), unpublished study

Το αρχαίο θέατρο της Λάρισας, Πρακτικά του Α΄ Ιστορικού-Αρχαιολογικού Συμποσίου, Larissa, pp. 162–185 (1985)

Tziafalias, A.: Το έργο της ΙΕ΄ ΕΠΚΑ Λάρισας, Το έργο των Εφορείων Αρχαιοτήτων και Νεωτέρων Μνημείων του ΥΠΠΟ στη Θεσσαλία και την ευρύτερη περιοχή της (1990–1998),*1η Επιστημονική Συνάντηση*, Volos, pp. 91–96 (2000)

Protection and Preservation Strategies

The Royal Roads in Mexico: Engineering and Architectural Heritage

Andres A. Torres-Acosta[1](✉), Joel Bustamante-Altamirano[2], and Alicia Esparza-Carrillo[2]

[1] Instituto Tecnológico de Estudios Superiores de Monterrey, Campus Querétaro. Epigmenio González 500, San Pablo, 76130 Santiago de Querétaro, Querétaro, México
atorresa@tec.mx

[2] SATELSA SA de CV, Paseo de la Reforma No. 184, Queretaro, Queretaro, Mexico
joel.bustamante@satelsa.com.mx

Abstract. When central Mexico, was conquered by Spaniards, the Conquistadors immediately began the process of colonization for which communication channels were necessary to bring the population under the power of the Spanish Crown. Since one of Spain's principal interests was extraction of precious metals: silver and gold, new regions were colonized and mines developed. Roads were built to access these mine fields, supply them with inputs and export the products which simultaneously led to growth in other productive sectors. Two main, perpendicular communication axes were opened during the Viceroyalty period. The first was opened from east to west from Veracruz port (Gulf of Mexico coast) to Mexico City and then to Acapulco port (Pacific Ocean). A north-south axis connected Mexico City with mining regions to the North and to the South. This investigation shows the methodology used to georeference a part of the Camino Real de Tierra Adentro, in the Queretaro State. It started with old cartography review (XVIII and XIX century maps); followed up with the direct walk of the route in almost 300-km long trip, obtaining direct information, a photographic survey of the route and face-to-face interview with people still living close to the Camino Real vestiges.

Keywords: Georeferenced information systems · Royal roads

1 Introduction

In Mexico, buildings with historical value are regulated by INAH (National Institute of Anthropology and History, by its acronym in Spanish), an institution directly dependent to the federal government of Mexico. This institute owns information regarding all the historical properties in the country and such information is deposited in an inventory generated by each INAH office (being 32 offices in total, one for each of the 32 states in Mexico). These might be private or state properties (any of the three levels: Municipal, State, or Federal) and this situation complicates the sharing of any information from them with other research centers, civil associations, or the three government levels, any of which could promote programs for use, restoration, or space adequacy.

I. Vayas and F. M. Mazzolani (Eds.): PROHITECH 2021, LNCE 209, pp. 1173–1184, 2022.
https://doi.org/10.1007/978-3-030-90788-4_90

The inventory generated by INAH includes several information fields such as location, sketches or architectural outlines, photos (no more than a dozen) and some historical details. Although some of the properties do not include specific geographic coordinates and the historical details are relegated to an observations field and can be sparse. The inventories generated by each of the 31+1 states are kept in each of their INAH's Regional Centers as well as in INAH's General Coordination of Historical Sites and Monuments (CGSMH, by its acronym in Spanish), located in Mexico City. Another challenge is the fact that in order to obtain this information it is necessary to submit a written application to the CGSMH which, in turn, sends the printed information (PDF files) via electronic mail, a disposition that makes its potential usage difficult.

So far there are three main challenges surrounding INAH's architectural heritage information. The first one is the difficulty in sharing information between institutions and individuals due to the type of ownership. The second one is related to the amount of information available for the researchers and professionals interested in historic property. The third one concerns the formats in which the information is currently available and its immediate readiness to the public. Nowadays, the organizations in charge of harboring the assets of cultural interest, have diverse typologies and capacities, which leaves them in the position of having to manage them in an effective way with very limited resources. The information they have is diversified, decentralized and, fundamentally, in paper support and, rarely, in digital support. In the latter's case, it is not always under standardized formats and safeguarded by one entity.

The second challenge is well depicted in the specific case of viceregal roads and their infrastructures since there are no inventories that provide information concerning their existence and location. The task of generating an inventory containing details such as location, physical state, and regional importance of the settlements that comprise these roads and the vestiges found along them is complicated, and more so if the property considered is a road with an extension of several hundreds or thousands of kilometers. This is due to the number of transformations undergone by the roads and the landscape surrounding them in the five centuries of their existence. The inventory of constructed heritage that the INAH currently has includes various types of constructions from different periods of Mexico's history, spanning from the pre-Columbian and viceregal periods to the independent period of the country (from Spain). In the inventories there tend to be more accurate and detailed information for properties from the latter period since their evidence tend to be in better shape and key subjects that could bear witness to those properties can be easily found.

However, in the case of pre-Columbian and Viceregal roads, their categorization and description can be complicated due to the fact that those are not structures with a single location (since they are extended through several states at once) and they possess elements that are part of a whole body, which is the case with bridges, haciendas, convents, ventas, inns and presidiums that, taken individually, would make up for a jigsaw puzzle of information. There is no archive that collects information regarding the existing vestiges from the viceregal roads, which is why their existence is not known in an official manner but rather by means of hearsay and accounts by the local population who may not know all the specific historical details.

1.1 The Inland Royal Road or Camino Real de Tierra Adentro (CRTA)

The Inland Royal Road (CRTA. by its acronym in Spanish) is one of the four main routes and Royal Roads built during Mexico's Viceroyalty and it's been functioning since the middle of the 16th century and continued to be used after the independence (1821) under the name of National Road. It has a length of almost 2,900 km (some says 2,600 km) and it crosses the present day eleven Mexican States: Sate of México, Hidalgo, Querétaro, Jalisco, Guanajuato, Aguascalientes, Zacatecas, San Luis Potosí, Coahuila, Durango, and Chihuahua. It is an important route not only because it connects the Vice regal Capital (Mexico City) to the northern and western regions but because this was the route used for the transportation of silver and gold from the mining districts to Mexico City and to the Pacific and Gulf of Mexico ports. Consumption goods and raw material were transported from different regions to Mexico City and then to the mining districts and other communities as well.

The CRTA has value as historic and cultural heritage because it is a main communication channel and trade route but it is also the backdrop and main actor in geographic, cultural, and social transformations of the country through almost five centuries. From the CRTA the history of the Vice regal communications, trade channels, transformations in constructive systems and materials of the roads, of short and long- term settlement processes and of the economic and cultural activities of the populations can be traced.

The influence of the CRTA in different spheres of human life is such that it has been a study subject in various disciplines, from different perspectives: settlement, engineering, architecture, economy, anthropology, demography, etc.

On August the 1st, 2010, the CRTA was declared a "Cultural Heritage Route" by the UNESCO, entering 60 sites, throughout the different States where CRTA passes, as Cultural Heritage. To obtain the declaratory, a guiding document or dossier was prepared by INAH in 2009 about the CRTA, its extension, infrastructure, movable and immovable property, etc.

While the cultural and historic value of CRTA was recognized and its heritage will benefit in many ways thanks to the declaration by the UNESCO by attracting researchers and professionals as well as possible projects with economic and social impact, there are still other royal roads that served important roles in the lives of the population throughout the centuries and these roads do not count with the same level of recognition, and while studied and researched, the preservation efforts for these roads is still lacking. By gathering accurate information in different areas about these roads, that is, by constructing the same kind of puzzle that is currently under works for the CRTA, the information could potentially attract projects aimed at harnessing the cultural and historic value of these roads for the economic and social benefit of the communities that surround the roads. Such a project could be the creation of "cultural routes" like the one planned for the CRTA.

2 The Royal Roads

Mexican pre Columbian societies didn't need actual roads because they ignored the usage of the wheel and track animals. The tamemes (person instructed since childhood for carrying goods using its back) were in charge of transporting any merchandise from one place to the other by foot. They could lift around 23 kg in their backs and could reach as far as Guatemala, from Tenochtitlan (actual Mexico City). For this kind of transport, a footpath was sufficient and the passage of the travelers over time was responsible for the gradual flattening and outlining of the roads, making them easier to transit.

However, the Mayans had, long before the conquest, opened real long-distance roads, the sacbé-ob (plural of sacbé, it comes from the Mayan sac: white and be: road), and they are preserved up to date, event though their specific function is unknown and the archeologists consider them rather as pilgrimage roads. There were some roads around Tenochtitlán that were part of the hydraulic system since they served as water dikes, and they were called calzadas.

When central Mexico, was conquered by a small group of Spaniards; the first incursion was led by Hernán Cortés, who left from La Antigua, near the present-day Veracruz port, and reached, in 1519, the most important city of the plateau: Tenochtitlán, a city bigger than any European city at that moment. From the fall of Tenochtitlán, in 1521, due in big part to the incidence of epidemics that decimated the population, the Spaniards carried on with their ventures of conquest. The Conquistadors opened horse paths made of dirt in order to reach the different indigenous populations with their horses and subdue them to the Spanish crown. In these early stages, they were content with merely broadening the existing pre-Columbian roads so that the horses and mules, then unknown in the New World, could transit.

The interest during the Viceroyalty for different products, substantially modified the network of exchange and drove the conquerors to open other roads to be able to travel to regions of difficult access. The efforts aimed for the discovery and exploitation of natural resources gave way to new channels of communications. During this period, the most prominent case was, without a doubt, the mining activity and mainly the extraction of silver. New regions were turned into colonies and road networks were shaped according to the locations of the mining reales or centers, with the purpose of exporting the products from the mines as well as benefitting from them. As part of this process, other activities (agriculture, ranching, certain types of manufacture needed for the daily life and trade) surfaced as they were indispensable to the continued exploitation at the mining sites since they provided consumption goods for the workers and the raw materials for the mining activities. As new seams were discovered or as mining centers were consolidated, population centers emerged and roads with their corresponding branches were opened around them.

From the 16th Century, when the conquerors settled in the different regions of New Spain, the New Kingdom of Galicia, New Biscay, New Santander, and the New Kingdom of León, new routes for the transport of merchandise were opened. Communications during the viceregal period were built based upon two perpendicular axes. The first one was outlined from Orient to Occident, between Veracruz Port- México- Acapulco/Navidad Ports and it constituted a colossal bridge between the Atlantic and Pacific oceans, known back then as the North Sea and the South Sea, respectively. The other axis, North-South, united the norther regions of the viceroyalty (septentrion) with México City as well as with Oaxaca and Guatemala in the South. It should be noted, however, that this axes were not travelled all year long. The Acapulco and Xalapa (Veracruz) fairs were celebrated once a year and it was just then that the merchants came together in those two places to acquire European and Asian goods. As for the conductas or coaches that went all the way to New Mexico, they travelled only biannually.

Once opened, these axes of communication during the colonial period were called Caminos Reales or Royal Roads, although this denomination is not entirely accurate since neither the King of Spain nor the Viceroy of the New Spain provided any resources to open them or give them maintenance.

In total, four main channels were built and they were known as:

Royal Road of México-Veracruz or "Viceroys Royal Road" (Fig. 1)
Royal Road of México-Acapulco or "Royal Road of the Orient" (Fig. 2)
Royal Road of México-Santa Fe or "Interior Land Royal Road" (Figs. 3 and 4) and
Royal Road of México-Guatemala or "Royal Road to the Salazones (salted meat)" (Fig. 5).

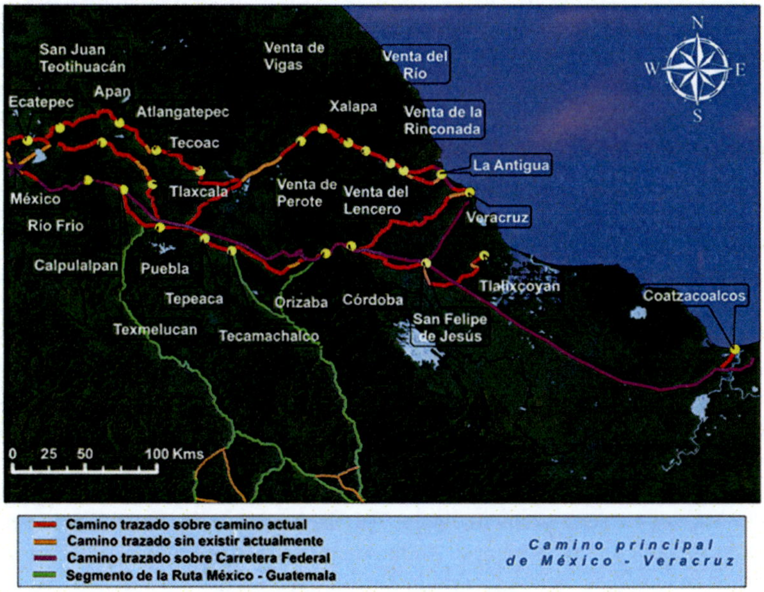

Fig. 1. Road México - Veracruz

Military campaigns, missionaries, merchants, stockmen, and expeditionary forces streamed from these roads, carrying seeds, devotions, beliefs, customs, migration, culinary traditions, rituals, ideas, and knowledge that came from different places and they established in the New Spain and its provinces, a complex economic and cultural system that resulted in the building of cities, villages, baserris, haciendas, ranches, presidiums, hospitals, convents, and mining centers from where the precious metals, that for a long time supported the economy of the old continent, were extracted.

In these Royal Roads trails/footpaths or transversal roads that communicated the remote colonial settlements to the main roads came together. The ones that stand out amongst them are the ones that communicated San Luis Potosí with Nuevo León and Coahuila with Texas (to the orient) and Mexico City with Toluca, Valladolid, Guadalajara with the Pacific Ocean in Colima, Nayarit, Sinaloa, part of Sonora and Baja California (to the west). These transversal roads were also of great importance during the Viceroyalty and especially at the end of the 18th century when the monopoly of the Veracruz and Acapulco ports was liberated.

The modifications applied to these roads with the passage of time were mainly due to the needs of the population. In the beginning, the means of transportation were the same people that travelled the roads (tamemes and armies) then, after the arrival of the conquerors, the usage of cargo animals (horses, mules, oxen, etc.) were responsible for the radical modifications that the outline and the constructive system of the roads underwent.

In a similar manner, this modernization in the field of land communications did not only modify the outline of the roads but also their outfitting. In the beginning, the pre-Columbian roads were transited by people, and so, the roads or footpaths were gradually flattened and shaped by the passage of said people and it didn't matter that they travelled through routes that were not straight or with pronounced slopes. With the arrival of horses and cargo mules, the roads had to be widened and with subtler slopes so that the cargo animals could go through easily, thus ensuring the safety of the travelers and their valued loads, especially when this loads came from the mining centers. Where obstacles or difficult terrains were unavoidable the roads' outfitting needed to include structures such as bridges to ease the passage. Constructions for the lodging of travelers were also needed since some trips were so long that months were needed to complete them. This outfitting included mesones, ventas and inns, which were required only after ordinances were created by Hernán Cortés himself for the new road of Mexico-Veracruz, that passed through Xalapa and Perote and required that such establishments be constructed at distances no greater than 40–50 leagues (1 Spanish league = 4.82 km, a day in the road in normal conditions) from each other.

The precious metals extracted at the mining sites had to be moved from the mines in the center and north of the Viceroyalty to the coasts of Veracruz thus, stronger vehicles and animals were needed, as well as the widening and reinforcement of the roads to avoid the sinking of such precious loads. This is the way that the road's construction systems went from simple paving or flattening of the surface to being reinforced with other materials such as pebbles, similar to the ones used in the roman roads at the start of the 1st or 2nd Century B.C. The bridges were reinforced as well and, since the lodging structures were not good enough to be used as shelter from possible attacks by bandits or natives seeking to steal the loads, fortresses, forts, and strongholds became a necessity in the middle of the XVI century. This structures that served as shelters were strategically placed along the roads according to the location of mining centers and

Fig. 2. Road México – Acapulco

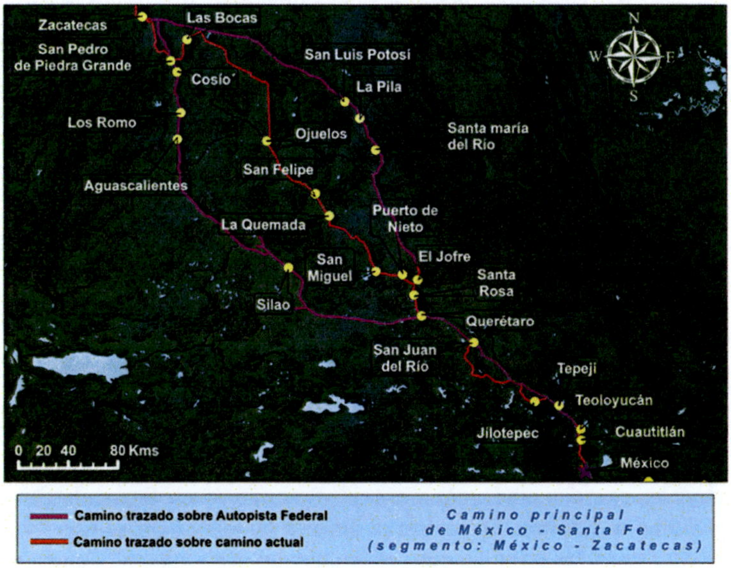

Fig. 3. México – Zacatecas

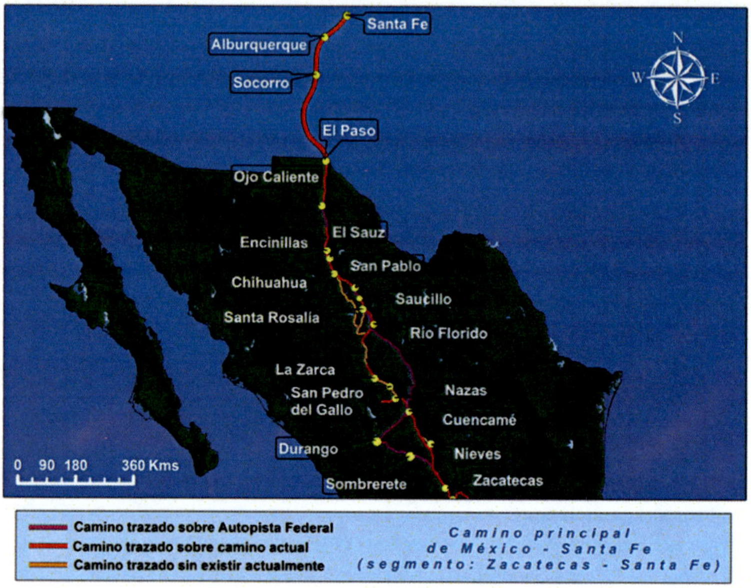

Fig. 4. Road Zacatecas – Santa Fe

supplying communities as well as close to the most common spots chosen for robbery due to the advantages that the same landscape gave to the robbers.

Even after Mexico's Independence War this routes were still used and known to the majority of the population since the influence of these communication channels in the spatial and social configuration was quite vast. The Royal Roads would remain during the 19th and some of the 20th century the guiding compass for most new communication paths that emerged after the Viceroyalty ended.

It is possible to know from travelers' diaries that even in the 19th century, these roads were still referred to as "the royal roads" by the local population even when the roads went by different names, such as "Camino Nacional" or National Road.

In modern times, transformations in the roads or highways are still ongoing as an updating process to the road network in the country. In these updates, new roads are frequently outlined, the number of lanes are widened in the existing roads, the materials initially used for the construction are modified and/or replaced, and the original outlines are modified in order to facilitate mobility and improve transportation time. The communities connected through this network of roads often transform according to the changes in both the transit of the people and the consumption goods through the roads. The influence of one over the other is such that the communities that increasingly become more isolated from the main roads usually suffer severe depopulation while the population living in communities nearer to the new outline of a road usually increases due to the change in the physiognomy of the nearby environment. Communities and communication infrastructure (such as bridges) emerge and vanish in different historical periods, making it difficult to pinpoint the outline of these roads using resources of a single historical period. This is why it is important to evaluate the transformations of the communication routes by means of the cartographic or documental evidence form travelers available.

In Mexico the safeguard of its Viceregal routes has only been undertaken for the CRTA, with the execution of isolated projects with scarce technological repercussion by several higher-education institutions and research centers in the country (Regional Centers of the INAH, for example).

Fig. 5. Road México – Oaxaca/Chiapas

3 Methodology

3.1 Documentary Research

The review of documents relating to roads in general: documents in the AGN (Nation's General Archive, from its acronym in Spanish), INAH's libraries, etc. These documents can include travelers' memoirs, technical details in engineering, construction reports, newspaper libraries, etc.

A key source of information are the records and documents (diaries, pamphlets, traveler's guides and reports) produced by the travelers that journeyed through the roads in different periods of time and with different objectives. From this records it is possible to obtain geographic and demographic information from the roads, their surroundings, the natural and cultural landscape in different historical periods, as well as descriptions of the state of the roads, the materials used for their construction, the development of secondary branches that stem from the main roads, means of transportations used, lodging structures along the road, quantity and type of provisions that could be found by travelers along the roads and many more aspects.

Another valuable source are the statistic and administrative memoirs generated by the government ministries for each state. Many of these documents can be found online in the digital libraries of higher education institutions. In this official documents there are records of construction and restoration processes in the roads with dates, budgets, labor force and materials used. There can also be found records of crop fields, ranches, haciendas, dams and constructions in general surrounding the roads, their owners and the state in which they can be found. The location and name of the sentries and the tolls charged in different points along the routes as well as the interaction of the built structures and the population with the natural landscape and the climate.

3.2 Cartographic Research

Nowadays there is not a nationwide program for digital dating and the georeferencing of road networks and associated monuments in Mexico, to record the Viceregal routes. Only isolated projects with scarce technological repercussion by several higher-education institutions, research centers, or history institutions (i.e. Regional INAH Centers) has being undertaken for the CRTA, in few of the States where this road pass through.

It is a first level priority to count with an accurate and updated survey for the location of these roads, based on their digitalization with information and georeferenced positioning system (GPS) technologies and thus, have an inventory of the historical sites and monuments that come together in these roads. This is the reason behind the interest in the execution of this project in pursuit of survey these Vice regal roads. This, in turn, will open the doors to business in Mexico for the partners involved in the project (INAH Querétaro, ITESM, UAQ, SATELSA), with a broadening into providing these types of services for other Iberoamerican countries (including Spain); the ones named "Reales" (main) and the transversal, thus encouraging the creation of cultural itineraries such as the one being undertaken in the CRTA's stretch in the state of Querétaro.

3.3 Perform the Georeferenced Survey

Nowadays there is not a nationwide program for digital dating and the georeferencing of road networks and associated monuments in Mexico, to record the viceregal routes. Only isolated projects with scarce technological repercussion by several higher-education institutions, research centers, or history institutions (i.e. Regional INAH Centers) has being undertaken for the CRTA, in few of the States where this road pass through.

It is a first level priority to count with an accurate and updated survey for the location of these roads, based on their digitalization with information and georeferenced positioning system (GPS) technologies and thus, have an inventory of the historical sites and monuments that come together in these roads. This is the reason behind the interest in the execution of this project in pursuit of survey these viceregal roads. This, in turn, will open the doors to business in Mexico for the partners involved in the project (INAH Querétaro, ITESM, UAQ, SATELSA), with a broadening into providing these types of services for other Iberoamerican countries (including Spain); the ones named "Reales" (main) and the transversal, thus encouraging the creation of cultural itineraries such as the one being undertaken in the CRTA's stretch in the state of Querétaro.

3.4 Produce a Cartographical Model with the Obtained Information

SATELSA already has several years working in the spatial analysis of these viceregal and nineteenth century roads with the guidance of cartographic information from the National Institute of Statistics and Geography (INEGI) and from the Ministry of Communications and Transport (SCT). The spatial analyses that, to this date, have been executed are all in a bureau level, based on maps that stem from the literature and the routes travelled of some travelers (Figs. 1, 2, 3, 4 and 5).

4 Conclusions

All of these roads could be visualized as historical-cultural itineraries because they contribute to illustrate the social notions from their corresponding historical period, and thus, they could help illustrate a contemporary social notion of the value of the cultural heritage as a non-offshoring resource for a sustainable social and economic development. These roads are the engine whose historical guiding thread has generated- or is still generating- the creation of immovable property and movable cultural heritage of Mexico, in particular, and of the humanity, in general, that is to say, that these routes are not only a sum of many elements joined together by it (cities and historical sites, cultural landscapes, etc.), but they rather incorporate a historical spirit that can be tangible or intangible, that agglutinates said element in a unique and singular "whole".

To be able to locate them with higher accuracy in contrast to how they are nowadays known, could bring a higher level of recognition to the communities by which these roads thread and in this way encourage a sense of propriety in these communities regarding their history and their tangible and intangible heritage, in turn, promoting the creation of cultural itineraries that could sustain their economies.

This work will synthesize and graphically present all the information external to the developed acquisition system. A study in the incorporation of additional information to the GIS (historical cartography, antique images, commentaries, voice notes and illustrations) is proposed. All of this information could also be georeferenced and looked up using the GIS tool, thus achieving the integration of all the available data and having a global and detailed view of the road studied.

References

Torres-Acosta, A.A., Bustamante, J., Cramaussel, C.: Road Stories. Mexico's Viceroyalty Period, Routes-Roads, vol. 367, pp. 86–91, July–September 2015. ISSN: 0004-556 X

Torres-Acosta, A.A., Gradilla-Hernández, L.A.: Road Stories. Cortes' Route to the Conquest of Tenochtitlán, Routes-Roads, vol. 366, pp. 88–93, April–June 2015. ISSN: 0004-556 X

Torres-Acosta, A.A.: Road Stories. Pre-Hispanic Roads in Mexico: Mexican Calzada and Mayan Sacbé-oob, Routes-Roads, vol. 365, pp. 90–93, January–March 2015. ISSN: 0004-556X

Torres Acosta, A.A., y Bustamante Altamirano, J., Mendoza Pérez, I.N., y Serna Cerrillo, L.L.: Uso de las geotecnologías de información para la ubicación del camino real de Tierra Adentro en Querétaro, México, Boletín de Monumentos Históricos, Tercera Época, vol. 29, pp.180–191. Instituto Nacional de Antropología e Historia, septiemnre-diciembre (2013). ISSN: 0188-4638

Torres Acosta, A.A., y Bustamante Altamirano, J.: El Antiguo Camino a Guatemala en la época virreinal. Relatos e Historias en México **5**(55), 80–85 (2013)

Mejía González, E.J., y Torres Acosta, A.A.: Trotamundos Gemelli Careri en Nueva España 1697. Relatos e Historias en México **5**(50), 26–33 (2012)

Torres Acosta, A.A., y Bustamante Altamirano, J.: El Antiguo Camino de Acapulco. Relatos e Historias en México **4**(46), 80–84 (2012)

Morán Miranda, C.A., Torres Acosta, A.A., Bustamante Altamirano, J.: Viajeros del siglo XVII en Veracruz," Vías Terrestres, Número 17, Año 3, mayo-junio. Asociación Mexicana de Vías Terrestres A.C., pp. 21–24 (2012)

Torres Acosta, A.A., y Bustamante Altamirano, J.: La Ruta de la Plata Camino de Tierra Adentro. Relatos e Historias en México **4**(44), 46–51 (2012)

Torres Acosta, A.A., y Bustamante Altamirano, J.: Camino Real de México a Veracruz. Relatos e Historias en México **4**(43), 65–69 (2012)

Morán Miranda, C.A., Torres Acosta, A.A.: Viajeros del siglo XVI en Veracruz. Vías Terrestres, Número 16, Año 3, marzo-abril, Asociación Mexicana de Vías Terrestres A.C., pp. 26–28 (2012)

Torres Acosta, A.A., y Bustamante Altamirano, J.: Caminos de Nueva España: Rutas y formas de transporte en la colonia. Relatos e Historias en México **4**(42), 62–67 (2012)

Castillo Colmenares, L., Torres Acosta, A.A.: Paseos coloniales por Oaxaca: crónicas de Thomas Gage. Vías Terrestres, Número 15, Año 2, enero-febrero, Asociación Mexicana de Vías Terrestres A.C., pp. 31–35 (2012)

Torres Acosta, A.A., y Bustamante Altamirano, J.: Historia de los Caminos en México. Tomo I, Elaboración de Mapas, Coordinadores: Ovalle Favela, H.S. y Cal y Mayor Leach, P., Colegio de Ingenieros Civiles de México AC, 2ª Edición, Banco Nacional de Obras y Servicios Públicos SNC, México, DF, México (2011). ISBN 978-607-9561-80-2

Morán Miranda, C.A., Torres Acosta, A.A.: Paseos decimonónicos: recorrido de Guillermo Prieto por los caminos de Querétaro, Vías Terrestres, Número 12, Año 2, julio-agosto, Asociación Mexicana de Vías Terrestres A.C., pp. 16–20 (2011)

Castillo Colmenares, L., Torres Acosta, A.A.: Paseos coloniales por Oaxaca, Vías Terrestres, Número 11, Año 2, mayo-junio, Asociación Mexicana de Vías Terrestres A.C., pp. 32–36 (2011)

Torres Acosta, A., Cramaussel Ballet, Ch., Bustamante Altamirano, J.: Los caminos de la historia: época virreinal, Vías Terrestres, Número 10, Año 2, marzo-abril, Asociación Mexicana de Vías Terrestres A.C., pp. 28–31 (2011)

Torres Acosta, A., Gradilla Hernández, L.A.: Los caminos de la historia: Conquista de México-Tenochtitlán, Vías Terrestres, Número 9, Año 2, enero-febrero, Asociación Mexicana de Vías Terrestres A.C. pp. 22–26 (2011)

Torres Acosta, A., Cramaussel Ballet, Ch., Gradilla Hernández, L.A.: Los Senderos y Caminos del México Antiguo, Vías Terrestres, Asociación Mexicana de Vías Terrestres A.C., Número 8, Año 2, noviembre-diciembre, pp. 26–29 (2010)

Conversion of an Existing Reinforced Concrete Building into a New Office Building

Marina Traykova[1](✉) and Petar Doychinov[2]

[1] Department of Reinforced Concrete Structures, University of Architecture, Civil Engineering and Geodesy, Sofia, Bulgaria
marina5261@abv.bg
[2] SM Engineering Ltd, Sofia, Bulgaria

Abstract. We are all open to the idea of preserving the old buildings. The renovation of the heritage structures is a topic of great importance worldwide. It results from the need to improve the structures of the existing heritage in accordance with the requirements of the new conditions for sustainable development and preserve it for future generations. Rehabilitation and strengthening of cultural and historic heritage buildings play an important role in sustainability policies. The applied rehabilitation and strengthening should be determined respecting the results to the integrity and authenticity of the cultural and historic heritage buildings. The construction sector is increasingly in need of strengthening and restoring existing buildings that have exceeded their service life but are still in good condition or have historical significance for the country. The understanding of the performance of the structure and the estimation of the actual conditions is basic for the decision making for the final solutions for rehabilitation and strengthening.

The research in the paper aims to offer option for the seismic assessment and strengthening of the structure of a building for gravity loading and seismic actions. Extending the service life and preserving the architectural interior and the exterior of the building have been the guiding principles in the research and the proposed solutions for rehabilitation and strengthening.

Keywords: Heritage buildings · Rehabilitation · Strengthening · Reinforced concrete

1 Introduction

There are many old buildings which are in very unfavorable conditions (earthquake, progressive deterioration of the material used for construction of building, etc.) as compared to the time they have been built. All these buildings were built before the development of current seismic codes, or without complying with current seismic codes. Such hazardous existing buildings have to be rehabilitated for safety of life and property. The amounts of scientific researches and innovations on strengthening methods continuously increase. A major problem in the design process is the seismic activity of the site (Kostov et al. 2006). Older design concepts are mostly focused on the effects of gravity loads and they do not dedicate enough attention to provide the lateral resistance and ductility. Most of

I. Vayas and F. M. Mazzolani (Eds.): PROHITECH 2021, LNCE 209, pp. 1185–1195, 2022.
https://doi.org/10.1007/978-3-030-90788-4_91

the historic buildings provide limited ductility when subjected to seismic action. For this reason, it is very important to protect them by taking measures from structural aspect to ensure their sustainable durability and development (Traykova 2015).

The paper presents the building "B", part of a group of buildings-historical monument in the center of Sofia, Bulgaria, former printing factory, emblematic of the city and an exclusive example of the architectural style in the 50ties of XX century (Fig. 1).

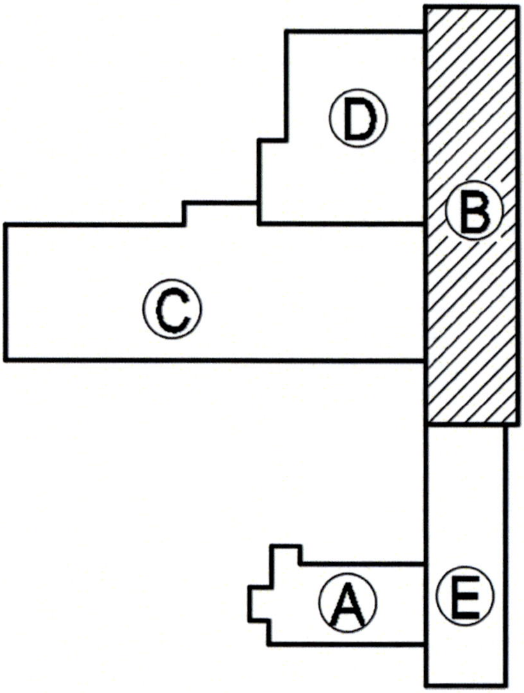

Fig. 1. Building plan

The construction period of the buildings is between 1949–1954. The area of the buildings is around 50 000 m^2. The structure of the buildings is a monolithic reinforced concrete frame structure consisting of reinforced concrete columns, transverse and longitudinal beams and reinforced concrete slabs. The new design aims to transform the building into a new office building according to the contemporary concepts for such type of buildings.

The paper tries to give a brief study about earthquake resistant design and the methodology about the seismic evolution and rehabilitation of an existing heritage structure. It is a continuation of other studies in Bulgaria, some of them in the area of the rehabilitation of industrial buildings (Traykov 2012), (Traykov 2017). The objective of the research was to develop an economic and easily applicable rehabilitation method. It also provides certain aspects of computer software modeling against the seismic loads and shows the necessity of seismic upgrading in a reinforced concrete building. The seismic evaluation determines the vulnerable and weak components and deficiency of the building during an earthquake. The seismic rehabilitation process aims to improve seismic performance and correct the deficiencies by increasing strength, stiffness or deformation capacity.

Preliminary investigations are prerequisites in the development of rehabilitation projects. Preliminary investigations were conducted to help in determining the future works. The conceptual design is based on a preliminary realized on-site investigation concerning details of the structure, grade of materials, damages.

2 Objectives

The main objectives of the present study are as below:

- To investigate the effects of earthquake forces on the building and to study the performance of the existing structure. Usually is better to start with the simplest realistic model and then, if necessary, develop a model that reflects more structural features and complexity.
- To identify the seismic deficiency of the structure
- To check the capacity of the structure under gravity loads
- To study the effectiveness of retrofitting strategies for this type of structures
- To propose effective structural solution for strengthening.

3 Building Modeling

3.1 Description of the Structure- as Built

The building consists of 14 axes in the longitudinal direction, located over 6 m and 3 axes in the transverse direction - over 6 and 8 m. There is an expansion joint on axis 8 – 5 cm. There are 5 above ground levels with a design height of 4.40 m and one basement with a design height of 3.30 m (Fig. 2).

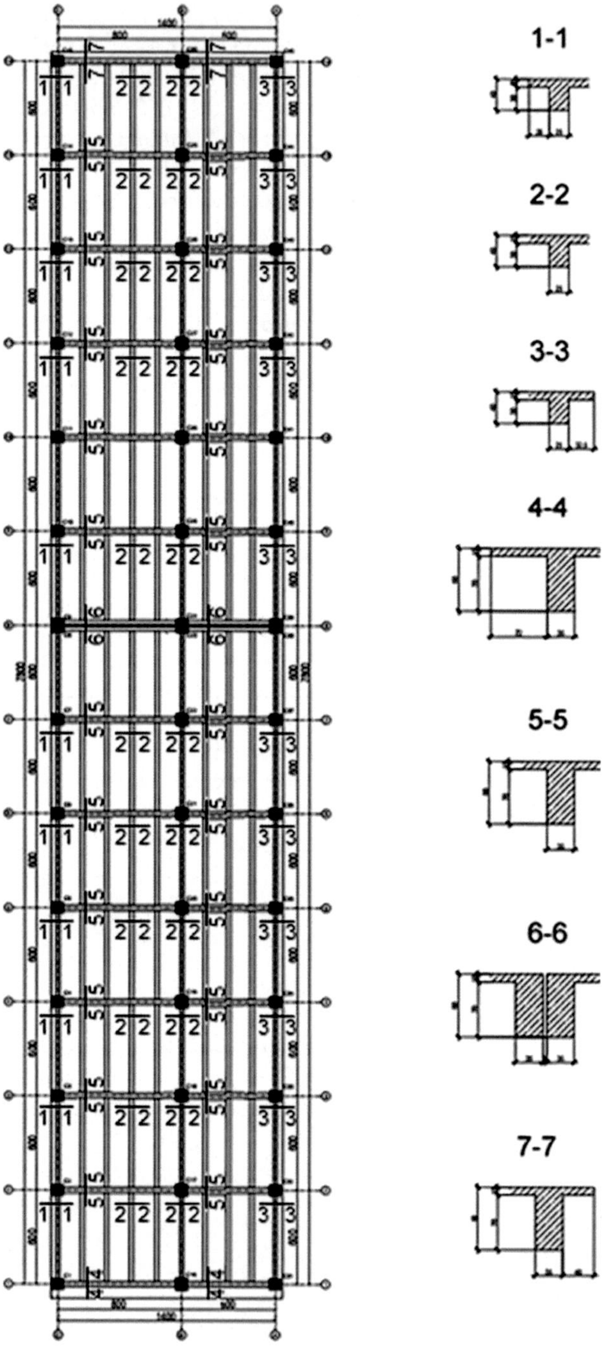

Fig. 2. Building "B"-as built

The structure of the buildings is a monolithic reinforced concrete frame structure consisting of reinforced concrete columns, transverse and longitudinal beams and reinforced concrete slabs.

The cross section of the columns starts from 60/70 cm in the basement and ends with 40/40 cm at the last level. The main beams' cross-section is 35/80 cm and an axial spacing of 8.00 m and the secondary beams with a cross-section of 25/40 cm and an axial spacing of 1.60 m. The thickness of the floor slabs is 10 cm.

After testing, the compressive strength of the concrete was found to be in class C16/20. The concrete cover survey revealed the following:

- for slabs in 48% of cases it is higher than the normative / 1.5 cm /, and in the rest 52% it is less.
- for beams in 23% of cases it is higher than the normative / 2,0 cm /, and in the rest 78% is less.
- for columns in 75% of cases it is higher than the normative / 2.5 cm /, and in the rest 25% it is less.

The corrosion test for reinforcement shows that there is a 10% chance of surface corrosion. The type of reinforcement is class S235 (A1).

The foundation is solved with single three-level concrete foundations with the size of the main plane in plan 250/260 cm. The height of the steps is 55 cm for the first step, 73 cm for the second step and 80 cm for the third step.

3.2 Seismic Strengthening and Numerical Modeling for Earthquake

The building is considered seismically unsecured according to the Bulgarian standards, it is envisaged that new vertical load-bearing elements - reinforced concrete shear walls will be designed and executed. The lack of evidence for the connection between the columns and the beams led to the solution of the structure for seismic actions with new RC shear walls. An important influence for the choice of the shear walls is the preservation of the storey drifts (displacements) of the two sections of block "B" in the limits of 5 cm (Fig. 3). The considered variant is the most favorable for the performance and the future function of the building and gives the possibility to prevent the "pounding effect" between the adjacent sections. The seismic effects and energy dissipation will be fully relied on the walls. Existing columns and beams are mainly dimensioned for vertical loads due to current design rules during the time of the construction of the building. The provided information is not enough for the detailing of the joints between the beams and columns.

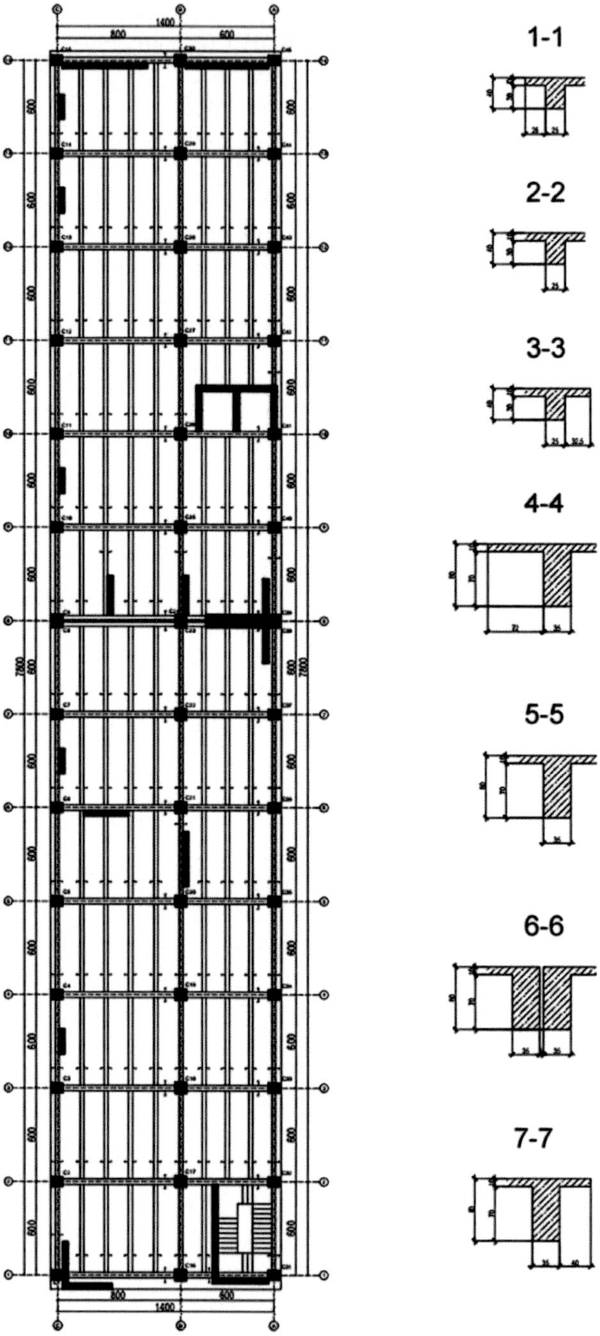

Fig. 3. Building "B"-new formwork plan

According to (EC8) for the behavior factor "q" of the newly designed wall structural system is accepted q = 3. Due to the fact that the building is classified as irregular in height, according to (EC8) the coefficient is reduced by 0.8.

The building is founded in soil type C, for the city of Sofia, where the value of seismic acceleration is agr = 0,23 g.

The SAP2000 software was used to design the design model:

- reinforced concrete walls are modeled with shell elements;
- an infinitely rigid diaphragm is modeled at each floor slab;

The seismic impact components are combined in accordance with the requirements of (EC8) using the SAP2000 standard procedure as follows:

- Structural response is calculated by applying the CQC rule to combine the effects of seismic impact by forms;
- The maximum value of each effect of a structural impact due to the two horizontal seismic components is determined by the SRSS of the effects of each horizontal component.

The results of the analysis show the period, shape and frequency of vibration of the building (Fig. 4).

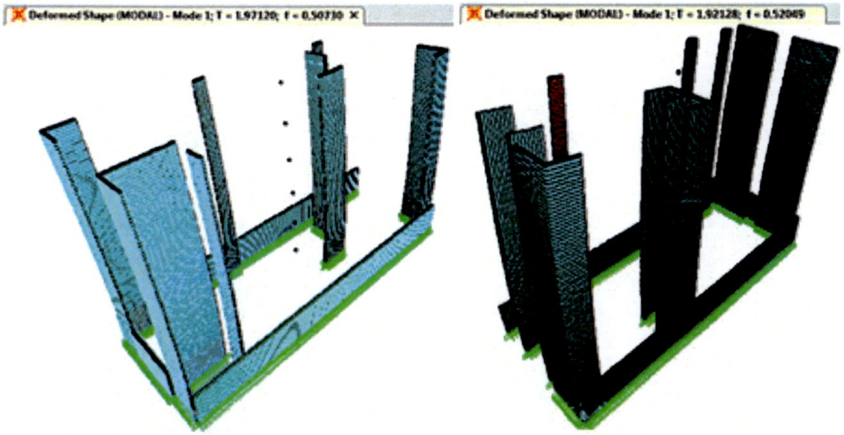

Structure between axes 1 to 8 ***Structure between axes 8 to 14***

Fig. 4. First deformed shape, period (T)s and frequencies (f) Hz

A special investigation of the existing expansion joint between the two parts of the building is provided. The wall system is designed so that maximum displacements at the top of the buildings are limited and consistent with existing joint between them. The results of the analysis show that the requirements are fulfilled.

Expressed according to (EC 8) and based on the coefficients of behavior of the two parts of the building and the elastic displacements at the top, the minimum required joint width is less than 50 mm (the existing joint between the two existing sections). The check shows that the width of the existing joint is sufficient and the provided solution with the new system of shear walls is well defined.

3.3 Strengthening for Gravity Loads

The floor slabs are calculated using the "elastic system" method as a one-way spanning reinforced concrete floor slab system. The beams are modeled in the SAP2000 program, looking for the most unfavorable combination and arrangement of loads.

Specific checks for the columns and their foundations for vertical loads are investigated.

After the calculations are carried out, the decision is to completely strengthen the load-bearing structure - columns, beams, slabs and some of the concrete single pad footing of the building. In order to preserve the aesthetic appearance and functions of the building, the strengthening of the structural elements is designed based on non - metallic reinforcement of fiber reinforced polymers (FRP) (Chardakova et al. 2015) (Figs. 5 and 6). The foundations are strengthen by additional "reinforced concrete jacket".

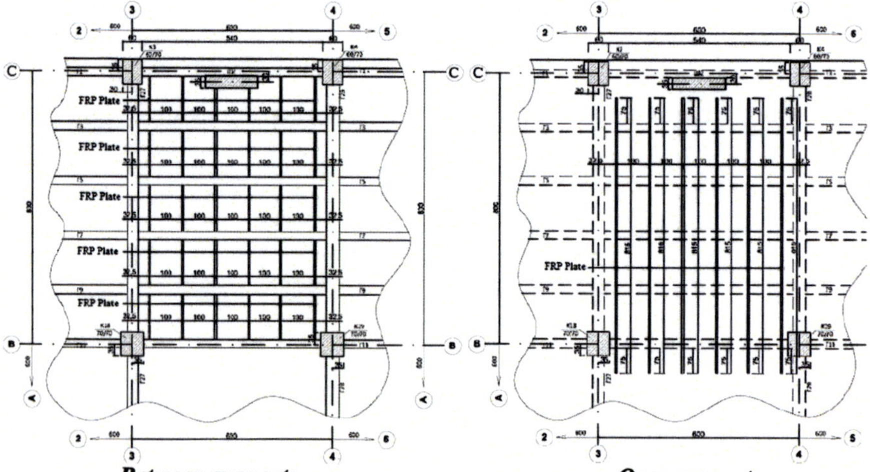

Between supports *Over supports*

Fig. 5. Strengthening of slabs

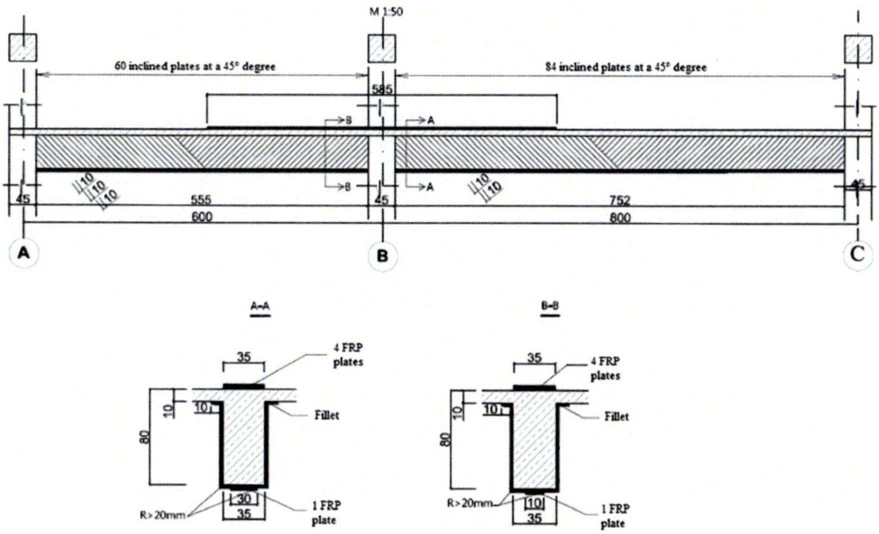

Fig. 6. Strengthening of beams

Calculation and construction of the new reinforced concrete shear walls (Fig. 7) that dissipate seismic energy are performed in accordance with all Eurocode 8 requirements for medium ductility (DCM). The new foundations are solved with single reinforced concrete foundations.

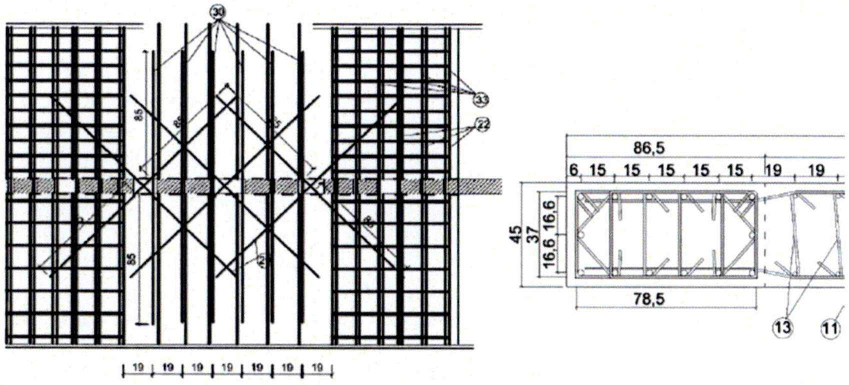

Fig. 7. Strengthening of shear wall

The project envisages the design and the construction of a new staircase Fig. 3. For this purpose it is necessary to cut the structural elements - beams and slabs and to make openings in all slabs in height of the building. This leads to a change in the static scheme of the cut elements and the necessecity of their redesign and strengthening.

4 General Conclusions and Recommendations

The main objectives of the present study are as below: following the analysis, a solution was adopted to provide to the structure the seismic performance according to current contemporary requirements. The design of new shear walls is the best solution for the structure. The adopted variant has very good behavior and gives limited displacements of the floors. This fact is important because of the fixed size of the seismic expansion joint. The strengthening for gravity loads has the minimum contribution to the total loading. The effect on the existing foundations is minimal. The new elements have new foundations.

Based on the analysis and the design, the following conclusions and recommendations could be presented:

1) The contemporary trends in the construction sector include the increasing rate of reuse of existing structures compared with construction of new structures.
2) Conservation, strengthening, restoration and eventual rehabilitation of built heritage require a multidisciplinary approach based on solid theoretical considerations
3) The social and cultural value of heritage is regarded as a powerful tool in sustainable development, economic regeneration and creation of sustainable communities
4) Preliminary investigations are necessary to acquire detailed and valuable knowledge about heritage, to assess the condition, to understand all alterations, decay phenomena, and environment, in order to determine the most viable option for rehabilitation.
5) Older design concepts mostly focused on the effects of gravity loads and they did not dedicate enough attention to provide the lateral resistance and ductility.
6) One of the main problems in the case of retrofitting is related to the accepted degree of the interventions and the necessity to reach the safety of the new buildings.
7) Evaluation of the risk, in case of lack of interventions for upgrading the performance of the building.
8) The project for historical buildings must take into account the position of the building, the type of structure and used materials, the constructive system, the damages in the structural elements, etc. Sometimes the change of those factors may make the project infeasible.

References

BDS EN 1998-1: Design of structures for earthquake resistance-Part 1: General rules, seismic actions and rules for buildings (EC 8)

Chardakova, T., Vuzgechev, K., Traykova, M.: Strengthening of structural elements – slabs, beams, columns. Numerical Examples. Nova Kampania, Sofia (2015). ISBN: 978-619-90368-1-5

Kostov, M., Kaneva, A., Vaseva, E., Stefanov, D., Varbanov, G., Koleva, N.: An advanced approach to earthquake risk scenarios of Sofia. In: International Conference on Civil Engineering Design and Construction, Varna, Bulgaria, pp. 245–251 (2006)

Sap 2000 – Computational program

Traykov, A.: Rehabilitation & repair of the structural elements of industrial buildings & facilities. In: 14th International Conference and Exhibition "Structural Faults & Repair", Edinburgh, Scotland (2012)

Traykov, A. (contributor): Rehabilitation of the complex reinforced concrete shell roof structure of an industrial building. In: Pelke, E., Bruhwiler, E. (eds.) Engineering history and heritage structures - Viewpoints and approaches, Structural Engineering Document, vol. 15, pp. 67–70. IABSE (Zurich) (2017). ISBN 978-3-85748-154-3

Traykova, M.: Study of the problems in the seismic retrofitting of existing and historical buildings. In: IABSE Conference, Presentation for the WG7, Geneva (2015)

The 100 Years of Bauhaus - The Neufert Building in Weißwasser, Germany

Hartmut Pasternak[1](✉), Thomas Krausche[1], and Natasa Zivaljevic-Luxor[2]

[1] IPP, Cottbus/Braunschweig & BTU, Cottbus, Germany
h.pasternak@web.de, i-p-p@gmx.de
[2] National Heritage Foundation, Belgrade, Serbia

Abstract. The 1919 marks a 100 years since Walter Gropius founded Bauhaus, one of the most influential design schools of all times. Numerous buildings (and artefacts) still exist today and some of them are considered heritage, submitted to architectural and urban conservation. Within Bauhaus concept, which seem to be initiated already with Fagus factory (designed between 1911 and 1913), there was, inter alia, an emphasis on industrial building design and standardization. In this article, we revised industrial building complex designed by architect Ernst Neufert in Weißwasser (North Saxony – 140 km south of Berlin) regarding challenges of structural conservation. It consists of six-story warehouse and a single-story loading bay. The warehouse and dispatch center (which were originally used by glass manufacturer), have both a load-bearing steel frame and a brick infill. The visible steel frame gives a rudiment rhythm and order to the façade. The interior is shaped by pattern defined by steel columns, resting on the beams with intervening strained system ceilings out of stone iron. We discussed the state of this industrial complex, a project of renovation and the form in the context of its relevance for the legacy of its author who is widely known for his book *Architects' Data* and for legacy of Bauhaus, in general.

Keywords: Neufert · Industrial heritage · Bauhaus · Steel building · Standardization · Normative rules

1 Introduction

The industrial complex, warehouse and dispatch center, in Weißwasser (Fig. 1) was designed by architect Ernst Neufert in the period while he was working on the book *Architects' Data* for which he is widely known. The complex is located in the countryside of North Saxony, 140 km south of Berlin, in a town named Weißwasser, close to the current border with Poland. The building was designed and built in 1936 for glass manufacturer Vereinigte Lausitzer Glaswerke (VLG, in English: United Lusatia Glassworks). Neufert became the resident architect there in 1935/1936, and in the same year came out the first of many editions of his *Architects' Data,* which has been published since that time worldwide, translated into numerous languages. This book is *de facto* accepted as most comprehensive and best known normative book for architects. Neufert was educated in Bauhaus, which was apparently of great importance for his professional development and legacy.

I. Vayas and F. M. Mazzolani (Eds.): PROHITECH 2021, LNCE 209, pp. 1196–1205, 2022.
https://doi.org/10.1007/978-3-030-90788-4_92

Fig. 1. The Neufert building in Weißwasser

2 The Architect Ernst Neufert and Normative Rules

Ernst Neufert (1900–1986) was born in Freyburg, on the Unstrut, near Weimar. He was educated in Baugewerksschule in Weimar (1918–1919) and immediately after in Bauhaus first generation of students (1919–1920). Before that, during his teenage years, Neufert worked as bricklayer apprentice, which seems to be important for his interest in contemporary architectural standardization and normative rules. Neufert came to VLG upon invitation of the former fellow-student from Bauhaus Wilhelm Wagenfeld who became the artistic director of VLG in 1935. During his Weißwasser period, Neufert designed many buildings (various housing, office, and factory buildings) in the town itself and in nearby Tschernitz and Kamenz.

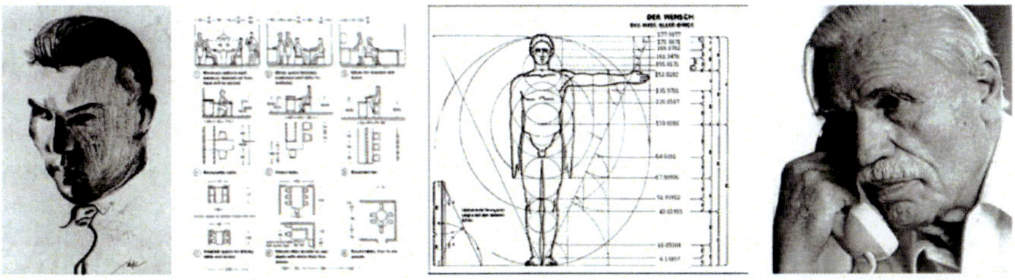

Fig. 2. Ernst Neufert as young man (left) author: Gyula Pap, 1930, samples of normative rules from his book Architects' Data (middle) (Neufert et al. 2002), and photo of himself as elderly man (right)

After the Bauhaus was closed, as well as two other schools where he was employed (Bauhochschule, known as "the other Bauhaus", also in Weimar, where he worked 1926–1928, and private school for art and architecture founded by Johannes Itten in Berlin where he worked in 1933–1934), being a family man, Neufert chose a safe job in the countryside in those politically turbulent times. Neufert used that time wisely: he worked as architectural practitioner and on shaping research on normative rules in architecture into a book. The book made him famous both in Europe and in USA already after the first edition (Fig. 2). From 1939, Neufert worked with Albert Speer, Hitler's General Building Inspector on typology, standardization and rationalization of Berlin's housing.

Fig. 3. Examples of Neufert's work in cooperation with his son Peter (Prigge 1999)

In the last phase of his career, Ernst Neufert was professor of architecture at the Technische Hochschule Darmstadt (1945–1965) and, at the same time, he kept working as freelance architecture practitioner. In this period, with his son Peter he designed several industrial buildings (Fig. 3). In his entire career, industrial building design, standardization and normative rules remained main foci.

3 Impact of Bauhaus

3.1 Legacy Regarding Architectural Design and Standardization

Bauhaus was founded the same year as German Workers' Party (soon after renamed into National Socialist Party) i.e. they were both established immediately after the World War I, the biggest and the most devastating war the World had seen by then. Germany was heavily impoverished, and in response to lack of resources, famine, under employment, in need to meet basic needs of population, minimum standards became one of imperatives of Bauhaus. A year after the National Socialist came to power, in 1933 they established one-party rule by changing Weimar Constitution, the Constitution of German Reich according to which Germany was democratic parliamentary republic, and that marked the end of what had name Weimar Republic (1919–1933). Bauhaus was not at any point isolated from those events, not the least for being originally located in Weimar, which, has also an obvious symbolic connotation.

Gropius and his partner in architectural studio Adolf Meyer were apprentices of Peter Behrens, and it is generally considered that "all started" from the design for AEG Turbine factory (Fig. 4). Industrial building design was among main foci in Bauhaus.

Fig. 4. AEG Turbine Factory in Berlin by Peter Behrens studio (left) Doris Antony, Berlin, Germany; Fagus pencil case Factory in Alfeld on the Leine by W. Gropius/A. Meyer in (middle) Carsten Janssen (Jaeggi 1998): and Van Nellefabriek in Rotterdam by Johannes-Andreas Brinkman and Leendert-Cornelis van der Vlugt.

When Walter Gropius changed location of Bauhaus in 1925, he chose Dessau which, by coincidence or not, had reputation of industrial town. While Bauhaus was led by Hannes Meyer, a Marxist, in 1928–1930, social ideals including minimum standards and affordable housing became prevailingly important over aesthetics. However, need for standardization was widely recognized. Ernst May's huge projects in Russia (20 cities in 3 years, the utopian city for 1 million citizens and more) were, de facto, standardization in practice. Standardization of kitchen by Margarete Schutte-Lihotzky (who was part of May's team and friend of Bruno Taut) has had enormous impact on home design ever since. The Dessau period ended in 1932, and for one semester Bauhaus existed in Berlin – situated in an old telephone factory (again a factory) in Berlin-Steglitz, but it was ultimately closed on the August 10th 1933, while directed by Mies van der Rohe who was, by common opinion, mostly interested in aesthetic. It all proves what is commonly known: industrial architecture was highly relevant of for Bauhaus. Further, it implicated preferences in materials used for designs, among which metal was often favored, and there are numerous proofs of that from design of artifacts to architecture. Last, but not least among key impacts of Bauhaus to Neufert's work, due to Bauhaus Neufert was acquainted with many prominent architects and historical figures of his time and he successfully cooperated with number of them: Walter Gropuis, Adolf Meyer, (1919–1920, 1921–1925), Ricardo Magdalena (1920–1921), Paul Klee, Wassillie Kandinsky, Georg Muche, Otto Bartning (1926), Frenk Loyd Right (1936), Albert Speer (1939–1944) and more. Neufert is said to have met Antonio Gaudi in 1921, during his sabbatical in Spain as well as to be impressed by his approach to architecture. It indicates that Neufert was driven by creative individuals and had interest in diversity.

3.2 Legacy Regarding Artefacts

Bauhaus artefacts contributed immensely to its fame, and among them Wagenfeld lamp stands justly (Fig. 5). Wilhelm Wagenfeld (1900–1990) studied with Laszlo Maholy-Nagy at Bauhaus in 1924. His career (also) had hard impact from political side, which caused several twists, however he worked at United Lausitzer Glass in Weißwasser from 1935–1947, except the period while he was involved in combat of the WWII (which he mostly served as military prisoner in Russia). Some of Wagenfeld designs are produced until today, and they won several awards. His design for United Lausitzer Glass won recognized internationally: it won prize of Exposition Internationale des Arts et Techniques dans la Vie Moderne (International Exposition of Art and Technology in Modern Life) in 1937, and a prize at the 1940 Milan Triennial VII. Wagenfeld was among those students of Bauhaus who changed forever common attitude to everyday household objects. They become affordable and beautiful at the same time. They were made of cheap materials and created by technics suitable for mass industrial production, and they spread the glory of Bauhaus worldwide.

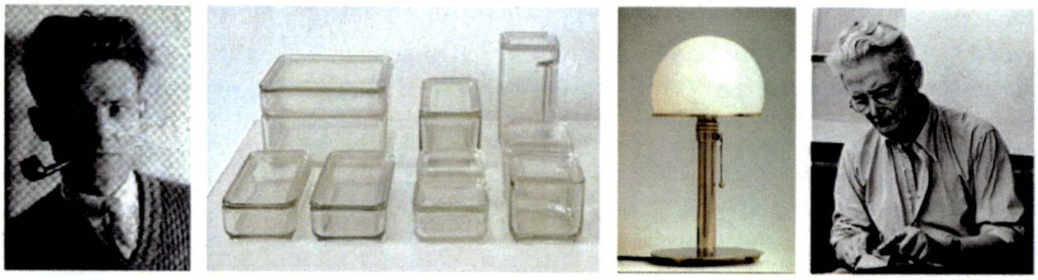

Fig. 5. Wilhelm Wagenfeld and his designs (Wagenfeld 1948), (Scheiffele 1994)

4 The Industrial Complex

4.1 Specifications

The building complex consists of a six-story warehouse and a single-story loading bay. The building construction (Fig. 6) is essentially formed by a steel structure and brickwork infills (12 cm wall thickness) in the outer walls, which is still largely original today, as well as between the cross bars exciting steel stone ceilings. The steel structure is composed of columns arranged in a regular pattern (HEA profiles) and floor-to-ceiling hinges connected to them as floor-to-floor units. The individual ceiling panel of the steel

Fig. 6. Warehouse during assembly (www.neufertbau.de)

stone ceilings are resting on the lower flange of the steel cross bars. The warehouse is accessed by a staircase and two larger freight elevators (in practice out of service) and has about 700 m^2 of storage space per floor.

The functionality of the supporting steel skeleton is not only visible in the façade, but also serves as a basic system of order. The interior is characterized by a steel column grid, on which beams with interlocked system ceilings ("stone-iron ceilings") rest with shingled brick formwork. The steel columns are tapered - for reasons of rationality - visible on the upper floors the steel frames of the east and west façades were hung so that the building could actually be "pulled up" like a construction kit. The loading hall, which adjoins the warehouse at right angles, has a direct functional connection and takes up the formal idiom of the warehouse. Noteworthy is the off-center column, which increases the functionality of the hall and which found its way into Neufert's Architects Data. From the loading hall there is direct access to the two freight elevators, here are loading ramps for the truck delivery of glassware and a loading ramp for shipping by rail.

4.2 Current State

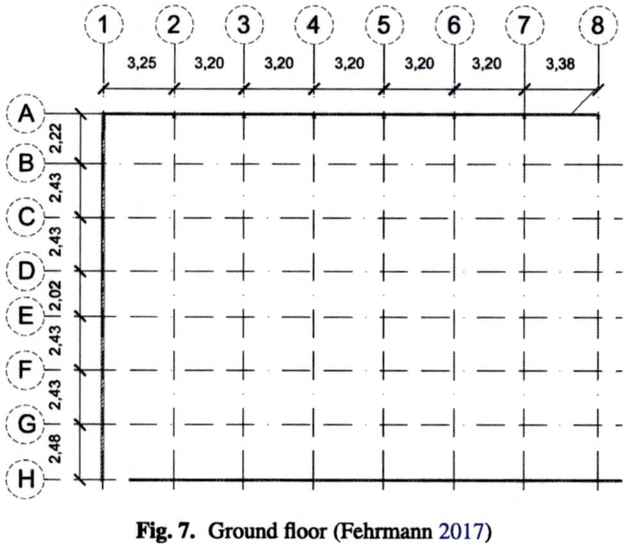

Fig. 7. Ground floor (Fehrmann 2017)

The building was erected as a warehouse in the years 1935 to 1937. It was partly damaged in 1945. Those war damages affected the building in axes 1–5/C-H in levels 3–6 (Figs. 7 and 8) and they were repaired in 1954/1955. The outer brick walls and the ceiling, including the steel structure, were completely rebuilt. Being vacant for years and neglected by changing owners coupled with the widespread image of "obsolete" (simple storage without purpose, in addition to bad condition). The current state of the construction requires urgent action: the roof and the roof drainage have been mostly

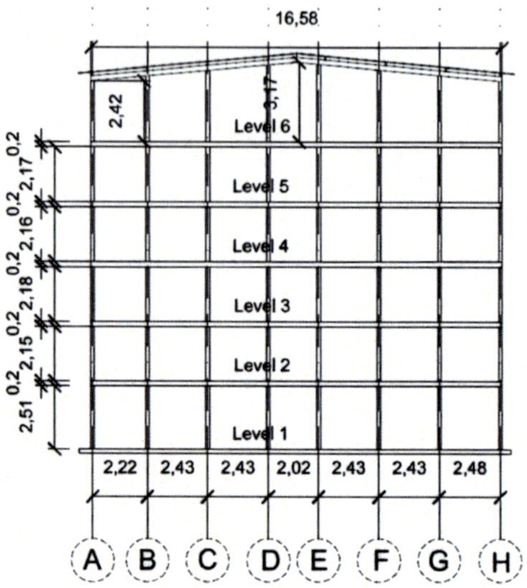

Fig. 8. Cross-section (Fehrmann 2017)

destroyed which leads to large-scale drenching on all floors. Part of the vertical glass window band on the south facade has collapsed, many windows on the east and west front are broken. Some props also lack the force-fit, so that now there is a risk that parts of the building go out.

4.3 Renovation

The destroyed roof and the windows have to be fixed urgently to keep moisture and other weather conditions outside. After fixing of the outer cover, the whole steel skeleton structure is to be checked. A damage from corrosion is largely visibly (Fig. 9) and it will continue to progress unless counter action is provided. Most columns and bars will need a surface blasting with new coating.

Fig. 9. Outer steel skelleton with corrosion damage

Fig. 10. Column in the building with corrosion damage

The connections, which have been already damaged due to the expansion of corrosion, must be completely rebuilt (Fig. 10). Those damaged connections already cause many problems in other parts of the building like the façade. The bolted joints are no longer effective and will jeopardize the carrying capacity. The bolts will get additionally to the planed tensile and shear forces also bending forces which weakens the connection and changes the original transfer of loads. To refurbish the inner columns joints, it is not possible to have an outer scaffold construction. For this reason, an inner auxiliary construction is necessary and should be arranged among each other in every floor to support the ceiling. To take the load off the columns, a counterweight presses the ceiling beams upwards is required, while fixing the joints (Fig. 11).

The preservation requires comprehensive intervention and finding new purpose for the building. Providing new purpose, which could be most efficient preservation, is action constrained by very small story height (2.15 m) and small distance between columns

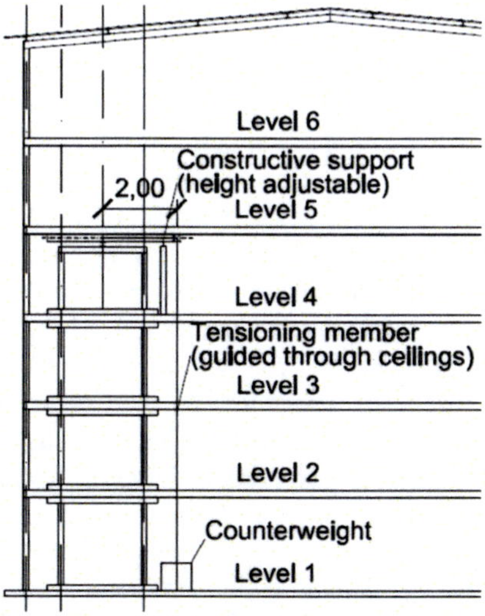

Fig. 11. Cross section with counterweight presses and supports during fixing

(2.02–2.43 m), according to contemporary standards. Windows are immediately under the ceiling, and parapet is high and in disproportion of interior, serving the rhythm of the façade. Some efforts for using building with new purpose were made in the past. Behind the building, a swimming pool and other details were added, in attempt to change the industrial complex into cultural hub.

5 Outlook, Heritage Value and Preservation Prospects

Currently, an association - Neufert-Bau Weißwasser e.V. - is collecting funds for refurbishment, therefore the future use is not clear yet. There are several evident problems which make reuse difficult: ceiling is too low for contemporary standards; windows are too small and position of windows seem to be in favor of façade, shaped by means which are as modest as possible. The building itself has primarily historical value because one of the chapters of the epic story of Bauhaus and its students was related to it. A small town Weißwasser in Germany is to be remembered as the place where one of the best known architectural books was written. It certainly has memorial value (regarding history of construction for early contemporary application of modular system), scientific value (for a constant application of normative rules) and more. In ideal case, the building would be reused with original purpose – as a warehouse. It could be beneficial to keep part of the building open for public and present it as museum. It could bring additional means which could cover at least some maintenance cost. Such model of management (which was promoted by its Canadian authors under the name *Economuseum*) may give good results in a countryside as such, far from areas which grossly attract tourists.

6 Discussion

Bauhaus gave enormous contribution to analyses of minimum standards in architectural design from its beginning. Under Hannes Meyer, social aspects, standardization and affordable housing became even more important. Concerns of social impact of how we built already had history in Germany, which can easily be traced (at least) to Deutscher Werkbund and Herman Muthesius. Having that in mind, Neufert had significant share in Bauhaus legacy. In the Weißwasser phase, it seems that Ernst Neufert experimented with application of his ergonometric normative rules and possibilities of different material. The design of the factory proves dedication to minimum standards as well as to Bauhaus design principles within low budget and using simple materials, which were at disposal locally. Besides brick and concrete, Neufert used metal and glass. His design keeps promoting genuine values and preferences thought at Bauhaus. We can easily recognize order, rationality, outlook determined by structural elements, respect for functionality, use of materials suitable for industrial production etc. Generalizing from historical distance, one can notice two streams of architects within Bauhaus according to their interest in normative rules vs. aesthetic, and Neufert definitely belongs among architects who rely on standardization and typology in architecture. However, compared to some other architects of the 20th c, e.g. Christopher Alexander, Neufert is more into standardization than typology, and between standardization and optimization, more into the second. Neufert's approach brings greatness by modesty; he does not seem to be imposing or overambitious, but rather comprehensive, practical and into optimization. It seems that in his case, normative rules are there to assist and serve, not to lead or constrain the individuality, or to falsely simplify architectural design. As consequence, Neufert's own designs differ among each other and reveal personal development and permanent search, which deserves full respect from practitioners, in addition to gratitude from countless generations of architectural students who used his Architects' Data to enter the world of Architecture. Therefore, the didactic value of Architects' Data for architectural students can be hardly overestimated, or its value as reference for practitioners.

7 Conclusion

The Neufert building in Weißwasser is an industrial building in the formal language of the "classical" modern response of the New Objectivity of the 1930s. It was planned and built as a central warehouse and logistics center of the "Vereinigte Lausitzer Glaswerke" (VLG), for the storage of glass products, among other designs by Wilhelm Wagenfeld - famous for the Wagenfeld lamp, as well as to the envelope production warehouse rail. Bauhaus, was founded in 1919 by Walter Gropius in Weimar as art and craft school and today, on its 100th anniversary, it is celebrated worldwide as *alma mater* of European 20th c avant-garde design. Nowadays, many Bauhaus designs, from households to industrial buildings, are considered contemporary design classics, and such achievement of Bauhaus is most surprising considering its short existence (all three phases of its existence together lasted less than 15 years). Neufert gave exceptional contribution to dissemination of knowledge regarding minimal standards based on architectural analyses. His Architects' Data set reference for all normative rules in 20c. His industrial complex design in Weißwasser proves that his written works and his architectural design were consistent. In regard to his overall legacy in architecture, we could not help association to Bach's contribution to music: predictable but in spite of that beautiful and diverse. Therefore, we conclude the world of architecture would be impoverished without legacy of Ernst Neufert and the effort should made to preserve his industrial complex in Weißwasser as standing testimony of that.

References

Neufert, E., Neufert, P., Baiche, B., Walliman, N.: Architects' Data, 3rd edn. Wiley-Blackwell, New York (2002).ISBN 978-0-632-05771-9

Prigge, W. ed.: Ernst Neufert. Normierte Baukultur im 20. Jahrhundert. Edition Bauhaus Dessau. Frankfurt am Main: Campus Verlag (1999). ISBN 3-593-36256-2

Jaeggi, A.: Fagus. Industriekultur zwischen Werkbund und Bauhaus. Verlag Jovis, Berlin (1998)

Weber, H.: Walter Gropius und das Faguswerk. Verlag Georg D.W Callway München, Munich (1961)

Wagenfeld, W.: Wesen und Gestalt der Dinge um uns, Worpsweder Verlag (1948)

Scheiffele, W.: Wilhelm Wagenfeld und die moderne Glasindustrie Verlag Gerd Hatje, Stuttgart (1994)

www.neufertbau.de

Fehrmann, K.: Approval and Execution Planning. Weißwasser, Germany (2017)

60th Anniversary of the Salvation of the Temples of Abu Simbel

Natasa Zivaljevic-Luxor[1](✉) and Hartmut Pasternak[2]

[1] National Heritage Foundation, Belgrade, Serbia
[2] Brandenburg Universtity of Technology (BTU), Cottbus, Germany
Hartmut.Pasternak@b-tu.de

Abstract. In 2020 we mark 60 years since the beginning of international cooperation on salvation of the two temples of Abu Simbel – Great Temple dedicated to Ramses II and the Small temple dedicated to his wife Nefertari. It was one of most demanding and most famous heritage preservation action ever undertaken; therefore it became a reference for all large heritage projects aftermath.

In 1960 Egypt began to build huge Aswan High Dam on The River Nile, by far the biggest structure ever built in its history of state already worldwide famous for huge structures, and it was clear that artificial lake will endanger two temples of Abu Simbel located at the bank on Nile. Therefore state of Egypt asked UNESCO for help and three years later the two temples were re-erected at a site 64 m higher and 180 m further inland from the original location.

The temples had to be dissembled, transported and assembled again. Originally, temples were curved out the rocky hill, and to replace it, the artificial concrete dome was built over the top of each temple. A 55 m wide curved scaffold was created for the cylindrical part of the protective dome. The spherical part of the dome framework was formed by half-arches made of SL 15 girders with a span of approx. 19 m. The requirements of the clients for minimal deformations taking into account the influence of the bolt clearance and the assembly and displacement problems could be kept. The measured deformations showed a very good agreement with the calculated ones.

Recalling memory of this great venture, in this paper we describe why and how it was done, and argue the current state, emerging protection issues and its symbolic meaning.

Keywords: Abu Simbel · Egypt · Heritage · Protection · Preservation · Cultural landscape · Large-span structure

1 Introduction

In 1960 Egypt with the help of former Soviet Union's experts built High Aswan Dam. It was clear, that the dam will create an artificial lake with a surface 60 m above the present floor of the monuments. In this situation the governments of Egypt and Sudan turned to the UNESCO with the request for assistance to protect the two temples located on the western bank of Nile, taking advantage of its water surface as a mirror.

I. Vayas and F. M. Mazzolani (Eds.): PROHITECH 2021, LNCE 209, pp. 1206–1215, 2022.
https://doi.org/10.1007/978-3-030-90788-4_93

In 1960 the Director of UNESCO made an International appeal for the salvation of those monuments located in the Nubian dessert. Three years later Egypt and the "Joint Venture Abu Simbel" signed a contract. In 1963 the project cost was estimated to 25M US$ (Martini 1968), which would worth today close to 210 million US $.

2 The Ancient History

The main facts and features of Abu Simbel temples are very well known and therefore we only briefly recall them here. Situated near the southern border of Egypt with Sudan, the Temples was hewn out of a cliff beside the Nile around 1260 B.C. The Great Temple was built for glorious Ramses II who belonged to the 19th Dynasty, and the Small temple, situated north from it, was dedicated to Nefertari, his most beloved wife. It is believed that the monuments were erected on the 30th anniversary of the pharaoh's reign. It is also believed that the main axe of the temple was carefully designed to allow sun rays to illuminate the back wall of the temple (which nowadays happens twice each year, on the 22nd of October and the 22nd of February) and that those dates were not randomly chosen but related to pharaoh's life.

The most famous features of the Great Temple are the four colossal sitting statues of Ramses II, each 22 m high. The façade of Small temple has six statues, each 10 m high. The four of them represent the pharaoh and the two his wife.

It is believed 30,000 workers over 30 years were employed on the construction sites (Schliephacke 1968).

3 The Contemporary History

The temple was covered by sand and forgotten, until it was rediscovered by Jean-Luis Burckhardt in 1813, who had it dig out in cooperation with Giovanni Belzoni in 1817.

However, those ancient monuments became part of another "plot" in mid of the 20th c, during Presidency of Gamal Abdel Nasser (1954–1970) in Egypt. President Nasser proved his political skills and successfully came out of Suez crises following nationalization of the Suez Canal. Among his many projects (e.g. he initiated pan-Arab unity, implemented numerous reforms, liberated Egypt from imperialists, co-founded Non-aligned Movement etc.), at a peak of his popularity he initiated construction of Aswan High Damn (there was also Aswan Low Damn built in 1902). The aim was to put under control oscillation of annual flooding of Nile, which in the same rhythm affected Egyptian economy. Even without taking his other projects into account, it is obvious that President Nasser ambition was at least equal to ambition of a pharaoh.

The historical moment was in favor of President Nasser's grand plans. Second WW left most of the world devastated and mistrustful and newly established United Nations were working on rebuilding international cooperation. Recalling that basic mission of UNESCO (established in 1946) was to **build peace through international cooperation in Education, the Sciences and Culture,** this project was a perfect opportunity for all parties to show themselves in best light: British decision-makers, Russian experts, UNESCO as supervisor, German companies and worldwide financial and other support, monitoring included. The project nevertheless contributed to the glory of President

Nasser, whose name was given to the new artificial lake. One can hardly argue his deeds did not justify that honor.

Unquestionably, the relocation of Abu Simbel temples became a milestone of heritage protection. The realization of such huge internationally coordinated project added several cultural layers to already most valuable monuments of Abu Simbel.

4 Removal and Rebuilding of the Temples

Three main proposals were discussed (Jurecka 1965):

- Protection of the temple complex in its existing location by an 80 m high dam
  It was not possible to prevent water from entering completely. Furthermore, natural lighting of the temples would no longer be possible due to the protective dam situated to the East.
- Elevation of the temples as a whole. Presses of 265000 t for the Great Temple and 55000t for the Small Temple would have been necessary.
- Disassembling the monuments and re-assembling them in a higher location.

It was decided to use the 3rd proposal and rebuild both temples under a reinforced concrete protection dome (to keep image of the original landscape).

The member firms who joined the "Joint Venture Abu Simbel" (1 Martini 1968):

- HOCHTIEF, Germany, General Manager
- ATLAS, Egypt
- GRANDS TRAVAUX DE MARSEILLE, France
- IMPREGILO Impresit-Girola-Lodigiani S.p.a, Italy
- SENTAB and SKANSKA CEMENTGJUTERIET, Sweden

The whole salvage work is accomplished in four main actions:

- Construction of the cofferdam
- Excavation of rock above the temples and removal of rock to expose ceilings and walls
- Cutting, transportation and storage of elements
- Re-erection at the new site

At first, a 360 m long cofferdam (Fig. 1) had to be erected to shut-off the temples from the rising Nile water. Behind this dam, the dismantling work could be safely carried out and the parts of ceilings and walls transported away. The lower bulkhead measured max. 25 m in length, and the upper 12 m, all together ca. 11 000 m^2.

Fig. 1. Cofferdam and Great Temple (Desroches-Noblecourta and Gerster 1968)

For the excavation above the temples (Fig. 2) the sandstone was chipped with rippers a rock chisels. Because walls and ceilings inside the temples show cracks, the sandstone has to be strengthened by injecting resin (Fig. 3). 23.5t of resin were used. Around 90% of all temple parts had to be secured so that they could survive transport and reconstruction without damage.

The rock adjacent to the frescoes was removed with hand tools to avoid vibrations. Finally, ca. 80 cm thick blocks from walls and ceilings were received.

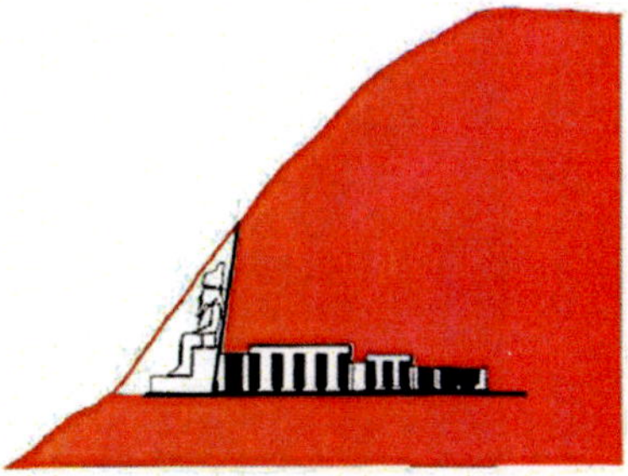

Fig. 2. Scheme of the initial situation (Schliephacke 1968)

Fig. 3. Injecting resin (Desroches-Noblecourta and Gerster 1968)

The next step was the cutting of the walls and ceilings into blocks and dismantling of the colossal statues at the facades (Fig. 4). Therefore diamond saws, chainsaws, wire saws and bucksaws were used. The lifting hooks were fixed in the stone blocks with resin. To shift the blocks and to lift them lower-loader trucks, two guy-rope derricks of 20 resp. 30 ton capacity were set-up at each temple (Fig. 5). Ceiling and wall parts should not be heavier than 20 tons, façade figures not heavier than 30 tons. After dismantling and before transportation to the two storage places (with a total area of 44 000 m^2), every single block was numbered and embedded in sand to protect the reliefs and statues.

Fig. 4. Cutting lines on the façade (Schliephacke 1968)

Fig. 5. Removal of the Ramses Temple with Derrick crane (Schliephacke 1968, Desroches-Noblecourta and Gerster 1968)

The temples will be re-erected at a site 64 m higher and 180 m inland from the original location. After the four colossal statues of Ramses were installed, the construction of the protective dome began.

The cylindrical part of the protection dome of the Great Temple has a span of ca. 55 m and an arch rise of ca. 19 m. The contractor Hünnebeck GmbH used scaffold girders SL 15 girders (Preinfalck 1968) (Fig. 6). Three sections, each 2.5 m wide, were build. Later the formwork was moved. The spherical part of the dome framework was formed by half-arches made of SL 15 girders. These half-arches had a tieback to make the support conditions simpler. The load at the apex was 20 t/sqm. The requirements of the clients for minimal deformations taking into account the influence of the bolt clearance and the assembly and displacement problems could be kept. The measured deformations showed a very good agreement with the calculated ones (Scheer 1969).

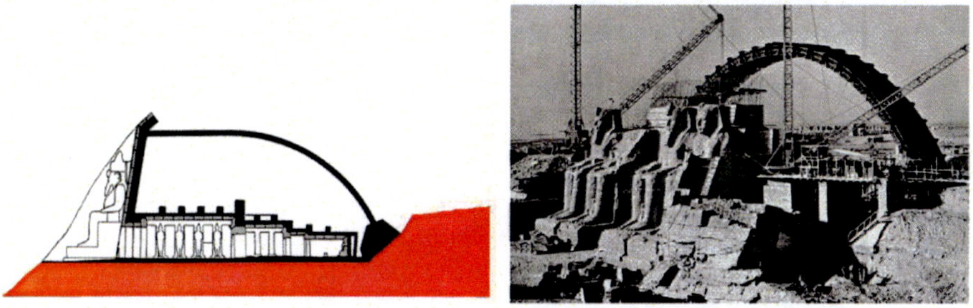

Fig. 6. Re-erection at the new construction site (Schliephacke 1968)

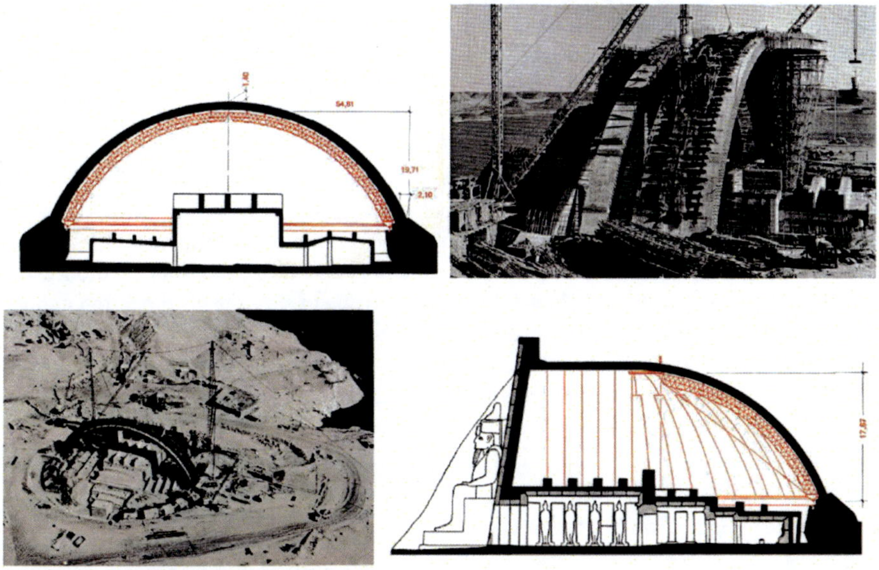

Fig. 7. Semi-finished concrete dome above the Great Temple (Scheer 1969, Schliephacke 1968)

The Small Temple consists of a cylindrical and spherical part, too. The cylindrical part has a span of ca. 21 m and an arch rise of ca. 6.6 m.

Step by step a concrete dome was erected (Fig. 7) and covered with stones to recreate the original landscape (Fig. 8).

Fig. 8. Great temple covered by stones (Schliephacke 1968, Desroches-Noblecourta and Gerster 1968)

5 Discussion: Legacy of Relocation of Abu Simbel

5.1 Legacy of Technical Intervention

After 60 years the concept of salvage may be recognized as satisfactory. The infrastructure nowadays is much better (a highway exists); the derrick cranes would be replaced by tower cranes which have easier handling; the mechanization and digitalization would have additional impact.

The relocation was part of a much bigger project – the Aswan High Dam. The lake changed the microclimate of the designated part of Nubian desert, increasing humidity for more than 10%. Although the average value stays relatively low (in all months roughly under 40%, annual average 29%) (https://weather-and-climate.com/average-monthly-Humidity-perc,abu-simbel-aswan-governorate-eg,Egypt), it should not be forgotten that long life of ancient Egyptian buildings comes great deal from low humidity of a dessert climate, which keeps process of decomposing very slow. Many of materials used in the projects were never used before in such amount on those latitude. It implies that preservation of relocated and changed monument (due to process of relocation and use of new materials) in new climate conditions are entirely new task.

The technical solution was well chosen considering the alternative. Technical solution applied on Abu Simbel monuments was not a perfect solution proving that nothing in real life usually is. However it sufficiently close to perfection and it was executed successfully.

5.2 Legacy Regarding Heritage Preservation

The deep analyses of numerous aspects of legacy regarding heritage preservation, in this case, taking into account change in criteria between 1960s and today, based on internationally adopted rules on heritage protection, overcome the size of this paper (Fig. 9).

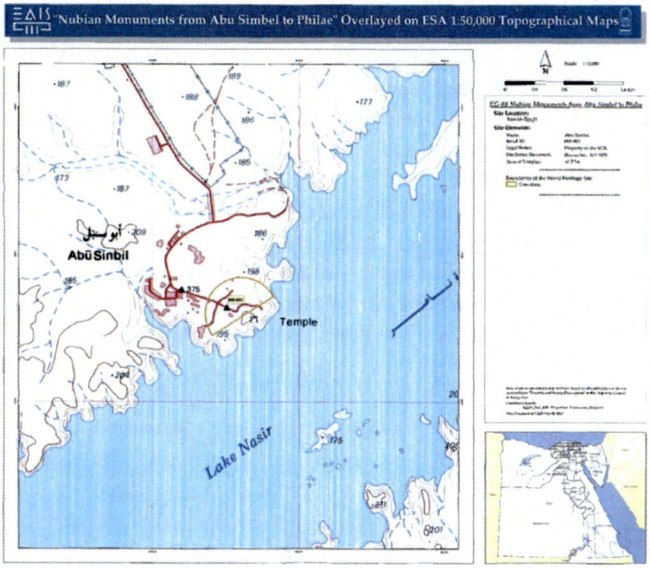

Fig. 9. Map "Nubian Monuments from Abu Simbel to Philae" 1:50000 (https://whc.unesco.org/en/list/88/multiple=1&unique_number=94)

The Abu Simbel temples was inscribed in the World Heritage List in 1979, seven years after it was established (https://whc.unesco.org/en/decisions/5732). The decision was based on criteria (i) (iii) (vi), meaning that its Outstanding Universal Value was recognized based on:

- To represent a masterpiece of human creative genius: the temples of Abu Simbel "…unanimously recognized as masterpieces of creative spirit of man… The success of campaign which was launched throughout the world by UNESCO, sufficed to prove…that these monuments, in fact perceived as relating to cultural heritage of mankind."
- To bear a unique or at least exceptional testimony to a cultural tradition or to a civilization which is living or which has disappeared; that is, in words of ICOMOS, the monuments were recognized as "…a unique series of prestigious monuments dating from 15th century B.C. to the second century A.D. which are, at once, both rare and ancient…";
- (vi) To be directly or tangibly associated with events or living traditions, with ideas, or with beliefs, with artistic and literary works of outstanding universal significance. (The Committee considers that this criterion should preferably be used in conjunction with other criteria) In this case, it meant: "The Open Air Museum of Nubia and

Aswan brings together cultural properties closely associated with the unfolding of a long sequence of Egyptian pharaonic history."

Since that time, the protection of temples came in international focus only once in 1999, due to the construction of structure which would protect them from noise and light, which was rapidly stopped upon requests from heritage experts by Egyptian authority (https://whc.unesco.org/archive/1999/whc-99-conf204-15e.pdf).

Relocation of monument is considered extreme instrument for protection, and the two temples of Abu Simbel set an anthological example proving it can be done, if necessary, even in the most complex cases.

The relocation became symbol of international cooperation and symbol of reconciliation after the Second WW, helping to establish UNESCO as the highest international authority regarding heritage preservation. This new symbolic value enriched those ancient monuments which were inevitably changed in the process of their salvation, to the level which was recognized by experts and incorporated into values for which the temples were inscribed into WHL.

There question remains unanswered if it would be possible to successfully fundraise internationally for such project nowadays. We can only speculate that the funds would be provided more probably from state of Egypt, private donors, Global Heritage fund etc. rather than through solidarity from UN state members, for a simple reason that nowadays common, universal and indivisible heritage of mankind is seen as such in cultural sense but not necessarily in economic, as well.

6 Coda

It rarely happens that a monument gets so many layers of symbolic meaning at this one. Beyond what was mentioned previously regarding heritage and engineering issues, we would like to remind of its wider sociological and economic importance. The Aswan High Dam did change economy of Egypt. It also became a symbol of raise of free, modern Egypt capable of building partnership with other nations on equal terms, at the same time remaining fully proud of its exceptional history. It became also a symbol of political power of a man who rose from humble origin to level of pharaoh, which was a privilege impossible in the past.

One of the most important requirements of contemporary heritage preservation is its **reversibility**, which is respected to great extent in this case. Hypothetically, if that would ever been required and executed, many layers from more recent history would inevitably be lost, and they are living testimony of unorthodox actions in epic times. Preservation of Abu Simbel temples is not over and new, challenging actions will be required in the future.

References

Martini, K.-H.: Abu Simbel, Wirtschaftsreport Düsseldorf (1968)

Schliephacke, H.: Abu Simbel - eine zweimalige künstlerische und technische Leistung Sonderdruck der Hünnebeck GmbH (1968)

Jurecka, W.: Die Bergung der nubischen Baudenkmäler. Bauingenieur **40**, 147–154 (1965)
Desroches-Noblecourt, C., Gerster, G.: Die Welt rettet Abu Sibel. A.F. Koska (1968)
Preinfalck, E.: Freitragende Bogenlehrgerüste in Abu Simbel. Bauingenieur **43**, 57–62 (1968)
Scheer, J.: Weitgespannte Bogenlehrgerüste für die Temple in Abu Simbel. Bauingenieur **44**, 128–132 (1969)

Rehabilitation of 1920s Housing Settlements: Examples from Cologne and Frankfurt

Zeren Önsel Atala(✉)

Department of Architecture, Istanbul Technical University, Taşkışla Taksim 34437, Istanbul, Turkey

Abstract. The concept of urban planning in Germany has become important for both urban officials and citizens since the end of the 19th century. Rapidly growing cities with the industrial revolution and the increasing urban population, has led to the transformation of medieval cities. Between 1925–1929 urban planners in many German cities acted as the propagandists of the new order. In these new housing settlements with basic facilities such as small shops and yet attached to the main cities, one could enjoy the new lifestyle: apartments with central heating and hot water systems, fresh air and light which gives the impression of spaciousness to the small housing units. Housing settlements constructed between 1920–1930 in Cologne and new housing settlements located in Frankfurt also known as results of New Frankfurt project, are good examples to reflect the main planning attitude of the era mentioned above. The renovation of the settlements raises the question of how modern living requirements and energy-efficiency can be agreed with the preservation of the original architectural and urban design principles. This paper aims to discuss the intervention techniques used in rehabilitation of 1920s modernist buildings through housing settlements located in Cologne and Frankfurt.

Keywords: Housing settlement · New Frankfurt · Renovation · Energy efficiency

1 Introduction

High-density residential areas built in the late 19th century as a consequence of rapid industrialization in Germany and attitudes of real estate brokers have been criticized frequently. These developments in urban level had influences on political level as a tension between the national and local governments. Many definitions and theories were developed in order to clarify how a city should be designed. According to anti-urban ideologies, the new city was culturally evil, and it should be better to turn to the medieval city. As Choay (1969) states that there were also theoreticians who argued for the improvement of cities due to Baron Haussmann's Garden City guidelines. In Germany Stübben and Baumeister were the promoters of this Garden City approach. Airflow, low density, green areas and adequate water and sewage systems were the main requirements for a healthy city. However, at the beginning of 20th century and prior to World War I, city planning was realized by wealthy elite sector and national government was not generally included in the housing improvements. Mullin (1982) argues that the elite controlled

I. Vayas and F. M. Mazzolani (Eds.): PROHITECH 2021, LNCE 209, pp. 1216–1227, 2022.
https://doi.org/10.1007/978-3-030-90788-4_94

housing supply and it was totally within the domain of private enterprise. Even though few urban planning applications were made between 1918–1925, many utopias, theories and manifestos were produced and published. Theories of the period had four main concerns about city fabric: the importance of streets in urban planning, the role of urban design in cultural development, the need of low-density residential areas and the need to unite man and nature. These discussions were derived from the Garden City theory and yet were applied slightly different than pre-war period. Instead of destruction of old cities in order to widen the roads, the new settlements were constructed in outer skirts of existent city center. The other important development of the period was the nationalization of local urban planning theories. Unhealthy city life was forced the production of new urban planning policies and the role of art, architecture, technology and design in society was concerned by a group of theoreticians and artists such as Walter Gropius, Adolf Behne, Bruno Taut and Max Taut who designed to develop a society where the needs of everyday life can be met and where the changes of post-war period should be reflected.

2 First Applications and New Frankfurt Project

One of the first two application examples of these theoretical concepts is the Georgsgarten Siedlung in Celle designed by Otto Haesler in 1923 and the other one is the housing settlement in Magdeburg restored by Bruno Taut the same year.

Between 1925–1929 urban planners in many German cities, acted as the propagandists of the new order. In these new housing settlements with basic facilities such as small shops and yet attached to the main cities, one could enjoy the new lifestyle. As Tafuri (1976) indicates, apartments with central heating and hot water systems, fresh air and light give the impression of spaciousness to the small housing units.

Housing settlements constructed between 1920–1930 in Cologne, are good examples to reflect the main planning attitude of the era mentioned above. All of them were constructed by a GAG, a semi-private cooperative housing company in that time as workers housing units.

Born in 1886 and died in 1970, Ernst May is the German planner architect who applied urban design techniques to city of Frankfurt during Weimar Period between 1925–1930. Even if he spent only five years on planning of housing settlements in Frankfurt, he is known as the creator of the large *Siedlungen* in the city and as the head of the New Frankfurt (Mohr 2011). Ernst May who could question traditional approaches was charged by Ludwig Landmann, the major supported by political parties such as the DP, the SDP and the Zentrum, with the planning and implementation of new housing projects in Frankfurt. May was able to develop compact and semi-independent housing units. The settlement was equipped with playgrounds, schools, theatres and common washing areas.

By autumn of 1925 as a response to housing crisis in Frankfurt, a housing construction program was developed and a master plan, which was later revised in 1926 and completed in 1930, prepared showing the existent built area and the area of development. This initiative encompassed the construction of affordable public housing and modern amenities throughout the city (Fig. 1). New Frankfurt was the name given to this public housing program and at the same time to the magazine published from 1926 to 1931 mentioning topics such as tendencies in architecture, art, housing and education. The program was carried out under the control of Ernst May, known as the head of the New Frankfurt, by various architects and designers such as Max Cetto, Martin Elsaesser, Walter Gropius, Ferdinand Kramer, Adolf Meyer, Leberecht Migge, Margarete Schütte Lihotzky, Mart Stam and Bruno Taut. As the location for new housing settlements, the north and northwest part of the existing ring road was selected. Main issue was to maximize the sunlight, airflow and create space to allow gardening. For that reason, a modification of street-oriented planning with generous interior gardens was used. Even if he tried to construct housing as inexpensively as possible, May's efforts were mostly to find the minimum dwelling based on biological requirements. A prefabrication system and a factory were developed in order to produce the standardized units with the intent to cut costs to the absolute minimum.

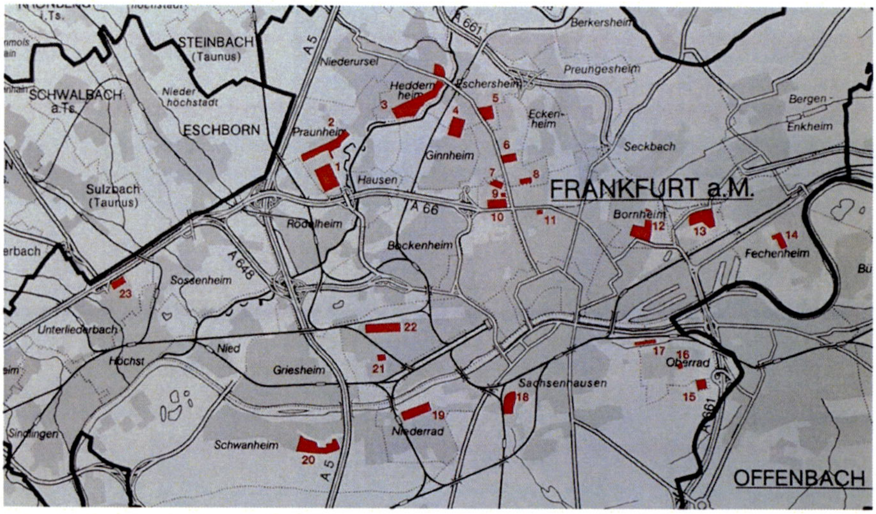

Fig. 1. Bickendorf Siedlung in 1925, Riphahn (1996).

At the core of this transformation were about 10000 kitchens designed by Grete Schütte-Lihotzky and constructed as an integral element of the new dwelling units. As Mohr (2011) states, three future focal points of *Siedlungen* were developed in 1925: the Niddatal project, the Riederwaldkolonie and Bruchfeldsstrasse in Niederrad. Two Siedlungen forms the spine of the complex: the Praunheim and Römerstadt, built between 1927–1928. Römerstadt situated in Nidda Valley is the case where Ernst May reflects his artistic design capabilities. The curved building with shops at the northern entrance with its rounded ends is a result of May's design principle. Not only building typologies but also the urban planning decisions as curved connecting roads; public and private greens are completely new forms of design.

New Frankfurt development became known throughout the world of architecture through the 2nd CIAM Congress realized in Frankfurt in 1929 where the problem of

Minimum Dwelling-*das Existenzminimum* was discussed. Four lectures held by Gropius, Bourgeois, Schmidt and Le Corbusier during the congress were on various aspects of minimum dwelling (Mumford 2002). One of the main out-put of the congress was the exhibition The Minimum Dwelling Unit with the same-scale comparative plans prepared by May's associates of minimum apartment plans from 26 European cities and USA. The selected examples were a demonstration of rational/economical solutions, with particular attention to the best relation between constructed area and number of beds made available. The bed, as a place for each person to sleep, was seen as the spatial/area reference to the relative size of the apartment.

3 Housing Settlements of Wilhelm Riphahn in Cologne

The Riehl Naumann settlement constructed between 1927–1929 was designed by architects Fritz Fuss, Otto Scheib and Manfred Faber who worked for GAG. The settlement, included 450 apartments, was realized for the workers of Cologne's northern parts industries.

The Bickendof I is the first *Siedlung* of Cologne and was realized by GAG after World War I in 1921, even if the design was the 1st prize of a competition organized in 1913. The architect of the settlement, Wilhelm Riphahn, was affected by Garden City theory. He designed single houses with back and front gardens where one could get fresh air and be with nature.

The Bickendorf II settlement constructed between 1922–1938 and the architect was again Wilhelm Riphahn. Riphahn used the three main design elements in this settlement: light, fresh air and green-*Licht, Luft, Bäumchen.* Front gardens, inner courtyards, windows at corners, balconies and loggias are the main characteristics of apartment blocks (Fig. 2). Colored façades can be considered as an influence of Bruno Taut on Riphahn's design characteristics.

Fig. 2. Bickendorf Siedlung in 1925, Riphahn (1996).

The Höhenberg Germania settlement is the last housing settlement designed by Wilhelm Riphahn in Cologne. It has architecturally more qualified apartment blocks.

The good spatial relationship of blocks with street is gained with the use of not linear street patterns and private-semi-private spaces: front gardens and courtyards (Fig. 3). In addition to the apartment blocks, even if it had some changes, also the church was designed by Riphahn.

Fig. 3. Höhenberg Germania Siedlung, Hagspiel (1996).

On the other hand, the blue courtyard-*Blauer Hof* designed in Germania Siedlung by Wilhelm Riphahn and Caspar Maria Grod between 1926–1927 has similarities to the settlement designed by Bruno Taut and Martin Wagner in Berlin-*Hufeisensiedlung*. The attached apartment blocks are situated around a semi-private U-shaped courtyard with one block on each side projecting towards the same courtyard. The name blue is given to the space because of the blue color used for the façade of apartment blocks.

4 Housing Settlements of Ernst May in Frankfurt

The Praunheim settlement, included 450 apartments, was realized for the workers of Cologne's northern parts industries. It is a settlement located adjacent to Römerstadt and built in three phases. The first phase containing 173 units was completed in 1926, the second phase with 569 units was built between 1927–1928 and the third phase with 699 units in 1929. Other than housing units it has 13 shops, 1 heating unit, 2 washing units, 1 church, 1 bar-restaurant and 1 big playing ground for children. Praunheim is the first Siedlung planned within the New Frankfurt program and it consists of single-family houses. 322 of the housing units belonging to ABG Frankfurt Holding were planned and still they are rented apartments.

Römerstadt settlement is situated on the north side of the river Rhine, on the site of one of the largest Roman towns. The Römerstadt, built between 1927–1928 consists of 1182 rented apartments, a high school-*Geschwister Scholl Schule*, a kindergarten and a religious center. The architectural elements like flat roofs and colored walls are also used in this project. Each building was painted with soft hues of red, gold and blue, as well as white. Each unit consists of a separate kitchen, a bath, a water closet and closets. In addition to standardization for housing units' and kitchen design, some patterns were designed too for garden layouts. Leberecht Migge who is known with his early revolutionary ideas on the self-sufficient garden is involved in this housing settlement project as well. As Haney (2011) states, he planned a large percentage of vegetable gardens and planting areas along the streets. He prepared also a report on the green planning for the whole Frankfurt metropolitan area. Today in the Ernst May House in Römerstadt the Ernst May Society-*Ernst-May-Gesellschaft* documents the architect's work http://ernst-may-gesellschaft.de.

Westhausen housing settlement is separated by Praunheim by west cemetery and consists of 1116 rented apartments. In every housing units a kitchen, a bath and a wash closet are included. The orientation of the housing rows is in such way that each unit could get a maximum amount of sunlight. Open-ended courtyards allow air circulation and safer pedestrian movement. The residents were mostly working-class families in 1930s and today most of the first tenants still reside in Westhausen (Fig. 4).

Fig. 4. Aerial view of Westhausen, Ernst May Society Archive.

Constructed between 1926–1927 the Niederrad housing settlement consists of 654 rented 1 to 4-bedroom apartments. All units are equipped with a separate kitchen, a bathroom, central heating, central radio transmission system, roof gardens and front gardens. Other than housing units, there are also shops, a library, a kindergarten, a social center and a washhouse. As Dreysee (2000) asserts also a high school was planned but never constructed. The enclosed complex of row houses situated in Breubergstrasse, known as Zig-Zag Houses, is the only example with a monumental view within the whole New Frankfurt program.

5 Rehabilitation of 1920s Housing Settlements and Energy Efficiency

Contrary to most other European and overseas countries, the conservation of monuments in Germany is not centrally organized. Instead the protection and conservation of monuments and sites are the responsibility of each individual State-*Land*, which itself pursues this task in co-operation with the towns and communities on a regional and local level. ICOMOS Germany (2001) clearly states that the State Conservation Offices-*Landesämter für Denkmalpflege* are responsible for the conservation of monuments and sites as well as for parks and archaeological sites on a regional level.

The renewal of the settlements, which are sometimes listed or like in Berlin are part of the World Heritage Site, raises the question of how modern living requirements and the requirements of the energy-efficiency can be agreed with the preservation of the original architectural and urban design principles. More than that, it is also a challenge to keep the costs portable and get the consent of the residents.

One of the aims of European Energy and Climate Policy is to reduce the consumption of energy and to increase energy efficiency. New building elements of the existing buildings should at least meet the current requirements of the energy efficiency provisions.

In the article for historically or architecturally valuable buildings, §24/1, of the Energy Saving Ordinance in Germany-*EnEV* it is stated that the cultural assets have priority over the energy efficiency measures. So, when the energy efficiency measures negatively affect the substance or appearance of a monument, the modernization decisions are rejected by the State Conservation Office.

The sustainable usability of monuments should also be sought over the will for substantial preservation of architectural heritage. Duda (2010) argues that this is achieved, especially for the listed residential buildings only if users are not forced to give up modern comforts.

The sustainable usability of monuments should also be sought over the will for substantial preservation of architectural heritage. This is achieved, especially for the listed residential buildings only if users are not forced to give up modern comforts.

In order to have energy efficient restoration with limited environmental impact and maximum comfort and quality of use, different methodologies and approaches are adopted. According to Duda (2010), architecturally and economically the safest and sensible way of carrying out a renovation is window renovation and exterior wall insulation at the same time and in monuments often comes only internal insulation in question, while building elements and physical boundary conditions are to be considered accurate.

There are lots of publicity, standard and study on how to reduce energy consumption in new buildings. Although religious institutions and buildings with historical or architectural importance are not subject to the same regulations as the new buildings it is a challenge to produce appropriate details for building elements with minimum loss of original architectural character of historic buildings.

6 Case Studies

6.1 Riehl Naumann Siedlung-Cologne

The renovation of the Naumann settlement from the 1920s in Cologne-Riehl by GAG Immobilien AG Cologne was awarded the special prize, Heritage Protection in Residential, in the competition for the German Developers' Prize due to the particularly careful design elaboration. Since 2008 the apartments are renovated and refurbished. The project takes place in ten phases and thanks to the gradual modernization of the different areas, the tenants are affected as little as possible (Fig. 5a).

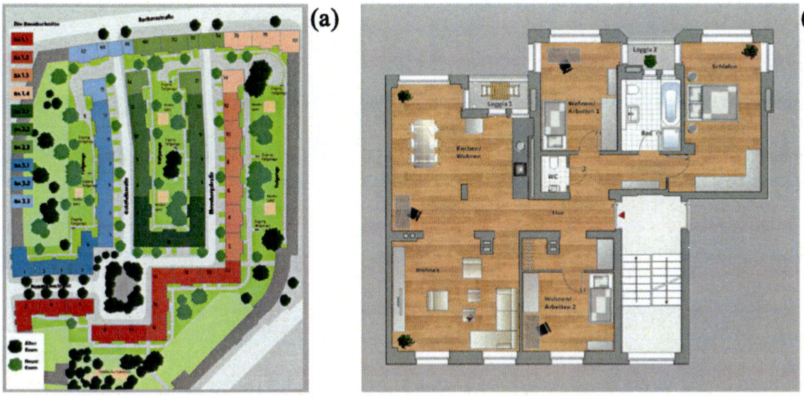
(a) (b)

Fig. 5. Site plan showing the ten phases of the project (a) and one of the new apartment plans (b).

The restoration work includes refurbishments from the open space design to facade detail. Thanks to the consultation between the clients, the architect and the city conservator Cologne it was possible to create new floor plans with open, flexible space planning and a more versatile apartment design (Fig. 5b).

Through the construction of underground car parks, the tenants had the opportunity to get the old quality open space again. The apartments are thermally insulated to modern standards and new water and sewage systems are inserted (Fig. 6a and 6b). In addition to the renovation of the roof trusses 105 new apartments are added to the roofs and 44 apartments to the formerly storage basements (Fig. 7a and 7b). So, the significant financial expenses offset by a significant growth of the residential space in the roof.

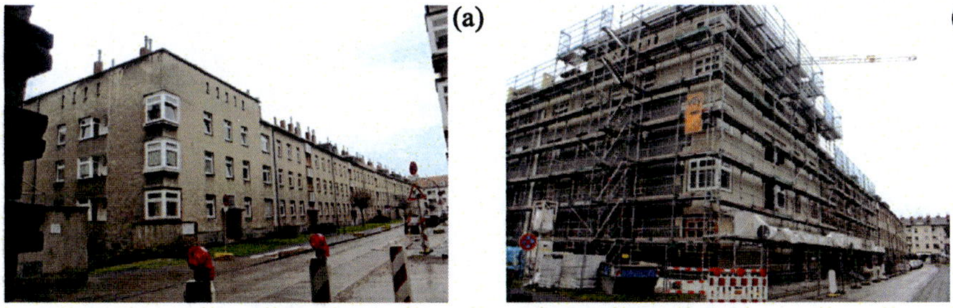
(a) (b)

Fig. 6. Before (a) and during the renovation (b).

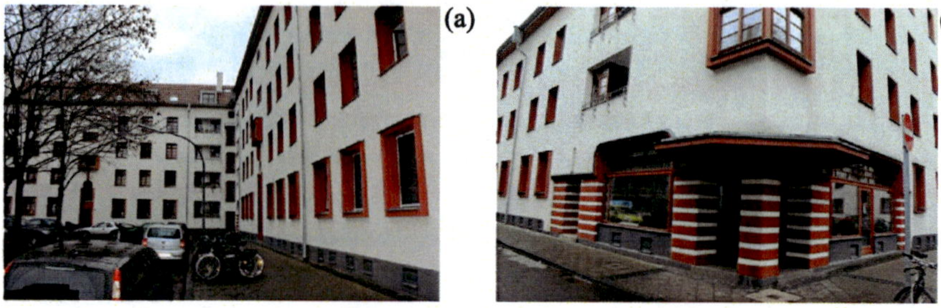
(a) (b)

Fig. 7. Renovated building with added apartments (a) and a view of a corner shop (b).

6.2 Westhausen Siedlung-Frankfurt

The conservation project prepared for Westhausen Settlement is focused on the exterior design of the apartment blocks, the gardens and all outdoor facilities. A guide has been compiled in order to get the overall picture of the settlement in the long term and to remain committed to the original condition as possible. Rohe (2015) explains that in order to avoid the inconsistent picture of various individual interventions, it is recommended that the roof and outer walls should only be restored in housing groups (Fig. 8).

Fig. 8. Westhausen housing settlement in 2015.

A package of measures is designed to support the owners in the renovation of their homes. It offers design details, material proposals and accurate color options. According to the Monument Protection Act of State of Hesse, any change is subject to the authorization of the State Office for the Preservation of Historical Monuments of Hesse. The

State Office has the possibility to promote restoration measures. According to the Protection Act, the additional costs caused by the restoration of historic monuments could be balanced via tax amortization benefit.

Werner Architekten realized the renovation project regarding the skin of the buildings. As Werner (2007) states, the facades of the listed buildings from the 1920s are marked by smooth surfaces with almost flush windows. This view should be maintained and the requirements of the current energy saving regulations are complied with. To meet all requirements, vacuum insulation panels-*VIP* were used (Fig. 9a). In order to reduce costs, conventional vacuum insulation panels were used only in exterior wall insulation. Taking thermal bridges into account, the outer spaces with a low head height, such as under door canopies are covered with insulation panels made of resol-hard foam.

6.3 Römerstadt Siedlung-Frankfurt

Römerstadt settlement is a cultural monument according to the Article No. 2 of the Monument Protection Act of State of Hesse and is therefore subject to the provisions of Law. The preservation and restoration of cultural monuments is in principle under the responsibility of the owners, in this case ABG Frankfurt Holding. According to the Protection Act, the owners are obliged to obtain a permit issued before the renovation work starts.

The State Office gave an advisory report for the protection of Römerstadt settlement in 1988 and the houses No. 15-33 are renovated according to that report. Kaiser (2015) informs that the report is still valid but has been updated several times according to the new technical requirements and standards regarding thermal insulation. The restoration work is carried out by ABG Frankfurt Holding and is supervised by the architects and construction managers of the same company.

The financing of the decommissioning of the Ernst-May House situated within Römerstadt was carried out by membership fees and donations (Fig. 9b). Grants and consultancy received the Ernst May Society also from the city of Frankfurt and its Monuments Office, the State Conservation Office Hesse, the German Foundation for Monument Protection and numerous sponsors. The dismantling of the garden has been realized through the active support of the Office of Green Space (Bausteine für ein klimagerechtes Frankfurt:Passivhäuser 2012).

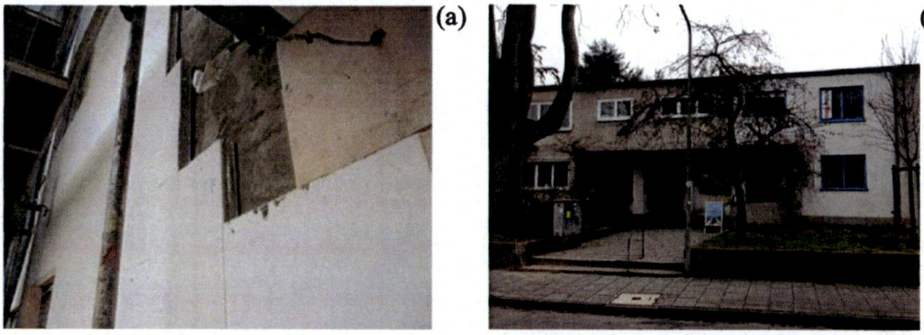
(a) (b)

Fig. 9. One of the VIP implementations (a) and Ernst-May House (a).

7 Evaluation and Conclusion

In determining the conservation values of modern architectural heritage, it is both important and difficult to construct a criterion system that will create common ground for local structure or area categories and different situations and places. It is not the right approach to make a classification emphasizing one value superior to another. The categories of values to be created for this heritage should not be brought together in a hierarchical order, but rather should be side by side categories of the same importance. For example, in most social housing settlements, the continuation of the designer's original idea appears to be a priority over the preservation of the original material or construction system.

As Tostões (2011) states that modern movement is often mistakenly related to a style, but in fact it has always shown great concern with such issues, seeking for efficiency and economy and an accurate use of materials.

Since the restoration and renovation works of the housing settlements discussed within this paper are of minimum budget and the average size of the apartments is very small, the intervention decisions compromise on some of the original architectural features. The construction of parking areas with their ramps designed for access under courtyards reduce the visual effect of the original façade configuration or the use of insulation materials to the outer walls causes the loss of brick ornaments in most of the cases.

However, the interventions are focused to satisfy the minimum insulation standards, to use the thin insulation thicknesses in order to maintain the position of the windows and to preserve the building proportions and components. The housing settlements demonstrate new and experimental forms of urban life and urban design that have been developed in the 20th century and the conservation approaches implemented for these areas are developed in a way that protects these specific planning schemes as stated in the Madrid-New Delhi Document (2017).

Acknowledgements. The paper has been developed from the on-going Ph.D. thesis titled "1920–1960 Arası Dönemde Planlanan Çok Katmanlı Batı Anadolu Kentlerinin Modern Mimarlık Mirası Değerlerinin Korunması" under the supervision of Asst. Prof. Dr. S. Yıldız Salman, within the Istanbul Technical University, Department of Architecture, Restoration Program and is a part of the research study carried out in 2015 with Prof. Dr. Helmut Bott, within the University of Stuttgart, Department of Architecture and Urban Planning during the Ph.D. period.

References

Bausteine für ein klimagerechtes Frankfurt: Passivhäuser, Green Buildings und zukunftsweisende Sanierungsvorhaben, Darmstadt, September 2012

Choay, F.: The Modern City: Planning in the Nineteenth Century. George Braziller, New York (1969)

Dreysee, D.W.: May Siedlungen: Architekturführer durch Acht Siedlungen des Neuen Frankfurt 1926–1930, Buchhandlung Walter König (2000)

Duda, E.: Möglichkeiten der energetischen Sanierung im Denkmal: am Beispiel eines Wohnhauses in der Hufeisensiedlung Berlin-Britz. Siedlungen der Moderne und ihre Zukunft – Kulturerbe in Berlin und Ost-Europa. Bautec Berlin (2010)

http://ernst-may-gesellschaft.de

Hagspiel, W.: Wilhem Riphahn-ein Kölner Baumeister. Neue Werkkunst. Gebr., Köln/Mann Verlag, Berlin (1996)

Haney, D.: Birds and fishes versus potatoes and cabbages: Max Bromme and Leberecht Migge's attitudes towards green space planning in the New Frankfurt. Ernst May 1886–1970, Prestel, 70 (2011)

ICOMOS Germany, ICOMOS World Report on Monuments and Sites in Danger 2001: Heritage @ Risk. http://www.icomos.org/risk/2001/germ2001.htm

Mohr, C.: The New Frankfurt: housing construction and the city 1925–1930. Ernst May 1886–1970, Prestel, pp. 51–53 (2011)

Mullin, J.R.: Ideology, planning theory and the German City in the inter-war years. Town Plann. Rev. **53**(2), 115–130 (1982)

Mumford, E.P.: The CIAM Discourse on Urbanism, 1928–1960. MIT Press, Cambridge (2002)

Personal email interview with Kaiser, Heike, Stadt Frankfurt am Main -Der Magistrat-Denkmalamt, July 2015

Personal email interview with Rohe, Johanna, KC Vertrieb Bestandsimmobilien und Liegenschaftsverwaltung, July 2015

Riphahn, W.: Neue Werkkunst. Gebr., Köln/Mann Verlag, Berlin (1996)

Tostões, A.: Modern and Sustainable. Docomomo Int. J. **44**, 2–4 (2011)

Werner, J.: Konzepte zur energetischen Sanierung von erhaltenswerten Gebäuden in Frankfurt am Main. Energetische Sanierung von erhaltenswerten Gebäuden in Frankfurt (2007)

Aventicum

Protection of Roman Monuments in Switzerland

Michail Zotos(✉)

Department of History of Art and Faculty of Humanities, University of Geneva, 24 rue Général-Dufour, 1211 Genève, Switzerland

Abstract. The archaeological site of Aventicum preserves several vestiges of an ancient agglomeration developed during the Roman years in the Vaud canton in Switzerland. The privileged location of the city onto the most important crossroads of the Swiss plateau favored the development of the capital of Aventicum, as the largest city in the region. At the end of the 1st century AD, the city became a Roman colony disposing of a rampart, a theater, an amphitheater, a temple of the imperial cult, and several thermal complexes. The roman baths were discovered in 1954 during the construction of a silo. The excavations revealed some parts of the walls, the floors and the pillars of the building. The Roman baths had a classic layout with three separated spaces (Frigidarium, Tepidarium, and Caldarium) of different temperatures. The most important finding was the "hypocaust", the underground heating system that used steam to heat the floor. The hypocaust was constructed of tiles dimensioned 20 × 20 × 5 cm. The tiles formed pillars to support the floor and let the hot air pass through in order to heat the room. A few years after the site discovery (1957–58), the fragility of the vestiges forced the construction of a protective hangar. The hangar is formed of wooden frames supporting a two slopes roof, and it is founded on concrete blocks. Nevertheless, this construction protects the site inadequately. The rain and the snow can penetrate and affect the vestiges. In addition, the underground water table and the capillarity affect the masonry. The tiles of the hypocaust are severely touched by the humidity and the temperature's differences. A new construction is proposed to protect the vestiges and facilitate the visit of the archaeological site. This is an example of ancient remnants in a very important Roman site for the Swiss territory that helps us to precise the preservation constraints and the restoration techniques in a cold climate.

Keywords: Roman baths · Archaeological excavation · Aventicum · Shelter

1 The Avenches City from Antiquity to Our days

The archaeological site of Aventicum preserves several vestiges of an ancient agglomeration developed during the Roman years in the Vaud canton in Switzerland. The privileged

I. Vayas and F. M. Mazzolani (Eds.): PROHITECH 2021, LNCE 209, pp. 1228–1245, 2022.
https://doi.org/10.1007/978-3-030-90788-4_95

situation of Avenches region, which stands on the crossroads of important communication routes[1], has favored the establishment of agglomerations since prehistoric times (Fig. 1a). There is evidence of an uninterrupted activity in the region since the Neolithic era[2].

The Aventicum city seems to have been founded in 15 BC, after the annexation of the Swiss territory to the Roman Empire during the Augustinian era. In the following years, the city expanded rapidly and it was organized according to an orthogonal network of streets. In the Tibetan era, the construction of a forum in the center of the city is an inevitable proof of an augmented commercial activity. Towards the end of the 1st century, Aventicum city reached its peak and disposed of a theater, an amphitheater, a temple of imperial worship, and several thermal complexes (Fig. 1b). The entire city is protected by an extended rampart.

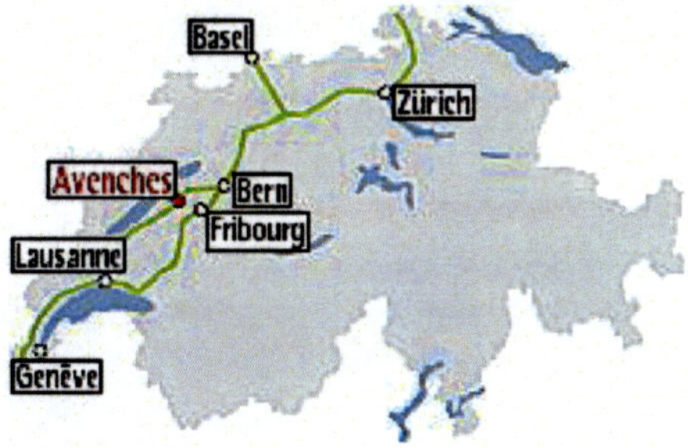

Fig. 1a. Avenches is on the road that connects Geneva to Bern and Zurich Source: http://www.avenches.ch/fr/acces/situationavenches

After the center of the region moved south to Lausanne in the 16th century, the former Roman capital started to decline. The ancient buildings have been dismantled to be used as materials sources to build churches, castles, or fountains. The medieval city has been restrained up to the hill in the southern part of the ancient Aventicum (Fig. 1c).

During the 16th century, more and more intellectuals were getting interested in the ancient history of Avenches and especially in its ancient inscriptions. Several travelers passed from Avenches in search of epigraphs. During the 18th century, the promotion

[1] (Castella 2001): «Deux grands axes routiers, figurant sur la Table de Peutinger, traversent l'agglomération. L'un relie les vallées du Rhône et de la Saône à la région rhénane, en passant notamment par Beçanson, Yverdon-Les-Bains et Augst, l'autre relie l'Italie à la Germanie, en passant par le col du Grand-Saint-Bernard, la région lémanique et le vicus de Moudon notamment. La navigation permet quant à elle le transport régional de marchandises sur les lacs de Neuchâtel, Bienne et jusqu'au Rhin par le bassin de l'Aar...»

[2] (Martin Pruvot 2006): «...des vestiges datés du Néolithique (environ 5000 à 2300 av. J.C.) au Premier âge du Fer (environ 800 à 450 av. J.C.) sont attestés, par exemple, au lieu-dit En Chaplix, à environ un kilomètre au nord d'Avenches... plusieurs aménagements funéraires et cultuels antérieures, datés de la Tène finale, ont étés découverts au pieds de la colline d'Avenches à l'ouest et dans la plaine du Lavoëx.» Chaplix, à environ un kilomètre au nord d'Avenches... plusieurs aménagements funéraires et cultuels antérieurs, datés de la Tène finale, ont été découverts au pied de la colline d'Avenches à l'ouest et dans la plaine du Lavoëx.»

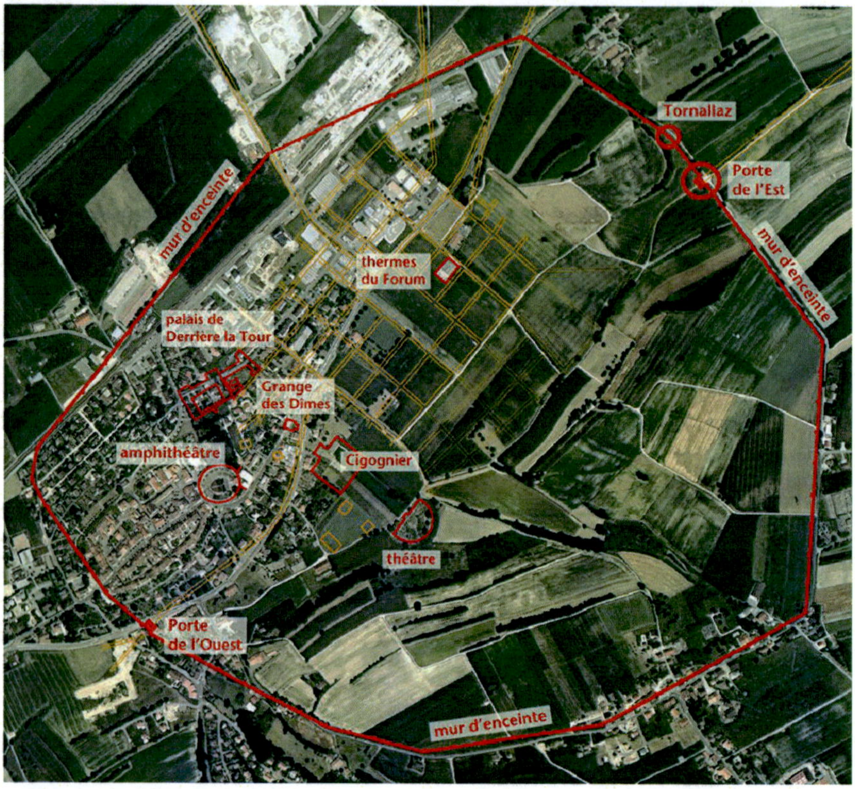

Fig. 1b. Aerial photo showing the most important monuments of the Aventicum archaeological site. The baths of "Perruet" are indicated as "thermes du Forum" on the northeast of the city. Source: http://www.aventicum.org/index.php/fr/site-monuments

of archeology and the interest in the Roman past led to a precise representation of Aventicum as well as the first in-depth studies of the remnants in situ[3]. The summer of 1727, Johann Caspar Hagenbuch, composed one of the first images of the ancient Aventicum, drawing the archaeological remains already in situ[4].

Moving onto the 18th century several Roman monuments were unearthed and highlighted thanks to the great interest of Erasmus Ritter and Earl of Northampton. In 1786, they carried out various excavations inside the walls of the ancient city.

Moreover, during the 19th century, the protection of ancient remnants became increasingly important and more and more systematic. In June 1822, the Vaud canton

[3] (Meylan Krause 2004) p. 55: « Les trouvailles romaines, qui jusqu'ici finissaient le plus souvent chez les collectionneurs et les érudits, commencent à être considérées non plus seulement comme des témoins du passé mais comme des outils de recherche capables d'enrichir la connaissance de l'histoire et de combler les lacunes des sources écrites».

[4] (Fontannaz) p. 24.

Fig. 1c. An illustration of Avenches during medieval times. Source: http://www.antiqbook.com/books/bookinfo.phtml?nr=1353462771&l=en&seller=alt

delegated for the first time curators of Antiquities. For the north region of the canton, including Avenches, François-Rodolphe de Dompierre was appointed. In 1824, the "Musée d'Avenches" was founded in order to accommodate the great number of the antiquities found in the region. The Pro Aventico Association created in 1885 continues until today to stimulate the public interest in favor of the ancient ruins of Avenches[5]. In 1898, the law on "the conservation of monuments and works of art having a historical or artistic interest" introduced the term of historical monuments and its measures of protection. Whereas in 1912, the Swiss Civil Code declared any archaeological object found in the underground as a national property (Fig. 1d).

Fig. 1d. From left to right "Sanctuaire du Cigognier", "Amphithéâtre", "Théâtre" of Aventicum.

2 The Thermal Baths of "Perruet"

In Roman times, the thermal baths of "Perruet on the northeast of the forum (Fig. 1b) were the most important of the city[6]. The spa complex covered an entire block (the insula 29) of the ancient city (Fig. 2a). It was equipped with three main rooms which corresponded to the bathing ritual starting from the cold room: the frigidarium (F), continuing to the temperate room: the tepidarium (T), and ending at the hot room: the caldarium (C). These three rooms placed on the same axis are framed by large stone walls founded

[5] Statuts de l'Association Pro Aventico.
[6] (Castella et al. 2015) p. 77.

on a dense network of oak stilts. Around them, there were all the auxiliaries such as a cloakroom (B), a sauna (S) and an underground corridor (CH) which gave access to the furnaces (p) which heated the main rooms. The complex also had outdoor spaces such as a sports field (P) and a courtyard with a swimming pool (N).

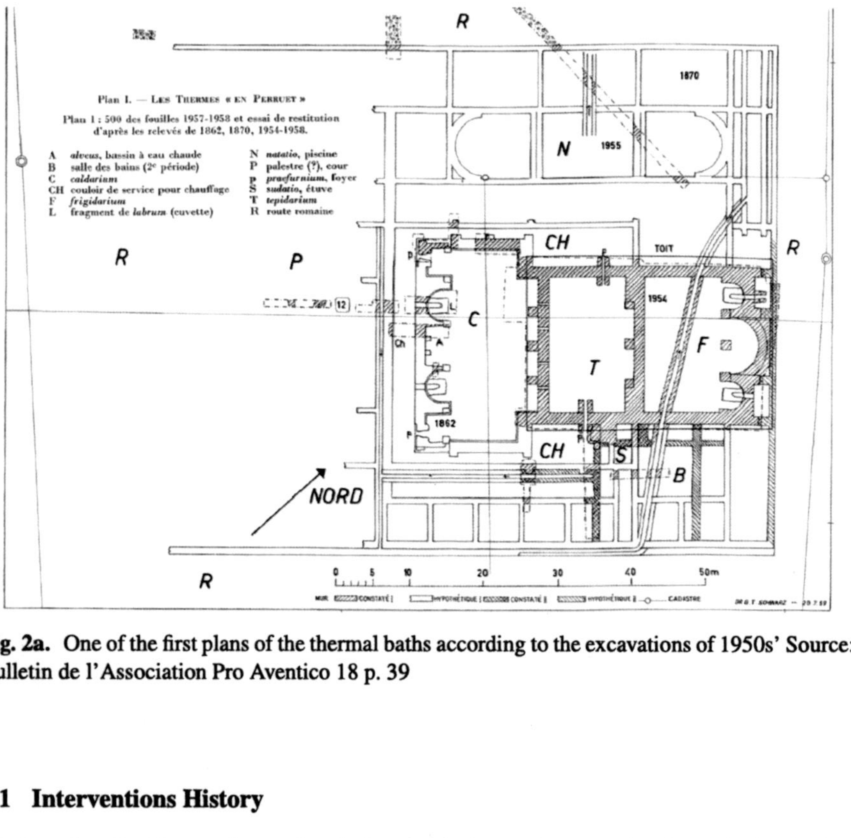

Fig. 2a. One of the first plans of the thermal baths according to the excavations of 1950s' Source: Bulletin de l'Association Pro Aventico 18 p. 39

2.1 Interventions History

The location "en Perruet" was already known for its ancient remains from the beginning of the 19th century. In 1804, for example it was reported the presence of monolithic columns which were later destroyed or dismantled during the construction of the municipal roads[7]. The importance of the location "en Perruet" has been confirmed via the discovery of several vestiges during the 19th century, such as a part of a golden sculpture or a mosaic with an inscription on it.

[7] (Martin Pruvot 2006) En 1804, l'architecte Aubert Parent observe deux colonnes monolithiques dessuées au milieu d'une région marécageuse.

The "Perruet" thermal baths were brought out during the construction of a silo in 1953. The first excavations were carried out by J. Bourquin from December 1953 to June 1954 (Fig. 2b). The excavations are carried out in two stages: firstly from November to December 1953, and secondly from March to June 1954[8]. The archaeological survey made it possible to reconstruct a typical section of the hypocaust (floor slabs, pillars, layers of mortar, slabs in marble) (Fig. 2c) and convey several pieces of information concerning the foundations (chain stilts), the coating (marble slabs), and the decoration (stucco) of the walls of Roman baths.

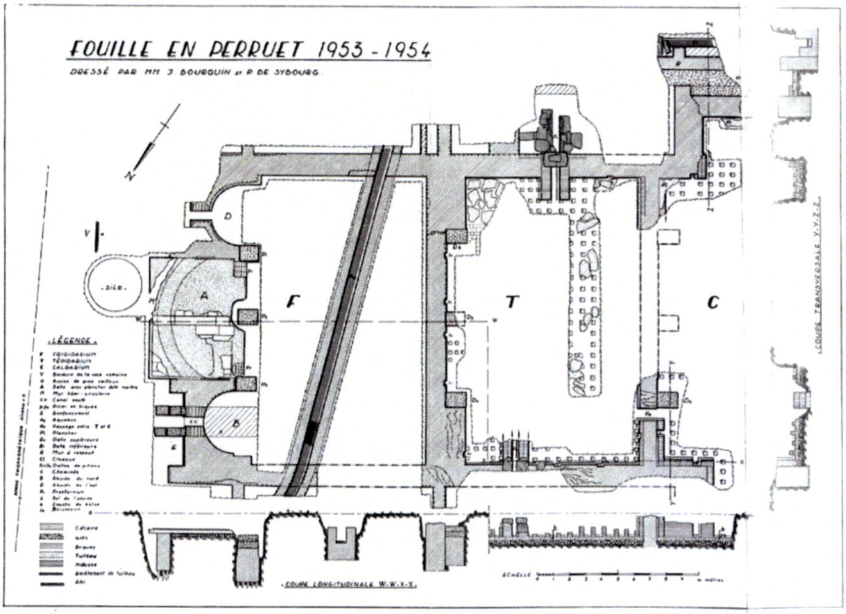

Fig. 2b. The first plan of the excavation by the architect M. Pierre de Sybourg. Source: BPA 14_1954

Such remarkable but fragile remains necessitate rapid interventions. The State of Vaud, therefore, bought the plot where the Roman baths are located and, despite the constrained budget, it supported, during the winter of 1957–58, the construction of a protection hangar to cover and protect the ancient remains (Fig. 2d). The gable roof of the hangar was covered with corrugated pieces of prefabricated asbestos panels posed on wooden rafters. In the eastern part of the roof, there was a skylight covered with a translucent plate. The substructure of the roof was built without an intermediate support using the laminated wood frame technique. Six (6) wood trusses, which cover a length

8 « Enfin, la fouille étant conduite avant tout avec l'idée de chercher à acquérir des notions topographiques les plus étendues avec le moins de frais possible, ne fut poursuivie qu'aux endroits les plus utiles, laissant entre eux des parties inexplorées, parce que suffisamment compréhensives.» (Bourquin 1955) p. 108.

Fig. 2c. The typical section of the hypocaust. Source: BPA 14_1954 PLATE XIV, Fig. 23

of about 20m, were installed at the edges of the excavation. The wooden trusses were based on concrete blocks found in about 3m depth and formed a square sole at the bottom (2.00 × 2.40 m). At their upper level, the concrete blocks form a pillar that exceeds 1.82m above ground level. A part of the ancient walls of the tepidarium had been destroyed during the construction of the foundations.

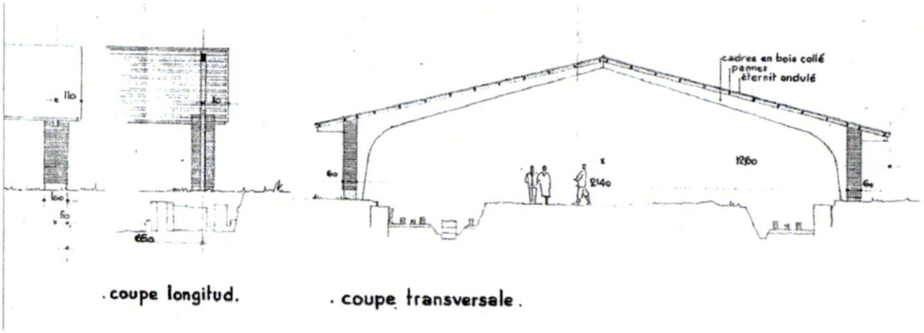

Fig. 2d. Longitudinal and transverse sections of the shelter from 1956. Source: Archives of SMRA

After the installation of the roof, the archaeological surveys continued. G. Theodor Schwarz succeeded M.J. Bourquin as the director of the excavations. The second archaeological campaign revealed the two main rooms of the Roman baths (Frigidarium, Caldarium) in their full extend[9](Fig. 2e).

Fig. 2e. The archaeological site under the newly constructed roof Source: Archives of SMRA

However, the problems of flooding[10], that continued to affect the old vestiges, have led to the launch of a new project by the Lonchamp + Froidevaux architects in the spring of 1995[11]. The project had two objectives. Its pivotal goal was to solve the problems caused by the slick groundwater, while it also focused on optimizing the evacuation of surface water. Both groundwater and surface water had an impact on the good preservation of the Roman remains. The intervention included the construction of a peripheral drainage network as well as the construction of a concrete wall forming a frame around the excavation. This framing concrete wall of 1.90m height was based on a foundation of 1m. The foundation of the wall was interrupted in the junctions with the ancient walls, so as not to ravage the archaeological vestiges (Fig. 2f). During the

[9] (Schwarz GT 1958) "The aim was to fully expose the parts protected by the roof and to fix approximately the extent of all the baths".

[10] In 1992 already, Philippe Bridel proposed a project to modify the hangar; to protect more efficiently the ancient remains which were in a concerning state of conservation.

[11] Cf. Ph. Bridel, Fondation pro Aventico, rehabilitation project for the Perruet thermal baths, 19.07.1995 (internal report). 95% of the pills are destroyed and 80% of the tiles in the area are fragmented.

construction of the wall the excavation was flooded due to the very high groundwater level. The engineer Arthur Spagnol proposed two solutions to lower the water table; a sump equipped with a pump with float or the Well-point system. After 15 days of installing the sump at the southeast corner of the field, the land became dry enough to start working again. This situation made even more evident the flooding problem on the site and the need for a drain system to evacuate the water outside the archaeological site. The new drain system also used a part of the ancient channels of the roman baths. The drainage channels were filled with layers of gravel, pebbles and a concrete soil to accommodate the drains.

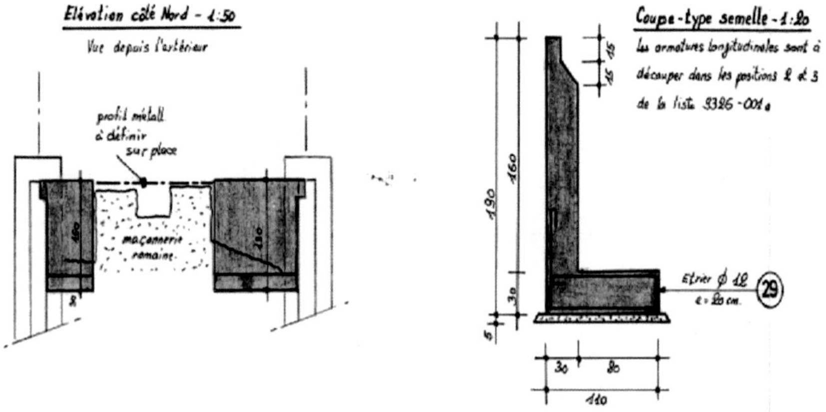

Fig. 2f. The detailed drawings for the perimeter wall. Source: SMRA Archives

In parallel with the architectural project of the Thermes, the archaeologist Philippe Bridel prepared, in July 1995, a global project defining the general directions for the restoration of the monument. This major project which was carried out from 1995 to 2005 included many restoration works on some very important elements of the ancient building. The heating system of the Tepidarium made of terracotta tiles was severely affected by the flooding and the humidity. A great amount of these tiles were replaced by new ones of similar composition and color. A part of the bath's original marble floor preserved on the northwest part of the Frigidarium which was collapsing because of the exposure to the rain and the snow was repositioned and consolidated. The brick soil of the aqueduct which traverses the Frigidarium was reconstructed. The outer walls of the ancient building were also partially reconstructed and restored. These walls conserved on their interior facing a brick coating which was protecting them from the high temperatures developed into the hypocaust. This part of the walls was in a very bad condition due to its function during antiquity but also due to the humidity. Some of the bricks were replaced and the bonds were consolidated.

After 2005, the interventions on the monument concerned mainly the maintenance of the vestiges and the cleaning of the archaeological site (Fig. 2g).

Fig. 2g. The protective hangar. Source: photo by Michail Zotos, 02/10/2015

2.2 The Current State of the Archaeological Site

The ancient masonry of the thermal baths is preserved today at a height of 1.50 m above ground level. Only two out of the three main rooms of the Roman baths, (the frigidarium and the tepidarium) are visible to the visitors of the archaeological site. The Roman vestiges include the wall foundations of these two rooms, a part of the floor, and the heating system (hypocaust). Since its revelation, the archaeological site has undergone a series of interventions (protective roof, concrete frame around the excavation, drainage, restorations) in order to ensure its maintenance. However, it is evident that the roman vestiges need more measures of protection. Fissures, loss of material, chromatic alterations, biological deposits, and colonization of microorganisms affect the conservation of the roman vestiges. All the walls present some degradations, both in their lower and upper parts. There are some green and black scabs, erosions and cracks in certain stones as well as mortar dissolutions. The tiles of the hypocaust present cracks and deformation, fragmentation, encrusting, algae or mold (Fig. 2h).

The monument today is threatened by a series of pathological problems affecting its form and figure. One of the major problems is the underground water. Avenches is part of the Swiss plateau, a region characterized by a moderate climate. The annual precipitation level is around 1000 mm, and the average temperature is slightly below 10 °C. Hydrology played an important role in the formation of the underground. The Aar sub-basin, a tributary of the Rhine, irrigates a significant part of the territory where Avenches is located. The soft subsoil facilitates the formation of large reservoirs for groundwater. Around the old thermal baths, there are flood deposits which include gravelly and salty sands. Groundwater is found at shallow depths (2–3 m) and only 1 m above the level of

Fig. 2h. Tepidarium, tiles cracked due to the frost and the humidity. Source: photo by Michail Zotos, 02/10/2015

the excavation while it can rise from time to time (Fig. 2i). Consequently, the monument is highly affected by the capillary rise. Almost all the walls are damp at their base. One can perceive how disastrous this situation is for the thermal baths by thinking that the water present in the masonry freezes quickly during the winter. The absence of a waterproofing membrane in the ground aggravates the situation.

Moreover, the roof and the concrete frame protect the vestiges insufficiently, since they do not prevent the penetration of rain and snow in the archaeological site. Therefore, the rainwater is concentrated on the walls and it penetrates in the gaps of the Roman masonry (Fig. 2i). The vestiges are affected by frost and humidity almost all year round; some large puddles remain several days after the thunderstorms. The water infiltrating into the ancient masonry freezes when the temperature falls, damaging the yellow stones, the limestone, and the terracotta tiles. The rainwater also draws off the walls causing the growth of green molds. The outer casing of the excavation is not waterproof enough so that the meteoric water outside the archaeological site can infiltrate the concrete and create green crusts on the interior facings of the peripheral walls (Fig. 2j).

Another problem for the archaeological site is the flora and the fauna since animals, especially birds, can easily access it. The birds that build their nests into the archaeological site move the vestiges and leave impurities. The wind brings grains and dust favors the growth of vegetation so that the site must be weeded twice per year.

The problems of the highlighting and the comprehension of the vestiges are also present. The hangar, built in 1954, does not correspond to the character of the archaeological site. Additionally, the access is not well distinguished and the visitors' paths are not well indicated. The visitors can only observe the vestiges from far away and

Fig. 2i. The flooding of the tepidarium in 2001. Source: Archives de SMRA

Fig. 2j. The tops of the walls remain wet for a long period of year. Source: Archives of SMRA

they cannot "get into" the excavation in order to understand the function of the ancient building and its connection to the roman city of Aventicum (Fig. 2k).

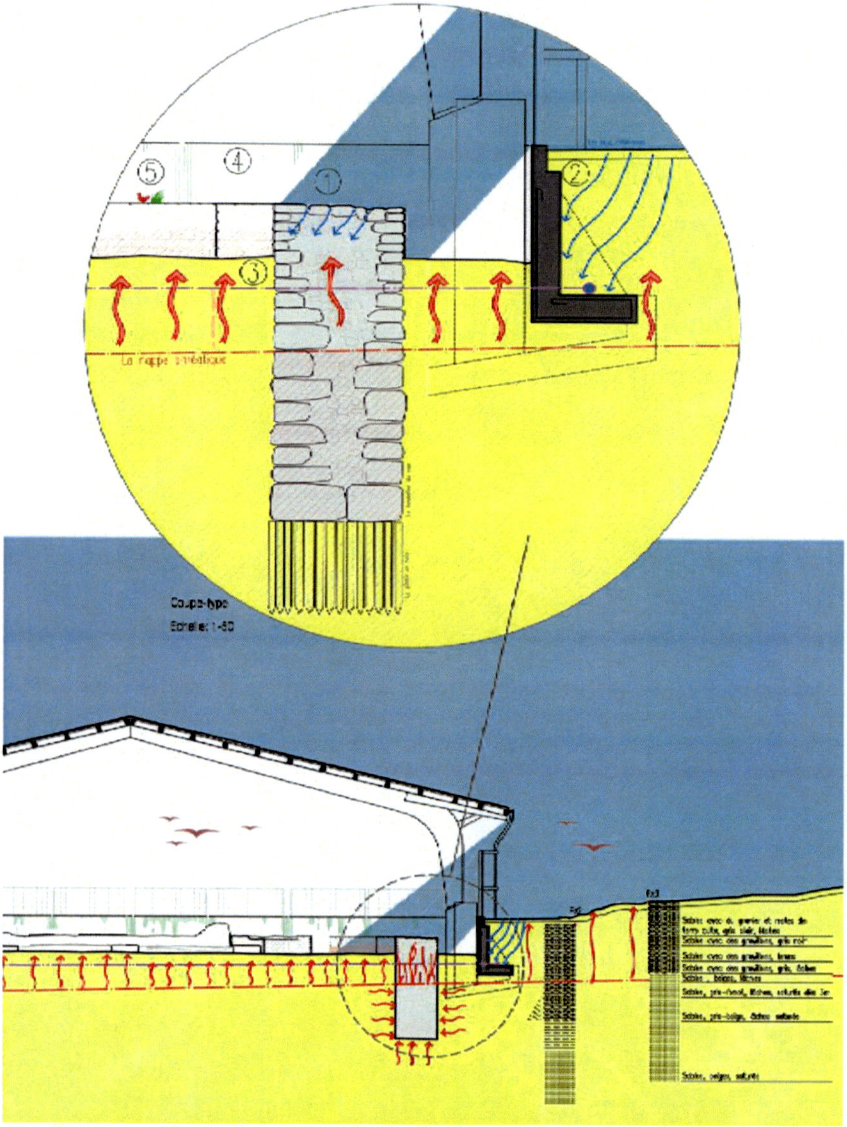

Fig. 2k. Main causes of degradation: 1. Water and snow from precipitation on walls and floors: Damage by moisture and frost 2. Infiltration of the meteoric water: Presentation of dark crusts on the concrete walls 3. Capillary rise of the water table: Presentation of humidity to the lower parts of the walls and the floor 4. Rain penetration into the excavation: Presentation of mold and green scabs on the concrete wall 5. Fauna and flora: Penetration of birds and other animals. Grains of plants which are brought by the wind.

2.3 The Proposal

After carrying out the diagnosis of the current building and identifying the preservation problems a new architectural project was proposed. The main objective of the new building has been concentrated on the alteration of the causes that damage the archaeological vestiges through a minimum and reversible intervention, which promotes the better presentation of the monument to the public. The intervention will have 6 axes (Fig. 21).

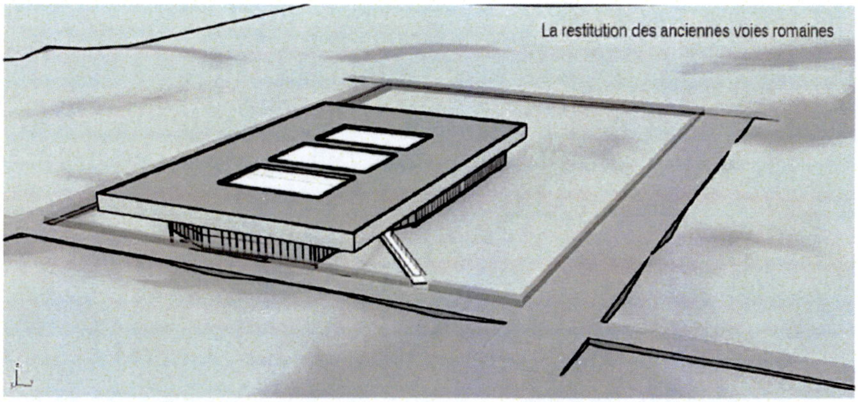

Fig. 21. The new thermal baths project in Perruet, the restoration of the old Roman roads around the thermal baths and the planting of shrubs on the edges of the islet. Source: computer generated image by Michail Zotos

2.3.1 The Camouflage of the Peripheral Wall

The concrete wall around the current excavation creates a very distinctive border and gives a false impression about the real extent of the ancient building as it frames only the two of the three main rooms of the Roman baths. The creation of slopes would hide the concrete walls and would give the impression of a work in progress so as to show that the survey has not finished and the ancient building is not fully unearthed. In the western part of the excavation the concrete wall could be fully demolished and give its place to a reconstruction of the Caldarium's floor so as to make more evident the continuity of the building.

2.3.2 The New Roof

The new roof consists of a simple rectangular metal construction, located approximately 6m above the excavation, covering its entire extent. The new construction will give thus the third dimension of the excavated space by signaling it from afar. Furthermore, three skylights in the dimensions of the bath's main rooms (frigidarium, tepidarium) will adequately illuminate all the important elements of the excavation. This new roof is supported by round metal columns and reuses the old concrete foundations of the former hangar in order to have the minimum impact on the archaeological vestiges (Fig. 2m).

Fig. 2m. A representation of the interior of the Perruet pavilion. Source: computer generated image by Michail Zotos

2.3.3 The Visitor's Paths

The project also includes the redesigning of the visit paths thus to get a closer look at the roman vestiges and a better understanding of the shape and the function of the thermal baths. A footbridge over the vestiges will cross the main rooms (frigidarium, tepidarium, caldarium) following the main bathing itinerary. This new construction will be suspended from the roof in order to prevent the excavation of foundations in the middle of the thermal baths (Fig. 2n).

Fig. 2n. The perforated steel floor will allow a look at the remains that lie just below. Source: computer generated image by Michail Zotos

2.3.4 An Extension of the Excavation

The visiting itinerary could be extended to the service corridor that surrounded the Tepidarium and the Caldarium in order to make the function and the heating system of the building more understandable (Fig. 2o).

Fig. 2o. A sketch of the architectural proposition. Source: Michail Zotos

2.3.5 A Glass Wall

The flooding problems of the monument require the best protection against the wind, the rain, and the snow but also the prevention of the animals' access. A glass wall suspended from the roof with no impact on the vestiges and allowing the air circulation can provide the needed isolation for the archaeological site.

2.3.6 Waterproofing

There is a need for a waterproofing membrane all around the excavation that will protect the remains from rising capillaries. A lifting pump could be used to lower the level of the underground water (Fig. 2p).

Fig. 2p. The section of the new construction Source: Michail Zotos

3 Conclusions

In general, the interventions with new buildings in archaeological sites raise several questions. A new intervention on an archaeological site aims to present the vestiges to the public, to facilitate their understanding and to upgrade them without ever distorting their meaning. The architects must adequately answer the question of the minimum impact on the archaeological site and the reversibility of their intervention. The challenge of providing the most efficient protection of the ancient vestiges against the elements of nature must also be taken into consideration. The ancient remains have usually suffered from a series of additional damage due to natural conditions (rain, snow, underground water, frost, animals) but also from human factors rendering them even more vulnerable.

The maintenance requirements of an archaeological site are quite difficult to fulfill; especially in Switzerland, where snow, wind, and rain can easily damage the vestiges. In an attempt to meet some restoration requirements, it is often necessary to build a shelter to protect the fragile remains. In Avenches, the needs of protection led very early, to the construction of a shelter for the excavation. Only two years after their discovery, the thermal baths of "Perruet" were protected and accessible to the public. This immediate reaction demonstrates the great interest in the protection but also the promotion of the ancient Aventicum (Fig. 2q).

Fig. 2q 3D representation of the new roof Source: Michail Zotos

References

Blanc, P., Frey-Kupper, S., Frei-Stolba, R.: Aux origines de la capitale des Helvètes. Avenches, capitale des Helvètes, pp. 6–11 (2001)

Bögli, H.: Aventicum. La ville romaine et le musée. Guides archéologiques de la Suisse, vol. 19 (1989)

Bourquin, J.: Fouilles de Pro Aventico: les thermes de Perruet. Bulletin de l'Association Pro Aventico 16 (1955)

Bridel, P.: Conservation et mise en valeur de vestiges d'époque romaine. Quelques exemples de Suisse romande. La Valorisation des sites archéologiques: actes du colloque international de Martigny, pp. 19–33 (2012)

Bridel, P., Bigovic, S.: La vie des monuments. Bulletin de l'Association Pro Aventico 49 (2007)

Bridel, P., Kaufmann, M.: La vie des monuments. Bulletin de l'Association Pro Aventico **42**, 279–284 (2001)

Castella, D.: Territoire et voies de communication. Avenches, capitale des Helvètes, pp. 15–19 (2001)

Castella, D., Blanc, P., Flück, M., Hufschmid, Th., Meylan Krause, M.-F.: Aventicum: une capitale romaine. Avenches: Association Pro Aventico (2015)

Faccani, G.: Tempel, Kirche, Friedhof und Holzgebäude. Bulletin de l'Association Pro Aventico **46**, 7–65 (2004)

Filgis, M.N.: Vor 225 Jahren entdeckt, unter Stahl und Glas präsentiert. Denkmalpflege in Baden-Württemberg (2009)

Frisch, E.C., Zingg, S., Diethelm, A.: Construire en acier: concevoir et construire. Steel doc 01, 34 (2006)

Hufschmid, T., Terrapon, N.: Le site et les monuments en 2013. Bulletin de l'association Pro Aventico 55 (2014)

Martin Pruvot, C.: L'insula 19 à Avenches. Cahiers d'archéologie romande **103**, 78 (2006)

Meylan Krause, M.-F.: Aventicum, ville en vues. (A. P. Aventico, Ed.) Fribourg (2004)

Morel, J., Blanc, P., Bridel, P., Chevalley, C., Christe, F., Dal Bianco, J.-P.: Chronique des fouilles archéologiques 1995. Bulletin de l'Association Pro Aventico 37 (1995)

Schwarz, G.T.: Avenches - Les thermes "En Perruet". Fouilles de l'hiver 1957/58. La Suisse Primitive, 2(Notices sur la préhistoire et l'archéologie suisses) (1958)

Schwarz, T.G.: Fouilles de 1957/58 aux thermes "en Perruet". Bulletin de l'Association Pro Aventico 18 (1961)

Wible, F.: Avant-propos. La valorisation des sites archéologiques. Actes du colloque international de Martigny. Cahiers d'archéologie romande 134 (2012)

The ARCHYTAS Intelligent Decision-Support System for the Protection of Monumental Structures

Dimitrios Vamvatsikos[1(✉)], Michalis Fragiadakis[1], Ioannis-Orestis Georgopoulos[2], Vlasis K. Koumousis[1], Demetris Koutsoyiannis[1], Alexandros Manetas[3], Vasileios E. Melissianos[1], Christos Papadopoulos[4], Konstantinos E. Papanikolopoulos[2], and Eleni-Eva Toumpakari[5]

[1] National Technical University of Athens, Heroon Polytechneiou 9, Athens, Greece
divamva@mail.ntua.gr
[2] AETMON IKE, Aiolou 102, Athens, Greece
[3] iTEAM S.A., Leontariou 10-12, Pallini, Greece
[4] DOMOS E.E., Ellanikou 38-40, Athens, Greece
[5] Directorate of Restoration of Ancient Monuments, Ministry of Culture and Sports, 12 Karytsi Sq., Athens, Greece

Abstract. The ARCHYTAS platform is based on using (i) reliable mechanical models and damage thresholds for assessing structural performance (ii) a network of sensors for updating the model parameters, (iii) detailed estimates of earthquake and flood hazard at the sites of interest and (iv) a state-of-art approach for multi-hazard risk assessment that can deliver accurate pre/trans/post-event evaluation of the risk at multiple geographically distributed cultural heritage sites. The core of the proposed system comprises a cloud-deployed computational platform, where data obtained from on-site measuring systems is processed, critical environmental actions are identified and flags are raised to provide alerts on the predicted monument structural condition. The decision-support system is fully uncertainty-aware, employing the concept of the mean annual frequency of limit-state exceedance under specified confidence levels to offer monument-specific courses of action based on the convolution of the current state of the monument (as determined by its best-estimate fragility, and updated by current or past measurements) and the predicted, recorded or evolving hazard. All-in-all, the platform can assist the relevant authorities to prioritize inspection, maintenance and rehabilitation actions before or after events subject to limited available resources.

Keywords: Cultural heritage · Seismic hazard · Flood hazard · Sensor network · Uncertainty · Risk assessment

1 Introduction

The protection of cultural heritage is a difficult and ever-evolving task as authorities attempt to tackle the steady onslaught of extreme natural hazards and continuous weathering deterioration. At present, most Mediterranean countries have their share of monumental structures whose structural condition is relatively poor and are in danger of

I. Vayas and F. M. Mazzolani (Eds.): PROHITECH 2021, LNCE 209, pp. 1246–1255, 2022.
https://doi.org/10.1007/978-3-030-90788-4_96

sustaining non-recoverable damage or even losing their structural integrity. Protecting endangered monuments from environmental actions, such as the dominant seismic and pluvial flooding hazards that are prevalent around the Mediterranean Basin is a challenge that only becomes more daunting when tackled within severe budgetary constraints. Finding ways to maximize the impact of every euro spent is of paramount importance, especially within a crisis environment. Such structures are priceless and, consequently, the application of novel engineering solutions needs to be handled with care, involving archeologists, restoration specialists, architects and structural engineers to ensure optimal outcomes. All the above elements highlight the need for an intelligent decision-support platform for the prioritization of rehabilitation actions, before or after a damaging event takes place. The ARCHYTAS project aims to develop such a software & hardware system that will offer the data required for reliable decision-making by relevant stakeholders. Its theoretical foundations and conceptual architecture are presented as applied to monuments of classical antiquity in Greece.

2 Platform Architecture

The ARCHYTAS platform is the core of the intelligent decision-support system for the protection of monumental structures and consists of four conceptual entities:

- End-user: The end-user of the platform is the Ministry of Culture and Sports, where certified personnel can access the platform, handle the data, and assess the potential risks to the monuments.
- Sensors: The sensors are installed in the monuments and provide the data to the web-cloud to perform the risk calculations.
- Computational models: Hazard and vulnerability computational models run offline and generate data that is stored in the web-cloud to be used for risk assessment.
- Web-Cloud Middleware: The core of the platform is hosted in the web-cloud, where sensor data are stored, risk assessment calculations are performed, the platform's website and warning engine are provided with data.

The conceptual architecture of the platform is illustrated in Fig. 1, where the modules are presented with green color, the sensors with orange, and the database with yellow. The platform architecture supports the seamless flow of data from sensors and computational models to the end-user, as well as the interaction with the end-user.

2.1 Sensor Data Module

The Sensor Data Module is a physical entity and includes the Logger and the Gateway. The module collects the data from the sensors and sends it to the webserver located in the web-cloud that is part of the Middleware. The Data Logger collects data from the sensors, saves it and forwards it to the Data Gateway. Usually, one recorder per monument is used and more than one sensor is connected to it. The Data Gateway receives the sensor data and transmits it to the Web Server located in the web-cloud. The Data Gateway can be part of the Data Logger or a separate physical entity (hardware). The data is transmitted asynchronously either wirelessly, or via a landline.

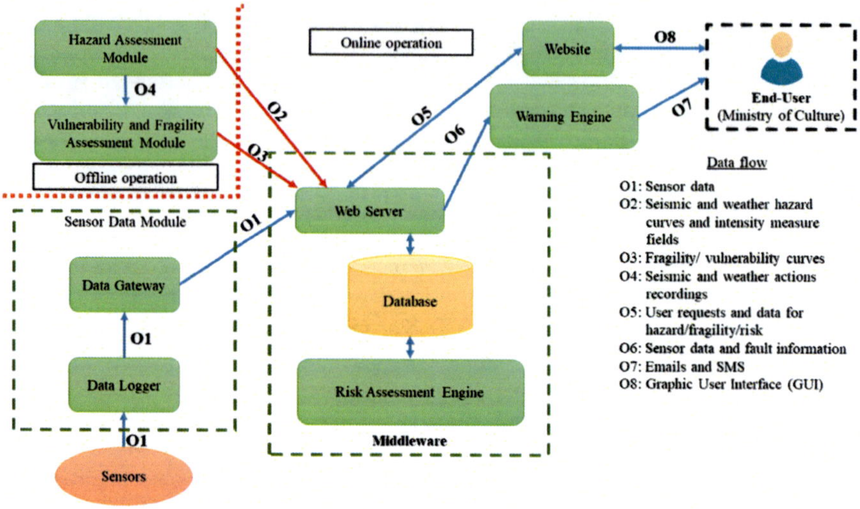

Fig. 1. ARCHYTAS conceptual system architecture

2.2 Hazard, Fragility and Vulnerability Assessment Modules

The hazard assessment for each site is carried out in the Hazard Assessment Module (operating offline), which supplies the Middleware with the (seismic and weather) hazard curves and the intensity measure fields, which are stored in the Database for use by the Risk Assessment Engine. Seismic hazard assessment is carried out with the OpenQuake (Pagani et al. 2014) open-source platform and the European seismic model SHARE (Woessner et al. 2015) to produce the seismic hazard curve for the site of the monument, the seismic intensity measure fields. Then, the appropriate accelerograms are selected using state-of-the-art tools and procedures. Accelerograms are transferred to the Vulnerability and Fragility Assessment Module, while the seismic hazard curves and the intensity measures fields are transferred to the Middleware. For weather hazards such as wind, rain, temperature, etc., local measurements from national meteorological stations and stochastic models are used. The weather hazard curves are transferred to the Vulnerability and Fragility Assessment Module, while the weather hazard data is transferred to the Middleware.

The assessment of the monument fragility is critical for calculating the risk. Fragility presents the possibility of exceeding a predetermined level of damage for a given value of the intensity measure. The Vulnerability and Fragility Assessment Module operates offline. The module supplies the Middleware with the structure's fragility curves, which are stored in the Database to be used by the Risk Assessment Engine.

2.3 Web-Cloud Middleware

The Middleware is the core of the platform and is located in the web-cloud. It comprises a Web Server that interacts with the rest of the platform modules and part, the Database and the Risk Assessment Engine. The Webserver provides the required functionality, in the form of RESTful web services, to (i) the Sensor Data Module that records the measurements from the sensors installed in the monuments, (ii) the system's Website where information for the monuments and measurements are presented, and (iii) the Warning Engine. The Web Server also receives data to be stored in the Database by the Hazard Assessment Module and the Vulnerability and Fragility Assessment Module.

The Database is part of the Middleware, where sensor data, the fragility curves of the monuments, the hazard curves and the intensity measure fields are stored, as well as the results obtained from the Risk Assessment Engine.

The Risk Assessment Engine is located in the cloud as being part of the Middleware and interacts with the Database for data exchange:

- Seismic and weather hazard curves produced in the Hazard Assessment Module;
- Intensity measures fields generated in the Hazard Assessment Module;
- Fragility curves produced in the Vulnerability and Fragility Assessment Module;
- Sensor data sent from the Sensor Data Module.

After performing the seismic and/or weather risk calculations, results are sent to the Database in order to be transmitted to the Website or the Warning Engine via the Web Server. The calculation of the risk is carried out by taking into account the effects of the hazard and the fragility. The result is the calculation of the mean annual frequency of exceeding a predefined limit state.

The mean annual frequency of exceedance of a limit state indicates the risk level of the monument. The ranges of this mean annual frequency are illustrated with color indications: green for low damage estimation, orange for medium damage estimation and need for on-site inspection, red for severe damage estimation and need for immediate visit to the monument. This color scale enables the End-User to immediately prioritize its needs and optimize the allocation of available financial and human resources.

2.4 User Interaction Modules

The Website is the central end-user interface to the ARCHYTAS platform. The Website receives End-User requests via GUI and transmits them to the Web Server and receives the hazard, vulnerability and risk data for the monuments from the Web server to be presented to the End-User. The Warning Engine sends event alerts to End-User via email and/or short message system (SMS) after receiving sensor data and fault information from the Web Server.

The End-User of the telemetry control platform is the Directorate for the Restoration of Ancient Monuments of the Ministry of Culture and Sport as the authority responsible for monitoring the status of the monuments. The End-User accesses the platform via the Website and receives notifications of emergency events from the Warning Engine.

3 Case Studies

3.1 Horologion of Andronikos Kyrristos

The Horologion, also known as the tower of the Winds, is an octagonal clocktower with 3.2m sides and a conical roof built in the first half of the 1st century B.C. by the astronomer Andronicos, from Kyrrhos, Macedonia. It resides in the Roman Agora of Athens and comprises sundials and a water clock. It is entirely composed of marble blocks, fitted without the use of mortar joints, and it suffers from lateral spreading (rotation) of the top of walls with subsequent lowering of the roof (Fig. 2). Due to the obvious risk of loss of structural integrity, it has been designated as "very vulnerable" by the authorities. The monitoring system comprises extensometers and inclinometers (Fig. 3), which will be complemented by the addition of accelerometers and a local weather station that will transmit data directly to the Web-Cloud Middleware.

Fig. 2. The laterally spreading roof of the Horologion of Andronikos Kyrristos

Fig. 3. Sensors currently installed at the Horologion: Extensometer (left) and inclinometer (right)

3.2 Temple of Aphaia

The Temple of Aphaia stands on a hill at the north-east side of the island of Aigina, within view of Athens. It is a Doric temple constructed from porous stone (tuffa) and erected circa 500BC. It comprises both monolithic and multi-drum columns, some standing alone and others connected by architraves (Fig. 4). Originally, plastering protected the porous stone of the columns, yet this has mostly disappeared over the millennia. Due to its exposure to the elements, it has sustained significant damage, with visible cracks in some architraves and columns, some requiring temporary shoring until a permanent solution is found (Fig. 5). The monitoring system will only comprise accelerometers.

Fig. 4. North view of the Temple of Aphaia, Aegina Island, Greece

Fig. 5. Signs of visible damage in the columns of the Temple of Aphaia (top) and temporary local support measures for a cracked lintel (bottom)

4 Hazard and Risk Assessment Framework

Probabilistic representations of hazard will be adopted to capture the natural randomness of environmental stressors such as earthquakes, pluvial floods and weathering of the stone. In the case of earthquakes, Probabilistic Seismic Hazard Analysis (PSHA) will be employed (Cornell 1968; Baker 2008). Specifically, the event-based approach of PSHA will be adopted to offer scenarios of seismic intensity occurring due to a single seismic event over multiple sites, as shown in Fig. 6. This allows us to capture the spatial variability and correlation of ground motion, as well as the cotemporaneous nature of losses at multiple monuments subject to the same event, as these will most stretch the resources for any post-event restoration.

Risk assessment is based on the concept of Performance-Based Earthquake Engineering, as proposed by Cornell and Krawinkler (2000), and adopted by the Pacific

Earthquake Engineering Research (PEER) Center:

$$\lambda(DV) = \int\int\int G(DV|DM)|dG(DM|EDP)||dG(EDP|IM)||d\lambda(IM)| \quad (1)$$

DV is a scalar or vector of Decision Variables, e.g., monetary cost, downtime or human casualties, to be used by decision makers to undertake restoration actions. *DM* is the Damage Measure, representing the level of damage and typically discretized into distinct damage states of structural and non-structural elements. The Engineering Demand Parameter (*EDP*) characterizes the structural response in engineering terms (e.g., strains, stresses, displacements, etc.) while *IM* is the intensity measure whose mean annual frequency is estimated by PSHA. $G(x)$ is the complementary cumulative distribution function of x, and $\lambda(y)$ offers the mean annual frequency of y. In essence, this is the embodiment of the total probability theorem that allows us to combine the results of multiple seismic hazard scenarios (Fig. 6) with the fragility assessment results to determine the risk of one or more monumental structures in terms understandable by non-engineer stakeholders.

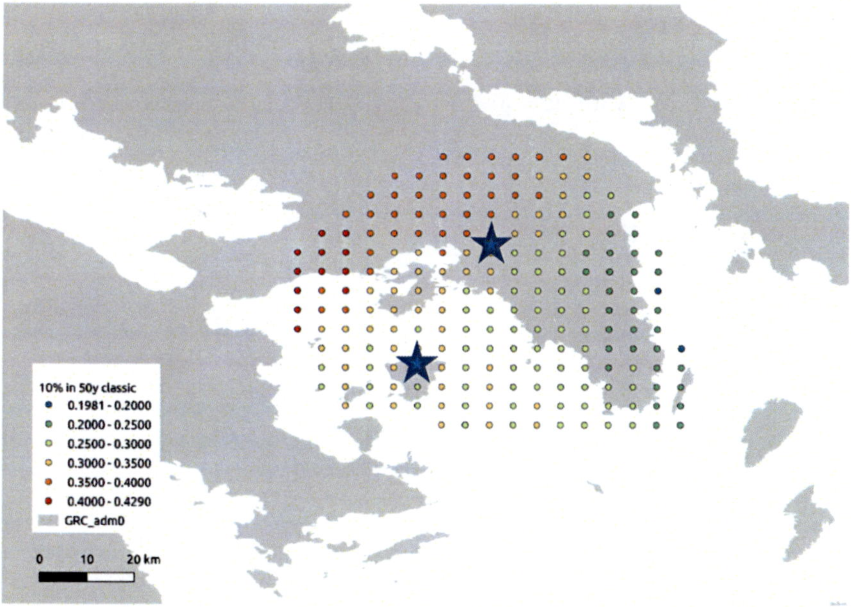

Fig. 6. Peak ground acceleration (PGA) with a return period of 475 years in the region of Attica based on the SHARE model (Woessner et al. 2015). The location of the case studies is indicated by two stars

5 Structural Modeling

Monuments of classical antiquity, like the two case studies presented, are mainly prone to rocking under seismic excitation. Both 2D and 3D models have been proposed to capture such a behavior. The simplest approach for modeling the rocking motion of a rigid body has been proposed by Housner (1963). Since then, many authors have recommended more elaborate models to capture rocking and/or sliding using either equivalent non-rocking oscillators or explicit solutions of the rocking/sliding motion. These range from single-degree-of-freedom systems, comprehensively presented in Fig. 7 by Diamantopoulos and Fragiadakis (2019), up to large-scale multi-degree-of-freedom finite element or discrete element models (Dasiou et al. 2009; Psycharis et al. 2013), as shown in Fig. 8. Each modeling choice represents a different compromise in terms of computational cost, modeling complexity, ease of convergence and accuracy, whose pros and cons need careful consideration for each case study. Selecting the simplest possible model with an acceptable level of fidelity is key for efficiently running the time-history analyses required for accurate fragility assessment.

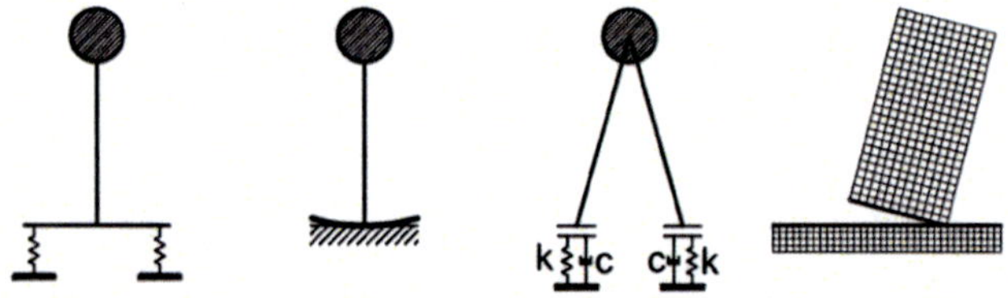

Fig. 7. Alternative models of rigid or flexible rocking bodies (Diamantopoulos and Fragiadakis 2019)

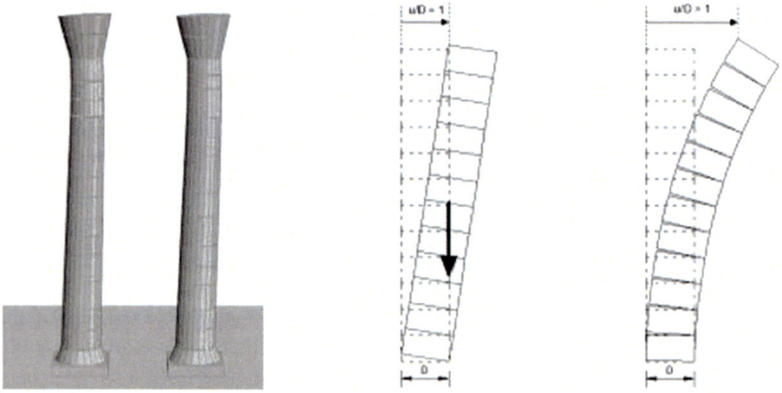

Fig. 8. Discrete element models of multi-drum columns (Psycharis et al. 2013)

6 Conclusions

The ARCHYTAS platform is a viable approach for supporting decision-making for pre-event or post-event rehabilitation actions for cultural-heritage sites. It supports multiple monuments at the same, or at spatially distributed sites. It allows integrating models of different levels of complexity and fidelity within a homogeneous flexible risk assessment framework with true multi-hazard capabilities. In the ever-present fiscal constraints and under a changing climate, it represents a way forward to achieve efficient resource management for protecting our cultural heritage.

Acknowledgements. This research has been co-financed by the European Regional Development Fund of the European Union and Greek national funds through the Operational Program Competitiveness, Entrepreneurship and Innovation, under the call RESEARCH – CREATE – INNOVATE (project code: T1EDK-00956), project: "ARCHYTAS: Archetypal telemetry and decision support system for the protection of monumental structures".

References

Baker, J.W.: An introduction to probabilistic seismic hazard analysis. Version 1.3 (2008). https://web.stanford.edu/~bakerjw/Publications/Baker_(2008)_Intro_to_PSHA_v1_3.pdf

Cornell, C.A.: Engineering seismic risk analysis. Bull. Seismol. Soc. Am. **58**(5), 1583–1606 (1968)

Cornell, C.A., Krawinkler, H.: Progress and challenges in seismic performance assessment. PEER Center News **3**(2), 1–4 (2000)

Dasiou, M.E., Psycharis, I.N., Vayas, I.: Verification of numerical models used for the analysis of ancient temples, Protection of Historical Buildings. In: Proceedings of International Conference on Protection of Historical Buildings, PROHITECH 2009, Rome, Italy (2009)

Diamantopoulos, S., Fragiadakis, M.: Seismic response assessment of rocking systems using single degree-of-freedom oscillators. Earthquake Eng. Struct. Dynam. **48**(7), 1–20 (2019)

Pagani, M., et al.: OpenQuake engine: an open hazard (and risk) software for the global earthquake model. Seismol. Res. Lett. **85**(3), 692–702 (2014)

Psycharis, I.N., Fragiadakis, M., Stefanou, I.: Seismic reliability assessment of classical columns subjected to near-fault ground motions. Earthquake Eng. Struct. Dynam. **42**(14), 2061–2079 (2013)

Woessner, J., Danciu, L., Giardini, D., The SHARE consortium: The 2013 European Seismic Hazard Model: key components and results. Bull. Earthq. Eng. **13**(12), 3553–3596 (2015)

H-BIM – Innovative and Digital Tools to Improve the Management of the Existing Buildings

Christian Musella(✉), Milena Serra, Antonio Salzano, Costantino Menna, and Domenico Asprone

Università Degli Studi Di Napoli Federico II, via Claudio, 21, 80125 Napoli, NA, Italy
christian.musella@unina.it

Abstract. With the introduction of the BIM methodology, the digitalization is consolidating mostly the process of new building design defining new standardized working procedures in order to optimize the data flow. On the contrary, pertaining to existing buildings, the rules are still dictated by the traditional methodology leading to fragmented and disaggregated information flow. However, some tools have been recently developed and are quickly upgrading to meet the specific aim to pursue progress, dynamism and experimentation in order to stay in tune with market demands. One common issue, is due to the fact that the documents are paper, hard to find and numerous so that the digitalization represents the best real solution to improve their management.

In this regard, the heart of the informative flow is represented by the Common Data Environment (CDE), a cloud storage in which the documents are collected once uploaded and easily manageable with specific platforms. By way of example, a case study is developed in the paper in which the management process of the H-BIM of the XV century's structure placed in the old town of Naples is carried out in a digital way, through the usBIM.platform (released by ACCA software company) using specific tools, like links, markers and tags. The goal of the paper is to define both a new organization for the data archive, in which the documents are structured and easily traceable, both a new way of conceiving the BIM model that it is thought as an informative vehicle and a key access to the information rather than a mere geometric representation.

Keywords: H-BIM · Information management · Masonry buildings · Cultural heritage · Common Data Environment · #TagBIM

1 Introduction

The Building Information Modelling BIM technologies (Penttilä 2006) are progressively getting used in the Architecture, Engineering and Construction AEC industry (Takim et al. 2013), increasingly conditioning the daily work method of professionals. These new operating methodologies are consolidating more in the field of design of new buildings, under the impulse of some European countries (Grzyl et al. 2017) that see in the

I. Vayas and F. M. Mazzolani (Eds.): PROHITECH 2021, LNCE 209, pp. 1256–1267, 2022.
https://doi.org/10.1007/978-3-030-90788-4_97

new buildings the greatest field of application, through standardizing workflows that allow managing in a better way the exchange of information among the various actors involved in each contract. According to the open BIM practices, the information is placed in a known architecture, handled by software that refer to open standards such as IFC (Sustainability of Digital Formats: Planning for Library of Congress Collections 2016), an open format promoted by buildingSMART international (buildingSMARTInternational, n.d.), which guarantee the information exchange even through a Common Data Environment (CDE). In the design of new buildings, it is quite simple to pursue these objectives: in traditional workflows, a team of technicians with different skills meets to develop a design solution and summarizes all the information produced in a detailed and structured way in an informative model, which constitute the digital twin of the asset. This, in fact, allows the use of digital models in various cases, since the same information can be used several times for different BIM uses. Existing assets, instead, give a completely different panorama, in particular for the operations of Facility Management (FM) that involve them. In order to improve the management of the Historical Building Information Modelling H-BIM (Logothetis et al. 2015) a new digital methodology have been developed, involving new technologies that are nowadays strongly establishing on the market. It consists in an approach that overcome the traditional working methods encouraging communication in a multi-disciplinary team thanks to the revision of the customary steps. In fact, each of the different professional figures involved in the process works simultaneously to the others, taking care of different aspects of the same project so that all the activities are to be planned in advance in a detailed and schedule. Moreover, the activities accurately reprogrammed and carried out in a digital way allow the organization, time and costs result anticipated, consequently the financial compensation of the whole process can be easily determined.

2 The Complex Combination of BIM and Existing Buildings

In matter to existing buildings, due to the complexity of the intervention, a dedicated workflow has been developed. Differently from the design of new buildings, all the phases are often difficult to organize, monitor and achieve; in fact, the first difficulties already appear in the knowledge phase of the building in which the main building data is collected and design and construction choices are retraced. In this phase the historical documentation, if available, is digitalized, so that the information is not structured to be available in a timely, updated and reliable manner. Following, in order to establish the real condition of the heritage, those data collected need to be revised and raised through in-situ surveys, geometric reliefs (manual or digital), such as tests, inspections, thermographs and geo-radar. Moreover, the BIM model of the construction can be difficult to build due to disparate and irregular geometries of elements rather widespread in historical buildings. Given the complexity of the whole process, thus it is necessary to develop a workflow for a structured working phases and data, to obtain the benefits of the BIM also for the management of existing constructions (Khadda and Srour 2016). More in detail, the methodology performed allows to generate and manage information of buildings from H-BIM models that are not only three-dimensional representations, but dynamic models of data derived from a parametric model of the objects of the heritage. It is a parameterization based on the Object Oriented logic (Liu 2001), which allows first the problem-free implementation and then the simple and automatic data extraction, thus creating new databases available to the entire supply chain operating in the sector. The workflow has been made up for the purpose and implemented in a dedicate platform called usBIM.platform (Piattaforma BIM (Collaboration Platform) - usBIM.platform, n.d.), developed by the Italian company ACCA software, that is a tool created and developed to support the control of the exchange of information among a project team, and applied in this specific case for the management of an existing building.

3 Palazzo Penne: Application on a Real Case Study

The application of the new methodology proposed, has been experienced in occasion of Palazzo Penne's renovation, a renaissance masonry palace located in the historical center of Naples and built in 1406. The renovation project was elaborated a few years ago in the traditional way so that the interventions will have to start soon.

Indeed, a digital way of information management has been simulated, having available all the documents elaborated in the several phases of the real case study so that all the information produced have been taken into account. The building is an example of civil habitation of XV century, one of the few constructions that did not collapsed after the devastating earthquake of 1456. Several changes of the building have undergone several time, as witnessed by the walled arches and the slabs built with different construction systems, such as wooden, steel, and reinforced concrete. The complex has been dived into three Minimal Unit of Intervention (MUI) that is a part of a construction consisting in uniform structural unit subject to the same retrofit intervention. Moreover, the number of levels results variable even for each MUI, because of the presence of lofts that raise the complexity of the building. Actually, the difficulties acknowledged in Palazzo Penne are rather common in the cultural heritage, so that it can be considered representative of a big amount of buildings (Fig. 1).

Fig. 1. Palazzo Penne's main façade

Indeed, due to the complexity and the importance of this construction, a workflow developed taking into account all the difficulties in its management can result an accurate and transparent example for the information management. The aim of this process is to create adequate guidelines to improve the management of the existing buildings.

4 A Workflow for the H-BIM Management

Because of the complexity of the task, a dedicated workflow has been engineered for the improvement of the assets management. All the steps of the traditionally performed have been deployed but in a digital way. Thanks to the use of a collaborative platform, it has been given rise to new methodology that enables multiple applications. In fact, it results a simple tool that can be deployed quickly and on a large scale, so that all the operations that goes from the determination of the stat-of-art to inspections and subsequent actions to be taken, make up a continuous information stream that does never stops. Thanks to the automation of the processes, the possibilities of making mistakes are drastically reduced, thus raising the results and reducing the costs. The steps for the H-BIM management are summarized in Fig. 2.

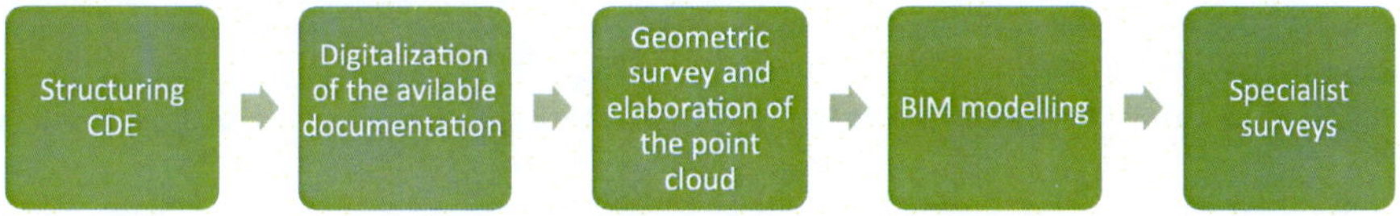

Fig. 2. H-BIM management

More in detail, all the specific activity to carry out in each step have been summarized in flow charts called "workflows", which should be established well in advance, before starting any operation. As a way of example, the workflows for the geometric survey, the BIM modelling and the damage assessment (one of the activities to carry out during the specialist surveys) are reported in Fig. 3.

In particular, the interactions among the different actors involved are managed making clear the information exchange. Having a definite road map, in fact, make the progress status of the project clear to all the team members that always know when to perform their tasks. Moreover, they provide support in case of delays or unforeseen events, to take some actions aimed at bringing the project in the pre-arranged path.

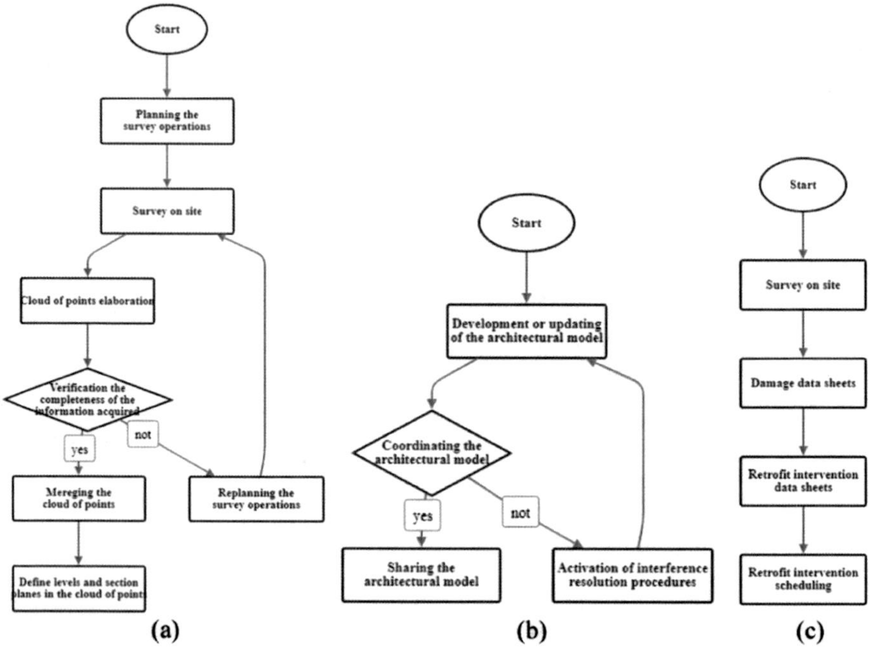

Fig. 3. Workflow for the geometric survey (a), the BIM modelling (b) and the damage assessment (c)

5 Digital and Smart Archive in the CDE

For the performance of all these operations, an IT structure has been established and identified in the CDE, which is the unique working environment in which all the information is shared among the team members. This instrument allows the complete stream of information and it is smartly accessible thanks to some platform tools.

More in detail, the CDE includes folders according to a Structured Query Language SQL (Fortier and Prichard 1994) database and it is organized according to the process phases, i.e. State-of-art, Survey, Investigation and tests and the Retrofit interventions. Each area has been sub-divided in Work in Progress, Shared, Published and Archive, i.e. the four processing states of the files within the CDE, according to ISO 19650-1 (Fig. 4).

In order to complete the settings of the CDE in usBIM.platform, the roles that each user have to endorse have been defined, taking into account both the technical roles (such as the structural engineer, the architect, etc.) and the informative ones (BIM modeler, BIM coordinator and BIM manager). It follows that a connection was established between each user and the tasks noted in the workflow that to be performed, so that each activity is defined and well-organized.

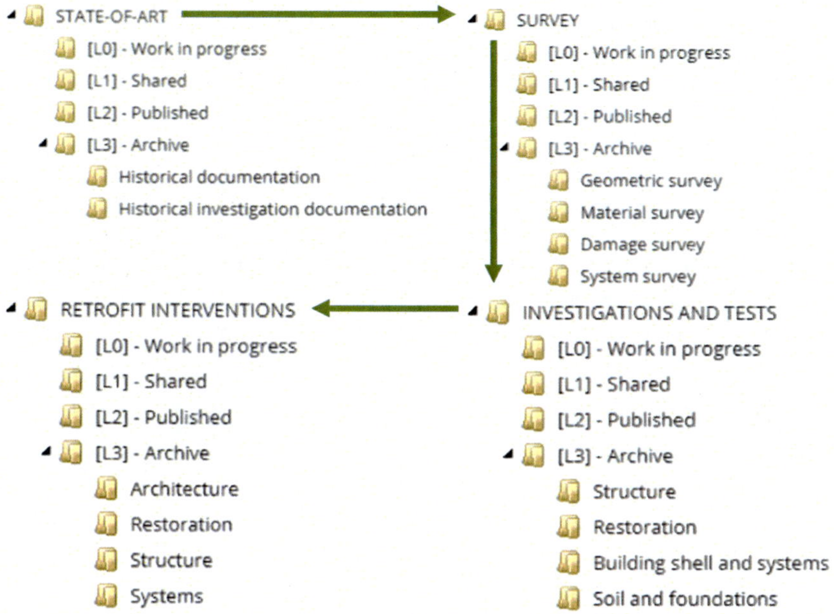

Fig. 4. Information flow in the Common Data Environment

Moreover, other important platform features are the definition of limitations to the access to each folder, ensuring the information security and confidentiality, and the historical storage of all the operation carried out. After establishing the CDE, a careful research of paper documentation, such as old reports, plants, elevations and sections was carried out.

It follows that a connection was established between each user and the tasks noted in the workflow that to be performed, so that each activity is defined and well-organized. Moreover, other important platform features are the definition of limitations to the access to each folder, ensuring the information security and confidentiality, and the historical storage of all the operation carried out.

After establishing the CDE, a careful research of paper documentation, such as old reports, plants, elevations and sections was carried out. Together with the documentation produced for the renovation project, they have been digitalized to move the paper archives into digital ones. It resulted that in the platform have been uploaded:

- 97 documental report, including the damage data sheets, report for the state of art, report for the design project reports for the investigations;
- 30 multimedia documents, including material relief drawings, damage relief drawings;
- 39 graphic documents, including plants, elevations and sections;
- 5 digital models, including BIM models and 3D cloud of points.

A total of 171 documents have been placed in the platform so it has to be noticed that even after a structuring in a comfortable way the CDE, it could still be difficult the research of the documents both for the team members and for the customer. For this reason, some specific tools called #TagBIM have been introduced. They allow filter operations among documents thanks to markers pointed to the documents, to draw attention to information in a quick and smart way for the dynamic recovery of each file. Therefore, the #TagBIM, according to the Not only SQL (NoSQL), make possible transversal pathways within the different folders regardless of the folder structure. Indeed, in order to facilitate the research of the documents, they have been defined referring to the document year, type and nature, to the discipline involved and to the document content. They have been allocated to all documents of CDE as showed in Fig. 5.

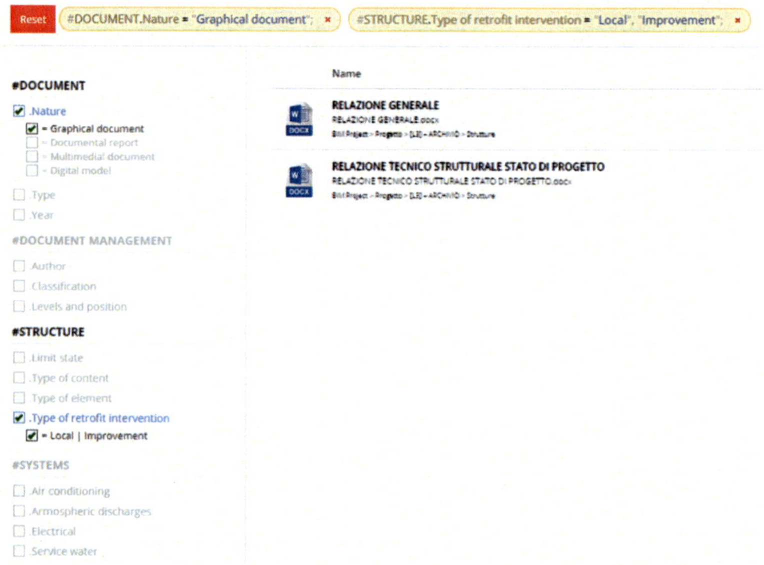

Fig. 5. #TagBIM for the documents

6 H-BIM Modelling and Management

Regarding to the geometry relief, with the traditional method, especially for historical construction as Palazzo Penne, the data derived are difficult to manage due to the complexity of the building itself. This operation, indeed, is simplified thanks to the support given by modern technologies for the elaboration of point clouds (Wong et al. 2014). In this regard, laser-scanning (Randall 2011) and photogrammetric techniques (Dore and Murphy 2013) are new ways of making surveys that meet the digital requirements and are little by little expanding their field of application. The point cloud derived from the digital relief represents a digital copy of the building in the form of a 3D model. The laser-scanner technique has been used to acquire Palazzo Penne's point cloud and from it, bi-dimensional drawings have been obtained, as sections, plans and elevations. The graphics have been used as reference for the creation of the BIM Model in Edificius, the BIM authoring software. However, many difficulties have been encountered modelling the building as BIM authoring software have been developed for the design of new buildings, so that even the modelling operation is difficult to catty out for an existing one. The reason for that can be identified in the uniqueness and complexity of each element that hardly fit the concept of "parametric object" and even because of the absence of a dedicated BIM library for specific objects. Indeed, modelling Palazzo Penne, significant deficiencies have been identified about specific objects that characterize the historical buildings, such as wall openings that have been closed, arches and vaults. Nevertheless, an accurate reflection showed that it is not always necessary to model the objects in all the details. In fact, some elements, such as doors, windows and portals have been chosen among the objects already existing in the library that mostly matched the real ones in terms of shape and typological features. In this regard, the procedure implemented in usBIM.platform takes into account an important concept linked to BIM modelling: the model elaboration has to be commensurated with its use; the use and the Level of Detail of the objects modelling (LOD according to ISO 19650-2) are closely linked. Famous, in fact, is the motto "begin with the end in mind" which aim to state this concept. In this regard, the platform tools make possible to store information in different options so that, all the data and features of the objects resulting hard and heavy for the model, are made available in alternative way and no loss of information is ensured.

After the geometric survey, each specialist member can perform the investigations of its own competence and integrate the information collected pertaining the state-of-art of the construction on the BIM model. Hence, the structural engineer generates additional files regarding materials tests and cracks and collapses suffered by the building while the architect report and drawings pertaining the degradations. These files have been uploaded to the CDE and structured through the properly #TagsBIM.

Then, they have been introduced in the BIM model in its IFC format, making it not merely a geometric representation, but a real store of information in different form, including elements, parameters of BIM objects or alternative solutions. Because not all the data can be coded according to the IFC format specifications, some dedicated platform tools have resulted very useful allowing the data storage on the model and making them easily available. In particular, thanks to the online viewer usbim.browser directly connected to the platform, data, file and models even in different format, **are directly linked in the CDE.**

More in detail, the damage sheets, the degradation sheets and the mesh model of some specific elements impossible to model (like the front door or the façade ashlars with the coats of arms of d'Angiò and Penne) have been linked directly to the object to which they refer, placing the localization of the information directly on the model and so enabling advanced search capabilities. Thanks to the point cloud, it is even possible to deduce the measurements of objects not contained into the BIM model (Fig. 6).

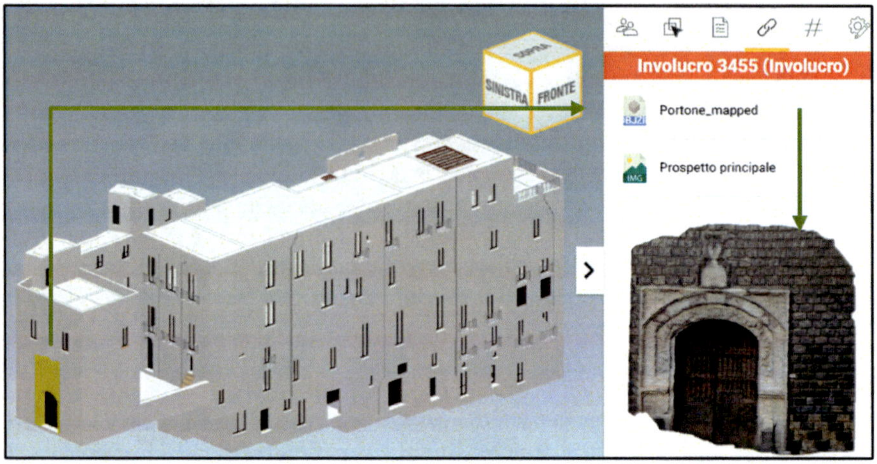

Fig. 6. Links among documents in the Common Data Environment

Moreover, even definite pointers allow data referencing on the model, favouring a quick and easy navigation and visualization of the files. With reference to the survey activity, the structural engineering has referenced photos of the endoscopy of the walls with its identification number while the architect photos of the degradations inspected, in order to have a visualization of the damage that has not been modelled directly (Fig. 7). This enables optimizing analysis phases and the scheduling of maintenance interventions downstream.

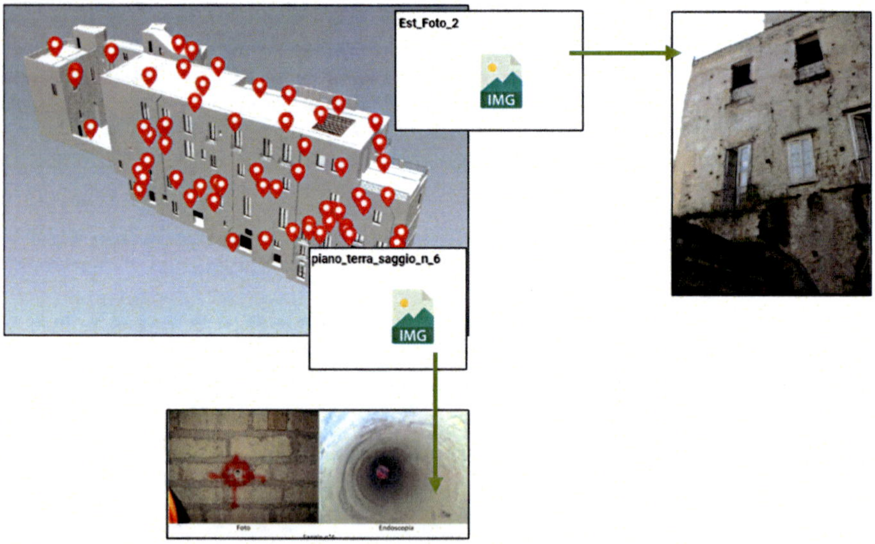

Fig. 7. Geolocation of the inspections on the model

In addition, the #TagBIM have been implemented even into the object of the model, a tool particularly useful to the user to quickly browse through it and view the objects and the content of CDE connected. In particular, the #TagBIM implemented pertain the damages inspected, the investigations and the retrofit interventions scheduled. Checking the options offered by the different #TagBIM, all this information can be crossed facilitating the research activities, as shown in Fig. 8.

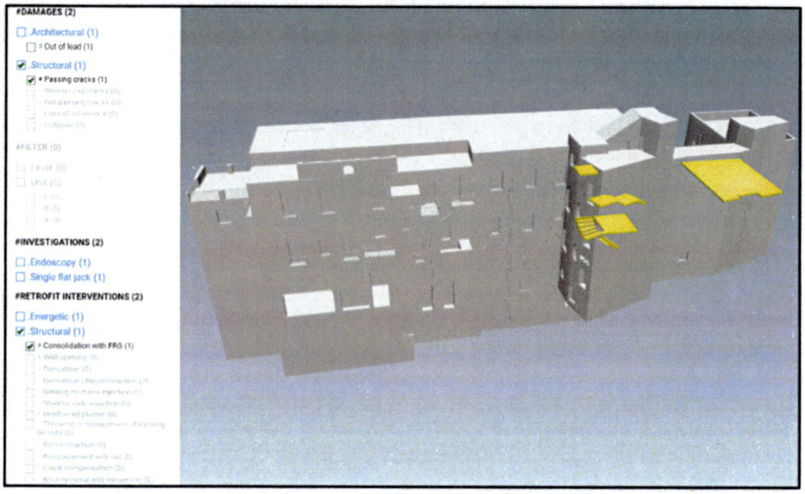

Fig. 8. #TagBIM of the model's objects

With these tools the information can be "googled" directly from an online browser and find the files and the documents linked to their respective objects.

7 Conclusions

In the steps below, it has been illustrated the potentialities of tools currently existing and their implementation for the H-BIM management. In particular, the strong point of the methodology explained consists in the fact that all the operations have been performed in a model in IFC format, the standard format introduced by buildingSMART that ensure that the openBIM principals have been taken into account, regardless of the software used for modelling. In the case study shown, it can be noticed how the model uses are completely different comparing to the common application for the design of new buildings. In this case, in fact, the model, after the uploading in the management platform, has been used as a storage for all the information produced in the renovation process of an ancient masonry building. The BIM model, in fact, rather than a mere 3D representation of the building, has become the key access to all the information of the CDE, dealing a huge amounts of data previously fragmented in the files. Thanks to pointers and #tagBIM, all the information has been centralized on the BIM model that in this way represents a point of reference for all the team members and constantly support their activities. In fact, as the process goes on, the model is always enriched with new information input in a structured manner, in order to always be traceable. As in the traditional process the use a paper archive represents a significant source of loss of time, thanks to the digital instruments the process is improved, allowing to convey the right information at the right time and to the right people. This purpose is also achieved thanks to the implementation of standardized workflows, which introduce the quality principles in the management of the design of retrofit intervention phase. The BIM methodology, if correctly implemented, can indeed represent an important turning point even for the management of existing building, new frontier of digitization, overcoming the currently widely recognized limits of classical methodologies.

Acknowledgements. The activity presented in the paper is part of the research grant BIM ReCulT (Il metodo BIM per il Recupero del patrimonio CulTurale).

References

Logothetis, S., Delinasiou, A., Stylianidis, E.: Building information modelling for cultural heritage: a review. In: 25th International CIPA Symposium 2015, Taipei, Taiwan, pp. 177–183 (2015)

buildingSMART International: Retrieved from buildingSMART - The Home of BIM (n.d.). https://www.buildingsmart.org/about/

Dore, C., Murphy, M.: Grammars using historic building information modelling. In: International Archives of the Photogrammetry, Remote Sensing and Spatial Information Sciences, Volume XL-5/W1, 2013 3D-ARCH 2013 - 3D Virtual Reconstruction and Visualization of Complex Architectures. Semi-Automatic Modelling of Building Façades with Shape, Trento, Italy, pp. 57–64 (2013)

Fortier, P., Prichard, J.: Concept for a real-time structured database query language (RT-SQL). Ann. Rev. Autom. Program. **27**, 167–174 (1994)

Grzyl, B., Miszewska-UrbaĚska, E., Apollo, M.: Building Information Modelling as an opportunity and risk for stakeholders involved in construction investment process. Procedia Eng. **196**, 1026–1033 (2017)

Khadda, M., Srour, I.: Using BIM to retrofit existing buildings. Procedia Eng. **145**, 1526–1533 (2016)

Liu, M.: Pluto: an object-oriented logic programming language. In: Proceedings 39th International Conference and Exhibition on Technology of Object-Oriented Languages and Systems, TOOLS 39, Santa Barbara, CA, USA, USA. IEEE (2001)

Penttilä, H.: Describing the changes in architectural information technology to understand design complexity and free-form architectural expression. ITcon **11**, 395–408 (2006)

Piattaforma BIM (Collaboration Platform) - usBIM.platform. Retrieved from BIM Collaboration Software | usBIM.platform | ACCA software (n.d.). https://www.acca.it/bim-collaboration-software

Randall, T.: Construction engineering requirements for integrating laser scanning technology and building information modeling. J. Constr. Eng. Manage. **137**(10), 797–805 (2011)

Sustainability of Digital Formats: Planning for Library of Congress Collections. Retrieved from Industry Foundation Classes (IFC), Clear Text Family (2016). https://www.loc.gov/preservation/digital/formats/fdd/fdd000447.shtml

Takim, R., Harris, M., Nawawi, A.: Building Information Modeling (BIM): a new paradigm for quality of life within Architectural, Engineering and Construction (AEC) industry. Procedia – Soc. Behav. Sci. **101**, 23–32 (2013)

Wong, J., Wang, X., Li, H., Chan, G., Li, H.: A review of cloud-based BIM technology in the construction sector. J. Inf. Technol. Constr. **19**, 281–287 (2014)

Giving Depth to the Traces of Nossa Senhora da Piedade da Caparica

Jesse Rafeiro[1](✉), Ana Tomé[2], and João Luís Inglês Fontes[3]

[1] Carleton Immersive Media Studio, Carleton University, 1125 Colonel By Drive, Ottawa, ON K1S5B6, Canada
jrafeiro@cims.carleton.ca

[2] Instituto Superior Técnico, Universidade de Lisboa, Av. Rovisco Pais, 1, 1049-001 Lisboa, Portugal
anatome@tecnico.ulisboa.pt

[3] Faculdade de Ciências Sociais e Humanas, Universidade NOVA de Lisboa, Avenida de Berna, 26-C / 1069-061, Lisboa, Portugal
joaofontes@fcsh.unl.pt

Abstract. Following an extensive renovation project in the 1950s and a lack of documentation throughout its life, the Convent of the Capuchos of Caparica (1558) today is in a state of confusion about the original configuration of its spaces. Since 1558, generations of Capuchos lived in the convent until the religious orders of Portugal were extinguished in 1834. Following this, it was abandoned for over one-hundred years until the renovation of the 1950s that inserted many new spaces. Presently, the convent serves the community of Almada as a space for recitals, concerts, wedding receptions and other social events that although fruitful to the community, mask its original conception as a place of silence and solitary contemplation for a small group of Franciscan friars known across Portugal as Capuchos. The research project described in this paper employs an interdisciplinary, mixed-methodological research approach to virtual reconstruction to provide the scientific validation for an HBIM reconstruction of the past condition of the convent prior to the 1950s. The approach combines historic text analysis and visualization with photogrammetric surveying of details and spaces found in other convents across the region as a means to visualize the traces that today are no longer present in Caparica.

Keywords: Virtual reconstruction · Heritage documentation · Heritage photogrammetry · Digital modelling · Building Information Modelling (HBIM) · 3D printing · Franciscan convents

1 Introduction

Following an extensive renovation project in the 1950s and a lack of documentation throughout its life (contemporary and historic), the Convent of Nossa Senhora da Piedade da Caparica today is in a state of confusion about the original configuration of its spaces. Presently, the convent serves the community of Almada as a space for recitals, concerts,

I. Vayas and F. M. Mazzolani (Eds.): PROHITECH 2021, LNCE 209, pp. 1268–1284, 2022.
https://doi.org/10.1007/978-3-030-90788-4_98

wedding receptions and other social events that although fruitful to the community, mask its original conception as a place of silence and solitary contemplation for a small group of Franciscan friars known across Portugal as Capuchos. A more thorough understanding of the building's history is necessary as the convent seeks to better understand its past and continued role within the municipality of Almada.

Since 1558, generations of Capuchos lived in the convent until the religious orders of Portugal were extinguished in 1834. Following this, it was abandoned for over one-hundred years until the renovation project of the 1950s that inserted many new programs and spaces unrelated to the activity of its past. To achieve better knowledge about the original condition of the building without undergoing an extensive and invasive archeological investigation would not be possible. Even still, such an investigation could leave us with an incomplete picture. Because of this, and because no record drawings or descriptions of the building exist prior to the renovation, the research project described in this paper employs an interdisciplinary, mixed-methodological research approach to virtual reconstruction to provide the scientific validation for an eventual HBIM reconstruction of a speculative past condition of the convent prior to the 1950s.

The approach combines historic text analysis and visualization with photogrammetric surveying of details and spaces found in other convents of the Arrábida province across the region. A primary text for the research was the 17th century *Estatutos da Província de Santa Maria da Arrábida* that outlines rules to the construction of all Capuchos convents in the Arrábida province. It includes descriptions of required spaces and strict dimensions to maintain consistent austerity across the order. By visualizing the spaces and relations of the ideal convent as described in the statutes, a comparison was also made between the arrangement of specific instances of convents across the same period, region and Franciscan ideology. Through this, the remaining traces of the present could be linked to the past. Another method of comparison for the study involved the use of photogrammetric surveys to acquire models of common details and spaces as seen between other Arrábid convents in Alferrara, Arrábida and Sintra–a "Franciscan triangle" around Lisbon–as a means to compare and visualize traces of the convent that today are no longer present. Following these studies, the paper concludes by outlining a future HBIM reconstruction that will identify building elements according to various sources of information (text, drawing, site visit) and levels of reliability (certainty or speculation) as recorded throughout the research.

2 History of the Franciscan Orders in Portugal

While the Franciscan communities studied throughout this research focus on a particular group of Arrábid convents in Portugal in the 16th-19th century it should be briefly discussed, that the roots of this particular form of Franciscan life began in Italy and moved accordingly across other regions in Europe. According to Vasconcelos et al. (2016), the Franciscans were first established in Portugal in 1217 following the General Chapter that sent Franciscan friars out of Italy, leading to the creation of several provinces, among them, one in Spain, which at the time, covered all Christian Kingdoms of the Iberian

Peninsula.[1] In their study, they describe that in the middle of the 14th century two consistent tendencies had emerged under the Franciscans across Europe: the Conventual and the Observance. This last group, which share roots to the Capuchos of Caparica in this study, entered Portugal in 1392 with friars from the province of Santiago who vouched to follow more strictly, the rules set by St. Francis for austerity and poverty in everyday life and building, and established their communities in rural locations in favor of prayer and contemplation. In sequence, in 1517 Pope Leo X officially decreed the division of Franciscans into two separate orders. As a result of this decision, two separate provinces emerged in Portugal: the Regular Observance (27 houses headquartered in Lisbon) and the Conventuals (22 houses headquartered in Porto).[2] By 1584, the Conventual Order was definitively suppressed in sequence of a decree of Pope Pio V in 1567. Thereafter, the Observant way of life was imposed to all Franciscan houses, organized into two groups: the Regular Observance and the Strict Observance.[3] The Regular Observance included the provinces of Portugal (founded 1517) and the Algarve (founded 1532–1533) while the Strict Observance included Piedade (founded 1517–1518), Arrábida (founded 1552–1560) and Santo António (founded 1565–1568).[4] The Strict Observance (or the Most Strict and Regular Observance), in this new formulation, aspired to return to a more austere version of Franciscan life. It was these friars that later took the name Capuchos as identified by their pointed hoods.[5]

3 History of Nossa Senhora da Piedade da Caparica

The Nossa Senhora da Piedade da Caparica, was a convent following the strictest observance of Franciscan orders in the Province of Arrábida. In hopes of spreading, Arrábid friars began to seek new settlements across the region, in areas apart from major cities in order to ensure the retreat of friars but close enough for them to develop their pastoral activities and ask for alms. Only hermitages like Arrábida and Santa Cruz de Sintra were completely apart from populated areas such villages and cities. As reported in the exposition catalogue of the convent, *O Convento dos Capuchos - Vida, Memória, Identidade: Catálogo da Exposição*, in 1550 Lourenço de Sousa initiated the first community in Almada, by inviting a small group of friars from the strict observance of Arrábida to inhabit his farm (Quinta da Conceição) in Murfacém.[6] By 1558, this would become a more official settlement in Caparica, given assistance and land from Lourenço Pires Távora, a diplomat and servant of the monarchy, to provide a more suitable environment for the group of friars who sought a more secluded location to carry out worship and that could be self-sustaining for the basic necessities of life for an independent community.[7] There, with a new location they would find the conditions for the full realization of their way of life, including the means of water gathering to support life and in turn building a

[1] Vasconcelos e Sousa et al. (2016).
[2] Vasconcelos e Sousa et al. (2016).
[3] Vasconcelos e Sousa et al. (2016).
[4] Vasconcelos e Sousa et al. (2016).
[5] Caeiro and Fontes (2013).
[6] Caeiro and Fontes (2013).
[7] Caeiro and Fontes (2013).

convent with the typical spaces of cloister, chapel, dormitories, refectory, kitchen, among other spaces, outbuildings, and a perimeter wall to enclose and privatize the community. Such provisions would have been detailed, as will be described further in a later section, in accordance with the *Estatutos da Província de Santa Maria da Arrábida.*

According to various historic documents uncovered and recounted, the convent experienced several adjustments and extensions throughout the sixteenth to eighteenth centuries. The following adjustments are recounted chronologically:[8]

1560: the dormitories, refectory, church and other spaces were adjusted

1560–1598: improvements to the convent's water supply.

1618: the dormitories were rebuilt along with some other spaces.

1630: the original church was enlarged, adding the choir and the porch.

1630: the current façade was installed with the chapel enlargement.

1630: mention of work in the kitchen and pantry.

1755–1759: earthquake damage caused unknown works lasting about 4 years.

The catalogue also provides the following list of known aspects of the building as found across disparate historic texts:[9]

There were likely confessionals in the wall, allowing access from the galilee and into the choir.

A small tower with bell and a clock to provide the rhythms of daily life.

The refectory housed six fixed wooden tables in stonework pillars.

The upper floor housed the cells, connected by a corridor extending from the west side to the choir.

A space connected to the pantry, that housed a small stone mill and two wine-carrying kites.

Anywhere between 12–19 inhabitants lived in the convent throughout any given time. In the first decades of the 19th century, for example, the community included 29 members.

Several more speculative spaces are also discussed briefly in the catalogue:

The library, registry office, the Chapel of the Lord Jesus of Health, possibly located on the right of the galilee as in other convents.

Most of the information from Caeiro and Fontes (2013) is given by the chronicles of the Province, mainly the one of Friar António da Piedade (1737),[10] since the archive of the convent had disappeared after 1834.

In 1834, an imperial decree ordered the extinction of religious orders in Portugal, with many convents, including Caparica, being closed. It is speculated that within the

[8] Caeiro and Fontes (2013).

[9] Caeiro and Fontes (2013).

[10] Piedade (1737).

space of closure for over 100 years, many informal settlements of farmers could have used or modified the space until eventually, the Municipality of Almada purchased the convent, reconstructing it, without documentation over the span of two years ending in 1952. Major interventions would have included the new roofs, the grandiose landscape interventions, the omission of walls to open spaces as in the case of the original dwelling cells. A new chapel, outside the main building was erected over the old hermitage dedicated to S. Peter and many ornaments, art, tiles were brought to beautify the interior spaces and exterior gardens to promote public access and use.[11]

Much of this intervention was of course, far out of keeping with the secluded Capuchos community that originally resided there and little documentation other than a few newspaper articles with exterior photos can be found regarding the as found condition of the building prior to this 1950 intervention. This also includes the absence of any drawings prior to the 1950s intervention. Following this, other public functions were added, including a 1976 temporary primary school, which introduced many modern features of lighting, heating with fireplaces, sanitation facilities and security as well as the large interior spaces that were opened became spaces to host concerts and exhibitions with a new conception of the convent as space of arts and especially to host major Musical Activities.[12] In 1984 the Municipality of Almada installed a Museum which again would have seen adjustments to the facilities and functionality of the space including adaptation of the rooms and inclusion of spaces for administrative functions.[13]

In 2000, another major restoration of the convent took place with the aim of repairing old mistakes and to prevent foreseeable damages in the future. These included corrections of wall pathologies and recovering volumes and traces of the original building – such as the revealing of stonework in the ceiling of the refectory. Other adjustments were the replacement plaster surfaces, roofs, frames, sills and trimmings; along with stone repair and cleansing and waterproofing that attempted to achieve more closely, the primitive spirit of austerity, simplicity and poverty that had marked its construction.[14] Following the permanent exhibition and publication of the catalogue in 2013, the location of the ministry window, between the kitchen and refectory was found.

4 Estatutos da Província de Santa Maria da Arrábida

The *Estatutos da Província de Santa Maria da Arrábida* was a document that provided a common set of rules governing the construction of Arrábid convents and codes of conduct for the communal life of the friars who inhabited them.[15] The rules provided directives to the maintenance of a life of strict observance to the rules of St. Francis that upheld poverty, silent contemplation, communitarian life and pastoral care. Although we know the original version of the *Estatutos da Província de Santa Maria da Arrábida* is lost, a second and refined version, approved first by the Pope in May and by the Arrábida

[11] Piedade (1737).

[12] Piedade (1737).

[13] Piedade (1737).

[14] Piedade (1737).

[15] Estatutos da Provincia de Santa Maria da Arrábida da mais perfeyta Observancia de nosso Seraphico Padre S. Francisco (1698).

Chapter in July of 1697 which provides its closest approximation.[16] This account of history, we know through the chronicles of Fr. Anthony, who reports on this piece of history in *Espelho de Penitentes, Chronica da Provincia de Santa Maria da Arrabida* (1737), a document which contains a reference to the original document and its posterior publications.[17]

A common message throughout the *Estatutos da Província de Santa Maria da Arrábida*, whether in dictating practices of daily life or in the architectural rules, was to live with as little as possible, using the least amount and least expensive materials available. As such, the rules contain various chapters on separate aspects of life, dictating political hierarchies within the community, the procedures of masses, practices of the chorus, rituals of fasting, rules of the wardrobe and the ornamentation of spaces among others. Beyond the internal community, the text also describes the necessary relations to the external public life, suggesting that the location of the buildings should not be too far or too near to the village or city and making other provisions, such as the necessity of a perimeter wall around the convent to make clear the separation of their territory from the city. Beyond these ways of life, the text also dictates architectural aspects of the convents, providing specific planar dimensions concerning a series of conventional spaces: the church, chorus, sacristy, the chapter, kitchen, refectory, pantry, dormitories, library and hostel.

Despite outlining these overall programs and dimensions, however, it appears that there was much leeway in the interpretation of how these spaces could be arranged within each individual convent since the programmatic adjacencies are not described. While the spaces identified within the rules can, for the most part, be seen across all the convent buildings, the relations of spaces may differ according to specific needs of a community or in the necessary adjustments of the spaces to the landscape as seen across the planning of the convents (Fig. 1).[18,19,20,21]

In a first effort, a series of translations of the original text were produced. The measurement systems, using palms and fingers as units of measurement in the document, were converted into modern dimensions using a conversion chart produced by Oliveira Marques.[22] The translations led to a series of digital volumes whose images, along with the text were recorded into a physical document that served as the basis to insert field notes as the visits across the other Arrábid convents were made. When an observation was made in one of the neighbouring sites, notes and references could be added to concretize the findings as a way of record-keeping and means for asking futures research questions to other experts in the study (Fig. 2).

---

16 Caeiro and Fontes (2013).

17 Piedade (1737).

18 Convento de Nossa Senhora da Conceição da Província da Arrábida de Alferrara/Convento de São Francisco de Alferrara/Convento de Capuchos de Alferrara 2011).

19 Convento de Santa Maria da Arrábida/Convento da Arrábida (2011).

20 Convento dos Capuchos/Convento de Santa Cruz da Serra de Sintra/Convento da Cortiça (2011).

21 Convento dos Capuchos de Nossa Senhora da Piedade/Igreja de Nossa Senhora da Conceição/Centro Cultural Costa Azul (2011).

22 Oliveira Marques (1963)

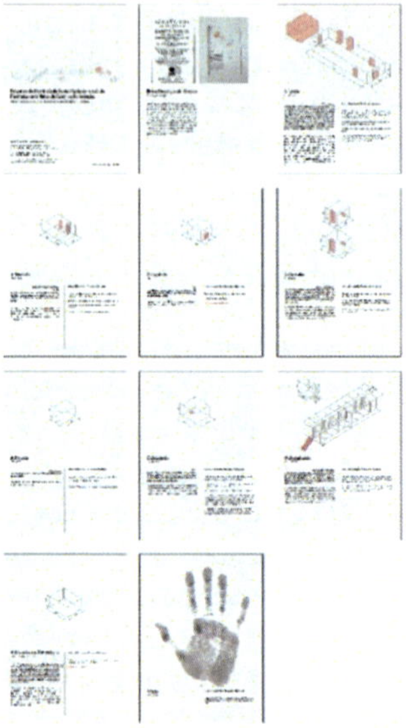

Fig. 1. Pages of the illustrated *Estatutos da Provincia de Santa Maria da Arrábida.*

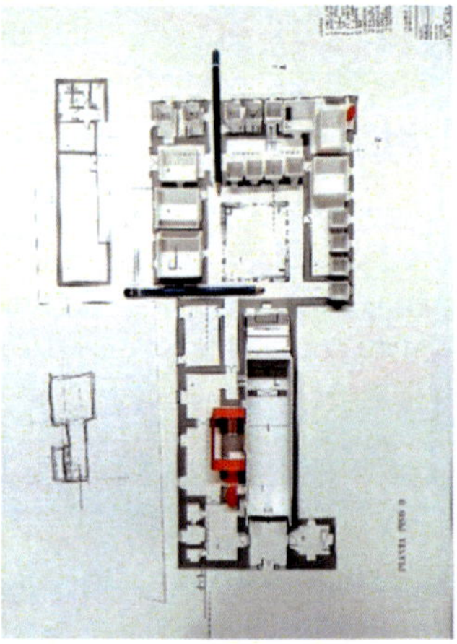

Fig. 2. Physical reconstruction of Caparica 1558 using the 3D printed models of the *Estatutos da Província de Santa Maria da Arrábida.*

Secondly, the series of digital models of the main spaces described and dimensioned by the statutes were constructed as 3D printed models in order to produce and visualize an ideal Capuchin convent as both a tool to help determine the original arrangement of spaces in Caparica, to better understand the other convents in the study and as a final public dissemination artifact. The hypothesis of this exercise was that the physical model could allow a more inclusive platform of exchange between architects, historians and the general public in proposing hypothetical arrangements of the spaces as the project continues over time and as a way of demonstrating visually, the gaps of information about the building. The 3D printed objects allowed physical manipulation and easy rearrangement of the volumes on top of a series of 2D plan drawings printed at the same scale as the 3D printed objects, of each of the convents studied in order to compare dimensions between the statutes and the real conditions. The study was produced also to allow a better means of proposing hypotheses about how the arrangement of the pre 1950 spaces of Caparica could have been.

In order to differentiate between the dimensionless and the dimensioned volumes written in the text, elements of the model were printed in different colours. The red pieces are spaces not defined by specific dimensions but that are mentioned as supporting spaces to other programs. The white pieces are the spaces described by actual dimensions in the original text of the Estatutos.

5 Case Studies of Other Convents (Alferrara, Sintra, Arrábida)

The provisions contained in the statutes were an important departure point to the study of Caparica and the neighboring convents by providing a common set of spaces and dimensions that could support evidence for highlight missing or altered spaces. By investigating the relationship first, between the ideal convent as described in the statutes and real convents in sites such as Sintra, Alferrara and Arrábida, as documented in drawings, photogrammetry and site visits, it was possible to ascertain what common grounds existed or not between the convents as a way to speculate on the missing elements in Caparica (Fig. 3).

Fig. 3. Map of sites studied throughout the research.

5.1 Alferrara

The Capuchos convent in Alferrara today remains one of the closest approximations to what the convent of Caparica could have looked like and how the spaces of the convent were organized prior to the 1950 restoration. Like in Caparica, aspects of the experiences and activities of the monks are not easily known, with little record keeping across history. As far as it is known however, the building has not undergone any major renovations since the extinctions of the order in 1834, other than the addition of a series of woodworks and treatments of the windows as a first phase of an eventual renovation.[23] The major difference of the convent in Alferrara, which also followed the rules of the statutes closely, was the location of the spaces in close proximity to a steep terrain on a narrow plot of land which caused a completely different configuration to appear between the cloister and the church than is seen in Caparica and Arrábida. This decision, it is assumed, was due to the importance of having adequate and natural waterlines for the site as a crucial element in the selection of all Arrábid convents.[24] Other aspects, however, remain the same between both convents, such as the extension of the façade in the 17th century with a Serliana entrance and choir above the entrance on the same level as the dormitories placed on the second level (Fig. 4).

Fig. 4. 3D prints comparing the ministry window of Alferrara (left) and Sintra (right).

What this building allows us to see, in its state of ruin, is material aspects, the remains of azuelos tiles, ceiling frescos, stonework and other detailing that could have been present in the convent of Caparica as well. The ministry window between the refectory and the kitchen is another example of how a similar detail could have looked within Caparica, but that today, we see only a trace through the infill of a masonry wall. We also see evidence of side altars and their materiality which allows some level of verification that the altars in Caparica are in fact likely kept in their original size and dimension.

5.2 Sintra

The Convent of the Capuchos in Sintra, also known as the "Convent of the Cork" and the Convent of Santa Cruz da Serra de Sintra, among others, was founded in 1560 by D.

[23] Pessoa et al. (2017).

[24] Pessoa et al. (2017).

Álvaro de Castro, a counselor of the State of King Sebastião. Like the other convents, it was also abandoned in 1834 as a result of the extinction of the religious orders but was soon acquired by the first Viscount of Monserrate Francis Cook in 1873, and later in 1949, by the Portuguese State.[25] Today the convent is part of the Cultural Landscape of Sintra, and since 1995 was classified by UNESCO as a World Heritage Site. As part of this continuous ownership, the building remains in a well-maintained state with much of its woodwork and cork-work, still intact.

Unlike the other convents in the study, the convent in Sintra was built without the intention of public access by any means which severely altered the arrangement and proportions of some spaces such as the chapel and choir. This aspect also resulted in the omission of other features that would normally appear in other convents such as confessionals. Furthermore, it is known, more than the other convents for its higher degree of poverty, as evident in the close relationship of the building to the natural terrain and due to the smallness of the spaces and corridors that were the result of a less exuberant use of materials. In this convent, the difference between human-made and natural is often confused as the building literally situated within the rocks, used the natural formations of the interior spaces in the rock to guide the planning of the spaces. This led to a much more extreme arrangement in-situ with the landscape that is evident in the many levels and spaces between levels that can't be found in the other convents.

Although the convent in Sintra was the result of an entirely different agenda than the others studied, the evidence of the statutes as a guiding aspect for the convent is evident. Despite having no spaces for public functions other than those of the community of friars, the essential ingredients of the statutes are present. When measuring the distance of the sink and the chimney, it was discovered that the same correspondence mentioned in the statutes is articulated. Furthermore, expressions such as the detailing of the cells, the ministry window and the seat in the refectory for the recital of the lessons of the table were other aspects of comparison that were no longer visible in Caparica. These traces, as captured by photogrammetry on the site visit, add to the understanding of how the inhabitants of Caparica could have dwelled.

5.3 Arrábida

The Convento de Santa Maria da Arrábida acted as the seed for other Capuchos communities of the region at the period of history. The land on which the convent sits was provided by the first Duke of Aveiro, D. João de Lencastre to Friar Martinho de Santa Maria, who founded the original convent in 1542.[26] Originally, the Memória Shrine, which existed on the land prior to the original convent was also the destination of major pilgrimages and the place where the first four Arrábida friars lived for two years in dormitories that were carved into the rocky hillside.[27] Although today, the convent is defined by two sections, the old and the new, the growth of the convent would have been perceived more organically at the time, with the same basic lifestyle sought and

[25] Convento dos Capuchos/Convento de Santa Cruz da Serra de Sintra/Convento da Cortiça (2011).

[26] Pereira and Benito (2006).

[27] Pereira and Benito (2006).

maintained by an expanded community who sought a higher degree of isolation.[28] What today is known as the new convent, thus was simply a series of additions to make the isolation and autonomy of the community life possible. Much unlike the convents in Caparica and in Alferrara, the spaces of the convent appear much more intricately connected to the high slope of the landscape, with a more disperse allocation of rooms than being contained within one volume. The convent also maintains much of the woodwork intact, allowing us to see unique architectural elements such as the revolving wood door of the ministry window and the detail of a swinging plank of wood built into the wall of the refectory (both unique details to this convent), used to signal the activities of the day through a series of coded knocks. Other spaces such as the sacristy are well kept, showing the woodworking of the cabinets and details such as the holy water font (Fig. 5).

Fig. 5. Examples of photogrammetric models made during site visits. Top row: Alferrara, middle row: Sintra, bottom row: Arrábida.

6 Principles of Virtual Reconstruction and a Strategy for HBIM

As in any case of virtual reconstruction, and especially considering the speculative nature of many of the aspects of reconstruction involved in Caparica, it was necessary to consult existing frameworks and principles for the validation of digital cultural heritage and virtual reconstruction projects. Three documents that helped support the framework of our methodology were: UNESCO's Charter on the Preservation of digital heritage (2003), The London Charter (Charter 2012) and The Principles of Seville (2011). Considering

[28] Pereira and Benito (2006).

the range of datasets, and overlapping historical periods, an HBIM strategy to properly organize and present the information was necessary. Identifying datasets, historical phases, levels of speculation.

6.1 Accessibility of Information Across Time

In recognition of a new legacy of digital heritage, the UNESCO Charter on the Preservation of Digital Heritage provides principles to maintain, increase, disseminate, preserve and make accessible, the documentation of heritage worldwide.[29] As such, the UNESCO charter was written as a direct response to the challenges that a digital era poses on digital heritage documentation and the possibilities of loss that could result from the rapid movement and uncareful use of technology in the field. Factors they identified include the obsolescence of hardware and software, a lack of frameworks for preserving data, accounting for aspects of resource uncertainty, the responsibility of maintaining and storing information and the overall absence of legislative support to govern these aspects. The main purpose of preserving digital heritage as stated by the UNESCO document, is long-term accessibility and limited restrictions to the access of knowledge produced. As such, the Charter advocates that measures should be made throughout the entire process, from content creation to the aftermath of any digital artifact to ensure access to the digital information.

Such a practice necessitates a schematic system and set of procedures to accompany the workflow in order to provide authentic and stable digital objects. One aspect of this activity in the case of Caparica has been the use of BIM itself which, through its IFC compatibility provides the strongest grounds today for the use and operation of the data across multiple software platforms. Also, the production of physical scale models and 2D graphic materials that can supplement for scenarios in which the access to the software is made difficult to the public, has been part of the ongoing consideration within the methodology.

6.2 Uncertainty

The London Charter provides a set of principles that provide guidelines to ensure the rigour of the processes involved computer-based visualizations of heritage.[30] The principles advocate a range of practices aimed to ensure, among many aspects, the intellectual integrity of visualizations, the reliability of information, the thorough documentation of methods and decisions, the sustainability of the output for future generations as well as provisions for easy public accessibility. One area of significance for the many states of uncertainty involved in the reconstruction of Caparica, is the degree to which aspects of the visualization can be labelled certain. The London Charter provides that a clear differentiation between the different conditions of the existing (as-found) state, an evidence-based reconstruction and a hypothetical reconstruction should be made evident.

---

29 United Nations Educational Scientific and Cultural Organization (2003).

30 London Charter (2012).

It is evident that within these cases of uncertainty, it is necessary to identify clearly, the extent and conditions that create these various levels of probability and speculation. Similar outlooks on the need for uncertainty representation have been expressed from the cultural heritage research community. A research framework for cultural heritage research proposed in 2012 aimed to extend predictive rendering to include a temporal component as well as address uncertainty arguing it is a necessary consideration for heritage modellers.[31] A summit on Digital Tools for the Humanities at the University of Virginia in Charlottesville, Virginia, on September 28–30, 2005 for example expressed the need for an, "integrated suite of tools," to collect, evaluate, date and represent the distinction between certainty and uncertainty.[32] Koller et al. (2012) also vouch for an open source of scientifically authenticated 3D reconstructions, suggesting there is an unfulfilled need for techniques to clearly communicate the uncertainty in cultural heritage visualizations and that the lack of clarity in these uncertainties is one of the issues interfering with this realization.[33]

One archaeological case study involving the representation of uncertainty is the Wesleyan-Brown Monastic Archaeology project. The project is a multidisciplinary collaboration aimed at understanding and reconstructing monastic life at the abbey of Saint-Jean-des-Vignes, Soissons, during the medieval and early modern periods. Visual solutions for the representation of uncertainty in this case consisted of the use of dotted lines, varying line weights, degrees of transparency or color-coding, wireframe, texture abstraction and non-photo-realistic representation.[34] Another virtual reconstruction project in 2006 was presented for the hunting residence of Mary of Hungary, at Mariemont that was no longer existent. For the reconstruction, authors relied heavily on surveys of the original construction translated into words and a significant quantity of iconographic source material to produce the reconstruction. In this case uncertainty is represented as a reconstruction showing every building part and its corresponding degree of certainty, the levels of uncertainty in all of the modeled elements were made transparent and a scale from green to red was used to depict the degrees of uncertainty wherein green represented the highest degree of certainty, red for hypothesis and yellow or orange being used to represent levels of uncertainty between the two extremes.[35] This example, while expressing varying degrees of uncertainty clearly, is a difficult model to comprehend through 2D images unless the user has fluency with the modelling software and or is able to navigate throughout the model space.

Another case study, the digital reconstruction of the Vinohrady Synagogue in Prague demolished in 1951 is one example where reliance on BIM software to model and represent architectural cultural heritage was explored. The software used in this reconstruction was Graphisoft ArchiCAD. It was described as a valuable approach to visualizing reconstruction assumptions as a 3D representation. Use of the tool *Renovation Filter* was explored to show or hide objects according to the filter selected although its intentional use is meant for contemporary renovations which limited the modellers to three types of

[31] Happa et al. (2012).
[32] Report on Summit Accomplishments (2005).
[33] Koller et al. (2009).
[34] Bonde et al. (2009).
[35] Vandevyvere et al. (2006).

fixed properties (existing, to be demolished and new). The users suggest potential for the tool to develop for more "fine-grained 'tagging', e.g. to show different building stages."[36] In other research, such aspects of uncertainty in modelled elements has been achieved in a HBIM through the adoption of custom project parameters that can be defined for all modelled elements.[37] This project, using Revit 2019 found a solution to the limitations of phasing from the example above, building upon custom project parameters and project views that can be used in 2D visualizations for public dissemination. Building on the research in this paper that mostly dealt with as-built conditions as opposed to a virtual reconstruction of a speculative past condition of a building, the project parameters of an HBIM of the convent at Caparica would be more intimately related to the levels of certainty that each element of the BIM is given. For this, the necessity to link objects to their data source is a necessary element in the scientific validation of the BIM.

Similar to the London Charter, the Principles of Seville, aim to provide basic principles to help oversee the range of practices conducted within the broad and increasingly expansive field of digital heritage.[38] The guidelines expand further upon the principles already set forth within the London Charter with the specific aim of improving its implementation into the specific domain of archaeological heritage. One aspect of this is providing clearer distinctions between relevant fields of virtual archeology and virtual reconstruction. For the work in Caparica, virtual reconstruction was the most appropriate definition provided by the Seville document, which describes it as using techniques of digital visualization to demonstrate a building or object in a condition of the past that can be based upon, "available physical evidence of these buildings or objects, scientifically-reasonable comparative inferences and in general all studies carried out by archaeologists and other experts in relation to archaeological and historical science."[39] As such, the fourth principle found in Seville: authenticity, expands upon the notions of speculation and probability highlighted previously in the London Charter, suggesting that the scientific basis of archaeology in principle are not exact and should aim to openly provide for numerous alternatives of any virtual interpretation in cases where the same scientific validity could allow more than one possibility.[40] In the case of Caparica, where many possible configurations could exist for a given time period, an HBIM methodology that can construct equally plausible histories should be considered. Specifically, the project phasing tool of Revit can be adopted to represent multiple variants of the same historic period while still maintaining the functionality of custom parameters for labelling aspects of certainty as described above (Fig. 6).

36 Boeykens et al. (2012).
37 Chow and Fai (2017).
38 The Seville Principles (2011).
39 The Seville Principles (2011).
40 The Seville Principles (2011).

Fig. 6. HBIM of 1630 facade produced from point clouds from laser scanning and photogrammetry surveys in Revit 2019.

7 Conclusions

Beyond the main agenda of a HBIM reconstruction of the pre-1950s convent, the overall methodology of the study has prompted a range of interrelated research outputs helpful to historians and the public in understanding the history of the convent in Caparica and the Arrábid history in a broader sense. From the results of the study of the statutes and physically reconstructing two phases of the convent, several gaps in the knowledge of the building were revealed, including cases where spaces were much larger than dictated by the statutes or where large unexplained spaces after the 1630s were identified. The missing gaps identified provide a specific set of research questions that can be addressed visually to the appropriate experts involved in future studies. The photogrammetry and 3D print models of the "ideal convent" and the specific details captured during site visits can serve as a means to educate the public on the way of life of the Capuchos as a whole and elements of the past life of the convent in Caparica. The research has contributed to a means of scientifically validating the elements of speculation and certainty for an eventual HBIM reconstruction that will use the various sources of information gathered throughout the process. As seen in this paper, a range of datasets were used throughout the research including historic text, historic photographs, photogrammetric models, and drawings that can be used to verify an element in the BIM according to its level of certainty through the use of custom project parameters. This process provides a necessary means of record keeping, to ensure all elements of speculation are sourced according to the guidelines of computer-based visualization as recited in the principles of UNESCO, the London Charter and The Seville Principles as a means of a faithful historical evolution and transfiguration of the built heritage.

Acknowledgements. The activity presented in this paper was made possible by funding from the "New Paradigm / New Tools for Architectural Heritage in Canada" training program funded by the Social Sciences and Humanities Research Council (SSHRC).

References

Caeiro, N., Fontes, J.: O Convento dos Capuchos - Vida, Memória, Identidade: Catálogo da Exposição. Câmara Municipal de Almada, Almada (2013)

Chow, L., Fai, S.: Developing verification systems for building information models of heritage buildings with heterogeneous datasets. The International Archives of the Photogrammetry, Remote Sensing and Spatial Information Sciences, Volume XLII-2/W5, 2017 26th International CIPA Symposium (2017)

Convento de Nossa Senhora da Conceição da Província da Arrábida de Alferrara/Convento de São Francisco de Alferrara / Convento de Capuchos de Alferrara. Sistema de Informação para o Património Arquitetónico (2011). http://www.monumentos.gov.pt/site/app_pagesuser/SIPA.aspx?id=10470

Convento de Santa Maria da Arrábida / Convento da Arrábida. Sistema de Informação para o Património Arquitetónico (2011). http://www.monumentos.gov.pt/site/app_pagesuser/SIPA.aspx?id=3447

Convento dos Capuchos/Convento de Santa Cruz da Serra de Sintra/Convento da Cortiça: Sistema de Informação para o Património Arquitetónico (2011). http://www.monumentos.gov.pt/Site/APP_PagesUser/SIPA.aspx?id=6410

Convento dos Capuchos de Nossa Senhora da Piedade / Igreja de Nossa Senhora da Conceição/Centro Cultural Costa Azul. Sistema de Informação para o Património Arquitetónico (2011). http://www.monumentos.gov.pt/Site/APP_PagesUser/SIPA.aspx?id=20051

Koller, D., Frischer, B., Humphreys, G.: Research challenges for digital archives of 3D cultural heritage models. ACM J. Comput. Cult. Heritage **2**(3), 7 (2009)

Estatutos da Provincia de Santa Maria da Arrábida da mais perfeyta Observancia de nosso Seraphico Padre S. Francisco. Na Officina de Miguel Deslandes, Lisboa (1698)

Vandevyvere, H., Neuckermans, H., De Jonge, K.: Digital historic reconstruction: case studies of an interdisciplinary task. Int. J. Des. Sci. Technol. **13**(1) (2006)

Happa, J., Bashford-Rogers, T., Wilkie, A., Artusi, A., Debattista, K., Chalmers, A.: Cultural heritage predictive rendering. Predictive Rendering Comput. Graph. Forum **31**(6) (2012)

London Charter: London Charter for the Computer-based Visualisation of Cultural Heritage (2012). http://www.londoncharter.org/introduction.html

Oliveira Marques, A.H.: Pesos e medidas. In: Serrão, J. (ed.) Dictionário de História de Portugal, vol. V, pp. 67–72. Livraria Figueirinhas, Porto (1963)

Pereira, P., Benito, P.: Convento da Arrábida: a Porta do Céu. Fundação Oriente, Lisboa (2006)

Pessoa, F., Rodrigues Delgado J., Aleixo, S., Pina, T., Mestre V.: Cadernos de Alferrara. Associação de Municípios da Região de Setúbal (2017)

Piedade, A.: Espelho de Penitentes e Chronica da Provincia de Santa Maria da Arrábida, Tomo Primeiro. Lisboa Occidental: Na Officina de Joseph Antonio Da Sylva (1737)

Report on Summit Accomplishments. Summit on Digital Tools for the Humanities (2005)

Boeykens, S., Himpe, C., Martens, B.: A case study of using BIM in historical reconstruction – The Vinohrady synagogue in Prague. eCAADe, Edition 30 (2012)
Bonde, S., Maines, C., Mylonas, E., Flanders, J.: The virtual monastery: re-presenting time, human movement, and uncertainty at saint-Jean-des-Vignes, Soissons. Visual Resources **25**(4) (2009)
The Seville Principles: International Principles of Virtual Archaeology (2011). http://smartheritage.com/seville-principles/seville-principles
United Nations Educational Scientific and Cultural Organization: The UNESCO Charter on the Preservation of Digital Heritage (2003). http://portal.unesco.org/en/ev.php-URL_ID=17721&URL_DO=DO_TOPIC&URL_SECTION=201.html
Vasconcelos e Sousa, B., Pina, I., Andrade, Ferraz de Oliveira Silva Santos, M. L. , 2016.Vasconcelos e Sousa, B., Pina, I., Andrade, Ferraz de Oliveira Silva Santos, M.L.: Ordens Religiosas em Portugal: das Origens a Trento. Guia Histórico. Third edn. Lisboa: Livros Horizonte (2016)
Walsh, M., Bernardello, R.A.: Heritage Visualisation and Potential Speculative Reconstructions in Digital Space: The Medieval Church of St. Anne in Famagusta, Cyprus. Disegnarecon **11**(21) (2018)

The Bulgarian Architectural Complex at the Golden Horn in Constantinople

Marina Traykova[1](✉) and Blagovesta Ivanova[2]

[1] University of Architecture, Civil Engineering and Geodesy, Sofia, Bulgaria
marina5261@abv.bg
[2] University of Structural Engineering and Architecture "Lyuben Karavelov", 175 Suhodolska Street, Sofia, Bulgaria
bg_fineart@abv.bg

Abstract. The investigation discusses the history of construction of the Orthodox Christian architectural complex St Stefan in Constantinople, comprising a steel church and a convent. The paper pays special attention to the third stage of the creating of the complex: the steel structure of prefabricate elements and the ornamented cast-iron cladding manufactured and fitted by the Vienna-based company Rudolph Philipp Waagner in 1896–1898 on a design of Turkish architect Hovsep Aznavur. The church is a monument of cultural heritage of Bulgaria and Turkey. The study is a result of in situ work and of an analysis of unpublished records kept in the Central State Archives in Sofia.

The paper deals with some structural features of the church. The application of the steel for structural purposes is an important step in the structural design. The concept of the structure is analyzed and some general conclusions are presented based on the comparison of the church with similar steel structures from this period.

Comparisons are made between the St Stefan Church, the steel churches in the other countries as Latvia, the Philippines, Peru, Chile, Mexico, Costa Rica, and Russia that are a product of the industrial development in the 19th century and now are the rare architectural monuments.

Keywords: Orthodox Christian architectural complex St Stefan in Constantinople · Prefabricate elements · Similar steel structures from the late of 19th c

1 Introduction

The construction of the St Stefan church had three periods in its historical development. The first period begins in 1849. It is the time of the timber structure. This period coincides with the beginning of the reforms in the Ottoman Empire known as the Tanzimât Reform Era.

The second period overlaps with the mid-19th century and the aftermath of the Crimean War. This is also the time of struggle for the independence of the Bulgarian church which was destroyed as an independent institution when Bulgaria was conquered

I. Vayas and F. M. Mazzolani (Eds.): PROHITECH 2021, LNCE 209, pp. 1285–1293, 2022.
https://doi.org/10.1007/978-3-030-90788-4_99

by the Ottoman Empire. This is the period of the masonry church. The architects who designed the church are Gaspare and Giuseppe Fossati from Morcote. They are of Swiss-Italian origin. They have worked in the Ottoman capital on the project of the Russian embassy and the restoration of the grate byzantine church St Sofia, transformed at the time of the Ottoman Empire to a mosque.

The duration of the third period in the life of the church edifice is between 1870 and 1898. The assembling of the St Stefan church took place during the Liberation (1878) period and before the Independence (1908) of Bulgaria. The building is very important for the history of Bulgarian Orthodox Exarchate which at that time is in Istanbul (Fig. 1).

Fig. 1. The map of Eastern Europe in 1878. https://en.wikipedia.org/wiki/Treaty_of_Berlin_(1878)

The used steel structures of that period have the connection with the synchronic processes of building of the modern temporary structures. Many steel churches have been executed in different points of the world. They are the temples of the people migrating to the "new lands" of the Latin America or Philippines. Some of the public buildings (railway stations, markets, industrial buildings, etc.) during this period are with steel structures as well. These trends are presented in the creation and the function of the Cristal palace (1851, cast-iron and plate-glass structure of Joseph Paxton), the Gallery of the Machines in Paris (1889) assembled of iron, steel and glass by the French architect Charles Louis Ferdinand Duter. These examples related to the Modern époque are part of the time of the application of the new principles for space organization and modern functions of the buildings.

Studying St Stefan Church, so far, it is important to analyse the question considering its construction as a part of a bigger architectural complex. The steel church was planned as part of a larger complex also including a new presbytery (convent) and a school, as it was meant to become a spiritual and educational centre for the Bulgarians living in the capital of the Ottoman Empire. The presbytery was, also, designed by Hovsep Aznavur, in the late 19th century (1892 and 1894), with Neoclassical bearing walls. Unfortunately, the school and the new presbytery were not built due to lack of money. The second main topic is related to the construction of the church, which was erected as a temporary edifice in the embankment, placed between the Byzantine fortress wall and the Golden horn in the third quarter of the 19th c.

2 Vienna Steel Factory Rudolph Philipp Waagner and Style Eclecticism of the Époque

The steel elements for the structure and outer wall decoration of the St Stefan church are produced by the famous Vienna-based company Rudolph Philipp Waagner in the 1890s, according to Hovsep Aznavur's projects. This period is related to the first steps in the development of the steel architecture in Europe. The style of the church was on the borderline between historicism and the innovative tendencies of the new form-formation and the ensuing new use of metal.

From a structural and artistic point of view, the St Stefan Church integrates all the achievements of the European architecture from the very first use of cast-iron structures in the 18th century till the representation of the aesthetic properties of the material. The steel shows the advantages as a structural material but as well as a traditional material for decorative features as garlands, angels and typical Christian symbols. This came as a result of attempting to hide the structure behind metal walls which was done in the spirit of historicism and decorated in the style of Classicism.

The creation of St Stefan coincides with the end of the period of historicism in architecture when steel was a material *per se,* when its advantages are highlighted in combination with glass, and when new lighter and transparent forms have appeared, as a beginning of the Modernism. Adopting the new mastery in manufacturing and assembling of steel structures, as well as employing a new form-formation after the advent of metals as new building materials, e.g. The Cristal Palace in London (1851)|, reveals that the structural solutions for the St Stefan church is an accomplished phase of their utilization (Fig. 2).

Fig. 2. The control assembling of the St Stefan church in the Vienna factory yard, 1894. The photo has the encryption on the on the back in German: "Rudolph Philipp Waagner Design Office Iron and Enamel Works" and was taken by the architect Yanaki Shamarjiev from Vienna as he marked in his report about the assembling of the constructions. (CSA, Fund, 166 K, Inv. 1, Archival unit 120, Sheet 20).

3 Basic Information for the Steel and the Steel Profiles of the Structure

The significant point in relation to the profiles used in the structure is the guarantees for the quality of the metal. It was requested by the Bulgarian state architects, who perform the technical supervision of the quality of the steel. Details of its certification are provided. The certificate was issued in Vienna in 1894 by the Royal Imperial Technical Museum of Craft Technology, and four pieces of tensile-tested steel are the basis for this certificate.

Attached are two tables for profiles and flat metal (probably considered steel sheets). The length of the specimens is 30 cm for the standard test and 20 cm for the test for the ductile behavior. The results shown (yield strength, fracture limit, relative elongation, etc.) give rise to the definition of the steel type as well as its classification as manganese steel (with manganese alloying element), subsequently confirmed by X-ray analysis. (Kalchevska and Ganev 2008). The certificate is one of the most important documents concerning the characteristics of the materials. Four bars are tested in the laboratory for grading of construction materials and machineries. (Ganev, Hristov and Barcheva 2009, 221–225; Ivanova 2018, 275, 281, 158; Ivanova and Ganev 2009) (Fig. 3).

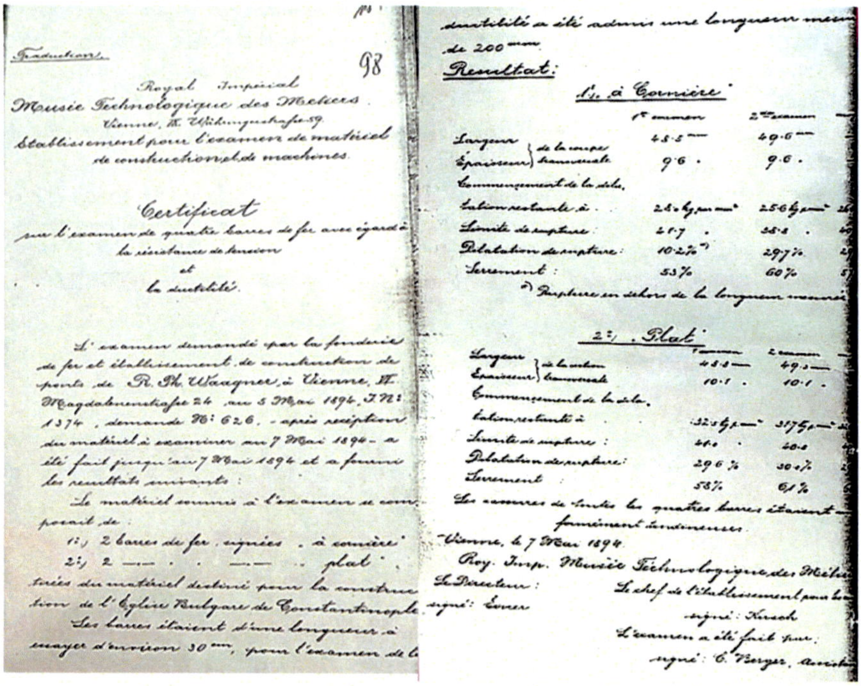

Traduction.

98

Royal Impérial
Musée Technologique des Métiers.
Vienne, IX. Währingerstrasse 59.
Établissement pour l'examen de matériel de construction et de machines.

Certificat
sur l'examen de quatre barres de fer avec égard à la résistance de tension et la ductilité.

L'examen demandé par la fonderie de fer et établissement de construction de ponts de R. Ph. Waagner, à Vienne, IV. Magdalenenstrasse 24, au 5 Mai 1894, J. N: 1374, demande N: 626, – après réception du matériel à examiner au 7 Mai 1894 – a été fait jusqu'au 7 Mai 1894 et a fourni les résultats suivants:

Le matériel soumis à l'examen se composait de:

1:) 2 barres de fer, signées, à cornière
2:) 2 — " — " — " — plat

tirées du matériel destiné pour la construction de l'Église Bulgare de Constantinople.

Les barres étaient d'une longueur à essayer d'environ 30 cm, pour l'examen de la ductilité a été admis une longueur mesurée de 200 mm.

Résultat:

1:) à Cornière

	1er examen	2me examen
Largeur } de la coupe	45.5 mm	49.6 mm
Épaisseur } transversale	9.6 "	9.6 "
Commencement de la dilatation,		
tension restante à	28.0 kg par mm²	25.6 kg p. mm²
Limite de rupture	41.7 "	38.8 "
Dilatation de rupture:	10.2 %*)	29.7 %
Serrement:	53 %	60 %

*) Rupture en dehors de la longueur mesurée

2:) Plat

	1 examen	2 examen
Largeur } de la section	45.8 —	49.5 —
Épaisseur } transversale	10.1 "	10.1 "
Commencement de la dilatation,		
tension restante à	32.5 kg p. mm²	31.7 kg p. mm²
Limite de rupture:	44.0 "	40.8
Dilatation de rupture:	29.6 %	30.0 %
Serrement:	58 %	61 %

Les cassures de toutes les quatres barres étaient fermement tendineuses.

Vienne, le 7 Mai 1894.

Roy. Imp. Musée Technologique des Métiers
Le Directeur: signé: Exner
Le chef de l'établissement pour la [illegible] signé: Kirsch
L'examen a été fait par: signé: C. Berger, assistant

Fig. 3. The Protocol from the Royal Impérial Musée Technologique des Métiers. (Central State Archive (CSA), Funds 166 K, Inv. 1, Archival unit 120, Sheet 26–41. Vienne, IX Währingerstrass 59. Establishment pour l'examen de materiel de construction et de machines. Certificat. 7 Mai 1894.)

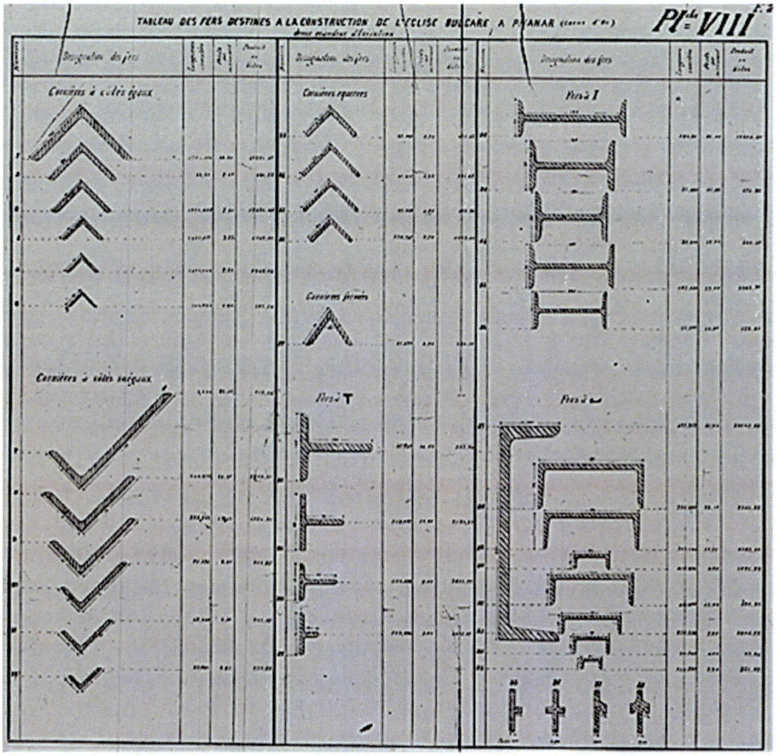

Fig. 4. Table of the profiles used in the structure. (Tableau des fers destines a la construction de l'Eglise Bulgare a Phanar.) (CSA, F. 246 K, inv. 3, a. u. 340, sh. 24–25; 284 K, F. 321 K, inv. 1, a. u. 473, sh. 174–189 (drawings of all projects), sh. 143–144)

The table of the profiles used in the structure gives an information concerning the type of the profile, the geometry, the total length of the used profiles (Fig. 5), the total weight of the different types. The presented table clarifies many specific features (Fig. 4).

Fig. 5. Structure at the level of the church choir of the St Stefan church. Photographer Blagovesta Ivanova, 29 September 2018. Semicilindrical vault upper the choir of the St Stefan church. Photographer Blagovesta Ivanova, 29 September 2018.

4 The Geotechnical Tests and the Control of the Embankment Terrain

The issue with the terrain perpetuated as the most complex one over the years, for the land borders the waters of the Golden Horn. The field surveys were carried out in the 70^{ies} and the 80^{ies} of XIX century by Istanbul architects and engineers of Italian and Greek descent. They work together with Bulgarian engineers and architects from the Ministry of Construction in Sofia.

In 1876 it was found that at the foundations of the unbuilt masonry church there was a crack, which was due to the slip of the terrain to the water of the Golden Horn. According to Istanbul engineer George D. Stampa, the landing of the terrain is due to the not so solid and short piles, which are subject to decay and poorly executed top structure. The architect recommends that, during the reconstruction, the construction of a solid quay of large stone blocks is first to strengthen the terrain. Georges Stampa does not propose a new site for the new temple. He worked on the option of laying it on the old stone foundations after their strengthening and partly on new ones. Stampa recommends his variant for the construction materials to be used in the new temple. According to him, the part of the temple that will rest on the new foundations should be made of cut stone and the part that will be laid on the crack should be of hollow bricks and use timber vaults for lightening the self weight of the structure.

In 1882, after a delay in the on-site investigation, the architect Perpigniani and the architects - engineers Alphonse Cingria and Frangoulis Eustratius Favoulis E. Mavrosordato, decided to make new field drilling. They examine the foundations and the cracks in them, the condition of the piles heads and the iron grid on them.

Singria and Mavrocord note that they have found that the piles are of unsatisfactory diameter, the masonry on the foundations of the temple is cracked along its entire length parallel to the sea, proving sliding and settlements. They believe that it is necessary to open an excavation along the foundations. The place where they mark on the situational plan of the old foundations is not preserved. The final opinion of the three architects is, for the above reasons, not to use the remains of the masonry of the Fosati brothers' church and to remove them. From here it appears that they do not indicate a place for the future temple and their opinion is for the complete removal of the old foundation and for the new strengthening of the terrain.

The Bulgarian engineer Georgi Mikhailovich, who make the control at the Ministry of Construction in Sofia in 1882, noted that the weight of the temple depends on the plan of the approved project, and the foundation of the building should stand on solid ground, which corresponds to the weight of the designed structure. Its condition must be determined by drilling or excavation. The load should be determined by the weight of the temple and the quality of the terrain. According to Mikhailovich, the specified location of the new temple will determine how much of the old foundations must be demolished, but this cannot be done before drilling is carried out.

The damaged foundations of the masonry church from the 50^{ies} of the XIX century and the lack of funds do not allow the execution of a new deep foundation. For this reason, the decision to build a lighter temple is taken just to prevent the overloading of the terrain. This determines the use of new steel structure.

Istanbul architects and engineers suggest the use of piles with a grillage top structure. Bulgarian specialists and Hovsep Aznavour also propose the use of iron sleepers and "jacks", if, as a result of a general settlement, the stone base eventually loses its level, it will be regulated by a system of jacks connected to the frame at different points.

This concept of the load-bearing structure is innovative. The idea of the jacks to maintain the stability and the balance is used. All specialists agree on the proposal that the compromised foundations should not be used.

Just before the new construction began, the foundations were made with timber piles, and those from the previous construction were removed.

Some actual investigations concerning the foundations are made as following: prof. Venkov 1985 (special geological profile of the site), and eng. Gilbert Wiplinger 1989 (proposal for drainage system around the church) (Scientific Archive of Institute of for immovable cultural heritage of Bulgaria; Tapan and Kuruyazici 1991) (Fig. 6).

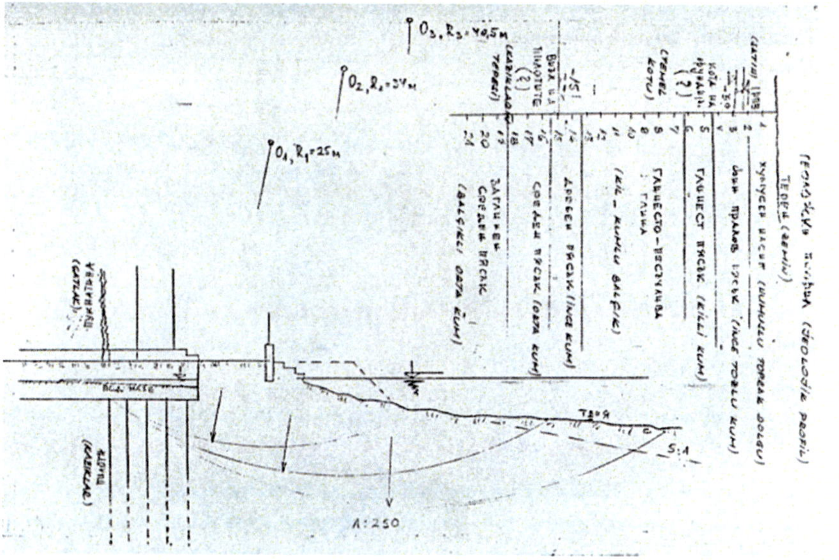

Fig. 6. The Geological Survey of prof. Venkov, published in: Kuruyazici, H., M. Tapan. Sveti Stefan Bulgar Kilisesi. Istanbul, 2009, s. 94–95.

According to Benjamin Mouton, in the 1980s and 1990s, dredging was carried out at the bottom of the Golden Horn. He argued for the need to initially strengthen the site in the area of decay, after which he recommended that the behavior of the church should be monitored and if necessary, to strengthen the foundations of the temple. With regard to conservation and restoration activities, arch. Mutton believes that work is needed to strengthen the foundations.

In 1985, in the basement of the church to the south wall, on proposal of Prof. Venkov, a study was conducted to investigate the depth of the foundations, the water level, the diameter and the condition of the piles' heads in front of the site of the main crack. (SAIICH, T-I-1254-115/12.2001).

The 2002 Geological Survey confirms prof. Vesselin Venkov's study of the composition of the terrain, namely that it consists of sand and clay and the piles are of insufficient length to prevent slipping. With the latest foundations done in 2005, the inclination of the building and the cracking of the structure with visible traces on the facade metal panels were stopped. (Ivanova 2018, 365–372.)

At that time, 192 piles were placed in the terrain to improve its bearing capacity around the building and 90 more in the most vulnerable part of the sea using the jet-grout method. It is known that the application of such technology is based on the principle of injecting solutions into the base to fill gaps in the loss of timber piles. (SAПCH, T-2-2-I ПСД).

Some actual investigations concerning the foundations are made by Immo Boyken (Boyken 1989), and H. Kuruyazigi-M. Tapan (Kuruyazigi end Tapan 2009).

5 Conclusions and Recommendations

The analysis in this paper could be summarized in the following conclusions and recommendations:

1) All projects dealing with heritage structures should be preceded by a detailed investigation of the historical artefacts.
2) The decision for the interventions is related with the actual condition of the structure but the original structural project is basic for the final project.
3) This historic building provides an opportunity for a truly international team aimed at safeguarding a significant heritage building, which brought together the industrial skills and efforts of several countries in implementing a technically complicated project.

References

Boyken, I.: Vorgefertigte Eisenkirchen. Jahrbuch des Sonderforschungsbereiches 315, Erhalten historisch bedeutsamer Bauwerke 1988, Ernst& und Sohn, Berlin, pp. 191–201 (1989)

Central State Archive of Bulgaria (CSA): Fund, 166 K, inv. 1, archival unit 120, sh. 20

Central State Archive (CSA): Funds 166 K, Inv. 1, Archival unit 120, Sheet 26–41

Ganev, R., Hristov, I., Barcheva, M.: Ганев, Р., Ив. Христов, М. Барчева. Протокол от 7 май 1895 г. за изпитване на стомана на опън. Journal of the Technical University Sofia, branch Plovdiv "Fundamental Sciences and Applications", vol. 14 (2009). Proceedings of the International Conference. Engineering, Technologies and Systems. Plovdiv, pp. 221–225 (2009)

Ivanova, B.l., Ganev, R.: Иванова, Бл., Р. Ганев. "Желязото" и стоманата в храма "Св. Стефан" в Истанбул. *Паметници, реставрация, музеи*, бр. 1 – 2, с. 19–26. [The iron and the steel in the church St Stefan in Istanbul. *Monuments, restoration, museums*, 1, 19–26] (2009)

Ivanova, B.l.: Иванова, Бл. Геотехнически проучвания и архитектурни проекти за църквата "Св. Стефан" в Истанбул от 70-те – 90-те години на XIX Geotechnical researches and architectural projects for the St Stefan church in Istanbul from 1870 to 1890-ies.]. International Conference of Civil Engineerin, Design and Construction, Science and Practice), 13–15 September, Varna Free University, "Chernorizetz Hrabar", Varna, Bulgaria, 67–72 (2012)

Ivanova, B.l.: Иванова, Бл. Българският архитектурен комплекс на Златния рог. Изд. "Парадигма". [Ivanova, Bl. Bulgarian Architectural complex on the Golden horn. Sofia Paradgma ed.] (2018)

Kalchevska, K., Ganev, R.: Калчевска, К., Р. Ганев. Нови изследвания на покритията и стоманата от конструкцията на българската черква „Св. Стефан" в Истанбул. Паметници, реставрация, музеи. 2008, бр. 1–2. Kalchevska, K., Ganev, R.: New investigations of the coverings and the steel of Bulgarian church "St. Stefan" in Istanbul, Monuments, Restoration, Museums, 1–2, 39–44 (2008)

Kuruyazici, H., Tapan, M.: Sveti Stefan Bulgar Kilisesi, Istanbul (2009)

SAПCH: Scientific Archive of Institute of for immovable cultural heritage of Bulgaria, the St Stefan church in Istanbul, T-I-1254-115/12.2001 г. The expert report of eng. prof. Vesselin Venkov

SAПCH: Scientific Archive of Institute of for immovable cultural heritage of Bulgaria, the St Stefan church in Istanbul, T-2-1-I. The expert report of arch. Benjamin Mouton

SAПCH: T-2–2-I ПСД

Tapan, M., Kuruyazici, H.: Sveti Stefan Bulgar Kilisesi/ Fener-Istanbul. Yapi, N **121**, 67–76 (1991)

Proposal for Project Management for the Intervention of Buildings with Historical and Architectural Heritage Declaration in Costa Rica

Jose Bulgarelli-Bolaños(✉)

School of Architecture and Urbanism, Costa Rica Institute of Technology (TEC), Barrio Amón, av. 9, c. 5, San José, Costa Rica
jpbulgarelli@tec.ac.cr

Abstract. This communication sets out the map of the proposed process for the professional management framework for the intervention projects of heritage buildings in Costa Rica, which guides the technical direction of this kind of projects. To achieve the objective of designing said model, it was proposed to diagnose the status of the management of heritage projects in the country; the processes of the model and the management tools were formulated, considering the diagnosis, the guide of the Project Management Institute and the suggested by the HBIM protocols. The Costa Rican State owns approximately 80% of the historical buildings, but it does not have institutional protocols about intervention processes so that the country does not have standards minimums that guarantee the use of the resources invested in these projects. Therefore, the model intends to standardize the processes of heritage interventions in Costa Rica and offers new professional management tools for the star-up and planning phases, witch present particularities compared to those used in the management of new construction. In addition, this article presents the new research project of the School of Architecture and Urbanism of the Costa Rica Institute of Technology, as a continuation of the work here exposed.

Keywords: Cultural property preservation · Planning methods · Intervention protocols · Project management

1 Introduction

This article is part of an investigation as part of the Final Graduation Project to opt for the Master's Degree in Project Management, of the Academic Area of Project Management of the Costa Rica Institute of Technology (TEC), which sought, from the base of local and international knowledge of the construction industry, to propose a framework for the management of projects of patrimonial intervention in Costa Rica, focused on the phases of initiation and planning (Bulgarelli-Bolaños 2019). In this article the objectives of the research are ex-posted, the methodology used is described, the results of the research are exposed, specifically what is related to the process map of intervention projects

I. Vayas and F. M. Mazzolani (Eds.): PROHITECH 2021, LNCE 209, pp. 1294–1311, 2022.
https://doi.org/10.1007/978-3-030-90788-4_100

in heritage buildings, as well as the particular tools of this type of projects for the contemplated phases and ends with the conclusions of the research. However, to begin with, the context in which the research was formulated is explained.

Intervention projects in heritage buildings deal with more challenges in their planning, design and construction than new construction projects. In addition, Basu, quoted by Warr and Farnsworth (2016), estimated that preserving and adapting an old property costs on average between 10% and 20% more per square foot when compared to new construction (Warr and Farnsworth 2016). These authors point out that both time and costs are subject to unforeseen conditions, materials that are usually difficult to acquire, and labor-intensive processes (Warr and Farnsworth 2016). On the other hand, Jordan-Palomar et al. (2018), indicate that something that contributes to the dispersion of information, and therefore to the increase in costs, in heritage restoration projects is the disjointed use of divergent protocols and techniques. In addition, Sorli Rojo and Azorín López (2014) state that on many occasions there has been a lack of coordination between the different professionals in charge of architectural restoration projects and some mimetic actions. As a consequence of the uncoordinated professional work between the different professionals in these projects, Jordan-Palomar et al. (2018) point out: the economic losses and mistrust in the management of the project, in turn provoke uncertainties as to the costs and schedules for the real estate developers when developing a restoration or historical intervention (Jordan-Palomar et al. 2018).

The problem detected in Costa Rica is that the public interest character of the interventions in patrimonial properties; reflected in the second article of Law N 7555, Law of Architectural Historical Heritage of Costa Rica and, in the fourth article of the Regulation for said Law; complicates the management of architectural restoration projects in this type of buildings. In the country the concept of intervention refers to any action that involves research, conservation, protection, restoration, rehabilitation, maintenance, dissemination and education in favor of architectural historical heritage (Regulation to Law 7555 2005). The complexity of the technical, legal and financial management of heritage intervention projects, when compared to new construction projects, puts at risk the country's architectural heritage, since it discourages the owners of buildings with heritage declarations to comply with their duty to protect it, the work falling into a state of deterioration and high risk, or causing the loss of the building. In addition, this complexity also affects the owners of buildings without a declaration, but with architectural historical value, since new construction projects are more profitable than heritage intervention projects, which encourages the demolition of the work to be replaced by another that guarantees better income.

At the technical level, the challenge of intervening in a heritage building can be high due to several factors, among them: a) the materials with which the building was constructed are no longer manufactured or are very difficult to obtain; b) the construction techniques are very specialized or completely unknown; c) the structural and physical characteristics of the building make it difficult to install new systems to guarantee the safety of the users; and d) the risks of the project are proportional to the state of the work to be intervened. On the other hand, the law obliges the owners of properties with a patrimonial declaration, as explained by Vargas (2007), to maintain and protect the property and to allow the inspections deemed necessary by the *Ministerio de Cultura y*

Juventud[1] (MCJ by its Spanish acronym). In addition, all persons involved in this type of project must respect the scope of the Law and its Regulations, which translates into the civil responsibility of the owners and those in charge of managing any process related to architectural heritage. Additionally, patrimony properties must adapt to accessibility and safety regulations, complying with the provisions of the Law on Equal Opportunities for Persons with Disabilities, Law 7600, and the Costa Rican Electrical Code for the Safety of Life and Property (Electrical Code) in force since 2012. This is complex, according to Artavia (2017), given the delicacy required for an intervention in a heritage building, since the project must ensure that it does not interfere with its historical fabric.

With respect to the financing of intervention projects in heritage buildings in Costa Rica, as indicated by Vargas (2007), in 2001 the Tax Simplification and Efficiency Law was enacted, with which the incentives of Law Number 7555 (articles 13 and 22) were eliminated, where deductibles were declared for income tax purposes, donations and investments, as well as improvements made in favor of the national patrimony. This discourages private owners to protect proprieties with patrimonial value. In addition, the "Save Our Architectural Historical Heritage" contest, which has existed since 1997, is the only way for these owners to have access to public funds for interventions. According to information from the *Centro de Investigación y Conservación del Patrimonio Cultural*[2] (CICPC or *Centro de Patrimonio*[3]) of the 387 structures with patrimonial declaration in Costa Rica, 334 are buildings, of which nearly 80% are property of the State (Central Government or Local Governments), through some of its entities. However, according to data from the Ministerio de Cultura y Juventud (2018) exposed the Satellite Account of Culture of Costa Rica (Sicultura), specifically in the production account of architectural activities, during the years 2016 and 2017 the production in this sector was approximately 252 million dollar, while according to data obtained from the *Sistema Integrado de Compras Públicas*[4] (SICOP), the State, through several entities[5] belonging to the Central Government and Local Governments, awarded through different contracts during those same years, approximately 2 million dollar, through the classification code 72101507 (building maintenance service) to conservation work, restoration or maintenance of buildings with patrimonial declaratory.

That means that about 0.81% of the production of the architectural sector in Costa Rica, during the period 2016 and 2017, was generated through projects of patrimonial intervention, whose sponsor was the State. This data, despite the fact that it does not have information from private sponsors, along with those from Sicultura, represents a valuable indicator that demonstrates: a) the existence of a sector dedicated to this type of projects in the country; b) that public money has been invested in these projects and that, therefore, the adequate management of resources must be guaranteed; and finally, c) that more and

[1] Ministry of Culture and Youth.

[2] Centre for Research and Conservation of Cultural Heritage.

[3] Heritage Center.

[4] Integrated Public Procurement System.

[5] Ministry of Culture and Youth, Costa Rican Art Museum, Costa Rican Center for Film Production, National Museum of Costa Rica, City Hall of Moravia, City Hall of San José, City Hall of Alajuela, City Hall of San Isidro de Heredia, Costa Rican Post Office and Ministry of Foreign Affairs and Worship (SICOP 2016–2017).

better indicators are needed that allow a traceability of the activities within the incipient sector. The above, as well as the data generated by the JMF (2018), serve as a basis for justifying the need for more and more standardized and professionalized management processes of heritage interventions. Irrespective of the type of owner, Eduarte-Ramírez (2015) points out that Costa Rica needs to seek external funds in order to carry out, in a responsible manner, actions aimed at the conservation of heritage. But few are the projects that can accede to external international funds, the previous thing forces each owner, public or private, to look for sources of financing in the local scope, with the difficulty that the request of a credit for this type of projects presents the problem that the only thing that can be mortgaged is the lot and not the construction. This is because the credit policies cover buildings that comply with certain characteristics present in the current regulations, which in general the great majority of architectural heritage in Costa Rica cannot comply with, since many buildings were built under another legal framework; therefore, to the perception of the owners, mainly private, the real estate becomes a lien that prevents them from improving the physical conditions of the place.

Finally, in this type of projects, the list of interested parties increases and diversifies, comparing them with other construction projects, since the buildings are part of the country's cultural heritage. In the case of Costa Rica, those involved can be classified into groups such as: authorities, civil society organizations, academies, the media, the construction industry and owners. Each with its own degree of power, urgency and legitimacy in the intervention. An interesting concept for managing stakeholders in such projects is the Heritage Building Information Modeling (HBIM). Asjana (2018) describes it as the inclusion of information from the historic property to the BIM model, which is applied inversely to the new construction process, since it starts from an existing construction with a life cycle of many years and, in the case of Costa Rica, with legal protection whose intervention represents a civil and criminal liability for those stakeholders who design and execute the planned on the heritage object. Now, Bacci et al. (2019) describe in a more general way the HBIM because they define it as a methodology of digitalization of existing artefacts with artistic and historical importance, not only architectural objects, and considers the HBIM as an approach where multiple disciplines such as: geomatics, restoration, history and data collection are interchanged to manage different activities and knowledge. What the authors agree is that the HBIM protocols improve the execution of intervention projects by clarifying processes, unifying information and therefore improving communication between the stakeholders involved in heritage projects through the use of digital tools; this is why it is considered the integration of such protocols within the management processes in heritage interventions in Costa Rica.

The foregoing points to the need for Costa Rica to properly administer processes, efficiently manage resources and direct them to the people involved in heritage intervention projects. The proposed management model, as a result of the research, is discussed in depth below, according to the structure described at the beginning of this section.

2 Objectives

The general objective is presented below, as well as two of the three specific project objectives addressed in this article.

2.1 General Objective

To design a framework for the management of real estate intervention projects with a declaration of architectural historical heritage in Costa Rica, to guide the technical direction of the start-up and planning phases, considering the good practices of the construction industry.

2.2 Specific Objectives

- To diagnose the state of the management of projects of intervention of real estate with patrimonial declaration, by means of the analysis of the criterion of national and international experts, as reference for the proposal.
- To formulate the tools of a model, based on the results of the diagnosis, the guide of the PMI and the suggested by the HBIM protocols, that guide the professional management in the phases of beginning and planning of heritage buildings interventions in Costa Rica.

3 Method

Depending on the objectives of the research, the individual experience of experts, or subjects of information, and their own experience around heritage intervention projects was important; therefore, according to the criterion of classification by design, the research was qualitative phenomenological interpretative, collecting and analyzing the necessary data in order to design the model. The information subjects were seven architecture professionals with experience in Costa Rica, Panama, Guatemala, Mexico and Spain. The criteria for the selection of these subjects were:

- Participation in some project of patrimonial intervention in Costa Rica, either from the direction of the project, or as representative authority of the State.
- Access to subjects and information they may provide.
- Professionals specialized in some process of patrimonial intervention: preservation, conservation or restoration.

Now, considering the classification by purpose, the work is catalogued as an applied research, whose subclassification is framed in what is known as a feasible project, because the management model that was proposed to solve the problem of the complexity of the intervention projects of heritage buildings in Costa Rica. The research considered two categories of analysis, the first corresponds to stakeholders, which was divided into internal and external. The second was called requirements, whose subcategories were: legal, financial and technical. The three products proposed were: a diagnosis, the tools of the intervention project model and the development of some particular tools of the intervention master plan of the Chapel of the Old House Gonzalez Feo. The development of the first product consisted in consulting seven experts, with experience in five countries, including Costa Rica, through a questionnaire and a semi-structured interview, where the categories of analysis described were explored. The diagnosis did not include the analysis of particular projects, since what was sought was to know the methodological frameworks of management that are currently being implemented in the national and international context.

The second product contains the suggested minimum necessary tools, adapted to the national context, according to the methodological framework proposed by the PMI, for the management of intervention projects in heritage buildings, specifically for the start-up and planning processes. In addition, the model's tools are adapted to the HBIM prototype guidelines. Execution, monitoring and control process groups and closure were not considered. This product was not intended to be: a) a new financing model, c) a particular methodology applicable only to a specific building, d) a plan to put a building into use, or c) a maintenance plan. The last product included the development of management tools for the start-up and planning process groups specific to intervention projects in heritage buildings.

4 Results

4.1 Process Map for Intervention Projects in Heritage Buildings in Costa Rica

Considering the information provided by the experts, the literature on projects and on the HBIM protocols, mainly those formulated by Jordan-Palomar et al. (2018) and Asjana (2018), a process map describing how this type of projects could be executed in the country was proposed. Figure 1 illustrates the cycle´s life of the Projects of Intervention of Heritage Buildings in Costa Rica, below is described the map of processes of each of the phases of that cycle for this type of projects in the country.

4.1.1 Process Map, Start-Up Phase

To start a project of intervention in heritage buildings is necessary to have an officialization that guarantees the necessary resources for the implementation of this. The first question that must be answered is: do you have the authorization to start the project? If the answer is negative, the integration and interested processes must be executed. With respect to the integration process, what is developed is the Constitution Act of the Project of Intervention in Heritage Buildings, where in addition to the officialization of equipment and resources, contains a record of assumptions, including possible uses of the property and contemplates the necessary elements that address the implementation of the HBIM models.

Three-dimensional (3D) models are integrated to this type of projects because they have proven to be an effective instrument in heritage interventions, hence the importance of determining the form of collaboration of HBIM models, specifying the technical requirements for developing 3D, the form of naming files and folders, as well as clarifying which uses of the BIM platform are primary and which secondary. Finally, this document

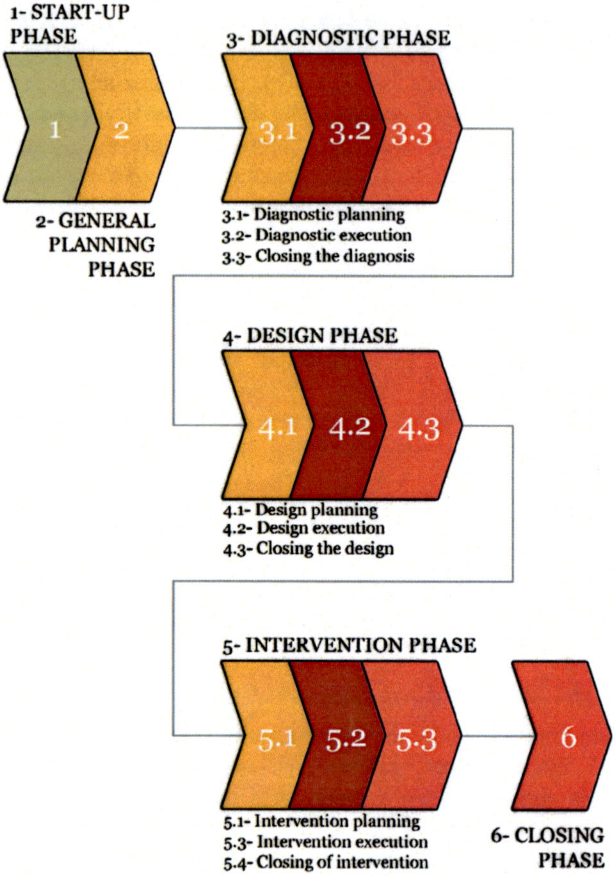

Fig. 1. Cycle's life of the projects of intervention of heritage buildings in Costa Rica

should also consider different possible uses of the property once it has been restored. In order to elaborate this product, it is possible to contemplate as entries the documents of this type of business, agreements, environmental factors of the organization and its assets.

With respect to the process of integration of the interested parties, from the identification of these it must elaborate a clear registry of the people who will be involved in the process. In this document, according to the PMI, information related to project stakeholders is compiled such as: interests, participation, interdependencies, influence and possible impact on the success of the project. It is also recommended to classify them according to the Mitchell Typology (typology proposed by Mitchell, Agle and Woods, cited by Rivera and Malaver 2011) where global strategies and specific actions can be determined that the project team can use to address each stakeholder, according to the degree of power, urgency and legitimacy of these in the project. The entries to elaborate this registry are the Constitution Act of the Project and those described for the previous deliverable.

When both products are available, what proceeds is to prepare a pre-diagnosis of the property, this document is a general description of the state of the property, depending on the data available from the property and at least one visit by the project management or experts who collaborate in this. The pre-diagnosis will be used as the basis for the planning stage; therefore, in addition to general data on the work, it contemplates the description of: a) the values for which it was declared historical-architectural heritage, b) the property (it can be accompanied by graphic elements), c) architectural style, d) constructive system, e) injuries detected; and f) necessary analysis to be carried out. The three tools described are the basis for planning the heritage intervention.

4.1.2 Process Map, Planning Phase

In this phase, the General Director Plan of the Project of Patrimonial Intervention is developed, where the different subsidiary plans that describe how the project will be executed are integrated. This plan is "general" because of the type of project, where it is necessary for the planning to be gradual; therefore, each phase must be planned in detail starting from the information obtained from the previous process. As inputs, the deliverables of the previous stage are contemplated, the outputs of the processes of the different areas of knowledge raised by the PMI, the environmental factors of the company and the assets of the processes of the organization.

4.1.3 Process Map, Diagnostic Phase

Once the Master Plan has been drawn up, a stylistic, historical and documentary investigation of the building is carried out. For this process it is necessary to have a server which will store the information of the different HBIM models; therefore, if you do not have one, you proceed to the acquisition of the equipment and the creation of the server. If it is not possible to have a physical server, it is necessary to enable a virtual hosting server such as Google Drive, Dropbox or Skydrive, so that the technicians involved have access to the information. When the server is available, the central model and the generation of subproyec-tos are created, that is to say "a kind" of template for the HBIM models. With the information obtained from the researches described above, an initial HBIM mass model is elaborated, with a level of development or LOD 100[6] and base plans of the project. Both the model, the plans and the texts, products of the different investigations, form the first deliverable of the phase: the documentary diagnosis.

The next step is to ask whether or not you have the technological equipment necessary to develop a virtual survey of the property. If you don't have it and you can't acquire it, what proceeds is to make measurements on site and carry out tastings, with this information you proceed to generate what is necessary for a rectified HBIM model whose level of development, recommended by the literature, must be LOD 200. If you have the equipment, you proceed to collect information by means of a laser scan and thermal imaging. With this data, as in the previous case, the HBIM materials are generated, as

[6] LOD 100 represents a conceptual a-dimensional model (Brumana et al. 2018).

well as the library of construction details, the HBIM families with stylistic elements and a historical evolution of the building; whose final result is a reference HBIM model from point clouds; equally the expected LOD is 200.

Once the rectified or reference HBIM Model is developed, a damage survey is carried out, which is a detailed record of the injuries presented by the property and their causes. If necessary, laboratory tests can be used to obtain more detailed information on the composition of the materials used or on the causes of the injuries found. The elaboration of the models described above, and the removal of damages can be done sequentially, if there are not many staff within the team, or in parallel if there are enough human resources to do so simultaneously. Once we have the information of the models and injuries of the property, we proceed to elaborate the existing HBIM Condition Model, with a development level LOD 300[7], from which the plans of the current condition are derived.

With the previous model and the set of plans a document is elaborated that describes in detail the current condition of the patrimonial real estate. This document is the second product of this phase and is the physical diagnosis. From the first and second deliverable, contemplating the possible uses exposed in the Constitution Act of the Project; the use that best adapts to the architectural and structural conditions is defined, the previous conforms the last deliverable: the proposal of use.

4.1.4 Process Map, Design Phase

Once the diagnostic stage has been completed, it is necessary to ask: is an intervention required to implement the proposed use? If the answer is positive, we continue with the elaboration of a consulting contract that specifies what is -but of this stage according to the requirements, needs and expectations of the owner and future users of the property. For this process, a HBIM model is generated that unifies the information with a LOD 300; this model contains the type of intervention that is intended to be carried out. From the model, a set of architectural plans is elaborated containing architectural plans, elevations, sections and architectural details, as well as area tables and symbology if required to understand the project as a whole. With this information together with the plans and models, both of the current state and of the survey of damages, a preliminary project is elaborated.

This pre-project is presented to the CICPC and to the institutions involved in the process, in principle to *Bomberos*[8] and the *Ministerio de Salud*[9] (MS), but it could also involve the *Secretaría Técnica Nacional Ambiental*[10] (SETENA) or some service company, depending on the size of the project. The intention of presenting the an-project to the authorities, prior to the application for construction permits, is to foresee reprocesses before involving more professionals in the project. Depending on the size of the project and the complexity of the intervention, more than one meeting with the authorities will

[7] LOD 300 represents a three-dimensional model in the executive design phase, applied to the design simulation solutions (Brumana et al. 2018).

[8] Firefighters Office.

[9] Ministry of Health.

[10] National Environmental Technical Secretariat.

be necessary to correct possible errors in the initial proposal. The important thing about this part of the process is to convene those people who have the authority to make decisions within the institutions involved, to ensure that the changes made will be approved at the time of applying for the respective permits.

Once the comments of the authorities have been corrected, the different HBIM models are elaborated: the architectural model, the structural model and the one that contemplates the electrical and mechanical installations. The construction plans are made up of architectural, structural, electrical and mechanical plans, as well as the respective graphic representations necessary for the project to be understood. The deliverable of the proposal will be complete when the technical specifications are incorporated. This second deliverable is the one that is presented, together with the form of declared buildings, to the Heritage Centre, which may or may not authorise the project. If it is not authorized, it is processed to reformulate the proposal from the preliminary project; if it is authorized with condition, the observations are corrected, and the proposal is presented again.

Once the CICPC's authorization is obtained, the proposal is presented to the *Colegio Federado de Ingenieros y Arquitectos de Costa Rica*[11] (CFIA) through the system called *Administración de Proyectos de Construcción*[12] (APC), where the contract for professional services is elaborated under the modality of patrimonial projects. In this system, the necessary documents are uploaded such as: CICPC authorization, soil studies, etc. The CFIA sends the proposal to the institutions involved in the process; if they have any observations, the next step is to correct them. When the authorities approve the project, the CFIA evaluates it, so that the corresponding pa-go is carried out. Once the payment has been made, the CFIA seals the plans. These plans are presented either through the APC or physically to the co-responding municipality, which reviews the project and if it has observations, these are corrected. When the municipality accepts the project, the Labor Risk Policy is paid to the *Instituto Nacional de Seguros*[13] (INS) and then the municipal taxes are paid. Once this is done, the third deliverable of this phase is obtained; that is to say, the construction permits.

4.1.5 Process Map, Intervention Phase

This phase begins with the preparation of the Construction Contract and the awarding of the same. Once the construction company is known, we proceed with the execution of the work, which includes the construction of witnesses. Parallel to this, an HBIM model is elaborated (LOD 400[14]) and "as built" plans, which will serve as the basis for the preparation of an intervention report and findings that are part of this deliverable. If the inspection indicates that the work does not conform to what was planned, it goes back to the previous process; once the work has been approved by the inspection, it is indicated that the building has been intervened and the building is presented to the client. It is

[11] Federated College of Engineers and Architects of Costa Rica.

[12] Construction Project Management System.

[13] National Insurance Institute.

[14] LOD 350–400 represents the model implemented for the construction phase (Brumana et al. 2018).

necessary to clarify it is very common in these projects the appearance of unforeseen; however, the more rigorous and detailed the two previous phases would have been, the less unforeseen are expected. It is necessary to carry out a documentary investigation at the CICPC to determine the percentage, in terms of costs and times, of unforeseen events in this type of project and to be able to contemplate it when planning a project of patrimonial intervention. If the client receives as the project proceeds to the last phase of the process.

4.1.6 Mapa De Proceso, Fase De Cierre

Once we have received the agreement of the client or owner of the property, we elaborate a HBIM model of conservation with a LOD 500[15], this is part of the HBIM management plan and building maintenance. In addition, the restoration is presented to the community, by means of an official act, where the intervention is explained; this helps to disseminate and give value to the building, which generates an awareness of the conservation of the architectural historical heritage. The last deliverable is the signature of the closing act of the project.

4.2 Management Tools for Heritage Intervention Projects in Costa Rica for the Start-Up and Planning Phases

The tools proposed in the model are a reference for heritage projects, which does not mean that all instruments should be applied in all interventions. The use or discard of any of these documents, and even the incorporation of others, will depend on the complexity of the project that involves aspects such as: the size of the work, level of deterioration, type of intervention, etc. The tools of the model exposed here are those that present particularities for projects of intervention of real estate with declaration of architectonic historical patrimony in Costa Rica, in the phases of beginning and planning, when compared with the projects of new work. The reader will notice that these are simple tools, but that during the investigation it was concluded that these are not elaborated and that, therefore, this affects the planning of the projects of pa-trimonial intervention in Costa Rica, since during the execution many unforeseen and reprocesses are generated that could be known before the intervention if some countries of registry of information like those that contemplates what is proposed here are fulfilled. The tools considered are:

- Acts of constitution of the project of intervention in patrimonial buildings
- Pre-diagnosis of the intervention project in heritage buildings
- Acts of the scope of the project of intervention in heritage buildings.

[15] LOD 500 represents the as-built updating after the construction phase. The concept of the LOD 500 richness needed to be anticipated to detect the damaged structures, it's characterized by a detailed level of geometry enriched by stratigraphic data (model) and texturing (information) here required before the management phase respect to the new construction (Brumana et al. 2018).

4.2.1 Acts of Constitution of the Project of Intervention in Patrimonial Buildings

With this document, which forms part of the start-up phase, it is with which the project is authorized and authority is given to the project director to use the organization's resources in the execution of project activities. Below are the minimum elements that a template of the project constitution act should present for the intervention of heritage real estate. The first section of the minutes is the so-called "General Information" (GI), where the basic data of the project are presented. The first section of the GI is the "Project Code", where it is assigned by the organization a code with which the project can be identified and tracked in the future, in this codification is considered as reference criteria the type of project, the acronym of the name with which the property is recognized to intervene, the number of the project within the organization and the year of beginning of the project. For the first reference criterion, the following clarification is recommended:

- IP = Intervention project. A project is classified as a IP when it is not known exactly the type of intervention that will be executed in the furniture or when it is necessary to apply different types of interventions to ensure the good condition of the building.
- PP = Preservation project. A project is classified as a PP when the context or environmental conditions of the building are involved, to ensure that the living conditions of the building are maintained, without altering its aesthetic, artistic, historical or technical values.
- CP = Conservation Project. A project is classified as CP when the building is involved, experiencing the least amount of alterations possible, which must be technically justified, to ensure the habitability conditions of the building, without altering its aesthetic, artistic, historical or technical values.
- RP = Restoration projects. A project is classified as RP when the property is partially or totally intervened to return the perceptible features of the property to a previous state, without altering its aesthetic, artistic, historical or technical values.
- RCP = Reconstruction Project. A project is classified as an RCP when it partially or totally rebuilds a building that has been destroyed for some cause and the community considers it important to regain the building respecting the aesthetic, artistic, historical or technical values of the building in one of its previous states.

For example, if the project is an intervention project in the *Antigua Casa Gonzalez Feo*[16], is the third intervention of the organization in that particular residence and begins in 2020, the proposed codification will be: IP_CGF_003_2020. The next removed from the GI are the name of the project that will be recognized by the organization, the owner, the type of project with a brief description of this, and the geographical location of the project indicating: district, canton and province. The sixth section of the GI is the project team, this should indicate, more than the project director, who integrate the teams: diagnostic, consulting, HBIM coordination and intervention team. The last sections of the GI are the record of the date of preparation of the document and the date of approval of the project indicating year, month and day.

The following sections of the Articles of Incorporation are considered to be:

[16] Old Gonzalez Feo House.

- Justification of the project, where the origin of the project is detailed in a clear way specifying if it is a problem, a need or an idea to solve.
- The project's objectives, indicating what it is intended to do, how it is going to be done and what it is going to be done for.
- Description of the project, which refers to the product, service or expected end result of the intervention.
- Intervention criteria specify which criteria of patrimonial intervention support the project objectives, as well as the expected product.
- Life cycle, an image of the life cycle of the project is included with the required phases.
- Preliminary scope description: a description of the phases and main deliverables of the intervention project is elaborated. This description should involve Revit modeling and documentation, including levels of development of HBIM models.

The eighth section of the Constitution Act refers to the collaboration of HBIM models, where three elements are mainly described. The first are the coordinates of the models, where the points of origin are specified, which are based on architectural models. Each model must be located according to this orientation for the correct coordination with all the disciplines of the project, for example: "origin to origin" by Revit. Following this, the information exchange protocol is specified, the platform for the storage of the models that each consultant will deliver is specified, as well as the periodicity of each delivery, for review and respective assembly of analysis. Finally, the modeling units are specified, i.e. the metric system to be used.

In the following section the technical requirements of the HBIM models are specified, in the first section of this section the experience in HBIM software is indicated, where each consultant will have to revise his internal work protocol and make the necessary adjustments for the delivery of each model in its respective phase, before starting a project under HBIM standards. The following section refers to the information communication platform, indicating how the exchange of information will be and the tools to be used for there to be adequate communication between the parties. The third element is the specification of the hardware to be used and finally the software or software for: modeling and documentation, coordination, cloud exchange, optional and virtual meetings.

In the tenth section the files are named, that is to say a codification is proposed to each file, the first part of the code is the project to which it belongs, then the type of file, if it is a two-dimensional plane (P) or if it is a model (M), in addition it is recommended to indicate and program and the version of the used program[17]. What you run then is to indicate the level or location in the case of plans, followed by the discipline[18] to which the file corresponds, then the authorship placing the code of the person according to the team and finally the version of the file. For example, continuing with the intervention project in the old Casa González Feo, if reference is made to the architectural model elaborated by Pedro Jiménez, the proposed coding is: IP_CGF_003_2020_M(R2019)_A_PJ_001.

[17] Autocad 2019 = (A2019); Revit 2019 = (R2019).

[18] A = Architecture, S = Structural, M = Mechanical, E = Electrical, F = Fire, A/C = Air conditioning.

In the tenth section the files are named, that is to say a codification is proposed to each file, the first part of the code is the project to which it belongs, then the type of file, if it is a two-dimensional plane (P) or if it is a model (M), in addition, it is recommended to indicate and program and the version of the program used. What corresponds then is to indicate the level or location in the case of planes, followed is indicated the disci-pline to which the file corresponds, then the authorship placing the code of the person according to the equipment and finally the version of the file. For example, they continue with the intervention project in the old Casa González Feo, if reference is made to the architectural model elaborated by Pedro Jiménez, the coding proposed is: IP_CGF_003_2020_M(R2019)_A_PJ_001.

In the section: "Denomination of work folders", it is recommended to name the mother folder with the code of the project, followed by the subsidiary folders, these can be: 01-MODELS, 02-CAD PLANS, 03-PLANS PDF, 04-BUILDING PLANS and 06-DOCUMENTS. Then, inside each subsidiary folder it is recommended to name the internal folders with the name of the different disciplines involved in the process. The twelfth section clarifies the uses of the BIM platform, i.e. it classifies which are primary and which are secondary. In the following section a table is elaborated with the estimated terms of the phases of the project, depending on the complexity of this can be presented in weeks or months. Subsequently, in the costs section is placed the projection of the amount or estimated range of money that will be required for each phase.

The fifteenth section records the assumptions and risks, understanding the assumption as a positive situation or factor that is very likely to materialize or is necessary for the project, while irrigation refers to a negative factor that should be avoided. The above is mounted on a table and each of these is classified with the letter "S" if it is an assumption or "R" if it is a risk and a consecutive number. The following section indicates what the uses of the property could be once it has been restored, whether or not a business idea was previously contemplated. Finally, in this tool, the signatures of the sponsor, the client or user and the project director are recorded.

4.2.2 Pre-diagnosis of the Intervention Project in a Heritage Building

This document describes in a general way the state of the building that will be intervened. It is recommended to prepare it at the beginning of the project and the objective of the tool is that, from a visit to the site, sufficient information is recorded to decide which analyses will be necessary in the diagnostic stage; in addition, it provides sufficient material to determine the type of intervention to be carried out and the intervention criteria. The necessary data in the pre-diagnosis are listed below. In the first section of this tool is placed the same general information that was determined in the Constitution of the Project, with the indication of who are responsible for preparing the document.

The following section is called "Characteristics of the building" in which details of the work are specified in seven sections which are: number of the declaration of the building, the author of the work which indicates the person responsible for designing and building the building or the part that you want to intervene. This is followed by a general description that involves the physical description of the building and its immediate context, including the area, levels, rooms and any element that allows the dimension of the project to be understood. It then describes the predominant architectural style or styles in the work, as well as the construction system or systems. Finally, in this section we add the available graphics of the building, architectural plans, elevations, photography and models.

The last two sections of this tool are referred to as "Injury Description" and "Necessary Analysis Description". The first part details the damage found in the property during the visit and the second part specifies the analyses that must be carried out in order to have enough information to be able to design the intervention. This tool is important because it provides the necessary information to plan and quote the value of the diagnosis and the formal investigation of the state of the property.

4.2.3 Act of the Scope of the Project for the Intervention of a Heritage Building

This is a tool typical of the planning phase, in this stage the actions and processes are executed to establish the total scope of the project, the objectives are defined and the line of action required to achieve them is developed. In the management of the scope it is defined, in a detailed way, both what is included and what is not in the product and in the project, considering all the requirements, needs and expectations of those involved. All the work is established in order to follow the proposed objective, that is, the scope of the project and then the scope of the product is detailed, establishing the characteristics of the product.

The first section of this tool is where the general information of the project is recorded, specifically the project code, on behalf of the project, the name of the owner either physical or legal person and the date of approval of the project which is when the Constitution Act of the project is signed, recording year, month and day. The following section is called "strategic foundation of the project", where it is indicated if the project obeys a need of the organization, a market order, a legal requirement, a technological advance, a social impact or any strategic aspect with which the organization justifies the project. It details why it is linked to any of the aspects described above.

The third section corresponds to the registration of the project team, indicating the name of who will lead the project and the names of collaborators by teams. A table is drawn up for each type of team, indicating the code of the person, which can be the initials of the name and surname, the profession, the full name, the role and responsibilities. The recommended tables are those corresponding to the management team, the diagnostic team, the consulting team, the HBIM coordination team and the intervention team. The following section shows how the organization of the project is structured by roles, including the sponsor and the client. The following sections are vital because in the fifth section you elaborate a detailed description of the scope of the product and in the sixth section of the project, indicating what is included according to the needs, requirements and expectations of the project stakeholders.

In the following section it is recommended to elaborate a matrix of project deliverables for each of its phases, which should include a brief description of the product to be delivered, the approval criteria and the person responsible for approval. The eighth section specifies the levels of development (LOD) of the HBIM models, it is necessary to specify through tables the level of detail of each model to be used in each phase of the project. After this it is indicated which are the exclusions of the project and of the product; that is to say, all that does not form part of the scope. Finally, the signatures of the sponsor, client or user and project director are recorded.

5 Next Steps

The above is a professional management framework for real estate intervention projects composed of three by-products: diagnosis of the state of management, model with tools for this type of projects in the initial and planning stages and development of instruments that are part of a master plan of asset intervention. However, for the model to be of practical use in Costa Rica, it is necessary to elaborate the management tools for the execution and closing phases, as well as for the pending control and follow-up processes. On the other hand, it is essential to create HBIM tools for the programmed conservation after the intervention, since the adequate follow-up given to the property is the way to safeguard its correct state and justify the investment made.

The School of Architecture and Urbanism and the School of Engineering in Construction of the TEC, together with the Department of Architecture of the University of Bologna; the Department of Graphic Expression and the University Institute of Architecture and Construction Sciences of the University of Seville; and the CICPC, intend to develop the research: "Programmed management and conservation tools in interventions of patrimony properties in Costa Rica", through the analysis of documentary information and previous experiences based on the criteria of national and international experts. The collaborative work between these institutions aims to integrate HBIM methodologies, with that suggested by the PMI to manage, administer and direct the phases of execution, closure and programmed conservation of projects of intervention of heritage real estate in Costa Rica.

The products to be obtained at the end of the project will be a protocol of the execution, closure and conservation phases programmed in the life cycle of the intervention projects of heritage real estate. In addition, management tools will be produced for these phases and two scientific articles will be published in indexed journals. These products will serve as input to all those professionals in the construction industry working in the field of architectural heritage conservation.

6 Conclusions

The MCJ data, presented at the beginning of this paper, allow us to establish that there is a sector in the cultural design industry dedicated to heritage intervention projects, a sector in which the State has directed public money to make possible interventions in the properties it owns and that, therefore, the adequate management of resources must be guaranteed. However, it is necessary to support from academia, or from other fields of research, to the tool of Sicultura with more and better indicators that allow a traceability of the activities within the sector of the interventions to the buildings with declaration of historical-architectural heritage in Costa Rica; with the objective of professionalizing and standardizing the processes more and more.

For intervention projects in heritage buildings, six phases are suggested within their life cycle: start, planning, diagnosis, design, intervention and closure; with phases three, four and five being what is known in PMBoK® (Projecto Managemente Institute, Inc, 2017) as the execution phase. For the type of project it is proposed that the planning should be gradual; therefore, before each execution phase it should be planned starting from the information obtained from the previous process. The diagnosis was considered to be the critical phase of the process because, according to the data obtained from the different analyses of this process, the heritage building is better known and, therefore, precise decisions can be made on the design and intervention, thus minimizing the risk of unforeseen events.

The good practices recommended for IBIP projects are a) research; b) follow due process, which implicitly entails technically justifying each decision taken, systematizing the results of the analyses and documenting each approval by the competent authorities; and lastly, c) consider the timing of the materials, whether they are to be prepared on-site or if they are to be tested prior to the intervention since normally the time of these materials are longer than those used in new construction.

One of the main lessons learned in the processes of intervention projects in heritage buildings is to communicate strategically, because not all the information has to reach everyone and not in the same way. This is because heritage projects are of public interest, their stakeholders are very diverse and do not necessarily have sufficient technical preparation to understand the decisions made at some point in the process.

Finally, in Costa Rica, technology has been used mainly as a means of graphic communication by those involved; however, according to publications made in 2018, the HBIM protocols have shown that the use of technology improves communication between those involved in the various processes, reducing reprocesses, increasing savings in costs and time. Therefore, it is recommended to incorporate these protocols in the projects of intervention in pa-trimoniales buildings.

Acknowledgements. The activity presented in the paper is part of the scholarship as a special research assistant granted by the Postgraduate Direction of TEC and the link with the research Amon_RA (https://www.tec.ac.cr/proyectos/amon-ra).

References

Artavia, S.: Heritage buildings face the challenge of becoming safe sites. La Nación (2017). https://www.nacion.com/ciencia/patrimonio/edificios-patrimoniales-encaran-el-reto-de-convertirse-en-sitios-seguros/YZIJWWIV2JFVTF4ZUOLNWYEFV4/story/

Asjana, J.: Development of a protocol for the historical building information modelling (HBIM) model applied to the case study of the San Juan church of the hospital in the city of Valencia. (Master's thesis). Universitat Politécnica de Valencia, Valencia, Spain (2018)

Bulgarelli-Bolaños, J.: Professional management model of intervention projects with patrimonial declaration in Costa Rica. Case study: intervention plan of the Chapel of the Antigua Casa González Feo (Master's thesis). Costa Rica Institute of Technology, San José, Costa Rica (2019)

Bacci, G., et al.: HBIM methodologies for the architectural restoration. The case of the ex-church of San Quirico All'Olivo in Lucca, Tuscany. International Archives of the Photogrammetry, Remote Sensing & Spatial Information Sciences (2019)

Brumana, R., et al.: Generative HBIM modelling to embody complexity (LOD, LOG, LOA, LOI): surveying, preservation, site intervention—the Basilica di Collemaggio (L'Aquila). Appl. Geomatics **10**(4), 545–567 (2018)
Eduarte-Ramírez, A.C.: Estado de la conservación-restauración en Costa Rica y nuevas perspectivas. In: La conservación del patrimonio cultural en Costa Rica, pp. 39–51. Aracne Editrice (2015)
Jordan-Palomar, I., Tzortzopoulos, P., García-Valldecabres, J., Pellicer, E.: Protocol to manage heritage-building interventions using heritage building information modelling (HBIM). Sustainability **10**(4), 908 (2018)
Ministerio de Cultura y Juventud. Costa Rica Culture Satellite Account: Methodology and Results (2018). https://si.cultura.cr/cuenta-satelite-cultura.html
Muñoz, S.: Contemporary restoration theory. Editorial Síntesis S.A., Madrid, Spain (2003)
Project Management Institute, Inc.: The Guide to the Basics of Project Management, Sixth Edition (PMBOK® Guide). Project Management Institute, Inc., Chicago, USA (2017)
Regulations to Law 7555 "Architectural Historical Heritage Law" N° 32749. Diario Oficial La Gaceta (2005). http://www.pgrweb.go.cr/scij/Busqueda/Normativa/Normas/nrm_texto_completo.aspx?param1=NRTC&nValor1=1&nValor2=55888&nValor3=66234&strTipM=TC
Rivera, H., Malaver, M.: The organisation: stakeholders and social responsibility. Documento de investigación (97), 1–34 (2011)
Sistema Integrado de Compras Públicas. File Listing. Costa Rica (2016 – 2017). https://www.sicop.go.cr/moduloPcont/pcont/ctract/es/CE_CEJ_ESQ001.jsp
Sorli Rojo, Á., Azorín Lopez, V.: Conservation and restoration of buildings in Spain through IETcc publications. Jornadas Internacionales Conmemorativas del 80 Aniversario del IETCC: Conservar, rehabilitar, innovar (2014). ISBN 978-84-7292-382-9
Vargas, C.: The Architectural Historical Heritage in Costa Rica and its Legal Guardianship. Centro de Investigación y Conservación del Patrimonio Cultural, San José (2007)
Warr, R., Farnsworth, C.: Perceptions of historic projects in the construction industry. In: Construction Research Congress 2016, pp. 1223–1233 (2016). https://doi.org/10.1061/9780784479827.123

Earth Architecture in Çatalhöyük, Konya, Türkiye

Selin Duran(✉) and Yesim Kamile Aktuglu

Faculty of Architecture, Department of Architecture, Dokuz Eylül University, İzmir, Türkiye
selin.duran@ogr.deu.edu.tr, yesim.aktuglu@deu.edu.tr

Abstract. Çatalhöyük is one of the oldest and important city in the world because of the use of earth in construction of buildings. It is located at Konya. Çatalhöyük was a very large Neolithic and Chalcolithic protocity settlement in southern Anatolia. It was existed from approximately 9000 years ago. It was inscribed as a UNESCO World Heritage Site. All of the ruins of the buildings are constructed with adobe that earth is stabilized with straw. There are two archeological areas here. They have lots of layers that constructed with adobe. The houses that non-used were filled with raw earth and then new ones were constructed on old ones' wall. In this paper, all of the ruins will be examined through the photos of today, in 2021 and their construction styles, socioeconomical situations for in the city. The adobe ruins (walls and buildings) will be disgussed through past's and today's situations.

Keywords: Adobe · Çatalhöyük · Earth buildings · Raw earth · World Heritage Site

1 Introduction

Earth; is one of the most important building materials that have been used since ancient times. The factors that using this building material a lot are that soil and water (including additives used) can be easily obtained from or near the construction area and can be prepared practically in the construction area (Naumann 2007).

In the Early Neolithic Age in Anatolia (after 7000 BC), the first settlements in Çatalhöyük show the use of adobe. The adobe used were prepared with clay soil mixed with chopped straw or with ungroomed sandy and clay soil (Naumann 2007).

In this study, the usage of adobe as a building material will be examined with examples from history and adobe settlements in Çatalhöyük will be documented with 2021 photographs.

2 What is Adobe? How It Was Used in Anatolia?

In The Encyclopedia Americana (1829), "adobe is sun-dried bricks made of natural clays, especially made in the arid southwestern regions of America and Mexico. Clay

I. Vayas and F. M. Mazzolani (Eds.): PROHITECH 2021, LNCE 209, pp. 1312–1316, 2022.
https://doi.org/10.1007/978-3-030-90788-4_101

mixed with grass or straw is molded to the desired shapes and sizes and dried under the sun for a week or two. It is used in the construction of a mortar consisting of the same material. Adobe, which was used as an early construction material in Assyria and Egypt, is still used in Japan and China."

Adobe is a building material obtained by pouring the soil containing clay in appropriate proportions into molds (Fig. 1) which are blended with water and additives and kneaded and soaked and dried in dry and open air (Duran 2016).

Fig. 1. Molding of adobe in Konya (Duran 2016).

In Troy (layer 1), clay soil mixed with crushed straw was used in adobe bricks, and a similar, finer-grained soil was used as mortar connecting adobes. In Boğazköy, clay soil was kneaded with chopped animal feed and tiny stones. The adobe bricks were soaked with some kind of slime milk before they were placed and placed without mortar. The main characteristics of the excavations in Hacilar are the presence of rectangular adobe structures with red plastered floors (Mellart 2002; Naumann 2007).

3 Çatalhöyük

In the Early Neolithic Age in Anatolia (after 7000 BC), the first settlements in Çatalhöyük show the use of adobe (Fig. 2). The adobes used were prepared with clay soil mixed with chopped straw or with ungroomed sandy and clay soil (Naumann 2007).

With Eriha in Jordan, Çatalhöyük takes its place among the first attempts of urban development of mankind. The mound was discovered in 1958 by the appearance of adobe walls, broken bones, etc., marked by southwest winds, although most of them were covered with grass on a carcass side consisting of J.Mellart, A.Hall and D.French. It was noticed that it continued to a height of 15 m (Mellart 2002).

Vertical stratification is observed in the settlement (Fig. 3).

Fig. 2. Northern excavation area, Çatalhöyük, Konya (Duran 2021).

Fig. 3. Vertical stratification, Southern excavation area, Çatalhöyük, Konya (Duran 2021).

Surrounding each structure with its own walls as well as the walls of other structures makes these structures more durable than a single structure (Fig. 4). Continuous plastering of the walls from inside and outside has also provided the structures to have a long life against abrasion. The walls, floors and ceilings are plastered with white, thin and sticky clay, known as white earth. Thirty to sixty plaster layers were found in the buildings. It was understood that structures with more layers were used for longer periods. The worn and dangerous structures were demolished by first taking wooden beams carrying flat roofs and roofs and then wooden posts. The average life of a building is determined as 100–120 years (Mellart 2002).

After 5600 BC this settlement was abandoned for an unknown reason and a new settlement was established in the West. It has been lived here for at least 700 years and it was abandoned for an unknown reason without any sign of destruction or attack (Mellart 2002).

Fig. 4. Side by side units, Southern excavation area, Çatalhöyük, Konya (Duran 2021).

In Çatalhöyük, all buildings were built with sun-dried rectangular adobes, reeds and plasters. Since there were no stones, six rows of adobe were used below the base level. Straw-dipped adobe (good condition adobe with unswept sand dipped in layer III) and a black mortar with ash and bone debris were used as mortar (Mellart 2002).

The area of the buildings varies between 11,25 m^2–48 m^2 depending on the need. Mostly (6 m * 4.5 m) they have usable area. As the old walls are used as a base and new structures are constructed in this way, a similar settlement plan emerges in different building levels. Depending on the needs, as a result of activities such as merging and separation in new buildings, building floor plans may differ (Mellart 2002).

The flat roofs are formed by laying straw bundles and laying a thick layer of mud on the two main beams on the walls. The entrance to the house is one of the characteristic features of Çatalhöyük architecture possible from the staircase to the roof. It is thought that the defense requirement may have played an active role in the development of this architecture (Mellart 2002).

4 Conclusions

Adobe is a building material that has been used since the first settled life. In addition, it is possible to see living adobe houses that have users in many of the rural settlements in Central Anatolia. There are traces of adobe settlement dating back to 9000 years in Çatalhöyük. Today in Konya, covered with steel systems are protected and functioned as an open air museum. The transportation to the region is done by a private vehicle and is located in a remote area away from the city settlements. In the prototype of the researched houses, experimental houses were built and exhibited reflecting the socio-cultural structure of the period. In a closed museum, Çatalhöyük history and research findings are exhibited. The excavations at Çatalhöyük are still continuing today. A new settlement is unearthed next to the northern excavation area.

References

Duran, S.: Akşehir' de Kerpiç Malzemeli Yapıların ve Binaların Taşıyıcı Sistem Açısından İrdelenmesi ve Günümüz Koşullarında Değerlendirilmesi. Master's thesis. Supervisor, Prof. Dr. Yeşim Kamile Aktuğlu. Dokuz Eylül University, Faculty of Architecture, Department of Architecture, İzmir (2016)

Duran, S.: Personal photograph archive. Çatalhöyük, Konya (2021)

Encyclopedia Americana Corporation: The encyclopedia Americana, USA, Americana Corporation (1829)

Mellart, J.: Anadolu'da bir neolitik kent, İstanbul: Yapı Kredi Yayınları (2002)

Naumann, R.: Eski Anadolu mimarlığı (5. press). (B. Maral, translation.). Ankara: Ankara Türk Tarih Kurumu. (Original work date of issue 1975) (2007)

The Historic French Wharf in Lavrio: Documentation, Conservation, Reuse, Revitalization

Irene Efesiou(✉), Nikolaos Belavilas, Andreas Georgopoulos, Maria Balodimou, and N. Athanasopoulou

NTUA, Athens, Greece
irenefe@otenet.gr

Abstract. The Historic Wharf is situated at the Port of Lavrio in Attica Greece and was constructed in 1887–1888 It is a technologically advanced Industrial construction, a significant landmark, a listed monument since 1995. The people of Lavrio are very concerned about the degradation of the monument, because they always considered it as a valuable symbol of their recent history and demand it's conservation. The documentation, conservation and reuse project of the wharf was a very difficult and challenging project that needed intensive interdisciplinary collaboration.

The geometric documentation was executed using contemporary methods of data acquisition and processing and was part of an interdisciplinary approach concerning historical, architectural, structural and mechanical documentation. The architectural reuse proposal aims to conserve and revitalize the Wharf, reactivating it as a "living" cultural monument open to society.

The stone walls and the metal structure will be conserved and other parts will be reconstructed. The circulation of people on the metal part will be restricted to modern corridors, following the rails where the wagons used to circulate. The architectural proposal also emphasises sustainability and viability, proposing changes to the surrounding urban fabric.

Keywords: Metal wharf · Lavrio · Restoration · Interdisciplinary · Industrial cultural heritage · Rail wagons

1 Introduction

The work presented in this paper concerns the documentation, conservation and reuse of the historic French wharf in Lavrio. It is part of an innovative research program (2018–2019) developed by the National Technical University of Athens in collaboration and sponsorship of the Municipality of Lavreotiki.

The wharf is a significant landmark for the area and the people of Lavrio. Although it is a listed monument since 1995, the wharf suffers from serious problems and damages due to the absence of protection and maintenance for many years.

The documentation, conservation, reuse project of the wharf was a very difficult and challenging project that needed intensive and continuous interdisciplinary collaboration.

I. Vayas and F. M. Mazzolani (Eds.): PROHITECH 2021, LNCE 209, pp. 1317–1343, 2022.
https://doi.org/10.1007/978-3-030-90788-4_102

Experts from the School of Architectural Engineering, School of Civil Engineering, School of Rural and Surveying Engineering, School of Mechanical Engineering, School of Mining and Metallurgical Engineering and the Post graduate Program "Protection of Monuments" collaborated creatively for more than a year in order to analyse and propose innovative and compatible conservation methods and reuse - revitalisation design strategies. NTUA Laboratories that took place in the research program were: Architectural Technology, Urban Environment, Photogrammetry, Steel structures, Soil mechanics-geotechnical engineering, Statics and Aseismic research, Mining and environmental technology, Dynamics and Structures laboratories.

In this paper we will focus on the Geometric and Architectural documentation as well as the Architectural conservation and reuse design project.

2 The Historic French Wharf in Lavrio. General Data

The Historic Wharf of the French mining company "Compagnie Française des Mines du Laurium" (CFML) is situated at the Port of Lavrio in Attica Greece and was built in 1887–1888 by "Société Anonyme des Anciens Établissements Cail, Paris".

The wharf was the coastal part of a larger port installation comprising warehouses connected by a network of railway tracks for the circulation of wagons transporting minerals. The minerals were brought from the industrial complex of CFML, stored near the wharf and transported with wagons to the incoming/outgoing ships and boats (Figs. 1, 2 and 3).

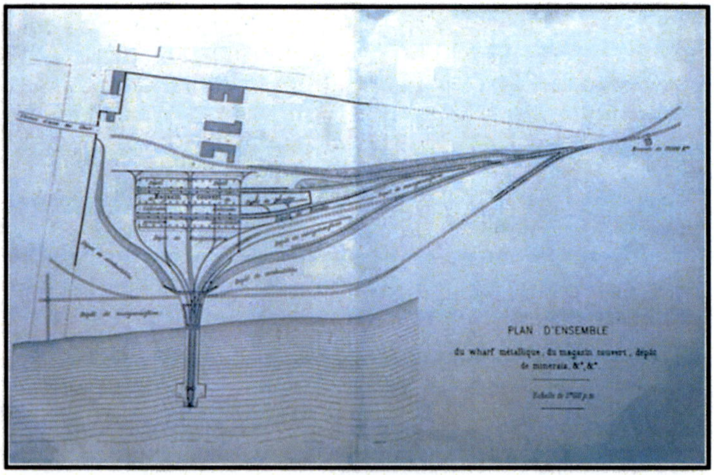

Fig. 1. Plan of the wharf and the port installations, transportation and storage facilities of The French Societe des Mines de Laurium in 1889 (Source: M. Marmanis archive)

Fig. 2. Historic photo of the west side of the wharf with the two revolving cranes (Source: M. Marmanis archive)

Fig. 3. Historic photo of the west side of the wharf and of the storage buildings after WW2 (Source: Man. Markoulis archive)

Fig. 4. The French wharf in 1980 (Source: A. Papaioannou Professor NTUA archive)

In 1950 the transportation method of the minerals to and from the ships changed causing the limitation of the wharf's function. The gradual closing of the CFML Company during the 1980's leads the wharf to complete obsolescence. Since then the wharf has been left to corrode and deteriorate, without any maintenance. This degradation caused deep grief to the habitants of Lavrio, who always considered it as a valuable symbol of their recent History and always demanded the conservation of the wharf (Figs. 4, 5, 6, 7 and 8).

Fig. 5. Photo of the west facade of the Historic French Wharf (2019)

Fig. 6. Photo of the east facade of the Historic French Wharf (2019)

Fig. 7. Photo of the destroyed warehouses (2019)

Today most of the warehouses are destroyed and the wharf stands completely isolated from the mainland, cut off by a heavy traffic road that leads to the town of Lavrio and the new port. The vehicles heading towards the new port pass under the horizontal platform causing serious problems to the iron beams of the structure due to vibration.

Fig. 8. Photo of the north facade of the French Wharf and the destroyed storage buildings (2019)

The pedestrian route coming from the new port stops abruptly on the stone wall of the wharf (Figs. 9 and 10).

Fig. 9. View from the East (2019)

In order to analyse and understand the architecture, construction, structure, defects, damages and other characteristics of the wharf, a detailed geometrical documentation and survey was required.

Fig. 10. North wall (2019)

3 Geometric Documentation of the Lavrio Metal Wharf

For the geometric documentation of the Wharf in Lavrio contemporary methods of data acquisition and processing were employed. The required documentation products were a detailed (scale 1:20) orthophoto of the uppermost surface of the wharf, a general orthophoto and plan of the greater area (scale 1:100) and, indicatively, the orthophotos of the supporting structure of the wharf, including exterior and interior facades of the bridge and of one façade of a wall from the remaining installations.

The establishment of a precise geodetic network with suitable permanent markings for its points was the first necessary action. This would ensure the georeference of all subsequent measurements and drawings to a common reference system. The network was measured with geodetic methods preferred to satellite positioning for greater internal accuracy. However, the points' coordinates were referred to the Greek Geodetic Reference System (EGAS'87, GGRS'87), using a minimum constraints method so as not to introduce distortions due to the projection scale. The sketch in Fig. 11 (image Google Earth ©) shows the positions of the network points marked with permanent marking and in particular a circular bolt of 2 cm in diameter, with the initials "NTUA".

The detailed drawings included orthophotos and were both horizontal and vertical ones. It was considered appropriate to produce two horizontal drawings. The first one covers the wider area of the Wharf, from the waterfront and the Wharf to its parallel road, north of the rest of the French Company's buildings. This project was decided to be at a scale of 1:200 with the related resolution and accuracy (Fig. 12). The second drawing includes only the metal wharf and was decided to be of very high resolution and precision, so its scale was decided to be 1:20 (Fig. 13).

The vertical drawings produced include the three (3) exterior facades of the stone-built section of the Wharf (East, West and North), two (2) vertical sections in the tunnel,

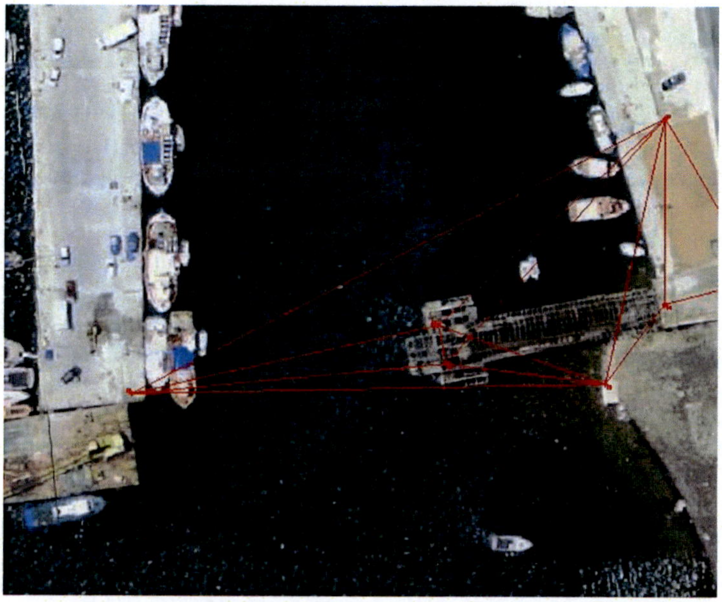

Fig. 11. Image showing positions of the network points marked

Fig. 12. Orthophoto of the greater area

and the West side of the only wall still standing in the northernmost part of the greater area.

For the production of the above deliverables, a combination of modern topographic and photogrammetric methods was used, in order to better provide the necessary details. Image-based modelling was mainly employed in order to acquire full coloured textures of the structures. For the detailed products, images were acquired using a full-frame DSLR camera from the ground as well as from above using an octocopter UAV drone. In addition, images for the general overview were acquired via a DJI Phantom 4 Pro UAV. Ground control points were determined in key positions to accurately scale the 3D models and georeference them to a common reference system and relate the image-based products to supporting survey measurements for the detailed documentation of the steel structure.

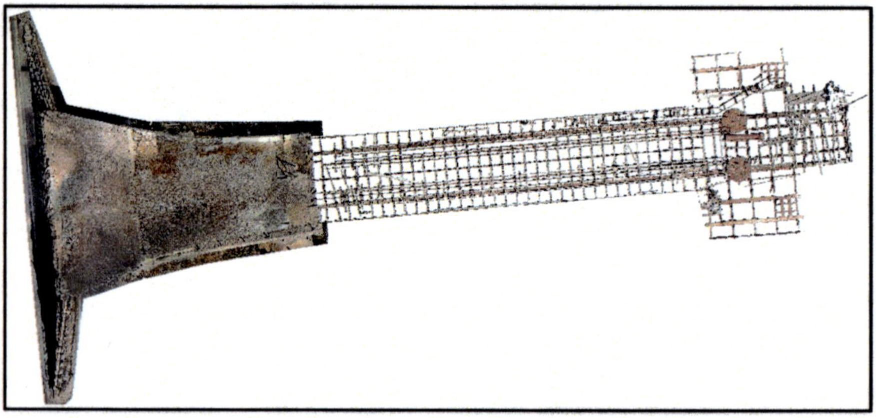

Fig. 13. The detailed orthophoto of the upper surface (1:20)

The images were suitably processed in Agisoft Metashape© software and each step was thoroughly examined for its accuracy and reliability, as the method has no self-control incorporated yet. The orthophotos produced were projected onto pre-defined planes, according to the specifications and the desired final documentation drawings. Special care was given to the filtering of unnecessary information, especially for the detailed orthophoto of the upper surface of the steel structure. The resulting orthophotos formed the necessary base for the production of the final drawings. Samples of the orthophotos are presented in the following images (Figs. 14 and 15).

Fig. 14. Orthophoto of the north facade of the supporting structure (1:50)

Fig. 15. Orthophoto of the tunnel (1:50).

4 Architectural-Constructional Documentation of the Lavrio Metal Wharf

The members of NTUA's group for the architectural-constructional survey of the monument were the architects N. Belavilas, I.Efesiou, M.Balodimou, N.Athanasopoulou, V.Savatianou and civil engineer V. Pougatioti (Fig. 16).

The wharf consists of two parts: a metal structure on the south and a stone structure to the north. The stone part is composed by a retaining wall, that used to hold the elevated ground providing the horizontal circulation of wagons to/from the storage installations and a trapezoid heavy stone structure. These two elements are separated by CFML's wagons covered passage. Today the passage is used by automobiles and tourist buses heading to the new port.

The metal structure consists of a linear longitudinal horizontal platform, based on circular section iron columns with technologically innovative foundation piles. CFML wagons used to circulate on the platform's iron rails in order to load minerals on the ships and unload all other elements needed. This platform has a length of 48,4 m width of 16,70 m and a height over sea level of about 8 m (Fig. 17).

The geometric documentation was combined with historical, urban, architectural, structural and mechanical documentation in order to discover all the secrets of the wharf's former function, understand the structural organization and pathology of the industrial monument and thus design an adequate and creative reuse-conservation project.

This interdisciplinary documentation procedure enabled the production of the very complex and sophisticated outline and detailed drawings which were the basis for the architectural reuse project in full accordance with the structural conservation project.

Fig. 16. Orthophoto of the wharf [Lab of Photogrammetry NTUA]

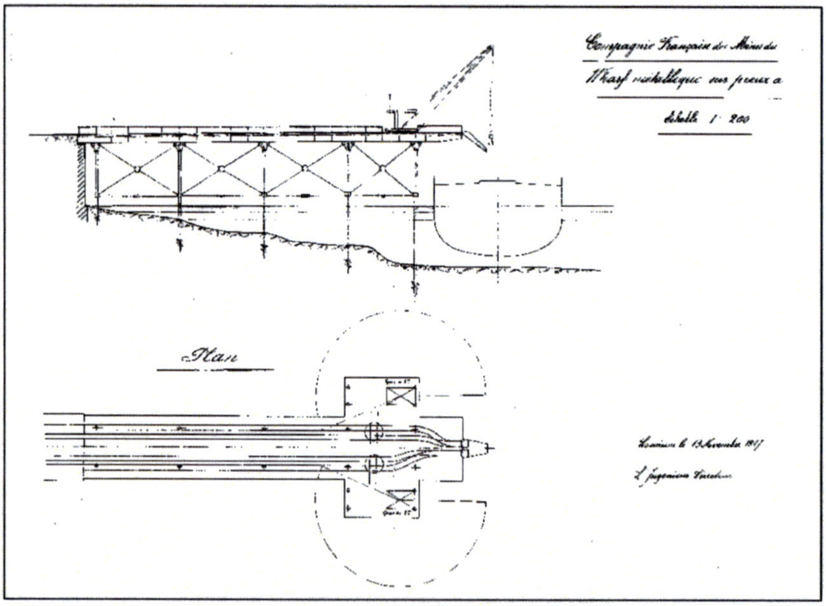

Fig. 17. Historic Drawing of the west facade and the plan of the metal part of the wharf, with the two revolving cranes 13 November 1917 (Source: M. Markoulis archive)

During the first in situ visits the architects made hand drawn sketches, in order to measure the details and fully survey the structure's load bearing system. The in situ work and in situ interdisciplinary collaboration is indispensable for the survey of monuments (Figs. 18, 19 and 20).

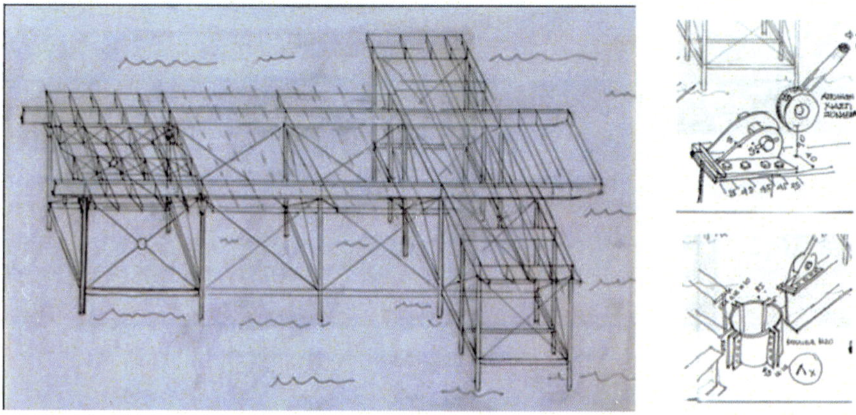

Fig. 18. In situ Sketches of the metal structure in order to survey and measure in detail

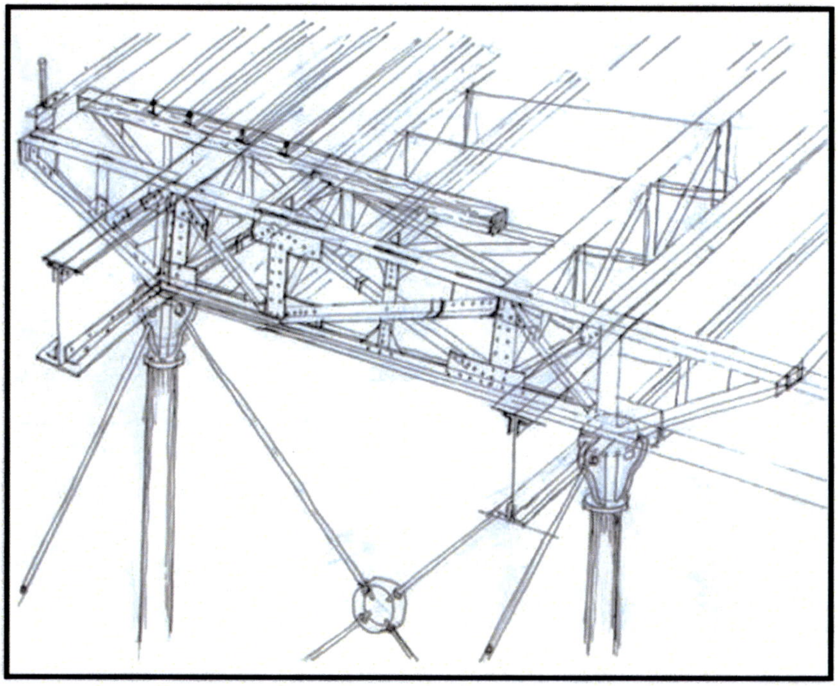

Fig. 19. In situ Sketches of the load bearing system

Fig. 20. 3D drawing of the load bearing system

According to the analysis conducted by NTUA's Mineral technology lab, the metal part of the wharf was constructed with cast iron vertical columns and a system of horizontal trussed girders made of steel sections with riveted connections. It is a rather typical metal structure of the time, with some technologically advanced details (Figs. 21 and 22).

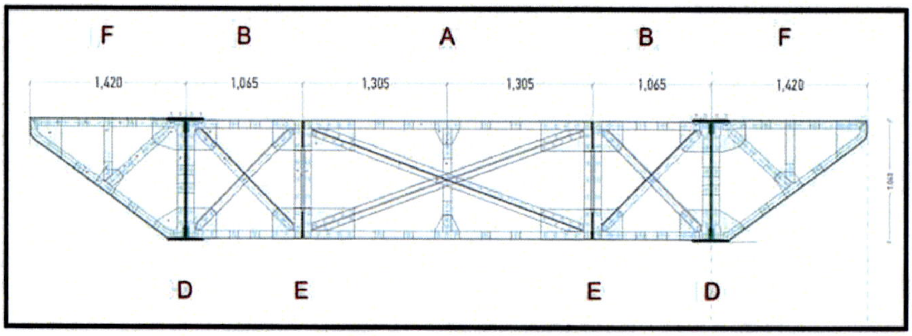

Fig. 21. Section of the horizontal load bearing system

The architectural survey drawings were conceived and designed step by step continuously interpreting the structural system according to the details and vice-versa (Figs. 23, 24, 25, 26, 27, 28, 29 and 30).

Fig. 22. Section and 3D drawing of the horizontal load bearing system of the wharf based on circular section cast iron columns with capitals

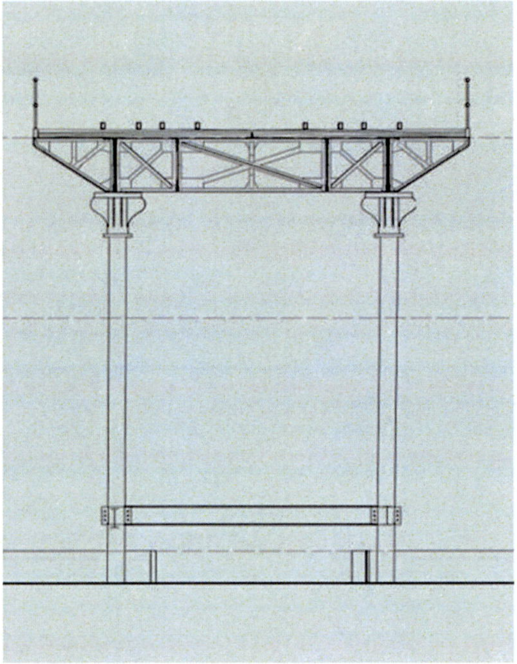

Fig. 23. Typical section of metal structure

Today the monument faces many problems and damages due to absence of maintenance for more than 50 years and careless repairs, conversions, additions, maritime accidents etc. Visiting the wharf is forbidden because it is currently unsafe with very narrow and dangerous staircases, with no handrails on the platform, with cracks on the stone, extended corrosion of many iron parts - especially on the interface of the columns with the sea water and many other damages and defects. Immediate safety measures must be taken and the monument must be properly conserved in order to become once more

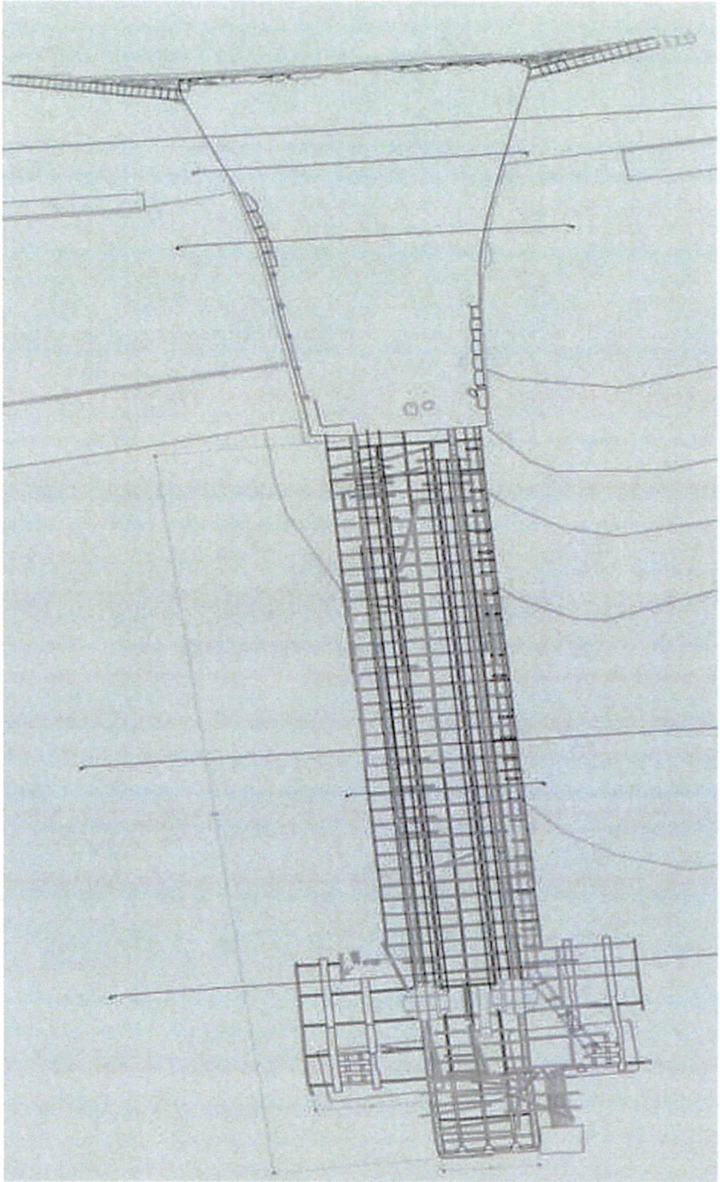

Fig. 24. Plan of the Wharf

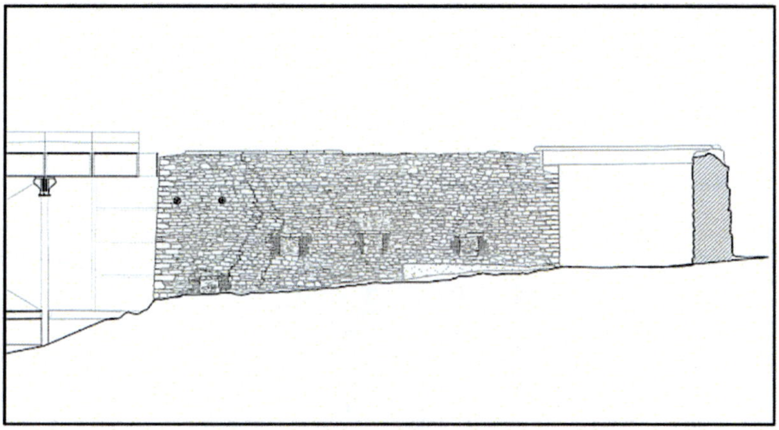

Fig. 25. Elevation of east stone wall

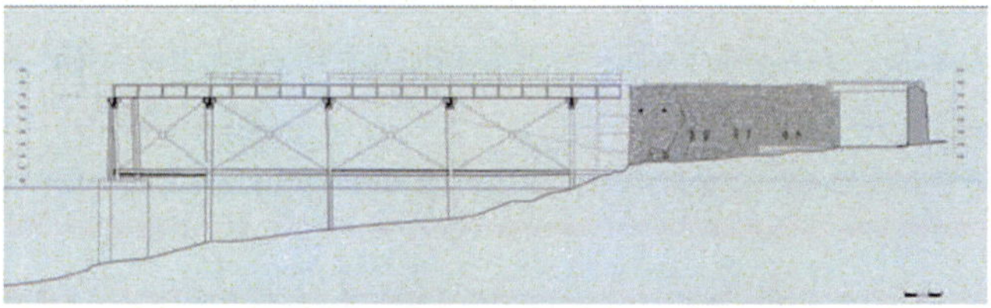

Fig. 26. East Elevation of the Wharfl

Fig. 27. Facade of East Wall

Fig. 28. Western entrance of the tunnel

Fig. 29. Stair to the West of the Wharf

Fig. 30. Details of the metal structure

a vivid, a "living" industrial landmark linked with the waterfront and the everyday life of the people of Lavrio, a safe,educational and recreational place for children, tourists, people with special needs and disabilities (Fig. 31).

Fig. 31. Perspective drawing of part of the metal structure.

5 Architectural Design Project for the Conservation-Reuse-Revitalization of the Metal Wharf

5.1 The Significance of the Wharf as an Industrial Monument and the Principles of Its Conservation and Restoration

The French Wharf as an industrial monument is part of Lavrio's cultural heritage and must be protected, preserving its authenticity and historical integrity. It encompasses a variety of values such as historical, technological, architectural and social. It is a landmark, as well as a record of the past industrial activity in Lavrio.

The restoration of the French Wharf is significant for Lavrio as it embodies great social value as a remnant of the everyday life of the past.

The architectural proposal offers a new use – making the monument accessible - in an effort to highlight its old use.

The architectural conservation- reuse proposal consists of drawings and a technical report concerning the restoration of the monument and its nearby surroundings, and focusing on the design of an environmentally sustainable project. The proposed new use of the wharf and the visitor's circulation on it attempts to highlight the historical function of the monument. It also gives general guidelines for the urban development and the vehicle circulation in the wider surrounding areas, which are indispensable elements for the revitalization of the wharf.

5.2 Restoring the Relation Between the French Wharf and the Surrounding Area

In order to reuse and revitalize the wharf, general interventions must be carried out in the area. There is no use in restoring the monument without including alterations in the master plan of Lavrion, organizing a seafront pedestrian route- a promenade from Lavrion to the new port, in which the wharf will play an important role. The vehicle circulation will be diverted on the north of the wharf, letting the pedestrians pass under it.

The future connection of the wharf with a cultural-leisure complex created in the ex CFML mining storage area and the existing listed buildings must also be planned via an elevated footbridge passing over the highway. This cultural-leisure complex organized by the Municipality of Lavrion will give new life to the seafront and the wharf will become a vital part of it (Fig. 32).

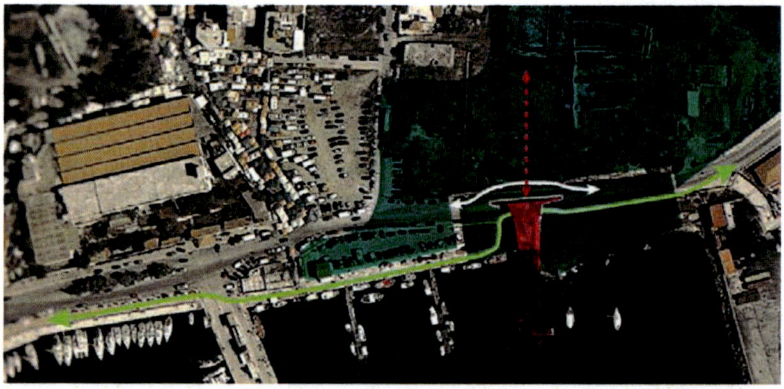

Fig. 32. Master plan depicting the restoration of the monument, the public access, the revitalization of the surrounding environment, the proposal for the pedestrian road towards the port and the town center, as well as the connection of the wharf with the area of the warehouses via a pedestrian bridge.

5.3 Description of the Wharfs Restoration Proposal

The new use of the monument will play an important role for its revitalization and its future existence. Thus the monument is planned be functionally integrated in the everyday life of the people of Lavrion and be open to visitors. A key component of the architectural project is the preservation of the wharf's iron rails and their replacement, where missing. Visitors will circulate along the rails, using a pathway located over the tracks used for the mining wagons.

It must be noted that the proposal for the restoration was made in collaboration with the relevant NTUA Civil Engineering study groups and the Metal Stuctures Laboratory.

Based on the documentation, it is chosen to restore the monument to its original phase - until the 1950s - while retaining certain features of the second phase.

A key component of the architectural proposal is the preservation of the rails and their replacement where missing. Visitors will circulate on foot along the rails instead of ore trolleys (Fig. 33).

Fig. 33. **Transportation** of minerals from CFML storage area to the wharf with mining wagons (Source: Kaloupakis collection)

Regarding the loads from visitors on the wooden deck of the quay, laboratory studies and calculations showed that it was not possible to allow free access everywhere, without very large scale reinforcements and replacements of original material. It was decided that the deck loads will be distributed over the column area, as it was previously foreseen for the circulation of mining wagons. This limitation was a major challenge for the architectural design proposal (Figs. 34 and 35).

Thus public access is restricted to two parallel paths above the two rows of metallic columns supporting the length of the wharf. The limited access and circulation of visitors will be on specially designed pathways of 1.10m wide. These pathways are new metal structures that lay directly on the existing metal bearing structure. The wooden beams of the floor on which the rails are placed are also restored - as the entire metal part of the wharf. The floor of these pathways is elevated so that the new structure remains on top

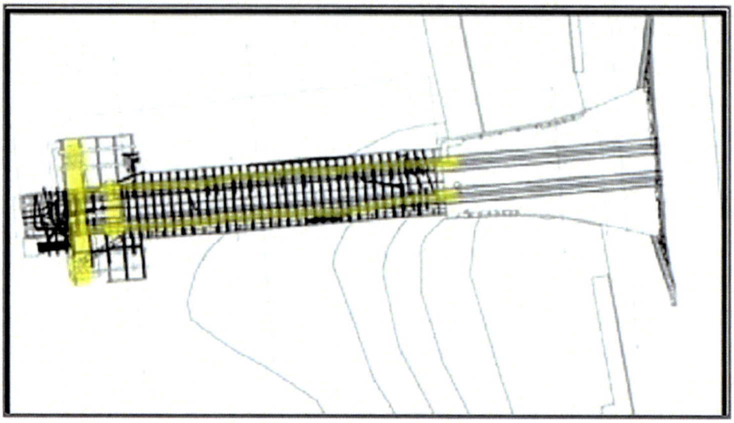

Fig. 34. The circulation of visitors on the pathways of the metal Wharf

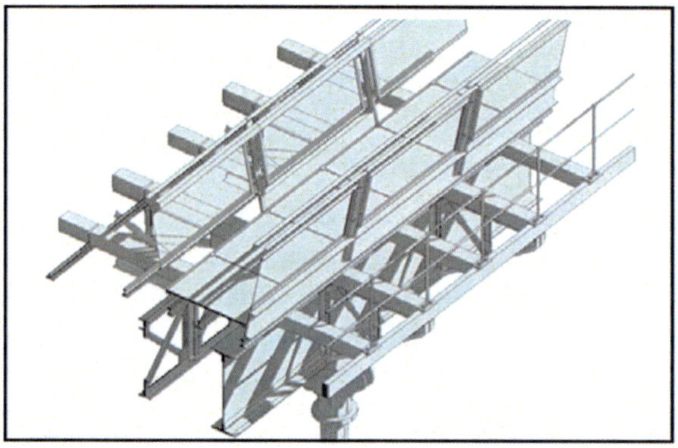

Fig. 35. Detail of the specially designed pathways of 1,10 m for the limited access and circulation of visitors

respecting the underlying authentic or restored elements of the wharf. The floor is made of metal grids in order to reduce wind stress and make the structure and continuity of the rails visible. The pathway parapets are made out of perforated metallic surfaces divided in units with a length of 2 m, having also a variety in the size of their perforations in order to give the optical impression of different densities on the facade (Figs. 36, 37 and 38).

Fig. 36. The floor of the corridors is made of metal grids in order to reduce wind stress and make the structure and continuity of the rails visible.

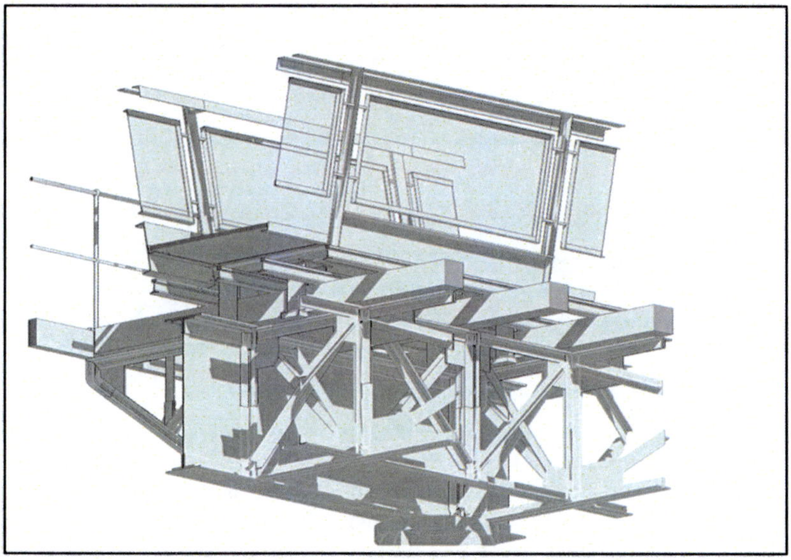

Fig. 37. New metal structures for the pathways.

This intervention, although defined and resulting from structural and functional needs, is intended to preserve the memory of the monument's operation in the past. The denser optical side panels create an impression to anyone looking at the wharf from far away that wagons are moving on the rails. In order to design the metal pathways for the visitors, research and documentation was carried out for the shape and main characteristics of mining wagons from other mining facilities in Greece and France. This intervention on the monument is carefully designed in order to be distinct from the historic construction and is also completely reversible (Fig. 39, 40 and 41).

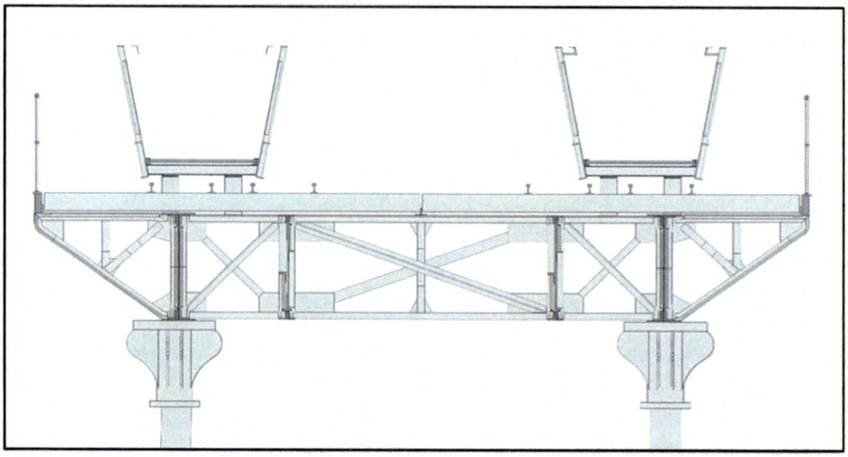

Fig. 38. The new pathways are elevated and distinct in relation with the historic metal construction.

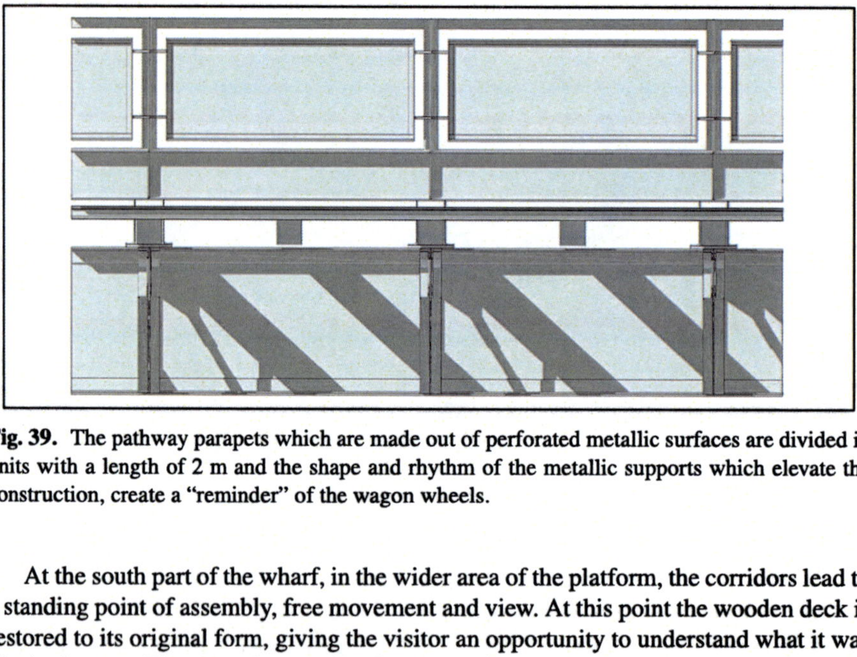

Fig. 39. The pathway parapets which are made out of perforated metallic surfaces are divided in units with a length of 2 m and the shape and rhythm of the metallic supports which elevate the construction, create a "reminder" of the wagon wheels.

At the south part of the wharf, in the wider area of the platform, the corridors lead to a standing point of assembly, free movement and view. At this point the wooden deck is restored to its original form, giving the visitor an opportunity to understand what it was like in the past. The more dense rails, the rotary plates located there, and the location of the crane as it used to be, complete the image of the loading site (Figs. 42 and 43).

Fig. 40. Wagons and workers on the Wharf during the 19th c. (Source: Marmanis Collection)

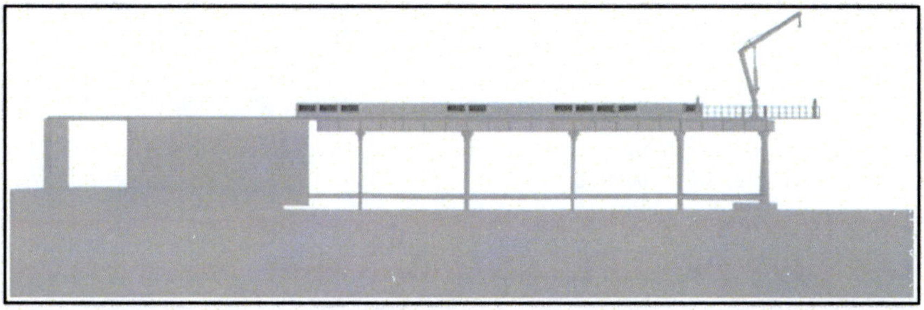

Fig. 41. Design of the proposal with the parapets placed randomly on the new corridor on the metallic part of the wharf.

The metal pathways stop when they reach the wharf's masonry structure. The wagon rails continue until the northern retaining wall. In the future a pedestrian bridge will symbolically extend the line of the wagon rails towards the ex FCML warehouses, which will become a cultural-leisure area. Visitors are protected by iron hand rails constructed according to the authentic design and placed on the east and west side. In order to ascend the wharf's upper platform, a new linear safety staircase is designed, parallel to the east historic staircase (Fig. 44).

On the east side of the wharf the metallic pier was locally destroyed in order to place the mechanism and the belt for the automated transportation of the products to the ships. The part of the pier that was damaged for the above reason will be repaired, but in remembrance of this newer historic phase the diagonal axis of transportation will be preserved.

Fig. 42. Historic photo in which the wooden deck that covered the metallic part of the wharf can be seen (Source: Manthos collection)

Fig. 43. View towards the south part of the Wharf (Computer generated image)

Accessible circulation on the upper platform of the wharf is assured by the use of a specially designed elevator (a vertical platform lift), mounted on the east side of the stone wall and powered by solar energy. When not in use, the platform will be lowered and hidden underground in a specially designed enclosed space.

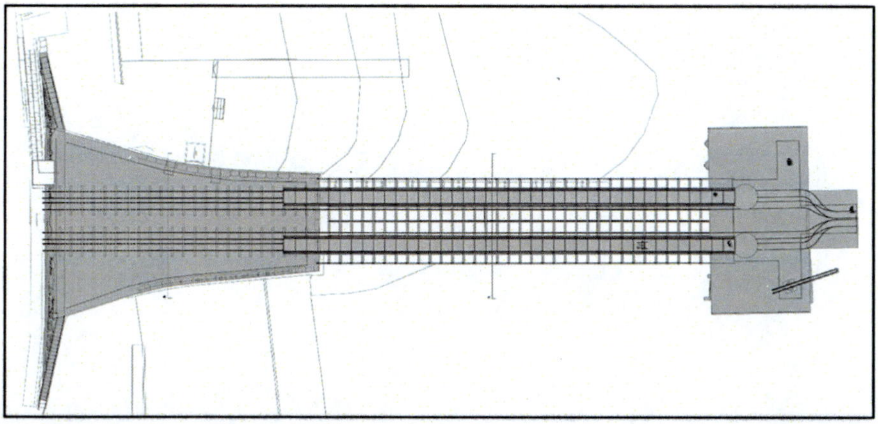

Fig. 44. On all the upper surface of the wharf, all the mechanic parts are restored and at least one crane is placed back at its original place, in order to give an educational dimension to the monument.

The upper deck will be fully accessible.

The new construction materials will be compatible with the existing ones and resistant to sea water corrosion. Stainless steel will be used where necessary, in concrete reinforcements, in handrails, in rails, grids, perforated sheets, vertical platform lift etc.). Exceptionally, where possible, preoxidised CORTEN type metal parts may be used.

Efforts will be made to preserve and repair as many authentic parts of the wharf as possible. Before removing an item, a detailed survey and cataloguing will take place in order to reposition it in the exact place, without losing any historic data.

The existing metallic elements after adequate conservation will be coated with anti-rust protection paints, with appropriate high quality durable colors, that will refer to the monuments rust color of the past. All parts that must be replaced partially or entirely, will be recognizable by slight morphological differences, without altering the historic image of the monument.

The whole metal wharf will be shielded by cathodic protection.

The methods of repairing and restoring the Wharf are described in detail in a report compiled by the Laboratory of Steel Structures and Soil-Foundations teams (Figs. 45 and 46).

Fig. 45. View of the restored monument (Computer generated image)

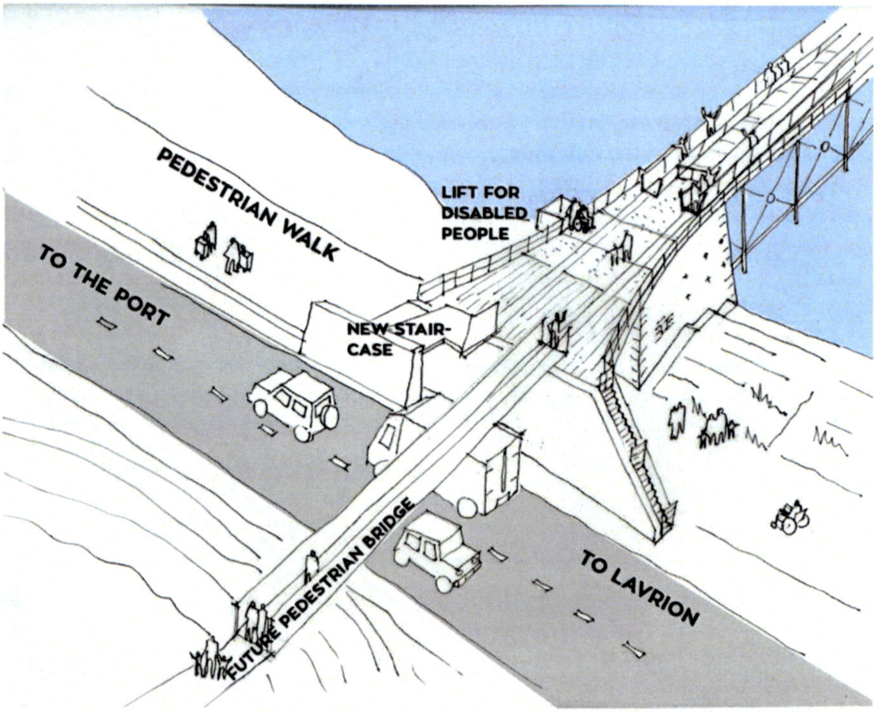

Fig. 46. Sketch of the proposal for the circulation in the surrounding area

6 Conclusions

The French Wharf of Lavrio is a valuable and unique industrial heritage monument, a landmark for the local community that needs immediate rescue measures, and the implementation of an innovative environmentally sustainable conservation-reuse project.

This project must be immediately implemented by the Municipality of Lavrion in order to save the monument from the risk of total destruction and the surrounding area from the risk of fatal degradation.

Interdisciplinary collaboration, carried out by the different Engineering Schools and expert Laboratories of N.T.U.A. has managed to survey and propose innovative solutions to the technical problems concerning the restoration project. The architectural project apart from the conservation-reuse proposal and the constructional details also underlines a more visionary approach for the upgrading of the surrounding area with the wharf as the "epicenter" and focus.

The Protection of industrial monuments is a multi-level scientific, social, economic, political issue, but above all it is a fundamental step in order to understand, evaluate, preserve and highlight the 19th–20th century Cultural Heritage of Greece and Europe.

References

Αρχιτεκτονικά Αποτυπώματα βιομηχανικής Αρχαιολογίας στο Λαύριο. Από την τεκμηρίωση στην επανάχρηση./Architectural footprints of Industrial Archaeology in Lavrion. From documentation to Reuse. Edition: National Technical University of Athens/School of Architecture (2010)

Collette, Q.: Riveted Connections in Historical Metal Structures (1840–1940). Hot-driven Rivets: Technology, Design and Experiments (2014). https://doi.org/10.13140/2.1.3157.2801

Collette, Q., Wouters, I., Lauriks, L., Verswijver, K.: Riveted connections in historical iron and steel structures: one turbulent century of technological, structural and geometrical considerations (1840–1940). Matériaux Techniques **100**, 137–154 (2012). https://doi.org/10.1051/mattech/2012023

Construire en acier: structures, enveloppes, assemblages et composants. Le Moniteur, Paris (1993)

Δερμάτης Γ. ΛΑΥΡΕΙΟ ΤΟ ΜΑΥΡΟ ΦΩΣ Η μεταλλευτική και μεταλλουργική βιομηχανία στο Λαύρειο 1860–1917 Ελληνική και Ευρωπαϊκή Διάσταση Εκδοση Τεχνολογικού Πάρκου Λαυρίου Ε.Μ.Π.2003

Δούση, Μ.: Η αρχιτεκτονική και τα δομικά συστήματα των ιστορικών σιδηρών κατασκευών. 18ος - 20ός αιώνας. Αθήνα: University Studio Press (2005)

Riba Stirling Prize: Hastings Pier ByDrmm 2017 24 October. https://www.architectsjournal.co.uk/buildings/riba-stirling-prize-hastings-pier-by-drmm/10024451.article

Schulitz, H.C., Habermann, K.J., Sobek, W.: Steel Contruction Manual. Birkhauser Edition-Detail (2012)

Contemporary Challenges of Residential Buildings from Mid-twentieth Century: For a Renewed Consciousness of Heritage Preservation and Energy Sustainability

Luciana Rocha(✉) and Rui Fernandes Póvoas

Center for Studies in Architecture and Urbanism (CEAU-FAUP), Via Panorâmica S/N, 4150-564 Porto, Portugal
lrocha@arq.up.pt

Abstract. This paper seeks to reflect on the relationship between preventive measures for the protection of built heritage and the implementation of sustainability strategies. The research aims to contribute to the safeguarding of the built heritage and the improvement of construction conditions, promoting the greater thermal comfort and energy efficiency of buildings. Through both the analysis of a specific programme, the 'Improvement Plan' for the city of Porto (1956–1966), including its subsequent extension (1966–1972), and the in-depth study of specific cases that were the subject of recent interventions, the aim is to identify and characterise the actions undertaken and their impacts in terms of preserving the cultural value of the buildings, as well as on the respective improvements at the level of sustainability. This research also seeks to demonstrate the importance of the user's experience in developing intervention strategies to improve living conditions and, in this way, to promote the effective involvement of the population in the maintenance of housing.

Keywords: Housing · Heritage preservation · Thermal comfort · Energy sustainability

1 Introduction

This paper is part of a broader ongoing research, based on the comparative analysis of case studies, both with and without recent intervention actions. The study seeks to establish a relationship between the identification and characterisation of the actions undertaken and the definition and application of new intervention strategies, in order to contribute to the safeguarding of architectural heritage, even current housing, and to the improvement of the thermal comfort and energy efficiency of buildings.

The research work focused on the collection of data from the original project, contemporary interventions and the current state of building complexes. The analysis of the original project relied on the documentary corpus available at various Portuguese

I. Vayas and F. M. Mazzolani (Eds.): PROHITECH 2021, LNCE 209, pp. 1344–1358, 2022.
https://doi.org/10.1007/978-3-030-90788-4_103

Institutions (IHRU[1], SIPA[2], LNEC[3], municipalities, etc.) and compiled in the database developed by the research project 'Mapping Public Housing'[4], referring to the architectural design: plans, cross sections, elevations, photographs, location, urban insertion, authorship, relevant dates or facts, the public entity or department promoting the project, and bibliographical references, also complemented with primary sources such as archival research; the collection of data on intervention projects was based on interviews with the architects; and the diagnosis of the current state of these building complexes was based on the collection of data *in situ* (visits to the buildings and photographic records). This paper analyses three case studies – the ***Cerco*** **housing complex** (1963); the ***Engenheiro Machado Vaz*** **housing complex** (1966); and the ***Falcão*** **housing complex** (1973) – located in the eastern part of the city of Porto, in Campanhã, and built under the scope of the 'Improvement Plan' for the city of Porto, with different approaches and constructive solutions and different levels of intervention, which consequently impacted on the formal and functional aspects of the original constructions.

The Cerco housing complex, originally built in 1963, comprised 804 dwellings in 32 buildings of four or five storeys, with two access systems (vertical and gallery) corresponding to different types of solutions. The neighbourhood designed by Vasco Mendes and Miguel Reimão Pinto was enlarged in 1991, with two new buildings and 88 more dwellings. This housing complex is located between Rua do Cerco do Porto, Rua Vila Nova de Foz Côa and Rua Peso da Régua, and occupies an area of some considerable size[5]. The ***Engenheiro Machado Vaz*** **housing complex** is located near the Contumil railway station. The original project, designed by Vasco Mendes, has 272 dwellings in 13 multifamily housing buildings of four storeys, with vertical accesses. This plan forms a grid composed mainly of groups of buildings positioned parallel to each other and at right angles or at an oblique angle to the street[6].

The ***Falcão housing complex*** is close to the Jardim da Corujeira. The project was divided into two phases: the first, designed by Rui Paixão, with 224 dwellings contained within 9 multifamily housing buildings with a vertical system of access; and the second, by Florêncio de Carvalho, with 160 dwellings in 6 buildings. This article mainly analyses

---

[1] Instituto de Habitação e Reabilitação Urbana (Institute for Housing and Urban Renewal).

[2] Sistema de Informação para o Património Arquitetónico (Architectural Heritage Information System).

[3] Laboratório Nacional de Engenharia Civil (National Laboratory for Civil Engineering).

[4] **Mapping Public Housing** is a project carried out within Atlas da Casa [AdC], one of the research groups of the Centre for Studies in Architecture and Urbanism [CEAU] at the Faculty of Architecture of the University of Porto [FAUP]. Further bibliographical references about the neighbourhoods can be found on the MdH online database/Project List: [https://mappingpublichousing.up.pt/en/].

[5] Data taken from the MdH online database/Project List: [https://mappingpublichousing.up.pt/en/].

[6] Data taken from the MdH online database/Project List: [https://mappingpublichousing.up.pt/en/]. The intervention project refers to 316 dwellings and 16 blocks, with three of these being from the older *São Roque da Lameira* housing complex.

the actions resulting from the implementation of the general intervention project on the buildings built in the first phase[7].

The methodology for this research thus results in the interpretation and comparison of the data gathered in order to substantiate the main potentialities and limitations of the complexes and their interventions, as well as to identify the main measures needed to improve the comfort of the dwellings.

In the next phase, this analysis will be further complemented by the development of sensitivity studies *in situ* for a complete evaluation of the effect of the different intervention levels in improving the thermal comfort of buildings.

2 Learning from the Past. A General Overview

By analysing the historical framework of the ensembles – focusing on a general overview of the 'Improvement Plan' for the city of Porto and on the architectural and constructive system of the collective housing from mid-twentieth century – one can understand the social, economic and political constrains with impact on the constructions standards of the buildings and recognize the original projects characteristics and their potentialities and limitations. Furthermore, this analysis may also clarify how have those buildings subsisted throughout time and what are their major needs to improve their living conditions.

2.1 The 'Improvement Plan' for the City of Porto

The 'Improvement Plan' for the city of Porto (Plano de Melhoramentos: CMP 1966) (Fig. 1), approved by Decree-Law No. 40616, of 28 May, 1956 (MOP 1956), had as its main objective to help solve the housing problem of the city's unhealthy neighbourhoods. Furthermore, this housing programme also presented an opportunity for expanding the city in keeping with the urbanisation plans developed by the municipality. Thus, the 'Improvement Plan' envisaged the construction of a minimum of 6000 dwellings within ten years and the implementation of an urbanisation programme that, in addition to the creation of expansion areas, sought to demolish the insalubrious neighbourhoods and proceed to their respective rehabilitation, as well as to develop the urbanisation of the *Campo Alegre* area and its access infrastructures[8].

Following the implementation of this plan, Decree-Law No. 47443, of 30 December, 1966, approved its further expansion, envisaging the construction of more than 3000 new dwellings by 1972 (MOP 1966).

[7] Data taken from the MdH online database/Project List: [https://mappingpublichousing.up.pt/en/]. The Project Brief developed by the intervention team refers to 231 dwellings, but includes 7 dwellings located in semi-basements not represented in the original documents (Calapez and Sousa 2016).

[8] Further information about the 'Improvement Plan' at Pereira (2016) and Queirós (2016).

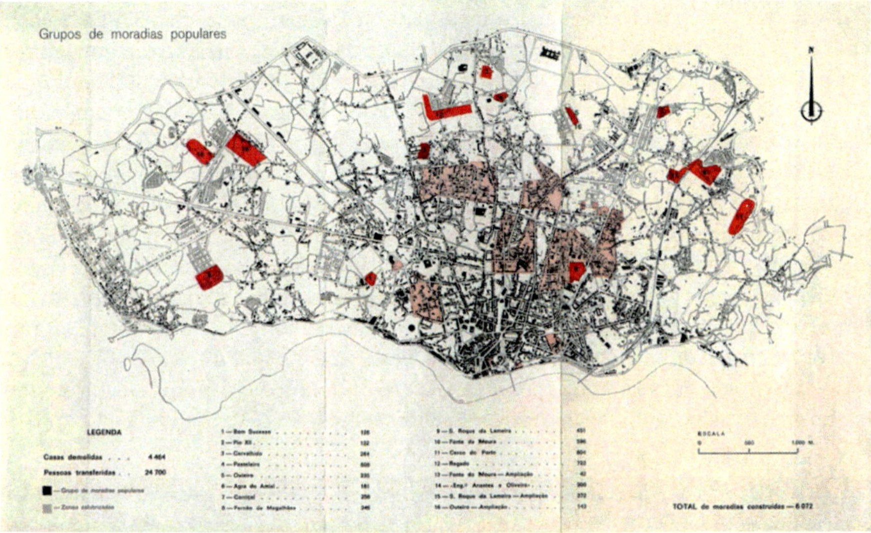

Fig. 1. 'Improvement Plan' for the city of Porto. Housing complexes location (CMP 1966, 17)

2.2 Architectural Analysis

The residential complexes built under the 'Improvement Plan' for the city of Porto now constitute an architectural heritage that, although it does not display any particularly exceptional architectural quality, reveals evident signs of modernity, as well as some innovative formal and functional characteristics. This common language is highlighted mainly by the urban concept of the whole development, but also by the experimental use of different access systems, typologies and architectural languages.

These complexes follow common principles of urban design, being characterised by lots of a significant size, with parallel buildings, green areas, footpaths and roads and various connections with the adjacent urban structure (Fig. 3).

The most common access systems consist of vertical communication cores composed of a staircase providing access to two dwellings (left-right access, e.g. *Falcão* and *Machado Vaz*) or connecting to a distribution gallery with access to multiple dwellings (access by gallery, e.g. *Cerco,* Fig. 4).

The experimentation with the access systems derived mainly from the diversity of solutions found in these galleries and the advantages that were associated with this circulation area, due to the economy of means (access to several apartments through a single vertical communication core) and to the multifunctional character of this space, which served as an area both for socialisation and transition. In the *Falcão* housing complex, the staircase also acquired exceptional prominence in the architectural language of the development as a whole.

The various housing typologies resulted from the necessary reduction in the areas of the dwellings, the consequent compactness of their available space and the rationalisation of their functional programme. The living area, which served as the central common space, acquired a multifunctional character, either by connecting to the kitchen area (e.g.

Machado Vaz) or by incorporating the circulation areas (e.g. *Cerco,* Fig. 3). The kitchen tends to be a minimal, rational and functional space and either appears as an autonomous area or is incorporated into the common space. This typological experimentation resulted in a gradual improvement in the design of the domestic space, in what amounted to a progressively less typified process, as can be seen in the case of the *Falcão* housing complex.

The architectural language of these complexes derived mainly from the rationalisation of the design and construction process, which was necessary in order to comply with the assumptions of low-cost construction. However, the absence of a hierarchy between the façades (front/rear) and the incorporation of grilles into the façades to hide the service areas are clear signs of modernity. Some buildings are also notable for the significant contribution that they make to the overall design of the complex. The *Falcão* housing complex, for example, combines the use of traditional materials, such as stone or bricks in the side walls, with new application techniques (Fig. 2).

Fig. 2. Architectural language. *Falcão* housing complex. 1973. ©Arquivo Histórico Municipal do Porto

2.3 The Constructive System for Collective Housing in the Mid-twentieth Century

The buildings constructed in Portugal in the mid-twentieth century belong to a period of transition characterised by the progressive use of reinforced concrete in current housing construction.

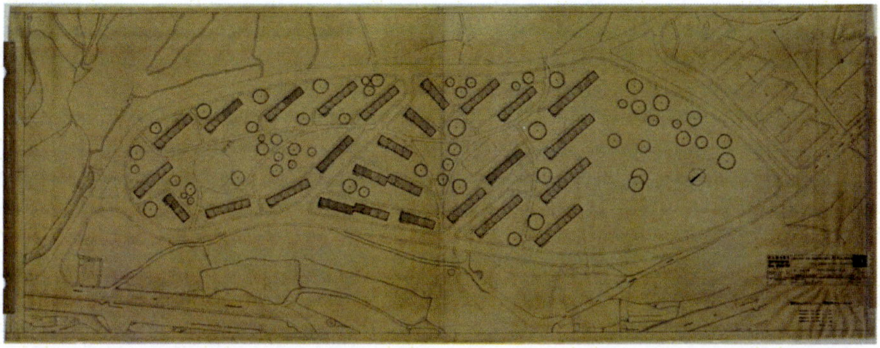

Fig. 3. Urban concept. *Cerco* housing complex. ©Arquivo Histórico Municipal do Porto

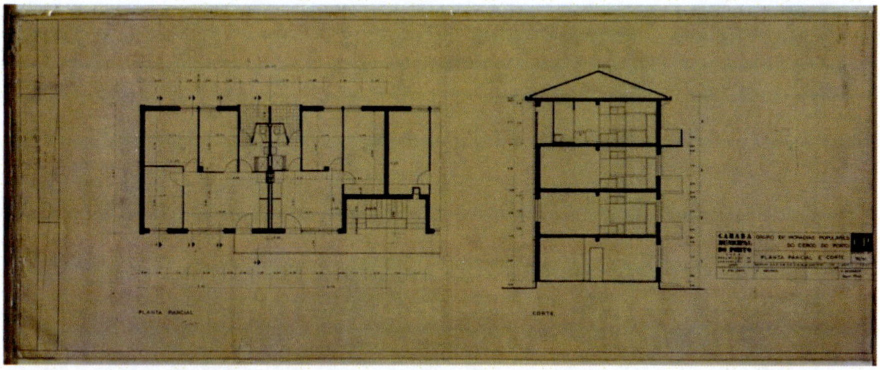

Fig. 4. Typologies and access systems. *Cerco* housing complex. ©Arquivo Histórico Municipal do Porto

A recent study by Vale and Abrantes (2010, 2012), seeking to characterise housing construction in Porto from the second quarter of the twentieth century onwards, refers to common alterations in the construction system involving the use of horizontal and vertical supporting elements. The floors of early twentieth-century buildings, which typically consisted of a wooden structure supported by party walls, increasingly gave rise to reinforced concrete slabs, especially after the 1940s. This change was mainly due to requests to use incombustible materials in view of the fire risk (Vale and Abrantes 2012, 3). The first slabs were solid and of lower thickness, with small spans between walls or support beams. Later, slabs of greater thickness were introduced, lightened through the use of perforated ceramic elements (Vale and Abrantes 2012, 4).

The walls consisted of supporting elements made of stone masonry, mostly granite, concrete blocks and/or brick, sometimes complemented with concrete pillars. These elements were progressively used for reinforcement and bracing, although the exclusive use of a reinforced concrete portico structure, replacing the resistant masonry wall, only occurred in exceptional cases (Vale and Abrantes 2012, 5).

The roofs are mostly sloping, tiled and with a wooden structure, although some exceptions have been identified containing flat reinforced concrete roofs (Vale and Abrantes 2012, 4).

The case studies display similar original building systems, consisting of exterior walls in stone masonry, concrete slabs and sloping tiled roofs (e.g. *Cerco*, Fig. 5). The roof structure, at least in the *Engenheiro Machado Vaz* housing complex, seems to be composed of brick walls, supporting beams and a substructure for the roof tiles. These neighbourhood buildings also have a granite base in the exterior walls (Fig. 6).

The *Falcão* housing complex, built at a later date, differs in the use of both double external walls and distinct building materials, with stone masonry in the side walls and brick in the remaining exterior walls. These buildings also seem to already incorporate a portico structure.

Fig. 5. *Cerco* housing complex. Construction phase. 1961. ©Arquivo Histórico Municipal do Porto

Fig. 6. *Machado Vaz* housing complex. Construction. 1965. ©Arquivo Histórico Municipal do Porto

3 Contemporary Interventions. Analysis of the Case Studies

The Porto Municipal Council, through Domus Social E.M.[9], has undertaken several interventions to promote the rehabilitation of social housing in the municipality. These recent interventions, mainly taking place during the last decade, have mostly focused on the general maintenance of the buildings, through the treatment of the façades, roofs and common spaces. Nevertheless, there are also some cases with more profound interventions and/or essential actions made in the common exterior areas.

This research is based on the specific analysis of three case studies, each involving different types of general interventions and approaches in terms of architectural features

[9] Domus Social, EM, is a housing and maintenance company run by the Porto Municipal Council, formed in 2000 under the scope of Article 19 of Law No. 50, of 31 August, 2012. Further information at: www.domussocial.pt.

and constructive details. These cases display different types of intervention in the envelope (façade and roof), resulting in distinct levels of preservation and transformation, with obvious consequences for the balance between architectural heritage protection and thermal comfort adjustments.

3.1 Reinterpretation of the Materiality Through a Second Skin Façade. The Case of the Cerco Housing Complex

The general intervention in the *Cerco* housing complex, commissioned by Domus Social E.M., consists of three execution phases with different architects: the intervention in buildings with 'left-right' access by architect Virgínio Moutinho[10] (completed); the intervention in buildings with distribution galleries by architect José Gigante[11] (ongoing); and the project for the surrounding public space by architect José Manuel Soares (under development).

This research focuses specifically on the project for the buildings with distribution galleries because of the distinctive characteristics of the intervention (Fig. 7). This project involves rehabilitating the constructive envelope (walls, exterior window frames and roofs) and common accesses (stairs and galleries), although the intervention is also compatible with a possible subsequent typological reconfiguration of the dwellings, already presented in a previous study.

The creation of simple exterior walls in stone masonry involves the application of a new layer of ceramic bricks over thermal insulation (extruded polystyrene). This option is designed to increase the thermal performance of the existing wall and improve the comfort of the dwellings by preventing condensation on the interior surfaces of the building (Gigante 2017, 4). Furthermore, the durability and strength of the ceramic bricks should ensure a better performance over time, without the need for any significant maintenance works (Gigante 2017, 5).

The project also proposes the replacement of the original exterior wooden window frames (already damaged or illegally altered) with double-glazed aluminium window frames, aluminium blinds and self-adjusting ventilation grilles.

The intervention on the roofs includes the reconstruction or reconfiguration of the structure, the replacement of the ceramic roof tiles (including new ventilation tiles), the application of thermal insulation over the floor slab, and the installation of casings, capping and other zinc elements and drop tubes (Gigante 2017, 5).

The intervention in the common areas includes the replacement of infrastructural network systems. Furthermore, the project contemplates the demolition of the illegally enclosed areas on the ground floor and their replacement with delimited yards, as well as the application of metal structures to the laundry windows to aid the drying process.

[10] The Virgínio Moutinho intervention also included the buildings from the extension to the *Cerco* housing complex, built in 1991.

[11] This project is being implemented in keeping with the Hygrothermal Study made by the engineer Vasco Peixoto de Freitas.

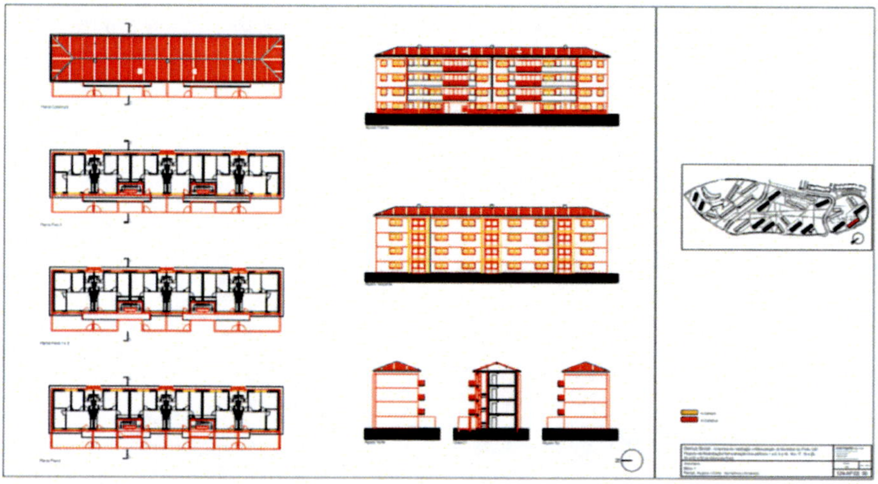

Fig. 7. *Cerco* housing complex (ongoing). Intervention project - Building 1. Courtesy José Gigante

3.2 Comfort Improvement with the External Thermal Insulation Composite System (ETICS). The Case of the Engenheiro Machado Vaz Housing Complex

The *Engenheiro Machado Vaz* housing complex was the subject of a general intervention completed in 2017 (Fig. 8). This intervention was based on the architectural project developed by the architect Nuno Abrantes and the specialities project developed by the engineer Vítor Abrantes. This project focused mainly on the façades, roofs and common areas, seeking to solve the existing problems of visible damage, but also improving the comfort of dwellings, promoting the energy efficiency of buildings and the adequate ventilation of interior compartments. The project also included the exterior areas, such as common public spaces and a small playing field.

One of the main interventions on the façades included the application of a new coating material, the External Thermal Insulation Composite System (ETICS), with benefits for the thermal comfort of the dwellings. This solution had no structural implications and also allowed for the preservation of the existing granite base (Figs. 9a and 9b). Other interventions involved the replacement of the existing window frames and the introduction of new frames (made of anodised aluminium) on the balconies, clothes washing areas and staircases in order to close off these spaces. The window frames include ventilation systems designed to reinforce the circulation of air in the interior spaces. These interventions include the main alterations that were made to the architectural language of the complex as a whole, particularly in the clothes washing areas, originally marked out on the façade with horizontal strips of cement, and now replaced with glass bricks and window frames.

The most prominent alteration to the roof involved the replacement of the structure and the installation of solar panels, while, in the common areas, in addition to the general maintenance of the coating materials (floors and walls), the intervention included the reformulation of the common infrastructures for water supply, the drainage of rainwater, and the electricity, gas and telecommunication systems.

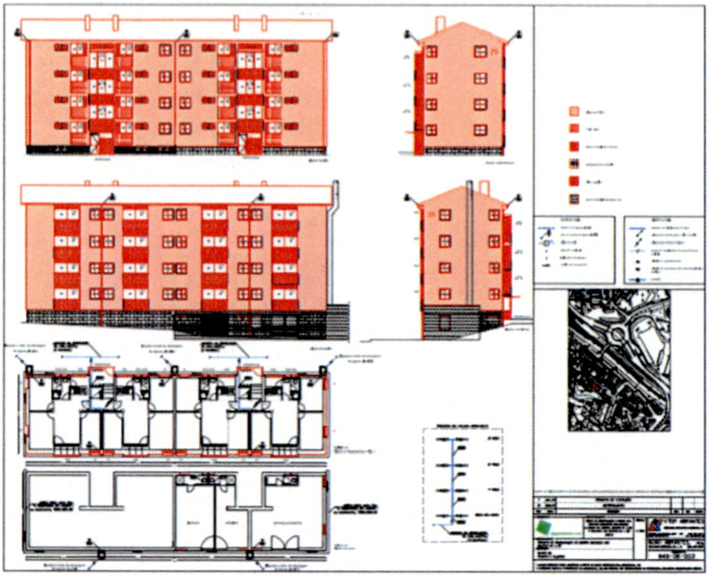

Fig. 8. *Machado Vaz* housing complex. Intervention project. Courtesy Nuno Abrantes

(a) (b)

Fig. 9. a and b *Machado Vaz* housing complex. General views. 2018. ©Luciana Rocha

3.3 Preservation of the Building Materials with Specific Adjustments. The Case of the Falcão Housing Complex

The first phase of the *Falcão* housing complex was marked by its expressive use of exterior materials, resulting in a vernacular architecture with some modern features, combining granite (on the side walls) with ceramic bricks (the predominant material) and exposed concrete (the guards for the stairs, the lines of the floor slabs visible in the façade, the grilles of the clothes washing areas) (Figs. 11a and 11b).

The intervention in this neighbourhood took place in 2017 and the project was developed by Castro Calapez Arquitectos Lda., under the coordination of the architect Paulo Calapez (Fig. 10). As in the previous cases, this project focused mainly on the treatment of the façades, roofs and common areas. However, this intervention was marked by its attempt to preserve as much as possible of the exterior walls. As such, the interventions mainly corresponded to the treatment and maintenance of existing materials (including the removal of elements illegally affixed to the façades), the replacement of the exterior wooden window frames with anodised aluminium frames with interior blinds and the installation of a new aluminium structure in the clothes washing areas. One of the main changes corresponded to the closure of the staircase and the common accesses through the introduction of iron window frames in the intermediate levels and the closure of the main entrances with a partial wall made from similar ceramic bricks, intended for the installation of the mailboxes. The most notable interventions in the common areas were the maintenance and recovery of the existing materials and the reinforcement of the common infrastructures. The dwellings also benefited from the installation of new ventilation ducts and extraction fans in the kitchens, clothes washing areas and bathrooms. The intervention on the roofs included the replacement of the ceramic tiles, the application of thermal insulation and the treatment of specific features, such as the eaves and the ridges of the roof, and the installation of solar panels.

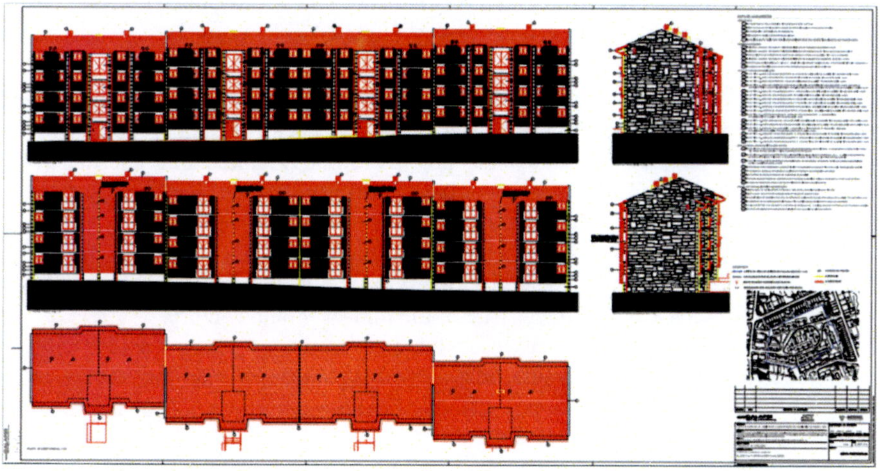

Fig. 10. *Falcão* housing complex. Intervention project. Courtesy Paulo Calapez

(a) (b)

Fig. 11. a and b *Falcão* housing complex. General views. 2018. ©Luciana Rocha

4 An Interpretative Analysis. Architectural and Constructive Challenges

This article is based on the specific analysis of three case studies, involving housing complexes where interventions have recently been made, with the aim of identifying the specific actions that were taken and analysing the formal (architectural) and functional (constructive) impact of these solutions, paying particular attention to the relationship between built heritage preservation and housing comfort improvement.

The case studies consist of multifamily housing complexes built under the scope of the 'Improvement Plan' for the city of Porto, based on similar architectural and construction principles and characteristic of the mid-twentieth century.

The general interventions focused on the constructive envelope (façade, window frames and roof) and the common spaces (accesses), resulting from the need to solve technical issues (infrastructures and visible damage) and, above all, to improve the thermal comfort of the dwellings.

The analysis reveals that these housing complexes present a high level of adaptability and tolerance to transformation. Despite the different types of intervention and the varying degrees of change in the architectural language of the buildings, the solutions have tended to follow an essentially similar overall approach.

The strategy involving the highest degree of change (*Cerco*) introduces significant alterations in the language of the complex, namely by applying a new layer of ceramic bricks, different from the original façade. However, this proposal has advantages in terms of the durability and strength of materials and also seeks to combine the thermal insulation of the walls with space ventilation strategies that have a significant potential for improving the energy performance of the dwellings. In contrast, the strategy that involved the fewest transformations (*Falcão*) gave importance to the preservation of the exterior walls. Thus, the potential improvements result mainly from the other changes, namely to the exterior window frames (replacing the windows and closing off the staircase) and the roof (thermal insulation inserted over the roof slab). This solution was important in terms of recognising the building's architectural features, but it may nevertheless result in a poorer performance in terms of thermal comfort.

The intervention at the *Engenheiro Machado Vaz* housing complex is an intermediate strategy, since it seeks to reproduce the language of the existing façade, although with the application of a new exterior coating material and with several alterations to the exterior frames. This solution preserves some characteristic features of the buildings such as the granite base and the tiled roofs, but it mainly seeks to adapt the buildings to the comfort needs of the dwellings.

The questions that now arise relate to the performance of these solutions over time, in view of the patterns of use demonstrated by the inhabitants of these complexes: are the most transformative strategies really rewarding in terms of thermal efficiency? Or can the same results be achieved with less significant interventions? The future development of sensitivity studies *in situ* will make it possible to analyse the performance and the real benefits of different strategies, and thus to develop tools to achieve compromise solutions between preservation and transformation.

Acknowledgements. This paper was developed with the support of Center for Studies in Architecture and Urbanism at Faculty of Architecture of the University of Porto (CEAU-FAUP) and funded by the Foundation for Science and Technology (FCT, Lisboa, Portugal), under the project UIDB/00145/2020.

References

Calapez, P., Sousa, S.: Projecto de execução. Memória Descritiva. Tomo 1 – Projecto para a reabilitação do "envelope" do Bairro do Falcão, Blocos 1 a 9, Porto (2016)

Câmara Municipal do Porto: Plano de Melhoramentos, 1956–1966, CMP, Porto (1966)

Gigante, J.: Anteprojecto Bloco 1. Memória Descritiva, Projecto de reabilitação/Remodelação dos edifícios 1 a 3, 5 a 10, 16 e 17, 19 e 20, 30 a 32 e 33 do Cerco do Porto, Projecto Geral de Arquitectura, Porto (2017)

MINISTÉRIO DAS OBRAS PÚBLICAS (MOP 1956): Decreto-Lei N. 40616. Diário da República, I Série, No. 108, pp. 629–633, 28 May 1956

MINISTÉRIO DAS OBRAS PÚBLICAS (MOP 1966): DECRETO-LEI N. 47443. Diário da República, I Série, No. 302, pp. 2358–2359, 30 December 1966

Pereira, V.B.: A Habitação Social na Transformação da Cidade. Sobre a génese e efeitos do "Plano de Melhoramentos para a Cidade do Porto" de 1956. Edições Afrontamento, Porto (2016). 978-972-36-1478-7

Queirós, J.: O 'Plano de melhoramentos para a cidade do Porto' de 1956: Enquadramento político-social e elenco de realizações. In: Pereira, V.B. (ed.) O Estado, a habitação e a questão social na cidade do Porto, Edições Afrontamento, Porto, pp. 37–64 (2016)

Vale, C., Abrantes, V.: Os processos de licenças de obra na cidade do Porto. Caracterização construtiva do património edificado entre 1911 e 1940, no eixo da Boavista, In: Reabilitar 2010 - Encontro Nacional Conservação e Reabilitação de Estruturas, LNEC, Lisbon (2010)
Vale, C., Abrantes, V.: Entre tradição construtiva e modernidade arquitectónica. Caracterização construtiva da habitação corrente da cidade do Porto no segundo quartel do século XX, In: PATORREB 2012 – 4° Congreso de patología y rehabilitación de edifícios, Santiago de Compostela (2012)

DOCK SHEDS in VALENCIA PORT. Historical Memory of *Tinglados* in Valencia Port, Damages Registered in *Tinglado* #2 and Recent Intervention

Ernesto Fenollosa(✉), Verónica Llopis, Iván Cabrera, Ricardo Perelló, and Adolfo Alonso

Department of Mechanics of the Continuous Medium and Theory of Structures, Universitat Politècnica de València, Cno de Vera, s/n, 46022 Valencia, Spain
efenollo@mes.upv.es

Abstract. Sheds (Spa. *Tinglados*) located next to the docks of Valencia Port, are heritage buildings originally intended for storage of goods. They are formed by a repetitive metal module between masonry constructions. By their composition, rhythm, monumentality and ornamentation they have become an interesting example of the Valencian industrial architecture. Nowadays they are cataloged like Building of Local Relevance. Since their construction, these buildings have suffered the vicissitudes of history. For example, they were bombed during the Spanish Civil War and later two of the initial six buildings were demolished. In addition, they show the damages caused by the aggressiveness of maritime environment and the condition of various events held in its proximity. There have been several projects that have enabled their reconstruction, restoration of structural and decorative elements and reform to adapt to new functional requirements. Tinglado #2, which had not been intervened since 1984, came to present an image of worrying degradation mainly due to corrosion of the structure and roof, cracking and spalling of items of masonry facades and the presence of improper facilities. This paper will explain the form, function, construction and structure of this interesting industrial architecture. It will review the effect of the passage of history and time, will describe the extent of damages achieved in Tinglado #2 and will show the intervention to which it has recently been subjected.

Keywords: Tinglado · Shed · Industrial architecture · Heritage · Dock · Valencia · Restoration

I. Vayas and F. M. Mazzolani (Eds.): PROHITECH 2021, LNCE 209, pp. 1359–1372, 2022.
https://doi.org/10.1007/978-3-030-90788-4_104

1 Introduction

Sheds named as *Tinglados*[1], located in the Valencia Port, have achieved more than 100 years of presence in the port landscape of the dock.

Hermosilla defined them like "one of the examples more interesting of the Valencian industrial architecture in which, the industrial and functional character is combined with the artistic and monumental" (Hermosilla 2007). The industrial character was the result of the rise of the industrial revolution and a common characteristic among the public buildings of the time, such as markets and railway stations. The monumentality is evidenced by the repetitive rhythm of the main module as by the fine buildings arranged like endings of the volume (Fig. 1). Its construction provided a new facade to the dock, changing the landscape of the port.

Fig. 1. Facade of Tinglado #2 and #3. "Project for the reform and expansion of covered warehouses for merchandise in sections 1, 5, 6 and 7. Fuster, J. (1911). General Archive. Port Authority of Valencia.

Tinglados of the Port of Valencia are buildings initially destined for the storage of goods. Although 6 sheds were built, today there are only three left (#2, #4 and #5). All of them are cataloged as Local Relevance Assets (BRL Nos 11.12.10, 11.12.08 and 11.12.09) and they are included in the Catalog of Buildings and Places which are protected by the General Urban Plan of Valencia (Latorre Hernández 2010).

Their general description according to the index card of the named Catalog is: "*Tinglados are composed by three parts: two prismatic volumes at the ends and a succession of uneven decks on slender supports in the center. The composition is classic, with symmetry in the layout of the masonry buildings. In the intermediate space, large decks are built with a rhythm of large spans according to its use as a warehouse*".

2 The Origin of *Tinglados*

During the second half of the nineteenth century, agricultural production experienced a high growth and the export of products became the specialty of the port, which required infrastructure improvements. Exports were multiplied by 25 and the need of new sheds

[1] According to the Dictionary of the Royal Academy of the Spanish Language, Tinglado comes from the old French word "Tingle". Of the different meanings, in our case, the term refers to "Shed". The VOX Spanish Language Dictionary defines it as "Covered place to contain controlled goods". Valencia's sheds were initially known as "*Deposito*" in the Projects designed by Fuster. They adopted the name *Tinglados* from the "Memories about the works" written by Vilar in 1942.

to keep the products became increasingly important. Although the construction of these sheds was previously considered, it is in the Manuel Maese project (1895), where they acquire true importance (Maese 1895). This project included the construction of 5 closed buildings with different dimensions but similar typology. These are lengthened plan buildings with a slight curvature following the curved shape of the dock. They were designed with a metal structure following a repetitive module both on the facade and on the floor plan. All of them included a series of features and ornaments that responded to the commercial architecture of their time, the use of materials such as steel and naked brick and decoration inspired by various activities commercial (agriculture, fishing, etc.). Despite several attempts to tender the works, various causes such as the high price of iron, prevented the execution of the project (Carreres and Domenech 2012).

In 1910, the engineer José María Fuster, Director of the works council of the Valencia's Port, carried out a study of the increase in port activities with a detailed analysis of export products and he proposed the placement of four sheds (Fuster 1910). As Fig. 2 shows, it is a simple building that preserves the modular construction, with a series of gable roofs made with steel structure, scarce ornamentation and without the massive volumes. In the same document, the engineer defined its geometry increasing the initial distance between axes from 12 to 15 m, designing the roof slope and describing its materials and skylights characteristics.

Fig. 2. Facade of Tinglados #4 and #5. "Project of four warehouses in the 1st, 5th, 6th and 7th sections". Fuster, J. (1910). General Archive. Port Authority of Valencia.

Due to the increase in the flow of exports, in 1911 a modification of the 1910 project was made. The area of the sheds, which he called "*Depósitos*", is extended in anticipation of the future Port necessities. At the end of the year 1911, the shed #3 is divided into the final sheds #4 and #5 and the construction of two other new deposits is decided (the current sheds #3 and #6 in the position of the initial shed #4). In this way, the final general plan (Fig. 3) is prepared with the current location of the sheds, where they already adopt a rectangular shape plan.

In the final design of the sheds #4 and #5, the engineer added 3 masonry volumes[2], two at the ends and one in the middle of the building (Fig. 4). In the Shed #2, the central volume was not added due its smaller length (Fig. 1). These volumes have a more monumental character thanks to its greater height with respect to the rest of the warehouse, its

[2] *"The facade of the masonry side buildings is configured as a triumphal arch with two masts at the ends and two cast columns in the entire height of the facade"* Text from the Catalog index card.

composition as a triumphal arch and its profuse decoration. Fuster attributes to Federico G. de Membrillera[3] the inclusion of these changes in the Tinglados final design.

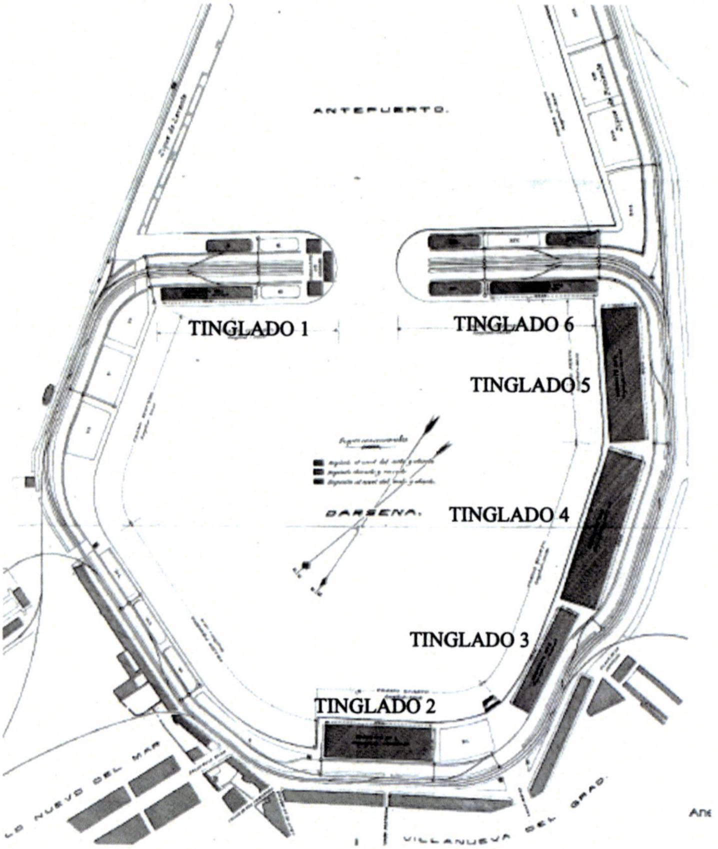

Fig. 3. General Plan of the Sheds in the dock. "Project for the reform and extension of warehouses for merchandise in the 1st, 5th, 6th and 7th sections". Fuster, J. 1911. General Archive. Port Authority of Valencia.

According to Hermosilla, the buildings designed by Membrillera were clearly eclectic with Neo-Baroque ornament. The work of the locksmith and glass, designed on the front and side part of the masonry volumes, stands out.

[3] Federico G. de Membrillera was deputy director of the works council of the Valencia Port.

Fig. 4. Facade of Tinglados #4 and #5. "Project for the reform and extension of warehouses for merchandise in the 1st, 5th, 6th and 7th sections". Fuster, J. (1911). General Archive. Port Authority of Valencia.

3 Construction Works

The construction of Tinglados #1, #4, #5 and #6 was awarded in April 1911 to the company "Land and Maritime Machinist Society" (Fuster 1914). The provisional delivery of the works took place in May 1913, after several delays due to the difficulty of harmonizing the works with the commercial activities (Fig. 5a). However, the complementary works finished later, beginning the full operation of the buildings in August.

The steel structure of the sheds #2 and #3 (Fig. 5b) was awarded to the company "Material Society for Railways and Constructions" in 1912. Its construction begins late due to interference with works of the masonry volumes, so the delivery of these sheds is not made until 1914.

(a) (b)

Fig. 5. a) Tinglados #4 and #5. b) Tinglado #3 in construction. "Report on the status and progress of the works during the year 1913". Unpublished report. Fuster, J. (1914). General Archive. Port Authority of Valencia.

Simultaneously the following complementary works are carried out (Fuster 1914):

- Streets' lighting with metallic filament lamps, which allows to increase loading and unloading works at night;
- Sewerage system, extending the urban sewerage network to the sheds area

- The replacement of the stone's pavement of the sheds 4 and 5 by hydraulic concrete, a work of high difficulty due to the huge commercial movement that had already reached the Port.

4 Structural and Constructive Characterization. Ornamentation

The most representative aspect of the sheds is its naked steel structure, designed following the patterns of the industrial architecture of the time (Benévolo 1999). Its geometry conforms to a rectangular grid of 15 m span in its longitudinal direction and 13.5 m in its transverse side, on whose axes the supports are located. This strict unit provides order and it configures the overall volume of the warehouse.

Primary trusses, of 13.5 m span, were arranged in the transverse direction to the main façade. They are trusses Long-type of 1.10 m edge, made with parallel chords, double diagonal bars and vertical posts.

Secondary trusses, of 15.0 m span are arranged orthogonally to the first ones. These are Pratt-type trusses with slope upper chord (Fig. 6a), which provides the characteristic exterior image of the sheds. Its lower chord has an unusual elliptical shape, vertical posts every 1.5 m and diagonals pulled. Four secondary trusses are arranged for each transverse module, with 4.50 m between them. So, two trusses are supported on the slender columns and other two intermediate on the main trusses. Due to the shape of their chords, the edge of secondary trusses is variable, with an average value of 1.80 m.

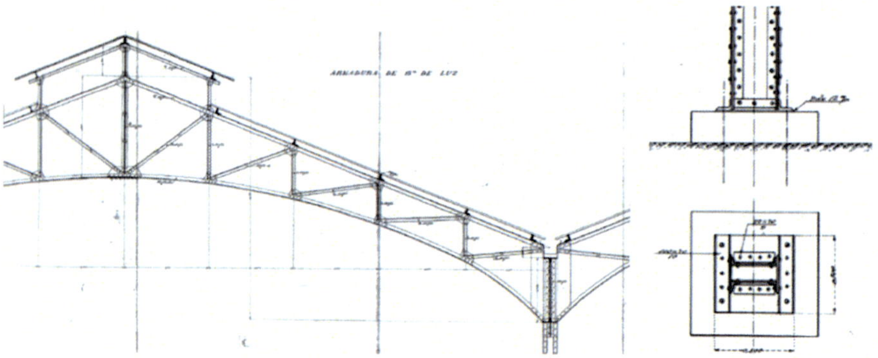

(a) (b)

Fig. 6. a) Pratt-type steel trusses b) Drawing of the columns. "Project for the reform and extension of warehouses for merchandise in the 1st, 5th, 6th and 7th sections". Fuster, J. (1911). General Archive. Port Authority of Valencia.

Roof purlins, with 0.75 m between them are supported on the trusses of sloped chords, so one is placed on the truss joints and the other in the middle of the bar. In addition, the structure has a bracing system formed by minor trusses and X-shape bars.

Angular section profiles have been used in trusses and IPN - UPN in purlins while the original columns are formed by plates and angular sections profiles (Fig. 6b). Some of the supports have been replaced H-shaped profiles, due to the damages caused by the passing of time and its use. The original unions between the pieces are riveted, however welding was used in the parts intervened in the 80 s.

Ornamentation focuses on masonry volumes (Fig. 7). It was used elements such as sculptures, ceramics and lights, whose modernist style improves the unity of the work with respect to the project. The inspiration of these elements was obtained from industrial elements such as anvils, hammers and cogwheels, as well as others related to the sea such as fish, anchors and fishing tackle or vegetables such as fruits and garlands.

Fig. 7. Ornamental elements of the facade of Tinglado #3. Own source.

Two types of natural elements are used in baseboard of decorative tiles (Fig. 8). Oranges and bunches of grapes are represented as a symbol of Valencian regionalism at the end of the s. XIX. And on the other hand, fish and algae remembering the original fishing activity of the place where warehouses are located.

Fig. 8. Detail of the ornamentation with industrial elements and baseboard of decorative tiles. Own source.

According to the Catalog index card: "*In the original project there was a mixture of ornaments from the Viennese Secession with other elements from the II French Empire. The construction works preserved the compositive structure of the project although majority of the II Empire ornamentation was replaced by another in more accordance with industrial Spain*".

5 Tinglados' Vicissitudes

The first important intervention was the replacement of the galvanized iron sheet roof of Tinglado #4 carried out between 1934 and 1935. It was damaged by corrosion due to the aggressive environment in which these buildings are located (Vilar 1942).

The Spanish civil war, 1936–39, was the event that most significantly affected the sheds (Fig. 9a). Roofs, structure and facades of all sheds, as well as other port buildings, were damaged due to the multiple bombings that suffered the city of Valencia. Tinglado #6 was practically destroyed due to being in the most bombed area, while the least damaged was Tinglado #2, since not containing war material it did not represent a priority objective for air raid. After the war, during the 40 s, the port buildings were repaired or rebuilt, depending on the deterioration suffered. Due to the works, last one lost a part of the ornamentation, several stained-glass windows were removed, and the wooden lattices were replaced by iron sheets. A mobile crane of 3 Tm was located between the Tinglado #2 and the dock, so it lost a module of its volume and part of the façade (Fig. 9b). After these works, the sheds recovered their original function as warehouses.

(a) (b)

Fig. 9. Tinglado #2. a) Damages caused by the bombing. b) Reconstruction carried on the 40 s. Photographies of 1982. Port Authority of Valencia.

The next significant event was the demolition of Tinglado #1 at the end of the 60 s, and Tinglado #3 in 1980 to get the space that required the construction of a New Maritime Station. In 1982, the "Tinglado 4 Restoration Project" was carried on, to use it as a car park for users of the future maritime station (Ronda 1982).

In 1984, the "Tinglado 2 Restoration Project" is carried on, whose objective is to improve and recover the original shed aspect to integrate it with dignity in the environment of the Maritime Station (Ronda 1984). This project included the demolition of the annexed constructions and other elements added to the original, the replacement of the roof, the extension of 4.5 m of the warehouse towards the dock, the restoration of all the ornamental elements and the replacement of the baseboard of decorative tiles, the stained glass and other wooden elements lost during the Spanish civil war.

Tinglado #5 is restored between 1993 and 1994 (Ronda 1989) with the same characteristics and objectives and Tinglado #6 is demolished in 1999 (Ronda 1999).

Tinglados stopped being used as warehouses at the end of the 90 s, being used since then in various activities and events. For example, Tinglado #2 began to be used as a free space for diverse public activities use such as skating.

In 2003, Valencia city is chosen as the venue for the 32nd American Cup to be held in 2007. For this reason the environment of the dock was transformed again: it began to be called "Marina Real Juan Carlos I", a new channel of communication with the sea was opened, the building "Veles e Vents" (designed by D. Chipperfield as symbol of the competition) was built and headquarters for teams were raised. Some of them were located next to Tinglado #2 following the edge of the water but others were located in front of the main facade of Tinglado #4 and #5, leaving them isolated from the dock (Fig. 10a). In this event, Tinglados were used to accommodate spectators who followed the regattas through huge screens.

A year later, the first Formula 1 European Grand Prix was held, once again affecting the surrounding area. Specifically, Tinglado #4 and #5 were reformed to adapt to the new necessities (Manzanares et al. 2008). Inside of them, an intermediate floor was built that served as paddock for pilots and the ground floor was adapted for boxes and services. Fortunately, all the added elements were designed on a respectful way with the original decoration. They were used for this purpose between 2008 and 2012 (Fig. 10b). Nowadays they keep the works made although there is a lack of use.

(a) (b)

Fig. 10. a) Dock Aerial view during the America's Cup. http://www.granjotitichallenge.com/es/puerto/marina-real-juan-carlos-i b) Tinglado #4 and #5 during the European Grand Prix of Formula 1. http://www.f1aldia.com/16185/valencia-street-circuit-espera-regresar-f1-2014/

In 2013, Valencia city council approved the Urban Special Plan for the "Real Marina Juan Carlos I". Acording to the Plan, the main objectives are: "Renovation, reform, rehabilitation, recovery, reconstruction, restoration, reuse, recycling… terms initially conceived and used to talk about interventions on the built heritage…" (Lozano 2013). This especially affects the sheds, the old boathouse, the Clock building, the commercial docks, the old customs, the old port authority and the health building. The Urban Planning assigns the damaged Tinglado #2 a use as "Food", specifically a Gourmet Market.

6 Tinglado #2: Current Status

As we have explained, in Tinglado #4 and #5, restoration works have been carried out throughout his life, the last ones recently. However, in Tinglado #2 the last maintenance works were carried out in 1984. At present, it shows a progressive deterioration due to the inevitable passage of time, (it was built over 100 years ago), the aggressiveness of its marine environment, and the regular use as a warehouse of goods with continuous circulation of vans and little lorries. Some of the damages may be a risk for people who use the shed as a leisure space or for skating. In May 2013, Ms. Rita Barberá[4] Mayor of Valencia at that time, expressed her concern about the situation of serious deterioration of this historic building of the city. She urged to Mr. Pablo Landero, director of the "Marina Real Juan Carlos I" Consortium, to tender the works of its restoration before summer that year (García 2013).

Corrosion of steel elements is one of the most serious damages and it is due to the marine environment and, locally, due to the droppings of the pigeons and the urine of the dogs. It is observed both in the columns and in the steel profiles of the roof, being especially aggressive in the downpipes and rainwater collection gutters which are made of sheet metal (Fig. 11a). The perforation of the gutters gets the filtration of water, which runs through the metal structure, aggravating the problem. The damages observed in many supports are the breakage of rivets due to the swelling produced by the corrosion and dents produced by impacts of vehicles and machinery (Fig. 11b). Peeling of paint due to corrosion is widespread throughout the building (Alvarado 2015).

Passing of time and the absence of maintenance and replacement of several pieces have driven deterioration and poor appearance in the facade elements. Probably the problem will have been aggravated due the deformations of the structure produced by temperature variations and by the corrosion of the metallic elements used for its anchoring. The masonry facade and its decorative elements suffered cracks and fissures producing various landslides that forced the provision of safety nets (Fig. 12a). In all the facades there were losses of coating, broken glass that has not been replaced, and in the front walls and the absence of some pieces of decorative tiles (Lluch 2018).

The different events held in it have required the installation of additional elements such as antennas and lighting that have been arranged in a disorganized way using the shed structure. Although they do not represent a direct risk, the presence of ducts and boxes damage the building aesthetics. The American Cup championship in the Valencia port required the construction of the headquarters for the different teams next to the sheds, covering part of its facades, hiding its architectural value and generating residual spaces. Some of these temporary constructions have remained in time affecting the sheds environment (Fig. 12b).

The building damages reached its critical point in 2014, causing its partial closure. Concern about the serious situation of Tinglado #2, led the Consortium of Marina Real

[4] Ms. Rita Barberá (1948–2016) was Mayor of the city council of Valencia between 1991 and 2015 and senator from 2015 until her death.

(a) (b)

Fig. 11. Tinglado #2 a) Detail of the roof. b) Detail of a column. Own source.

Fig. 12. Tinglado #2. Facade to *Ingeniero Manuel Soto* avenue. Boxes in front of the dock. Own source.

Juan Carlos I to call a contest for the rehabilitation of the building to recovery a historic building and its privileged environment for the city (Magrinyá and Maza 2005). The resolution of the contest awarded the first prize to Arkitera S.L.P. which in 2015 presented the intervention project supported in the three following premises (Arkitera 2015):

- Minimal intervention to guarantee the maintenance of the aesthetic, material, historical and cultural documentation of the monument.
- All historical additions should be preserved, as they are testimonies of the experiences of the piece and historical-cultural documents.

- Any new intervention carried out will have to be able to eliminate without damaging the original element.

The intervention in the building was carried out in two phases. The works of the first phase, consisting of structural and functional repair, were carried out among May and October 2018. The concrete maintenance interventions in the steel structure consisted of cleaning, replacement of rivets and fire and corrosion protection paint. Structural repair included the provision of bracing elements as well as the reinforcement of beams, supports and walls. Finally, the baseboards were cleaned up, the ornamental elements, moldings and stained glass were restored, overflow channel were executed and the waste water facility was improved.

The works of the second phase were adjudicated by the City Council in June 2019, (Moreno 2019) and they are currently in progress (Fig. 13). This phase consists of the replacement of the metal plates of the roof, the adaptation of the rainwater facility and the restoration of the ceramic decorative elements of the masonry buildings. The electrical transformation center was moved, the current restrooms are being replaced with adapted toilets, and the bases of the supports are being repaired. In addition, the lights will be replaced to enhance the monumentality of the building and a broadband "Wi-Fi" network will be installed to attract the public with the aim of making the building the place of temporary events and activities of neighborhood and cultural associations.

Fig. 13. Tinglado #2. Facade to the dock (January 10, 2020). Own source.

7 Conclusions

Tinglados of the Valencia Port, located next to the dock, are an interesting example of the industrial architecture of the late XIX century. They are included in the Catalog of Buildings and Places which are protected by the General Urban Plan of Valencia to promote their conservation.

The most representative aspect of the sheds is its naked steel structure. Its geometry conforms to a rectangular grid of 15 m span in its longitudinal direction and 13.5 m in its transverse side. Primary trusses, of 13.5 m span and parallel chords, were arranged in the transverse direction to the main façade.

Secondary trusses, of 15.0 m span are Pratt-type trusses with slope upper chord which provides the characteristic exterior image of the sheds. Its lower chord has an unusual elliptical shape.

Since its construction they have suffered numerous damages caused by the passage of time and the aggressiveness of their maritime environment. They have undergone several reconstructions to recover the damages caused by the civil war and other interventions in order to adapt them to the new functional necessities.

The dock's environment has also been altered by different events such as the 32nd edition of the América's Cup of sailing and the great prizes of Formula 1 held in Valencia. Although the first one left isolated Tinglados #4 and #5 with respect to the dock, the second one served as a reason for their restoration and reform.

On the other hand, no maintenance works had been carried out in Tinglado #2 for more than 30 years. This shed was progressively deteriorating until reaching a worrying state both for the safety of the users and for the integrity of the building.

In May 2018, rehabilitation works were initiated. They include, among others, the structural reinforcement, the replacement of the roof plate and the improvement of the drainage facility. So, a historic building and its privileged surroundings will be recovered for the citizens.

Acknowledgments. To the Port Authority of Valencia, for allowing access to the historical documentation of its General Archives.

References

Alvarado, J.D.: Los Tinglados del Puerto de Valencia: análisis histórico y evaluación estructural. (Tinglados in Valencia Port: historical analysis and structural evaluation). Master Thesis. Universitat Politècnica de València (2015)

Arkitera S.L.P.: Tinglado #2 de la Marina Real Juan Carlos I, Puerto de Valencia (2015). http://www.arkitera.es/portfolio/tinglado-2-marina-real-juan-carlos-i-puerto-valencia/

Benévolo, L.: History of modern architecture. 8th revised and expanded edition. Gustavo Gili. Barcelona (1999).

Carreres Rodríguez, M., Doménech García, S.: Los tinglados del puerto de valencia. Una obra centenaria. (Tinglados of Valencia Port. A centenary buildings). Ed. Consellería de Infraestructuras, Territorio y Medio Ambiente. Cátedra Demetrio Ribes UVEG-CITMA (2012). 978-84-482-5670-8

Fuster, J.: Proyecto de cuatro depósitos cubiertos en los tramos 1°, 5°, 6° y 7°. (Project of four sheds in Sect. 1°, 5°, 6° and 7°) General Archive. Port Authority of Valencia (1910)

Fuster, J.: Memoria sobre el estado y progreso de las obras durante el año 1913. (Report on the status and progress of the works during the year 1913). Unpublished report. General Archive. Port Authority of Valencia (1914)

García, H.: Barberá reclama prioridad para el Tinglado #2. (Barberá claims priority for Tinglado #2). Levante. El Mercantil Valenciano, Valencia (2013). http://www.levante-emv.com/valencia/2013/05/01/barbera-reclama-prioridad-tinglado-2/994201.html

Hermosilla, J.: Historia del puerto de Valencia. (History of the Valencia Port). Valencia: Universitat de València (2007)

Latorre Hernández, M.: Fichas BRL 11.12.10, 11.12.08 y 11.12.09. Catálogo de Bienes y Espacios Protegidos. Revisión simplificada del Plan General de Valencia. (Index card BRL 11.12.10, 11.12.08 y 11.12.09. Catalog of Buildings and Places Protected. General Urban Plan of Valencia) (2010). http://www.valencia.es/revisionpgou/catalogo/urbano/11.10%20TINGLADO_2%20DEL%20PUERTO%20DE%20VALENCIA_firmado.pdf. https://www.valencia.es/revisionpgou/catalogo/urbano/11.08%20TINGLADO_4%20DEL%20PUERTO%20DE%20VALENCIA_firmado.pdf. http://www.valencia.es/revisionpgou/catalogo/urbano/11.09%20TINGLADO_5%20DEL%20PUERTO%20DE%20VALENCIA_firmado.pdf

Lluch, M.: Informe técnico de la ornamentación del Tinglado #2 del puerto de Valencia. (Technical report on the ornamentation of the Tinglado #2 of the Valencia Port). Final Degree Project. Universitat Politècnica de València (2018)

Lozano, J.M.: Plan Especial Marina Real Juan Carlos I. (Special Urban Plan for Marina Real Juan Carlos I). Valencia City Council. Port Authority of Valencia (2013)

Maese, M.: Puerto de Valencia. Proyecto de obras de distribución y servicio de los muelles. (Valencia Port. Project to distribution and service works of docks). Anales de la Revista de Obras Públicas, Cuaderno 3°, Madrid (1985)

Magrinyá F., Maza, G.: Tinglados de Bar-cel-ona: la incorporación de espacios del puerto a la ciudad (1981–2002). (Tinglados of Bar-cel-ona: the incorporation of spaces from the port to the city (1981–2002)). Scripta Nova, vol. 9, no. 193 (2005)

Manzanares, J., Salas, J., Machi, J.: Modificado Proyecto Constructivo de Arquitectura del Tinglado 4 del Circuito de F1 de Valencia. (Modified Construction Project of Tinglado #4 of the Formula 1 Circuit of Valencia). Generalitat Valenciana (2008)

Moreno, P.: La reforma del Tinglado 2 se centrará en la cubierta, los cimientos y la decoración. (The reform of Tinglado #2 will focus on the roof, the foundations and the decoration). Las Provincias (2019) https://www.lasprovincias.es/valencia-ciudad/reforma-tinglado-centrara-20190628002521-ntvo.html

Ronda, J.: Proyecto de restauración del cabecero norte y naves anexas del tinglado n°4. (Restoration project of the north part and annexed sections of Tinglado n° 4). General Archive, Port Authority of Valencia (1982)

Ronda, J.: Proyecto de restauración del cabecero norte y naves anexas del tinglado n°2. Restoration project of the north part and annexed sections of Tinglado n ° 2). General Archive, Port Authority of Valencia (1984)

Ronda, J.: Proyecto de restauración del tinglado n°5. (Restoration project of Tinglado n°5). General Archive, Port Authority of Valencia (1989)

Ronda, J.: Proyecto de demolición del tinglado n°6 en el transversal de poniente del puerto de Valencia. (Demolition project of Tinglado n°6 in the west area of the Valencia Port). General Archive, Port Authority of Valencia (1999)

Vilar, J.: Memoria sobre el estado y progreso de las obras desde el 1° de enero de 1933 hasta el 31 de diciembre de 1941. (Report on the status and progress of the works from January 1, 1933 to December 31, 1941). Unpublished report. General Archive, Port Authority of Valencia (1942)

A Simplified Procedure for Seismic Risk Assessment of Existing Italian School Buildings

Gianfranco De Matteis(✉) and Pasquale Bencivenga

Department of Architecture and Industrial Design, University of Campania "Luigi Vanvitelli", Abbazia San Lorenzo ad Septimum, Via San Lorenzo, 81031 Aversa, Italy
{gianfranco.dematteis,pasquale.bencivenga}@unicampania.it

Abstract. Seismic risk assessment of school building heritage is a relevant topic in Italy, as most of them were built several years ago and were not designed using anti-seismic criteria. In addition, a large portion of existing school structures is characterized by a bad state of conservation, mainly due to the lack of maintenance and structure deterioration. Seismic risk evaluation of these buildings has to take into account the combination of seismic hazard, structural vulnerability and exposure considering the mainly characteristics. In the last decades, to this aim, many methods have been introduced, which differ from each other by the refinement of the applied methodology of analysis. At national level, in order to provide a tool to solve this problem, a specific database called "Sistema Nazionale delle Anagrafi dell'Edilizia Scolastica (SNAES)", which consists of a census of the entire school building heritage, has been developed. Such database contains many parameters related to the consistency, degradation state and functionality.

In this paper, starting from the existing methods and considering the parameters and the information actually present in the school building files of SNAES, a simplified and fast methodology for assessing the seismic risk of both masonry and reinforced concrete schools is provided. This procedure can be applied to a large number of structures and in this paper it is analyzed with reference to a sample of school buildings located in the Caserta Province.

Keywords: Risk assessment · School buildings · Seismic vulnerability · Seismic exposure · Seismic hazard

1 Introduction

The Italian school building heritage is constituted by many public "historical" structures, which may be characterized by inadequate structural safety in case of seismic events. From the data emerged by a recent analysis, more than 63% of the relevant buildings, in fact, were built before 1975 and often require urgent maintenance (46.8% of the sample) (XIX Rapporto di Legambiente sulla qualità dell'edilizia scolastica, delle strutture e dei servizi 2018). Moreover, this building heritage does not comply with safety standards imposed by seismic regulations. More than 41% of the school buildings are located in highly prone seismic areas (namely seismic zone 1 and 2). It is highlighted that only 14.2% of all the structures are designed according to seismic criteria and the seismic

I. Vayas and F. M. Mazzolani (Eds.): PROHITECH 2021, LNCE 209, pp. 1373–1390, 2022.
https://doi.org/10.1007/978-3-030-90788-4_105

structural vulnerability has been produced on 32.9% of the sample. Analyzing the funding lines of the last five years, it appears that only 4.4% of the interventions concerned the seismic upgrading or energy efficiency of buildings located in seismic areas, with a consequent estimated timeline that could allow the achievement of security goal for all the school buildings in more than 100 years. This long time, obviously, is not compatible with social needs.

Also for this reason, the instrument called "Sistema Nazionale delle Anagrafi dell'Edilizia Scolastica" (SNAES) has been proposed in Italy. It is still incomplete and inaccurate, and therefore unable to provide effective information for the buildings safety. In order to improve such a system, in Regione Campania, a specific research project named "Potenziamento e analisi critica dell'Anagrafe dell'Edilizia Scolastica della Regione Campania" has been developed in 2018 launched, aiming at increasing the dissemination of the tool among provinces and municipalities (De Matteis et al. 2018, De Matteis and Bencivenga 2019).

2 The School Building Heritage in the Caserta District

The school building heritage of the province of Caserta, similarly to the national level, is characterized by various critical issues. It is possible to have a clear picture of the situation by the "Sistema Nazionale delle Anagrafi dell'Edilizia Scolastica" (SNAES). This system has been activated on 1st December 2014, despite it had already been introduced in 1996 (Norme per l'edilizia scolastica, Legge 11 gennaio 1996, n. 23). It consists in a digital platform, which is continuously updated by filling in appropriate forms information on the functionality and conditions of the building.

Starting from the 781 school buildings located in the territory of Caserta (information reported on the SNAES and updated to November 2019), the sample analyzed in this paper, corresponding to the 27% of the stock, is composed by 211 school buildings: in particular 141 school are characterized by a masonry structure and the remaining 70 are reinforced concrete buildings.

Seismic risk evaluation of a building requires the assessment of three fundamental factors: vulnerability, hazard and exposure. Therefore, in the following paper, the possibility to define a fast methodology for the seismic risk assessment is investigated, considering the following main tasks:

1. Identification of the main features of the building in order to characterize the structural vulnerability;
2. Simplified assessment of exposure and hazard.

3 Seismic Vulnerability Assessment

3.1 General

In the existing literature there are specific methods for seismic vulnerability assessment of existing buildings, which can be grouped into two main categories. The first one is characterized by adopting a semi-empirical statistical-observational approach (Zuccaro and Cacace 2007, Zuccaro and Cacace 2015, Dolce et al. 2004), which is essentially based on the detection of building performance especially after significant earthquakes. By observing the behavior of existing buildings, a number of vulnerability factors may be defined, which are differentiated for masonry and reinforced concrete buildings. Starting from these vulnerability factors, taking into account their different influence on the seismic response of the buildings, global indices, based on semi-empirical formulations of vulnerability as function of seismic intensity, can be introduced.

The second category is referred to the adoption of more refined analytical approaches, such as the direct evaluation of seismic capacity of single or groups of buildings (Dolce and Moroni 2005, Gattesco et al. 2012). Obviously, even though it is more reliable, such a method requires a more specific knowledge of the structure, including geometrical and mechanical parameters.

In this paper, a simplified method related to the first category (observational based method) has been developed, based only on the data included in the S.N.A.E.S. forms. This methodology has been subdivided into the three main steps described in the following.

3.2 Vulnerability Assessment – Step 1

The seismic vulnerability of a structure is a measure of the possible damage occurred by an earthquake with a predefined seismic intensity. Therefore, it is related to the intrinsic structural capacity of the building itself. With reference to the examined sample made up of 141 school buildings, a preliminary vulnerability assessment was conducted based on a rapid methodology present in the literature, which is based on the use of a classification matrix which considers only two data corresponding to the vertical and horizontal structural system (Di Pasquale et al. 2000).

According to this matrix, which represents the first step of the vulnerability assessment, it is possible to assign to each building one among 5 different vulnerability classes, as provided in Fig. 1: High (H); Medium-High (MH); Medium (M); Medium-Low (ML); Low (L).

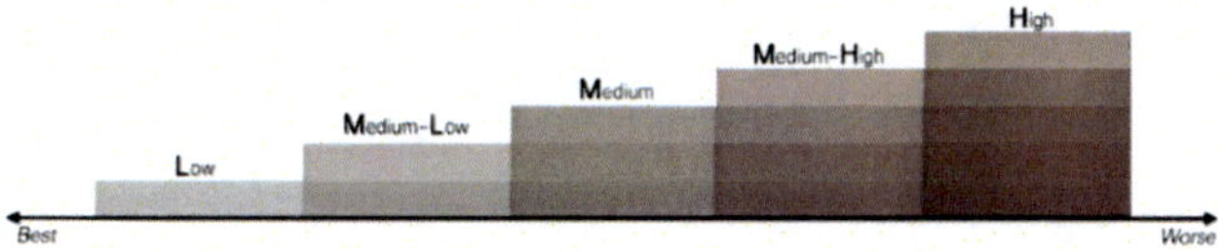

Fig. 1. Vulnerability assessment: step 1

In order to account for both the horizontal and vertical structural systems, Table 1 has been assumed to associate the basic vulnerability class to the examined structure for traditional buildings and Table 2 for monumental buildings.

The first matrix shows that the most vulnerable buildings, corresponding to those in the first and second line, are characterized by vaulted structures and wood slabs. It is apparent that the seismic vulnerability of buildings decreases passing from the first column to the last one and from the first line to the last one. This aspect is due to the fact that the last structural vertical and horizontal systems are typically characterized by more accurate technical solutions.

Table 1. Definition of basic seismic vulnerability classes of traditional masonry buildings

Structural horizontal system	Structural vertical system			
	Mixed masonry - r.c. structures	Sack masonry	Bearing masonry	Tufa masonry
Vaulted structures	H	H	H	H
Wooden slabs	H	H	MH	MH
Steel slabs	MH	MH	M	M
Reinforced concrete slabs	M	MH	M	ML
Other	M	MH	M	ML

Table 2. Definition of basic seismic vulnerability classes of monumental masonry buildings

Structural horizontal system	Structural vertical system			
	Mixed masonry - r.c. structures	Sack masonry	Bearing masonry	Tufa masonry
Vaulted structures	MH	M	ML	ML
Wooden slabs	M	MH	M	M
Steel slabs	M	MH	M	M
Reinforced concrete slabs	M	MH	MH	M

Table 3. Definition of basic seismic vulnerability classes of reinforced concrete buildings

Structural resisting system	Vertical loads designed (Zone 4)	Low seismic forces designed (Zone 3)	Medium seismic forces designed (Zone 2)	High seismic forces designed (Zone 1)
Frames – columns and high beams, with light cladding	H	MH	MH	M
Frames – columns and high beams, with heavy cladding	MH (M)	M (ML)	ML	ML
Walls	ML	L	L	L

Similarly to this procedure, for the sample of 70 reinforced concrete buildings, a matrix (Table 3) that considers two data only, namely the structural resisting system and seismicity level has been proposed.

3.3 Vulnerability Assessment – Step 2

Once basic seismic vulnerability has been assessed by Tables 1, 2 and 3, in order to refine the obtained basic classification, additional criteria are considered to either modify or confirm the vulnerability class deduced by Tables 1, 2 and 3. Such additional criteria have been established considering the information included in the SNAES forms (conservation status, designed forces considered at the time of construction and changes that these have undergone over time, number of floors, age of construction, planimetric and elevation regularity,… etc.). For each of these parameters, a specific score is provided. Furthermore, three different intervals have been defined so that the sum of the scores obtained considering all the parameters can determine the effect on the previous vulnerability classification by having:

- a positive class change (increase of vulnerability);
- no class change;
- a negative class change (decrease of vulnerability).

The selected additional parameters that can influence the vulnerability of the building are the following:

- Seismic zone; if the building is located in a territory whose seismic classification has changed since it was built, it will be more vulnerable, as it has been designed without taking into account the current earthquake hazard. The actual seismic zone has been considered referring to O.P.C.M. n° 3274/2003. In Tables 4 and 5 is indicated +3, + 2, +1, +0 stand for buildings located in seismic zones actually increased of three, two, one or zero classes. Furthermore >2003 indicates schools realized or designed after the new classification.
- Number of structural storeys; higher buildings are more vulnerable, as they are subjected to seismic amplification at upper storeys and larger forces at lower storeys.
- Type of roofing system; this parameter is considered only in terms of morphology (flat, pitched, mixed): in fact, in the SNAES forms, any additional technical information is not provided.
- Type of interventions 1; in this classification are included interventions of: enlargement, superelevation (EN./S.E.), restructuring (RE.) and extraordinary maintenance (E.M.). Except for the first two, which can increase the seismic vulnerability if not realized correctly, the structural improvement provided by restructuring or maintenance interventions can contribute to reduce the seismic vulnerability.
- Conservation status of vertical bearing structures and masonry, slabs, stairs and roofing; the conservation state of structural elements may significantly influence the vulnerability of a building. In Table 4 and 5, T.R. and P.R. indicate that buildings need total and partial replacement respectively, O.M. stands for ordinary maintenance and N.I. express that no interventions are necessary.

- Age of construction; this parameter reflects the development of construction techniques, according to the technical code requirement, and, starting from a preliminary classification (Calderoni et al. 2017), four different intervals have been defined;
- Planimetric configuration; each building is classified according to the ratio B/L and to the projections on each side;
- Altimetric configuration; this parameter takes into account the regularity of stiffness, masses and the distribution of the openings for each storey.

The table was organized in such a way that, for each of the above parameters, 4 different conditions are identified, which are characterized by a different and decreasing effect on seismic vulnerability, passing from class I to class IV. Moreover, for each parameter, a "weight" has been associated to take into account the different importance on the global seismic structural behavior of the building. A summary scheme of the conditions defined for each parameter is provided in Table 4 and in Table 5 for masonry structures and reinforced concrete buildings, respectively. In such tables, it is evident that buildings characterized by higher vulnerability are those having the above parameters classified in "Class I", while a lower vulnerability is associated to parameters classified in "Class IV".

In order to provide a quantitative measure to the above additional parameters, a specific numerical coefficient has been associated to each vulnerability class, as shown in Table 6 for masonry structures and in Table 7 for reinforced concrete buildings. Such numerical coefficients have been assigned in such a way to define numerical intervals that allow to establish a positive or negative class change.

By applying Table 6 and 7, the lower and upper limits of the vulnerability parameter (P_V), given by the sum of the products of the coefficients of the "class" column multiplied by the corresponding "weight" are between −6,00 and +36,90 for masonry buildings and from −6,70 to +28,20 for reinforced concrete structures.

Such a range has been normalized in order to have a variation range of the vulnerability parameter (P_V) [0; 1]. Within such a range, for masonry buildings the parameter P_V should be in the interval (0,29 ÷ 0,45) in order to do not have any variation of vulnerability classes; similarly the interval (0,18 ÷ 0,27) is valid for reinforced concrete school buildings.

Therefore, for masonry structures the following situation has been defined (Fig. 2):

- If P_V ∊ *[0; 0,29[* the building has a negative class change (decrease of vulnerability);
- If P_V ∊ *[0,29; 0,45]* the building has not any vulnerability class change;
- If P_V ∊*]0,45; 1,00]* the building has a positive class change (increase of vulnerability).

As in the previous case, for reinforced concrete buildings the following classification has been provided too (Fig. 3):

- If P_V ∊ *[0; 0,18[* the building has a negative class change (decrease of vulnerability);
- If P_V ∊ *[0,18; 0,27]* the building has not any vulnerability class change;
- If P_V ∊*]0,27; 1,00]* the building has a positive class change (increase of vulnerability).

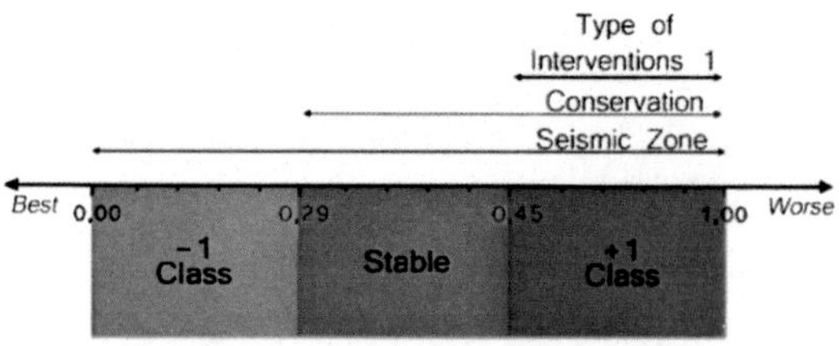

Fig. 2. Vulnerability assessment: step 2 (masonry structures)

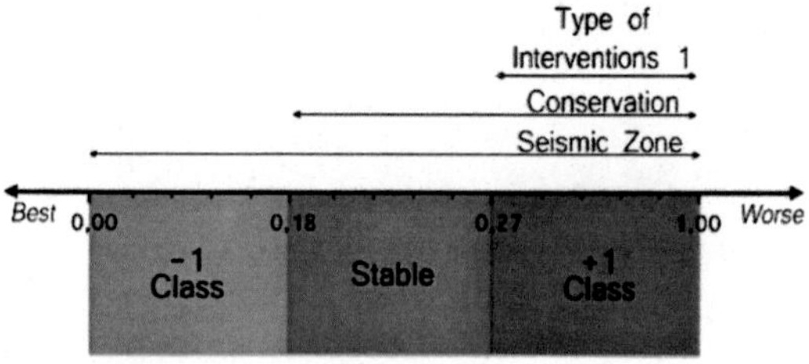

Fig. 3. Vulnerability assessment: step 2 (r.c. buildings)

3.4 Vulnerability Assessment – Step 3

The vulnerability class obtained in the previous step can be further modified only in presence of structural interventions included in a specific category named "Type of interventions 2". This parameter includes seismic interventions as: upgrading, improvement and local repair, aimed at increasing the seismic structural capacity and, therefore, reducing the seismic vulnerability. The following assumptions have been made (Fig. 4):

- In case of seismic local repair intervention, the building undergoes a class change obtaining a minimum vulnerability class M;
- In case of seismic improvement interventions, the building undergoes a class change obtaining a minimum vulnerability class ML;
- In case of seismic upgrading interventions, the building undergoes a class change obtaining a vulnerability class L.

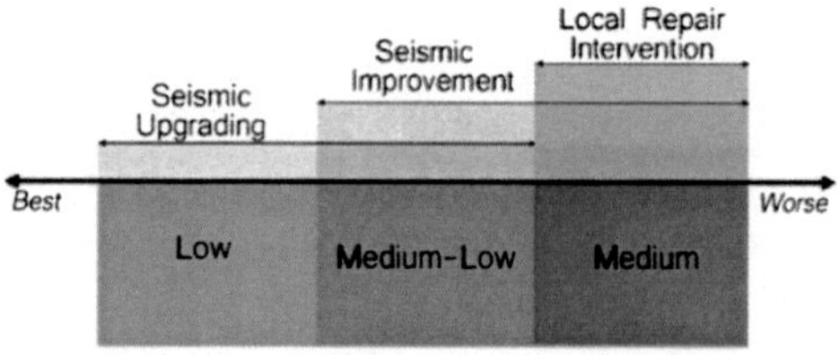

Fig. 4. Vulnerability assessment: step 3

Table 4. Vulnerability conditions and weights related to each additional parameter (for masonry structures)

Parameter	Vulnerability conditions				Weight
	Class I	Class II	Class III	Class IV	
Seismic zone	+3	+2	+1	+0	3,00
Number of storeys	N.4	N.3	N.2	N.1	1,00
Type of roofing system	Pitched	Mixed	Plane	X	1,00
Type of interventions 1	EN./S.E.	N.I.	E.M.	R.E	1,00
Conservation status: vertical bearing structures and masonry	T.R.	P.R.	O.M.	N.I.	0,80
Conservation status: slabs	T.R.	P.R.	O.M.	N.I.	0,80
Conservation status: stairs	T.R.	P.R.	O.M.	N.I.	0,80
Conservation status: roof	T.R.	P.R.	O.M.	N.I.	0,80
Age of construction	<1900	1901 ÷ 1937	1938 ÷ 1960	>1961	0,50
Planimetric configuration	Irregular [B/L > 4]	Almost irregular [all projections > 25%]	Almost regular [at least one projection < 25%]	Regular [B/L < 4]	0,80
Elevation configuration	Irregular stiffness, masses and distribution of openings	Irregular stiffness, masses and regular distribution of openings	Regular stiffness, masses and irregular distribution of openings	Regular stiffness, masses and distribution of openings	0,80

Table 5. Vulnerability conditions and weights related to each additional parameter (for r.c. buildings)

Parameter	Vulnerability conditions				Weight
	Class I	Class II	Class III	Class IV	
Seismic zone	+2	+1	+0	>2003	3,00
Number of storeys	N.4	N.3	N.2	N.1	0,50
Type of roofing system	Pitched	Mixed	Plane	X	0,30
Type of interventions 1	EN./S.E.	N.I.	EM	RE	1,00
Conservation status: vertical bearing structures and masonry	T.R.	P.R.	O.M.	N.I.	0,90
Conservation status: slabs	T.R.	P.R.	O.M.	N.I.	0,90
Conservation status: stairs	T.R.	P.R.	O.M.	N.I.	0,90
Conservation status: roof	T.R.	P.R.	O.M.	N.I.	0,90

Table 6. Numerical coefficients associated with the vulnerability classes and weights of each parameter (for masonry structures)

Parameter	Vulnerability conditions				Weight
	Class I	Class II	Class III	Class IV	
Seismic zone	3	2	1	0	3,00
Number of storeys	3	2	1	0	1,00
Type of roofing system	3	2	1	x	1,00
Type of interventions 1	3 (<2009)	0	−1	−3 (<2003)	1,00
	1 (>2009)			−6 (>2003)	
Conservation status: vertical bearing structures and masonry	3	2	1	0	0,80
Conservation status: slabs	3	2	1	0	0,80
Conservation status: stairs	3	2	1	0	0,80
Conservation status: roof	3	2	1	0	0,80
Age of construction	3	2	1	0	0,50
Planimetric configuration	3	2	1	0	0,80
Elevation configuration	3	2	1	0	0,80

Furthermore, to define in numerical terms the capacity of the structure for each obtained vulnerability category, Table 8 has been proposed. In such a table the corresponding resisting P.G.A. values (structural capacity) are provided as a function of the determined vulnerability class and of the seismic classification of the territory referring to the year of construction. It should be noted that in the case of a building that was subsequently adapted or improved seismically, it is necessary to refer to the seismic zone relative to the intervention period.

Table 7. Numerical coefficients associated with the vulnerability classes and weights of each parameter (for r.c. buildings)

Parameter	Vulnerability conditions			
	Class I	Class II	Class III	Class IV
Seismic zone	3	2	1	0
Number of structural storeys	3	2	1	0
Type of roofing system	3	2	1	x
Type of interventions 1	3 (before 2009)	0	−1	−3 (before 2003)
	1 (after 2009)			−6 (after 2003)
Conservation status: vertical bearing structures and masonry	3	2	1	0
Conservation status: slabs	3	2	1	0
Conservation status: stairs	3	2	1	0
Conservation status: roof	3	2	1	0

Table 8. Vulnerability classes and related P.G.A. values

Seismic vulnerability class	H	MH	M	ML	L
Zone 1 (P.G.A.)	0,125	0,150	0,200	0,225	0,250
Zone 2 (P.G.A.)	0,100	0,125	0,150	0,175	0,200
Zone 3 (P.G.A.)	0,050	0,075	0,100	0,125	0,150
Zone 4 (P.G.A.)	0,025	0,030	0,035	0,040	0,050

Finally, in Table 9 a simple value is associated to each vulnerability class by defining the corresponding vulnerability coefficient C_V.

Table 9. Vulnerability coefficients C_V

Vulnerability class	H	MH	M	ML	L
Vulnerability coefficient C_V [–]	0,90	0,70	0,50	0,30	0,10

3.5 Application to School Buildings of Caserta District

The proposed method has been applied to determine the vulnerability class of the analyzed stock of 211 school buildings for the district of Caserta. The obtained result is shown in Fig. 5. It is evident that the majority of the analyzed school buildings (34% of the sample) is characterized by a medium-high vulnerability class and 29% by a medium vulnerability class. Only two school buildings may be identified with a low vulnerability. Furthermore the 25% of the sample is representative of the vulnerability class H and the remaining part of the sample (11%) corresponds to the class ML.

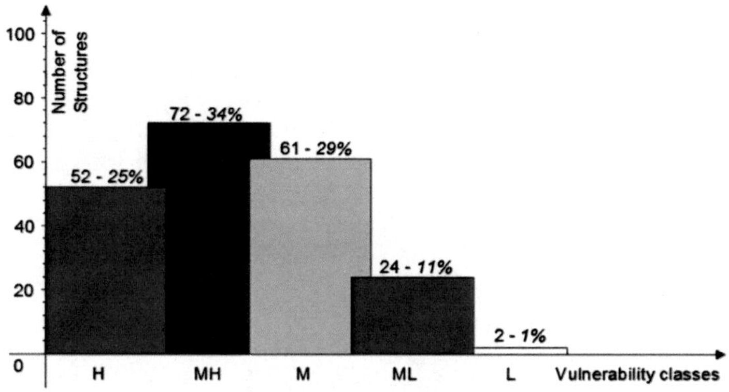

Fig. 5. Vulnerability classes determined for the sample of 211 school buildings of the Caserta district.

4 Hazard Assessment

4.1 The Proposed Method

The seismic hazard represents a measure of the destructive capacity of an earthquake and is linked to the frequency of this aleatory phenomenon, as well as to the geological characteristics of the area in which the event manifests. In this way, the knowledge of the seismic hazard of a site is a fundamental tool to predict the severity of the expected earthquakes. For the specific study case, the hazard (H) has been defined according to the expected peak ground acceleration (PGA), which has been assumed as the one having a probability of exceedance equal to 10% every 50 years. Such a value, considering that school buildings belong to use class III ("important buildings"), corresponds to an earthquake return period $T_R = 712$ years. The corresponding PGA value can be determined from the seismic hazard map of the national territory according to the geographic coordinates of each school building. Then, this value may be normalized, considering a maximum of PGA equal to 0.35g, i.e. the maximum value established in NTC2008. Hence, a numerical coefficient (C_H) can be obtained, which represents the hazard of the site where the school building is located.

4.2 Application to School Buildings of Caserta District

The proposed method has been applied to determine the vulnerability class of the analyzed stock of 211 school buildings for the district of Caserta (Fig. 6). Hazard coefficient values ranging between 0.853 (maximum) and 0.279 (minimum) have been obtained.

Table 10. Hazard classes and hazard coefficient C_H.

Hazard class	H	MH	M	ML	L
Hazard coefficient C_H [–]	$C_H > 0,80$	$0,60 < C_H \leq 0,80$	$0,40 < C_H \leq 0,60$	$0,20 < C_H \leq 0,40$	$C_H \leq 0,20$

For the sake of simplicity, the hazard values have been grouped into five different hazard classes according to Table 10.

It is evident that the majority of school buildings, namely 185 buildings corresponding to 86% of the whole sample, are equally subdivided in vulnerability classes M and ML.

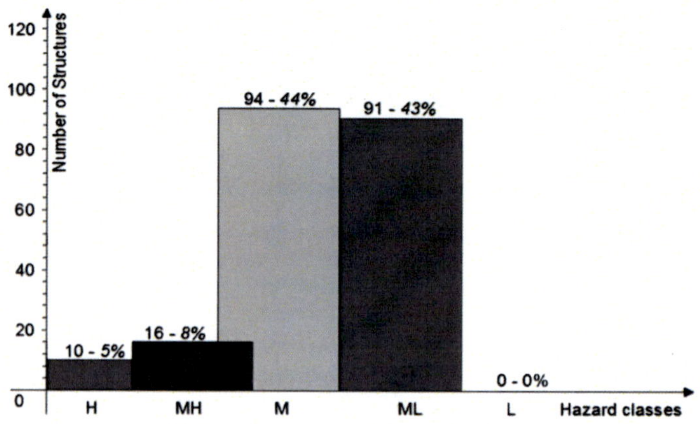

Fig. 6. Hazard classes determined for the sample of 211 school buildings of the Caserta district.

5 Exposure Assessment

5.1 The Proposed Method

Generally, the exposure is associated to the nature, the quality and quantity of the goods exposed to the risk. Therefore, the exposure takes into account the quantification of the number of artefacts (buildings, infrastructures, etc.), of the strategical functions and of the number of personnel that presumably might be involved in the seismic event, in relation to their reaction ability. With reference to the exposure parameter, school buildings are classified as "important buildings", as they are usually subject to considerable crowding (Eurocode 8, 2005).

The exposure is generally related to a functional and a user component: in particular it can be assumed that the exposure coefficient (C_E) is obtained as a product of two indices, namely the user index (I_U) and the function index (I_F) (Ferrini 1998, Polidoro 2010). It is significant to note that while the function index (I_F) is obtained from qualitative considerations on the functions performed in the analyzed structure, the user index

(I_U) is provided by quantitative data concerning the users and number of operators in the buildings. In the specific study case, referring only to school buildings, it seems reasonable to assume that exposure is not dependent on the function index.

The user index (I_F) is, in turn, depending on the behavioral capacity of the users into the building, the period of use of the structure, the building crowding index and the storeys number.

Summarizing, the exposure coefficient (C_E) can be determined by the following symbolic equation:

$$C_E = Is \cdot I_D \cdot I_P \quad (1)$$

where:

- I_S is the index that takes into account the behavioral ability of users, that is related to the reaction capacity, dependent mainly on the age of students, physical conditions and freedom of movement, and also on the period of use of the school building, which is simply the number of weekly hours during which users attend the school activities (Table 11).
- I_D is the user density that is a function of the ratio between the user number (N_U) and the total area of the various floors of the building (S_{TOT}). For each value of this ratio is possible to associate a given coefficient, as shown in Table 12.

Table 11. School index

Type of school	Is [–]
Preschool	1,0
Primary school	0,80
Secondary school grade I	0,70
Secondary school grade II	0,60
Comprehensive institute	0,85

Table 12. Density index.

N_U/S_{TOT}	I_D [–]
$N_U/S_{TOT} < 0{,}08$	0,75
$0{,}08 \leq N_U/S_{TOT} < 0{,}16$	0,85
$0{,}16 \leq N_U/S_{TOT} < 0{,}24$	0,90
$0{,}24 \leq N_U/S_{TOT} < 0{,}32$	0,95
$N_U/S_{TOT} \geq 0{,}32$	1,00

Table 13. Floor number index.

Number of storeys	N_f [–]
1	0,60
2	0,75
3	0,85
4	0,95
≥5	1,00

- I_P is the index relative to the number of storeys, since it is related to the evacuation easiness, without taking into account the characteristics of the users; it is defined according to Table 13.

For the sake of simplicity, the exposure risk values have been grouped into five different exposure classes according to Table 14.

Table 14. Exposure classes and exposure coefficient C_E.

Exposure class	H	MH	M	ML	L
Exposure coefficient C_E [–]	$C_E > 0,80$	$0,60 < C_E \leq 0,80$	$0,40 < C_E \leq 0,60$	$0,20 < C_E \leq 0,40$	$C_E \leq 0,20$

5.2 Application to School Buildings of Caserta District

The proposed method has been applied to determine the seismic risk of the analyzed stock of 211 school buildings for the district of Caserta (Fig. 7). It is noticeable that the majority of school buildings, namely 173 buildings corresponding to 82% of the whole sample, is characterized by a Medium Hazard class.

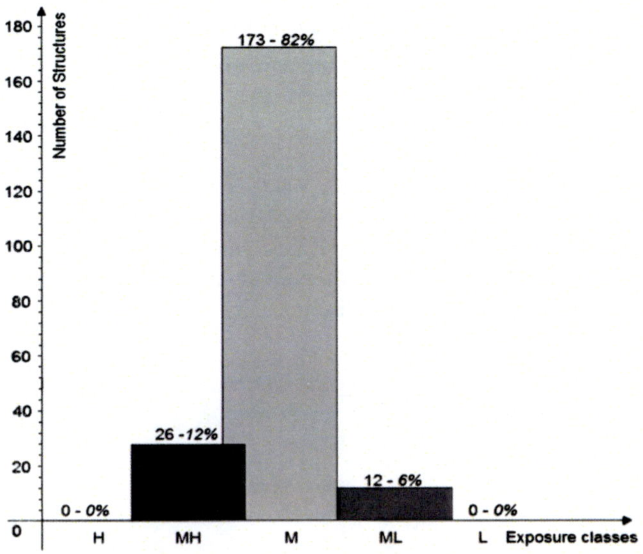

Fig. 7. Seismic exposure determined for the sample of 211school buildings of the Caserta district.

6 Seismic Risk Assessment

6.1 The Proposed Method

Once the numerical values of the various factors concurring to define seismic vulnerability, hazard and exposure have been obtained, the seismic risk (R) can be consequently evaluated by combining the above coefficients C_V, C_P, C_E. As an homogeneous class of buildings is considered, in order to highlight the differences among the buildings, a simplified additive model can be applied to combine the above risk factors rather than a combination model based on their product; then, to account for the different importance of the above three coefficients, a weight equal to 0,4 has been assigned to the vulnerability coefficient C_V, while a weight equal to 0,3 has been considered for the other two coefficients C_H and C_E.

Therefore, the seismic risk R is evaluated by applying the following combination formula:

$$R = 0,4 \cdot C_V + 0,3 \cdot C_H + 0,3 \cdot C_E \tag{2}$$

For the sake of synthesis, the results got by the above application have been grouped defining five different classes of seismic risk (Table 15).

6.2 Application to School Buildings of Caserta District

The proposed method has been applied to determine the seismic risk of the analyzed stock of 211 school buildings for the district of Caserta and the results are shown in Fig. 8. The obtained values range between a minimum of 0,33 to a maximum of 0,72.

Table 15. Seismic risk classes

Risk class	H	MH	M	ML	L
Risk coefficient R [–]	R > 0,80	0,60 < R ≤ 0,80	0,40 < R ≤ 0,60	0,20 < R ≤ 0,40	R ≤ 0,20

The school building representative of the greatest seismic risk is located in seismic zone 1, characterized by a high seismic action, a medium-high vulnerability and exposure: in fact, it hosts a preschool with a medium user density.

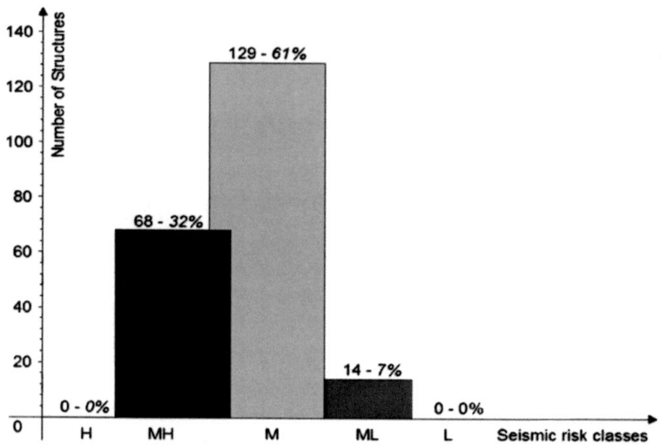

Fig. 8. Seismic risk classes determined for the sample of 211 school buildings in the Caserta district.

The school building characterized by the lowest seismic risk is located in seismic zone 2, built in the last fifteen years and therefore characterized by low features of vulnerability and medium characteristics of exposure and hazard. Furthermore, the majority of the analyzed school buildings, namely 129 school buildings corresponding to 61% of the whole sample, is classified M; 32% of the school buildings present a medium seismic risk class MH; only 14 school building structures can be classified in the ML class.

7 Conclusions

In this paper, a simplified method to define seismic vulnerability, seismic hazard and exposure of masonry and reinforced concrete school buildings has been proposed. It is based on the structural and typological parameters which are included in the "Sistema Nazionale delle Anagrafi dell'Edilizia Scolastica" (SNAES). Starting from a seismic vulnerability value provided by a preliminary matrix that depends only on the vertical and horizontal structural systems, using additional specific parameters taken from the existing database, it is possible to define specific vulnerability classes.

In order to prove the actual possibility of application of such a methodology, it has been applied to evaluate the seismic risk of the school building heritage of the province of Caserta. Therefore, starting from the 781 schools of the Caserta district (information reported on the SNAES and updated to November 2019), a stock of buildings composed by 211 schools has been analyzed.

The obtained results show that the five classes of vulnerability are not homogeneously represented by these buildings. In fact, the majority of the analyzed schools (34% of the sample) is characterized by a medium-high vulnerability class, 29% are classifiable in the medium class and only two buildings may be identified with a low vulnerability. Furthermore the 25% of the sample is representative of the vulnerability class H and the remaining part of the sample (11%) corresponds to the class ML. Analogously, in terms of seismic risk, which accounts also for seismic hazard and exposure, 61% of the whole sample of the analyzed schools have been classified with a Medium seismic risk, while 32% with a Medium-High seismic risk, while only 14 school building structures are characterized by a Medium-Low seismic risk class.

In the whole, the obtained results confirm that the school building heritage present many critical issues that should be solved in order to mitigate seismic risk. The proposed simplified methodology seems to represent a useful tool to categorize the seismic risk of schools at large territorial scale in order to define effective intervention priorities.

Acknowledgments. The present study has been developed within the regional project "Potenziamento e analisi critica dell'Anagrafe dell'Edilizia scolastica della Regione Campania" financed by Regione Campania.

References

Calderoni, B., Sandoli, A., Cordasco, E.A.: Valutazione speditiva della vulnerabilita' sismica dei centri urbani italiani: classificazione tipologica strutturale degli edifici esistenti in muratura ed in c.a. Structural (2017). https://doi.org/10.12917/Stru210.09. 210 – marzo/aprile 2017 – paper 09 – ISSN 2282-3794

D.M. 14 gennaio 2008: Norme tecniche per le costruzioni, Consiglio Superiore dei Lavori Pubblici, Gazzetta Ufficiale della Repubblica Italiana (2008)

De Matteis, G., Bencivenga, P., Guarino, F., Nola, C.: A simplified analysis for seismic risk assessment of reinforced concrete school buildings. In: World Heritage and Knowledge. Le Vie dei Mercanti - XVI International Forum, Naples, IT, 14–16 June (2018). ISBN: 978-88-492-3633-0, ID 194

De Matteis G., Bencivenga, P.: A simplified procedure for seismic risk assessment of masonry school buildings. In: XVIII Convegno ANIDIS, 16–19 September, Ascoli Piceno, IT (2019). ISBN: 978-88-3339-256-1

Di Pasquale, G., Dolce, M., Martinelli, A.: Censimento di vulnerabilità a campione dell'edilizia corrente dei centri abitati nelle Regioni Abruzzo, Basilicata, Calabria, Campania, Molise, Puglia e Sicilia Roma (2000)

Dolce, M., et al.: Valutazione della vulnerabilità sismica di edifici scolastici della Provincia di Potenza. In: XI Congresso Anidis, 24–29 January, Genova, IT (2004)

Dolce, M., Moroni, C.: La valutazione della vulnerabilità e del rischio sismico degli edifici pubblici mediante le procedure VC (vulnerabilità c.a.) e VM (vulnerabilità muratura). In: Atti del Dipartimento di Strutture, Geotecnica, Geologia Applicata all'Ingegneria della Università della Basilicata, vol. 4 (2005)

Eurocode 8.: Design of structures for earthquake resistance – Part 3: Assessment and retrofitting of buildings (2005)
Ferrini, M.: Progetto terremoto in Garfagnana e Lunigiana. Un modello di calcolo per la determinazione di un indice di esposizione di edifici pubblici strategici (1998)
Gattesco, N., Franceschinis, R., Zornini, F.: Stima della vulnerabilità sismica delle strutture in c.a. mediante procedura semplificata. In: XV Convegno ANIDIS, 30 June–4 July, Padova, IT (2012)
Legambiente: Ecosistema scuola - XIX Rapporto di Legambiente sulla qualità dell'edilizia scolastica, delle strutture e dei servizi (2018)
Norme per l'edilizia scolastica, Legge 11 gennaio 1996, n. 23 (1996)
Ordinanza 3274 della Presidenza del Consiglio dei Ministri: Primi elementi in materia di criteri generali per la classificazione sismica del territorio nazionale e di normative tecniche per le costruzioni in zona sismica (2003)
Ordinanza 3519 del Presidente del Consiglio dei Ministri: Criteri generali per l'individuazione delle zone sismiche e per la formazione e l'aggiornamento degli elenchi delle stesse zone (2006)
Polidoro, B.: La valutazione della vulnerabilità sismica: il caso di Pettino (AQ). Tesi di Laurea Magistrale in Ingegneria Strutturale e Geotecnica, Università degli Studi di Napoli "Federico II" (2010)
Zuccaro, G., Cacace, F.: Proposta per una procedura di valutazione semplificata di rischio sismico a scala nazionale dell'edilizia scolastica. In: Proceedings of National Congress of Italian National Association of Earthquake Engineering, 2007, Pisa, IT (2007)
Zuccaro, G., Cacace, F.: Seismic vulnerability assessment based on typological characteristics. The first level procedure SAVE. Soil Dynamics and Earthquake Engineering (2015)

Limit Analysis for Triumphal Arches in Masonry Churches: Case Study from L'Aquila 2009 and Central Italy 2016–17 Earthquakes

Mattia Zizi, Valentina Corlito, Daniela Cacace, and Gianfranco De Matteis(✉)

Department of Architecture and Industrial Design (DADI), University of Campania "Luigi Vanvitelli", San Lorenzo Abbey, 81030 Aversa, Italy
gianfranco.dematteis@unicampania.it

Abstract. The protection and conservation of historical monumental assets is a priority for contemporary societies. As it is widely recognized, historical masonry buildings (and in particular masonry churches) are unable to adequately withstand horizontal actions and generally suffer heavy damage if struck by a seismic event. Although a correct evaluation of their seismic capacity is necessary in order to preventively plan retrofitting interventions, the complexity of the constructions and the uncertainties in material modelling make it difficult to fully understand the behaviour of masonry churches. In this context, the macro-element approach is largely applied and confirmed by the partial collapses observed after the last seismic events. In this paper, a simplified procedure based on the limit analysis is introduced to evaluate the seismic capacity of triumphal arches, considered as part of the most vulnerable macro-elements in masonry churches. Triumphal arches from churches affected by the L'Aquila (2009) and Central-Italy (2016–17) earthquakes have been investigated by means of a linear kinematic approach, defining the possible positions of the non-dissipative hinges through step-by-step numerical analyses. The obtained results, finally, have been compared with the damage scenario observed in the considered triumphal arches, aimed at evaluating the effectiveness of the proposed procedure.

Keywords: Masonry churches · Seismic vulnerability · Triumphal arches · Limit analysis · Linear kinematic analysis

1 Introduction

The Italian architectural heritage includes a great number of masonry historical buildings of considerable value, characterized by evident architectural and structural differences. Despite the heterogeneity, the earthquake always represents the main cause of damage and loss of these assets. The events occurred in Italy in the last decades (i.e. Umbria and Marche 1997, L'Aquila 2009, Emilia 2012, Central Italy, 2016–17) stressed the need to define suitable preventive procedures aimed at managing the seismic risk in the whole national territory.

I. Vayas and F. M. Mazzolani (Eds.): PROHITECH 2021, LNCE 209, pp. 1391–1407, 2022.
https://doi.org/10.1007/978-3-030-90788-4_106

As far as masonry churches are concerned, the last seismic events stressed that the seismic behaviour of these kind of structures is constantly repeated, regardless the different features of the detected buildings.

In accordance with the post-earthquake evidences, a widely recognized approach, based on the subdivision of these structures into homogeneous parts (macro-elements) and on the identification of the recurring collapse mechanisms, has been defined. This approach is also suggested by the Italian Guidelines of Cultural Heritage, which proposes the macro-element/mechanism combination given in Table 1 (Ministry of Heritage and Cultural Activities 2011).

Table 1. Abacus of macro-elements and mechanisms for churches according to Guidelines of Cultural Heritage (2011).

	Mechanisms		Mechanisms
Out-of-plane	1. Façade overturning	Dome	14. Dome
	10. Transept overturning		15. Lantern
	16. Apse overturning	Roof system	19. Top of the nave
Façade	2. Top of the façade		20. Top of transept
	3. Plan of the façade		21. Top of the apse
	4. Narthex	Bell tower	27. Bell tower
Lateral walls	6. Shear mech. walls		28. Bell cell
	11. Shear mech. transept	Arches & Vaults	5. Transversal response
	17. Shear mech. apse		8. Central nave vaults
	25. Irregularities		9. Aisles vaults
Chapels	22. Out-of-plane		12. Transept vaults
	23. Shear mechanisms		13. Triumphal arches
Columns	7. Longitudinal response		18. Apse vaults
			24. Chapel vaults

The surveys carried out by the research group on a population of 132 masonry churches (64 three-nave churches; 68 one-nave churches) heavily damaged by the L'Aquila (2009) and Central Italy (2016–17) seismic events, respectively, evidenced the high fragility of the detected structures, allowing to identify the occurred failure mechanisms and the recurrent structural vulnerabilities.

In particular, the classification of the observed damage has been carried out taking into account the typical mechanisms described in Table 1, according to the observation criteria introduced by Grunthal (1998).

The damage analyses recently conducted by the authors put into evidence that curved structures in a masonry church are among the most vulnerable elements in case of earthquake occurrence (De Matteis et al. 2019, De Matteis and Zizi 2019).

Indeed, as it is shown in Fig. 1, in the both stocks of detected churches, the collapse mechanisms related to the macro-element "Arches & Vaults" were all characterized by the highest level of average damage. Moreover, the damage analyses also returned that the collapse mechanism "Triumphal arches" exhibited a significant value of mean damage $\mu_{d,o}$, which resulted equal to 1.953 and 1.944 for L'Aquila and Central Italy samples of churches, respectively.

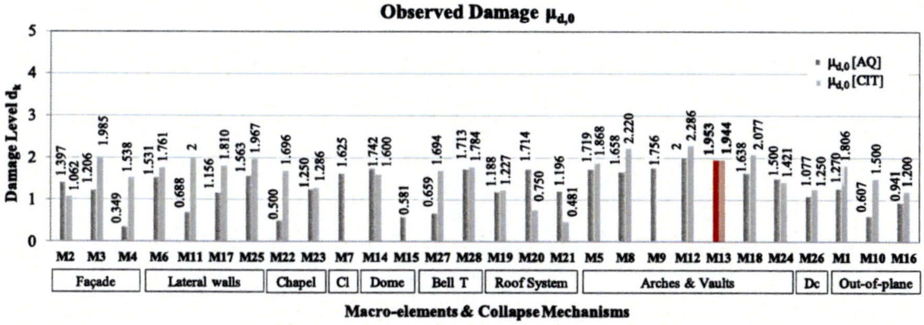

Fig. 1. Values of the mean damage $\mu_{d,o}$ defined for the possible collapse mechanisms in the detected stock of churches.

The high fragility of the triumphal arches is further demonstrated in Fig. 2, where a comparison between the possible and the activated mechanisms is shown for the analysed churches. Of course, the presence of each structural element has been considered as enough to make possible the activation of the associated collapse mechanisms.

In particular, for M13 (i.e. Triumphal arches) the presence of the macro-element was recorded in 67% and 79% of the cases, for L'Aquila and Central Italy churches, respectively, and the mechanism was activated in the 67% and 72% of cases, with a presence/activation frequency equal to 100% and 91%, respectively. Therefore, it is possible to state that if this structural element is present, there is a very high percentage of activation of the relevant collapse mechanism.

Based on the surveyed evidences, the present paper focuses on the assessment of the seismic behaviour of the triumphal arches, not neglecting, from a typological point of view, the various differences found in this structural element, mainly related to the possible presence of a diaphragm system (Fig. 3).

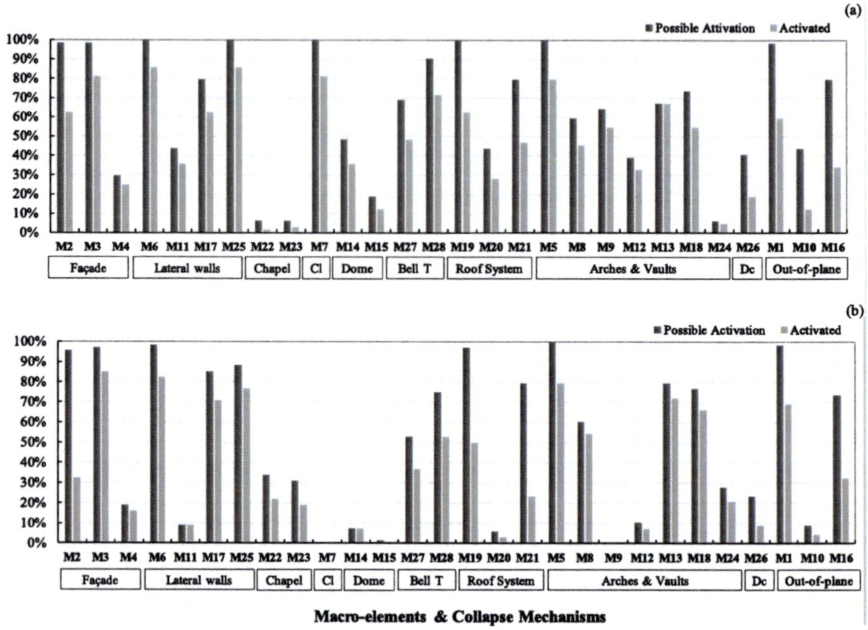

Fig. 2. Frequencies of possible collapse mechanisms and activations detected in the analysed stock of churches in the aftermath of the (a) 2009 L'Aquila and (b) 2016–17 Central Italy earthquakes.

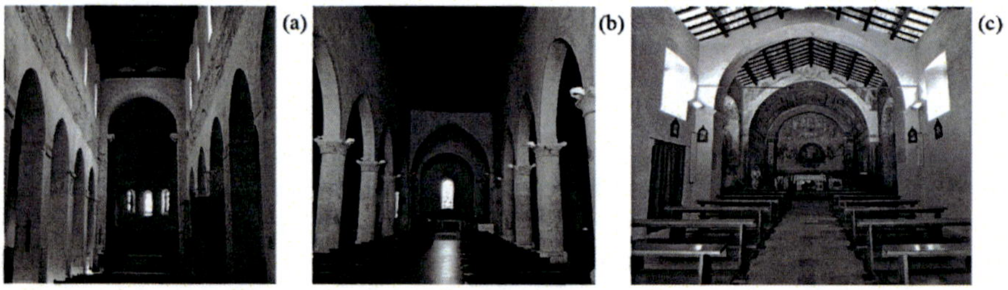

Fig. 3. Triumphal arches with diaphragm system detected in the churches of: (a) San Pelino in Corfinio (AQ); (b) Santa Maria della Tomba in Sulmona (AQ); (c) Santa Maria delle Grazie in Sant'Anatolia di Narco (PG).

The stress distribution and cracking pattern evaluation of a diaphragm arch could result more complicated than of a stand-alone arch having its own constant thickness.

This is because, if in the arch the thrust line is necessarily confined within the thickness of the ashlars, in the diaphragm arch the definition of the resistant system could result different and more complex. This becomes even more true in very stiff structures, where the masonry totally replaces the static functions of the arch (Cangi 2012).

The possible collapse mechanisms detectable in a triumphal arch are deeply influenced by the interaction with the other structural elements. In particular, given the high contribution provided by the naves, the apse, the transversal or adjacent walls (when detected in churches) in inhibiting the development of out-of-plane (or I MODE) mechanisms, the typical collapses of triumphal arches can be mainly related to the in-plane (or II MODE) mechanisms.

In the light of the above, and considering the wide range of variables involved in the issue, it is clear that the definition of the structural behaviour of this macro-element is a problem of no simple solution, especially when it is subjected to horizontal actions and simplified approaches are considered. In this regard, it is worth noticing that several studies have been carried out in the literature aimed at evaluating the seismic response by means of simplified procedure of curved structures, although mainly referred to stand-alone arches (De Luca et al. 2004, Huerta 2009, De Matteis and Mazzolani 2010, Alexakis and Markis 2015, Dimitri and Tornabene 2015, Gattesco and Boem 2019, Cacace et al. 2019).

In the following, four triumphal arches from churches affected by the L'Aquila (2009) and Central-Italy (2016–17) earthquakes have been investigated by means of limit analyses with a kinematic approach. Considering the difficulties related to the definition of the possible kinematic chains, step-by-step numerical analyses have been preliminarily performed with the aim of defining the possible positions of the non-dissipative hinges. The obtained collapse mechanisms and corresponding capacity (in terms of spectral accelerations) have been compared with the actual activations or collapses observed in the considered triumphal arches in the aftermath of aforementioned seismic events, as well as with current Italian code provisions.

2 Case Studies

2.1 Abbey of S. Felice e Mauro in Sant'Anatolia di Narco (PG)

The church of S. Felice e Mauro [SFM] (Fig. 4), which dates back to the 12th century, is part of the homonymous abbey complex placed in Sant'Anatolia di Narco (Perugia District, Umbria Region) and belongs to the competence territory of the Archdiocese of Spoleto-Norcia. Nowadays, the church still presents the native Romanesque features, with no significant signs of transformation due to subsequent interventions. In particular, the church is characterized by a one-nave aisleless layout with an upraised presbytery and an apsidal termination. The church was repaired and retrofitted after the 1997 earthquake occurred in Umbria and Marche. After the recent seismic event of 2016–17, a significant damage was surveyed, especially in the façade, apse and triumphal arch.

2.2 Church of S. Sebastiano in Esanatoglia (MC)

The Church of S. Sebastiano [SSB] (Fig. 5) is placed in the historic centre of Esanatoglia, in the Diocese of Camerino (Macerata District, Marche Region). The construction of the church can be traced back to the beginning of the 13th century, although many later interventions can be recognized. Moreover, due to the damages occurred after the Umbria and Marche earthquakes (1997), the church was subjected to a significant restoration and consolidation intervention. The church is characterized by a one-nave aisleless plan, with an external narthex and an apsidal termination. After the Norcia seismic event (30th October 2016), the church revealed a moderate damage state, which mainly interested the vaults and the top of the flat façade.

2.3 Church of S. Maria della Tomba in Sulmona (AQ)

The church of Santa Maria della Tomba [SMT] (Fig. 6) is located in the town of Sulmona, in the Sulmona-Valva Diocese (in L'Aquila District, Abruzzi Region). The church was already documented in the 13th century as a reference point of the surrounding villages. It has a flat façade (typical of the L'Aquila District) with a Gothic portal and a rose window dated back to 15th century. The church has a simple basilica plan covered by timber trusses, without transept and dome. The church was not damaged by the 6th April 2009 earthquake.

2.4 Church of S. Eusanio Martire in S. Eusanio Forconese (AQ)

The church of Sant'Eusanio Martire [SEU] in Sant'Eusanio Forconese (Fig. 7) is located in the Diocese of L'Aquila (in L'Aquila District, Abruzzi Region). A primitive structure of the [SEU] church was probably built in the 3rd century. Nowadays only few remains of the ancient Christian basilica can be detected. The new church was built in the 12th century and was transformed in the 17th with the addition of a transept and a dome. In 1703 the church was damaged by the earthquake and subsequently renewed in the Baroque style. The original medieval structure can be recognized in the external part of the church. The structure was also heavily damaged after the L'Aquila 2009 seismic event. Serious damage and collapses involved the top of the façade, the roof system, the vaults, the lateral walls and, partially, the decorative system. The church was declared unsafe and included in the list of 45 monuments to be restored by the Ministry for Cultural Heritage.

3 The Implemented Limit Analysis Procedure

3.1 General

As confirmed by the partial collapses observed in the aftermath of the last seismic events, a masonry building can be considered as an assembly of statically determined elements: in case of good masonry quality and if the connections between these elements are absent (or not adequate), the global behaviour of the whole structure can be compromised, so that damage can occur with the loss of equilibrium of rigid blocks.

This is particularly true in masonry churches, where the plan-height complexity, the typological configuration and the historical stratifications may imply the presence of remarkable internal disconnections.

The analysis of local rigid-body mechanisms considers each structural element composed of a certain number of non-deformable wall blocks, connected to each other and to

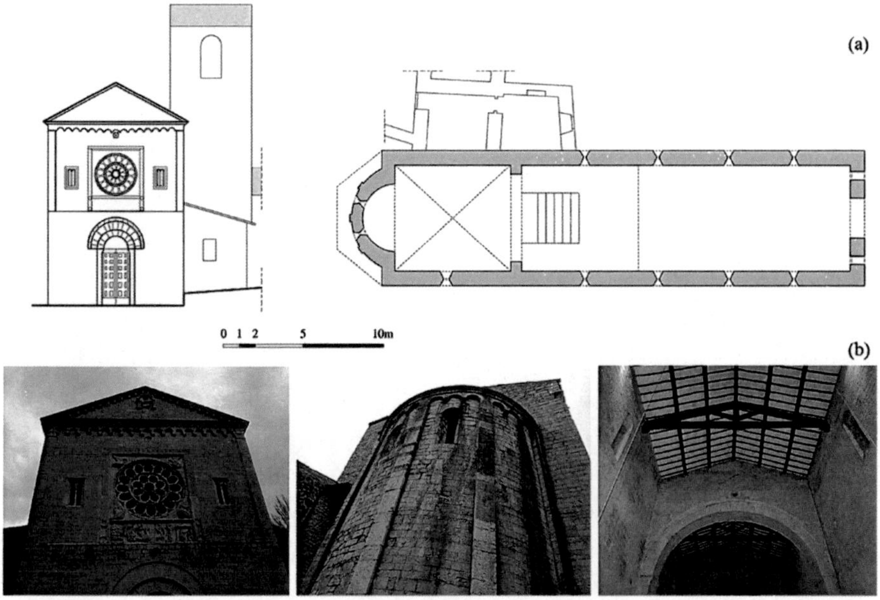

Fig. 4. [SFM] Abbey in Sant'Anatolia di Narco (PG). (a) Geometric and (b) photographic survey.

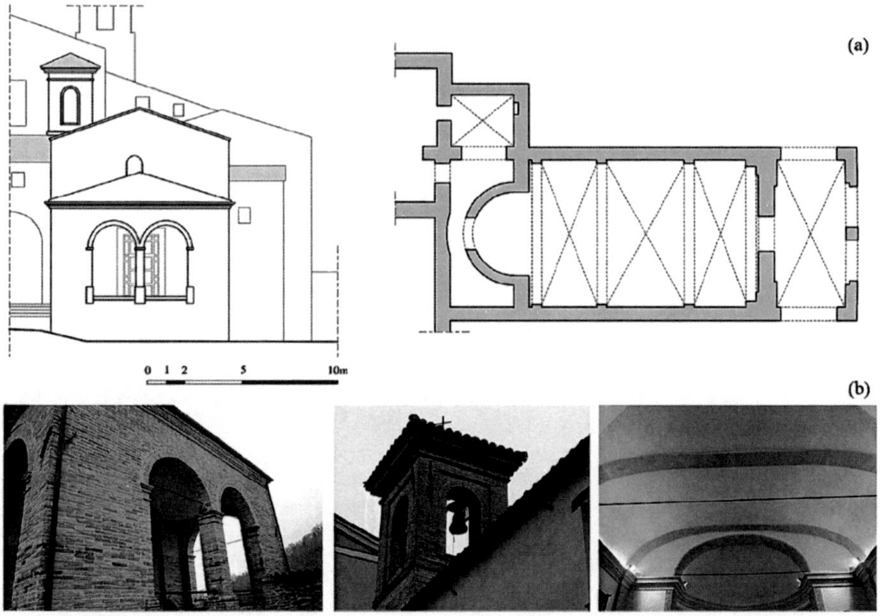

Fig. 5. [SSB] church in Esanatoglia (MC). (a) Geometric and (b) photographic survey.

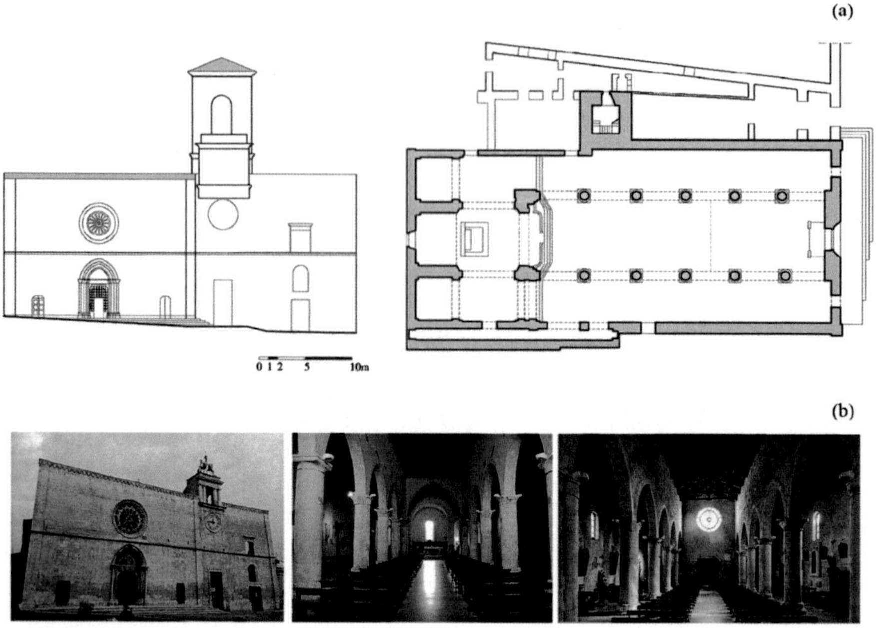

Fig. 6. [SMT] church in Sulmona (AQ). (a) Geometric and (b) photographic survey.

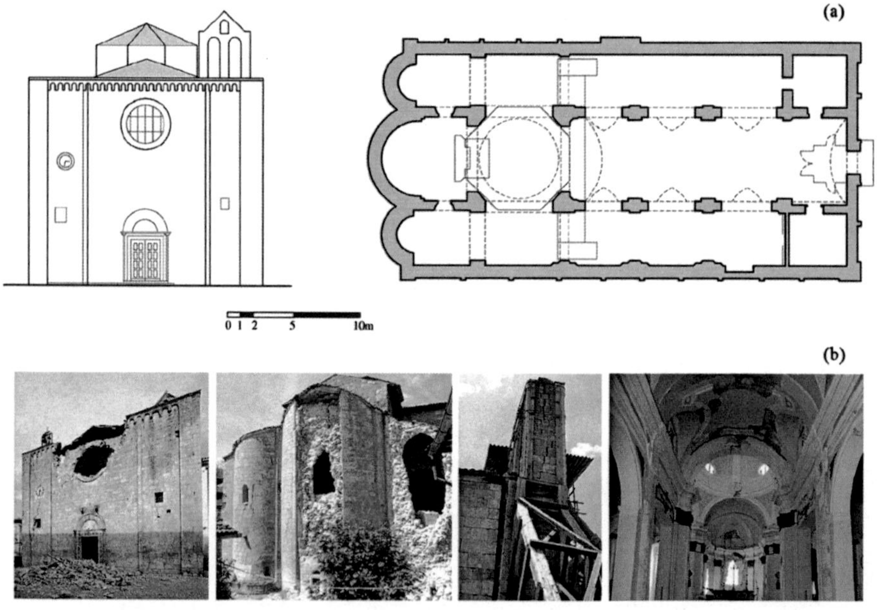

Fig. 7. S. [SEU] church in S. Eusanio Forconese (AQ). (a) Geometric and (b) photographic survey.

the rest of construction through internal and external constraints. The local analyses of rigid bodies are based on the Heyman's (1995) hypotheses of the mechanical behaviour of masonry: *(i)* infinite compressive strength; *(ii)* zero tensile strength; *(iii)* sliding failure cannot occur.

The verification of local mechanisms is conducted through the methods of limit analysis with a kinematic approach, according to the following phases: *(i)* identification of possible collapse mechanisms; *(ii)* definition of the value of the horizontal load multiplier α that involves the activation of the specific mechanism; *(iii)* security check, through the comparison between the seismic demand and the structural capacity, evaluated by means of the linear and non-linear kinematic approach (Doherty et al. 2002, D'Ayala 2005, Ministry of Heritage and Cultural Activities 2011).

These criteria can be extended to monumental buildings by recognizing kinematic chains with a single degree of freedom (SDOF). In particular, the analysis of the local mechanisms for masonry churches can be conducted assuming the macro-elements approach identified by the Italian Cultural Heritage Guidelines (Ministry of Heritage and Cultural Activities 2011).

3.2 Identification of the Collapse Mechanism

The identification of the real collapse mechanism should provide an iterative process that involves all the statically admissible distributions of the hinges and the consequently possible block subdivisions. In particular, the hinges represent discontinuities which provoke the subdivision of the considered structural elements in rigid blocks and the definition of a kinematic chain. Among all the detected possible collapse mechanisms, the most reliable is the one returning the lowest structural capacity.

Since the block definition depends both on the position of the hinges and on the possible crack directions, such an iterative procedure may assume a significant relevance for the triumphal arches: a meticulous approach should provide the evaluation of the horizontal load multiplier α for each statically admissible position of hinges, to which can correspond uncountable possible directions of the cracks.

Therefore, in order to avoid computational efforts, as well as to evaluate reliable positions of hinges, simplified step-by-step numerical analyses are proposed in this study.

In particular, four FE Models have been implemented in *Midas Gen* by adopting a macro-modelling technique for each analysed triumphal arch. Since the in-plane behaviour is investigated, bi-dimensional plate elements have been used, with a thickness of 1 m.

The masonry has been modelled as a linear material, with a density of $\rho = 2000$ kg/m^3 and an Elastic Modulus E = 2000 MPa, both typical of a limestone masonry. Fixed constraints have been introduced for the base nodes. Hence, a gravity vertical acceleration and a small initial uniform horizontal acceleration (0.02 *g*) have been applied to the masonry elements.

The step-by-step procedure provides the increment of the horizontal acceleration until the occurrence in the masonry mesh of a tensile stress higher than 0.3 MPa, which has been assumed as conventional limit tensile strength. Once the conventional stress has been overcome, a subdivision has been introduced in the mesh in order to simulate the cracking process. In particular, the direction of the crack has been considered starting from the tensile stress concentrations toward the most compressed zones, considered as the possible position of the hinge. Such an operation has been repeated until that an admissible kinematic chain has been obtained.

Once the step-by-step analyses have been implemented for each detected triumphal arch, both the infinitesimal vertical and horizontal virtual displacements of the rigid blocks have been evaluated, according to the graphical method.

3.3 The Limit Analysis with a Kinematic Approach

The analyses of the identified kinematic motions are developed according to the limit analysis with the linear kinematic approach. For each analysed triumphal arch and for each identified mechanism, the horizontal load multiplier α_0 is defined according to Eq. (1) by applying the Principle of Virtual Works.

$$\alpha_0\left(\sum_{i=1}^{n} P_i \cdot \delta_{x,i} + \sum_{j=n+1}^{n+m} P_j \cdot \delta_{x,j}\right) - \sum_{i=1}^{n} P_i \cdot \delta_{y,i} - \sum_{h=1}^{o} F_h \cdot \delta_h = L_{fi} \tag{1}$$

In the above equation, n is the number of the rigid blocks, P_i is their self-weight, $\delta_{x,i}$ and $\delta_{y,i}$ are the corresponding horizontal and vertical virtual displacements according to the kinematic chain, P_j represents the weight of the m element not directly involved in the mechanism that generates horizontal forces on the blocks and δ_{xj} the corresponding virtual displacement, F_h is the horizontal external force acting on the blocks and δ_h is the corresponding displacements. L_{fi} is the internal work, which has been assumed null.

Since in the conducted study P_j, F_h and L_{fi} have been neglected, the load multipliers can be determined according to the Eq. (2).

$$\alpha_0 = \frac{\sum_{i=1}^{n} P_i \cdot \delta_{y,i}}{\sum_{i=1}^{n} P_i \cdot \delta_{x,i}} \tag{2}$$

The spectral acceleration $a_0{}^*$ that activates the identified collapse mechanisms, related to the specific values of α_0, is therefore calculated according to Italian current code (Superior Council of the Public Works 2019) as reported in Eq. (3):

$$a_0^* = \frac{a_0 \cdot g}{FC \cdot e^*} \tag{3}$$

where g is the gravity acceleration, FC is the confidence factor related to the level of knowledge (LC) of the investigated structures, and e^* is the portion of mass participating in the considered mechanism evaluated according to Eq. (4).

$$e^* = \frac{\left(\sum_{i+1}^{n+m} P_i \cdot \delta_{x,i}\right)^2}{\sum_{i+1}^{n+m} P_i \cdot \sum_{i+1}^{n+m} P_i \cdot \delta_{x,i}^2} \tag{4}$$

It is worth noticing that if the compressive strength of the masonry is neglected in the calculations, the current Italian code suggest adopting the highest value of the confidence factor, which corresponds to the minimum knowledge level LC1 (Superior Council of the Public Works 2019). Hence, in the conducted calculation FC = 1.35 has been assumed.

4 Application of the Procedure to the Case Studies

4.1 Definition of the Kinematic Chains

According to the step-by-step numerical analyses described in Sect. 3.2 and the graphical method, the kinematic chains and the corresponding both horizontal and vertical displacements have been determined.

In Figs. 8, 9, 10 and 11 the block subdivisions obtained by the FE step-by-step analyses, and the kinematic chains are shown for the four investigated triumphal arches.

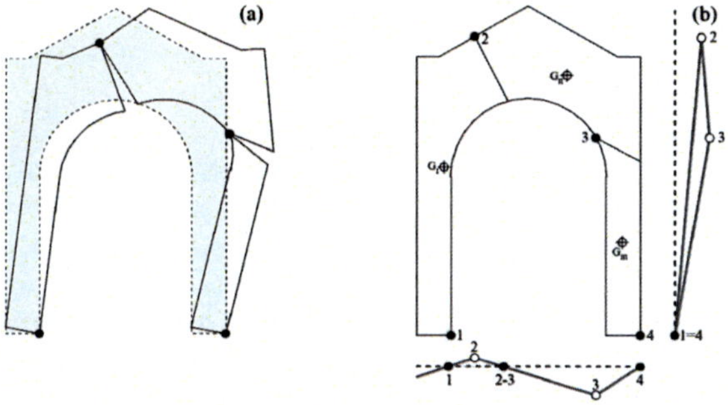

Fig. 8. Block subdivision (a) and kinematic chain (b) for the triumphal arch of S. Felice and Mauro Abbey [SFM]

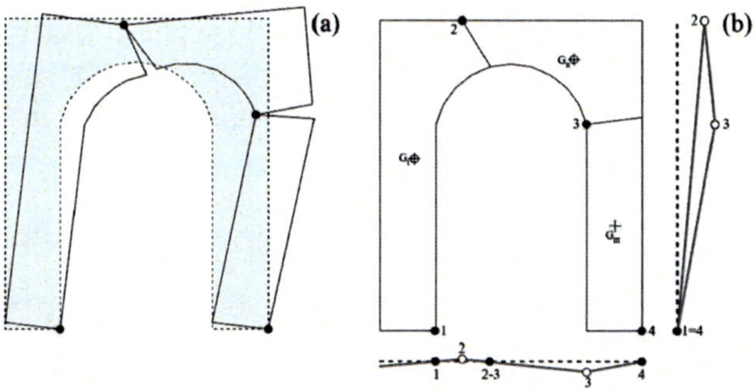

Fig. 9. Block subdivision (a) and kinematic chain (b) for the triumphal arch of S. Sebastiano church [SSB]

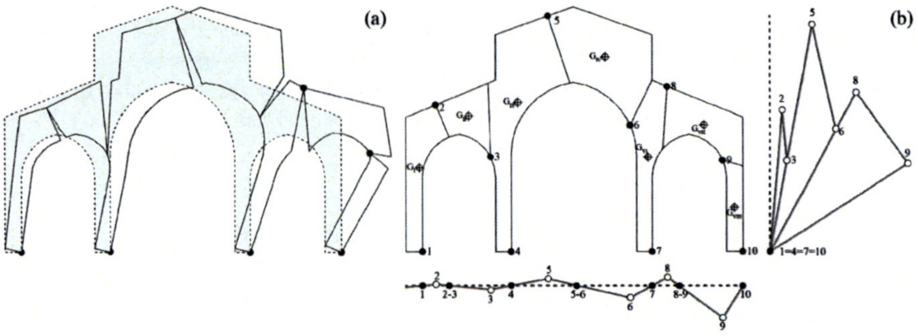

Fig. 10. Block subdivision (a) and kinematic chain (b) for the triumphal arch of S. Maria della Tomba church [SMT]

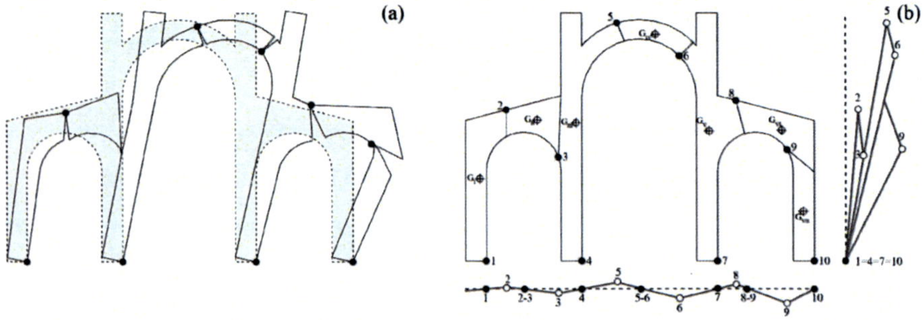

Fig. 11. Block subdivision (a) and kinematic chain (b) for the triumphal arch of S. Eusanio Martire church [SEU]

4.2 Results of the Limit Analyses

Once the kinematic chains [K, TR] have been defined for each considered triumphal arch, the linear kinematic analysis described in Sect. 3.3 has been performed.

The seismic capacities of the investigated triumphal arches, defined in terms of spectral acceleration a_0^*, are summarised in Table 2, where also the values of the horizontal load multiplier (α_0) and the portions of the mass participating in the considered mechanisms (e^*) are reported.

Table 2. Main results of the linear kinematic analysis.

	α_0 [g]	e^* [%]	a_0^* [g]
K, TR $_{[SFM]}$	0.148	90%	0.121
K, TR $_{[SSB]}$	0.226	90%	0.185
K, TR $_{[SMT]}$	0.098	68%	0.106
K, TR $_{[SEU]}$	0.110	82%	0.099

According to the results shown in Table 2, the single-span triumphal arch of [SSB] church presents the highest value of the horizontal load multiplier and the highest structural capacity in terms of spectral acceleration ($a_0^* = 0.185\ g$). On the contrary, the triumphal arches of the [SMT] and [SEU] three-nave churches show a higher vulnerability, being their capacity around 0.1 g. Finally, the analysis performed for the [SFM] church returned an intermediate value of the activating spectral acceleration ($a_0^* = 0.121\ g$).

Qualitatively, the results show that three-span triumphal arches have a worse seismic capacity than one-span ones.

Nevertheless, the obtained evidences cannot be considered explanatory of the seismic capacity, since they are strongly affected by the geometrical features, e.g. the width of the piers and the thickness of the arches.

Therefore, in order to check the reliability of the model, in the following a comparison of the estimated capacities with the seismic inputs recorded during the recent events have been carried out. Moreover, further considerations have been made regarding the seismic demands arisen from the current Italian code application.

4.3 Comparison Between Capacities and Demands

In order to corroborate the obtained results and to check the reliability of the proposed procedure, a comparison with the real damage exhibited by the churches in the aftermath of the L'Aquila (2009) and Central Italy (2016–2017) earthquakes has been carried out, both in terms of collapse mechanisms and seismic input.

In Fig. 12, pictures of the damage states of the considered triumphal arches are shown.

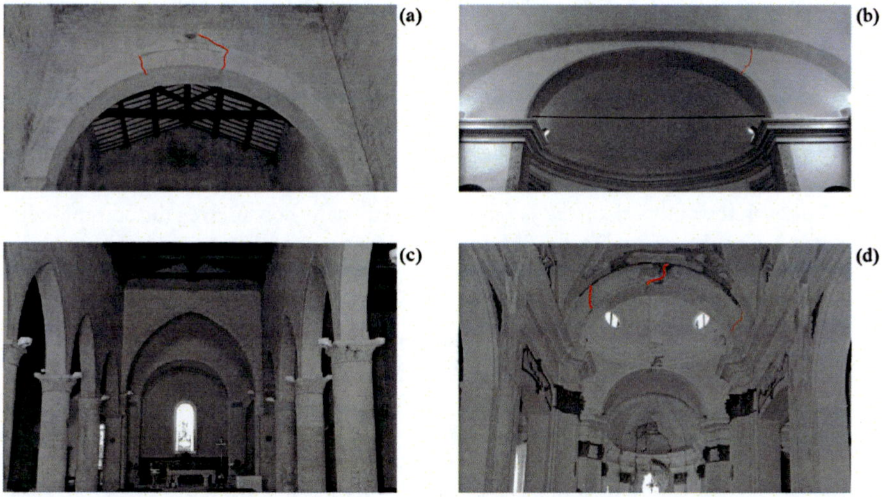

Fig. 12. Damage state of the investigated triumphal arches after the considered seismic events: (a) S. Felice and Mauro [SFM], (b) S. Sebastiano [SSB], (c) S. Maria della Tomba [SMT] and (d) S. Eusanio Martire [SEU] churches.

As it is possible to observe, the [SFM] arch showed an activation of the collapse mechanism very similar to that obtained in the step-by-step procedure, with a significant crack near the mid-span of the arch, which provoked a partial detachment of the keystone blocks (Fig. 12a).

With regard to the [SSB] church, it is worth to note the presence of a transversal tie road, which has been neglected in the conducted calculations. However, after the seismic event, the triumphal arch exhibited a very slight damage concentrated in the keystone proximity (Fig. 12b), corroborating the possible collapse mechanism evaluated with the numerical procedure.

On the contrary, the triumphal arch of [SEU] church developed a three-hinge collapse mechanism, which is more typical in case of heavy vertical loads than in case of earthquake (Fig. 12d). In this case, indeed, the presence of a heavy dome (not considered in the conducted simplified analyses) could have returned to the activation of a different collapse mechanism.

Finally, the [SMT] church did not exhibit the activation of any collapse mechanism and resulted undamaged in the aftermath of the 2009 L'Aquila earthquake (Fig. 12c).

According to the disposition of the current Italian Technical Standards (Ministry of Infrastructures and Transports 2018), for a structure (or a portion thereof) lying directly on the ground, the SLD (Damage Limit State) safety check could be considered satisfied in terms of the spectral acceleration $a_0{}^*$ if the following inequality of Eq. (5) is respected.

$$a_0^* \geq a_g(P_{VR}) \cdot S \tag{5}$$

In the above equation, $a_g(P_{VR})$ is the spectral seismic acceleration, as a function of the probability of exceeding the chosen limit state, S is the subsoil coefficient, which is evaluated as the product between the stratigraphy (S_s) and the topography (S_T) coefficients.

Therefore, the comparison has been carried by considering the PGA values recorded in the corresponding sites during the mentioned seismic events, according to the shake maps provided by the INGV (Faenza and Michelini 2011, Gruppo di Lavoro INGV sul terremoto in centro Italia 2016).

Given the uncertainties in defining the stratigraphic coefficient S_s, a range of the S factor has been considered, accounting for the real topographic categories, which has led to assume $S_T = 1.4$ for the [SSB] church and $S_T = 1.2$ for the other cases.

Moreover, the comparison has been also carried out by considering the current Italian code provisions, by assuming a reference life of 50 years and an exceeding probability of 63%, which is typical of the damage limit state (*SLD*).

The seismic evaluation has been therefore based on the definition of an I_S safety index, obtained as a ratio *Cap/Dem* between the seismic capacity of the structure (with $Cap = a_0{}^*$) and the seismic demand as function of the reference limit state (with $Dem = a_g(P_{VR}) \cdot S$, in the case of mechanisms at the level of the foundation). The check is satisfied when $Is \geq 1$.

In particular, the safety indexes have been determined considering the variation range of the seismic demand related to the uncertainties in defining the subsoil coefficient *S*.

The obtained outcomes are summarised in Table 3, where *Dem* represents the seismic demand, the subscripts *SI* and *SLD* mean seismic input and limit state of damage, respectively, and *min* and *max* are the limits of the considered ranges.

As shown in Table 3, the obtained outcomes seem to return a good correspondence for the considered cases.

Table 3. Comparison between capacity and demands of the investigated triumphal arches.

	a_0^* [g]	$Dem_{,SI}$ [g]	$Dem_{,SLD}$[g]	$I_{S(SI,min)}$	$I_{S(SI,max)}$	$I_{S(SLD,min)}$	$I_{S(SID,max)}$
K, TR [SFM]	0.121	0.096–0.172	0.115–0.207	0.70	1.26	0.58	1.05
K, TR [SSB]	0.185	0.168–0.302	0.112–0.202	0.61	1.10	0.92	1.65
K, TR [SMT]	0.106	0.048–0.086	0.112–0.218	1.23	2.21	0.49	0.87
K, TR [SEU]	0.099	0.144–0.259	0.125–0.225	0.38	0.69	0.44	0.79

In particular, the [SMT] triumphal arch experienced a seismic demand $Dem,_{SI}[g]$ lower than the obtained structural capacity, as it was demonstrated by the undamaged state surveyed in the aftermath of the L'Aquila seismic event: in this case the safety indexes resulted greater than 1, regardless of the quality of the subsoil.

On the contrary, the estimated capacity of the [SEU] arch was inadequate to face the seismic demand $Dem,_{SI}[g]$ arisen from the L'Aquila earthquake, returning an I_S index smaller than 1, also for the upper bound of the range. The significant seismic damage observed on this triumphal arch corroborates the obtained results.

A particular case is represented by the [SSB] church, for which the obtained range of I_S could be ascertained by the slight damage surveyed after the Central-Italy seismic events, despite the presence of the transversal tie road has been neglected.

Finally, as far as the [SFM] Abbey is concerned, the analysis revealed a high probability that the capacity of the triumphal arch was exceeded by the demand induced by the 30th October 2016 seismic event (epicenter in Norcia). Nevertheless, according to the revealed damage, the *Cap/Dem* ration should return a value close to 1, since a slight activation of the collapse mechanism has been observed in the detected arch.

If the demands arisen from the code provisions are considered, it is clear that, except for the [SSB] church, all the triumphal arches are affected by a significant vulnerability, since the safety indexes returned values smaller than 1, especially for bad subsoil categories.

5 Conclusions

In this paper, a simplified procedure for the structural assessment of triumphal arches, based on a combined numerical and analytical analyses, is proposed. The suggested procedure has been applied to real cases of triumphal arches affected by the L'Aquila (2009) or the Central Italy (2016–17) earthquakes. In order to corroborate the reliability of such a proposed methodology, the obtained outcomes have been compared with the real damages and the seismic inputs experienced by the considered churches.

According to the obtained evidences some concluding remarks can be stated. Above all, the adopted procedure revealed a good agreement in a both qualitative and quantitative points of view, despite the assumed approximations and simplifications. In particular, the collapse mechanisms obtained from the step-by-step analyses showed a good correspondence with the real crack patterns exhibited by the damaged churches. Moreover, the values of the safety index obtained from the analytical procedure resulted to be confirmed by the damage states surveyed in the detected triumphal arches.

Based on the above outcomes, it is clear that the proposed methodology could be considered a valid alternative to the more complex iterative processes for the definition of non-dissipative hinges, representing a valid starting point for a wider application on real cases, as well as for the implementation of vulnerability models for existing masonry churches.

Acknowledgements. The present study has been developed within the agreement DPC-ReLUIS 2019-2021 and as part of a scholarship funded in the framework of the National Operational Program ESF-ESFR Research and Innovation (PON RI 2014-2020), Action I.1 related to Innovative Industrial Ph.D. (Project Code – CUP - B29D17001610006).

References

Alexakis, H., Makris, N.: Limit equilibrium analysis of masonry arches. Arch. Appl. Mech. **85**(9–10), 1363–1381 (2015). https://doi.org/10.1007/s00419-014-0963-6

Cacace, D., Rouhi, J., Zizi, M., De Matteis, G.: Vulnerabilità sismica di archi in muratura: uno studio parametrico con approccio cinematico. In: XVIII Convegno ANIDIS – L'Ingegneria sismica in Italia, 15–19 September, Ascoli Piceno, Italy (2019)

Cangi, G.: Manuale del Recupero Strutturale e Antisismico (II edizione). Edizioni DEI, Tipografia del Genio Civile di Roma 2005/ IIed. 2012 (2012). (In Italian)

D'Ayala, D.: Force and displacement based vulnerability assessment for traditional buildings. Bull. Earthq. Eng. **3**, 235–265 (2005)

De Luca, A., Giordano, A., Mele, E.: A semplified procedure for assessing the seismic capacity of masonry arches. Eng. Struct. **26**, 1915–1929 (2004)

De Matteis, G., Mazzolani, F.M.: The Fossanova church: seismic vulnerability assessment by numeric and physical testing. Int. J. Archit. Herit. **4**, 222–245 (2010)

De Matteis, G., Zizi, M.: Seismic damage prediction of masonry churches by a PGA-based approach. Int. J. Archit. Herit. **13**(7), 1165–1179 (2019)

De Matteis, G., Brando, G., Corlito, V.: Predictive model for seismic vulnerability assessment of churches based on the 2009 L'Aquila earthquake. Bull. Earthq. Eng. **17**(9), 4909–4936 (2019). https://doi.org/10.1007/s10518-019-00656-7

Dimitri, R., Tornabene, F.: A parametric investigation of the seismic capacity for masonry arches and portals of different shapes. Eng. Fail. Anal. **52**, 1–34 (2015)

Doherty, K., Griffith, M.C., Lam, N., Wilson, J.: Displacement-based seismic analysis for out-of-plane bending of unreinforced masonry walls. Earthq. Eng. Struct. Dyn. **31**, 833–850 (2002)

Faenza, L., Michelini, A.: Regression analysis of MCS intensity and ground motion spectral accelerations (SAs) in Italy. Geophys. J. Int. **186**, 1415–1430 (2011). https://doi.org/10.1111/j.1365-246X.2011.05125.x

Gattesco, N., Boem, I.: Review of experimental tests and numerical study on masonry vaults reinforced through fiber-reinforced mortar coating. Bull. Earthq. Eng. **17**(7), 4027–4048 (2019). https://doi.org/10.1007/s10518-019-00619-y

Grunthal, G.: European Macroseismic Scale, Centre Européen de Géodynamique et de Séismologie, Luxembourg, vol. 15 (1998)

Gruppo di Lavoro INGV sul terremoto in centro Italia: Rapporto di sintesi sul Terremoto in centro Italia Mw 6.5 del 30 ottobre 2016 (2016). https://doi.org/10.5281/zenodo.166019

Heyman, J.: The Stone Skeleton: Structural Engineering of Masonry Architecture. Cambridge University Press, Cambridge (1995)

Huerta, S.: The debate about the structural behaviour of Gothic Vaults: From Viollet-le-Duc to Heyman. In: 3D International Congress on Construction History, Cottbus, Germany, 20–24 May (2009)

Ministry of Heritage and Cultural Activities: Circular n.26, Italian Code for protection of cultural heritage. Linee Guida per la valutazione e la riduzione del rischio Sismico del patrimonio culturale con riferimento alle norme tecniche per le costruzioni (2011). (In Italian)

Ministry of Infrastructures and Transports: Aggiornamento delle "Norme tecniche per le costruzioni" (2018). (In Italian)

Superior Council of the Public Works: Istruzioni per l'applicazione dell'«Aggiornamento delle "Norme tecniche per le costruzioni"» di cui al decreto ministeriale 17 gennaio 2018" (2019). (In Italian)

Check for updates

Reshaping the Space. The Larger Goals of Historic Preservation and Protection

Ioan Andreescu[1,2](✉)

[1] Faculty of Architecture and Urbanism, "Politehnica" University Timisoara, Str. Traian Lalescu 2A, Timisoara, Romania
[2] "Andreescu&Gaivoronschi" Design Firm, St. George sq. 4, ap.10, Timisoara, Romania

Abstract. Historic structures are not only a part of our cultural identity, but also a component of our daily life. Most of these heritage presences – objects, buildings, landscapes – are embedded in our habitat, in our cities and are influencing the way we are foreseeing the future. The strategies and methodologies developed for the material conservation of historic structures must be integrated with the projects for tomorrow. It is not only about blending old and new, but about integrating the re-shaped, re-informed elements of the past into an evolutionary process. Our office, "Andreescu&Gaivoronschi", has been involved in heritage projects from conserving singular objects to large-scale urban landscaping and landscaping designs. We confronted challenges regarding methodologies, proceedings, material research. During our work we discovered the important, less evident parts of heritage: geometry, visual connectivity, meaning; we also became aware of the benefits of using contemporary art. The goal was to imagine new and meaningful roles for the remnants of the paSt. The paper presents our quest and findings, following a string of projects organized according to scale.

Keywords: Heritage · Re-informing · Conservation · Geometry · Landscape · Art

1 Introduction

The heritage is not only a component of our cultural identity but, also, an important part of our daily life. Objects, works of art, buildings, urban landscapes, "cultural" landscapes are constituents of our environment and subjects for our visions of the future.

As architects we have at our disposal a large array of theoretical and practical tools in order to asses and conserve the material imprint of the past, but this activity must be connected with our planning for the world to come. The flow of time cannot be stopped and our main task is to manage a blend of old and new elements in order to reshape, re-use, re-inform the remnants of our past into a continuous process (Gausa et al. 2003).

During the years, our office "Andreescu & Gaivoronschi" has produced a large number of projects, from object and building conservation to large scale urban and landscaping interventions.

We were confronted with challenges regarding research and working methodologies, operational proceedings, but our goal was always the same – to find a new role for our

I. Vayas and F. M. Mazzolani (Eds.): PROHITECH 2021, LNCE 209, pp. 1408–1428, 2022.
https://doi.org/10.1007/978-3-030-90788-4_107

heritage, a new dimension for our inherited space, a better connection between – past, present and future.

An important part of our projects is situated in the historic city of Timisoara, in Western Romania (Fig. 1). Its territory is a reclaimed marshland and like many Central-European cities, it has a central area formed by a core - the Citadel - satellites and an in-between connecting tissue. The actual shape of the city is the result of a succession of different planning operations between mid XVIIIth and mid XXth century.

Intervening at various scales in different parts of the city we learned about the importance of the visible and invisible parts of the heritage: the geometry, the visual connections, the atmosphere or "ambiente". We also became aware of the importance of using works of art both ancient or contemporary, in order to infuse meaning and visual consistency to our interventions.

Our experiences and discoveries will be presented along a string of selected projects organized according to their scale – from building conservation/restoration to landscape interventions.

(a) (b)

Fig. 1. Timisoara urban map 1913. (a) Aerial photography of the city Centre (b)

2 Beginnings

Even before the start of our office, together with my partner arch. Vlad Gaivoronschi, we investigated the complex issues connected with architecture, memory, art and the city, mainly by participating in architecture contests.

Two contest projects explored a park occupying a corner of the inner city: "The theatre for Tours" (Prague 1983) (Fig. 2) and "The Civic Centre" (1985) (Fig. 3). The first was a project of ideas and the second was dedicated to a real problem - edifying a new city center in the near proximity of the existing core. Both projects dealt with the power of memory, of art and society. The Centre was selected for the development, and received the National Prize in 1986, but was never implemented.

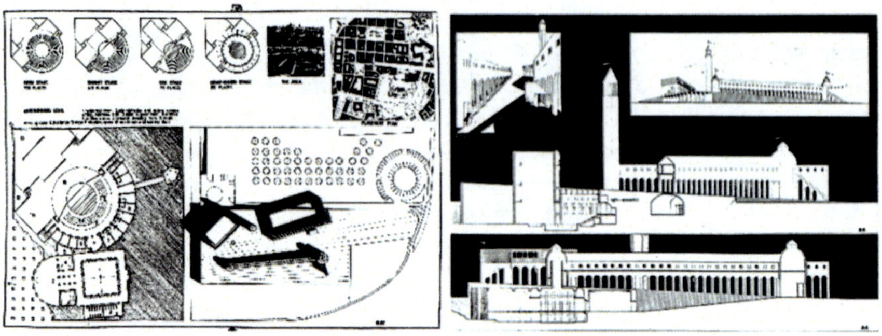

Fig. 2. Theatre for tours, 1983

(a) (b)

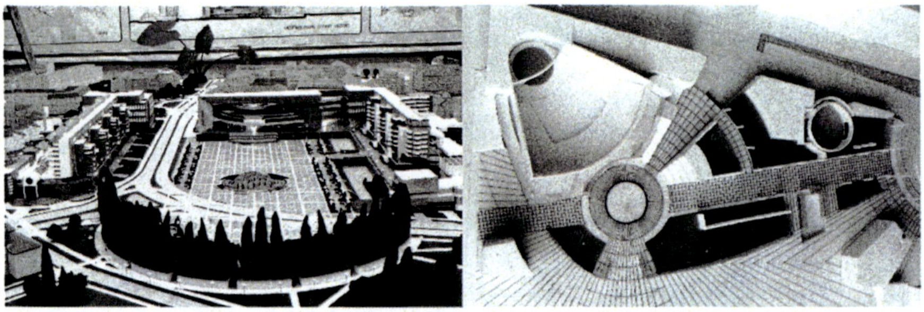

Fig. 3. The Civic Centre, 1985. Model photos: (a) general view, (b) urban detail

More interesting is the project for "A Poetic Dwelling" – "A Marine House" (1986) (Fig. 4), proposing a submerged house, floating on high seas. It was a modernized atrium house able to submerge into the sea, a metaphor for the fascinating relationship between form, technology and environment; it was to be a prescient image for our continuous exploration of the merger between project and context.

Fig. 4. A Poetic Dwelling, 1986, "A Marine House"

In 1993, proposing a project for the remodeling of the historic "St. George Square" (Fig. 5), in Timisoara, we uncovered an amazing fact - folding the geometric traces of the surrounding facades over the square surface, a perfect geometric layout appeared, imbued with symbolism. In the 1880's, a liberal, progressive City Council tore down the Jesuit Church and College, reshaped the place and "signed" the operation with this "invisible" geometric seal. It was a lesson for future projects about the value of "hidden geometry" in historic sites.

(a) (b)

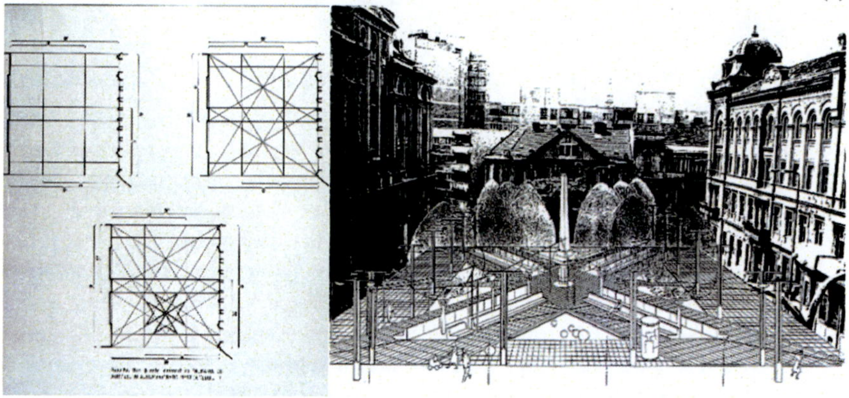

Fig. 5. St. George Square Timişoara, 1993, (a) Geometric Survey, (b) Proposal

3 Restauration. Conservation

An important part of any intervention on existing structures is the conservation of the material substance, the so called "consistenza materica", especially the surface, which is the vector of the "image" (Brandi 1996). Restoration could be the main purpose of the project or an important part in a re-modelling or re-informing design.

The inner city of Timişoara contains hundreds of buildings displaying beautiful details and ornamental patterns. Most of them have been modified or re-built during the XIXth century or the early XXth century. I agree with P. Philipot who affirmed that it is a must to preserve the image of the premodern city from the late XIXth century and the early XXth, without proposing "unimportant periods" (Philippot and Mora 1987).

In the early 1990's we addressed the problem in a traditional way, either during the assessment or the intervention itself, using skilled masters in plaster, gypsum, wood at the V. Alecsandri no.5 façade restoration in Timişoara (Fig. 6).

Fig. 6. Historic façade restauration, 5 Alecsandri St., Timişoara, 1998, (a) views, (b) details

But in 2015 the situation was different. Skilled labor was in scant supply and replacement materials offered by specialized firms were not appropriate in many cases. The complex project for the restoration and reconversion of the "Moorish Synagogue" (1869) (Fig. 7a), in the Citadel, led us to use high-tech fabrication techniques in order to replace missing facade tiles: we used a "Linear motion Central System for Digital Photogrammetric Survey", in order to collect data, then we generated a 3D element solid textural model and printed it with a **Projet** device (using a power infiltrant & CMYK binders) (Fig. 7b). The result was stunning, a perfect reproduction was issued, but raising the thorny problem of authenticity (Gheorghiu and Andreescu 2015).

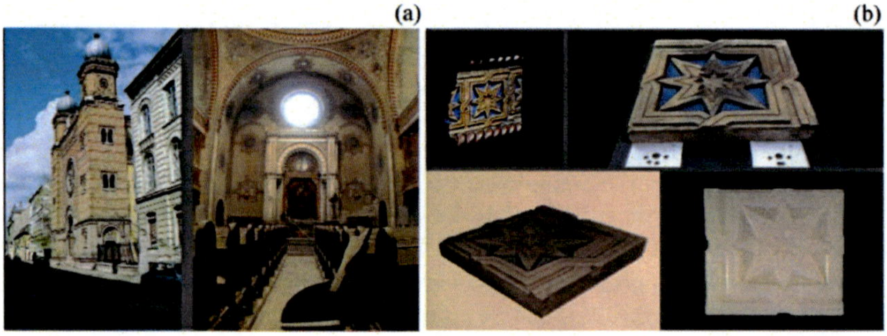

Fig. 7. The Moorish Synagoge Timişoara (a) exterior, interior view, (b) printing of the façade tile

A challenging situation presented itself when, during the site works for the office development City Business Center - CBC (2006–2016), "The Gate to the Center", the lock system of the former Vauban fortress moat was discovered (2015).

It was a well-preserved complex structure, composed of brick and stone culverts with stone wave breakings walls and gutters, and intake and exit pools covered with oak planks, everything founded on oak piles. After many studies and evaluations, the decision was made by the Heritage Committee to preserve and integrate the main lock ensemble (cca. 450 m^2) within the urban plaza created by our project team, by relocating the ensemble in order to fit the CBC general outlook.

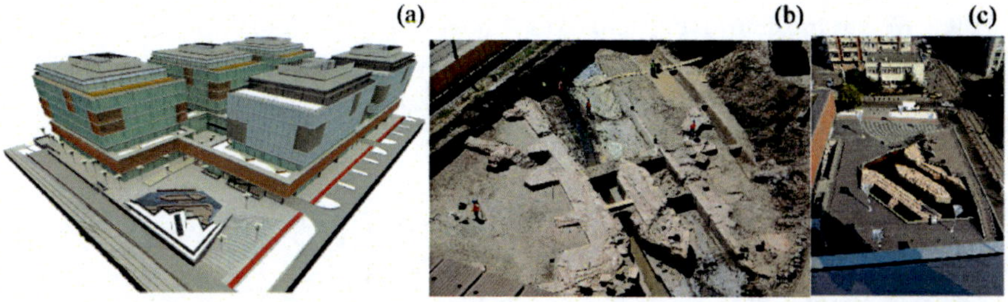

Fig. 8. City Business Centre, Timişoara, 2006–2016 (a) integration into the project, (b) relocation of the lock, (c) the final outlook

Dismantling the brickwork and the planks, treating the materials and relocating them into a basin, under controlled layers of water was a complex, but rewarding task (Andreescu et al. 2015a).

The integration of this precious hydrotechnical device from the early XVIIIth century opened an interesting opportunity to mix old and new architecture inside the new urban pattern (Fig. 8).

Working with historic wood structures was a constant preoccupation, since the buildings from the inner city presented complex roofing structures from the late XVIIIth century to the late XIXth century. Assessment and study of various roofing structures revealed more complex issues than the technical aspects. Roof builders from the past used consistently the so called "sacred geometries" – until the beginning of the XXth century.

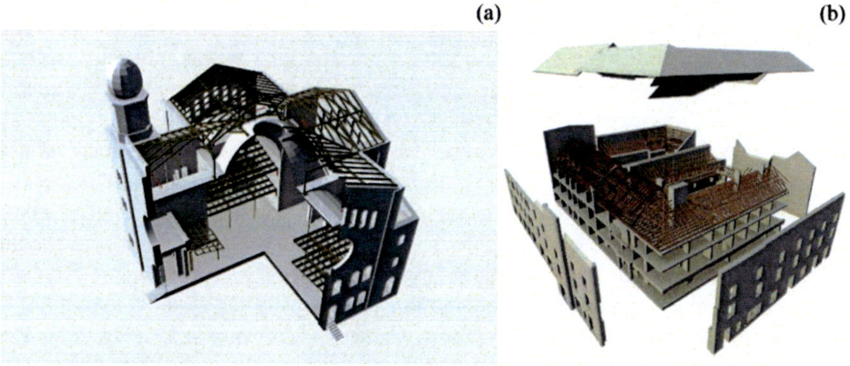

Fig. 9. Geometry and structure (a) The Synagogue, Timişoara, (b) Historic building St. George sq. 4, Timişoara

Various geometric patterns were used in relation to key events in Craft Guilds history, advancement and decline, so the geometric analysis is an important tool for dating those complex structures. Assessment of some of these roofing configurations (like the Synagogue) (Fig. 9) proved that coherent "sacred geometry" patterns were consistent with the best structural behaviors including anti-seismic response (Andreescu et al.

2016a). We put this knowledge into work, organizing a late XVIIIth roof as a working space for our office. Minimum interventions were required and the structure works perfectly in the changed circumstances (Fig. 10).

(a) (b)

Fig. 10. Restored attic St. George sq. 4 Timişoara, 2013 (a) interior view, (b) geometric analysis

4 Introducing New Functions into Historic Buildings

The next step, both in scale and complexity, was to address valuable historic buildings in order to accommodate new functions: office space, cultural centers, conference, retail. Conserving the authentic materials and the original outlook was a cultural priority. In order not to violently disturb the inner organization of the houses we adopted a specific procedure – the "**augmentative approach**" (Andreescu et al. 2016b). The potentially disturbing new functions (like auditoriums) were inserted in the available residual open space, avoiding street visibility. Where necessary, transparent atrium-like connecting volumes were provided enhancing the connectivity. Timisoara is situated into a seismic area – some of the benefits of this approach is to connect the old buildings to the new infills in order to enhance the response to horizontal forces.

Among the most interesting projects in this category are the Hotel "Savoy" (2005–2006) (Fig. 11), and the "Christian Youth Centre" (2017-onwards) (Fig. 12). Both projects are based on exceptional modernist structures from the early 1930's, situated in protected areas in the close proximity of the historic core.

In both cases a similar strategy has been used: the bulk of the new functions, (hotel rooms and car park, respectively the Conference Center have been organized in the back yard, being connected to the existing building by a spectacular atrium space).

(a) (b)

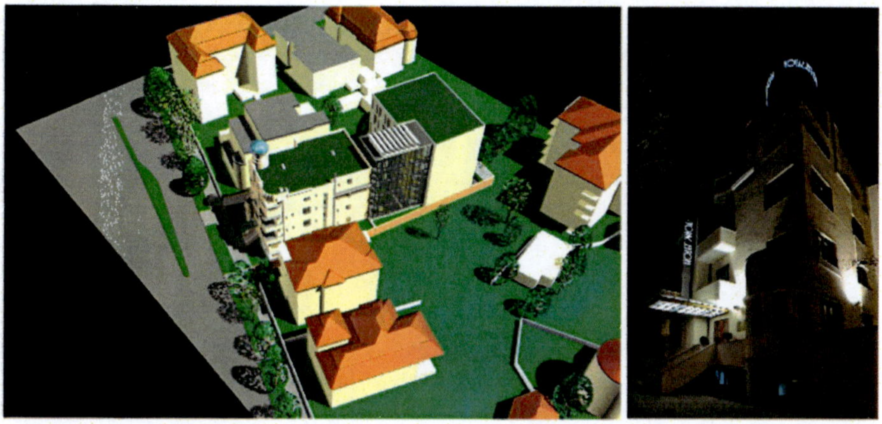

Fig. 11. Hotel Savoy Timişoara, 2005–2006, (a) general view, (b) main entrance – night view

(a) (b) (c)

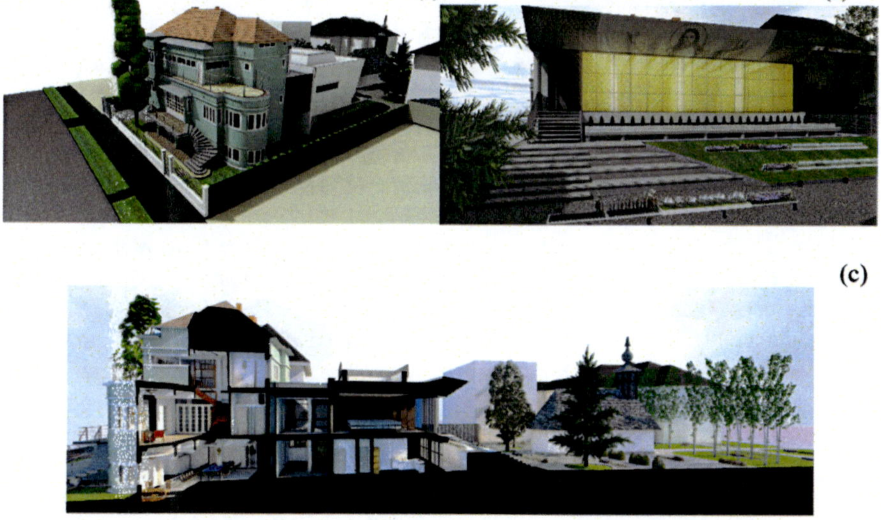

Fig. 12. Christian Youth Center, Timişoara, 2017-onwards (a) general view, (b) garden elevation, (c) perspective cross-section

The "Youth Center" location (Fig. 12), offered an interesting urban opportunity: it is linked back-to-back to the Archbishop's Residence and garden, which contains a precious wooden Chapel. After completion the separating wall will be torn down, and the garden will be extended by a gentle slope towards the Conference Center. So, the new building will have a façade of its own supporting an exceptional stained-glass wall by the artist Orăvitzan, dedicated to the "Divine Light" (Andreescu et al. 2019).

In both cases, intensive attention was given to the preservation and enhancement of the plastic and spatial characteristics of the early modern structures. Flooring, decoration, fine tracery was either restored or reproduced identically. A great contribution was realized by using both historic and contemporary works of art, some of them dedicated to the projects.

We consistently used high quality works of art in restoration and re-use, re-information of existing structures, because they are offering both a visual identity and high quality to the project and perfect connectors between old and new parts.

5 Urban Configurations

The city continues to be the main stage of the social, economic, cultural and political evolution. It is one of the most complex creations since the beginning of civilization.

The association between urban form and architecture is essential, the physical dimension of the city, being an important topic for both the urban studies and architectural practice.

Being confronted to interventions in sensitive urban areas we soon discovered that urban conservation is rather a legend. Cities are evolving and a proper balance has to be realized between preserving elements and characteristics of the urban morphology and creating new connections, and contemporary infills in the process of re-use and development.

Our city – Timisoara – evolved around a massive nuclear Citadel and its historic satellites. The destruction of the fortification was followed by a succession of masterplans between 1893 and the late XXth century aiming at the remodeling of the in-between spaces and the future expansion.

Some of the in-between areas are containing several of the most elaborate urban spaces in the city, created or remodeled in the critical years between late XIXth and mid XXth century. We had the privilege to work on two of them: The "Victoria Plaza" (2009–2010), and the "Hay Market" (1995–2011). Our objective in both cases was somewhat similar – to consolidate the main urban characteristics of the place and upgrade it to a new level of activity and presence.

"Victoria Plaza", a monumental esplanade expanding the city core to the south-west, connects two iconic buildings: "The Opera House" (1875, reconditioned 1923) and the Orthodox Cathedral (1938-41). Most of the front buildings are spectacular Sezession structures, build between 1903–1915. In the 1920's it became the real center of the city, and it still is today.

Our proposed intervention was rather restraint, improving connectivity, conserving surfaces, proposing new lighting and urban furnishing, adapted to the actual functions (Fig. 13).

(a)

(b) (c)

Fig. 13. Victoria Plaza, Timişoara (a) aerial view, (b) proposal general view, (c) proposal axial view

The intervention on the "Hay Market" was more radical. The "Hay Market" ("Badea Cârţan" Market) connecting one of the satellites to the main east-west historic road, was a declining space, unable to exploit its exceptional urban form and position. In 1995 the City decided to act, realizing a structure intendent to function also as a support for community activities. A caterpillar like steel and tensile membrane was realized covering 90% of the market (Andreescu et al. 2015b). It was scaled in order to fit into the urban ensemble; it was opened laterally in order to allow the view of the historic facades (Fig. 14). The building received the ECCS – European Design for Steel Structure award in 1996) (Andreescu et al. 2015b).

In 2010, a high-quality membrane replaced the original yellow one and a transparent line of small shops were built along the edge. The low silhouette of the shops, helped the maintain the panoramic views of the historic fronts, allowing a spectacular visual connectivity. The covered market was an instant success, becoming a hub for the entire city. At the community's request a new church was built in the small plaza adjoining the market; the city decided to extend the development of the river banks in order to include the market area.

Fig. 14. The "Hay Market", Timişoara, 1996–2011, (a) general view 1996, (b) community church 2006, (c) night view 2010, (d) interior view 2010

The most challenging urban "conservation" intervention our studio undertook, was a project for the re-using and modernization of the historic core of the spectacular Spa Town of "Herculane Baths" (Fig. 15), in the mountains of South Banat (2011). The first edification was in 102 AD, by the emperor Trajan; after the rebuilding, starting from 1736, the town displayed a magnificent center, composed by two plazas situated on the opposing banks of the river and connected by a historic bridge. The first plaza (1736–1840) contains hotels, a casino, bathhouse and a neo-classical church. The second "esplanade" (1870–1900) – crossed by the river contains an elevated casino, the big "roman" bath edifice, hotels and the intendent villa.

The aim of the project was to re-use the magnificent buildings without severe alterations and – most important – to create an elegant "ambiente" as a contemporary counterpoint for the famed atmosphere à la "Belle Époque", when Emperor Franz Josef considered Herculane like "the most beautiful resort of the continent" (Fig. 16). Unfortunately, due to a disagreement between the investment fund and local vested interest the project was stopped and the beautiful plazas were sunk in disrepair; some enthusiastic NGO's are trying to save bits and pieces.

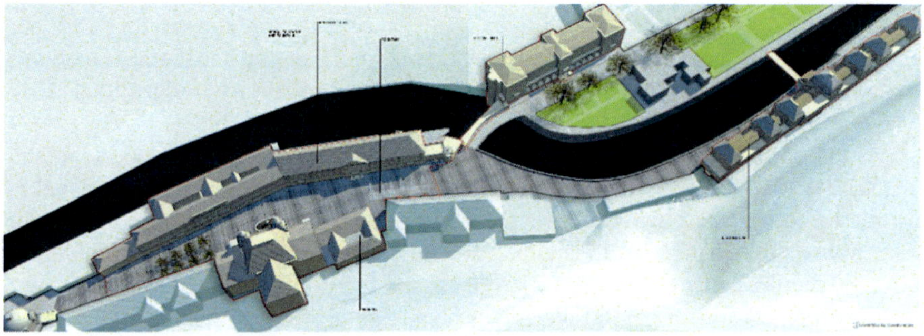

Fig. 15. Herculane Baths, Spa City, 2011–2012, proposal: general view

(a) (b)

Fig. 16. Herculane Baths, Spa City, main plaza proposal 2011–2012, (a) general view, (b) street view

6 Large-Scale Urban Interventions

In recent years we were confronted with several large-scale urban interventions situated in the so-called in-between spaces connecting the City core of Timisoara with the former historic satellites.

At the beginning of the XXth century, successive Master Plans were proposed in order to transform these connecting areas into metropolitan districts a la Otto Wagner. Complex geometric patterns, consistent urban plazas and representative public buildings were created but none of the schemes was fully implemented, and today some of these central areas lost their coherence, quality and became subjected to massive redevelopment.

Based on our previous studies ("St. George Square", "Victoria Plaza", "Hay Market", etc.) we imagined a design strategy of "reprogramming", "urban recycling" and "reactivating" these areas in order to generate complex relationships between existing and proposed elements.

Integrating geometric and historic data created the conceptual support for developing landscaping operating systems of vistas, panoramas and viewing corridors, as they were first used in the planning proceedings by The Greater London Authority (2004, 2011) (Mayor of London Assembly 2011).

We first applied this approach for the design of an important waterfront development – the ISHO – 2016-onwards, situated east of the city core between the river Bega and the historic east-west thoroughfare, near the crossing with the main historic north-south road; where the "Hay Market" is situated. The development contains cca. 50000 m^2 office space, a five-star hotel, 1200 apartments and retail, offices etc.

The urban complexity of the site and its challenging geometry inspired the planning strategy: to create a pattern of axes, viewing lines, visual and physical connections. This pattern influenced the organizing of the interior landscape, the typological configuration of the buildings and the structure of apartments (Fig. 17).

The site was divided in three east-west stripes: along the main road, along the riverside and between them. The road side lane contains massive structures: office, hotel, high-rise apartment buildings, the middle lane contains parks and landscaped areas, the riverside lane aligns U-shaped low-rise buildings oriented towards the river banks and a 21-stories tower containing premium apartments, acting as a signal. The strips are intersected by transversal crossing, either visual, pedestrian or destined for vehicles. Two bridges – the main traffic bridge and a secondary pedestrian bridge will be realized. The ISHO development, now nearing completion acts as an effective hub connecting the city core with the Fabric satellite, the eastward avenue to the riverbank, and is visually integrating an important historic landmark from the south bank – The "Millennium" Church (Fig. 18).

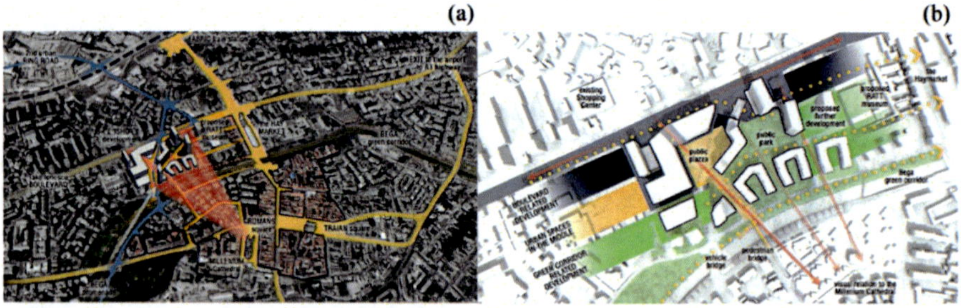

Fig. 17. The ISHO Development, Timişoara, 2016-onwards, organizing principles: (a) a hub of connections, (b) axes and stripes

Fig. 18. The ISHO Development, Timişoara, 2017-onwards: (a) integrating the "Millennium" Church, (b) river view, (c) UAV survey, (d) U-shaped building

The other major important development was the intervention in the "Old Slaughterhouse" district in order to create a mixed-use development: "The Zenit Views" (Fig. 18, 48500 m^2, 900 apartments, retail, schools, services) (Andreescu and Gaivoronschi 2019). The district was a part of a south-east in-between area connecting the city core with satellites. The plot was configured by the 1901-03 "Systematization Plan" in order to contain the "Slaughterhouse" (1904-06), a monumental neo-Romanesque complex designed by a noted architect – Laslo Szekely.

The area is situated between the high-density northern part of the district and a low-density "garden city" in the south, bordered by a semi-circular green area.

The "Slaughterhouse" complex was organized around the main water-tower, acting both as the center of an elegant radial system of axis connecting the main elements: the northern area, the main road, the "Slaughterhouse" pavilions, the park and the residential area. The tower played also as major accent along the east-west road. The elegance and complexity of the geometric pattern, along with the central role of the tower inspired the project's logo "Zenit Views" (Fig. 19).

(a) (b)

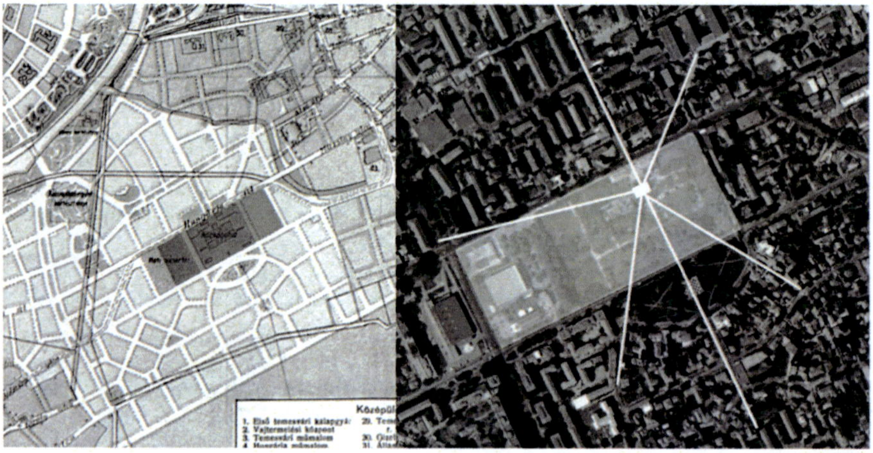

Fig. 19. Slaughterhouse district urbanization, Timişoara: (a) Systematization Plan Detail, 1901-03, (b) Identification of the main axes

Reviving and developing the geometric pattern's regulating potential, we obtained the instrument for organizing, rhythmically, the project's elements, either new or old. As for the ISHO, the area was divided in three strips:

- the northern strip containing the existing buildings, monumental statues, completed with new additions in order to form an entrance Agora.
- the mid-strip, containing high-rise apartment buildings (10–14 stories), with diminishing upper floors and with the building's heights attenuate towards the center in order to favor the view of the Tower.
- the south strip with lower height apartment buildings, dispersed in an alternate rhythm in order to connect the development to the park.

The old buildings were supposed to be restored with limited alterations, and the new ones were imagined as contemporary elements, using the in-situ surface finishing: brickwork, stone, plaster, metal.

The patterned composition sustains a variety of public spaces. Unfortunately, the developer chose a simpler scheme (Fig. 20).

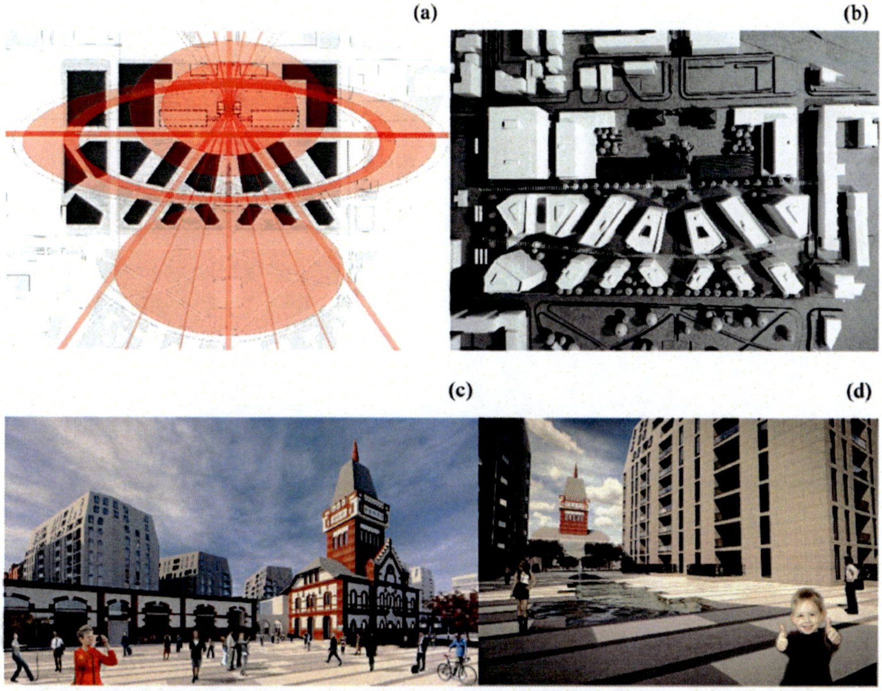

Fig. 20. The "Zenit Views" Development Proposal, Timişoara, 2019, (a) geometric patterns, (b) general organization, (c) agora view, (d) south view towards the tower

7 Landscape Interventions

The ultimate challenge in large scale intervention projects is the historic landscape, or "the designed landscape", a complex cultural product representing the artistic, social and political values of an environment.

Those landscapes are telling stories, are fabulous connecting tools both physically and visually, they have hidden organizing mechanisms and were and are able to support multiple interventions in different times.

One of the most important characteristics of a "cultural" landscape is its hidden geometry, the not so-visible ordering structure.

A few years ago, we were able to uncover the hidden structure of a late XVIIIth century park, the Banloc Manor Gardens, 50 km south of Timisoara. The park was created in the mid XVIIIth century as a formal landscape, modified according to English fashion at the end of the XVIIIth century, and slowly evolving until the mid XXth century when it fell into decay.

In 1995, after a careful traditional topographical survey we proposed a project for the park restoration, some twenty years later we started a new exploration using three layers of investigation (Andreescu et al. 2014):

- an UAV (unmanned aerial vehicle) survey of the park (2014) resulting in high resolution 2D Georeferenced Ortophotos and 3D Digital Elevation Models
- a selection of historic maps, resulting in geo-referential topographic material
- a geometric study of the garden's tracing patterns, using formulas specific for the period (late XVIIIth, early XIXth century)

The amazing accuracy of the orthophotos geo-referenced map, superposed with geo-referential and corrected historic maps, proved to be invaluable tool for research (Fig. 21). Every small detail was made visible. The final coordination with the geometric analysis produced the final touch – the "phantom" XVIIIth century park emerged and all its successive alterations became visible (Fig. 22).

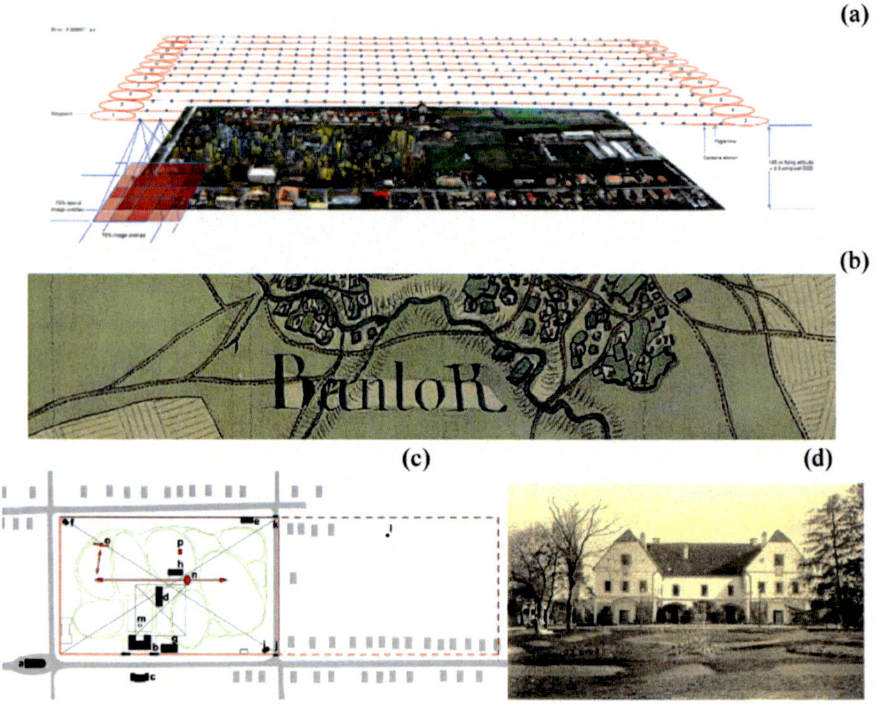

Fig. 21. The Banloc Castle Survey and Assessment, 1995–2014: (a) UAV survey, (b) historic maps, (c) geometric analysis, (d) garden view

Cultural landscapes could be a powerful immersive tool, a blend of natural, historic, cultural elements able to sustain an engaging personal experience; they are perfect platforms for connecting and interpreting multiple elements.

(a) (b)

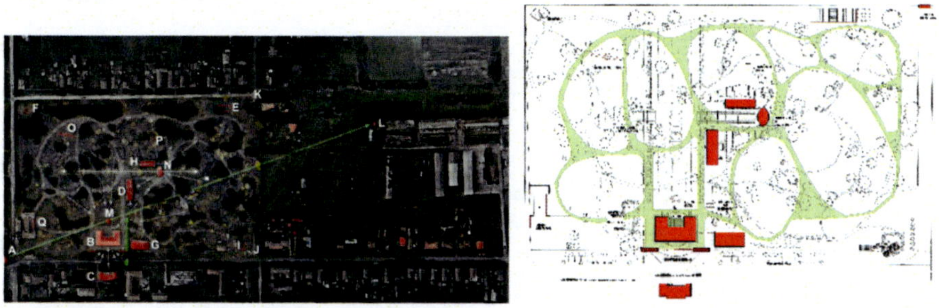

Fig. 22. The Banloc Castle Park Analysis, 1995, 2014: (a) park configuration hypothesis (1840), (b) superposition of the proposals (1995–2014)

In 2007–2011 we were implicated in a large-scale project for organizing the mobility and visiting infrastructure for the "Iron Gates National Park" (124 000 ha), situated in south Banat (Fig. 23), along the most dramatic sequence of the Danube river. The park has also historic and cultural relevance, containing important historic and archeological remnants: the Trajan bridge (102–106 AD) fortifications, old villages etc.

(a) (b) (c)

Fig. 23. The Iron Gates National Park Visiting Centre, 2007–2011, (a) river view, (b) ascending to the belvedere, (c) night view

Our task was to create a visiting center and an open-air theatre in the most spectacular viewing point of the riverbank, deploying a spectacular promenade along the shore.

We have chosen to merge the Visitor Center, the Theater and the Belvedere platform in a unique sculptural project – a double-shell open and floating arena, a distinct reminder of our project – The (floating) "Poetic House" from 1985 (Fig. 24).

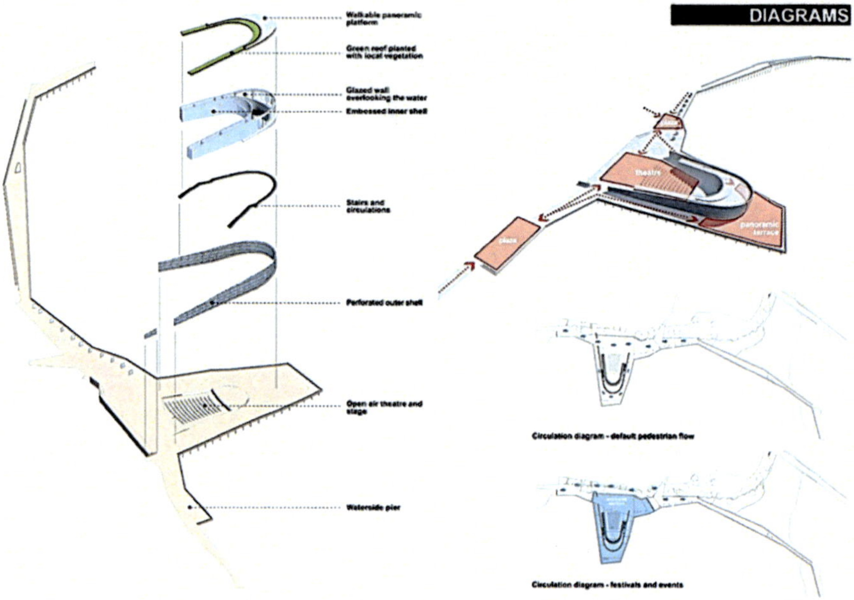

Fig. 24. The Iron Gates National Park Visiting Centre, 2007–2011, compositional diagrams

This apparently straightforward, but very complex structure was configured in collaboration with the late sculptor Ştefan Călărăşan – 1947–2017. The white concrete outer skin was imprinted with the artist's "Runes", an obscure alphabet connected to distant traces and cultures, evocative of the site's powerful historic and natural heritage (Fig. 24). The interplay of light and shadows during day and night, the spectacular views to and from the arena, the sound of music and the public rumor – all was coordinated in order to produce and immersive experience (Fig. 25).

Unfortunately, shortly before the inception of the works, the project was abandoned due to political infighting.

Fig. 25. The Iron Gates National Park Visiting Centre, 2007–2011, riverbank view

8 Conclusions

A "pure" conservation and/or restoration project is a rare occurrence. Objects, houses, places, city sections, urban landscapes are in a continuous development process. Architects and urbanists have today at their disposition a multitude of rules and internationally promoted regulations, together with legal instruments starting with the **1972 World Heritage Convention**. A growing body of experience and practices are also at hand. But the whole system is outrun by the various evolutions affecting the contemporary world and its architectural and urban components. There is a gap between rules and theories and the practical reality (Bondarin and van Oers 2012). Innovative approaches and changes have to take place. Starting with the **2005 Vienna Memorandum on "World Heritage and Contemporary Architecture"** – the gap is now closing, contemporary architectural design being considered an appropriate tool of intervention in the built historic environment, including open spaces, new constructions, additions to or extensions of historic buildings and sites and conversions.

Our experience at the Andreescu&Gaivoronschi office, allowed us to draw the following conclusion:

- Every conservation, restoration or (re)conversion project has to be based on a careful consideration about the balance between new and old elements. They have to be compatible, well adapted to each other and forming a consistent synthesis. New methods and technologies must be used with responsibility, considering their perceptive impact and evolution in time.
- One of the most important characteristic of a "cultural" object and/or landscape is its generative geometric pattern. During the entire history, geometric patterns were used to configure formal, spatial and territorial shapes, and served as a container for meaning and significations. They are perfect tools for exploring both the history and the meaning of a building or a landscape. Identified and used with intelligence they could offer consistency and signification to the project.
- Collaboration with contemporary artists is essential for the success of a conservation or restructuration project of any scale. Art works embedded into the project contribute to enhancing the living and working conditions and are expressions of our aims and aspirations, to be transmitted to the future.

Developing a comprehensive vision about the impact and the future evolution of the conservation/restoration project is the main prerequisite for its success.

Acknowledgements. The activity presented in the paper is based on the studies and projects elaborated by the "Andreescu&Gaivoronschi" architectural design firm; special acknowledgements to arch. Vlad Gaivoronschi Ph.D., senior partner and to arch. Dinu Dan – Razvan. Gratitude to our artistic collaborations: Silviu Orăvitzan and Ştefan Călărăşan (1947–2017).

References

Andreescu, I., Gaivoronschi, V., Lacatusu, A.: Urban Regeneration and its surprise due to the discovery of XVIIIth century architectural artefacts. In: RE-COND' 15 International Conference Re-evaluating Contemporary Designs in Historical Context, 22–24 July, Istanbul, Turkey (2015a)

Andreescu, I., Gaivoronschi, V., Micsa, O.: The Badea Cartan covered market. A case study into the evolution of a modern infill project in a historic area. In: RE-COND' 15 International Conference Re-evaluating Contemporary Designs in Historical Context, 22–24 July, Istanbul, Turkey (2015b)

Andreescu, I., Gaivoronschi, V., Mosoarca, M.: Old and new–the complex problem of integrating new functions into old building. Procedia Eng. **161**, 1103–1108 (2016)

Andreescu, I., Keller, A., Mosoarca, M.: Complex assessment of roof structures. Procedia Eng. **161**, 1204–1210 (2016)

Andreescu, I., Mosoarca, M., Dinu, D.-R.: Reshaping the villa – complex intervention in a 1930's structure. In: Aguilar, R., Torrealva, D., Moreira, S., Pando, M.A., Ramos, L.F. (eds.) Structural Analysis of Historical Constructions. RB, vol. 18, pp. 2314–2322. Springer, Cham (2019). https://doi.org/10.1007/978-3-319-99441-3_248

Andreescu, I., Gaivoronschi, V.: The reborn structure. Timisoara old slaughterhouse development. In: LivenARCH VI-2019: Replacing Architecture, 25–28 September, Trabzon, Turkey (2019)

Andreescu, I., Gaivoronschi, V., Micle, D., Vuicin, L.: The Banloc castle. The rebirth of a landscape park. In: PROHITECH' 14 International Conference on Protection of Historical Constructions, 7–9 May, Antalya, Turkey (2014)

Brandi C.: Teoria Restaurarii (Teoria del Restaura), Editura Meridiane Bucureşti, pp. 37–38 (1996)

Bondarin, F., van Oers, R.: The Historic Urban Landscape. Managing Heritage in an Urban Century, pp. 15–20. Willey-Blackwell (2012)

Gheorghiu, A., Andreescu, I.: Synthetic of architectural details using 3D modelling and 3D printing. In: RE-COND' 15 International Conference Re-evaluating Contemporary Designs in Historical Context, 22–24 July, Istanbul, Turkey (2015)

Gausa, M., Guallart, V., Müller, W., Soriano, F., Porras, F., Morales, J.: The Metapolis Dictionary of Advanced Architecture: City, Technology and Society in the Information Age. ACTAR, Barcelona, pp. 520–522 (2003)

Mayor of London Assembly: London Management Framework, UK (2011). https://www.london.gov.uk/what-we-do/planning/implementing-london-plan/planning-guidance-and-practice-notes/london-view-management. LVMF March 2012 part 1.pdf, retrieved 22.07.2019

Philippot, P., Mora, L.E.P.: Il restauro degli intonaci colorati in architettura: l'esempio di Roma e la questione di metodo, Bolletino d'Arte, supplemento al n. 35–36, Italia, parte II, pp. 114–121 (1987)

Large-Scale Analysis of Masonry Bell Towers in Naples

Daniela Cacace, Corrado Chisari, and Gianfranco De Matteis(✉)

Department of Architecture and Industrial Design (DADI), University of Campania "Luigi Vanvitelli", Via San Lorenzo, 81031 Aversa, Italy
{daniela.cacace,corrado.chisari, gianfranco.dematteis}@unicampania.it

Abstract. Masonry bell towers often represent a distinctive element of the Italian landscape. Their different typological, constructive, and structural peculiarities, along with the great exposure given by their relevant artistic value, make them both structurally and seismically vulnerable. Since the conservation of masonry bell towers requires a multidisciplinary approach, the PREVENT project (Integrated PRocedure for assEssing and improVing the resiliENce of existing masonry bell Towers on a territorial scale) aims at defining a two-level procedure for assessment and enhancement, at both a territorial and architectural scale.

In this paper, the preliminary results that emerged from the first census carried out on a surveyed population of 56 masonry bell towers in the city of Naples (Italy), are presented. A classification has been made from a geometric-typological point of view, in order to frame their structural peculiarities. Also, preliminary assessments of the dynamic behaviour of a sub-sample of 40 masonry tower bell towers (i.e. excluding the bell-gables and double tower types) have been carried out. Further analyses will allow the elaboration of "resilience" indexes, as synthetic indicators of the architectural, historic, and cultural value, as well as of the structural safety, of the considered bell towers.

Keywords: Masonry bell towers · Structural assessment · Seismic vulnerability · Large-scale study · Naples

1 Introduction

Italy is often recognized as the "country of bell towers": almost every city, neighbourhood, religious building, from the majestic basilicas and cathedrals, up to the minor churches, is equipped with a bell tower, carrying a symbolic-identity function, as well as historical, artistic and architectural value. To date, there is no exhaustive mapping of their presence on the Italian territory. The seismic events that hit our country, in more or less recent years, underlined the great vulnerability of these historical structures, which reported a high frequency of large damage and even collapse (De Matteis and Zizi 2019). Their weakness against horizontal loads is partly due to their construction peculiarities (heterogeneity of the masonry, slenderness, etc.), partly to the lack of maintenance that is often reserved to them, being frequently abandoned, inoperative or even inaccessible.

I. Vayas and F. M. Mazzolani (Eds.): PROHITECH 2021, LNCE 209, pp. 1429–1442, 2022.
https://doi.org/10.1007/978-3-030-90788-4_108

The study of a historical masonry building, mainly aimed at its conservation and enhancement, requires a multidisciplinary approach, as recommended by ICOMOS/ISCARSAH (2003) and encouraged by Italian Guidelines (2011). However, recent scientific literature has seen little evidence and application examples in this respect, when referred to the topic of bell towers preservation. The bibliographic studies carried out, in fact, underline the need for the development of new large-scale methodologies, which include the architectural and structural survey, the evaluation of the state of conservation, the monitoring and the rehabilitation, not only aimed at the single-building scale. This last aspect becomes crucial for the calibration of fast-assessment procedures, which allow the implementation of adequate intervention policies aimed at mitigating the structural risk and at the architectural and functional enhancement of the bell towers.

In this context, the PREVENT (Integrated PRocedure for assEssing and improVing the resiliENce of existing masonry bell Towers at territorial scale) project, funded under the VALERE program (VAnviteLli per la RicErca) of the University of Campania "Luigi Vanvitelli" aims at the integrated and multidisciplinary study of historical masonry bell towers, conducted on a territorial scale. The phase of data collection, preparatory to the acquisition of the necessary knowledge on historical bell towers, has been carried out in a territory so stratified and heterogeneous as that of the city of Naples (Italy).

The first outcomes of these preliminary analyses, which will lay the foundations for the set-up of the procedure, are reported in the following sections, where, for simplicity, the term "bell tower" generally refers to both each kind of structure bearing a bell (i.e., including bell-gables), and those built with a tall shaft, while the term "belfry" always indicates the part of a bell tower where the bells are housed.

2 Italian Bell Towers and Bell-Gables: Brief Historical Notes and Key Features

Although there are no univocal bibliographical sources regarding the invention of bell towers in primitive churches, it is ascertained that, after the period of persecutions, it was customary among Christian communities to use the sound of a bell to call the worshippers to religious functions (Sarnelli et al. 1686). This bell (sometimes more than one) was placed at a certain height, such as to favour the propagation of sound.

Over the centuries the bells were gradually positioned on towers built by monastic buildings, until similar turreted structures, in which the large clocks were often coupled to the bells, were used to adorn civil buildings, especially during the Lordships.

In the Early Renaissance architectural treatises, where geometric fundamental ratios and instructions for civil and sacred buildings were collected to disseminate the so-called *rules of thumb*, few references are addressed to the bell tower as a religious structure.

In Leon Battista Alberti's *De re aedificatoria* (trans. 1546), as a non-exhaustive example of this evidence, the Florentine architect mentions the *vedette*, as ornamental towers standing out over the great Italian cities, to symbolise the honour and splendour of noblest families.

It is interesting to report here some construction rules given by Alberti for these towers, in Ch. 5 of book VIII, which are summarised in Table 1. According to him, the towers, which he divided into cylindrical or square, can be slender or stocky. In particular, for the construction of slender cylindrical towers, the height (H) must be four times the diameter (d) of the base; on the contrary, for stocky cylindrical towers, the height is equal to three times the base diameter. As for the slender square towers, according to Alberti, the base (B) must be equal to 1/6 of the height, while, in the case thicker square towers, the base must be a quarter of the height.

In addition to the slenderness ratio, Alberti also identifies a relationship between the minimum thickness (t) of the walls and the height of the structure itself: for every 10 *cubits* in height, he proposes a minimum wall thickness of 1 *foot*. Although these units of measurement were affected by a strong regionalism, being a Renaissance cubit equal to about 6 palms and being a foot equal to 4 palms, this ratio can be summarised as 60 cubits every 4 feet, that is a thickness equal to 1/15 of the height.

Table 1. Leon Battista Alberti's fundamental ratios for the construction of *vedette* civil towers.

Typology	Geometrical ratio
Cylindrical slender tower	$H = 4 * d$
Cylindrical stocky tower	$H = 3 * d$
Square slender tower	$B = H/6$
Square stocky tower	$B = H/4$
All	$t/H = 1/15$

A first real series of instructions about the construction of the bell tower as a space for sacred architecture is provided by San Carlo Borromeo (Borromei 2000). In Ch. XXVI of his work, the Milanese Cardinal makes a description of the bell tower, in the aftermath of the Counter-Reformation, which effectively reflects the general Italian style of many of these structures, apart from other local influences (for example, Oriental ones in the Venetian areas or some areas of Southern Italy).

In particular, he conceives the bell tower as an appendix of the church: isolated on each side and preferably square in shape, the most suitable height for this kind of structure must not overhang that of the church. Without limiting the number of floors, he prescribes that the first and the last one are always covered by a masonry vault, while the intermediate ones are made of wooden horizontal slabs; at each floor there will be elongated windows on all sides, and larger openings in the belfry; the roof, preferably reached through masonry or wooden spiral stairs, is pyramidal or domed, and the bells are at least two for the parish churches, up to five or seven for the greater cathedrals. It is interesting to note how the concept of a bell-gable is then introduced, indicated as a cheaper solution, in the form of hanging arches raised on the church façade, to which the bells can be suspended.

A similar geometric prescription of a bell tower construction is provided by the last (for the sake of brevity) reference reported here. Indeed, Francesco Milizia and Cipriani (1813), in Ch. IV, Book I of his theoretical treatise *Principj di Architettura civile*, affirms that, in the harmony of the façade of a building, a bell tower must have two architectural orders (i.e. marked by stringcourses) above a high base, and must end with a pyramidal roof. In Ch. XIII of Book II, he suggests for each bell tower the creation of a solid base and a square or polygonal shape, possibly tapered upwards.

It is worth noticing that Milizia advises against building isolated bell towers, as well as placing them in a corner, suggesting a better positioning in the back of the church or, where necessary, creating a double tower on the church façade. However, the scholar generally demonstrates a scarce consideration of these tall towers because of their predisposition to collapse due to lightning, earthquakes, and shaking of bells.

3 Structural Survey of Neapolitan Bell Towers

3.1 Overview

The historical centre of the city of Naples, identified by the ancient Greek-Roman road layout still visible, is one of the largest in Italy, so rich in history and monuments that it has been a UNESCO heritage site since 1995. Although an in-depth description of its evolution cannot be offered here, it is important to underline the high stratification in terms of construction periods that can be read in this territory, essentially due to the fact that, as it is well known, Naples is a layered city, where the urban space has grown inside the various perimeters of city walls that followed one another over the centuries.

Furthermore, the geomorphological characteristics of the territory, which alternates promontories with small hills, almost without flat areas, make the historical and cultural heritage of Naples a highly heterogeneous object of study.

Within the data collection phase, 56 bell towers have been selected in an area that includes a large part of the Historical Centre, extending in some cases to the surrounding quarters (Fig. 1). The survey consisted of the study of the identified bell towers according to a so-called level 0 of knowledge, based on visual inspections and/or on the study of archival documents, to be reported in a data collection tool.

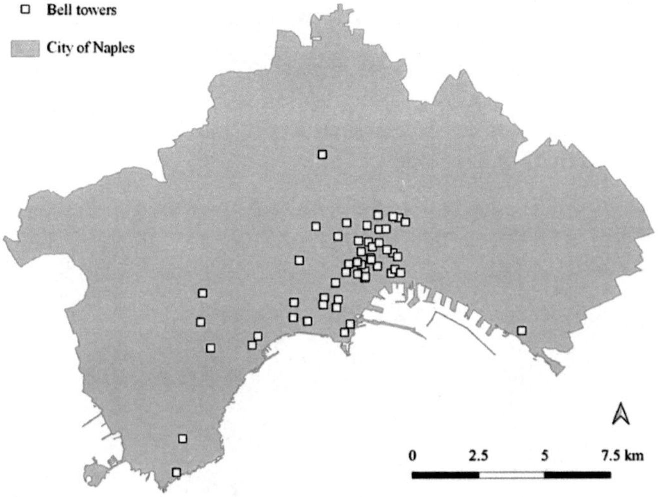

ID	Name	ID	Name
A01	M. S.ma del Carmine	I55	S. Orsola a Chiaia
A02	S. Domenico Soriano	I56	S. Caterina da Siena
A07	S. M. della Pazienza	I57	S. Lucia a Mare
B10	S. Pasquale a Chiaia	I58	S. M. della Catena
B12	S. M. di Piedigrotta	J61	S. Domenico Maggiore
B13	S. M. in Portico	J63	S. Pietro a Majella
B15	S.mo Redentore	J64	S. Chiara
D04	S. Vitale Martire	K06	S. Rocco
D17	S. Nicola da Tolentino	K65	Pietatella a Carbonara
D21	S. M. Ognibene	K66	S. M. del Carmine
D22	S.ma Trinità degli Spagnoli	K67	Gesù delle Monache
D24	Basilica dello Spirito Santo	K68	Monte dei Poveri
E27	S.S. Annunziata Maggiore	K69	S. Gioacchino a Pontenuovo
E29	S. Giorgio Maggiore	K71	S. Giovanni Battista
E32	S. Eligio Maggiore	K72	S. Giuseppe dei Ruffi
E33	S. Croce e Purgatorio al Mercato	K73	S. Lorenzo Maggiore
E34	S. M. la Scala	K75	S. Anna a Marconiglie
E36	S. M. Stella Maris	K76	S. Antonio Abate
F39	Cappella Pappacoda	K79	S. M. della Fede
F41	S. Girolamo delle Monache	K82	S. M. Maggiore alla Pietrasanta
F44	S. M. la Nova	K83	S. M. Regina Coeli
G45	S. Luigi Gonzaga	K84	S. Gregorio Armeno
G46	S. M. del Faro	K85	Santi Apostoli
G48	S. Antonio a Posillipo	L88	S. M. del Rosario alle Pigne
G49	S. Strato a Posillipo	L90	S. M. dei Miracoli
H50	S. M. degli Angeli alle croci	L93	S. M. della Sanità
I52	S. Carlo alle Mortelle	O08	S. Giovanni Battista
I54	S. Anna di Palazzo	X01	S. M. Immacolata di Lourdes

Fig. 1. List of the 56 censed bell towers in the city of Naples with the alphanumeric ID as referred to hereinafter.

3.2 Typological Classification of the Reference Sample

Most of the historical Neapolitan bell towers analysed here have generally articulated shapes, showy stringcourses on the façades (which hide the masonry and do not always reflect the internal structure), superstructures and decorative stuccoes, with large openings in the belfry (Fig. 2a, b, c).

(a) (b) (c)

Fig. 2. Examples of bell towers in the sample: S. M. in Portico (a), S. Luigi Gonzaga (b), S. M. del Faro (c).

This tendency was expected, since the construction (or re-construction) period of almost the 50% of the sample can be placed between the 17th and 18th centuries (Fig. 3a), thus denoting the predominance of a Baroque architectural style (or re-styling, excluding many renowned examples of stylistic restoration, such as the church of S. Chiara), which, in fact, has found in Naples one of its greatest expressions in that decades. The 84% of the surveyed elements have a tower-shaped structural type, with a square or rectangular shaft, rarely octagonal (Figs. 3b and 3c), mostly flat roofs, less pyramidal or bulbous (Fig. 3d).

The irregularity of the territory under examination, as asserted in Sect. 3.1, has been confirmed by the fact that more than 1/3 of the selected samples have supports qualitatively defined as "irregular" (placed on a slope, subjected to subsequent changes at the level of the base, etc.).

The average height of the tower structures is about 30.00 m (Fig. 4a): since a very small percentage of bell towers are isolated (Fig. 4b), the concept of "emerging height" (H_{emer}) has been introduced, as the part of the bell tower not embedded and not bounded, thus emerging with respect to the surrounding buildings.

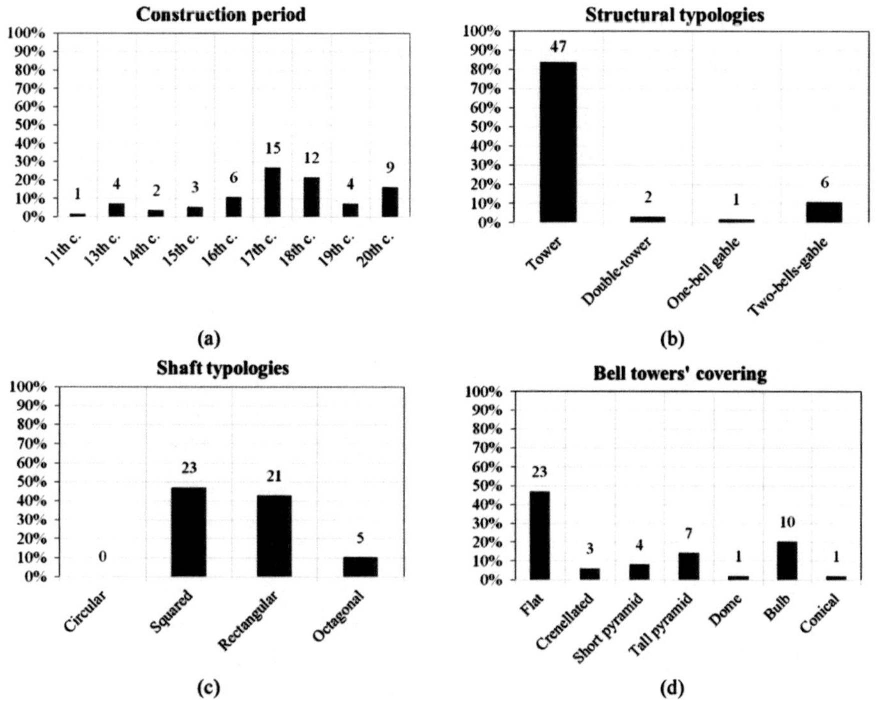

Fig. 3. Typological features of the 56 bell towers sample (a), (b) and of the 49 tower-shaped sub-sample (c), (d).

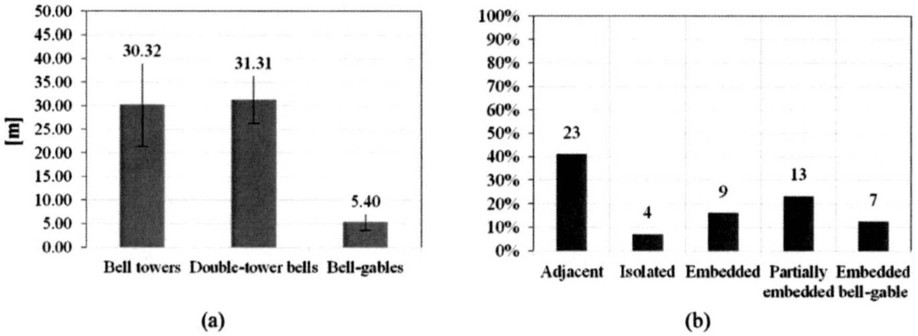

Fig. 4. Average height in the sample ± standard deviation (a). Position of the bell towers with respect to the church (b).

The total height H_{tot}, divided into H_{emer} and H_{embed} (embedded height) of the 49 bell towers sub-sample is shown in Fig. 5a, where, for isolated or totally embedded towers, $H_{emer} = H_{tot}$ and $H_{embed} = H_{tot}$, respectively. Moreover, the graph of Fig. 5b represents the relationship between H_{tot} and the base measured on the minimum length (L_{min}) for each of these towers: the two dashed lines indicate the two limits for stocky (4) and slender (6) towers suggested by Leon Battista Alberti, as described in Sect. 2 and Table 1. It is possible to appreciate how, according to this parameter, many of the bell towers analysed can be qualitatively defined as slender.

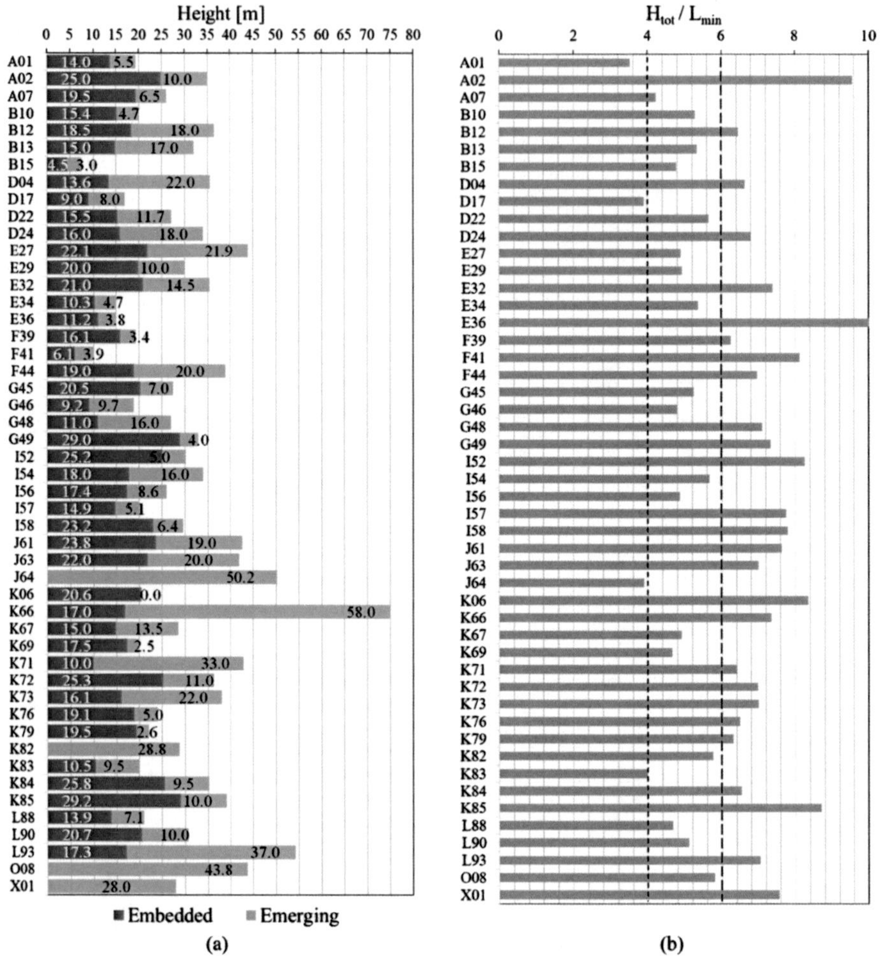

Fig. 5. H_{emer} and H_{embed} of the 49 tower-shaped sub-sample (a). Correspondence to Alberti's H/B rule for stocky (H/B = 4) or slender (H/B = 6) towers (b).

3.3 Preliminary Considerations on Seismic Vulnerability

The phase of geometrical knowledge of the sample has been followed by the appraisal of the structural features which should be considered for the vulnerability assessment of the shaft and the belfry, respectively. As mentioned in Sect. 3.2, most of the bell towers analysed have a stylistic conformation that hides the pattern of the masonry, as the façades are entirely covered by plasters. Although in some cases it was possible to overcome this problem by taking the information from historical photographs or by observing missing parts of plaster on the walls, in 21% of cases it was not possible to identify the masonry arrangement (Figs. 6a and 6b).

Intrinsic information on the vulnerability of these structures has been provided by the fact that access to the bell towers is often forbidden, both for mere functional reasons, but above all due to inability, inoperability, and lack of safety conditions. For this reason, information on the internal structure could not be collected for almost half of the sample (Fig. 6c). At the belfry level, however, some information on the arrangement of the bells has been acquired at a certain distance and is shown in Fig. 6d.

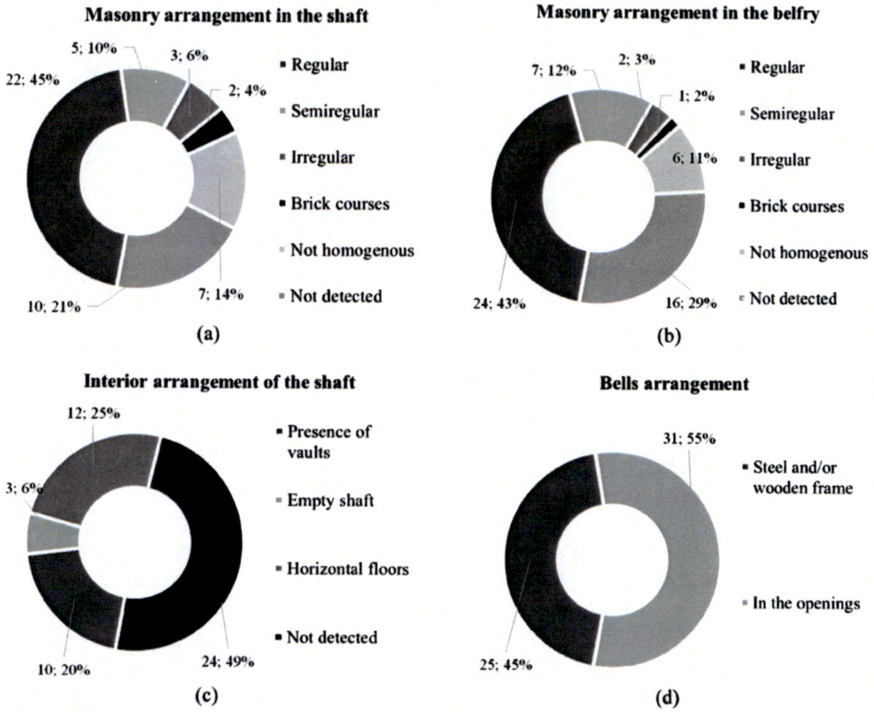

Fig. 6. Masonry arrangement in the shaft (a) and in the belfry, (b). Interior arrangement of the shaft (c). Position of the bells (d).

In the same way, it has been possible to collect information regarding the maintenance status of the structures: in particular, an attempt has been made to define a level of severity of the crack pattern, when present (Fig. 7a and 7b), and the presence / absence of anti-seismic devices (Fig. 7c and 7d).

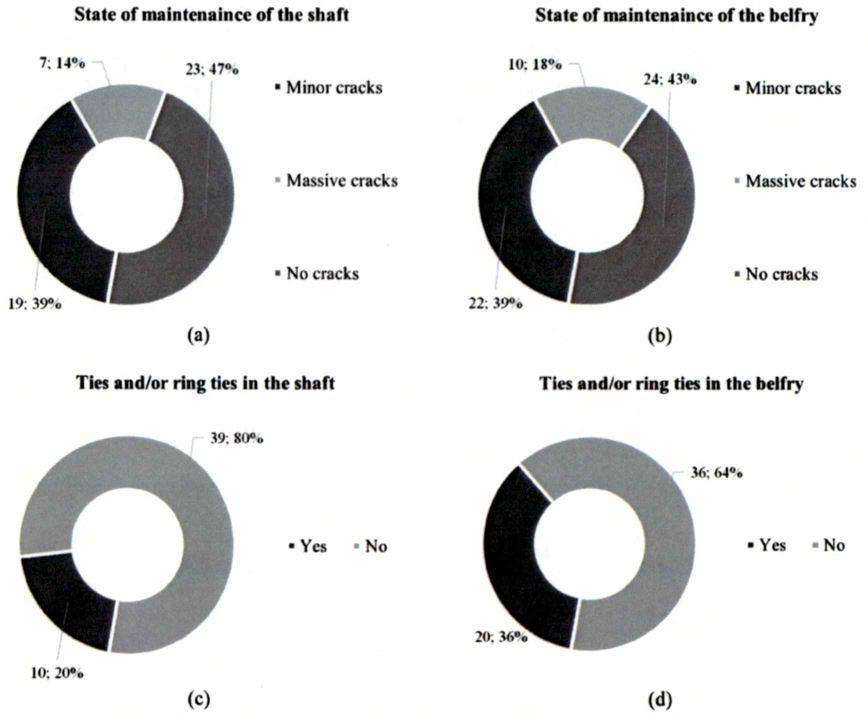

Fig. 7. Presence/absence of cracks in the shaft and in the belfry (a), (b). Presence/absence of anti-seismic devices in the shaft and in the belfry (c), (d).

At this point of the research, it emerged, from an intersection of the collected data with the bibliographic ones, that the bell towers dating back in the 20th century have often been constructed using a mixed structure (i.e. masonry, with some reinforced concrete parts, especially staircases). For this reason, 9 bell towers pertaining to that period have been excluded from the study, thus a new sub-sample of 40 bell towers has been set.

The slenderness of the new sub-sample is shown in the graph of Fig. 8a, as the ratio between H_{tot} and L_{min}, with most of the values ranging between 4.00 and 7.00. Furthermore, despite the large scale at which the surveys have been carried out, it has been possible, for 33 bell towers in the sub-sample, to estimate the average thickness (t) of the walls, which is related to the H_{tot} in the graph in Fig. 8b. If these last data are compared with Alberti's prescriptions for civil towers (see Sect. 2, Table 1), the value recommended by the scholar $1/15 = 0.066$ represents an upper bound since in the great majority of the cases smaller values have been found.

As pointed out in the first sections, there are only few studies in the literature on large-scale vulnerability assessment of historical masonry bell towers. Among those, the method called Vulnerability of Towers (VULNeT) (Sepe et al. 2008), which evaluates the seismic vulnerability of slender structures (in particular towers) by collecting qualitative parameters, according to two different levels of accuracy, has been applied to the 40 bell towers sub-sample.

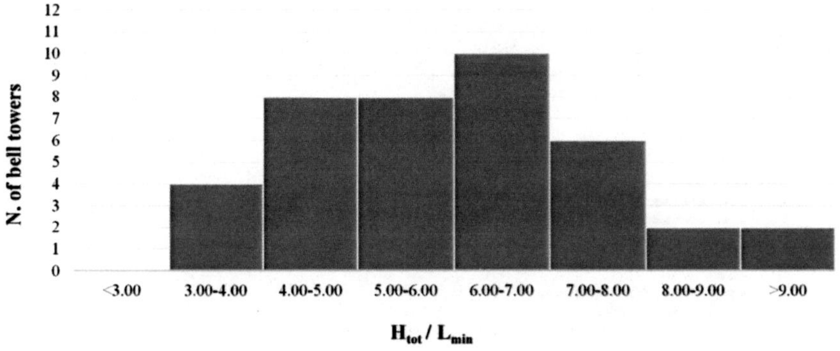

(a)

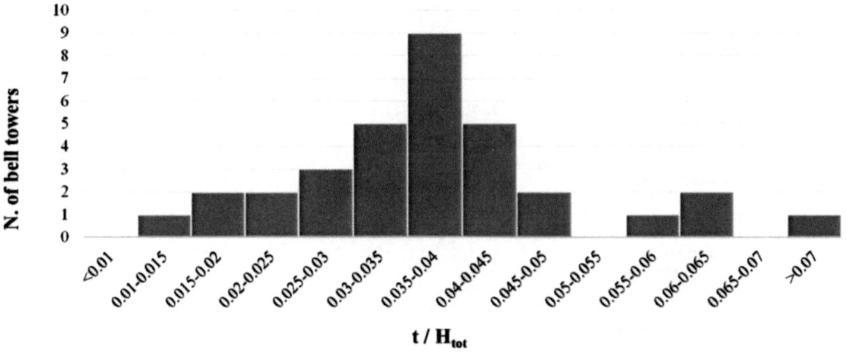

(b)

Fig. 8. Slenderness values in the 40 bell towers sub-sample (a). Slenderness of the walls with respect to the total height of the structures (b).

The formula used for Level 1-VULNeT survey form considers three parameters:

- usability level (*u*), ranging between 0 and 3, where 0 means "utilised" and 3 means "not utilised";
- conservation level (*c*), ranging between 0 and 1, where 0 means "good" and 1 means "very bad";
- local MCS intensity (*Mercalli-Cancani-Sieberg* scale) experienced after construction or restoration date (*s*), ranging between 6 and 11.

In this way, the seismic vulnerability index Iv_1 is defined by Eq. 1:

$$Iv_1 = \frac{\alpha \cdot (u + u_0) \cdot (c + c_0)}{(s + s_0)} \tag{1}$$

where *u*, *c* and *s* are collected in the survey form, while the symbols α, u_0, c_0 and s_0 are constant (equal to 0.1, 4, 2, and 0, respectively).

The graph of Fig. 9 provides the Iv_1 values for the 40 bell towers sub-sample, normalized to 100 (Iv_{1_norm}).

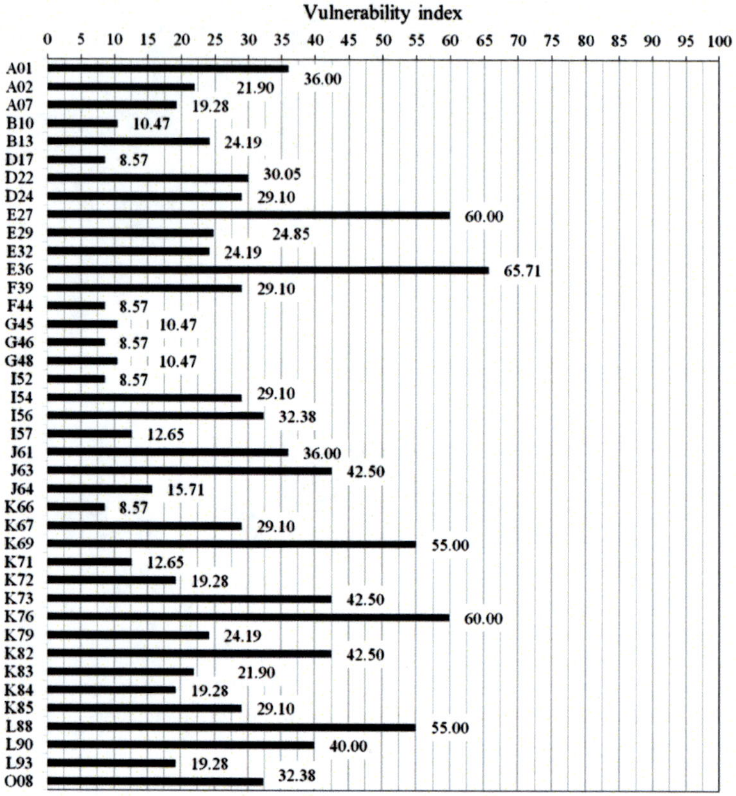

Fig. 9. Iv_{1_norm} values of the 40 bell towers sub-sample (procedure by Sepe et al. 2008).

As already highlighted in the original study, although this procedure is useful for fast vulnerability assessment procedures, to establish intervention priorities, it does not take into account some intrinsic factors of vulnerability, such as the slenderness of the structures. As an example, the bell tower of S. M. del Carmine (K66) which, with L_{min} equal to about 10 m, measures 75 m in height (including the non-structural parts), shows the lowest Iv_{1_norm} value among those registered: this is due to the fact that an important conservative restoration of this bell tower has just been completed at the date of this research work.

Diaferio et al. (2018) proposed approximated formulations for the estimation of the fundamental frequencies of bounded (Eq. 2) and isolated (Eq. 3) masonry bell towers. These respectively reads:

$$f_1(L_{min}, H_{eff}, H) = 14.61 L_{min}^{-0.254} H_{eff}^{-0.341} H^{-0.216} \quad (2)$$

$$f_1(L_{min}, H) = 208.54\, L_{min}^{0.55}\, H^{-1.73} \quad (3)$$

The relationship between the fundamental periods (obtained as inverse of the frequencies of Eq. 2 and 3) of the 40 towers of the population herein analysed and the pseudo-acceleration elastic spectra given by Italian Code NTC (2018) for different soil types is displayed in Fig. 10. The range of the most probable fundamental periods is expected between 0.3 and 0.7 s, in a zone of the elastic design spectrum which depending on the soil type maybe in the descending branch (soil type A, rock soil with shear wave velocity $V_s > 800$ m/s) or on the plateau (soil type D, deposits of loose-to-medium cohesionless soil, $V_s < 180$ m/s). This highlights the importance of correct soil characterisation for the prediction of the maximum actions expected on the bell tower.

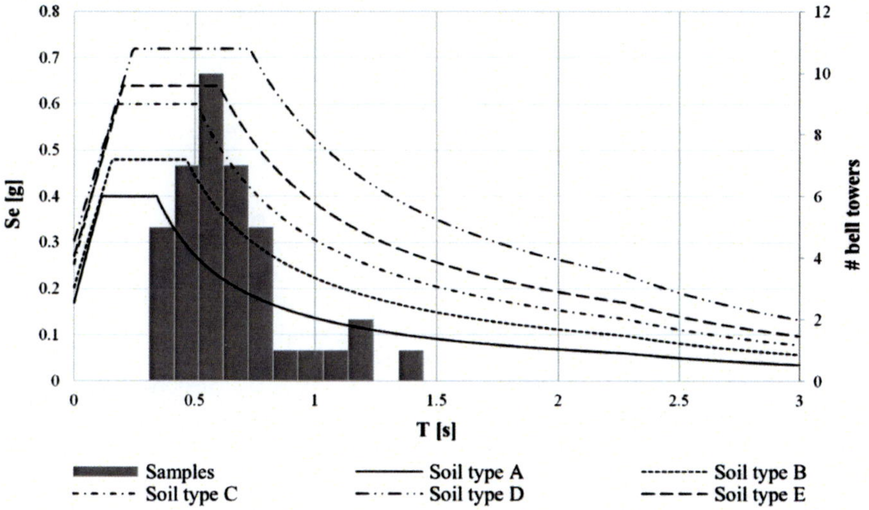

Fig. 10. Relationship of the estimated fundamental periods and the elastic spectra for different soil types according to NTC 2018.

4 Conclusions

In this paper, the preliminary results of the data collection phase for PREVENT project, aimed at assessing the structural vulnerability of bell towers in Naples historic centre are presented. 56 bell towers have been visually surveyed and classified according to the construction period and architectural and structural features. The large majority of bell towers was built between 17th and 18th century, has squared cross-section and interacts significantly with the adjacent buildings, with only 4 isolated towers in the samples. Preliminary vulnerability assessment has been performed by means of literature formulations in which very few parameters related to usability, state of conservation and maximum local intensity experienced during the bell towers' history are considered. Basic geometric information has been collected and is being exploited in ongoing research activities to estimate the vulnerability of Neapolitan bell towers at territorial and architectural scale.

Acknowledgements. The activity presented in the paper has been developed within the project "Integrated PRocedure for assEssing and improVing the resiliENce of existing masonry bell Towers on a territorial scale" – PREVENT, funded under the VALERE 2019 program at the University of Campania "Luigi Vanvitelli". C.C. is funded by MUR (Ministry of University and Research) through PON FSE 2014-2020 program (project AIM1879349-2).

References

Alberti, L.B.: Dieci Libri De L'Architettvra Di Leon Battista De Gli Alberti Fiorentino [...], La commodità, l'utilità, la neceßità, e la dignità di tale opera [...] (P. Lauro, Trans.), Appresso Vincenzo Vaugris, In Vinegia (1546)

Borromei, C.: Instructionum fabricae et supellectilis ecclesiasticae. Libri II (1577). In: Della Torre, S., Marinelli, M. (eds.) Libreria Editrice Vaticana (2000)

De Matteis, G., Zizi, M.: Seismic damage prediction of masonry churches by a PGA-based approach. Int. J. Archit. Herit. **13**(7), 1165–1179 (2019). https://doi.org/10.1080/15583058.2019.1597215

Diaferio, M., Foti, D., Potenza, F.: Prediction of the fundamental frequencies and modal shapes of historic masonry towers by empirical equations based on experimental data. Eng. Struct. **156**, 433–442 (2018). https://doi.org/10.1016/j.engstruct.2017.11.061

ICOMOS/ISCARSAH Committee: Recommendations for the analysis, conservation and structural restoration of architectural heritage (2003). www.icomos.org

Milizia, F.: Principj di Architettura Civile di Francesco Milizia: terza edizione Veneta, riveduta, amendata, ed accresciuta di figure disegnate ed incise in Roma da Gio. Battista Cipriani Sanese, vol. 3. Remondini, Bassano (1813)

Ministry of Heritage and Cultural Activities, Circular n. 26, Italian Code for Protection of Cultural Heritage: Linee Guida per la valutazione e la riduzione del rischio Sismico del patrimonio culturale con riferimento alle norme tecniche per le costruzioni. Prot 10953 of 02/12/2010 (updated to 2011). [in Italian]

NTC2018 – Norme tecniche per le costruzioni: Decree of the Minister of the Infrastructures n. 42 of Feb. 20th, 2018. [in Italian]

Sarnelli, P., Evora, J., Bulifon, A., Roselli, G., Convento di Santa Maria in Aracoeli: Biblioteca. Antica basilicografia di Pompeo Sarnelli dottor della S. teologia, e delle leggi, ... Dedicata all'eminentiss. ... Vincenzo Maria ... cardinale Orsini ... a spese di Antonio Bulifon, libraro di S.E. (1686)

Sepe, V., Speranza, E., Viskovic, A.: A method for large-scale vulnerability assessment of historic towers. Struct. Control Health Monit. **15**, 389–415 (2008). https://doi.org/10.1002/stc.243

Seismic Structural Evaluation of Candi Siva, Prambanan World Heritage Temple, by Introducing Muography

Toshikazu Hanazato[1,2](✉), Hiroyuki Tanaka[3], Yasufumi Uekita[4], and Yoyok Subroto[5]

[1] Graduate School of Engineering, Mie University, Kurimamachiya-cho 1577, Tsu-shi, Mie 5148507, Japan
hanazato@arch.mie-u.ac.jp

[2] Research Institute for Engineering, Kanagawa University, Rokkakubashi 3-27-1, Kanagawa-ku, Yokohama 2218686, Japan

[3] Earthquake Research Institute, The University of Tokyo, Yayoi 1-1-1, Bunkyo-ku, Tokyo 1130032, Japan

[4] Faculty of Art and Design, The University of Tsukuba, Tenno-dai 1-1-1, Tsukuba, Ibaragi 3058571, Japan

[5] Faculty of Engineering, Gadjamada University, Jl. Grafika No.2, Kampus UGM, Yogyakarta 55281, Indonesia

Abstract. Muon cosmic-ray can penetrate rocks and soils, which give us projection of the path's density. It indicated that muography technology would be useful for non-destructive survey of inner conditions of massive masonry heritage structures. In order to survey the inner structural condition, this muon detecting technology was successfully introduced to conduct the seismic safety assessment of the Prambanan World Heritage Temple, Indonesia, damaged by the Central Java Earthquake of 2006. The muography showed the volume of the hidden space. The stone masonry monuments of the Prambanan Temple were reconstructed by introducing RCC modern technique in 20 centuries. It was needed in the restoration project to make it clear, in particular, whether the inner RCC structure was damaged or not. On the basis of the muography, 3-D finite element model of the structure was made to evaluate the earthquake-induced stress. As dynamic soil-structure interaction would affect the seismic response of such massive masonry structures, the soil-structure model was employed for the analysis. The seismic response analysis showed the inner RCC frame would not be structurally damaged. The present study gave them helpful knowledge for the structural restoration.

Keywords: Muography · Structural survey · Masonry · Seismic analysis · Monitoring

1 Introduction

Muon cosmic-ray can penetrate rocks and soils and give us projection of the path' density, therefore, muography technology has been successfully developed in the geological

I. Vayas and F. M. Mazzolani (Eds.): PROHITECH 2021, LNCE 209, pp. 1443–1454, 2022.
https://doi.org/10.1007/978-3-030-90788-4_109

field for disaster prevention of volcano explosion (Tanaka et al. 2010). Furthermore, it was utilized to survey the inner condition of a blast furnace in a steel mill during its operation time. On the other hand, non-destructive tests are required, in general, to conduct structural survey of heritage structures with cultural and historical values. In particular, when World Heritage Monuments are surveyed, we have to follow this principle strictly. There are a number of World Cultural Heritages of masonry in seismic regions in the World. When their seismic safety is assessed, seismic structural survey is conducted by employing non-destructive tests. Considering that muography technology can be useful for structural survey of massive masonry structures as a non-destructive test, we installed the muon detecting system at the Prambanan Temples, World Cultural Heritage in Indonesia and monitored the muon cosmic-ray for 5 months. Here, the Prambanan Temples of stone masonry structures were severely damaged by Central Java Earthquake of 2006. We have been successfully involved in architectural and structural survey project conducted by an international and interdisciplinary team. The damaged masonry monuments have been restored after the earthquake, however, restoration work of Candi Siva, the oldest and highest monument of the Prambanan Temples, was not started yet, as its inner structural condition was unknown. If the inner structural conditions are revealed, 3-D finite element model is available for seismic structural diagnosis of such massive masonry structures. The scope of the present paper is to describe this challenge of non-destructive test utilizing muography technology for the Prambanan restoration project and to demonstrate applicability of this advanced technology to structural survey of World Cultural Heritages of masonry. The muon data obtained at the site indicated that the monument must have inner chambers that had been unknown. The date also indicated their sizes and locations. This information will be useful to provide analysis model for seismic evaluation.

2 Prambanan World Heritage Temple

The heritage structures, of the local andesite blocks, composing Prambanan Temples originated in the 9th century. Although they had been ruined during their long histories, the structures were reconstructed in the 20th century (Candi Siva was reconstructed first by Dutch engineers in the middle of 20th century; the other buildings were reconstructed by Indonesian engineers during the end of 1980s and beginning of 1990s using their technology). In 1991, those architectural heritages were registered as World Heritage by UNESCO. Photo 1 shows an overview of the Prambanan Temples, taken by a radio control helicopter in February, 2007.

The stone monuments were reconstructed by introducing inner concrete structures composed of reinforced concrete frames and infill rubble materials. The reinforced concrete frame was employed to confine infill concrete frames and infill rubble materials (See Fig. 1).The reinforced concrete frame was employed to confine infill rubble material structure for ensuring structural stability. The decorative stones of "ratna" and "stupa" were connected to the inner concrete structure by use of metal anchor bolts. Structural issues to assess the seismic safety concerning possibility of structural damage of the inner concrete structure therefore rose.

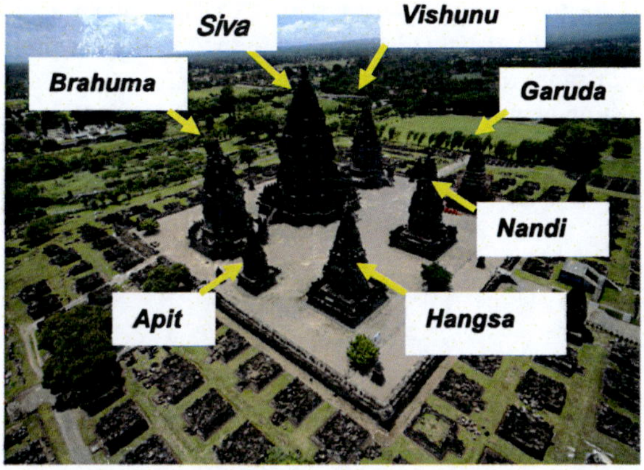

Photo 1. Overview of the Prambanan Temple

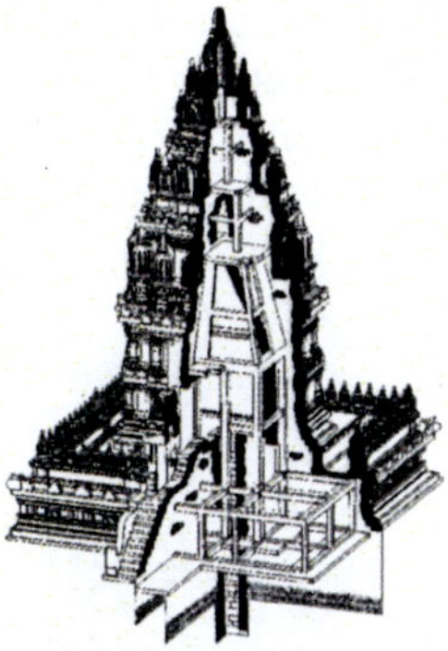

Fig. 1. Inner structure

A devastating earthquake of magnitude (M_W) 6.4 struck Central Java, Indonesia, having affected various structural types of architectural heritages in and around the historic city of Yogyakarta which locates 25 km from the epicenter. A traditional timber building in Royal Palace of Sultan, Kraton, constructed in the unique Java architectural style, totally collapsed. The historical stone masonry in the former royal palace, Taman Sari, also suffered severe damage. The earthquake affected Koda Gede, a historic area where a number of traditional types of masonry and timber buildings had been maintained and preserved with historic townscape. The damage that attracted worldwide concern immediately after the earthquake was to World Heritage Prambanan Temple compounds. Features of damage to Prambanan Temple Compounds are outlined as follows:

Photo 2. Damage of decorative stones

Of the six main temples composing Prambanan Temple, Candi Brahma and Candi Garuda suffered apparently the most severe damage. Many ratna and stupa that decorated the nave roof fell, shown in Photo 2. The upper part of the roof of Candi Garuda appeared seriously unstable. Significant relative displacement and horizontal openings between the stone elements were found at the freestanding gate on the platform in front of the entrance. Candi Siva, the largest monument in Prambanan Temple, suffered serious cracks at its base and wall of the nave, shown in Photo 3.

Photo 3. Cracks produced at entrance

In response to the Indonesian Government request just after the earthquake, an interdisciplinary expert team was sent from Japanese Government to support them to propose the restoration plan (National Institute for Cultural Heritage, Japan Center for International Cooperation in Conservation National Research Institute for Cultural Properties, Tokyo 2008). Following that governmental project, the scientific survey was started as international cooperation research for the structural restoration of the damage heritage monuments.

Photo 4. Candi Siva

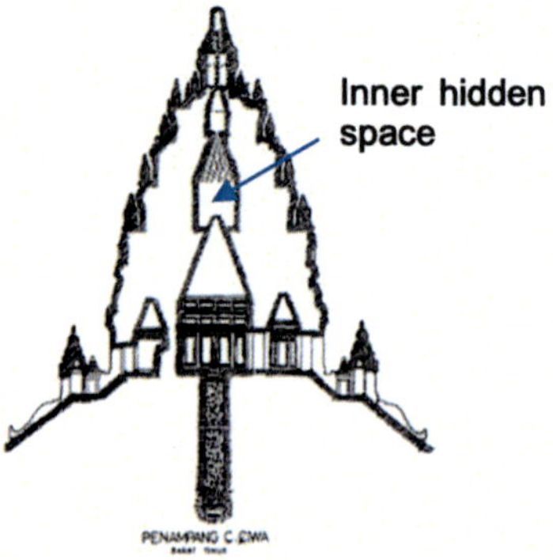

Fig. 2. Vertical section of Candi Siva

As mentioned above, the Prambanan Temple was composed of existing 8 monuments. Among those monuments, Candi Siva (See Photo 4), the highest one, was first reconstructed 1950's. However, since there found no design drawing of Candi Siva, the inner structural condition was actually uncertain, which would require the structural survey to perform the structural analysis. In Fig. 2 showing an old picture (vertical section) kept by the Archeological Site Office of Yogyakarta Special Province, there was hidden space in the internal structure of Candi Siva. It was required to conduct the non-destructive survey for such World Heritage Monuments, therefore, we employed the muography technology to know the size of the inner hidden space which would affect the dynamic response behaviors to the earthquake.

3 Muon Radiography Monitoring

The method of cosmic-ray muon radiography was a new radiographic method that has been developed by Earthquake Research Institute, The University of Tokyo. Muons are part of our total natural radiation expose and harmless to the human body. The measured count rate is 1 muon/cm^2/min for vertical detection at sea level: One or two muons strike on our palm every second. Muon cosmic-ray can penetrate rocks and soils and give us projection of the path's density, A high energy muon can go through long length of rock before stopping — the higher the energy, the more rock it can penetrate. Muography, based on the absorption of these muons in matter, is in some sense similar to X-ray radiography and is suitable for investigating huge objects such as volcanoes. Therefore, applications of muon radiography have been considered for archeology and geophysics.

Figure 3 shows the schematic view of muon monitoring system that we installed in Candi Siva. The muon detector was composed of 24 plastic scintillators and Photomultiplier Tubes (PMT). When muon penetrated the upper PMT and lower PMT, it enabled us to catch the direction of muon (See Fig. 4). This detector had an effective area of 1764 cm^2 and an angular interval of 70/1000 rad (70 mrad). Figure 5 shows the horizontal section of the Candi Siva and installed equipment. The monitoring system was located at the NW corner in the room. The center of the detector was placed 0.4 m from north wall, 0.7 m from west wall and 0.5 m from the floor level.

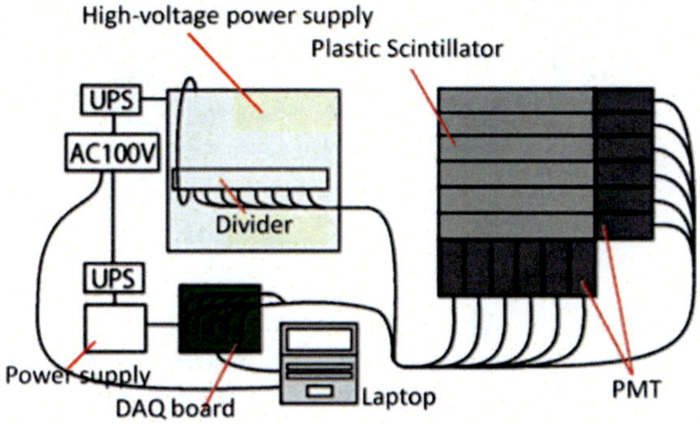

Fig. 3. Schematic view of detector system

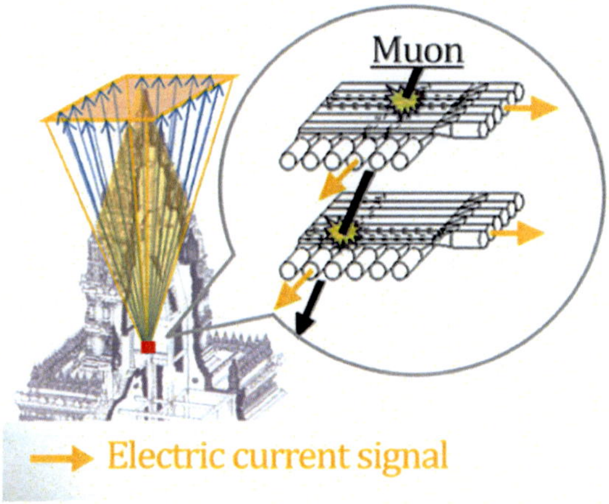

Fig. 4. Muon radiography monitoring for survey of Candi Siva

We installed the muon detector at the northwest of the central room of Candi Siva, shown in Fig. 5. The measurement period was between Oct. 18, 2013 and Feb. 13, 2014. In this period, we rotated the detector 180° around the vertical axis on Dec. 17, 2013 in order to correct the detector characteristics.

In our first simulation, the following four models were assumed: 0-volume model, 0.5-volume model, 1-volume model, 2-volume model, 3-volume model and 4-volume

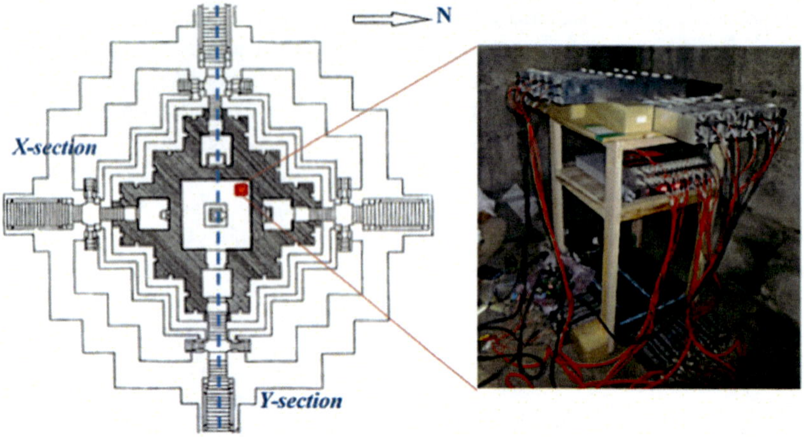

Fig. 5. Horizontal section of Candi Siva and location of muon detector

model. Hence, "1-volume" means that there was an inner room with same size of the size illustrated in the drawing of Candi Siva, shown in Fig. 2. For each model, a cubic grid with a size of 50 × 50 × 50 cm^3 was assumed for the simulation. In order to cancel the detector characteristics, the ratios of the data before to after the rotation of the equipment were calculated, shown in Table 1. In the table, the horizontal axis denotes X-direction (unit: mrad), while the vertical axis correspond to Y-direction (unit: mrad). In case the ratio exceeds 1.0, it means muon flux was larger than the flux from opposite side of the direction of the "before rotation".

Table 1. Ratio of recorded data before and after the rotation

y¥x	−210	−140	−70	0	70	140	210
280	1.1425	1.0073	0.913	1.0211	1.2113	1.1555	0.9793
210	1.2745	1.1059	1.1938	1.0607	1.1289	1.1108	1.2147
140	1.2842	1.0848	0.9849	0.9191	1.0228	0.9957	0.895
70	1.0495	1.1311	0.9892	0.9134	1.0501	0.9735	0.9052
0	0.9711	0.9601	0.9513	1.053	1.1058	0.9496	0.8646
−70	1.0245	1.0304	1.0931	0.9665	0.9242	0.906	0.7883
−140	1.1671	1.1317	1.0707	0.9809	0.9316	0.9178	0.8876
−210	1.1121	0.9902	1.0478	0.9854	0.9136	0.8578	0.7743
−280	0.9845	1.0155	0.9519	0.99	1.0318	0.8791	0.8488

Utilizing the assumed models from 0.5-volume to 4-volume, the simulation was performed. In this simulation, the pass length was calculated (See Fig. 4), furthermore, the bulk density ρwas calculated as;

$$\rho = \frac{\rho_1 \cdot V_1 + \rho_2 \cdot V_2}{V_1 + V_2} \tag{1}$$

Here, ρ_1 and ρ_2 denote density of stone material (2000 kg/m^3) and 1.18 kg/m^3, respectively. V_1 and V_2 denote bulk of stone material and inner space in one segment region, respectively. Comparison of the measurement and the simulation indicated that 2-volume model would be the most probable model. Table 2 shows the simulation of 2-Volume model. Figure 6 describes the confidence levels evaluated from comparison of the measurement and the simulation. Figure 7 displays the density profile of the upper half of Candi Siva, given by the muography.

Table 2. Simulation of 2-Volume model

y¥x	-210	-140	-70	0	70	140	210
280	1.318	1.066	0.942	0.842	1.013	1.12	1
210	1.728	1.489	1.158	0.78	0.985	1.132	0.962
140	1.709	1.547	1.066	0.657	0.836	0.971	0.827
70	1.33	1.149	0.877	0.683	0.909	1.058	0.94
0	0.843	0.726	0.779	1	1.284	1.378	1.186
-70	1.063	0.946	1.1	1.465	1.14	0.87	0.752
-140	1.209	1.03	1.197	1.521	0.938	0.646	0.585
-210	1.04	0.883	1.015	1.282	0.864	0.672	0.579
-280	1	0.892	0.988	1.188	1.062	0.938	0.759

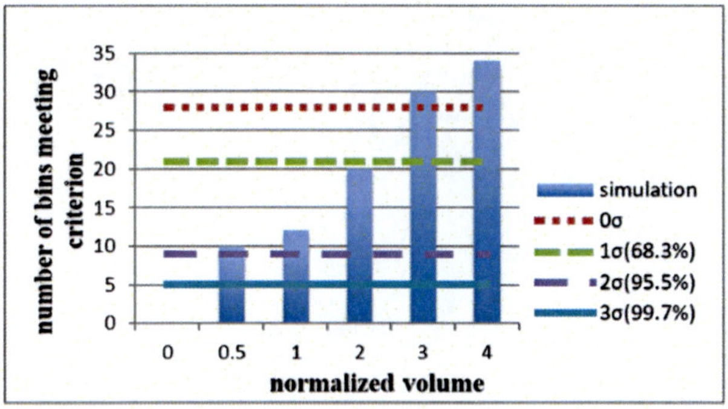

Fig. 6. Confidence level of simulation model with measurement

4 Seismic Response Analysis

Seismic response analysis was conducted by utilizing the structural model identified by the muographical technology. It should be pointed out that dynamic behaviors of such massive masonry structure would essentially affected by the soil-structure interaction. The authors conducted the seismic response analysis employing the lumped masses model where the soil springs were utilized (Nakanani et al. 2012) Their analysis demonstrated that the simulation was in good agreement with the earthquake monitoring record, and that the soli-structure interaction should be taken into account for seismic evaluation. However, as the lumped masses model was a simplified model, it might be difficult to discuss the earthquake damage of the inner structure of RCC. In the present study, 3-D finite element model was employed to discuss whether the inner structure was

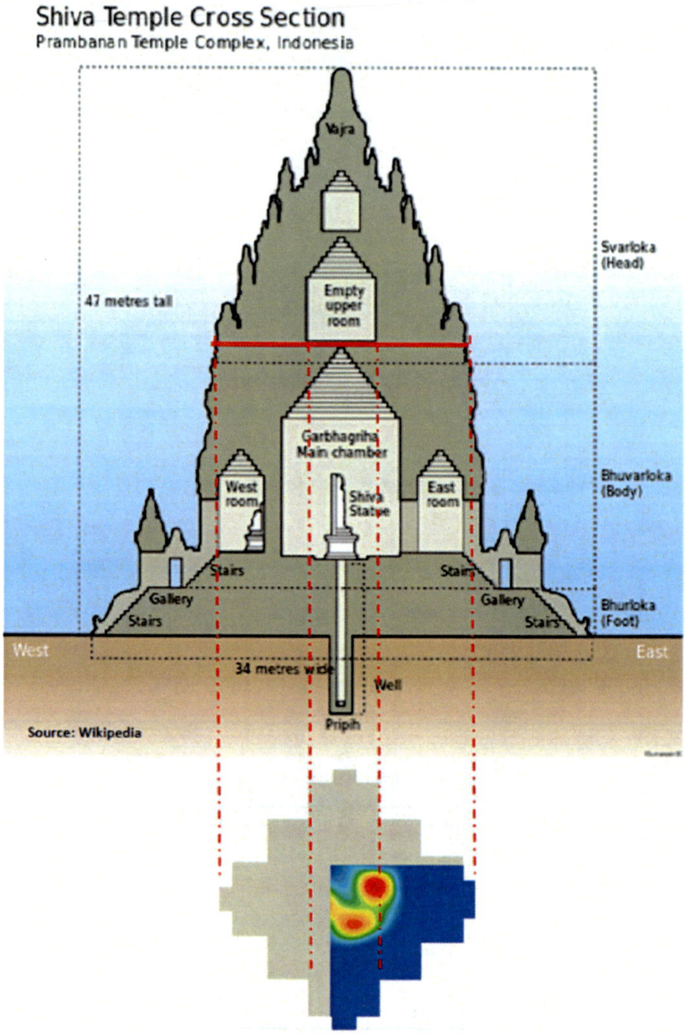

Fig. 7. Density profile of upper half structure of Candi Siva

damaged or intact by the Central Java Earthquake of 2006. In order to take in to account the soil-structure interaction, the analysis model of the soil-structure system was made, shown in Fig. 8. The muography indicated the hidden inner space would be 2-volume model, described in the previous chapter, 2-volume model was utilized in the present analysis (See Fig. 9). As the inner RCC frame, it was assumed the columns had a square section of 25 cm for the upper structure, 35 cm for the lower structure. The RCC frames was modeled by the beam elements. The natural frequency of the analysis model was 1.9 Hz, which was consistent with the past study (Nakatani et al. 2012).

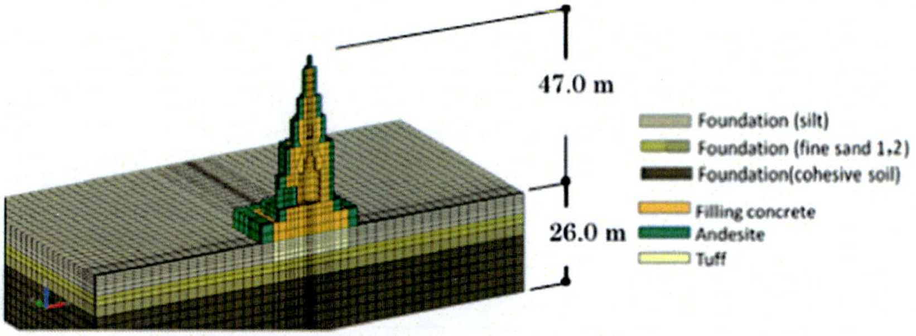

Fig. 8. Analysis model with geological condition

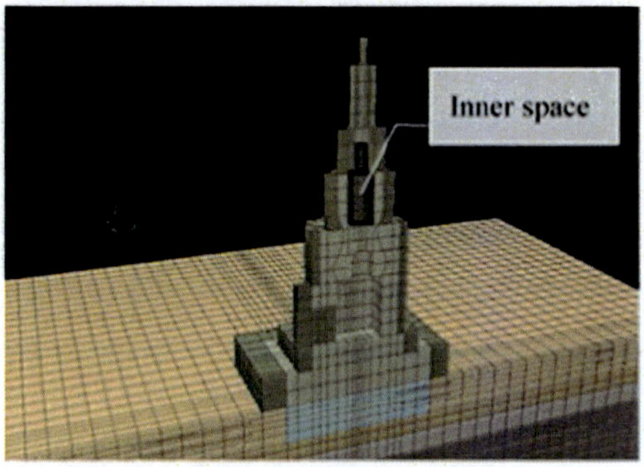

Fig. 9. Analysis model with inner space

Since the Central Java Earthquake of 2006 were not recorded both near the Prambanan Temple and in the epicentral area, the record of the Athens Earthquake at the site SPLA 1 was utilized as the input seismic motion. This earthquake occurred on September 7th, 1999 near Athens, Greece, which was the near-field earthquake with magnitude of 5.7 being similar to the Central Java Earthquake, as well as it brought significant damage to historical masonry buildings. The transverse component of the peak velocity was 12.4 kine (cm/s), and the peak acceleration was 217 gal (cm/s^2). This ground motion level could be accepted, as it was estimated that the peak velocity of around 10 kine (cm/s) at the site of the Prambanan Temples on the basis of empirical equation of attenuation with distance. Figures 10 show the acceleration waveform and the acceleration response spectrum of the input ground motion.

As analysis results, the peak compressive stress and tensile stress of the stone structure are shown in Figs. 11 and 12, respectively. Furthermore, those peak stresses distribution around the inner space are shown in Figs. 13 and 14. The compressive strength of the andesite stone was 13,600 kN/m^2, while the peak compressive stress was calculated to be 2215 kN/m^2. Provided that the tensile strength of the andesite stone was 1/10 of the compressive strength, the tensile strength of the stone would be 1,360 kN/m^2. The peak tensile stress was calculated to be 1,267 kN/m^2. The analysis results showed that the peak

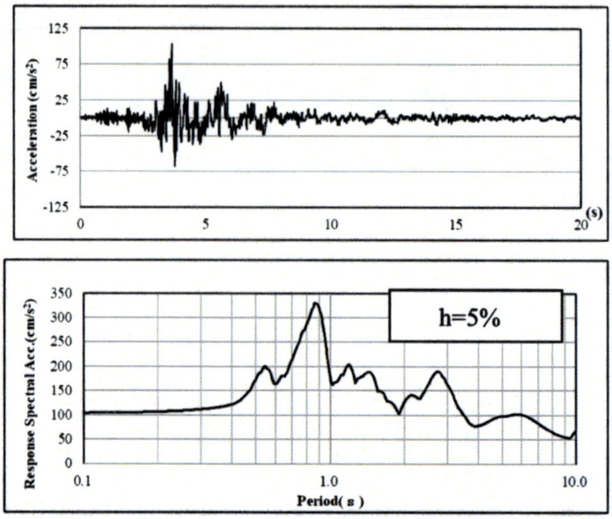

Fig. 10. Acceleration wave form and response spectrum of input ground motion

stresses, compressive was enough lower than the strength. On the other hand, the peak tensile stress was a little lower than the strength. It suggested the outer visible cracks were produced by the Central Java Earthquake of 2006 (See Photo 3). However, the peak stress around the inner space was rather lower than the strength, which indicated that the inner stone structure might be subjected to no damage. The peak bending moment of the inner RCC frame, induced by the earthquake, was evaluated to be 12.6 kNm which was lower than the short-term allowable bending moment of 30.8 kNm. This analysis result indicated that the damage might be caused to the RCC frame during that earthquake.

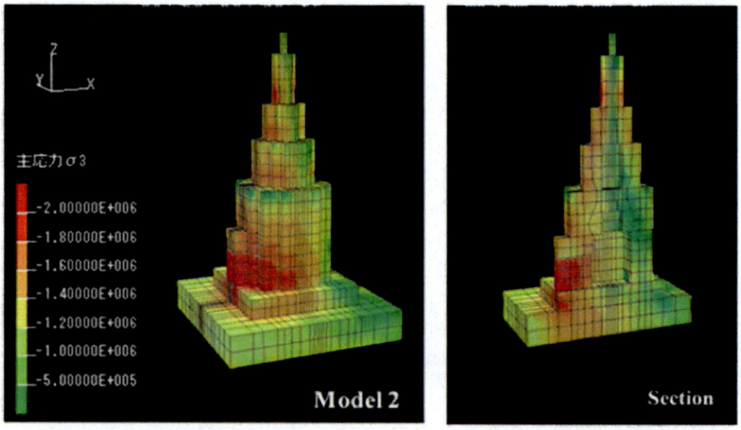

Fig. 11. Peak compressive stress evaluated by analysis

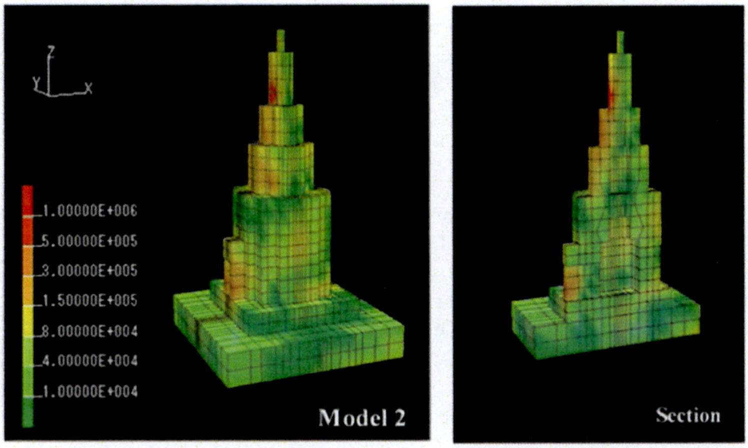

Fig. 12. Peak tensile stress evaluated by analysis

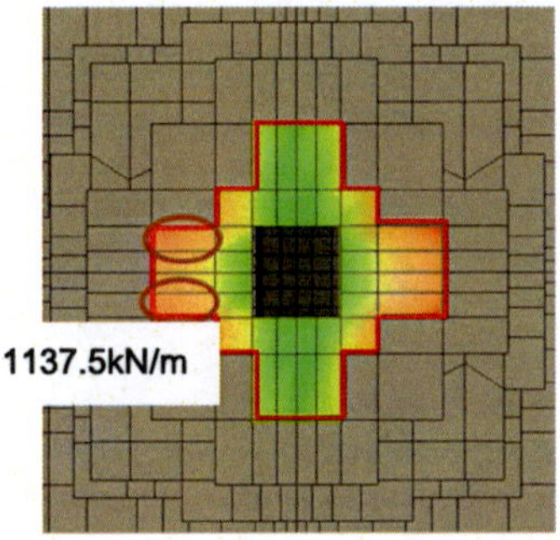

Fig. 13. Peak compressive stress around inner space

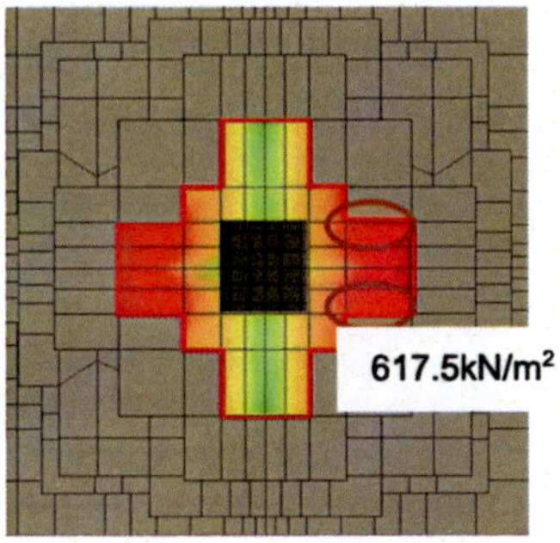

Fig. 14. Peak tensile stress around inner space

5 Conclusions

The muography technology was successfully applied to conduct the non-destructive survey of the Prambanan World Heritage Monument of masonry. It made it clear the size of the inner invisible hidden space in the stone structure. Taking into account this outcome by the muography, seismic response analysis was conducted by employing 3-D finite element model of soil-structure system. The analysis demonstrated that no significant damage might be caused to the inner structure, both the stone and RCC frame, by the Central Java Earthquake of 2006. This knowledge was reported for the restoration project conducted by Indonesian Government.

Acknowledgements. The authors wish to express their sincere gratitude to the government of Indonesia Ministry of Culture and Tourism Directorate General of History and Archaeology, the preservation office of Archaeological Site Yogyakarta Special District, the staff responsible for field office for their generosity and help, without which this experiment would have been realized. The present research had been financially supported by Grant-Aid-for Scientific Research, provided by Japanese Government.

References

Tanaka, H.K.M., et al.: Three-dimensional computational axial tomography scan of a volcano with cosmic ray muon radiograph. J. Geophys. Res. **115**, B12332 (2010)

National Institute for Cultural Heritage, Japan Center for International Cooperation in Conservation National Research Institute for Cultural Properties, Tokyo: Research Cooperation on Conservational Restoration of the Cultural Heritage damaged by Central Jawa Earthquake -Reconstructive Supporting Report on the Prambanan World Heritage Site-, pp. 35–71 (2008)

Nakatani, A., et al.: Seismic structural performance of the Prambanan World Heritage temple damaged by the Central Java Earthquake of 2006 in Indonesia. In: Proceedings of 12 WCEE, Paper No. 0681 (2012)

Statically Analysis and Retrofitting of the Old Timber Roof Structures for the Storage Heritage Building in Kustendil, Bulgaria

Petar Grekov[1], Anton Gorolomov[1], Doncho Partov[1](✉), Nikolay Tuleshkov[1], Miloš Drdácky[2], and Gaetano Pianese[3]

[1] University of Structural Engineering and Architecture, VSU "L.Karavelov", Street Suhodolska 175, 1373 Sofia, Bulgaria
partov@vsu.bg
[2] Ústav Teoretické a Aplikované Mechaniky AV ČR, Prague, Czech Republic
[3] Milano TU, Milan, Italy

Abstract. The building is a masterpiece of the industrial architecture in the thirties of XX century. The original design is the work of the architect Rudolf Fischer. The construction of the building is realized in the period 1924–1925. The combination of different construction materials: masonry, reinforced concrete and timber, gives a unique structure. This contribution introduces a structural investigation, using MFE software, into the stability and efficiency of the design of the roof truss of the storage heritage building in Kustendil. A two dimensional model was used in numerical analysis to discuss and highlight structural sense of Austrian architect R. Fischer in the period of 1925–1925. The modeling of timber construction and the final conclusions are basic for the decision concerning the possible interventions in the building and the possibility to extend the service life of the roof structure.

Keywords: Masterpiece · Industrial architecture · Intervention · Structural investigation

1 Introduction

The building is a model of industrial architecture in the thirties of the twentieth century, which still dominates with its volume in the modern urban structure in the town of Kyustendil. It was designed as a tobacco warehouse by the Austrian architect Rudolf Fischer, who lived in Kyustendil in the period 1909–1934. The building was built in the period 1924–1925. The plan scheme is a one-dimensional structure dissected by the rhythm of the columns. The scheme is symmetrical with an axis of symmetry defined by the projection middle part, covering the stairwells and the entrance of the building. The building is included in the list of buildings declared as cultural monuments from the period 1878–1944, approved by the director of NIPC in the town of Kyustendil" with letter № 2765/17.07.1979. An immovable cultural property with a category of "local significance" was declared after a meeting with document: № 6/10.03.1997 and approval, and was published in SJ, with No.: 25/04.03.1998. After a meeting with document: № RD

I. Vayas and F. M. Mazzolani (Eds.): PROHITECH 2021, LNCE 209, pp. 1455–1467, 2022.
https://doi.org/10.1007/978-3-030-90788-4_110

09-29/29.01.2008, the proposal of the director of RIM Kyustendil to change the building to “national significance” is accepted. The urban situation of the building complex of the Regional Historical Museum of Kyustendil is extremely favorable. The terrain is located near the central part of the city, east of the attractive pedestrian “green” axis, connecting the central square spaces with the station square/railway station and bus station. The architectural volume of the central building is clearly expressed in the silhouettes on the main pedestrian axis - from the west, park and square space - from the north and street - from the east, as well as from the tall buildings of the city or in the panoramic view from the southern hill “Hisarlaka”. The facades of the building are flat and treated equally on all four sides (Fig. 1). The windows are in pairs alternating with flesh. The height of the windows in the different floors is different with a certain gradation. The attic space is illuminated with dormers along the long axis and with windows along the narrow fronts (Fig. 1).

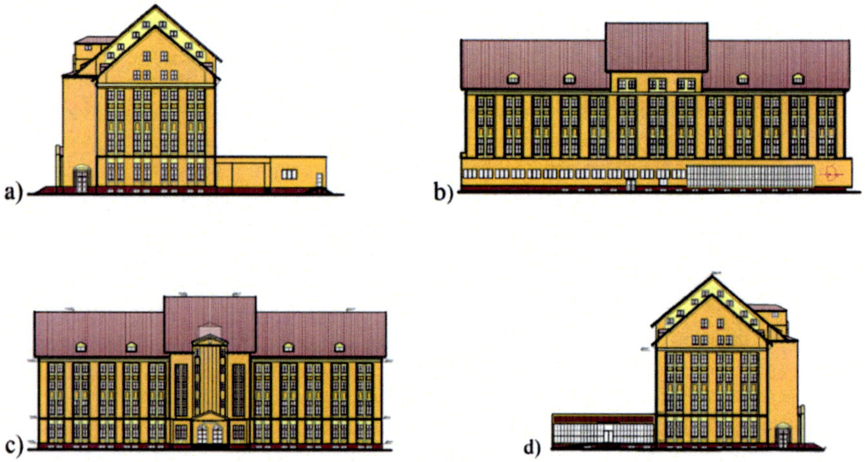

Fig. 1. East elevation-a), Nord elavation-b), South elevation-c), West elavation-d)

The height of the building consists of a semi-dug basement, three main floors, a mezzanine and attics. The presence of a high brick superstructure favors the use of the attic space. The building is designed as one main volume, with a rectangular base and five main floors, basement and attic floor: Basement, elevation −2.92, with: $H_{light} = 2.75$ m; First floor, elevation −0.00, with: $H_{light} = 3.96$ m; Second floor, elevation: +4.13, with: $H_{liht} = 3.88$ m; Third floor, elevation +8.26, with: $H_{light} = 3{,}50$ m; Fourth floor, elevation + 12.01, with $H_{light} = 1.95$ m (installation); Fifth floor, elevation +14.21, with: $H_{light} = 2.68$ m; Sixth and seventh attic floor: elevation: +17.14 and elevation: +19.73 are from wooden structure (Fig. 2.).

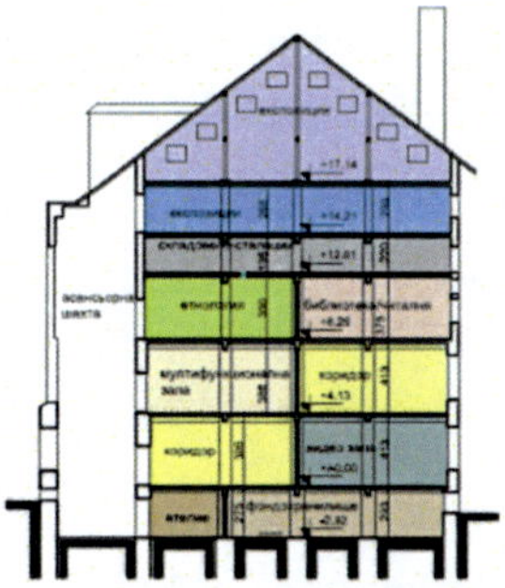

Fig. 2. Cross section of the Museum building

The building in plan has rectangular dimensions with a width of 18.1 m and a length of 59.79 m and a built-up area of approximately 1082 m^2. In the transverse direction, the columns of the building are located in three wheel bases with dimensions: 4.60 + 4.00 + 4.00 + 4.60 = 17.200 m, which is the light spans of the building. In the longitudinal direction, in the ground floor the columns are arranged as follows: 5.00 + 4.40 + 4.40 + 4.40 + 4.10 (Fig. 3, Fig. 4 and Fig. 5).

Fig. 3. Horizontal section: over Ground floor

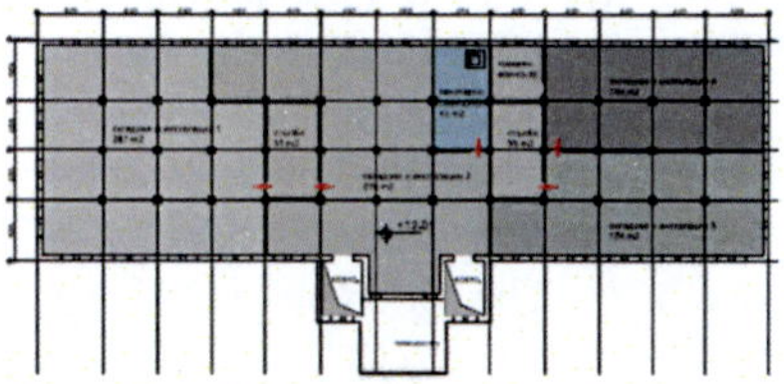

Fig. 4. Horizontal section: over 4- floor

a b

Fig. 5. **a, b** General view of the building and the south-east facade

2 Description of the Constructive System of Building

The construction of the building is massive with external load-bearing brick walls 80 cm thick in the basement, 60 cm on the first and second floor, 45 cm on the third and fourth floor, 40 cm on the fifth, sixth and seventh floors. The columns are from reinforce concrete with dimensions 78/78 cm in the basement, and from brick with dimensions 64/64 cm on the first and second floor, 50/50 cm on the third and fourth floor (Fig. 6a). In the transverse direction, the columns of the building are located in three wheelbases with dimensions: 4.60 + 4.00 + 4.00 + 4.60 = 17,200 m, which is the light spans of the building. In the longitudinal direction, in the first section the columns are arranged as follows: 5.00 + 4.40 + 4.40 + 4.40 + 4.10 m. The mezzanine structures at elevation: +(−) 0.00 (slab above the basement) and at elevation: +4.13 m (slab above the first floor) are a beam system with monolithic reinforced concrete slabs and beams, which lie on brick (respectively) RC columns (Fig. 6).

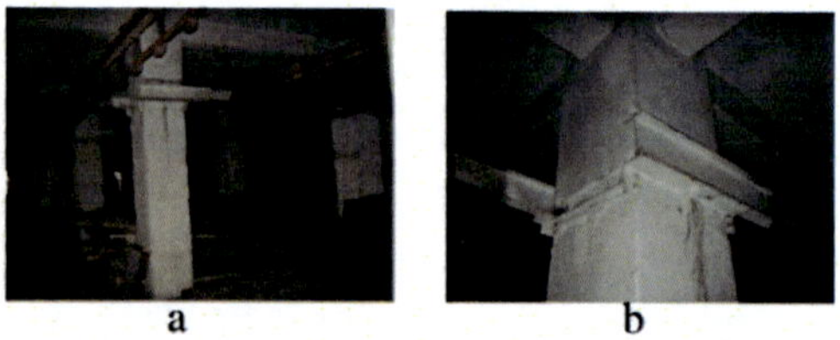

Fig. 6. **a, b** Distribution of the columns in the basement of the building-a), View on the floor and column in the first level of building-b)

At elevation 8.26 m (slab above the second floor), as well as at elevations +12.01 (slab above the third floor) and +14.21 m (slab above the fourth floor) the structures are made of wooden beams with a caratavan ceiling. The wooden beams are oriented in the longitudinal direction of the building and are supported in the transverse axes of the building on steel double T-beams, which in turn lie on the columns of the building (Fig. 7a, b, c). The wooden beams have different sizes varying in the range: 16/18, 18/20, 17/17, 20/20, 17/18, 18/18. The 7 timber beams are located in the end and 5 in the intermediate axes of the building. The steel beams on the second floor are № 30 and on the third and fourth are № 22 and № 24. The floors are made of wooden sheets, placed on a joist structure of the ridge (Fig. 7).

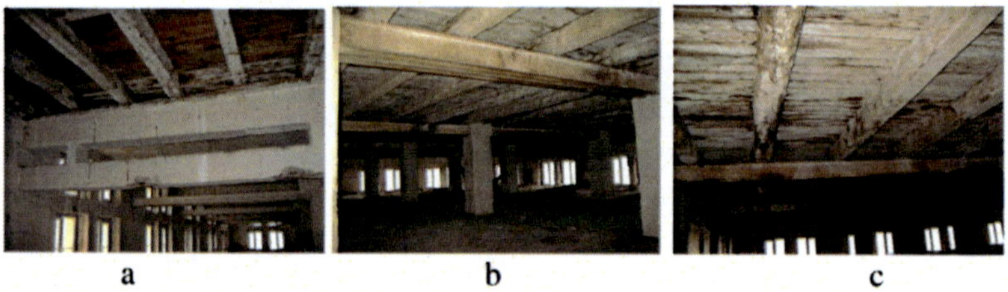

Fig. 7. **a, b, c** Supporting of the wooden beams on steel double T- beams; Supporting of the steel beams on the columns of the building.

a) 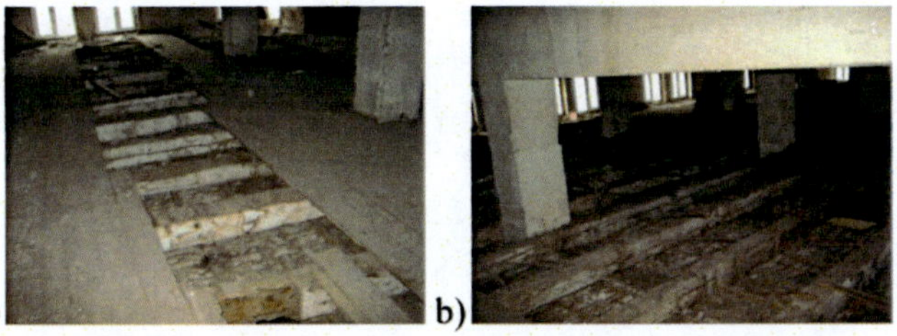 b)

Fig. 8. **a, b** View on the beams in the floor construction

View inside on the floor construction is shown in the Fig. 8. The floor construction of the elevation: +17.14 m (ceiling slab) is also made of wooden joists, lying on steel beams located in the transverse axes of the building, which rest on the brick (respectively. Concrete columns) (Fig. 9).

a b

Fig. 9. **a, b** Floor construction of the elevation: +17.14 m (ceiling slab), made of wooden beams, lying on steel beams located in the transverse axes of the building, which are supported by the brick (respectively: RC columns).

3 Description of the Constructive System of Timber Roof of Building

The roof structure of the museum in Kyustendil is a gable roof with alpine type and height to the ridge: H = 10.10 m from the level of the ceiling slab. The base of the isosceles configuration of the gable roof is light in size between the enclosing walls: 17,220 m. The supporting structure of the roof is made entirely of wood (Fig. 10). The roof structure is made entirely of wood - made of shaped wood. The roof is gabled with a steep slope covered with marseille tiles on a wooden cladding.

The main load carrying elements are: binder, columns-with capitals (from diagonal supporting struts), or slopes to each columns, purlins in the longitudinal direction of the building (Fig. 11) of the three-storey roof structure and also the presence of kings on the third level of the structure (Fig. 11).

The roof construction is characterized by the presence of inclined beams at the beginning and end of the main spine of the structure, which start from the floor beam above the wooden floor, pass through the two-piece construction of pliers, located above

a b

Fig. 10. **a, b** The drawing of the roof truss structure- and View on the timber roof

a b c

Fig. 11. **a, b, c** Binder columns (kings) anchored on a binder beam-a); capitals in the longitudinal direction, slopes to each column-b) Pliers and kings at each level of the three-storey structure of the roof-c)

the masonry wall and at the level of the first floor joist, reach second level of the structure and, resting on the side at the top of the column of the second level (Fig. 12a, b, c). The inclined beam (pos. 5A), which is supported at its lower end in the main row (pos. 1), is supported at its upper end in the two end columns (pos. 10). These inclined beams in all wheel-bases are: 15/17 cm. It is gripped in turn with pliers (pos.6) to the roof rib. This small pliers is made of two planks with dimensions: 8/15 cm. Above this pliers on the inclined beam is located a little column (pos. 7) with dimensions 14/14 cm. Above it is a longitudinal purlin (pos. 9) with dimensions 10/10 cm, under which a pillow (pos. 8) with dimensions 10/10 cm is mounted.

a) b) c)

Fig. 12. **a, b, c** Masonry wall attics and inclined beams at the beginning and end of the main spine of the structure

The roof ribs resting on the little masonry attics have an independent slope (Fig. 13). A 4 cm thick wooden cladding is nailed on them, on which Marseille tiles are arranged. The ribs are located at approximately equal distances from each other, varying in the range of: 40–50–60–70 cm. There are usually 7 ribs in each field.

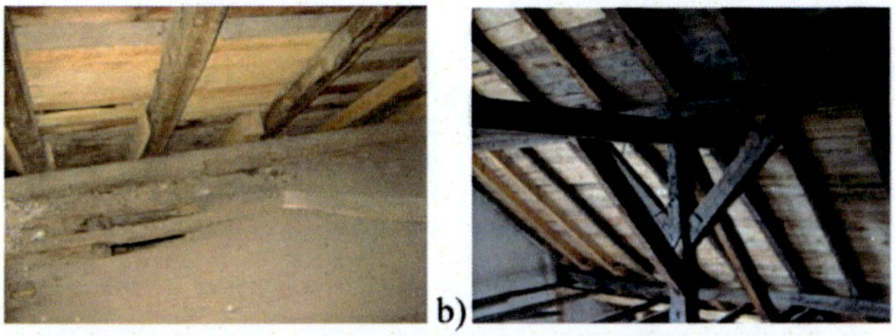

a) b)

Fig. 13. **a, b** Second floor of the roof; ribs supported by pliers

The three-storey wooden structure of the roof starts from a main wooden beam (pos. 1.-binder), lying on the bricks, respectively on the concrete columns (Fig. 14). The binder of the roof truss is supported continuously on the beams of the floor joists of the ceiling, which in turn are supported on the transverse steel double T beams with a height of 22 cm. This method of support presupposes the calculation of the binder as a continuous beam, which must be checked for bending obtained from the positive and negative moments obtained in it (in the continuous beam) and also for axial forces obtained in the binder during its participation in the roof truss structure. It consists of separate single-span beams with dimensions: 20/22 cm, fast connected each other. On them in the places of the massive columns, wooden columns (pos. 2-kings) with dimensions 20/22 cm are mounted (Fig. 15).

Fig. 14. Supporting the binder on the floor joist; supporting the floor joist from the ceiling on the steel beams; Supporting a column (king) from the bar truss on a brick column

Fig. 15. Connection between binder and column (king)

The wooden columns are 3 in the first level and are at a distance of 400 cm from each other in the transverse direction (Fig. 16).

Fig. 16. Floor joists of the first new of the spatial structure

Fig. 17. Connection between column, purlins and pliers on the first level of the roof

Purlins with dimensions: 18/20 cm (pos. 3) are laid on them in the longitudinal direction. Above the longitudinally oriented purlins, in the transverse direction are mounted wooden pliers (pos. 4) with dimensions: 2 × 8/15 cm, which are supported by braces (pos. 5) in the transverse direction (Fig. 17). The diagonal struts have the dimensions of their cross sections: 14/16 (pos. 5). The pliers (pos. 4) surround the columns of the second level (pos. 10-2 pcs., and pos. 11-1 pcs.), on the side, as the connection between them is with a bolted connection (Fig. 18 and Fig. 19). The pliers (pos.4), which are made of planks with dimensions: 2 × 8/15, grip each other at both ends, the inclined beams to the main supporting structure as well as the corresponding roof ribs, which are located in this axis.

Fig. 18. Connection between column, purlins, pliers and struts on the second level of the roof

Parallel to the pliers on the capitals a wooden beam with circular cross section has been developed: round wood: Φ12/12 cm; as well as with rectangular section: 14/15 cm; 12/15; 16/15; 18/15 cm. Half of this wooden platform is plastered (Fig. 19a, b, c).

The clear distance between this floor structure and the lower level is equal to 225 cm. Above this floor structure at a distance of 170 cm from it is located a second level - trimmer joists, parallel to the horizontal pliers - pos. 13(2 × 8/15). Above the central king of the second level, longitudinally on the roof is a purlin situate: (pos. 12) with dimensions 18/20 cm. At both ends of the pliers (pos. 13) are located longitudinally on the roof purlins: (pos. 15) with dimensions 16/19 cm (Fig. 19). On these purlins, located next to the contour of the roof, the roof ribs rest laterally. The third level of the roof structure consists of a central "king" (pos. 16) with dimensions 20/22 cm, which at the top supports a ridge purlin: (pos. 21) with dimensions: 18/20 cm Fig. 20). The central king lies at his lower end on the capital, passing through the pliers - pos.13. Under the

a b c

Fig. 19. **a, b, c** View on the second level of the roof structure

ridge purlin there is a pliers (pos. 20) with dimensions: 2 × 8/15 cm, which covers the roof ribs, which fall into its axis and is connected to them with a bolted joint. In the longitudinal direction this "king" is reinforced with two diagonal struts: (pos. 24) with dimensions 14/14 cm.

Fig. 20. Central column (king), roof purlin in the longitudinal direction, diaonal slopes to the central column, pliers under the roof of the three-storey roof structure

At a distance of 195 cm from the central king are located on both sides two hanging kings (pos. 17) with dimensions: 20/22 cm. They are supported in the pliers below them (pos. 13), passing through the joint formed by the two planks: 2 × 8/15 cm (Fig. 21). The connection between them is realized with a bolted connection BM20. At its upper end this king (pos. 17) supports a longitudinal purlin: (pos. 19) with dimensions: 16/19 cm. Pos. 17 is supported laterally with pliers: (pos. 18) with dimensions: 2 × 8/15 cm.

Fig. 21. Central column (king), side hanging columns (kings), roof purlins in the longitudinal direction, diagonal slopes to the central column, pliers under the roof of the three-storey roof structure on both levels

4 Computational 3D Model of the Truss Construction of the Building

4.1 Load Model

Loads are adopted pursuant to Ordinance № 3/2004 years. "Basics of designing the structure of the buildings and the impacts on them" for the region of Kyustendil as follows: snow region III with value 130 daN/m^2 and wind to area II with value 32 daN/m^2. Computing values of these values are achieved by multiplying the normative values of the two impacts with load factor: = 1.50. Studied are all occurring in wooden frames species components. Static analysis of the various structural elements of the roof done according to the methods of structural mechanics and verified with FEM program "SAP2000". Static schemes of frames and roof are modeled as plane joint-rod systems. All fixtures in the wooden structure of the roof are filled with bolt-screws. Dimensions of the different components of the roof frame of strength, stability and deformability are carried out according to EC5, 2007. The quality of the wood, when testing the structural elements of the roof frame is adopted as a second category (SI-II). On the Fig. 22 is demonstrated the Computational 3D model of the truss construction of the museums building. The drawings of the main truss construction is on the Fig. 23 Cross sections of the constructive element of the timber truss system is the Table 1. As additional information is important to know that: length of the truss is: 17200 mm, and high: 10010 mm. The length of the roof is: 59,79 m, width: 18,10 m and high: 10,10 m.

Fig. 22. Computational 3D model of the truss construction of the roof

Fig. 23. The drawings of the truss construction of the roof

Table 1. Cross sections of individual structural elements

Cross-section	b (cm)	h (cm)
Pos. 1-binder	20	22
Pos. 2-column	20	22
Pos. 3-purlin	18	20

(continued)

Table 1. (*continued*)

Cross-section	b (cm)	h (cm)
Pos. 4-pliers	2 × 8	15
Pos. 5/5A-incline beam	14/15	16/17
Pos. 6-pliers	8	15
Pos. 7-column	14	14
Pos. 8-pilow	10	10
Pos. 9-purlin	10	10
Pos. 10-column	20	22
Pos. 11-column	20	22
Pos. 12-purlin	18	20
Pos. 13-pliers	2 × 8	15
Pos. 14 diagonal bar	14	16
Pos. 15-purlin	16	19
Pos. 16-colum	20	22
Pos. 17-column	20	22
Pos. 18-pliers	8	15
Pos. 19 -purlin	16	19
Pos. 20-pliers	2 × 8	15
Pos. 21-purlin	20	20
Pos. 22-rib	14	14
Pos. 23-timber sheets	17	17
Pos. 24-inclined supports bar	16	16 r

5 Calculation of the Construction According to the Theory of the Second Order

In order to be able to comment on the reserves of the truss supporting structure, we performed a full calculation links with SAP2000 software, according to theory I. and II. order. The combinations of loads and values are given in next text as follows:

5.1 Load Combinations (According: EC5, 2007)

1.35 DL + 1.5 SL; 1.35DL + 1.5SL + 0.75WL; 1.35DL + 1.5WL (pressure) + 0.75SL; 1.0DL + 1.5WL (suction); 1.35DL + 0.75WL.

5.2 Load Values

a) Covering: 0.70 (kN/m^2); b) Snow: 1.30 (kN/m^2); c) Wind (pressure): 0,32/−0.32 (kN/m^2); d)-Wind (suction): −1.00/−1.00 (kN/m^2).

Due to the age of the construction were created extraordinary load values, which in this case can count on a 100-year return Scheme of external load: dead load, snow and wind load and are given in Fig. 24.

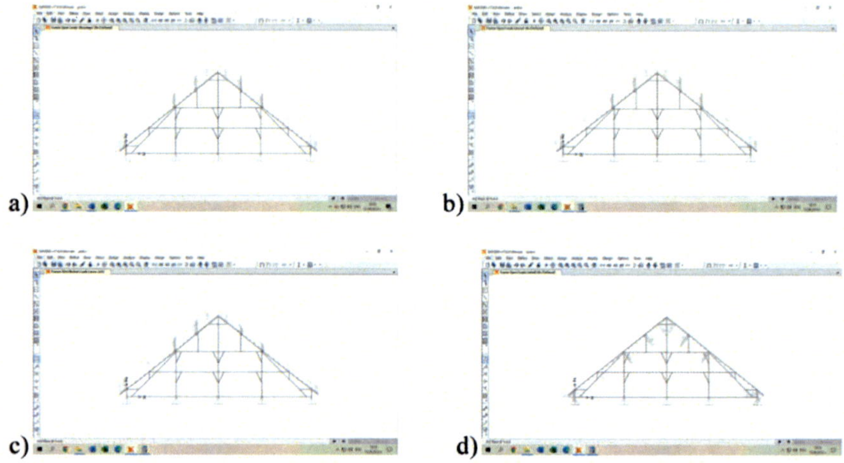

Fig. 24. **a, b, c, d** Scheme of external load are given

Static scheme and courses of internal forces for dead load M, Q, N values according to theory of second order is shown in Fig. 25a, b, c, d.

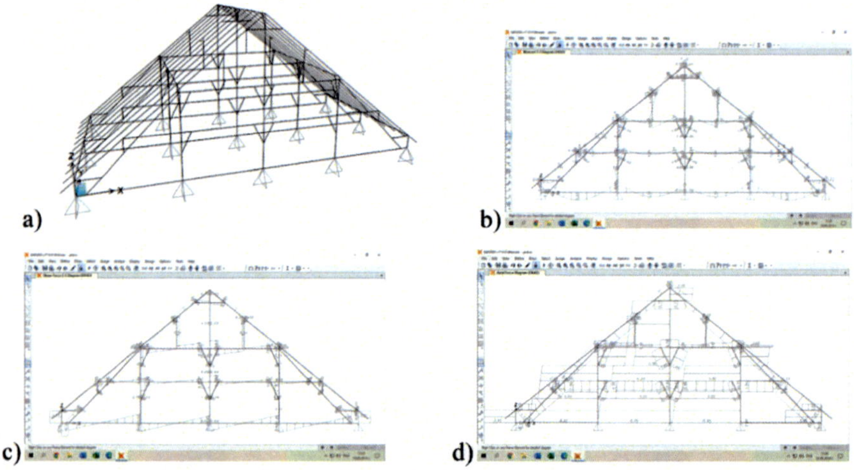

Fig. 25. **a, b, c, d** Static scheme, Diagrams of bending moments-M, shear forces (Q) and normal forces (N), according to theory II. Order

6 Conclusion

The calculation verified that if there is no further significant damage to the timber construction by biotic influences, there is no danger of the collapse of the truss of the building in Kustendil. If we think about the constructional feeling of the arch. R. Fischer, we must state that this was considerably advanced. Although they had no modern computing resources, they did comprehensively combine various requirements for the functionality of the work, including some protection against biotic influences. From today's point of view, the building withstood all the effects of a load with a 95-year return and still despite some damage to the truss does not threaten its immediate collapse.

Acknowledgement. The research was supported within the research project: "Investigation of behavior between historic Timber Roof Structures and Masonry Building, due to load of Wind and Earthquake using MFE", contract number 2/2020-22, granted by the research Fund of Ministry of Education.

References

Drdácký, M.: Historic roofs and timber frames, state of the art studies, European Research on Cultural Heritage. In: Proceedings of the ARCCHIP Workshops supported from the EC 5th FP project, ARIADNE 18, Prague, vol. 4, pp. 313–531 (2006)

Kuklik, P.: History of timber structures. In: Handbook 1-Timber Structures, Leonardo da Vinci Pilot Project, CZ/06/B/F/PP/168007, Education Materials for Designing and Testing of Timber Structures-TEMTIS, pp.1–14 (2008)

Lorenco, P.B.: Selected case studies for ancient Portuguese timber structures. In: Workshop on Earthquake Engineering on Timber Structures, Coibra, Portugal, pp. 1–9 (2006)

Straka, B., Novotný, M., Krupicová, J., Šmak, M., Šuhajda, K., Vejpustek, Z.: Konstrukce šikmých střech. Grada Publishing (2013)

Tampone, G.: Mechanical failures of the timber structural systems. In: Proceedings of ICOMOS IWC - XVI International Symposium, From Material to Structure - Mechanical Behaviour and Failures of the Timber Structures, Florence, Venice and Vicenza, pp. 1–21 (2007)

Vinař, J., Kufner, V.: Historické Krovy, Konstrukce a Statika. Grada Publishing, Praha (2004)

Author Index

I. Vayas and F. M. Mazzolani (Eds.): PROHITECH 2021, LNCE 209, pp. 1469–1472, 2022.
https://doi.org/10.1007/978-3-030-90788-4

S

T

U

V

Y

Z